Beginning and Intermediate Algebra

Fifth Edition

John Tobey

North Shore Community College
Danvers, Massachusetts

Jeffrey Slater

North Shore Community College
Danvers, Massachusetts

Jamie Blair

Orange Coast College
Costa Mesa, California

Jennifer Crawford

Normandale Community College
Bloomington, Minnesota

PEARSON

Boston Columbus Indianapolis New York San Francisco
Amsterdam Cape Town Dubai London Madrid Milan Munich Paris Montréal Toronto
Delhi Mexico City São Paulo Sydney Hong Kong Seoul Singapore Taipei Tokyo

Editorial Director, Mathematics: *Christine Hoag*
Editor in Chief: *Michael Hirsch*
Senior Acquisition Editor: *Rachel Ross*
Editorial Assistant: *Megan Tripp*
Development Editor: *Elaine Page*
Project Management Team Lead: *Christina Lepre*
Project Manager: *Ron Hampton*
Program Management Team Lead: *Karen Wernholm*
Program Manager: *Beth Kaufman*
Interior Design: *Tamara Newnam*
Interior Design Supervision/Cover Design: *Barbara T. Atkinson*
Design Manager: *Andrea Nix*
Manager, Course Production: *Ruth Berry*
Senior Multimedia Producer: *Vicki Dreyfus*
Media Producer: *Nicholas Sweeny*
Executive Content Manager: *Rebecca Williams (MathXL)*
Associate Content Manager: *Eric Gregg (MathXL)*
Senior Content Developer: *John Flanagan (TestGen)*
Senior Product Marketing Manager: *Jennifer Edwards*
Marketing Assistant: *Alexandra Habashi*
Senior Author Support/Technology Specialist: *Joe Vetere*
Procurement Manager: *Carol Melville*
Rights Manager: *Gina Cheselka*
Rights Project Manager: *Martha Shethar*
Production Management, Composition, and Art: *Integra*
Cover Images: *Illustration by Amy DeVoogd*

Library of Congress Cataloging-in-Publication Data

Tobey, John,
Beginning and Intermediate Algebra—5th edition / John Tobey, Jr., North Shore Community College, Jeffrey Slater, North Shore Community College, Jamie Blair, Orange Coast College, Jennifer Crawford, Normandale Community College.—5th edition.
pages cm
ISBN 0-13-417364-3
1. Mathematics—Textbooks. I. Slater, Jeffrey, II. Blair, Jamie.
III. Crawford, Jennifer IV. Title.
QA152.3T64 2017
512.9—dc23 2015007596

Photo credits are located on page C-1 and represent an extension of this copyright.

www.pearsonhighered.com

0-13-417364-3 (Student Edition paperback)
978-0-13-417364-1 (Student Edition paperback)

20 2020

This book is dedicated to my husband, Nate.
Thank you for your support and patience while I worked.
You were the voice of reason countless times, exactly when I needed it.

Contents

Preface

TO THE INSTRUCTOR

Developmental mathematics course structures, trends, and dynamics continue to evolve and change, as **course redesign trends** continue to evolve and change, including the introduction of **new pathways-type courses.** Developmental mathematics instructors are increasingly challenged with helping their students **navigate career-oriented math tracks (including non-STEM and STEM pathways),** plus helping students think about **selecting a major** and **work-force readiness.** To help instructors on this front, with this revision of *Beginning and Intermediate Algebra* you'll find a **new emphasis on, and integration of, Career Explorations** throughout the text and MyMathLab course.

Additionally, the program retains its hallmark characteristics that have always made the text so easy to learn and teach from, including its building-block organization. Each section is written to stand on its own, and every homework set is completely self-testing. Exercises are paired and graded and are of varying levels and types to ensure that all skills and concepts are covered. As a result, the text offers students an effective and proven learning program suitable for a variety of course formats—including lecture-based classes; computer-lab based or hybrid classes; discussion-oriented, activity-driven classes; modular and/or self-paced programs; and distance-learning, online programs.

We have visited and listened to teachers across the country and have incorporated a number of suggestions into this edition to help you with the particular learning-delivery system at your school. The following pages describe the key changes in this fifth edition.

WHAT'S NEW IN THE FIFTH EDITION?

New Career Explorations Interactions for Students

Each chapter begins with a **Career Opportunities** feature that enables students to personally investigate possible future career options while putting the math into context. Students are asked simple, interactive questions prompting them to consider employment opportunities that perhaps they had never thought possible.

Then, the students are directed to the corresponding **Career Exploration Problems** where they can actually solve problems that help them visualize what work would be like in that career field. This feature opens up possibilities for personal success in future employment.

The Career Exploration Problems are also assignable in MyMathLab, allowing this feature to be seamlessly integrated with the technology. The problems help to foster active learning and better understanding of the math concepts.

New Guided Learning Videos

Faculty have asked for specific interactive videos that will clearly show each step of the **key concepts** of each chapter. With this revision, you'll find a new series of **Guided Learning Videos** that show in a powerful, interactive way **how to solve the most important types of problems contained in each chapter.** For student ease, icons throughout the eText indicate where the videos are available. The eText is clickable, opening the videos on the spot. Plus, a new ***Video Workbook with the Math Coach*** allows students to take notes and practice by studying and solving problems.

Expanded Video Program 

In addition to the new Guided Learning Videos with icons throughout the eText, objective-level video clips have also been added to the MyMathLab course with accompanying icons throughout the eText. These video additions expand upon an already complete video lecture series available in MyMathLab. Students and instructors will also find complete Section Lecture Videos, Math Coach Videos, and Chapter Test Prep Videos.

- **The Math Coach** has been expanded within the MyMathLab course, with even more stepped-out, guided Math Coach problems assignable in MyMathLab. Within the text, following each Chapter Test, the **Math Coach** provides students with a personal office-hour experience by walking them through problems step-by-step and pointing out some helpful hints to keep them from making common errors on test problems. For additional help, students can also watch the authors work through these problems on the accompanying Math Coach videos in the MyMathLab course. Instructors can also assign the Math Coach problems in MyMathLab and use the companion *Video Workbook with the Math Coach* for additional practice and to serve as the foundation for a course notebook.
- Fifteen percent of the exercises throughout the text have been refreshed.
- Real-world application problems have been updated throughout the text.
- **New Use Math to Save Money Animations** have been added to the MyMathLab course. The animations expand upon a favorite feature from the text, allowing students to put the math they just learned into context. These newly created animations are set to music and depict real-life scenarios and real-life people using math to cut costs and spend less. To ensure that students watch and understand the animations, there are accompanying Use Math to Save Money homework assignments available in MyMathLab, which are prebuilt for instructor convenience.

Additionally, we've created an even stronger connection between the approach that is used to teach the concepts in the text, and the media assets and assignable exercises within the accompanying MyMathLab course.

To make sure you and your students are getting the most out of the text *and* the MyMathLab course, see the following MyMathLab feature descriptions.

BREAKTHROUGH
To improving results

Get the most out of MyMathLab®

MyMathLab is the world's leading online resource for teaching and learning mathematics. MyMathLab helps students and instructors improve results and provides engaging experiences and personalized learning for each student so learning can happen in any environment. Plus, MyMathLab offers flexible and time-saving course-management features to allow instructors to easily manage their classes while remaining in complete control, regardless of course format.

Personalized Support for Students

- MyMathLab comes with many learning resources–eText, animations, videos, and more–all designed to support your students as they progress through their course.
- The Adaptive Study Plan acts as a personal tutor, updating in real time based on student performance to provide personalized recommendations on what to work on next. With the new Companion Study Plan assignments, instructors can now assign the Study Plan as a prerequisite to a test or quiz, helping to guide students through concepts they need to master.
- Personalized Homework allows instructors to create homework assignments tailored to each student's specific needs by focusing on just the topics students have not yet mastered.

Used by nearly 4 million students each year, the MyMathLab and MyStatLab family of products delivers consistent, measurable gains in student learning outcomes, retention, and subsequent course success.

www.mymathlab.com

Resources for Success

MyMathLab® Online Course

Beginning and Intermediate Algebra by Tobey/Slater/Blair/Crawford

(access code required)

MyMathLab is available to accompany Pearson's market-leading text offerings. To give students a consistent tone, voice, and teaching method, each text's approach is tightly integrated throughout the accompanying MyMathLab course, making learning the material as seamless as possible.

New Career Explorations Interactions

A new integration of Career Explorations has been added throughout the text and MyMathLab course in an interactive format that engages students and gets them thinking about future career possibilities. Each chapter starts with a **Career Opportunities** feature that puts the math into context and ends with multiple **Career Exploration Problems** that are also assignable in MyMathLab!

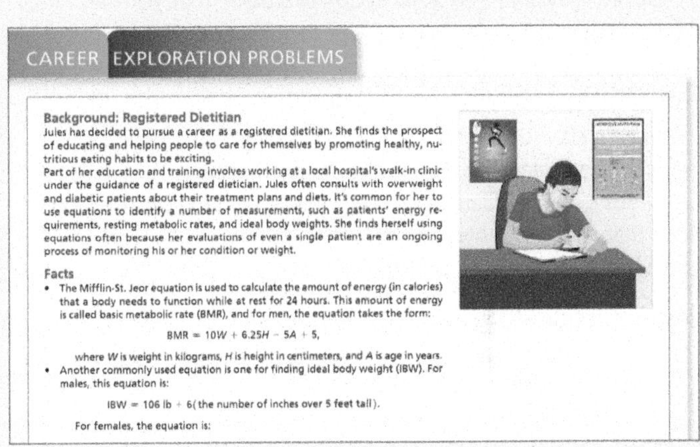
CAREER EXPLORATION PROBLEMS

Background: Registered Dietitian

Jules has decided to pursue a career as a registered dietitian. She finds the prospect of educating and helping people to care for themselves by promoting healthy, nutritious eating habits to be exciting.

Part of her education and training involves working at a local hospital's walk-in clinic under the guidance of a registered dietician. Jules often consults with overweight and diabetic patients about their treatment plans and diets. It's common for her to use equations to identify a number of measurements, such as patients' energy requirements, resting metabolic rates, and ideal body weights. She finds herself using equations often because her evaluations of even a single patient are an ongoing process of monitoring his or her condition or weight.

Facts

- The Mifflin-St. Jeor equation is used to calculate the amount of energy (in calories) that a body needs to function while at rest for 24 hours. This amount of energy is called basic metabolic rate (BMR), and for men, the equation takes the form:

$$BMR = 10W + 6.25H - 5A + 5,$$

where W is weight in kilograms, H is height in centimeters, and A is age in years.
- Another commonly used equation is one for finding ideal body weight (IBW). For males, this equation is:

$$IBW = 106 \text{ lb} + 6(\text{the number of inches over 5 feet tall}).$$

For females, the equation is:

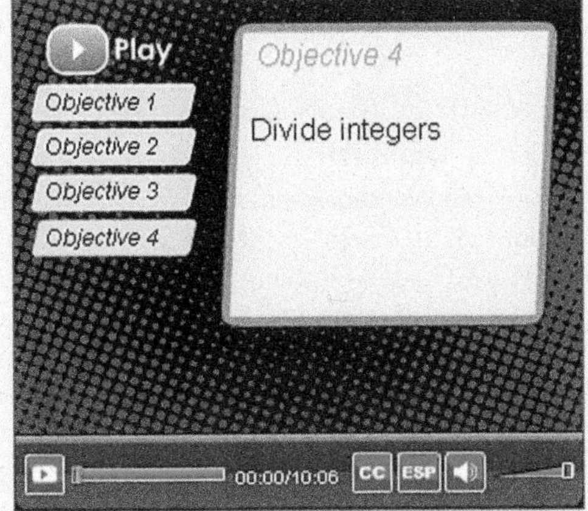

New Guided Learning Videos, Objective-Level Video Clips, and Video Workbook

New Guided Learning Videos show in a powerful, interactive way how to solve the most important types of problems in each chapter. Icons throughout the eText indicate where videos are available. The eText is clickable, opening videos on the spot. Plus, a new *Video Workbook with the Math Coach* ties it all together and provides opportunity for extra practice.

New Use Math to Save Money Animations

These newly created animations, which have been added to the MyMathLab course, are set to music and depict real-life scenarios in which people use math to cut costs and spend less. Accompanying Use Math to Save Money homework assignments are available in MyMathLab to help further students' understanding.

Resources for Success

With MyMathLab, students and instructors get a robust course-delivery system, the full Tobey/Slater/Blair/Crawford eText, and many assignable exercises and media assets. Additionally, MyMathLab also houses these additional instructor and student resources, making the entire set of resources available in one easy-to-access online location.

Instructor Resources

Annotated Instructor's Edition

This version of the text includes answers to all exercises presented in the book, as well as helpful teaching tips. This resource is available as a hardcopy textbook that you can request through your Pearson sales representative.

Learning Catalytics™ Integration

Generate class discussion, guide your lecture, and promote peer-to-peer learning with real-time analytics. MyMathLab now provides Learning Catalytics—an interactive student-response tool that uses students' smartphones, tablets, or laptops to engage them in more sophisticated tasks and thinking.

Instructors can

- Pose a variety of open-ended questions that help students develop critical-thinking skills.
- Monitor responses to find out where students are struggling.
- Use real-time data to adjust instructional strategy and try other ways of engaging students during class.
- Manage student interactions by automatically grouping students for discussion, teamwork, and peer-to-peer learning.

Instructor's Solutions Manual

The *Instructor's Solutions Manual* is available for download from the Pearson Instructor Resource Center or within the MyMathLab course, and it includes detailed, step-by-step solutions to the even-numbered section exercises as well as solutions to every exercise (odd and even) in the Classroom Quiz, mid-chapter reviews, chapter reviews, chapter tests, cumulative tests, and practice final.

Instructor's Resource Manual with Tests and Mini Lectures

Also available for download from the Pearson Instructor Resource Center and within the MyMathLab course, the *Instructor's Resource Manual* includes a mini lecture for each text section, two short group activities per chapter, three forms of additional practice exercises, two pretests, six tests, and two final exams for every chapter, both free response and multiple choice, as well as two cumulative tests for every even numbered chapter. The *Instructor's Resource Manual* also contains the answers to all items.

PowerPoint Lecture Slides

Available through www.pearsonhighered.com and in MyMathLab, these fully editable lecture slides include definitions, key concepts, and examples for use in a lecture setting.

TestGen

TestGen® (www.pearsoned.com/testgen) enables instructors to build, edit, print, and administer tests using a computerized bank of questions developed to cover all the objectives of the text. TestGen is algorithmically based, allowing instructors to create multiple but equivalent versions of the same question or test with the click of a button. Instructors can also modify test bank questions or add new questions. The software and test bank are available for download from Pearson's Instructor Resource Center.

Student Resources

Student Solutions Manual

The *Student Solutions Manual* provides worked-out solutions to all odd-numbered section exercises, even and odd exercises in the Quick Quiz, mid-chapter reviews, chapter reviews, chapter tests, Math Coach, and cumulative reviews. Instructors have the option to make an electronic version available to students within the MyMathLab course, or students can purchase it separately in printed form.

New Video Workbook with the Math Coach

The new *Video Workbook with the Math Coach* expands upon the popular *Math Coach* workbook format and is correlated with the new Guided Learning Videos to serve as a video note-taking and practice guide for students. It is available to students in electronic form within the MyMathLab course, and students can also purchase it separately in printed form.

Student Success Module in MyMathLab

This new interactive module is available in the left-hand navigation of MyMathLab and includes videos, activities, and post-tests for these three student-success areas:

- **Math-Reading Connections**, including topics such as "Using Word Clues" and "Looking for Patterns."
- **Study Skills**, including topics such as "Time Management" and "Preparing for and Taking Exams."
- **College Success**, including topics such as "College Transition" and "Online Learning."

Instructors can assign these videos and/or activities as media assignments, along with prebuilt post-tests to make sure students learn and understand how to improve their skills in these areas. Instructors can integrate these assignments with their traditional MyMathLab homework assignments to incorporate student success topics into their course, as they deem appropriate.

Diagnostic Pretest: Beginning and Intermediate Algebra

Follow the directions for each problem. Simplify each answer.

Chapter 0

1. $\frac{7}{8} + \frac{2}{3} - \frac{1}{4}$

2. $5\frac{2}{7} + 2\frac{1}{14}$

3. $\frac{15}{18} \times \frac{36}{25}$

4. $5\frac{1}{4} \div 4\frac{3}{8}$

5. $2.3 + 7.522 + 0.088$

6. 81.4×0.05

7. $0.2496 \div 0.12$

8. What is 4% of 120.8?

9. 55 is what percent of 220?

Chapter 1

10. Add. $-3 + (-4) + (12)$

11. Subtract. $-20 - (-23)$

12. Combine. $5x - 6xy - 12x - 8xy$

13. Evaluate. $2x^2 - 3x - 4$ when $x = -3$.

14. Remove the grouping symbols. $2 - 3\{5 + 2[x - 4(3 - x)]\}$

15. Evaluate. $-3(2 - 6)^2 + (-12) \div (-4)$

Chapter 2

In questions 16–19, solve each equation for x.

16. $40 + 2x = 60 - 3x$

17. $7(3x - 1) = 5 + 4(x - 3)$

18. $\frac{2}{3}x - \frac{3}{4} = \frac{1}{6}x + \frac{21}{4}$

19. $\frac{4}{5}(3x + 4) = 20$

20. Solve for x. $-16x \geq 9 + 2x$

21. Solve for x and graph the result. $42 - 18x < 48x - 24$

−2 −1 0 1 2 3

22. The length of a rectangle is 7 meters longer than the width. The perimeter is 46 meters. Find the dimensions.

23. The drama club put on a play for Thursday, Friday, and Saturday nights. The total attendance for the three nights was 6210. Thursday night had 300 fewer people than Friday night. Saturday night had 510 more people than Friday night. How many people came each night?

24. Two men travel in separate trucks. They each travel a distance of 225 miles on a country road. Art travels at exactly 60 mph and Lester travels at 50 mph. How much time did the trip take each man? (Use the formula distance = rate · time or $d = rt$.)

25. Each of the equal angles of an isosceles triangle is twice as large as the third angle. What is the measure of each angle?

1. ______
2. ______
3. ______
4. ______
5. ______
6. ______
7. ______
8. ______
9. ______
10. ______
11. ______
12. ______
13. ______
14. ______
15. ______
16. ______
17. ______
18. ______
19. ______
20. ______
21. ______
22. ______
23. ______
24. ______
25. ______

26. ______

27. ______

28. ______

29. ______

30. ______

31. ______

32. ______

33. ______

34. ______

35. ______

36. ______

37. ______

Chapter 3

26. Graph $y = 2x - 4$.

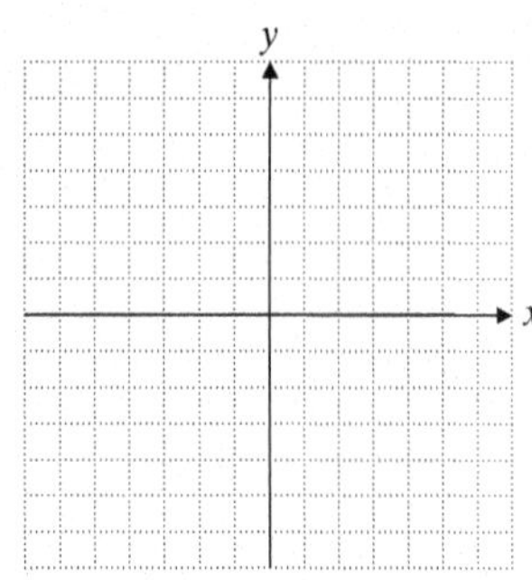

27. Graph $3x + 4y = -12$.

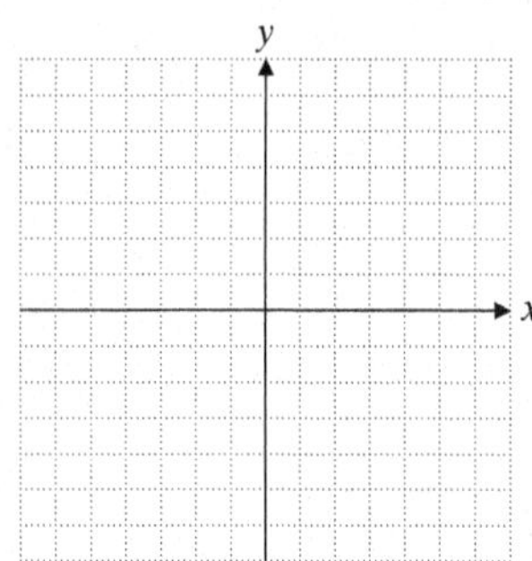

28. What is the slope of a line passing through $(6, -2)$ and $(-3, 4)$?

29. If $f(x) = 2x^2 - 3x + 1$, find $f(3)$.

30. Graph the region. $y \geq -\frac{1}{3}x + 2$

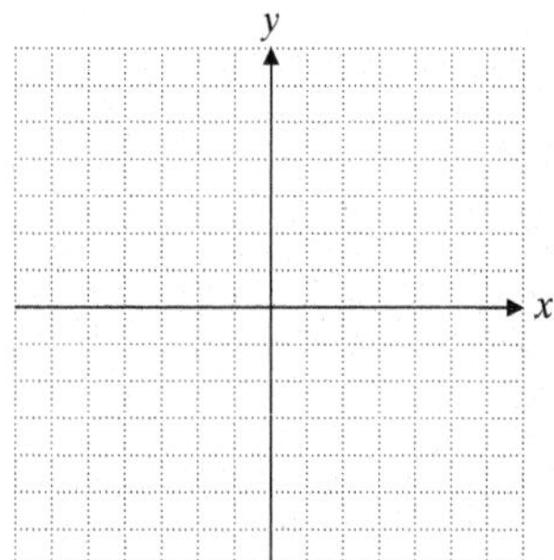

31. Find an equation of the line with a slope of $\frac{3}{5}$ that passes through the point $(-1, 3)$.

Chapter 4

Solve the following:

32.
$$\begin{aligned} 3x + 5y &= 30 \\ 5x + 3y &= 34 \end{aligned}$$

33.
$$\begin{aligned} 2x - y + 2z &= 8 \\ x + y + z &= 7 \\ 4x + y - 3z &= -6 \end{aligned}$$

34. A speedboat can travel 90 miles with the current in 2 hours. It can travel upstream 105 miles against the current in 3 hours. How fast is the boat in still water? How fast is the current?

35. Graph the system.
$$\begin{aligned} x - y &\leq -42 \\ 2x + y &\leq 0 \end{aligned}$$

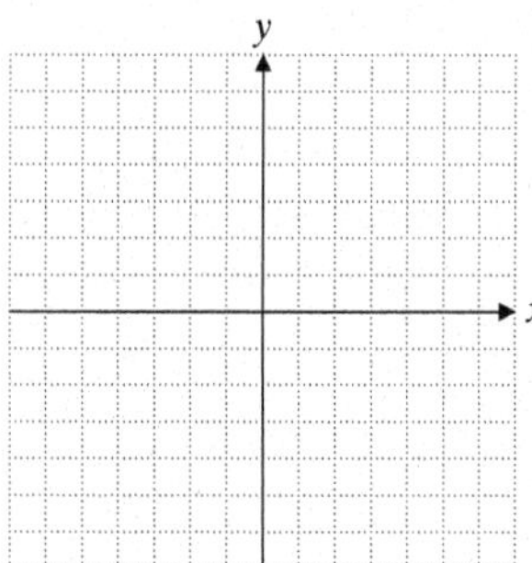

Chapter 5

36. Multiply. $(-2xy^2)(-4x^3y^4)$

37. Divide. $\dfrac{36x^5y^6}{-18x^3y^{10}}$

38. Raise to the indicated power. $(-2x^3y^4)^5$

39. Evaluate. $(-3)^{-4}$

40. Multiply. $(3x^2 + 2x - 5)(4x - 1)$

41. Divide. $(x^3 + 6x^2 - x - 30) \div (x - 2)$

Chapter 6

Factor completely.

42. $5x^2 - 5$

43. $x^2 - 12x + 32$

44. $8x^2 - 2x - 3$

45. $3ax - 8b - 6a + 4bx$

Solve for x.

46. $16x^2 - 24x + 9 = 0$

47. $\frac{x^2 + 8x}{5} = -3$

Chapter 7

48. Simplify. $\frac{x^2 + 3x - 18}{2x - 6}$

49. Multiply. $\frac{6x^2 - 14x - 12}{6x + 4} \cdot \frac{x + 3}{2x^2 - 2x - 12}$

50. Divide and simplify. $\frac{x^2}{x^2 - 4} \div \frac{x^2 - 3x}{x^2 - 5x + 6}$

51. Add. $\frac{3}{x^2 - 7x + 12} + \frac{4}{x^2 - 9x + 20}$

52. Solve for x. $2 - \frac{5}{2x} = \frac{2x}{x + 1}$

53. Simplify. $\frac{3 + \frac{1}{x}}{\frac{9}{x} + \frac{3}{x^2}}$

Chapter 8

Assume that all expressions under radicals represent nonnegative numbers.

54. Multiply and simplify. $(\sqrt{3} + \sqrt{2x})(\sqrt{7} - \sqrt{2x^3})$

55. Rationalize the denominator. $\frac{3\sqrt{x} + \sqrt{y}}{\sqrt{x} - \sqrt{y}}$

56. Solve and check your solutions. $2\sqrt{x - 1} = x - 4$

Chapter 9

57. Solve for x. $x^2 - 2x - 4 = 0$

58. Solve for x. $x^4 - 12x^2 + 20 = 0$

59. Graph $f(x) = (x - 2)^2 + 3$. Label the vertex.

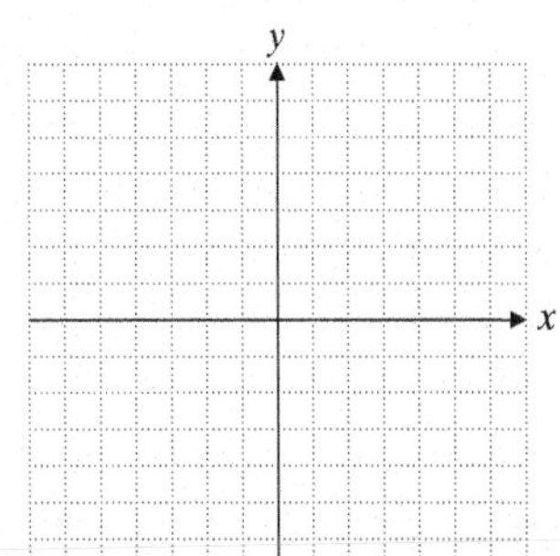

38. ______
39. ______
40. ______
41. ______
42. ______
43. ______
44. ______
45. ______
46. ______
47. ______
48. ______
49. ______
50. ______
51. ______
52. ______
53. ______
54. ______
55. ______
56. ______
57. ______
58. ______
59. ______

60. ______

61. ______

62. ______

63. ______

64. ______

65. ______

66. ______

67. ______

68. ______

69. ______

70. ______

71. ______

72. ______

60. Solve for x. $\left|3\left(\frac{2}{3}x - 4\right)\right| \leq 12$

61. Solve for y. $|3y - 2| + 5 = 8$

Chapter 10

62. Write in standard form the equation of the circle with center at $(5, -2)$ and a radius of 6.

63. Write in standard form the equation of the ellipse whose center is at $(0, 0)$ and whose intercepts are at $(3, 0)$, $(-3, 0)$, $(0, 4)$, and $(0, -4)$.

64. Solve the following nonlinear system of equations. $\begin{aligned} x^2 + 4y^2 &= 9 \\ x + 2y &= 3 \end{aligned}$

Chapter 11

65. If $f(x) = 2x^2 - 3x + 4$, find $f(a + 2)$.

66. Graph on one axis $f(x) = |x + 3|$ and $g(x) = |x + 3| - 3$.

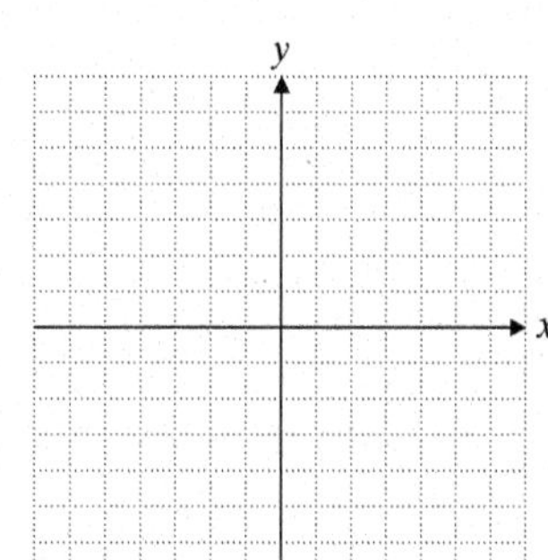

67. If $f(x) = \frac{3}{x + 2}$ and $g(x) = 3x^2 - 1$, find $g[f(x)]$.

68. If $f(x) = -\frac{1}{2}x - 5$, find $f^{-1}(x)$.

Chapter 12

69. Find y if $\log_5 125 = y$.

70. Find b if $\log_b 4 = \frac{2}{3}$.

71. What is $\log 10{,}000$?

72. Solve for x. $\log_6(5 + x) + \log_6 x = 2$

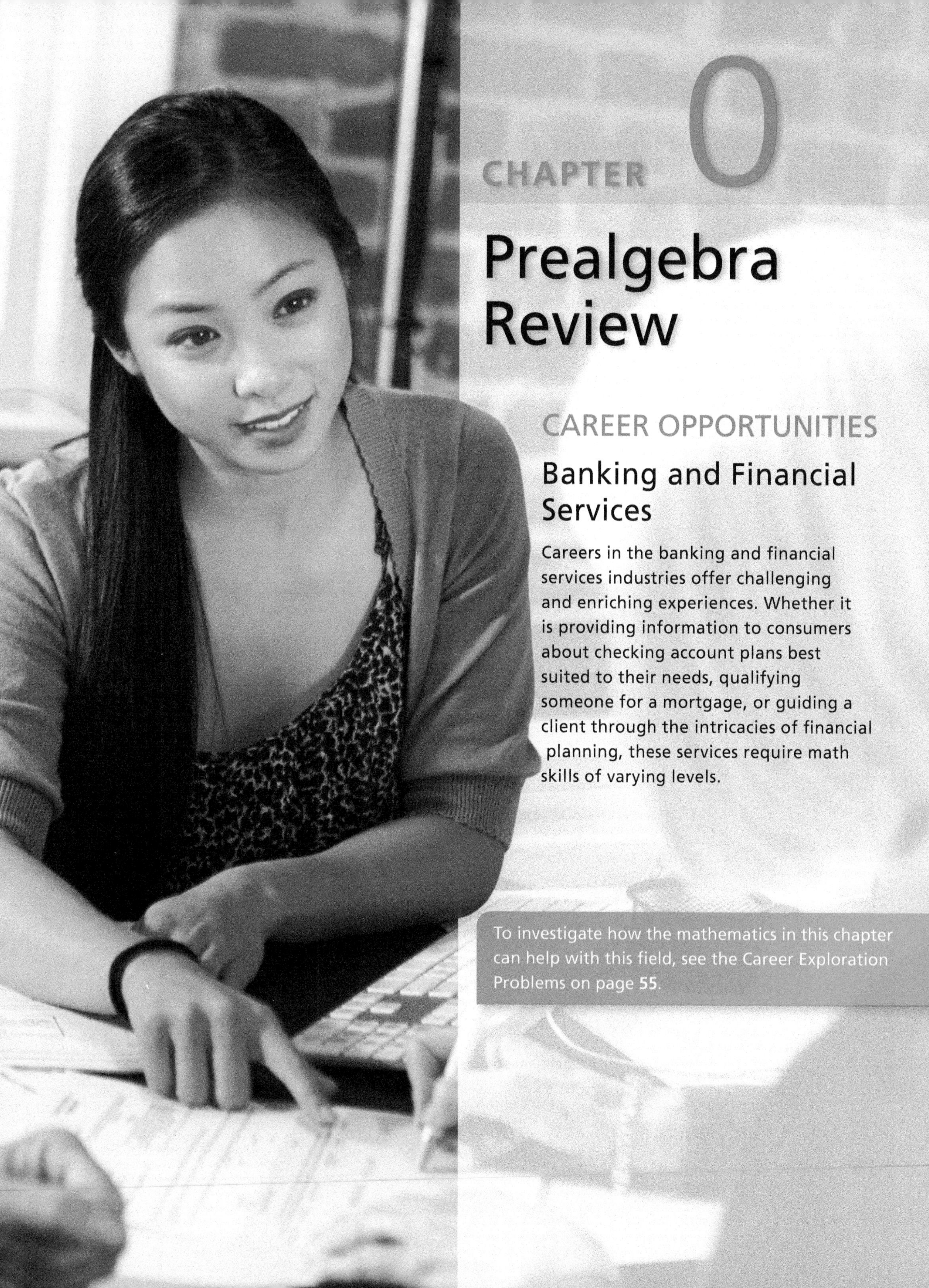

CHAPTER 0

Prealgebra Review

CAREER OPPORTUNITIES

Banking and Financial Services

Careers in the banking and financial services industries offer challenging and enriching experiences. Whether it is providing information to consumers about checking account plans best suited to their needs, qualifying someone for a mortgage, or guiding a client through the intricacies of financial planning, these services require math skills of varying levels.

To investigate how the mathematics in this chapter can help with this field, see the Career Exploration Problems on page **55**.

0.1 Simplifying Fractions

Student Learning Objectives

After studying this section, you will be able to:

1. Understand basic mathematical definitions.
2. Simplify fractions to lowest terms using prime numbers.
3. Convert between improper fractions and mixed numbers.
4. Change a fraction to an equivalent fraction with a given denominator.

Chapter 0 is designed to give you a mental "warm-up." In this chapter you'll be able to step back a bit and tone up your math skills. This brief review of prealgebra will increase your math flexibility and give you a good running start into algebra.

1 Understanding Basic Mathematical Definitions 

Whole numbers are the set of numbers 0, 1, 2, 3, 4, 5, 6, 7, ... They are used to describe whole objects, or entire quantities.

Fractions are a set of numbers that are used to describe parts of whole quantities. In the object shown in the figure there are four equal parts. The *three* of the *four* parts that are shaded are represented by the fraction $\frac{3}{4}$. In the fraction $\frac{3}{4}$ the number 3 is called the **numerator** and the number 4, the **denominator.**

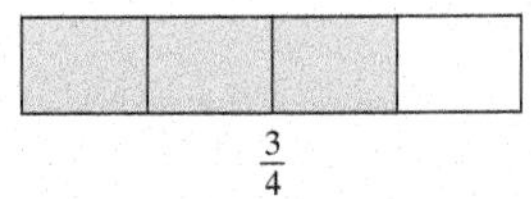

$\frac{3}{4}$

3 ← *Numerator* is on the top
4 ← *Denominator* is on the bottom

The *denominator* of a fraction shows the number of equal parts in the whole and the *numerator* shows the number of these parts being talked about or being used.

Numerals are symbols we use to name numbers. There are many different numerals that can be used to describe the same number. We know that $\frac{1}{2} = \frac{2}{4}$. The fractions $\frac{1}{2}$ and $\frac{2}{4}$ both describe the same number.

Usually, we find it more useful to use fractions that are simplified. A fraction is considered to be in **simplest form** or **reduced form** when the numerator (top) and the denominator (bottom) have no common divisor other than 1, and the denominator is greater than 1.

$\frac{1}{2}$ is in simplest form.

$\frac{2}{4}$ is *not* in simplest form, since the numerator and the denominator can both be divided by 2.

If you get the answer $\frac{2}{4}$ to a problem, you should state it in simplest form, $\frac{1}{2}$. The process of changing $\frac{2}{4}$ to $\frac{1}{2}$ is called **simplifying** or **reducing** the fraction.

2 Simplifying Fractions to Lowest Terms Using Prime Numbers

Natural numbers or **counting numbers** are the set of whole numbers excluding 0. Thus the natural numbers are the numbers 1, 2, 3, 4, 5, 6, ...

When two or more numbers are multiplied, each number that is multiplied is called a **factor.** For example, when we write $3 \times 7 \times 5$, each of the numbers 3, 7, and 5 is called a factor.

Prime numbers are natural numbers greater than 1 whose only natural number factors are 1 and themselves. The number 5 is prime. The only natural number factors of 5 are 5 and 1.

$$5 = 5 \times 1$$

The number 6 is not prime. The natural number factors of 6 are 3 and 2 or 6 and 1.

$$6 = 3 \times 2 \qquad 6 = 6 \times 1$$

The first 15 prime numbers are

2, 3, 5, 7, 11, 13, 17, 19, 23, 29, 31, 37, 41, 43, 47.

Any natural number greater than 1 either is prime or can be written as the product of prime numbers. For example, we can take each of the numbers 12, 30, 14, 19, and 29 and either indicate that they are prime or, if they are not prime, write them as the product of prime numbers. We write as follows:

$12 = 2 \times 2 \times 3$ $\quad 30 = 2 \times 3 \times 5$ $\quad 14 = 2 \times 7$

19 is a prime number. 29 is a prime number.

To reduce a fraction, we use prime numbers to factor the numerator and the denominator. Write each part of the fraction (numerator and denominator) as a product of prime numbers. Note any *factors* that appear in both the *numerator* (top) and *denominator* (bottom) of the fraction. If we divide numerator and denominator by these values we will obtain an equivalent fraction in *simplest form.* When the new fraction is simplified, it is said to be in **lowest terms.** Throughout this text, to *simplify* a fraction will always mean to write the fraction in lowest terms.

Example 1 Simplify each fraction.

(a) $\frac{14}{21}$ **(b)** $\frac{15}{35}$ **(c)** $\frac{20}{70}$

Solution

(a) $\frac{14}{21} = \frac{\cancel{7} \times 2}{\cancel{7} \times 3} = \frac{2}{3}$ We factor 14 and factor 21. Then we divide numerator and denominator by 7.

(b) $\frac{15}{35} = \frac{\cancel{5} \times 3}{\cancel{5} \times 7} = \frac{3}{7}$ We factor 15 and factor 35. Then we divide numerator and denominator by 5.

(c) $\frac{20}{70} = \frac{2 \times \cancel{2} \times \cancel{5}}{7 \times \cancel{2} \times \cancel{5}} = \frac{2}{7}$ We factor 20 and factor 70. Then we divide numerator and denominator by both 2 and 5. □

Student Practice 1 Simplify each fraction.

(a) $\frac{10}{16}$ **(b)** $\frac{24}{36}$ **(c)** $\frac{36}{42}$

Sometimes when we simplify a fraction, all the prime factors in the top (numerator) are divided out. When this happens, we must remember that a 1 is left in the numerator.

Example 2 Simplify each fraction.

(a) $\frac{7}{21}$ **(b)** $\frac{15}{105}$

Solution

(a) $\frac{7}{21} = \frac{\cancel{7} \times 1}{\cancel{7} \times 3} = \frac{1}{3}$ **(b)** $\frac{15}{105} = \frac{\cancel{5} \times \cancel{3} \times 1}{7 \times \cancel{5} \times \cancel{3}} = \frac{1}{7}$ □

Student Practice 2 Simplify each fraction.

(a) $\frac{4}{12}$ **(b)** $\frac{25}{125}$ **(c)** $\frac{73}{146}$

If all the prime numbers in the bottom (denominator) are divided out, we do not need to leave a 1 in the denominator, since we do not need to express the answer as a fraction. The answer is then a whole number and is not usually expressed as a fraction.

Example 3 Simplify each fraction.

(a) $\frac{35}{7}$ **(b)** $\frac{70}{10}$

Solution

(a) $\frac{35}{7} = \frac{5 \times \cancel{7}}{\cancel{7} \times 1} = 5$ **(b)** $\frac{70}{10} = \frac{7 \times \cancel{5} \times \cancel{2}}{\cancel{5} \times \cancel{2} \times 1} = 7$ □

Student Practice 3 Simplify each fraction.

(a) $\frac{18}{6}$ **(b)** $\frac{146}{73}$ **(c)** $\frac{28}{7}$

Sometimes the fraction we use represents how many of a certain thing are successful. For example, if a baseball player was at bat 30 times and achieved 12 hits, we could say that he had a hit $\frac{12}{30}$ of the time. If we reduce the fraction, we could say he had a hit $\frac{2}{5}$ of the time.

Example 4 Cindy got 48 out of 56 questions correct on a test. Write this as a fraction in simplest form.

Solution Express as a fraction in simplest form the number of correct responses out of the total number of questions on the test.

$$48 \text{ out of } 56 \rightarrow \frac{48}{56} = \frac{2 \times 3 \times \cancel{2} \times \cancel{2} \times \cancel{2}}{7 \times \cancel{2} \times \cancel{2} \times \cancel{2}} = \frac{6}{7}$$

Cindy answered the questions correctly $\frac{6}{7}$ of the time. □

Student Practice 4 The major league pennant winner in 1917 won 56 games out of 154 games played. Express as a fraction in simplest form the number of games won in relation to the number of games played.

The number *one* can be expressed as $1, \frac{1}{1}, \frac{2}{2}, \frac{6}{6}, \frac{8}{8}$, and so on, since

$$1 = \frac{1}{1} = \frac{2}{2} = \frac{6}{6} = \frac{8}{8}.$$

We say that these numerals are *equivalent ways* of writing the number *one* because they all express the same quantity even though they appear to be different.

Sidelight: The Multiplicative Identity

When we simplify fractions, we are actually using the fact that we can multiply any number by 1 without changing the value of that number. (Mathematicians call the number 1 the **multiplicative identity** because it leaves any number it multiplies with the same identical value as before.)

Let's look again at one of the previous examples.

$$\frac{14}{21} = \frac{7 \times 2}{7 \times 3} = \frac{7}{7} \times \frac{2}{3} = 1 \times \frac{2}{3} = \frac{2}{3}$$

So we see that

$$\frac{14}{21} = \frac{2}{3}$$

When we simplify fractions, we are using this property of multiplying by 1.

3 Converting Between Improper Fractions and Mixed Numbers

If the numerator is less than the denominator, the fraction is a **proper fraction.** A proper fraction is used to describe a quantity smaller than a whole.

Fractions can also be used to describe quantities larger than a whole. The following figure shows two bars that are equal in size. Each bar is divided into 5 equal pieces. The first bar is shaded completely. The second bar has 2 of the 5 pieces shaded.

The shaded-in region can be represented by $\frac{7}{5}$ since 7 of the pieces (each of which is $\frac{1}{5}$ of a whole box) are shaded. The fraction $\frac{7}{5}$ is called an improper fraction. An **improper fraction** is one in which the numerator is larger than or equal to the denominator.

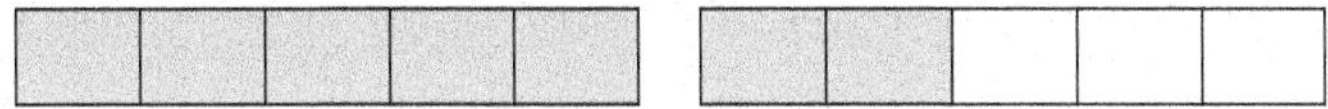

The shaded-in region can also be represented by 1 whole added to $\frac{2}{5}$ of a whole, or $1 + \frac{2}{5}$. This is written as $1\frac{2}{5}$. The fraction $1\frac{2}{5}$ is called a mixed number. A **mixed number** consists of a whole number added to a proper fraction (the numerator is smaller than the denominator). The addition is understood but not written. When we write $1\frac{2}{5}$, it represents $1 + \frac{2}{5}$. The numbers $1\frac{7}{8}$, $2\frac{3}{4}$, $8\frac{1}{3}$, and $126\frac{1}{10}$ are all mixed numbers. From the preceding figure it seems clear that $\frac{7}{5} = 1\frac{2}{5}$. This suggests that we can change from one form to the other without changing the value of the fraction.

From a picture it is easy to see how to *change improper fractions to mixed numbers*. For example, suppose we start with the fraction $\frac{11}{3}$ and represent it by the following figure (where 11 of the pieces, each of which is $\frac{1}{3}$ of a box, are shaded). We see that $\frac{11}{3} = 3\frac{2}{3}$, since 3 whole boxes and $\frac{2}{3}$ of a box are shaded.

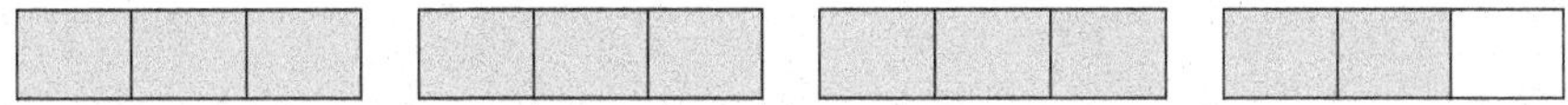

Changing Improper Fractions to Mixed Numbers You can follow the same procedure without a picture. For example, to change $\frac{11}{3}$ to a mixed number, we can do the following:

$$\frac{11}{3} = \frac{3}{3} + \frac{3}{3} + \frac{3}{3} + \frac{2}{3}$$ Use the rule for adding fractions (which is discussed in detail in Section 0.2).

$$= 1 + 1 + 1 + \frac{2}{3}$$ Write 1 in place of $\frac{3}{3}$, since $\frac{3}{3} = 1$.

$$= 3 + \frac{2}{3}$$ Write 3 in place of $1 + 1 + 1$.

$$= 3\frac{2}{3}$$ Use the notation for mixed numbers.

Now that you know how to change improper fractions to mixed numbers and why the procedure works, here is a shorter method.

> **TO CHANGE AN IMPROPER FRACTION TO A MIXED NUMBER**
>
> 1. Divide the denominator into the numerator.
> 2. The quotient is the whole-number part of the mixed number.
> 3. The remainder from the division will be the numerator of the fraction. The denominator of the fraction remains unchanged.

We can write the fraction as a division statement and divide. The arrows show how to write the mixed number.

$$\frac{7}{5} \qquad 5\overline{)7}\ \text{quotient } 1,\ 7-5=2$$

Whole-number part → 1 in $1\frac{2}{5}$

Numerator of fraction ← 2 (Remainder)

Thus, $\frac{7}{5} = 1\frac{2}{5}$.

$$\frac{11}{3} \qquad \begin{array}{r} 3 \\ 3\overline{)11} \\ \underline{9} \\ 2 \end{array}$$

Whole-number part → $3\frac{2}{3}$ ← Numerator of fraction

Remainder

Thus, $\frac{11}{3} = 3\frac{2}{3}$.

Sometimes the remainder is 0. In this case, the improper fraction changes to a whole number.

Example 5 Change to a mixed number or to a whole number.

(a) $\frac{7}{4}$ **(b)** $\frac{15}{3}$

Solution

(a) $\frac{7}{4} = 7 \div 4 \qquad \begin{array}{r} 1 \\ 4\overline{)7} \\ \underline{4} \\ 3 \end{array}$ Remainder

Thus $\frac{7}{4} = 1\frac{3}{4}$.

(b) $\frac{15}{3} = 15 \div 3 \qquad \begin{array}{r} 5 \\ 3\overline{)15} \\ \underline{15} \\ 0 \end{array}$ Remainder

Thus $\frac{15}{3} = 5$. □

Student Practice 5 Change to a mixed number or to a whole number.

(a) $\frac{12}{7}$ **(b)** $\frac{20}{5}$

Changing Mixed Numbers to Improper Fractions It is not difficult to see how to change mixed numbers to improper fractions. Suppose that you wanted to write $2\frac{2}{3}$ as an improper fraction.

$$2\frac{2}{3} = 2 + \frac{2}{3} \qquad \text{The meaning of mixed number notation}$$

$$= 1 + 1 + \frac{2}{3} \qquad \text{Since } 1 + 1 = 2$$

$$= \frac{3}{3} + \frac{3}{3} + \frac{2}{3} \qquad \text{Since } 1 = \frac{3}{3}$$

When we draw a picture of $\frac{3}{3} + \frac{3}{3} + \frac{2}{3}$, we have this figure:

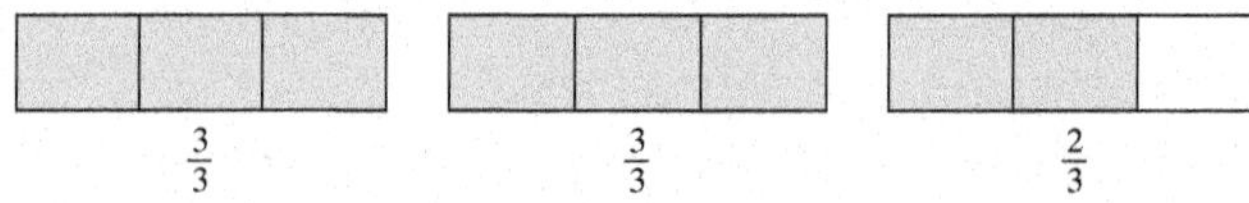

If we count the shaded parts, we see that

$$\frac{3}{3} + \frac{3}{3} + \frac{2}{3} = \frac{8}{3}. \quad \text{Thus} \quad 2\frac{2}{3} = \frac{8}{3}.$$

Now that you have seen how this change can be done, here is a shorter method.

TO CHANGE A MIXED NUMBER TO AN IMPROPER FRACTION

1. Multiply the whole number by the denominator.
2. Add this to the numerator. The result is the new numerator. The denominator does not change.

Example 6 Change to an improper fraction.

(a) $3\frac{1}{7}$ **(b)** $5\frac{4}{5}$

Solution

(a) $3\frac{1}{7} = \frac{(3 \times 7) + 1}{7} = \frac{21 + 1}{7} = \frac{22}{7}$

(b) $5\frac{4}{5} = \frac{(5 \times 5) + 4}{5} = \frac{25 + 4}{5} = \frac{29}{5}$ □

Student Practice 6 Change to an improper fraction.

(a) $3\frac{2}{5}$ **(b)** $1\frac{3}{7}$ **(c)** $2\frac{6}{11}$ **(d)** $4\frac{2}{3}$

4 Changing a Fraction to an Equivalent Fraction with a Given Denominator

A fraction can be changed to an equivalent fraction with a different denominator by multiplying both numerator and denominator by the same number.

$$\frac{5}{6} = \frac{5 \times 2}{6 \times 2} = \frac{10}{12} \quad \text{and} \quad \frac{3}{7} = \frac{3 \times 3}{7 \times 3} = \frac{9}{21} \text{ so}$$

$$\frac{5}{6} \text{ is equivalent to } \frac{10}{12} \quad \text{and} \quad \frac{3}{7} \text{ is equivalent to } \frac{9}{21}.$$

We often multiply in this way to obtain an equivalent fraction with a *particular denominator.*

Example 7 Find the missing numerator.

(a) $\frac{3}{5} = \frac{?}{25}$ **(b)** $\frac{4}{7} = \frac{?}{14}$ **(c)** $\frac{2}{9} = \frac{?}{36}$

Solution

(a) $\frac{3}{5} = \frac{?}{25}$ Observe that we need to multiply the denominator by 5 to obtain 25. So we multiply the numerator 3 by 5 also.

$\frac{3 \times 5}{5 \times 5} = \frac{15}{25}$ The desired numerator is 15.

(b) $\frac{4}{7} = \frac{?}{14}$ Observe that $7 \times 2 = 14$. We need to multiply the numerator by 2 to get the new numerator.

$\frac{4 \times 2}{7 \times 2} = \frac{8}{14}$ The desired numerator is 8.

(c) $\frac{2}{9} = \frac{?}{36}$ Observe that $9 \times 4 = 36$. We need to multiply the numerator by 4 to get the new numerator.

$\frac{2 \times 4}{9 \times 4} = \frac{8}{36}$ The desired numerator is 8. □

Student Practice 7 Find the missing numerator.

(a) $\frac{3}{8} = \frac{?}{24}$ **(b)** $\frac{5}{6} = \frac{?}{30}$ **(c)** $\frac{2}{7} = \frac{?}{56}$

0.1 Exercises MyMathLab®

Verbal and Writing Skills, Exercises 1–4

1. In the fraction $\frac{12}{13}$, what number is the numerator?

2. In the fraction $\frac{13}{17}$, what number is the denominator?

3. What is a factor? Give an example.

4. Give some examples of the number 1 written as a fraction.

5. Draw a diagram to illustrate $2\frac{2}{3}$.

6. Draw a diagram to illustrate $3\frac{3}{4}$.

Simplify each fraction.

7. $\frac{9}{15}$ **8.** $\frac{20}{24}$ **9.** $\frac{12}{36}$ **10.** $\frac{8}{48}$ **11.** $\frac{60}{12}$ **12.** $\frac{72}{18}$

13. $\frac{24}{36}$ **14.** $\frac{32}{64}$ **15.** $\frac{30}{85}$ **16.** $\frac{33}{55}$ **17.** $\frac{42}{54}$ **18.** $\frac{63}{81}$

Change to a mixed number.

19. $\frac{17}{6}$ **20.** $\frac{19}{5}$ **21.** $\frac{47}{5}$ **22.** $\frac{54}{7}$ **23.** $\frac{38}{7}$ **24.** $\frac{41}{6}$

25. $\frac{41}{2}$ **26.** $\frac{25}{3}$ **27.** $\frac{32}{5}$ **28.** $\frac{79}{7}$ **29.** $\frac{111}{9}$ **30.** $\frac{124}{8}$

Change to an improper fraction or whole number.

31. $3\frac{1}{5}$ **32.** $4\frac{2}{5}$ **33.** $6\frac{3}{5}$ **34.** $5\frac{1}{12}$ **35.** $1\frac{2}{9}$ **36.** $1\frac{5}{6}$

37. $8\frac{3}{7}$ **38.** $6\frac{2}{3}$ **39.** $24\frac{1}{4}$ **40.** $10\frac{1}{9}$ **41.** $\frac{72}{9}$ **42.** $\frac{78}{6}$

Find the missing numerator.

43. $\frac{3}{8} = \frac{?}{64}$ **44.** $\frac{5}{9} = \frac{?}{54}$ **45.** $\frac{3}{5} = \frac{?}{35}$

46. $\frac{5}{9} = \frac{?}{45}$ **47.** $\frac{4}{13} = \frac{?}{39}$ **48.** $\frac{13}{17} = \frac{?}{51}$

49. $\frac{3}{7} = \frac{?}{49}$ **50.** $\frac{10}{15} = \frac{?}{60}$ **51.** $\frac{3}{4} = \frac{?}{20}$

52. $\frac{7}{8} = \frac{?}{40}$ **53.** $\frac{35}{40} = \frac{?}{80}$ **54.** $\frac{45}{50} = \frac{?}{100}$

Applications

Solve.

55. ***Women's Professional Basketball*** During the 2014 WNBA basketball season, Maya Moore of the Minnesota Lynx scored 799 points in 34 games. Express as a mixed number in simplified form how many points she averaged per game.

56. ***Kentucky Derby Nominations*** In 2014, 424 horses were nominated to compete in the Kentucky Derby. Only 20 horses were actually chosen to compete in the Derby. What simplified fraction shows what portions of the nominated horses actually competed?

57. ***Income Tax*** Last year, my parents had a combined income of $64,000. They paid $13,200 in federal income taxes. What simplified fraction shows how much my parents spent on their federal taxes?

58. ***Employment*** A large employment agency was able to find jobs within 6 months for 1400 people out of 2420 applicants who applied at one of its branches. What simplified fraction shows what portion of applicants gained employment?

Trail Mix *The following chart gives recipes for two trail mix blends.*

The Rocking R trail mix

Premium blend **High-energy blend**

Sunflower seeds
Nuts
Raisins
1 1 1 1

Sunflower seeds
Nuts
1 1 1 1 1

59. What fractional part of the premium blend is nuts?

60. What fractional part of the high-energy blend is raisins?

61. What fractional part of the premium blend is not sunflower seeds?

62. What fractional part of the high-energy blend does not contain nuts?

College Enrollment *The following chart provides statistics about the total enrollment of male and female students in U.S. colleges for specific years during the period from 1984 to 2014.*

63. What fractional part of the number of students enrolled in 2014 are female?

64. What fractional part of the number of students enrolled in 1994 are male?

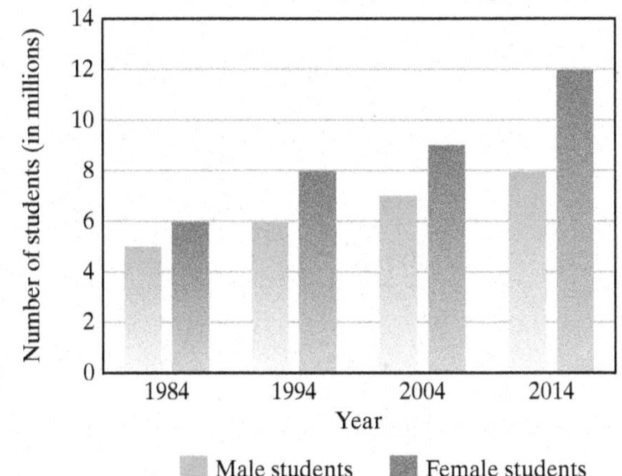

Source: Digest of Education statistics, 2014

Quick Quiz 0.1

1. Simplify. $\frac{84}{92}$

2. Write as an improper fraction. $6\frac{9}{11}$

3. Write as a mixed number. $\frac{103}{21}$

4. **Concept Check** Explain in your own words how to change a mixed number to an improper fraction.

0.2 Adding and Subtracting Fractions

Student Learning Objectives

After studying this section, you will be able to:

1. Add or subtract fractions with a common denominator.
2. Use prime factors to find the least common denominator of two or more fractions.
3. Add or subtract fractions with different denominators.
4. Add or subtract mixed numbers.

1 Adding or Subtracting Fractions with a Common Denominator

If fractions have the same denominator, the numerators may be added or subtracted. The denominator remains the same.

> **TO ADD OR SUBTRACT TWO FRACTIONS WITH A COMMON DENOMINATOR**
>
> 1. Add or subtract the numerators.
> 2. Keep the same (common) denominator.
> 3. Simplify the answer whenever possible.

Example 1 Add the fractions. Simplify your answer whenever possible.

(a) $\frac{5}{7}+\frac{1}{7}$ **(b)** $\frac{2}{3}+\frac{1}{3}$ **(c)** $\frac{1}{8}+\frac{3}{8}+\frac{2}{8}$ **(d)** $\frac{3}{5}+\frac{4}{5}$

Solution

(a) $\frac{5}{7}+\frac{1}{7}=\frac{5+1}{7}=\frac{6}{7}$

(b) $\frac{2}{3}+\frac{1}{3}=\frac{2+1}{3}=\frac{3}{3}=1$

(c) $\frac{1}{8}+\frac{3}{8}+\frac{2}{8}=\frac{1+3+2}{8}=\frac{6}{8}=\frac{3}{4}$

(d) $\frac{3}{5}+\frac{4}{5}=\frac{3+4}{5}=\frac{7}{5}$ or $1\frac{2}{5}$ □

Student Practice 1 Add the fractions. Simplify your answer whenever possible.

(a) $\frac{3}{6}+\frac{2}{6}$ **(b)** $\frac{3}{11}+\frac{8}{11}$ **(c)** $\frac{1}{8}+\frac{2}{8}+\frac{1}{8}$ **(d)** $\frac{5}{9}+\frac{8}{9}$

Example 2 Subtract the fractions. Simplify your answer whenever possible.

(a) $\frac{9}{11}-\frac{2}{11}$ **(b)** $\frac{5}{6}-\frac{1}{6}$

Solution

(a) $\frac{9}{11}-\frac{2}{11}=\frac{9-2}{11}=\frac{7}{11}$

(b) $\frac{5}{6}-\frac{1}{6}=\frac{5-1}{6}=\frac{4}{6}=\frac{2}{3}$ □

Student Practice 2 Subtract the fractions. Simplify your answer whenever possible.

(a) $\frac{11}{13}-\frac{6}{13}$ **(b)** $\frac{8}{9}-\frac{2}{9}$

Although adding and subtracting fractions with the same denominator is fairly simple, most problems involve fractions that do not have a common denominator. Fractions and mixed numbers such as halves, fourths, and eighths are often used. To add or subtract such fractions, we begin by finding a common denominator.

2 Using Prime Factors to Find the Least Common Denominator of Two or More Fractions

Before you can add or subtract fractions, they must have the same denominator. To save work, we select the smallest possible common denominator. This is called the **least common denominator** or LCD (also known as the *lowest common denominator*).

The LCD of two or more fractions is the smallest whole number that is exactly divisible by each denominator of the fractions.

Example 3 Find the LCD. $\frac{2}{3}$ and $\frac{1}{4}$

Solution The numbers are small enough to find the LCD by inspection. The LCD is 12, since 12 is exactly divisible by 4 and by 3. There is no smaller number that is exactly divisible by 4 and 3. □

 **Student Practice 3** Find the LCD. $\frac{1}{8}$ and $\frac{5}{7}$

In some cases, the LCD cannot easily be determined by inspection. If we write each denominator as the product of prime factors, we will be able to find the LCD. We will use $(\cdot)$ to indicate multiplication. For example, $30 = 2 \cdot 3 \cdot 5$. This means $30 = 2 \times 3 \times 5$.

PROCEDURE TO FIND THE LCD USING PRIME FACTORS

1. Write each denominator as the product of prime factors.
2. The LCD is a product containing each different factor.
3. If a factor occurs more than once in any one denominator, the LCD will contain that factor repeated the greatest number of times that it occurs in any one denominator.

Example 4 Find the LCD of $\frac{5}{6}$ and $\frac{1}{15}$ using the prime factor method.

Solution

$6 = 2 \cdot 3$ — Write each denominator as the product of prime factors.

$15 = \quad 3 \cdot 5$

$\downarrow \ \downarrow \ \downarrow$

$\text{LCD} = 2 \cdot 3 \cdot 5$ — The LCD is a product containing each different prime factor.

$\text{LCD} = 2 \cdot 3 \cdot 5 = 30$ — The different factors are 2, 3, and 5, and each factor appears at most once in any one denominator. □

Student Practice 4 Find the LCD of $\frac{8}{35}$ and $\frac{6}{15}$ using the prime factor method.

Great care should be used to determine the LCD in the case of repeated factors.

Example 5 Find the LCD of $\frac{4}{27}$ and $\frac{5}{18}$.

Solution

$27 = 3 \cdot 3 \cdot 3$ — Write each denominator as the product of prime factors. We observe that the factor 3 occurs three times in the factorization of 27.

$18 = \quad 3 \cdot 3 \cdot 2$

$\downarrow \ \downarrow \ \downarrow \ \downarrow$

$\text{LCD} = 3 \cdot 3 \cdot 3 \cdot 2$

$\text{LCD} = 3 \cdot 3 \cdot 3 \cdot 2 = 54$

The LCD is a product containing each different factor. The factor 3 *occurred most* in the factorization of 27, where it occurred *three* times. Thus the LCD will be the product of *three* 3s and *one* 2. □

Student Practice 5 Find the LCD of $\frac{5}{12}$ and $\frac{7}{30}$.

Example 6 Find the LCD of $\frac{5}{12}, \frac{1}{15}$, and $\frac{7}{30}$.

Solution

$$
\begin{aligned}
12 &= 2\cdot 2\cdot 3 \\
15 &= \quad\;\; 3\cdot 5 \\
30 &= \quad 2\cdot 3\cdot 5 \\
&\;\;\downarrow\;\;\downarrow\;\;\downarrow\;\;\downarrow \\
\text{LCD} &= 2\cdot 2\cdot 3\cdot 5 \\
\text{LCD} &= 2\cdot 2\cdot 3\cdot 5 = 60
\end{aligned}
$$

Write each denominator as the product of prime factors. Notice that the only repeated factor is 2, which occurs twice in the factorization of 12.

The LCD is the product of each different factor, with the factor 2 appearing twice since it occurred twice in one denominator. □

 **Student Practice 6** Find the LCD of $\frac{2}{27}, \frac{1}{18}$, and $\frac{5}{12}$.

3 Adding or Subtracting Fractions with Different Denominators

Before you can add or subtract them, fractions must have the same denominator. Using the LCD will make your work easier. First you must find the LCD. Then change each fraction to a fraction that has the LCD as the denominator. Sometimes one of the fractions will already have the LCD as the denominator. Once all the fractions have the same denominator, you can add or subtract. Be sure to simplify the fraction in your answer if this is possible.

TO ADD OR SUBTRACT FRACTIONS THAT DO NOT HAVE A COMMON DENOMINATOR

1. Find the LCD of the fractions.
2. Change each fraction to an equivalent fraction with the LCD for a denominator.
3. Add or subtract the fractions.
4. Simplify the answer whenever possible.

Let us return to the two fractions of Example 3. We have previously found that the LCD is 12.

Example 7 Bob picked $\frac{2}{3}$ of a bushel of apples on Monday and $\frac{1}{4}$ of a bushel of apples on Tuesday. How much did he pick in total?

Solution To solve this problem we need to add $\frac{2}{3}$ and $\frac{1}{4}$, but before we can do so, we must change $\frac{2}{3}$ and $\frac{1}{4}$ to fractions with the same denominator. We change each fraction to an equivalent fraction with a common denominator of 12, the LCD.

$$\frac{2}{3} = \frac{?}{12} \qquad \frac{2\times 4}{3\times 4} = \frac{8}{12} \quad \text{so} \quad \frac{2}{3} = \frac{8}{12}$$

$$\frac{1}{4} = \frac{?}{12} \qquad \frac{1\times 3}{4\times 3} = \frac{3}{12} \quad \text{so} \quad \frac{1}{4} = \frac{3}{12}$$

Then we rewrite the problem with common denominators and add.

$$\frac{2}{3} + \frac{1}{4} = \frac{8}{12} + \frac{3}{12} = \frac{8+3}{12} = \frac{11}{12}$$

In total Bob picked $\frac{11}{12}$ of a bushel of apples. □

Student Practice 7 Carol planted corn in $\frac{5}{7}$ of the farm fields at the Old Robinson Farm. Connie planted soybeans in $\frac{1}{8}$ of the farm fields. What fractional part of the farm fields of the Old Robinson Farm was planted in corn or soybeans?

Sometimes one of the denominators is the LCD. In such cases the fraction that has the LCD for the denominator will not need to be changed. If every other denominator divides into the largest denominator, the largest denominator is the LCD.

Example 8 Find the LCD and then add. $\frac{3}{5} + \frac{7}{20} + \frac{1}{2}$

Solution We can see by inspection that both 5 and 2 divide exactly into 20. Thus 20 is the LCD. Now add.

$$\frac{3}{5} + \frac{7}{20} + \frac{1}{2}$$

We change $\frac{3}{5}$ and $\frac{1}{2}$ to equivalent fractions with a common denominator of 20, the LCD.

$$\frac{3}{5} = \frac{?}{20} \qquad \frac{3 \times 4}{5 \times 4} = \frac{12}{20} \quad \text{so} \quad \frac{3}{5} = \frac{12}{20}$$

$$\frac{1}{2} = \frac{?}{20} \qquad \frac{1 \times 10}{2 \times 10} = \frac{10}{20} \quad \text{so} \quad \frac{1}{2} = \frac{10}{20}$$

Then we rewrite the problem with common denominators and add.

$$\frac{3}{5} + \frac{7}{20} + \frac{1}{2} = \frac{12}{20} + \frac{7}{20} + \frac{10}{20} = \frac{12 + 7 + 10}{20} = \frac{29}{20} \text{ or } 1\frac{9}{20}$$ □

Student Practice 8 Find the LCD and then add.

$$\frac{4}{5} + \frac{6}{25} + \frac{1}{50}$$

Now we turn to examples where the selection of the LCD is not so obvious. In Examples 9 through 11 we will use the prime factorization method to find the LCD.

Example 9 Add. $\frac{7}{18} + \frac{5}{12}$

Solution First we find the LCD.

$$18 = 3 \cdot 3 \cdot 2$$

$$12 = \phantom{3 \cdot{}} 3 \cdot 2 \cdot 2$$

$$\downarrow \; \downarrow \; \downarrow \; \downarrow$$

$$\text{LCD} = 3 \cdot 3 \cdot 2 \cdot 2 = 36$$

Now we change $\frac{7}{18}$ and $\frac{5}{12}$ to equivalent fractions that have the LCD.

$$\frac{7}{18} = \frac{?}{36} \qquad \frac{7 \times 2}{18 \times 2} = \frac{14}{36}$$

$$\frac{5}{12} = \frac{?}{36} \qquad \frac{5 \times 3}{12 \times 3} = \frac{15}{36}$$

Now we add the fractions.

$$\frac{7}{18} + \frac{5}{12} = \frac{14}{36} + \frac{15}{36} = \frac{29}{36}$$ This fraction cannot be simplified. □

Student Practice 9 Add.

$$\frac{1}{49} + \frac{3}{14}$$

Example 10 Subtract. $\frac{25}{48} - \frac{5}{36}$

Solution First we find the LCD.

$$48 = 2 \cdot 2 \cdot 2 \cdot 2 \cdot 3$$
$$36 = \quad\quad 2 \cdot 2 \cdot 3 \cdot 3$$
$$\text{LCD} = 2 \cdot 2 \cdot 2 \cdot 2 \cdot 3 \cdot 3 = 144$$

Now we change $\frac{25}{48}$ and $\frac{5}{36}$ to equivalent fractions that have the LCD.

$$\frac{25}{48} = \frac{?}{144} \qquad \frac{25 \times 3}{48 \times 3} = \frac{75}{144}$$
$$\frac{5}{36} = \frac{?}{144} \qquad \frac{5 \times 4}{36 \times 4} = \frac{20}{144}$$

Now we subtract the fractions.

$$\frac{25}{48} - \frac{5}{36} = \frac{75}{144} - \frac{20}{144} = \frac{55}{144}$$ This fraction cannot be simplified. □

Student Practice 10 Subtract.

$$\frac{1}{12} - \frac{1}{30}$$

Example 11 Combine. $\frac{1}{5} + \frac{1}{6} - \frac{3}{10}$

Solution First we find the LCD.

$$5 = 5$$
$$6 = \quad 2 \cdot 3$$
$$10 = 5 \cdot 2$$
$$\text{LCD} = 5 \cdot 2 \cdot 3 = 30$$

Now we change $\frac{1}{5}, \frac{1}{6},$ and $\frac{3}{10}$ to equivalent fractions that have the LCD for a denominator.

$$\frac{1}{5} = \frac{?}{30} \qquad \frac{1 \times 6}{5 \times 6} = \frac{6}{30}$$
$$\frac{1}{6} = \frac{?}{30} \qquad \frac{1 \times 5}{6 \times 5} = \frac{5}{30}$$
$$\frac{3}{10} = \frac{?}{30} \qquad \frac{3 \times 3}{10 \times 3} = \frac{9}{30}$$

Now we combine the three fractions.

$$\frac{1}{5} + \frac{1}{6} - \frac{3}{10} = \frac{6}{30} + \frac{5}{30} - \frac{9}{30} = \frac{2}{30} = \frac{1}{15}$$

Note the important step of simplifying the fraction to obtain the final answer. □

Student Practice 11 Combine.

$$\frac{2}{3} + \frac{3}{4} - \frac{3}{8}$$

4 Adding or Subtracting Mixed Numbers

If your addition or subtraction problem has mixed numbers, change them to improper fractions first and then combine (add or subtract). As a convention in this book, if the original problem contains mixed numbers, express the result as a mixed number rather than as an improper fraction.

Example 12 Combine. Simplify your answer whenever possible.

(a) $5\frac{1}{2} + 2\frac{1}{3}$ **(b)** $2\frac{1}{5} - 1\frac{3}{4}$ **(c)** $1\frac{5}{12} + \frac{7}{30}$

Solution

(a) First we change the mixed numbers to improper fractions.

$$5\frac{1}{2} = \frac{5 \times 2 + 1}{2} = \frac{11}{2} \qquad 2\frac{1}{3} = \frac{2 \times 3 + 1}{3} = \frac{7}{3}$$

Next we change each fraction to an equivalent form with the common denominator of 6.

$$\frac{11}{2} = \frac{?}{6} \qquad \frac{11 \times 3}{2 \times 3} = \frac{33}{6}$$

$$\frac{7}{3} = \frac{?}{6} \qquad \frac{7 \times 2}{3 \times 2} = \frac{14}{6}$$

Finally, we add the two fractions and change our answer to a mixed number.

$$\frac{33}{6} + \frac{14}{6} = \frac{47}{6} = 7\frac{5}{6}$$

Thus $5\frac{1}{2} + 2\frac{1}{3} = 7\frac{5}{6}$.

(b) First we change the mixed numbers to improper fractions.

$$2\frac{1}{5} = \frac{2 \times 5 + 1}{5} = \frac{11}{5} \qquad 1\frac{3}{4} = \frac{1 \times 4 + 3}{4} = \frac{7}{4}$$

Next we change each fraction to an equivalent form with the common denominator of 20.

$$\frac{11}{5} = \frac{?}{20} \qquad \frac{11 \times 4}{5 \times 4} = \frac{44}{20}$$

$$\frac{7}{4} = \frac{?}{20} \qquad \frac{7 \times 5}{4 \times 5} = \frac{35}{20}$$

Now we subtract the two fractions.

$$\frac{44}{20} - \frac{35}{20} = \frac{9}{20}$$

Thus $2\frac{1}{5} - 1\frac{3}{4} = \frac{9}{20}$.

Note: It is not necessary to use these exact steps to add and subtract mixed numbers. If you know another method and can use it to obtain the correct answers, it is all right to continue to use that method throughout this chapter.

(c) Now we add $1\frac{5}{12} + \frac{7}{30}$.

The LCD of 12 and 30 is 60. Why? Change the mixed number to an improper fraction. Then change each fraction to an equivalent form with a common denominator.

$$1\frac{5}{12} = \frac{17 \times 5}{12 \times 5} = \frac{85}{60} \qquad \frac{7 \times 2}{30 \times 2} = \frac{14}{60}$$

Continued on next page

Then add the fractions, simplify, and write the answer as a mixed number.

$$\frac{85}{60} + \frac{14}{60} = \frac{99}{60} = \frac{33}{20} = 1\frac{13}{20}$$

Thus $1\frac{5}{12} + \frac{7}{30} = 1\frac{13}{20}$. □

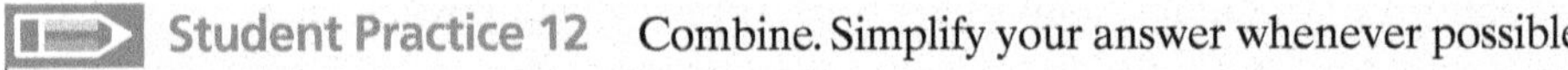

Student Practice 12 Combine. Simplify your answer whenever possible.

(a) $1\frac{2}{3} + 2\frac{4}{5}$ **(b)** $5\frac{1}{4} - 2\frac{2}{3}$

▲ **Example 13** Manuel is enclosing a triangle-shaped exercise yard for his new dog. He wants to determine how many feet of fencing he will need. The sides of the yard measure $20\frac{3}{4}$ feet, $15\frac{1}{2}$ feet, and $18\frac{1}{8}$ feet. What is the perimeter of (total distance around) the triangle?

Solution ***Understand the problem.*** Begin by drawing a picture.

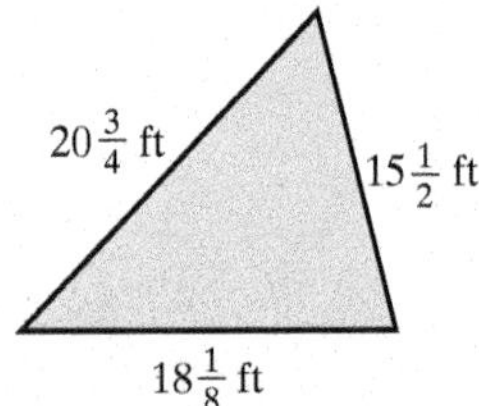

We want to add up the lengths of all three sides of the triangle. This distance around the triangle is called the **perimeter.**

$$20\frac{3}{4} + 15\frac{1}{2} + 18\frac{1}{8} = \frac{83}{4} + \frac{31}{2} + \frac{145}{8}$$

$$= \frac{166}{8} + \frac{124}{8} + \frac{145}{8} = \frac{435}{8} = 54\frac{3}{8}$$

He will need $54\frac{3}{8}$ feet of fencing. □

▲ **Student Practice 13** Find the perimeter of a rectangle with sides of $4\frac{1}{5}$ cm and $6\frac{1}{2}$ cm. Begin by drawing a picture. Label the picture by including the measure of *each* side.

STEPS TO SUCCESS Faithful Class Attendance Is Well Worth It.

If you attend a traditional mathematics class that meets one or more times each week:

Get started in the right direction. Make a personal commitment to attend class every day, beginning with the first day of class. Teachers and students all over the country have discovered that faithful class attendance and good grades go together.

The vital content of class. What goes on in class is designed to help you learn more quickly. Each day significant information is given that will truly help you to understand concepts. There is no substitute for this firsthand learning experience.

Meet a friend. You will soon discover that other students are also coming to class every single class period. It is easy to strike up a friendship with students like you who have this common commitment. They will usually be available to answer a question after class and give you an additional source of help when you encounter difficulty.

Making it personal: Write down what you think is the most compelling reason to come to every class meeting. Make that commitment and see how much it helps you. ▼

0.2 Exercises MyMathLab®

Verbal and Writing Skills, Exercises 1 and 2

1. Explain why the denominator 8 is the least common denominator of $\frac{3}{4}$ and $\frac{5}{8}$.

2. What must you do before you add or subtract fractions that do not have a common denominator?

Find the LCD (least common denominator) of each set of fractions. Do not combine the fractions; only find the LCD.

3. $\frac{4}{9}$ and $\frac{5}{12}$

4. $\frac{21}{30}$ and $\frac{17}{20}$

5. $\frac{7}{10}$ and $\frac{1}{4}$

6. $\frac{5}{18}$ and $\frac{1}{24}$

7. $\frac{5}{18}$ and $\frac{7}{54}$

8. $\frac{5}{16}$ and $\frac{7}{48}$

9. $\frac{1}{15}$ and $\frac{4}{21}$

10. $\frac{11}{12}$ and $\frac{7}{20}$

11. $\frac{17}{40}$ and $\frac{13}{60}$

12. $\frac{7}{30}$ and $\frac{8}{45}$

13. $\frac{2}{5}, \frac{3}{8}$, and $\frac{5}{12}$

14. $\frac{1}{7}, \frac{3}{14}$, and $\frac{9}{35}$

15. $\frac{5}{6}, \frac{9}{14}$, and $\frac{17}{26}$

16. $\frac{3}{8}, \frac{5}{12}$, and $\frac{11}{42}$

17. $\frac{1}{2}, \frac{1}{18}$, and $\frac{13}{30}$

18. $\frac{5}{8}, \frac{3}{14}$, and $\frac{11}{16}$

Combine. Be sure to simplify your answer whenever possible.

19. $\frac{3}{8} + \frac{2}{8}$

20. $\frac{3}{11} + \frac{5}{11}$

21. $\frac{5}{14} - \frac{1}{14}$

22. $\frac{11}{15} - \frac{2}{15}$

23. $\frac{5}{12} + \frac{5}{8}$

24. $\frac{3}{20} + \frac{13}{15}$

25. $\frac{5}{7} - \frac{2}{9}$

26. $\frac{4}{5} - \frac{3}{7}$

27. $\frac{1}{3} + \frac{2}{5}$

28. $\frac{3}{8} + \frac{1}{3}$

29. $\frac{5}{9} + \frac{5}{12}$

30. $\frac{2}{15} + \frac{7}{10}$

31. $\frac{11}{15} - \frac{31}{45}$

32. $\frac{21}{12} - \frac{23}{24}$

33. $\frac{16}{24} - \frac{1}{6}$

34. $\frac{13}{15} - \frac{1}{5}$

35. $\frac{3}{8} + \frac{4}{7}$

36. $\frac{7}{4} + \frac{5}{9}$

37. $\frac{2}{3} + \frac{7}{12} + \frac{1}{4}$

38. $\frac{4}{7} + \frac{7}{9} + \frac{1}{3}$

39. $\frac{5}{30} + \frac{3}{40} + \frac{1}{8}$

40. $\frac{1}{12} + \frac{3}{14} + \frac{4}{21}$

41. $\frac{1}{3} + \frac{1}{12} - \frac{1}{6}$

42. $\frac{1}{5} + \frac{2}{3} - \frac{11}{15}$

43. $\frac{5}{36} + \frac{7}{9} - \frac{5}{12}$

44. $\frac{5}{24} + \frac{3}{8} - \frac{1}{3}$

45. $4\frac{1}{3} + 3\frac{2}{5}$

46. $3\frac{1}{8} + 2\frac{1}{6}$

47. $1\frac{5}{24} + \frac{5}{18}$

48. $6\frac{2}{3} + \frac{3}{4}$

49. $7\frac{1}{6} - 2\frac{1}{4}$

50. $7\frac{2}{5} - 3\frac{3}{4}$

51. $8\frac{5}{7} - 2\frac{1}{4}$

52. $7\frac{8}{15} - 2\frac{3}{5}$

53. $2\frac{1}{8} + 3\frac{2}{3}$

54. $3\frac{1}{7} + 4\frac{1}{3}$

55. $11\frac{1}{7} - 6\frac{5}{7}$

56. $12\frac{1}{3} - 5\frac{2}{3}$

57. $3\frac{5}{12} + 5\frac{7}{12}$

58. $9\frac{12}{13} + 9\frac{1}{13}$

Mixed Practice

59. $\frac{7}{8} + \frac{1}{12}$

60. $\frac{19}{30} + \frac{3}{10}$

61. $3\frac{3}{16} + 4\frac{3}{8}$

62. $5\frac{2}{3} + 7\frac{2}{5}$

63. $\frac{16}{21} - \frac{2}{7}$

64. $\frac{15}{24} - \frac{3}{8}$

65. $5\frac{1}{5} - 2\frac{1}{2}$

66. $6\frac{1}{3} - 4\frac{1}{4}$

67. $25\frac{2}{3} - 6\frac{1}{7}$

68. $45\frac{3}{8} - 26\frac{1}{10}$

69. $1\frac{1}{6} + \frac{3}{8}$

70. $1\frac{2}{3} + \frac{5}{18}$

71. $8\frac{1}{4} + 3\frac{5}{6}$

72. $7\frac{3}{4} + 6\frac{2}{5}$

73. $36 - 2\frac{4}{7}$

74. $28 - 3\frac{5}{8}$

Applications

75. ***Inline Skating*** Nancy and Sarah meet three mornings a week to skate. They skated $8\frac{1}{4}$ miles on Monday, $10\frac{2}{3}$ miles on Wednesday, and $5\frac{3}{4}$ miles on Friday. What was their total distance for those three days?

76. ***Marathon Training*** Paco and Eskinder are training for the Boston Marathon. Their coach gave them the following schedule: a medium run of $10\frac{1}{2}$ miles on Thursday, a short run of $5\frac{1}{4}$ miles on Friday, a rest day on Saturday, and a long run of $18\frac{2}{3}$ miles on Sunday. How many miles did they run over these four days?

77. ***Restaurant Management*** The manager of a Boston restaurant must have his staff replace unsafe and rusted knives and replace tables and chairs in the dining area on Monday when the restaurant is closed. He has scheduled the staff for $15\frac{1}{2}$ hours of work. He estimates it will take $3\frac{2}{3}$ hours to replace the unsafe and rusted knives. He estimates it will take $9\frac{1}{4}$ hours to replace the tables and chairs in the dining area. In the time remaining, he wants them to wash the front windows. How much time will be available for washing the front windows?

78. ***Aquariums*** Carl bought a 20-gallon aquarium. He put $17\frac{3}{4}$ gallons of water into the aquarium, but it looked too low, so he added $1\frac{1}{4}$ more gallons of water. He then put in the artificial plants and the gravel but now the water was too high, so he siphoned off $2\frac{2}{3}$ gallons of water. How many gallons of water are now in the aquarium?

To Think About

Carpentry *Carpenters use fractions in their work. The picture below is a diagram of a spice cabinet. The symbol " means inches. Use the picture to answer exercises 79 and 80.*

79. Before you can determine where the cabinet will fit, you need to calculate the height, A, and the width, B. Don't forget to include the $\frac{1}{2}$-inch thickness of the wood where needed.

80. Look at the close-up of the drawer. The width is $4\frac{9}{16}''$. In the diagram, the width of the opening for the drawer is $4\frac{5}{8}''$. What is the difference?

Why do you think the drawer is smaller than the opening?

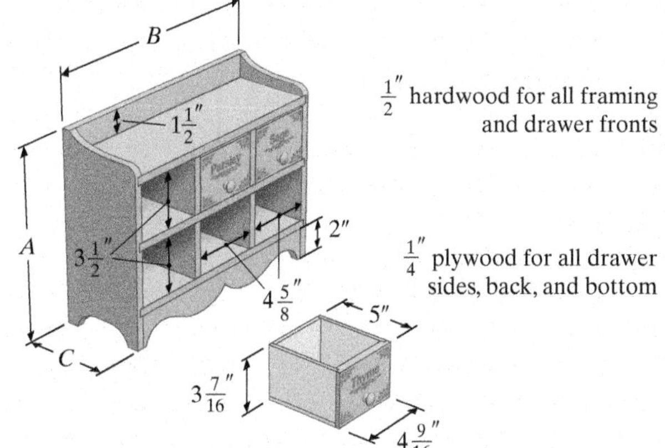

81. ***Facilities Management*** The Falmouth Country Club maintains the putting greens with a grass height of $\frac{7}{8}$ inch. The grass on the fairways is maintained at a height of $2\frac{1}{2}$ inches. How much must the mower blade be lowered by a person mowing the fairways if that person will be using the same mowing machine on the putting greens?

82. ***Facilities Management*** The director of facilities maintenance at the club in Exercise 81 discovered that due to slippage in the adjustment lever, the lawn mower actually cuts the grass $\frac{1}{16}$ of an inch too long or too short on some days. What is the maximum height that the fairway grass could be after being mowed with this machine? What is the minimum height that the putting greens could be after being mowed with this machine?

Cumulative Review

83. **[0.1.2]** Simplify. $\frac{36}{44}$

84. **[0.1.3]** Change to an improper fraction. $26\frac{3}{5}$

Quick Quiz 0.2 *Perform the operations indicated. Simplify your answers whenever possible.*

1. $\frac{3}{4} + \frac{1}{2} + \frac{5}{12}$

2. $2\frac{3}{5} + 4\frac{14}{15}$

3. $6\frac{1}{9} - 3\frac{5}{6}$

4. **Concept Check** Explain how you would find the LCD of the fractions $\frac{4}{21}$ and $\frac{5}{18}$.

0.3 Multiplying and Dividing Fractions

Student Learning Objectives

After studying this section, you will be able to:

1 Multiply fractions, whole numbers, and mixed numbers.

2 Divide fractions, whole numbers, and mixed numbers.

1 Multiplying Fractions, Whole Numbers, and Mixed Numbers

Multiplying Fractions During a recent snowstorm, the runway at Beverly Airport was plowed. However, the plow cleared only $\frac{3}{5}$ of the width and $\frac{2}{7}$ of the length. What fraction of the total runway area was cleared? To answer this question, we need to multiply $\frac{3}{5} \times \frac{2}{7}$.

The answer is that $\frac{6}{35}$ of the total runway area was cleared.

The multiplication rule for fractions states that to multiply two fractions, we multiply the two numerators and multiply the two denominators.

> **TO MULTIPLY ANY TWO FRACTIONS**
>
> 1. Multiply the numerators.
> 2. Multiply the denominators.

Example 1 Multiply.

(a) $\frac{3}{5} \times \frac{2}{7}$ **(b)** $\frac{1}{3} \times \frac{5}{4}$ **(c)** $\frac{7}{3} \times \frac{1}{5}$ **(d)** $\frac{6}{5} \times \frac{2}{3}$

Solution

(a) $\frac{3}{5} \times \frac{2}{7} = \frac{3 \cdot 2}{5 \cdot 7} = \frac{6}{35}$

(b) $\frac{1}{3} \times \frac{5}{4} = \frac{1 \cdot 5}{3 \cdot 4} = \frac{5}{12}$

(c) $\frac{7}{3} \times \frac{1}{5} = \frac{7 \cdot 1}{3 \cdot 5} = \frac{7}{15}$

(d) $\frac{6}{5} \times \frac{2}{3} = \frac{6 \cdot 2}{5 \cdot 3} = \frac{12}{15} = \frac{4}{5}$

Note that we must simplify this fraction. □

Student Practice 1 Multiply.

(a) $\frac{2}{7} \times \frac{5}{11}$ **(b)** $\frac{1}{5} \times \frac{7}{10}$ **(c)** $\frac{9}{5} \times \frac{1}{4}$ **(d)** $\frac{8}{9} \times \frac{3}{10}$

It is possible to avoid having to simplify a fraction as the last step. In many cases we can divide by a value that appears as a factor in both a numerator and a denominator. Often it is helpful to write the numbers as products of prime factors in order to do this.

Example 2 Multiply.

(a) $\frac{3}{5} \times \frac{5}{7}$ **(b)** $\frac{4}{11} \times \frac{5}{2}$ **(c)** $\frac{15}{8} \times \frac{10}{27}$

Solution

(a) $\frac{3}{5} \times \frac{5}{7} = \frac{3 \cdot 5}{5 \cdot 7} = \frac{3 \cdot \overset{1}{\cancel{5}}}{7 \cdot \underset{1}{\cancel{5}}} = \frac{3}{7}$ Note that here we divided numerator and denominator by 5.

If we factor each number, we can see the common factors.

(b) $\frac{4}{11} \times \frac{5}{2} = \frac{2 \cdot \overset{1}{\cancel{2}}}{11} \times \frac{5}{\underset{1}{\cancel{2}}} = \frac{10}{11}$ **(c)** $\frac{15}{8} \times \frac{10}{27} = \frac{\overset{1}{\cancel{3}} \cdot 5}{2 \cdot 2 \cdot \underset{1}{\cancel{2}}} \times \frac{5 \cdot \overset{1}{\cancel{2}}}{\underset{1}{\cancel{3}} \cdot 3 \cdot 3} = \frac{25}{36}$

After dividing out common factors, the resulting multiplication problem involves smaller numbers and the answers are in simplified form. □

 **Student Practice 2** Multiply.

(a) $\frac{3}{5} \times \frac{4}{3}$ **(b)** $\frac{9}{10} \times \frac{5}{12}$

Sidelight: Dividing Out Common Factors

Why does this method of dividing out a value that appears as a factor in both numerator and denominator work? Let's reexamine one of the examples we solved previously.

$$\frac{3}{5} \times \frac{5}{7} = \frac{3 \cdot 5}{5 \cdot 7} = \frac{3 \cdot \overset{1}{\cancel{5}}}{7 \cdot \underset{1}{\cancel{5}}} = \frac{3}{7}$$

Consider the following steps and reasons.

$\frac{3}{5} \times \frac{5}{7} = \frac{3 \cdot 5}{5 \cdot 7}$ Definition of multiplication of fractions.

$= \frac{3 \cdot 5}{7 \cdot 5}$ Change the order of the factors in the denominator, since $5 \cdot 7 = 7 \cdot 5$. This is called the commutative property of multiplication.

$= \frac{3}{7} \cdot \frac{5}{5}$ Definition of multiplication of fractions.

$= \frac{3}{7} \cdot 1$ Write 1 in place of $\frac{5}{5}$, since 1 is another name for $\frac{5}{5}$.

$= \frac{3}{7}$ $\frac{3}{7} \cdot 1 = \frac{3}{7}$, since any number can be multiplied by 1 without changing the value of the number.

Think about this concept. It is an important one that we will use again when we discuss rational expressions.

Multiplying a Fraction by a Whole Number Whole numbers can be named using fractional notation. 3, $\frac{9}{3}$, $\frac{6}{2}$, and $\frac{3}{1}$ are ways of expressing the number *three*. Therefore,

$$3 = \frac{9}{3} = \frac{6}{2} = \frac{3}{1}.$$

When we multiply a fraction by a whole number, we merely express the whole number as a fraction whose denominator is 1 and follow the multiplication rule for fractions.

Example 3 Multiply.

(a) $7 \times \frac{3}{5}$ **(b)** $\frac{3}{16} \times 4$

Solution

(a) $7 \times \frac{3}{5} = \frac{7}{1} \times \frac{3}{5} = \frac{21}{5}$ or $4\frac{1}{5}$ **(b)** $\frac{3}{16} \times 4 = \frac{3}{16} \times \frac{4}{1} = \frac{3}{4 \cdot \cancel{4}} \times \frac{\cancel{4}}{1} = \frac{3}{4}$

Notice that in **(b)** we did not use *prime* factors to factor 16. We recognized that $16 = 4 \cdot 4$. This is a more convenient factorization of 16 for this problem. Choose the factorization that works best for each problem. If you cannot decide what is best, factor into primes. □

 **Student Practice 3** Multiply.

(a) $4 \times \frac{2}{7}$ **(b)** $12 \times \frac{3}{4}$

Multiplying Mixed Numbers When multiplying mixed numbers, we first change them to improper fractions and then follow the multiplication rule for fractions.

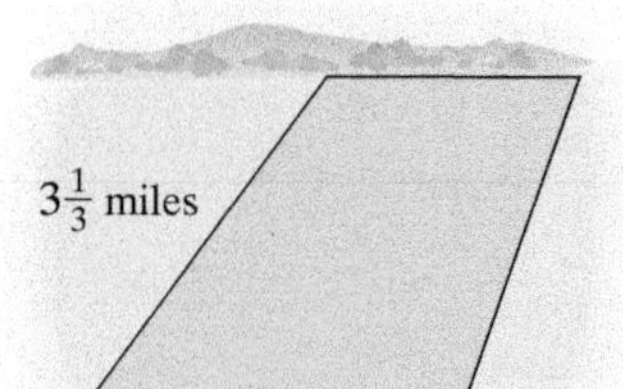

▲ **Example 4** How do we find the area of a rectangular field $3\frac{1}{3}$ miles long and $2\frac{1}{2}$ miles wide?

Solution To find the area, we multiply length times width.

$$3\frac{1}{3} \times 2\frac{1}{2} = \frac{10}{3} \times \frac{5}{2} = \frac{\cancel{2} \cdot 5}{3} \times \frac{5}{\cancel{2}} = \frac{25}{3} = 8\frac{1}{3}$$

The area is $8\frac{1}{3}$ square miles. □

▲ **Student Practice 4** Delbert Robinson has a farm with a rectangular field that measures $5\frac{3}{5}$ miles long and $3\frac{3}{4}$ miles wide. What is the area of that field?

Example 5 Multiply. $2\frac{2}{3} \times \frac{1}{4} \times 6$

Solution

$$2\frac{2}{3} \times \frac{1}{4} \times 6 = \frac{8}{3} \times \frac{1}{4} \times \frac{6}{1} = \frac{\cancel{4} \cdot 2}{\cancel{3}} \times \frac{1}{\cancel{4}} \times \frac{2 \cdot \cancel{3}}{1} = \frac{4}{1} = 4$$ □

Student Practice 5 Multiply.

$$3\frac{1}{2} \times \frac{1}{14} \times 4$$

2 Dividing Fractions, Whole Numbers, and Mixed Numbers

Dividing Fractions To divide two fractions, we invert the second fraction (that is, the divisor) and then multiply the two fractions.

TO DIVIDE TWO FRACTIONS

1. Invert the second fraction (that is, the divisor).
2. Now multiply the two fractions.

Example 6 Divide.

(a) $\frac{1}{3} \div \frac{1}{2}$ **(b)** $\frac{2}{5} \div \frac{3}{10}$ **(c)** $\frac{2}{3} \div \frac{7}{5}$

Solution

(a) $\frac{1}{3} \div \frac{1}{2} = \frac{1}{3} \times \frac{2}{1} = \frac{2}{3}$ Note that we always invert the *second* fraction.

(b) $\frac{2}{5} \div \frac{3}{10} = \frac{2}{5} \times \frac{10}{3} = \frac{2}{\cancel{5}} \times \frac{\cancel{5} \cdot 2}{3} = \frac{4}{3}$ or $1\frac{1}{3}$ **(c)** $\frac{2}{3} \div \frac{7}{5} = \frac{2}{3} \times \frac{5}{7} = \frac{10}{21}$ □

Student Practice 6 Divide.

(a) $\frac{2}{5} \div \frac{1}{3}$ **(b)** $\frac{12}{13} \div \frac{4}{3}$

Dividing a Fraction and a Whole Number The process of inverting the second fraction and then multiplying the two fractions should be done very carefully when one of the original values is a whole number. Remember, a whole number such as 2 is equivalent to $\frac{2}{1}$.

Example 7 Divide.

(a) $\frac{1}{3} \div 2$ **(b)** $5 \div \frac{1}{3}$

Solution

(a) $\frac{1}{3} \div 2 = \frac{1}{3} \div \frac{2}{1} = \frac{1}{3} \times \frac{1}{2} = \frac{1}{6}$

(b) $5 \div \frac{1}{3} = \frac{5}{1} \div \frac{1}{3} = \frac{5}{1} \times \frac{3}{1} = \frac{15}{1} = 15$ □

Student Practice 7 Divide.

(a) $\frac{3}{7} \div 6$ **(b)** $8 \div \frac{2}{3}$

Sidelight: Number Sense

Look at the answers to the problems in Example 7. In part (a), you will notice that $\frac{1}{6}$ is less than the original number $\frac{1}{3}$. Does this seem reasonable? Let's see. If $\frac{1}{3}$ is divided by 2, it means that $\frac{1}{3}$ will be divided into two equal parts. We would expect that each part would be less than $\frac{1}{3}$. $\frac{1}{6}$ is a reasonable answer to this division problem.

In part **(b)**, 15 is greater than the original number 5. Does this seem reasonable? Think of what $5 \div \frac{1}{3}$ means. It means that 5 will be divided into thirds. Let's think of an easier problem. What happens when we divide 1 into thirds? We get *three* thirds. We would expect, therefore, that when we divide 5 into thirds, we would get 5×3 or 15 thirds. 15 is a reasonable answer to this division problem.

Complex Fractions Sometimes division is written in the form of a **complex fraction** with one fraction in the numerator and one fraction in the denominator. It is best to write this in standard division notation first; then complete the problem using the rule for division.

Example 8 Divide.

(a) $\dfrac{\frac{3}{7}}{\frac{3}{5}}$ **(b)** $\dfrac{\frac{2}{9}}{\frac{5}{7}}$

Solution

(a) $\dfrac{\frac{3}{7}}{\frac{3}{5}} = \frac{3}{7} \div \frac{3}{5} = \frac{\cancel{3}}{7} \times \frac{5}{\cancel{3}} = \frac{5}{7}$

(b) $\dfrac{\frac{2}{9}}{\frac{5}{7}} = \frac{2}{9} \div \frac{5}{7} = \frac{2}{9} \times \frac{7}{5} = \frac{14}{45}$ □

Student Practice 8 Divide.

(a) $\dfrac{\frac{3}{11}}{\frac{5}{7}}$ **(b)** $\dfrac{\frac{12}{5}}{\frac{8}{15}}$

Sidelight: Invert and Multiply

Why does the method of "invert and multiply" work? The division rule really depends on the property that any number can be multiplied by 1 without changing the value of the number. Let's look carefully at an example of division of fractions:

$$\frac{2}{5} \div \frac{3}{7} = \frac{\frac{2}{5}}{\frac{3}{7}}$$ We can write the problem using a complex fraction.

$$= \frac{\frac{2}{5}}{\frac{3}{7}} \times 1$$ We can multiply by 1, since any number can be multiplied by 1 without changing the value of the number.

$$= \frac{\frac{2}{5}}{\frac{3}{7}} \times \frac{\frac{7}{3}}{\frac{7}{3}}$$ We write 1 in the form $\frac{\frac{7}{3}}{\frac{7}{3}}$, since any nonzero number divided by itself equals 1. We choose this value as a multiplier because it will help simplify the denominator.

$$= \frac{\frac{2}{5} \times \frac{7}{3}}{\frac{3}{7} \times \frac{7}{3}}$$ Definition of multiplication of fractions.

$$= \frac{\frac{2}{5} \times \frac{7}{3}}{1} = \frac{2}{5} \times \frac{7}{3}$$ The product in the denominator equals 1.

Thus we have shown that $\frac{2}{5} \div \frac{3}{7}$ is equivalent to $\frac{2}{5} \times \frac{7}{3}$ and have shown justification for the "invert and multiply rule."

Dividing Mixed Numbers This method for division of fractions can be used with mixed numbers. However, we first must change the mixed numbers to improper fractions and then use the rule for dividing fractions.

Example 9 Divide.

(a) $2\frac{1}{3} \div 3\frac{2}{3}$ **(b)** $\frac{2}{3\frac{1}{2}}$

Solution

(a) $2\frac{1}{3} \div 3\frac{2}{3} = \frac{7}{3} \div \frac{11}{3} = \frac{7}{\cancel{3}} \times \frac{\cancel{3}}{11} = \frac{7}{11}$

(b) $\frac{2}{3\frac{1}{2}} = 2 \div 3\frac{1}{2} = \frac{2}{1} \div \frac{7}{2} = \frac{2}{1} \times \frac{2}{7} = \frac{4}{7}$ ☐

Student Practice 9 Divide.

(a) $1\frac{2}{5} \div 2\frac{1}{3}$ **(b)** $4\frac{2}{3} \div 7$

(c) $\frac{1\frac{1}{5}}{1\frac{2}{7}}$

Example 10 A chemist has 96 fluid ounces of a solution. She pours the solution into test tubes. Each test tube holds $\frac{3}{4}$ fluid ounce. How many test tubes can she fill?

Solution We need to divide the total number of ounces, 96, by the number of ounces in each test tube, $\frac{3}{4}$.

$$96 \div \frac{3}{4} = \frac{96}{1} \div \frac{3}{4} = \frac{96}{1} \times \frac{4}{3} = \frac{\cancel{3} \cdot 32}{1} \times \frac{4}{\cancel{3}} = \frac{128}{1} = 128$$

She will be able to fill 128 test tubes.

Check: Pause for a moment to think about the answer. Does 128 test tubes filled with solution seem like a reasonable answer? Did you perform the correct operation? □

Student Practice 10 A chemist has 64 fluid ounces of a solution. He wishes to fill several jars, each holding $5\frac{1}{3}$ fluid ounces. How many jars can he fill?

Sometimes when solving word problems involving fractions or mixed numbers, it is helpful to solve the problem using simpler numbers first. Once you understand what operation is involved, you can go back and solve using the original numbers in the word problem.

Example 11 A car traveled 301 miles on $10\frac{3}{4}$ gallons of gas. How many miles per gallon did it get?

Solution Use simpler numbers: 300 miles on 10 gallons of gas. We want to find out how many miles the car traveled on 1 gallon of gas. You may want to draw a picture.

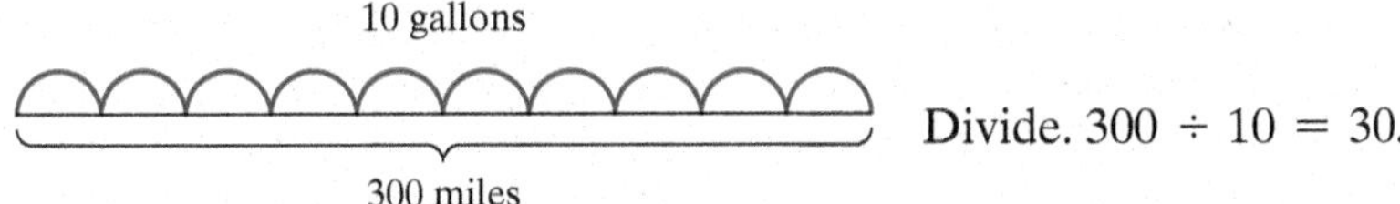

Divide. $300 \div 10 = 30$.

Now use the original numbers given in the problem.

$$301 \div 10\frac{3}{4} = \frac{301}{1} \div \frac{43}{4} = \frac{301}{1} \times \frac{4}{43} = \frac{1204}{43} = 28$$

The car got 28 miles per gallon. □

Student Practice 11 A car traveled 126 miles on $5\frac{1}{4}$ gallons of gas. How many miles per gallon did it get?

STEPS TO SUCCESS Doing Homework for Each Class Is Critical.

Many students in the class ask the question, "Is homework really that important? Do I actually have to do it?"

You learn by doing. It really makes a difference. Mathematics involves mastering a set of skills that you learn by practicing, not by watching someone else do it. Your instructor may make solving a mathematics problem look very easy, but for you to learn the necessary skills, you must practice them over and over.

The key to success is practice. Learning mathematics is like learning to play a musical instrument, to type, or to play a sport. No matter how much you watch someone else do mathematical calculations, no matter how many books you read on "how to" do it, and no matter how easy it appears to be, the key to success in mathematics is practice on each homework set.

Do each kind of problem. Some exercises in a homework set are more difficult than others. Some stress different concepts. Usually you need to work at least all the odd-numbered problems in the exercise set. This allows you to cover the full range of skills in the problem set. Remember, the more exercises you do, the better you will become in your mathematical skills.

Making it personal: Write down your personal reason for why you think doing the homework in each section is very important for success. Which of the three points given do you find is the most convincing? ▼

0.3 Exercises MyMathLab®

Verbal and Writing Skills, Exercises 1 and 2

1. Explain in your own words how to multiply two mixed numbers.

2. Explain in your own words how to divide two proper fractions.

Multiply. Simplify your answer whenever possible.

3. $\frac{28}{5} \times \frac{6}{35}$

4. $\frac{5}{7} \times \frac{28}{15}$

5. $\frac{17}{18} \times \frac{3}{5}$

6. $\frac{17}{26} \times \frac{13}{34}$

7. $\frac{4}{5} \times \frac{3}{10}$

8. $\frac{3}{11} \times \frac{5}{7}$

9. $\frac{24}{25} \times \frac{5}{2}$

10. $\frac{15}{24} \times \frac{8}{9}$

11. $\frac{7}{12} \times \frac{8}{28}$

12. $\frac{6}{21} \times \frac{9}{18}$

13. $\frac{6}{35} \times 5$

14. $\frac{2}{21} \times 15$

15. $9 \times \frac{2}{5}$

16. $\frac{8}{11} \times 3$

Divide. Simplify your answer whenever possible.

17. $\frac{8}{5} \div \frac{8}{3}$

18. $\frac{13}{9} \div \frac{13}{7}$

19. $\frac{3}{7} \div 3$

20. $\frac{7}{8} \div 4$

21. $10 \div \frac{5}{7}$

22. $18 \div \frac{2}{9}$

23. $\frac{6}{14} \div \frac{3}{8}$

24. $\frac{8}{12} \div \frac{5}{6}$

25. $\frac{7}{24} \div \frac{9}{8}$

26. $\frac{9}{28} \div \frac{4}{7}$

27. $\dfrac{\frac{7}{8}}{\frac{3}{4}}$

28. $\dfrac{\frac{5}{6}}{\frac{10}{13}}$

29. $\dfrac{\frac{5}{6}}{\frac{7}{9}}$

30. $\dfrac{\frac{3}{4}}{\frac{11}{12}}$

31. $1\frac{3}{7} \div 6\frac{1}{4}$

32. $4\frac{1}{2} \div 3\frac{3}{8}$

33. $3\frac{1}{3} \div 2\frac{1}{2}$

34. $5\frac{1}{2} \div 3\frac{3}{4}$

35. $6\frac{1}{2} \div \frac{3}{4}$

36. $\frac{1}{4} \div 1\frac{7}{8}$

37. $\dfrac{15}{2\frac{2}{5}}$

38. $\dfrac{18}{4\frac{1}{2}}$

39. $\dfrac{\frac{2}{3}}{1\frac{1}{4}}$

40. $\dfrac{\frac{5}{6}}{2\frac{1}{2}}$

Mixed Practice *Perform the proper calculations. Simplify your answer whenever possible.*

41. $\frac{4}{7} \times \frac{21}{2}$

42. $\frac{12}{18} \times \frac{9}{2}$

43. $\frac{5}{14} \div \frac{2}{7}$

44. $\frac{5}{6} \div \frac{11}{18}$

45. $10\frac{3}{7} \times 5\frac{1}{4}$

46. $10\frac{2}{9} \div 2\frac{1}{3}$

47. $25 \div \frac{5}{8}$

48. $15 \div 1\frac{2}{3}$

49. $6 \times 4\frac{2}{3}$

50. $6\frac{1}{2} \times 12$

51. $2\frac{1}{2} \times \frac{1}{10} \times \frac{3}{4}$

52. $2\frac{1}{3} \times \frac{2}{3} \times \frac{3}{5}$

53. **(a)** $\frac{1}{15} \times \frac{25}{21}$
(b) $\frac{1}{15} \div \frac{25}{21}$

54. **(a)** $\frac{1}{6} \times \frac{24}{15}$
(b) $\frac{1}{6} \div \frac{24}{15}$

55. **(a)** $\frac{2}{3} \div \frac{12}{21}$
(b) $\frac{2}{3} \times \frac{12}{21}$

56. **(a)** $\frac{3}{7} \div \frac{21}{25}$
(b) $\frac{3}{7} \times \frac{21}{25}$

Applications

57. ***Shirt Manufacturing*** A denim shirt at the Gap requires $2\frac{3}{4}$ yards of material. How many shirts can be made from $71\frac{1}{2}$ yards of material?

58. ***Pullover Manufacturing*** A fleece pullover requires $1\frac{5}{8}$ yards of material. How many fleece pullovers can be made from $29\frac{1}{4}$ yards of material?

▲ **59.** ***Window Construction*** Jesse needs to find the area of a large window he has been hired to build so that he can order enough glass. The window measures $11\frac{1}{3}$ feet long and 12 feet wide. What is the area of this window?

▲ **60.** ***Gardening*** Sara must find the area of her flower garden so that she can determine how much fertilizer to purchase. What is the area of her rectangular garden, which measures 15 feet long and $10\frac{1}{5}$ feet wide?

Cumulative Review *In exercises 61 and 62, find the missing numerator.*

61. **[0.1.4]** $\frac{11}{15} = \frac{?}{75}$

62. **[0.1.4]** $\frac{7}{9} = \frac{?}{63}$

Quick Quiz 0.3 *Perform the operations indicated. Simplify answers whenever possible.*

1. $\frac{7}{15} \times \frac{25}{14}$

2. $3\frac{1}{4} \times 4\frac{1}{2}$

3. $3\frac{3}{10} \div 2\frac{1}{2}$

4. **Concept Check** Explain the steps you would take to perform the calculation $3\frac{1}{4} \div 2\frac{1}{2}$.

0.4 Using Decimals

Student Learning Objectives

After studying this section, you will be able to:

1. Understand the meaning of decimals.
2. Change a fraction to a decimal.
3. Change a decimal to a fraction.
4. Add and subtract decimals.
5. Multiply decimals.
6. Divide decimals.
7. Multiply and divide a decimal by a multiple of 10.

1 Understanding the Meaning of Decimals 

We can express a part of a whole as a fraction or as a decimal. A **decimal** is another way of writing a fraction whose denominator is 10, 100, 1000, and so on.

$$\frac{3}{10} = 0.3 \qquad \frac{5}{100} = 0.05 \qquad \frac{172}{1000} = 0.172 \qquad \frac{58}{10{,}000} = 0.0058$$

The period in decimal notation is known as the **decimal point.** The number of digits in a number to the right of the decimal point is known as the number of **decimal places** of the number. The place value of decimals is shown in the following chart.

Hundred-thousands	Ten-thousands	Thousands	Hundreds	Tens	Ones	← Decimal point	Tenths	Hundredths	Thousandths	Ten-thousandths	Hundred-thousandths
100,000	10,000	1000	100	10	1	•	$\frac{1}{10}$	$\frac{1}{100}$	$\frac{1}{1000}$	$\frac{1}{10{,}000}$	$\frac{1}{100{,}000}$

Example 1 Write each of the following decimals as a fraction or mixed number. State the number of decimal places. Write out in words the way the number would be spoken.

(a) 0.6 **(b)** 0.29 **(c)** 0.527 **(d)** 1.38 **(e)** 0.00007

Solution

Decimal Form	*Fraction Form*	*Number of Decimal Places*	*The Words Used to Describe the Number*
(a) 0.6	$\frac{6}{10}$	one	six tenths
(b) 0.29	$\frac{29}{100}$	two	twenty-nine hundredths
(c) 0.527	$\frac{527}{1000}$	three	five hundred twenty-seven thousandths
(d) 1.38	$1\frac{38}{100}$	two	one and thirty-eight hundredths
(e) 0.00007	$\frac{7}{100{,}000}$	five	seven hundred-thousandths

Student Practice 1 State the number of decimal places. Write each decimal as a fraction or mixed number and in words.

(a) 0.9 **(b)** 0.09 **(c)** 0.731 **(d)** 1.371 **(e)** 0.0005

You have seen that a given fraction can be written in several different but equivalent ways. There are also several different equivalent ways of writing the decimal form of a fraction. The decimal 0.18 can be written in the following equivalent ways:

$$\text{Fractional form: } \frac{18}{100} = \frac{180}{1000} = \frac{1800}{10{,}000} = \frac{18{,}000}{100{,}000}$$

$$\text{Decimal form: } 0.18 = 0.180 = 0.1800 = 0.18000$$

Thus we see that *any number of terminal zeros may be added to the right-hand side of a decimal* without changing its value.

$$0.13 = 0.1300 \qquad 0.162 = 0.162000$$

Similarly, *any number of terminal zeros may be removed from the right-hand side of a decimal* without changing its value.

2 Changing a Fraction to a Decimal

A fraction can be changed to a decimal by dividing the denominator into the numerator.

Example 2 Write each of the following fractions as a decimal.

(a) $\frac{3}{4}$ **(b)** $\frac{21}{20}$ **(c)** $\frac{1}{8}$ **(d)** $\frac{3}{200}$

Solution

(a) $\frac{3}{4} = 0.75$ since

```
   0.75
4)3.00
  2 8
    20
    20
     0
```

(b) $\frac{21}{20} = 1.05$ since

```
     1.05
20)21.00
   20
    1 00
    1 00
       0
```

(c) $\frac{1}{8} = 0.125$ since

```
   0.125
8)1.000
    8
    20
    16
     40
     40
      0
```

(d) $\frac{3}{200} = 0.015$ since

```
     0.015
200)3.000
     2 00
     1 000
     1 000
         0
```

□

Student Practice 2 Write each of the following fractions as a decimal.

(a) $\frac{3}{8}$ **(b)** $\frac{7}{200}$ **(c)** $\frac{33}{20}$

Sometimes division yields an infinite repeating decimal. We use three dots to indicate that the pattern continues forever. For example,

$$\frac{1}{3} = 0.3333\ldots \qquad \begin{array}{r} 0.333 \\ 3\overline{)1.000} \\ \underline{9} \\ 10 \\ \underline{9} \\ 10 \\ \underline{9} \\ 1 \end{array}$$

An alternative notation is to place a bar over the repeating digit(s):

$$0.3333\ldots = 0.\overline{3} \qquad 0.575757\ldots = 0.\overline{57}$$

Example 3 Write each fraction as a decimal.

(a) $\frac{2}{11}$ **(b)** $\frac{5}{6}$

Solution

(a) $\frac{2}{11} = 0.181818\ldots$ or $0.\overline{18}$

$$\begin{array}{r} 0.1818 \\ 11\overline{)2.0000} \\ \underline{1\,1} \\ 90 \\ \underline{88} \\ 20 \\ \underline{11} \\ 90 \\ \underline{88} \\ 2 \end{array}$$

(b) $\frac{5}{6} = 0.8333\ldots$ or $0.8\overline{3}$

$$\begin{array}{r} 0.8333 \\ 6\overline{)5.0000} \\ \underline{4\,8} \\ 20 \\ \underline{18} \\ 20 \\ \underline{18} \\ 20 \\ \underline{18} \\ 2 \end{array}$$

Note that the 8 does not repeat. Only the digit 3 is repeating. □

Student Practice 3 Write each fraction as a decimal.

(a) $\frac{1}{6}$ **(b)** $\frac{5}{11}$

Calculator

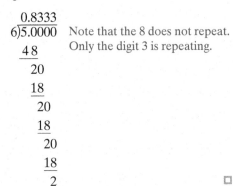

Fraction to Decimal

You can use a calculator to change $\frac{3}{5}$ to a decimal.

Enter:

3 [÷] 5 [=]

The display should read

[0.6]

Try the following.

(a) $\frac{17}{25}$ **(b)** $\frac{2}{9}$

(c) $\frac{13}{10}$ **(d)** $\frac{15}{19}$

Sometimes division must be carried out to many places in order to observe the repeating pattern. This is true in the following example:

$\frac{2}{7} = 0.285714285714285714\ldots$ This can also be written as $\frac{2}{7} = 0.\overline{285714}$.

It can be shown that the denominator determines the maximum number of decimal places that might repeat. So $\frac{2}{7}$ must repeat in the seventh decimal place or sooner.

3 Changing a Decimal to a Fraction 

To convert from a decimal to a fraction, merely write the decimal as a fraction with a denominator of 10, 100, 1000, 10,000, and so on, and simplify the result when possible.

Example 4 Write each decimal as a fraction and simplify whenever possible.

(a) 0.2 **(b)** 0.35 **(c)** 0.516 **(d)** 0.74 **(e)** 0.138 **(f)** 0.008

Solution

(a) $0.2 = \frac{2}{10} = \frac{1}{5}$ **(b)** $0.35 = \frac{35}{100} = \frac{7}{20}$

(c) $0.516 = \frac{516}{1000} = \frac{129}{250}$ **(d)** $0.74 = \frac{74}{100} = \frac{37}{50}$

(e) $0.138 = \frac{138}{1000} = \frac{69}{500}$ **(f)** $0.008 = \frac{8}{1000} = \frac{1}{125}$ □

Student Practice 4 Write each decimal as a fraction and simplify whenever possible.

(a) 0.8 **(b)** 0.88 **(c)** 0.45 **(d)** 0.148 **(e)** 0.612 **(f)** 0.016

All repeating decimals can also be converted to fractional form. In practice, however, repeating decimals are usually rounded to a few places. It will not be necessary, therefore, to learn how to convert $0.\overline{033}$ to $\frac{11}{333}$ for this course.

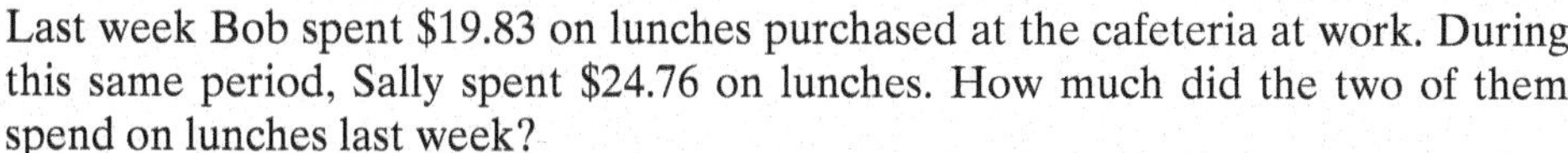

4 Adding and Subtracting Decimals

Last week Bob spent $19.83 on lunches purchased at the cafeteria at work. During this same period, Sally spent $24.76 on lunches. How much did the two of them spend on lunches last week?

Adding and subtracting decimals is similar to adding and subtracting whole numbers, except that it is necessary to line up decimal points. To perform the operation 19.83 + 24.76, we line up the numbers in column form and add the digits:

$$\begin{array}{r} 19.83 \\ +\ 24.76 \\ \hline 44.59 \end{array}$$

Thus Bob and Sally spent $44.59 on lunches last week.

ADDITION AND SUBTRACTION OF DECIMALS

1. Write in column form and line up the decimal points.
2. Add or subtract the digits.

Example 5 Add or subtract.

(a) 3.6 + 2.3 **(b)** 127.32 − 38.48 **(c)** 3.1 + 42.36 + 9.034 **(d)** 5.0006 − 3.1248

Solution

(a) $\begin{array}{r} 3.6 \\ +\ 2.3 \\ \hline 5.9 \end{array}$ **(b)** $\begin{array}{r} 127.32 \\ -\ 38.48 \\ \hline 88.84 \end{array}$ **(c)** $\begin{array}{r} 3.1 \\ 42.36 \\ +\ 9.034 \\ \hline 54.494 \end{array}$ **(d)** $\begin{array}{r} 5.0006 \\ -\ 3.1248 \\ \hline 1.8758 \end{array}$ □

Student Practice 5 Add or subtract.

(a) 3.12 + 5.08 **(b)** 152.003 − 136.118

(c) 1.1 + 3.16 + 5.123 **(d)** 1.0052 − 0.1234

Sidelight: Adding Zeros to the Right-Hand Side of the Decimal

When we added fractions, we had to have common denominators. Since decimals are really fractions, why can we add them without having common denominators? Actually, we have to have common denominators to add any fractions, whether they are in decimal form or fraction form. However, sometimes the notation does not show this. Let's examine Example 5**(c)**.

Original Problem

$$\begin{array}{r} 3.1 \\ 42.36 \\ +\ 9.034 \\ \hline 54.494 \end{array}$$

We are adding the three numbers:

$$3\tfrac{1}{10} + 42\tfrac{36}{100} + 9\tfrac{34}{1000}$$

$$3\tfrac{100}{1000} + 42\tfrac{360}{1000} + 9\tfrac{34}{1000}$$

$3.100 + 42.360 + 9.034$ This is the new problem.

Original Problem	***New Problem***
$\begin{array}{r} 3.1 \\ 42.36 \\ +\ 9.034 \\ \hline 54.494 \end{array}$	$\begin{array}{r} 3.100 \\ 42.360 \\ +\ 9.034 \\ \hline 54.494 \end{array}$

We notice that the results are the same. The only difference is the notation. We are using the property that any number of zeros may be added to the right-hand side of a decimal without changing its value.

This shows the convenience of adding and subtracting fractions in decimal form. Little work is needed to change the decimals so that they have a common denominator. All that is required is to add zeros to the right-hand side of the decimal (and we usually do not even write out that step except when subtracting).

As long as we line up the decimal points, we can add or subtract any decimal fractions.

In the following example we will find it useful to add zeros to the right-hand side of the decimal.

Example 6 Perform the following operations.

(a) $1.0003 + 0.02 + 3.4$ **(b)** $12 - 0.057$

Solution We will add zeros so that each number shows the same number of decimal places.

(a)
$$\begin{array}{r} 1.0003 \\ 0.0200 \\ +\ 3.4000 \\ \hline 4.4203 \end{array}$$

(b)
$$\begin{array}{r} 12.000 \\ -\ 0.057 \\ \hline 11.943 \end{array}$$

□

Student Practice 6 Perform the following operations.

(a) $0.061 + 5.0008 + 1.3$ **(b)** $18 - 0.126$

5 Multiplying Decimals 

MULTIPLICATION OF DECIMALS

To multiply decimals, you first multiply as with whole numbers. To determine the position of the decimal point, you count the total number of decimal places in the two numbers being multiplied. This will determine the number of decimal places that should appear in the answer.

Example 7 Multiply. 0.8×0.4

Solution

$$\begin{array}{rl} 0.8 & \text{(one decimal place)} \\ \times\, 0.4 & \text{(one decimal place)} \\ \hline 0.32 & \text{(two decimal places)} \end{array}$$

Student Practice 7 Multiply. 0.5×0.3

Note that you will often have to add zeros to the left of the digits obtained in the product so that you obtain the necessary number of decimal places.

Example 8 Multiply. 0.123×0.5

Solution

$$\begin{array}{rl} 0.123 & \text{(three decimal places)} \\ \times\quad 0.5 & \text{(one decimal place)} \\ \hline 0.0615 & \text{(four decimal places)} \end{array}$$

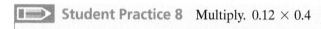

Student Practice 8 Multiply. 0.12×0.4

Here are some examples that involve more decimal places.

Example 9 Multiply.

(a) 2.56×0.003

(b) 0.0036×0.008

Solution

(a)
$$\begin{array}{rl} 2.56 & \text{(two decimal places)} \\ \times\, 0.003 & \text{(three decimal places)} \\ \hline 0.00768 & \text{(five decimal places)} \end{array}$$

(b)
$$\begin{array}{rl} 0.0036 & \text{(four decimal places)} \\ \times\ 0.008 & \text{(three decimal places)} \\ \hline 0.0000288 & \text{(seven decimal places)} \end{array}$$

Student Practice 9 Multiply.

(a) 1.23×0.005

(b) 0.003×0.00002

Sidelight: Counting the Number of Decimal Places

Why do we count the number of decimal places? The rule really comes from the properties of fractions. If we write the problem in Example 8 in fraction form, we have

$$0.123 \times 0.5 = \frac{123}{1000} \times \frac{5}{10} = \frac{615}{10{,}000} = 0.0615.$$

6 Dividing Decimals 

When discussing division of decimals, we frequently refer to the three primary parts of a division problem. Be sure you know the meaning of each term.

The **divisor** is the number you divide into another.
The **dividend** is the number to be divided.
The **quotient** is the result of dividing one number by another.

In the problem 6 ÷ 2 = 3 we represent each of these terms as follows:

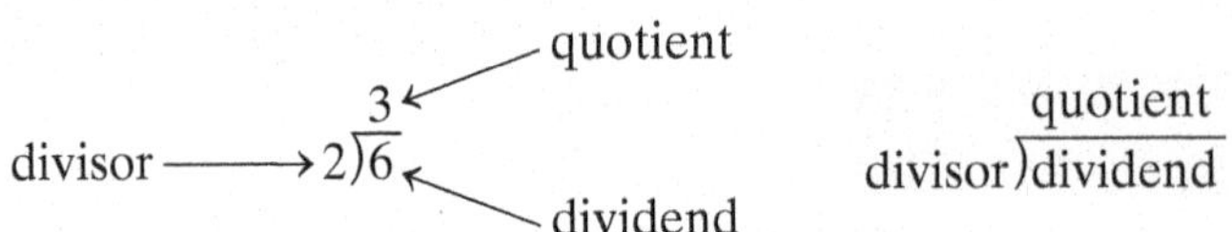

When dividing two decimals, count *the number of decimal places* in the divisor. Then *move the decimal point to the right* that *same number of places* in both *the divisor* and *the dividend.* Mark that position with a caret (‸). Finally, perform the division. Be sure to line up the decimal point in the quotient with the position indicated by the caret.

Example 10 Four friends went out for lunch. The total bill, including tax, was \$32.68. How much did each person pay if they shared the cost equally?

Solution To answer this question, we must calculate 32.68 ÷ 4.

```
    8.17
 4)32.68
   32
    06
     4
     28
     28
      0
```

Since there are no decimal places in the divisor, we do not need to move the decimal point. We must be careful, however, to place the decimal point in the quotient directly above the decimal point in the dividend.

Thus 32.68 ÷ 4 = 8.17, and each friend paid \$8.17. □

Student Practice 10 Sally Keyser purchased 6 boxes of paper for a laser printer. The cost was \$31.56. There was no tax since she purchased the paper for a charitable organization. How much did she pay for each box of paper?

Note that sometimes we will need to place extra zeros in the dividend in order to move the decimal point the required number of places.

Example 11 Divide. 16.2 ÷ 0.027

Solution

```
0.027‸)16.200‸
```

There are **three** decimal places in the divisor, so we move the decimal point **three places** to **the right** in the **divisor** and **dividend** and mark the new position by a caret. Note that we must add two zeros to 16.2 in order to do this.

three decimal places

```
           600.
0.027‸)16.200‸
        16 2
          000
```

Now perform the division as with whole numbers. The decimal point in the answer is directly above the caret.

Thus 16.2 ÷ 0.027 = 600. □

Student Practice 11 Divide. 1800. ÷ 0.06

Special care must be taken to line up the digits in the quotient. Note that sometimes we will need to place zeros in the quotient after the decimal point.

Example 12 Divide. 0.04288 ÷ 3.2

Solution

$3.2_{\wedge}\overline{)0.0_{\wedge}4288}$ There is **one** decimal place in the divisor, so we move the decimal point **one place** to **the right** in the **divisor** and **dividend** and mark the new position by a caret.

one decimal place

$$
\begin{array}{r}
0.0134 \\
3.2_{\wedge}\overline{)0.0_{\wedge}4288} \\
\underline{32} \\
108 \\
\underline{96} \\
128 \\
\underline{128} \\
0
\end{array}
$$

Now perform the division as for whole numbers. The decimal point in the answer is directly above the caret. Note the need for the initial zero after the decimal point in the answer.

Thus 0.04288 ÷ 3.2 = 0.0134. □

Student Practice 12 Divide. 0.01764 ÷ 4.9

Sidelight: Dividing Decimals by Another Method

Why does this method of dividing decimals work? Essentially, we are using the steps we used in Section 0.1 to change a fraction to an equivalent fraction by multiplying both the numerator and denominator by the same number. Let's reexamine Example 12.

$0.04288 \div 3.2 = \dfrac{0.04288}{3.2}$ Write the original problem using fraction notation.

$= \dfrac{0.04288 \times 10}{3.2 \times 10}$ Multiply the numerator and denominator by 10. Since this is the same as multiplying by 1, we are not changing the fraction.

$= \dfrac{0.4288}{32}$ Write the result of multiplication by 10.

$= 0.4288 \div 32$ Rewrite the fraction as an equivalent problem with division notation.

Notice that we have obtained a new problem that is the same as the problem in Example 12 when we moved the decimal one place to the right in the divisor and dividend. We see that the reason we can move the decimal point as many places as necessary to the right in the divisor and dividend is that this is the same as multiplying the numerator and denominator of a fraction by a power of 10 to obtain an equivalent fraction.

7 Multiplying and Dividing a Decimal by a Multiple of 10

When multiplying by 10, 100, 1000, and so on, a simple rule may be used to obtain the answer. For every zero in the multiplier, move the decimal point one place to the right.

Example 13 Multiply.

(a) 3.24×10 **(b)** 15.6×100 **(c)** 0.0026×1000

Solution

(a) $3.24 \times 10 = 32.4$ One zero—move decimal point one place to the right.

(b) $15.6 \times 100 = 1560$ Two zeros—move decimal point two places to the right.

(c) $0.0026 \times 1000 = 2.6$ Three zeros—move decimal point three places to the right. □

Student Practice 13 Multiply.

(a) 0.0016×100 **(b)** 2.34×1000 **(c)** $56.75 \times 10{,}000$

The reverse rule is true for division. When dividing by 10, 100, 1000, 10,000, and so on, move the decimal point one place to the left for every zero in the divisor.

Example 14 Divide.

(a) $52.6 \div 10$ **(b)** $0.0038 \div 100$ **(c)** $5936.2 \div 1000$

Solution

(a) $\frac{52.6}{10} = 5.26$ Move decimal point one place to the left.

(b) $\frac{0.0038}{100} = 0.000038$ Move decimal point two places to the left.

(c) $\frac{5936.2}{1000} = 5.9362$ Move decimal point three places to the left. □

Student Practice 14 Divide.

(a) $\frac{5.82}{10}$ **(b)** $123.4 \div 1000$ **(c)** $\frac{0.00614}{10{,}000}$

STEPS TO SUCCESS Do You Realize How Valuable Friendship Is?

In a math class a friend is a person of fantastic value. Robert Louis Stevenson once wrote "A friend is a gift you give yourself." This is especially true when you take a mathematics class and make a friend in the class. You will find that you enjoy sitting together and drawing support and encouragement from each other. You may want to exchange phone numbers or e-mail addresses. You may want to study together or review together before a test.

How do you get started? Try talking to the students seated around you. Ask someone for help about something you did not understand in class. Take the time to listen to them and their interests and concerns. You may discover you have a lot in common. If the first few people you talk to seem uninterested, try sitting in a different part of the room and talk to those students who are seated around you in the new location. Don't force a friendship on anyone but just look for chances to open up a good channel of communication.

Making it personal: What do you think is the best way to make a friend of someone in your class? Which of the given suggestions do you find the most helpful? Will you take the time to reach out to someone in your class this week and try to begin a new friendship? ▼

0.4 Exercises MyMathLab®

Verbal and Writing Skills, Exercises 1–4

1. A decimal is another way of writing a fraction whose denominator is ______________________.

2. We write 0.42 in words as ______________________.

3. When dividing 7432.9 by 1000 we move the decimal point _______ places to the _________.

4. When dividing 96.3 by 10,000 we move the decimal point _______ places to the _________.

Write each fraction as a decimal.

5. $\frac{7}{8}$ **6.** $\frac{18}{25}$ **7.** $\frac{3}{15}$ **8.** $\frac{9}{15}$ **9.** $\frac{7}{11}$ **10.** $\frac{1}{6}$

Write each decimal as a fraction in simplified form.

11. 0.8 **12.** 0.5 **13.** 0.25 **14.** 0.35 **15.** 0.625 **16.** 0.775

17. 0.06 **18.** 0.08 **19.** 3.4 **20.** 4.8 **21.** 5.5 **22.** 6.25

Add or subtract.

23. $1.71 + 0.38$ **24.** $4.64 + 0.23$ **25.** $2.5 + 3.42 + 4.9$ **26.** $6.31 + 4.2 + 8.5$

27. $46.03 + 215.1 + 0.078$ **28.** $33.01 + 0.38 + 175.401$ **29.** $147.18 - 15.39$ **30.** $121.52 - 79.85$

31. $6.0054 - 2.0257$ **32.** $5.0032 - 3.0036$ **33.** $125.43 - 2.8$ **34.** $212.54 - 3.6$

Multiply or divide.

35. 7.21×4.2 **36.** 6.12×3.4 **37.** 0.04×0.08 **38.** 6.32×1.31

39. 4.23×0.025 **40.** 3.84×0.0017 **41.** $58{,}200 \times 0.0015$ **42.** $23{,}000 \times 0.0042$

43. $3.616 \div 64$ **44.** $12.6672 \div 39$ **45.** $7.9728 \div 3.02$ **46.** $6.519 \div 2.05$

47. $0.5230 \div 0.002$ **48.** $0.031 \div 0.005$ **49.** $0.03048 \div 0.06$ **50.** $0.00855 \div 0.09$

Multiply or divide by moving the decimal point.

51. 3.45×1000 **52.** 1.36×1000 **53.** $0.76 \div 100$ **54.** $175{,}318 \div 1000$

55. $7.36 \times 10{,}000$ **56.** $0.00243 \times 100{,}000$ **57.** $73{,}892 \div 100{,}000$ **58.** $3.52 \div 1000$

59. 0.1498×100 **60.** $85.54 \times 10{,}000$ **61.** $1.931 \div 100$ **62.** $96.12 \div 10{,}000$

Mixed Practice *Perform the calculations indicated.*

63. 54.8×0.15

64. 8.252×0.005

65. $13.75 + 2.55 + 0.078$

66. $1.109 + 0.088 + 16.4$

67. $0.05724 \div 0.027$

68. $77.136 \div 0.003$

69. 0.7683×1000

70. $25.62 \times 10{,}000$

71. $56.37 - 4.29$

72. $14.3 - 0.68$

73. $153.7 \div 100$

74. $0.58 \div 1000$

Applications

75. ***Measurement*** While mixing solutions in her chemistry lab, Mia needed to change the measured data from pints to liters. There is 0.4732 liter in one pint and the original measurement was 5.5 pints. What is the measured data in liters?

76. ***Mileage of Hybrid Cars*** In order to minimize fuel costs, Chris Smith purchased a used Honda Civic Hybrid that averages 44 miles per gallon in the city. The gas tank holds 13.2 gallons of gas. How many miles can Chris drive the car in the city on a full tank of gas?

77. ***Wages*** Harry has a part-time job at Stop and Shop. He earns \$9 an hour. He requested enough hours of work each week so that he could earn at least \$185 a week. How many hours will he have to work to achieve his goal? By how much will he exceed his earning goal of \$185 per week?

78. ***Drinking Water*** The EPA standard for safe drinking water is a maximum of 1.3 milligrams of copper per liter of water. A water testing firm found 6.8 milligrams of copper in a 5-liter sample drawn from Jim and Sharon LeBlanc's house. Is the water safe or not? By how much does the amount of copper exceed or fall short of the maximum allowed?

Cumulative Review *Perform each operation. Simplify all answers.*

79. **[0.3.2]** $3\frac{1}{2} \div 5\frac{1}{4}$

80. **[0.3.1]** $\frac{3}{8} \cdot \frac{12}{27}$

81. **[0.2.3]** $\frac{12}{25} + \frac{9}{20}$

82. **[0.2.4]** $1\frac{3}{5} - \frac{1}{2}$

Quick Quiz 0.4 *Perform the calculations indicated.*

1. $8.0567 - 2.3489$

2. 58.7×0.06

3. $4.608 \div 0.16$

4. **Concept Check** Explain how you would place the decimal points when performing the calculation $0.252 \div 0.0035$.

Use Math to Save Money

Did You Know? Managing your debts can save you money over time.

Live Debt Free

Understanding the Problem:

Tracy and Max are in debt.

- Each of their three credit cards is maxed out to the limit of $8000.
- They owe $12,000 in hospital bills and $2000 for their car loan.
- After borrowing from friends, they still owe $100 to one friend and $300 to another.

Before they can save for a much-needed vacation, they must pay off their debts.

Making a Plan:

Tracy and Max come up with a plan for tackling their debt.

Step 1: They list all of their debts, ordered from smallest to largest.
Task 1: Complete Step 1 for Tracy and Max. Remember that there are three credit cards.

Step 2: They make minimum monthly payments on each debt.

- Each credit card has a minimum monthly payment of $25.
- The hospital expects a payment of $50 per month.
- The monthly car payment is $200.
- Tracy and Max arrange to pay $20 per month for each loan from friends.

Task 2: What is the total amount of their minimum monthly payments?

Step 3: They pay off their three smallest debts first.

Task 3: What are their three smallest debts?

Finding a Solution:

Step 4: Tracy and Max decide to eliminate any unnecessary spending until the three smallest debts are paid off. By doing this, they can pay off the two smallest debts in only two months while still making the minimum payments on the other debts.
Task 4: What is the total amount of the minimum monthly payments for the two smallest debts?

Step 5: Each month, Tracy and Max take the amount that they would have used to pay the two smallest debts and apply it toward the third smallest, all while paying the minimum monthly payments on the remaining debts.

Task 5: How many more months will it take Tracy and Max to pay off the third smallest debt if they follow Step 5? Round your answer to the nearest whole number when you perform division operations.

Step 6: After the three smallest debts are paid off, Tracy and Max take the money that they would have spent per month to pay those debts and use it on the principal of the remaining debts. To pay the debts more quickly, Tracy and Max decide to stop using credit cards for new purchases. After a few years of careful budgeting, not using credit cards, and paying more than the minimum payment on their credit card debt, they finally pay off their debts.
Task 6: Besides avoiding credit cards for new purchases, can you think of other ways that Tracy and Max could have budgeted their money and paid off the debt more quickly?

Applying the Situation to Your Life:

Debt counselors often provide this simple, practical plan for people in debt:

- Arrange debts in order.
- Pay off the smallest debt first.
- Let the consequences of paying off the smallest debt help you to pay off the rest of the debts more quickly.
- Avoid unnecessary spending and incorporate budgeting strategies into your daily life.

0.5 Percents, Rounding, and Estimating

Student Learning Objectives

After studying this section, you will be able to:

1 Change a decimal to a percent.

2 Change a percent to a decimal.

3 Find the percent of a given number.

4 Find the missing percent when given two numbers.

5 Use rounding to estimate.

1 Changing a Decimal to a Percent

A **percent** is a fraction that has a denominator of 100. When you say "sixty-seven percent" or write 67%, you are just expressing the fraction $\frac{67}{100}$ in another way. The word *percent* is a shortened form of the Latin words *per centum,* which means "by the hundred." In everyday use, percent means per one hundred.

Russell Camp owns 100 acres of land in Montana. 49 of the acres are covered with trees. The rest of the land is open fields. We say that 49% of his land is covered with trees.

It is important to see that 49% means 49 parts out of 100 parts. It can also be written as a fraction, $\frac{49}{100}$, or as a decimal, 0.49. Understanding the meaning of the notation allows you to change from one form to another. For example,

$$49\% = 49 \text{ out of } 100 \text{ parts} = \frac{49}{100} = 0.49.$$

Similarly, you can express a fraction with denominator 100 as a percent or a decimal.

$$\frac{11}{100} \text{ means 11 parts out of 100 or } 11\%. \quad \text{So } \frac{11}{100} = 11\% = 0.11.$$

Now that we understand the concept, we can use some quick procedures to change from a decimal to a percent, and vice versa.

CHANGING A DECIMAL TO A PERCENT

1. Move the decimal point two places to the right.
2. Add the % symbol.

Sometimes percents are less than 1%. When we follow the procedure above, we see that 0.01 is 1%. Thus we would expect 0.001 to be less than 1%. 0.001 = 0.1% or one-tenth (0.1) of a percent.

Example 1 Change to a percent.

(a) 0.0364 **(b)** 0.0008 **(c)** 0.4

Solution We move the decimal point two places to the right and add the % symbol.

(a) 0.0364 = 3.64% **(b)** 0.0008 = 0.08% **(c)** 0.4 = 0.40 = 40% □

Student Practice 1 Change to a percent.

(a) 0.92 **(b)** 0.0736 **(c)** 0.7 **(d)** 0.0003

Percents can be greater than 100%. Since 1 = 1.00 = 100%, we would expect 1.5 to be greater than 100%. In fact, 1.5 = 150%.

Example 2 Change to a percent.

(a) 1.48 **(b)** 2.938 **(c)** 4.5

Solution We move the decimal point two places to the right and add the % symbol.

(a) 1.48 = 148% **(b)** 2.938 = 293.8% **(c)** 4.5 = 4.50 = 450% □

Student Practice 2 Change to a percent.

(a) 3.04 **(b)** 5.186 **(c)** 2.1

2 Changing a Percent to a Decimal 

In this procedure we move the decimal point to the left and remove the % symbol.

> **CHANGING A PERCENT TO A DECIMAL**
>
> 1. Move the decimal point two places to the left.
> 2. Remove the % symbol.

Example 3 Change to a decimal.

(a) 4% **(b)** 0.6% **(c)** 254.8%

Solution First we move the decimal point two places to the left. Then we remove the % symbol.

(a) 4% = 4.% = 0.04

↑

The unwritten decimal point is understood to be here.

(b) 0.6% = 0.006

(c) 254.8% = 2.548 □

 **Student Practice 3** Change to a decimal.

(a) 7% **(b)** 9.3% **(c)** 131% **(d)** 0.04%

Calculator

Percent to Decimal

You can use a calculator to change 52% to a decimal. If your calculator has a [%] key, do the following:

Enter: 52 [%]

The display should read:

[0.52]

If your calculator does not have a [%] key, divide the number by 100.

Enter: 52 [÷] 100 [=]

Try the following:

(a) 46% **(b)** 137%

(c) 9.3% **(d)** 6%

Note: The calculator divides by 100 when the percent key is pressed.

3 Finding the Percent of a Given Number 

How do we find 60% of 20? Let us relate it to problems we did in Section 0.3 involving multiplication of fractions.

Consider the following problem.

What	is	$\frac{3}{5}$	of	20?
↓	↓	↓	↓	↓
[?]	=	$\frac{3}{5}$	×	20

$$\boxed{?} = \frac{3}{\cancel{5}_1} \times \cancel{20}^4 = 12 \quad \text{The answer is 12.}$$

Since a percent is really a fraction, a percent problem is solved similarly to the way a fraction problem is solved. Since $\frac{3}{5} = \frac{3 \cdot 20}{5 \cdot 20} = \frac{60}{100} = 60\%$, we could write the problem what is $\frac{3}{5}$ of 20? as

What	is	60%	of	20?	Replace $\frac{3}{5}$ with 60%.
↓	↓	↓	↓	↓	
[?]	=	60%	×	20	

[?] = 0.60 × 20

[?] = 12.0 The answer is 12.

Thus we have developed the following rule.

> **FINDING THE PERCENT OF A NUMBER**
>
> To find the percent of a number, change the percent to a decimal and multiply the number by the decimal.

Calculator

Finding the Percent of a Number

You can use a calculator to find 12% of 48.

Enter: 12 [%] [×] 48 [=]

The display should read

[5.76]

If your calculator does not have a [%] key, do the following:

Enter: 0.12 [×] 48 [=]

What is 54% of 450?

Example 4 Find.

(a) 10% of 36 **(b)** 2% of 350 **(c)** 182% of 12 **(d)** 0.3% of 42

Solution

(a) 10% of 36 $= 0.10 \times 36 = 3.6$ **(b)** 2% of 350 $= 0.02 \times 350 = 7$

(c) 182% of 12 $= 1.82 \times 12 = 21.84$ **(d)** 0.3% of 42 $= 0.003 \times 42 = 0.126$ □

Student Practice 4 Find.

(a) 18% of 50 **(b)** 4% of 64 **(c)** 156% of 35

There are many real-life applications for finding the percent of a number. When you go shopping in a store, you may find sale merchandise marked 35% off. This means that the sale price is 35% off the regular price. That is, 35% of the regular price is subtracted from the regular price to get the sale price.

Example 5 A store is having a sale of 35% off the retail price of all sofas. Melissa wants to buy a particular sofa that normally sells for $595.

(a) How much will Melissa save if she buys the sofa on sale?

(b) What will the purchase price be if Melissa buys the sofa on sale?

Solution

(a) To find 35% of $595 we will need to multiply 0.35×595.

$$\begin{array}{r} 595 \\ \times\ 0.35 \\ \hline 2975 \\ 1785 \\ \hline 208.25 \end{array}$$

Thus Melissa will save $208.25 if she buys the sofa on sale.

(b) The purchase price is the difference between the original price and the amount saved.

$$\begin{array}{r} 595.00 \\ -\ 208.25 \\ \hline 386.75 \end{array}$$

If Melissa buys the sofa on sale, she will pay $386.75. □

Student Practice 5 John received a 4.2% pay raise at work this year. He had previously earned $38,000 per year.

(a) What was the amount of his pay raise in dollars?

(b) What is his new salary?

4 Finding the Missing Percent When Given Two Numbers

Recall that we can write $\frac{3}{4}$ as $\frac{75}{100}$ or 75%. If we were asked the question, "What percent is 3 of 4?" we would say 75%. This gives us a procedure for finding what percent one number is of a second number.

FINDING THE MISSING PERCENT

1. Write a fraction with the two numbers. The number *after* the word *of* is always the denominator, and the other number is the numerator.
2. Simplify the fraction (if possible).
3. Change the fraction to a decimal.
4. Express the decimal as a percent.

Example 6 What percent of 24 is 15?

Solution This can be solved quickly as follows.

Step 1 $\frac{15}{24}$ Write the relationship as a fraction. The number after "of" is 24, so 24 is the denominator.

Step 2 $\frac{15}{24} = \frac{5}{8}$ Simplify the fraction (when possible).

Step 3 $= 0.625$ Change the fraction to a decimal.

Step 4 $= 62.5\%$ Change the decimal to percent. □

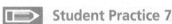

Student Practice 6 What percent of 148 is 37?

The question in Example 6 can also be written as "15 is what percent of 24?" To answer the question, we begin by writing the relationship as $\frac{15}{24}$. Remember that "of 24" means 24 will be the denominator.

MC **Example 7**

(a) What percent of 16 is 3.8? **(b)** \$150 is what percent of \$120?

Solution

(a) What percent of 16 is 3.8?

$\frac{3.8}{16}$ Write the relationship as a fraction.

You can divide to change the fraction to a decimal and then change the decimal to a percent.

$$16\overline{)3.8000} = 0.2375 \rightarrow 23.75\%$$

(b) \$150 is what percent of \$120?

$\frac{150}{120} = \frac{5}{4}$ Reduce the fraction whenever possible to make the division easier.

$$4\overline{)5.00} = 1.25 \rightarrow 125\%$$

□

Student Practice 7

(a) What percent of 48 is 24?

(b) 4 is what percent of 25?

Example 8 Marcia made 29 shots on goal during the last high school field hockey season. She actually scored a goal 8 times. What percent of her total shots were goals? Round your answer to the nearest whole percent.

Solution Marcia scored a goal 8 times out of 29 tries. We want to know what percent of 29 is 8.

Step 1 $\frac{8}{29}$ Express the relationship as a fraction. The number after the word *of* is 29, so 29 appears in the denominator.

Step 2 $\frac{8}{29} = \frac{8}{29}$ Note that this fraction cannot be reduced.

Step 3 $= 0.2758\ldots$ The decimal equivalent of the fraction has many digits.

Step 4 $= 27.58\ldots\%$ Change the decimal to a percent, rounded to the nearest whole percent.

$\approx 28\%$ (The $\approx$ symbol means approximately equal to.)

Therefore, Marcia scored a goal approximately 28% of the time she made a shot on goal. □

Calculator

Finding the Percent Given Two Numbers

You can use a calculator to find a missing percent.

What percent of 95 is 19?

1. Enter as a fraction.

19 [÷] 95

2. Change to a percent.

19 [÷] 95 [×] 100 [=]

The display should read

[20]

This means 20%.

What percent of 625 is 250?

Student Practice 8 Roberto scored a basket 430 times out of 1256 attempts during his high school basketball career. What percent of the time did he score a basket? Round your answer to the nearest whole percent.

5 Using Rounding to Estimate

Before we proceed in this section, we will take some time to be sure you understand the idea of rounding a number. You will probably recall the following simple rule from your previous mathematics courses.

ROUNDING A NUMBER

If the first digit to the right of the round-off place is

1. less than 5, we make no change to the digit in the round-off place.
2. 5 or more, we increase the digit in the round-off place by 1.

To illustrate, 4689 rounded to the nearest hundred is 4700. Rounding 233,987 to the nearest ten thousand, we obtain 230,000. We will now use our experience in rounding as we discuss the general area of estimation.

Estimation is the process of finding an approximate answer. It is not designed to provide an exact answer. Estimation will give you a rough idea of what the answer might be. For any given problem, you may choose to estimate in many different ways.

ESTIMATION BY ROUNDING

1. Round each number so that there is one nonzero digit.
2. Perform the calculation with the rounded numbers.

Example 9 Use estimation by rounding to find the product. 5368×2864

Solution

Step 1 Round 5368 to 5000.
Round 2864 to 3000.

Step 2 Multiply.

$$5000 \times 3000 = 15{,}000{,}000$$

An estimate of the product is 15,000,000. □

Student Practice 9 Use estimation by rounding to find the product.
$128{,}621 \times 378$

▲ **Example 10** The four walls of a college classroom are $22\frac{1}{4}$ feet long and $8\frac{3}{4}$ feet high. A painter needs to know the area of these four walls in square feet. Since paint is sold in gallons, an estimate will do. Use estimation by rounding to approximate the area of the four walls.

Solution

Step 1 Round $22\frac{1}{4}$ feet to 20 feet.
Round $8\frac{3}{4}$ feet to 9 feet.

Step 2 Multiply 20×9 to obtain an estimate of the area of one wall.
Multiply $20 \times 9 \times 4$ to obtain an estimate of the area of all four walls.

$$20 \times 9 \times 4 = 720$$

Our estimate for the painter is 720 square feet of wall space. □

▲ **Student Practice 10** Mr. and Mrs. Ramirez need to carpet two rooms of their house. One room measures $12\frac{1}{2}$ feet by $9\frac{3}{4}$ feet. The other room measures $11\frac{1}{4}$ feet by $18\frac{1}{2}$ feet. Use estimation by rounding to approximate the number of square feet (square footage) in these two rooms.

Example 11 Won Lin has a small compact car in Honolulu. He drove 396.8 miles in his car and used 8.4 gallons of gas.

(a) Estimate the number of miles he gets per gallon.

(b) Estimate how much he will pay for fuel to drive 2764 miles in the next month if gasoline usually costs $\$3.59\frac{9}{10}$ per gallon.

Solution

(a) Round 396.8 miles to 400 miles. Round 8.4 gallons to 8 gallons. Now divide.

$$8\overline{)400}\quad = 50$$

Won Lin's car gets about 50 miles per gallon.

(b) We will need to use the information we found in part (a) to determine how many gallons of gasoline Won Lin will use. Round 2764 miles to 3000 miles and divide 3000 miles by 50 miles per gallon.

$$50\overline{)3000}\quad = 60$$

Won Lin will use about 60 gallons of gas for traveling next month.

To estimate the cost, we need to ask ourselves, "What kind of an estimate are we looking for?" It may be sufficient to round $\$3.59\frac{9}{10}$ to \$4.00 and multiply.

$$60 \times \$4.00 = \$240.00$$

Keep in mind that this is a broad estimate. You may want an estimate that will be closer to the exact answer. In that case round $\$3.59\frac{9}{10}$ to \$3.60 and multiply.

$$60 \times \$3.60 = \$216.00$$

Student Practice 11 Roberta drove 422.8 miles in her truck in Alaska and used 19.3 gallons of gas. Assume that gasoline costs $\$3.69\frac{9}{10}$ per gallon.

(a) Estimate the number of miles she gets per gallon.

(b) Estimate how much it will cost her to drive this winter if she drives 3862 miles.

STEPS TO SUCCESS What Happens When You Read a Mathematics Textbook?

Reading a math book can give you amazing insight. Always take the time at the start of a homework section by reading the section(s) assigned in your textbook.

Remember this book was written to help you become successful in this mathematics class. However, you have to read it if you want to take advantage of all its benefits.

Always read your book with a pen or a highlighter in hand. Especially watch for helpful hints and suggestions as you look over the sample explanations. Underline any step that you think is hard or not clear to you. Put question marks by words you do not understand. You may want to write down a list of these in your notebook. Ask your instructor about items you do not understand.

When you come to a sample example, make your mind work it through step by step. Underline steps that you think are especially important.

Make sure you are thinking about what you are reading. If your mind is wandering, get up and get a drink of water or walk around the room—anything to help you get your mind back on track.

Making it personal: Look over the given suggestions. Pick one or two ideas that you think will be the most helpful to you. Use them this week as you read the book carefully before doing homework assignments. ▼

0.5 Exercises MyMathLab®

Verbal and Writing Skills, Exercises 1 and 2

1. When you write 19%, what do you really mean? Describe the meaning in your own words.

2. When you try to solve a problem like "What percent of 80 is 30?" how do you know if you should write the fraction as $\frac{80}{30}$ or as $\frac{30}{80}$?

Change to a percent.

3. 0.79 **4.** 0.54 **5.** 0.568 **6.** 0.063

7. 0.076 **8.** 0.046 **9.** 2.39 **10.** 7.49

11. 3.6 **12.** 5.7 **13.** 3.672 **14.** 8.674

Change to a decimal.

15. 3% **16.** 6% **17.** 0.4% **18.** 0.62%

19. 250% **20.** 175% **21.** 7.4% **22.** 8.7%

23. 0.52% **24.** 0.1% **25.** 100% **26.** 200%

Find the following.

27. What is 8% of 65? **28.** What is 7% of 69? **29.** What is 10% of 130?

30. What is 25% of 600? **31.** What is 112% of 65? **32.** What is 154% of 270?

33. 36 is what percent of 24? **34.** 49 is what percent of 28? **35.** What percent of 340 is 17?

36. What percent of 35 is 28? **37.** 30 is what percent of 500? **38.** 48 is what percent of 600?

39. 80 is what percent of 200? **40.** 75 is what percent of 30?

Applications

41. ***Exam Grades*** Dave took an exam with 80 questions. He answered 68 correctly. What was his grade for the exam? Write the grade as a percent.

42. ***Assembly Line Supervisor*** Ken Thompson is supervising the production line. He conducted a random inspection of 440 units as they came off the assembly line. A total of 11 of those units were defective. What percent of the units were defective?

43. ***Tipping*** Diana and Russ ate a meal costing \$32.80 when they went out to dinner. If they want to leave the standard 15% tip for their server, how much will they tip and what will their total bill be?

44. ***Startup Business Success Rate*** The Knoxville Better Business Bureau examined the records of 350 new businesses that started up last year in Knoxville, TN. It discovered that there was a failure rate of 28%. If that statistic is true, how many of those 350 new businesses in Knoxville failed?

45. ***Food Budget*** The Gonzalez family has a combined monthly income of \$1850. Their food budget is \$380 per month. What percent of their monthly income is budgeted for food? (Round your answer to the nearest whole percent.)

46. ***Survey*** A college survey found that 137 out of 180 students had a grandparent not born in the United States. What percent of the students had a grandparent not born in the United States? (Round your answer to the nearest hundredth of a percent.)

47. ***Gift Returns*** Last Christmas season, the Jones Mill Outlet Store chain calculated that they sold 36,000 gift items. If they assume that there will be a 1.5% return rate after the holidays, how many gifts can they expect to be returned?

48. ***Rent*** The total cost of a downtown Boston apartment, including utilities, is \$3690 per month. Sara's share is 17% of the total cost each month. What is her monthly cost?

49. ***Computer Hardware Sales*** Abdul sells computers for a local computer outlet. He gets paid \$450 per month plus a commission of 3.8% on all the computer hardware he sells. Last year he sold \$780,000 worth of computer hardware.
   (a) What was his sales commission on the \$780,000 worth of hardware he sold?
   (b) What was his annual salary (that is, his monthly pay and commission combined)?

50. ***Medical Supply Sales*** Bruce sells medical supplies on the road. He logged 18,600 miles in his car last year. He declared 65% of his mileage as business travel.
   (a) How many miles did he travel on business last year?
   (b) If his company reimburses him 31 cents for each mile traveled, how much should he be paid for travel expenses?

In exercises 51–62, use estimation by rounding to approximate the value. ***Do not find the exact value.***

51. 586×421

52. 729×688

53. 3547×4693

54. 8192×5984

55. $14 + 73 + 80 + 21 + 56$

56. $318 + 494 + 613 + 243$

57. $41\overline{)829{,}346}$

58. $16\overline{)5{,}846{,}213}$

59. $\dfrac{2714}{31{,}500}$

60. $\dfrac{53{,}610}{786}$

61. Find 17% of \$21,365.85.

62. Find 4.9% of \$9321.88.

Applications

In exercises 63–66, use estimation by rounding to approximate the answer. ***Do not find the exact value.***

63. ***Checkout Sales*** A typical customer at the local Piggly Wiggly supermarket spends approximately \$82 at the checkout register. The store keeps four registers open, and each handles 22 customers per hour. *Estimate* the amount of money the store receives in one hour.

64. ***Weekend Spending Money*** The Westerly Credit Union has found that on Fridays, the average customer withdraws \$85 from the ATM for weekend spending money. Each ATM averages 19 customers per hour. There are five machines. *Estimate* the amount of money withdrawn by customers in one hour.

65. ***Gas Mileage*** Rod's trip from Salt Lake City to his cabin in the mountains was 117.7 miles. If his car used 3.8 gallons of gas for the trip, *estimate* the number of miles his car gets to the gallon.

66. ***Bookstore Manager*** Betty is transferring her stock of books from one store to another. If her car can transport 430 books per trip and she has 11,900 books to move, *estimate* the number of trips it will take.

Cumulative Review

67. **[0.4.6]** ***Gas Mileage*** Dan took a trip in his Ford Taurus. At the start of the trip his car odometer read 68,459.5 miles. At the end of the trip his car odometer read 69,229.5 miles. He used 35 gallons of gas on the trip. How many miles per gallon did his car achieve?

68. **[0.4.4]** ***Rainfall*** In Hilo, Hawaii, Brad spent three months working in a restaurant for a summer job. In June he observed that 4.6 inches of rain fell. In July the rainfall was 4.5 inches and in August it was 2.9 inches. What was the average monthly rainfall that summer for those three months in Hilo?

Quick Quiz 0.5

1. What is 114% of 85?

2. 63 is what percent of 420?

3. Use estimation by rounding to find the quotient. $34{,}987\overline{)567{,}238}$

4. **Concept Check** Explain how you would change 0.0078 to a percent.

0.6 Using the Mathematics Blueprint for Problem Solving

Student Learning Objective

After studying this section, you will be able to:

1 Use the Mathematics Blueprint to solve real-life problems.

1 Using the Mathematics Blueprint to Solve Real-Life Problems

When a builder constructs a new home or office building, he or she often has a blueprint. This accurate drawing shows the basic form of the building. It also shows the dimensions of the structure to be built. This blueprint serves as a useful reference throughout the construction process.

Similarly, when solving real-life problems, it is helpful to have a "mathematics blueprint." This is a simple way to organize the information provided in the word problem in a chart or in a graph. You can record the facts you need to use. You can determine what it is you are trying to find and how you can go about actually finding it. You can record other information that you think will be helpful as you work through the problem.

As we solve real-life problems, we will use three steps.

Step 1. ***Understand the problem.*** Here we will read through the problem. We draw a picture if it will help, and use the Mathematics Blueprint as a guide to assist us in thinking through the steps needed to solve the problem.

Step 2. ***Solve and state the answer.*** We will use arithmetic or algebraic procedures along with problem-solving strategies to find a solution.

Step 3. ***Check.*** We will use a variety of techniques to see if the answer in step 2 is the solution to the word problem. This will include estimating to see if the answer is reasonable, repeating our calculation, and working backward from the answer to see if we arrive at the original conditions of the problem.

▲**Example 1** Nancy and John want to install wall-to-wall carpeting in their living room. The floor of the rectangular living room is $11\frac{2}{3}$ feet wide and $19\frac{1}{2}$ feet long. How much will it cost if the carpet is $18.00 per square yard?

Solution

1. ***Understand the problem.*** First, read the problem carefully. Drawing a sketch of the living room may help you see what is required. The carpet will cover the floor of the living room, so we need to find the area. Now we fill in the Mathematics Blueprint.

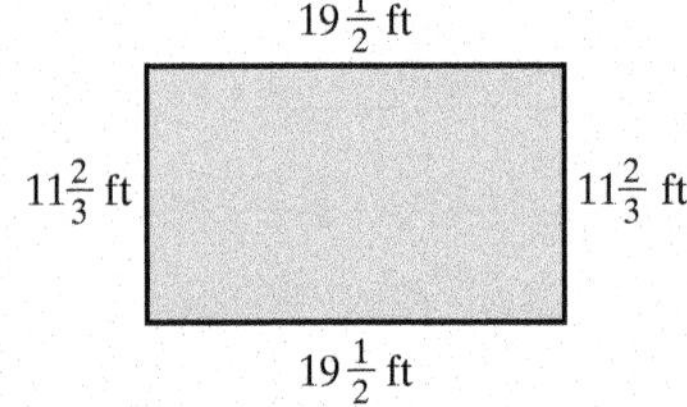

Mathematics Blueprint for Problem Solving

Gather the Facts	What Am I Solving For?	What Must I Calculate?	Key Points to Remember
The living room measures $11\frac{2}{3}$ ft by $19\frac{1}{2}$ ft. The carpet costs $18.00 per square yard.	**(a)** the area of the room in square feet **(b)** the area of the room in square yards **(c)** the cost of the carpet	**(a)** Multiply $11\frac{2}{3}$ ft by $19\frac{1}{2}$ ft to get area in square feet. **(b)** Divide the number of square feet by 9 to get the number of square yards. **(c)** Multiply the number of square yards by $18.00.	There are 9 square feet, 3 feet × 3 feet, in 1 square yard; therefore, we must divide the number of square feet by 9 to obtain square yards.

2. ***Solve and state the answer.***

(a) To find the area of a rectangle, we multiply the length times the width.

$$11\frac{2}{3} \times 19\frac{1}{2} = \frac{35}{3} \times \frac{39}{2}$$
$$= \frac{455}{2} = 227\frac{1}{2}$$

A minimum of $227\frac{1}{2}$ square feet of carpet will be needed. We say a minimum because some carpet may be wasted in cutting. Carpet is sold by the square yard. We will want to know the amount of carpet needed in square yards.

(b) To determine the area in square yards, we divide $227\frac{1}{2}$ by 9. ($9 \text{ ft}^2 = 1 \text{ yd}^2$)

$$227\frac{1}{2} \div 9 = \frac{455}{2} \div \frac{9}{1}$$
$$= \frac{455}{2} \times \frac{1}{9} = \frac{455}{18} = 25\frac{5}{18}$$

A minimum of $25\frac{5}{18}$ square yards of carpet will be needed.

(c) Since the carpet costs \$18.00 per square yard, we will multiply the number of square yards needed by \$18.00.

$$25\frac{5}{18} \times 18 = \frac{455}{18} \times \frac{18}{1} = 455$$

The carpet will cost a minimum of \$455.00 for this room.

"*Remember to estimate. It will save you time and money!*"

3. *Check.* We will estimate to see if our answers are reasonable.

(a) We will estimate by rounding each number to the nearest 10.

$$11\frac{2}{3} \times 19\frac{1}{2} \longrightarrow 10 \times 20 = 200$$

This is close to our answer of $227\frac{1}{2} \text{ ft}^2$. Our answer is reasonable. ✓

(b) We will estimate by rounding to the nearest hundred and ten, respectively.

$$227\frac{1}{2} \div 9 \longrightarrow 200 \div 10 = 20$$

This is close to our answer of $25\frac{5}{18} \text{ yd}^2$. Our answer is reasonable. ✓

(c) We will estimate by rounding each number to the nearest 10.

$$25\frac{5}{18} \times 18 \longrightarrow 30 \times 20 = 600$$

This is close to our answer of \$455. Our answer seems reasonable. □

▲ **Student Practice 1** Jeff went to help Abby pick out wall-to-wall carpet for her new house. Her rectangular living room measures $16\frac{1}{2}$ feet by $10\frac{1}{2}$ feet. How much will it cost to carpet the room if the carpet costs \$20 per square yard?

Mathematics Blueprint for Problem Solving

Gather the Facts	What Am I Solving For?	What Must I Calculate?	Key Points to Remember

To Think About: *Example 1 Follow-Up* Assume that the carpet in Example 1 comes in a standard width of 12 feet. How much carpet will be wasted if it is laid out on the living room floor in one strip that is $19\frac{1}{2}$ feet long? How much carpet will be wasted if it is laid in two sections side by side that are each $11\frac{2}{3}$ feet long? Assuming you have to pay for wasted carpet, what is the minimum cost to carpet the room?

Example 2 The following chart shows the 2015 sales of Micropower Computer Software for each of the four regions of the United States. Use the chart to answer the following questions (round all answers to the nearest whole percent):

(a) What percent of the sales personnel are assigned to the Northeast?
(b) What percent of the volume of sales is attributed to the Northeast?
(c) What percent of the sales personnel are assigned to the Southeast?
(d) What percent of the volume of sales is attributed to the Southeast?
(e) Which of these two regions of the country has sales personnel that appear to be more effective in terms of the volume of sales?

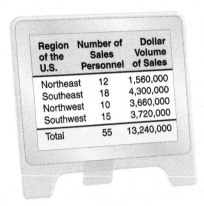

Region of the U.S.	Number of Sales Personnel	Dollar Volume of Sales
Northeast	12	1,560,000
Southeast	18	4,300,000
Northwest	10	3,660,000
Southwest	15	3,720,000
Total	55	13,240,000

Solution

1. ***Understand the problem.*** We will only need to deal with figures from the Northeast region and the Southeast region.

Mathematics Blueprint for Problem Solving

Gather the Facts	What Am I Solving For?	What Must I Calculate?	Key Points to Remember
Personnel: 12 Northeast 18 Southeast 55 total Sales Volume: \$1,560,000 NE \$4,300,000 SE \$13,240,000 Total	**(a)** the percent of the total personnel in the Northeast **(b)** the percent of the total sales in the Northeast **(c)** the percent of the total personnel in the Southeast **(d)** the percent of the total sales in the Southeast **(e)** compare the percentages from the two regions	**(a)** 12 is what percent of 55? Divide. 12 ÷ 55 **(b)** 1,560,000 is what percent of 13,240,000? 1,560,000 ÷ 13,240,000 **(c)** 18 ÷ 55 **(d)** 4,300,000 ÷ 13,240,000	We do not need to use the numbers for the Northwest or the Southwest.

2. ***Solve and state the answer.***

(a) $\frac{12}{55} = 0.21818\ldots$

$\approx 22\%$

(b) $\frac{1{,}560{,}000}{13{,}240{,}000} = \frac{156}{1324} \approx 0.1178$

$\approx 12\%$

(c) $\frac{18}{55} = 0.32727\ldots$

$\approx 33\%$

(d) $\frac{4{,}300{,}000}{13{,}240{,}000} = \frac{430}{1324} \approx 0.3248$

$\approx 32\%$

(e) We notice that the Northeast sales force, 22% of the total personnel, made only 12% of the sales. The percent of the sales compared to the percent of the sales force is about half (12% of 24% would be half) or 50%. The Southeast sales force, 33% of the total personnel, made 32% of the sales. The percent of sales compared to the percent of the sales force is close to 100%. We must be cautious here. *If there are no other significant factors,* it would appear that the Southeast sales force is more effective. (There may be other significant factors affecting sales, such as a recession in the Northeast, new and inexperienced sales personnel, or fewer competing companies in the Southeast.)

3. ***Check.*** You may want to use a calculator to check the division in step 2, or you may use estimation.

(a) $\frac{12}{55} \rightarrow \frac{10}{60} \approx 0.17$
$= 17\% \checkmark$

(b) $\frac{1{,}560{,}000}{13{,}240{,}000} \rightarrow \frac{1{,}600{,}000}{13{,}000{,}000} \approx 0.12$
$= 12\% \checkmark$

(c) $\frac{18}{55} \rightarrow \frac{20}{60} \approx 0.33$
$= 33\% \checkmark$

(d) $\frac{4{,}300{,}000}{13{,}240{,}000} \rightarrow \frac{4{,}300{,}000}{13{,}000{,}000} \approx 0.33$
$= 33\% \checkmark$ □

Student Practice 2 Using the chart for Example 2, answer the following questions. (Round all answers to the nearest whole percent.)

(a) What percent of the sales personnel are assigned to the Northwest?

(b) What percent of the sales volume is attributed to the Northwest?

(c) What percent of the sales personnel are assigned to the Southwest?

(d) What percent of the sales volume is attributed to the Southwest?

(e) Which of these two regions of the country has sales personnel that appear to be more effective in terms of volume of sales?

Mathematics Blueprint for Problem Solving

Gather the Facts	What Am I Solving For?	What Must I Calculate?	Key Points to Remember

To Think About: *Example 2 Follow-Up* Suppose in 2015 the number of sales personnel (55) increased by 60%. What would the new number of sales personnel be? Suppose in 2016 the number of sales personnel decreased by 60% from the number of sales personnel in 2015. What would the new number be? Why is this number not 55, since we have increased the number by 60% and then decreased the result by 60%? Explain.

0.6 Exercises

Applications *Use the Mathematics Blueprint for Problem Solving to help you solve each of the following exercises.*

▲ **1.** ***Mustang Parking Area*** Richard Penta recently restored a 1965 Mustang Coupe. He will park it on a new parking area in his backyard that uses concrete pavers. The parking area is rectangular and measures $7\frac{1}{3}$ yards by $4\frac{1}{3}$ yards. The concrete pavers each measure 1 square foot and cost $4.50 each. How much will the pavers cost that he will need to complete his parking area?

▲ **2.** ***Carpentry Estimates*** Samuel is a carpenter, and he recently completed an estimate to build a new deck for Russ and Norma Camp. The deck measures $22\frac{1}{2}$ feet by $12\frac{1}{2}$ feet. He estimated the new decking would cost $8.00 per square foot. What are the estimated costs of the decking for this new deck?

▲ **3.** ***Swimming Pool Estimates*** Wally and Mike install swimming pools in the summer in Manchester-by-the-Sea. They are building a swimming pool for Kevin Baird that measures $14\frac{1}{2}$ feet wide and $23\frac{1}{2}$ feet long.

(a) At the outside edge of the pool is a protective fence. How many feet of fencing would be needed to surround the pool area?

(b) Kevin discovers that he can buy a packaged box of 90 feet of fencing for $859 or he can buy cut-to-order fencing that costs $11.20 per foot. Which type of fencing should he buy? How much money does he save?

▲ **4.** ***Swimming Pool Maintenance*** Wally and Mike recommend to their customers that they add the proper amount of granular chlorinator to their swimming pool each day.

(a) The swimming pool that they built in exercise 3 has an interior measurement of $12\frac{1}{2}$ feet wide by $20\frac{1}{2}$ feet long. The pool is 5 feet deep. How much water (in cubic feet) will it take to fill the pool? Round your answer to the nearest cubic foot.

(b) If each cubic foot of water is 7.5 gallons, how many gallons of water will the swimming pool hold? Round your answer to the nearest 1000 gallons.

(c) If every 5000 gallons should have 4 ounces of granular chlorinator added to the pool, how many ounces of chlorinator should be added to this pool each day?

Exercise Training *The following directions are posted on the wall at the gym.*

Beginning exercise training schedule

On day 1, each athlete will begin the morning as follows:

Jog	$1\frac{1}{2}$ miles
Walk	$1\frac{3}{4}$ miles
Rest	$2\frac{1}{2}$ minutes
Walk	1 mile

5. Betty's athletic trainer told her to follow the beginning exercise training schedule on day 1. On day 2, she is to increase all distances and times by $\frac{1}{3}$ that of day 1. On day 3, she is to increase all distances and times by $\frac{1}{3}$ that of day 2. What will be her training schedule on day 3?

6. Melinda's athletic trainer told her to follow the beginning exercise training schedule on day 1. On day 2, she is to increase all distances and times by $\frac{1}{3}$ that of day 1. On day 3, she is to once again increase all distances and times by $\frac{1}{3}$ that of day 1. What will be her training schedule on day 3?

To Think About *Refer to exercises 5 and 6 in working exercises 7–10.*

7. Who will have a more demanding schedule on day 3, Betty or Melinda? Why?

8. If Betty kept up the same type of increase day after day, how many miles would she be jogging on day 5?

9. If Melinda kept up the same type of increase day after day, how many miles would she be jogging on day 7?

10. Which athletic trainer would appear to have the best plan for training athletes if they used this plan for 14 days? Why?

11. ***Snickers Bars*** Franklin Clarence Mars invented the Snickers Bar and introduced it in 1930. More than 15 million Snickers Bars are produced each day. Each week approximately 6,300,000 pounds of this popular bar are produced.

(a) In 1 day, how many pounds of Snickers Bars are produced?

(b) What percent of the total weekly pounds of Snickers Bars are produced in 1 day? Round to the nearest hundredth of a percent.

12. ***Egg Weight*** Chicken eggs are classified by weight per dozen eggs. Large eggs weigh 24 ounces per dozen and medium eggs weigh 21 ounces per dozen.

(a) If you do not include the shell, which is 12% of the total weight of an egg, how many ounces of eggs do you get from a dozen large eggs? From a dozen medium eggs?

(b) At a local market, large eggs sell for \$1.79 a dozen and medium eggs for \$1.39 a dozen. If you do not include the shell, which is a better buy, large or medium eggs?

Family Budgets *For the following exercises, use the chart below.*

Michael and Dianne have been married for two years. Michael has graduated from community college. Dianne is still attending community college. Michael works full time and Dianne part time.

Rent	22%	Clothing	7%
Food	29%	Medical	8%
Utilities	8%	Savings	4%
Entertainment	6%	Loan and credit payments	8%
Cell phones	4%	Charitable contributions	4%

13. Their combined annual income from their two jobs is \$68,500.

(a) If 29% of their salary is withheld for various taxes, how much money do Michael and Dianne have available for their budget?

(b) How much of their take-home pay is available for food?

14. They want to pay off one large credit card account.

(a) They determine that if they eat out less and cook more meals at home, they can save 20% of their food costs. If they achieve that goal, what is their new annual food budget in dollars?

(b) If they use all that savings to pay down their credit card, how much would be available to reduce their debt?

Paycheck Stub *Use the following information from a paycheck stub to solve exercises 15–18.*

TOBEY & SLATER INC. 5000 Stillwell Avenue Queens, NY 10001	Check Number	Payroll Period From Date	Payroll Period To Date	Pay Date
	495885	10-31-15	11-30-15	12-01-15

Name	Social Security No.	I.D. Number	File Number	Rate/Salary	Department	MS	DEP	Res
Fred J. Gilliani	012-34-5678	01	1379	1150.00	0100	M	5	NY

	Current	Year to Date		Current	Year to Date
GROSS	1 , 150 . 00	6 , 670 . 00	STATE	67 . 76	388 . 45
FEDERAL	138 . 97	781 . 07	LOCAL	5 . 18	30 . 04
FICA	87 . 98	510 . 28	DIS-SUI	. 00	. 00
W-2 GROSS		6 , 670 . 00	NET	790 . 47	4 , 960 . 16

Earnings						Special Deductions		
No.	Type	Hours	Rate	Amount	Dept/Job No.	No.	Description	Amount
96	REGULAR			1,150 . 00	0100	82	Retirement	12 . 56
						75	Medical	36 . 28
						56	Union Dues	10 . 80

Gross pay is the pay an employee receives for his or her services before deductions. Net pay is the pay the employee actually gets to take home. You may round each amount to the nearest whole percent for exercises 15–18.

15. What percent of Fred's gross pay is deducted for federal, state, and local taxes?

16. What percent of Fred's gross pay is deducted for retirement and medical?

17. What percent of Fred's gross pay does he actually get to take home?

18. What percent of Fred's deductions are special deductions?

Quick Quiz 0.6

▲ **1.** The college courtyard has 525 square yards of space. It is being paved with squares made of Vermont granite stone. Each stone has an area of 1.5 square yards. How many stones will be required to pave the courtyard?

2. Melinda is paid \$16,200 per year. She is also paid a sales commission of 4% of the value of her sales. Last year she sold \$345,000 worth of products. What percent of her total income was her commission? Round to the nearest whole percent.

▲ **3.** A carpenter is building a fence for a rectangular yard that measures 40 feet by 95 feet. The fence material costs \$4.50 per foot. If he makes a fence large enough to surround the yard, how much will the fence cost?

4. Concept Check Hank knows that 1 kilometer $\approx$ 0.62 mile. Explain how Hank could find out how many miles he traveled on a 12-kilometer trip in Mexico.

Background: Financial Counselor

As a college student, Sybelle had an internship with the federal government at a local Social Security Administration office. Upon graduation, she was offered a full-time position as a financial counselor. Today, she is meeting with Cora, who wants to understand how monthly retirement benefits are calculated.

Facts

- Cora is 62 years old in 2015.
- Full retirement age is 67 years old.
- At age 66, the percent reduction in monthly benefits is about 6.7%.
- At age 65, the percent reduction in monthly benefits is about 13.3%.
- Cora's average monthly earnings are $4200.
- The guidelines used for calculating a monthly benefit are as follows:

  90% of the first $826 of average monthly earnings
  +32% of the average monthly earnings between $826 and $4980
  +15% of the earnings over $4980

Tasks

1. Sybelle must explain to Cora the calculations involved in determining the monthly benefit she'll receive at the full retirement age of 67. What does Sybelle calculate Cora's monthly benefit to be?

2. Sybelle also wants to express Cora's monthly benefit as a percent of her average monthly pre-retirement earnings. What is this percentage?

3. Finally, Sybelle wants to calculate Cora's monthly benefit if her client retires early.
   (a) By what amount would Cora's monthly benefit be reduced if she began collecting it at age 66?

   (b) By what amount would it be reduced if she began collecting it at age 65?

Chapter 0 Organizer

Topic and Procedure	Examples	You Try It
Simplifying fractions, p. 2 1. Write the **numerator** and **denominator** as products of prime factors. 2. Use the basic rule of fractions that $\frac{a \times c}{b \times c} = \frac{a}{b}$ for any factor that appears in both the numerator and the denominator. 3. Multiply the remaining factors for the numerator and for the denominator separately.	Simplify. **(a)** $\frac{15}{25} = \frac{\cancel{5} \cdot 3}{\cancel{5} \cdot 5} = \frac{3}{5}$ **(b)** $\frac{36}{48} = \frac{\cancel{2} \cdot \cancel{2} \cdot 3 \cdot \cancel{3}}{\cancel{2} \cdot \cancel{2} \cdot 2 \cdot 2 \cdot \cancel{3}} = \frac{3}{4}$ **(c)** $\frac{26}{39} = \frac{2 \cdot \cancel{13}}{3 \cdot \cancel{13}} = \frac{2}{3}$	1. Simplify. **(a)** $\frac{18}{27}$ **(b)** $\frac{45}{60}$ **(c)** $\frac{34}{85}$
Changing improper fractions to mixed numbers, p. 5 1. Divide the denominator into the numerator to obtain the whole-number part of the mixed number. 2. The remainder from the division will be the numerator of the fraction. 3. The denominator remains unchanged.	Change to a mixed number. **(a)** $\frac{14}{3} = 4\frac{2}{3}$ since $3\overline{)14}$ gives 4: $14 - 12 = 2$ **(b)** $\frac{19}{8} = 2\frac{3}{8}$ since $8\overline{)19}$ gives 2: $19 - 16 = 3$	2. Change to a mixed number. **(a)** $\frac{21}{5}$ **(b)** $\frac{37}{7}$
Changing mixed numbers to improper fractions, p. 6 1. Multiply the whole number by the denominator and add the result to the numerator. This will yield the new numerator. 2. The denominator does not change.	Change to an improper fraction. **(a)** $4\frac{5}{6} = \frac{(4 \times 6) + 5}{6} = \frac{24 + 5}{6} = \frac{29}{6}$ **(b)** $3\frac{1}{7} = \frac{(3 \times 7) + 1}{7} = \frac{21 + 1}{7} = \frac{22}{7}$	3. Change to an improper fraction. **(a)** $2\frac{2}{5}$ **(b)** $6\frac{1}{9}$
Changing fractions to equivalent fractions with a given denominator, p. 7 1. Divide the original denominator into the new denominator. This result is the value that we use for multiplication. 2. Multiply the numerator and the denominator of the original fraction by that value.	Find the missing numerator. $\frac{4}{7} = \frac{?}{21}$ $7\overline{)21}$ = 3 ← Use this to multiply $\frac{4 \times 3}{7 \times 3} = \frac{12}{21}$.	4. Find the missing numerator. $\frac{3}{8} = \frac{?}{40}$
Finding the LCD (least common denominator) of two or more fractions, p. 11 1. Write each denominator as the product of prime factors. 2. The LCD is a product containing each different factor. 3. If a factor occurs more than once in any one denominator, the LCD will contain that factor repeated the greatest number of times that it occurs in any one denominator.	**(a)** Find the LCD of each pair of fractions. $\frac{4}{15}$ and $\frac{3}{35}$ $15 = 5 \cdot 3 \quad 35 = 5 \cdot 7 \quad \text{LCD} = 3 \cdot 5 \cdot 7 = 105$ **(b)** $\frac{11}{18}$ and $\frac{7}{45}$ $\quad 18 = 3 \cdot 3 \cdot 2$ (factor 3 appears twice) $45 = 3 \cdot 3 \cdot 5$ (factor 3 appears twice) $\text{LCD} = 2 \cdot 3 \cdot 3 \cdot 5 = 90$	5. Find the LCD of each pair of fractions. **(a)** $\frac{3}{10}$ and $\frac{7}{12}$ **(b)** $\frac{17}{25}$ and $\frac{9}{20}$
Adding and subtracting fractions that do not have a common denominator, p. 12 1. Find the LCD. 2. Change each fraction to an equivalent fraction with the LCD for a denominator. 3. Add or subtract the fractions and simplify the answer if possible.	Perform the operation indicated. **(a)** $\frac{3}{8} + \frac{1}{3} = \frac{3 \cdot 3}{8 \cdot 3} + \frac{1 \cdot 8}{3 \cdot 8} = \frac{9}{24} + \frac{8}{24} = \frac{17}{24}$ **(b)** $\frac{11}{12} - \frac{1}{4} = \frac{11}{12} - \frac{1 \cdot 3}{4 \cdot 3} = \frac{11}{12} - \frac{3}{12} = \frac{8}{12} = \frac{2}{3}$	6. Perform the operation indicated. **(a)** $\frac{1}{2} + \frac{1}{9}$ **(b)** $\frac{19}{24} - \frac{3}{8}$
Adding and subtracting mixed numbers, p. 15 1. Change the mixed numbers to improper fractions. 2. Follow the rules for adding and subtracting fractions. 3. If necessary, change your answer to a mixed number.	**(a)** Add. $1\frac{2}{3} + 1\frac{3}{4} = \frac{5}{3} + \frac{7}{4} = \frac{5 \cdot 4}{3 \cdot 4} + \frac{7 \cdot 3}{4 \cdot 3}$ $= \frac{20}{12} + \frac{21}{12} = \frac{41}{12} = 3\frac{5}{12}$ **(b)** Subtract. $2\frac{1}{4} - 1\frac{3}{4} = \frac{9}{4} - \frac{7}{4} = \frac{2}{4} = \frac{1}{2}$	7. **(a)** Add. $2\frac{2}{3} + 3\frac{1}{9}$ **(b)** Subtract. $3\frac{5}{6} - 1\frac{1}{3}$

Topic and Procedure	Examples	You Try It
Multiplying fractions, p. 20 **1.** If there are no common factors, multiply the numerators. Then multiply the denominators. **2.** If possible, write the numerators and denominators as the product of prime factors. Use the basic rule of fractions to divide out any value that appears in both a numerator and a denominator. Multiply the remaining factors in the numerator. Multiply the remaining factors in the denominator.	Multiply. **(a)** $\frac{3}{7} \times \frac{2}{13} = \frac{6}{91}$ **(b)** $\frac{6}{15} \times \frac{35}{91} = \frac{2 \cdot \cancel{3}}{\cancel{3} \cdot \cancel{5}} \times \frac{\cancel{5} \cdot \cancel{7}}{\cancel{7} \cdot 13} = \frac{2}{13}$ **(c)** $3 \times \frac{5}{8} = \frac{3}{1} \times \frac{5}{8} = \frac{15}{8}$ or $1\frac{7}{8}$	8. Multiply. **(a)** $\frac{5}{11} \times \frac{2}{3}$ **(b)** $\frac{7}{10} \times \frac{15}{21}$ **(c)** $6 \times \frac{3}{5}$
Dividing fractions, p. 22 **1.** Invert the second fraction. **2.** Multiply the fractions.	Divide. **(a)** $\frac{4}{7} \div \frac{11}{3} = \frac{4}{7} \times \frac{3}{11} = \frac{12}{77}$ **(b)** $\frac{5}{9} \div \frac{5}{7} = \frac{\cancel{5}}{9} \times \frac{7}{\cancel{5}} = \frac{7}{9}$	9. Divide. **(a)** $\frac{3}{14} \div \frac{2}{5}$ **(b)** $\frac{7}{12} \div \frac{7}{5}$
Multiplying and dividing mixed numbers, pp. 21, 22, 24 **1.** Change each mixed number to an improper fraction. **2.** Use the rules for multiplying or dividing fractions. **3.** If necessary, change your answer to a mixed number.	**(a)** Multiply. $2\frac{1}{4} \times 3\frac{3}{5} = \frac{9}{4} \times \frac{18}{5}$ $= \frac{3 \cdot 3}{2 \cdot \cancel{2}} \times \frac{\cancel{2} \cdot 3 \cdot 3}{5} = \frac{81}{10} = 8\frac{1}{10}$ **(b)** Divide. $1\frac{1}{4} \div 1\frac{1}{2} = \frac{5}{4} \div \frac{3}{2} = \frac{5}{2 \cdot \cancel{2}} \times \frac{\cancel{2}}{3} = \frac{5}{6}$	10. **(a)** Multiply. $1\frac{5}{6} \times 2\frac{1}{4}$ **(b)** Divide. $3\frac{1}{2} \div 1\frac{3}{4}$
Changing fractional form to decimal form, p. 29 Divide the denominator into the numerator.	Write as a decimal. $\frac{5}{8} = 0.625$ since $8\overline{)5.000}$ = 0.625	11. Write as a decimal. $\frac{7}{8}$
Changing decimal form to fractional form, p. 30 **1.** Write the decimal as a fraction with a denominator of 10, 100, 1000, and so on. **2.** Simplify the fraction, if possible.	Write as a fraction. **(a)** $0.37 = \frac{37}{100}$ **(b)** $0.375 = \frac{375}{1000} = \frac{3}{8}$	12. Write as a fraction. **(a)** 0.29 **(b)** 0.175
Adding and subtracting decimals, p. 31 **1.** Carefully line up the decimal points as indicated for addition and subtraction. (Extra zeros may be added to the right-hand side of the decimals if desired.) **2.** Add or subtract the appropriate digits.	**(a)** Add. 1.236 + 7.825 1.236 + 7.825 9.061 **(b)** Subtract. 2 − 1.32 2.00 − 1.32 0.68	13. **(a)** Add. 2.338 + 6.195 **(b)** Subtract. 6 − 2.54
Multiplying decimals, p. 32 **1.** First multiply the digits. **2.** Count the total number of decimal places in the numbers being multiplied. This number determines the number of decimal places in the answer.	Multiply. **(a)** 0.9 (one place) × 0.7 (one place) 0.63 (two places) **(b)** 0.009 (three places) × 0.07 (two places) 0.00063 (five places)	14. Multiply. **(a)** 1.5 × 0.9 **(b)** 5.12 × 0.67
Dividing decimals, p. 33 **1.** Count the number of decimal places in the divisor. **2.** Move the decimal point to the right the same number of places in both the divisor and dividend. **3.** Mark that position with a caret (‸). **4.** Perform the division. Line up the decimal point in the quotient with the position indicated by the caret.	Divide. 7.5 ÷ 0.6 Move decimal point one place to the right. $0.6_\wedge\overline{)7.5_\wedge0}$ = 12.5 6 15 12 30 30 0 Therefore, 7.5 ÷ 0.6 = 12.5	15. Divide. 9.25 ÷ 0.5

Topic and Procedure	Examples	You Try It
Changing a decimal to a percent, p. 40 **1.** Move the decimal point two places to the right. **2.** Add the % symbol.	Write as a percent. **(a)** 0.46 = 46% **(b)** 0.002 = 0.2% **(c)** 0.013 = 1.3% **(d)** 1.59 = 159% **(e)** 0.0007 = 0.07%	16. Write as a percent. **(a)** 0.52 **(b)** 0.008 **(c)** 1.86 **(d)** 0.077 **(e)** 0.0009
Changing a percent to a decimal, p. 41 **1.** Move the decimal point two places to the left. **2.** Remove the % symbol.	Write as a decimal. **(a)** 49% = 0.49 **(b)** 59.8% = 0.598 **(c)** 180% = 1.8 **(d)** 0.13% = 0.0013	17. Write as a decimal. **(a)** 28% **(b)** 7.42% **(c)** 165% **(d)** 0.25%
Finding a percent of a number, p. 41 **1.** Convert the percent to a decimal. **2.** Multiply the decimal by the number.	Find 12% of 86. 12% = 0.12 $0.12 \times 86 = 10.32$ Therefore, 12% of 86 is 10.32.	18. Find 15% of 92.
Finding what percent one number is of another number, p. 42 **1.** Place the number after the word *of* in the denominator. **2.** Place the other number in the numerator. **3.** If possible, simplify the fraction. **4.** Change the fraction to a decimal. **5.** Express the decimal as a percent.	**(a)** What percent of 8 is 7? $\frac{7}{8} = 0.875 = 87.5\%$ **(b)** 42 is what percent of 12? $\frac{42}{12} = \frac{7}{2} = 3.5 = 350\%$	19. **(a)** What percent of 12 is 10? Round to the nearest tenth of a percent. **(b)** 50 is what percent of 40?
Estimation, p. 44 **1.** Round each number so that there is one nonzero digit. **2.** Perform the calculation with the rounded numbers.	Estimate the number of square feet in a room that is 22 feet long and 13 feet wide. Assume that the room is rectangular. We round 22 to 20. We round 13 to 10. To find the area of a rectangle, we multiply length times width. $20 \times 10 = 200$ We estimate that there are 200 square feet in the room.	20. Estimate the number of square feet in a rectangular game room that measures 27 feet long by 11.5 feet wide.
Problem solving, p. 48 In solving a real-life problem, you may find it helpful to complete the following steps. You will not use all of the steps all of the time. Choose the steps that best fit the conditions of the problem. **1.** ***Understand the problem.*** **(a)** Read the problem carefully. **(b)** Draw a picture if this helps you. **(c)** Use the Mathematics Blueprint for Problem Solving. **2.** ***Solve and state the answer.*** **3.** ***Check.*** **(a)** Estimate to see if your answer is reasonable. **(b)** Repeat your calculation. **(c)** Work backward from your answer. Do you arrive at the original conditions of the problem?	Susan is installing wall-to-wall carpeting in her $10\frac{1}{2}$-ft-by-12-ft bedroom. How much will it cost at $20 a square yard? **1.** ***Understand the problem.*** We need to find the area of the room in square yards. Then we can find the cost. **2.** ***Solve and state the answer.*** Area: $10\frac{1}{2} \times 12 = \frac{21}{2} \times \frac{12}{1} = 126 \text{ ft}^2$ $126 \div 9 = 14 \text{ yd}^2$ Cost: $14 \times 20 = \$280$ The carpeting will cost $280. **3.** ***Check.*** Estimate: $10 \times 10 = 100 \text{ ft}^2$ $100 \div 10 = 10 \text{ yd}^2$ $10 \times 20 = \$200$ Our answer is reasonable. ✓	21. Wayne is installing new tile in his basement, which measures 15 feet by $21\frac{3}{4}$ feet. How much will the tile cost at $4.25 per square yard? Round to the nearest cent.

Chapter 0 Review Problems

Section 0.1

In exercises 1–4, simplify.

1. $\frac{36}{48}$ **2.** $\frac{15}{50}$ **3.** $\frac{36}{82}$ **4.** $\frac{18}{30}$

5. Write $7\frac{1}{8}$ as an improper fraction. **6.** Write $\frac{34}{5}$ as a mixed number. **7.** Write $\frac{80}{3}$ as a mixed number.

Find the missing numerator.

8. $\frac{5}{8} = \frac{?}{24}$ **9.** $\frac{1}{7} = \frac{?}{35}$ **10.** $\frac{3}{5} = \frac{?}{75}$ **11.** $\frac{2}{5} = \frac{?}{55}$

Section 0.2

Combine.

12. $\frac{3}{5} + \frac{1}{4}$ **13.** $\frac{7}{12} + \frac{5}{8}$ **14.** $\frac{7}{20} - \frac{1}{12}$ **15.** $\frac{7}{10} - \frac{4}{15}$

16. $3\frac{1}{6} + 2\frac{3}{5}$ **17.** $2\frac{7}{10} + 3\frac{3}{4}$ **18.** $6\frac{2}{9} - 3\frac{5}{12}$ **19.** $3\frac{1}{15} - 1\frac{3}{20}$

Section 0.3

Multiply.

20. $6 \times \frac{5}{11}$ **21.** $2\frac{1}{3} \times 4\frac{1}{2}$ **22.** $16 \times 3\frac{1}{8}$ **23.** $\frac{4}{7} \times 5$

Divide.

24. $\frac{3}{8} \div 6$ **25.** $\frac{\frac{8}{3}}{\frac{5}{9}}$ **26.** $\frac{15}{16} \div 6\frac{1}{4}$ **27.** $2\frac{6}{7} \div \frac{10}{21}$

Section 0.4

Combine.

28. $1.634 + 3.007 + 2.560$ **29.** $24.831 - 17.094$ **30.** $47.251 - 17.69$ **31.** $1.9 + 2.53 + 0.006$

Multiply.

32. 0.007×5.35 **33.** 362.341×1000 **34.** $2.6 \times 0.03 \times 1.02$ **35.** $2.51 \times 100 \times 0.5$

Divide.

36. $71.32 \div 1000$ **37.** $0.523 \div 0.4$ **38.** $1.35 \div 0.015$ **39.** $4.186 \div 2.3$

40. Write as a decimal: $\frac{3}{8}$. **41.** Write as a fraction in simplified form: 0.36.

Section 0.5

In exercises 42–45, write each percent as a decimal.

42. 1.4% **43.** 36.1% **44.** 0.02% **45.** 125.3%

In exercises 46–49, write each decimal in percent form.

46. 0.0025 **47.** 0.325 **48.** 0.9 **49.** 0.1

50. What is 30% of 400?

51. Find 7.2% of 55.

52. 76 is what percent of 80?

53. What percent of 1250 is 750?

54. ***Cell Phones*** 80% of Del Mar Community College students have a cell phone. If there are 16,850 students in the college, how many students have cell phones?

55. ***Math Deficiency*** In a given university, 720 of the 960 freshmen had a math deficiency. What percent of the class had a math deficiency?

In exercises 56–61, estimate. Do not find an exact value.

56. $234{,}897 \times 1{,}936{,}112$

57. $357 + 923 + 768 + 417$

58. $634{,}318 - 284{,}000$

59. $21\frac{1}{5} - 8\frac{4}{5} - 1\frac{2}{3}$

60. Find 18% of $56,297.

61. $12{,}482 \div 389$

62. ***Salary*** Estimate Carmen's salary for the week if she earns $8.35 per hour and worked 38.5 hours.

63. ***Vacation Cost*** Estimate the weekly amount owed by each of five families who are renting a resort that costs $3875 per week.

Section 0.6

Solve. You may use the Mathematics Blueprint for Problem Solving.

Gas Mileage *A six-passenger Piper Cub airplane has a gas tank that holds 240 gallons. Use this information to answer exercises 64 and 65.*

64. When flying at cruising speed, the plane averages $7\frac{2}{3}$ miles per gallon. How far can the plane fly at cruising speed? If the pilot never plans to fly more than 80% of his maximum cruising distance, what is the longest trip he would plan to fly?

65. When flying at maximum speed, the plane averages $6\frac{1}{4}$ miles per gallon. How far can the plane fly at maximum speed? If the pilot never plans to fly more than 70% of his maximum flying distance when flying at full speed, what is the longest trip he would plan to fly at full speed?

▲ **66.** ***Carpeting*** Mr. and Mrs. Carr are installing wall-to-wall carpeting in a room that measures $12\frac{1}{2}$ ft by $9\frac{2}{3}$ ft. How much will it cost if the carpet is $26.00 per square yard?

67. ***Sales Commission*** Mike sells sporting goods on an 8% commission. During the first week in July, he sold goods worth $5785. What was his commission for the week?

68. ***Car Loan*** Dick and Ann Wright purchased a new car. They took out a loan of $9214.50 to help pay for the car and paid the rest in cash. They paid off the loan with payments of $225 per month for four years. How much more did they pay back than the amount of the car loan? (This is the amount of interest they were charged for the car loan.)

69. ***Wages*** Kevin works as an overnight stocker at Target. He is paid $7.50 an hour for a 40-hour week. For any additional time he gets paid 1.5 times the normal rate. Last week he worked 49 hours. How much did he get paid last week?

How Am I Doing? Chapter 0 Test

MATH COACH MyMathLab® YouTube

After you take this test, read through the Math Coach on pages 62–63. Math Coach videos are available via MyMathLab and YouTube. Step-by-step test solutions on the Chapter Test Prep Videos are also available via MyMathLab and YouTube. (Search "TobeyBegInterAlg" and click on "Channels.")

In exercises 1 and 2, simplify.

1. $\frac{16}{18}$

2. $\frac{48}{36}$

3. Write as an improper fraction. $6\frac{3}{7}$

4. Write as a mixed number: $\frac{105}{9}$

In exercises 5–12, perform the operations indicated. Simplify answers whenever possible.

5. $\frac{2}{3} + \frac{5}{6} + \frac{3}{8}$

6. $1\frac{1}{8} + 3\frac{3}{4}$

MC **7.** $3\frac{2}{3} - 2\frac{5}{6}$

8. $\frac{5}{7} \times \frac{28}{15}$

9. $\frac{7}{4} \div \frac{1}{2}$

MC **10.** $5\frac{3}{8} \div 2\frac{3}{4}$

11. $2\frac{1}{2} \times 3\frac{1}{4}$

12. $\frac{\frac{7}{8}}{\frac{1}{4}}$

In exercises 13–18, perform the operations indicated.

13. $1.6 + 3.24 + 9.8$

14. $7.0046 - 3.0149$

15. 32.8×0.04

16. 0.07385×1000

MC **17.** $12.88 \div 0.056$

18. $26{,}325.9 \div 100$

19. Write as a percent. 0.073

20. Write as a decimal. 196.5%

21. What is 3.5% of 180?

MC **22.** 39 is what percent of 650?

23. A 4-inch stack of computer chips is on the table. Each computer chip is $\frac{2}{9}$ of an inch thick. How many computer chips are in the stack?

In exercises 24–25, estimate. Round each number to one nonzero digit. Then calculate.

24. $52{,}344\overline{)4{,}678{,}987}$

25. $285.36 + 311.85 + 113.6$

Solve. You may use the Mathematics Blueprint for Problem Solving.

26. Allison is paid $14,000 per year plus a sales commission of 3% of the value of her sales. Last year she sold $870,000 worth of products. What percent of her total income was her commission? Round to the nearest whole percent.

▲ **27.** Fred and Melinda are laying wall tile in the kitchen. Each tile covers $3\frac{1}{2}$ square inches of space. They plan to cover 210 square inches of wall space. How many tiles will they need?

1. ☐	2. ☐
3. ☐	4. ☐
5. ☐	6. ☐
7. ☐	8. ☐
9. ☐	10. ☐
11. ☐	12. ☐
13. ☐	14. ☐
15. ☐	16. ☐
17. ☐	18. ☐
19. ☐	20. ☐
21. ☐	22. ☐
23. ☐	
24. ☐	25. ☐
26. ☐	
27. ☐	

Total Correct: ☐

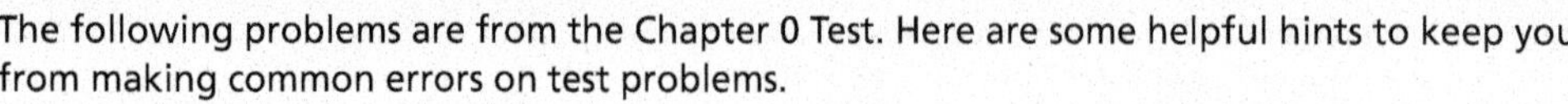

Mastering the skills you need to do well on the test.

The following problems are from the Chapter 0 Test. Here are some helpful hints to keep you from making common errors on test problems.

Chapter 0 Test, Problem 7 Subtract. $3\frac{2}{3} - 2\frac{5}{6}$

HELPFUL HINT First change the mixed numbers to improper fractions. Next find the LCD of the two denominators. Then change the fractions to an equivalent form with the LCD as the common denominator before subtracting.

Did you change $3\frac{2}{3}$ to $\frac{11}{3}$ and $2\frac{5}{6}$ to $\frac{17}{6}$?

Yes ______ No ______

If you answered No, stop and change the two mixed numbers to improper fractions.

Did you find the LCD to be 6?

Yes ______ No ______

If you answered No, consider how to find the LCD of the two fractions. Once the fractions are written as equivalent fractions with 6 as the denominator, the two like fractions can be subtracted.

If you answered Problem 7 incorrectly, go back and rework the problem using these suggestions.

Chapter 0 Test, Problem 10 Divide. $5\frac{3}{8} \div 2\frac{3}{4}$

HELPFUL HINT Be sure to change the mixed numbers to improper fractions before dividing.

Did you change $5\frac{3}{8}$ to $\frac{43}{8}$ and $2\frac{3}{4}$ to $\frac{11}{4}$ before doing any other steps?

Yes ______ No ______

If you answered No, stop and change the two mixed numbers to improper fractions.

Next did you change the division to multiplication to obtain $\frac{43}{8} \times \frac{4}{11}$?

Yes ______ No ______

If you answered No, stop and make this change.

Did you simplify the product?

Yes ______ No ______

If you answered No, try dividing a 4 from the second numerator and first denominator before multiplying. The product will be an improper fraction that can be converted to a mixed number.

Now go back and rework the problem using these suggestions.

Need more help? Watch the MATH COACH videos in MyMathLab® or on YouTube™.

Chapter 0 Test, Problem 17 Divide. $12.88 \div 0.056$

HELPFUL HINT Be careful as you move the decimal point in the divisor to the right. Make sure that the resulting divisor is an integer. Then move the decimal point in the dividend the same number of places to the right. Add zeros if necessary.

Did you move the decimal point in the divisor three places to the right to get 56?

Yes _____ No _____

If you answered No, stop and perform this step first.

Did you move the decimal point in the dividend three places to the right and add one zero to get 12880?

Yes _____ No _____

If you answered No, perform this step now. Be careful of calculation errors as you perform the division.

If you answered Problem 17 incorrectly, go back and rework the problem using these suggestions.

Chapter 0 Test, Problem 22 39 is what percent of 650?

HELPFUL HINT Write a fraction with the two numbers. The number after the word *of* is always the denominator, and the other number is the numerator.

Did you write the fraction $\frac{39}{650}$?

Yes _____ No _____

If you answered No, stop and perform this step.

Did you simplify the fraction to $\frac{3}{50}$ before changing the fraction to a decimal?

Yes _____ No _____

If you answered No, consider that simplifying the fraction first makes the division step a little easier. Be sure to place the decimal point correctly in your quotient.

Did you change the quotient from a decimal to a percent?

Yes _____ No _____

If you answered No, stop and perform this final step.

Now go back and rework the problem using these suggestions.

CHAPTER 1

Real Numbers and Variables

CAREER OPPORTUNITIES

Electrical Engineer

When we think of electrical work, mathematics isn't necessarily the first thing that comes to mind. Instead, we envision electricians running cable through walls, wiring outlets, and installing appliances. In reality, though, the field is broader than this. Electrical engineers and their assistants use quite a bit of math to create, install, or repair electrical systems. They perform calculations using ratios, formulas, and geometry to perform a wide range of tasks.

To investigate how the mathematics in this chapter can help with this field, see the Career Exploration Problems on page **122**.

1.1 Adding Real Numbers

1 Identifying Different Types of Numbers

Let's review some of the basic terms we use to talk about numbers.

Whole numbers are numbers such as 0, 1, 2, 3, 4, ...
Integers are numbers such as . . . , −3, −2, −1, 0, 1, 2, 3, ...
Rational numbers are numbers such as $\frac{3}{2}, \frac{5}{7}, -\frac{3}{8}, -\frac{4}{13}, \frac{6}{1}$, and $-\frac{8}{2}$.

Rational numbers can be written as one integer divided by another integer (as long as the denominator is not zero!). Integers can be written as fractions ($3 = \frac{3}{1}$, for example), so we can see that all integers are rational numbers. Rational numbers can be expressed in decimal form. For example, $\frac{3}{2} = 1.5$, $-\frac{3}{8} = -0.375$, and $\frac{1}{3} = 0.333\ldots$ or $0.\overline{3}$. It is important to note that rational numbers in decimal form are either terminating decimals or repeating decimals.

Irrational numbers are numbers that cannot be expressed as one integer divided by another integer. The numbers π, $\sqrt{2}$, and $\sqrt[3]{7}$ are irrational numbers.

Irrational numbers can be expressed in decimal form. The decimal form of an irrational number is a nonterminating, nonrepeating decimal. For example, $\sqrt{2} = 1.414213\ldots$ can be carried out to an infinite number of decimal places with no repeating pattern of digits.

Finally, **real numbers** are all the rational numbers and all the irrational numbers.

Student Learning Objectives

After studying this section, you will be able to:

1. Identify different types of numbers.
2. Use real numbers in real-life situations.
3. Add real numbers with the same sign.
4. Add real numbers with different signs.
5. Use the addition properties for real numbers.

Example 1 Classify as an integer, a rational number, an irrational number, and/or a real number.

(a) 5 (b) $-\dfrac{1}{3}$ (c) 2.85 (d) $\sqrt{2}$ (e) 0.777. . .

Solution Make a table. Check off the description of the number that applies.

	Number	Integer	Rational Number	Irrational Number	Real Number
(a)	5	✓	✓		✓
(b)	$-\frac{1}{3}$		✓		✓
(c)	2.85		✓		✓
(d)	$\sqrt{2}$			✓	✓
(e)	0.777 . . .		✓		✓

Student Practice 1 Classify as an integer, a rational number, an irrational number, and/or a real number.

(a) $-\dfrac{2}{5}$ (b) 1.515151 . . . (c) −8 (d) π

Any real number can be pictured on a **number line.**

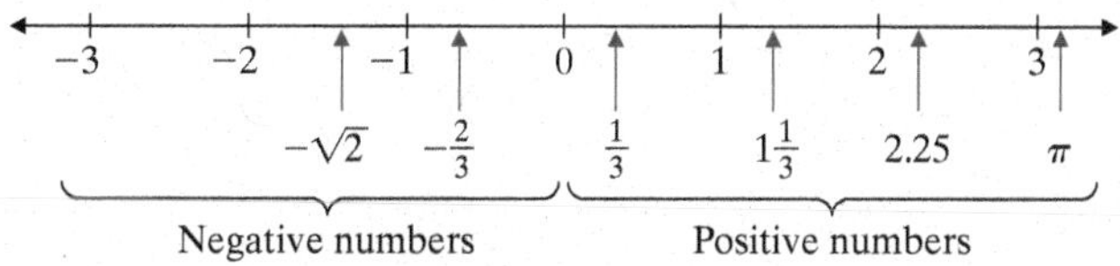

Positive numbers are to the right of 0 on a number line.
Negative numbers are to the left of 0 on a number line.

The **real numbers** include the positive numbers, the negative numbers, and zero.

2 Using Real Numbers in Real-Life Situations 

We often encounter practical examples of number lines that include positive and negative rational numbers. For example, we can tell by reading the accompanying thermometer that the temperature is 20° below 0. From the stock market report, we see that the stock opened at 36 and closed at 34.5, so the net change for the day was −1.5.

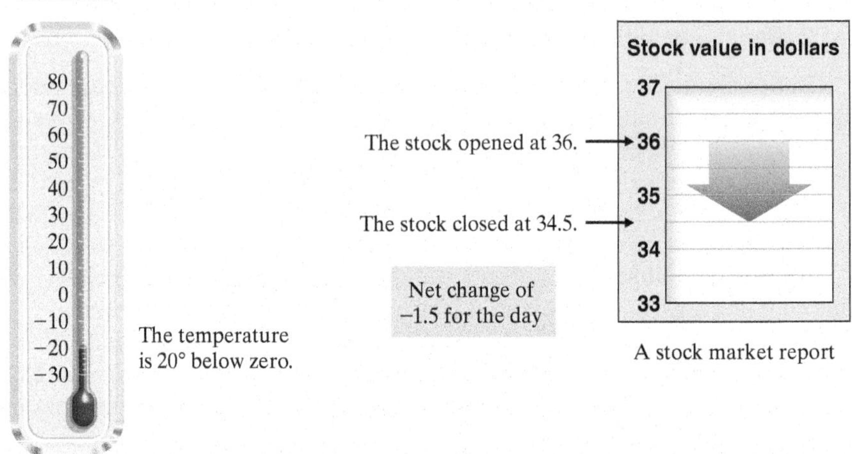

In the following example we use real numbers to represent real-life situations.

Example 2 Use a real number to represent each situation.

(a) A temperature of 128.6°F below zero is recorded at Vostok, Antarctica.

(b) The Himalayan peak K2 rises 29,064 feet above sea level.

(c) The Dow gains 10.24 points.

(d) An oil drilling platform extends 328 feet below sea level.

Solution A key word can help you to decide whether a number is positive or negative.

(a) 128.6°F *below* zero is −128.6.

(b) 29,064 feet *above* sea level is +29,064.

(c) A *gain* of 10.24 points is +10.24.

(d) 328 feet *below* sea level is −328. □

Student Practice 2 Use a real number to represent each situation.

(a) A population growth of 1259

(b) A depreciation of $763

(c) A wind-chill factor of minus 10° F

In everyday life we consider positive numbers the opposite of negative numbers. For example, a gain of 3 yards in a football game is the opposite of a loss of 3 yards; a check written for $2.16 on a checking account is the opposite of a deposit of $2.16.

Each positive number has an opposite negative number. Similarly, each negative number has an opposite positive number. **Opposite numbers,** also called **additive inverses,** have the same magnitude but different signs and can be represented on a number line.

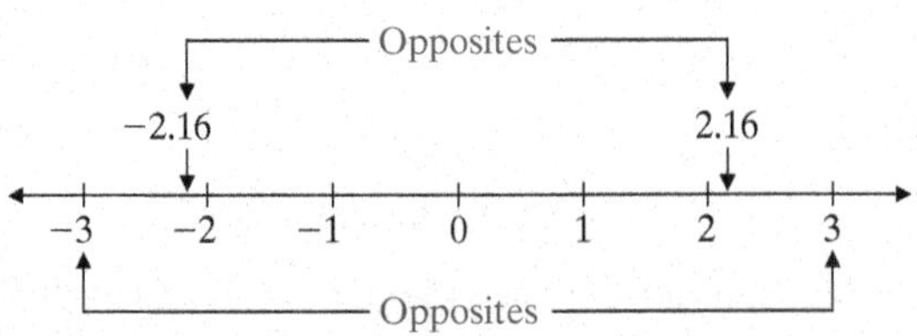

Example 3 Find the additive inverse (that is, the opposite).

(a) -7 **(b)** $\frac{1}{4}$ **(c)** A temperature rise of 40°

Solution

(a) The opposite of -7 is $+7$.

(b) The opposite of $\frac{1}{4}$ is $-\frac{1}{4}$.

(c) The opposite of $+40°$ is $-40°$. □

Student Practice 3 Find the additive inverse (the opposite).

(a) $\frac{2}{5}$ **(b)** -1.92 **(c)** A loss of 12 yards on a football play

3 Adding Real Numbers with the Same Sign 

To use a real number, we need to be clear about its sign. When we write the number three as $+3$, the sign indicates that it is a positive number. The positive sign can be omitted. If someone writes three (3), it is understood that it is a positive three $(+3)$. To write a negative number such as negative three (-3), we must include the sign.

A concept that will help us add and subtract real numbers is the idea of absolute value. The **absolute value** of a number is the distance between that number and zero on a number line. The absolute value of 3 is written $|3|$.

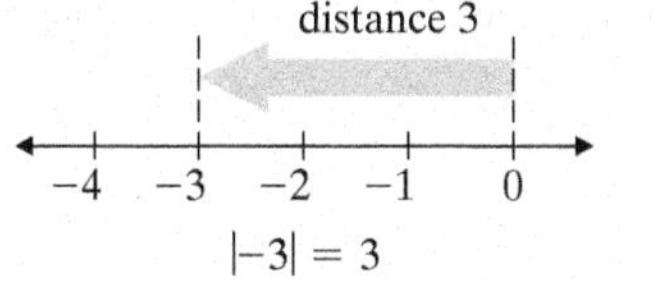

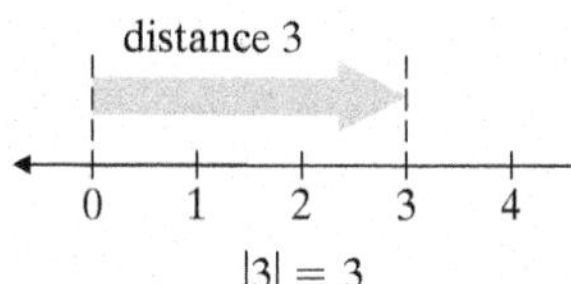

Distance is always a positive number regardless of the direction of travel. This means that the absolute value of any number will be a positive value or zero. We place the symbols | and | around a number to mean the absolute value of the number.

The distance from 0 to 3 is 3, so $|3| = 3$. This is read "the absolute value of 3 is 3."

The distance from 0 to -3 is 3, so $|-3| = 3$. This is read "the absolute value of -3 is 3."

Some other examples are

$$|-22| = 22, \quad |5.6| = 5.6, \quad \text{and} \quad |0| = 0.$$

Thus, the absolute value of a number can be thought of as the magnitude of the number, without regard to its sign.

Example 4 Find the absolute value.

(a) $|-4.62|$ **(b)** $\left|\frac{3}{7}\right|$ **(c)** $\left|\frac{0}{5}\right|$

Solution

(a) $|-4.62| = 4.62$ **(b)** $\left|\frac{3}{7}\right| = \frac{3}{7}$ **(c)** $\left|\frac{0}{5}\right| = 0$ □

Student Practice 4 Find the absolute value.

(a) $|-7.34|$ **(b)** $\left|\frac{5}{8}\right|$ **(c)** $\left|\frac{0}{2}\right|$

Now let's look at addition of real numbers when the two numbers have the same sign. Suppose that you are keeping track of your checking account at a local

bank. When you make a deposit of 5 dollars, you record it as +5. When you write a check for 4 dollars, you record it as −4, as a debit. (If you do not have a checking account but have a debit card, think of depositing 5 dollars in your account and then making a purchase of 4 dollars.)

Situation 1: *Total Deposit* You made a deposit of 20 dollars on one day and a deposit of 15 dollars the next day. You want to know the total value of your deposits.

Your record for situation 1.

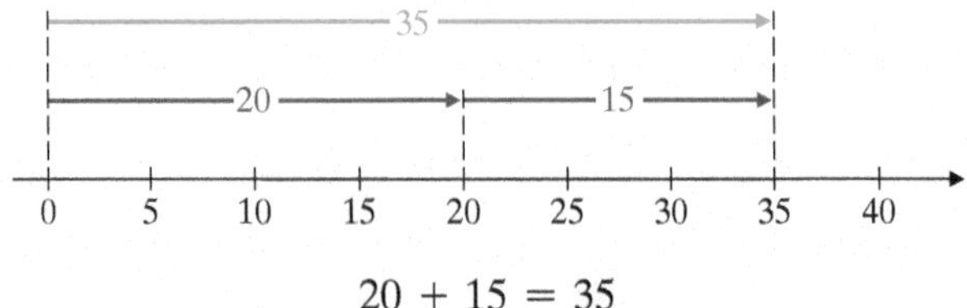

$$20 + 15 = 35$$

The amount of the deposit on the first day added to the amount of the deposit on the second day is the total of the deposits made over the two days.

Situation 2: *Total Debit* You write a check for 25 dollars to pay one bill and two days later write a check for 5 dollars. You want to know the total value of the debits to your account for the two checks.

Your record for situation 2.

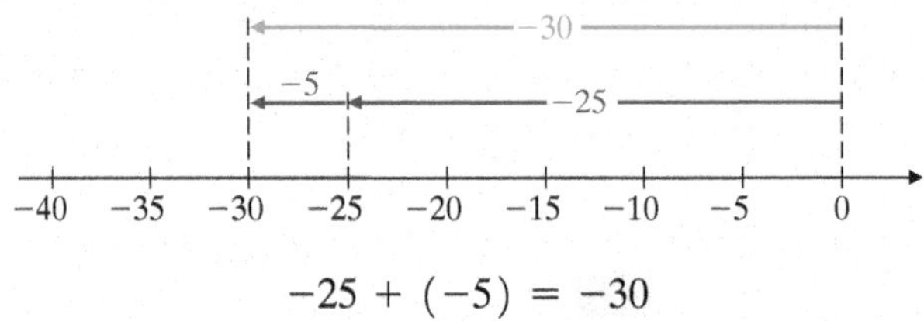

$$-25 + (-5) = -30$$

The value of the first check added to the value of the second check is the total debit to your account.

In each situation we found that we added the absolute value of each number. (That is, we added the numbers without regarding their sign.) The answer always contained the sign that was common to both numbers.

We will now state these results as a formal rule.

ADDITION RULE FOR TWO NUMBERS WITH THE SAME SIGN

To add two numbers with the same sign, add the absolute values of the numbers and use the common sign in the answer.

Example 5 Add.

(a) $14 + 16$ **(b)** $-8 + (-7)$

Solution

(a) $14 + 16$

$14 + 16 = 30$ Add the absolute values of the numbers.

$14 + 16 = +30$ Use the common sign in the answer. Here the common sign is the + sign.

(b) $-8 + (-7)$

$8 + 7 = 15$ Add the absolute values of the numbers.

$-8 + (-7) = -15$ Use the common sign in the answer. Here the common sign is the − sign. □

Student Practice 5 Add.

(a) $37 + 19$ **(b)** $-23 + (-35)$

Example 6 Add. $\frac{2}{3} + \frac{1}{7}$

Solution

$$\frac{2}{3} + \frac{1}{7}$$

$$\frac{14}{21} + \frac{3}{21}$$ Change each fraction to an equivalent fraction with a common denominator of 21.

$$\frac{14}{21} + \frac{3}{21} = +\frac{17}{21} \text{ or } \frac{17}{21}$$ Add the absolute values of the numbers. Use the common sign in the answer. Note that if no sign is written, the number is understood to be positive. □

Student Practice 6 Add.

$$-\frac{3}{5} + \left(-\frac{4}{7}\right)$$

Example 7 Add. $-4.2 + (-3.94)$

Solution

$-4.2 + (-3.94)$

$4.20 + 3.94 = 8.14$ Add the absolute values of the numbers.

$-4.20 + (-3.94) = -8.14$ Use the common sign in the answer. □

Student Practice 7 Add. $-12.7 + (-9.38)$

The rule for adding two numbers with the same sign can be extended to more than two numbers. If we add more than two numbers with the same sign, the answer will have the sign common to all.

Example 8 Add. $-7 + (-2) + (-5)$

Solution

$-7 + (-2) + (-5)$ We are adding three real numbers, all with the same sign. We begin by adding the first two numbers.

$= -9 + (-5)$ Add $-7 + (-2) = -9$.

$= -14$ Add $-9 + (-5) = -14$.

Of course, this can be shortened by adding the three numbers without regard to sign and then using the common sign for the answer. □

Student Practice 8 Add. $-7 + (-11) + (-33)$

4 Adding Real Numbers with Different Signs

What if the signs of the numbers you are adding are different? Let's consider our checking account again to see how such a situation might occur.

Situation 3: *Net Increase* You made a deposit of 30 dollars on one day. On the next day you write a check for 25 dollars. You want to know the result of your two transactions.

Calculator

Negative Numbers

To enter a negative number on most scientific calculators, find the key marked [+/−]. To enter the number −2, press the key 2 and then the key +/−. The display should read

[−2]

To find −32 + (−46), enter

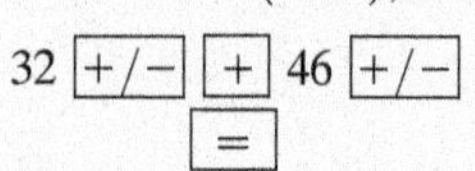

The display should read

[− 78]

Try the following.

(a) −256 + 184

(b) 94 + (−51)

(c) − 18 + (−24)

(d) −6 + (−10) + (−15)

Note: The [+/−] key changes the sign of a number from + to − or − to +. On some calculators the negative sign must be entered first, followed by the number.

Ans:
(a) −72 **(b)** 43
(c) −42 **(d)** −31

Your record for situation 3.

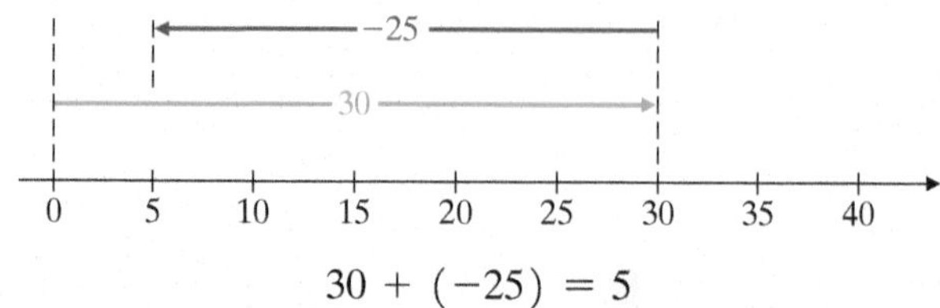

$$30 + (-25) = 5$$

A positive 30 for the deposit added to a negative 25 for the check, which is a debit, gives a net increase of 5 dollars in the account.

Situation 4: *Net Decrease* You made a deposit of 10 dollars on one day. The next day you write a check for 40 dollars. You want to know the result of your two transactions.

Your record for situation 4.

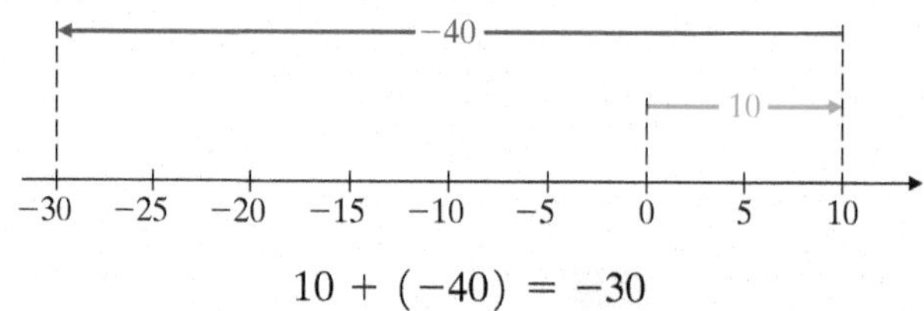

$$10 + (-40) = -30$$

A positive 10 for the deposit added to a negative 40 for the check, which is a debit, gives a net decrease of 30 dollars in the account.

The result is a negative thirty (−30), because the check was larger than the deposit. If you did not have at least 30 dollars in your account at the start of *situation 4*, you have overdrawn your account.

What do we observe from *situations 3 and 4*? In each case, first we found the difference of the absolute values of the two numbers. Then the sign of the result was always the sign of the number with the greater absolute value. Thus, in *situation 3*, 30 is larger than 25. The sign of 30 is positive. The sign of the answer (5) is positive. In *situation 4*, 40 is larger than 10. The sign of 40 is negative. The sign of the answer (−30) is negative.

We will now state these results as a formal rule.

ADDITION RULE FOR TWO NUMBERS WITH DIFFERENT SIGNS

1. Find the difference between the larger absolute value and the smaller absolute value.
2. Give the answer the sign of the number having the larger absolute value.

Example 9 Add. $8 + (-7)$

Solution

$8 + (-7)$	We are to add two numbers with different signs.
$8 - 7 = 1$	Find the difference between the two absolute values, which is 1.
$+8 + (-7) = +1$ or 1	The answer will have the sign of the number with the larger absolute value. That number is +8. Its sign is **positive,** so the answer will be +1.

□

Student Practice 9 Add. $-9 + 15$

5 Using the Addition Properties for Real Numbers 

It is useful to know the following three properties of real numbers.

1. *Addition is commutative.*
   This property states that if two numbers are added, the result is the same no matter which number is written first. The order of the numbers does not affect the result.

$$3 + 6 = 6 + 3 = 9$$
$$-7 + (-8) = (-8) + (-7) = -15$$
$$-15 + 3 = 3 + (-15) = -12$$

2. *Addition of zero to any given number will result in that given number again.*

$$0 + 5 = 5$$
$$-8 + 0 = -8$$

3. *Addition is associative.*
   This property states that if three numbers are added, it does not matter which two numbers are grouped by parentheses and added first.

$$3 + (5 + 7) = (3 + 5) + 7$$
$$3 + (12) = (8) + 7$$
$$15 = 15$$

First combine numbers inside parentheses; then combine the remaining numbers. The results are the same no matter which numbers are grouped and added first.

We can use these properties along with the rules we have for adding real numbers to add three or more numbers. We go from left to right, adding two numbers at a time.

Example 10 Add. $\frac{2}{15} + \left(-\frac{8}{15}\right) + \frac{1}{15}$

Solution

$-\frac{6}{15} + \frac{1}{15}$ Add $\frac{2}{15} + (-\frac{8}{15}) = -\frac{6}{15}$. The answer is negative since the larger of the two absolute values is negative.

$= -\frac{1}{3}$ Add $-\frac{6}{15} + \frac{1}{15} = -\frac{5}{15} = -\frac{1}{3}$. The answer is negative since the larger of the two absolute values is negative. □

Student Practice 10 Add.

$$-\frac{5}{12} + \frac{7}{12} + \left(-\frac{11}{12}\right)$$

Sometimes the numbers being added have the same signs; sometimes the signs are different. When adding three or more numbers, you may encounter both situations.

Example 11 Add. $-1.8 + 1.4 + (-2.6)$

Solution

$-0.4 + (-2.6)$ Add $-1.8 + 1.4$. We take the difference of 1.8 and 1.4 and use the sign of the number with the larger absolute value.

$= -3.0$ Add $-0.4 + (-2.6) = -3.0$. The signs are the same; we add the absolute values of the numbers and use the common sign. □

Student Practice 11 Add. $-6.3 + (-8.0) + 3.5$

If many real numbers are added, it is often easier to add numbers with the same sign in a column format. Remember that addition is commutative and associative; therefore, real numbers can be added *in any order*. You do *not* need to combine the first two numbers as your first step.

Example 12 Add. $-8 + 3 + (-5) + (-2) + 6 + 5$

Solution

-8		$+3$	
-5	All the signs are the same.	$+6$	All the signs are the same.
$\underline{-2}$	Add the three negative	$\underline{+5}$	Add the three positive
-15	numbers to obtain -15.	$+14$	numbers to obtain $+14$.

Add the two results.

$$-15 + 14 = -1$$

The answer is negative because the number with the larger absolute value is negative. □

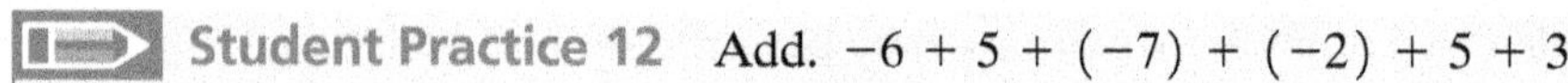

Student Practice 12 Add. $-6 + 5 + (-7) + (-2) + 5 + 3$

A word about notation: The only time we really need to show the sign of a number is when the number is negative, for example -3. The only time we need to show parentheses when we add real numbers is when we have two different signs preceding a number. For example, $-5 + (-6)$.

Example 13 Add.

(a) $2.8 + (-1.3)$

(b) $-\frac{2}{5} + \left(-\frac{3}{4}\right)$

Solution

(a) $2.8 + (-1.3) = 1.5$

(b) $-\frac{2}{5} + \left(-\frac{3}{4}\right) = -\frac{8}{20} + \left(-\frac{15}{20}\right) = -\frac{23}{20}$ or $-1\frac{3}{20}$ □

Student Practice 13 Add.

(a) $-2.9 + (-5.7)$

(b) $\frac{2}{3} + \left(-\frac{1}{4}\right)$

1.1 Exercises

Verbal and Writing Skills, Exercises 1–10

Check off any description of the number that applies.

	Number	Integer	Rational Number	Irrational Number	Real Number
1.	23				
2.	$-\frac{4}{5}$				
3.	π				
4.	2.34				
5.	−6.666...				

	Number	Integer	Rational Number	Irrational Number	Real Number
6.	$-\frac{7}{9}$				
7.	−2.3434...				
8.	14				
9.	$\sqrt{2}$				
10.	3.232232223...				

Use a real number to represent each situation.

11. Jules Verne wrote a book with the title *20,000 Leagues under the Sea.*

12. The value of the dollar is up \$0.07 with respect to the yen.

13. Ramona lost $37\frac{1}{2}$ pounds on Weight Watchers.

14. The scouts hiked from sea level to the top of a 3642-foot-high mountain.

15. The temperature rises 7°F.

16. The lowest point in Australia is Lake Eyre at 52 feet below sea level.

Find the additive inverse (opposite).

17. 8

18. $-\frac{3}{7}$

19. −2.73

20. 85.4

Find the absolute value.

21. $|-1.3|$

22. $|-5.9|$

23. $\left|\frac{5}{6}\right|$

24. $\left|\frac{3}{11}\right|$

Add.

25. $-8 + (-7)$

26. $-14 + (-3)$

27. $-20 + (-30)$

28. $(-17) + (-23)$

29. $-\frac{7}{20} + \frac{13}{20}$

30. $-\frac{3}{7} + \left(-\frac{2}{7}\right)$

31. $-\frac{2}{13} + \left(-\frac{5}{13}\right)$

32. $-\frac{3}{16} + \frac{5}{16}$

33. $-\frac{2}{5} + \frac{3}{7}$

34. $-\frac{2}{7} + \frac{3}{14}$

35. $-10.3 + (-8.9)$

36. $-5.4 + (-12.8)$

37. $0.6 + (-0.2)$

38. $-0.8 + 0.5$

39. $-5.26 + (-8.9)$

40. $-6.48 + (-3.7)$

41. $-8 + 5 + (-3)$

42. $5 + (-9) + (-2)$

43. $-2 + (-8) + 10$

44. $-8 + 7 + (-15)$

45. $-\frac{3}{10}+\frac{3}{4}$

46. $-\frac{3}{8}+\frac{11}{24}$

47. $-14+9+(-3)$

48. $-18+10+(-5)$

Mixed Practice *Add.*

49. $8+(-11)$

50. $15+(-26)$

51. $-83+142$

52. $-114+186$

53. $-\frac{4}{9}+\frac{5}{6}$

54. $-\frac{3}{5}+\frac{2}{3}$

55. $-\frac{1}{10}+\frac{1}{2}$

56. $-\frac{2}{3}+\left(-\frac{1}{4}\right)$

57. $5.18+(-7.39)$

58. $8.33+(-14.2)$

59. $4+(-8)+16$

60. $38+(-15)+(-6)$

61. $26+(-19)+12+(-31)$

62. $-16+12+(-26)+15$

63. $17.85+(-2.06)+0.15$

64. $28.37+4.08+(-16.98)$

Applications

65. ***Profit/Loss*** Holly paid \$47 for a vase at an estate auction. She resold it to an antiques dealer for \$214. What was her profit or loss?

66. ***Temperature*** When we skied at Jackson Hole, Wyoming, yesterday, the temperature at the summit was −12°F. Today when we called the ski report, the temperature had risen 7°F. What is the temperature at the summit today?

67. ***Home Equity Line of Credit*** Ramon borrowed \$2300 from his home equity line of credit to pay off his car loan. He then borrowed another \$1500 to pay to have his kitchen repainted. Represent how much Ramon owed on his home equity line of credit as a real number.

68. ***Time Change*** During the winter, New York City is on Eastern Standard Time (EST). Melbourne, Australia, is 15 hours ahead of New York. If it is 11 P.M. in Melbourne, what time is it in New York?

69. ***Football*** During the Homecoming football game, Quentin lost 15 yards, gained 3 yards, and gained 21 yards in three successive running plays. What was his total gain or loss?

70. ***School Fees*** Wanda's financial aid account at school held \$643.85. She withdrew \$185.50 to buy books for the semester. Does she have enough left in her account to pay the \$475.00 registration fee for the next semester? If so, how much extra money does she have? If not, how much is she short?

71. ***Butterfly Population*** The population of a particular butterfly species was 8000. Twenty years later there were 3000 fewer. Today, there are 1500 fewer. Study the graph to the right. What is the new population?

72. ***Credit Card Balance*** Aaron owes \$258 to a credit card company. He makes a purchase of \$32 with the card and then makes a payment of \$150 on the account. How much does he still owe?

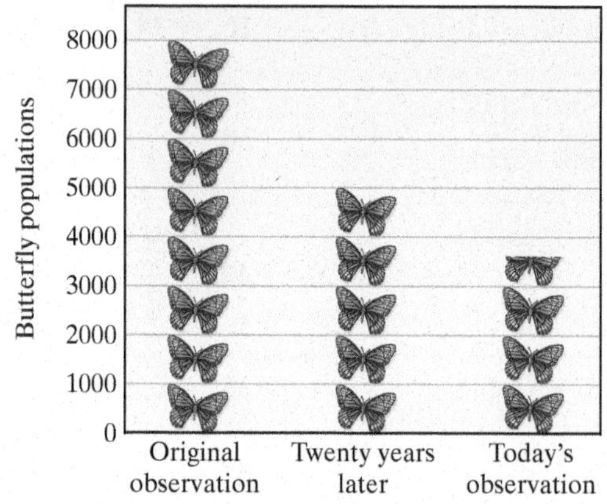

Profit/Loss *During the first five months of 2015, a regional Midwest airline posted profit and loss figures for each month of operation, as shown in the accompanying bar graph.*

73. For the first three months of 2015, what was the total earnings of the airline?

74. For the first five months of 2015, what was the total earnings for the airline?

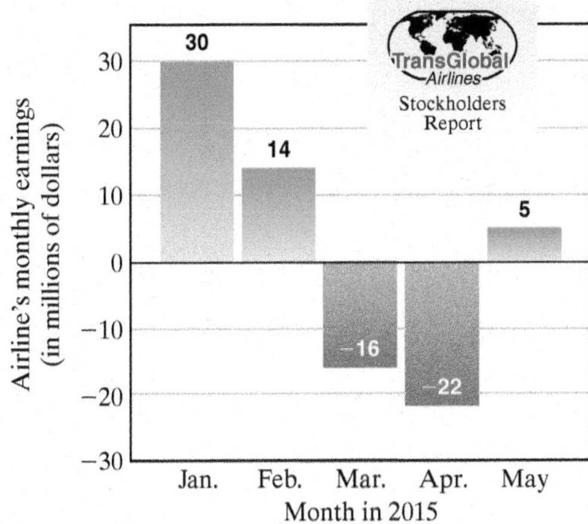

To Think About

75. What number must be added to -13 to get 5?

76. What number must be added to -18 to get 10?

Cumulative Review *Perform the calculations indicated.*

77. **[0.2.3]** $\frac{15}{16} + \frac{1}{4}$

78. **[0.3.1]** $\frac{3}{7} \times \frac{14}{9}$

79. **[0.2.3]** $\frac{2}{15} - \frac{1}{20}$

80. **[0.3.2]** $2\frac{1}{2} \div 3\frac{2}{5}$

81. **[0.4.4]** $0.72 + 0.8$

82. **[0.4.4]** $1.63 - 0.98$

83. **[0.4.5]** 1.63×0.7

84. **[0.4.6]** $0.208 \div 0.8$

Quick Quiz 1.1 *Add.*

1. $-18 + (-16)$

2. $-2.7 + 8.6 + (-5.4)$

3. $-\frac{5}{6} + \frac{7}{24}$

4. **Concept Check** Explain why when you add two negative numbers, you always obtain a negative number, but when you add one negative number and one positive number, you may obtain zero, a positive number, or a negative number.

1.2 Subtracting Real Numbers

Student Learning Objective

After studying this section, you will be able to:

1 Subtract real numbers with the same or different signs.

1 Subtracting Real Numbers with the Same or Different Signs

So far we have developed the rules for adding real numbers. We can use these rules to subtract real numbers. Let's look at a bank account situation to see how.

Situation 5: *Subtract a Deposit and Add a Debit* You have a balance of 20 dollars in your debit card account. The bank calls you and says that a deposit of 5 dollars that belongs to another account was erroneously added to your account. They say they will correct the account balance to 15 dollars. You want to keep track of what's happening to your account.

Your record for situation 5.

$$20 - (+5) = 15$$

From your present balance *subtract* the *deposit* to get the new balance.

This equation shows what needs to be done to your account. The bank tells you that because the error happened in the past, they cannot "take it away." However, they can add to your account a debit of 5 dollars. Here is the equivalent addition.

$$20 + (-5) = 15$$

To your present balance *add* a *debit* to get the new balance.

We see that subtracting a positive 5 has the same effect as adding a negative 5.

SUBTRACTION OF REAL NUMBERS

To subtract real numbers, add the opposite of the second number (that is, the number you are subtracting) to the first.

The rule tells us to do three things when we subtract real numbers. First, change subtraction to addition. Second, replace the second number by its opposite. Third, add the two numbers using the rules for addition of real numbers.

Example 1 Subtract. $6 - (-2)$

Solution $6 \quad - \quad (-2)$

Change subtraction to addition.

Write the opposite of the second number.

$= 6 \quad + \quad (+2)$

Add the two real numbers with the same sign.

$= 8$

Student Practice 1 Subtract. $9 - (-3)$

Example 2 Subtract. $-8 - (-6)$

Solution

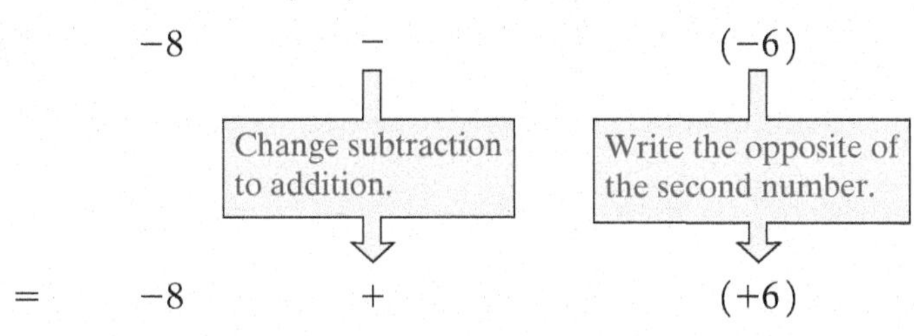

 **Student Practice 2** Subtract. $-12 - (-5)$

Example 3 Subtract.

(a) $\frac{3}{7} - \frac{6}{7}$ **(b)** $-\frac{7}{18} - \left(-\frac{1}{9}\right)$

Solution

(a) $\frac{3}{7} - \frac{6}{7} = \frac{3}{7} + \left(-\frac{6}{7}\right)$ Change the subtraction problem to one of adding the opposite of the second number. We note that the problem has two fractions with the same denominator.

$= -\frac{3}{7}$ Add two numbers with different signs.

(b) $-\frac{7}{18} - \left(-\frac{1}{9}\right) = -\frac{7}{18} + \frac{1}{9}$ Change subtracting to adding the opposite.

$= -\frac{7}{18} + \frac{2}{18}$ Change $\frac{1}{9}$ to $\frac{2}{18}$ since LCD $= 18$.

$= -\frac{5}{18}$ Add two numbers with different signs. □

Student Practice 3 Subtract.

(a) $\frac{5}{9} - \frac{7}{9}$ **(b)** $-\frac{5}{21} - \left(-\frac{3}{7}\right)$

> **Calculator**
>
> **More with Negative Numbers**
>
> Subtract. Remember to use the [+/−] key to change the sign of a number from + to − or from − to +.
>
> **(a)** $-18 - (-24)$
>
> **(b)** $-6 + (-10) - (-15)$
>
> **Ans:**
>
> **(a)** 6 **(b)** −1

Example 4 Subtract. $-5.2 - (-5.2)$

Solution

$-5.2 - (-5.2) = -5.2 + 5.2$ Change the subtraction problem to one of adding the opposite of the second number.

$= 0$ Add two numbers with different signs. □

Student Practice 4 Subtract. $-17.3 - (-17.3)$

Example 4 illustrates what is sometimes called the **additive inverse property.** When you add two real numbers that are opposites of each other, you will obtain zero. Examples of this are the following:

$$5 + (-5) = 0 \qquad -186 + 186 = 0 \qquad -\frac{1}{8} + \frac{1}{8} = 0.$$

Example 5 Subtract.

(a) $-8 - 2$ **(b)** $23 - 28$ **(c)** $5 - (-3)$ **(d)** $\frac{1}{4} - 8$

Solution To subtract, we add the opposite of the second number to the first.

(a) $-8 - 2 = -8 + (-2) = -10$

In a similar fashion we have

(b) $23 - 28 = 23 + (-28) = -5$

(c) $5 - (-3) = 5 + 3 = 8$

(d) $\frac{1}{4} - 8 = \frac{1}{4} + (-8) = \frac{1}{4} + \left(-\frac{32}{4}\right) = -\frac{31}{4}$ or $-7\frac{3}{4}$ □

Student Practice 5 Subtract.

(a) $-21 - 9$ **(b)** $17 - 36$ **(c)** $12 - (-15)$ **(d)** $\frac{3}{5} - 2$

Example 6 A satellite is recording radioactive emissions from nuclear waste buried 3 miles below sea level. The satellite orbits Earth at 98 miles above sea level. How far is the satellite from the nuclear waste?

Solution We want to find the difference between +98 miles and −3 miles. This means we must subtract −3 from 98.

$$98 - (-3) = 98 + 3 = 101$$

The satellite is 101 miles from the nuclear waste. □

Student Practice 6 A helicopter is directly over a sunken vessel. The helicopter is 350 feet above sea level. The vessel lies 186 feet below sea level. How far is the helicopter from the sunken vessel?

STEPS TO SUCCESS What Are the Absolute Essentials to Succeed in This Course?

Students who are successful in this course find there are six absolute essentials.
Here they are:

1. Attend every class session.
2. Read the textbook for every assigned section.
3. Take notes in class.
4. Do the assigned homework for every class.
5. Get help immediately when you need assistance.
6. Review what you are learning.

Making it personal: Which of the six suggestions above is the one you have the greatest trouble actually doing? Will you make a personal commitment to doing that one thing consistently for the next two weeks? You will be amazed at the results. ▼

If you are in an online class or a nontraditional class:

Students in an online class or a self-paced class find it really helps to spread the homework and the reading over five different days each week.

Making it personal: Will you try to do your homework for five days each week over the next two weeks? You will be amazed at the results. Write out a study plan for the next two weeks. ▼

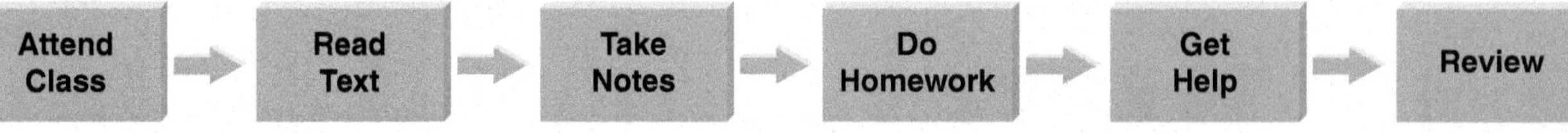

1.2 Exercises MyMathLab®

Verbal and Writing Skills, Exercises 1 and 2

1. Explain in your own words how you would perform the necessary steps to find $-8 - (-3)$.

2. Explain in your own words how you would perform the necessary steps to find $-10 - (-15)$.

Subtract by adding the opposite.

3. $27 - 49$
4. $23 - 57$
5. $19 - 23$
6. $8 - 19$
7. $-14 - (-3)$
8. $-17 - (-13)$
9. $-52 - (-60)$
10. $-48 - (-80)$
11. $0 - (-5)$
12. $0 - (-7)$
13. $-18 - (-18)$
14. $-24 - (-24)$
15. $-17 - (-20)$
16. $-11 - (-19)$
17. $\frac{2}{5} - \frac{4}{5}$
18. $\frac{2}{9} - \frac{7}{9}$
19. $\frac{3}{4} - \left(-\frac{3}{5}\right)$
20. $-\frac{2}{3} - \frac{1}{4}$
21. $-\frac{3}{4} - \frac{5}{6}$
22. $-\frac{7}{10} - \frac{10}{15}$
23. $-0.6 - 0.3$
24. $-0.9 - 0.5$
25. $2.64 - (-1.83)$
26. $-0.03 - 0.06$

Mixed Practice *Calculate.*

27. $\frac{3}{5} - 4$
28. $\frac{5}{6} - 3$
29. $-\frac{2}{3} - 4$
30. $-\frac{1}{6} - 5$
31. $34 - 87$
32. $19 - 76$
33. $-25 - 48$
34. $-74 - 11$
35. $2.3 - (-4.8)$
36. $8.4 - (-2.7)$
37. $8 - \left(-\frac{3}{4}\right)$
38. $\frac{2}{3} - (-6)$
39. $\frac{5}{6} - 7$
40. $9 - \frac{2}{3}$
41. $-\frac{2}{7} - \frac{4}{5}$
42. $-\frac{5}{6} - \frac{1}{5}$
43. $-135 - (-126.5)$
44. $-97.6 - (-146)$
45. $\frac{1}{5} - 6$
46. $\frac{2}{7} - (-3)$
47. $4.5 - (-1.56)$
48. $5.2 - (-3.88)$
49. $-3 - 2.047$
50. $-1.043 - 4$
51. Subtract -9 from -2.
52. Subtract -12 from 20.
53. Subtract 13 from -35.

One Step Further *Change each subtraction operation to "adding the opposite." Then combine the numbers.*

54. $9 + (-7) - 5$

55. $7 + (-6) - 3$

56. $-18 + 12 - (-6)$

57. $-13 + 12 - (-1)$

58. $-23 - (-12) - (-4) + 17$

59. $16 + (-20) - (-15) - 1$

60. $-8.3 - (-2.6) + 1.9$

61. $-7.8 - (-5.2) + 3.7$

Applications

62. ***Sea Rescue*** A rescue helicopter is 300 feet above sea level. The captain has located an ailing submarine directly below it that is 126 feet below sea level. How far is the helicopter from the submarine?

63. ***Sea Rescue*** Suppose a rescue helicopter 600 feet above sea level descends 300 feet to search for debris or any other signs of a submarine in trouble. When the captain does not see anything, he ascends 200 feet and stays at that altitude while he tries to get a radio signal from the submarine. A few moments later he speaks to the captain and is informed that the submarine is 126 feet below sea level. How far is the helicopter from the submarine?

64. ***Debit Card Account Balance*** Yesterday Jackie had \$156 in her debit card account. Today her account reads "balance −\$37." Find the difference in these two amounts.

Cumulative Review *In exercises 65–69, perform the operations indicated.*

65. **[1.1.4]** $-37 + 16$

66. **[1.1.3]** $-37 + (-14)$

67. **[1.1.3]** $-3 + (-6) + (-10)$

68. **[1.1.4]** ***Temperature*** On a winter morning, Alisa noticed that the outside temperature was −5°F. By the afternoon the temperature had risen 20°F. What was the afternoon temperature?

69. **[0.3.1]** ***Hiking*** Sean and Khalid went hiking in the Blue Ridge Mountains. During their $8\frac{1}{3}$-mile hike, $\frac{4}{5}$ of the distance was covered with snow. How many miles were snow covered?

Quick Quiz 1.2 *Subtract.*

1. $-8 - (-15)$

2. $-1.3 - 0.6$

3. $\frac{5}{8} - \left(-\frac{2}{7}\right)$

4. **Concept Check** Explain the different results that are possible when you start with a negative number and then subtract a negative number.

1.3 Multiplying and Dividing Real Numbers

1 Multiplying Real Numbers

Student Learning Objectives

After studying this section, you will be able to:

1. Multiply real numbers.
2. Use the multiplication properties for real numbers.
3. Divide real numbers.

We are familiar with the meaning of multiplication for positive numbers. For example, $5 \times 90 = 450$ might mean that you receive five weekly checks of 90 dollars each and you gain \$450. Let's look at a situation that corresponds to $5 \times (-90)$. What might that mean?

Situation 6: *Checking an Account Balance* You write a check for five weeks in a row to pay your weekly room rent of 90 dollars. If you do not have a checking account and instead have a debit card account, a similar result would occur if you were charged \$90 each week for five weeks. You want to know the total impact on your account balance.

Your record for situation 6.

$(+5)$	$\times$	(-90)	$=$	-450
The number of checks you have written	times	negative 90, the value of each check that was a debit to your account,	gives	negative 450 dollars, a net debit to your account

Note that a multiplication symbol is not needed between the $(+5)$ and the (-90) because the two sets of parentheses indicate multiplication. The multiplication $(5)(-90)$ is the same as repeated addition of five (-90)s. Note that 5 multiplied by -90 can be written as $5(-90)$ or $(5)(-90)$.

$$(-90) + (-90) + (-90) + (-90) + (-90) = -450$$

This example seems to show that a positive number multiplied by a negative number is negative.

What if the negative number is the one that is written first? If $(5)(-90) = -450$ then $(-90)(5) = -450$ by the commutative property of multiplication. This is an example showing that *when two numbers with different signs* (one positive, one negative) *are multiplied, the result is negative.*

But what if both numbers are negative? Consider the following situation.

Situation 7: *Renting a Room* Last year at college you rented a room at 90 dollars per week for 36 weeks, which included two semesters and summer school. This year you will not attend the summer session, so you will be renting the room for only 30 weeks. Thus the number of weekly rental checks will be six less than last year. You are making out your budget for this year. You want to know the financial impact of renting the room for six fewer weeks.

Your record for situation 7.

(-6)	$\times$	(-90)	$=$	540
The difference in the number of checks this year compared to last year is -6, which is negative to show a decrease,	times	-90, the value of each check paid out,	gives	$+540$ dollars. The product is positive, because your financial situation will be 540 dollars better this year.

You could check that the answer is positive by calculating the total rental expenses.

	Dollars in rent last year	$(36)(90)$	$= 3240$
(subtract)	Dollars in rent this year	$-(30)(90)$	$= -2700$
	Extra dollars available this year		$= +540$

This agrees with our previous answer: $(-6)(-90) = +540$.

Note that -6 times -90 can be written as $-6(-90)$ or $(-6)(-90)$.

In this situation it seems reasonable that a negative number times a negative number yields a positive answer. We already know from arithmetic that a positive number times a positive number yields a positive answer. Thus we might see the general rule that *when two numbers with the same sign* (both positive or both negative) *are multiplied, the result is positive.*

We will now state our rule.

> **MULTIPLICATION OF REAL NUMBERS**
>
> To multiply two real numbers with **the same sign,** multiply the absolute values.
> The sign of the result is **positive.**
>
> To multiply two real numbers with **different signs,** multiply the absolute values.
> The sign of the result is **negative.**

Example 1 Multiply.

(a) $(3)(6)$ **(b)** $\left(-\frac{5}{7}\right)\left(-\frac{2}{9}\right)$ **(c)** $-4(8)$ **(d)** $\left(\frac{2}{7}\right)(-3)$

Solution

(a) $(3)(6) = 18$

(b) $\left(-\frac{5}{7}\right)\left(-\frac{2}{9}\right) = \frac{10}{63}$

When multiplying two numbers with the same sign, the result is a positive number.

(c) $-4(8) = -32$

(d) $\left(\frac{2}{7}\right)(-3) = \left(\frac{2}{7}\right)\left(-\frac{3}{1}\right) = -\frac{6}{7}$

When multiplying two numbers with different signs, the result is a negative number.

Student Practice 1 Multiply.

(a) $(-6)(-2)$ **(b)** $(7)(9)$ **(c)** $\left(-\frac{3}{5}\right)\left(\frac{2}{7}\right)$ **(d)** $\left(\frac{5}{6}\right)(-7)$

To multiply more than two numbers, multiply two numbers at a time.

Example 2 Multiply. $(-4)(-3)(-2)$

Solution

$(-4)(-3)(-2) = (+12)(-2)$ We begin by multiplying the first two numbers, (-4) and (-3). The signs are the same. The answer is positive 12.

$= -24$ Now we multiply $(+12)$ and (-2). The signs are different. The answer is negative 24.

Student Practice 2 Multiply. $(-5)(-2)(-6)$

Example 3 Multiply.

(a) $-3(-1.5)$ **(b)** $\left(-\frac{1}{2}\right)(-1)(-4)$ **(c)** $-2(-2)(-2)(-2)$

Solution Multiply two numbers at a time. See if you find a pattern.

(a) $-3(-1.5) = 4.5$ Be sure to place the decimal point in your answer.

(b) $\left(-\frac{1}{2}\right)(-1)(-4) = +\frac{1}{2}(-4) = -2$

(c) $-2(-2)(-2)(-2) = +4(-2)(-2) = -8(-2) = +16$ or 16

What kind of answer would we obtain if we multiplied five negative numbers? If you guessed "negative," you probably see the pattern. □

Student Practice 3 Determine the sign of the product. Then multiply to check.

(a) $-2(-3)$ **(b)** $(-1)(-3)(-2)$

(c) $-4\left(-\frac{1}{4}\right)(-2)(-6)$

When you multiply two or more nonzero real numbers:

1. The result is always **positive** if there is an **even** number of negative signs.
2. The result is always **negative** if there is an **odd** number of negative signs.

2 Using the Multiplication Properties for Real Numbers 

For convenience, we will list the properties of multiplication.

1. *Multiplication is commutative.*
   This property states that if two real numbers are multiplied, the order of the numbers does not affect the result. The result is the same no matter which number is written first.
   $$(5)(7) = (7)(5) = 35, \quad \left(\frac{1}{3}\right)\left(\frac{2}{7}\right) = \left(\frac{2}{7}\right)\left(\frac{1}{3}\right) = \frac{2}{21}$$
2. *Multiplication of any real number by zero will result in zero.*
   $$(5)(0) = 0, \quad (-5)(0) = 0, \quad (0)\left(\frac{3}{8}\right) = 0, \quad (0)(0) = 0$$
3. *Multiplication of any real number by 1 will result in that same number.*
   $$(5)(1) = 5, \quad (1)(-7) = -7, \quad (1)\left(-\frac{5}{3}\right) = -\frac{5}{3}$$
4. *Multiplication is associative.*
   This property states that if three real numbers are multiplied, it does not matter which two numbers are grouped by parentheses and multiplied first.

   $2 \times (3 \times 4) = (2 \times 3) \times 4$ First multiply the numbers in parentheses. Then multiply the remaining numbers.

   $2 \times (12) = (6) \times 4$ The results are the same no matter which numbers are grouped and multiplied first.

   $24 = 24$

3 Dividing Real Numbers

What about division? Any division problem can be rewritten as a multiplication problem.

We know that $20 \div 4 = 5$ because $4(5) = 20$.
Similarly, $-20 \div (-4) = 5$ because $-4(5) = -20$.

In both division problems the answer is positive 5. Thus we see that *when you divide two numbers with the same sign* (both positive or both negative), *the answer is positive.* What if the signs are different?

We know that $-20 \div 4 = -5$ because $4(-5) = -20$.
Similarly, $20 \div (-4) = -5$ because $-4(-5) = 20$.

In these two problems the answer is negative 5. So we have reasonable evidence to see that *when you divide two numbers with different signs* (one positive and one negative), *the answer is negative.*

We will now state our rule for division.

DIVISION OF REAL NUMBERS

To divide two real numbers with **the same sign,** divide the absolute values. The sign of the result is **positive.**
To divide two real numbers with **different signs,** divide the absolute values. The sign of the result is **negative.**

Example 4 Divide.

(a) $12 \div 4$ **(b)** $(-25) \div (-5)$ **(c)** $\dfrac{-36}{18}$ **(d)** $\dfrac{42}{-7}$

Solution

(a) $12 \div 4 = 3$

(b) $(-25) \div (-5) = 5$

When dividing two numbers with the same sign, the result is a positive number.

(c) $\dfrac{-36}{18} = -2$

(d) $\dfrac{42}{-7} = -6$

When dividing two numbers with different signs, the result is a negative number.

□

Student Practice 4 Divide.

(a) $-36 \div (-2)$ **(b)** $49 \div 7$ **(c)** $\dfrac{50}{-10}$ **(d)** $\dfrac{-39}{13}$

Example 5 Divide. **(a)** $-36 \div 0.12$ **(b)** $-2.4 \div (-0.6)$

Solution

(a) $-36 \div 0.12$ Look at the problem to determine the sign. When dividing two numbers with different signs, the result will be a negative number.

We then divide the absolute values.

$$\begin{array}{r} 3\ 00. \\ 0.12_{\wedge}\overline{)36.00_{\wedge}} \\ \underline{36} \\ 00 \end{array}$$

Thus $-36 \div 0.12 = -300$. The answer is a negative number.

(b) $-2.4 \div (-0.6)$ Look at the problem to determine the sign. When dividing two numbers with the same sign, the result will be positive.

We then divide the absolute values.

$$0.6_{\wedge}\overline{)2.4_{\wedge}} = 4.$$
$$\underline{2\,4}$$

Thus $-2.4 \div (-0.6) = 4$. The answer is a positive number. □

Student Practice 5 Divide.

(a) $-12.6 \div (-1.8)$ **(b)** $0.45 \div (-0.9)$

Note that the rules for multiplication and division are the same. When you **multiply** or **divide** two numbers with the **same** sign, you obtain a **positive** number. When you **multiply** or **divide** two numbers with **different** signs, you obtain a **negative** number.

Example 6 Divide. $-\frac{12}{5} \div \frac{2}{3}$

Solution

$= \left(-\frac{12}{5}\right)\left(\frac{3}{2}\right)$ Divide two fractions. We invert the second fraction and multiply by the first fraction.

$= \left(-\frac{\overset{6}{\cancel{12}}}{5}\right)\left(\frac{3}{\underset{1}{\cancel{2}}}\right)$

$= -\frac{18}{5}$ or $-3\frac{3}{5}$ The answer is negative since the two numbers divided have different signs. □

Student Practice 6 Divide.

$$-\frac{5}{16} \div \left(-\frac{10}{13}\right)$$

Note that division can be indicated by the symbol $\div$ or by the fraction bar —. $\frac{2}{3}$ means $2 \div 3$.

Example 7 Divide. **(a)** $\dfrac{\frac{7}{8}}{-21}$ **(b)** $\dfrac{-\frac{2}{3}}{-\frac{7}{13}}$

Solution

(a) $\dfrac{\frac{7}{8}}{-21}$

$= \frac{7}{8} \div \left(-\frac{21}{1}\right)$ Change -21 to a fraction. $-21 = -\frac{21}{1}$

$= \frac{\overset{1}{\cancel{7}}}{8}\left(-\frac{1}{\underset{3}{\cancel{21}}}\right)$ Change the division to multiplication. Divide out common factor.

$= -\frac{1}{24}$ Simplify.

Continued on next page

(b) $\dfrac{-\dfrac{2}{3}}{-\dfrac{7}{13}} = -\dfrac{2}{3} \div \left(-\dfrac{7}{13}\right) = -\dfrac{2}{3}\left(-\dfrac{13}{7}\right) = \dfrac{26}{21}$ or $1\dfrac{5}{21}$ □

Student Practice 7 Divide.

(a) $\dfrac{-12}{-\dfrac{4}{5}}$ **(b)** $\dfrac{-\dfrac{2}{9}}{\dfrac{8}{13}}$

1. *Division of 0 by any nonzero real number gives 0 as a result.*

$$0 \div 5 = 0, \quad 0 \div \frac{2}{3} = 0, \quad \frac{0}{5.6} = 0, \quad \frac{0}{1000} = 0$$

You can divide zero by 5, $\frac{2}{3}$, 5.6, 1000, or any number (except 0).

2. *Division of any real number by 0 is* **undefined.**

$$7 \div 0 \qquad \frac{64}{0}$$
$$\uparrow \qquad \uparrow$$

Neither of these operations is possible. **Division by zero is undefined.**

You may be wondering why division by zero is undefined. Let us think about it for a minute. We said that $7 \div 0$ is undefined. Suppose there were an answer. Let us call the answer a. So we assume for a minute that $7 \div 0 = a$. Then it would have to follow that $7 = 0(a)$. But this is impossible. Zero times any number is zero. So we see that if there were such a number, it would contradict known mathematical facts. Therefore there is no number a such that $7 \div 0 = a$. Thus we conclude that division by zero is undefined.

When combining two numbers, it is important to be sure you know which rule applies. Think about the concepts in the following chart. See if you agree with each example.

Operation	**Two Real Numbers with the Same Sign**	**Two Real Numbers with Different Signs**
Addition	Result may be positive or negative. $9 + 2 = 11$ $-5 + (-6) = -11$	Result may be positive or negative. $-3 + 7 = 4$ $4 + (-12) = -8$
Subtraction	Result may be positive or negative. $15 - 6 = 15 + (-6) = 9$ $-12 - (-3) = -12 + 3 = -9$	Result may be positive or negative. $-12 - 3 = -12 + (-3) = -15$ $5 - (-6) = 5 + 6 = 11$
Multiplication	Result is always positive. $9(3) = 27$ $-8(-5) = 40$	Result is always negative. $-6(12) = -72$ $8(-3) = -24$
Division	Result is always positive. $150 \div 6 = 25$ $-72 \div (-2) = 36$	Result is always negative. $-60 \div 10 = -6$ $30 \div (-6) = -5$

Example 8 The Hamilton-Wenham Generals recently analyzed the 48 plays their team made while in possession of the football during their last game. The bar graph illustrates the number of plays made in each category. The team statistician prepared the following chart indicating the average number of yards gained or lost during each type of play.

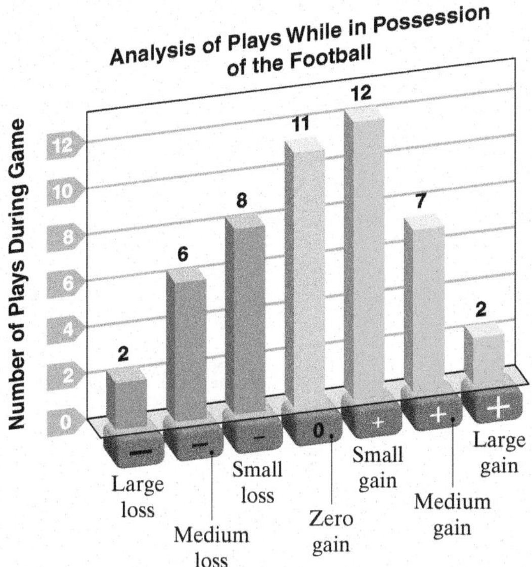

Type of Play	Average Yards Gained or Lost for Play
Large gain	+25
Medium gain	+15
Small gain	+5
Zero gain	0
Small loss	−5
Medium loss	−10
Large loss	−15

(a) How many yards were lost by the Generals in the plays that were considered small losses?

(b) How many yards were gained by the Generals in the plays that were considered small gains?

(c) If the total yards gained in small gains were combined with the total yards lost in small losses, what would be the result?

Solution

(a) We multiply the number of small losses by the average number of total yards lost on each small loss:

$$8(-5) = -40.$$

The team lost approximately 40 yards with plays that were considered small losses.

(b) We multiply the number of small gains by the average number of yards gained on each small gain:

$$12(5) = 60.$$

The team gained approximately 60 yards with plays that were considered small gains.

(c) We combine the results for **(a)** and **(b)**:

$$-40 + 60 = 20.$$

A total of 20 yards was gained during the plays that were small losses and small gains. □

Student Practice 8 Using the information provided in Example 8, answer the following:

(a) How many yards were lost by the Generals in the plays that were considered medium losses?

(b) How many yards were gained by the Generals in the plays that were considered medium gains?

(c) If the total yards gained in medium gains were combined with the total yards lost in medium losses, what would be the result?

1.3 Exercises MyMathLab®

Verbal and Writing Skills, Exercises 1 and 2

1. Explain in your own words the rule for determining the correct sign when multiplying two real numbers.

2. Explain in your own words the rule for determining the correct sign when multiplying three or more real numbers.

Multiply. Be sure to write your answer in the simplest form.

3. $8(-5)$

4. $9(-9)$

5. $0(-12)$

6. $0(136)$

7. $14(3.5)$

8. $7.5(8)$

9. $(-1.32)(-0.2)$

10. $(-2.3)(-0.11)$

11. $1.8(-2.5)$

12. $(3.4)(-2.2)$

13. $\left(\frac{3}{8}\right)(-4)$

14. $(5)\left(-\frac{7}{10}\right)$

15. $\left(-\frac{3}{5}\right)\left(-\frac{15}{11}\right)$

16. $\left(-\frac{4}{9}\right)\left(-\frac{3}{5}\right)$

17. $\left(\frac{12}{13}\right)\left(\frac{-5}{24}\right)$

18. $\left(\frac{14}{17}\right)\left(-\frac{3}{28}\right)$

Divide.

19. $0 \div (-9)$

20. $0 \div (-13)$

21. $-48 \div (-8)$

22. $-64 \div 8$

23. $-120 \div (-8)$

24. $-180 \div (-4)$

25. $156 \div (-13)$

26. $-0.6 \div 0.3$

27. $-9.1 \div 0.07$

28. $8.1 \div (-0.03)$

29. $0.54 \div (-0.9)$

30. $-7.2 \div 8$

31. $-6.3 \div 7$

32. $\frac{2}{7} \div \left(-\frac{3}{5}\right)$

33. $\left(-\frac{1}{5}\right) \div \left(\frac{2}{3}\right)$

34. $\left(-\frac{5}{6}\right) \div \left(-\frac{7}{18}\right)$

35. $-\frac{5}{7} \div \left(-\frac{3}{28}\right)$

36. $\left(-\frac{4}{9}\right) \div \left(-\frac{8}{15}\right)$

37. $\left(-\frac{7}{12}\right) \div \left(-\frac{5}{6}\right)$

38. $\dfrac{12}{-\frac{2}{5}}$

39. $\dfrac{-6}{-\frac{3}{7}}$

40. $\dfrac{-\frac{3}{8}}{-\frac{2}{3}}$

41. $\dfrac{\frac{-2}{3}}{\frac{8}{15}}$

42. $\dfrac{\frac{9}{2}}{-3}$

43. $\dfrac{\frac{8}{3}}{-4}$

Multiply. You may want to determine the sign of the product before you multiply.

44. $-6(2)(-3)(4)$

45. $-1(-2)(-3)(4)$

46. $-3(2)(-1)(-2)(5)$ **47.** $-2(4)(3)(-1)(-3)$ **48.** $-6(2)(-3)(0)(-9)$

49. $-3(0)\left(\frac{1}{3}\right)(-4)(2)$ **50.** $-2(0.14)(-3)(0.5)$ **51.** $25(-0.04)(-0.3)(-1)$

52. $\left(\frac{3}{7}\right)\left(-\frac{2}{3}\right)\left(-\frac{5}{3}\right)$ **53.** $\left(-\frac{4}{5}\right)\left(-\frac{6}{7}\right)\left(-\frac{1}{3}\right)$

54. $\left(-\frac{2}{3}\right)\left(-\frac{1}{4}\right)\left(\frac{3}{5}\right)\left(-\frac{2}{7}\right)$ **55.** $\left(-\frac{3}{4}\right)\left(-\frac{7}{15}\right)\left(-\frac{8}{21}\right)\left(-\frac{5}{9}\right)$

Mixed Practice *Take a minute to review the chart before Example 8. Be sure that you can remember the sign rules for each operation. Then do exercises 56–65. Perform the calculations indicated.*

56. $-5 - (-2)$ **57.** $-36 \div (-4)$ **58.** $-4(-8)$ **59.** $12 + (-8)$

60. $(-30) \div 5$ **61.** $8 - (-9)$ **62.** $-6 + (-3)$ **63.** $6(-12)$

64. $18 \div (-18)$ **65.** $-37 \div 37$

Applications

66. ***Stock Trading*** During one day in October 2015, the value of one share of Alamo Group stock opened at \$50.46. At the end of the day, it closed at \$47.24. If you owned 90 shares, what was your profit or loss that day?

67. ***Equal Contributions*** Ed, Ned, Ted, and Fred went camping. They each contributed an equal share of money toward food. Fred did the shopping. When he returned from the store, he had \$17.60 left. How much money did Fred give back to each person?

68. ***Student Loans*** Ramon will pay \$6480 on his student loan over the next three years. If \$180 is automatically deducted from his bank account each month to pay the loan off, how much does he still owe after one year?

69. ***Car Payments*** Muriel will pay the Volkswagen dealer a total of \$14,136 to be paid in 60 equal monthly installments. What is her monthly bill?

Football *The Beverly Panthers recently analyzed the 37 plays their team made while in possession of the football during their last game. The team statistician prepared the following chart indicating the number of plays in each category and the average number of yards gained or lost during each type of play. Use this chart to answer exercises 70–75.*

Type of Play	Number of Plays	Average Yards Gained or Lost per Play
Large gain	1	+25
Medium gain	6	+15
Small gain	4	+5
Zero gain	5	0
Small loss	10	−5
Medium loss	7	−10
Large loss	4	−15

70. How many yards were lost by the Panthers in the plays that were considered small losses?

71. How many yards were gained by the Panthers in the plays that were considered small gains?

72. If the total yards gained in small gains were combined with the total yards lost in small losses, what would be the result?

73. How many yards were lost by the Panthers in the plays that were considered medium losses?

74. How many yards were gained by the Panthers in the plays that were considered medium gains?

75. If the total yards gained in medium gains were combined with the total yards lost in medium losses, what would be the result?

Cumulative Review

76. **[1.1.5]** $-17.4 + 8.31 + 2.40$

77. **[1.1.5]** $-\frac{3}{4} + \left(-\frac{2}{3}\right) + \left(-\frac{5}{12}\right)$

78. **[1.2.1]** $-47 - (-32)$

79. **[1.2.1]** $-37 - 51$

Quick Quiz 1.3 *Perform the operations indicated.*

1. $\left(-\frac{3}{8}\right)(5)$

2. $-4(3)(-5)(-2)$

3. $-2.4 \div (-0.6)$

4. **Concept Check** Explain how you can determine the sign of the answer if you multiply several negative numbers.

STEPS TO SUCCESS Getting the Greatest Value from Your Homework

Read the textbook first before doing the homework. Take some time to read the text and study the examples. Try working out the Student Practice problems. You will be amazed at the amount of understanding you will obtain by studying the book before jumping into the homework exercises.

Take your time. Read the directions carefully. Be sure you understand what is being asked. Check your answers with those given in the back of the textbook. If your answer is incorrect, study similar examples in the text. Then redo the problem, watching for errors.

Make a schedule. You will need to allow two hours outside of class for each hour of actual class time. Make a weekly schedule of the times you have class. Now write down the times each day you will devote to doing math homework. Then write down the times you will spend doing homework for your other classes. If you have a job be sure to write down all your work hours.

Making it personal: Write down your own schedule of class, work, and study time. ▼

Sunday	Monday	Tuesday	Wednesday	Thursday	Friday	Saturday

1.4 Exponents

1 Writing Numbers in Exponent Form

In mathematics, we use exponents as a way to abbreviate repeated multiplication.

Long Notation		Exponent Form
$2 \cdot 2 \cdot 2 \cdot 2 \cdot 2 \cdot 2$	=	2^6

There are two parts to exponent notation: (1) the **base** and (2) the **exponent.** The **base** tells you what number is being multiplied and the **exponent** tells you how many times this number is used as a factor. (A *factor,* you recall, is a number being multiplied.)

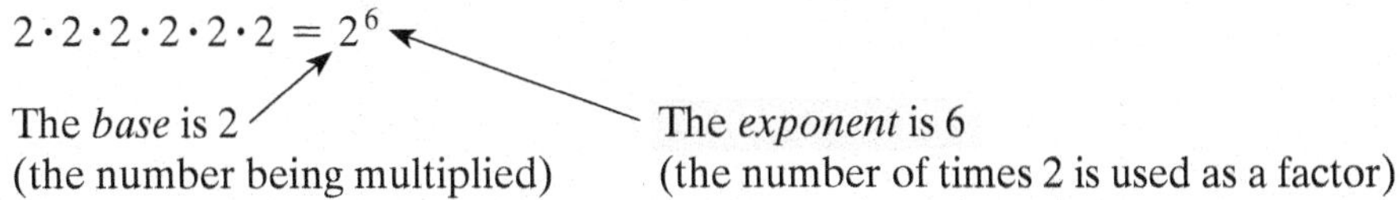

The *base* is 2 (the number being multiplied)   The *exponent* is 6 (the number of times 2 is used as a factor)

If the base is a *positive* real number, the exponent appears to the right and slightly above the level of the number as in, for example, 5^6 and 8^3. If the base is a *negative* real number, then parentheses are used around the number and the exponent appears outside the parentheses. For example, $(-2)(-2)(-2) = (-2)^3$.

In algebra, if we do not know the value of a number, we use a letter to represent the unknown number. We call the letter a **variable.** This is quite useful in the case of exponents. Suppose we do not know the value of a number, but we know the number is multiplied by itself several times. We can represent this with a variable base and a whole-number exponent. For example, when we have an unknown number, represented by the variable x, and this number occurs as a factor four times, we have

$$(x)(x)(x)(x) = x^4.$$

Likewise, if an unknown number, represented by the variable w, occurs as a factor five times, we have

$$(w)(w)(w)(w)(w) = w^5.$$

Example 1 Write in exponent form.

(a) $9(9)(9)$ **(b)** $13(13)(13)(13)$ **(c)** $-7(-7)(-7)(-7)(-7)$
(d) $-4(-4)(-4)(-4)(-4)(-4)$ **(e)** $(x)(x)$ **(f)** $(y)(y)(y)$

Solution

(a) $9(9)(9) = 9^3$ **(b)** $13(13)(13)(13) = 13^4$

(c) The -7 is used as a factor five times. The answer must contain parentheses. Thus $-7(-7)(-7)(-7)(-7) = (-7)^5$.

(d) $-4(-4)(-4)(-4)(-4)(-4) = (-4)^6$

(e) $(x)(x) = x^2$ **(f)** $(y)(y)(y) = y^3$ □

Student Practice 1 Write in exponent form.

(a) $6(6)(6)(6)$ **(b)** $-2(-2)(-2)(-2)(-2)$
(c) $108(108)(108)$ **(d)** $-11(-11)(-11)(-11)(-11)(-11)$
(e) $(w)(w)(w)$ **(f)** $(z)(z)(z)(z)$

If the base has an exponent of 2, we say the base is **squared.**

If the base has an exponent of 3, we say the base is **cubed.**

If the base has an exponent greater than 3, we say the base is raised **to the (exponent)-th power.**

x^2 is read "x squared."

y^3 is read "y cubed."

3^6 is read "three to the sixth power" or simply "three to the sixth."

Student Learning Objectives

After studying this section, you will be able to:

1. Write numbers in exponent form.
2. Evaluate numerical expressions that contain exponents.

2 Evaluating Numerical Expressions That Contain Exponents

Example 2 Evaluate.

(a) 2^5 **(b)** $2^3 + 4^4$

Solution

(a) $2^5 = (2)(2)(2)(2)(2) = 32$

(b) First we evaluate each power.

$2^3 = 8 \qquad 4^4 = 256$

Then we add. $8 + 256 = 264$ □

Student Practice 2 Evaluate.

(a) 3^5 **(b)** $2^2 + 3^3$

Calculator

Exponents

You can use a calculator to evaluate 3^5. Press the following keys:

3 y^x 5 =

The display should read

243

Try the following.

(a) 4^6 **(b)** $(0.2)^5$

(c) 18^6 **(d)** 3^{12}

Ans:

(a) 4096 **(b)** 0.00032

(c) 34,012,224 **(d)** 531,441

The steps needed to raise a number to a power are slightly different on some calculators.

If the base is negative, be especially careful in determining the sign. Notice the following:

$$(-3)^2 = (-3)(-3) = +9 \qquad (-3)^3 = (-3)(-3)(-3) = -27$$

From Section 1.3 we know that when you multiply two or more real numbers, first you multiply their absolute values.

- The result is positive if there is an even number of negative signs.
- The result is negative if there is an odd number of negative signs.

SIGN RULE FOR EXPONENTS

Suppose that a number is written in exponent form and the base is negative. The result is **positive** if the exponent is **even.** The result is **negative** if the exponent is **odd.**

Be careful how you read expressions with exponents and negative signs.

$(-3)^4$ means $(-3)(-3)(-3)(-3)$ or $+81$.

-3^4 means $-(3)(3)(3)(3)$ or -81.

Example 3 Evaluate.

(a) $(-2)^3$ **(b)** $(-4)^6$ **(c)** -3^6 **(d)** $-(5^4)$

Solution

(a) $(-2)^3 = -8$ The answer is negative since the base is negative and the exponent 3 is odd.

(b) $(-4)^6 = +4096$ The answer is positive since the exponent 6 is even.

(c) $-3^6 = -729$ The negative sign is not contained in parentheses. Thus we find 3 raised to the sixth power and then take the negative of that value.

(d) $-(5^4) = -625$ The negative sign is outside the parentheses. □

Student Practice 3 Evaluate.

(a) $(-3)^3$ **(b)** $(-2)^6$ **(c)** -2^4 **(d)** $-(6^3)$

Example 4 Evaluate.

(a) $\left(\frac{1}{2}\right)^4$ (b) $(0.2)^4$ (c) $\left(\frac{2}{5}\right)^3$

(d) $(3)^3(2)^5$ (e) $2^3 - 3^4$

Solution

(a) $\left(\frac{1}{2}\right)^4 = \left(\frac{1}{2}\right)\left(\frac{1}{2}\right)\left(\frac{1}{2}\right)\left(\frac{1}{2}\right) = \frac{1}{16}$

(b) $(0.2)^4 = (0.2)(0.2)(0.2)(0.2) = 0.0016$

(c) $\left(\frac{2}{5}\right)^3 = \left(\frac{2}{5}\right)\left(\frac{2}{5}\right)\left(\frac{2}{5}\right) = \frac{8}{125}$

(d) First we evaluate each power.
$3^3 = 27 \quad 2^5 = 32$

Then we multiply. $(27)(32) = 864$

(e) $2^3 - 3^4 = 8 - 81 = -73$ □

Student Practice 4 Evaluate.

(a) $\left(\frac{1}{3}\right)^3$

(b) $(0.3)^4$

(c) $\left(\frac{3}{2}\right)^4$

(d) $(3)^4(4)^2$

(e) $4^2 - 2^4$

STEPS TO SUCCESS Helping Your Accuracy

It is easy to make a mistake. But here are five ways to cut down on errors. Look over each one and think about how each suggestion can help you.

1. Work carefully and take your time. Do not rush through a problem just to get it done.
2. Concentrate on the problem. Sometimes your mind starts to wander. Then you get careless and will likely make a mistake.
3. Check your problem. Be sure you copied it correctly from the book.
4. Check your computations from step to step. Did you do each step correctly?
5. Check your final answer. Does it work? Is it reasonable?

Making it personal: Look over these five suggestions. Which one do you think will help you the most? Write down how you can use this suggestion to help you personally as you try to improve your accuracy. ▼

1.4 Exercises MyMathLab®

Verbal and Writing Skills, Exercises 1–6

1. Explain in your own words how to evaluate 3^4.

2. Explain in your own words how to evaluate 9^2.

3. Explain how you would determine whether $(-5)^3$ is negative or positive.

4. Explain how you would determine whether $(-2)^5$ is negative or positive.

5. Explain the difference between $(-2)^4$ and -2^4. What answers do you obtain when you evaluate the expressions?

6. Explain the difference between $(-3)^4$ and -3^4. What answers do you obtain when you evaluate the expressions?

Write in exponent form.

7. $(6)(6)(6)(6)(6)$

8. $(8)(8)(8)(8)(8)(8)$

9. $(w)(w)$

10. $(x)(x)(x)(x)$

11. $(p)(p)(p)(p)$

12. $(r)(r)(r)(r)(r)(r)(r)$

13. $(3q)(3q)(3q)$

14. $(2w)(2w)(2w)(2w)(2w)$

Evaluate.

15. 3^3

16. 9^2

17. 3^4

18. 7^3

19. 6^3

20. 12^2

21. $(-3)^3$

22. $(-2)^3$

23. $(-4)^2$

24. $(-5)^4$

25. -5^2

26. -4^2

27. $\left(\frac{1}{4}\right)^2$

28. $\left(\frac{3}{4}\right)^2$

29. $\left(\frac{2}{5}\right)^3$

30. $\left(\frac{1}{3}\right)^3$

31. $(2.1)^2$

32. $(1.5)^2$

33. $(0.2)^5$

34. $(0.7)^3$

35. $(-16)^2$

36. $(-7)^4$

37. -16^2

38. -7^4

Evaluate.

39. $5^3 + 6^2$

40. $6^2 + 2^3$

41. $10^2 - 11^2$

42. $5^3 - 2^2$

43. $(-4)^2 - (12)^2$

44. $14^2 - (-6)^2$

45. $2^5 - (-3)^2$

46. $7^2 - (-2)^4$

47. $(-5)^3(-2)^3$

Cumulative Review *Evaluate.*

48. **[1.1.5]** $(-11) + (-13) + 6 + (-9) + 8$

49. **[1.3.3]** $\frac{3}{4} \div \left(-\frac{9}{20}\right)$

50. **[1.2.1]** $-17 - (-9)$

51. **[1.3.1]** $(-2.1)(-1.2)$

52. **[0.5.3]** Amanda decided to invest her summer job earnings of $1600. At the end of the year she earned 6% on her investment. How much money did Amanda have at the end of the year?

Quick Quiz 1.4 *Evaluate.*

1. $(-4)^4$

2. $(1.8)^2$

3. $\left(\frac{3}{4}\right)^3$

4. **Concept Check** Explain the difference between $(-2)^6$ and -2^6. How do you decide if the answers are positive or negative?

1.5 The Order of Operations

Student Learning Objective

After studying this section, you will be able to:

1 Use the order of operations to simplify numerical expressions.

1 Using the Order of Operations to Simplify Numerical Expressions

It is important to know *when* to do certain operations as well as how to do them. For example, to simplify the expression $2 - 4 \cdot 3$, should we subtract first or multiply first?

Also remember that multiplication can be written several ways. Thus $4 \cdot 3$, 4×3, $4(3)$, and $(4)(3)$ all indicate that we are multiplying 4 times 3.

The following list will assist you. It tells which operations to do first: the correct **order of operations.** You might think of it as a *list of priorities.*

ORDER OF OPERATIONS FOR NUMBERS

Follow this order of operations:

Do first	1. Do all operations inside parentheses.
	2. Raise numbers to a power.
	3. Multiply and divide numbers from left to right.
Do last	4. Add and subtract numbers from left to right.

Let's return to the problem $2 - 4 \cdot 3$. There are no parentheses or numbers raised to a power, so the first thing we do is multiply. Then we subtract since this comes last on our list.

$2 - 4 \cdot 3 = 2 - 12$ Follow the order of operations by first multiplying $4 \cdot 3 = 12$.

$= -10$ Combine $2 - 12 = -10$.

Example 1 Evaluate. $8 \div 2 \cdot 3 + 4^2$

Solution

$8 \div 2 \cdot 3 + 4^2 = 8 \div 2 \cdot 3 + 16$ Evaluate $4^2 = 16$ because the highest priority in this problem is raising to a power.

$= 4 \cdot 3 + 16$ Next multiply and divide from left to right. We have $8 \div 2 = 4$.

$= 12 + 16$ $4 \cdot 3 = 12$

$= 28$ Finally, add. □

Student Practice 1 Evaluate. $25 \div 5 \cdot 6 + 2^3$

Note: Multiplication and division have equal priority. We do not do multiplication first. Rather, we work from left to right, doing any multiplication or division that we encounter. Similarly, addition and subtraction have equal priority.

Example 2 Evaluate. $(-3)^3 - 2^4$

Solution The highest priority is to raise the expressions to the appropriate powers.

$(-3)^3 - 2^4 = -27 - 16$ In $(-3)^3$ we are cubing the number -3 to obtain -27. Be careful; -2^4 is not $(-2)^4$! We raise 2 to the fourth power.

$= -43$ The last step is to add and subtract from left to right. □

Student Practice 2 Evaluate. $(-4)^3 - 2^6$

Example 3 Evaluate. $2 \cdot (2-3)^3 + 6 \div 3 + (8-5)^2$

Solution

$2 \cdot (2-3)^3 + 6 \div 3 + (8-5)^2$

$= 2 \cdot (-1)^3 + 6 \div 3 + 3^2$ Combine the numbers inside the parentheses. Note that we need parentheses for -1 because of the negative sign, but they are not needed for 3.

$= 2 \cdot (-1) + 6 \div 3 + 9$ Next, raise to a power.

$= -2 + 2 + 9$ Next, multiply and divide from left to right.

$= 0 + 9$

$= 9$ Finally, add and subtract from left to right.

Student Practice 3 Evaluate.
$6 - (8-12)^2 + 8 \div 2$

Example 4 Evaluate. $\left(-\frac{1}{5}\right)\left(\frac{1}{2}\right) - \left(\frac{3}{2}\right)^2$

Solution The highest priority is to raise $\frac{3}{2}$ to the second power.

$$\left(\frac{3}{2}\right)^2 = \left(\frac{3}{2}\right)\left(\frac{3}{2}\right) = \frac{9}{4}$$

$$\left(-\frac{1}{5}\right)\left(\frac{1}{2}\right) - \left(\frac{3}{2}\right)^2 = \left(-\frac{1}{5}\right)\left(\frac{1}{2}\right) - \frac{9}{4}$$

$= -\frac{1}{10} - \frac{9}{4}$ Next we multiply.

$= -\frac{1 \cdot 2}{10 \cdot 2} - \frac{9 \cdot 5}{4 \cdot 5}$ We need to write each fraction as an equivalent fraction with the LCD of 20.

$= -\frac{2}{20} - \frac{45}{20}$

$= -\frac{47}{20}$ or $-2\frac{7}{20}$ Add.

Student Practice 4 Evaluate.

$$\left(-\frac{1}{7}\right)\left(-\frac{14}{5}\right) + \left(-\frac{1}{2}\right) \div \left(\frac{3}{4}\right)^2$$

Calculator

Order of Operations

Use your calculator to evaluate $3 + 4 \cdot 5$. Enter

3 [+] 4 [×] 5 [=]

If the display is [23], the correct order of operations is built in. If the display is not 23, you will need to modify the way you enter the problem. You should use

4 [×] 5 [+] 3 [=]

Try $6 + 3 \cdot 4 - 8 \div 2$.

Ans:
14

1.5 Exercises MyMathLab®

Verbal and Writing Skills, Exercises 1–4

Game Points *You have lost a game of UNO and are counting the points left in your hand. You announce that you have three fours and six fives.*

1. Write this as a number expression.

2. How many points do you have in your hand?

3. What answer would you get for the number expression if you simplified it by

(a) performing the operations from left to right?

(b) following the order of operations?

4. Which procedure in exercise 3 gives the correct number of total points?

Evaluate.

5. $(7 - 9)^2 \div 2 \times 5$

6. $(5 - 8)^2 \div 3 \times 6$

7. $9 + 4(5 + 2 - 8)$

8. $13 + 2(-8 + 6 - 3)$

9. $8 - 2^3 \cdot 5 + 3$

10. $7 - 3^2 \cdot 4 + 5$

11. $4 + 42 \div 3 \cdot 2 - 8$

12. $7 + 36 \div 12 \cdot 3 - 14$

13. $3 \cdot 5 + 7 \cdot 3 - 5 \cdot 3$

14. $2 \cdot 6 + 5 \cdot 3 - 7 \cdot 4$

15. $8 - 5(2)^3 \div (-8)$

16. $11 - 3(4)^2 \div (-6)$

17. $3(5 - 7)^2 - 6(3)$

18. $-2(3 - 6)^2 - (-2)$

19. $5 \cdot 6 - (3 - 5)^2 + 8 \cdot 2$

20. $(-5)^2 + (15 \div 3) + 4 \cdot 2$

21. $\frac{1}{2} \div \frac{2}{3} + 6 \cdot \frac{1}{4}$

22. $\frac{5}{6} \div \frac{2}{3} - 6 \cdot \left(\frac{1}{2}\right)^2$

23. $0.8 + 0.3(0.6 - 0.2)^2$

24. $0.05 + 1.4 - (0.5 - 0.7)^3$

25. $\frac{3}{8}\left(-\frac{1}{6}\right) - \frac{7}{8} + \frac{1}{2}$

26. $\frac{1}{2} \div \frac{4}{5} - \frac{3}{4}\left(\frac{5}{6}\right)$

Mixed Practice

27. $(3 - 7)^2 \div 8 + 3$

28. $(5 - 6)^2 \cdot 7 - 4$

29. $\left(\frac{3}{4}\right)^2(-16) + \frac{4}{5} \div \frac{-8}{25}$

30. $\left(2\frac{4}{7}\right) \div \left(-1\frac{1}{5}\right)$

31. $-2.6 - (-1.8)(2.3) + (4.1)^2$

32. $5.15 + 4.2 \div (-0.3) - (3.5)^2$

33. $\left(\frac{2}{3}\right)^2 + \frac{1}{2} - \left(\frac{1}{3} - \frac{3}{4}\right) + \left(-\frac{1}{2}\right)^2$

34. $\left(\frac{1}{2}\right)^3 + \frac{1}{4} - \left(\frac{2}{3} - \frac{1}{6}\right) + \left(-\frac{1}{3}\right)^2$

Applications *Often in golf, it becomes important to keep track of eagles, bogies, pars, and birdies. Consider the following example during a year when Tiger Woods won the Buick Invitational. The following scorecard shows the results of Round 4. There are 18 holes per round and the table shows the number of times Woods got each type of score.*

Score on a Hole	Number of Times the Score Occurred
Eagle (−2)	1
Birdie (−1)	5
Par (0)	10
Bogey (+1)	2

35. Write his score as the sum of eagles, birdies, pars, and bogeys.

36. What was his final score for the round, when compared to par?

37. What answer do you get if you do the arithmetic left to right?

38. Explain why the answers in exercises 36 and 37 do not match.

Cumulative Review *Simplify.*

39. **[1.4.2]** $(0.5)^3$

40. **[1.2.1]** $-\frac{3}{4} - \frac{5}{6}$

41. **[1.4.2]** -1^{20}

42. **[0.3.2]** $3\frac{3}{5} \div 6\frac{1}{4}$

Quick Quiz 1.5 *Evaluate.*

1. $7 - 3^4 + 2 - 5$

2. $(0.3)^2 - 4.2(-4) + 0.07$

3. $(7 - 9)^4 + 22 \div (-2) + 6$

4. **Concept Check** Explain in what order you would perform the calculations to evaluate the expression? $4 - (-3 + 4)^3 + 12 \div (-3)$.

Use Math to Save Money

Did You Know? Having a budget can help you control how you spend and save your money.

Time to Budget

Understanding the Problem:

One way to improve your financial situation is to learn to manage the money you have with a budget. A budget can maximize your efforts to ensure you have enough money to cover your fixed expenses, as well as your variable expenses.

Michael is a teacher. One of Michael's goals is to go back to school to earn his master's degree in education. He knows that this will not only help him further his career but also provide better financial stability in the long run. His net monthly income is currently $2500. So, Michael knows he'll need to put himself on a budget for a period of time in order to save money to go back to school.

Making a Plan:

Michael needs to budget his expenses to help control his spending and to save for college. He wants to see how long it will take to save up the needed money.

Step 1: Research shows that consumer credit counseling services recommend allocating the following percentages for each category of the monthly budget:

Housing	25%
Transportation	10%
Savings	5%
Utilities	5%
Debt Payments	20%
Food	15%
Misc.	20%

Task 1: If Michael follows this plan, how much will he put away in savings at the end of one year?

Task 2: If Michael follows this plan, how much will he spend on food by the end of one year?

Step 2: After investigating several schools in his area, Michael chooses to attend a state college that offers a one-year program for the degree he wishes to pursue. Michael will need $3000 for tuition and fees the first semester, and $450 for textbooks.

Task 3: What is the total cost for books and fees, and textbooks for one semester?

Finding a Solution:

Step 3: By controlling his spending Michael is able to save for his master's degree.

Task 4: How many months will Michael have to save to pay for one semester at this college? Round to the nearest month.

Task 5: If Michael can increase his savings to 10% of his monthly budget by making cuts in other areas, how long would he have to save to pay for one semester? Round to the nearest month.

Step 4: Once Michael earns his advanced degree, his salary will increase on the following schedule:

- Year 1 an additional $4200
- Year 2 an additional $4800
- Year 3 an additional $5200
- Year 4 an additional $5500
- Year 5 an additional $5800

Task 6: If Michael goes back to his original budget, how much will he be saving with his new income after five years?

Task 7: How much will he have available for misc. spending?

Applying the Situation to Your Life:

Having a budget can help you control your spending.

Task 8: Do you have a budget?

Task 9: How would you adjust Michael's budget to fit your needs?

Task 10: You may not be planning to get a master's degree but probably you are thinking of saving up for some important purchase. Are you thinking of buying a new car that is more fuel-efficient? Are you planning a special trip? How could you adjust Michael's budget to help you with your goal?

1.6 Using the Distributive Property to Simplify Algebraic Expressions

Student Learning Objective

After studying this section, you will be able to:

1. Use the distributive property to simplify algebraic expressions.

1 Using the Distributive Property to Simplify Algebraic Expressions

As we learned previously, we use letters called *variables* to represent unknown numbers. If a number is multiplied by a variable, we do not need any symbol between the number and variable. Thus, to indicate $(2)(x)$, we write $2x$. To indicate $3 \cdot y$, we write $3y$. If one variable is multiplied by another variable, we place the variables next to each other. Thus, $(a)(b)$ is written ab. We use exponent form if an unknown number (a variable) is used several times as a factor. Thus, $x \cdot x \cdot x = x^3$. Similarly, $(y)(y)(y)(y) = y^4$.

In algebra, we need to be familiar with several definitions. We will use them throughout the remainder of this book. Take some time to think through how each of these definitions is used.

An **algebraic expression** is a quantity that contains numbers and variables, such as $a + b$, $2x - 3$, and $5ab^2$. In this chapter we will be learning rules about adding and multiplying algebraic expressions. A **term** is a number, a variable, a product, or a quotient of numbers and variables. 17, x, $5xy$, and $22xy^3$ are all examples of terms. We will refer to terms when we discuss the distributive property.

An important property of algebra is the **distributive property.** We can state it in an equation as follows:

DISTRIBUTIVE PROPERTY

For all real numbers *a*, *b*, and *c*,

$$a(b + c) = ab + ac.$$

A numerical example shows that it does seem reasonable.

$$5(3 + 6) = 5(3) + 5(6)$$
$$5(9) = 15 + 30$$
$$45 = 45$$

We can use the distributive property to multiply any term by the sum of two or more terms. In Section 0.1, we defined the word *factor*. Two or more algebraic expressions that are multiplied are called **factors.** Consider the following examples of multiplying algebraic expressions.

Example 1 Multiply.

(a) $5(a + b)$ **(b)** $-3(3x + 2y)$

Solution

(a) $5(a + b) = 5a + 5b$ Multiply the factor $(a + b)$ by the factor 5.

(b) $-3(3x + 2y) = -3(3x) + (-3)(2y)$ Multiply the factor $(3x + 2y)$ by the factor (-3).

$= -9x - 6y$ □

Student Practice 1 Multiply.

(a) $3(x + 2y)$ **(b)** $-2(a + 3b)$

If the parentheses are preceded by a negative sign, we consider this to be the product of (-1) and the expression inside the parentheses.

Example 2 Multiply. $-(a - 2b)$

Solution

$$-(a - 2b) = (-1)(a - 2b) = (-1)(a) + (-1)(-2b) = -a + 2b$$

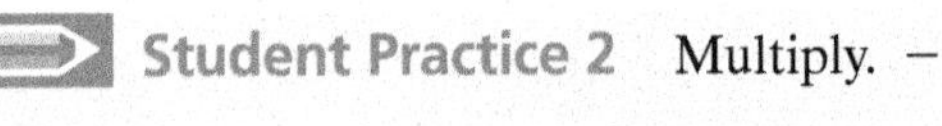

Student Practice 2 Multiply. $-(-3x + y)$

In general, we see that in all these examples we have multiplied each term of the expression in the parentheses by the expression in front of the parentheses.

Example 3 Multiply.

(a) $\frac{2}{3}(x^2 - 6x + 8)$ **(b)** $1.4(a^2 + 2.5a + 1.8)$

Solution

(a)
$$\begin{aligned} \frac{2}{3}(x^2 - 6x + 8) &= \left(\frac{2}{3}\right)(1x^2) + \left(\frac{2}{3}\right)(-6x) + \left(\frac{2}{3}\right)(8) \\ &= \frac{2}{3}x^2 + (-4x) + \frac{16}{3} \\ &= \frac{2}{3}x^2 - 4x + \frac{16}{3} \end{aligned}$$

(b)
$$\begin{aligned} 1.4(a^2 + 2.5a + 1.8) &= 1.4(1a^2) + (1.4)(2.5a) + (1.4)(1.8) \\ &= 1.4a^2 + 3.5a + 2.52 \end{aligned}$$

Student Practice 3 Multiply.

(a) $\frac{3}{5}(a^2 - 5a + 25)$ **(b)** $2.5(x^2 - 3.5x + 1.2)$

There are times we multiply a variable by itself and use exponent notation. For example, $(x)(x) = x^2$ and $(x)(x)(x) = x^3$. In other cases there will be numbers and variables multiplied at the same time.

We will see problems like $(2x)(x) = (2)(x)(x) = 2x^2$. Some expressions will involve multiplication of more than one variable. We will see problems like $(3x)(xy) = (3)(x)(x)(y) = 3x^2y$. There will be times when we use the distributive property and all of these methods will be used. For example,

$$\begin{aligned} 2x(x - 3y + 2) &= 2x(x) + (2x)(-3y) + (2x)(2) \\ &= 2x^2 + (-6)(xy) + 4(x) \\ &= 2x^2 - 6xy + 4x. \end{aligned}$$

We will discuss this type of multiplication of variables with exponents in more detail in Section 5.1. At that point we will expand these examples and other similar examples to develop the general rule for multiplication $(x^a)(x^b) = x^{a+b}$.

Example 4 Multiply. $-2x(3x + y - 4)$

Solution

$$\begin{aligned} -2x(3x + y - 4) &= -2(x)(3)(x) + (-2)(x)(y) + (-2)(x)(-4) \\ &= -2(3)(x)(x) + (-2)(xy) + (-2)(-4)(x) \\ &= -6x^2 - 2xy + 8x \end{aligned}$$

Student Practice 4 Multiply. $-4x(x - 2y + 3)$

The distributive property can also be presented with the a on the right.

$$(b + c)a = ba + ca$$

The a is "distributed" over the b and c inside the parentheses.

Example 5 Multiply. $(2x^2 - x)(-3)$

Solution

$$(2x^2 - x)(-3) = 2x^2(-3) + (-x)(-3)$$
$$= -6x^2 + 3x$$

Student Practice 5 Multiply.

$$(3x^2 - 2x)(-4)$$

▲ **Example 6** A farmer has a rectangular field that is 300 feet wide. One portion of the field is $2x$ feet long. The other portion of the field is $3y$ feet long. Use the distributive property to find an expression for the area of this field.

Solution First we draw a picture of a field that is 300 feet wide and $2x + 3y$ feet long.

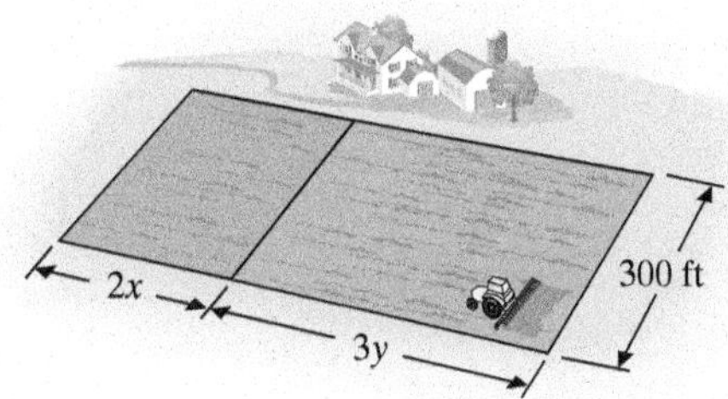

To find the area of the field, we multiply the width times the length.

$$300(2x + 3y) = 300(2x) + 300(3y) = 600x + 900y$$

Thus the area of the field in square feet is $600x + 900y$.

▲ **Student Practice 6** A farmer has a rectangular field that is 400 feet wide. One portion of the field is $6x$ feet long. The other portion of the field is $9y$ feet long. Use the distributive property to find an expression for the area of this field.

STEPS TO SUCCESS Review a Little Every Day

Successful students find that review is not something you do the night before the test. Take time to review a little each day. When you are learning new material, take a little time to look over the concepts previously learned in the chapter. By this continual review you will find the pressure is reduced to prepare for a test. You need time to think about what you have learned and make sure you really understand it. This will help to tie together the different topics in the chapter. A little review of each idea and each kind of problem will enable you to feel confident. You will think more clearly and have less tension when it comes to test time.

Making it personal: Which of these suggestions is the one you most need to follow? Write down what you need to do to improve in this area. ▼

1.6 Exercises MyMathLab®

Verbal and Writing Skills, Exercises 1–6

In exercises 1 and 2, complete each sentence by filling in the blank.

1. A ______________ is a symbol used to represent an unknown number.

2. When we write an expression with numbers and variables such as $7x$, it indicates that we are ______________ 7 by x.

3. Explain in your own words how we multiply a problem like $(4x)(x)$.

4. Explain in your own words why you think the property $a(b + c) = ab + ac$ is called the distributive property. What does "distribute" mean?

5. Does the following distributive property work?

$$a(b - c) = ab - ac$$

Why or why not? Give an example.

6. Susan tried to use the distributive property and wrote

$$-5(x + 3y - 2) = -5x - 15y - 10.$$

What did she do wrong?

Multiply. Use the distributive property.

7. $5(2x - 5y)$

8. $6(3x - 6y)$

9. $-2(4a - 3b)$

10. $-3(2a - 4b)$

11. $3(3x + y)$

12. $8(5x + y)$

13. $8(-m - 3n)$

14. $10(-2m - n)$

15. $-(x - 3y)$

16. $-(-3y + x)$

17. $-9(9x - 5y + 8)$

18. $-3(4x + 8 - 6y)$

19. $2(-5x + y - 6)$

20. $3(2x - 6y - 5)$

21. $\frac{5}{6}(12x^2 - 24x + 18)$

22. $\frac{2}{3}(-27a^4 + 9a^2 - 21)$

23. $\frac{x}{5}(x + 10y - 4)$ $\left(\textit{Hint: } \frac{x}{5} = \frac{1}{5}x\right)$

24. $\frac{y}{3}(3y - 4x - 6)$ $\left(\textit{Hint: } \frac{y}{3} = \frac{1}{3}y\right)$

25. $5x(x + 2y + z)$

26. $3a(2a + b - c)$

27. $(-4.5x + 5)(-3)$

28. $(-3.2x + 5)(-4)$

29. $(6x + y - 1)(3x)$

30. $(2x - 2y + 6)(3x)$

Mixed Practice *Multiply. Use the distributive property.*

31. $(3x + 2y - 1)(-xy)$

32. $(5a - 3b - 1)(-ab)$

33. $(-a - 2b + 4)5ab$

34. $(2a - b - 5)3ab$

35. $\frac{1}{4}(8a^2 - 16a - 5)$

36. $\frac{1}{3}(-15a^2 - 21a + 4)$

37. $-0.3x(-1.2x^2 - 0.3x + 0.5)$
Hint: $x(x^2) = x(x)(x) = x^3$

38. $-0.6q(1.2q^2 + 2.5r - 0.7s)$
Hint: $q(q^2) = q(q)(q) = q^3$

39. $0.4q(-3.3q^2 - 0.7r - 10)$
Hint: $q(q^2) = q(q)(q) = q^3$

Applications

▲ **40.** ***Geometry*** Gary Roswell owns a large rectangular field where he grows corn and wheat. The width of the field is 850 feet. The portion of the field where corn grows is $10x$ feet long and the portion where wheat grows is $7y$ feet long. Use the distributive property to find an expression for the area of this field.

▲ **41.** ***Geometry*** Kathy Maris has a rectangular field that is 800 feet wide. One portion of the field is $5x$ feet long. The other portion of the field is $14y$ feet long. Use the distributive property to find an expression for the area of this field.

To Think About

▲ **42.** ***Athletic Field*** The athletic field at Gordon College is $2x$ feet wide. It used to be 1800 feet long. An old, rundown shed was torn down, making the field $3y$ feet longer. Use the distributive property to find an expression for the area of the new field.

▲ **43.** ***Airport Runway*** The runway at Beverly Airport is $4x$ feet wide. Originally, the airport was supposed to have a 3000-foot-long runway. However, some of the land was wetland, so a runway could not be built on all of it. Therefore, the length of the runway was decreased by $2y$ feet. Use the distributive property to find an expression for the area of the final runway.

Cumulative Review *In exercises 44–48, evaluate.*

44. [1.1.5] $-18 + (-20) + 36 + (-14)$

45. [1.4.2] $(-2)^6$

46. [1.2.1] $-27 - (-41)$

47. [1.5.1] $25 \div 5(2) + (-6)$

48. [1.5.1] $(12 - 10)^2 + (-3)(-2)$

Quick Quiz 1.6 *Multiply. Use the distributive property.*

1. $5(-3a - 7b)$

2. $-2x(x - 4y + 8)$

3. $-3ab(4a - 5b - 9)$

4. Concept Check Explain how you would multiply to obtain the answer for $(-\frac{3}{7})(21x^2 - 14x + 3)$.

1.7 Combining Like Terms

Student Learning Objectives

After studying this section, you will be able to:

1. Identify like terms.
2. Combine like terms.

1 Identifying Like Terms 

We can add or subtract quantities that are *like quantities*. This is called **combining** like quantities.

$$5 \text{ inches} + 6 \text{ inches} = 11 \text{ inches}$$
$$20 \text{ square inches} - 16 \text{ square inches} = 4 \text{ square inches}$$

However, we cannot combine things that are not the same.

$$16 \text{ square inches} - 4 \text{ inches} \quad \text{(Cannot be done!)}$$

Similarly, in algebra we can **combine like terms.** This means to add or subtract like terms. Remember, we cannot combine terms that are not the same. Recall that a *term* is a number, a variable, a product, or a quotient of numbers and variables. **Like terms** are terms that have identical variables and exponents. In other words, like terms must have exactly the same letter parts.

Example 1 List the like terms of each expression.

(a) $5x - 2y + 6x$ **(b)** $2x^2 - 3x - 5x^2 - 8x$

Solution

(a) $5x$ and $6x$ are like terms. These are the only like terms in this expression.

(b) $2x^2$ and $-5x^2$ are like terms.

$-3x$ and $-8x$ are like terms.

Note that x^2 and x are not like terms. □

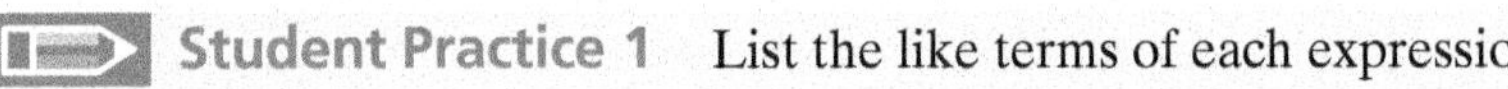

Student Practice 1 List the like terms of each expression.

(a) $5a + 2b + 8a - 4b$ **(b)** $x^2 + y^2 + 3x - 7y^2$

Do you really understand what a term is? A term is a number, a variable, a product, or a quotient of numbers and variables. Terms are the parts of an algebraic expression separated by plus or minus signs. The sign in front of the term is considered part of the term.

2 Combining Like Terms 

It is important to know how to combine like terms. Since

$$4 \text{ inches} + 5 \text{ inches} = 9 \text{ inches},$$

we would expect in algebra that $4x + 5x = 9x$.

Why is this true? Let's take a look at the distributive property.

Like terms may be added or subtracted by using the distributive property:

$$ab + ac = a(b + c) \quad \text{and} \quad ba + ca = (b + c)a.$$

For example,

$$-7x + 9x = (-7 + 9)x = 2x$$
$$5x^2 + 12x^2 = (5 + 12)x^2 = 17x^2.$$

Example 2 Combine like terms.

(a) $-4x^2 + 8x^2$ **(b)** $5x + 3x + 2x$

Solution

(a) Notice that each term contains the factor x^2. Using the distributive property, we have

$$-4x^2 + 8x^2 = (-4 + 8)x^2 = 4x^2.$$

(b) Note that each term contains the factor x. Using the distributive property, we have

$$5x + 3x + 2x = (5 + 3 + 2)x = 10x.$$

□

 **Student Practice 2** Combine like terms.

(a) $16y^3 + 9y^3$ **(b)** $5a + 7a + 4a$

In this section, the direction *simplify* means to remove parentheses and/or combine like terms.

Example 3 Simplify. $5a^2 - 2a^2 + 6a^2$

Solution

$$5a^2 - 2a^2 + 6a^2 = (5 - 2 + 6)a^2 = 9a^2$$

□

 **Student Practice 3** Simplify. $-8y^2 - 9y^2 + 4y^2$

After doing a few problems, you will find that it is not necessary to write out the step of using the distributive property. We will omit this step for the remaining examples in this section.

Example 4 Simplify.

(a) $5.6a + 2b + 7.3a - 6b$

(b) $3x^2y - 2xy^2 + 6x^2y$

(c) $2a^2b + 3ab^2 - 6a^2b^2 - 8ab$

Solution

(a) $5.6a + 2b + 7.3a - 6b = 12.9a - 4b$ We combine the a terms and the b terms separately.

(b) $3x^2y - 2xy^2 + 6x^2y = 9x^2y - 2xy^2$ **Note:** x^2y and xy^2 are not like terms because of different powers.

(c) $2a^2b + 3ab^2 - 6a^2b^2 - 8ab$ These terms cannot be combined; there are no like terms in this expression. □

Student Practice 4 Simplify.

(a) $1.3x + 3a - 9.6x + 2a$

(b) $5ab - 2ab^2 - 3a^2b + 6ab$

(c) $7x^2y - 2xy^2 - 3x^2y^2 - 4xy$

The two skills in this section that a student must practice are identifying like terms and correctly adding and subtracting like terms. If a problem involves many terms, you may find it helpful to rearrange the terms so that like terms are together.

Example 5 Simplify. $3a - 2b + 5a^2 + 6a - 8b - 12a^2$

Solution There are three pairs of like terms.

$$\underbrace{3a + 6a}_{a \text{ terms}} \underbrace{- 2b - 8b}_{b \text{ terms}} \underbrace{+ 5a^2 - 12a^2}_{a^2 \text{ terms}}$$ You can rearrange the terms so that like terms are together, making it easier to combine them.

$$= 9a - 10b - 7a^2$$ Combine like terms.

Because of the commutative property, the order of terms in an answer to this problem is not significant. These three terms can be rearranged in a different order. $-10b + 9a - 7a^2$ and $-7a^2 + 9a - 10b$ are also correct. Later, we will learn the preferred way to write the answer. □

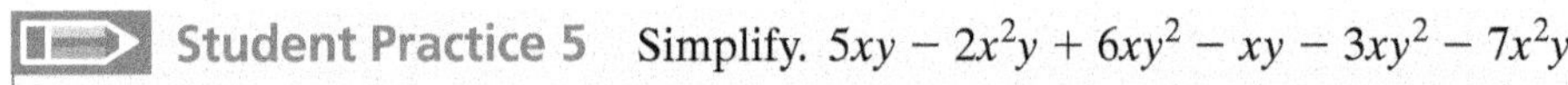

Student Practice 5 Simplify. $5xy - 2x^2y + 6xy^2 - xy - 3xy^2 - 7x^2y$

Use extra care with fractional values.

Example 6 Simplify. $\frac{3}{4}x^2 - 5y - \frac{1}{8}x^2 + \frac{1}{3}y$

Solution We need the least common denominator for the x^2 terms, which is 8. Change $\frac{3}{4}$ to eighths by multiplying the numerator and denominator by 2.

$$\frac{3}{4}x^2 - \frac{1}{8}x^2 = \frac{3 \cdot 2}{4 \cdot 2}x^2 - \frac{1}{8}x^2 = \frac{6}{8}x^2 - \frac{1}{8}x^2 = \frac{5}{8}x^2$$

The least common denominator for the y terms is 3. Change -5 to thirds.

$$-\frac{5}{1}y + \frac{1}{3}y = \frac{-5 \cdot 3}{1 \cdot 3}y + \frac{1}{3}y = \frac{-15}{3}y + \frac{1}{3}y = -\frac{14}{3}y$$

Thus, our solution is $\frac{5}{8}x^2 - \frac{14}{3}y$. □

Student Practice 6 Simplify.

$$\frac{1}{7}a^2 - \frac{5}{12}b + 2a^2 - \frac{1}{3}b$$

Example 7 Simplify. $6(2x + 3xy) - 8x(3 - 4y)$

Solution First remove the parentheses; then combine like terms.

$$6(2x + 3xy) - 8x(3 - 4y) = 12x + 18xy - 24x + 32xy$$ Use the distributive property.

$$= -12x + 50xy$$ Combine like terms. □

Student Practice 7 Simplify. $5a(2 - 3b) - 4(6a + 2ab)$

1.7 Exercises MyMathLab®

Verbal and Writing Skills, Exercises 1–6

1. Explain in your own words the mathematical meaning of the word *term*.

2. Explain in your own words the mathematical meaning of the phrase *like terms*.

3. Explain which terms are like terms in the expression $5x - 7y - 8x$.

4. Explain which terms are like terms in the expression $12a - 3b - 9a$.

5. Explain which terms are like terms in the expression $7xy - 9x^2y - 15xy^2 - 14xy$.

6. Explain which terms are like terms in the expression $-3a^2b - 12ab + 5ab^2 + 9ab$.

Combine like terms.

7. $-16x^2 - 15x^2$

8. $-14x^3 - 21x^3$

9. $5a^3 - 7a^2 + a^3$

10. $2b^3 + 8b^2 - 9b^3$

11. $3x + 2y - 8x - 7y$

12. $5x - 9b - 6x - 5b$

13. $1.3x - 2.6y + 5.8x - 0.9y$

14. $3.1a - 0.2b - 0.8a + 5.3b$

15. $1.6x - 2.8y - 3.6x - 5.9y$

16. $1.9x - 2.4b - 3.8x - 8.2b$

17. $3p - 4q + 2p + 3 + 5q - 21$

18. $6x - 5y - 3y + 7 - 11x - 5$

19. $2ab + 5bc - 6ac - 2ab$

20. $7ab - 3bc - 12ac + 8ab$

21. $2x^2 - 3x - 5 - 7x + 8 - x^2$

22. $5x + 7 - 6x^2 + 6 - 11x + 4x^2$

23. $2y^2 - 8y + 9 - 12y^2 - 8y + 3$

24. $3y^2 + 9y - 12 - 4y^2 - 6y + 2$

25. $\frac{1}{3}x - \frac{2}{3}y - \frac{2}{5}x + \frac{4}{7}y$

26. $\frac{2}{5}s - \frac{3}{8}t - \frac{4}{15}s - \frac{5}{12}t$

27. $\frac{3}{4}a^2 - \frac{1}{3}b - \frac{1}{5}a^2 - \frac{1}{2}b$

28. $\frac{2}{5}y - \frac{3}{4}x^2 - \frac{1}{3}y + \frac{7}{8}x^2$

29. $3rs - 8r + s - 5rs + 10r - s$

30. $-rs + 10s + 5r - rs + 6s - 2r$

31. $4xy + \frac{5}{4}x^2y + \frac{3}{4}xy + \frac{3}{4}x^2y$

32. $\frac{3}{7}ab - \frac{2}{7}a^2b + 2ab + \frac{9}{7}a^2b$

Simplify. Use the distributive property to remove parentheses; then combine like terms.

33. $5(2a - b) - 3(5b - 6a)$

34. $8(3x - 2y) + 4(3y - 5x)$

35. $-3b(5a - 3b) + 4(-3ab - 5b^2)$

36. $3a(2a - 3b) - 3(-5a^2 + ab)$

37. $6(c - 2d^2) - 2(4c - d^2)$

38. $-4(3cd + 2c^2) + 2c(5d - c)$

39. $3(4 - x) - 2(-4 - 7x)$

40. $5(6 - x) - 2(9 - 11x)$

To Think About

▲ **41.** ***Fencing in a Pool*** Mr. Jimenez has a pool behind his house that needs to be fenced in. The backyard is an odd quadrilateral shape. The four sides are $3a$, $2b$, $4a$, and $7b$ in length. How much fencing (the length of the perimeter) would he need to enclose the pool?

▲ **42.** ***Framing a Masterpiece*** The new Degas masterpiece purchased by the Museum of Fine Arts in Boston needs to be reframed. If the rectangular picture measures $6x - 3$ wide by $8x - 7$ high, what is the perimeter of the painting?

▲ **43.** ***Geometry*** A rectangle is $5x - 10$ feet long and $2x + 6$ feet wide. What is the perimeter of the rectangle?

▲ **44.** ***Geometry*** A triangle has sides of length $3a + 8$ meters, $5a - b$ meters, and $12 - 2b$ meters. What is the perimeter of the triangle?

Cumulative Review *Evaluate.*

45. **[1.2.1]** $-\frac{3}{4} - \frac{1}{3}$

46. **[1.3.1]** $\left(\frac{2}{3}\right)\left(-\frac{9}{16}\right)$

47. **[1.1.4]** $\frac{4}{5} + \left(-\frac{1}{25}\right) + \left(-\frac{3}{10}\right)$

48. **[1.3.3]** $\left(\frac{5}{7}\right) \div \left(-\frac{14}{3}\right)$

Quick Quiz 1.7 *Combine like terms.*

1. $3xy - \frac{2}{3}x^2y - \frac{5}{6}xy + \frac{7}{3}x^2y$

2. $8.2a^2b + 5.5ab^2 - 7.6a^2b - 9.9ab^2$

3. $2(3x - 5y) - 2(-7x - 4y)$

4. **Concept Check** Explain how you would remove parentheses and then combine like terms to obtain the answer for $1.2(3.5x - 2.2y) - 4.5(2.0x + 1.5y)$.

1.8 Using Substitution to Evaluate Algebraic Expressions and Formulas

1 Evaluating an Algebraic Expression for a Specified Value

You will use the order of operations to **evaluate** variable expressions. Suppose we are asked to evaluate

$$6 + 3x \text{ for } x = -4.$$

In general, x represents some unknown number. Here we are told x has the value -4. We can replace x with -4. Use parentheses around -4. Note that we always put replacement values in parentheses.

$$6 + 3(-4) = 6 + (-12) = -6$$

When we replace a variable by a particular value, we say we have **substituted** the value for the variable. We then evaluate the expression (that is, find a value for it).

Student Learning Objectives

After studying this section, you will be able to:

1. Evaluate an algebraic expression for a specified value.
2. Evaluate a formula by substituting values.

Example 1 Evaluate $\frac{2}{3}x - 5$ for $x = -6$.

Solution

$\frac{2}{3}x - 5 = \frac{2}{3}(-6) - 5$ Substitute -6 for x. Be sure to enclose the -6 in parentheses.

$= -4 - 5$ Multiply $\left(\frac{2}{3}\right)\left(-\frac{6}{1}\right) = -4$.

$= -9$ Combine. ☐

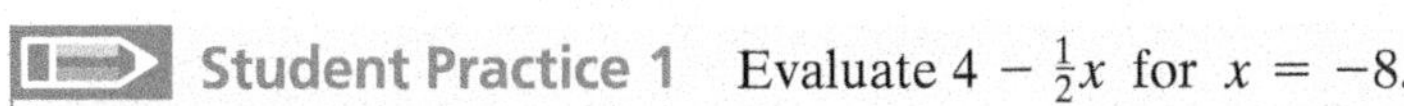

Student Practice 1 Evaluate $4 - \frac{1}{2}x$ for $x = -8$.

Compare parts **(a)** and **(b)** in the next example. The two parts illustrate that you must be careful what value you raise to a power. *Note:* In part **(b)** we will need parentheses within parentheses. To avoid confusion, we use brackets [ ] to represent the outside parentheses.

Example 2 Evaluate for $x = -3$.

(a) $2x^2$ **(b)** $(2x)^2$

Solution

(a) Here the value x is squared.

$2x^2 = 2(-3)^2$

$= 2(9)$ First square -3.

$= 18$ Then multiply.

(b) Here the value $(2x)$ is squared.

$(2x)^2 = [2(-3)]^2$

$= (-6)^2$ First multiply the numbers inside the brackets.

$= 36$ Then square -6. ☐

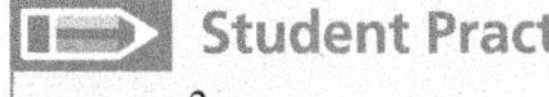

Student Practice 2 Evaluate for $x = -3$.

(a) $4x^2$ **(b)** $(4x)^2$

Carefully study the solutions to Example 2**(a)** and Example 2**(b)**. You will find that taking the time to see *how* and *why* they are different is a good investment of time.

Example 3 Evaluate $x^2 + 3x$ for $x = -4$.

Solution

$x^2 + 3x = (-4)^2 + 3(-4)$ Replace each x by -4 in the original expression.

$= 16 + (3)(-4)$ Raise to a power.

$= 16 - 12$ Multiply.

$= 4$ Finally, we add the opposite of 12. □

Student Practice 3

Evaluate $2x^2 - 3x$ for $x = -2$.

Example 4 Evaluate $x^3 + 2xy - 3x + 1$ for $x = 2$ and $y = -\frac{1}{4}$.

Solution

$x^3 + 2xy - 3x + 1 = (2)^3 + 2(2)\left(-\frac{1}{4}\right) - 3(2) + 1$ Replace x by 2 and y by $-\frac{1}{4}$.

$= 8 + (-1) - 6 + 1$ Calculate $2^3 = 8$, and multiply $2(2)\left(-\frac{1}{4}\right) = 4\left(-\frac{1}{4}\right) = -1$. Also, multiply $-3(2) = -6$.

$= 8 + (-1) + (-6) + 1$ Change subtracting to adding the opposite.

$= 9 + (-7)$ Find the sum of the positive numbers and find the sum of the negative numbers.

$= 2$ We take the difference of 9 and 7 and use the sign of the number with the larger absolute value. □

Student Practice 4

Evaluate $6a + 4ab^2 - 5$ for $a = -\frac{1}{6}$ and $b = 3$.

2 Evaluating a Formula by Substituting Values

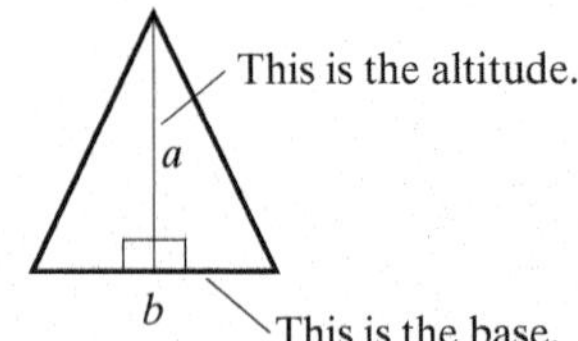

We can *evaluate a formula* by substituting values for the variables. For example, the area of a triangle can be found using the formula $A = \frac{1}{2}ab$, where b is the length of the base of the triangle and a is the altitude of the triangle (see figure). If we know values for a and b, we can substitute those values into the formula to find the area. The units for area are *square units.*

Because some of the examples and exercises in this section involve geometry, it may be helpful to review this topic.

The following information is very important. If you have forgotten some of this material (or if you have never learned it), please take the time to learn it completely now. Throughout the entire book we will be using this information in solving applied problems.

Perimeter is the distance around a plane figure. Perimeter is measured in linear units (inches (in.), feet (ft), centimeters (cm), miles (mi)). **Area** is a measure of the amount of surface in a region. Area is measured in square units (square inches (in.^{2}), square feet (ft^2), square centimeters (cm^2)).

In our sketches we will show angles of 90° by using a small square (∟). This indicates that the two lines are at right angles. All angles that measure 90° are called **right angles.** An **altitude** is perpendicular to the base of a figure. That is, the altitude forms a right angle with the base. The small corner square in a sketch helps us identify the altitude of the figure.

The following box provides a handy guide to some facts and formulas you will need to know. Use it as a reference when solving word problems involving geometric figures.

GEOMETRIC FORMULAS: TWO-DIMENSIONAL FIGURES

A **parallelogram** is a four-sided figure with opposite sides parallel. In a parallelogram, opposite sides are equal and opposite angles are equal.

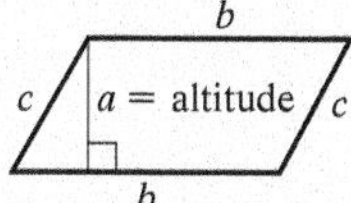

Perimeter = the sum of all four sides

Area = ab

A **rectangle** is a parallelogram with all interior angles measuring 90°.

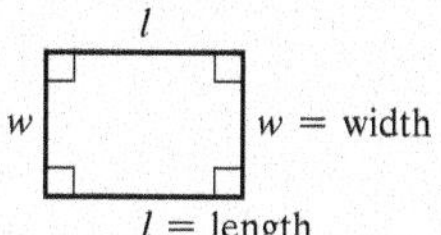

Perimeter = $2l + 2w$

Area = lw

A **square** is a rectangle with all four sides equal.

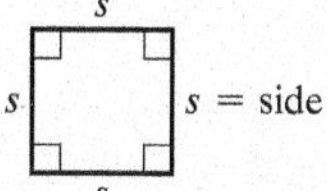

Perimeter = $4s$

Area = s^2

A **trapezoid** is a four-sided figure with two sides parallel. The parallel sides are called the *bases* of the trapezoid.

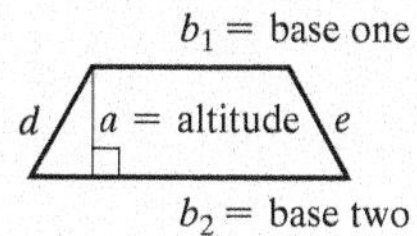

Perimeter = the sum of all four sides

Area = $\frac{1}{2}a(b_1 + b_2)$

A **triangle** is a closed plane figure with three sides.

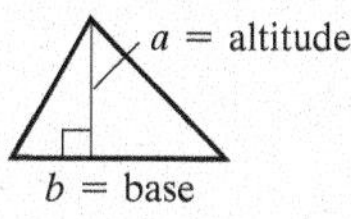

Perimeter = the sum of all three sides

Area = $\frac{1}{2}ab$

A **circle** is a plane curve consisting of all points at an equal distance from a given point called the center. **Circumference** is the distance around a circle.

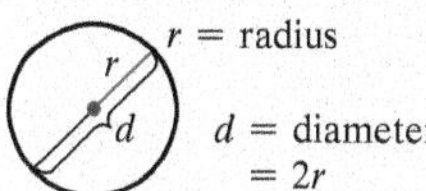

Circumference = $2\pi r$ or πd

Area = πr^2

π (the number *pi*) is a constant associated with circles. It is an irrational number that is approximately 3.141592654. We usually use 3.14 as a sufficiently accurate approximation. Thus we write $\pi \approx 3.14$ for most of our calculations involving π.

▲ **Example 5** Find the area of a triangle with a base of 16 centimeters (cm) and an altitude of 12 centimeters (cm).

Solution Substitute $a = 12$ cm and $b = 16$ cm in $A = \frac{1}{2}ab$.

$$A = \frac{1}{2}(12\text{ cm})(16\text{ cm})$$
$$= (6)(16)(\text{cm})^2 = 96 \text{ square centimeters}$$

The area of the triangle is 96 square centimeters or 96 cm^2. □

▲ **Student Practice 5** Find the area of a triangle with an altitude of 3 meters and a base of 7 meters.

The area of a circle is given by $A = \pi r^2$. We will use 3.14 as an approximation for the *irrational number* π.

▲ **Example 6** Find the area of a circle if the radius is 2 inches.

Solution

$A = \pi r^2 \approx (3.14)(2 \text{ inches})^2$ Write the formula and substitute the given values for the letters.

$\approx (3.14)(4)(\text{in.})^2$ Raise to a power. Then multiply.

≈ 12.56 square inches or 12.56 in.^2 □

▲ **Student Practice 6** Find the area of a circle if the radius is 3 meters.

The formula $C = \frac{5}{9}(F - 32)$ allows us to find the Celsius temperature if we know the Fahrenheit temperature. That is, we can substitute a value for F in degrees Fahrenheit into the formula to obtain a temperature C in degrees Celsius.

Example 7 What is the Celsius temperature when the Fahrenheit temperature is $F = -22°$?

Solution Use the formula.

$$\begin{aligned} C &= \frac{5}{9}(F - 32) \\ &= \frac{5}{9}[(-22) - 32] && \text{Substitute } -22 \text{ for } F \text{ in the formula.} \\ &= \frac{5}{9}(-54) && \text{Combine the numbers inside the brackets.} \\ &= (5)(-6) && \text{Simplify.} \\ &= -30 && \text{Multiply.} \end{aligned}$$

The temperature is $-30°$ Celsius or $-30°$C. □

Student Practice 7 What is the Celsius temperature when the Fahrenheit temperature is $F = 68°$? Use the formula $C = \frac{5}{9}(F - 32)$.

When driving in Canada or Mexico, we must observe speed limits posted in kilometers per hour. A formula that converts r (miles per hour) to k (kilometers per hour) is $k \approx 1.61r$. Note that this is an approximation.

Example 8 You are driving on a highway in Mexico. It has a posted maximum speed of 100 kilometers per hour. You are driving at 61 miles per hour. Are you exceeding the speed limit?

Solution Use the formula.

$$\begin{aligned} k &\approx 1.61r \\ &\approx (1.61)(61) && \text{Replace } r \text{ by 61.} \\ &\approx 98.21 && \text{Multiply the numbers.} \end{aligned}$$

You are driving at approximately 98 kilometers per hour. You are not exceeding the speed limit. □

Student Practice 8 You are driving behind a heavily loaded truck on a Canadian highway. The highway has a posted minimum speed of 65 kilometers per hour. When you travel at exactly the same speed as the truck ahead of you, you observe that the speedometer reads 35 miles per hour. Assuming that your speedometer is accurate, determine whether the truck is violating the minimum speed law.

1.8 Exercises MyMathLab®

Evaluate.

1. $-3x + 5$ for $x = 4$

2. $-5x - 6$ for $x = 6$

3. $\frac{2}{5}y - 8$ for $y = -10$

4. $\frac{5}{6}y - 5$ for $y = -6$

5. $5x + 10$ for $x = \frac{1}{2}$

6. $5x + 15$ for $x = -\frac{3}{2}$

7. $2 - 4x$ for $x = 7$

8. $3 - 5x$ for $x = 8$

9. $3.5 - 2x$ for $x = 2.4$

10. $6.3 - 3x$ for $x = 2.3$

11. $9x + 13$ for $x = -\frac{3}{4}$

12. $5x + 7$ for $x = -\frac{2}{3}$

13. $x^2 - 3x$ for $x = -2$

14. $x^2 + 3x$ for $x = 4$

15. $5y^2$ for $y = -1$

16. $8y^2$ for $y = -1$

17. $-3x^3$ for $x = 2$

18. $-5x^3$ for $x = 3$

19. $-5x^2$ for $x = -2$

20. $-2x^2$ for $x = -3$

21. $2x^2 + 3x$ for $x = -3$

22. $18 + 3x^2$ for $x = -3$

23. $(2x)^2 + x$ for $x = 3$

24. $2 - x^2$ for $x = -2$

25. $2 - (-x)^2$ for $x = -2$

26. $2x - 3x^2$ for $x = -4$

27. $10a + (4a)^2$ for $a = -2$

28. $9a - (2a)^2$ for $a = -3$

29. $4x^2 - 6x$ for $x = \frac{1}{2}$

30. $5 - 9x^2$ for $x = \frac{1}{3}$

31. $x^3 - 7y + 3$ for $x = 2$ and $y = 5$

32. $a^3 + 2b - 4$ for $a = 1$ and $b = -4$

33. $\frac{1}{2}a^2 - 3b + 9$ for $a = -4$ and $b = \frac{2}{3}$

34. $\frac{1}{3}x^2 + 4y - 5$ for $x = -3$ and $y = \frac{3}{4}$

35. $2r^2 + 3s^2 - rs$ for $r = -1$ and $s = 3$

36. $-r^2 + 5rs + 4s^2$ for $r = -2$ and $s = 3$

37. $a^3 + 2abc - 3c^2$ for $a = 5, b = 9$, and $c = -1$

38. $a^3 - 2ab + 2c^2$ for $a = 3, b = 2$, and $c = -4$

39. $\dfrac{a^2 + ab}{3b}$ for $a = -1$ and $b = -2$

40. $\dfrac{x^2 - 2xy}{2y}$ for $x = -2$ and $y = -3$

Applications

▲ **41.** ***Geometry*** A sign is made in the shape of a parallelogram. The base measures 22 feet. The altitude measures 16 feet. What is the area of the sign?

▲ **42.** ***Geometry*** A field is shaped like a parallelogram. The base measures 92 feet. The altitude measures 54 feet. What is the area of the field?

▲ **43.** ***TV Parts*** A square support unit in a television is made with a side measuring 3 centimeters. A new model being designed for next year will have a larger square with a side measuring 3.2 centimeters. By how much will the area of the square be increased?

▲ **44.** ***Computer Chips*** A square computer chip for last year's computer had a side measuring 23 millimeters. This year the computer chip has been reduced in size. The new square chip has a side of 20 millimeters. By how much has the area of the chip decreased?

▲ **45.** ***Carpentry*** A carpenter cut out a small trapezoid as a wooden support for a front step. It has an altitude of 4 inches. One base of the trapezoid measures 9 inches and the other base measures 7 inches. What is the area of this support?

▲ **46.** ***Cable Television Technician*** Dan is a Comcast cable technician. He works on a signal tower that has a small trapezoid frame on the top of the tower. The frame has an altitude of 9 inches. One base of the trapezoid is 20 inches and the other base measures 17 inches. What is the area of this small trapezoidal frame?

▲ **47.** ***Geometry*** Bradley Palmer State Park has a triangular piece of land on the border. The altitude of the triangle is 400 feet. The base of the triangle is 280 feet. What is the area of this piece of land?

▲ **48.** ***Roofing*** The ceiling in the Madisons' house has a leak. The roofer exposed a triangular region that needs to be sealed and then reroofed. The region has an altitude of 14 feet. The base of the region is 19 feet. What is the area of the region that needs to be reroofed?

▲ **49.** ***Geometry*** The radius of a circular tablecloth is 3 feet. What is the area of the tablecloth? (Use $\pi \approx 3.14$.)

50. ***Landscape Architect Assistant*** Lexi works as an assistant for a landscape architect in West Chicago, Illinois. The center of the flower garden at a new estate has a circular concrete platform for an elaborate water fountain. The circular platform has a diameter of 12 feet. What will be the area of the circular platform? (Use $\pi \approx 3.14$.)

Temperature *For exercises 51 and 52, use the formula* $C = \frac{5}{9}(F - 32)$ *to find the Celsius temperature.*

51. Dry ice is solid carbon dioxide. Dry ice does not melt, it goes directly from the solid state to the gaseous state. Dry ice changes from a solid to a gas at −109.3°F. What is this temperature in Celsius?

52. ***Winter Jogging*** Jenny went jogging in January on a trail in Minneapolis when the weather outside was −12 degrees Celsius. Her outside winter jogging coat is rated for −3 degrees Fahrenheit. Is the coat going to be warm enough? What is the outside winter jogging coat rated for in degrees Celsius?

Solve.

▲ **53.** ***Sail Dimensions*** Find the total cost of making a triangular sail that has a base dimension of 12 feet and an altitude of 20 feet if the price for making the sail is $19.50 per square foot.

▲ **54.** ***Replacement Window Technician*** Caleb works for a company that installs energy-efficient replacement windows. He installed a new semicircular window in a large dining room. The owners want the window coated with a layer of sunblocking material. The semicircular window has a radius of 14 inches. The sunblock coating costs $1.15 per square inch to apply. What is the total cost of coating the window to the nearest cent? (Use $\pi \approx 3.14$.)

55. ***Temperature Extremes on the Moon*** The results of measurements made by NASA with the Lunar Orbiter show that the coldest temperature recorded on the surface of the moon was −238°C. The warmest temperature recorded on the surface of the moon was 123°C. What are the corresponding temperatures in degrees Fahrenheit? (Use the formula $F = \frac{9}{5}C + 32$.)

56. ***Tour de France*** A recently completed Tour de France was 3642 kilometers long and was completed in 20 stages. What was the average length of each stage in miles? Use the formula $r \approx 0.62k$, where r is the number of miles and k is the number of kilometers. Round to the nearest tenth of a mile.

Cumulative Review *In exercises 57 and 58, simplify.*

57. **[1.5.1]** $(-2)^4 - 4 \div 2 - (-2)$

58. **[1.7.2]** $3(x - 2y) - (x^2 - y) - (x - y)$

Quick Quiz 1.8 *Evaluate.*

1. $2x^2 - 4x - 14$ for $x = -2$

2. $5a - 6b$ for $a = \frac{1}{2}$ and $b = -\frac{1}{3}$

3. $x^3 + 2x^2y + 5y + 2$ for $x = -2$ and $y = 3$

4. **Concept Check** Explain how you would find the area of a circle if you know its diameter is 12 meters.

STEPS TO SUCCESS What Is the Best Way to Review Before a Test?

Here is what students have found.

1. Read over your textbook again. Make a list of any terms, rules, or formulas you need to know for the exam. Make sure you understand all of them.
2. Go over your notes. Look back at your homework and quizzes. Redo the problems you missed. Make sure you can get the right answer.
3. Practice some of each type of problem covered in the chapter(s) you are to be tested on.
4. At the end of the chapter are special sections you need to complete. Study each part of the Chapter Organizer and do the You Try It problems. Be sure to do the Chapter Review Problems.
5. When you think you are ready take the How Am I Doing? Chapter Test. Make sure you study the Math Coach notes right after the test.
6. Get help with concepts that are giving you difficulty. Don't be afraid to ask for help. Teachers, tutors, friends in class, and other friends are ready to help you. Don't wait. Get help now.

Making it personal: Which of the six steps do you most need to follow? Take some time today to start doing these things. These methods of review have helped thousands of students! They can help you—NOW! ▼

1.9 Grouping Symbols

Student Learning Objective

After studying this section, you will be able to:

1 Simplify algebraic expressions by removing grouping symbols.

1 Simplifying Algebraic Expressions by Removing Grouping Symbols

Many expressions in algebra use **grouping symbols** such as parentheses, brackets, and braces. Sometimes expressions are inside other expressions. Because it can be confusing to have more than one set of parentheses, brackets and braces are also used. How do we know what to do first when we see an expression like $2[5 - 4(a + b)]$?

To simplify the expression, we start with the innermost grouping symbols. Here it is a set of parentheses. We first use the distributive property to multiply.

$$2[5 - 4(a + b)] = 2[5 - 4a - 4b]$$

We use the distributive law again.

$$= 10 - 8a - 8b$$

There are no like terms, so this is our final answer.

Notice that we started with two sets of grouping symbols, but our final answer has none. So we can say we *removed* the grouping symbols. Of course, we didn't just take them away; we used the distributive property and the rules for real numbers to simplify as much as possible. Although simplifying expressions like this involves many steps, we sometimes say "remove parentheses" as a shorthand direction. Sometimes we say "simplify."

Remember to remove the innermost grouping symbols first. Keep working from the inside out.

Example 1 Simplify. $3[6 - 2(x + y)]$

Solution We want to remove the innermost parentheses first. Therefore, we first use the distributive property to simplify $-2(x + y)$.

$$3[6 - 2(x + y)] = 3[6 - 2x - 2y] \quad \text{Use the distributive property.}$$
$$= 18 - 6x - 6y \quad \text{Use the distributive property again.}$$

Student Practice 1 Simplify. $5[4x - 3(y - 2)]$

You recall that a negative sign in front of parentheses is equivalent to having a coefficient of negative 1. You can write the -1 and then multiply by -1 using the distributive property.

$$-(x + 2y) = -1(x + 2y) = -x - 2y$$

Notice that this has the effect of removing the parentheses. Each term in the result now has its sign changed.

Similarly, a positive sign in front of parentheses can be viewed as multiplication by $+1$.

$$+(5x - 6y) = +1(5x - 6y) = 5x - 6y$$

If a grouping symbol has a positive or negative sign in front, we mentally multiply by $+1$ or -1, respectively.

Fraction bars are also considered grouping symbols. In exercises 39 and 40 of Section 1.8, your first step was to simplify expressions above and below the fraction bars. Later in this book we will encounter further examples requiring the same first step. This type of operation will have some similarities to the operation of removing parentheses.

Example 2 Simplify. $-2[3a - (b + 2c) + (d - 3e)]$

Solution

$= -2[3a - b - 2c + d - 3e]$ Remove the two innermost sets of parentheses. Since one is not inside the other, we remove both sets at once.

$= -6a + 2b + 4c - 2d + 6e$ Now we remove the brackets by multiplying each term by -2. □

Student Practice 2 Simplify. $-3[2a - (3b - c) + 4d]$

Example 3 Simplify. $2[3x - (y + w)] - 3[2x + 2(3y - 2w)]$

Solution

$= 2[3x - y - w] - 3[2x + 6y - 4w]$ In each set of brackets, remove the inner parentheses.

$= 6x - 2y - 2w - 6x - 18y + 12w$ Remove each set of brackets by multiplying by the appropriate number.

$= -20y + 10w$ or $10w - 20y$ Combine like terms: $6x - 6x = 0x = 0$. □

Student Practice 3 Simplify. $3[4x - 2(1 - x)] - [3x + (x - 2)]$

You can always simplify problems with many sets of grouping symbols by the method shown. Essentially, you just keep removing one level of grouping symbols at each step. Finally, at the end you combine any like terms remaining if possible.

Sometimes it is possible to combine like terms at each step.

MC **Example 4** Simplify. $-3\{7x - 2[x - (2x - 1)]\}$

Solution

$= -3\{7x - 2[x - 2x + 1]\}$ Remove the inner parentheses by multiplying each term within the parentheses by -1.

$= -3\{7x - 2[-x + 1]\}$ Combine like terms by combining $+x - 2x$.

$= -3\{7x + 2x - 2\}$ Remove the brackets by multiplying each term within them by -2.

$= -3\{9x - 2\}$ Combine the x terms.

$= -27x + 6$ Remove the braces by multiplying each term by -3. □

Student Practice 4

Simplify. $-2\{5x - 3[2x - (3 - 4x)]\}$

STEPS TO SUCCESS How Important Is the Quick Quiz?

At the end of each exercise set is a Quick Quiz. Please be sure to do that. You will be amazed how it helps you in proving that you have mastered the homework. Make sure you do that each time as a necessary part of your homework. It will really help you gain confidence in what you have learned. The Quick Quiz is essential.

Here is a fun way you can use the Quick Quiz. Ask a friend in the class if he or she will "quiz you" about 5 minutes before class by asking you to work out (without using your book) the solution to one of the problems on the Quick Quiz. Tell your friend you can do the same thing for him or for her. This will force you to "be ready for anything" when you come to class. You will be amazed how this little trick will keep you sharp and ready for class.

Making it personal: Which of these suggestions do you find most helpful? Use those suggestions as you do each section.

1.9 Exercises MyMathLab®

Verbal and Writing Skills, Exercises 1–4

1. Rewrite the expression $-3x - 2y$ using a negative sign and parentheses.

2. Rewrite the expression $-x + 5y$ using a negative sign and parentheses.

3. To simplify expressions with grouping symbols, we use the ____________ property.

4. When an expression contains many grouping symbols, remove the ____________ grouping symbols first.

Simplify. Remove grouping symbols and combine like terms.

5. $8x - 4(x - 3y)$

6. $-6y - 2(x - 5y)$

7. $5(c - 3d) - (3c + d)$

8. $6(2c - d) - (4c + d)$

9. $-3(x + 3y) + 2(2x + y)$

10. $-4(x + 5y) + 3(6y - 3x)$

11. $2x[4x^2 - 2(x - 3)]$

12. $4y[-3y^2 + 2(4 - y)]$

13. $2[5(x + y) - 2(3x - 4y)]$

14. $-3[2(3a + b) - 5(a - 2b)]$

15. $[10 - 4(x - 2y)] + 3(2x + y)$

16. $[-5(-x + 3y) - 12] - 4(2x - 3)$

17. $5[3a - 2a(3a + 6b) + 6a^2]$

18. $3[x - y(3x + y) + y^2]$

19. $6a(2a^2 - 3a - 4) - a(a - 2)$

20. $7b(3b^2 - 2b - 5) - 2b(4 - b)$

21. $3a^2 - 4[2b - 3b(b + 2)]$

22. $2b^2 - 3[5b + 2b(2 - b)]$

23. $5b + \{-[3a + 2(5a - 2b)] - 1\}$

24. $3a - \{-2[a - 4(3a + b)] - 2\}$

25. $2\{3x^2 + 5[2x - (3 - x)]\}$

26. $4\{4x^2 + 2[3x - (4 - x)]\}$

27. $-4\{3a^2 - 2[4a^2 - (b + a^2)]\}$

28. $-3\{x^2 - 5[x - (x - 3x^2)]\}$

Cumulative Review

29. **[1.8.2]** ***Melting Point of Gold*** The melting point of pure gold is 1064.18°C. Use $F = 1.8C + 32$ to find the melting point of pure gold in degrees Fahrenheit. Round to the nearest hundredth of a degree.

▲ **30.** **[1.8.2]** ***Geometry*** Use 3.14 as an approximation for π to compute the area covered by a circular irrigation system with radial arm of length 380 feet. Use $A = \pi r^2$.

31. **[1.8.2]** ***Dog Weight*** An average Great Dane weighs between 120 and 150 pounds. Express the range of weight for a Great Dane in kilograms. Use the formula $k = 0.45p$ (where k = kilograms, p = pounds).

32. **[1.8.2]** ***Dog Weight*** An average Miniature Pinscher weighs between 9 and 14 pounds. Express the range of weight for a Miniature Pinscher in kilograms. Use the formula $k = 0.45p$ (where k = kilograms, p = pounds).

Quick Quiz 1.9 *Simplify.*

1. $2[3x - 2(5x + y)]$

2. $3[x - 3(x + 4) + 5y]$

3. $-4\{2a + 2[2ab - b(1 - a)]\}$

4. **Concept Check** Explain how you would simplify the following expression to combine like terms whenever possible. $3\{2 - 3[4x - 2(x + 3) + 5x]\}$

CAREER EXPLORATION PROBLEMS

Background: Electrical Engineering Assistant

Dan Silva is employed as an electrical engineering assistant. In order to finish a bid for a new client, Dan's manager has given him a few problems to solve involving the use of math and formulas.

Facts

- Each of two conductors carries 12A (amps) and has a voltage drop of 1.8V (volts). The formula used to calculate power loss is $P = I \times E$, where P is the power loss in watts (W), I is the current, and E is the voltage.
- A third conductor has a voltage drop of 3% in a 240V circuit carrying a current of 24A.
- An electrical circuit supply is 120V and its resistance is 200 ohms. The formula for current flow is $I = \frac{E}{R}$, where I is the current flow, E is the voltage, and R is the resistance.
- A circuit box has the dimensions 4 inches by 4 inches by $1\frac{1}{2}$ inches, and volume (V) = length $(L) \times$ width $(W) \times$ height (H).

Tasks

1. Dan must calculate the power loss of each conductor that carries 12A (amps) and has a voltage drop of 1.8V (volts) along with the total power loss. What are these amounts?

2. Dan needs to calculate the power loss in watts for the conductor having a voltage drop of 3% in the 240V circuit carrying a current of 24A. What is this amount?

3. Dan must determine the amount of current flow in amps (A) in the circuit where the supply is 120V and the resistance is 200 ohms using the formula $I = \frac{E}{R}$, where I is the current flow, E is the voltage, and R is the resistance. What is this amount?

4. Dan's final calculation involves a little geometry. He needs to determine the total space inside seven 4-by-4-by-$1\frac{1}{2}$-inch circuit boxes. What is this total volume?

Chapter 1 Organizer

Topic and Procedure	Examples	You Try It
Absolute value, p. 67 The absolute value of a number is the distance between that number and zero on the number line. The absolute value of any number will be positive or zero.	Find the absolute value. **(a)** $\lvert 3 \rvert = 3$ **(b)** $\lvert -2 \rvert = 2$ **(c)** $\lvert 0 \rvert = 0$ **(d)** $\left\lvert -\frac{5}{6} \right\rvert = \frac{5}{6}$ **(e)** $\lvert -1.38 \rvert = 1.38$	1. Find the absolute value. **(a)** $\lvert 5 \rvert$ **(b)** $\lvert -1 \rvert$ **(c)** $\lvert 0.5 \rvert$ **(d)** $\left\lvert -\frac{1}{4} \right\rvert$ **(e)** $\lvert -4.57 \rvert$
Adding real numbers with the same sign, p. 68 If the signs are the same, add the absolute values of the numbers. Use the common sign in the answer.	Add. $-3 + (-7) = -10$	2. Add. $-10 + (-4)$
Adding real numbers with different signs, p. 70 If the signs are different: **1.** Find the difference between the larger and the smaller absolute values. **2.** Give the answer the sign of the number having the larger absolute value.	Add. **(a)** $(-7) + 13 = 6$ **(b)** $7 + (-13) = -6$	3. Add. **(a)** $(-5) + 11$ **(b)** $5 + (-11)$
Adding several real numbers, p. 71 When adding several real numbers, separate them into two groups by sign. Find the sum of all the positive numbers and the sum of all the negative numbers. Combine these two subtotals by the method described above.	Add. $-7 + 6 + 8 + (-11) + (-13) + 22$ $\begin{array}{rr} -7 & +6 \\ -11 & +8 \\ \underline{-13} & \underline{+22} \\ -31 & +36 \end{array}$ $-31 + 36 = 5$ The answer is positive since 36 is positive.	4. Add. $-4 + 1 + (-8) + 12 + (-3) + 5$
Subtracting real numbers, p. 76 Add the opposite of the second number.	Subtract. $-3 - (-13) = -3 + (+13) = 10$	5. Subtract. $-8 - (-7)$
Multiplying and dividing real numbers, pp. 82, 84 **1.** If the two numbers have the same sign, multiply (or divide) the absolute values. The result is positive. **2.** If the two numbers have different signs, multiply (or divide) the absolute values. The result is negative.	Multiply or divide. **(a)** $-5(-3) = +15$ **(b)** $-36 \div (-4) = +9$ **(c)** $28 \div (-7) = -4$ **(d)** $-6(3) = -18$	6. Multiply or divide. **(a)** $9(-6)$ **(b)** $24 \div (-3)$ **(c)** $-48 \div (-8)$ **(d)** $-3(-7)$
Exponent form, p. 91 The base tells you what number is being multiplied. The exponent tells you how many times this number is used as a factor.	Evaluate. **(a)** $2^5 = 2 \cdot 2 \cdot 2 \cdot 2 \cdot 2 = 32$ **(b)** $4^3 = 4 \cdot 4 \cdot 4 = 64$ **(c)** $(-3)^4 = (-3)(-3)(-3)(-3) = 81$	7. Evaluate. **(a)** 3^4 **(b)** $(1.5)^2$ **(c)** $\left(\frac{1}{2}\right)^4$
Raising a negative number to a power, p. 92 When the base is negative, the result is positive for even exponents and negative for odd exponents.	Evaluate. **(a)** $(-3)^3 = -27$ **(b)** $(-2)^4 = 16$	8. Evaluate. **(a)** $(-2)^3$ **(b)** $(-4)^4$
Order of operations, p. 96 Remember the proper order of operations: **1.** Perform operations inside parentheses. **2.** Raise to powers. **3.** Multiply and divide from left to right. **4.** Add and subtract from left to right.	Evaluate. $3(5 + 4)^2 - 2^2 \cdot 3 \div (9 - 2^3)$ $= 3 \cdot 9^2 - 4 \cdot 3 \div (9 - 8)$ $= 3 \cdot 81 - 12 \div 1$ $= 243 - 12 = 231$	9. Evaluate. $4^2 + 2(6 - 3)^3 - (5 - 2)^2 \div 3$
Removing parentheses, p. 101 Use the distributive property to remove parentheses. $a(b + c) = ab + ac$	Multiply. **(a)** $3(5x + 2) = 15x + 6$ **(b)** $-4(x - 3y) = -4x + 12y$	10. Multiply. **(a)** $4(2a - 3)$ **(b)** $-5(5x - 1)$
Combining like terms, p. 106 Combine terms that have identical variables and exponents.	Simplify. $7x^2 - 3x + 4y + 2x^2 - 8x - 9y = 9x^2 - 11x - 5y$	11. Simplify. $9a^2 - 10a + 3ab + 7a - 12a^2 + 5ab$

Topic and Procedure	Examples	You Try It
Substituting into variable expressions, p. 111 1. Replace each variable by the numerical value given for it. 2. Follow the order of operations in evaluating the expression.	Evaluate. $2x^3 + 3xy + 4y^2$ for $x = -3$ and $y = 2$. $2(-3)^3 + 3(-3)(2) + 4(2)^2$ $= 2(-27) + 3(-3)(2) + 4(4)$ $= -54 - 18 + 16$ $= -56$	12. Evaluate. $6x^2 - xy + 3y^2$ for $x = 4$ and $y = -1$.
Using formulas, p. 112 1. Replace the variables in the formula by the given values. 2. Evaluate the expression. 3. Label units carefully.	Find the area of a circle with radius 4 feet. Use $A = \pi r^2$, with π as approximately 3.14. $A \approx (3.14)(4 \text{ ft})^2$ $\approx (3.14)(16 \text{ ft}^2)$ $\approx 50.24 \text{ ft}^2$ The area of the circle is approximately 50.24 square feet.	13. Find the area of a trapezoid whose altitude is 50 feet and whose bases are 40 feet and 60 feet.
Removing grouping symbols, p. 118 1. Remove innermost grouping symbols first. 2. Then remove remaining innermost grouping symbols. 3. Continue until all grouping symbols are removed. 4. Combine like terms.	Simplify. $5\{3x - 2[4 + 3(x - 1)]\}$ $= 5\{3x - 2[4 + 3x - 3]\}$ $= 5\{3x - 8 - 6x + 6\}$ $= 15x - 40 - 30x + 30$ $= -15x - 10$	14. Simplify. $4\{9x - [2(x + 3) - 8]\}$

Chapter 1 Review Problems

Section 1.1

Add.

1. $-6 + (-2)$

2. $-12 + 7.8$

3. $5 + (-2) + (-12)$

4. $3.7 + (-1.8)$

5. $\frac{1}{2} + \left(-\frac{5}{6}\right)$

6. $-\frac{3}{11} + \left(-\frac{1}{22}\right)$

7. $\frac{3}{4} + \left(-\frac{1}{12}\right) + \left(-\frac{1}{2}\right)$

8. $\frac{2}{15} + \frac{1}{6} + \left(-\frac{4}{5}\right)$

Section 1.2

Add or subtract.

9. $5 - (-3)$

10. $-2 - (-15)$

11. $-30 - (+3)$

12. $8 - (-1.2)$

13. $-\frac{7}{8} + \left(-\frac{3}{4}\right)$

14. $-\frac{3}{8} + \frac{5}{6}$

15. $-20.8 - 1.9$

16. $-151 - (-63)$

Section 1.3

Multiply or divide.

17. $87 \div (-29)$

18. $-10.4 \div (-0.8)$

19. $\dfrac{-24}{-\frac{3}{4}}$

20. $-\frac{2}{3} \div \left(-\frac{4}{5}\right)$

21. $\frac{5}{7} \div \left(-\frac{5}{25}\right)$

22. $-6(3)(4)$

23. $-1(-4)(-3)(-5)$

24. $(-5)\left(-\frac{1}{2}\right)(4)(-3)$

Section 1.4

Evaluate.

25. $(-3)^5$

26. $(-2)^6$

27. $(-5)^4$

28. $\left(-\frac{2}{3}\right)^3$

29. -9^2

30. $(0.6)^2$

31. $\left(\frac{5}{6}\right)^2$

32. $\left(\frac{3}{4}\right)^3$

Section 1.5

Simplify using the order of operations.

33. $5(-4) + 3(-2)^3$

34. $8 \div 0.4 + 0.1 \times (0.2)^2$

35. $(3 - 6)^2 + (-12) \div (-3)(-2)$

Section 1.6

Use the distributive property to multiply.

36. $7(-3x + y)$

37. $3x(6 - x + 3y)$

38. $-(7x^2 - 3x + 11)$

39. $(2xy + x - y)(-3y^2)$

Section 1.7

Combine like terms.

40. $3a^2b - 2bc + 6bc^2 - 8a^2b - 6bc^2 + 5bc$

41. $9x + 11y - 12x - 15y$

42. $4x^2 - 13x + 7 - 9x^2 - 22x - 16$

43. $-x + \frac{1}{2} + 14x^2 - 7x - 1 - 4x^2$

Section 1.8

Evaluate for the given value of the variables.

44. $7x - 6$ for $x = -7$

45. $7 - \frac{3}{4}x$ for $x = 8$

46. $x^2 + 3x - 4$ for $x = -3$

47. $-x^2 + 5x - 9$ for $x = 3$

48. $2x^3 - x^2 + 6x + 9$ for $x = -1$

49. $b^2 - 4ac$ for $a = -1$, $b = 5$, and $c = -2$

50. $\frac{mMG}{r^2}$ for $m = -4, M = 15, G = -1$, and $r = -2$

Solve.

51. ***Simple Interest*** Find the simple interest on a loan of \$6000 at an annual interest rate of 18% per year for $\frac{3}{4}$ of a year. Use $I = prt$, where p = principal, r = rate per year (in decimal form), and t = time in years.

52. ***Medication*** The label of a medication warns the user that it must be stored at a temperature between 20°C and 25°C. What is this temperature range in degrees Fahrenheit? Use the formula $F = \frac{9C + 160}{5}$.

▲ **53.** ***Sign Painting*** How much will it cost to paint a circular sign with a radius of 4 feet if the painter charges \$1.50 per square foot? Use $A = \pi r^2$, where π is approximately 3.14.

54. ***Profit*** Find the daily profit P at a furniture factory if the initial cost of setting up the factory is C = \$1200, rent R = \$300, and sale price of furniture S = \$56. Use the profit formula $P = 180S - R - C$.

▲ **55.** ***Parking Lot Sealer*** A parking lot is in the shape of a trapezoid. The altitude of the trapezoid is 200 feet, and the bases of the trapezoid are 300 feet and 700 feet. What is the area of the parking lot? If the parking lot had a sealer applied that costs \$2 per square foot, what was the cost of the amount of sealer needed for the entire parking lot?

▲ **56.** ***Signal Paint*** The Green Mountain Telephone Company has a triangular signal tester at the top of a communications tower. The altitude of the triangle is 3.8 feet and the base is 5.5 feet. What is the area of the triangular signal tester? If the signal tester was painted with a special metallic surface paint that costs \$66 per square foot, what was the cost of the amount of paint needed to paint one side of the triangle?

Section 1.9

Simplify.

57. $5x - 7(x - 6)$

58. $3(x - 2) - 4(5x + 3)$

59. $2[3 - (4 - 5x)]$

60. $-3x[x + 3(x - 7)]$

61. $2xy^3 - 6x^3y - 4x^2y^2 + 3(xy^3 - 2x^2y - 3x^2y^2)$

62. $-5(x + 2y - 7) + 3x(2 - 5y)$

63. $-(a + 3b) + 5[2a - b - 2(4a - b)]$

64. $-5\{2a - [5a - b(3 + 2a)]\}$

65. $-3\{2x - [x - 3y(x - 2y)]\}$

66. $2\{3x + 2[x + 2y(x - 4)]\}$

Mixed Practice

Simplify the following.

67. $-6.3 + 4$

68. $4 + (-8) + 12$

69. $-\frac{2}{3} - \frac{4}{5}$

70. $-\frac{7}{8} - \left(-\frac{3}{4}\right)$

71. $3 - (-4) + (-8)$

72. $-1.1 - (-0.2) + 0.4$

73. $\left(-\frac{9}{10}\right)\left(-2\frac{1}{4}\right)$

74. $3.6 \div (-0.45)$

75. $-14.4 \div (-0.06)$

76. $(-8.2)(3.1)$

77. ***Jeopardy*** A Jeopardy quiz show contestant began the second round (Double Jeopardy) with \$400. She buzzed in on the first two questions, answering a \$1000 question correctly but then giving the incorrect answer to a \$800 question. What was her score?

Simplify the following.

78. $(-0.3)^4$

79. -0.5^4

80. $9(5) - 5(2)^3 + 5$

81. $3.8x - 0.2y - 8.7x + 4.3y$

82. Evaluate $\frac{2p + q}{3q}$ for $p = -2$ and $q = 3$.

83. Evaluate $\frac{4s - 7t}{s}$ for $s = -3$ and $t = -2$.

84. ***Dog Body Temperature*** The normal body temperature of a dog is 38.6°C. Your dog has a temperature of 101.1°F. Does your dog have a fever? Use the formula $F = \frac{9}{5}C + 32$.

85. $-7(x - 3y^2 + 4) + 3y(4 - 6y)$

86. $-2\{6x - 3[7y - 2y(3 - x)]\}$

How Am I Doing? Chapter 1 Test

MATH COACH MyMathLab® YouTube

After you take this test read through the Math Coach on pages 128–129. Math Coach videos are available via MyMathLab and YouTube. Step-by-step test solutions on the Chapter Test Prep Videos are also available via MyMathLab and YouTube. (Search "TobeyBegInterAlg" and click on "Channels.")

Simplify.

1. $-2.5 + 6.3 + (-4.1)$ **2.** $-5 - (-7)$ **3.** $\left(-\frac{2}{3}\right)(7)$

4. $-5(-2)(7)(-1)$ **5.** $-12 \div (-3)$ **6.** $-1.8 \div (0.6)$

7. $(-4)^3$ **8.** $(1.6)^2$ MC **9.** $\left(\frac{2}{3}\right)^4$

10. $(0.2)^2 - (2.1)(-3) + 0.46$ MC **11.** $3(4 - 6)^3 + 12 \div (-4) + 2$

12. $-5x(x + 2y - 7)$ **13.** $-2ab^2(-3a - 2b + 7ab)$

14. $6ab - \frac{1}{2}a^2b + \frac{3}{2}ab + \frac{5}{2}a^2b$ **15.** $2.3x^2y - 8.1xy^2 + 3.4xy^2 - 4.1x^2y$

16. $3(2 - a) - 4(-6 - 2a)$ **17.** $5(3x - 2y) - (x + 6y)$

In questions 18–20, evaluate for the values of the variables indicated.

18. $x^3 - 3x^2y + 2y - 5$ for $x = 3$ and $y = -4$

MC **19.** $3x^2 - 7x - 11$ for $x = -3$

20. $2a - 3b$ for $a = \frac{1}{3}$ and $b = -\frac{1}{2}$

21. If you are traveling 60 miles per hour on a highway in Canada, how fast are you traveling in kilometers per hour? (Use $k = 1.61r$, where r = rate in miles per hour and k = rate in kilometers per hour.)

▲ **22.** A field is in the shape of a trapezoid. The altitude of the trapezoid is 120 feet and the bases of the trapezoid are 180 feet and 200 feet. What is the area of the field?

▲ **23.** Jeff Slater's garage has a triangular roof support beam. The support beam is covered with a sheet of plywood. The altitude of the triangular region is 6.8 feet and the base is 8.5 feet. If the triangular piece of plywood was painted with paint that cost \$0.80 per square foot, what was the cost of the amount of paint needed to coat one side of the triangle?

▲ **24.** You wish to apply blacktop sealer to your driveway but do not know how much to buy. If your rectangular driveway measures 60 feet long by 10 feet wide, and a can of blacktop sealer claims to cover 200 square feet, how many cans should you buy?

Simplify.

25. $3[x - 2y(x + 2y) - 3y^2]$ MC **26.** $-3\{a + b[3a - b(1 - a)]\}$

1. ______
2. ______
3. ______
4. ______
5. ______
6. ______
7. ______
8. ______
9. ______
10. ______
11. ______
12. ______
13. ______
14. ______
15. ______
16. ______
17. ______
18. ______
19. ______
20. ______
21. ______
22. ______
23. ______
24. ______
25. ______
26. ______

Total Correct: ______

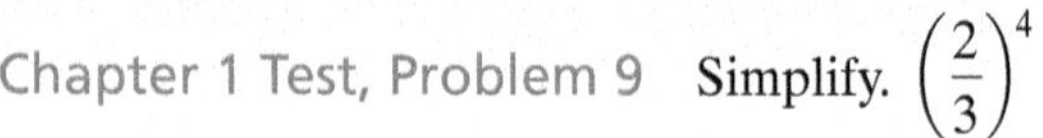

Mastering the skills you need to do well on the test.

The following problems are from the Chapter 1 Test. Here are some helpful hints to keep you from making common errors on test problems.

Chapter 1 Test, Problem 9 Simplify. $\left(\frac{2}{3}\right)^4$

HELPFUL HINT Always write out the repeated multiplication.

- If a number is raised to the fourth power, then we write out the multiplication with that number appearing as a factor a total of four times.
- When the base is a fraction, this means that we must multiply the numerators and then multiply the denominators.

Did you rewrite the problem as $\left(\frac{2}{3}\right)\left(\frac{2}{3}\right)\left(\frac{2}{3}\right)\left(\frac{2}{3}\right)$?

Yes ______ No ______

If you answered No to this question, please complete this step now.

Did you multiply $2 \times 2 \times 2 \times 2$ in the numerator and $3 \times 3 \times 3 \times 3$ in the denominator?

Yes ______ No ______

If you answered No, make this correction and complete the calculations.

If you answered Problem 9 incorrectly, go back and rework the problem using these suggestions.

Chapter 1 Test, Problem 11 Simplify. $3(4 - 6)^3 + 12 \div (-4) + 2$

HELPFUL HINT First, do all operations inside parentheses. Second, raise numbers to a power. Then multiply and divide numbers from left to right. As the last step, add and subtract numbers from left to right.

Did you first combine $4 - 6$ to obtain -2?

Yes ______ No ______

If you answered No, perform the operation inside the parentheses first.

Next, did you calculate $(-2)^3 = -8$ *before* multiplying by 3?

Yes ______ No ______

If you answered No, go back and evaluate the exponential expression.

Did you multiply and divide *before* adding?

Yes ______ No ______

If you answered No, remember that multiplication and division must be performed before addition and subtraction once the operations inside the parentheses are done and exponents are evaluated.

Now go back and rework the problem using these suggestions.

Need more help? Watch the MATH COACH videos in MyMathLab® or on YouTube™.

Chapter 1 Test, Problem 19 Evaluate $3x^2 - 7x - 11$ for $x = -3$.

HELPFUL HINT When you replace a variable by a particular value, place parentheses around that value. Then use the order of operations to evaluate the expression.

Did you first rewrite the expression as $3(-3)^2 - 7(-3) - 11$, using parentheses to complete the substitution?

Yes ______ No ______

If you answered No, please go back and perform that substitution step using parentheses around the specified value.

Did you next raise -3 to the second power to obtain $3(9) - 7(-3) - 11$?

Yes ______ No ______

If you answered No, review the order of operations and complete this step.

Remember that multiplication must be performed before addition and subtraction.

If you answered Problem 19 incorrectly, go back and rework this problem using these suggestions.

Chapter 1 Test, Problem 26 Simplify. $-3\{a + b[3a - b(1 - a)]\}$

HELPFUL HINT Work from the inside out. Remove the innermost symbols, ( ), first. Then remove the next level of innermost symbols, [ ]. Finally remove the outermost symbols, { }. Be careful to avoid sign errors.

Did you first obtain the expression $-3\{a + b[3a - b + ab]\}$ when you removed the innermost parentheses?

Yes ______ No ______

If you answered No, go back to the original problem and use the distributive property to remove the innermost grouping symbol, the parentheses.

Did you next obtain the expression $-3\{a + 3ab - b^2 + ab^2\}$ when you removed the brackets?

Yes ______ No ______

If you answered No, use the distributive property to distribute the b on the outside of the bracket, [ ], to each term inside the bracket.

Finally, use the distributive property in the last step to distribute the -3 across all of the terms inside the outermost brackets, { }. Be careful with the +/− signs.

Now go back and rework the problem using these suggestions.

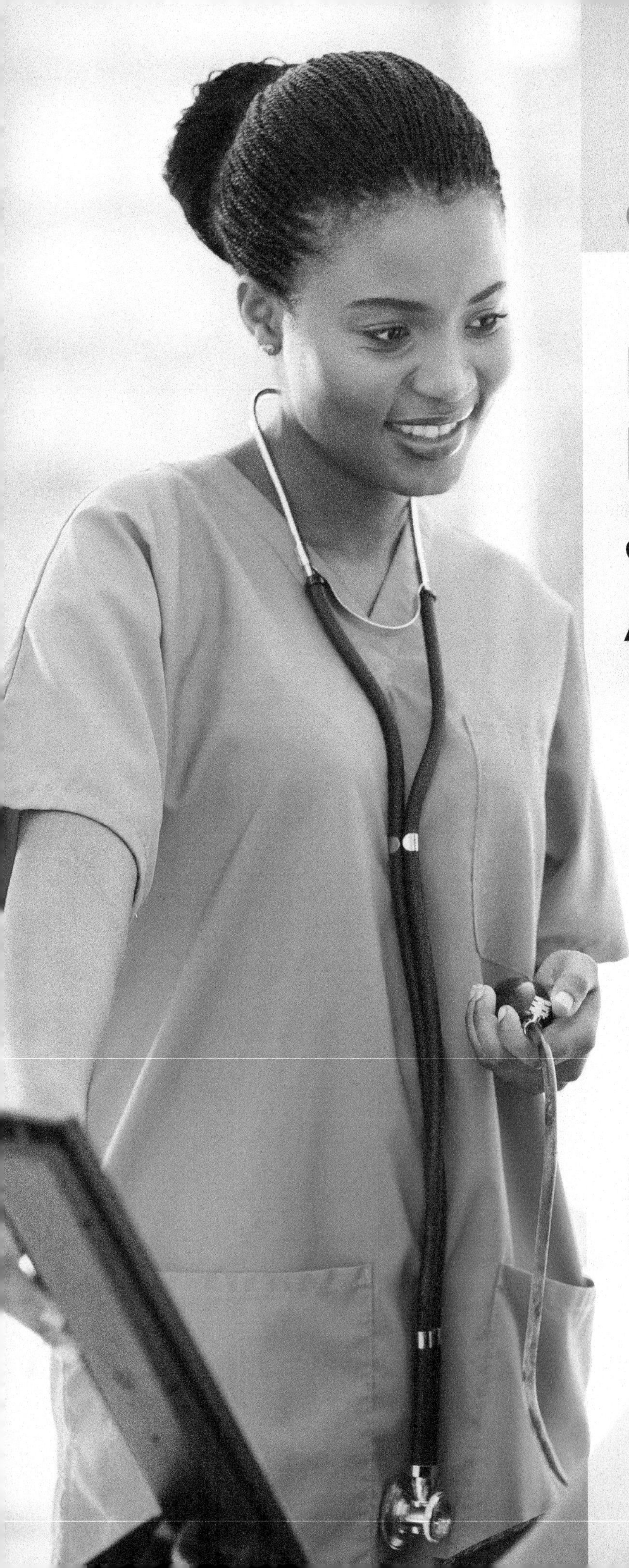

CHAPTER 2

Equations, Inequalities, and Applications

CAREER OPPORTUNITIES

Dietitian, Nutritionist

Dietitians and nutritionists design food programs that promote healthy eating habits and lifestyles. They play important roles in assisting health care professionals by evaluating patients' nutritional needs and implementing individualized dietary plans. The mathematics covered in this chapter can be applied to the development of treatment plans for maintaining healthy, balanced lifestyles.

To investigate how the mathematics in this chapter can help with this field, see the Career Exploration Problems on page **193**.

2.1 The Addition Principle of Equality

1 Using the Addition Principle to Solve Equations of the Form $x + b = c$

Student Learning Objective

After studying this section, you will be able to:

1 Use the addition principle to solve equations of the form $x + b = c$.

When we use an equals sign $(=)$, we are indicating that two expressions are equal in value. Such a statement is called an **equation.** For example, $x + 5 = 23$ is an equation. A **solution** of an equation is a number that when substituted for the variable makes the equation true. Thus 18 is a solution of $x + 5 = 23$ because $18 + 5 = 23$. Equations that have exactly the same solutions are called **equivalent equations.** By following certain procedures, we can often transform an equation to a simpler, equivalent one that has the form $x =$ some number. Then this number is a solution of the equation. The process of finding all solutions of an equation is called **solving the equation.**

One of the first procedures used in solving equations has an everyday application. Suppose that we place a 10-kilogram box on one side of a seesaw and a 10-kilogram stone on the other side. If the center of the box is the same distance from the balance point as the center of the stone, we would expect the seesaw to balance. The box and the stone do not look the same, but their weights are equal. If we add a 2-kilogram lead weight to the center of each object at the same time, the seesaw would still balance. The weights are still equal.

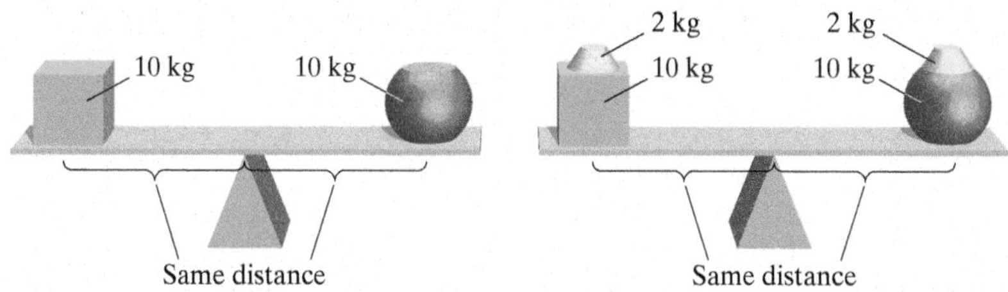

There is a similar principle in mathematics. We can state it in words as follows.

THE ADDITION PRINCIPLE

If the same number is added to both sides of an equation, the results on both sides are equal in value.

We can restate it in symbols this way.

For real numbers a, b, and c, if $a = b$, then $a + c = b + c$.

Here is an example.

$$\text{If } 3 = \frac{6}{2}, \text{ then } 3 + 5 = \frac{6}{2} + 5.$$

Since we added the same amount, 5, to both sides, the sides remain equal to each other.

$$3 + 5 = \frac{6}{2} + 5$$

$$8 = \frac{6}{2} + \frac{10}{2}$$

$$8 = \frac{16}{2}$$

$$8 = 8$$

We can use the addition principle to solve certain equations.

Example 1 Solve for x. $x + 16 = 20$

Solution

$x + 16 + (-16) = 20 + (-16)$	Use the addition principle to add -16 to both sides.
$x + 0 = 4$	Simplify.
$x = 4$	The value of x is 4.

We have just found a solution of the equation. A **solution** is a value for the variable that makes the equation true. We then say that the value 4 in our example **satisfies** the equation. We can easily verify that 4 is a solution by substituting this value into the original equation. This step is called **checking** the solution.

Check. $x + 16 = 20$

$4 + 16 \stackrel{?}{=} 20$

$20 = 20$ ✓

When the same value appears on both sides of the equals sign, we call the equation an **identity.** Because the two sides of the equation in our check have the same value, we know that the original equation has been solved correctly. We have found a solution, and since no other number makes the equation true, it is the only solution. □

Student Practice 1 Solve for x and check your solution.

$x + 14 = 23$

Notice that when you are trying to solve these types of equations, you must add a particular number to both sides of the equation. What is the number to choose? Look at the number that is on the same side of the equation with x, that is, the number added to x. Then think of the number that is **opposite in sign.** This is called the **additive inverse** of the number. The additive inverse of 16 is -16. The additive inverse of -3 is 3. The number to add to both sides of the equation is precisely this additive inverse.

It does not matter which side of the equation contains the variable. The x-term may be on the right or left. In the next example the x-term will be on the right.

Example 2 Solve for x. $14 = x - 3$

Solution

$14 + 3 = x - 3 + 3$	Notice that -3 is being added to x in the original equation. Add 3 to both sides, since 3 is the additive inverse of -3. This will eliminate the -3 on the right and isolate x.
$17 = x + 0$	Simplify.
$17 = x$	The value of x is 17.

Check.

$14 = x - 3$	
$14 \stackrel{?}{=} 17 - 3$	Replace x by 17.
$14 = 14$ ✓	Simplify. It checks. The solution is 17.

□

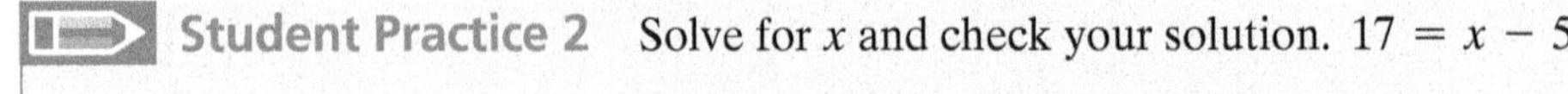

Student Practice 2 Solve for x and check your solution. $17 = x - 5$

Before you add a number to both sides, you should always simplify the equation. The following example shows how combining numbers by addition—separately, on both sides of the equation—simplifies the equation.

Example 3 Solve for x. $1.5 + 0.2 = 0.3 + x + 0.2$

Solution

$$1.5 + 0.2 = 0.3 + x + 0.2$$

$$1.7 = x + 0.5 \quad \text{Simplify by adding.}$$

$$1.7 + (-0.5) = x + 0.5 + (-0.5) \quad \text{Add the value } -0.5 \text{ to both sides, since } -0.5 \text{ is the additive inverse of } 0.5.$$

$$1.2 = x \quad \text{Simplify. The value of } x \text{ is } 1.2.$$

Check.

$$1.5 + 0.2 = 0.3 + x + 0.2$$

$$1.5 + 0.2 \stackrel{?}{=} 0.3 + 1.2 + 0.2 \quad \text{Replace } x \text{ by } 1.2 \text{ in the original equation.}$$

$$1.7 = 1.7 \; \checkmark \quad \text{It checks.}$$

□

 **Student Practice 3** Solve for x and check your solution.

$$0.5 - 1.2 = x - 0.3$$

In Example 3 we added -0.5 to each side. You could subtract 0.5 from each side and get the same result. In Chapter 1 we discussed how subtracting a 0.5 is the same as adding a negative 0.5. Do you see why?

Just as it is possible to add the same number to both sides of an equation, it is also possible to subtract the same number from both sides of an equation. This is so because any subtraction problem can be rewritten as an addition problem. For example, $1.7 - 0.5 = 1.7 + (-0.5)$. Thus the addition principle tells us that we can subtract the same number from both sides of the equation.

We can determine whether a value is the solution to an equation by following the same steps used to check an answer. Substitute the value to be tested for the variable in the original equation. We will obtain an identity if the value is the solution.

Example 4 Is 10 the solution to the equation $-15 + 2 = x - 3$? If it is not, find the solution.

Solution We substitute 10 for x in the equation and see if we obtain an identity.

$$-15 + 2 = x - 3$$

$$-15 + 2 \stackrel{?}{=} 10 - 3$$

$$-13 \neq 7 \quad \text{The values are not equal. The statement is not an identity.}$$

Thus, 10 is not the solution.

Now we take the original equation and solve to find the solution.

$$-15 + 2 = x - 3$$

$$-13 = x - 3 \quad \text{Simplify by adding.}$$

$$-13 + 3 = x - 3 + 3 \quad \text{Add 3 to both sides. 3 is the additive inverse of } -3.$$

$$-10 = x$$

Check. Replace x with -10 and verify that it is a solution. □

Look at the solution above. Notice that if you make a sign error you could obtain 10 as the solution instead of -10. As we saw above, 10 is *not* a solution. We must be especially careful to write the correct sign for each number when solving equations.

Student Practice 4 Is -2 the solution to the equation $x + 8 = -22 + 6$? If it is not, find the solution.

Example 5 Find the value of x that satisfies the equation.

$$\frac{1}{5} + x = -\frac{1}{10} + \frac{1}{2}$$

Solution To be combined, the fractions must have common denominators. The least common denominator (LCD) of the fractions is 10.

$$\frac{1 \cdot 2}{5 \cdot 2} + x = -\frac{1}{10} + \frac{1 \cdot 5}{2 \cdot 5}$$ Change each fraction to an equivalent fraction with a denominator of 10.

$$\frac{2}{10} + x = -\frac{1}{10} + \frac{5}{10}$$ This is an equivalent equation.

$$\frac{2}{10} + x = \frac{4}{10}$$ Simplify by adding.

$$\frac{2}{10} + \left(-\frac{2}{10}\right) + x = \frac{4}{10} + \left(-\frac{2}{10}\right)$$ Add the additive inverse of $\frac{2}{10}$ to each side. You could also say that you are subtracting $\frac{2}{10}$ from each side.

$$x = \frac{2}{10}$$ Add the fractions.

$$x = \frac{1}{5}$$ Simplify the answer.

Check. We substitute $\frac{1}{5}$ for x and see if we obtain an identity.

$$\frac{1}{5} + x = -\frac{1}{10} + \frac{1}{2}$$ Write the original equation.

$$\frac{1}{5} + \frac{1}{5} \stackrel{?}{=} -\frac{1}{10} + \frac{1}{2}$$ Substitute $\frac{1}{5}$ for x.

$$\frac{2}{5} \stackrel{?}{=} -\frac{1}{10} + \frac{5}{10}$$

$$\frac{2}{5} \stackrel{?}{=} \frac{4}{10}$$

$$\frac{2}{5} = \frac{2}{5} \checkmark$$ It checks. □

Student Practice 5 Find the value of x that satisfies the equation.

$$\frac{1}{20} - \frac{1}{2} = x + \frac{3}{5}$$

STEPS TO SUCCESS Taking Good Notes in Every Class Session

Don't copy down everything the teacher says. You will get overloaded with facts. Instead write down the important ideas and examples as the instructor lectures. Be sure to include any helpful hints or suggestions that your instructor gives you. You will be amazed at how easily you forget these if you do not write them down.

Be an active listener. Keep your mind on what the instructor is saying. Be ready with questions whenever you do not understand something. Stay alert in class. Keep your mind on mathematics.

Preview the lesson material. Before class, glance over the topics that will be covered. You can take better notes if you know the topics of the lecture ahead of time.

Look back at your notes. Try to review them the same day sometime after class. You will find the content of your notes easier to understand if you read them again within a few hours of class.

Making it personal: Which of these suggestions are the most helpful to you? How can you improve your skills in note taking? ▼

2.1 Exercises MyMathLab®

Verbal and Writing Skills, Exercises 1–6 *In exercises 1–3, fill in each blank with the appropriate word.*

1. When we use the ________ sign, we indicate two expressions are ________ in value.

2. If the ________ ________ is added to both sides of an equation, the results on each side are equal in value.

3. The _______ of an equation is a value of the variable that makes the equation true.

4. What is the additive inverse of -20?

5. Why do we add the additive inverse of a to each side of $x + a = b$ to solve for x?

6. What is the additive inverse of a?

Solve for x. Check your answers.

7. $x + 14 = 21$

8. $x + 15 = 21$

9. $20 = 9 + x$

10. $23 = 8 + x$

11. $x - 3 = 14$

12. $x - 13 = 4$

13. $0 = x + 5$

14. $0 = x + 9$

15. $x - 6 = -19$

16. $x - 11 = -13$

17. $-12 + x = 50$

18. $-16 + x = 47$

19. $3 + 5 = x - 7$

20. $8 - 2 = x + 5$

21. $32 - 17 = x - 6$

22. $32 - 11 = x - 4$

23. $4 + 8 + x = 6 + 6$

24. $19 - 3 + x = 10 + 6$

25. $18 - 7 + x = 7 + 9 - 5$

26. $3 - 17 + 8 = 8 + x - 3$

27. $-12 + x - 3 = 15 - 18 + 9$

28. $-19 + x - 7 = 20 - 42 + 10$

In exercises 29–36, determine whether the given solution is correct. If it is not, find the solution.

29. Is $x = 5$ the solution to $-7 + x = 2$?

30. Is $x = 7$ the solution to $-13 + x = 4$?

31. Is -6 the solution to $-11 + 5 = x + 8$?

32. Is -9 the solution to $-13 - 4 = x - 8$?

33. Is -33 the solution to $x - 23 = -56$?

34. Is -8 the solution to $-39 = x - 47$?

35. Is 35 the solution to $15 - 3 + 20 = x - 3$?

36. Is -12 the solution to $x + 8 = 12 - 19 + 3$?

Find the value of x that satisfies each equation.

37. $2.5 + x = 0.7$

38. $8.2 + x = 3.2$

39. $12.5 + x - 8.2 = 4.9$

40. $4.3 + x - 2.6 = 3.4$

41. $x - \frac{1}{4} = \frac{3}{4}$

42. $x + \frac{1}{3} = \frac{2}{3}$

43. $\frac{2}{3} + x = \frac{1}{6} + \frac{1}{4}$

44. $\frac{2}{5} + x = \frac{1}{2} - \frac{3}{10}$

Mixed Practice *Solve for x.*

45. $3 + x = -12 + 8$

46. $12 + x = -7 + 20$

47. $5\frac{1}{6} + x = 8$

48. $3\frac{3}{4} + x = 9$

49. $\frac{3}{14} - \frac{2}{7} = x - \frac{1}{2}$

50. $\frac{3}{16} - \frac{1}{4} = x - \frac{3}{8}$

51. $1.6 + x - 3.2 = -2 + 5.6$

52. $1.8 + x - 4.6 = -3 + 4.2$

53. $x - 18.225 = 1.975$

54. $x - 10.012 = -16.835$

Cumulative Review *Simplify by combining like terms.*

55. **[1.7.2]** $x + 3y - 5x - 7y + 2x$

56. **[1.7.2]** $y^2 + y - 12 - 3y^2 - 5y + 16$

Quick Quiz 2.1 *Solve for the variable.*

1. $x - 4.7 = 9.6$

2. $-8.6 + x = -12.1$

3. $3 - 12 + 7 = 8 + x - 2$

4. **Concept Check** Explain how you would check to verify whether $x = 3.8$ is the solution to $-1.3 + 1.6 + 3x = -6.7 + 4x + 3.2$.

2.2 The Multiplication Principle of Equality 

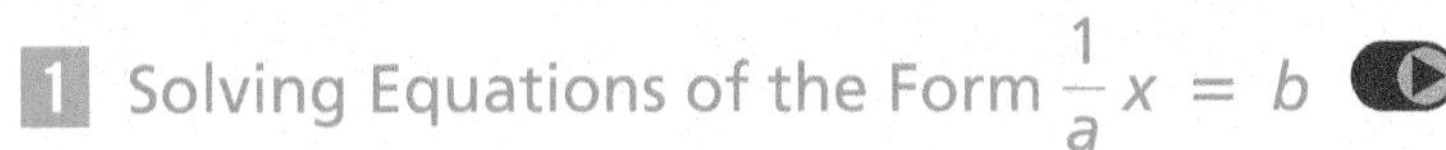

1 Solving Equations of the Form $\frac{1}{a}x = b$

Student Learning Objectives

After studying this section, you will be able to:

1. Solve equations of the form $\frac{1}{a}x = b$.
2. Solve equations of the form $ax = b$.

The addition principle allows us to add the same number to both sides of an equation. What would happen if we multiplied each side of an equation by the same number? For example, what would happen if we multiplied each side of an equation by 3?

To answer this question, let's return to our simple example of the box and the stone on a balanced seesaw. If we triple the weight on each side (that is, multiply the weight on each side by 3), the seesaw would still balance. The weight values of both sides remain equal.

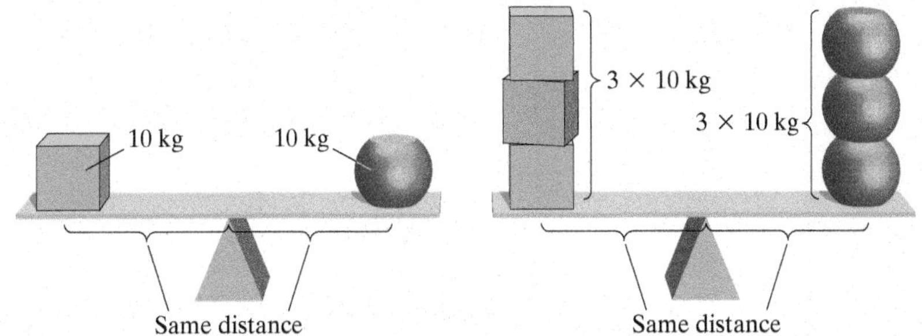

In words we can state this principle thus.

> **MULTIPLICATION PRINCIPLE**
>
> **If both sides of an equation are multiplied by the same nonzero number, the results on both sides are equal in value.**

In symbols we can restate the multiplication principle this way.

For real numbers a, b, and c with $c \neq 0$, if $a = b$, then $ca = cb$.

Let us look at an equation where it would be helpful to multiply each side by 3.

Example 1 Solve for x. $\frac{1}{3}x = -15$

Solution We know that $(3)(\frac{1}{3}) = 1$. We will multiply each side of the equation by 3 because we want to isolate the variable x.

$$3\left(\frac{1}{3}x\right) = 3(-15) \quad \text{Multiply each side of the equation by 3 since } (3)(\tfrac{1}{3}) = 1.$$

$$\left(\frac{3}{1}\right)\left(\frac{1}{3}\right)(x) = -45$$

$$1x = -45 \quad \text{Simplify.}$$

$$x = -45 \quad \text{The solution is } -45.$$

Check. $\frac{1}{3}(-45) \stackrel{?}{=} -15$ Substitute -45 for x in the original equation.

$-15 = -15$ ✓ It checks. □

Student Practice 1 Solve for x.

$$\frac{1}{8}x = -2$$

Note that $\frac{1}{5}x$ can be written as $\frac{x}{5}$. To solve the equation $\frac{x}{5} = 3$, we could multiply each side of the equation by 5. Try it. Then check your solution.

2 Solving Equations of the Form $ax = b$

We can see that using the multiplication principle to multiply each side of an equation by $\frac{1}{2}$ is the same as dividing each side of the equation by 2. Thus, it would seem that the multiplication principle would allow us to divide each side of the equation by any nonzero real number. Is there a real-life example of this idea?

Let's return to our simple example of the box and the stone on a balanced seesaw. Suppose that we were to cut the two objects in half (so that the amount of weight of each was divided by 2). We then return the objects to the same places on the seesaw. The seesaw would still balance. The weight values of both sides remain equal.

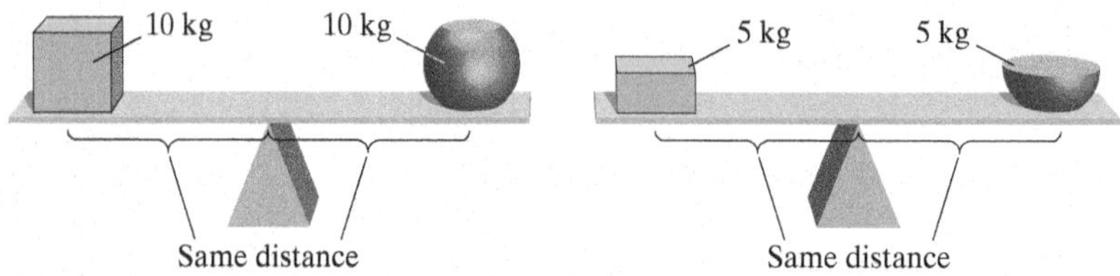

In words we can state this principle thus.

> **DIVISION PRINCIPLE**
>
> If both sides of an equation are divided by the same nonzero number, the results on both sides are equal in value.

Note: We put a restriction on the number by which we are dividing. We cannot divide by zero. We say that expressions like $\frac{2}{0}$ are not defined. Thus we restrict our divisor to *nonzero* numbers. We can restate the division principle this way.

$$\text{For real numbers } a, b, \text{ and } c \quad \text{where } c \neq 0, \quad \text{if } a = b, \quad \text{then } \frac{a}{c} = \frac{b}{c}.$$

Example 2 Solve for x. $5x = 125$

Solution

$$\frac{5x}{5} = \frac{125}{5} \quad \text{Divide both sides by 5.}$$

$$x = 25 \quad \text{Simplify. The solution is 25.}$$

Check.

$$5x = 125$$

$$5(25) \stackrel{?}{=} 125 \quad \text{Replace } x \text{ by 25.}$$

$$125 = 125 \checkmark \quad \text{It checks.}$$

 **Student Practice 2** Solve for x. $9x = 72$

For equations of the form $ax = b$ (a number multiplied by x equals another number), we solve the equation by choosing to divide both sides by a particular number. What is the number to choose? We look at the side of the equation that contains x. We notice the number that is multiplied by x. We divide by that number. The division principle tells us that we can still have a true equation provided that we divide by that number *on both sides* of the equation.

The solution to an equation may be a proper fraction or an improper fraction.

Example 3 Solve for x. $9x = 60$

Solution

$$\frac{9x}{9} = \frac{60}{9} \quad \text{Divide both sides by 9.}$$

$$x = \frac{20}{3} \text{ or } 6\frac{2}{3} \quad \text{Simplify. The solution is } \frac{20}{3} \text{ or } 6\frac{2}{3}.$$

Check.

$$9x = 60$$

$$\overset{3}{\cancel{9}}\left(\frac{20}{\cancel{3}}\right) \stackrel{?}{=} 60 \quad \text{Replace } x \text{ by } \frac{20}{3}.$$

$$60 = 60 \checkmark \quad \text{It checks.}$$

□

 **Student Practice 3** Solve for x.

$$6x = 50$$

In Examples 2 and 3 we *divided by the number multiplied by x.* This procedure is followed regardless of whether the sign of that number is positive or negative. In equations of the form $ax = b$, a is a number multiplied by x. The **coefficient** of x is a. A coefficient is a multiplier.

Sidelight As you work through the exercises in this book, you will notice that the solutions of equations can be integers, fractions, or decimals. Recall from page 65 that a **terminating decimal** is one that has a definite number of digits. Unless directions state that the solution should be rounded, the decimal form of a solution should be given only if it is a terminating decimal. Is the decimal form of the solution in Example 3 a terminating decimal?

Example 4 Solve for x. $-3x = 48$

Solution

$$\frac{-3x}{-3} = \frac{48}{-3} \quad \text{Divide both sides by the coefficient } -3.$$

$$x = -16 \quad \text{The solution is } -16.$$

Check. Can you check this solution? □

 **Student Practice 4** Solve for x. $-27x = 54$

The coefficient of x may be 1 or -1. You may have to rewrite the equation so that the coefficient 1 or -1 is obvious. With practice you may be able to recognize the coefficient without actually rewriting the equation.

Example 5 Solve for x. $-x = -24$.

Solution

$$-1x = -24 \quad \text{Rewrite the equation. } -1x \text{ is the same as } -x. \text{ Now the coefficient } -1 \text{ is obvious.}$$

$$\frac{-1x}{-1} = \frac{-24}{-1} \quad \text{Divide both sides by } -1.$$

$$x = 24 \quad \text{The solution is 24.}$$

Check. Can you check this solution? □

 **Student Practice 5** Solve for x. $-x = 36$

The variable can be on either side of the equation.

Example 6 Solve for x. $-78 = 5x - 8x$

Solution

$$-78 = 5x - 8x \quad \text{There are like terms on the right side.}$$

$$-78 = -3x \quad \text{Combine like terms.}$$

$$\frac{-78}{-3} = \frac{-3x}{-3} \quad \text{Divide both sides by } -3.$$

$$26 = x \quad \text{The solution is 26.}$$

Check. The check is left to you. □

Student Practice 6 Solve for x.

$$-51 = 3x - 9x$$

There is a mathematical concept that unites what we have learned in this section. The concept uses the idea of a multiplicative inverse. For any nonzero number a, the **multiplicative inverse** of a is $\frac{1}{a}$. Likewise, for any nonzero number a, the multiplicative inverse of $\frac{1}{a}$ is a. So to solve an equation of the form $ax = b$, we say that we need to multiply each side by the multiplicative inverse of a. Thus to solve $5x = 45$, we would multiply each side of the equation by the multiplicative inverse of 5, which is $\frac{1}{5}$. In similar fashion, if we wanted to solve the equation $\frac{1}{6}x = 4$, we would multiply each side of the equation by the multiplicative inverse of $\frac{1}{6}$, which is 6. In general, all the problems we have covered so far in this section can be solved by multiplying both sides of the equation by the multiplicative inverse of the coefficient of x.

Example 7 Solve for x. $31.2 = 6.0x - 0.8x$

Solution

$31.2 = 6.0x - 0.8x$ There are like terms on the right side.

$31.2 = 5.2x$ Combine like terms.

$\dfrac{31.2}{5.2} = \dfrac{5.2x}{5.2}$ Divide both sides by 5.2 (which is the same as multiplying both sides by the multiplicative inverse of 5.2).

$6 = x$ The solution is 6.

Note: Be sure to place the decimal point in the quotient directly above the caret (‸) when performing the division.

$$\begin{array}{r} 6. \\ 5.2_{\wedge}\overline{)\,31.2_{\wedge}} \\ \underline{31\ 2} \\ 0 \end{array}$$

Check. The check is left to you. □

Student Practice 7 Solve for x. $16.2 = 5.2x - 3.4x$

STEPS TO SUCCESS HELP! When Do I Get It? Where Do I Get It?

Getting the right kind of help at the right time can be the key ingredient to being successful in Beginning Algebra. When you have made every effort to learn the math on your own and you still need assistance, it is time to ask for help.

When should you ask for help? Ask as soon as you discover that you don't understand something. Don't wait until the night before a test. When you try the homework and have trouble and you are unable to clear up the difficulty in the next class period, *that is the time to seek help immediately.*

Where do you go for help? The best source is your instructor. Make an appointment to see your instructor and explain exactly where you are having trouble. If that is not possible, use the tutoring services at your college, visit the mathematics lab, watch the videos, use MyMathLab and let it help you, or call the 1-800 phone number for phone tutoring. You may also want to talk with a classmate to see if you can help one another.

Making it personal: Look over these suggestions. Write down the one that you think would work best for you and then use it this week. Make it a priority in your life to get help immediately in this math course whenever you do not understand something. We all need a little help from time to time. ▼

2.2 Exercises MyMathLab®

Verbal and Writing Skills, Exercises 1–4

In exercises 1–4, fill in the blank with the appropriate number.

1. To solve the equation $6x = -24$, divide each side of the equation by _____.

2. To solve the equation $-7x = 56$, divide each side of the equation by _____.

3. To solve the equation $\frac{1}{7}x = -2$, multiply each side of the equation by _____.

4. To solve the equation $\frac{1}{9}x = 5$, multiply each side of the equation by _____.

Solve for x. Be sure to simplify your answer. Check your solution.

5. $\frac{1}{8}x = 6$

6. $\frac{1}{5}x = 12$

7. $\frac{1}{2}x = -15$

8. $\frac{1}{9}x = -8$

9. $\frac{x}{5} = 16$

10. $\frac{x}{12} = -7$

11. $-3 = \frac{x}{5}$

12. $\frac{x}{6} = -2$

13. $13x = 52$

14. $15x = 60$

15. $56 = 7x$

16. $46 = 2x$

17. $-16 = 6x$

18. $-35 = 21x$

19. $1.5x = 75$

20. $2x = 0.36$

21. $-15 = -x$

22. $32 = -x$

23. $-112 = 16x$

24. $-108 = -18x$

25. $0.4x = 0.08$

26. $2.5x = 0.5$

27. $-3.9x = -15.6$

28. $-4.7x = -14.1$

Determine whether the given solution is correct. If not, find the correct solution.

29. Is 7 the solution for $-3x = 21$?

30. Is 8 the solution for $5x = -40$?

31. Is -15 the solution for $-x = 15$?

32. Is -8 the solution for $-11x = 88$?

Mixed Practice *Find the value of the variable that satisfies the equation.*

33. $7y = -0.21$

34. $-6y = 2.16$

35. $-56 = -21t$

36. $26 = -39t$

37. $4.6y = -3.22$

38. $-2.8y = -3.08$

39. $4x + 3x = 21$

40. $5x + 4x = 36$

41. $2x - 7x = 20$

42. $3x - 9x = 18$

43. $\frac{1}{4}x = -9$

44. $\frac{1}{5}x = -4$

45. $12 - 19 = -7x$

46. $24 - 27 = -9x$

47. $8m = -14 + 30$

48. $8x = 26 - 50$

49. $\frac{3}{4}x = 63$

50. $\frac{5}{6}x = 40$

51. $-2.5133x = 26.38965$

52. $-5.42102x = -45.536568$

Cumulative Review *Simplify.*

53. **[1.7.2]** $-3y(2x + y) + 5(3xy - y^2)$

54. **[1.9.1]** $-\{2(x - 3) + 3[x - (2x - 5)]\}$

55. **[0.6.1]** ***Humpback Whales*** During the summer months, a group of humpback whales gather at Stellwagen Bank, near Gloucester, Massachusetts, to feed. When they return to the Caribbean for winter, they will lose up to 25% of their body weight in blubber. If a humpback whale weighs 30 tons after feeding at Stellwagen Bank, how much will it weigh after losing 25% of its body weight?

56. **[0.6.1]** ***Earthquakes*** In an average year worldwide, there are 20 earthquakes of magnitude 7 on the Richter scale. If next year is predicted to be an exceptional year and the number of earthquakes of magnitude 7 is expected to increase by 35%, how many earthquakes of magnitude 7 can be expected?

Quick Quiz 2.2 *Solve for the variable.*

1. $2.5x = -95$

2. $-3.9x = -54.6$

3. $7x - 12x = 60$

4. **Concept Check** Explain how you would check to verify whether $x = 36\frac{2}{3}$ is the solution to $-22 = -\frac{3}{5}x$.

2.3 Using the Addition and Multiplication Principles Together

1 Solving Equations of the Form $ax + b = c$

Jenny Crawford scored several goals in field hockey during April. Her teammates scored three more than five times the number she scored for a total of 18 goals scored in April. How many did Jenny score? To solve this problem, we need to solve the equation $5x + 3 = 18$.

To solve an equation of the form $ax + b = c$, we must use both the addition principle and the multiplication or division principle.

Student Learning Objectives

After studying this section, you will be able to:

1. Solve equations of the form $ax + b = c$.
2. Solve equations with the variable on both sides of the equation.
3. Solve equations with parentheses.

Example 1 Solve for x in the equation $5x + 3 = 18$ to determine how many goals Jenny scored. Check your solution.

Solution We first want to isolate the variable term.

$$5x + 3 = 18$$

$$5x + 3 + (-3) = 18 + (-3)$$ Use the addition principle to add -3 to both sides.

$$5x = 15$$ Simplify.

$$\frac{5x}{5} = \frac{15}{5}$$ Use the division principle to divide both sides by 5.

$$x = 3$$ The solution is 3. Thus Jenny scored 3 goals.

Check. $5(3) + 3 \stackrel{?}{=} 18$

$$15 + 3 \stackrel{?}{=} 18$$

$$18 = 18 \checkmark$$ It checks.

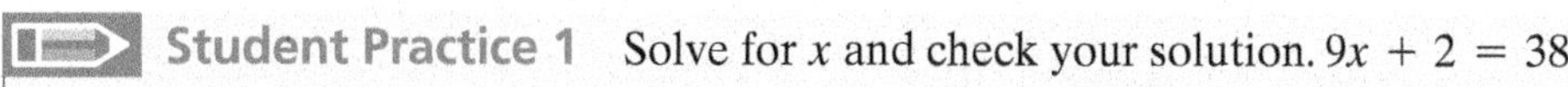

Student Practice 1 Solve for x and check your solution. $9x + 2 = 38$

2 Solving Equations with the Variable on Both Sides of the Equation

In some cases the variable appears on both sides of the equation. We would like to rewrite the equation so that all the terms containing the variable appear on one side. To do this, we apply the addition principle to the variable term.

Example 2 Solve for x. $9x = 6x + 15$

Solution

$$9x + (-6x) = 6x + (-6x) + 15$$ Add $-6x$ to both sides. Notice $6x + (-6x)$ eliminates the variable on the right side.

$$3x = 15$$ Combine like terms.

$$\frac{3x}{3} = \frac{15}{3}$$ Divide both sides by 3.

$$x = 5$$ The solution is 5.

Check. The check is left to the student.

Student Practice 2 Solve for x. $13x = 2x - 66$

In many problems the variable terms and constant terms appear on both sides of the equation. You will want to get all the variable terms on one side and all the constant terms on the other side.

Example 3 Solve for x and check your solution. $9x + 3 = 7x - 2$

Solution First we want to isolate the variable term.

$9x + (-7x) + 3 = 7x + (-7x) - 2$ Add $-7x$ to both sides of the equation.

$2x + 3 = -2$ Combine like terms.

$2x + 3 + (-3) = -2 + (-3)$ Add -3 to both sides.

$2x = -5$ Simplify.

$\frac{2x}{2} = \frac{-5}{2}$ Divide both sides by 2.

$x = -\frac{5}{2} \text{ or } -2\frac{1}{2}$ The solution is $-\frac{5}{2}$ or $2\frac{1}{2}$.

Check. $9x + 3 = 7x - 2$

$9\left(-\frac{5}{2}\right) + 3 \stackrel{?}{=} 7\left(-\frac{5}{2}\right) - 2$ Replace x by $-\frac{5}{2}$.

$-\frac{45}{2} + 3 \stackrel{?}{=} -\frac{35}{2} - 2$ Simplify.

$-\frac{45}{2} + \frac{6}{2} \stackrel{?}{=} -\frac{35}{2} - \frac{4}{2}$ Change to equivalent fractions with a common denominator.

$-\frac{39}{2} = -\frac{39}{2}$ ✓ It checks. The solution is $-\frac{5}{2}$. □

 **Student Practice 3** Solve for x and check your solution.

$$3x + 2 = 5x + 2$$

In our next example we will study equations that need simplifying before any other steps are taken. Where it is possible, you should first collect like terms on one or both sides of the equation. The variable terms can be collected on the right side or the left side. In this example we will collect all the x-terms on the right side.

Example 4 Solve for y. $5y + 26 - 6 = 9y + 12y$

Solution $5y + 26 - 6 = 9y + 12y$

$5y + 20 = 21y$ Combine like terms.

$5y + (-5y) + 20 = 21y + (-5y)$ Add $-5y$ to both sides.

$20 = 16y$ Combine like terms.

$\frac{20}{16} = \frac{16y}{16}$ Divide both sides by 16.

$\frac{5}{4} \text{ or } 1\frac{1}{4} = y$ Don't forget to reduce the resulting fraction.

Check. The check is left to the student. □

Student Practice 4 Solve for z.

$$-z + 8 - z = 3z + 10 - 3$$

Do you really need all these steps? No. As you become more proficient you will be able to combine or eliminate some of these steps. However, it is best to write each step in its entirety until you are consistently obtaining the correct solution. It is much better to show every step than to take a lot of shortcuts and possibly obtain a wrong answer. This is a section of the algebra course where working neatly and accurately will help you—both now and as you progress through the course.

3 Solving Equations with Parentheses

The equations that you just solved are simpler versions of equations that we will now discuss. These equations contain parentheses. If the parentheses are first removed, the problems then become just like those encountered previously. We use the distributive property to remove the parentheses.

Example 5 Solve for x and check your solution.

$$4(x + 1) - 3(x - 3) = 25$$

Solution

$4(x + 1) - 3(x - 3) = 25$

$4x + 4 - 3x + 9 = 25$ Multiply by 4 and −3 to remove parentheses. Be careful of the signs. Remember that $(-3)(-3) = 9$.

After removing the parentheses, it is important to combine like terms on each side of the equation. Do this before going on to isolate the variable.

$x + 13 = 25$ Combine like terms.

$x + 13 - 13 = 25 - 13$ Subtract 13 from both sides to isolate the variable.

$x = 12$ The solution is 12.

Check.

$4(12 + 1) - 3(12 - 3) \stackrel{?}{=} 25$ Replace x by 12.

$4(13) - 3(9) \stackrel{?}{=} 25$ Combine numbers inside parentheses.

$52 - 27 \stackrel{?}{=} 25$ Multiply.

$25 = 25$ ✓ Simplify. It checks. □

 **Student Practice 5** Solve for x and check your solution.

$$4x - (x + 3) = 12 - 3(x - 2)$$

Example 6 Solve for x. $3(-x - 7) = -2(2x + 5)$

Solution

$3(-x - 7) = -2(2x + 5)$

$-3x - 21 = -4x - 10$ Remove parentheses. Watch the signs carefully.

$-3x + 4x - 21 = -4x + 4x - 10$ Add $4x$ to both sides.

$x - 21 = -10$ Simplify.

$x - 21 + 21 = -10 + 21$ Add 21 to both sides.

$x = 11$ The solution is 11.

Check. The check is left to the student. □

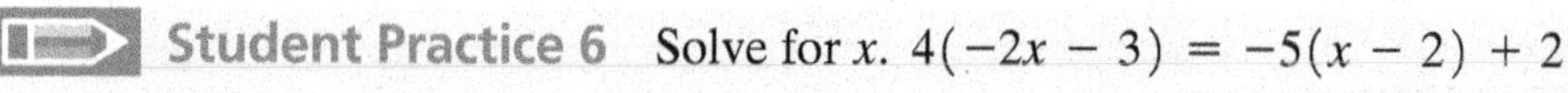

Student Practice 6 Solve for x. $4(-2x - 3) = -5(x - 2) + 2$

In problems that involve decimals, great care should be taken. In some steps you will be multiplying decimal quantities, and in other steps you will be adding them.

Example 7 Solve for x. $0.3(1.2x - 3.6) = 4.2x - 16.44$

Solution

$0.3(1.2x - 3.6) = 4.2x - 16.44$	
$0.36x - 1.08 = 4.2x - 16.44$	Remove parentheses.
$0.36x - 0.36x - 1.08 = 4.2x - 0.36x - 16.44$	Subtract $0.36x$ from both sides.
$-1.08 = 3.84x - 16.44$	Combine like terms.
$-1.08 + 16.44 = 3.84x - 16.44 + 16.44$	Add 16.44 to both sides.
$15.36 = 3.84x$	Simplify.
$\frac{15.36}{3.84} = \frac{3.84x}{3.84}$	Divide both sides by 3.84.
$4 = x$	The solution is 4.

Check. The check is left to the student. □

Student Practice 7 Solve for x.

$$0.3x - 2(x + 0.1) = 0.4(x - 3) - 1.1$$

Example 8 Solve for z and check.

$$2(3z - 5) + 2 = 4z - 3(2z + 8)$$

Solution

$2(3z - 5) + 2 = 4z - 3(2z + 8)$	
$6z - 10 + 2 = 4z - 6z - 24$	Remove parentheses.
$6z - 8 = -2z - 24$	Combine like terms.
$6z + 2z - 8 = -2z + 2z - 24$	Add $2z$ to both sides.
$8z - 8 = -24$	Simplify.
$8z - 8 + 8 = -24 + 8$	Add 8 to both sides.
$8z = -16$	Simplify.
$\frac{8z}{8} = \frac{-16}{8}$	Divide both sides by 8.
$z = -2$	Simplify. The solution is -2.

Check.

$2[3(-2) - 5] + 2 \stackrel{?}{=} 4(-2) - 3[2(-2) + 8]$	Replace z by -2.
$2[-6 - 5] + 2 \stackrel{?}{=} -8 - 3[-4 + 8]$	Multiply.
$2[-11] + 2 \stackrel{?}{=} -8 - 3[4]$	Simplify.
$-22 + 2 \stackrel{?}{=} -8 - 12$	
$-20 = -20$ ✓	It checks. □

Student Practice 8 Solve for z and check.

$$5(2z - 1) + 7 = 7z - 4(z + 3)$$

STEPS TO SUCCESS Keep Trying! Do Not Quit!

Math opens doors. We live in a highly technical world that depends more and more on mathematics. You cannot afford to give up. Dropping mathematics may prevent you from entering an interesting career field. Understanding mathematics can open new doors for you.

Practice daily. Learning mathematics requires time and effort. You will find that regular study and daily practice are necessary. This will help your level of academic success and lead you toward a mastery of mathematics. It may open a path for you to a good paying job. Do not quit! You can do it!

Making it personal: Which of these statements do you find most helpful? What things can you do to help you master mathematics? Don't let yourself quit or let up on your studies. Your work now will help your financial success in the future! ▼

2.3 Exercises MyMathLab®

Find the value of the variable that satisfies each equation in exercises 1–22. Check your solution.

1. $3x + 23 = 50$

2. $4x + 7 = 35$

3. $4x - 11 = 13$

4. $5x - 9 = 36$

5. $7x - 18 = -46$

6. $8x - 15 = -47$

7. $-4x + 17 = -35$

8. $-6x + 25 = -83$

9. $2x + 3.2 = 9.4$

10. $4x + 4.6 = 9.2$

11. $\frac{1}{4}x + 6 = 13$

12. $\frac{1}{2}x + 1 = 7$

13. $\frac{1}{3}x + 5 = -4$

14. $\frac{1}{8}x - 3 = -9$

15. $8x = 48 + 2x$

16. $5x = 22 + 3x$

17. $-6x = -27 + 3x$

18. $-7x = -26 + 6x$

19. $44 - 2x = 6x$

20. $21 - 5x = 7x$

21. $54 - 2x = -8x$

22. $72 - 4x = -12x$

In exercises 23–26, determine whether the given solution is correct. If it is not, find the solution.

23. Is 2 the solution for $2y + 3y = 12 - y$?

24. Is 4 the solution for $5y + 2 = 6y - 6 + y$?

25. Is 11 a solution for $7x + 6 - 3x = 2x - 5 + x$?

26. Is -12 a solution for $9x + 2 - 5x = -8 + 5x - 2$?

Solve for the variable. You may move the variable terms to the right or to the left.

27. $14 - 2x = -5x + 11$

28. $8 - 3x = 7x + 8$

29. $x - 6 = 8 - x$

30. $-x + 12 = -4 + x$

31. $0.6y + 0.8 = 0.1 - 0.1y$

32. $1.1y + 0.3 = -1.3 + 0.3y$

33. $5x - 9 = 3x + 23$

34. $9x - 5 = 7x + 43$

To Think About, Exercises 35 and 36

For exercises 35 and 36, first combine like terms on each side of the equation. Then solve for y by collecting all the y-terms on the left. Then solve for y by collecting all the y-terms on the right. Which approach is better?

35. $-3 + 10y + 6 = 15 + 12y - 18$

36. $7y + 21 - 5y = 5y - 7 + y$

Remove the parentheses and solve for the variable. Check your solution.

37. $5(x + 3) = 35$

38. $7(x + 3) = 28$

39. $5(4x - 3) + 8 = -2$

40. $4(2x + 1) - 7 = 6 - 5$

41. $7x - 3(5 - x) = 10$

42. $8x - 2(4 - x) = 14$

43. $0.5x - 0.3(2 - x) = 4.6$

44. $0.4x - 0.2(3 - x) = 1.8$

45. $4(a - 3) + 2 = 2(a - 5)$

46. $6(a + 3) - 2 = -4(a - 4)$

47. $-2(x + 3) + 4 = 3(x + 4) + 2$

48. $-3(x + 5) + 2 = 4(x + 6) - 9$

49. $-3(y - 3y) + 4 = -4(3y - y) + 6 + 13y$

50. $2(4x - x) + 6 = 2(2x + x) + 8 - x$

Mixed Practice *Solve for the variable.*

51. $5.7x + 3 = 4.2x - 3$

52. $4x - 3.1 = 5.3 - 3x$

53. $5z + 7 - 2z = 32 - 2z$

54. $8 - 7z + 2z = 20 + 5z$

55. $-0.3a + 1.4 = -1.2 - 0.7a$

56. $-0.7b + 1.6 = -1.7 - 1.5b$

57. $6x + 8 - 3x = 11 - 12x - 13$

58. $4 - 7x - 13 = 8x - 3 - 5x$

59. $-3.5x + 1.3 = -2.7x + 1.5$

60. $1.4x - 0.8 = 1.2x - 0.2$

61. $5(4 + x) = 3(3x - 1) - 9$

62. $5(2x - 3) = 3(3x + 2) - 17$

63. $-1.7x + 4.4 + 5x = 0.3x - 0.1$

64. $6x - 3.7 - 1.2x = 0.8x + 1.1$

Cumulative Review *Evaluate using the correct order of operations. (Be careful to avoid sign errors.)*

65. [1.5.1] $(-6)(-8) + (-3)(2)$

66. [1.5.1] $(-3)^3 + (-20) \div 2$

67. [1.5.1] $5 + (2 - 6)^2$

68. [0.6.1] ***Investments*** On May 1, 2015, Marcella owned three different stocks: Motorola, Barnes & Noble, and CVS. Her portfolio contained the following:

35 shares of Motorola valued at $9.11 per share,
16 shares of Barnes & Noble valued at $22.70 per share, and
5 shares of CVS valued at $100.46 per share.

Find the market value of Marcella's stock holdings on May 1, 2015.

69. [0.6.1] ***Employee Discount*** Marvin works at Best Buy and gets a 10% discount on anything he buys from the store. A GPS navigation system that Marvin wishes to purchase costs $899 and is on sale at a 20% discount.

(a) What is the sale price if Marvin has a total discount of 30%? (Disregard sales tax.)

(b) What is the price if Marvin gets a 10% discount on the 20% sale price? (Disregard sales tax.)

Quick Quiz 2.3 *Solve for the variable.*

1. $7x - 6 = -4x - 10$

2. $-3x + 6.2 = -5.8$

3. $2(3x - 2) = 4(5x + 3)$

4. Concept Check Explain how you would solve the equation $3(x - 2) + 2 = 2(x - 4)$.

2.4 Solving Equations with Fractions

Student Learning Objective

After studying this section, you will be able to:

1 Solve equations with fractions.

1 Solving Equations with Fractions

Equations with fractions can be rather difficult to solve. This difficulty is simply due to the extra care we usually have to use when computing with fractions. The actual equation-solving procedures are the same, with fractions or without. To avoid unnecessary work, we transform the given equation with fractions to an equivalent equation that does not contain fractions. How do we do this? We multiply each side of the equation by the least common denominator of all the fractions contained in the equation. We then use the distributive property so that the LCD is multiplied by each term of the equation.

Example 1 Solve for x. $\frac{1}{4}x - \frac{2}{3} = \frac{5}{12}x$

Solution First we find that the LCD = 12.

$12\left(\frac{1}{4}x - \frac{2}{3}\right) = 12\left(\frac{5}{12}x\right)$	Multiply both sides by 12.
$\left(\frac{12}{1}\right)\left(\frac{1}{4}\right)(x) - \left(\frac{12}{1}\right)\left(\frac{2}{3}\right) = \left(\frac{12}{1}\right)\left(\frac{5}{12}\right)(x)$	Use the distributive property.
$3x - 8 = 5x$	Simplify.
$3x + (-3x) - 8 = 5x + (-3x)$	Add $-3x$ to both sides.
$-8 = 2x$	Simplify.
$\frac{-8}{2} = \frac{2x}{2}$	Divide both sides by 2.
$-4 = x$	Simplify.

Check. The check is left to the student. □

 **Student Practice 1** Solve for x.

$$\frac{3}{8}x - \frac{3}{2} = \frac{1}{4}x$$

In Example 1 we multiplied both sides of the equation by the LCD. However, most students prefer to go immediately to the second step and multiply each term by the LCD. This avoids having to write out a separate step using the distributive property.

Example 2 Solve for x and check your solution. $\frac{x}{3} + 3 = \frac{x}{5} - \frac{1}{3}$

Solution

$15\left(\frac{x}{3}\right) + 15(3) = 15\left(\frac{x}{5}\right) - 15\left(\frac{1}{3}\right)$	The LCD is 15. Use the multiplication principle to multiply each term by 15.
$5x + 45 = 3x - 5$	Simplify.
$5x - 3x + 45 = 3x - 3x - 5$	Subtract $3x$ from both sides.
$2x + 45 = -5$	Combine like terms.
$2x + 45 - 45 = -5 - 45$	Subtract 45 from both sides.
$2x = -50$	Simplify.
$\frac{2x}{2} = \frac{-50}{2}$	Divide both sides by 2.
$x = -25$	The solution is -25.

Check. $\frac{-25}{3} + 3 \stackrel{?}{=} \frac{-25}{5} - \frac{1}{3}$

$$-\frac{25}{3} + \frac{9}{3} \stackrel{?}{=} -\frac{5}{1} - \frac{1}{3}$$

$$-\frac{16}{3} \stackrel{?}{=} -\frac{15}{3} - \frac{1}{3}$$

$$-\frac{16}{3} = -\frac{16}{3} \quad ✓$$

□

 **Student Practice 2** Solve for x and check your solution.

$$\frac{5x}{4} - 1 = \frac{3x}{4} + \frac{1}{2}$$

Example 3 Solve for x. $\frac{x + 2}{8} = \frac{x}{4} + \frac{1}{2}$

Solution

$\frac{x}{8} + \frac{2}{8} = \frac{x}{4} + \frac{1}{2}$ — First we rewrite the left side as two fractions. This is actually multiplying $\frac{1}{8}(x + 2) = \frac{x}{8} + \frac{2}{8}$.

$\frac{x}{8} + \frac{1}{4} = \frac{x}{4} + \frac{1}{2}$ — Once we write $\frac{x+2}{8}$ as two fractions we can simplify $\frac{2}{8} = \frac{1}{4}$.

$8\left(\frac{x}{8}\right) + 8\left(\frac{1}{4}\right) = 8\left(\frac{x}{4}\right) + 8\left(\frac{1}{2}\right)$ — We observe that the LCD is 8, so we multiply each term by 8.

$x + 2 = 2x + 4$ — Simplify.

$x - x + 2 = 2x - x + 4$ — Subtract x from both sides.

$2 = x + 4$ — Combine like terms.

$2 - 4 = x + 4 - 4$ — Subtract 4 from both sides.

$-2 = x$ — The solution is -2.

Check. The check is left to the student. □

CAUTION: We may *not* use slashes to divide out *part of an addition and subtraction* problem. We may use slashes if we are *multiplying* factors.

$\frac{x + \not{2}}{\not{8}} = \frac{x + 1}{4}$ THIS IS WRONG!

$\frac{x \cdot \not{2}}{\not{8}} = \frac{x}{4}$ THIS IS CORRECT.

Student Practice 3 Solve for x.

$$\frac{x + 6}{9} = \frac{x}{6} + \frac{1}{2}$$

If a problem contains both parentheses and fractions, it is best to remove the parentheses first. Many students find it is helpful to have a written procedure to follow in solving these more-involved equations.

PROCEDURE TO SOLVE EQUATIONS

1. Remove any parentheses.
2. If fractions exist, multiply all terms on both sides by the least common denominator of all the fractions.
3. Combine like terms if possible.
4. Add or subtract terms on both sides of the equation to get all terms with the variable on one side of the equation.
5. Add or subtract a constant value on both sides of the equation to get all terms not containing the variable on the other side of the equation.
6. Divide both sides of the equation by the coefficient of the variable.
7. Simplify the solution (if possible).
8. Check your solution.

Example 4 Solve for x and check your solution.

$$\frac{1}{3}(x-2)=\frac{1}{5}(x+4)+2$$

Solution

Step 1 $\frac{x}{3}-\frac{2}{3}=\frac{x}{5}+\frac{4}{5}+2$ Remove parentheses.

Step 2 Multiply by the LCD, 15.

$$15\left(\frac{x}{3}\right)-15\left(\frac{2}{3}\right)=15\left(\frac{x}{5}\right)+15\left(\frac{4}{5}\right)+15(2)$$

$5x-10=3x+12+30$ Simplify.

Step 3 $5x-10=3x+42$ Combine like terms.

Step 4 $5x-3x-10=3x-3x+42$ Subtract $3x$ from both sides.

$2x-10=42$ Simplify.

Step 5 $2x-10+10=42+10$ Add 10 to both sides.

$2x=52$ Simplify.

Step 6 $\frac{2x}{2}=\frac{52}{2}$ Divide both sides by 2.

Step 7 $x=26$ Simplify the solution.

Step 8 *Check.* $\frac{1}{3}(26-2)\stackrel{?}{=}\frac{1}{5}(26+4)+2$ Replace x by 26.

$\frac{1}{3}(24)\stackrel{?}{=}\frac{1}{5}(30)+2$ Combine values within parentheses.

$8\stackrel{?}{=}6+2$ Simplify.

$8=8$ ✓ The solution is 26. □

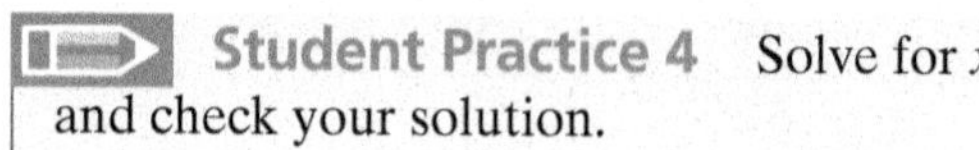

Student Practice 4 Solve for x and check your solution.

$$\frac{1}{2}(x+5)=\frac{1}{5}(x-2)+\frac{1}{2}$$

Remember that not every step will be needed in each problem. You can combine some steps as well, *as long as you are consistently obtaining the correct solution.* However, you are encouraged to write out every step as a way of helping you to avoid careless errors.

It is important to remember that when we write decimals, these numbers are really fractions written in a special way. Thus, $0.3=\frac{3}{10}$ and $0.07=\frac{7}{100}$. It is possible to take an equation containing decimals and to multiply each term by the appropriate value to obtain integer coefficients.

Example 5 Solve for x. $0.2(1-8x)+1.1=-5(0.4x-0.3)$

Solution

$0.2-1.6x+1.1=-2.0x+1.5$ Remove parentheses.

Next, we multiply each term by 10 to move the decimal point one place to the right.

$$10(0.2)-10(1.6x)+10(1.1)=10(-2.0x)+10(1.5)$$

$$2-16x+11=-20x+15$$

$-16x+13=-20x+15$ Simplify.

Student Practice 5 Solve for x.

$$2.8=0.3(x-2)+2(0.1x-0.3)$$

$$-16x + 20x + 13 = -20x + 20x + 15 \quad \text{Add } 20x \text{ to both sides.}$$
$$4x + 13 = 15 \quad \text{Simplify.}$$
$$4x + 13 - 13 = 15 - 13 \quad \text{Subtract 13 from both sides.}$$
$$4x = 2 \quad \text{Simplify.}$$
$$\frac{4x}{4} = \frac{2}{4} \quad \text{Divide both sides by 4.}$$
$$x = \frac{1}{2} \text{ or } 0.5 \quad \text{Simplify.}$$

Earlier we stated that the decimal form of a solution should only be given if it is a terminating decimal. You can decide which form of the answer you want to use in the check. Here we use 0.5.

Check.

$$0.2[1 - 8(0.5)] + 1.1 \stackrel{?}{=} -5[0.4(0.5) - 0.3]$$
$$0.2[1 - 4] + 1.1 \stackrel{?}{=} -5[0.2 - 0.3]$$
$$0.2[-3] + 1.1 \stackrel{?}{=} -5[-0.1]$$
$$-0.6 + 1.1 \stackrel{?}{=} 0.5$$
$$0.5 = 0.5 \quad ✓$$

□

To Think About: *Does Every Equation Have One Solution?* Actually, no. There are some rare cases where an equation has no solution at all. Suppose we try to solve the equation

$$5(x + 3) = 2x - 8 + 3x.$$

If we remove the parentheses and combine like terms, we have

$$5x + 15 = 5x - 8.$$

If we add $-5x$ to each side, we obtain

$$15 = -8.$$

Clearly this is impossible. There is no value of x for which these two numbers are equal. We would say this equation has **no solution.**

One additional surprise may happen. An equation may have an infinite number of solutions. Suppose we try to solve the equation

$$9x - 8x - 7 = 3 + x - 10.$$

If we combine like terms on each side, we have the equation

$$x - 7 = x - 7.$$

If we add $-x$ to each side, we obtain

$$-7 = -7.$$

Now this statement is always true, no matter what the value of x. We would say this equation has **an infinite number of solutions.**

In the To Think About exercises in this section, we will encounter some equations that have no solution or an infinite number of solutions.

2.4 Exercises MyMathLab®

In exercises 1–16, solve for the variable and check your answer.

1. $\frac{1}{6}x + \frac{2}{3} = -\frac{1}{2}$

2. $\frac{1}{3}x + \frac{5}{6} = \frac{1}{2}$

3. $\frac{2}{3}x = \frac{1}{15}x + \frac{3}{5}$

4. $\frac{4}{15}x + \frac{1}{5} = \frac{2}{3}x$

5. $\frac{x}{2} + \frac{x}{5} = \frac{7}{10}$

6. $\frac{x}{8} + \frac{x}{4} = -\frac{3}{4}$

7. $5 - \frac{1}{3}x = \frac{1}{12}x$

8. $15 - \frac{1}{2}x = \frac{1}{4}x$

9. $2 + \frac{y}{2} = \frac{3y}{4} - 3$

10. $\frac{x}{3} + 3 = \frac{5x}{6} + 2$

11. $\frac{x - 3}{5} = 1 - \frac{x}{3}$

12. $\frac{y - 5}{4} = 1 - \frac{y}{5}$

13. $\frac{x + 3}{4} = \frac{x}{2} + \frac{1}{6}$

14. $\frac{x - 2}{3} = \frac{x}{12} + \frac{5}{4}$

15. $0.6x + 5.9 = 3.8$

16. $-3.2x - 5.1 = 2.9$

Answer Yes or No for exercises 17–20.

17. Is 4 a solution to $\frac{1}{2}(y - 2) + 2 = \frac{3}{8}(3y - 4)$?

18. Is 2 a solution to $\frac{1}{5}(y + 2) = \frac{1}{10}y + \frac{3}{5}$?

19. Is $\frac{5}{8}$ a solution to $\frac{1}{2}\left(y - \frac{1}{5}\right) = \frac{1}{5}(y + 2)$?

20. Is $\frac{1}{2}$ a solution to $\frac{1}{3}\left(x - \frac{1}{4}\right) = \frac{1}{8} + \frac{1}{3}x$

Remove parentheses first. Then combine like terms. Solve for the variable.

21. $\frac{3}{4}(3x + 1) = 2(3 - 2x) + 1$

22. $\frac{1}{4}(3x + 1) = 2(2x - 4) - 8$

23. $2(x - 2) = \frac{2}{5}(3x + 1) + 2$

24. $2(x - 4) = \frac{5}{6}(x + 6) - 6$

25. $0.3x - 0.2(3 - 5x) = -0.5(x - 6)$

26. $0.2(x + 1) + 0.5x = -0.3(x - 4)$

27. $-8(0.1x + 0.4) - 0.9 = -0.1$

28. $0.6x + 1.5 = 0.3x - 0.6(2x + 5)$

Mixed Practice *Solve.*

29. $\frac{1}{3}(y + 2) = 3y - 5(y - 2)$

30. $\frac{1}{4}(y + 6) = 2y - 3(y - 3)$

31. $\frac{1 + 2x}{5} + \frac{4 - x}{3} = \frac{1}{15}$

32. $\frac{1 + 3x}{2} + \frac{2 - x}{3} = \frac{5}{6}$

33. $\frac{3}{4}(x - 2) + \frac{3}{5} = \frac{1}{5}(x + 1)$

34. $\frac{2}{3}(x + 4) = 6 - \frac{1}{4}(3x - 2) - 1$

35. $\frac{1}{3}(x - 2) = 3x - 2(x - 1) + \frac{16}{3}$

36. $\frac{1}{4}(x + 5) = 3x - 2(3 - x) - 7$

37. $\frac{4}{5}x - \frac{2}{3} = \frac{3x + 1}{2}$

38. $\frac{5}{12}x + \frac{1}{3} = \frac{2x - 3}{4}$

39. $0.3x - 0.2(5x - 1) = -0.4(x + 2)$

40. $0.7(x + 3) = 0.2(x - 5) + 0.1$

To Think About *Solve. Be careful to examine your work to see if the equation has a solution, no solution, or an infinite number of solutions.*

41. $-1 + 5(x - 2) = 12x + 3 - 7x$

42. $x + 3x - 2 + 3x = -11 + 7(x + 2)$

43. $9(x + 3) - 6 = 24 - 2x - 3 + 11x$

44. $7(x + 4) - 10 = 3x + 20 + 4x - 2$

45. $-3(4x - 1) = 5(2x - 1) + 8$

46. $11x - 8 = -4(x + 3) + 4$

47. $3(4x + 1) - 2x = 2(5x - 3)$

48. $5(-3 + 4x) = 4(2x + 4) + 12x$

Cumulative Review

49. **[0.3.1]** Multiply. $\left(-3\frac{1}{4}\right)\left(5\frac{1}{3}\right)$

50. **[0.3.2]** Divide. $5\frac{1}{2} \div 1\frac{1}{4}$

51. **[0.6.1]** ***Peregrine Falcons*** Peregrine falcons are the fastest birds on record, reaching horizontal speeds of 40–55 miles per hour. Also, they are one of the few animals in which the females are larger than the males. If female peregrine falcons are 30% larger than the males and males measure 440–750 grams, what is the weight range for females?

52. **[1.8.2]** ***Auditorium Seating*** The seating area of an auditorium is shaped like a trapezoid, with front and back sides parallel. The front of the auditorium measures 88 feet across, the back of the auditorium measures 150 feet across, and the auditorium is 200 feet from front to back. If each seat requires a space that is 2.5 feet wide by 3 feet deep, how many seats will the auditorium hold? (This will only be an approximation because of the angled side walls. Round off to the nearest whole number.)

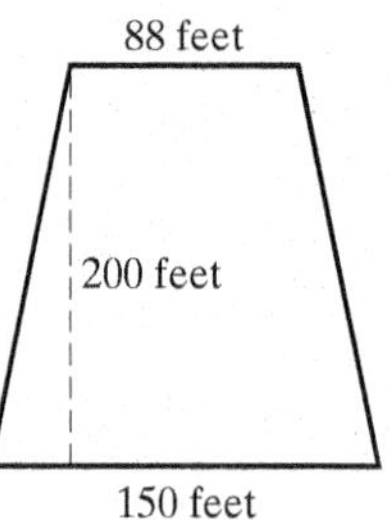

Quick Quiz 2.4 *Solve for the variable.*

1. $\frac{3}{4}x + \frac{5}{12} = \frac{1}{3}x - \frac{1}{6}$

2. $\frac{2}{3}x - \frac{3}{5} + \frac{7}{5}x + \frac{1}{3} = 1$

3. $\frac{2}{3}(x + 2) + \frac{1}{4} = \frac{1}{2}(5 - 3x)$

4. **Concept Check** Explain how you would solve the equation $\frac{x + 5}{6} = \frac{x}{2} + \frac{3}{4}$.

Use Math to Save Money

Did You Know? You may be paying more than you think.

See Through the Hidden Fees

Understanding the Problem:

Sam lives in California. Before taking a trip one weekend in June 2015 he needed to put gas in his car.

- On his street is a Shell gas station, which charged $4.55 per gallon for gas.
- There is also an ARCO gas station, which charged $4.43 per gallon for gas.
- The ARCO station also charged a single "ATM Transaction Fee" of $0.45.

Sam needed to decide which gas station to go to in order to pay the least amount for gas.

Making a Plan:

Sam needed to know how much gas would cost at each station to make a choice that would save him money.

Step 1: Sam needed to find out how much filling up his car at the gas station really cost after any hidden fees.

Task 1: Determine how much it cost Sam to buy one gallon of gas at each station.

Task 2: Determine how much it cost Sam to buy three gallons of gas at each station.

Task 3: Determine how much it cost Sam to buy four gallons of gas at each station.

Task 4: Determine how much it cost Sam to buy 10 gallons of gas at each station.

Step 2: Notice that when Sam bought more gas, it was less expensive to buy at the ARCO station than at the Shell station. The price at Shell rose faster than the price at ARCO.

Task 5: Find the number of gallons for which the cost was the same.

Making a Decision:

Step 3: Sam needed to take into consideration the price of gas and any fees involved when he made his decision about where to buy gas.

Task 6: Which station was less expensive if Sam only needed a small amount of gas, say less than four gallons?

Task 7: Which station was less expensive if Sam needed more than four gallons of gas?

Applying the Situation to Your Life:

Some gas stations have different prices depending on whether you pay with cash or a credit card.

Task 8: Does the station where you normally buy gas charge the same price for cash or credit?

Task 9: Do you know if the gas station charges an ATM transaction fee?

Task 10: Have the increases in gas prices caused you to change your driving habits? If so, please explain.

There is a simple plan to help you save money by avoiding hidden fees:

- Always check for any hidden fees.
- Find the real total cost for your purchases, including fees, taxes, shipping costs, and tips.
- Watch for different prices for cash or credit.

2.5 Translating English Phrases into Algebraic Expressions

Student Learning Objectives

After studying this section, you will be able to:

1. Translate English phrases into algebraic expressions.
2. Write an algebraic expression to compare two or more quantities.

1 Translating English Phrases into Algebraic Expressions

One of the most useful applications of algebra is solving word problems. One of the first steps in solving a word problem is translating the conditions of the problem into algebra. In this section we show you how to translate common English phrases into algebraic expressions. This process is similar to translating between languages like Spanish and French.

Several English phrases describe the operation of addition. If we represent an unknown number by the variable x, all of the following phrases can be translated into algebra as $x + 3$.

English Phrases Describing Addition	***Algebraic Expression***	***Diagram***
Three *more than* a number		
The *sum of* a number and three		
A number *increased by* three		
Three is *added to* a number.	$x + 3$	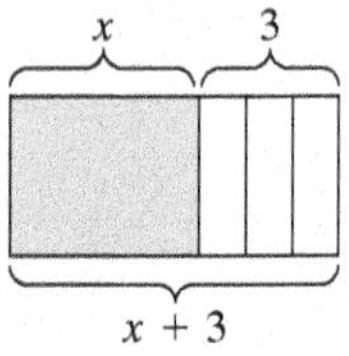 x; 3; $x + 3$
Three *greater than* a number		
A number *plus* three		

In a similar way we can use algebra to express English phrases that describe the operations of subtraction, multiplication, and division.

CAUTION: Since subtraction is not commutative, the order is essential. A number decreased by five is $x - 5$. It is not correct to say $5 - x$. Use extra care as you study each example. Make sure you understand the proper order.

English Phrases Describing Subtraction	***Algebraic Expression***	***Diagram***
A number *decreased by* four		
Four *less than* a number		
Four is *subtracted from* a number.		
Four *smaller than* a number	$x - 4$	x; 4; $x - 4$
Four *fewer than* a number		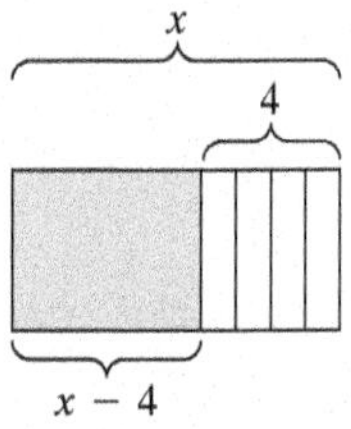
A number *diminished by* four		
A number *minus* 4		
The *difference between* a number and four		

English Phrases Describing Multiplication	***Algebraic Expression***	***Diagram***
Double a number		
Twice a number		
The *product of* two and a number	$2x$	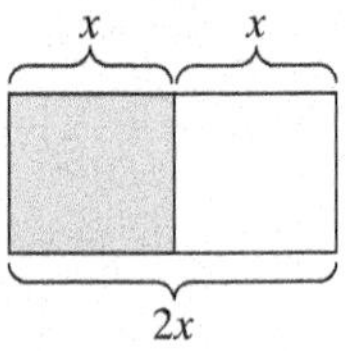 x; x; $2x$
Two *of* a number		
Two *times* a number		

Since division is not commutative, the order is essential. A number divided by 3 is $\frac{x}{3}$. It is not correct to say $\frac{3}{x}$. Use extra care as you study each example.

English Phrases Describing Division	*Algebraic Expression*	*Diagram*
A number *divided by* five *One-fifth of* a number The *quotient of* a number and five	$\frac{x}{5}$	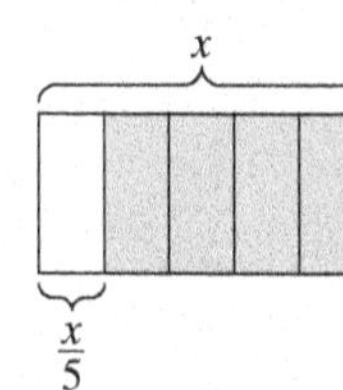

Often other words are used in English instead of the word *number*. We can use a variable, such as x, here also.

Example 1 Write each English phrase as an algebraic expression.

English Phrase	*Algebraic Expression*
(a) A *quantity* is increased by five.	$x + 5$
(b) Double the *value*	$2x$
(c) One-third of the *weight*	$\frac{x}{3}$ or $\frac{1}{3}x$
(d) Twelve more than an *unknown number*	$x + 12$

The variable or expression that follows the words *more than* technically comes before the plus sign. However, since addition is commutative, it also can be written after the plus sign.

(e) Seven less than a *number*	$x - 7$

Note that the algebraic expression for "seven less than a number" does not follow the order of the words in the English phrase. The variable or expression that follows the words *less than* always comes first.

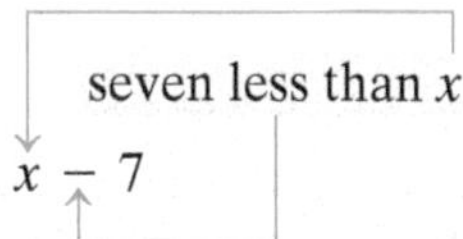

□

 **Student Practice 1** Write each English phrase as an algebraic expression.

(a) Four more than a number
(b) Triple a value
(c) Eight less than a number
(d) One-fourth of a height

More than one operation can be described in an English phrase. Sometimes parentheses must be used to make clear which operation is done first.

Example 2 Write each English phrase as an algebraic expression.

English Phrase	*Algebraic Expression*
(a) Seven more than double a number	$2x + 7$ *Note that these are **not** the same.*
(b) The value of the number is increased by seven and then doubled.	$2(x + 7)$ *Note that the word **then** tells us to add x and 7 before doubling.*
(c) One-half of the sum of a number and 3	$\frac{1}{2}(x + 3)$

□

Student Practice 2 Write each English phrase as an algebraic expression.

(a) Eight more than triple a number

(b) A number is increased by eight and then it is tripled.

(c) One-third of the sum of a number and 4

2 Writing an Algebraic Expression to Compare Two or More Quantities

Often in a word problem two or more quantities are described in terms of another. We will want to use a variable to represent one quantity and then write an algebraic expression using *the same variable* to represent the other quantity. Which quantity should we let the variable represent? We usually let the variable represent the quantity that is the basis of comparison: the quantity that the others are being *compared to.*

Example 3 Use a variable and an algebraic expression to describe the two quantities in the English sentence "Mike's salary is \$2000 more than Fred's salary."

Solution The two quantities that are being compared are Mike's salary and Fred's salary. Since Mike's salary is being *compared to* Fred's salary, we let the variable represent Fred's salary. The choice of the letter f helps us to remember that the variable represents Fred's salary.

Let f = Fred's salary.

Then $f + \$2000$ = Mike's salary, since Mike's salary is \$2000 *more than* Fred's. □

Student Practice 3 Use a variable and an algebraic expression to describe the two quantities in the English sentence "Marie works 17 hours per week less than Ann."

Example 4 The length of a rectangle is 3 meters shorter than twice the width. Use a variable and an algebraic expression to describe the length and the width. Draw a picture of the rectangle and label the length and width.

Solution The length of the rectangle is being *compared to* the width. Use the letter w for width.

Let w = the width.

$$\underbrace{\text{3 meters shorter than twice the width}}$$

Then $2w - 3$ = the length.

A picture of the rectangle is shown.

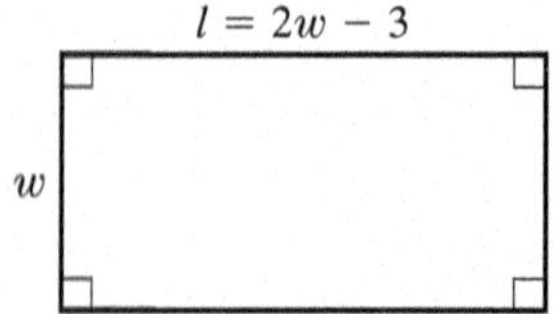

□

Student Practice 4 The length of a rectangle is 5 meters longer than double the width. Use a variable and an algebraic expression to describe the length and the width. Draw a picture of the rectangle and label the length and width.

Example 5 The measure of the first angle of a triangle is triple the measure of the second angle. The measure of the third angle of the triangle is 12° more than the measure of the second angle. Describe each angle measure algebraically. Draw a diagram of the triangle and label its parts.

Solution Since the first and third angle measures are described in terms of the second angle, we let the variable represent the number of degrees in the second angle.

Let $s =$ the number of degrees in the second angle.

Then $3s =$ the number of degrees in the first angle.

And $s + 12 =$ the number of degrees in the third angle.

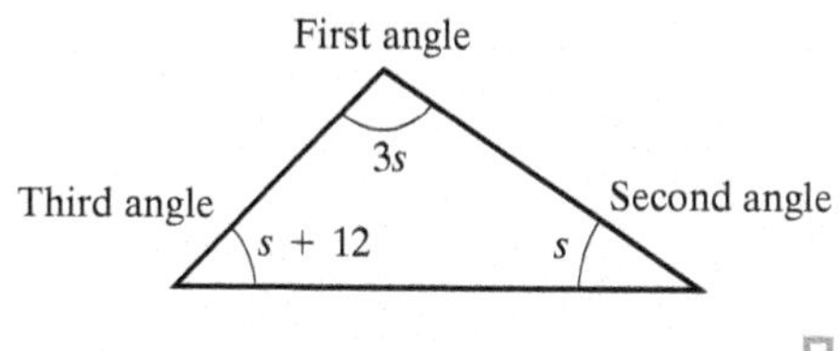

□

Student Practice 5 The measure of the first angle of a triangle is 16° less than the measure of the second angle. The measure of the third angle is double the measure of the second angle. Describe each angle measure algebraically. Draw a diagram of the triangle and label its parts.

Some comparisons will involve fractions.

Example 6 A theater manager was examining the records of attendance for last year. The number of people attending the theater in January was one-half of the number of people attending the theater in February. The number of people attending the theater in March was three-fifths of the number of people attending the theater in February. Use algebra to describe the attendance each month.

Solution What are we looking for? The *number of people* who attended the theater *each month*. The basis of comparison is February. That is where we begin.

Let $f =$ the number of people who attended in February.

Then $\frac{1}{2}f =$ the number of people who attended in January.

And $\frac{3}{5}f =$ the number of people who attended in March. □

Student Practice 6 The college dean noticed that in the spring the number of students on campus was two-thirds of the number of students on campus in the fall. She also noticed that in the summer the number of students on campus was one-fifth the number of students on campus in the fall. Use algebra to describe the number of students on campus in each of these three time periods.

2.5 Exercises MyMathLab®

Verbal and Writing Skills *Write an algebraic expression for each quantity. Let x represent the unknown value.*

1. eleven more than a number

2. the sum of a number and five

3. twelve less than a number

4. seven subtracted from a number

5. one-eighth of a quantity

6. one-sixth of a quantity

7. twice a quantity

8. triple a number

9. three more than half of a number

10. five more than one-third of a number

11. double a quantity increased by nine

12. ten times a number increased by 1

13. one-third of the sum of a number and seven

14. one-fourth of the sum of a number and 5

15. one-third of a number reduced by twice the same number

16. one-fifth of a number reduced by double the same number

17. five times a quantity decreased by eleven

18. four less than seven times a number

Write an algebraic expression for each of the quantities being compared.

19. ***Stock Value*** The value of a share of IBM stock one day was \$74.50 more than the value of a share of AT&T stock.

20. ***Investments*** One day in June 2015, the value of a share of Twitter stock was \$47.49 less than the value of a share of Target stock.

▲ **21.** ***Geometry*** The length of a rectangle is 7 inches more than double the width.

▲ **22.** ***Geometry*** The length of a rectangle is 3 meters more than triple the width.

23. ***Cookie Sales*** The number of boxes of cookies sold by Sarah was 43 fewer than the number of boxes of cookies sold by Keiko. The number of boxes of cookies sold by Imelda was 53 more than the number sold by Keiko.

24. ***April Rainfall*** The average April rainfall in Savannah, Georgia, is about 13 inches more than that of Burlington, Vermont. The average rainfall in Phoenix, Arizona, is about 28 inches less than that of Burlington.

▲ **25.** ***Geometry*** The measure of the first angle of a triangle is 25 degrees less than the measure of the second angle. The measure of the third angle of the triangle is triple the measure of the second angle.

▲ **26.** ***Geometry*** The measure of the second angle of a triangle is double the measure of the first angle. The measure of the third angle of the triangle is 30 degrees more than the measure of the first angle.

27. ***Exports*** The value of the exports of Japan was twice the value of the exports of Canada.

28. ***Text Messages*** The number of text messages Marisol received was three times the number received by her brother.

29. ***Concert Tickets*** The price of the All Star Concert tickets was one-half the price of the Summer on the Beach Concert tickets.

30. ***Technology Convention*** The attendance at the Innovative Technology convention last year was one-third the attendance at the convention this year.

To Think About

Kayak Rentals *The following bar graph depicts the number of people who rented sea kayaks at Essex Boat Rental during the summer of 2015. Use the bar graph to answer exercises 31 and 32.*

31. Write an expression for the number of men who rented sea kayaks at Essex Boat Rental in each age category. Start by using x for the number of men aged 16 to 24 who rented kayaks.

32. Write an expression for the number of women who rented sea kayaks at Essex Boat Rental. Start by using y for the number of women aged 25 to 34 who rented kayaks.

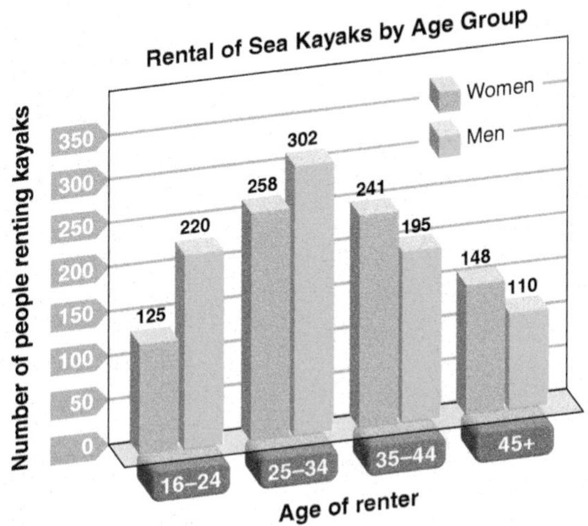

Cumulative Review *Solve for the variable.*

33. **[2.4.1]** $x + \frac{1}{2}(x - 3) = 9$

34. **[2.4.1]** $\frac{3}{5}x - 3(x - 1) = 9$

Quick Quiz 2.5 *Write an algebraic expression for each quantity. Let x represent the unknown value.*

1. Ten greater than a number

2. Five less than double a number

3. The measure of the first angle of a triangle is 15 degrees more than the measure of the second angle. The measure of the third angle of the triangle is five times the measure of the second angle. Write an algebraic expression for the measure of each of the three angles.

4. **Concept Check** Explain how you would decide whether to use $\frac{1}{3}(x + 7)$ or $\frac{1}{3}x + 7$ as the algebraic expression for the phrase "one-third of the sum of a number and seven."

2.6 Using Equations to Solve Word Problems

Student Learning Objectives

After studying this section, you will be able to:

1 Solve number problems.

2 Use the Mathematics Blueprint to solve applied word problems.

In Section 0.6 we introduced a simple three-step procedure to solve applied problems. You have had an opportunity to use that approach to solve word problems in Exercises 0.6 and in the Applications exercises in Chapter 1. Now we are going to focus our attention on solving applied problems that require the use of variables, translating English phrases into algebraic expressions, and setting up equations. The process is a little more involved. Some students find the following outline a helpful way to keep organized while solving such problems.

1. ***Understand the problem.***
   (a) Read the word problem carefully to get an overview.
   (b) Determine what information you will need to solve the problem.
   (c) Draw a sketch. Label it with the known information. Determine what needs to be found.
   (d) Choose a variable to represent one unknown quantity.
   (e) If necessary, represent other unknown quantities in terms of that very same variable.
2. ***Write an equation.***
   (a) Look for key words to help you to translate the words into algebraic symbols and expressions.
   (b) Use a given relationship in the problem or an appropriate formula to write an equation.
3. ***Solve and state the answer.***
4. ***Check.***
   (a) Check the solution in the original equation. Is the answer reasonable?
   (b) Be sure the solution to the equation answers the question in the word problem. You may need to do some additional calculations if it does not.

1 Solving Number Problems 

Example 1 Two-thirds of a number is eighty-four. What is the number?

Solution

1. ***Understand the problem.*** Draw a sketch.

   Let x = the unknown number.

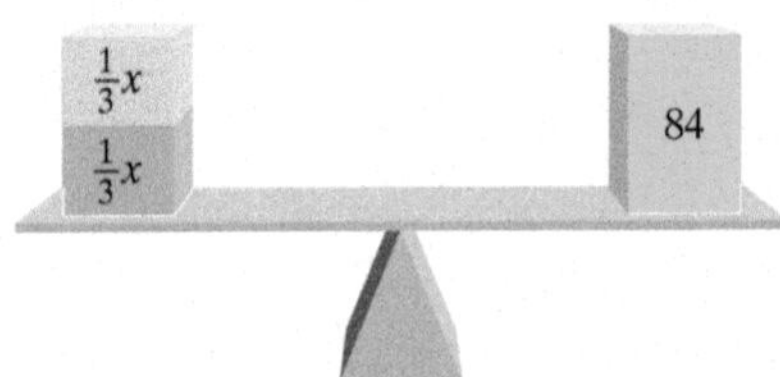

2. ***Write an equation.***

$$\underbrace{\text{Two-thirds of a number}}_{\frac{2}{3}x} \quad \underset{\downarrow}{\text{is}} \quad \underbrace{\text{eighty-four.}}_{84}$$

$$\frac{2}{3}x = 84$$

3. ***Solve and state the answer.***

$$\frac{2}{3}x = 84$$

$$3\left(\frac{2}{3}x\right) = 3(84) \quad \text{Multiply both sides of the equation by 3.}$$

$$2x = 252 \quad \text{Simplify.}$$

$$\frac{2x}{2} = \frac{252}{2} \quad \text{Divide both sides by 2.}$$

$$x = 126$$

The number is 126.

4. ***Check.*** Is two-thirds of 126 eighty-four?

$$\frac{2}{3}(126) \stackrel{?}{=} 84$$

$$84 = 84 \checkmark$$

□

Student Practice 1 Three-fourths of a number is negative eighty-one. What is the number?

Learning to solve problems like Examples 1 and 2 is a very useful skill. You will find that learning the remaining material in this text will be much easier if you can master the procedure used in these two examples.

Example 2 Five more than six times a quantity is three hundred five. Find the number.

Solution

1. ***Understand the problem.*** Read the problem carefully. You may not need to draw a sketch.

Let $x =$ the unknown quantity.

2. ***Write an equation.***

Five more than	six times a quantity	is	three hundred five.
5 +	$6x$	=	305

3. ***Solve and state the answer.*** You may want to rewrite the equation to make it easier to solve.

$$6x + 5 = 305$$

$$6x + 5 - 5 = 305 - 5 \quad \text{Subtract 5 from both sides.}$$

$$6x = 300 \quad \text{Simplify.}$$

$$\frac{6x}{6} = \frac{300}{6} \quad \text{Divide both sides by 6.}$$

$$x = 50$$

The quantity, or number, is 50.

4. ***Check.*** Is five more than six times 50 three hundred five?

$$6(50) + 5 \stackrel{?}{=} 305$$

$$300 + 5 \stackrel{?}{=} 305$$

$$305 = 305 \checkmark$$

□

Student Practice 2 Two less than triple a number is forty-nine. Find the number.

Example 3 The larger of two numbers is three more than twice the smaller. The sum of the numbers is thirty-nine. Find each number.

Solution

1. ***Understand the problem.*** The problem refers to *two* numbers. We must write an algebraic expression for *each number* before writing the equation. If we are comparing one quantity to others, we usually let the variable represent the quantity that the others are being *compared* to. In this case, the larger number is being compared to the smaller number. We want to use *one variable* to describe each number.

$$\text{Let } s = \text{the smaller number.}$$
$$\text{Then } \underbrace{2s + 3}_{\text{three more than twice the smaller number}} = \text{the larger number.}$$

2. ***Write an equation.*** The sum of the numbers is thirty-nine.

$$s + (2s + 3) = 39$$

3. ***Solve.***

$$s + (2s + 3) = 39$$
$$3s + 3 = 39 \quad \text{Combine like terms.}$$
$$3s = 36 \quad \text{Subtract 3 from each side.}$$
$$s = 12 \quad \text{Divide both sides by 3.}$$

4. ***Check.***

$$12 + [2(12) + 3] \stackrel{?}{=} 39$$
$$39 = 39 \quad ✓$$

The solution checks, but have we solved the word problem? We need to find *each* number. 12 is the smaller number. Substitute 12 into the expression $2s + 3$ to find the larger number.

$$2s + 3 = 2(12) + 3 = 27$$

The smaller number is 12. The larger number is 27. □

Student Practice 3 Consider two numbers. The second number is twelve less than triple the first number. The sum of the two numbers is twenty-four. Find each number.

We frequently encounter triangles in word problems. There are four important facts about triangles, which we list for convenient reference.

TRIANGLE FACTS

1. The sum of the interior angles of any triangle is 180°. That is,

   measure of $\angle A$ + measure of $\angle B$ + measure of $\angle C = 180°$.

2. An **equilateral** triangle is a triangle with three sides equal in length and three angles that measure 60° each.

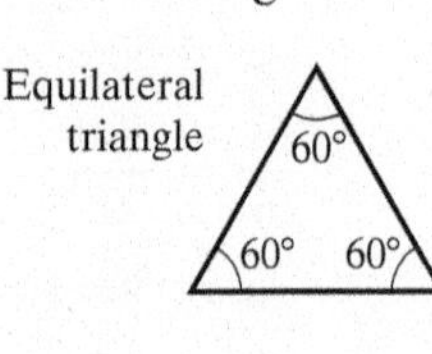

3. An **isosceles** triangle is a triangle with two equal sides. The two angles opposite the equal sides are also equal.

   measure of $\angle A$ = measure of $\angle B$

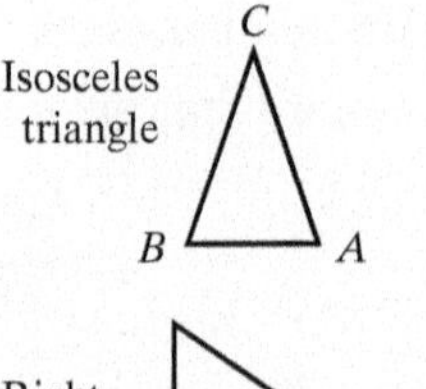

4. A **right** triangle is a triangle with one angle that measures 90°.

Example 4 The smallest angle of an isosceles triangle measures 24°. The other two angles are larger. What are the measurements of the other two angles?

Solution We know that in an isosceles triangle the measures of two angles are equal. We know that the sum of the measures of all three angles is 180°. Both of the larger angles must be different from 24°, therefore these two larger angles must be equal.

Let $x =$ the measure in degrees of each of the larger angles.

Then we can write
$$24° + x + x = 180°$$
$$24° + 2x = 180° \quad \text{Add like terms.}$$
$$2x = 156° \quad \text{Subtract 24 from each side.}$$
$$x = 78° \quad \text{Divide each side by 2.}$$

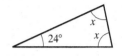

Thus the measure of each of the other two angles of the triangle must be 78°. □

Student Practice 4 The largest angle of an isosceles triangle measures 132°. The other two angles are smaller. What are the measurements of the other two angles?

Example 5 In a triangle, the measure of the first angle is 25° more than the measure of the second angle. The measure of the third angle is three times the measure of the second angle. What are the measurements of all three angles?

Solution The sum of the measures of all three angles of a triangle is 180°. We are comparing both the first and third angles to the second, so we let $x =$ the measure of the second angle. Then $x + 25° =$ the measure of the first angle, and $3x =$ the measure of the third angle.

Then we can write
$$x + 25° + x + 3x = 180°$$
$$5x + 25° = 180° \quad \text{Add like terms.}$$
$$5x = 155° \quad \text{Subtract 25° from each side.}$$
$$x = 31° \quad \text{Divide each side by 5.}$$

Thus, the measure of the second angle is 31°, the measure of the first angle is $31° + 25° = 56°$, and the measure of the third angle is $3(31°) = 93°$.

Check. Add all three angles to verify the sum is 180°. ✓ □

Student Practice 5 In a triangle, the measure of the second angle is five degrees more than twice the measure of the first angle. The measure of the third angle is half the measure of the first angle. What are the measurements of all three angles?

2 Using the Mathematics Blueprint to Solve Applied Word Problems

To facilitate understanding more involved word problems, we will use a Mathematics Blueprint similar to the one we used in Section 0.6. This format is a simple way to organize facts, determine what to set variables equal to, and select a method or approach that will assist you in finding the desired quantity. You will find using this form helpful, particularly in those cases when you read through a word problem and mentally say to yourself, "Now where do I begin?" You begin by responding to the headings of the blueprint. Soon a procedure for solving the problem will emerge.

Mathematics Blueprint for Problem Solving

Gather the Facts	Assign the Variable	Basic Formula or Equation	Key Points to Remember

Sometimes the relationship between two quantities is so well understood that we have developed a formula to describe that relationship. The following example shows how you can use a formula to solve a word problem.

Example 6 Two people traveled in separate cars. They each traveled a distance of 330 miles on an interstate highway. To maximize fuel economy, Fred traveled at exactly 50 mph. Sam traveled at exactly 55 mph. How much time did the trip take each person? (Use the formula distance = rate · time or $d = rt$.)

Solution

Mathematics Blueprint for Problem Solving

Gather the Facts	Assign the Variable	Basic Formula or Equation	Key Points to Remember
Each person drove 330 miles. Fred drove at 50 mph. Sam drove at 55 mph.	Time is the unknown quantity for each driver. Use subscripts to denote different values of t. t_f = Fred's time t_s = Sam's time	distance = (rate)(time) or $d = rt$	The time is expressed in hours.

We must find t in the formula $d = rt$. To simplify calculations we can solve for t before we substitute values:

$$\frac{d}{r} = \frac{\cancel{r}t}{\cancel{r}} \rightarrow \frac{d}{r} = t$$

Substitute the known values into the formula and solve for t.

$$\frac{d}{r} = t \qquad\qquad \frac{d}{r} = t$$

$$\frac{330}{50} = t_f \qquad\qquad \frac{330}{55} = t_s$$

$$6.6 = t_f \qquad\qquad 6 = t_s$$

It took Fred 6.6 hours to drive 330 miles. It took Sam 6 hours to drive 330 miles.

Check. Is this reasonable? Yes, you would expect Fred to take longer to drive the same distance because Fred was driving at a slower speed.

Note: You may wish to express 6.6 hours in hours and minutes. To change 0.6 hour to minutes, proceed as follows:

$$0.6\ \cancel{\text{hour}} \cdot \frac{60 \text{ minutes}}{1\ \cancel{\text{hour}}} = (0.6)(60) \text{ minutes} = 36 \text{ minutes}$$

Thus, Fred drove for 6 hours and 36 minutes. □

Student Practice 6 Sarah left the city to visit her aunt and uncle, who live in a rural area north of the city. She traveled the 220-mile trip in 4 hours. On her way home she took a slightly longer route, which measured 225 miles on the car odometer. The return trip took 4.5 hours. (Use the formula distance = rate · time or $d = rt$.)

(a) What was her average speed on the trip leaving the city?

(b) What was her average speed on the return trip?

(c) On which trip did she travel faster and by how much?

Example 7 A teacher told Melinda that she had a course average of 78 based on her six math tests. When she got home, Melinda found five of her tests. She had scores of 87, 63, 79, 71, and 95 on the five tests. She could not find her sixth test. What score did she obtain on that test? (Use the formula that an average = the sum of scores ÷ the number of scores.)

Solution

Mathematics Blueprint for Problem Solving

Gather the Facts	Assign the Variable	Basic Formula or Equation	Key Points to Remember
Her five known test scores are 87, 63, 79, 71, and 95. Her course average is 78.	We do not know the score Melinda received on her sixth test. Let x = the score on the sixth test.	$\text{average} = \dfrac{\text{sum of scores}}{\text{number of scores}}$	Since there are six test scores, we will need to divide the sum by 6.

When you average anything, you total up the sum of all the values and then divide it by the number of values.

We now write the equation for the average of six items. This involves adding the scores for all the tests and dividing by 6.

$$\frac{87 + 63 + 79 + 71 + 95 + x}{6} = 78$$

$$\frac{395 + x}{6} = 78 \quad \text{Add the numbers in the numerator.}$$

$$\frac{395}{6} + \frac{x}{6} = 78 \quad \text{First we rewrite the left side as two fractions. This is actually multiplying } \frac{1}{6}(395 + x) = \frac{395}{6} + \frac{x}{6}.$$

$$6 \cdot \frac{395}{6} + 6 \cdot \frac{x}{6} = 6(78) \quad \text{Multiply both sides of the equation by 6 to remove the fraction.}$$

$$395 + x = 468 \quad \text{Simplify.}$$

$$x = 73 \quad \text{Subtract 395 from both sides to find } x.$$

Melinda's score on the sixth test was 73.

Check. To verify that this is correct, we check that the average of the six tests is 78.

$$\frac{87 + 63 + 79 + 71 + 95 + 73}{6} \stackrel{?}{=} 78$$

$$\frac{468}{6} \stackrel{?}{=} 78$$

$$78 = 78 \checkmark$$

The problem checks. We know that the score on the sixth test was 73. □

Student Practice 7 Barbara's math course has four tests and one final exam. The final exam counts as much as two tests. Barbara has test scores of 78, 80, 100, and 96 on her four tests. What grade does she need on the final exam if she wants to have a 90 average for the course? (Use the formula that an average = the sum of scores ÷ the number of scores.)

2.6 Exercises MyMathLab®

Solve. Check your solution.

1. What number minus 543 gives 718?

2. What number added to 74 gives 265?

3. A number divided by eight is 296. What is the number?

4. A number divided by nine is 189. What is the number?

5. Seventeen greater than a number is 199. Find the number.

6. Three times a number is one. What is the number?

7. A number is doubled and then increased by seven. The result is ninety-three. What is the original number?

8. Eight times a number is decreased by thirty-two. The result is one hundred twenty. Find the original number.

9. When eighteen is reduced by two-thirds of a number, the result is 12. Find the number.

10. Twice a number is increased by one-third the same number. The result is 42. Find the number.

11. Eight less than triple a number is the same as five times the same number. Find the number.

12. Ten less than double a number is the same as seven times the number. Find the number.

13. A number, half of that number, and one-third of that number are added. The result is 22. What is the number?

14. A number, twice that number, and one-third of that number are added. The result is 20. What is the number?

Applications *Solve. Check to see if your answer is reasonable.*

15. ***Tablet Cases*** Lester orders supplies for the Tech Warehouse and noticed that the number of black tablet cases sold this year was three times the number of red cases. If he orders 120 black tablet cases and expects the same sales pattern, how many red cases should he order?

16. ***Inventory*** The college store maintains an inventory of baseball T-shirts and sweatshirts that have the team logo. Due to demand, the number of baseball T-shirts kept in inventory is four times the number of baseball sweatshirts. If the store has 164 T-shirts, how many sweatshirts are in inventory at the college store?

17. ***Geometry*** The smallest angle of an isosceles triangle measures 36°. The other two angles are larger. What is the measure of each of the other two angles?

18. ***TV Antenna*** Part of a cable TV antenna is an isosceles triangle. The largest angle measures 146°. The other two angles are smaller. What are the measurements of the other two angles in the triangular part of the cable TV antenna?

19. ***Parking Lot*** A triangular region of the community college parking lot was measured. The measure of the first angle of this triangle is double the measure of the second angle. The measure of the third angle is 19° greater than double the measure of the first angle. What are the measurements of all three angles?

20. ***Lobster Pot*** A lobster pot used in Gloucester, Massachusetts, has a small triangular piece of wood at the point where the lobster enters. The measure of the first angle of this triangle is triple the measure of the second angle. The measure of the third angle of the triangle is 30° less than double the measure of the first angle. What are the measurements of all three angles?

21. ***One-Day Sale*** Raquelle went to Weller's Department Store's one-day sale. She bought two blouses for \$38 each and a pair of shoes for \$49. She also wanted to buy some jewelry. Each item of jewelry was bargain priced at \$11.50 each. If she brought \$171 with her, how many pieces of jewelry could she buy?

22. ***Sales Clerk Bonus*** Samuel is a sales clerk in a major department store and earns \$11 per hour. Last week the store had a credit card promotion offering a bonus of \$2.50 for each store credit card application the sales clerks obtained from customers. Samuel worked 30 hours during the promotional week, and his paycheck for that week was \$377.50 before deductions. How many credit card applications did Samuel obtain during the promotional week?

23. ***Running Time*** A track star in Arizona ran the 100-meter event in six track meets. Her average time for all six meets was 11.8 seconds. She could not find the record for her running time in the last meet, but she had all the other ones: 11.7 seconds, 11.6 seconds, 12 seconds, 12.1 seconds, 11.9 seconds. What was her running time for her last meet?

24. ***Algebra Exams*** Leo has taken four of the five exams in his algebra class. His scores on the four exams were 94, 89, 92, and 88. The instructor told students they must have at least a 90 as their average test score for all five exams to get a grade of A in the class. What is the lowest score Leo can get on the fifth exam to earn an A in his algebra class?

25. ***In-Line Skating*** Two in-line skaters, Nell and Kristin, start from the same point and skate in the same direction. Nell skates at 12 miles per hour and Kristin skates at 14 miles per hour. If they can keep up that pace for 2.5 hours, how far apart will they be at the end of that time?

26. ***Truck Drivers*** Allan and Tiana are transporting cars on their trucks from the same pick-up point and taking them to the same dealership. Tiana is traveling at 55 mph, while Allan's speed is a little slower, 45 mph, since he has a larger load. If they both continue at the same speed, how far apart will they be after traveling for 3.5 hours?

27. ***Travel Routes*** Nella drove from Albuquerque, New Mexico, to the Garden of the Gods rock formation in Colorado Springs. It took her six hours to travel 312 miles over the mountain road. She came home on the highway. On the highway she took five hours to travel 320 miles. How fast did she travel using the mountain route? How much faster (in miles per hour) did she travel using the highway route?

28. ***Road Trip*** Ester and her roommate MaryAnn decided to visit a friend at her beach home. MaryAnn had to return a few days earlier than Ester so they took separate cars. They both left their apartment at the same time for the 420-mile trip to the beach house. MaryAnn's car is not fuel efficient so she traveled at50 mph to save money on gas, while Ester drove 60 mph. How much time did the trip take each person? How much longer did it take MaryAnn to travel to the beach house?

29. ***Grading System*** Danielle's chemistry professor has a complicated grading system. Each lab counts once and each test counts as two labs. At the end of the semester there is a final lab that counts as three regular labs. Danielle received scores of 84, 81, and 93 on her labs and scores of 89 and 94 on her tests. What score must she get on the final lab to receive an A (an A is 90)?

30. ***Life Expectancy*** Just before Sara's presentation on the top seven countries' life expectancies, she lost a few note cards for her speech, and therefore she did not have the data for Spain and Sweden. She remembered they both had the exact same age for life expectancy. Sara used the following data from her note cards to calculate the missing information. Japan: 82.6 years; Hong Kong: 82.2 years; Iceland: 81.8 years; Switzerland: 81.7 years; Australia 81.2 years; the average age for all 7 countries: 81.6 years. What age did Sara calculate as the life expectancy for Spain and Sweden? Round your answer to the nearest tenth.

To Think About

31. ***Cricket Chirps*** In warmer climates, approximate temperature predictions can be made by counting the number of chirps of a cricket during a minute. The Fahrenheit temperature decreased by forty is equivalent to one-fourth of the number of cricket chirps.

(a) Write an equation for this relationship.

(b) Approximately how many chirps per minute should be recorded if the temperature is 90°F?

(c) If a person recorded 148 cricket chirps in a minute, what would be the Fahrenheit temperature according to this formula?

Cumulative Review *Simplify.*

32. **[1.6.1]** $5x(2x^2 - 6x - 3)$

33. **[1.6.1]** $-2a(ab - 3b + 5a)$

34. **[1.7.2]** $7x - 3y - 12x - 8y + 5y$

35. **[1.7.2]** $5x^2y - 7xy^2 - 8xy - 9x^2y$

Quick Quiz 2.6

1. A number is tripled and then decreased by 15. The result is 36. What is the number?

2. The largest angle of an isosceles triangle measures 70°. The other two angles are smaller. What are the measurements of the other two angles?

3. James scored 84, 89, 73, and 80 on four tests. What must he score on the next test to have an 80 average in the course?

4. **Concept Check** Explain how you would set up an equation to solve the following problem.

Phil purchased two shirts for \$23 each and then purchased several pairs of socks. The socks were priced at \$0.75 per pair. How many pairs of socks did he purchase if the total cost was \$60.25?

2.7 Solving Word Problems: The Value of Money and Percents 

The problems we now present are frequently encountered in business. They deal with money: buying, selling, and renting items; earning and borrowing money; and the value of collections of stamps or coins. Many applications require an understanding of the use of percents and decimals. Review Sections 0.4 and 0.5 if you are weak in these skills.

Student Learning Objectives

After studying this section, you will be able to:

1. Solve problems involving periodic rate charges.
2. Solve percent problems.
3. Solve investment problems involving simple interest.
4. Solve coin problems.

1 Solving Problems Involving Periodic Rate Charges

Example 1 A business executive rented a car. The Supreme Car Rental Agency charged \$39 per day and \$0.28 per mile. The executive rented the car for two days and the total rental cost was computed to be \$176. How many miles did the executive drive the rented car?

Solution

1. ***Understand the problem.*** How do you calculate the cost of renting a car?
   total cost = per-day cost + mileage cost
   What is known?
   It cost a total of \$176 to rent the car for two days, the per-day cost is \$39 and the mileage cost is \$0.28 per mile.
   What do you need to find?
   The number of miles the car was driven.
   Choose a variable:
   Let m = the number of miles driven in the rented car.
2. ***Write an equation.*** Use the relationship for calculating the total cost.

$$\begin{array}{ccccc} \text{per-day cost} & + & \text{mileage cost} & = & \text{total cost} \\ (39)(2) & + & (0.28)m & = & 176 \end{array}$$

3. ***Solve and state the answer.***

$$\begin{aligned} (39)(2) + (0.28)(m) &= 176 & & \\ 78 + 0.28m &= 176 & & \text{Simplify the equation.} \\ 0.28m &= 98 & & \text{Subtract 78 from both sides.} \\ \frac{0.28m}{0.28} &= \frac{98}{0.28} & & \text{Divide both sides by 0.28.} \\ m &= 350 & & \text{Simplify.} \end{aligned}$$

   The executive drove 350 miles.
4. ***Check.*** Does this seem reasonable? If he drove the car 350 miles in two days, would it cost \$176?

$$\begin{gathered} \left(\begin{array}{c}\text{cost of \$39 per day} \\ \text{for 2 days}\end{array}\right) + \left(\begin{array}{c}\text{cost of \$0.28 per mile} \\ \text{for 350 miles}\end{array}\right) \stackrel{?}{=} \text{total cost of \$176} \\ (\$39)(2) + (350)(\$0.28) \stackrel{?}{=} \$176 \\ \$78 + \$98 \stackrel{?}{=} \$176 \\ \$176 = \$176 \ \checkmark \end{gathered}$$

□

Student Practice 1 Alfredo wants to rent a truck to move to Florida. He has determined that the cheapest rental rates for a truck of the correct size are from a local company that will charge him \$25 per day and \$0.20 per mile. He has not yet completed an estimate of the mileage of the trip, but he knows that he will need the truck for three days. He has allowed \$350 in his moving budget for the truck. How far can he travel for a rental cost of exactly \$350?

2 Solving Percent Problems

Many applied situations require finding a percent of an unknown number. If we want to find 23% of \$400, we multiply 0.23 by 400: $0.23(400) = 92$. If we want to find 23% of an unknown number, we can express this using algebra by writing $0.23n$, where n represents the unknown number.

Example 2 A sofa was marked with the following sign: "The price of this sofa has been reduced by 23%. You can save \$138 if you buy now." What was the original price of the sofa?

Solution

1. ***Understand the problem.***

Let $s =$ the original price of the sofa.

Then $0.23s =$ the amount of the price reduction, which is \$138.

2. ***Write an equation and solve.***

$$0.23s = 138 \quad \text{Write the equation.}$$

$$\frac{0.23s}{0.23} = \frac{138}{0.23} \quad \text{Divide each side of the equation by 0.23.}$$

$$s = 600 \quad \text{Simplify.}$$

The original price of the sofa was \$600.

3. ***Check.*** Is \$600 a reasonable answer? ✓ Does 23% of \$600 = \$138? ✓ □

Student Practice 2 John earns a commission of 38% of the cost of every set of chef's knives that he sells. Last year he earned \$17,100 in commissions. What was the cost of the chef's knives that he sold last year?

Example 3 Hector received a pay raise this year. The raise was 6% of last year's salary. This year he will earn \$15,900. What was his salary last year before the raise?

Solution

1. ***Understand the problem.*** What do we need to find?

Hector's salary last year.

What do we know?

Hector received a 6% pay raise and now earns \$15,900.

What does this mean?

Reword the problem: *This year's salary of* \$15,900 *is* 6% *more than last year's salary.*

Choose a variable:

Let $x =$ Hector's salary last year.

Then $0.06x =$ the amount of the raise.

2. ***Write an equation and solve.***

Last year's salary	+	the amount of his raise	=	this year's salary	
x	+	$0.06x$	=	15,900	Write the equation.
$1.00x$	+	$0.06x$	=	15,900	Rewrite x as $1.00x$.
		$1.06x$	=	15,900	Combine like terms.
		$\frac{1.06x}{1.06}$	=	$\frac{15{,}900}{1.06}$	Divide by 1.06.
		x	=	15,000	Simplify.

Thus Hector's salary was $15,000 last year before the raise.

3. ***Check.*** Does it seem reasonable that Hector's salary last year was $15,000? The check is up to you. □

Student Practice 3 The price of Betsy's new car is 7% more than the price of a similar model last year. She paid $19,795 for her car this year. What would a similar model have cost last year?

3 Solving Investment Problems Involving Simple Interest

Interest is a charge for borrowing money or an income from investing money. Interest rates affect our lives. They affect the national economy and they affect a consumer's ability to borrow money for big purchases. For these reasons, a student of mathematics should be able to solve problems involving interest.

There are two basic types of interest: simple and compound. **Simple interest** is computed by multiplying the amount of money borrowed or invested (which is called the *principal*) times the rate of interest times the period of time over which it is borrowed or invested (usually measured in years unless otherwise stated).

$$\text{Simple interest} = \text{principal} \times \text{rate} \times \text{time}$$
$$I = prt$$

You often hear of banks offering a certain interest rate *compounded* quarterly, monthly, weekly, or daily. In **compound interest** the amount of interest is added to the amount of the original principal at the end of each time period, so future interest is based on the sum of both principal and previous interest. Most financial institutions use compound interest in their transactions.

However, all examples and exercises in this chapter will involve **simple interest.**

Example 4 Find the interest on $3000 borrowed at a simple interest rate of 18% for one year.

Solution

$I = prt$	The simple interest formula
$I = (3000)(0.18)(1)$	Substitute the values of the variables: principal = 3000, the rate = 18% = 0.18, the time = one year.
$I = 540$	Simplify.

Thus the interest charge for borrowing $3000 for one year at a simple interest rate of 18% is $540. □

Student Practice 4 Find the interest on $7000 borrowed at a simple interest rate of 12% for one year.

Calculator

Interest

You can use a calculator to find simple interest. Find the interest on $450 invested at an annual rate of 6.5% for 15 months. Notice that the time is in months. Since the interest formula $I = prt$ is in years, you need to change 15 months to years by dividing 15 by 12.

Enter: 15 [÷] 12 [=]

Display: [1.25]

Leave this on the display and multiply as follows:

1.25 [×] 450 [×] 6.5 [%] [=]

The display should read [36.5625] which would round to $36.56.

Try finding the simple interest in the following.

(a) $9516 invested at 12% for 30 months

(b) $593 borrowed at 8% for 5 months

Ans:

(a) $2854.80

(b) $19.77

Now we apply this concept to a word problem about investments.

Student Practice 5 A woman invested her savings of \$8000 in two accounts that each calculate interest only once per year. She placed one amount in a special notice account that yields 9% annual interest. The remainder she placed in a tax-free All-Savers account that yields 7% annual interest. At the end of the year, she had earned \$630 in interest from the two accounts together. How much had she invested in each account?

Example 5 A woman invested an amount of money in two accounts for one year. She invested some at 8% simple interest and the rest at 6% simple interest. Her total amount invested was \$1250. At the end of the year she had earned \$86 in interest. How much money had she invested in each account?

Solution The simple interest formula is $I = prt$.

1. ***Understand the problem.***
   What do we know? We have $t = 1$ year, so we can calculate $I = p \times r \times 1$ or $I = pr$: *Interest = amount invested × rate.*

   What do we need to find? *How much money was invested at* 8% *and at* 6%.

   Choose a variable: Let x = amount invested at 8%.
   Then we know:

   amount invested at 8% + amount invested at 6% = \$1250
   x + amount invested at 6% = \$1250
   amount invested at 6% = \$1250 − x

   We let $(1250 - x)$ = *amount invested* at 6%.

2. ***Write an equation and solve.***
   We use $I = pr$ (interest = amount invested × rate)

interest earned at 8%	+	interest earned at 6%	=	total interest earned during the year
$0.08x$	+	$0.06(1250 - x)$	=	86

Note: Be sure you write $(1250 - x)$ for the amount of money invested at 6%. Students often write it backwards by mistake. It is *not* correct to use $(x - 1250)$ instead of $(1250 - x)$. The order of the terms is very important.

Solve and state the answer.

$$0.08x + 75 - 0.06x = 86 \quad \text{Remove parentheses.}$$
$$0.02x + 75 = 86 \quad \text{Combine like terms.}$$
$$0.02x = 11 \quad \text{Subtract 75 from both sides.}$$
$$\frac{0.02x}{0.02} = \frac{11}{0.02} \quad \text{Divide both sides by 0.02.}$$
$$x = 550 \quad \text{The amount invested at 8\% simple interest is \$550.}$$
$$1250 - x = 1250 - 550 = 700 \quad \text{The amount invested at 6\% simple interest is \$700.}$$

Check. Are these values reasonable? Yes. Do the amounts equal \$1250?

$$\$550 + \$700 \stackrel{?}{=} \$1250$$
$$\$1250 = \$1250 \checkmark$$

Would these amounts earn \$86 interest in one year invested at the specified rates?

$$0.08(\$550) + 0.06(\$700) \stackrel{?}{=} \$86$$
$$\$44 + \$42 \stackrel{?}{=} \$86$$
$$\$86 = \$86 \checkmark$$

□

4 Solving Coin Problems

Coin problems provide an unmatched opportunity to use the concept of *value.* We must make a distinction between *how many coins* there are and the *value* of the coins.

Consider the next example. Here we know the *value* of some coins, but do not know *how many* we have.

Example 6 When Bob got out of math class, he wanted to buy a coffee at the snack bar. He had exactly enough dimes and quarters to buy a coffee that would cost \$2.55. He had one fewer quarter than he had dimes. How many coins of each type did he have?

Solution

1. ***Understand the problem.***

$$\text{Let } d = \text{the number of dimes.}$$
$$\text{Then } d - 1 = \text{the number of quarters.}$$

The total value of the coins was \$2.55. How can we represent the value of the dimes and the value of the quarters? Think.

Each dime is worth \$0.10.
5 dimes are worth $(5)(0.10) = 0.50$.
d dimes are woth $(d)(0.10) = 0.10d$.

Each quarter is worth \$0.25.
8 quarters are worth $(8)(0.25) = 2.00$.
$(d - 1)$ quarters are worth:
$(d - 1)(0.25) = 0.25(d - 1)$.

2. ***Write an equation and solve.*** Now we can write an equation for the total value.

$$(\text{value of dimes}) + (\text{value of quarters}) = \$2.55$$
$$0.10d + 0.25(d - 1) = 2.55$$
$$0.10d + 0.25d - 0.25 = 2.55 \quad \text{Remove parentheses.}$$
$$0.35d - 0.25 = 2.55 \quad \text{Combine like terms.}$$
$$0.35d = 2.80 \quad \text{Add 0.25 to both sides.}$$
$$\frac{0.35d}{0.35} = \frac{2.80}{0.35} \quad \text{Divide both sides by 0.35.}$$
$$d = 8 \quad \text{Simplify.}$$
$$d - 1 = 8 - 1 = 7$$

Thus Bob had eight dimes and seven quarters.

3. ***Check.*** Is this answer reasonable? Yes. Does Bob have one fewer quarter than he has dimes?

$$8 - 7 \stackrel{?}{=} 1$$
$$1 = 1 \checkmark$$

Are eight dimes and seven quarters worth \$2.55?

$$8(\$0.10) + 7(\$0.25) \stackrel{?}{=} \$2.55$$
$$\$0.80 + \$1.75 \stackrel{?}{=} \$2.55$$
$$\$2.55 = \$2.55 \checkmark$$

□

Student Practice 6 Ginger has five more quarters than dimes. She has \$5.10 in change. If she has only quarters and dimes, how many coins of each type does she have?

Example 7 Michele and her two children returned from the grocery store with \$2.80 in change. She had twice as many quarters as nickels. She had two more dimes than nickels. How many nickels, dimes, and quarters did she have?

Solution

Mathematics Blueprint for Problem Solving

Gather the Facts	Assign the Variable	Basic Formula or Equation	Key Points to Remember
Michele had \$2.80 in change. She had twice as many quarters as nickels. She had two more dimes than nickels.	x = number of nickels. $2x$ = number of quarters. $x + 2$ = number of dimes. $0.05x$ = value of the nickels. $0.25(2x)$ = value of the quarters. $0.10(x + 2)$ = value of the dimes.	The value of the nickels + the value of the dimes + the value of the quarters = \$2.80.	Don't add the number of coins to get \$2.80. You must add the values of the coins!

$$(\text{value of nickels}) + (\text{value of dimes}) + (\text{value of quarters}) = \$2.80$$

$$0.05x + 0.10(x + 2) + 0.25(2x) = 2.80$$

Solve.

$$0.05x + 0.10x + 0.20 + 0.50x = 2.80 \quad \text{Remove parentheses.}$$

$$0.65x + 0.20 = 2.80 \quad \text{Combine like terms.}$$

$$0.65x = 2.60 \quad \text{Subtract 0.20 from both sides.}$$

$$\frac{0.65x}{0.65} = \frac{2.60}{0.65} \quad \text{Divide both sides by 0.65.}$$

$$x = 4 \quad \text{Simplify. Michele had four nickels.}$$

$$2x = 2(4) = 8 \quad \text{She had eight quarters.}$$

$$x + 2 = 4 + 2 = 6 \quad \text{She had six dimes.}$$

When Michele left the grocery store she had four nickels, eight quarters, and six dimes.

Check. Is the answer reasonable? Yes. Did Michele have twice as many quarters as nickels?

$$(4)(2) \stackrel{?}{=} 8 \qquad 8 = 8 \checkmark$$

Did she have two more dimes than nickels?

$$4 + 2 \stackrel{?}{=} 6 \qquad 6 = 6 \checkmark$$

Do four nickels, eight quarters, and six dimes have a value of \$2.80?

$$4(\$0.05) + 8(\$0.25) + 6(\$0.10) \stackrel{?}{=} \$2.80$$

$$\$0.20 + \$2.00 + \$0.60 \stackrel{?}{=} \$2.80$$

$$\$2.80 = \$2.80 \checkmark$$

Student Practice 7 A young boy told his friend that he had twice as many nickels as dimes in his pocket. He also said that he had four more quarters than dimes. He said that he had \$2.35 in change in his pocket. Can you determine how many nickels, dimes, and quarters he had?

2.7 Exercises MyMathLab®

Applications

Solve. All problems involving interest refer to simple interest.

1. ***Assistant Manager*** Clyde works as a part-time assistant manager for the Barkley Coffeehouse to earn money to help pay for his college expenses. He makes $11.50 per hour. The coffeehouse sells whole bean coffee, mugs, and other coffee products. Clyde earns $0.50 for each of these items he sells. Last week he worked 30 hours and earned $364. How many coffee products did he sell last week?

2. ***Waitress*** Kayleigh must earn $385 a week in order to cover her expenses while attending nursing school. Due to her past experience, she was hired at a high-end restaurant and scheduled weekends, which resulted in higher tip earnings. The manager told her the pay is $8.75 per hour plus tips, which average $8 per table. If Kayleigh works 12 hours a week, how many tables would she have to serve in order to make $385 per week?

3. ***Tax Accountant*** Eli Moran is an accountant at J & R Tax Prep, and during tax season he works overtime to meet the demand. He is paid $32 per hour for the first 40 hours and $48 per hour for each hour in the week worked over 40 hours. To save for a nice vacation, Eli's goal is to earn $1520 per week. How many hours overtime per week will he need to achieve his goal?

4. ***Construction Apprentice*** Anna is an apprentice sheet metal worker for a construction company. When the company falls behind schedule on a job, she works overtime. She is paid $20 per hour for the first 40 hours and $30 per hour for each hour in the week worked over 40 hours. Anna is saving money for a new truck, so she would like to work enough overtime to earn $980 per week. How many hours overtime per week will she need to achieve this goal?

5. ***Layaway*** Mrs. Peterson's triplets all want to attend the local private academy, where they are required to wear uniforms. Mrs. Peterson knows that the only way she can afford these uniforms is to shop early, put the uniforms on layaway, and pay a little every week. The total uniform cost for the three girls is $1817.75. If she put down a deposit of $600.00 and she could afford $105.00 per week, how long would it take her to pay off the uniforms? (Round to nearest whole number.)

6. ***Party Costs*** The Swedish Chef Catering Company charges $50.00 for setup, $85.00 for cleanup, and the special menu that Judy Carter ordered cost $23.50 per person. If Judy Carter's bill came to $558.00, how many people came to the party?

7. ***Camera Sale*** The camera Melissa wanted for her birthday is on sale at 28% off the usual price. The amount of the discount is $100.80. What was the original price of the camera?

8. ***Work Force*** The number of women working full-time in Springfield has risen 12% this year. This means 216 more women have full-time jobs. What was the number of women working full-time last year?

9. ***Salary*** The cost of living last year went up 3%. Fortunately, Alice Swanson got a 3% raise in her salary from last year. This year she is earning $22,660. How much did she make last year?

10. ***Stock Profit*** A speculator bought stocks and later sold them for $5136, making a profit of 7%. How much did the stocks cost him?

11. ***Investments*** Katerina Lubov inherited some money and invested it at 6% simple interest. At the end of the year, the total amount of her original principal and the interest was $12,720. How much did she originally invest?

12. ***Investments*** Robert Campbell invested some money at 8% simple interest. At the end of the year, the total amount of his original principal and the interest was $7560. How much did he originally invest?

13. ***Trust Fund*** Mr. and Mrs. Wright set up a trust fund for their children last year. Part of it is earning 7% simple interest per year while the rest of it is earning 5% simple interest per year. They placed \$5000 in the trust fund. In one year the trust fund has earned \$310. How much did they invest at each interest rate?

14. ***Savings Bonds*** Anne and Michael invested \$8000 last year in tax-free bonds. Some of the bonds earned 8% simple interest while the rest earned 6% simple interest. At the end of the year, they had earned \$580 in interest. How much did they invest at each interest rate?

15. ***Mutual Funds*** Plymouth Rock Bank invested \$400,000 last year in mutual funds. The conservative fund earned 8% simple interest. The growth fund earned 12% simple interest. At the end of the year, the bank had earned \$38,000 from these mutual funds. How much did it invest in each fund?

16. ***Mutual Funds*** Millennium Securities last year invested \$600,000 in mutual funds. The international fund earned 11% simple interest. The high-tech fund earned 7% simple interest. At the end of the year, the company had earned \$50,000 in interest. How much did it invest in each fund?

17. ***Investments*** Dave Horn invested half of his money at 5%, one-third of his money at 4%, and the rest of his money at 3.5%. If his total annual investment income was \$530, how much had he invested?

18. ***Investments*** When Karin Schumann won the lottery, she decided to invest half her winnings in a mutual fund paying 4% interest, one-third of it in a credit union paying 4.5% interest, and the rest in a bank CD paying 3% interest. If her annual investment income was \$2400, how much money had she invested?

19. ***Coin Bank*** Little Melinda has nickels and quarters in her bank. She has four fewer nickels than quarters. She has \$3.70 in the bank. How many coins of each type does she have?

20. ***Coin Change*** Reggie's younger brother had several coins when he returned from his paper route. He had a total of \$5.35 in dimes and quarters. He had six more quarters than he had dimes. How many of each coin did he have?

21. ***Coin Change*** A newspaper carrier has \$3.75 in change. He has three more quarters than dimes but twice as many nickels as quarters. How many coins of each type does he have?

22. ***Coin Change*** Tim Whitman has \$4.50 in change in his desk drawer to use on the vending machines downstairs. He has four more quarters than dimes. He has three times as many nickels as dimes. How many coins of each type does he have?

23. ***Office Manager*** Huy is an office manager for a small law firm. One of his responsibilities is to order supplies. He noticed that in the past, the number of boxes of white paper ordered was double the number of boxes of beige paper. He must spend no more than \$112.50 on paper so that when tax and shipping are added he stays within the office budget. If a box of white paper cost \$3.50 and a box of beige paper cost \$4.25, how many boxes of each can Huy order and stay within his \$112.50 budget?

24. ***Ticket Donation*** Mario's Marionettes donated free tickets for its show to the local Boys & Girls Club. It claimed that the ticket value was \$176.75. A child's ticket cost \$5.50 and an adult's ticket cost \$8.75. If the number of children's tickets was three times the number of adults' tickets, how many adults and children got to attend the show for free?

25. ***Paper Currency*** Jim and Amy sold many pieces of furniture and made \$1380 during their garage sale. They had sixteen more \$10 bills than \$50 bills. They had one more than three times as many \$20 bills as \$50 bills. How many of each denomination did they have?

26. ***Paper Currency*** Roberta Burgess came home with \$325 in tips from two nights on her job as a waitress. She had \$20 bills, \$10 bills, and \$5 bills. She discovered that she had three times as many \$5 bills as she had \$10 bills. She also found that she had 4 fewer \$20 bills than she had \$10 bills. How many of each denomination did she have?

27. ***Salary*** Madelyn Logan is an office furniture dealer who earns an \$18,000 base salary. She also earns a 4% commission on sales. How much must she sell to earn a total of \$55,000?

28. ***Animal Shelter*** The local animal shelter accepts abandoned cats and dogs. They usually receive twice as many cats as dogs. They estimate that 80% of the cats and 60% of the dogs that come in need some kind of medical treatment. If they treated 286 animals last year, how many cats and dogs did they take in?

Cumulative Review *Perform the operations in the proper order.*

29. **[1.5.1]** $5(3) + 6 \div (-2)$

30. **[1.5.1]** $5(-3) - 2(12 - 15)^2 \div 9$

Evaluate for $a = -1$ and $b = 4$.

31. **[1.8.1]** $a^2 - 2ab + b^2$

32. **[1.8.1]** $a^3 + ab^2 - b - 5$

Quick Quiz 2.7

1. Charles is renting a large copy machine for his company and has been given a budget of \$1250. The machine he wants to rent will require a \$224 installation fee and a \$114-per-month rental contract. How many months will Charles be able to rent the copy machine with his current budget?

2. Last year the cost of Marcia's health plan went up 7%. The cost for coverage for Marcia and her family this year is \$12,412. What did it cost last year before the increase went into effect?

3. Walter and Barbara invested \$5000 in two CDs last year. One of the CDs paid 4% interest. The other one paid 5% interest. At the end of one year they earned \$228 in simple interest. How much did they invest at each interest rate?

4. **Concept Check** Explain how you would set up an equation to solve the following problem.

Robert has \$2.55 in change consisting of nickels, dimes, and quarters. He has twice as many dimes as quarters. He has one more nickel than he has quarters. How many of each coin does he have?

2.8 Solving Inequalities in One Variable

Student Learning Objectives

After studying this section, you will be able to:

1. Interpret inequality statements.
2. Graph an inequality on a number line.
3. Translate English phrases into algebraic statements.
4. Solve and graph an inequality.

1 Interpreting Inequality Statements

We frequently speak of one value being greater than or less than another value. We say that "5 is less than 7" or "9 is greater than 4." These relationships are called **inequalities.** We can write inequalities in mathematics by using symbols. We use the symbol $<$ to represent the words "**is less than.**" We use the symbol $>$ to represent the words "**is greater than.**"

Statement in Words	*Statement in Algebra*
5 is less than 7.	$5 < 7$
9 is greater than 4.	$9 > 4$

Note: "5 is less than 7" and "7 is greater than 5" have the same meaning. Similarly, $5 < 7$ and $7 > 5$ have the same meaning. They represent two equivalent ways of describing the same relationship between the two numbers 5 and 7.

We can better understand the concept of inequality if we examine a number line.

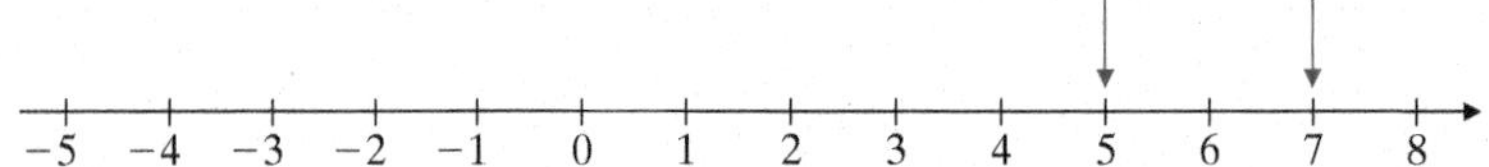

We say that one number is greater than another if it is to the right of the other on the number line. Thus $7 > 5$, since 7 is to the right of 5.

What about negative numbers? We can say "-1 is greater than -3" and write it in symbols as $-1 > -3$ because we know that -1 lies to the right of -3 on the number line.

Example 1 In each statement, replace the question mark with the symbol $<$ or $>$.

(a) $3 ? -1$ **(b)** $-2 ? 1$ **(c)** $-3 ? -4$ **(d)** $0 ? 3$ **(e)** $-3 ? 0$

Solution

(a) $3 > -1$ Use $>$, since 3 is to the right of -1 on the number line.

(b) $-2 < 1$ Use $<$, since -2 is to the left of 1. (Or equivalently, we could say that 1 is to the right of -2.)

(c) $-3 > -4$ Note that -3 is to the right of -4.

(d) $0 < 3$ Use $<$, since 0 is to the left of 3.

(e) $-3 < 0$ Note that -3 is to the left of 0. □

Student Practice 1 In each statement, replace the question mark with the symbol $<$ or $>$.

(a) $7 ? 2$ **(b)** $-2 ? -4$ **(c)** $-1 ? 2$ **(d)** $-8 ? -5$ **(e)** $0 ? -2$ **(f)** $5 ? -3$

2 Graphing an Inequality on a Number Line

Sometimes we will use an inequality to express the relationship between a variable and a number. $x > 3$ means that x could have the value of *any number* greater than 3.

Any number that makes an inequality true is called a **solution** of the inequality. The set of all numbers that make the inequality true is called the **solution set.** A picture that represents all of the solutions of an inequality is called a **graph** of the inequality.

The inequality $x > 3$ can be graphed on a number line as follows:

Case 1

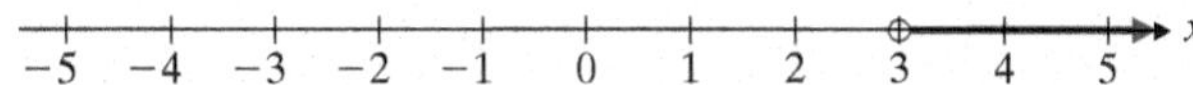

Note that all of the points to the right of 3 are shaded. The open circle at 3 indicates that *we do not include* the point for the number 3.

Sometimes a variable will be either less than or equal to a certain number. In the statement "x is less than or equal to -1," we are implying that x could have the value of -1 or any number less than -1. We write this as $x \leq -1$. We graph it as follows:

Case 2

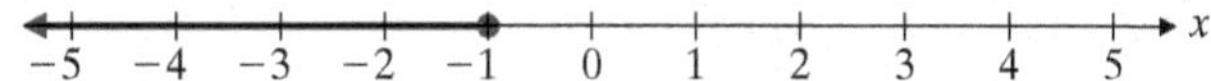

Note that the closed circle at -1 indicates that *we do include* the point for the number -1.

Be careful you do not confuse ○———► with ●———►. It is important to decide if you need an open circle or a closed one. Case 1 uses an open circle, and Case 2 uses a closed circle.

Example 2 State each mathematical relationship in words and then graph it.

(a) $x < -1$ **(b)** $x \geq -2$ **(c)** $-3 < x$ **(d)** $x \leq -\frac{1}{2}$

Solution

(a) We state that "x is less than -1."

$x < -1$

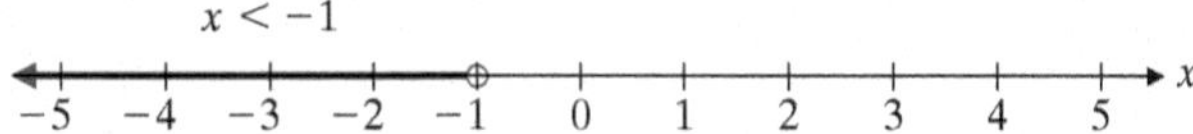

(b) We state that "x is greater than or equal to -2."

$x \geq -2$

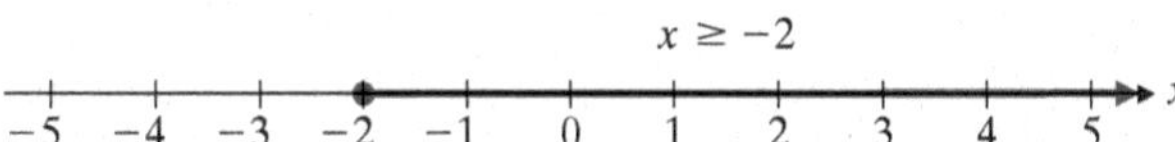

(c) We can state that "-3 is less than x" or, equivalently, that "x is greater than -3." Be sure you see that $-3 < x$ is equivalent to $x > -3$. Although both statements are correct, we *usually write the variable first* in a simple inequality containing a variable and a numerical value.

$x > -3$

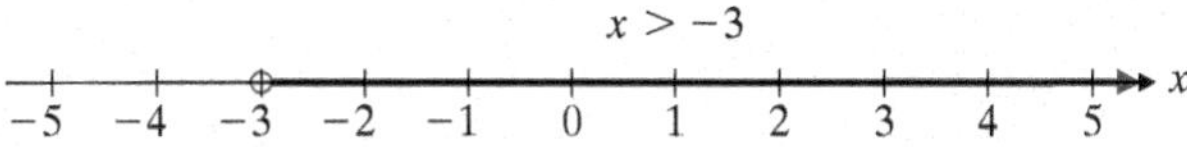

(d) We state that "x is less than or equal to $-\frac{1}{2}$."

$x \leq -\frac{1}{2}$

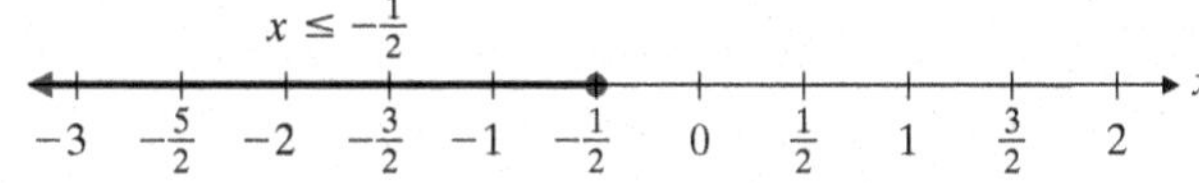

□

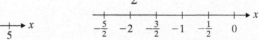

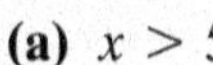 **Student Practice 2** State each mathematical relationship in words and then graph it on the given number line.

(a) $x > 5$

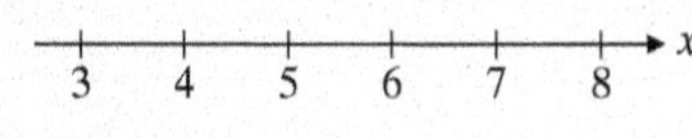

(b) $x \leq -2$

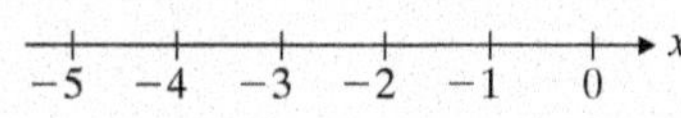

(c) $3 > x$

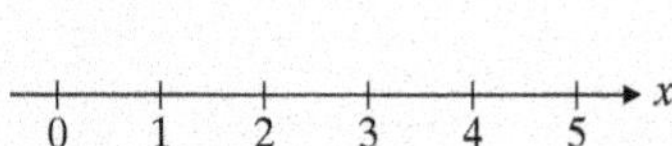

(d) $x \geq -\frac{3}{2}$

To Think About: *Comparing Results* What is the difference between the graphs in Example 2(a) and Case 2 above? Why are these graphs different?

3 Translating English Phrases into Algebraic Statements 

We can translate many everyday situations into algebraic statements with an unknown value and an inequality symbol. This is the first step in solving word problems using inequalities.

Example 3 Translate each English statement into an algebraic statement.

(a) The police on the scene said that the car was traveling more than 80 miles per hour. (Use the variable s for speed.)

(b) The owner of the trucking company said that the payload of a truck must never exceed 4500 pounds. (Use the variable p for payload.)

Solution

(a) Since the speed must be greater than 80, we have $s > 80$.

(b) If the payload of the truck can never exceed 4500 pounds, then the payload must be always less than or equal to 4500 pounds. Thus we write $p \leq 4500$. □

Student Practice 3 Translate each English statement into an algebraic statement.

(a) During the drying cycle, the temperature inside the clothes dryer must never exceed 180° Fahrenheit. (Use the variable t for temperature.)

(b) The bank loan officer said that the total consumer debt incurred by Wally and Mary must be less than $15,000 if they want to qualify for a mortgage to buy their first home. (Use the variable d for debt.)

4 Solving and Graphing an Inequality 

When we **solve an inequality,** we are finding *all* the values that make it true. To solve an inequality, we simplify it to the point where we can clearly see all possible values for the variable. We've solved equations by adding, subtracting, multiplying by, and dividing by a particular value on both sides of the equation. Here we perform similar operations with inequalities with one important exception. We'll show some examples so that you can see how these operations can be used with inequalities just as with equations.

We will first examine the pattern that occurs when we perform these operations *with a positive value* on both sides of an inequality.

Original Inequality	Operations with a Positive Number	New Inequality
$4 < 6$	Add 2 to both sides.	$6 < 8$
	Subtract 2 from both sides.	$2 < 4$
	Multiply both sides by 2.	$8 < 12$
	Divide both sides by 2.	$2 < 3$

Notice that the inequality symbol remains the same when these operations are performed with a positive value.

Now let us examine what happens when we perform these operations *with a negative value.*

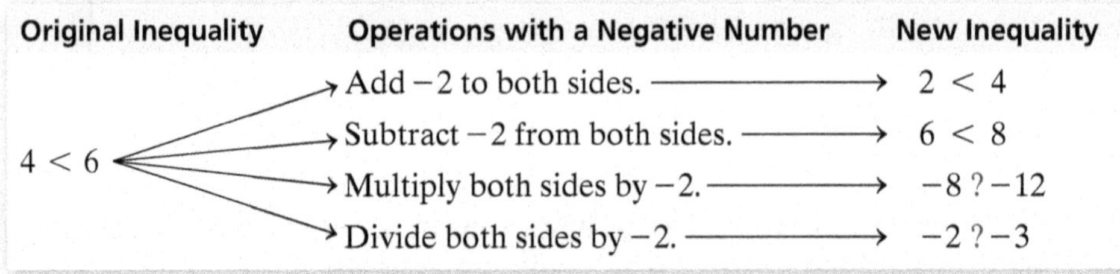

Original Inequality	Operations with a Negative Number	New Inequality
$4 < 6$	Add -2 to both sides.	$2 < 4$
	Subtract -2 from both sides.	$6 < 8$
	Multiply both sides by -2.	$-8 \; ? -12$
	Divide both sides by -2.	$-2 \; ? -3$

What happens to the inequality sign when we multiply both sides by a negative number? Since -8 is to the right of -12 on a number line, we know that the new inequality should be $-8 > -12$ if we want the statement to remain true. Notice how we reverse the direction of the inequality from $<$ (less than) to $>$ (greater than) when we multiply by a negative value. Thus we have the following.

$$4 < 6 \longrightarrow \text{Multiply both sides by } -2. \longrightarrow -8 > -12$$

The same thing happens when we divide by a negative number. The inequality is reversed from $<$ to $>$. We know this since -2 is to the right of -3 on a number line.

$$4 < 6 \longrightarrow \text{Divide both sides by } -2. \longrightarrow -2 > -3$$

Similar reversals take place in the next example.

Example 4 Perform the given operations and write the new inequalities.

Original Inequality		New Inequality
(a) $-2 < -1$	$\longrightarrow$ Multiply both sides by -3.	$\longrightarrow$ $6 > 3$
(b) $0 > -4$	$\longrightarrow$ Divide both sides by -2.	$\longrightarrow$ $0 < 2$
(c) $8 \geq 4$	$\longrightarrow$ Divide both sides by -4.	$\longrightarrow$ $-2 \leq -1$

Solution Notice that we perform the arithmetic with signed numbers just as we always do. But the new inequality signs are reversed (from those of the original inequalities).

Whenever both sides of an inequality are multiplied or divided by a negative quantity, the direction of the inequality is reversed. □

Student Practice 4 Perform the given operations and write the new inequalities.

(a) $7 > 2$	Multiply each side by -2.
(b) $-3 < -1$	Multiply each side by -1.
(c) $-10 \geq -20$	Divide each side by -10.
(d) $-15 \leq -5$	Divide each side by -5.

PROCEDURE FOR SOLVING INEQUALITIES

You may use the same procedures to solve inequalities that you used to solve equations *except* that the direction of an inequality is *reversed* if you *multiply* or *divide* both sides *by a negative number.*

It may be helpful to think over quickly what we have discussed here. The inequalities remain the same when we add a number to both sides or subtract a number from both sides of the inequality. The inequalities remain the same when we multiply both sides by a positive number or divide both sides by a positive number.

However, if we *multiply* both sides of an inequality by a *negative number* or if we *divide* both sides of an inequality by a *negative number*, then *the inequality is reversed.*

Example 5 Solve and graph. $3x + 7 \geq 13$

Solution

$$3x + 7 - 7 \geq 13 - 7 \quad \text{Subtract 7 from both sides.}$$

$$3x \geq 6 \quad \text{Simplify.}$$

$$\frac{3x}{3} \geq \frac{6}{3} \quad \text{Divide both sides by 3.}$$

$$x \geq 2$$

Simplify. Note that the direction of the inequality is not changed, since we have divided by a positive number.

The graph is as follows:

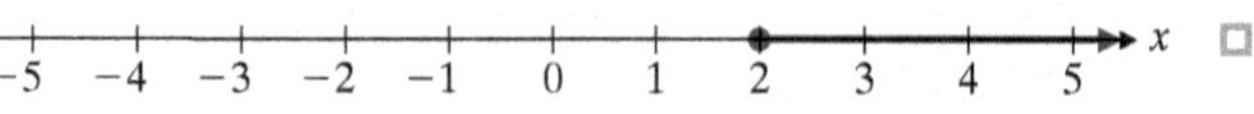

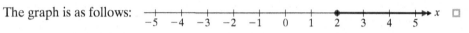

Student Practice 5 Solve and graph. $8x - 2 < 3$

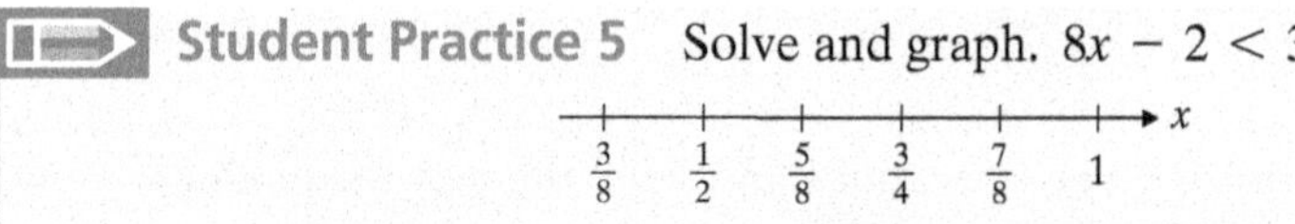

Example 6 Solve and graph. $5 - 3x > 7$

Solution

$$5 - 5 - 3x > 7 - 5 \quad \text{Subtract 5 from both sides.}$$

$$-3x > 2 \quad \text{Simplify.}$$

$$\frac{-3x}{-3} < \frac{2}{-3}$$

Divide by -3 and **reverse the inequality** since we are dividing by a negative number.

$$x < -\frac{2}{3} \quad \text{Note the direction of the inequality.}$$

The graph is as follows:

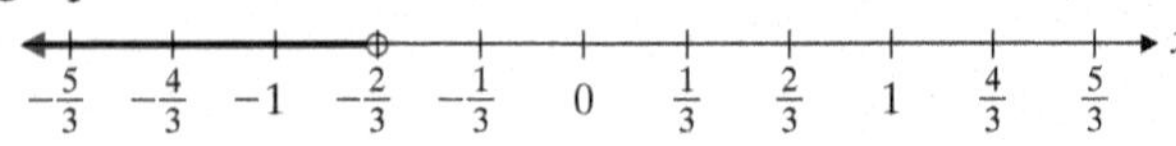

Student Practice 6 Solve and graph.

$$4 - 5x > 7$$

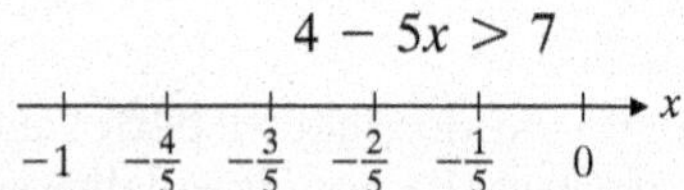

Just like equations, some inequalities contain parentheses and fractions. The initial steps to solve these inequalities will be the same as those used to solve equations with parentheses and fractions. When the variable appears on both sides of the inequality, it is advisable to collect the x-terms on the left side of the inequality symbol.

Example 7 Solve and graph. $-\frac{13x}{2} \leq \frac{x}{2} - \frac{15}{8}$

Solution

$$8\left(\frac{-13x}{2}\right) \leq 8\left(\frac{x}{2}\right) - 8\left(\frac{15}{8}\right)$$

Multiply all terms by LCD $= 8$. We do **not** reverse the direction of the inequality symbol since we are multiplying by a positive number.

$$-52x \leq 4x - 15 \quad \text{Simplify.}$$

$$-52x - 4x \leq 4x - 15 - 4x \quad \text{Subtract } 4x \text{ from both sides.}$$

$$-56x \leq -15 \quad \text{Combine like terms.}$$

$$\frac{-56x}{-56} \geq \frac{-15}{-56}$$

Divide both sides by -56. We **reverse** the direction of the inequality when we divide both sides by a negative number.

$$x \geq \frac{15}{56}$$

The graph is as follows:

Student Practice 7 Solve and graph.

$$\frac{1}{2}x + 3 < \frac{2}{3}x$$

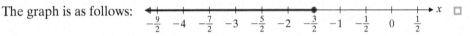

16 17 18 19 20 21 x

Example 8 Solve and graph. $\frac{1}{3}(3 - 2x) \le -4(x + 1)$

Solution $\frac{1}{3}(3 - 2x) \le -4(x + 1)$

$1 - \frac{2x}{3} \le -4x - 4$ Remove parentheses. $\frac{3}{3} = 1$.

$3(1) - 3\left(\frac{2x}{3}\right) \le 3(-4x) - 3(4)$ Multiply all terms by LCD $= 3$.

$3 - 2x \le -12x - 12$ Simplify.

$3 - 2x + 12x \le -12x + 12x - 12$ Add $12x$ to both sides.

$3 + 10x \le -12$ Combine like terms.

$3 - 3 + 10x \le -12 - 3$ Subtract 3 from both sides.

$10x \le -15$ Simplify.

$\frac{10x}{10} \le \frac{-15}{10}$ Divide both sides by 10. Since we are dividing by a **positive** number, the inequality is **not** reversed.

$x \le -\frac{3}{2}$

The graph is as follows:

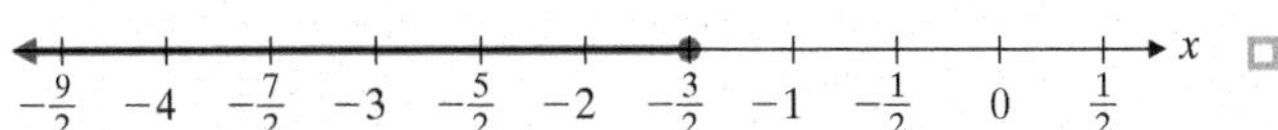

Student Practice 8 Solve and graph.

$$\frac{1}{2}(3 - x) \le 2x + 5$$

−2 $-\frac{9}{5}$ $-\frac{8}{5}$ $-\frac{7}{5}$ $-\frac{6}{5}$ −1 x

CAUTION: The most common error students make when solving inequalities is forgetting to reverse the direction of the inequality symbol when multiplying or dividing both sides of the inequality by a negative number.

Normally when you solve inequalities, you solve for x by putting the variables on the left side. If you solve by placing the variables on the right side, you will end up with statements like $3 > x$. This is equivalent to $x < 3$. It is wise to express your answer with the variables on the left side.

Example 9 A hospital director has determined that the costs of operating one floor of the hospital for an eight-hour shift must never exceed \$3000. An expression for the cost of operating one floor of the hospital is $200n + 1200$, where n is the number of nurses. This expression is based on an estimate of \$1200 in fixed costs and a cost of \$200 per nurse for an eight-hour shift. Solve the inequality $200n + 1200 \le 3000$ to determine the number of nurses that may be on duty on this floor during an eight-hour shift if the director's cost control measure is followed.

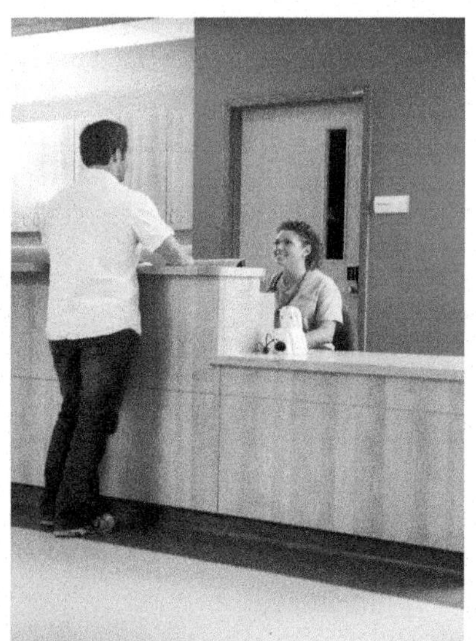

Solution

$200n + 1200 \le 3000$ The inequality we must solve.

$200n + 1200 - 1200 \le 3000 - 1200$ Subtract 1200 from each side.

$200n \le 1800$ Simplify.

$\frac{200n}{200} \le \frac{1800}{200}$ Divide each side by 200.

$n \le 9$

The number of nurses on duty on this floor during an eight-hour shift must always be less than or equal to nine.

Student Practice 9 The company president of Staywell, Inc., wants the monthly profits never to be less than \$2,500,000. He has determined that an expression for monthly profit is $2000n - 700,000$. In the expression, n is the number of exercise machines manufactured each month. The profit on each machine is \$2000, and the $-\$700,000$ in the expression represents the fixed costs of running the company.

Solve the inequality $2000n - 700,000 \ge 2,500,000$ to find how many machines must be made and sold each month to satisfy these financial goals.

STEPS TO SUCCESS Why Do We Need to Study Mathematics?

Students often question the value of studying math. They see little real use for it in their everyday lives. Knowing math can help in various ways.

Get a good job. Mathematics is often the key that opens the door to a better-paying job or just to get a job if you are unemployed. In our technological world, people use mathematics daily. Many vocational and professional areas—such as business, statistics, economics, psychology, finance, computer science, chemistry, physics, engineering, electronics, nuclear energy, banking, quality control, nursing, medical technician, and teaching—require a certain level of expertise in mathematics. Those who want to work in these fields must be able to function at a given mathematical level. Those who cannot will not be able to enter these job areas.

Save money. These are challenging financial times. We are all looking for ways to save money. The more mathematics you learn, the more you will be able to find ways to save money. Several suggestions for saving money are given in this book. Be sure to read over each one and think how it might apply to your life.

Make decisions. Should I buy a car or lease one? Should I buy a house or rent? Should I drive to work or take public transportation? What career field should I pick if I want to increase my chances of getting a good job? Mathematics will help you to think more clearly and make better decisions.

Making it personal: Which of these three paragraphs is most relevant to you? Write down what you think is the most important reason why you should study mathematics.

2.8 Exercises MyMathLab®

Verbal and Writing Skills, Exercises 1 and 2

1. Is the statement $5 > -6$ equivalent to the statement $-6 < 5$? Why?

2. Is the statement $-8 < -3$ equivalent to the statement $-3 > -8$? Why?

Replace the ? by < or >.

3. $8 \ ? \ -6$

4. $-10 \ ? \ 6$

5. $0 \ ? \ -8$

6. $-8 \ ? \ 0$

7. $-4 \ ? \ -2$

8. $-5 ? -8$

9. **(a)** $-7 \ ? \ 2$
(b) $2 \ ? \ -7$

10. **(a)** $-5 \ ? \ 11$
(b) $11 \ ? \ -5$

11. **(a)** $15 \ ? \ -15$
(b) $-15 \ ? \ 15$

12. **(a)** $-17 \ ? \ 17$
(b) $17 \ ? \ -17$

13. $\frac{1}{3} \ ? \ \frac{9}{10}$

14. $\frac{4}{6} \ ? \ \frac{7}{9}$

15. $\frac{7}{8} \ ? \ \frac{25}{31}$

16. $\frac{9}{11} \ ? \ \frac{41}{53}$

17. $-6.6 \ ? \ -8.9$

18. $-4.2 \ ? \ -7.3$

19. $-4.2 \ ? \ 3.5$

20. $-3.7 \ ? \ 3.7$

21. $-\frac{10}{3} \ ? \ -3$

22. $-5 \ ? \ -\frac{29}{4}$

23. $-\frac{5}{8} \ ? \ -\frac{3}{5}$

24. $-\frac{2}{3} \ ? \ -\frac{1}{2}$

Graph each inequality on the number line.

25. $x > 7$

−1 0 1 2 3 4 5 6 7 8 9 x

26. $x < 1$

−5 −4 −3 −2 −1 0 1 2 3 4 5 x

27. $x \geq -6$

−9 −8 −7 −6 −5 −4 −3 −2 −1 0 1 x

28. $x \leq -2$

−5 −4 −3 −2 −1 0 1 2 3 4 5 x

29. $x < -\frac{1}{4}$

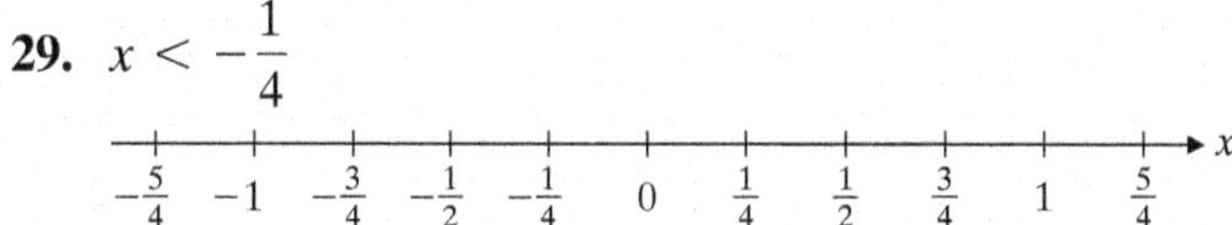

30. $x \leq -\frac{3}{2}$

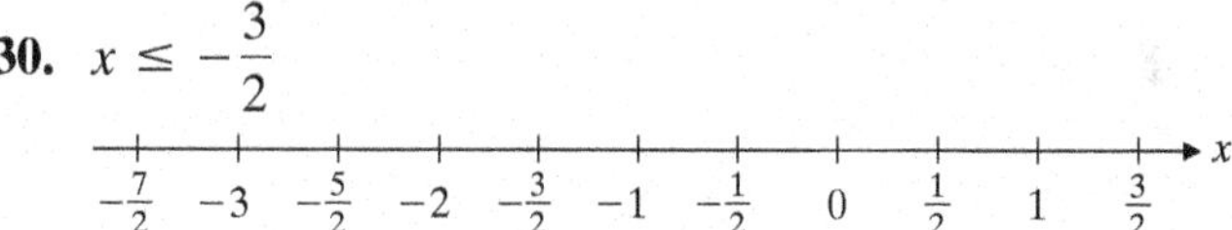

31. $x \leq -5.3$

−5.8 −5.7 −5.6 −5.5 −5.4 −5.3 −5.2 −5.1 −5.0 −4.9 −4.8 x

32. $x > -3.5$

−4.1 −4.0 −3.9 −3.8 −3.7 −3.6 −3.5 −3.4 −3.3 −3.2 −3.1 x

33. $25 < x$

−10 −5 0 5 10 15 20 25 30 35 40 x

34. $35 \geq x$

20 25 30 35 40 45 50 55 60 65 70 x

Translate each graph to an inequality using the variable x.

35.

$-\frac{5}{3}$ $-\frac{4}{3}$ −1 $-\frac{2}{3}$ $-\frac{1}{3}$ 0 $\frac{1}{3}$ $\frac{2}{3}$ 1 $\frac{4}{3}$ $\frac{5}{3}$ x

36.

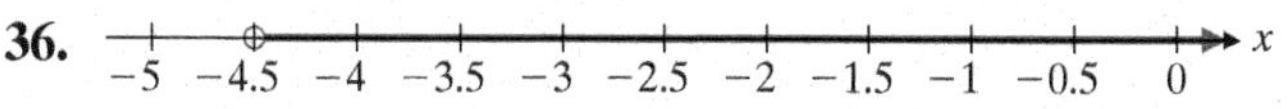

37.

−25 −24 −23 −22 −21 −20 −19 −18 −17 −16 −15 x

38.

$-\frac{3}{2}$ −1 $-\frac{1}{2}$ 0 $\frac{1}{2}$ 1 $\frac{3}{2}$ 2 $\frac{5}{2}$ 3 $\frac{7}{2}$ x

39. 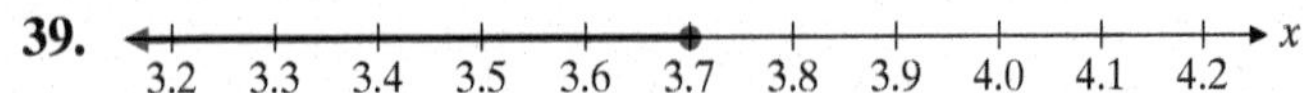

40. 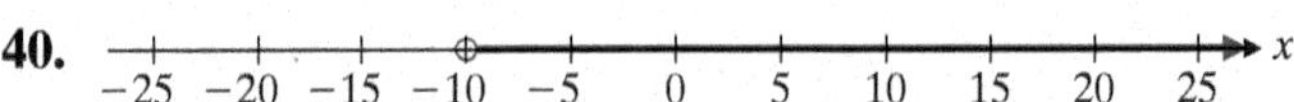

Translate each English statement into an inequality.

41. ***Full-Time Student*** At Normandale Community College the number of credits a student takes per semester must not be less than 12 to be considered full time. (Use the variable c for credits.)

42. ***BMI Category*** A person is considered underweight if his or her BMI (body mass index) measurement is smaller than 18.5. (Use the variable B for BMI.)

43. ***Height Limit*** In order to ride the roller coaster at the theme park, your height must be at least 48 inches. (Use h for height.)

44. ***Boxing Category*** To box in the featherweight category, your weight must not exceed 126 pounds. (Use w for weight.)

To Think About, Exercises 45 and 46

45. Suppose that the variable x must satisfy *all* of these conditions.

$$x \leq 2, \quad x > -3, \quad x < \frac{5}{2}, \quad x \geq -\frac{5}{2}$$

Graph on a number line the region that satisfies all of the conditions.

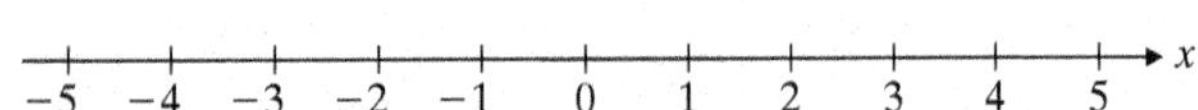

46. Suppose that the variable x must satisfy *all* of these conditions.

$$x < 4, \quad x > -4, \quad x \leq \frac{7}{2}, \quad x \geq -\frac{9}{2}$$

Graph on a number line the region that satisfies all of the conditions.

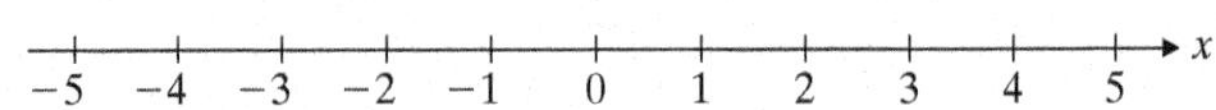

Solve and graph the result.

47. $x + 7 \leq 4$

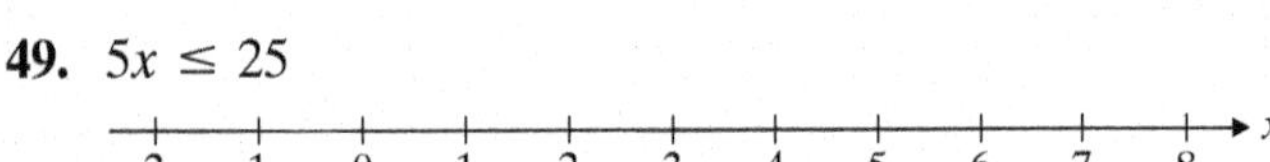

48. $x - 5 < -3$

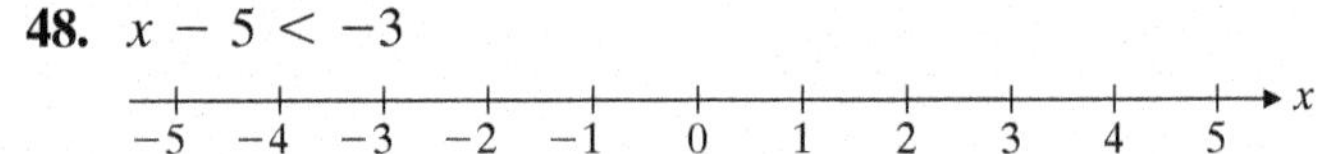

49. $5x \leq 25$

50. $6x \geq -42$

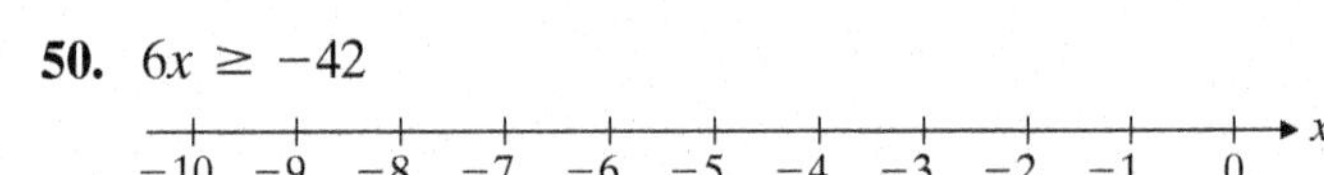

51. $-2x < 18$

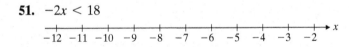

52. $-7x < 28$

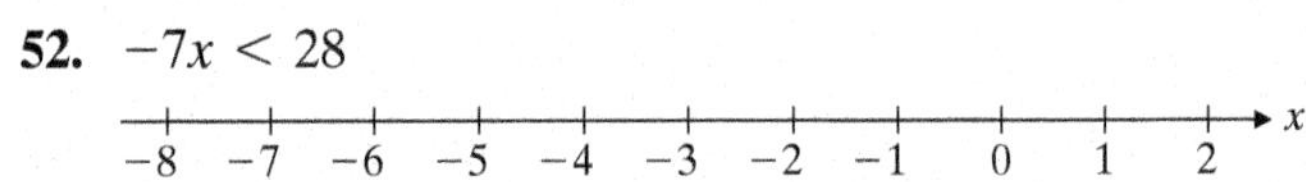

53. $\frac{1}{2}x \geq 4$

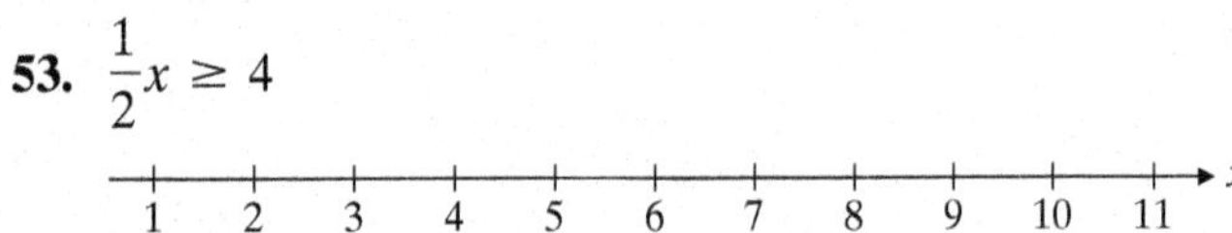

54. $\frac{1}{3}x \leq 2$

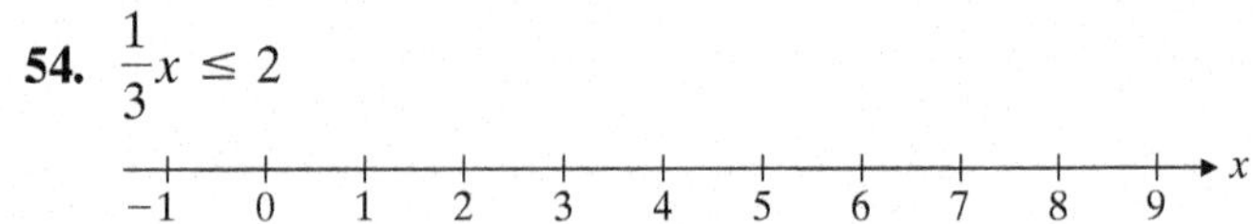

55. $-\frac{1}{4}x > 3$

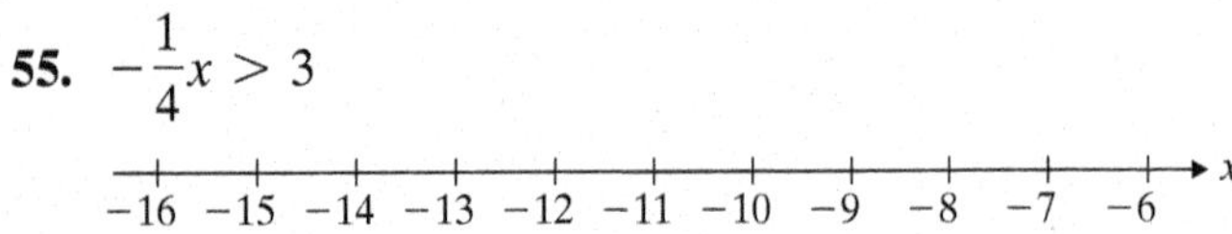

56. $-\frac{1}{5}x < 10$

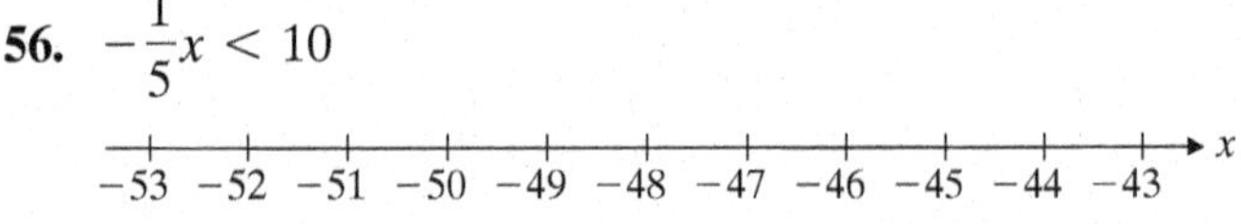

57. $8 - 5x > 13$

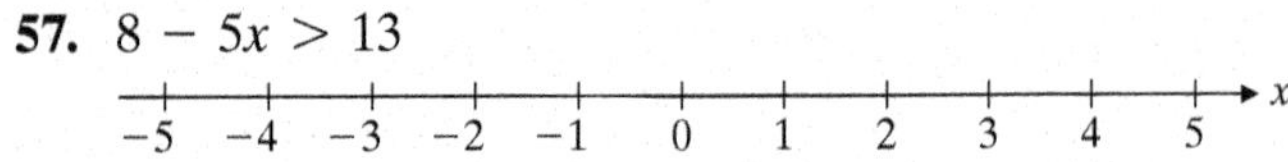

58. $9 - 4x \leq 21$

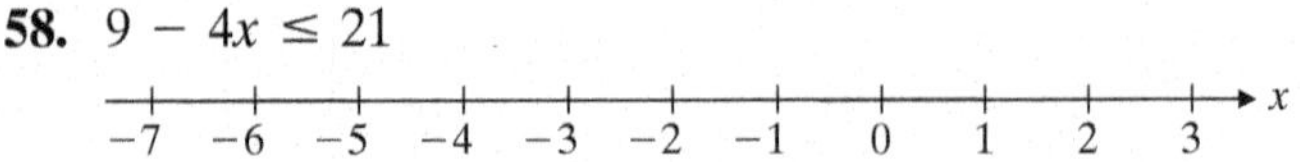

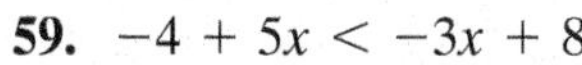

59. $-4 + 5x < -3x + 8$

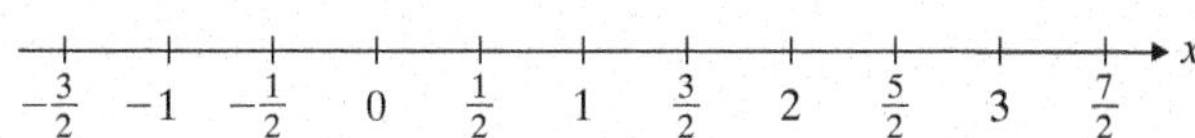

60. $-6 - 4x < 1 - 6x$

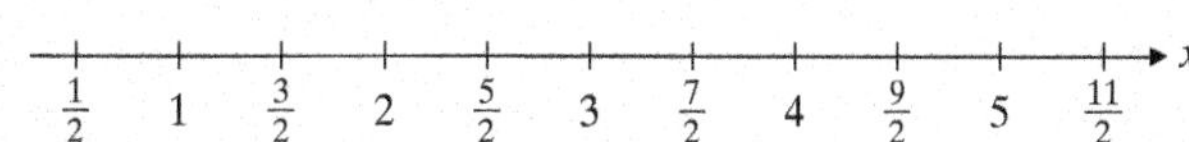

61. $\frac{5x}{6} - 5 > \frac{x}{6} - 9$

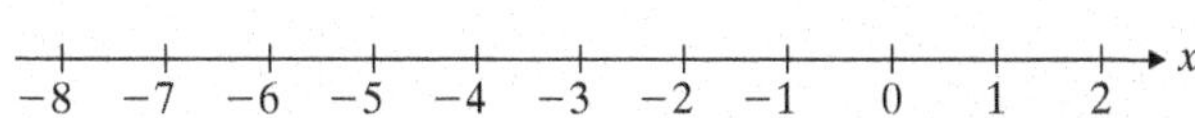

62. $\frac{x}{4} - 2 < \frac{3x}{4} + 5$

-21 -20 -19 -18 -17 -16 -15 -14 -13 -12 -11 x

63. $2(3x + 4) > 3(x + 3)$

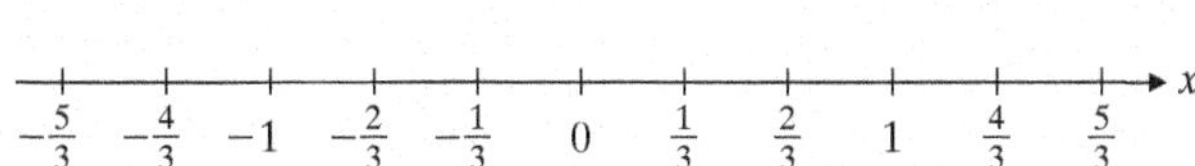

64. $5(x - 3) \leq 2(x - 3)$

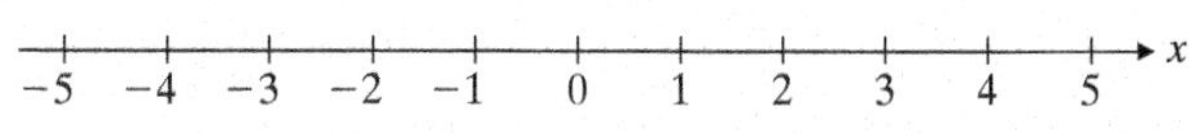

Verbal and Writing Skills *In exercises 65 and 66, answer the questions in your own words.*

65. Add -2 to both sides of the inequality $5 > 3$. What is the result? Why is the direction of the inequality not reversed?

66. Divide -3 into both sides of the inequality $-21 > -29$. What is the result? Why is the direction of the inequality reversed?

Mixed Practice *Solve. Collect the variable terms on the left side of the inequality.*

67. $5x + 2 > 8x - 7$

68. $7x + 8 < 12x - 2$

69. $6x - 2 \geq 4x + 6$

70. $9x - 8 \leq 7x + 4$

71. $0.3(x - 1) < 0.1x - 0.5$

72. $0.4(2 - x) + 0.6 > 0.2(x - 2)$

73. $3 + 5(2 - x) \geq -3(x + 5)$

74. $9 - 3(2x - 1) \leq 4(x + 2)$

75. $\frac{x + 6}{7} - \frac{3}{7} > \frac{x + 3}{2}$

76. $\frac{3x + 5}{4} - \frac{7}{12} > -\frac{x}{6}$

Applications

77. ***Course Average*** To pass a course with a B grade, a student must have an average of 80 or greater. A student's grades on three tests are 75, 83, and 86. Solve the inequality $\frac{75 + 83 + 86 + x}{4} \geq 80$ to find what score the student must get on the next test to get a B average or better.

78. ***Payment Options*** Sharon sells very expensive European sports cars. She may choose to receive \$10,000.00 per month or 8% of her sales as payment for her work. Solve the inequality $0.08x > 10{,}000$ to find how much she needs to sell to make the 8% offer a better deal.

79. ***Elephant Weight*** The average African elephant weighs 268 pounds at birth. During its first three weeks, a baby elephant will usually gain about 4 pounds per day. Assuming that growth rate, solve the inequality $268 + 4x \geq 300$ to find how many days it will be until a baby elephant weighs at least 300 pounds.

80. ***Car Loan*** Rennie is buying a used car that costs \$4500. The deal called for a \$600 down payment, and payments of \$260 monthly. He wants to know whether he can pay off the car within a year. Solve the inequality $600 + 260x \geq 4500$ to find out the minimum number of months it will take to pay off the car.

Cumulative Review

81. **[0.5.3]** Find 16% of 38.

82. **[0.5.4]** 18 is what percent of 120?

83. **[0.5.4]** ***Percent Accepted*** For the most coveted graduate study positions, only 16 out of 800 students are accepted. What percent are accepted?

84. **[0.5.4]** Write the fraction $\frac{3}{8}$ as a percent.

Quick Quiz 2.8

1. Graph $x \leq -3.5$ on the given number line.

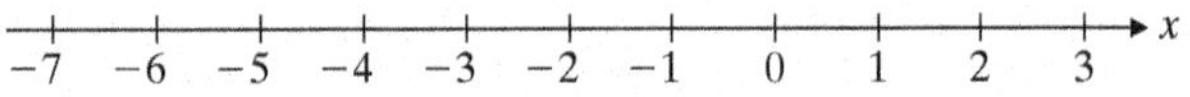

−7 −6 −5 −4 −3 −2 −1 0 1 2 3
x

Solve and graph the result for each of the following.

2. $-12 + 4x \leq 2x$

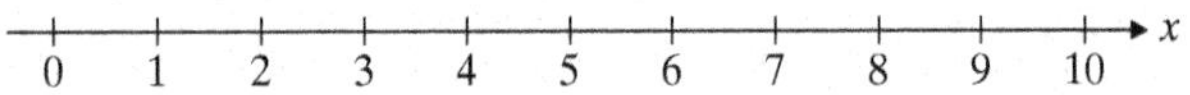

0 1 2 3 4 5 6 7 8 9 10
x

3. $\frac{x}{2} - 1 < \frac{3}{2}x + 4$

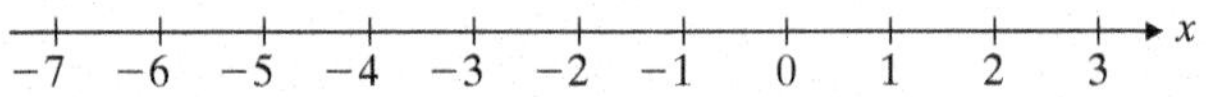

−7 −6 −5 −4 −3 −2 −1 0 1 2 3
x

4. **Concept Check** Explain the difference between $12 < x$ and $x > 12$. Would the graphs of these inequalities be the same or different?

CAREER EXPLORATION PROBLEMS

Background: Registered Dietitian

Jules has decided to pursue a career as a registered dietitian. She finds the prospect of educating and helping people to care for themselves by promoting healthy, nutritious eating habits to be exciting.

Part of her education and training involves working at a local hospital's walk-in clinic under the guidance of a registered dietician. Jules often consults with overweight and diabetic patients about their treatment plans and diets. It's common for her to use equations to identify a number of measurements, such as patients' energy requirements, resting metabolic rates, and ideal body weights. She finds herself using equations often because her evaluations of even a single patient are an ongoing process of monitoring his or her condition or weight.

Facts

- The Mifflin-St. Jeor equation is used to calculate the amount of energy (in calories) that a body needs to function while at rest for 24 hours. This amount of energy is called basic metabolic rate (BMR), and for men, the equation takes the form:

$$\text{BMR} = 10W + 6.25H - 5A + 5,$$

where W is weight in kilograms, H is height in centimeters, and A is age in years.

- Another commonly used equation is one for finding ideal body weight (IBW). For males, this equation is:

$$\text{IBW} = 106 \text{ lb} + 6(\text{the number of inches over 5 feet tall}).$$

For females, the equation is:

$$\text{IBW} = 100 \text{ lb} + 5(\text{the number of inches over 5 feet tall}).$$

Tasks

1. Jules wants to solve the Mifflin-St. Jeor equation for both weight (W) and height (H). What are her solutions?

2. Jules wants to determine the height of a male whose ideal weight is 190 pounds. What should she do to determine this?

3. She now wants to identify her own ideal weight, her IBW. She is 5 feet 5 inches tall. What is her ideal weight?

STEPS TO SUCCESS Brainstorming with Yourself to Find Greater Success When You Take a Test

Do you sometimes get a test back and feel that you could have done a lot better? Most of us have had that happen. Here are some great ideas to improve your test performance.

When you get a graded test back, take some time within the next four hours to look it over carefully and then quickly write some suggestions to yourself. Do some brainstorming. What do you think you could have done to improve your test score?

1. Did you study too much the night before the test and not adequately prepare as you were going through the chapter?
2. Did you neglect to do some of the homework assignments?
3. Did you fail to get help with some of the problems that you did not understand or could not do?
4. Did you neglect to review the Chapter Organizer or work the Chapter Review problems in the textbook?

Making it personal: Think about how you did on the last test you took in this course. Which of the four questions listed here is the most relevant to the way you studied? What can you do to make sure your performance is better on the next test?

Chapter 2 Organizer

Topic and Procedure | **Examples** | **You Try It**

Solving equations without parentheses or fractions, p. 143

1. On both sides of the equation, combine like terms if possible.
2. Add or subtract terms on both sides of the equation in order to get all terms with the variable on one side of the equation.
3. Add or subtract a value on both sides of the equation to get all terms not containing the variable on the other side of the equation.
4. Divide both sides of the equation by the coefficient of the variable.
5. If possible, simplify the solution.
6. Check your solution by substituting the obtained value into the original equation.

Examples

Solve for x.

$$\begin{aligned} 5x + 2 + 2x &= -10 + 4x + 3 \\ 7x + 2 &= -7 + 4x \\ 7x - 4x + 2 &= -7 + 4x - 4x \\ 3x + 2 &= -7 \\ 3x + 2 - 2 &= -7 - 2 \\ 3x &= -9 \\ \frac{3x}{3} &= \frac{-9}{3} \\ x &= -3 \end{aligned}$$

Check:

Is -3 the solution of $5x + 2 + 2x = -10 + 4x + 3$?

$$\begin{aligned} 5(-3) + 2 + 2(-3) &\stackrel{?}{=} -10 + 4(-3) + 3 \\ -15 + 2 + (-6) &\stackrel{?}{=} -10 + (-12) + 3 \\ -13 - 6 &\stackrel{?}{=} -22 + 3 \\ -19 &= -19 \checkmark \end{aligned}$$

You Try It

1. Solve for x.
$-8x - 1 + x = 13 - 6x - 2$

Solving equations with parentheses and/or fractions, pp. 145, 150

1. Remove any parentheses.
2. Simplify, if possible.
3. If fractions exist, multiply all terms on both sides by the least common denominator of all the fractions.
4. Now follow the steps for solving an equation without parentheses or fractions.

Examples

Solve for y.

$$\begin{aligned} 5(3y - 4) &= \frac{1}{4}(6y + 4) - 48 \\ 15y - 20 &= \frac{3}{2}y + 1 - 48 \\ 15y - 20 &= \frac{3}{2}y - 47 \\ 2(15y) - 2(20) &= 2\left(\frac{3}{2}y\right) - 2(47) \\ 30y - 40 &= 3y - 94 \\ 30y - 3y - 40 &= 3y - 3y - 94 \\ 27y - 40 &= -94 \\ 27y - 40 + 40 &= -94 + 40 \\ 27y &= -54 \\ \frac{27y}{27} &= \frac{-54}{27} \\ y &= -2 \end{aligned}$$

You Try It

2. Solve for y.
$\frac{1}{3}(y + 5) = \frac{1}{4}(5y - 8)$

Solving number problems, p. 164

1. Use a variable to describe the quantity that the other quantities are being compared to.
2. Write an expression in terms of that variable for each of the quantities.

Examples

One number is one-third of a second number. The sum of the two numbers is twelve. Find each number. The problem refers to two numbers, so we must write an expression for both numbers. Let x represent the second number since the first number is described in terms of the second.

$$\begin{aligned} x &= \text{the second number} \\ \frac{1}{3}x &= \text{the first number} \end{aligned}$$

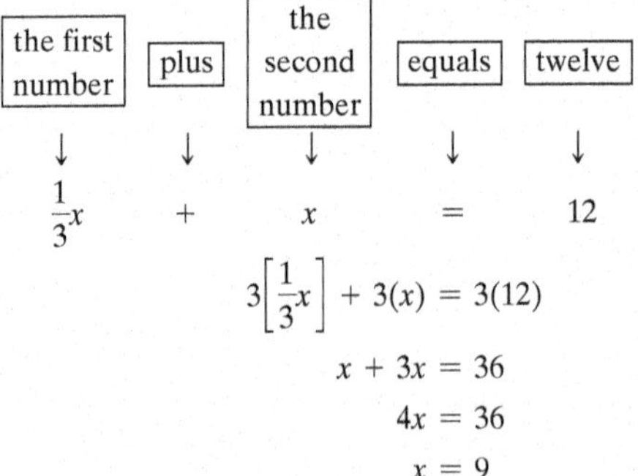

$$\begin{aligned} 3\left[\frac{1}{3}x\right] + 3(x) &= 3(12) \\ x + 3x &= 36 \\ 4x &= 36 \\ x &= 9 \end{aligned}$$

Since x = *the second number*, then 9 is the value of the second number. Now $\frac{1}{3}x$ = *the first number*, so we have $\frac{1}{3}(9) = 3$, the value of the first number.

You Try It

3. The smaller of two numbers is one-fifth the larger. The sum of the two numbers is sixty. Find each number.

Topic and Procedure | **Examples** | **You Try It**

Money and percents, pp. 175–178

Let $x =$ one amount and the total $- x =$ the other amount. Find the percent of each amount.

Gina saved some money for college. She invested \$2400 for one year and earned \$225 in simple interest. She invested part of it at 12% and the rest of it at 9%. How much did she invest at each rate?

1. ***Understand the problem.***
   We want to find each amount that was invested.
   interest earned at 12% + interest earned at 9% = total interest of \$225

   We let x represent one quantity of money.
   Let $x =$ amount of money invested at 12%.

   We started with \$2400. If we invest x at 12%, we still have $(2400 - x)$ left.
   Then $2400 - x =$ the amount of money invested at 9%.

   Interest $= prt$

   Interest at 12% $\quad I_1 = 0.12x$

   Interest at 9% $\quad I_2 = 0.09(2400 - x)$

2. ***Write an equation.***
   $I_1 + I_2 = 225.$ $\quad 0.12x + 0.09(2400 - x) = 225$

3. ***Solve.***

$$0.12x + 216 - 0.09x = 225$$
$$0.03x + 216 = 225$$
$$0.03x = 9$$
$$\frac{0.03x}{0.03} = \frac{9}{0.03}$$
$$x = 300$$

\$300 was invested at 12%.
$2400 - x = 2400 - 300 = 2100$
\$2100 was invested at 9%.

4. ***Check.***
   Is this reasonable? Yes.
   Are the conditions of the problem satisfied?
   Does the total amount invested equal \$2400?
   Will \$2100 at 9% and \$300 at 12% yield \$225 in interest?

$$2100 + 300 \stackrel{?}{=} 2400$$
$$2400 = 2400 \quad ✓$$
$$0.09(2100) + 0.12(300) \stackrel{?}{=} 225$$
$$189 + 36 \stackrel{?}{=} 225$$
$$225 = 225 \quad ✓$$

4. Caleb saved some money from his summer job. He invested \$3600 for one year and earned \$132 in simple interest. He invested part of it at 2% and the rest of it at 4%. How much did he invest at each rate?

Solving inequalities, p. 187

1. Follow the steps for solving an equation up to the multiplication or division step.
2. If you divide or multiply both sides of the inequality by a *positive number,* the **direction** of the inequality **is not reversed.**
3. If you divide or multiply both sides of the inequality by a *negative number,* the **direction** of the inequality **is reversed.**

Solve for x and graph your solution.

$$\frac{1}{2}(3x - 2) \leq -5 + 5x - 3$$ First remove parentheses and simplify.

$$\frac{3}{2}x - 1 \leq -8 + 5x$$

$$2\left(\frac{3}{2}x\right) - 2(1) \leq 2(-8) + 2(5x)$$ Now multiply each term by 2.

$$3x - 2 \leq -16 + 10x$$
$$3x - 10x - 2 \leq -16 + 10x - 10x$$
$$-7x - 2 \leq -16$$
$$-7x - 2 + 2 \leq -16 + 2$$
$$-7x \leq -14$$

$$\frac{-7x}{-7} \geq \frac{-14}{-7}$$ When we divide both sides by a negative number, the inequality is reversed.

$$x \geq 2$$

Graph of the solution:

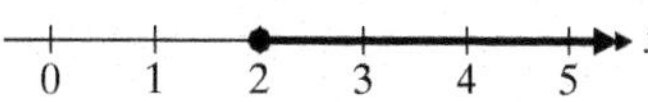

5. Solve for x and graph your solution.
   $4 + 3x - 5 \geq \frac{1}{3}(10x + 1)$

Chapter 2 Review Problems

Sections 2.1–2.3

Solve for the variable.

1. $3x + 2x = -35$

2. $x - 19 = -29 + 7$

3. $18 - 10x = 63 + 5x$

4. $x - (0.5x + 2.6) = 17.6$

5. $3(x - 2) = -4(5 + x)$

6. $12 - 5x = -7x - 2$

7. $2(3 - x) = 1 - (x - 2)$

8. $4(x + 5) - 7 = 2(x + 3)$

9. $3 = 2x + 5 - 3(x - 1)$

10. $2(5x - 1) - 7 = 3(x - 1) + 5 - 4x$

Section 2.4

Solve for the variable.

11. $\frac{3}{4}x - 3 = \frac{1}{2}x + 2$

12. $1 = \frac{5x}{6} + \frac{2x}{3}$

13. $\frac{7x}{5} = 5 + \frac{2x}{5}$

14. $\frac{7x - 3}{2} - 4 = \frac{5x + 1}{3}$

15. $\frac{3x - 2}{2} + \frac{x}{4} = 2 + x$

16. $\frac{-3}{2}(x + 5) = 1 - x$

17. $-0.2(x + 1) = 0.3(x + 11)$

18. $1.2x - 0.8 = 0.8x + 0.4$

19. $3.2 - 0.6x = 0.4(x - 2)$

20. $\frac{1}{3}(x - 2) = \frac{x}{4} + 2$

21. $\frac{3}{4} - \frac{2}{3}x = \frac{1}{3}x + \frac{3}{4}$

22. $-\frac{8}{3}x - 8 + 2x - 5 = -\frac{5}{3}$

23. $\frac{1}{6} + \frac{1}{3}(x - 3) = \frac{1}{2}(x + 9)$

24. $\frac{1}{7}(x + 5) - \frac{3}{7} = \frac{1}{2}(x + 3)$

Section 2.5

Write an algebraic expression. Use the variable x to represent the unknown value.

25. 19 more than a number

26. two-thirds of a number

27. half a number

28. 18 less than a number

29. triple the sum of a number and 4

30. twice a number decreased by three

Write an expression for each of the quantities compared. Use the letter specified.

31. ***Workforce*** The number of working people is four times the number of retired people. The number of unemployed people is one-half the number of retired people. (Use the letter r.)

▲ **32.** ***Geometry*** The length of a rectangle is 5 meters more than triple the width. (Use the letter w.)

▲ **33.** ***Geometry*** A triangle has three angles A, B, and C. The number of degrees in angle A is double the number of degrees in angle B. The number of degrees in angle C is 17 degrees less than the number of degrees in angle B. (Use the letter b.)

34. ***Class Size*** There are 29 more students in biology class than in algebra. There are one-half as many students in geology class as in algebra. (Use the letter a.)

Section 2.6

Solve.

35. Fourteen is taken away from triple a number. The result is -5. What is the number?

36. When twice a number is reduced by seven, the result is -21. What is the number?

37. ***Age*** Jon is twice as old as his cousin David. If Jon is 32, how old is David?

38. ***Biology Score*** Zach took four biology tests and got scores of 83, 86, 91, and 77. He needs to complete one more test. If he wants to get an average of 85 on all five tests, what score does he need on his last test?

39. ***Speed Rates*** Two cars left San Francisco and drove to San Diego, 800 miles away. One car was driven at 60 miles an hour. The other car was driven at 65 miles an hour. How long did it take each car to make the trip? Round your answers to the nearest tenth.

▲ **40.** ***Geometry*** The measure of the second angle of a triangle is three times the measure of the first angle. The measure of the third angle is 12 degrees less than twice the measure of the first. Find the measure of each of the three angles.

Section 2.7

41. ***Electric Bill*** The electric bill at Jane's house this month was $71.50. The charge is based on a flat rate of $25 per month plus a charge of $0.15 per kilowatt-hour of electricity used. How many kilowatt-hours of electricity were used?

42. ***Car Rental*** Abel rented a car from Sunshine Car Rentals. He was charged $39 per day and $0.25 per mile. He rented the car for three days and paid a rental cost of $187. How many miles did he drive the car?

43. ***Simple Interest*** Boxell Associates has $7400 invested at 5.5% simple interest. After money was withdrawn, $242 was earned for one year on the remaining funds. How much was withdrawn?

44. ***CD Player*** Jamie bought a new CD player for her car. When she bought it, she found that the price had been decreased by 18%. She was able to buy the CD player for $36 less than the original price. What was the original price?

45. ***Investments*** Peter and Shelly invested $9000 for one year. Part of it was invested at 12% and the remainder was invested at 8%. At the end of one year the couple had earned exactly $1000 in simple interest. How much did they invest at each rate?

46. ***Investments*** A man invested $5000 in a savings bank. He placed part of it in a checking account that earns 4.5% and the rest in a regular savings account that earns 6%. His total annual income from this investment in simple interest was $270. How much was invested in each account?

47. ***Coin Change*** Mary has \$3.75 in nickels, dimes, and quarters. She has three more quarters than dimes. She has twice as many nickels as quarters. How many of each coin does she have?

48. ***Tip Jar*** After the morning rush at the neighborhood coffee shop, the tip jar contained \$9.80 in coins. There were two more quarters than nickels. There were three fewer dimes than nickels. How many coins of each type were there?

Section 2.8

Solve each inequality and graph the result.

49. $9 + 2x \leq 6 - x$

−5 −4 −3 −2 −1 0 1 2 3 4 5 x

50. $2x - 3 + x > 5(x + 1)$

−7 −6 −5 −4 −3 −2 −1 0 1 2 3 x

51. $-x + 4 < 3x + 16$

−5 −4 −3 −2 −1 0 1 2 3 4 5 x

52. $8 - \frac{1}{3}x \leq x$

−1 0 1 2 3 4 5 6 7 8 9 x

53. $7 - \frac{3}{5}x > 4$

−3 −2 −1 0 1 2 3 4 5 6 7 x

54. $-4x - 14 < 4 - 2(3x - 1)$

7 8 9 10 11 12 13 14 15 16 17 x

55. $3(x - 2) + 8 < 7x + 14$

−5 −4 −3 −2 −1 0 1 2 3 4 5 x

Use an inequality to solve.

56. ***Wages*** Julian earns \$15 per hour as a plasterer's assistant. His employer determines that the current job allows him to pay \$480 in wages to Julian. What are the maximum number of hours that Julian can work on this job? (*Hint:* Use $15h \leq 480$.)

57. ***Hiring a Substitute*** The cost of hiring a substitute elementary teacher for a day is \$110. The school's budget for substitute teachers is \$2420 per month. What is the maximum number of times a substitute teacher may be hired during a month? (*Hint:* Use $110n \leq 2420$.)

Mixed Practice

Solve for the variable.

58. $10(2x + 4) - 13 = 8(x + 7) - 3$

59. $-9x + 15 - 2x = 4 - 3x$

60. $-2(x - 3) = -4x + 3(3x + 2)$

61. $\frac{1}{2} + \frac{5}{4}x = \frac{2}{5}x - \frac{1}{10} + 4$

Solve each inequality and graph the result.

62. $5 - \frac{1}{2}x > 4$

−5 −4 −3 −2 −1 0 1 2 3 4 5 x

63. $2(x - 1) \geq 3(2 + x)$

−10 −9 −8 −7 −6 −5 −4 −3 −2 −1 0 x

64. $\frac{1}{3}(x + 2) \leq \frac{1}{2}(3x - 5)$

$\frac{12}{7}$ $\frac{13}{7}$ 2 $\frac{15}{7}$ $\frac{16}{7}$ $\frac{17}{7}$ $\frac{18}{7}$ $\frac{19}{7}$ $\frac{20}{7}$ 3 $\frac{22}{7}$ x

65. $4(2 - x) - (-5x + 1) \geq -8$

−18 −17 −16 −15 −14 −13 −12 −11 −10 −9 −8 x

How Am I Doing? Chapter 2 Test

MATH COACH MyMathLab® YouTube

After you take this test read through the Math Coach on pages 201–202. Math Coach videos are available via MyMathLab and YouTube. Step-by-step test solutions on the Chapter Test Prep Videos are also available via MyMathLab and YouTube. (Search "TobeyBegInterAlg" and click on "Channels.")

Solve for the variable.

1. $3x + 5.6 = 11.6$

2. $9x - 8 = -6x - 3$

3. $2(2y - 3) = 4(2y + 2)$

4. $\frac{1}{7}y + 3 = \frac{1}{2}y$

5. $4(7 - 4x) = 3(6 - 2x)$

MC **6.** $0.8x + 0.18 - 0.4x = 0.3(x + 0.2)$

7. $\frac{2y}{3} + \frac{1}{5} - \frac{3y}{5} + \frac{1}{3} = 1$

8. $3 - 2y = 2(3y - 2) - 5y$

9. $5(20 - x) + 10x = 165$

10. $5(x + 40) - 6x = 9x$

11. $-2(2 - 3x) = 76 - 2x$

MC **12.** $20 - (2x + 6) = 5(2 - x) + 2x$

In questions 13–17, solve for x.

13. $2x - 3 = 12 - 6x + 3(2x + 3)$

14. $\frac{1}{3}x - \frac{3}{4}x = \frac{1}{12}$

15. $\frac{3}{5}x + \frac{7}{10} = \frac{1}{3}x + \frac{3}{2}$

16. $\frac{15x - 2}{28} = \frac{5x - 3}{7}$

MC **17.** $\frac{2}{3}(x + 8) + \frac{3}{5} = \frac{1}{5}(11 - 6x)$

Solve and graph the inequality.

18. $3(x - 2) \geq 5x$

Number line: -5, -4, -3, -2, -1, 0, x

MC **19.** $2 - 7(x + 1) - 5(x + 2) < 0$

Number line: $-\frac{3}{2}$, $-\frac{5}{4}$, -1, $-\frac{3}{4}$, $-\frac{1}{2}$, $-\frac{1}{4}$, x

20. $5 + 8x - 4 < 2x + 13$

Number line: 0, 1, 2, 3, 4, 5, x

21. $\frac{1}{4}x + \frac{1}{16} \leq \frac{1}{8}(7x - 2)$

Number line: -1, $-\frac{1}{2}$, 0, $\frac{1}{2}$, 1, $\frac{3}{2}$, x

1. ______
2. ______
3. ______
4. ______
5. ______
6. ______
7. ______
8. ______
9. ______
10. ______
11. ______
12. ______
13. ______
14. ______
15. ______
16. ______
17. ______
18. ______
19. ______
20. ______
21. ______

22.

23.

24.

25.

26.

27.

28.

29.

30.

Total Correct:

Solve.

22. A number is doubled and then decreased by 11. The result is 59. What is the original number?

23. The sum of one-half of a number, one-ninth of the number, and one-twelfth of the number is twenty-five. Find the original number.

24. One number is six less than three times a second number. The sum of the two numbers is twenty-two. Find each number.

25. Jerome and Steven traveled in separate cars from their apartment to a ski resort 300 miles away. Steven wanted to get there early, so he traveled the maximum speed limit, 60 mph. Jerome just bought a new car, so he drove a little slower at exactly 50 mph. If they both left at the same time, how much sooner did Steven arrive at the ski resort than Jerome?

26. The measure of the first angle in a triangle is triple the measure of the second angle. The measure of the third angle is 10 degrees more than the second angle. What is the measure of each angle?

27. Raymond has a budget of $1940 to rent a computer for his company office. The computer company he wants to rent from charges $200 for installation and service as a one-time fee. Then they charge $116 per month rental for the computer. How many months will Raymond be able to rent a computer with this budget?

28. Last year the yearly tuition at Westmont College went up 8%. This year's charge for tuition for the year is $34,560. What was it last year before the increase went into effect?

29. Franco invested $4000 in money market funds. Part was invested at 14% interest, the rest at 11% interest. At the end of each year the fund company pays interest. After one year he earned $482 in simple interest. How much was invested at each interest rate?

30. Mary has $3.50 in change. She has twice as many nickels as quarters. She has one fewer dime than she has quarters. How many of each coin does she have?

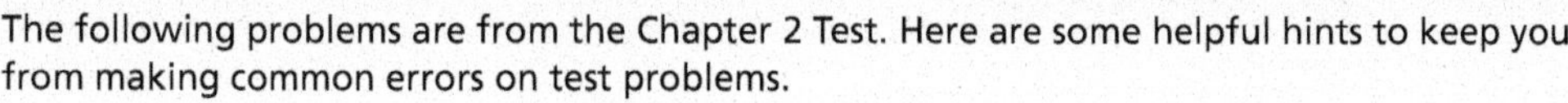

Mastering the skills you need to do well on the test.

The following problems are from the Chapter 2 Test. Here are some helpful hints to keep you from making common errors on test problems.

Chapter 2 Test, Problem 6 Solve for the variable. $0.8x + 0.18 - 0.4x = 0.3(x + 0.2)$

HELPFUL HINT After removing parentheses, it might be most helpful for you to multiply both sides of the equation by 100 in order to obtain a simpler, equivalent equation without decimals. Check to make sure that you did not make any errors in calculations before solving the equation.

Did you remove the parentheses to get the equation $0.8x + 0.18 - 0.4x = 0.3x + 0.06$?

Yes ______ No ______

If you answered No, go back and use the distributive property to remove the parentheses. Be careful to place the decimal point in the correct location when multiplying 0.3 and 0.2 together.

Did you multiply each term of the equation by 100 to move the decimal point two places to the right to get the equivalent equation $80x + 18 - 40x = 30x + 6$?

Yes ______ No ______

If you answered No, stop and carefully complete this step before solving the equation. Remember that you may need to add a 0 to the end of a term in order to move the decimal point two places to the right.

If you answered Problem 6 incorrectly, go back and rework the problem using these suggestions.

Chapter 2 Test, Problem 12 Solve for the variable. $20 - (2x + 6) = 5(2 - x) + 2x$

HELPFUL HINT Slowly complete the necessary steps to remove each set of parentheses before doing any other steps. Be careful to avoid sign errors.

Did you obtain the equation $20 - 2x - 6 = 10 - 5x + 2x$ after removing each set of parentheses?

Yes ______ No ______

If you answered No, go back and carefully use the distributive property to remove each set of parentheses. Locate any mistakes you have made and make a note of the type of error discovered.

Did you combine like terms to get the equation $14 - 2x = 10 - 3x$?

Yes ______ No ______

If you answered No, stop and perform that step correctly.

Now go back and rework the problem using these suggestions.

Need more help? Watch the MATH COACH videos in MyMathLab® or on YouTube™.

Chapter 2 Test, Problem 17 Solve for x. $\frac{2}{3}(x + 8) + \frac{3}{5} = \frac{1}{5}(11 - 6x)$

HELPFUL HINT Remove the parentheses first. This is the most likely place to make a mistake. Next, carefully show every step of your work as you multiply each fraction by the LCD. Be sure to check your work.

Did you remove each set of parentheses to obtain the equation $\frac{2}{3}x + \frac{16}{3} + \frac{3}{5} = \frac{11}{5} - \frac{6}{5}x$?

Yes ______ No ______

If you answered No, stop and carefully redo your steps of multiplication, showing every part of your work.

Did you identify the LCD as 15 and then multiply each term by 15 to get $10x + 80 + 9 = 33 - 18x$?

Yes ______ No ______

If you answered No, stop and write out your steps slowly.

If you answered Problem 17 incorrectly, go back and rework the problem using these suggestions.

Chapter 2 Test, Problem 19 Solve and graph the inequality. $2 - 7(x + 1) - 5(x + 2) < 0$

HELPFUL HINT Be sure to remove parentheses and combine any like terms on each side of the inequality before solving for the variable. Always verify the following:

1. Did you multiply or divide by a negative number? If so, did you reverse the inequality symbol?
2. In the graph, is your choice of an open circle or closed circle correct?

Did you remove parentheses to get $2 - 7x - 7 - 5x - 10 < 0$? Did you combine like terms to obtain the inequality $-15 - 12x < 0$? Next, did you add 15 to both sides of the inequality?

Yes ______ No ______

If you answered No to any of these questions, stop now and perform those steps.

Did you remember to reverse the inequality symbol in the last step?

Yes ______ No ______

If you answered No, please review the rules for when to reverse the inequality symbol and then go back and perform this step.

Did you use an open circle in your number line graph? Is your arrow pointing to the right?

Yes ______ No ______

If you answered No to either question, please review the rules for how to graph an inequality involving the $<$ or $>$ inequality symbols.

Now go back and rework the problem using these suggestions.

Need more help? Look for section examples marked with MG to review.

CHAPTER 3

Graphing and Functions

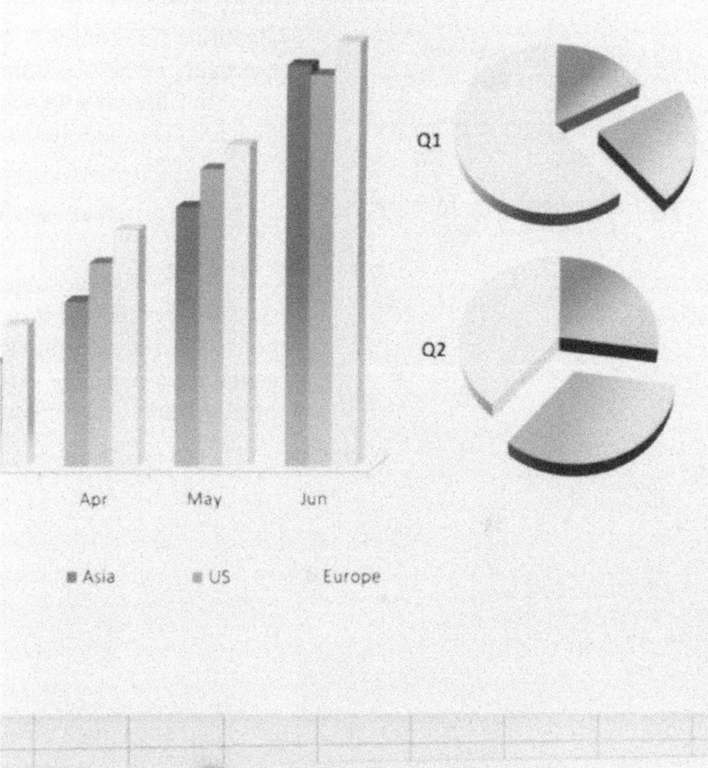

CAREER OPPORTUNITIES

Economist

Economists play an important role throughout society by studying the production, supply, demand, and exchange of goods and services. Identifying trends and their possible impact, these professionals help shape not only business decision making but also public policy. Many economists are employed in businesses and academic institutions. Though their job titles differ, they all share a strong background in mathematics, and they use the math presented in this chapter to assist their research, analysis, and forecasting.

To investigate how the mathematics in this chapter can help with this field, see the Career Exploration Problems on page **259**.

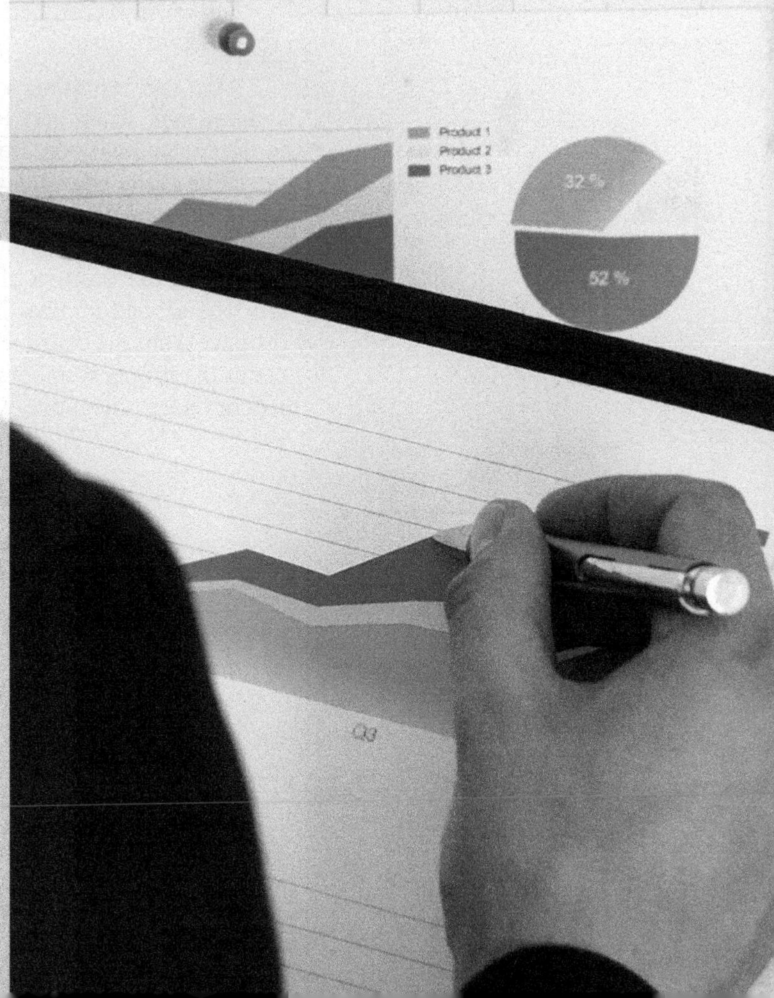

3.1 The Rectangular Coordinate System

Student Learning Objectives

After studying this section, you will be able to:

1. Plot a point, given the coordinates.
2. Determine the coordinates of a plotted point.
3. Find ordered pairs for a given linear equation.

1 Plotting a Point, Given the Coordinates

Often we can better understand an idea if we see a picture. This is the case with many mathematical concepts, including those relating to algebra. We can illustrate algebraic relationships with drawings called **graphs.** Before we can draw a graph, however, we need a frame of reference.

In Chapter 1 we showed that any real number can be represented on a number line. Look at the following number line. The arrow indicates the positive direction.

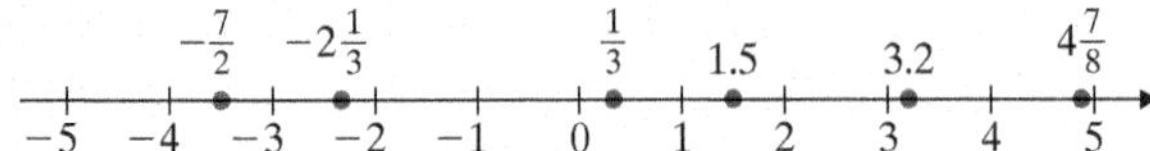

To form a **rectangular coordinate system,** we draw a second number line vertically. We construct it so that the 0 point on each number line is exactly at the same place. We refer to this location as the **origin.** The horizontal number line is often called the ***x*-axis.** The vertical number line is often called the ***y*-axis.** Arrows show the positive direction for each axis.

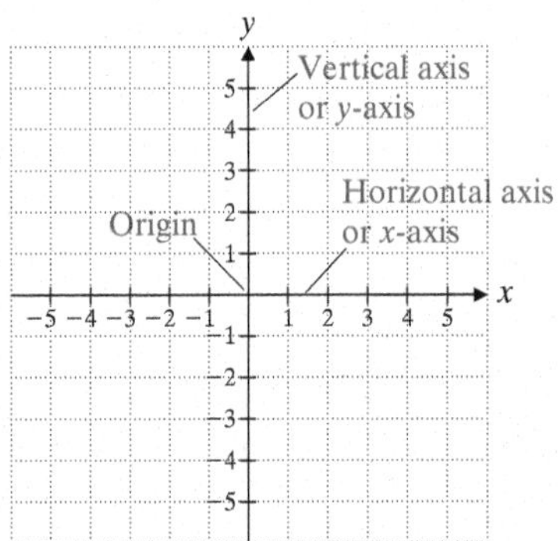

We can represent a point in this rectangular coordinate system by using an **ordered pair** of numbers. For example, (5, 2) is an ordered pair that represents a point in the rectangular coordinate system. The numbers in an ordered pair are often referred to as the **coordinates** of the point. The first number is called the ***x*-coordinate** and it represents the distance from the origin measured along the horizontal or *x*-axis. If the *x*-coordinate is positive, we count the proper number of squares to the right (that is, in the positive direction). If the *x*-coordinate is negative, we count to the left. The second number in the pair is called the ***y*-coordinate** and it represents the distance from the origin measured along the *y*-axis. If the *y*-coordinate is positive, we count the proper number of squares upward (that is, in the positive direction). If the *y*-coordinate is negative, we count downward.

(5, 2)
x-coordinate ⌟ ⌞ *y*-coordinate

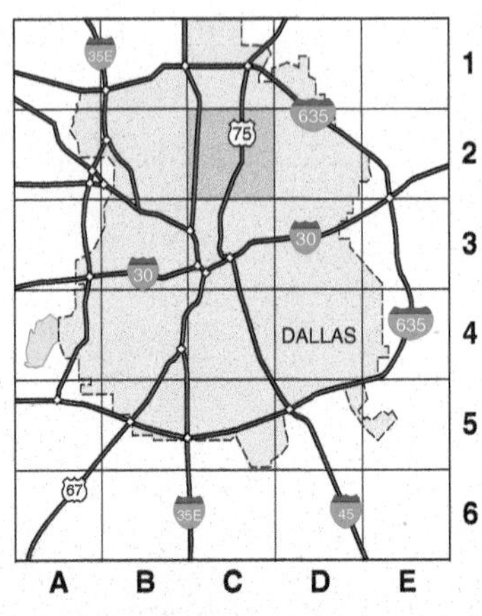

Suppose the directory for the map on the left indicated that you would find a certain street in the region C2. To find the street you would first scan across the horizontal scale until you found section C; from there you would scan up the map until you hit section 2 along the vertical scale. As we will see in the next example, plotting a point in the rectangular coordinate system is much like finding a street on a map with grids.

Example 1 Plot the point (5, 2) on a rectangular coordinate system. Label this point as *A*.

Solution Since the *x*-coordinate is 5, we first count 5 units to the right on the *x*-axis. Then, because the *y*-coordinate is 2, we count 2 units up from the point where we stopped on the *x*-axis. This locates the point corresponding to (5, 2). We mark this point with a dot and label it *A*.

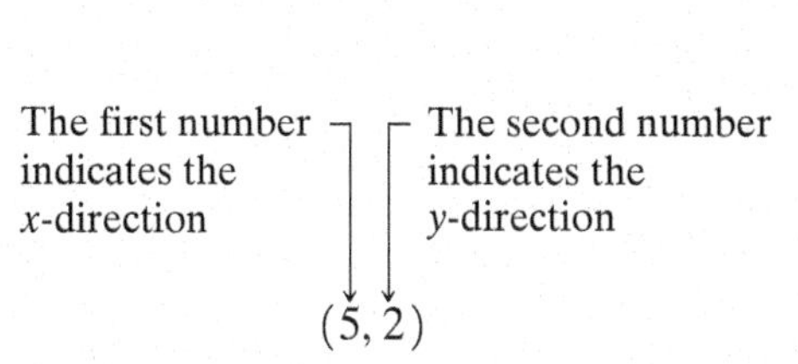

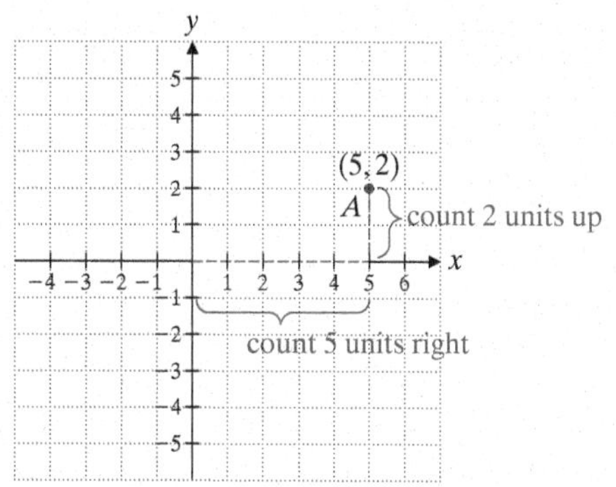

Student Practice 1 Plot the point (3, 4) on the preceding rectangular coordinate system. Label this point as B.

It is important to remember that the first number in an ordered pair is the x-coordinate and the second number is the y-coordinate. The ordered pairs (5, 2) and (2, 5) represent different points.

Example 2 Use the rectangular coordinate system to plot each point. Label the points F, G, and H, respectively.

(a) $(-5, -3)$ **(b)** $(2, -6)$ **(c)** $(-6, 2)$

Solution

(a) $(-5, -3)$ Notice that the x-coordinate, -5, is negative. On the coordinate grid, negative x-values appear to the left of the origin. Thus, we will begin by counting 5 squares to the left, starting at the origin. Since the y-coordinate, -3, is negative, we will count 3 units down from the point where we stopped on the x-axis.

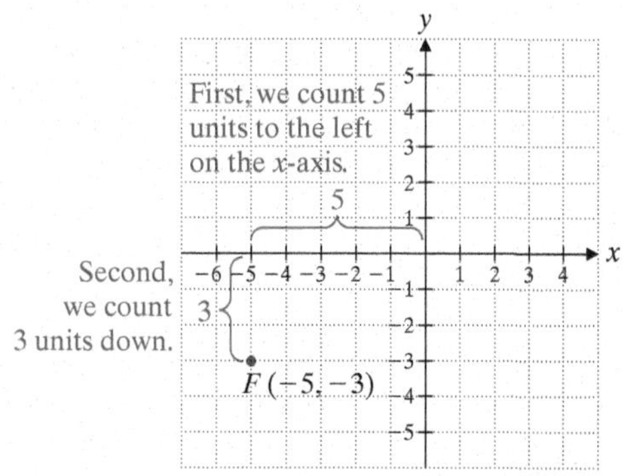

(b) $(2, -6)$ The x-coordinate is positive. Begin by counting 2 squares to the right of the origin. Then count down because the y-coordinate is negative.

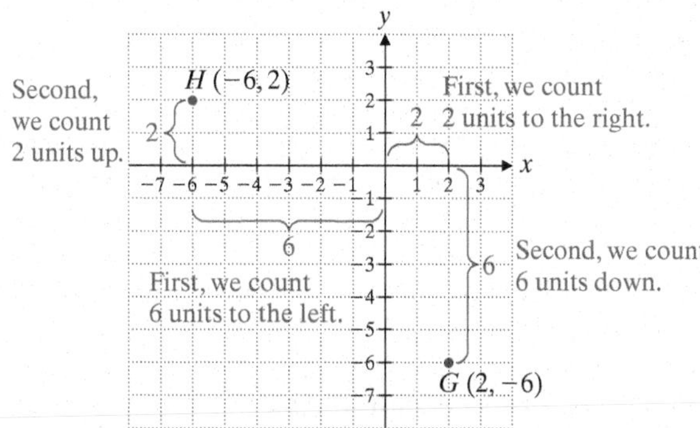

(c) $(-6, 2)$ The x-coordinate is negative. Begin by counting 6 squares to the left of the origin. Then count up because the y-coordinate is positive.

Continued on next page

Student Practice 2

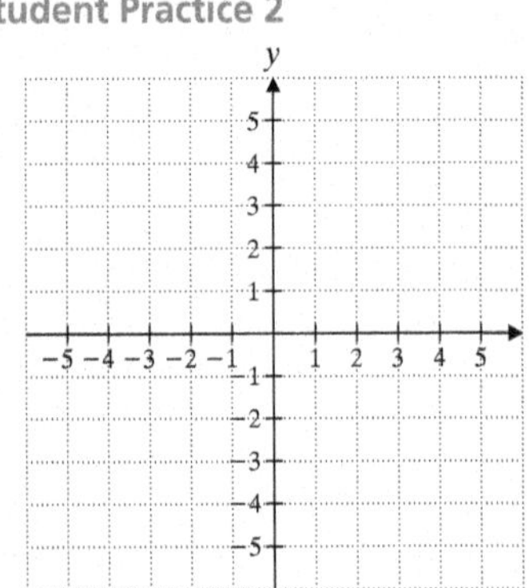

Student Practice 2 Use the rectangular coordinate system in the margin to plot each point. Label the points I, J, and K, respectively.

(a) $(-2, -4)$ **(b)** $(-4, 5)$ **(c)** $(4, -2)$

Example 3 Plot the following points.

F: $(0, 5)$ G: $\left(3, \frac{3}{2}\right)$ H: $(-6, 4)$ I: $(-3, -4)$

J: $(-4, 0)$ K: $(2, -3)$ L: $(6.5, -7.2)$

Solution These points are plotted in the figure.

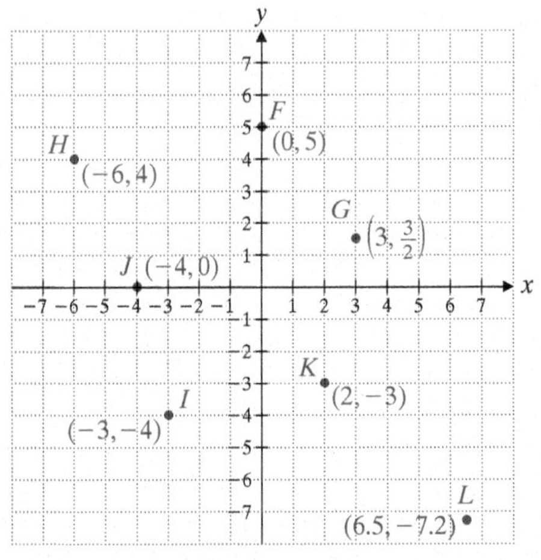

Note: When you are plotting decimal values like $(6.5, -7.2)$, plot the point halfway between 6 and 7 in the x-direction (for the 6.5) and at your best approximation of -7.2 in the y-direction.

Student Practice 3

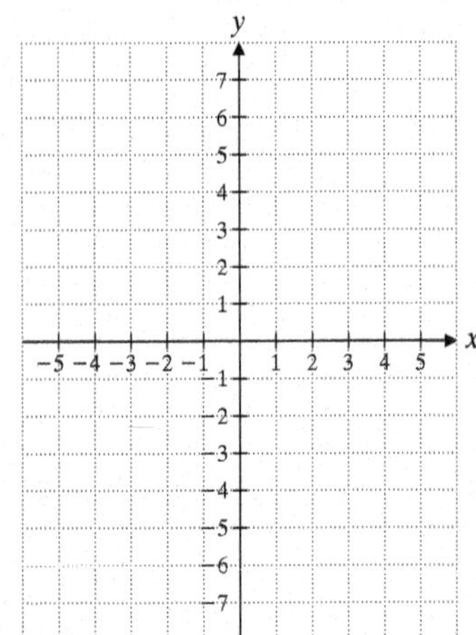

Student Practice 3 Plot the following points. Label each point with both the letter and the ordered pair. Use the coordinate system provided in the margin.

A: $(3, 7)$; B: $(0, -6)$; C: $(3, -4.5)$; D: $\left(-\frac{7}{2}, 2\right)$

2 Determining the Coordinates of a Plotted Point

Sometimes we need to find the coordinates of a point that has been plotted. First, we count the units we need on the x-axis to get as close as possible to the point. Next we count the units up or down that we need to go from the x-axis to reach the point.

Example 4 What ordered pairs of numbers represent point A and point B on the graph?

Solution To find point A, we move along the x-axis until we get as close as possible to A, ending up at the number 5. Thus we obtain 5 as the first number of the ordered pair. Then we count 4 units upward on a line parallel to the y-axis to reach A. So we obtain 4 as the second number of the ordered pair. Thus point A is represented by the ordered pair (5, 4). We use the same approach to find point B: $(-5, -3)$.

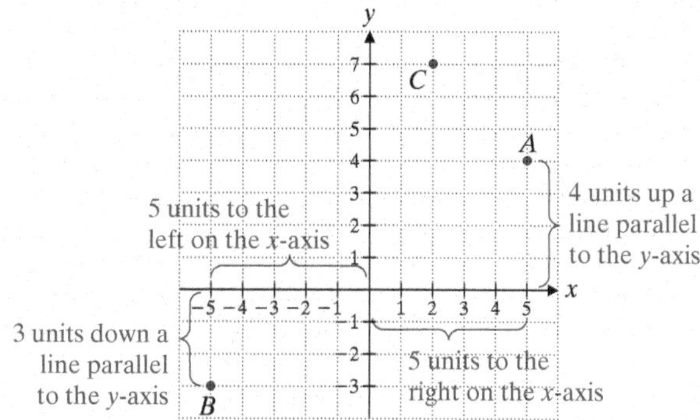

Student Practice 4 What ordered pair of numbers represents point C on the graph in Example 4?

In examining data from real-world situations, we often find that plotting data points shows useful trends. In such cases, it is often necessary to use a different scale, one that displays only positive values.

Example 5 The number of motor vehicle accidents in millions is recorded in the following table for the years 2000 to 2012.

(a) Plot points that represent this data on the given coordinate system.

(b) What trends are apparent from the plotted data?

Number of Years Since 2000	Number of Motor Vehicle Accidents (in Millions)
0	13
2	18
4	11
6	10
8	10
10	11
12	10

Source: 2012 Traffic Safety Facts

Solution

(a)

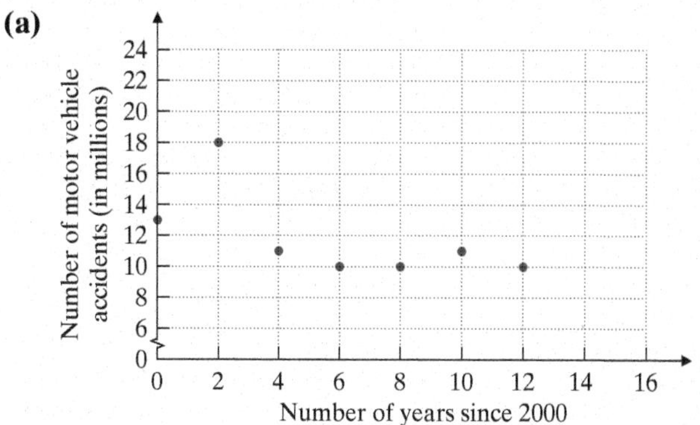

(b) From 2000 to 2002, there was a *significant* increase in the number of accidents. From 2002 to 2004, there was a *significant* decrease in the number of accidents. From 2004 to 2012, the number of accidents was relatively stable. □

Student Practice 5 The number of motor vehicle deaths in thousands is recorded in the following table for the years 1980 to 2010.

(a) Plot points that represent this data on the given coordinate system.

(b) What trends are apparent from the plotted data?

Continued on next page

Number of Years Since 1980	Number of Motor Vehicle Deaths (in Thousands)
0	51
5	44
10	45
15	42
20	42
25	44
30	33

Source: 2012 Statistical Abstract; 2012 Traffic Safety Facts

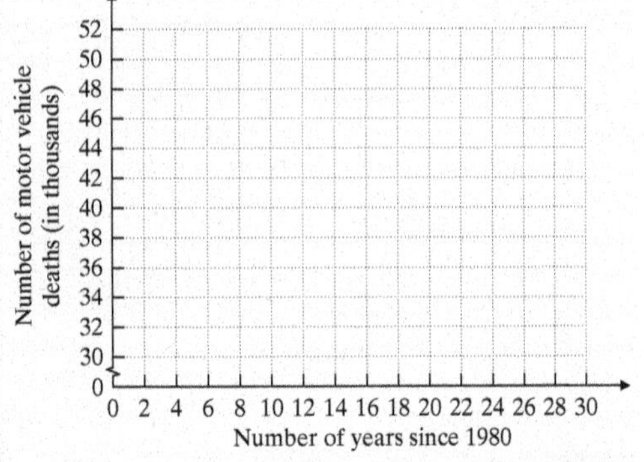

3 Finding Ordered Pairs for a Given Linear Equation

An equation such as $x + 2 = 8$ is called a **linear equation in one variable.** It is called *linear* because the exponent on x is understood to be 1.

In earlier chapters we found the solutions to linear equations with one variable. For example, the solution to $x + 2 = 8$ is $x = 6$. Now, consider the equation $x + y = 8$. This equation has *two variables* and is called a **linear equation in two variables.** A solution to $x + y = 8$ consists of a pair of numbers, one for x and one for y, whose sum is 8. When we find solutions to $x + y = 8$, we are answering the question: "The sum of what two numbers equals 8?" Since $6 + 2 = 8$, the pair of numbers $x = 6$ and $y = 2$ or $(6, 2)$ is a solution.

$$x + y = 8 \quad \text{The sum of what two numbers equals 8?}$$
$$6 + 2 = 8 \quad x = 6 \text{ and } y = 2 \text{ or } (6, 2) \text{ is a solution.}$$

Now, $3 + 5 = 8$, so the pair of numbers $x = 3$ and $y = 5$ or $(3, 5)$ is also a solution to the equation $x + y = 8$. Can you find another pair of numbers whose sum is 8? Of course you can: $(7, 1)$, $(6\frac{1}{2}, 1\frac{1}{2})$, $(0, 8)$, $(4, 4)$ and so on. In fact, there are *infinitely* many pairs of numbers with a sum of 8 and thus infinitely many ordered pairs that represent solutions to $x + y = 8$. It is important that you realize that not just any ordered pair of numbers is a solution to $x + y = 8$. Only ordered pairs whose sum is 8 are solutions to $x + y = 8$.

We now state the definition for a linear equation in two variables.

> A **linear equation in two variables** is an equation that can be written in the form $Ax + By = C$ where A, B, and C are real numbers but A and B are not *both* zero.

Replacement values for x and y that make *true mathematical statements* of the equation are called *truth values;* and an ordered pair of these truth values is called a **solution.**

Example 6 Are the following ordered pairs solutions to the equation $3x + 2y = 5$?

(a) $(-1, 4)$ **(b)** $(2, 2)$

Solution We replace values for x and y in the equation to see if we obtain a true statement.

(a)

$$3x + 2y = 5$$
$$3(-1) + 2(4) \stackrel{?}{=} 5 \quad \text{Replace } x \text{ with } -1 \text{ and } y \text{ with } 4.$$
$$-3 + 8 \stackrel{?}{=} 5$$
$$5 = 5 \checkmark \quad \text{True statement}$$

The ordered pair $(-1, 4)$ is a solution to $3x + 2y = 5$ because when we replace x by -1 and y by 4, we obtain a true statement.

(b) $3x + 2y = 5$

$3(2) + 2(2) \stackrel{?}{=} 5$ Replace x with 2 and y with 2.

$6 + 4 \stackrel{?}{=} 5$

$10 = 5$ False statement

The ordered pair $(2, 2)$ is *not* a solution to $3x + 2y = 5$ because when we replace x by 2 and y by 2, we obtain a *false* statement. □

Student Practice 6 Are the following ordered pairs solutions to the equation $3x + 2y = 5$?

(a) $(3, -1)$ **(b)** $\left(2, -\frac{1}{2}\right)$

Example 7

(a) Is $(-1, 5)$ a solution to the equation $x + y = 4$?

(b) List two other ordered pairs of numbers that are solutions to the equation $x + y = 4$.

Solution

(a) The ordered pair $(-1, 5)$ is a solution if, when we replace x with -1 and y with 5 in the equation $x + y = 4$, we get a true statement.

Check: $x = -1$ and $y = 5$, or $(-1, 5)$ $x + y = 4$

$-1 + 5 = 4,\quad 4 = 4$ ✓

Since $4 = 4$ is true, $(-1, 5)$ is a solution of $x + y = 4$.

(b) There are infinitely many solutions, so answers may vary. We can choose any two numbers whose sum is 4.

$x = 0$ and $y = 4$, or $(0, 4)$ $x + y = 4$

$0 + 4 = 4,\quad 4 = 4$ ✓

$x = 2$ and $y = 2$, or $(2, 2)$ $x + y = 4$

$2 + 2 = 4,\quad 4 = 4$ ✓

Therefore $(0, 4)$ and $(2, 2)$ are solutions to $x + y = 4$. □

Student Practice 7

(a) Is $(-2, 13)$ a solution to the equation $x + y = 11$?

(b) List three other ordered pairs of numbers that are solutions to the equation $x + y = 11$.

If one value of an ordered-pair solution to a linear equation is known, the other can be quickly obtained. To do so, we replace the proper variable in the equation by the known value. Then, using the methods learned in Chapter 2, we solve the resulting equation for the other variable.

Example 8 Find the missing coordinate to complete the following ordered-pair solutions to the equation $2x + 3y = 15$.

(a) $(0, ?)$ **(b)** $(?, 1)$

Solution

(a) For the ordered pair $(0, ?)$, we know that $x = 0$. Replace x by 0 in the equation and solve for y.

Continued on next page

$$\begin{aligned} 2x + 3y &= 15 \\ 2(0) + 3y &= 15 \quad \text{Replace } x \text{ with 0.} \\ 0 + 3y &= 15 \quad \text{Simplify.} \\ y &= 5 \quad \text{Divide both sides by 3.} \end{aligned}$$

Thus we have the ordered pair (0, 5).

(b) For the ordered pair (?, 1), we *do not know* the value of x. However, we do know that $y = 1$. So we start by replacing the variable y by 1. We will end up with an equation with one variable, x. We can then solve for x.

$$\begin{aligned} 2x + 3y &= 15 \\ 2x + 3(1) &= 15 \quad \text{Replace } y \text{ with 1.} \\ 2x + 3 &= 15 \quad \text{Simplify.} \\ 2x &= 12 \quad \text{Isolate the variable term.} \\ x &= 6 \quad \text{Solve for } x. \end{aligned}$$

Thus we have the ordered pair (6, 1). □

Student Practice 8 Find the missing coordinate to complete the following ordered-pair solutions to the equation $3x - 4y = 12$.

(a) (0, ?) **(b)** (?, 3) **(c)** (?, −6)

Sometimes it is convenient to first solve for y, then find the missing coordinates of the ordered-pair solutions.

Example 9 Find the missing coordinates to complete the ordered-pair solutions to $3x - 2y = 6$.

(a) (2, ?) **(b)** (4, ?)

Solution Since we are given the values for x and must find the values for y in each ordered pair, solving the equation for y first simplifies the calculations.

$$3x - 2y = 6$$

$$-2y = 6 - 3x \quad \text{We want to isolate the term containing } y \text{, so we subtract } 3x \text{ from both sides.}$$

$$\frac{-2y}{-2} = \frac{6 - 3x}{-2} \quad \text{Divide both sides by the coefficient of } y.$$

$$y = \frac{6}{-2} + \frac{-3x}{-2} \quad \text{Change subtracting to adding the opposite: } 6 - 3x = 6 + (-3x). \text{ Then rewrite the fraction on the right side as two fractions.}$$

$$y = \frac{3}{2}x - 3 \quad \text{Simplify and reorder the terms on the right.}$$

Now we find the values of y by replacing x with 2 and 4.

(a) $y = \frac{3}{\cancel{2}}(\cancel{2}) - 3$
$y = 3 - 3$
$y = 0$

(b) $y = \frac{3}{\cancel{2}}(\overset{2}{\cancel{4}}) - 3$
$y = 6 - 3$
$y = 3$

Thus we have the ordered pairs (2, 0) and (4, 3). □

Student Practice 9 Find the missing coordinates to complete the ordered-pair solutions to $8 - 2y + 3x = 0$.

(a) (0, ?) **(b)** (2, ?)

To Think About: *An Alternate Method* Find the missing coordinates in Example 9 using the equation in the form $3x - 2y = 6$. Which form do you prefer? Why?

3.1 Exercises MyMathLab®

Verbal and Writing Skills, Exercises 1–6

Unless otherwise indicated, assume each grid line represents one unit.

1. What is the x-coordinate of the origin?

2. What is the y-coordinate of the origin?

3. Explain why (5, 1) is referred to as an *ordered* pair of numbers.

4. Explain how you would locate the point (4, 3) on graph paper.

5. Explain why the ordered pairs (2, 7) and (7, 2) do not represent the same point on a graph.

6. The equation $x + y = 10$ has how many ordered-pair solutions?

7. Plot the following points.

J: $(-4, 3.5)$ K: $(6, 0)$ L: $(5, -6)$
M: $(0, -4)$ N: $(3, 4)$ P: $(-6, 5)$

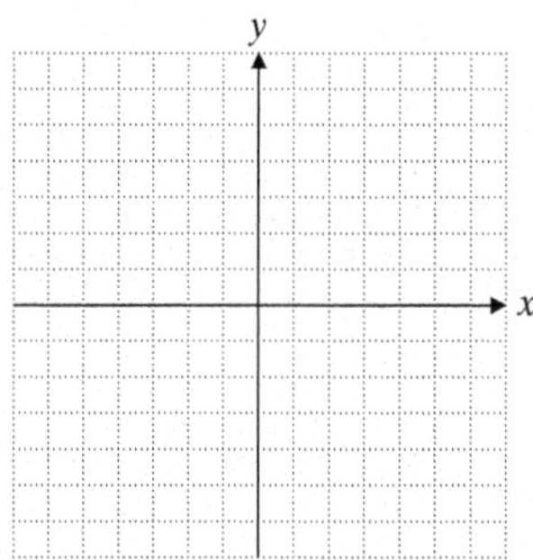

8. Plot the following points.

R: $(-3, 0)$ S: $(3.5, 4)$ T: $(-2, -2.5)$
V: $(0, 5)$ W: $(3, 0)$ X: $(2, -4)$

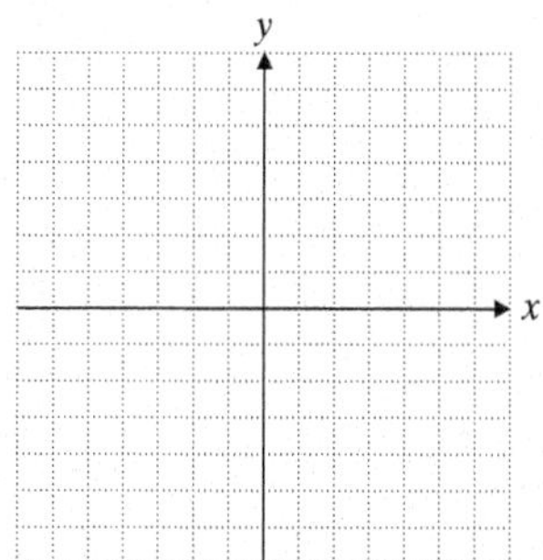

Consider the points plotted on the graph at right.

9. Give the coordinates for points R, S, X, and Y.

10. Give the coordinates for points T, V, W, and Z.

In exercises 11 and 12, six points are plotted in each figure. List all the ordered pairs needed to represent the points.

11.

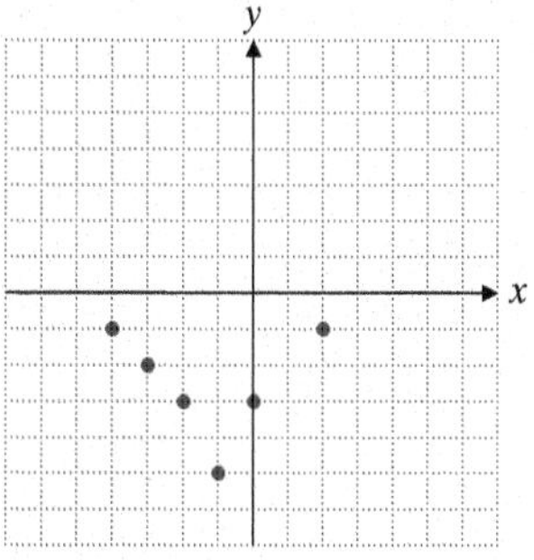

12.

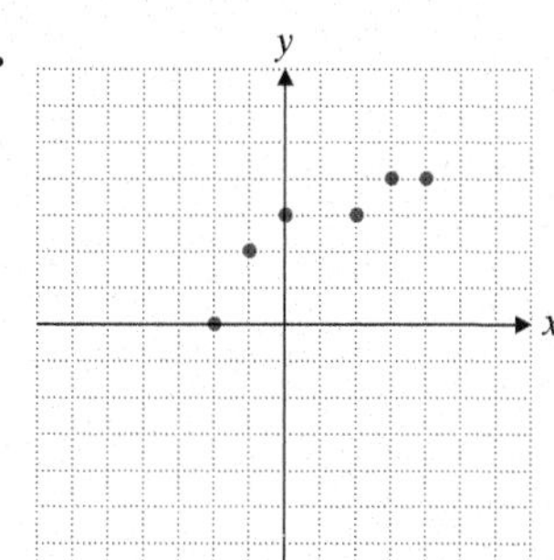

Using Road Maps *The map below shows a portion of New York, Connecticut, and Massachusetts. Like many maps used in driving or flying, it has horizontal and vertical grid markers for ease of use. For example, Newburgh, New York, is located in grid B3. Use the grid labels to indicate the locations of the following cities.*

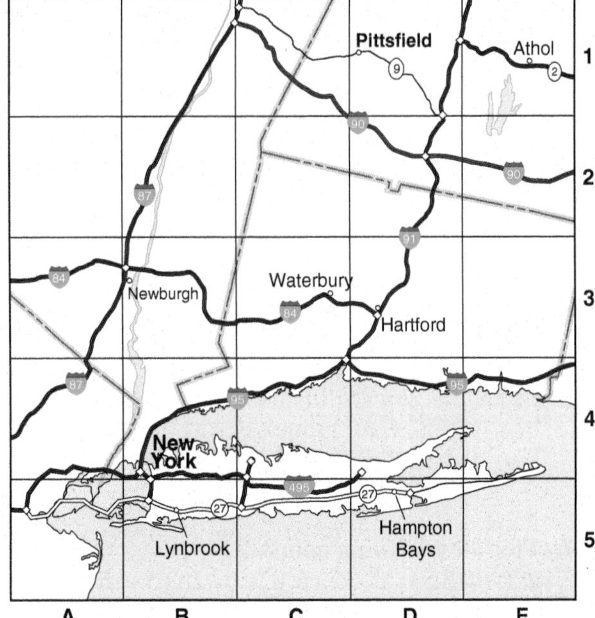

13. Lynbrook, New York

14. Hampton Bays, New York

15. Athol, Massachusetts

16. Pittsfield, Massachusetts

17. Hartford, Connecticut

18. Waterbury, Connecticut

19. ***DVD Movie Shipments*** According to a large movie DVD manufacturer, the number of DVDs shipped by the manufacturer decreased significantly from 2008 to 2015. The number of DVDs shipped during these years is recorded in the following table and is measured in thousands. For example, 803 in the second column means 803 thousand, or 803,000.

(a) Plot points that represent this data on the given rectangular coordinate system.

(b) What trends are apparent from the plotted data?

Number of Years Since 2008	Number of DVDs Shipped (in Thousands)
0	803
1	746
2	767
3	705
4	615
5	511
6	368
7	293

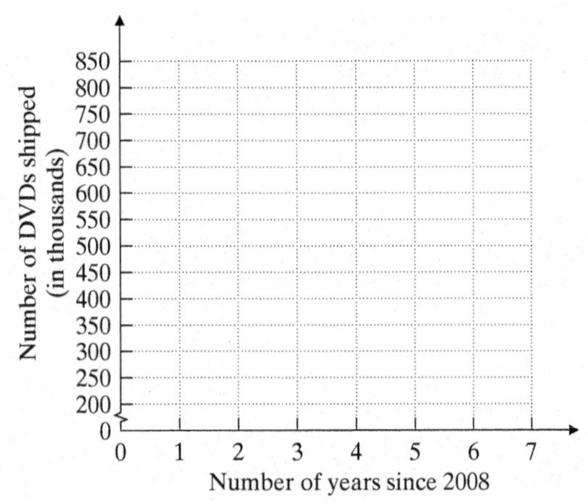

20. ***Customers Visiting a Restaurant Chain*** The number of customers visiting a large restaurant chain per year in the United States for selected years starting in 1990 is recorded in the following table. The number of customers is measured in millions. For example, 15 in the second column means 15 million, or 15,000,000 customers.

(a) Plot points that represent the data on the given rectangular coordinate system.

(b) What trends are apparent from the plotted data?

Number of Years Since 1990	Customers Visiting Restaurant Chain per year (in Millions of customers)
0	15
5	20
10	21
15	22
20	22
25	23

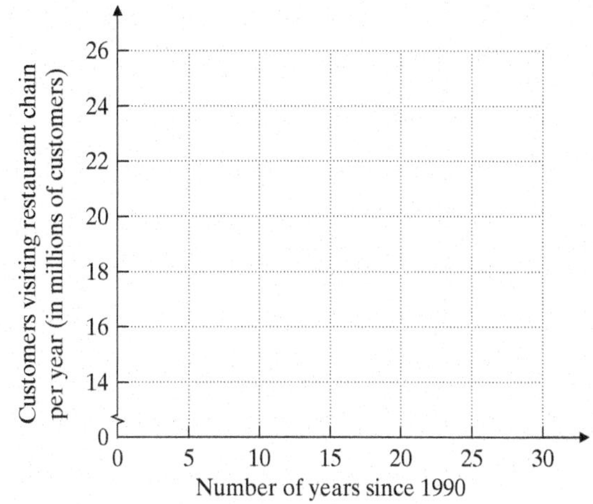

21. ***Buying Books Online*** A large university's bookstore decided to offer students the option of buying their books online. A review of online sales at the end of the fourth year indicated that online sales have increased at a significant rate. The following chart records the amount spent each year for Year 1 through Year 4.

Year	Total Amount Spent (in Millions of Dollars)
1	3.2
2	3.6
3	3.9
4	4.2

(a) Plot points that represent the data on the given rectangular coordinate system.

(b) Based on your graph, estimate the amount students will spend buying books online in year 5.

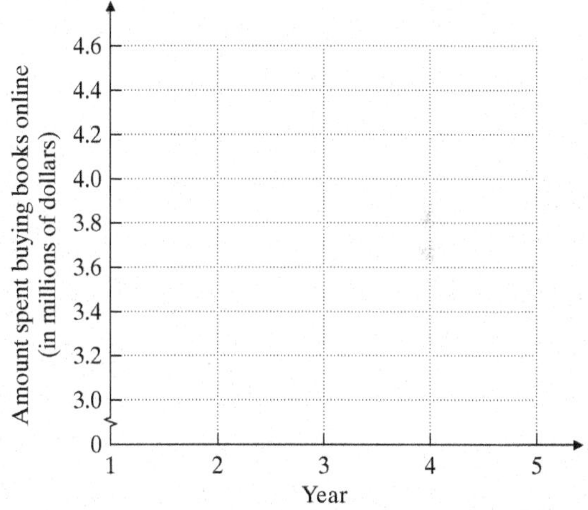

22. ***University Debit Card Usage*** Students at a university can use a prepaid debit card to make all their purchases on campus, including food and other essentials. Since the university began this program in 2012, the amount of money spent using the university's debit card has increased at a significant rate. The following chart records the amount spent each year from 2012 to 2015.

Year	Total Amount Spent (in Thousands of Dollars)
2012	53.9
2013	67.3
2014	80.9
2015	95.3

(a) Plot points that represent the data on the given rectangular coordinate system.

(b) Based on your graph, estimate the amount students spend using their debit cards in 2016.

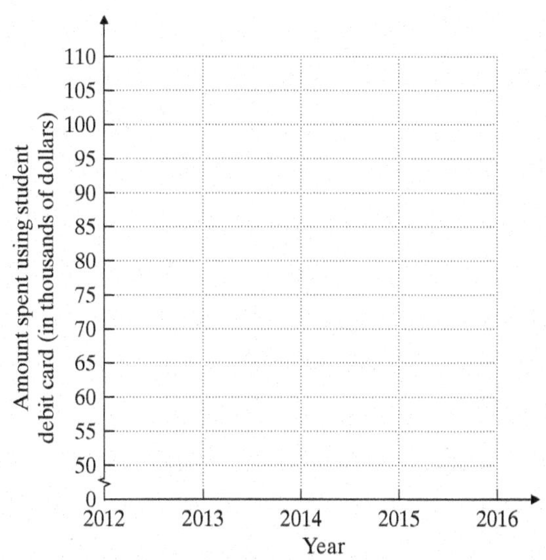

Are the following ordered pairs solutions to the equation $2x - y = 6$?

23. $(1, 0)$

24. $(6, 2)$

25. $(2, -2)$

26. $(0, -6)$

To Think About, Exercises 27 and 28

27. After an exam, Jon and Syed discussed their answers to the following exam question. Which of the following is an ordered-paired solution to the equation $2x - 2y = 4$, $(3, 5)$ or $(5, 3)$? Jon said the answer was $(5, 3)$, and Syed stated that $(3, 5)$ is the correct answer. Who is right? Why?

28. On an exam, students were asked to write an ordered-pair solution for the equation $x + y = 18$. Damien, Jennifer, and Tanya compared their answers after the exam and discovered they each had different answers. Is it possible for each of them to be correct? Explain.

For each of the following: ***(a)*** *Solve for y;* ***(b)*** *Find the missing coordinate to complete the ordered pair.*

29. **(a)** $y - 4 = -\frac{2}{3}x$

(b) $(6, \ )$

30. **(a)** $y + 9 = \frac{3}{8}x$

(b) $(8, \ )$

31. **(a)** $4x + y = 11$

(b) $(2, \ )$

32. **(a)** $5x + y = 3$

(b) $(2, \ )$

33. **(a)** $2x + 3y = 6$

(b) $(3, \ )$

34. **(a)** $3y - 5x = 9$

(b) $(6, \ )$

Find the missing coordinate to complete the ordered-pair solution to the given linear equation.

35. $y = 4x + 7$
(a) $(0, \ )$
(b) $(2, \ )$

36. $y = 6x + 5$
(a) $(0, \ )$
(b) $(3, \ )$

37. $y + 6x = 5$
(a) $(-1, \ )$
(b) $(3, \ )$

38. $y + 4x = 9$
(a) $(-2, \ )$
(b) $(5, \ )$

39. $3x - 4y = 11$
(a) $(-3, \ )$
(b) $(\ , 1)$

40. $5x - 2y = 9$
(a) $(7, \ )$
(b) $(\ , -7)$

41. $3x + 2y = -6$
(a) $(-2,\ )$
(b) $(\ , 3)$

42. $5x + 4y = 10$
(a) $(-2,\ )$
(b) $(\ , 0)$

43. $y - 1 = \frac{2}{7}x$
(a) $(7,\ )$
(b) $\left(\ , \frac{5}{7}\right)$

44. $x - 2 = \frac{5}{6}y$
(a) $(7,\ )$
(b) $\left(\ , \frac{6}{5}\right)$

45. $3x + \frac{1}{2}y = 7$
(a) $(\ , 2)$
(b) $\left(\frac{3}{2},\ \right)$

46. $3x + \frac{1}{4}y = 11$
(a) $(\ , 8)$
(b) $\left(\frac{13}{4},\ \right)$

Cumulative Review

▲ **47.** **[1.8.2]** ***Circular Swimming Pool*** The circular pool at the hotel where Bob and Linda stayed in Orlando, Florida, has a radius of 19 yards. What is the area of the pool? (Use $\pi \approx 3.14$.)

48. **[2.6.1]** A number is doubled and then decreased by three. The result is twenty-one. What is the original number?

Quick Quiz 3.1

1. Plot and label the following points.
$A: (3, -4)$
$B: (-6, -2)$
$C: (0, 5)$
$D: (-3, 6)$

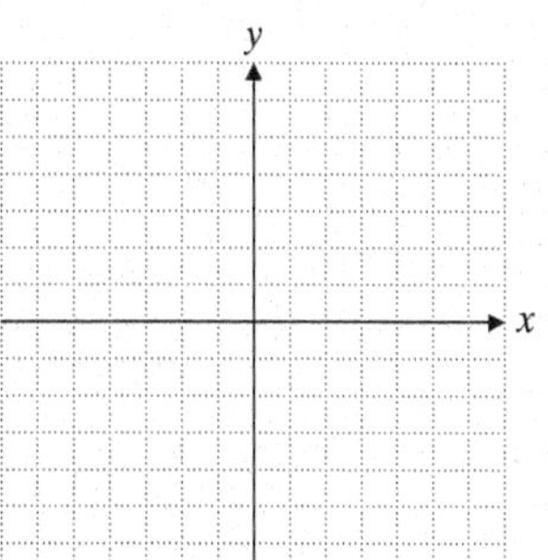

Find the missing coordinate to complete the ordered-pair solution to the given linear equation.

2. $y = -5x - 7$
(a) $(-2,\ )$
(b) $(3,\ )$
(c) $(0,\ )$

3. $4x - 3y = -12$
(a) $(3,\ )$
(b) $(\ , -8)$
(c) $(\ , 10)$

4. **Concept Check** Explain how you would find the missing coordinate to complete the ordered-pair solution to the equation $2.5x + 3y = 12$ if the ordered pair was of the form $(\ , -6)$.

3.2 Graphing Linear Equations

Student Learning Objectives

After studying this section, you will be able to:

1. Graph a linear equation by plotting three ordered pairs.
2. Graph a straight line by plotting its intercepts.
3. Graph horizontal and vertical lines.

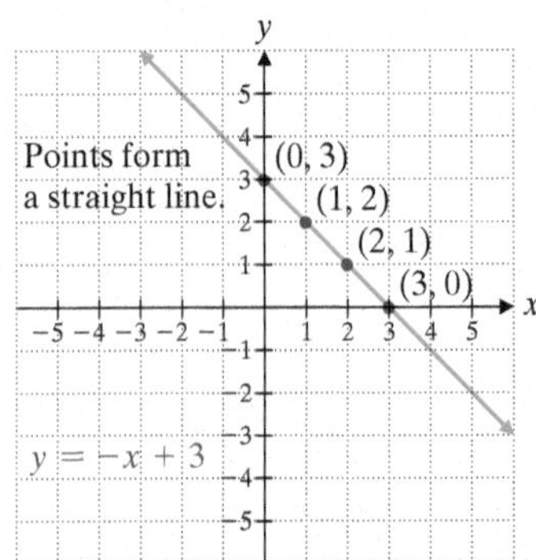

1 Graphing a Linear Equation by Plotting Three Ordered Pairs

We have seen that a solution to a linear equation in two variables is an ordered pair. The graph of an ordered pair is a point. Thus we can graph an equation by graphing the points corresponding to its ordered-pair solutions.

A linear equation in two variables has an infinite number of ordered-pair solutions. We can see that this is true by noting that we can substitute any number for x in the equation and solve it to obtain a y-value. For example, if we substitute $x = 0, 1, 2, 3, \ldots$ into the equation $y = -x + 3$ and solve for y, we obtain the ordered-pair solutions $(0, 3), (1, 2), (2, 1), (3, 0), \ldots$. (If desired, substitute these values into the equation to convince yourself.) If we plot these points on a rectangular coordinate system, we notice that they fall on a straight line, as illustrated in the margin.

It turns out that all of the points corresponding to the ordered-pair solutions of $y = -x + 3$ lie on this line, and the line extends forever in both directions. A similar statement can be made about any linear equation in two variables.

The graph of any linear equation in two variables is a straight line.

From geometry, we know that two points determine a line. Thus to graph a linear equation in two variables, we need to graph only two ordered-pair solutions of the equation and then draw the line that passes through them. Having said this, we recommend that you use three points to graph a line. Two points will determine where the line is. The third point verifies that you have drawn the line correctly. For ease in plotting, it is better if the ordered pairs contain integers.

TO GRAPH A LINEAR EQUATION

1. Find three ordered pairs that are solutions to the equation.
2. Plot the points.
3. Draw a line through the points.

Example 1 Find three ordered pairs that satisfy $y = -2x + 4$. Then graph the resulting straight line.

Solution Since we can choose any value for x, we choose numbers that are convenient. To organize the results, we will make a table of values. We will let $x = 0$, $x = 1$, and $x = 3$. We write these numbers under x in our table of values. For each of these x-values, we find the corresponding y-value in the equation $y = -2x + 4$.

$$y = -2x + 4 \qquad y = -2x + 4 \qquad y = -2x + 4$$
$$y = -2(0) + 4 \qquad y = -2(1) + 4 \qquad y = -2(3) + 4$$
$$y = 0 + 4 \qquad y = -2 + 4 \qquad y = -6 + 4$$
$$y = 4 \qquad y = 2 \qquad y = -2$$

We record these results by placing each y-value in the table next to its corresponding x-value. Keep in mind that these values represent ordered pairs, each of which is a solution to the equation. If we plot these ordered pairs and connect the three points, we

Table of Values

x	y
0	4
1	2
3	−2

Student Practice 1 Find three ordered pairs that satisfy $y = -3x - 1$. Then graph the resulting straight line. Use the given coordinate system.

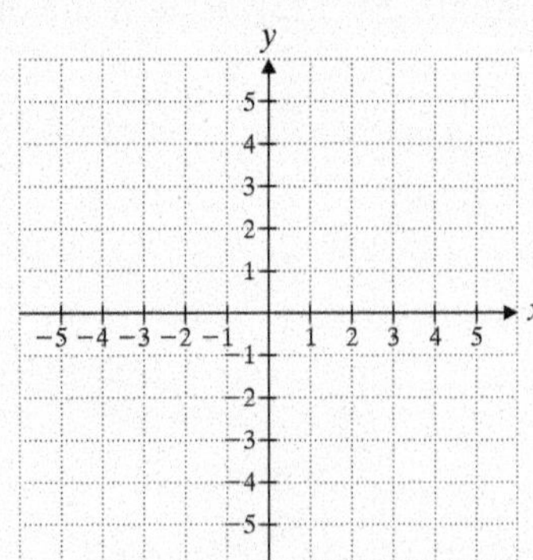

get a straight line that is the graph of the equation $y = -2x + 4$. The graph of the equation is shown in the figure below.

Find the point $(2, 0)$ on the graph. Is it on the line? Check to verify that it is a solution to $y = -2x + 4$. Find another solution by *locating* another point on the line.

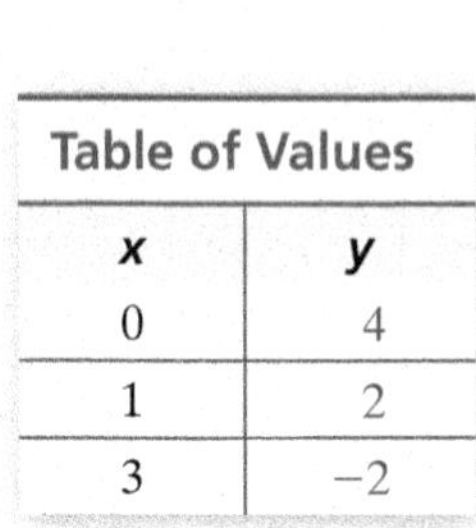

Table of Values	
x	y
0	4
1	2
3	−2

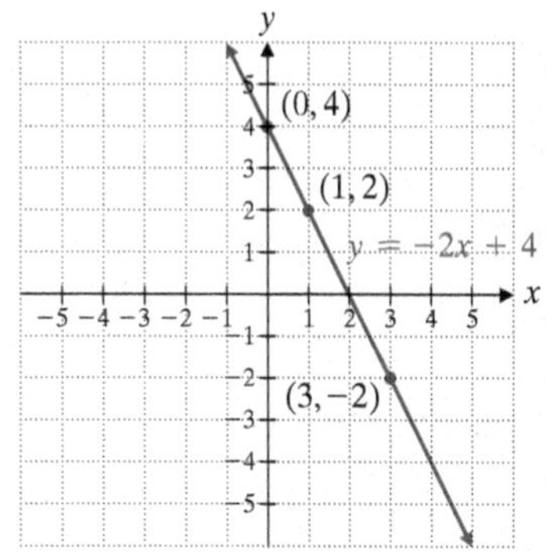

Sidelight: Determining Values for Ordered Pairs

Why can we choose any value for either x or y when we are finding ordered pairs that are solutions to an equation?

The numbers for x and y that are solutions to the equation come in pairs. When we try to find a pair that fits, we have to start with some number. Usually, it is an x-value that is small and easy to work with. Then we must find the value for y so that the pair of numbers (x, y) is a solution to the given equation.

Example 2 Graph $5x - 4y + 2 = 2$.

Solution First, we simplify the equation $5x - 4y + 2 = 2$ by subtracting 2 from each side.

$$5x - 4y + 2 - 2 = 2 - 2$$
$$5x - 4y = 0$$

Table of Values	
x	y
0	0

Since we are free to choose any value of x, $x = 0$ is a natural choice. Calculate the value of y when $x = 0$ and place the results in the table of values.

$$5(0) - 4y = 0$$
$$-4y = 0 \quad \text{Remember: Any number times 0 is 0.}$$
$$y = 0 \quad \text{Since } -4y = 0,\ y \text{ must equal 0.}$$

Now let's see what happens when $x = 1$.

$$5(1) - 4y = 0$$
$$5 - 4y = 0$$
$$-4y = -5$$
$$y = \frac{-5}{-4} \text{ or } \frac{5}{4} \quad \text{This is not an easy number to graph.}$$

A better choice for a replacement of x is a number that is divisible by 4. Let's see why. Let $x = 4$ and let $x = -4$.

$$5(4) - 4y = 0 \qquad\qquad 5(-4) - 4y = 0$$
$$20 - 4y = 0 \qquad\qquad -20 - 4y = 0$$
$$-4y = -20 \qquad\qquad -4y = 20$$
$$y = \frac{-20}{-4} \text{ or } 5 \qquad\qquad y = \frac{20}{-4} \text{ or } -5$$

Continued on next page

Student Practice 2

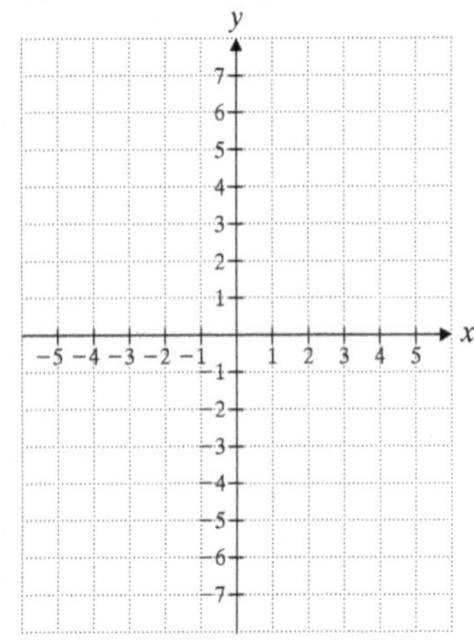

Now we can put these numbers into our table of values and graph the line.

Table of Values

x	y
0	0
4	5
−4	−5

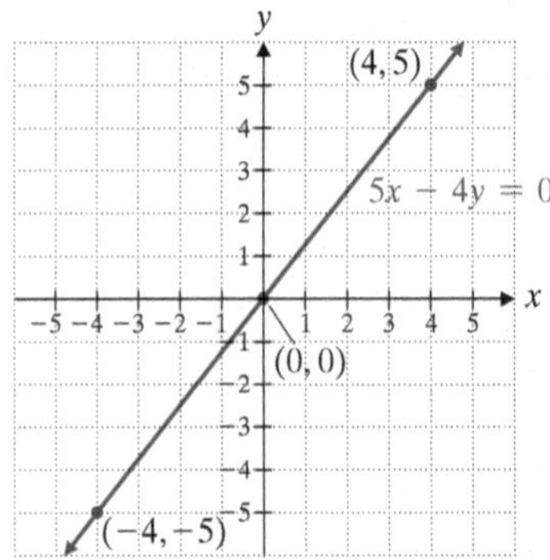

Student Practice 2 Graph $7x + 3 = -2y + 3$ on the coordinate system in the margin.

To Think About: *An Alternative Approach* In Example 2, when we picked the value of 1 for x, we found that the corresponding value for y was a fraction. To avoid fractions, we can solve the equation for the variable y *first,* then choose the values for x.

$$5x - 4y = 0 \quad \text{We must isolate } y.$$

$$-4y = -5x \quad \text{Subtract } 5x \text{ from each side.}$$

$$\frac{-4y}{-4} = \frac{-5x}{-4} \quad \text{Divide each side by } -4.$$

$$y = \frac{5}{4}x$$

Now let $x = -4$, $x = 0$, and $x = 4$, and find the corresponding values of y. Explain why you would choose multiples of 4 as replacements for x in this equation. Graph the equation and compare it to the graph in Example 2.

In the previous two examples we began by picking values for x. We could just as easily have chosen values for y.

2 Graphing a Straight Line by Plotting Its Intercepts 

What values should we pick for x and y? Which points should we use for plotting? For many straight lines it is easiest to pick the two *intercepts*. Some lines have only one intercept. We will discuss these separately.

The **x-intercept** of a line is the point where the line crosses the x-axis; it has the form $(a, 0)$. The **y-intercept** of a line is the point where the line crosses the y-axis; it has the form $(0, b)$.

INTERCEPT METHOD OF GRAPHING

To graph an equation using intercepts, we:

1. Find the x-intercept by letting $y = 0$ and solving for x.
2. Find the y-intercept by letting $x = 0$ and solving for y.
3. Find one additional ordered pair so that we have three points with which to plot the line.

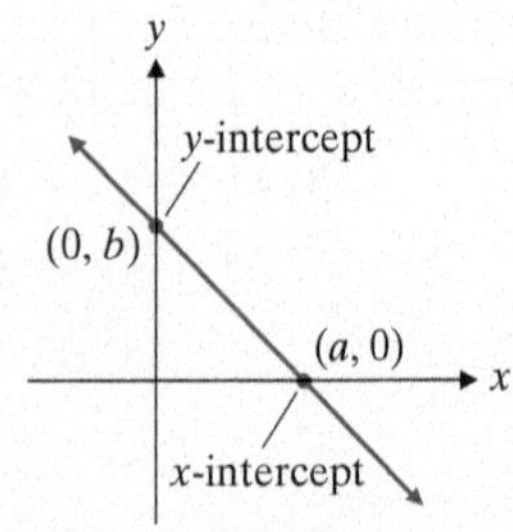

Example 3 Complete **(a)** and **(b)** for the equation $5y - 3x = 15$.

(a) State the x- and y-intercepts.

(b) Use the intercept method to graph.

Solution Substitute values in the equation $5y - 3x = 15$.

(a) Let $y = 0$.

$5(0) - 3x = 15$ Replace y by 0.

$-3x = 15$ Simplify.

$x = -5$ Divide both sides by -3.

x	y
−5	0

x-intercept

The ordered pair $(-5, 0)$ is the x-intercept.

Let $x = 0$.

$5y - 3(0) = 15$ Replace x by 0.

$5y = 15$ Simplify.

$y = 3$ Divide both sides by 5.

x	y
0	3

y-intercept

The ordered pair $(0, 3)$ is the y-intercept.

(b) We find another ordered pair to have a third point and then graph.

Let $y = 6$.

$5(6) - 3x = 15$ Replace y by 6.

$30 - 3x = 15$ Simplify.

$-3x = -15$ Subtract 30 from both sides.

$x = \frac{-15}{-3}$ or 5

The ordered pair is (5, 6).
Our table of values is

x	y
−5	0
0	3
5	6

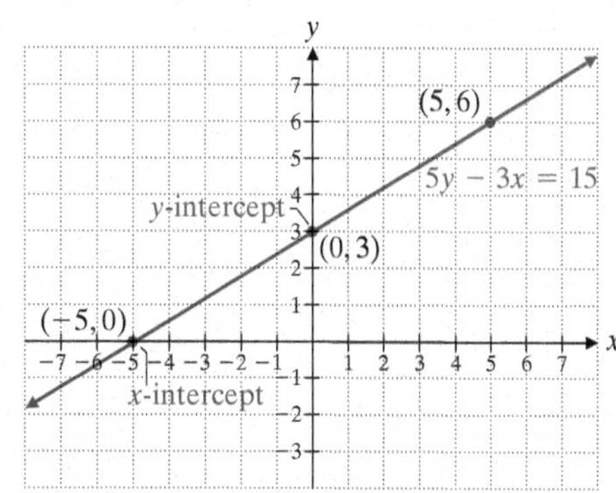

CAUTION: The three points on the graph must form a straight line. If the three points do not form a straight line, you made a calculation error. □

Student Practice 3 Use the intercept method to graph $2y - x = 6$. Use the given coordinate system.

Student Practice 3

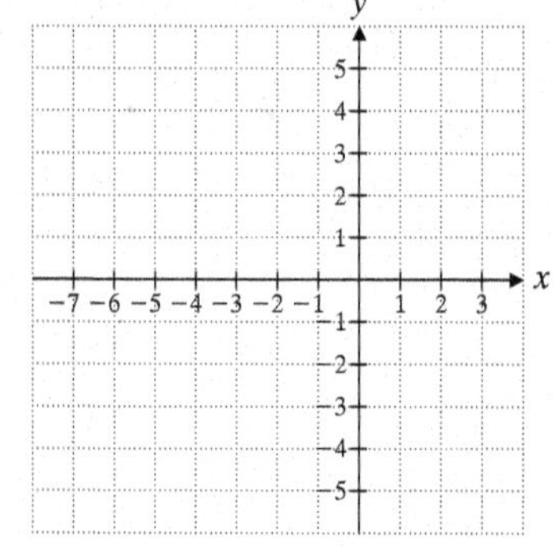

To Think About: *Lines That Go Through the Origin* Can you draw all straight lines by the intercept method? Not really. Some straight lines may go through the origin and have only one intercept. If a line goes through the origin, it will have an equation of the form $Ax + By = 0$, where $A \neq 0$ or $B \neq 0$ or both. Refer to Example 2. When we simplified the equation, we obtained $5x - 4y = 0$. Notice that the graph goes through the origin, and thus there is only one intercept. In such cases you should plot two additional points besides the origin.

3 Graphing Horizontal and Vertical Lines 

You will notice that the x-axis is a horizontal line. It is the line $y = 0$, since for any value of x, the value of y is 0. Try a few points. The points $(1, 0)$, $(3, 0)$, and $(-2, 0)$ all lie on the x-axis. Any horizontal line will be parallel to the x-axis. Lines such as $y = 5$ and $y = -2$ are horizontal lines.

What does $y = 5$ mean? It means that for any value of x, y is 5. Likewise $y = -2$ means that for any value of x, $y = -2$.

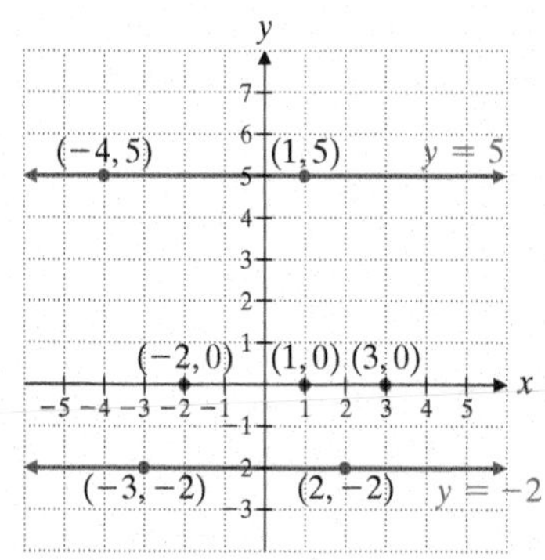

How can we recognize the equation of a line that is horizontal, that is, parallel to the x-axis?

> If the graph of an equation is a straight line that is parallel to the x-axis (that is, a horizontal line), the equation will be of the form $y = b$, where b is some real number.

Example 4 Graph $y = -3$.

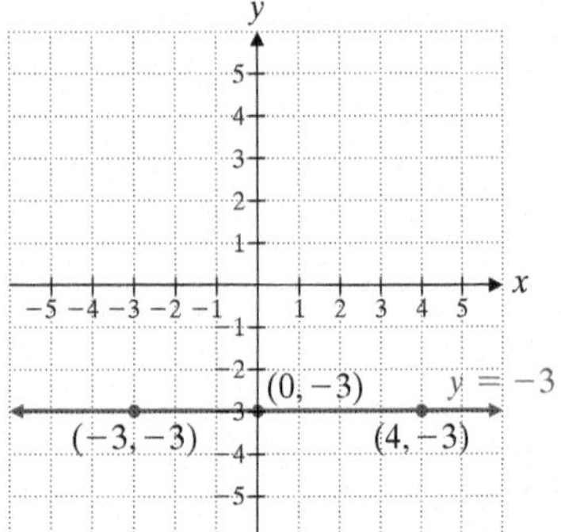

Solution A solution to $y = -3$ is any ordered pair that has y-coordinate -3. The x-coordinate can be any number as long as y is -3.

$(0, -3)$, $(-3, -3)$, and $(4, -3)$ are solutions to $y = -3$ since all the **y-values are -3.**

Since the y-coordinate of every point on the line is -3, it is easy to see that the horizontal line will be 3 units below the x-axis.

Note: You could write the equation $y = -3$ as $0x + y = -3$. Then it is clear that for any value of x, you will always obtain $y = -3$. Try it and see. □

Student Practice 4 Graph $y = 2$ on the given coordinate system.

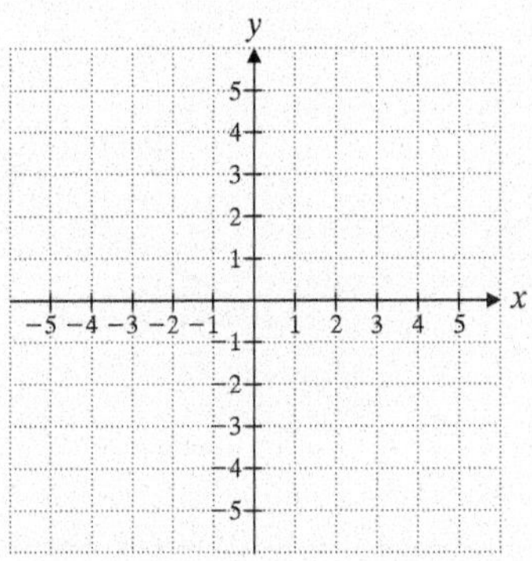

Notice that the y-axis is a vertical line. This is the line $x = 0$, since for any y, x is 0. Try a few points. The points $(0, 2)$, $(0, -3)$, and $(0, \frac{1}{2})$ all lie on the y-axis. Any vertical line will be parallel to the y-axis. Lines such as $x = 2$ and $x = -3$ are vertical lines.

Think of what $x = 2$ means. It means that for any value of y, x is 2. The graph of $x = 2$ is a vertical line two units to the right of the y-axis.

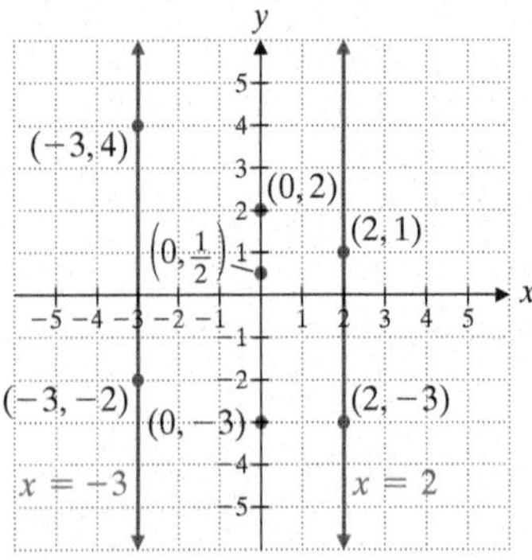

How can we recognize the equation of a line that is vertical, that is, parallel to the y-axis?

> If the graph of an equation is a straight line that is parallel to the y-axis (that is, a vertical line), the equation will be of the form $x = a$, where a is some real number.

Example 5 Graph $2x + 1 = 11$.

Solution Notice that there is only one variable, x, in the equation. This is an indication that we can simplify the equation to the form $x = a$.

$$2x + 1 = 11 \quad x \text{ is the only variable in the equation.}$$
$$2x = 10 \quad \text{Solve for } x.$$
$$x = 5$$

Since the x-coordinate of every point on this line is 5, we can see that the vertical line will be 5 units to the right of the y-axis.

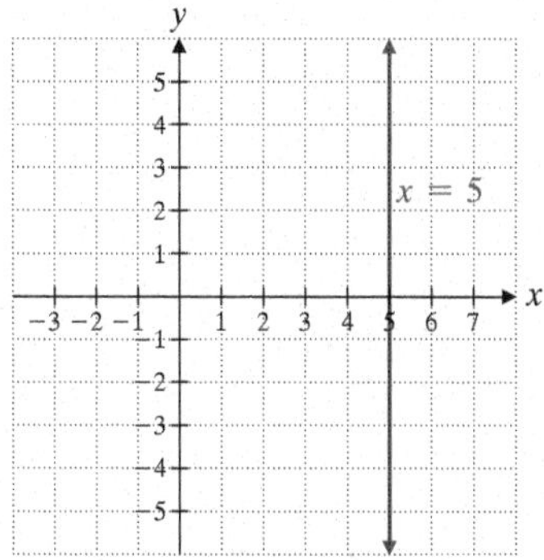

□

Student Practice 5 Graph $3x + 1 = -8$ on the following coordinate system.

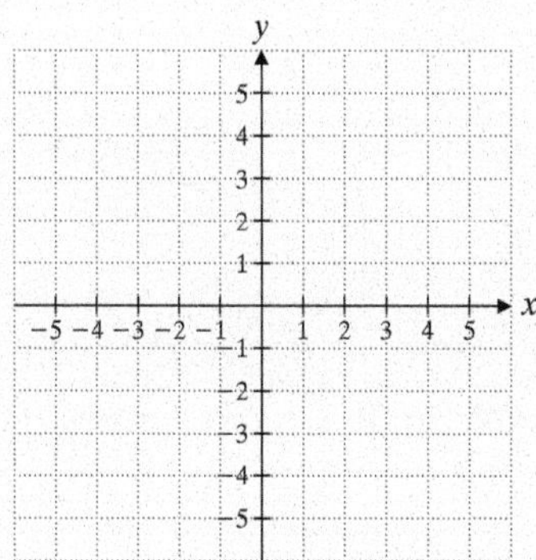

3.2 Exercises MyMathLab®

Verbal and Writing Skills, Exercises 1–4

Unless otherwise indicated, assume each grid line represents one unit.

1. Is the point $(-2, 5)$ a solution to the equation $2x + 5y = 0$? Why or why not?

2. The graph of a linear equation in two variables is a __________ __________.

3. The x-intercept of a line is the point where the line crosses the __________.

4. The graph of the equation $y = b$ is a __________ line.

Complete the ordered pairs so that each is a solution of the given linear equation. Then graph the equation by plotting each solution and connecting the points by a straight line.

5. $y = x - 4$
$(0, \quad)$
$(2, \quad)$
$(4, \quad)$

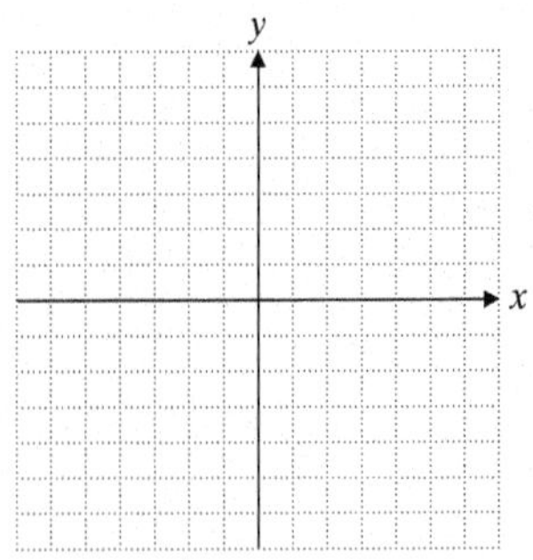

6. $y = x + 4$
$(-1, \quad)$
$(0, \quad)$
$(1, \quad)$

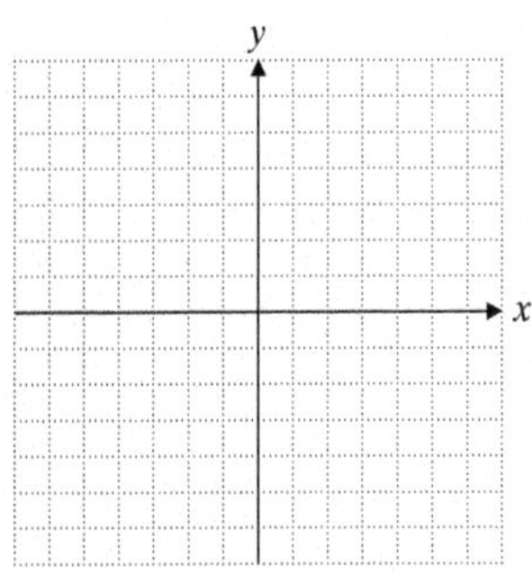

7. $y = -2x + 1$
$(0, \quad)$
$(-2, \quad)$
$(1, \quad)$

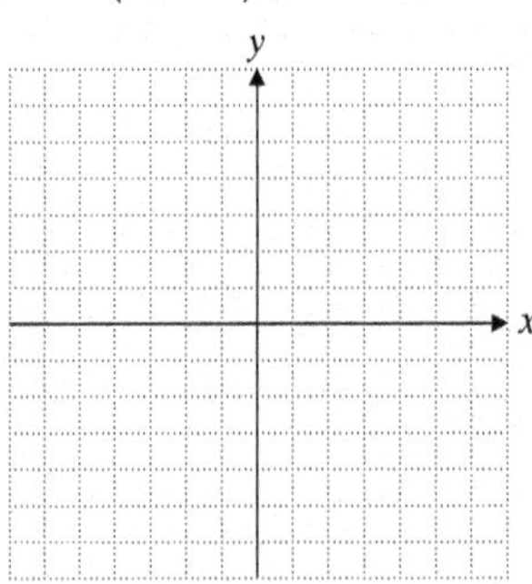

8. $y = -3x - 4$
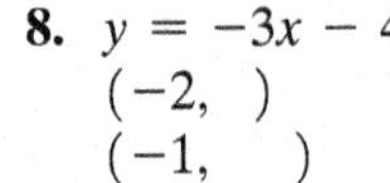
$(-2, \quad)$
$(-1, \quad)$
$(0, \quad)$

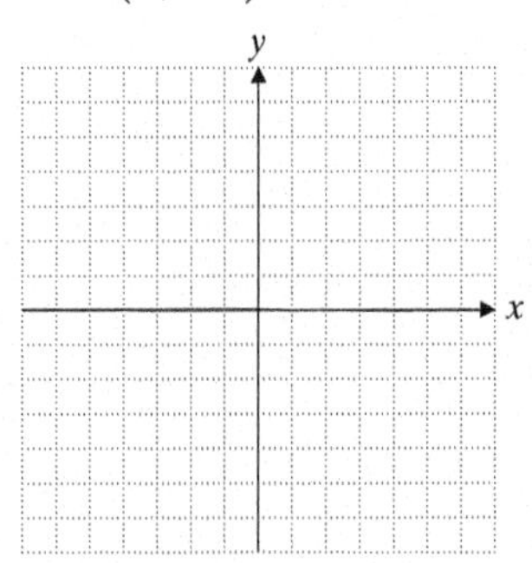

9. $y = 3x - 1$
$(0, \quad)$
$(2, \quad)$
$(-1, \quad)$

10. $y = -2x + 3$
$(0, \quad)$
$(2, \quad)$
$(4, \quad)$

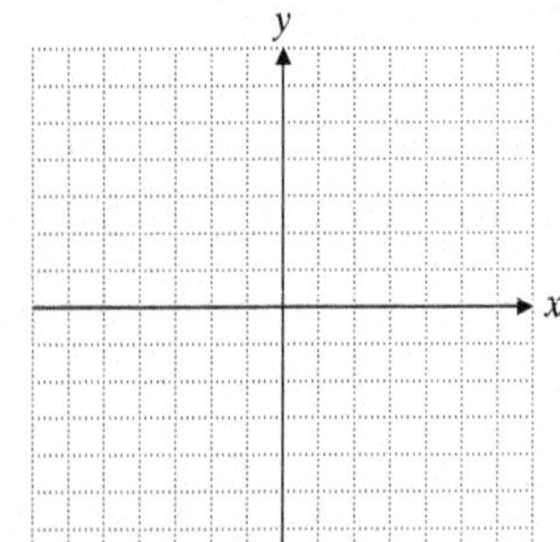

11. $y = 2x - 5$
$(0, \quad)$
$(2, \quad)$
$(4, \quad)$

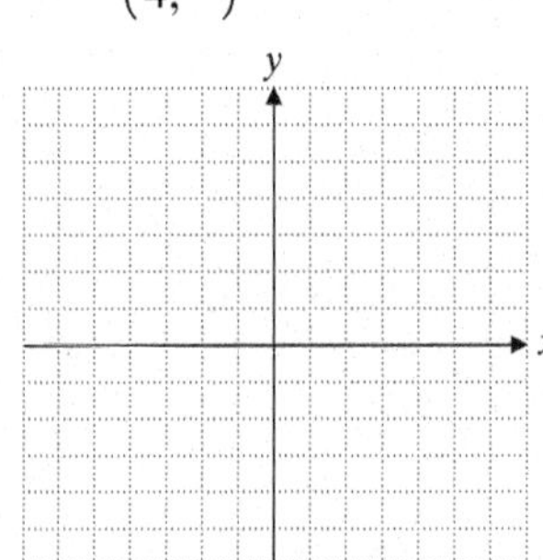

12. $y = 3x + 2$
$(-1, \quad)$
$(0, \quad)$
$(1, \quad)$

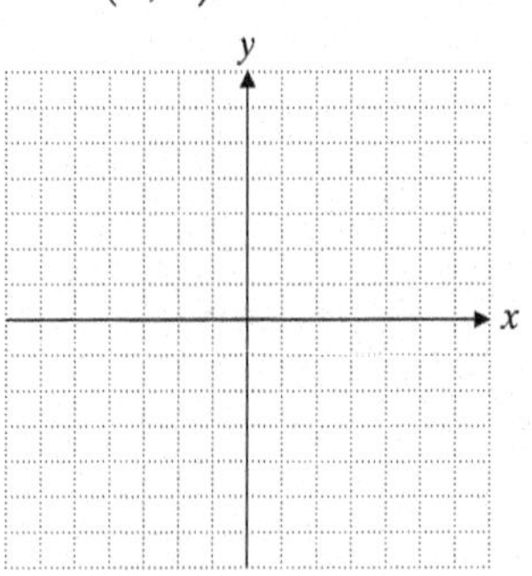

Graph each equation by plotting three points and connecting them. Use a table of values to organize the ordered pairs.

13. $y = -x + 3$

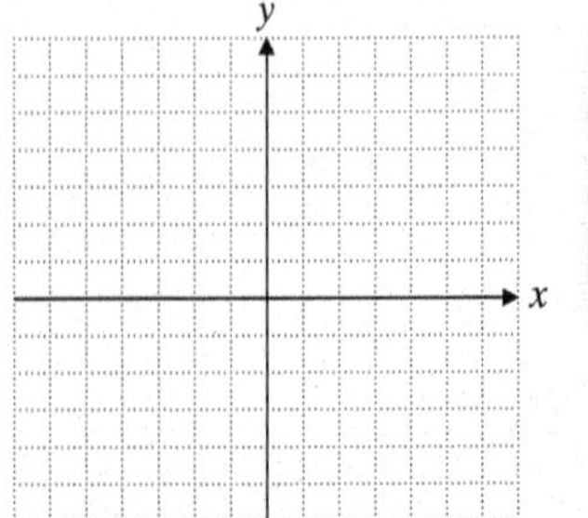

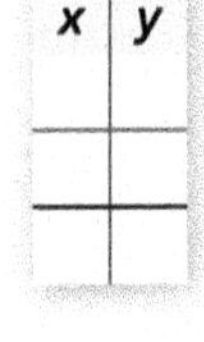

x	y

14. $y = -3x + 2$

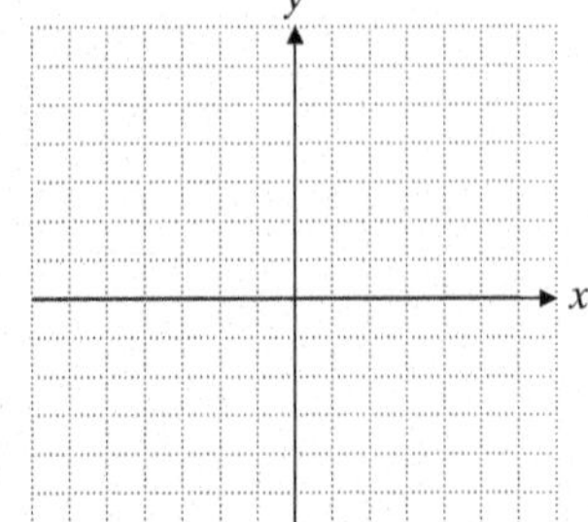

x	y

15. $3x - 2y = 0$

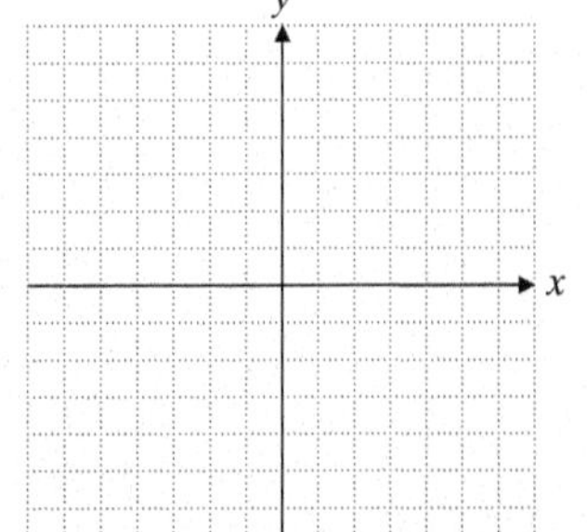

x	y

16. $2y - 5x = 0$

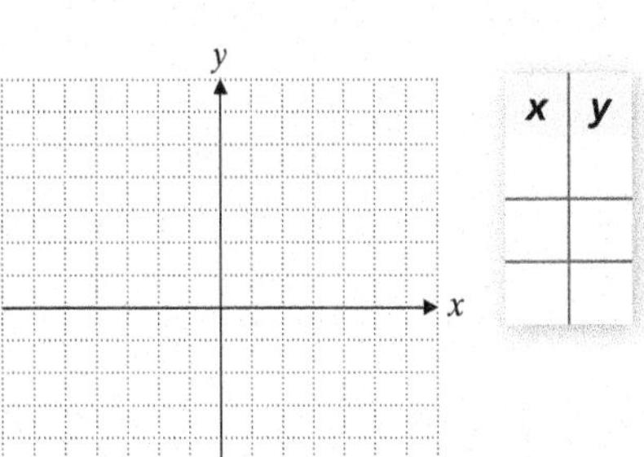

17. $y = -\frac{3}{4}x + 3$

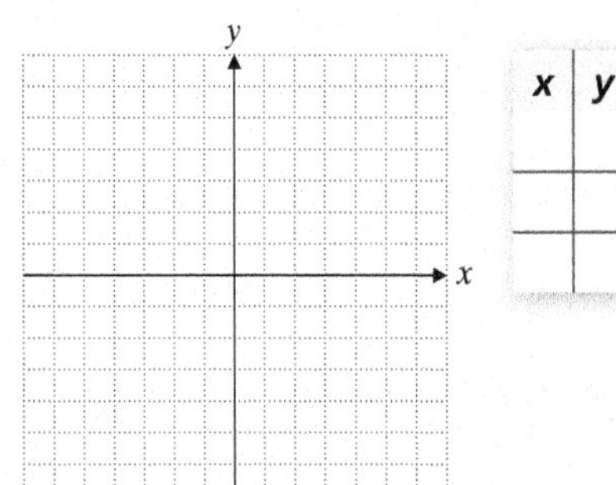

18. $y = \frac{2}{3}x + 2$

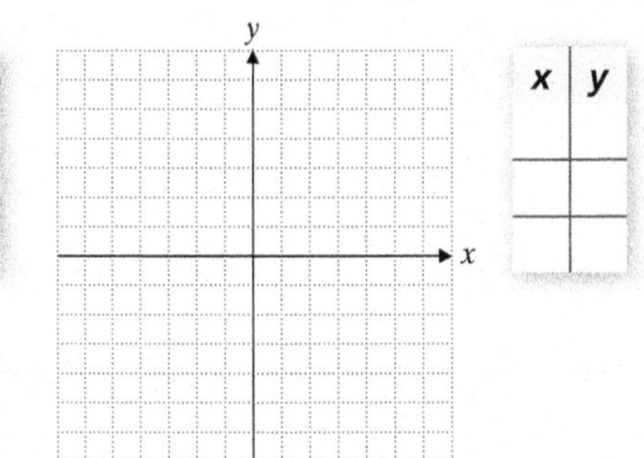

19. $4x + 6 + 3y = 18$

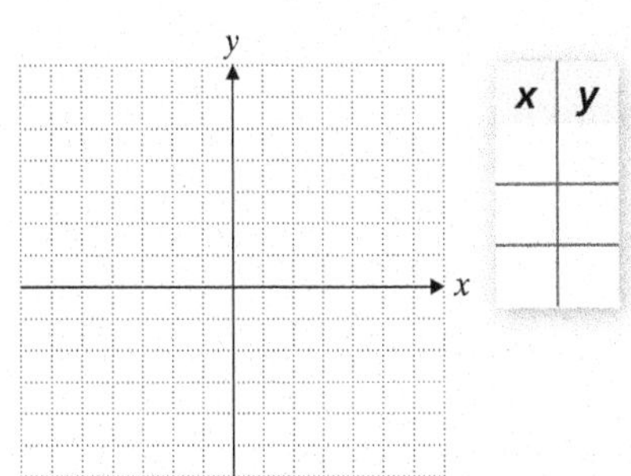

20. $-2x + 3 + 4y = 15$

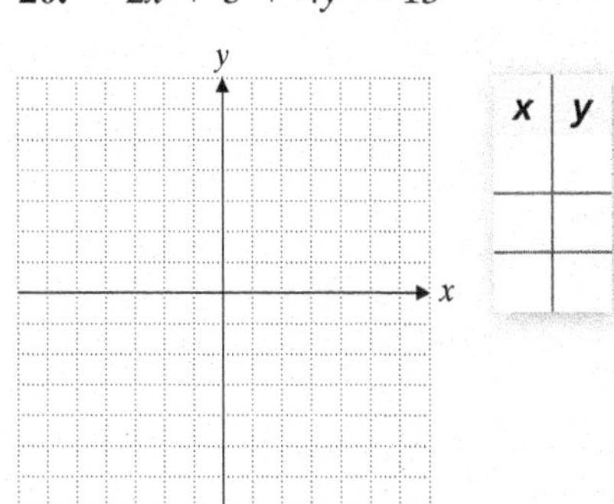

*For exercises 21–24: (**a**) state the x- and y-intercepts; (**b**) use the intercept method to graph.*

21. $y = 6 - 2x$

(a) x-intercept: ____
y-intercept: ____

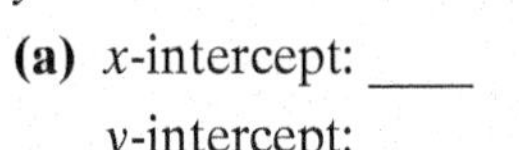

(b)

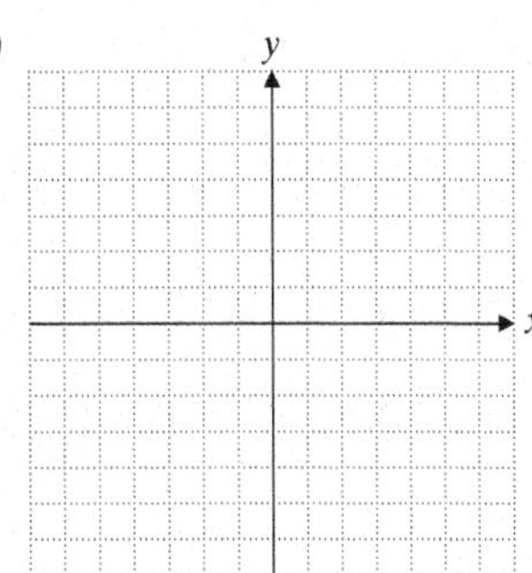

22. $y = 4 - 2x$

(a) x-intercept: ____
y-intercept: ____

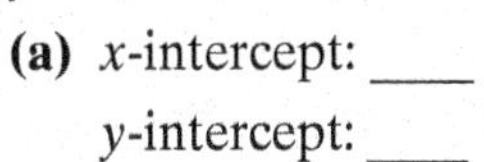

(b)

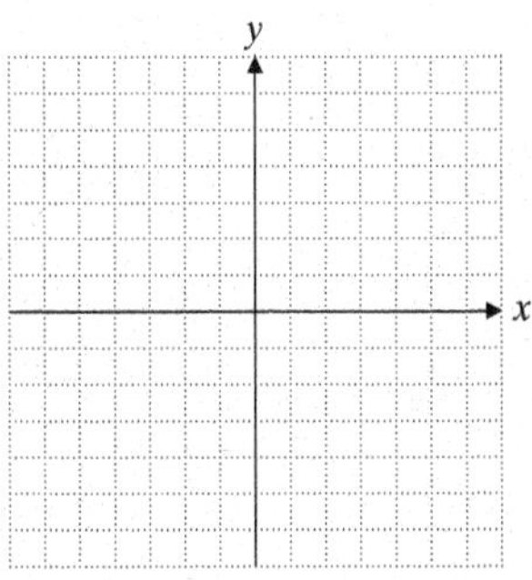

23. $x + 3 = 6y$

(a) x-intercept: ______
y-intercept: ______

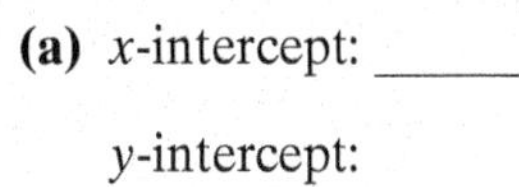

(b)

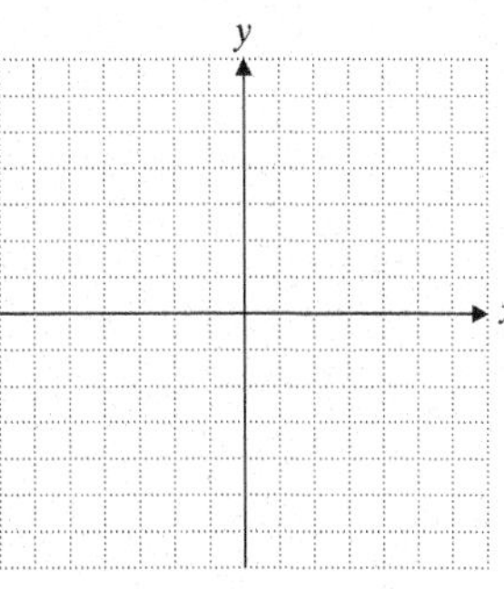

24. $x - 6 = 2y$

(a) x-intercept: ____
y-intercept: _____

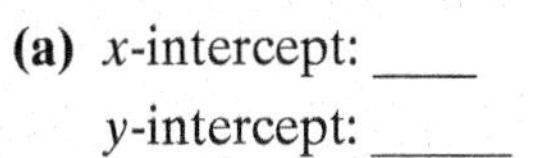

(b)

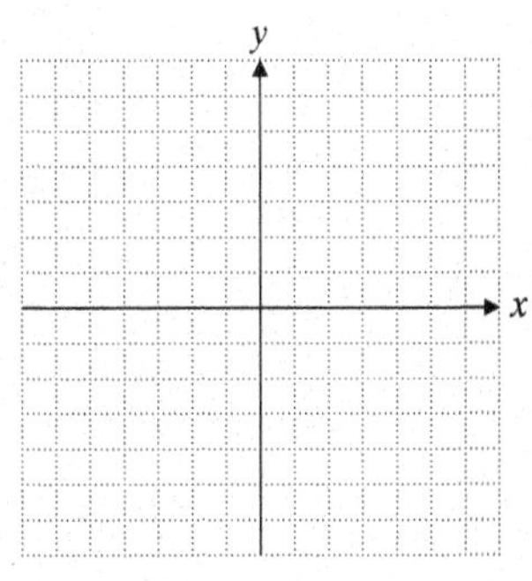

Graph the equation. Be sure to simplify the equation before graphing it.

25. $x = 4$

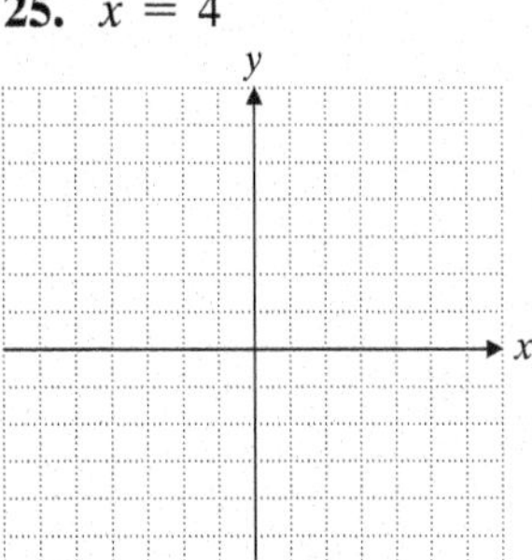

26. $y = -4$

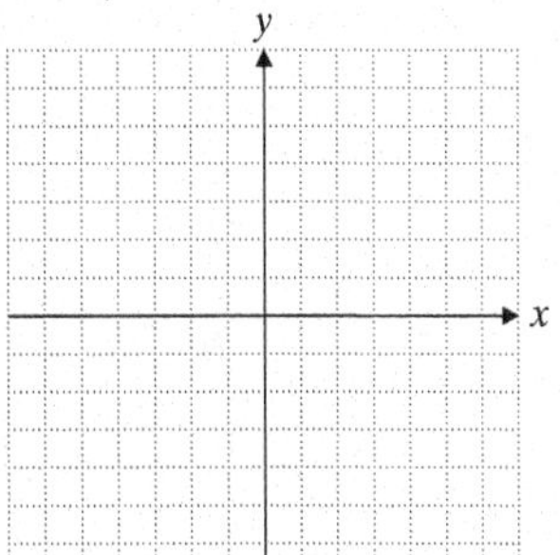

27. $y - 2 = 3y$

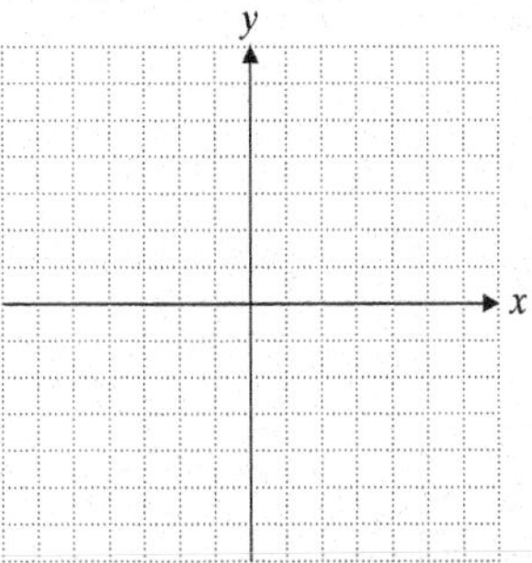

28. $3x - 4 = -13$

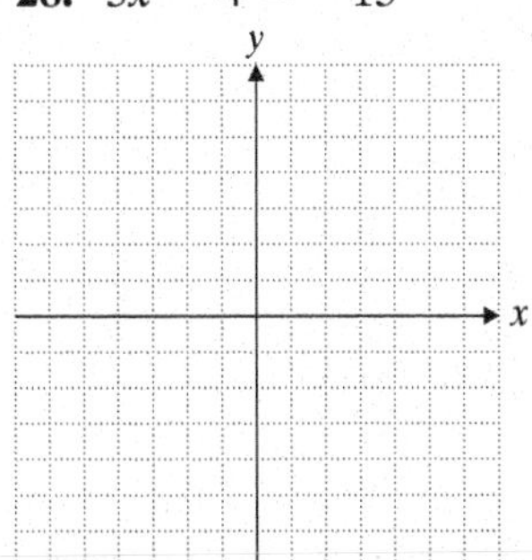

Mixed Practice *Graph.*

29. $2x + 5y - 2 = -12$

30. $3x - 4y - 5 = -17$

31. $2x + 9 = 5x$

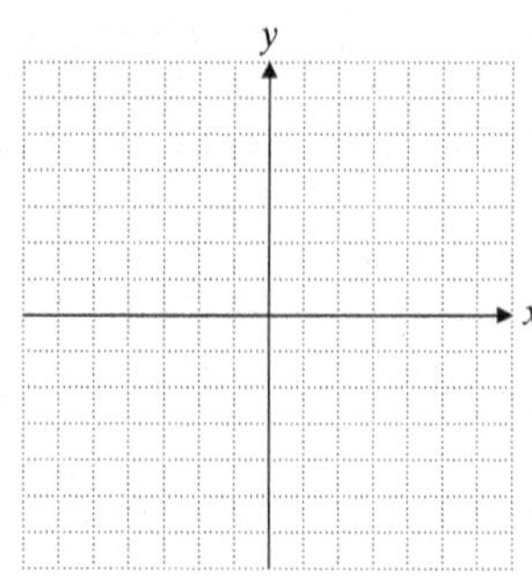

32. $3y + 1 = 7$

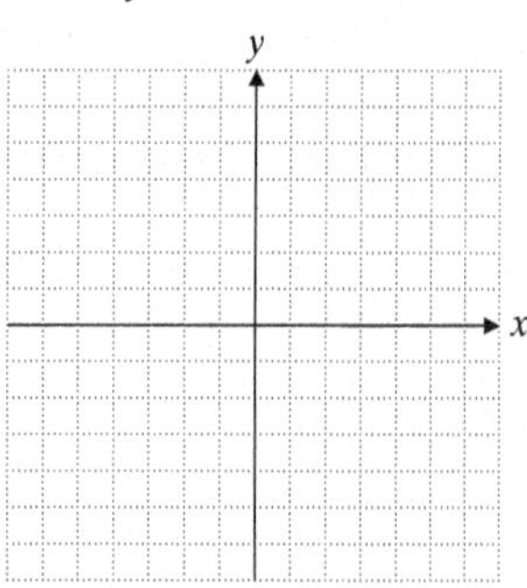

Applications

33. ***Cross-Country Skiing*** The number of calories burned by an average person while cross-country skiing is given by the equation $C = 8m$, where m is the number of minutes. (*Source:* National Center for Health Statistics.) Graph the equation for $m = 0, 15, 30, 45, 60,$ and 75.

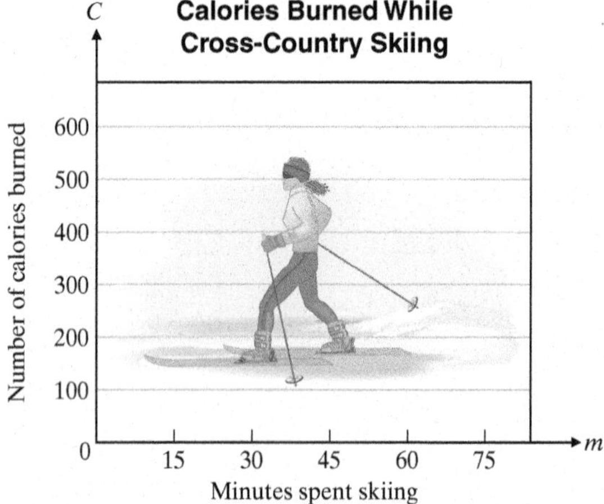

34. ***Calories Burned While Jogging*** The number of calories burned by an average person while jogging is given by the equation $C = \frac{28}{3}m$, where m is the number of minutes. (*Source:* National Center for Health Statistics.) Graph the equation for $m = 0, 15, 30, 45, 60,$ and 75.

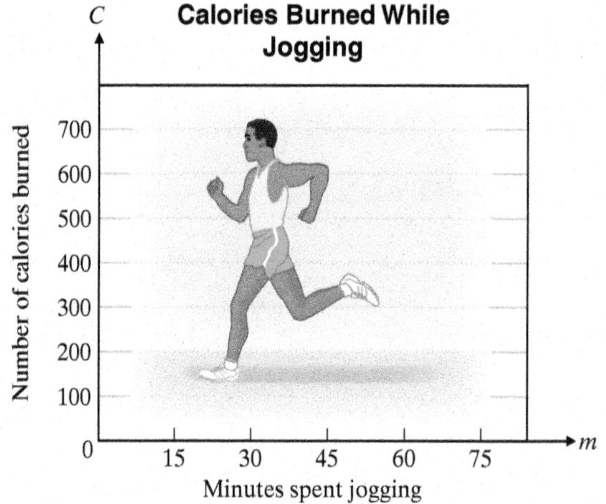

35. ***Foreign Students in the United States*** The number of foreign students enrolled in college in the United States can be approximated by the equation $S = 11t + 395$, where t stands for the number of years since 1990, and S is the number of foreign students (in thousands). (*Source:* www.opendoors.iienetwork.org.) Graph the equation for $t = 0, 4, 8, 16$.

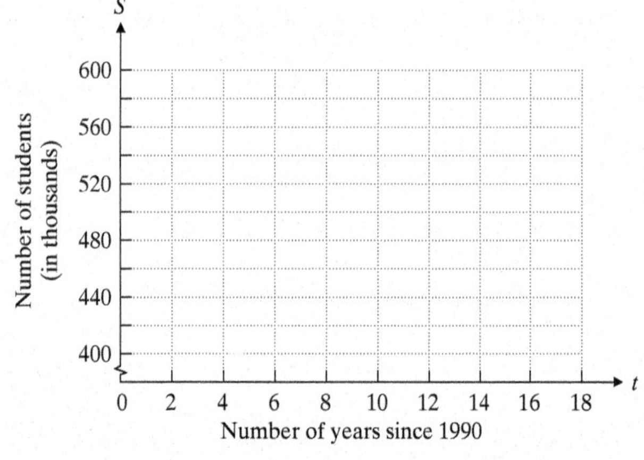

36. ***Food and Beverage Sales*** The amount of money spent at a popular amusement park on food and beverages can be approximated by the equation $S = 16.8t + 223$, where S is the sales in thousands of dollars and t is the number of years since 1995. Graph the equation for $t = 0, 6, 10, 14, 18$.

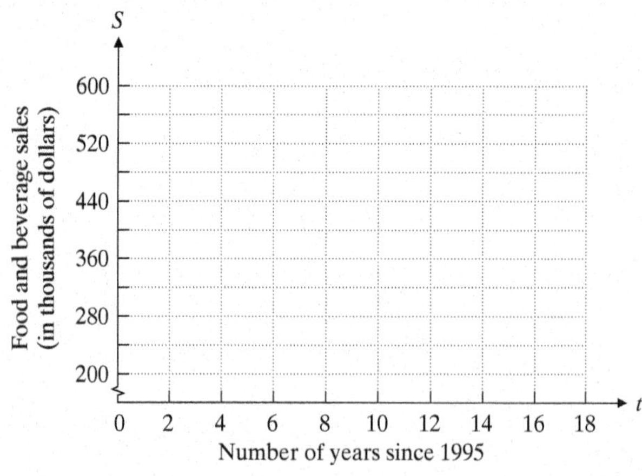

Cumulative Review

37. **[2.3.3]** Solve. $2(x + 3) + 5x = 3x - 2$

38. **[2.8.4]** Solve and graph on a number line. $4 - 3x \leq 18$

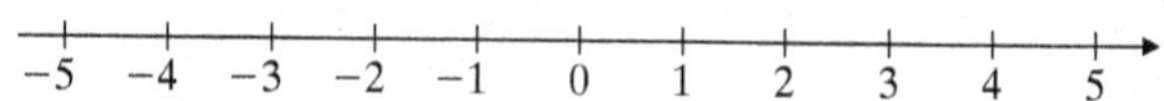

Quick Quiz 3.2

1. Graph $4y + 1 = x + 9$.

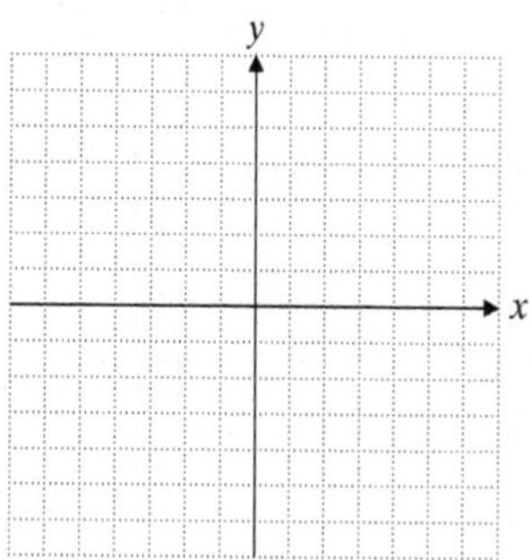

2. Graph $3y = 2y + 4$.

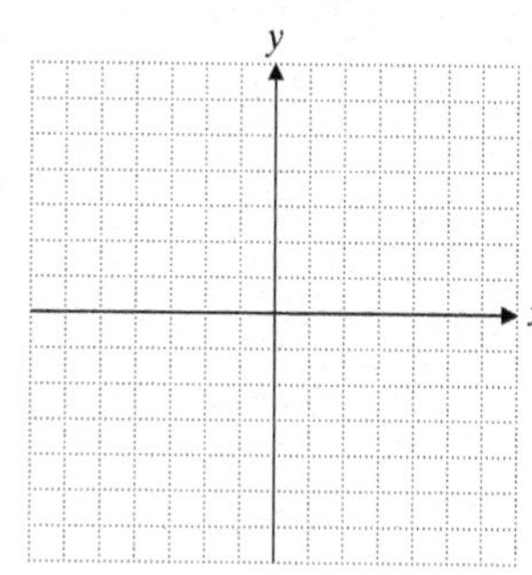

3. Complete **(a)** and **(b)** for the equation $y = -2x + 4$.

(a) State the x- and y-intercepts.

(b) Use the intercept method to graph.

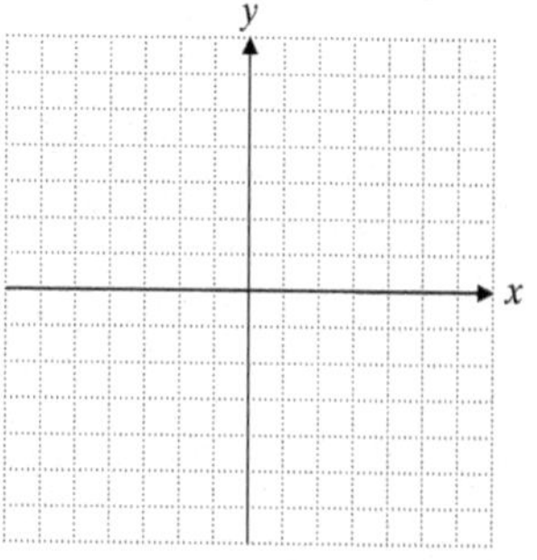

4. **Concept Check** In graphing the equation $3y - 7x = 0$, what is the most important ordered pair to obtain before drawing a graph of the line? Why is that ordered pair so essential to drawing the graph?

3.3 The Slope of a Line

Student Learning Objectives

After studying this section, you will be able to:

1. Find the slope of a line given two points on the line.
2. Find the slope and y-intercept of a line given its equation.
3. Write the equation of a line given the slope and y-intercept.
4. Graph a line using the slope and y-intercept.
5. Find the slopes of parallel and perpendicular lines.

1 Finding the Slope of a Line Given Two Points on the Line

We often use the word *slope* to describe the incline (the steepness) of a hill. A carpenter or a builder will refer to the *pitch* or *slope* of a roof. The slope is the change in the vertical distance (the rise) compared to the change in the horizontal distance (the run) as you go from one point to another point along the roof. If the change in the vertical distance is greater than the change in the horizontal distance, the slope will be steep. If the change in the horizontal distance is greater than the change in the vertical distance, the slope will be gentle.

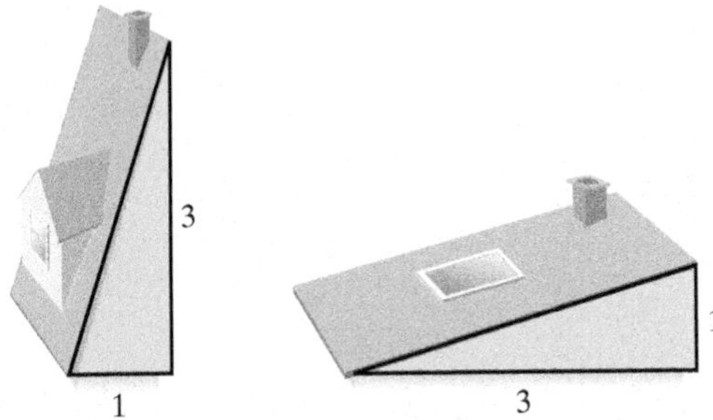

In a coordinate plane, the **slope** of a straight line is defined by the change in y divided by the change in x.

$$\text{slope} = \frac{\text{change in } y}{\text{change in } x}$$

Consider the line drawn through points A and B in the figure. If we measure the change from point A to point B in the x-direction and the y-direction, we will have an idea of the steepness (or the slope) of the line. From point A to point B, the change in y-values is from 2 to 4, a *change of* 2. From point A to point B, the change in x-values is from 1 to 5, a *change of* 4. Thus

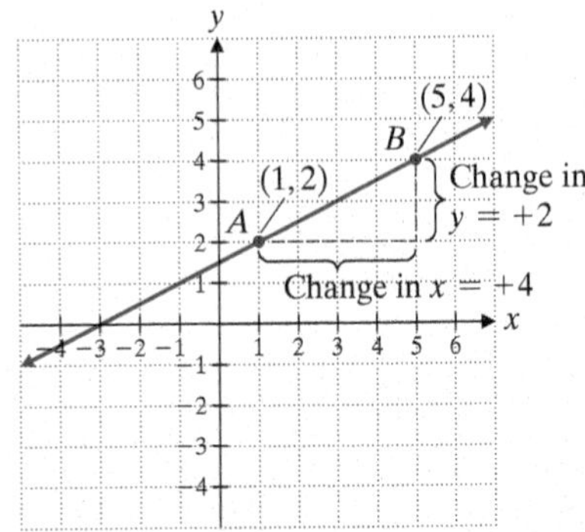

$$\text{slope} = \frac{\text{change in } y}{\text{change in } x} = \frac{2}{4} = \frac{1}{2}.$$

Informally, we can describe this move as the rise over the run: slope $= \dfrac{\text{rise}}{\text{run}}$. We now state a more formal (and more frequently used) definition.

(x₂, y₂) / Change in y (rise) / (x₁, y₁) / Change in x (run)

DEFINITION OF SLOPE OF A LINE

The **slope** of any *nonvertical* straight line that contains the points with coordinates (x_1, y_1) and (x_2, y_2) is defined by the difference ratio

$$\text{slope} = m = \frac{y_2 - y_1}{x_2 - x_1} \quad \text{where } x_2 \neq x_1.$$

The use of subscripted terms such as x_1, x_2, and so on, is just a way of indicating that the first x-value is x_1 and the second x-value is x_2. Thus (x_1, y_1) are the coordinates of the first point and (x_2, y_2) are the coordinates of the second point. The letter m is commonly used for the slope.

Example 1 Find the slope of the line that passes through (2, 0) and (4, 2).

Solution Let (2,0) be the first point (x_1, y_1) and (4,2) be the second point (x_2, y_2).

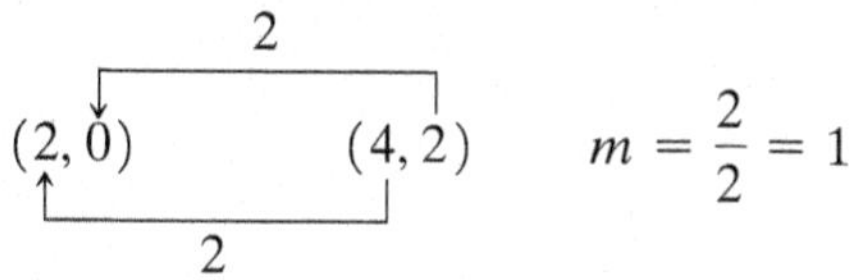

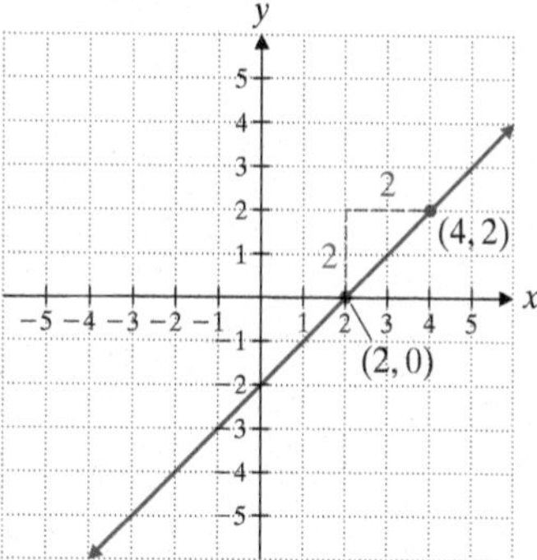

$$\text{slope} = m = \frac{y_2 - y_1}{x_2 - x_1} = \frac{2 - 0}{4 - 2} = \frac{2}{2} = 1.$$

Note that the slope of the line will be the same if we let (4, 2) be the first point (x_1, y_1) and (2, 0) be the second point (x_2, y_2).

−2

(4, 2) (2, 0)

−2

$$m = \frac{y_2 - y_1}{x_2 - x_1} = \frac{0 - 2}{2 - 4} = \frac{-2}{-2} = 1$$

Thus, given two points, it does not matter which you call (x_1, y_1) and which you call (x_2, y_2). □

CAUTION: Be careful, however, not to put the x's in one order and the y's in another order when finding the slope from two points on a line.

Student Practice 1 Find the slope of the line that passes through (6, 1) and (−4, −1).

It is a good idea to have some concept of what different slopes mean. In downhill skiing, a very gentle slope used for teaching beginning skiers might drop 1 foot vertically for each 10 feet horizontally. The slope would be $\frac{1}{10}$. The speed of a skier on a hill with such a gentle slope would be only about 6 miles per hour.

A triple diamond slope for experts might drop 11 feet vertically for each 10 feet horizontally. The slope would be $\frac{11}{10}$. The speed of a skier on such an expert trail would be in the range of 60 miles per hour.

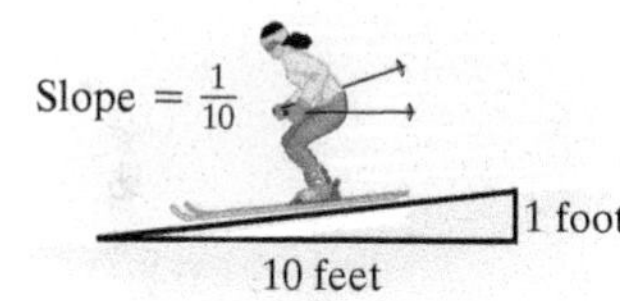

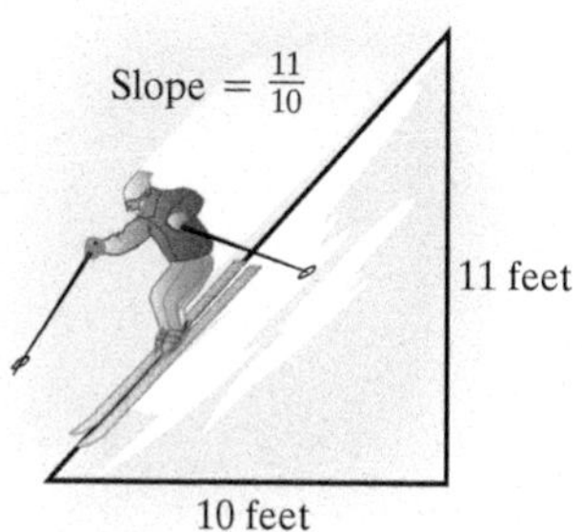

It is important to see how positive and negative slopes affect the graphs of lines.

POSITIVE SLOPE

$m = \frac{5-2}{4-1} = \frac{3}{3} = 1$

(4, 5)

y is increasing from 2 to 5 as you go from left to right

(1, 2)

NEGATIVE SLOPE

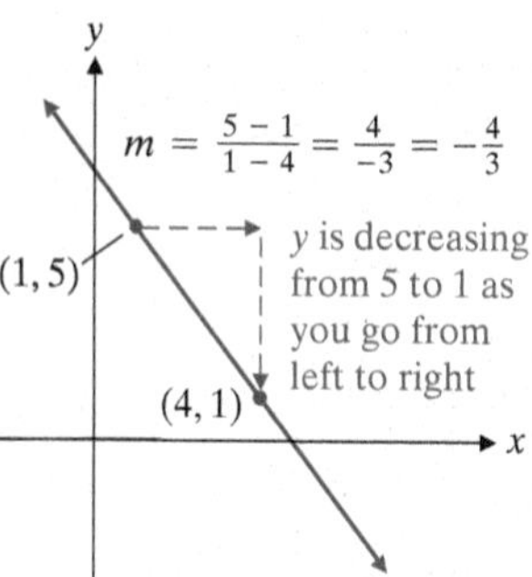

1. If the y-values increase as you go from left to right, the slope of the line is positive.
2. If the y-values decrease as you go from left to right, the slope of the line is negative.

Example 2 Find the slope of the line that passes through $(-3, 2)$ and $(2, -4)$.

Solution Let $(-3, 2)$ be (x_1, y_1) and $(2, -4)$ be (x_2, y_2).

$$m = \frac{y_2 - y_1}{x_2 - x_1} = \frac{-4 - 2}{2 - (-3)} = \frac{-4 - 2}{2 + 3} = \frac{-6}{5} = -\frac{6}{5}$$

The slope of this line is negative. We would expect this, since the y-value decreased from 2 to -4 as the x-value increased. What does the graph of this line look like? Plot the points and draw the line to verify. □

Student Practice 2 Find the slope of the line that passes through $(2, 0)$ and $(-1, 1)$.

To Think About: *Using the Slope to Describe a Line* Describe the line in Student Practice 2 by looking at its slope. Then verify by drawing the graph.

Example 3 Find the slope of the line that passes through the given points.

(a) $(0, 2)$ and $(5, 2)$ **(b)** $(-4, 0)$ and $(-4, -4)$

Solution

(a) Take a moment to look at the y-values. What do you notice? What does this tell you about the line? Now calculate the slope.

$(0, 2)$ and $(5, 2)$ $\quad m = \frac{2 - 2}{5 - 0} = \frac{0}{5} = 0$

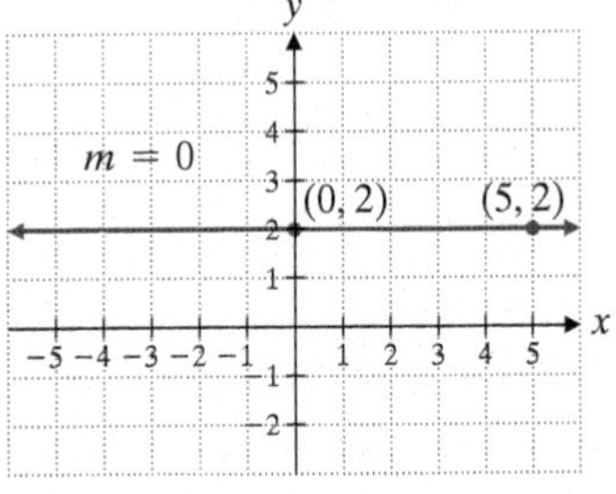

Since any two points on a horizontal line will have the same y-value, the slope of a horizontal line is 0.

(b) Take a moment to look at the x-values. What do you notice? What does this tell you about the line? Now calculate the slope.

$(-4, 0)$ and $(-4, -4)$ $\quad m = \frac{-4 - 0}{-4 - (-4)} = \frac{-4}{0}$

Recall that division by 0 is undefined. The slope of a vertical line is undefined. We say that a vertical line has **no slope.**

Notice in our definition of slope that $x_2 \neq x_1$. Thus it is not appropriate to use the formula for slope for the points in part **(b)**. We did so to illustrate what would happen if $x_2 = x_1$. We get an impossible situation, $\frac{y_2 - y_1}{0}$. Now you can see why we include the restriction $x_2 \neq x_1$ in our definition. □

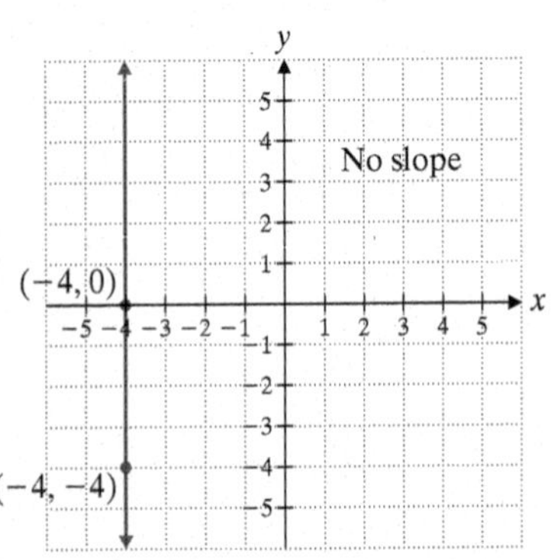

Student Practice 3 Find the slope of the line that passes through the given points.

(a) $(-5, 6)$ and $(-5, 3)$ **(b)** $(-7, -11)$ and $(3, -11)$

> **SLOPE OF A STRAIGHT LINE**
>
> 1. Positive slope
>   Line goes upward to the right
>
> 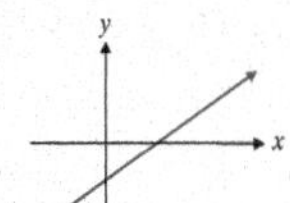
>
>
> *Lines with positive slopes go upward* as you go from left to right.
>
> 2. Negative slope
>   Line goes downward to the right
>
> 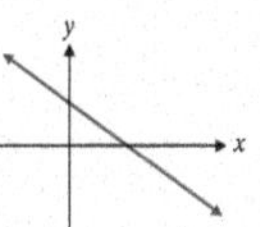
>
>
> *Lines with negative slopes go downward* as you go from left to right.
>
> 3. Zero slope
>   Horizontal line
>
> Horizontal lines have a slope of 0.
>
> 4. Undefined slope
>   Vertical line
>
> 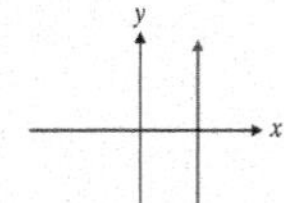
>
>
> A vertical line is said to have undefined slope. The slope of a vertical line is not defined. In other words, a vertical line has no slope.

2 Finding the Slope and *y*-Intercept of a Line Given Its Equation

Recall that the equation of a line is a linear equation in two variables. This equation can be written in several different ways. A very useful form of the equation of a straight line is the slope–intercept form. This form can be derived in the following way. Suppose that a straight line with slope m crosses the y-axis at a point $(0, b)$. Consider any other point on the line and label the point (x, y). Then we have the following.

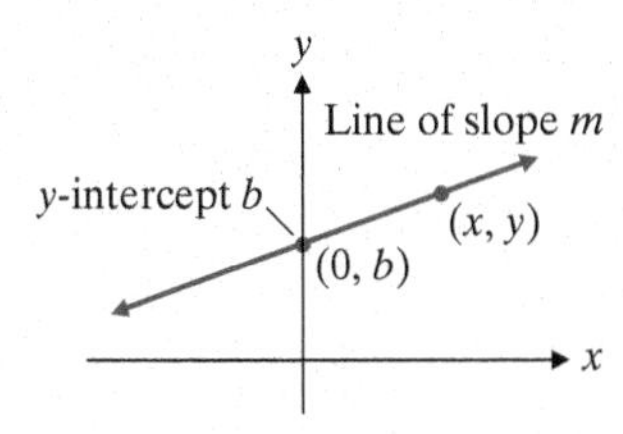

$$\frac{y_2 - y_1}{x_2 - x_1} = m \qquad \text{Definition of slope.}$$

$$\frac{y - b}{x - 0} = m \qquad \text{Substitute } (0, b) \text{ for } (x_1, y_1) \text{ and } (x, y) \text{ for } (x_2, y_2).$$

$$\frac{y - b}{x} = m \qquad \text{Simplify.}$$

$$y - b = mx \qquad \text{Multiply both sides by } x.$$

$$y = mx + b \qquad \text{Add } b \text{ to both sides.}$$

This form of a linear equation immediately reveals the slope of the line, m, and the y-coordinate of the point where the line intercepts (crosses) the y-axis, b.

> **SLOPE–INTERCEPT FORM OF A LINE**
>
> The slope–intercept form of the equation of the line that has slope m and y-intercept $(0, b)$ is given by
>
> $$y = mx + b.$$

By using algebraic operations, we can write any linear equation in slope–intercept form and use this form to identify the slope and the y-intercept of the line.

Example 4 What is the slope and the y-intercept of the line $5x + 3y = 2$?

Solution We want to solve for y and get the equation in the form $y = mx + b$. We need to isolate the y-variable.

$$5x + 3y = 2$$

$$3y = -5x + 2 \qquad \text{Subtract } 5x \text{ from both sides.}$$

$$y = \frac{-5x + 2}{3} \qquad \text{Divide both sides by 3.}$$

Continued on next page

$$y = -\frac{5}{3}x + \frac{2}{3}$$ Using the property $\frac{a+b}{c} = \frac{a}{c} + \frac{b}{c}$, write the right-hand side as two fractions.

$m = -\frac{5}{3}$ and $b = \frac{2}{3}$ The *slope* is $-\frac{5}{3}$. The *y-intercept* is $\left(0, \frac{2}{3}\right)$.

Note: We write the y-intercept as an ordered pair of the form $(0, b)$. □

Student Practice 4 What is the slope and the y-intercept of the line $4x - 2y = -5$?

3 Writing the Equation of a Line Given the Slope and y-Intercept

If we know the slope of a line and the y-intercept, we can write the equation of the line, $y = mx + b$.

Example 5 Find an equation of the line with slope $\frac{2}{5}$ and y-intercept $(0, -3)$.

(a) Write the equation in slope–intercept form, $y = mx + b$.

(b) Write the equation in the form $Ax + By = C$.

Solution

(a) We are given that $m = \frac{2}{5}$ and $b = -3$. Thus we have the following.

$$y = mx + b$$
$$y = \frac{2}{5}x + (-3)$$
$$y = \frac{2}{5}x - 3$$

(b) To write the equation in the form $Ax + By = C$, let us first clear the equation of fractions so that A, B, and C are integers. Then we move the x-term to the left side.

$$5y = 5\left(\frac{2x}{5}\right) - 5(3)$$ Multiply each term by 5.

$$5y = 2x - 15$$ Simplify.

$$-2x + 5y = -15$$ Subtract $2x$ from each side.

$$2x - 5y = 15$$ Multiply each term by -1. The form $Ax + By = C$ is usually written with A as a positive integer. □

Student Practice 5 Find an equation of the line with slope $-\frac{3}{7}$ and y-intercept $(0, \frac{2}{7})$.

(a) Write the equation in slope–intercept form.

(b) Write the equation in the form $Ax + By = C$.

4 Graphing a Line Using the Slope and y-Intercept

If we know the slope of a line and the y-intercept, we can draw the graph of the line.

Example 6 Graph the line with slope $m = \frac{2}{3}$ and y-intercept $(0, -3)$. Use the given coordinate system.

Solution Recall that the y-intercept is the point where the line crosses the y-axis. We need a point to start the graph with. So we plot the point $(0, -3)$ on the y-axis.

Graphing Calculator

Graphing Lines

You can graph a line given in the form $y = mx + b$ using a graphing calculator. For example, to graph $y = 2x + 4$, enter the right-hand side of the equation in the Y = editor of your calculator and graph. Choose an appropriate window to show all the intercepts. The following window is -10 to 10 by -10 to 10.
Display:

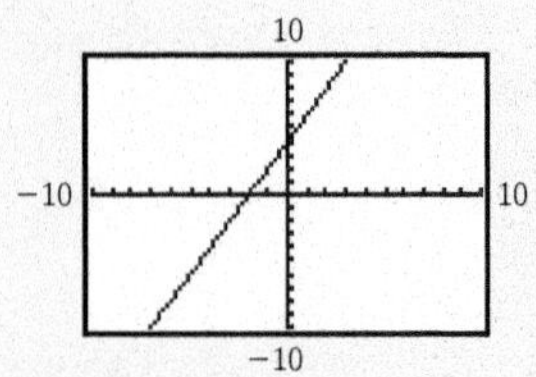

Try graphing other equations given in slope–intercept form.

Recall that slope $= \frac{\text{rise}}{\text{run}}$. Since the slope of this line is $\frac{2}{3}$, we will go up (rise) 2 units and go over (run) to the right 3 units from the point $(0, -3)$. Look at the figure to the right. This is the point $(3, -1)$. Plot the point. Draw a line that connects the two points $(0, -3)$ and $(3, -1)$.

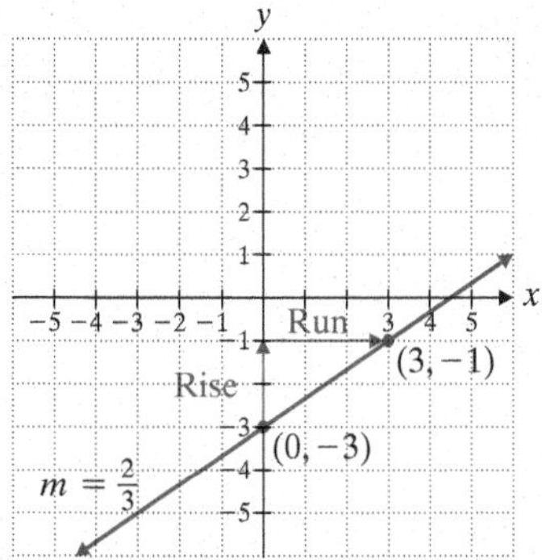

This is the graph of the line with slope $\frac{2}{3}$ and y-intercept $(0, -3)$. □

Student Practice 6 Graph the line with slope $= \frac{3}{4}$ and y-intercept $(0, -1)$. Use the coordinate system in the margin.

Student Practice 6

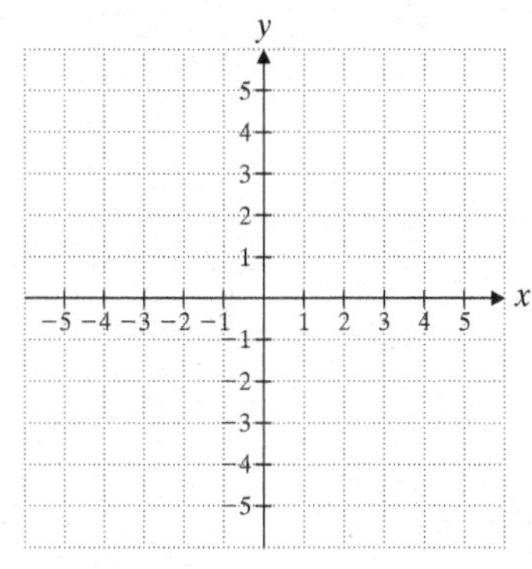

Let's summarize the process used in Example 6. If we know both the slope m and the y-intercept b, we can graph the equation without performing any calculations as follows.

Step 1. Plot the y-intercept point on the graph and then begin at this point.

Step 2. Using the slope, $\frac{\text{rise}}{\text{run}}$, we *rise* in the y-direction. We move up if the number is positive and down if the number is negative. We *run* in the x-direction. We move right if the number is positive and left if the number is negative.

Step 3. Plot this point and draw a line through the two points.

Example 7 Graph the equation $y = -\frac{1}{2}x + 4$. Use the following coordinate system.

Solution

Step 1. Since $b = 4$, we plot the y-intercept $(0, 4)$.

Step 2. Using $m = -\frac{1}{2}$ or $\frac{-1}{2}$, we begin at $(0, 4)$ and go down 1 unit and to the right 2 units.

Step 3. This is the point $(2, 3)$. Plot this point and draw a line that connects the points $(0, 4)$ and $(2, 3)$.

This is the graph of the equation $y = -\frac{1}{2}x + 4$.

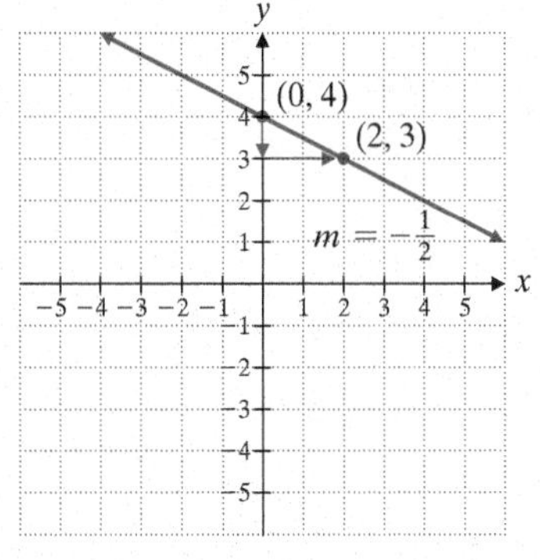

Note: The slope $-\frac{1}{2} = \frac{-1}{2} = \frac{1}{-2}$. Therefore, we get the same line if we write the slope as $\frac{1}{-2}$. Try it. □

Student Practice 7 Graph the equation $y = -\frac{2}{3}x + 5$. Use the coordinate system in the margin.

Student Practice 7

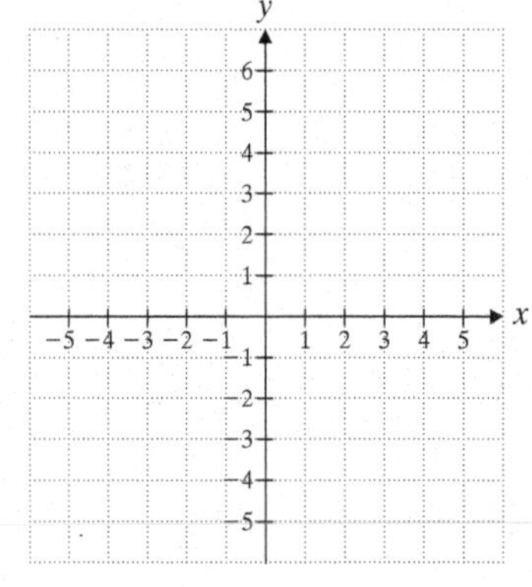

Graphing Using Different Methods In this section and in Section 3.2 we learned the following three ways to graph a linear equation:

1. Plot any 3 points.
2. Find the intercepts.
3. Find the slope and y-intercept.

The following Graphing Organizer summarizes each method and notes the advantages and disadvantages of each.

Graphing Organizer

Method 1: Find Any Three Ordered Pairs

Graph $2y - x = 2$ by plotting three points.

Choose any value for either x or y.

x	y
1	
2	
−4	

Substitute to find the other value.

$2y - \mathbf{1} = 2$ $\quad$ $2y - \mathbf{2} = 2$ $\quad$ $2y - (\mathbf{-4}) = 2$

$2y = 3$ $\quad$ $2y = 4$ $\quad$ $2y + 4 = 2$

$y = \mathbf{\frac{3}{2}}$ $\quad$ $y = \mathbf{2}$ $\quad$ $2y = -2$

$y = \mathbf{-1}$

x	y
1	$\frac{3}{2}$
2	2
−4	−1

Plot points and draw line through points.

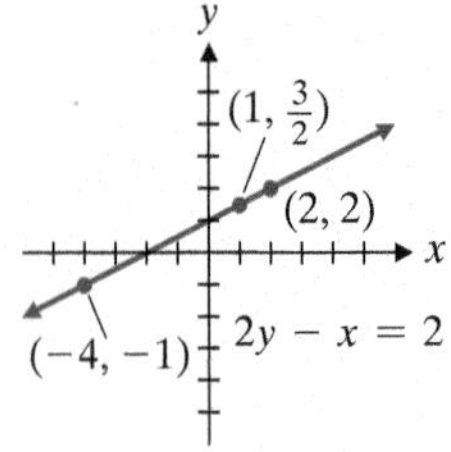

Advantage

1. Method is easy to remember: just choose any point for either x or y then solve for the other value.

Disadvantages

1. It may be difficult to avoid fractions.
2. Whether you start with a value for x or y, solving for the other value can be difficult.

You Try It

Graph $-3x + 2 = y$. Find and label three points.

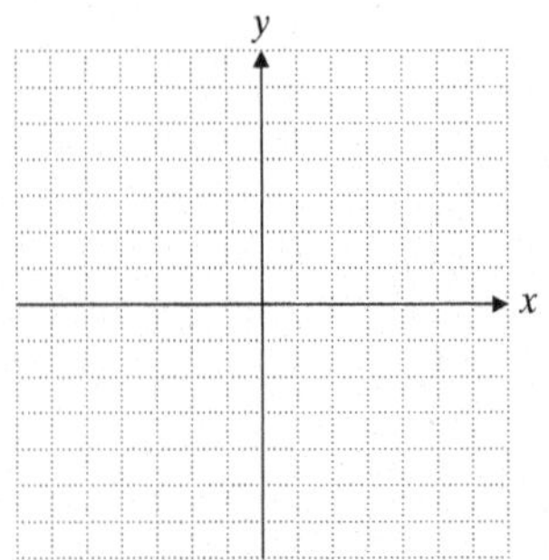

Method 2: Find the x- and y-Intercepts and One Additional Ordered Pair

Graph $2y - x = 2$. State the x- and y-intercepts.

Choose $x = 0$ and $y = 0$, and any other value for either x or y.

x	y
0	
	0
−1	

Notice we are still choosing 3 values, but to get the intercepts we must pick $x = 0$ and $y = 0$.

Substitute to find the other value.

$2y - \mathbf{0} = 2$ $\quad$ $2(\mathbf{0}) - x = 2$ $\quad$ $2y - (\mathbf{-1}) = 2$

$2y = 2$ $\quad$ $-x = 2$ $\quad$ $2y + 1 = 2$

$y = \mathbf{1}$ $\quad$ $x = \mathbf{-2}$ $\quad$ $2y = 1$

$y = \mathbf{\frac{1}{2}}$

x	y
0	1
−2	0
−1	$\frac{1}{2}$

y-intercept: $(0, 1)$
x-intercept: $(-2, 0)$

Plot points and draw line through points.

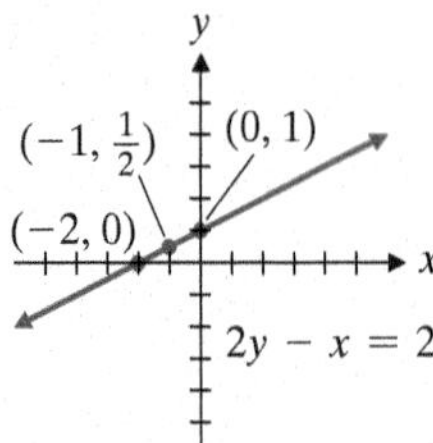

Advantage

1. When we substitute 0 into an equation it is easy to calculate the value of the other variable.

Disadvantage

1. It may be difficult to avoid fractions.

You Try It

Graph $-3x + 2 = y$. Label the x- and y-intercepts.

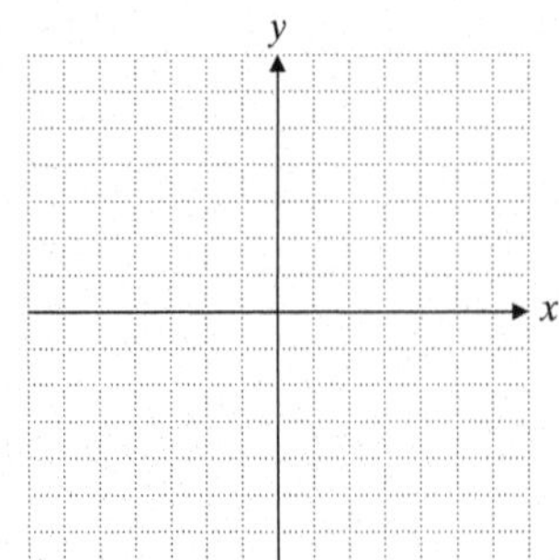

Method 3: Write the Equation in Slope–Intercept Form, $y = mx + b$

Graph $2y - x = 2$. State m and b.

Solve for y.

$$2y = x + 2$$
$$y = \frac{1}{2}x + 1$$

Once we solve for y we can choose any method to graph.

Option 1. Use the slope–intercept method, $y = \frac{1}{2}x + 1$.

Identify b: $b = 1$; y-intercept: $(0, 1)$.

Identify the slope, m: $m = \frac{1}{2}$

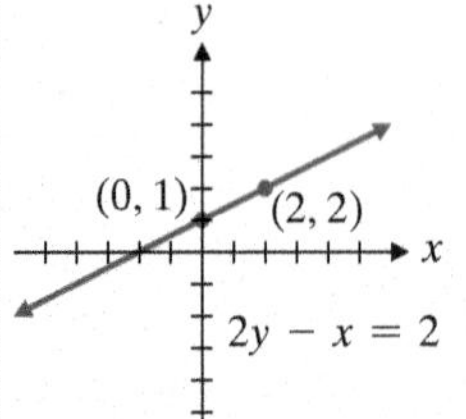

Start at b, move up 1 unit, then move right 2 units. Plot this point and draw a line through the two points.

Option 2. We can also choose to find intercepts or any three ordered pairs when equation is in the form $y = mx + b$.

$$y = \frac{1}{2}x + 1$$

Substitute $x = 2, 0, -2$ and find y.

x	y
2	2
0	1
−2	0

Plot these points on the coordinate planes in columns 1 and 2. You should get the same line. Why?

Advantages

1. When using the slope–intercept method, we can avoid doing any calculations.
2. When we find ordered pairs, the slope–intercept form makes it easy to see how to avoid fractions.

$y = \frac{1}{2}x + 1$ $\quad$ Choosing a multiple of 2 for x will avoid getting a fraction for y.

Disadvantage

1. It may be difficult to solve for y.

You Try It

Graph $-3x + 2 = y$. State m and b.

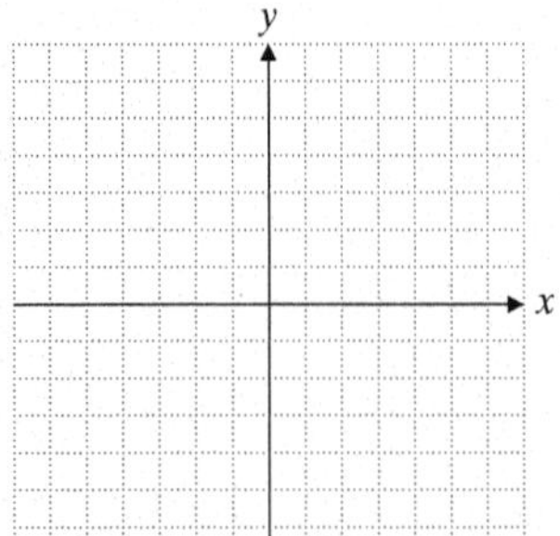

5 Finding the Slopes of Parallel and Perpendicular Lines

Parallel lines are two straight lines that never touch. Look at the parallel lines in the figure. Notice that the slope of line a is -3 and the slope of line b is also -3. Why do you think the slopes must be equal? What would happen if the slope of line b were -1? Graph it and see.

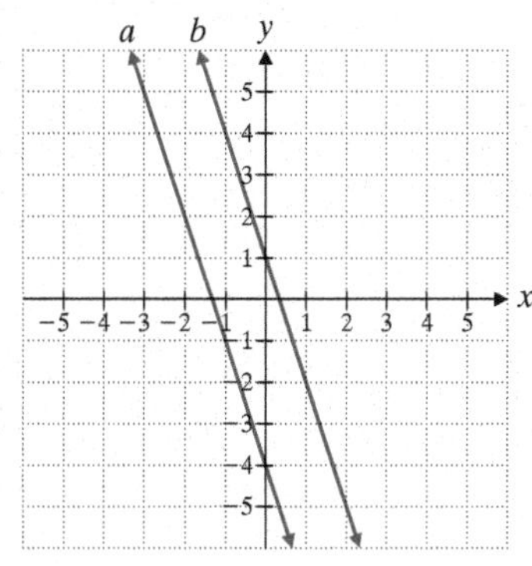

Graphing Calculator

Graphing Lines

You can graph a line given in the form $y = mx + b$ using a graphing calculator. For example, to graph $y = 2x + 4$, enter the right-hand side of the equation in the Y = editor of your calculator and graph. Choose an appropriate window to show all the intercepts. The following window is -10 to 10 by -10 to 10.
Display:

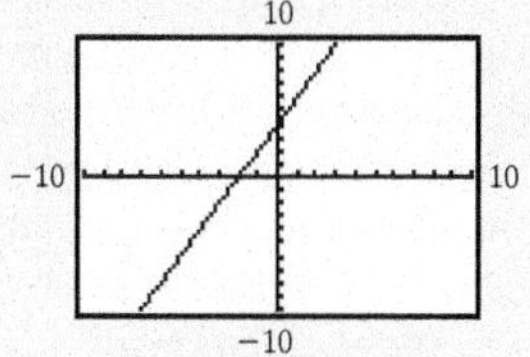

Try graphing other equations given in slope–intercept form.

PARALLEL LINES

Parallel lines are two straight lines that never touch.
Parallel lines have the same slope but different *y*-intercepts.

$m_1 = m_2$

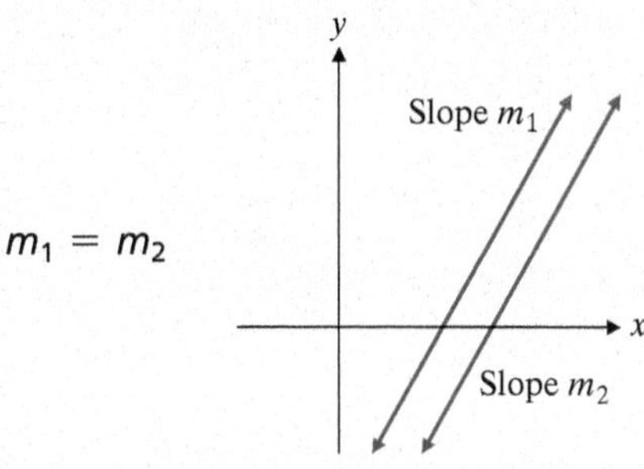

Perpendicular lines are two lines that meet at a 90° angle. Look at the perpendicular lines in the figure at the right. The slope of line c is -3. The slope of line d is $\frac{1}{3}$. Notice that

$$(-3)\left(\frac{1}{3}\right) = \left(-\frac{3}{1}\right)\left(\frac{1}{3}\right) = -1.$$

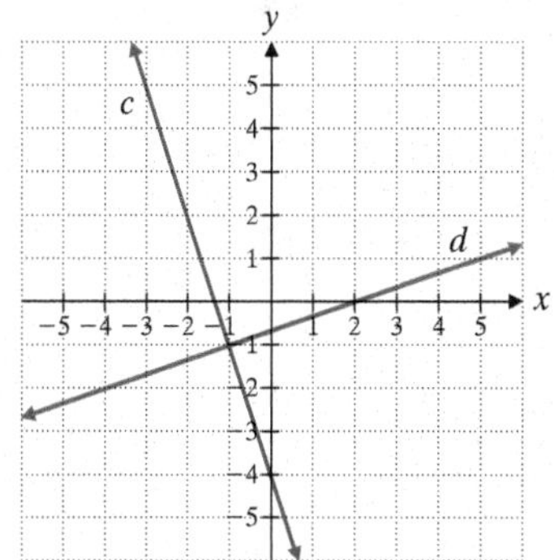

You may wish to draw several pairs of perpendicular lines to determine whether the product of their slopes is always -1.

PERPENDICULAR LINES

Perpendicular lines are two lines that meet at a 90° angle.
Perpendicular lines have slopes whose product is -1. If m_1 and m_2 are slopes of perpendicular lines, then

$$m_1 m_2 = -1 \quad or \quad m_1 = -\frac{1}{m_2}.$$

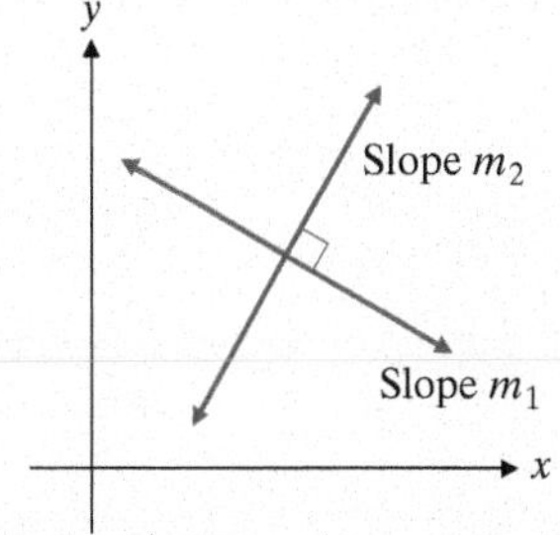

Example 8 Line h has a slope of $-\frac{2}{3}$.

(a) If line f is parallel to line h, what is its slope?

(b) If line g is perpendicular to line h, what is its slope?

Solution

(a) Parallel lines have the same slope. Line f has a slope of $-\frac{2}{3}$.

(b) Perpendicular lines have slopes whose product is -1.

$$m_1 m_2 = -1$$

$$-\frac{2}{3}m_2 = -1 \qquad \text{Substitute } -\frac{2}{3} \text{ for } m_1.$$

$$\left(-\frac{3}{2}\right)\left(-\frac{2}{3}\right)m_2 = -1\left(-\frac{3}{2}\right) \qquad \text{Multiply both sides by } -\frac{3}{2}.$$

$$m_2 = \frac{3}{2}$$

Thus line g has a slope of $\frac{3}{2}$. □

Graphing Calculator

Graphing Parallel Lines

If two equations are in the form $y = mx + b$, then it will be obvious that they are parallel because the slope will be the same. On a graphing calculator, graph both of these equations:

$$y = -2x + 6$$
$$y = -2x - 4$$

Use a window of -10 to 10 for both x and y.

Display:

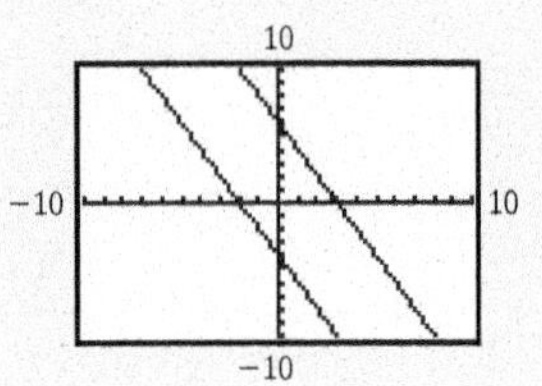

 **Student Practice 8** Line h has a slope of $\frac{1}{4}$.

(a) If line j is parallel to line h, what is its slope?

(b) If line k is perpendicular to line h, what is its slope?

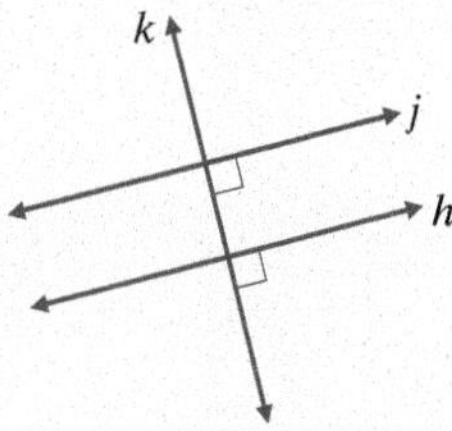

Example 9 The equation of line l is $y = -2x + 3$.

(a) What is the slope of a line that is parallel to line l?

(b) What is the slope of a line that is perpendicular to line l?

Solution

(a) Looking at the equation, we can see that the slope of line l is -2. The slope of a line that is parallel to line l is -2.

(b) Perpendicular lines have slopes whose product is -1.

$$m_1 m_2 = -1$$

$$(-2)m_2 = -1 \qquad \text{Substitute } -2 \text{ for } m_1.$$

$$m_2 = \frac{1}{2} \qquad \text{Because } (-2)\left(\frac{1}{2}\right) = -1.$$

The slope of a line that is perpendicular to line l is $\frac{1}{2}$. □

 **Student Practice 9** The equation of line n is $y = \frac{2}{3}x - 4$.

(a) What is the slope of a line that is parallel to line n?

(b) What is the slope of a line that is perpendicular to line n?

3.3 Exercises MyMathLab®

Verbal and Writing Skills, Exercises 1 and 2

1. Can you find the slope of the line passing through $(5, -12)$ and $(5, -6)$? Why or why not?

2. Can you find the slope of the line passing through $(6, -2)$ and $(-8, -2)$? Why or why not?

Find the slope of the straight line that passes through the given pair of points.

3. $(2, 3)$ and $(5, 9)$

4. $(1, 4)$ and $(9, 12)$

5. $(5, 10)$ and $(6, 5)$

6. $(5, 3)$ and $(2, 15)$

7. $(-2, 1)$ and $(3, 4)$

8. $(7, 4)$ and $(-2, 8)$

9. $(-6, -5)$ and $(2, -7)$

10. $(-8, -3)$ and $(4, -9)$

11. $(-3, 0)$ and $(0, -4)$

12. $(0, 5)$ and $(5, 3)$

13. $(5, -1)$ and $(-7, -1)$

14. $(-1, 5)$ and $(-1, 7)$

15. $\left(\frac{3}{4}, -4\right)$ and $(2, -8)$

16. $\left(\frac{5}{3}, -2\right)$ and $(3, 6)$

Find the slope and the y-intercept.

17. $y = 8x + 9$

18. $y = 2x + 10$

19. $3x + y - 4 = 0$

20. $8x + y + 7 = 0$

21. $y = -\frac{8}{7}x + \frac{3}{4}$

22. $y = \frac{5}{3}x - \frac{4}{5}$

23. $y = -6x$

24. $y = 4x$

25. $y = -2$

26. $y = 7$

27. $7x - 3y = 4$

28. $9x - 4y = 18$

Write the equation of the line in slope–intercept form.

29. $m = \frac{3}{5}$, y-intercept $(0, 3)$

30. $m = \frac{2}{3}$, y-intercept $(0, 5)$

31. $m = 4$, y-intercept $(0, -5)$

32. $m = 2$, y-intercept $(0, -1)$

33. $m = -1$, y-intercept $(0, 0)$

34. $m = 1$, y-intercept $(0, 0)$

35. $m = -\frac{5}{4}$, y-intercept $\left(0, -\frac{3}{4}\right)$

36. $m = -4$, y-intercept $\left(0, \frac{1}{2}\right)$

Graph the line $y = mx + b$ for the given values.

37. $m = \frac{3}{4}, b = -4$

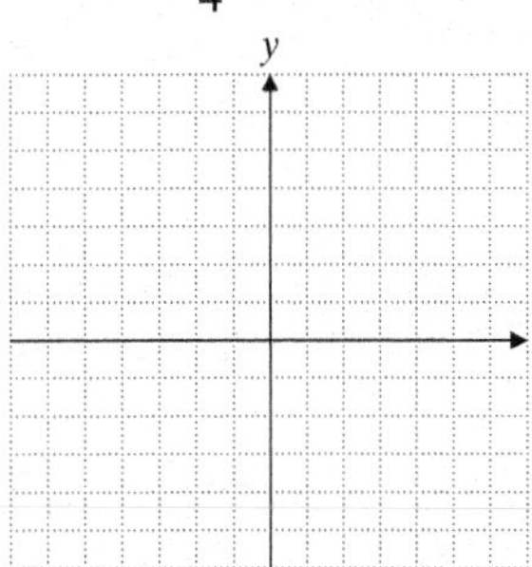

38. $m = \frac{1}{3}, b = -2$

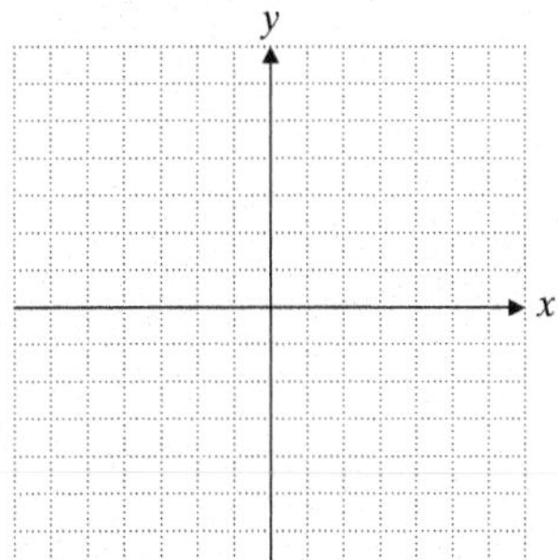

39. $m = -\frac{5}{3}, b = 2$

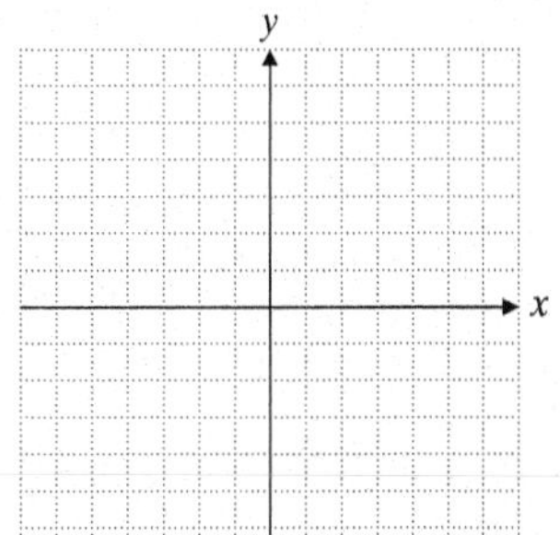

40. $m = -\frac{3}{2}, b = 4$

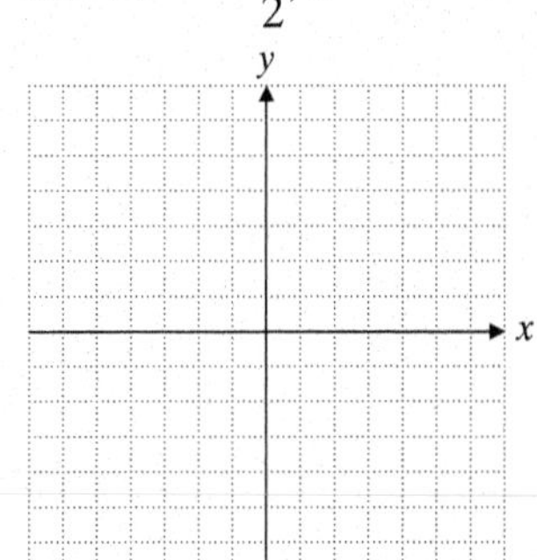

In exercises 41–46, graph the line. You may use any method: $y = mx + b$, intercepts, or finding any three points.

41. $y = \frac{2}{3}x + 2$

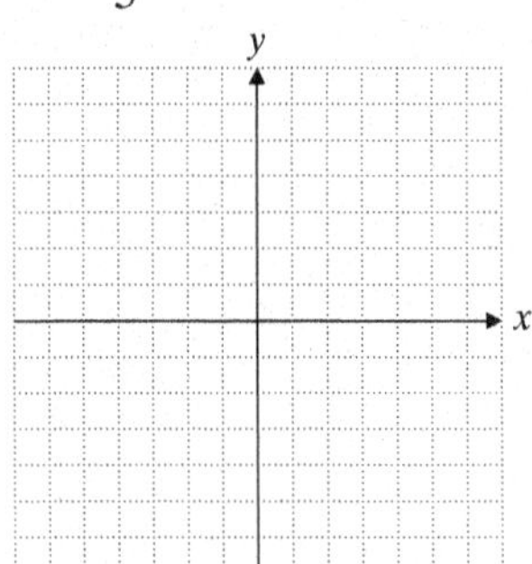

42. $y = \frac{3}{4}x + 1$

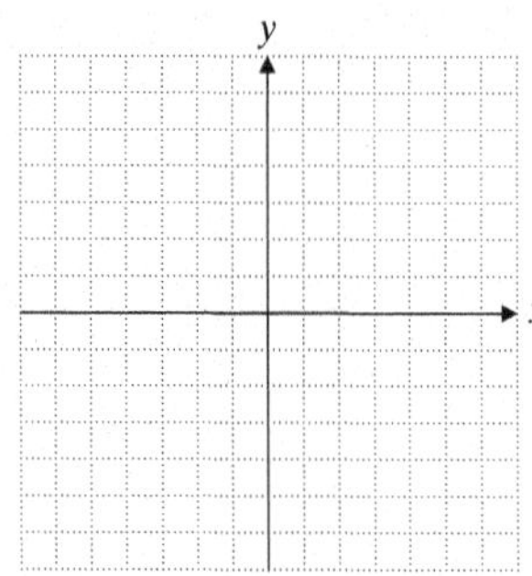

43. $y + 2x = 3$

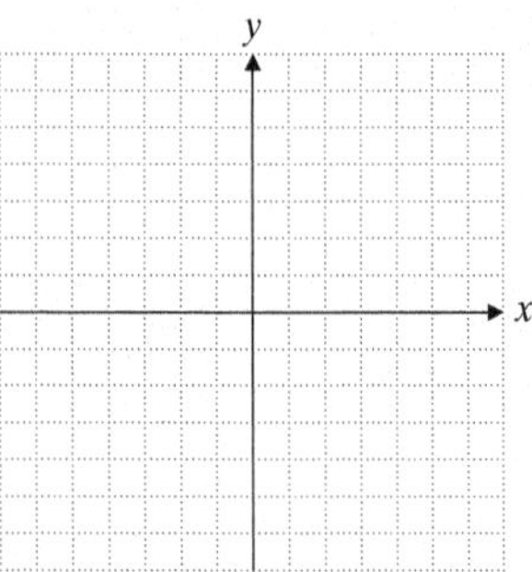

44. $y + 4x = 5$

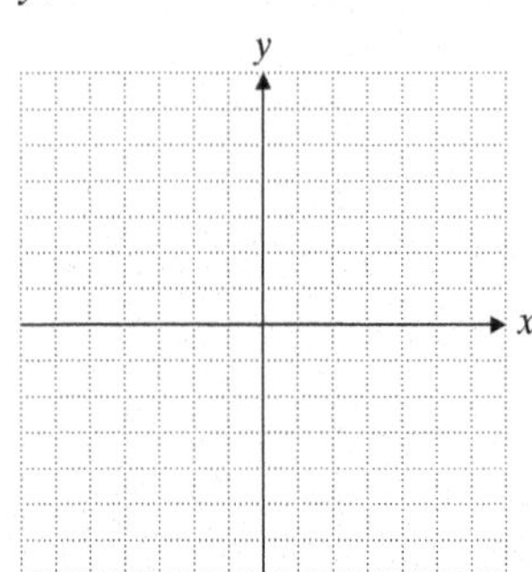

45. $y = 2x$

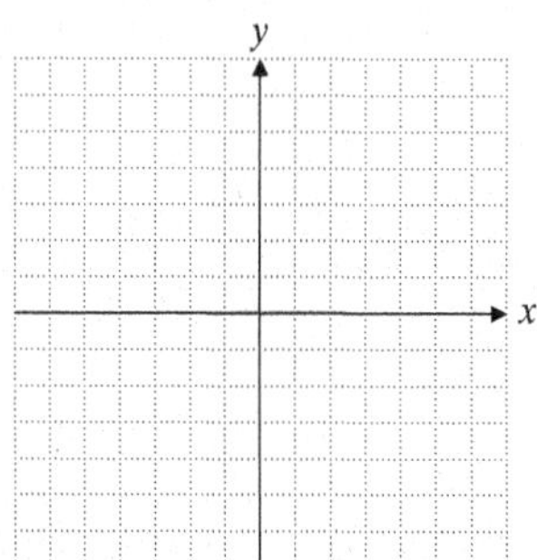

46. $y = 3x$

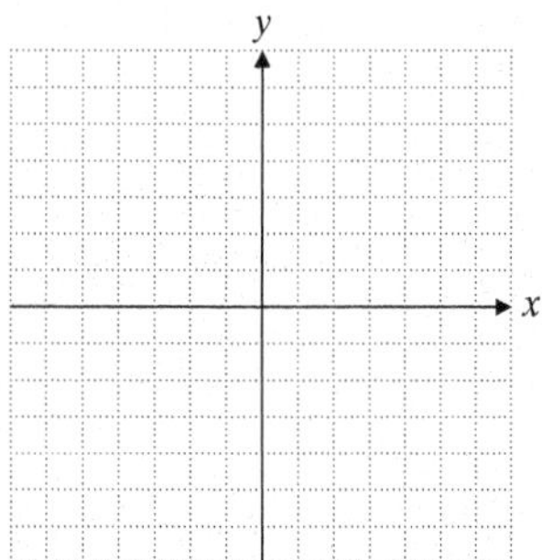

47. A line has a slope of $\frac{5}{6}$.

(a) What is the slope of a line parallel to it?

(b) What is the slope of a line perpendicular to it?

48. A line has a slope of $\frac{11}{4}$.

(a) What is the slope of a line parallel to it?

(b) What is the slope of a line perpendicular to it?

49. A line has a slope of -8.

(a) What is the slope of a line parallel to it?

(b) What is the slope of a line perpendicular to it?

50. A line has a slope of $-\frac{5}{8}$.

(a) What is the slope of a line parallel to it?

(b) What is the slope of a line perpendicular to it?

51. The equation of a line is $y = \frac{2}{3}x + 6$.

(a) What is the slope of a line parallel to it?

(b) What is the slope of a line perpendicular to it?

52. The equation of a line is $y = -\frac{3}{4}x - 2$.

(a) What is the slope of a line parallel to it?

(b) What is the slope of a line perpendicular to it?

To Think About

53. Do the points $(3, -4)$, $(18, 6)$, and $(9, 0)$ all lie on the same line? If so, what is the equation of the line?

54. Do the points $(2, 1)$, $(-3, -2)$, and $(7, 4)$ lie on the same line? If so, what is the equation of the line?

55. ***Cell Phone Accessory Sales*** A distribution center projected that one of the fastest-selling products in the United States between 2010 and 2020 will be cell phone accessories. During this 10-year period, the number of cell phone accessories sold can be approximated by the equation $y = 5(7x + 125)$, where x is the number of years since 2010 and y is the number of accessories sold in thousands.

(a) Write the equation in slope–intercept form.

(b) Find the slope and the y-intercept.

(c) In this specific equation, what is the meaning of the slope? What does it indicate?

56. ***Solar Energy*** Essex Solar Company executives projected that one of the fastest-growing sources of home energy in the United States between 2014 and 2024 will be solar energy. During this 10-year period, the number of homes using solar energy can be approximated by the equation $y = \frac{1}{10}(31x + 620)$, where x is the number of years since 2014 and y is the number of homes in thousands.

(a) Write the equation in slope–intercept form.

(b) Find the slope and the y-intercept.

(c) In this specific equation, what is the meaning of the slope? What does it indicate?

Cumulative Review *Solve for x and graph the solution.*

57. [2.8.4] $\frac{1}{4}x + 3 > \frac{2}{3}x + 2$

58. [2.8.4] $\frac{1}{2}(x + 2) \le \frac{1}{3}x + 5$

Quick Quiz 3.3

1. Find the slope of the straight line that passes through the points $(-2, 5)$ and $(-6, 3)$.

2. (a) Find the slope and y-intercept of $6x + 2y - 4 = 0$.
(b) Graph.

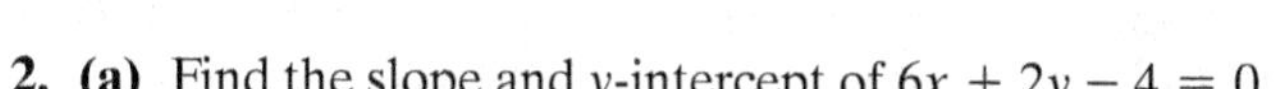

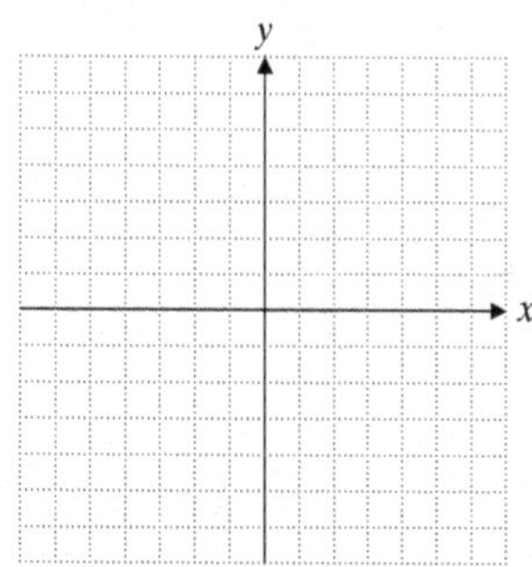

3. Write an equation of the line in slope–intercept form that passes through $(0, -5)$ and has a slope of $-\frac{5}{7}$.

4. Concept Check Consider the formula for slope: $m = \frac{y_2 - y_1}{x_2 - x_1}$. Explain why we substitute the y coordinates in the numerator.

Use Math to Save Money

Did You Know? Saving only 5% of your income can go a long way toward replacing your income in retirement.

Retirement Planning

Understanding the Problem:

It is never too early to begin thinking about your retirement. Louis began investing money in his retirement account at the age of 25 with the goal of retiring at the age of 65. He has a job where he gets paid $42,000 per year. His company has a retirement plan where he can contribute a percentage of his salary and have it invested in a retirement account that has an average annual return of 8%.

Making a Plan:

Louis is interested in what his monthly income will be at retirement if he saves 5% of his earnings for the next 40 years. To calculate his projected income at retirement, Louis needs to determine what the projected balance of his retirement account will be. He uses the retirement formula

$$FV = PMT \times \frac{(1 + I)^N - 1}{I}$$

with the variables defined as follows:

- *FV* is the future value of the retirement account.
- *PMT* is the monthly payment contributed to the retirement account.
- *I* is the interest rate in decimal notation divided by 12, which is the amount of interest being earned each month.
- *N* is the number of payments made into the account, which is 12 times the number of years the account is contributed to.

Step 1: Louis calculates what his monthly payment (*PMT*) will be if he saves 5% of his earnings.

Task 1: How much does Louis earn per month?

Task 2: What is 5% of his monthly income? This value will be the PMT.

Step 2: Using $I = 0.08 / 12 \approx 0.0067$ and $N = 12 \times 40 = 480$, Louis substitutes these values into the retirement formula to solve for *FV*.

Task 3: Substitute and solve for FV. Round your answer to the nearest dollar.

Finding a Solution:

Louis now knows how much he can expect to have in his retirement account when he retires. He has been told that to make sure his money lasts throughout his retirement, he should withdraw 4% of his account balance the first year and then increase that amount to keep up with inflation every year thereafter.

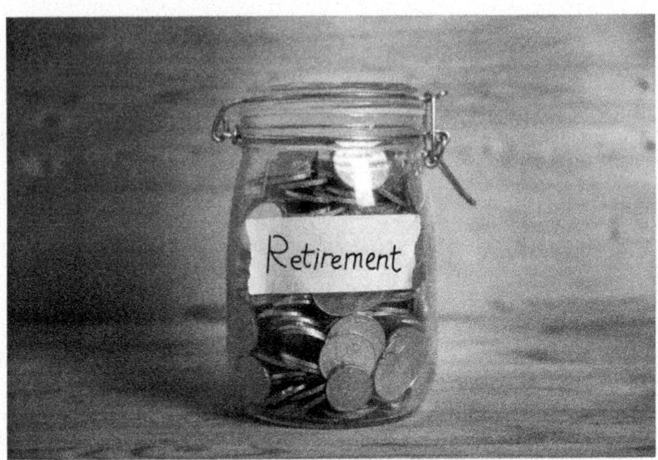

Step 3: Louis calculates his monthly income based on a withdrawal of 4% the first year. His goal is to contribute just enough to his retirement fund so that his retirement income is equal to his current working income.

Task 4: How much can Louis withdraw the first year?

Task 5: What will Louis's monthly income be the first year?

Task 6: Calculate the difference between Louis's current monthly income and his projected monthly income in retirement.

Task 7: Does Louis need to change the percentage he is contributing to his retirement plan to reach his goal? If so, should he increase or decrease the amount?

Applying the Situation to Your Life:

These are simplified calculations for projecting income in retirement. Other factors like inflation, which are more complicated to calculate, need to be taken into consideration. Inflation will reduce the value of money over time so that in 40 years, $42,000 may not be enough money to support the same lifestyle. You should also expect your yearly income to increase over time, so that the amount you are saving will increase. In addition there will be other sources of income in retirement like Social Security and pensions. The above calculations will give you a good idea of how much you will have in retirement, based on your current earnings and the expected number of years until retirement.

3.4 Writing the Equation of a Line

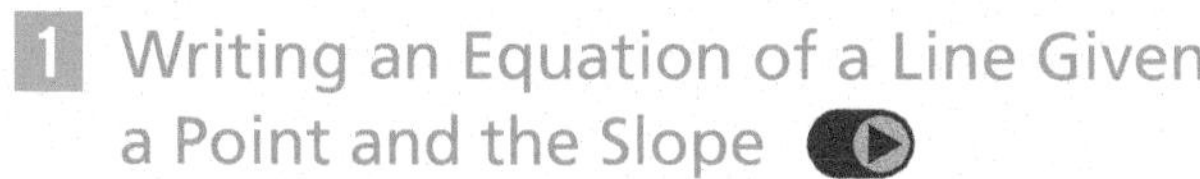

1 Writing an Equation of a Line Given a Point and the Slope

Student Learning Objectives

After studying this section, you will be able to:

1. Write an equation of a line given a point and the slope.
2. Write an equation of a line given two points.
3. Write an equation of a line given the graph of the line.

If we know the slope of a line and the y-intercept, we can write the equation of the line in slope–intercept form. Sometimes we are given the slope and a point on the line. We use the information to find the y-intercept. Then we can write the equation of the line.

It may be helpful to summarize our approach.

TO FIND AN EQUATION OF A LINE GIVEN A POINT AND THE SLOPE

1. Substitute the given values of x, y, and m into the equation $y = mx + b$.
2. Solve for b.
3. Use the values of b and m to write the equation in the form $y = mx + b$.

Example 1 Find an equation of the line that passes through $(-3, 6)$ with slope $-\frac{2}{3}$.

Solution We are given the values $m = -\frac{2}{3}$, $x = -3$, and $y = 6$, and we must find b.

$$y = mx + b$$

$$6 = \left(-\frac{2}{3}\right)(-3) + b \quad \text{Substitute known values.}$$

$$6 = 2 + b \quad \text{Solve for } b.$$

$$4 = b \quad \text{Now use the values for } b \text{ and } m \text{ to write the equation in slope–intercept form.}$$

An equation of the line is $y = -\frac{2}{3}x + 4$. □

Student Practice 1 Find an equation of the line that passes through $(-8, 12)$ with slope $-\frac{3}{4}$.

2 Writing an Equation of a Line Given Two Points

We can use our procedure when we are *given two points* if we first find the slope m and then find b. Recall from Section 3.3 that if we are given two points on a line, we can find the slope m using the formula $m = \frac{y_2 - y_1}{x_2 - x_1}$. In the next example we will use this information to find the values of m and b so that we can write an equation in the form $y = mx + b$.

MG **Example 2** Find an equation of the line that passes through $(2, 5)$ and $(6, 3)$.

Solution We must find both m and b. We first find m using the formula for slope. Then we proceed as in Example 1 to find b.

$$m = \frac{y_2 - y_1}{x_2 - x_1}$$

$$m = \frac{3 - 5}{6 - 2} \quad \text{Substitute } (x_1, y_1) = (2, 5) \text{ and } (x_2, y_2) = (6, 3) \text{ into the formula.}$$

$$= \frac{-2}{4} = -\frac{1}{2}$$

Student Practice 2 Find an equation of the line that passes through $(3, 5)$ and $(-1, 1)$.

Continued on next page

Now that we know $m = -\frac{1}{2}$, we can find b. Choose either point, say (2, 5), to substitute into $y = mx + b$ as in Example 1.

$$5 = -\frac{1}{2}(2) + b \quad \text{Substitute known values into } y = mx + b.$$
$$5 = -1 + b$$
$$6 = b$$

Use the values for b and m to write the equation.

An equation of the line is $y = -\frac{1}{2}x + 6$.

Note: We could have substituted the slope and the other point, (6, 3), into the slope–intercept form and arrived at the same answer. Try it. □

As we saw in Examples 1 and 2, once we find m and b we can write the equation of a line. Now, if we are given the graph of a line, can we write the equation of this line? That is, from the graph can we find m and b? We will look at this in Example 3.

3 Writing an Equation of a Line Given the Graph of the Line

Example 3 What is the equation of the line in the figure at the right?

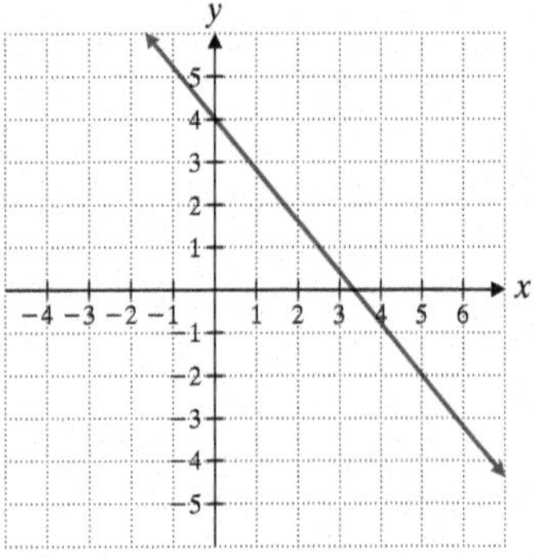

Solution First, look for the y-intercept. The line crosses the y-axis at (0, 4). Thus $b = 4$.

Second, find the slope.

$$m = \frac{\text{change in } y}{\text{change in } x} = \frac{\text{rise}}{\text{run}}$$

Look for another point on the line. We choose (5, −2). Count the number of vertical units from 4 to −2 (rise). Count the number of horizontal units from 0 to 5 (run).

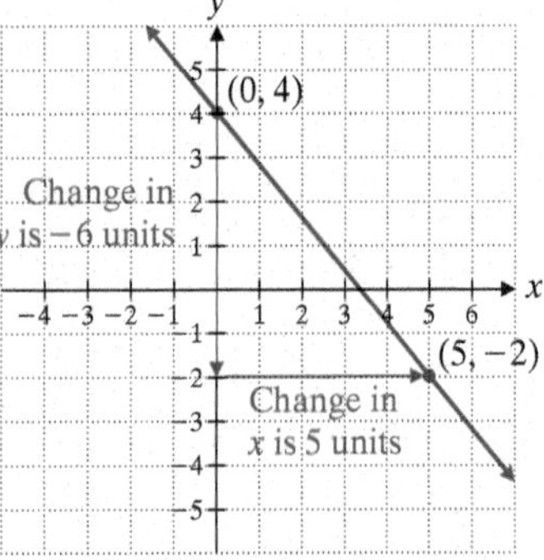

$$m = \frac{-6}{5}$$

Now, using $m = -\frac{6}{5}$ and $b = 4$, we can write an equation of the line.

$$y = mx + b$$
$$y = -\frac{6}{5}x + 4$$
□

Student Practice 3

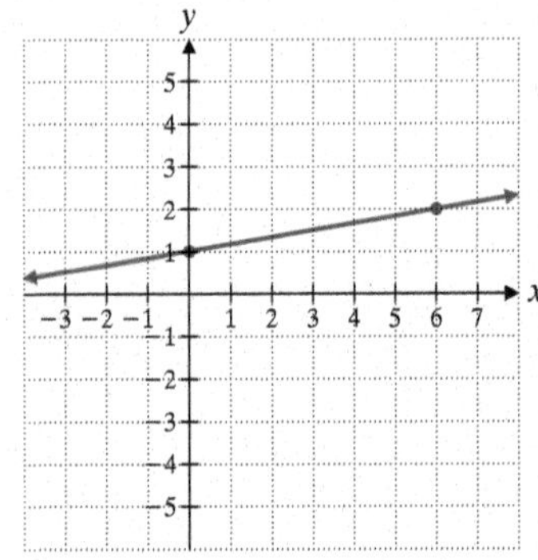

Student Practice 3 What is the equation of the line in the figure at the left?

3.4 Exercises MyMathLab®

Find an equation of the line that has the given slope and passes through the given point.

1. $m = 3, (-2, 0)$

2. $m = 5, (4, -1)$

3. $m = -2, (3, 5)$

4. $m = -4, (1, 1)$

5. $m = -3, \left(\frac{1}{2}, 2\right)$

6. $m = -4, \left(1, \frac{1}{5}\right)$

7. $m = \frac{1}{4}, (4, 5)$

8. $m = \frac{2}{3}, (3, -2)$

Write an equation of the line passing through the given points.

9. $(3, -12)$ and $(-4, 2)$

10. $(-1, 8)$ and $(0, 5)$

11. $(2, -6)$ and $(-1, 6)$

12. $(-1, 1)$ and $(5, 7)$

13. $(3, 5)$ and $(-1, -15)$

14. $(-1, -19)$ and $(2, 2)$

15. $\left(1, \frac{5}{6}\right)$ and $\left(3, \frac{3}{2}\right)$

16. $(2, 0)$ and $\left(\frac{3}{2}, \frac{1}{2}\right)$

Mixed Practice, Exercises 17–20

17. Find an equation of the line with a slope of -3 that passes through the point $(-1, 3)$.

18. Find an equation of the line with a slope of $-\frac{1}{2}$ that passes through the point $(6, -2)$.

19. Find an equation of the line that passes through $(2, -3)$ and $(-1, 6)$.

20. Find an equation of the line that passes through $(1, -8)$ and $(2, -14)$.

Write an equation of each line.

21.

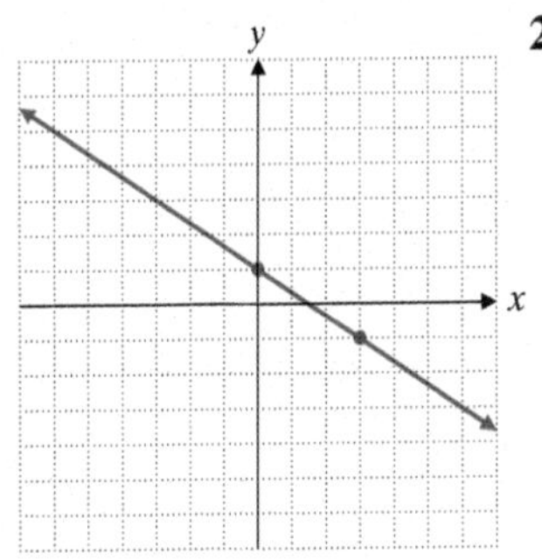

22.

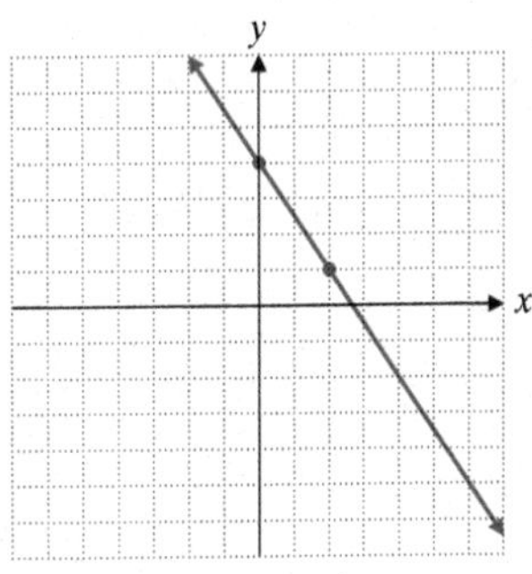

23.

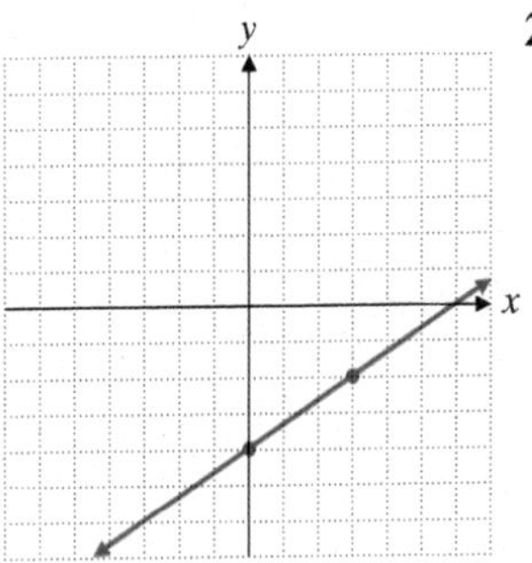

24.

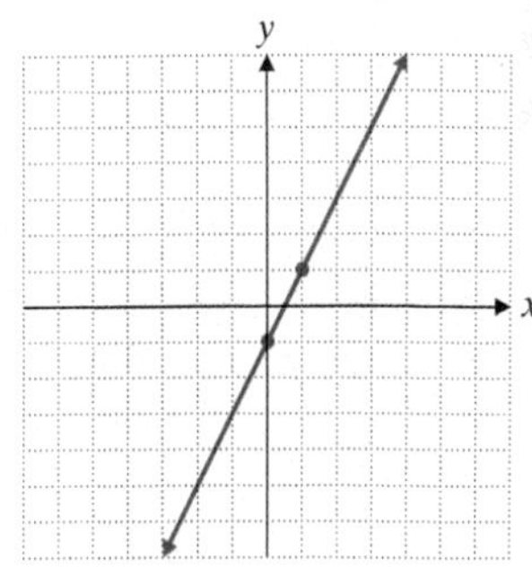

25.

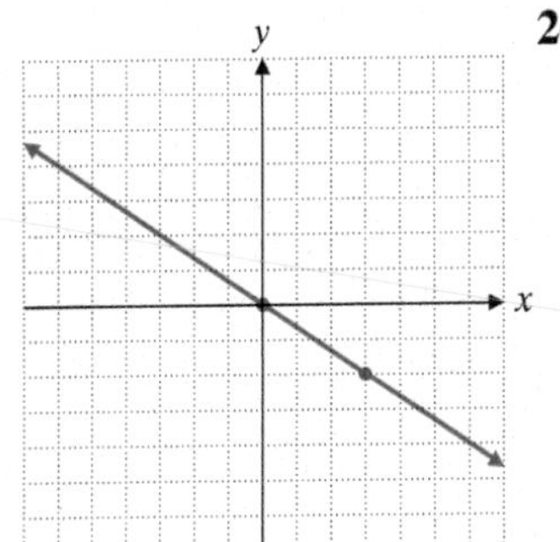

26.

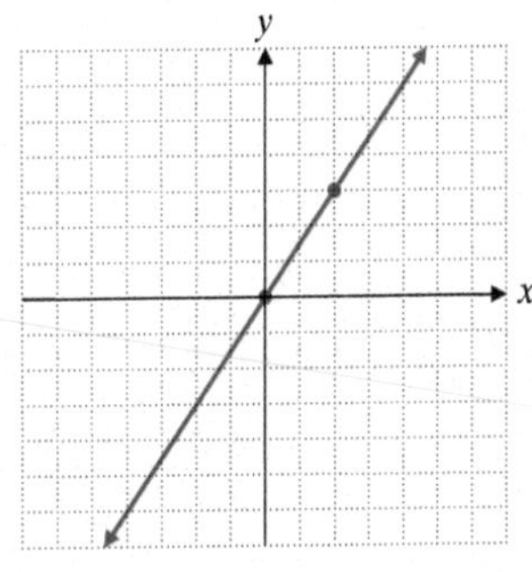

27.

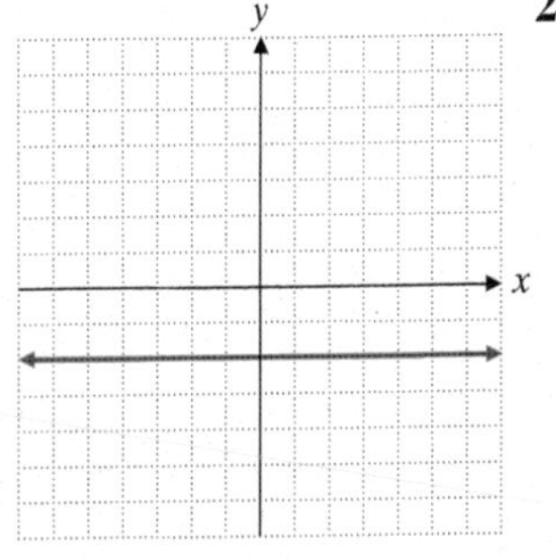

28.

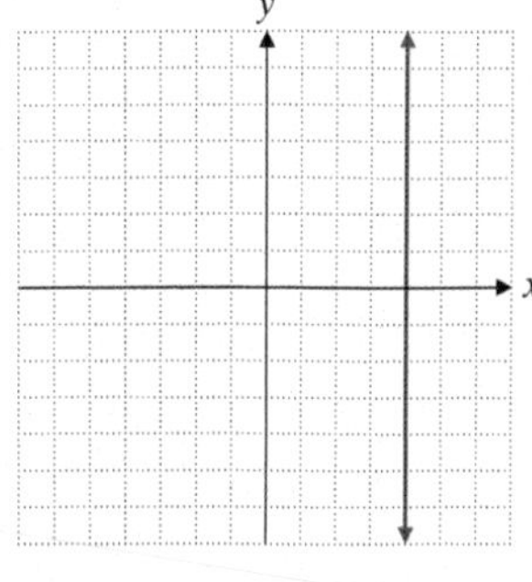

To Think About, Exercises 29–36 *Find an equation of the line that fits each description.*

29. Passes through $(7, -5)$ and has zero slope

30. Passes through (9, 3) and has undefined slope

31. Passes through $(4, -6)$ and is perpendicular to the x-axis

32. Passes through $(-7, 11)$ and is perpendicular to the y-axis

33. Passes through $(0, 5)$ and is parallel to $y = \frac{1}{3}x + 4$

34. Passes through $(0, 5)$ and is perpendicular to $y = \frac{1}{3}x + 4$

35. Passes through (2, 3) and is perpendicular to $y = 2x - 9$

36. Passes through (2, 9) and is parallel to $y = 5x - 3$

37. ***Population Growth*** The growth of the population of a large city during the period from 1990 to 2018 can be approximated by an equation of the form $y = mx + b$, where x is the number of years since 1990 and y is the population measured in thousands. Find the equation if two ordered pairs that satisfy it are (0, 227) and (10, 251).

38. ***Home Equity Loans*** The amount of debt outstanding on home equity loans at a bank during the period from 2000 to 2015 can be approximated by an equation of the form $y = mx + b$, where x is the number of years since 2000 and y is the debt measured in millions of dollars. Find the equation if two ordered pairs that satisfy it are (1, 280) and (6, 500).

Cumulative Review

39. **[2.8.4]** Solve. $10 - 3x > 14 - 2x$

40. **[2.8.4]** Solve. $2x - 3 \geq 7x - 18$

41. **[0.6.1]** ***Basketball Sneakers*** A pair of basketball sneakers sells for \$80. The next week the sneakers go on sale for 15% off. The third week there is a coupon in the newspaper offering a 10% discount off the second week's price. How much would you have paid for the sneakers during the third week if you had used the coupon?

42. **[2.7.1]** ***Cell Phone Costs*** Dave and Jane Wells have a cell phone. The plan they subscribe to costs \$50 per month. It includes 200 free minutes of calling time. Each minute after the 200 is charged at the rate of \$0.21 per minute. Last month their cell phone bill was \$68.90. How many total minutes did they use their cell phone during the month?

Quick Quiz 3.4

1. Write an equation in slope–intercept form of the line that passes through the point $(3, -5)$ and has a slope of $\frac{2}{3}$.

2. Write an equation in slope–intercept form of the line that passes through the points $(-2, 7)$ and $(-4, -5)$.

3. Write an equation of the line that passes through $(4, 5)$ and $(4, -2)$. What is the slope of this line?

4. **Concept Check** How would you find an equation of the line that passes through $(-2, -3)$ and has zero slope?

3.5 Graphing Linear Inequalities

In Section 2.8 we discussed inequalities in one variable. Look at the inequality $x < -2$ (x is less than -2). Some of the solutions to the inequality are -3, -5, and $-5\frac{1}{2}$. In fact all numbers to the left of -2 on a number line are solutions. The graph of the inequality is given in the following figure. Notice that the open circle at -2 indicates that -2 is *not* a solution.

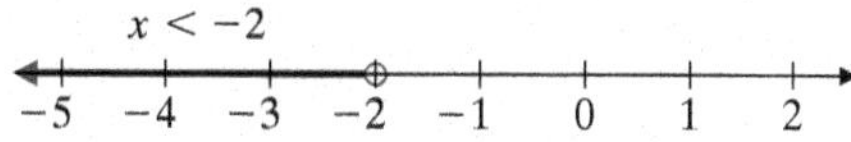

Now we will extend our discussion to consider linear inequalities in two variables.

Student Learning Objective

After studying this section, you will be able to:

1 Graph linear inequalities in two variables.

1 Graphing Linear Inequalities in Two Variables

Consider the inequality $y \geq x$. The solution of the inequality is the set of all possible ordered pairs that when substituted into the inequality will yield a true statement. Which ordered pairs will make the statement $y \geq x$ true? Let's try some.

$(0, 6)$	$(-2, 1)$	$(1, -2)$	$(3, 5)$	$(4, 4)$
$6 \geq 0$, true	$1 \geq -2$, true	$-2 \geq 1$, false	$5 \geq 3$, true	$4 \geq 4$, true

$(0, 6)$, $(-2, 1)$, $(3, 5)$, and $(4, 4)$ are solutions to the inequality $y \geq x$. In fact, every point at which the y-coordinate is greater than or equal to the x-coordinate is a solution to the inequality. This is shown by the solid line and the shaded region in the graph at the right.

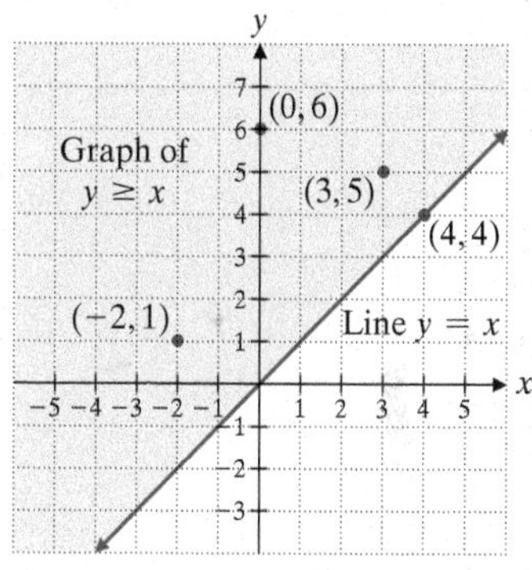

Is there an easier way to graph a linear inequality in two variables? It turns out that we can graph such an inequality by first graphing the associated linear equation and then testing one point that is not on that line. That is, we can change the inequality symbol to an equals sign and graph the equation. If the inequality symbol is $\geq$ or $\leq$, we use a solid line to indicate that the points on the line are included in the solution of the inequality. If the inequality symbol is $>$ or $<$, we use a dashed line to indicate that the points on the line are not included in the solution of the inequality. Then we test one point that is not on the line. If the point is a solution to the inequality, we shade the region on the side of the line that includes the point. If the point is not a solution, we shade the region on the other side of the line.

Example 1 Graph $5x + 3y > 15$.

Solution We begin by graphing the line $5x + 3y = 15$. You may use any method discussed previously to graph the line. Since there is no equals sign in the inequality, we will draw a *dashed line* to indicate that the line is *not* part of the solution set.

Look for a test point. The easiest point to test is $(0, 0)$. Substitute $(0, 0)$ for (x, y) in the inequality.

$$5x + 3y > 15$$
$$5(0) + 3(0) > 15$$
$$0 > 15 \quad \text{false}$$

$(0, 0)$ is *not* a solution. Shade the region on the side of the line that does *not* include $(0, 0)$.

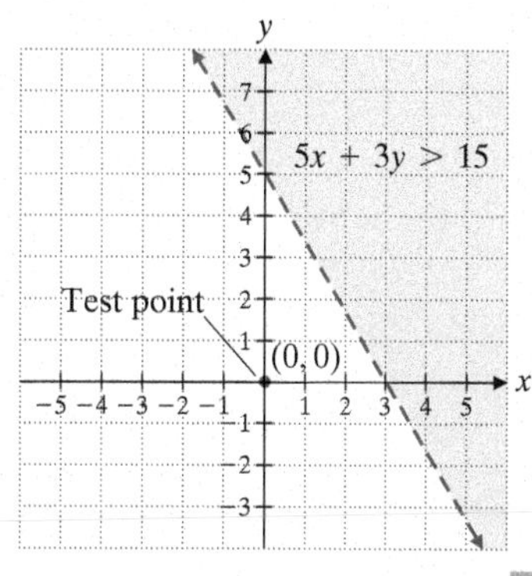

□

Student Practice 1

Graph $x - y \geq -10$.

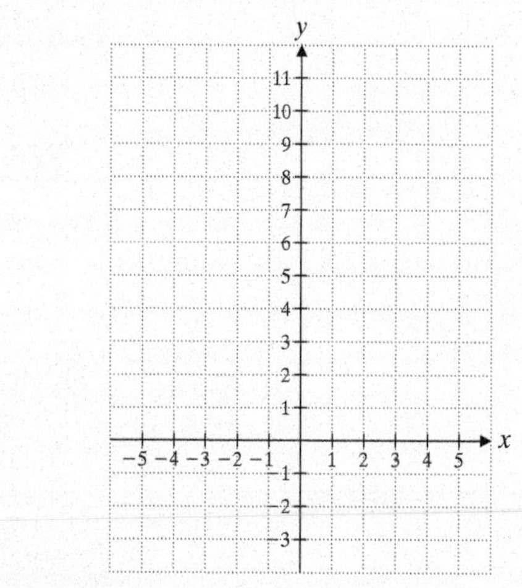

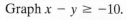

GRAPHING LINEAR INEQUALITIES

1. Replace the inequality symbol by an equality symbol. Graph the line.
   (a) The line will be solid if the inequality is $\geq$ or $\leq$.
   (b) The line will be dashed if the inequality is $>$ or $<$.
2. Test the point (0, 0) in the inequality if (0, 0) does not lie on the graphed line in step 1.
   (a) If the inequality is true, shade the region on the side of the line that includes (0, 0).
   (b) If the inequality is false, shade the region on the side of the line that does not include (0, 0).
3. If the point (0, 0) is a point on the line, choose another test point and proceed accordingly.

Example 2 Graph $2y \leq -3x$.

Solution

Step 1 Graph $2y = -3x$. Since $\leq$ is used, the line will be a solid line.

Step 2 We see that the line passes through $(0, 0)$.

Step 3 Choose another test point. We will choose $(-3, -3)$.

$$2y \leq -3x$$
$$2(-3) \leq -3(-3)$$
$$-6 \leq 9 \quad \text{true}$$

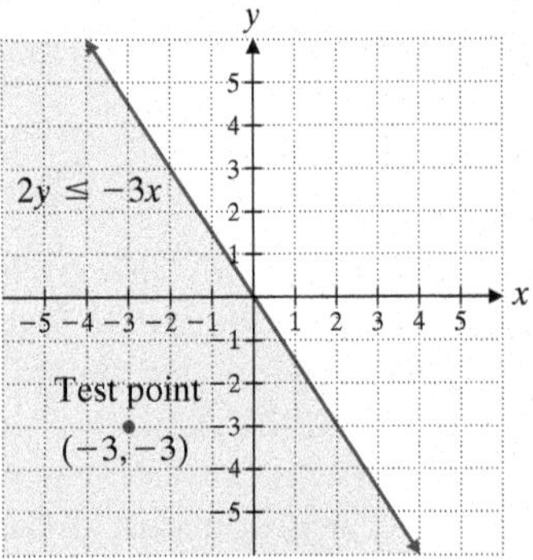

Shade the region that includes $(-3, -3)$, that is, the region below the line. □

Student Practice 2

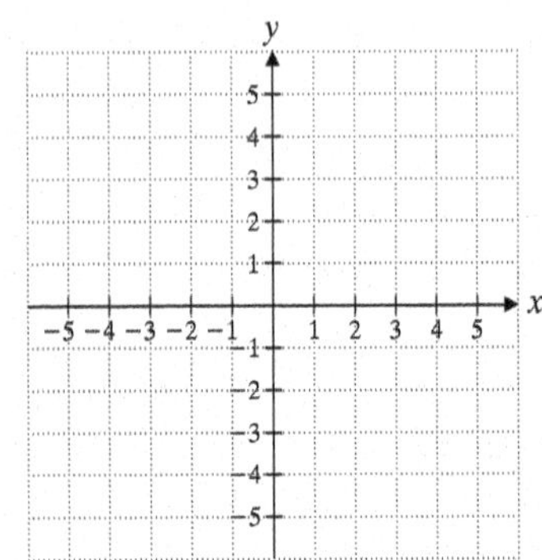

Student Practice 2 Graph $2y > x$ on the coordinate system in the margin.

If we are graphing the inequality $x < -2$ on a coordinate plane, the solution will be a region. Notice that this is very different from the solution of $x < -2$ on a number line discussed earlier.

Example 3 Graph $x < -2$.

Solution

Step 1 Graph $x = -2$. Since $<$ is used, the line will be dashed.

Step 2 Test $(0, 0)$ in the inequality.

$$x < -2$$
$$0 < -2 \quad \text{false}$$

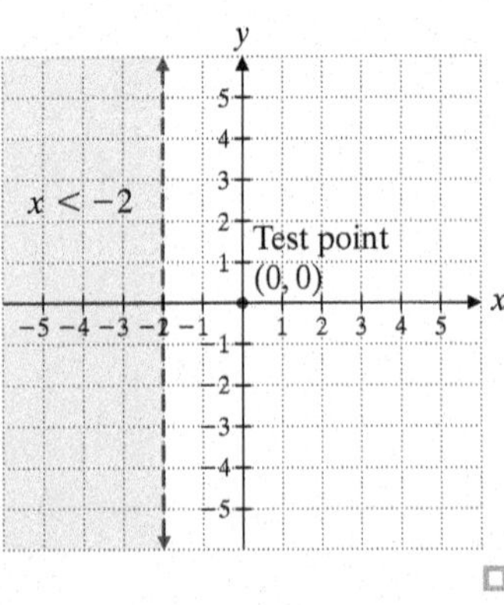

Shade the region that does not include (0, 0), that is, the region to the left of the line $x = -2$. Observe that every point in the shaded region has an x-value that is less than -2. □

Student Practice 3

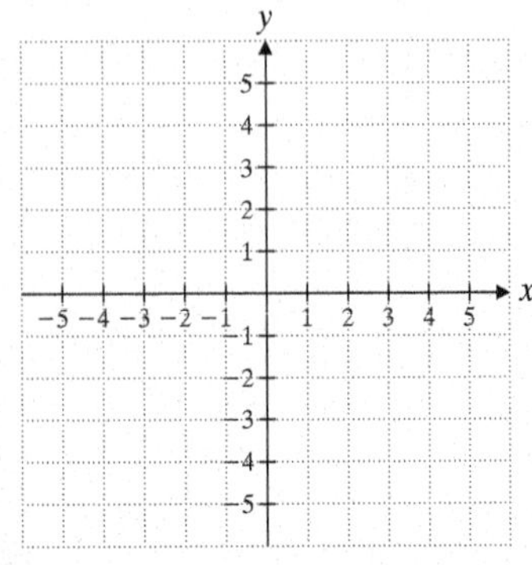

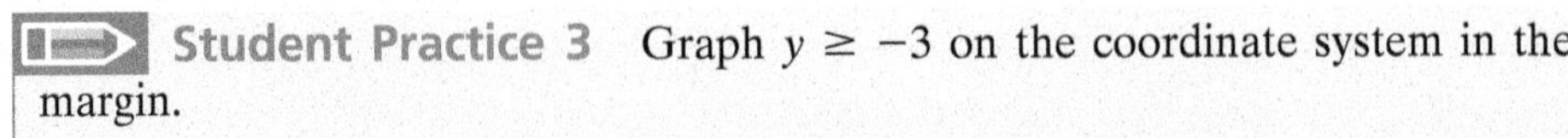

Student Practice 3 Graph $y \geq -3$ on the coordinate system in the margin.

3.5 Exercises MyMathLab®

Verbal and Writing Skills, Exercises 1 and 2

1. Does it matter what point you use as your test point? Justify your response.

2. Explain when to use a solid line or a dashed line when graphing a linear inequality in two variables.

Graph the region described by the inequality.

3. $y \geq 4x$

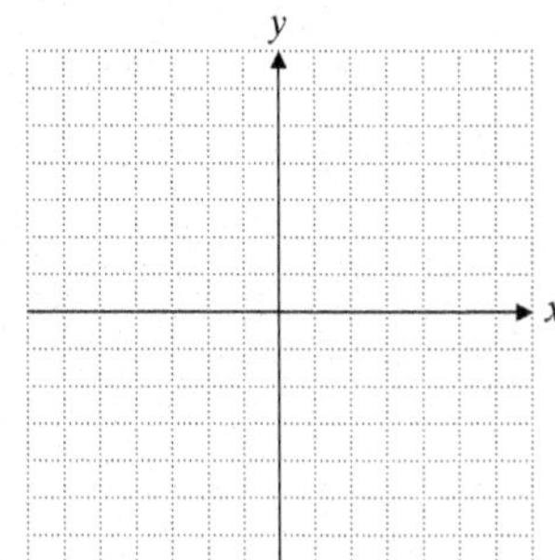

4. $y \leq -2x$

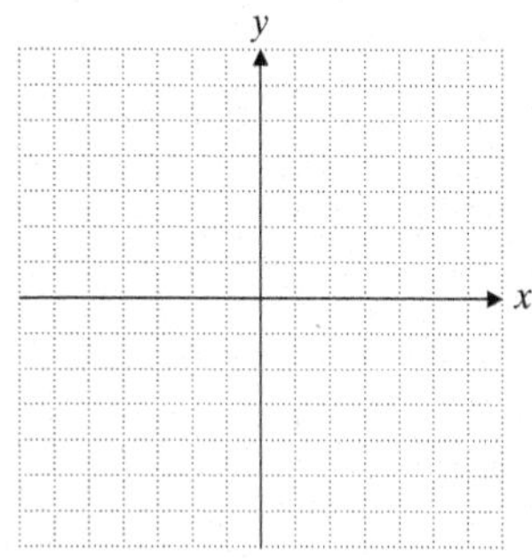

5. $2x - 3y < 6$

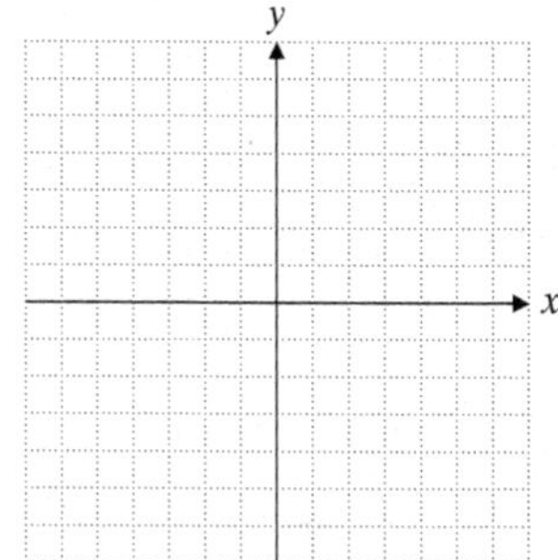

6. $3x + 2y < -6$

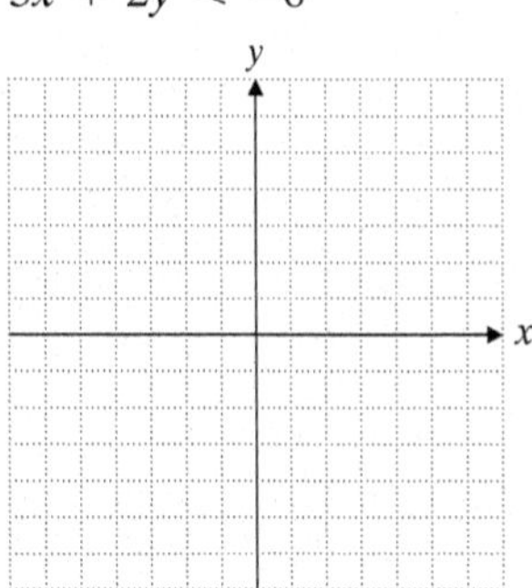

7. $2x - y \geq 3$

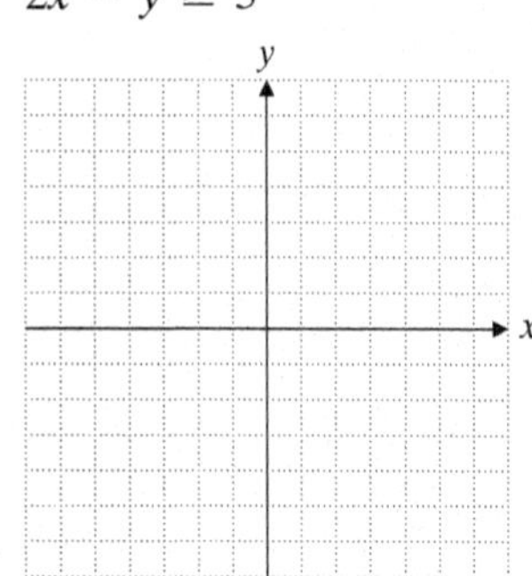

8. $3x - y \geq 4$

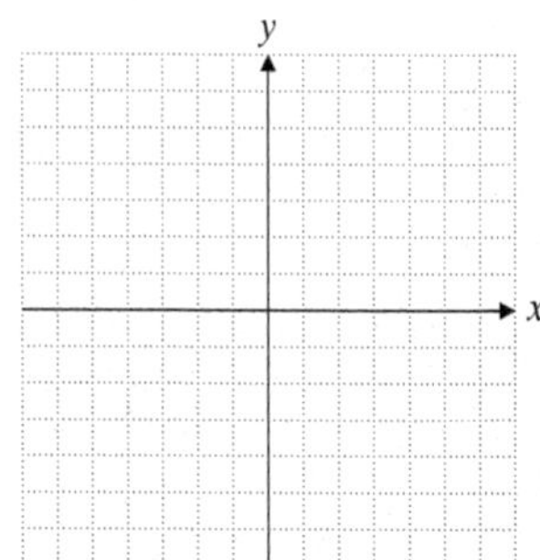

9. $y < 2x - 4$

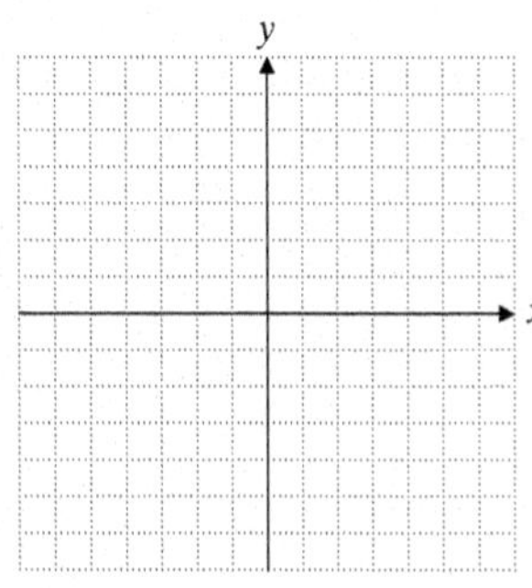

10. $y > 1 - 3x$

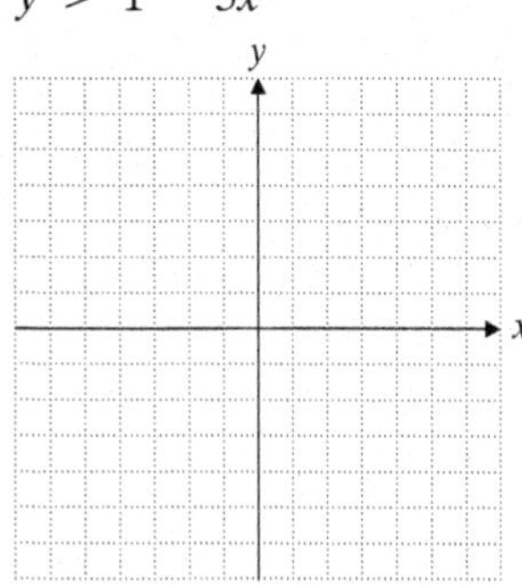

11. $y < -\frac{1}{2}x$

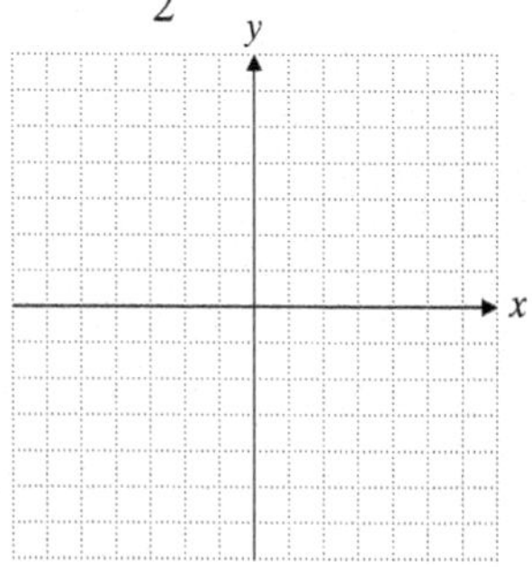

12. $y > \frac{1}{5}x$

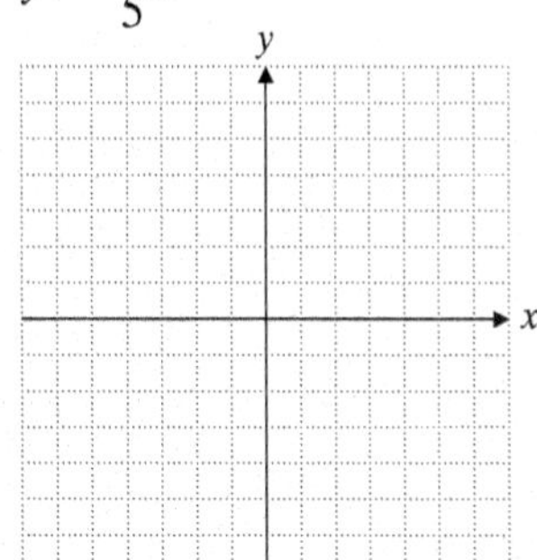

13. $x \geq 2$

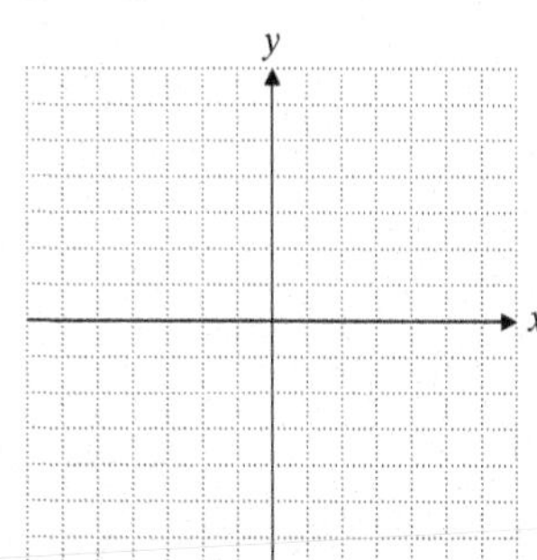

14. $y \leq -2$

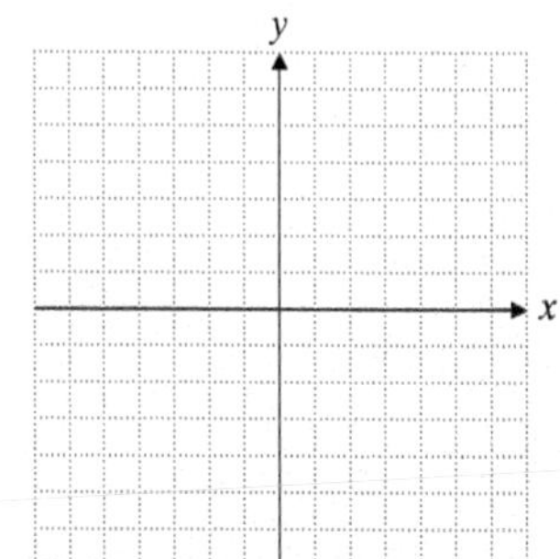

15. $2x - 3y + 6 \geq 0$

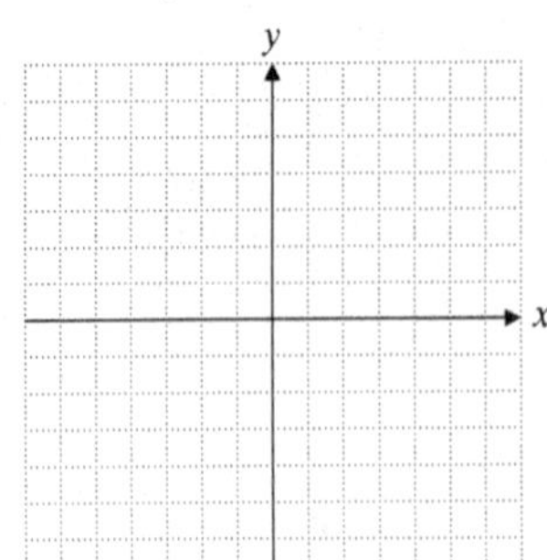

16. $3x + 4y - 8 \leq 0$

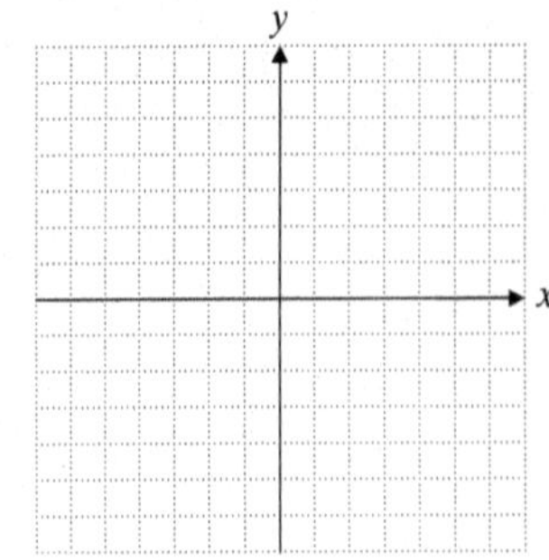

17. $x > -2y$

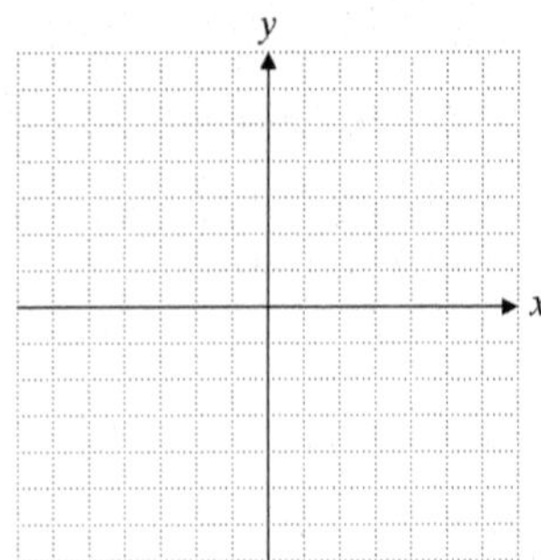

18. $x < -3y$

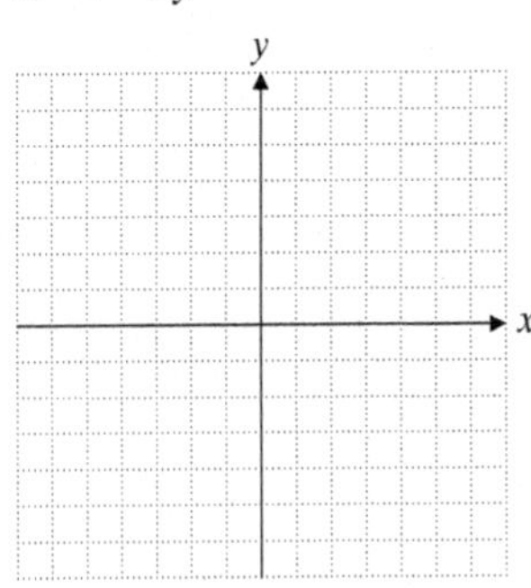

19. $2x > 3 - y$

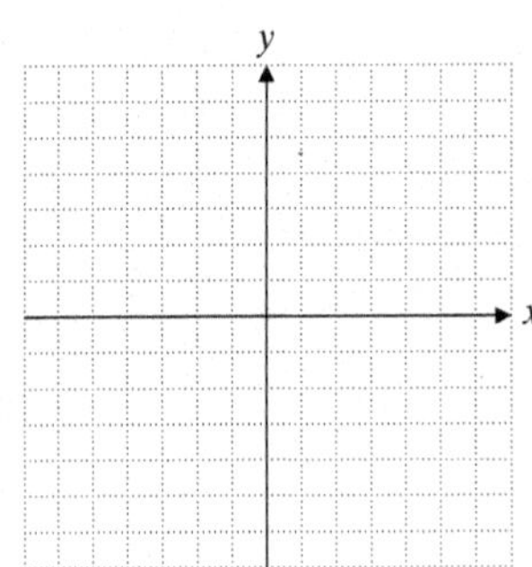

20. $x > 4 - y$

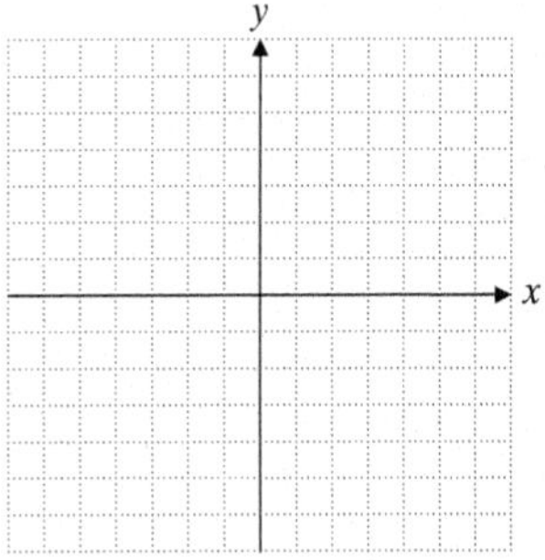

21. $2x \geq -3y$

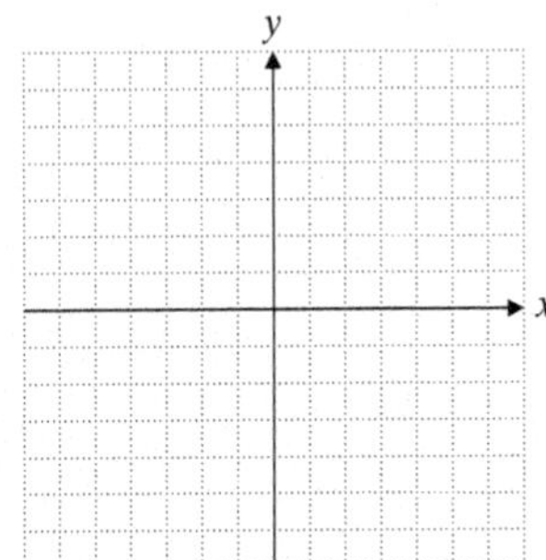

22. $3x \leq -2y$

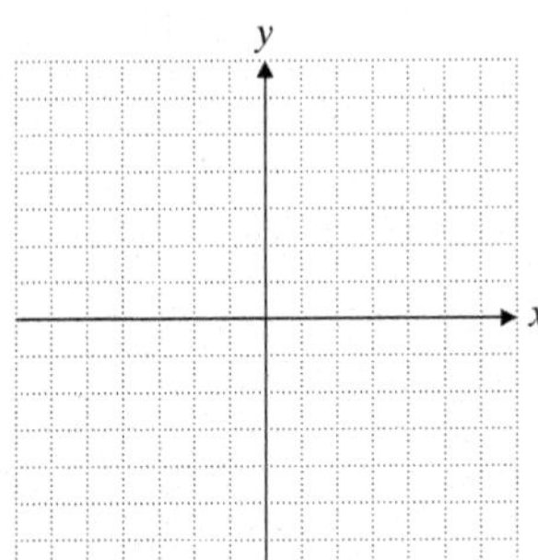

Cumulative Review *Perform the operations in the proper order.*

23. [1.5.1] $6(2) + 10 \div (-2)$

24. [1.5.1] $3(-3) + 2(12 - 15)^2 \div 9$

Evaluate for $x = -2$ and $y = 3$.

25. [1.8.1] $2x^2 + 3xy - 2y^2$

26. [1.8.1] $x^3 - 5x^2 + 3y - 1$

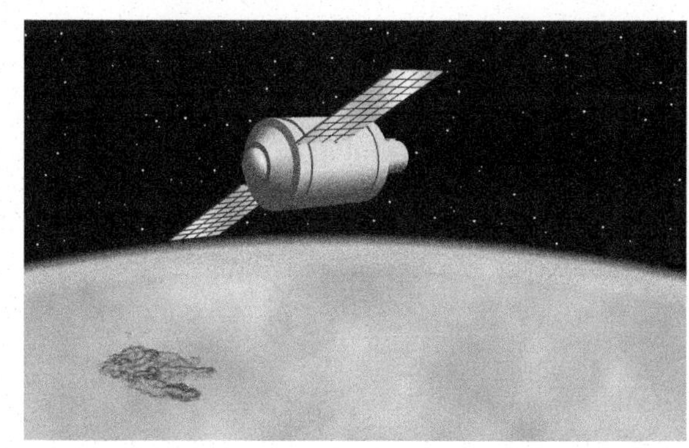

***Satellite Parts* [0.5.3]** *Brian sells high-tech parts to satellite communications companies. In his negotiations he originally offers to sell one company 200 parts for a total of \$22,400. However, after negotiations, he offers to sell that company the same parts at a 15% discount if the company agrees to sign a purchasing contract for 200 additional parts at some future date.*

27. What is the average cost per part if the parts are sold at the discounted price?

28. How much will the total bill be for the 200 parts at the discounted price?

Quick Quiz 3.5

1. When you graph the inequality $y > -3x + 1$, should you use a solid line for a boundary or a dashed line for a boundary? Why?

2. Graph the region described by $3y \leq -7x$.

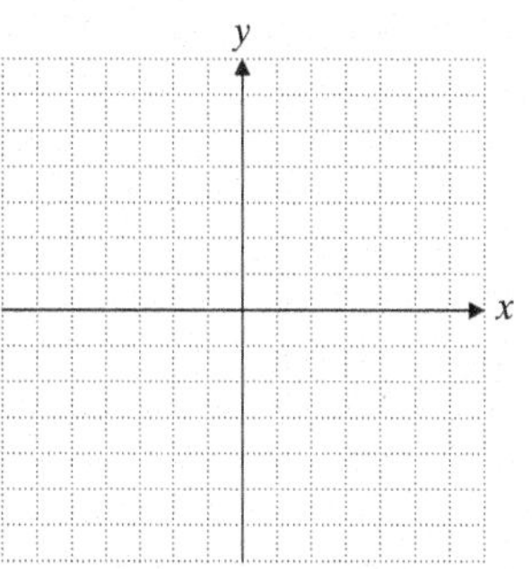

3. Graph the region described by $-5x + 2y > -3$.

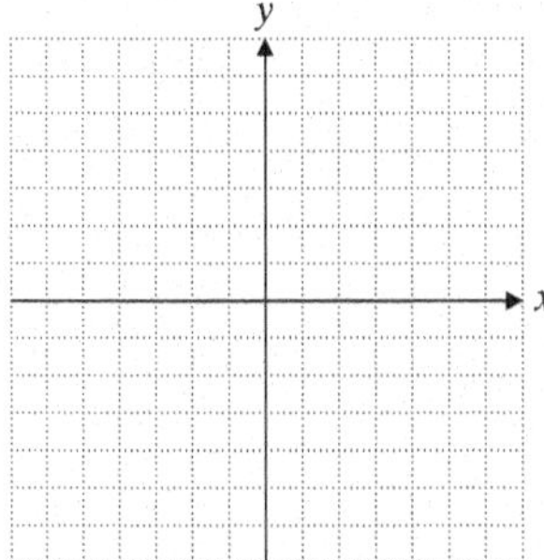

4. Concept Check Explain how you would determine if you should shade the region above the line or below the line if you were to graph the inequality $y > -3x + 4$ using $(0, 0)$ as a test point.

3.6 Functions

Student Learning Objectives

After studying this section, you will be able to:

1. Understand the meanings of a relation and a function.
2. Graph simple nonlinear equations.
3. Determine whether a graph represents a function.
4. Use function notation.

1 Understanding the Meanings of a Relation and a Function

Thus far you have studied linear equations in two variables. You have seen that such an equation can be represented by a table of values, by the algebraic equation itself, or by a graph.

x	0	1	3
y	4	1	−5

$y = -3x + 4$

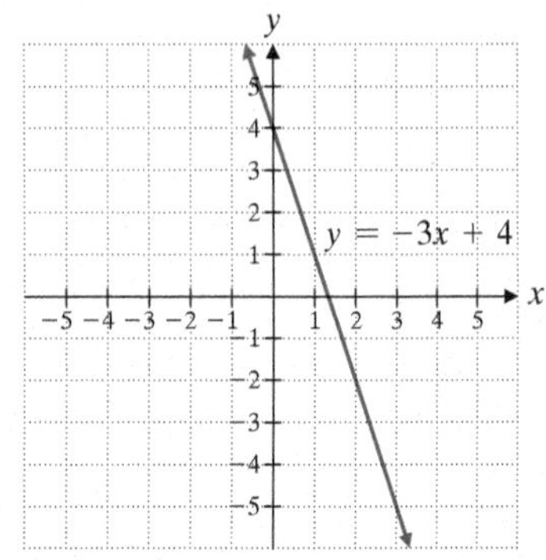

The solutions to the linear equation are all the ordered pairs that satisfy the equation (make the equation true). They are all the points that lie on the graph of the line. These ordered pairs can be represented in a table of values. Notice the relationship between the ordered pairs. We can choose any value for x. But once we have chosen a value for x, the value of y is determined. For the preceding equation $y = -3x + 4$, if x is 0, then y must be 4. We say that x is the **independent variable** and that y is the **dependent variable**.

Mathematicians call such a pairing of two values a *relation*.

DEFINITION OF A RELATION

A **relation** is any set of ordered pairs.

All the *first* coordinates in all of the ordered pairs of the relation make up the **domain** of the relation. All the *second* coordinates in all of the ordered pairs make up the **range** of the relation. Notice that the definition of a relation is very broad. Some relations cannot be described by an equation. These relations may simply be a set of discrete ordered pairs.

Example 1 State the domain and range of the relation.

$$\{(5, 7), (9, 11), (10, 7), (12, 14)\}$$

Solution The domain consists of all the first coordinates in the ordered pairs.

Domain

$$\{(5, 7), (9, 11), (10, 7), (12, 14)\}$$

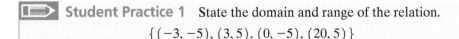

Range

The range consists of all the second coordinates in the ordered pairs. We usually list the values of a domain or range from smallest to largest.

The domain is $\{5, 9, 10, 12\}$.

The range is $\{7, 11, 14\}$. We list 7 only once. □

Student Practice 1 State the domain and range of the relation.

$$\{(-3, -5), (3, 5), (0, -5), (20, 5)\}$$

Some relations have the special property that no two different ordered pairs have the same first coordinate. Such relations are called **functions**. The relation $y = -3x + 4$ is a function. If we substitute a value for x, we get just one value for y. Thus no two ordered pairs will have the same x-coordinate and different y-coordinates.

DEFINITION OF A FUNCTION

A **function** is a relation in which no two different ordered pairs have the same first coordinate.

Example 2 Determine whether the relation is a function.

(a) $\{(3, 9), (4, 16), (5, 9), (6, 36)\}$ **(b)** $\{(7, 8), (9, 10), (12, 13), (7, 14)\}$

Solution

(a) Look at the ordered pairs. No two ordered pairs have the same first coordinate. Thus this set of ordered pairs defines a function. Note that the ordered pairs (3, 9) and (5, 9) have the same second coordinate, but the relation is still a function. It is the first coordinates that cannot be the same.

(b) Look at the ordered pairs. Two different ordered pairs, (7, 8) and (7, 14), have the same first coordinate. Thus this relation is *not* a function. □

Student Practice 2 Determine whether the relation is a function.

(a) $\{(-5, -6), (9, 30), (-3, -3), (8, 30)\}$

(b) $\{(60, 30), (40, 20), (20, 10), (60, 120)\}$

A functional relationship is often what we find when we analyze two sets of data. Look at the following table of values, which compares Celsius temperature with Fahrenheit temperature. Is there a relationship between degrees Fahrenheit and degrees Celsius? Is the relation a function?

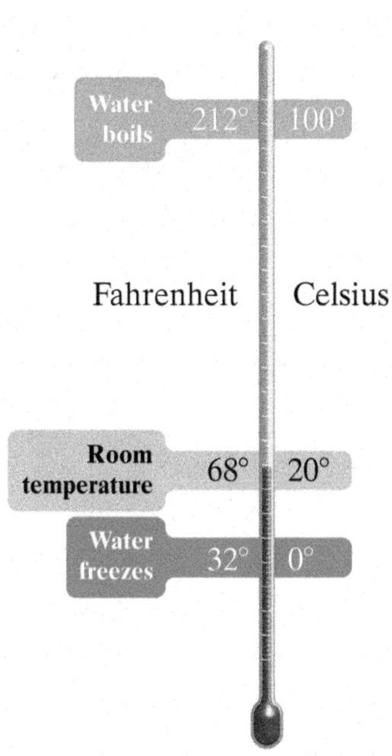

Temperature

°F	23	32	41	50
°C	−5	0	5	10

Since every Fahrenheit temperature produces a unique Celsius temperature, we would expect this to be a function. We can verify our assumption by looking at the formula $C = \frac{5}{9}(F - 32)$ and its graph. The formula is a linear equation, and its graph is a line with slope $\frac{5}{9}$ and y-intercept at about -17.8. The relation is a function. In the equation given here, notice that the *dependent variable* is C, since the value of C *depends* on the value of F. We say that F is the *independent variable*. The *domain* can be described as the set of possible values of the independent variable. The *range* is the set of corresponding values of the dependent variable. Scientists believe that the coldest temperature possible is approximately −273°C. In Fahrenheit, they call this temperature **absolute zero**. Thus,

Domain = {all possible Fahrenheit temperatures from absolute zero to infinity}

Range = {all corresponding Celsius temperatures from −273°C to infinity}.

Example 3 Each of the following tables contains some data pertaining to a relation. Determine whether the relation suggested by the table is a function. If it is a function, identify the domain and range.

(a) Circle

Radius	1	2	3	4	5
Area	3.14	12.56	28.26	50.24	78.5

(b) $4000 Loan at 8% for a Minimum of One Year

Time (yr)	1	2	3	4	5
Interest	$320	$665.60	$1038.85	$1441.96	$1877.31

Solution

(a) Looking at the table, we see that no two different ordered pairs have the same first coordinate. The area of a circle is a function of the length of the radius.

Next we need to identify the independent variable to determine the domain. Sometimes it is easier to identify the dependent variable. Here we notice that the area of the circle depends on the length of the radius. Thus radius is the independent variable. Since a negative length does not make sense, the radius cannot be a negative number. Although only integer radius values are listed in the table, the radius of a circle can be any nonnegative real number.

Domain = {all nonnegative real numbers}
Range = {all nonnegative real numbers}

(b) No two different ordered pairs have the same first coordinate. Interest is a function of time.

Since the amount of interest paid on a loan depends on the number of years (term of the loan), interest is the dependent variable and time is the independent variable. Negative numbers do not apply in this situation. Although the table includes only integer values for the time, the length of a loan in years can be any real number that is greater than or equal to 1.

Domain = {all real numbers greater than or equal to 1}
Range = {all positive real numbers greater than or equal to $320} □

Student Practice 3 Determine whether the relation suggested by the table is a function. If it is a function, identify the domain and the range.

(a) 28 Mpg at $4.16 per Gallon

Distance	0	28	42	56	70
Cost	$0	$4.16	$6.24	$8.32	$10.40

(b) Store's Inventory of Shirts

Number of Shirts	5	10	5	2	8
Price of Shirt	$20	$25	$30	$45	$50

To Think About: *Is It a Function?* Look at the following bus schedule. Determine whether the relation is a function. Which is the independent variable? Explain your choice.

Bus Schedule

Bus Stop	Main St.	8th Ave.	42nd St.	Sunset Blvd.	Cedar Lane
Time	7:00	7:10	7:15	7:30	7:39

2 Graphing Simple Nonlinear Equations 

Thus far in this chapter we have graphed linear equations in two variables. We now turn to graphing a few nonlinear equations. We will need to plot more than three points to get a good idea of what the graph of a nonlinear equation will look like.

Example 4 Graph $y = x^2$.

Solution Begin by constructing a table of values. We select values for x and then determine by the equation the corresponding values of y. We will include negative values for x as well as positive values. We then plot the ordered pairs and connect the points with a smooth curve.

x	$y = x^2$	y
-2	$y = (-2)^2 = 4$	4
-1	$y = (-1)^2 = 1$	1
0	$y = (0)^2 = 0$	0
1	$y = (1)^2 = 1$	1
2	$y = (2)^2 = 4$	4

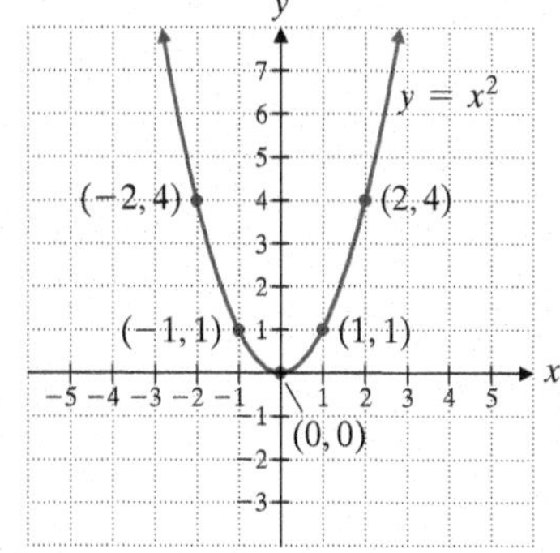

This type of curve is called a *parabola*. □

Student Practice 4 Graph $y = x^2 - 2$ on the coordinate system in the margin.

Student Practice 4

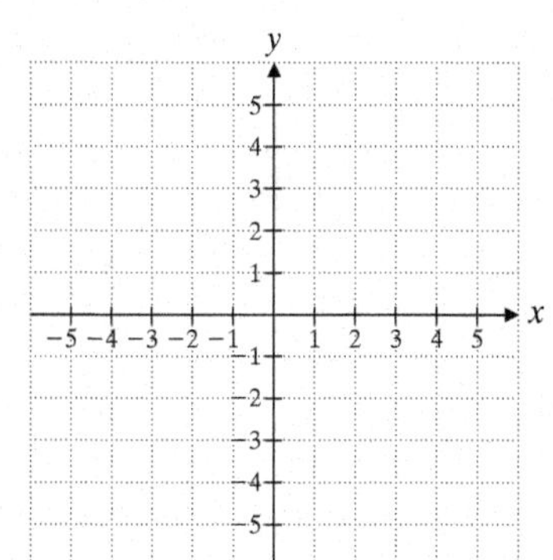

Some equations are solved for x. Usually, in those cases we pick values of y and then obtain the corresponding values of x from the equation.

Example 5 Graph $x = y^2 + 2$.

Solution Since the equation is solved for x, we start by picking a value of y. We will find the value of x from the equation in each case. We will select $y = -2$ first. Then we substitute it into the equation to obtain x. For convenience in graphing, we will repeat the y column at the end so that it is easy to write the ordered pairs (x, y).

y	$x = y^2 + 2$	x	y
-2	$x = (-2)^2 + 2 = 4 + 2 = 6$	6	-2
-1	$x = (-1)^2 + 2 = 1 + 2 = 3$	3	-1
0	$x = (0)^2 + 2 = 0 + 2 = 2$	2	0
1	$x = (1)^2 + 2 = 1 + 2 = 3$	3	1
2	$x = (2)^2 + 2 = 4 + 2 = 6$	6	2

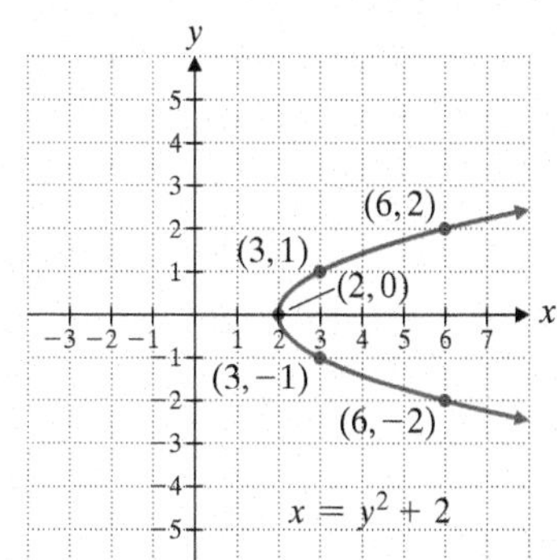

□

Student Practice 5 Graph $x = y^2 - 1$ on the coordinate system in the margin.

Student Practice 5

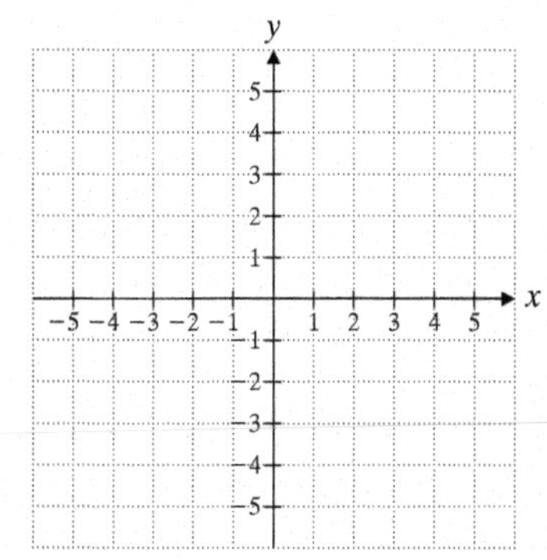

If the equation involves fractions with variables in the denominator, we must use extra caution. Remember that you may never divide by zero.

Example 6 Graph $y = \frac{4}{x}$.

Solution It is important to note that x cannot be zero because division by zero is not defined. $y = \frac{4}{0}$ is not allowed! Observe that when we draw the graph we get two separate branches that do not touch.

x	$y = \frac{4}{x}$	y
-4	$y = \frac{4}{-4} = -1$	-1
-2	$y = \frac{4}{-2} = -2$	-2
-1	$y = \frac{4}{-1} = -4$	-4
0	We cannot divide by zero.	There is no value.
1	$y = \frac{4}{1} = 4$	4
2	$y = \frac{4}{2} = 2$	2
4	$y = \frac{4}{4} = 1$	1

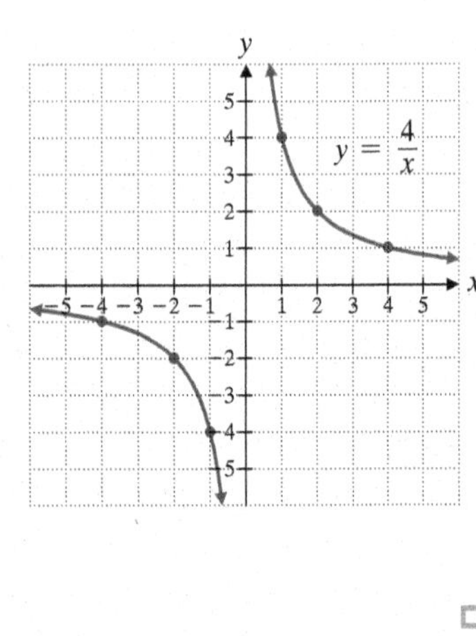

□

Student Practice 6

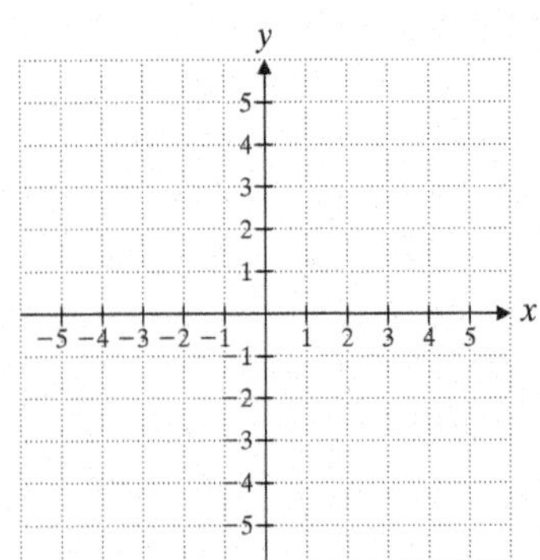

Student Practice 6 Graph $y = \frac{6}{x}$ on the coordinate system in the margin.

Graphing Calculator

Graphing Nonlinear Equations

You can graph nonlinear equations solved for y using a graphing calculator. For example, graph $y = x^2 - 2$ on a graphing calculator using an appropriate window. Display:

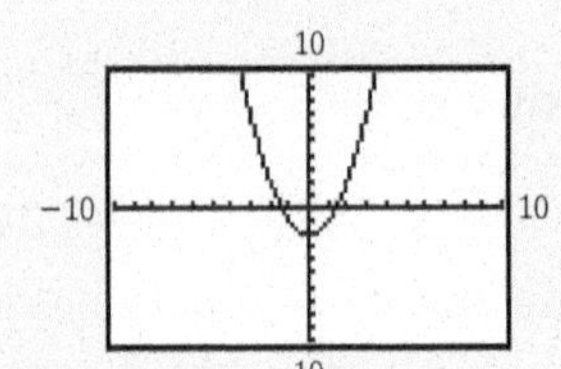

Try graphing $y = \frac{5}{x}$.

3 Determining Whether a Graph Represents a Function

Can we tell whether a graph represents a function? Recall that a function cannot have two different ordered pairs with the same first coordinate. That is, each value of x must have a separate, unique value of y. Look at the graph below of the function $y = x^2$. Each x-value has a unique y-value. Now look at the graph of $x = y^2 + 2$ below. At $x = 6$ there are two y-values, 2 and -2. In fact, for every x-value greater than 2, there are two y-values. $x = y^2 + 2$ is not a function.

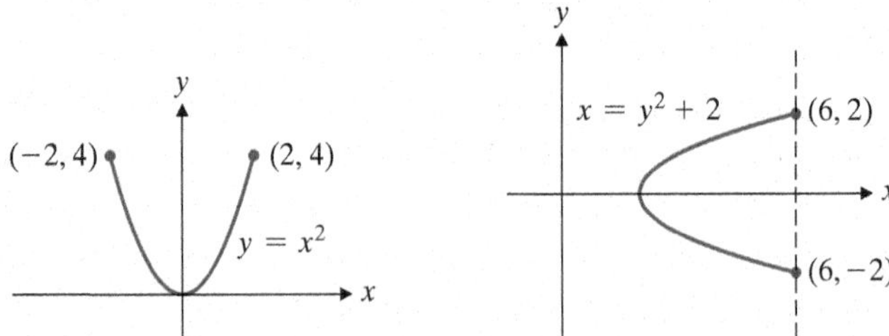

Observe that we can draw a vertical line through $(6, 2)$ and $(6, -2)$. Any graph that is not a function will have at least one region in which a vertical line will cross the graph more than once.

> **VERTICAL LINE TEST**
>
> If a vertical line can intersect the graph of a relation more than once, the relation is not a function. If no such line can be drawn, then the relation is a function.

Example 7 Determine whether each of the following is the graph of a function.

(a)

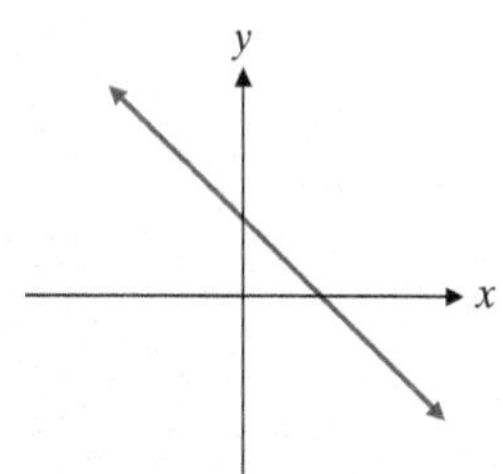

(b)

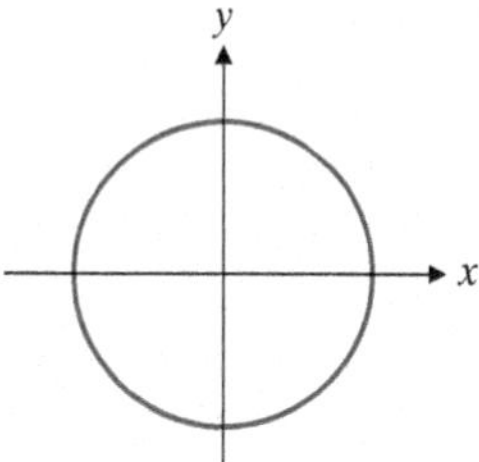

(c)

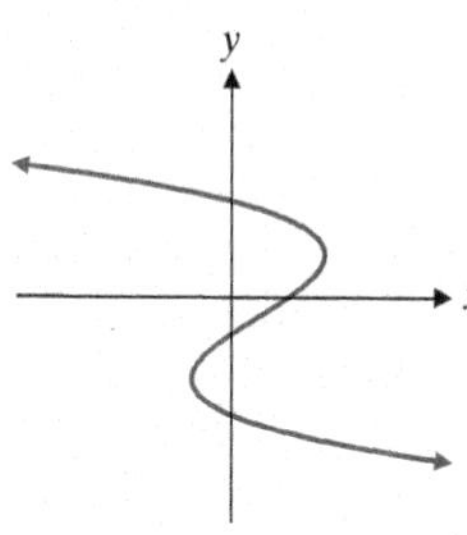

Solution

(a) The graph of the straight line is a function. Any vertical line will cross this straight line in only one location.

(b) and **(c)** Each of these graphs is not the graph of a function. In each case there exists a vertical line that will cross the curve in more than one place.

(a)

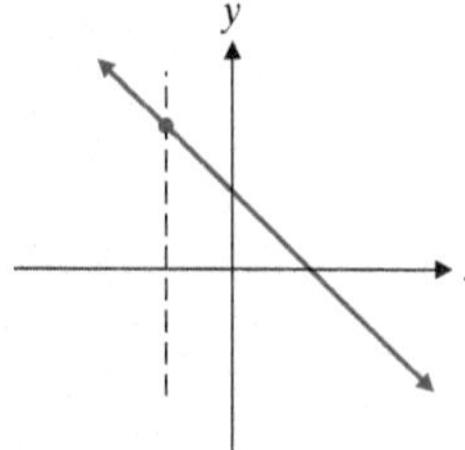

(b)

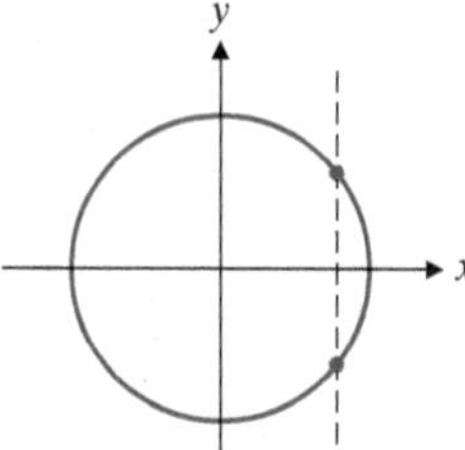

(c)

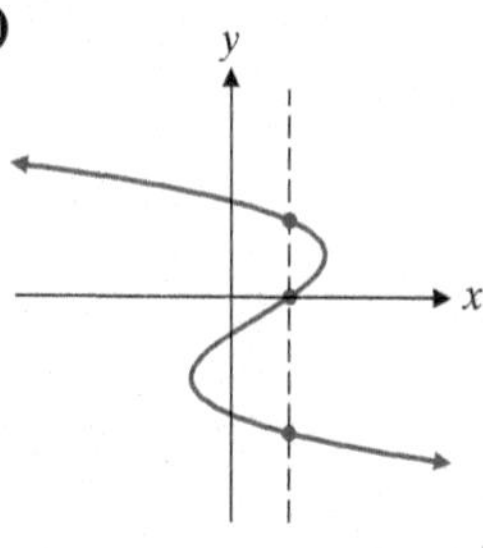

□

Student Practice 7 Determine whether each of the following is the graph of a function.

(a)

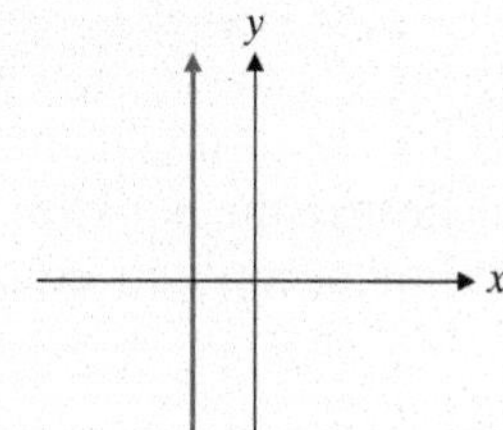

(b)

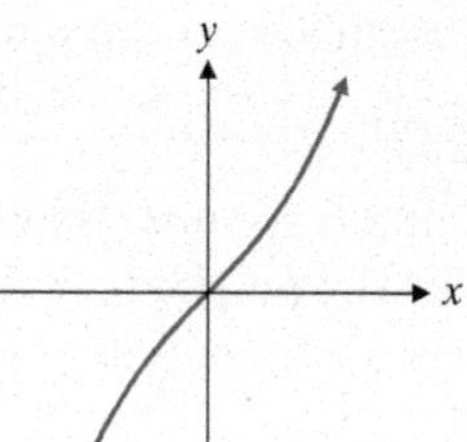

(c)

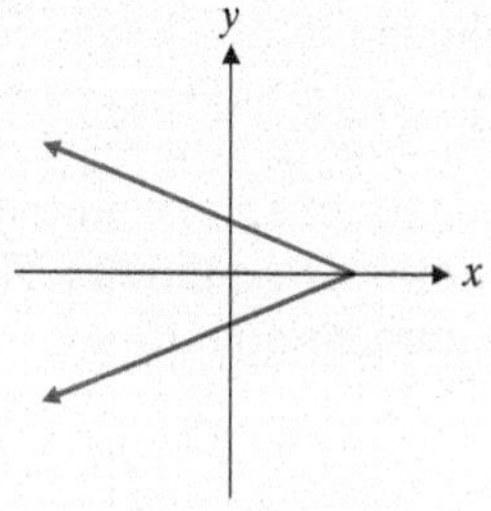

4 Using Function Notation 

We have seen that an equation like $y = 2x + 7$ is a function. For each value of x, the equation assigns a unique value to y. We could say, "y is a function of x." If we name the function f, this statement can be symbolized by using the **function notation** $y = f(x)$. Many times we avoid using the y-variable completely and write the function as $f(x) = 2x + 7$.

CAUTION: Be careful. The notation $f(x)$ does not mean f multiplied by x.

Example 8 If $f(x) = 3x^2 - 4x + 5$, find each of the following.

(a) $f(-2)$ **(b)** $f(4)$ **(c)** $f(0)$

Solution

(a) $f(-2) = 3(-2)^2 - 4(-2) + 5 = 3(4) - 4(-2) + 5$
$= 12 + 8 + 5 = 25$

(b) $f(4) = 3(4)^2 - 4(4) + 5 = 3(16) - 4(4) + 5$
$= 48 - 16 + 5 = 37$

(c) $f(0) = 3(0)^2 - 4(0) + 5 = 3(0) - 4(0) + 5 = 0 - 0 + 5 = 5$ □

Student Practice 8

If $f(x) = -2x^2 + 3x - 8$, find each of the following.

(a) $f(2)$

(b) $f(-3)$

(c) $f(0)$

When evaluating a function, it is helpful to place parentheses around the value that is being substituted for x. Taking the time to do this will minimize sign errors in your work.

Some functions are useful in medicine, anthropology, and forensic science. For example, the approximate length of a man's femur (thigh bone) is given by the function $f(x) = 0.53x - 17.03$, where x is the height of the man in inches. If a man is 68 inches tall, $f(68) = 0.53(68) - 17.03 = 19.01$. A man 68 inches tall would have a femur length of approximately 19.01 inches.

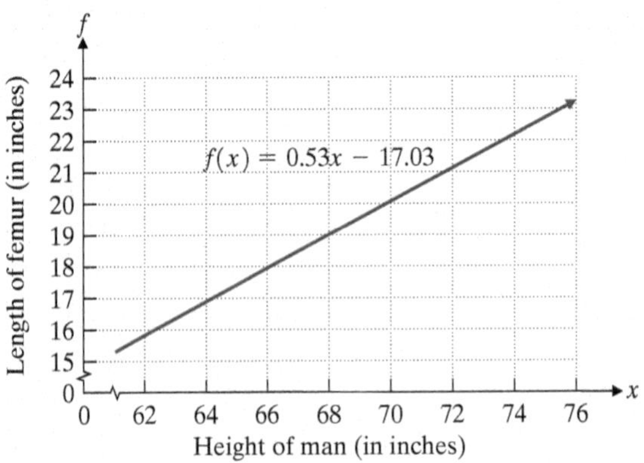

Source: www.nsbri.org

Sidelight: Linear Versus Nonlinear

When graphing, keep the following facts in mind.

The graph of a linear equation will be a straight line.
The graph of a nonlinear equation will *not* be a straight line.

The graphs in Examples 4, 5, and 6 are *not* straight lines because the equations are *not* linear equations. The graph shown above *is* a straight line because its equation *is* a linear equation.

3.6 Exercises MyMathLab®

Verbal and Writing Skills, Exercises 1–6

1. What are the three ways you can describe a function?

2. What is the difference between a function and a relation?

3. The domain of a function is the set of ____________ ____________ of the ____________ variable.

4. The range of a function is the set of ____________ ____________ of the ____________ variable.

5. How can you tell whether a graph is the graph of a function?

6. Without drawing a graph, how could you tell if the equation $x = y^2$ is a function or not?

(a) *Find the domain and range of the relation.* **(b)** *Determine whether the relation is a function.*

7. $\left\{\left(\frac{3}{7}, 4\right), \left(3, \frac{3}{7}\right), \left(-3, \frac{3}{7}\right), \left(\frac{3}{7}, -1\right)\right\}$

8. $\left\{\left(\frac{2}{3}, -4\right), (-4, 5), \left(\frac{2}{3}, 2\right), (5, -4)\right\}$

9. $\{(6, 2.5), (3, 1.5), (0, 0.5)\}$

10. $\{(6, 1), (7, 8), (5, 1)\}$

11. $\{(12, 1), (14, 3), (1, 12), (9, 12)\}$

12. $\{(7.2, 8), (7.3, 6), (9, 5.8), (4, 6)\}$

13. $\{(3, 75), (5, 95), (3, 85), (7, 100)\}$

14. $\{(85, 3), (95, 11), (110, 15), (110, 20)\}$

Graph the equation.

15. $y = x^2 + 3$

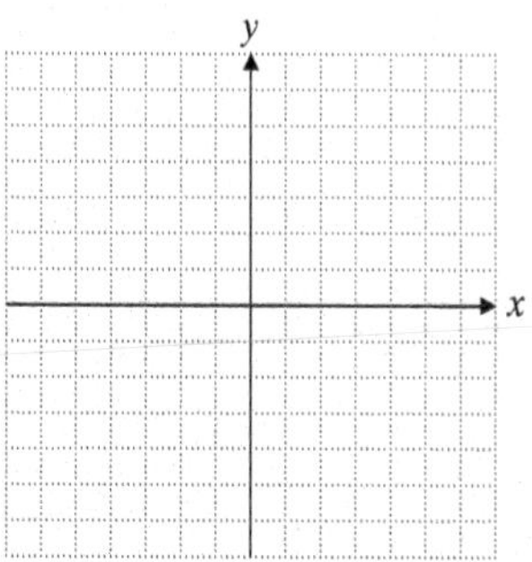

16. $y = x^2 - 1$

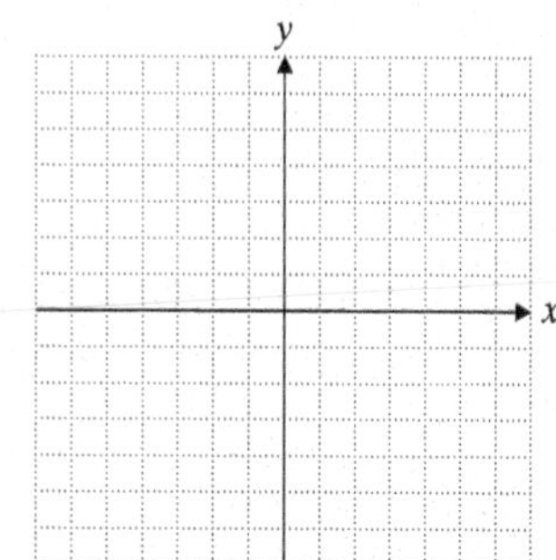

17. $y = 2x^2$

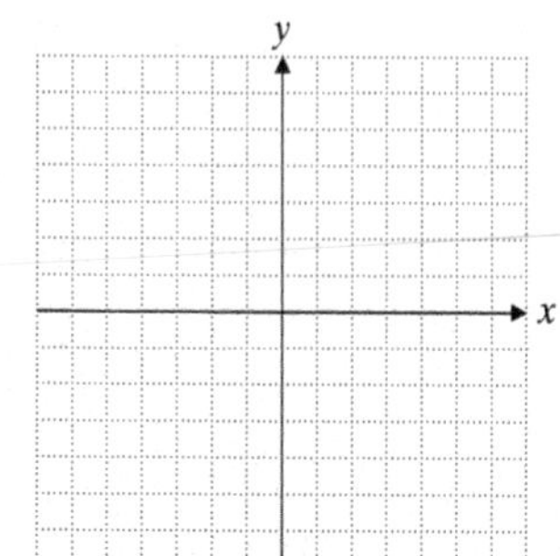

18. $y = \frac{1}{2}x^2$

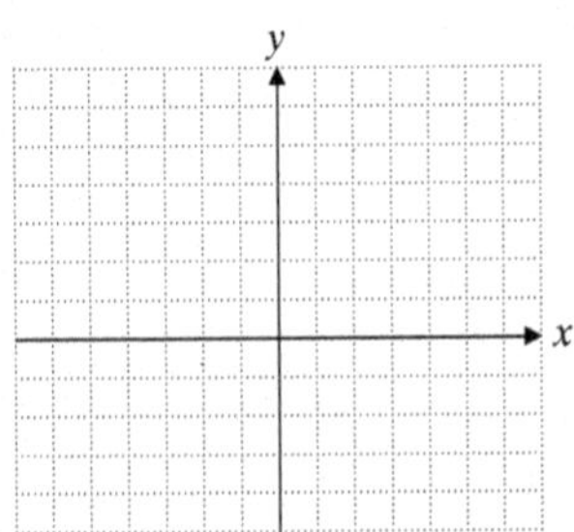

19. $x = -2y^2$

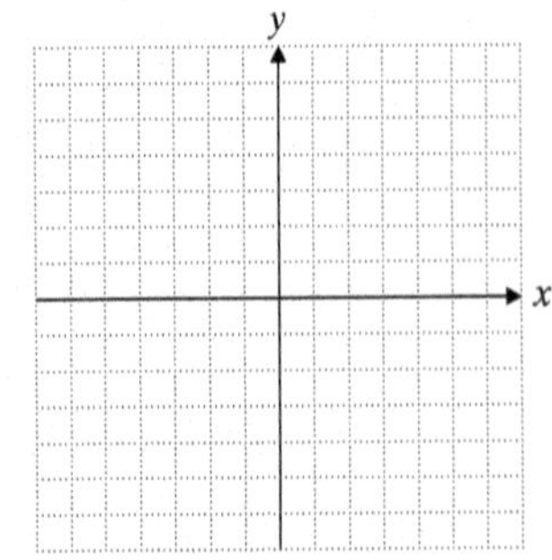

20. $x = \frac{1}{2}y^2$

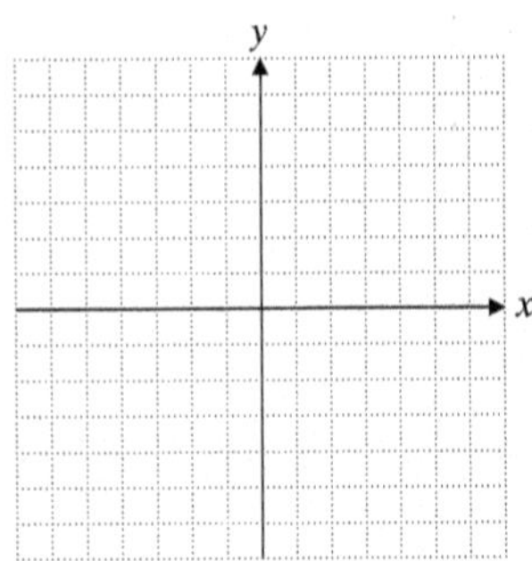

21. $x = y^2 - 4$

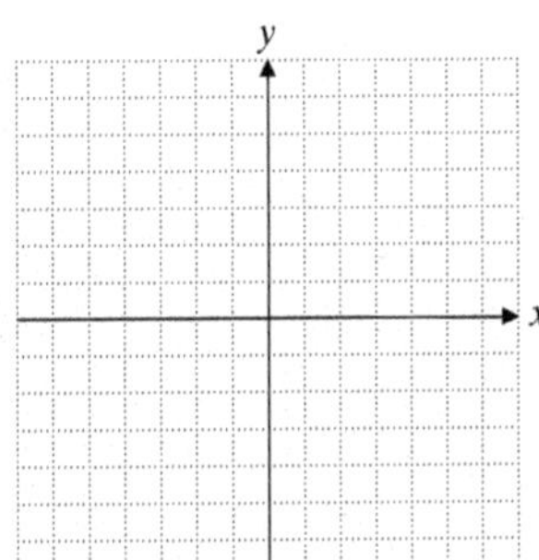

22. $x = 2y^2$

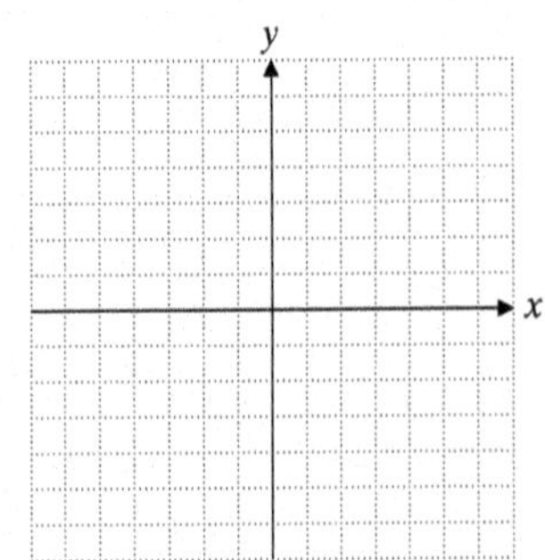

23. $y = \frac{2}{x}$

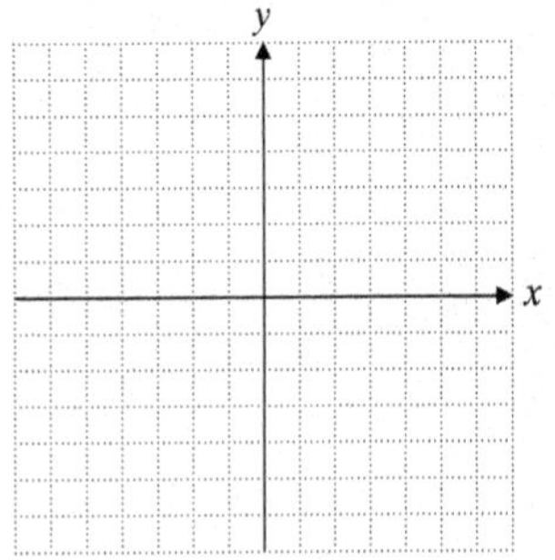

24. $y = -\frac{2}{x}$

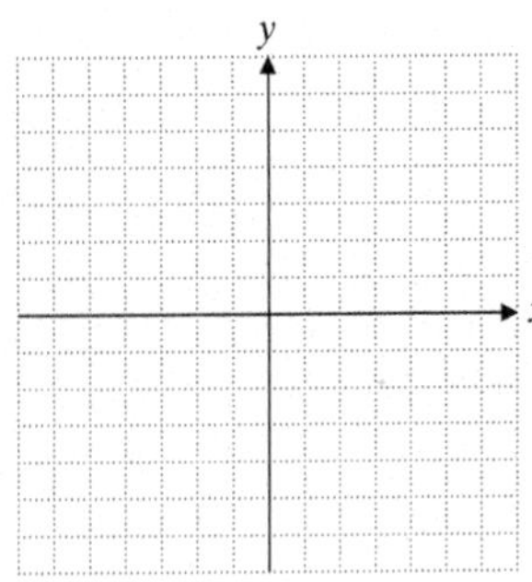

25. $y = \frac{4}{x^2}$

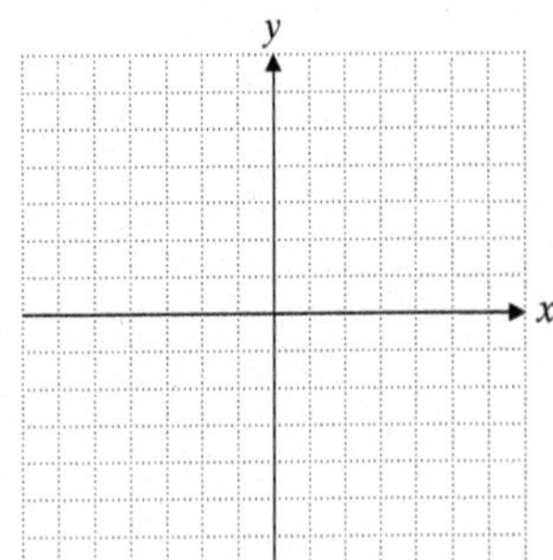

26. $y = -\frac{6}{x^2}$

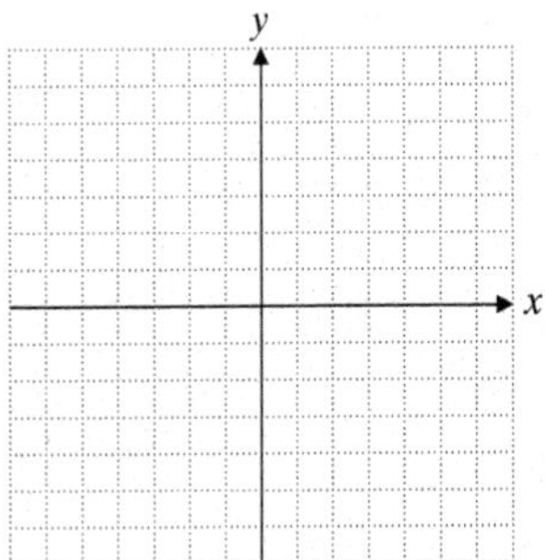

27. $x = (y + 1)^2$

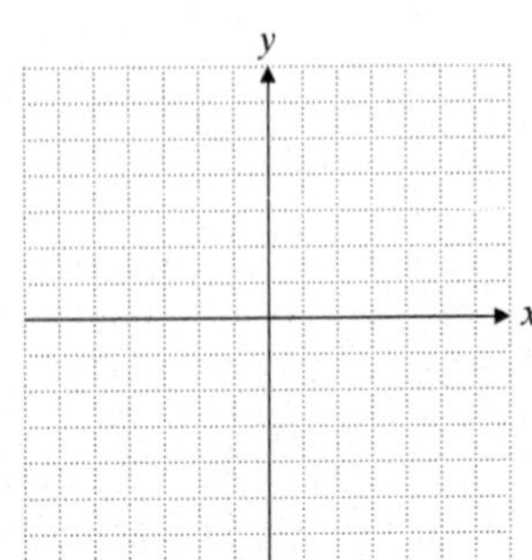

28. $y = (x - 3)^2$

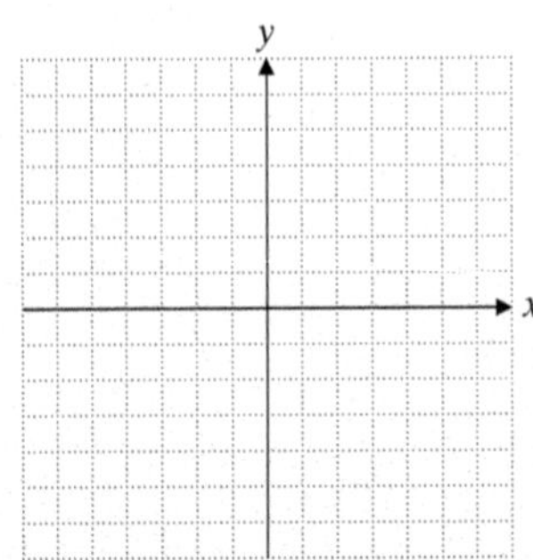

29. $y = \dfrac{4}{x - 2}$

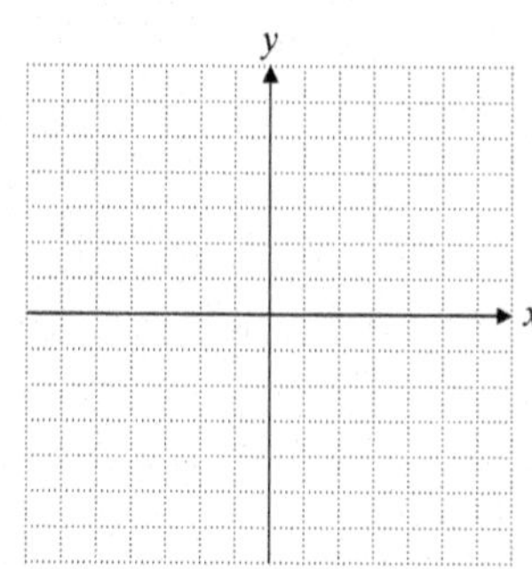

30. $x = \dfrac{2}{y + 1}$

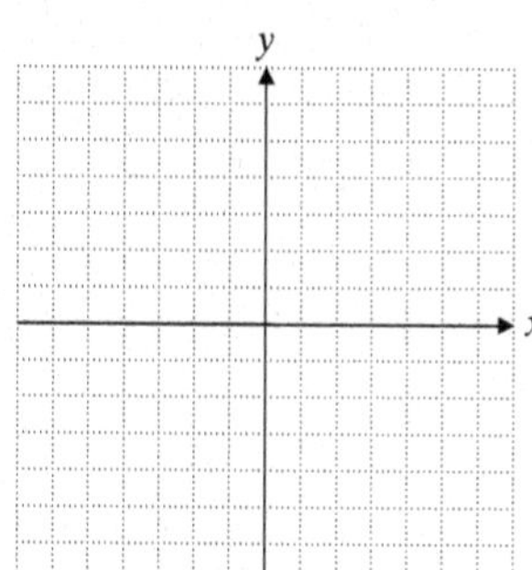

Determine whether each relation is a function.

31.

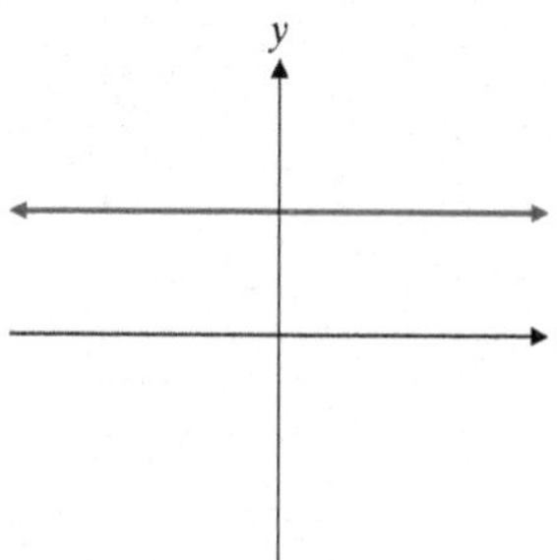

32.

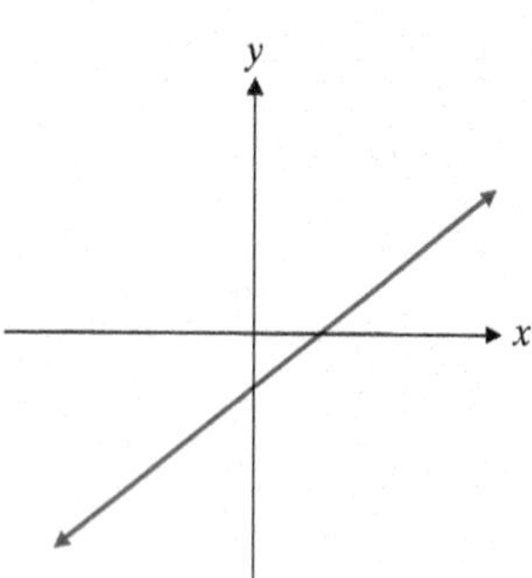

33.

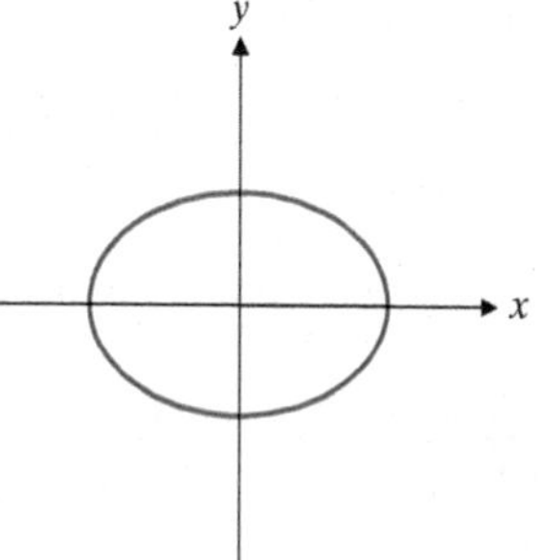

34.

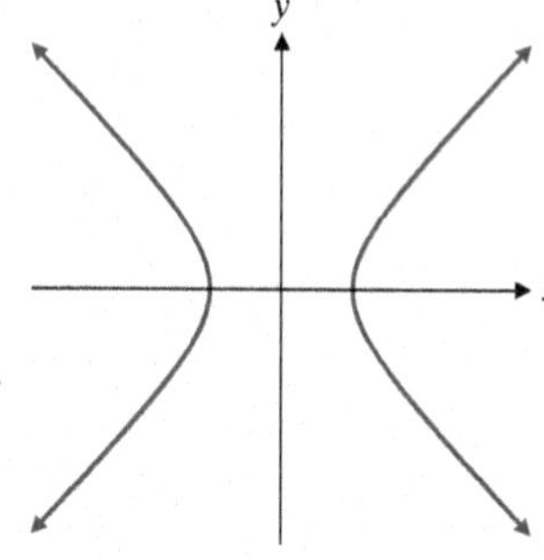

35.

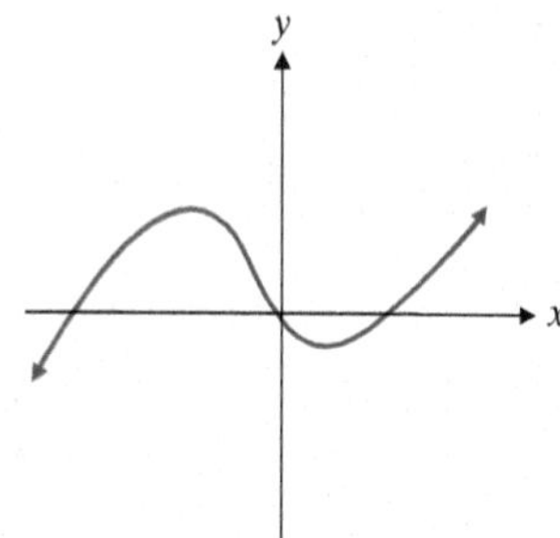

36.

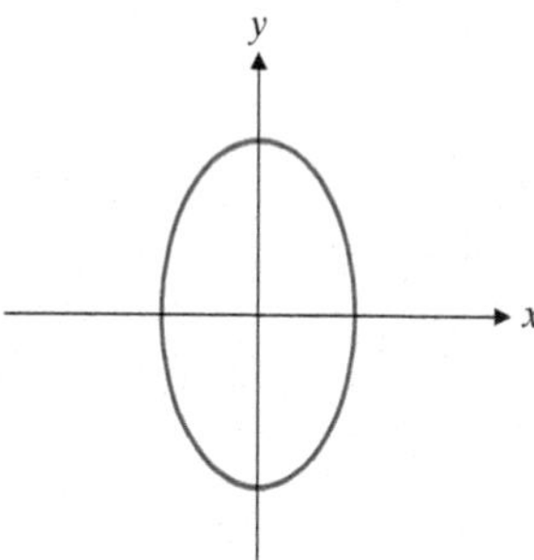

37.

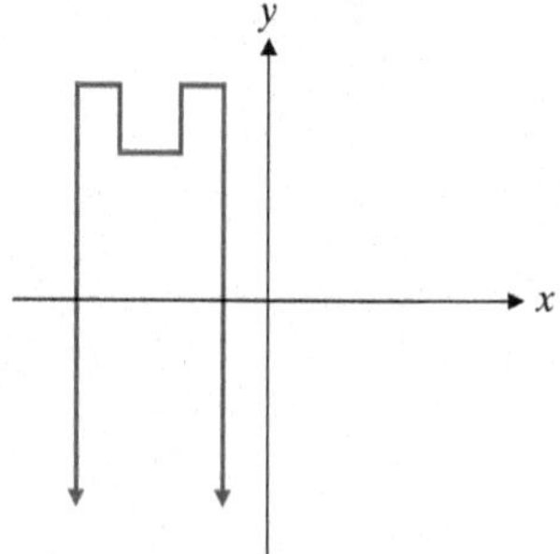

38.

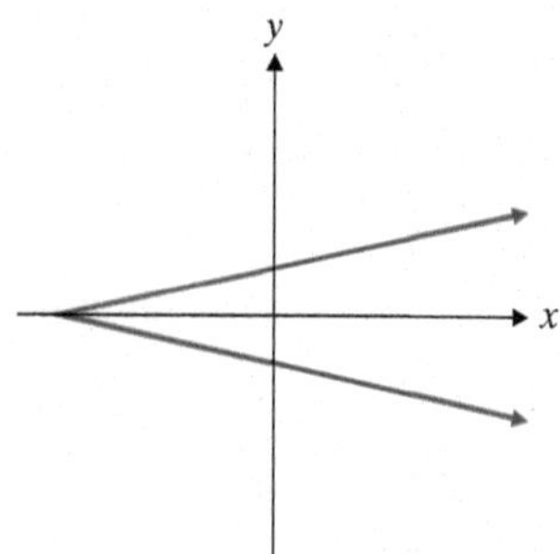

Given the following functions, find the indicated values.

39. $f(x) = 2 - 3x$
(a) $f(-8)$ **(b)** $f(0)$ **(c)** $f(2)$

40. $f(x) = 5 - 2x$
(a) $f(-3)$ **(b)** $f(1)$ **(c)** $f(2)$

41. $f(x) = 2x^2 - x + 3$
(a) $f(0)$ **(b)** $f(-3)$ **(c)** $f(2)$

42. $f(x) = -x^2 - 2x + 3$
(a) $f(-1)$ **(b)** $f(2)$ **(c)** $f(0)$

Applications

43. ***Pet Ownership*** A market research company predicted that pet ownership will increase in the United States during the period 2014 to 2024. The approximate number of pet owners measured in thousands could be predicted by the function $f(x) = 0.02x^2 + 0.08x + 31.6$, where x is the number of years since 2014. Find $f(0)$, and predict $f(4)$, and $f(10)$. Graph the function. What pattern do you observe?

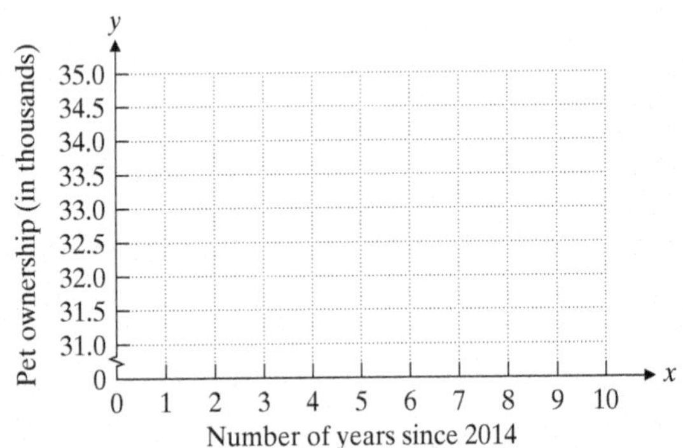

44. ***Kentucky Population*** During a population growth period in Kentucky, from 1995 to 2005, the approximate population of the state measured in thousands could be predicted by the function $f(x) = -0.64x^2 + 30.2x + 3860$, where x is the number of years since 1995. (*Source:* www.census.gov.) Find $f(0), f(5)$, and $f(10)$. Use the function to predict the population in 2015 and 2017. Graph the function. What pattern do you observe?

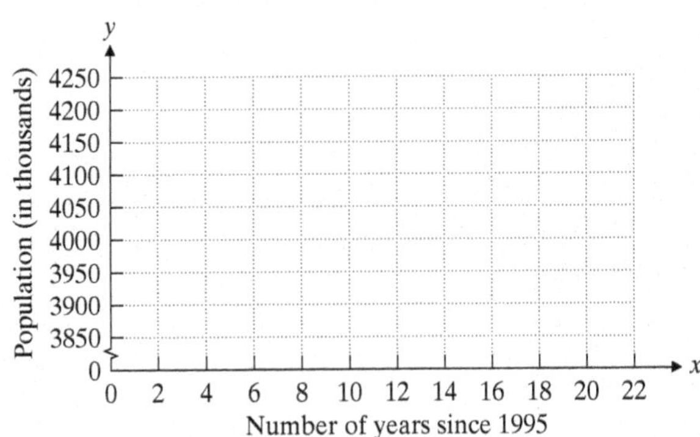

Cumulative Review *Simplify.*

45. **[1.6.1]** $-4x(2x^2 - 3x + 8)$

46. **[1.6.1]** $5a(ab + 6b - 2a)$

47. **[1.7.2]** $-7x + 10y - 12x - 8y - 2$

48. **[1.7.2]** $3x^2y - 6xy^2 + 7xy + 6x^2y$

Quick Quiz 3.6

1. Is this relation a function? $\{(5, 7), (7, 5), (5, 5)\}$ Why?

2. For $f(x) = 3x^2 - 4x + 2$:

(a) Find $f(-3)$. **(b)** Find $f(4)$.

3. For $g(x) = \dfrac{7}{x - 3}$:

(a) Find $g(2)$. **(b)** Find $g(-5)$.

4. **Concept Check** In the relation $\{(3, 4), (5, 6), (3, 8), (2, 9)\}$, why is there a different number of elements in the domain than in the range?

CAREER EXPLORATION PROBLEMS

Background: Economist

The U.S. Bureau of Labor Statistics collects and analyzes data related to economic and social issues important to the American public's interests. Salina, an economist for the bureau, is working on a project involving the consumer price index (CPI), a value indicating the change in the price of a defined set of goods and services over time. She's examining the CPI and its trends over the past decade to see if it's possible to predict future CPI values.

Facts

Salina has assembled data about the consumer price index over the past ten years. See Table 1.

Table 1
Consumer Price Index 2005–2014

Year	CPI
2005	195.30
2006	201.60
2007	207.34
2008	215.30
2009	214.54
2010	218.06
2011	224.94
2012	229.59
2013	232.96
2014	236.74

Source: U.S. Bureau of Labor Statistics

Tasks

1. Salina wants to display the data she's assembled in graphical form to get a general sense of when the CPI was increasing or decreasing. Record the approximate location of each year's data point on the line graph outline below.

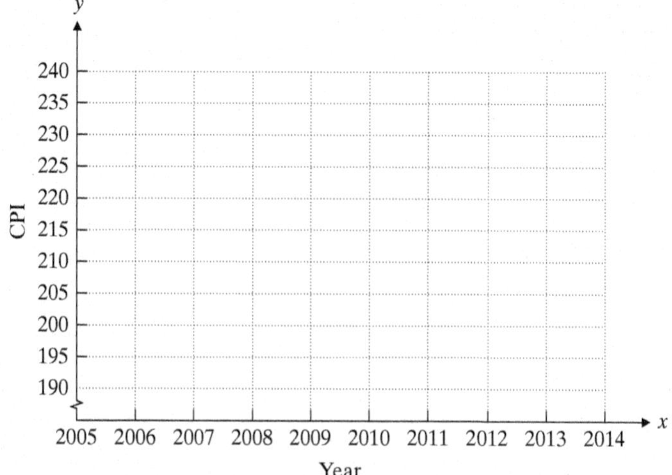

Source: U.S. Bureau of Labor Statistics

2. Salina now computes the average rate of change, given by slope, for three time periods: 2005–2008, 2008–2009, and 2010–2014. (Use the precise data points from the table, and round the final values to the nearest hundredth when necessary.) What is the average rate of change for each of these three periods?

3. Now, Salina wants to write the equation of the line that connects the first and last data points to describe the relation of CPI to year. What is this equation?

4. Salina uses the equation she found to approximate the CPI for two future years, if trends continue. How much will the set of goods and services purchased for \$236.74 in 2014 cost in 2016? How much will it cost in 2020?

Chapter 3 Organizer

Topic and Procedure	Examples	You Try It

Graphing straight lines by plotting three ordered pairs, p. 216

An equation of the form

$$Ax + By = C$$

has a graph that is a straight line.

To graph such an equation, plot any three points; two give the line and the third checks it.

Graph $x + 2y = 4$.

x	y
0	2
2	1
−2	3

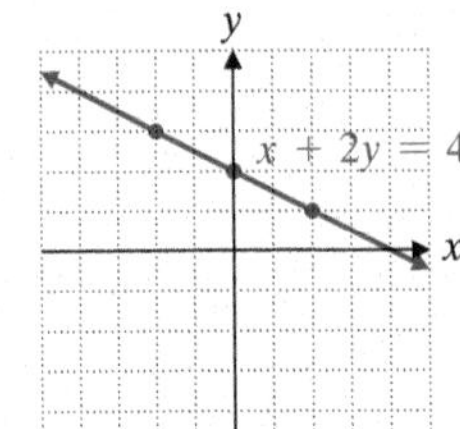

1. Graph $2x - y = 6$.

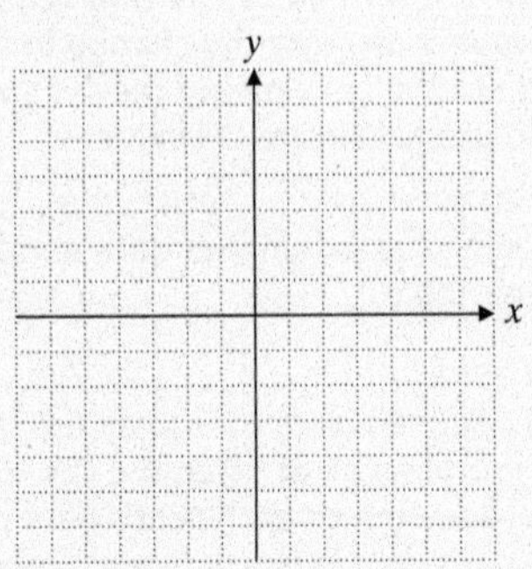

Graphing straight lines by plotting intercepts, p. 218

1. Let $x = 0$ to find the y-intercept.
2. Let $y = 0$ to find the x-intercept.
3. Find another ordered pair to have a third point.

For the equation $3x + 2y = 6$:

(a) State the intercepts.

(b) Graph using the intercept method.

x	y
0	3
2	0
4	−3

y-intercept: $(0, 3)$
x-intercept: $(2, 0)$

2. For the equation $-2x + 4y = -8$:

(a) State the intercepts.

(b) Graph using the intercept method.

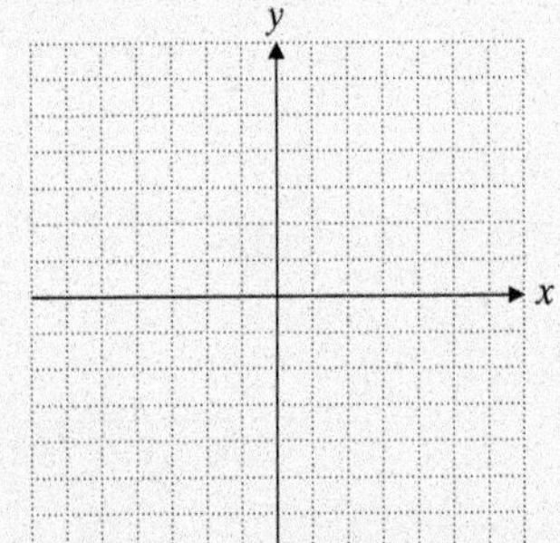

Graphing horizontal and vertical lines, pp. 220–221

The graph of $x = a$, where a is a real number, is a vertical line.

The graph of $y = b$, where b is a real number, is a horizontal line.

Graph each line.

(a) $x = 3$

(b) $y = -4$

(a)

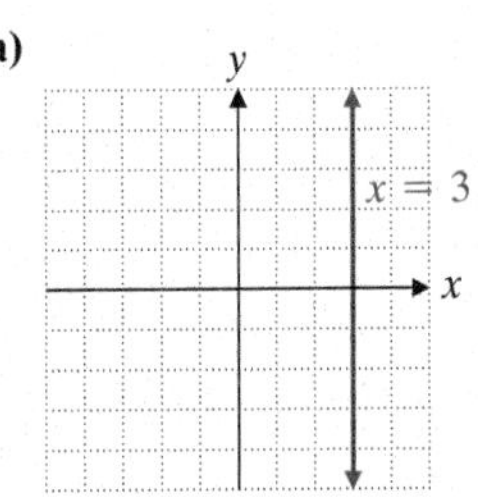

(b)

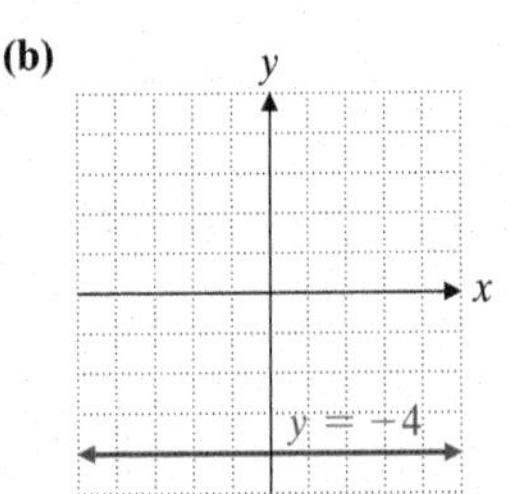

3. Graph each line.

(a) $x = -2$

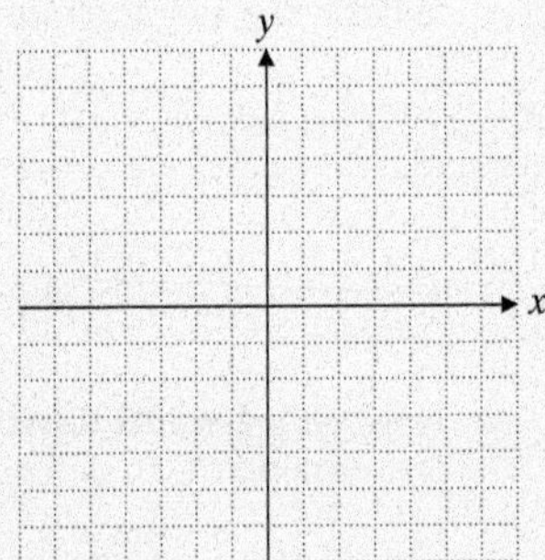

(b) $y = 5$

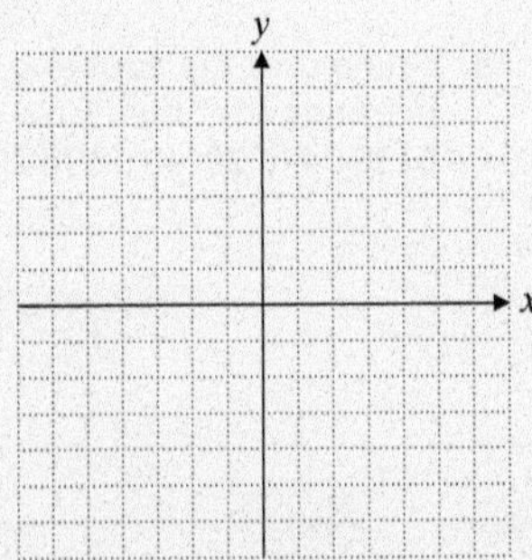

Topic and Procedure	Examples	You Try It
Finding the slope given two points, p. 226 Nonvertical lines passing through distinct points (x_1, y_1) and (x_2, y_2) have slope $m = \dfrac{y_2 - y_1}{x_2 - x_1}$. The slope of a horizontal line is 0. The slope of a vertical line is undefined.	What is the slope of the line through (2, 8) and (5, 1)? $m = \dfrac{1-8}{5-2} = -\dfrac{7}{3}$	4. What is the slope of the line through $(-2, 5)$ and $(0, 1)$?
Finding the slope and y-intercept of a line given the equation, p. 229 **1.** Rewrite the equation in the form $y = mx + b$. **2.** The slope is m. **3.** The y-intercept is $(0, b)$.	Find the slope and y-intercept. $3x - 4y = 8$ $-4y = -3x + 8$ $y = \frac{3}{4}x - 2$ The slope is $\frac{3}{4}$. The y-intercept is $(0, -2)$.	5. Find the slope and y-intercept. $5x + 3y = 9$
Finding an equation of a line given the slope and y-intercept, p. 230 The slope–intercept form of the equation of a line is $y = mx + b$. The slope is m and the y-intercept is $(0, b)$.	Find an equation of the line with y-intercept $(0, 7)$ and with slope $m = 3$. $y = 3x + 7$	6. Find an equation of the line with y-intercept $(0, -3)$ and with slope $m = 2$.
Graphing a line using slope and y-intercept, p. 230 **1.** Plot the y-intercept. **2.** Starting from $(0, b)$, plot a second point using the slope. slope $= \dfrac{\text{rise}}{\text{run}}$ **3.** Draw a line that connects the two points.	Graph $y = -4x + 1$. First plot the y-intercept at $(0, 1)$. Slope $= -4$ or $\frac{-4}{1}$ 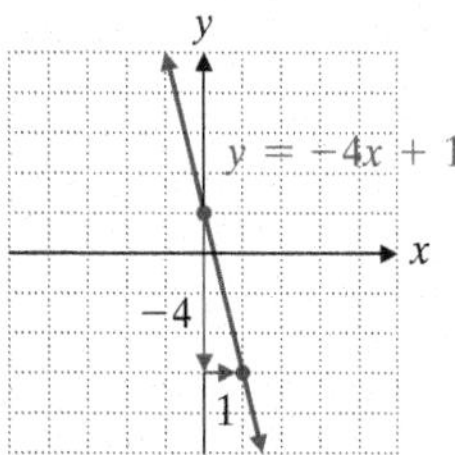	7. Graph $y = 3x - 4$. 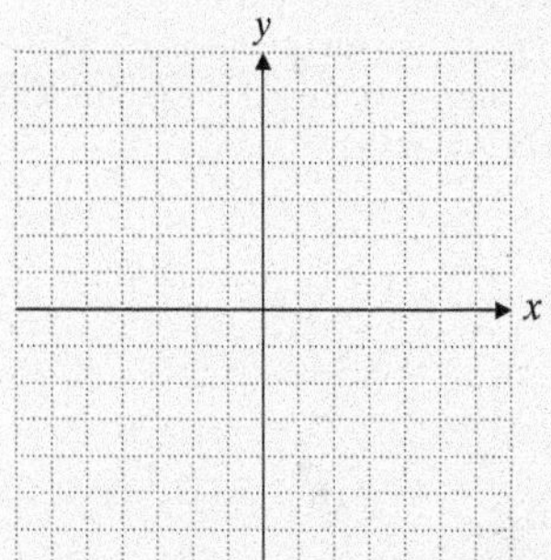
Finding the slopes of parallel and perpendicular lines, p. 233 Parallel lines have the same slope. Perpendicular lines have slopes whose product is -1.	Line q has a slope of 2. The slope of a line parallel to q is 2. The slope of a line perpendicular to q is $-\frac{1}{2}$.	8. Line q has a slope of 3. **(a)** What is the slope of a line parallel to line q? **(b)** What is the slope of a line perpendicular to line q?
Finding an equation of the line through a point with a given slope, p. 239 **1.** Write the slope–intercept form of the equation of a line: $y = mx + b$. **2.** Find m (if not given). **3.** Substitute the known values into the equation $y = mx + b$. **4.** Solve for b. **5.** Use the values of m and b to write the slope–intercept form of the equation.	Find an equation of the line through $(3, 2)$ with slope $m = \frac{4}{5}$. $y = mx + b \quad 2 = \frac{4}{5}(3) + b$ $2 = \frac{12}{5} + b$ $-\frac{2}{5} = b$ An equation is $y = \frac{4}{5}x - \frac{2}{5}$.	9. Find an equation of the line through $(1, 3)$ with slope $m = -\frac{1}{2}$.
Finding an equation of the line through two points, p. 239 **1.** Find the slope. **2.** Use the procedure for finding the equation of a line when given a point and the slope.	Find an equation of the line through $(3, 2)$ and $(13, 10)$. $m = \dfrac{y_2 - y_1}{x_2 - x_1} = \dfrac{10-2}{13-3} = \dfrac{8}{10} = \dfrac{4}{5}$ We are given the point $(3, 2)$. $y = mx + b \quad 2 = \frac{4}{5}(3) + b$ $2 = \frac{12}{5} + b$ $-\frac{2}{5} = b$ The equation is $y = \frac{4}{5}x - \frac{2}{5}$.	10. Find an equation of the line through $(3, 2)$ and $(-1, 0)$.

Topic and Procedure	Examples	You Try It

Graphing linear inequalities, pp. 243–244

1. Graph as if it were an equation.
   If the inequality symbol is $>$ or $<$, use a dashed line.
   If the inequality symbol is $\geq$ or $\leq$, use a solid line.
2. Look for a test point. The easiest test point is $(0, 0)$, unless the line passes through $(0, 0)$. In that case, choose another test point.
3. Substitute the coordinates of the test point into the inequality.
4. If it is a true statement, shade the region on the side of the line containing the test point. If it is a false statement, shade the region on the side of the line that does *not* contain the test point.

Example:

Graph $y \geq 3x + 2$.
Graph the line $y = 3x + 2$. Use a solid line.

Test $(0, 0)$. $\quad 0 \geq 3(0) + 2$

$0 \geq 2$ false

Shade the region that does not contain $(0, 0)$.

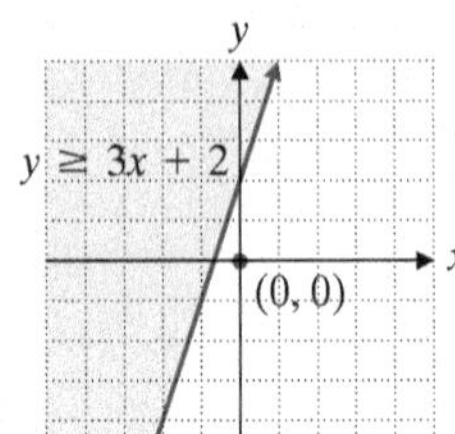

You Try It:

11. Graph $y \leq 2x - 4$.

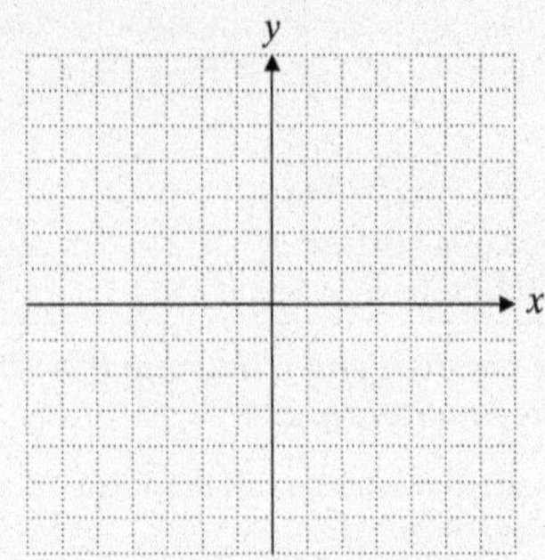

Determining whether a relation is a function, p. 249

A function is a relation in which no two different ordered pairs have the same first coordinate.

Example:

Is this relation a function?

$\{(5, 7), (3, 8), (5, 10)\}$

It is *not* a function since $(5, 7)$ and $(5, 10)$ are two different ordered pairs with the same x-coordinate, 5.

You Try It:

12. Is this relation a function?
$\{(2, -1), (3, 0), (-1, 4), (0, 4)\}$

Graphing nonlinear equations, p. 251

Make a table of values to find several ordered pairs that lie on the graph. Plot the ordered pairs and connect the points with a smooth curve.

Example:

Graph $y = x^2 - 5$.

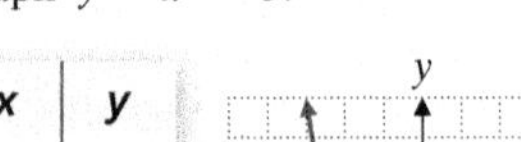

x	y
−2	−1
−1	−4
0	−5
1	−4
2	−1

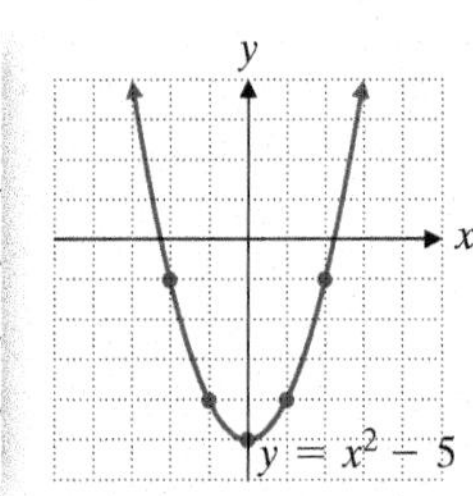

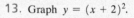

You Try It:

13. Graph $y = (x + 2)^2$.

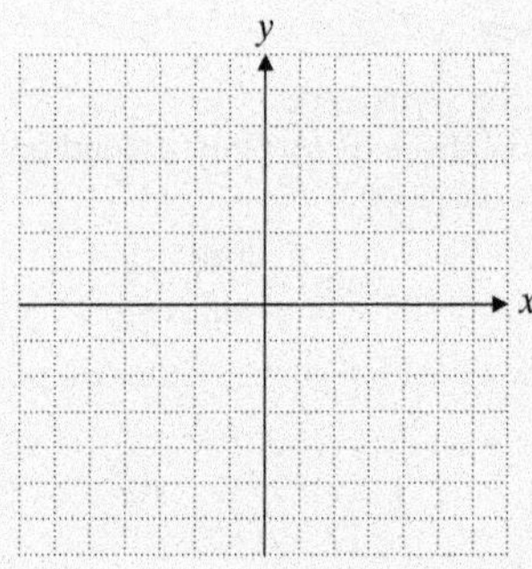

Determining whether a graph represents a function, pp. 252–253

If a vertical line can intersect the graph of a relation more than once, the relation is not a function. If no such line exists, the relation is a function.

Example:

Does this graph represent a function?

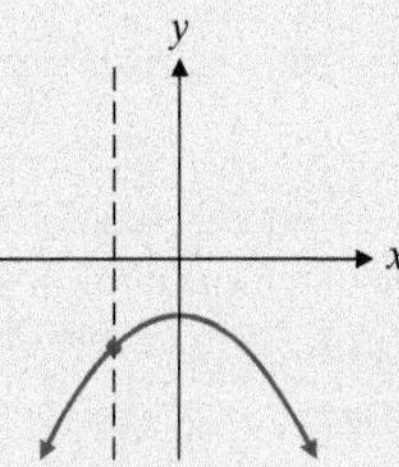

Yes. Any vertical line will intersect it at most once.

You Try It:

14. Does this graph represent a function?

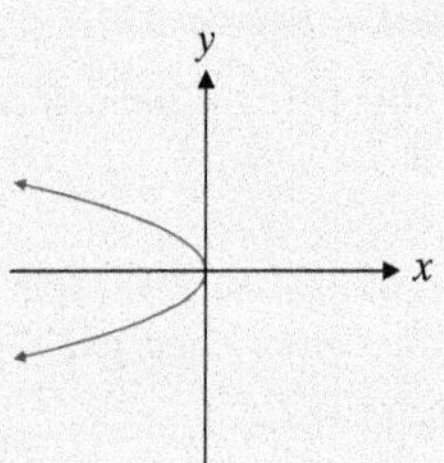

Evaluating a function, p. 254

To evaluate a function, substitute the given value of x into the expression.

Example:

If $f(x) = -3x + 8$, find $f(3)$ and $f(-4)$.

(a) $f(3) = -3(3) + 8 = -9 + 1 = -1$

(b) $f(-4) = -3(-4) + 8 = 12 + 8 = 20$

You Try It:

15. If $f(x) = x^2 - x$, find
(a) $f(5)$ and
(b) $f(-2)$.

Chapter 3 Review Problems

Section 3.1

1. Plot and label the following points.

$A: (2, -3)$
$B: (-1, 0)$
$C: (3, 2)$
$D: (-2, -3)$

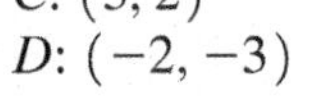

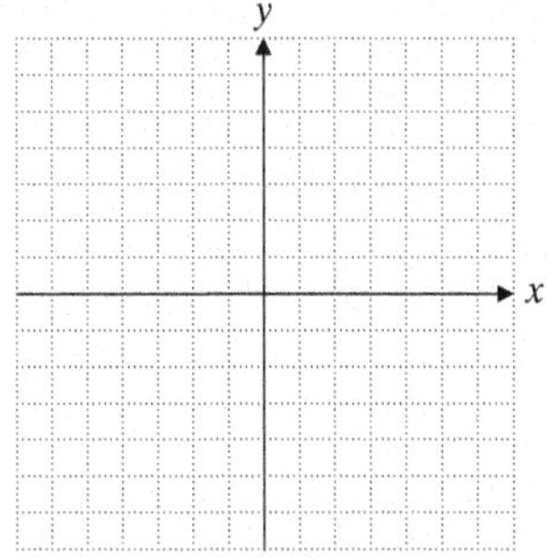

2. Give the coordinates of each point.

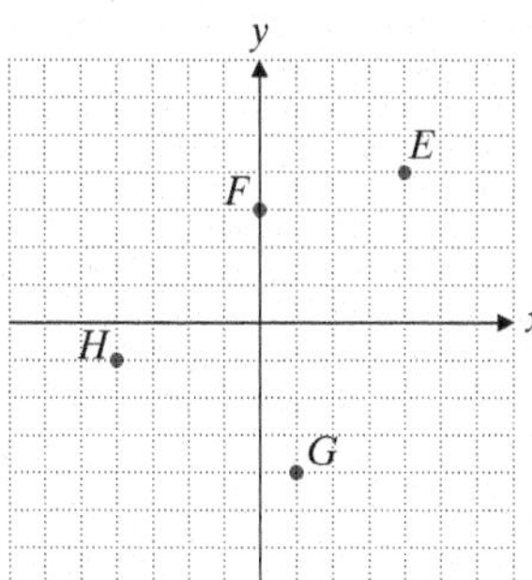

Complete the ordered pairs so that each is a solution to the given equation.

3. $y = 7 - 3x$
(a) $(0, \ \ )$ **(b)** $(\ \ , 10)$

4. $2x + 5y = 12$
(a) $(1, \ \ )$ **(b)** $(\ \ , 4)$

5. $x = 6$
(a) $(\ \ , -1)$ **(b)** $(\ \ , 3)$

Section 3.2

6. Graph $5y + x = -15$.

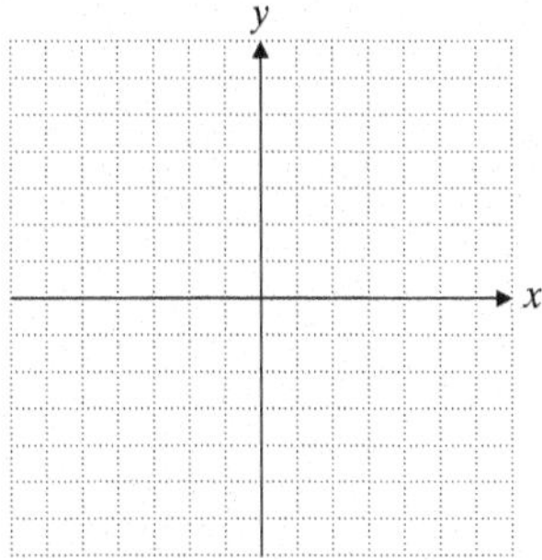

7. Graph $2y + 4x = -8 + 2y$.

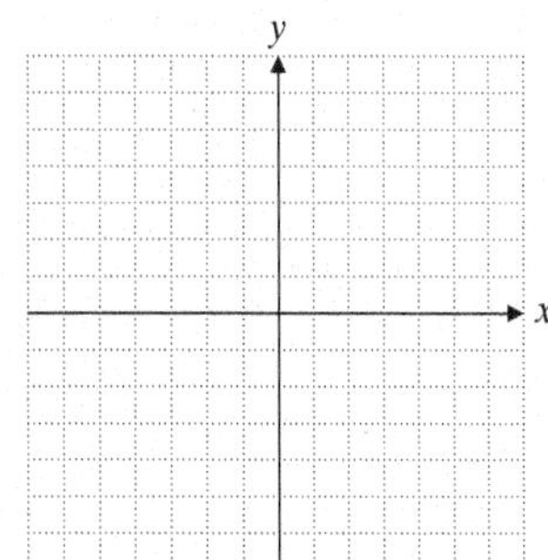

8. Graph $3y = 2x + 6$ and label the intercepts.

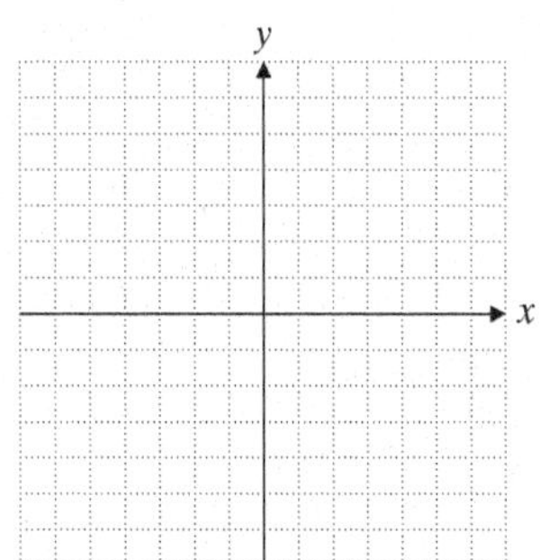

Section 3.3

9. Find the slope of the line passing through $(5, -3)$ and $\left(2, -\frac{1}{2}\right)$.

10. The equation of a line is $y = \frac{3}{5}x - 2$. What is the slope of a line perpendicular to that line?

11. Find the slope and y-intercept of the line $9x - 11y + 15 = 0$.

12. Write an equation of the line with slope $-\frac{1}{2}$ and y-intercept $(0, 3)$.

13. Graph $y = -\dfrac{1}{2}x + 3$.

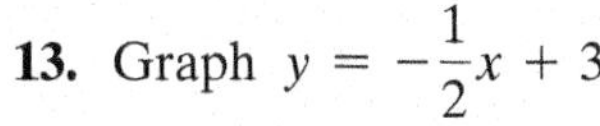

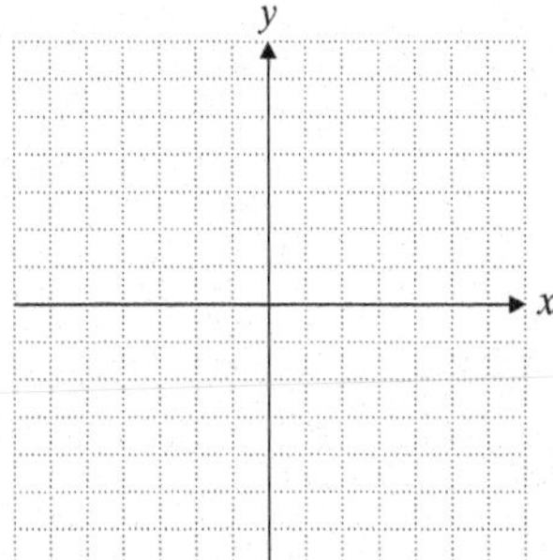

14. Graph $2x - 3y = -12$.

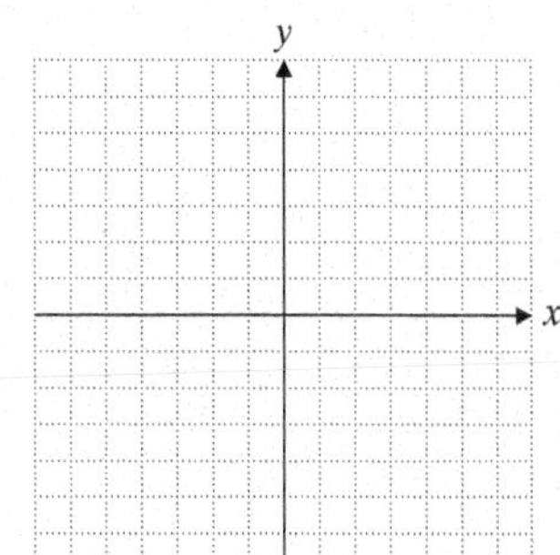

15. Graph $y = -2x$.

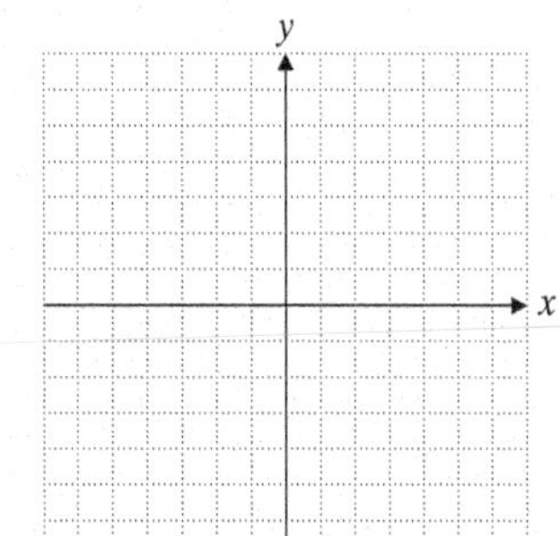

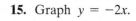

Section 3.4

16. Write an equation of the line passing through $(3, -4)$ having a slope of -6.

17. Write an equation of the line passing through $(-1, 4)$ having a slope of $-\frac{1}{3}$.

18. Write an equation of the line passing through $(2, 5)$ having a slope of 1.

19. Write an equation of the line passing through $(3, 7)$ and $(-6, 7)$.

Write an equation of the graph.

20.

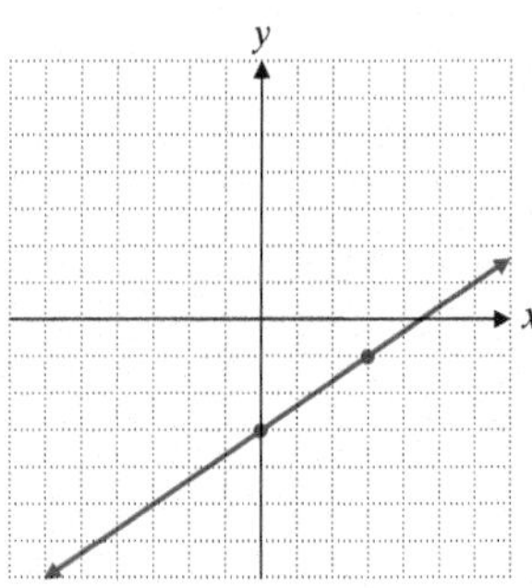

21.

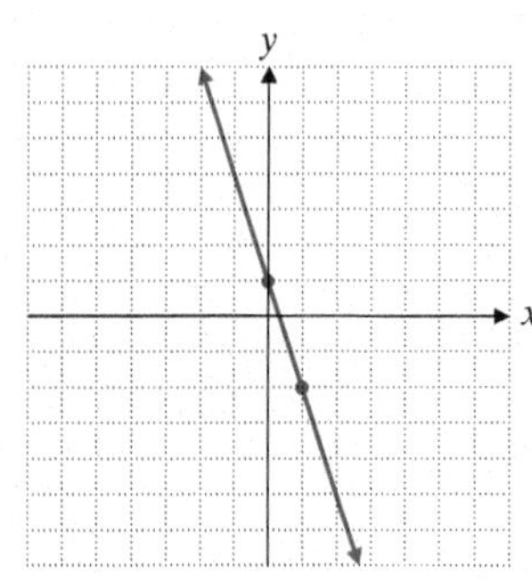

22.

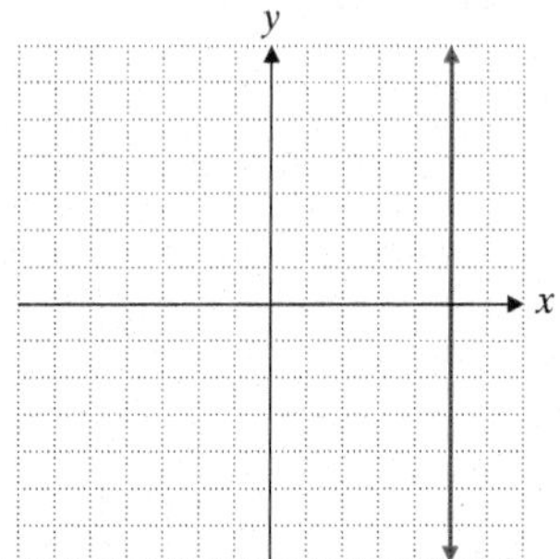

Section 3.5

23. Graph $y < \frac{1}{3}x + 2$.

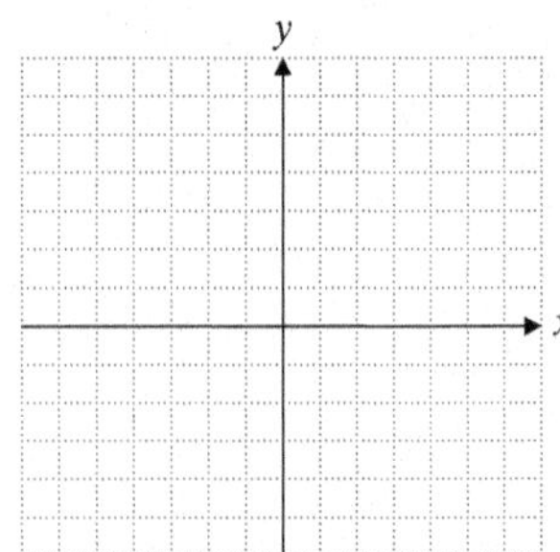

24. Graph $3y + 2x \geq 12$.

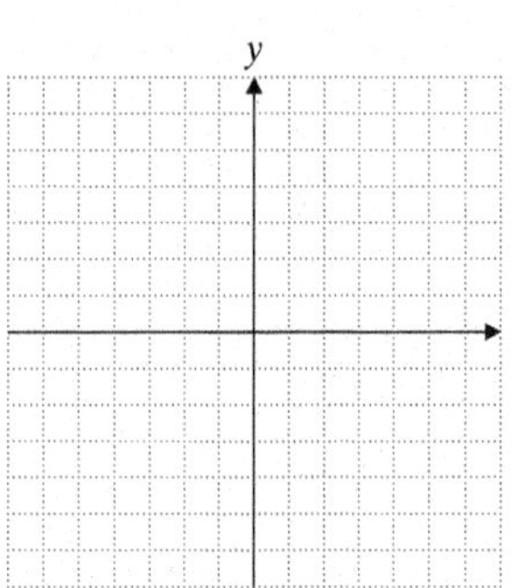

25. Graph $x \leq 2$.

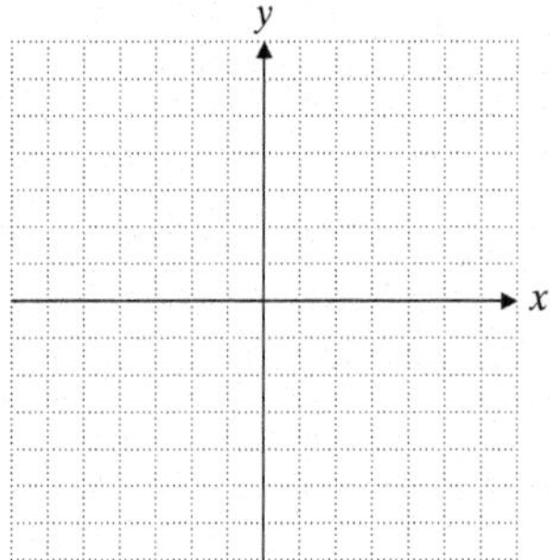

Section 3.6

Determine the domain and range of the relation. Determine whether the relation is a function.

26. $\{(5, -6), (-6, 5), (-5, 5), (-6, -6)\}$

27. $\{(2, -3)(5, -3)(6, 4)(-2, 4)\}$

In exercises 28–30, determine whether the graphs represent a function.

28.

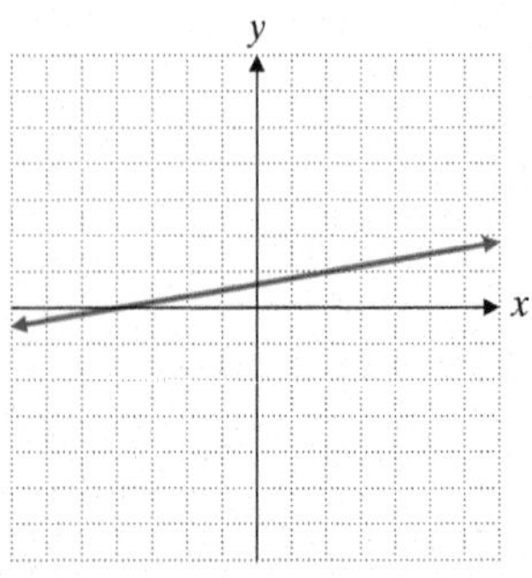

29.

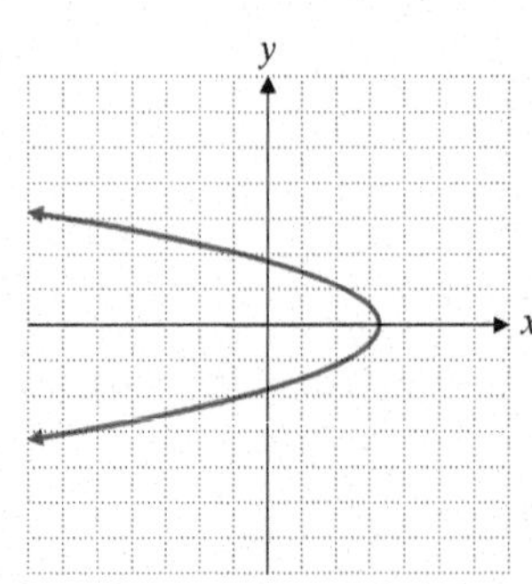

30.

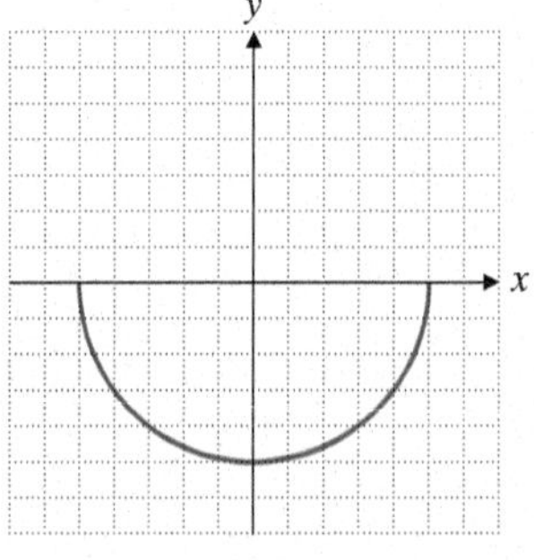

31. Graph $y = x^2 - 5$.

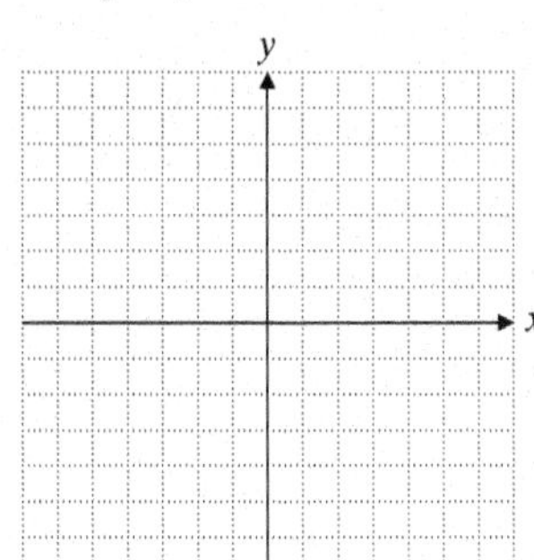

32. Graph $x = y^2 + 3$.

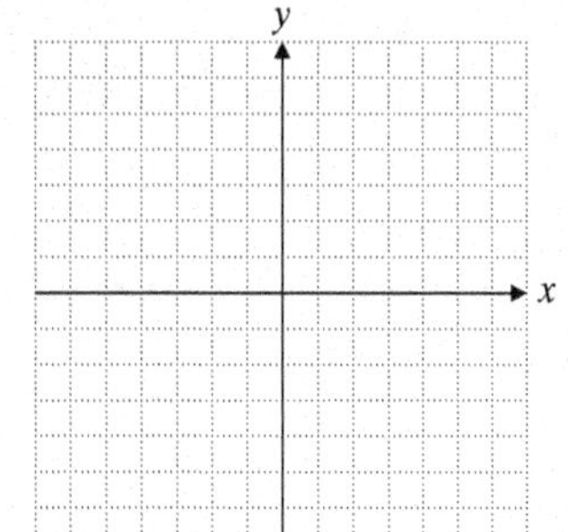

33. Graph $y = (x - 3)^2$.

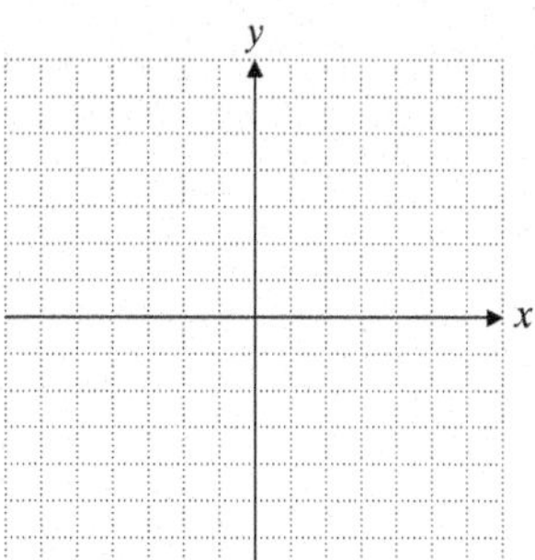

Given the following functions, find the indicated values.

34. $f(x) = 7 - 6x$

(a) $f(0)$ **(b)** $f(-4)$

35. $g(x) = -2x^2 + 3x + 4$

(a) $g(-1)$ **(b)** $g(3)$

36. $f(x) = \dfrac{2}{x + 4}$

(a) $f(-2)$ **(b)** $f(6)$

37. $f(x) = x^2 - 2x + \dfrac{3}{x}$

(a) $f(-1)$ **(b)** $f(3)$

Mixed Practice

38. Graph $5x + 3y = -15$.

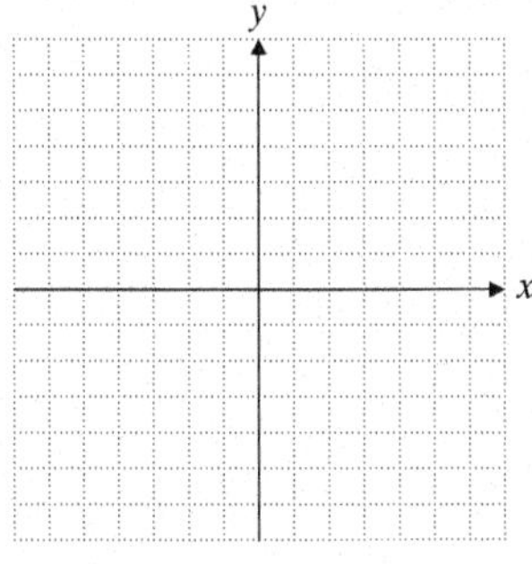

39. Graph $y = \frac{3}{4}x - 3$.

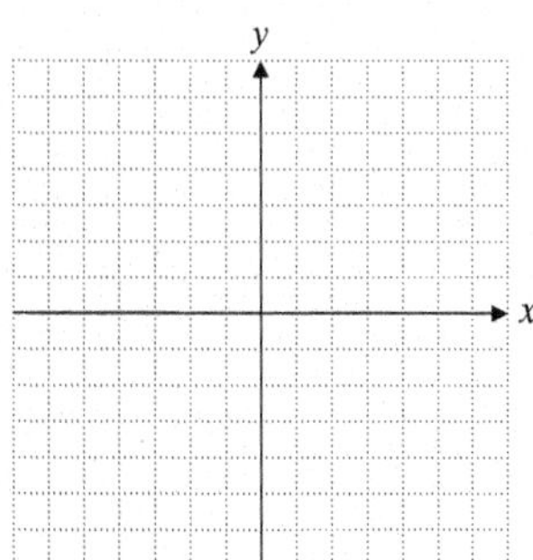

40. Graph $y < -2x + 1$.

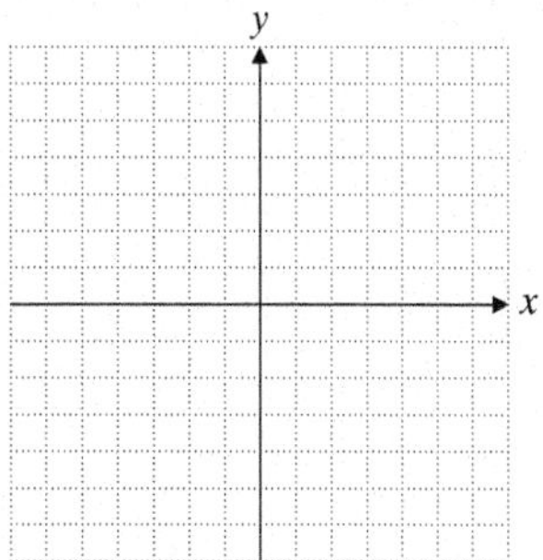

41. Find the slope of the line through $(2, -7)$ and $(-3, -5)$.

42. What is the slope and the y-intercept of the line $7x + 6y - 10 = 0$?

43. Write an equation of the line that passes through $(3, -5)$ and has a slope of $\frac{2}{3}$.

44. Write an equation of the line that passes through $(-1, 4)$ and $(2, 1)$.

Applications

Monthly Electric Bill *Russ and Norma Camp found that their monthly electric bill could be calculated by the equation $y = 30 + 0.09x$. In this equation y represents the amount of the monthly bill in dollars and x represents the number of kilowatt-hours used during the month.*

45. What would be their monthly bill if they used 2000 kilowatt-hours of electricity?

46. What would be their monthly bill if they used 1600 kilowatt-hours of electricity?

47. Write the equation in the form $y = mx + b$, and determine the numerical value of the y-intercept. What is the significance of this y-intercept? What does it tell us?

48. If the equation is placed in the form $y = mx + b$, what is the numerical value of the slope? What is the significance of this slope? What does it tell us?

49. If Russ and Norma have a monthly bill of \$147, how many kilowatt-hours of electricity did they use?

50. If Russ and Norma have a monthly bill of \$246, how many kilowatt-hours of electricity did they use?

Job Losses in Manufacturing *Economists in the Labor Department are concerned about the continued job loss in the manufacturing industry. The number of people employed in manufacturing jobs in the United States can be predicted by the equation $y = -269x + 17{,}020$, where x is the number of years since 1994 and y is the number of employees in* ***thousands*** *in the manufacturing industry. Use this data to answer the following questions. (Source: www.bls.gov)*

51. How many people were employed in manufacturing in 1994? In 2008? In 2014?

52. Use your answers for 51 to draw a graph of the equation $y = -269x + 17{,}020$.

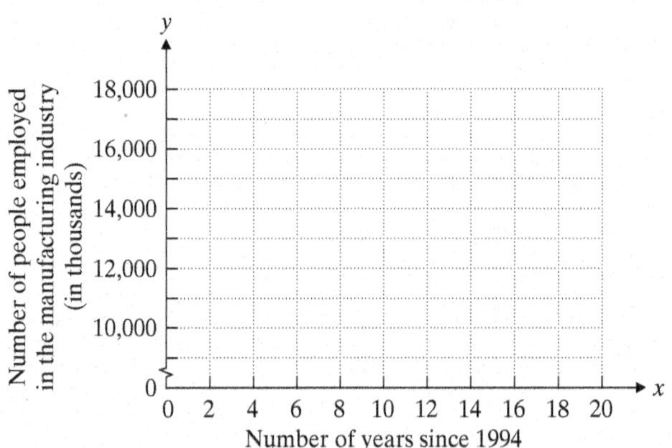

53. What is the slope of this equation? What is its significance?

54. What is the y-intercept of this equation? What is its significance?

55. Use the equation to predict in what year the number of manufacturing jobs will be 11,102,000.

56. Use the equation to predict in what year the number of manufacturing jobs will be 10,295,000.

How Am I Doing? Chapter 3 Test

After you take this test read through the Math Coach on pages 269–270. Math Coach videos are available via MyMathLab and YouTube. Step-by-step test solutions on the Chapter Test Prep Videos are also available via MyMathLab and YouTube. (Search "TobeyBegInterAlg" and click on "Channels.")

1. Plot and label the following points.

B: $(6, 1)$ C: $(-4, -3)$

D: $(-3, 0)$ E: $(5, -2)$

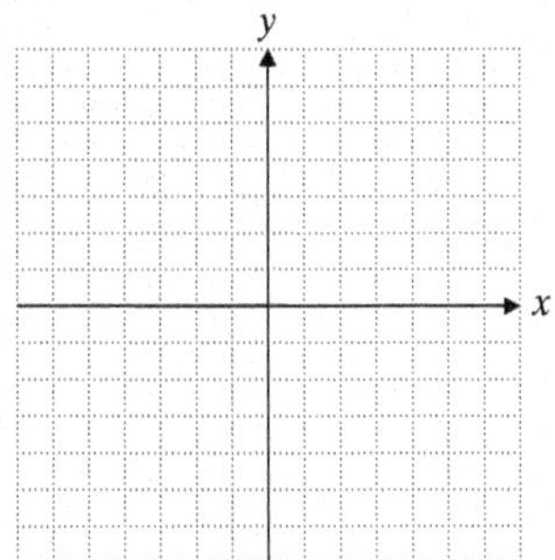

2. Graph the line $6x - 3 = 5x - 2y$.

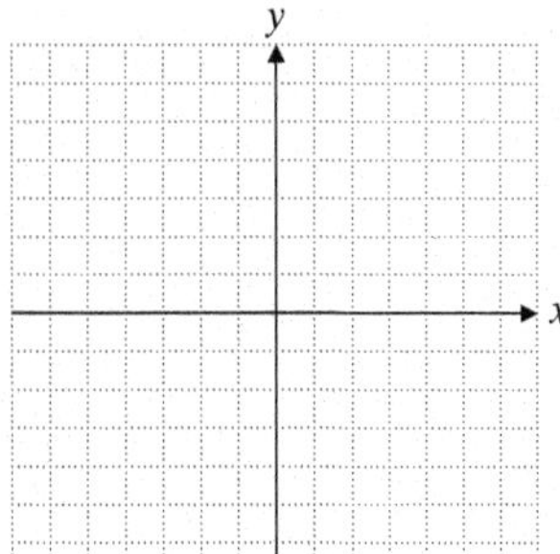

3. Graph $-3x + 9 = 6x$.

MC **4.** Graph $y = \frac{2}{3}x - 4$.

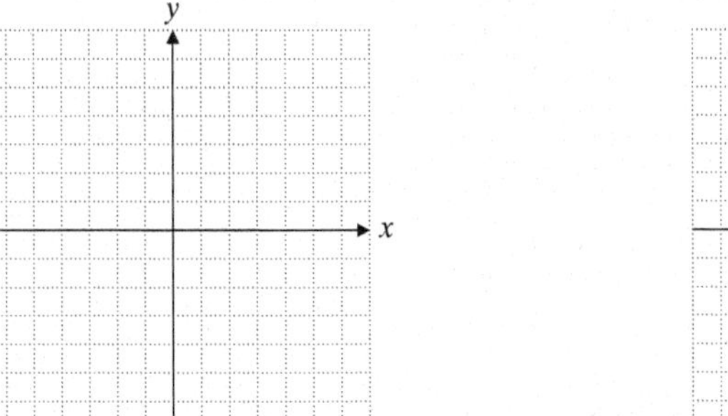

5. Complete **(a)** and **(b)** for the equation $4x + 2y = -8$.

(a) State the x- and y-intercepts.

(b) Use the intercept method to graph.

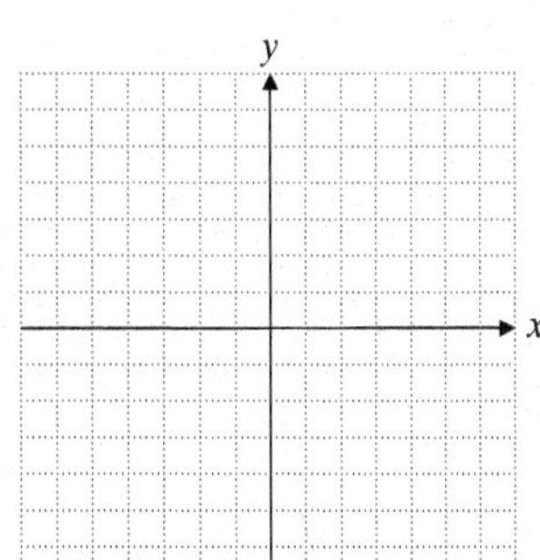

6. Find the slope of the line that passes through $(8, 6)$ and $(-3, -5)$.

7. What is the slope and the y-intercept of the line $3x + 2y - 5 = 0$?

1. ______

2. ______

3. ______

4. ______

5. (a) ______

(b) ______

6. ______

7. ______

8. ______

9. (a) ______

(b) ______

10. ______

11. ______

12. ______

13. ______

14. ______

15. ______

16. (a) ______

(b) ______

Total Correct: ______

8. Find an equation of the line with slope $\frac{3}{4}$ and y-intercept $(0, -6)$.

9. (a) Write an equation for the line f that passes through $(4, -2)$ and has a slope of $\frac{1}{2}$.

(b) What is the slope of a line perpendicular to line f?

MC 10. Find the equation for the line passing through $(5, -4)$ and $(-3, 8)$.

11. Graph the region described by $4y \leq 3x$.

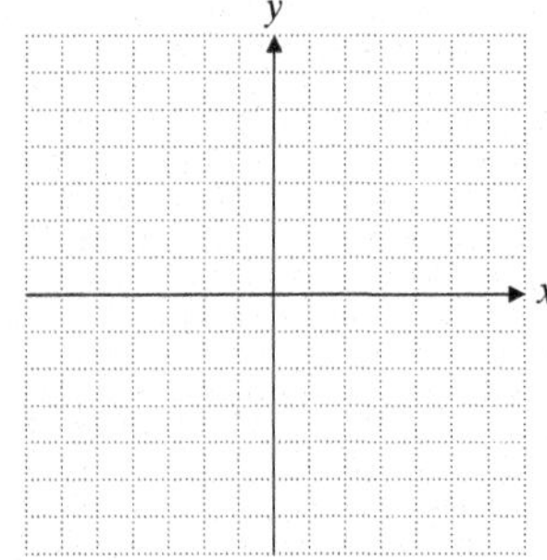

MC 12. Graph the region described by $-3x - 2y > 10$.

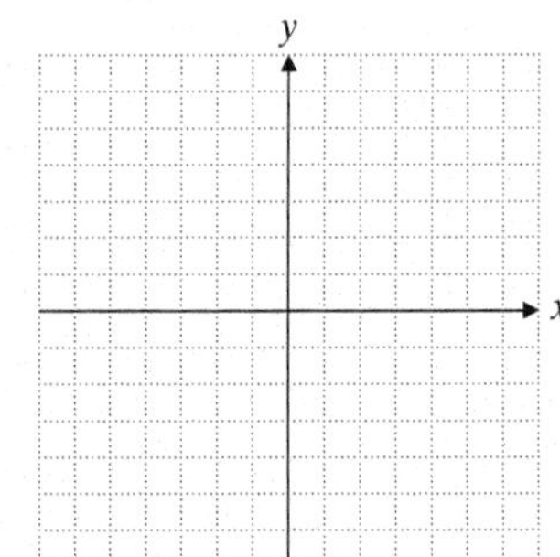

13. Is this relation a function? $\{(2, -8), (3, -7), (2, 5)\}$ Why?

14. Look at the relation graphed below. Is this relation a function? Why?

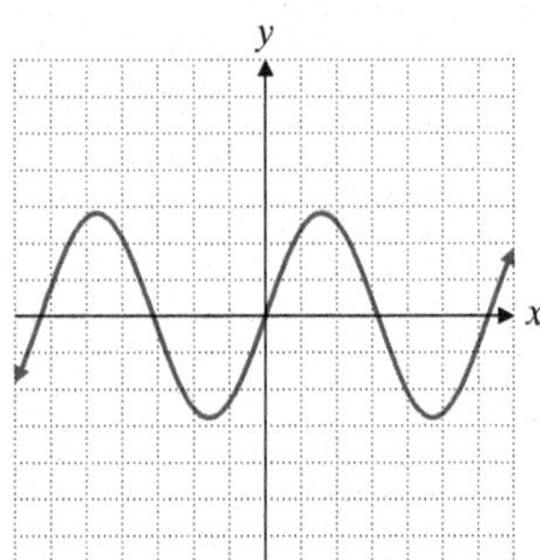

15. Graph $y = 2x^2 - 3$.

x	y
−2	
−1	
0	
1	
2	

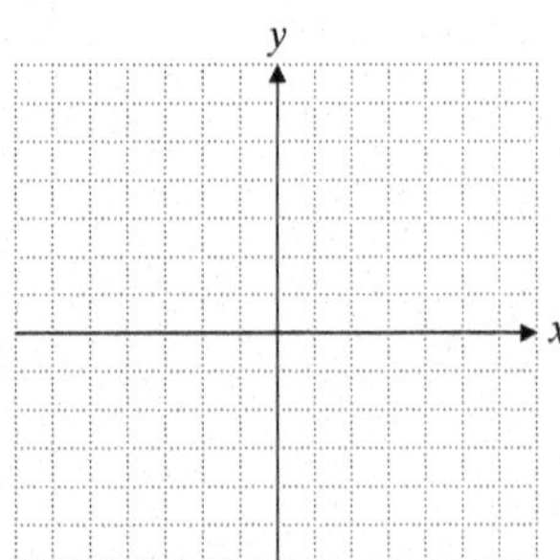

MC 16. For $f(x) = -x^2 - 2x - 3$:

(a) Find $f(0)$.

(b) Find $f(-2)$.

MATH COACH

Mastering the skills you need to do well on the test.

The following problems are from the Chapter 3 Test. Here are some helpful hints to keep you from making common errors on test problems.

Chapter 3 Test, Problem 4 Graph $y = \frac{2}{3}x - 4$.

> **HELPFUL HINT** Find three ordered pairs that are solutions to the equation. Plot those 3 points. Then draw a line through the points. When the equation is solved for *y* and there are fractional coefficients on *x*, it is sometimes a good idea to choose values for *x* that result in integer values for *y*. This will make graphing easier.

Choosing values for x that result in integer values for y means choosing 0 or multiples of 3 since 3 is the denominator of the fractional coefficient on x. When we multiply by these values, the result becomes an integer.

Did you choose values for x that result in values for y that are not fractions?

Yes ______ No ______

If you answered No, try using $x = 0$, $x = 3$, and $x = 6$. Solve the equation for y in each case to find the y-coordinate. Remember that it will make the graphing process easier if you choose values for x that will clear the fraction from the equation.

Did you plot the three points and connect them with a line?

Yes ______ No ______

If you answered No, go back and complete this step.

If you feel more comfortable using the slope–intercept method to solve this problem, simply identify the y-intercept from the equation in $y = mx + b$ form, and then use the slope m to find two other points.

If you answered Problem 4 incorrectly, go back and rework the problem using these suggestions.

Chapter 3 Test, Problem 10 Find the equation for the line passing through $(5, -4)$ and $(-3, 8)$.

> **HELPFUL HINT** When given two points (x_1, y_1) and (x_2, y_2), we can find the slope *m* using the slope formula $m = \frac{y_2 - y_1}{x_2 - x_1}$. Then you can substitute *m* into the equation $y = mx + b$ along with the coordinates of one of the points to find *b*, the *y*-intercept.

When you substituted the points into the slope formula to find m, did you obtain either $m = \frac{8 - (-4)}{-3 - 5}$ or $m = \frac{-4 - 8}{5 - (-3)}$?

Yes ______ No ______

If you answered No, check your work to make sure you substituted the points correctly. Be careful to avoid any sign errors.

Need more help? Watch the MATH COACH videos in MyMathLab® or on YouTube™.

Did you use $m = -\frac{3}{2}$ and either of the points given when you substituted the values into the equation $y = mx + b$ to find the value of b?

Yes ______ No ______

If you answered No, stop and make a careful substitution for $m = -\frac{3}{2}$ and either $x = 5$ and $y = -4$ or $x = -3$ and $y = 8$. See if you can solve the resulting equation for b.

Now go back and rework the problem using these suggestions.

Chapter 3 Test, Problem 12 Graph the region described by $-3x - 2y > 10$.

> **HELPFUL HINT** First graph the equation $-3x - 2y = 10$. Determine if the line should be solid or dashed. Then pick a test point to see if it satisfies the inequality $-3x - 2y > 10$. If the test point satisfies the inequality, shade the side of the line on which the point lies. If the test point does not satisfy the inequality, shade the opposite side of the line.

Examine your work. Does the line $-3x - 2y = 10$ pass through the point $(0, -5)$?

Yes ______ No ______

If you answered No, substitute $x = 0$ into the equation and solve for y. Check the calculations for each of the points you plotted to find the graph of this equation.

Did you draw a solid line?

Yes ______ No ______

If you answered Yes, look at the inequality symbol. Remember that we only use a solid line with the symbols $\leq$ and $\geq$. A dashed line is used for $<$ and $>$.

Did you shade the area above the dashed line?

Yes ______ No ______

If you answered Yes, stop now and use $(0, 0)$ as a test point and substitute it into the inequality $-3x - 2y > 10$. Then use the Helpful Hint to determine which side to shade.

If you answered Problem 12 incorrectly, go back and rework the problem using these suggestions.

Chapter 3 Test, Problem 16(a) and 16(b)

For $f(x) = -x^2 - 2x - 3$: **(a)** Find $f(0)$. **(b)** Find $f(-2)$.

> **HELPFUL HINT** Replace x with the number indicated. It is a good idea to place parentheses around the value to avoid any sign errors. Then use the order of operations to evaluate the function in each case.

Did you replace x with 0 and write

$$f(0) = -(0)^2 - 2(0) - 3?$$

Yes ______ No ______

If you answered No, take time to go over your steps one more time, remembering that 0 times any number is 0.

Did you replace x with -2 and write

$$f(-2) = -(-2)^2 - 2(-2) - 3?$$

Yes ______ No ______

If you answered No, go over your steps again, remembering to place parentheses around -2. Note that $(-2)^2 = 4$ and therefore $-(-2)^2 = -4$.

Now go back and rework the problem again using these suggestions.

Need more help? Look for section examples marked with MG to review.

Cumulative Test for Chapters 0–3

This test provides a comprehensive review of the key objectives for Chapters 0–3.

1. Divide. $2.4\overline{)8.856}$

2. Add. $\frac{3}{8} + \frac{5}{12} + \frac{1}{2}$

3. Add. $1.386 + (-2.9)$

4. Subtract. $9 - 0.48$

5. Multiply. -12.04×0.72

6. Simplify.
$-4(4x - y + 5) - 3(6x - 2y)$

7. Evaluate. $12 - 3(2 - 4) + 12 \div 4$

8. Multiply. $(-3)(-5)(-1)(2)(-1)$

9. Combine like terms. $3st - 8s^2t + 12st^2 + s^2t - 5st + 2st^2$

10. Simplify. $(5x)^2$

11. Simplify. $2\{3x - 4[5 - 3y(2 - x)]\}$

12. Solve for x. $-2(x + 5) = 4x - 15$

13. Solve for x. $\frac{1}{3}(x + 5) = 2x - 5$

14. Solve for y. $\frac{2y}{3} - \frac{1}{4} = \frac{1}{6} + \frac{y}{4}$

Graph the inequality.

15. $\frac{1}{2}(x - 5) \geq x - 4$

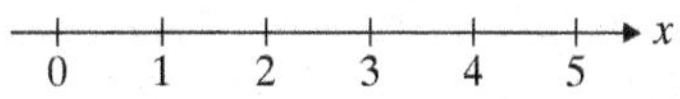

▲ **16.** Find the area of a circle with radius 3 inches.

▲ **17.** Find the area of a triangle with an altitude of 13 meters and a base of 25 meters. What is the cost to construct such a triangle out of sheet metal that costs $4.50 per square meter?

18. When a number is doubled and then increased by 15, the result is 1. Find the number.

19. Ian has a budget of $2000 to spend on a company retirement luncheon for his supervisor at the company lodge. The catering company charges a $250 flat fee to cover glassware, linens, dishes, and silverware. The company charges $30 per person for the meal. How many office staff can Ian invite to the retirement luncheon and stay within his budget?

1. ______
2. ______
3. ______
4. ______
5. ______
6. ______
7. ______
8. ______
9. ______
10. ______
11. ______
12. ______
13. ______
14. ______
15. ______
16. ______
17. ______
18. ______
19. ______

20. ______

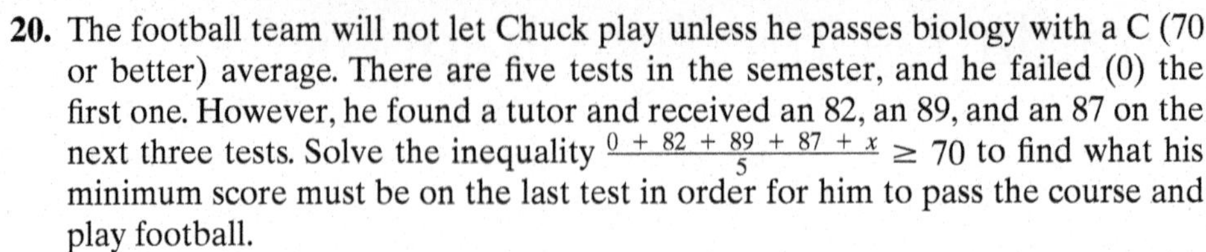

20. The football team will not let Chuck play unless he passes biology with a C (70 or better) average. There are five tests in the semester, and he failed (0) the first one. However, he found a tutor and received an 82, an 89, and an 87 on the next three tests. Solve the inequality $\frac{0 + 82 + 89 + 87 + x}{5} \geq 70$ to find what his minimum score must be on the last test in order for him to pass the course and play football.

21. ______

21. Find the slope of the line through $(-8, -3)$ and $(11, -3)$.

22. Find the equation of the line that passes through $(3, -5)$ with slope $-\frac{2}{3}$.

22. ______

23. Graph $y = \frac{2}{3}x - 4$.

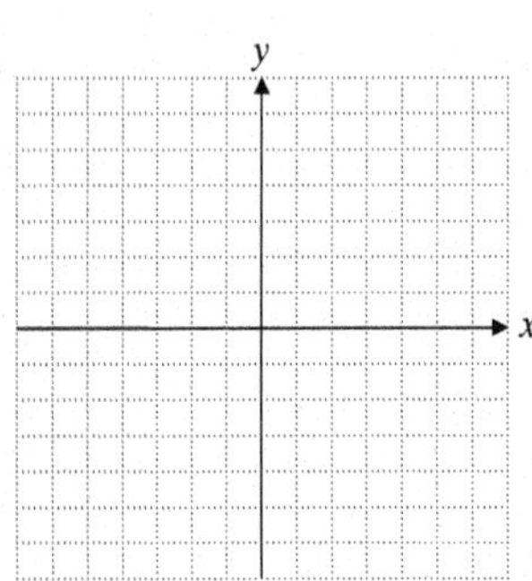

23. ______

24. Graph $3x + 8 = 5x$.

24. ______

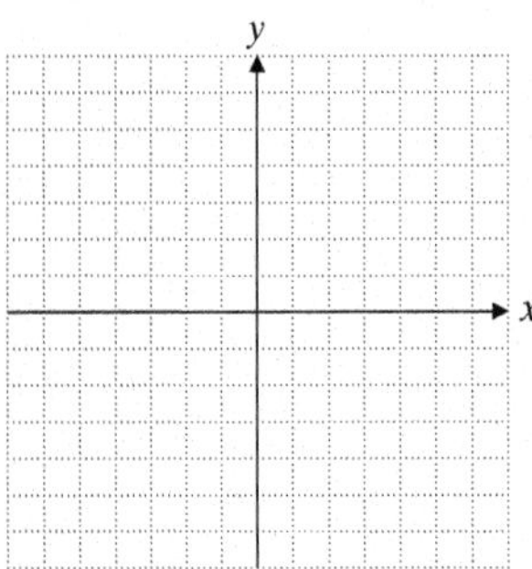

25. ______

25. Graph the region $2x + 5y \leq -10$.

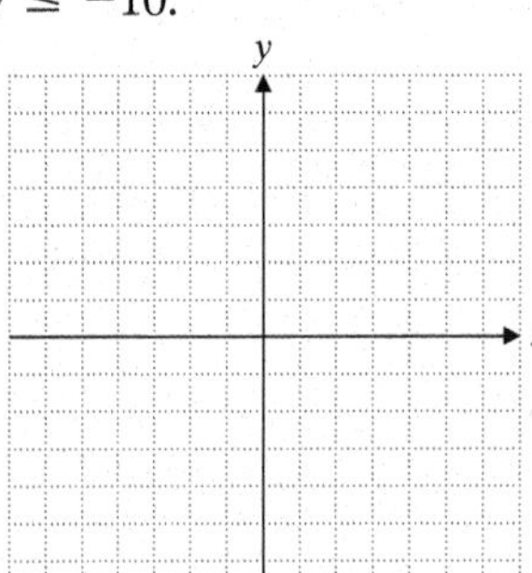

26. (a) ______

26. For $f(x) = 3x^2 - 2x + \frac{4}{x}$:

(b) ______

(a) Find $f(-1)$. **(b)** Find $f(2)$.

CHAPTER 4

Systems of Linear Equations and Inequalities

CAREER OPPORTUNITIES

Accountant

Accountants use their expertise to prepare financial documents and help manage the financial activities of businesses and individuals. Maintaining accurate records of transactions and financial operations helps them to evaluate a business' practices and advise it on how to best manage its expenses and revenues in a manner leading to increased profitability.

To investigate how the mathematics in this chapter can help with this field, see the Career Exploration Problems on page **315**.

4.1 Systems of Linear Equations in Two Variables

Student Learning Objectives

After studying this section, you will be able to:

1. Determine whether an ordered pair is a solution to a system of two linear equations.
2. Solve a system of two linear equations by the graphing method.
3. Solve a system of two linear equations by the substitution method.
4. Solve a system of two linear equations by the addition (elimination) method.
5. Identify systems of linear equations that do not have a unique solution.
6. Choose an appropriate method to solve a system of linear equations algebraically.

1 Determining Whether an Ordered Pair Is a Solution to a System of Two Linear Equations

In Chapter 3 we found that a linear equation containing two variables, such as $4x + 3y = 12$, has an unlimited number of ordered pairs (x, y) that satisfy it. For example, $(3, 0)$, $(0, 4)$, and $(-3, 8)$ all satisfy the equation $4x + 3y = 12$. We call *two* linear equations in two unknowns a **system of two linear equations in two variables.** Many such systems have exactly one solution. A **solution to a system** of two linear equations in two variables is an *ordered pair* that is a solution to *each* equation.

Example 1 Determine whether $(3, -2)$ is a solution to the following system.

$$x + 3y = -3$$
$$4x + 3y = 6$$

Solution We will begin by substituting $(3, -2)$ into the first equation to see whether the ordered pair is a solution to the first equation.

$$3 + 3(-2) \stackrel{?}{=} -3$$
$$3 - 6 \stackrel{?}{=} -3$$
$$-3 = -3 \checkmark$$

Likewise, we will determine whether $(3, -2)$ is a solution to the second equation.

$$4(3) + 3(-2) \stackrel{?}{=} 6$$
$$12 - 6 \stackrel{?}{=} 6$$
$$6 = 6 \checkmark$$

Since $(3, -2)$ is a solution to each equation in the system, it is a solution to the system itself.

Student Practice 1 Determine whether $(-3, 4)$ is a solution to the following system.

$$2x + 3y = 6$$
$$3x - 4y = 7$$

It is important to remember that we cannot confirm that a particular ordered pair is in fact the solution to a system of two equations unless we have checked to see whether the solution satisfies both equations. Merely checking one equation is not sufficient. Determining whether an ordered pair is a solution to a system of equations requires that we verify that the solution satisfies *both* equations. □

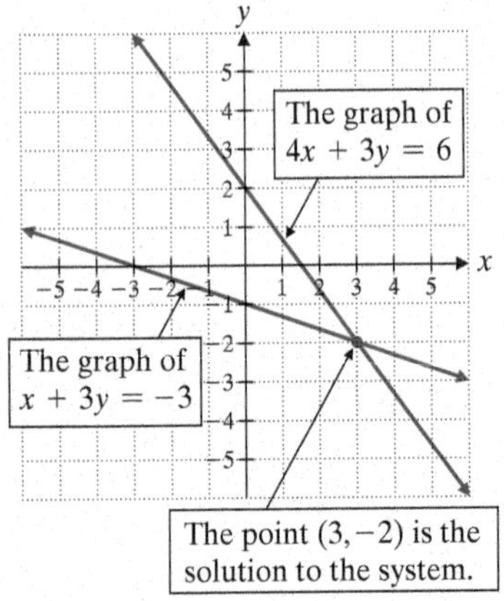

2 Solving a System of Two Linear Equations by the Graphing Method

We can verify the solution to a system of linear equations by graphing each equation. If the lines intersect, the system has a unique solution. The point of intersection lies on both lines. Thus, it is a solution to each equation and the solution to the system. We will illustrate this by graphing the equations in Example 1. Notice that the coordinates of the point of intersection are $(3, -2)$. The solution to the system is $(3, -2)$.

This example shows that we can find the solution to a system of linear equations by graphing each line and determining the point of intersection.

Example 2 Solve this system of equations by graphing.

$$2x + 3y = 12$$
$$x - y = 1$$

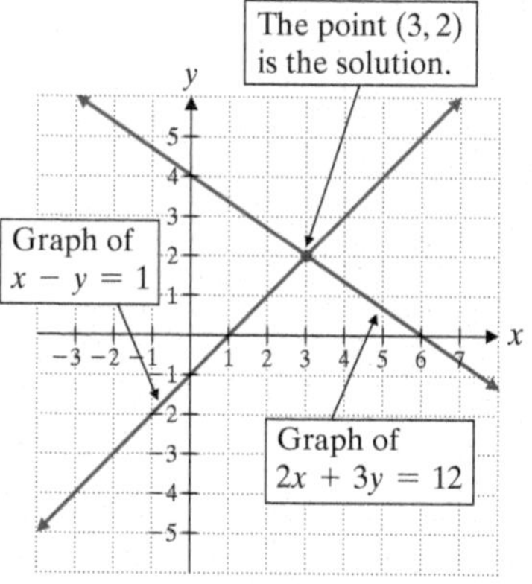

Solution Using the methods that we developed in Chapter 3, we graph each line and determine the point at which the two lines intersect. The graph is to the left.

Finding the solution by the graphing method does not always lead to an accurate result, however, because it involves visual estimation of the point of intersection. Also, our plotting of one or more of the lines could be off slightly. Thus, we verify that our answer is correct by substituting $x = 3$ and $y = 2$ into the system of equations.

$$2x + 3y = 12 \qquad\qquad x - y = 1$$
$$2(3) + 3(2) \stackrel{?}{=} 12 \qquad\qquad 3 - 2 \stackrel{?}{=} 1$$
$$12 = 12 \ ✓ \qquad\qquad 1 = 1 \ ✓$$

Thus, we have verified that the solution to the system is (3, 2). □

Student Practice 2 Solve this system of equations by graphing. Check your solution.

$$3x + 2y = 10$$
$$x - y = 5$$

Student Practice 2

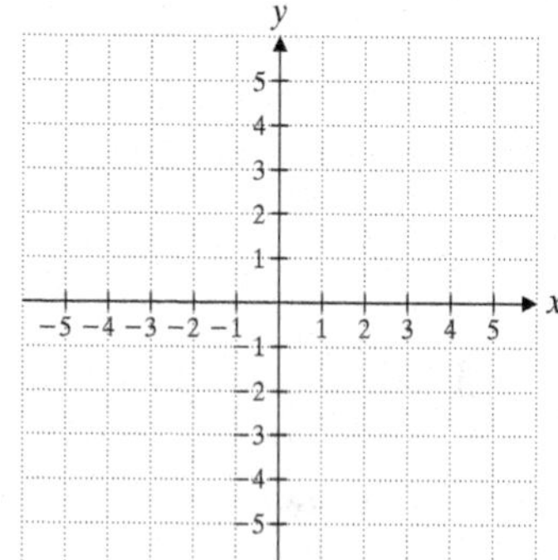

Many times when we graph a system, we find that the two lines intersect at one point. However, it is possible for a given system to have as its graph two parallel lines. In such a case there is no solution because there is no point that lies on both lines (i.e., no ordered pair that satisfies both equations). Such a system of equations is said to be **inconsistent.** Another possibility is that when we graph each equation in the system, we obtain one line. In such a case there is an infinite number of solutions. Any point (i.e., any ordered pair) that lies on the first line will also lie on the second line. A system of equations in two variables is said to have **dependent equations** if it has infinitely many solutions. We will discuss these situations in more detail after we have developed algebraic methods for solving systems of equations.

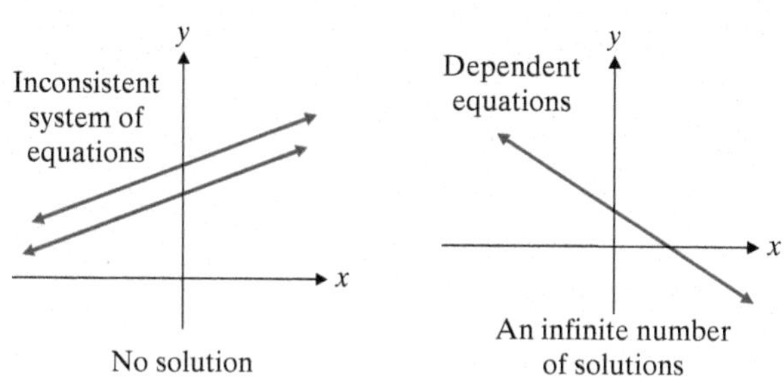

3 Solving a System of Two Linear Equations by the Substitution Method

One algebraic method of solving a system of linear equations in two variables is the **substitution method.** To use this method, we choose one equation and solve for one variable. It is usually best to solve for a variable that has a coefficient of +1 or −1, if possible. This will help us avoid introducing fractions. When we solve for one variable, we obtain an expression that contains the other variable. We *substitute* this expression into the second equation. Then we have one equation with one unknown, which we can easily solve. Once we know the value of this variable, we can substitute it into one of the original equations to find the value of the other variable.

Example 3 Find the solution to the following system of equations. Use the substitution method.

$$x + 3y = -7 \quad \textbf{(1)}$$
$$4x + 3y = -1 \quad \textbf{(2)}$$

Solution We can work with equation **(1)** or equation **(2).** Let's choose equation **(1)** because x has a coefficient of 1. Now let us solve for x. This gives us equation **(3).**

$$x = -7 - 3y \quad \textbf{(3)} \quad \text{We solve } x + 3y = -7 \text{ for } x.$$

Now we substitute this expression for x into equation **(2)** and solve the equation for y.

$$4x + 3y = -1 \quad \textbf{(2)}$$
$$4(-7 - 3y) + 3y = -1 \quad \text{Replace } x \text{ with } -7 - 3y.$$
$$-28 - 12y + 3y = -1 \quad \text{Simplify.}$$
$$-28 - 9y = -1 \quad \text{Now we solve for } y.$$
$$-9y = -1 + 28$$
$$-9y = 27$$
$$y = -3$$

Now we substitute $y = -3$ into equation **(1)** or **(2)** to find x. Let's use **(1):**

$$x + 3(-3) = -7$$
$$x - 9 = -7$$
$$x = -7 + 9$$
$$x = 2$$

Therefore, our solution is the ordered pair $(2, -3)$.
Check. We must verify the solution in both of the *original* equations.

$$x + 3y = -7 \quad \textbf{(1)} \qquad\qquad 4x + 3y = -1 \quad \textbf{(2)}$$
$$2 + 3(-3) \stackrel{?}{=} -7 \qquad\qquad 4(2) + 3(-3) \stackrel{?}{=} -1$$
$$2 - 9 \stackrel{?}{=} -7 \qquad\qquad 8 - 9 \stackrel{?}{=} -1$$
$$-7 = -7 \checkmark \qquad\qquad -1 = -1 \checkmark$$

Student Practice 3 Use the substitution method to solve this system.

$$2x - y = 7$$
$$3x + 4y = -6$$

We summarize the substitution method here.

HOW TO SOLVE A SYSTEM OF TWO LINEAR EQUATIONS BY THE SUBSTITUTION METHOD

1. Choose one of the two equations and solve for one variable in terms of the other variable.
2. Substitute the expression from step 1 into the *other* equation.
3. You now have one equation with one variable. Solve this equation for that variable.
4. Substitute this value for the variable into one of the original or equivalent equations to obtain a value for the second variable.
5. Check the solution in both original equations.

Graphing Calculator

Solving Systems of Linear Equations

We can solve systems of equations graphically by using a graphing calculator. For example, to solve the system of equations in Example 2, first rewrite each equation in slope–intercept form.

$$y = -\frac{2}{3}x + 4$$
$$y = x - 1$$

Then graph $y_1 = -\frac{2}{3}x + 4$ and $y_2 = x - 1$ on the same screen.

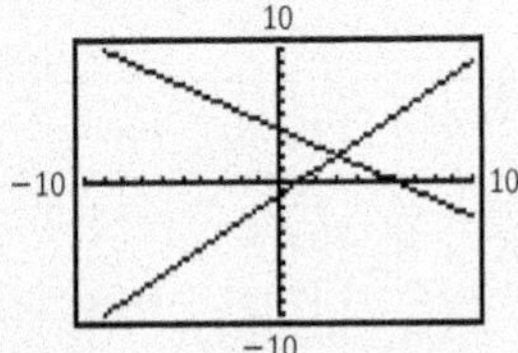

Next, use the Zoom and Trace features to find the intersection of the two lines.
Some graphing calculators have a command to find and calculate the intersection.

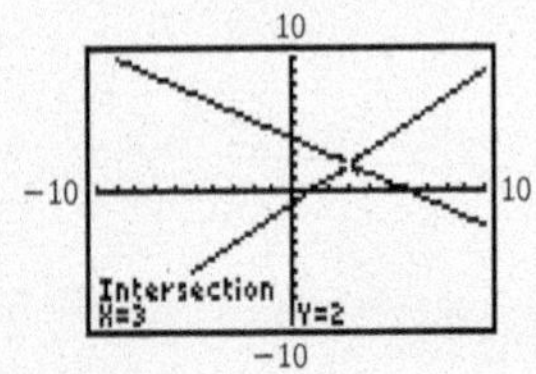

Try to find the solution to:

$$y_1 = -3x + 9$$
$$y_2 = 4x - 5$$

Optional Graphing Calculator Note. Before solving the system in Example 3 with a graphing calculator, you will first need to solve each equation for y. Equation **(1)** can be written as $y_1 = -\frac{1}{3}x - \frac{7}{3}$ or as $y_1 = \frac{-x-7}{3}$. Likewise, equation **(2)** can be written as $y_2 = -\frac{4}{3}x - \frac{1}{3}$ or as $y_2 = \frac{-4x-1}{3}$.

Example 4 Solve the following system of equations.

$$\frac{1}{2}x - \frac{1}{4}y = -\frac{3}{4} \quad \textbf{(1)}$$
$$3x - 2y = -6 \quad \textbf{(2)}$$

Solution First clear equation **(1)** of fractions by multiplying each term by 4.

$$4\left(\frac{1}{2}x\right) - 4\left(\frac{1}{4}y\right) = 4\left(-\frac{3}{4}\right) \quad \textbf{(1)} \quad \text{Multiply each term of the equation by the LCD.}$$
$$2x - y = -3 \quad \textbf{(3)}$$

We now have an equivalent system that does not contain fractions:

$$2x - y = -3 \quad \textbf{(3)}$$
$$3x - 2y = -6 \quad \textbf{(2)}$$

Now follow the five-step procedure.

Step 1 Let's solve equation **(3)** for y. We select this because the y-variable has a coefficient of -1.

$$2x - y = -3 \quad \text{Subtract } 2x \text{ from both sides.}$$
$$-y = -3 - 2x \quad \text{Multiply each term by } -1.$$
$$y = 3 + 2x$$

Step 2 Substitute this expression for y into equation **(2)**.

$$3x - 2(3 + 2x) = -6$$

Step 3 Solve this equation for x.

$$3x - 6 - 4x = -6 \quad \text{Remove parentheses.}$$
$$-6 - x = -6$$
$$-x = -6 + 6$$
$$-x = 0$$
$$x = 0 \quad \text{Multiply each term by } -1.$$

Step 4 To find the value of y, the easiest equation to use is our solution for y from step 1. Substitute $x = 0$ into this equation.

$$y = 3 + 2x$$
$$y = 3 + 2(0)$$
$$y = 3$$

So our solution is $(0, 3)$.

Step 5 We must verify the solution in both original equations.

$$\frac{1}{2}x - \frac{1}{4}y = -\frac{3}{4} \quad \textbf{(1)}$$
$$\frac{0}{2} - \frac{3}{4} \stackrel{?}{=} -\frac{3}{4}$$
$$-\frac{3}{4} = -\frac{3}{4} \quad \checkmark$$

$$3x - 2y = -6 \quad \textbf{(2)}$$
$$3(0) - 2(3) \stackrel{?}{=} -6$$
$$-6 = -6 \quad \checkmark$$

□

Student Practice 4 Use the substitution method to solve this system.

$$\frac{1}{2}x + \frac{2}{3}y = 1$$
$$\frac{1}{3}x + y = -1$$

4 Solving a System of Two Linear Equations by the Addition (Elimination) Method

Another way to solve a system of two linear equations in two variables is to add the two equations so that a variable is eliminated. This technique is called the **addition method** or the **elimination method.** We usually have to multiply one or both of the equations by suitable factors so that we obtain opposite coefficients on one variable (either x or y) in the equations.

Example 5 Solve the following system by the addition method.

$$5x + 8y = -1 \quad \textbf{(1)}$$
$$3x + y = 7 \quad \textbf{(2)}$$

Solution We can eliminate either the x- or the y-variable. Let's choose y. We multiply equation **(2)** by -8.

$$-8(3x) + (-8)(y) = -8(7)$$
$$-24x - 8y = -56 \quad \textbf{(3)}$$

We now add equations **(1)** and **(3).**

$$\begin{aligned} 5x + 8y &= -1 \quad \textbf{(1)} \\ -24x - 8y &= -56 \quad \textbf{(3)} \\ \hline -19x \qquad &= -57 \end{aligned}$$

We solve for x.

$$x = \frac{-57}{-19} = 3$$

Now we substitute $x = 3$ into either of the original equations **(1)** or **(2)** or equivalent equation **(3).** We will use equation **(2).**

$$3(3) + y = 7$$
$$9 + y = 7$$
$$y = -2$$

Our solution is $(3, -2)$.

Check.

$$5x + 8y = -1 \quad \textbf{(1)} \qquad\qquad 3x + y = 7 \quad \textbf{(2)}$$
$$5(3) + 8(-2) \stackrel{?}{=} -1 \qquad\qquad 3(3) + (-2) \stackrel{?}{=} 7$$
$$15 + (-16) \stackrel{?}{=} -1 \qquad\qquad 9 + (-2) \stackrel{?}{=} 7$$
$$-1 = -1 \checkmark \qquad\qquad 7 = 7 \checkmark$$

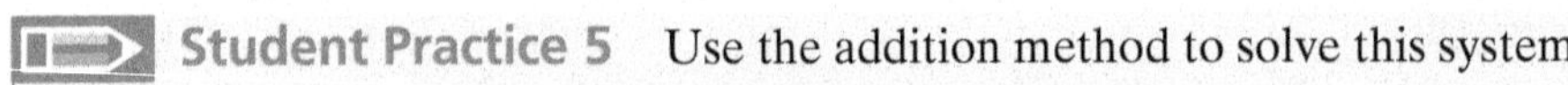

Student Practice 5 Use the addition method to solve this system.

$$-3x + y = 5$$
$$2x + 3y = 4$$

For convenience, we summarize the addition method on the next page.

HOW TO SOLVE A SYSTEM OF TWO LINEAR EQUATIONS BY THE ADDITION (ELIMINATION) METHOD

1. Arrange each equation in the form $ax + by = c$. (Remember that a, b, and c can be any real numbers.)
2. Multiply one or both equations by appropriate numbers so that the coefficients of one of the variables are opposites.
3. Add the two equations from step 2 so that one variable is eliminated.
4. Solve the resulting equation for the remaining variable.
5. Substitute this value into one of the original or equivalent equations and solve to find the value of the other variable.
6. Check the solution in both of the original equations.

MG Example 6 Solve the following system by the addition method.

$$\frac{x}{4} + \frac{y}{6} = -\frac{2}{3} \quad \textbf{(1)}$$
$$\frac{x}{5} + \frac{y}{2} = \frac{1}{5} \quad \textbf{(2)}$$

Solution Clear equation **(1)** of fractions by multiplying each term by 12.

$$12\left(\frac{x}{4}\right) + 12\left(\frac{y}{6}\right) = 12\left(-\frac{2}{3}\right)$$
$$3x + 2y = -8 \quad \textbf{(3)}$$

Clear equation **(2)** of fractions by multiplying each term by 10.

$$10\left(\frac{x}{5}\right) + 10\left(\frac{y}{2}\right) = 10\left(\frac{1}{5}\right)$$
$$2x + 5y = 2 \quad \textbf{(4)}$$

We now have an equivalent system that does not contain fractions.

$$3x + 2y = -8 \quad \textbf{(3)}$$
$$2x + 5y = 2 \quad \textbf{(4)}$$

To eliminate the variable x, we multiply equation **(3)** by 2 and equation **(4)** by -3. We now have the following equivalent system.

$$\begin{array}{rcl} 6x + 4y &=& -16 \\ -6x - 15y &=& -6 \\ \hline -11y &=& -22 \quad \text{Add the equations.} \\ y &=& 2 \quad \text{Solve for } y. \end{array}$$

Since the original equations contain fractions, it will be easier to find the value of x if we use one of the equivalent equations. Substitute $y = 2$ into equation **(3)**.

$$3x + 2(2) = -8$$
$$3x + 4 = -8$$
$$3x = -12$$
$$x = -4$$

The solution to the system is $(-4, 2)$.

Check. Verify that this solution is correct.

Note. We could have easily eliminated the variable y by multiplying equation **(3)** by 5 and equation **(4)** by -2. Try it. Is the solution the same? Why? □

Student Practice 6

Use the addition (elimination) method to solve this system.

$$\frac{x}{4} + \frac{y}{5} = \frac{23}{20}$$
$$\frac{7}{15}x - \frac{y}{5} = 1$$

> **Graphing Calculator**
>
> **Identifying Systems of Linear Equations**
>
> If there is a concern as to whether or not a system of equations has no solution, how can this be determined quickly using a graphing calculator? What is the most important thing to look for on the graph? If the equations are dependent, how can we be sure of this by looking at the display of a graphing calculator? Why? Determine whether there is one solution, no solution, or an infinite number of solutions for each of the following systems.
>
> **1.** $y_1 = -2x + 1$
> $y_2 = -6x + 3$
>
> **2.** $y_1 = 3x + 6$
> $y_2 = 3x + 2$
>
> **3.** $y_1 = x - 3$
> $y_2 = -2x + 6$

5 Identifying Systems of Linear Equations That Do Not Have a Unique Solution

So far we have examined only those systems that have one solution. But other systems must also be considered. These systems can best be illustrated with graphs. In general, the system of equations

$$ax + by = c$$
$$dx + ey = f$$

may have one solution, no solution, or an infinite number of solutions.

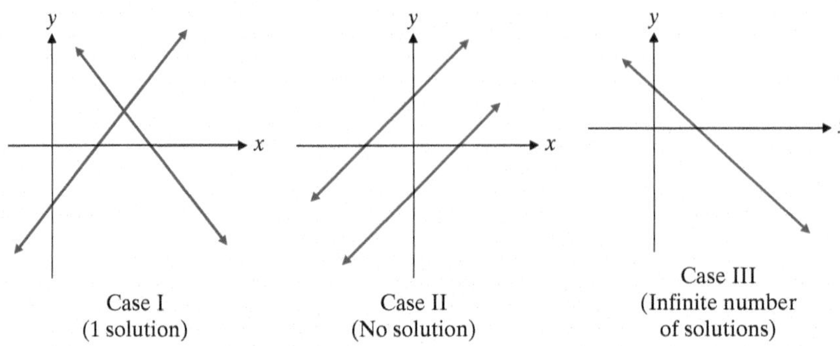

Case I (1 solution) Case II (No solution) Case III (Infinite number of solutions)

Case I: *One solution.* The two graphs intersect at one point, which is the solution. We say that the equations are **independent.** It is a **consistent system** of equations. There is a point (an ordered pair) *consistent* with both equations.

Case II: *No solution.* The two graphs are parallel and so do not intersect. We say that the system of equations is **inconsistent** because there is no point consistent with both equations.

Case III: *An infinite number of solutions.* The graphs of each equation yield the same line. Every ordered pair on this line is a solution to both of the equations. We say that the equations are **dependent.**

Example 7 If possible, solve the system.

$$2x + 8y = 16 \quad \textbf{(1)}$$
$$4x + 16y = -8 \quad \textbf{(2)}$$

Solution To eliminate the variable y, we'll multiply equation **(1)** by -2.

$$-2(2x) + (-2)(8y) = (-2)(16)$$
$$-4x - 16y = -32 \quad \textbf{(3)}$$

We now have the following equivalent system.

$$-4x - 16y = -32 \quad \textbf{(3)}$$
$$4x + 16y = -8 \quad \textbf{(2)}$$

When we add equations **(3)** and **(2),** we get

$$0 = -40,$$

which, of course, is false. Thus, we conclude that this system of equations is inconsistent, and **there is no solution.** Therefore, equations **(1)** and **(2)** do not intersect, as we can see on the graph to the right.

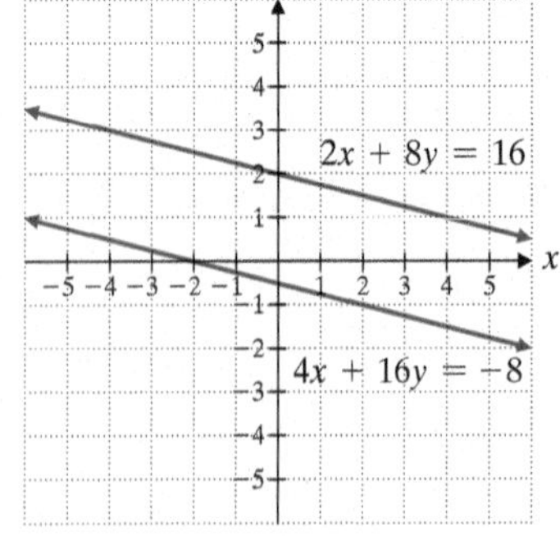

If we had used the substitution method to solve this system, we still would have obtained a false statement. When you try to solve an inconsistent system of linear equations by any method, you will always obtain a mathematical statement that is not true. □

Student Practice 7 If possible, solve the system.

$$4x - 2y = 6$$
$$-6x + 3y = 9$$

Example 8 If possible, solve the system.

$$\begin{aligned} 0.5x - 0.2y &= 1.3 \quad \textbf{(1)} \\ -1.0x + 0.4y &= -2.6 \quad \textbf{(2)} \end{aligned}$$

Solution Although we could work directly with the decimals, it is easier to multiply each equation by the appropriate power of 10 (10, 100, and so on) so that the coefficients of the new system are integers. Therefore, we will multiply equations **(1)** and **(2)** by 10 to obtain the following equivalent system.

$$\begin{aligned} 5x - 2y &= 13 \quad \textbf{(3)} \\ -10x + 4y &= -26 \quad \textbf{(4)} \end{aligned}$$

We can eliminate the variable y by multiplying each term of equation **(3)** by 2.

$$\begin{aligned} 10x - 4y &= 26 \quad \textbf{(5)} \\ -10x + 4y &= -26 \quad \textbf{(4)} \\ \hline 0 &= 0 \end{aligned}$$

Add the equations.

This statement is always true; it is an **identity.** Hence, the two equations are dependent, and there is an infinite number of solutions. Any solution satisfying equation **(1)** will also satisfy equation **(2).** For example, (3, 1) is a solution to equation **(1).** (Prove this.) Hence, it must also be a solution to equation **(2).** (Prove it). Thus, the equations actually describe the same line, as you can see on the graph.

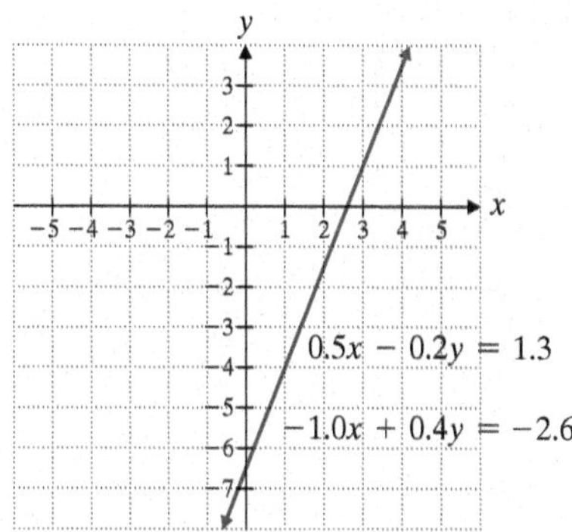

□

Student Practice 8 If possible, solve the system.

$$\begin{aligned} 0.3x - 0.9y &= 1.8 \\ -0.4x + 1.2y &= -2.4 \end{aligned}$$

6 Choosing an Appropriate Method to Solve a System of Linear Equations Algebraically

At this point we will review the algebraic methods for solving systems of linear equations and discuss the advantages and disadvantages of each method.

Method	Advantage	Disadvantage
Substitution	Works well if one or more variables has a coefficient of 1 or −1.	Often becomes difficult to use if no variable has a coefficient of 1 or −1.
Addition	Works well if equations have fractional or decimal coefficients. Works well if no variable has a coefficient of 1 or −1.	None

Example 9 Select a method and solve each system of equations.

(a) $x + y = 3080$
$2x + 3y = 8740$

(b) $5x - 2y = 19$
$-3x + 7y = 35$

Solution

(a) Since there are x- and y-values that have coefficients of 1, we will select the substitution method.

$x + y = 3080$
$y = 3080 - x$ Solve the first equation for y.
$2x + 3(3080 - x) = 8740$ Substitute the expression into the second equation.
$2x + 9240 - 3x = 8740$ Remove parentheses.
$-1x = -500$ Simplify.
$x = 500$ Divide each side by -1.

Substitute $x = 500$ into the first equation.

$x + y = 3080$
$500 + y = 3080$
$y = 3080 - 500$
$y = 2580$ Simplify.

The solution is (500, 2580).

Check. Verify that this solution is correct.

(b) Because none of the x- and y-variables has a coefficient of 1 or -1, we select the addition method. We choose to eliminate the y-variable. Thus, we would like the coefficients of y to be -14 and 14.

$7(5x) - 7(2y) = 7(19)$ Multiply each term of the first equation by 7.
$2(-3x) + 2(7y) = 2(35)$ Multiply each term of the second equation by 2.
$35x - 14y = 133$ We now have an equivalent system of equations.
$-6x + 14y = 70$
$29x = 203$ Add the two equations.
$x = 7$ Divide each side by 29.

Substitute $x = 7$ into one of the original equations. We will use the first equation.

$5(7) - 2y = 19$
$35 - 2y = 19$ Solve for y.
$-2y = -16$
$y = 8$

The solution is (7, 8).

Check. Verify that this solution is correct. □

Student Practice 9 Select a method and solve each system of equations.

(a) $3x + 5y = 1485$
$x + 2y = 564$

(b) $7x + 6y = 45$
$6x - 5y = -2$

To Think About: *Two Linear Equations in Two Variables* Now is a good time to look back over what we have learned. When you graph a system of two linear equations, what possible kinds of graphs will you obtain?

What will happen when you try to solve a system of two linear equations using algebraic methods? How many solutions are possible in each case? The following chart may help you to organize your answers to these questions.

Graph	Number of Solutions	Algebraic Interpretation
(6, −3) Two lines intersect at one point	**One unique solution**	You obtain one value for x and one value for y. For example, $x = 6, \quad y = -3.$
Parallel lines	**No solution**	You obtain an equation that is inconsistent with known facts. For example, $0 = 6.$ The system of equations is inconsistent.
Lines coincide	**Infinite number of solutions**	You obtain an equation that is always true. For example, $8 = 8.$ The equations are dependent.

4.1 Exercises MyMathLab®

Verbal and Writing Skills, Exercises 1–4

1. Explain what happens when a system of two linear equations is inconsistent. What effect does it have in obtaining a solution? What would the graph of such a system look like?

2. Explain what happens when a system of two linear equations has dependent equations. What effect does it have in obtaining a solution? What would the graph of such a system look like?

3. How many possible solutions can a system of two linear equations in two unknowns have?

4. When you have graphed a system of two linear equations in two unknowns, how do you determine the solution of the system?

Determine whether the given ordered pair is a solution to the system of equations.

5. $\left(\frac{5}{2}, 2\right)$ $\quad 2x - 8 = y - 5$
$\quad 4x - 3y = 4$

6. $\left(3, -\frac{1}{3}\right)$ $\quad 3(y + 4) - x = 8$
$\quad 9y = x - 6$

Solve the system of equations by graphing. Check your solution.

7. $3x + y = 2$
$2x - y = 3$

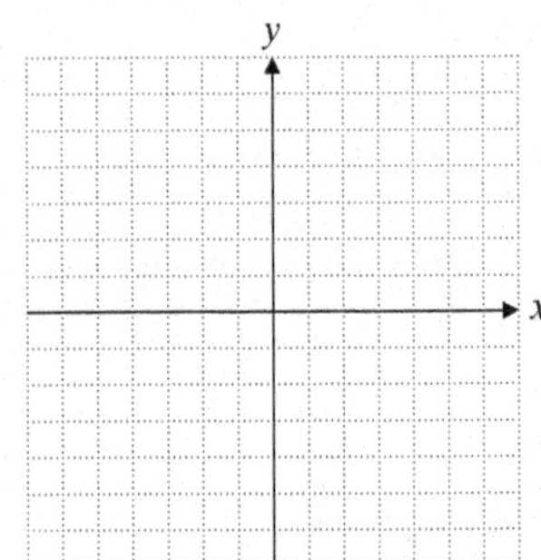

8. $-x + y = 5$
$x + 2y = 4$

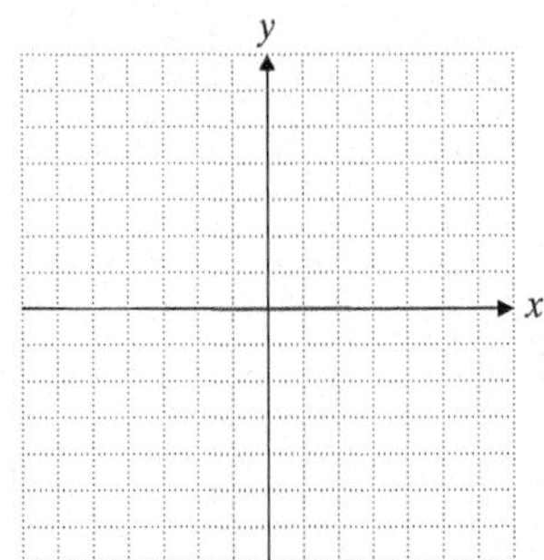

9. $3x - 2y = 6$
$4x + y = -3$

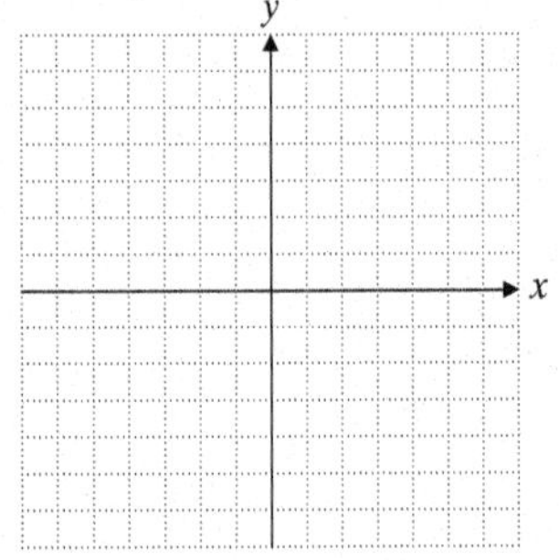

10. $3x - y = 5$
$2x - 3y = -6$

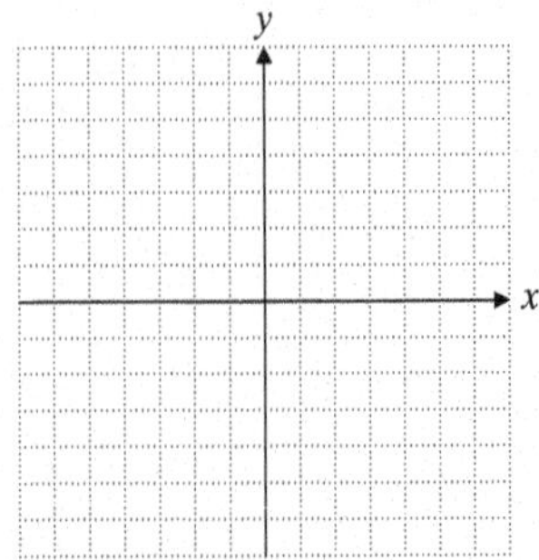

11. $y = -x + 3$
$x + y = -\frac{2}{3}$

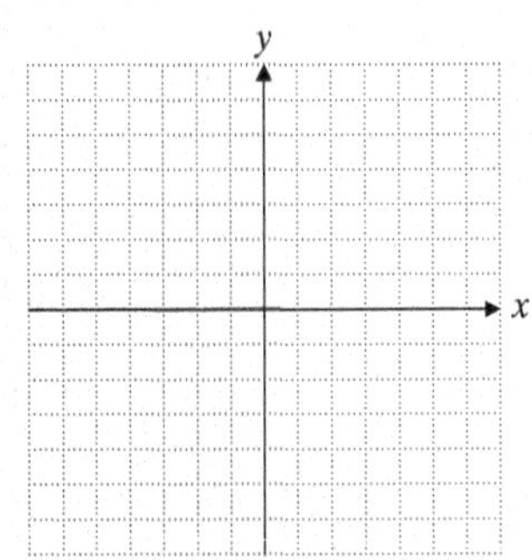

12. $y = \frac{1}{2}x + 2$
$-x + 2y = -8$

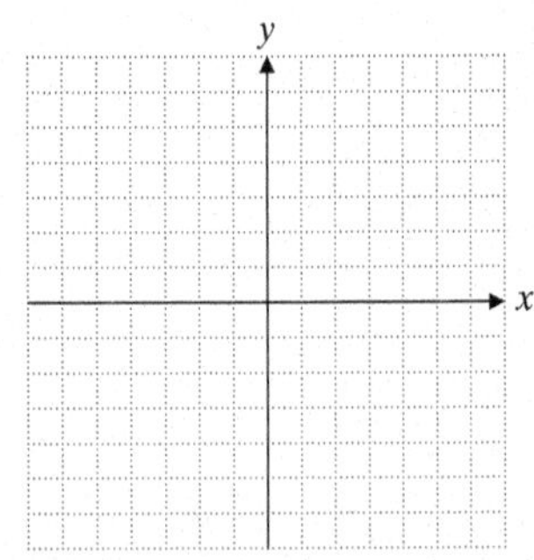

13. $y = -2x + 5$
$3y + 6x = 15$

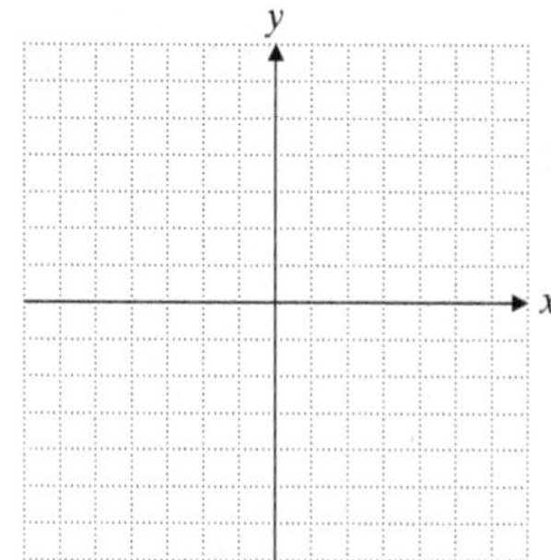

14. $x - 3 = 2y + 1$
$y - \frac{x}{2} = -2$

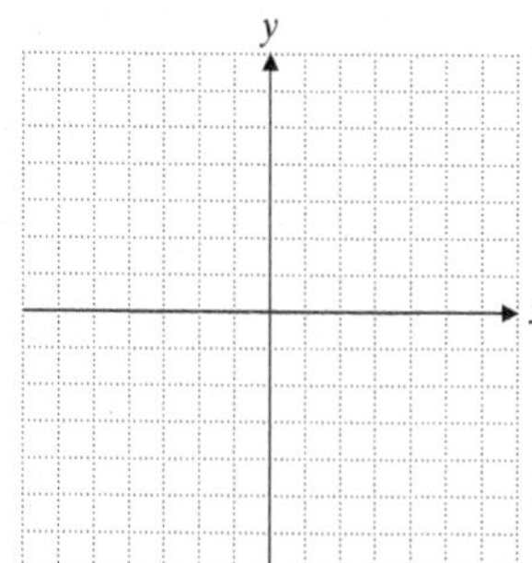

Find the solution to each system by the substitution method. Check your answers for exercises 15–18.

15. $3x + 4y = 14$
$x + 2y = -2$

16. $3x - 5y = 7$
$x - 4y = -7$

17. $-x + 3y = -8$
$2x - y = 6$

18. $10x + 3y = 8$
$2x + y = 2$

19. $2x - \frac{1}{2}y = -3$
$\frac{x}{5} + 2y = \frac{19}{5}$

20. $x - \frac{3}{2}y = 1$
$2x - 7y = 10$

21. $\frac{1}{2}x - \frac{1}{8}y = 3$
$\frac{2}{3}x + \frac{3}{4}y = 4$

22. $x + \frac{1}{4}y = 1$
$\frac{3}{8}x - y = -4$

Find the solution to each system by the addition (elimination) method. Check your answers for exercises 23–26.

23. $9x + 2y = 2$
$3x + 5y = 5$

24. $12x - 5y = -7$
$4x + 2y = 5$

25. $3s + 3t = 10$
$4s - 9t = -4$

26. $5s + 9t = 6$
$4s - 3t = 15$

27. $\frac{7}{2}x + \frac{5}{2}y = -4$
$3x + \frac{2}{3}y = 1$

28. $\frac{4}{3}x - y = 4$
$\frac{3}{4}x - y = \frac{1}{2}$

29. $1.6x + 1.5y = 1.8$
$0.4x + 0.3y = 0.6$

30. $2.5x + 0.6y = 0.2$
$0.5x - 1.2y = 0.7$

Mixed Practice

If possible, solve each system of equations. Use any method. If there is not a unique solution to a system, state a reason.

31. $7x - y = 6$
$3x + 2y = 22$

32. $8x - y = 17$
$4x + 3y = 33$

33. $3x + 4y = 8$
$5x + 6y = 10$

34. $4x + 5y = 22$
$9x + 2y = -6$

35. $2x + y = 4$
$\frac{2}{3}x + \frac{1}{4}y = 2$

36. $2x + 3y = 16$
$5x - \frac{3}{4}y = 7$

37. $0.2x = 0.1y - 1.2$
$2x - y = 6$

38. $0.1x - 0.6 = 0.3y$
$0.3x + 0.1y + 2.2 = 0$

39. $5x - 7y = 12$
$-10x + 14y = -24$

40. $-3x + 5y = 1$
$6x - 10y = 3$

41. $0.4x + 0.1y = -0.9$
$-0.1x + 0.2y = 1.8$

42. $5x = -y + 4$
$0.5y = -2.5x - 3$

43. $\frac{5}{3}b = \frac{1}{3} + a$
$9a - 15b = 2$

44. $a - 3b = \frac{3}{4}$
$\frac{a}{3} = \frac{1}{4} + b$

45. $\frac{2}{3}x - y = 4$
$2x - \frac{3}{4}y = 21$

46. $\frac{4}{3}x + y = 8$
$2x - \frac{1}{8}y = 25$

47. $3.2x - 1.5y = -3$
$0.7x + y = 2$

48. $3x - 0.2y = 1$
$1.1x + 0.4y = -2$

49. $3 - (2x + 1) = y + 6$
$x + y + 5 = 1 - x$

50. $2(y - 3) = x + 3y$
$x + 2 = 3 - y$

To Think About

51. ***Bathroom Tile*** Wayne Burton is having some tile replaced in his bathroom. He has obtained an estimate from two tile companies. Old World Tile gave an estimate of \$200 to remove the old tile and \$50 per hour to place new tile on the wall. Modern Bathroom Headquarters gave an estimate of \$300 to remove the old tile and \$30 per hour to place new tile on the wall.

(a) Create a cost equation for each company where y is the total cost of the tile work and x is the number of hours of labor used to install new tile. Write a system of equations.

(b) Graph the two equations using the values $x = 0, 4$, and 8.

(c) Determine from your graph how many hours of installing new tile will be required for the two companies to cost the same.

(d) Determine from your graph which company costs less to remove old tile and to install new tile if the time needed to install new tile is 6 hours.

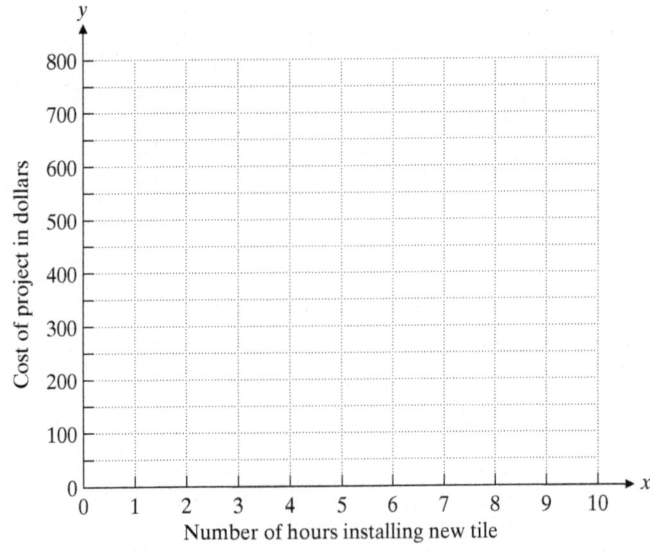

Optional Graphing Calculator Problems

On a graphing calculator, graph each system of equations on the same set of axes. Find the point of intersection to the nearest hundredth.

52. $y_1 = -0.81x + 2.3$
$y_2 = 1.6x + 0.8$

53. $y_1 = -1.7x + 3.8$
$y_2 = 0.7x - 2.1$

54. $5.86x + 6.22y = -8.89$
$-2.33x + 4.72y = -10.61$

55. $0.5x + 1.1y = 5.5$
$-3.1x + 0.9y = 13.1$

Cumulative Review

56. **[1.3.3]** ***Winter Road Salt*** Nine million tons of salt are applied to American highways for road deicing each year. The cost of buying and applying the salt totals \$200 million. How much is this per pound? (Recall that 1 ton = 2000 pounds.) Round your answer to the nearest cent.

57. **[2.6.2]** ***City Parking Space*** Four-fifths of the automobiles that enter the city of Boston during rush hour will have to park in private or municipal parking lots. If there are 273,511 private or municipal lot spaces filled by cars entering the city during rush hour every morning, how many cars enter the city during rush hour? Round your answer to the nearest car.

Quick Quiz 4.1

1. Solve using the substitution method.
$5x - 3y = 14$
$2x - y = 6$

2. Solve using the addition method.
$6x + 7y = 26$
$5x - 2y = 6$

3. Solve by any method.
$\frac{2}{3}x + \frac{3}{5}y = -17$
$\frac{1}{2}x - \frac{1}{3}y = -1$

4. **Concept Check** Explain what happens when you go through the steps to solve the following system of equations. Why does this happen?
$6x - 4y = 8$
$-9x + 6y = -12$

4.2 Systems of Linear Equations in Three Variables

Student Learning Objectives

After studying this section, you will be able to:

1. Determine whether an ordered triple is the solution to a system of three linear equations in three variables.
2. Find the solution to a system of three linear equations in three variables if none of the coefficients is zero.
3. Find the solution to a system of three linear equations in three variables if some of the coefficients are zero.

1 Determining Whether an Ordered Triple Is the Solution to a System of Three Linear Equations in Three Variables

We are now going to study **systems of three linear equations in three variables** (unknowns). A **solution** to a system of three linear equations in three unknowns is an **ordered triple** of real numbers (x, y, z) that satisfies each equation in the system.

Example 1 Determine whether $(2, -5, 1)$ is a solution to the following system.

$$\begin{aligned} 3x + y + 2z &= 3 \\ 4x + 2y - z &= -3 \\ x + y + 5z &= 2 \end{aligned}$$

Solution How can we determine whether $(2, -5, 1)$ is a solution to this system? We will substitute $x = 2$, $y = -5$, and $z = 1$ into each equation. If a true statement occurs each time, $(2, -5, 1)$ is a solution to each equation and hence, a solution to the system. For the first equation:

$$\begin{aligned} 3(2) + (-5) + 2(1) &\stackrel{?}{=} 3 \\ 6 - 5 + 2 &\stackrel{?}{=} 3 \\ 3 &= 3 \quad \checkmark \end{aligned}$$

For the second equation:

$$\begin{aligned} 4(2) + 2(-5) - 1 &\stackrel{?}{=} -3 \\ 8 - 10 - 1 &\stackrel{?}{=} -3 \\ -3 &= -3 \quad \checkmark \end{aligned}$$

For the third equation:

$$\begin{aligned} 2 + (-5) + 5(1) &\stackrel{?}{=} 2 \\ 2 - 5 + 5 &\stackrel{?}{=} 2 \\ 2 &= 2 \quad \checkmark \end{aligned}$$

Since we obtained three true statements, the ordered triple $(2, -5, 1)$ is a solution to the system. □

Student Practice 1 Determine whether $(3, -2, 2)$ is a solution to this system.

$$\begin{aligned} 2x + 4y + z &= 0 \\ x - 2y + 5z &= 17 \\ 3x - 4y + z &= 19 \end{aligned}$$

To Think About: *Graphs in Three Variables* Can we graph an equation in three variables? How? What would the graph look like? What would the graph of the system in Example 1 look like? Describe the graph of the solution. At the end of this section we will show you how the graphs might look.

2 Finding the Solution to a System of Three Linear Equations in Three Variables If None of the Coefficients Is Zero

One way to solve a system of three equations with three variables is to obtain from it a system of two equations in two variables; in other words, we eliminate one variable from both equations. We can then use the methods of Section 4.1 to solve the resulting system. You can find the third variable (the one that was eliminated) by substituting the two variables that you have found into one of the original equations.

Example 2 Find the solution to (that is, solve) the following system of equations.

$$\begin{aligned} -2x + 5y + z &= 8 \quad \textbf{(1)} \\ -x + 2y + 3z &= 13 \quad \textbf{(2)} \\ x + 3y - z &= 5 \quad \textbf{(3)} \end{aligned}$$

Solution Let's eliminate z because it can be done easily by adding equations **(1)** and **(3)**.

$$\begin{aligned} -2x + 5y + z &= 8 \quad \textbf{(1)} \\ x + 3y - z &= 5 \quad \textbf{(3)} \\ \hline -x + 8y \quad &= 13 \quad \textbf{(4)} \end{aligned}$$

Now we need to choose a *different pair* from the original system of equations and once again eliminate the same variable. In other words, we have to use equations **(1)** and **(2)** or equations **(2)** and **(3)** and eliminate z. Let's multiply each term of equation **(3)** by 3 (and call it equation **(6)**) and add the result to equation **(2)**.

$$\begin{aligned} -x + 2y + 3z &= 13 \quad \textbf{(2)} \\ 3x + 9y - 3z &= 15 \quad \textbf{(6)} \\ \hline 2x + 11y \quad &= 28 \quad \textbf{(5)} \end{aligned}$$

We now can solve the resulting system of two linear equations.

$$\begin{aligned} -x + 8y &= 13 \quad \textbf{(4)} \\ 2x + 11y &= 28 \quad \textbf{(5)} \end{aligned}$$

Multiply each term of equation **(4)** by 2.

$$\begin{aligned} -2x + 16y &= 26 \\ 2x + 11y &= 28 \\ \hline 27y &= 54 \quad \text{Add the equations.} \\ y &= 2 \quad \text{Solve for } y. \end{aligned}$$

Substituting $y = 2$ into equation **(4)**, we have the following:

$$\begin{aligned} -x + 8(2) &= 13 \\ -x &= -3 \\ x &= 3. \end{aligned}$$

Now substitute $x = 3$ and $y = 2$ into one of the original equations (any one will do) to solve for z. Let's use equation **(1)**.

$$\begin{aligned} -2x + 5y + z &= 8 \\ -2(3) + 5(2) + z &= 8 \\ -6 + 10 + z &= 8 \\ z &= 4 \end{aligned}$$

The solution to the system is (3, 2, 4).

Check. Verify that (3, 2, 4) satisfies *each* of the three *original* equations. □

Student Practice 2 Solve this system.

$$\begin{aligned} x + 2y + 3z &= 4 \\ 2x + y - 2z &= 3 \\ 3x + 3y + 4z &= 10 \end{aligned}$$

Here's a summary of the procedure that we just used.

HOW TO SOLVE A SYSTEM OF THREE LINEAR EQUATIONS IN THREE VARIABLES

1. Use the addition method to eliminate any variable from any pair of equations. (The choice of variable is arbitrary.)
2. Use appropriate steps to eliminate the *same variable* from a *different pair* of equations. (If you don't eliminate the same variable, you will still have three unknowns.)
3. Solve the resulting system of two equations in two variables.
4. Substitute the values obtained in step 3 into one of the three original equations. Solve for the remaining variable.
5. Check the solution in all of the original equations.

It is helpful to write all equations in the form $Ax + By + Cz = D$ before using this five-step method.

3 Finding the Solution to a System of Three Linear Equations in Three Variables If Some of the Coefficients Are Zero

If a system of three linear equations in three variables contains one or more equations of the form $Ax + By + Cz = D$, where one of the values of A, B, or C is zero, then we will slightly modify our approach to solving the system. We will select one equation that contains only two variables. Then we will take the remaining system of two equations and eliminate the variable that is missing in the equation that we selected.

Example 3 Solve the system.

$$
\begin{aligned}
4x + 3y + 3z &= 4 \quad \textbf{(1)}\\
3x + 2z &= 2 \quad \textbf{(2)}\\
2x - 5y &= -4 \quad \textbf{(3)}
\end{aligned}
$$

Solution Note that equation **(2)** has no y-term and equation **(3)** has no z-term. Obviously, that makes our work easier. Let's work with equations **(2)** and **(1)** to obtain an equation that contains only x and y.

Step 1 Multiply equation **(1)** by 2 and equation **(2)** by -3 to obtain the following system.

$$
\begin{aligned}
8x + 6y + 6z &= 8 \quad \textbf{(4)}\\
-9x - 6z &= -6 \quad \textbf{(5)}\\
\hline
-x + 6y &= 2 \quad \textbf{(6)}
\end{aligned}
$$

Step 2 This step is already done, since equation **(3)** has no z-term.

Step 3 Now we can solve the system formed by equations **(3)** and **(6)**.

$$
\begin{aligned}
2x - 5y &= -4 \quad \textbf{(3)}\\
-x + 6y &= 2 \quad \textbf{(6)}
\end{aligned}
$$

If we multiply each term of equation **(6)** by 2, we obtain the system

$$\begin{aligned} 2x - 5y &= -4 \\ -2x + 12y &= 4 \\ \hline 7y &= 0 \quad \text{Add.} \\ y &= 0. \quad \text{Solve for } y. \end{aligned}$$

Substituting $y = 0$ in equation **(6),** we find the following:

$$\begin{aligned} -x + 6(0) &= 2 \\ -x &= 2 \\ x &= -2. \end{aligned}$$

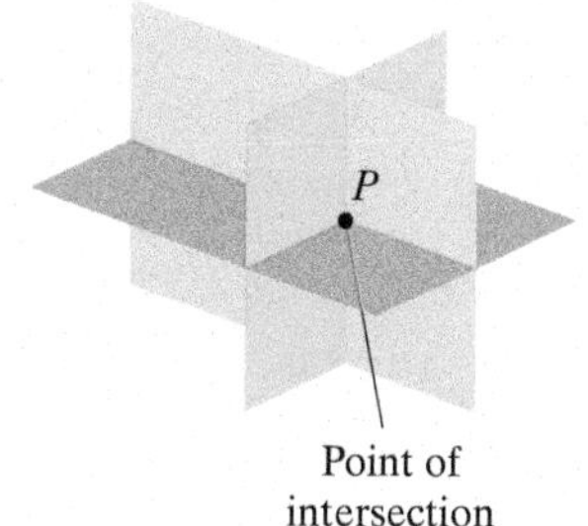

(a)

Step 4 To find z, we substitute $x = -2$ and $y = 0$ into one of the original equations containing z. Since equation **(2)** has only two variables, let's use it.

$$\begin{aligned} 3x + 2z &= 2 \\ 3(-2) + 2z &= 2 \\ 2z &= 8 \\ z &= 4 \end{aligned}$$

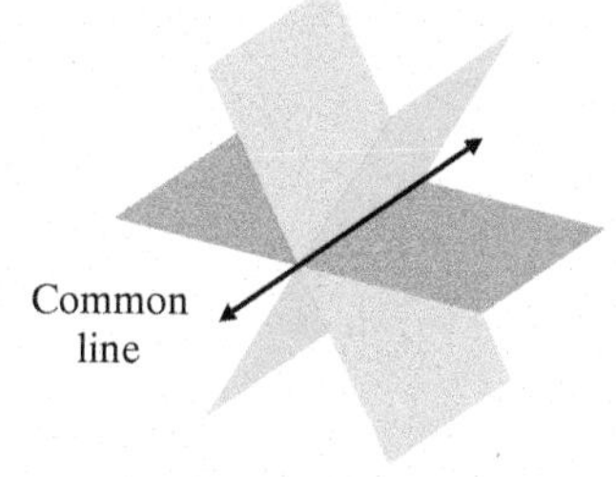

(b)

The solution to the system is $(-2, 0, 4)$.

Check. Verify this solution by substituting these values into equations **(1), (2),** and **(3).** □

 **Student Practice 3** Solve the system.

$$\begin{aligned} 2x + y + z &= 11 \\ 4y + 3z &= -8 \\ x - 5y &= 2 \end{aligned}$$

(c)

A linear equation in three variables is a plane in three-dimensional space. A system of three linear equations in three variables is three planes. The solution to the system is the set of points at which all three planes intersect. There are three possible results. The three planes may intersect at one point. (See figure **(a)** in the margin.) This point is described by an ordered triple of the form (x, y, z) and lies in each plane.

(d)

The three planes may intersect at a line. (See figure **(b)** in the margin.) In this case the system has an infinite number of solutions; that is, all the points on the line are solutions to the system.

Finally, all three planes may not intersect at any points. It may mean that all three planes never share any point of intersection. (See figure **(c)** in the margin.) It may mean that two planes intersect. (See figures **(d)** and **(e)** in the margin.) In all such cases there is no solution to the system of equations.

(e)

4.2 Exercises MyMathLab®

1. Determine whether $(2, 1, -4)$ is a solution to the system.

$$\begin{aligned} 2x - 3y + 2z &= -7 \\ x + 4y - z &= 10 \\ 3x + 2y + z &= 4 \end{aligned}$$

2. Determine whether $(-2, 3, 0)$ is a solution to the system.

$$\begin{aligned} x + 4y - 9z &= 1 \\ -3x - y + 5z &= -1 \\ 4x - 2y + 11z &= 3 \end{aligned}$$

3. Determine whether $(-1, 5, 1)$ is a solution to the system.

$$\begin{aligned} 3x + 2y - z &= 6 \\ x - y - 2z &= -8 \\ 4x + y + 2z &= 5 \end{aligned}$$

4. Determine whether $(1, 2, 4)$ is a solution to the system.

$$\begin{aligned} 3x + y + z &= 9 \\ x + 2y + 2z &= 13 \\ -4x + 3y + z &= 6 \end{aligned}$$

Solve each system.

5. $$\begin{aligned} x + y + 2z &= 0 \\ 2x - y - z &= 1 \\ x + 2y + 3z &= 1 \end{aligned}$$

6. $$\begin{aligned} 2x + 2y + z &= -6 \\ x - y + 3z &= 1 \\ x + 4y + z &= 8 \end{aligned}$$

7. $$\begin{aligned} x + 3y - z &= -5 \\ -2x - y + z &= 8 \\ x - y + 3z &= 3 \end{aligned}$$

8. $$\begin{aligned} -4x + 2y + z &= 1 \\ x - y + 3z &= -5 \\ 3x + y - 4z &= 10 \end{aligned}$$

9. $$\begin{aligned} 8x - 5y + z &= 15 \\ 3x + y - z &= -7 \\ x + 4y + z &= -3 \end{aligned}$$

10. $$\begin{aligned} -x + 2y + 3z &= 0 \\ 3x - 2y - 2z &= 5 \\ 4x + y - 3z &= -9 \end{aligned}$$

11. $$\begin{aligned} x + 4y - z &= -5 \\ -2x - 3y + 2z &= 5 \\ x - \frac{2}{3}y + z &= \frac{11}{3} \end{aligned}$$

12. $$\begin{aligned} x - 4y + 4z &= -1 \\ -x + \frac{y}{2} - \frac{5}{2}z &= -3 \\ -x + 3y - z &= 5 \end{aligned}$$

13. $$\begin{aligned} 2x \qquad + 2z &= -7 + 3y \\ \frac{3}{2}x + y + \frac{1}{2}z &= 2 \\ x + 4y \qquad &= 10 + z \end{aligned}$$

14. $$\begin{aligned} 3y - z &= 9 - x \\ -x + 2y &= 15 + 3z \\ \frac{1}{3}x + y - \frac{4}{3}z &= 9 \end{aligned}$$

15. $$\begin{aligned} 0.2a + 0.1b + 0.2c &= 0.1 \\ 0.3a + 0.2b + 0.4c &= -0.1 \\ 0.6a + 1.1b + 0.2c &= 0.3 \end{aligned}$$

16. $$\begin{aligned} -0.1a + 0.2b + 0.3c &= 0.1 \\ 0.2a - 0.6b + 0.3c &= 0.5 \\ 0.3a - 1.2b - 0.4c &= -0.4 \end{aligned}$$

Find the solution for each system of equations. Round your answers to five decimal places.

17.
$$\begin{aligned} x - 4y + 4z &= -3.72186 \\ -x + 3y - z &= 5.98115 \\ 2x - y + 5z &= 7.93645 \end{aligned}$$

18.
$$\begin{aligned} 4x + 2y + 3z &= 9 \\ 9x + 3y + 2z &= 3 \\ 2.987x + 5.027y + 3.867z &= 18.642 \end{aligned}$$

Solve each system.

19.
$$\begin{aligned} x - y &= 5 \\ 2y - z &= 1 \\ 3x + 3y + z &= 6 \end{aligned}$$

20.
$$\begin{aligned} -x + 4z &= -3 \\ -y + 2z &= 5 \\ 3x + y + 2z &= 4 \end{aligned}$$

21.
$$\begin{aligned} -y + 2z &= 1 \\ x + y + z &= 2 \\ -x + 3z &= 2 \end{aligned}$$

22.
$$\begin{aligned} -2x + y - 3z &= 0 \\ -2y - z &= -1 \\ x + 2y - z &= 5 \end{aligned}$$

23.
$$\begin{aligned} x - 2y + z &= 0 \\ -3x - y &= -6 \\ y - 2z &= -7 \end{aligned}$$

24.
$$\begin{aligned} 2x - 4z &= -1 \\ 5x + y - 2z &= 1 \\ 2y + 8z &= 4 \end{aligned}$$

25.
$$\begin{aligned} a - \frac{b}{2} - 2c &= -3 \\ 3a - b &= -12 \\ 2a + \frac{3b}{2} + 2c &= 3 \end{aligned}$$

26.
$$\begin{aligned} \frac{-a}{3} - \frac{2b}{3} + c &= -10 \\ a + b - \frac{c}{3} &= 7 \\ b - c &= 13 \end{aligned}$$

Try to solve the system of equations. Explain your result in each case.

27. $\begin{aligned} 2x + y &= -3 \\ 2y + 16z &= -18 \\ -7x - 3y + 4z &= 6 \end{aligned}$

28. $\begin{aligned} x - 2y + 5z &= 4 \\ -9x + 3y + 12z &= 9 \\ 3x - y - 4z &= 5 \end{aligned}$

29. $\begin{aligned} 3x + 3y - 3z &= -1 \\ 4x + y - 2z &= 1 \\ -2x + 4y - 2z &= -8 \end{aligned}$

30. $\begin{aligned} 2x + y - 4z &= -1 \\ 8x - 2y + 3z &= 2 \\ -10x - 5y + 20z &= 5 \end{aligned}$

To Think About

Solve each system.

31. $\begin{aligned} a &= 8 + 3b - 2c \\ 4a + 2b - 3c &= 10 \\ c &= 10 + b - 2a \end{aligned}$

32. $\begin{aligned} a &= c - b \\ 3a - 2b + 6c &= 1 \\ c &= 4 - 3b - 7a \end{aligned}$

Cumulative Review

33. [2.3.3] Solve for x. $-2(3x - 4) + 12 = 6x - 4$

34. [3.3.1] Find the slope of the line that passes through $(-1, 6)$ and $(4, -3)$.

35. [3.4.2] Find the equation in slope-intercept form of the line that passes through $(1, 4)$ and $(-2, 3)$.

36. [3.4.4] Find the slope of a line that is perpendicular to $y = -\frac{2}{3}x + 4$.

Quick Quiz 4.2 *Solve each system.*

1. $\begin{aligned} 4x - y + 2z &= 0 \\ 2x + y + z &= 3 \\ 3x - y + z &= -2 \end{aligned}$

2. $\begin{aligned} x + 2y + 2z &= -1 \\ 2x - y + z &= 1 \\ x + 3y - 6z &= 7 \end{aligned}$

3. $\begin{aligned} 4x - 2y + 6z &= 0 \\ 6y + 3z &= 3 \\ x + 2y - z &= 5 \end{aligned}$

4. Concept Check Explain how you would eliminate the variable z and obtain two equations with only the variables x and y in the following system.

$$\begin{aligned} 2x + 4y - 2z &= -22 \\ 4x + 3y + 5z &= -10 \\ 5x - 2y + 3z &= 13 \end{aligned}$$

Use Math to Save Money

Did You Know? How much you get paid at your job is not how much you get to spend because of expenses.

Determining Your Earnings

Understanding the Problem:

Your real hourly wage is the amount you make per hour after taxes and other work-related expenses. When calculating real hourly wage, we cannot just look at how much a job pays an hour or how much the paycheck will be. We need to consider work-related expenses as well as extra time spent for the job.

Sharon is not currently employed. She is considering two different positions and needs to determine which job she should take. If she takes a job, she estimates she'll have the following weekly expenses: daycare, $175; after-school program, $75; clothing, $25; other work-related expenses, $50, as well as $0.35 per mile for transportation.

Making a Plan:

Sharon needs to determine what the income and expenses are for each position.

Step 1: First she calculates what her income would be for each position.

Job 1: Donivan Tech is offering her a full-time management position at a yearly salary of $50,310. Sharon estimates that she will work an average of 45 hours per week when she considers the extra work commitment required for managers. The commute of 45 miles a day will be about five hours a week in driving. She estimates that payroll deductions will be 15% of her gross salary.

Job 2: J & R Financial Group is offering her a part-time position as an assistant manager. The work schedule would be 6 hours per day, 5 days a week at an hourly rate of $20.50. The commute of 5 miles a day will be about one hour a week in driving. Due to the reduced work schedule, she calculates that payroll deductions will only be 12%, and daycare, after-school program, and other expenses will all be reduced by $25 per week.

Task 1: What is the weekly pay, before deductions or expenses, at Job 1? At Job 2?

Step 2: She needs to calculate what her expenses would be for each position.

Task 2: How much are weekly payroll deductions at Job 1? At Job 2?

Task 3: What are the weekly job-related expenses at Job 1? At Job 2?

Finding a Solution

Step 3: Using weekly pay, payroll deductions and job-related expenses, Sharon can determine what her real hourly wage will be. Then she'll be able to compare the two jobs while taking into consideration all expenses.

Task 4: How much will Sharon have left to spend every week after subtracting deductions and expenses from the weekly pay at Job 1? At Job 2?

Task 5: Considering the total time spent working and commuting, what is the real hourly wage at Job 1? At Job 2?

Step 4: Sharon can now decide which job to take.

Task 6: If Sharon is deciding on a job based only on real hourly wage, which job should she take?

Task 7: If Sharon is deciding on a job based on how much she will have left to spend each week after deductions and expenses, which job should she take?

Applying the Situation to Your Life:

When comparing jobs, one needs to determine the real hourly wage because what you earn is not as important as what you actually get to spend. Work-related expenses and time spent commuting have an impact on the real hourly rate. Taking these into consideration will help you make an informed decision.

4.3 Applications of Systems of Linear Equations

Student Learning Objectives

After studying this section, you will be able to:

1 Solve applications requiring the use of a system of two linear equations in two unknowns.

2 Solve applications requiring the use of a system of three linear equations in three unknowns.

1 Solving Applications Requiring the Use of a System of Two Linear Equations in Two Unknowns

We will now examine how a system of linear equations can assist us in solving applied problems.

Example 1 For the paleontology lecture on campus, advance tickets cost \$5 and tickets at the door cost \$6. The ticket sales this year came to \$4540. The department chairman wants to raise prices next year to \$7 for advance tickets and \$9 for tickets at the door. He said that if exactly the same number of people attend next year, the ticket sales at these new prices will total \$6560. If he is correct, how many tickets were sold in advance this year? How many tickets were sold at the door?

Solution

1. ***Understand the problem.*** Since we are looking for the number of tickets sold, we let

$$x = \text{number of tickets bought in advance and}$$
$$y = \text{number of tickets bought at the door.}$$

2. ***Write a system of two equations in two unknowns.*** If advance tickets cost \$5, then the total sales will be $5x$; similarly, total sales of door tickets will be $6y$. Since the total sales of both types of tickets was \$4540, we have

$$5x + 6y = 4540.$$

By the same reasoning, we have

$$7x + 9y = 6560.$$

Thus, our system is as follows:

$$5x + 6y = 4540 \quad \textbf{(1)}$$
$$7x + 9y = 6560 \quad \textbf{(2)}$$

3. ***Solve the system of equations and state the answer.*** We multiply each term of equation **(1)** by -3 and each term of equation **(2)** by 2 to obtain the following equivalent system.

$$\begin{aligned} -15x - 18y &= -13{,}620 \\ \underline{14x + 18y} &= \underline{13{,}120} \\ -x &= -500 \end{aligned}$$

Therefore, $x = 500$. Substituting $x = 500$ into equation **(1),** we have the following:

$$\begin{aligned} 5(500) + 6y &= 4540 \\ 6y &= 2040 \\ y &= 340 \end{aligned}$$

Thus, 500 advance tickets and 340 door tickets were sold.

4. ***Check.*** We need to check our answers. Do they seem reasonable?

Would 500 advance tickets at \$5 and 340 door tickets at \$6 yield \$4540?

$$\begin{aligned} 5(500) + 6(340) &\stackrel{?}{=} 4540 \\ 2500 + 2040 &\stackrel{?}{=} 4540 \\ 4540 &= 4540 \checkmark \end{aligned}$$

Would 500 advance tickets at \$7 and 340 door tickets at \$9 yield \$6560?

$$\begin{aligned} 7(500) + 9(340) &\stackrel{?}{=} 6560 \\ 3500 + 3060 &\stackrel{?}{=} 6560 \\ 6560 &= 6560 \checkmark \end{aligned}$$

Student Practice 1 Coach Perez purchased baseballs at \$6 each and bats at \$21 each last week for the college baseball team. The total cost of the purchase was \$318. This week he noticed that the same items are on sale. Baseballs are now \$5 each and bats are \$17. He found that if he made the same purchase this week, it would cost only \$259. How many baseballs and how many bats did he buy last week?

Example 2 An electronics firm makes two types of switching devices. Type A takes 4 minutes to make and requires \$3 worth of materials. Type B takes 5 minutes to make and requires \$5 worth of materials. When the production manager reviewed the latest batch, he found that it took 35 hours to make these switches with a materials cost of \$1900. How many switches of each type were produced for this latest batch?

Solution

1. ***Understand the problem.*** We are given a lot of information, but the major concern is to find out how many of the type A devices and the type B devices were produced. This becomes our starting point to define the variables we will use. Let

$$A = \text{the number of type } A \text{ devices produced and}$$
$$B = \text{the number of type } B \text{ devices produced.}$$

2. ***Write a system of two equations.*** How should we construct the equations? What relationships exist between our variables (or unknowns)? According to the problem, the devices are related by time and by cost. So we set up one equation in terms of time (minutes in this case) and one in terms of cost (dollars). Each type A took 4 minutes to make, each type B took 5 minutes to make, and the total time was $60(35) = 2100$ minutes. Each type A used \$3 worth of materials, each type B used \$5 worth of materials, and the total materials cost was \$1900. We can gather this information in a table. Making a table will help us form the equations.

	Type A Devices	Type B Devices	Total
Number of Minutes	$4A$	$5B$	2100
Cost of Materials	$3A$	$5B$	1900

$$4A + 5B = 2100$$
$$3A + 5B = 1900$$

Therefore, we have the following system.

$$4A + 5B = 2100 \quad \textbf{(1)}$$
$$3A + 5B = 1900 \quad \textbf{(2)}$$

3. ***Solve the system of equations and state the answer.*** Multiplying equation **(2)** by -1 and adding the equations, we find the following:

$$\begin{aligned} 4A + 5B &= 2100 \\ -3A - 5B &= -1900 \\ \hline A \quad &= 200 \end{aligned}$$

Substituting $A = 200$ into equation **(1),** we have the following:

$$800 + 5B = 2100$$
$$5B = 1300$$
$$B = 260$$

Thus, 200 type A devices and 260 type B devices were produced.

Continued on next page

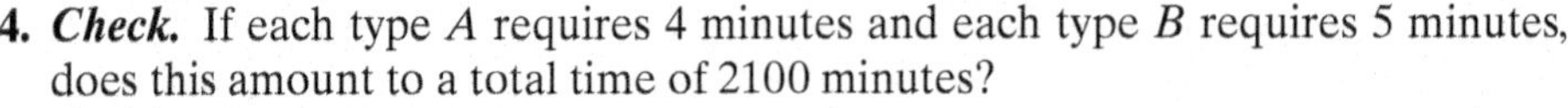

4. *Check.* If each type A requires 4 minutes and each type B requires 5 minutes, does this amount to a total time of 2100 minutes?

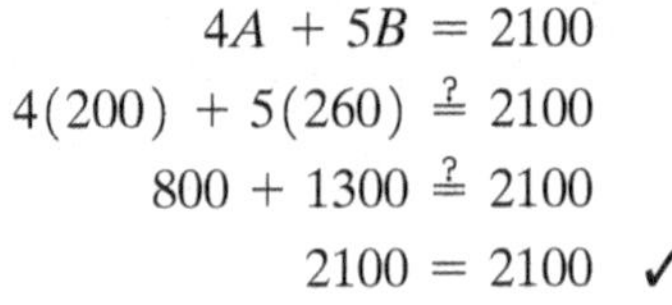

$$4A + 5B = 2100$$
$$4(200) + 5(260) \stackrel{?}{=} 2100$$
$$800 + 1300 \stackrel{?}{=} 2100$$
$$2100 = 2100 \checkmark$$

If each type A costs \$3 and each type B costs \$5, does this amount to a total cost of \$1900?

$$3A + 5B = 1900$$
$$3(200) + 5(260) \stackrel{?}{=} 1900$$
$$600 + 1300 \stackrel{?}{=} 1900$$
$$1900 = 1900 \checkmark$$

□

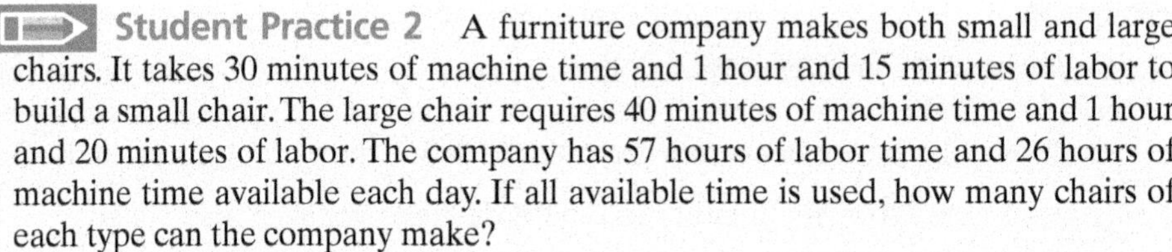

Student Practice 2 A furniture company makes both small and large chairs. It takes 30 minutes of machine time and 1 hour and 15 minutes of labor to build a small chair. The large chair requires 40 minutes of machine time and 1 hour and 20 minutes of labor. The company has 57 hours of labor time and 26 hours of machine time available each day. If all available time is used, how many chairs of each type can the company make?

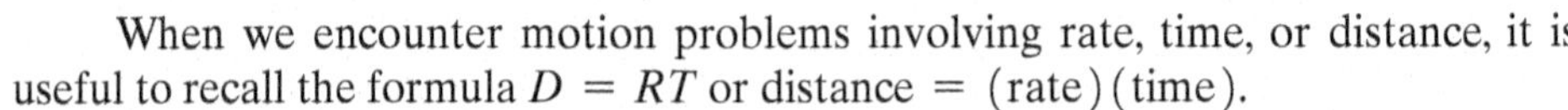

When we encounter motion problems involving rate, time, or distance, it is useful to recall the formula $D = RT$ or distance = (rate)(time).

Example 3 An airplane travels between two cities that are 1500 miles apart. The trip against the wind takes 3 hours. The return trip with the wind takes $2\frac{1}{2}$ hours. What is the speed of the plane in still air (in other words, how fast would the plane travel if there were no wind)? What is the speed of the wind?

Solution

1. ***Understand the problem.*** Our unknowns are the speed of the plane in still air and the speed of the wind.

Let

x = the speed of the plane in still air and

y = the speed of the wind.

Let's make a sketch to help us see how these speeds are related to one another. When we travel against the wind, the wind is slowing us down. Since the wind speed opposes the plane's speed in still air, we must subtract: $x - y$.

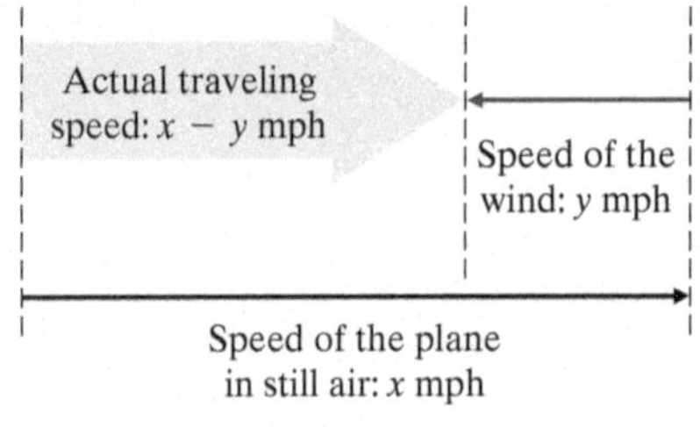

When we travel with the wind, the wind is helping us travel forward. Thus, the wind speed is added to the plane's speed in still air, and we add: $x + y$.

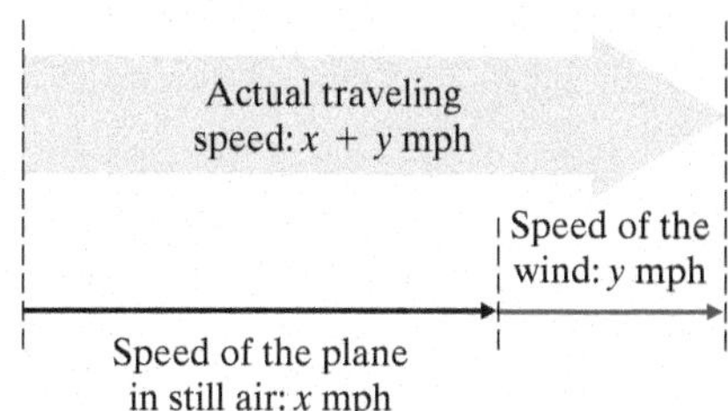

2. ***Write a system of two equations.*** To help us write our equations, we organize the information in a chart. The chart will be based on the formula $RT = D$, which is (rate)(time) = distance.

	R	$\cdot$ T	= D
Flying against the wind	$x - y$	3	1500
Flying with the wind	$x + y$	2.5	1500

Using the rows of the chart, we obtain a system of equations.

$$(x - y)(3) = 1500$$
$$(x + y)(2.5) = 1500$$

If we remove the parentheses, we will obtain the following system.

$$3x - 3y = 1500 \quad \textbf{(1)}$$
$$2.5x + 2.5y = 1500 \quad \textbf{(2)}$$

3. ***Solve the system of equations and state the answer.*** It will be helpful to clear equation **(2)** of decimal coefficients. Although we could multiply each term by 10, doing so will result in large coefficients on x and y. For this equation, multiplying by 2 is a better choice.

$$3x - 3y = 1500 \quad \textbf{(1)}$$
$$5x + 5y = 3000 \quad \textbf{(3)}$$

If we multiply equation **(1)** by 5 and equation **(3)** by 3, we will obtain the following system.

$$\begin{aligned} 15x - 15y &= 7500 \\ 15x + 15y &= 9000 \\ \hline 30x &= 16{,}500 \\ x &= 550 \end{aligned}$$

Substituting this result in equation **(1),** we obtain the following:

$$\begin{aligned} 3(550) - 3y &= 1500 \\ 1650 - 3y &= 1500 \\ -3y &= -150 \\ y &= 50 \end{aligned}$$

Thus, the speed of the plane in still air is 550 miles per hour, and the speed of the wind is 50 miles per hour. □

Student Practice 3 An airplane travels west from city A to city B against the wind. It takes 3 hours to travel 1950 kilometers. On the return trip the plane travels east from city B to city C, a distance of 1600 kilometers in a time of 2 hours. On the return trip the plane travels with the wind. What is the speed of the plane in still air? What is the speed of the wind?

Graphing Calculator

Exploration

A visual interpretation of two equations in two unknowns is sometimes helpful. Study Example 3. Graph the two equations.

$$3x - 3y = 1500$$
$$2.5x + 2.5y = 1500$$

What is the significance of the point of intersection? If you were an air traffic controller, how would you interpret the linear equation $3x - 3y = 1500$? Why would this be useful? How would you interpret $2.5x + 2.5y = 1500$? Why would this be useful?

2 Solving Applications Requiring the Use of a System of Three Linear Equations in Three Unknowns

Example 4 A trucking firm has three sizes of trucks. The biggest truck holds 10 tons of gravel, the next size holds 6 tons, and the smallest holds 4 tons. The firm has a contract to provide fifteen trucks to haul 104 tons of gravel. To reduce fuel costs, the firm's manager wants to use two more of the fuel-efficient 10-ton trucks than the 6-ton trucks. How many trucks of each type should she use?

Student Practice 4 A factory uses three machines to wrap boxes for shipment. Machines A, B, and C working together can wrap 260 boxes in 1 hour. If machine A runs 3 hours and machine B runs 2 hours, they can wrap 390 boxes. If machine B runs 3 hours and machine C runs 4 hours, 655 boxes can be wrapped. How many boxes per hour can each machine wrap?

Solution

1. ***Understand the problem.*** Since we need to find three things (the numbers of 10-ton trucks, 6-ton trucks, and 4-ton trucks), it would be helpful to have three variables. Let

x = the number of 10-ton trucks used,

y = the number of 6-ton trucks used, and

z = the number of 4-ton trucks used.

2. ***Write a system of three equations.*** We know that fifteen trucks will be used; hence, we have the following:

$$x + y + z = 15 \quad \textbf{(1)}$$

How can we get our second equation? Well, we also know the *capacity* of each truck type, and we know the total tonnage to be hauled. The first type of truck hauls 10 tons, the second type 6 tons, and the third type 4 tons, and the total tonnage is 104 tons. Hence, we can write the following:

$$10x + 6y + 4z = 104 \quad \textbf{(2)}$$

We still need one more equation. What other given information can we use? The problem states that the manager wants to use two more 10-ton trucks than 6-ton trucks. Thus, we have the following:

$$x = 2 + y \quad \textbf{(3)}$$

(We could also have written $x - y = 2$.) Hence, our system of equations is as follows:

$$\begin{aligned} x + y + z &= 15 \quad \textbf{(1)} \\ 10x + 6y + 4z &= 104 \quad \textbf{(2)} \\ x - y &= 2 \quad \textbf{(3)} \end{aligned}$$

3. ***Solve the system of equations and state the answer.*** Equation **(3)** doesn't contain the variable z. Let's work with equations **(1)** and **(2)** to eliminate z. First, we multiply equation **(1)** by -4 and add it to equation **(2)**.

$$\begin{aligned} -4x - 4y - 4z &= -60 \quad \textbf{(4)} \\ 10x + 6y + 4z &= 104 \quad \textbf{(2)} \\ \hline 6x + 2y &= 44 \quad \textbf{(5)} \end{aligned}$$

Make sure you understand how we got equation **(5)**. Dividing each term of equation **(5)** by 2 and adding to equation **(3)** gives the following:

$$\begin{aligned} 3x + y &= 22 \quad \textbf{(6)} \\ x - y &= \ \ 2 \quad \textbf{(3)} \\ \hline 4x \qquad &= 24 \\ x &= \ \ 6 \end{aligned}$$

For $x = 6$, equation **(3)** yields the following:

$$\begin{aligned} 6 - y &= 2 \\ 4 &= y \end{aligned}$$

Now we substitute the known x- and y-values into equation **(1)**.

$$\begin{aligned} 6 + 4 + z &= 15 \\ z &= \ \ 5 \end{aligned}$$

Thus, the manager needs six 10-ton trucks, four 6-ton trucks, and five 4-ton trucks.

4. ***Check.*** The check is left to the student. □

STEPS TO SUCCESS Why Do We Have to Learn to Solve Word Problems?

Many students ask that question. It is a fair question, and it deserves an honest answer.

Applications or word problems are the very life of mathematics! They are the reason for doing mathematics. They teach you how to put into use the mathematical skills you have developed.

Almost all branches of mathematics are studied because they solve problems in real life. You will find in this textbook that studying the Use Math to Save Money problems will help you manage your finances. Many decisions in daily life can be made more easily if mathematics is used to evaluate the possible options.

How can I learn to solve word problems more easily? Many students ask that question also. The key to success is practice. Make yourself do as many problems as you can. You may not be able to do them all correctly at first, but keep trying. If you cannot solve a problem, try another one. Ask for help from your teacher or someone in the tutoring lab. Ask other classmates how they solved the problem. Soon you will see great progress in your own problem-solving ability.

Making it personal: Which of these statements best answers the question "Why do we have to learn to solve word problems?" What suggestion do you think is the most helpful to help you to learn to solve word problems more easily? Write the suggestion in your own words. Start to apply it today to help you with the homework in Exercises 4.3. Remember, much of life is like a word problem. ▼

4.3 Exercises MyMathLab®

Applications

Use a system of two linear equations to solve exercises 1–22.

1. The sum of 2 numbers is 87. If twice the smaller number is subtracted from the larger number, the result is 12. Find the two numbers.

2. The difference between two numbers is 11. If triple the smaller is added to double the larger, the result is 42. Find the two numbers.

3. ***Temporary Employment Agency*** An employment agency specializing in temporary construction help pays heavy equipment operators $140 per day and general laborers $90 per day. If thirty-five people were hired and the payroll was $3950, how many heavy equipment operators were employed? How many laborers?

4. ***Broadway Ticket Prices*** A Broadway performance of *The Lion King* had a paid attendance of 387 people. Mezzanine tickets cost $70 and orchestra tickets cost $95. Ticket sales receipts totaled $30,965. How many tickets of each type were sold?

5. ***Amtrak Train Tickets*** Ninety-eight passengers rode an Amtrak train from Boston to Denver. Tickets for regular coach seats cost $120. Tickets for sleeper car seats cost $290. The receipts for the trip totaled $19,750. How many passengers purchased each type of ticket?

6. ***Farming Operations*** The Mulder Farm has 600 acres of land allotted for raising corn and wheat. The cost to cultivate corn is $40 per acre. The cost to cultivate wheat is $25 per acre. The Mulders have $18,000 available to cultivate these crops. How many acres of each crop should the Mulders plant?

7. ***Computer Training Coordinator*** Reggie is in charge of scheduling computer training for the employees at his company. The human resources department is to be trained to use new spreadsheet software and a new payroll program. Experienced employees can learn the spreadsheet in 3 hours and the payroll in 4 hours. New employees need 4 hours to learn the spreadsheet and 7 hours to learn the payroll. The company can afford to pay for 115 hours of spreadsheet instruction and 170 hours of payroll instruction. How many of each type of employee can Reggie send to the training?

8. ***Radar Detector Manufacturing*** Ventex makes auto radar detectors. Ventex has found that its basic model requires 3 hours of manufacturing for the inside components and 2 hours for the housing and controls. Its advanced model requires 5 hours to manufacture the inside components and 3 hours for the housing and controls. This week, the production division has available 1050 hours for producing inside components and 660 hours for housing and controls. How many detectors of each type can be made?

9. ***Farm Management*** David manages the family farm. He has several packages of fertilizer for his new grain crop. The old packages contain 50 pounds of long-term-growth supplement and 60 pounds of weed killer. The new packages contain 65 pounds of long-term-growth supplement and 45 pounds of weed killer. Using past experience, David estimates that he needs 3125 pounds of long-term-growth supplement and 2925 pounds of weed killer for the fields. How many old packages of fertilizer and how many new packages of fertilizer should he use?

10. ***Hospital Dietician*** Geneva is a dietician in a hospital. She has two prepackaged mixtures of vitamin additives available for patients. Mixture 1 contains 5 grams of vitamin C and 3 grams of niacin; mixture 2 contains 6 grams of vitamin C and 5 grams of niacin. On an average day she needs 87 grams of niacin and 117 grams of vitamin C. How many packets of each mixture will Geneva need?

11. ***Coffee and Snack Expenses*** On Monday, Luis picked up five scones and six large coffees for the office staff. He paid $25.15. On Tuesday, Rachel picked up four scones and seven large coffees for the office staff. She paid $24.30. What is the cost of one scone? What is the cost of one large coffee?

12. ***Advertising Printer*** A local department store is preparing four-color sales brochures to insert into the *Salem Evening News*. Sam is the advertising printer it has hired. Sam has a fixed charge to set up the printing of the brochure and a specific per-copy amount for each brochure printed. He quoted a price of $1350 for printing five thousand brochures and a price of $1750 for printing seven thousand brochures. What is the fixed charge to set up the printing? What is the per-copy cost that Sam charges for printing a brochure?

13. ***Airspeed*** Against the wind a small plane flew 210 miles in 1 hour and 10 minutes. The return trip took only 50 minutes. What was the speed of the wind? What was the speed of the plane in still air?

14. ***Aircraft Operation*** Against the wind, a commercial aircraft in South America flew 630 miles in 3 hours and 30 minutes. With a tailwind, the return trip took 3 hours. What was the speed of the wind? What was the speed of the plane in still air?

15. ***Fishing and Boating*** Don Williams uses his small motorboat to go 8 miles upstream to his favorite fishing spot. Against the current, the trip takes $\frac{2}{3}$ hour. With the current, the trip takes $\frac{1}{2}$ hour. How fast can the boat travel in still water? What is the speed of the current?

16. ***Canoe Trip*** It look Lance and Ivan 6 hours to travel 33 miles downstream by canoe on the Mississippi River. The next day they traveled for 8 hours upstream for 20 miles. What was the rate of the current? What was their average speed in still water?

17. ***Basketball*** In 1962, Wilt Chamberlain set an NBA single-game scoring record. He scored 64 times for a total of 100 points. He made no 3-point shots (these were not part of the professional basketball game until 1979), but made several free throws worth 1 point each and several regular shots worth 2 points each. How many free throws did he make? How many 2-point shots did he make?

18. ***Basketball*** Blake Griffin of the Los Angeles Clippers scored 40 points in a recent basketball game. He scored no 3-point shots, but made a number of 2-point field goals and 1-point free throws. He scored 24 times. How many 2-point shots did he make? How many free throws did he make?

19. ***International Messaging Charges*** Nick traveled out of the country for a month. He purchased a plan from his cell phone company that would allow him to send text messages to friends and family back home. The company charges $0.20 for a regular text message and $0.50 for a multimedia text message. Nick's bill for the month was $87, and he sent a total of 315 messages. How many regular text messages and how many multimedia text messages did he send?

20. ***Office Supply Manager*** A new catalog from an office supply company shows that some of its prices will increase next month. This month, copier paper costs $6.50 per ream and printer cartridges cost $38.00. Mateo is in charge of ordering supplies for the office. If he submits his order this month, the cost will be $651. Next month, when paper will cost $7.00 per ream and the cartridges will cost $40, his order would cost $690. How many reams of paper and how many printer cartridges are in his order?

21. ***Highway Department Purchasing*** This year the state highway department in Montana purchased 256 identical cars and 183 identical trucks for official use. The purchase price was $5,791,948. Due to a budget shortfall, next year the department plans to purchase only 64 cars and 107 trucks. It will be charged the same price for each car and for each truck. Next year it plans to spend $2,507,612. How much does the department pay for each car and for each truck?

22. ***Concert Ticket Prices*** A recent concert at Gordon College had a paid audience of 987 people. Advance tickets were $9.95 and tickets at the door were $12.95. A total of $10,738.65 was collected in ticket sales. How many of each type of ticket were sold?

Use a system of three linear equations to solve exercises 23–31.

23. ***Bookstore Supplies*** Johanna bought 15 items at the college bookstore. The items cost a total of $23.00. The pens cost $0.50 each, the notebooks were $3.00 each, and the highlighters cost $1.50 each. She bought 2 more notebooks than highlighters. How many of each item did she buy?

24. ***Basketball*** In November 2012, Jack Taylor of Grinnell College set the all-time NCAA record for points scored in a single game. He made 59 shots for a total of 138 points. He made several free throws worth 1 point each, several regular shots worth 2 points each, and several 3-point shots. He made two more 3-point shots than 2-point shots. How many of each type of shot did he make?

25. ***Holiday Concert*** A total of three hundred people attended a holiday concert at the city theater. The admission prices were $12 for adults, $9 for high school students, and $5 for children. The ticket sales totaled $2460. The concert coordinator suggested that next year they raise the prices to $15 for adults, $10 for high school students, and $6 for children. She said that if the same number of people attend next year, the ticket sales at the higher prices will total $2920. How many adults, high school students, and children attended this year?

26. ***CPR Training*** A college conducted a CPR training class for students, faculty, and staff. Faculty were charged $10, staff were charged $8, and students were charged $2 to attend the class. A total of four hundred people came. The receipts for all who attended totaled $2130. The college president remarked that if he had charged faculty $15 and staff $10 and let students come free, the receipts this year would have been $2425. How many students, faculty, and staff came to the CPR training class?

27. ***City Subway Token Costs*** A total of twelve thousand passengers normally ride the green line of the MBTA during the morning rush hour. The token prices for a ride are $1.05 for junior and high school students, $2.10 for adults, and $1.05 for senior citizens, and the revenue from these riders is $23,100. If the token prices were raised to $1.25 for junior and high school students and $2.40 for adults, and the senior citizen price were unchanged, the expected revenue from these riders would be $26,340. How many riders in each category normally ride the green line during the morning rush hour?

28. ***Commission Sales*** The owner of Danvers Ford found that he sold a total of 520 cars, Escapes, and Explorers last year. He paid the sales staff a commission of \$100 for every car, \$200 for every Escape, and \$300 for every Explorer sold. The total of these commissions last year was \$87,000. In the coming year he is contemplating an increase so that the commission will be \$150 for every car and \$250 for every Escape, with no change in the commission for Explorer sales. If the sales are the same this year as they were last year, the commissions will total \$106,500. How many vehicles in each category were sold last year?

29. ***Pizza Costs*** The Hamilton House of Pizza delivered twenty pepperoni pizzas to Gordon College on the first night of final exams. The cost of these pizzas totaled \$233. A small pizza costs \$8 and contains 3 ounces of pepperoni. A medium pizza costs \$11 and contains 4 ounces of pepperoni. A large pizza costs \$15 and contains 5 ounces of pepperoni. The owner of the pizza shop used 5 pounds 2 ounces of pepperoni in making these twenty pizzas. How many pizzas of each size were delivered to Gordon College? (Recall that there are 16 ounces in 1 pound.)

30. ***Roast Beef Sandwich Costs*** One of the favorite meeting places for local college students is Nick's Roast Beef in Beverly, Massachusetts. Last night from 8 P.M. to 9 P.M. Nick served twenty-four roast beef sandwiches. He sliced 15 pounds 8 ounces of roast beef to make these sandwiches and collected \$131 for them. The medium roast beef sandwich has 6 ounces of beef and costs \$4. The large roast beef sandwich has 10 ounces of beef and costs \$5. The extra large roast beef sandwich has 14 ounces of beef and costs \$7. How many of each size of roast beef sandwich did Nick sell from 8 P.M. to 9 P.M.?

31. ***Packing Fruit*** Florida Fruits packs three types of gift boxes containing oranges, grapefruit, and tangerines. Box A contains 12 oranges, 5 grapefruit, and 3 tangerines. Box B contains 10 oranges, 6 grapefruit, and 4 tangerines. Box C contains 5 oranges, 8 grapefruit, and 5 tangerines. The shipping manager has available 91 oranges, 63 grapefruit, and 40 tangerines. How many gift boxes of each type can she prepare?

Cumulative Review *Solve for the variable indicated.*

32. [2.4.1] Solve for x. $\frac{1}{3}(4 - 2x) = \frac{1}{2}x - 3$

33. [2.4.1] Solve for x. $0.06x + 0.15(0.5 - x) = 0.04$

34. [2.3.3] Solve for y. $2(y - 3) - (2y + 4) = -6y$

35. [2.3.3] Solve for x. $4(3x + 1) = -6 + x$

Quick Quiz 4.3

1. A plane flew 1200 miles with a tailwind in 2.5 hours. The return trip against the wind took 3 hours. Find the speed of the wind. Find the speed of the plane in still air.

2. A man needed a car in Alaska to travel to fishing sites. He found a company that rented him a car but charged him a daily rental fee and a mileage charge. He rented the car for 8 days, drove 300 miles, and was charged \$355. Next week his friend rented the same car for 9 days, drove 260 miles, and was charged \$380. What was the daily rental charge? How much did the company charge per mile?

3. Three friends went on a trip to Uganda. They went to the same store and purchased the exact same items. Nancy purchased 3 drawings, 2 carved elephants, and 1 set of drums for \$55. John purchased 2 drawings, 3 carved elephants, and 1 set of drums for \$65. Steve purchased 4 drawings, 3 carved elephants, and 2 sets of drums for \$85. What was the price of each drawing, carved elephants, and set of drums?

4. Concept Check Explain how you would set up two equations using the information in problem 1 above if the plane flew 1500 miles instead of 1200 miles.

4.4 Systems of Linear Inequalities

1 Graphing a System of Linear Inequalities

We learned how to graph a linear inequality in two variables in Section 3.5. We call two linear inequalities in two variables a **system of linear inequalities in two variables.** We now consider how to graph such a system. The solution to a system of inequalities is the intersection of the solution sets of the individual inequalities of the system.

Student Learning Objective

After studying this section, you will be able to:

1 Graph a system of linear inequalities.

Example 1 Graph the solution of the system.

$$y \leq -3x + 2$$
$$-2x + y \geq -1$$

Solution

In this example, we will first graph each inequality separately. The graph of $y \leq -3x + 2$ is the region on and below the line $y = -3x + 2$.

The graph of $-2x + y \geq -1$ consists of the region on and above the line $-2x + y = -1$.

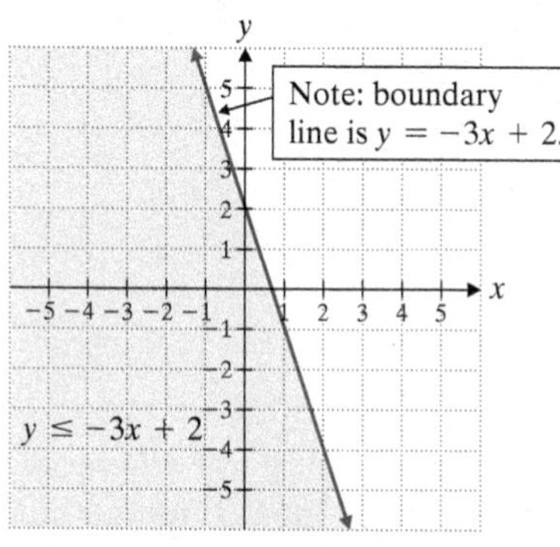

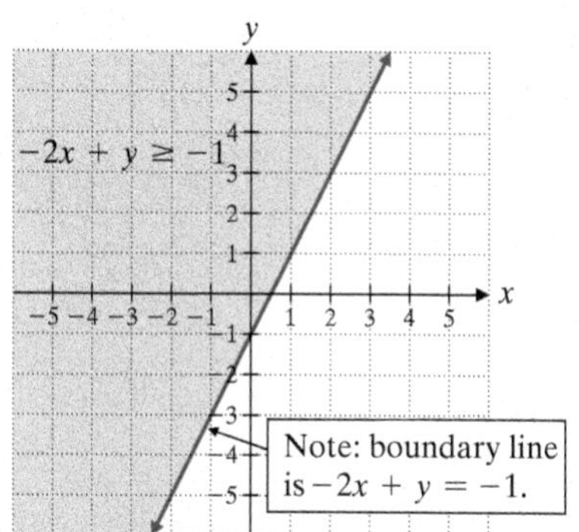

We will now place these graphs on one rectangular coordinate system. The darker shaded region is the intersection of the two graphs. Thus, the solution to the system of two inequalities is the darker shaded region and its boundary lines.

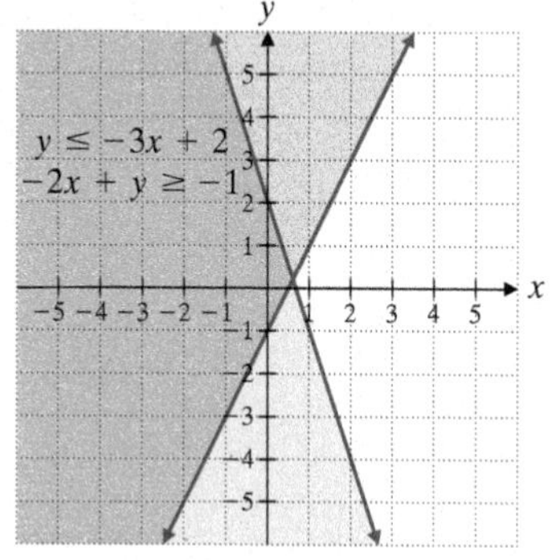

Student Practice 1 Graph the solution of the system.

$$-2x + y \leq -3$$
$$x + 2y \geq 4$$

Student Practice 1

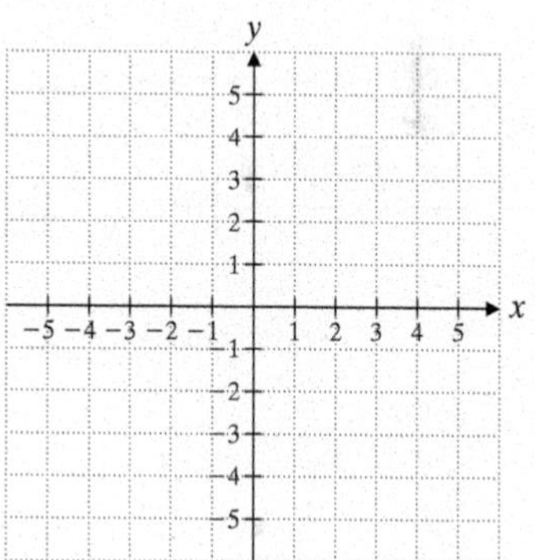

Usually we sketch the graphs of the individual inequalities on one set of axes. We will illustrate that concept with the following example.

Example 2 Graph the solution of the system.

$$y < 4$$
$$y > \frac{3}{2}x - 2$$

Solution The graph of $y < 4$ is the region below the line $y = 4$. It does not include the line since we have the $<$ symbol. Thus, we use a dashed line to indicate that the boundary line is not part of the answer. The graph of $y > \frac{3}{2}x - 2$ is the region above the line $y = \frac{3}{2}x - 2$. Again, we use the dashed line to indicate that the boundary line is not part of the answer. The final solution is the darker shaded region. The solution does *not* include the dashed boundary lines.

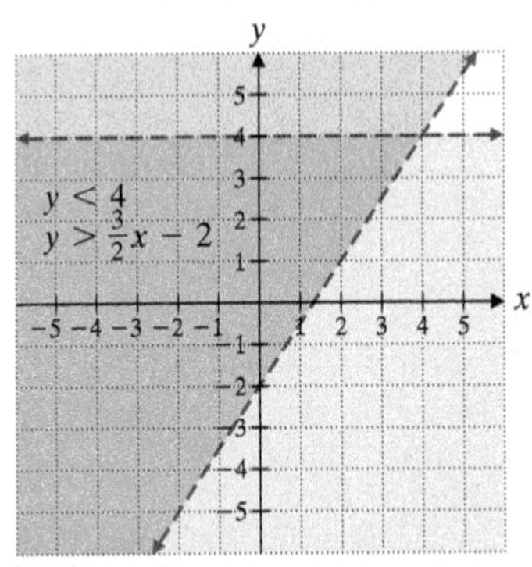

□

Student Practice 2

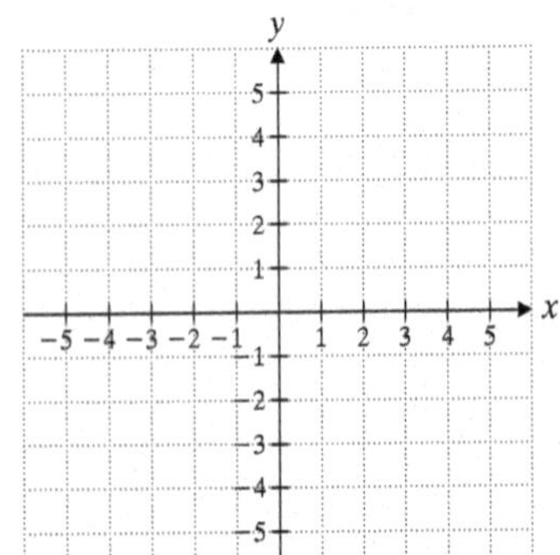

Student Practice 2 Graph the solution of the system.

$$y > -1$$
$$y < -\frac{3}{4}x + 2$$

There are times when we require the exact location of the point where two boundary lines intersect. In these cases the boundary points are labeled on the final sketch of the solution.

Example 3 Graph the solution of the following system of inequalities. Find the coordinates of any points where boundary lines intersect.

$$x + y \leq 5$$
$$x + 2y \leq 8$$
$$x \geq 0$$
$$y \geq 0$$

Solution The graph of $x + y \leq 5$ is the region on and below the line $x + y = 5$. The graph of $x + 2y \leq 8$ is the region on and below the line $x + 2y = 8$. We solve the system containing the equations $x + y = 5$ and $x + 2y = 8$ to find that their point of intersection is $(2, 3)$. The graph of $x \geq 0$ is the y-axis and all the region to the right of the y-axis. The graph of $y \geq 0$ is the x-axis and all the region above the x-axis. Thus, the solution to the system is the shaded region and its boundary lines. There are four points where boundary lines intersect.

These points are called the **vertices** of the solution. Thus, the vertices of the solution are (0, 0), (0, 4), (2, 3), and (5, 0).

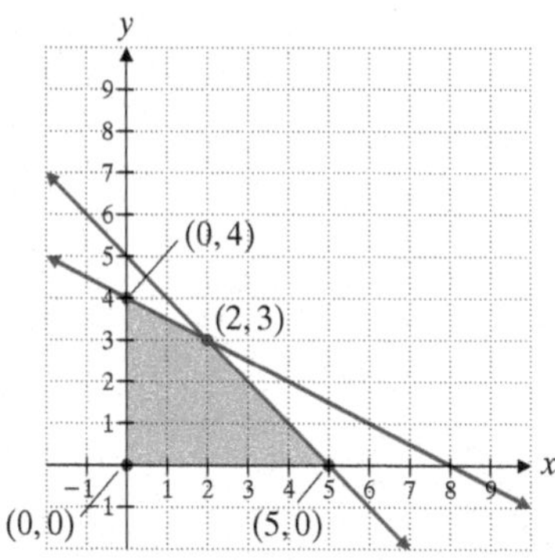

Student Practice 3 Graph the solution to the system of inequalities. Find the vertices of the solution.

$$x + y \leq 6$$
$$3x + y \leq 12$$
$$x \geq 0$$
$$y \geq 0$$

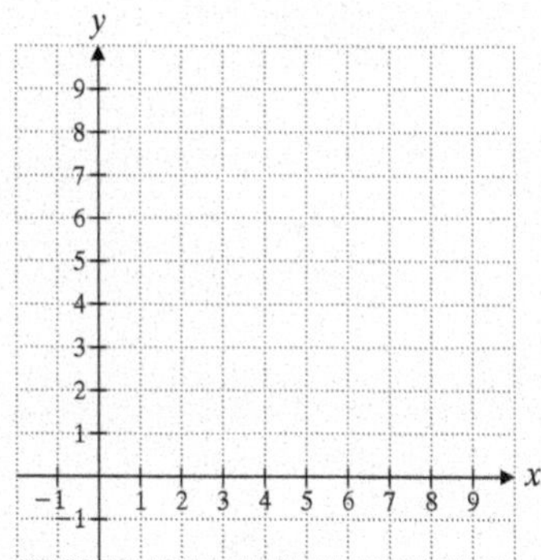

Graphing Calculator

Graphing Systems of Linear Inequalities

On some graphing calculators, you can graph a system of linear inequalities. To graph the system in Example 1, first rewrite it as:

$$y \leq -3x + 2$$
$$y \geq 2x - 1.$$

Enter each expression into the Y = editor of your graphing calculator and then select the appropriate direction for shading.

Display:

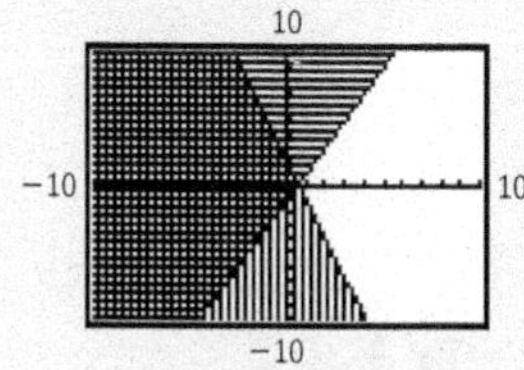

Notice that one inequality is shaded vertically and the other is shaded horizontally. The intersection is the solution. Note also that the graphing calculator will not indicate whether or not the boundary of the solution region is included in the solution.

STEPS TO SUCCESS Look Ahead to See What Is Coming

You will find that learning new material is much easier if you know what is coming. Take a few minutes at the end of your study time to glance over the next section of the book. If you quickly look over the topics and ideas in this new section, it will help you get your bearings when the instructor presents new material. Students find that when they preview new material it enables them to see what is coming. It helps them to be able to grasp new ideas much more quickly.

Making it personal: Do this right now: Look ahead to the next section of the book. Glance over the ideas and concepts. Write down a couple of facts about the next section.

4.4 Exercises MyMathLab®

Verbal and Writing Skills, Exercises 1–4

1. In the graph of the system $y > 3x + 1$ and $y < -2x + 5$, would the boundary lines be solid or dashed? Why?

2. In the graph of the system $y \geq -6x + 3$ and $y \leq -4x - 2$, would the boundary lines be solid or dashed? Why?

3. Stephanie wanted to know if the point $(3, -4)$ lies in the region that is a solution for $y < -2x + 3$ and $y > 5x - 3$. How could she determine if this is true?

4. John wanted to know if the point $(-5, 2)$ lies in the region that is a solution for $x + 2y < 3$ and $-4x + y > 2$. How could he determine if this is true?

Graph the solution of each of the following systems.

5. $y \geq 2x - 2$
$x + y \leq 4$

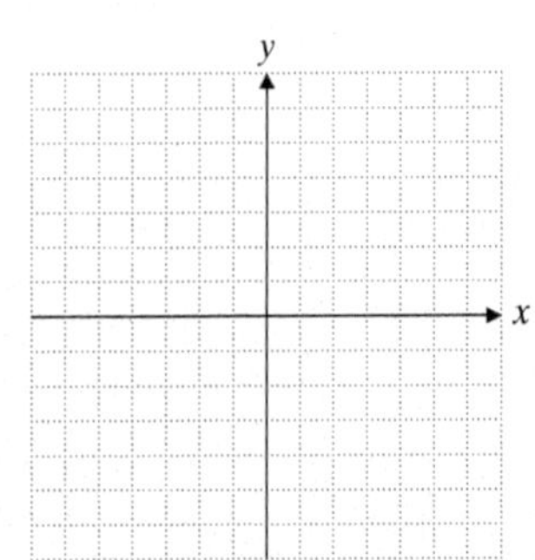

6. $y \geq x - 5$
$x + y \geq 3$

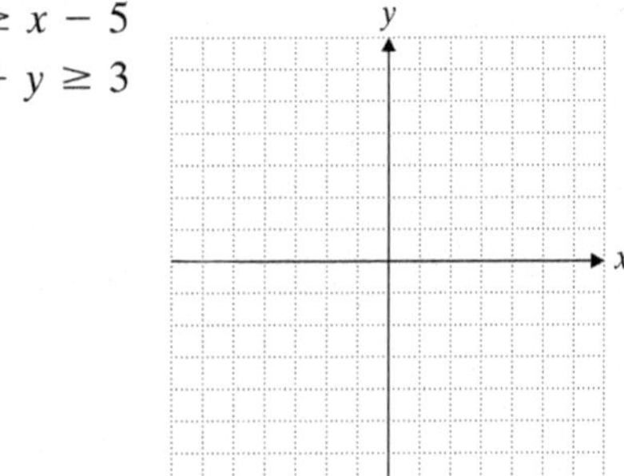

7. $y \geq -2x$
$y \geq 3x + 5$

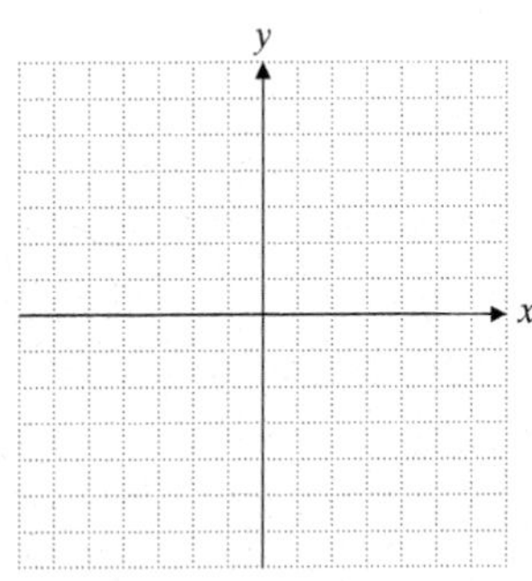

8. $y \geq -x$
$y \leq -3x + 4$

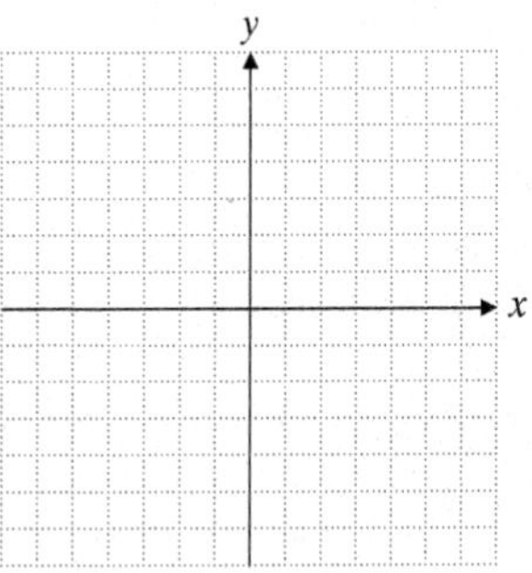

9. $y \geq 2x - 3$
$y \leq \frac{2}{3}x$

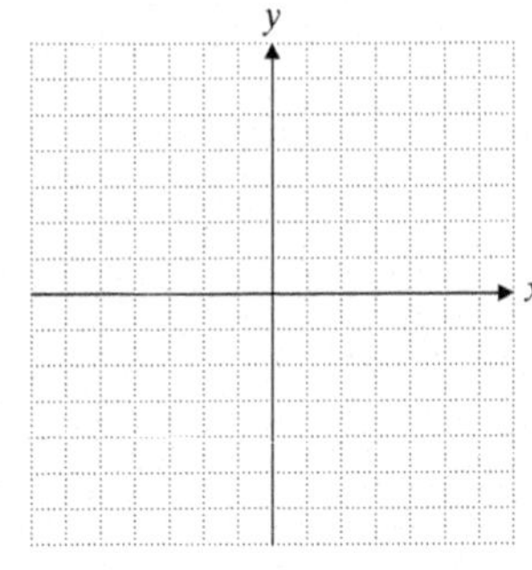

10. $y \leq \frac{1}{2}x + 3$
$y \geq -\frac{1}{2}x$

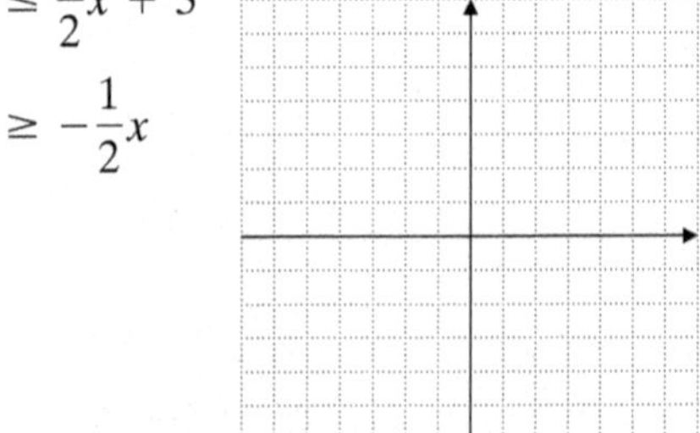

11. $x - y \geq -1$
$-3x - y \leq 4$

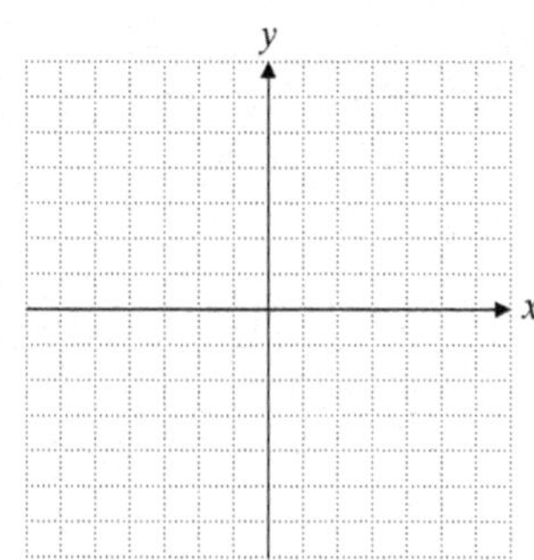

12. $x + y \leq 5$
$-3x + 4y \geq -8$

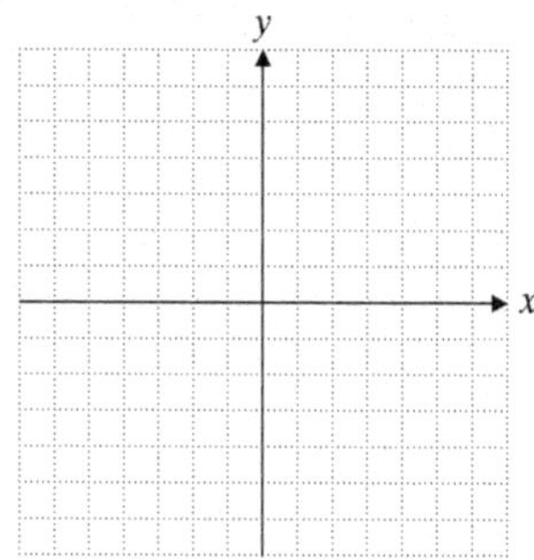

13. $x + 2y < 6$
$y < 3$

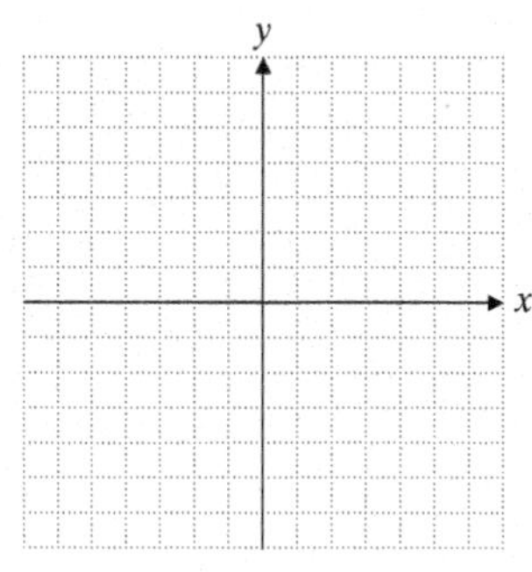

14. $-2x + y < -4$
$y > -4$

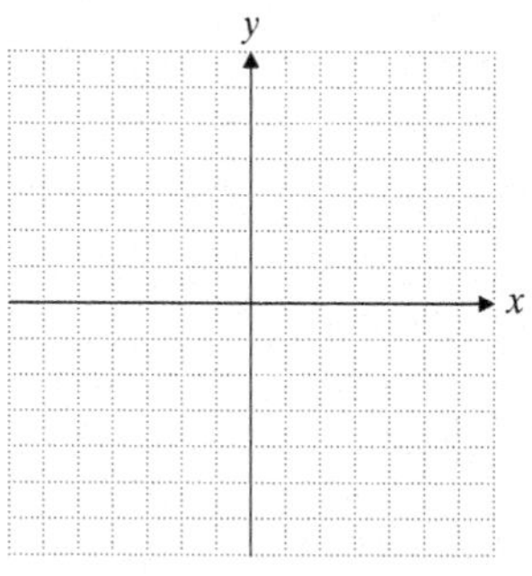

15. $y < 4$
$x > -2$

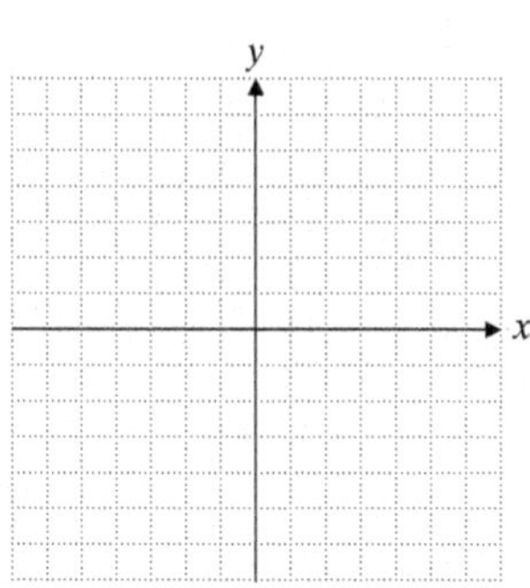

16. $y > -5$
$x < 3$

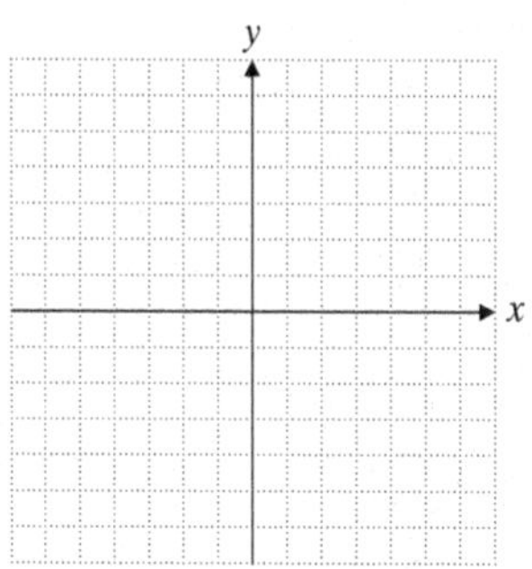

Mixed Practice

Graph the solution of each of the following systems.

17. $x - 4y \geq -4$
$3x + y \leq 3$

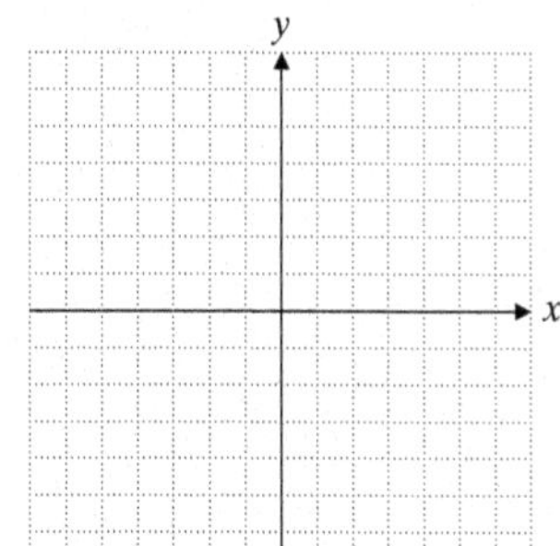

18. $5x - 2y \leq 10$
$x - y \geq -1$

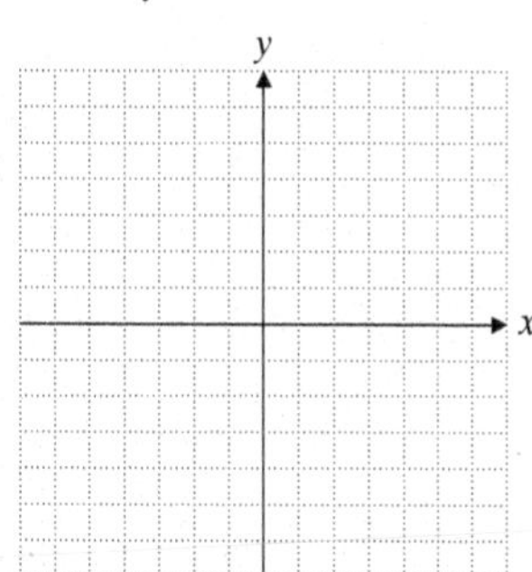

19. $3x + 2y < 6$
$3x + 2y > -6$

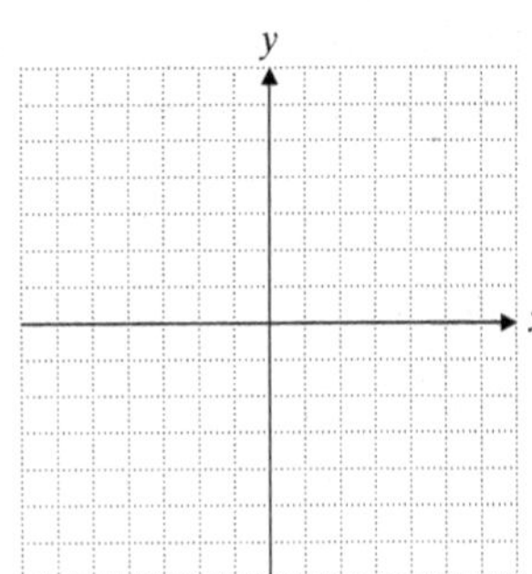

20. $2x - y < 2$
$2x - y > -2$

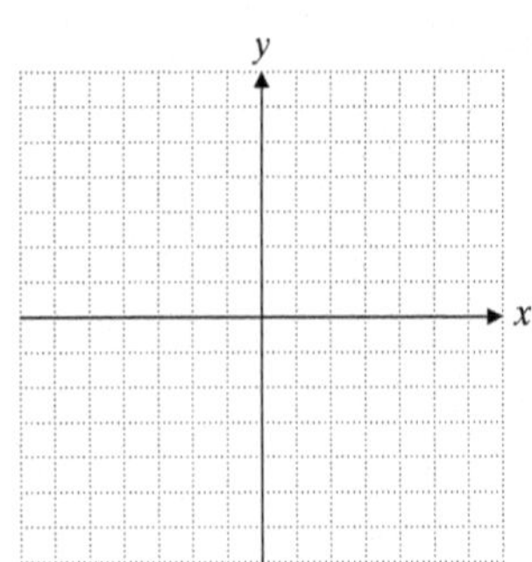

Graph the solution of the following systems of inequalities. Find the vertices of the solution.

21. $x + y \leq 5$
$2x - y \geq 1$

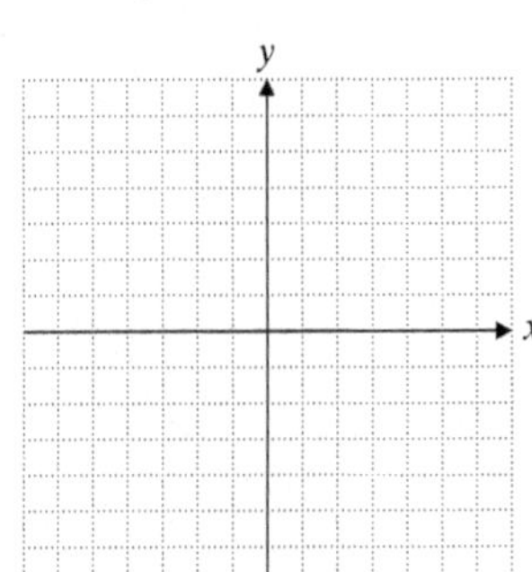

22. $-x + y \geq 3$
$y + 2x \leq 6$

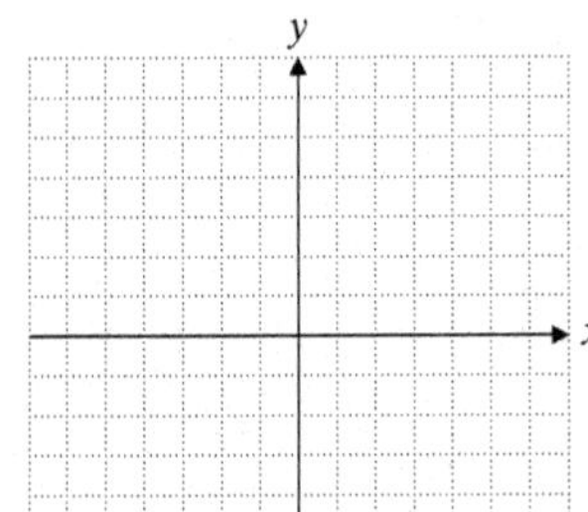

23. $x + 4y < -20$
$y \leq x$

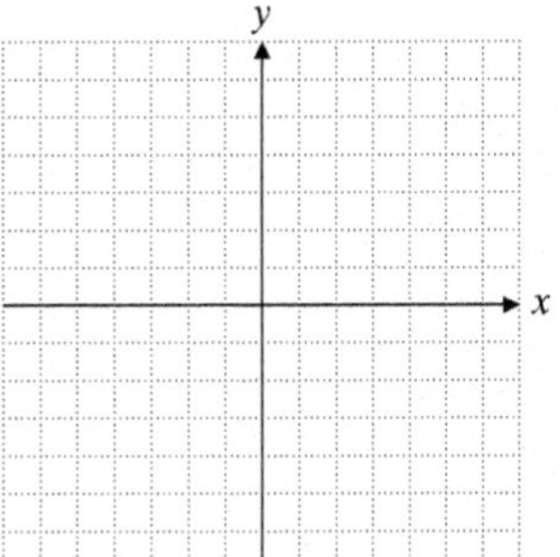

24. $x + 2y \leq 4$
$y < -x$

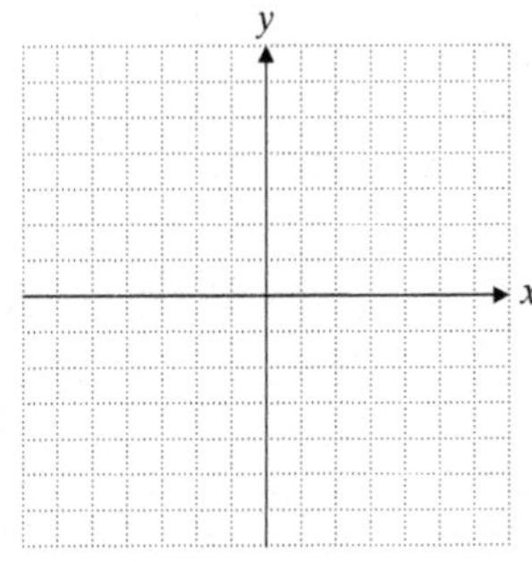

25. $y \leq x$
$x + y \geq 1$
$x \leq 5$

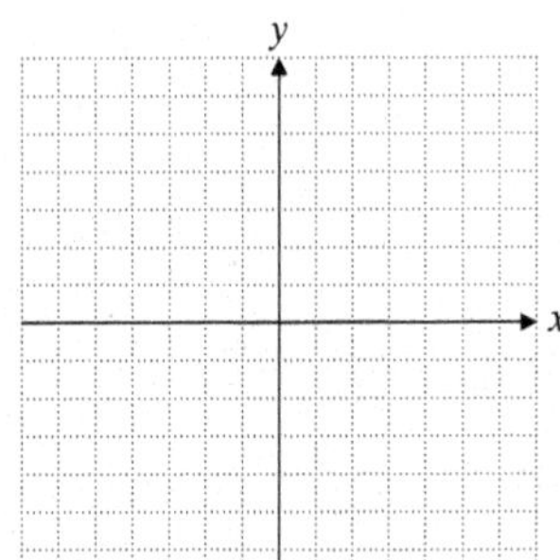

26. $x + y \leq 2$
$x - y \leq 2$
$x \geq -3$

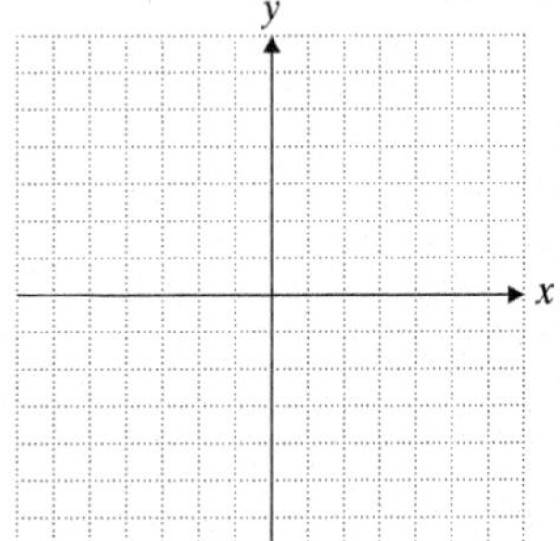

To Think About

Graph the region determined by each of the following systems.

27. $y \leq 3x + 6$
$4y + 3x \leq 3$
$x \geq -2$
$y \geq -3$

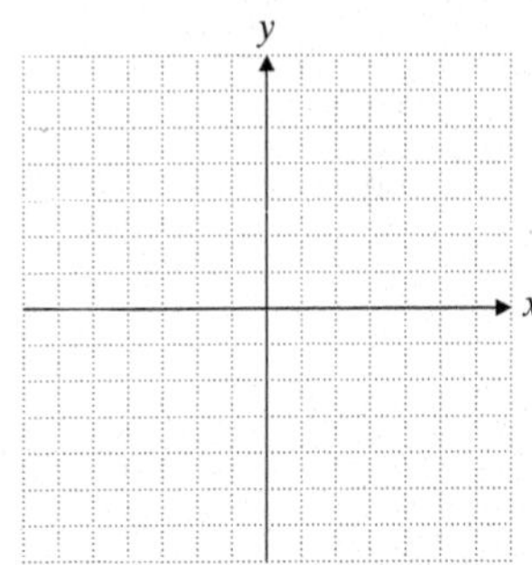

28. $-x + y \leq 100$
$x + 3y \leq 150$
$x \geq -80$
$y \geq 20$
(*Hint:* Use a scale of each square $= 20$ units on both axes.)

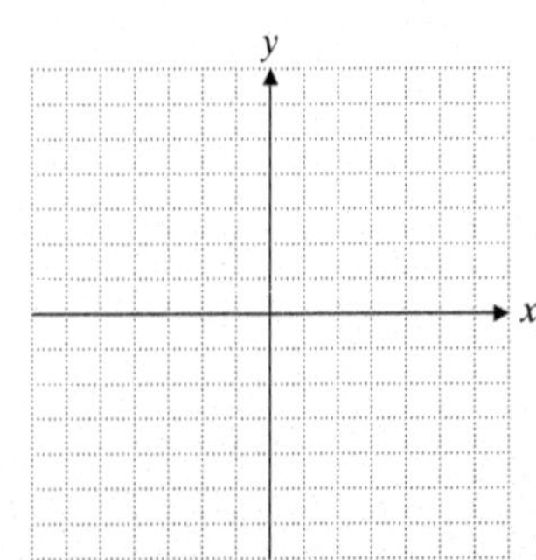

Applications

Hint: In exercises 29 and 30, if the coordinates of a boundary point contain fractions, it is wise to obtain the point of intersection algebraically rather than graphically.

29. ***Hospital Staffing Levels*** The equation that represents the proper level of medical staffing in the cardiac care unit of a local hospital is $N \le 2D$, where N is the number of nurses on duty and D is the number of doctors on duty. In order to control costs, the equation $4N + 3D \le 20$ is appropriate. The number of doctors and nurses on duty at any time cannot be negative, so $N \ge 0$ and $D \ge 0$.

(a) Graph the region satisfying all of the medical requirements for the cardiac care unit. Use the following special graph grid, where D is measured on the horizontal axis and N is measured on the vertical axis.

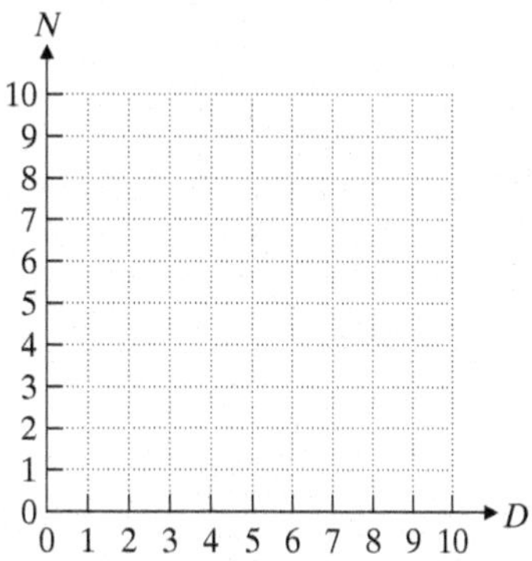

(b) If there are three doctors and two nurses on duty in the cardiac care unit, are all of the medical requirements satisfied?

(c) If there is one doctor and four nurses on duty in the cardiac care unit, are all of the medical requirements satisfied?

30. ***Traffic Control*** The equation that represents the proper traffic control and emergency vehicle response availability in the city of Salem is $P + 3F \le 18$, where P is the number of police cars on active duty and F is the number of fire trucks that have left the firehouse and are involved in a response to a call. In order to comply with staffing limitations, the equation $4P + F \le 28$ is appropriate. The number of police cars on active duty and the number of fire trucks that have left the firehouse cannot be negative, so $P \ge 0$ and $F \ge 0$.

(a) Graph the regions satisfying all of the availability and staffing limitation requirements for the city of Salem. Use the following special graph grid, where P is measured on the horizontal axis and F is measured on the vertical axis.

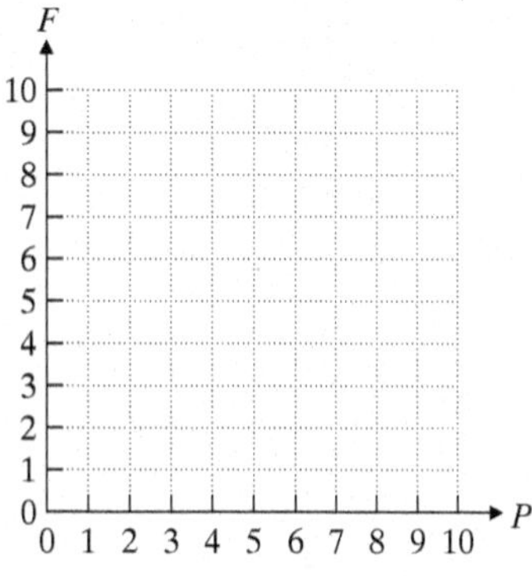

(b) If four police cars are on active duty and four fire trucks have left the firehouse in response to a call, are all of the requirements satisfied?

(c) If two police cars are on active duty and six fire trucks have left the firehouse in response to a call, are all of the requirements satisfied?

Cumulative Review

31. [1.8.1] Evaluate for $x = 2, y = -1$.
$-3x^2y - x^2 + 5y^2$

32. [1.9.1] Simplify. $2x - 2[y + 3(x - y)]$

33. **[4.3.1]** ***Driving Range Receipts*** Golf Galaxy Indoor Driving Range took in \$6050 on one rainy day and six sunny days. The next week the driving range took in \$7400 on four rainy days and three sunny days. What is the average amount of money taken in at Golf Galaxy on a rainy day? On a sunny day?

34. **[4.3.2]** ***Office Lunch Expenses*** Two weeks ago, Raquel took the office lunch order to Kramer's Deli and bought 3 chicken sandwiches, 2 side salads, and 3 sodas for \$27.75. One week ago it was Gerry's turn to take the trip to Kramer's. He bought 3 chicken sandwiches, 4 side salads, and 4 sodas for \$36.75. This week Sheryl made the trip. She purchased 4 chicken sandwiches, 3 side salads, and 5 sodas for \$39.75. What is the cost of one chicken sandwich? Of one side salad? Of one soda?

Quick Quiz 4.4

1. In graphing the inequality $3x + 5y < 15$, do you shade the region above the line or below the line?

2. In graphing the following region, should the boundary lines be dashed lines or solid lines?

$$4x + 5y \geq 20$$
$$2x - 3y \geq 6$$

3. Graph the solution of the following system of linear inequalities.

$$3x + 2y > 6$$
$$x - 2y < 2$$

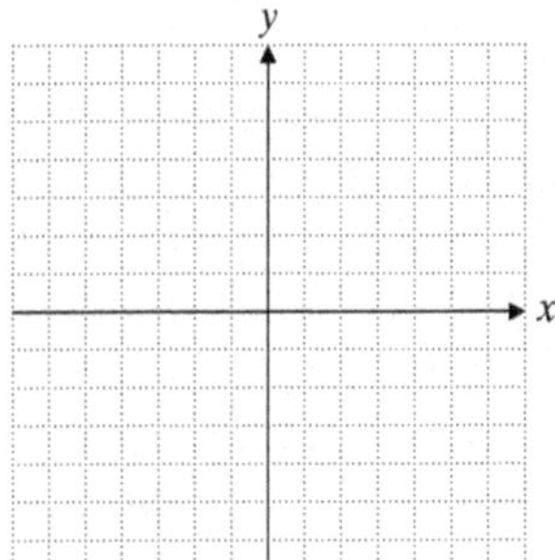

4. **Concept Check** Explain how you would graph the region described by the following.

$$y > x + 2$$
$$x < 3$$

Background: Accountant

Pierre works as an accountant for a major accounting firm and is responsible for managing the services provided to many of the firm's technology-based clients. Periodic audits of expenses and revenues are completed and provided to the firm's clients along with advisory information that helps the businesses manage their finances.

Pierre's accounting firm was recently hired by RealTek, a software-development company. They've asked Pierre to evaluate their revenues and expenses to help identify whether or not a revised pricing structure is needed for a software suite RealTek has recently developed.

Facts

- Currently RealTek sells their new software for $19.95 per unit with free shipping and handling.
- Current costs of production are $7.50 per unit.
- Costs for rent and other fixed expenses are $19,920 monthly.
- RealTek is considering increasing the software's selling price to $22.95 per unit to include shipping and handling, lowering its production costs to $2.95 per unit. Fixed expenses will remain at $19,920 monthly.
- RealTek is also considering moving to a smaller office location thereby decreasing its rental costs and overall fixed expenses to $11,205 monthly. If they relocate their office, RealTek will keep the selling price of the software as $19.95 per unit and production costs will also remain at $7.50 per unit.

Tasks

1. Pierre computes RealTek's break-even point when the sale price per unit is $19.95, monthly fixed costs are $19,920, and the production costs are $7.50 per unit. How many units of software must be sold monthly in order for RealTek to meet their break-even point?

2. Pierre now determines the break-even point for RealTek's new price structure. In this scenario, the sale price per unit is $22.95, the production costs are $2.95 per unit, and fixed expenses will remain the same. How many units of software must be sold monthly in order for RealTek to meet their break-even point under the new price structure?

3. Pierre feels it would be wise for RealTek to move to a smaller office location that doesn't cost as much monthly and won't affect RealTek's business operations to any significant extent. How many units of software would need to be sold monthly in order for RealTek to meet their break-even point if they relocate?

Chapter 4 Organizer

Topic and Procedure	Examples	You Try It
Finding a solution to a system of equations by the graphing method, p. 274 1. Graph the first equation. 2. Graph the second equation. 3. Approximate from your graph the coordinates of the point where the two lines intersect, if they intersect at one point. 4. If the lines are parallel, there is no solution. If the lines coincide, there is an infinite number of solutions.	Solve by graphing. $x + y = 6$ $2x - y = 6$ Graph each line. The solution is (4, 2). 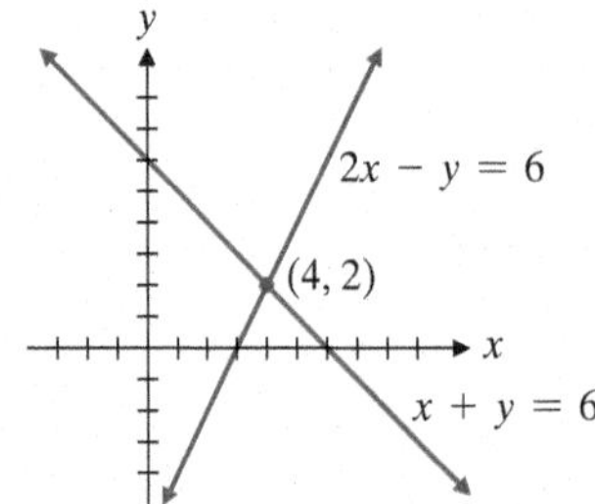 	1. Solve by graphing. $-x + 3y = 6$ $2x - 3y = -9$ 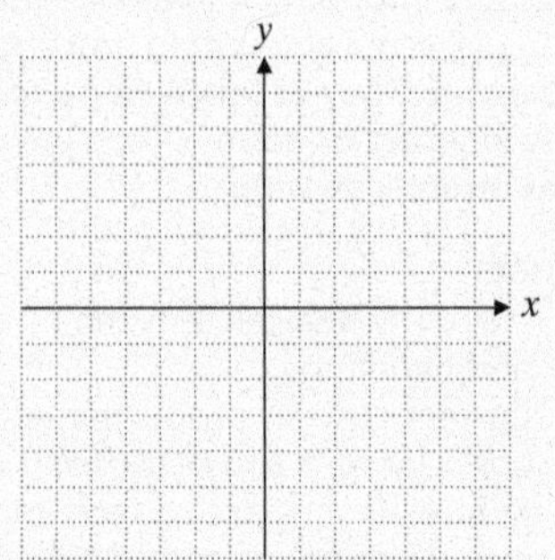
Solving a system of two linear equations by the substitution method, p. 275 The substitution method is most appropriate when *at least one variable has a coefficient of* 1 *or* −1. 1. Solve for one variable in one of the equations. 2. In the other equation, replace that variable with the expression you obtained in step 1. 3. Solve the resulting equation. 4. Substitute the numerical value you obtain for the variable into one of the original or equivalent equations and solve for the other variable. 5. Check the solution in both original equations.	Solve. $2x + y = 11$ **(1)** $x + 3y = 18$ **(2)** $y = 11 - 2x$ from equation **(1)**. Substitute into **(2)**. $x + 3(11 - 2x) = 18$ $x + 33 - 6x = 18$ $-5x = -15$ $x = 3$ Substitute $x = 3$ into $2x + y = 11$. $2(3) + y = 11$ $y = 5$ The solution is (3, 5).	2. Solve by the substitution method. $-3x + 4y = -10$ $x - 5y = 7$
Solving a system of two linear equations by the addition (or elimination) method, p. 278 The addition method is most appropriate when the variables *all have coefficients other than* 1 *or* −1. 1. Arrange each equation in the form $ax + by = c$. 2. Multiply one or both equations by appropriate numerical values so that when the two resulting equations are added, one variable is eliminated. 3. Solve the resulting equation. 4. Substitute the numerical value you obtain for the variable into one of the original or equivalent equations. 5. Solve this equation to find the other variable.	Solve. $2x + 3y = 5$ **(1)** $-3x - 4y = -2$ **(2)** Multiply equation **(1)** by 3 and **(2)** by 2. $6x + 9y = 15$ $-6x - 8y = -4$ $y = 11$ Substitute $y = 11$ into equation **(1)**. $2x + 3(11) = 5$ $2x + 33 = 5$ $2x = -28$ $x = -14$ The solution is (−14, 11).	3. Solve by the addition method. $4x - 5y = 5$ $-3x + 7y = 19$
Inconsistent system of equations, p. 280 If there is *no solution* to a system of linear equations, the system of equations is inconsistent. When you try to solve an inconsistent system algebraically, you obtain an equation that is not true, such as $0 = 5$.	Solve. $4x + 3y = 10$ **(1)** $-8x - 6y = 5$ **(2)** Multiply equation **(1)** by 2 and add to **(2)**. $8x + 6y = 20$ $-8x - 6y = 5$ $0 = 25$ But $0 \neq 25$. Thus, there is no solution. The system of equations is inconsistent.	4. Solve. $2x + 12y = 3$ $x + 6y = 8$
Dependent equations, p. 280 If there is an *infinite number of solutions* to a system of linear equations, at least one pair of equations is dependent. When you try to solve a system that contains dependent equations, you will obtain an equation that is always true (such as $0 = 0$ or $3 = 3$). These equations are called *identities*.	Attempt to solve the system. $x - 2y = -5$ **(1)** $-3x + 6y = 15$ **(2)** Multiply equation **(1)** by 3 and add to **(2)**. $3x - 6y = -15$ $-3x + 6y = 15$ $0 = 0$ There is an infinite number of solutions. The equations are dependent.	5. Solve. $x - y = 2$ $-4x + 4y = -8$

Topic and Procedure	Examples	You Try It
Solving a system of three linear equations by algebraic methods, p. 288 If there is one solution to a system of three linear equations in three unknowns, it may be obtained in the following manner. **1.** Choose two equations from the system. **2.** Multiply one or both of the equations by the appropriate constants so that by adding the two equations together, one variable can be eliminated. **3.** Choose a *different* pair of the three original equations and eliminate the *same* variable using the procedure of step 2. **4.** Solve the system formed by the two equations resulting from steps 2 and 3 to find values for both variables. **5.** Substitute the values obtained in step 4 into one of the original three equations to find the value of the third variable.	Solve. $2x - y - 2z = -1$ **(1)** $x - 2y - z = 1$ **(2)** $x + y + z = 4$ **(3)** Add equations **(2)** and **(3)** to eliminate z. $2x - y = 5$ **(4)** Multiply equation **(3)** by 2 and add to **(1)**. $4x + y = 7$ **(5)** Add equations **(4)** and **(5)**. $2x - y = 5$ $4x + y = 7$ $6x = 12$ $x = 2$ Substitute $x = 2$ into equation **(5)**. $4(2) + y = 7$ $y = -1$ Substitute $x = 2, y = -1$ into equation **(3)**. $2 + (-1) + z = 4$ $z = 3$ The solution is $(2, -1, 3)$.	6. Solve. $x + 2y - z = -13$ $-2x + 3y - 2z = -8$ $x - y + z = 3$
Graphing the solution to a system of inequalities in two variables, p. 307 **1.** Determine the region that satisfies each individual inequality. **2.** Shade the common region that satisfies all the inequalities.	Graph. $3x + 2y \leq 10$ $-x + 2y \geq 2$ **1.** $3x + 2y \leq 10$ can be graphed more easily as $y \leq -\frac{3}{2}x + 5$. We draw a solid line and shade the region below it. $-x + 2y \geq 2$ can be graphed more easily as $y \geq \frac{1}{2}x + 1$. We draw a solid line and shade the region above it. **2.** The common region is shaded. 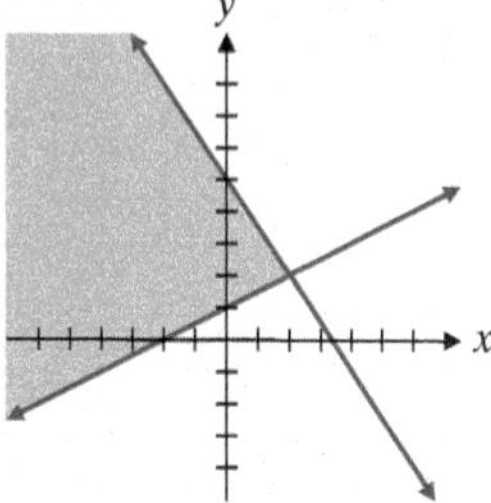	7. Graph. $x - 3y < 6$ $2x + y > 5$ 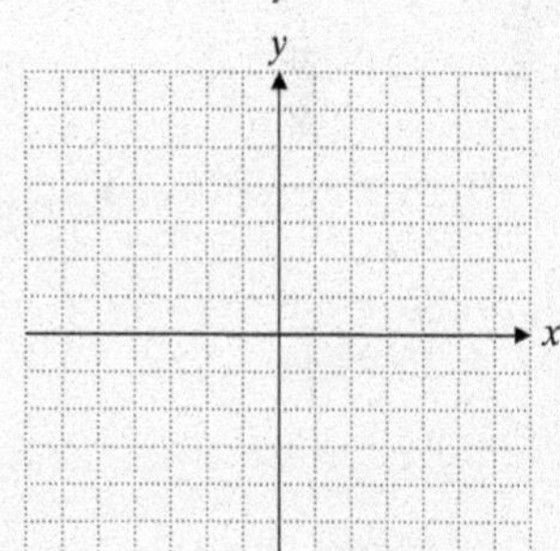

Chapter 4 Review Problems

Solve the following systems by graphing.

1. $x + 2y = 8$
$x - y = 2$

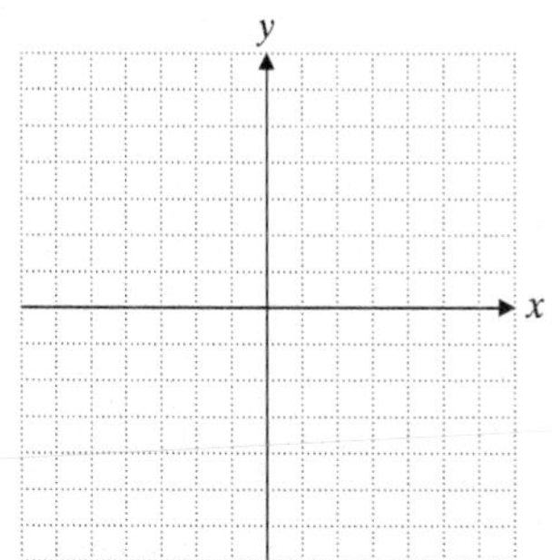

2. $2x + y = 6$
$3x + 4y = 4$

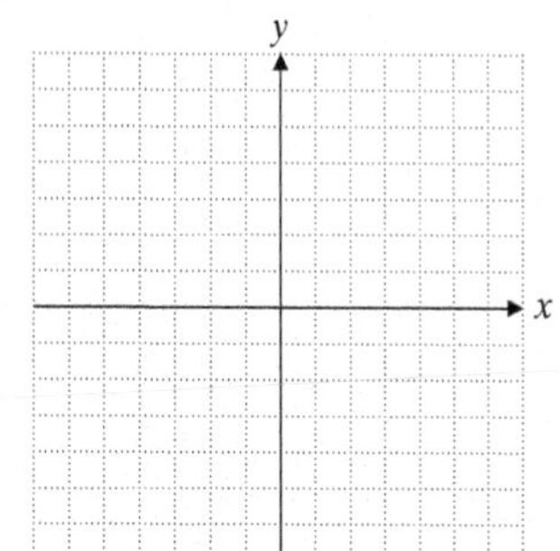

Solve the following systems by substitution.

3. $\begin{aligned} 3x - 2y &= -9 \\ 2x + y &= 1 \end{aligned}$

4. $\begin{aligned} 4x + 5y &= 2 \\ 3x - y &= 11 \end{aligned}$

5. $\begin{aligned} 3x - 4y &= -12 \\ x + 2y &= -4 \end{aligned}$

Solve the following systems by addition.

6. $\begin{aligned} -2x + 5y &= -12 \\ 3x + y &= 1 \end{aligned}$

7. $\begin{aligned} 7x - 4y &= 2 \\ 6x - 5y &= -3 \end{aligned}$

8. $\begin{aligned} 5x + 2y &= 3 \\ 7x + 5y &= -20 \end{aligned}$

Solve by any appropriate method. If there is no unique solution, state why.

9. $\begin{aligned} x &= 3 - 2y \\ 3x + 6y &= 8 \end{aligned}$

10. $\begin{aligned} x + 5y &= 10 \\ y &= 2 - \frac{1}{5}x \end{aligned}$

11. $\begin{aligned} 5x - 2y &= 15 \\ 3x + y &= -2 \end{aligned}$

12. $\begin{aligned} x + \frac{1}{3}y &= 1 \\ \frac{1}{4}x - \frac{3}{4}y &= -\frac{9}{4} \end{aligned}$

13. $\begin{aligned} 3a + 8b &= 0 \\ 9a + 2b &= 11 \end{aligned}$

14. $\begin{aligned} x + 3 &= 3y + 1 \\ 1 - 2(x - 2) &= 6y + 1 \end{aligned}$

15. $\begin{aligned} 10(x + 1) - 13 &= -8y \\ 4(2 - y) &= 5(x + 1) \end{aligned}$

16. $\begin{aligned} 0.2x - 0.1y &= 0.8 \\ 0.1x + 0.3y &= 1.1 \end{aligned}$

Solve by an appropriate method.

17. $\begin{aligned} 3x - 2y - z &= 3 \\ 2x + y + z &= 1 \\ -x - y + z &= -4 \end{aligned}$

18. $\begin{aligned} -2x + y - z &= -7 \\ x - 2y - z &= 2 \\ 6x + 4y + 2z &= 4 \end{aligned}$

19. $\begin{aligned} 2x + 5y + z &= 3 \\ x + y + 5z &= 42 \\ 2x + y &= 7 \end{aligned}$

20. $\begin{aligned} x + 2y + z &= 5 \\ 3x - 8y &= 17 \\ 2y + z &= -2 \end{aligned}$

21. $\begin{aligned} -3x - 4y + z &= -4 \\ x + 6y + 3z &= -8 \\ 5x + 3y - z &= 14 \end{aligned}$

22. $\begin{aligned} 3x + 2y &= 7 \\ 2x + 7z &= -26 \\ 5y + z &= 6 \end{aligned}$

Use a system of linear equations to solve each of the following exercises.

23. ***Commercial Airline*** A plane flies 720 miles against the wind in 3 hours. The return trip with the wind takes only $2\frac{1}{2}$ hours. Find the speed of the wind. Find the speed of the plane in still air.

24. ***Football*** During the 2015 Pro Bowl between the American Football Conference and the National Football Conference, 54 points were scored in touchdowns and field goals. The total number of touchdowns (6 points each) and field goals (3 points each) was 10. How many touchdowns and how many field goals were scored?

25. ***Temporary Help Expenses*** When the circus came to town last year, they hired general laborers at \$70 per day and mechanics at \$90 per day. They paid \$1950 for this temporary help for one day. This year they hired exactly the same number of people of each type, but they paid \$80 for general laborers and \$100 for mechanics for the one day. This year they paid \$2200 for temporary help. How many general laborers did they hire? How many mechanics did they hire?

26. ***Children's Theater Prices*** A total of 330 tickets were sold for the opening performance of *Frog and Toad*. Children's admission tickets were \$8, and adults' tickets were \$13. The ticket receipts for the performance totaled \$3215. How many children's tickets were sold? How many adults' tickets were sold?

27. ***Baseball Equipment*** A baseball coach bought two hats, five shirts, and four pairs of pants for \$177. His assistant purchased one hat, one shirt, and two pairs of pants for \$66. The next week the coach bought two hats, three shirts, and one pair of pants for \$81. What was the cost of each item?

28. ***Math Exam Scores*** Jess, Chris, and Nick scored a combined total of 249 points on their last math exam. Jess's score was 20 points higher than Chris's score. Twice Nick's score was 6 more than the sum of Chris's and Jess's scores. What did each of them score on the exam?

29. ***Food Costs*** Four jars of jelly, three jars of peanut butter, and five jars of honey cost \$32.50. Two jars of jelly, two jars of peanut butter, and one jar of honey cost \$14.80. Three jars of jelly, four jars of peanut butter, and two jars of honey cost \$27.00. Find the cost of each item.

30. ***Transportation Logistics*** A church youth group is planning a trip to Mount Washington. A total of 127 people need rides. The church has available buses that hold forty passengers, and several parents have volunteered station wagons that hold eight passengers or sedans that hold five passengers. The youth leader is planning to use nine vehicles to transport the people. One parent said that if they didn't use any buses, tripled the number of station wagons, and doubled the number of sedans, they would be able to transport 126 people. How many buses, station wagons, and sedans are they planning to use if they use nine vehicles?

Mixed Practice, Exercises 31–40

Solve by any appropriate method.

31. $$\begin{aligned} x - y &= 1 \\ 5x + y &= 7 \end{aligned}$$

32. $$\begin{aligned} 2x + 5y &= 4 \\ 5x - 7y &= -29 \end{aligned}$$

33. $$\begin{aligned} \frac{x}{2} - 3y &= -6 \\ \frac{4}{3}x + 2y &= 4 \end{aligned}$$

34. $$\begin{aligned} \frac{x}{2} - y &= -12 \\ x + \frac{3}{4}y &= 9 \end{aligned}$$

35. $\begin{aligned} 7(x+3) &= 2y+25 \\ 3(x-6) &= -2(y+1) \end{aligned}$

36. $\begin{aligned} 0.3x - 0.4y &= 0.9 \\ 0.2x - 0.3y &= 0.4 \end{aligned}$

37. $\begin{aligned} 1.2x - y &= 1.6 \\ x + 1.5y &= 6 \end{aligned}$

38. $\begin{aligned} x - \frac{y}{2} + \frac{1}{2}z &= -1 \\ 2x \phantom{- \frac{y}{2}} + \frac{5}{2}z &= -1 \\ \frac{3}{2}y + 2z &= 1 \end{aligned}$

39. $\begin{aligned} x - 4y + 4z &= -1 \\ 2x - y + 5z &= -3 \\ x - 3y + z &= 4 \end{aligned}$

40. $\begin{aligned} x - 2y + z &= -5 \\ 2x + z &= -10 \\ y - z &= 15 \end{aligned}$

Graph the solution of each of the following systems of linear inequalities.

41. $\begin{aligned} y &\geq -\frac{1}{2}x - 1 \\ -x + y &\leq 5 \end{aligned}$

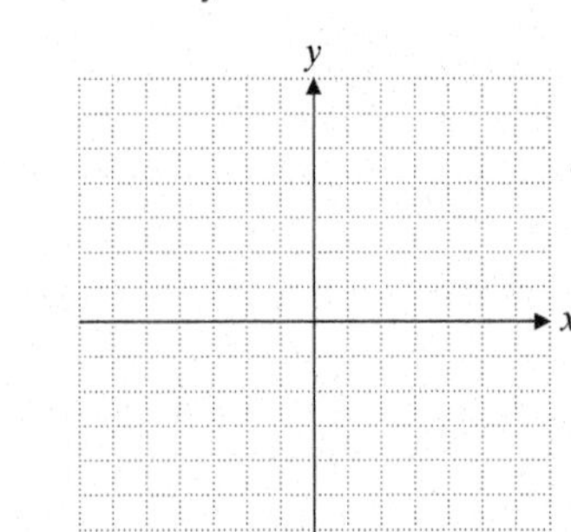

42. $\begin{aligned} -2x + 3y &< 6 \\ y &> -2 \end{aligned}$

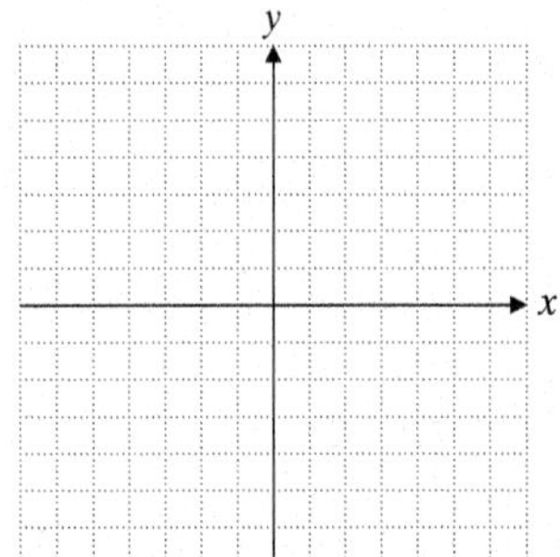

43. $\begin{aligned} x + y &> 1 \\ 2x - y &< 5 \end{aligned}$

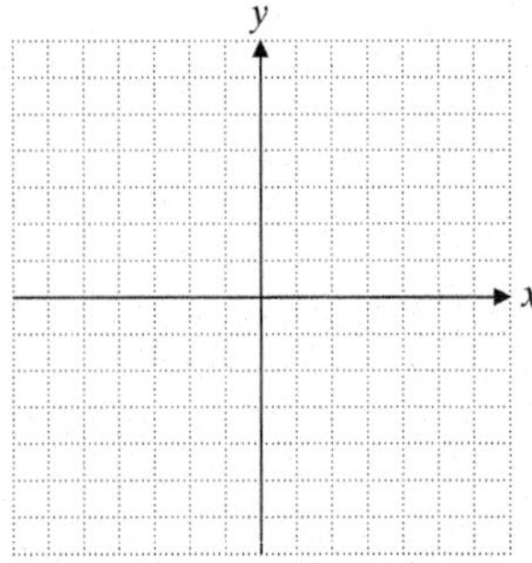

44. $\begin{aligned} x + y &\geq 4 \\ y &\leq x \\ x &\leq 6 \end{aligned}$

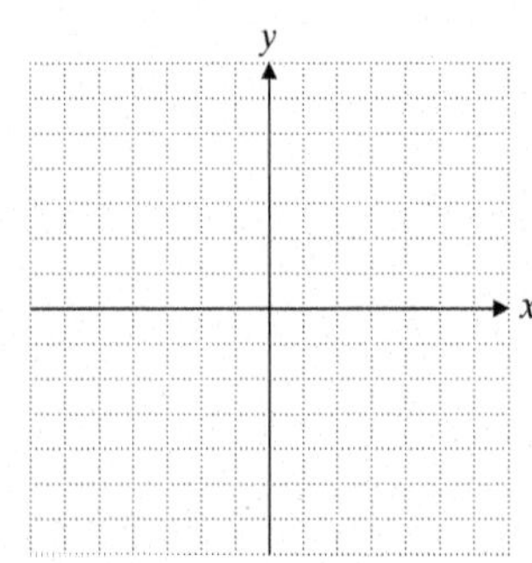

How Am I Doing? Chapter 4 Test

MATH COACH   MyMathLab®   YouTube

After you take this test read through the Math Coach on pages 322–323. Math Coach videos are available via MyMathLab and YouTube. Step-by-step test solutions in the Chapter Test Prep Videos are also available via MyMathLab and YouTube. (Search "TobeyBegInterAlg" and click on "Channels.")

Solve each system of equations. If there is no solution to the system, give a reason.

1. Solve using the substitution method.

$$\begin{aligned} x - y &= 3 \\ 2x - 3y &= -1 \end{aligned}$$

2. Solve using the addition method.

$$\begin{aligned} 3x + 2y &= 1 \\ 5x + 3y &= 3 \end{aligned}$$

In exercises 3–9, solve using any method.

3. $$\begin{aligned} 5x - 3y &= 3 \\ 7x + y &= 25 \end{aligned}$$

4. $$\begin{aligned} \frac{1}{4}a - \frac{3}{4}b &= -1 \\ \frac{1}{3}a + b &= \frac{5}{3} \end{aligned}$$

MC **5.** $$\begin{aligned} \frac{1}{3}x + \frac{5}{6}y &= 2 \\ \frac{3}{5}x - y &= -\frac{7}{5} \end{aligned}$$

6. $$\begin{aligned} 8x - 3y &= 5 \\ -16x + 6y &= 8 \end{aligned}$$

MC **7.** $$\begin{aligned} 3x + 5y - 2z &= -5 \\ 2x + 3y - z &= -2 \\ 2x + 4y + 6z &= 18 \end{aligned}$$

8. $$\begin{aligned} 3x + 2y &= 0 \\ 2x - y + 3z &= 8 \\ 5x + 3y + z &= 4 \end{aligned}$$

9. $$\begin{aligned} x + 5y + 4z &= -3 \\ x - y - 2z &= -3 \\ x + 2y + 3z &= -5 \end{aligned}$$

Use a system of linear equations to solve the following exercises.

10. A plane flew 1000 miles with a tailwind in 2 hours. The return trip against the wind took $2\frac{1}{2}$ hours. Find the speed of the wind and the speed of the plane in still air.

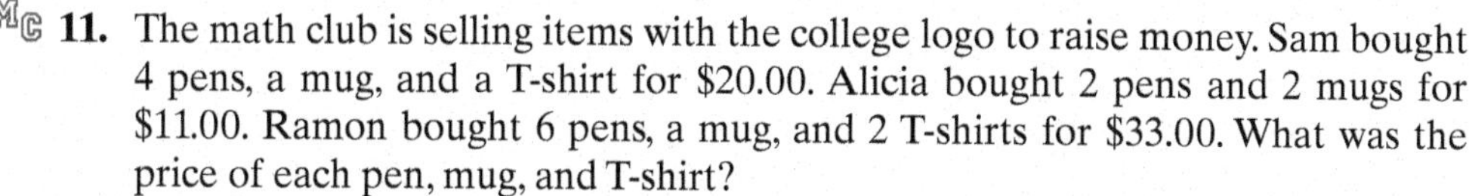

MC **11.** The math club is selling items with the college logo to raise money. Sam bought 4 pens, a mug, and a T-shirt for \$20.00. Alicia bought 2 pens and 2 mugs for \$11.00. Ramon bought 6 pens, a mug, and 2 T-shirts for \$33.00. What was the price of each pen, mug, and T-shirt?

12. Sue Miller had to move some supplies to Camp Cedarbrook for the summer camp program. She rented a Portland Rent-A-Truck in April for 5 days and drove 150 miles. She paid \$180 for the rental in April. Then in May she rented the same truck again for 7 days and drove 320 miles. She paid \$274 for the rental in May. How much does Portland Rent-A-Truck charge for a daily rental of the truck? How much does it charge per mile?

Solve the following systems of linear inequalities by graphing.

13. $$\begin{aligned} x + 2y &\leq 6 \\ -2x + y &\geq -2 \end{aligned}$$

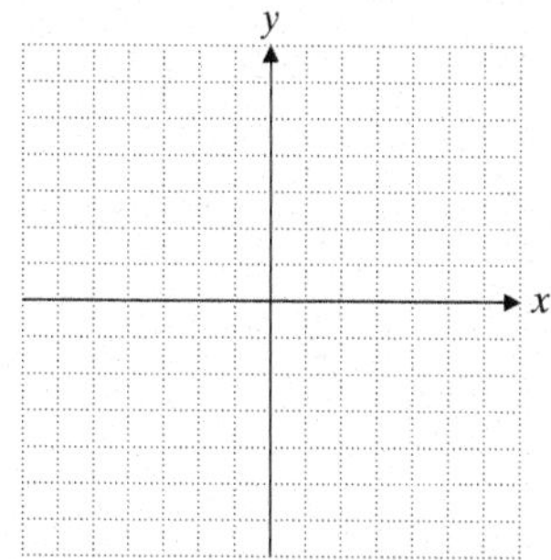

MC **14.** $$\begin{aligned} 3x + y &> 8 \\ x - 2y &> 5 \end{aligned}$$

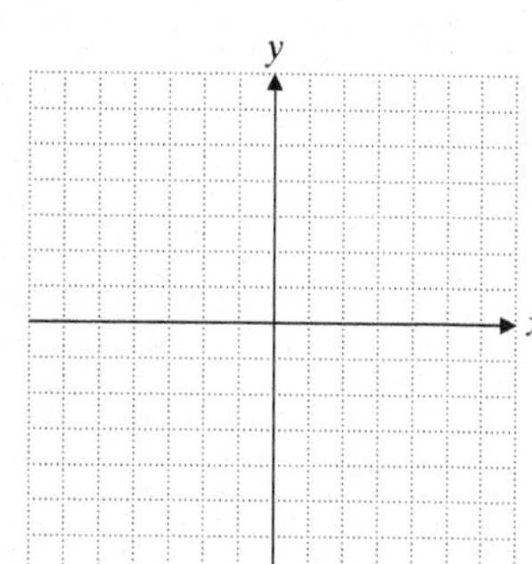

1. ______
2. ______
3. ______
4. ______
5. ______
6. ______
7. ______
8. ______
9. ______
10. ______
11. ______
12. ______
13. ______
14. ______

Total Correct: ______

MATH COACH

Mastering the skills you need to do well on the test.

The following problems are from the Chapter 4 Test. Here are some helpful hints to keep you from making common errors on test problems.

Chapter 4 Test, Problem 5

Solve the system.
$$\frac{1}{3}x + \frac{5}{6}y = 2$$
$$\frac{3}{5}x - y = -\frac{7}{5}$$

HELPFUL HINT First clear fractions by multiplying the first equation by its LCD and multiplying the second equation by its LCD. Then choose the appropriate method to solve this system.

Did you multiply the first equation by the LCD, 6? Did you obtain the equation $2x + 5y = 12$?

Yes ______ No ______

If you answered No, consider why 6 is the LCD and perform the multiplication again. Remember to multiply each term of the equation by the LCD.

Did you multiply the second equation by 5? Did you obtain the equation $3x - 5y = -7$?

Yes ______ No ______

If you answered No, go back and perform those multiplications again. Be careful as you carry out the calculations.

Did you use the addition (elimination) method to complete the solution?

Yes ______ No ______

If you answered No, consider why this method might make the solution easiest to find.

If you answered Problem 5 incorrectly, go back and rework the problem using these suggestions.

Chapter 4 Test, Problem 7

Solve the system.

(1) $3x + 5y - 2z = -5$
(2) $2x + 3y - z = -2$
(3) $2x + 4y + 6z = 18$

HELPFUL HINT Try to eliminate one of the variables from the original first and second equations. Then eliminate this same variable from the original second and third equations. The result should be a system of two equations in two variables.

Did you choose the variable z to eliminate?

Yes ______ No ______

If you answered No, consider why this might be the easiest variable to eliminate out of the three.

Did you multiply equation (2) by -2 and add the result to equation (1) to obtain the equation $-x - y = -1$?

Yes ______ No ______

If you answered No, stop and perform those operations using only original equations (2) and (1).

Next did you multiply the original equation (2) by 6 and add the result to equation (3) to obtain the equation $14x + 22y = 6$?

Yes ______ No ______

If you answered No, stop and perform those operations using only original equations (2) and (3).

Make sure that after performing these steps, your result is a system of two linear equations in two variables. Once you solve this system in two variables, remember to go back to one of the original equations and substitute in the resulting values for x and y to find the value of z.

Now go back and rework the problem using these suggestions.

Need more help? Watch the MATH COACH videos in MyMathLab® or on YouTube™.

Chapter 4 Test, Problem 11 The math club is selling items with the college logo to raise money. Sam bought 4 pens, a mug, and a T-shirt for $20.00. Alicia bought 2 pens and 2 mugs for $11.00. Ramon bought 6 pens, a mug, and 2 T-shirts for $33.00. What was the price of each pen, mug, and T-shirt?

HELPFUL HINT Read through the word problem carefully. Specify what each variable represents. Then construct appropriate equations for Sam, Alicia, and Ramon. Solve the resulting system of three linear equations in three variables.

Did you realize that this problem requires a system of three equations in three variables?

Yes ______ No ______

Did you let x = the price of pens, y = the price of mugs, and z = the price of T shirts?

Yes ______ No ______

If you answered No to either question, go back and read the problem carefully again. Consider that you have three items with unknown prices. Consider that information is provided for three people—Sam, Alicia, and Ramon.

Did you write Sam's equation as $4x + y + z = 20$?

Yes ______ No ______

If you answered No, reread the second sentence in the word problem. Apply the information using the variables listed. Now translate this information into a linear equation in three variables.

Did you write Alicia's equation as $2x + 2y = 11$?

Yes ______ No ______

If you answered No, reread the third sentence in the word problem. Notice that we do not use the variable z because Alicia did not buy any T-shirts.

Now see if you can write a third equation for Ramon that totals to $33. Then solve the resulting system.

If you answered Problem 11 incorrectly, go back and rework the problem using these suggestions.

Chapter 4 Test, Problem 14 Solve the system of linear inequalities by graphing.

$$3x + y > 8$$
$$x - 2y > 5$$

HELPFUL HINT Remember to use dashed lines when the inequality symbols are $>$ or $<$. Take one inequality at a time and graph the border line. Then shade above or below each border line based on test points. The intersection of the two shaded regions is the solution to the system.

First graph the border line $3x + y = 8$. Did you see that this line passes through $(3, -1)$ and $(1, 5)$? Did you see that the line should be dashed?

Yes ______ No ______

If you answered No to any of these questions, stop and examine the first inequality again. Notice the inequality symbol used. Then carefully substitute $x = 3$ into the equation and solve for y. Substitute $x = 1$ into the equation and solve for y. Now you have two ordered pairs to use when graphing the border line.

Now you must decide to shade above or below the first border line. Did you substitute $(0, 0)$ into $3x + y > 8$ and decide to shade above the line?

Yes ______ No ______

If you answered No, remember that $3(0) + 0$ is not greater than 8. So we do not shade on the side of the line that contains $(0, 0)$. We must shade above the line.

Follow this procedure for the second border line. Remember that your solution is the intersection of the two shaded regions.

Now go back and rework the problem using these suggestions.

CHAPTER 5

Exponents and Polynomials

CAREER OPPORTUNITIES

Biotechnology Professionals

The biotechnology industry features a wide range of careers. Using the most recent advances made in biology and technology, such jobs deal with important issues in the areas of pharmaceuticals, farming, food production, genetics, disease prevention, and the environment, to name just a few. The mathematics seen in this chapter is used by biotechnology professionals in the research, development, and problem solving that lead to better, safer products and services for the entire world.

To investigate how the mathematics in this chapter can help with this field, see the Career Exploration Problems on page **368**.

5.1 The Rules of Exponents

1 Using the Product Rule to Multiply Exponential Expressions with Like Bases

Student Learning Objectives

After studying this section, you will be able to:

1 Use the product rule to multiply exponential expressions with like bases.

2 Use the quotient rule to divide exponential expressions with like bases.

3 Raise exponential expressions to a power.

Recall that x^2 means $x \cdot x$. That is, x appears as a factor two times. The 2 is called the **exponent.** The **base** is the variable x. The expression x^2 is called an **exponential expression.** What happens when we multiply $x^2 \cdot x^2$? Is there a pattern that will help us form a general rule?

$$(2^2)(2^3) = \overbrace{(2 \cdot 2)(2 \cdot 2 \cdot 2)}^{5\text{ twos}} = 2^5$$ The exponent means 2 occurs 5 times as a factor.

$$(3^3)(3^4) = \overbrace{(3 \cdot 3 \cdot 3)(3 \cdot 3 \cdot 3 \cdot 3)}^{7\text{ threes}} = 3^7$$ Notice that $3 + 4 = 7$.

$$(x^3)(x^5) = \overbrace{(x \cdot x \cdot x)(x \cdot x \cdot x \cdot x \cdot x)}^{8\ x\text{'s}} = x^8$$ The sum of the exponents is $3 + 5 = 8$.

$$(y^4)(y^2) = \overbrace{(y \cdot y \cdot y \cdot y)(y \cdot y)}^{6\ y\text{'s}} = y^6$$ The sum of the exponents is $4 + 2 = 6$.

We can state the pattern in words and then use variables.

THE PRODUCT RULE

To multiply two exponential expressions that have the same base, keep the base and *add the exponents.*

$$x^a \cdot x^b = x^{a+b}$$

Be sure to notice that this rule applies only to expressions that have the *same base.* Here x represents the base, while the letters a and b represent the exponents that are added.

It is important that you apply this rule even when an exponent is 1. Every variable that does not have a written exponent is understood to have an exponent of 1. Thus $x = x^1$, $y = y^1$, and so on.

Example 1 Multiply. **(a)** $x^3 \cdot x^6$ **(b)** $m \cdot m^5$

Solution

(a) $x^3 \cdot x^6 = x^{3+6} = x^9$

(b) $m \cdot m^5 = m^{1+5} = m^6$ Note that the exponent of the first m is 1. □

 **Student Practice 1** Multiply. **(a)** $a^7 \cdot a^5$ **(b)** $w^{10} \cdot w$

Example 2 Simplify, if possible. **(a)** $y^5 \cdot y^{11}$ **(b)** $2^3 \cdot 2^5$ **(c)** $x^6 \cdot y^8$

Solution

(a) $y^5 \cdot y^{11} = y^{5+11} = y^{16}$

(b) $2^3 \cdot 2^5 = 2^{3+5} = 2^8$ Note that the base does not change! Only the exponent changes.

(c) $x^6 \cdot y^8$ The rule for multiplying exponential expressions does not apply since the bases are not the same. This cannot be simplified. □

Student Practice 2 Simplify, if possible.

(a) $x^3 \cdot x^9$ **(b)** $3^7 \cdot 3^4$ **(c)** $a^3 \cdot b^2$

We can now look at multiplying expressions such as $(2x^5)(3x^6)$.

The number 2 in $2x^5$ is called the **numerical coefficient.** Recall that a numerical coefficient is a number that is multiplied by a variable. When we multiply two expressions such as $2x^5$ and $3x^6$, we first multiply the numerical coefficients; we multiply the variables with exponents separately.

As you do the following problems, keep in mind the rule for multiplying expressions with exponents and the rules for multiplying signed numbers.

Example 3 Multiply.

(a) $(2x^5)(3x^6)$ **(b)** $(-5x^3)(x^6)$ **(c)** $(-6x)(-4x^5)$

Solution

(a)
$$\begin{aligned}(2x^5)(3x^6) &= (2\cdot 3)(x^5\cdot x^6)\\ &= 6(x^5\cdot x^6) \\ &= 6x^{11}\end{aligned}$$

Multiply the numerical coefficients.

Use the rule for multiplying expressions with exponents. Add the exponents.

(b) Every variable that does not have a visible numerical coefficient is understood to have a numerical coefficient of 1. Thus x^6 has a numerical coefficient of 1.

$$(-5x^3)(x^6) = (-5x^3)(1x^6) = (-5\cdot 1)(x^3\cdot x^6) = -5x^9$$

(c) $(-6x)(-4x^5) = (-6)(-4)(x^1\cdot x^5) = 24x^6$ Remember that x has an exponent of 1. □

Student Practice 3 Multiply.

(a) $(7a^8)(a^4)$ **(b)** $(3y^2)(-2y^3)$ **(c)** $(-4x^3)(-5x^2)$

Problems of this type may involve more than one variable or more than two factors.

Example 4 Multiply. $(5ab)\left(-\frac{1}{3}a\right)(9b^2)$

Solution
$$\begin{aligned}(5ab)\left(-\tfrac{1}{3}a\right)(9b^2) &= (5)\left(-\tfrac{1}{3}\right)(9)(a\cdot a)(b\cdot b^2)\\ &= -15a^2b^3\end{aligned}$$
□

Student Practice 4 Multiply.

$$(2xy)\left(-\tfrac{1}{4}x^2y\right)(6xy^3)$$

2 Using the Quotient Rule to Divide Exponential Expressions with Like Bases

Frequently, we must divide exponential expressions. Because division by zero is undefined, in all problems in this chapter we assume that the denominator of any variable expression is not zero. We'll look at division in three separate parts.

Suppose that we want to simplify $x^5 \div x^2$. We could do the division the long way.

$$\frac{x^5}{x^2} = \frac{(x)(x)(x)\cancel{(x)}\cancel{(x)}}{\cancel{(x)}\cancel{(x)}} = x^3$$

Here we are using the arithmetical property of reducing fractions (see Section 0.1). When the same factor appears in both numerator and denominator, that factor can be removed.

A simpler way is to *subtract the exponents*. Notice that the base remains the same.

THE QUOTIENT RULE (PART 1)

$\frac{x^a}{x^b} = x^{a-b}$ Use this form if the larger exponent is in the numerator and $x \neq 0$.

Example 5 Divide.

(a) $\frac{2^{16}}{2^{11}}$ **(b)** $\frac{x^5}{x^3}$ **(c)** $\frac{y^{16}}{y^7}$

Solution

(a) $\frac{2^{16}}{2^{11}} = 2^{16-11} = 2^5$ Note that the base does *not* change.

(b) $\frac{x^5}{x^3} = x^{5-3} = x^2$ **(c)** $\frac{y^{16}}{y^7} = y^{16-7} = y^9$ □

Student Practice 5 Divide.

(a) $\frac{10^{13}}{10^7}$ **(b)** $\frac{x^{11}}{x}$ **(c)** $\frac{y^{18}}{y^8}$

Now we consider the situation where the larger exponent is in the denominator. Suppose that we want to simplify $x^2 \div x^5$.

$$\frac{x^2}{x^5} = \frac{\cancel{(x)}\cancel{(x)}}{\cancel{(x)}\cancel{(x)}(x)(x)(x)} = \frac{1}{x^3}$$

THE QUOTIENT RULE (PART 2)

$\frac{x^a}{x^b} = \frac{1}{x^{b-a}}$ Use this form if the larger exponent is in the denominator and $x \neq 0$.

Example 6 Divide.

(a) $\frac{12^{17}}{12^{20}}$ **(b)** $\frac{b^7}{b^9}$ **(c)** $\frac{x^{20}}{x^{24}}$

Solution

(a) $\frac{12^{17}}{12^{20}} = \frac{1}{12^{20-17}} = \frac{1}{12^3}$ Note that the base does *not* change.

(b) $\frac{b^7}{b^9} = \frac{1}{b^{9-7}} = \frac{1}{b^2}$ **(c)** $\frac{x^{20}}{x^{24}} = \frac{1}{x^{24-20}} = \frac{1}{x^4}$ □

Student Practice 6 Divide.

(a) $\frac{c^3}{c^4}$ **(b)** $\frac{10^{31}}{10^{56}}$ **(c)** $\frac{z^{15}}{z^{21}}$

When there are numerical coefficients, use the rules for dividing signed numbers to reduce fractions to lowest terms.

Example 7 Divide.

(a) $\frac{5x^5}{25x^7}$ **(b)** $\frac{-12y^8}{4y^3}$ **(c)** $\frac{-16x^7}{-24x^8}$

Solution

(a) $\frac{5x^5}{25x^7} = \frac{1}{5x^{7-5}} = \frac{1}{5x^2}$ **(b)** $\frac{-12y^8}{4y^3} = -3y^{8-3} = -3y^5$

(c) $\frac{-16x^7}{-24x^8} = \frac{2}{3x^{8-7}} = \frac{2}{3x}$ □

Student Practice 7 Divide.

(a) $\frac{-7x^7}{-21x^9}$ **(b)** $\frac{15x^{11}}{-3x^4}$ **(c)** $\frac{23b^8}{46b^9}$

You have to work very carefully if two or more variables are involved. Treat the coefficients and each variable separately.

Example 8 Divide.

(a) $\frac{a^3b^2}{5ab^6}$ **(b)** $\frac{-3x^2y^5}{12x^6y^8}$

Solution

(a) $\frac{a^3b^2}{5ab^6} = \frac{a^2}{5b^4}$ **(b)** $\frac{-3x^2y^5}{12x^6y^8} = -\frac{1}{4x^4y^3}$ □

Student Practice 8 Divide.

(a) $\frac{r^7s^9}{s^{10}}$ **(b)** $\frac{12x^5y^6}{-24x^3y^8}$

Suppose that a given base appears with the same exponent in the numerator and denominator of a fraction. In this case we can use the fact that *any nonzero number divided by itself is* 1.

Example 9 Divide.

(a) $\frac{7^6}{7^6}$ **(b)** $\frac{3x^5}{x^5}$

Solution

(a) $\frac{7^6}{7^6} = 1$ **(b)** $\frac{3x^5}{x^5} = 3\left(\frac{x^5}{x^5}\right) = 3(1) = 3$ □

Student Practice 9 Divide.

(a) $\frac{10^7}{10^7}$ **(b)** $\frac{12a^4}{2a^4}$

Do you see that if we had subtracted exponents when simplifying $\frac{x^6}{x^6}$ we would have obtained x^0 in Example 9? So we can surmise that any number (except 0) to the 0 power equals 1. We can write this fact as a separate rule.

THE QUOTIENT RULE (PART 3)

$$\frac{x^a}{x^a} = x^0 = 1 \quad \text{if } x \neq 0 \quad (0^0 \text{ remains undefined}).$$

To Think About: *What Is 0 to the 0 Power?* What about 0^0? Why is it undefined? $0^0 = 0^{1-1}$. If we use the quotient rule, $0^{1-1} = \frac{0}{0}$. Since division by zero is undefined, we must agree that 0^0 is undefined.

Example 10 Divide.

(a) $\dfrac{4x^0y^2}{8^0y^5z^3}$ (b) $\dfrac{5b^2c}{10b^2c^3}$

Solution

(a) $\dfrac{4x^0y^2}{8^0y^5z^3} = \dfrac{4(1)y^2}{(1)y^5z^3} = \dfrac{4y^2}{y^5z^3} = \dfrac{4}{y^3z^3}$ (b) $\dfrac{5b^2c}{10b^2c^3} = \dfrac{1b^0}{2c^2} = \dfrac{(1)(1)}{2c^2} = \dfrac{1}{2c^2}$ □

Student Practice 10 Divide.

(a) $\dfrac{-20a^3b^8c^4}{28a^3b^7c^5}$ (b) $\dfrac{5x^0y^6}{10x^4y^8}$

We can combine all three parts of the quotient rule we have developed.

THE QUOTIENT RULE

$\dfrac{x^a}{x^b} = x^{a-b}$ Use this form if the larger exponent is in the numerator and $x \neq 0$.

$\dfrac{x^a}{x^b} = \dfrac{1}{x^{b-a}}$ Use this form if the larger exponent is in the denominator and $x \neq 0$.

$\dfrac{x^a}{x^a} = x^0 = 1$ if $x \neq 0$.

We can combine the product rule and the quotient rule to simplify algebraic expressions that involve both multiplication and division.

Example 11 Simplify. $\dfrac{(8x^2y)(-3x^3y^2)}{-6x^4y^3}$

Solution

$$\frac{(8x^2y)(-3x^3y^2)}{-6x^4y^3} = \frac{-24x^5y^3}{-6x^4y^3} = 4x$$ □

Student Practice 11 Simplify.

$$\frac{(-6ab^5)(3a^2b^4)}{16a^5b^7}$$

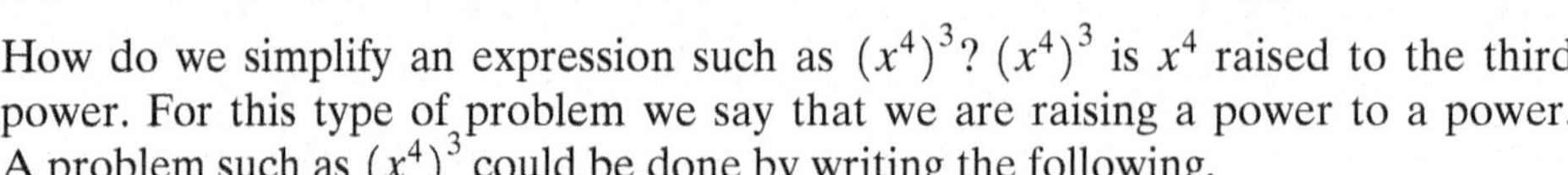

3 Raising Exponential Expressions to a Power

How do we simplify an expression such as $(x^4)^3$? $(x^4)^3$ is x^4 raised to the third power. For this type of problem we say that we are raising a power to a power. A problem such as $(x^4)^3$ could be done by writing the following.

$$(x^4)^3 = x^4 \cdot x^4 \cdot x^4 \quad \text{By definition}$$
$$= x^{12} \quad \text{By adding exponents}$$

Notice that when we add the exponents we get $4 + 4 + 4 = 12$. This is the same as multiplying 4 by 3. That is, $4 \cdot 3 = 12$. This process can be summarized by the following rule.

RAISING A POWER TO A POWER

To raise a power to a power, keep the same base and multiply the exponents.

$$(x^a)^b = x^{ab}$$

Recall what happens when you raise a negative number to a power. $(-1)^2 = 1$. $(-1)^3 = -1$. In general,

$$(-1)^n = \begin{cases} +1 & \text{if } n \text{ is even} \\ -1 & \text{if } n \text{ is odd.} \end{cases}$$

Example 12 Simplify.

(a) $(x^3)^5$ **(b)** $(2^7)^3$ **(c)** $(-1)^8$

Solution

(a) $(x^3)^5 = x^{3 \cdot 5} = x^{15}$ **(b)** $(2^7)^3 = 2^{7 \cdot 3} = 2^{21}$ **(c)** $(-1)^8 = +1$

Note that in both parts **(a)** and **(b)** the base does not change. □

Student Practice 12 Simplify.

(a) $(a^4)^3$ **(b)** $(10^5)^2$ **(c)** $(-1)^{15}$

Here are two rules involving products and quotients that are very useful: the product raised to a power rule and the quotient raised to a power rule. We'll illustrate each with an example.

If a product in parentheses is raised to a power, the parentheses indicate that *each factor* must be raised to that power.

$$(xy)^2 = x^2y^2 \qquad (xy)^3 = x^3y^3$$

RAISING A PRODUCT TO A POWER

$$(xy)^a = x^ay^a$$

Example 13 Simplify.

(a) $(ab)^8$ **(b)** $(3x)^4$ **(c)** $(-2x^2)^3$

Solution

(a) $(ab)^8 = a^8b^8$ **(b)** $(3x)^4 = 3^4x^4 = 81x^4$

(c) $(-2x^2)^3 = (-2)^3 \cdot (x^2)^3 = -8x^6$ □

Student Practice 13 Simplify.

(a) $(3xy)^3$ **(b)** $(yz)^{37}$ **(c)** $(-3x^3)^2$

If a fractional expression within parentheses is raised to a power, the parentheses indicate that both numerator and denominator must be raised to that power.

$$\left(\frac{x}{y}\right)^5 = \frac{x^5}{y^5} \qquad \left(\frac{x}{y}\right)^2 = \frac{x^2}{y^2} \qquad \text{if } y \neq 0$$

RAISING A QUOTIENT TO A POWER

$$\left(\frac{x}{y}\right)^a = \frac{x^a}{y^a} \qquad \text{if } y \neq 0$$

Example 14 Simplify.

(a) $\left(\frac{x}{y}\right)^5$ **(b)** $\frac{(7w^2)^3}{(2w)^4}$

Solution

(a) $\left(\frac{x}{y}\right)^5 = \frac{x^5}{y^5}$ **(b)** $\frac{(7w^2)^3}{(2w)^4} = \frac{7^3w^6}{2^4w^4} = \frac{343w^6}{16w^4} = \frac{343w^2}{16}$ □

Student Practice 14 Simplify.

(a) $\left(\frac{x}{5}\right)^3$ **(b)** $\frac{(4a)^2}{(ab)^6}$

Many expressions can be simplified by using the previous rules involving exponents. Be sure to take particular care to determine the correct sign, especially if there is a negative numerical coefficient.

Example 15 Simplify. $\left(\frac{-3x^2z^0}{y^3}\right)^4$

Solution

$\left(\frac{-3x^2z^0}{y^3}\right)^4 = \left(\frac{-3x^2}{y^3}\right)^4$ Simplify inside the parentheses first. Note that $z^0 = 1$.

$= \frac{(-3)^4x^8}{y^{12}}$ Apply the rules for raising a power to a power. Notice that we wrote $(-3)^4$ and not -3^4. We are raising -3 to the fourth power.

$= \frac{81x^8}{y^{12}}$ Simplify the coefficient: $(-3)^4 = +81$. □

Student Practice 15 Simplify.

$$\left(\frac{-2x^3y^0z}{4xz^2}\right)^5$$

We list here the rules of exponents we have discussed in Section 5.1.

$$x^a \cdot x^b = x^{a+b}$$

$$\frac{x^a}{x^b} = \begin{cases} x^{a-b} & \text{if} \quad a > b \quad x \neq 0 \\ \dfrac{1}{x^{b-a}} & \text{if} \quad b > a \quad x \neq 0 \\ x^0 = 1 & \text{if} \quad a = b \quad x \neq 0 \end{cases}$$

$$(x^a)^b = x^{ab}$$

$$(xy)^a = x^a y^a$$

$$\left(\frac{x}{y}\right)^a = \frac{x^a}{y^a} \qquad y \neq 0$$

STEPS TO SUCCESS Why Does Reviewing Make Such a Big Difference?

Students are often amazed that reviewing makes such a huge difference in helping them to learn. It is one of the most powerful tools that a math student can use.

Mathematics involves learning concepts one step at a time. Then the concepts are put together in a chapter. At the end of the chapter you need to know each of these concepts. Therefore, to succeed in each chapter you need to be able to put together all the pieces of each chapter. Reviewing each section and reviewing at the end of each chapter are the amazing tools that help you to master the mathematical concepts.

As you review, if you find you cannot work out a problem, be sure to study the examples and the Student Practice problems very carefully. Then things will become more clear.

Making it personal: Start with this section. Do each Cumulative Review Problem at the end of the chapter. Check your answer for each one in the back of the book. If you miss any problem, then go back to the appropriate section of the book for help.

For example if you miss problem 97 in Section 5.1, notice that it is coded [1.3.3]. This means to go back to Section 1.3 of the book and look at Objective 3. There you will find similar examples that explain this kind of problem. Which of these suggestions do you find most helpful? Write out your plan and begin using it today. ▼

5.1 Exercises MyMathLab®

Verbal and Writing Skills, Exercises 1–6

1. Write in your own words the product rule for exponents.

2. To be able to use the rules of exponents, what must be true of the bases?

3. If the larger exponent is in the denominator, the quotient rule states that $\frac{x^a}{x^b} = \frac{1}{x^{b-a}}$ if $x \neq 0$. Provide an example to show why this is true.

In exercises 4 and 5, identify the numerical coefficient, the base(s), and the exponent(s).

4. $-5xy^3$

5. $6x^{11}y$

6. Evaluate **(a)** $3x^0$ and **(b)** $(3x)^0$. **(c)** Why are the results different?

Write in simplest exponent form.

7. $2 \cdot 2 \cdot a \cdot a \cdot a \cdot b$

8. $6 \cdot x \cdot x \cdot x \cdot x \cdot y$

9. $(-5)(x)(y)(z)(y)(x)(x)(z)$

10. $(-4)(b)(b)(a)(b)(c)(c)(b)(c)$

Multiply. Leave your answer in exponent form.

11. $(7^4)(7^6)$

12. $(3^2)(3^3)$

13. $(8^9)(8^{12})$

14. $(3^7)(3^8)$

15. $x^4 \cdot x^8$

16. $a^8 \cdot a^{12}$

17. $t^{15} \cdot t$

18. $w^{18} \cdot w$

Multiply.

19. $-5x^4(4x^2)$

20. $6x^2(-9x^3)$

21. $(5x)(10x^2)$

22. $(-2x^3)(-5x)$

23. $(2xy^3)(9x^2y^5)$

24. $(-3a^2b)(7ab^4)$

25. $\left(\frac{2}{5}xy^3\right)\left(\frac{1}{3}x^2y^2\right)$

26. $\left(\frac{4}{5}x^5y\right)\left(\frac{15}{16}x^2y^4\right)$

27. $(1.1x^2z)(-2.5xy)$

28. $(2.3wx^4)(-3.5xy^4)$

29. $(8a)(2a^3b)(0)$

30. $(5ab)(2a^2)(0)$

31. $(-16x^2y^4)(-5xy^3)$

32. $(-12x^4y)(-7x^5y^3)$

33. $(-8x^3y^2)(3xy^5)$

34. $(9x^2y^6)(-11x^3y^3)$

35. $(-2x^3y^2)(0)(-3x^4y)$ **36.** $(-4x^8y^2)(13y^3)(0)$ **37.** $(8a^4b^3)(-3x^2y^5)$ **38.** $(-4wz^4)(-9x^2y^3)$

39. $(2x^2y)(-3y^3z^2)(5xz^4)$ **40.** $(3ab)(5a^2c)(-2b^2c^3)$

Divide. Leave your answer in exponent form. Assume that all variables in any denominator are nonzero.

41. $\dfrac{y^{12}}{y^5}$ **42.** $\dfrac{x^{11}}{x^4}$ **43.** $\dfrac{y^5}{y^8}$ **44.** $\dfrac{b^{20}}{b^{23}}$

45. $\dfrac{11^{18}}{11^{30}}$ **46.** $\dfrac{8^9}{8^{12}}$ **47.** $\dfrac{2^{17}}{2^{10}}$ **48.** $\dfrac{7^{18}}{7^9}$

49. $\dfrac{a^{13}}{4a^5}$ **50.** $\dfrac{b^{16}}{5b^{13}}$ **51.** $\dfrac{x^7}{y^9}$ **52.** $\dfrac{x^5}{y^4}$

53. $\dfrac{48x^5y^3}{24xy^3}$ **54.** $\dfrac{45a^4b^3}{15a^4b^2}$ **55.** $\dfrac{16x^5y}{-32x^2y^3}$ **56.** $\dfrac{-36x^3y^7}{72x^5y}$

57. $\dfrac{1.8f^4g^3}{54f^2g^8}$ **58.** $\dfrac{3.1s^5t^3}{62s^8t}$ **59.** $\dfrac{(-17x^5y^4)(5y^6)}{-5xy^7}$ **60.** $\dfrac{(-6xy)(10x^5y^2)}{-4x^6y}$

61. $\dfrac{8^0x^2y^3}{16x^5y}$ **62.** $\dfrac{2^3x^5y^3}{2^0x^3y^7}$ **63.** $\dfrac{18a^6b^3c^0}{24a^5b^3}$ **64.** $\dfrac{12a^7b^8}{16a^3b^8c^0}$

Simplify.

65. $(x^2)^6$ **66.** $(w^5)^8$ **67.** $(x^3y)^5$ **68.** $(ab^3)^4$

69. $(rs^2)^6$ **70.** $(m^3n^2)^5$ **71.** $(3a^3b^2c)^3$ **72.** $(2x^4yz^3)^2$

73. $(-3a^4)^2$ **74.** $(-2a^5)^4$ **75.** $\left(\dfrac{x}{2m^4}\right)^7$ **76.** $\left(\dfrac{p^5}{6x}\right)^5$

77. $\left(\frac{5x}{7y^2}\right)^2$ **78.** $\left(\frac{3b^3}{2a^4}\right)^3$ **79.** $(-3a^2b^3c^0)^4$ **80.** $(-a^3b^0c^4)^5$

81. $(-2x^3y^0z)^3$ **82.** $(-4xy^0z^4)^3$ **83.** $\frac{(3x)^5}{(3x^2)^3}$ **84.** $\frac{(4y)^4}{(4y^5)^2}$

Mixed Practice

85. $(-5a^2b^3)^2(ab)$ **86.** $(-2ab^3)^5(a^2b)$ **87.** $\left(\frac{7}{a^5}\right)^2$

88. $\left(\frac{4}{x^6}\right)^3$ **89.** $\left(\frac{2x}{y^3}\right)^4$ **90.** $\frac{18a^3}{(3a)(2b)}$

91. $\frac{(10ac^3)(7a)}{40b}$ **92.** $\frac{7x^3y^6}{35x^4y^8}$ **93.** $\frac{11x^7y^2}{33x^8y^3}$

Cumulative Review *Simplify.*

94. [1.2.1] $-3 - 8$

95. [1.1.5] $-17 + (-32) + (-24) + 27$

96. [1.3.1] $\left(-\frac{3}{5}\right)\left(-\frac{2}{15}\right)$

97. [1.3.3] $-\frac{5}{4} \div \frac{5}{16}$

Amazon Rainforest *In 2012, the size of the Amazon rainforest was 7,760,000 km^2. Over half the rainforest, 4,966,400 km^2, lies in Brazil. (Source:* rainforests.mongabay.com*) Round your answers to the nearest tenth of a percent.*

98. [0.5.3] What percent of the Amazon rainforest lay in Brazil in 2012?

99. [0.5.3] In 2013, 7500 km^2 of the Amazon rainforest in Brazil were lost. This number would have been much higher if conservation efforts were not in place to slow the deforestation rate. Through these efforts, the goal is to lose at most 5250 km^2 of the rainforest for each of the years 2014, 2015, 2016, and 2017. What percent of the Amazon rainforest in Brazil will be lost from 2013 through 2017?

Quick Quiz 5.1 *Simplify the following.*

1. $(2x^2y^3)(-5xy^4)$

2. $\frac{-28x^6y^6}{35x^3y^8}$

3. $(-3x^3y^5)^4$

4. Concept Check Explain the steps you would need to follow to simplify the expression.

$$\frac{(4x^3)^2}{(2x^4)^3}$$

5.2 Negative Exponents and Scientific Notation 

Student Learning Objectives

After studying this section, you will be able to:

1. Use negative exponents.
2. Use scientific notation.

1 Using Negative Exponents

If n is an integer and $x \neq 0$, then x^{-n} is defined as follows:

> **DEFINITION OF A NEGATIVE EXPONENT**
>
> $$x^{-n} = \frac{1}{x^n} \quad \text{where } x \neq 0$$

Example 1 Write with positive exponents.

(a) y^{-3} **(b)** z^{-6} **(c)** w^{-1}

Solution

(a) $y^{-3} = \frac{1}{y^3}$ **(b)** $z^{-6} = \frac{1}{z^6}$ **(c)** $w^{-1} = \frac{1}{w^1} = \frac{1}{w}$ □

Student Practice 1 Write with positive exponents.

(a) x^{-12} **(b)** w^{-5} **(c)** z^{-2}

To evaluate a numerical expression with a negative exponent, first write the expression with a positive exponent. Then simplify.

Example 2 Evaluate.

(a) 3^{-2} **(b)** 2^{-5}

Solution

(a) $3^{-2} = \frac{1}{3^2} = \frac{1}{9}$ **(b)** $2^{-5} = \frac{1}{2^5} = \frac{1}{32}$ □

Student Practice 2 Evaluate.

(a) 4^{-3} **(b)** 2^{-4}

All the previously studied laws of exponents are true for any integer exponent. These laws are summarized in the following box.

> **LAWS OF EXPONENTS WHERE $x, y, \neq 0$**
>
> **The Product Rule**
>
> $$x^a \cdot x^b = x^{a+b}$$
>
> **The Quotient Rule**
>
> $$\frac{x^a}{x^b} = x^{a-b} \quad \text{Use if } a > b. \qquad \frac{x^a}{x^b} = \frac{1}{x^{b-a}} \quad \text{Use if } a < b.$$
>
> **Power Rules**
>
> $$(xy)^a = x^a y^a, \qquad (x^a)^b = x^{ab}, \qquad \left(\frac{x}{y}\right)^a = \frac{x^a}{y^a}$$

By using the definition of a negative exponent and the properties of fractions, we can derive two more helpful properties of exponents.

PROPERTIES OF NEGATIVE EXPONENTS WHERE $x, y, \neq 0$

$$\frac{1}{x^{-n}} = x^n \qquad \frac{x^{-m}}{y^{-n}} = \frac{y^n}{x^m}$$

Example 3 Simplify. Write the expression with no negative exponents.

(a) $\frac{1}{x^{-6}}$ **(b)** $\frac{x^{-3}y^{-2}}{z^{-4}}$ **(c)** $x^{-2}y^3$

Solution

(a) $\frac{1}{x^{-6}} = x^6$ **(b)** $\frac{x^{-3}y^{-2}}{z^{-4}} = \frac{z^4}{x^3y^2}$ **(c)** $x^{-2}y^3 = \frac{y^3}{x^2}$ □

Student Practice 3 Simplify. Write the expression with no negative exponents.

(a) $\frac{3}{w^{-4}}$ **(b)** $\frac{x^{-6}y^4}{z^{-2}}$ **(c)** $x^{-6}y^{-5}$

Example 4 Simplify. Write the expression with no negative exponents.

(a) $(3x^{-4}y^2)^{-3}$ **(b)** $\frac{x^2y^{-4}}{x^{-5}y^3}$

Solution

(a) $(3x^{-4}y^2)^{-3} = 3^{-3}x^{12}y^{-6}$ We use the power to a power rule: $3^{1(-3)} = 3^{-3}$; $x^{-4(-3)} = x^{12}$; $y^{2(-3)} = y^{-6}$.

$= \frac{x^{12}}{3^3y^6} = \frac{x^{12}}{27y^6}$ We rewrite the expression so that only positive exponents appear, then simplify.

(b) $\frac{x^2y^{-4}}{x^{-5}y^3} = \frac{x^2x^5}{y^4y^3} = \frac{x^7}{y^7}$ First rewrite the expression so that only positive exponents appear. Then simplify using the product rule. □

Student Practice 4 Simplify. Write the expression with no negative exponents.

(a) $(2x^4y^{-5})^{-2}$ **(b)** $\frac{y^{-3}z^{-4}}{y^2z^{-6}}$

2 Using Scientific Notation

One common use of negative exponents is in writing numbers in scientific notation. Scientific notation is most useful in expressing very large and very small numbers.

SCIENTIFIC NOTATION

A positive number is written in **scientific notation** if it is in the form $a \times 10^n$, where $1 \leq a < 10$ and n is an integer.

Example 5 Write in scientific notation.

(a) 4567 **(b)** 157,000,000

Solution

(a) $4567 = 4.567 \times 1000$ To change 4567 to a number that is greater than or equal to 1 but less than 10, we move the decimal point *three* places to the *left*. We must then multiply the number by a power of 10 so that we do not change the value of the number. Use 1000.

$= 4.567 \times 10^3$

Notice we moved the decimal point 3 places to the left, so we must multiply by 10^3.

(b) $157{,}000{,}000 = \underset{\text{8 places}}{1.57000000} \times \underbrace{100000000}_{\text{8 zeros}}$

$= 1.57 \times 10^8$

Student Practice 5 Write in scientific notation.

(a) 78,200 **(b)** 4,786,000

Numbers that are smaller than 1 will have a *negative power* of 10 if they are written in scientific notation.

Example 6 Write in scientific notation.

(a) 0.061 **(b)** 0.000052

Solution

(a) We need to write 0.061 as a number that is greater than or equal to 1 but less than 10. In which direction do we move the decimal point?

$0.061 = 6.1 \times 10^{-2}$ Move the decimal point 2 places to the *right*.

(b) $0.000052 = 5.2 \times 10^{-5}$ Why?

Student Practice 6 Write in scientific notation.

(a) 0.98 **(b)** 0.000092

The reverse procedure transforms scientific notation into ordinary decimal notation.

Example 7 Write in decimal notation.

(a) 1.568×10^2 **(b)** 7.432×10^{-3}

Solution

(a) $1.568 \times 10^2 = 1.568 \times 100 = 156.8$

Alternative Method

$1.568 \times 10^2 = 156.8$ The exponent 2 tells us to move the decimal point 2 places to the right.

(b) $7.432 \times 10^{-3} = 7.432 \times \frac{1}{1000} = 0.007432$

Alternative Method

$7.432 \times 10^{-3} = 0.007432$ The exponent -3 tells us to move the decimal point 3 places to the left.

Student Practice 7 Write in decimal notation.

(a) 1.93×10^6 **(b)** 8.562×10^{-5}

Calculator

Scientific Notation

Most scientific calculators can display only eight digits at one time. Numbers with more than eight digits are usually shown in scientific notation. 1.12 E 08 or 1.12 8 means 1.12×10^8. You can use a calculator to compute with large numbers by entering the numbers using scientific notation. For example,

$(7.48 \times 10^{24}) \times (3.5 \times 10^8)$

is entered as follows.

7.48 [EXP] 24 [×]
3.5 [EXP] 8 [=]

Display: [2.618 E 33]
or [2.618 33]

Note: Some calculators have an [EE] key instead of [EXP]

The distance light travels in one year is called a *light-year*. A light-year is a convenient unit of measure to use when investigating the distances between stars.

Example 8 A light-year is a distance of 9,460,000,000,000,000 meters. Write this number in scientific notation.

Solution 9,460,000,000,000,000 meters $= 9.46 \times 10^{15}$ meters □

Student Practice 8 Astronomers measure distances to faraway galaxies in parsecs. A parsec is a distance of 30,900,000,000,000,000 meters. Write this number in scientific notation.

To perform a calculation involving very large or very small numbers, it is usually helpful to write the numbers in scientific notation and then use the laws of exponents to do the calculation.

Example 9 Use scientific notation and the laws of exponents to find the following. Leave your answer in scientific notation.

(a) $(32{,}000{,}000)(3{,}500{,}000{,}000{,}000)$ **(b)** $\dfrac{0.00063}{0.021}$

Solution

(a) $(32{,}000{,}000)(3{,}500{,}000{,}000{,}000)$

$= (3.2 \times 10^7)(3.5 \times 10^{12})$ Write each number in scientific notation.

$= 3.2 \times 3.5 \times 10^7 \times 10^{12}$ Rearrange the order. Remember that multiplication is commutative.

$= 11.2 \times 10^{19}$ Multiply 3.2×3.5. Multiply $10^7 \times 10^{12}$.

$= 1.12 \times 10^{20}$ Rewrite 11.2 in scientific notation and combine powers of 10.

(b) $\dfrac{0.00063}{0.021} = \dfrac{6.3 \times 10^{-4}}{2.1 \times 10^{-2}}$ Write each number in scientific notation.

$= \dfrac{6.3}{2.1} \times \dfrac{10^{-4}}{10^{-2}}$ Rearrange the order. We are actually using the definition of multiplication of fractions.

$= \dfrac{6.3}{2.1} \times \dfrac{10^{2}}{10^{4}}$ Rewrite with positive exponents.

$= 3.0 \times 10^{-2}$ □

Student Practice 9 Use scientific notation and the laws of exponents to find the following. Leave your answer in scientific notation.

(a) $(56{,}000)(1{,}400{,}000{,}000)$ **(b)** $\dfrac{0.000111}{0.00000037}$

When we use scientific notation, we are often writing approximate numbers. We must include some zeros so that the decimal point can be properly located. However, all other digits except for these zeros are considered **significant digits.** The number 34.56 has four significant digits. The number 0.0049 has two significant digits. The zeros are considered placeholders. The number 634,000 has three significant digits (unless we have specific knowledge to the contrary). The zeros are considered placeholders. We sometimes round numbers to a specific number of significant digits. For example, 0.08746 rounded to two significant digits is 0.087. When we round 1,348,593 to three significant digits, we obtain 1,350,000.

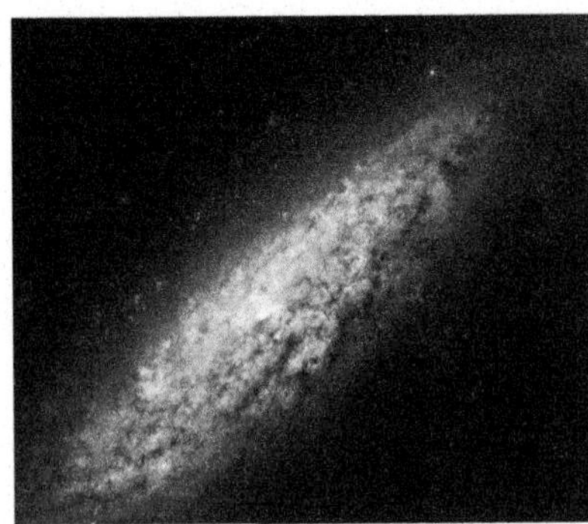

Example 10 The approximate distance from Earth to the star Polaris is 208 parsecs. A parsec is a distance of approximately 3.09×10^{13} kilometers. How long would it take a space probe traveling at 40,000 kilometers per hour to reach the star? Round to three significant digits.

Solution

1. ***Understand the problem.*** Recall that the distance formula is

$$\text{distance} = \text{rate} \times \text{time}.$$

We are given the distance and the rate. We need to find the time.
Let's take a look at the distance. The distance is given in parsecs, but the rate is given in kilometers per hour. We need to change the distance to kilometers. We are told that a parsec is approximately 3.09×10^{13} kilometers. That is, there are 3.09×10^{13} kilometers per parsec. We use this information to change 208 parsecs to kilometers.

$$208 \text{ parsecs} = \frac{(208 \text{ parsecs})(3.09 \times 10^{13} \text{ kilometers})}{1 \text{ parsec}} = 642.72 \times 10^{13} \text{ kilometers}$$

2. ***Write an equation.*** Use the distance formula.

$$d = r \times t$$

3. ***Solve the equation and state the answer.*** Substitute the known values into the formula and solve for the unknown, time.

$$642.72 \times 10^{13} \text{ km} = \frac{40{,}000 \text{ km}}{1 \text{ hr}} \times t$$

$$6.4272 \times 10^{15} \text{ km} = \frac{4 \times 10^4 \text{ km}}{1 \text{ hr}} \times t \quad \text{Change the numbers to scientific notation.}$$

Next, multiply both sides by the reciprocal of $\dfrac{4 \times 10^4 \text{ km}}{1 \text{ hr}}$.

$$6.4272 \times 10^{15} \text{ km} \times \frac{1 \text{ hr}}{4 \times 10^4 \text{ km}} = \frac{4 \times 10^4 \text{ km}}{1 \text{ hr}} \times \frac{1 \text{ hr}}{4 \times 10^4 \text{ km}} \times t$$

$$\frac{(6.4272 \times 10^{15} \cancel{\text{km}})(1 \text{ hr})}{4 \times 10^4 \cancel{\text{km}}} = t \quad \text{Simplify.}$$

$$1.6068 \times 10^{11} \text{ hr} = t$$

Rounding to three significant digits, we have

$$1.6068 \times 10^{11} \text{ hr} \approx 1.61 \times 10^{11} \text{ hr}.$$

4. ***Check.*** Unless you have had a great deal of experience working in astronomy, it would be difficult to determine whether this is a reasonable answer. You may wish to reread your analysis and redo your calculations as a check. □

Student Practice 10 The average distance from Earth to the distant star Betelgeuse is 159 parsecs. How many hours would it take a space probe to travel from Earth to Betelgeuse at a speed of 50,000 kilometers per hour? Round to three significant digits.

5.2 Exercises MyMathLab®

Simplify. Express your answer with positive exponents. Assume that all variables are nonzero.

1. x^{-4} **2.** y^{-5} **3.** 3^{-4} **4.** 2^{-4} **5.** $\frac{1}{y^{-8}}$ **6.** $\frac{1}{z^{-10}}$

7. $\frac{x^{-4}y^{-5}}{z^{-6}}$ **8.** $\frac{x^{-6}y^{-2}}{z^{-5}}$ **9.** a^3b^{-2} **10.** a^5b^{-8} **11.** $(2x^{-3})^{-3}$ **12.** $(4x^{-4})^{-2}$

13. $3x^{-2}$ **14.** $5y^{-7}$ **15.** $(3xy^2)^{-2}$ **16.** $(5x^3y)^{-3}$

Mixed Practice, Exercises 17–28

17. $\frac{3xy^{-2}}{z^{-3}}$ **18.** $\frac{4x^{-2}y^{-3}}{y^4}$ **19.** $\frac{(4xy)^{-1}}{(4xy)^{-2}}$ **20.** $\frac{(3a^2b)^{-2}}{(3a^2b)^{-3}}$

21. $a^{-1}b^3c^{-4}d$ **22.** $x^{-5}y^{-2}z^3$ **23.** $(8^{-2})(2^3)$ **24.** $(3^{-3})(9^2)$

25. $\left(\frac{3x^0y^2}{z^4}\right)^{-2}$ **26.** $\left(\frac{2a^3b^0}{c^2}\right)^{-3}$ **27.** $\frac{x^{-2}y^{-3}}{x^4y^{-2}}$ **28.** $\frac{a^3b^{-1}}{a^{-5}b^{-4}}$

Write in scientific notation.

29. 123,780 **30.** 5,786,100 **31.** 0.063 **32.** 0.0000871

33. 889,610,000,000 **34.** 7,652,000,000 **35.** 0.00000342 **36.** 0.00783

In exercises 37–42, write in decimal notation.

37. 3.02×10^5 **38.** 8.137×10^7 **39.** 4.7×10^{-4} **40.** 5.36×10^{-2} **41.** 9.83×10^5 **42.** 3.5×10^{-8}

43. ***Bamboo Growth*** The growth rate of some species of bamboo is 0.0000237 miles per hour. Write this in scientific notation.

44. ***Neptune*** Neptune is 2.793×10^9 miles from the sun. Write this in decimal notation.

45. ***Astronomical Unit*** The astronomical unit (AU) is a unit of length approximately equal to 1.496×10^8 km. Write this in decimal notation.

46. ***Gold Atom*** The average volume of an atom of gold is 0.0000000000000000000000001695 cubic centimeters. Write this in scientific notation.

Evaluate by using scientific notation and the laws of exponents. Leave your answer in scientific notation.

47. (42,000,000)(150,000,000)

48. (55,000,000,000)(16,000,000)

49. $\frac{(5{,}000{,}000)(16{,}000)}{8{,}000{,}000{,}000}$

50. (0.0075)(0.0000002)(0.001)

51. $(0.003)^4$

52. $(500{,}000)^4$

53. (150,000,000)(0.00005)(0.002)(30,000)

54. $\frac{(160{,}000)(0.0003)}{1600}$

Applications

National Debt *In May 2015, the national debt was about* 1.8×10^{13} *dollars.* (*Source:* www.factfinder.census.gov)

55. The Census Bureau estimates that in May 2015, the population of the United States was 3.21×10^8 people. If the national debt were evenly divided among every person in the country, how much debt would be assigned to each individual? Round to three significant digits.

56. The Census Bureau estimates that in May 2015, the number of people in the United States who were 18 years or older was approximately 2.82×10^8. If the national debt were evenly divided among every person 18 years or older in the country, how much would be assigned to each individual? Round to three significant digits.

57. ***Watch Hand*** The tip of a $\frac{1}{3}$-inch-long hour hand on a watch travels at a speed of 0.00000275 miles per hour. How far has it traveled in a day?

58. ***Neutron Mass*** The mass of a neutron is approximately 1.675×10^{-27} kilogram. Find the mass of 150,000 neutrons.

59. ***Mission to Pluto*** In January 2006, a spacecraft called *New Horizons* began its journey to Pluto. The trip was 3.5×10^9 miles long and took $9\frac{1}{2}$ years. How many miles did *New Horizons* travel per year? (*Source:* www.pluto.jhuapl.edu)

60. ***Molecules per Mole*** Avogadro's number says that there are approximately 6.02×10^{23} molecules/mole. How many molecules can one expect in 0.00483 mole?

61. ***Construction Costs*** In March 2015, the cost for construction of new private buildings was estimated at $\$5.46 \times 10^{10}$. In March 2005, the estimated cost for construction of new private buildings was $\$6.29 \times 10^{10}$. What was the percent decrease from March 2005 to March 2015?

62. ***Construction Costs*** In March 2015, the cost for construction of new public buildings was estimated at $\$1.81 \times 10^{11}$. In March 2005, the estimated cost for construction of new public buildings was $\$1.65 \times 10^{11}$. What was the percent increase from March 2005 to March 2015?

Cumulative Review *Simplify.*

63. **[1.2.1]** $-2.7 - (-1.9)$

64. **[1.4.2]** $(-1)^{33}$

65. **[1.1.4]** $-\frac{3}{4} + \frac{5}{7}$

Quick Quiz 5.2 *Simplify and write your answer with only positive exponents.*

1. $3x^{-3}y^2z^{-4}$

2. $\dfrac{4a^3b^{-4}}{8a^{-5}b^{-3}}$

3. Write in scientific notation. 0.00876

4. **Concept Check** Explain how you would simplify a problem like the following so that your answer had only positive exponents.

$$(4x^{-3}y^4)^{-3}$$

5.3 Fundamental Polynomial Operations

Student Learning Objectives

After studying this section, you will be able to:

1. Recognize polynomials and determine their degrees.
2. Add polynomials.
3. Subtract polynomials.
4. Evaluate polynomials to predict a value.

1 Recognizing Polynomials and Determining Their Degrees

A **polynomial** in x is the sum of a finite number of terms of the form ax^n, where a is any real number and n is a whole number. Usually these polynomials are written in descending powers of the variable, as in

$$5x^3 + 3x^2 - 2x - 5 \quad \text{and} \quad 3.2x^2 - 1.4x + 5.6.$$

A **multivariable polynomial** is a polynomial with more than one variable. The following are multivariable polynomials:

$$5xy + 8, \qquad 2x^2 - 7xy + 9y^2, \qquad 17x^3y^9$$

The **degree of a term** is the sum of the exponents of all of the variables in the term. For example, the degree of $7x^3$ is three. The degree of $4xy$ is two. The degree of $10x^4y^2$ is six.

The **degree of a polynomial** is the highest degree of all of the terms in the polynomial. For example, the degree of $5x^3 + 8x^2 - 20x - 2$ is three. The degree of $6xy - 4x^2y + 2xy^3$ is four. A polynomial consisting of a constant only is said to have degree 0.

There are special names for polynomials with one, two, or three terms.

A **monomial** has *one* term:

$$5a, \qquad 3x^3yz^4, \qquad 12xy$$

A **binomial** has *two* terms:

$$7x + 9y, \qquad -6x - 4, \qquad 5x^4 + 2xy^2$$

A **trinomial** has *three* terms:

$$8x^2 - 7x + 4, \qquad 2ab^3 - 6ab^2 - 15ab, \qquad 2 + 5y + y^4$$

Example 1 State the degree of the polynomial and whether it is a monomial, a binomial, or a trinomial.

(a) $5xy + 3x^3$ **(b)** $-7a^5b^2$ **(c)** $8x^4 - 9x - 15$

Solution

(a) This polynomial is of degree 3. It has two terms, so it is a binomial.

(b) The sum of the exponents is $5 + 2 = 7$. Therefore this polynomial is of degree 7. It has one term, so it is a monomial.

(c) This polynomial is of degree 4. It has three terms, so it is a trinomial. □

Student Practice 1 State the degree of the polynomial and whether it is a monomial, a binomial, or a trinomial.

(a) $-7x^5 - 3xy$ **(b)** $22a^3b^4$ **(c)** $-3x^3 + 3x^2 - 6x$

2 Adding Polynomials

We usually write a polynomial in x so that the exponents on x decrease from left to right. For example, the polynomial

$$5x^2 - 6x + 2$$

is said to be written in **decreasing order** since each exponent is decreasing as we move from left to right.

You can add, subtract, multiply, and divide polynomials. Let us take a look at addition. To add two polynomials, we add their like terms.

Example 2 Add. $(5x^2 - 6x - 12) + (-3x^2 - 9x + 5)$

Solution

$$\begin{aligned}(5x^2 - 6x - 12) + (-3x^2 - 9x + 5) &= [5x^2 + (-3x^2)] + [-6x + (-9x)] + [-12 + 5] \\ &= [(5 - 3)x^2] + [(-6 - 9)x] + [-12 + 5] \\ &= 2x^2 + (-15x) + (-7) \\ &= 2x^2 - 15x - 7\end{aligned}$$

Student Practice 2 Add. $(-8x^3 + 3x^2 + 6) + (2x^3 - 7x^2 - 3)$

The numerical coefficients of polynomials may be any real number. Thus polynomials may have numerical coefficients that are decimals or fractions.

Example 3 Add. $\left(\frac{1}{2}x^2 - 6x + \frac{1}{3}\right) + \left(\frac{1}{5}x^2 - 2x - \frac{1}{2}\right)$

Solution

$$\begin{aligned}\left(\tfrac{1}{2}x^2 - 6x + \tfrac{1}{3}\right) + \left(\tfrac{1}{5}x^2 - 2x - \tfrac{1}{2}\right) &= \left[\tfrac{1}{2}x^2 + \tfrac{1}{5}x^2\right] + [-6x + (-2x)] + \left[\tfrac{1}{3} + \left(-\tfrac{1}{2}\right)\right] \\ &= \left[\left(\tfrac{1}{2} + \tfrac{1}{5}\right)x^2\right] + [(-6 - 2)x] + \left[\tfrac{1}{3} + \left(-\tfrac{1}{2}\right)\right] \\ &= \left[\left(\tfrac{5}{10} + \tfrac{2}{10}\right)x^2\right] + [-8x] + \left[\tfrac{2}{6} - \tfrac{3}{6}\right] \\ &= \tfrac{7}{10}x^2 - 8x - \tfrac{1}{6}\end{aligned}$$

Student Practice 3 Add.

$$\left(-\tfrac{1}{3}x^2 - 6x - \tfrac{1}{12}\right) + \left(\tfrac{1}{4}x^2 + 5x - \tfrac{1}{3}\right)$$

Example 4 Add. $(1.2x^3 - 5.6x^2 + 5) + (-3.4x^3 - 1.2x^2 + 4.5x - 7)$

Solution Group like terms.

$$\begin{aligned}(1.2x^3 - 5.6x^2 + 5) + (-3.4x^3 - 1.2x^2 + 4.5x - 7) &= (1.2 - 3.4)x^3 + (-5.6 - 1.2)x^2 + 4.5x + (5 - 7) \\ &= -2.2x^3 - 6.8x^2 + 4.5x - 2\end{aligned}$$

Student Practice 4 Add.

$$(3.5x^3 - 0.02x^2 + 1.56x - 3.5) + (-0.08x^2 - 1.98x + 4)$$

3 Subtracting Polynomials

Recall that subtraction of real numbers can be defined as adding the opposite of the second number. Thus $a - b = a + (-b)$. That is, $3 - 5 = 3 + (-5)$. A similar method is used to subtract two polynomials.

To subtract two polynomials, change the sign of each term in the second polynomial and then add.

Example 5 Subtract. $(7x^2 - 6x + 3) - (5x^2 - 8x - 12)$

Solution We change the sign of each term in the second polynomial and then add.

$$\begin{aligned}(7x^2 - 6x + 3) - (5x^2 - 8x - 12) &= (7x^2 - 6x + 3) + (-5x^2 + 8x + 12)\\ &= (7 - 5)x^2 + (-6 + 8)x + (3 + 12)\\ &= 2x^2 + 2x + 15\end{aligned}$$

□

Student Practice 5 Subtract.

$$(5x^3 - 15x^2 + 6x - 3) - (-4x^3 - 10x^2 + 5x + 13)$$

As mentioned previously, polynomials may involve more than one variable. When subtracting polynomials in two variables, you will need to use extra care in determining which terms are like terms. For example, $6x^2y$ and $5x^2y$ are like terms. In a similar fashion, $3xy$ and $8xy$ are like terms. However, $7xy^2$ and $15x^2y^2$ are not like terms. Every exponent of every variable in the two terms must be the same if the terms are to be like terms. You will use this concept in Example 6.

Example 6 Subtract.

$$(-6x^2y - 3xy + 7xy^2) - (5x^2y - 8xy - 15x^2y^2)$$

Solution Change the sign of each term in the second polynomial and add. Look for like terms.

$$\begin{aligned}&(-6x^2y - 3xy + 7xy^2) + (-5x^2y + 8xy + 15x^2y^2)\\ &= (-6 - 5)x^2y + (-3 + 8)xy + 7xy^2 + 15x^2y^2\\ &= -11x^2y + 5xy + 7xy^2 + 15x^2y^2\end{aligned}$$

Nothing further can be done to combine these four terms. □

Student Practice 6 Subtract.

$$(x^3 - 7x^2y + 3xy^2 - 2y^3) - (2x^3 + 4xy - 6y^3)$$

4 Evaluating Polynomials to Predict a Value 

Sometimes polynomials are used to predict values. In such cases we need to **evaluate** the polynomial. We do this by substituting a known value for the variable and determining the value of the polynomial.

Example 7 Passenger cars sold in the United States have become more fuel efficient over the years due to regulations from Congress. The number of miles per gallon obtained by the average passenger car in the United States can be described by the polynomial

$$0.3x + 27.6,$$

where x is the number of years since 1985. (*Source:* www.rita.dot.gov.) Use this polynomial to estimate the number of miles per gallon obtained by the average passenger car in

(a) 1988 **(b)** 2020

Solution

(a) The year 1988 is three years later than 1985, so $x = 3$.

Thus the number of miles per gallon obtained by the average passenger car in 1988 can be estimated by evaluating $0.3x + 27.6$ when $x = 3$.

$$\begin{aligned} 0.3(3)+27.6 &= 0.9 + 27.6 \\ &= 28.5 \end{aligned}$$

We estimate that the average passenger car in 1988 obtained 28.5 miles per gallon.

(b) The year 2020 is 35 years after 1985, so $x = 35$.

Thus the estimated number of miles per gallon obtained by the average passenger car in 2020 can be predicted by evaluating $0.3x + 27.6$ when $x = 35$.

$$\begin{aligned} 0.3(35) + 27.6 &= 10.5 + 27.6 \\ &= 38.1 \end{aligned}$$

We therefore predict that the average passenger car in 2020 will obtain 38.1 miles per gallon. □

Student Practice 7 The number of miles per gallon obtained by the average light truck (less than 8500 lb) in the United States can be described by the polynomial $0.16x + 20.7$, where x is the number of years since 1985. (*Source:* www.rita.dot.gov.) Use this polynomial to estimate the number of miles per gallon obtained by the average truck in

(a) 1990 **(b)** 2025

5.3 Exercises MyMathLab®

Verbal and Writing Skills, Exercises 1–4

1. State in your own words a definition for a polynomial in x and give an example.

2. State in your own words a definition for a multivariable polynomial and give an example.

3. State in your own words how to determine the degree of a polynomial in x.

4. State in your own words how to determine the degree of a multivariable polynomial.

State the degree of the polynomial and whether it is a monomial, a binomial, or a trinomial.

5. $6x^3y$

6. $-9x^2y^3$

7. $20x^5 + 6x^3 - 7x$

8. $9x^3 - 10x^2 + 5$

9. $4x^2y^3 - 7x^3y^3$

10. $-7x^4y + 12x^5y^2$

Add.

11. $(6x - 11) + (-9x - 4)$

12. $(-2x + 19) + (5x - 6)$

13. $(6x^2 + 5x - 6) + (-8x^2 - 3x + 5)$

14. $(x^2 - 4x - 8) + (-x^2 - x + 1)$

15. $\left(\frac{1}{2}x^2 + \frac{1}{3}x - 4\right) + \left(\frac{1}{3}x^2 + \frac{1}{6}x - 5\right)$

16. $\left(\frac{1}{4}x^2 - \frac{2}{3}x - 10\right) + \left(-\frac{1}{3}x^2 + \frac{1}{9}x + 2\right)$

17. $(3.4x^3 - 7.1x + 3.4) + (2.2x^2 - 6.1x - 8.8)$

18. $(-4.6x^3 + 5.6x - 0.3) + (9.8x^2 + 4.5x - 1.7)$

Subtract.

19. $(2x - 19) - (-3x + 5)$

20. $(5x - 5) - (6x - 3)$

21. $\left(\frac{2}{5}x^2 - \frac{1}{2}x + 5\right) - \left(\frac{1}{3}x^2 - \frac{3}{7}x - 6\right)$

22. $\left(\frac{3}{8}x^2 - \frac{2}{3}x - 7\right) - \left(\frac{2}{3}x^2 - \frac{1}{2}x + 2\right)$

23. $(4x^3 + 3x) - (x^3 + x^2 - 5x)$

24. $(2x^3 - x^2 + 5) - (6x^3 + x^2 - x + 1)$

25. $(0.5x^4 - 0.7x^2 + 8.3) - (5.2x^4 + 1.6x + 7.9)$

26. $(1.3x^4 - 3.1x^3 + 6.3x) - (x^4 - 5.2x^2 + 6.5x)$

Perform the operations indicated.

27. $(8x + 2) + (x - 7) - (3x + 1)$

28. $(x - 5) - (3x + 8) - (5x - 2)$

29. $(-4x^2y^2 + 9xy - 3) + (8x^2y^2 - 5xy - 7)$

30. $(12x^2y - xy^2 + 5) + (-2x^2y + 5xy^2 - 8)$

31. $(3x^4 - 4x^2 - 18) - (2x^4 + 3x^3 + 6)$

32. $(3b^3 - 5b^2 - 7b) - (2b^3 + 3b - 5)$

Prisons *The number of prisoners held in federal and state prisons, measured in thousands, can be described by the polynomial* $-2.06x^2 + 77.82x + 743$. *The variable x represents the number of years since 1990.* (*Source:* www.ojp.usdoj.gov)

33. Estimate the number of prisoners in 1990.

34. Estimate the number of prisoners in 2005.

35. According to the polynomial, by how much did the prison population increase from 2002 to 2007?

36. According to the polynomial, by how much did the prison population decrease from 2008 to 2012?

Applications

▲ **37.** ***Geometry*** The lengths and the widths of the following three rectangles are labeled. Create a polynomial that describes the sum of the *area* of these three rectangles.

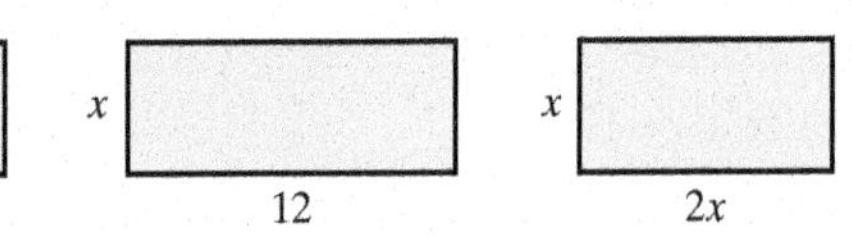

▲ **38.** ***Geometry*** The dimensions of the sides of the following figure are labeled. Create a polynomial that describes the *perimeter* of this figure.

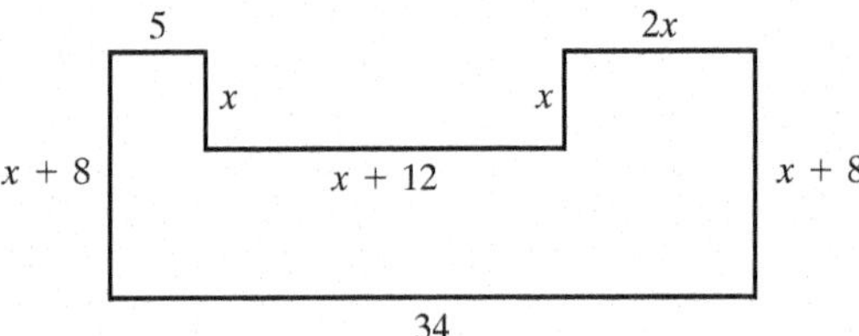

Cumulative Review

39. **[3.3.2]** Find the slope and y-intercept for the line $3y - 8x = 2$.

40. **[2.8.4]** Solve and graph. $\frac{5x}{7} - 4 > \frac{2x}{7} - 1$

−1 0 1 2 3 4 5 6 7 8 9 x

41. **[2.3.3]** Solve for x. $-2(x - 5) + 6 = 2^2 - 9 + x$

42. **[2.4.1]** Solve for x. $\frac{x}{6} + \frac{x}{2} = \frac{4}{3}$

Quick Quiz 5.3 *Combine.*

1. $(3x^2 - 5x + 8) + (-7x^2 - 6x - 3)$

2. $(2x^2 - 3x - 7) - (-4x^2 + 6x + 9)$

3. $(5x - 3) - (2x - 4) + (-6x + 7)$

4. **Concept Check** Explain how you would determine the degree of the following polynomial and how you would decide if it is a monomial, a binomial, or a trinomial.

$$2xy^2 - 5x^3y^4$$

Use Math to Save Money

Did You Know? With some simple changes in your shopping habits, you can save a lot on your grocery bill.

Saving at the Grocery Store

Understanding the Problem:

A rise in the cost of gas affects our budget with higher gas prices; it also affects our budget indirectly by causing the price of groceries to increase due to an increased cost in shipping and production. Jenny is a single mom with two children. Concerned with rising costs, she decided to track her grocery expenses for three months. In the first month she spent $450, in the second month she spent $425, and in the third month she spent $460.

Making a Plan:

By using coupons and buying store brand items, it is possible to reduce the amount of money spent at the grocery store. Jenny wants to incorporate these strategies into her shopping so she can save money.

Step 1: Jenny determines her monthly average spending on groceries so she'll have a baseline to use to calculate her savings.

Task 1: What are Jenny's average monthly grocery expenses?

Step 2: To help bring the monthly costs down, Jenny starts clipping coupons and cutting back on expensive treats. In the fourth month she saved her family 6% of their average expenses by using coupons and an additional $30 by cutting out treats.

Task 2: How much was the grocery bill the fourth month?

Step 3: Jenny is pleased with the results but knows she can save even more. The next month, she decides to choose store brand products when possible. She is happy to see that it saved her 20% of what she spent in the fourth month.

Task 3: How much was the grocery bill the fifth month?

Task 4: Compare Jenny's original average costs to her costs in the fifth month and then calculate the savings. Show it as a dollar amount and as a percentage of original costs, rounded to the nearest tenth of a percent.

Finding a Solution:

Jenny decides to make these changes permanent. These are substantial savings and will keep adding up as long as she sticks to her new shopping strategy.

Step 4: To give her family encouragement to accept the changes she's made in her shopping habits, Jenny calculates how much she will save in a year.

Task 5: If she continues to save the same amount every month, how much will she have saved at the end of one year?

Applying the Situation to Your Life:

Grocery shopping is an expense that everyone has. Incorporating these savings strategies into your own shopping will help you save money as well. Calculate how much you spend a month, and then calculate what you could save a month using the percentage of savings you solved for above. That figure should give you encouragement to implement these strategies in your own life. Even if you only use one of these methods, you will see savings. Some additional ways to save on groceries are to stock up on items when they are on sale and to buy less convenience food, instead making more meals from scratch.

5.4 Multiplying Polynomials

Student Learning Objectives

After studying this section, you will be able to:

1. Multiply a monomial by a polynomial.
2. Multiply two binomials.

1 Multiplying a Monomial by a Polynomial

We use the distributive property to multiply a monomial by a polynomial. Remember, the distributive property states that for real numbers a, b, and c,

$$a(b + c) = ab + ac.$$

Example 1 Multiply. $3x^2(5x - 2)$

Solution

$$\begin{aligned} 3x^2(5x - 2) &= 3x^2(5x) + 3x^2(-2) && \text{Use the distributive property.} \\ &= (3 \cdot 5)(x^2 \cdot x) + (3)(-2)x^2 \\ &= 15x^3 - 6x^2 \end{aligned}$$

Student Practice 1 Multiply. $4x^3(-2x^2 + 3x)$

Try to do as much of the multiplication as you can mentally.

Example 2 Multiply.

(a) $2x(x^2 + 3x - 1)$ **(b)** $-2xy^2(x^2 - 2xy - 3y^2)$

Solution

(a) $2x(x^2 + 3x - 1) = 2x^3 + 6x^2 - 2x$

(b) $-2xy^2(x^2 - 2xy - 3y^2) = -2x^3y^2 + 4x^2y^3 + 6xy^4$

Notice in part **(b)** that you are multiplying each term by the negative expression $-2xy^2$. This will change the sign of each term in the product.

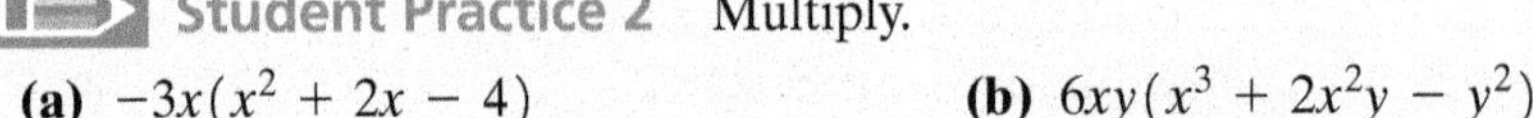

Student Practice 2 Multiply.

(a) $-3x(x^2 + 2x - 4)$ **(b)** $6xy(x^3 + 2x^2y - y^2)$

When we multiply by a monomial, the monomial may be on the right side.

Example 3 Multiply. $(x^2 - 2x + 6)(-2xy)$

Solution $(x^2 - 2x + 6)(-2xy) = -2x^3y + 4x^2y - 12xy$

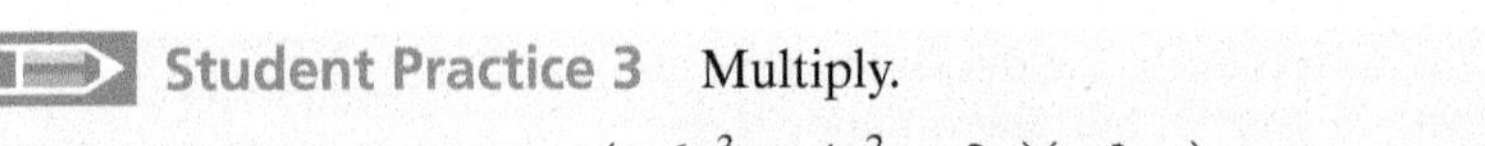

Student Practice 3 Multiply.

$$(-6x^3 + 4x^2 - 2x)(-3xy)$$

2 Multiplying Two Binomials

We can build on our knowledge of the distributive property and our experience with multiplying monomials to learn how to multiply two binomials. Let's suppose that we want to multiply $(x + 2)(3x + 1)$. We can use the distributive property. Since a can represent any quantity, let $a = x + 2$. Then let $b = 3x$ and $c = 1$. We now have the following.

$$\begin{aligned} a(b + c) &= \quad ab \quad + \quad ac \\ (x + 2)(3x + 1) &= (x + 2)(3x) + (x + 2)(1) \\ &= 3x^2 + 6x + x + 2 \\ &= 3x^2 + 7x + 2 \end{aligned}$$

Let's take another look at the original problem, $(x + 2)(3x + 1)$. This time we will assign a letter to each term in the binomials. That is, let $a = x$, $b = 2$, $c = 3x$, and $d = 1$. Using substitution, we have the following.

$$\begin{aligned}(x + 2)(3x + 1) &= (a + b)(c + d)\\ &= (a + b)c + (a + b)d\\ &= ac + bc + ad + bd\\ &= (x)(3x) + (2)(3x) + (x)(1) + (2)(1) \quad \text{By substitution}\\ &= 3x^2 + 6x + x + 2\\ &= 3x^2 + 7x + 2\end{aligned}$$

How does this compare with the preceding result?

The distributive property shows us *how* the problem can be done and *why* it can be done. In actual practice there is a memory device to help students remember the steps involved. It is often referred to as FOIL. The letters FOIL stand for the following.

F multiply the *F*irst terms
O multiply the *O*uter terms
I multiply the *I*nner terms
L multiply the *L*ast terms

The FOIL letters are simply a way to remember the four terms in the final product and how they are obtained. Let's return to our original problem.

$(x + 2)(3x + 1)$	F	Multiply the *first* terms to obtain $3x^2$.
$(x + 2)(3x + 1)$	O	Multiply the *outer* terms to obtain x.
$(x + 2)(3x + 1)$	I	Multiply the *inner* terms to obtain $6x$.
$(x + 2)(3x + 1)$	L	Multiply the *last* terms to obtain 2.

The result so far is $3x^2 + x + 6x + 2$. These four terms are the same four terms that we obtained when we multiplied using the distributive property. We can combine the like terms to obtain the final answer: $3x^2 + 7x + 2$. Now let's study the FOIL method in a few examples.

Example 4 Multiply. $(2x - 1)(3x + 2)$

Solution

First Last

$(2x - 1)(3x + 2)$

$-3x$ Inner

$+4x$ Outer

First + Outer + Inner + Last

$$\begin{aligned}&\quad F \qquad\; O \qquad\;\; I \qquad\; L\\ &= 6x^2 + 4x - 3x - 2\\ &= 6x^2 + x - 2 \quad \text{Combine like terms.}\end{aligned}$$

Notice that we combine the inner and outer terms to obtain the middle term. □

Student Practice 4 Multiply. $(5x - 1)(x - 2)$

Example 5 Multiply. $(3a - 2b)(4a - b)$

Solution

First Last

$(3a - 2b)(4a - b) \quad = 12a^2 - 3ab - 8ab + 2b^2$

Inner

Outer

$= 12a^2 - 11ab + 2b^2$ ◻

Student Practice 5 Multiply. $(8a - 5b)(3a - b)$

After you have done several problems, you may be able to combine the outer and inner products mentally.

In some problems the inner and outer products cannot be combined.

Example 6 Multiply. $(3x + 2y)(5x - 3z)$

Solution

First Last

$(3x + 2y)(5x - 3z) \quad = 15x^2 - 9xz + 10xy - 6yz$

Inner

Outer

Since there are no like terms, we cannot combine any terms. ◻

Student Practice 6 Multiply. $(3a + 2b)(2a - 3c)$

Example 7 Multiply. $(7x - 2y)^2$

Solution

$(7x - 2y)(7x - 2y)$ When we square a binomial, it is the same as multiplying the binomial by itself.

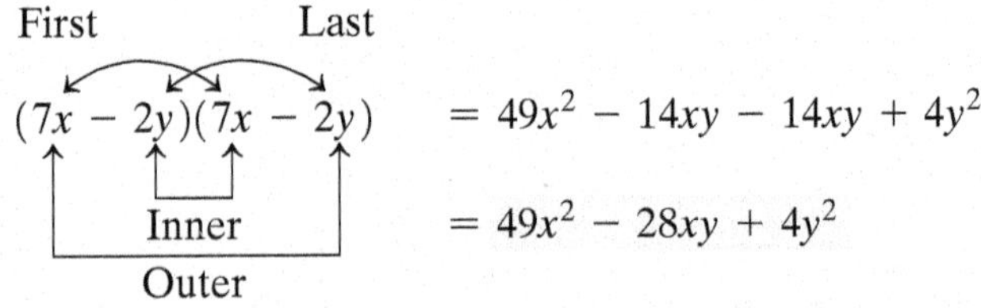

$= 49x^2 - 14xy - 14xy + 4y^2$

$= 49x^2 - 28xy + 4y^2$ ◻

Student Practice 7 Multiply. $(3x - 2y)^2$

We can multiply binomials containing exponents that are greater than 1. That is, we can multiply binomials containing x^2 or y^3, and so on.

Example 8 Multiply. $(3x^2 + 4y^3)(2x^2 + 5y^3)$

Solution $(3x^2 + 4y^3)(2x^2 + 5y^3) = 6x^4 + 15x^2y^3 + 8x^2y^3 + 20y^6$

$= 6x^4 + 23x^2y^3 + 20y^6$ □

Student Practice 8 Multiply. $(2x^2 + 3y^2)(5x^2 + 6y^2)$

▲ **Example 9** The width of a living room is $(x + 4)$ feet. The length of the room is $(3x + 5)$ feet. What is the area of the room in square feet?

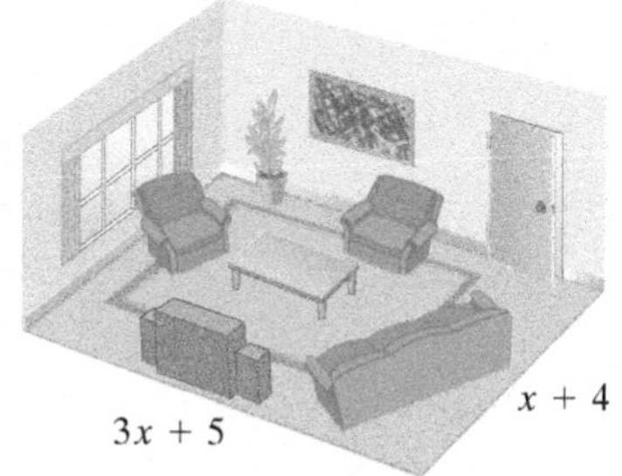

Solution

$$A = (\text{length})(\text{width}) = (3x + 5)(x + 4)$$
$$= 3x^2 + 12x + 5x + 20$$
$$= 3x^2 + 17x + 20$$

There are $(3x^2 + 17x + 20)$ square feet in the room. □

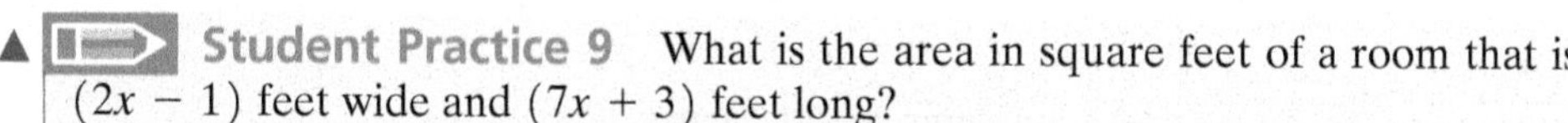

▲ **Student Practice 9** What is the area in square feet of a room that is $(2x - 1)$ feet wide and $(7x + 3)$ feet long?

STEPS TO SUCCESS Keeping Yourself on Schedule

The key to success is to keep on schedule. In a class where you determine your own pace, you will need to commit yourself to follow the suggested pace provided in your course materials. Check off each assignment as you do it so you can see your progress.

Make sure all your class materials are organized. Keep all course schedules and assignments where you can quickly find them. Review them often to be sure you are doing everything that you should.

Discipline yourself to follow the detailed schedule. Professor Tobey and Professor Slater have both taught online classes for several years. They have found that students usually succeed in the course if they do *every* suggested activity. This will give you a guideline for determining what you need to be successful.

Making it personal: Write all your assignment due dates, quiz and exam schedules in the space below. Then plan when you will complete each assignment and study for quizzes and exams. Then write this in a daily planner. ▼

5.4 Exercises MyMathLab®

Multiply.

1. $-2x(6x^3 - x)$

2. $5x(-3x^4 + 4x)$

3. $4x^2(6x - 1)$

4. $-4x^2(2x - 9)$

5. $2x^3(-2x^3 + 5x - 1)$

6. $5x^3(-x^2 + 3x + 2)$

7. $\frac{1}{2}(2x + 3x^2 + 5x^3)$

8. $\frac{2}{3}(4x + 6x^2 - 2x^3)$

9. $(2x^3 - 4x^2 + 5x)(-x^2y)$

10. $(3b^3 + 3b^2 - ab)(-5a^2)$

11. $(3x^3 + x^2 - 8x)(3xy)$

12. $(x^4 + 2x^3 - 8x)(xy)$

13. $(x^3 - 3x^2 + 5x - 2)(3x)$

14. $(-3x^3 + 2x^2 - 6x + 5)(5x)$

15. $(x^2y^2 - 6xy + 8)(-2xy)$

16. $(x^2y^2 + 5xy - 9)(-3xy)$

17. $(-7x^3 + 3x^2 + 2x - 1)(4x^2y)$

18. $(-4x^3 + 8x^2 - 9x - 2)(xy^2)$

19. $(3d^4 - 4d^2 + 6)(-2c^2d)$

20. $(-4x^3 + 6x^2 - 5x)(-7xy^2)$

21. $6x^3(2x^4 - x^2 + 3x + 9)$

22. $8x^3(-2x^4 + 3x^2 - 5x - 14)$

23. $-2x^3(8x^3 - 5x^2 + 6x)$

24. $-4x^6(x^2 - 3x + 5)$

Multiply. Try to do most of the exercises mentally without writing down intermediate steps.

25. $(x + 5)(x + 7)$

26. $(x + 6)(x + 3)$

27. $(x + 6)(x + 2)$

28. $(x + 9)(x + 3)$

29. $(x - 8)(x + 2)$

30. $(x + 3)(x - 6)$

31. $(x - 5)(x - 4)$

32. $(x - 6)(x - 5)$

33. $(5x - 2)(-4x - 3)$

34. $(7x + 1)(-2x - 3)$

35. $(2x - 5)(x + 3y)$

36. $(x - 3)(2x + 3y)$

37. $(5x + 2)(3x - y)$

38. $(3x + 4)(5x - y)$

39. $(4y + 1)(5y - 3)$

40. $(5y + 1)(6y - 5)$

41. $(5x^2 + 4y^3)(2x^2 + 3y^3)$

42. $(3a^2 + 4b^4)(2a^2 + b^4)$

To Think About

43. What is wrong with this multiplication?
$(x - 2)(-3) = 3x - 6$

44. What is wrong with this answer?
$-(3x - 7) = -3x - 7$

45. What is the missing term?
$(5x + 2)(5x + 2) = 25x^2 +$ __________ $+ 4$

46. Multiply the binomials and write a brief description of what is special about the result. $(5x - 1)(5x + 1)$

Mixed Practice *Multiply.*

47. $(4x - 3y)(5x - 2y)$

48. $(3b - 5c)(2b - 7c)$

49. $(7x - 2)^2$

50. $(3x - 7)^2$

51. $(4a + 2b)^2$

52. $(5a + 3b)^2$

53. $(0.2x + 3)(4x - 0.3)$

54. $(0.5x - 2)(6x - 0.2)$

55. $\left(\frac{1}{2}x + \frac{1}{3}\right)\left(\frac{1}{2}x - \frac{1}{4}\right)$

56. $\left(\frac{1}{3}x + \frac{1}{5}\right)\left(\frac{1}{3}x - \frac{1}{2}\right)$

57. $(2x^2 + 4y^3)(3x^2 + 2y^3)$

58. $(4x^3 + 2y^2)(2x^3 + 3y^2)$

Find the area of the rectangle.

▲ **59.** Rectangle with length $5x + 2$ and width $2x - 3$

▲ **60.** Rectangle with length $7x + 3$ and width $4x - 6$

Cumulative Review

61. **[2.3.3]** Solve for x. $3(x - 6) = -2(x + 4) + 6x$

62. **[2.3.3]** Solve for w. $3(w - 7) - (4 - w) = 11w$

63. **[2.7.4]** ***Paper Currency*** Heather returned from the bank with \$375. She had one more \$20 bill than she had \$10 bills. The number of \$5 bills she had was one less than triple the number of \$10 bills. How many of each denomination did she have?

Social Security *Every year, the U.S. government disburses billions of dollars in veterans benefits checks. The polynomial* $2.23x + 25$ *can be used to estimate the total value of veterans benefits checks sent in a year in billions of dollars, where* x *is the number of years since 2000. Round your answers to the nearest tenth. (Source: washingtontimes.com)*

64. **[5.3.4]** Estimate the total value of veterans benefits checks sent in 2000.

65. **[5.3.4]** Estimate the total value of veterans benefits checks sent in 2005.

66. **[5.3.4]** Estimate the total value of veterans benefits checks sent in 2015.

67. **[5.3.4]** Predict the total value of veterans benefits checks that will be sent in 2019.

Quick Quiz 5.4 *Multiply.*

1. $(2x^2y^2 - 3xy + 4)(4xy^2)$

2. $(2x + 3)(3x - 5)$

3. $(6a - 4b)(2a - 3b)$

4. **Concept Check** Explain how you would multiply $(7x - 3)^2$.

5.5 Multiplication: Special Cases

Student Learning Objectives

After studying this section, you will be able to:

1 Multiply binomials of the type $(a + b)(a - b)$.

2 Multiply binomials of the type $(a + b)^2$ and $(a - b)^2$.

3 Multiply polynomials with more than two terms.

1 Multiplying Binomials of the Type $(a + b)(a - b)$

The case when you multiply $(x + y)(x - y)$ is interesting and deserves special consideration. Using the FOIL method, we find

$$(x + y)(x - y) = x^2 - xy + xy - y^2 = x^2 - y^2.$$

Notice that the sum of the inner product and the outer product is zero. We see that

$$(x + y)(x - y) = x^2 - y^2.$$

This works in all cases when the binomials are the sum and difference of the same two terms. That is, in one factor the terms are added, while in the other factor the same two terms are subtracted.

$$\begin{aligned}(5a + 2b)(5a - 2b) &= 25a^2 - 10ab + 10ab - 4b^2 \\ &= 25a^2 - 4b^2\end{aligned}$$

The product is the difference of the squares of the terms. That is, $(5a)^2 - (2b)^2$ or $25a^2 - 4b^2$.

Many students find it helpful to memorize this equation.

MULTIPLYING BINOMIALS: A SUM AND A DIFFERENCE

$$(a + b)(a - b) = a^2 - b^2$$

You may use this relationship to find the product quickly in cases where it applies. The terms must be the same and there must be a sum and a difference.

Example 1 Multiply. $(7x + 2)(7x - 2)$

Solution

$$(7x + 2)(7x - 2) = (7x)^2 - (2)^2 = 49x^2 - 4$$

Check. Multiply the binomials using FOIL to verify that the sum of the inner and outer products is zero. □

Student Practice 1 Multiply. $(6x + 7)(6x - 7)$

Example 2 Multiply. $(5x - 8y)(5x + 8y)$

Solution

$$(5x - 8y)(5x + 8y) = (5x)^2 - (8y)^2 = 25x^2 - 64y^2$$ □

Student Practice 2 Multiply. $(3x - 5y)(3x + 5y)$

2 Multiplying Binomials of the Type $(a + b)^2$ and $(a - b)^2$

A second case that is worth special consideration is a binomial that is squared. Consider the following problem.

$$\begin{aligned}(3x + 2)^2 &= (3x + 2)(3x + 2) \\ &= 9x^2 + 6x + 6x + 4 \\ &= 9x^2 + 12x + 4\end{aligned}$$

If you complete enough problems of this type, you will notice a pattern. The answer always contains the square of the first term added to double the product of the first and last terms added to the square of the last term.

$3x$ is the first term	2 is the last term		Square the first term: $(3x)^2$		Double the product of the first and last terms: $2(3x)(2)$		Square the last term: $(2)^2$
↓	↓		↓		↓		↓
$(3x$ +	$2)^2$	=	$9x^2$	+	$12x$	+	4

We can show the same steps using variables instead of words.

$$(a + b)^2 = a^2 + 2ab + b^2$$

There is a similar formula for the square of a difference:

$$(a - b)^2 = a^2 - 2ab + b^2$$

We can use this formula to simplify $(2x - 3)^2$.

$$\begin{aligned}(2x - 3)^2 &= (2x)^2 - 2(2x)(3) + (3)^2 \\ &= 4x^2 - 12x + 9\end{aligned}$$

You may wish to multiply this product using FOIL to verify.

These two types of products, the square of a sum and the square of a difference, can be summarized as follows.

A BINOMIAL SQUARED

$$(a + b)^2 = a^2 + 2ab + b^2$$
$$(a - b)^2 = a^2 - 2ab + b^2$$

Example 3 Multiply.

(a) $(5y - 2)^2$ **(b)** $(8x + 9y)^2$

Solution

(a) $(5y - 2)^2 = (5y)^2 - (2)(5y)(2) + (2)^2$
$= 25y^2 - 20y + 4$

(b) $(8x + 9y)^2 = (8x)^2 + (2)(8x)(9y) + (9y)^2$
$= 64x^2 + 144xy + 81y^2$

Student Practice 3 Multiply.

(a) $(4a - 9b)^2$ **(b)** $(5x + 4)^2$

CAUTION: $(a + b)^2 \neq a^2 + b^2$! The two sides are not equal! Squaring the sum $(a + b)$ does not give $a^2 + b^2$! Beginning algebra students often make this error. Make sure you remember that when you square a binomial, there is always a *middle term.*

$$(a + b)^2 = a^2 + 2ab + b^2$$

Sometimes a numerical example helps you to see this.

$$(3 + 4)^2 \neq 3^2 + 4^2$$
$$7^2 \neq 9 + 16$$
$$49 \neq 25$$

Notice that what is missing on the right is $2ab = 2 \cdot 3 \cdot 4 = 24$.

3 Multiplying Polynomials with More Than Two Terms 

We used the distributive property to multiply two binomials $(a + b)(c + d)$, and we obtained $ac + ad + bc + bd$. We could also use the distributive property to multiply the polynomials $(a + b)$ and $(c + d + e)$, and we would then obtain $ac + ad + ae + bc + bd + be$. Let us see if we can find a direct way to multiply products such as $(3x - 2)(x^2 - 2x + 3)$. It can be done quickly using an approach similar to that used in arithmetic for multiplying whole numbers. Consider the following arithmetic problem.

128	
× 43	
384 ←	The product of 128 and 3
512 ←	The product of 128 and 4 moved one space to the left
5504 ←	The sum of the two partial products

Let us use a similar format to multiply the two polynomials. For example, multiply $(x^2 - 2x + 3)$ and $(3x - 2)$.

$x^2 - 2x + 3$	
$3x - 2$	This is often called **vertical multiplication.**
$-2x^2 + 4x - 6$ ←	The product $(x^2 - 2x + 3)(-2)$
$3x^3 - 6x^2 + 9x$ ←	The product $(x^2 - 2x + 3)(3x)$ moved one space to the left so that like terms are underneath each other
$3x^3 - 8x^2 + 13x - 6$ ←	The sum of the two partial products

Example 4 Multiply vertically. $(3x^3 + 2x^2 + x)(x^2 - 2x - 4)$

Solution

$3x^3 + 2x^2 + x$	We place one polynomial over the other.
$x^2 - 2x - 4$	
$-12x^3 - 8x^2 - 4x$ ←	The product $(3x^3 + 2x^2 + x)(-4)$
$-6x^4 - 4x^3 - 2x^2$ ←	The product $(3x^3 + 2x^2 + x)(-2x)$
$3x^5 + 2x^4 + x^3$ ←	The product $(3x^3 + 2x^2 + x)(x^2)$
$3x^5 - 4x^4 - 15x^3 - 10x^2 - 4x$ ←	The sum of the three partial products

Note that the answers for each partial product are placed so that like terms are underneath each other. □

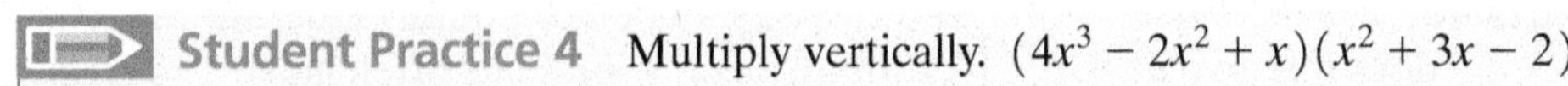

Student Practice 4 Multiply vertically. $(4x^3 - 2x^2 + x)(x^2 + 3x - 2)$

Alternative Method: **FOIL Horizontal Multiplication**

Some students prefer to do this type of multiplication using a horizontal format similar to the FOIL method. The following example illustrates this approach.

Example 5 Multiply horizontally. $(x^2 + 3x + 5)(x^2 - 2x - 6)$

Solution We will use the distributive property repeatedly.

$$(x^2 + 3x + 5)(x^2 - 2x - 6) = x^2(x^2 - 2x - 6) + 3x(x^2 - 2x - 6) + 5(x^2 - 2x - 6)$$
$$= \boxed{x^4 - 2x^3 - 6x^2} + \boxed{3x^3 - 6x^2 - 18x} + \boxed{5x^2 - 10x - 30}$$
$$= x^4 + x^3 - 7x^2 - 28x - 30$$

□

Student Practice 5 Multiply horizontally. $(2x^2 + 5x + 3)(x^2 - 3x - 4)$

Some problems may need to be done in two or more separate steps.

Example 6 Multiply. $(2x - 3)(x + 2)(x + 1)$

Solution We first need to multiply any two of the binomials. Let us select the first pair.

$$\underbrace{(2x - 3)(x + 2)}_{\text{Find this product first.}}(x + 1)$$

$$(2x - 3)(x + 2) = 2x^2 + 4x - 3x - 6$$
$$= 2x^2 + x - 6$$

Now we replace the first two factors with their resulting product.

$$\underbrace{(2x^2 + x - 6)}_{\text{First product}}(x + 1)$$

We then multiply again.

$$(2x^2 + x - 6)(x + 1) = (2x^2 + x - 6)x + (2x^2 + x - 6)1$$
$$= 2x^3 + x^2 - 6x + 2x^2 + x - 6$$
$$= 2x^3 + 3x^2 - 5x - 6$$

The vertical format of Example 4 is an alternative method for this type of problem.

$$\begin{array}{r} 2x^2 + x - 6 \\ x + 1 \\ \hline 2x^2 + x - 6 \\ 2x^3 + x^2 - 6x \\ \hline 2x^3 + 3x^2 - 5x - 6 \end{array}$$

Thus we have

$$(2x - 3)(x + 2)(x + 1) = 2x^3 + 3x^2 - 5x - 6.$$

Note that it does not matter which two binomials are multiplied first. For example, you could first multiply $(2x - 3)(x + 1)$ to obtain $2x^2 - x - 3$ and then multiply that product by $(x + 2)$ to obtain the same result. □

Student Practice 6 Multiply.
$(3x - 2)(2x + 3)(3x + 2)$
(*Hint:* Rearrange the factors.)

Sometimes we encounter a binomial raised to a third power. In such cases we would write out the binomial three times as a product and then multiply. So to evaluate $(3x + 4)^3$ we would first write $(3x + 4)(3x + 4)(3x + 4)$ and then follow the method of Example 6.

5.5 Exercises MyMathLab®

Verbal and Writing Skills, Exercises 1–4

1. In the special case of $(a + b)(a - b)$, a binomial times a binomial is a ______________.

2. Identify which of the following could be the answer to a problem using the formula for $(a + b)(a - b)$. Why?

(a) $9x^2 - 16$ **(b)** $4x^2 + 25$

(c) $9x^2 + 12x + 4$ **(d)** $x^4 - 1$

3. A student evaluated $(4x - 7)^2$ as $16x^2 + 49$. What is missing? State the correct answer.

4. The square of a binomial, $(a - b)^2$, always produces which of the following?

(a) binomial **(b)** trinomial

(c) four-term polynomial

Use the formula $(a + b)(a - b) = a^2 - b^2$ to multiply.

5. $(y - 7)(y + 7)$ **6.** $(x + 3)(x - 3)$ **7.** $(x - 9)(x + 9)$

8. $(x + 10)(x - 10)$ **9.** $(6x - 5)(6x + 5)$ **10.** $(5x + 2)(5x - 2)$

11. $(2x - 7)(2x + 7)$ **12.** $(4x + 1)(4x - 1)$ **13.** $(5x - 3y)(5x + 3y)$

14. $(6a + 5b)(6a - 5b)$ **15.** $(0.6x + 3)(0.6x - 3)$ **16.** $(3x - 0.8)(3x + 0.8)$

Use the formula for a binomial squared to multiply.

17. $(2y + 5)^2$ **18.** $(3x - 1)^2$ **19.** $(5x - 4)^2$

20. $(6x + 5)^2$ **21.** $(7x + 3)^2$ **22.** $(8x - 3)^2$

23. $(3x - 7)^2$ **24.** $(3x + 5y)^2$ **25.** $\left(\frac{2}{3}x + \frac{1}{4}\right)^2$

26. $\left(\frac{3}{4}x + \frac{1}{2}\right)^2$ **27.** $(9xy + 4z)^2$ **28.** $(7y - 3xz)^2$

Mixed Practice, Exercises 29–36 *Multiply. Use the special formula that applies.*

29. $(7x + 3y)(7x - 3y)$ **30.** $(12x - 5y)(12x + 5y)$

31. $(3c - 5d)^2$ **32.** $(6c - d)^2$

33. $(9a - 10b)(9a + 10b)$ **34.** $(11a + 6b)(11a - 6b)$ **35.** $(5x + 9y)^2$ **36.** $(4x + 8y)^2$

Use the distributive property to multiply.

37. $(x^2 - x + 5)(x - 3)$ **38.** $(x^2 + 4x + 2)(x - 5)$ **39.** $(2x + 1)(x^3 + 3x^2 - x + 4)$

40. $(3x - 1)(x^3 + x^2 - 4x - 2)$ **41.** $(a^2 - 3a + 2)(a^2 + 4a - 3)$ **42.** $(b^2 + 5b - 1)(b^2 - 4b + 1)$

43. $(x + 3)(x - 1)(3x - 8)$ **44.** $(x - 7)(x + 4)(2x - 5)$ **45.** $(2x - 5)(x - 1)(x + 3)$

46. $(3x - 2)(x + 6)(x - 3)$ **47.** $(a - 5)(2a + 3)(a + 5)$ **48.** $(b - 3)(b + 3)(2b - 1)$

To Think About

▲ **49.** Find the volume of this object.

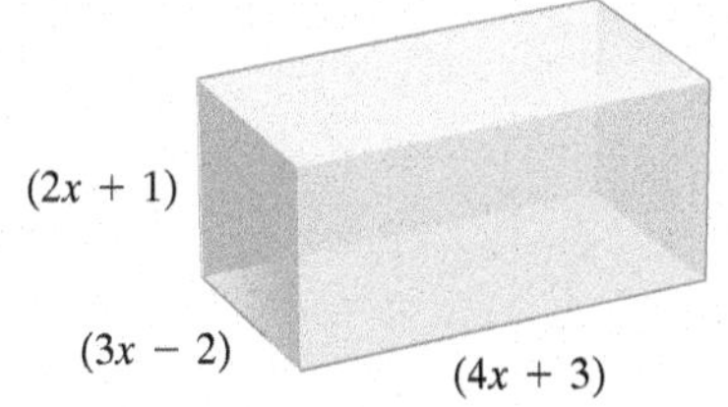

▲ **50.** ***Geometry*** The formula for the volume of a pyramid is $V = \frac{1}{3}Bh$, where B is the area of the base and h is the height. Find the volume of the following pyramid.

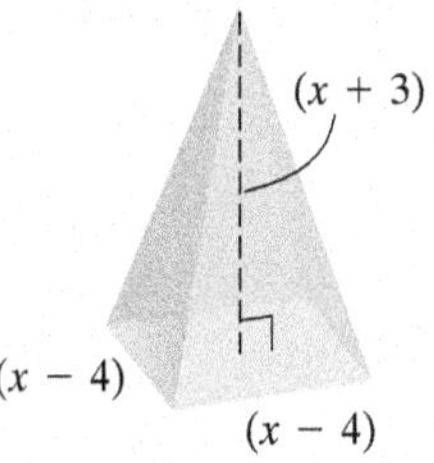

Cumulative Review

51. **[2.6.1]** One number is three more than twice a second number. The sum of the two numbers is 60. Find both numbers.

▲ **52.** **[2.6.1]** ***Room Dimensions*** The perimeter of a rectangular room measures 34 meters. The width is 2 meters more than half the length. Find the dimensions of the room.

Quick Quiz 5.5 *Multiply.*

1. $(7x - 12y)(7x + 12y)$

2. $(2x + 3)(x - 2)(3x + 1)$

3. $(3x - 2)(5x^3 - 2x^2 - 4x + 3)$

4. **Concept Check** Using the formula $(a + b)^2 = a^2 + 2ab + b^2$, explain how to multiply $(6x - 9y)^2$.

5.6 Dividing Polynomials

Student Learning Objectives

After studying this section, you will be able to:

1. Divide a polynomial by a monomial.
2. Divide a polynomial by a binomial.

1 Dividing a Polynomial by a Monomial

To divide a polynomial by a monomial, divide each term of the numerator by the denominator; then write the sum of the results. We are using the property of fractions that states that

$$\frac{a+b}{c} = \frac{a}{c} + \frac{b}{c}$$

DIVIDING A POLYNOMIAL BY A MONOMIAL

1. Divide each term of the polynomial by the monomial.
2. When dividing variables, use the property $\frac{x^a}{x^b} = x^{a-b}$.

Example 1 Divide. $\dfrac{8y^6 - 8y^4 + 24y^2}{8y^2}$

Solution $\dfrac{8y^6 - 8y^4 + 24y^2}{8y^2} = \dfrac{8y^6}{8y^2} - \dfrac{8y^4}{8y^2} + \dfrac{24y^2}{8y^2} = y^4 - y^2 + 3$ □

Student Practice 1 Divide.

$$\frac{15y^4 - 27y^3 - 21y^2}{3y^2}$$

2 Dividing a Polynomial by a Binomial

Division of a polynomial by a binomial is similar to long division in arithmetic. Notice the similarity in the following division problems.

Division of a three-digit number by a two-digit number	***Division of a polynomial by a binomial***
$\begin{array}{r} 32 \\ 21\overline{)672} \\ \underline{63} \\ 42 \\ \underline{42} \\ 0 \end{array}$	$\begin{array}{r} 3x + 2 \\ 2x + 1\overline{)6x^2 + 7x + 2} \\ \underline{6x^2 + 3x} \\ 4x + 2 \\ \underline{4x + 2} \\ 0 \end{array}$

DIVIDING A POLYNOMIAL BY A BINOMIAL

1. Place the terms of the polynomial and binomial in descending order. Insert a 0 for any missing term.
2. Divide the first term of the polynomial by the first term of the binomial. The result is the first term of the answer.
3. Multiply the first term of the answer by the binomial and subtract the result from the first two terms of the polynomial. Bring down the next term to obtain a new polynomial.
4. Divide the new polynomial by the binomial using the process described in step 2.
5. Continue dividing, multiplying, and subtracting until the degree of the remainder is less than the degree of the binomial divisor.
6. Write the remainder as the numerator of a fraction that has the binomial divisor as its denominator.

Example 2 Divide. $(x^3 + 5x^2 + 11x + 4) \div (x + 2)$

Solution

Step 1 The terms are arranged in descending order. No terms are missing.

Step 2 Divide the first term of the polynomial by the first term of the binomial. In this case, divide x^3 by x to get x^2.

$$\begin{array}{r} x^2 \\ x + 2 \overline{) x^3 + 5x^2 + 11x + 4} \end{array}$$

Step 3 Multiply x^2 by $x + 2$ and subtract the result from the first two terms of the polynomial, $x^3 + 5x^2$ in this case.

$$\begin{array}{rl} x^2 & \\ x + 2 \overline{) x^3 + 5x^2 + 11x + 4} & \\ \underline{x^3 + 2x^2} \quad \downarrow & \text{Subtract: } 5x^2 - 2x^2 = 3x^2 \\ 3x^2 + 11x & \text{Bring down the next term.} \end{array}$$

Step 4 Continue to use the step 2 process. Divide $3x^2$ by x. Write the resulting $3x$ as the next term of the answer.

$$\begin{array}{r} x^2 + 3x \\ x + 2 \overline{) x^3 + 5x^2 + 11x + 4} \\ \underline{x^3 + 2x^2} \\ 3x^2 + 11x \end{array}$$

Step 5 Continue multiplying, dividing, and subtracting until the degree of the remainder is less than the degree of the divisor. In this case, we stop when the remainder does not have an x.

$$\begin{array}{rl} x^2 + 3x + 5 & \\ x + 2 \overline{) x^3 + 5x^2 + 11x + 4} & \\ \underline{x^3 + 2x^2} & \\ 3x^2 + 11x & \\ \underline{3x^2 + 6x} & 3x(x + 2) = 3x^2 + 6x \\ 5x + 4 & \text{Bring down 4.} \\ \underline{5x + 10} & \text{Subtract } 4 - 10 = -6. \\ -6 & \text{The remainder is } -6. \end{array}$$

Step 6 The answer is $x^2 + 3x + 5 + \dfrac{-6}{x + 2}$.

Check. To check the answer, we multiply $(x + 2)(x^2 + 3x + 5)$ and add the remainder, -6.

$$\begin{aligned} (x + 2)(x^2 + 3x + 5) + (-6) &= x^3 + 5x^2 + 11x + 10 - 6 \\ &= x^3 + 5x^2 + 11x + 4 \end{aligned}$$

This is the original polynomial. It checks. □

Student Practice 2 Divide.

$$(x^3 + 10x^2 + 31x + 25) \div (x + 4)$$

Take great care with the subtraction step when negative numbers are involved.

Example 3 Divide. $(5x^3 - 24x^2 + 9) \div (5x + 1)$

Solution We must first insert $0x$ to represent the missing x-term. Then we divide $5x^3$ by $5x$.

$$\begin{array}{r|l} & x^2 \\ 5x + 1 & 5x^3 - 24x^2 + 0x + 9 \\ & \underline{5x^3 + x^2} \\ & -25x^2 \end{array}$$

Note that we are subtracting:
$-24x^2 - (+1x^2) = -24x^2 - 1x^2$
$= -25x^2$

Next we divide $-25x^2$ by $5x$.

$$\begin{array}{r|l} & x^2 - 5x \\ 5x + 1 & 5x^3 - 24x^2 + 0x + 9 \\ & \underline{5x^3 + x^2} \\ & -25x^2 + 0x \\ & \underline{-25x^2 - 5x} \\ & 5x \end{array}$$

Note that we are subtracting:
$0x - (-5x) = 0x + 5x = 5x$

Finally, we divide $5x$ by $5x$.

$$\begin{array}{r|l} & x^2 - 5x + 1 \\ 5x + 1 & 5x^3 - 24x^2 + 0x + 9 \\ & \underline{5x^3 + x^2} \\ & -25x^2 + 0x \\ & \underline{-25x^2 - 5x} \\ & 5x + 9 \\ & \underline{5x + 1} \\ & 8 \end{array}$$

$\longleftarrow$ The remainder is 8.

The answer is $x^2 - 5x + 1 + \dfrac{8}{5x + 1}$.

Check. To check, multiply $(5x + 1)(x^2 - 5x + 1)$ and add the remainder, 8.

$$\begin{aligned}(5x + 1)(x^2 - 5x + 1) + 8 &= 5x^3 - 24x^2 + 1 + 8 \\ &= 5x^3 - 24x^2 + 9\end{aligned}$$

This is the original polynomial. Our answer is correct. □

Student Practice 3 Divide.
$(2x^3 - x^2 + 1) \div (x - 1)$

Now we will perform the division by writing a minimum of steps. See if you can follow each step.

Example 4 Divide and check. $(12x^3 - 11x^2 + 8x - 4) \div (3x - 2)$

Solution

$$\begin{array}{r} 4x^2 - x + 2 \\ 3x - 2\overline{)12x^3 - 11x^2 + 8x - 4} \\ \underline{12x^3 - 8x^2} \\ -3x^2 + 8x \\ \underline{-3x^2 + 2x} \\ 6x - 4 \\ \underline{6x - 4} \\ 0 \end{array}$$

Check. $(3x - 2)(4x^2 - x + 2) = 12x^3 - 3x^2 + 6x - 8x^2 + 2x - 4$

$= 12x^3 - 11x^2 + 8x - 4$ Our answer is correct. □

 **Student Practice 4** Divide and check.

$$(20x^3 - 11x^2 - 11x + 6) \div (4x - 3)$$

STEPS TO SUCCESS Key Points to Help You While You Are Taking an Exam

You will be amazed to see how your performance on exams can improve just by using a good strategy while taking an exam. Look over the following suggestions, then apply them while taking your next exam. You will really see a difference!

Before You Begin the Exam

Arrive early. This will give you time to collect your thoughts and ready yourself.

Relax. After you get your exam, take two or three moderately deep breaths. You will feel your entire body begin to relax.

Look over the entire exam quickly. This will help you pace yourself and use your time wisely. If point values are given, plan to spend more time on questions that are worth more points. Write down formulas in a blank area of the test.

During the Test

Read directions carefully. Be sure to answer all questions clearly and ask your instructor about anything that is not clear.

Answer questions that are easiest first. This will help you build your confidence.

Keep track of time. Do not get bogged down with one question. Return to it after you complete the rest of the questions.

Check your work. Keep your work neat and organized so that you are less likely to make errors and can check your steps easily.

Stay calm. Stop and take a deep breath periodically. If others leave before you, don't worry. You are entitled to use the full amount of allotted time.

Making it personal: Which of these suggestions do you find most helpful? Can you list any other suggestions? Use these strategies on your next exam. ▼

5.6 Exercises MyMathLab®

Divide.

1. $\dfrac{25x^4 - 15x^2 + 20x}{5x}$

2. $\dfrac{20b^4 - 4b^3 + 16b^2}{4b}$

3. $\dfrac{3y^5 + 21y^3 - 9y^2}{3y^2}$

4. $\dfrac{10y^4 - 35y^3 + 5y^2}{5y^2}$

5. $\dfrac{81x^7 - 36x^5 - 63x^3}{9x^3}$

6. $\dfrac{28x^8 - 14x^6 + 35x^4}{7x^3}$

7. $(48x^7 - 54x^4 + 36x^3) \div 6x^3$

8. $(72x^8 - 45x^6 - 36x^3) \div 9x^3$

Divide. Check your answers for exercises 9–16 by multiplication.

9. $\dfrac{6x^2 + 13x + 5}{2x + 1}$

10. $\dfrac{12x^2 + 19x + 5}{3x + 1}$

11. $\dfrac{x^2 - 8x - 17}{x - 5}$

12. $\dfrac{x^2 - 9x - 5}{x - 3}$

13. $\dfrac{3x^3 - x^2 + 4x - 2}{x + 1}$

14. $\dfrac{2x^3 - 3x^2 - 3x + 6}{x + 1}$

15. $\dfrac{4x^3 + 4x^2 - 19x - 15}{2x + 5}$

16. $\dfrac{6x^3 + 11x^2 - 8x + 5}{2x + 5}$

17. $\dfrac{10x^3 + 11x^2 - 11x + 2}{5x - 2}$

18. $\dfrac{5x^3 - 28x^2 - 20x + 21}{5x - 3}$

19. $\dfrac{4x^3 + 3x + 5}{2x - 3}$

20. $\dfrac{8x^3 + 8x + 5}{2x - 1}$

21. $(y^3 - y^2 - 13y - 12) \div (y + 3)$

22. $(y^3 - 2y^2 - 26y - 4) \div (y + 4)$ **23.** $(y^4 - 9y^2 - 5) \div (y - 2)$ **24.** $(2y^4 + 3y^2 - 5) \div (y - 2)$

Cumulative Review

25. **[0.5.3]** ***Milk Prices*** In January 2008, the average price for a gallon of whole milk was \$3.87. By January 2010, the price had decreased by 16%. In January 2015, the price increased 16% from the 2010 price. What was the average price of a gallon of whole milk in January 2010? What was the price in January 2015? (*Source:* www.bls.gov)

26. **[2.6.1]** ***Page Numbers*** Thomas was assigned to read a special two-page case study in his psychology book. He can't remember the page numbers, but does remember that the pages are consecutive and the two numbers add to 341. What are the page numbers?

27. **[2.6.2]** ***Hurricane Pattern*** The National Hurricane Center has noticed an interesting pattern in the number of Atlantic hurricanes each year. Examine the bar graph, then answer the following questions. Round all answers to the nearest tenth.

(a) What was the mean number of hurricanes per year during the years 2003 to 2006?

(b) What was the mean number of hurricanes per year during the years 2007 to 2010?

(c) What was the mean number of hurricanes per year during the years 2011 to 2014?

(d) What was the percent decrease of the mean from the four-year period 2003 to 2006 to the four-year period 2007 to 2010?

(e) What was the percent decrease of the mean from the four-year period 2007 to 2010 to the four-year period 2011 to 2014?

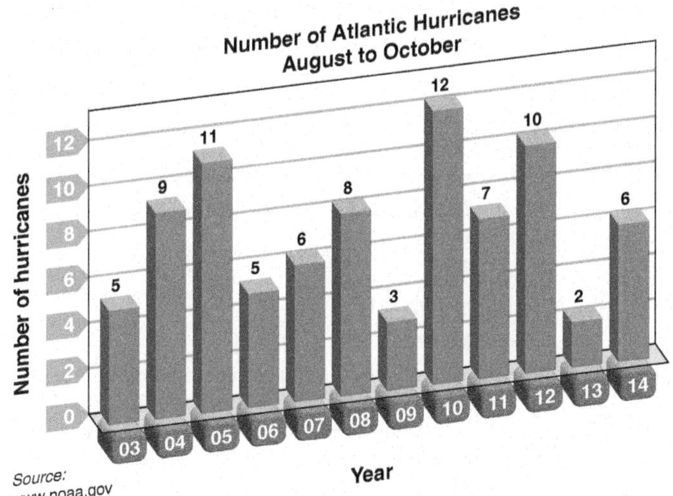

Quick Quiz 5.6 *Divide.*

1. $\dfrac{20x^5 - 64x^4 - 8x^3}{4x^2}$

2. $(8x^3 + 2x^2 - 19x - 6) \div (2x + 3)$

3. $(x^3 + 4x - 3) \div (x - 2)$

4. **Concept Check** Explain how you would check your answer to problem 3. Perform the check. Does your answer check?

CAREER EXPLORATION PROBLEMS

Background: Biological Technician

Jeannie is a biological technician working at a major biotechnology company. Jeannie's responsibilities are varied in her role assisting a team of scientists in the protocols of their experiments. She spends a good deal of time observing, recording, and calculating results from these experiments.

Facts

- One of the projects she's working on involves the study of how to best modify bacterial cells, and she needs to calculate a transformation efficiency value, a measure of the cell's ability to be transformed. Typically these values are calculated and expressed in scientific notation. After measuring and recording a first set of data, her final calculation appears as:

$$\text{Transformation efficiency} = 140 \text{ transformants} \div \left[\frac{250}{2{,}500{,}000}\right]$$

  A second set of data generates the following calculation:

$$\text{Transformation efficiency} = 164 \text{ transformants} \times \left[\frac{1}{5 \times 10^{-3}}\right]$$

  In her lab reports, Jeannie's measurements for cell density, absorption, and number of bacteria must be converted from scientific notation to decimal notation.

- Another project Jeannie is working on involves genetics. In it, she's writing polynomial expressions that describe the possible genetic makeup for the offspring of two parents. Each parent has half dominant genes, symbolized by X, and half recessive genes, symbolized by y. The binomial describing each parent's makeup is $(0.5X + 0.5y)$.

Tasks

1. What are the final transformation efficiency values expressed in scientific notation for the two sets of data Jeannie has measured and recorded?
2. What are the decimal equivalents of the following measurements expressed in scientific notation?
   (a) Cell density: 5.01×10^7
   (b) Absorption: 3.02×10^{-3}
   (c) Number of bacteria: $237 \times 10 \times 10^5$
3. Given that the binomial describing each parent's genetic makeup is $(0.5X + 0.5y)$, the polynomial that describes the genetic makeup of the offspring is the product of $(0.5X + 0.5y)^2$. What is this product?

Chapter 5 Organizer

Topic and Procedure	Examples	You Try It
Multiplying monomials, p. 325 $x^a \cdot x^b = x^{a+b}$ **1.** Multiply the numerical coefficients. **2.** Add the exponents of a given base.	Multiply. **(a)** $3^{12} \cdot 3^{15} = 3^{27}$ **(b)** $(-3x^2)(6x^3) = -18x^5$ **(c)** $(2ab)(4a^2b^3) = 8a^3b^4$	1. Multiply. **(a)** $2^9 \cdot 2^{14}$ **(b)** $(-8a^3)(-2a^5)$ **(c)** $(-ab^2)(3a^4b^2)$
Dividing monomials, p. 327 $(x \neq 0)$ $\frac{x^a}{x^b} = \begin{cases} x^{a-b} & \text{Use if } a \text{ is greater than } b. \\ \frac{1}{x^{b-a}} & \text{Use if } b \text{ is greater than } a. \end{cases}$ **1.** Divide or reduce the fraction created by the quotient of the numerical coefficients. **2.** Subtract the exponents of a given base.	Divide. **(a)** $\frac{16x^7}{8x^3} = 2x^4$ **(b)** $\frac{5x^3}{25x^5} = \frac{1}{5x^2}$ **(c)** $\frac{-12x^5y^7}{18x^3y^{10}} = -\frac{2x^2}{3y^3}$	2. Divide. **(a)** $\frac{21x^5}{3x^2}$ **(b)** $\frac{-3x}{9x^2}$ **(c)** $\frac{14ab^7}{28a^3b}$
Exponent of zero, p. 329 $x^0 = 1 \quad \text{if } x \neq 0$	Simplify. **(a)** $5^0 = 1$ **(b)** $\frac{x^6}{x^6} = 1$ **(c)** $w^0 = 1$ **(d)** $3x^0y = 3y$	3. Simplify. **(a)** 9^0 **(b)** m^0 **(c)** $\frac{a^5}{a^5}$ **(d)** $6ab^0$
Raising a power, product, or quotient to a power, pp. 330–331 $(x^a)^b = x^{ab}$ $(xy)^a = x^ay^a$ $\left(\frac{x}{y}\right)^a = \frac{x^a}{y^a} \quad (y \neq 0)$ **1.** Raise the numerical coefficient to the power outside the parentheses. **2.** Multiply the exponent outside the parentheses times all exponents inside the parentheses.	Simplify. **(a)** $(x^9)^3 = x^{27}$ **(b)** $(3x^2)^3 = 27x^6$ **(c)** $\left(\frac{2x^2}{y^3}\right)^3 = \frac{8x^6}{y^9}$ **(d)** $(-3x^4y^5)^4 = 81x^{16}y^{20}$ **(e)** $(-5ab)^3 = -125a^3b^3$	4. Simplify. **(a)** $(a^4)^5$ **(b)** $(2n^3)^2$ **(c)** $\left(\frac{3x^3}{y}\right)^3$ **(d)** $(-5s^2t^5)^2$ **(e)** $(-a^2b)^5$
Negative exponents, p. 336 If $x \neq 0$ and $y \neq 0$, then $x^{-n} = \frac{1}{x^n}, \quad \frac{1}{x^{-n}} = x^n, \quad \frac{x^{-m}}{y^{-n}} = \frac{y^n}{x^m}$	Write with positive exponents. **(a)** $x^{-6} = \frac{1}{x^6}$ **(b)** $\frac{1}{w^{-3}} = w^3$ **(c)** $\frac{w^{-12}}{z^{-5}} = \frac{z^5}{w^{12}}$ **(d)** $2^{-2} = \frac{1}{2^2} = \frac{1}{4}$	5. Rewrite with positive exponents. **(a)** a^{-3} **(b)** $\frac{1}{x^{-1}}$ **(c)** $\frac{m^{-9}}{n^{-6}}$ **(d)** 3^{-2}
Scientific notation, p. 337–338 A positive number is written in scientific notation if it is in the form $a \times 10^n$, where $1 \leq a < 10$ and n is an integer.	Write in scientific notation. **(a)** $2{,}568{,}000 = 2.568 \times 10^6$ **(b)** $0.0000034 = 3.4 \times 10^{-6}$	6. Write in scientific notation. **(a)** 386,400 **(b)** 0.000052
Performing calculations with numbers written in scientific notation, p. 339 Rearrange the order when multiplying. Use the definition of multiplication of fractions when dividing.	Multiply or divide. **(a)** $(5.2 \times 10^7)(1.8 \times 10^5)$ $= 5.2 \times 1.8 \times 10^7 \times 10^5$ $= 9.36 \times 10^{12}$ **(b)** $\frac{4.5 \times 10^6}{6 \times 10^3}$ $= \frac{4.5}{6} \times \frac{10^6}{10^3}$ $= 0.75 \times 10^3$ $= 7.5 \times 10^2$	7. Multiply or divide. **(a)** $(3.1 \times 10^6)(2.5 \times 10^4)$ **(b)** $\frac{3.8 \times 10^9}{1.25 \times 10^5}$
Adding polynomials, pp. 343–344 To add two polynomials, we add their like terms.	Add. $(-7x^3 + 2x^2 + 5) + (x^3 + 3x^2 + x)$ $= -6x^3 + 5x^2 + x + 5$	8. Add. $(x^4 - 5x^3 + 2x^2) + (-7x^4 + x^3 - x^2)$
Subtracting polynomials, p. 345 To subtract the polynomials, change all signs of the second polynomial and add the result to the first polynomial. $a - b = a + (-b)$	Subtract. $(5x^2 - 6) - (-3x^2 + 2) = (5x^2 - 6) + (3x^2 - 2)$ $= 8x^2 - 8$	9. Subtract. $(8 - x^2) - (5 + 2x^2)$

Topic and Procedure	Examples	You Try It
Multiplying a monomial by a polynomial, p. 350 Use the distributive property. $a(b + c) = ab + ac$ $(b + c)a = ba + ca$	Multiply. **(a)** $-5x(2x^2 + 3x - 4) = -10x^3 - 15x^2 + 20x$ **(b)** $(6x^3 - 5xy - 2y^2)(3xy) = 18x^4y - 15x^2y^2 - 6xy^3$	10. Multiply. **(a)** $-2a(3a^2 - 5a + 1)$ **(b)** $(-x^2 + 3xy - 3y^2)(4xy)$
Multiplying two binomials, pp. 350–351, 356 **1.** The product of the sum and difference of the same two terms: $(a + b)(a - b) = a^2 - b^2$ **2.** The square of a binomial: $(a + b)^2 = a^2 + 2ab + b^2$ $(a - b)^2 = a^2 - 2ab + b^2$ **3.** Use FOIL for other binomial multiplication. The middle terms can often be combined, giving a trinomial.	Multiply. **(a)** $(3x + 7y)(3x - 7y) = 9x^2 - 49y^2$ **(b)** $(3x + 7y)^2 = 9x^2 + 42xy + 49y^2$ $(3x - 7y)^2 = 9x^2 - 42xy + 49y^2$ **(c)** $(3x - 5)(2x + 7) = 6x^2 + 21x - 10x - 35$ $= 6x^2 + 11x - 35$	11. Multiply. **(a)** $(2a + 5b)(2a - 5b)$ **(b)** $(2a + 5b)^2$ **(c)** $(2a - 5b)^2$ **(d)** $(2a - b)(3a + 5b)$
Multiplying two polynomials, p. 358 To multiply two polynomials, multiply each term of one by each term of the other. This method is similar to the multiplication of many-digit numbers.	Multiply. $(3x^2 - 7x + 4)(3x - 1)$ Vertical method: $\begin{array}{r} 3x^2 - 7x + 4 \\ \times \qquad 3x - 1 \\ \hline -3x^2 + 7x - 4 \\ 9x^3 - 21x^2 + 12x \qquad \\ \hline 9x^3 - 24x^2 + 19x - 4 \end{array}$ Horizontal method: $(5x + 2)(2x^2 - x + 3)$ $= 10x^3 - 5x^2 + 15x + 4x^2 - 2x + 6$ $= 10x^3 - x^2 + 13x + 6$	12. Multiply. **(a)** $\begin{array}{r} 6x^2 - 5x + 3 \\ 2x + 1 \\ \hline \end{array}$ **(b)** $(x - 5)(3x^2 - 2x + 1)$
Multiplying three or more polynomials, p. 359 **1.** Multiply any two polynomials. **2.** Multiply the result by any remaining polynomials.	Multiply. $(2x + 1)(x - 3)(x + 4) = (2x^2 - 5x - 3)(x + 4)$ $= 2x^3 + 3x^2 - 23x - 12$	13. Multiply. $(x + 5)(x - 1)(3x + 2)$
Dividing a polynomial by a monomial, p. 362 **1.** Divide each term of the polynomial by the monomial. **2.** When dividing variables, use the property $\frac{x^a}{x^b} = x^{a-b}$.	Divide. $(15x^3 + 20x^2 - 30x) \div (5x)$ $= \frac{15x^3}{5x} + \frac{20x^2}{5x} - \frac{30x}{5x}$ $= 3x^2 + 4x - 6$	14. Divide. $(18a^3 - 9a^2 + 3a) \div (3a)$
Dividing a polynomial by a binomial, p. 362 **1.** Place the terms of the polynomial and binomial in descending order. Insert a 0 for any missing term. **2.** Divide the first term of the polynomial by the first term of the binomial. **3.** Multiply the first term of the answer by the binomial, and subtract the result from the first two terms of the polynomial. Bring down the next term to obtain a new polynomial. **4.** Divide the new polynomial by the binomial using the process described in step 2. **5.** Continue dividing, multiplying, and subtracting until the degree of the remainder is less than the degree of the binomial divisor. **6.** Write the remainder as the numerator of a fraction that has the binomial divisor as its denominator.	Divide. $(8x^3 + 2x^2 - 13x + 7) \div (4x - 1)$ $\begin{array}{r} 2x^2 + x - 3 \\ 4x - 1 \overline{)8x^3 + 2x^2 - 13x + 7} \\ \underline{8x^3 - 2x^2} \qquad\qquad \\ 4x^2 - 13x \qquad \\ \underline{4x^2 - x} \qquad \\ -12x + 7 \\ \underline{-12x + 3} \\ 4 \end{array}$ The answer is $2x^2 + x - 3 + \frac{4}{4x - 1}$.	15. Divide. $(3x^3 + 13x^2 - 13x + 2) \div (3x - 2)$

Chapter 5 Review Problems

Section 5.1

Simplify. In problems 1–12, leave your answer in exponent form.

1. $(-6a^2)(3a^5)$

2. $(5^{10})(5^{13})$

3. $(3xy^2)(2x^3y^4)$

4. $(2x^3y^4)(-7xy^5)$

5. $\dfrac{7^{15}}{7^{27}}$

6. $\dfrac{x^{12}}{x^{17}}$

7. $\dfrac{y^{30}}{y^{16}}$

8. $\dfrac{9^{13}}{9^{24}}$

9. $\dfrac{-15xy^2}{25x^6y^6}$

10. $\dfrac{-12a^3b^6}{18a^2b^{12}}$

11. $(x^3)^8$

12. $\dfrac{(2b^2)^4}{(5b^3)^6}$

13. $(-3a^3b^2)^2$

14. $(3x^3y)^4$

15. $\left(\dfrac{5ab^2}{c^3}\right)^2$

16. $\left(\dfrac{x^0y^3}{4w^5z^2}\right)^3$

Section 5.2

Simplify. Write with positive exponents.

17. $a^{-3}b^5$

18. m^8p^{-5}

19. $\dfrac{2x^{-6}}{y^{-3}}$

20. $(2x^{-5}y)^{-3}$

21. $(6a^4b^5)^{-2}$

22. $\dfrac{3x^{-3}}{y^{-2}}$

23. $\dfrac{4x^{-5}y^{-6}}{w^{-2}z^8}$

24. $\dfrac{3^{-3}a^{-2}b^5}{c^{-3}d^{-4}}$

Write in scientific notation.

25. 156,340,200,000

26. 179,632

27. 0.00092

28. 0.00000174

Write in decimal notation.

29. 1.2×10^5

30. 6.034×10^6

31. 2.5×10^{-1}

32. 4.32×10^{-5}

Perform the calculation indicated. Leave your answer in scientific notation.

33. $\dfrac{(28{,}000{,}000)(5{,}000{,}000{,}000)}{7000}$

34. $(3.12 \times 10^5)(2.0 \times 10^6)(1.5 \times 10^8)$

35. $\dfrac{(0.00078)(0.000005)(0.00004)}{0.002}$

36. ***Mission to Pluto*** The *New Horizons* spacecraft began its 3.5×10^9-mile journey to Pluto in January 2006. If the cost of this mission is \$0.20 per mile, find the total cost. (*Source*: www.pluto.jhuapl.edu)

37. ***Atomic Clock*** An atomic clock is based on the fact that cesium emits 9,192,631,770 cycles of radiation in one second. How many of these cycles occur in one day? Round to three significant digits.

38. ***Computer Speed*** Today's fastest modern computers can perform one operation in 1×10^{-11} second. How many operations can such a computer perform in 1 minute?

Section 5.3

Combine.

39. $(2.8x^2 - 1.5x + 3.4) + (2.7x^2 + 0.5x - 5.7)$

40. $(4x^3 - x^2 - x + 3) - (-3x^3 + 2x^2 + 5x - 1)$

41. $\left(\dfrac{3}{5}x^2y - \dfrac{1}{3}x + \dfrac{3}{4}\right) - \left(\dfrac{1}{2}x^2y + \dfrac{2}{7}x + \dfrac{1}{3}\right)$

42. $\dfrac{1}{2}x^2 - \dfrac{3}{4}x + \dfrac{1}{5} - \left(\dfrac{1}{4}x^2 - \dfrac{1}{2}x + \dfrac{1}{10}\right)$

43. $(x^2 - 9) - (4x^2 + 5x) + (5x - 6)$

Section 5.4

Multiply.

44. $(3x + 1)(5x - 1)$

45. $(7x - 2)(4x - 3)$

46. $(2x + 3)(10x + 9)$

47. $5x(2x^2 - 6x + 3)$

48. $(xy^2 + 5xy - 6)(-4xy^2)$

49. $(5a + 7b)(a - 3b)$

50. $(2x^2 - 3)(4x^2 - 5y)$

Section 5.5

Multiply.

51. $(4x + 3)^2$

52. $(a + 5b)(a - 5b)$

53. $(7x + 6y)(7x - 6y)$

54. $(5a - 2b)^2$

55. $(x^2 + 7x + 3)(4x - 1)$

56. $(x - 6)(2x - 3)(x + 4)$

Section 5.6

Divide.

57. $(12y^3 + 18y^2 + 24y) \div (6y)$

58. $(30x^5 + 35x^4 - 90x^3) \div (5x^2)$

59. $(16x^3 - 24x^2 + 32x) \div (4x)$

60. $(15x^2 + 11x - 14) \div (5x + 7)$

61. $(12x^2 - x - 63) \div (4x + 9)$

62. $(2x^3 - x^2 + 3x - 1) \div (x + 2)$

63. $(6x^2 + x - 9) \div (2x + 3)$

64. $(x^3 - x - 24) \div (x - 3)$

65. $(2x^3 - 3x + 1) \div (x - 2)$

Applications

Solve. Express your answer in scientific notation.

66. ***Fighting Obesity*** President Obama signed into law the Healthy Hunger-Free Kids Act of 2010, which includes 4.5×10^9 dollars to provide children with healthy food in schools and fight childhood obesity. If the population of the United States in 2010 was 3.1×10^8 people, how much per person did the United States spend to fight childhood obesity? Write your answer in dollars and cents. (*Source:* www.whitehouse.gov)

67. ***Population*** The population of Africa in 2020 is projected to be 1.25×10^9. The projected population of North America in 2020 is 5.97×10^8. What is the total population of the two countries projected to be? (*Hint*: First write 5.97×10^8 as 0.597×10^9 before you do the calculations.)

68. ***Electron Mass*** The mass of an electron is approximately 9.11×10^{-28} gram. Find the mass of 30,000 electrons.

69. ***Gray Whales*** During feeding season, gray whales eat 3.4×10^5 pounds of food per day. If the feeding period lasts for 140 days, how many pounds of food total will a gray whale consume?

To Think About

Find a polynomial that describes the shaded area.

▲ **70.**

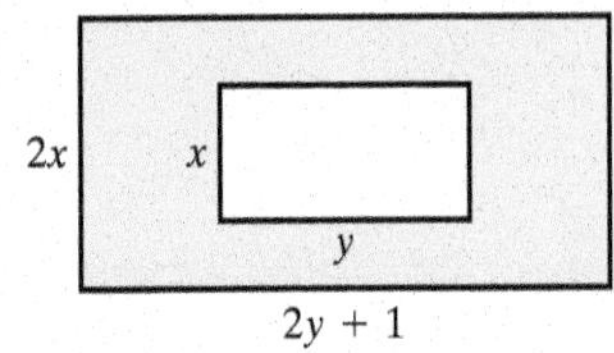

▲ **71.**

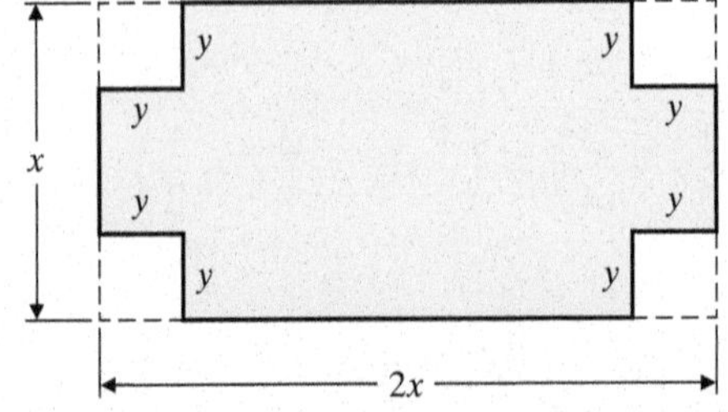

How Am I Doing? Chapter 5 Test

MATH COACH MyMathLab® YouTube

After you take this test read through the Math Coach on pages 374–375. Math Coach videos are available via MyMathLab and YouTube. Step-by-step test solutions on the Chapter Test Prep Videos are also available via MyMathLab and YouTube. (Search "TobeyBegInterAlg" and click on "Channels.")

Simplify. Leave your answer in exponent form.

1. $(3^{10})(3^{24})$

2. $\dfrac{25^{18}}{25^{34}}$

3. $(8^4)^6$

In questions 4–8, simplify.

4. $(-3xy^4)(-4x^3y^6)$

5. $\dfrac{-35x^8y^{10}}{25x^5y^{10}}$

6. $(-5xy^6)^3$

7. $\left(\dfrac{7a^7b^2}{3c^0}\right)^2$

MC **8.** $\dfrac{(3x^2)^3}{(6x)^2}$

9. Evaluate. 4^{-3}

In questions 10 and 11, simplify and write with only positive exponents.

10. $6a^{-4}b^{-3}c^5$

MC **11.** $\dfrac{3x^{-3}y^2}{x^{-4}y^{-5}}$

12. Write in scientific notation. 0.0005482

13. Write in decimal notation. 5.82×10^8

14. Multiply. Leave your answer in scientific notation.
$(4.0 \times 10^{-3})(3.0 \times 10^{-8})(2.0 \times 10^4)$

Combine.

15. $(2x^2 - 3x - 6) + (-4x^2 + 8x + 6)$

16. $(3x^3 - 4x^2 + 3) - (14x^3 - 7x + 11)$

Multiply.

17. $-7x^2(3x^3 - 4x^2 + 6x - 2)$

18. $(5x^2y^2 - 6xy + 2)(3x^2y)$

19. $(5a - 4b)(2a + 3b)$

MC **20.** $(3x + 2)(2x + 1)(x - 3)$

21. $(7x^2 + 2y^2)^2$

22. $(5s - 11t)(5s + 11t)$

23. $(3x - 2)(4x^3 - 2x^2 + 7x - 5)$

24. $(3x^2 - 5xy)(x^2 + 3xy)$

Divide.

25. $(15x^6 - 5x^4 + 25x^3) \div 5x^3$

26. $(8x^3 - 22x^2 - 5x + 12) \div (4x + 3)$

MC **27.** $(2x^3 - 6x - 36) \div (x - 3)$

Solve. Express your answer in scientific notation. Round to the nearest hundredth.

28. At the end of 2014, Saudi Arabia was estimated to have 2.6×10^{11} barrels of oil reserves. By one estimate, they have 69 years of reserves remaining. If they disbursed the oil in an equal amount each year, how many barrels of oil would they pump each year? (*Source:* saudiembassy.net)

29. A space probe is traveling from Earth to Pluto at a speed of 2.49×10^4 miles per hour. How far would this space probe travel in one week?

1. ______ ☐
2. ______ ☐
3. ______ ☐
4. ______ ☐
5. ______ ☐
6. ______ ☐
7. ______ ☐
8. ______ ☐
9. ______ ☐
10. ______ ☐
11. ______ ☐
12. ______ ☐
13. ______ ☐
14. ______ ☐
15. ______ ☐
16. ______ ☐
17. ______ ☐
18. ______ ☐
19. ______ ☐
20. ______ ☐
21. ______ ☐
22. ______ ☐
23. ______ ☐
24. ______ ☐
25. ______ ☐
26. ______ ☐
27. ______ ☐
28. ______ ☐
29. ______ ☐

Total Correct: ☐

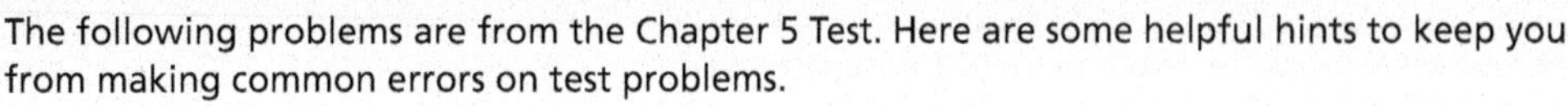

MATH COACH

Mastering the skills you need to do well on the test.

The following problems are from the Chapter 5 Test. Here are some helpful hints to keep you from making common errors on test problems.

Chapter 5 Test, Problem 8 Simplify. $\dfrac{(3x^2)^3}{(6x)^2}$

HELPFUL HINT Do the problem in three stages. First, use the power to a power rule to raise the numerator to the third power. Second, raise the denominator to the second power. Then divide the monomials using the rules of exponents. Be sure to simplify any fractions.

Did you use the power to a power rule to raise both 3^1 and x^2 to the third power in the numerator and both 6^1 and x^1 to the second power in the denominator?

Yes ______ No ______

If you answered No, stop and review the power to a power rule before completing these steps again.

Did you remember to simplify the fraction $\dfrac{27}{36}$?

Yes ______ No ______

Finally, did you remember to use the quotient rule to subtract the exponents in the x terms?

Yes ______ No ______

If you answered No to either of these questions, go back and examine your work carefully before completing these steps again.

If you answered Problem 8 incorrectly, go back and rework the problem using these suggestions.

Chapter 5 Test, Problem 11 Simplify and write with only positive exponents. $\dfrac{3x^{-3}y^2}{x^{-4}y^{-5}}$

HELPFUL HINT First, use the definition of a negative exponent to rewrite the expression using only positive exponents. Then use the rules for exponents to simplify the resulting expression.

Did you remove the negative exponents by rewriting the expression as $\dfrac{3x^4y^2y^5}{x^3}$?

Yes ______ No ______

If you answered No, review the definition of negative exponents in Section 5.2 and complete this step again.

Did you use the quotient rule to simplify the x terms and the product rule to simplify the y terms?

Yes ______ No ______

If you answered No, review the rules for exponents in Sections 5.1 and 5.2 and simplify the expression again.

Now go back and rework the problem using these suggestions.

Need more help? Watch the MATH COACH videos in MyMathLab® or on YouTube™.

Chapter 5 Test, Problem 20 Multiply. $(3x + 2)(2x + 1)(x - 3)$

HELPFUL HINT A good approach is to start by multiplying the first two binomials. Then multiply that result by the third binomial. Be careful to avoid sign errors when multiplying, and be careful to write down the correct exponent for each term.

Did you use the FOIL method to multiply the first two binomials and obtain $6x^2 + 7x + 2$?

Yes ______ No ______

If you answered No, stop and complete this step.

Did you multiply the result above by $(x - 3)$?

Yes ______ No ______

Did you multiply *each term* of $6x^2 + 7x + 2$ by x?

Yes ______ No ______

Did you multiply *each term* of $6x^2 + 7x + 2$ by -3?

Yes ______ No ______

If you answered No to any of these questions, go back and examine each step of the multiplication carefully. Be sure to write the correct exponent each time that you multiply. Be careful to avoid sign errors when multiplying by -3. Then combine like terms before writing your final answer.

If you answered Problem 20 incorrectly, go back and rework the problem using these suggestions.

Chapter 5 Test, Problem 27 Divide. $(2x^3 - 6x - 36) \div (x - 3)$

HELPFUL HINT Review the procedure for dividing a polynomial by a binomial in Section 5.6. Make sure you understand each step. Be sure you understand where the expression $0x^2$ came from in the dividend. Be careful with subtraction. Write out the subtraction steps to avoid sign errors.

Did you write the division problem in the form $x - 3\overline{)2x^3 + 0x^2 - 6x - 36}$?

Yes ______ No ______

If you answered No, remember that every power must be represented. We must use $0x^2$ as a placeholder so that we can perform our division.

When you carried out the first step of division, did you obtain $2x^2$ as the first part of your answer?

Yes ______ No ______

When you multiplied $x - 3$ by $2x^2$ and then subtracted, did you get the result $6x^2$?

Yes ______ No ______

If you answered No to these questions, stop and examine your first division step carefully. Make sure that you subtracted carefully too. Remember to write out the subtraction steps: $0x^2 - (-6x^2) = 6x^2$.

Next, did you bring down $-6x$ from the dividend to obtain $6x^2 - 6x$?

Yes ______ No ______

If you answered No, go back and look at the dividend again and see how to obtain this result.

Now go back and rework the problem using these suggestions.

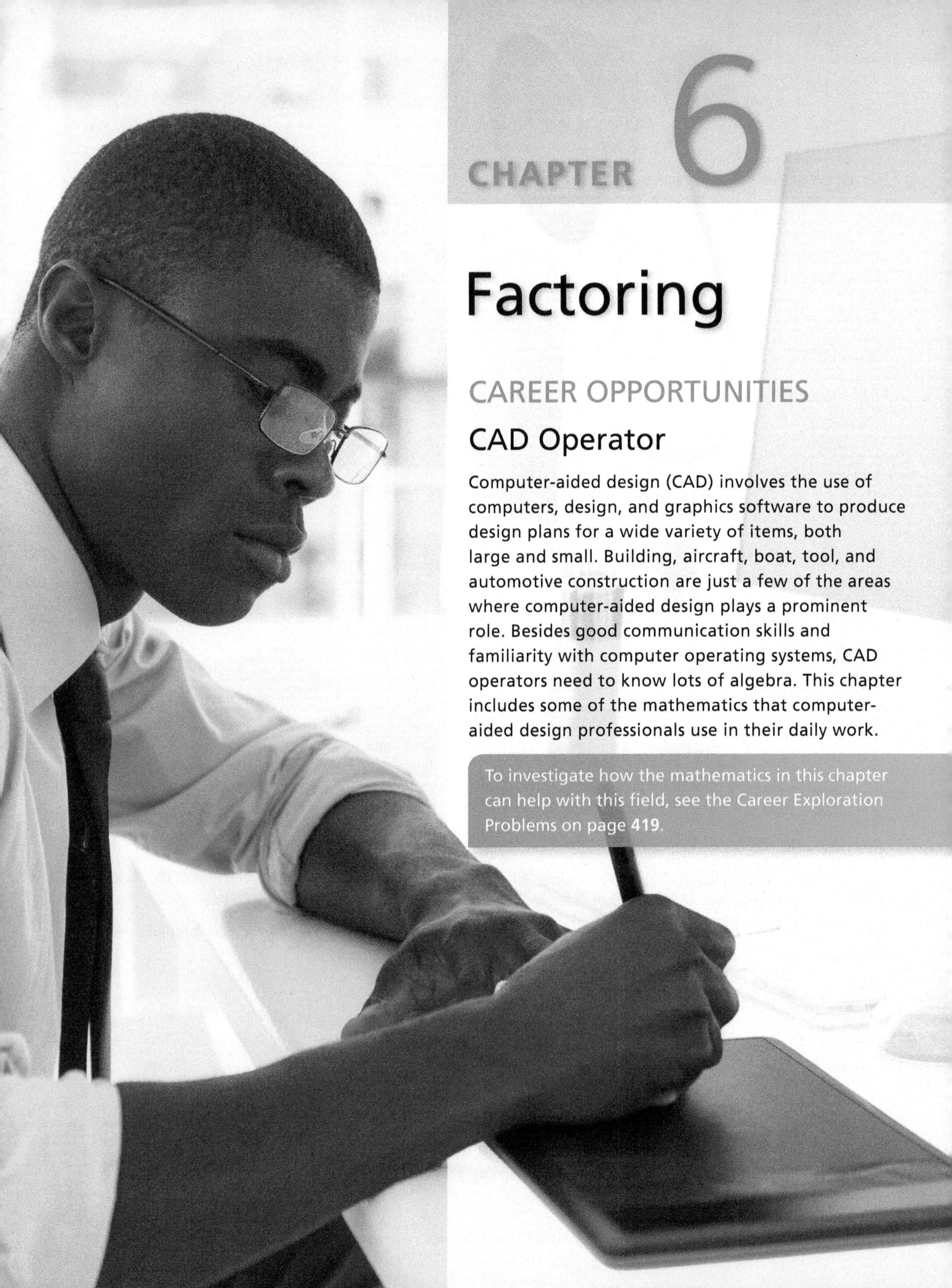

CHAPTER 6

Factoring

CAREER OPPORTUNITIES

CAD Operator

Computer-aided design (CAD) involves the use of computers, design, and graphics software to produce design plans for a wide variety of items, both large and small. Building, aircraft, boat, tool, and automotive construction are just a few of the areas where computer-aided design plays a prominent role. Besides good communication skills and familiarity with computer operating systems, CAD operators need to know lots of algebra. This chapter includes some of the mathematics that computer-aided design professionals use in their daily work.

To investigate how the mathematics in this chapter can help with this field, see the Career Exploration Problems on page **419**.

6.1 Removing a Common Factor

Student Learning Objective

After studying this section, you will be able to:

1 Factor polynomials whose terms contain a common factor.

1 Factoring Polynomials Whose Terms Contain a Common Factor

Recall that when two or more numbers, variables, or algebraic expressions are multiplied, each is called a **factor.**

$$3 \cdot 2 \qquad 3x^2 \cdot 5x^3 \qquad (2x-3)(x+4)$$

factor factor   factor factor   factor factor

When you are asked **to factor** a number or an algebraic expression, you are being asked, "What factors, when multiplied, will give that number or expression?"

For example, you can factor 6 as $3 \cdot 2$ since $3 \cdot 2 = 6$. You can factor $15x^5$ as $3x^2 \cdot 5x^3$ since $3x^2 \cdot 5x^3 = 15x^5$. Factoring is simply the reverse of multiplying. 6 and $15x^5$ are simple expressions to factor and can be factored in different ways.

The factors of the polynomial $2x^2 + 5x - 12$ are not so easy to recognize. Factoring a polynomial changes an addition and/or subtraction problem into a multiplication problem. In this chapter we will be learning techniques for finding the factors of a polynomial. We will begin with **common factors.**

Example 1 Factor. **(a)** $3x - 6y$ **(b)** $9x + 2xy$

Solution Begin by looking for a common factor, a factor that both terms have in common. Then rewrite the expression as a product.

(a) $3x - 6y = 3(x - 2y)$ This is true because $3(x - 2y) = 3x - 6y$.

(b) $9x + 2xy = x(9 + 2y)$ This is true because $x(9 + 2y) = 9x + 2xy$.

Some people find it helpful to think of factoring as the distributive property in reverse. When we write $3x - 6y = 3(x - 2y)$, we are doing the reverse of distributing the 3. This is the main point of this section. We are doing problems of the form $ca + cb = c(a + b)$. The common factor c becomes the first factor of our answer. □

Student Practice 1 Factor.

(a) $21a - 7b$ **(b)** $5xy + 8x$

When we factor, we begin by looking for the **greatest common factor.** For example, in the polynomial $48x - 16y$, a common factor is 2. We could factor $48x - 16y$ as $2(24x - 8y)$. *However, this is not complete.* To factor $48x - 16y$ completely, we look for the greatest common factor of 48 and of 16.

$$48x - 16y = 16(3x - y)$$

Example 2 Factor $24xy + 12x^2 + 36x^3$. Remember to remove the greatest common factor.

Solution We start by finding the greatest common factor of 24, 12, and 36. You may want to factor each number, or you may notice that 12 is a common factor. 12 is the greatest numerical common factor.

Notice also that x is a factor of each term. Thus, $12x$ is the greatest common factor.

$$24xy + 12x^2 + 36x^3 = 12x(2y + x + 3x^2)$$ □

Student Practice 2 Factor $12a^2 + 16ab^2 - 12a^2b$. Be careful to remove the greatest common factor.

FACTORING A POLYNOMIAL WITH COMMON FACTORS

1. Determine the greatest common numerical factor by asking, "What is the largest integer that will divide into the coefficients of all the terms?"
2. Determine the greatest common variable factor by first asking, "What variables are common to all the terms?" Then, for each variable that is common to all the terms, ask, "What is the largest exponent of the variable that is common to all the terms?"
3. The common factor(s) found in steps 1 and 2 are the first part of the answer.
4. After removing the common factor(s), what remains is placed in parentheses as the second factor.

Example 3 Factor. **(a)** $12x^2 - 18y^2$ **(b)** $x^2y^2 + 3xy^2 + y^3$

Solution

(a) Note that the largest integer that is common to both terms is 6 (not 3 or 2).

$$12x^2 - 18y^2 = 6(2x^2 - 3y^2)$$

(b) Although y is common to all of the terms, we factor out y^2 since 2 is the largest exponent of y that is common to all terms. We do not factor out x, since x is not common to all of the terms.

$$x^2y^2 + 3xy^2 + y^3 = y^2(x^2 + 3x + y)$$

□

 **Student Practice 3** Factor.

(a) $16a^3 - 24b^3$ **(b)** $r^3s^2 - 4r^4s + 7r^5$

Checking Is Very Important! You can check any factoring problem by multiplying the factors you obtain. The result must be the same as the original polynomial.

Example 4 Factor. $8x^3y + 16x^2y^2 - 24x^3y^3$

Solution We see that 8 is the largest integer that will divide evenly into the three numerical coefficients. We can factor x^2 out of each term. We can also factor y out of each term.

$$8x^3y + 16x^2y^2 - 24x^3y^3 = 8x^2y(x + 2y - 3xy^2)$$

Check.

$$8x^2y(x + 2y - 3xy^2) = 8x^3y + 16x^2y^2 - 24x^3y^3 \checkmark$$

□

 **Student Practice 4** Factor. $18a^3b^2c - 27ab^3c^2 - 45a^2b^2c^2$

Example 5 Factor. $9a^3b^2 + 9a^2b^2$

Solution We observe that both terms contain a common factor of 9. We can also factor a^2 and b^2 out of each term. Thus the greatest common factor is $9a^2b^2$.

$$9a^3b^2 + 9a^2b^2 = 9a^2b^2(a + 1)$$

CAUTION: Don't forget to include the 1 inside the parentheses in Example 5. The solution is wrong without it. You will see why if you try to check a result written without the 1.

□

Student Practice 5 Factor and check. $30x^3y^2 - 24x^2y^2 + 6xy^2$

Example 6 Factor. $3x(x - 4y) + 2(x - 4y)$

Solution Be sure you understand what are *terms* and what are *factors* of the polynomial in this example. There are two terms. The expression $3x(x - 4y)$ is the first term. The expression $2(x - 4y)$ is the second term. Each term is made up of two factors.

Observe that the binomial $(x - 4y)$ is a common factor of the terms. A common factor may be any type of polynomial. Thus we can factor out the common factor $(x - 4y)$.

$$3x(x - 4y) + 2(x - 4y) = (x - 4y)(3x + 2)$$ □

 **Student Practice 6** Factor. $3(a + 5b) + x(a + 5b)$

Example 7 Factor. $7x^2(2x - 3y) - (2x - 3y)$

Solution The common factor of the terms is $(2x - 3y)$. What happens when we factor out $(2x - 3y)$? What are we left with in the second term?

Recall that $(2x - 3y) = 1(2x - 3y)$.

Thus $7x^2(2x - 3y) - (2x - 3y)$

$= 7x^2(2x - 3y) - \mathbf{1}(2x - 3y)$ Rewrite the original expression with 1 as a factor.

$= (2x - 3y)(7x^2 - 1)$ Factor out $(2x - 3y)$. □

Student Practice 7 Factor. $8y(9y^2 - 2) - (9y^2 - 2)$

▲ **Example 8** A computer programmer is writing a program to find the total area of 4 circles. She uses the formula $A = \pi r^2$. The radii of the circles are a, b, c, and d, respectively. She wants the final answer to be in factored form with the value of π occurring only once, in order to minimize the rounding error. Write the total area of the 4 circles with a formula that has π occurring only once.

Solution For each circle, $A = \pi r^2$, where $r = a, b, c,$ or d.

We add the area of each of the 4 circles.

The total area is $\pi a^2 + \pi b^2 + \pi c^2 + \pi d^2$.

In factored form the total area $= \pi(a^2 + b^2 + c^2 + d^2)$. □

▲ **Student Practice 8** Use $A = \pi r^2$ to find the shaded area. The radius of the larger circle is b. The radius of the smaller circle is a. Write the shaded-area formula in factored form so that π appears only once.

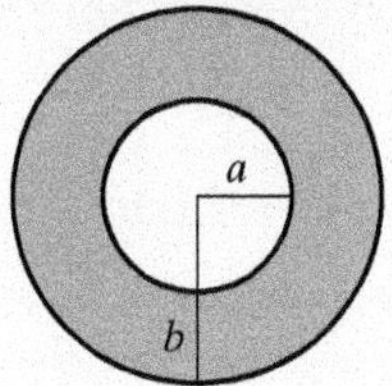

6.1 Exercises MyMathLab®

Verbal and Writing Skills, Exercises 1–4 *In exercises 1 and 2, write a word or words to complete each sentence.*

1. In the expression $3x^2 \cdot 5x^3$, $3x^2$ and $5x^3$ are called ______.

2. In the expression $3x^2 + 5x^3$, $3x^2$ and $5x^3$ are called ______.

3. We can factor $30a^4 + 15a^3 - 45a^2$ as $5a(6a^3 + 3a^2 - 9a)$. Is the factoring complete? Why or why not?

4. We can factor $4x^3 - 8x^2 + 20x$ as $4(x^3 - 2x^2 + 5x)$. Is the factoring complete?

Remove the greatest common factor. Check your answers for exercises 5–28 by multiplication.

5. $8a^2 + 8a$

6. $7b^2 + 7b$

7. $21ab - 14ab^2$

8. $24ab - 18ab^2$

9. $2\pi rh + 2\pi r^2$

10. $5a^2b^2 - 35ab$

11. $5x^3 + 25x^2 - 15x$

12. $8x^3 - 10x^2 - 14x$

13. $12ab - 28bc + 20ac$

14. $14xy + 21yz - 42xz$

15. $16x^5 + 24x^3 - 32x^2$

16. $36x^6 + 45x^4 - 18x^2$

17. $14x^2y - 35xy - 63x$

18. $40a^2 - 16ab - 24a$

19. $54x^2 - 45xy + 18x$

20. $48xy - 24y^2 + 40y$

21. $3xy^2 - 2ay + 5xy - 2y$

22. $2ab^3 + 3xb^2 - 5b^4 + 2b^2$

23. $24x^2y - 40xy^2$

24. $35abc^2 - 49ab^2c$

25. $7x^3y^2 + 21x^2y^2$

26. $8x^3y^2 + 32xy^2$

27. $16x^4y^2 - 24x^2y^2 - 8x^2y$

28. $18x^2y^2 - 12xy^3 + 6xy$

Hint: In exercises 29–42, refer to Examples 6 and 7.

29. $7a(x + 2y) - b(x + 2y)$

30. $6(3a + b) - z(3a + b)$

31. $3x(x - 4) - 2(x - 4)$

32. $5x(x - 7) + 3(x - 7)$

33. $6b(2a - 3c) - 5d(2a - 3c)$

34. $7x(3y + 5z) - 6t(3y + 5z)$

35. $7c(b - a^2) - 5d(b - a^2) + 2f(b - a^2)$

36. $2a(x^2 - y) - 5b(x^2 - y) + c(x^2 - y)$

37. $3a(ab - 4) - 5(ab - 4) - b(ab - 4)$

38. $4b(3ab - 2) + 2(3ab - 2) - 5a(3ab - 2)$

39. $4a^3(a - 3b) + (a - 3b)$

40. $2b^2(2a - 5b) + (2a - 5b)$

41. $(a + 2) - x(a + 2)$

42. $y(3x - 2) - (3x - 2)$

Use the figure below for exercises 43 and 44.

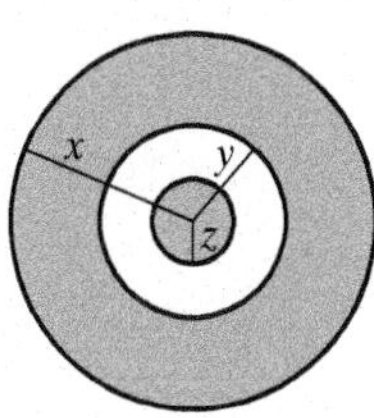

▲ **43.** Use $C = 2\pi r$ to find the circumference of each circle. The radii of the circles are x, y, and z, respectively. Write the sum of the circumferences of all three circles in factored form so π appears only once.

▲ **44.** Use $A = \pi r^2$ to find the shaded area. The radii of the circles are x, y, and z, respectively. Write the shaded-area formula in factored form so π appears only once.

Cumulative Review

Coffee Production *In a recent year, the top six coffee-producing countries of the world produced 6,400,000 metric tons of coffee beans. The percent of this 6.40 million metric tons produced by each of these countries is shown in the following graph. Use the graph to answer the following questions. (Source:* worldatlas.com)

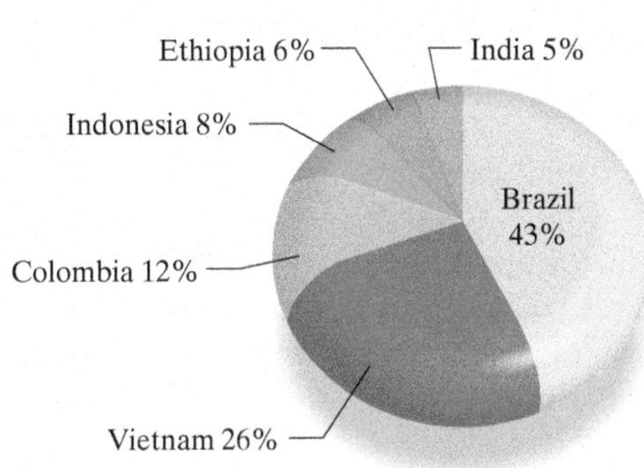

45. **[0.5.3]** How many metric tons of coffee were produced in Vietnam?

46. **[0.5.3]** How many metric tons of coffee were produced in Brazil?

47. **[0.6.1]** The population of Vietnam in 2015 was approximately 89,700,000 people. How many pounds of coffee were produced in Vietnam for each person? (Round your answer to the nearest whole number. One metric ton is about 2205 pounds.)

48. **[0.6.1]** The population of Brazil in 2015 was approximately 200,400,000 people. How many pounds of coffee were produced in Brazil for each person? (Round your answer to the nearest whole number. One metric ton is about 2205 pounds.)

Quick Quiz 6.1 *Remove the greatest common factor.*

1. $3x - 4x^2 + 2xy$

2. $20x^3 - 25x^2 - 5x$

3. $8a(a + 3b) - 7b(a + 3b)$

4. **Concept Check** Explain how you would remove the greatest common factor from the following polynomial.

$$36a^3b^2 - 72a^2b^3$$

6.2 Factoring by Grouping

Student Learning Objective

After studying this section, you will be able to:

1 Factor expressions with four terms by grouping.

1 Factoring Expressions with Four Terms by Grouping

A common factor of a polynomial can be a number, a variable, or an algebraic expression. Sometimes the polynomial is written so that it is easy to recognize the common factor. This is especially true when the common factor is enclosed by parentheses.

Example 1 Factor. $x(x - 3) + 2(x - 3)$

Solution Observe each term:

$$\underbrace{x(x-3)}_{\text{first term}} + \underbrace{2(x-3)}_{\text{second term}}$$

The common factor of the first and second terms is the quantity $(x - 3)$, so we have

$$x(x - 3) + 2(x - 3) = (x - 3)(x + 2).$$

Student Practice 1 Factor. $3y(2x - 7) - 8(2x - 7)$

Now let us face a new challenge. Think carefully. Try to follow this new idea. Suppose the polynomial in Example 1, $x(x - 3) + 2(x - 3)$, were written in the form $x^2 - 3x + 2x - 6$. (Note that this form is obtained by multiplying the factors of the first and second terms.) How would we factor a four-term polynomial like this?

In such cases we remove a common factor from the first two terms and a different common factor from the second two terms. That is, we would factor x from $x^2 - 3x$ and 2 from $2x - 6$.

$$x^2 - 3x + 2x - 6 = x(x - 3) + 2(x - 3)$$

Because the resulting terms have a common factor (the binomial enclosed by the parentheses), we would then proceed as we did in Example 1 to obtain the factored form $(x - 3)(x + 2)$. This procedure for factoring is often called **factoring by grouping.**

Example 2 Factor. $2x^2 + 3x + 6x + 9$

Solution

$$2x^2 + 3x \quad + \quad 6x + 9$$

Factor out a common factor of x from the first two terms. — Factor out a common factor of 3 from the second two terms.

$$\underbrace{x(2x + 3)} \qquad \underbrace{3(2x + 3)}$$

Note that the sets of parentheses in the two terms contain the same expression at this step.

The expression in parentheses is now a common factor of the terms. Now we finish the factoring.

$$\begin{aligned} 2x^2 + 3x + 6x + 9 &= x(2x + 3) + 3(2x + 3) \\ &= (2x + 3)(x + 3) \end{aligned}$$

Student Practice 2 Factor. $6x^2 - 15x + 4x - 10$

Example 3 Factor by grouping. $4x + 8y + ax + 2ay$

Solution

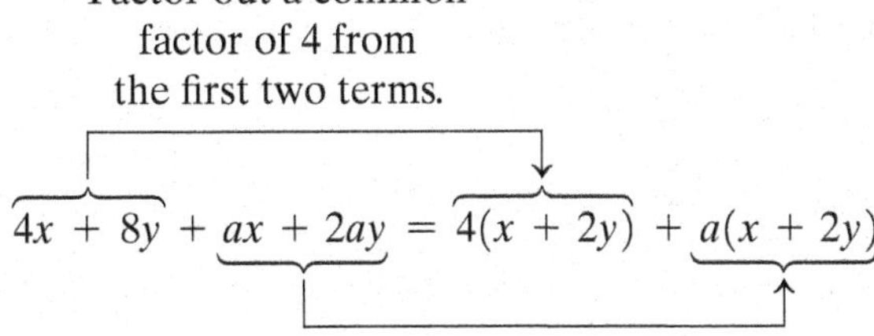

Factor out a common
factor of 4 from
the first two terms.
$4x + 8y + ax + 2ay = 4(x + 2y) + a(x + 2y)$
Factor out a common
factor of a from
the second two terms.

$4(x + 2y) + a(x + 2y) = (x + 2y)(4 + a)$ The common factor of the terms is the expression in parentheses, $x + 2y$. □

Student Practice 3 Factor by grouping. $ax + 2a + 4bx + 8b$

In some problems the terms are out of order. In this case, we have to rearrange the order of the terms first so that the first two terms have a common factor.

Example 4 Factor. $bx + 4y + 4b + xy$

Solution

$bx + 4y + 4b + xy = bx + 4b + xy + 4y$ Rearrange the terms so that the first two terms have a common factor.

$= b(x + 4) + y(x + 4)$ Factor out the common factor b from the first two terms and the common factor y from the second two terms.

$= (x + 4)(b + y)$ □

Student Practice 4 Factor. $6a^2 + 5bc + 10ab + 3ac$

Sometimes you will need to *factor out a negative common factor* from the second two terms to obtain two terms that contain the same parenthetical expression.

Example 5 Factor. $2x^2 + 5x - 4x - 10$

Solution

$2x^2 + 5x - 4x - 10 = x(2x + 5) - 4x - 10$ Factor out the common factor x from the first two terms.

$= x(2x + 5) - 2(2x + 5)$ Factor out the common factor -2 from the second two terms.

$= (2x + 5)(x - 2)$

CAUTION: If you factored out the common factor +2 in the second step, the two resulting terms would not contain the same parenthetical expression. If the expressions inside the two sets of parentheses are *not exactly the same*, you cannot express the polynomial as a product of two factors! □

Student Practice 5 Factor. $6xy + 14x - 15y - 35$

Example 6 Factor. $2ax - a - 2bx + b$

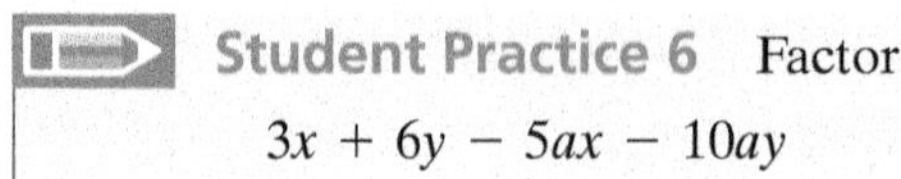

Student Practice 6 Factor.

$3x + 6y - 5ax - 10ay$

Solution

$2ax - a - 2bx + b$

$= a(2x - 1) - b(2x - 1)$ Factor out the common factor a from the first two terms. Factor out the common factor $-b$ from the second two terms.

$= (2x - 1)(a - b)$ Since the two resulting terms contain the same parenthetical expression, we can complete the factoring. □

CAUTION: Many students find that they make a factoring error in the first step of problems like Example 6. When factoring out $-b$, be sure to check your signs carefully: $-2bx + b = -b(2x - 1)$.

Example 7 Factor and check your answer. $8ad + 21bc - 6bd - 28ac$

Solution We observe that the first two terms do not have a common factor.

$8ad + 21bc - 6bd - 28ac$

$= 8ad - 6bd - 28ac + 21bc$ Rearrange the order using the commutative property of addition.

$= 2d(4a - 3b) - 7c(4a - 3b)$ Factor out the common factor $2d$ from the first two terms and the common factor $-7c$ from the last two terms.

$= (4a - 3b)(2d - 7c)$ Factor out the common factor $(4a - 3b)$.

To check, we multiply the two binomials using the FOIL procedure.

$(4a - 3b)(2d - 7c) = 8ad - 28ac - 6bd + 21bc$

$= 8ad + 21bc - 6bd - 28ac$ ✓ Rearrange the order of the terms. This is the original problem. Thus it checks. □

Student Practice 7 Factor and check your answer.

$10ad + 27bc - 6bd - 45ac$

STEPS TO SUCCESS I Don't Really Need to Actually Read This Book, Do I?

Successful students find that they can get a lot of benefit from reading the book. Here are some suggestions they have made after they completed the course:

1. This book was written by faculty who have learned how to help you succeed in doing math problems. So take the time to read over the assigned section of the book. You will be amazed by the understanding you will acquire.
2. Read your textbook with a paper and pen handy. As you come across a new definition or a new idea, underline it. As you come across a suggestion that helps you understand things better, be sure to underline it.
3. Interact with the book. When you see something you don't understand put a big question mark next to it. Try reading it again. Find a similar example in the book and go over the steps. If that doesn't help, ask your instructor or a classmate to explain the procedure or idea.
4. Use your pen to follow each step that is worked out for each example. Underline the steps that you think are especially challenging. This will help you when you do the homework.

Making it personal: Which of these four suggestions do you think you would benefit from the most? Put a check mark by that suggestion. Now write down what you need to do in order to take full advantage of your textbook. Make a commitment to doing this for your next homework assignment. ▼

6.2 Exercises MyMathLab®

Verbal and Writing Skills, Exercises 1 and 2

1. To factor $3x^2 - 6xy + 5x - 10y$, we must first remove a common factor of $3x$ from the first two terms. What do we do with the last two terms? What should we get for the answer?

2. To factor $5x^2 + 15xy - 2x - 6y$, we must first remove a common factor of $5x$ from the first two terms. What do we do with the last two terms? What should we get for the answer?

Factor by grouping. Check your answers for exercises 3–26.

3. $ab - 4a + 6b - 24$

4. $xy - 2x + 7y - 14$

5. $x^3 - 4x^2 + 3x - 12$

6. $x^3 - 8x^2 + 3x - 24$

7. $2ax + 6bx - ay - 3by$

8. $4y - 12x - 3yw + 9xw$

9. $3ax + bx - 6a - 2b$

10. $ad + 3a - d^2 - 3d$

11. $5a + 12bc + 10b + 6ac$

12. $4u^2 + v + 4uv + u$

13. $6c - 12d + cx - 2dx$

14. $xy + 5x - 5y - 25$

15. $y^2 - 2y - 3y + 6$

16. $ax - 3a - 2bx + 6b$

17. $54 - 6y + 9y - y^2$

18. $35 - 5a + 7a - a^2$

19. $6ax - y + 2ay - 3x$

20. $6tx + r - 3t - 2rx$

21. $2x^2 + 8x - 3x - 12$

22. $3y^2 - y + 9y - 3$

23. $t^3 - t^2 + t - 1$

24. $x^2 - 2x - xy + 2y$

25. $6x^2 + 15xy^2 + 8xw + 20y^2w$

26. $10xw + 14x^2 + 25wy^2 + 35xy^2$

To Think About

27. Although $6a^2 - 12bd - 8ad + 9ab = 6(a^2 - 2bd) - a(8d - 9b)$ is true, it is not the correct solution to the problem "Factor $6a^2 - 12bd - 8ad + 9ab$." Explain. Can this expression be factored?

28. Tim was trying to factor $5x^2 - 3xy - 10x + 6y$. In his first step he wrote down $x(5x - 3y) + 2(-5x + 3y)$. Was he doing the problem correctly? What is the answer?

Cumulative Review

29. **[1.3.3]** Divide. $\frac{6}{7} \div \left(-\frac{2}{5}\right)$

30. **[1.1.5]** Add. $-\frac{2}{3} + \frac{4}{5}$

31. **[5.1.2]** Simplify. $\frac{-5a^2b^8}{25ab^{10}}$

32. **[5.5.2]** Multiply. $(2x - 5)^2$

33. **[2.6.1]** ***Salaries in the Pharmacy Industry*** In 2015, the average annual salary of a pharmacist in the United States was \$22,000 more than triple the average annual salary of a pharmacy technician. The two average annual salaries totaled \$162,000. Find the average salary of a pharmacist and the average salary of a pharmacy technician in 2015. (*Source:* www.pharmacist.com)

34. **[2.7.2]** ***Oil Production in North America*** In 2011, in North America 17,000 thousand barrels of oil were produced each day. By 2013, this figure had increased by 13% in North America. By 2015, this figure had increased by 11% from the amount produced in 2013. How many thousand barrels of oil were produced each day in North America in 2015?

Quick Quiz 6.2 *Factor by grouping.*

1. $7ax + 12a - 14x - 24$

2. $2xy^2 - 15 + 6x - 5y^2$

3. $10xy - 3x + 40by - 12b$

4. **Concept Check** Explain how you would factor the following polynomial.

$$10ax + b^2 + 2bx + 5ab$$

6.3 Factoring Trinomials of the Form $x^2 + bx + c$

1 Factoring Polynomials of the Form $x^2 + bx + c$

Suppose that you wanted to factor $x^2 + 5x + 6$. After some trial and error you *might* obtain $(x + 2)(x + 3)$, or you might get discouraged and not get an answer. If you did get these factors, you could check this answer by the FOIL method.

$$\begin{aligned}(x + 2)(x + 3) &= x^2 + 3x + 2x + 6 \\ &= x^2 + 5x + 6\end{aligned}$$

But trial and error can be a long process. There is another way. Let's look at the preceding equation again.

$$(x + 2)(x + 3) = \overset{F}{x^2} + \underbrace{\overset{O}{3x} + \overset{I}{2x}} + \overset{L}{6}$$

$$= x^2 \quad + 5x \quad + 6$$

The first thing to notice is that the product of the first terms in the factors gives the first term of the polynomial. That is, $x \cdot x = x^2$.

The first term is the product of these terms.

$$x^2 + 5x + 6 \quad = \quad (x + 2)(x + 3)$$

The next thing to notice is that the sum of the products of the outer and inner terms in the factors produces the middle term of the polynomial. That is, $(x \cdot 3) + (2 \cdot x) = 3x + 2x = 5x$. Thus we see that the sum of the second terms in the factors, $2 + 3$, gives the coefficient of the middle term, 5.

Finally, note that the product of the last terms of the factors gives the last term of the polynomial. That is, $2 \cdot 3 = 6$.

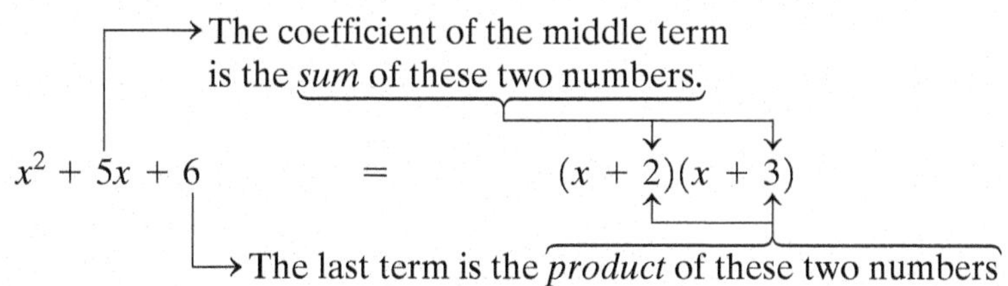

The coefficient of the middle term
is the sum of these two numbers.
x² + 5x + 6 = (x + 2)(x + 3)
The last term is the product of these two numbers

Let's summarize our observations in general terms and then try a few examples.

Student Learning Objectives

After studying this section, you will be able to:

1. Factor polynomials of the form $x^2 + bx + c$.
2. Factor polynomials that have a common factor and a factor of the form $x^2 + bx + c$.

FACTORING TRINOMIALS OF THE FORM $x^2 + bx + c$

1. The answer will be of the form $(x + m)(x + n)$.
2. m and n are numbers such that:
   (a) When you multiply them, you get the last term, which is c.
   (b) When you add them, you get the coefficient of the middle term, which is b.

Example 1 Factor. $x^2 + 7x + 12$

Solution The answer will be of the form $(x + m)(x + n)$. We want to find the two numbers, m and n, that you can multiply to get 12 and add to get 7. The numbers are 3 and 4.

$$x^2 + 7x + 12 = (x + 3)(x + 4)$$

□

Student Practice 1 Factor. $x^2 + 8x + 12$

Example 2 Factor. $x^2 + 12x + 20$

Solution We want two numbers that have a product of 20 and a sum of 12. The numbers are 10 and 2.

$$x^2 + 12x + 20 = (x + \underline{10})(x + \underline{2})$$

Note: If you cannot think of the numbers in your head, write down the possible factors whose product is 20.

Product	Sum
$1 \cdot 20 = 20$	$1 + 20 = 21$
$2 \cdot 10 = 20$	$2 + 10 = 12$ ←
$4 \cdot 5 = 20$	$4 + 5 = 9$

Then select the pair whose sum is 12. Select this pair. □

Student Practice 2 Factor. $x^2 + 17x + 30$

You may find that it is helpful to first list all the factors whose product is 30.

So far we have factored only trinomials of the form $x^2 + bx + c$, where b and c are positive numbers. The same procedure applies if b is a negative number and c is positive. Because m and n have a positive product and a negative sum, they must both be negative.

Example 3 Factor. $x^2 - 8x + 15$

Solution We want two numbers that have a product of $+15$ and a sum of -8. They must be negative numbers since the sign of the middle term is negative and the sign of the last term is positive.

the sum $-5 + (-3)$

$$x^2 - 8x + 15 = (x - 5)(x - 3)$$

the product $(-5)(-3)$

Think: $(-5)(-3) = +15$ *and* $-5 + (-3) = -8$.

Multiply using FOIL to check. □

Student Practice 3 Factor. $x^2 - 11x + 18$

Example 4 Factor. $x^2 - 9x + 14$

Solution We want two numbers whose product is 14 and whose sum is -9. The numbers are -7 and -2. So

$$x^2 - 9x + 14 = (x - 7)(x - 2) \text{ or } (x - 2)(x - 7).$$ □

Student Practice 4 Factor. $x^2 - 11x + 24$

All the examples so far have had a positive last term. What happens when the last term is negative? If the last term is a negative number, one of the numbers m or n must be a positive number and the other must be a negative number. Why? The product of a positive number and a negative number is negative.

Example 5 Factor. $x^2 - 3x - 10$

Solution We want two numbers whose product is -10 and whose sum is -3. The two numbers are -5 and $+2$.

$$x^2 - 3x - 10 = (x - 5)(x + 2)$$ □

Student Practice 5 Factor. $x^2 - 5x - 24$

What if we made a sign error and *incorrectly* factored the trinomial $x^2 - 3x - 10$ as $(x + 5)(x - 2)$? We could detect the error immediately since the sum of $+5$ and -2 is 3. We need a sum of -3!

Example 6 Factor and check your answer. $y^2 + 10y - 24$

Solution The two numbers whose product is -24 and whose sum is $+10$ are $+12$ and -2.

$$y^2 + 10y - 24 = (y + 12)(y - 2)$$

CAUTION: It is very easy to make a sign error in these problems. Make sure that you mentally multiply your answer by FOIL to obtain the original expression. Check each sign carefully.

Check. $(y + 12)(y - 2) = y^2 - 2y + 12y - 24 = y^2 + 10y - 24$ ✓ □

Student Practice 6 Factor $y^2 + 17y - 60$.
Multiply your answer to check.

Example 7 Factor. $x^2 - 16x - 36$

Solution We want two numbers whose product is -36 and whose sum is -16.

List all the possible factors of 36 (without regard to sign). Find the pair that has a difference of 16. We are looking for a difference because the signs of the factors are different.

Factors of 36	*The Difference Between the Factors*
36 and 1	35
18 and 2	16 ← This is the value we want.
12 and 3	9
9 and 4	5
6 and 6	0

Once we have picked the pair of numbers (18 and 2), it is not difficult to find the signs. For the coefficient of the middle term to be -16, we will have to add the numbers -18 and $+2$.

$$x^2 - 16x - 36 = (x - 18)(x + 2)$$ □

Student Practice 7 Factor. $x^2 - 7x - 60$
You may find it helpful to list the pairs of numbers whose product is 60.

At this point you should work several problems to develop your factoring skills. This is one section where you really need to drill by doing many problems.

Feel a little confused about the signs? If you do, you may find these facts helpful.

FACTS ABOUT FACTORING TRINOMIALS OF THE FORM $x^2 + bx + c$

The *two numbers m* and *n* will have the *same sign* if the last term of the polynomial is *positive*.

1. They will both be *positive* if the *coefficient* of the *middle* term is *positive*. $x^2 + bx + c = (x \quad m)(x \quad n)$, $x^2 + 5x + 6 = (x + 2)(x + 3)$
2. They will both be *negative* if the *coefficient* of the *middle* term is *negative*. $x^2 - 5x + 6 = (x - 2)(x - 3)$

The two numbers *m* and *n* will have *opposite signs* if the last term is *negative*.

1. The *larger* of the absolute values of the two numbers will be given a plus sign if the coefficient of the *middle term* is *positive*. $x^2 + 6x - 7 = (x + 7)(x - 1)$
2. The larger of the absolute values of the two numbers will be given a negative sign if the coefficient of the *middle term* is *negative*. $x^2 - 6x - 7 = (x - 7)(x + 1)$

Do not memorize these facts; rather, try to understand the pattern.

Sometimes the exponent of the first term of the polynomial will be greater than 2. If the exponent is an even power, it is a square. For example, $x^4 = (x^2)(x^2)$. Likewise, $x^6 = (x^3)(x^3)$.

Example 8 Factor. $y^4 - 2y^2 - 35$

Solution Think: $y^4 = (y^2)(y^2)$ This will be the first term of each set of parentheses.

$(y^2 \quad)(y^2 \quad)$

$(y^2 + \quad)(y^2 - \quad)$ The last term of the polynomial is negative. Thus the signs of m and n will be different.

$(y^2 + 5)(y^2 - 7)$ Now think of factors of 35 whose difference is 2.

Multiply using FOIL to check. □

Student Practice 8 Factor. $a^4 + a^2 - 42$

2 Factoring Polynomials That Have a Common Factor and a Factor of the Form $x^2 + bx + c$

Some factoring problems require two steps. Often we must first factor out a common factor from each term of the polynomial. Once this is done, we may find that the other factor is a trinomial that can be factored using the methods previously discussed in this section.

Example 9 Factor. $2x^2 + 36x + 160$

Solution

$$2x^2 + 36x + 160 = 2(x^2 + 18x + 80)$$ First factor out the common factor 2 from each term of the polynomial.

$$= 2(x + 8)(x + 10)$$ Then factor the remaining polynomial.

The final answer is $2(x + 8)(x + 10)$. *Be sure to list all parts of the answer.*

Check. $2(x + 8)(x + 10) = 2(x^2 + 18x + 80) = 2x^2 + 36x + 160$ ✓

Thus we are sure that the answer is $2(x + 8)(x + 10)$. □

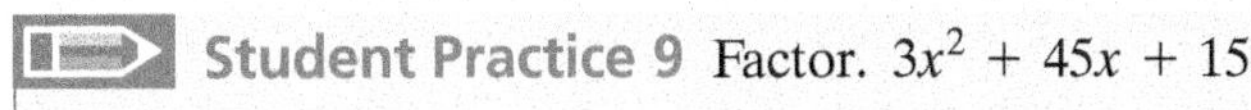

Student Practice 9 Factor. $3x^2 + 45x + 150$

Example 10 Factor. $3x^2 + 9x - 162$

Solution

$$3x^2 + 9x - 162 = 3(x^2 + 3x - 54)$$ First factor out the common factor 3 from each term of the polynomial.

$$= 3(x - 6)(x + 9)$$ Then factor the remaining polynomial.

The final answer is $3(x - 6)(x + 9)$. *Be sure you include the 3.*

Check. $3(x - 6)(x + 9) = 3(x^2 + 3x - 54) = 3x^2 + 9x - 162$ ✓

Thus we are sure that the answer is $3(x - 6)(x + 9)$. □

Student Practice 10

Factor. $4x^2 - 8x - 140$

CAUTION: Don't forget the common factor!

It is quite easy to forget to look for a greatest common factor as the first step of factoring a trinomial. Therefore, it is a good idea to examine your final answer in any factoring problem and ask yourself, "Can I factor out a common factor from any binomial contained inside a set of parentheses?" Often you will be able to see a common factor at that point if you missed it in the first step of the problem.

▲ **Example 11** Find a polynomial in factored form for the shaded area in the figure.

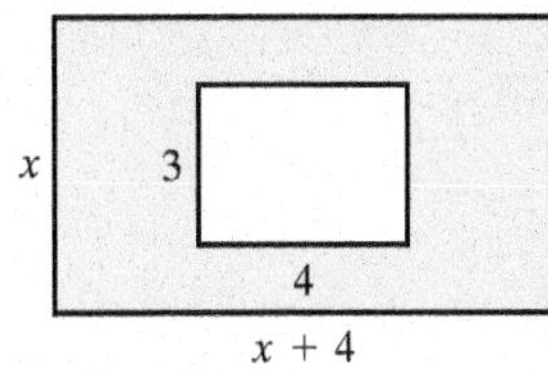

Solution To obtain the shaded area, we find the area of the larger rectangle and subtract from it the area of the smaller rectangle. Thus we have the following:

$$\text{shaded area} = x(x + 4) - (4)(3)$$
$$= x^2 + 4x - 12$$

Now we factor this polynomial to obtain the shaded area $= (x + 6)(x - 2)$. □

▲ **Student Practice 11** Find a polynomial in factored form for the shaded area in the figure.

Student Practice 11

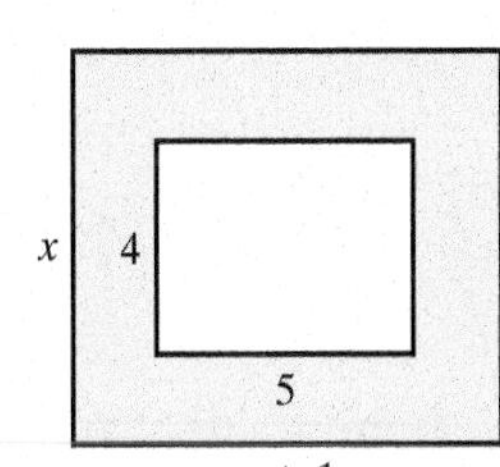

6.3 Exercises MyMathLab®

Verbal and Writing Skills, Exercises 1 and 2

Fill in the blanks.

1. To factor $x^2 + 5x + 6$, find two numbers whose ________ is 6 and whose ________ is 5.

2. To factor $x^2 + 5x - 6$, find two numbers whose ________ is -6 and whose ________ is 5.

Factor.

3. $x^2 + 8x + 16$

4. $x^2 + 13x + 42$

5. $x^2 + 12x + 35$

6. $x^2 + 10x + 21$

7. $x^2 - 4x + 3$

8. $x^2 - 8x + 15$

9. $x^2 - 11x + 28$

10. $x^2 - 13x + 12$

11. $x^2 + 5x - 24$

12. $x^2 + 5x - 36$

13. $x^2 - 13x - 14$

14. $x^2 - 6x - 16$

15. $x^2 + 2x - 35$

16. $x^2 + x - 12$

17. $x^2 - 2x - 24$

18. $x^2 - 11x - 26$

19. $x^2 + 15x + 36$

20. $x^2 + 15x + 44$

21. $x^2 - 10x + 24$

22. $x^2 - 13x + 42$

23. $x^2 + 13x + 30$

24. $x^2 + 9x + 20$

25. $x^2 - 6x + 5$

26. $x^2 - 15x + 54$

Mixed Practice *Look over your answers to exercises 3–26 carefully. Be sure that you are clear on your sign rules. Exercises 27–42 contain a mixture of all the types of problems in this section. Make sure you can do them all. Check your answers by multiplication.*

27. $a^2 + 6a - 16$

28. $a^2 - 10a + 24$

29. $x^2 - 12x + 32$

30. $x^2 - 6x - 27$

31. $x^2 + 4x - 21$

32. $x^2 - 9x + 18$

33. $x^2 + 15x + 56$

34. $x^2 + 20x + 99$

35. $y^2 + 4y - 45$

36. $x^2 + 12x - 45$

37. $x^2 + 9x - 36$

38. $x^2 - 13x + 36$

39. $x^2 - 2xy - 15y^2$

40. $x^2 - 3xy - 18y^2$

41. $x^2 - 16xy + 63y^2$

42. $x^2 + 19xy + 48y^2$

In exercises 43–54, first factor out the greatest common factor from each term. Then factor the remaining polynomial. Refer to Examples 9 and 10.

43. $4x^2 + 24x + 20$

44. $4x^2 + 28x + 40$

45. $6x^2 + 18x + 12$

46. $6x^2 + 24x + 18$

47. $5x^2 - 30x + 25$

48. $2x^2 - 20x + 32$

49. $3x^2 - 6x - 72$

50. $3x^2 - 18x - 48$

51. $7x^2 + 21x - 70$

52. $4x^2 - 8x - 60$

53. $5x^2 - 35x + 30$

54. $7x^2 - 35x + 42$

▲ **55.** ***Geometry*** Find a polynomial in factored form for the shaded area. Both figures are rectangles with dimensions as labeled.

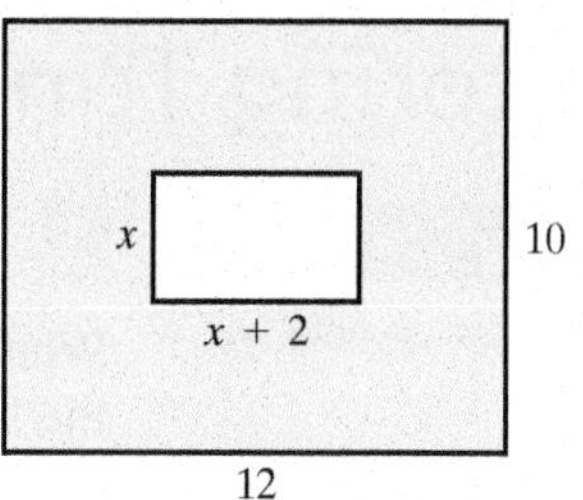

▲ **56.** ***Geometry*** How much larger is the perimeter of the large rectangle than the perimeter of the small rectangle?

Cumulative Review

57. **[5.1.1]** Multiply. $(9ab^3)(2a^5b^6c^0)$

58. **[5.1.3]** Simplify. $(-5y^6)^2$

59. **[5.2.1]** Simplify. Write the expression with no negative exponents. $\dfrac{x^4y^{-3}}{x^{-2}y^5}$

60. **[5.4.2]** Multiply. $(2x + 3y)(4x - 2y)$

61. **[2.6.2]** ***Travel Speed*** A new car that maintains a constant speed travels from Watch Hill, Rhode Island, to Greenwich, Connecticut, in 2 hours. A train, traveling 20 mph faster, makes the trip in $1\frac{1}{2}$ hours. How far is it from Watch Hill to Greenwich? (*Hint:* Let $c =$ the car's speed. First find the speed of the car and the speed of the train. Then be sure to answer the question.)

62. **[0.6.1]** ***Salary*** Carla works in an electronics store. She is paid $600 per month plus 4% commission on her sales. If Carla's sales for the month are $80,000, how much will she earn for the month?

Average Temperature *The equation $T = 19 + 2M$ has been used by some meteorologists to predict the monthly average temperature for the small island of Menorca off the coast of Spain during the first 6 months of the year. The variable T represents the average monthly temperature measured in degrees Celsius. The variable M represents the number of months since January.*

63. **[1.8.2]** What is the average temperature of Menorca during the month of April?

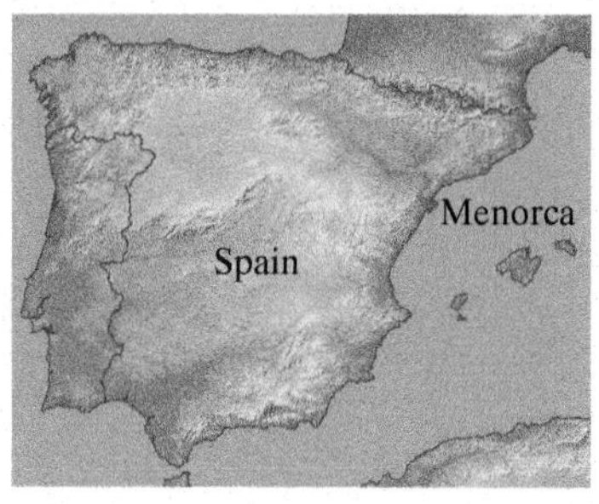

64. **[1.8.2]** During what month is the average temperature 29°C?

Quick Quiz 6.3 *Factor completely.*

1. $x^2 + 17x + 70$

2. $x^2 - 14x + 48$

3. $2x^2 - 4x - 96$

4. **Concept Check** Explain how you would completely factor $4x^2 - 4x - 120$.

6.4 Factoring Trinomials of the Form $ax^2 + bx + c$

Student Learning Objectives

After studying this section, you will be able to:

1. Factor a trinomial of the form $ax^2 + bx + c$ by the trial-and-error method.
2. Factor a trinomial of the form $ax^2 + bx + c$ by the grouping method.
3. Factor a trinomial of the form $ax^2 + bx + c$ after a common factor has been factored out of each term.

1 Using the Trial-and-Error Method

When the coefficient of the x^2-term in a trinomial of the form $ax^2 + bx + c$ is not 1, the trinomial is more difficult to factor. Several possibilities must be considered.

Example 1 Factor. $2x^2 + 5x + 3$

Solution In order for the coefficient of the x^2-term of the polynomial to be 2, the coefficients of the x-terms in the factors must be 2 and 1.

$$\text{Thus } 2x^2 + 5x + 3 = (2x \qquad)(x \qquad).$$

In order for the last term of the polynomial to be 3, the constants in the factors must be 3 and 1.

Since all signs in the polynomial are positive, we know that each factor in parentheses will contain only positive signs. However, we still have two possibilities. They are as follows:

$$(2x + 1)(x + 3)$$
$$(2x + 3)(x + 1)$$

We check them by multiplying by the FOIL method.

$$(2x + 1)(x + 3) = 2x^2 + 7x + 3 \quad \text{Wrong middle term}$$
$$(2x + 3)(x + 1) = 2x^2 + 5x + 3 \quad \text{Correct middle term}$$

Thus the correct answer is

$$(2x + 3)(x + 1) \quad \text{or} \quad (x + 1)(2x + 3).$$ □

Student Practice 1 Factor. $2x^2 + 7x + 5$

Some problems have many more possibilities.

Example 2 Factor. $4x^2 - 13x + 3$

Solution

The Different Factorizations of 4 Are:	*The Factorization of 3 Is:*
(2)(2)	(1)(3)
(1)(4)	

Let us list the possible factoring combinations and compute the middle term by the FOIL method. Note that the signs of the constants in both factors will be negative. Why?

Possible Factors	*Middle Term*	*Correct?*
$(2x - 3)(2x - 1)$	$-8x$	No
$(4x - 3)(x - 1)$	$-7x$	No
$(4x - 1)(x - 3)$	$-13x$	Yes

The correct answer is $(4x - 1)(x - 3)$ or $(x - 3)(4x - 1)$.
This method is called the **trial-and-error method.** □

Student Practice 2 Factor. $9x^2 - 64x + 7$

Example 3 Factor. $3x^2 - 2x - 8$

Solution

Factorization of 3	*Factorizations of 8*
(3)(1)	(8)(1)
	(4)(2)

Let us list only one-half of the possibilities. We'll let the constant in the first factor of each product be positive.

Possible Factors	*Middle Term*	*Correct Factors?*
$(x + 8)(3x - 1)$	$+23x$	No
$(x + 1)(3x - 8)$	$-5x$	No
$(x + 4)(3x - 2)$	$+10x$	No
$(x + 2)(3x - 4)$	$+2x$	No (but only because the sign is wrong)

So we just *reverse* the signs of the constants in the factors.

	Middle Term	*Correct Factors?*
$(x - 2)(3x + 4)$	$-2x$	Yes

The correct answer is $(x - 2)(3x + 4)$ or $(3x + 4)(x - 2)$. □

 **Student Practice 3** Factor. $3x^2 - x - 14$

Example 4 Factor. $6x^2 + 5x - 4$

Solution

Factorizations of 6	*Factorizations of 4*
(6)(1)	(4)(1)
(2)(3)	(2)(2)

We list a few of the possibilities (without regard to the sign) starting with $(2x)(3x)$ for the first term of each product.

$(2x \quad 4)(3x \quad 1) \qquad (2x \quad 1)(3x \quad 4) \qquad (2x \quad 2)(3x \quad 2)$

Look at the options below. Using the FOIL method, which option would yield a middle term $+5x$ when adding the outer and inner products?

$(2x \quad 4)(3x \quad 1)$	$(2x \quad 1)(3x \quad 4)$	$(2x \quad 2)(3x \quad 2)$
$12x$	$3x$	$6x$
$2x$	$8x$	$4x$
No possible sum of $+5x$	$5x$ ✓ $+5x$ is possible with $-3x$ and $+8x$	No possible sum of $+5x$

The middle possibility yields $+5x$ if we make the $3x$ negative. Once we have picked the pair of numbers, it is not difficult to find the signs and write the factors $(2x - 1)(3x + 4)$.

Now check your answer using FOIL to be sure you obtain the original expression. Pay particular attention to the last term; it must be -4.

Check. $(2x - 1)(3x + 4) = 6x^2 + 5x - 4$ ✓

Note: If you cannot find the factors using $(2)(3)$ for the coefficient of x^2, then try the same method using $(6)(1)$. □

Student Practice 4 Factor. $10x^2 + 7x - 6$

It takes a good deal of practice to readily factor problems of this type. The more problems you do, the more proficient you will become. The following method may help you factor more quickly.

2 Using the Grouping Method

One way to factor a trinomial of the form $ax^2 + bx + c$ is to write it with four terms and factor by grouping, as we did in Section 6.2. For example, the trinomial $2x^2 + 13x + 20$ can be written as $2x^2 + 5x + 8x + 20$. Using the methods of Section 6.2, we factor it as follows.

$$\begin{aligned} 2x^2 + 5x + 8x + 20 &= x(2x + 5) + 4(2x + 5) \\ &= (2x + 5)(x + 4) \end{aligned}$$

We can factor all factorable trinomials of the form $ax^2 + bx + c$ in this way. We will use the following procedure.

GROUPING METHOD FOR FACTORING TRINOMIALS OF THE FORM $ax^2 + bx + c$

1. Obtain the grouping number ac.
2. Find the two numbers whose product is the grouping number and whose sum is b.
3. Use those numbers to write bx as the sum of two terms.
4. Factor by grouping.
5. Multiply to check.

Example 5 Factor by grouping. $3x^2 - 2x - 8$

Solution

1. The grouping number is $(3)(-8) = -24$.
2. We want two numbers whose product is -24 and whose sum is -2. They are -6 and 4.
3. We write $-2x$ as the sum $-6x + 4x$.
4. Factor by grouping.

$$\begin{aligned} 3x^2 - 6x + 4x - 8 &= 3x(x - 2) + 4(x - 2) \\ &= (x - 2)(3x + 4) \end{aligned}$$

□

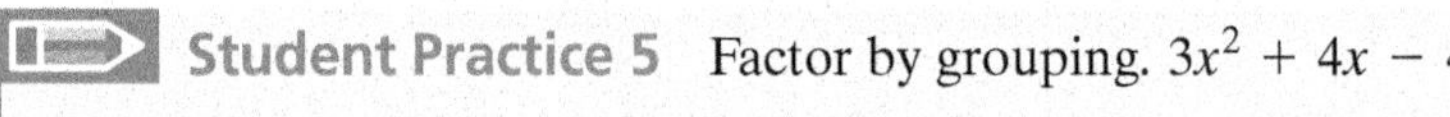
Student Practice 5 Factor by grouping. $3x^2 + 4x - 4$

Example 6 Factor by grouping. $4x^2 - 13x + 3$

Solution

1. The grouping number is $(4)(3) = 12$.
2. The factors of 12 are $(12)(1)$ or $(4)(3)$ or $(6)(2)$. Note that the middle term of the polynomial is negative. Thus we choose the numbers -12 and -1 because their product is still 12 and their sum is -13.
3. We write $-13x$ as the sum $-12x + (-1x)$ or $-12x - 1x$.
4. Factor by grouping.

$$\begin{aligned} 4x^2 - 13x + 3 &= 4x^2 - 12x - 1x + 3 \\ &= 4x(x - 3) - 1(x - 3) \end{aligned}$$

Remember to factor out -1 from the last two terms so that both sets of parentheses contain the same expression.

$$= (x - 3)(4x - 1)$$

□

Student Practice 6 Factor by grouping. $9x^2 - 64x + 7$

To factor polynomials of the form $ax^2 + bx + c$, use the method, either trial-and-error or grouping, that works best for you.

3 Factoring Out a Common Factor

Some problems require first factoring out the greatest common factor and then factoring the trinomial by one of the two methods of this section.

Example 7 Factor. $9x^2 + 3x - 30$

Solution

$9x^2 + 3x - 30 = 3(3x^2 + 1x - 10)$ We first factor out the common factor 3 from each term of the trinomial.

$= 3(3x - 5)(x + 2)$ We then factor the trinomial by the grouping method or by the trial-and-error method. ◻

Student Practice 7 Factor. $8x^2 + 8x - 6$

Be sure to remove the greatest common factor as the very first step.

Example 8 Factor. $32x^2 - 40x + 12$

Solution

$32x^2 - 40x + 12 = 4(8x^2 - 10x + 3)$ We first factor out the greatest common factor 4 from each term of the trinomial.

$= 4(2x - 1)(4x - 3)$ We then factor the trinomial by the grouping method or by the trial-and-error method. ◻

Student Practice 8 Factor. $24x^2 - 38x + 10$

STEPS TO SUCCESS Look Ahead to See What Is Coming

You will find that learning new material is much easier if you know what is coming. Take a few minutes at the end of your study time to glance over the next section of the book. If you quickly look over the topics and ideas in this new section, it will help you get your bearings when the instructor presents new material. Students find that when they preview new material, it enables them to see what is coming. It helps them to be able to grasp new ideas much more quickly.

Making it personal: Do this right now. Look ahead to the next section of the book. Glance over the ideas and concepts. Write down a couple of facts about the next section. ▼

6.4 Exercises MyMathLab®

Factor by the trial-and-error method. Check your answers using FOIL.

1. $4x^2 + 21x + 5$ **2.** $3x^2 + 13x + 12$ **3.** $5x^2 + 7x + 2$ **4.** $4x^2 + 5x + 1$

5. $4x^2 + 5x - 6$ **6.** $5x^2 + 9x - 2$ **7.** $2x^2 - 5x - 3$ **8.** $2x^2 - x - 6$

Factor by the grouping method. Check your answers by using FOIL.

9. $9x^2 + 9x + 2$ **10.** $4x^2 + 11x + 6$ **11.** $15x^2 - 34x + 15$ **12.** $10x^2 - 29x + 10$

13. $2x^2 + 3x - 20$ **14.** $6x^2 + 11x - 10$ **15.** $8x^2 + 10x - 3$ **16.** $5x^2 - 34x - 7$

Factor by any method.

17. $6x^2 - 5x - 6$ **18.** $3x^2 - 13x - 10$ **19.** $10x^2 + 3x - 1$ **20.** $6x^2 + x - 5$

21. $7x^2 - 5x - 18$ **22.** $9x^2 - 22x - 15$ **23.** $9y^2 - 13y + 4$ **24.** $5y^2 - 11y + 2$

25. $5a^2 - 13a - 6$ **26.** $2a^2 + 5a - 12$ **27.** $14x^2 + 17x - 6$ **28.** $32x^2 + 36x - 5$

29. $15x^2 + 4x - 4$ **30.** $8x^2 - 11x + 3$ **31.** $12x^2 + 28x + 15$ **32.** $24x^2 + 17x + 3$

33. $12x^2 - 16x - 3$ **34.** $12x^2 + x - 6$ **35.** $3x^4 - 14x^2 - 5$ **36.** $4x^4 + 8x^2 - 5$

37. $2x^2 + 11xy + 15y^2$ **38.** $15x^2 + 28xy + 5y^2$ **39.** $5x^2 + 16xy - 16y^2$ **40.** $12x^2 + 11xy - 5y^2$

Factor by first factoring out the greatest common factor. See Examples 7 and 8.

41. $10x^2 - 25x - 15$ **42.** $20x^2 - 25x - 30$ **43.** $6x^3 + 9x^2 - 60x$ **44.** $6x^3 - 16x^2 - 6x$

Mixed Practice *Factor.*

45. $5x^2 + 3x - 2$ **46.** $6x^2 + x - 2$ **47.** $12x^2 - 38x + 20$ **48.** $30x^2 - 26x + 4$

49. $12x^3 - 20x^2 + 3x$ **50.** $12x^3 - 13x^2 + 3x$ **51.** $8x^2 + 24x - 14$ **52.** $12x^2 + 26x - 10$

Cumulative Review

53. [3.3.1] Find the slope of the line that passes through $(-1, 6)$ and $(2, 4)$.

54. [2.4.1] Solve. $\frac{x}{3} - \frac{x}{5} = \frac{7}{15}$

Overseas Travelers *The double-bar graph shows the top states visited by foreign travelers (excluding travelers from Canada and Mexico) for last year and five years ago.*

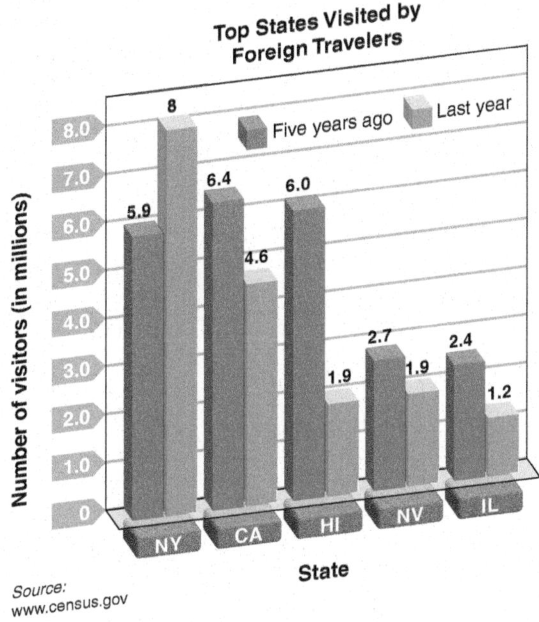

55. [0.6.1] (a) How many overseas travelers visited these five states last year?

(b) What percent of these travelers visited Hawaii (HI)? Round to the nearest tenth.

56. [0.6.1] Of the overseas visitors to these five states five years ago, what percent traveled to California (CA)? Round to the nearest tenth.

57. [0.6.1] (a) Find the difference in the number of visitors to California (CA) last year versus five years ago.

(b) This decrease is what percent of the visitors from five years ago? Round to the nearest tenth.

58. [0.6.1] (a) Find the difference in the number of visitors to New York (NY) last year versus five years ago.

(b) This increase is what percent of the visitors from five years ago? Round to the nearest tenth.

Quick Quiz 6.4 *Factor by any method.*

1. $12x^2 + 16x - 3$

2. $10x^2 - 21x + 9$

3. $6x^3 - 3x^2 - 30x$

4. Concept Check Explain how you would factor $10x^3 + 18x^2y - 4xy^2$.

Use Math to Save Money

Did You Know? Even if you always pay on time, having a high debt-to-credit ratio will hurt your credit score.

Get Out of Debt and Improve Your Credit Score

Understanding the Problem:

Megan would like to purchase a house in a few years, but currently her credit score is not very high. This will affect her ability to get a mortgage. She needs to find a way to improve her credit score. Part of her problem is that she has almost maxed out her credit cards.

- Credit card 1 has a current balance of $5100 and a credit limit of $5500.
- Credit card 2 has a current balance of $3800 and a credit limit of $4000.
- Credit card 3 has a current balance of $3200 and a credit limit of $3500.

Making a Plan:

Megan talks to a credit counselor and is told that even though she has always paid her bills on time, she has a high debt-to-credit ratio, and this is lowering her credit score. (To calculate this ratio, first total the current balances on all the credit cards. Next, total the credit limits on all the credit cards. Then divide the total current balance by the total credit limit. Finally, convert this value to a percentage.) Lowering this percentage will increase her credit score. Getting this percentage down to less than 50% would be good. It would be even better for her credit score if she could get this percentage down to less than 33%.

Step 1: Megan and her counselor total up her balances and come up with a plan for Megan to reduce her debt over the next couple of years. Reducing her balances without reducing her credit limits will lower her debt-to-credit ratio.

Task 1: What is the total current balance (or total debt) on Megan's cards?

Task 2: What is her total credit limit?

Step 2: Now they can calculate her debt-to-credit ratio.

Task 3: Calculate Megan's debt-to-credit ratio as a percentage. Round to the nearest percent.

Finding a Solution:

They decide that Megan should try to reduce her debt-to-credit ratio to 50% over the next two years.

Step 3: They calculate how much she needs to pay off to achieve this goal.

Task 4: What amount of debt would be 50% of her total credit limit?

Task 5: How much does she need to pay off to reach that amount?

Step 4: Megan is going to have to cut some spending to reach this goal in two years, but she knows it is going to be worth her efforts. As she pays down her debt, a greater percentage of her payments will go toward the principal, so she will be paying down her balances faster as time goes on. She wonders if she will be able to reduce her debt-to-credit ratio to 33% by the end of the third year.

Task 6: How much would she need to pay off in the third year to reach 33%?

Applying the Situation to Your Life:

Most people know that paying their bills on time is very important and that it has a positive impact on their credit score. What many people don't know is that having a high debt-to-credit ratio will lower their score. Paying down the debt is a great way to fix this problem. Some people think they should cancel their credit cards after they pay them off, but keeping them open increases the total credit limit. This reduces the debt-to-credit ratio, which will boost the credit score. You should close your accounts only if you are having trouble controlling your spending.

6.5 Special Cases of Factoring 

As we proceed in this section you will be able to reduce the time it takes you to factor polynomials by quickly recognizing and factoring two special types of polynomials: the difference of two squares and perfect-square trinomials.

Student Learning Objectives

After studying this section, you will be able to:

1. Recognize and factor expressions of the type $a^2 - b^2$ (difference of two squares).
2. Recognize and factor expressions of the type $a^2 + 2ab + b^2$ (perfect-square trinomial).
3. Recognize and factor expressions that require factoring out a common factor and then using a special-case formula.

1 Factoring the Difference of Two Squares

Recall the formula from Section 5.5:

$$(a + b)(a - b) = a^2 - b^2.$$

In reverse form we can use it for factoring.

DIFFERENCE OF TWO SQUARES

$$a^2 - b^2 = (a + b)(a - b)$$

We can state it in words in this way:

"The difference of two squares can be factored into the sum and difference of those values that were squared."

Example 1 Factor. $9x^2 - 1$

Solution We see that the polynomial is in the form of the difference of two squares. $9x^2$ is a square and 1 is a square. So using the formula we can write the following.

$$9x^2 - 1 = (3x + 1)(3x - 1) \quad \text{Because } 9x^2 = (3x)^2 \text{ and } 1 = (1)^2$$

Student Practice 1 Factor. $64x^2 - 1$

Example 2 Factor. $25x^2 - 16$

Solution Again we use the formula for the difference of squares.

$$25x^2 - 16 = (5x + 4)(5x - 4) \quad \text{Because } 25x^2 = (5x)^2 \text{ and } 16 = (4)^2$$

Student Practice 2 Factor. $36x^2 - 49$

Sometimes the polynomial contains two variables.

Example 3 Factor. $4x^2 - 49y^2$

Solution We see that

$$4x^2 - 49y^2 = (2x + 7y)(2x - 7y).$$

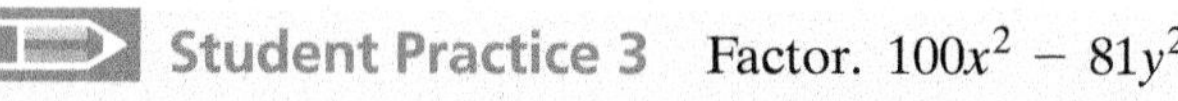

Student Practice 3 Factor. $100x^2 - 81y^2$

CAUTION: Please note that the difference of two squares formula only works if the last term is negative. So if Example 3 had been to factor $4x^2 + 49y^2$, we would **not** have been able to factor the problem. We will examine this in more detail in Section 6.6.

Some problems may involve more than one step.

Example 4 Factor. $81x^4 - 1$

Solution We see that

$81x^4 - 1 = (9x^2 + 1)(9x^2 - 1)$. Because $81x^4 = (9x^2)^2$ and $1 = (1)^2$

Is the factoring complete? No. We can factor $9x^2 - 1$.

$81x^4 - 1 = (9x^2 + 1)(3x + 1)(3x - 1)$ Because $(9x^2 - 1) = (3x + 1)(3x - 1)$ □

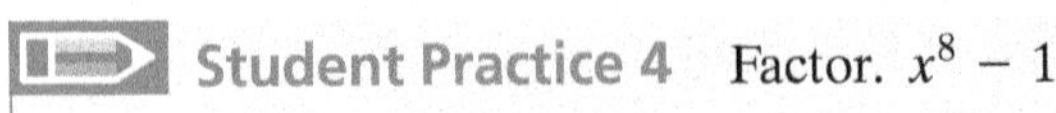

Student Practice 4 Factor. $x^8 - 1$

2 Factoring Perfect-Square Trinomials

There is a formula that will help us to very quickly factor certain trinomials called **perfect-square trinomials.** Recall from Section 5.5 the formulas for binomials squared.

$$(a + b)^2 = a^2 + 2ab + b^2$$
$$(a - b)^2 = a^2 - 2ab + b^2$$

We can use these two equations in reverse form for factoring.

PERFECT-SQUARE TRINOMIALS

$$a^2 + 2ab + b^2 = (a + b)^2$$
$$a^2 - 2ab + b^2 = (a - b)^2$$

A perfect-square trinomial is a trinomial that is the result of squaring a binomial. How can we recognize a perfect-square trinomial?

1. The first and last terms are *perfect squares.*
2. The middle term is twice the product of the values whose squares are the first and last terms.

Example 5 Factor. $x^2 + 6x + 9$

Solution This is a perfect-square trinomial.

1. The first and last terms are perfect squares because $x^2 = (x)^2$ and $9 = (3)^2$.
2. The middle term, $6x$, is twice the product of x and 3.

Since $x^2 + 6x + 9$ is a perfect-square trinomial, we can use the formula

$$a^2 + 2ab + b^2 = (a + b)^2$$

with $a = x$ and $b = 3$. So we have

$$x^2 + 6x + 9 = (x + 3)^2.$$ □

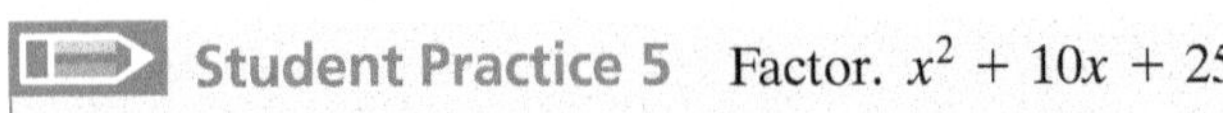

Student Practice 5 Factor. $x^2 + 10x + 25$

Example 6 Factor. $4x^2 - 20x + 25$

Solution This is a perfect-square trinomial.

Note that $20x = 2(2x \cdot 5)$. Also note the negative sign.

Thus we have the following.

$4x^2 - 20x + 25 = (2x - 5)^2$ Since $a^2 - 2ab + b^2 = (a - b)^2$ □

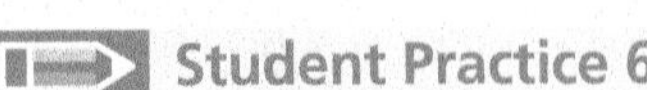

Student Practice 6

Factor. $25x^2 - 30x + 9$

A polynomial may have more than one variable and its exponents may be higher than 2. The same principles apply.

Example 7 Factor.

(a) $49x^2 + 42xy + 9y^2$ **(b)** $36x^4 - 12x^2 + 1$

Solution

(a) This is a perfect-square trinomial. Why?

$49x^2 + 42xy + 9y^2 = (7x + 3y)^2$ Because $49x^2 = (7x)^2$, $9y^2 = (3y)^2$, and $42xy = 2(7x \cdot 3y)$

(b) This is a perfect-square trinomial. Why?

$36x^4 - 12x^2 + 1 = (6x^2 - 1)^2$ Because $36x^4 = (6x^2)^2$, $1 = (1)^2$, and $12x^2 = 2(6x^2 \cdot 1)$ □

Student Practice 7 Factor.

(a) $25x^2 + 60xy + 36y^2$ **(b)** $64x^6 - 48x^3 + 9$

Be *careful*. Some polynomials appear to be perfect-square trinomials but are not. They were factored in other ways in Section 6.4.

Example 8 Factor. $49x^2 + 35x + 4$

Solution This is *not* a perfect-square trinomial! Although the first and last terms are perfect squares since $(7x)^2 = 49x^2$ and $(2)^2 = 4$, the middle term, $35x$, is not double the product of 2 and $7x$. $35x \neq 28x$! So we must factor by trial and error or by grouping to obtain

$$49x^2 + 35x + 4 = (7x + 4)(7x + 1).$$ □

Student Practice 8 Factor. $9x^2 + 15x + 4$

3 Factoring Out a Common Factor Before Using a Special-Case Formula

For some polynomials, we first need to factor out the greatest common factor. Then we will find an opportunity to use the difference of two squares formula or one of the perfect-square trinomial formulas.

Example 9 Factor. $12x^2 - 48$

Solution We see that the greatest common factor is 12.

$12x^2 - 48 = 12(x^2 - 4)$ First we factor out 12.

$= 12(x + 2)(x - 2)$ Then we use the difference of two squares formula, $a^2 - b^2 = (a + b)(a - b)$. □

Student Practice 9 Factor. $20x^2 - 45$

Look carefully at Example 10. Can you identify the greatest common factor?

Example 10 Factor. $24x^2 - 72x + 54$

Solution

$24x^2 - 72x + 54 = 6(4x^2 - 12x + 9)$ First we factor out the greatest common factor, 6.

$= 6(2x - 3)^2$ Then we use the perfect-square trinomial formula, $a^2 - 2ab + b^2 = (a - b)^2$. □

Student Practice 10 Factor. $75x^2 - 60x + 12$

6.5 Exercises MyMathLab®

Factor by using the difference of two squares formula.

1. $100x^2 - 1$

2. $36x^2 - 1$

3. $81x^2 - 16$

4. $49x^2 - 25$

5. $x^2 - 49$

6. $9x^2 - 64$

7. $25x^2 - 81$

8. $49x^2 - 4$

9. $x^2 - 25$

10. $x^2 - 36$

11. $1 - 16x^2$

12. $1 - 64x^2$

13. $16x^2 - 49y^2$

14. $25x^2 - 81y^2$

15. $36x^2 - 169y^2$

16. $4x^2 - 9y^2$

17. $100x^2 - 81$

18. $16a^2 - 25$

19. $25a^2 - 81b^2$

20. $9x^2 - 49y^2$

Factor by using the perfect-square trinomial formula.

21. $9x^2 + 6x + 1$

22. $25x^2 + 10x + 1$

23. $y^2 - 10y + 25$

24. $y^2 - 12y + 36$

25. $36x^2 - 60x + 25$

26. $9x^2 - 42x + 49$

27. $49x^2 + 28x + 4$

28. $25x^2 + 30x + 9$

29. $x^2 + 14x + 49$

30. $x^2 + 8x + 16$

31. $25x^2 - 40x + 16$

32. $64x^2 - 16x + 1$

33. $81x^2 + 36xy + 4y^2$

34. $36x^2 + 60xy + 25y^2$

35. $9x^2 - 30xy + 25y^2$

36. $49x^2 - 28xy + 4y^2$

Mixed Practice *Factor by using either the difference of two squares or the perfect-square trinomial formula.*

37. $16a^2 + 72ab + 81b^2$

38. $169a^2 + 26ab + b^2$

39. $49x^2 - 42xy + 9y^2$

40. $9x^2 - 30xy + 25y^2$

41. $64x^2 + 80x + 25$

42. $16x^2 + 56x + 49$

43. $144x^2 - 1$

44. $16x^2 - 121$

45. $x^4 - 16$

46. $x^4 - 1$

47. $9x^4 - 24x^2 + 16$

48. $64x^4 - 48x^2 + 9$

To Think About, Exercises 49–52

49. In Example 4, first we factored $81x^4 - 1$: $(9x^2 + 1)(9x^2 - 1)$. Then we factored $9x^2 - 1$: $(3x + 1)(3x - 1)$. Show why you cannot factor $9x^2 + 1$.

50. What two numbers could replace the b in $36x^2 + bx + 49$ so that the resulting trinomial would be a perfect square? (*Hint*: One number is negative.)

51. What value could you give to c so that $16y^2 - 24y + c$ would become a perfect-square trinomial? Is there only one answer or more than one?

52. Jerome says that he can find two values of b so that $100x^2 + bx - 9$ will be a perfect square. Kesha says there is only one that fits, and Larry says there are none. Who is correct and why?

Factor by first looking for a greatest common factor. See Examples 9 and 10.

53. $16x^2 - 36$

54. $27x^2 - 75$

55. $147x^2 - 3y^2$

56. $16y^2 - 100x^2$

57. $16x^2 - 16x + 4$

58. $45x^2 - 60x + 20$

59. $98x^2 + 84x + 18$

60. $50x^2 + 80x + 32$

Mixed Practice *Factor. Be sure to look for common factors first.*

61. $x^2 + 16x + 63$

62. $x^2 - 3x - 40$

63. $2x^2 + 5x - 3$

64. $15x^2 - 11x + 2$

65. $12x^2 - 27$

66. $16x^2 - 36$

67. $9x^2 + 42x + 49$

68. $9x^2 + 30x + 25$

69. $36x^2 - 36x + 9$

70. $6x^2 + 60x + 150$

71. $2x^2 - 32x + 126$

72. $2x^2 - 34x + 140$

Cumulative Review

73. **[5.6.2]** Divide. $(x^3 + x^2 - 2x - 11) \div (x - 2)$

74. **[5.6.2]** Divide. $(6x^3 + 11x^2 - 11x - 20) \div (3x + 4)$

Iguana Diet *The green iguana can reach a length of 6 feet and weigh up to 18 pounds. Of the basic diet of the iguana, 40% should consist of greens such as lettuce, spinach, and parsley; 35% should consist of bulk vegetables such as broccoli, zucchini, and carrots; and 25% should consist of fruit.*

75. **[0.6.1]** If a certain iguana weighing 150 ounces has a daily diet equal to 2% of its body weight, compose a diet for it in ounces that will meet the iguana's one-day requirement for nutrition.

76. **[0.6.1]** If another iguana weighing 120 ounces has a daily diet equal to 3% of its body weight, compose a diet for it in ounces that will meet the iguana's one-day requirement for nutrition.

Quick Quiz 6.5 *Factor the following.*

1. $49x^2 - 81y^2$

2. $9x^2 - 48x + 64$

3. $162x^2 - 200$

4. **Concept Check** Explain how to factor the polynomial $24x^2 + 120x + 150$.

6.6 A Brief Review of Factoring

Student Learning Objectives

After studying this section, you will be able to:

1. Identify and factor any polynomial that can be factored.
2. Determine whether a polynomial is prime.

1 Identifying and Factoring Polynomials

Often the various types of factoring problems are all mixed together. We need to be able to identify each type of polynomial quickly. The following table summarizes the information we have learned about factoring.

Many polynomials require more than one factoring method. When you are asked to factor a polynomial, it is expected that you will factor it completely. Usually, the first step is factoring out a common factor; then the next step will become apparent.

Carefully go through each example in the following **Factoring Organizer.** Be sure you understand each step that is involved.

Factoring Organizer

Number of Terms in the Polynomial	Identifying Name and/or Formula	Example
A. Any number of terms	**Common factor** The terms have a common factor consisting of a number, a variable, or both.	$2x^2 - 16x = 2x(x - 8)$ $3x^2 + 9y - 12 = 3(x^2 + 3y - 4)$ $4x^2y + 2xy^2 - wxy + xyz = xy(4x + 2y - w + z)$
B. Two terms	**Difference of two squares** First and last terms are perfect squares. $a^2 - b^2 = (a + b)(a - b)$	$16x^2 - 1 = (4x + 1)(4x - 1)$ $25y^2 - 9x^2 = (5y + 3x)(5y - 3x)$
C. Three terms	**Perfect-square trinomial** First and last terms are perfect squares. $a^2 + 2ab + b^2 = (a + b)^2$ $a^2 - 2ab + b^2 = (a - b)^2$	$25x^2 - 10x + 1 = (5x - 1)^2$ $16x^2 + 24x + 9 = (4x + 3)^2$
D. Three terms	**Trinomial of the form $x^2 + bx + c$** It starts with x^2. The constants of the two factors are numbers whose product is c and whose sum is b.	$x^2 - 7x + 12 = (x - 3)(x - 4)$ $x^2 + 11x - 26 = (x + 13)(x - 2)$ $x^2 - 8x - 20 = (x - 10)(x + 2)$
E. Three terms	**Trinomial of the form $ax^2 + bx + c$** It starts with ax^2, where a is any number but 0 or 1.	Use trial-and-error or the grouping method to factor $12x^2 - 5x - 2$. **1.** The grouping number is -24. **2.** The two numbers whose product is -24 and whose sum is -5 are 3 and -8. **3.** $12x^2 - 5x - 2 = 12x^2 + 3x - 8x - 2$ $= 3x(4x + 1) - 2(4x + 1)$ $= (4x + 1)(3x - 2)$
F. Four terms	**Factor by grouping** Rearrange the order if the first two terms do not have a common factor.	$wx - 6yz + 2wy - 3xz = wx + 2wy - 3xz - 6yz$ $= w(x + 2y) - 3z(x + 2y)$ $= (x + 2y)(w - 3z)$

Example 1 Factor. These example problems are mixed.

(a) $25x^3 - 10x^2 + x$ **(b)** $20x^2y^2 - 45y^2$ **(c)** $15x^2 - 3x^3 + 18x$

Solution

(a) $25x^3 - 10x^2 + x = x(25x^2 - 10x + 1)$ Factor out the common factor x.

$= x(5x - 1)^2$ The other factor is a perfect-square trinomial.

(b) $20x^2y^2 - 45y^2 = 5y^2(4x^2 - 9)$ Factor out the common factor $5y^2$.

$= 5y^2(2x + 3)(2x - 3)$ The other factor is a difference of squares.

(c) $15x^2 - 3x^3 + 18x = -3x^3 + 15x^2 + 18x$ Rearrange the terms in descending order of powers of x.

$= -3x(x^2 - 5x - 6)$ Factor out the common factor $-3x$.

$= -3x(x - 6)(x + 1)$ Factor the trinomial. □

Student Practice 1 Factor. Be careful. These practice problems are mixed.

(a) $3x^2 - 36x + 108$ **(b)** $9x^4y^2 - 9y^2$

(c) $12x - 9 - 4x^2$

Example 2 Factor. $ax^2 - 9a + 2x^2 - 18$

Solution We factor by grouping since there are four terms.

$ax^2 - 9a + 2x^2 - 18 = a(x^2 - 9) + 2(x^2 - 9)$ Factor out the common factor a from the first two terms and 2 from the second two terms.

$= (a + 2)(x^2 - 9)$ Factor out the common factor $(x^2 - 9)$.

$= (a + 2)(x - 3)(x + 3)$ Factor $x^2 - 9$ using the difference of two squares formula. □

Student Practice 2 Factor. $5x^3 - 20x + 2x^2 - 8$

2 Determining Whether a Polynomial Is Prime

Not all polynomials can be factored using the methods in this chapter. If we cannot factor a polynomial by elementary methods, we will identify it as a **prime** polynomial. If, after you have mastered the factoring techniques in this chapter, you encounter a polynomial that you cannot factor with these methods, you should feel comfortable enough to say, "The polynomial cannot be factored with the methods in this chapter, so it is prime," rather than "I can't do it—I give up!"

Example 3 Factor, if possible.

(a) $x^2 + 6x + 12$ **(b)** $25x^2 + 4$

Solution

(a) The factors of 12 are

$$(1)(12) \text{ or } (2)(6) \text{ or } (3)(4).$$

None of these pairs add up to 6, the coefficient of the middle term. Thus the problem *cannot be factored* by the methods of this chapter. It is prime.

(b) We have a formula to factor the difference of two squares. There is no way to factor the sum of two squares. That is, $a^2 + b^2$ *cannot be factored.*

Thus $25x^2 + 4$ is prime. □

Student Practice 3 Factor, if possible.

(a) $x^2 - 9x - 8$ **(b)** $25x^2 + 82x + 4$

6.6 Exercises MyMathLab®

Review the six basic types of factoring in the Factoring Organizer on page 406. Each of the six types is included in exercises 1–12. Factor if possible. Check your answer by multiplying.

1. $3x^2 - 6xy + 5x$

2. $8x^2 + 5xy - 7x$

3. $16x^2 - 25y^2$

4. $36x^2 - 49y^2$

5. $x^2 + 64$

6. $25x^2 + 80xy + 64y^2$

7. $x^2 + 8x + 15$

8. $x^2 + 15x + 54$

9. $15x^2 + 7x - 2$

10. $6x^2 + 13x - 5$

11. $ax - 3cx + 3ay - 9cy$

12. $bx - 2dx + 5by - 10dy$

Mixed Practice *Factor, if possible. Be sure to factor completely. Always factor out the greatest common factor first, if one exists.*

13. $y^2 + 14y + 49$

14. $y^2 + 16y + 64$

15. $4x^2 - 12x + 9$

16. $16x^2 + 25$

17. $2x^2 - 11x + 12$

18. $3x^2 - 10x + 8$

19. $x^2 - 3xy - 70y^2$

20. $x^2 - 6xy - 16y^2$

21. $ax - 5a + 3x - 15$

22. $by + 7b - 6y - 42$

23. $16x - 4x^3$

24. $8y^2 + 10y - 12$

25. $2x^2 + 3x - 36$

26. $7x^2 + 3x - 2$

27. $3xyz^2 - 6xyz - 9xy$

28. $-16x^2 - 2x - 32x^3$

29. $3x^2 + 6x - 105$

30. $7x^2 + 21x - 70$

31. $5x^3y^3 - 10x^2y^3 + 5xy^3$

32. $2x^4y - 12x^3y + 18x^2y$

33. $7x^2 - 2x^4 + 4$

34. $14x^2 - x^3 + 32x$

35. $6x^2 - 3x + 2$

36. $4x^3 + 8x^2 - 60x$

37. $5x^4 - 5x^2 + 10x^3y - 10xy$

38. $2a^4b - 8a^2b + 3a^2b - 12b$

Remove the greatest common factor first. Then continue to factor, if possible.

39. $5x^2 + 10xy - 30y$

40. $6x^2 + 30x - 54y$

41. $30x^3 + 3x^2y - 6xy^2$

42. $48x^3 + 20x^2y - 8xy^2$

43. $30x^2 - 38x + 12$

44. $15x^2 - 35x + 10$

To Think About

45. A polynomial that cannot be factored by the methods of this chapter is called ______.

46. A binomial of the form $x^2 - d$ can be quickly factored or identified as prime. If it can be factored, what is true of the number d?

Cumulative Review

47. [2.7.2] ***Independent Contractor*** When Dave Barry decided to leave the company and work as an independent contractor, he took a pay cut of 14%. He earned \$24,080 this year. What did he earn in his previous job?

48. [0.6.1] ***Antiviral Drug*** A major pharmaceutical company is testing a new, powerful antiviral drug. It kills 13 strains of virus every hour. If there are presently 294 live strains of virus in the test container, how many live strains were there 6 hours ago?

49. [4.1.6] Solve the system.

$$\frac{1}{2}x - y = 7$$
$$-3x + 2y = -22$$

50. [3.3.4] Graph. $y = 2x + 3$

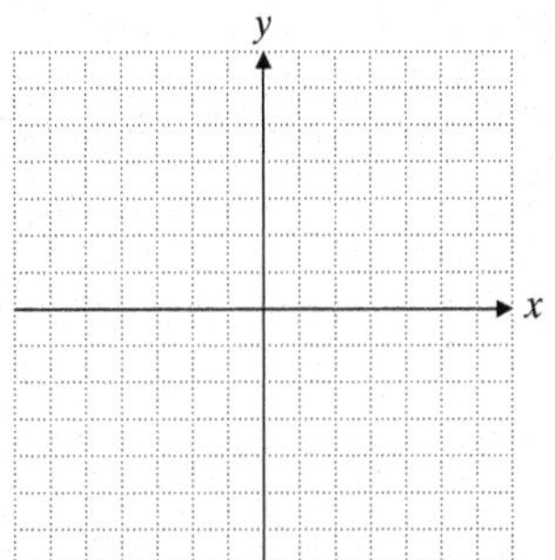

Quick Quiz 6.6 *Completely factor the following, if possible.*

1. $6x^2 - 17x + 12$

2. $60x^2 - 9x - 6$

3. $25x^2 + 49$

4. Concept Check Explain how to completely factor $2x^2 + 6xw - 5x - 15w$.

6.7 Solving Quadratic Equations by Factoring

Student Learning Objectives

After studying this section, you will be able to:

1 Solve quadratic equations by factoring.

2 Use quadratic equations to solve applied problems.

1 Solving Quadratic Equations by Factoring

In Chapter 2, we learned how to solve linear equations such as $3x + 5 = 0$ by finding the root (or value of x) that satisfied the equation. Now we turn to the question of how to solve equations like $3x^2 + 5x + 2 = 0$. Such equations are called **quadratic equations.** A quadratic equation is a polynomial equation in one variable that contains a variable term of degree 2 and no terms of higher degree.

The *standard form* of a quadratic equation is $ax^2 + bx + c = 0$, where a, b, and c are real numbers and $a \neq 0$.

In this section, we will study quadratic equations in standard form, where a, b, and c are integers.

Many quadratic equations have two real number solutions (also called **real roots**). But how can we find them? The most direct approach is the factoring method. This method depends on a very powerful property.

ZERO FACTOR PROPERTY

If $a \cdot b = 0$, then $a = 0$ or $b = 0$.

Notice the word *or* in the zero factor property. When we make a statement in mathematics using this word, we intend it to mean *one or the other or both*. Therefore, the zero factor property states that if the product $a \cdot b$ is zero, then a can equal zero or b can equal zero or *both a and b* can equal zero. We can use this principle to solve quadratic equations. Before you start, make sure that the equation is in standard form.

SOLVING A QUADRATIC EQUATION

1. Make sure the equation is set equal to zero.
2. Factor, if possible, the quadratic expression that equals zero.
3. Set each factor containing a variable equal to zero.
4. Solve the resulting equations to find each root.
5. Check each root.

Example 1 Solve the equation to find the two roots and check. $3x^2 + 5x + 2 = 0$

Solution

$$3x^2 + 5x + 2 = 0 \quad \text{The equation is in standard form.}$$

$$(3x + 2)(x + 1) = 0 \quad \text{Factor the quadratic expression.}$$

$$3x + 2 = 0 \qquad x + 1 = 0 \quad \text{Set each factor equal to 0.}$$

$$3x = -2 \qquad x = -1 \quad \text{Solve the equations to find the two roots.}$$

$$x = -\frac{2}{3}$$

The two roots (that is, solutions) are $-\frac{2}{3}$ and -1.

Check. We can determine if the two numbers $-\frac{2}{3}$ and -1 are solutions to the equation. Substitute $-\frac{2}{3}$ for x in the *original equation*. If an identity results, $-\frac{2}{3}$ is a solution. Do the same for -1.

$$
\begin{aligned}
3x^2 + 5x + 2 &= 0 \\
3\left(-\frac{2}{3}\right)^2 + 5\left(-\frac{2}{3}\right) + 2 &\stackrel{?}{=} 0 \\
3\left(\frac{4}{9}\right) + 5\left(-\frac{2}{3}\right) + 2 &\stackrel{?}{=} 0 \\
\frac{4}{3} - \frac{10}{3} + 2 &\stackrel{?}{=} 0 \\
\frac{4}{3} - \frac{10}{3} + \frac{6}{3} &\stackrel{?}{=} 0 \\
0 &= 0 \quad \checkmark
\end{aligned}
\qquad
\begin{aligned}
3x^2 + 5x + 2 &= 0 \\
3(-1)^2 + 5(-1) + 2 &\stackrel{?}{=} 0 \\
3(1) + 5(-1) + 2 &\stackrel{?}{=} 0 \\
3 - 5 + 2 &\stackrel{?}{=} 0 \\
-2 + 2 &\stackrel{?}{=} 0 \\
0 &= 0 \quad \checkmark
\end{aligned}
$$

Thus $-\frac{2}{3}$ and -1 are both roots of the equation $3x^2 + 5x + 2 = 0$. □

Student Practice 1 Solve the equation by factoring to find the two roots and check. $10x^2 - x - 2 = 0$

Example 2 Solve the equation to find the two roots. $2x^2 + 13x - 7 = 0$

Solution

$2x^2 + 13x - 7 = 0$		The equation is in standard form.
$(2x - 1)(x + 7) = 0$		Factor.
$2x - 1 = 0$	$x + 7 = 0$	Set each factor equal to 0.
$2x = 1$	$x = -7$	Solve the equations to find the two roots.
$x = \frac{1}{2}$		

The two roots are $\frac{1}{2}$ and -7.

Check. If $x = \frac{1}{2}$, then we have the following.

$$
\begin{aligned}
2\left(\frac{1}{2}\right)^2 + 13\left(\frac{1}{2}\right) - 7 &= 2\left(\frac{1}{4}\right) + 13\left(\frac{1}{2}\right) - 7 \\
&= \frac{1}{2} + \frac{13}{2} - \frac{14}{2} = 0 \quad \checkmark
\end{aligned}
$$

If $x = -7$, then we have the following.

$$
\begin{aligned}
2(-7)^2 + 13(-7) - 7 &= 2(49) + 13(-7) - 7 \\
&= 98 - 91 - 7 = 0 \quad \checkmark
\end{aligned}
$$

Thus $\frac{1}{2}$ and -7 are both roots of the equation $2x^2 + 13x - 7 = 0$. □

Student Practice 2 Solve the equation to find the two roots.

$$3x^2 + 11x - 4 = 0$$

If the quadratic equation $ax^2 + bx + c = 0$ has no visible constant term, then $c = 0$. All such quadratic equations can be solved by factoring out a common factor and then using the zero factor property to obtain two solutions that are real numbers.

Example 3 Solve the equation to find the two roots. $7x^2 - 3x = 0$

Solution

$7x^2 - 3x = 0$	The equation is in standard form. Here $c = 0$.
$x(7x - 3) = 0$	Factor out the common factor.
$x = 0 \quad 7x - 3 = 0$	Set each factor equal to 0 by the zero factor property.
$7x = 3$	Solve the equations to find the two roots.
$x = \frac{3}{7}$	

The two roots are 0 and $\frac{3}{7}$.

Check. Verify that 0 and $\frac{3}{7}$ are the roots of $7x^2 - 3x = 0$.

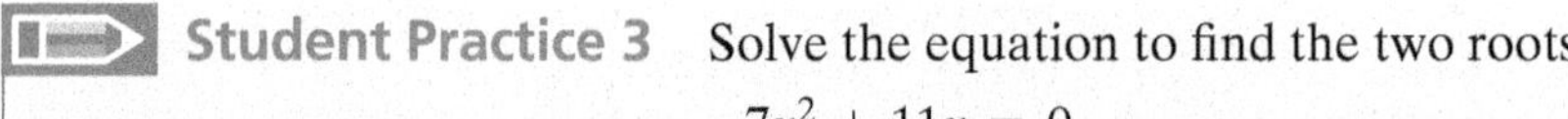

Student Practice 3 Solve the equation to find the two roots.

$$7x^2 + 11x = 0$$

If the quadratic equation is not in standard form, we use the same basic algebraic methods we studied in Sections 2.1–2.3 to place the terms on one side and zero on the other so that we can use the zero factor property.

Example 4 Solve. $x^2 = 12 - x$

Solution

$x^2 = 12 - x$	The equation is not in standard form.
$x^2 + x - 12 = 0$	Add x and -12 to both sides of the equation so that the right side is equal to zero; we can now factor.
$(x - 3)(x + 4) = 0$	Factor.
$x - 3 = 0 \quad x + 4 = 0$	Set each factor equal to 0 by the zero factor property.
$x = 3 \quad\quad x = -4$	Solve the equations for x.

Check. If $x = 3$: $(3)^2 \stackrel{?}{=} 12 - 3$; $9 \stackrel{?}{=} 12 - 3$; $9 = 9$ ✓

If $x = -4$: $(-4)^2 \stackrel{?}{=} 12 - (-4)$; $16 \stackrel{?}{=} 12 + 4$; $16 = 16$ ✓

Both roots check.

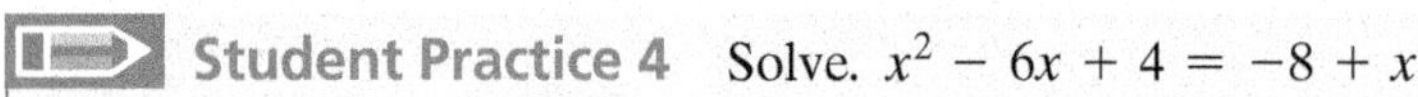

Student Practice 4 Solve. $x^2 - 6x + 4 = -8 + x$

Example 5 Solve. $\dfrac{x^2 - x}{2} = 6$

Solution We must first clear the fractions from the equation.

$2\left(\dfrac{x^2 - x}{2}\right) = 2(6)$	Multiply each side by 2.
$x^2 - x = 12$	Simplify.
$x^2 - x - 12 = 0$	Place in standard form.
$(x - 4)(x + 3) = 0$	Factor.
$x - 4 = 0 \quad x + 3 = 0$	Set each factor equal to zero.
$x = 4 \quad\quad x = -3$	Solve the equations for x.

The check is left to the student.

Student Practice 5 Solve.

$$\frac{2x^2 - 7x}{3} = 5$$

2 Using Quadratic Equations to Solve Applied Problems

Certain types of word problems—for example, some geometry applications—lead to quadratic equations. We'll show how to solve such word problems in this section.

It is particularly important to check the apparent solutions to the quadratic equation with conditions stated in the word problem. Often a particular solution to the quadratic equation will be eliminated by the conditions of the word problem.

▲ **Example 6** Carlos lives in Mexico City. He has a rectangular brick walkway in front of his house. The length of the walkway is 3 meters longer than twice the width. The area of the walkway is 44 square meters. Find the length and width of the rectangular walkway.

Solution

1. ***Understand the problem.***
   Draw a picture.

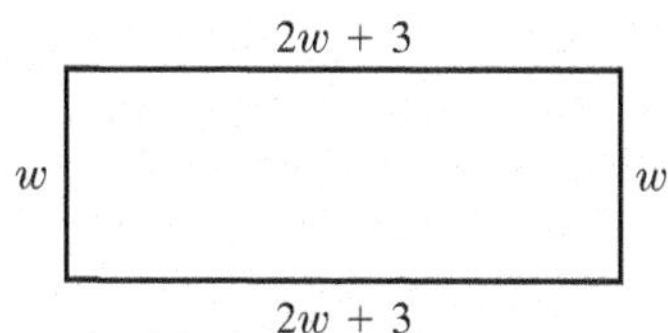

Let $w =$ the width in meters.
Then $2w + 3 =$ the length in meters.

2. ***Write an equation.***

$$\text{area} = (\text{width})(\text{length})$$
$$44 = w(2w + 3)$$

3. ***Solve and state the answer.***

$44 = w(2w + 3)$

$44 = 2w^2 + 3w$ Remove parentheses.

$0 = 2w^2 + 3w - 44$ Put in standard form.

$0 = (2w + 11)(w - 4)$ Factor.

$2w + 11 = 0 \qquad w - 4 = 0$ Set each factor equal to 0.

$2w = -11 \qquad w = 4$ Simplify and solve.

$w = -5\frac{1}{2}$

Although $-5\frac{1}{2}$ is a solution to the quadratic equation, it is not a valid solution to the word problem. It would not make sense to have a rectangle with a negative number as a width.

Since $w = 4$, the width of the walkway is 4 meters. The length is $2w + 3$, so we have $2(4) + 3 = 8 + 3 = 11$.

Thus the length of the walkway is 11 meters.

4. ***Check.*** Is the length 3 meters more than twice the width?

$$11 \stackrel{?}{=} 3 + 2(4) \qquad 11 = 3 + 8 \ ✓$$

Is the area of the rectangle 44 square meters?

$$4 \times 11 \stackrel{?}{=} 44 \qquad 44 = 44 \ ✓$$

□

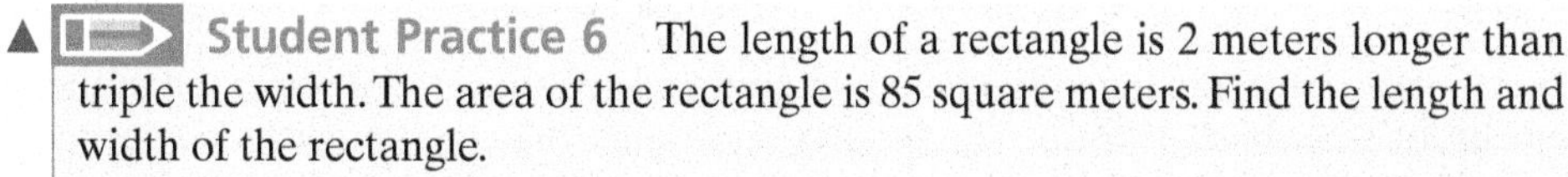

▲ **Student Practice 6** The length of a rectangle is 2 meters longer than triple the width. The area of the rectangle is 85 square meters. Find the length and width of the rectangle.

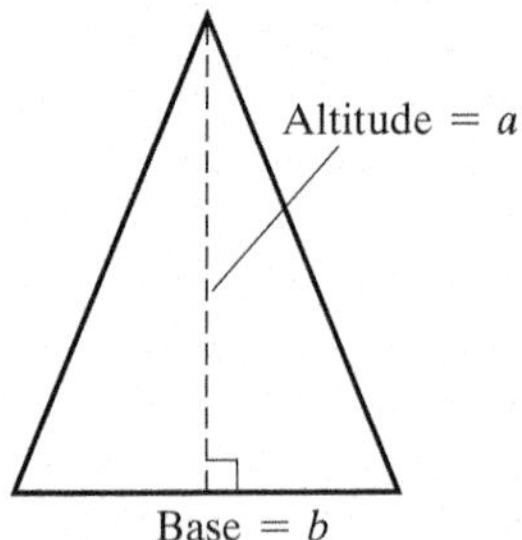

▲ **Example 7** The top of a local cable television tower has several small triangular reflectors. The area of each triangle is 49 square centimeters. The altitude of each triangle is 7 centimeters longer than the base. Find the altitude and the base of one of the triangles.

Solution

Let b = the length of the base in centimeters

$b + 7$ = the length of the altitude in centimeters

To find the area of a triangle, we use

$$\text{area} = \frac{1}{2}(\text{altitude})(\text{base}) = \frac{1}{2}ab = \frac{ab}{2}.$$

$$\frac{ab}{2} = 49 \quad \text{Write an equation.}$$

$$\frac{(b+7)(b)}{2} = 49 \quad \text{Substitute the expressions for altitude and base.}$$

$$\frac{b^2 + 7b}{2} = 49 \quad \text{Simplify.}$$

$$b^2 + 7b = 98 \quad \text{Multiply each side of the equation by 2.}$$

$$b^2 + 7b - 98 = 0 \quad \text{Place the quadratic equation in standard form.}$$

$$(b - 7)(b + 14) = 0 \quad \text{Factor.}$$

$$b - 7 = 0 \qquad b + 14 = 0 \quad \text{Set each factor equal to zero.}$$

$$b = 7 \qquad b = -14 \quad \text{Solve the equations for } b.$$

We cannot have a base of -14 centimeters, so we reject the negative answer. The only possible solution is 7. So the base is 7 centimeters. The altitude is $b + 7 = 7 + 7 = 14$. The altitude is 14 centimeters.

The triangular reflector has a base of 7 centimeters and an altitude of 14 centimeters.

Check. When you do the check, answer the following two questions.

1. Is the altitude 7 centimeters longer than the base?
2. Is the area of a triangle with a base of 7 centimeters and an altitude of 14 centimeters actually 49 square centimeters? □

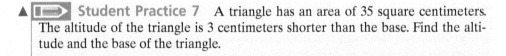

▲ **Student Practice 7** A triangle has an area of 35 square centimeters. The altitude of the triangle is 3 centimeters shorter than the base. Find the altitude and the base of the triangle.

Many problems in the sciences require the use of quadratic equations. You will study these in more detail if you take a course in physics or calculus in college. Often a quadratic equation is given as part of the problem.

When an object is thrown upward, its height (S) in meters is given, approximately, by the quadratic equation

$$S = -5t^2 + vt + h.$$

The letter h represents the initial height in meters. The letter v represents the initial velocity of the object thrown. The letter t represents the time in seconds starting from the time the object is thrown.

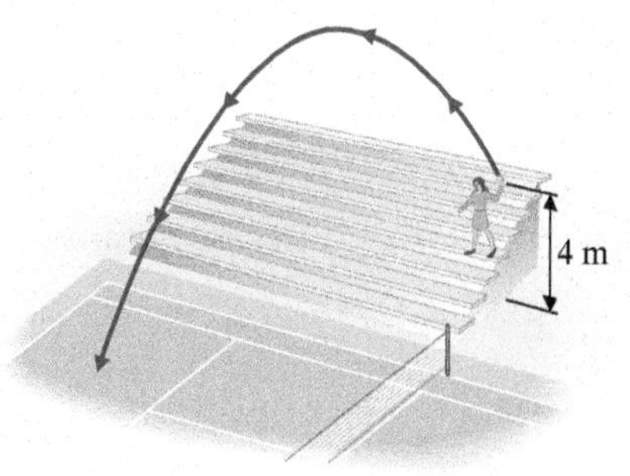

Example 8 A tennis ball is thrown upward with an initial velocity of 8 meters/second. Suppose that the initial height above the ground is 4 meters. At what time t will the ball hit the ground?

Solution In this case $S = 0$ since the ball will hit the ground. The initial upward velocity is $v = 8$ meters/second. The initial height is 4 meters, so $h = 4$.

$S = -5t^2 + vt + h$ Write an equation.

$0 = -5t^2 + 8t + 4$ Substitute all values into the equation.

$5t^2 - 8t - 4 = 0$ Isolate the terms on the left side. (Most students can factor more readily if the squared variable is positive.)

$(5t + 2)(t - 2) = 0$ Factor.

$5t + 2 = 0 \quad t - 2 = 0$ Set each factor equal to 0.

$5t = -2 \quad t = 2$ Solve the equations for t.

$$t = -\frac{2}{5}$$

We want a positive time t in seconds; thus we do not use $t = -\frac{2}{5}$.

Therefore, the ball will strike the ground 2 seconds after it is thrown.

Check. Verify the solution. □

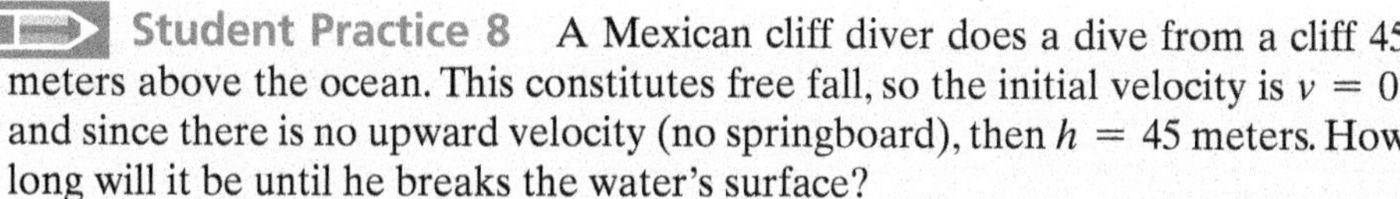

Student Practice 8 A Mexican cliff diver does a dive from a cliff 45 meters above the ocean. This constitutes free fall, so the initial velocity is $v = 0$, and since there is no upward velocity (no springboard), then $h = 45$ meters. How long will it be until he breaks the water's surface?

STEPS TO SUCCESS Be Involved.

If you are in a traditional class:

Don't just sit on the sidelines of the class and watch. Take part in the classroom discussion. People learn mathematics best through active participation. Whenever you are not clear about something, ask a question. Usually your questions will be helpful to other students in the room. When the teacher asks for suggestions, be sure to contribute your own ideas. Sit near the front where you can see and hear well. This will help you to focus on the material being covered.

Making it personal: Which of the suggestions above is the one you most need to follow? Write down what you need to do to improve in this area. ▼

If you are in an online class or a nontraditional class:

Be sure to e-mail the teacher. Talk to the tutor on duty. Ask questions. Think about concepts. Make your mind interact with the textbook. Be mentally involved. This active mental interaction is the key to your success.

Making it personal: Which of the suggestions above is the one you most need to follow? Write down what you need to do to improve in this area. ▼

6.7 Exercises MyMathLab®

Using the factoring method, solve for the roots of each quadratic equation. Be sure to place the equation in standard form before factoring. Check your answers.

1. $x^2 - 4x - 12 = 0$

2. $x^2 - x - 12 = 0$

3. $x^2 + 14x + 24 = 0$

4. $x^2 - 6x - 40 = 0$

5. $2x^2 - 7x + 6 = 0$

6. $3x^2 - 17x + 10 = 0$

7. $6x^2 - 13x = -6$

8. $10x^2 + 19x = 15$

9. $x^2 + 13x = 0$

10. $6x^2 - x = 0$

11. $8x^2 = 72$

12. $9x^2 = 81$

13. $5x^2 + 3x = 8x$

14. $3x^2 - x = 4x$

15. $6x^2 = 16x - 8$

16. $24x^2 = -10x + 4$

17. $(x - 5)(x + 2) = -4(x + 1)$

18. $(x - 5)(x + 4) = 2(x - 5)$

19. $9x^2 + x + 1 = -5x$

20. $4x^2 - 5x + 25 = 15x$

21. $\frac{x^2}{4} + \frac{5x}{4} + 2 = 2$

22. $\frac{x^2}{6} + \frac{2x}{3} + 1 = 1$

23. $\frac{x^2 + 10x}{8} = -2$

24. $\frac{x^2 + 5x}{4} = 9$

25. $\frac{10x^2 - 25x}{12} = 5$

26. $\frac{12x^2 - 4x}{5} = 8$

To Think About

27. Why can an equation in standard form with $c = 0$ (that is, an equation of the form $ax^2 + bx = 0$) always be solved?

28. Martha solved $(x + 3)(x - 2) = 14$ as follows:

$$x + 3 = 14 \quad \text{or} \quad x - 2 = 14$$
$$x = 11 \quad \text{or} \quad x = 16$$

Josette said this had to be wrong because these values do not check. Explain what is wrong with Martha's method.

Applications

▲ **29.** ***Geometry*** The area of a rectangular garden is 140 square meters. The width is 3 meters longer than one-half of the length. Find the length and the width of the garden.

▲ **30.** ***Geometry*** The area of a triangular sign is 33 square meters. The base of the triangle is 1 meter less than double the altitude. Find the altitude and the base of the sign.

Forming Groups *Suppose the number of students in a mathematics class is x. The teacher insists that each student participate in group work each class. The number of possible groups is:*

$$G = \frac{x^2 - 3x + 2}{2}$$

31. The class has 13 students. How many possible groups are there?

32. Four students withdraw from the class in exercise 31. How many fewer groups can be formed?

33. A teacher claims each student could be in 36 different groups. How many students are there?

34. The teacher wants each student to participate in a different group each day. There are 45 class days in the semester. How many students must be in the class?

Falling Object *Use the following information for exercises 35 and 36. When an object is thrown upward, its height (S), in meters, is given (approximately) by the quadratic equation*

$$S = -5t^2 + vt + h,$$

where $v =$ the upward initial velocity in meters/second,
$t =$ the time of flight in seconds, and
$h =$ the height above level ground from which the object is thrown.

35. Johnny is standing on a platform 6 meters high and throws a ball straight up as high as he can at a velocity of 13 meters/second. At what time t will the ball hit the ground? How far from the ground is the ball after 2 seconds have elapsed from the time of the throw? (Assume that the ball is 6 meters from the ground when it leaves Johnny's hand.)

36. You are standing on the edge of a cliff near Acapulco, overlooking the ocean. The place where you stand is 180 meters from the ocean. You drop a pebble into the water. ("Dropping" the pebble implies that there is no initial velocity, so $v = 0$.) How many seconds will it take to hit the water? How far has the pebble dropped after 3 seconds?

Internal Phone Calls *The technology and communication office of a local company has set up a new telephone system so that each employee has a separate telephone and extension number. They discovered that if there are x people in the office and each person talks to everyone else in the office by phone, the number of different calls (T) that take place is described by the equation* $T = 0.5(x^2 - x)$, *where x is the number of people in the office. Use this information to answer exercises 37 and 38.*

37. If 70 people are presently employed at the office, how many different telephone calls can be made between these 70 people?

38. On the day after Thanksgiving, only a small number of employees were working at the office. It has been determined that on that day, a total of 105 different phone calls could have been made from people working in the office to other people working in the office. How many people worked on the day after Thanksgiving?

The same equation given above is used in the well-known "handshake problem." If there are n people at a party and each person shakes hands with every other person once, the number of handshakes that take place among the n people is $H = 0.5(n^2 - n)$.

39. Barry is hosting a holiday party and has invited 16 friends. If all his friends attend the party and everyone, including Barry, shakes hands with everyone else at the party, how many handshakes will take place?

40. At a mathematics conference, 55 different handshakes took place among the organizers during the opening meeting. How many organizers were there?

Cumulative Review *Simplify.*

41. **[5.1.1]** $(2x^2y^3)(-5x^3y)$

42. **[5.1.1]** $(3a^4b^5)(4a^6b^8)$

43. **[5.1.2]** $\dfrac{21a^5b^{10}}{-14ab^{12}}$

44. **[5.1.2]** $\dfrac{18x^3y^6}{54x^8y^{10}}$

Quick Quiz 6.7 *Solve.*

1. $15x^2 - 8x + 1 = 0$

2. $4 + x(x - 2) = 7$

3. $4x^2 = 9x + 9$

4. **Concept Check** Explain how you would solve the following problem: A rectangle has an area of 65 square feet. The length of the rectangle is 3 feet longer than double the width. Find the length and the width of the rectangle.

CAREER EXPLORATION PROBLEMS

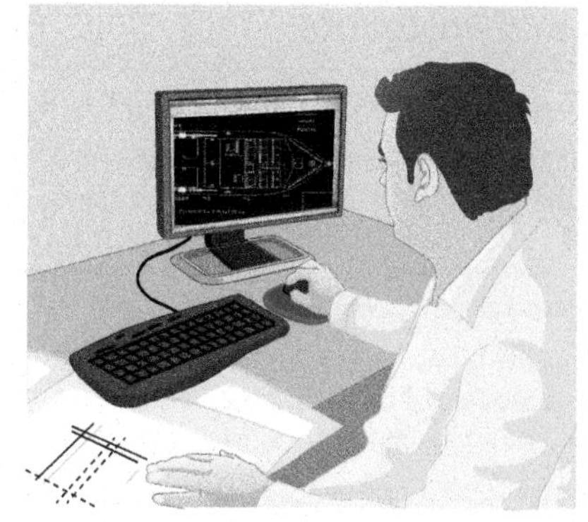

Background: CAD Operator

Mauricio is a skilled CAD operator who has worked with boats since he was a child. As a recent college graduate with a degree in computer-aided design technology, he's accepted a job with a nationally known sail maker and marine supply manufacturer. He feels incredibly lucky to be putting his skills to use in a career he loves.

As he begins his workweek, Mauricio is presented with two new job orders. The first is to create a construction design drawing with dimensions for a boat cover based on general specifications. The second project is to create a design for a racing sailboat's mainsail.

Facts

- The boat storage cover has two pieces, one rectangular and the other triangular. Mauricio examines the following information: For the rectangular piece, 126 square feet of material is needed, and its length has to be 3 feet shorter than 3 times its width. For the triangular piece, 10 square feet of material is needed. The height of the triangular section is 1 foot less than the length of its base. Mauricio begins to determine the dimensions of both pieces, writing a program that he'll be able to use for similar orders.

- The second order Mauricio needs to complete is a construction design drawing with the specifications of the racing boat's mainsail. He knows that the material used for these sails is very expensive and is measured in square yards, so ultimately, he'll have to convert to those units of measurement. The general information given is that the total area for the constructed sail is 132 square feet. Regulations require that the sail's height, when fully assembled, be 2 feet less than double its base.

Tasks

1. Mauricio must write a program to calculate dimensions for new orders of boat storage covers. What are the dimensions of the rectangular piece (length and width) and the triangular piece (base and height) for this initial storage cover? (Note: Each dimension must be a positive number, so eliminate any negative solutions.)

2. Mauricio must complete a construction design drawing with the specifications for the racing boat's mainsail. What are the dimensions (base and height) of the mainsail?

3. Because the material used for sail construction is measured in square yards, Mauricio must convert the measurements for the sail's area. What is the sail's area measured in square yards?

Chapter 6 Organizer

Topic and Procedure	Examples	You Try It
Common factor, p. 377 Factor out the largest common factor from each term.	Factor. **(a)** $2x^2 - 2x = 2x(x - 1)$ **(b)** $3a^2 + 3ab - 12a = 3a(a + b - 4)$ **(c)** $8x^4y - 24x^3 = 8x^3(xy - 3)$	1. Factor. **(a)** $5a^2 - 15a$ **(b)** $4x^2 - 8xy + 4x$ **(c)** $6x^4 - 18x^2$
Four terms. Factor by grouping, p. 382 Rearrange the terms if necessary so that the first two terms have a common factor. Then factor out the common factors. $ax + ay - bx - by = a(x + y) - b(x + y)$ $= (x + y)(a - b)$	Factor by grouping. $2ax^2 + 21 + 14x^2 + 3a$ $= 2ax^2 + 14x^2 + 3a + 21$ $= 2x^2(a + 7) + 3(a + 7)$ $= (a + 7)(2x^2 + 3)$	2. Factor by grouping. $3ax^2 - 12x^2 - 8 + 2a$
Trinomials of the form $x^2 + bx + c$, p. 387 Factor trinomials of the form $x^2 + bx + c$ by finding two numbers that have a product of c and a sum of b. If each term of the trinomial has a common factor, factor it out as the first step.	Factor. **(a)** $x^2 - 18x + 77 = (x - 7)(x - 11)$ **(b)** $x^2 + 7x - 18 = (x + 9)(x - 2)$ **(c)** $5x^2 - 10x - 40 = 5(x^2 - 2x - 8)$ $= 5(x - 4)(x + 2)$	3. Factor. **(a)** $x^2 + 9x + 18$ **(b)** $x^2 + 2x - 35$ **(c)** $3x^2 - 9x - 12$
Trinomials of the form $ax^2 + bx + c$, where $a \neq 0$ and $a \neq 1$, p. 394 Factor trinomials of the form $ax^2 + bx + c$ by the grouping method or by the trial-and-error method.	Factor. $6x^2 + 11x - 10$ Grouping number $= -60$ Two numbers whose product is -60 and whose sum is $+11$ are $+15$ and -4. $6x^2 + 15x - 4x - 10 = 3x(2x + 5) - 2(2x + 5)$ $= (2x + 5)(3x - 2)$	4. Factor. $8x^2 + 6x - 9$
Special cases ***Difference of two squares, p. 401*** ***Perfect-square trinomials, p. 402*** If you recognize the special cases, you will be able to factor quickly. $a^2 - b^2 = (a + b)(a - b)$ $a^2 + 2ab + b^2 = (a + b)^2$ $a^2 - 2ab + b^2 = (a - b)^2$	Factor. **(a)** $25x^2 - 36y^2 = (5x + 6y)(5x - 6y)$ **(b)** $16x^4 - 1 = (4x^2 + 1)(2x + 1)(2x - 1)$ **(c)** $25x^2 + 10x + 1 = (5x + 1)^2$ **(d)** $49x^2 - 42xy + 9y^2 = (7x - 3y)^2$	5. Factor. **(a)** $9x^2 - 16y^2$ **(b)** $81x^4 - 1$ **(c)** $16a^2 + 24a + 9$ **(d)** $4x^2 - 20xy + 25y^2$
Multistep factoring, p. 406 Many problems require two or three steps of factoring. Always try to factor out the greatest common factor as the first step.	Factor completely. **(a)** $3x^2 - 21x + 36 = 3(x^2 - 7x + 12)$ $= 3(x - 4)(x - 3)$ **(b)** $2x^3 - x^2 - 6x = x(2x^2 - x - 6)$ $= x(2x + 3)(x - 2)$ **(c)** $25x^3 - 49x = x(25x^2 - 49)$ $= x(5x + 7)(5x - 7)$ **(d)** $8x^2 - 24x + 18 = 2(4x^2 - 12x + 9)$ $= 2(2x - 3)^2$	6. Factor completely. **(a)** $4x^2 + 4x - 24$ **(b)** $3x^3 + 7x^2 + 2x$ **(c)** $9x^3 - 64x$ **(d)** $48x^2 - 24x + 3$
Prime polynomials, p. 407 A polynomial that is not factorable is called prime.	**(a)** $x^2 + y^2$ is prime. **(b)** $x^2 + 5x + 7$ is prime.	7. Explain why each polynomial is prime. **(a)** $x^2 + 4$ **(b)** $x^2 + x + 2$
Solving quadratic equations by factoring, p. 410 **1.** Write as $ax^2 + bx + c = 0$. **2.** Factor. **3.** Set each factor equal to 0. **4.** Solve the resulting equations.	Solve. $3x^2 + 5x = 2$ $3x^2 + 5x - 2 = 0$ $(3x - 1)(x + 2) = 0$ $3x - 1 = 0 \quad x + 2 = 0$ $x = \frac{1}{3} \quad x = -2$	8. Solve. $2x^2 - x = 3$

Topic and Procedure	Examples	You Try It
Using quadratic equations to solve applied problems, p. 413 Some word problems, like those involving the product of two numbers, area, and formulas with a squared variable, can be solved using the factoring methods we have shown.	The length of a rectangle is 4 less than three times the width. Find the length and width if the area is 55 square inches. Let w = width. Then $3w - 4$ = length. $55 = w(3w - 4)$ $55 = 3w^2 - 4w$ $0 = 3w^2 - 4w - 55$ $0 = (3w + 11)(w - 5)$ $w = -\frac{11}{3}$ or $w = 5$ $-\frac{11}{3}$ is not a valid solution. Thus width = 5 inches and length = 11 inches.	9. The length of a rectangle is 3 more than twice the width. Find the length and width if the area is 90 square feet.

Chapter 6 Review Problems

Section 6.1

Factor out the greatest common factor.

1. $12x^3 - 20x^2y$

2. $10x^3 - 35x^3y$

3. $24x^3y - 8x^2y^2 - 16x^3y^3$

4. $3a^3 + 6a^2 - 9ab + 12a$

5. $2a(a + 3b) - 5(a + 3b)$

6. $15x^3y + 6xy^2 + 3xy$

Section 6.2

Factor by grouping.

7. $2ax + 5a - 8x - 20$

8. $a^2 - 4ab + 7a - 28b$

9. $x^2y + 3y - 2x^2 - 6$

10. $30ax - 15ay + 42x - 21y$

11. $15x^2 - 3x + 10x - 2$

12. $30w^2 - 18w + 5wz - 3z$

Section 6.3

Factor completely. Be sure to factor out the greatest common factor as your first step.

13. $x^2 + 6x - 27$

14. $x^2 + 9x - 10$

15. $x^2 + 14x + 48$

16. $x^2 + 8xy + 15y^2$

17. $x^4 + 13x^2 + 42$

18. $x^4 - 2x^2 - 35$

19. $6x^2 + 30x + 36$

20. $2x^2 - 28x + 96$

Section 6.4

Factor completely. Be sure to factor out the greatest common factor as your first step.

21. $4x^2 + 7x - 15$

22. $12x^2 + 11x - 5$

23. $2x^2 - x - 3$

24. $3x^2 + 2x - 8$

25. $20x^2 + 48x - 5$

26. $20x^2 + 21x - 5$

27. $6x^2 + 4x - 10$

28. $6x^2 - 4x - 10$

29. $4x^2 - 26x + 30$

30. $4x^2 - 20x - 144$

31. $12x^2 + xy - 6y^2$

32. $6x^2 + 5xy - 25y^2$

Section 6.5

Factor these special cases. Be sure to factor out the greatest common factor.

33. $49x^2 - y^2$

34. $16x^2 - 36y^2$

35. $y^2 - 36x^2$

36. $9y^2 - 25x^2$

37. $36x^2 + 12x + 1$

38. $25x^2 - 20x + 4$

39. $16x^2 - 24xy + 9y^2$

40. $49x^2 - 28xy + 4y^2$

41. $2x^2 - 32$

42. $3x^2 - 27$

43. $28x^2 + 140x + 175$

44. $72x^2 - 192x + 128$

Section 6.6

If possible, factor each polynomial completely. If a polynomial cannot be factored, state that it is prime.

45. $4x^2 - 9y^2$

46. $x^2 + 13x - 30$

47. $9x^2 - 9x - 4$

48. $50x^3y^2 + 30x^2y^2 - 10x^2y^2$

49. $3x^2 - 18x + 27$

50. $25x^3 - 60x^2 + 36x$

51. $4x^2 - 13x - 12$

52. $3x^3a^3 - 11x^4a^2 - 20x^5a$

53. $12a^2 + 14ab - 10b^2$

54. $121a^2 + 66ab + 9b^2$

55. $7a - 7 - ab + b$

56. $3x^3 - 3x + 5yx^2 - 5y$

Mixed Practice

If possible, factor each polynomial completely. If a polynomial cannot be factored, state that it is prime.

57. $18b - 42 + 3bc - 7c$

58. $10b + 16 - 24x - 15bx$

59. $5xb - 35x + 4by - 28y$

60. $x^4 - 81y^{12}$

61. $6x^4 - x^2 - 15$

62. $28yz - 16xyz + x^2yz$

63. $12x^3 + 17x^2 + 6x$

64. $12w^2 - 12w + 3$

65. $4y^3 + 10y^2 - 6y$

66. $9x^4 - 144$

67. $x^2 - 6x + 12$

68. $8x^2 - 19x - 6$

69. $8y^5 + 4y^3 - 60y$

70. $16x^4y^2 - 56x^2y + 49$

71. $2ax + 5a - 10b - 4bx$

72. $2x^3 - 9 + x^2 - 18x$

Section 6.7

Solve the following equations by factoring.

73. $x^2 + x - 20 = 0$

74. $2x^2 + 11x - 6 = 0$

75. $7x^2 = 15x + x^2$

76. $5x^2 - x = 4x^2 + 12$

77. $2x^2 + 9x - 5 = 0$

78. $x^2 + 11x + 24 = 0$

79. $x^2 + 14x + 45 = 0$

80. $5x^2 = 7x + 6$

81. $3x^2 + 6x = 2x^2 - 9$

82. $4x^2 + 9x - 9 = 0$

83. $5x^2 - 11x + 2 = 0$

Solve.

▲ **84.** ***Geometry*** The area of a triangle is 25 square inches. The altitude is 5 inches longer than the base. Find the length of the base and the altitude.

▲ **85.** ***Geometry*** The area of a rectangle is 30 square feet. The length of the rectangle is 4 feet shorter than double the width. Find the length and width of the rectangle.

86. ***Rocket Height*** The height in feet that a model rocket attains is given by $h = -16t^2 + 80t + 96$, where t is the time measured in seconds. How many seconds will it take until the rocket finally reaches the ground? (*Hint:* At ground level $h = 0$.)

87. ***Output Power*** An electronic technician is working with a 100-volt electric generator. The output power of the generator is given by the equation $p = -5x^2 + 100x$, where x is the amount of current measured in amperes and p is measured in watts. The technician wants to find the value for x when the power is 480 watts. Can you find the two answers?

How Am I Doing? Chapter 6 Test

MATH COACH MyMathLab® YouTube

After you take this test read through the Math Coach on pages 425–426. Math Coach videos are available via MyMathLab and YouTube. Step-by-step test solutions on the Chapter Test Prep Videos are also available via MyMathLab and YouTube. (Search "TobeyBegInterAlg" and click on "Channels.")

If possible, factor each polynomial completely. If a polynomial cannot be factored, state that it is prime.

1. $x^2 + 12x - 28$

MC **2.** $16x^2 - 81$

3. $10x^2 + 27x + 5$

MC **4.** $9a^2 - 30a + 25$

5. $7x - 9x^2 + 14xy$

MC **6.** $10xy + 15by - 8x - 12b$

7. $6x^3 - 20x^2 + 16x$

8. $5a^2c - 11ac + 2c$

9. $81x^2 - 100$

10. $9x^2 - 15x + 4$

11. $20x^2 - 45$

12. $36x^2 + 1$

13. $3x^3 + 11x^2 + 10x$

14. $60xy^2 - 20x^2y - 45y^3$

15. $81x^2 - 1$

16. $81y^4 - 1$

17. $2ax + 6a - 5x - 15$

18. $aw^2 - 8b + 2bw^2 - 4a$

MC **19.** $3x^2 - 3x - 90$

20. $2x^3 - x^2 - 15x$

Solve.

21. $x^2 + 14x + 45 = 0$

22. $14 + 3x(x + 2) = -7x$

23. $2x^2 + x - 10 = 0$

24. $x^2 - 3x - 28 = 0$

Solve using a quadratic equation.

▲ **25.** The park service is studying a rectangular piece of land that has an area of 91 square miles. The length of this piece of land is 1 mile shorter than double the width. Find the length and width of this rectangular piece of land.

1. ____
2. ____
3. ____
4. ____
5. ____
6. ____
7. ____
8. ____
9. ____
10. ____
11. ____
12. ____
13. ____
14. ____
15. ____
16. ____
17. ____
18. ____
19. ____
20. ____
21. ____
22. ____
23. ____
24. ____
25. ____

Total Correct: ____

MATH COACH

Mastering the skills you need to do well on the test.

The following problems are from the Chapter 6 Test. Here are some helpful hints to keep you from making common errors on test problems.

Chapter 6 Test, Problem 2 Factor completely. $16x^2 - 81$

HELPFUL HINT It is important to learn the difference-of-two-squares formula: $a^2 - b^2 = (a + b)(a - b)$. Remember that the numerical values in both terms will be perfect squares. The first ten perfect squares are 1, 4, 9, 16, 25, 36, 49, 64, 81, and 100.

Did you remember that $(4x)^2 = 16x^2$?

Yes _____ No _____

Did you remember that $9^2 = 81$?

Yes _____ No _____

If you answered No to these questions, stop and review the list of the first ten perfect squares. Consider that $4^2 = 16$ and $x \cdot x = x^2$.

Do you see how $16x^2 - 81$ can be factored using the formula $a^2 - b^2$?

Yes _____ No _____

If you answered No, stop and review the difference-of-two-squares formula again. Make sure that one set of parentheses contains a + sign and the other set of parentheses contains a − sign.

If you answered Problem 2 incorrectly, go back and rework the problem using these suggestions.

Chapter 6 Test, Problem 4 Factor completely. $9a^2 - 30a + 25$

HELPFUL HINT Remember the perfect-square-trinomial formula: $a^2 - 2ab + b^2 = (a - b)^2$. You must verify two things to determine if you can use this formula:

(1) The numerical values in the first term and the last term must be perfect squares.

(2) The middle term must equal "twice the product of the values whose squares are the first and last terms."

Did you remember that $(3a)^2 = 9a^2$?

Yes _____ No _____

Did you remember that $5^2 = 25$?

Yes _____ No _____

If you answered No to these questions, stop and review the first ten perfect squares. Consider that $3^2 = 9$ and $a \cdot a = a^2$.

Do you see how $9a^2 - 30a + 25$ can be factored using the formula $(a - b)^2$?

Yes _____ No _____

If you answered No, check to see if the middle term, $30a$, equals twice the product of $3a$ and 5.

Now go back and rework the problem using these suggestions.

Need more help? Watch the MATH COACH videos in MyMathLab® or on YouTube™.

Chapter 6 Test, Problem 6 Factor completely. $10xy + 15by - 8x - 12b$

HELPFUL HINT Look for common factors first. We can find the greatest common factor of the first two terms and factor. Then we can find the greatest common factor of the second two terms and factor. Make sure that you obtain the *same binomial factor* for each step. Be careful with +/− signs.

Did you identify $5y$ as the greatest common factor of the first two terms: $10xy + 15by$?

Yes ______ No ______

Did you identify 4 as the greatest common factor of the second two terms: $-8x - 12b$?

Yes ______ No ______

If you answered Yes to these questions, then you obtained $(2x + 3b)$ in the first term and $(-2x - 3b)$ in the second term. These are *not* the same binomial factor. Stop and consider how to get the same binomial factor of $(2x + 3b)$.

In your final answer, is the binomial factor of $(2x + 3b)$ listed once?

Yes ______ No ______

If you answered No, remember that to factor completely, we must remove all common factors from both terms. Stop now and complete this step.

If you answered Problem 6 incorrectly, go back and rework the problem using these suggestions.

Chapter 6 Test, Problem 19 Factor completely. $3x^2 - 3x - 90$

HELPFUL HINT Look for the greatest common factor of all three terms as your *first step*. Don't forget to include this common factor as part of your answer. Always check your final product to make sure that it matches the original polynomial. Do this by multiplying.

Did you obtain $(3x - 18)(x + 5)$ or $(3x + 15)(x - 6)$ as your answer?

Yes ______ No ______

If you answered Yes, then you forgot to factor out the greatest common factor 3 as your first step.

Do you see how to factor $x^2 - x - 30$?

Yes ______ No ______

If you answered No, remember that we are looking for two numbers with a product of -30 and a sum of -1.

Be sure to double-check your final answer to be sure there are no common factors and include 3 in your final answer.

Now go back and rework the problem using these suggestions.

Need more help? Look for section examples marked with MC to review.

Cumulative Test for Chapters 0–6

This test provides a comprehensive review of the key objectives for Chapters 0–6.

In questions 1–3, simplify.

1. $\frac{11}{15} - \frac{7}{10}$

2. $-\frac{5}{3} + \frac{1}{2} + \frac{5}{6}$

3. $\left(-4\frac{1}{2}\right) \div \left(5\frac{1}{4}\right)$

4. Find 6% of 1842.5

5. A national walkout of 11,904 employees of the VBM Corp. occurred last month. This was 96% of the total number of employees. How many employees does VBM have?

6. Evaluate $3a^2 + ab - 4b^2$ for $a = 5$ and $b = -1$.

7. Simplify.
$7x(3x - 4) - 5x(2x - 3) - (3x)^2$

In questions 8–10, solve.

8. $\frac{2}{3}x + 6 = 4(x - 11)$

9. $7x - 3(4 - 2x) = 14x - (3 - x)$

10. $2 - 5x > 17$

Solve by any method. If there is not one solution to a system, state why.

11. $\frac{1}{3}x - \frac{1}{2}y = 4$
$\frac{3}{8}x - \frac{1}{4}y = 7$

12. $10x - 5y = 45$
$3x - 8y = 7$

13. A boat trip 6 miles upstream takes 3 hours. The return takes 90 minutes. How fast is the stream?

14. Find the slope of the line passing through $(6, -1)$ and $(-4, -2)$.

15. Find an equation of the line that passes through (2, 5) with slope $-\frac{3}{4}$.

1. ______
2. ______
3. ______
4. ______
5. ______
6. ______
7. ______
8. ______
9. ______
10. ______
11. ______
12. ______
13. ______
14. ______
15. ______

16.

17.

18.

19.

20.

21.

22.

23.

24.

25.

26.

27.

28.

16. Graph the line $4x - 8y = 10$. Plot at least three points.

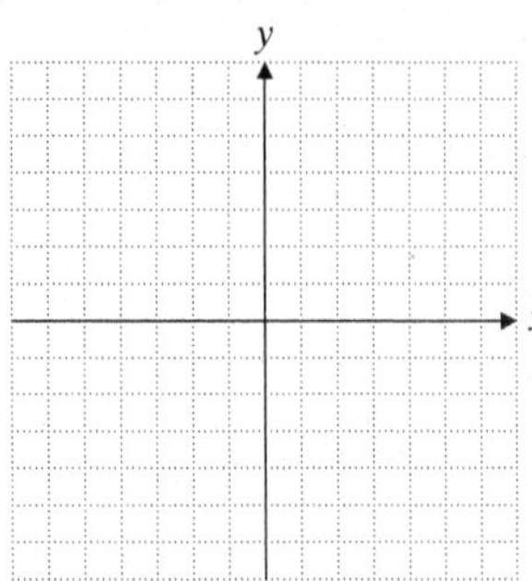

17. Multiply. $(4x - 5)(5x + 1)$

18. Multiply. $(3x - 5)^2$

In questions 19 and 20, simplify.

19. $(-4x^4y^5)(5xy^3)$

20. $(-2x^3y^2z^0)^4$

21. Write with only positive exponents. $\dfrac{9x^{-3}y^{-4}}{w^2z^{-8}}$

22. Write in scientific notation. 0.00056

Factor each polynomial completely. If a polynomial cannot be factored, state that it is prime.

23. $121x^2 - 64y^2$

24. $4x + 120 - 80x^2$

25. $16x^3 + 40x^2 + 25x$

26. $16x^4 - b^4$

27. $2ax - 4bx + 3a - 6b$

28. Solve: $x^2 + 5x - 36 = 0$

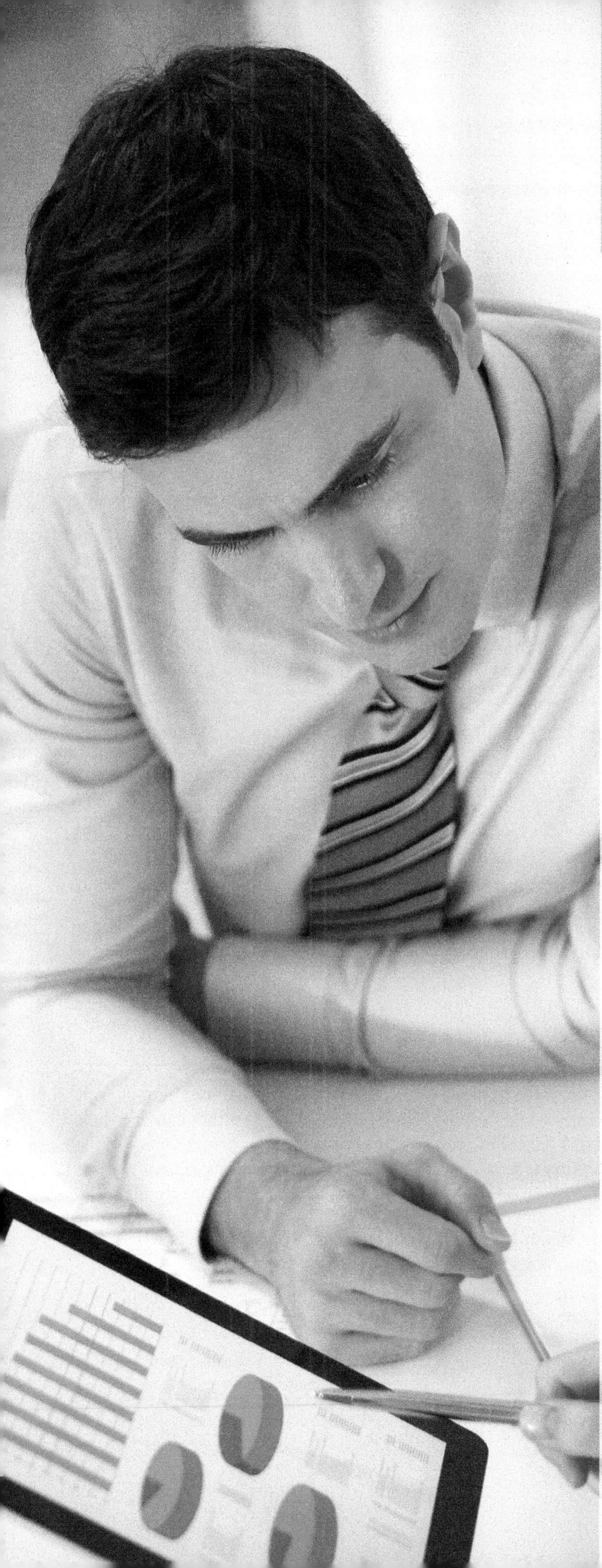

CHAPTER 7

Rational Expressions and Equations

CAREER OPPORTUNITIES

Operations Analyst

The effective management of the logistics and daily operations of a business is vital to its success. Teams of operations management and logistics professionals are the individuals who accomplish this. From predicting average costs to identifying the time required for manufacturing goods, these professionals use math every day to simulate the best practices needed for creating efficient operations.

To investigate how the mathematics in this chapter can help with this field, see the Career Exploration Problems on page **469**.

7.1 Simplifying Rational Expressions

Student Learning Objective

After studying this section, you will be able to:

1 Simplify rational expressions by factoring.

Recall that a rational number is a number that can be written as one integer divided by another integer, such as $3 \div 4$ or $\frac{3}{4}$. We usually use the word *fraction* to mean $\frac{3}{4}$. We can extend this idea to algebraic expressions. A **rational expression** is a polynomial divided by another polynomial, such as

$$(3x + 2) \div (x + 4) \quad \text{or} \quad \frac{3x + 2}{x + 4}.$$

The last fraction is sometimes also called a **fractional algebraic expression.** There is a special restriction for all fractions, including fractional algebraic expressions: The denominator of the fraction cannot be 0. For example, in the expression

$$\frac{3x + 2}{x + 4},$$

the denominator cannot be 0. Therefore, the value of x cannot be -4. The following important restriction will apply throughout this chapter. We state it here to avoid having to mention it repeatedly throughout this chapter.

RESTRICTION

The denominator of a rational expression cannot be zero. Any value of the variable that would make the denominator zero is not allowed.

We have discovered that fractions can be simplified (or reduced) in the following way.

$$\frac{15}{25} = \frac{3 \cdot \cancel{5}}{5 \cdot \cancel{5}} = \frac{3}{5}$$

This is sometimes referred to as the **basic rule of fractions** and can be stated as follows.

BASIC RULE OF FRACTIONS

For any rational expression $\frac{a}{b}$ and any polynomials a, b, and c (where $b \neq 0$ and $c \neq 0$),

$$\frac{ac}{bc} = \frac{a}{b}.$$

We will examine several examples where a, b, and c are real numbers, as well as more involved examples where a, b, and c are polynomials. In either case we shall make extensive use of our factoring skills in this section.

One essential property is revealed by the basic rule of fractions: If the numerator and denominator of a given fraction are multiplied by the same nonzero quantity, an equivalent fraction is obtained. The rule can be used two ways. You can start with $\frac{ac}{bc}$ and end with the equivalent fraction $\frac{a}{b}$. Or, you can start with $\frac{a}{b}$ and end with the equivalent fraction $\frac{ac}{bc}$. In this section we focus on the first process.

Example 1 Reduce. $\frac{21}{39}$

Solution $\frac{21}{39} = \frac{7 \cdot \cancel{3}}{13 \cdot \cancel{3}} = \frac{7}{13}$ Use the rule $\frac{ac}{bc} = \frac{a}{b}$. Let $c = 3$ because 3 is the greatest common factor of 21 and 39. □

Student Practice 1 Reduce.

$$\frac{28}{63}$$

1 Simplifying Rational Expressions by Factoring

The process of reducing fractions shown in Example 1 is sometimes called *dividing out* common factors. When you do this, you are **simplifying the fraction.** Remember, only **factors** of both the numerator and the denominator can be divided out. To apply the basic rule of fractions, it is usually helpful if the numerator and denominator of the fraction are completely factored. You will need to use your factoring skills from Chapter 6 to accomplish this step.

Example 2 Simplify. $\dfrac{4x + 12}{5x + 15}$

Solution

$$\frac{4x + 12}{5x + 15} = \frac{4(x + 3)}{5(x + 3)}$$ Factor 4 from the numerator. Factor 5 from the denominator.

$$= \frac{4\cancel{(x + 3)}}{5\cancel{(x + 3)}}$$ Apply the basic rule of fractions.

$$= \frac{4}{5}$$ □

Student Practice 2 Simplify.

$$\frac{12x - 6}{14x - 7}$$

Example 3 Simplify. $\dfrac{x^2 + 9x + 14}{x^2 - 4}$

Solution

$$= \frac{(x + 7)(x + 2)}{(x - 2)(x + 2)}$$ Factor the numerator. Factor the denominator.

$$= \frac{(x + 7)\cancel{(x + 2)}}{(x - 2)\cancel{(x + 2)}}$$ Apply the basic rule of fractions.

$$= \frac{x + 7}{x - 2}$$ □

CAUTION: Do not try to remove terms that are added. In Example 3, do not try to remove the x^2 in the top and the x^2 in the bottom of the fraction $\frac{x^2 + 9x + 14}{x^2 - 4}$. The basic rule of fractions applies only to quantities that are factors of both numerator and denominator.

Student Practice 3 Simplify.

$$\frac{4x^2 - 9}{2x^2 - x - 3}$$

Some problems may involve more than one step of factoring. Always remember to factor out the greatest common factor as the first step, if it is possible to do so.

Example 4 Simplify. $\dfrac{x^3 - 9x}{x^3 + x^2 - 6x}$

Solution

$$= \frac{x(x^2 - 9)}{x(x^2 + x - 6)}$$ Factor out the greatest common factor from the polynomials in the numerator and the denominator.

$$= \frac{\cancel{x}\cancel{(x + 3)}(x - 3)}{\cancel{x}\cancel{(x + 3)}(x - 2)}$$ Factor each polynomial and apply the basic rule of fractions.

$$= \frac{x - 3}{x - 2}$$ □

Student Practice 4 Simplify.

$$\frac{x^3 - 16x}{x^3 - 2x^2 - 8x}$$

When you are simplifying, be on the lookout for the special situation where *a factor in the denominator is the opposite of a factor in the numerator.* In such a case you should factor a negative number from one of the factors so that it becomes equivalent to the other factor and can be divided out. Look carefully at the following two examples.

Example 5 Simplify. $\dfrac{5x - 15}{6 - 2x}$

Solution Notice that the variable term in the numerator, $5x$, and the variable term in the denominator, $-2x$, *are opposite in sign.* Likewise, the numerical terms -15 and 6 *are opposite in sign.* Factor out a negative number from the denominator.

$$\frac{5x - 15}{6 - 2x} = \frac{5\,(x - 3)}{-2\,(-3 + x)}$$ Factor 5 from the numerator.
Factor -2 from the denominator.
Note that $(x - 3)$ and $(-3 + x)$ are equivalent since $(+x - 3) = (-3 + x)$.

$$= \frac{5\,\cancel{(x - 3)}}{-2\,\cancel{(-3 + x)}}$$ Apply the basic rule of fractions.

$$= -\frac{5}{2}$$

Note that $\frac{5}{-2}$ is not considered to be in simplest form. We usually avoid leaving a negative number in the denominator. Therefore, to simplify, give the result as $-\frac{5}{2}$ or $\frac{-5}{2}$. □

 **Student Practice 5** Simplify.

$$\frac{8x - 20}{15 - 6x}$$

Example 6 Simplify. $\dfrac{2x^2 - 11x + 12}{16 - x^2}$

Solution

$$= \frac{(x - 4)(2x - 3)}{(4 - x)(4 + x)}$$ Factor the numerator and the denominator.
Observe that $(x - 4)$ and $(4 - x)$ are opposites.

$$= \frac{(x - 4)(2x - 3)}{-1(-4 + x)(4 + x)}$$ Factor -1 out of $(+4 - x)$ to obtain $-1(-4 + x)$.

$$= \frac{\cancel{(x - 4)}(2x - 3)}{-1\cancel{(-4 + x)}(4 + x)}$$ Apply the basic rule of fractions since $(x - 4)$ and $(-4 + x)$ are equivalent.

$$= \frac{2x - 3}{-1(4 + x)}$$

$$= -\frac{2x - 3}{4 + x}$$ □

Student Practice 6 Simplify.

$$\frac{4x^2 + 3x - 10}{25 - 16x^2}$$

After doing Examples 5 and 6, you will notice a pattern. Whenever the factor in the numerator and the factor in the denominator are opposites, the value -1 results. We could actually make this a property.

> For all monomials A and B where $A \neq B$, it is true that
> $$\frac{A - B}{B - A} = -1.$$

You may use this property when reducing fractions if it is helpful to you. Otherwise, you may use the factoring method shown in Examples 5 and 6.

Some problems will involve two or more variables. In such cases, you will need to factor carefully and make sure that each set of parentheses contains the correct terms.

Example 7 Simplify. $\dfrac{x^2 - 7xy + 12y^2}{2x^2 - 7xy - 4y^2}$

Solution

$$= \frac{(x - 4y)(x - 3y)}{(2x + y)(x - 4y)}$$ Factor the numerator. Factor the denominator.

$$= \frac{\cancel{(x - 4y)}(x - 3y)}{(2x + y)\cancel{(x - 4y)}}$$ Apply the basic rule of fractions.

$$= \frac{x - 3y}{2x + y}$$ □

Student Practice 7 Simplify.

$$\frac{x^2 - 8xy + 15y^2}{2x^2 - 11xy + 5y^2}$$

Example 8 Simplify. $\dfrac{6a^2 + ab - 7b^2}{36a^2 - 49b^2}$

Solution

$$= \frac{(6a + 7b)(a - b)}{(6a + 7b)(6a - 7b)}$$ Factor the numerator. Factor the denominator.

$$= \frac{\cancel{(6a + 7b)}(a - b)}{\cancel{(6a + 7b)}(6a - 7b)}$$ Apply the basic rule of fractions.

$$= \frac{a - b}{6a - 7b}$$ □

Student Practice 8 Simplify.

$$\frac{25a^2 - 16b^2}{10a^2 + 3ab - 4b^2}$$

7.1 Exercises MyMathLab®

Simplify.

1. $\dfrac{4x - 24y}{x - 6y}$

2. $\dfrac{3x + 4y}{9x + 12y}$

3. $\dfrac{6x + 18}{x^2 + 3x}$

4. $\dfrac{25x - 30}{5x^3 - 6x^2}$

5. $\dfrac{9x^2 + 6x + 1}{1 - 9x^2}$

6. $\dfrac{16x^2 + 8x + 1}{1 - 16x^2}$

7. $\dfrac{3a^2b(a - 2b)}{6ab^2}$

8. $\dfrac{6a^2b}{9a^2b^2(a + 3b)}$

9. $\dfrac{x^2 + x - 2}{x^2 - x}$

10. $\dfrac{x^2 + x - 12}{x^2 - 3x}$

11. $\dfrac{x^2 - 3x - 10}{3x^2 + 5x - 2}$

12. $\dfrac{4x^2 - 10x + 6}{2x^2 + x - 3}$

13. $\dfrac{x^2 + 4x - 21}{x^3 - 49x}$

14. $\dfrac{x^3 - 3x^2 - 40x}{x^2 + 10x + 25}$

15. $\dfrac{3x^2 - 11x - 4}{x^2 + x - 20}$

16. $\dfrac{x^2 - 9x + 18}{2x^2 - 9x + 9}$

17. $\dfrac{3x^2 - 8x + 5}{4x^2 - 5x + 1}$

18. $\dfrac{3y^2 - 8y - 3}{3y^2 - 10y + 3}$

19. $\dfrac{5x^2 - 27x + 10}{5x^2 + 3x - 2}$

20. $\dfrac{2x^2 + 4x - 6}{x^2 + 3x - 4}$

Mixed Practice *Take some time to review exercises 1–20 before you proceed with exercises 21–30.*

21. $\dfrac{12 - 3x}{5x^2 - 20x}$

22. $\dfrac{20 - 4ab}{a^2b - 5a}$

23. $\dfrac{2x^2 - 7x - 15}{25 - x^2}$

24. $\dfrac{49 - x^2}{2x^2 - 9x - 35}$

25. $\dfrac{(3x + 4)^2}{9x^2 + 9x - 4}$

26. $\dfrac{12x^2 - 11x - 5}{(4x - 5)^2}$

27. $\dfrac{3x^2 + 13x - 10}{20 - x - x^2}$

28. $\dfrac{2x^2 + 15x + 18}{42 + x - x^2}$

29. $\dfrac{a^2 + ab - 6b^2}{3a^2 + 8ab - 3b^2}$

30. $\dfrac{a^2 + 2ab - 15b^2}{4a^2 - 13ab + 3b^2}$

Cumulative Review *Multiply.*

31. **[5.5.2]** $(3x - 7)^2$

32. **[5.5.1]** $(7x + 6y)(7x - 6y)$

33. **[5.4.2]** $(2x + 3)(x - 4)$

34. **[5.5.3]** $(2x + 3)(x - 4)(x - 2)$

35. **[1.7.2]** Simplify. $\dfrac{2a^2}{7} + \dfrac{3b}{2} + 3a^2 - \dfrac{3b}{4}$

36. **[1.3.3]** Divide. $\dfrac{-35}{12} \div \dfrac{5}{14}$

37. **[0.3.2]** ***Dividing Acreage*** David and Connie Swensen wish to divide $4\frac{7}{8}$ acres of farmland into three equal-size house lots. What will be the acreage of each lot?

38. **[0.3.1]** ***Roasting a Turkey*** Ron and Mary Larson are planning to cook a $17\frac{1}{2}$-pound turkey. The directions suggest a cooking time of 22 minutes per pound for turkeys that weigh between 16 and 20 pounds. How many hours and minutes should they allow for an approximate cooking time?

Quick Quiz 7.1 *Simplify.*

1. $\dfrac{x^3 + 3x^2}{x^3 - 2x^2 - 15x}$

2. $\dfrac{6 - 2ab}{ab^2 - 3b}$

3. $\dfrac{8x^2 + 6x - 5}{16x^2 + 40x + 25}$

4. **Concept Check** Explain why it is important to completely factor both the numerator and the denominator when simplifying

$$\frac{x^2y - y^3}{x^2y + xy^2 - 2y^3}$$

7.2 Multiplying and Dividing Rational Expressions

Student Learning Objectives

After studying this section, you will be able to:

1. Multiply rational expressions.
2. Divide rational expressions.

1 Multiplying Rational Expressions 

To multiply two rational expressions, we multiply the numerators and multiply the denominators. As before, the denominators cannot equal zero.

> For any two rational expressions $\frac{a}{b}$ and $\frac{c}{d}$ where $b \neq 0$ and $d \neq 0$,
>
> $$\frac{a}{b} \cdot \frac{c}{d} = \frac{ac}{bd}.$$

Simplifying or reducing fractions *prior to multiplying them* usually makes the computations easier to do. Leaving the reducing step until the end makes the simplifying process longer and increases the chance for error. This long approach should be avoided.

As an example, let's do the same problem two ways to see which one is easier. Let's simplify the following problem by multiplying first and then reducing the result.

$$\frac{5}{7} \times \frac{49}{125}$$

$\frac{5}{7} \times \frac{49}{125} = \frac{245}{875}$ Multiply the numerators and multiply the denominators.

$= \frac{7}{25}$ Reduce the fraction. (*Note:* It can take a bit of trial and error to discover how to reduce it.)

Compare this with the following method, where we reduce the fractions prior to multiplying them.

$$\frac{5}{7} \times \frac{49}{125}$$

$\frac{5}{7} \times \frac{7 \cdot 7}{5 \cdot 5 \cdot 5}$ **Step 1.** It is easier to factor first. We factor the numerator and denominator of the second fraction.

$= \frac{\cancel{5}}{\cancel{7}} \times \frac{\cancel{7} \cdot 7}{\cancel{5} \cdot 5 \cdot 5} = \frac{7}{25}$ **Step 2.** Then we apply the basic rule of fractions to divide out the common factors 5 and 7 that appear in both a numerator and in a denominator.

A similar approach can be used with the multiplication of rational expressions. We first factor the numerator and denominator of each fraction. Then we divide out any factor that is common to a numerator and a denominator. Finally, we multiply the remaining numerators and the remaining denominators.

Example 1 Multiply. $\frac{x^2 - x - 12}{x^2 - 16} \cdot \frac{2x^2 + 7x - 4}{x^2 - 4x - 21}$

Solution

$\frac{(x-4)(x+3)}{(x-4)(x+4)} \cdot \frac{(x+4)(2x-1)}{(x+3)(x-7)}$ Factoring is always the first step.

$= \frac{\cancel{(x-4)}\cancel{(x+3)}}{\cancel{(x-4)}\cancel{(x+4)}} \cdot \frac{\cancel{(x+4)}(2x-1)}{\cancel{(x+3)}(x-7)}$ Apply the basic rule of fractions. (Three pairs of factors divide out.)

$= \frac{2x-1}{x-7}$ The final answer. □

Student Practice 1 Multiply.

$$\frac{6x^2 + 7x + 2}{x^2 - 7x + 10} \cdot \frac{x^2 + 3x - 10}{2x^2 + 11x + 5}$$

In some cases, it may take several steps to factor a given numerator or denominator. You should always check for the *greatest common factor* as your first step.

Example 2 Multiply. $\dfrac{x^4 - 16}{x^3 + 4x} \cdot \dfrac{2x^2 - 8x}{4x^2 + 2x - 12}$

Solution

$= \dfrac{(x^2 + 4)(x^2 - 4)}{x(x^2 + 4)} \cdot \dfrac{2x(x - 4)}{2(2x^2 + x - 6)}$ Factor each numerator and denominator. Factoring out the greatest common factor first is very important.

$= \dfrac{(x^2 + 4)(x + 2)(x - 2)}{x(x^2 + 4)} \cdot \dfrac{2x(x - 4)}{2(x + 2)(2x - 3)}$ Factor again where possible.

$= \dfrac{\cancel{(x^2 + 4)}\cancel{(x + 2)}(x - 2)}{\cancel{x}\cancel{(x^2 + 4)}} \cdot \dfrac{\cancel{2}\cancel{x}(x - 4)}{\cancel{2}\cancel{(x + 2)}(2x - 3)}$ Divide out factors that appear in both a numerator and a denominator. (There are four such pairs of factors.)

$= \dfrac{(x - 2)(x - 4)}{2x - 3}$ or $\dfrac{x^2 - 6x + 8}{2x - 3}$ Write the answer as one fraction. (Usually, if there is more than one factor in a numerator or denominator, the answer is left in factored form.)

 **Student Practice 2** Multiply.

$$\frac{2y^2 - 6y - 8}{y^2 - y - 2} \cdot \frac{y^2 - 5y + 6}{2y^2 - 32}$$

2 Dividing Rational Expressions 

For any two fractions $\frac{a}{b}$ and $\frac{c}{d}$, the operation of division can be performed by inverting the second fraction and multiplying it by the first fraction. When we invert a fraction, we are finding its *reciprocal.* Two numbers are **reciprocals** of each other if their product is 1. The reciprocal of $\frac{3}{5}$ is $\frac{5}{3}$. The reciprocal of 7 is $\frac{1}{7}$. The reciprocal of $\frac{a}{b}$ is $\frac{b}{a}$. Sometimes people state the rule for dividing fractions this way: "To divide two fractions, keep the first fraction unchanged and multiply by the reciprocal of the second fraction."

> The definition for division of fractions is
>
> $$\frac{a}{b} \div \frac{c}{d} = \frac{a}{b} \cdot \frac{d}{c}.$$

This property holds whether *a, b, c,* and *d* are polynomials or numerical values. (It is assumed, of course, that no denominator is zero.)

In the first step for dividing two rational expressions, invert the second fraction and rewrite the quotient as a product. Then follow the procedure for multiplying rational expressions.

MG **Example 3** Divide. $\dfrac{6x + 12y}{2x - 6y} \div \dfrac{9x^2 - 36y^2}{4x^2 - 36y^2}$

Solution

$= \dfrac{6x + 12y}{2x - 6y} \cdot \dfrac{4x^2 - 36y^2}{9x^2 - 36y^2}$ Invert the second fraction and write the problem as the product of two fractions.

$= \dfrac{6(x + 2y)}{2(x - 3y)} \cdot \dfrac{4(x^2 - 9y^2)}{9(x^2 - 4y^2)}$ Factor each numerator and denominator.

Continued on next page

 **Student Practice 3** Divide.

$$\frac{x^2 + 5x + 6}{x^2 + 8x} \div \frac{2x^2 + 5x + 2}{2x^2 + x}$$

$$= \frac{(3)(2)(x + 2y)}{2(x - 3y)} \cdot \frac{(2)(2)(x + 3y)(x - 3y)}{(3)(3)(x + 2y)(x - 2y)}$$ Factor again where possible.

$$= \frac{\cancel{(3)}\cancel{(2)}\cancel{(x + 2y)}}{\cancel{2}\cancel{(x - 3y)}} \cdot \frac{(2)(2)(x + 3y)\cancel{(x - 3y)}}{\cancel{(3)}(3)\cancel{(x + 2y)}(x - 2y)}$$ Divide out factors that appear in both a numerator and a denominator.

$$= \frac{(2)(2)(x + 3y)}{3(x - 2y)}$$ Write the result as one fraction.

$$= \frac{4(x + 3y)}{3(x - 2y)}$$ Simplify.

Although it is correct to write this answer as $\frac{4x + 12y}{3x - 6y}$, it is customary to leave the answer in factored form to ensure that the final answer is simplified. □

A polynomial that is not in fraction form can be written as a fraction if you give it a denominator of 1.

Example 4 Divide. $\frac{15 - 3x}{x + 6} \div (x^2 - 9x + 20)$

Solution

Note that $x^2 - 9x + 20$ can be written as $\frac{x^2 - 9x + 20}{1}$.

$$= \frac{15 - 3x}{x + 6} \cdot \frac{1}{x^2 - 9x + 20}$$ Invert and multiply.

$$= \frac{-3(-5 + x)}{x + 6} \cdot \frac{1}{(x - 5)(x - 4)}$$ Factor where possible. Note that we had to factor -3 from the first numerator so that it would have a factor in common with the second denominator.

$$= \frac{-3\cancel{(-5 + x)}}{x + 6} \cdot \frac{1}{\cancel{(x - 5)}(x - 4)}$$ Divide out the common factor. $(-5 + x)$ is equivalent to $(x - 5)$.

$$= \frac{-3}{(x + 6)(x - 4)}$$ The final answer. Note that the answer can be written in several equivalent forms.

or $-\frac{3}{(x + 6)(x - 4)}$ or $\frac{3}{(x + 6)(4 - x)}$ □

Student Practice 4 Divide.

$$\frac{x + 3}{x - 3} \div (9 - x^2)$$

CAUTION: It is logical to assume that the exercises in Section 7.2 have at least one common factor that can be divided out. Therefore, if after factoring, you do not observe any common factors, you should be somewhat suspicious. In such cases, it would be wise to double-check your factoring steps for errors.

7.2 Exercises MyMathLab®

Verbal and Writing Skills, Exercises 1 and 2

1. Before multiplying rational expressions, we should always first try to

____

______________________________________.

2. Division of two rational expressions is done by keeping the first fraction unchanged and then

____

______________________________________.

Multiply.

3. $\frac{4x + 12}{x - 4} \cdot \frac{x^2 + x - 20}{x^2 + 6x + 9}$

4. $\frac{3x - 6}{x + 5} \cdot \frac{x^2 + 6x + 5}{2x^2 - 8}$

5. $\frac{24x^3}{4x^2 - 36} \cdot \frac{2x^2 + 6x}{16x^2}$

6. $\frac{2x^2}{6x + 15} \cdot \frac{4x^3 + 26x^2 + 40x}{10x^3}$

7. $\frac{x^2 + 3x - 10}{x^2 + x - 20} \cdot \frac{x^2 - 3x - 4}{x^2 + 4x + 3}$

8. $\frac{x^2 - 2x - 35}{x^2 - 5x - 6} \cdot \frac{x^2 - 2x - 24}{x^2 - 3x - 28}$

Divide.

9. $\frac{x + 6}{x - 8} \div \frac{x + 5}{x^2 - 6x - 16}$

10. $\frac{x - 4}{x + 9} \div \frac{x - 7}{x^2 + 13x + 36}$

11. $(5x + 4) \div \frac{25x^2 - 16}{5x^2 + 11x - 12}$

12. $\frac{4x^2 - 25}{4x^2 - 20x + 25} \div (4x + 10)$

13. $\frac{3x^2 + 12xy + 12y^2}{x^2 + 4xy + 3y^2} \div \frac{4x + 8y}{x + y}$

14. $\frac{5x^2 + 10xy + 5y^2}{x^2 + 5xy + 6y^2} \div \frac{3x + 3y}{x + 2y}$

Mixed Practice *Perform the operation indicated.*

15. $\frac{(x + 5)^2}{3x^2 - 7x + 2} \cdot \frac{x^2 - 4x + 4}{x + 5}$

16. $\frac{3x^2 - 10x - 8}{(4x + 5)^2} \cdot \frac{4x + 5}{(x - 4)^2}$

17. $\frac{x^2 + x - 30}{10 - 2x} \div \frac{x^2 + 4x - 12}{5x + 15}$

18. $\frac{x^2 + 4x - 12}{16 - 4x^2} \div \frac{x^2 - 3x - 54}{x^2 + 10x + 16}$

19. $\frac{y^2 + 4y - 12}{y^2 + 2y - 24} \cdot \frac{y^2 - 16}{y^2 + 2y - 8}$

20. $\frac{4y^2 + 13y + 3}{12y^2 - y - 1} \cdot \frac{6y^2 + y - 1}{2y^2 + 7y + 3}$

21. $\frac{x^2 + 10x + 24}{2x^2 + 13x + 6} \cdot \frac{2x^2 - 5x - 3}{x^2 + 5x - 24}$

22. $\frac{x^2 + x - 12}{3x^2 - 8x - 3} \cdot \frac{3x^2 - 14x - 5}{x^2 + 2x - 8}$

Cumulative Review

23. [2.3.2] Solve. $6x^2 + 3x - 18 = 5x - 2 + 6x^2$

24. [1.2.1] Perform the operations indicated. $\frac{3}{4} + \frac{1}{2} - \frac{4}{7}$

▲ **25. [2.5.2]** ***Golden Gate Bridge*** The Golden Gate Bridge has a total length (including approaches) of 8981 feet and a road width of 90 feet. The width of the sidewalk is 10.5 feet. (The sidewalk spans the entire length of the bridge.) Assume it would cost $\$x$ per square foot to resurface the road or the sidewalk. Write an expression for how much more it would cost to resurface the road than the sidewalk.

▲ **26. [2.6.1]** ***Garden Design*** Harold Rafton planted a square garden bed. George Avis also planted a garden that was 2 feet less in width but 3 feet longer in length than Harold's garden. If the area of George's garden was 36 square feet, find the dimensions of each garden.

Quick Quiz 7.2

1. Multiply. $\frac{2x - 10}{x - 4} \cdot \frac{x^2 + 5x + 4}{x^2 - 4x - 5}$

2. Multiply. $\frac{3x^2 - 13x - 10}{3x^2 + 2x} \cdot \frac{x^2 - 25x}{x^2 - 25}$

3. Divide. $\frac{2x^2 - 18}{3x^2 + 3x} \div \frac{x^2 + 6x + 9}{x^2 + 4x + 3}$

4. Concept Check Explain how you would divide $\frac{21x - 7}{9x^2 - 1} \div \frac{1}{3x + 1}$.

7.3 Adding and Subtracting Rational Expressions

1 Adding and Subtracting Rational Expressions with a Common Denominator

Student Learning Objectives

After studying this section, you will be able to:

1. Add and subtract rational expressions with a common denominator.
2. Determine the LCD for rational expressions with different denominators.
3. Add and subtract rational expressions with different denominators.

If rational expressions have the same denominator, they can be combined in the same way as arithmetic fractions. The numerators are added or subtracted and the denominator remains the same.

ADDING RATIONAL EXPRESSIONS

For any rational expressions $\frac{a}{b}$ and $\frac{c}{b}$,

$$\frac{a}{b} + \frac{c}{b} = \frac{a + c}{b} \quad \text{where } b \neq 0.$$

Example 1 Add. $\dfrac{5a}{a + 2b} + \dfrac{6a}{a + 2b}$

Solution

$$\frac{5a}{a + 2b} + \frac{6a}{a + 2b} = \frac{5a + 6a}{a + 2b} = \frac{11a}{a + 2b}$$

Note that the denominators are the same. Only add the numerators. Keep the same denominator.

Do not change the denominator. □

Student Practice 1 Add.

$$\frac{2s + t}{2s - t} + \frac{s - t}{2s - t}$$

SUBTRACTING RATIONAL EXPRESSIONS

For any rational expressions $\frac{a}{b}$ and $\frac{c}{b}$,

$$\frac{a}{b} - \frac{c}{b} = \frac{a - c}{b} \quad \text{where } b \neq 0.$$

Example 2 Subtract. $\dfrac{3x}{(x + y)(x - 2y)} - \dfrac{8x}{(x + y)(x - 2y)}$

Solution

$$\frac{3x}{(x + y)(x - 2y)} - \frac{8x}{(x + y)(x - 2y)} = \frac{3x - 8x}{(x + y)(x - 2y)}$$

Write as one fraction.

$$= \frac{-5x}{(x + y)(x - 2y)}$$

Simplify. □

Student Practice 2 Subtract.

$$\frac{b}{(a - 2b)(a + b)} - \frac{2b}{(a - 2b)(a + b)}$$

2 Determining the LCD for Rational Expressions with Different Denominators

How do we add or subtract rational expressions when the denominators are not the same? First we must find the **least common denominator** (LCD). You need to be clear on how to find the least common denominator and how to add and subtract fractions from arithmetic before you attempt this section. Review Sections 0.1 and 0.2 if you have any questions about this topic.

> **HOW TO FIND THE LCD OF TWO OR MORE RATIONAL EXPRESSIONS**
>
> 1. Factor each denominator completely.
> 2. The LCD is a product containing each *different factor.*
> 3. If a factor occurs more than once in any one denominator, the LCD will contain that factor repeated the greatest number of times that it occurs in any one denominator.

Example 3 Find the LCD. $\dfrac{5}{2x-4}, \dfrac{6}{3x-6}$

Solution Factor each denominator.

$$2x - 4 = 2(x - 2) \qquad 3x - 6 = 3(x - 2)$$

The different factors are 2, 3, and $(x - 2)$. Since no factor appears more than once in any one denominator, the LCD is the product of these three factors.

$$\text{LCD} = (2)(3)(x - 2) = 6(x - 2)$$

Student Practice 3 Find the LCD.

$$\frac{7}{6x+21}, \frac{13}{10x+35}$$

Example 4 Find the LCD.

(a) $\dfrac{5}{12ab^2c}, \dfrac{13}{18a^3bc^4}$ **(b)** $\dfrac{8}{x^2-5x+4}, \dfrac{12}{x^2+2x-3}$

Solution If a factor occurs more than once in any one denominator, the LCD will contain that factor repeated the greatest number of times that it occurs in any one denominator.

(a)

$$12ab^2c = 2\cdot2\cdot3\cdot a\cdot b\cdot b\cdot c$$
$$18a^3bc^4 = 2\cdot3\cdot3\cdot a\cdot a\cdot a\cdot b\cdot c\cdot c\cdot c\cdot c$$
$$\text{LCD} = 2\cdot2\cdot3\cdot3\cdot a\cdot a\cdot a\cdot b\cdot b\cdot c\cdot c\cdot c\cdot c$$
$$\text{LCD} = 2^2\cdot3^2\cdot a^3\cdot b^2\cdot c^4 = 36a^3b^2c^4$$

(b)

$$x^2 - 5x + 4 = (x - 4)(x - 1)$$
$$x^2 + 2x - 3 = (x - 1)(x + 3)$$
$$\text{LCD} = (x - 4)(x - 1)(x + 3)$$

Student Practice 4 Find the LCD.

(a) $\dfrac{3}{50xy^2z}, \dfrac{19}{40x^3yz}$ **(b)** $\dfrac{2}{x^2+5x+6}, \dfrac{6}{3x^2+5x-2}$

3 Adding and Subtracting Rational Expressions with Different Denominators

If two rational expressions have different denominators, we first change them to equivalent rational expressions with the least common denominator. Then we add or subtract the numerators and keep the common denominator.

Example 5 Add. $\frac{5}{xy} + \frac{2}{y}$

Solution The denominators are different. We must find the LCD. The two factors are x and y. We observe that the LCD is xy.

$\frac{5}{xy} + \frac{2}{y} = \frac{5}{xy} + \frac{2}{y} \cdot \frac{x}{x}$ Multiply the second fraction by $\frac{x}{x}$.

$= \frac{5}{xy} + \frac{2x}{xy}$ Now each fraction has a common denominator of xy.

$= \frac{5 + 2x}{xy}$ Write the sum as one fraction. □

Student Practice 5 Add.

$$\frac{7}{a} + \frac{3}{abc}$$

Example 6 Add. $\frac{3x}{x^2 - y^2} + \frac{5}{x + y}$

Solution We factor the first denominator so that $x^2 - y^2 = (x + y)(x - y)$. Thus, the factors of the denominators are $(x + y)$ and $(x - y)$. We observe that the LCD $= (x + y)(x - y)$.

$\frac{3x}{(x + y)(x - y)} + \frac{5}{(x + y)} \cdot \frac{x - y}{x - y}$ Multiply the second fraction by $\frac{x - y}{x - y}$.

$= \frac{3x}{(x + y)(x - y)} + \frac{5x - 5y}{(x + y)(x - y)}$ Now each fraction has a common denominator of $(x + y)(x - y)$.

$= \frac{3x + 5x - 5y}{(x + y)(x - y)}$ Write the sum of the numerators over the common denominator.

$= \frac{8x - 5y}{(x + y)(x - y)}$ Combine like terms. □

Student Practice 6 Add.

$$\frac{2a - b}{a^2 - 4b^2} + \frac{2}{a + 2b}$$

It is important to remember that the LCD is the smallest algebraic expression into which each denominator can be divided. For rational expressions the LCD must contain *each factor* that appears in any denominator. If the factor is repeated, the LCD must contain that factor the greatest number of times that it appears in any one denominator.

In many cases, the denominators in an addition or subtraction problem are not in factored form. You must factor each denominator to determine the LCD. Combine like terms in the numerator; then determine whether that final numerator can be factored. If so, you may be able to simplify the fraction.

Example 7 Add. $\dfrac{5}{x^2 - y^2} + \dfrac{3x}{x^3 + x^2y}$

Solution

$\dfrac{5}{x^2 - y^2} + \dfrac{3x}{x^3 + x^2y}$

$= \dfrac{5}{(x + y)(x - y)} + \dfrac{3x}{x^2(x + y)}$ Factor the two denominators. Observe that the LCD is $x^2(x + y)(x - y)$.

$= \dfrac{5}{(x + y)(x - y)} \cdot \dfrac{x^2}{x^2} + \dfrac{3x}{x^2(x + y)} \cdot \dfrac{x - y}{x - y}$ Multiply each fraction by the appropriate expression to obtain a common denominator of $x^2(x + y)(x - y)$.

$= \dfrac{5x^2}{x^2(x + y)(x - y)} + \dfrac{3x^2 - 3xy}{x^2(x + y)(x - y)}$

$= \dfrac{5x^2 + 3x^2 - 3xy}{x^2(x + y)(x - y)}$ Write the sum of the numerators over the common denominator.

$= \dfrac{8x^2 - 3xy}{x^2(x + y)(x - y)}$ Combine like terms.

$= \dfrac{x\,(8x - 3y)}{x^2\,(x + y)(x - y)}$ Divide out the common factor x in the numerator and denominator and simplify.

$= \dfrac{8x - 3y}{x(x + y)(x - y)}$ □

 **Student Practice 7** Add.

$$\frac{7a}{a^2 + 2ab + b^2} + \frac{4}{a^2 + ab}$$

It is very easy to make a sign mistake when subtracting two fractions. You will find it helpful to place parentheses around the numerator of the second fraction so that you will not forget to subtract the entire numerator.

MC **Example 8** Subtract. $\dfrac{3x + 4}{x - 2} - \dfrac{x - 3}{2x - 4}$

Solution Factor the second denominator.

$= \dfrac{3x + 4}{x - 2} - \dfrac{x - 3}{2(x - 2)}$ Observe that the LCD is $2(x - 2)$.

$= \dfrac{2}{2} \cdot \dfrac{3x + 4}{x - 2} - \dfrac{x - 3}{2(x - 2)}$ Multiply the first fraction by $\frac{2}{2}$ so that the resulting fraction will have the common denominator.

$= \dfrac{2(3x + 4) - (x - 3)}{2(x - 2)}$ Write the indicated subtraction as one fraction. Note the parentheses around $x - 3$.

$= \dfrac{6x + 8 - x + 3}{2(x - 2)}$ Remove the parentheses in the numerator.

$= \dfrac{5x + 11}{2(x - 2)}$ Combine like terms. □

Student Practice 8 Subtract.

$$\frac{x + 7}{3x - 9} - \frac{x - 6}{x - 3}$$

Sidelight: Alternate Method

To avoid making errors when subtracting two fractions, some students prefer to change subtraction to addition of the opposite of the second fraction. In other words, we use the property that $\frac{a}{b} - \frac{c}{b} = \frac{a}{b} + \frac{-c}{b}$.

Let's revisit Example 8 to see how to use this method.

$\frac{3x+4}{x-2} - \frac{(x-3)}{2(x-2)}$ Insert parentheses around $x - 3$.

$= \frac{2}{2} \cdot \frac{3x+4}{x-2} + \frac{-(x-3)}{2(x-2)}$ Change subtraction to addition of the opposite, and multiply the first fraction by $\frac{2}{2}$.

$= \frac{2(3x+4) + -(x-3)}{2(x-2)}$

$= \frac{6x + 8 + (-x) + 3}{2(x-2)}$ Apply distributive property to simplify the numerator.

$= \frac{5x+11}{2(x-2)}$ Combine like terms.

Try each method and see which one helps you avoid making errors.

Example 9 Subtract and simplify. $\frac{8x}{x^2-16} - \frac{4}{x-4}$

Solution

$\frac{8x}{x^2-16} - \frac{4}{x-4}$

$= \frac{8x}{(x+4)(x-4)} + \frac{-4}{x-4}$ Factor the first denominator. Use the property that $\frac{a}{b} - \frac{c}{b} = \frac{a}{b} + \frac{-c}{b}$.

$= \frac{8x}{(x+4)(x-4)} + \frac{-4}{x-4} \cdot \frac{x+4}{x+4}$ Multiply the second fraction by $\frac{x+4}{x+4}$.

$= \frac{8x + (-4)(x+4)}{(x+4)(x-4)}$ Write the sum of the numerators over the common denominator.

$= \frac{8x - 4x - 16}{(x+4)(x-4)}$ Remove parentheses in the numerator.

$= \frac{4x - 16}{(x+4)(x-4)}$ Combine like terms. Note that the numerator can be factored.

$= \frac{4\cancel{(x-4)}}{(x+4)\cancel{(x-4)}}$ Since $(x - 4)$ is a *factor* of the numerator *and* the denominator, we may divide out the common factor.

$= \frac{4}{x+4}$ □

Student Practice 9 Subtract and simplify.

$$\frac{x-2}{x^2-4} - \frac{x+1}{2x^2+4x}$$

7.3 Exercises MyMathLab®

Verbal and Writing Skills, Exercises 1 and 2

1. Suppose two rational expressions have denominators of $(x + 3)(x + 5)$ and $(x + 3)^2$. Explain how you would determine the LCD.

2. Suppose two rational expressions have denominators of $(x - 4)^2(x + 7)$ and $(x - 4)^3$. Explain how you would determine the LCD.

Perform the operation indicated. Be sure to simplify.

3. $\dfrac{3x + 2}{5 + 2x} + \dfrac{x}{2x + 5}$

4. $\dfrac{x + 6}{3x + 8} + \dfrac{4}{8 + 3x}$

5. $\dfrac{3x}{x + 3} - \dfrac{x + 5}{x + 3}$

6. $\dfrac{4x + 8}{x - 6} - \dfrac{x + 6}{x - 6}$

7. $\dfrac{8x + 3}{5x + 7} - \dfrac{6x + 10}{5x + 7}$

8. $\dfrac{7x + 6}{5x + 2} - \dfrac{2x + 3}{5x + 2}$

Find the LCD. Do not combine fractions.

9. $\dfrac{10}{3a^2b^3}, \dfrac{8}{ab^2}$

10. $\dfrac{12}{5a^2}, \dfrac{9}{a^3}$

11. $\dfrac{5}{18x^2y^5}, \dfrac{7}{30x^3y^3}$

12. $\dfrac{9}{14xy^3}, \dfrac{14}{35x^4y^2}$

13. $\dfrac{9}{2x - 6}, \dfrac{5}{9x - 27}$

14. $\dfrac{12}{3x + 12}, \dfrac{5}{5x + 20}$

15. $\dfrac{8}{x + 3}, \dfrac{15}{x^2 - 9}$

16. $\dfrac{5}{x^2 - 4}, \dfrac{3}{x + 2}$

17. $\dfrac{7}{3x^2 + 14x - 5}, \dfrac{4}{9x^2 - 6x + 1}$

18. $\dfrac{4}{2x^2 - 9x - 35}, \dfrac{3}{4x^2 + 20x + 25}$

Add.

19. $\dfrac{7}{ab} + \dfrac{3}{b}$

20. $\dfrac{8}{cd} + \dfrac{9}{d}$

21. $\dfrac{3}{x + 7} + \dfrac{8}{x^2 - 49}$

22. $\dfrac{5}{x^2 - 2x + 1} + \dfrac{3}{x - 1}$

23. $\dfrac{4y}{y + 1} + \dfrac{y}{y - 1}$

24. $\dfrac{5}{y - 3} + \dfrac{2}{y + 3}$

25. $\dfrac{6}{5a} + \dfrac{5}{3a + 2}$

26. $\dfrac{2}{3a - 5} + \dfrac{3}{4a}$

27. $\dfrac{2}{3xy} + \dfrac{1}{6yz}$

28. $\dfrac{5}{4xy} + \dfrac{5}{12yz}$

Subtract.

29. $\dfrac{5x + 6}{x - 3} - \dfrac{x - 2}{2x - 6}$

30. $\dfrac{7x + 3}{x - 4} - \dfrac{x - 3}{2x - 8}$

31. $\dfrac{3x}{x^2 - 25} - \dfrac{2}{x + 5}$

32. $\dfrac{7x}{x^2 - 9} - \dfrac{6}{x + 3}$

33. $\dfrac{a + 3b}{2} - \dfrac{a - b}{5}$

34. $\dfrac{2b - a}{4} - \dfrac{a + 3b}{3}$

35. $\dfrac{8}{2x - 3} - \dfrac{6}{x + 2}$

36. $\dfrac{5}{3x + 2} - \dfrac{2}{x - 4}$

37. $\dfrac{x}{x^2 + 2x - 3} - \dfrac{x}{x^2 - 5x + 4}$

38. $\dfrac{1}{x^2 - 2x} - \dfrac{5}{x^2 - 4x + 4}$

39. $\dfrac{3}{x^2 + 9x + 20} + \dfrac{1}{x^2 + 10x + 24}$

40. $\dfrac{2}{x^2 - 3x - 10} + \dfrac{5}{x^2 - 2x - 15}$

41. $\dfrac{3x - 8}{x^2 - 5x + 6} + \dfrac{x + 2}{x^2 - 6x + 8}$

42. $\dfrac{3x + 5}{x^2 + 4x + 3} + \dfrac{-x + 5}{x^2 + 2x - 3}$

Mixed Practice *Add or subtract.*

43. $\dfrac{6x}{y - 2x} - \dfrac{5x}{2x - y}$

44. $\dfrac{8b}{2b - a} - \dfrac{3b}{a - 2b}$

45. $\dfrac{3y}{8y^2 + 2y - 1} - \dfrac{5y}{2y^2 - 9y - 5}$

46. $\dfrac{3x}{x^2 - 5x - 6} - \dfrac{x + 5}{x^2 + 3x + 2}$

47. $\dfrac{2x}{2x^2 - 5x - 3} + \dfrac{3}{x - 3}$

48. $\dfrac{4x}{3x^2 - 5x - 2} + \dfrac{5}{x - 2}$

Cumulative Review

49. **[2.4.1]** Solve. $\dfrac{1}{3}(x - 2) + \dfrac{1}{2}(x + 3) = \dfrac{1}{4}(3x + 1)$

50. **[2.3.3]** Solve. $4.8 - 0.6x = 0.8(x - 1)$

51. **[2.8.4]** Solve for x. $x - \dfrac{1}{5}x > \dfrac{1}{2} + \dfrac{1}{10}x$

52. **[5.1.3]** Simplify. $(3x^3y^4)^4$

53. **[2.8.4]** ***Commuting Costs*** A single subway fare costs \$2.75. A monthly unlimited-ride subway pass costs \$90. How many days per month would you have to use the subway to go to work (assume one subway fare to get to work and one subway fare to get back home) in order for it to be cheaper to buy a monthly subway pass?

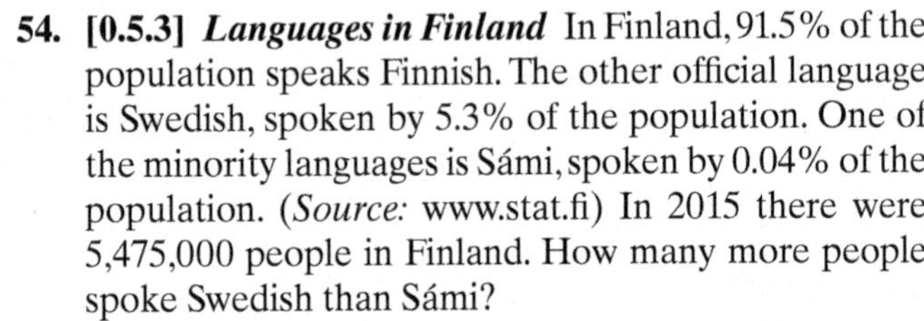

54. **[0.5.3]** ***Languages in Finland*** In Finland, 91.5% of the population speaks Finnish. The other official language is Swedish, spoken by 5.3% of the population. One of the minority languages is Sámi, spoken by 0.04% of the population. (*Source:* www.stat.fi) In 2015 there were 5,475,000 people in Finland. How many more people spoke Swedish than Sámi?

Quick Quiz 7.3 *Perform the operation indicated. Simplify.*

1. $\dfrac{3}{x^2 - 2x - 8} + \dfrac{2}{x - 4}$

2. $\dfrac{2x + y}{xy} - \dfrac{b - y}{by}$

3. $\dfrac{2}{x^2 - 9} + \dfrac{3}{x^2 + 7x + 12}$

4. **Concept Check** Explain how to find the LCD of the fractions $\dfrac{3}{10xy^2z}, \dfrac{6}{25x^2yz^3}$.

Use Math to Save Money

Did You Know? A lower credit score could cost you hundreds of dollars a year in higher finance fees.

Credit Card Follies

Understanding the Problem:

Stores tempt customers with the promise of "15% off all purchases made today" if the customer opens and puts the purchases on a store credit card. Adam is thinking about applying for one. He has a lot of shopping to do, and saving 15% on his $500 purchase is hard to resist. He hesitates, though, because his friend Matt is telling him that in some cases, people lose money on the deal.

Making a Plan:

Opening up a store credit card can cost you in two different ways. The first way is through the high interest rates store cards usually charge. The second way is not as straightforward. Opening a store credit account can lower your credit score because it increases your debt liability. If you need to apply for credit elsewhere, you will pay a higher interest rate due to your lower score. You need to weigh these costs against the benefit of the one-time savings.

Step 1: Adam calculates how much he will save today if he opens a store credit account.

Task 1: How much will Adam save today?

Step 2: The card has an annual interest rate of 25%. Using simple interest, Adam calculates that he would pay about $62.50 in finance fees if he carried the balance for six months. (*Note:* $500 \times 0.25 \times \frac{1}{2} = 62.50$)

Task 2: How much would Adam pay in interest if he carried the balance for a year?

Step 3: Adam is surprised at how much he could end up paying in finance charges. Matt then tells him other ways it could cost him. His credit score would go down due to his opening a new account. Also, since the store card has a $1000 limit, the $500 purchase would be 50% of the available credit. The higher the percentage of available credit you are using, the lower your credit score. Matt tells Adam that if he needs to put the purchase on a card, he should use his existing card, which has a $5000 limit.

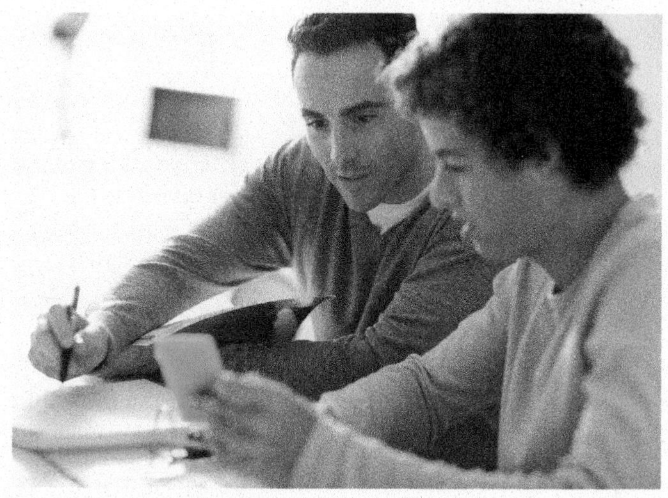

Task 3: What percentage of available credit would Adam be using if he put the purchase on his existing card?

Step 4: Even though Adam isn't planning on applying for a large loan next year, it is still a possibility. What if his car broke down and he decided to finance a new one? An increase in the interest rate due to a lower credit score could cost him hundreds of dollars over the life of the loan.

Task 4: If he paid $9 more a month because of a higher interest rate, how much more would he pay over the life of a four-year auto loan?

Finding a Solution:

Adam decides that the risk of higher future fees is not worth the one-time savings on his purchase. He declines the offer and puts the purchase on his existing card.

Applying the Situation to Your Life:

The next time you are tempted to sign up for a store credit card, you will have a better idea of the costs associated with that decision. That way, you can weigh the choices and make an informed decision.

7.4 Simplifying Complex Rational Expressions

Student Learning Objectives

After studying this section, you will be able to:

1. Simplify complex rational expressions by adding or subtracting in the numerator and denominator.
2. Simplify complex rational expressions using the LCD.

1 Simplifying Complex Rational Expressions by Adding or Subtracting in the Numerator and Denominator

A **complex rational expression** (also called a **complex fraction**) has a fraction in the numerator or in the denominator or both.

$$\frac{3+\frac{2}{x}}{\frac{x}{7}+2} \qquad \frac{\frac{x}{y}+1}{2} \qquad \frac{\frac{a+b}{3}}{\frac{x-2y}{4}}$$

The bar in a complex rational expression is both a grouping symbol and a symbol for division.

$$\frac{\frac{a+b}{3}}{\frac{x-2y}{4}} \quad \text{is equivalent to} \quad \left(\frac{a+b}{3}\right) \div \left(\frac{x-2y}{4}\right)$$

We need a procedure for simplifying complex rational expressions.

PROCEDURE TO SIMPLIFY A COMPLEX RATIONAL EXPRESSION: ADDING AND SUBTRACTING

1. Add or subtract so that you have a single fraction in the numerator and in the denominator.
2. Divide the fraction in the numerator by the fraction in the denominator. This is done by inverting the fraction in the denominator and multiplying it by the numerator.

Example 1 Simplify. $\dfrac{\frac{1}{x}}{\frac{2}{y^2}+\frac{1}{y}}$

Solution

Step 1 Add the two fractions in the denominator.

$$\frac{\frac{1}{x}}{\frac{2}{y^2}+\frac{1}{y}\cdot\frac{y}{y}} = \frac{\frac{1}{x}}{\frac{2+y}{y^2}}$$

Step 2 Divide the fraction in the numerator by the fraction in the denominator.

$$\frac{1}{x} \div \frac{2+y}{y^2} = \frac{1}{x}\cdot\frac{y^2}{2+y} = \frac{y^2}{x(2+y)}$$ □

Student Practice 1

Simplify. $\dfrac{\frac{1}{a}+\frac{1}{a^2}}{\frac{2}{b^2}}$

A complex rational expression may contain two or more fractions in the numerator and the denominator.

Example 2 Simplify. $\dfrac{\dfrac{1}{x} + \dfrac{1}{y}}{\dfrac{3}{a} - \dfrac{2}{b}}$

Solution We observe that the LCD of the fractions in the numerator is xy. The LCD of the fractions in the denominator is ab.

$$= \dfrac{\dfrac{1}{x}\cdot\dfrac{y}{y} + \dfrac{1}{y}\cdot\dfrac{x}{x}}{\dfrac{3}{a}\cdot\dfrac{b}{b} - \dfrac{2}{b}\cdot\dfrac{a}{a}}$$ Multiply each fraction by the appropriate value to obtain common denominators.

$$= \dfrac{\dfrac{y + x}{xy}}{\dfrac{3b - 2a}{ab}}$$ Add the two fractions in the numerator. Subtract the two fractions in the denominator.

$$= \dfrac{y + x}{xy}\cdot\dfrac{ab}{3b - 2a}$$ Invert the fraction in the denominator and multiply it by the numerator.

$$= \dfrac{ab(y + x)}{xy(3b - 2a)}$$ Write the answer as one fraction. □

Student Practice 2

Simplify. $\dfrac{\dfrac{1}{a} + \dfrac{1}{b}}{\dfrac{x}{2} - \dfrac{5}{y}}$

For some complex rational expressions, factoring may be necessary to determine the LCD and to combine fractions.

Example 3 Simplify. $\dfrac{\dfrac{1}{x^2 - 1} + \dfrac{2}{x + 1}}{x}$

Solution We need to factor $x^2 - 1$.

$$= \dfrac{\dfrac{1}{(x + 1)(x - 1)} + \dfrac{2}{x + 1}\cdot\dfrac{x - 1}{x - 1}}{x}$$ The LCD for the fractions in the numerator is $(x + 1)(x - 1)$.

$$= \dfrac{\dfrac{1 + 2x - 2}{(x + 1)(x - 1)}}{x}$$ Add the two fractions in the numerator.

$$= \dfrac{2x - 1}{(x + 1)(x - 1)}\cdot\dfrac{1}{x}$$ Simplify the numerator. Invert the fraction in the denominator and multiply.

$$= \dfrac{2x - 1}{x(x + 1)(x - 1)}$$ Write the answer as one fraction. □

Student Practice 3

Simplify. $\dfrac{\dfrac{x}{x^2 + 4x + 3} + \dfrac{2}{x + 1}}{x + 1}$

When simplifying complex rational expressions, always check to see if the final fraction can be reduced or simplified.

Example 4 Simplify. $\dfrac{\dfrac{3}{a+b}-\dfrac{3}{a-b}}{\dfrac{5}{a^2-b^2}}$

Solution The LCD of the two fractions in the numerator is $(a + b)(a - b)$.

$$= \frac{\dfrac{3}{a+b}\cdot\dfrac{a-b}{a-b}-\dfrac{3}{a-b}\cdot\dfrac{a+b}{a+b}}{\dfrac{5}{a^2-b^2}}$$

$$= \frac{\dfrac{3a-3b}{(a+b)(a-b)}-\dfrac{3a+3b}{(a+b)(a-b)}}{\dfrac{5}{a^2-b^2}}$$

Study carefully how we combine the two fractions in the numerator. Do you see how we obtain $-6b$?

$$= \frac{\dfrac{-6b}{(a+b)(a-b)}}{\dfrac{5}{(a+b)(a-b)}}$$ Factor $a^2 - b^2$ as $(a + b)(a - b)$.

$$= \frac{-6b}{(a+b)(a-b)}\cdot\frac{(a+b)(a-b)}{5}$$ Since $(a + b)(a - b)$ are factors in both a numerator and a denominator, they may be divided out.

$$= \frac{-6b}{5} \text{ or } -\frac{6b}{5}$$ □

Student Practice 4 Simplify.

$$\frac{\dfrac{6}{x^2-y^2}}{\dfrac{1}{x-y}+\dfrac{3}{x+y}}$$

2 Simplifying Complex Rational Expressions Using the LCD

There is another way to simplify complex rational expressions: Multiply the numerator and denominator of the complex fraction by the least common denominator of all the denominators appearing in the complex fraction.

PROCEDURE TO SIMPLIFY A COMPLEX RATIONAL EXPRESSION: MULTIPLYING BY THE LCD

1. Determine the LCD of all individual denominators occurring in the numerator and denominator of the complex rational expression.
2. Multiply both the numerator and the denominator of the complex rational expression by the LCD.
3. Simplify, if possible.

Example 5 Simplify by multiplying by the LCD. $\dfrac{\dfrac{5}{ab^2} - \dfrac{2}{ab}}{3 - \dfrac{5}{2a^2b}}$

Solution The LCD of all the denominators in the complex rational expression is $2a^2b^2$.

$$= \frac{2a^2b^2\left(\dfrac{5}{ab^2} - \dfrac{2}{ab}\right)}{2a^2b^2\left(3 - \dfrac{5}{2a^2b}\right)}$$ Multiply the numerator and denominator by the LCD.

$$= \frac{2a^2b^2\left(\dfrac{5}{ab^2}\right) - 2a^2b^2\left(\dfrac{2}{ab}\right)}{2a^2b^2(3) - 2a^2b^2\left(\dfrac{5}{2a^2b}\right)}$$ Multiply each term by $2a^2b^2$.

$$= \frac{10a - 4ab}{6a^2b^2 - 5b}$$ Simplify. □

Student Practice 5 Simplify by multiplying by the LCD.

$$\frac{\dfrac{2}{3x^2} - \dfrac{3}{y}}{\dfrac{5}{xy} - 4}$$

So that you can compare the two methods, we will redo Example 4 by multiplying by the LCD.

Example 6 Simplify by multiplying by the LCD. $\dfrac{\dfrac{3}{a+b} - \dfrac{3}{a-b}}{\dfrac{5}{a^2 - b^2}}$

Solution The LCD of all individual fractions contained in the complex fraction is $(a + b)(a - b)$.

$$= \frac{(a+b)(a-b)\left(\dfrac{3}{a+b}\right) - (a+b)(a-b)\left(\dfrac{3}{a-b}\right)}{(a+b)(a-b)\left(\dfrac{5}{(a+b)(a-b)}\right)}$$ Multiply each term by the LCD.

$$= \frac{3(a-b) - 3(a+b)}{5}$$ Simplify.

$$= \frac{3a - 3b - 3a - 3b}{5}$$ Remove parentheses.

$$= -\frac{6b}{5}$$ Simplify. □

Student Practice 6 Simplify by multiplying by the LCD.

$$\frac{\dfrac{6}{x^2 - y^2}}{\dfrac{7}{x-y} + \dfrac{3}{x+y}}$$

7.4 Exercises MyMathLab®

Simplify.

1. $\dfrac{\dfrac{5}{x}}{\dfrac{4}{x}+\dfrac{3}{x^2}}$

2. $\dfrac{\dfrac{8}{x}}{\dfrac{5}{x^2}+\dfrac{6}{x}}$

3. $\dfrac{\dfrac{4}{a}+\dfrac{1}{b}}{\dfrac{5}{ab}}$

4. $\dfrac{\dfrac{3}{a}+\dfrac{5}{b}}{\dfrac{4}{ab}}$

5. $\dfrac{\dfrac{x}{6}-\dfrac{1}{3}}{\dfrac{2}{3x}+\dfrac{5}{6}}$

6. $\dfrac{\dfrac{x}{6}-\dfrac{1}{3}}{\dfrac{5}{6x}+\dfrac{1}{2}}$

7. $\dfrac{\dfrac{7}{5x}-\dfrac{1}{x}}{\dfrac{3}{5}+\dfrac{2}{x}}$

8. $\dfrac{\dfrac{8}{x}-\dfrac{2}{3x}}{\dfrac{2}{3}+\dfrac{5}{x}}$

9. $\dfrac{\dfrac{5}{x}+\dfrac{3}{y}}{3x+5y}$

10. $\dfrac{\dfrac{1}{x}+\dfrac{1}{y}}{x+y}$

11. $\dfrac{4-\dfrac{1}{x^2}}{2+\dfrac{1}{x}}$

12. $\dfrac{1-\dfrac{36}{x^2}}{1-\dfrac{6}{x}}$

13. $\dfrac{\dfrac{2}{x+6}}{\dfrac{2}{x-6}-\dfrac{2}{x^2-36}}$

14. $\dfrac{\dfrac{9}{x^2-1}}{\dfrac{4}{x-1}-\dfrac{2}{x+1}}$

15. $\dfrac{a+\dfrac{3}{a}}{\dfrac{a^2+2}{3a}}$

16. $\dfrac{x+\dfrac{4}{x}}{\dfrac{x^2+3}{4x}}$

17. $\dfrac{\dfrac{3}{x-3}}{\dfrac{1}{x^2-9}+\dfrac{2}{x+3}}$

18. $\dfrac{\dfrac{4}{x+5}}{\dfrac{2}{x-5}-\dfrac{1}{x^2-25}}$

19. $\dfrac{\dfrac{3}{x-1}+4}{\dfrac{3}{x-1}-4}$

20. $\dfrac{\dfrac{2x}{x-2}+3}{\dfrac{3x}{x-2}+4}$

To Think About, Exercises 21 and 22

21. Consider the complex fraction $\dfrac{\dfrac{4}{x+3}}{\dfrac{5}{x}-1}$. What values are not allowable replacements for the variable x?

22. Consider the complex fraction $\dfrac{\dfrac{5}{x-2}}{\dfrac{6}{x}+1}$. What values are not allowable replacements for the variable x?

Simplify.

23. $\dfrac{x+5y}{x-6y} \div \left(\dfrac{1}{5y} - \dfrac{1}{x+5y}\right)$

24. $\left(\dfrac{1}{x+2y} - \dfrac{1}{x-y}\right) \div \dfrac{2x-4y}{x^2-3xy+2y^2}$

Cumulative Review

25. [3.3.2] Find the slope and y-intercept of the line $5x + 6y = 8$.

26. [2.8.4] Solve and graph. $7 + x < 11 + 5x$

−5 −4 −3 −2 −1 0 1 2 3 4 5 x

27. [2.6.1] When nine is subtracted from double a number, the result is the same as one-half of the same number. What is the number?

28. [2.7.2] Isabella received a pay raise this year. The raise was 5% of last year's salary. This year she will earn $25,200. What was her salary last year before the raise?

Quick Quiz 7.4 *Simplify.*

1. $\dfrac{\dfrac{a}{4b} - \dfrac{1}{3}}{\dfrac{5}{4b} - \dfrac{4}{a}}$

2. $\dfrac{a+b}{\dfrac{1}{a} + \dfrac{1}{b}}$

3. $\dfrac{\dfrac{10}{x^2-25}}{\dfrac{3}{x+5} + \dfrac{2}{x-5}}$

4. **Concept Check** To simplify the following complex fraction, explain how you would add the two fractions in the numerator.

$$\dfrac{\dfrac{7}{x-3} + \dfrac{15}{2x-6}}{\dfrac{2}{x+5}}$$

7.5 Solving Equations Involving Rational Expressions

Student Learning Objectives

After studying this section, you will be able to:

1. Solve equations involving rational expressions that have solutions.
2. Determine whether an equation involving rational expressions has no solution.

1 Solving Equations Involving Rational Expressions That Have Solutions

In Section 2.4 we developed procedures to solve linear equations containing fractions whose denominators are numbers. In this section we use a similar approach to solve equations containing fractions whose denominators are polynomials. It would be wise for you to review Section 2.4 briefly *before you begin this section.* It will be especially helpful to carefully study Examples 1 and 2 in that section.

TO SOLVE AN EQUATION CONTAINING RATIONAL EXPRESSIONS

1. Determine the LCD of all the denominators.
2. Multiply each term of the equation by the LCD.
3. Solve the resulting equation.
4. Check your solution. Exclude from your solution any value that would make the LCD equal to zero.

Example 1 Solve for x and check your solution. $\frac{5}{x} + \frac{2}{3} = -\frac{3}{x}$

Solution

$$3x\left(\frac{5}{x}\right) + 3x\left(\frac{2}{3}\right) = 3x\left(-\frac{3}{x}\right)$$ Observe that the LCD is $3x$. Multiply each term by $3x$.

$$15 + 2x = -9$$

$$2x = -9 - 15$$ Subtract 15 from both sides.

$$2x = -24$$

$$x = -12$$ Divide both sides by 2.

Check.

$$\frac{5}{-12} + \frac{2}{3} \stackrel{?}{=} -\frac{3}{-12}$$ Replace each x by -12.

$$-\frac{5}{12} + \frac{8}{12} \stackrel{?}{=} \frac{3}{12}$$

$$\frac{3}{12} = \frac{3}{12} \checkmark$$ It checks. □

Student Practice 1 Solve for x and check your solution.

$$\frac{3}{x} + \frac{4}{5} = -\frac{2}{x}$$

Example 2 Solve and check. $\frac{6}{x+3} = \frac{3}{x}$

Solution

Observe that the LCD $= x(x + 3)$.

$$x(x+3)\left(\frac{6}{x+3}\right) = x(x+3)\left(\frac{3}{x}\right)$$ Multiply both sides by $x(x + 3)$.

$$6x = 3(x + 3)$$ Simplify. Do you see how this is done?

$$6x = 3x + 9$$ Remove parentheses.

$$3x = 9$$ Subtract $3x$ from both sides.

$$x = 3$$ Divide both sides by 3.

Check. $\dfrac{6}{3+3} \stackrel{?}{=} \dfrac{3}{3}$ Replace each x by 3.

$\dfrac{6}{6} = \dfrac{3}{3}$ It checks. ✓ □

Student Practice 2 Solve and check.

$$\frac{6}{2x+1} = \frac{2}{x+2}$$

It is sometimes necessary to factor denominators before the correct LCD can be determined.

Example 3 Solve and check. $\dfrac{3}{x+5} - 1 = \dfrac{4-x}{2x+10}$

Solution

$\dfrac{3}{x+5} - 1 = \dfrac{4-x}{2(x+5)}$ Factor $2x + 10$. We determine that the LCD is $2(x + 5)$.

$2(x+5)\left(\dfrac{3}{x+5}\right) - 2(x+5)(1) = 2(x+5)\left[\dfrac{4-x}{2(x+5)}\right]$ Multiply each term by the LCD.

$2(3) - 2(x+5) = 4 - x$ Simplify.

$6 - 2x - 10 = 4 - x$ Remove parentheses.

$-2x - 4 = 4 - x$ Combine like terms.

$-4 = 4 + x$ Add $2x$ to both sides.

$-8 = x$ Subtract 4 from both sides.

Check. $\dfrac{3}{-8+5} - 1 \stackrel{?}{=} \dfrac{4-(-8)}{2(-8)+10}$ Replace each x in the original equation by -8.

$\dfrac{3}{-3} - 1 \stackrel{?}{=} \dfrac{4+8}{-16+10}$

$-1 - 1 \stackrel{?}{=} \dfrac{12}{-6}$

$-2 = -2$ ✓ It checks. The solution is -8. □

Student Practice 3

Solve and check.

$$\frac{x-1}{x^2-4} = \frac{2}{x+2} + \frac{4}{x-2}$$

2 Determining Whether an Equation Involving Rational Expressions Has No Solution

Equations containing rational expressions sometimes appear to have solutions when in fact they do not. By this we mean that the "solutions" we get by using completely correct methods are, in actuality, not solutions.

In the case where a value makes a denominator in the equation equal to zero, we say it is not a solution to the equation. Such a value is called an **extraneous solution.** An extraneous solution is an apparent solution that does *not* satisfy the original equation. If all of the apparent solutions of an equation are extraneous solutions, we say that the equation has **no solution.** It is important that you check all apparent solutions in the original equation.

Example 4 Solve and check. $\frac{y}{y-2} - 4 = \frac{2}{y-2}$

Solution

Observe that the LCD is $y - 2$.

$$(y-2)\left(\frac{y}{y-2}\right) - (y-2)(4) = (y-2)\left(\frac{2}{y-2}\right) \quad \text{Multiply each term by } (y-2).$$

$$y - 4(y-2) = 2 \quad \text{Simplify. Do you see how this is done?}$$

$$y - 4y + 8 = 2 \quad \text{Remove parentheses.}$$

$$-3y + 8 = 2 \quad \text{Combine like terms.}$$

$$-3y = -6 \quad \text{Subtract 8 from both sides.}$$

$$\frac{-3y}{-3} = \frac{-6}{-3} \quad \text{Divide both sides by } -3.$$

$$y = 2 \quad \text{2 is only an apparent solution.}$$

This equation has no solution.

Why? We can see immediately that $y = 2$ is not a solution of the original equation. When we substitute 2 for y in a denominator, the denominator is equal to zero and the expression is undefined.

Check. $\frac{y}{y-2} - 4 = \frac{2}{y-2}$ Suppose that you try to check the apparent solution by substituting 2 for y.

$$\frac{2}{2-2} - 4 \stackrel{?}{=} \frac{2}{2-2}$$

$$\frac{2}{0} - 4 = \frac{2}{0}$$ This does not check since you do not obtain a real number when you divide by zero.

These expressions are not defined.

There is no such number as $2 \div 0$. We see that 2 does *not* check. This equation has **no solution.** □

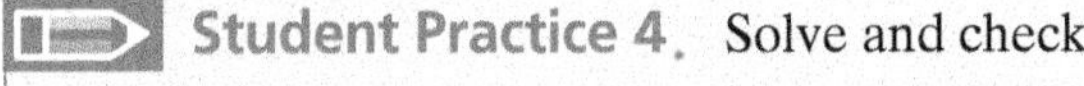

Student Practice 4. Solve and check.

$$\frac{2x}{x+1} = \frac{-2}{x+1} + 1$$

7.5 Exercises MyMathLab®

Solve and check exercises 1–16.

1. $\frac{7}{x} + \frac{3}{4} = \frac{-2}{x}$

2. $\frac{8}{x} + \frac{2}{5} = \frac{-2}{x}$

3. $\frac{3}{x} - \frac{5}{4} = \frac{1}{2x}$

4. $\frac{1}{3x} + \frac{5}{6} = \frac{2}{x}$

5. $\frac{5x + 3}{3x} = \frac{7}{3} - \frac{9}{x}$

6. $\frac{2x + 3}{4x} - \frac{1}{x} = \frac{3}{2}$

7. $\frac{x + 5}{3x} = \frac{1}{2}$

8. $\frac{x - 4}{5x} = \frac{3}{10}$

9. $\frac{6}{3x - 5} = \frac{3}{2x}$

10. $\frac{3}{x + 4} = \frac{2}{x}$

11. $\frac{2}{2x + 5} = \frac{4}{x - 4}$

12. $\frac{3}{x + 5} = \frac{3}{3x - 2}$

13. $\frac{2}{x} + \frac{x}{x + 1} = 1$

14. $\frac{5}{2} = 3 + \frac{2x + 7}{x + 6}$

15. $\frac{85 - 4x}{x} = 7 - \frac{3}{x}$

16. $\frac{63 - 2x}{x} = 2 - \frac{5}{x}$

Mixed Practice *Solve and check. If there is no solution, say so.*

17. $\frac{1}{x + 4} - 2 = \frac{3x - 2}{x + 4}$

18. $\frac{2}{x + 5} - 1 = \frac{3x - 4}{x + 5}$

19. $\frac{2}{x - 6} - 5 = \frac{2(x - 5)}{x - 6}$

20. $5 - \frac{x}{x + 3} = \frac{3}{3 + x}$

21. $\frac{2}{x + 1} - \frac{1}{x - 1} = \frac{2x}{x^2 - 1}$

22. $\frac{8x}{4x^2 - 1} = \frac{3}{2x + 1} + \frac{3}{2x - 1}$

23. $\frac{x + 2}{x^2 - x - 12} = \frac{1}{x + 3} - \frac{1}{x - 4}$

24. $\frac{x + 3}{x^2 - 3x - 10} = \frac{2}{x - 5} - \frac{2}{x + 2}$

25. $\frac{2x}{x + 4} - \frac{8}{x - 4} = \frac{2x^2 + 32}{x^2 - 16}$

26. $\frac{4x}{x + 3} - \frac{12}{x - 3} = \frac{4x^2 + 36}{x^2 - 9}$

27. $\dfrac{4}{x^2 - 1} + \dfrac{7}{x + 1} = \dfrac{5}{x - 1}$

28. $\dfrac{9}{9x^2 - 1} + \dfrac{1}{3x + 1} = \dfrac{2}{3x - 1}$

29. $\dfrac{x + 11}{x^2 - 5x + 4} + \dfrac{3}{x - 1} = \dfrac{5}{x - 4}$

30. $\dfrac{6}{x - 3} = \dfrac{-5}{x - 2} + \dfrac{-5}{x^2 - 5x + 6}$

To Think About *In each of the following equations, what values are not allowable replacements for the variable x? Do not solve the equation.*

31. $\dfrac{5x}{x + 6} - \dfrac{2x}{3x + 1} = \dfrac{2}{3x^2 + 19x + 6}$

32. $\dfrac{4x}{x - 3} - \dfrac{3x}{2x - 1} = \dfrac{6}{2x^2 - 5x - 3}$

Cumulative Review

33. **[6.4.1]** Factor. $8x^2 - 2x - 1$

34. **[2.3.3]** Solve. $5(x - 2) = 8 - (3 + x)$

▲ **35.** **[2.6.1]** ***Geometry*** The perimeter of a rectangular computer monitor is 44 inches. The length is 8 inches less than twice the width. Find the dimensions of the monitor.

36. **[3.6.1]** Determine the domain and range of the relation. Is the relation a function? $\{(7, 3), (2, 2), (-2, 0), (2, -2), (7, -3)\}$

Quick Quiz 7.5 *Solve. If there is no solution, say so.*

1. $\dfrac{3}{4x} - \dfrac{5}{6x} = 2 - \dfrac{1}{2x}$

2. $\dfrac{x}{x - 1} - \dfrac{2}{x} = \dfrac{1}{x - 1}$

3. $\dfrac{6}{x^2 - 2x - 8} + \dfrac{5}{x + 2} = \dfrac{1}{x - 4}$

4. **Concept Check** Explain how to find the LCD for the following equation. Do not solve the equation.

$$\frac{x}{x^2 - 9} + \frac{2}{3x - 9} = \frac{5}{2x + 6} + \frac{3}{2x^2 - 18}$$

7.6 Ratio, Proportion, and Other Applied Problems

Student Learning Objectives

After studying this section, you will be able to:

1. Solve problems involving ratio and proportion.
2. Solve problems involving similar triangles.
3. Solve distance problems involving rational expressions.
4. Solve work problems.

1 Solving Problems Involving Ratio and Proportion

A **ratio** is a comparison of two quantities. You may be familiar with ratios that compare miles to hours or miles to gallons. A ratio is often written as a quotient in the form of a fraction. For example, the ratio of 7 to 9 can be written as $\frac{7}{9}$.

A **proportion** is an equation that states that two ratios are equal. For example,

$$\frac{7}{9} = \frac{21}{27}, \quad \frac{2}{3} = \frac{10}{15}, \quad \text{and} \quad \frac{a}{b} = \frac{c}{d} \quad \text{are proportions.}$$

Let's take a closer look at the last proportion. We can see that the LCD of the fractional equation is bd.

$$(b\,d)\frac{a}{b} = (b\,d)\frac{c}{d} \qquad \text{Multiply each side by the LCD and simplify.}$$

$$da = bc$$

$$ad = bc \qquad \text{Since multiplication is commutative, } da = ad.$$

Thus we have proved the following.

> **THE PROPORTION EQUATION**
>
> $$\text{If } \frac{a}{b} = \frac{c}{d}, \text{ then } ad = bc$$
>
> for all real numbers a, b, c, and d, where $b \neq 0$ and $d \neq 0$.

This is sometimes called **cross multiplying.** It can be applied only if you have *one* fraction and nothing else on each side of the equation.

Example 1 Michael took 5 hours to drive 245 miles on the turnpike. At the same rate, how many hours will it take him to drive a distance of 392 miles?

Solution

1. ***Understand the problem.*** Let $x =$ the number of hours it will take to drive 392 miles. If 5 hours are needed to drive 245 miles, then x hours are needed to drive 392 miles.
2. ***Write an equation.*** We can write this as a proportion. Compare time to distance in each ratio.

$$\begin{array}{lcl} \text{Time} \longrightarrow & \dfrac{5 \text{ hours}}{245 \text{ miles}} = \dfrac{x \text{ hours}}{392 \text{ miles}} & \longleftarrow \text{Time} \\ \text{Distance} \longrightarrow & & \longleftarrow \text{Distance} \end{array}$$

3. ***Solve and state the answer.***

$$5(392) = 245x \qquad \text{Cross-multiply.}$$

$$\frac{1960}{245} = x \qquad \text{Divide both sides by 245.}$$

$$8 = x$$

It will take Michael 8 hours to drive 392 miles.

4. ***Check.*** Is $\frac{5}{245} = \frac{8}{392}$? Do the computation and see. □

Student Practice 1 It took Brenda 8 hours to drive 420 miles. At the same rate, how long would it take her to drive 315 miles?

Example 2 If $\frac{3}{4}$ inch on a map represents an actual distance of 20 miles, how long is the distance represented by $4\frac{1}{8}$ inches on the same map?

Solution Let x = the distance represented by $4\frac{1}{8}$ inches.

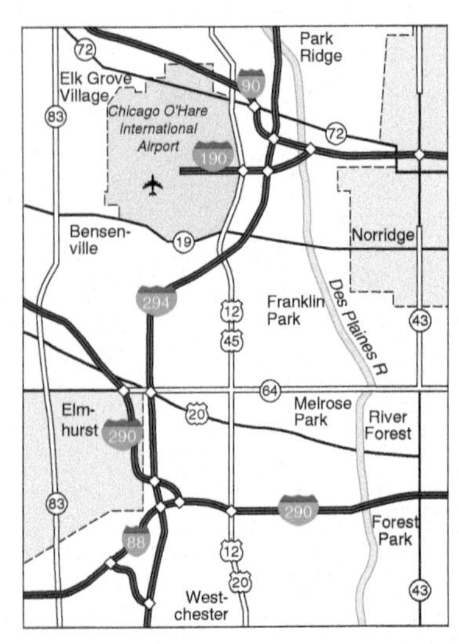

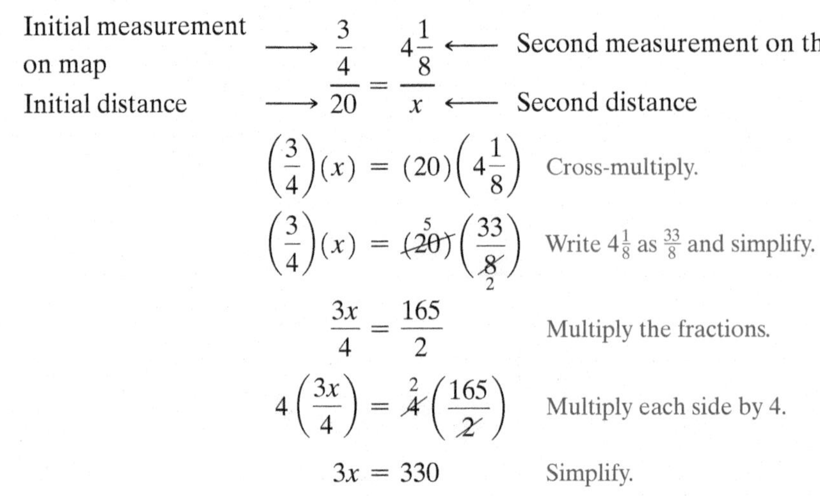

Initial measurement on map $\longrightarrow$ / Initial distance $\longrightarrow$ $\quad \dfrac{\frac{3}{4}}{20} = \dfrac{4\frac{1}{8}}{x}$ $\longleftarrow$ Second measurement on the map / $\longleftarrow$ Second distance

$$\left(\frac{3}{4}\right)(x) = (20)\left(4\frac{1}{8}\right) \quad \text{Cross-multiply.}$$

$$\left(\frac{3}{4}\right)(x) = \overset{5}{\cancel{(20)}}\left(\frac{33}{\underset{2}{\cancel{8}}}\right) \quad \text{Write } 4\tfrac{1}{8} \text{ as } \tfrac{33}{8} \text{ and simplify.}$$

$$\frac{3x}{4} = \frac{165}{2} \quad \text{Multiply the fractions.}$$

$$4\left(\frac{3x}{4}\right) = \overset{2}{\cancel{4}}\left(\frac{165}{\cancel{2}}\right) \quad \text{Multiply each side by 4.}$$

$$3x = 330 \quad \text{Simplify.}$$

$$x = 110 \quad \text{Divide both sides by 3.}$$

$4\frac{1}{8}$ inches on the map represents an actual distance of 110 miles. □

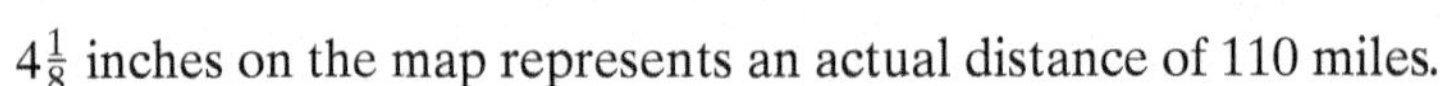

Student Practice 2 If $\frac{5}{8}$ inch on a map represents an actual distance of 30 miles, how long is the distance represented by $2\frac{1}{2}$ inches on the same map?

2 Solving Problems Involving Similar Triangles

Similar triangles are triangles that have the same shape but may be different sizes. For example, if you draw a triangle on a sheet of paper, place the paper in a photocopy machine, and make a copy that is reduced by 25%, you would create a triangle that is similar to the original triangle. The two triangles will have the same shape. The corresponding sides of the triangles will be proportional. The corresponding angles of the triangles will be equal.

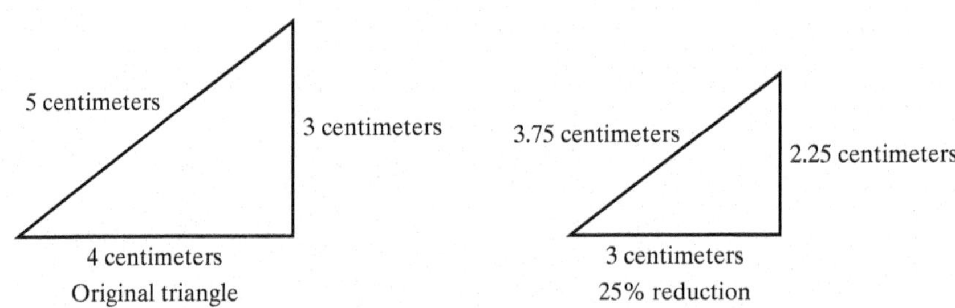

You can use the proportion equation to show that the corresponding sides of the triangles above are proportional. In fact, you can use the proportion equation to find an unknown length of a side of one of two similar triangles.

▲ **Example 3** A ramp is 32 meters long and rises up 15 meters. A ramp at the same angle is 9 meters long. How high does the second ramp rise?

Solution To answer this question, we find the length of side x in the following two similar triangles.

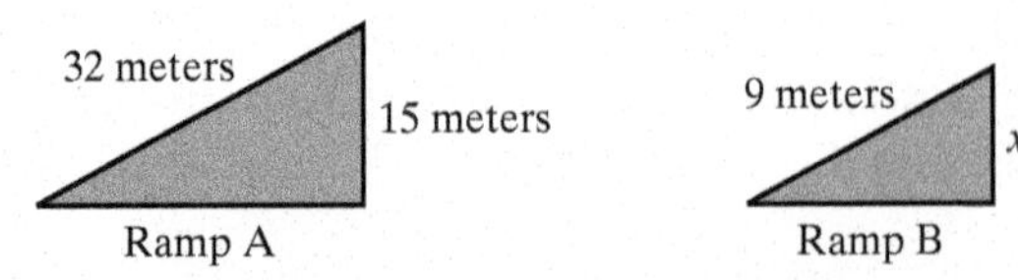

$$\begin{array}{lcl} \text{Length of ramp A} \longrightarrow & \dfrac{32}{9} = \dfrac{15}{x} & \longleftarrow \text{Rise of ramp A} \\ \text{Length of ramp B} \longrightarrow & & \longleftarrow \text{Rise of ramp B} \end{array}$$

$$32x = (9)(15) \quad \text{Cross-multiply.}$$

$$32x = 135 \quad \text{Simplify.}$$

$$x = \frac{135}{32} \quad \text{Divide both sides by 32.}$$

$$\text{or} \quad x = 4\frac{7}{32}$$

The ramp rises $4\frac{7}{32}$ meters high. □

▲ **Student Practice 3** Triangle C is similar to Triangle D. Find the length of side x. Express your answer as a mixed number.

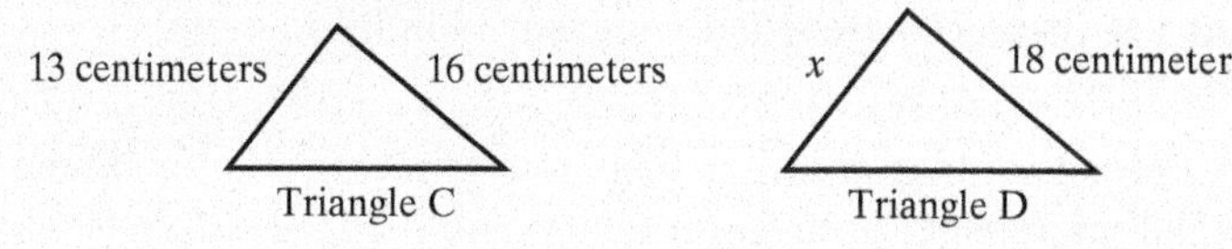

We can also use similar triangles for indirect measurement—for instance, to find the height of an object that is too tall to measure using standard measuring devices. When the sun shines on two vertical objects at the same time, the shadows and the objects form similar triangles.

▲ **Example 4** A woman who is 5 feet tall casts a shadow that is 8 feet long. At the same time of day, a building casts a shadow that is 72 feet long. How tall is the building?

Solution

1. ***Understand the problem.*** First we draw a sketch. We do not know the height of the building, so we call it x.

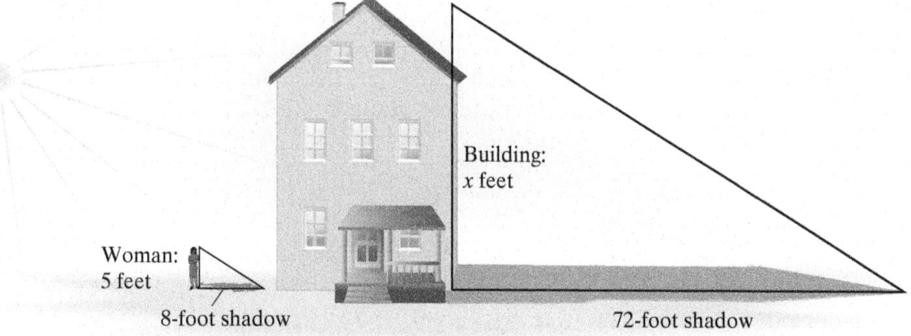

2. ***Write an equation and solve.***

$$\begin{array}{lcl} \text{Height of woman} \longrightarrow & \dfrac{5}{8} = \dfrac{x}{72} & \longleftarrow \text{Height of building} \\ \text{Length of woman's shadow} \longrightarrow & & \longleftarrow \text{Length of building's shadow} \end{array}$$

$$(5)(72) = 8x \quad \text{Cross-multiply.}$$

$$360 = 8x$$

$$45 = x$$

The height of the building is 45 feet. □

▲ **Student Practice 4** A man who is 6 feet tall casts a shadow that is 7 feet long. At the same time of day, a large flagpole casts a shadow that is 38.5 feet long. How tall is the flagpole?

In problems such as Example 4, we are assuming that the building and the person are standing exactly perpendicular to the ground. In other words, each triangle is assumed to be a right triangle. In other similar triangle problems, if the triangles are not right triangles, you must be careful that the corresponding angles in the two triangles are the same.

Les Ailes
Parce que vous
n'êtes pas un oiseau

3 Solving Distance Problems Involving Rational Expressions

Some distance problems are solved using equations with rational expressions. We will need the formula Distance = Rate × Time, $D = RT$, which we can write in the form $T = \frac{D}{R}$. In the United States, distances are usually measured in miles. In most other countries, distances are usually measured in kilometers.

Example 5 A French commuter airline flies from Paris to Avignon. Plane A flies at a speed that is 50 kilometers per hour faster than plane B. Plane A flies 500 kilometers in the amount of time that plane B flies 400 kilometers. Find the speed of each plane.

Solution

1. ***Understand the problem.*** Let $s =$ the speed of plane B in kilometers per hour. Then $s + 50 =$ the speed of plane A in kilometers per hour.
Make a simple table for D, R, and T.

	D	R	$T = \frac{D}{R}$
Plane A	500	$s + 50$	?
Plane B	400	s	?

Since $T = \frac{D}{R}$, for each plane we divide the expression for D by the expression for R and write it in the table in the column for time.

	D	R	$T = \frac{D}{R}$
Plane A	500	$s + 50$	$\frac{500}{s + 50}$
Plane B	400	s	$\frac{400}{s}$

2. ***Write an equation and solve.*** Each plane flies the same amount of time. That is, the time for plane A equals the time for plane B.

$$\frac{500}{s + 50} = \frac{400}{s}$$

You can solve this equation using the methods in Section 7.5 or you may cross-multiply. Here we will cross-multiply.

$$500s = (s + 50)(400) \quad \text{Cross-multiply.}$$
$$500s = 400s + 20{,}000 \quad \text{Remove parentheses.}$$
$$100s = 20{,}000 \quad \text{Subtract } 400s \text{ from each side.}$$
$$s = 200 \quad \text{Divide each side by 100.}$$

Plane B travels 200 kilometers per hour.

Since $s + 50 = 200 + 50 = 250$, plane A travels 250 kilometers per hour. □

Student Practice 5 Two European freight trains traveled toward Paris for the same amount of time. Train A traveled 180 kilometers, while train B traveled 150 kilometers. Train A traveled 10 kilometers per hour faster than train B. What was the speed of each train?

4 Solving Work Problems 

Some applied problems involve the length of time needed to do a job. These problems are often referred to as work problems.

Example 6 Reynaldo can sort a huge stack of mail on an old sorting machine in 9 hours. His brother Carlos can sort the same amount of mail using a newer sorting machine in 8 hours. How long would it take them to do the job working together? Express your answer in hours and minutes. Round to the nearest minute.

Solution

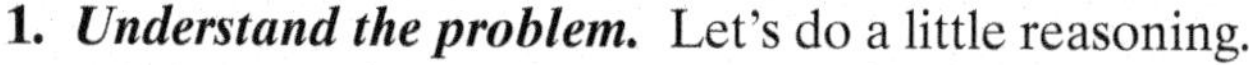

1. ***Understand the problem.*** Let's do a little reasoning.

   If Reynaldo can do the job in 9 hours, then in *1 hour* he could do $\frac{1}{9}$ of the job.

   If Carlos can do the job in 8 hours, then in *1 hour* he could do $\frac{1}{8}$ of the job.

   Let $x =$ the number of hours it takes Reynaldo and Carlos to do the job together. In *1 hour* together they could do $\frac{1}{x}$ of the job.

2. ***Write an equation and solve.*** The amount of work Reynaldo can do in 1 hour plus the amount of work Carlos can do in 1 hour must be equal to the amount of work they could do together in 1 hour.

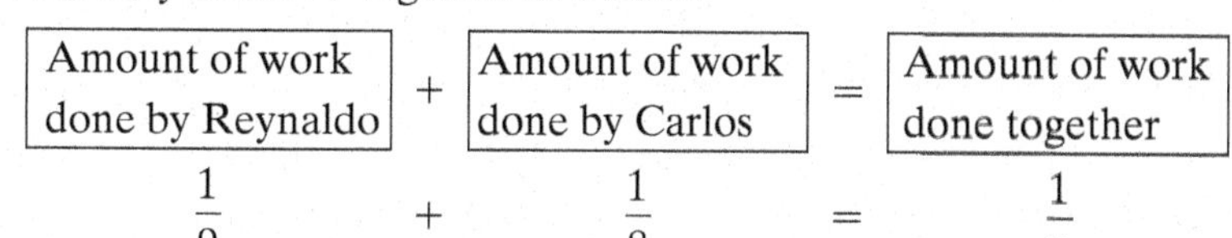

Amount of work done by Reynaldo
+
Amount of work done by Carlos
=
Amount of work done together

$$\frac{1}{9} + \frac{1}{8} = \frac{1}{x}$$

Let us solve for x. We observe that the LCD is $72x$.

$$72x\left(\frac{1}{9}\right) + 72x\left(\frac{1}{8}\right) = 72x\left(\frac{1}{x}\right) \quad \text{Multiply each term by the LCD.}$$

$$8x + 9x = 72 \quad \text{Simplify.}$$

$$17x = 72 \quad \text{Combine like terms.}$$

$$x = \frac{72}{17} \quad \text{Divide each side by 17.}$$

$$x = 4\frac{4}{17}$$

To change $\frac{4}{17}$ of an hour to minutes, we multiply.

$$\frac{4}{17}\ \cancel{\text{hour}} \times \frac{60 \text{ minutes}}{1\ \cancel{\text{hour}}} = \frac{240}{17} \text{ minutes, which is approximately 14.118 minutes}$$

To the nearest minute this is 14 minutes. Thus doing the job together will take 4 hours and 14 minutes. □

Student Practice 6 John Tobey and Dave Wells obtained night custodian jobs at a local factory while going to college part-time. Using the buffer machine, John can buff all the floors in the building in 6 hours. Dave takes a little longer and can do all the floors in the building in 7 hours. Their supervisor bought another buffer machine. How long will it take John and Dave to do all the floors in the building working together, each with his own machine? Express your answer in hours and minutes. Round to the nearest minute.

> **Calculator**
>
> **Reciprocals**
>
> You can find $\frac{1}{x}$ for any value of x on a scientific calculator by using the key labeled $\boxed{x^{-1}}$ or the key labeled $\boxed{1/x}$. For example, to find $\frac{1}{9}$, we use $9\ \boxed{x^{-1}}$ or $9\ \boxed{1/x}$. The display will read 0.11111111. Therefore we can solve Example 6 as follows:
>
> $9\ \boxed{x^{-1}}\ \boxed{+}\ 8\ \boxed{x^{-1}}\ \boxed{=}$
>
> The display will read 0.2361111. Thus we have obtained the equation $0.2361111 = \frac{1}{x}$. Now, this is equivalent to $x = \frac{1}{0.2361111}$. (Do you see why?) Thus we enter $0.2361111\ \boxed{x^{-1}}$, and the display reads 4.2352943. If we round to the nearest hundredth, we have $x \approx 4.24$ hours, which is approximately equal to our answer of $4\frac{4}{17}$ hours.

7.6 Exercises MyMathLab®

Solve.

1. $\dfrac{5}{11} = \dfrac{8}{x}$

2. $\dfrac{7}{14} = \dfrac{x}{9}$

3. $\dfrac{x}{17} = \dfrac{12}{5}$

4. $\dfrac{18}{x} = \dfrac{7}{3}$

5. $\dfrac{9.1}{8.4} = \dfrac{x}{6}$

6. $\dfrac{3}{x} = \dfrac{12.5}{3.2}$

7. $\dfrac{7}{x} = \dfrac{40}{130}$

8. $\dfrac{x}{12} = \dfrac{7}{3}$

Applications *Use a proportion to answer exercises 9–14.*

9. ***Exchange Rates*** Robyn spent two months traveling in New Zealand. The day she arrived, the exchange rate was 1.3 New Zealand dollars per U.S. dollar.

(a) If she exchanged \$500 U.S. dollars when she arrived, how many New Zealand dollars did she receive?

(b) Three days later the exchange rate of the New Zealand dollar was 1.15 New Zealand dollars per U.S. dollar. How much less money would she have received had she waited three days to exchange her money?

10. ***Exchange Rates*** Sean spent a semester studying in Germany. On the day he arrived in Berlin, the exchange rate for the euro was 0.77 euro per U.S. dollar. Sean converted \$350 to euros that day.

(a) How many euros did Sean receive for his \$350?

(b) On his way home Sean decided to spend a week in London. He had €200 (200 euros) that he wanted to change into British pounds. If the exchange rate was 0.83 British pounds per euro, how many pounds did he receive?

11. ***Speed Units*** Alfonse and Melinda are taking a drive in Mexico. They know that a speed of 100 kilometers per hour is approximately equal to 62 miles per hour. They are now driving on a Mexican road that has a speed limit of 90 kilometers per hour. How many miles per hour is the speed limit? Round to the nearest mile per hour.

12. ***Baggage Weight*** Dick and Anne took a trip to France. Their suitcases were weighed at the airport and the weight recorded was 39 kilograms. If 50 kilograms is equivalent to 110 pounds, how many pounds did their suitcases weigh? Round to the nearest pound.

13. ***Map Scale*** On a map the distance between two mountains is $3\frac{1}{2}$ inches. The actual distance between the mountains is 136 miles. Russ is camped at a location that on the map is $\frac{3}{4}$ inch from the base of the mountain. How many miles is he from the base of the mountain? Round to the nearest mile.

14. ***Map Scale*** John, Stephanie, Stella, Nathaniel, and Josiah are taking a trip from Denver to Pueblo. The scale on the AAA map of Colorado is approximately $\frac{3}{4}$ inch to 15 miles. If the distance from Denver to Pueblo measures 5.5 inches on the map, how far apart are the two cities?

Geometry *Triangles A and B are similar. Use them to answer exercises 15–18. Leave your answers as fractions.*

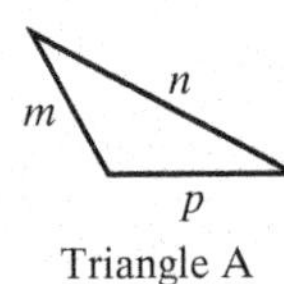

Triangle A

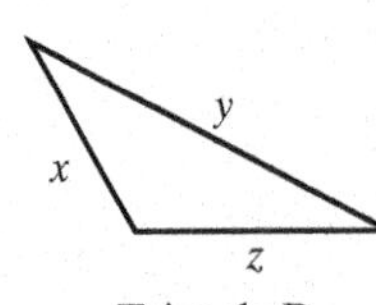

Triangle B

▲ **15.** If $x = 20$ in., $y = 29$ in., and $m = 13$ in., find the length of side n.

▲ **16.** If $p = 12$ in., $m = 16$ in., and $z = 20$ in., find the length of side x.

▲ **17.** If $x = 175$ meters, $n = 40$ meters, and $m = 35$ meters, find the length of side y.

▲ **18.** If $z = 18$ cm, $y = 25$ cm, and $n = 9$ cm, find the length of side p.

Geometry *Just as we have discussed similar triangles, other geometric shapes can be similar. Similar geometric shapes will have sides that are proportional. Quadrilaterals abcd and ghjk are similar. Use them to answer exercises 19–22. Leave your answers as fractions.*

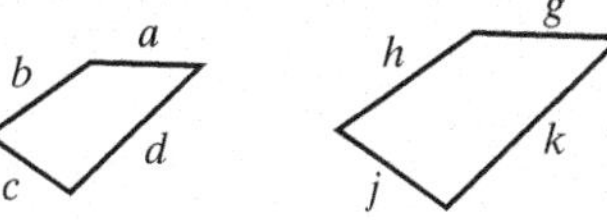

▲ **19.** If $a = 5$ ft, $d = 8$ ft, and $g = 7$ ft, find the length of side k.

▲ **20.** If $j = 12$ in., $k = 14$ in., and $c = 9$ in., find the length of side d.

▲ **21.** If $b = 20$ m, $h = 24$ m, and $d = 32$ m, find the length of side k.

▲ **22.** If $a = 20$ cm, $d = 24$ cm, and $k = 30$ cm, find the length of side g.

Use a proportion to solve.

▲ **23.** ***Geometry*** A rectangle whose width-to-length ratio is approximately 5 to 8 is called a **golden rectangle** and is said to be pleasing to the eye. Using this ratio, what should the length of a rectangular picture be if its width is to be 30 inches?

▲ **24.** ***Shadows*** Samantha is 5.5 feet tall and notices that she casts a shadow of 9 feet. At the same time, the new sculpture at the local park casts a shadow of 20 feet. How tall is the sculpture? Round your answer to the nearest foot.

▲ **25.** ***Floral Displays*** Floral designers often create arrangements where the flower height to container height ratio is 5 to 3. The FIU Art Museum wishes to create a floral display for the opening of a new show. They know they want to use an antique Chinese vase from their collection that is 13 inches high. How tall will the entire flower arrangement be if they use this standard ratio? (Round your answer to the nearest inch.)

▲ **26.** ***Securing Wires*** A wire line helps to secure a radio transmission tower. The wire measures 23 meters from the tower to the ground anchor pin. The wire is secured 14 meters up on the tower. If a second wire is secured 130 meters up on the tower and is extended from the tower at the same angle as the first wire, how long would the second wire need to be to reach an anchor pin on the ground? Round to the nearest meter.

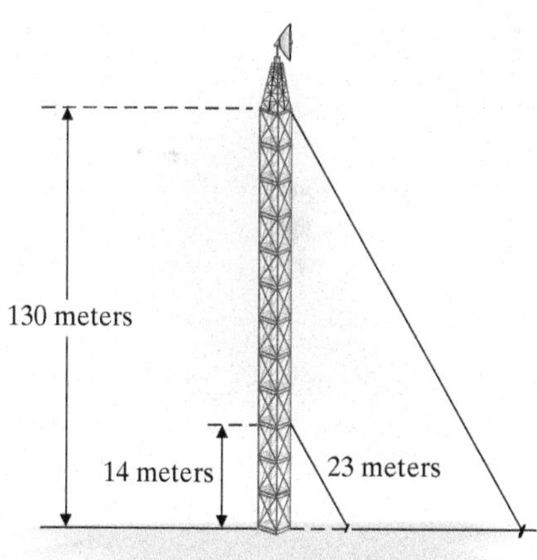

27. ***Acceleration*** Ben Hale is driving his new Toyota Camry on Interstate 90 at 45 miles per hour. He accelerates at the rate of 3 miles per hour every 2 seconds. How fast will he be traveling after accelerating for 11 seconds?

28. ***Braking*** Tim Newitt is driving a U-Haul truck to Chicago. He is driving at 55 miles per hour and has to hit the brakes because of heavy traffic. His truck slows at the rate of 2 miles per hour for every 3 seconds. How fast will he be traveling 10 seconds after he hits the brakes? Round to the nearest tenth.

29. ***Flight Speeds*** A Montreal commuter airliner travels 40 kilometers per hour faster than the television news helicopter over the city. The commuter airliner travels 1250 kilometers during the same time that the television news helicopter travels only 1050 kilometers. How fast does the commuter airliner fly? How fast does the television news helicopter fly?

30. ***Driving Speeds*** Jenny drove to Dallas while Mary drove to Houston in the same amount of time. Jenny drove 225 miles, while Mary drove 175 miles. Jenny traveled 12 miles per hour faster than Mary on her trip. What was the average speed in miles per hour for each woman?

31. ***Fluff Containers*** Marshmallow fluff comes in only two sizes, a $7\frac{1}{2}$-oz glass jar and a 16-oz plastic tub. At the local Stop and Shop, the 16-oz tub costs \$2.19 and the $7\frac{1}{2}$-oz jar costs \$1.29.

(a) How much does marshmallow fluff in the glass jar cost per ounce? (Round your answer to the nearest cent.)

(b) How much does marshmallow fluff in the plastic tub cost per ounce? (Round your answer to the nearest cent.)

(c) If the marshmallow fluff company decided to add a third size, a 40-oz bucket, how much would the price be if it was at the same unit price as the 16-oz plastic tub? (*Hint*: Set up a proportion. Do *not* use your answer from part **(b)**. Round your answer to the nearest cent.)

32. ***Green Tea*** Won Ling is a Chinese tea importer in Boston's Chinatown. He charges \$12.25 for four sample packs of his famous green tea. The packs are in the following sizes: 25 grams, 40 grams, 50 grams, and 60 grams.

(a) How much is Won Ling charging per gram for his green tea?

(b) How much would you pay for a 60-gram pack if he were willing to sell that one by itself?

(c) How much would you pay for an 800-gram package of green tea if it cost the same amount per gram?

33. ***Raking Leaves*** When all the leaves have fallen at Fred and Suzie's house in Concord, New Hampshire, Suzie can rake the entire yard in 6 hours. When Fred does it alone, it takes him 8 hours. How long would it take them to rake the yard together? Round to the nearest minute.

34. ***Meal Preparation*** To celebrate Diwali, a major Indian festival, Deepak and Alpa host a party each year for all their friends. Deepak can decorate and prepare all the food in 6 hours. Alpa takes 5 hours to decorate and prepare the food. How long would it take if they decorated and made the food together? Round your answer to the nearest minute.

Cumulative Review

35. **[5.2.2]** Write in scientific notation. 0.000892465

36. **[5.2.2]** Write in decimal notation. 6.83×10^9

37. **[5.2.1]** Write with positive exponents. $\dfrac{x^{-3}y^{-2}}{z^4w^{-8}}$

38. **[5.2.1]** Evaluate. $\left(\dfrac{2}{3}\right)^{-3}$

Quick Quiz 7.6

1. Solve for x. $\dfrac{16.5}{2.1} = \dfrac{x}{7}$

▲ **2.** While hiking in the White Mountains, Phil, Melissa, Noah, and Olivia saw a tall tree that cast a shadow 34 feet long. They observed at the same time that a 6-foot-tall person cast a shadow that was 8.5 feet long. How tall is the tree?

3. Last week Jeff Slater noted that 164 of the 205 flights to Chicago's O'Hare Airport flying out of Logan Airport departed on time. During the next week 215 flights left Logan Airport for Chicago's O'Hare Airport. If the same ratio holds, how many of those flights would he expect to depart on time?

4. **Concept Check** Mike found that his car used 18 gallons of gas to travel 396 miles. He needs to take a trip of 450 miles and wants to know how many gallons of gas it will take. He set up the equation $\frac{18}{x} = \frac{450}{396}$. Explain what error he made and how he should correctly solve the problem.

CAREER EXPLORATION PROBLEMS

Background: Operations Manager

Brian is a member of the operations management team of Fitness Equipment Engineering. As a member of the team, he's responsible for identifying the cost and time associated with manufacturing new equipment.

Facts

- Recently, a new treadmill with state-of-the-art features has been approved for manufacture. Fixed daily costs are \$22,000, and the cost of producing each treadmill is \$350.
- The production facility has two assembly lines, line A and a newer, faster robotic line, line B. The production cycle for a single shipment of new treadmills using assembly line A is 18 hours, whereas the production cycle for a single shipment of the new treadmills using line B is 14 hours.

Tasks

1. Brian wants to create a model that will identify the average daily cost when n treadmills are produced. What form should this model take?

2. Assembly line A is slated to be used to produce the new treadmills, but Brian proposes that both assembly lines be used for the production cycle of the single shipment of new treadmills. His supervisor wants to know how long the production cycle will be if the two lines are used simultaneously. Rounded to the nearest hour, how long is the production cycle for the single shipment using both assembly lines?

3. Brian now wants to determine the average cost per hour for producing 50 treadmills and 100 treadmills, given that a workday is 8 hours.

Chapter 7 Organizer

Topic and Procedure	Examples	You Try It
Simplifying rational expressions, p. 430 1. Factor the numerator and denominator. 2. Divide out any factor common to both the numerator and denominator.	Simplify. $\dfrac{36x^2 - 16y^2}{18x^2 + 24xy + 8y^2}$ $= \dfrac{4(3x + 2y)(3x - 2y)}{2(3x + 2y)(3x + 2y)} = \dfrac{2(3x - 2y)}{3x + 2y}$	1. Simplify. $\dfrac{6x^2 - 12x - 90}{3x^2 - 27}$
Multiplying rational expressions, p. 436 1. Factor all numerators and denominators. 2. Simplify the resulting rational expression as described above.	Multiply. $\dfrac{x^2 - y^2}{x^2 + 2xy + y^2} \cdot \dfrac{x^2 + 4xy + 3y^2}{x^2 - 4xy + 3y^2}$ $= \dfrac{(x + y)(x - y)}{(x + y)(x + y)} \cdot \dfrac{(x + y)(x + 3y)}{(x - y)(x - 3y)}$ $= \dfrac{x + 3y}{x - 3y}$	2. Multiply. $\dfrac{x^2 - 4xy - 5y^2}{2x^2 - 9xy - 5y^2} \cdot \dfrac{4x^2 - y^2}{4x^2 - 4xy + y^2}$
Dividing rational expressions, p. 437 1. Invert the second fraction and rewrite the problem as a product. 2. Multiply the rational expressions.	Divide. $\dfrac{14x^2 + 17x - 6}{x^2 - 25} \div \dfrac{4x^2 - 8x - 21}{x^2 + 10x + 25}$ $= \dfrac{(2x + 3)(7x - 2)}{(x + 5)(x - 5)} \cdot \dfrac{(x + 5)(x + 5)}{(2x - 7)(2x + 3)}$ $= \dfrac{(7x - 2)(x + 5)}{(x - 5)(2x - 7)}$	3. Divide. $\dfrac{2x^2 + 3x - 20}{8x + 8} \div \dfrac{x^2 - 16}{4x^2 - 12x - 16}$
Adding rational expressions, p. 441 1. If the denominators differ, factor them and determine the least common denominator (LCD). 2. Use multiplication to change each fraction into an equivalent one with the LCD as the denominator. 3. Add the numerators; put the answer over the LCD. 4. Simplify as needed.	Add. $\dfrac{x - 1}{x^2 - 4} + \dfrac{x - 1}{3x + 6}$ $= \dfrac{x - 1}{(x + 2)(x - 2)} + \dfrac{x - 1}{3(x + 2)}$ LCD $= 3(x + 2)(x - 2)$ $= \dfrac{x - 1}{(x + 2)(x - 2)} \cdot \dfrac{3}{3} + \dfrac{x - 1}{3(x + 2)} \cdot \dfrac{x - 2}{x - 2}$ $= \dfrac{3x - 3 + x^2 - 3x + 2}{3(x + 2)(x - 2)}$ $= \dfrac{x^2 - 1}{3(x + 2)(x - 2)}$ $= \dfrac{(x + 1)(x - 1)}{3(x + 2)(x - 2)}$	4. Add. $\dfrac{x + 2}{2x + 6} + \dfrac{x}{x^2 - 9}$
Subtracting rational expressions, p. 441 Move the subtraction sign to the numerator of the second fraction. Add. Simplify if possible. $\dfrac{a}{b} - \dfrac{c}{b} = \dfrac{a}{b} + \dfrac{-c}{b}$	Subtract. $\dfrac{5x}{x - 2} - \dfrac{3x + 4}{x - 2}$ $= \dfrac{5x}{x - 2} + \dfrac{-(3x + 4)}{x - 2}$ $= \dfrac{5x - 3x - 4}{x - 2}$ $= \dfrac{2x - 4}{x - 2}$ $= \dfrac{2(x - 2)}{(x - 2)} = 2$	5. Subtract. $\dfrac{9x}{x + 3} - \dfrac{3x - 18}{x + 3}$
Simplifying complex rational expressions, p. 450 1. Add the two fractions in the numerator. 2. Add the two fractions in the denominator.	Simplify. $\dfrac{\dfrac{x}{x^2 - 4} + \dfrac{1}{x + 2}}{\dfrac{3}{x + 2} - \dfrac{4}{x - 2}}$ $= \dfrac{\dfrac{x}{(x + 2)(x - 2)} + \dfrac{1}{x + 2} \cdot \dfrac{x - 2}{x - 2}}{\dfrac{3}{x + 2} \cdot \dfrac{x - 2}{x - 2} - \dfrac{4}{x - 2} \cdot \dfrac{x + 2}{x + 2}}$	6. Simplify. $\dfrac{\dfrac{x}{x - 3} + \dfrac{2}{x + 3}}{\dfrac{1}{x - 3} + \dfrac{3}{x^2 - 9}}$

Topic and Procedure	Examples	You Try It
3. Divide the fraction in the numerator by the fraction in the denominator. This is done by inverting the fraction in the denominator and multiplying by the numerator. 4. Simplify.	$= \dfrac{\dfrac{x + x - 2}{(x+2)(x-2)}}{\dfrac{3x - 6 - 4x - 8}{(x+2)(x-2)}}$ $= \dfrac{2x-2}{(x+2)(x-2)} \div \dfrac{-x-14}{(x+2)(x-2)}$ $= \dfrac{2(x-1)}{(x+2)(x-2)} \cdot \dfrac{(x+2)(x-2)}{-x-14}$ $= \dfrac{2(x-1)}{-x-14}$ or $-\dfrac{2(x-1)}{x+14}$ or $\dfrac{-2(x-1)}{x+14}$	
Solving equations involving rational expressions, p. 456 1. Determine the LCD of all denominators. 2. Note what values will make the LCD equal to 0. These are excluded from your solutions. 3. Multiply each side by the LCD, distributing as needed. 4. Solve the resulting polynomial equation. 5. Check. Be sure to exclude those values found in step 2.	Solve. $\dfrac{3}{x-2} = \dfrac{4}{x+2}$ LCD $= (x-2)(x+2)$. $(x-2)(x+2)\dfrac{3}{x-2} = \dfrac{4}{x+2}(x-2)(x+2)$ $3(x+2) = 4(x-2)$ $3x + 6 = 4x - 8$ $-x = -14$ $x = 14$ *Check:* $\dfrac{3}{14-2} \stackrel{?}{=} \dfrac{4}{14+2}$ $\dfrac{3}{12} \stackrel{?}{=} \dfrac{4}{16}$ $\dfrac{1}{4} = \dfrac{1}{4}$ ✓	7. Solve. $\dfrac{5x}{x^2-16} = \dfrac{5}{x+4}$
Solving applied problems with proportions, p. 461 1. Organize the data. 2. Write a proportion equating the respective parts. Let x represent the value that is not known. 3. Solve the proportion.	Renee can make five cherry pies with 3 cups of flour. How many cups of flour does she need to make eight cherry pies? $\dfrac{5 \text{ cherry pies}}{3 \text{ cups flour}} = \dfrac{8 \text{ cherry pies}}{x \text{ cups flour}}$ $\dfrac{5}{3} = \dfrac{8}{x}$ $5x = 24$ $x = \dfrac{24}{5}$ $x = 4\dfrac{4}{5}$ $4\frac{4}{5}$ cups of flour are needed for eight cherry pies.	▲ 8. On the blueprint of Rob and Amy's new family room addition, the room measures 2 inches wide by 3 inches long. The actual width of the family room will be 10.5 feet. How long will the actual room be?

Chapter 7 Review Problems

Section 7.1

Simplify.

1. $\dfrac{bx}{bx - by}$

2. $\dfrac{4x - 4y}{5y - 5x}$

3. $\dfrac{x^3 - 4x^2}{x^3 - x^2 - 12x}$

4. $\dfrac{2x^2 + 7x - 15}{25 - x^2}$

5. $\dfrac{2x^2 - 2xy - 24y^2}{2x^2 + 5xy - 3y^2}$

6. $\dfrac{4 - y^2}{3y^2 + 5y - 2}$

7. $\dfrac{5x^3 - 10x^2}{25x^4 + 5x^3 - 30x^2}$

8. $\dfrac{16x^2 - 4y^2}{4x - 2y}$

Section 7.2

Multiply or divide.

9. $\dfrac{2x^2 + 6x}{3x^2 - 27} \cdot \dfrac{x^2 + 3x - 18}{4x^2 - 4x}$

10. $\dfrac{y^2 + 8y + 16}{5y^2 + 20y} \div \dfrac{y^2 + 7y + 12}{2y^2 + 5y - 3}$

11. $\dfrac{6y^2 + 13y - 5}{9y^2 + 3y} \div \dfrac{4y^2 + 20y + 25}{12y^2}$

12. $\dfrac{3xy^2 + 12y^2}{2x^2 - 11x + 5} \div \dfrac{2xy + 8y}{8x^2 + 2x - 3}$

13. $\dfrac{x^2 - 5xy - 24y^2}{2x^2 - 2xy - 24y^2} \cdot \dfrac{4x^2 + 4xy - 24y^2}{x^2 - 10xy + 16y^2}$

14. $\dfrac{2x^2 + 10x + 2}{8x - 8} \cdot \dfrac{3x - 3}{4x^2 + 20x + 4}$

Section 7.3

Add or subtract.

15. $\dfrac{6}{y + 2} + \dfrac{2}{3y}$

16. $3 + \dfrac{2}{x + 1} + \dfrac{1}{x}$

17. $\dfrac{7}{x + 2} + \dfrac{3}{x - 4}$

18. $\dfrac{2}{x^2 - 9} + \dfrac{x}{x + 3}$

19. $\dfrac{x}{y} + \dfrac{3}{2y} + \dfrac{1}{y + 2}$

20. $\dfrac{4}{a} + \dfrac{2}{b} + \dfrac{3}{a + b}$

21. $\dfrac{3x + 1}{3x} - \dfrac{1}{x}$

22. $\dfrac{x + 4}{x + 2} - \dfrac{1}{2x}$

23. $\dfrac{27}{x^2 - 81} + \dfrac{3}{2(x + 9)}$

24. $\dfrac{1}{x^2 + 7x + 10} - \dfrac{x}{x + 5}$

Section 7.4

Simplify.

25. $\dfrac{\dfrac{4}{3y} - \dfrac{2}{y}}{\dfrac{1}{2y} + \dfrac{1}{y}}$

26. $\dfrac{\dfrac{5}{x} + \dfrac{1}{2x}}{\dfrac{x}{4} + x}$

27. $\dfrac{w - \dfrac{4}{w}}{1 + \dfrac{2}{w}}$

28. $\dfrac{1 - \dfrac{w}{w - 1}}{1 + \dfrac{w}{1 - w}}$

29. $\dfrac{1 + \dfrac{1}{y^2 - 1}}{\dfrac{1}{y + 1} - \dfrac{1}{y - 1}}$

30. $\dfrac{\dfrac{1}{y} + \dfrac{1}{x + y}}{1 + \dfrac{2}{x + y}}$

31. $\dfrac{\dfrac{1}{a+b} - \dfrac{1}{a}}{b}$

32. $\dfrac{\dfrac{2}{a+b} - \dfrac{3}{b}}{\dfrac{1}{a+b}}$

Section 7.5

Solve for the variable. If there is no solution, say so.

33. $\dfrac{8a-1}{6a+8} = \dfrac{3}{4}$

34. $\dfrac{8}{a-3} = \dfrac{12}{a+3}$

35. $\dfrac{2x-1}{x} - \dfrac{1}{2} = -2$

36. $\dfrac{5}{4} - \dfrac{1}{2x} = \dfrac{1}{x} + 2$

37. $\dfrac{7}{8x} - \dfrac{3}{4} = \dfrac{1}{4x} + \dfrac{1}{2}$

38. $\dfrac{3}{y-3} = \dfrac{3}{2} + \dfrac{y}{y-3}$

39. $\dfrac{3x}{x^2-4} - \dfrac{2}{x+2} = -\dfrac{4}{x-2}$

40. $\dfrac{3y-1}{3y} - \dfrac{6}{5y} = \dfrac{1}{y} - \dfrac{4}{15}$

41. $\dfrac{y+18}{y^2-16} = \dfrac{y}{y+4} - \dfrac{y}{y-4}$

42. $\dfrac{4}{x^2-1} = \dfrac{2}{x-1} + \dfrac{2}{x+1}$

43. $\dfrac{3y+1}{y^2-y} - \dfrac{3}{y-1} = \dfrac{4}{y}$

44. $\dfrac{3}{y-2} + \dfrac{4}{3y+2} = \dfrac{1}{2-y}$

Section 7.6

Solve.

45. $\dfrac{x}{4} = \dfrac{7}{10}$

46. $\dfrac{8}{5} = \dfrac{2}{x}$

47. $\dfrac{33}{10} = \dfrac{x}{8}$

48. $\dfrac{16}{x} = \dfrac{24}{9}$

49. $\dfrac{13.5}{0.6} = \dfrac{360}{x}$

50. $\dfrac{2\frac{1}{2}}{3\frac{1}{4}} = \dfrac{7}{x}$

Use a proportion to answer each question.

▲ 51. ***Paint Needs*** Five gallons of paint will cover 240 square feet. How many gallons of paint will be needed to cover 400 square feet? Round to the nearest tenth of a gallon.

52. ***Recipe Ratios*** Aunt Lexie uses 3 pounds of sugar to make 100 cookies. How many cookies can she make with 5 pounds of sugar? Round to the nearest whole cookie.

53. ***Map Scale*** On a map of Texas, the distance between El Paso and Dallas is 4 inches. The actual distance between these cities is 640 miles. Houston and Dallas are 1.5 inches apart on the same map. How many miles apart are Houston and Dallas?

54. ***Travel Speeds*** A train travels 180 miles in the same time that a car travels 120 miles. The speed of the train is 20 miles per hour faster than the speed of the car. Find the speed of the train and the speed of the car.

▲ **55.** ***Shadows*** Mary takes a walk across a canyon in New Mexico. She stands 5.75 feet tall and her shadow is 3 feet long. At the same time, the shadow from the peak of the canyon wall casts a shadow that is 95 feet long. How tall is the peak of the canyon? Round to the nearest foot.

▲ **56.** ***Shadows*** A flagpole that is 8 feet tall casts a shadow of 3 feet. At the same time of day, a tall office building in the city casts a shadow of 450 feet. How tall is the office building?

57. ***Window Cleaning*** As part of their spring cleaning routine, Tina and Mathias wash all of their windows, inside and out. Tina can do this job in 4 hours. When Mathias washes the windows, it takes him 6 hours. How long would it take them if they worked together?

58. ***Plowing Fields*** Sally runs the family farm in Boone, Iowa. She can plow the fields of the farm in 20 hours. Her daughter Brenda can plow the fields of the farm in 30 hours. If they have two identical tractors, how long would it take Brenda and Sally to plow the fields of the farm if they worked together?

Mixed Practice

Perform the operation indicated. Simplify.

59. $\dfrac{a^2 + 2a - 8}{6a^2 - 3a^3}$

60. $\dfrac{4a^3 + 20a^2}{2a^2 + 13a + 15}$

61. $\dfrac{x^2 - y^2}{x^2 + 4xy + 3y^2} \cdot \dfrac{x^2 + xy - 6y^2}{x^2 + xy - 2y^2}$

62. $\dfrac{x}{x + 3} + \dfrac{9x + 18}{x^2 + 3x}$

63. $\dfrac{x - 30}{x^2 - 5x} + \dfrac{x}{x - 5}$

64. $\dfrac{a + b}{ax + ay} - \dfrac{a + b}{bx + by}$

65. $\dfrac{\dfrac{5}{3x} + \dfrac{2}{9x}}{\dfrac{3}{x} + \dfrac{8}{3x}}$

66. $\dfrac{\dfrac{4}{5y} - \dfrac{8}{y}}{y + \dfrac{y}{5}}$

67. $\dfrac{x - 3y}{x + 2y} \div \left(\dfrac{2}{y} - \dfrac{12}{x + 3y}\right)$

68. $\dfrac{7}{x + 2} = \dfrac{4}{x - 4}$

69. $\dfrac{2x - 1}{3x - 8} = \dfrac{5}{8}$

70. $2 + \dfrac{4}{b - 1} = \dfrac{4}{b^2 - b}$

How Am I Doing? Chapter 7 Test

MATH COACH MyMathLab® YouTube

After you take this test read through the Math Coach on pages 476–477. Math Coach videos are available via MyMathLab and YouTube. Step-by-step test solutions on the Chapter Test Prep Videos are also available via MyMathLab and YouTube. (Search "TobeyBegInterAlg" and click on "Channels.")

Perform the operation indicated. Simplify.

1. $\dfrac{2ac + 2ad}{3a^2c + 3a^2d}$

2. $\dfrac{8x^2 - 2x^2y^2}{y^2 + 4y + 4}$

3. $\dfrac{x^2 + 2x}{2x - 1} \cdot \dfrac{10x^2 - 5x}{12x^3 + 24x^2}$

4. $\dfrac{x + 2y}{12y^2} \cdot \dfrac{4y}{x^2 + xy - 2y^2}$

MC 5. $\dfrac{2a^2 - 3a - 2}{a^2 + 5a + 6} \div \dfrac{a^2 - 5a + 6}{a^2 - 9}$

6. $\dfrac{1}{a^2 - a - 2} + \dfrac{3}{a - 2}$

7. $\dfrac{x - y}{xy} - \dfrac{a - y}{ay}$

MC 8. $\dfrac{3x}{x^2 - 3x - 18} - \dfrac{x - 4}{x - 6}$

9. $\dfrac{\dfrac{x}{3y} - \dfrac{1}{2}}{\dfrac{4}{3y} - \dfrac{2}{x}}$

MC 10. $\dfrac{\dfrac{6}{b} - 4}{\dfrac{5}{bx} - \dfrac{10}{3x}}$

11. $\dfrac{2x^2 + 3xy - 9y^2}{4x^2 + 13xy + 3y^2}$

12. $\dfrac{1}{x + 4} - \dfrac{2}{x^2 + 6x + 8}$

In questions 13–18, solve for x. Check your answers. If there is no solution, say so.

13. $\dfrac{4}{3x} - \dfrac{5}{2x} = 5 - \dfrac{1}{6x}$

MC 14. $\dfrac{x - 3}{x - 2} = \dfrac{2x^2 - 15}{x^2 + x - 6} - \dfrac{x + 1}{x + 3}$

15. $3 - \dfrac{7}{x + 3} = \dfrac{x - 4}{x + 3}$

16. $\dfrac{3}{3x - 5} = \dfrac{7}{5x + 4}$

17. $\dfrac{9}{x} = \dfrac{13}{5}$

18. $\dfrac{9.3}{2.5} = \dfrac{x}{10}$

19. A random check of America West air flights last month showed that 113 of the 150 flights checked arrived on time. If the inspectors check 200 flights next month, how many can be expected to be on time? (Round your answer to the nearest whole number.)

20. In northern Michigan the Gunderson family heats their home with firewood. They used \$100 worth of wood in 25 days. Mr. Gunderson estimates that he needs to burn wood at that rate for about 92 days during the winter. If that is so, how much will the 92-day supply of wood cost?

▲ 21. A hiking club is trying to construct a rope bridge across a canyon. A 6-foot construction pole held upright casts a 7-foot shadow. At the same time of day, a tree at the edge of the canyon casts a shadow that exactly covers the distance that is needed for the rope bridge. The tree is exactly 87 feet tall. How long should the rope bridge be? Round to the nearest foot.

1. ☐
2. ☐
3. ☐
4. ☐
5. ☐
6. ☐
7. ☐
8. ☐
9. ☐
10. ☐
11. ☐
12. ☐
13. ☐
14. ☐
15. ☐
16. ☐
17. ☐
18. ☐
19. ☐
20. ☐
21. ☐

Total Correct: ☐

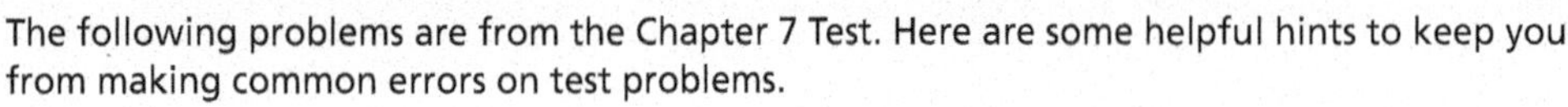

Mastering the skills you need to do well on the test.

The following problems are from the Chapter 7 Test. Here are some helpful hints to keep you from making common errors on test problems.

Chapter 7 Test, Problem 5 Simplify. $\dfrac{2a^2 - 3a - 2}{a^2 + 5a + 6} \div \dfrac{a^2 - 5a + 6}{a^2 - 9}$

HELPFUL HINT The operation of division is performed by inverting the *second fraction* and multiplying it by the first fraction. This step should be done first, before you begin factoring.

Did you keep the first fraction the same as it is written and then multiply it by $\dfrac{a^2 - 9}{a^2 - 5a + 6}$?

Yes _____ No _____

If you answered No, stop and make this correction to your work.

Were you able to factor each expression so that you obtained $\dfrac{(2a + 1)(a - 2)}{(a + 2)(a + 3)} \cdot \dfrac{(a + 3)(a - 3)}{(a - 2)(a - 3)}$?

Yes _____ No _____

If you answered No, go back and check each factoring step to see if you can get the same result. Complete the problem by dividing out any common factors.

If you answered Problem 5 incorrectly, go back and rework the problem using these suggestions.

Chapter 7 Test, Problem 8 Simplify. $\dfrac{3x}{x^2 - 3x - 18} - \dfrac{x - 4}{x - 6}$

HELPFUL HINT First factor the trinomial in the denominator so that you can determine the LCD of these two fractions. Then multiply by what is needed in the second fraction so that it becomes an equivalent fraction with the LCD as the denominator. When subtracting, it is a good idea to place brackets around the second numerator to avoid sign errors.

Did you factor the first denominator into $(x - 6)(x + 3)$?

Yes _____ No _____

Did you then determine that the LCD is $(x - 6)(x + 3)$?

Yes _____ No _____

If you answered No to these questions, stop and review how to factor the trinomial in the first denominator and how to find the LCD when working with polynomials as denominators.

Did you multiply the numerator and denominator of the second fraction by $(x + 3)$ and place brackets around the this product to obtain

$$\frac{3x}{(x - 6)(x + 3)} - \frac{[(x - 4)(x + 3)]}{(x - 6)(x + 3)}?$$

Yes _____ No _____

Did you obtain $3x - x^2 - x - 12$ in the numerator?

Yes _____ No _____

If you answered Yes, you forgot to distribute the negative sign. Rework the problem and make this correction.

As your final step, remember to combine like terms and then factor the new numerator before dividing out common factors.

Now go back and rework the problem using these suggestions.

Need more help? Watch the MATH COACH videos in MyMathLab® or on YouTube™.

Chapter 7 Test, Problem 10 Simplify. $\dfrac{\frac{6}{b} - 4}{\frac{5}{bx} - \frac{10}{3x}}$

HELPFUL HINT There are two ways to simplify this expression:

1. combine the numerators and the denominators separately, or
2. multiply the numerator and denominator of the complex fraction by the LCD of all the denominators.

Consider both methods and choose the one that seems easiest to you. We will show the steps of the first method for this particular problem.

Did you multiply 4 by $\frac{b}{b}$ and then subtract $\frac{6}{b} - \frac{4b}{b}$?

Yes ______ No ______

If you answered No, remember that 4 can be written as $\frac{4}{1}$, and to find the LCD, you must multiply numerator and denominator by the variable b.

In the denominator of the complex fraction, did you obtain $3bx$ as the LCD and multiply the first fraction by $\frac{3}{3}$ and the second fraction by $\frac{b}{b}$ before subtracting the two fractions?

Yes ______ No ______

If you answered No, stop and carefully change these two fractions into equivalent fractions with $3bx$ as the denominator. Then subtract the numerators and keep the common denominator.

Were you able to rewrite the problem as follows: $\frac{6 - 4b}{b} \div \frac{15 - 10b}{3bx}$?

Yes ______ No ______

If you answered No, examine your steps carefully. Once you have one fraction in the numerator and one fraction in the denominator, you can rewrite the division as multiplication by inverting the fraction in the denominator.

If you answered Problem 10 incorrectly, go back and rework the problem using these suggestions.

Chapter 7 Test, Problem 14 Solve for x. $\dfrac{x - 3}{x - 2} = \dfrac{2x^2 - 15}{x^2 + x - 6} - \dfrac{x + 1}{x + 3}$

HELPFUL HINT First factor any denominators that need to be factored so that you can determine the LCD of all the denominators. Verify that you are solving an equation, then multiply *each term* of the equation by the LCD. Solve the resulting equation. Check your solution.

Did you factor $x^2 + x - 6$ to get $(x + 3)(x - 2)$ and identify the LCD as $(x + 3)(x - 2)$?

Yes ______ No ______

If you answered No, go back and complete these steps again.

Did you notice that we are solving an equation and that we can multiply each term of the equation by the LCD and then multiply the binomials to obtain the equation $x^2 - 9 = (2x^2 - 15) - (x^2 - x - 2)$?

Yes ______ No ______

If you answered No, look at the step $-[(x + 1)(x - 2)]$.

After you multiplied the binomials, did you distribute the negative sign and change the sign of all terms inside the grouping symbols?

Yes ______ No ______

If you answered No, review how to multiply binomials and subtract polynomial expressions. Then combine like terms and solve the equation for x.

Now go back and rework the problem using these suggestions.

Need more help? Look for section examples marked with MC to review.

CHAPTER 8

Rational Exponents and Radicals

CAREER OPPORTUNITIES

Surveyor

Surveyors are an essential part of property development and real estate transactions. Using technology, legal records, and mapping systems, land surveyors develop detailed documents identifying the boundaries and characteristics of properties on both land and water. Their work requires precise measurement and accurate computation. The mathematics in this chapter is a vital component in producing properly completed surveys that support real estate purchases, construction, and development activity.

To investigate how the mathematics in this chapter can help with this field, see the Career Exploration Problems on page 532.

8.1 Rational Exponents

1 Simplifying Expressions with Rational Exponents

Before studying this section, you may need to review Sections 5.1 and 5.2. For convenience, we list the rules of exponents that we learned there.

$$x^m x^n = x^{m+n} \qquad x^{-n} = \frac{1}{x^n} \qquad x^0 = 1 \qquad (xy)^n = x^n y^n$$

$$\frac{x^m}{x^n} = x^{m-n} \qquad \frac{x^{-n}}{y^{-m}} = \frac{y^m}{x^n} \qquad (x^m)^n = x^{mn} \qquad \left(\frac{x}{y}\right)^n = \frac{x^n}{y^n}$$

Recall the steps that you would take to simplify an expression such as $\left(\frac{x^3y^6}{-8x}\right)(2y^2)$.

$$\left(\frac{x^3y^6}{-8x}\right)(2y^2) = \frac{2x^3y^6y^2}{-8x^1} = \frac{x^{3-1}y^{6+2}}{-4} = -\frac{x^2y^8}{4}$$

To ensure that you understand these rules, study Example 1 carefully and work Student Practice 1. If you have any difficulties, we also refer you to Appendix A.1, which contains additional review and practice of these rules.

Example 1 Simplify. $\left(\frac{5xy^{-3}}{2x^{-4}y}\right)^{-2}$

Solution

$$\left(\frac{5xy^{-3}}{2x^{-4}y}\right)^{-2} = \frac{(5xy^{-3})^{-2}}{(2x^{-4}y)^{-2}} \qquad \left(\frac{x}{y}\right)^n = \frac{x^n}{y^n}$$

$$= \frac{5^{-2}x^{-2}(y^{-3})^{-2}}{2^{-2}(x^{-4})^{-2}y^{-2}} \qquad (xy)^n = x^n y^n$$

$$= \frac{5^{-2}x^{-2}y^6}{2^{-2}x^8y^{-2}} \qquad (x^m)^n = x^{mn}$$

$$= \frac{5^{-2}}{2^{-2}} \cdot \frac{x^{-2}}{x^8} \cdot \frac{y^6}{y^{-2}}$$

$$= \frac{2^2}{5^2} \cdot x^{-2-8} \cdot y^{6+2} \qquad \frac{x^{-n}}{y^{-m}} = \frac{y^m}{x^n}; \frac{x^m}{x^n} = x^{m-n}$$

$$= \frac{4}{25}x^{-10}y^8$$

The answer can also be written as $\frac{4y^8}{25x^{10}}$. Explain why. □

Student Practice 1 Simplify.

$$\left(\frac{3x^{-2}y^4}{2x^{-5}y^2}\right)^{-3}$$

Sidelight: **Example 1 Follow-Up** Deciding when to use the rule $\frac{x^{-n}}{y^{-m}} = \frac{y^m}{x^n}$ is entirely up to you. In Example 1, we could have begun by writing

$$\left(\frac{5xy^{-3}}{2x^{-4}y}\right)^{-2} = \left(\frac{5x \cdot x^4}{2y \cdot y^3}\right)^{-2} = \left(\frac{5x^5}{2y^4}\right)^{-2}.$$

Complete the steps to simplify this expression.

Likewise, in the fourth step in Example 1, we could have written

$$\frac{5^{-2}x^{-2}y^6}{2^{-2}x^8y^{-2}} = \frac{2^2y^6y^2}{5^2x^8x^2}.$$

Student Learning Objectives

After studying this section, you will be able to:

1. Simplify expressions with rational exponents.
2. Add expressions with rational exponents.
3. Factor expressions with rational exponents.

Complete the steps to simplify this expression. Are the two answers the same as the answer in Example 1? Why or why not?

We generally begin to simplify a rational expression with exponents by raising a power to a power because sometimes negative powers become positive. The order in which you use the rules of exponents is up to you. Work carefully. Keep track of your exponents and where you are as you simplify the rational expression.

These rules for exponents can also be extended to include rational exponents—that is, exponents that are fractions. As you recall, rational numbers are of the form $\frac{a}{b}$, where a and b are integers and b does not equal zero. We will write fractional exponents using diagonal lines. Thus, we will write $\frac{5}{6}$ as $5/6$, and if a fractional exponent contains a variable such as $\frac{a}{b}$, we will write it as a/b. For now we restrict the base to *positive* real numbers. Later we will talk about negative bases.

Example 2 Simplify.

(a) $(x^{2/3})^4$ **(b)** $\dfrac{x^{5/6}}{x^{1/6}}$ **(c)** $x^{2/3} \cdot x^{-1/3}$ **(d)** $5^{3/7} \cdot 5^{2/7}$

Solution We will not write out every step or every rule of exponents that we use. You should be able to follow the solutions.

(a) $(x^{2/3})^4 = x^{(2/3)(4/1)} = x^{8/3}$

(b) $\dfrac{x^{5/6}}{x^{1/6}} = x^{5/6-1/6} = x^{4/6} = x^{2/3}$

(c) $x^{2/3} \cdot x^{-1/3} = x^{2/3-1/3} = x^{1/3}$

(d) $5^{3/7} \cdot 5^{2/7} = 5^{3/7+2/7} = 5^{5/7}$ □

Student Practice 2 Simplify.

(a) $(x^4)^{3/8}$ **(b)** $\dfrac{x^{3/7}}{x^{2/7}}$ **(c)** $x^{-7/5} \cdot x^{4/5}$

Sometimes fractional exponents will not have the same denominator. Remember that you need to change the fractions to equivalent fractions with the same denominator when the rules of exponents require you to add or to subtract them.

Example 3 Simplify. Express your answers with positive exponents only.

(a) $(2x^{1/2})(3x^{1/3})$ **(b)** $\dfrac{18x^{1/4}y^{-1/3}}{-6x^{-1/2}y^{1/6}}$

Solution

(a) $(2x^{1/2})(3x^{1/3}) = 6x^{1/2+1/3} = 6x^{3/6+2/6} = 6x^{5/6}$

(b)
$$\begin{aligned}\frac{18x^{1/4}y^{-1/3}}{-6x^{-1/2}y^{1/6}} &= -3x^{1/4-(-1/2)}y^{-1/3-1/6}\\ &= -3x^{1/4+2/4}y^{-2/6-1/6}\\ &= -3x^{3/4}y^{-3/6}\\ &= -3x^{3/4}y^{-1/2}\\ &= -\frac{3x^{3/4}}{y^{1/2}}\end{aligned}$$
□

Student Practice 3 Simplify. Express your answers with positive exponents only.

(a) $(-3x^{1/4})(2x^{1/2})$ **(b)** $\dfrac{13x^{1/12}y^{-1/4}}{26x^{-1/3}y^{1/2}}$

Example 4 Multiply and simplify. $-2x^{5/6}(3x^{1/2} - 4x^{-1/3})$

Solution We will need to be very careful when we add the exponents for x as we use the distributive property. Study each step of the following example. Be sure you understand each operation.

$$\begin{aligned}
-2x^{5/6}(3x^{1/2} - 4x^{-1/3}) &= -6x^{5/6+1/2} + 8x^{5/6-1/3} \\
&= -6x^{5/6+3/6} + 8x^{5/6-2/6} \\
&= -6x^{8/6} + 8x^{3/6} \\
&= -6x^{4/3} + 8x^{1/2}
\end{aligned}$$

Student Practice 4 Multiply and simplify. $-3x^{1/2}(2x^{1/4} + 3x^{-1/2})$

Sometimes we can use the rules of exponents to simplify numerical values raised to rational powers.

Example 5 Evaluate.

(a) $(25)^{3/2}$ **(b)** $(27)^{2/3}$

Solution

(a) $(25)^{3/2} = (5^2)^{3/2} = 5^{2/1 \cdot 3/2} = 5^3 = 125$

(b) $(27)^{2/3} = (3^3)^{2/3} = 3^{3/1 \cdot 2/3} = 3^2 = 9$

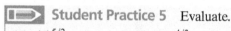

Student Practice 5 Evaluate.

(a) $(4)^{5/2}$ **(b)** $(27)^{4/3}$

2 Adding Expressions with Rational Exponents

Adding expressions with rational exponents may require several steps. Sometimes this involves removing negative exponents. For example, to add $2x^{-1/2} + x^{1/2}$, we begin by writing $2x^{-1/2}$ as $\frac{2}{x^{1/2}}$. This is a rational expression. Recall that to add rational expressions we need to have a common denominator. Take time to look at the steps needed to write $2x^{-1/2} + x^{1/2}$ as one term.

Example 6 Write as one fraction with positive exponents. $2x^{-1/2} + x^{1/2}$

Solution

$$2x^{-1/2} + x^{1/2} = \frac{2}{x^{1/2}} + \frac{x^{1/2} \cdot x^{1/2}}{x^{1/2}} = \frac{2}{x^{1/2}} + \frac{x^1}{x^{1/2}} = \frac{2 + x}{x^{1/2}}$$

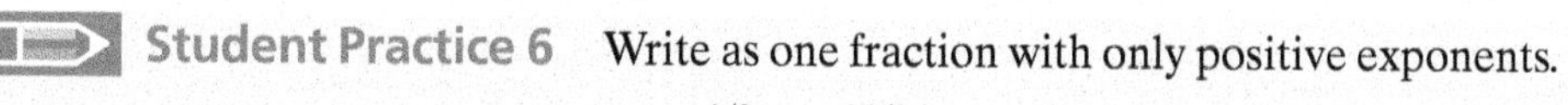

Student Practice 6 Write as one fraction with only positive exponents.

$3x^{1/3} + x^{-1/3}$

3 Factoring Expressions with Rational Exponents 

To factor expressions, we need to be able to recognize common factors. If the terms of the expression contain exponents, we look for the same exponential factor in each term. For example, in the expression $6x^5 + 4x^3 - 8x^2$, the common factor of each term is $2x^2$. Thus, we can factor out the common factor $2x^2$ from each term. The expression then becomes $2x^2(3x^3 + 2x - 4)$.

We do exactly the same thing when we factor expressions with rational exponents. The key is to identify the exponent of the common factor. In the expression $6x^{3/4} + 4x^{1/2} - 8x^{1/4}$, the common factor is $2x^{1/4}$. Thus, we factor the expression $6x^{3/4} + 4x^{1/2} - 8x^{1/4}$ as $2x^{1/4}(3x^{1/2} + 2x^{1/4} - 4)$. We do not always need to factor out the greatest common factor. In the following examples we simply factor out a common factor.

Example 7 Factor out the common factor $2x$. $2x^{3/2} + 4x^{5/2}$

Solution

We rewrite the exponent of each term so that we can see that each term contains the factor $2x$ or $2x^{2/2}$.

$$\begin{aligned} 2x^{3/2} + 4x^{5/2} &= 2x^{2/2+1/2} + 4x^{2/2+3/2} \\ &= 2(x^{2/2})(x^{1/2}) + 4(x^{2/2})(x^{3/2}) \\ &= 2x(x^{1/2} + 2x^{3/2}) \end{aligned}$$

□

Student Practice 7 Factor out the common factor $4y$. $4y^{3/2} - 8y^{5/2}$

For convenience we list here the properties of exponents that we have discussed in this section, as well as the property $x^0 = 1$.

When x and y are **positive real numbers** and a and b are **rational numbers:**

$$x^m x^n = x^{m+n} \qquad \frac{x^m}{x^n} = x^{m-n} \qquad x^0 = 1$$

$$x^{-n} = \frac{1}{x^n} \qquad \frac{x^{-n}}{y^{-m}} = \frac{y^m}{x^n}$$

$$(x^m)^n = x^{mn} \qquad (xy)^n = x^n y^n \qquad \left(\frac{x}{y}\right)^n = \frac{x^n}{y^n}$$

8.1 Exercises MyMathLab®

Remember that additional review and practice are available in Sections 5.1 and 5.2 as well as Appendix A.1.
Simplify. Express your answer with positive exponents.

1. $\left(\dfrac{3xy^{-1}}{z^2}\right)^4$

2. $\left(\dfrac{3xy^{-2}}{x^3}\right)^2$

3. $\left(\dfrac{2a^2b}{3b^{-1}}\right)^3$

4. $\left(\dfrac{-a^2b}{2b^{-1}}\right)^3$

5. $\left(\dfrac{2x^2}{y}\right)^{-3}$

6. $\left(\dfrac{5y^2}{x^3}\right)^{-3}$

7. $\left(\dfrac{3xy^{-2}}{y^3}\right)^{-2}$

8. $\left(\dfrac{5x^{-2}y}{x^4}\right)^{-2}$

9. $(x^{3/4})^2$

10. $(x^{5/6})^3$

11. $(y^{12})^{2/3}$

12. $(y^4)^{3/2}$

13. $\dfrac{x^{3/5}}{x^{1/5}}$

14. $\dfrac{y^{6/7}}{y^{3/7}}$

15. $\dfrac{x^{8/9}}{x^{2/9}}$

16. $\dfrac{x^{11/12}}{x^{5/12}}$

17. $\dfrac{a^2}{a^{1/4}}$

18. $\dfrac{x^5}{x^{1/2}}$

19. $x^{1/7} \cdot x^{3/7}$

20. $a^{3/4} \cdot a^{1/4}$

21. $a^{3/8} \cdot a^{1/2}$

22. $b^{2/5} \cdot b^{2/15}$

23. $y^{3/5} \cdot y^{-1/10}$

24. $y^{7/8} \cdot y^{-1/2}$

Write each expression with positive exponents.

25. $x^{-3/4}$

26. $x^{-1/4}$

27. $a^{-5/6}b^{1/3}$

28. $4a^{-3/8}b^{1/2}$

29. $6^{-1/2}$

30. $5^{-1/8}$

31. $3a^{-1/3}$

32. $-3y^{-5/4}$

Evaluate or simplify the numerical expressions.

33. $(27)^{5/3}$

34. $(16)^{3/4}$

35. $(4)^{3/2}$

36. $(9)^{3/2}$

37. $(-8)^{5/3}$

38. $(-27)^{5/3}$

39. $(-27)^{2/3}$

40. $(-64)^{2/3}$

Mixed Practice, Exercises 41–64 *Simplify and express your answers with positive exponents. Evaluate or simplify the numerical expressions.*

41. $(x^{1/4}y^{-1/3})(x^{3/4}y^{1/2})$

42. $(x^{1/4}y^{1/2})(x^{3/8}y^{-1/2})$

43. $(7x^{1/3}y^{1/4})(-2x^{1/4}y^{-1/6})$

44. $(8x^{-1/5}y^{1/3})(-3x^{-1/4}y^{1/6})$

45. $6^2 \cdot 6^{-2/3}$

46. $11^{1/2} \cdot 11^3$

47. $\dfrac{2x^{1/5}}{x^{-1/2}}$

48. $\dfrac{3y^{2/3}}{y^{-1/4}}$

49. $\dfrac{-20x^2y^{-1/5}}{5x^{-1/2}y}$

50. $\dfrac{12x^{-2/3}y}{-6xy^{-3/4}}$

51. $\left(\dfrac{8a^2b^6}{a^{-1}b^3}\right)^{1/3}$

52. $\left(\dfrac{36a^{-1}b^3}{a^{-7}b}\right)^{1/2}$

53. $(-4x^{1/4}y^{5/2}z^{1/2})^2$

54. $(4x^{2/3}y^{-1/2}z^{3/5})^3$

55. $x^{2/3}(x^{4/3} - x^{1/5})$

56. $y^{-2/3}(y^{2/3} + y^{3/2})$

57. $m^{7/8}(m^{-1/2} + 2m)$

58. $n^{1/6}(n^{5/12} + 3n)$

59. $(8)^{-1/3}$

60. $(100)^{-1/2}$

61. $(64)^{-2/3}$

62. $(16)^{-5/4}$

63. $(81)^{3/4} + (25)^{1/2}$

64. $(4)^{3/2} + 121^{1/2}$

Write each expression as one fraction with positive exponents.

65. $3y^{1/2} + y^{-1/2}$

66. $2y^{1/3} + y^{-2/3}$

67. $x^{-1/3} + 6^{4/3}$

68. $5^{-1/4} + x^{-1/2}$

Factor out the common factor 2a.

69. $10a^{5/4} - 4a^{8/5}$

70. $6a^{4/3} - 8a^{3/2}$

Factor out the common factor 3x.

71. $12x^{4/3} - 3x^{5/2}$

72. $18x^{5/3} - 6x^{11/8}$

To Think About

73. What is the value of a if $x^a \cdot x^{1/4} = x^{-1/8}$?

74. What is the value of b if $x^b \div x^{1/3} = x^{-1/12}$?

Applications

Radius and Volume of a Sphere *The radius needed to create a sphere with a given volume V can be approximated by the equation* $r = 0.62(V)^{1/3}$. *Find the radius of the spheres with the following volumes.*

▲ **75.** 27 cubic meters

▲ **76.** 125 cubic meters

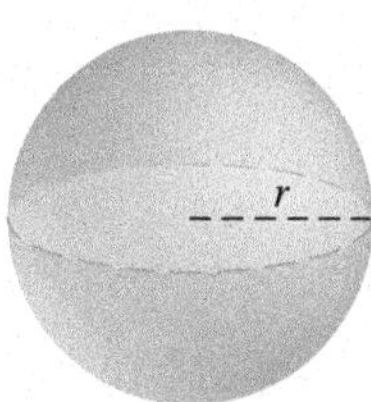

Radius and Volume of a Cone *The radius required for a cone to have a volume V and a height h is given by the equation* $r = \left(\dfrac{3V}{\pi h}\right)^{1/2}$. *Find the necessary radius to have a cone with the properties below. Use* $\pi \approx 3.14$.

▲ **77.** $V = 314$ cubic feet and $h = 12$ feet.

▲ **78.** $V = 3140$ cubic feet and $h = 30$ feet.

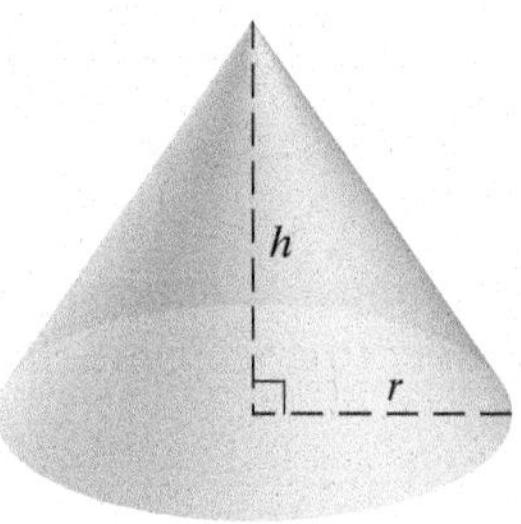

Cumulative Review

79. **[2.4.1]** Solve for x. $-4(x + 1) = \dfrac{1}{3}(3 - 2x)$

80. **[2.4.1]** Solve for x. $\dfrac{2}{3}x + 4 = -\dfrac{1}{2}(x - 1)$

Quick Quiz 8.1 *Simplify.*

1. $(-4x^{2/3}y^{1/4})(3x^{1/6}y^{1/2})$

2. $\dfrac{16x^4}{8x^{2/3}}$

3. $(25x^{1/4})^{3/2}$

4. Concept Check Explain how you would simplify the following.

$$x^{9/10} \cdot x^{-1/5}$$

8.2 Radical Expressions and Functions 

Student Learning Objectives

After studying this section, you will be able to:

1 Evaluate radical expressions and functions.

2 Change radical expressions to expressions with rational exponents.

3 Change expressions with rational exponents to radical expressions.

4 Evaluate higher-order radicals containing variable radicands that represent any real number (including negative real numbers).

1 Evaluating Radical Expressions and Functions

The **square root** of a number is a value that when multiplied by itself is equal to the original number. That is, since $3 \cdot 3 = 9$, 3 is a square root of 9. But $(-3) \cdot (-3) = 9$, so -3 is also a square root. We call the positive square root the **principal square root.**

The symbol $\sqrt{\ }$ is called a **radical sign.** We use it to denote positive square roots (and positive higher-order roots also). A negative square root is written $-\sqrt{\ }$. Thus, we have the following:

$$\sqrt{9} = 3 \qquad -\sqrt{9} = -3$$
$$\sqrt{64} = 8 \qquad (\text{because } 8 \cdot 8 = 64)$$
$$\sqrt{121} = 11 \qquad (\text{because } 11 \cdot 11 = 121)$$

Because $\sqrt{9} = \sqrt{3 \cdot 3} = \sqrt{3^2} = 3$, we can say the following:

DEFINITION OF SQUARE ROOT

If x is a nonnegative real number, then $\sqrt{x}$ is the *nonnegative* (or principal) *square root* of x; in other words, $(\sqrt{x})^2 = x$.

Note that x must be *nonnegative.* Why? Suppose we want to find $\sqrt{-36}$. We must find a real number that when multiplied by itself gives -36. Is there one? No, because

$$6 \cdot 6 = 36 \quad \text{and}$$
$$(-6)(-6) = 36.$$

So there is no real number that we can square to get -36.

We call $\sqrt[n]{x}$ a **radical expression.** The $\sqrt{\ }$ symbol is the radical sign, the x is the **radicand,** and the n is the **index** of the radical. When no number for n appears in the radical expression, it is understood that 2 is the index, which means that we are looking for the square root. For example, in the radical expression $\sqrt{25}$, with no number given for the index n, we take the index to be 2. Thus, $\sqrt{25}$ is the principal square root of 25.

We can extend the notion of square root to **higher-order roots,** such as cube roots, fourth roots, and so on. A **cube root** of a number is a value that when cubed is equal to the original number. The index n of the radical is 3, and the radical used is $\sqrt[3]{\ }$. Similarly, a **fourth root** of a number is a value that when raised to the fourth power is equal to the original number. The index n of the radical is 4, and the radical used is $\sqrt[4]{\ }$. Thus, we have the following:

$$\sqrt[3]{27} = 3 \qquad \text{because } 3 \cdot 3 \cdot 3 = 3^3 = 27$$
$$\sqrt[4]{81} = 3 \qquad \text{because } 3 \cdot 3 \cdot 3 \cdot 3 = 3^4 = 81$$
$$\sqrt[5]{32} = 2 \qquad \text{because } 2 \cdot 2 \cdot 2 \cdot 2 \cdot 2 = 2^5 = 32$$
$$\sqrt[3]{-64} = -4 \qquad \text{because } (-4)(-4)(-4) = (-4)^3 = -64$$
$$\sqrt[6]{729} = 3 \qquad \text{because } 3 \cdot 3 \cdot 3 \cdot 3 \cdot 3 \cdot 3 = 3^6 = 729$$

You should be able to see a pattern here.

$$\sqrt[3]{27} = \sqrt[3]{3^3} = 3$$
$$\sqrt[4]{81} = \sqrt[4]{3^4} = 3$$
$$\sqrt[5]{32} = \sqrt[5]{2^5} = 2$$
$$\sqrt[3]{-64} = \sqrt[3]{(-4)^3} = -4$$
$$\sqrt[6]{729} = \sqrt[6]{3^6} = 3$$

In these cases, we see that $\sqrt[n]{x^n} = x$. We now give the following definition.

DEFINITION OF HIGHER-ORDER ROOTS

1. If x is a *nonnegative* real number, then $\sqrt[n]{x}$ is a nonnegative nth root and has the property that

$$(\sqrt[n]{x})^n = x.$$

2. If x is a *negative* real number, then
   (a) When n is an *odd integer,* $\sqrt[n]{x}$ has the property that $(\sqrt[n]{x})^n = x$.
   (b) When n is an *even integer,* $\sqrt[n]{x}$ is *not* a real number.

Example 1 If possible, find the root of each negative number. If there is no real number root, say so.

(a) $\sqrt[3]{-216}$ **(b)** $\sqrt[5]{-32}$ **(c)** $\sqrt[4]{-16}$ **(d)** $\sqrt[6]{-64}$

Solution

(a) $\sqrt[3]{-216} = \sqrt[3]{(-6)^3} = -6$ because $(-6)^3 = -216$.

(b) $\sqrt[5]{-32} = \sqrt[5]{(-2)^5} = -2$ because $(-2)^5 = -32$.

(c) $\sqrt[4]{-16}$ is not a real number because n is even and x is negative.

(d) $\sqrt[6]{-64}$ is not a real number because n is even and x is negative. □

Student Practice 1 If possible, find the roots. If there is no real number root, say so.

(a) $\sqrt[3]{216}$ **(b)** $\sqrt[5]{32}$ **(c)** $\sqrt[3]{-8}$ **(d)** $\sqrt[4]{-81}$

In Section 3.6 we saw that an equation like $y = 2x + 7$ is a function. For each value of x, the equation assigns a unique value to y. We could say, "y is a function of x." If we name the function f, this statement can be symbolized by using the **function notation** $y = f(x)$. Many times we avoid using the y-variable completely and write the function as $f(x) = 2x + 7$.

Because the symbol $\sqrt{x}$ represents exactly one real number for all real numbers x that are nonnegative, we can use it to define the **square root function** $f(x) = \sqrt{x}$. This function has a domain of all real numbers x that are greater than or equal to zero.

We can use function notation when we determine the *function value of y* for specific values of x, as shown in Example 2 below.

Example 2 Find the indicated function values of the function $f(x) = \sqrt{2x + 4}$. Round your answers to the nearest tenth when necessary.

(a) $f(-2)$ **(b)** $f(6)$ **(c)** $f(3)$

Solution

(a) $f(-2) = \sqrt{2(-2) + 4} = \sqrt{-4 + 4} = \sqrt{0} = 0$ The square root of zero is zero.

(b) $f(6) = \sqrt{2(6) + 4} = \sqrt{12 + 4} = \sqrt{16} = 4$

(c) $f(3) = \sqrt{2(3) + 4} = \sqrt{6 + 4} = \sqrt{10} \approx 3.2$ We use a calculator to approximate $\sqrt{10}$. □

Student Practice 2

Find the indicated values of the function $f(x) = \sqrt{4x - 3}$. Round your answers to the nearest tenth when necessary.

(a) $f(3)$ **(b)** $f(4)$ **(c)** $f(7)$

For the function $f(x) = \sqrt{x}$, we know that x cannot be negative. Therefore, the domain of this function is the set of all real numbers x that are greater than or equal to 0.

Example 3 Find the domain of the function. $f(x) = \sqrt{3x - 6}$

Solution We know that the expression $3x - 6$ must be nonnegative. That is, $3x - 6 \geq 0$.

$$3x - 6 \geq 0$$
$$3x \geq 6$$
$$x \geq 2$$

Thus, the domain is all real numbers x where $x \geq 2$. □

Student Practice 3 Find the domain of the function. $f(x) = \sqrt{0.5x + 2}$

Frequently, we are given a function in the form of an equation and are asked to graph it. Each value of the function $f(x)$, often labeled y, corresponds to a value in the domain, often labeled x. This correspondence is the ordered pair (x, y), or $(x, (f(x))$. The graph of the function is the graph of the ordered pairs.

Example 4 Graph the function $f(x) = \sqrt{x + 2}$. Use the values $f(-2)$, $f(-1)$, $f(0)$, $f(1)$, $f(2)$, and $f(7)$. Round your answers to the nearest tenth when necessary.

Solution We show the table of values here.

x	$f(x)$
−2	0
−1	1
0	1.4
1	1.7
2	2
7	3

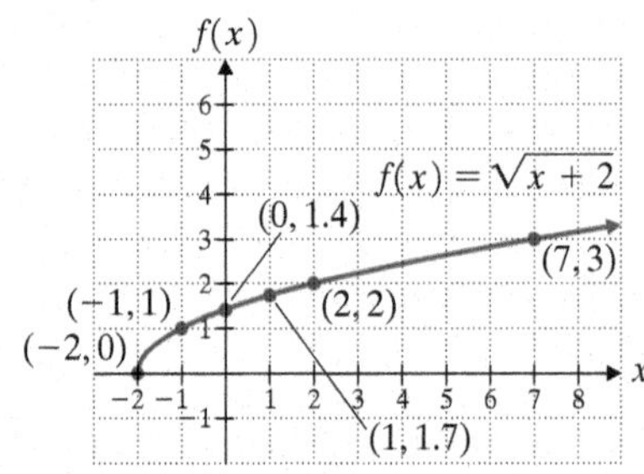

□

Student Practice 4

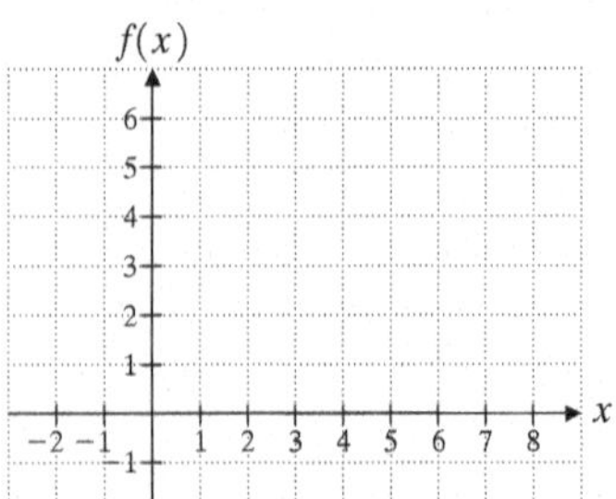

Student Practice 4 Graph the function $f(x) = \sqrt{3x - 9}$. Use the values $f(3)$, $f(4)$, $f(5)$, and $f(6)$.

2 Changing Radical Expressions to Expressions with Rational Exponents

Now we want to extend our definition of roots to rational exponents. By the laws of exponents we know that

$$x^{1/2} \cdot x^{1/2} = x^{1/2+1/2} = x^1 = x.$$

Since $x^{1/2}x^{1/2} = x$, $x^{1/2}$ must be a square root of x. That is, $x^{1/2} = \sqrt{x}$. Is this true? By the definition of square root, $(\sqrt{x})^2 = x$. Does $(x^{1/2})^2 = x$? Using the law of exponents we have

$$(x^{1/2})^2 = x^{(1/2)(2)} = x^1 = x.$$

We conclude that

$$x^{1/2} = \sqrt{x}.$$

In the same way we can write the following:

$$x^{1/3} \cdot x^{1/3} \cdot x^{1/3} = x \qquad x^{1/3} = \sqrt[3]{x}$$
$$x^{1/4} \cdot x^{1/4} \cdot x^{1/4} \cdot x^{1/4} = x \qquad x^{1/4} = \sqrt[4]{x}$$
$$\vdots \qquad \vdots$$
$$\underbrace{x^{1/n} \cdot x^{1/n} \cdot \cdots \cdot x^{1/n}}_{n \text{ factors}} = x \qquad x^{1/n} = \sqrt[n]{x}$$

Therefore, we are ready to define fractional exponents in general.

DEFINITION

If n is a positive integer and x is a nonnegative real number, then

$$x^{1/n} = \sqrt[n]{x}.$$

Example 5 Change to rational exponents and simplify. Assume that all variables are nonnegative real numbers.

(a) $\sqrt[4]{x^4}$ **(b)** $\sqrt[5]{(32)^5}$

Solution

(a) $\sqrt[4]{x^4} = (x^4)^{1/4} = x^{4/4} = x^1 = x$

(b) $\sqrt[5]{(32)^5} = (32^5)^{1/5} = 32^{5/5} = 32^1 = 32$ □

Student Practice 5 Change to rational exponents and simplify. Assume that all variables are nonnegative real numbers.

(a) $\sqrt[3]{x^3}$ **(b)** $\sqrt[4]{y^4}$

Example 6 Replace all radicals with rational exponents.

(a) $\sqrt[3]{x^2}$ **(b)** $(\sqrt[5]{w})^7$

Solution

(a) $\sqrt[3]{x^2} = (x^2)^{1/3} = x^{2/3}$

(b) $(\sqrt[5]{w})^7 = (w^{1/5})^7 = w^{7/5}$ □

Student Practice 6 Replace all radicals with rational exponents.

(a) $\sqrt[4]{x^3}$ **(b)** $\sqrt[5]{(xy)^7}$

Example 7 Evaluate or simplify. Assume that all variables are nonnegative.

(a) $\sqrt[5]{32x^{10}}$ **(b)** $\sqrt[3]{125x^9}$ **(c)** $(16x^4)^{3/4}$

Solution

(a) $\sqrt[5]{32x^{10}} = (2^5x^{10})^{1/5} = 2x^2$

(b) $\sqrt[3]{125x^9} = (5^3x^9)^{1/3} = 5x^3$

(c) $(16x^4)^{3/4} = (2^4x^4)^{3/4} = 2^3x^3 = 8x^3$ □

Student Practice 7 Evaluate or simplify. Assume that all variables are nonnegative.

(a) $\sqrt[4]{81x^{12}}$ **(b)** $\sqrt[3]{27x^6}$ **(c)** $(32x^5)^{3/5}$

3 Changing Expressions with Rational Exponents to Radical Expressions

Sometimes we need to change an expression with rational exponents to a radical expression. This is especially helpful because the value of the radical form of an expression is sometimes more recognizable. For example, because of our experience with radicals, we know that $\sqrt{25} = 5$. It is not as easy to see that $25^{1/2} = 5$. Therefore, we simplify expressions with rational exponents by first rewriting them as radical expressions. Recall that

$$x^{1/n} = \sqrt[n]{x}.$$

Again, using the laws of exponents, we know that

$$x^{m/n} = (x^m)^{1/n} = (x^{1/n})^m,$$

when x is nonnegative. We can make the following general definition.

DEFINITION

For positive integers m and n and any real number x for which $x^{1/n}$ is defined,

$$x^{m/n} = \left(\sqrt[n]{x}\right)^m = \sqrt[n]{x^m}.$$

If it is also true that $x \neq 0$, then

$$x^{-m/n} = \frac{1}{x^{m/n}} = \frac{1}{\left(\sqrt[n]{x}\right)^m} = \frac{1}{\sqrt[n]{x^m}}.$$

Example 8 Change to radical form.

(a) $(xy)^{5/7}$ **(b)** $w^{-2/3}$ **(c)** $3x^{3/4}$ **(d)** $(3x)^{3/4}$

Solution

(a) $(xy)^{5/7} = \sqrt[7]{(xy)^5} = \sqrt[7]{x^5y^5}$ or $(xy)^{5/7} = \left(\sqrt[7]{xy}\right)^5$

(b) $w^{-2/3} = \dfrac{1}{w^{2/3}} = \dfrac{1}{\sqrt[3]{w^2}}$ or $w^{-2/3} = \dfrac{1}{w^{2/3}} = \dfrac{1}{\left(\sqrt[3]{w}\right)^2}$

(c) $3x^{3/4} = 3\sqrt[4]{x^3}$ or $3x^{3/4} = 3\left(\sqrt[4]{x}\right)^3$

(d) $(3x)^{3/4} = \sqrt[4]{(3x)^3} = \sqrt[4]{27x^3}$ or $(3x)^{3/4} = \left(\sqrt[4]{3x}\right)^3$ □

Graphing Calculator

Rational Exponents

To evaluate Example 9(a) on a graphing calculator we use

125 [^] [(] 2 [÷] 3 [)] [ENTER]

Note the need to include the parentheses around the quantity $\frac{2}{3}$. Try each part of Example 9 on your graphing calculator.

Student Practice 8 Change to radical form.

(a) $(xy)^{3/4}$ **(b)** $y^{-1/3}$ **(c)** $(2x)^{4/5}$ **(d)** $2x^{4/5}$

Example 9 Change to radical form and evaluate.

(a) $125^{2/3}$ **(b)** $(-16)^{5/2}$ **(c)** $144^{-1/2}$

Solution

(a) $125^{2/3} = \left(\sqrt[3]{125}\right)^2 = (5)^2 = 25$

(b) $(-16)^{5/2} = \left(\sqrt{-16}\right)^5$; however, $\sqrt{-16}$ is not a real number. Thus, $(-16)^{5/2}$ is not a real number.

(c) $144^{-1/2} = \dfrac{1}{144^{1/2}} = \dfrac{1}{\sqrt{144}} = \dfrac{1}{12}$ □

Student Practice 9 Change to radical form and evaluate.

(a) $8^{2/3}$ (b) $(-8)^{4/3}$ (c) $100^{-3/2}$

4 Evaluating Higher-Order Radicals Containing Variable Radicands That Represent Any Real Number (Including Negative Real Numbers)

We now give a definition of higher-order radicals that works for all radicals, no matter what their signs are.

DEFINITION

For all real numbers x (including negative real numbers),

$\sqrt[n]{x^n} = |x|$ when n is an *even* positive integer, and

$\sqrt[n]{x^n} = x$ when n is an *odd* positive integer.

Example 10 Evaluate; x may be any real number.

(a) $\sqrt[3]{(-2)^3}$ (b) $\sqrt[4]{(-2)^4}$ (c) $\sqrt[5]{x^5}$ (d) $\sqrt[6]{x^6}$

Solution

(a) $\sqrt[3]{(-2)^3} = -2$ because the index is odd

(b) $\sqrt[4]{(-2)^4} = |-2| = 2$ because the index is even

(c) $\sqrt[5]{x^5} = x$ because the index is odd

(d) $\sqrt[6]{x^6} = |x|$ because the index is even ☐

Student Practice 10 Evaluate; y and w may be any real numbers.

(a) $\sqrt[5]{(-3)^5}$ (b) $\sqrt[4]{(-5)^4}$ (c) $\sqrt[4]{w^4}$ (d) $\sqrt[7]{y^7}$

Example 11 Simplify. Assume that x and y may be any real numbers.

(a) $\sqrt{49x^2}$ (b) $\sqrt[4]{81y^{16}}$ (c) $\sqrt[3]{27x^6y^9}$

Solution

(a) We observe that the index is an even positive number. We will need the absolute value. $\sqrt{49x^2} = 7|x|$

(b) Again, we need the absolute value. $\sqrt[4]{81y^{16}} = 3|y^4|$
Since we know that $3y^4$ is positive (anything to the fourth power will be positive), we can write $3|y^4|$ without the absolute value symbol. Thus, $\sqrt[4]{81y^{16}} = 3y^4$.

(c) The index is an odd integer. The absolute value is never needed in such a case. $\sqrt[3]{27x^6y^9} = \sqrt[3]{3^3(x^2)^3(y^3)^3} = 3x^2y^3$ ☐

Student Practice 11 Simplify. Assume that x and y may be any real numbers.

(a) $\sqrt{36x^2}$ (b) $\sqrt[4]{16y^8}$ (c) $\sqrt[3]{125x^3y^6}$

8.2 Exercises MyMathLab®

Verbal and Writing Skills, Exercises 1–4

1. In a simple sentence, explain what a square root is.

2. In a simple sentence, explain what a cube root is.

3. Give an example to show why the cube root of a negative number is a negative number.

4. Give an example to show why it is not possible to find a real number that is the square root of a negative number.

Evaluate if possible.

5. $\sqrt{100}$

6. $\sqrt{64}$

7. $\sqrt{16} + \sqrt{81}$

8. $\sqrt{36} + \sqrt{121}$

9. $-\sqrt{\frac{1}{9}}$

10. $-\sqrt{\frac{4}{25}}$

11. $\sqrt{-36}$

12. $\sqrt{-49}$

13. $\sqrt{0.25}$

14. $\sqrt{0.0036}$

For the given function, find the indicated function values. Find the domain of each function. Round your answers to the nearest tenth when necessary. You may use a calculator as needed.

15. $f(x) = \sqrt{4x + 12};\ f(-3), f(-1), f(1), f(3)$

16. $f(x) = \sqrt{2x - 9};\ f(5), f(6), f(7), f(8)$

17. $f(x) = \sqrt{0.5x - 5};\ f(10), f(12), f(18), f(20)$

18. $f(x) = \sqrt{2.5x + 10};\ f(2), f(4), f(6), f(8)$

Graph each of the following functions. Plot at least four points for each function.

19. $f(x) = \sqrt{x - 1}$

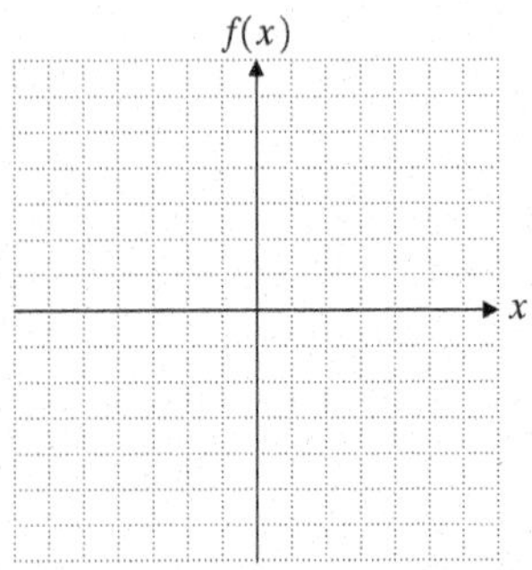

20. $f(x) = \sqrt{x - 3}$

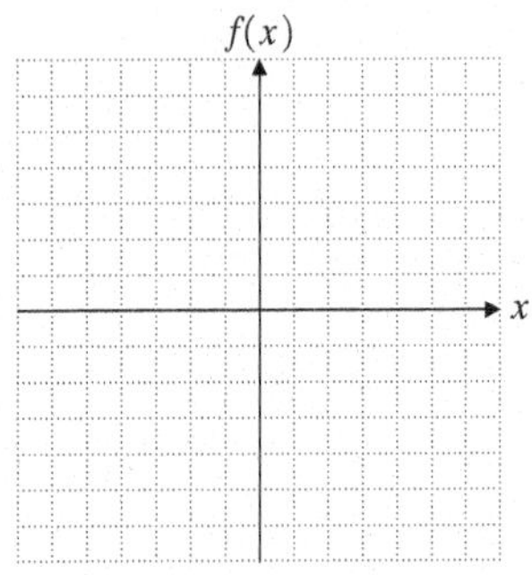

21. $f(x) = \sqrt{3x + 9}$

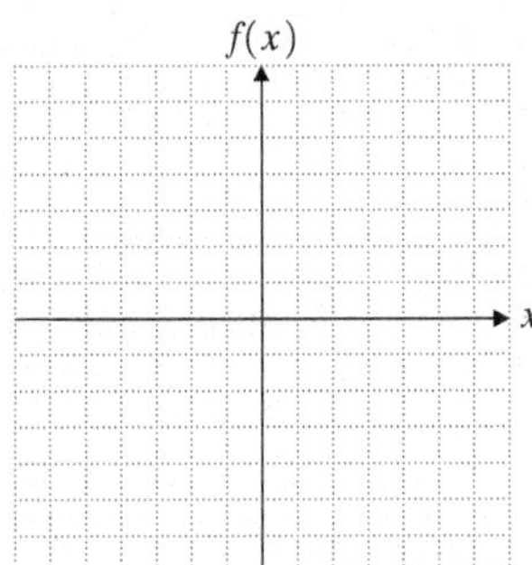

22. $f(x) = \sqrt{2x + 4}$

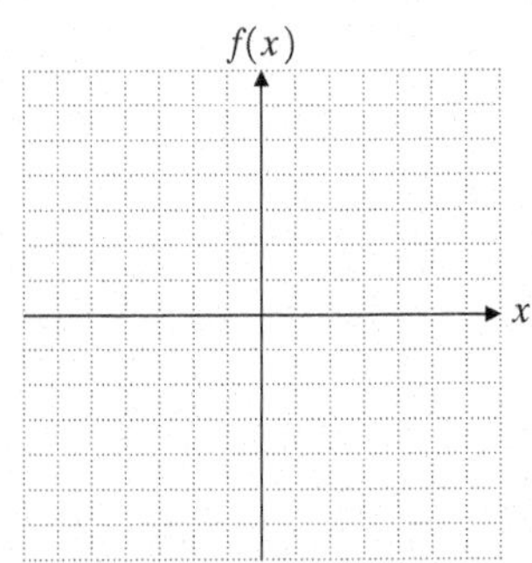

Evaluate if possible.

23. $\sqrt[3]{64}$

24. $\sqrt[3]{27}$

25. $\sqrt[3]{-1000}$

26. $\sqrt[3]{-125}$

27. $\sqrt[4]{16}$

28. $\sqrt[4]{625}$

29. $\sqrt[4]{81}$

30. $-\sqrt[6]{64}$

31. $\sqrt[5]{(8)^5}$

32. $\sqrt[6]{(9)^6}$

33. $\sqrt[8]{5^8}$

34. $\sqrt[7]{11^7}$

35. $\sqrt[3]{-\frac{1}{8}}$

36. $\sqrt[3]{-\frac{8}{27}}$

For exercises 37–78, assume that variables represent positive real numbers. Replace all radicals with rational exponents.

37. $\sqrt[3]{y}$

38. $\sqrt{a}$

39. $\sqrt[5]{m^3}$

40. $\sqrt[6]{b^5}$

41. $\sqrt[4]{2a}$

42. $\sqrt[5]{4b}$

43. $\sqrt[7]{(a + b)^3}$

44. $\sqrt[9]{(a - b)^5}$

45. $\sqrt{\sqrt[3]{x}}$

46. $\sqrt[5]{\sqrt{y}}$

47. $(\sqrt[6]{3x})^5$

48. $(\sqrt[5]{2x})^3$

Simplify.

49. $\sqrt[6]{(12)^6}$ **50.** $\sqrt[5]{(-11)^5}$ **51.** $\sqrt[3]{x^{12}y^3}$ **52.** $\sqrt[3]{x^6y^{15}}$

53. $\sqrt{36x^8y^4}$ **54.** $\sqrt{64x^6y^{10}}$ **55.** $\sqrt[4]{16a^8b^4}$ **56.** $\sqrt[4]{81a^{12}b^{20}}$

Change to radical form.

57. $y^{4/7}$ **58.** $x^{5/6}$ **59.** $7^{-2/3}$ **60.** $5^{-3/5}$

61. $(a + 5b)^{3/4}$ **62.** $(3a + b)^{2/7}$ **63.** $(-x)^{3/5}$ **64.** $(-y)^{5/7}$

65. $(2xy)^{3/5}$ **66.** $(3ab)^{2/7}$

Mixed Practice, Exercises 67–78 *Evaluate or simplify. Assume that variables represent positive real numbers.*

67. $9^{3/2}$ **68.** $8^{2/3}$ **69.** $\left(\frac{4}{25}\right)^{1/2}$ **70.** $\left(\frac{1}{16}\right)^{3/4}$

71. $\left(\frac{1}{8}\right)^{-1/3}$ **72.** $\left(\frac{25}{9}\right)^{-1/2}$ **73.** $(64x^4)^{-1/2}$ **74.** $(100a^{12})^{-1/2}$

75. $\sqrt{121x^4}$ **76.** $\sqrt{49x^8}$ **77.** $\sqrt{144a^6b^{24}}$ **78.** $\sqrt{81a^{18}b^2}$

Simplify. Assume that the variables represent any real number.

79. $\sqrt{25x^2}$ **80.** $\sqrt{100x^2}$ **81.** $\sqrt[3]{-8x^6}$ **82.** $\sqrt[3]{-27x^9}$ **83.** $\sqrt[4]{x^4y^{24}}$

84. $\sqrt[4]{x^{16}y^{40}}$ **85.** $\sqrt[4]{a^{12}b^4}$ **86.** $\sqrt[4]{a^4b^{20}}$ **87.** $\sqrt{25x^{12}y^4}$ **88.** $\sqrt{64x^6y^{10}}$

Cumulative Review

89. **[3.3.2]** Solve for x. $-5x + 2y = 6$

90. **[3.3.2]** Solve for y. $x = \frac{2}{3}y + 4$

Quick Quiz 8.2 *Simplify. Assume that variables represent positive real numbers.*

1. $\left(\frac{4}{25}\right)^{3/2}$

2. $\sqrt[3]{-64}$

3. $\sqrt{121x^{10}y^{12}}$

4. Concept Check Explain how you would simplify the following.

$$\sqrt[4]{81x^8y^{16}}$$

8.3 Simplifying, Adding, and Subtracting Radicals

Student Learning Objectives

After studying this section, you will be able to:

1. Simplify a radical by using the product rule.
2. Add and subtract like radical terms.

1 Simplifying a Radical by Using the Product Rule

When we simplify a radical, we want to get an equivalent expression with the smallest possible quantity in the radicand. We can use the product rule for radicals to simplify radicals.

> **PRODUCT RULE FOR RADICALS**
>
> For all nonnegative real numbers *a* and *b* and positive integers *n*,
>
> $$\sqrt[n]{a}\,\sqrt[n]{b} = \sqrt[n]{ab}.$$

You should be able to derive the product rule from your knowledge of the laws of exponents. We have

$$\sqrt[n]{a}\,\sqrt[n]{b} = a^{1/n}b^{1/n} = (ab)^{1/n} = \sqrt[n]{ab}.$$

Throughout the remainder of this chapter, assume that all variables in any radicand represent nonnegative numbers, unless a specific statement is made to the contrary.

Example 1 Simplify. $\sqrt{32}$

Solution 1: $\sqrt{32} = \sqrt{16 \cdot 2} = \sqrt{16}\sqrt{2} = 4\sqrt{2}$

Solution 2: $\sqrt{32} = \sqrt{4 \cdot 8} = \sqrt{4}\sqrt{8} = 2\sqrt{8} = 2\sqrt{4 \cdot 2} = 2\sqrt{4}\sqrt{2} = 4\sqrt{2}$

Although we obtained the same answer both times, the first solution is much shorter. You should try to use the largest factor that is a perfect square when you use the product rule. □

Student Practice 1 Simplify. $\sqrt{20}$

Example 2 Simplify. $\sqrt{48}$

Solution $\sqrt{48} = \sqrt{16}\sqrt{3} = 4\sqrt{3}$ □

Student Practice 2 Simplify. $\sqrt{27}$

Example 3 Simplify.

(a) $\sqrt[3]{16}$ **(b)** $\sqrt[3]{-81}$

Solution

(a) $\sqrt[3]{16} = \sqrt[3]{8}\sqrt[3]{2} = 2\sqrt[3]{2}$

(b) $\sqrt[3]{-81} = \sqrt[3]{-27}\sqrt[3]{3} = -3\sqrt[3]{3}$ □

Student Practice 3 Simplify.

(a) $\sqrt[3]{24}$ **(b)** $\sqrt[3]{-108}$

Example 4 Simplify. $\sqrt[4]{48}$

Solution $\sqrt[4]{48} = \sqrt[4]{16}\sqrt[4]{3} = 2\sqrt[4]{3}$ □

Student Practice 4 Simplify. $\sqrt[4]{64}$

Example 5 Simplify.

(a) $\sqrt{27x^3y^4}$ **(b)** $\sqrt[3]{16x^4y^3z^6}$

Solution

(a) $\sqrt{27x^3y^4} = \sqrt{9 \cdot 3 \cdot x^2 \cdot x \cdot y^4} = \sqrt{9x^2y^4}\sqrt{3x}$ Factor out the perfect squares.

$= 3xy^2\sqrt{3x}$

(b) $\sqrt[3]{16x^4y^3z^6} =$

$\sqrt[3]{8 \cdot 2 \cdot x^3 \cdot x \cdot y^3 \cdot z^6} = \sqrt[3]{8x^3y^3z^6}\sqrt[3]{2x}$ Factor out the perfect cubes. Why is z^6 a perfect cube?

$= 2xyz^2\sqrt[3]{2x}$ □

Student Practice 5 Simplify.

(a) $\sqrt{45x^6y^7}$ **(b)** $\sqrt[3]{27a^7b^8c^9}$

2 Adding and Subtracting Like Radical Terms

Only like radicals can be added or subtracted. Two radicals are **like radicals** if they have the same radicand and the same index. $2\sqrt{5}$ and $3\sqrt{5}$ are like radicals. $2\sqrt{5}$ and $2\sqrt{3}$ are not like radicals; $2\sqrt{5}$ and $2\sqrt[3]{5}$ are not like radicals. When we combine radicals, we combine like terms by using the distributive property.

Example 6 Combine. $2\sqrt{5} + 3\sqrt{5} - 4\sqrt{5}$

Solution $2\sqrt{5} + 3\sqrt{5} - 4\sqrt{5} = (2 + 3 - 4)\sqrt{5} = 1\sqrt{5} = \sqrt{5}$ □

Student Practice 6 Combine. $19\sqrt{xy} + 5\sqrt{xy} - 10\sqrt{xy}$

Sometimes when you simplify radicands, you may find you have like radicals.

Example 7 Combine. $5\sqrt{3} - \sqrt{27} + 2\sqrt{48}$

Solution

$$\begin{aligned} 5\sqrt{3} - \sqrt{27} + 2\sqrt{48} &= 5\sqrt{3} - \sqrt{9}\sqrt{3} + 2\sqrt{16}\sqrt{3} \\ &= 5\sqrt{3} - 3\sqrt{3} + 2(4)\sqrt{3} \\ &= 5\sqrt{3} - 3\sqrt{3} + 8\sqrt{3} \\ &= 10\sqrt{3} \end{aligned}$$

Student Practice 7 Combine.

$$4\sqrt{2} - 5\sqrt{50} - 3\sqrt{98}$$

Example 8 Combine. $6\sqrt{x} + 4\sqrt{12x} - \sqrt{75x} + 3\sqrt{x}$

Solution

$$\begin{aligned} &6\sqrt{x} + 4\sqrt{12x} - \sqrt{75x} + 3\sqrt{x} \\ &= 6\sqrt{x} + 4\sqrt{4}\sqrt{3x} - \sqrt{25}\sqrt{3x} + 3\sqrt{x} \\ &= 6\sqrt{x} + 8\sqrt{3x} - 5\sqrt{3x} + 3\sqrt{x} \\ &= 6\sqrt{x} + 3\sqrt{x} + 8\sqrt{3x} - 5\sqrt{3x} \\ &= 9\sqrt{x} + 3\sqrt{3x} \end{aligned}$$

Student Practice 8 Combine.

$$4\sqrt{2x} + \sqrt{18x} - 2\sqrt{125x} - 6\sqrt{20x}$$

Example 9 Combine. $2\sqrt[3]{81x^3y^4} + 3xy\sqrt[3]{24y}$

Solution

$$\begin{aligned} 2\sqrt[3]{81x^3y^4} + 3xy\sqrt[3]{24y} &= 2\sqrt[3]{27x^3y^3}\sqrt[3]{3y} + 3xy\sqrt[3]{8}\sqrt[3]{3y} \\ &= 2(3xy)\sqrt[3]{3y} + 3xy(2)\sqrt[3]{3y} \\ &= 6xy\sqrt[3]{3y} + 6xy\sqrt[3]{3y} \\ &= 12xy\sqrt[3]{3y} \end{aligned}$$

Student Practice 9 Combine.

$$3x\sqrt[3]{54x^4} - 3\sqrt[3]{16x^7}$$

8.3 Exercises MyMathLab®

Simplify. Assume that all variables are nonnegative real numbers.

1. $\sqrt{8}$

2. $\sqrt{12}$

3. $\sqrt{18}$

4. $\sqrt{75}$

5. $\sqrt{28}$

6. $\sqrt{54}$

7. $\sqrt{50}$

8. $\sqrt{80}$

9. $\sqrt{9x^2}$

10. $\sqrt{36x^4}$

11. $\sqrt{40a^6b^7}$

12. $\sqrt{150a^4b^9}$

13. $\sqrt{90x^3yz^4}$

14. $\sqrt{32x^2y^5z^4}$

15. $\sqrt[3]{8}$

16. $\sqrt[3]{125}$

17. $\sqrt[3]{40}$

18. $\sqrt[3]{128}$

19. $\sqrt[3]{54a^2}$

20. $\sqrt[3]{32n}$

21. $\sqrt[3]{27a^5b^9}$

22. $\sqrt[3]{125a^6b^2}$

23. $\sqrt[3]{40x^{12}y^{13}}$

24. $\sqrt[3]{108x^5y^{17}}$

25. $\sqrt[4]{81kp^{23}}$

26. $\sqrt[4]{16k^{12}p^{18}}$

27. $\sqrt[5]{-32x^5y^6}$

28. $\sqrt[5]{-243x^4y^{10}}$

To Think About, Exercises 29 and 30

29. $\sqrt[4]{1792} = a\sqrt[4]{7}$. What is the value of a?

30. $\sqrt[3]{3072} = b\sqrt[3]{6}$. What is the value of b?

Combine.

31. $4\sqrt{5} + 8\sqrt{5}$

32. $3\sqrt{13} + 7\sqrt{13}$

33. $4\sqrt{3} + \sqrt{7} - 5\sqrt{7}$

34. $2\sqrt{6} + \sqrt{2} - 5\sqrt{6}$

35. $3\sqrt{32} - \sqrt{2}$

36. $\sqrt{90} - \sqrt{10}$

37. $4\sqrt{12} + \sqrt{27}$

38. $5\sqrt{75} + \sqrt{48}$

39. $\sqrt{8} + \sqrt{50} - 2\sqrt{72}$

40. $\sqrt{45} + \sqrt{80} - 3\sqrt{20}$

41. $\sqrt{48} - 2\sqrt{27} + \sqrt{12}$

42. $-4\sqrt{18} + \sqrt{98} - \sqrt{32}$

43. $-2\sqrt{24} + 3\sqrt{6} + \sqrt{54}$

44. $2\sqrt{7} + 2\sqrt{63} - 4\sqrt{28}$

Combine. Assume that all variables represent nonnegative real numbers.

45. $3\sqrt{48x} - 2\sqrt{12x}$

46. $5\sqrt{27x} - 4\sqrt{75x}$

47. $5\sqrt{2x} + 2\sqrt{18x} + 2\sqrt{32x}$

48. $4\sqrt{50x} + 3\sqrt{2x} + \sqrt{72x}$

49. $\sqrt{44} - 3\sqrt{63x} + 4\sqrt{28x}$

50. $\sqrt{20x} - \sqrt{72x} + \sqrt{45x}$

51. $\sqrt{200x^3} - x\sqrt{32x}$

52. $a\sqrt{48a} + \sqrt{27a^3}$

53. $\sqrt[3]{16} + 3\sqrt[3]{54}$

54. $\sqrt[3]{128} - 4\sqrt[3]{16}$

55. $4\sqrt[3]{x^4y^3} - 3\sqrt[3]{xy^5}$

56. $2y\sqrt[3]{27x^3} - 3\sqrt[3]{125x^3y^3}$

To Think About

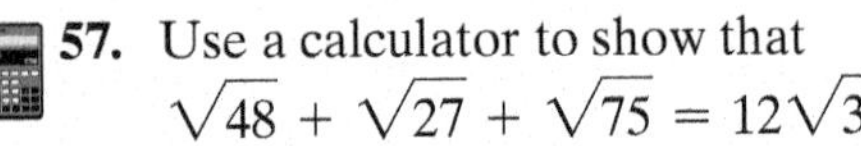

57. Use a calculator to show that $\sqrt{48} + \sqrt{27} + \sqrt{75} = 12\sqrt{3}$.

58. Use a calculator to show that $\sqrt{98} + \sqrt{50} + \sqrt{128} = 20\sqrt{2}$.

Applications

Electricians *Electricians approximate the amount of current in amps I (amperes) drawn by an appliance in the home using the formula*

$$I = \sqrt{\frac{P}{R}},$$

where P is the power measured in watts and R is the resistance measured in ohms. In exercises 59 and 60, round your answers to three decimal places.

59. What is the amount of current drawn by a refrigerator if $P = 825$ watts and $R = 15$ ohms?

60. What is the amount of current drawn by a clothes dryer if $P = 3600$ watts and $R = 12$ ohms?

Period of a Pendulum *The* **period** *of a pendulum is the amount of time it takes the pendulum to make one complete swing back and forth. If the length of the pendulum L is measured in feet, then its period T measured in seconds is given by the formula*

$$T = 2\pi\sqrt{\frac{L}{32}}.$$

Use $\pi \approx 3.14$ *for exercises 61–62.*

61. Find the period of a pendulum if its length is 8 feet.

62. How much longer is the period of a pendulum consisting of a person swinging on a 30-foot rope than that of a person swinging on a 10-foot rope? Round your answer to the nearest hundredth.

Cumulative Review *Factor completely.*

63. **[6.5.3]** $16x^3 - 56x^2y + 49xy^2$

64. **[6.5.3]** $81x^2y - 25y$

65. **[5.5.1]** $(2x + 3)(2x - 3)$

66. **[5.5.2]** $(2x + 3)^2$

Quick Quiz 8.3 *Simplify.*

1. $\sqrt{120x^7y^8}$

2. $\sqrt[3]{16x^{15}y^{10}}$

3. Combine.
$2\sqrt{75} + 3\sqrt{48} - 4\sqrt{27}$

4. **Concept Check** Explain how you would simplify the following.

$$\sqrt[4]{16x^{13}y^{16}}$$

8.4 Multiplying and Dividing Radicals

Student Learning Objectives

After studying this section, you will be able to:

1. Multiply radical expressions.
2. Divide radical expressions.
3. Simplify radical expressions by rationalizing the denominator.

1 Multiplying Radical Expressions

We use the product rule for radicals to multiply radical expressions. Recall that $\sqrt[n]{a}\sqrt[n]{b} = \sqrt[n]{ab}$.

Example 1 Multiply. $(3\sqrt{2})(5\sqrt{11x})$

Solution $(3\sqrt{2})(5\sqrt{11x}) = (3)(5)\sqrt{2 \cdot 11x} = 15\sqrt{22x}$ □

Student Practice 1 Multiply. $(-4\sqrt{2})(-3\sqrt{13x})$

Example 2 Multiply. $\sqrt{6x}(\sqrt{3} + \sqrt{2x} + \sqrt{5})$

Solution

$$\begin{aligned}\sqrt{6x}(\sqrt{3} + \sqrt{2x} + \sqrt{5}) &= (\sqrt{6x})(\sqrt{3}) + (\sqrt{6x})(\sqrt{2x}) + (\sqrt{6x})(\sqrt{5}) \\ &= \sqrt{18x} + \sqrt{12x^2} + \sqrt{30x} \\ &= \sqrt{9}\sqrt{2x} + \sqrt{4x^2}\sqrt{3} + \sqrt{30x} \\ &= 3\sqrt{2x} + 2x\sqrt{3} + \sqrt{30x}\end{aligned}$$ □

Student Practice 2 Multiply. $\sqrt{2x}(\sqrt{5} + 2\sqrt{3x} + \sqrt{8})$

To multiply two binomials containing radicals, we can use the distributive property. Most students find that the FOIL method is helpful in remembering how to find the four products.

Example 3 Multiply. $(\sqrt{2} + 3\sqrt{5})(2\sqrt{2} - \sqrt{5})$

Solution By FOIL:

$$\begin{aligned}(\sqrt{2} + 3\sqrt{5})(2\sqrt{2} - \sqrt{5}) &= 2\sqrt{4} - \sqrt{10} + 6\sqrt{10} - 3\sqrt{25} \\ &= 4 + 5\sqrt{10} - 15 \\ &= -11 + 5\sqrt{10}\end{aligned}$$

By the distributive property:

$$\begin{aligned}(\sqrt{2} + 3\sqrt{5})(2\sqrt{2} - \sqrt{5}) &= (\sqrt{2} + 3\sqrt{5})(2\sqrt{2}) - (\sqrt{2} + 3\sqrt{5})(\sqrt{5}) \\ &= (\sqrt{2})(2\sqrt{2}) + (3\sqrt{5})(2\sqrt{2}) - (\sqrt{2})(\sqrt{5}) - (3\sqrt{5})(\sqrt{5}) \\ &= 2\sqrt{4} + 6\sqrt{10} - \sqrt{10} - 3\sqrt{25} \\ &= 4 + 5\sqrt{10} - 15 \\ &= -11 + 5\sqrt{10}\end{aligned}$$ □

Student Practice 3 Multiply. $(\sqrt{7} + 4\sqrt{2})(2\sqrt{7} - 3\sqrt{2})$

Example 4 Multiply. $(7 - 3\sqrt{2})(4 - \sqrt{3})$

Solution

$$(7 - 3\sqrt{2})(4 - \sqrt{3}) = 28 - 7\sqrt{3} - 12\sqrt{2} + 3\sqrt{6}$$ □

Student Practice 4 Multiply. $(2 - 5\sqrt{5})(3 - 2\sqrt{2})$

Example 5 Multiply. $(\sqrt{7} + \sqrt{3x})^2$

Solution

Method 1: We can use the FOIL method or the distributive property.

$$\begin{aligned}(\sqrt{7} + \sqrt{3x})(\sqrt{7} + \sqrt{3x}) &= \sqrt{49} + \sqrt{21x} + \sqrt{21x} + \sqrt{9x^2}\\ &= 7 + 2\sqrt{21x} + 3x\end{aligned}$$

Method 2: We could also use the Chapter 5 formula.

$$(a + b)^2 = a^2 + 2ab + b^2,$$

where $a = \sqrt{7}$ and $b = \sqrt{3x}$. Then

$$\begin{aligned}(\sqrt{7} + \sqrt{3x})^2 &= (\sqrt{7})^2 + 2\sqrt{7}\sqrt{3x} + (\sqrt{3x})^2\\ &= 7 + 2\sqrt{21x} + 3x\end{aligned}$$

□

Student Practice 5 Multiply, using the approach that seems easiest to you.

$$(\sqrt{5x} + \sqrt{10})^2$$

Example 6 Multiply.

(a) $\sqrt[3]{3x}(\sqrt[3]{x^2} + 3\sqrt[3]{4y})$

(b) $(\sqrt[3]{2y} + \sqrt[3]{4})(2\sqrt[3]{4y^2} - 3\sqrt[3]{2})$

Solution

(a)
$$\begin{aligned}\sqrt[3]{3x}(\sqrt[3]{x^2} + 3\sqrt[3]{4y}) &= (\sqrt[3]{3x})(\sqrt[3]{x^2}) + 3(\sqrt[3]{3x})(\sqrt[3]{4y})\\ &= \sqrt[3]{3x^3} + 3\sqrt[3]{12xy}\\ &= x\sqrt[3]{3} + 3\sqrt[3]{12xy}\end{aligned}$$

(b)
$$\begin{aligned}(\sqrt[3]{2y} + \sqrt[3]{4})(2\sqrt[3]{4y^2} - 3\sqrt[3]{2}) &= 2\sqrt[3]{8y^3} - 3\sqrt[3]{4y} + 2\sqrt[3]{16y^2} - 3\sqrt[3]{8}\\ &= 2(2y) - 3\sqrt[3]{4y} + 2\sqrt[3]{8}\sqrt[3]{2y^2} - 3(2)\\ &= 4y - 3\sqrt[3]{4y} + 4\sqrt[3]{2y^2} - 6\end{aligned}$$

□

Student Practice 6 Multiply.

(a) $\sqrt[3]{2x}(\sqrt[3]{4x^2} + 3\sqrt[3]{y})$

(b) $(\sqrt[3]{7} + \sqrt[3]{x^2})(2\sqrt[3]{49} - \sqrt[3]{x})$

2 Dividing Radical Expressions

We can use the laws of exponents to develop a rule for dividing two radicals.

$$\sqrt[n]{\frac{a}{b}} = \left(\frac{a}{b}\right)^{1/n} = \frac{a^{1/n}}{b^{1/n}} = \frac{\sqrt[n]{a}}{\sqrt[n]{b}}$$

This quotient rule is very useful. We now state it more formally.

QUOTIENT RULE FOR RADICALS

For all nonnegative real numbers *a*, all positive real numbers *b*, and positive integers *n*,

$$\frac{\sqrt[n]{a}}{\sqrt[n]{b}} = \sqrt[n]{\frac{a}{b}}.$$

Sometimes it will be best to change $\sqrt[n]{\frac{a}{b}}$ to $\frac{\sqrt[n]{a}}{\sqrt[n]{b}}$, whereas at other times it will be best to change $\frac{\sqrt[n]{a}}{\sqrt[n]{b}}$ to $\sqrt[n]{\frac{a}{b}}$. To use the quotient rule for radicals, you need to have good number sense. You should know your squares up to 15^2 and your cubes up to 5^3.

Example 7 Divide.

(a) $\frac{\sqrt{48}}{\sqrt{3}}$ **(b)** $\sqrt[3]{\frac{125}{8}}$ **(c)** $\frac{\sqrt{28x^5y^3}}{\sqrt{7x}}$

Solution

(a) $\frac{\sqrt{48}}{\sqrt{3}} = \sqrt{\frac{48}{3}} = \sqrt{16} = 4$

(b) $\sqrt[3]{\frac{125}{8}} = \frac{\sqrt[3]{125}}{\sqrt[3]{8}} = \frac{5}{2}$

(c) $\frac{\sqrt{28x^5y^3}}{\sqrt{7x}} = \sqrt{\frac{28x^5y^3}{7x}} = \sqrt{4x^4y^3} = 2x^2y\sqrt{y}$ □

Student Practice 7 Divide.

(a) $\frac{\sqrt{75}}{\sqrt{3}}$ **(b)** $\sqrt[3]{\frac{27}{64}}$ **(c)** $\frac{\sqrt{54a^3b^7}}{\sqrt{6b^5}}$

3 Simplifying Radical Expressions by Rationalizing the Denominator

Recall that to simplify a radical we want to get the smallest possible quantity in the radicand. Whenever possible, we find the square root of a perfect square. Thus, to simplify $\sqrt{\frac{7}{16}}$ we have

$$\sqrt{\frac{7}{16}} = \frac{\sqrt{7}}{\sqrt{16}} = \frac{\sqrt{7}}{4}.$$

Notice that the denominator does not contain a square root. The expression $\frac{\sqrt{7}}{4}$ is in simplest form.

Let's look at $\sqrt{\frac{16}{7}}$. We have

$$\sqrt{\frac{16}{7}} = \frac{\sqrt{16}}{\sqrt{7}} = \frac{4}{\sqrt{7}}.$$

Notice that the denominator contains a square root. If an expression contains a square root in the denominator, it is not considered to be simplified. How can we rewrite $\frac{4}{\sqrt{7}}$ as an equivalent expression that does not contain the $\sqrt{7}$ in the denominator? Since $\sqrt{7}\sqrt{7} = 7$, we can multiply the numerator and the denominator by the radical in the denominator.

$$\frac{4}{\sqrt{7}} \cdot \frac{\sqrt{7}}{\sqrt{7}} = \frac{4\sqrt{7}}{\sqrt{49}} = \frac{4\sqrt{7}}{7}$$

This expression is considered to be in simplest form. We call this process rationalizing the denominator.

Rationalizing the denominator is the process of transforming a fraction with one or more radicals in the denominator into an equivalent fraction without a radical in the denominator.

Example 8 Simplify by rationalizing the denominator. $\frac{3}{\sqrt{2}}$

Solution

$$\frac{3}{\sqrt{2}} = \frac{3}{\sqrt{2}} \cdot \frac{\sqrt{2}}{\sqrt{2}} \quad \text{since } \frac{\sqrt{2}}{\sqrt{2}} = 1$$

$$= \frac{3\sqrt{2}}{\sqrt{4}} \quad \text{product rule for radicals}$$

$$= \frac{3\sqrt{2}}{2}$$

□

Student Practice 8 Simplify by rationalizing the denominator.

$$\frac{7}{\sqrt{3}}$$

We can rationalize the denominator either before or after we simplify the denominator.

Example 9 Simplify. $\frac{3}{\sqrt{12x}}$

Solution

Method 1: First we simplify the radical in the denominator, and then we multiply in order to rationalize the denominator.

$$\frac{3}{\sqrt{12x}} = \frac{3}{\sqrt{4}\sqrt{3x}} = \frac{3}{2\sqrt{3x}} \cdot \frac{\sqrt{3x}}{\sqrt{3x}} = \frac{\cancel{3}\sqrt{3x}}{2(\cancel{3}x)} = \frac{\sqrt{3x}}{2x}$$

Method 2: We can multiply numerator and denominator by a value that will make the radicand in the denominator a perfect square (i.e., rationalize the denominator).

$$\frac{3}{\sqrt{12x}} = \frac{3}{\sqrt{12x}} \cdot \frac{\sqrt{3x}}{\sqrt{3x}}$$

$$= \frac{3\sqrt{3x}}{\sqrt{36x^2}} \quad \text{since } \sqrt{12x}\sqrt{3x} = \sqrt{36x^2}$$

$$= \frac{3\sqrt{3x}}{6x} = \frac{\sqrt{3x}}{2x}$$

□

Student Practice 9 Simplify.

$$\frac{8}{\sqrt{20x}}$$

If a radicand contains a fraction, it is not considered to be simplified. We can use the quotient rule for radicals and then rationalize the denominator to simplify the radical. We have already rationalized denominators when they contain square roots. Now we will rationalize denominators when they contain radical expressions that are cube roots or higher-order roots.

Example 10 Simplify. $\sqrt[3]{\dfrac{2}{3x^2}}$

Solution

Method 1: $\sqrt[3]{\dfrac{2}{3x^2}} = \dfrac{\sqrt[3]{2}}{\sqrt[3]{3x^2}}$ quotient rule for radicals

$= \dfrac{\sqrt[3]{2}}{\sqrt[3]{3x^2}} \cdot \dfrac{\sqrt[3]{9x}}{\sqrt[3]{9x}}$ Multiply the numerator and denominator by an appropriate value so that the new radicand in the denominator will be a perfect cube.

$= \dfrac{\sqrt[3]{18x}}{\sqrt[3]{27x^3}}$ Observe that we can evaluate the cube root in the denominator.

$= \dfrac{\sqrt[3]{18x}}{3x}$

Method 2: $\sqrt[3]{\dfrac{2}{3x^2}} = \sqrt[3]{\dfrac{2}{3x^2} \cdot \dfrac{9x}{9x}}$

$= \sqrt[3]{\dfrac{18x}{27x^3}}$

$= \dfrac{\sqrt[3]{18x}}{\sqrt[3]{27x^3}}$

$= \dfrac{\sqrt[3]{18x}}{3x}$ □

Student Practice 10 Simplify.

$\sqrt[3]{\dfrac{6}{5x}}$

If the denominator of a radical expression contains a sum or difference with radicals, we multiply the numerator and denominator by the *conjugate* of the denominator. For example, the conjugate of $x + \sqrt{y}$ is $x - \sqrt{y}$; similarly, the conjugate of $x - \sqrt{y}$ is $x + \sqrt{y}$. What is the conjugate of $3 + \sqrt{2}$? It is $3 - \sqrt{2}$. How about $\sqrt{11} + \sqrt{xyz}$? It is $\sqrt{11} - \sqrt{xyz}$.

CONJUGATES

The expressions $a + b$ and $a - b$, where a and b represent any algebraic term, are called **conjugates.** Each expression is the conjugate of the other expression.

Multiplying by conjugates is simply an application of the formula

$$(a + b)(a - b) = a^2 - b^2.$$

For example,

$$\left(\sqrt{x} + \sqrt{y}\right)\left(\sqrt{x} - \sqrt{y}\right) = \left(\sqrt{x}\right)^2 - \left(\sqrt{y}\right)^2 = x - y.$$

Example 11 Simplify. $\dfrac{5}{3+\sqrt{2}}$

Solution

$$\frac{5}{3+\sqrt{2}} = \frac{5}{3+\sqrt{2}} \cdot \frac{3-\sqrt{2}}{3-\sqrt{2}}$$ Multiply the numerator and denominator by the conjugate of $3+\sqrt{2}$.

$$= \frac{15-5\sqrt{2}}{3^2-(\sqrt{2})^2}$$

$$= \frac{15-5\sqrt{2}}{9-2} = \frac{15-5\sqrt{2}}{7}$$ □

Student Practice 11 Simplify.

$$\frac{4}{2+\sqrt{5}}$$

Example 12 Simplify. $\dfrac{\sqrt{7}+\sqrt{3}}{\sqrt{7}-\sqrt{3}}$

Solution The conjugate of $\sqrt{7}-\sqrt{3}$ is $\sqrt{7}+\sqrt{3}$.

$$\frac{\sqrt{7}+\sqrt{3}}{\sqrt{7}-\sqrt{3}} \cdot \frac{\sqrt{7}+\sqrt{3}}{\sqrt{7}+\sqrt{3}} = \frac{\sqrt{49}+2\sqrt{21}+\sqrt{9}}{(\sqrt{7})^2-(\sqrt{3})^2}$$

$$= \frac{7+2\sqrt{21}+3}{7-3}$$

$$= \frac{10+2\sqrt{21}}{4}$$

$$= \frac{2(5+\sqrt{21})}{2 \cdot 2}$$

$$= \frac{\cancel{2}(5+\sqrt{21})}{\cancel{2} \cdot 2}$$

$$= \frac{5+\sqrt{21}}{2}$$ □

Student Practice 12 Simplify.

$$\frac{\sqrt{11}+\sqrt{5}}{\sqrt{11}-\sqrt{5}}$$

8.4 Exercises MyMathLab®

Multiply and simplify. Assume that all variables represent nonnegative numbers.

1. $\sqrt{5}\sqrt{7}$
2. $\sqrt{11}\sqrt{6}$
3. $(4\sqrt{7})(-2\sqrt{3})$
4. $(-4\sqrt{5})(-2\sqrt{3})$
5. $(3\sqrt{10})(-4\sqrt{2})$
6. $(-5\sqrt{5})(-6\sqrt{6})$
7. $(-3\sqrt{y})(\sqrt{5x})$
8. $(3\sqrt{x})(-6\sqrt{y})$
9. $(3x\sqrt{2x})(-2\sqrt{10xy})$
10. $(4\sqrt{3a})(a\sqrt{6ab})$
11. $5\sqrt{a}(3\sqrt{b}-5)$
12. $-\sqrt{a}(5+4\sqrt{b})$
13. $-3\sqrt{a}(\sqrt{2b}+2\sqrt{5})$
14. $4\sqrt{x}(\sqrt{3y}-3\sqrt{6})$
15. $-\sqrt{a}(\sqrt{a}-2\sqrt{b})$
16. $-2\sqrt{ab}(5\sqrt{a}-\sqrt{ab})$
17. $7\sqrt{x}(2\sqrt{3}-5\sqrt{x})$
18. $3\sqrt{y}(4\sqrt{6}+11\sqrt{y})$
19. $(3-\sqrt{2})(8+\sqrt{2})$
20. $(\sqrt{6}+3)(\sqrt{6}-1)$
21. $(2\sqrt{3}+\sqrt{2})(2\sqrt{3}-4\sqrt{2})$
22. $(3\sqrt{3}+\sqrt{5})(\sqrt{3}-2\sqrt{5})$
23. $(\sqrt{7}+4\sqrt{5x})(2\sqrt{7}+3\sqrt{5x})$
24. $(\sqrt{6}+3\sqrt{3y})(5\sqrt{6}+2\sqrt{3y})$
25. $(\sqrt{3}+2\sqrt{2})(\sqrt{5}+\sqrt{3})$
26. $(3\sqrt{5}+\sqrt{3})(\sqrt{2}+2\sqrt{5})$
27. $(\sqrt{5}-2\sqrt{6})^2$
28. $(\sqrt{3}+4\sqrt{7})^2$
29. $(9-2\sqrt{b})^2$
30. $(3\sqrt{a}+4)^2$
31. $(\sqrt{3x+4}+3)^2$
32. $(\sqrt{2x+1}-2)^2$
33. $(\sqrt[3]{x^2})(3\sqrt[3]{4x}-4\sqrt[3]{x^5})$
34. $(2\sqrt[3]{x})(\sqrt[3]{4x^2}-\sqrt[3]{14x})$
35. $(\sqrt[3]{3}+\sqrt[3]{2})(\sqrt[3]{9}-\sqrt[3]{4})$
36. $(\sqrt[3]{4}-\sqrt[3]{6})(\sqrt[3]{2}+\sqrt[3]{36})$

Divide and simplify. Assume that all variables represent positive numbers.

37. $\sqrt{\frac{49}{25}}$

38. $\sqrt{\frac{16}{36}}$

39. $\sqrt{\frac{12x}{49y^6}}$

40. $\sqrt{\frac{27a^4}{64x^2}}$

41. $\sqrt[3]{\frac{8x^5y^6}{27}}$

42. $\sqrt[3]{\frac{125a^3b^4}{64}}$

43. $\frac{\sqrt[3]{5y^8}}{\sqrt[3]{27x^3}}$

44. $\frac{\sqrt[3]{5y^{10}}}{\sqrt[3]{216x^3}}$

Simplify by rationalizing the denominator.

45. $\frac{3}{\sqrt{2}}$

46. $\frac{5}{\sqrt{7}}$

47. $\sqrt{\frac{4}{3}}$

48. $\sqrt{\frac{25}{2}}$

49. $\frac{1}{\sqrt{5y}}$

50. $\frac{1}{\sqrt{3x}}$

51. $\frac{\sqrt{14a}}{\sqrt{2y}}$

52. $\frac{\sqrt{24x}}{\sqrt{3y}}$

53. $\frac{\sqrt{2}}{\sqrt{6x}}$

54. $\frac{\sqrt{5y}}{\sqrt{10x}}$

55. $\frac{x}{\sqrt{5} - \sqrt{2}}$

56. $\frac{a}{\sqrt{10} + \sqrt{2}}$

57. $\frac{2y}{\sqrt{6} + \sqrt{5}}$

58. $\frac{3x}{\sqrt{10} - \sqrt{2}}$

59. $\frac{\sqrt{y}}{\sqrt{6} + \sqrt{2y}}$

60. $\frac{\sqrt{y}}{\sqrt{3y} + \sqrt{5}}$

61. $\frac{\sqrt{5} + \sqrt{3}}{\sqrt{5} - \sqrt{3}}$

62. $\frac{\sqrt{11} - \sqrt{5}}{\sqrt{11} + \sqrt{5}}$

63. $\frac{\sqrt{3x} - 2\sqrt{y}}{\sqrt{3x} + \sqrt{y}}$

64. $\frac{\sqrt{x} + \sqrt{y}}{\sqrt{x} - 2\sqrt{y}}$

Mixed Practice *Simplify each of the following.*

65. $2\sqrt{32} - \sqrt{72} + 3\sqrt{18}$

66. $\sqrt{45} - 2\sqrt{125} - 3\sqrt{20}$

67. $(3\sqrt{2} - 5\sqrt{3})(\sqrt{2} + 2\sqrt{3})$

68. $(5\sqrt{6} - 3\sqrt{2})(\sqrt{6} + 2\sqrt{2})$

69. $\dfrac{9}{\sqrt{8x}}$

70. $\dfrac{1}{\sqrt{18x}}$

71. $\dfrac{\sqrt{5} + 1}{\sqrt{5} + 2}$

72. $\dfrac{\sqrt{3} + 1}{2\sqrt{3} - 1}$

To Think About

73. A student rationalized the denominator of $\dfrac{\sqrt{6}}{2\sqrt{3} - \sqrt{2}}$ and obtained $\dfrac{\sqrt{3} + 3\sqrt{2}}{5}$. Find a decimal approximation of each expression. Are the decimals equal? Did the student do the work correctly?

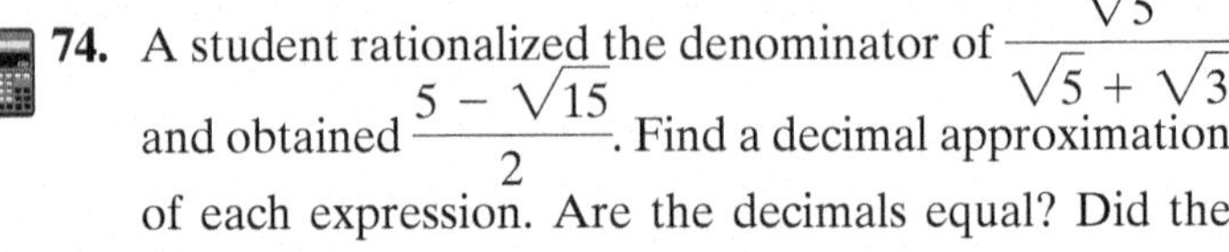

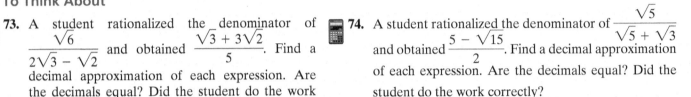

74. A student rationalized the denominator of $\dfrac{\sqrt{5}}{\sqrt{5} + \sqrt{3}}$ and obtained $\dfrac{5 - \sqrt{15}}{2}$. Find a decimal approximation of each expression. Are the decimals equal? Did the student do the work correctly?

In calculus, students are sometimes required to rationalize the numerator of an expression. In this case the numerator will not have a radical in the answer. Rationalize the numerator in each of the following:

75. $\dfrac{\sqrt{2} + 3\sqrt{5}}{4}$

76. $\dfrac{\sqrt{6} - 3\sqrt{3}}{7}$

Applications

Fertilizer Costs *The cost of fertilizing a lawn is $1.25 per square yard. Find the cost to fertilize each of the triangular lawns in exercises 77 and 78. Round your answers to the nearest cent.*

▲ **77.** The base of the triangle is $(12 + \sqrt{5})$ yards, and the altitude is $\sqrt{80}$ yards.

▲ **78.** The base of the triangle is $(14 + \sqrt{6})$ yards, and the altitude is $\sqrt{96}$ yards.

▲ **79.** ***Pacemaker Control Panel*** A medical doctor has designed a pacemaker that has a rectangular control panel. This rectangle has a width of $(\sqrt{x} + 3)$ millimeters and a length of $(\sqrt{x} + 5)$ millimeters. Find the area in square millimeters of this rectangle.

▲ **80.** ***FBI Listening Device*** An FBI agent has designed a secret listening device that has a rectangular base. The rectangle has a width of $(\sqrt{x} + 7)$ centimeters and a length of $(\sqrt{x} + 11)$ centimeters. Find the area in square centimeters of this rectangle.

Cumulative Review

81. **[4.1.6]** Solve the system.

$$\begin{aligned} -2x + 3y &= 21 \\ 3x + 2y &= 1 \end{aligned}$$

82. **[4.2.2]** Solve the system.

$$\begin{aligned} 2x + 3y - z &= 8 \\ -x + 2y + 3z &= -14 \\ 3x - y - z &= 10 \end{aligned}$$

Income of U.S. Households *The bar graph below shows the percentages of U.S. households that were in varying income ranges in 2013. In that year, there were about 123,000,000 households in the United States. Use the graph to answer exercises 83–85.*

83. [1.1.3] What percent of U.S. households earned \$49,999 or less?

84. [0.5.3] How many households earned \$150,000 or more?

85. [0.5.3] How many households earned \$49,999 or less?

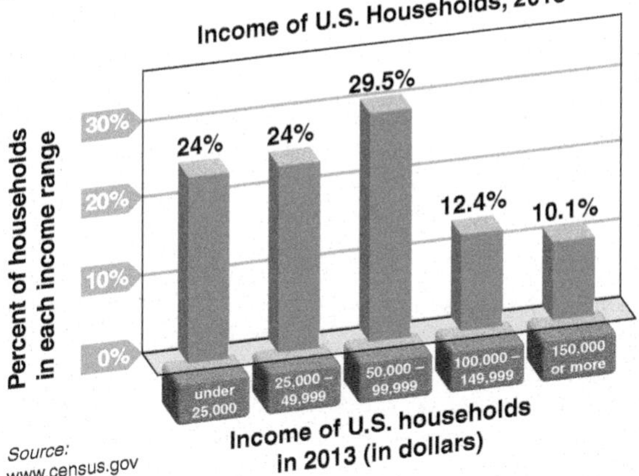

Quick Quiz 8.4

1. Multiply and simplify. $(2\sqrt{3} - \sqrt{5})(3\sqrt{3} + 2\sqrt{5})$

Rationalize the denominator.

2. $\dfrac{9}{\sqrt{3x}}$

3. $\dfrac{1 + 2\sqrt{5}}{4 - \sqrt{5}}$

4. Concept Check Explain how you would rationalize the denominator of the following.

$$\frac{5\sqrt{3} - 3\sqrt{2}}{3\sqrt{2} - 2\sqrt{3}}$$

Use Math to Save Money

Did You Know? Balancing your checkbook can help you keep track of what you are spending.

Balancing Your Finances

Understanding the Problem:

One of the first steps in saving money is to determine your current spending trends. The first step in that process is learning to balance your finances.

Terry balanced his checkbook once a month when he received his bank statement. Below is a table that records the deposits Terry made for the month of May. The beginning balance for May was $300.50.

Date	Deposit
May 1	$200.00
May 3	$150.50
May 10	$120.25
May 25	$50.00
May 28	$25.00

Keeping a Record of Checks:

Terry needs to know if he is depositing enough money to cover his monthly expenses. Below is a table that records each check Terry wrote for the month of May.

Date	Check Number	Checks
May 2	102	$238.50
May 6	103	$75.00
May 12	104	$200.00
May 28	105	$28.56
May 30	106	$36.00

Finding the Facts:

Step 1: Terry needs to know how much he deposits into the bank every month.

Task 1: Determine how much Terry deposited in the bank in May.

Step 2: Terry needs to know how much he spends each month.

Task 2: Determine the total amount of the checks Terry wrote in May.

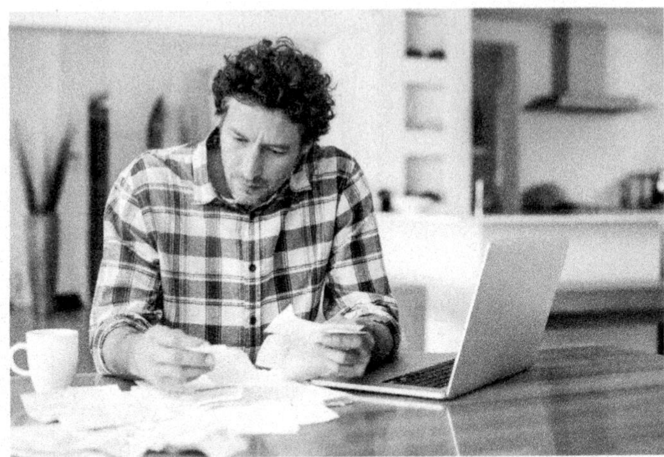

Task 3: Based on the given information, will Terry be able to cover all his expenses for the month of May?

Making a Decision

Step 3: Terry needs to know if he can continue to spend money at the same rate or if he needs to cut back on his spending.

Task 4: Assuming all the checks cleared for May, what would Terry's balance be at the beginning of June?

Task 5: If Terry continues these spending habits, what will happen?

Applying the Situation to Your Life:

Knowing your monthly income and spending habits can help you to save. Balance your checkbook and monitor your spending habits each month, and try to cut out unnecessary expenses. If you are charged ATM fees for making withdrawals from your checking account with an ATM card be sure to subtract those costs from your checkbook balance. If you pay monthly fees to the bank for the cost of your checking account be sure to subtract those costs from your checkbook balance.

Task 6: How often do you balance your checkbook?

Task 7: Do you have any unnecessary expenses that can be cut out?

8.5 Radical Equations

1 Solving a Radical Equation by Squaring Each Side Once

Student Learning Objectives

After studying this section, you will be able to:

1 Solve a radical equation that requires squaring each side once.

2 Solve a radical equation that requires squaring each side twice.

A **radical equation** is an equation with a variable in one or more of the radicals. $3\sqrt{x} = 8$ and $\sqrt{3x - 1} = 5$ are radical equations. We solve radical equations by raising each side of the equation to the appropriate power. In other words, we square both sides if the radicals are square roots, cube both sides if the radicals are cube roots, and so on. Once we have done this, solving for the unknown becomes routine.

Sometimes after we square each side, we obtain a quadratic equation. In this case we collect all terms on one side and use the zero factor method that we developed in Section 6.7. After solving the equation, *always* check your answers to see whether extraneous solutions have been introduced.

We will now generalize this rule because it is very useful in higher-level mathematics courses.

RAISING EACH SIDE OF AN EQUATION TO A POWER

If $y = x$, then $y^n = x^n$ for all natural numbers n.

Example 1 Solve. $\sqrt{2x + 9} = x + 3$

Solution

$(\sqrt{2x + 9})^2 = (x + 3)^2$ Square each side.

$2x + 9 = x^2 + 6x + 9$ Simplify.

$0 = x^2 + 4x$ Collect all terms on one side.

$0 = x(x + 4)$ Factor.

$x = 0$ or $x + 4 = 0$ Set each factor equal to zero.

$x = 0$ $\quad x = -4$ Solve for x.

Check.

For $x = 0$: $\sqrt{2(0) + 9} \stackrel{?}{=} 0 + 3$; $\sqrt{9} \stackrel{?}{=} 3$; $3 = 3$ ✓

For $x = -4$: $\sqrt{2(-4) + 9} \stackrel{?}{=} -4 + 3$; $\sqrt{1} \stackrel{?}{=} -1$; $1 \neq -1$

Therefore, 0 is the only solution to this equation. □

Graphing Calculator

Solving Radical Equations

On a graphing calculator Example 1 can be solved in two ways. Let $y_1 = \sqrt{2x + 9}$ and let $y_2 = x + 3$. Use your graphing calculator to determine where y_1 intersects y_2. What value of x do you obtain? Now let $y = \sqrt{2x + 9} - x - 3$ and find the value of x when $y = 0$. What value of x do you obtain? Which method seems more efficient?

Use the method above that you found most efficient to solve the following equations, and round your answers to the nearest tenth.

$\sqrt{x + 9.5} = x - 2.3$

$\sqrt{6x + 1.3} = 2x - 1.5$

Student Practice 1 Solve and check your solution(s). $\sqrt{3x - 8} = x - 2$

As you begin to solve more complicated radical equations, it is important to make sure that one radical expression is alone on one side of the equation. This is often referred to as **isolating the radical term.**

Example 2 Solve. $\sqrt{10x + 5} - 1 = 2x$

Solution

$\sqrt{10x + 5} = 2x + 1$ Isolate the radical term.

$(\sqrt{10x + 5})^2 = (2x + 1)^2$ Square each side.

$10x + 5 = 4x^2 + 4x + 1$ Simplify.

$0 = 4x^2 - 6x - 4$ Collect all terms on one side.

$0 = 2(2x^2 - 3x - 2)$ Factor out the common factor.

$0 = 2(2x + 1)(x - 2)$ Factor completely.

$2x + 1 = 0$ or $x - 2 = 0$ Set each factor equal to zero.

$2x = -1$ $\quad x = 2$ Solve for x.

$x = -\frac{1}{2}$

Continued on next page

Student Practice 2 Solve and check your solution(s). $-4 + \sqrt{x + 4} = x$

Check.

$x = -\frac{1}{2}$: $\sqrt{10\left(-\frac{1}{2}\right) + 5} - 1 \stackrel{?}{=} 2\left(-\frac{1}{2}\right)$

$$\sqrt{-5 + 5} - 1 \stackrel{?}{=} -1$$
$$\sqrt{0} - 1 \stackrel{?}{=} -1$$
$$-1 = -1 \checkmark$$

$x = 2$: $\sqrt{10(2) + 5} - 1 \stackrel{?}{=} 2(2)$

$$\sqrt{25} - 1 \stackrel{?}{=} 4$$
$$5 - 1 \stackrel{?}{=} 4$$
$$4 = 4 \checkmark$$

Both answers check, so $-\frac{1}{2}$ and 2 are roots of the equation. □

2 Solving a Radical Equation by Squaring Each Side Twice

In some exercises, we must square each side twice in order to remove all the radicals. It is important to isolate at least one radical before squaring each side.

Example 3 Solve. $\sqrt{5x + 1} - \sqrt{3x} = 1$

Solution

$\sqrt{5x + 1} = 1 + \sqrt{3x}$	Isolate one of the radicals.
$(\sqrt{5x + 1})^2 = (1 + \sqrt{3x})^2$	Square each side.
$5x + 1 = (1 + \sqrt{3x})(1 + \sqrt{3x})$	
$5x + 1 = 1 + 2\sqrt{3x} + 3x$	
$2x = 2\sqrt{3x}$	Isolate the remaining radical.
$x = \sqrt{3x}$	Divide each side by 2.
$(x)^2 = (\sqrt{3x})^2$	Square each side.
$x^2 = 3x$	
$x^2 - 3x = 0$	Collect all terms on one side.
$x(x - 3) = 0$	Factor.
$x = 0$ or $x - 3 = 0$	Solve for x.
$x = 3$	

Check.

$x = 0$: $\sqrt{5(0) + 1} - \sqrt{3(0)} \stackrel{?}{=} 1$

$$\sqrt{1} - \sqrt{0} \stackrel{?}{=} 1$$
$$1 = 1 \checkmark$$

$x = 3$: $\sqrt{5(3) + 1} - \sqrt{3(3)} \stackrel{?}{=} 1$

$$\sqrt{16} - \sqrt{9} \stackrel{?}{=} 1$$
$$1 = 1 \checkmark$$

Both answers check. The solutions are 0 and 3. □

Student Practice 3 Solve and check your solution(s).

$$\sqrt{2x + 5} - 2\sqrt{2x} = 1$$

We will now formalize the procedure for solving radical equations.

PROCEDURE FOR SOLVING RADICAL EQUATIONS

1. Perform algebraic operations to obtain one radical by itself on one side of the equation.
2. If the equation contains square roots, square each side of the equation. Otherwise, raise each side to the appropriate power for third- and higher-order roots.
3. Simplify, if possible.
4. If the equation still contains a radical, repeat steps 1 to 3.
5. Collect all terms on one side of the equation.
6. Solve the resulting equation.
7. Check all apparent solutions. Solutions to radical equations must be verified.

Example 4 Solve. $\sqrt{2y+5} - \sqrt{y-1} = \sqrt{y+2}$

Solution

$$\left(\sqrt{2y+5} - \sqrt{y-1}\right)^2 = \left(\sqrt{y+2}\right)^2$$

$$\left(\sqrt{2y+5} - \sqrt{y-1}\right)\left(\sqrt{2y+5} - \sqrt{y-1}\right) = y + 2$$

$$2y + 5 - 2\sqrt{(y-1)(2y+5)} + y - 1 = y + 2$$

$$-2\sqrt{(y-1)(2y+5)} = -2y - 2$$

$$\sqrt{(y-1)(2y+5)} = y + 1 \quad \text{Divide each side by } -2.$$

$$\left(\sqrt{2y^2 + 3y - 5}\right)^2 = (y+1)^2 \quad \text{Square each side.}$$

$$2y^2 + 3y - 5 = y^2 + 2y + 1$$

$$y^2 + y - 6 = 0 \quad \text{Collect all terms on one side.}$$

$$(y+3)(y-2) = 0$$

$$y = -3 \quad \text{or} \quad y = 2$$

Check. Verify that 2 is a valid solution but −3 is not a valid solution. □

 **Student Practice 4** Solve and check your solution(s).

$$\sqrt{y-1} + \sqrt{y-4} = \sqrt{4y-11}$$

8.5 Exercises MyMathLab®

Verbal and Writing Skills, Exercises 1 and 2

1. Before squaring each side of a radical equation, what step should be taken first?

2. Why do we have to check the solutions when we solve radical equations?

Solve each radical equation. Check your solution(s).

3. $\sqrt{8x + 1} = 5$

4. $\sqrt{5x - 4} = 6$

5. $\sqrt{4x - 3} - 3 = 0$

6. $2 = 7 - \sqrt{2x + 3}$

7. $y + 1 = \sqrt{5y - 1}$

8. $\sqrt{y + 10} = y - 2$

9. $2x = \sqrt{11x + 3}$

10. $4x = \sqrt{6x + 1}$

11. $2 = 5 + \sqrt{2x + 1}$

12. $12 + \sqrt{4x + 5} = 7$

13. $y - \sqrt{y - 3} = 5$

14. $\sqrt{2y - 6} + 3 = y$

15. $y = \sqrt{y + 3} - 3$

16. $y = \sqrt{2y + 9} + 3$

17. $x - 2\sqrt{x - 3} = 3$

18. $2\sqrt{2x + 2} + 1 = x + 4$

19. $\sqrt{3x^2 - x} = x$

20. $\sqrt{5x^2 - 3x} = 2x$

21. $\sqrt[3]{2x + 3} = 2$

22. $\sqrt[3]{3x - 1} = 5$

23. $\sqrt[3]{4x - 1} = 3$

24. $\sqrt[3]{3 - 5x} = 2$

Solve each radical equation. This will usually involve squaring each side twice. Check your solutions.

25. $\sqrt{x + 4} = 1 + \sqrt{x - 3}$

26. $\sqrt{x + 5} + 1 = \sqrt{x + 18}$

27. $\sqrt{9x + 1} = 1 + \sqrt{7x}$

28. $2\sqrt{x + 4} = 1 + \sqrt{2x + 9}$

29. $\sqrt{x + 6} = 1 + \sqrt{x + 2}$

30. $\sqrt{3x + 1} - \sqrt{x - 4} = 3$

31. $\sqrt{3x + 13} = 1 + \sqrt{x + 4}$

32. $\sqrt{8x + 17} = \sqrt{9x + 7} + 1$

33. $\sqrt{2x + 9} - \sqrt{x + 1} = 2$

34. $\sqrt{2x + 6} = \sqrt{7 - 2x} + 1$

35. $\sqrt{4x + 6} = \sqrt{x + 1} - \sqrt{x + 5}$

36. $\sqrt{3x + 4} + \sqrt{x + 5} = \sqrt{7 - 2x}$

37. $2\sqrt{x} - \sqrt{x - 5} = \sqrt{2x - 2}$

38. $\sqrt{3 - 2\sqrt{x}} = \sqrt{x}$

Optional Graphing Calculator Problems

Solve for x. Round your answer to four decimal places.

39. $x = \sqrt{4.28x - 3.15}$

40. $\sqrt[3]{5.62x + 9.93} = 1.47$

Applications

41. ***Police Officer*** Police officers often use a formula that relates a car's speed with the length of the skid marks it leaves after the brakes are applied. When a car traveling on wet pavement at a speed V in miles per hour stops suddenly, it will produce skid marks of length S feet according to the formula $V = 2\sqrt{3S}$.

   (a) Solve the equation for S.

   (b) Use your result from **(a)** to find the length of the skid mark S if the car is traveling at 30 miles per hour

▲42. ***Flight Data Recorder*** The volume V of a steel container inside a flight data recorder is defined by the equation

$$x = \sqrt{\frac{V}{5}},$$

where x is the sum of the length and the width of the container in inches and the height of the container is 5 inches.

   (a) Solve the equation for V.

   (b) Use the result from **(a)** to find the volume of the container whose length and width total 3.5 inches.

Stopping Distance *Recently an experiment was conducted relating the speed a car is traveling and the stopping distance. In this experiment, a car is traveling on dry pavement at a constant rate of speed. From the instant that a driver recognizes the need to stop, the number of feet it takes for him or her to stop the car is recorded. For example, for a driver traveling at 50 miles per hour, it requires a stopping distance of 190 feet. In general, the stopping distance x in feet is related to the speed of the car y in miles per hour by the equation*

$$0.11y + 1.25 = \sqrt{3.7625 + 0.22x}.$$

(Source: www.nhtsa.gov)

43. Solve this equation for x.

44. Use your answer from exercise 43 to find what the stopping distance x would have been for a car traveling at $y = 60$ miles per hour.

To Think About

45. The solution to the equation

$$\sqrt{x^2 - 3x + c} = x - 2$$

is $x = 3$. What is the value of c?

46. The solution to the equation

$$\sqrt{x + a} - \sqrt{x} = -4$$

is $x = 25$. What is the value of a?

Cumulative Review *Simplify.*

47. [8.1.1] $(4^3x^6)^{2/3}$

48. [8.1.1] $(2^{-3}x^{-6})^{1/3}$

49. [8.2.2] $\sqrt[3]{-216x^6y^9}$

50. [8.2.2] $\sqrt[5]{-32x^{15}y^5}$

51. [4.3.1] ***Mississippi Queen*** The Mississippi Queen is a steamboat that travels up and down the Mississippi River at a cruising speed of 12 miles per hour in still water. After traveling 4 hours downstream with the current, it takes 5 hours to get upstream against the current and return to its original starting point. What is the speed of the Mississippi River's current?

52. [4.3.1] ***Veterinarian Costs*** The Concord Veterinary Clinic saw a total of 28 dogs and cats in one day for a standard annual checkup. The checkup cost for a cat is \$55, and the checkup cost for a dog is \$68. The total amount charged that day for checkups was \$1748. How many dogs were examined? How many cats?

Quick Quiz 8.5 *Solve and check your solutions.*

1. $\sqrt{5x - 4} = x$

2. $x = 3 - \sqrt{2x - 3}$

3. $4 - \sqrt{x - 4} = \sqrt{2x - 1}$

4. Concept Check When you try to solve the equation $2 + \sqrt{x + 10} = x$, you obtain the values $x = -1$ and $x = 6$. Explain how you would determine if either of these values is a solution of the radical equation.

8.6 Complex Numbers

1 Simplifying Expressions Involving Complex Numbers

Until now we have not been able to solve an equation such as $x^2 = -4$ because there is no *real* number that satisfies this equation. However, this equation *does* have a nonreal solution. This solution is an *imaginary number.*

We define a new number:

$$i = \sqrt{-1} \text{ or } i^2 = -1$$

Now let us use the product rule

$$\sqrt{-a} = \sqrt{-1}\sqrt{a} \quad \text{and see if it is valid.}$$

If $x^2 = -4$, $x = \sqrt{-4}$. Then $\sqrt{-4} = \sqrt{4(-1)} = \sqrt{4}\sqrt{-1} = \sqrt{4} \cdot i = 2i$.

Thus, one solution to the equation $x^2 = -4$ is $2i$. Let's check it.

$$x^2 = -4$$
$$(2i)^2 \stackrel{?}{=} -4$$
$$4i^2 \stackrel{?}{=} -4$$
$$4(-1) \stackrel{?}{=} -4$$
$$-4 = -4 \checkmark$$

The value $-2i$ is also a solution. You should verify this.

Now we formalize our definitions and give some examples of imaginary numbers.

Student Learning Objectives

After studying this section, you will be able to:

1. Simplify expressions involving complex numbers.
2. Add and subtract complex numbers.
3. Multiply complex numbers.
4. Evaluate complex numbers of the form i^n.
5. Divide two complex numbers.

DEFINITION OF IMAGINARY NUMBERS

The **imaginary number** ***i*** is defined as follows:

$$i = \sqrt{-1} \quad \textit{and} \quad i^2 = -1.$$

The set of imaginary numbers consists of numbers of the form bi, where b is a real number and $b \neq 0$.

DEFINITION

For all positive real numbers a,

$$\sqrt{-a} = \sqrt{-1}\sqrt{a} = i\sqrt{a}.$$

Example 1 Simplify.

(a) $\sqrt{-36}$ **(b)** $\sqrt{-17}$

Solution

(a) $\sqrt{-36} = \sqrt{-1}\sqrt{36} = (i)(6) = 6i$

(b) $\sqrt{-17} = \sqrt{-1}\sqrt{17} = i\sqrt{17}$ □

Student Practice 1 Simplify. **(a)** $\sqrt{-49}$ **(b)** $\sqrt{-31}$

To avoid confusing $\sqrt{17i}$ with $\sqrt{17}i$, we write the i before the radical. That is, we write $i\sqrt{17}$.

Example 2 Simplify. $\sqrt{-45}$

Solution

$$\sqrt{-45} = \sqrt{-1}\sqrt{45} = i\sqrt{45} = i\sqrt{9}\sqrt{5} = 3i\sqrt{5}$$ □

Student Practice 2 Simplify. $\sqrt{-98}$

The rule $\sqrt{a}\sqrt{b} = \sqrt{ab}$ requires that $a \geq 0$ and $b \geq 0$. Therefore, we cannot use our product rule when the radicands are negative unless we first use the definition of $\sqrt{-1}$. Notice that

$$\sqrt{-1} \cdot \sqrt{-1} = i \cdot i = i^2 = -1.$$

Example 3 Multiply. $\sqrt{-16} \cdot \sqrt{-25}$

Solution

First we must use the definition $\sqrt{-1} = i$. Thus, we have the following:

$$\begin{aligned}(\sqrt{-16})(\sqrt{-25}) &= (i\sqrt{16})(i\sqrt{25}) \\ &= i^2(4)(5) \\ &= -1(20) \qquad i^2 = -1 \\ &= -20\end{aligned}$$ □

 **Student Practice 3** Multiply. $\sqrt{-8} \cdot \sqrt{-2}$

Now we formally define a complex number.

DEFINITION OF COMPLEX NUMBER

A number that can be written in the form $a + bi$, where a and b are real numbers, is a **complex number.** We say that a is the **real part** and bi is the **imaginary part.**

Under this definition, every real number is also a complex number. For example, the real number 5 can be written as $5 + 0i$. Therefore, 5 is a complex number. In a similar fashion, the imaginary number $2i$ can be written as $0 + 2i$. So $2i$ is a complex number. Thus, the set of complex numbers includes the set of real numbers and the set of imaginary numbers.

DEFINITION

Two complex numbers $a + bi$ and $c + di$ are equal if and only if $a = c$ and $b = d$.

This definition means that two complex numbers are equal if and only if their real parts are equal *and* their imaginary parts are equal.

Graphing Calculator

Complex Numbers

Some graphing calculators, such as the TI-84, have a complex number mode. If your graphing calculator has this capability, you will be able to use it to do complex number operations. First you must use the Mode command to transfer selection from "Real" to "Complex" or "$a + bi$." To verify your status, try to find $\sqrt{-7}$ on your graphing calculator. If you obtain an approximate answer of "2.645751311 i," then your calculator is operating in the complex number mode. If you obtain "ERROR: NONREAL ANSWER," then your calculator is not operating in the complex number mode.

Example 4 Find the real numbers x and y if $x + 3i\sqrt{7} = -2 + yi$.

Solution By our definition, the real parts must be equal, so x must be -2; the imaginary parts must also be equal, so y must be $3\sqrt{7}$. □

Student Practice 4 Find the real numbers x and y if

$$-7 + 2yi\sqrt{3} = x + 6i\sqrt{3}.$$

2 Adding and Subtracting Complex Numbers 

ADDING AND SUBTRACTING COMPLEX NUMBERS

For all real numbers a, b, c, and d,

$$(a + bi) + (c + di) = (a + c) + (b + d)i \quad \text{and}$$
$$(a + bi) - (c + di) = (a - c) + (b - d)i.$$

In other words, to combine complex numbers we add (or subtract) the real parts, and we add (or subtract) the imaginary parts.

Example 5 Subtract. $(6 - 2i) - (3 - 5i)$

Solution

$$(6 - 2i) - (3 - 5i) = (6 - 3) + [-2 - (-5)]i = 3 + (-2 + 5)i = 3 + 3i$$ □

Student Practice 5 Subtract. $(3 - 4i) - (-2 - 18i)$

3 Multiplying Complex Numbers

As we might expect, the procedure for multiplying complex numbers is similar to the procedure for multiplying polynomials. We will see that the complex numbers obey the associative, commutative, and distributive properties.

Example 6 Multiply. $(7 - 6i)(2 + 3i)$

Solution Use FOIL.

$$\begin{aligned}(7 - 6i)(2 + 3i) &= (7)(2) + (7)(3i) + (-6i)(2) + (-6i)(3i)\\ &= 14 + 21i - 12i - 18i^2\\ &= 14 + 21i - 12i - 18(-1)\\ &= 14 + 21i - 12i + 18\\ &= 32 + 9i\end{aligned}$$ □

 **Student Practice 6** Multiply. $(4 - 2i)(3 - 7i)$

Example 7 Multiply. $3i(4 - 5i)$

Solution Use the distributive property.

$$\begin{aligned} 3i(4 - 5i) &= (3)(4)i + (3)(-5)i^2 \\ &= 12i - 15i^2 \\ &= 12i - 15(-1) \\ &= 15 + 12i \end{aligned}$$

Student Practice 7 Multiply. $-2i(5 + 6i)$

4 Evaluating Complex Numbers of the Form in i^n

How would you evaluate i^n, where n is any positive integer? We look for a pattern. We have defined

$$i^2 = -1.$$

We could write

$$i^3 = i^2 \cdot i = (-1)i = -i.$$

We also have the following:

$$\begin{aligned} i^4 &= i^2 \cdot i^2 = (-1)(-1) = +1 \\ i^5 &= i^4 \cdot i = (+1)i = +i \end{aligned}$$

We notice that $i^5 = i$. Let's look at i^6.

$$i^6 = i^4 \cdot i^2 = (+1)(-1) = -1$$

We begin to see a pattern that starts with i and repeats itself for i^5. Will $i^7 = -i$? Why or why not?

VALUES OF i^n

$i = i$	$i^5 = i$	$i^9 = i$
$i^2 = -1$	$i^6 = -1$	$i^{10} = -1$
$i^3 = -i$	$i^7 = -i$	$i^{11} = -i$
$i^4 = +1$	$i^8 = +1$	$i^{12} = +1$

We can use this pattern to evaluate powers of i.

Example 8 Evaluate.

(a) i^{36} **(b)** i^{27}

Solution

(a) $i^{36} = (i^4)^9 = (1)^9 = 1$

(b) $i^{27} = (i^{24+3}) = (i^{24})(i^3) = (i^4)^6(i^3) = (1)^6(-i) = -i$

This suggests a quick method for evaluating powers of i. Divide the exponent by 4. i^4 raised to any power will be 1. Then use the first column of the values of i^n chart above to evaluate the remainder.

Student Practice 8 Evaluate.

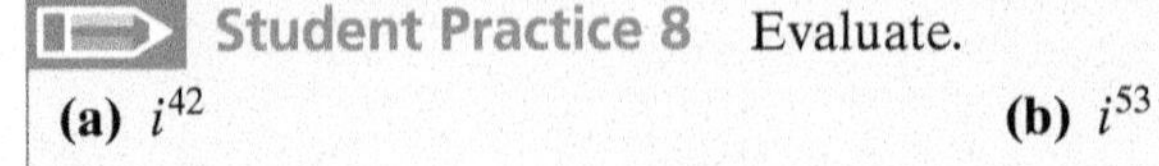

(a) i^{42} **(b)** i^{53}

5 Dividing Two Complex Numbers

The complex numbers $a + bi$ and $a - bi$ are called **conjugates.** The product of two complex conjugates is always a real number.

$$\begin{aligned}(a + bi)(a - bi) &= a^2 - abi + abi - b^2i^2 \\ &= a^2 - b^2(-1) \\ &= a^2 + b^2\end{aligned}$$

When dividing two complex numbers, we want to remove any expression involving i from the denominator. So we multiply the numerator and denominator by the conjugate of the denominator. This is just what we did when we rationalized the denominators of radical expressions.

Example 9 Divide. $\dfrac{7 + i}{3 - 2i}$

Solution

$$\begin{aligned}\frac{7 + i}{3 - 2i} \cdot \frac{3 + 2i}{3 + 2i} = \frac{21 + 14i + 3i + 2i^2}{9 - 4i^2} &= \frac{21 + 17i + 2(-1)}{9 - 4(-1)} \\ &= \frac{21 + 17i - 2}{9 + 4} \\ &= \frac{19 + 17i}{13} \quad \text{or} \quad \frac{19}{13} + \frac{17}{13}i\end{aligned}$$

Student Practice 9 Divide.

$$\frac{4 + 2i}{3 + 4i}$$

Graphing Calculator

Complex Operations

When we perform complex number operations on a graphing calculator, the answer will usually be displayed as an approximate value in decimal form. Try Example 9 on your graphing calculator by entering $(7 + i) \div (3 - 2i)$. You should obtain an approximate answer of $1.461538462 + 1.307692308i$.

Example 10 Divide. $\dfrac{3 - 2i}{4i}$

Solution

The conjugate of $0 + 4i$ is $0 - 4i$ or simply $-4i$.

$$\begin{aligned}\frac{3 - 2i}{4i} \cdot \frac{-4i}{-4i} = \frac{-12i + 8i^2}{-16i^2} &= \frac{-12i + 8(-1)}{-16(-1)} \\ &= \frac{-8 - 12i}{16} = \frac{\cancel{4}(-2 - 3i)}{\cancel{4} \cdot 4} \\ &= \frac{-2 - 3i}{4} \quad \text{or} \quad -\frac{1}{2} - \frac{3}{4}i\end{aligned}$$

Student Practice 10 Divide.

$$\frac{5 - 6i}{-2i}$$

8.6 Exercises MyMathLab®

Verbal and Writing Skills, Exercises 1–4

1. Does $x^2 = -9$ have a real number solution? Why or why not?

2. Describe a complex number and give an example.

3. Are the complex numbers $2 + 3i$ and $3 + 2i$ equal? Why or why not?

4. Describe in your own words how to add or subtract complex numbers.

Simplify. Express in terms of i.

5. $\sqrt{-25}$

6. $\sqrt{-100}$

7. $\sqrt{-28}$

8. $\sqrt{-80}$

9. $\sqrt{-\frac{25}{4}}$

10. $\sqrt{-\frac{49}{64}}$

11. $-\sqrt{-81}$

12. $-\sqrt{-36}$

13. $2 + \sqrt{-3}$

14. $5 + \sqrt{-7}$

15. $-2.8 + \sqrt{-16}$

16. $8.1 + \sqrt{-144}$

17. $-3 + \sqrt{-24}$

18. $-2 - \sqrt{-40}$

19. $(\sqrt{-5})(\sqrt{-2})$

20. $(\sqrt{-7})(\sqrt{-3})$

21. $(\sqrt{-36})(\sqrt{-4})$

22. $(\sqrt{-25})(\sqrt{-9})$

Find the real numbers x and y.

23. $x - 3i = 5 + yi$

24. $x - 9i = 8 + yi$

25. $1.3 - 2.5yi = x - 5i$

26. $3.4 - 0.8i = 2x - yi$

27. $23 + yi = 17 - x + 3i$

28. $2 + x - 11i = 19 + yi$

Perform the addition or subtraction.

29. $(1 + 8i) + (-6 + 3i)$

30. $(-10 - 4i) + (7 + i)$

31. $\left(-\frac{3}{2} + \frac{1}{2}i\right) + \left(\frac{5}{2} - \frac{3}{2}i\right)$

32. $\left(\frac{2}{3} - \frac{1}{3}i\right) + \left(\frac{10}{3} + \frac{4}{3}i\right)$

33. $(2.8 - 0.7i) - (1.6 - 2.8i)$

34. $(6.5 + 7.2i) - (2.3 + 4.9i)$

Multiply and simplify your answers. Place in i notation before doing any other operations.

35. $(2i)(7i)$ **36.** $(6i)(3i)$ **37.** $(-7i)(6i)$

38. $(i)(-3i)$ **39.** $(2 + 3i)(2 - i)$ **40.** $(5 - 2i)(1 + 4i)$

41. $9i - 3(-2 + i)$ **42.** $15i - 5(4 + i)$ **43.** $2i(5i - 6)$

44. $4i(7 - 2i)$ **45.** $\left(\frac{1}{2} + i\right)^2$ **46.** $\left(\frac{1}{5} - i\right)^2$

47. $(i\sqrt{3})(i\sqrt{7})$ **48.** $(i\sqrt{2})(i\sqrt{6})$

49. $(3 + \sqrt{-2})(4 + \sqrt{-5})$ **50.** $(2 + \sqrt{-3})(6 + \sqrt{-2})$

Evaluate.

51. i^{17} **52.** i^{21} **53.** i^{24} **54.** i^{16}

55. i^{46} **56.** i^{83} **57.** i^{47} **58.** i^{10}

59. $i^{30} + i^{28}$ **60.** $i^{32} + i^{42}$ **61.** $i^{100} - i^{7}$ **62.** $2i^{80} - 2i^{7}$

Divide.

63. $\frac{2 + i}{3 - i}$ **64.** $\frac{4 + 2i}{2 - i}$ **65.** $\frac{i}{1 + 4i}$

66. $\frac{-3i}{2 + 5i}$ **67.** $\frac{5 - 2i}{6i}$ **68.** $\frac{6 + 9i}{2i}$

69. $\frac{2}{i}$ **70.** $\frac{-5}{i}$ **71.** $\frac{7}{5 - 6i}$

72. $\frac{3}{4 + 2i}$ **73.** $\frac{5 - 2i}{3 + 2i}$ **74.** $\frac{3 + 4i}{6 - i}$

Mixed Practice *Simplify.*

75. $\sqrt{-98}$

76. $\sqrt{-72}$

77. $(8 - 5i) - (-1 + 3i)$

78. $(-4 + 6i) - (3 - 4i)$

79. $(3i - 1)(5i - 3)$

80. $(2i + 7)(7 - i)$

81. $\dfrac{2 - 3i}{2 + i}$

82. $\dfrac{1 + i}{1 - i}$

Applications *The impedance Z in an alternating current circuit (like the one used in your home and in your classroom) is given by the formula* $Z = \dfrac{V}{I}$*, where V is the voltage and I is the current.*

83. Find the value of Z if $V = 3 + 2i$ and $I = 3i$.

84. Find the value of Z if $V = 4 + 2i$ and $I = -3i$.

Cumulative Review

85. **[2.6.2]** ***Factory Production*** A grape juice factory produces juice in three different types of containers. $x + 3$ hours per week are spent on producing juice in glass bottles. $2x - 5$ hours per week are spent on producing juice in cans. $4x + 2$ hours per week are spent on producing juice in plastic bottles. If the factory operates 105 hours per week, how much time is spent producing juice in each type of container?

86. **[2.7.2]** ***Donation of Computers*** Citizens Bank has decided to donate its older personal computers to the Boston Public Schools. Each computer donated is worth \$120 in tax-deductible dollars to the bank. In addition, the computer company supplying the bank with its new computers gives a 7% rebate to any customer donating used computers to schools. If sixty new computers are purchased at a list price of \$1850 each and sixty older computers are donated to the Boston Public Schools, what is the net cost to the bank for this purchase?

Quick Quiz 8.6 *Simplify.*

1. $(6 - 7i)(3 + 2i)$

2. $\dfrac{4 + 3i}{1 - 2i}$

3. i^{33}

4. **Concept Check** Explain how you would simplify the following.

$$(3 + 5i)^2$$

8.7 Variation

Student Learning Objectives

After studying this section, you will be able to:

1. Solve problems requiring the use of direct variation.
2. Solve problems requiring the use of inverse variation.
3. Solve problems requiring the use of joint or combined variation.

1 Solving Problems Using Direct Variation

Many times in daily life we observe how a change in one quantity produces a change in another. If we order one large pepperoni pizza, we pay \$8.95. If we order two large pepperoni pizzas, we pay \$17.90. For three large pepperoni pizzas, it is \$26.85. The change in the number of pizzas we order results in a corresponding change in the price we pay.

Notice that the price we pay for each pizza stays the same. That is, each pizza costs \$8.95. The number of pizzas changes, and the corresponding price of the order changes. From our experience with functions and with equations, we see that the cost of the order is $y = \$8.95x$, where the price y depends on the number of pizzas x. We see that the variable y is a constant multiple of x. The two variables are said to *vary directly*. That is, y varies directly with x. We write a general equation that represents this idea as follows: $y = kx$.

When we solve problems using direct variation, we usually are not given the value of the constant of variation k. This is something that we must find. Usually all we are given is a point of reference. That is, we are given the value of y for a specific value of x. Using this information, we can find k.

Example 1 The time of a pendulum's period varies directly with the square root of its length. If the pendulum is 1 foot long when the time is 0.2 second, find the time if the length is 4 feet.

Solution Let t = the time and L = the length.

We then have the equation

$$t = k\sqrt{L}.$$

We can evaluate k by substituting $L = 1$ and $t = 0.2$ into the equation.

$$t = k\sqrt{L}$$
$$0.2 = k(\sqrt{1})$$
$$0.2 = k \qquad \text{because } \sqrt{1} = 1$$

Now we know the value of k and can write the equation more completely.

$$t = 0.2\sqrt{L}$$

When $L = 4$, we have the following:

$$t = 0.2\sqrt{4}$$
$$t = (0.2)(2)$$
$$t = 0.4 \text{ second}$$

Student Practice 1 The maximum speed of a racing car varies directly with the square root of the horsepower of the engine. If the maximum speed of a car with 256 horsepower is 128 miles per hour, what is the maximum speed of a car with 225 horsepower?

2 Solving Problems Using Inverse Variation 

In some cases when one variable increases, another variable decreases. For example, as you increase your speed while driving to a particular location, the time it takes to arrive decreases. If one variable is a constant multiple of the reciprocal of the other, the two variables are said to *vary inversely*.

Example 2 If y varies inversely with x and $y = 12$ when $x = 5$, find the value of y when $x = 14$.

Solution If y varies inversely with x, we can write the equation $y = \frac{k}{x}$. We can find the value of k by substituting the values $y = 12$ and $x = 5$.

$$12 = \frac{k}{5}$$

$$60 = k$$

We can now write the equation

$$y = \frac{60}{x}.$$

To find the value of y when $x = 14$, we substitute 14 for x in the equation.

$$y = \frac{60}{14}$$

$$y = \frac{30}{7}$$

□

Student Practice 2 If y varies inversely with x and $y = 45$ when $x = 16$, find the value of y when $x = 36$.

Example 3 The amount of light from a light source varies inversely with the square of the distance to the light source. If an object receives 6.25 lumens when the light source is 8 meters away, how much light will the object receive if the light source is 4 meters away?

Solution Let L = the amount of light and d = the distance to the light source.

Since the amount of light varies inversely with the *square of the distance* to the light source, we have

$$L = \frac{k}{d^2}.$$

Substituting the known values of $L = 6.25$ and $d = 8$, we can find the value of k.

$$6.25 = \frac{k}{8^2}$$

$$6.25 = \frac{k}{64}$$

$$400 = k$$

We are now able to write a more specific equation,

$$L = \frac{400}{d^2}.$$

We will use this to find L when $d = 4$ meters.

$$L = \frac{400}{4^2}$$

$$L = \frac{400}{16}$$

$$L = 25 \text{ lumens}$$

Check. Does this answer seem reasonable? Would we expect to have more light if we move closer to the light source? ✓ □

Student Practice 3 The weight that can safely be supported on top of a cylindrical column varies inversely with the square of its height. If a 7.5-ft column can support 2 tons, how much weight can a 3-ft column support?

3 Solving Problems Using Joint or Combined Variation

Sometimes a quantity depends on the variation of two or more variables. This is called joint or **combined variation.**

Example 4 y varies directly with x and z and inversely with d^2. When $x = 7$, $z = 3$, and $d = 4$, the value of y is 20. Find the value of y when $x = 5$, $z = 6$, and $d = 2$.

Solution We can write the equation

$$y = \frac{kxz}{d^2}.$$

To find the value of k, we substitute into the equation $y = 20$, $x = 7$, $z = 3$, and $d = 4$.

$$20 = \frac{k(7)(3)}{4^2}$$

$$20 = \frac{21k}{16}$$

$$320 = 21k$$

$$\frac{320}{21} = k$$

Now we substitute $\frac{320}{21}$ for k into our original equation.

$$y = \frac{\frac{320}{21}xz}{d^2} \quad \text{or} \quad y = \frac{320xz}{21d^2}$$

Continued on next page

We use this equation to find y for the known values of x, z, and d. We want to find y when $x = 5$, $z = 6$, and $d = 2$.

$$y = \frac{320(5)(6)}{21(2)^2} = \frac{9600}{84}$$

$$y = \frac{800}{7}$$

□

Student Practice 4 y varies directly with z and w^2 and inversely with x. $y = 20$ when $z = 3$, $w = 5$, and $x = 4$. Find y when $z = 4$, $w = 6$, and $x = 2$.

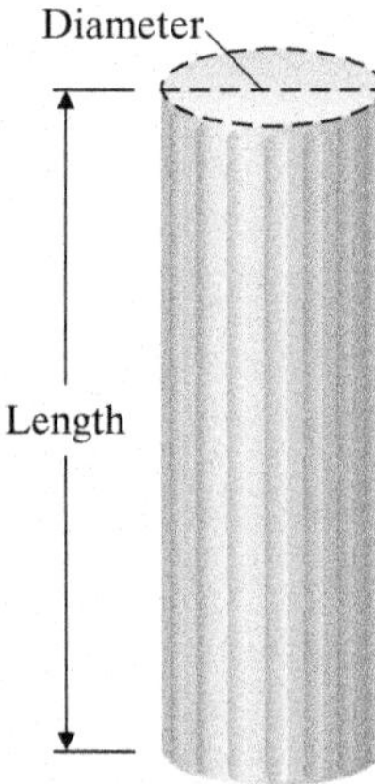

Many applied problems involve joint variation. For example, a cylindrical concrete column has a safe load capacity that varies directly with the diameter raised to the fourth power and inversely with the square of its length.

Therefore, if d = diameter and l = length, the equation would be of the form

$$y = \frac{kd^4}{l^2}.$$

8.7 Exercises MyMathLab®

Verbal and Writing Skills, Exercises 1–4

1. Give an example in everyday life of direct variation and write an equation as a mathematical model.

2. The general equation $y = kx$ means that y varies ________ with x. k is called the ________ of variation.

3. If y varies inversely with x, we write the equation ________.

4. Write a mathematical model for the following situation: The strength of a rectangular beam varies directly with its width and the square of its depth.

Round all answers to the nearest tenth unless otherwise directed.

5. If y varies directly with x and $y = 20$ when $x = 25$, find y when $x = 65$.

6. If y varies directly with x and $y = 30$ when $x = 9$, find y when $x = 90$.

7. ***Pressure on a Submarine*** A marine biology submarine was searching the waters for blue whales at 50 feet below the surface, where it experienced a pressure of 21 pounds per square inch (psi). If the pressure of water on a submerged object varies directly with its distance beneath the surface, how much pressure would the submarine experience if it had to dive to 170 feet?

8. ***Spring Stretching*** The distance a spring stretches varies directly with the weight of the object hung on the spring. If a 20-pound weight stretches a spring 8 inches, how far will a 50-pound weight stretch this spring?

9. ***Stopping Distance*** A car's stopping distance varies directly with the square of its speed. A car that is traveling 30 miles per hour can stop in 40 feet. What distance will it take to stop if it is traveling 60 miles per hour?

10. ***Time of Fall in Gravitation*** When an object is dropped, the distance it falls in feet varies directly with the square of the duration of the fall in seconds. An apple that falls from a tree falls 1 foot in $\frac{1}{4}$ second. How far will it fall in 1 second? How far will it fall in 2 seconds?

11. If y varies inversely with the square of x, and $y = 5$ when $x = 3$, find y when $x = 0.3$.

12. If y varies inversely with the square root of x, and $y = 3.5$ when $x = 0.16$, find y when $x = 0.3$.

13. ***Gasoline Prices*** Last summer the price of gasoline changed frequently. One station owner noticed that the number of gallons he sold each day seemed to vary inversely with the price per gallon. If he sold 2800 gallons when the price was \$3.10, how many gallons could he expect to sell if the price fell to \$2.90? Round your answer to the nearest gallon.

14. ***Weight of an Object*** The weight of an object on Earth's surface varies inversely with the square of its distance from the center of Earth. Suppose a person weighs 230 pounds on Earth's surface. This is approximately 4000 miles from the center of Earth. How much would the person weigh 4500 miles from the center of Earth?

15. ***Beach Cleanup*** Every year on Earth Day, a group of volunteers pick up garbage at Hidden Falls Park. The time it takes to clean the park varies inversely with the number of people picking up garbage. Last year, 39 volunteers took 6 hours to clean the park. If 60 volunteers come to pick up garbage this year, how long will it take to clean the park?

16. ***Emptying a Tank*** The time required for a pump to empty a tank varies inversely with the rate of pumping. If the pump can empty a tank in 2.5 hours at a rate of 400 gallons per minute, how long will it take the same pump to empty the tank at a rate of 500 gallons per minute?

17. ***Construction Worker*** Keith is a construction worker and knows that the weight that can be safely supported by a 2- by 6-inch support beam varies inversely with its length. He finds that a support beam that is 8 feet long will support 900 pounds. Find the weight that can be safely supported by a beam that is 18 feet long.

18. ***Satellite Orbit Speed*** The speed that is required to maintain a satellite in a circular orbit around Earth varies directly with the square root of the distance of the satellite from the center of Earth. We will assume that the radius of Earth is approximately 4000 miles. A satellite that is 100 miles above the surface of Earth is orbiting at approximately 18,000 miles per hour. What speed would be necessary for the satellite to orbit 500 miles above the surface of Earth? Round to the nearest mile per hour.

19. ***Pool Maintenance Worker*** Michelle works for Aqua City Pools as a maintenance worker. She knows that the amount of time it takes to fill a whirlpool tub is inversely proportional to the square of the radius of the pipe used to fill it. She has a pipe of radius 2.5 inches that can fill a tub in 6 minutes. How long will it take the tub to fill if a pipe of radius 3.5 inches is used?

20. ***Wind Generator*** The force on a blade of a wind generator varies jointly with the product of the blade's area and the square of the wind velocity. The force of the wind is 20 pounds when the area is 3 square feet and the velocity is 30 feet per second. Find the force when the area is increased to 5 square feet and the velocity is reduced to 25 feet per second.

Cumulative Review *Solve each of the following equations or word problems.*

21. **[6.7.1]** $3x^2 - 8x + 4 = 0$

22. **[6.7.1]** $4x^2 = -28x + 32$

23. **[2.7.2]** ***Sales Tax*** In 2015, the state with the highest sales tax was Tennessee with 9.45%. Donny shopped in Nashville, Tennessee, and bought an amplifier for his stereo that cost \$503.47 after tax. What was the original price of the amplifier?

24. **[7.6.1]** ***Tennis Courts*** It takes 7.5 gallons of white paint to properly paint lines on three tennis courts. How much paint is needed to paint twenty-two tennis courts?

Quick Quiz 8.7

1. If y varies inversely with x and $y = 9$ when $x = 3$, find the value of y when $x = 6$.

2. Suppose y varies directly with x and inversely with the square of z. $y = 6$ when $x = 3$ and $z = 5$. Find y when $x = 6$ and $z = 10$.

3. The distance a pickup truck requires to stop varies directly with the square of the speed. A new Ford pickup truck traveling on dry pavement can stop in 80 feet when traveling at 50 miles per hour. What distance will the truck take to stop if it is traveling at 65 miles per hour?

4. **Concept Check** If y varies directly with the square root of x and $y = 50$ when $x = 5$, explain how you would find the constant of variation k.

CAREER EXPLORATION PROBLEMS

Background: Surveyor

Clarisse is an associate surveyor employed by Cliffside Surveying & Mapping. In addition to the field work she does, her duties include checking the data and computations her team has compiled to verify its accuracy. After doing this, Clarisse enters the data into a spreadsheet that will be used to compile Cliffside's final set of diagrams and report.

Clarisse has three current assignments. The first is surveying a stretch of shoreline abutting an apartment complex so that extensive repairs to its seawall can be made. A second project involves verifying the property lines of an office park so excavation for a parking lot and new lighting can then begin. Her third project is determining the boundaries between city property and private property around a planned roadway expansion.

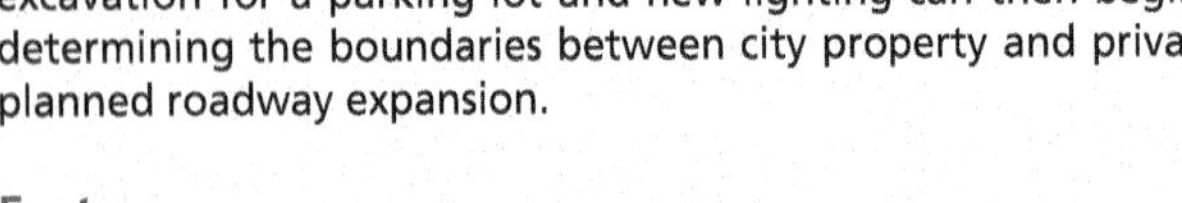

Facts

- The shoreline's measured distance is made up of two sections. The first is the hypotenuse of a right triangle whose legs have been measured to be 170 feet and 38 feet. The second section is the leg of a right triangle whose hypotenuse and second leg are measured as 84 feet and 27 feet, respectively.
- The volume of earth being removed from the office park area is determined by the formula $V = \frac{1}{2}hl(a + b)$, where $hl = 20\sqrt{3}\,\text{yd}^2$, $a = 6\sqrt{50}$ yd, and $b = 7\sqrt{8}$ yd.
- The amount of light the parking lot surface will receive is to be 3900 lumens from an LED source of 50 watts mounted atop light poles 25 feet high.
- The endpoint of a section of roadway measuring 40.5 feet is 13 feet below the starting point of the measured length.

Tasks

1. Clarisse needs to approximate to the nearest foot, what the distance of the shoreline is so materials and equipment can be put in place to begin the seawall repair. What are the lengths of the two sections and the overall length rounded to the nearest foot?

2. Looking over the data for the parking lot excavation project, Clarisse detects an error in the computation for the volume of the earth being excavated. It is given as approximately $3083\,\text{yd}^3$, so she decides to recalculate using the following formula:

   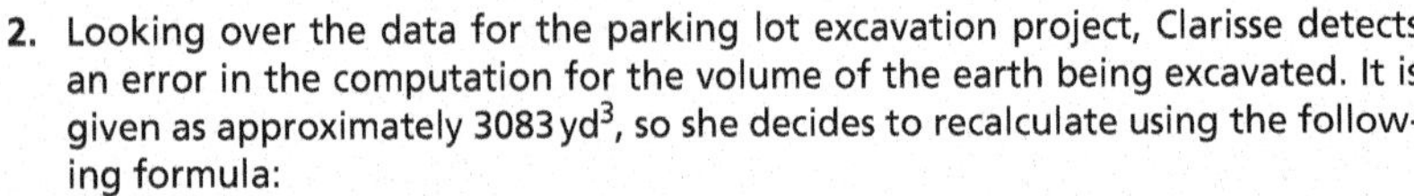

   $V = \frac{1}{2}hl(a + b)$, where $hl = 20\sqrt{3}\,\text{yd}^2$, $a = 6\sqrt{50}$ yd, and $b = 7\sqrt{8}$ yd.

   What is her result, rounded to the nearest cubic yard?

3. Clarisse knows that the intensity of light from its source varies inversely as the square of its distance from the source. The intensity of light is measured in lumens, L, and the distance from the source is expressed as d. The project manager wants her to recompute how much light the parking lot area would receive if the 50-watt LED source was mounted atop light poles 20 feet and 30 feet high.

4. Clarisse sees that the horizontal distance for the stretch of roadway measuring 40.5 feet whose endpoint is 13 feet below the starting point hasn't been calculated. What is this value, rounded to the nearest tenth of a foot?

Chapter 8 Organizer

Topic and Procedure	Examples	You Try It
Raising a variable with an exponent to a power, p. 479 $(x^m)^n = x^{mn}$ $(xy)^n = x^n y^n$ $\left(\frac{x}{y}\right)^n = \frac{x^n}{y^n}, \quad y \neq 0$	Simplify. **(a)** $(x^{-1/2})^{-2/3} = x^{1/3}$ **(b)** $(3x^{-2}y^{-1/2})^{2/3} = 3^{2/3}x^{-4/3}y^{-1/3}$ **(c)** $\left(\frac{4x^{-2}}{3^{-1}y^{-1/2}}\right)^{1/4} = \frac{4^{1/4}x^{-1/2}}{3^{-1/4}y^{-1/8}}$	1. Simplify. **(a)** $(x^{-2/5})^{-1/2}$ **(b)** $(2a^{-3}b^{-1/4})^{1/3}$ **(c)** $\left(\frac{5x^{-3}}{4^{-2}y^2}\right)^{1/4}$
Multiplication of variables with rational exponents, p. 480 $x^m x^n = x^{m+n}$	Multiply. $(3x^{1/5})(-2x^{3/5}) = -6x^{4/5}$	2. Multiply. $(-a^{1/2})(4a^{1/3})$
Division of variables with rational exponents, p. 480 $\frac{x^m}{x^n} = x^{m-n}, \quad x \neq 0$	Divide. $\frac{-16x^{3/20}}{24x^{5/20}} = -\frac{2x^{-1/10}}{3}$	3. Divide. $\frac{12x^{7/12}}{-2x^{1/3}}$
Removing negative exponents, p. 480 $x^{-n} = \frac{1}{x^n}, \quad x \text{ and } y \neq 0$ $\frac{x^{-n}}{y^{-m}} = \frac{y^m}{x^n}$	Write with positive exponents. **(a)** $3x^{-4} = \frac{3}{x^4}$ **(b)** $\frac{2x^{-6}}{5y^{-8}} = \frac{2y^8}{5x^6}$ **(c)** $4^{-2} = \frac{1}{4^2} = \frac{1}{16}$	4. Write with positive exponents. **(a)** $5a^{-3}$ **(b)** $\frac{3a^{-2}}{6a^{-6}}$ **(c)** 2^{-5}
Multiplication of expressions with rational exponents, p. 481 Add exponents whenever expressions with the same base are multiplied.	Multiply. $x^{2/3}(x^{1/3} - x^{1/4}) = x^{3/3} - x^{2/3+1/4} = x - x^{11/12}$	5. Multiply. $x^{1/2}(x^{2/3} - x^{1/2})$
Zero exponent, p. 482 $x^0 = 1 \quad (\text{if } x \neq 0)$	Simplify. $(3x^{1/2})^0 = 1$	6. Simplify. $(-5x^{2/3})^0$
Higher-order roots, p. 487 If x is a nonnegative real number, $\sqrt[n]{x}$ is a nonnegative n th root and has the property that $(\sqrt[n]{x})^n = x.$ If x is a negative real number, and n is an odd integer, then $\sqrt[n]{x}$ has the property that $(\sqrt[n]{x})^n = x.$ If x is a negative real number, and n is an even integer, $\sqrt[n]{x}$ is not a real number.	Simplify. **(a)** $\sqrt[3]{27} = 3$ because $3^3 = 27$. **(b)** $\sqrt[5]{-32} = -2$ because $(-2)^5 = -32$. **(c)** $\sqrt[4]{-16}$ is *not* a real number.	7. Simplify. **(a)** $\sqrt[4]{16}$ **(b)** $\sqrt[5]{-1}$ **(c)** $\sqrt[6]{-64}$
Rational exponents and radicals, p. 490 For positive integers m and n and any real number x for which $x^{1/n}$ is defined, $x^{m/n} = (\sqrt[n]{x})^m = \sqrt[n]{x^m}.$ If it is also true that $x \neq 0$, then $x^{-m/n} = \frac{1}{x^{m/n}} = \frac{1}{(\sqrt[n]{x})^m} = \frac{1}{\sqrt[n]{x^m}}.$	**(a)** Write as a radical. $x^{3/7} = \sqrt[7]{x^3}$ or $(\sqrt[7]{x})^3$, $3^{1/5} = \sqrt[5]{3}$ **(b)** Write as an expression with a fractional exponent. $\sqrt[3]{w^4} = w^{4/3}$ **(c)** Evaluate. $25^{3/2} = (\sqrt{25})^3 = (5)^3 = 125$	8. **(a)** Write as a radical. $x^{4/5}$ **(b)** Write as an expression with a fractional exponent. $\sqrt[4]{v^9}$ **(c)** Evaluate. $27^{4/3}$
Higher-order roots and absolute value, p. 491 $\sqrt[n]{x^n} = \lvert x \rvert$ when n is an even positive integer. $\sqrt[n]{x^n} = x$ when n is an odd positive integer.	Simplify. Assume x can be any real number. **(a)** $\sqrt[6]{x^6} = \lvert x \rvert$ **(b)** $\sqrt[5]{x^5} = x$	9. Simplify. Assume x can be any real number. **(a)** $\sqrt[4]{x^4}$ **(b)** $\sqrt[7]{y^7}$
Evaluation of higher-order roots, p. 491 Use exponent notation.	Simplify. $\sqrt[5]{-32x^{15}} = \sqrt[5]{(-2)^5 x^{15}}$ $= [(-2)^5 x^{15}]^{1/5} = (-2)^1 x^3 = -2x^3$	10. Simplify. $\sqrt[3]{-64m^{18}}$

Topic and Procedure	Examples	You Try It
Simplification of radicals with the product rule, p. 494 For nonnegative real numbers a and b and positive integers n, $\sqrt[n]{a}\,\sqrt[n]{b} = \sqrt[n]{ab}$.	Simplify when $x \geq 0, y \geq 0$. **(a)** $\sqrt{75x^3} = \sqrt{25x^2}\sqrt{3x}$ $= 5x\sqrt{3x}$ **(b)** $\sqrt[3]{16x^5y^6} = \sqrt[3]{8x^3y^6}\,\sqrt[3]{2x^2}$ $= 2xy^2\,\sqrt[3]{2x^2}$	11. Simplify when $x \geq 0, y \geq 0$. **(a)** $\sqrt{24x^5}$ **(b)** $\sqrt[3]{54r^4s^9}$
Combining radicals, p. 495 Simplify radicals and combine them if they have the same index and the same radicand.	Combine. $2\sqrt{50} - 3\sqrt{98} = 2\sqrt{25}\sqrt{2} - 3\sqrt{49}\sqrt{2}$ $= 2(5)\sqrt{2} - 3(7)\sqrt{2}$ $= 10\sqrt{2} - 21\sqrt{2} = -11\sqrt{2}$	12. Combine. $3\sqrt{72} + 4\sqrt{18}$
Multiplying radicals, p. 500 1. Multiply coefficients outside the radical and then multiply the radicands. 2. Simplify your answer.	Multiply. **(a)** $(2\sqrt{3})(4\sqrt{5}) = 8\sqrt{15}$ **(b)** $2\sqrt{6}(\sqrt{2} - 3\sqrt{12}) = 2\sqrt{12} - 6\sqrt{72}$ $= 2\sqrt{4}\sqrt{3} - 6\sqrt{36}\sqrt{2}$ $= 4\sqrt{3} - 36\sqrt{2}$ **(c)** $(\sqrt{2} + \sqrt{3})(2\sqrt{2} - \sqrt{3})$ Use the FOIL method. $= 2\sqrt{4} - \sqrt{6} + 2\sqrt{6} - \sqrt{9}$ $= 4 + \sqrt{6} - 3$ $= 1 + \sqrt{6}$	13. Multiply. **(a)** $(\sqrt{6})(3\sqrt{5})$ **(b)** $3\sqrt{3}(2\sqrt{6} - \sqrt{15})$ **(c)** $(\sqrt{3} - \sqrt{5})(2\sqrt{3} + \sqrt{5})$
Simplifying quotients of radicals with the quotient rule, p. 501 For nonnegative real numbers a, positive real numbers b, and positive integers n, $\sqrt[n]{\frac{a}{b}} = \frac{\sqrt[n]{a}}{\sqrt[n]{b}}$.	Simplify. $\sqrt[3]{\frac{5}{27}} = \frac{\sqrt[3]{5}}{\sqrt[3]{27}} = \frac{\sqrt[3]{5}}{3}$	14. Simplify. $\sqrt[4]{\frac{3}{16}}$
Rationalizing denominators, p. 502 Multiply numerator and denominator by a value that eliminates the radical in the denominator.	Rationalize the denominator. **(a)** $\frac{2}{\sqrt{7}} = \frac{2}{\sqrt{7}} \cdot \frac{\sqrt{7}}{\sqrt{7}} = \frac{2\sqrt{7}}{7}$ **(b)** $\frac{3}{\sqrt{5} + \sqrt{2}} = \frac{3}{\sqrt{5} + \sqrt{2}} \cdot \frac{\sqrt{5} - \sqrt{2}}{\sqrt{5} - \sqrt{2}}$ $= \frac{3\sqrt{5} - 3\sqrt{2}}{(\sqrt{5})^2 - (\sqrt{2})^2} = \frac{3\sqrt{5} - 3\sqrt{2}}{5 - 2}$ $= \frac{3\sqrt{5} - 3\sqrt{2}}{3} = \sqrt{5} - \sqrt{2}$	15. Rationalize the denominator. **(a)** $\frac{3}{\sqrt{6}}$ **(b)** $\frac{4}{\sqrt{2} - \sqrt{3}}$
Solving radical equations, p. 511 1. Perform algebraic operations to obtain one radical by itself on one side of the equation. 2. If the equation contains square roots, square each side of the equation. Otherwise, raise each side to the appropriate power for third- and higher-order roots. 3. Simplify, if possible. 4. If the equation still contains a radical, repeat steps 1 to 3. 5. Collect all terms on one side of the equation. 6. Solve the resulting equation. 7. Check all apparent solutions. Solutions to radical equations must be verified.	Solve. $x = \sqrt{2x + 9} - 3$ $x + 3 = \sqrt{2x + 9}$ $(x + 3)^2 = (\sqrt{2x + 9})^2$ $x^2 + 6x + 9 = 2x + 9$ $x^2 + 6x - 2x + 9 - 9 = 0$ $x^2 + 4x = 0$ $x(x + 4) = 0$ $x = 0$ or $x = -4$ *Check.* $x = 0$: $0 \stackrel{?}{=} \sqrt{2(0) + 9} - 3$ $0 \stackrel{?}{=} \sqrt{9} - 3$ $0 = 3 - 3$ $x = -4$: $-4 \stackrel{?}{=} \sqrt{2(-4) + 9} - 3$ $-4 \stackrel{?}{=} \sqrt{1} - 3$ $-4 \neq -2$ The only solution is 0.	16. Solve. $5 + \sqrt{3x - 11} = x$

Topic and Procedure	Examples	You Try It
Simplifying imaginary numbers, p. 517 Use $i = \sqrt{-1}$, $i^2 = -1$, and $\sqrt{-a} = \sqrt{-1}\sqrt{a}$ for $a \geq 0$.	Simplify. **(a)** $\sqrt{-16} = \sqrt{-1}\sqrt{16} = 4i$ **(b)** $\sqrt{-18} = \sqrt{-1}\sqrt{18} = i\sqrt{9}\sqrt{2} = 3i\sqrt{2}$	17. Simplify. **(a)** $\sqrt{-100}$ **(b)** $\sqrt{-24}$
Adding and subtracting complex numbers, p. 519 Combine real parts and imaginary parts separately.	Add or subtract. **(a)** $(5 + 6i) + (2 - 4i) = 7 + 2i$ **(b)** $(-8 + 3i) - (4 - 2i) = -8 + 3i - 4 + 2i$ $= -12 + 5i$	18. Add or subtract. **(a)** $(8 + i) + (3 - 7i)$ **(b)** $(4 + 3i) - (5 + i)$
Multiplying complex numbers, p. 519 Use the FOIL method and $i^2 = -1$.	Multiply. $(5 - 6i)(2 - 4i) = 10 - 20i - 12i + 24i^2$ $= 10 - 32i + 24(-1)$ $= 10 - 32i - 24$ $= -14 - 32i$	19. Multiply. $(1 - 3i)(3 + 2i)$
Raising i to a power, p. 520 $i^1 = i$ $i^2 = -1$ $i^3 = -i$ $i^4 = 1$	Evaluate. $i^{27} = i^{24} \cdot i^3$ $= (i^4)^6 \cdot i^3$ $= (1)^6(-i)$ $= -i$	20. Evaluate. i^{36}
Dividing complex numbers, p. 521 Multiply the numerator and denominator by the conjugate of the denominator.	Divide. $\frac{5 + 2i}{4 - i} = \frac{5 + 2i}{4 - i} \cdot \frac{4 + i}{4 + i} = \frac{20 + 5i + 8i + 2i^2}{4^2 - i^2}$ $= \frac{20 + 13i + 2(-1)}{16 - (-1)}$ $= \frac{20 + 13i - 2}{16 + 1}$ $= \frac{18 + 13i}{17}$ or $\frac{18}{17} + \frac{13}{17}i$	21. Divide. $\frac{4 - 5i}{2 + i}$
Direct variation, p. 525 If y varies directly with x, there is a constant of variation k such that $y = kx$. After k is determined, other values of y or x can easily be computed.	y varies directly with x. When $x = 2$, $y = 7$. $y = kx$ $7 = k(2)$ Substitute. $k = \frac{7}{2}$ Solve. $y = \frac{7}{2}x$ What is y when $x = 18$? $y = \frac{7}{2}x = \frac{7}{2} \cdot 18 = 63$	22. If y varies directly with the square of x, and $y = 16$ when $x = 2$, what is y when $x = -3$?
Inverse variation, p. 525 If y varies inversely with x, the constant k is such that $y = \frac{k}{x}$.	y varies inversely with x. When x is 5, y is 12. What is y when x is 30? $y = \frac{k}{x}$ $12 = \frac{k}{5}$ Substitute. $k = 60$ Solve. $y = \frac{60}{x}$ Substitute. When $x = 30$, $y = \frac{60}{30} = 2$.	23. If w varies inversely with z, and $w = 5$ when $z = 0.5$, what is w when $z = 5$?

Chapter 8 Review Problems

In all exercises assume that the variables represent positive real numbers unless otherwise stated. Simplify using only positive exponents in your answers.

1. $(3xy^{1/2})(5x^2y^{-3})$

2. $(16a^6b^5)^{1/2}$

3. $3^{1/2} \cdot 3^{1/6}$

4. $\dfrac{6x^{2/3}y^{1/10}}{12x^{1/6}y^{-1/5}}$

5. $(2x^{-1/5}y^{1/10}z^{4/5})^{-5}$

6. $\left(\dfrac{49a^3b^6}{a^{-7}b^4}\right)^{1/2}$

7. $\left(\dfrac{8a^4}{a^{-2}}\right)^{1/3}$

8. $(4^{5/3})^{6/5}$

9. Combine as one fraction containing only positive exponents. $2x^{1/3} + x^{-2/3}$

10. Factor out the common factor $3x$ from $6x^{3/2} - 9x^{1/2}$.

In exercises 11–36, assume that all variables represent nonnegative real numbers. Evaluate, if possible.

11. $-\sqrt{16}$

12. $\sqrt[5]{-32}$

13. $-\sqrt{\dfrac{1}{25}}$

14. $\sqrt{0.04}$

15. $\sqrt[4]{-256}$

16. $\sqrt[3]{-\dfrac{1}{8}}$

17. $64^{2/3}$

18. $125^{4/3}$

Simplify.

19. $\sqrt{49x^4y^{10}z^2}$

20. $\sqrt[3]{64a^{12}b^{30}}$

21. $\sqrt[3]{-8a^{12}b^{15}c^{21}}$

22. $\sqrt{49x^{22}y^2}$

Replace radicals with rational exponents.

23. $\sqrt[5]{a^2}$

24. $\sqrt{2b}$

25. $\sqrt[3]{5a}$

26. $(\sqrt[5]{xy})^7$

Change to radical form.

27. $m^{1/2}$

28. $y^{3/5}$

29. $(3z)^{2/3}$

30. $(2x)^{3/7}$

Evaluate or simplify.

31. $16^{3/4}$

32. $(-27)^{2/3}$

33. $\left(\dfrac{1}{9}\right)^{1/2}$

34. $(0.49)^{1/2}$

35. $\left(\dfrac{1}{36}\right)^{-1/2}$

36. $(25a^2b^4)^{3/2}$

Combine where possible.

37. $\sqrt{50} + 2\sqrt{32} - \sqrt{8}$

38. $\sqrt{28} - 4\sqrt{7} + 5\sqrt{63}$

39. $2\sqrt{12} - \sqrt{48} + 5\sqrt{75}$

40. $\sqrt{125x^3} + x\sqrt{45x}$

41. $2\sqrt{32x} - 5x\sqrt{2} + \sqrt{18x}$

42. $3\sqrt[3]{16} - 4\sqrt[3]{54}$

Multiply and simplify.

43. $(5\sqrt{12})(3\sqrt{6})$

44. $(-2\sqrt{15})(4x\sqrt{3})$

45. $3\sqrt{x}\,(2\sqrt{8x} - 3\sqrt{48})$

46. $\sqrt{3a}\,(4 - \sqrt{21a})$

47. $2\sqrt{7b}\,(\sqrt{ab} - b\sqrt{3bc})$

48. $(5\sqrt{2} + \sqrt{3})(\sqrt{2} - 2\sqrt{3})$

49. $(2\sqrt{5} - 3\sqrt{6})^2$

50. $(\sqrt[3]{2x} + \sqrt[3]{6})(\sqrt[3]{4x^2} - \sqrt[3]{y})$

51. Let $f(x) = \sqrt{4x + 16}$.
(a) Find $f(12)$.
(b) What is the domain of $f(x)$?

52. Let $f(x) = \sqrt{36 - 3x}$.
(a) Find $f(9)$.
(b) What is the domain of $f(x)$?

Rationalize the denominator and simplify the expression.

53. $\sqrt{\dfrac{6y^2}{x}}$

54. $\dfrac{3}{\sqrt{5y}}$

55. $\dfrac{3\sqrt{7x}}{\sqrt{21x}}$

56. $\dfrac{2}{\sqrt{6} - \sqrt{5}}$

57. $\dfrac{\sqrt{x}}{3\sqrt{x} + \sqrt{y}}$

58. $\dfrac{\sqrt{5}}{\sqrt{7} - 3}$

59. $\dfrac{2\sqrt{3} + \sqrt{6}}{\sqrt{3} + 2\sqrt{6}}$

60. $\dfrac{2xy}{\sqrt[3]{16xy^5}}$

61. Simplify. $\sqrt{-16} + \sqrt{-45}$

62. Find x and y. $2x - 3i + 5 = yi - 2 + \sqrt{6}$

Simplify by performing the operation indicated.

63. $(-12 - 6i) + (3 - 5i)$

64. $(2 - i) - (12 - 3i)$

65. $(5 - 2i)(3 + 3i)$

66. $(6 - 2i)^2$

67. $2i(3 + 4i)$

68. $3 - 4(2 + i)$

69. Evaluate. i^{34}

70. Evaluate. i^{65}

Divide.

71. $\dfrac{7 - 2i}{3 + 4i}$

72. $\dfrac{5 - 2i}{1 - 3i}$

73. $\dfrac{4 - 3i}{5i}$

74. $\dfrac{12}{3 - 5i}$

Solve and check your solution(s).

75. $\sqrt{3x - 2} = 5$

76. $\sqrt[3]{3x - 1} = 2$

77. $\sqrt{2x + 1} = 2x - 5$

78. $1 + \sqrt{3x + 1} = x$

79. $\sqrt{3x + 1} - \sqrt{2x - 1} = 1$

80. $\sqrt{7x + 2} = \sqrt{x + 3} + \sqrt{2x - 1}$

Round all answers to the nearest tenth.

81. If y varies directly with x and $y = 11$ when $x = 4$, find the value of y when $x = 6$.

82. ***Saturated Fat Intake*** The maximum amount of saturated fat a person should consume varies directly with the number of calories consumed. A person who consumes 2000 calories per day should have a maximum of 18 grams of saturated fat per day. For a person who consumes 2500 calories, what is the maximum amount of saturated fat he or she should consume?

83. ***Time of Falling Object*** The time it takes a falling object to drop a given distance varies directly with the square root of the distance traveled. A steel ball takes 2 seconds to drop a distance of 64 feet. How many seconds will it take to drop a distance of 196 feet?

84. If y varies inversely with x and $y = 8$ when $x = 3$, find the value of y when $x = 48$.

85. Suppose that y varies directly with x and inversely with the square of z. $y = 20$ when $x = 10$ and $z = 5$. Find y when $x = 8$ and $z = 2$.

▲ 86. ***Capacity of a Cylinder*** The capacity of a cylinder varies directly with the height and the square of the radius. A cylinder with a radius of 3 centimeters and a height of 5 centimeters has a capacity of approximately 135 cubic centimeters. What is the capacity of a cylinder with a height of 9 centimeters and a radius of 4 centimeters?

How Am I Doing? Chapter 8 Test

MATH COACH MyMathLab® YouTube

After you take this test read through the Math Coach on pages 541–542. Math Coach videos are available via MyMathLab and YouTube. Step-by-step test solutions in the Chapter Test Prep Videos are also available via MyMathLab and YouTube. (Search "TobeyBegInterAlg" and click on "Channels.")

Simplify.

1. $(2x^{1/2}y^{1/3})(-3x^{1/3}y^{1/6})$

2. $\dfrac{7x^3}{4x^{3/4}}$

3. $(8x^{1/3})^{3/2}$

4. Evaluate. $\left(\dfrac{4}{9}\right)^{3/2}$

5. Evaluate. $\sqrt[5]{-32}$

Evaluate.

6. $8^{-2/3}$

MC **7.** $16^{5/4}$

Simplify. Assume that all variables are nonnegative.

8. $\sqrt{75a^4b^9}$

9. $\sqrt{49a^4b^{10}}$

10. $\sqrt[3]{54m^3n^5}$

Combine where possible.

11. $3\sqrt{24} - \sqrt{18} + \sqrt{50}$

MC **12.** $\sqrt{40x} - \sqrt{27x} + 2\sqrt{12x}$

Multiply and simplify.

13. $(-3\sqrt{2y})(5\sqrt{10xy})$

14. $2\sqrt{3}\,(3\sqrt{6} - 5\sqrt{2})$

15. $(5\sqrt{3} - \sqrt{6})(2\sqrt{3} + 3\sqrt{6})$

1. ______
2. ______
3. ______
4. ______
5. ______
6. ______
7. ______
8. ______
9. ______
10. ______
11. ______
12. ______
13. ______
14. ______
15. ______

16. ______

17. ______

18. ______

19. ______

20. ______

21. ______

22. ______

23. ______

24. ______

25. ______

26. ______

27. ______

28. ______

29. ______

30. ______

Total Correct: ______

Rationalize the denominator.

16. $\dfrac{30}{\sqrt{5x}}$

17. $\sqrt{\dfrac{xy}{3}}$

MC **18.** $\dfrac{1 + 2\sqrt{3}}{3 - \sqrt{3}}$

Solve and check your solution(s).

19. $\sqrt{3x - 2} = x$

MC **20.** $5 + \sqrt{x + 15} = x$

21. $5 - \sqrt{x - 2} = \sqrt{x + 3}$

Simplify by using the properties of complex numbers.

22. $(8 + 2i) - 3(2 - 4i)$

23. $i^{18} + \sqrt{-16}$

24. $(3 - 2i)(4 + 3i)$

25. $\dfrac{2 + 5i}{1 - 3i}$

26. $(6 + 3i)^2$

27. i^{43}

28. If y varies inversely with x and $y = 9$ when $x = 2$, find the value of y when $x = 6$.

29. Suppose y varies directly with x and inversely with the square of z. When $x = 8$ and $z = 4$, then $y = 3$. Find y when $x = 5$ and $z = 6$.

30. A car's stopping distance varies directly with the square of its speed. A car traveling on pavement can stop in 30 feet when traveling at 30 miles per hour. What distance will the car take to stop if it is traveling at 50 miles per hour?

MATH COACH

Mastering the skills you need to do well on the test.

The following problems are from the Chapter 8 Test. Here are some helpful hints to keep you from making common errors on test problems.

Chapter 8 Test, Problem 7 Evaluate. $16^{5/4}$

HELPFUL HINT First see if the numerical base can be written in exponent form. If there are two alternate ways to rewrite this number, use the form that involves the smallest possible number as the base and has the largest exponent.

Did you write 16 as 2^4?

Yes _____ No _____

If you answered No, think of the different ways that 16 can be written in exponent form: $16^1, 4^2$, or 2^4.

Your choice is most likely between 4^2 and 2^4. The problem works out more easily with the smallest possible base, which is 2. See if you can redo this step correctly.

Do you see that if we use the rule $(x^m)^n = x^{mn}$, we obtain the expression $(2^4)^{5/4} = 2^{4/1 \cdot 5/4}$?

Yes _____ No _____

If you answered No, please review the rule about raising a power to a power and see if you can obtain the right result. Now simplify this expression further and evaluate the resulting exponential expression.

If you answered Problem 7 incorrectly, go back and rework the problem using these suggestions.

Chapter 8 Test, Problem 12 Combine where possible. $\sqrt{40x} - \sqrt{27x} + 2\sqrt{12x}$

HELPFUL HINT Remember to simplify each radical expression first. For this problem, remember to keep the variable *x* inside the radical.

Did you simplify $\sqrt{40x}$ to $2\sqrt{10x}$?

Yes _____ No _____

Did you simplify $\sqrt{27x}$ to $3\sqrt{3x}$?

Yes _____ No _____

If you answered No to either question, remember that $\sqrt{40x} = \sqrt{4} \cdot \sqrt{10x}$ and $\sqrt{27x} = \sqrt{9} \cdot \sqrt{3x}$. You can take the square roots of both 4 and 9. Try to do that part of the problem again and then simplify to see if you obtain the same results.

Did you simplify $2\sqrt{12x}$ to $4\sqrt{3x}$?

Yes _____ No _____

If you answered No, remember that $2\sqrt{12x} = 2 \cdot \sqrt{4} \cdot \sqrt{3x}$. Go back and see if you can obtain the correct answer to that step. Remember that in your final step, you can only add radical expressions if they have the same radicand.

Now go back and rework the problem using these suggestions.

Chapter 8 Test, Problem 18 Rationalize the denominator. $\dfrac{1 + 2\sqrt{3}}{3 - \sqrt{3}}$

HELPFUL HINT You will need to multiply both the numerator and the denominator by the conjugate of the denominator. Do this very carefully using the FOIL method.

Did you multiply the numerator and the denominator by $(3 + \sqrt{3})$?

Yes _____ No _____

If you answered No, review the definition of conjugate. Note that the conjugate of $3 - \sqrt{3}$ is $3 + \sqrt{3}$.

Did you multiply the binomials in the denominator to get $9 + 3\sqrt{3} - 3\sqrt{3} - \sqrt{9}$?

Yes _____ No _____

Did you multiply the binomials in the numerator to get $3 + \sqrt{3} + 6\sqrt{3} + 2\sqrt{9}$?

Yes _____ No _____

If you answered No to either question, go back over each step of multiplying the binomial expressions using the FOIL method. Be careful in writing which numbers go outside the radical and which ones go inside the radical. Now simplify each expression.

If you answered Problem 18 incorrectly, go back and rework the problem using these suggestions.

Chapter 8 Test, Problem 20 Solve and check your solution(s). $5 + \sqrt{x + 15} = x$

HELPFUL HINT Always isolate a radical term on one side of the equation before squaring each side.

Did you isolate the radical term first to obtain the equation $\sqrt{x + 15} = x - 5$?

Yes _____ No _____

After squaring each side, did you obtain $x + 15 = x^2 - 10x + 25$?

Yes _____ No _____

If you answered No to either question, review the process for solving a radical equation. Remember that $(\sqrt{x + 15})^2 = x + 15$ and $(x - 5)^2 = x^2 - 10x + 25$. Go back and try to complete these steps again.

Did you collect all like terms on one side to obtain the equation $0 = x^2 - 11x + 10$?

Yes _____ No _____

If you answered No, remember to add $-x - 15$ to both sides of the equation.

Now factor the quadratic equation and set each factor equal to 0 to find the solution(s). Make sure to check your possible solutions and discard any solution that does not check when substituted back into the original equation.

If you answered Problem 20 incorrectly, go back and rework the problem using these suggestions.

Need more help? Look for section examples marked with MG to review.

CHAPTER 9

Quadratic Equations and Inequalities

CAREER OPPORTUNITIES

Sales Manager

Sales managers work collaboratively with sales teams and clients to build relationships that lead to increased sales, positive customer relations, and greater profits for the business. Because purchasing decisions are data driven, the math involved in the sales process is a crucial part of initiating and completing transactions between buyers and sellers. The mathematics in this chapter is applied by sales managers to increase both their staff's effectiveness and their customer's purchasing decisions.

To investigate how the mathematics in this chapter can help with this field, see the Career Exploration Problems on page 610.

9.1 Quadratic Equations

Student Learning Objectives

After studying this section, you will be able to:

1. Solve quadratic equations by the square root property.
2. Solve quadratic equations by completing the square.

1 Solving Quadratic Equations by the Square Root Property

Recall from Section 6.7 that an equation written in the form $ax^2 + bx + c = 0$, where a, b, and c are real numbers and $a \neq 0$, is called a **quadratic equation.** Recall also that we call this the **standard form** of a quadratic equation. We have previously solved quadratic equations using the zero factor property. This has allowed us to factor the left side of an equation such as $x^2 - 7x + 12 = 0$ and obtain $(x - 3)(x - 4) = 0$ and then solve to find that $x = 3$ and $x = 4$. In this chapter we develop new methods of solving quadratic equations. The first method is often called the **square root property.**

> **THE SQUARE ROOT PROPERTY**
>
> If $x^2 = a$, then $x = \pm\sqrt{a}$ for all real numbers a.

The notation $\pm\sqrt{a}$ is a shorthand way of writing "$+\sqrt{a}$ or $-\sqrt{a}$." The symbol $\pm$ is read "plus or minus." We can justify this property by using the zero factor property. If we write $x^2 = a$ in the form $x^2 - a = 0$, we can factor it to obtain $(x + \sqrt{a})(x - \sqrt{a}) = 0$ and thus, $x = -\sqrt{a}$ or $x = +\sqrt{a}$. This can be written more compactly as $x = \pm\sqrt{a}$.

Example 1 Solve and check. $x^2 - 36 = 0$

Solution If we add 36 to each side, we have $x^2 = 36$.

$$x = \pm\sqrt{36}$$
$$x = \pm 6$$

Thus, the roots are 6 and -6.

Check.

$$(6)^2 - 36 \stackrel{?}{=} 0 \qquad (-6)^2 - 36 \stackrel{?}{=} 0$$
$$36 - 36 \stackrel{?}{=} 0 \qquad 36 - 36 \stackrel{?}{=} 0$$
$$0 = 0 \checkmark \qquad 0 = 0 \checkmark$$

 **Student Practice 1** Solve and check. $x^2 - 121 = 0$

Example 2 Solve. $x^2 = 48$

Solution

$$x = \pm\sqrt{48} = \pm\sqrt{16 \cdot 3}$$
$$x = \pm 4\sqrt{3}$$

The roots are $4\sqrt{3}$ and $-4\sqrt{3}$.

 **Student Practice 2** Solve. $x^2 = 18$

Example 3 Solve and check. $3x^2 + 2 = 77$

Solution

$$3x^2 + 2 = 77$$
$$3x^2 = 75$$
$$x^2 = 25$$
$$x = \pm\sqrt{25}$$
$$x = \pm 5$$

The roots are 5 and -5.

Check.

$$3(5)^2 + 2 \stackrel{?}{=} 77 \qquad 3(-5)^2 + 2 \stackrel{?}{=} 77$$
$$3(25) + 2 \stackrel{?}{=} 77 \qquad 3(25) + 2 \stackrel{?}{=} 77$$
$$75 + 2 \stackrel{?}{=} 77 \qquad 75 + 2 \stackrel{?}{=} 77$$
$$77 = 77 \checkmark \qquad 77 = 77 \checkmark$$

□

Student Practice 3 Solve. $5x^2 + 1 = 46$

Sometimes we obtain roots that are complex numbers.

Example 4 Solve and check. $4x^2 = -16$

Solution

$$x^2 = -4$$
$$x = \pm\sqrt{-4}$$
$$x = \pm 2i \qquad \text{Simplify using } \sqrt{-1} = i.$$

The roots are $2i$ and $-2i$.

Check.

$$4(2i)^2 \stackrel{?}{=} -16 \qquad 4(-2i)^2 \stackrel{?}{=} -16$$
$$4(4i^2) \stackrel{?}{=} -16 \qquad 4(4i^2) \stackrel{?}{=} -16$$
$$4(-4) \stackrel{?}{=} -16 \qquad 4(-4) \stackrel{?}{=} -16$$
$$-16 = -16 \checkmark \qquad -16 = -16 \checkmark$$

□

Student Practice 4 Solve and check. $3x^2 = -27$

Example 5 Solve. $(4x - 1)^2 = 5$

Solution

$$4x - 1 = \pm\sqrt{5}$$
$$4x = 1 \pm \sqrt{5}$$
$$x = \frac{1 \pm \sqrt{5}}{4}$$

The roots are $\frac{1 + \sqrt{5}}{4}$ and $\frac{1 - \sqrt{5}}{4}$. □

Student Practice 5 Solve. $(2x + 3)^2 = 7$

2 Solving Quadratic Equations by Completing the Square

Often, a quadratic equation cannot be factored (or it may be difficult to factor). So we use another method of solving the equation, called **completing the square.** When we complete the square, we are changing the polynomial to a perfect-square trinomial. The form of the equation then becomes $(x + d)^2 = e$.

We already know that

$$(x + d)^2 = x^2 + 2dx + d^2.$$

Notice three things about the quadratic expression on the right-hand side of the equation above.

1. The coefficient of the quadratic term (x^2) is 1.
2. The coefficient of the linear (x) term is $2d$.
3. The constant term (d^2) is the square of *half* the coefficient of the linear term.

For example, in the perfect-square trinomial $x^2 + 6x + 9$, the coefficient of the linear term is 6 and the constant term is $\left(\frac{6}{2}\right)^2 = (3)^2 = 9$.

For the perfect-square trinomial $x^2 - 10x + 25$, the coefficient of the linear term is -10 and the constant term is $\left(\frac{-10}{2}\right)^2 = (-5)^2 = 25$.

What number n makes the trinomial $x^2 + 12x + n$ a perfect-square trinomial?

$$n = \left(\frac{12}{2}\right)^2 = 6^2 = 36$$

Hence, the trinomial $x^2 + 12x + 36$ is a perfect-square trinomial and can be written as $(x + 6)^2$.

Now let's solve some equations.

Example 6 Solve by completing the square and check. $x^2 + 6x + 1 = 0$

Solution

Step 1 First we rewrite the equation in the form $ax^2 + bx = c$ by subtracting 1 from each side of the equation. Thus, we obtain

$$x^2 + 6x = -1.$$

Step 2 We verify that the coefficient of the quadratic term (x^2) is 1. If the coefficient of the x^2 term were not 1, you would divide each term of the equation by the coefficient.

Step 3 We want to complete the square on $x^2 + 6x$. That is, we want to add a constant term to $x^2 + 6x$ so that we get a perfect-square trinomial. We do this by taking half the coefficient of x and squaring it.

$$\left(\frac{6}{2}\right)^2 = 3^2 = 9$$

Adding 9 to $x^2 + 6x$ gives the perfect-square trinomial $x^2 + 6x + 9$, which we factor as $(x + 3)^2$. But we cannot add 9 to the left side of our equation unless we also add 9 to the right side. (Why?) We now have

$$x^2 + 6x + 9 = -1 + 9$$

Step 4 Now we factor the left side.

$$(x + 3)^2 = 8$$

Then we use the square root property.

$$x + 3 = \pm\sqrt{8}$$
$$x + 3 = \pm 2\sqrt{2}$$

Step 5 Next we solve for x by subtracting 3 from each side of the equation.

$$x = -3 \pm 2\sqrt{2}$$

The roots are $-3 + 2\sqrt{2}$ and $-3 - 2\sqrt{2}$.

Step 6 We *must* check our solution in the *original* equation (not the perfect-square trinomial we constructed).

$$\begin{aligned} x^2 + 6x + 1 &= 0 \\ (-3 + 2\sqrt{2})^2 + 6(-3 + 2\sqrt{2}) + 1 &\stackrel{?}{=} 0 \\ 9 - 12\sqrt{2} + 8 - 18 + 12\sqrt{2} + 1 &\stackrel{?}{=} 0 \\ 18 - 18 - 12\sqrt{2} + 12\sqrt{2} &\stackrel{?}{=} 0 \\ 0 &= 0 \ \checkmark \end{aligned}$$

$$\begin{aligned} x^2 + 6x + 1 &= 0 \\ (-3 - 2\sqrt{2})^2 + 6(-3 - 2\sqrt{2}) + 1 &\stackrel{?}{=} 0 \\ 9 + 12\sqrt{2} + 8 - 18 - 12\sqrt{2} + 1 &\stackrel{?}{=} 0 \\ 18 - 18 + 12\sqrt{2} - 12\sqrt{2} &\stackrel{?}{=} 0 \\ 0 &= 0 \ \checkmark \end{aligned}$$

□

Student Practice 6 Solve by completing the square. $x^2 + 8x + 3 = 0$

Let us summarize for future reference the six steps we use to solve a quadratic equation by completing the square.

COMPLETING THE SQUARE

1. Put the equation in the form $ax^2 + bx = c$.
2. If $a \neq 1$, divide each term of the equation by a.
3. Square half of the numerical coefficient of the linear term. Add the result to both sides of the equation.
4. Factor the left side; then take the square root of both sides of the equation.
5. Solve the resulting equation for x.
6. Check the solutions in the original equation.

Notice that step 2 is used when the coefficient of x^2 is not 1, as in Example 7.

Example 7 Solve by completing the square. $3x^2 - 8x + 1 = 0$

Solution

$$3x^2 - 8x = -1 \quad \text{Subtract 1 from each side.}$$

$$\frac{3x^2}{3} - \frac{8x}{3} = -\frac{1}{3} \quad \text{Divide each term by 3. (Remember that the coefficient of the quadratic term must be 1.)}$$

$$x^2 - \frac{8}{3}x + \frac{16}{9} = -\frac{1}{3} + \frac{16}{9} \quad \text{Square half of } \frac{8}{3}\text{. Add the result to both sides of the equation.}$$

$$\left(x - \frac{4}{3}\right)^2 = \frac{13}{9}$$

$$x - \frac{4}{3} = \pm\sqrt{\frac{13}{9}}$$

$$x - \frac{4}{3} = \pm\frac{\sqrt{13}}{3}$$

$$x = \frac{4}{3} \pm \frac{\sqrt{13}}{3}$$

$$x = \frac{4 \pm \sqrt{13}}{3}$$

Check. For $x = \dfrac{4 + \sqrt{13}}{3}$,

$$3\left(\frac{4 + \sqrt{13}}{3}\right)^2 - 8\left(\frac{4 + \sqrt{13}}{3}\right) + 1 \stackrel{?}{=} 0$$

$$\frac{16 + 8\sqrt{13} + 13}{3} - \frac{32 + 8\sqrt{13}}{3} + 1 \stackrel{?}{=} 0$$

$$\frac{16 + 8\sqrt{13} + 13 - 32 - 8\sqrt{13}}{3} + 1 \stackrel{?}{=} 0$$

$$\frac{29 - 32}{3} + 1 \stackrel{?}{=} 0$$

$$-\frac{3}{3} + 1 \stackrel{?}{=} 0$$

$$-1 + 1 = 0 \quad ✓$$

See whether you can check the solution $\dfrac{4 - \sqrt{13}}{3}$. □

Student Practice 7 Solve by completing the square. $2x^2 + 4x + 1 = 0$

9.1 Exercises MyMathLab®

Solve the equations by using the square root property. Express any complex numbers using i notation.

1. $x^2 = 100$

2. $x^2 = 49$

3. $4x^2 - 12 = 0$

4. $2x^2 - 20 = 0$

5. $2x^2 - 80 = 0$

6. $5x^2 - 40 = 0$

7. $x^2 = -81$

8. $x^2 = -25$

9. $x^2 + 16 = 0$

10. $x^2 + 121 = 0$

11. $(x - 3)^2 = 12$

12. $(x + 3)^2 = 18$

13. $(x + 9)^2 = 21$

14. $(x - 8)^2 = 23$

15. $(2x + 1)^2 = 7$

16. $(4x + 1)^2 = 6$

17. $(4x - 3)^2 = 36$

18. $(5x - 2)^2 = 25$

19. $(2x + 5)^2 = 49$

20. $(7x + 1)^2 = 64$

21. $2x^2 - 9 = 0$

22. $4x^2 - 7 = 0$

Solve the equations by completing the square. Simplify your answers.

23. $x^2 + 10x + 5 = 0$

24. $x^2 + 6x + 2 = 0$

25. $x^2 - 8x = 17$

26. $x^2 - 12x = 4$

27. $x^2 - 14x = -48$

28. $x^2 - 16x = -48$

29. $\frac{x^2}{2} + \frac{3}{2}x = 4$

30. $\frac{x^2}{3} - \frac{5}{3}x = 1$

31. $2y^2 + 10y = -11$

32. $7x^2 + 4x - 5 = 0$

33. $3x^2 + 10x - 2 = 0$

34. $5x^2 + 4x - 3 = 0$

Mixed Practice, Exercises 35–44 *Solve the equations by any method. Simplify your answers. Express any complex numbers using i notation.*

35. $x^2 + 4x - 6 = 0$

36. $x^2 + 6x + 2 = 0$

37. $\frac{x^2}{2} - x = 4$

38. $\frac{x^2}{3} - 4x = -9$

39. $3x^2 + 1 = x$

40. $2x^2 + 1 = -x$

41. $x^2 + 2 = x$

42. $x^2 + 5 = 4x$

43. $2x^2 + 2 = 3x$

44. $3x^2 + 8x + 3 = -3$

45. Check the solution $x = -1 + \sqrt{6}$ in the equation $x^2 + 2x - 5 = 0$.

46. Check the solution $x = 2 + \sqrt{3}$ in the equation $x^2 - 4x + 1 = 0$.

Applications *Volume of a Box* *The sides of the box shown are labeled with the dimensions in feet.*

▲ **47.** What is the value of x if the volume of the box is 200 cubic feet?

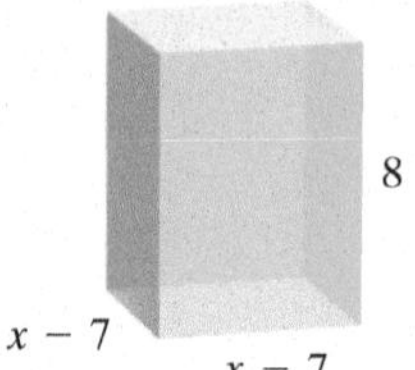

▲ **48.** What is the value of x if the volume of the box is 1800 cubic feet?

Basketball *The time a basketball player spends in the air when shooting a basket is called the "hang time." The vertical leap L measured in feet is related to the hang time t measured in seconds by the equation* $L = 4t^2$.

49. Kadour Ziani is a member of Slam Nation, a group that performs slam dunk shows. Ziani holds the vertical leap world record, with a leap of 5 feet. Find the hang time for this leap. Round to the nearest hundredth.

50. In 2013, DJ Stephens of the University of Memphis had a vertical leap of 3.8 feet. This was the highest vertical leap ever recorded by the NBA.

Cumulative Review *Evaluate the expressions for the given values.*

51. [1.8.1] $\sqrt{b^2 - 4ac}$; $b = 4, a = 3, c = -4$

52. [1.8.1] $\sqrt{b^2 - 4ac}$; $b = -5, a = 2, c = -3$

53. [1.8.1] $5x^2 - 6x + 8$; $x = -2$

54. [1.8.1] $2x^2 + 3x - 5$; $x = -3$

Quick Quiz 9.1

1. Solve by using the square root property.
$(4x - 3)^2 = 12$

2. Solve by completing the square. $x^2 - 8x = 28$

3. Solve by completing the square. $2x^2 + 10x = -11$

4. Concept Check Explain how you would decide what to add to each side of the equation to complete the square for the following equation.

$$x^2 + x = 1$$

9.2 The Quadratic Formula and Solutions to Quadratic Equations

1 Solving a Quadratic Equation by Using the Quadratic Formula

The last method we'll study for solving quadratic equations is the **quadratic formula.** This method works for *any* quadratic equation.

The quadratic formula is developed from completing the square. We begin with the **standard form** of the quadratic equation.

$$ax^2 + bx + c = 0$$

To complete the square, we want the equation to be in the form $x^2 + dx = e$. Thus, we divide by a and subtract the constant term $\frac{c}{a}$ from each side.

$$\frac{ax^2}{a} + \frac{b}{a}x + \frac{c}{a} = 0$$

$$x^2 + \frac{b}{a}x = -\frac{c}{a}$$

Now we complete the square by adding $\left(\frac{b}{2a}\right)^2$ to each side.

$$x^2 + \frac{b}{a}x + \left(\frac{b}{2a}\right)^2 = -\frac{c}{a} + \left(\frac{b}{2a}\right)^2$$

We factor the left side and write the right side as one fraction.

$$\left(x + \frac{b}{2a}\right)^2 = \frac{b^2 - 4ac}{4a^2}$$

Now we use the square root property.

$$x + \frac{b}{2a} = \pm\sqrt{\frac{b^2 - 4ac}{4a^2}}$$

We solve for x and simplify.

$$x = -\frac{b}{2a} \pm \sqrt{\frac{b^2 - 4ac}{4a^2}}$$

$$x = \frac{-b \pm \sqrt{b^2 - 4ac}}{2a}$$

This is the quadratic formula.

QUADRATIC FORMULA

For all equations $ax^2 + bx + c = 0$ where $a \neq 0$

$$x = \frac{-b \pm \sqrt{b^2 - 4ac}}{2a}.$$

Student Learning Objectives

After studying this section, you will be able to:

1. Solve a quadratic equation by using the quadratic formula.
2. Use the discriminant to determine the nature of the roots of a quadratic equation.
3. Write a quadratic equation given the solutions of the equation.

Example 1 Solve by using the quadratic formula. $x^2 + 8x = -3$

Solution The standard form is $x^2 + 8x + 3 = 0$. We substitute $a = 1, b = 8$, and $c = 3$.

$$x = \frac{-b \pm \sqrt{b^2 - 4ac}}{2a}$$

$$x = \frac{-8 \pm \sqrt{8^2 - 4(1)(3)}}{2(1)}$$

$$x = \frac{-8 \pm \sqrt{64 - 12}}{2} = \frac{-8 \pm \sqrt{52}}{2} = \frac{-8 \pm \sqrt{4}\sqrt{13}}{2}$$

$$x = \frac{-8 \pm 2\sqrt{13}}{2} = \frac{\cancel{2}(-4 \pm \sqrt{13})}{\cancel{2}}$$

$$x = -4 \pm \sqrt{13}$$

□

Student Practice 1 Solve by using the quadratic formula.

$$x^2 + 5x = -1 + 2x$$

Example 2 Solve by using the quadratic formula. $3x^2 - x - 2 = 0$

Solution Here $a = 3, b = -1$, and $c = -2$.

$$x = \frac{-b \pm \sqrt{b^2 - 4ac}}{2a}$$

$$x = \frac{-(-1) \pm \sqrt{(-1)^2 - 4(3)(-2)}}{2(3)}$$

$$x = \frac{1 \pm \sqrt{1 + 24}}{6} = \frac{1 \pm \sqrt{25}}{6}$$

$$x = \frac{1 + 5}{6} = \frac{6}{6} \quad \text{or} \quad x = \frac{1 - 5}{6} = -\frac{4}{6}$$

$$x = 1 \qquad\qquad x = -\frac{2}{3}$$

□

Student Practice 2 Solve by using the quadratic formula.

$$2x^2 + 7x + 6 = 0$$

Example 3 Solve by using the quadratic formula. $2x^2 - 48 = 0$

Solution This equation is equivalent to $2x^2 + 0x - 48 = 0$. Therefore, we know that $a = 2, b = 0$, and $c = -48$.

$$x = \frac{-b \pm \sqrt{b^2 - 4ac}}{2a}$$

$$x = \frac{-0 \pm \sqrt{(0)^2 - 4(2)(-48)}}{2(2)}$$

$$x = \frac{\pm\sqrt{384}}{4} \quad \text{This is not simplified.}$$

$$x = \frac{\pm\sqrt{64}\sqrt{6}}{4} = \frac{\pm 8\sqrt{6}}{4}$$

$$x = \pm 2\sqrt{6}$$ □

Student Practice 3 Solve by using the quadratic formula.

$$2x^2 - 26 = 0$$

Example 4 A small company that manufactures canoes makes a daily profit p according to the equation $p = -100x^2 + 3400x - 26{,}196$, where p is measured in dollars and x is the number of canoes made per day. Find the number of canoes that must be made each day to produce a zero profit for the company. Round your answer to the nearest whole number.

Solution Since $p = 0$, we are solving the equation $0 = -100x^2 + 3400x - 26{,}196$.

In this case we have $a = -100$, $b = 3400$, and $c = -26{,}196$. Now we substitute these into the quadratic formula.

$$x = \frac{-b \pm \sqrt{b^2 - 4ac}}{2a}$$

$$x = \frac{-3400 \pm \sqrt{(3400)^2 - 4(-100)(-26{,}196)}}{2(-100)}$$

We will use a calculator to assist us with computation in this problem.

$$x = \frac{-3400 \pm \sqrt{11{,}560{,}000 - 10{,}478{,}400}}{-200}$$

$$x = \frac{-3400 \pm \sqrt{1{,}081{,}600}}{-200}$$

$$x = \frac{-3400 \pm 1040}{-200}$$

We now obtain two answers.

$$x = \frac{-3400 + 1040}{-200} = \frac{-2360}{-200} = 11.8 \approx 12$$

$$x = \frac{-3400 - 1040}{-200} = \frac{-4440}{-200} = 22.2 \approx 22$$

A zero profit is obtained when approximately 12 canoes are produced or when approximately 22 canoes are produced. Actually a slight profit of $204 is made when these numbers of canoes are produced. The discrepancy is due to the rounding error that occurs when we approximate. By methods that we will learn later in this chapter, the maximum profit is produced when 17 canoes are made at the factory. We will investigate exercises of this kind later. □

Student Practice 4 A company that manufactures modems makes a daily profit p according to the equation $p = -100x^2 + 4800x - 52{,}559$, where p is measured in dollars and x is the number of modems made per day. Find the number of modems that must be made each day to produce a zero profit for the company. Round your answer to the nearest whole number.

When an equation contains fractions or rational expressions, eliminate them by multiplying each term by the LCD. Then rewrite the equation in standard form before using the quadratic formula.

Example 5 Solve by using the quadratic formula.

$$\frac{2x}{x+2} = 1 - \frac{3}{x+4}$$

Solution The LCD is $(x + 2)(x + 4)$.

$$\frac{2x}{x+2} = 1 - \frac{3}{x+4}$$

$$\frac{2x}{\cancel{x+2}}\cancel{(x+2)}(x+4) = 1(x+2)(x+4) - \frac{3}{\cancel{x+4}}(x+2)\cancel{(x+4)}$$

$$2x(x+4) = (x+2)(x+4) - 3(x+2)$$

$$2x^2 + 8x = x^2 + 6x + 8 - 3x - 6$$

$$2x^2 + 8x = x^2 + 3x + 2$$

$$x^2 + 5x - 2 = 0$$ Now we have an equation that is quadratic.

Now the equation is in standard form, and we can use the quadratic formula with $a = 1$, $b = 5$, and $c = -2$.

$$x = \frac{-5 \pm \sqrt{5^2 - 4(1)(-2)}}{2(1)} = \frac{-5 \pm \sqrt{25 + 8}}{2}$$

$$x = \frac{-5 \pm \sqrt{33}}{2}$$ □

Student Practice 5 Solve by using the quadratic formula.

$$\frac{1}{x} + \frac{1}{x-1} = \frac{5}{6}$$

Some quadratic equations will have solutions that are not real numbers. You should use i notation to simplify the solutions that are nonreal complex numbers.

Example 6 Solve and simplify your answer. $8x^2 - 4x + 1 = 0$

Solution $a = 8$, $b = -4$, and $c = 1$.

$$x = \frac{-(-4) \pm \sqrt{(-4)^2 - 4(8)(1)}}{2(8)}$$

$$x = \frac{4 \pm \sqrt{16 - 32}}{16} = \frac{4 \pm \sqrt{-16}}{16}$$

$$x = \frac{4 \pm 4i}{16} = \frac{4(1 \pm i)}{16} = \frac{1 \pm i}{4}$$ □

Student Practice 6 Solve by using the quadratic formula.

$$2x^2 - 4x + 5 = 0$$

You may have noticed that complex roots come in pairs. In other words, if $a + bi$ is a solution of a quadratic equation, its conjugate $a - bi$ is also a solution.

2 Using the Discriminant to Determine the Nature of the Roots of a Quadratic Equation

So far we have used the quadratic formula to solve quadratic equations that had two real roots. Sometimes the roots were rational, and sometimes they were irrational. We have also solved equations like Example 6 with nonreal complex numbers as solutions. Such solutions occur when the expression $b^2 - 4ac$, the radicand in the quadratic formula, is negative.

$$x = \frac{-b \pm \sqrt{b^2 - 4ac}}{2a}$$

The expression $b^2 - 4ac$ is called the **discriminant.** Depending on the value of the discriminant and whether the discriminant is positive, zero, or negative, the roots of the quadratic equation will be rational, irrational, or complex. We summarize the types of solutions in the following table.

If the Discriminant $b^2 - 4ac$ is:	Then the Quadratic Equation $ax^2 + bx + c = 0$, where a, b, and c are integers, $a \neq 0$, will have:
A positive number that is also a perfect square	Two different rational solutions Such an equation can be solved by factoring.
A positive number that is not a perfect square	Two different irrational solutions
Zero	One rational solution
Negative	Two complex solutions containing i They will be complex conjugates.

Example 7 What type of solutions does the equation $2x^2 - 9x - 35 = 0$ have? Do not solve the equation.

Solution $a = 2, b = -9$, and $c = -35$. Thus,

$$b^2 - 4ac = (-9)^2 - 4(2)(-35) = 361.$$

Since the discriminant is positive, the equation has two real roots.

Since $(19)^2 = 361$, 361 is a perfect square. Thus, the equation has two different rational solutions. This type of quadratic equation can always be factored. □

Student Practice 7 Use the discriminant to find what type of solutions the equation $9x^2 + 12x + 4 = 0$ has. Do not solve the equation.

Example 8 Use the discriminant to determine the type of solutions each of the following equations has.

(a) $3x^2 - 4x + 2 = 0$ **(b)** $5x^2 - 3x - 5 = 0$

Solution

(a) Here $a = 3, b = -4$, and $c = 2$. Thus,

$$b^2 - 4ac = (-4)^2 - 4(3)(2)$$
$$= 16 - 24 = -8$$

Since the discriminant is negative, the equation will have two complex solutions containing i.

(b) Here $a = 5, b = -3$, and $c = -5$. Thus,

$$b^2 - 4ac = (-3)^2 - 4(5)(-5)$$
$$= 9 + 100 = 109$$

Since this positive number is not a perfect square, the equation will have two different irrational solutions. □

Student Practice 8 Use the discriminant to determine the type of solutions each of the following equations has.

(a) $x^2 - 4x + 13 = 0$

(b) $9x^2 + 6x - 7 = 0$

3 Writing a Quadratic Equation Given the Solutions of the Equation

By using the zero factor property in reverse, we can find a quadratic equation that contains two given solutions. To illustrate, if 3 and 7 are the two solutions, then we could write the equation $(x - 3)(x - 7) = 0$, and therefore, a quadratic equation that has these two solutions is $x^2 - 10x + 21 = 0$.

This answer is not unique. Any constant multiple of $x^2 - 10x + 21 = 0$ would also have roots of 3 and 7. Thus, $2x^2 - 20x + 42 = 0$ also has roots of 3 and 7.

Example 9 Find a quadratic equation whose roots are 5 and -2.

Solution First we write the two equations.

$$x = 5 \qquad\qquad x = -2$$
$$x - 5 = 0 \quad \text{and} \quad x + 2 = 0$$
$$(x - 5)(x + 2) = 0$$
$$x^2 - 3x - 10 = 0$$

□

Student Practice 9 Find a quadratic equation whose roots are -10 and -6.

Example 10 Find a quadratic equation whose solutions are $3i$ and $-3i$.

Solution First we write the two equations.

$$x = 3i \qquad\qquad x = -3i$$
$$x - 3i = 0 \quad \text{and} \quad x + 3i = 0$$
$$(x - 3i)(x + 3i) = 0$$
$$x^2 + 3ix - 3ix - 9i^2 = 0$$
$$x^2 - 9(-1) = 0 \qquad \text{Use } i^2 = -1.$$
$$x^2 + 9 = 0$$

□

Student Practice 10 Find a quadratic equation whose solutions are $2i\sqrt{3}$ and $-2i\sqrt{3}$.

9.2 Exercises MyMathLab®

Verbal and Writing Skills, Exercises 1–4

1. How is the quadratic formula used to solve a quadratic equation?

2. The discriminant in the quadratic formula is the expression ________________.

3. If the discriminant in the quadratic formula is zero, then the quadratic equation will have ________________ solution(s).

4. If the discriminant in the quadratic formula is a perfect square, then the quadratic equation will have ________________ solution(s).

Solve by the quadratic formula. Simplify your answers.

5. $x^2 + x - 5 = 0$

6. $x^2 + 3x - 2 = 0$

7. $2x^2 + 3x - 3 = 0$

8. $4x^2 + x - 1 = 0$

9. $x^2 = \frac{2}{3}x$

10. $\frac{5}{2}x^2 = x$

11. $3x^2 - x - 2 = 0$

12. $7x^2 + 4x - 3 = 0$

13. $4x^2 + 3x - 2 = 0$

14. $6x^2 - 2x - 1 = 0$

15. $4x^2 + 1 = 7$

16. $5x^2 - 4 = 4$

Simplify each equation. Then solve by the quadratic formula. Simplify your answers and use i notation for nonreal complex numbers.

17. $2x(x + 3) - 3 = 4x - 2$

18. $5 + 3x(x - 2) = 4$

19. $x(x + 3) - 2 = 3x + 7$

20. $3(x^2 - 12) - 2x = 2x(x - 1)$

21. $(x - 2)(x + 1) = \frac{2x + 3}{2}$

22. $3x(x + 1) = \frac{7x + 1}{3}$

23. $\frac{1}{x + 2} + \frac{1}{x} = \frac{1}{3}$

24. $2y = \frac{1}{y} + \frac{5}{2}$

25. $\frac{1}{12} + \frac{1}{y} = \frac{2}{y + 2}$

26. $\frac{1}{4} + \frac{6}{y + 2} = \frac{6}{y}$

27. $x(x + 4) = -12$

28. $x^2 = 3(x - 3)$

29. $2x^2 + 11 = 0$

30. $3x^2 = -7$

31. $3x^2 - 8x + 7 = 0$

32. $3x^2 - 4x + 6 = 0$

Use the discriminant to find what type of solutions (two rational, two irrational, one rational, or two nonreal complex) each of the following equations has. Do not solve the equation.

33. $3x^2 + 4x = 2$

34. $5x^2 - 10x = -5$

35. $2x^2 + 10x + 8 = 0$

36. $2x^2 - 9x + 4 = 0$

37. $9x^2 + 4 = 12x$

38. $6x^2 - 2 = -7x$

39. $7x(x - 1) + 15 = 10$

40. $x^2 - 3(x - 8) = 2x$

Write a quadratic equation having the given solutions.

41. 13, 2

42. 9, 4

43. $-7, -6$

44. $-8, -5$

45. $4i, -4i$

46. $7i, -7i$

47. $3, -\frac{5}{2}$

48. $-\frac{3}{4}, 2$

Solve for x by using the quadratic formula. Approximate your answers to four decimal places.

49. $0.162x^2 + 0.094x - 0.485 = 0$

50. $20.6x^2 - 73.4x + 41.8 = 0$

Applications

51. ***Business Owner*** Anja is the owner of a company that manufactures mountain bikes. The company makes a daily profit p according to the equation $p = -100x^2 + 4800x - 54{,}351$, where p is measured in dollars and x is the number of mountain bikes made per day. Find the number of mountain bikes that must be made each day to produce a zero profit for the company. Round your answer to the nearest whole number.

52. ***Business Manager*** Rick is the manager of a company that manufactures sport parachutes. The company makes a daily profit p according to the equation $p = -100x^2 + 4200x - 39{,}476$, where p is measured in dollars and x is the number of parachutes made per day. Find the number of parachutes that must be made each day to produce a zero profit for the company. Round your answer to the nearest whole number.

Cumulative Review *Simplify.*

53. **[1.7.2]** $9x^2 - 6x + 3 - 4x - 12x^2 + 8$

54. **[1.7.2]** $3y(2 - y) + \frac{1}{5}(10y^2 - 15y)$

Quick Quiz 9.2 *Solve each of the following equations. Use i notation for any complex numbers.*

1. $11x^2 - 9x - 1 = 0$

2. $\frac{3}{4} + \frac{5}{4x} = \frac{2}{x^2}$

3. $(x + 2)(x + 1) + (x - 4)^2 = 9$

4. **Concept Check** Explain how you would determine if $2x^2 - 6x + 3 = 3$ has two rational, two irrational, one rational, or two nonreal complex solutions.

9.3 Equations That Can Be Transformed into Quadratic Form

Student Learning Objectives

After studying this section, you will be able to:

1 Solve equations of degree greater than 2 that can be transformed into quadratic form.

2 Solve equations with fractional exponents that can be transformed into quadratic form.

1 Solving Equations of Degree Greater than 2

Some higher-order equations can be solved by writing them in the form of quadratic equations. An equation is **quadratic in form** if we can substitute a linear term for the variable and get an equation of the form $ay^2 + by + c = 0$.

Example 1 Solve. $x^4 - 13x^2 + 36 = 0$

Solution Let $y = x^2$. Then $y^2 = x^4$. Thus, we obtain a new equation and solve it as follows:

$y^2 - 13y + 36 = 0$ Replace x^2 by y and x^4 by y^2.

$(y - 4)(y - 9) = 0$ Factor.

$y - 4 = 0$ or $y - 9 = 0$ Solve for y.

$y = 4$ $\quad y = 9$ These are *not* the roots to the original equation. We must replace y by x^2.

$x^2 = 4$ $\quad x^2 = 9$

$x = \pm\sqrt{4}$ $\quad x = \pm\sqrt{9}$

$x = \pm 2$ $\quad x = \pm 3$

Thus, there are *four* solutions to the original equation: $x = +2$, $x = -2$, $x = +3$, and $x = -3$. Check these values to verify that they are solutions.

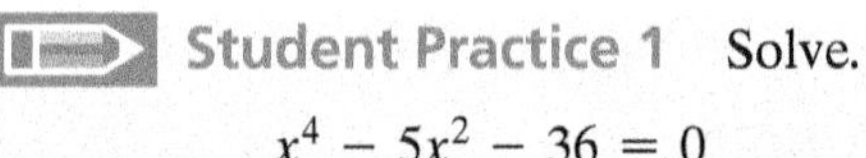

Student Practice 1 Solve.

$x^4 - 5x^2 - 36 = 0$

Example 2 Solve for all real roots. $2x^6 - x^3 - 6 = 0$

Solution Let $y = x^3$. Then $y^2 = x^6$. Thus, we have the following:

$2y^2 - y - 6 = 0$ Replace x^3 by y and x^6 by y^2.

$(2y + 3)(y - 2) = 0$ Factor.

$2y + 3 = 0$ or $y - 2 = 0$ Solve for y.

$y = -\frac{3}{2}$ $\quad y = 2$

$x^3 = -\frac{3}{2}$ $\quad x^3 = 2$ Replace y by x^3.

$x = \sqrt[3]{-\frac{3}{2}}$ $\quad x = \sqrt[3]{2}$ Take the cube root of each side of the equation.

$x = \frac{\sqrt[3]{-12}}{2}$ Simplify $\sqrt[3]{-\frac{3}{2}}$ by rationalizing the denominator.

Check these solutions.

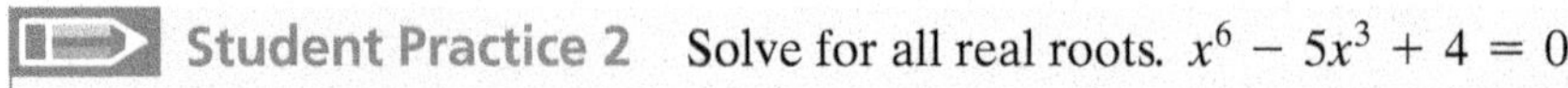

Student Practice 2 Solve for all real roots. $x^6 - 5x^3 + 4 = 0$

2 Solving Equations with Fractional Exponents

Example 3 Solve and check your solutions. $x^{2/3} - 3x^{1/3} + 2 = 0$

Solution Let $y = x^{1/3}$. Then $y^2 = x^{2/3}$.

$$y^2 - 3y + 2 = 0 \quad \text{Replace } x^{1/3} \text{ by } y \text{ and } x^{2/3} \text{ by } y^2.$$

$$(y - 2)(y - 1) = 0 \quad \text{Factor.}$$

$$y - 2 = 0 \quad \text{or} \quad y - 1 = 0$$

$$y = 2 \qquad y = 1 \quad \text{Solve for } y.$$

$$x^{1/3} = 2 \qquad x^{1/3} = 1 \quad \text{Replace } y \text{ by } x^{1/3}.$$

$$(x^{1/3})^3 = (2)^3 \qquad (x^{1/3})^3 = (1)^3 \quad \text{Cube each side of the equation.}$$

$$x = 8 \qquad x = 1$$

Check.

$x = 8$: $(8)^{2/3} - 3(8)^{1/3} + 2 \stackrel{?}{=} 0$ $\qquad$ $x = 1$: $(1)^{2/3} - 3(1)^{1/3} + 2 \stackrel{?}{=} 0$

$(\sqrt[3]{8})^2 - 3(\sqrt[3]{8}) + 2 \stackrel{?}{=} 0$ $\qquad$ $(\sqrt[3]{1})^2 - 3(\sqrt[3]{1}) + 2 \stackrel{?}{=} 0$

$(2)^2 - 3(2) + 2 \stackrel{?}{=} 0$ $\qquad$ $1 - 3 + 2 \stackrel{?}{=} 0$

$4 - 6 + 2 \stackrel{?}{=} 0$ $\qquad$ $0 = 0$ ✓

$0 = 0$ ✓ ◻

 **Student Practice 3** Solve and check your solutions.

$$3x^{4/3} - 5x^{2/3} + 2 = 0$$

The exercises that appear in this section are somewhat difficult to solve. Part of the difficulty lies in the fact that the equations have different numbers of solutions. A fourth-degree equation like the one in Example 1 has four different solutions. Whereas a sixth-degree equation such as the one in Example 2 has only two solutions, some sixth-degree equations will have as many as six solutions. Although the equation that we examined in Example 3 has only two solutions, other equations with fractional exponents may have one solution or even no solution at all. It is good to take some time to carefully examine your work to determine that you have obtained the correct number of solutions.

A graphing program on a computer such as TI Interactive, Derive, or Maple can be very helpful in determining or verifying the solutions to these types of problems. Of course a graphing calculator can be most helpful, particularly in verifying the value of a solution and the number of solutions.

Optional Graphing Calculator Exploration: If you have a graphing calculator, verify the solutions for Example 3 by graphing the equation

$$y = x^{2/3} - 3x^{1/3} + 2.$$

Determine from your graph whether the curve does in fact cross the x-axis (that is, $y = 0$) when $x = 1$ and $x = 8$. You will have to carefully select the window so that you can see the behavior of the curve clearly. For this equation a useful window is $[-1, 12]$ by $[-2, 2]$. Remember that with most graphing calculators, you will need to surround the exponents with parentheses.

Example 4 Solve and check your solutions. $2x^{1/2} = 5x^{1/4} + 12$

Solution

$$2x^{1/2} - 5x^{1/4} - 12 = 0 \quad \text{Place in standard form.}$$

$$2y^2 - 5y - 12 = 0 \quad \text{Replace } x^{1/4} \text{ by } y \text{ and } x^{1/2} \text{ by } y^2.$$

$$(2y + 3)(y - 4) = 0 \quad \text{Factor.}$$

$$2y + 3 = 0 \quad \text{or} \quad y - 4 = 0$$

$$y = -\frac{3}{2} \qquad y = 4 \quad \text{Solve for } y.$$

$$x^{1/4} = -\frac{3}{2} \qquad x^{1/4} = 4 \quad \text{Replace } y \text{ by } x^{1/4}.$$

$$(x^{1/4})^4 = \left(-\frac{3}{2}\right)^4 \qquad (x^{1/4})^4 = (4)^4 \quad \text{Solve for } x.$$

$$x = \frac{81}{16} \qquad x = 256$$

Continued on next page

Check.

$$x = \frac{81}{16}:\quad 2\left(\frac{81}{16}\right)^{1/2} - 5\left(\frac{81}{16}\right)^{1/4} - 12 \stackrel{?}{=} 0 \qquad x = 256:\quad 2(256)^{1/2} - 5(256)^{1/4} - 12 \stackrel{?}{=} 0$$

$$2\left(\frac{9}{4}\right) - 5\left(\frac{3}{2}\right) - 12 \stackrel{?}{=} 0 \qquad 2(16) - 5(4) - 12 \stackrel{?}{=} 0$$

$$\frac{9}{2} - \frac{15}{2} - 12 \stackrel{?}{=} 0 \qquad 32 - 20 - 12 \stackrel{?}{=} 0$$

$$-15 \neq 0 \qquad 0 = 0 \checkmark$$

$\frac{81}{16}$ is extraneous and not a valid solution. The only valid solution is 256. □

 **Student Practice 4** Solve and check your solutions.

$$3x^{1/2} = 8x^{1/4} - 4$$

Example 5 Solve and check your solutions. $x^{-2} = 5x^{-1} + 14$

Solution

$$x^{-2} - 5x^{-1} - 14 = 0 \qquad \text{Place in standard form.}$$

$$y^2 - 5y - 14 = 0 \qquad \text{Replace } x^{-1} \text{ by } y \text{ and } x^{-2} \text{ by } y^2.$$

$$(y - 7)(y + 2) = 0 \qquad \text{Factor.}$$

$$y - 7 = 0 \quad \text{or} \quad y + 2 = 0$$

$$y = 7 \qquad y = -2 \qquad \text{Solve for } y.$$

$$x^{-1} = 7 \qquad x^{-1} = -2 \qquad \text{Replace } y \text{ by } x^{-1}.$$

$$(x^{-1})^{-1} = (7)^{-1} \qquad (x^{-1})^{-1} = (-2)^{-1} \qquad \text{Solve for } x.$$

$$x = \frac{1}{7} \qquad x = -\frac{1}{2}$$

Check.

$$x = \frac{1}{7}:\quad \left(\frac{1}{7}\right)^{-2} - 5\left(\frac{1}{7}\right)^{-1} - 14 \stackrel{?}{=} 0 \qquad x = -\frac{1}{2}:\quad \left(-\frac{1}{2}\right)^{-2} - 5\left(-\frac{1}{2}\right)^{-1} - 14 \stackrel{?}{=} 0$$

$$49 - 5(7) - 14 \stackrel{?}{=} 0 \qquad 4 - 5(-2) - 14 \stackrel{?}{=} 0$$

$$0 = 0 \checkmark \qquad 0 = 0 \checkmark$$

□

 **Student Practice 5** Solve and check your solutions.

$$2x^{-2} + x^{-1} = 6$$

Although we have covered just five basic examples here, this substitution technique can be extended to other types of equations. In each case we substitute y for an appropriate expression in order to obtain a quadratic equation. The following table lists some substitutions that would be appropriate.

If You Want to Solve:	Then You Would Use the Substitution:	Resulting Equation
$x^4 - 13x^2 + 36 = 0$	$y = x^2$	$y^2 - 13y + 36 = 0$
$2x^6 - x^3 - 6 = 0$	$y = x^3$	$2y^2 - y - 6 = 0$
$x^{2/3} - 3x^{1/3} + 2 = 0$	$y = x^{1/3}$	$y^2 - 3y + 2 = 0$
$6(x - 1)^{-2} + (x - 1)^{-1} - 2 = 0$	$y = (x - 1)^{-1}$	$6y^2 + y - 2 = 0$
$(2x^2 + x)^2 + 4(2x^2 + x) + 3 = 0$	$y = 2x^2 + x$	$y^2 + 4y + 3 = 0$
$\left(\frac{1}{x - 1}\right)^2 + \frac{1}{x - 1} - 6 = 0$	$y = \frac{1}{x - 1}$	$y^2 + y - 6 = 0$
$2x - 5x^{1/2} + 2 = 0$	$y = x^{1/2}$	$2y^2 - 5y + 2 = 0$

Exercises like these appear in the exercise set.

9.3 Exercises MyMathLab®

Solve. Express any nonreal complex numbers with i notation.

1. $x^4 - 9x^2 + 20 = 0$

2. $x^4 + 6x^2 - 27 = 0$

3. $x^4 + x^2 - 12 = 0$

4. $x^4 - 2x^2 - 8 = 0$

5. $4x^4 - x^2 - 3 = 0$

6. $2x^4 - x^2 - 1 = 0$

In exercises 7–10, find all valid real roots for each equation.

7. $x^6 - 7x^3 - 8 = 0$

8. $x^6 - 3x^3 - 4 = 0$

9. $x^6 - 3x^3 = 0$

10. $x^6 + 27x^3 = 0$

Solve for real roots.

11. $x^8 = 17x^4 - 16$

12. $x^8 = 5x^4$

13. $3x^8 + 13x^4 = 10$

14. $3x^8 - 10x^4 = 8$

Solve for real roots.

15. $x^{2/3} + x^{1/3} - 12 = 0$ **16.** $x^{2/3} + 2x^{1/3} - 8 = 0$ **17.** $12x^{2/3} + 5x^{1/3} - 2 = 0$ **18.** $2x^{2/3} - 7x^{1/3} - 4 = 0$

19. $2x^{1/2} - 5x^{1/4} - 3 = 0$ **20.** $3x^{1/2} + 13x^{1/4} - 10 = 0$ **21.** $4x^{1/2} + 11x^{1/4} - 3 = 0$

22. $2x^{1/2} - x^{1/4} - 1 = 0$ **23.** $x^{2/5} - x^{1/5} - 2 = 0$ **24.** $2x^{2/5} - 7x^{1/5} + 6 = 0$

Mixed Practice *In each exercise make an appropriate substitution in order to obtain a quadratic equation. Find the complex values for x.*

25. $x^6 - 5x^3 = 14$ **26.** $x^6 + 2x^3 = 15$

27. $(x^2 + 3x)^2 - 2(x^2 + 3x) - 8 = 0$

28. $(x^2 + 4x)^2 + 8(x^2 + 4x) + 15 = 0$

29. $x - 5x^{1/2} + 6 = 0$

30. $x - 3x^{1/2} - 10 = 0$

31. $x^{-2} + 3x^{-1} = 0$

32. $3x^{-2} + x^{-1} = 0$

33. $x^{-2} = 3x^{-1} + 10$

34. $x^{-2} = -7x^{-1} - 6$

To Think About *Solve. Find all valid real roots for each equation.*

35. $15 - \frac{2x}{x - 1} = \frac{x^2}{x^2 - 2x + 1}$

36. $4 - \frac{x^3 + 1}{x^3 + 6} = \frac{x^3 - 3}{x^3 + 2}$

Cumulative Review

37. **[4.1.6]** Solve the system. $\begin{aligned} 2x + 3y &= 5 \\ -5x - 2y &= 4 \end{aligned}$

38. **[7.4.1]** Simplify. $\dfrac{5 + \frac{2}{x}}{\frac{7}{3x} - 1}$

Multiply and simplify.

39. **[8.4.1]** $3\sqrt{2}(\sqrt{5} - 2\sqrt{6})$

40. **[8.4.1]** $(\sqrt{2} + \sqrt{6})(3\sqrt{2} - 2\sqrt{5})$

41. **[1.3.3]** ***Salary and Educational Attainment*** How much greater is the median weekly salary of a man than a woman at each of the four levels of educational attainment shown in the graph? Express your answers to the nearest tenth of a percent.

42. **[1.3.3]** ***Salary and Educational Attainment*** Approximately how many extra weeks would an average woman with some college or an associate's degree have to work to earn the same annual earnings as a male counterpart who worked for 52 weeks? Round your answer to the nearest tenth of a week.

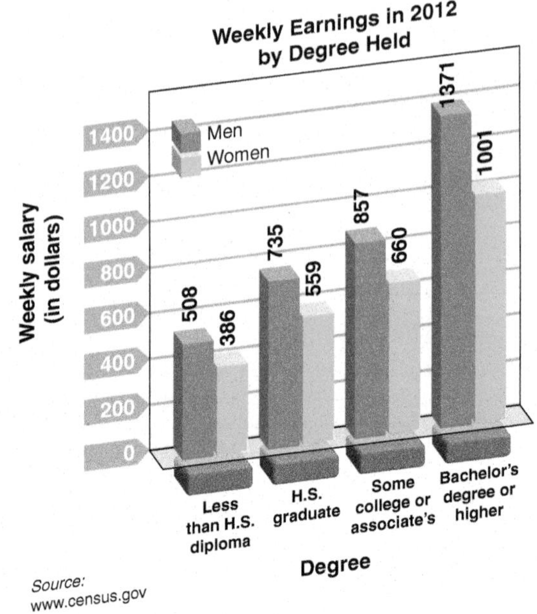

Quick Quiz 9.3 *Solve for any valid real roots.*

1. $x^4 - 18x^2 + 32 = 0$

2. $2x^{-2} - 3x^{-1} - 20 = 0$

3. $x^{2/3} - 4x^{1/3} - 5 = 0$

4. **Concept Check** Explain how you would solve the following. $x^8 - 6x^4 = 0$

Use Math to Save Money

Did You Know? You can save money over the course of a semester by choosing the right meal plan.

Choosing a Meal Plan

Understanding the Problem:

Jake is a college student who lives in a college dormitory. He is going over the meal plans and calculating his food budget for the semester. He is interested in choosing a meal plan that will cost him the least while still meeting his needs.

Making a Plan:

First he needs to compare the prices and options for the meal plans. Then he needs to determine how many meals he will eat in the dining halls so he can determine which plan is right for him.

Step 1. There are three dining plans that cover the 15-week semester.

- The Gold plan costs $2205 a semester and covers 21 meals a week.
- The Silver plan costs $1764 a semester and covers 16 meals a week, with extra meals costing $10.
- The Bronze plan costs $1386 a semester and covers 12 meals a week, with extra meals costing $10.

Which plan is best for the following situations:
Task 1: Jake plans on eating 20 meals per week in the dining halls?
Task 2: Jake plans on eating 18 meals per week in the dining halls?
Task 3: Jake plans on eating 16 meals per week in the dining halls?
Task 4: Jake plans on eating 14 meals per week in the dining halls?

Step 2. Jake also has the option of buying "dining dollars" to use at other eating establishments on campus. With dining dollars Jake will get 10% off all food purchases.

How much will Jake save over the course of the semester by using dining dollars in the following situations:

Task 5: if he spends $20 a week at these establishments?

Task 6: if he spends $40 a week at these establishments?

Task 7: if he spends $60 a week at these establishments?

Finding a Solution:

Step 3. Jake plans on eating 17 meals a week in the dining halls and spending $30 a week at other establishments.

Task 8: Which plan should he buy?
Task 9: How much will Jake spend on food for the semester?

Applying the Situation to Your Life:

You can do these same calculations for your own circumstances.

- See what plans are offered.
- Determine how often you will eat in the dining halls.
- Decide which plan would be right for you.
- Calculate how much you will need to spend on food for the semester.

9.4 Formulas and Applications

Student Learning Objectives

After studying this section, you will be able to:

1. Solve a quadratic equation containing several variables.
2. Solve problems requiring the use of the Pythagorean Theorem.
3. Solve applied problems requiring the use of a quadratic equation.

1 Solving a Quadratic Equation Containing Several Variables

In mathematics, physics, and engineering, we must often solve an equation for a variable in terms of other variables. Recall in Chapter 3 we solved equations such as $2y + 3x = 6$ for y in order to write the equation in slope-intercept form $(y = mx + b)$. We now examine several cases where the variable that we are solving for is squared. If the variable we are solving for is squared, and there is no other term containing that variable, then the equation can be solved using the square root property as illustrated in Example 1.

▲ **Example 1** The surface area of a sphere is given by $A = 4\pi r^2$. Solve this equation for r. (You do not need to rationalize the denominator.)

Solution

$$A = 4\pi r^2$$

$$\frac{A}{4\pi} = r^2$$

$$\pm\sqrt{\frac{A}{4\pi}} = r \quad \text{Use the square root property.}$$

$$\pm\frac{1}{2}\sqrt{\frac{A}{\pi}} = r \quad \text{Simplify.}$$

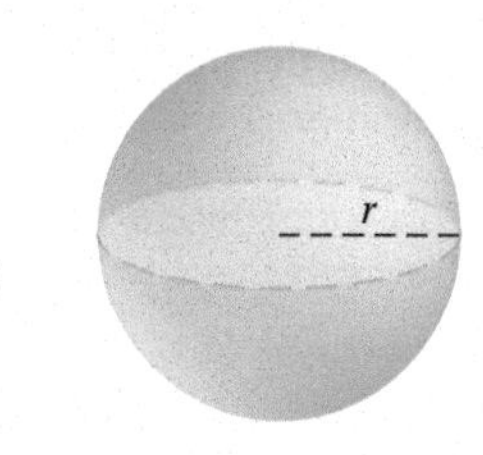

Since the radius of a sphere must be a positive value, we use only the principal root.

$$r = \frac{1}{2}\sqrt{\frac{A}{\pi}}$$

▲ **Student Practice 1** The volume of a cylindrical cone is $V = \frac{1}{3}\pi r^2 h$. Solve this equation for r. (You do not need to rationalize the denominator.)

Some quadratic equations containing many variables can be solved for one variable by factoring.

Example 2 Solve for y. $y^2 - 2yz - 15z^2 = 0$

Solution

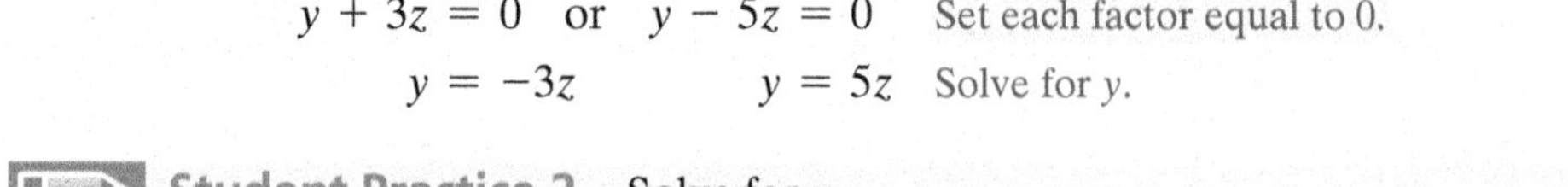

$$(y + 3z)(y - 5z) = 0 \quad \text{Factor.}$$

$$y + 3z = 0 \quad \text{or} \quad y - 5z = 0 \quad \text{Set each factor equal to 0.}$$

$$y = -3z \qquad y = 5z \quad \text{Solve for } y.$$

Student Practice 2 Solve for y.

$$2y^2 + 9wy + 7w^2 = 0$$

Sometimes the quadratic formula is required in order to solve the equation.

Example 3 Solve for x. $2x^2 + 3wx - 4z = 0$

Solution We use the quadratic formula where the variable is considered to be x and the letters w and z are considered constants. Thus, $a = 2$, $b = 3w$, and $c = -4z$.

$$x = \frac{-b \pm \sqrt{b^2 - 4ac}}{2a}$$

$$x = \frac{-3w \pm \sqrt{(3w)^2 - 4(2)(-4z)}}{2(2)} = \frac{-3w \pm \sqrt{9w^2 + 32z}}{4}$$

Note that this answer cannot be simplified any further. □

 **Student Practice 3** Solve for y.

$$3y^2 + 2fy - 7g = 0$$

▲ **Example 4** The formula for the curved surface area S of a right circular cone of altitude h and with base of radius r is $S = \pi r\sqrt{r^2 + h^2}$. Solve for r^2.

Solution

$$S = \pi r\sqrt{r^2 + h^2}$$

$$\frac{S}{\pi r} = \sqrt{r^2 + h^2} \qquad \text{Isolate the radical.}$$

$$\frac{S^2}{\pi^2 r^2} = r^2 + h^2 \qquad \text{Square both sides.}$$

$$\frac{S^2}{\pi^2} = r^4 + h^2r^2 \qquad \text{Multiply each term by } r^2.$$

$$0 = r^4 + h^2r^2 - \frac{S^2}{\pi^2} \qquad \text{Subtract } \frac{S^2}{\pi^2} \text{ from both sides.}$$

This equation is quadratic in form. If we let $y = r^2$, then we have

$$0 = y^2 + h^2y - \frac{S^2}{\pi^2}.$$

By the quadratic formula we have the following:

$$y = \frac{-h^2 \pm \sqrt{(h^2)^2 - 4(1)\left(-\frac{S^2}{\pi^2}\right)}}{2(1)}$$

$$y = \frac{-h^2 \pm \sqrt{\frac{\pi^2h^4}{\pi^2} + \frac{4S^2}{\pi^2}}}{2} \qquad \text{Obtain a common denominator under the radical.}$$

$$y = \frac{-h^2 \pm \frac{1}{\pi}\sqrt{\pi^2h^4 + 4S^2}}{2} \qquad \text{Factor out } \frac{1}{\pi^2} \text{ from both terms under the radical. } \sqrt{\frac{1}{\pi^2}} = \frac{1}{\pi} \text{ is written outside the radical.}$$

$$y = \frac{-\pi h^2 \pm \sqrt{\pi^2h^4 + 4S^2}}{2\pi} \qquad \text{Multiply the rational expression by } \frac{\pi}{\pi}.$$

Since $y = r^2$, we have

$$r^2 = \frac{-\pi h^2 + \sqrt{\pi^2h^4 + 4S^2}}{2\pi}.$$

Since the radius of a cone must be a positive value, we know $r^2 > 0$ and we use only the principal root. □

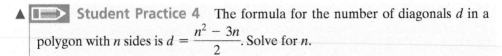

▲ **Student Practice 4** The formula for the number of diagonals d in a polygon with n sides is $d = \frac{n^2 - 3n}{2}$. Solve for n.

2 Solving Problems Requiring the Use of the Pythagorean Theorem

A very useful formula is the Pythagorean Theorem for right triangles.

PYTHAGOREAN THEOREM

If c is the length of the longest side of a right triangle and a and b are the lengths of the other two sides, then $a^2 + b^2 = c^2$.

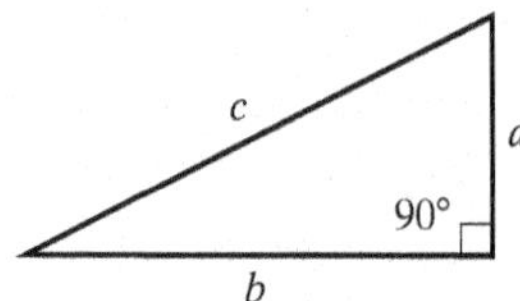

The longest side of a right triangle is called the **hypotenuse.** The other two sides are called the **legs** of the triangle.

Example 5

(a) Solve the Pythagorean Theorem $a^2 + b^2 = c^2$ for a.
(b) Find the value of a if $c = 13$ and $b = 5$.

Solution

(a) $a^2 = c^2 - b^2$ Subtract b^2 from each side.

$a = \pm\sqrt{c^2 - b^2}$ Use the square root property.

a, b, and c must be positive numbers because they represent lengths, so we use only the positive root, $a = \sqrt{c^2 - b^2}$.

(b) $a = \sqrt{c^2 - b^2}$

$a = \sqrt{(13)^2 - (5)^2} = \sqrt{169 - 25} = \sqrt{144} = 12$

Thus, $a = 12$. □

Student Practice 5

(a) Solve the Pythagorean Theorem for b.
(b) Find the value of b if $c = 26$ and $a = 24$.

There are many practical uses for the Pythagorean Theorem. For hundreds of years, it was used in land surveying and in ship navigation. So, for hundreds of years students learned it in public schools and were shown applications in surveying and navigation. In almost any situation where you have a right triangle and you know two sides and need to find the third side, you can use the Pythagorean Theorem.

See if you can see how it is used in Example 6. Notice that it is helpful right at the beginning of the problem to draw a picture of the right triangle and label information that you know.

▲ **Example 6** The perimeter of a triangular piece of land is 12 miles. One leg of the triangle is 1 mile longer than the other leg. Find the length of each boundary of the land if the triangle is a right triangle.

Solution

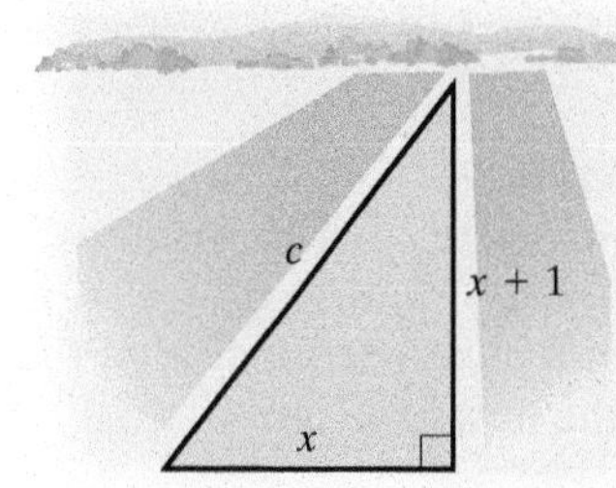

1. ***Understand the problem.*** Draw a picture of the piece of land and label the sides of the triangle.
2. ***Write an equation.*** We can use the Pythagorean Theorem. First, we want only one variable in our equation. (Right now, both c and x are not known.)

   We are given that the perimeter is 12 miles, so

$$x + (x + 1) + c = 12.$$

Thus,

$$c = -2x + 11.$$

By the Pythagorean Theorem,

$$x^2 + (x + 1)^2 = (-2x + 11)^2.$$

3. ***Solve the equation and state the answer.***

$$x^2 + (x + 1)^2 = (-2x + 11)^2$$
$$x^2 + x^2 + 2x + 1 = 4x^2 - 44x + 121$$
$$0 = 2x^2 - 46x + 120$$
$$0 = x^2 - 23x + 60$$

By the quadratic formula, we have the following:

$$x = \frac{-(-23) \pm \sqrt{(-23)^2 - 4(1)(60)}}{2(1)}$$
$$x = \frac{23 \pm \sqrt{289}}{2}$$
$$x = \frac{23 \pm 17}{2}$$
$$x = \frac{40}{2} = 20 \quad \text{or} \quad x = \frac{6}{2} = 3.$$

The answer $x = 20$ cannot be right because the perimeter (the sum of *all* the sides) is only 12. The only answer that makes sense is $x = 3$. Thus, the sides of the triangle are $x = 3$, $x + 1 = 3 + 1 = 4$, and $-2x + 11 = -2(3) + 11 = 5$. The longest boundary of this triangular piece of land is 5 miles. The other two boundaries are 4 miles and 3 miles.

Notice that we could have factored the quadratic equation instead of using the quadratic formula. $x^2 - 23x + 60 = 0$ can be written as $(x - 20)(x - 3) = 0$.

4. ***Check.***

Is the perimeter 12 miles?
$5 + 4 + 3 = 12$ ✓

Is one leg 1 mile longer than the other?
$4 = 3 + 1$ ✓ □

▲ **Student Practice 6** The perimeter of a triangular piece of land is 30 miles. One leg of the triangle is 7 miles shorter than the other leg. Find the length of each boundary of the land if the triangle is a right triangle.

3 Solving Applied Problems Requiring the Use of a Quadratic Equation

Many types of area problems can be solved with quadratic equations as shown in the next two examples.

▲ **Example 7** The radius of an old circular pipe under a roadbed is 10 feet. Designers want to replace it with a smaller pipe and have decided they can use one with a cross-sectional area that is 36π square feet smaller. What should the radius of the new pipe be?

Solution First we need the formula for the area of a circle,

$$A = \pi r^2,$$

where A is the area and r is the radius. The area of the cross section of the old pipe is as follows:

$$\begin{aligned} A_{old} &= \pi(10)^2 \\ &= 100\pi \end{aligned}$$

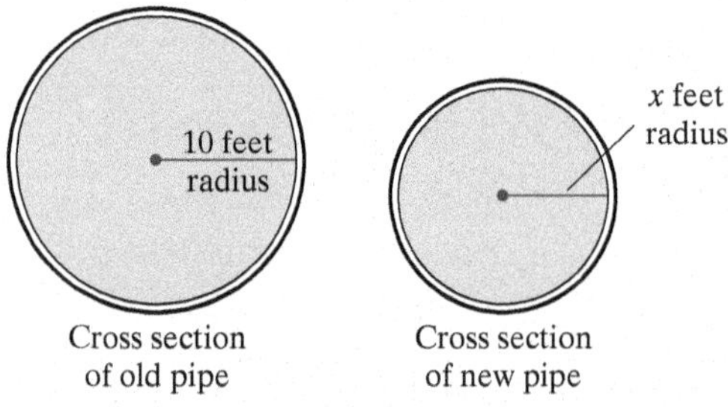

Let $x =$ the radius of the new pipe.

$$(\text{area of old pipe}) - (\text{area of new pipe}) = 36\pi$$

$$100\pi - \pi x^2 = 36\pi$$

$64\pi = \pi x^2$ Add πx^2 to each side and subtract 36π from each side.

$\frac{64\pi}{\pi} = \frac{\pi x^2}{\pi}$ Divide each side by π.

$64 = x^2$

$\pm 8 = x$ Use the square root property.

Since the radius must be positive, we select $x = 8$. The radius of the new pipe is 8 feet. Check to verify this solution. □

▲ **Student Practice 7** Redo Example 7 when the radius of the pipe under the roadbed is 6 feet and the designers want to replace it with a pipe that has a cross-sectional area that is 45π square feet larger. What should the radius of the new pipe be?

▲ **Example 8** A triangular sign marks the edge of some rocks in Rockport Harbor. The sign has an area of 35 square meters. Find the base and altitude of this triangular sign if the base is 3 meters shorter than the altitude.

Solution The area of a triangle is given by

$$A = \frac{1}{2}ab.$$

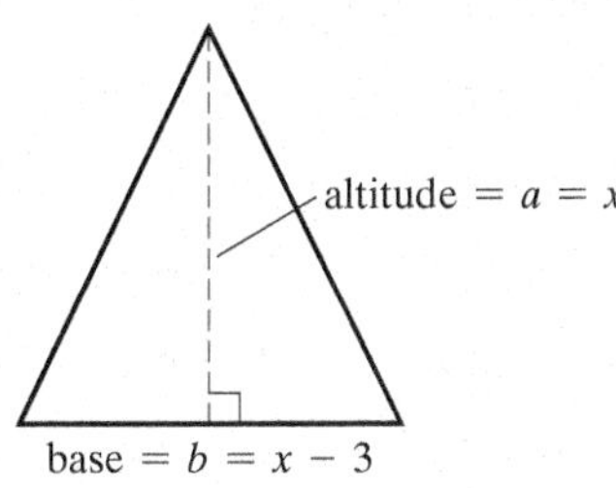

Let x = the length in meters of the altitude. Then $x - 3$ = the length in meters of the base.

$35 = \frac{1}{2}x(x - 3)$	Replace A (area) by 35, a (altitude) by x, and b (base) by $x - 3$.
$70 = x(x - 3)$	Multiply each side by 2.
$70 = x^2 - 3x$	Use the distributive property.
$0 = x^2 - 3x - 70$	Subtract 70 from each side.
$0 = (x - 10)(x + 7)$	
$x = 10 \quad \text{or} \quad x = -7$	

The length of a side of a triangle must be a positive number, so we disregard -7. Thus,

$$\text{altitude} = x = 10 \text{ meters and}$$
$$\text{base} = x - 3 = 7 \text{ meters.}$$

The check is left to the student. □

▲ **Student Practice 8** The length of a rectangle is 3 feet shorter than twice the width. The area of the rectangle is 54 square feet. Find the dimensions of the rectangle.

We will now examine a word problem that requires the use of the formula distance = (rate)(time) or $d = rt$.

Example 9 When Barbara was training for a bicycle race, she rode a total of 135 miles on Monday and Tuesday. On Monday she rode for 75 miles in the rain. On Tuesday she rode 5 miles per hour faster because the weather was better. Her total cycling time for the 2 days was 8 hours. Find her speed for each day.

Solution We can find each distance. If Barbara rode 75 miles on Monday and a total of 135 miles during the 2 days, then she rode $135 - 75 = 60$ miles on Tuesday.

Let $x =$ the cycling rate in miles per hour on Monday. Since Barbara rode 5 miles per hour faster on Tuesday, $x + 5 =$ the cycling rate in miles per hour on Tuesday.

Since distance divided by rate is equal to time $\left(\frac{d}{r} = t\right)$, we can determine that the time Barbara cycled on Monday was $\frac{75}{x}$ and the time she cycled on Tuesday was $\frac{60}{x+5}$.

Day	Distance	Rate	Time
Monday	75	x	$\frac{75}{x}$
Tuesday	60	$x + 5$	$\frac{60}{x+5}$
Total	135	(not used)	8

Since the total cycling time was 8 hours, we have the following:

$$\text{time cycling Monday} + \text{time cycling Tuesday} = 8 \text{ hours}$$
$$\frac{75}{x} + \frac{60}{x+5} = 8$$

The LCD of this equation is $x(x + 5)$. Multiply each term by the LCD.

$$\cancel{x}(x+5)\left(\frac{75}{\cancel{x}}\right) + x(\cancel{x+5})\left(\frac{60}{\cancel{x+5}}\right) = x(x+5)(8)$$

$$75(x + 5) + 60x = 8x(x + 5)$$
$$75x + 375 + 60x = 8x^2 + 40x$$
$$0 = 8x^2 - 95x - 375$$
$$0 = (x - 15)(8x + 25)$$
$$x - 15 = 0 \quad \text{or} \quad 8x + 25 = 0$$
$$x = 15 \qquad\qquad x = -\frac{25}{8}$$

We disregard the negative answer. The cyclist did not have a negative rate of speed—unless she was moving backward! Thus, $x = 15$. So Barbara's rate of speed on Monday was 15 mph, and her rate of speed on Tuesday was $x + 5 = 15 + 5 = 20$ mph. □

Student Practice 9 Carlos traveled in his car at a constant speed on a secondary road for 150 miles. Then he traveled 10 mph faster on a better road for 240 miles. If Carlos drove for 7 hours, find the car's speed for each part of the trip.

9.4 Exercises MyMathLab®

Solve for the variable specified. Assume that all other variables are nonzero. You do not need to rationalize the denominators.

1. $S = 16t^2$; for t

2. $E = mc^2$; for c

3. $S = 9\pi r^2$; for r

4. $A = \frac{1}{2}r^2\theta$; for r

5. $3N = \frac{2}{5}ax^2$; for x

6. $V = \frac{\pi r^2 h}{3}$; for r

7. $4(y^2 + w) - 5 = 7R$; for y

8. $4x^2 + 1 = 3A$; for x

9. $Q = \frac{3mwM^2}{2c}$; for M

10. $H = \frac{5abT^2}{7k}$; for T

11. $V = \pi(r^2 + R^2)h$; for r

12. $M = a(b^2 + c^2)$; for b

13. $7bx^2 - 3ax = 0$; for x

14. $2x^2 - 5ax = 0$; for x

15. $P = EI - RI^2$; for I

16. $A = P(1 + r)^2$; for r

17. $9w^2 + 5tw - 2 = 0$; for w

18. $6w^2 - sw + 4 = 0$; for w

19. $S = 2\pi rh + \pi r^2$; for r

20. $B = 3abx^2 - 5x$; for x

21. $(a + 1)x^2 + 5x + 2w = 0$; for x

22. $(b - 2)x^2 - 3x + 5y = 0$; for x

In exercises 23–28, use the Pythagorean Theorem to find the unknown side(s).

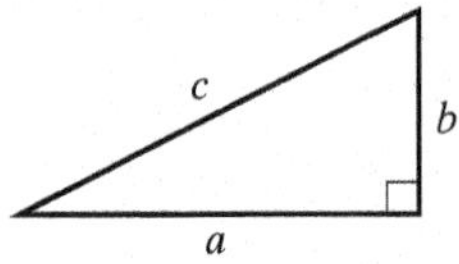

▲ **23.** $c = 6, a = 4$; find b

▲ **24.** $b = 5, c = 10$; find a

▲ **25.** $c = \sqrt{34}, b = \sqrt{19}$; find a

▲ **26.** $c = \sqrt{21}, a = \sqrt{5}$; find b

▲ **27.** $c = 12, b = 2a$; find b and a

▲ **28.** $c = 15, a = 2b$; find b and a

Applications

▲ **29.** ***Brace for a Shelf*** A brace for a shelf has the shape of a right triangle. Its hypotenuse is 10 inches long and the two legs are equal in length. How long are the legs of the triangle?

▲ **30.** ***Tent*** The two sides of a tent are equal in length and meet to form a right triangle. If the width of the floor of the tent is 8 feet, what is the length of the two sides?

▲ **31.** ***College Parking Lot*** Knox College is creating a new rectangular parking lot. The length is 0.07 mile longer than the width and the area of the parking lot is 0.026 square mile. Find the length and width of the parking lot.

▲ **32.** ***Computer Chip*** A high-tech company is producing a new rectangular computer chip. The length is 0.3 centimeter less than twice the width. The chip has an area of 1.04 square centimeters. Find the length and width of the chip.

▲ **33.** ***Barn*** The area of a rectangular wall of a barn is 126 square feet. Its length is 4 feet longer than twice its width. Find the length and width of the wall of the barn.

▲ **34.** ***Tennis Court*** The area of a doubles tennis court is 312 square yards. Its length is 2 yards longer than twice its width. Find the length and width of the tennis court.

▲ **35.** ***Triangular Flag*** The area of a triangular flag is 72 square centimeters. Its altitude is 2 centimeters longer than twice its base. Find the lengths of the altitude and the base.

▲ **36.** ***Children's Playground*** A children's playground is triangular in shape. Its altitude is 2 yards shorter than its base. The area of the playground is 60 square yards. Find the base and altitude of the playground.

37. ***Driving Speed*** Roberto drove at a constant speed in a rainstorm for 225 miles. He took a break, and the rain stopped. He then drove 150 miles at a speed that was 5 miles per hour faster than his previous speed. If he drove for 8 hours, find the car's speed for each part of the trip.

38. ***Driving Speed*** Benita traveled at a constant speed on a gravel road for 40 miles. She then traveled 10 miles per hour faster on a paved road for 150 miles. If she drove for 4 hours, find the car's speed for each part of the trip.

39. ***Commuter Traffic*** Bob drove from home to work at 50 mph. After work the traffic was heavier, and he drove home at 45 mph. His driving time to and from work was 1 hour and 16 minutes. How far does he live from his job?

40. ***Commercial Trucker*** Tranh drove his heavily loaded truck from the company warehouse to a delivery point at 35 mph. He unloaded the truck and drove back to the warehouse at 45 mph. The total trip took 5 hours and 20 minutes. How far is the delivery point from the warehouse?

Incarcerated Adults *The number of incarcerated adults N (measured in thousands) in the United States can be approximated by the quadratic equation* $N = -3.9x^2 + 70.9x + 1944$, *where x is the number of years since 2000. For example, when* $x = 1$, $N = 2001$. *This tells us that in 2001 there were approximately 2,011,000 incarcerated adults in the United States. Use this equation to answer the following questions.* (*Source:* www.bjs.ojp.usdoj.gov)

41. In 2010, the number of incarcerated adults peaked. How many adults were incarcerated in that year?

42. How many incarcerated adults does the equation predict there will be in the year 2018?

43. In what year was the number of incarcerated adults 2,130,000?

44. Determine the year in which the number of incarcerated adults is expected to be 1,802,000.

Cumulative Review *Rationalize the denominators.*

45. **[8.4.3]** $\dfrac{4}{\sqrt{3x}}$

46. **[8.4.3]** $\dfrac{5\sqrt{6}}{2\sqrt{5}}$

47. **[8.4.3]** $\dfrac{3}{\sqrt{x} + \sqrt{y}}$

48. **[8.4.3]** $\dfrac{2\sqrt{3}}{\sqrt{3} - \sqrt{6}}$

Quick Quiz 9.4

1. Solve for y.

$$H = \frac{5ab}{y^2}$$

2. Solve for z.

$$6z^2 + 7yz - 5w = 0$$

▲ **3.** The area of a rectangular field is 175 square yards. The length of the field is 4 yards longer than triple the width of the field. What is the width of the field? What is the length of the field?

▲ **4.** ***Concept Check*** In a right triangle the hypotenuse measures 12 meters. One of the legs of the triangle is three times as long as the other. Explain how you would find the length of each leg.

9.5 Quadratic Functions

1 Finding the Vertex and Intercepts of a Quadratic Function

In Section 3.6 we graphed functions such as $p(x) = x^2$ and $q(x) = (x + 2)^2$. We will now study quadratic functions in more detail.

DEFINITION OF A QUADRATIC FUNCTION

A **quadratic function** is a function of the form $f(x) = ax^2 + bx + c$, where a, b, and c are real numbers and $a \neq 0$.

Graphs of quadratic functions written in this form will be parabolas opening upward if $a > 0$ or downward if $a < 0$. The **vertex** of a parabola is the lowest point on a parabola opening upward or the highest point on a parabola opening downward. The vertex will occur at $x = \frac{-b}{2a}$. To find the y-value, or $f(x)$, when $x = \frac{-b}{2a}$, we find $f\left(\frac{-b}{2a}\right)$. Therefore, we can say that a quadratic function has its vertex at $\left(\frac{-b}{2a}, f\left(\frac{-b}{2a}\right)\right)$.

It is helpful to know the x-intercepts and the y-intercept when graphing a quadratic function.

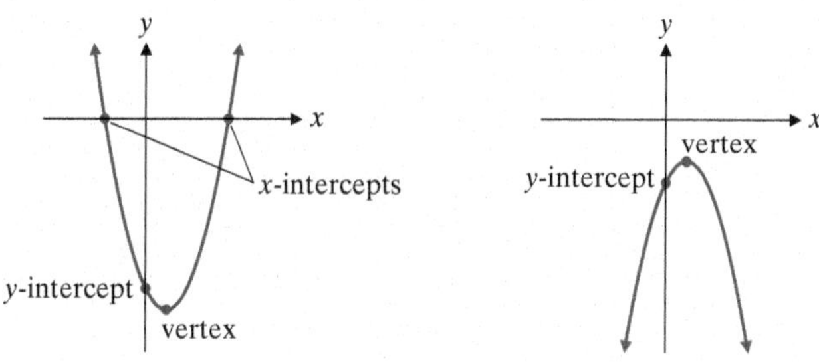

A quadratic function will always have exactly one y-intercept. However, it may have zero, one, or two x-intercepts. Why?

WHEN GRAPHING QUADRATIC FUNCTIONS OF THE FORM $f(x) = ax^2 + bx + c, a \neq 0$

1. The coordinates of the vertex are $\left(\frac{-b}{2a}, f\left(\frac{-b}{2a}\right)\right)$.
2. The y-intercept is at $f(0)$.
3. The x-intercepts (if they exist) occur where $f(x) = 0$. They can always be found with the quadratic formula and can sometimes be found by factoring.

Since we may replace y by $f(x)$, the graph is equivalent to the graph of $y = ax^2 + bx + c$.

Student Learning Objectives

After studying this section, you will be able to:

1. Find the vertex and intercepts of a quadratic function.
2. Graph a quadratic function.

Example 1 Find the coordinates of the vertex and the intercepts of the quadratic function $f(x) = x^2 - 8x + 15$.

Solution For this function $a = 1$, $b = -8$, and $c = 15$.

Step 1 The vertex occurs at $x = \frac{-b}{2a}$. Thus,

$$x = \frac{-(-8)}{2(1)} = \frac{8}{2} = 4.$$

The vertex has an x-coordinate of 4. To find the y-coordinate, we evaluate $f(4)$.

$$f(4) = 4^2 - 8(4) + 15 = 16 - 32 + 15 = -1$$

Thus, the vertex is $(4, -1)$.

Step 2 The y-intercept is at $f(0)$. We evaluate $f(0)$ to find the y-coordinate when x is 0.

$$f(0) = 0^2 - 8(0) + 15 = 15$$

The y-intercept is $(0, 15)$.

Step 3 If there are x-intercepts, they will occur when $f(x) = 0$—that is, when $x^2 - 8x + 15 = 0$. We solve for x.

$$(x - 5)(x - 3) = 0$$

$$x - 5 = 0 \qquad x - 3 = 0$$

$$x = 5 \qquad x = 3$$

Thus, we conclude that the x-intercepts are $(5, 0)$ and $(3, 0)$.

We list these four important points of the function in table form.

Name	x	f(x)
Vertex	4	−1
y-intercept	0	15
x-intercept	5	0
x-intercept	3	0

□

Student Practice 1 Find the coordinates of the vertex and the intercepts of the quadratic function $f(x) = x^2 - 6x + 5$.

Graphing Calculator

Finding the x-intercepts and the Vertex

To find the intercepts of the quadratic function $f(x) = x^2 - 4x + 3$, graph $y = x^2 - 4x + 3$ on a graphing calculator using an appropriate window.
Display:

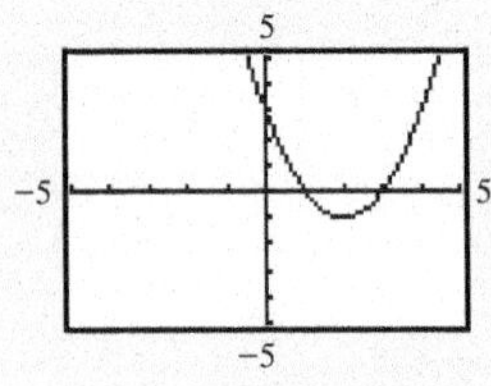

Next you can use the Trace and Zoom features or zero command of your calculator to find the x-intercepts.
You can also use the Trace and Zoom features to determine the vertex.
Some calculators have a feature that will calculate the maximum or minimum point on the graph. Use the feature that calculates the minimum point to find the vertex of $f(x) = x^2 - 4x + 3$.
Display:

5
-5 5
Minimum X=2 Y=-1
-5

Thus, the vertex is $(2, -1)$.

2 Graphing a Quadratic Function

It is helpful to find the vertex and the intercepts of a quadratic function before graphing it.

Example 2 Find the vertex and the intercepts, and then graph the function $f(x) = x^2 + 2x - 4$.

Solution Here $a = 1, b = 2$, and $c = -4$. Since $a > 0$, the parabola opens *upward*.

Step 1 We find the vertex.

$$x = \frac{-b}{2a} = \frac{-2}{2(1)} = \frac{-2}{2} = -1$$

$$f(-1) = (-1)^2 + 2(-1) - 4 = 1 + (-2) - 4 = -5$$

The vertex is $(-1, -5)$.

Step 2 We find the y-intercept. The y-intercept is at $f(0)$.

$$f(0) = (0)^2 + 2(0) - 4 = -4$$

The y-intercept is $(0, -4)$.

Step 3 We find the x-intercepts. The x-intercepts occur when $f(x) = 0$. Thus, we solve $x^2 + 2x - 4 = 0$ for x. We cannot factor this equation, so we use the quadratic formula.

$$x = \frac{-b \pm \sqrt{b^2 - 4ac}}{2a} = \frac{-2 \pm \sqrt{2^2 - 4(1)(-4)}}{2(1)} = \frac{-2 \pm \sqrt{20}}{2} = -1 \pm \sqrt{5}$$

To aid our graphing, we will approximate the value of x to the nearest tenth by using a square root table or a scientific calculator.

1 [+/−] [+] 5 [√] [=] 1.236068

$x \approx 1.2$

1 [+/−] [−] 5 [√] [=] −3.236068

$x \approx -3.2$

The x-intercepts are approximately $(-3.2, 0)$ and $(1.2, 0)$.

We have found that the vertex is $(-1, -5)$; the y-intercept is $(0, -4)$; and the x-intercepts are approximately $(-3.2, 0)$ and $(1.2, 0)$. We connect these points by a smooth curve to graph the parabola.

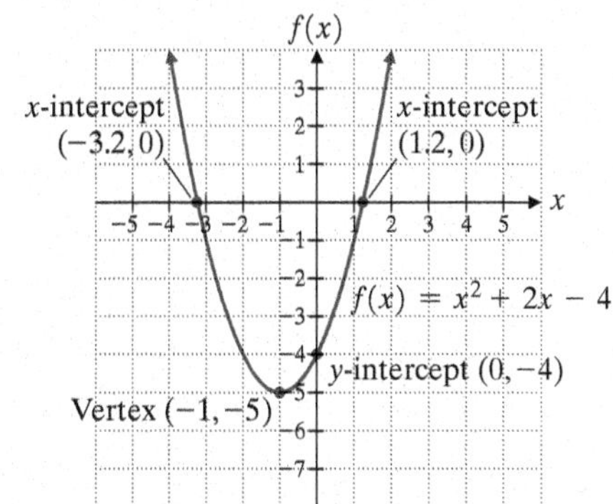

□

Student Practice 2 Find the vertex and the intercepts, and then graph the function $g(x) = x^2 - 2x - 2$.

Student Practice 2

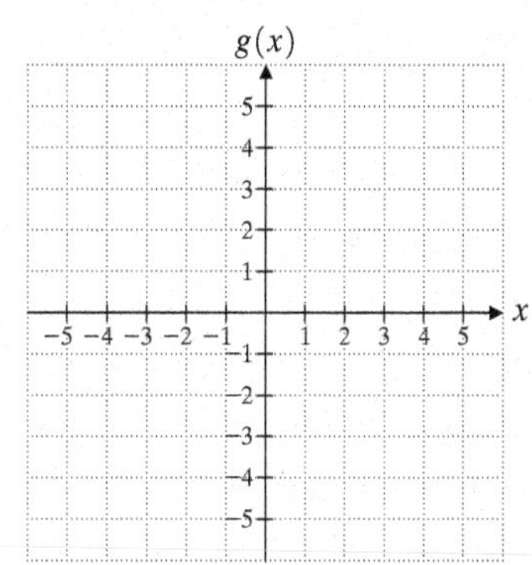

Example 3 Find the vertex and the intercepts, and then graph the function $f(x) = -2x^2 + 4x - 3$.

Solution Here $a = -2, b = 4$, and $c = -3$. Since $a < 0$, the parabola opens *downward*.

The vertex occurs at $x = \frac{-b}{2a}$.

$$x = \frac{-4}{2(-2)} = \frac{-4}{-4} = 1$$

$$f(1) = -2(1)^2 + 4(1) - 3 = -2 + 4 - 3 = -1$$

The vertex is $(1, -1)$.

The y-intercept is at $f(0)$.

$$f(0) = -2(0)^2 + 4(0) - 3 = -3$$

The y-intercept is $(0, -3)$.

If there are any x-intercepts, they will occur when $f(x) = 0$. We use the quadratic formula to solve $-2x^2 + 4x - 3 = 0$ for x.

$$x = \frac{-4 \pm \sqrt{4^2 - 4(-2)(-3)}}{2(-2)} = \frac{-4 \pm \sqrt{-8}}{-4}$$

Because $\sqrt{-8}$ yields an imaginary number, there are no real roots. Thus, there are no x-intercepts for the graph of the function. That is, the graph does not intersect the x-axis.

We know that the parabola opens *downward*. Thus, the vertex is a maximum value at $(1, -1)$. Since this graph has no x-intercepts, we will look for three additional points to help us in drawing the graph. We try $f(2), f(3)$, and $f(-1)$.

$$f(2) = -2(2)^2 + 4(2) - 3 = -8 + 8 - 3 = -3$$

$$f(3) = -2(3)^2 + 4(3) - 3 = -18 + 12 - 3 = -9$$

$$f(-1) = -2(-1)^2 + 4(-1) - 3 = -2 - 4 - 3 = -9$$

We plot the vertex, the y-intercept, and the points $(2, -3)$, $(3, -9)$, and $(-1, -9)$.

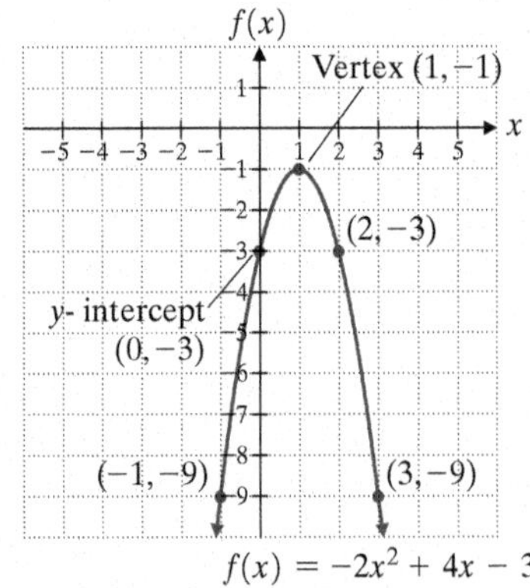

Student Practice 3 Find the vertex and the intercepts, and then graph the function $g(x) = -2x^2 - 8x - 6$.

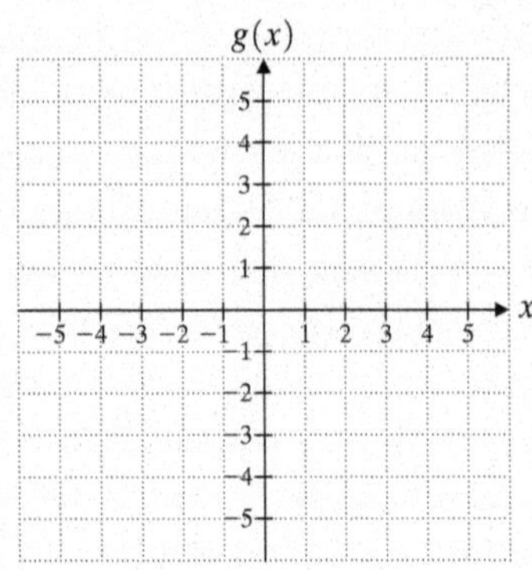

9.5 Exercises MyMathLab®

Find the coordinates of the vertex, the y-intercept, and the x-intercepts (if any exist) of each of the following quadratic functions. When necessary, approximate the x-intercepts to the nearest tenth.

1. $f(x) = x^2 - 8x + 15$

2. $f(x) = x^2 + 2x - 24$

3. $g(x) = -x^2 - 8x + 9$

4. $g(x) = -x^2 + 4x + 32$

5. $p(x) = 3x^2 + 12x + 3$

6. $p(x) = 2x^2 - 4x - 1$

7. $r(x) = -3x^2 - 2x - 6$

8. $f(x) = -2x^2 + 3x - 2$

9. $f(x) = 2x^2 + 2x - 4$

10. $f(x) = 5x^2 + 2x - 3$

In each of the following exercises, find the vertex, the y-intercept, and the x-intercepts (if any exist), and then graph the function.

11. $f(x) = x^2 - 6x + 8$

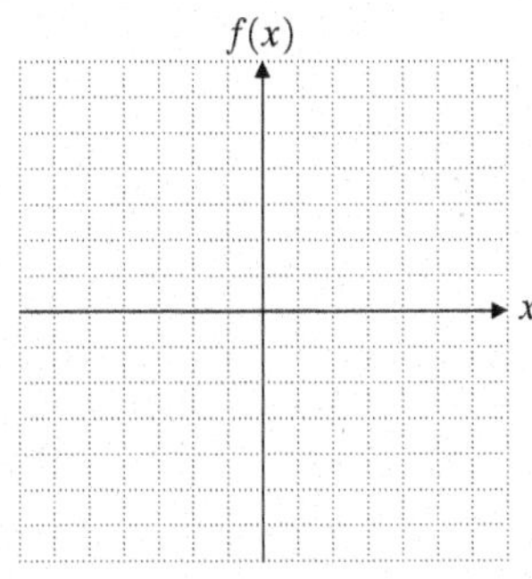

12. $f(x) = x^2 + 6x + 8$

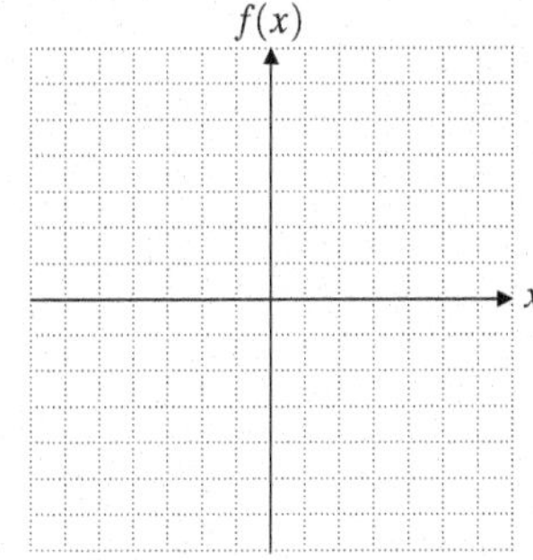

13. $g(x) = x^2 + 2x - 8$

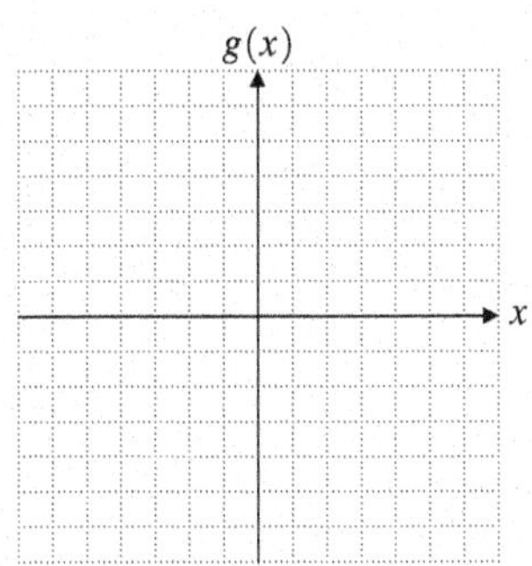

14. $g(x) = x^2 - 2x - 8$

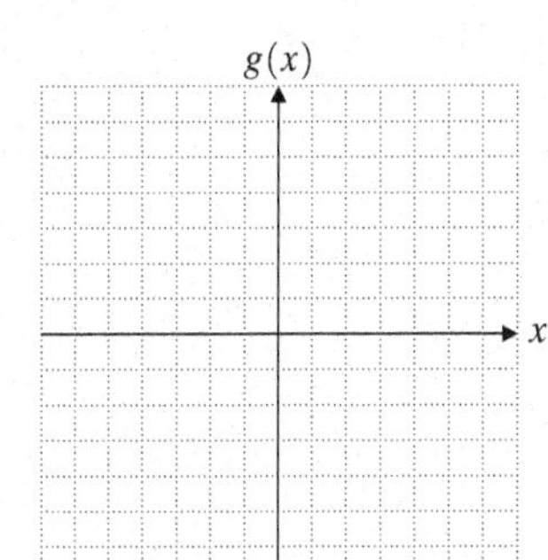

15. $p(x) = -x^2 + 8x - 12$

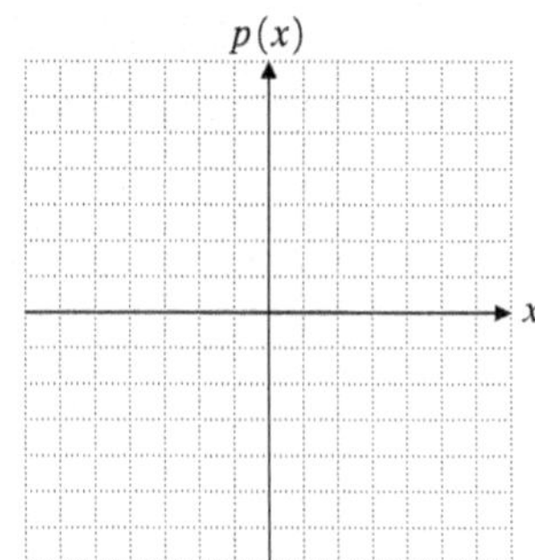

16. $p(x) = -x^2 - 8x - 12$

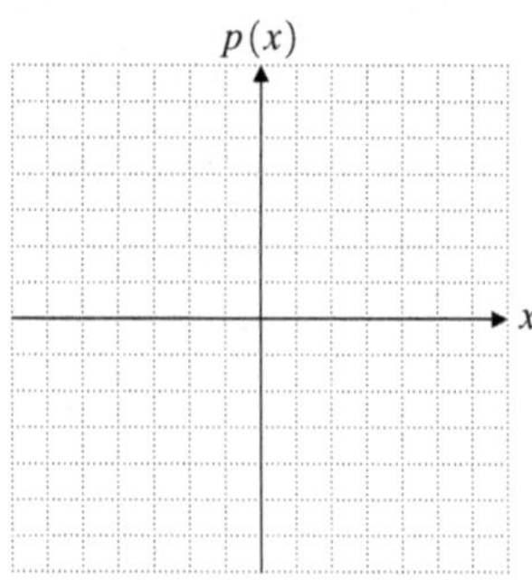

17. $r(x) = 3x^2 + 6x + 4$

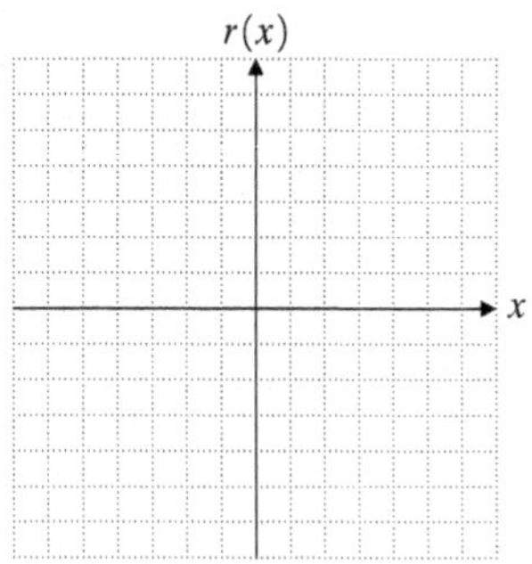

18. $r(x) = -3x^2 + 6x - 4$

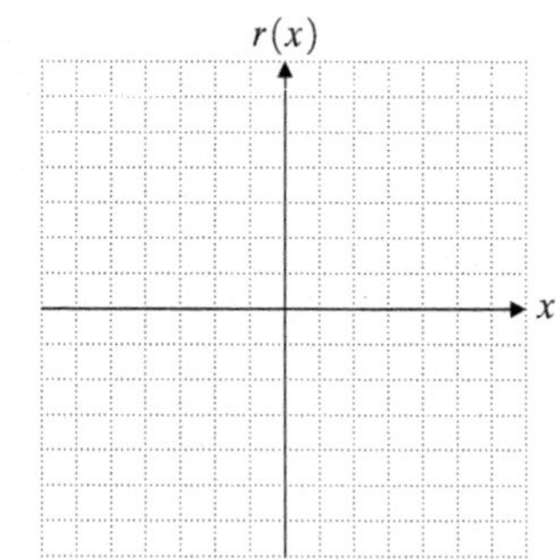

19. $f(x) = x^2 - 6x + 5$

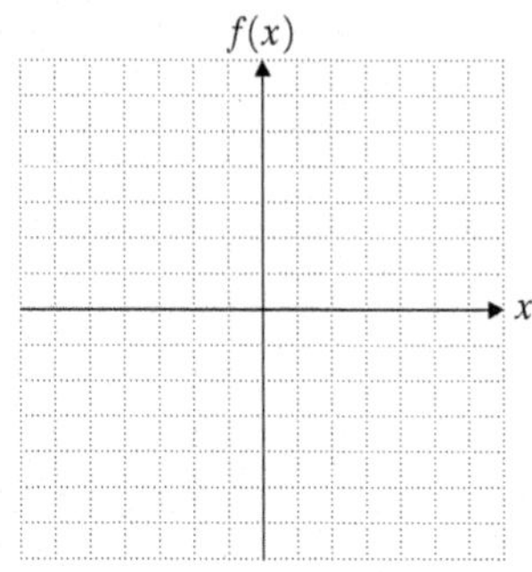

20. $g(x) = 2x^2 - 2x + 1$

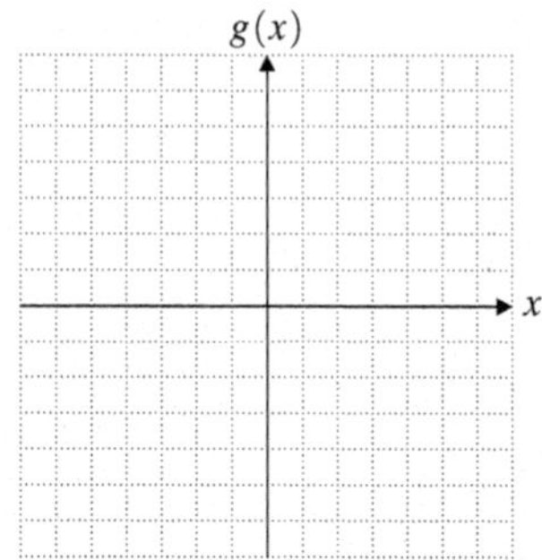

21. $f(x) = x^2 - 4x + 4$

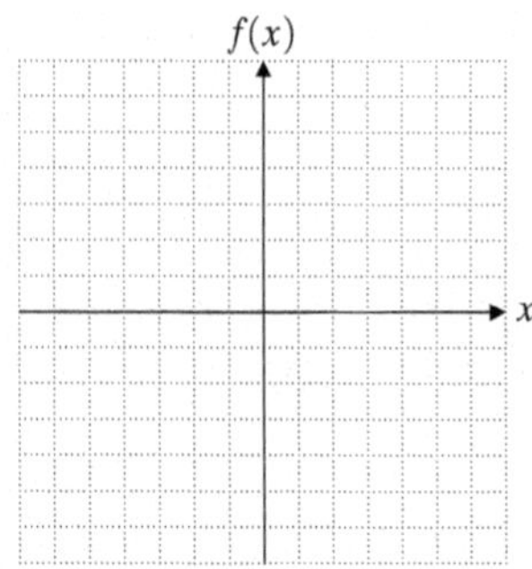

22. $g(x) = -x^2 + 6x - 9$

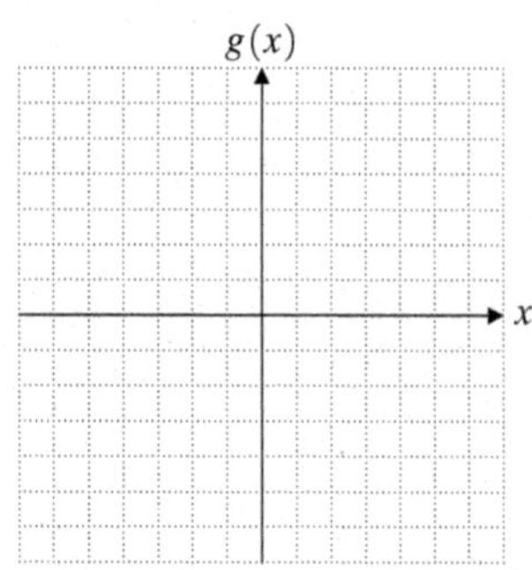

23. $f(x) = x^2 - 4$

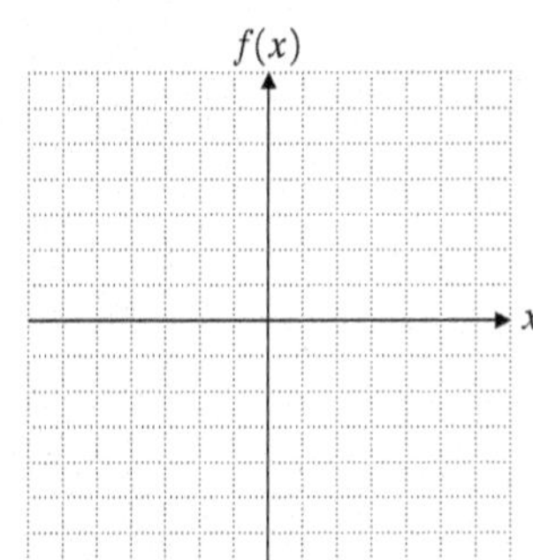

24. $r(x) = -x^2 + 1$

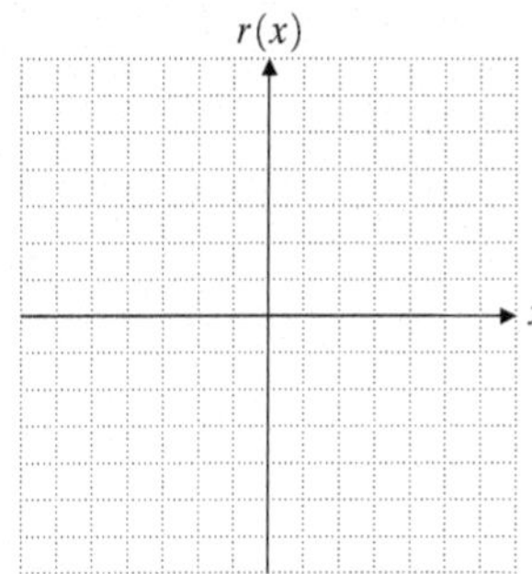

Applications

Boating *Some leisure activities such as boating are expensive. People with larger incomes are more likely to participate in such an activity. The number of people N (measured in thousands) who participate in boating (motor and power boating) can be described by the function* $N(x) = 1.99x^2 - 6.76x + 1098.4$, *where x is the mean income (measured in thousands) and* $x \geq 10$. *Use this information to answer exercises 25–30.* (*Source:* www.census.gov)

25. Find $N(10), N(30), N(50), N(70)$, and $N(90)$.

26. Use the results of exercise 25 to graph the function from $x = 10$ to $x = 90$. You may use the graph grid provided at the bottom of the page.

27. Find $N(60)$ from your graph. Explain what $N(60)$ means.

28. Find $N(60)$ from the equation for $N(x)$. Compare your answers for exercises 27 and 28.

29. Use your graph to find the value of x when $N(x)$ is equal to 10,000. Explain what this means.

30. Use your graph to find the value of x when $N(x)$ is equal to 4000. Explain what this means.

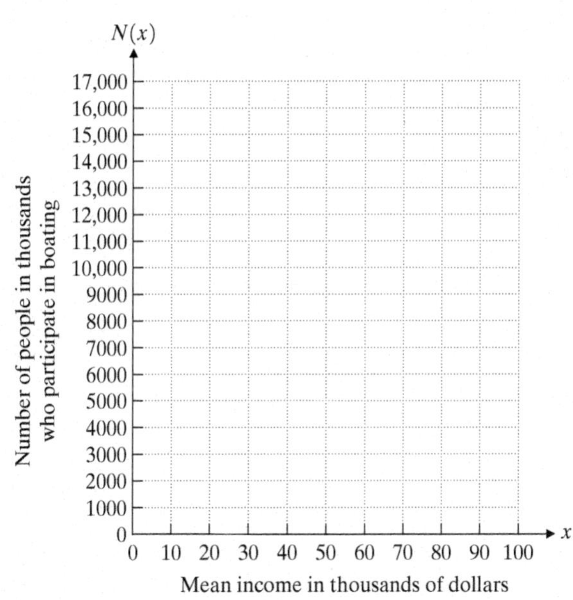

Company Owner *The daily profit P in dollars of Pine Tree Picnic Tables in the summer months is described by the function* $P(x) = -5x^2 + 300x - 4000$, *where x is the number of tables that are manufactured in one day. Scott, the owner, wants to be prepared for the summer. He is studying the graph of this function so he can determine how many employees to hire and have enough supplies on hand to make tables. Use this information to answer exercises 31–36.*

31. Find $P(18)$, $P(20)$, $P(28)$, $P(35)$, and $P(42)$.

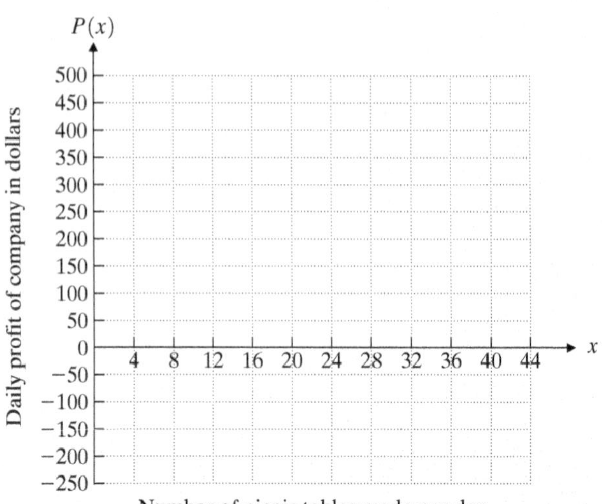

32. Use the results of exercise 31 to graph the function from $x = 18$ to $x = 42$.

33. The maximum profit of the company occurs at the vertex of the parabola. How many picnic tables should be made per day in order to obtain the maximum profit for the company? What is the maximum profit?

34. How many picnic tables per day should be made in order to obtain a daily profit of \$420? Why are there two answers to this question?

35. How many picnic tables are made per day if the company has a daily profit of zero dollars?

36. How many picnic tables are made per day if the company has a daily *loss* of \$105?

37. ***Softball*** Susan throws a softball upward into the air at a speed of 32 feet per second from a 40-foot platform. The height of the ball after t seconds is given by the function $d(t) = -16t^2 + 32t + 40$. What is the maximum height of the softball? How many seconds does it take to reach the ground after first being thrown upward? (Round your answer to the nearest tenth.)

38. ***Baseball*** Henry is standing on a platform overlooking a baseball stadium. It is 160 feet above the playing field. When he throws a baseball upward at 64 feet per second, the distance d from the baseball to the ground after t seconds is given by the function $d(t) = -16t^2 + 64t + 160$. What is the maximum height of the baseball if he throws it upward? How many seconds does it take until the ball finally hits the ground? (Round your answer to the nearest tenth.)

Optional Graphing Calculator Problems

39. Graph $y = 2.3x^2 - 5.4x - 1.6$. Find the x-intercepts to the nearest tenth.

40. Graph $y = -4.6x^2 + 7.2x - 2.3$. Find the x-intercepts to the nearest tenth.

To Think About

41. A graph of a quadratic equation of the form $y = ax^2 + bx + c$ passes through the points $(0, -10)$, $(3, 41)$, and $(-1, -15)$. What are the values of a, b, and c?

42. A graph of a quadratic equation of the form $y = ax^2 + bx + c$ has a vertex of $(1, -25)$ and passes through the point $(0, -24)$. What are the values of a, b, and c?

Cumulative Review *Solve each system.*

43. **[4.2.2]**
$$\begin{aligned} 3x - y + 2z &= 12 \\ 2x - 3y + z &= 5 \\ x + 3y + 8z &= 22 \end{aligned}$$

44. **[4.2.2]**
$$\begin{aligned} 7x + 3y - z &= -2 \\ x + 5y + 3z &= 2 \\ x + 2y + z &= 1 \end{aligned}$$

Quick Quiz 9.5 *Consider the function* $f(x) = -2x^2 - 4x + 6$.

1. Find the vertex.

2. Find the y-intercept and the x-intercepts.

3. Using the above points, graph the function.

$f(x)$

x

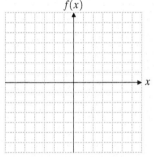

4. ***Concept Check*** Explain how you would find the vertex of $f(x) = 4x^2 - 9x - 5$.

9.6 Compound and Quadratic Inequalities

Student Learning Objectives

After studying this section, you will be able to:

1. Graph compound inequalities that use *and*.
2. Graph compound inequalities that use *or*.
3. Solve compound inequalities and graph their solutions.
4. Solve a factorable quadratic inequality in one variable.
5. Solve a nonfactorable quadratic inequality in one variable.

1 Graphing Compound Inequalities That Use *and*

In Section 2.8 we learned how to solve and graph inequalities such as $2x + 3 < 11$. In this section we will discuss inequalities that consist of two inequalities connected by the word *and* or the word *or*. They are called **compound inequalities.**

The solution of a compound inequality using the connective ***and*** includes all the numbers that make both parts true at the same time.

Example 1 Graph the values of x where $7 < x$ *and* $x < 12$.

Solution We read the inequality starting with the variable. Thus, we graph all values of x, where x is greater than 7 and where x is less than 12. All such values must be between 7 and 12. Numbers that are greater than 7 and less than 12 can be written as $7 < x < 12$.

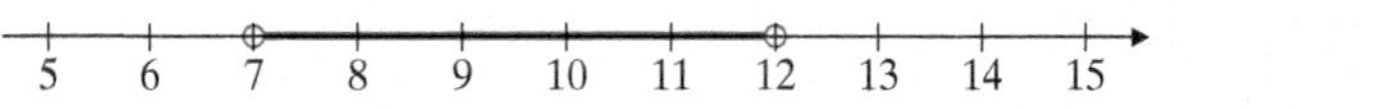

Student Practice 1 Graph the values of x where $-8 < x$ *and* $x < -2$.

−11 −10 −9 −8 −7 −6 −5 −4 −3 −2 −1

Example 2 Graph the values of x where $-6 \leq x \leq 2$.

Solution Here we have that x is greater than or equal to -6 and that x is less than or equal to 2. We remember to include the points -6 and 2 since the inequality symbols contain the equals sign.

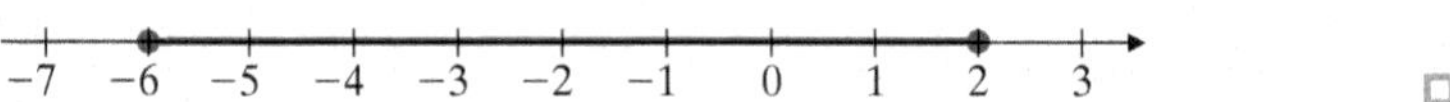

Student Practice 2 Graph the values of x where $-1 \leq x \leq 5$.

−4 −3 −2 −1 0 1 2 3 4 5 6

Example 3 Graph the salary range (s) of the full-time employees of Tentron Corporation. Each person earns at least \$190 weekly, but not more than \$800 weekly.

Solution "At least \$190" means that the weekly salary of each person is greater than or equal to \$190 weekly. We write $s \geq \$190$. "Not more than" means that the weekly salary of each person is less than or equal to \$800. We write $s \leq \$800$. Thus, s may be between 190 and 800 and may include those endpoints.

Student Practice 3 Graph the weekly salary range of a person who earns at least \$200 per week, but never more than \$950 per week.

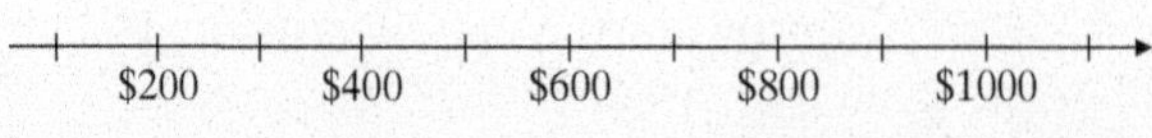

2 Graphing Compound Inequalities That Use *or*

The solution of a compound inequality using the connective ***or*** includes all the numbers that are solutions of either of the two inequalities.

Example 4 Graph the region where $x < 3$ *or* $x > 6$.

Solution Notice that a solution to this inequality need not be in both regions at the same time.

Read the inequality as "x is less than 3 or x is greater than 6." Thus, x can be less than 3 or x can be greater than 6. This includes all values to the left of 3 as well as all values to the right of 6 on a number line. We shade these regions.

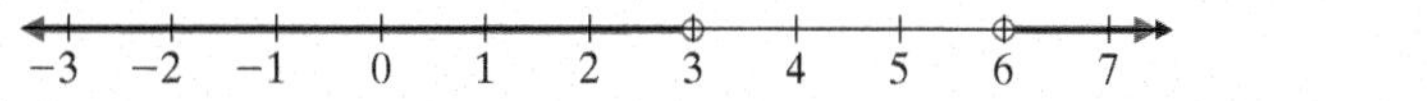

Student Practice 4 Graph the region where $x < 8$ *or* $x > 12$.

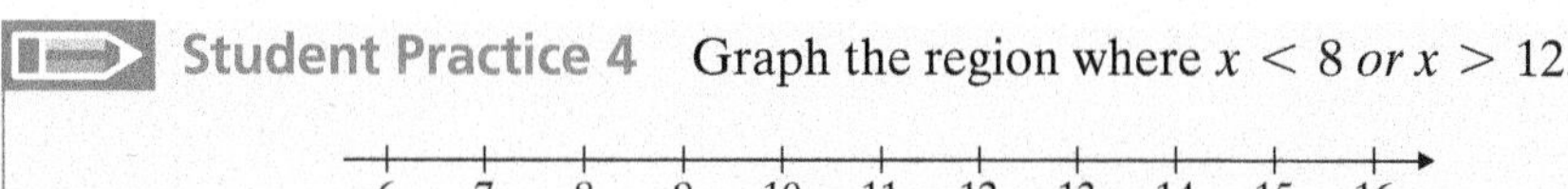

Example 5 Graph the region where $x > -2$ *or* $x \le -5$.

Solution Note the shaded circle at -5 and the open circle at -2.

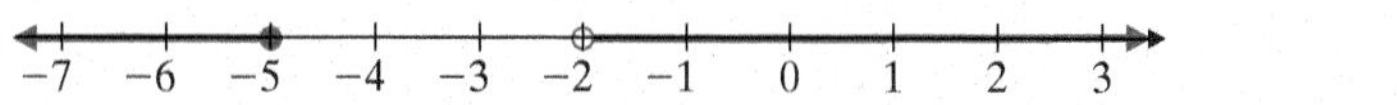

Student Practice 5 Graph the region where $x \le -6$ *or* $x > 3$.

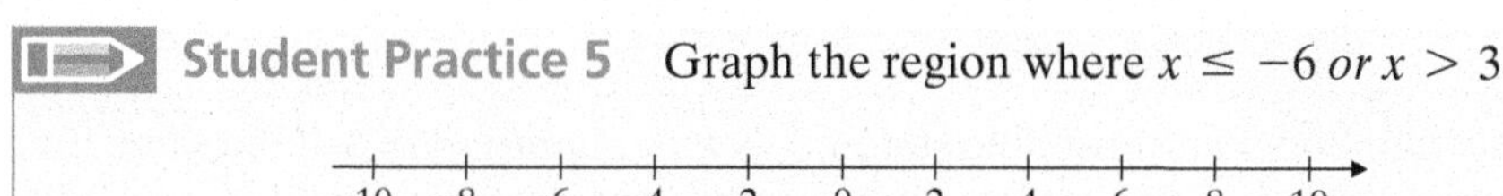

Example 6 Male applicants for the state police force in Fred's home state are ineligible for the force if they are shorter than 60 inches or taller than 76 inches. Graph the range of rejected applicants' heights.

Solution Each rejected applicant's height h will be less than 60 inches $(h < 60)$ or will be greater than 76 inches $(h > 76)$.

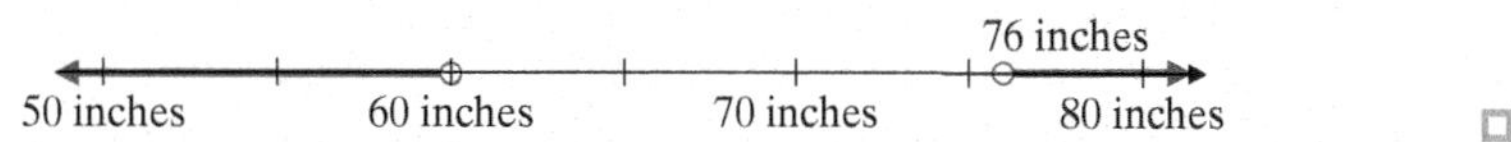

Student Practice 6 Female applicants are ineligible if they are shorter than 56 inches or taller than 70 inches. Graph the range of rejected applicant's heights.

50 in. 60 in. 70 in. 80 in. 90 in.

3 Solving Compound Inequalities and Graphing Their Solutions

When asked to solve a more complicated compound inequality for x, we normally solve each individual inequality separately.

Example 7 Solve for x and graph the compound solution.

$$3x + 2 > 14 \text{ or } 2x - 1 < -7$$

Solution We solve each inequality separately.

$$\begin{aligned} 3x + 2 &> 14 & \text{or} \quad 2x - 1 &< -7 \\ 3x &> 12 & 2x &< -6 \\ x &> 4 & x &< -3 \end{aligned}$$

The solution is $x < -3$ *or* $x > 4$.

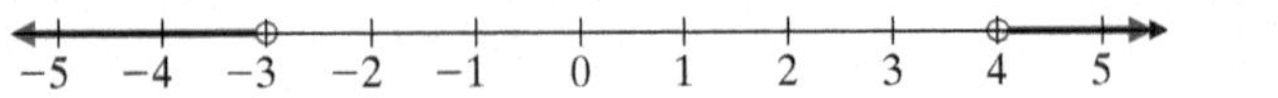

□

Student Practice 7 Solve for x and graph the compound solution.
$3x - 4 < -1$ *or* $2x + 3 > 13$

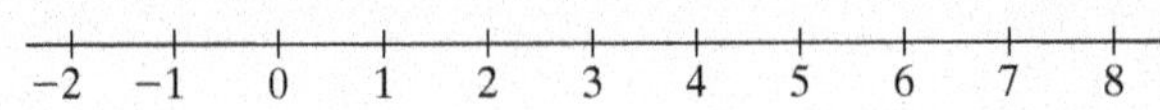

Example 8 Solve and graph. $2x + 5 \le 11$ *and* $-3x > 18$

Solution We solve each inequality separately.

$$\begin{aligned} 2x + 5 &\le 11 & \text{and} \quad -3x &> 18 \\ 2x &\le 6 & \frac{-3x}{-3} &< \frac{18}{-3} \\ x &\le 3 & x &< -6 \end{aligned}$$

Division by a negative number reverses the inequality symbol.

The separate solutions are $x < -6$ *and* $x \le 3$.

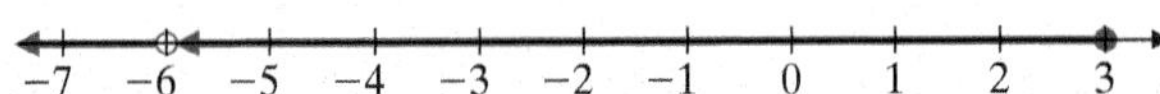

The only numbers that satisfy the statements $x \le 3$ *and* $x < -6$ at the same time are $x < -6$. Thus, $x < -6$ is the solution to the compound inequality.

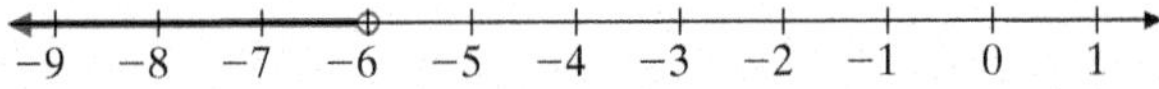

□

Student Practice 8 Solve and graph. $-2x + 3 < -7$ *and* $7x - 1 > -15$

Example 9 Solve. $-3x - 2 < -5$ *and* $4x + 6 < -12$

Solution We solve each inequality separately.

$$\begin{aligned} -3x - 2 &< -5 & \text{and} \quad 4x + 6 &< -12 \\ -3x &< -3 & 4x &< -18 \\ \frac{-3x}{-3} &> \frac{-3}{-3} & \frac{4x}{4} &< \frac{-18}{4} \\ x &> 1 & x &< -4\frac{1}{2} \end{aligned}$$

Now, clearly it is impossible for one number to be greater than 1 *and* at the same time be less than $-4\frac{1}{2}$.

Thus, there is *no solution*. We can express this by the notation $\varnothing$, which is the **empty set.** Or we can just state, "There is no solution."

□

Student Practice 9 Solve. $-3x - 11 < -26$ *and* $5x + 4 < 14$

4 Solving a Factorable Quadratic Inequality in One Variable

We will now solve quadratic inequalities such as $x^2 - 2x - 3 > 0$ and $2x^2 + x - 15 < 0$. A **quadratic inequality** has the form $ax^2 + bx + c < 0$ (or replace $<$ by $>$, $\le$, or $\ge$), where a, b, and c are real numbers and $a \neq 0$. We use our knowledge of solving quadratic equations to solve quadratic inequalities.

Let's solve the inequality $x^2 - 2x - 3 > 0$. We want to find the two points where the expression on the left side is equal to zero. We call these the **boundary points.** To do this, we replace the inequality symbol by an equals sign and solve the resulting equation.

$$x^2 - 2x - 3 = 0$$

$$(x + 1)(x - 3) = 0 \quad \text{Factor.}$$

$$x + 1 = 0 \quad \text{or} \quad x - 3 = 0 \quad \text{Zero factor property}$$

$$x = -1 \qquad\qquad x = 3$$

These two solutions form boundary points that divide a number line into three segments.

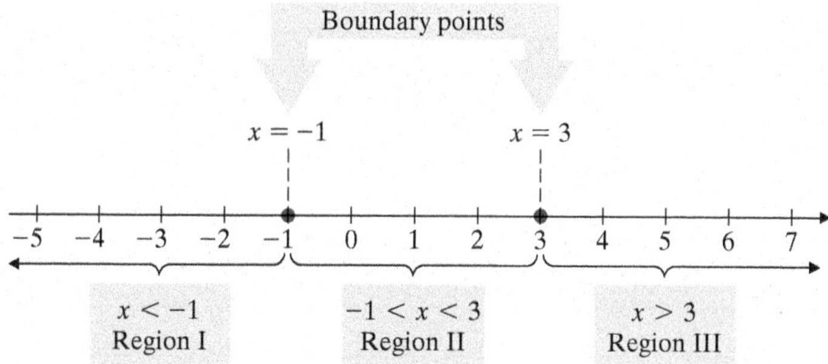

All values of x in a given segment produce results that are greater than zero, or all values of x in a given segment produce results that are less than zero.

To solve the quadratic inequality, we pick an arbitrary test point in each region and then substitute it into the inequality to determine whether it satisfies the inequality. If one point in a region satisfies the inequality, then *all* points in the region satisfy the inequality. We will test three values of x in the expression $x^2 - 2x - 3$.

$x < -1$, *Region I*: A sample point is $x = -2$.

$$(-2)^2 - 2(-2) - 3 = 4 + 4 - 3 = 5 > 0$$

$-1 < x < 3$, *Region II*: A sample point is $x = 0$.

$$(0)^2 - 2(0) - 3 = 0 + 0 - 3 = -3 < 0$$

$x > 3$, *Region III*: A sample point is $x = 4$.

$$(4)^2 - 2(4) - 3 = 16 - 8 - 3 = 5 > 0$$

Thus, we see that $x^2 - 2x - 3 > 0$ when $x < -1$ or $x > 3$. No points in Region II satisfy the inequality. The graph of the solution is shown next.

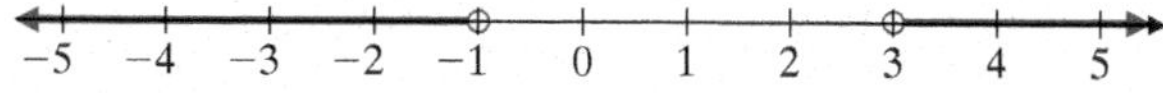

We summarize our method.

SOLVING A QUADRATIC INEQUALITY

1. Replace the inequality symbol by an equals sign. Solve the resulting equation to find the boundary points.
2. Use the boundary points to separate a number line into three distinct regions.
3. Evaluate the quadratic expression at a test point in each region.
4. Determine which regions satisfy the original conditions of the quadratic inequality.

Example 10 Solve and graph $x^2 - 10x + 24 > 0$.

Solution

1. We replace the inequality symbol by an equals sign and solve the resulting equation.

$$x^2 - 10x + 24 = 0$$

$$(x - 4)(x - 6) = 0$$

$$x - 4 = 0 \quad \text{or} \quad x - 6 = 0$$

$$x = 4 \qquad\qquad x = 6$$

2. We use the boundary points to separate a number line into distinct regions.

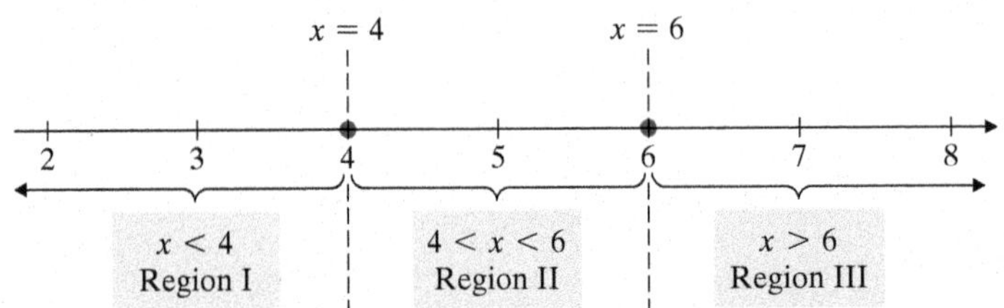

3. We evaluate the quadratic expression at a test point in each of the regions.

$$x^2 - 10x + 24$$

$x < 4$, *Region I*: We pick the test point $x = 1$.

$$(1)^2 - 10(1) + 24 = 1 - 10 + 24 = 15 > 0$$

$4 < x < 6$, *Region II*: We pick the test point $x = 5$.

$$(5)^2 - 10(5) + 24 = 25 - 50 + 24 = -1 < 0$$

$x > 6$, *Region III*: We pick the test point $x = 7$.

$$(7)^2 - 10(7) + 24 = 49 - 70 + 24 = 3 > 0$$

4. We determine which regions satisfy the original conditions of the quadratic inequality.

$$x^2 - 10x + 24 > 0 \text{ when } x < 4 \text{ or when } x > 6.$$

The graph of the solution is shown next.

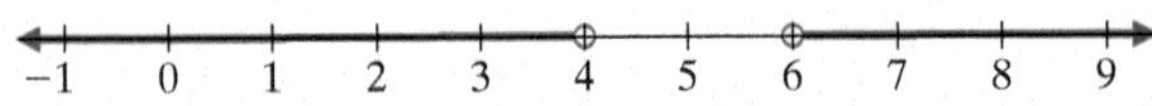

Student Practice 10

−3 −2 −1 0 1 2 3 4 5

Student Practice 10 Solve and graph $x^2 - 2x - 8 < 0$.

Example 11 Solve and graph $2x^2 + x - 6 \le 0$.

Solution We replace the inequality symbol by an equals sign and solve the resulting equation.

$$2x^2 + x - 6 = 0$$
$$(2x - 3)(x + 2) = 0$$
$$2x - 3 = 0 \quad \text{or} \quad x + 2 = 0$$
$$2x = 3 \qquad\qquad x = -2$$
$$x = \frac{3}{2} = 1.5$$

We use the boundary points to separate a number line into distinct regions. The boundary points are $x = -2$ and $x = 1.5$. Now we arbitrarily pick a test point in each region.

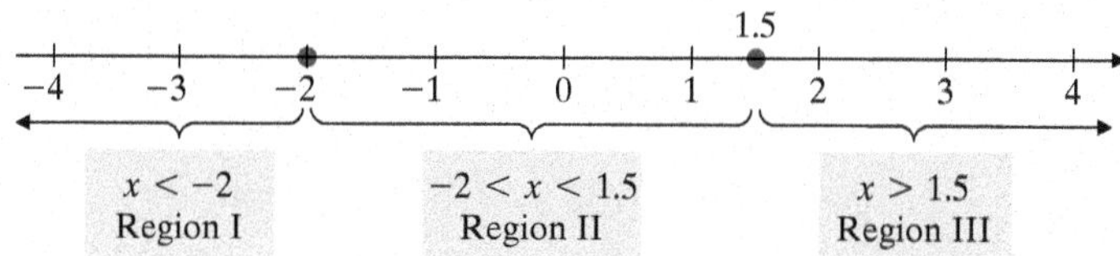

We will pick 3 values of x for the polynomial $2x^2 + x - 6$.

Region I: First we need an x-value less than -2. We pick $x = -3$.

$$2(-3)^2 + (-3) - 6 = 18 - 3 - 6 = 9 > 0$$

Region II: Next we need an x-value between -2 and 1.5. We pick $x = 0$.

$$2(0)^2 + (0) - 6 = 0 + 0 - 6 = -6 < 0$$

Region III: Finally we need an x-value greater than 1.5. We pick $x = 2$.

$$2(2)^2 + (2) - 6 = 8 + 2 - 6 = 4 > 0$$

Since our inequality is $\le$ and not just $<$, we need to include the boundary points. Thus, $2x^2 + x - 6 \le 0$ when $-2 \le x \le 1.5$. The graph of our solution is shown next.

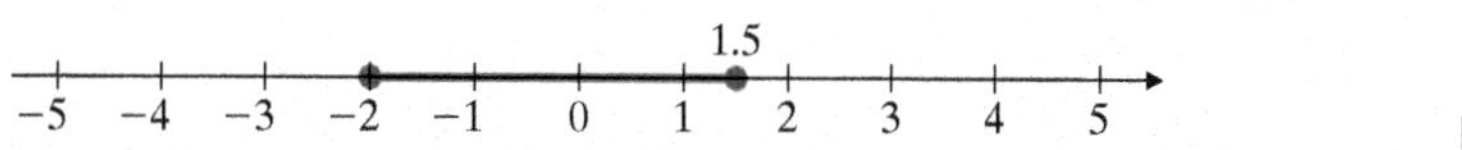

□

Student Practice 11 Solve and graph $3x^2 - x - 2 \ge 0$.

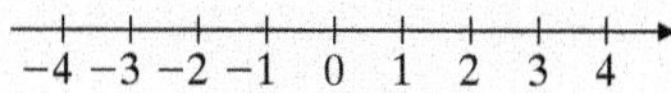

Graphing Calculator

Solving Quadratic Inequalities

To solve Example 11 on a graphing calculator, graph $y = 2x^2 + x - 6$, zoom in on the two x-intercepts, and use the Trace feature to find the value of x where y changes from a negative value to a positive value or zero. You can use the Zero command if your calculator has it. Can you verify from your graph that the solution to Example 11 is $-2 \le x \le 1.5$? Use your graphing calculator and this method to solve the exercises below. Round your answers to the nearest hundredth.

1. $3x^2 - 3x - 60 \ge 0$
2. $2.1x^2 + 4.3x - 29.7 > 0$
3. $15.3x^2 - 20.4x + 6.8 \ge 0$
4. $16.8x^2 > -16.8x - 35.7$

If you are using a graphing calculator and you think there is no solution to a quadratic inequality, how can you *verify* this from the graph? Why is this possible?

5 Solving a Nonfactorable Quadratic Inequality in One Variable

If the quadratic expression in a quadratic inequality cannot be factored, then we will use the quadratic formula to obtain the boundary points.

Example 12 Solve and graph $x^2 + 4x > 6$. Round your answer to the nearest tenth.

Solution First we write $x^2 + 4x - 6 > 0$. Because we cannot factor $x^2 + 4x - 6$, we use the quadratic formula to find the boundary points.

$$x = \frac{-4 \pm \sqrt{4^2 - 4(1)(-6)}}{2(1)} = \frac{-4 \pm \sqrt{16 + 24}}{2}$$

$$= \frac{-4 \pm \sqrt{40}}{2} = \frac{-4 \pm 2\sqrt{10}}{2} = -2 \pm \sqrt{10}$$

Using a calculator, we find the following:

$$-2 + \sqrt{10} \approx -2 + 3.162 \approx 1.162 \text{ or about } 1.2$$

$$-2 - \sqrt{10} \approx -2 - 3.162 \approx -5.162 \text{ or about } -5.2$$

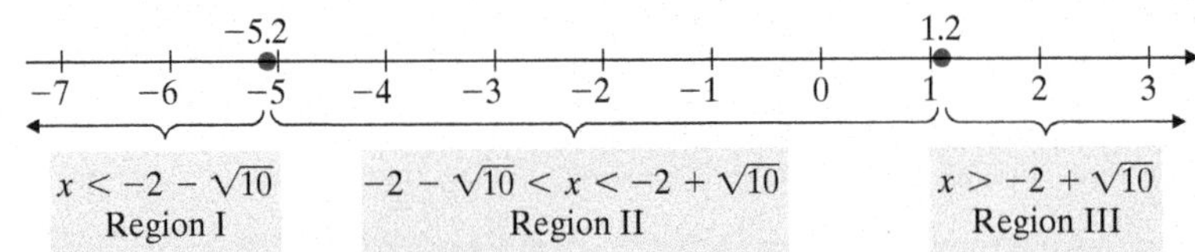

We will see where $x^2 + 4x - 6 > 0$.

Region I: Test $x = -6$.

$$(-6)^2 + 4(-6) - 6 = 36 - 24 - 6 = 6 > 0$$

Region II: Test $x = 0$.

$$(0)^2 + 4(0) - 6 = 0 + 0 - 6 = -6 < 0$$

Region III: Test $x = 2$.

$$(2)^2 + 4(2) - 6 = 4 + 8 - 6 = 6 > 0$$

Thus, $x^2 + 4x > 6$ when $x^2 + 4x - 6 > 0$, and this occurs when $x < -2 - \sqrt{10}$ or $x > -2 + \sqrt{10}$. Rounding to the nearest tenth, our answer is

$$x < -5.2 \quad \text{or} \quad x > 1.2.$$

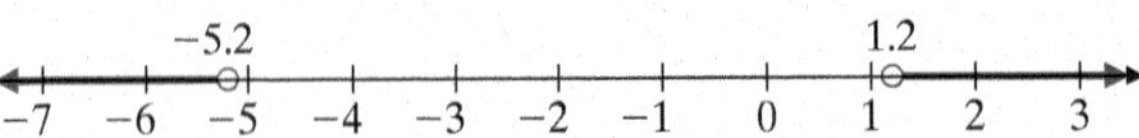

□

Student Practice 12 Solve and graph $x^2 + 2x < 7$. Round your answer to the nearest tenth.

Student Practice 12

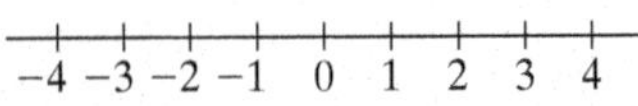

9.6 Exercises MyMathLab®

Graph the values of x that satisfy the conditions given.

1. $3 < x$ *and* $x < 8$

2. $5 < x$ *and* $x < 10$

3. $-4 < x$ *and* $x < 2$

4. $-5 < x$ *and* $x < -1$

5. $7 < x < 9$

6. $3 < x < 5$

7. $-2 < x \leq \frac{1}{2}$

8. $-\frac{7}{2} \leq x < 2$

9. $x > 8$ *or* $x < 2$

10. $x \geq 2$ *or* $x \leq 1$

11. $x \leq -\frac{5}{2}$ *or* $x > 4$

12. $x < 0$ *or* $x > \frac{9}{2}$

13. $x \leq -10$ *or* $x \geq 40$

14. $x \leq -6$ *or* $x \geq 2$

Solve for x and graph your results.

15. $2x + 3 \leq 5$ *and* $x + 1 \geq -2$

16. $4x - 1 < 7$ *and* $x \geq -1$

17. $4x - 6 > 4$ *or* $x + 4 < -2$

18. $x + 1 \geq 5$ *or* $x + 5 < 2.5$

19. $x < 8$ *and* $x > 10$

20. $x < 6$ *and* $x > 9$

Applications *Express as a compound inequality.*

21. ***Toothpaste*** A tube of toothpaste is not properly filled if the amount of toothpaste t in the tube is more than 11.2 ounces or less than 10.9 ounces.

22. ***Clothing Standards*** The width of a seam on a pair of blue jeans is unacceptable if it is narrower than 10 millimeters or wider than 12 millimeters.

23. ***Interstate Highway Travel*** The number of cars c driving over Interstate 91 during the evening hours in January is always at least 5000, but never more than 12,000.

24. ***Campsite Capacity*** The number of campers c at a campsite during the Independence Day weekend is always at least 490, but never more than 2000.

Temperature Conversion *Solve the following application problems by using the formula* $C = \frac{5}{9}(F - 32)$. *Round to the nearest tenth.*

25. When visiting Montreal this spring, Marcos had been advised that the temperature could range from −20°C to 11°C. Find an inequality that represents the range in Fahrenheit temperatures.

26. The temperature in Sao Paulo, Brazil, during January can range from 16°C to 24°C. Find an inequality that represents the range in Fahrenheit temperatures.

Exchange Rates *At one point in 2015, the exchange rate for converting American dollars into Japanese yen was* $Y = 119(d - 5)$. *In this equation, d is the number of American dollars, Y is the number of yen, and $5 represents a onetime fee that banks sometimes charge for currency conversion. Use the equation to solve the following problems. (Round answers to the nearest cent.)*

27. Frank is traveling to Tokyo, Japan, for two weeks, and he has been advised to have between 30,000 yen and 35,000 yen for spending money for each week he is there. Including the conversion charge, write an inequality that represents the number of American dollars he will need to exchange at the bank for this two-week period.

28. Carrie is traveling to Osaka, Japan, for three weeks. Her friend told her she should plan to have between 23,000 yen and 28,000 yen for spending money for each week she is there. Including the conversion charge, write an inequality that represents the number of American dollars she will need to exchange at the bank for the three-week period.

Mixed Practice *Solve each compound inequality.*

29. $x - 3 > -5$ *and* $2x + 4 < 8$

30. $x - 2 < 9$ *and* $x + 3 < 6$

31. $-6x + 5 \geq -1$ *and* $2 - x \leq 5$

32. $5x + 6 \geq -9$ *and* $10 - x \geq 3$

33. $4x - 3 < -11$ *or* $7x + 2 \geq 23$

34. $5x + 1 < 1$ *or* $3x - 9 > 9$

35. $-0.3x + 1 \geq 0.2x$ *or* $-0.2x + 0.5 > 0.7$

36. $-0.3x - 0.4 \geq 0.1x$ *or* $0.2x + 0.3 \leq -0.4x$

37. $\frac{5x}{2} + 1 \geq 3$ *and* $x - \frac{2}{3} \geq \frac{4}{3}$

38. $\frac{5x}{3} - 2 < \frac{14}{3}$ *and* $3x + \frac{5}{2} < -\frac{1}{2}$

39. $2x + 5 < 3$ *and* $3x - 1 > -1$

40. $6x - 10 < 8$ *and* $2x + 1 > 9$

41. $2x - 3 \geq 7$ *and* $5x - 8 \leq 2x + 7$

42. $7x + 2 \geq 11x + 14$ *and* $x + 9 \geq 6$

To Think About *Solve the compound inequality.*

43. $\frac{1}{4}(x + 2) + \frac{1}{8}(x - 3) \leq 1$ *and* $\frac{3}{4}(x - 1) > -\frac{1}{4}$

44. $\frac{x - 4}{6} - \frac{x - 2}{9} \leq \frac{5}{18}$ *or* $-\frac{2}{5}(x + 3) < -\frac{6}{5}$

Verbal and Writing Skills, Exercises 45 and 46

45. When solving a quadratic inequality, why is it necessary to find the boundary points?

46. What is the difference between solving an exercise like $ax^2 + bx + c > 0$ and an exercise like $ax^2 + bx + c \geq 0$?

Solve and graph.

47. $x^2 + x - 12 < 0$

48. $x^2 + x - 12 > 0$

49. $x^2 \leq 25$

50. $x^2 - 16 \geq 0$

51. $2x^2 + x - 3 < 0$

52. $4x^2 + 7x - 2 < 0$

Solve.

53. $x^2 + x - 20 > 0$

54. $x^2 + 6x - 27 > 0$

55. $4x^2 \le 11x + 3$

56. $4x^2 - 5 \le -8x$

57. $6x^2 - 5x > 6$

58. $3x^2 + 13x > -4$

59. $-2x + 30 \ge x(x + 5)$

Hint: Put variables on the right and zero on the left in your first step.

60. $55 - x^2 \ge 6x$

Hint: Put variables on the right and zero on the left in your first step.

61. $x^2 - 2x \ge -1$

62. $x^2 + 25 \ge 10x$

63. $x^2 - 4x \le -4$

64. $x^2 - 6x \le -9$

Solve each of the following quadratic inequalities if possible. Round your answer to the nearest tenth.

65. $x^2 - 2x > 5$

66. $2x^2 - 4x > 3$

67. $2x^2 - 2x < 3$

68. $4x^2 < 2x + 3$

69. $2x^2 \ge x^2 - 4$

70. $5x^2 \ge 4x^2$

71. $5x^2 \le 4x^2 - 1$

72. $x^2 - 1 \le -17$

Applications

Projectile Flight *In exercises 73 and 74, a projectile is fired vertically with an initial velocity of 640 feet per second. The distance s in feet above the ground after t seconds is given by the equation* $s = -16t^2 + 640t$.

73. For what range of time t (measured in seconds) will the height s be greater than 6000 feet?

74. For what range of time t (measured in seconds) will the height s be less than 4800 feet?

Business Analyst *One part of the job of a business analyst is to study the relationship between profit made and numbers of products or services sold. In exercises 75 and 76, the profit of a manufacturing company is determined by the number of units x sold each day according to the given equation.* ***(a)*** *Find when the profit is greater than zero.* ***(b)*** *Find the daily profit when 50 units are sold.* ***(c)*** *Find the daily profit when 60 units are sold.*

75. Profit $= -10(x^2 - 200x + 1800)$

76. Profit $= -20(x^2 - 320x + 3600)$

Cumulative Review

Cost of a Cruise *The Circle Line Cruise is a 2-hour, 24-mile cruise around southern Manhattan (New York City). The charge for adults is \$18; children 12 and under, \$10; and seniors over 62, \$16. For the 3-hour, 35-mile cruise around all of Manhattan, the charge for adults is \$22; children 12 and under, \$12; and seniors over 62, \$19. The Yoffa family has come to New York for their family reunion and is planning family activities. The family has ten adults, fourteen children under 12, and five senior members.*

77. **[1.3.3]** What would it cost for all of the family to take the 2-hour trip? The 3-hour trip?

78. **[4.3.2]** Six people do not take the cruise. If the rest of the family takes a 2-hour cruise, it will cost \$314. If the rest of the family takes a 3-hour cruise, it will cost \$380. How many adults, how many children, and how many senior members plan to take a cruise?

Quick Quiz 9.6 *Solve.*

1. $3x + 2 < 8$ *and* $3x > -16$

2. $x - 7 \leq -15$ *or* $2x + 3 \geq 5$

3. $x^2 - 7x + 6 > 0$

4. **Concept Check** Explain what happens when you solve the inequality $x^2 + 2x + 8 > 0$.

9.7 Absolute Value Equations and Inequalities

Student Learning Objectives

After studying this section, you will be able to:

1. Solve absolute value equations of the form $|ax + b| = c$.
2. Solve absolute value equations of the form $|ax + b| + c = d$.
3. Solve absolute value equations of the form $|ax + b| = |cx + d|$.
4. Solve absolute value inequalities of the form $|ax + b| < c$.
5. Solve absolute value inequalities of the form $|ax + b| > c$.

1 Solving Absolute Value Equations of the Form $|ax + b| = c$

From Section 1.1, you know that the absolute value of a number x can be pictured as the distance between 0 and x on a number line. Let's look at a simple absolute value equation, $|x| = 4$, and draw a picture.

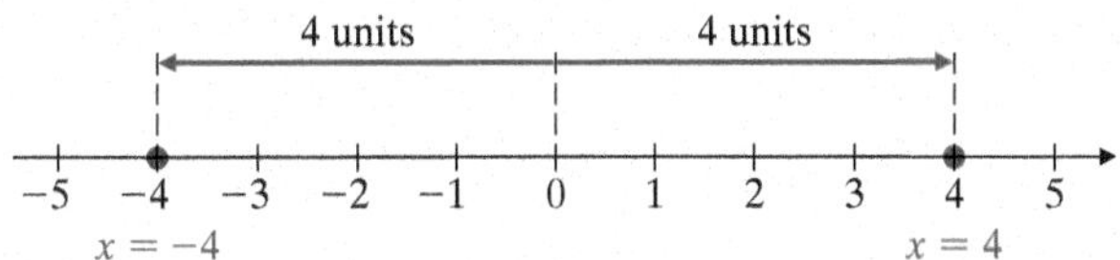

Thus, the equation $|x| = 4$ has two solutions, $x = 4$ or $x = -4$. Let's look at another example.

$$\text{If} \quad |x| = \frac{2}{3},$$

$$\text{then} \quad x = \frac{2}{3} \quad \text{or} \quad x = -\frac{2}{3},$$

$$\text{because} \quad \left|\frac{2}{3}\right| = \frac{2}{3} \quad \text{and} \quad \left|-\frac{2}{3}\right| = \frac{2}{3}.$$

We can solve these relatively simple absolute value equations by using the following definition of absolute value.

$$|x| = \begin{cases} x, & \text{if } x \geq 0 \\ -x, & \text{if } x < 0 \end{cases}$$

Now let's take a look at a more complicated absolute value equation: $|ax + b| = c$.

> The solutions of an equation of the form $|ax + b| = c$, where $a \neq 0$ and c is a positive number, are those values that satisfy
>
> $$ax + b = c \quad or \quad ax + b = -c.$$

Example 1 Solve and check your solutions. $|2x + 5| = 11$

Solution Using the rule established in the box, we have the following:

$$\begin{aligned} 2x + 5 &= 11 \quad \text{or} \quad & 2x + 5 &= -11 \\ 2x &= 6 & 2x &= -16 \\ x &= 3 & x &= -8 \end{aligned}$$

The two solutions are 3 and −8.

Check. **if $x = 3$**

$$\begin{aligned} |2x + 5| &= 11 \\ |2(3) + 5| &\stackrel{?}{=} 11 \\ |6 + 5| &\stackrel{?}{=} 11 \\ |11| &\stackrel{?}{=} 11 \\ 11 &= 11 \checkmark \end{aligned}$$

if $x = -8$

$$\begin{aligned} |2x + 5| &= 11 \\ |2(-8) + 5| &\stackrel{?}{=} 11 \\ |-16 + 5| &\stackrel{?}{=} 11 \\ |-11| &\stackrel{?}{=} 11 \\ 11 &= 11 \checkmark \end{aligned}$$

□

Student Practice 1 Solve and check your solutions. $|3x - 4| = 23$

Example 2 Solve and check your solutions. $\left|\frac{1}{2}x - 1\right| = 5$

Solution The solutions of the given absolute value equation must satisfy

$$\frac{1}{2}x - 1 = 5 \quad \text{or} \quad \frac{1}{2}x - 1 = -5.$$

If we multiply each term of both equations by 2, we obtain the following:

$$\begin{aligned} x - 2 &= 10 \quad \text{or} \quad & x - 2 &= -10 \\ x &= 12 & x &= -8 \end{aligned}$$

Check. **if $x = 12$** **if $x = -8$**

$$\begin{aligned} \left|\frac{1}{2}(12) - 1\right| &\stackrel{?}{=} 5 & \left|\frac{1}{2}(-8) - 1\right| &\stackrel{?}{=} 5 \\ |6 - 1| &\stackrel{?}{=} 5 & |-4 - 1| &\stackrel{?}{=} 5 \\ |5| &\stackrel{?}{=} 5 & |-5| &\stackrel{?}{=} 5 \\ 5 &= 5 \checkmark & 5 &= 5 \checkmark \end{aligned}$$

□

Student Practice 2 Solve and check your solutions.

$$\left|\frac{2}{3}x + 4\right| = 2$$

2 Solving Absolute Value Equations of the Form $|ax + b| + c = d$

Notice that in each of the previous examples the absolute value expression is on one side of the equation and a positive real number is on the other side of the equation. What happens when we encounter an equation of the form $|ax + b| + c = d$?

Example 3 Solve $|3x - 1| + 2 = 5$ and check your solutions.

Solution First we will rewrite the equation so that the absolute value expression is alone on one side of the equation.

$$\begin{aligned} |3x - 1| + 2 - 2 &= 5 - 2 \\ |3x - 1| &= 3 \end{aligned}$$

Now we solve $|3x - 1| = 3$.

$$\begin{aligned} 3x - 1 &= 3 \quad \text{or} \quad & 3x - 1 &= -3 \\ 3x &= 4 & 3x &= -2 \\ x &= \frac{4}{3} & x &= -\frac{2}{3} \end{aligned}$$

Check. **if $x = \frac{4}{3}$** **if $x = -\frac{2}{3}$**

$$\begin{aligned} \left|3\left(\frac{4}{3}\right) - 1\right| + 2 &\stackrel{?}{=} 5 & \left|3\left(-\frac{2}{3}\right) - 1\right| + 2 &\stackrel{?}{=} 5 \\ |4 - 1| + 2 &\stackrel{?}{=} 5 & |-2 - 1| + 2 &\stackrel{?}{=} 5 \\ |3| + 2 &\stackrel{?}{=} 5 & |-3| + 2 &\stackrel{?}{=} 5 \\ 3 + 2 &\stackrel{?}{=} 5 & 3 + 2 &\stackrel{?}{=} 5 \\ 5 &= 5 \checkmark & 5 &= 5 \checkmark \end{aligned}$$

□

Student Practice 3 Solve and check your solutions. $|2x + 1| + 3 = 8$

3 Solving Absolute Value Equations of the Form $|ax + b| = |cx + d|$

Let us now consider the possibilities for a and b if $|a| = |b|$.

Suppose $a = 5$; then $b = 5$ or -5.
If $a = -5$, then $b = 5$ or -5.

> To generalize, if $|a| = |b|$, then $a = b$ or $a = -b$.

We now apply this property to solve more complex equations.

Example 4 Solve and check. $|3x - 4| = |x + 6|$

Solution The solutions of the given equation must satisfy

$$3x - 4 = x + 6 \quad \text{or} \quad 3x - 4 = -(x + 6).$$

Now we solve each equation in the normal fashion.

$$\begin{aligned} 3x - 4 &= x + 6 & \text{or} \quad 3x - 4 &= -x - 6 \\ 3x - x &= 4 + 6 & 3x + x &= 4 - 6 \\ 2x &= 10 & 4x &= -2 \\ x &= 5 & x &= -\frac{1}{2} \end{aligned}$$

We will check each solution by substituting it into the *original equation*.

Check. **if $x = 5$** **if $x = -\frac{1}{2}$**

$$\begin{aligned} |3(5) - 4| &\stackrel{?}{=} |5 + 6| \\ |15 - 4| &\stackrel{?}{=} |11| \\ |11| &\stackrel{?}{=} |11| \\ 11 &= 11 \checkmark \end{aligned}$$

$$\begin{aligned} \left|3\left(-\frac{1}{2}\right) - 4\right| &\stackrel{?}{=} \left|-\frac{1}{2} + 6\right| \\ \left|-\frac{3}{2} - 4\right| &\stackrel{?}{=} \left|-\frac{1}{2} + 6\right| \\ \left|-\frac{3}{2} - \frac{8}{2}\right| &\stackrel{?}{=} \left|-\frac{1}{2} + \frac{12}{2}\right| \\ \left|-\frac{11}{2}\right| &\stackrel{?}{=} \left|\frac{11}{2}\right| \\ \frac{11}{2} &= \frac{11}{2} \checkmark \end{aligned}$$

Student Practice 4 Solve and check. $|x - 6| = |5x + 8|$

To Think About: *Two Other Absolute Value Equations* Explain how you would solve an absolute value equation of the form $|ax + b| = 0$. Give an example. Does $|3x + 2| = -4$ have a solution? Why or why not?

4 Solving Absolute Value Inequalities of the Form $|ax + b| < c$

We begin by looking at $|x| < 3$. What does this mean? The inequality $|x| < 3$ means that x is less than 3 units from 0 on a number line. We draw a picture.

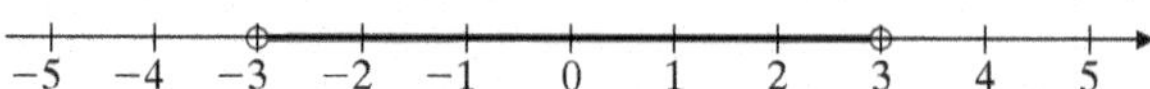

This picture shows all possible values of x such that $|x| < 3$. We see that this occurs when $-3 < x < 3$. We conclude that $|x| < 3$ and $-3 < x < 3$ are equivalent statements.

DEFINITION OF $|x| < a$

If a is a positive real number and $|x| < a$, then $-a < x < a$.

Example 5 Solve. $|x| \leq 4.5$

Solution The inequality $|x| \leq 4.5$ means that x is less than or equal to 4.5 units from 0 on a number line. We draw a picture.

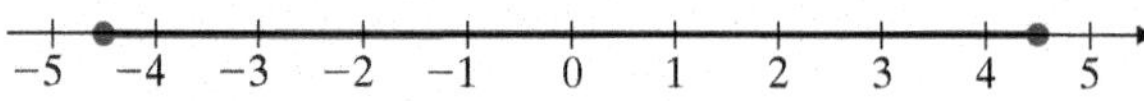

Thus, the solution is $-4.5 \leq x \leq 4.5$. □

Student Practice 5 Solve and graph. $|x| < 2$

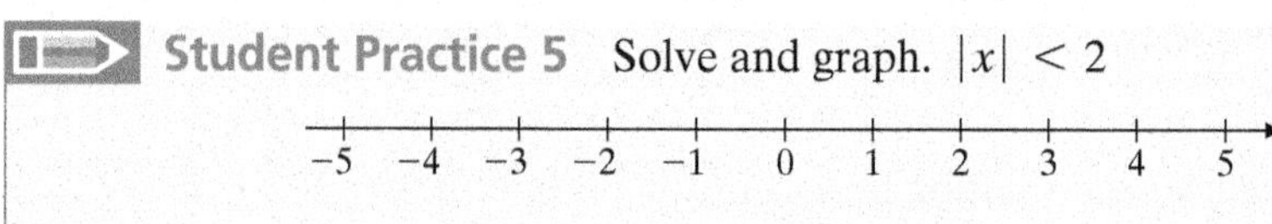

This same technique can be used to solve more complicated inequalities.

Example 6 Solve and graph the solution. $|x + 5| \leq 10$

Solution We want to find the values of x that make $-10 \leq x + 5 \leq 10$ a true statement. We need to solve the compound inequality.

To solve this inequality, we subtract 5 from each part.

$$-10 - 5 \leq x + 5 - 5 \leq 10 - 5$$
$$-15 \leq x \leq 5$$

Thus, the solution is $-15 \leq x \leq 5$. We graph this solution.

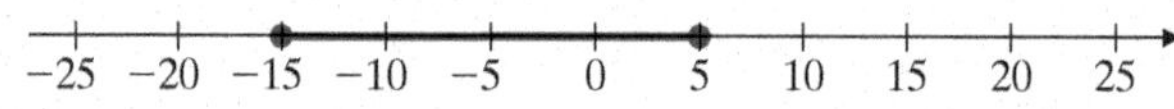

□

Student Practice 6 Solve and graph the solution. (*Hint:* Choose a convenient scale.) $|x - 6| < 15$

Example 7 Solve and graph the solution. $|2(x - 1) + 4| < 8$.

Solution First we simplify the expression within the absolute value.

$$|2x - 2 + 4| < 8$$
$$|2x + 2| < 8$$
$$-8 < 2x + 2 < 8 \quad \text{If } |x| < a, \text{ then } -a < x < a.$$
$$-8 - 2 < 2x + 2 - 2 < 8 - 2 \quad \text{Subtract 2 from each part.}$$
$$-10 < 2x < 6 \quad \text{Simplify.}$$
$$\frac{-10}{2} < \frac{2x}{2} < \frac{6}{2} \quad \text{Divide each part by 2.}$$
$$-5 < x < 3$$

□

 **Student Practice 7** Solve and graph the solution. $|2 + 3(x - 1)| < 20$

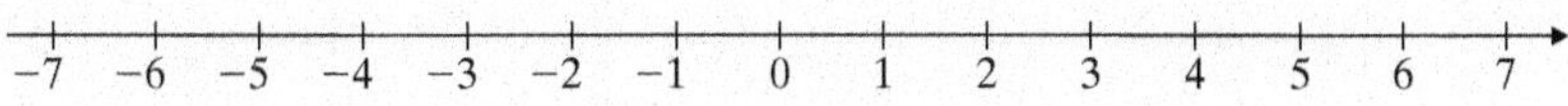

5 Solving Absolute Value Inequalities of the Form $|ax + b| > c$

Now consider $|x| > 3$. What does this mean? This inequality $|x| > 3$ means that x is greater than 3 units from 0 on a number line. We draw a picture.

This picture shows all possible values of x such that $|x| > 3$. This occurs when $x < -3$ or when $x > 3$. (Note that a solution can be either in the region to the left of -3 on the number line or in the region to the right of 3 on the number line.) We conclude that the expression $|x| > 3$ and the expression $x < -3$ or $x > 3$ are equivalent statements.

DEFINITION OF $|x| > a$

If a is a positive real number and $|x| > a$, then $x < -a$ or $x > a$.

Example 8 Solve and graph the solution. $|x| \geq 5\frac{1}{4}$

Solution The inequality $|x| \geq 5\frac{1}{4}$ means that x is $5\frac{1}{4}$ or more units from 0 on a number line. We draw a picture.

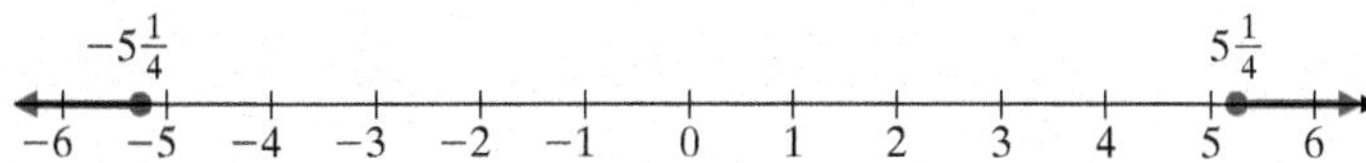

Thus, the solution is $x \leq -5\frac{1}{4}$ *or* $x \geq 5\frac{1}{4}$. □

 **Student Practice 8** Solve and graph the solution. $|x| > 2.5$.

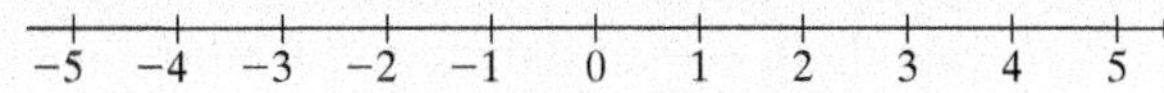

This same technique can be used to solve more complicated inequalities.

Example 9 Solve and graph the solution. $|-3x + 6| > 18$

Solution We want to find the values of x that make $-3x + 6 > 18$ or $-3x + 6 < -18$ a true statement. We need to solve the compound inequality. By definition, we have the following compound inequality.

We will solve each inequality separately

$$-3x + 6 > 18 \qquad \textit{or} \qquad -3x + 6 < -18$$

$$-3x > 12 \qquad\qquad -3x < -24$$

$$\frac{-3x}{-3} < \frac{12}{-3} \longleftarrow \text{Division by a negative number reverses the inequality sign.} \longrightarrow \frac{-3x}{-3} > \frac{-24}{-3}$$

$$x < -4 \qquad\qquad x > 8$$

The solution is $x < -4$ *or* $x > 8$.

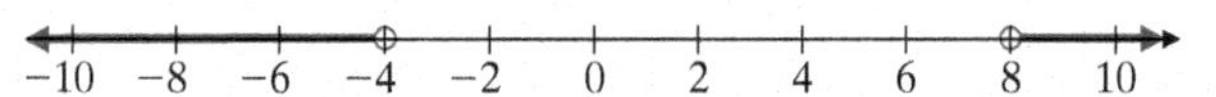

Student Practice 9 Solve and graph the solution. $|-5x - 2| > 13$.

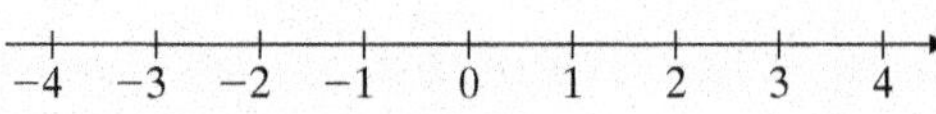

Example 10 Solve and graph the solution. $\left|3 - \frac{2}{3}x\right| \geq 5$

Solution By definition, we have the following compound inequality.

$$3 - \frac{2}{3}x \geq 5 \qquad \textit{or} \qquad 3 - \frac{2}{3}x \leq -5$$

$$3(3) - 3\left(\frac{2}{3}x\right) \geq 3(5) \qquad 3(3) - 3\left(\frac{2}{3}x\right) \leq 3(-5)$$

$$9 - 2x \geq 15 \qquad 9 - 2x \leq -15$$

$$-2x \geq 6 \qquad -2x \leq -24$$

$$\frac{-2x}{-2} \leq \frac{6}{-2} \qquad \frac{-2x}{-2} \geq \frac{-24}{-2}$$

$$x \leq -3 \qquad x \geq 12$$

The solution is $x \leq -3$ *or* $x \geq 12$.

Student Practice 10 Solve and graph the solution.

$$\left|4 - \frac{3}{4}x\right| \geq 5$$

Example 11 When a new car transmission is built, the diameter d of the transmission must not differ from the specified standard s by more than 0.37 millimeter. The engineers express this requirement as $|d - s| \leq 0.37$. If the standard s is 216.82 millimeters for a particular car, find the limits of d.

Solution

$$|d - s| \leq 0.37$$

$$|d - 216.82| \leq 0.37 \quad \text{Substitute the known value of } s.$$

$$-0.37 \leq d - 216.82 \leq 0.37 \quad \text{If } |x| \leq a, \text{ then } -a \leq x \leq a.$$

$$-0.37 + 216.82 \leq d - 216.82 + 216.82 \leq 0.37 + 216.82$$

$$216.45 \leq d \leq 217.19$$

Thus, the diameter of the transmission must be at least 216.45 millimeters, but not greater than 217.19 millimeters. □

 **Student Practice 11** The diameter d of a transmission must not differ from the specified standard s by more than 0.37 millimeter. This is written as $|d - s| \leq 0.37$. Solve to find the allowed limits of d for a truck transmission for which the standard s is 276.53 millimeters.

SUMMARY OF ABSOLUTE VALUE EQUATIONS AND INEQUALITIES

It may be helpful to review the key concepts of absolute value equations and inequalities that we have covered in this section. For real numbers a, b, and c, where $a \neq 0$ and $c > 0$, we have the following:

Absolute value form of the equation or inequality	Equivalent form without the absolute value	Type of solution obtained	Graphed form of the solution on a number line
$\lvert ax + b \rvert = c$	$ax + b = c$ or $ax + b = -c$	Two distinct numbers: m and n	m n
$\lvert ax + b \rvert < c$	$-c < ax + b < c$	The set of numbers between the two numbers m and n: $m < x < n$	m n
$\lvert ax + b \rvert > c$	$ax + b < -c$ or $ax + b > c$	The set of numbers less than m or the set of numbers greater than n: $x < m$ or $x > n$	m n

STEPS TO SUCCESS Helping Your Accuracy

It is easy to make a mistake. But here are five ways to cut down on errors. Look over each one and think about how each suggestion can help you.

1. Work carefully, and take your time. Do not rush through a problem just to get it done.
2. Concentrate on the problem. Sometimes your mind starts to wander. Then you get careless and will likely make a mistake.
3. Check your problem. Be sure you copied it correctly from the book.
4. Check your computations from step to step. Did you do each step correctly?
5. Check your final answer. Does it work? Is it reasonable?

Making it personal: Look over these five suggestions. Which one do you think will help you the most? Write down how you can use this suggestion to help you personally as you try to improve your accuracy. ▼

9.7 Exercises MyMathLab®

Verbal and Writing Skills, Exercises 1–4

1. The equation $|x| = b$, where b is a positive number, will always have how many solutions? Why?

2. The equation $|x| = b$ might have only one solution. How could that happen?

3. To solve an equation like $|x + 7| - 2 = 8$, what is the first step that must be done? What will be the result?

4. To solve an equation like $|3x - 1| + 5 = 14$, what is the first step that must be done? What will be the result?

Solve each absolute value equation. Check your solutions.

5. $|x| = 30$

6. $|x| = 14$

7. $|x + 7| = 15$

8. $|x + 6| = 13$

9. $|2x - 5| = 13$

10. $|4x - 7| = 9$

11. $|1.8 - 0.4x| = 1$

12. $|2.4 - 0.8x| = 2$

13. $|x + 2| - 1 = 7$

14. $|x + 3| - 4 = 8$

15. $\left|\frac{1}{2} - \frac{3}{4}x\right| + 1 = 3$

16. $\left|\frac{2}{3} - \frac{1}{2}x\right| - 2 = -1$

17. $\left|2 - \frac{2}{3}x\right| - 3 = 5$

18. $\left|5 - \frac{7}{2}x\right| + 1 = 11$

19. $\left|\frac{1 - 3x}{2}\right| = \frac{4}{5}$

20. $\left|\frac{2x - 1}{4}\right| = \frac{1}{3}$

Solve each absolute value equation.

21. $|x + 4| = |2x - 1|$

22. $|x - 7| = |3x - 1|$

23. $\left|\frac{x - 1}{2}\right| = |2x + 3|$

24. $\left|\frac{2x + 3}{3}\right| = |x + 4|$

25. $|1.5x - 2| = |x - 0.5|$

26. $|2.2x + 2| = |1 - 2.8x|$

27. $|3 - x| = \left|\frac{x}{2} + 3\right|$

28. $\left|\frac{2x}{5} + 1\right| = |1 - x|$

Mixed Practice *Solve each equation, if possible. Check your solutions.*

29. $|3(x + 4)| + 2 = 14$

30. $|4(x - 1)| + 5 = 15$

31. $\left|\frac{8x}{5} - 2\right| = 0$

32. $\left|\frac{3}{4}x + 9\right| = 0$

33. $\left|\frac{4}{3}x - \frac{1}{8}\right| = -5$

34. $\left|\frac{1}{2}x - \frac{3}{8}\right| = -1$

35. $\left|\frac{3x - 1}{3}\right| = \frac{2}{5}$

36. $\left|\frac{5x + 1}{2}\right| = \frac{3}{4}$

Solve and graph the solutions.

37. $|x| \leq 8$

38. $|x| < 6$

39. $|x + 4.5| < 5$

40. $|x + 6| < 3.5$

Solve for x.

41. $|x - 3| \le 5$

42. $|x - 8| \le 12$

43. $|3x + 1| \le 10$

44. $|2x + 3| \le 11$

45. $|0.5 - 0.1x| < 1$

46. $|0.6 - 0.3x| < 9$

47. $\left|\frac{1}{4}x + 2\right| < 6$

48. $\left|\frac{1}{3}x + 4\right| < 7$

49. $\left|\frac{2}{3}(x - 2)\right| < 4$

50. $\left|\frac{3}{4}(x + 1)\right| < 2$

51. $\left|\frac{3x - 2}{4}\right| < 3$

52. $\left|\frac{5x - 3}{2}\right| < 4$

53. $|x| > 5$

54. $|x| \ge 7$

55. $|x + 2| > 5$

56. $|x + 4| > 7$

57. $|x - 1| \ge 2$

58. $|x - 6| \ge 4$

59. $|4x - 7| \ge 9$

60. $|6x - 5| \ge 7$

61. $|6 - 0.1x| > 5$

62. $|0.5 - 0.1x| > 6$

63. $\left|\frac{1}{5}x - \frac{1}{10}\right| > 2$

64. $\left|\frac{1}{4}x - \frac{3}{8}\right| > 1$

Mixed Practice *Solve for x.*

65. $\left|\frac{1}{3}(x - 2)\right| < 5$

66. $\left|\frac{2}{5}(x - 2)\right| \le 4$

67. $|4x + 7| < 13$

68. $|2x + 3| < 5$

69. $|3 - 8x| > 19$

70. $|4 - 3x| > 4$

Applications

Manufacturing Standards *In a certain company, the measured thickness m of a helicopter blade must not differ from the standard s by more than 0.12 millimeter. The manufacturing engineer expresses this as* $|m - s| \le 0.12$.

71. Find the limits of m if the standard s is 18.65 millimeters.

72. Find the limits of m if the standard s is 17.48 millimeters.

Manufacturing Standards *Cell phone cases have dimension requirements to ensure the phone will fit properly in the case. The manufacturing engineer has written the specification that the new length n of the case can differ from the previous length p by only 0.03 centimeter or less. The inequality is* $|n - p| \le 0.03$.

73. Find the limits of the new length of an iPhone 5 case if the previous length was 17.78 centimeters.

74. Find the limits of the new length of an iPhone 6 case if the previous length was 19.8 centimeters.

Cumulative Review

75. [5.2.1] Simplify. $(3x^{-3}yz^{0})\left(\frac{5}{3}x^{4}y^{2}\right)$

76. [5.2.1] Simplify. $(3x^{4}y^{-3})^{-2}$

Quick Quiz 9.7 *Solve for x.*

1. $\left|\frac{2}{3}x + 1\right| - 3 = 5$

2. $|8x - 4| \leq 20$

3. $|5x + 2| > 7$

4. Concept Check Explain what happens when you try to solve for x. $|7x + 3| < -4$

STEPS TO SUCCESS Getting the Greatest Value from Your Homework

Read the textbook first before doing the homework. Take some time to read the text and study the examples. Try working out the Student Practice problems. You will be amazed at the amount of understanding you will obtain by studying the book before jumping into the homework exercises.

Take your time. Read the directions carefully. Be sure you understand what is being asked. Check your answers with those given in the back of the textbook. If your answer is incorrect, study similar examples in the text. Then redo the problem, watching for errors.

Make a schedule. You will need to allow two hours outside of class for each hour of actual class time. Make a weekly schedule of the times you have class. Now write down the times each day you will devote to doing math homework. Then write down the times you will spend doing homework for your other classes. If you have a job be sure to write down all your work hours.

Making it personal: Write down your own schedule of class, work, and study time. ▼

Sunday	Monday	Tuesday	Wednesday	Thursday	Friday	Saturday

CAREER EXPLORATION PROBLEMS

Background: Sales Manager

Edgar, a regional sales manager for a large pharmaceutical company, always provides his sales teams and their clients with the most accurate information about drug pricing, use, and sales so they can make informed presentations and decisions.

This week, Edgar is examining data that models the over-the-counter sales of a competitor's popular allergy medication that became available for purchase midway through 2010. Edgar's company recently received approval from the Food and Drug Administration (FDA) to sell an allergy medication over the counter that had previously required a prescription.

Facts

- The sales model for the competitor's allergy medication over the period from 2010 through 2015, in millions of dollars, is $S(t) = -0.153t^2 + 12.11t + 41.06$, where t represents the number of years since 2010.
- The sales of Edgar's company's prescription allergy medication, in millions of dollars, over the same period 2010 through 2015 is expressed as $P(t) = -1.48t^2 + 7.02t + 63.5$. Again, t represents the number of years since 2010.
- Over-the-counter sales for Edgar's company's newly approved allergy medication will begin in 2015. The model for projected annual profits from its sale, in millions of dollars, is $N(t) = -0.83t^2 + 4.15t + 47.25$, where t represents the number of years after 2015.

Tasks

1. Edgar wants to use the sales model for the competitor's over-the-counter allergy medication to determine its use in the years 2010 and 2015. How are these values calculated?

2. Edgar now wants to calculate the sales figures for his own company's prescription medication in the years 2010 and 2015. Which model does he use, what amounts does it generate, and what might his interpretation of the values be?

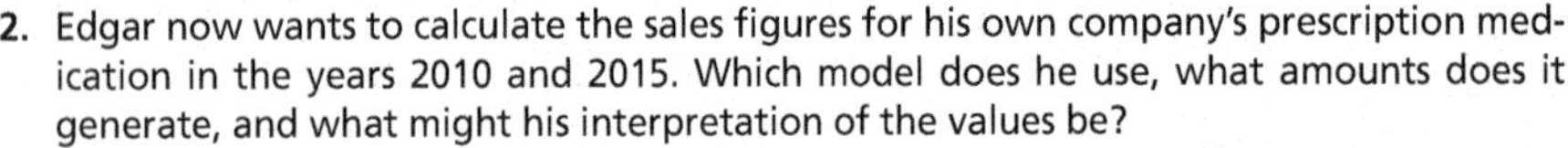

3. Edgar wants to determine at what point in time sales for the two medications were equal. How can he determine this?

4. Edgar is now interested in forecasting for his sales team when profits from sales of the newly approved medication are projected to be at a maximum along with what the maximum profit will be.

Chapter 9 Organizer

Topic and Procedure	Examples	You Try It
Solving a quadratic equation by using the square root property, p. 544 If $x^2 = a$, then $x = \pm\sqrt{a}$.	Solve. $2x^2 - 50 = 0$ $2x^2 = 50$ $x^2 = 25$ $x = \pm\sqrt{25}$ $x = \pm 5$	1. Solve. $3x^2 - 60 = 0$
Solving a quadratic equation by completing the square, p. 545 1. Rewrite the equation in the form $ax^2 + bx = c$. 2. If $a \neq 1$, divide each term of the equation by a. 3. Square half of the numerical coefficient of the linear term. Add the result to both sides of the equation. 4. Factor the left side; then take the square root of both sides of the equation. 5. Solve the resulting equation for x. 6. Check the solutions in the original equation.	Solve. $2x^2 - 4x - 1 = 0$ $2x^2 - 4x = 1$ $\frac{2x^2}{2} - \frac{4x}{2} = \frac{1}{2}$ $x^2 - 2x + \_\_\_\_ = \frac{1}{2} + \_\_\_\_$ $x^2 - 2x + 1 = \frac{1}{2} + 1$ $(x - 1)^2 = \frac{3}{2}$ $x - 1 = \pm\sqrt{\frac{3}{2}}$ $x - 1 = \pm\frac{\sqrt{6}}{2}$ $x = 1 \pm \frac{\sqrt{6}}{2} = \frac{2 \pm \sqrt{6}}{2}$	2. Solve. $2x^2 + 6x - 3 = 0$
Solve a quadratic equation by using the quadratic formula, p. 551 If $ax^2 + bx + c = 0$, where $a \neq 0$, $x = \frac{-b \pm \sqrt{b^2 - 4ac}}{2a}$. 1. Rewrite the equation in standard form. 2. Determine the values of a, b, and c. 3. Substitute the values of a, b, and c into the formula. 4. Simplify the result to obtain the values of x. 5. Any imaginary solutions to the quadratic equation should be simplified by using the definition $\sqrt{-a} = i\sqrt{a}$, where $a > 0$.	Solve. $2x^2 = 3x - 2$ $2x^2 - 3x + 2 = 0$ $a = 2, b = -3, c = 2$ $x = \frac{-(-3) \pm \sqrt{(-3)^2 - 4(2)(2)}}{2(2)}$ $x = \frac{3 \pm \sqrt{9 - 16}}{4}$ $x = \frac{3 \pm \sqrt{-7}}{4}$ $x = \frac{3 \pm i\sqrt{7}}{4}$	3. Solve. $x^2 - 8x = 5$
Placing a quadratic equation in standard form, p. 553 A quadratic equation in standard form is an equation of the form $ax^2 + bx + c = 0$, where a, b, and c are real numbers and $a \neq 0$. It is often necessary to remove parentheses and clear away fractions by multiplying each term of the equation by the LCD to obtain the standard form.	Rewrite in quadratic form. $\frac{2}{x - 3} + \frac{x}{x + 3} = \frac{5}{x^2 - 9}$ $(x + 3)(x - 3)\left[\frac{2}{x - 3}\right] + (x + 3)(x - 3)\left[\frac{x}{x + 3}\right]$ $= (x + 3)(x - 3)\left[\frac{5}{(x + 3)(x - 3)}\right]$ $2(x + 3) + x(x - 3) = 5$ $2x + 6 + x^2 - 3x = 5$ $x^2 - x + 1 = 0$	4. Rewrite in quadratic form. $\frac{5}{x - 1} + \frac{3}{x + 2} = 2$

Topic and Procedure	Examples	You Try It
Equations that can be transformed into quadratic form, p. 560 If the exponents are positive: 1. Find the variable with the smallest exponent. Let this quantity be replaced by y. 2. Continue to make substitutions for the remaining variable terms based on the first substitution. (You should be able to replace the variable with the largest exponent by y^2.) 3. Solve the resulting equation for y. 4. Reverse the substitution used in step 1. 5. Solve the resulting equation for x. 6. Check your solution in the *original* equation.	Solve. $x^{2/3} - x^{1/3} - 2 = 0$ Let $y = x^{1/3}$. Then $y^2 = x^{2/3}$. $y^2 - y - 2 = 0$ $(y-2)(y+1) = 0$ $y = 2$ or $y = -1$ $x^{1/3} = 2$ $\quad$ $x^{1/3} = -1$ $(x^{1/3})^3 = 2^3$ $\quad$ $(x^{1/3})^3 = (-1)^3$ $x = 8$ $\quad$ $x = -1$ *Check.* $x = 8$: $(8)^{2/3} - (8)^{1/3} - 2 \stackrel{?}{=} 0$ $2^2 - 2 - 2 \stackrel{?}{=} 0$ $4 - 4 = 0$ ✓ $x = -1$: $(-1)^{2/3} - (-1)^{1/3} - 2 \stackrel{?}{=} 0$ $(-1)^2 - (-1) - 2 \stackrel{?}{=} 0$ $1 + 1 - 2 = 0$ ✓ Both 8 and -1 are solutions.	4. Solve. $x + 3x^{1/2} - 4 = 0$
Solving quadratic equations containing two or more variables, p. 568 Treat the letter to be solved for as a variable, but treat all other letters as constants. Solve the equation by factoring, by using the square root property, or by using the quadratic formula.	Solve for x. **(a)** $6x^2 - 11xw + 4w^2 = 0$ By factoring: $(3x - 4w)(2x - w) = 0$ $3x - 4w = 0$ or $2x - w = 0$ $x = \frac{4w}{3}$ $\quad$ $x = \frac{w}{2}$ **(b)** $4x^2 + 5b = 2w^2$ Using the square root property: $4x^2 = 2w^2 - 5b$ $x^2 = \frac{2w^2 - 5b}{4}$ $x = \pm\sqrt{\frac{2w^2 - 5b}{4}} = \pm\frac{1}{2}\sqrt{2w^2 - 5b}$ **(c)** $2x^2 + 3xz - 10z = 0$ By the quadratic formula, with $a = 2, b = 3z, c = -10z$: $x = \frac{-3z \pm \sqrt{9z^2 + 80z}}{4}$	6. Solve for x. **(a)** $2x^2 - 5xy - 3y^2 = 0$ **(b)** $x^2 + 4y^2 = 4a$ **(c)** $4x^2 - 2xw = 9w$
The Pythagorean Theorem, p. 570 In any right triangle, if c is the length of the hypotenuse and a and b are the lengths of the two legs, then $c^2 = a^2 + b^2$.	Find a if $c = 7$ and $b = 5$. $49 = a^2 + 25$ $49 - 25 = a^2$ $24 = a^2$ $\sqrt{24} = a$ $2\sqrt{6} = a$ (triangle: c, b, a)	7. Find b if $c = 10$ and $a = 5$. 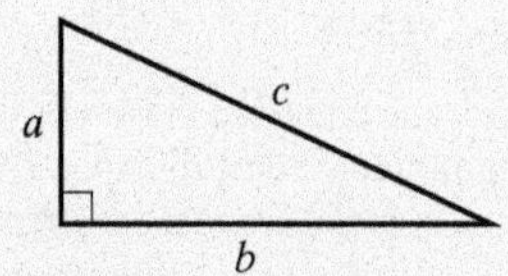

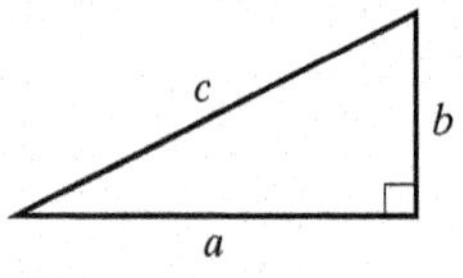

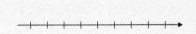

Topic and Procedure	Examples	You Try It
Graphing quadratic functions, p. 579 Graph quadratic functions of the form $f(x) = ax^2 + bx + c$ with $a \neq 0$ as follows: **1.** Find the vertex at $\left(\frac{-b}{2a}, f\left(\frac{-b}{2a}\right)\right)$. **2.** Find the y-intercept, which occurs at $f(0)$. **3.** Find the x-intercepts if they exist. Solve $f(x) = 0$ for x. **4.** Connect the points with a smooth curve.	Graph. $f(x) = x^2 + 6x + 8$ Vertex: $x = \frac{-6}{2} = -3$ $f(-3) = (-3)^2 + 6(-3) + 8 = -1$ The vertex is $(-3, -1)$. Intercepts: $f(0) = (0)^2 + 6(0) + 8 = 8$ The y-intercept is $(0, 8)$. $x^2 + 6x + 8 = 0$ $(x + 2)(x + 4) = 0$ $x = -2, x = -4$ The x-intercepts are $(-2, 0)$ and $(-4, 0)$. 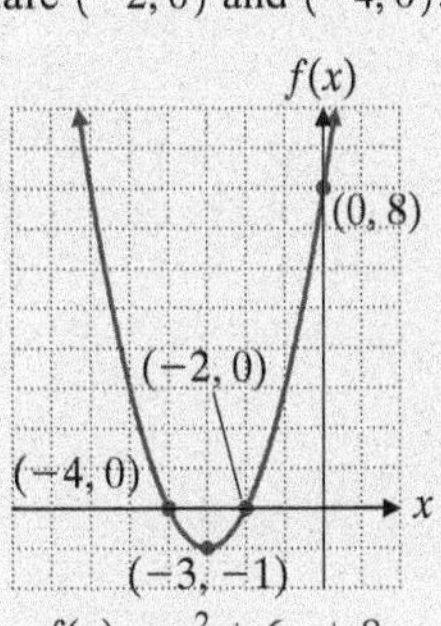 	**8.** Graph. $f(x) = x^2 + 3x - 4$ 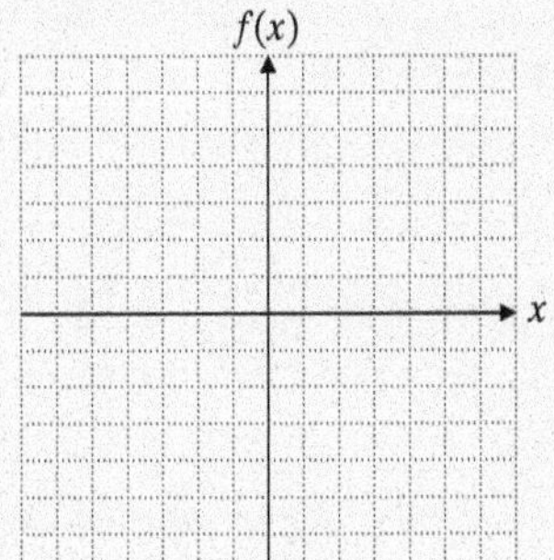
Solving compound inequalities containing* and*, p. 588 The solution is the desired region containing all values of x that meet both conditions.	Graph the values of x satisfying $x + 6 > -3$ *and* $2x - 1 < -4$. $x + 6 - 6 > -3 - 6$ *and* $2x - 1 + 1 < -4 + 1$ $x > -9$ *and* $x < -1.5$ −1.5 −9 −8 −7 −6 −5 −4 −3 −2 −1	**9.** Graph the values of x satisfying $x + 7 > -1$ *and* $3x + 4 < 10$.
Solving compound inequalities containing* or*, p. 589 The solution is the desired region containing all values of x that meet either of the two conditions.	Graph the values of x satisfying $-3x + 1 \leq 7$ *or* $3x + 1 \leq -11$. $-3x + 1 - 1 \leq 7 - 1$ *or* $3x + 1 - 1 \leq -11 - 1$ $-3x \leq 6$ *or* $3x \leq -12$ $\frac{-3x}{-3} \geq \frac{6}{-3}$ *or* $\frac{3x}{3} \leq \frac{-12}{3}$ $x \geq -2$ *or* $x \leq -4$  	**10.** Graph the values of x satisfying $5x + 2 \leq -8$ *or* $4x - 3 \geq 9$.
Solving quadratic inequalities in one variable, p. 591 **1.** Replace the inequality symbol by an equals sign. Solve the resulting equation to find the boundary points. **2.** Use the boundary points to separate the number line into distinct regions. **3.** Evaluate the quadratic expression at a test point in each region. **4.** Determine which regions satisfy the original conditions of the quadratic inequality.	Solve and graph. $3x^2 + 5x - 2 > 0$ **1.** $3x^2 + 5x - 2 = 0$ $(3x - 1)(x + 2) = 0$ $3x - 1 = 0$ or $x + 2 = 0$ $x = \frac{1}{3}$ $x = -2$ Boundary points are -2 and $\frac{1}{3}$. **2.** 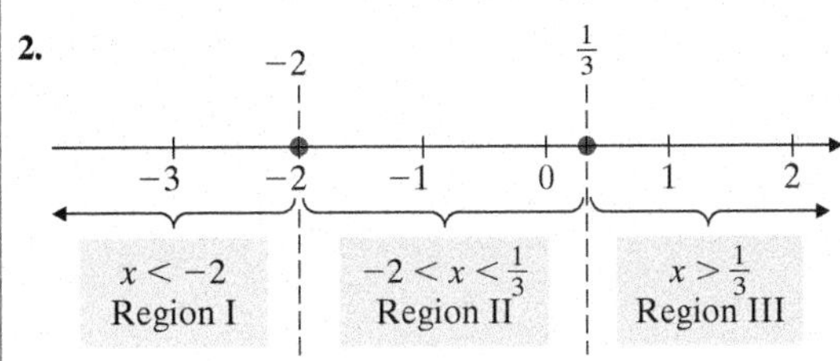 	**11.** Solve and graph. $2x^2 - 3x - 9 < 0$ 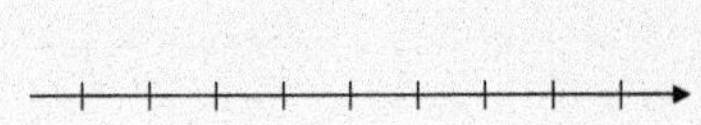

Topic and Procedure	Examples	You Try It
	3. $3x^2 + 5x - 2$ *Region I:* Pick $x = -3$. $3(-3)^2 + 5(-3) - 2 = 27 - 15 - 2 = 10 > 0$ *Region II:* Pick $x = 0$. $3(0)^2 + 5(0) - 2 = 0 + 0 - 2 = -2 < 0$ *Region III:* Pick $x = 3$. $3(3)^2 + 5(3) - 2 = 27 + 15 - 2 = 40 > 0$ **4.** We know that the expression is greater than zero (that is, $3x^2 + 5x - 2 > 0$) when $x < -2$ or $x > \frac{1}{3}$. 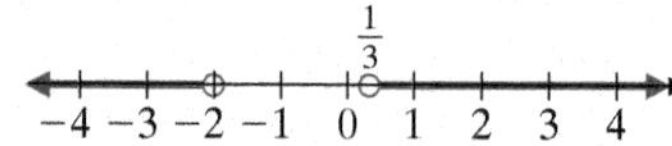	
Absolute value equations, p. 599 To solve an equation that involves an absolute value, we rewrite the absolute value equation as two separate equations without the absolute value. We solve each equation. If $\lvert ax + b \rvert = c$ where $c > 0$, then $ax + b = c$ or $ax + b = -c$.	Solve for x. $\lvert 4x - 1 \rvert = 17$ $4x - 1 = 17$ or $4x - 1 = -17$ $4x = 17 + 1$ $4x = -17 + 1$ $4x = 18$ $4x = -16$ $x = \frac{18}{4}$ $x = \frac{-16}{4}$ $x = \frac{9}{2}$ $x = -4$	12. Solve for x. $\lvert 3x + 5 \rvert = 11$
Solving absolute value inequalities involving < or ≤, p. 602 Let a be a positive real number. If $\lvert x \rvert < a$, then $-a < x < a$. If $\lvert x \rvert \le a$, then $-a \le x \le a$.	Solve and graph. $\lvert 3x - 2 \rvert < 19$ $-19 < 3x - 2 < 19$ $-19 + 2 < 3x - 2 + 2 < 19 + 2$ $-17 < 3x < 21$ $-\frac{17}{3} < \frac{3x}{3} < \frac{21}{3}$ $-5\frac{2}{3} < x < 7$ $-5\frac{2}{3}$ 7 −8 −6 −4 −2 0 2 4 6 8	13. Solve and graph. $\lvert 2x + 7 \rvert < 17$
Solving absolute value inequalities involving > or ≥, p. 603 Let a be a positive real number. If $\lvert x \rvert > a$, then $x < -a$ or $x > a$. If $\lvert x \rvert \ge a$, then $x \le -a$ or $x \ge a$.	Solve and graph. $\left\lvert \frac{1}{3}(x - 2) \right\rvert \ge 2$ $\frac{1}{3}(x - 2) \le -2$ *or* $\frac{1}{3}(x - 2) \ge 2$ $\frac{1}{3}x - \frac{2}{3} \le -2$ $\frac{1}{3}x - \frac{2}{3} \ge 2$ $x - 2 \le -6$ $x - 2 \ge 6$ $x \le -6 + 2$ $x \ge 6 + 2$ $x \le -4$ *or* $x \ge 8$ 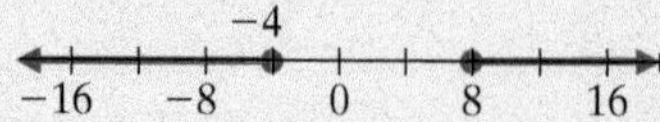	14. Solve and graph. $\left\lvert \frac{1}{4}(x + 8) \right\rvert > 1$

Chapter 9 Review Problems

Solve each of the following exercises by the specified method. Simplify all answers.

Solve by the square root property.

1. $6x^2 = 24$

2. $(x + 8)^2 = 81$

Solve by completing the square.

3. $x^2 + 8x + 13 = 0$

4. $4x^2 - 8x + 1 = 0$

Solve by the quadratic formula.

5. $x^2 - 4x - 2 = 0$

6. $3x^2 - 8x + 4 = 0$

Solve by any appropriate method and simplify your answers. Express any nonreal complex solutions using i notation.

7. $4x^2 - 12x + 9 = 0$

8. $x^2 - 14 = 5x$

9. $6x^2 - 23x = 4x$

10. $2x^2 = 5x - 1$

11. $5x^2 - 10 = 0$

12. $3x^2 - 2x = 15x - 10$

13. $6x^2 + 12x - 24 = 0$

14. $7x^2 + 24 = 5x^2$

15. $3x^2 + 5x + 1 = 0$

16. $2x(x - 4) - 4 = -x$

17. $9x(x + 2) + 2 = 12x$

18. $\frac{4}{5}x^2 + x + \frac{1}{5} = 0$

19. $y + \frac{5}{3y} + \frac{17}{6} = 0$

20. $\frac{15}{y^2} - \frac{2}{y} = 1$

21. $y(y + 1) + (y + 2)^2 = 4$

22. $\frac{2x}{x + 3} + \frac{3x - 1}{x + 1} = 3$

Determine the nature of the solution for each of the following quadratic equations. Do not solve the equation. Find the discriminant in each case and determine whether the equation has one rational solution, two rational solutions, two irrational solutions, or two non real complex solutions.

23. $4x^2 - 5x - 3 = 0$

24. $2x^2 - 7x + 6 = 0$

25. $25x^2 - 20x + 4 = 0$

Write a quadratic equation having the given numbers as solutions.

26. $5, -5$

27. $3i, -3i$

28. $-\frac{1}{4}, -\frac{3}{2}$

Solve for any valid real roots.

29. $x^4 - 6x^2 + 8 = 0$

30. $2x^6 - 5x^3 - 3 = 0$

31. $x^{2/3} - 3 = 2x^{1/3}$

32. $1 + 4x^{-8} = 5x^{-4}$

Solve for the variable specified. Assume that all radical expressions obtained have a positive radicand.

33. $3M = \frac{2A^2}{N}$; for A

34. $3t^2 + 4b = t^2 + 6ay$; for t

35. $yx^2 - 3x - 7 = 0$; for x

36. $20d^2 - xd - x^2 = 0$; for d

37. $2y^2 + 4ay - 3a = 0$; for y

38. $AB = 3x^2 + 2y^2 - 4x$; for x

Use the Pythagorean Theorem to find the unknown side. Assume that c is the length of the hypotenuse of a right triangle and that a and b are the lengths of the legs. Leave your answer as a radical in simplified form.

▲ **39.** $a = 3\sqrt{2}, b = 2$; find c

▲ **40.** $c = 16, b = 4$; find a

▲ **41.** ***Airplane Flight*** A plane is 6 miles away from an observer and exactly 5 miles above the ground. The plane is directly above a car. How far is the car from the observer? Round your answer to the nearest tenth of a mile.

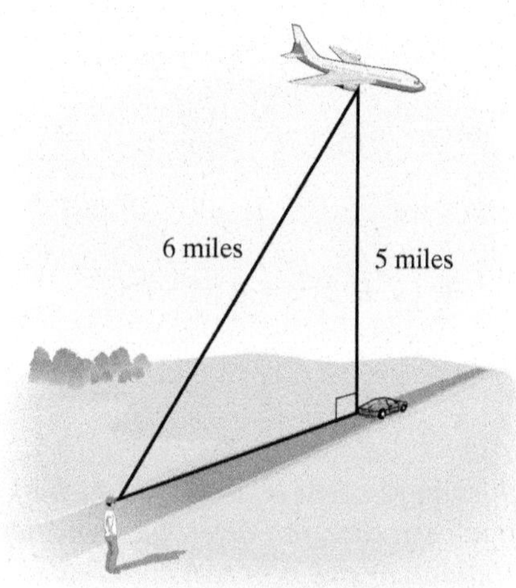

▲ **42.** ***Geometry*** The area of a triangle is 70 square centimeters. Its altitude is 6 meters longer than twice the length of the base. Find the lengths of the altitude and base.

▲ **43.** ***Geometry*** The area of a rectangle is 203 square meters. Its length is 1 meter longer than four times its width. Find the length and width of the rectangle.

44. ***Cruise Ship*** A cruise ship left port and traveled 80 miles at a constant speed. Then for 10 miles it traveled 10 miles per hour slower while circling an island before stopping. The trip took 5 hours. Find the ship's speed for each part of the trip.

45. ***Car Travel in the Rain*** Jessica drove at a constant speed for 200 miles. Then it started to rain. So for the next 90 miles she traveled 5 miles per hour slower. The entire trip took 6 hours of driving time. Find her speed for each part of the trip.

▲ **46.** ***Garden Walkway*** Mr. and Mrs. Gomez are building a rectangular garden that is 10 feet by 6 feet. Around the outside of the garden, they will build a brick walkway. They have 100 square feet of brick. How wide should they make the brick walkway? Round your answer to the nearest tenth of a foot.

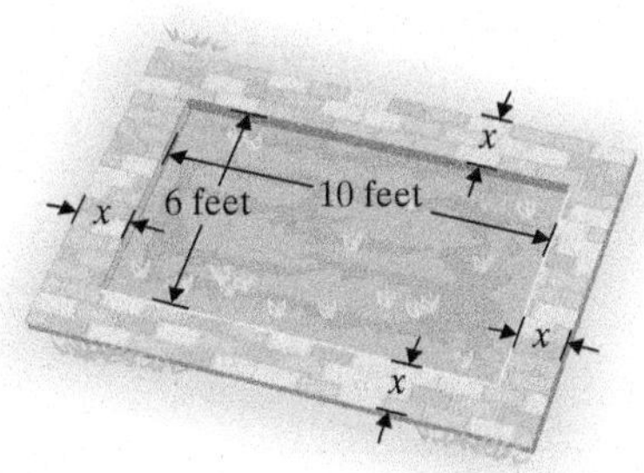

▲ **47.** ***Swimming Pool*** The local YMCA is building a new Olympic-sized pool of 50 meters by 25 meters. The builders want to make a walkway around the pool with a nonslip concrete surface. They have enough material to make 76 square meters of nonslip concrete surface. How wide should the walkway be?

Find the vertex and the intercepts of the following quadratic functions.

48. $g(x) = -x^2 + 6x - 11$

49. $f(x) = x^2 + 10x + 25$

In each of the following exercises, find the vertex, the y-intercept, and the x-intercepts (if any exist) and then graph the function.

50. $f(x) = x^2 + 6x + 5$

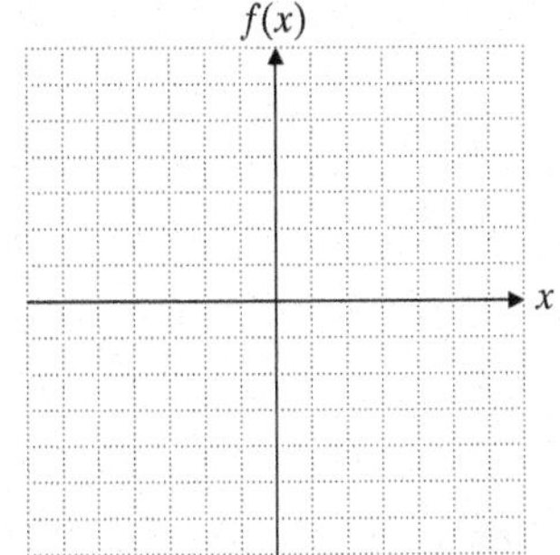

51. $f(x) = -x^2 + 6x - 5$

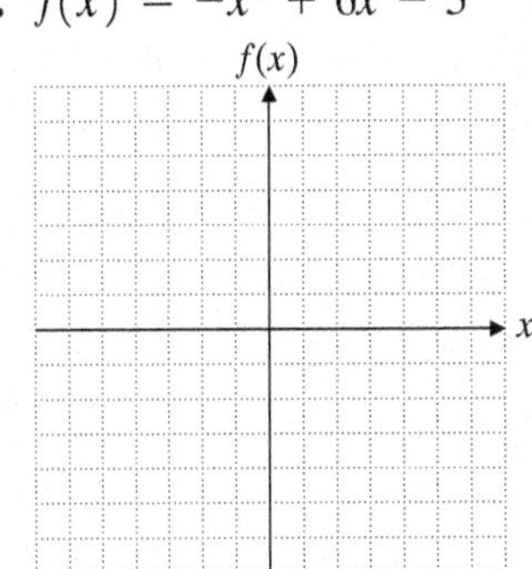

52. *Rocket Flight* A model rocket is launched upward from a platform 40 feet above the ground. The height of the rocket h is given at any time t in seconds by the function $h(t) = -16t^2 + 400t + 40$. Find the maximum height of the rocket. How long will it take the rocket to go through its complete flight and then hit the ground? (Assume that the rocket does *not* have a parachute.) Round your answer to the nearest tenth.

53. *Revenue for a Store* A salesman for an electronics store finds that in 1 month he can sell $(1200 - x)$ compact disc players that each sell for x dollars. Write a function for the revenue. What is the price x that will result in the maximum revenue for the store?

Graph the values of x that satisfy the conditions given.

54. $-3 \le x < 2$

55. $-8 \le x \le -4$

56. $x < -2$ *or* $x \ge 5$

57. $x > -5$ *and* $x < -1$

58. $x > -8$ *and* $x < -3$

59. $x + 3 > 8$ *or* $x + 2 < 6$

Solve for x.

60. $x - 2 > 7$ *or* $x + 3 < 2$

61. $x + 3 > 8$ *and* $x - 4 < -2$

62. $-1 < x + 5 < 8$

63. $0 \le 5 - 3x \le 17$

64. $2x - 7 < 3$ *and* $5x - 1 \ge 8$

65. $4x - 2 < 8$ *or* $3x + 1 > 4$

Solve and graph your solutions.

66. $x^2 + 7x - 18 < 0$

67. $x^2 - 9x + 20 > 0$

Solve each of the following if possible. Approximate, if necessary, any irrational solutions to the nearest tenth.

68. $2x^2 - x - 6 \le 0$

69. $3x^2 - 13x + 12 \le 0$

70. $9x^2 - 4 > 0$

71. $4x^2 - 8x \le 12 + 5x^2$

72. $x^2 + 13x > 16 + 7x$

73. $3x^2 - 12x > -11$

74. $4x^2 + 12x + 9 < 0$

Solve for x.

75. $|2x - 7| + 4 = 5$

76. $\left|\frac{2}{3}x - \frac{1}{2}\right| \le 3$

77. $|2 - 5x - 4| > 13$

Solve for x.

78. $|x + 7| < 15$

79. $|x + 9| < 18$

80. $\left|\frac{1}{2}x + 2\right| < \frac{7}{4}$

81. $|2x - 1| \ge 9$

82. $|3x - 1| \ge 2$

83. $|2(x - 5)| \ge 2$

How Am I Doing? Chapter 9 Test

MATH COACH MyMathLab® YouTube

After you take this test read through the Math Coach on pages 621–622. Math Coach videos are available via MyMathLab and YouTube. Step-by-step test solutions in the Chapter Test Prep Videos are also available via MyMathLab and YouTube. (Search "TobeyBegInterAlg" and click on "Channels.")

Solve the quadratic equations and simplify your answers. Use i notation for any complex numbers.

1. $8x^2 + 9x = 0$

2. $6x^2 - 3x = 1$

3. $\frac{3x}{2} - \frac{8}{3} = \frac{2}{3x}$

4. $x(x - 3) - 30 = 5(x - 2)$

5. $7x^2 - 4 = 52$

MC **6.** $\frac{2x}{2x + 1} - \frac{6}{4x^2 - 1} = \frac{x + 1}{2x - 1}$

7. $2x^2 - 6x + 5 = 0$

8. $2x(x - 3) = -3$

Solve for any valid real roots.

MC **9.** $x^4 - 11x^2 + 18 = 0$

10. $3x^{-2} - 11x^{-1} - 20 = 0$

11. $x^{2/3} - 3x^{1/3} - 4 = 0$

Solve for the variable specified.

12. $B = \frac{xyw}{z^2}$; for z

MC **13.** $5y^2 + 2by + 6w = 0$; for y

1. ______
2. ______
3. ______
4. ______
5. ______
6. ______
7. ______
8. ______
9. ______
10. ______
11. ______
12. ______
13. ______

14.

15.

16.

17.

18.

19.

20.

21.

22.

23.

24.

25

26.

Total Correct:

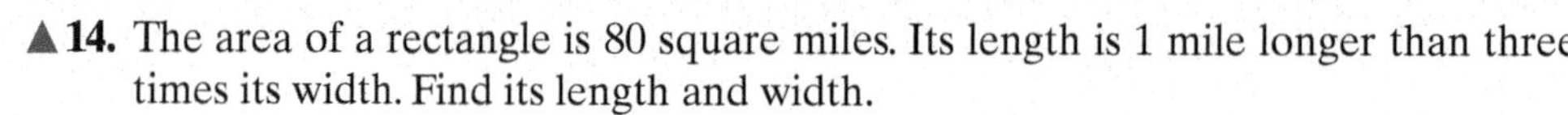

▲ **14.** The area of a rectangle is 80 square miles. Its length is 1 mile longer than three times its width. Find its length and width.

▲ **15.** Find the hypotenuse of a right triangle if the lengths of its legs are 6 and $2\sqrt{3}$.

16. Shirley and Bill paddled a canoe at a constant speed for 6 miles. They rested, had lunch, and then paddled 1 mile per hour faster for an additional 3 miles. The travel time for the entire trip was 4 hours. How fast did they paddle during each part of the trip?

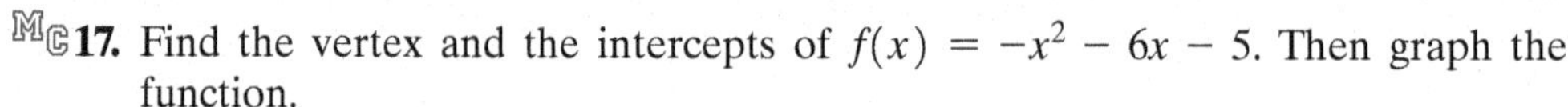

MC **17.** Find the vertex and the intercepts of $f(x) = -x^2 - 6x - 5$. Then graph the function.

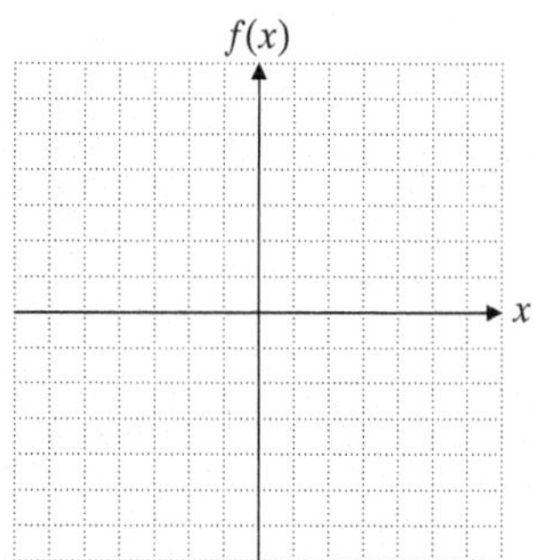

Find the values of x that satisfy the given conditions.

18. $-11 < 2x - 1 \leq -3$

19. $x - 4 \leq -6$ *or* $2x + 1 \geq 3$

Solve.

20. $2x^2 + 3x \geq 27$

21. $x^2 - 5x - 14 < 0$

22. Use a calculator to approximate to the nearest tenth the solution to $x^2 + 3x - 7 > 0$.

Solve for x.

23. $|5x - 2| = 37$

24. $\left|\frac{1}{2}x + 3\right| - 2 = 4$

Solve each absolute value inequality.

25. $|7x - 3| \leq 18$

26. $|3x + 1| > 7$

MATH COACH

Mastering the skills you need to do well on the test.

The following problems are from the Chapter 9 Test. Here are some helpful hints to keep you from making common errors on test problems.

Chapter 9 Test, Problem 6

Solve the quadratic equation. $\frac{2x}{2x+1} - \frac{6}{4x^2-1} = \frac{x+1}{2x-1}$

HELPFUL HINT If any denominators need to be factored, do that first. Then determine the LCD of all the denominators in the equation. Multiply each term of the equation by the LCD before solving for x.

Did you factor $4x^2 - 1$ as $(2x + 1)(2x - 1)$?

Yes ______ No ______

Did you identify the LCD to be $(2x + 1)(2x - 1)$?

Yes ______ No ______

If you answered No to these questions, review how to factor the difference of two squares and how to find the LCD of polynominal denominators.

Did you multiply the LCD by each term of the equation and remove parentheses to obtain $2x^2 - 5x - 7 = 0$?

Yes ______ No ______

Did you then use the quadratic formula and substitute for $a, b,$ and c to get $x = \frac{-(-5) \pm \sqrt{(-5)^2 - 4(2)(-7)}}{2(2)}$?

Yes ______ No ______

If you answered No to these questions, remember that your final equation should be in the form $ax^2 + bx + c = 0$. Then use $a = 2, b = -5,$ and $c = -7$ in the quadratic formula and simplify your result.

If you answered Problem 6 incorrectly, go back and rework the problem using these suggestions.

Chapter 9 Test, Problem 9

Solve for any valid real roots. $x^4 - 11x^2 + 18 = 0$

HELPFUL HINT Let $y = x^2$ and then $y^2 = x^4$. Write the new quadratic equation after these replacements have been made.

After making the necessary replacements, did you obtain the equation $y^2 - 11y + 18 = 0$?

Yes ______ No ______

Did you solve the quadratic equation using any method to result in $y = 9$ and $y = 2$?

Yes ______ No ______

If you answered No to these questions, stop and complete these steps again.

If $y = 9$ and $y = 2$, can you conclude that $x^2 = 9$ and $x^2 = 2$?

Yes ______ No ______

If you take the square root of each side of each equation, can you obtain $x = \pm 3$ and $x = \pm\sqrt{2}$?

Yes ______ No ______

If you answered No to these equations, remember that when you take the square root of each side of the equation there are two sign possibilities. Note that your final solution should consist of four values for x.

Now go back and rework the problem using these suggestions.

Chapter 9 Test, Problem 13

Solve for the variable specified. $5y^2 + 2by + 6w = 0$; for y

> **HELPFUL HINT** Think of the equation being written as $ay^2 + by + c = 0$. The quantities for a, b, or c may contain variables. Use the quadratic formula to solve for y.

Did you determine that $a = 5, b = 2b$, and $c = 6w$?

Yes ______ No ______

Did you substitute these values into the quadratic formula and simplify to obtain $y = \frac{-2b \pm \sqrt{4b^2 - 120w}}{10}$?

Yes ______ No ______

If you answered No to these questions, review the Helpful Hint again to make sure that you find the correct values for a, b, and c. Carefully substitute these values into the quadratic formula and simplify.

Were you able to simplify the radical expression by factoring out a 4 to obtain $\sqrt{4(b^2 - 30w)}$?

Yes ______ No ______

Did you simplify this expression further to get $2\sqrt{b^2 - 30w}$?

Yes ______ No ______

If you answered No to these questions, remember to always simplify radicals whenever possible. Make sure to write your solution as a simplified rational expression.

Now go back and rework the problem using these suggestions.

Chapter 9 Test, Problem 17

Find the vertex and the intercepts of $f(x) = -x^2 - 6x - 5$. Then graph the function.

> **HELPFUL HINT** When the function is written in $f(x) = ax^2 + bx + c$ form, we can find the vertex by using $x = \frac{-b}{2a}$ to find the x value of the vertex. We can solve for the intercepts using the substitutions $x = 0$ and $f(x) = 0$ to find the unknown coordinates. And, if $a < 0$, the graph is a parabola opening downward.

Do you see that $a = -1, b = -6$, and $c = -5$?

Yes ______ No ______

Did you use $x = \frac{-b}{2a}$ to discover that the vertex has an x-coordinate of -3?

Yes ______ No ______

If you answered No to these questions, notice that the function is written in $f(x) = ax^2 + bx + c$ form and review the formula: $x = \frac{-b}{2a}$. Substitute the resulting value for x into the original function to find the y-coordinate of the vertex point.

To find the y-intercept, did you substitute 0 for x into the original function to find the value for y?

Yes ______ No ______

After letting $f(x) = 0$ and substituting the values for a, b, and c into the quadratic formula, did you get the expression

$x = \frac{-(-6) \pm \sqrt{(-6)^2 - 4(-1)(-5)}}{2(-1)}$?

Yes ______ No ______

If you answered No to these questions, remember that the y-intercept will be an ordered pair in the form $(0, y)$, or in this case, $(0, f(0))$, and the x-intercept will be an ordered pair in the form $(x, 0)$. Be careful when substituting values for a, b, and c into the quadratic formula and remember to evaluate $\sqrt{16}$ as both 4 and -4. Simplify the expression to find the possible x-values.

Since $a < 0$, the parabola will open downward. Plot the vertex, x-intercept, and y-intercept points and connect these points with a curve to find the graph of the function.

If you answered Problem 17 incorrectly, go back and rework the problem using these suggestions.

Cumulative Test for Chapters 0–9

Approximately one-half of this test is based on the content of Chapters 0–8. The remainder is based on the content of Chapter 9.

1. Simplify. $(-3x^{-2}y^3)^4$

2. Combine like terms. $\frac{1}{2}a^3 - 2a^2 + 3a - \frac{1}{4}a^3 - 6a + a^2$

3. Solve for x. $\frac{1}{3}(x - 3) + 1 = \frac{1}{2}x - 2$

4. Graph. $6x - 3y = -12$

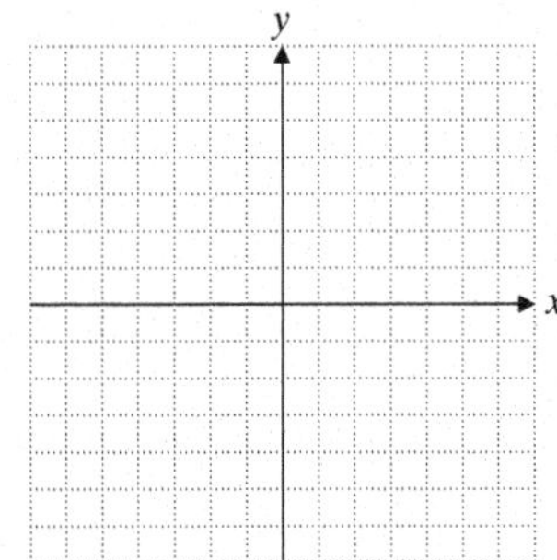

5. What is the slope of a line parallel to $2y + x = 8$?

6. Solve the system. $3x + 4y = -14$
$-x - 3y = 13$

7. Factor. $125x^3 - 27y^3$

8. Simplify. $\sqrt{72x^3y^6}$

9. Multiply. $(5 + \sqrt{3})(\sqrt{6} - \sqrt{2})$

10. Rationalize the denominator. $\frac{3x}{\sqrt{6}}$

Solve and simplify your answers. Use i notation for complex numbers

11. $3x^2 + 12x = 26x$

12. $12x^2 = 11x - 2$

13. $44 = 3(2x - 3)^2 + 8$

14. $3 - \frac{4}{x} + \frac{5}{x^2} = 0$

1. __________
2. __________
3. __________
4. __________
5. __________
6. __________
7. __________
8. __________
9. __________
10. __________
11. __________
12. __________
13. __________
14. __________

15. ______

Solve and check.

15. $\sqrt{6x + 12} - 2 = x$

16. $x^{2/3} + 9x^{1/3} + 18 = 0$

16. ______

Solve for y.

17. $2y^2 + 5wy - 7z = 0$

18. $3y^2 + 16z^2 = 5w$

17. ______

▲**19.** The hypotenuse of a right triangle is $\sqrt{38}$. One leg of the triangle is 5. Find the length of the other leg

18. ______

▲**20.** A triangle has an area of 45 square meters. The altitude is 3 meters longer than three times the length of the base. Find each dimension.

19. ______

Exercises 21 and 22 refer to the quadratic function $f(x) = -x^2 + 8x - 12$.

20. ______

21. Find the vertex and the intercepts of the function

21. ______

22. Graph the function.

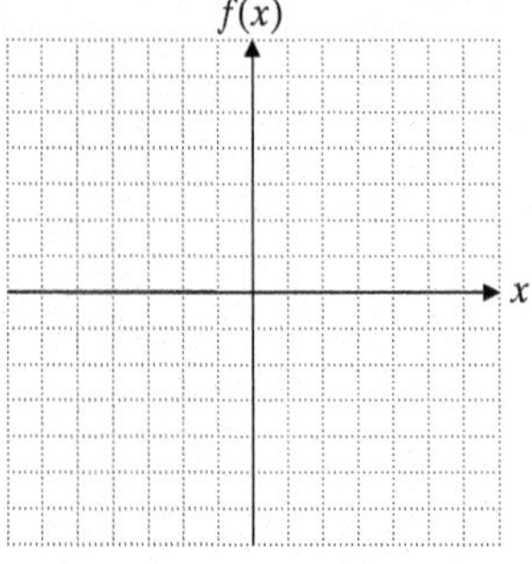

22. ______

23. ______

24. ______

Solve the following inequalities

23. $x + 5 \leq -4$ *or* $2 - 7x \leq 16$

24. $x^2 > -3x + 18$

25. ______

Solve for x.

25. $|3x + 1| = 16$

26. ______

Solve each absolute value inequality.

27. ______

26. $\left|\frac{1}{2}x + 2\right| \leq 8$

27. $|3x - 4| > 11$

CHAPTER 10

The Conic Sections

CAREER OPPORTUNITIES

Architect

The built environments in which we live, work, and relax are created by architects. Skyscrapers, shopping malls, houses, bridges, and tunnels all begin as architectural ideas that are transformed into reality. Architects use many mathematical tools to design structures that meet the aesthetic, structural, and economic requirements of the client.

To investigate how the mathematics in this chapter can help with this field, see the Career Exploration Problems on page **668**.

10.1 The Distance Formula and the Circle

Student Learning Objectives

After studying this section, you will be able to:

1. Find the distance between two points.
2. Find the center and radius of a circle and graph the circle if the equation is in standard form.
3. Write the equation of a circle in standard form given its center and radius.
4. Rewrite the equation of a circle in standard form.

In this chapter we'll talk about the equations and graphs of four special geometric figures–the circle, the parabola, the ellipse, and the hyperbola. These shapes are called **conic sections** because they can be formed by slicing a cone with a plane. The equation of any conic section is of degree 2.

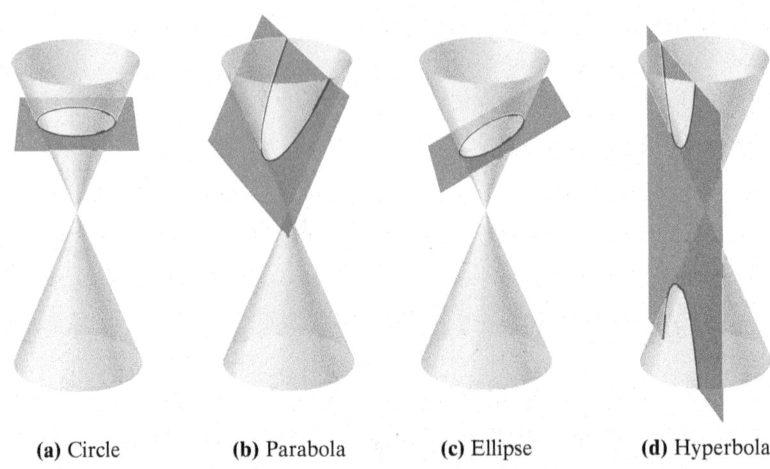

(a) Circle **(b)** Parabola **(c)** Ellipse **(d)** Hyperbola

Conic sections are an important and interesting subject. They are studied along with many other things in a branch of mathematics called *analytic geometry*. Conic sections can be found in applications of physics and engineering. Satellite transmission dishes have parabolic shapes; the orbits of planets are ellipses; the orbits of some comets are hyperbolas; the path of a ball, rocket, or bullet is a parabola (if we neglect air resistance).

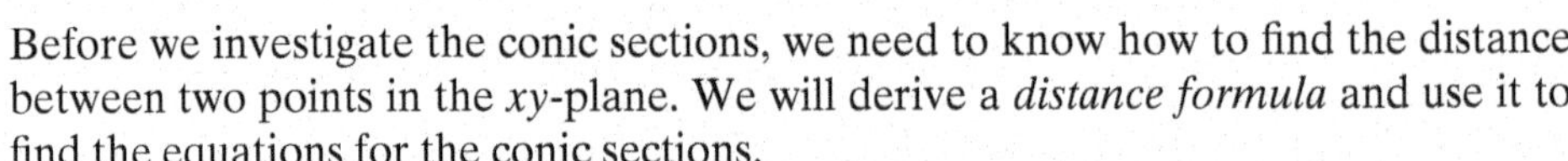

1 Finding the Distance Between Two Points

Before we investigate the conic sections, we need to know how to find the distance between two points in the *xy*-plane. We will derive a *distance formula* and use it to find the equations for the conic sections.

Recall from Chapter 1 that to find the distance between two points on a real number line, we simply find the absolute value of the difference of the values of the points. For example, the distance from -3 to 5 on the x-axis is

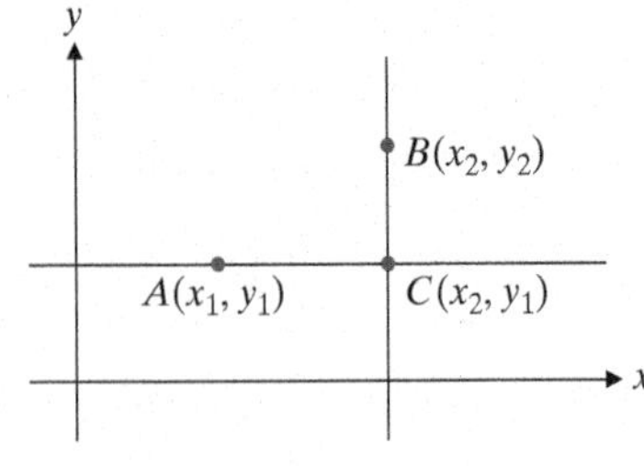

$$|5 - (-3)| = |5 + 3| = 8.$$

Remember that absolute value is another name for distance. We could have written

$$|-3 - 5| = |-8| = 8.$$

Similarly, the distance from -3 to 5 on the y-axis is

$$|5 - (-3)| = 8.$$

We use this simple fact to find the distance between two points in the xy-plane. Let $A(x_1, y_1)$ and $B(x_2, y_2)$ be points in a plane. First we draw a horizontal line through A, and then we draw a vertical line through B. (We could have drawn a horizontal line through B and a vertical line through A.) The lines intersect at point $C(x_2, y_1)$. Why are the coordinates (x_2, y_1)? The distance from A to C is $|x_2 - x_1|$ and from B to C, $|y_2 - y_1|$.

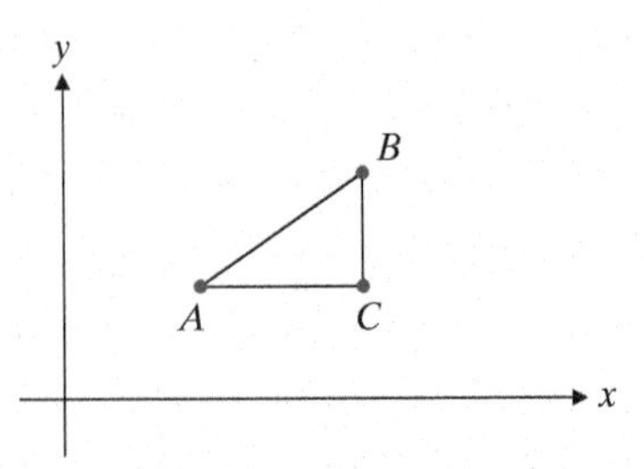

Now, if we draw a line from A to B, we have the right triangle ABC. We can use the Pythagorean Theorem to find the length (distance) of the line from A to B. By the Pythagorean Theorem,

$$(AB)^2 = (AC)^2 + (BC)^2.$$

Let's rename the distance AB as d. Then

$$d^2 = (|x_2 - x_1|)^2 + (|y_2 - y_1|)^2$$

$$d = \sqrt{(x_2 - x_1)^2 + (y_2 - y_1)^2}.$$

(Can you give a reason why we can drop the absolute value bars here?)

This is the **distance formula.**

> **DISTANCE FORMULA**
>
> The distance between two points (x_1, y_1) and (x_2, y_2) is
>
> $$d = \sqrt{(x_2 - x_1)^2 + (y_2 - y_1)^2}.$$

Example 1 Find the distance between $(3, -4)$ and $(-2, -5)$.

Solution To use the formula, we arbitrarily let $(x_1, y_1) = (3, -4)$ and $(x_2, y_2) = (-2, -5)$.

$$\begin{aligned} d &= \sqrt{(x_2 - x_1)^2 + (y_2 - y_1)^2} \\ &= \sqrt{(-2 - 3)^2 + [-5 - (-4)]^2} \\ &= \sqrt{(-5)^2 + (-5 + 4)^2} \\ &= \sqrt{(-5)^2 + (-1)^2} \\ &= \sqrt{25 + 1} = \sqrt{26} \end{aligned}$$

□

Student Practice 1 Find the distance between $(-6, -2)$ and $(3, 1)$.

The choice of which point is (x_1, y_1) and which point is (x_2, y_2) is up to you. We would obtain exactly the same answer in Example 1 if $(x_1, y_1) = (-2, -5)$ and if $(x_2, y_2) = (3, -4)$. Try it for yourself and see whether you obtain the same result.

2 Finding the Center and Radius of a Circle and Graphing the Circle

A **circle** is defined as the set of all points in a plane that are at a fixed distance from a point in that plane. The fixed distance is called the **radius,** and the point is called the **center** of the circle.

We can use the distance formula to find the equation of a circle. Let a circle of radius r have its center at (h, k). For any point (x, y) on the circle, the distance formula tells us that

$$\sqrt{(x - h)^2 + (y - k)^2} = r.$$

Squaring each side gives

$$(x - h)^2 + (y - k)^2 = r^2.$$

This is the equation of the circle with center at (h, k) and radius r.

> **Graphing Calculator**
>
> **Graphing Circles**
>
> A graphing calculator is designed to graph *functions.* In order to graph a circle, you need to separate it into two halves, each of which is a function. Thus, in order to graph the circle in Example 2 on the next page, first solve for y.
>
> $$(y - 3)^2 = 25 - (x - 2)^2$$
> $$y - 3 = \pm\sqrt{25 - (x - 2)^2}$$
> $$y = 3 \pm \sqrt{25 - (x - 2)^2}$$
>
> Now graph the two functions
>
> $$y_1 = 3 + \sqrt{25 - (x - 2)^2}$$
>
> (the upper half of the circle) and
>
> $$y_2 = 3 - \sqrt{25 - (x - 2)^2}$$
>
> (the lower half of the circle). To get a proper-looking circle, use a "square" window setting. Window settings will vary depending on the calculator. Display:
>
> 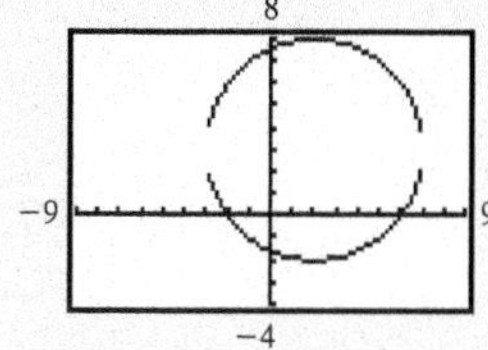
>
>
> Notice that due to limitations in the calculator, it is not a perfect circle and two gaps appear.

STANDARD FORM OF THE EQUATION OF A CIRCLE

The standard form of the equation of the circle with center at (h, k) and radius r is

$$(x - h)^2 + (y - k)^2 = r^2.$$

Example 2 Find the center and radius of the circle $(x - 2)^2 + (y - 3)^2 = 25$. Then sketch its graph.

Solution From the equation of a circle,

$$(x - h)^2 + (y - k)^2 = r^2,$$

we see that $(h, k) = (2, 3)$. Thus, the center of the circle is at $(2, 3)$. Since $r^2 = 25$, the radius of the circle is $r = 5$.

The graph of this circle is shown on the right.

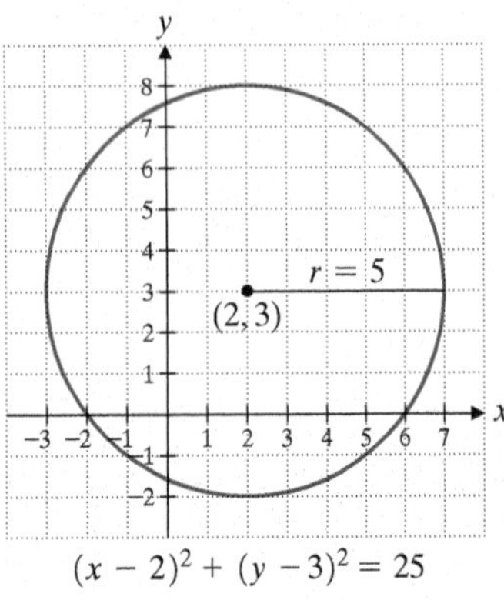

$(x - 2)^2 + (y - 3)^2 = 25$

Student Practice 2

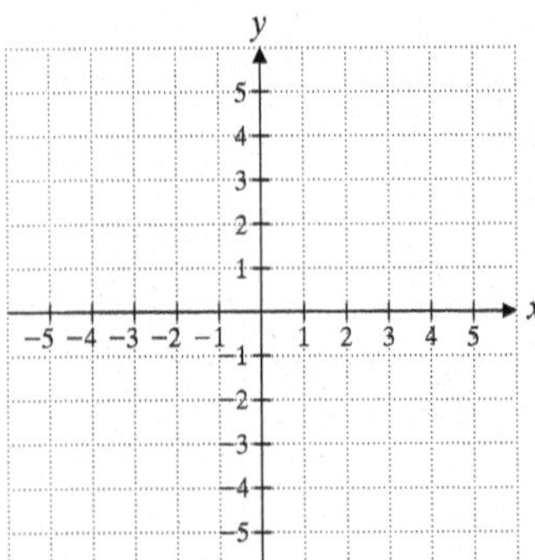

Student Practice 2 Find the center and radius of the circle

$$(x + 1)^2 + (y + 2)^2 = 9.$$

Then sketch its graph.

3 Writing the Equation of a Circle in Standard Form Given the Center and Radius

We can write the standard form of the equation of a specific circle if we are given the center and the radius. We use the definition of the standard form of the equation of a circle to write the equation we want.

Example 3 Write the equation of the circle with center $(-1, 3)$ and radius $\sqrt{5}$. Put your answer in standard form.

Solution We are given that $(h, k) = (-1, 3)$ and $r = \sqrt{5}$. Thus,

$$(x - h)^2 + (y - k)^2 = r^2$$

becomes the following:

$$[x - (-1)]^2 + (y - 3)^2 = (\sqrt{5})^2$$
$$(x + 1)^2 + (y - 3)^2 = 5$$

Be careful of the signs. It is easy to make a sign error in these steps.

Student Practice 3 Write the equation of the circle with center $(-5, 0)$ and radius $\sqrt{3}$. Put your answer in standard form.

4 Rewriting the Equation of a Circle in Standard Form

The standard form of the equation of a circle helps us sketch the graph of the circle. Sometimes the equation of a circle is not given in standard form, and we need to rewrite the equation.

Example 4 Write the equation of the circle $x^2 + 2x + y^2 + 6y = -6$ in standard form. Find the radius and center of the circle and sketch its graph.

Solution The standard form of the equation of a circle is

$$(x - h)^2 + (y - k)^2 = r^2.$$

If we multiply out the terms in the equation, we get

$$(x^2 - 2hx + h^2) + (y^2 - 2ky + k^2) = r^2.$$

Comparing this with the equation we were given,

$$(x^2 + 2x) + (y^2 + 6y) = -6,$$

suggests that we can complete the squares to put the equation in standard form.

$$x^2 + 2x + \_\_\_\_ + y^2 + 6y + \_\_\_\_ = -6$$

$$x^2 + 2x + 1 + y^2 + 6y + 9 = -6 + 1 + 9$$

$$x^2 + 2x + 1 + y^2 + 6y + 9 = 4$$

$$(x + 1)^2 + (y + 3)^2 = 4$$

Thus, the center is at $(-1, -3)$, and the radius is 2. The sketch of the circle is shown below.

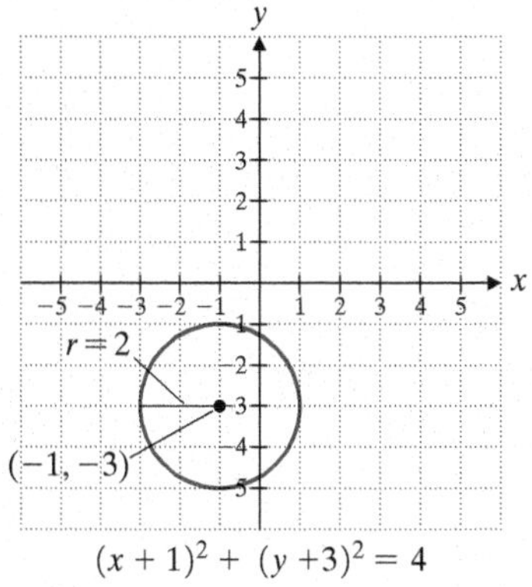

$(x + 1)^2 + (y + 3)^2 = 4$

Student Practice 4 Write the equation of the circle $x^2 + 4x + y^2 + 2y - 20 = 0$ in standard form. Find the radius and center of the circle and sketch its graph.

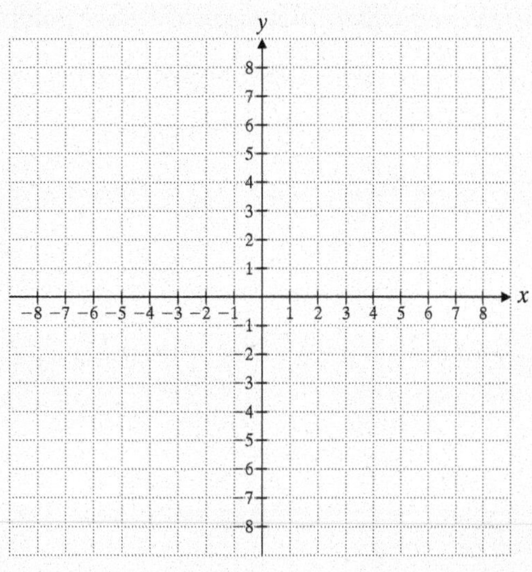

Graphing Calculator

Exploration

One way to graph the equation in Example 4 is to write it as a quadratic equation in y and then employ the quadratic formula.

$$y^2 + 6y + (x^2 + 2x + 6) = 0$$

$$ay^2 + by + c = 0$$

$$a = 1, b = 6, \text{ and}$$

$$c = x^2 + 2x + 6$$

$$y = \frac{-6 \pm \sqrt{36 - 4(1)(x^2 + 2x + 6)}}{2(1)}$$

$$y = \frac{-6 \pm \sqrt{12 - 8x - 4x^2}}{2}$$

Thus, we have the two halves of the circle.

$$y_1 = \frac{-6 + \sqrt{12 - 8x - 4x^2}}{2}$$

$$y_2 = \frac{-6 - \sqrt{12 - 8x - 4x^2}}{2}$$

We can graph these on one coordinate system to obtain the graph. Some graphing calculators have a feature for getting a background grid for your graph. If you have this feature, use it to find the coordinates of the center.

10.1 Exercises

Verbal and Writing Skills, Exercises 1–4

1. Explain how you would find the distance from -2 to 4 on the y-axis.

2. Explain how you would find the distance between $(3, -1)$ and $(-4, 0)$ in the xy-plane.

3. $(x - 1)^2 + (y + 2)^2 = 9$ is the equation of a circle. Explain how to determine the center and the radius of the circle.

4. $x^2 - 6x + y^2 - 2y = 6$ is the equation of a circle. Explain how you would rewrite the equation in standard form.

Find the distance between each pair of points. Simplify your answers.

5. $(1, 6)$ and $(2, 4)$

6. $(5, 4)$ and $(9, 3)$

7. $(-4, 3)$ and $(-2, 7)$

8. $(0, -2)$ and $(5, 3)$

9. $(4, -5)$ and $(-2, -13)$

10. $(-7, -1)$ and $(5, 8)$

11. $\left(\frac{5}{4}, -\frac{1}{3}\right)$ and $\left(\frac{1}{4}, -\frac{2}{3}\right)$

12. $\left(-1, \frac{1}{5}\right)$ and $\left(-\frac{1}{2}, \frac{11}{5}\right)$

13. $\left(\frac{1}{3}, \frac{3}{5}\right)$ and $\left(\frac{7}{3}, \frac{1}{5}\right)$

14. $\left(-\frac{1}{4}, \frac{1}{7}\right)$ and $\left(\frac{3}{4}, \frac{6}{7}\right)$

15. $(1.3, 2.6)$ and $(-5.7, 1.6)$

16. $(0.4, 1.8)$ and $(9.4, -1.2)$

Find the two values of the unknown coordinate so that the distance between the points is as given.

17. $(7, 2)$ and $(1, y)$; distance is 10

18. $(3, y)$ and $(3, -5)$; distance is 9

19. $(1.5, 2)$ and $(0, y)$; distance is 2.5

20. $\left(1, \frac{15}{2}\right)$ and $\left(x, -\frac{1}{2}\right)$; distance is 10

21. $(4, 7)$ and $(x, 5)$; distance is $\sqrt{5}$

22. $(1, 6)$ and $(3, y)$; distance is $2\sqrt{2}$

Applications, Exercises 23 and 24

Radar Detection *Use the following information to solve exercises 23 and 24. An airport is located at point O. A short-range radar tower is located at point R. The maximum range at which the radar can detect a plane is 4 miles from point R.*

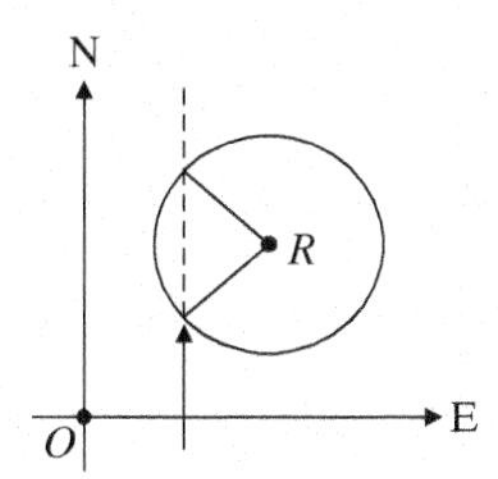

23. Assume that R is 5 miles east of O and 7 miles north of O. In other words, R is located at the point $(5, 7)$. An airplane is flying parallel to and 2 miles east of the north axis. (In other words, the plane is flying along the path $x = 2$.) What is the *shortest distance* north of the airport at which the plane can be detected by the radar tower at R? Round your answer to the nearest tenth of a mile.

24. Assume that R is 6 miles east of O and 6 miles north of O. In other words, R is located at the point $(6, 6)$. An airplane is flying parallel to and 4 miles east of the north axis. (In other words, the plane is flying along the path $x = 4$.) What is the *greatest distance* north of the airport at which the plane can still be detected by the radar tower at R? Round your answer to the nearest tenth of a mile.

Write in standard form the equation of the circle with the given center and radius.

25. Center $(-3, 7)$; $r = 6$

26. Center $(9, -4)$; $r = 5$

27. Center $(-2.4, 0)$; $r = \frac{3}{4}$

28. Center $\left(0, \frac{1}{2}\right)$; $r = \frac{1}{3}$

29. Center $\left(0, \frac{3}{8}\right)$; $r = \sqrt{3}$

30. Center $(-2.5, 0)$; $r = \sqrt{6}$

Give the center and radius of each circle. Then sketch its graph.

31. $x^2 + y^2 = 25$

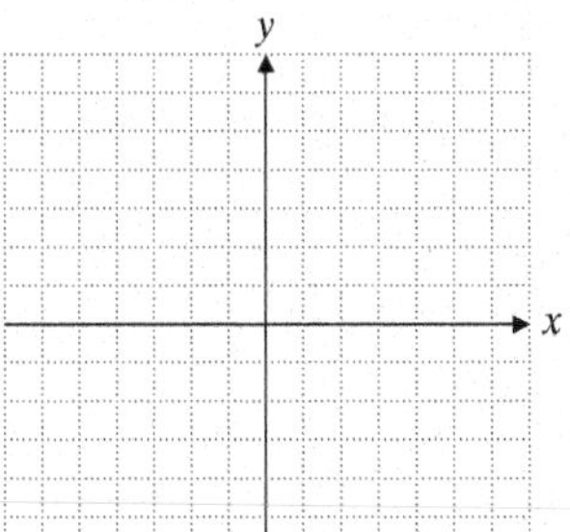

32. $x^2 + y^2 = 9$

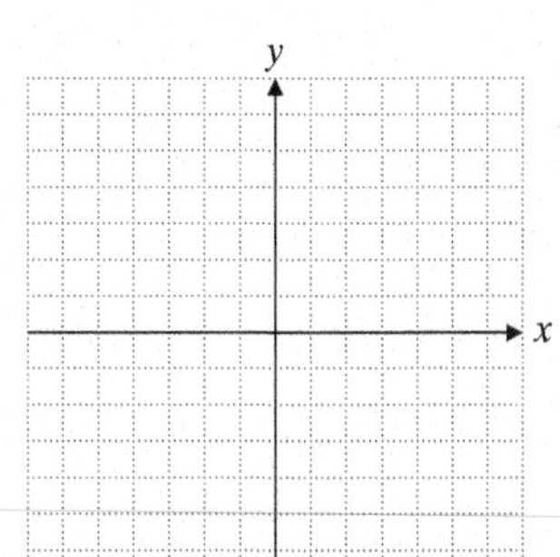

33. $(x - 5)^2 + (y - 3)^2 = 16$

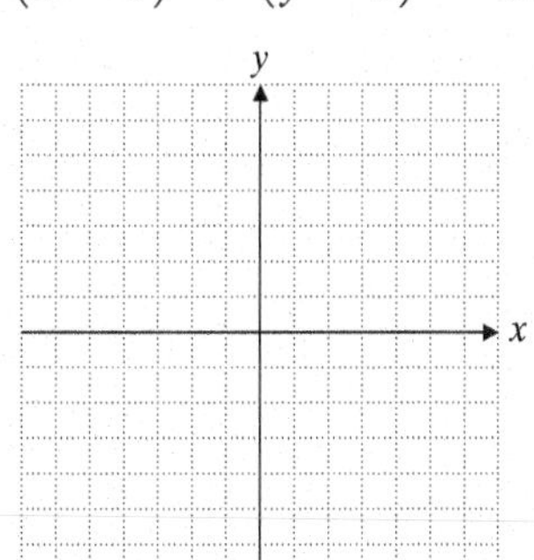

34. $(x-3)^2+(y-2)^2=4$

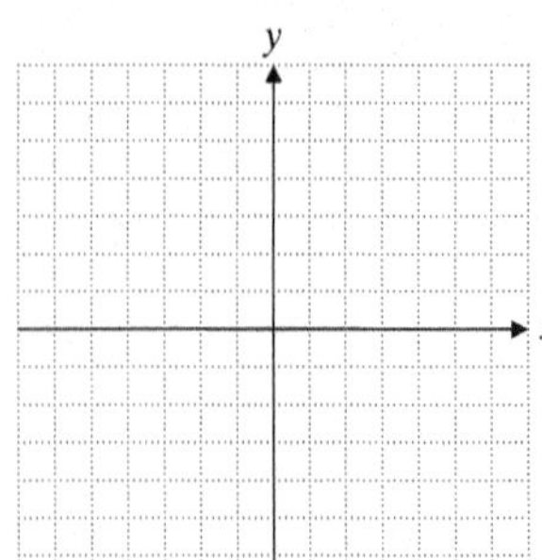

35. $(x+2)^2+(y-3)^2=25$

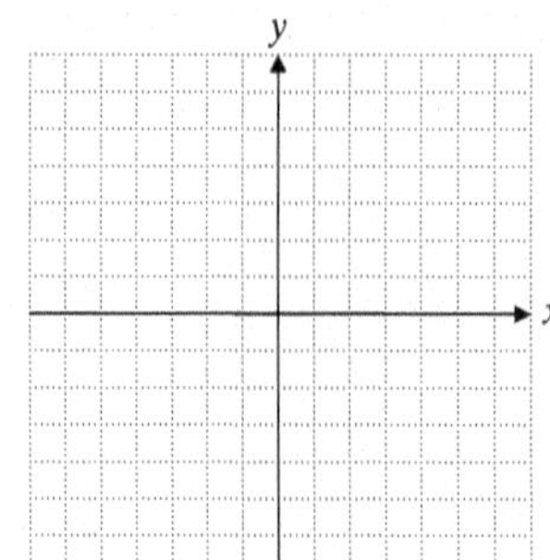

36. $\left(x-\frac{3}{2}\right)^2+(y+2)^2=9$

Rewrite each equation in standard form, using the approach of Example 4. Find the center and radius of each circle.

37. $x^2+y^2-4x+10y+4=0$

38. $x^2+y^2-8x+2y+8=0$

39. $x^2+y^2-10x+6y-2=0$

40. $x^2+y^2+12x+2y-27=0$

41. $x^2+y^2+3x-2=0$

42. $x^2+y^2-5x-1=0$

43. ***Ferris Wheels*** A Ferris wheel has a radius r of 44.2 feet. The height of the tower t is 55.8 feet. The distance d from the origin to the base of the tower is 61.5 feet. Find the standard form of the equation of the circle represented by the Ferris wheel.

44. ***Ferris Wheels*** The tallest Ferris wheel in the world is the High Roller in Las Vegas. This Ferris wheel has a radius r of 260 feet. The distance d from the origin to the base of the tower is 275 feet, and the height of the tower t is 290 feet. Find the standard form of the equation of the circle represented by the Ferris wheel.

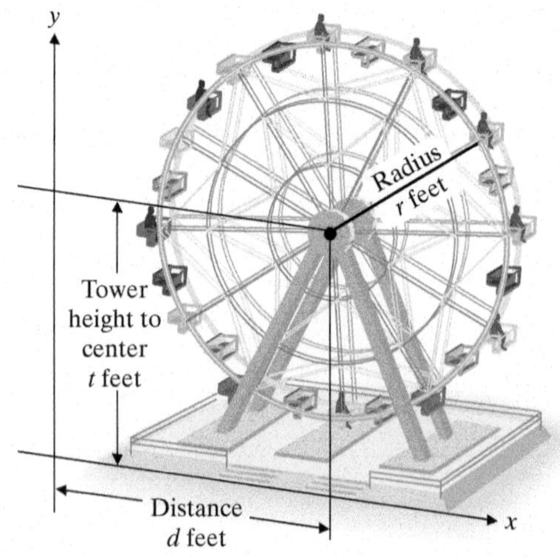

Optional Graphing Calculator Problems *Graph each circle with your graphing calculator.*

45. $(x - 5.32)^2 + (y + 6.54)^2 = 47.28$

46. $x^2 + 9.56x + y^2 - 7.12y + 8.9995 = 0$

Cumulative Review *Solve the quadratic equations by using the quadratic formula.*

47. [9.2.1] $4x^2 + 2x = 1$

48. [9.2.1] $5x^2 - 6x - 7 = 0$

▲ **49. [5.2.2]** ***Volcano Eruptions*** The 1980 eruptions of Mt. Saint Helens blew down or scorched 230 square miles of forest. A deposit of rock and sediments soon filled up a 20-square-mile area to an average depth of 150 feet. How many cubic feet of rock and sediments settled in this region?

50. [2.6.2] ***Volcano Eruptions*** Within a 15-mile radius north of Mt. Saint Helens, the blast of its 1980 eruption traveled at up to 670 miles per hour. If an observer 15 miles north of the volcano saw the blast and attempted to run for cover, how many seconds did he have to run before the blast reached his original location?

Quick Quiz 10.1

1. Find the distance between $(3, -4)$ and $(-2, -6)$.

2. Write in standard form the equation of the circle with center $(5, -6)$ and radius 7.

3. Rewrite the equation in standard form. Find the center and radius of the circle.

$$x^2 + 4x + y^2 - 6y + 4 = 0$$

4. Concept Check Explain how you would find the values of the unknown coordinate x if the distance between $(-6, 8)$ and $(x, 12)$ is 4.

10.2 The Parabola

Student Learning Objectives

After studying this section, you will be able to:

1. Graph vertical parabolas.
2. Graph horizontal parabolas.
3. Rewrite the equation of a parabola in standard form.

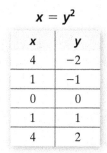

If we pass a plane through a cone so that the plane is parallel to but not touching a side of the cone, we form a parabola. A **parabola** is defined as the set of points that are the same distance from some fixed line (called the **directrix**) and some fixed point (called the **focus**) that is *not* on the line.

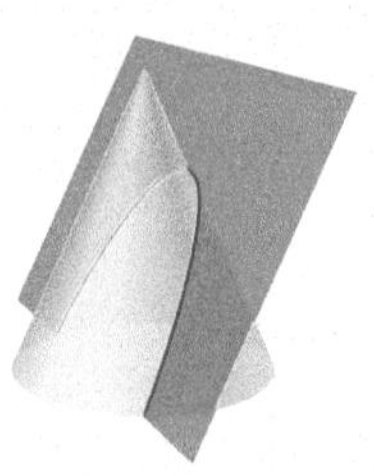

The shape of a parabola is a common one. For example, the cables that are used to support the weight of a bridge are often in the shape of a parabola.

The simplest form for the equation is one variable = (another variable)^2. That is, $y = x^2$ or $x = y^2$. We will make a table of values for each equation, plot the points, and draw a graph. For the first equation we choose values for x and find y. For the second equation we choose values for y and find x.

$y = x^2$

x	y
−2	4
−1	1
0	0
1	1
2	4

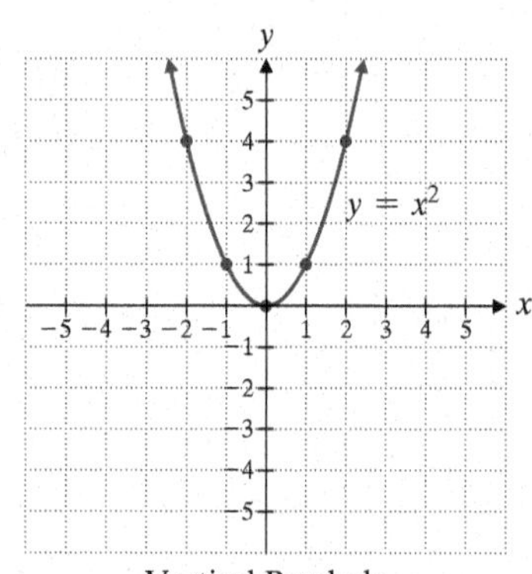

Vertical Parabola

$x = y^2$

x	y
4	−2
1	−1
0	0
1	1
4	2

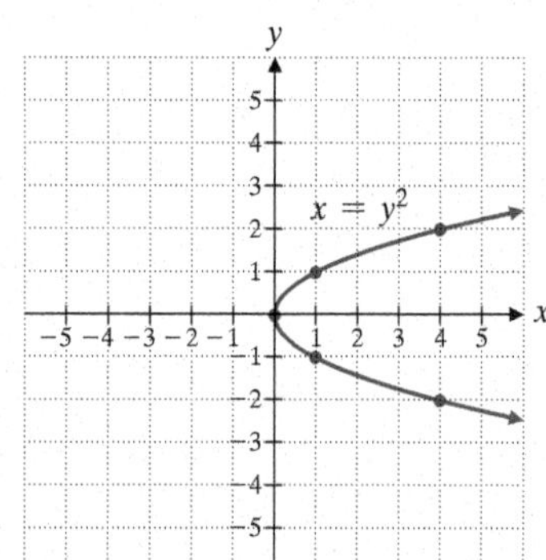

Horizontal Parabola

Notice that the graph of $y = x^2$ is symmetric about the y-axis. That is, if you folded the graph along the y-axis, the two parts of the curve would coincide. For this parabola, the y-axis is the **axis of symmetry.**

What is the axis of symmetry for the parabola $x = y^2$? Every parabola has an axis of symmetry. This axis can be *any* line; it depends on the location and orientation of the parabola in the rectangular coordinate system. The point at which the parabola crosses the axis of symmetry is the **vertex.** What are the coordinates of the vertex for $y = x^2$? For $x = y^2$?

1 Graphing Vertical Parabolas

Example 1 Graph $y = (x - 2)^2$. Identify the vertex and the axis of symmetry.

Solution We make a table of values. We begin with $x = 2$ in the middle of the table of values because $(2 - 2)^2 = 0$. That is, when $x = 2$, $y = 0$. We then fill in the x- and y-values above and below $x = 2$. We plot the points and draw the graph.

$y = (x - 2)^2$

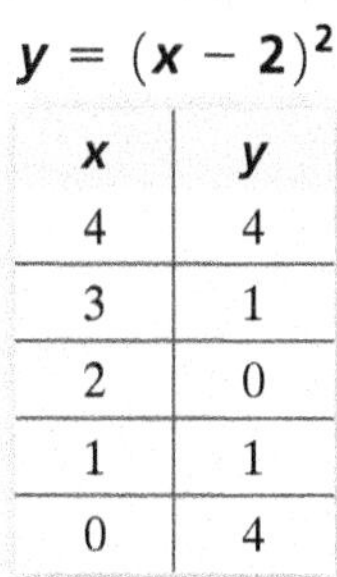

x	y
4	4
3	1
2	0
1	1
0	4

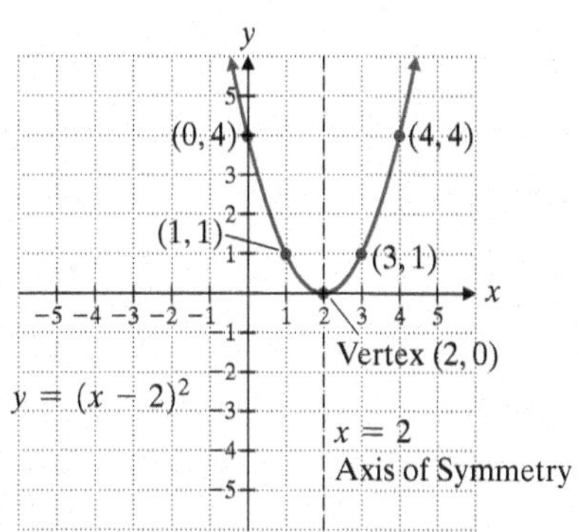

Student Practice 1

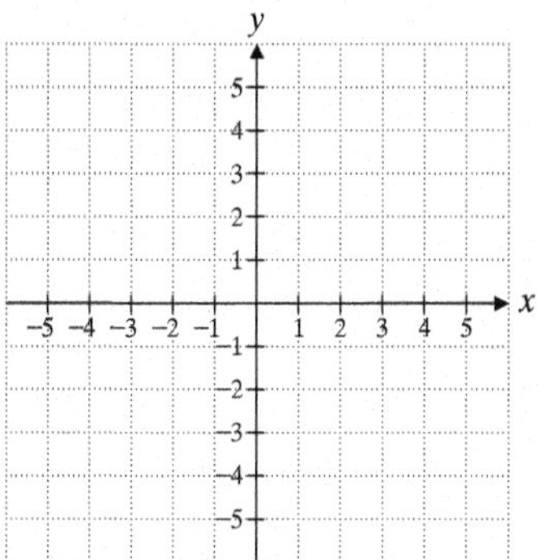

The vertex is $(2, 0)$, and the axis of symmetry is the line $x = 2$. □

Student Practice 1 Graph $y = -(x + 3)^2$. Identify the vertex and the axis of symmetry.

Example 2 Graph $y = (x - 2)^2 + 3$. Find the vertex, the axis of symmetry, and the y-intercept.

Solution This graph looks just like the graph of $y = x^2$, except that it is shifted 2 units to the right and 3 units up. The vertex is $(2, 3)$. The axis of symmetry is $x = 2$. We can find the y-intercept by letting $x = 0$ in the equation. We get

$$y = (0 - 2)^2 + 3 = 4 + 3 = 7.$$

Thus, the y-intercept is $(0, 7)$.

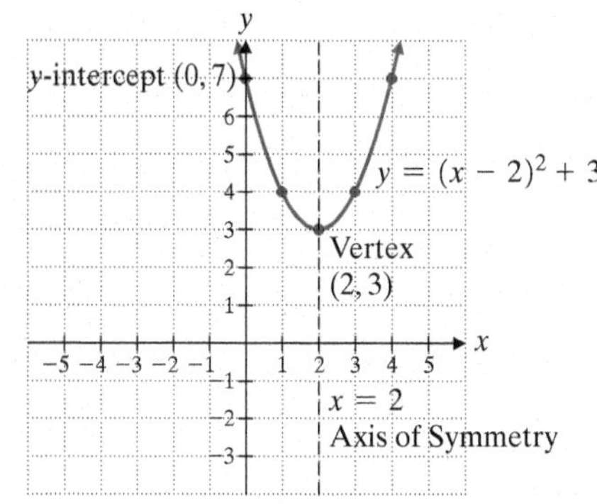

□

Student Practice 2 Graph the parabola $y = (x - 6)^2 + 4$.

Student Practice 2

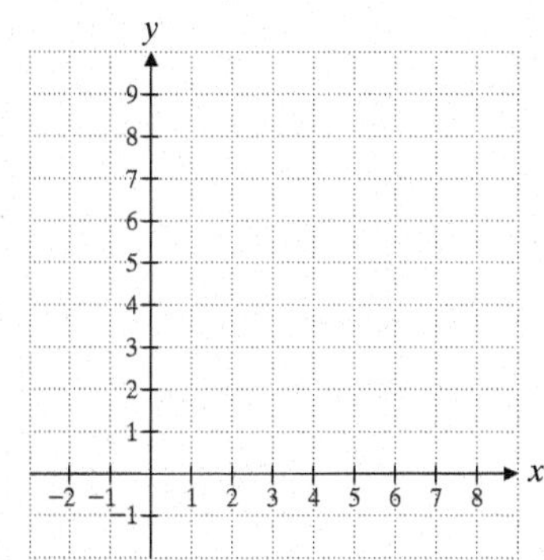

The examples we have studied illustrate the following properties of the standard form of the equation of a vertical parabola.

STANDARD FORM OF THE EQUATION OF A VERTICAL PARABOLA

1. The graph of $y = a(x - h)^2 + k$, where $a \neq 0$, is a vertical parabola.
2. The parabola opens upward ◡ if $a > 0$ and downward ◠ if $a < 0$.
3. The vertex of the parabola is (h, k).
4. The axis of symmetry is the line $x = h$.
5. The y-intercept is the point where the parabola crosses the y-axis (i.e., where $x = 0$).

We can use these properties as steps to graph a parabola. If we want greater accuracy, we should also plot a few other points.

Example 3 Graph $y = -\frac{1}{2}(x + 3)^2 - 1$.

Solution

Step 1 The equation has the form $y = a(x - h)^2 + k$, where $a = -\frac{1}{2}, h = -3$, and $k = -1$, so it is a vertical parabola.

$$y = a(x - h)^2 + k$$

$$y = -\frac{1}{2}[x - (-3)]^2 + (-1)$$

Step 2 $a < 0$; so the parabola opens downward.

Step 3 We have $h = -3$ and $k = -1$.
Therefore, the vertex of the parabola is $(-3, -1)$.

Step 4 The axis of symmetry is the line $x = -3$.

We plot a few points on either side of the axis of symmetry. We try $x = -1$ because $(-1 + 3)^2$ is 4 and $-\frac{1}{2}(4)$ is an integer. We avoid fractions. When $x = -1$, $y = -\frac{1}{2}(-1 + 3)^2 - 1 = -3$. Thus, the point is $(-1, -3)$. The image of this point on the other side of the axis of symmetry is $(-5, -3)$. We now try $x = 1$. When $x = 1$, $y = -\frac{1}{2}(1 + 3)^2 - 1 = -9$. Thus, the point is $(1, -9)$. The image of this point on the other side of the axis of symmetry is $(-7, -9)$.

Step 5 When $x = 0$, we have the following:

$$y = -\frac{1}{2}(0 + 3)^2 - 1$$
$$= -\frac{1}{2}(9) - 1$$
$$= -4.5 - 1 = -5.5$$

Thus, the y-intercept is $(0, -5.5)$.

The graph is shown on the right.

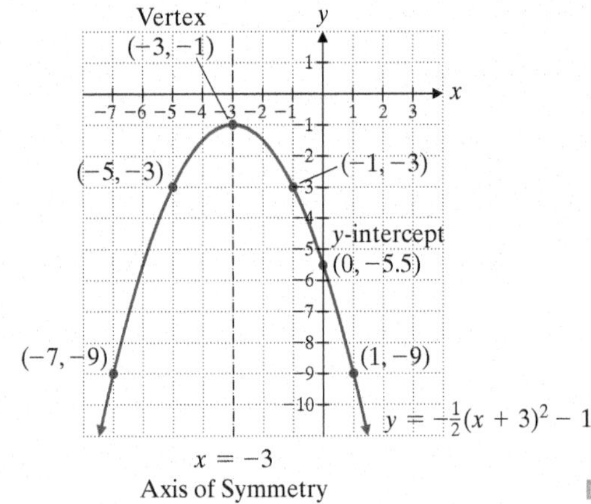

Student Practice 3

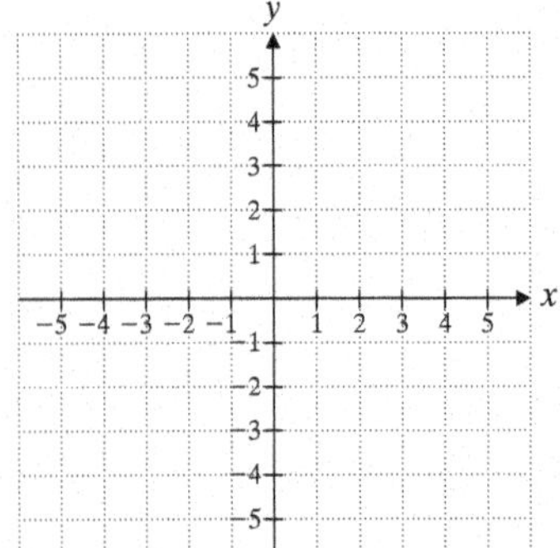

Student Practice 3 Graph $y = \frac{1}{4}(x - 2)^2 + 3$.

2 Graphing Horizontal Parabolas

Recall that the equation $x = y^2$, in which the squared term is the y-variable, describes a horizontal parabola. Horizontal parabolas open to the left or right. They are symmetric about the x-axis or about a line parallel to the x-axis. Let's now look at examples of horizontal parabolas.

Example 4 Graph $x = -2y^2$.

Solution Notice that the y-term is squared. This means that the parabola is horizontal. We make a table of values, plot points, and draw the graph. To make the table of values, we choose values for y and find x. We begin with $y = 0$.

$x = -2y^2$

x	y
−8	−2
−2	−1
0	0
−2	1
−8	2

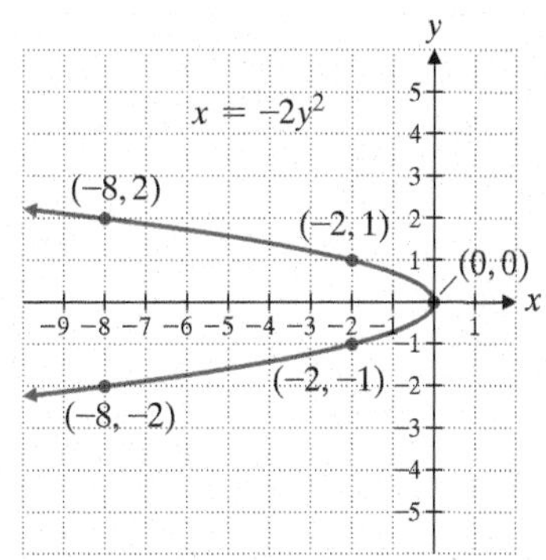

The parabola $x = -2y^2$ has its vertex at $(0, 0)$. The axis of symmetry is the x-axis. □

 **Student Practice 4** Graph the parabola $x = -2y^2 + 4$.

Student Practice 4

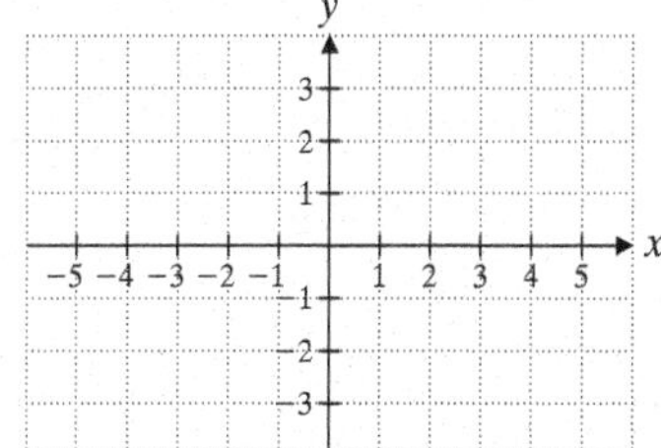

To Think About: *Example 4 Follow-up* Compare the graphs in Example 4 and Student Practice 4 to the graph of $x = y^2$. How are they different? How are they the same? What does the coefficient -2 in the equation $x = -2y^2$ do to the graph of the equation $x = y^2$? What does the constant 4 in $x = -2y^2 + 4$ do to the graph of $x = -2y^2$?

Now we can make the same type of observations for horizontal parabolas as we did for vertical ones.

STANDARD FORM OF THE EQUATION OF A HORIZONTAL PARABOLA

1. The graph of $x = a(y - k)^2 + h$, where $a \neq 0$, is a horizontal parabola.
2. The parabola opens to the right ⊂ if $a > 0$ and opens to the left ⊃ if $a < 0$.
3. The vertex of the parabola is (h, k).
4. The axis of symmetry is the line $y = k$.
5. The x-intercept is the point where the parabola crosses the x-axis (i.e., where $y = 0$).

Example 5 Graph $x = (y - 3)^2 - 5$. Find the vertex, the axis of symmetry, and the x-intercept.

Solution

Step 1 The equation has the form $x = a(y - k)^2 + h$, where $a = 1$, $k = 3$, and $h = -5$, so it is a horizontal parabola.

$$x = a(y - k)^2 + h$$

$$x = 1(y - 3)^2 + (-5)$$

Step 2 $a > 0$; so the parabola opens to the right.

Step 3 We have $k = 3$ and $h = -5$. Therefore, the vertex is $(-5, 3)$.

Step 4 The line $y = 3$ is the axis of symmetry.

We look for a few points on either side of the axis of symmetry. We will try y-values close to the vertex $(-5, 3)$. We try $y = 4$ and $y = 2$. When $y = 4$, $x = (4 - 3)^2 - 5 = -4$. When $y = 2$, $x = (2 - 3)^2 - 5 = -4$. Thus, the points are $(-4, 4)$ and $(-4, 2)$. (Remember to list the x-value first in a coordinate pair.) We try $y = 5$ and $y = 1$. When

Continued on next page

$y = 5, x = (5 - 3)^2 - 5 = -1$. When $y = 1$, $x = (1 - 3)^2 - 5 = -1$. Thus, the points are $(-1, 5)$ and $(-1, 1)$. You may prefer to find one point, graph it, and find its image on the other side of the axis of symmetry, as was done in Example 3.

Step 5 When $y = 0, x = (0 - 3)^2 - 5 = 9 - 5 = 4$. Thus, the x-intercept is $(4, 0)$.

We plot the points and draw the graph.

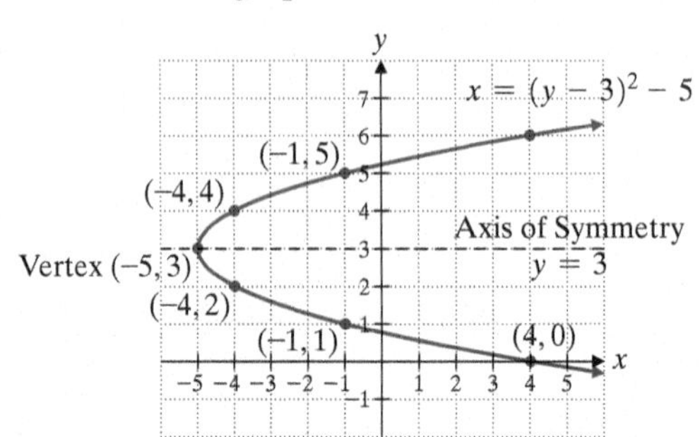

Notice that the graph also crosses the y-axis. You can find the y-intercepts by setting x equal to 0 and solving the resulting quadratic equation. Try it. □

Student Practice 5

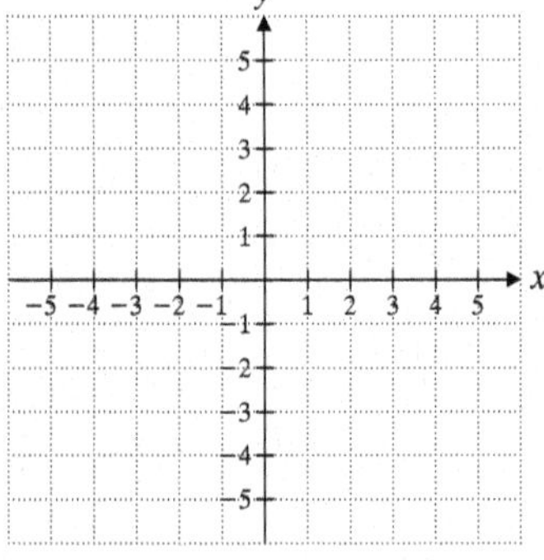

Student Practice 5 Graph the parabola $x = -(y + 1)^2 - 3$. Find the vertex, the axis of symmetry, and the x-intercept.

3 Rewriting the Equation of a Parabola in Standard Form

Example 6 Place the equation $x = y^2 + 4y + 1$ in standard form. Then graph it.

Solution Since the y-term is squared, we have a horizontal parabola. So the standard form is

$$x = a(y - k)^2 + h.$$

We must complete the square in order to write the equation in standard form.

$x = y^2 + 4y + \_\_\_\_ - \_\_\_\_ + 1$	Whatever we add to the right side we must also subtract from the right side.
$x = y^2 + 4y + \left(\frac{4}{2}\right)^2 - \left(\frac{4}{2}\right)^2 + 1$	Complete the square.
$x = (y^2 + 4y + 4) - 3$	Simplify.
$x = (y + 2)^2 - 3$	Standard form.

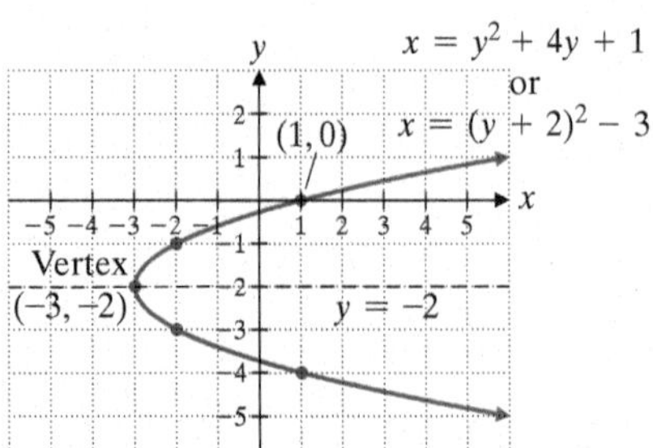

We see that $a = 1, k = -2$, and $h = -3$. Since a is positive, the parabola opens to the right. The vertex is $(-3, -2)$. The axis of symmetry is $y = -2$. If we let $y = 0$, we find that the x-intercept is $(1, 0)$. □

Student Practice 6 Place the equation $x = y^2 - 6y + 13$ in standard form and graph it.

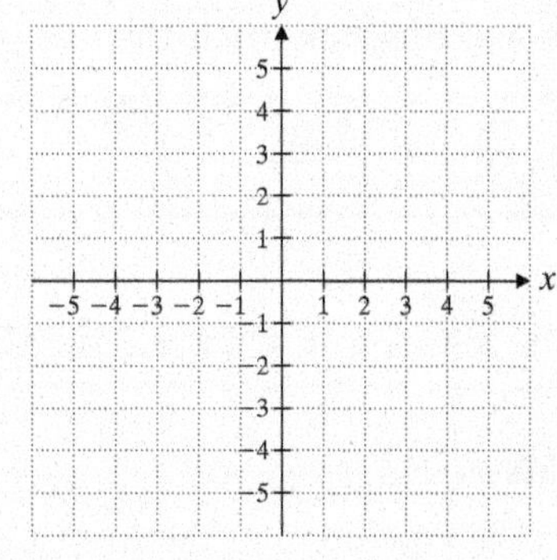

Example 7 Place the equation $y = 2x^2 - 4x - 1$ in standard form. Then graph it.

Solution This time the x-term is squared, so we have a vertical parabola. The standard form is

$$y = a(x - h)^2 + k.$$

We need to complete the square.

$$\begin{aligned} y &= 2(x^2 - 2x + \_\_\_) - \_\_\_ - 1 \\ &= 2[x^2 - 2x + (1)^2] - 2(1)^2 - 1 \\ &= 2(x - 1)^2 - 3 \end{aligned}$$

The parabola opens upward $(a > 0)$, the vertex is $(1, -3)$, the axis of symmetry is $x = 1$, and the y-intercept is $(0, -1)$.

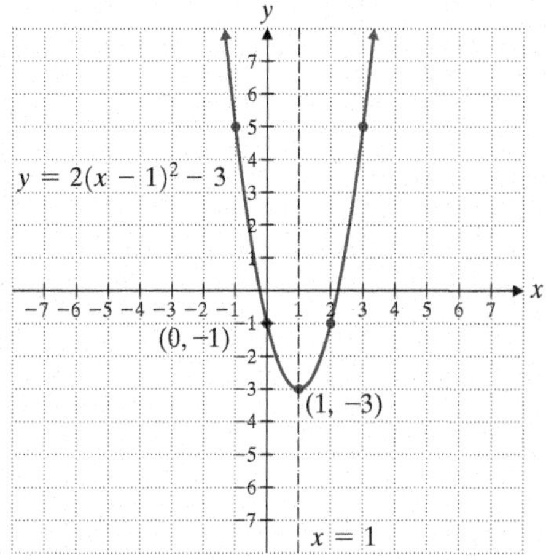

□

Student Practice 7 Place $y = 2x^2 + 8x + 9$ in standard form and graph it.

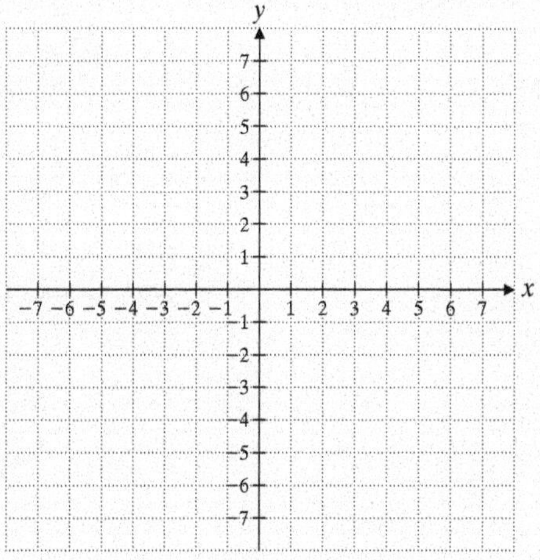

Graphing Calculator

Graphing Parabolas

Graphing horizontal parabolas such as the one in Example 6 on a graphing calculator requires dividing the curve into two halves. In this case the halves would be

$$y_1 = -2 + \sqrt{x + 3}$$

and

$$y_2 = -2 - \sqrt{x + 3}$$

Vertical parabolas can be graphed immediately on a graphing calculator. Why is this? How can you tell whether it is necessary to divide a curve into two halves?

10.2 Exercises MyMathLab®

Verbal and Writing Skills, Exercises 1–4

1. The graph of $y = x^2$ is symmetric about the ________. The graph of $x = y^2$ is symmetric about the ________.

2. Explain how to determine the axis of symmetry of the parabola $x = \frac{1}{2}(y + 5)^2 - 1$.

3. Explain how to determine the vertex of the parabola $y = 2(x - 3)^2 + 4$.

4. How does the coefficient -6 affect the graph of the parabola $y = -6x^2$?

Graph each parabola and label the vertex. Find the y-intercept.

5. $y = -4x^2$

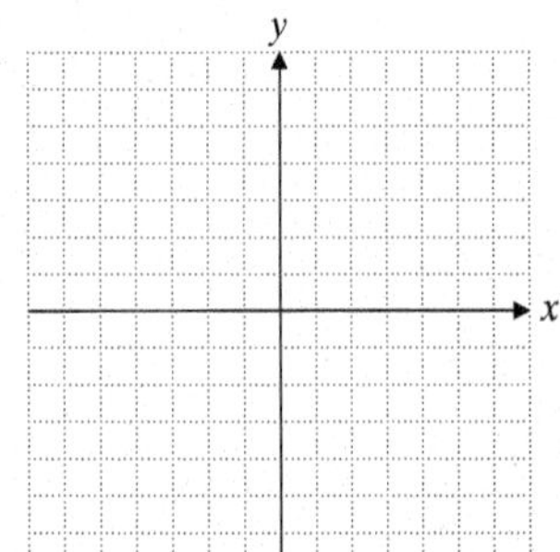

6. $y = -x^2$

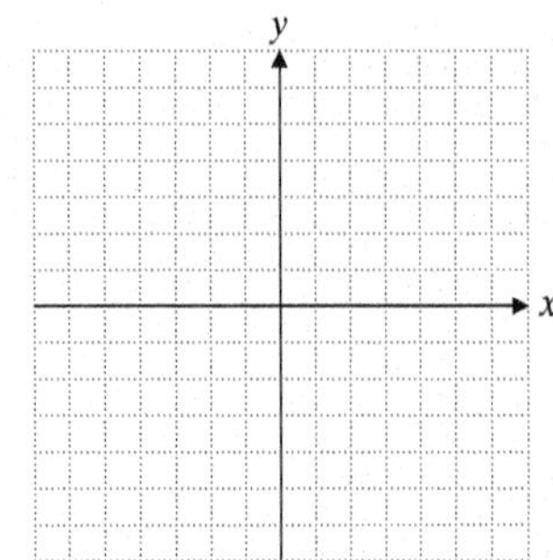

7. $y = x^2 - 2$

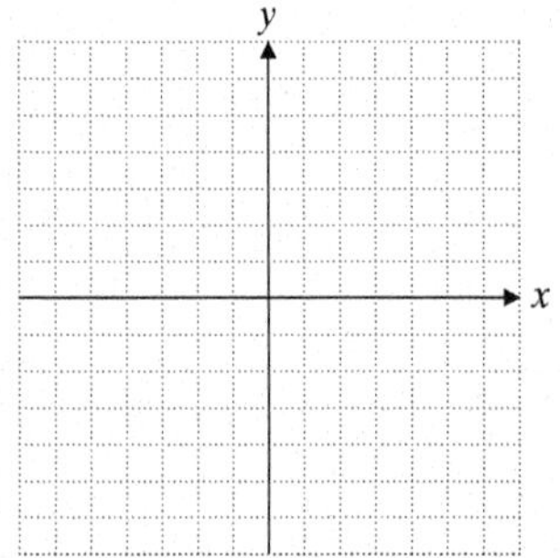

8. $y = x^2 + 3$

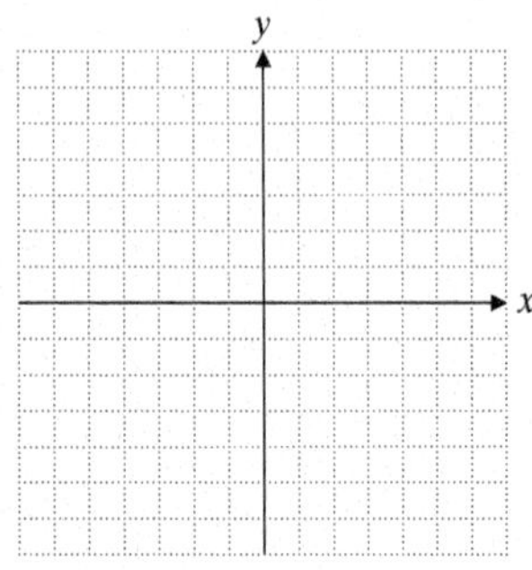

9. $y = \frac{1}{2}x^2 - 2$

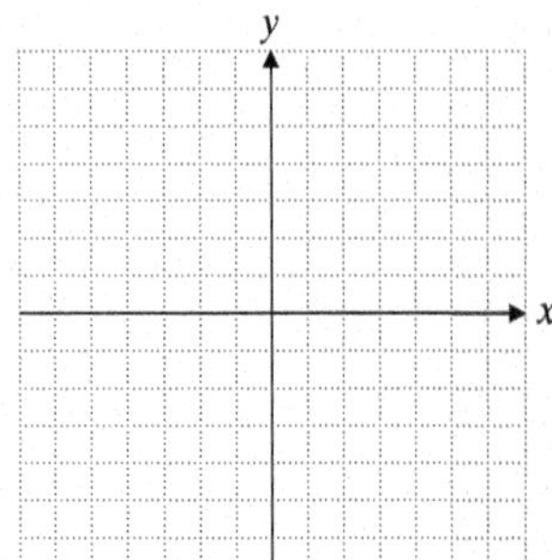

10. $y = \frac{1}{4}x^2 + 1$

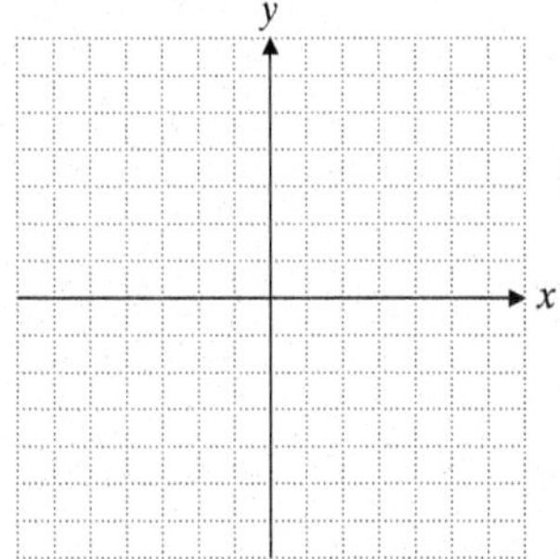

11. $y = (x - 3)^2 - 2$

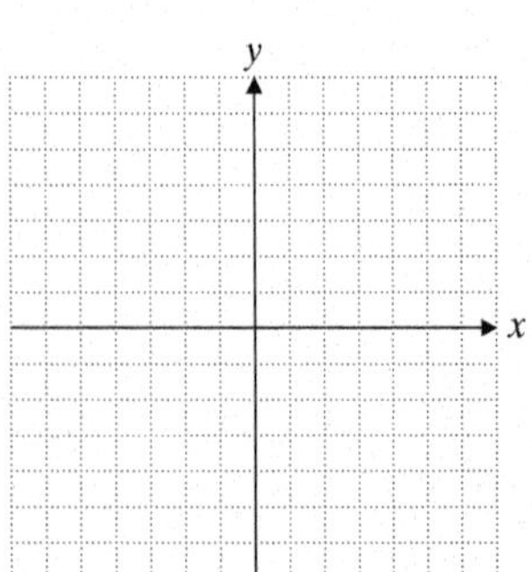

12. $y = (x + 2)^2 - 1$

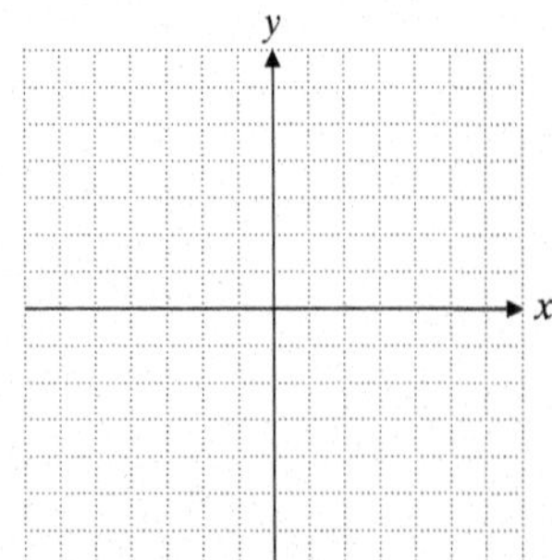

13. $y = 2(x - 1)^2 + \frac{3}{2}$

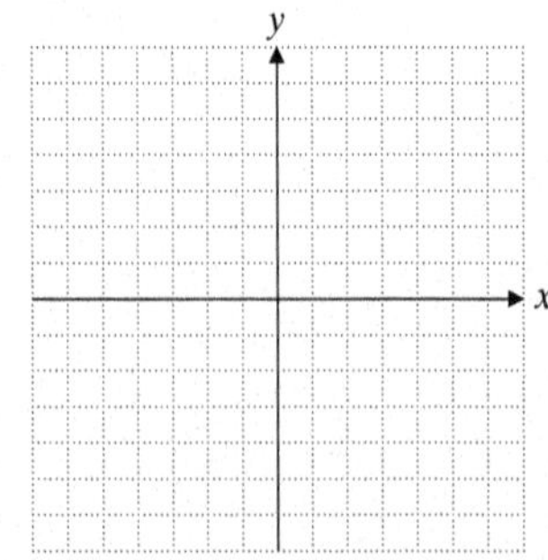

14. $y = 2(x - 2)^2 + \dfrac{5}{2}$

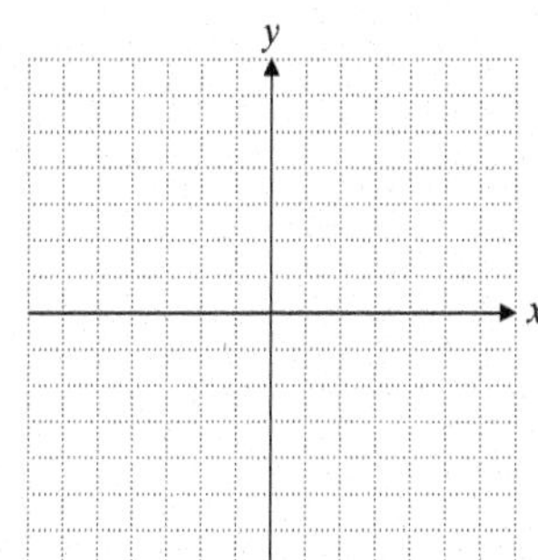

15. $y = -4\left(x + \dfrac{3}{2}\right)^2 + 5$

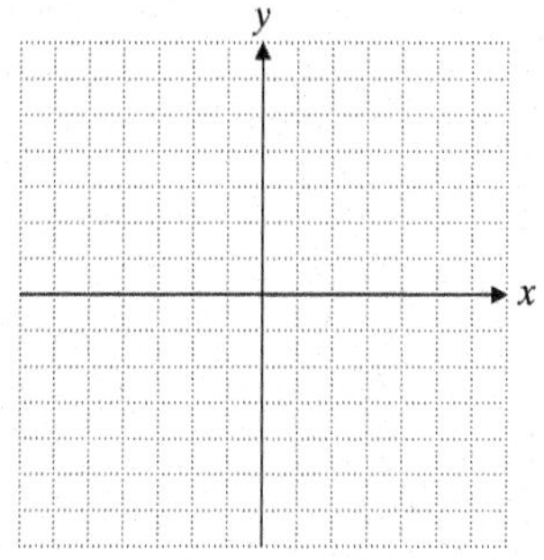

16. $y = -2\left(x + \dfrac{1}{2}\right)^2 - 1$

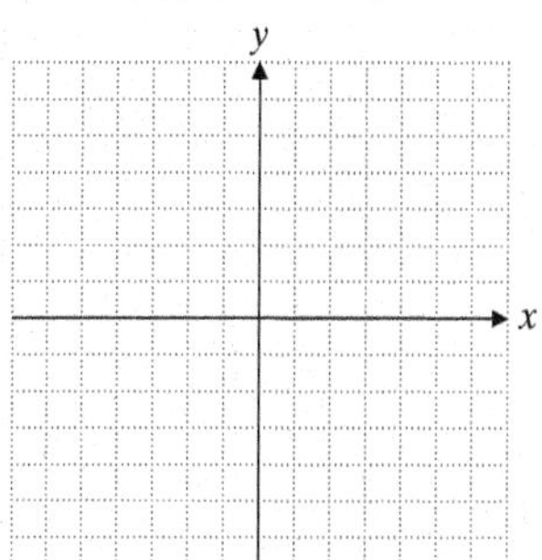

Graph each parabola and label the vertex. Find the x-intercept.

17. $x = \dfrac{1}{2}y^2$

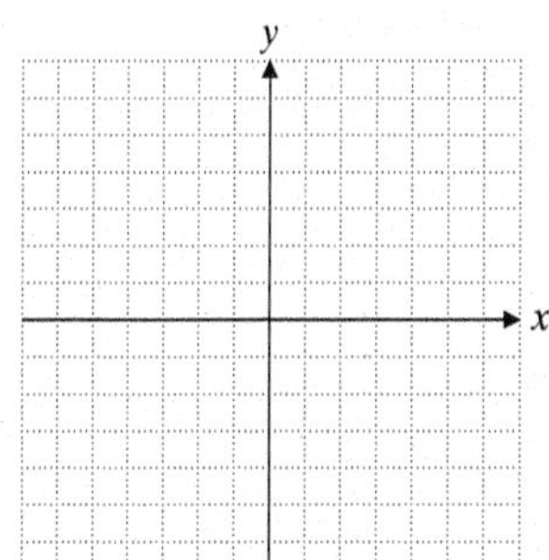

18. $x = \dfrac{2}{3}y^2$

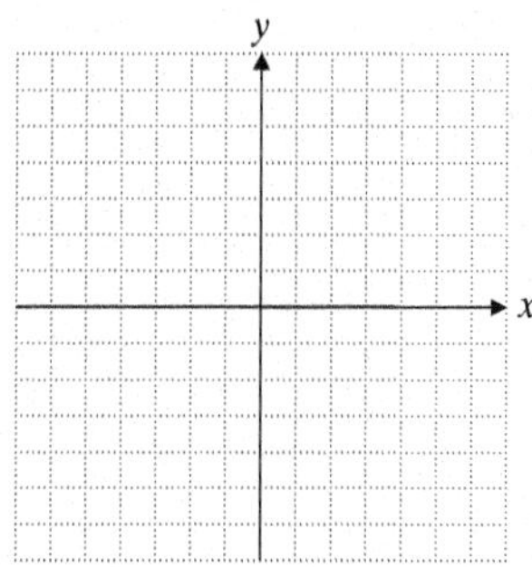

19. $x = \dfrac{1}{4}y^2 - 2$

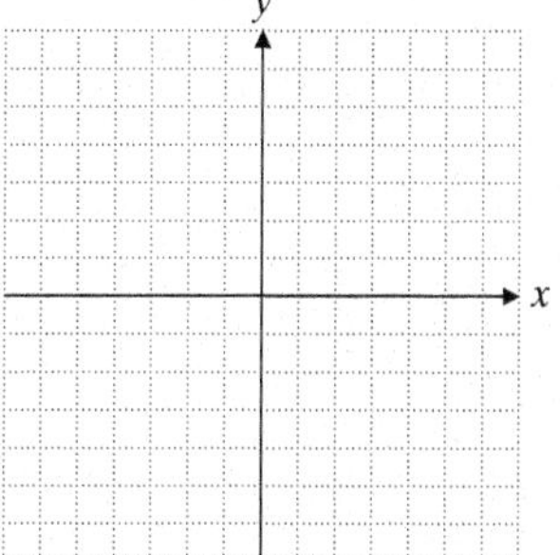

20. $x = \dfrac{1}{3}y^2 + 1$

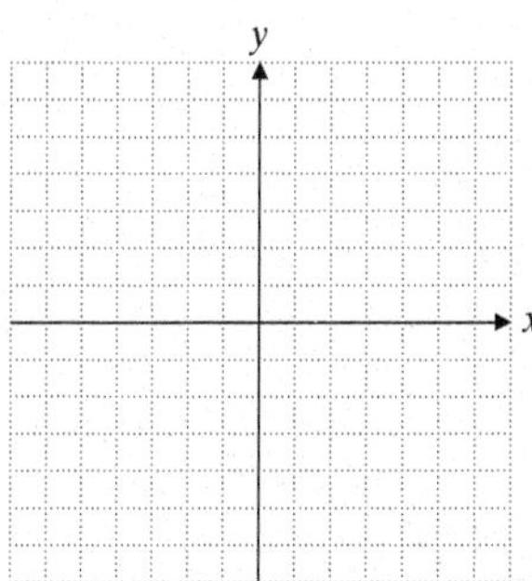

21. $x = -y^2 + 2$

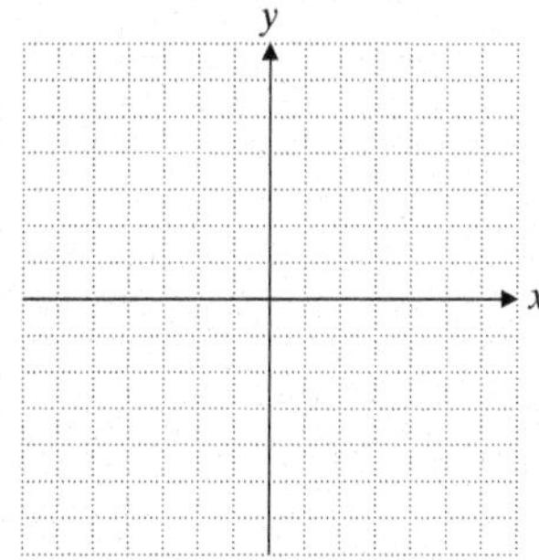

22. $x = -y^2 - 1$

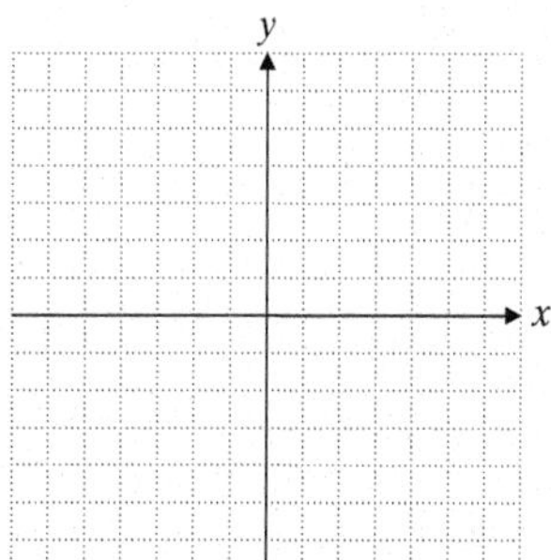

23. $x = (y - 2)^2 + 3$

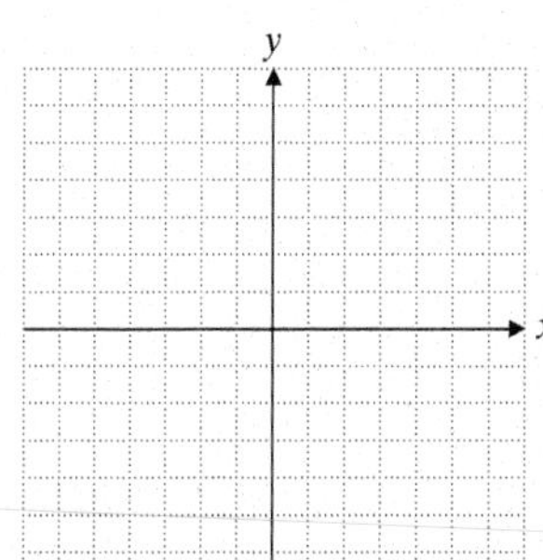

24. $x = (y - 4)^2 + 1$

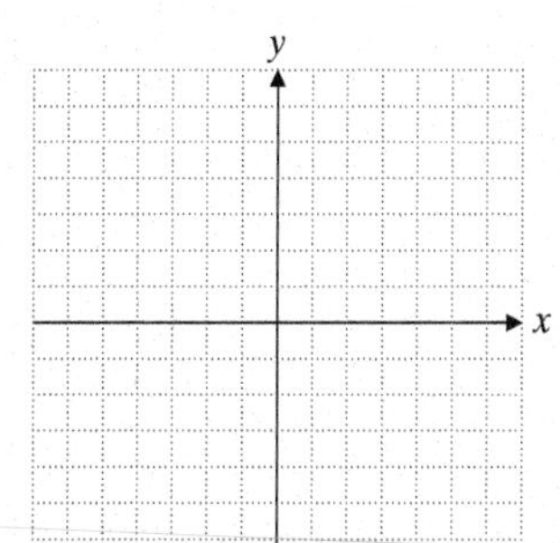

25. $x = -3(y + 1)^2 - 2$

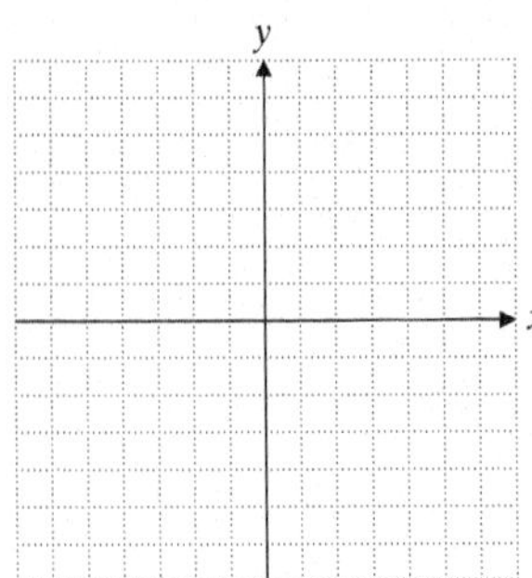

26. $x = -2(y + 3)^2 - 1$

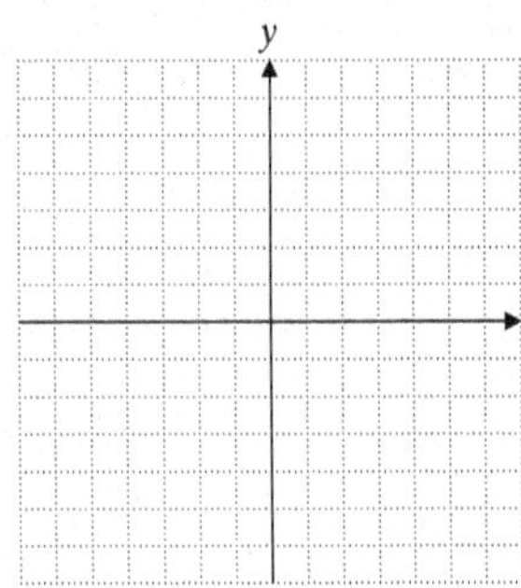

Rewrite each equation in standard form. Determine **(a)** *whether the parabola is horizontal or vertical,* **(b)** *the direction it opens, and* **(c)** *the vertex.*

27. $y = x^2 - 4x - 1$

28. $y = x^2 + 10x + 31$

29. $y = -2x^2 + 12x - 16$

30. $y = -3x^2 + 6x + 2$

31. $x = y^2 + 8y + 9$

32. $x = y^2 - 8y + 11$

Applications

Satellite Dishes *Modern satellite dishes intended for home television are generally between 18 and 31 inches in diameter and are less than 10 inches in depth. For exercises 33 and 34, find an equation of the form $y = ax^2$ that describes the outline of the satellite dish such that the bottom of the dish passes through $(0, 0)$ and has the given diameter and depth.*

33. The diameter is 20 inches, and the depth is 5 inches.

34. The diameter is 30 inches, and the depth is 9 inches.

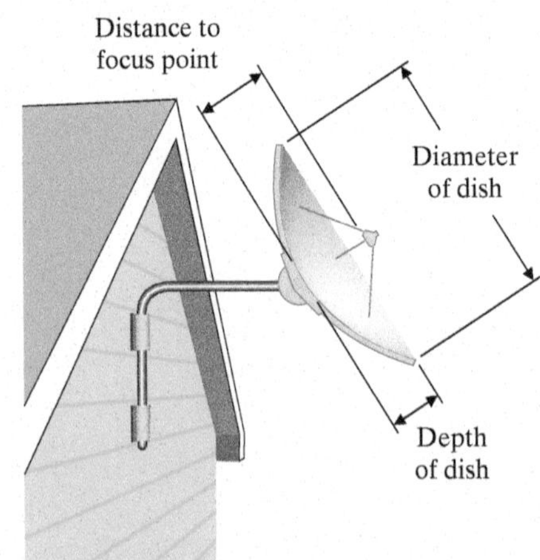

35. ***Satellite Dishes*** If the outline of a satellite dish is described by the equation $y = ax^2$, then the distance p from the center of the dish to the focus point of the dish is given by the equation $a = \frac{1}{4p}$. Find the distance p for the dish in exercise 33.

36. ***Satellite Dishes*** If the outline of a satellite dish is described by the equation $y = ax^2$, then the distance p from the center of the dish to the focus point of the dish is given by the equation $a = \frac{1}{4p}$. Find the distance p for the dish in exercise 34.

Optional Graphing Calculator Problems *Find the vertex and y-intercept of each parabola. Find the two x-intercepts.*

37. $y = 2x^2 + 6.48x - 0.1312$

38. $y = -3x^2 + 33.66x - 73.5063$

Applications *By writing a quadratic equation in the form $y = a(x - h)^2 + k$, we can find the maximum or minimum value of the equation and the value of x at which it occurs. Remember that the equation $y = a(x - h)^2 + k$ is a vertical parabola. For $a > 0$, the parabola opens upward. Thus, the y-coordinate of the vertex is the smallest (or minimum) value of the equation. Similarly, when $a < 0$, the parabola opens downward, so the y-coordinate of the vertex is the maximum value of the equation. Since the vertex occurs at (h, k), the maximum value of the equation occurs when $x = h$. Then*

$$y = -a(x - h)^2 + k = -a(0) + k = k.$$

For example, suppose the weekly profit of a manufacturing company in dollars is $P = -2(x - 45)^2 + 2300$ for x units manufactured. By looking at the equation, we see that the maximum profit per week is \$2300 and is attained when 45 units are manufactured. Use this approach for exercises 39–42.

39. ***Small Business Manager*** Deborah is the manager of a specialty watch company. The company's monthly profit equation is

$$P = -2x^2 + 280x + 35{,}200,$$

where x is the number of watches manufactured. At the next business meeting, Deborah wants to report what the maximum monthly profit could be. Find the maximum monthly profit and the number of watches that must be produced each month to attain the maximum profit.

40. ***Small Business Owner*** Frank is the owner of a small bicycle manufacturing company. The company's monthly profit equation is

$$P = -4x^2 + 640x + 3400,$$

where x is the number of bicycles manufactured. Frank wants to tell his employees what the maximum monthly profit could be. Find the maximum monthly profit and the number of bicycles that must be produced each month to attain the maximum profit.

41. ***Orange Grove Yield*** The effective yield from a grove of miniature orange trees is described by the equation $E = x(900 - x)$, where x is the number of orange trees per acre. What is the maximum effective yield? How many orange trees per acre should be planted to achieve the maximum yield?

42. ***Research Pharmacologist*** Linnea is a research pharmacologist, and she has determined that sensitivity S to a drug depends on the dosage d in milligrams, according to the equation $S = 650d - 2d^2$. What is the maximum sensitivity that will occur? What dosage will produce that maximum sensitivity?

Cumulative Review *Simplify.*

43. [8.3.2] $\sqrt{50x^3}$

44. [8.3.2] $\sqrt[3]{40x^3y^4}$

Add

45. [8.3.3] $\sqrt{98x} + x\sqrt{8} - 3\sqrt{50x}$

46. [8.3.3] $\sqrt[3]{16x^4} + 4x\sqrt[3]{2} - 8x\sqrt[3]{54}$

47. [0.5.3] ***Rose Bushes*** Sir George Tipkin of Sussex has a collection of eight large English rose bushes, each having approximately 1050 buds. In normal years this type of bush produces blooms from 73% of its buds. During years of drought this figure drops to 44%. During years of heavy rainfall the figure rises to 88%. How many blooms can Sir George expect on these bushes if there is heavy rainfall this year?

48. [0.5.3] ***Rose Bushes*** Last year Sir George had only six of the type of bushes described in exercise 45. It was a drought year, and he counted 2900 blooms. Using the bloom rates given in exercise 45, determine approximately how many buds appeared on each of these six bushes. (Round your answer to the nearest whole number.)

Quick Quiz 10.2

1. Find the vertex and the y-intercept of the parabola whose equation is $y = -3(x + 2)^2 + 5$.

2. Find the standard form of the parabola that opens to the right, has a vertex at the point (3, 2), and crosses the x-axis at (7, 0). Place your answer in the form $x = (y - k)^2 + h$.

3. Place the equation $y = 2x^2 - 12x + 12$ in standard form. Then graph it.

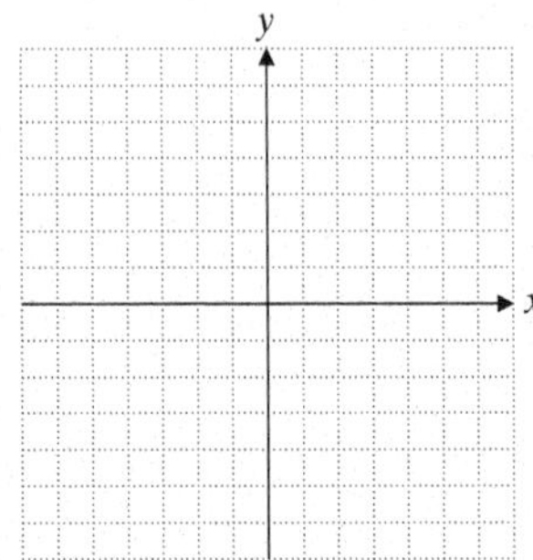

4. **Concept Check** Explain how you can tell which way the parabola opens for these equations: $y = 2x^2$, $y = -2x^2$, $x = 2y^2$, and $x = -2y^2$.

10.3 The Ellipse

Suppose a plane cuts a cone at an angle so that the plane intersects all sides of the cone. If the plane is not perpendicular to the axis of the cone, the conic section that is formed is called an ellipse.

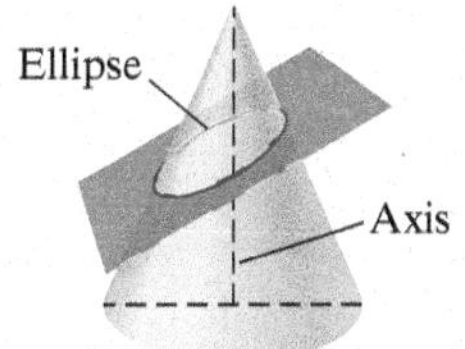

Student Learning Objectives

After studying this section, you will be able to:

1 Graph an ellipse whose center is at the origin.

2 Graph an ellipse whose center is at (h, k).

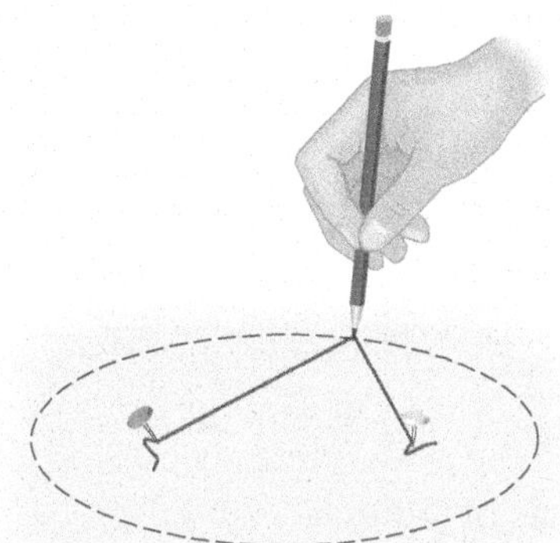

We define an **ellipse** as the set of points in a plane such that for each point in the set, the *sum* of its distances to two fixed points is constant. The fixed points are called **foci** (plural of *focus*).

We can use this definition to draw an ellipse using a piece of string tied at each end to a thumbtack. Place a pencil as shown in the drawing and draw the curve, keeping the pencil pushed tightly against the string. The two thumbtacks are the foci of the ellipse that results.

Examples of the ellipse can be found in the real world. The orbit of Earth (and each of the other planets) is approximately an ellipse with the sun at one focus. In the sketch at the right the sun is located at F and the other focus of the orbit of Earth is at F'.

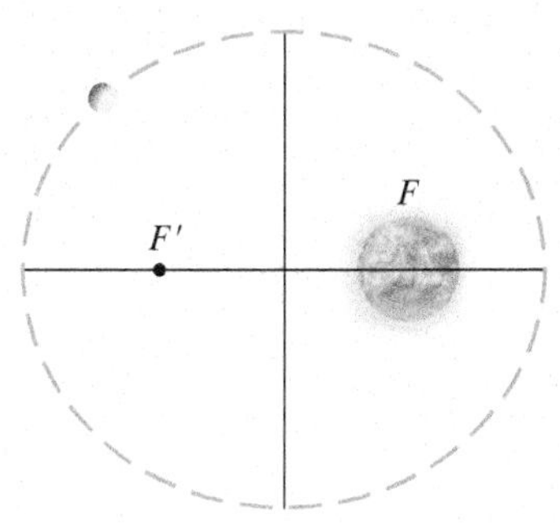

An elliptical surface has a special reflecting property. When sound, light, or some other object originating at one focus reaches the ellipse, it is reflected in such a way that it passes through the other focus. This property can be found in the United States Capitol in a famous room known as the Statuary Hall. If a person whispers at the focus of one end of this elliptically shaped room, a person at the other focus can easily hear him or her.

1 Graphing an Ellipse Whose Center Is at the Origin

The equation of an ellipse is similar to the equation of a circle. The standard form of the equation of an ellipse centered at the origin is given next.

> **STANDARD FORM OF THE EQUATION OF AN ELLIPSE WITH CENTER AT THE ORIGIN**
>
> An ellipse with center at the origin has the equation
>
> $$\frac{x^2}{a^2} + \frac{y^2}{b^2} = 1, \quad \text{where } a \text{ and } b > 0.$$
>
> The **vertices** of this ellipse are at $(a, 0)$, $(-a, 0)$, $(0, b)$, and $(0, -b)$.

To plot the ellipse, we need the x- and y-intercepts.

$$\frac{x^2}{a^2} + \frac{y^2}{b^2} = 1$$

Continued on next page

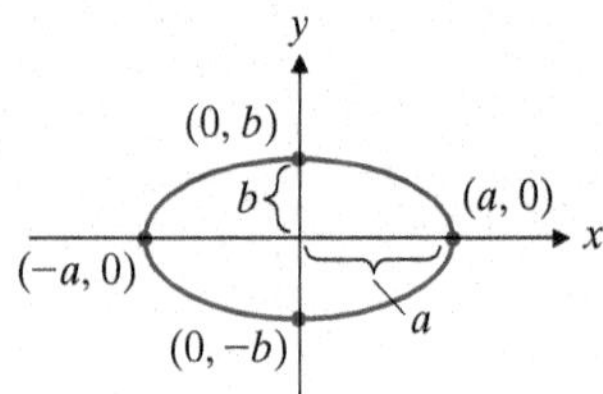

If $x = 0$, ***then*** $\dfrac{y^2}{b^2} = 1.$

$$y^2 = b^2$$
$$\pm\sqrt{y^2} = \pm\sqrt{b^2}$$
$$\pm y = \pm b \quad \text{or} \quad y = \pm b$$

If $y = 0$, ***then*** $\dfrac{x^2}{a^2} = 1.$

$$x^2 = a^2$$
$$\pm\sqrt{x^2} = \pm\sqrt{a^2}$$
$$\pm x = \pm a \quad \text{or} \quad x = \pm a$$

So the x-intercepts are $(a, 0)$ and $(-a, 0)$, and the y-intercepts are $(0, b)$ and $(0, -b)$ for an ellipse of the form $\dfrac{x^2}{a^2} + \dfrac{y^2}{b^2} = 1$.

A circle is a special case of an ellipse. If $a = b$, we get the following:

$$\frac{x^2}{a^2} + \frac{y^2}{a^2} = 1$$
$$x^2 + y^2 = a^2$$

This is the equation of a circle of radius a.

Example 1 Graph $x^2 + 3y^2 = 12$. Label the intercepts.

Solution Before we can graph this ellipse, we need to rewrite the equation in standard form.

$$\frac{x^2}{12} + \frac{3y^2}{12} = \frac{12}{12} \quad \text{Divide each side by 12.}$$
$$\frac{x^2}{12} + \frac{y^2}{4} = 1 \quad \text{Simplify.}$$

Thus, we have the following:

$$a^2 = 12 \quad \text{so} \quad a = 2\sqrt{3}$$
$$b^2 = 4 \quad \text{so} \quad b = 2$$

The x-intercepts are $(-2\sqrt{3}, 0)$ and $(2\sqrt{3}, 0)$, and the y-intercepts are $(0, 2)$ and $(0, -2)$. We plot these points and draw the ellipse.

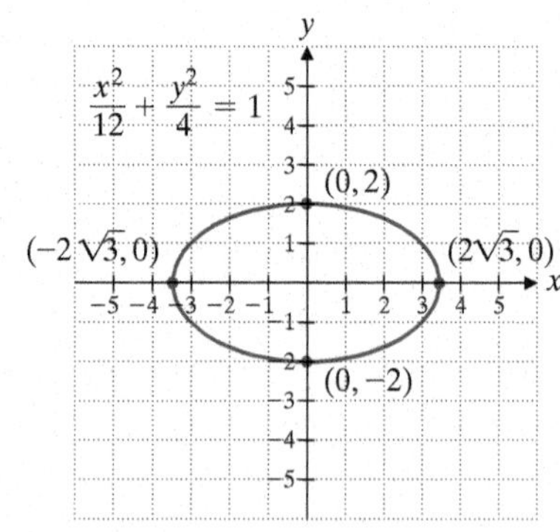

□

Student Practice 1

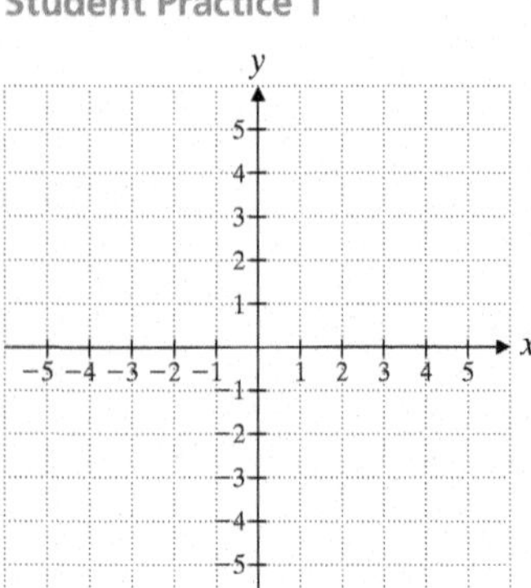

Student Practice 1 Graph $4x^2 + y^2 = 16$. Label the intercepts.

2 Graphing an Ellipse Whose Center Is at (h, k)

If the center of the ellipse is not at the origin but at some point whose coordinates are (h, k), then the standard form of the equation is changed.

STANDARD FORM OF THE EQUATION OF AN ELLIPSE WITH CENTER AT (h, k)

An ellipse with center at (h, k) has the equation

$$\frac{(x-h)^2}{a^2} + \frac{(y-k)^2}{b^2} = 1,$$

where a and $b > 0$.

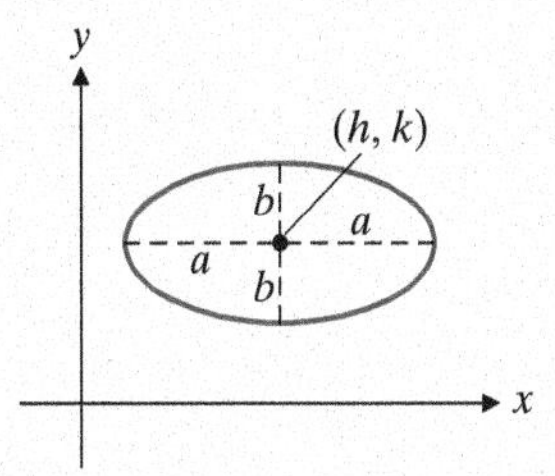

Note that a and b are *not* the x-intercepts and y-intercepts now. Why is this? Look at the sketch. You'll see that a is the horizontal distance from the center of the ellipse to a point on the ellipse. Similarly, b is the vertical distance. Hence, when the center of the ellipse is not at the origin, the ellipse may not cross either axis.

Graphing Calculator

Graphing Ellipses

In order to graph the ellipse in Example 2 on a graphing calculator, we first need to solve for y.

$$\frac{(y-6)^2}{4} = 1 - \frac{(x-5)^2}{9}$$

$$(y-6)^2 = 4\left[1 - \frac{(x-5)^2}{9}\right]$$

$$y = 6 \pm 2\sqrt{1 - \frac{(x-5)^2}{9}}$$

Is it necessary to break up the curve into two halves in order to graph the ellipse? Why or why not?

Example 2 Graph $\frac{(x-5)^2}{9} + \frac{(y-6)^2}{4} = 1.$

Solution The center of the ellipse is $(5, 6)$, $a = 3$, and $b = 2$. Therefore, we begin at $(5, 6)$. We plot points 3 units to the left, 3 units to the right, 2 units up, and 2 units down from $(5, 6)$. The points we plot are the *vertices* of the ellipse.

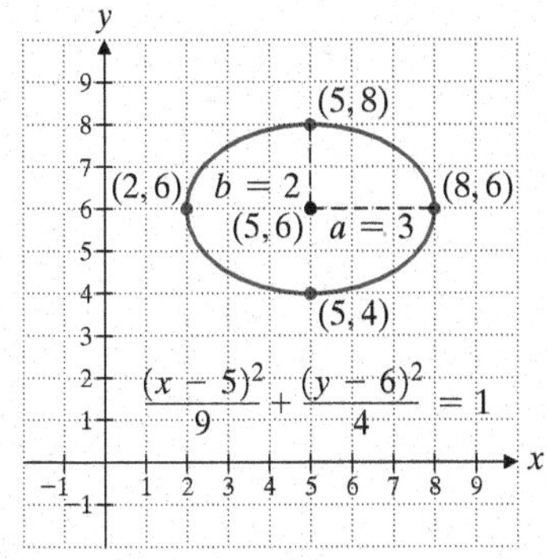

Student Practice 2 Graph

$$\frac{(x-2)^2}{16} + \frac{(y+3)^2}{9} = 1.$$

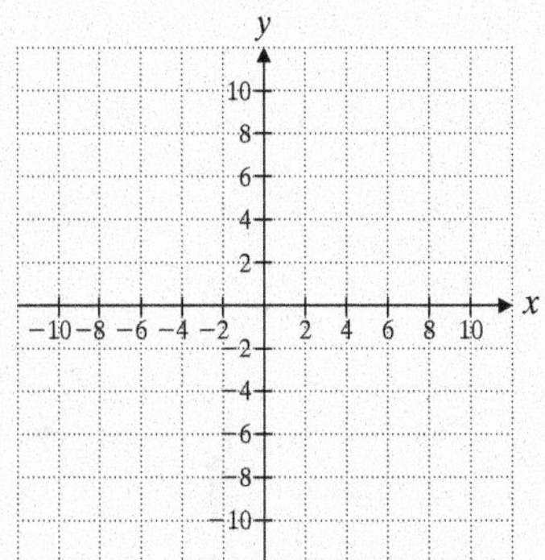

10.3 Exercises MyMathLab®

Verbal and Writing Skills, Exercises 1 and 2

1. Explain how to determine the center of the ellipse $\frac{(x+2)^2}{4} + \frac{(y-3)^2}{9} = 1$.

2. Explain how to determine the x- and y-intercepts of the ellipse $\frac{x^2}{9} + \frac{y^2}{16} = 1$.

Graph each ellipse. Label the intercepts. You may need to use a scale other than 1 square = 1 unit.

3. $\frac{x^2}{36} + \frac{y^2}{4} = 1$

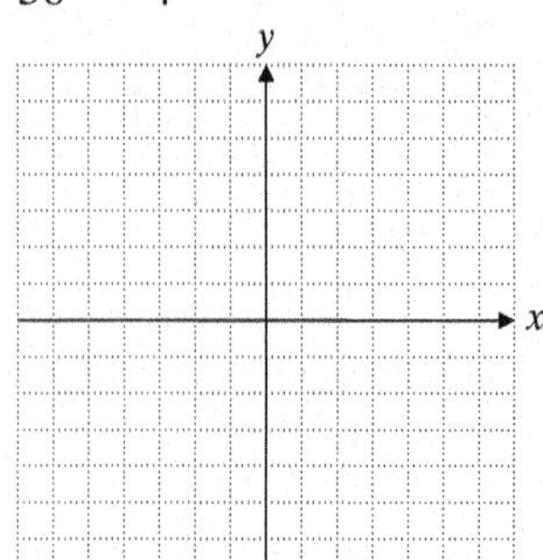

4. $\frac{x^2}{64} + \frac{y^2}{16} = 1$

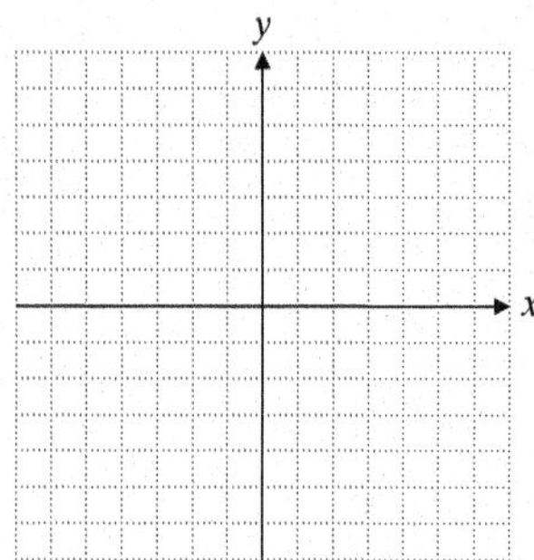

5. $\frac{x^2}{4} + \frac{y^2}{100} = 1$

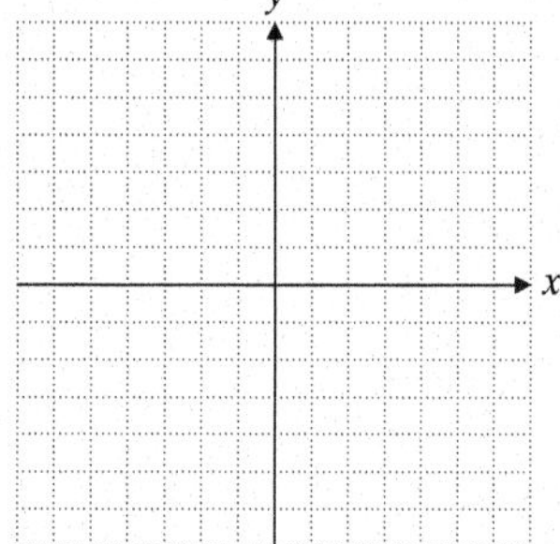

6. $\frac{x^2}{121} + \frac{y^2}{144} = 1$

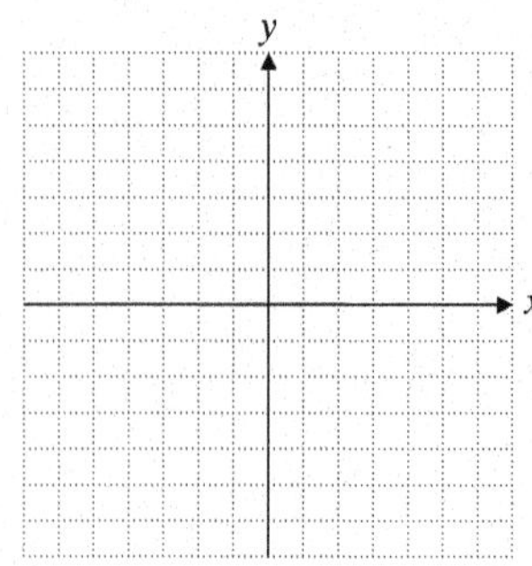

7. $4x^2 + y^2 - 36 = 0$

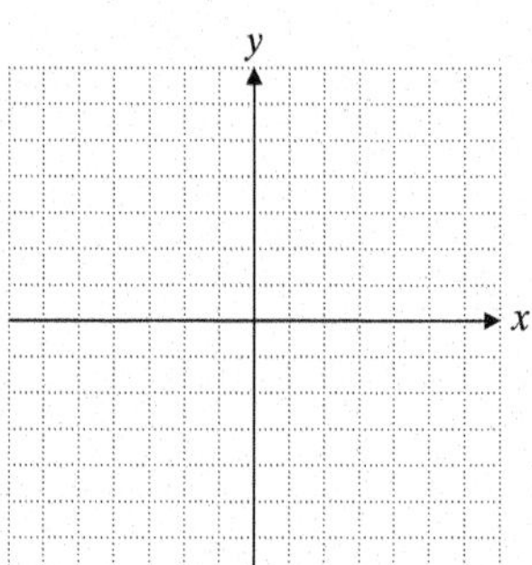

8. $4x^2 + y^2 - 4 = 0$

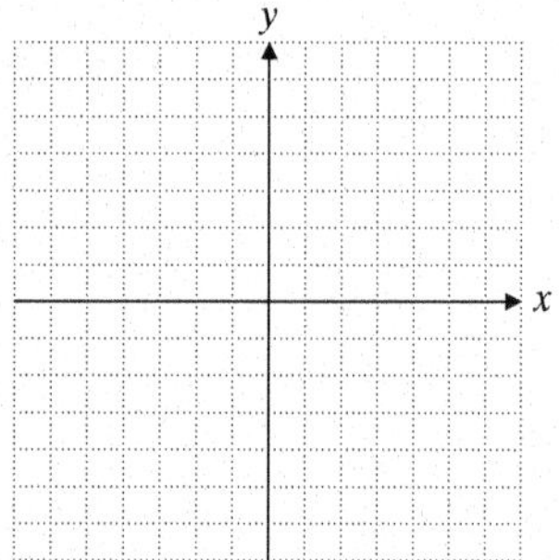

9. $x^2 + 9y^2 = 81$

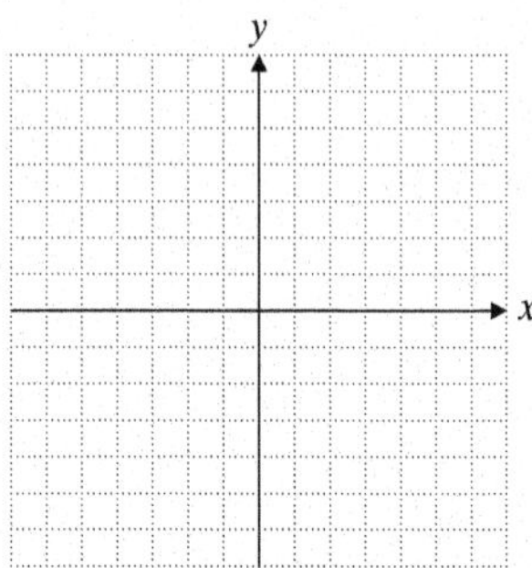

10. $x^2 + 4y^2 = 16$

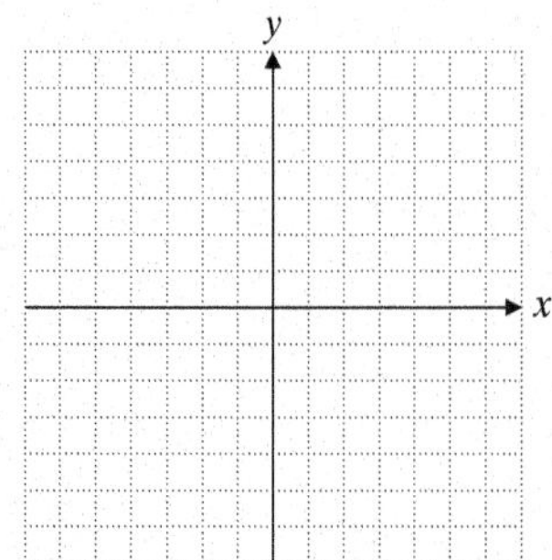

11. $x^2 + 12y^2 = 36$

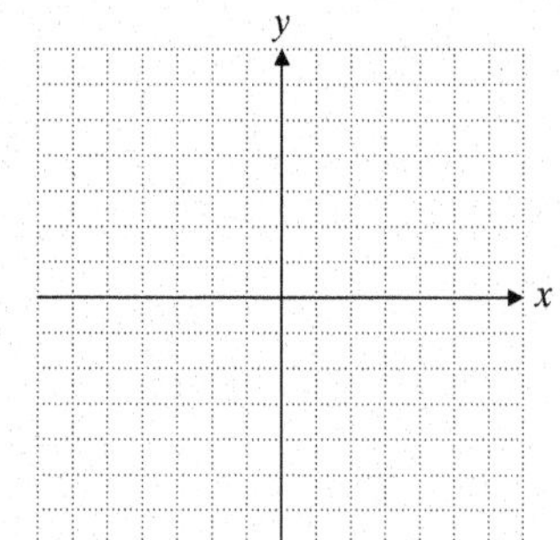

12. $2x^2 + 3y^2 = 18$

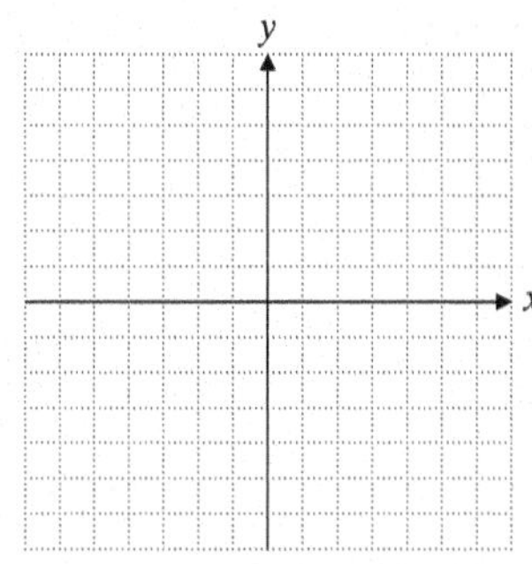

13. $\dfrac{x^2}{\frac{25}{4}} + \dfrac{y^2}{\frac{16}{9}} = 1$

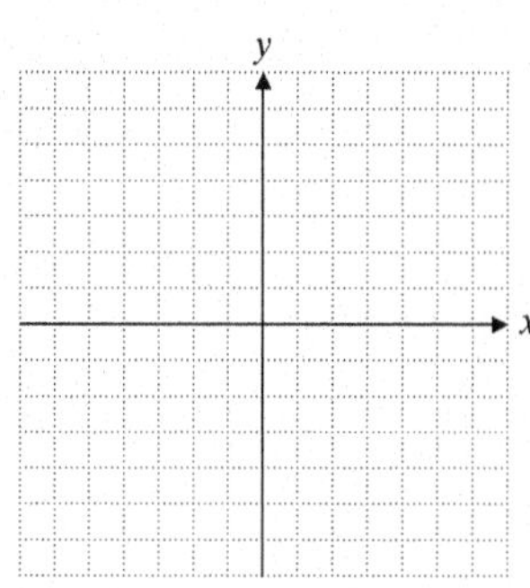

14. $\dfrac{x^2}{\frac{81}{4}} + \dfrac{y^2}{\frac{25}{16}} = 1$

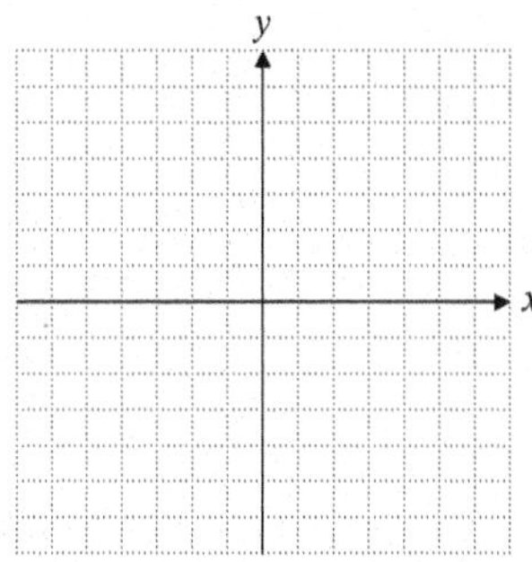

15. $121x^2 + 64y^2 = 7744$

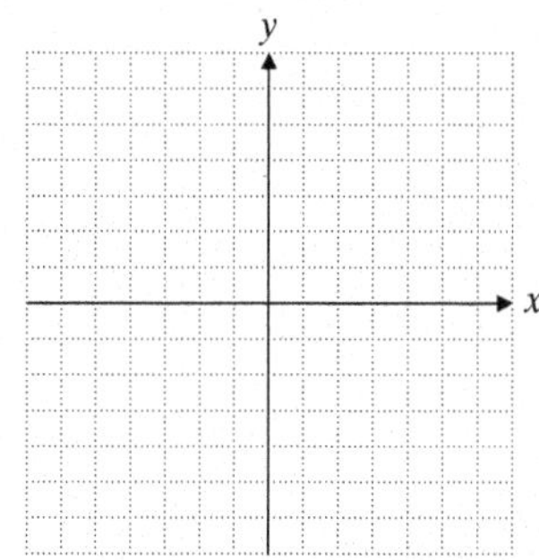

16. Write in standard form the equation of the ellipse with center at the origin, an x-intercept at $(-7, 0)$, and a y-intercept at $(0, 11)$.

17. Write in standard form the equation of the ellipse with center at the origin, an x-intercept at $(13, 0)$, and a y-intercept at $(0, -12)$.

18. Write in standard form the equation of the ellipse with center at the origin, an x-intercept at $(5\sqrt{2}, 0)$, and a y-intercept at $(0, -1)$.

19. Write in standard form the equation of the ellipse with center at the origin, an x-intercept at $(6, 0)$, and a y-intercept at $(0, 4\sqrt{3})$

Applications

20. ***Window Design*** The window shown in the sketch is in the shape of half of an ellipse. Find the equation for the ellipse if the center of the ellipse is at point $A = (0, 0)$.

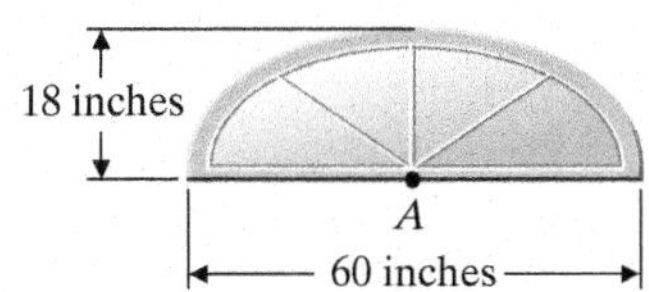

21. ***Orbit of Venus*** The orbit of Venus is an ellipse with the sun as one focus. If we say that the center of the ellipse is at the origin, an approximate equation for the orbit is

$$\frac{x^2}{5013} + \frac{y^2}{4970} = 1,$$

where x and y are measured in millions of miles. Find the largest possible distance across the ellipse. Round your answer to the nearest million miles.

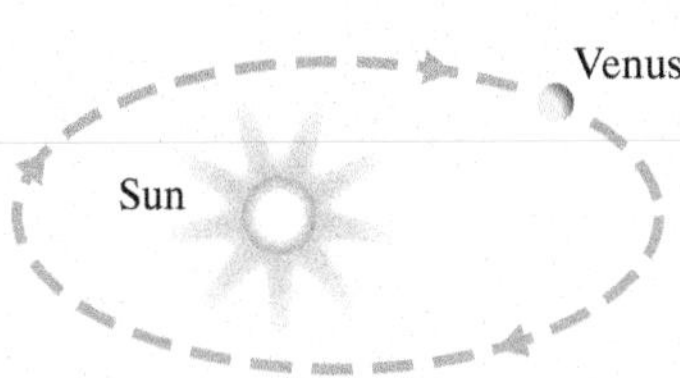

Graph each ellipse. Label the center. You may need to use a scale other than 1 square = 1 unit.

22. $\dfrac{(x-7)^2}{4}+\dfrac{(y-6)^2}{9}=1$

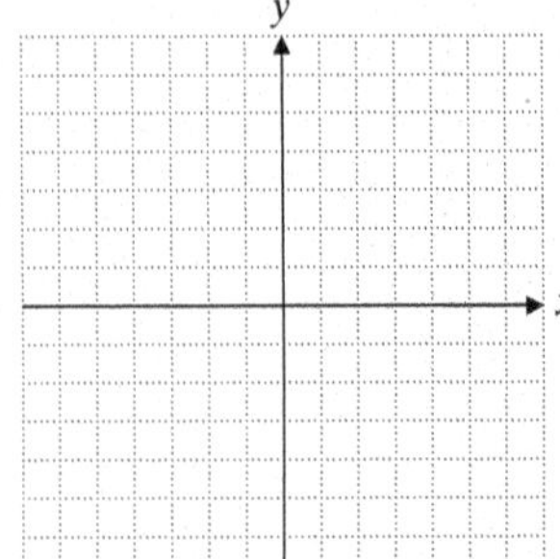

23. $\dfrac{(x-5)^2}{9}+(y-2)^2=1$

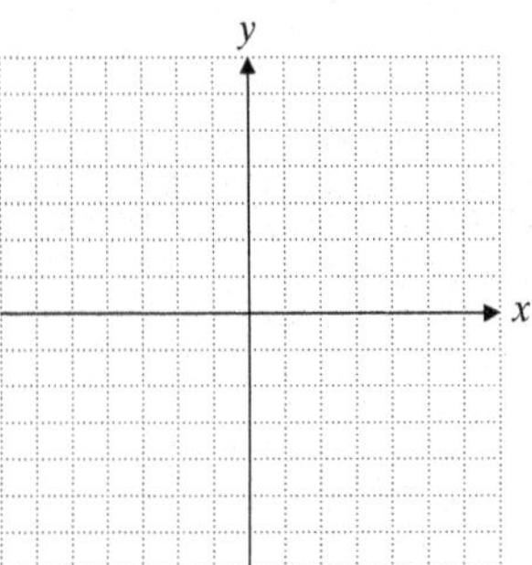

24. $\dfrac{(x+2)^2}{49}+\dfrac{y^2}{25}=1$

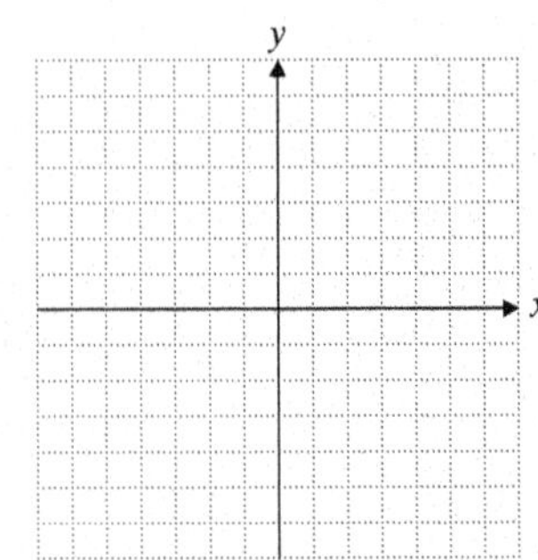

25. $\dfrac{x^2}{25}+\dfrac{(y-4)^2}{16}=1$

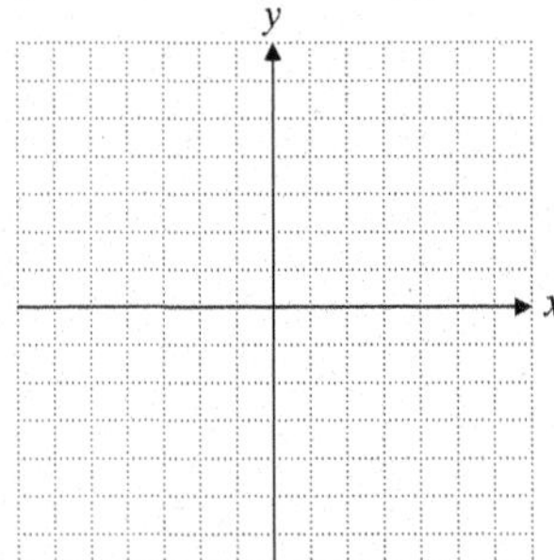

26. $(x+3)^2+\dfrac{(y+1)^2}{36}=1$

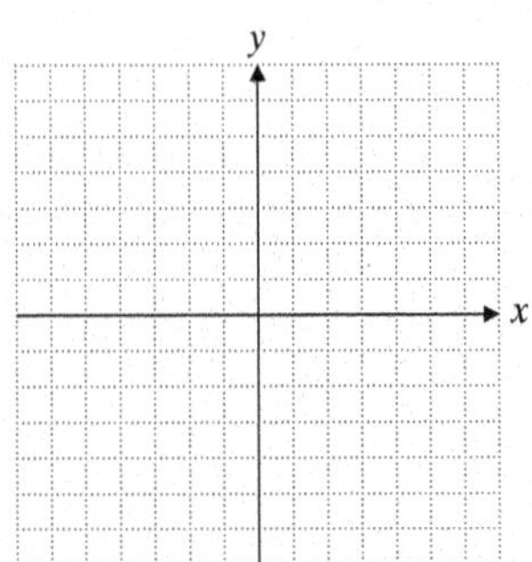

27. $\dfrac{(x+5)^2}{16}+\dfrac{(y+2)^2}{36}=1$

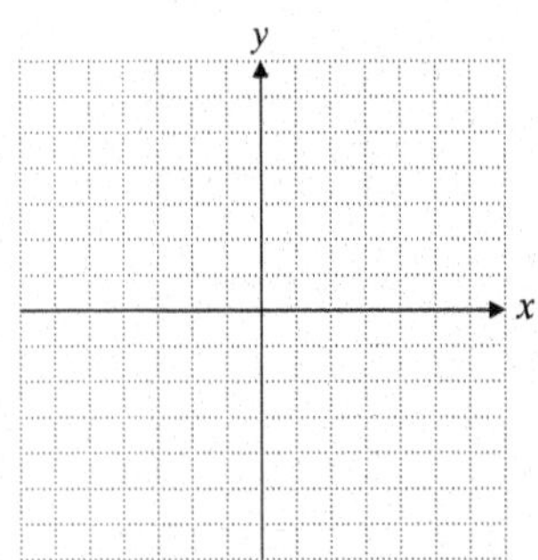

28. Write in standard form the equation of the ellipse whose vertices are $(-6, 2)$, $(0, 2)$, $(-3, -4)$, and $(-3, 8)$.

29. Write in standard form the equation of the ellipse whose vertices are $(5, -1)$, $(5, 3)$, $(4, 1)$, and $(6, 1)$.

30. For what values of a does the ellipse

$$\frac{(x+a)^2}{49} + \frac{(y-3)^2}{4} = 1$$

pass through the point $(-4, 3)$?

31. ***Pet Exercise Area*** Bob's backyard is a rectangle 40 meters by 60 meters. He uses this backyard as an exercise area for his dog. He drove two posts into the ground and fastened a rope to each post, passing the rope through the metal ring on his dog's collar. When the dog pulls on the rope while running, its path is an ellipse. (See the figure.) If the dog can just reach all four sides of the rectangle, find the equation of the elliptical path.

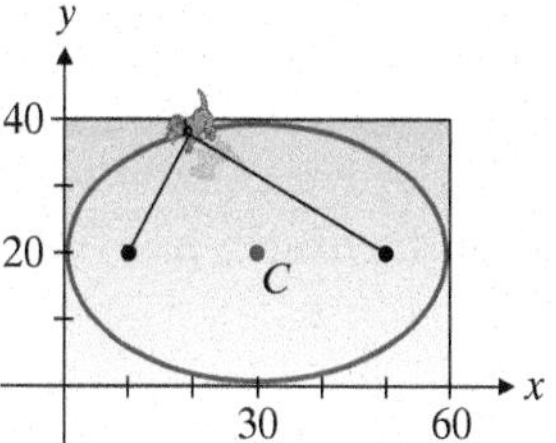

Optional Graphing Calculator Problems *Find the four intercepts, rounded to four decimal places, for each ellipse.*

32. $\frac{x^2}{12} + \frac{y^2}{19} = 1$

33. $\frac{(x-3.6)^2}{14.98} + \frac{(y-5.3)^2}{28.98} = 1$

To Think About *The area enclosed by the ellipse* $\frac{x^2}{a^2} + \frac{y^2}{b^2} = 1$ *is given by the equation* $A = \pi ab$. *Use the value* $\pi \approx 3.1416$ *to find an approximate value for the following answer.*

34. ***Mirror Designer*** Barry works for a company that frames mirrors of all shapes. He is designing an oval mirror that has an outer boundary in the shape of an ellipse. The width of the mirror is 20 inches, and the length of the mirror is 45 inches. Barry will charge the customer according to how many square inches the mirror is. What is the area of the mirror? Round your answer to the nearest tenth.

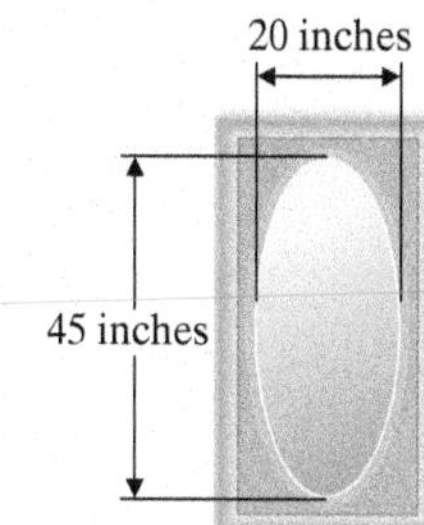

Cumulative Review

35. [8.4.3] Rationalize the denominator. $\dfrac{5}{\sqrt{2x} - \sqrt{y}}$

36. [8.4.1] Multiply and simplify.
$(2\sqrt{3} + 4\sqrt{2})(5\sqrt{6} - \sqrt{2})$

37. [1.2.2] ***Empire State Building*** The Empire State Building was the tallest building in the world for many years. Construction began on March 17, 1930, and the framework rose at a rate of 4.5 stories per week. How many weeks did it take to complete the framework for all 102 stories?

Quick Quiz 10.3

1. Write in standard form the equation of the ellipse with center at the origin, an x-intercept at $(5, 0)$, and a y-intercept at $(0, -6)$.

2. Write in standard form the equation of the ellipse with vertices at $(-7, -2)$, $(1, -2)$, $(-3, 1)$, and $(-3, -5)$.

3. Graph the ellipse. Label the center and the four vertices.
$$\frac{(x-2)^2}{9} + \frac{(y+1)^2}{25} = 1$$

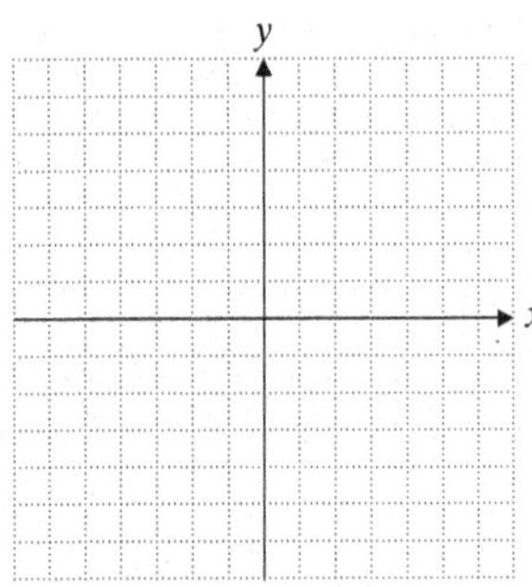

4. **Concept Check** Explain how you would find the four vertices of the ellipse $12x^2 + y^2 - 36 = 0$.

Use Math to Save Money

Did You Know? The government will pay interest on certain student loans while you are in college.

Choosing a Student Loan

Understanding the Problem:

Alicia needs a $10,000 college loan. A credit union offers a 20-year private loan with a 4.65% fixed rate. A subsidized federal loan provides a 20-year loan at a 6% fixed rate with the government paying interest during school and six months after graduation, for a total of 4.5 years. Which option is the better deal?

Making a Plan:

Step 1: Notice that the two interest rates are different for each loan. Also, the total number of payments is different because the government will pay for Alicia's interest for 4.5 years out of the 20-year life of the subsidized loan.

Task 1: What is the total amount that Alicia must pay back for each loan?

Task 2: What is surprising about the results of Task 1?

Step 2: The private loan requires that Alicia make more payments overall, and she must start making payments immediately. The subsidized loan allows for fewer payments, and she can wait until 6 months after graduation to begin making payments.

Task 3: What is the monthly payment for each loan?

Task 4: Which loan offers the lowest monthly payment?

Finding the Solution:

Step 3: Let's compare the loans.

Task 5: What are the values for the missing places in this table?

Type of Loan	Total Amount	Number of Monthly Payments	Monthly Payment Amount	When Payments Begin
Private loan				Immediately
Subsidized loan				6 months after graduation

Task 6: Which option provides the best deal? Explain your reasoning.

Applying the Solution to Your Life:

When choosing between the two loans, students must consider the total amount that they must repay, whether the government will pay interest while they are in school, and when they start making their monthly payments. To apply for a subsidized loan, students must fill out a Free Application for Federal Student Aid (FAFSA) form. Approval for subsidized loans is based on financial need, but you do not need a high credit score. On the other hand, a private loan from a bank or credit union will require a good credit score. Credit unions sometimes offer lower cost private loans than banks.
Make sure to check into all of your options before making a decision.

10.4 The Hyperbola

Student Learning Objectives

After studying this section, you will be able to:

1 Graph a hyperbola whose center is at the origin.

2 Graph a hyperbola whose center is at (h, k).

By cutting two branches of a cone by a plane as shown in the sketch, we obtain the two branches of a hyperbola. A comet moving with more than enough kinetic energy to escape the Sun's gravitational pull will travel in a hyperbolic path. Similarly, a rocket traveling with more than enough velocity to escape Earth's gravitational field will follow a hyperbolic path.

We define a **hyperbola** as the set of points in a plane such that for each point in the set, the absolute value of the *difference* of its distances to two fixed points (called **foci**) is constant.

1 Graphing a Hyperbola Whose Center Is at the Origin

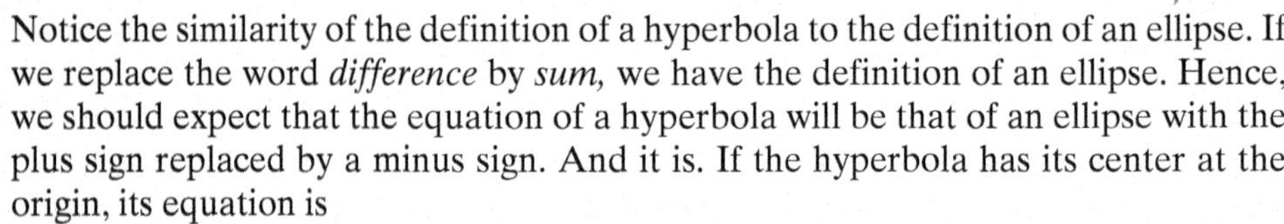

Notice the similarity of the definition of a hyperbola to the definition of an ellipse. If we replace the word *difference* by *sum,* we have the definition of an ellipse. Hence, we should expect that the equation of a hyperbola will be that of an ellipse with the plus sign replaced by a minus sign. And it is. If the hyperbola has its center at the origin, its equation is

$$\frac{x^2}{a^2} - \frac{y^2}{b^2} = 1 \quad \text{or} \quad \frac{y^2}{b^2} - \frac{x^2}{a^2} = 1.$$

A hyperbola has two branches. If the center of the hyperbola is at the origin and each branch has one x-intercept but no y-intercepts, the hyperbola is a horizontal hyperbola, and its **axis** is the x-axis. If the center of the hyperbola is at the origin and each branch has one y-intercept but no x-intercepts, the hyperbola is a vertical hyperbola, and its axis is the y-axis.

The points where the hyperbola intersects its axis are called the **vertices** of the hyperbola.

For hyperbolas centered at the origin, the vertices are also the intercepts.

STANDARD FORM OF THE EQUATION OF A HYPERBOLA WITH CENTER AT THE ORIGIN

Let a and b be any positive real numbers. A hyperbola with center at the origin and vertices $(-a, 0)$ and $(a, 0)$ has the equation

$$\frac{x^2}{a^2} - \frac{y^2}{b^2} = 1.$$

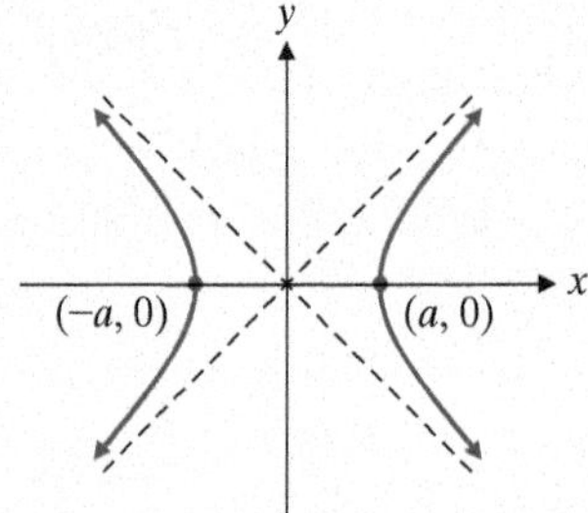

This is called a *horizontal hyperbola*.

A hyperbola with center at the origin and vertices $(0, b)$ and $(0, -b)$ has the equation

$$\frac{y^2}{b^2} - \frac{x^2}{a^2} = 1.$$

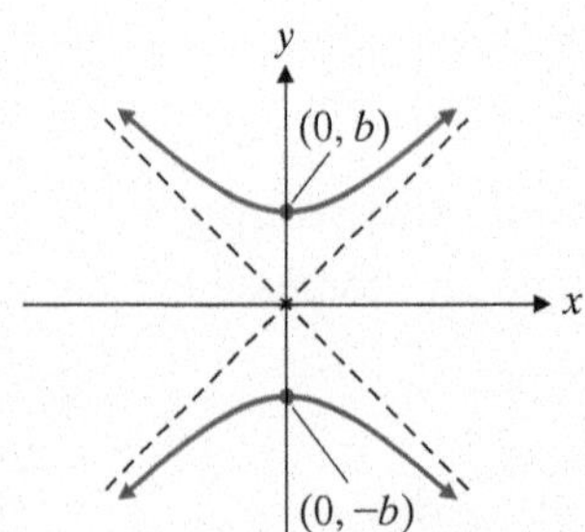

This is called a *vertical hyperbola*.

Notice that the two equations are slightly different. Be aware of this difference so that when you look at an equation you will be able to tell whether the hyperbola is horizontal or vertical.

Notice also the diagonal lines that we've drawn on the graphs of the hyperbolas. These lines are called **asymptotes.** The two branches of the hyperbola come increasingly closer to the asymptotes as the value of $|x|$ gets very large. By drawing the asymptotes and plotting the vertices, we can easily graph a hyperbola.

> **ASYMPTOTES OF HYPERBOLAS**
>
> The asymptotes of the hyperbolas $\frac{x^2}{a^2} - \frac{y^2}{b^2} = 1$ and $\frac{y^2}{b^2} - \frac{x^2}{a^2} = 1$ are
>
> $$y = \frac{b}{a}x \quad \text{and} \quad y = -\frac{b}{a}x.$$

Note that $\frac{b}{a}$ and $-\frac{b}{a}$ are the slopes of the asymptotes.

An easy way to find the asymptotes is to draw extended diagonals of the rectangle whose center is at the origin and whose corners are at (a, b), $(a, -b)$, $(-a, b)$, and $(-a, -b)$. (This rectangle is sometimes called the **fundamental rectangle.**) We draw the fundamental rectangle and the asymptotes with dashed lines because they are not part of the curve.

Example 1 Graph $\frac{x^2}{25} - \frac{y^2}{16} = 1$.

Solution The equation has the form $\frac{x^2}{a^2} - \frac{y^2}{b^2} = 1$, so it is a horizontal hyperbola. $a^2 = 25$, so $a = 5$; $b^2 = 16$, so $b = 4$. Since the hyperbola is horizontal, it has vertices at $(a, 0)$ and $(-a, 0)$ or $(5, 0)$ and $(-5, 0)$.

To draw the asymptotes, we construct a fundamental rectangle with corners at $(5, 4)$, $(5, -4)$, $(-5, 4)$, and $(-5, -4)$. We draw extended diagonals of the rectangle as the asymptotes. We construct each branch of the curve so that it passes through a vertex and gets closer to the asymptotes as it moves away from the origin.

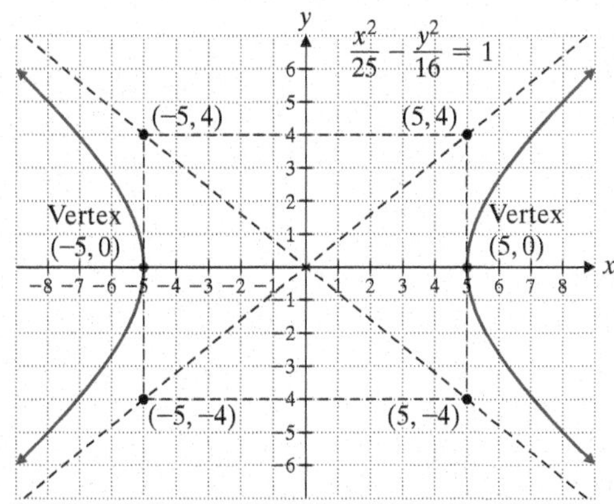

□

Student Practice 1

Graph $\frac{x^2}{16} - \frac{y^2}{25} = 1$.

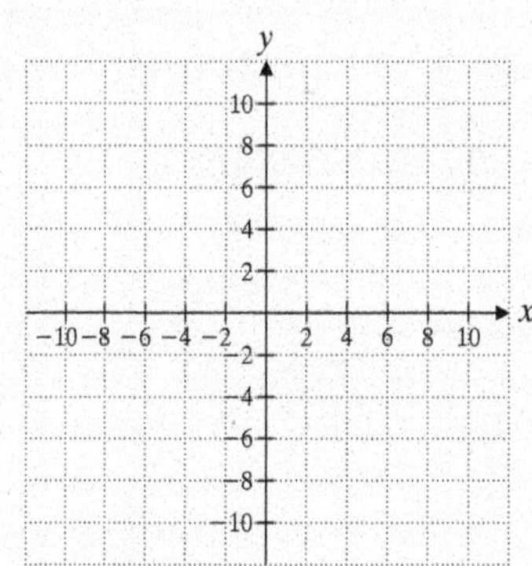

Example 2 Graph $4y^2 - 7x^2 = 28$.

Solution To find the vertices and asymptotes, we must rewrite the equation in standard form. Divide each term by 28.

$$\frac{4y^2}{28} - \frac{7x^2}{28} = \frac{28}{28}$$

$$\frac{y^2}{7} - \frac{x^2}{4} = 1$$

Thus, we have the standard form of a vertical hyperbola with center at the origin. Here $b^2 = 7$, so $b = \sqrt{7}$; $a^2 = 4$, so $a = 2$. The hyperbola has vertices at $(0, \sqrt{7})$ and $(0, -\sqrt{7})$. The fundamental rectangle has corners at $(2, \sqrt{7})$, $(2, -\sqrt{7})$, $(-2, \sqrt{7})$, and $(-2, -\sqrt{7})$. To aid us in graphing, we measure the distance $\sqrt{7}$ as approximately 2.6.

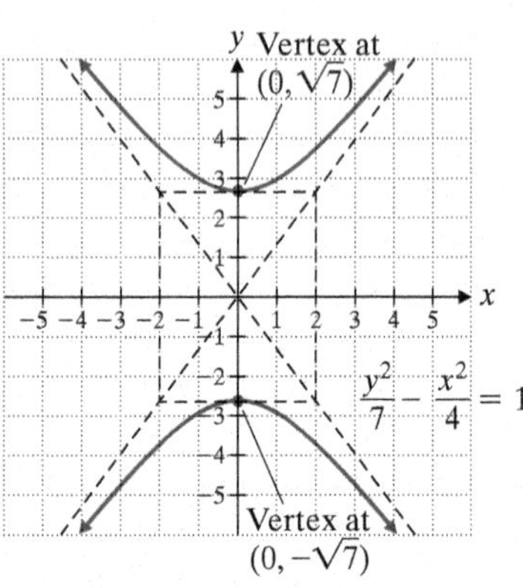

Student Practice 2

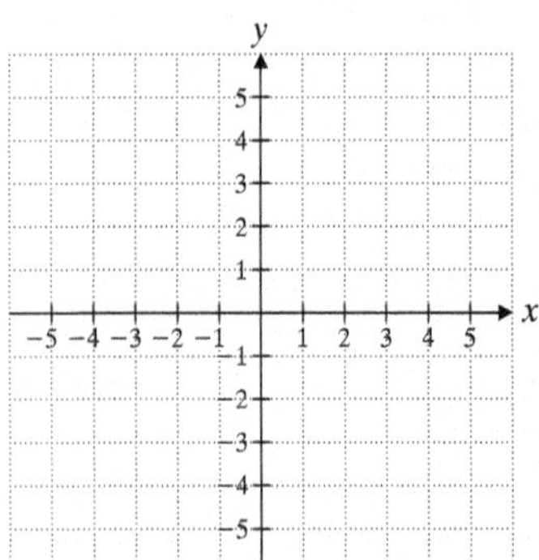

 **Student Practice 2** Graph $y^2 - 4x^2 = 4$.

2 Graphing a Hyperbola Whose Center Is at (h, k)

If a hyperbola does not have its center at the origin but is shifted h units to the right or left and k units up or down, its equation is one of the following:

STANDARD FORM OF THE EQUATION OF A HYPERBOLA WITH CENTER AT (h, k)

Let a and b be any positive real numbers. A horizontal hyperbola with center at (h, k) and vertices $(h - a, k)$ and $(h + a, k)$ has the equation

$$\frac{(x - h)^2}{a^2} - \frac{(y - k)^2}{b^2} = 1.$$

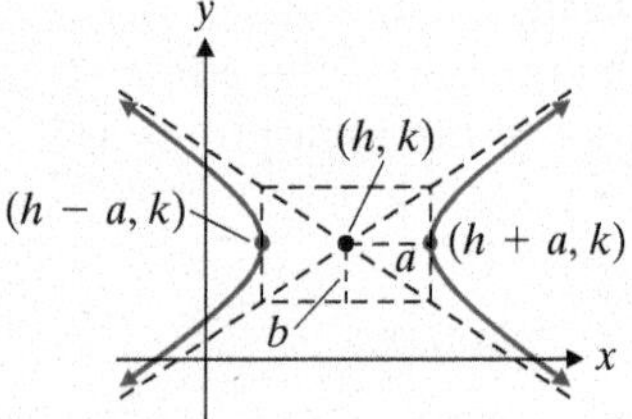

Horizontal Hyperbola

A vertical hyperbola with center at (h, k) and vertices $(h, k + b)$ and $(h, k - b)$ has the equation

$$\frac{(y - k)^2}{b^2} - \frac{(x - h)^2}{a^2} = 1.$$

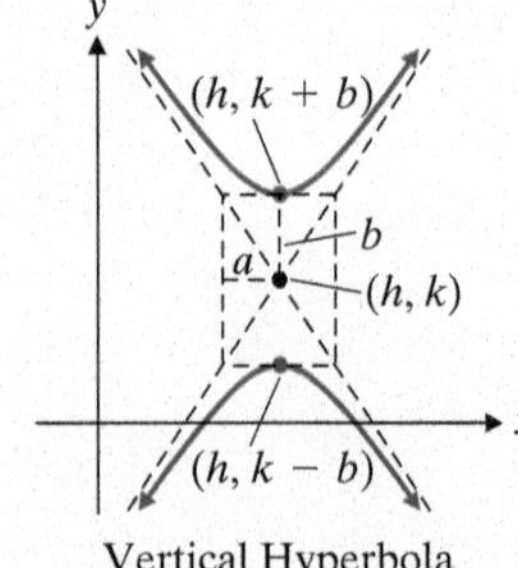

Vertical Hyperbola

Graphing Calculator

Graphing Hyperbolas

Graph on your graphing calculator the hyperbola in Example 2 using

$$y_1 = \frac{\sqrt{28 + 7x^2}}{2}$$

and

$$y_2 = -\frac{\sqrt{28 + 7x^2}}{2}.$$

Do you see how we obtained y_1 and y_2?

Example 3 Graph $\frac{(x-4)^2}{9} - \frac{(y-5)^2}{4} = 1$.

Solution The center is at $(4, 5)$, and the hyperbola is horizontal. We have $a = 3$ and $b = 2$, so the vertices are $(4 \pm 3, 5)$, or $(7, 5)$ and $(1, 5)$. We can sketch the hyperbola more readily if we can draw a fundamental rectangle. Using $(4, 5)$ as the center, we construct a rectangle $2a$ units wide and $2b$ units high. We then draw and extend the diagonals of the rectangle. The extended diagonals are the asymptotes for the branches of the hyperbola.

In this example, since $a = 3$ and $b = 2$, we draw a rectangle $2a = 6$ units wide and $2b = 4$ units high with a center at $(4, 5)$. We draw extended diagonals through the rectangle. From the vertex at $(7, 5)$, we draw a branch of the hyperbola opening to the right. From the vertex at $(1, 5)$, we draw a branch of the hyperbola opening to the left. The graph of the hyperbola is shown.

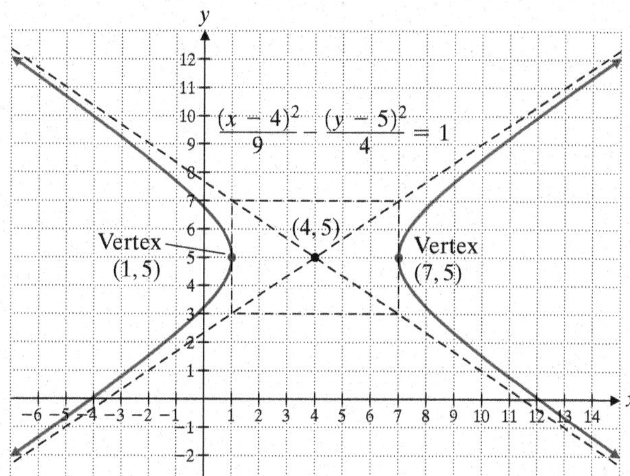

□

Student Practice 3 Graph $\frac{(y+2)^2}{9} - \frac{(x-3)^2}{16} = 1$.

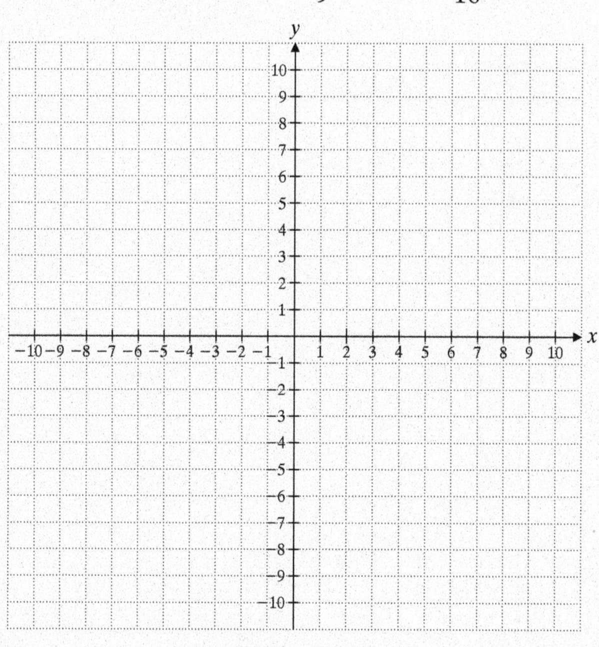

Graphing Calculator

Exploration

Graph the hyperbola in Example 3 using

$$y_1 = 5 + \sqrt{\frac{4(x-4)^2}{9} - 4}$$

and

$$y_2 = 5 - \sqrt{\frac{4(x-4)^2}{9} - 4}.$$

Do you see how we obtained y_1 and y_2?

10.4 Exercises MyMathLab®

Verbal and Writing Skills, Exercises 1–4

1. What is the standard form of the equation of a horizontal hyperbola centered at the origin?

2. What are the vertices of the hyperbola $\frac{y^2}{9} - \frac{x^2}{4} = 1$? Is this a horizontal hyperbola or a vertical hyperbola? Why?

3. Explain in your own words how you would draw the graph of the hyperbola $\frac{x^2}{16} - \frac{y^2}{4} = 1$.

4. Explain how you determine the center of the hyperbola $\frac{(x-2)^2}{4} - \frac{(y+3)^2}{25} = 1$.

Find the vertices and graph each hyperbola. If the equation is not in standard form, write it as such.

5. $\frac{x^2}{4} - \frac{y^2}{25} = 1$

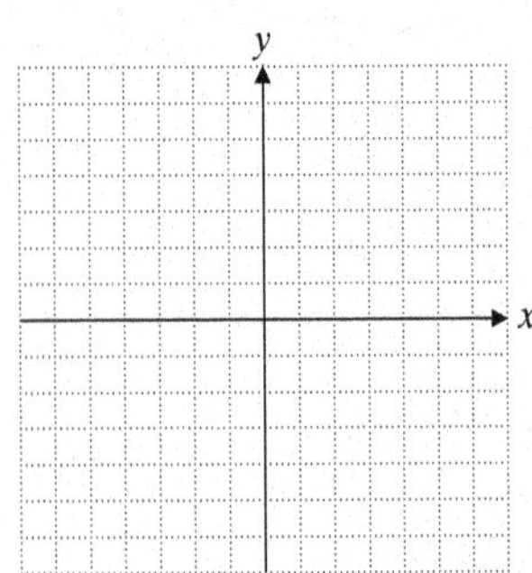

6. $\frac{x^2}{25} - \frac{y^2}{49} = 1$

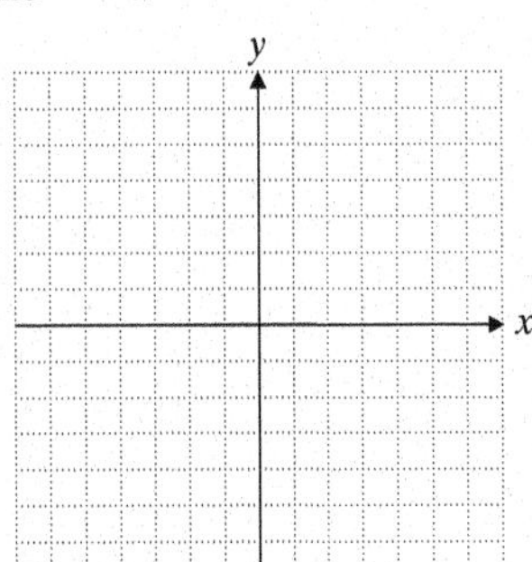

7. $\frac{y^2}{16} - \frac{x^2}{36} = 1$

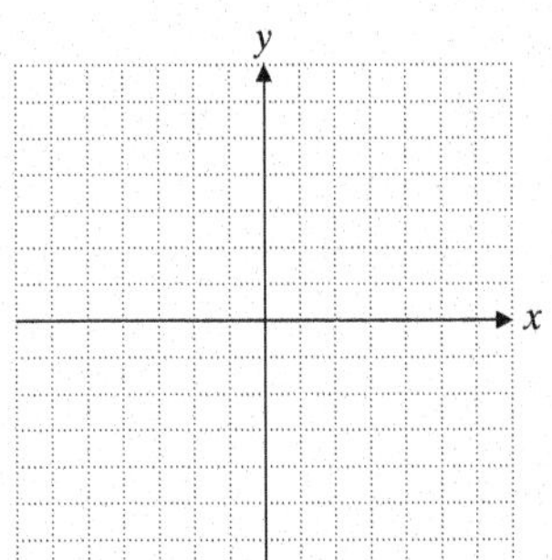

8. $\frac{y^2}{9} - \frac{x^2}{4} = 1$

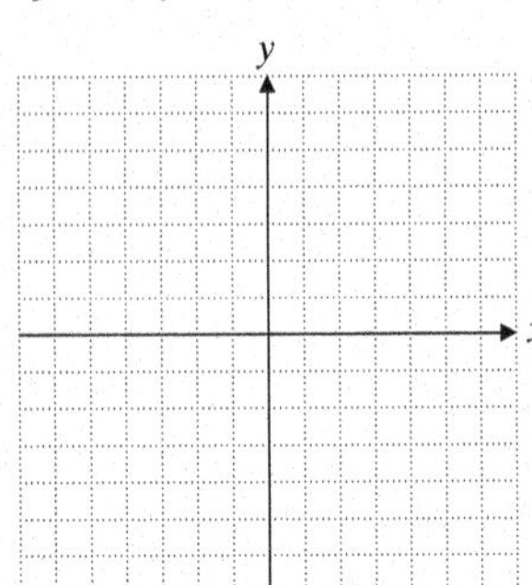

9. $4x^2 - y^2 = 64$

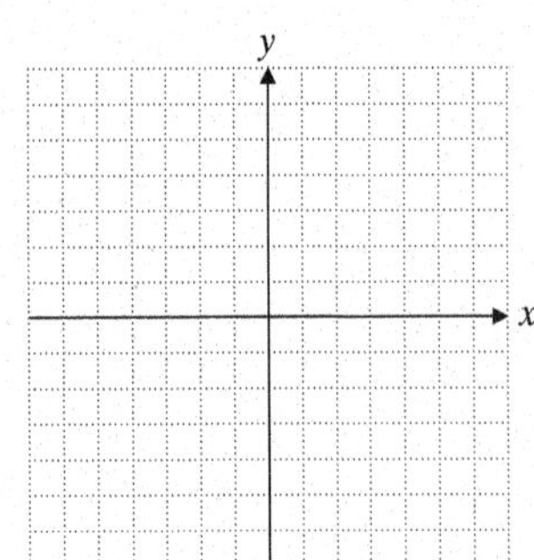

10. $x^2 - 4y^2 = 36$

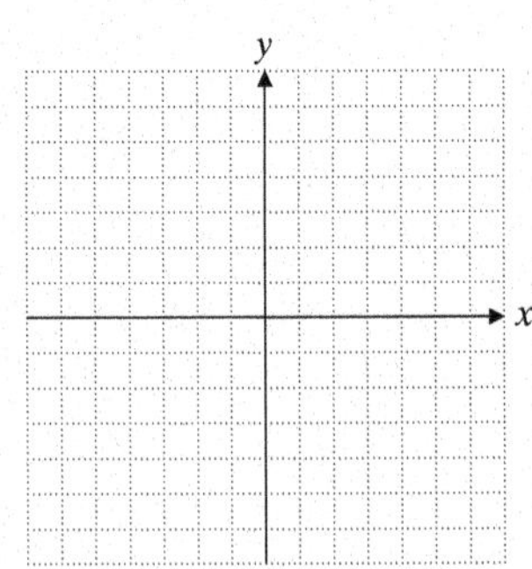

11. $8x^2 - y^2 = 16$

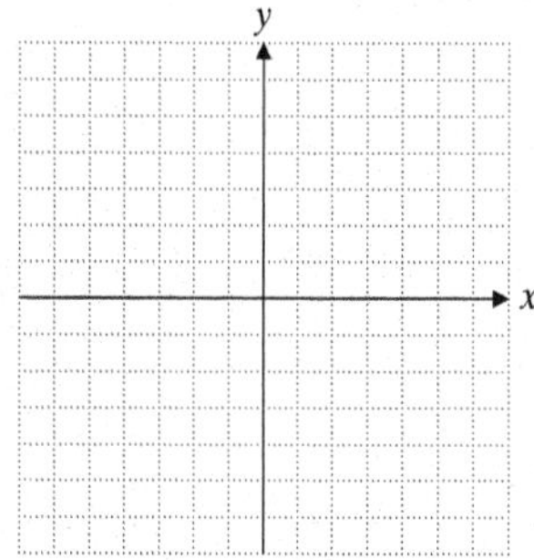

12. $9x^2 - 5y^2 = 45$

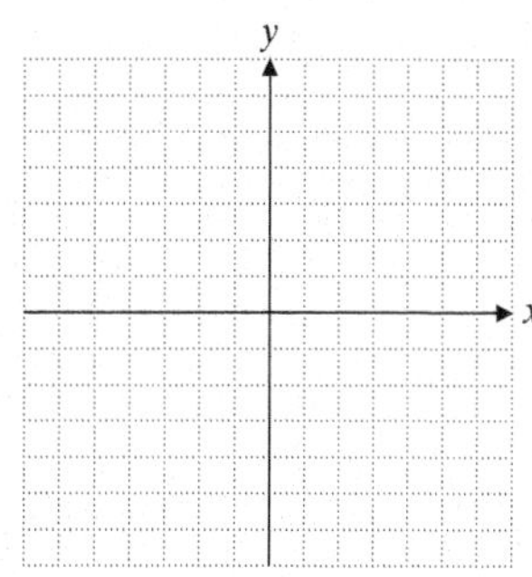

13. $4y^2 - 3x^2 = 48$

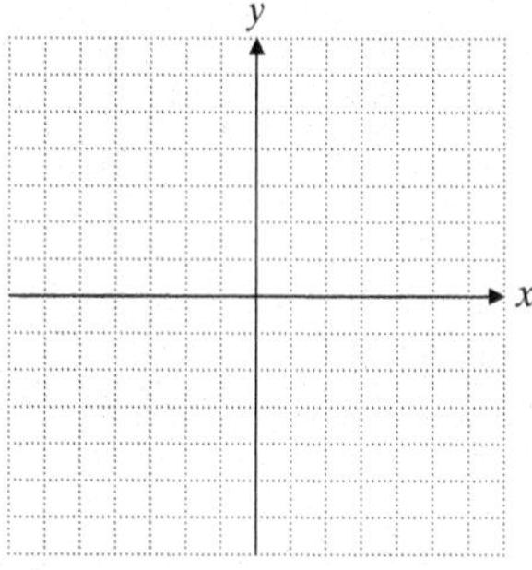

14. $8x^2 - 3y^2 = 24$

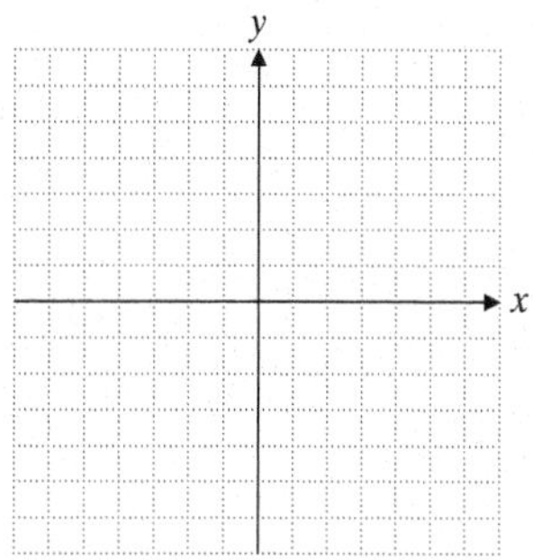

Write the standard form of the equation of the hyperbola with center at the origin and with the following vertices and asymptotes.

15. Vertices at $(3, 0)$ and $(-3, 0)$; asymptotes $y = \frac{4}{3}x$, $y = -\frac{4}{3}x$

16. Vertices at $(2, 0)$ and $(-2, 0)$; asymptotes $y = \frac{3}{2}x$, $y = -\frac{3}{2}x$

17. Vertices $(0, 11)$ and $(0, -11)$; asymptotes $y = \frac{11}{13}x$, $y = -\frac{11}{13}x$

18. Vertices $(0, \sqrt{15})$ and $(0, -\sqrt{15})$; asymptotes $y = \frac{\sqrt{15}}{4}x$, $y = -\frac{\sqrt{15}}{4}x$

Applications

19. ***Comet Orbits*** Some comets have orbits that are hyperbolic in shape with the sun at the focus of the hyperbola. A comet is heading toward Earth but then veers off as shown in the graph. It comes within 120 million miles of Earth. As it travels into the distance, it moves closer and closer to the line $y = 3x$ with Earth at the origin. Find the equation that describes the path of the comet.

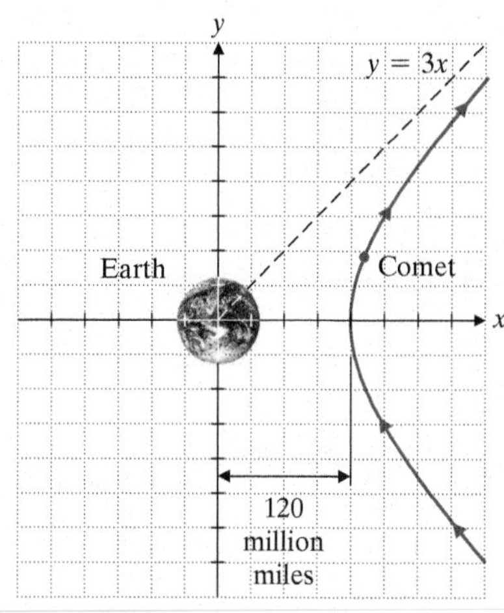

Scale on x-axis: each square is 30 million miles.
Scale on y-axis: each square is 90 million miles.

20. ***Rocket Path*** A rocket following the hyperbolic path shown in the graph turns rapidly at $(4, 0)$ and then moves closer and closer to the line $y = \frac{2}{3}x$ as the rocket gets farther from the tracking station at the origin. Find the equation that describes the path of the rocket if the center of the hyperbola is at $(0, 0)$.

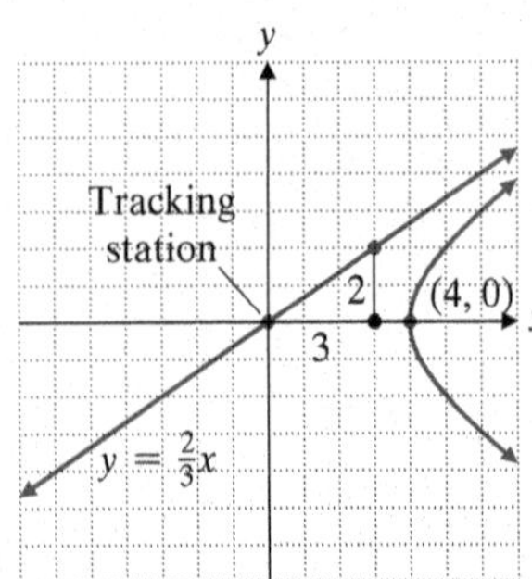

Find the center and then graph each hyperbola. You may want to use a scale other than 1 square = 1 unit.

21. $\dfrac{(x-1)^2}{4} - \dfrac{(y+2)^2}{9} = 1$

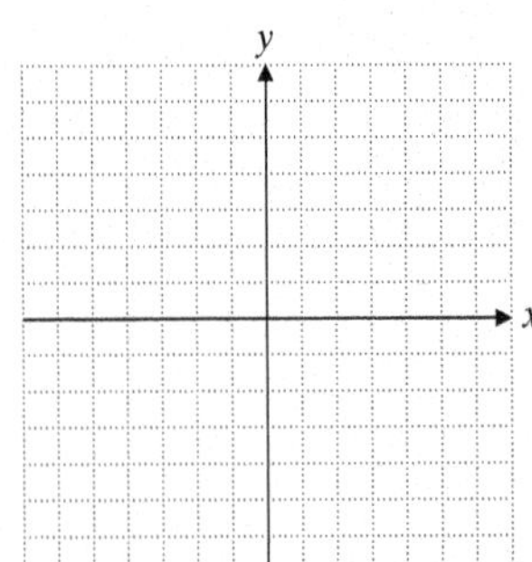

22. 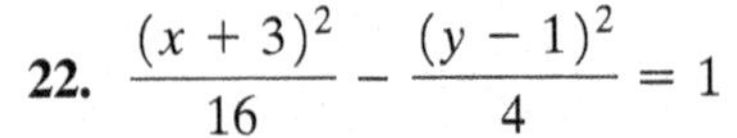 $\dfrac{(x+3)^2}{16} - \dfrac{(y-1)^2}{4} = 1$

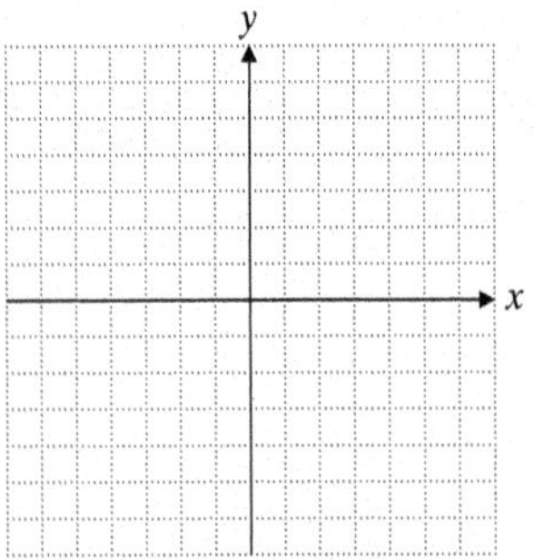

23. $\dfrac{(y+2)^2}{36} - \dfrac{(x+1)^2}{81} = 1$

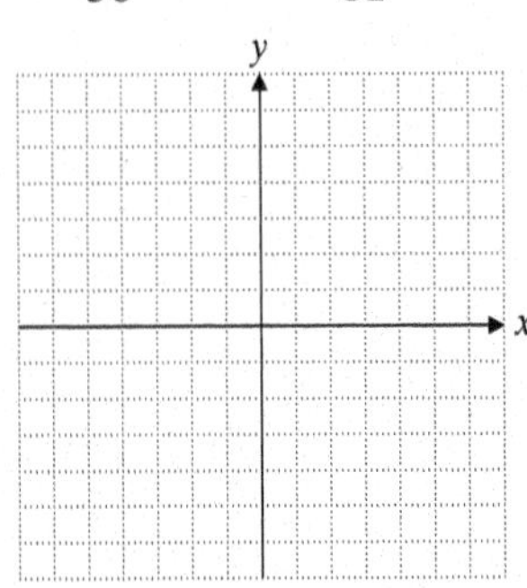

24. $\dfrac{(y+1)^2}{49} - \dfrac{(x+3)^2}{81} = 1$

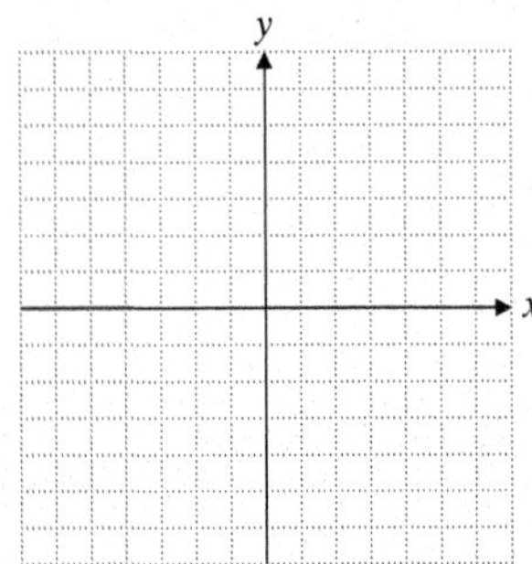

Find the center and the two vertices for each of the following hyperbolas.

25. $\dfrac{(x+6)^2}{7} - \dfrac{y^2}{3} = 1$

26. $\dfrac{x^2}{8} - \dfrac{(y-5)^2}{4} = 1$

27. A hyperbola's center is not at the origin. Its vertices are $(4, -14)$ and $(4, 0)$. One asymptote is $y = -\frac{7}{4}x$. Find the equation of the hyperbola.

28. A hyperbola's center is not at the origin. Its vertices are $(2, 0)$ and $(2, 6)$; one asymptote is $y = \frac{3}{2}x$. Find the equation of the hyperbola.

Optional Graphing Calculator Problems

29. For the hyperbola $8x^2 - y^2 = 16$, if $x = 3.5$, what are the two values of y?

30. For the hyperbola $x^2 - 12y^2 = 36$, if $x = 8.2$, what are the two values of y?

Cumulative Review *Combine.*

31. [7.3.3] $\dfrac{3}{x^2 - 5x + 6} + \dfrac{2}{x^2 - 4}$

32. [7.3.3] $\dfrac{2x}{5x^2 + 9x - 2} - \dfrac{3}{5x - 1}$

33. [0.5.3] ***Box Office*** For the weekend of April 10–12, 2015, two of the top ten movies in U.S. theaters were *Furious 7* and *Home*. Together, these two movies grossed \$78.1 million, which accounted for 63.3% of the total amount grossed by the top ten movies. What was the amount grossed by the top ten movies during that weekend? Round to the nearest tenth. (*Source:* www.boxofficemojo.com)

34. [0.5.3] ***Oil Consumption*** It is predicted that by 2050, global oil use will reach 105 million barrels per day. This will be a 19% increase in oil use from 2015. How many barrels of oil were used daily in 2015? Round your answer to the nearest tenth. (*Source:* www.energy.gov)

Quick Quiz 10.4

1. What are the two vertices of the hyperbola $36y^2 - 9x^2 = 36$

2. Find the equation in standard form of the hyperbola with center at the origin and vertices at $(4, 0)$ and $(-4, 0)$. One asymptote is $y = \dfrac{5}{4}x$.

3. Graph the hyperbola. Label the vertices.

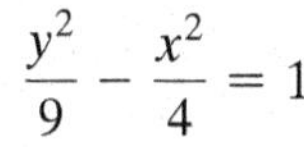

$$\frac{y^2}{9} - \frac{x^2}{4} = 1$$

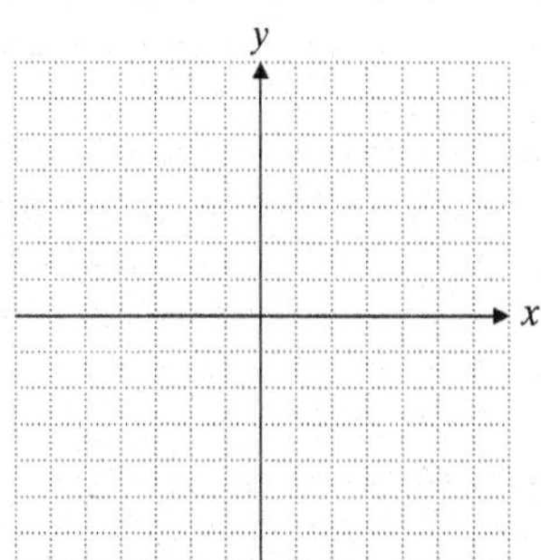

4. Concept Check Explain how you would find the equation of one of the asymptotes for the hyperbola $49x^2 - 4y^2 = 196$.

10.5 Nonlinear Systems of Equations

Student Learning Objectives

After studying this section, you will be able to:

1 Solve a nonlinear system by the substitution method.

2 Solve a nonlinear system by the addition method.

1 Solving a Nonlinear System by the Substitution Method

Any equation that is of second degree or higher is a **nonlinear equation.** In other words, the graph of the equation is not a straight line (which is what the word *nonlinear* means), and the equation can't be written in the form $y = mx + b$. A **nonlinear system of equations** includes at least one nonlinear equation.

The most frequently used method for solving a nonlinear system is the method of substitution. This method works especially well when one equation of the system is linear. A sketch can often be used to verify the solution(s).

Example 1 Solve the following nonlinear system and verify your answer with a sketch.

$$x + y - 1 = 0 \quad \textbf{(1)}$$
$$y - 1 = x^2 + 2x \quad \textbf{(2)}$$

Solution We'll use the substitution method.

$$y = -x + 1 \quad \textbf{(3)} \qquad \text{Solve for } y \text{ in equation } \textbf{(1)}.$$
$$(-x + 1) - 1 = x^2 + 2x \qquad \text{Substitute } \textbf{(3)} \text{ into equation } \textbf{(2)}.$$
$$-x + 1 - 1 = x^2 + 2x$$
$$0 = x^2 + 3x \qquad \text{Solve the resulting quadratic equation.}$$
$$0 = x(x + 3)$$
$$x = 0 \quad \text{or} \quad x = -3$$

Now substitute the values for x in the equation $y = -x + 1$.

For $x = -3$: $\quad y = -(-3) + 1 = +3 + 1 = 4$

For $x = 0$: $\quad y = -(0) \quad + 1 = +1 = 1$

Thus, the solutions of the system are $(-3, 4)$ and $(0, 1)$.

To sketch the system, we see that equation **(2)** describes a parabola. We can rewrite it in the form

$$y = x^2 + 2x + 1 = (x + 1)^2.$$

This is a parabola opening upward with its vertex at $(-1, 0)$. Equation **(1)** can be written as $y = -x + 1$, which is a straight line with slope $= -1$ and y-intercept $(0, 1)$.

A sketch shows the two graphs intersecting at $(0, 1)$ and $(-3, 4)$. Thus, the solutions are verified.

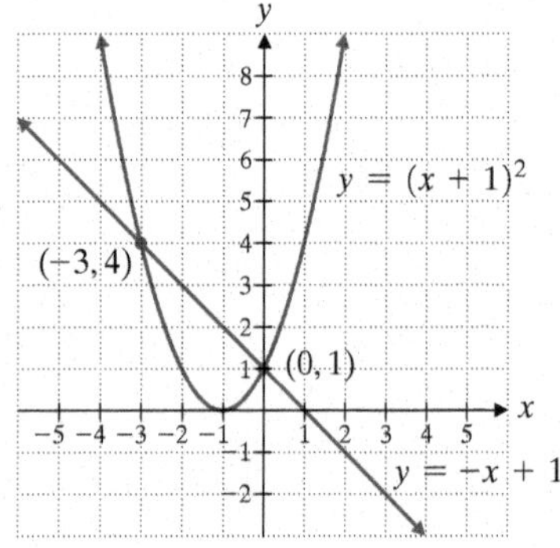

□

Student Practice 1 Solve the system.

$$\frac{x^2}{4} - \frac{y^2}{4} = 1$$
$$x + y + 1 = 0$$

Example 2 Solve the following nonlinear system and verify your answer with a sketch.

$$y - 2x = 0 \quad \textbf{(1)}$$

$$\frac{x^2}{4} + \frac{y^2}{9} = 1 \quad \textbf{(2)}$$

Solution

$$y = 2x \quad \textbf{(3)}$$ Solve equation **(1)** for y.

$$\frac{x^2}{4} + \frac{(2x)^2}{9} = 1$$ Substitute **(3)** into equation **(2)**.

$$\frac{x^2}{4} + \frac{4x^2}{9} = 1$$ Simplify.

$$36\left(\frac{x^2}{4}\right) + 36\left(\frac{4x^2}{9}\right) = 36(1)$$ Clear the fractions.

$$9x^2 + 16x^2 = 36$$

$$25x^2 = 36$$

$$x^2 = \frac{36}{25}$$

$$x = \pm\sqrt{\frac{36}{25}}$$

$$x = \pm\frac{6}{5} = \pm 1.2$$

For $x = +1.2$: $\quad y = 2(1.2) = 2.4$.

For $x = -1.2$: $\quad y = 2(-1.2) = -2.4$.

Thus, the solutions are $(1.2, 2.4)$ and $(-1.2, -2.4)$.

We recognize $\frac{x^2}{4} + \frac{y^2}{9} = 1$ as an ellipse with center at the origin and vertices $(0, 3)$, $(0, -3)$, $(2, 0)$, and $(-2, 0)$. When we rewrite $y - 2x = 0$ as $y = 2x$, we recognize it as a straight line with slope 2 passing through the origin. The sketch shows that the points of intersection at $(1.2, 2.4)$ and $(-1.2, -2.4)$ seem reasonable.

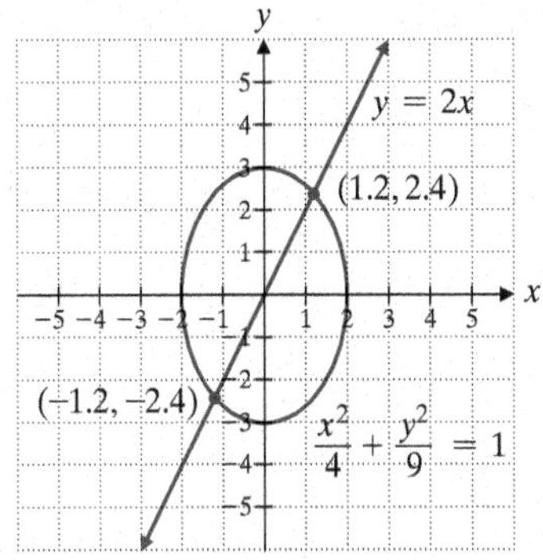

Graphing Calculator

Solving Nonlinear Systems

Use a graphing calculator to solve the following system. Round your answers to the nearest tenth.

$$30x^2 + 256y^2 = 7680$$
$$3x + y - 40 = 0$$

First we will need to obtain the equations

$$y_1 = \frac{\sqrt{7680 - 30x^2}}{16},$$

$$y_2 = -\frac{\sqrt{7680 - 30x^2}}{16},$$

and

$$y_3 = 40 - 3x$$

and graph them to approximate the solutions. Be sure that your window includes enough of the graphs to find the points of intersection.

Student Practice 2 Solve the system. Verify your answer with a sketch.

$$2x - 9 = y$$
$$xy = -4$$

Student Practice 2

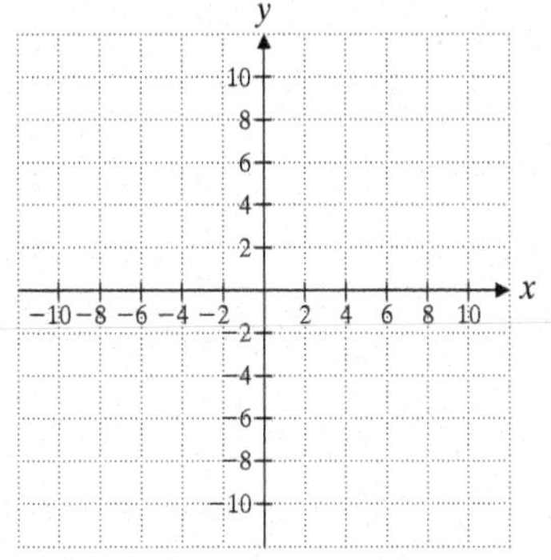

2 Solving a Nonlinear System by the Addition Method

Sometimes a system may be solved more readily by adding the equations together. It should be noted that some systems have no solution.

Example 3 Solve the system.

$$4x^2 + y^2 = 1 \quad \textbf{(1)}$$
$$x^2 + 4y^2 = 1 \quad \textbf{(2)}$$

Solution Although we could use the substitution method, it is easier to use the addition method because neither equation is linear.

$$\begin{aligned} -16x^2 - 4y^2 &= -4 \\ x^2 + 4y^2 &= 1 \\ \hline -15x^2 \qquad &= -3 \end{aligned}$$

Multiply equation **(1)** by -4 and add to equation **(2)**.

$$x^2 = \frac{-3}{-15}$$
$$x^2 = \frac{1}{5}$$
$$x = \pm\sqrt{\frac{1}{5}}$$

If $x = +\sqrt{\frac{1}{5}}$, then $x^2 = \frac{1}{5}$. Substituting this value into equation **(2)** gives

$$\frac{1}{5} + 4y^2 = 1$$
$$4y^2 = \frac{4}{5}$$
$$y^2 = \frac{1}{5}$$
$$y = \pm\sqrt{\frac{1}{5}}$$

Similarly, if $x = -\sqrt{\frac{1}{5}}$, then $y = \pm\sqrt{\frac{1}{5}}$. It is important to determine exactly how many solutions a nonlinear system of equations actually has. In this case, we have four solutions. When x is negative, there are two values for y. When x is positive, there are two values for y. If we rationalize each expression, the four solutions are $\left(\frac{\sqrt{5}}{5}, \frac{\sqrt{5}}{5}\right)$, $\left(\frac{\sqrt{5}}{5}, -\frac{\sqrt{5}}{5}\right)$, $\left(-\frac{\sqrt{5}}{5}, \frac{\sqrt{5}}{5}\right)$, and $\left(-\frac{\sqrt{5}}{5}, -\frac{\sqrt{5}}{5}\right)$. □

Student Practice 3 Solve the system.

$$x^2 + y^2 = 12$$
$$3x^2 - 4y^2 = 8$$

10.5 Exercises MyMathLab®

Solve each of the following systems by the substitution method. Graph each equation to verify that the answer seems reasonable.

1. $y^2 = 2x$
$y = -2x + 2$

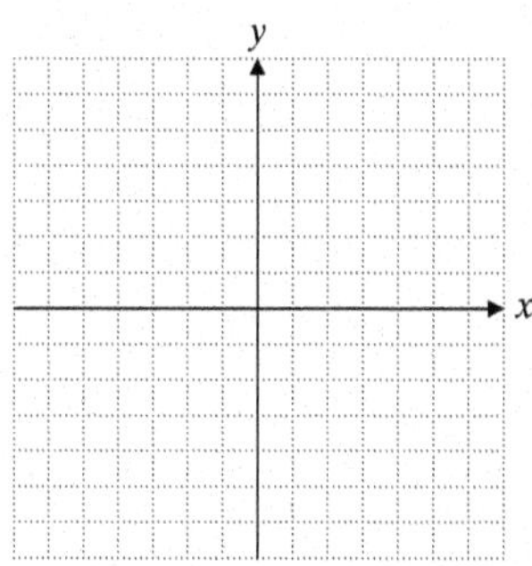

2. $y = x^2$
$y = x + 2$

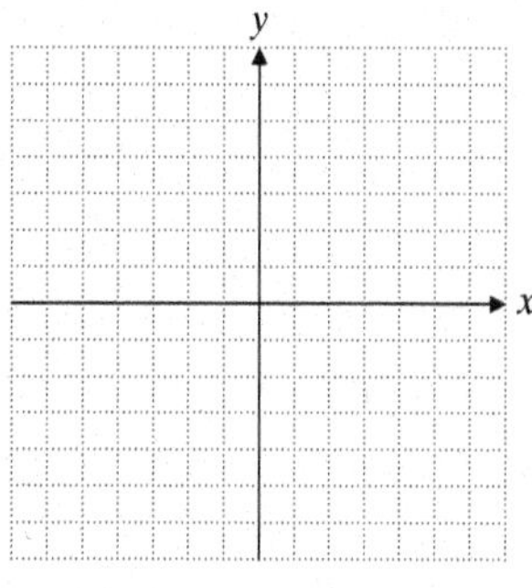

3. $x + 2y = 0$
$x^2 + 4y^2 = 32$

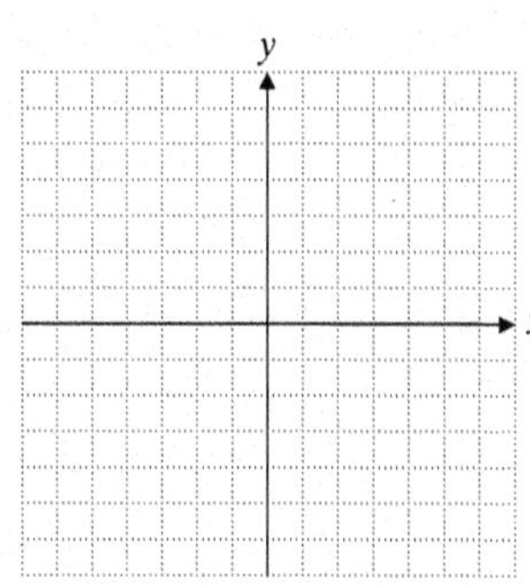

4. $y - 4x = 0$
$4x^2 + y^2 = 20$

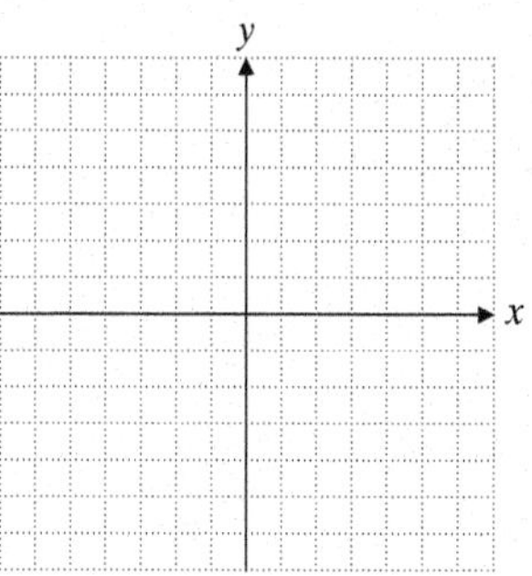

Solve each of the following systems by the substitution method.

5. $\frac{x^2}{1} - \frac{y^2}{3} = 1$
$x + y = 1$

6. $y = (x + 3)^2 - 3$
$2x - y + 2 = 0$

7. $x^2 + y^2 - 16 = 0$
$2y = x - 4$

8. $x^2 + y^2 - 4 = 0$
$3y = -x + 2$

9. $x^2 + 2y^2 = 4$
$y = -x + 2$

10. $2x^2 + 3y^2 = 27$
$y = x + 3$

11. $\frac{x^2}{4} - \frac{y^2}{4} = 1$
$x + y - 4 = 0$

12. $y^2 - x^2 = 3$
$y = 2x$

Solve each of the following systems by the addition method. Graph each equation to verify that the answer seems reasonable.

13. $2x^2 - 5y^2 = -2$
$3x^2 + 2y^2 = 35$

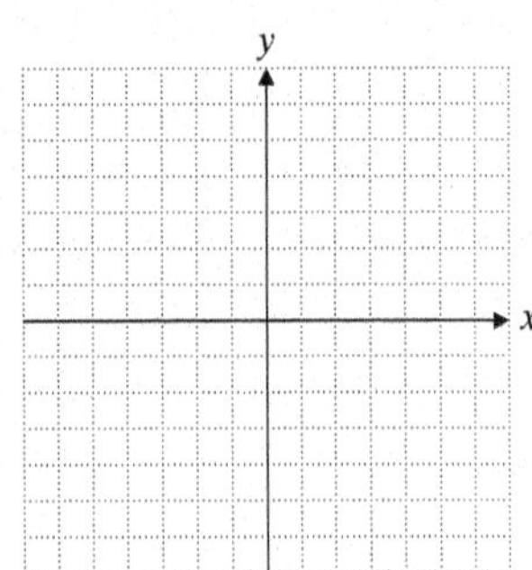

14. $2x^2 - 3y^2 = 5$
$3x^2 + 4y^2 = 16$

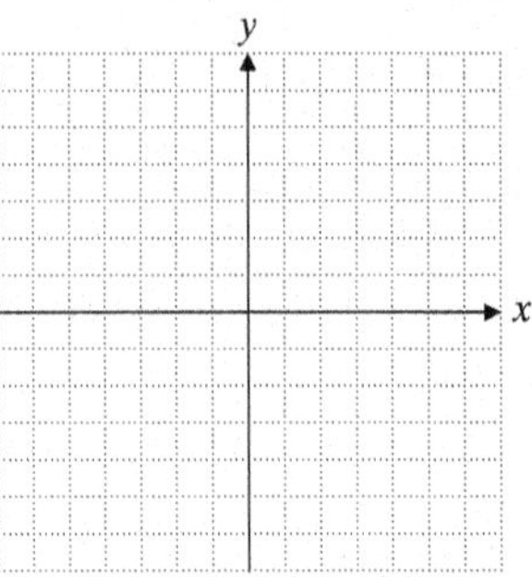

Solve each of the following systems by the addition method.

15. $x^2 + y^2 = 9$
$2x^2 - y^2 = 3$

16. $5x^2 - y^2 = 4$
$x^2 + 3y^2 = 4$

17. $x^2 + 2y^2 = 8$
$x^2 - y^2 = 1$

18. $x^2 + 6y^2 = 21$
$x^2 - 2y^2 = 5$

Mixed Practice *Solve each of the following systems by any appropriate method. If there is no real number solution, so state.*

19. $x^2 + y^2 = 7$
$\frac{x^2}{3} - \frac{y^2}{9} = 1$

20. $4x^2 + y^2 = 16$
$x^2 + y^2 = 16$

21. $xy = -2$
$x + 6y = -1$

22. $3xy = 6$
$x - 2y = -3$

23. $xy = -6$
$2x + y = -4$

24. $xy = 1$
$3x - y + 2 = 0$

25. $x + y = 5$
$x^2 + y^2 = 4$

26. $x^2 + y^2 = 1$
$x - y = 4$

Applications

27. ***Path of a Meteor*** The outline of Earth can be considered a circle with a radius of approximately 4000 miles. Thus, if we say that the center of Earth is located at (0, 0), then an equation for this circle is $x^2 + y^2 = 16{,}000{,}000$. Suppose an incoming meteor is approaching Earth in a hyperbolic path. The equation for the meteor's path is $25{,}000{,}000x^2 - 9{,}000{,}000y^2 = 2.25 \times 10^{14}$. Will the meteor strike Earth? Why or why not? If so, locate the point (x, y) where the meteor will strike Earth. Assume that x and y are both positive. Round your answer to the nearest ten.

28. ***Path of a Meteor*** Suppose that a second incoming meteor is approaching Earth in a hyperbolic path. The equation for this second meteor's path is $16{,}000{,}000x^2 - 25{,}000{,}000y^2 = 4.0 \times 10^{14}$. Will the second meteor strike Earth? Why or why not? If so, locate the point (x, y) where the meteor will strike Earth. Assume that x and y are both positive. Round your answer to the nearest ten.

Cumulative Review

29. **[5.6.2]** Divide. $(3x^3 - 8x^2 - 33x - 10) \div (3x + 1)$

30. **[7.1.1]** Simplify. $\dfrac{6x^4 - 24x^3 - 30x^2}{3x^3 - 21x^2 + 30x}$

Quick Quiz 10.5 *Solve each nonlinear system.*

1. $2x - y = 4$
$y^2 - 4x = 0$

2. $y - x^2 = -4$
$x^2 + y^2 = 16$

3. $(x + 2)^2 + (y - 1)^2 = 9$
$x = 2 - y$

4. **Concept Check** Explain how you would solve the following system.

$$y^2 + 2x^2 = 18$$
$$xy = 4$$

CAREER EXPLORATION PROBLEMS

Background: Architect

Faria works on a design team as an associate architect with the responsibility of researching design styles and their structural components. Her duties also include producing design drawings that identify design specifications and measurements along with their accompanying construction plans.

The projects Faria and her team are working on are the design of a bridge and tunnel over and under a river used for both commerce and recreation. The requirements her design team must meet are twofold: 1. allow for the safe passage of barges and sailboats on the river and 2. promote smooth traffic flow through a tunnel on one side of the river.

Facts

- The entrance to the tunnel is in the shape of a parabolic archway. Its maximum height from the center of the roadway is 30 feet. The width across the entire span of roadway is also 30 feet.
- The top of the archway of the tunnel entrance directly above the roadway's center is considered the parabola's vertex whose coordinates are (0, 0).
- The tunnel exit has the shape of half an ellipse. Its width is 80 feet, and its maximum height directly above the roadway's center is 30 feet.
- The center of the ellipse is located at the center of the roadway and its coordinates are (0, 0).
- The bridge over the river has an archway in the form of half an ellipse. The coordinates of this elliptical shape's center are (0, 0), and its equation is $\frac{x^2}{4800} + \frac{y^2}{2500} = 1$. Watercraft traveling the river must always keep a distance of 30 feet between them when traveling under the bridge.

Tasks

1. Faria's design drawing for the tunnel entrance includes the equation of the parabolic form that describes the outline of the archway. She expresses the equation in the form $y = ax^2$ and uses this equation to determine the vertical distance, y, from a point 8 feet to the right of the roadway's center to the top of the archway. What are the results of her calculations rounded to the nearest tenth? (Only positive values should be included.)

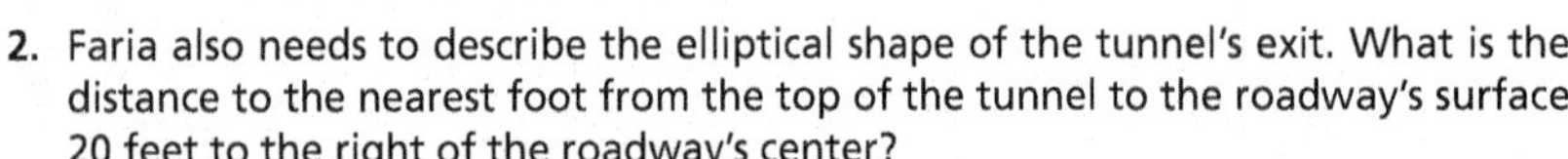

2. Faria also needs to describe the elliptical shape of the tunnel's exit. What is the distance to the nearest foot from the top of the tunnel to the roadway's surface 20 feet to the right of the roadway's center?

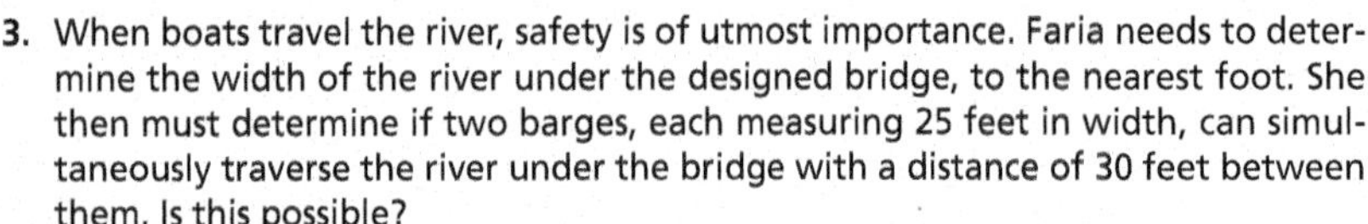

3. When boats travel the river, safety is of utmost importance. Faria needs to determine the width of the river under the designed bridge, to the nearest foot. She then must determine if two barges, each measuring 25 feet in width, can simultaneously traverse the river under the bridge with a distance of 30 feet between them. Is this possible?

Chapter 10 Organizer

Topic and Procedure	Examples	You Try It
Distance between two points, p. 627 The distance d between points (x_1, y_1) and (x_2, y_2) is $d = \sqrt{(x_2 - x_1)^2 + (y_2 - y_1)^2}$.	Find the distance between $(-6, -3)$ and $(5, -2)$. $d = \sqrt{[5 - (-6)]^2 + [-2 - (-3)]^2}$ $= \sqrt{(5 + 6)^2 + (-2 + 3)^2}$ $= \sqrt{121 + 1}$ $= \sqrt{122}$	1. Find the distance between $(7, -4)$ and $(3, -1)$.
Standard form of the equation of a circle, p. 628 The standard form of the equation of the circle with center at (h, k) and radius r is $(x - h)^2 + (y - k)^2 = r^2$.	Graph $(x - 3)^2 + (y + 4)^2 = 16$. Center at $(h, k) = (3, -4)$. Radius = 4. 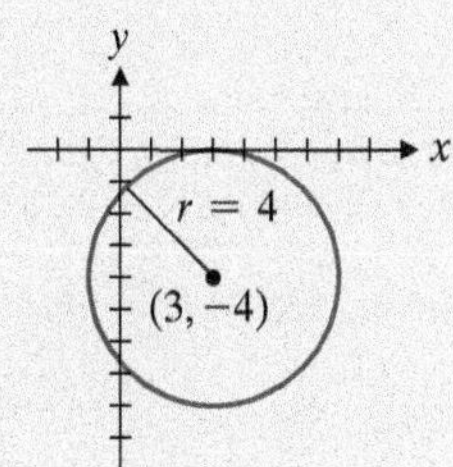	2. Graph $(x + 2)^2 + (y - 4)^2 = 9$. 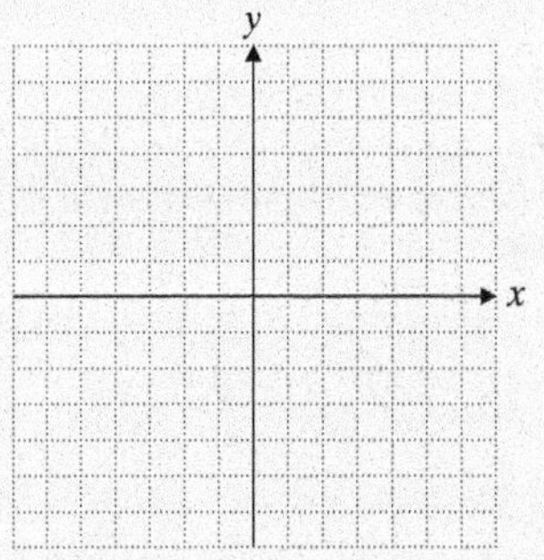
Standard form of the equation of a vertical parabola, p. 635 The equation of a vertical parabola with its vertex at (h, k) can be written in the form $y = a(x - h)^2 + k$. It opens upward if $a > 0$ and downward if $a < 0$. 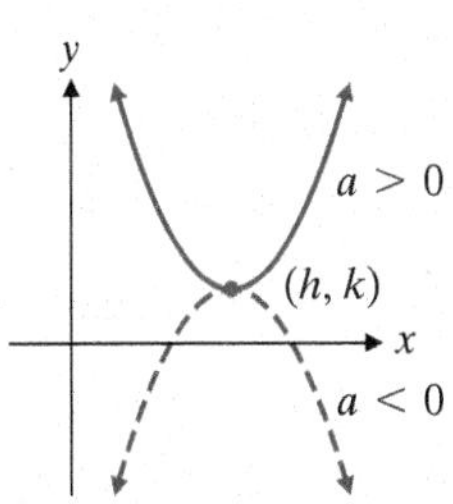	Graph $y = \frac{1}{2}(x - 3)^2 + 5$. $a = \frac{1}{2}$, so parabola opens upward. Vertex at $(h, k) = (3, 5)$. If $x = 0$, $y = 9.5$. 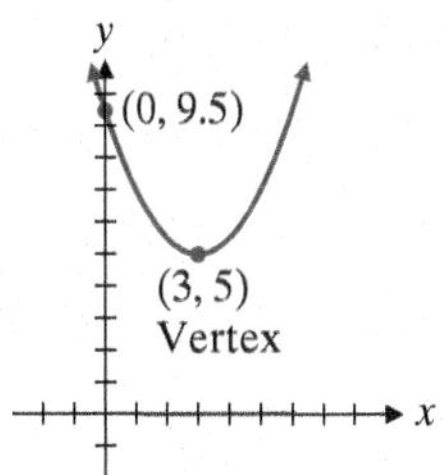	3. Graph $y = -2(x + 1)^2 + 3$. 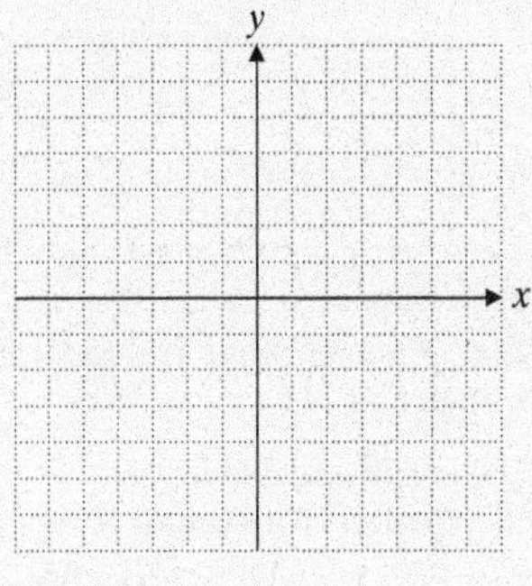
Standard form of the equation of a horizontal parabola, p. 637 The equation of a horizontal parabola with its vertex at (h, k) can be written in the form $x = a(y - k)^2 + h$. It opens to the right if $a > 0$ and to the left if $a < 0$. 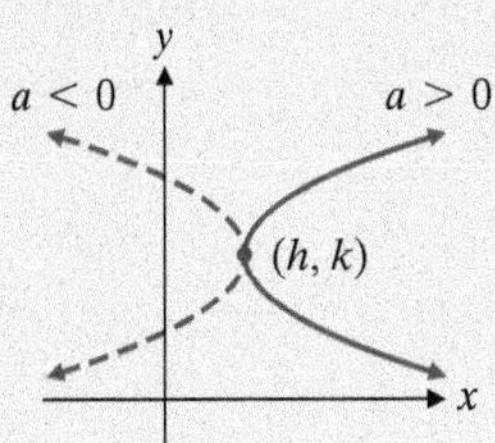	Graph $x = \frac{1}{3}(y + 2)^2 - 4$. $a = \frac{1}{3}$, so parabola opens to the right. Vertex at $(h, k) = (-4, -2)$. If $y = 0$, $x = -\frac{8}{3}$. 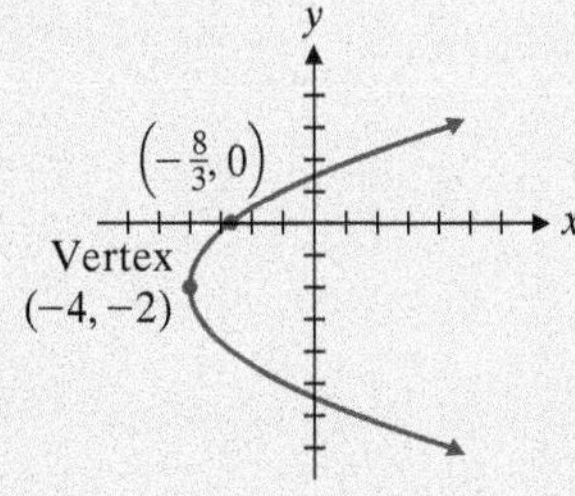	4. Graph $x = (y - 4)^2 - 1$. 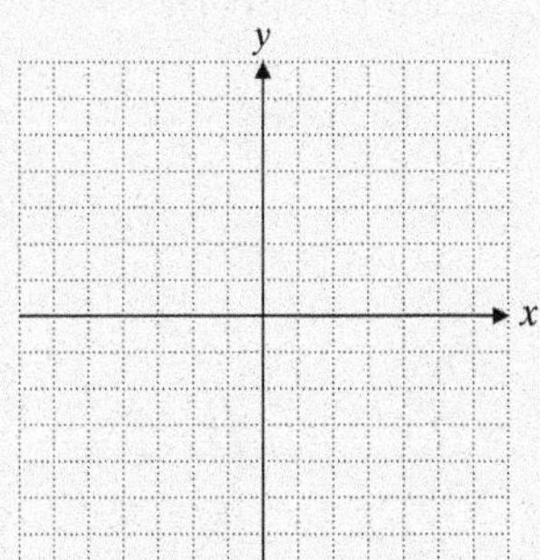

Topic and Procedure	Examples	You Try It
Standard form of the equation of an ellipse with center at (0, 0), p. 645 An ellipse with center at the origin has the equation $$\frac{x^2}{a^2} + \frac{y^2}{b^2} = 1,$$ where $a > 0$ and $b > 0$. 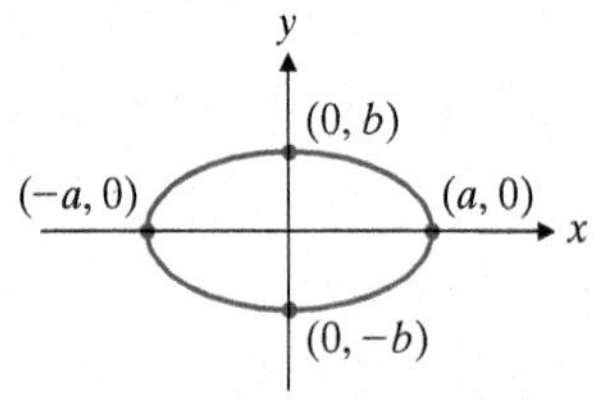 	Graph $\frac{x^2}{16} + \frac{y^2}{4} = 1$. $a^2 = 16, \quad a = 4; \quad b^2 = 4, \quad b = 2$ 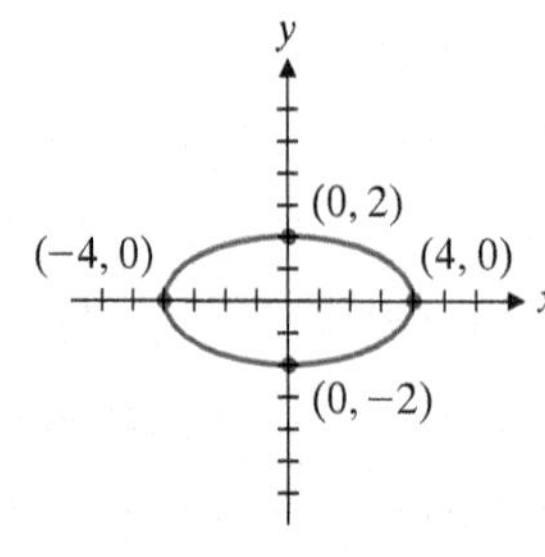 	5. Graph $\frac{x^2}{36} + y^2 = 1$. 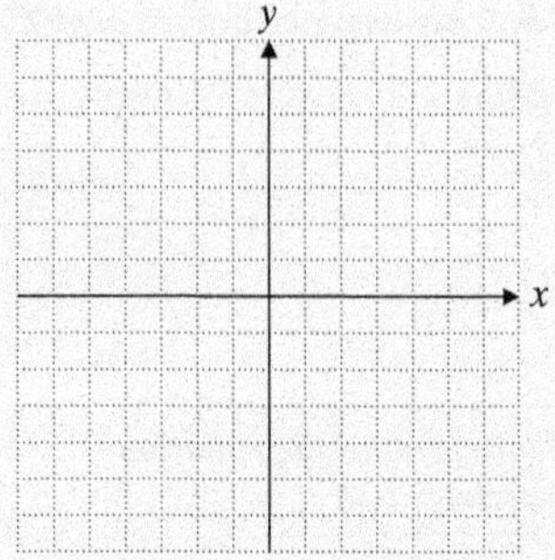
Standard form of an ellipse with center at (h, k), p. 647 An ellipse with center at (h, k) has the equation $$\frac{(x-h)^2}{a^2} + \frac{(y-k)^2}{b^2} = 1,$$ where $a > 0$ and $b > 0$. 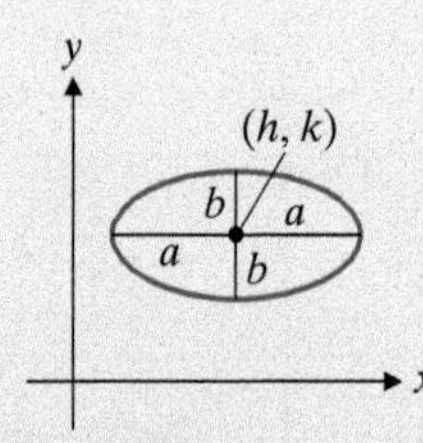 	Graph $\frac{(x+2)^2}{9} + \frac{(y+4)^2}{25} = 1$. $(h, k) = (-2, -4); \quad a = 3, \quad b = 5$ 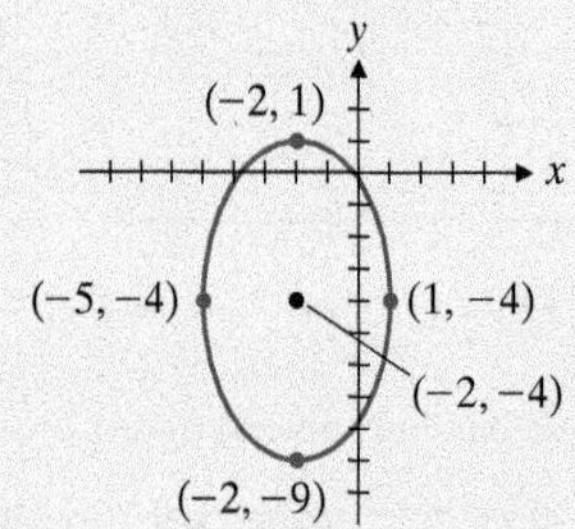 	6. Graph $\frac{(x-1)^2}{9} + \frac{(y+3)^2}{4} = 1$. 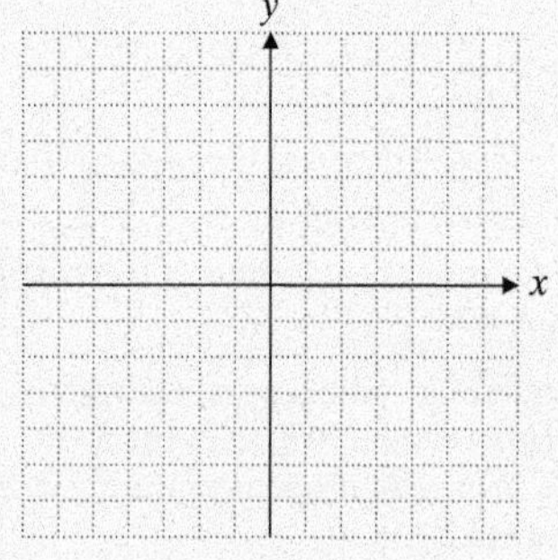
Standard form of a horizontal hyperbola with center at (0, 0), p. 654 Let a and b be positive real numbers. A horizontal hyperbola with center at the origin and vertices $(a, 0)$ and $(-a, 0)$ has the equation $$\frac{x^2}{a^2} - \frac{y^2}{b^2} = 1$$ and asymptotes $$y = \pm\frac{b}{a}x.$$ 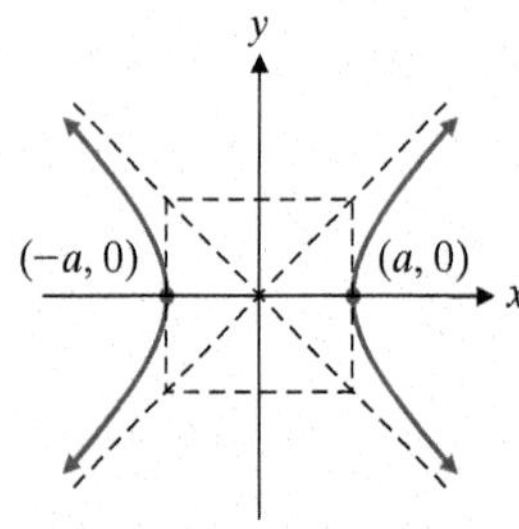 	Graph $\frac{x^2}{25} - \frac{y^2}{9} = 1$. $a = 5, \quad b = 3$ 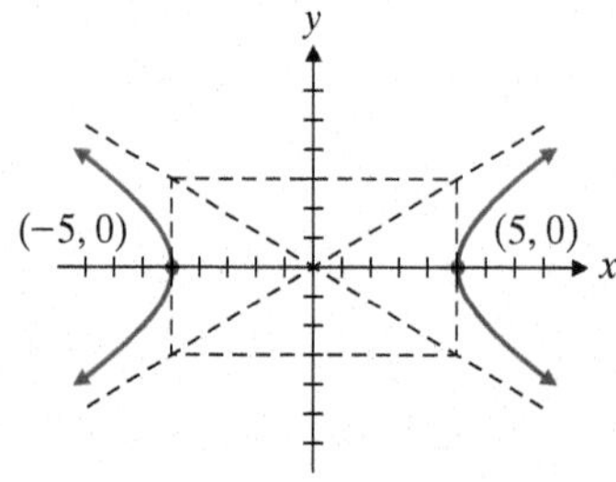 	7. Graph $\frac{x^2}{36} - \frac{y^2}{4} = 1$. 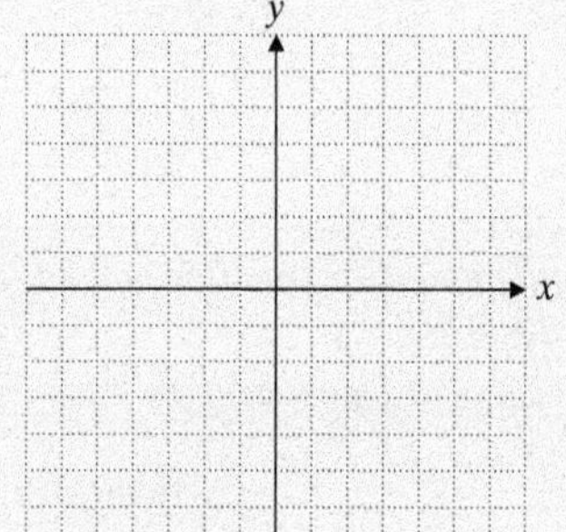

Topic and Procedure	Examples	You Try It
Standard form of a vertical hyperbola with center at (0, 0), p. 654 Let a and b be positive real numbers. A vertical hyperbola with center at the origin and vertices $(0, b)$ and $(0, -b)$ has the equation $$\frac{y^2}{b^2} - \frac{x^2}{a^2} = 1$$ and asymptotes $$y = \pm\frac{b}{a}x.$$ 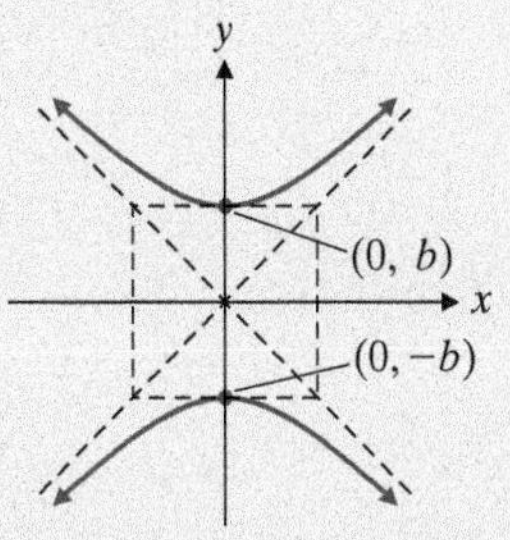	Graph $\frac{y^2}{9} - \frac{x^2}{4} = 1.$ $b = 3, \quad a = 2$ 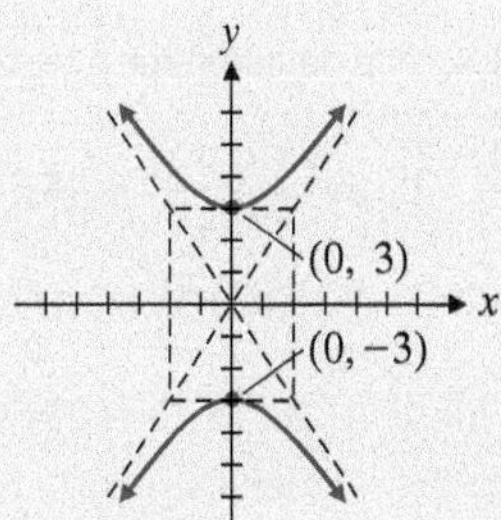	8. Graph $\frac{y^2}{16} - x^2 = 1.$ 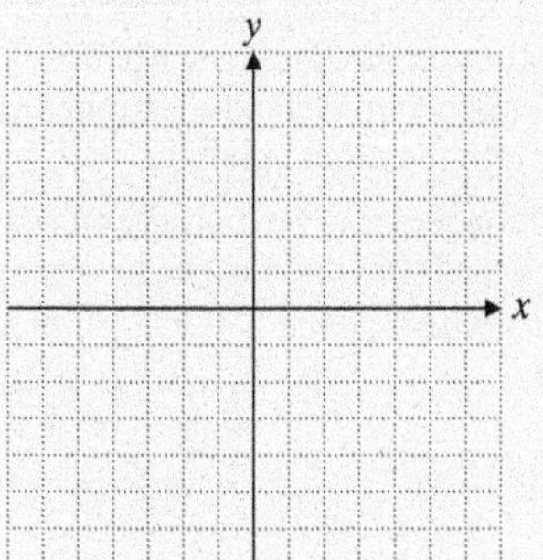
Standard form of a horizontal hyperbola with center at (h, k), p. 656 Let a and b be positive real numbers. A horizontal hyperbola with center at (h, k) and vertices $(h - a, k)$ and $(h + a, k)$ has the equation $$\frac{(x - h)^2}{a^2} - \frac{(y - k)^2}{b^2} = 1.$$ 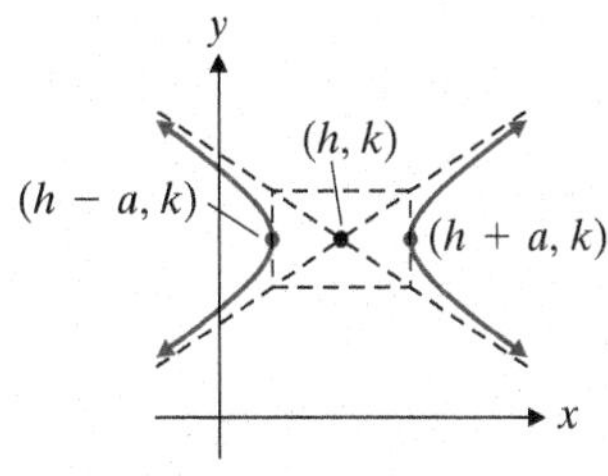	Graph $\frac{(x - 2)^2}{4} - \frac{(y - 3)^2}{25} = 1.$ Center at $(2, 3)$; $a = 2, b = 5$ 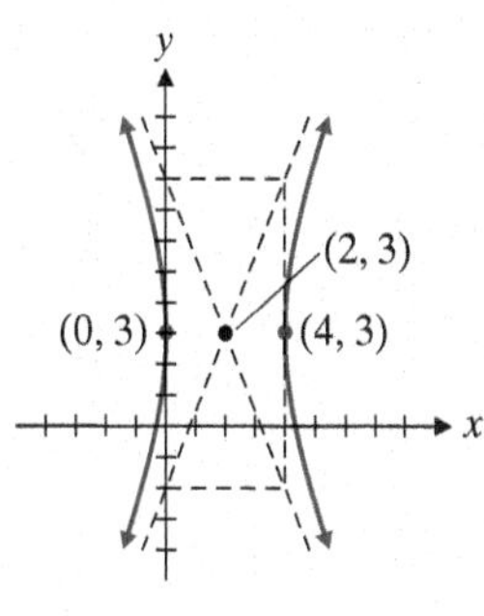	9. Graph $\frac{(x - 1)^2}{9} - \frac{(y + 2)^2}{16} = 1.$ 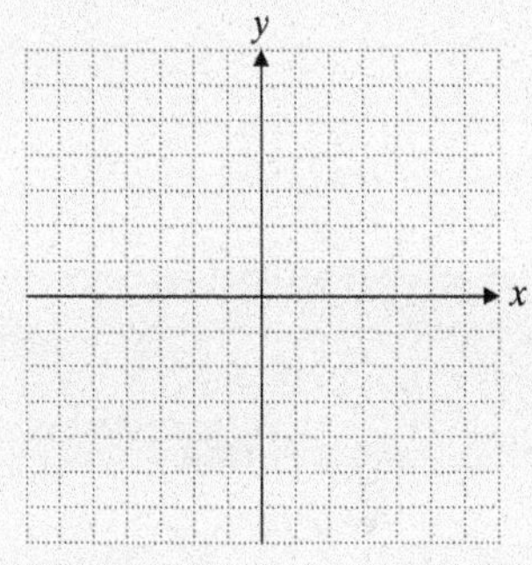
Standard form of a vertical hyperbola with center at (h, k), p. 656 Let a and b be positive real numbers. A vertical hyperbola with center at (h, k) and vertices $(h, k + b)$ and $(h, k - b)$ has the equation $$\frac{(y - k)^2}{b^2} - \frac{(x - h)^2}{a^2} = 1.$$ 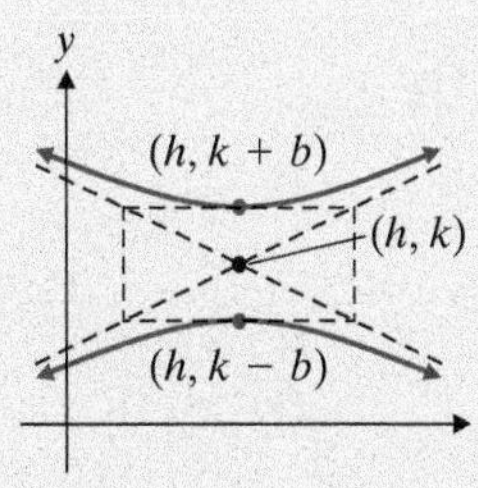	Graph $\frac{(y - 5)^2}{9} - \frac{(x - 4)^2}{4} = 1.$ Center at $(4, 5)$; $b = 3, a = 2$ 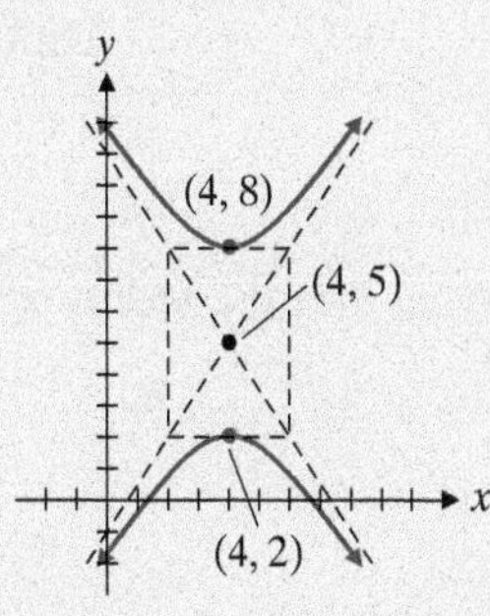	10. Graph $\frac{(y + 3)^2}{4} - \frac{(x - 2)^2}{4} = 1.$ 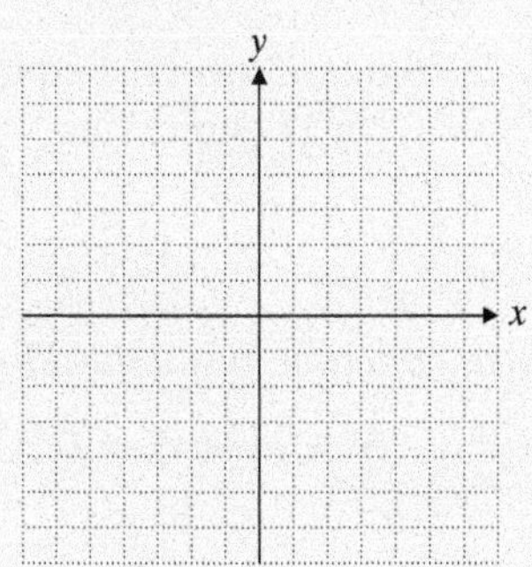

Topic and Procedure	Examples	You Try It
Nonlinear systems of equations, pp. 662–664 We can solve a nonlinear system by the substitution method or the addition method. In the substitution method we solve one equation for one variable and substitute that expression into the other equation. In the addition method, we multiply one or more equations by a numerical value and then add them together so that one variable is eliminated.	Solve by substitution. $2x^2 + y^2 = 18$ $xy = 4$ Solving the second equation for y, we have $y = \frac{4}{x}$. $2x^2 + \left(\frac{4}{x}\right)^2 = 18$ $2x^2 + \frac{16}{x^2} = 18$ $2x^4 + 16 = 18x^2$ $2x^4 - 18x^2 + 16 = 0$ $x^4 - 9x^2 + 8 = 0$ $(x^2 - 1)(x^2 - 8) = 0$ $x^2 - 1 = 0 \qquad x^2 - 8 = 0$ $x^2 = 1 \qquad x^2 = 8$ $x = \pm 1 \qquad x = \pm 2\sqrt{2}$ Since $xy = 4$, if $x = 1$, then $y = 4$. if $x = -1$, then $y = -4$. if $x = 2\sqrt{2}$, then $y = \sqrt{2}$. if $x = -2\sqrt{2}$, then $y = -\sqrt{2}$. The solutions are $(1, 4)$, $(-1, -4)$, $(2\sqrt{2}, \sqrt{2})$, and $(-2\sqrt{2}, -\sqrt{2})$.	11. Solve by substitution. $x^2 + 3y^2 = 12$ $y - 3x = -2$

Chapter 10 Review Problems

In exercises 1 and 2, find the distance between the points.

1. $(0, -6)$ and $(-3, 2)$

2. $(-7, 3)$ and $(-2, -1)$

3. Write in standard form the equation of the circle with center at $(-6, 3)$ and radius $\sqrt{15}$.

4. Write in standard form the equation of the circle with center at $(0, -7)$ and radius 5.

Rewrite each equation in standard form. Find the center and the radius of each circle.

5. $x^2 + y^2 + 2x - 6y + 5 = 0$

6. $x^2 + y^2 - 10x + 12y + 52 = 0$

Graph each parabola. Label its vertex and plot at least one intercept.

7. $x = \frac{1}{3}y^2$

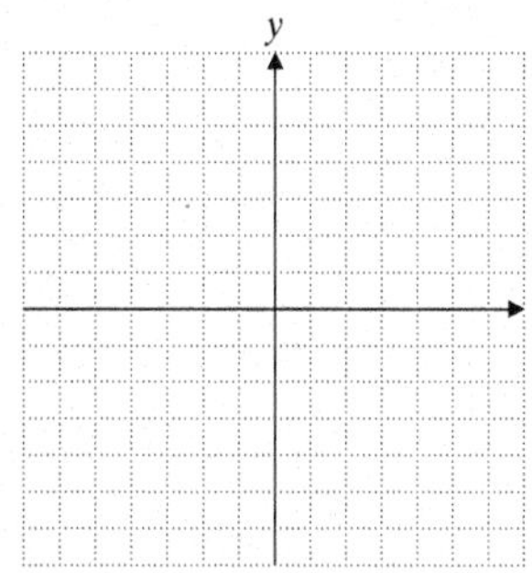

8. $x = \frac{1}{2}(y - 2)^2 + 4$

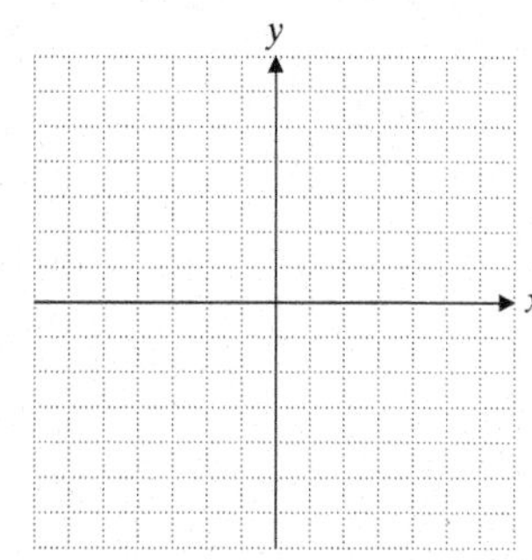

9. $y = -2(x + 1)^2 - 3$

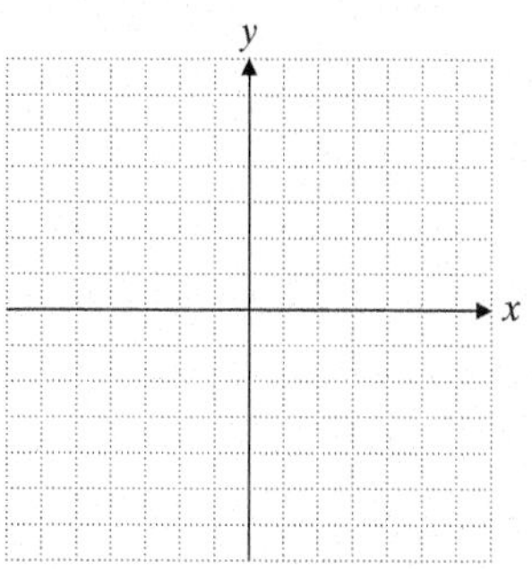

Rewrite each equation in standard form. Find the vertex and determine in which direction the parabola opens.

10. $x^2 + 6x = y - 4$

11. $x + 8y = y^2 + 10$

Graph each ellipse. Label its center and four other points.

12. $\frac{x^2}{4} + y^2 = 1$

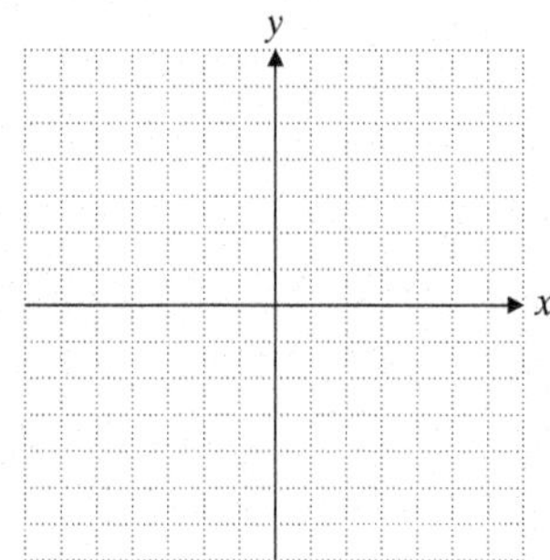

13. $16x^2 + y^2 - 32 = 0$

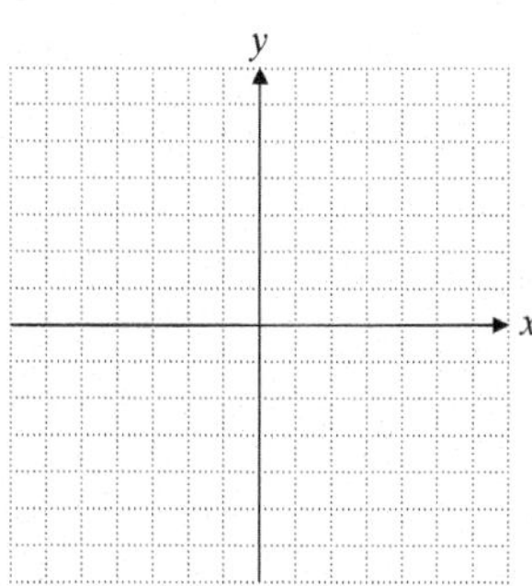

14. Determine the vertices and the center of the ellipse. $\frac{(x + 5)^2}{4} + \frac{(y + 3)^2}{25} = 1$

Find the center and vertices of each hyperbola and graph it.

15. $x^2 - 4y^2 - 16 = 0$

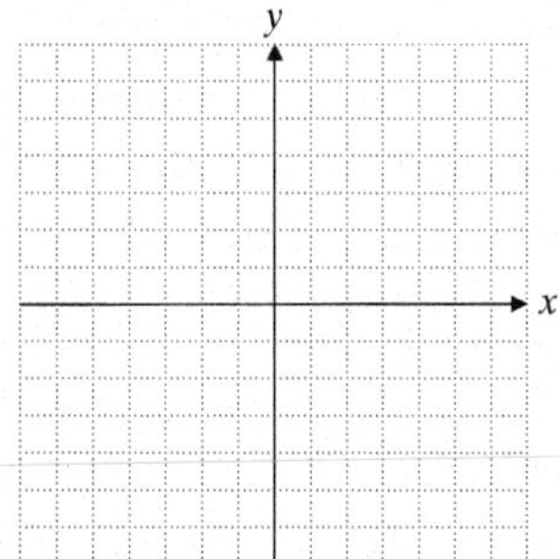

16. $3y^2 - x^2 = 27$

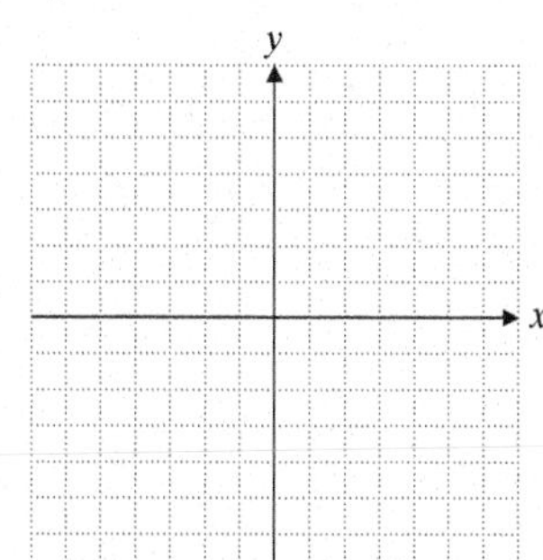

17. Determine the vertices and the center of the hyperbola. $\frac{(x-2)^2}{4} - \frac{(y+3)^2}{25} = 1$

Solve each nonlinear system. If there is no real number solution, so state.

18. $x^2 + y = 9$
$y - x = 3$

19. $x^2 + y^2 = 4$
$x + y = 2$

20. $2x^2 + y^2 = 17$
$x^2 + 2y^2 = 22$

21. $3x^2 - 4y^2 = 12$
$7x^2 - y^2 = 8$

22. $y = x^2 + 1$
$x^2 + y^2 - 8y + 7 = 0$

23. $2x^2 + y^2 = 18$
$xy = 4$

24. $y^2 - 2x^2 = 2$
$2y^2 - 3x^2 = 5$

25. $y^2 = 2x$
$y = \frac{1}{2}x + 1$

Applications

If the outline of a satellite dish (or other parabolic shape) is described by the equation $y = ax^2$, then the distance p from the center of the dish to the focus point of the dish is given by the equation $a = \frac{1}{4p}$. Use this information to answer exercises 26 and 27.

26. ***Searchlights*** The side view of an airport searchlight is shaped like a parabola. The center of the light source of the searchlight is located 2 feet from the base along the axis of symmetry, and the opening is 5 feet across. How deep should the searchlight be? Round your answer to the nearest hundredth.

27. ***Satellite Dishes*** The side view of a satellite dish on Jason and Wendy's house is shaped like a parabola. The signals that come from the satellite hit the surface of the dish and are then reflected to the point where the signal receiver is located. This point is the focus of the parabolic dish. The dish is 10 feet across at its opening and 4 feet deep at its center. How far from the center of the dish should the signal receiver be placed? Round your answer to the nearest hundredth.

How Am I Doing? Chapter 10 Test

MATH COACH MyMathLab® 

After you take this test read through the Math Coach on pages 677–678. Math Coach videos are available via MyMathLab and YouTube. Step-by-step test solutions in the Chapter Test Prep Videos are also available via MyMathLab and YouTube. (Search "TobeyBegInterAlg" and click on "Channels.")

1. Find the distance between $(-6, -8)$ and $(-2, 5)$.

Rewrite the equation in standard form. Identify the conic, find the center or vertex, plot at least one other point, and sketch the curve.

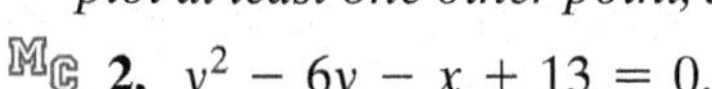

MC **2.** $y^2 - 6y - x + 13 = 0.$

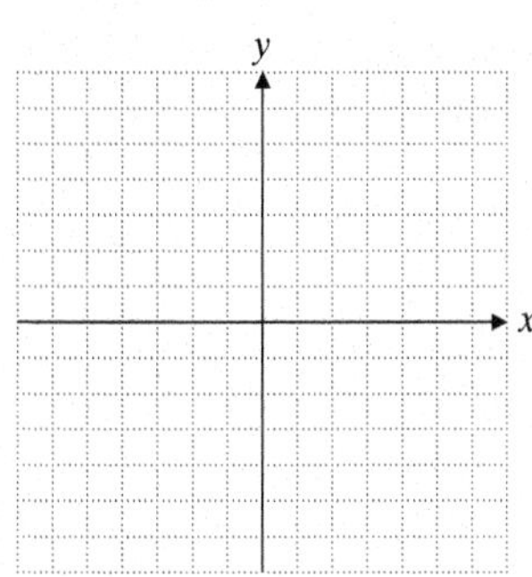

3. $x^2 + y^2 + 6x - 4y + 9 = 0$

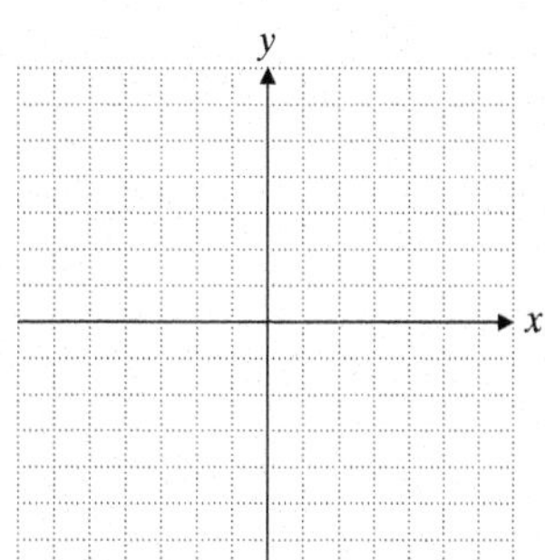

Identify and graph each conic section. Label the center and/or vertex as appropriate.

4. $\dfrac{x^2}{25} + \dfrac{y^2}{1} = 1$

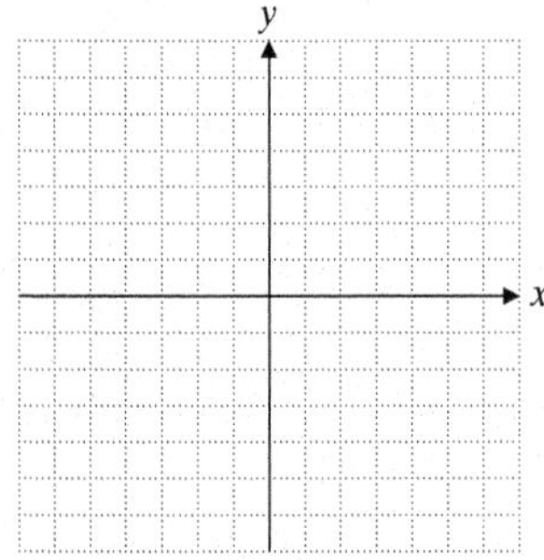

MC **5.** $\dfrac{x^2}{10} - \dfrac{y^2}{9} = 1$

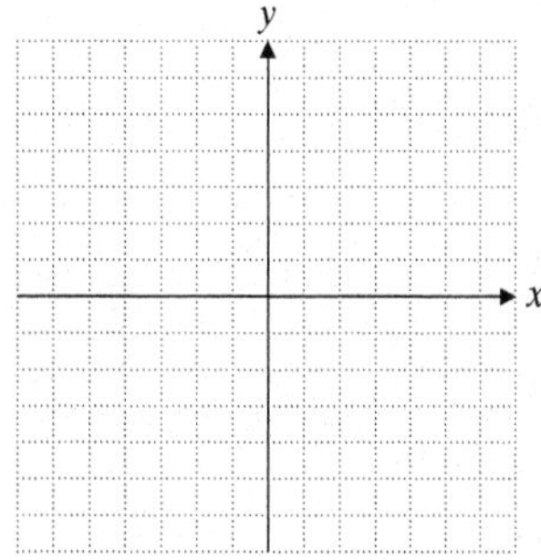

6. $y = -2(x + 3)^2 + 4$

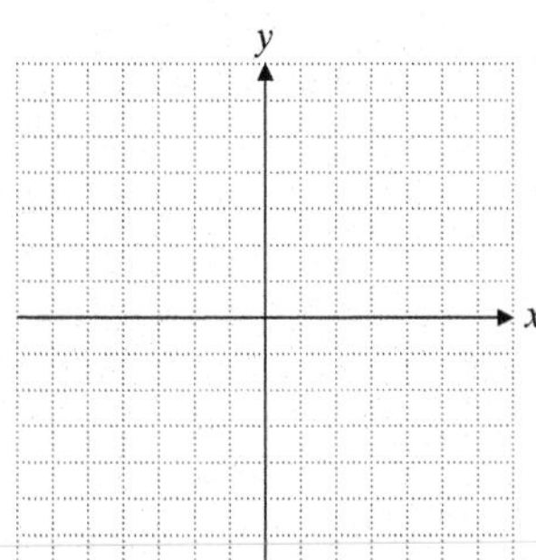

MC **7.** $\dfrac{(x + 2)^2}{16} + \dfrac{(y - 5)^2}{4} = 1$

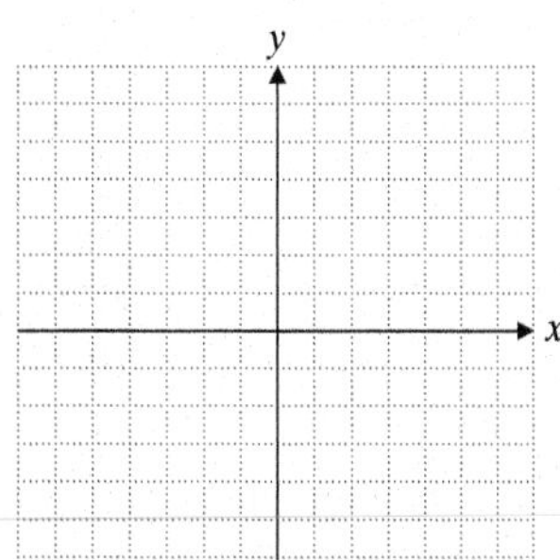

1. ______

2. ______

3. ______

4. ______

5. ______

6. ______

7. ______

8.

9.

10.

11.

12.

13.

14.

15.

16.

Total Correct:

8. $7y^2 - 7x^2 = 28$

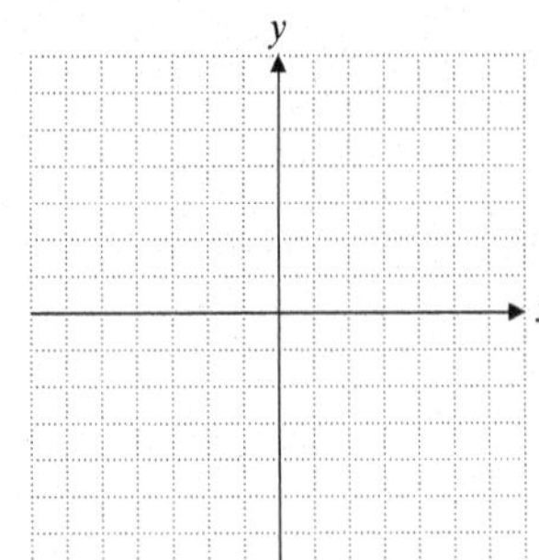

Find the standard form of the equation of each of the following:

9. Circle of radius $\sqrt{8}$ with its center at $(3, -5)$.

10. Ellipse with center at the origin, an x-intercept at $(3, 0)$, and a y-intercept at $(0, 5)$.

11. Parabola with its vertex at $(-7, 3)$ and that opens to the right. This parabola crosses the x-axis at $(2, 0)$. It is of the form $x = (y - k)^2 + h$.

12. Hyperbola with center at the origin, vertices at $(3, 0)$ and $(-3, 0)$. One asymptote is $y = \frac{5}{3}x$.

Solve each nonlinear system.

13. $-2x + y = 5$
$x^2 + y^2 - 25 = 0$

14. $x^2 + y^2 = 9$
$y = x - 3$

15. $4x^2 + y^2 - 4 = 0$
$9x^2 - 4y^2 - 9 = 0$

MC **16.** $x^2 + 2y^2 = 15$
$x^2 - y^2 = 6$

MATH COACH

Mastering the skills you need to do well on the test.

The following problems are from the Chapter 10 Test. Here are some helpful hints to keep you from making common errors on test problems.

Chapter 10 Test, Problem 2 Rewrite the equation in standard form. Identify the conic, find the center or vertex, plot at least one other point, and sketch the curve.

$$y^2 - 6y - x + 13 = 0$$

HELPFUL HINT If the equation of the parabola has a y^2 term, then it can be written in the standard form $x = a(y - k)^2 + h$. The vertex is (h, k), and the graph is a horizontal parabola.

Did you first rewrite the equation as $x = y^2 - 6y + 13$

Yes ______ No ______

Next did you complete the square to obtain $x = (y - 3)^2 + 4$

Yes ______ No ______

If you answered No to these questions, remember that when completing the square, we take half of -6 and square it, which results in 9. We must add both a positive and a negative 9 to the right side of the equation. Go back and try to complete these steps again.

Did you see that since $a = 1$ and $a > 0$, the parabola opens to the right?

Yes ______ No ______

Did you identify the vertex as $(4, 3)$?

Yes ______ No ______

If you answered No to these questions, review the helpful hint and other information about horizontal parabolas. Stop and perform this step again.

In your final answer, be sure to give the equation in standard form, list the vertex, identify the conic, and graph the equation.

If you answered Problem 2 incorrectly, go back and rework the problem using these suggestions.

Chapter 10 Test, Problem 5 Identify and graph the conic section. Label the center and/or vertex as appropriate. $\frac{x^2}{10} - \frac{y^2}{9} = 1$

HELPFUL HINT The standard form of a hyperbola with the center at the origin and vertices $(-a, 0)$ and $(a, 0)$ has the equation $\frac{x^2}{a^2} - \frac{y^2}{b^2} = 1$.

Did you realize that this equation represents a horizontal hyperbola with vertices at $(-\sqrt{10}, 0)$ and $(\sqrt{10}, 0)$?

Yes ______ No ______

If you answered No, review the rules for the standard form of an equation of a hyperbola with its center at the origin and read the Helpful Hint. Then remember that since $a^2 = 10$ and $a > 0$, $a = \sqrt{10}$. Likewise, if $b^2 = 9$ and $b > 0$, then $b = 3$.

Did you obtain a fundamental rectangle with corners at $(\sqrt{10}, 3)$, $(\sqrt{10}, -3)$, $(-\sqrt{10}, 3)$ and $(-\sqrt{10}, -3)$?

Yes ______ No ______

If you answered No, note that drawing a fundamental rectangle with corners at (a, b), $(a, -b)$, $(-a, b)$, and $(-a, -b)$ can help in creating the graph. The extended diagonals of the rectangle are the asymptotes.

In your final answer, remember to identify the conic section, create its graph, and label the center and the vertices.

Now go back and rework the problem using these suggestions.

Need more help? Watch the MATH COACH videos in MyMathLab® or on YouTube™.

Chapter 10 Test, Problem 7 Identify and graph the conic section. Label the center and/or vertex as appropriate. $\frac{(x+2)^2}{16} + \frac{(y-5)^2}{4} = 1$

HELPFUL HINT When an ellipse has a center that is not at the origin, the center has the coordinates (h, k) and the equation in standard form is $\frac{(x-h)^2}{a^2} + \frac{(y-k)^2}{b^2} = 1$, where both a and b are greater than zero.

Did you determine that the center of the ellipse is at $(-2, 5)$ with $a = 4$ and $b = 2$?

Yes ______ No ______

If you answered No, remember that the standard form of the equation involves $x - h$ and $y - k$, so you must be careful in determining the signs of h and k.

Did you start at the center and find points a units to the left, a units to the right, b units up, and b units down to plot the points $(-6, 5)$, $(2, 5)$, $(-2, 7)$, and $(-2, 3)$?

Yes ______ No ______

If you answered No, remember that to find these four points you need to find the following: $(h - a, k)$, $(h + a, k)$, $(h, k + b)$, and $(h, k - b)$.

Plot all four points and the center and label these on your graph. Then use the four points to make a sketch of the ellipse. Remember to identify the conic as an ellipse in your final answer.

If you answered Problem 7 incorrectly, go back and rework the problem using these suggestions.

Chapter 10 Test, Problem 16 Solve.

$$\begin{aligned} x^2 + 2y^2 &= 15 \\ x^2 - y^2 &= 6 \end{aligned}$$

HELPFUL HINT When two equations in a system have the form $ax^2 + by^2 = c$, where a, b, and c are real numbers, then it may be easiest to solve the system by the addition method.

If you multiplied the second equation by 2 and added the result to the first equation, do you get the equation $3x^2 = 27$?

Yes ______ No ______

Can you solve this equation for x to get $x = 3$ and $x = -3$?

Yes ______ No ______

If you answered No to these questions, remember that when you add the two equations together, the y^2 term adds to 0.

If you substitute $x = 3$ into the first equation, do you get the equation $9 + 2y^2 = 15$?

Yes ______ No ______

Can you solve this equation for y to get $y = \sqrt{3}$ and $y = -\sqrt{3}$?

Yes ______ No ______

If you answered No to these questions, try substituting $x = 3$ into the equation again and be careful to avoid calculation errors. Remember that you must also perform this same step with $x = -3$. Since x is squared, your results for y should be the same.

Your final answer should have a total of four possible ordered pair solutions to this system.

Now go back and rework this problem using these suggestions.

CHAPTER 11

Additional Properties of Functions

CAREER OPPORTUNITIES

Criminologist

Criminologists study criminal behavior and develop theories, programs, and techniques that make other criminal justice professionals more effective in their efforts at controlling crime and promoting its prevention. Analyzing crime data, patterns of behavior, crime scenes, and physical evidence to develop accurate profiles and programs involves the frequent use of math.

To investigate how the mathematics in this chapter can help with this field, see the Career Exploration Problems on page 712.

11.1 Function Notation

Student Learning Objectives

After studying this section, you will be able to:

1 Use function notation to evaluate expressions.

2 Use function notation to solve application exercises.

1 Using Function Notation to Evaluate Expressions

Function notation is useful in solving a number of interesting exercises. Suppose you wanted to skydive from an airplane. Your instructor tells you that you must wait 20 seconds before you pull the cord to open the parachute. How far will you fall in that time?

The approximate distance an object in free-fall travels when there is no initial downward velocity is given by the distance function $d(t) = 16t^2$, where time t is measured in seconds and distance $d(t)$ is measured in feet. (Neglect air resistance.)

How far will a person in free-fall travel in 20 seconds if he or she leaves the airplane with no downward velocity? We want to find the distance $d(20)$ where $t = 20$ seconds. To find the distance, we substitute 20 for t in the function $d(t) = 16t^2$.

$$d(20) = 16(20)^2 = 16(400) = 6400$$

Thus, a person in free-fall will travel approximately 6400 feet in 20 seconds.

Now suppose that person waits longer than he or she should to pull the parachute cord. How will this affect the distance he or she falls? Suppose the person falls for a seconds beyond the 20-second mark before pulling the parachute cord.

$$\begin{aligned} d(20 + a) &= 16(20 + a)^2 \\ &= 16(20 + a)(20 + a) \\ &= 16(400 + 40a + a^2) \\ &= 6400 + 640a + 16a^2 \end{aligned}$$

Thus, if the person waited 5 seconds too long before pulling the cord, he or she would fall the following distance.

$$\begin{aligned} d(20 + 5) &= 6400 + 640(5) + 16(5)^2 \\ &= 6400 + 3200 + 16(25) \\ &= 6400 + 3200 + 400 \\ &= 10{,}000 \text{ feet} \end{aligned}$$

This is 3600 feet farther than the person would have fallen in 20 seconds. Obviously, a delay of 5 seconds could have life or death consequences.

We now revisit a topic that we first discussed in Chapter 3, evaluating a function for particular values of the variable.

Example 1 If $g(x) = 5 - 3x$, find the following:

(a) $g(a)$ **(b)** $g(a + 3)$ **(c)** $g(a) + g(3)$

Solution

(a) $g(a) = 5 - 3a$

(b) $g(a + 3) = 5 - 3(a + 3) = 5 - 3a - 9 = -4 - 3a$

(c) This exercise requires us to find each addend separately. Then we add them together.

$$g(a) = 5 - 3a$$
$$g(3) = 5 - 3(3) = 5 - 9 = -4$$

Thus,
$$\begin{aligned} g(a) + g(3) &= (5 - 3a) + (-4) \\ &= 5 - 3a - 4 \\ &= 1 - 3a \end{aligned}$$

Notice that $g(a + 3) \neq g(a) + g(3)$. □

Student Practice 1 If $g(x) = \frac{1}{2}x - 3$, find the following:

(a) $g(a)$ **(b)** $g(a + 4)$ **(c)** $g(a) + g(4)$

To Think About: *Understanding Function Notation* Is $g(a + 4) = g(a) + g(4)$ in Student Practice 1? Why or why not?

Example 2 If $p(x) = 2x^2 - 3x + 5$, find the following:

(a) $p(-2)$ **(b)** $p(a)$

(c) $p(3a)$ **(d)** $p(a - 2)$

Solution

(a) $$\begin{aligned} p(-2) &= 2(-2)^2 - 3(-2) + 5 \\ &= 2(4) - 3(-2) + 5 \\ &= 8 + 6 + 5 \\ &= 19 \end{aligned}$$

(b) $p(a) = 2(a)^2 - 3(a) + 5 = 2a^2 - 3a + 5$

(c) $$\begin{aligned} p(3a) &= 2(3a)^2 - 3(3a) + 5 \\ &= 2(9a^2) - 3(3a) + 5 \\ &= 18a^2 - 9a + 5 \end{aligned}$$

(d) $$\begin{aligned} p(a - 2) &= 2(a - 2)^2 - 3(a - 2) + 5 \\ &= 2(a - 2)(a - 2) - 3(a - 2) + 5 \\ &= 2(a^2 - 4a + 4) - 3(a - 2) + 5 \\ &= 2a^2 - 8a + 8 - 3a + 6 + 5 \\ &= 2a^2 - 11a + 19 \end{aligned}$$

Student Practice 2

If $p(x) = -3x^2 + 2x + 4$, find the following:

(a) $p(-3)$

(b) $p(a)$

(c) $p(2a)$

(d) $p(a - 3)$

Example 3 If $r(x) = \frac{4}{x + 2}$, find

(a) $r(a + 3)$ **(b)** $r(a)$

(c) $r(a + 3) - r(a)$. Express this result as one fraction.

Solution

(a) $r(a + 3) = \frac{4}{a + 3 + 2} = \frac{4}{a + 5}$

(b) $r(a) = \frac{4}{a + 2}$

(c) $r(a + 3) - r(a) = \frac{4}{a + 5} - \frac{4}{a + 2}$

To express this as one fraction, we note that the LCD $= (a + 5)(a + 2)$.

$$\begin{aligned} r(a + 3) - r(a) &= \frac{4(a + 2)}{(a + 5)(a + 2)} - \frac{4(a + 5)}{(a + 2)(a + 5)} \\ &= \frac{4a + 8}{(a + 5)(a + 2)} - \frac{4a + 20}{(a + 5)(a + 2)} \\ &= \frac{4a + 8 - 4a - 20}{(a + 5)(a + 2)} \\ &= \frac{-12}{(a + 5)(a + 2)} \end{aligned}$$

Student Practice 3 If $r(x) = \dfrac{-3}{x + 1}$, find

(a) $r(a + 2)$ **(b)** $r(a)$

(c) $r(a + 2) - r(a)$. Express this result as one fraction.

Example 4 Let $f(x) = 3x - 7$. Find $\dfrac{f(x + h) - f(x)}{h}$.

Solution First

$$f(x + h) = 3(x + h) - 7 = 3x + 3h - 7$$

and

$$f(x) = 3x - 7.$$

So

$$\begin{aligned} f(x + h) - f(x) &= (3x + 3h - 7) - (3x - 7) \\ &= 3x + 3h - 7 - 3x + 7 \\ &= 3h. \end{aligned}$$

Therefore,

$$\frac{f(x + h) - f(x)}{h} = \frac{3h}{h} = 3.$$

□

Student Practice 4

Suppose that $g(x) = 2 - 5x$. Find $\dfrac{g(x + h) - g(x)}{h}$.

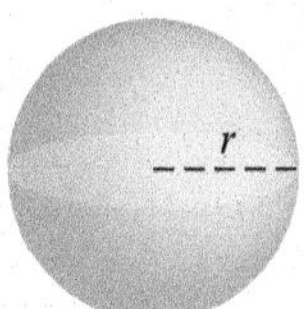

2 Using Function Notation to Solve Application Exercises

▲ **Example 5** The surface area of a sphere is given by $S = 4\pi r^2$ where r is the radius. If we use $\pi \approx 3.14$ as an approximation, this becomes $S = 4(3.14)r^2$, or $S = 12.56r^2$.

(a) Write the surface area of a sphere as a function of radius r.

(b) Find the surface area of a sphere with a radius of 3 centimeters.

(c) Suppose that an error is made and the radius of the sphere is calculated to be $(3 + e)$ centimeters. Find an expression for the surface area as a function of the error e.

(d) Evaluate the surface area for $r = (3 + e)$ centimeters when $e = 0.2$. Round your answer to the nearest hundredth of a centimeter. What is the difference in the surface area due to the error in measurement?

Solution

(a) $S(r) = 12.56r^2$

(b) $S(3) = 12.56(3)^2 = 12.56(9) = 113.04$ square centimeters

(c)
$$\begin{aligned} S(e) &= 12.56(3 + e)^2 \\ &= 12.56(3 + e)(3 + e) \\ &= 12.56(9 + 6e + e^2) \\ &= 113.04 + 75.36e + 12.56e^2 \end{aligned}$$

(d) If an error in measurement is made so that the radius is calculated to be $r = (3 + e)$ centimeters, where $e = 0.2$, we can use the function generated in part **(c)**.

$$\begin{aligned} S(e) &= 113.04 + 75.36e + 12.56e^2 \\ S(0.2) &= 113.04 + 75.36(0.2) + 12.56(0.2)^2 \\ &= 113.04 + 15.072 + 0.5024 \\ &= 128.6144 \end{aligned}$$

Rounding, we have $S = 128.61$ square centimeters.
Thus, if the radius of 3 centimeters was incorrectly calculated as 3.2 centimeters, the surface area would be approximately $128.61 - 113.04 = 15.57$ square centimeters too large. □

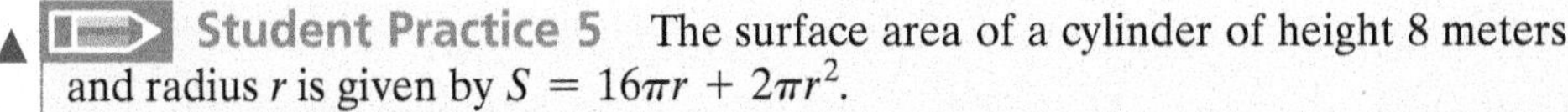

Student Practice 5 The surface area of a cylinder of height 8 meters and radius r is given by $S = 16\pi r + 2\pi r^2$.

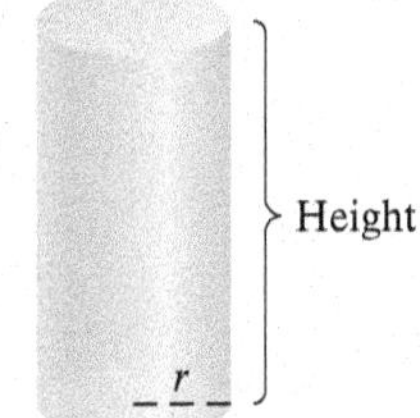

(a) Write the surface area of a cylinder of height 8 meters (using $\pi \approx 3.14$) and radius r as a function of r.

(b) Find the surface area if the radius is 2 meters.

(c) Suppose that an error is made and the radius is calculated to be $(2 + e)$ meters. Find an expression for the surface area as a function of the error e.

(d) Evaluate the surface area for $r = (2 + e)$ meters when $e = 0.3$. Round your answer to the nearest hundredth of a meter. What is the difference in the surface area due to the error in measurement?

11.1 Exercises MyMathLab®

For the function $f(x) = 3x - 5$, *find the following.*

1. $f\left(-\frac{2}{3}\right)$
2. $f(1.5)$
3. $f(a + 5)$
4. $f(t - 2)$

For the function $g(x) = \frac{1}{2}x - 3$, *find the following.*

5. $g(2) + g(a)$
6. $g(6) + g(b)$
7. $g(4a) - g(a)$
8. $g(-4b) + g(10b)$
9. $g(2a - 4)$
10. $g(3a + 1)$
11. $g(a^2) - g\left(\frac{2}{5}\right)$
12. $g(b^2) - g\left(\frac{4}{3}\right)$

If $p(x) = 3x^2 + 4x - 2$, *find the following.*

13. $p(-2)$
14. $p(-5)$
15. $p\left(\frac{1}{2}\right)$
16. $p\left(-\frac{1}{2}\right)$
17. $p(a + 1)$
18. $p(b - 1)$
19. $p\left(-\frac{2a}{3}\right)$
20. $p\left(\frac{1}{4}a\right)$

If $h(x) = \sqrt{x + 5}$, *find the following.*

21. $h(4)$
22. $h(-5)$
23. $h(7)$
24. $h(19)$
25. $h(a^2 - 1)$
26. $h(a^2 + 4)$
27. $h(-2b)$
28. $h(3b)$
29. $h(4a - 1)$
30. $h(8a - 1)$
31. $h(b^2 + b)$
32. $h(b^2 + b - 5)$

If $r(x) = \frac{7}{x - 3}$, *find the following and write your answers as one fraction.*

33. $r(7)$
34. $r(-4)$
35. $r(3.5)$
36. $r(-0.5)$
37. $r(a^2)$
38. $r(3b^2)$
39. $r(a + 2)$
40. $r(a - 3)$
41. $r\left(\frac{1}{3}\right) + r(-5)$
42. $r(0) + r(6)$

Find $\dfrac{f(x+h)-f(x)}{h}$ *for the following functions.*

43. $f(x) = 2x - 3$

44. $f(x) = 8 - 3x$

45. $f(x) = x^2 - x$

46. $f(x) = 3x^2$

Applications

47. ***Wind Generators*** A turbine wind generator produces P kilowatts of power for wind speed w (measured in miles per hour) according to the equation $P = 2.5w^2$.

(a) Write the number of kilowatts P as a function of w.

(b) Find the power in kilowatts when the wind speed is $w = 20$ miles per hour.

(c) Suppose that an error is made and the speed of the wind is calculated to be $(20 + e)$ miles per hour. Find an expression for the power as a function of error e.

(d) Evaluate the power for $w = (20 + e)$ miles per hour when $e = 2$.

▲ **48.** ***Geometry*** The area of a circle is $A = \pi r^2$.

(a) Write the area of a circle as the function of the radius r. Use $\pi \approx 3.14$.

(b) Find the area of a circle with a radius of 4.0 feet.

(c) Suppose that an error is made and the radius is calculated to be $(4 + e)$ feet. Find an expression for the area as a function of error e.

(d) Evaluate the area for $r = (4 + e)$ feet when $e = 0.4$. Round your answer to the nearest hundredth.

Lead Levels in Air *Because of the elimination of lead in automobile gasoline and increased use of emission controls in automobiles and industrial operations, the amount of lead in the air in the United States has shown a marked decrease. The percent of lead in the air p(x) expressed in terms of 1984 levels is given in the line graph. The variable x indicates the number of years since 1984. The function value p(x) indicates the amount of lead that remains in the air in selected regions of the United States, expressed as a percent of the amount of lead in the air in 1984.*

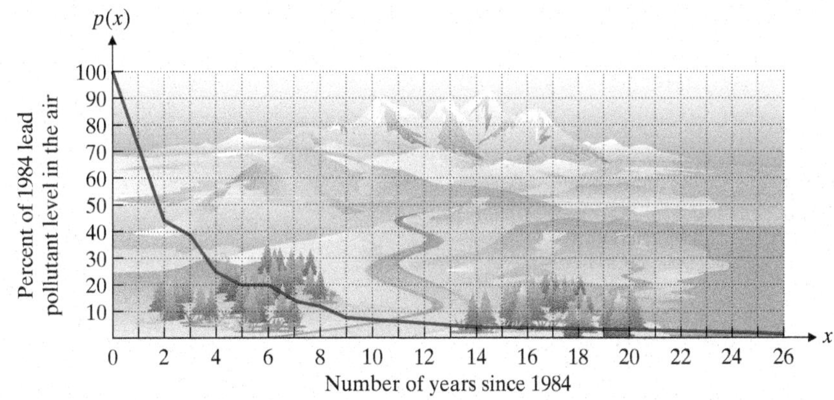

Source: www.epa.gov

49. If a new function were defined as $p(x) - 13$, what would happen to the function values associated with x? Find $p(3) - 13$.

50. If a new function were defined as $p(x + 2)$, what would happen to the function values associated with x? Find $p(x + 2)$ when $x = 4$.

If $f(x) = 3x^2 - 4.6x + 1.23$, *find each of the following function values to the nearest thousandth.*

51. $f(0.026a)$

52. $f(a + 2.23)$

Cumulative Review *Solve for x.*

53. **[7.5.1]** $\frac{7}{6} + \frac{5}{x} = \frac{3}{2x}$

54. **[7.5.1]** $\frac{1}{6} - \frac{2}{3x + 6} = \frac{1}{2x + 4}$

Use the formula for the volume of a sphere, $V = \frac{4}{3}\pi r^3$, *to complete exercises 55 and 56.*

▲ **55.** **[0.6.1, 1.8.2]** ***Planet Volume*** The diameter of the planet Mars is 4211 miles, while that of Earth is almost twice as large with a diameter of 7927 miles. How many times greater is the volume of Earth compared to the volume of Mars?

▲ **56.** **[0.6.1, 1.8.2]** ***Planet Volume*** The radius of the planet Neptune is 15,345 miles. The radius of Saturn (excluding its rings) is 37,366 miles. How many times greater is the volume of Saturn compared to the volume of Neptune?

Quick Quiz 11.1

1. If $f(x) = \frac{3}{5}x - 4$, find $f(a) - f(-3)$.

2. If $g(x) = 2x^2 - 3x + 4$, find $g\left(\frac{2}{3}a\right)$.

3. If $h(x) = \frac{3}{x + 4}$, find $h(a - 6)$.

4. Concept Check Explain how you would find $k(2a - 1)$ if $k(x) = \sqrt{3x + 1}$.

11.2 General Graphing Procedures for Functions 

1 Using the Vertical Line Test to Determine Whether a Graph Represents a Function

Student Learning Objectives

After studying this section, you will be able to:

1. Use the vertical line test to determine whether a graph represents a function.
2. Graph a function of the form $f(x+h)+k$ by means of horizontal and vertical shifts of the graph of $f(x)$.

Not every graph we observe is that of a function. By definition, a function must have no ordered pairs that have the same first coordinates and different second coordinates. A graph that includes the points $(4, 2)$ and $(4, -2)$, for example, would not be the graph of a function. Thus, the graph of $x = y^2$ would not be the graph of a function.

If any vertical line crosses a graph of a relation in more than one place, the relation is not a function. If no such line exists, the relation is a function.

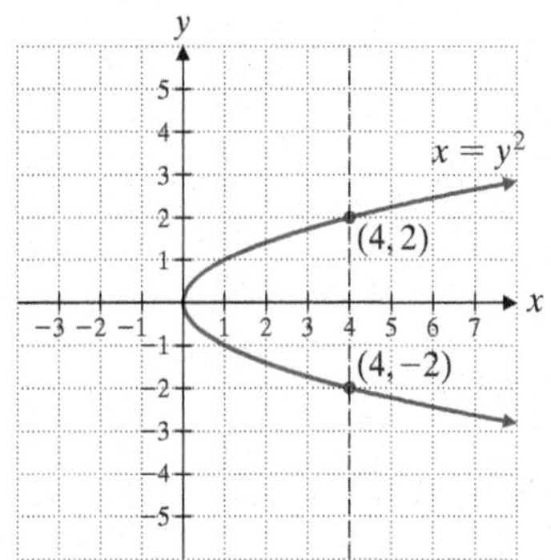

> **VERTICAL LINE TEST**
>
> If any vertical line intersects the graph of a relation more than once, the relation is not a function. If no such line exists, the relation is a function.

In the following sketches, we observe that the dashed vertical line crosses the curve of a function no more than once. The dashed vertical line crosses the curve of a relation that is *not* a function more than once.

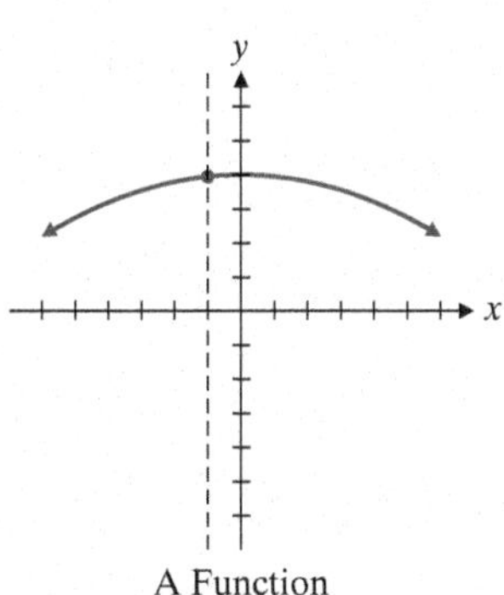

A Function

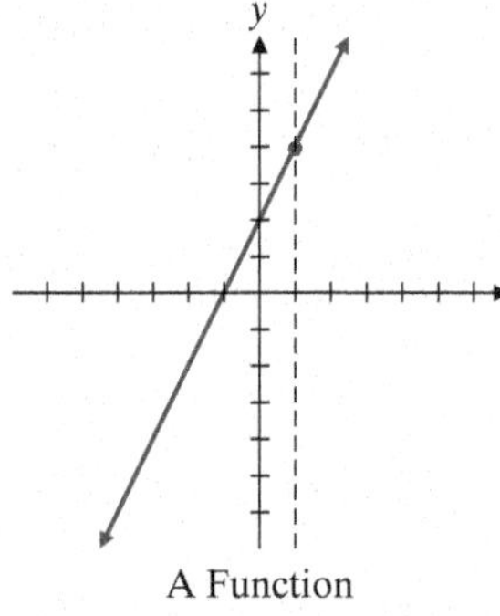

A Function

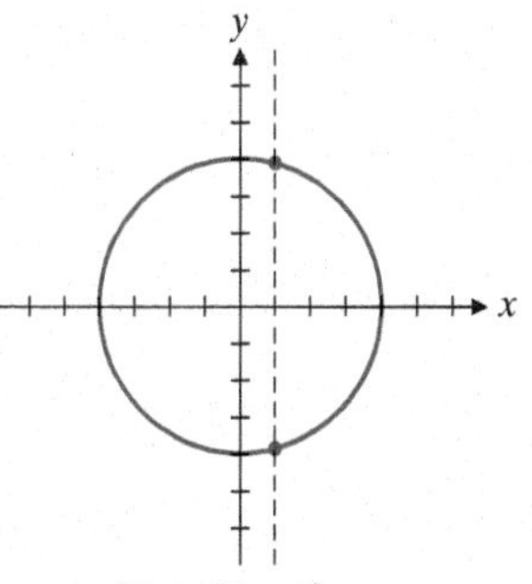

Not a Function

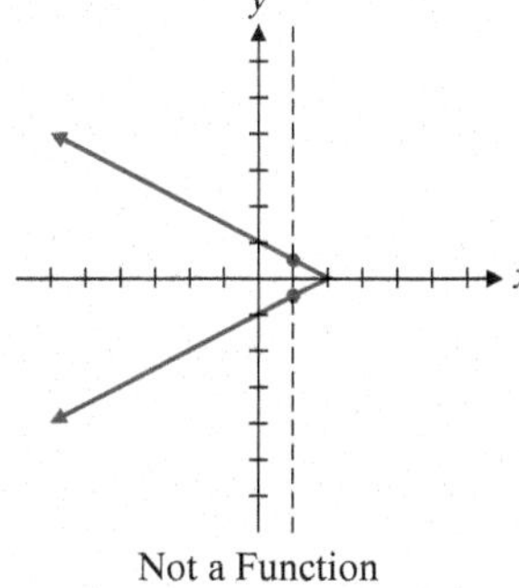

Not a Function

Example 1 Determine whether each of the following is the graph of a function.

(a)

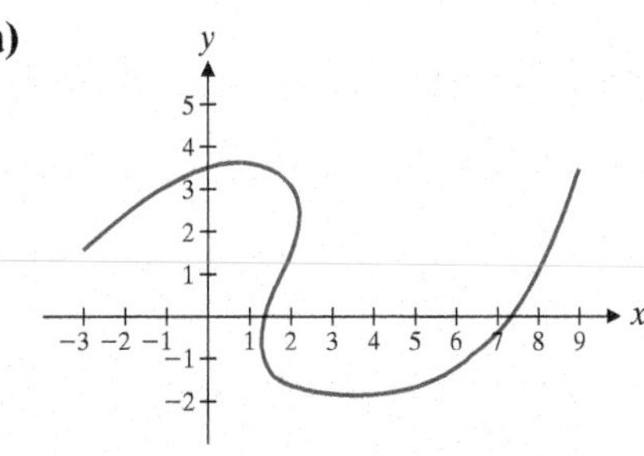

(b)

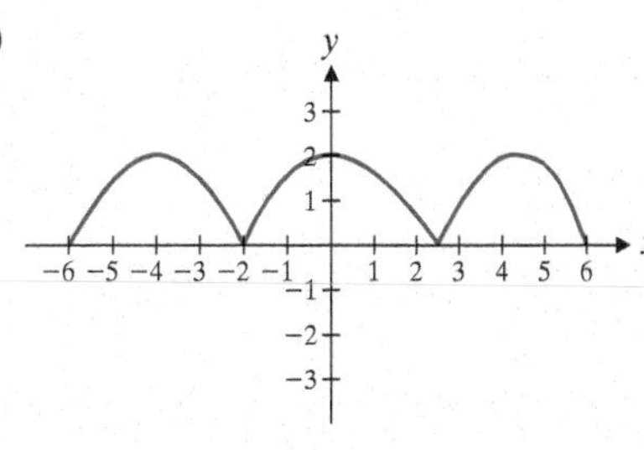

Continued on next page

Solution

(a) By the vertical line test, this relation is not a function.

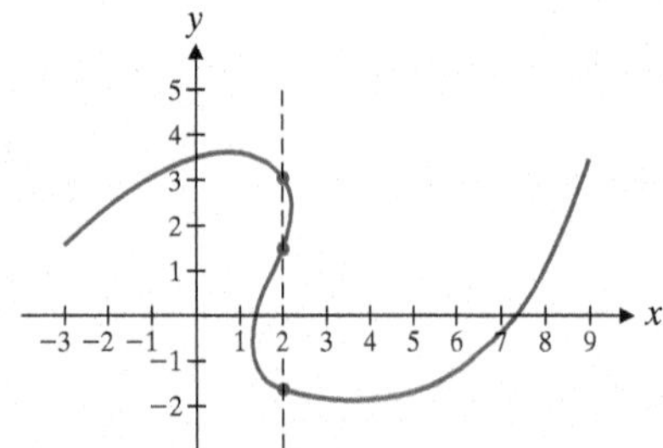

(b) By the vertical line test, this relation is a function.

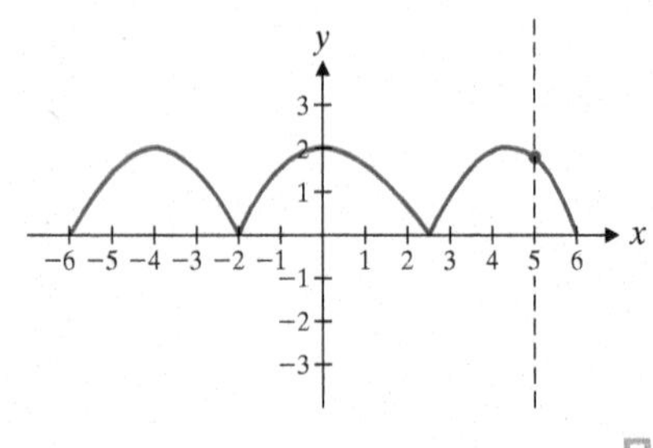

Student Practice 1 Does this graph represent a function? Why or why not?

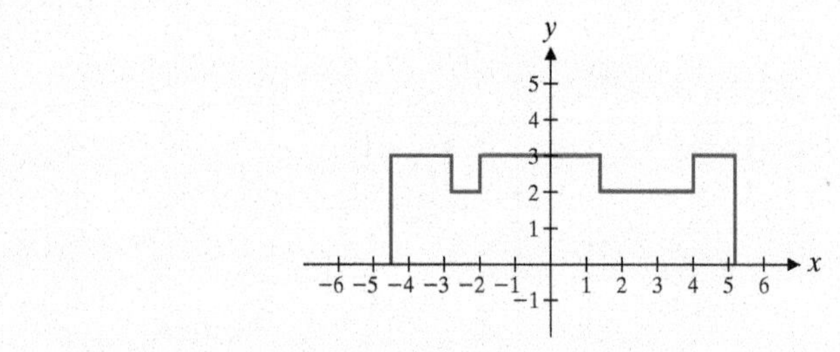

2 Graphing a Function of the Form $f(x + h) + k$ by Means of Horizontal and Vertical Shifts of the Graph of $f(x)$

The graphs of some functions are simple vertical shifts of the graphs of similar functions.

Example 2 Graph the functions on one coordinate plane.

$$f(x) = x^2 \quad \text{and} \quad h(x) = x^2 + 2$$

Solution First we make a table of values for $f(x)$ and for $h(x)$.

x	$f(x) = x^2$
−2	4
−1	1
0	0
1	1
2	4

x	$h(x) = x^2 + 2$
−2	6
−1	3
0	2
1	3
2	6

Now we graph each function on the same coordinate plane.

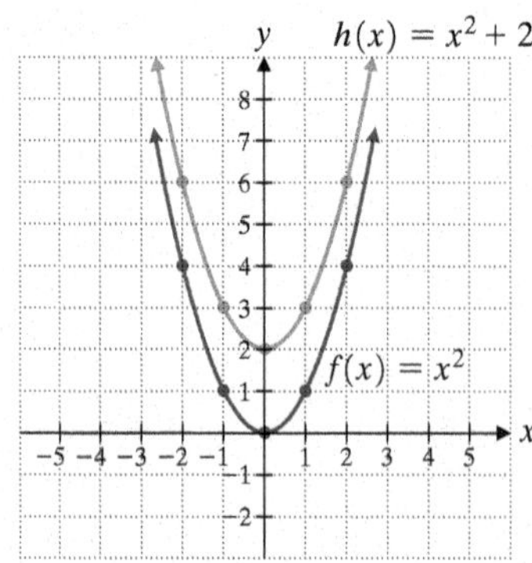

Notice that the graph of $h(x)$ is the graph of $f(x)$ moved 2 units upward. □

 **Student Practice 2** Graph the functions on one coordinate plane.

$$f(x) = x^2 \quad \text{and} \quad h(x) = x^2 - 5$$

Student Practice 2

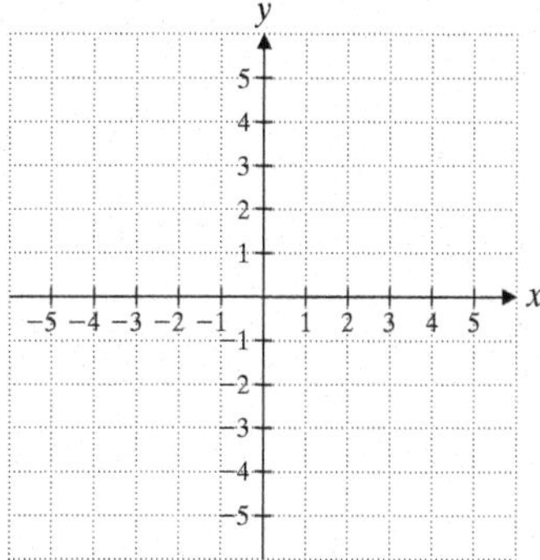

We have the following general summary.

> **VERTICAL SHIFTS**
>
> Suppose that k is a positive number.
>
> 1. To obtain the graph of $f(x) + k$, shift the graph of $f(x)$ up k units.
> 2. To obtain the graph of $f(x) - k$, shift the graph of $f(x)$ down k units.

Now we turn to the topic of horizontal shifts.

Graphing Calculator

Exploration

Most graphing calculators have an absolute value function (abs). Use this function to graph $f(x)$ and $p(x)$ from Example 3 on one coordinate plane.

Example 3 Graph the functions on one coordinate plane.

$$f(x) = |x| \quad \text{and} \quad p(x) = |x - 3|$$

Solution First we make a table of values for $f(x)$ and for $p(x)$.

x	$f(x) = \lvert x\rvert$
−2	2
−1	1
0	0
1	1
2	2
3	3
4	4

x	$p(x) = \lvert x - 3\rvert$
−2	5
−1	4
0	3
1	2
2	1
3	0
4	1

Now we graph each function on the same coordinate plane.

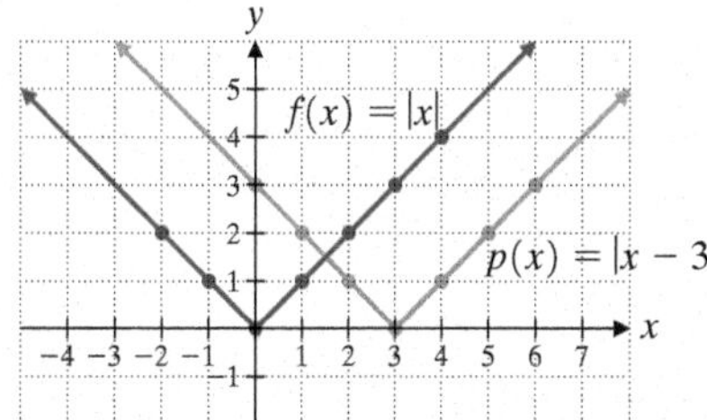

Notice that the graph of $p(x)$ is the graph of $f(x)$ shifted 3 units to the right. □

 **Student Practice 3** Graph the functions on one coordinate plane.

$$f(x) = |x| \quad \text{and} \quad p(x) = |x + 2|$$

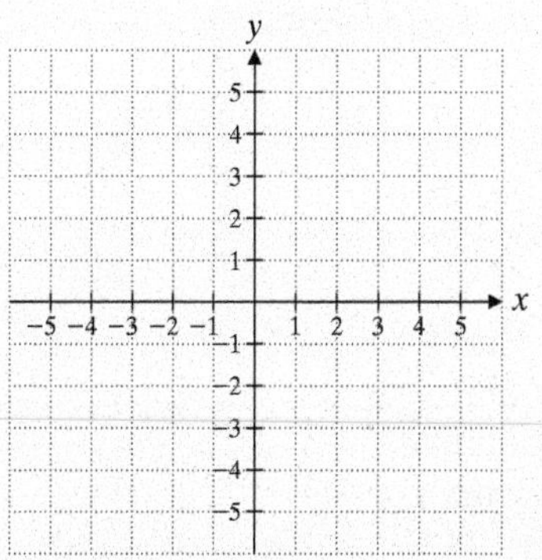

Now we can write the following general summary.

> **HORIZONTAL SHIFTS**
>
> Suppose that h is a positive number.
>
> 1. To obtain the graph of $f(x - h)$, shift the graph of $f(x)$ to the right h units.
> 2. To obtain the graph of $f(x + h)$, shift the graph of $f(x)$ to the left h units.

Some graphs will involve both horizontal and vertical shifts.

Example 4 Graph the functions on one coordinate plane.

$$f(x) = x^3 \quad \text{and} \quad h(x) = (x - 3)^3 - 2$$

Solution First we make a table of values for $f(x)$ and graph the function.

x	$f(x)$
−2	−8
−1	−1
0	0
1	1
2	8

Next we recognize that $h(x)$ will have a similar shape, but the curve will be shifted 3 units to the *right* and 2 units *downward*. We draw the graph of $h(x)$ using these shifts.

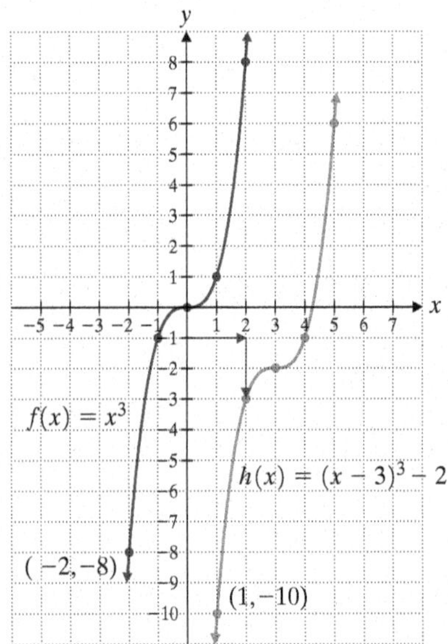

To Think About: *Shifting Points* The point $(-2, -8)$ has been shifted 3 units to the right and 2 units down to the point $(-2 + 3, -8 + (-2))$ or $(1, -10)$. The point $(-1, -1)$ is a point on $f(x)$. Use the same reasoning to find the image of $(-1, -1)$ on the graph of $h(x)$. Verify by checking the graphs. □

Student Practice 4 Graph the functions on one coordinate plane.

$$f(x) = x^3 \quad \text{and} \quad h(x) = (x + 4)^3 + 3$$

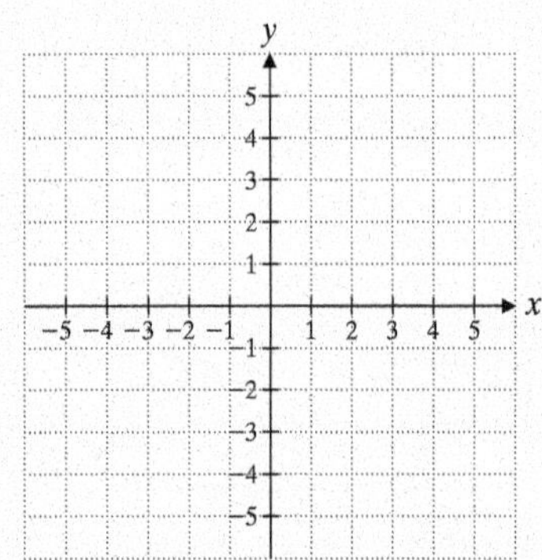

All the functions that we have sketched in this section so far have had a domain of all real numbers. Some functions have restricted domains.

Example 5 Graph the functions on one coordinate plane. State the domain of each function.

$$f(x) = \frac{4}{x} \quad \text{and} \quad g(x) = \frac{4}{x + 3} + 1$$

Solution First we make a table of values for $f(x)$. The domain of $f(x)$ is all real numbers, where $x \neq 0$. Note that $f(x)$ is not defined when $x = 0$ since we cannot divide by 0.

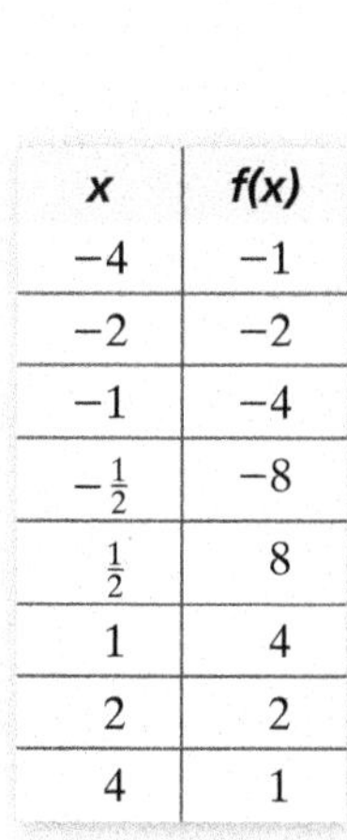

x	f(x)
−4	−1
−2	−2
−1	−4
$-\frac{1}{2}$	−8
$\frac{1}{2}$	8
1	4
2	2
4	1

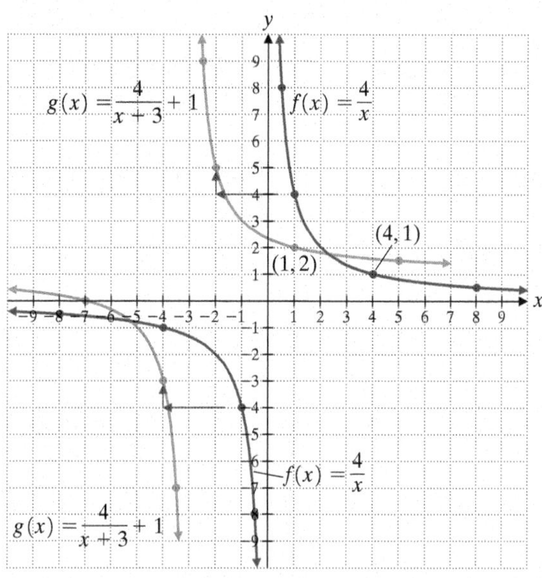

We draw $f(x)$ and note key points. From the equation, we see that the graph of $g(x)$ is 3 units to the left of and 1 unit above $f(x)$. We can find the image of each of the key points on $f(x)$ as a guide in graphing $g(x)$. For example, the image of $(4, 1)$ is $(4 - 3, 1 + 1)$ or $(1, 2)$.

Each point on $f(x)$ is shifted

$\Leftarrow$ 3 units left and
$\Uparrow$ 1 unit up

to form the graph of $g(x)$.

What is the domain of $g(x)$? Why? $g(x)$ contains the denominator $x + 3$. But $x + 3 \neq 0$. Therefore, $x \neq -3$. The domain of $g(x)$ is all real numbers, where $x \neq -3$. □

 **Student Practice 5** Graph the functions on one coordinate plane.

$$f(x) = \frac{2}{x} \quad \text{and} \quad g(x) = \frac{2}{x + 1} - 2$$

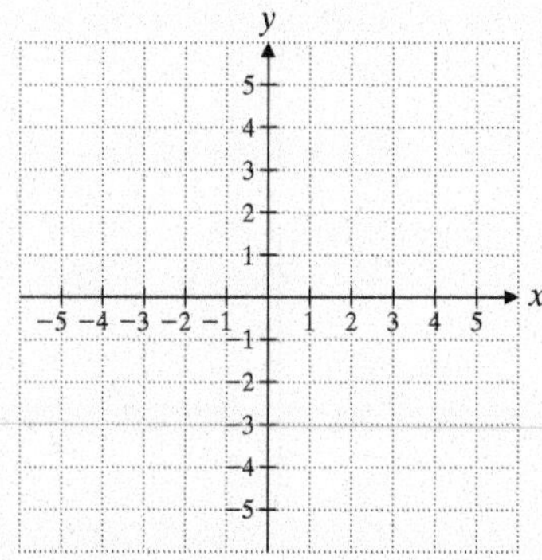

11.2 Exercises

Verbal and Writing Skills, Exercises 1–4

1. Does $f(x + 2) = f(x) + f(2)$? Why or why not? Give an example.

2. Explain what the vertical line test is and why it works.

3. To obtain the graph of $f(x) + k$ for $k > 0$, shift the graph of $f(x)$ ________ k units.

4. To obtain the graph of $f(x - h)$ for $h > 0$, shift the graph of $f(x)$ ____________ h units.

Determine whether or not each graph represents a function.

5.

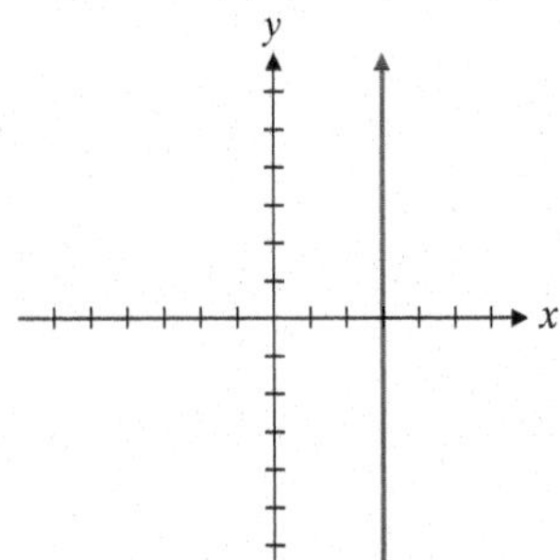

6.

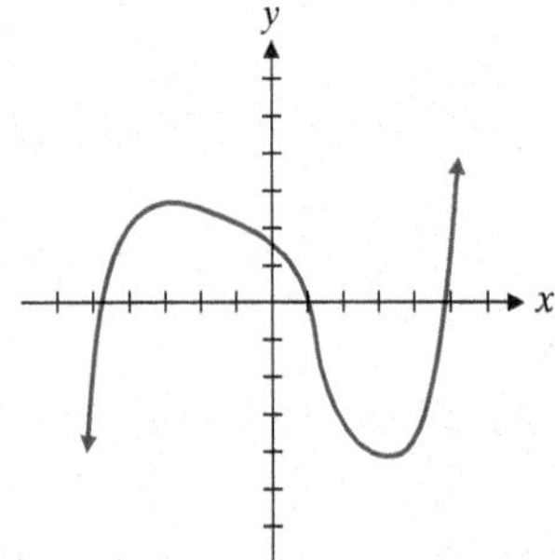

7.

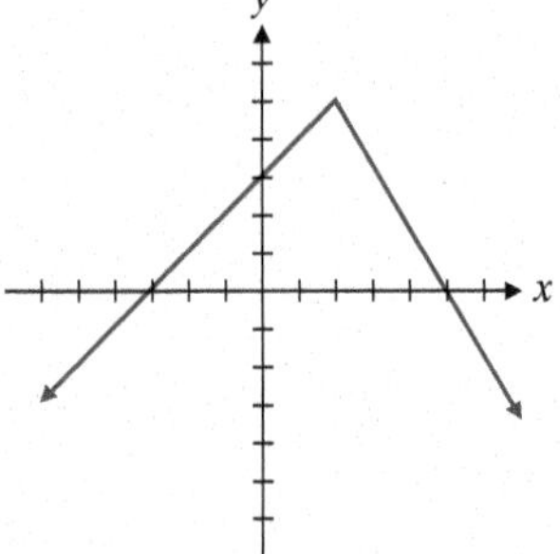

8.

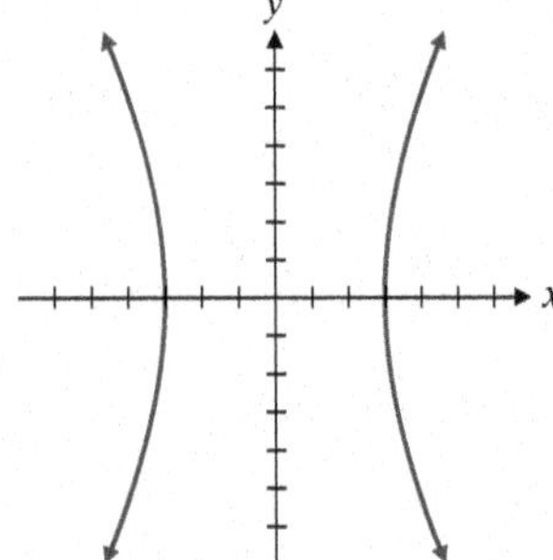

9.

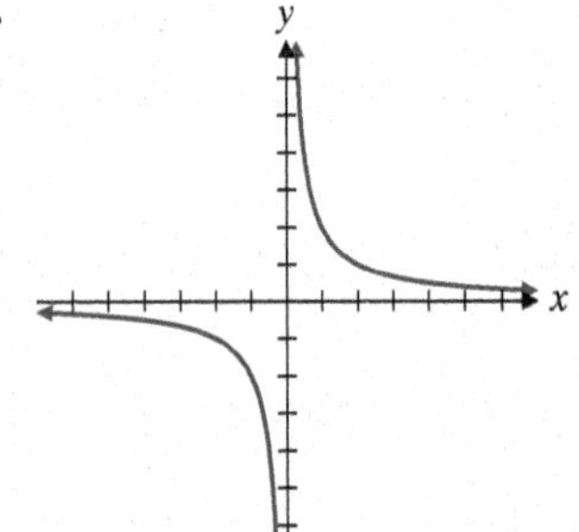

10.

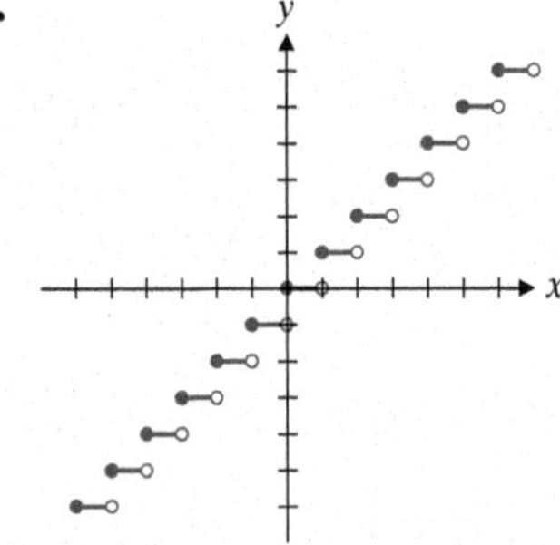

Hint: The open circle means that the function value does not exist at that point.

11.

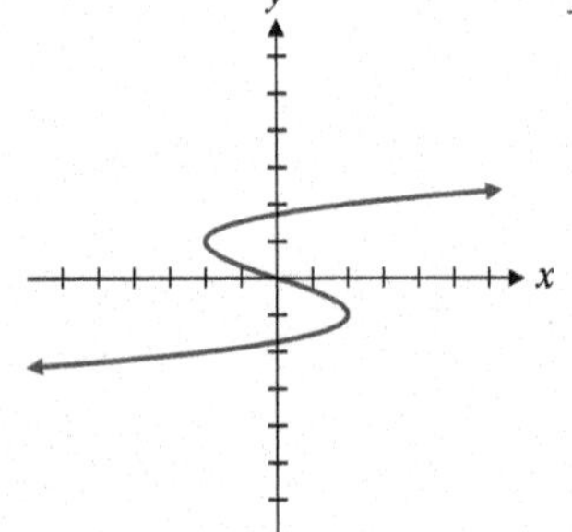

12.

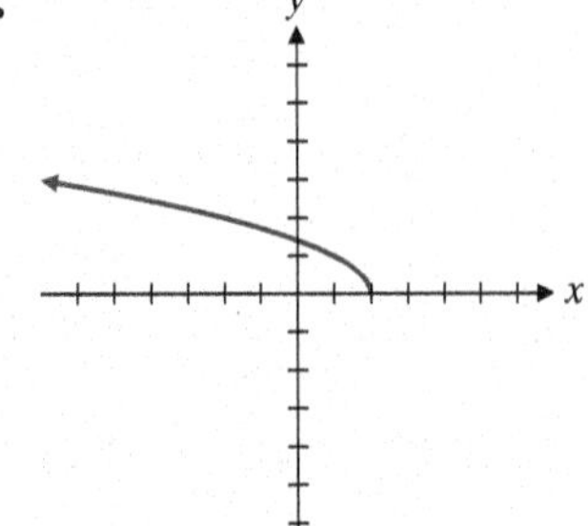

13.

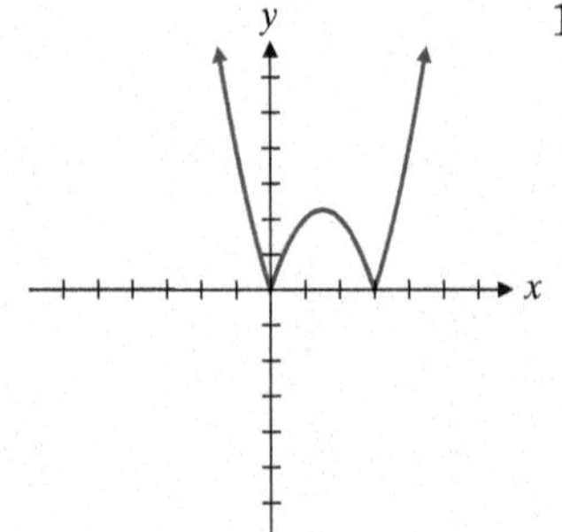

14.

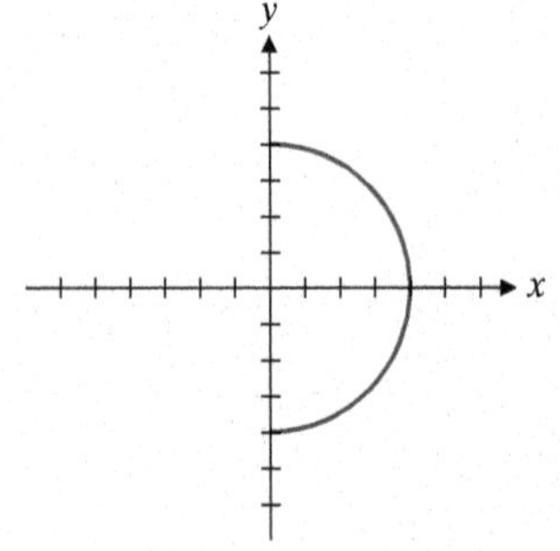

For each of exercises 15–28, graph the two functions on one coordinate plane.

15. $f(x) = x^2$
$h(x) = x^2 - 3$

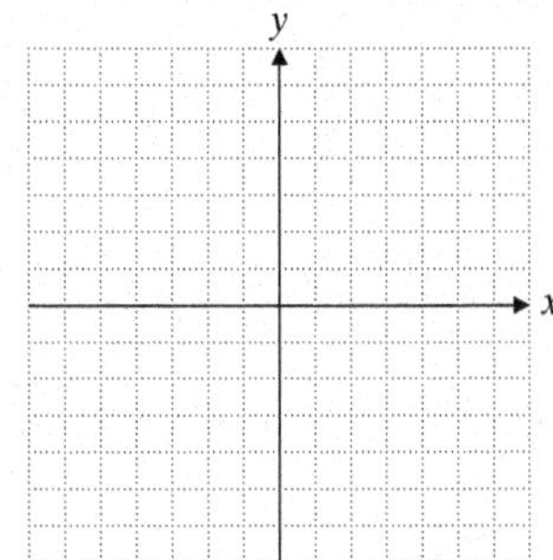

16. $f(x) = x^2$
$h(x) = x^2 + 3$

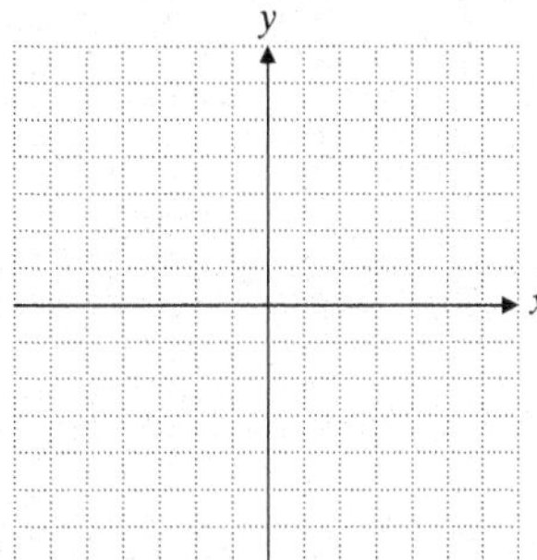

17. $f(x) = x^2$
$p(x) = (x + 3)^2$

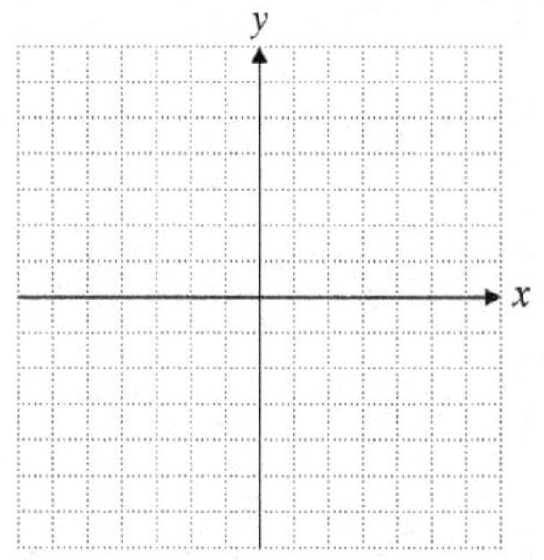

18. $f(x) = x^2$
$p(x) = (x - 2)^2$

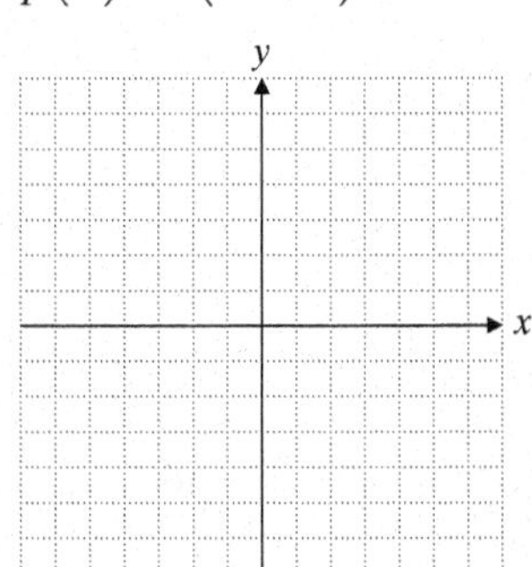

19. $f(x) = x^2$
$g(x) = (x - 2)^2 + 1$

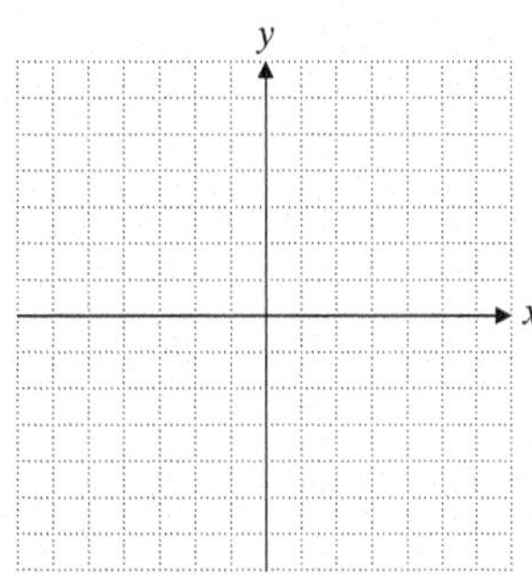

20. $f(x) = x^2$
$g(x) = (x + 1)^2 + 4$

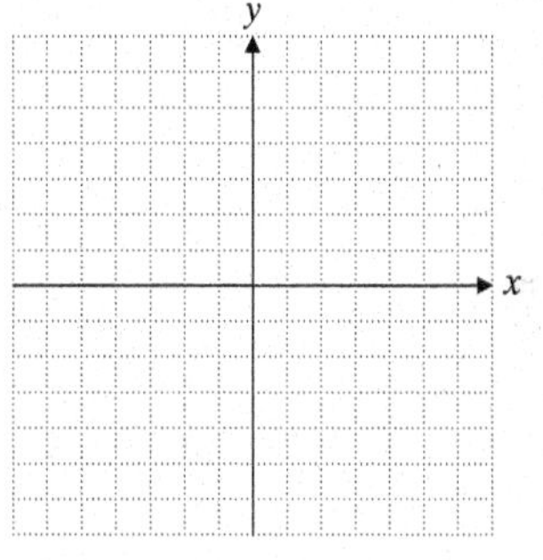

21. $f(x) = x^3$
$r(x) = x^3 - 1$

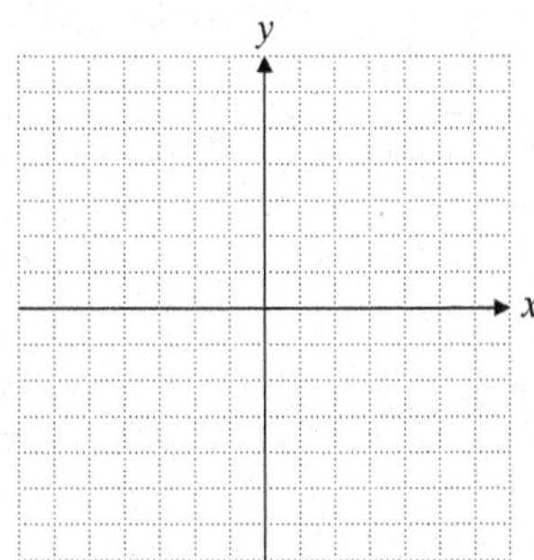

22. $f(x) = x^3$
$r(x) = x^3 + 2$

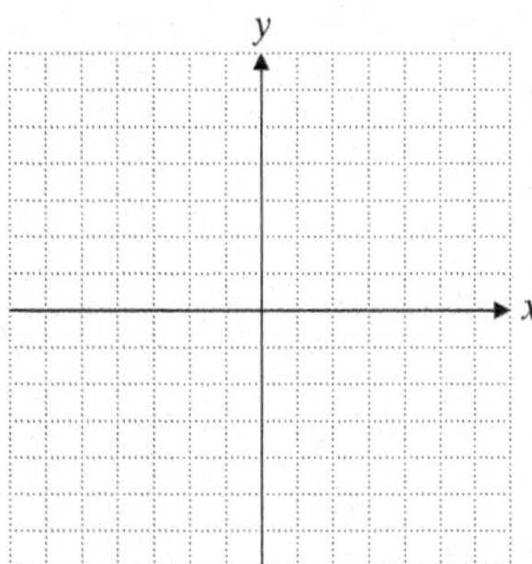

23. $f(x) = |x|$
$s(x) = |x + 4|$

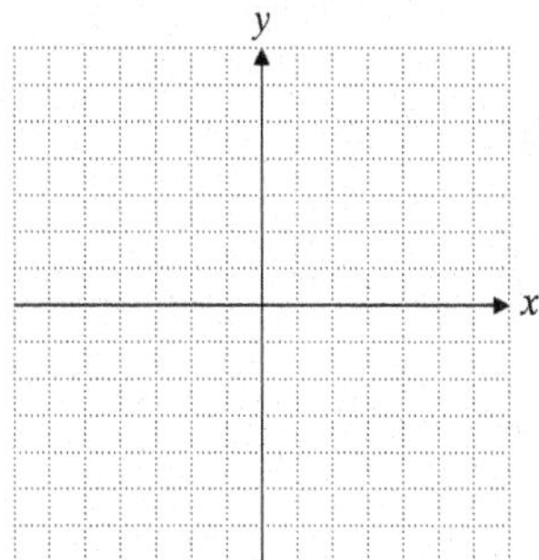

24. $f(x) = |x|$
$s(x) = |x - 5|$

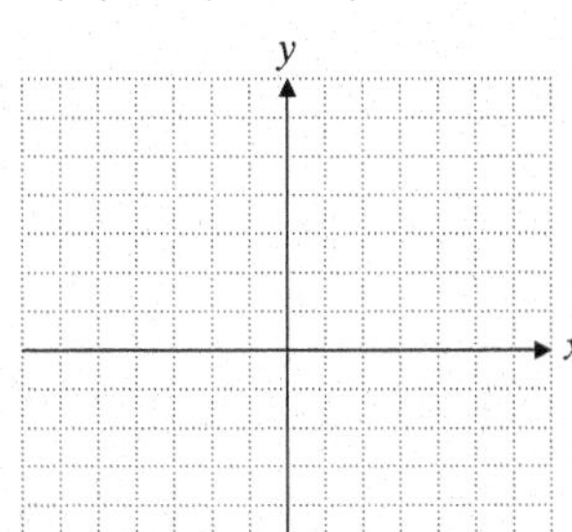

25. $f(x) = |x|$
$t(x) = |x - 3| - 4$

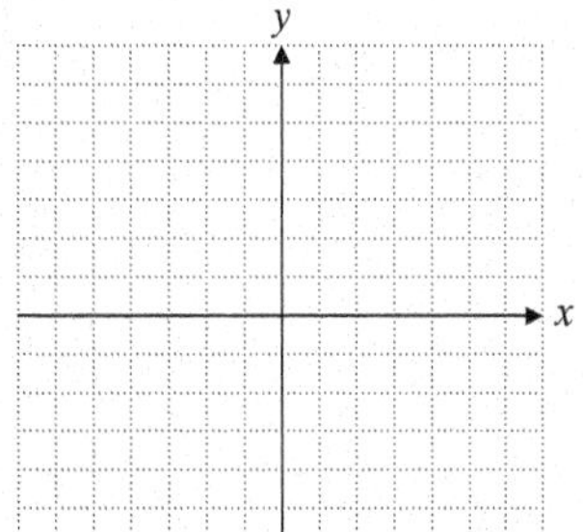

26. $f(x) = |x|$
$t(x) = |x + 2| - 5$

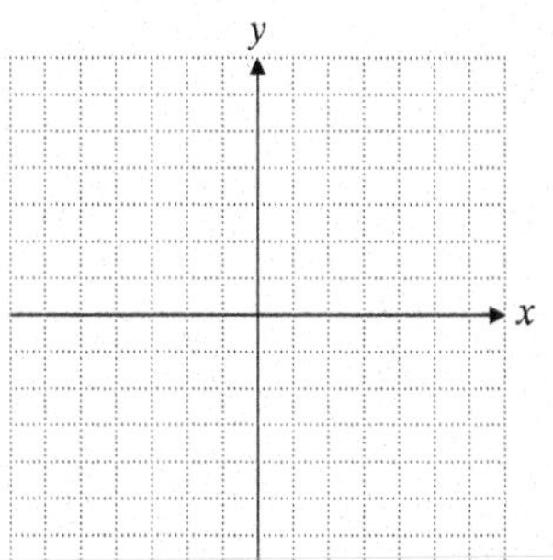

27. $f(x) = \dfrac{3}{x}$

$g(x) = \dfrac{3}{x} - 2$

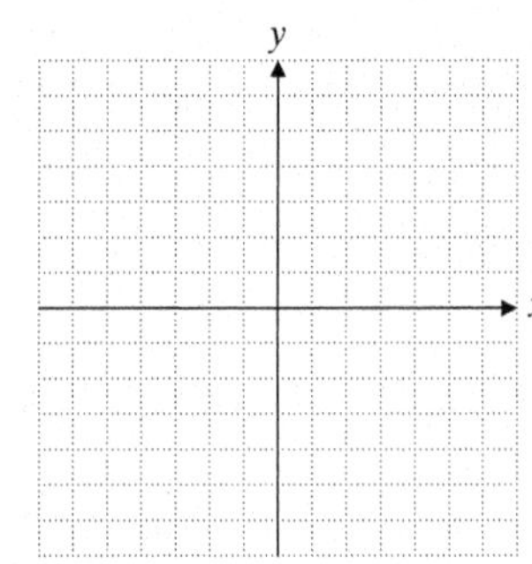

28. $f(x) = \dfrac{2}{x}$

$g(x) = \dfrac{2}{x} + 3$

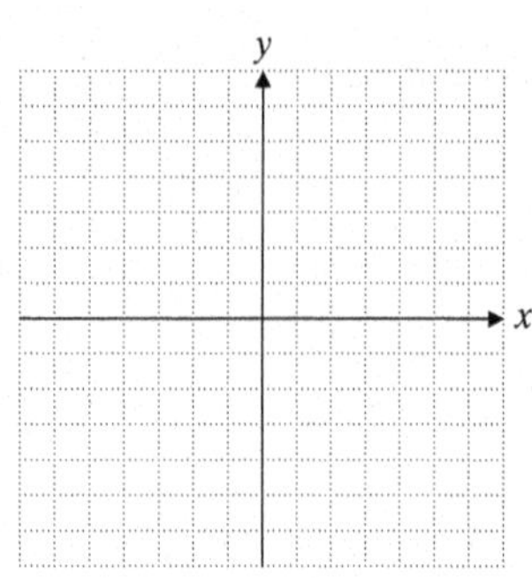

Cumulative Review *Simplify each expression. Assume that all variables are positive.*

29. [8.3.2] $\sqrt{12} + 3\sqrt{50} - 4\sqrt{27}$

30. [8.4.1] $(\sqrt{3x} - 1)^2$

31. [8.4.3] Rationalize the denominator. $\dfrac{\sqrt{5} - 2}{\sqrt{5} + 1}$

Quick Quiz 11.2

1. If $f(x) = x^2$ and $h(x) = (x - 3)^2$ were graphed on one coordinate plane, how would you describe where the graph of $h(x)$ is compared to the graph of $f(x)$?

2. If $f(x) = x^3$ and $g(x) = x^3 - 4$ were graphed on one coordinate plane, how would you describe where the graph of $g(x)$ is compared to the graph of $f(x)$?

3. Graph each pair of functions on one coordinate plane.

$f(x) = |x|, \quad k(x) = |x - 1| - 4$

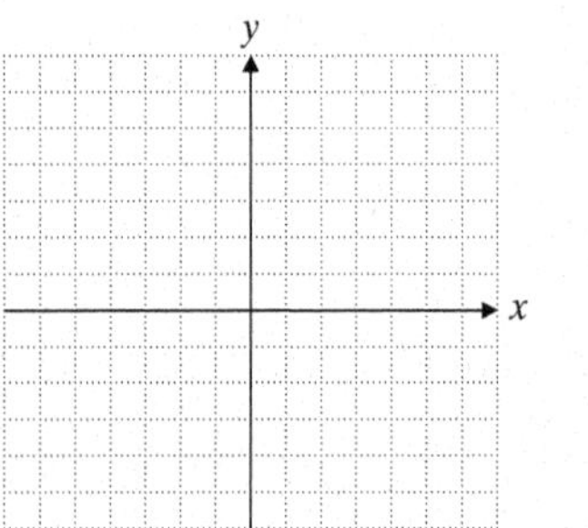

4. Concept Check Explain how you can use a vertical line to determine whether a graph represents a function.

Use Math to Save Money

Did You Know? You can save money by turning down your thermostat this winter.

Adjust the Thermostat

Understanding the Problem:

Mark heats his home with oil. He is interested in how much he needs to budget for the heating season and how changing the settings on his thermostat will affect that amount.

Making a Plan:

Mark needs to calculate the cost of his current oil usage and potential savings.

Step 1: During the winter of 2012 Mark paid $2.95 per gallon for home heating oil. The price has since increased to $3.55 per gallon during the winter of 2014.

Task 1: The oil company will only deliver 100 gallons of heating oil or more. Determine the cost of 100 gallons in December 2012 and November 2014.

Task 2: Mark knows he will need a 100-gallon heating oil delivery every month during the winter (November and December 2014 and January, February, March 2015). What can Mark expect to pay in home heating oil costs for the upcoming five months of winter?

Making a Decision

Step 2: Mark knows that for every 1 degree change in his thermostat setting, he can see a 2% change in his utility bills.

Task 3: Mark typically keeps his thermostat at 72°. How much will he save on his heating costs if he turns the thermostat down to 68° for the entire winter?

Task 4: How low would Mark have to set his thermostat to save one month's worth of heating costs ($355)?

Applying the Situation to Your Life:

If you lower your thermostat a few degrees, you too can realize savings on your heating bill. If you are not comfortable with the temperature set lower, you can still turn the heat down when you are away at work or school, as well as when you are sleeping. A programmable thermostat that will adjust the temperature depending on the time of day would make it easy to do that.

Task 5: At what temperature do you keep your thermostat?

Task 6: Calculate your savings if you lower the temperature of your thermostat during the winter.

11.3 Algebraic Operations on Functions

Student Learning Objectives

After studying this section, you will be able to:

1 Find the sum, difference, product, and quotient of two functions.

2 Find the composition of two functions.

1 Finding the Sum, Difference, Product, and Quotient of Two Functions

If f represents one function and g represents a second function, we can define new functions as follows:

THE SUM, DIFFERENCE, PRODUCT, AND QUOTIENT OF TWO FUNCTIONS

New functions can be formed by combining two given functions using algebraic operations.

Sum of Functions	$(f+g)(x) = f(x) + g(x)$
Difference of Functions	$(f-g)(x) = f(x) - g(x)$
Product of Functions	$(fg)(x) = f(x) \cdot g(x)$
Quotient of Functions	$\left(\frac{f}{g}\right)(x) = \frac{f(x)}{g(x)},\ g(x) \neq 0$

Example 1 Suppose that $f(x) = 3x^2 - 3x + 5$ and $g(x) = 5x - 2$.

(a) Find $(f+g)(x)$. **(b)** Evaluate $(f+g)(x)$ when $x = 3$.

Solution

(a)
$$\begin{aligned}(f+g)(x) &= f(x) + g(x)\\ &= (3x^2 - 3x + 5) + (5x - 2)\\ &= 3x^2 - 3x + 5 + 5x - 2\\ &= 3x^2 + 2x + 3\end{aligned}$$

(b) To evaluate $(f+g)(x)$ when $x = 3$, we write $(f+g)(3)$ and use the formula obtained in **(a)**.

$$\begin{aligned}(f+g)(x) &= 3x^2 + 2x + 3\\ (f+g)(3) &= 3(3)^2 + 2(3) + 3\\ &= 3(9) + 2(3) + 3\\ &= 27 + 6 + 3 = 36\end{aligned}$$

□

Student Practice 1 Given $f(x) = 4x + 5$ and $g(x) = 2x^2 + 7x - 8$, find the following:

(a) $(f+g)(x)$ **(b)** $(f+g)(4)$

Example 2 Given $f(x) = x^2 - 5x + 6$ and $g(x) = 2x - 1$, find the following:

(a) $(fg)(x)$ **(b)** $(fg)(-4)$

Solution

(a)
$$\begin{aligned}(fg)(x) &= f(x) \cdot g(x)\\ &= (x^2 - 5x + 6)(2x - 1)\\ &= 2x^3 - 10x^2 + 12x - x^2 + 5x - 6\\ &= 2x^3 - 11x^2 + 17x - 6\end{aligned}$$

(b) To evaluate $(fg)(x)$ when $x = -4$, we write $(fg)(-4)$ and use the formula obtained in **(a)**.

$$\begin{aligned}(fg)(x) &= 2x^3 - 11x^2 + 17x - 6\\ (fg)(-4) &= 2(-4)^3 - 11(-4)^2 + 17(-4) - 6\\ &= 2(-64) - 11(16) + 17(-4) - 6\\ &= -128 - 176 - 68 - 6\\ &= -378\end{aligned}$$

□

Student Practice 2 Given $f(x) = 3x + 2$ and $g(x) = x^2 - 3x - 4$, find the following:

(a) $(fg)(x)$ **(b)** $(fg)(2)$

When finding the quotient of a function, we must be careful to avoid division by zero. Thus, we always specify any values of x that must be eliminated from the domain.

Example 3 Given $f(x) = 3x + 1$, $g(x) = 2x - 1$, and $h(x) = 9x^2 + 6x + 1$, find the following:

(a) $\left(\frac{f}{g}\right)(x)$ **(b)** $\left(\frac{f}{h}\right)(x)$ **(c)** $\left(\frac{f}{h}\right)(-2)$

Solution

(a) $\left(\frac{f}{g}\right)(x) = \frac{3x + 1}{2x - 1}$

The denominator of the quotient can never be zero. Since $2x - 1 \neq 0$, we know that $x \neq \frac{1}{2}$.

(b) $\left(\frac{f}{h}\right)(x) = \frac{3x + 1}{9x^2 + 6x + 1} = \frac{3x + 1}{(3x + 1)(3x + 1)} = \frac{1}{3x + 1}$

Since $3x + 1 \neq 0$, we know that $x \neq -\frac{1}{3}$.

(c) To find $\left(\frac{f}{h}\right)(-2)$, we must evaluate $\left(\frac{f}{h}\right)(x)$ when $x = -2$.

$$\left(\frac{f}{h}\right)(x) = \frac{1}{3x + 1}$$

$$\left(\frac{f}{h}\right)(-2) = \frac{1}{(3)(-2) + 1} = \frac{1}{-6 + 1} = -\frac{1}{5}$$

□

Student Practice 3 Given $p(x) = 5x^2 + 6x + 1$, $h(x) = 3x - 2$, and $g(x) = 5x + 1$, find the following. (Be careful to find the two values for x in part b that must be eliminated from the domain.)

(a) $\left(\frac{g}{h}\right)(x)$ **(b)** $\left(\frac{g}{p}\right)(x)$ **(c)** $\left(\frac{g}{h}\right)(3)$

2 Finding the Composition of Two Functions

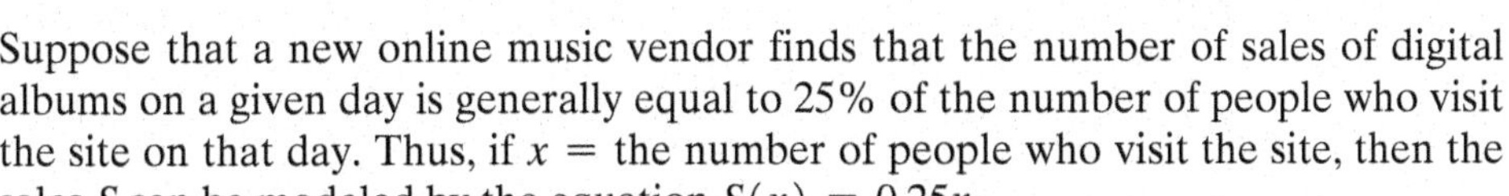

Suppose that a new online music vendor finds that the number of sales of digital albums on a given day is generally equal to 25% of the number of people who visit the site on that day. Thus, if $x =$ the number of people who visit the site, then the sales S can be modeled by the equation $S(x) = 0.25x$.

Suppose that the average album on the site sells for \$15. Then if $S =$ the number of albums sold on a given day, the income for that day can be modeled by the equation $P(S) = 15S$. Suppose that eight thousand people visit the site.

$$S(x) = 0.25x$$
$$S(8000) = 0.25(8000) = 2000$$

Thus, two thousand albums would be sold.

If two thousand albums were sold and the average price of an album is \$15, then we would have the following:

$$P(S) = 15S$$
$$P(2000) = 15(2000) = 30{,}000$$

That is, the income from the sales of albums would be \$30,000.

Continued on next page

Let us analyze the functions we have described and record a few values of x, $S(x)$, and $P(S)$.

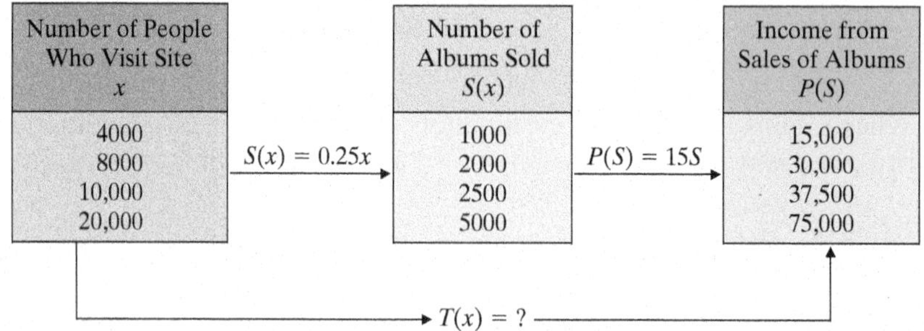

Is there a function $T(x)$ that describes the income from album sales as a function of x, the number of people who visit the site?

The number of albums sold is

$$S(x) = 0.25x.$$

Thus, $0.25x$ is the number of albums sold.

If we replace S in $P(S) = 15S$ by $S(x)$, we have

$$P[S(x)] = P(0.25x) = 15(0.25x) = 3.75x.$$

Thus, the formula $T(x)$ that describes the income in terms of the number of visitors is

$$T(x) = 3.75x.$$

Is this correct? Let us check by finding $T(20{,}000)$. From our table the result should be 75,000.

If $\quad T(x) = 3.75x,$

then $\quad T(20{,}000) = 3.75(20{,}000) = 75{,}000.$

Thus, we have found a function T that is the composition of the functions P and S: $T(x) = P[S(x)]$.

We now state a definition of the composition of one function with another.

The **composition** of the functions f and g, denoted $f \circ g$, is defined as follows: $(f \circ g)(x) = f[g(x)]$. The domain of $f \circ g$ is the set of all x-values in the domain of g such that $g(x)$ is in the domain of f.

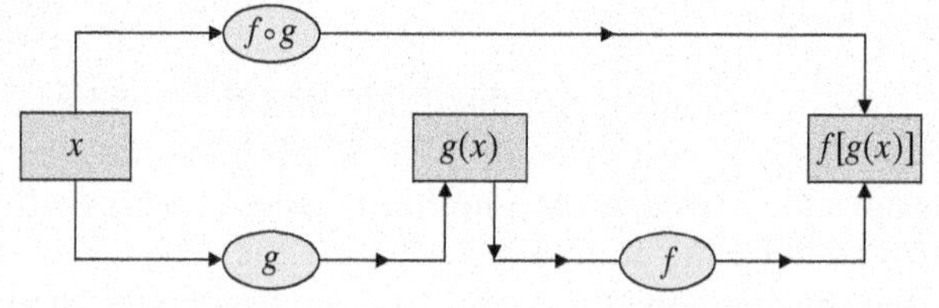

Example 4 Given $f(x) = 3x - 2$ and $g(x) = 2x + 5$, find $f[g(x)]$.

Solution

$f[g(x)]$	$= f(2x + 5)$	Substitute $g(x) = 2x + 5$.
	$= 3(2x + 5) - 2$	Apply the formula for $f(x)$.
	$= 6x + 15 - 2$	Remove parentheses.
	$= 6x + 13$	Simplify.

Student Practice 4 Given $f(x) = 2x - 1$ and $g(x) = 3x - 4$, find $f[g(x)]$.

In most situations $f[g(x)]$ and $g[f(x)]$ are not the same.

Example 5 Given $f(x) = \sqrt{x-4}$ with $x \geq 4$ and $g(x) = 3x + 1$, find the following:

(a) $f[g(x)]$ **(b)** $g[f(x)]$

Solution

(a) $f[g(x)] = f(3x+1)$ Substitute $g(x) = 3x + 1$.

$= \sqrt{(3x+1)-4}$ Apply the formula for $f(x)$.

$= \sqrt{3x+1-4}$ Remove parentheses.

$= \sqrt{3x-3}$ Simplify.

(b) $g[f(x)] = g\left(\sqrt{x-4}\right)$ Substitute $f(x) = \sqrt{x-4}$.

$= 3\left(\sqrt{x-4}\right) + 1$ Apply the formula for $g(x)$.

$= 3\sqrt{x-4} + 1$ Remove parentheses.

We note that $g[f(x)] \neq f[g(x)]$. □

Student Practice 5 Given $f(x) = 2x^2 - 3x + 1$ and $g(x) = x + 2$, find the following:

(a) $f[g(x)]$ **(b)** $g[f(x)]$

Graphing Calculator

Composition of Functions

You can evaluate the composition of functions on most graphing calculators by using the y-variable function (Y-VARS). To do Example 6 on most graphing calculators, you would use the following equations.

$$y_1 = \frac{1}{3x-4}$$

$$y_2 = 2(y_1)$$

To find the function value, you can use the TableSet command to let $x = 2$. Then enter Table and you will see displayed $y_1 = 0.5$, which represents $g(2) = 0.5$, and $y_2 = 1$, which represents $f[g(2)] = 1$.

MC **Example 6** Given $f(x) = 2x$ and $g(x) = \frac{1}{3x-4}$, $x \neq \frac{4}{3}$, find the following:

(a) $(f \circ g)(x)$

(b) $(f \circ g)(2)$

Solution

(a) $(f \circ g)(x) = f[g(x)] = f\left(\frac{1}{3x-4}\right)$ Substitute $g(x) = \frac{1}{3x-4}$.

$= 2\left(\frac{1}{3x-4}\right)$ Apply the formula for $f(x)$.

$= \frac{2}{3x-4}$ Simplify.

(b) $(f \circ g)(2) = \frac{2}{3(2)-4} = \frac{2}{6-4} = \frac{2}{2} = 1$ □

Student Practice 6

Given $f(x) = 3x + 1$ and $g(x) = \frac{2}{x-3}$, $x \neq 3$, find the following:

(a) $(g \circ f)(x)$

(b) $(g \circ f)(-3)$

11.3 Exercises MyMathLab®

For the following functions, find **(a)** $(f+g)(x)$, **(b)** $(f-g)(x)$, **(c)** $(f+g)(2)$, *and* **(d)** $(f-g)(-1)$.

1. $f(x) = -2x + 3,\ g(x) = 2 + 4x$

2. $f(x) = 2x + 1,\ g(x) = 2 - 3x$

3. $f(x) = 2x^2 - 1,\ g(x) = 4x + 1$

4. $f(x) = 3 - x,\ g(x) = x^2 - 5x + 4$

5. $f(x) = x^3 - \frac{1}{2}x^2 + x,\ g(x) = x^2 - \frac{x}{4} - 5$

6. $f(x) = 4.5x^2 - x + 1.2,\ g(x) = 1.5x^3 - 0.4x$

7. $f(x) = -5\sqrt{x+6}, x \ge -6,\ g(x) = 8\sqrt{x+6}, x \ge -6$

8. $f(x) = -4\sqrt{2-x},\ x \le 2,\ g(x) = -\sqrt{2-x}, x \le 2$

For the following functions, find **(a)** $(fg)(x)$ *and* **(b)** $(fg)(-3)$.

9. $f(x) = 2x - 3,\ g(x) = -2x^2 - 3x + 1$

10. $f(x) = x^2 - 3x + 2,\ g(x) = 1 - x$

11. $f(x) = \frac{2}{x^2},\ x \ne 0,\ g(x) = x^2 - x$

12. $f(x) = \frac{3x}{x+4}, x \ne -4,\ g(x) = \frac{x}{3}$

13. $f(x) = \sqrt{-2x+1},\ x \le \frac{1}{2},\ g(x) = -3x$

14. $f(x) = 4x,\ g(x) = \sqrt{3x+10}, x \ge -\frac{10}{3}$

For the following functions, find **(a)** $\left(\frac{f}{g}\right)(x)$ *and* **(b)** $\left(\frac{f}{g}\right)(2)$.

15. $f(x) = x - 6,\ g(x) = 3x$

16. $f(x) = 3x,\ g(x) = 4x - 1$

17. $f(x) = x^2 - 1,\ g(x) = x - 1$

18. $f(x) = x,\ g(x) = x^2 - 5x$

19. $f(x) = x^2 + 10x + 25,\ g(x) = x + 5$

20. $f(x) = 2x^2 + 5x - 12,\ g(x) = x + 4$

21. $f(x) = 4x - 1,\ g(x) = 4x^2 + 7x - 2$

22. $f(x) = 3x + 2,\ g(x) = 3x^2 - x - 2$

Let $f(x) = 3x + 2,\ g(x) = x^2 - 2x$, *and* $h(x) = \frac{x-2}{3}$. *Find the following:*

23. $(f-g)(x)$

24. $(g-f)(x)$

25. $\left(\frac{g}{h}\right)(x)$

26. $\left(\frac{g}{h}\right)(-2)$

27. $(fg)(-1)$

28. $(gh)(3)$

29. $\left(\frac{g}{f}\right)(-1)$

30. $\left(\frac{g}{f}\right)(x)$

Find $f[g(x)]$ for each of the following:

31. $f(x) = 2 - 3x, g(x) = 2x + 5$

32. $f(x) = 3x + 2, g(x) = 4x - 1$

33. $f(x) = 2x^2 + 5, g(x) = x - 1$

34. $f(x) = 9 + x^2, g(x) = x - 3$

35. $f(x) = 8 - 5x, g(x) = x^2 + 3$

36. $f(x) = 4 - x, g(x) = 1 - 3x + x^2$

37. $f(x) = \frac{7}{2x - 3}, g(x) = x + 2$

38. $f(x) = \frac{15}{1 - x}, g(x) = x - 5$

39. $f(x) = |x + 3|, g(x) = 2x - 1$

40. $f(x) = \left|-\frac{1}{2}x\right|, g(x) = 2x + 10$

Let $f(x) = x^2 + 2, g(x) = 3x + 5, h(x) = \frac{1}{x}$, and $p(x) = \sqrt{x - 1}$. Find each of the following:

41. $f[g(x)]$

42. $g[h(x)]$

43. $g[f(x)]$

44. $h[g(x)]$

45. $g[f(0)]$

46. $h[g(0)]$

47. $(p \circ f)(x)$

48. $(f \circ h)(x)$

49. $(g \circ h)(\sqrt{2})$

50. $(f \circ p)(x)$

51. $(p \circ f)(-3)$

52. $(f \circ p)\left(\frac{3}{2}\right)$

Applications

53. ***Temperature Scales*** Consider the Celsius function $C(F) = \frac{5F - 160}{9}$, which converts degrees Fahrenheit to degrees Celsius. A different temperature scale, called the Kelvin scale, is used by many scientists in their research. The Kelvin scale is similar to the Celsius scale, but it begins at absolute zero (the coldest possible temperature, which is around $-273°C$). To convert a Celsius temperature to a temperature on the Kelvin scale, we use the function $K(C) = C + 273$. Find $K[C(F)]$, which is the composite function that defines the temperature in Kelvins in terms of the temperature in degrees Fahrenheit.

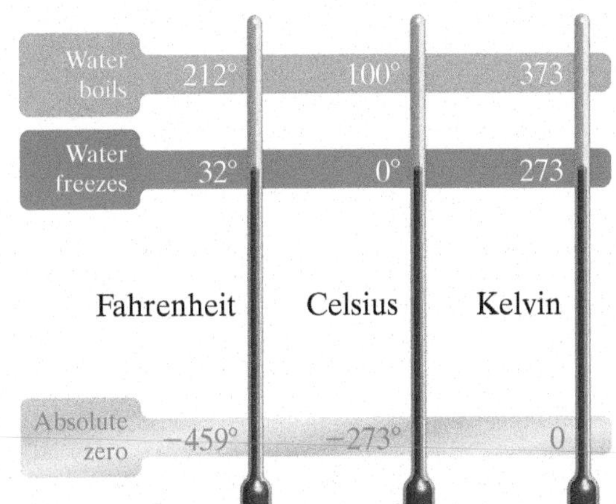

54. ***Business*** Suppose the dollar cost to produce n items in a factory is $c(n) = 5n + 4$. Furthermore, the number of items n produced in x hours is $n(x) = 3x$. Find $c[n(x)]$, which is the composite function that defines the dollar cost in terms of the number of hours of production x.

▲ **55.** ***Oil Slicks*** An oil tanker with a ruptured hull is leaking oil off the coast of Africa. There is no wind or significant current, so the oil slick is spreading in a circle whose radius is defined by the function $r(t) = 3t$, where t is the time in minutes since the tanker began to leak. The area of the slick for any given radius is approximately determined by the function $a(r) = 3.14r^2$, where r is the radius of the circle measured in feet. Find $a[r(t)]$, which is the composite function that defines the area of the oil slick in terms of the minutes t since the beginning of the leak. How large is the area after 20 minutes?

Cumulative Review *Factor each of the following:*

56. **[6.5.2]** $36x^2 - 12x + 1$

57. **[6.5.1]** $25x^4 - 1$

58. **[6.5.1]** $x^4 - 10x^2 + 9$

59. **[6.4.1]** $3x^2 - 7x + 2$

Quick Quiz 11.3

1. If $f(x) = 2x^2 - 4x - 8$ and $g(x) = -3x^2 + 5x - 2$, find $(f - g)(x)$.

2. If $f(x) = x^2 - 3$ and $g(x) = \frac{x - 4}{2}$, find $f[g(x)]$.

3. If $f(x) = x - 7$ and $g(x) = 2x - 5$, find $\left(\frac{g}{f}\right)(2)$.

4. **Concept Check** If $f(x) = 5 - 2x^2$ and $g(x) = 3x^2 - 5x + 1$, explain how you would find $(f - g)(-4)$.

11.4 Inverse of a Function

Americans driving in Canada or Mexico need to be able to convert miles per hour to kilometers per hour and vice versa.

Student Learning Objectives

After studying this section, you will be able to:

1. Determine whether a function is a one-to-one function.
2. Find the inverse function for a given function.
3. Graph a function and its inverse function.

If someone is driving at 55 miles per hour, how fast is he or she going in kilometers per hour?

Approximate Value in Miles per Hour	Approximate Value in Kilometers per Hour
35	56
40	64
45	72
50	80
55	88
60	96
65	104

A function f that converts from miles per hour to an approximate value in kilometers per hour is $f(x) = 1.6x$.

For example, $f(40) = 1.6(40) = 64$.

This tells us that 40 miles per hour is approximately equivalent to 64 kilometers per hour.

We can come up with a function that does just the opposite—that is, that converts kilometers per hour to an approximate value in miles per hour. This function is $f^{-1}(x) = 0.625x$.

For example, $f^{-1}(64) = 0.625(64) = 40$.

This tells us that 64 kilometers per hour is approximately equivalent to 40 miles per hour.

Miles per hour		**Kilometers per hour**
40 ⟶	$f(x) = 1.6x$	⟶ 64
40 ⟵	$f^{-1}(x) = 0.625x$	⟵ 64

We call a function f^{-1} that reverses the domain and range of a function f the **inverse function *f*.**

Most American cars have numbers showing kilometers per hour in smaller print on the car speedometer. Unfortunately, these numbers are usually hard to read. If we made a list of several function values of f and several inverse function values of f^{-1}, we could create a conversion scale like the one below that we could use if we should travel to Mexico or Canada with an American car.

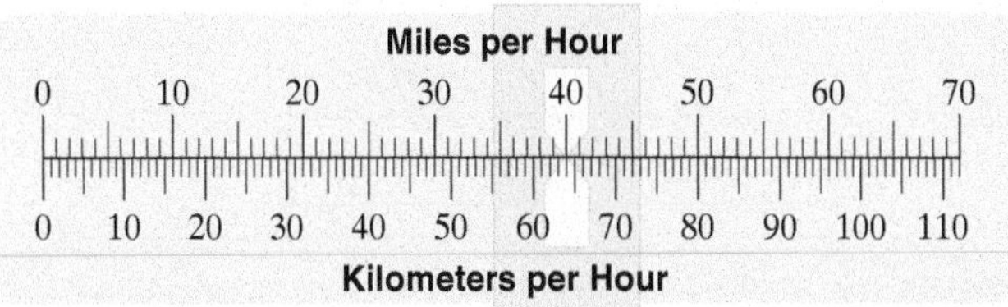

Continued on next page

The original function that we studied converts miles per hour to kilometers per hour. The corresponding inverse function converts kilometers per hour to miles per hour. How do we find inverse functions? Do all functions have inverse functions? These are questions we will explore in this section.

1 Determining Whether a Function Is a One-to-One Function

First we state that not all functions have inverse functions. To have an inverse that is a function, a function must be one-to-one. This means that for every value of y, there is only one value of x. Or, in the language of ordered pairs, no ordered pairs have the same second coordinate.

DEFINITION OF A ONE-TO-ONE FUNCTION

A **one-to-one function** is a function in which no ordered pairs have the same second coordinate.

To Think About: *Relationship of One-to-One and Inverses* Why must a function be one-to-one in order to have an inverse that is a function?

Example 1 Indicate whether the following are one-to-one functions.

(a) $M = \{(1, 3), (2, 7), (5, 8), (6, 12)\}$

(b) $P = \{(1, 4), (2, 9), (3, 4), (4, 18)\}$

Solution

(a) M is a function because no ordered pairs have the same first coordinate. M is also a one-to-one function because no ordered pairs have the same second coordinate.

(b) P is a function, but it is not one-to-one because the ordered pairs (1, 4) and (3, 4) have the same second coordinate. □

Student Practice 1

(a) Is the function $A = \{(-2, -6), (-3, -5), (-1, 2), (3, 5)\}$ one-to-one?

(b) Is the function $B = \{(0, 0), (1, 1), (2, 4), (3, 9), (-1, 1)\}$ one-to-one?

By examining the graph of a function, we can quickly tell whether it is one-to-one. If any horizontal line crosses the graph of a function in more than one place, the function is not one-to-one. If no such line exists, then the function is one-to-one.

HORIZONTAL LINE TEST

If any horizontal line intersects the graph of a function more than once, the function is not one-to-one. If no such line exists, the function is one-to-one.

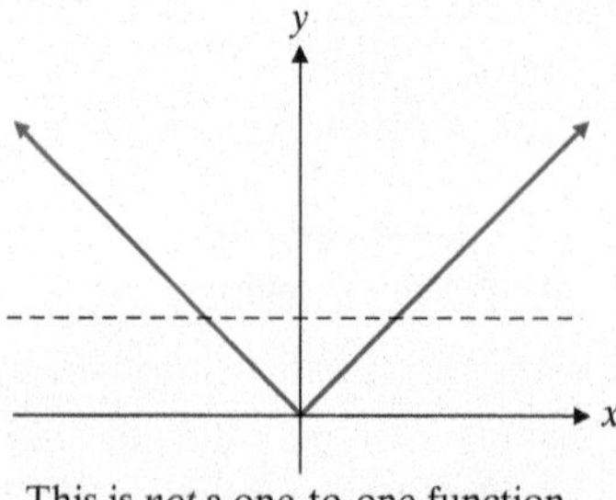

This is *not* a one-to-one function.

Example 2 Determine whether the functions graphed are one-to-one functions.

(a)

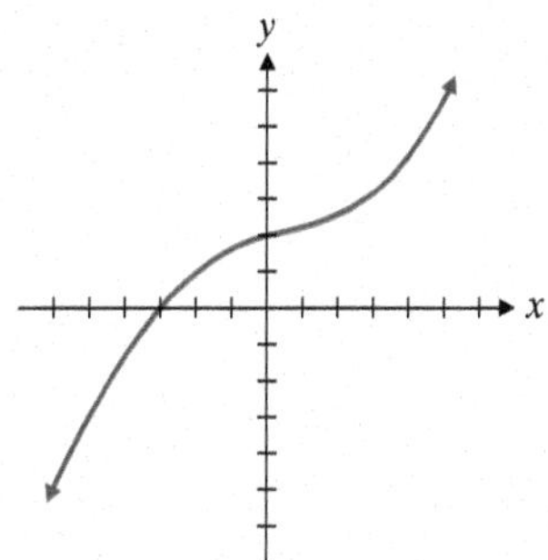

(b)

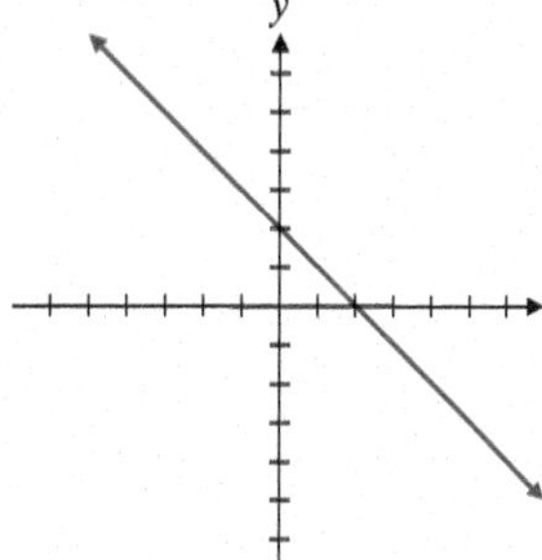

(c)

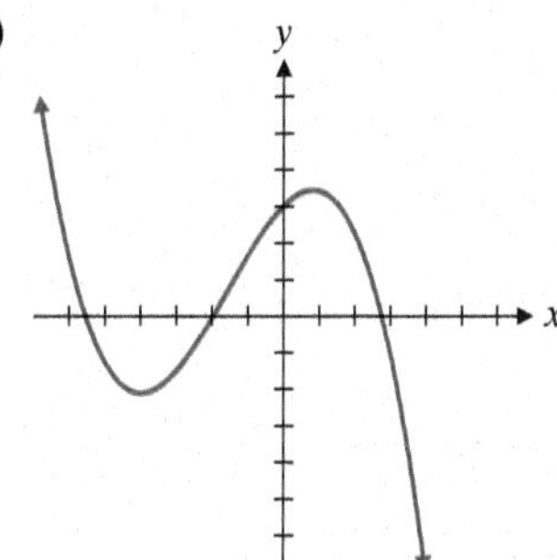

(d)

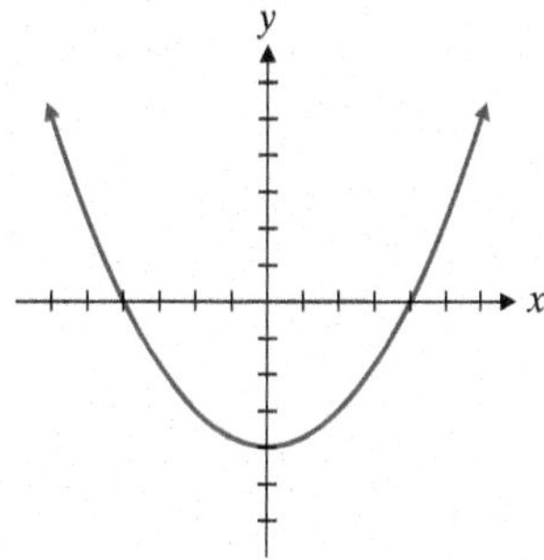

Solution The graphs of **(a)** and **(b)** represent one-to-one functions. Horizontal lines cross the graphs at most once.

(a)

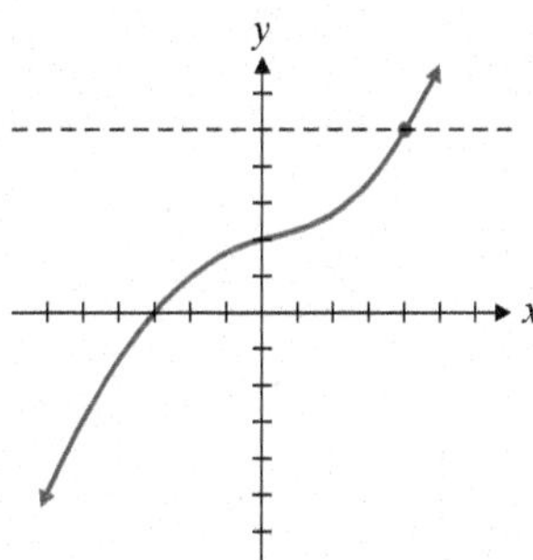

(b)

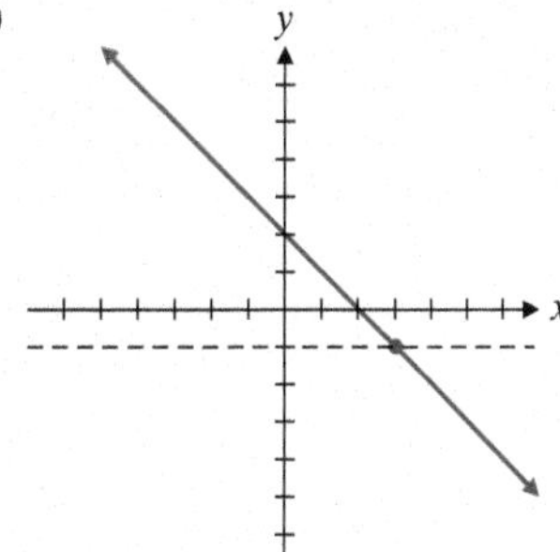

The graphs of **(c)** and **(d)** do not represent one-to-one functions. A horizontal line exists that crosses the graphs more than once.

(a)

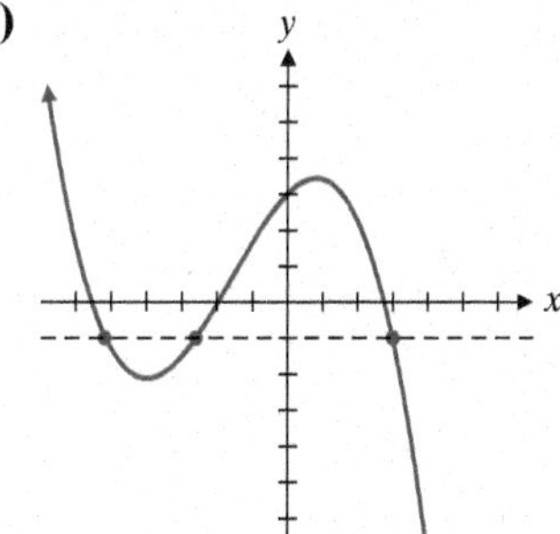

(b)

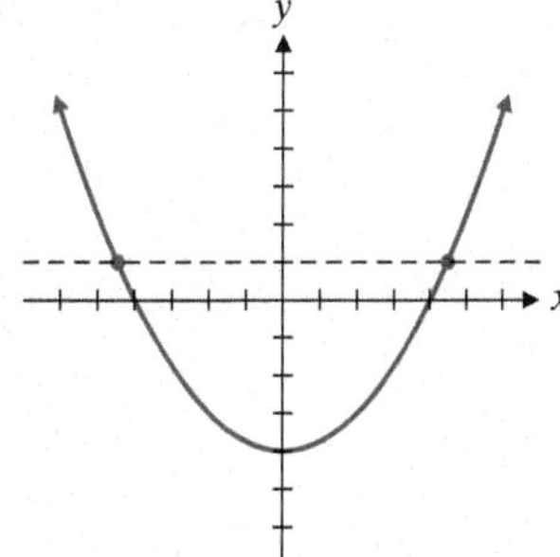

□

Student Practice 2 Do each of the following graphs of a function represent a one-to-one function? Why or why not?

(a)

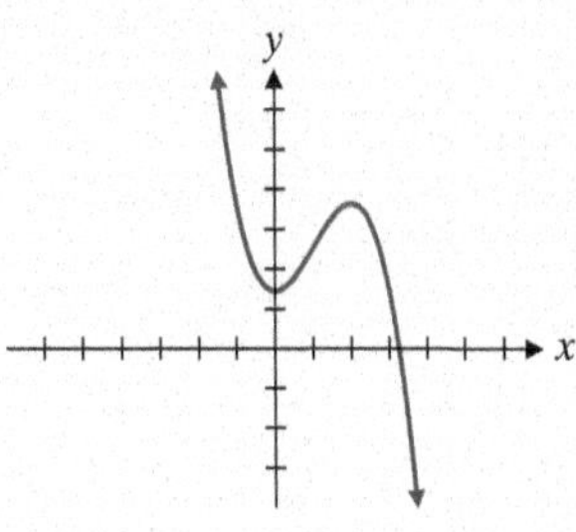

(b)

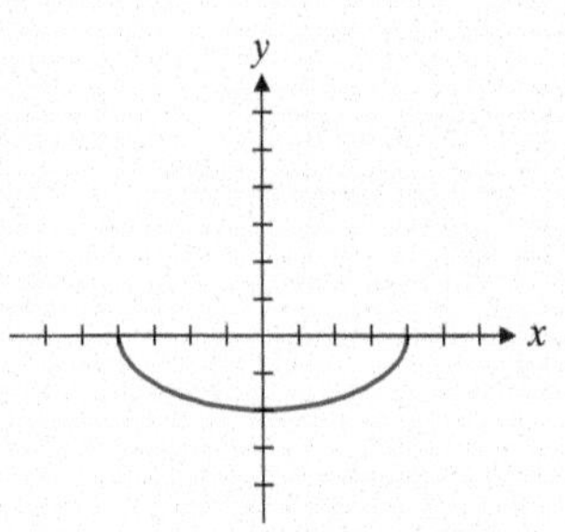

2 Finding the Inverse Function for a Given Function

How do we find the inverse of a function? If we have a list of ordered pairs, we simply interchange the coordinates of each ordered pair. In Example 1, we said that M has an inverse. What is it? Here is M again.

$$M = \{(1, 3), (2, 7), (5, 8), (6, 12)\}$$

The inverse of M, written M^{-1}, is

$$M^{-1} = \{(3, 1), (7, 2), (8, 5), (12, 6)\}.$$

Now do you see why a function must be one-to-one in order to have an inverse that is a function? Let's look at the function P from Example 1.

$$P = \{(1, 4), (2, 9), (3, 4), (4, 18)\}$$

If P had an inverse, it would be

$$P^{-1} = \{(4, 1), (9, 2), (4, 3), (18, 4)\}.$$

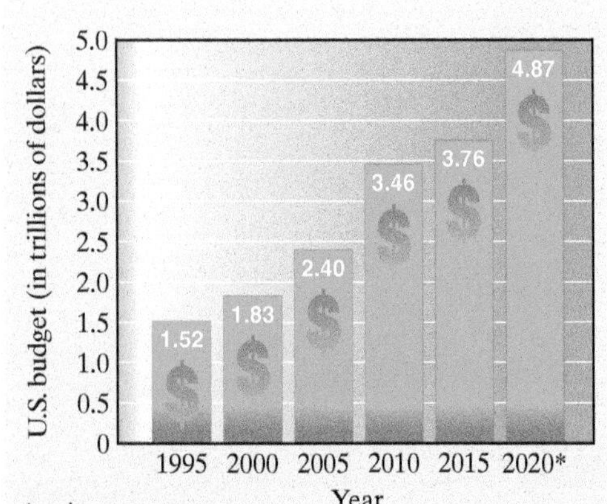

*estimate
Source: www.whitehouse.gov

But we have two ordered pairs with the same first coordinate. Therefore, P^{-1} is not a function (in other words, the inverse function does not exist).

A number of real-world situations are described by functions that have inverses. Consider the function that is defined by the ordered pairs (year, U.S. budget in trillions of dollars). Some function values are

$$F = \{(2020, 4.87), (2015, 3.76), (2010, 3.46), (2005, 2.40), (2000, 1.83), (1995, 1.52)\}.$$

In this case the inverse of the function is

$$F^{-1} = \{(4.87, 2020), (3.76, 2015), (3.46, 2010), (2.40, 2005), (1.83, 2000), (1.52, 1995)\}.$$

By the way, F^{-1} does *not* mean $\frac{1}{F}$. Here the -1 simply means "inverse."

Example 3 Determine the inverse function of the function

$$F = \{(6, 1), (12, 2), (13, 5), (14, 6)\}.$$

Solution The inverse function of F is

$$F^{-1} = \{(1, 6), (2, 12), (5, 13), (6, 14)\}.$$ □

Student Practice 3 Find the inverse of the one-to-one function.
$B = \{(1, 2), (7, 8), (8, 7), (10, 12)\}$

Suppose that a function is given in the form of an equation. How do we find the inverse? Since, by definition, we interchange the ordered pairs to find the inverse of a function, this means that the x-values of the function become the y-values of the inverse function and vice versa.

Four steps will help us find the inverse of a one-to-one function when we are given its equation.

FINDING THE INVERSE OF A ONE-TO-ONE FUNCTION

1. Replace $f(x)$ with y.
2. Interchange x and y.
3. Solve for y in terms of x.
4. Replace y with $f^{-1}(x)$.

Example 4 Find the inverse of $f(x) = 7x - 4$.

Solution

Step 1 $y = 7x - 4$ Replace $f(x)$ with y.

Step 2 $x = 7y - 4$ Interchange the variables x and y.

Step 3 $x + 4 = 7y$ Solve for y in terms of x.

$$\frac{x + 4}{7} = y$$

Step 4 $f^{-1}(x) = \dfrac{x + 4}{7}$ Replace y with $f^{-1}(x)$. □

 **Student Practice 4** Find the inverse of the function $g(x) = 4 - 6x$.

Let's see whether this technique works on a situation similar to the opening example in which we converted miles per hour to approximate values in kilometers per hour.

Example 5 Find the inverse function of $f(x) = \frac{9}{5}x + 32$, which converts Celsius temperature (x) into equivalent Fahrenheit temperature.

Solution

Step 1 $y = \dfrac{9}{5}x + 32$ Replace $f(x)$ with y.

Step 2 $x = \dfrac{9}{5}y + 32$ Interchange x and y.

Step 3 $5(x) = 5\left(\dfrac{9}{5}\right)y + 5(32)$ Solve for y in terms of x.

$$5x = 9y + 160$$

$$5x - 160 = 9y$$

$$\frac{5x - 160}{9} = \frac{9y}{9}$$

$$\frac{5x - 160}{9} = y$$

Step 4 $f^{-1}(x) = \dfrac{5x - 160}{9}$ Replace y with $f^{-1}(x)$.

Continued on next page

Graphing Calculator

Inverse Functions

If f and g are inverse functions, then their graphs will be symmetric about the line $y = x$. Use a square window setting when graphing. For Example 6, graph

$$y_1 = 3x - 2, y_2 = \frac{x+2}{3},$$

and $y_3 = x$.

Display:

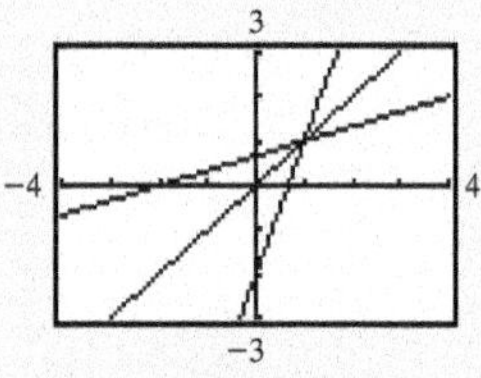

Note: Our inverse function $f^{-1}(x)$ will now convert Fahrenheit temperature to Celsius temperature. For example, suppose we wanted to know the Celsius temperature that corresponds to 86°F.

$$f^{-1}(86) = \frac{5(86) - 160}{9} = \frac{270}{9} = 30$$

This tells us that a temperature of 86°F corresponds to a temperature of 30°C. □

Student Practice 5 Find the inverse function of $f(x) = 0.75 + 0.55(x - 1)$, which gives the cost of a telephone call for any call over 1 minute if the telephone company charges 75 cents for the first minute and 55 cents for each minute thereafter. Here x = the number of minutes.

3 Graphing a Function and Its Inverse Function

The graph of a function and its inverse are symmetric about the line $y = x$. Why do you think that this is so?

Example 6 If $f(x) = 3x - 2$, find $f^{-1}(x)$. Graph f and f^{-1} on the same set of axes. Draw the line $y = x$ as a dashed line for reference.

Solution

$$f(x) = 3x - 2$$
$$y = 3x - 2$$
$$x = 3y - 2$$
$$x + 2 = 3y$$
$$\frac{x+2}{3} = y$$
$$f^{-1}(x) = \frac{x+2}{3}$$

Now we graph each line.

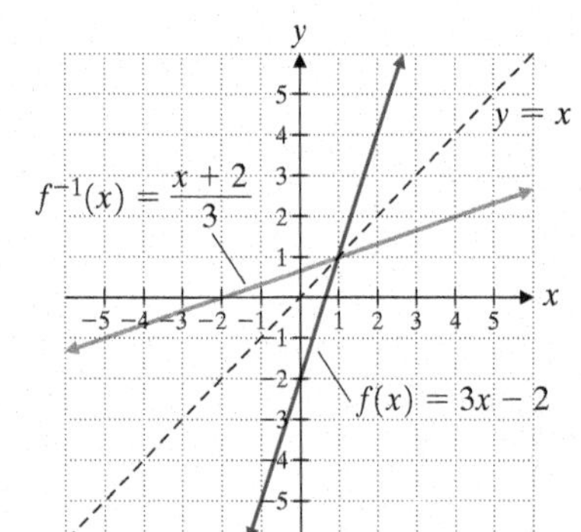

We see that the graphs of f and f^{-1} are symmetric about the line $y = x$. If we folded the graph paper along the line $y = x$, the graph of f would touch the graph of f^{-1}. Try it. Redraw the functions on a separate piece of graph paper. Fold the graph paper on the line $y = x$. □

Student Practice 6

If $f(x) = -\frac{1}{4}x + 1$, find $f^{-1}(x)$. Graph f and f^{-1} on the same coordinate plane. Draw the line $y = x$ as a dashed line for reference.

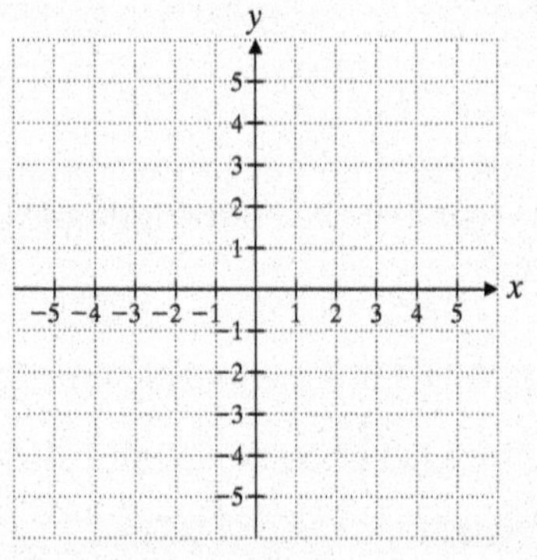

11.4 Exercises

Verbal and Writing Skills, Exercises 1–6

Complete the following:

1. A one-to-one function is a function in which no ordered pairs ______________________.

2. If any horizontal line intersects the graph of a function more than once, the function ______________.

3. The graphs of a function f and its inverse f^{-1} are symmetric about the line ________.

4. Do all functions have inverse functions? Why or why not?

5. Does the graph of a horizontal line represent a function? Why or why not? Does it represent a one-to-one function? Explain.

6. Does the graph of a vertical line represent a function? Why or why not? Does it represent a one-to-one function? Explain.

Indicate whether each function is one-to-one.

7. $B = \{(0, 1), (1, 0), (10, 0)\}$

8. $A = \{(-6, -2), (6, 2), (3, 4)\}$

9. $F = \{(\frac{2}{3}, 2), (3, -\frac{4}{5}), (-\frac{2}{3}, -2), (-3, \frac{4}{5})\}$

10. $C = \{(12, 3), (-6, 1), (6, 3)\}$

11. $E = \{(2, 3.5), (-1, 8), (10, 3.5), (0, -8)\}$

12. $F = \{(0, -3), (3, 2), (-2, 3), (-1, -2)\}$

Indicate whether each graph represents a one-to-one function.

13.

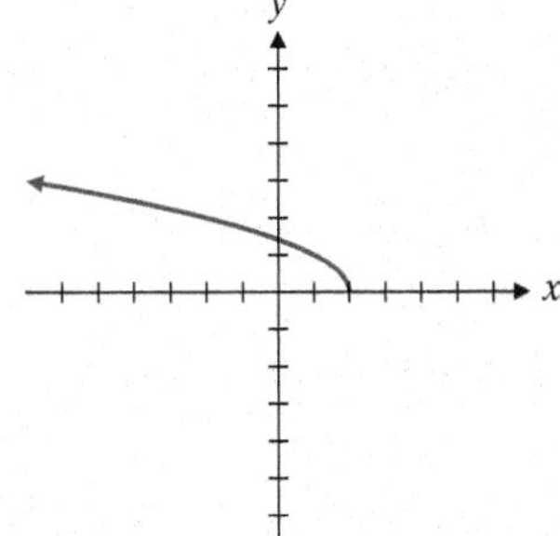

14.

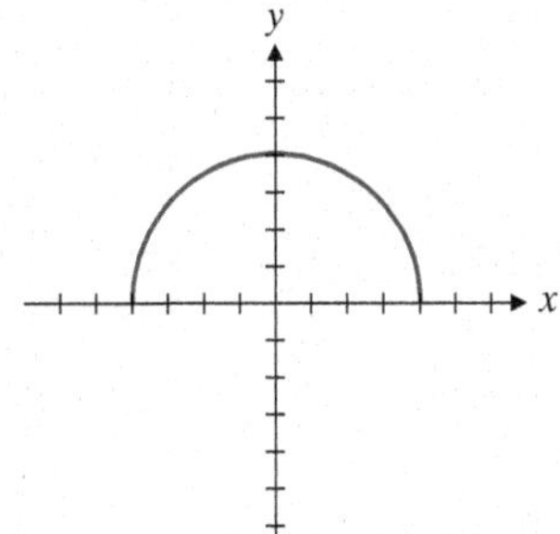

15.

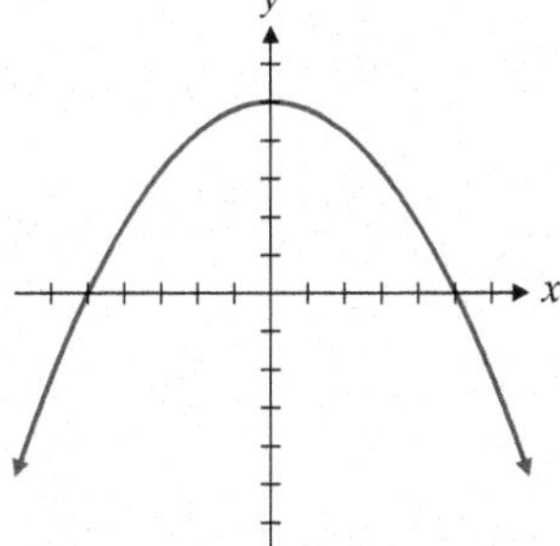

16.

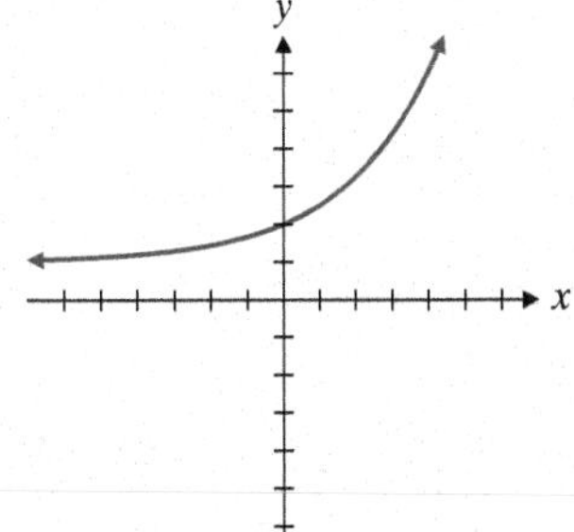

17.

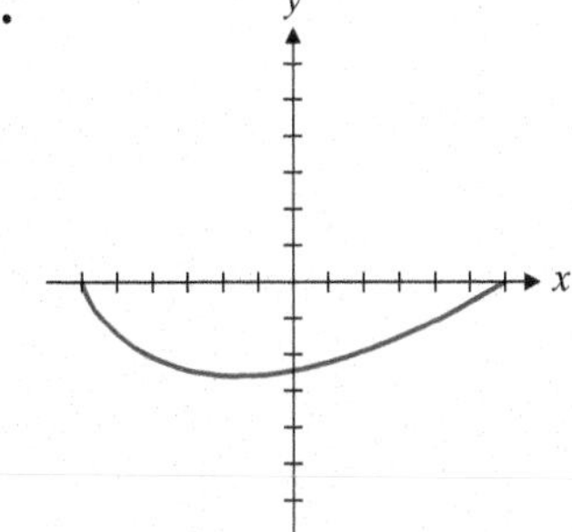

18.

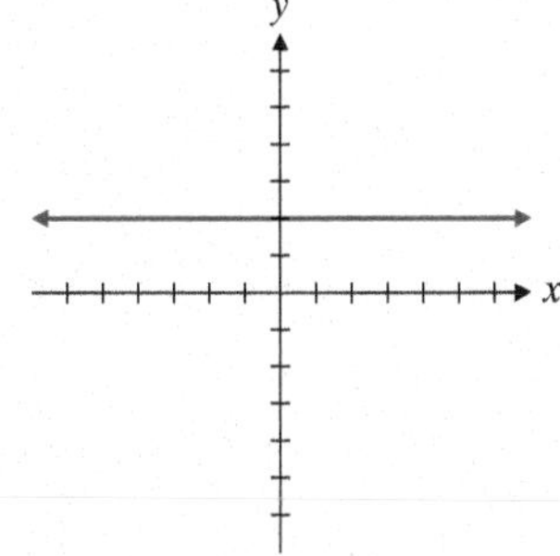

Find the inverse of each one-to-one function. Graph the function and its inverse on one coordinate plane.

19. $J = \{(8, 2), (1, 1), (0, 0), (-8, -2)\}$

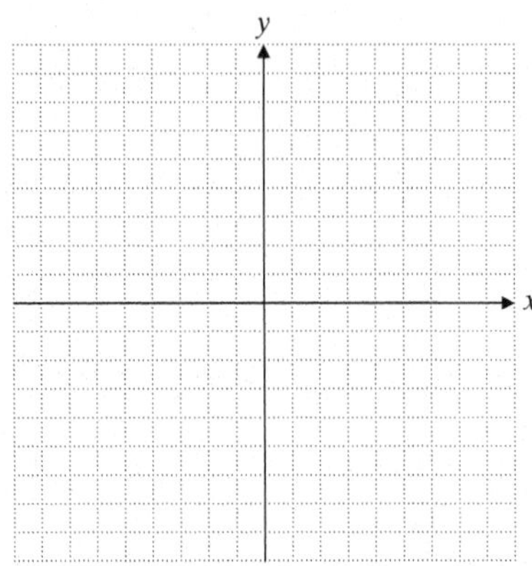

20. $K = \{(-7, 1), (6, 2), (3, -1), (2, 5)\}$

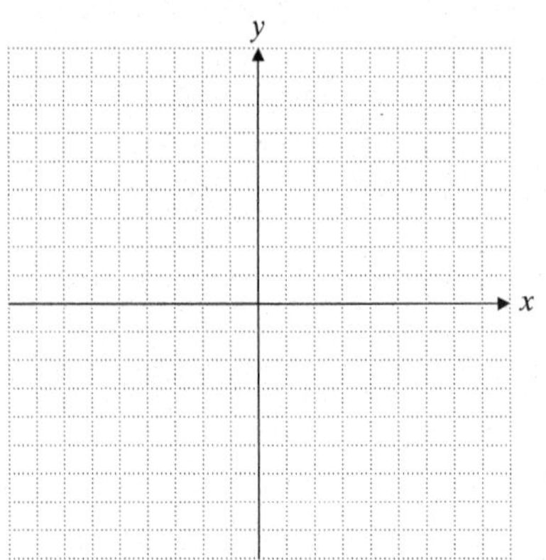

Find the inverse of each function.

21. $f(x) = 4x - 5$

22. $f(x) = 9 + 2x$

23. $f(x) = x^3 - 8$

24. $f(x) = 3 - x^3$

25. $f(x) = -\frac{4}{x}$

26. $f(x) = \frac{7}{x}$

27. $f(x) = \frac{4}{x - 5}$

28. $f(x) = \frac{-3}{5 + x}$

For the given function f and its inverse, find $f[f^{-1}(x)]$.

29. $f(x) = \frac{x - 3}{5}; f^{-1}(x) = 5x + 3$

30. $f(x) = \frac{x}{2} + 1; f^{-1}(x) = 2x - 2$

Find the inverse of each function. Graph the function and its inverse on one coordinate plane. Graph the line $y = x$ as a dashed line.

31. $g(x) = 2x + 5$

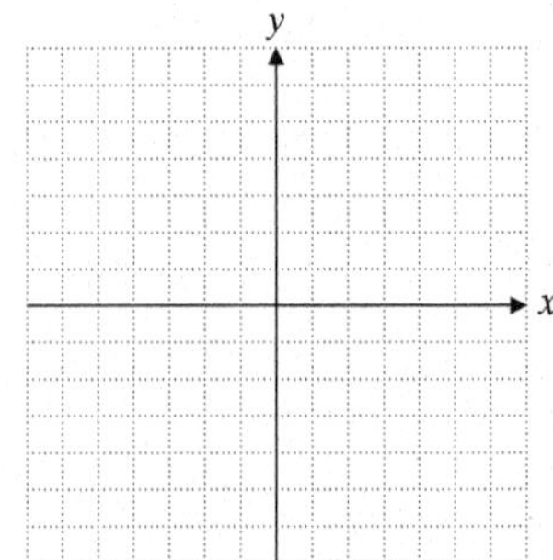

32. $f(x) = 3x + 4$

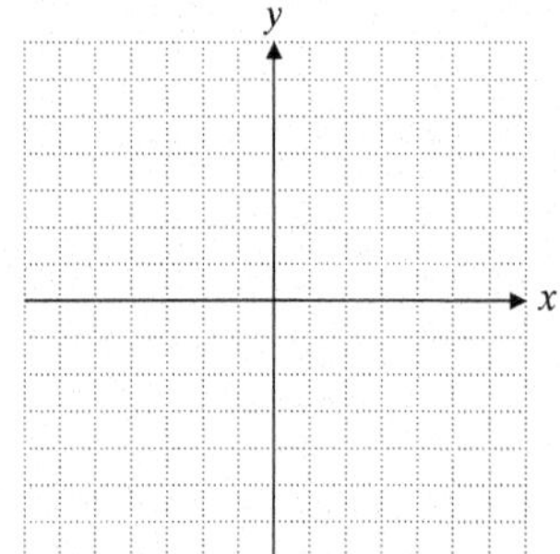

33. $h(x) = \frac{1}{2}x - 2$

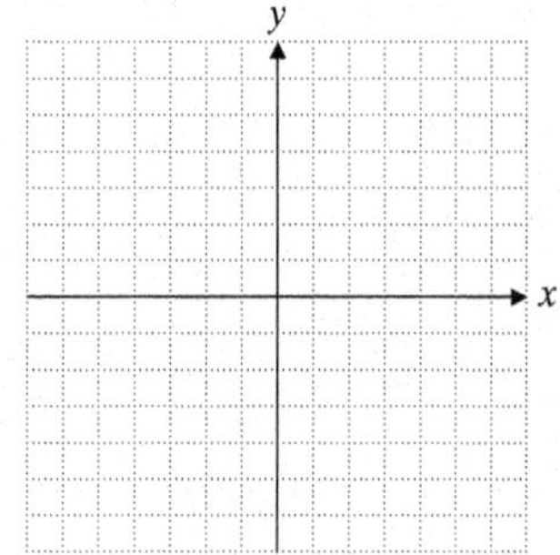

34. $p(x) = \frac{2}{3}x - 4$

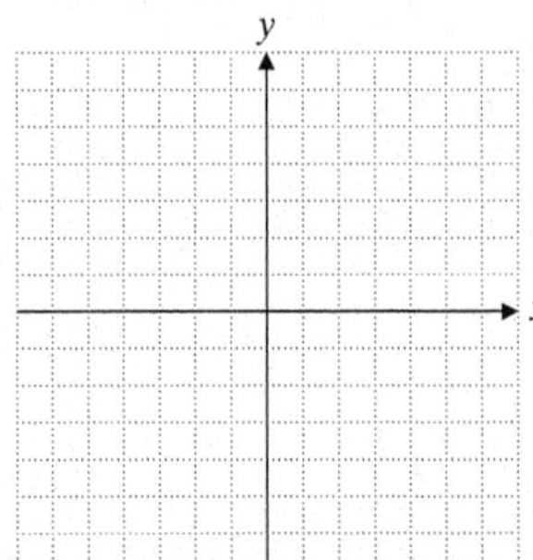

35. $r(x) = -3x - 1$

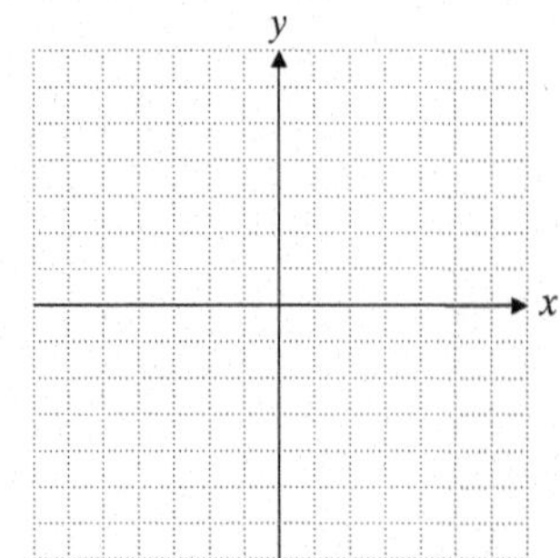

36. $k(x) = 3 - 2x$

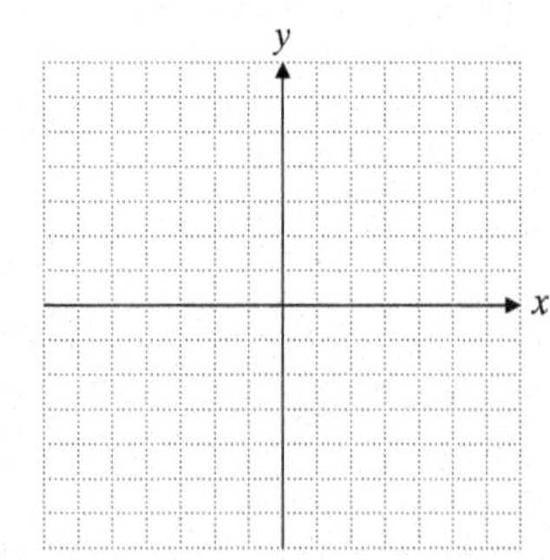

To Think About

37. Can you find an inverse function for the function $f(x) = 2x^2 + 3$? Why or why not?

38. Can you find an inverse function for the function $f(x) = |3x + 4|$? Why or why not?

For every function f and its inverse, f^{-1}, it is true that $f[f^{-1}(x)] = x$ and $f^{-1}[f(x)] = x$. Show that this is true for each pair of inverse functions.

39. $f(x) = 2x + \frac{3}{2}, f^{-1}(x) = \frac{1}{2}x - \frac{3}{4}$

40. $f(x) = -3x - 10, f^{-1}(x) = \frac{-x - 10}{3}$

Cumulative Review *Solve for x.*

41. **[8.5.1]** $\sqrt{20 - x} = x$

42. **[9.3.2]** $x^{2/3} + 7x^{1/3} + 12 = 0$

43. **[0.6.1]** ***Canadian Forests*** Forests are a dominant feature of Canada. Ten percent of all the world's forests lie in Canada. One out of every fifteen people in the labor force in Canada works in a job that relates to forests. If the labor force in Canada in 2013 was 19,446,000 people, how many people worked in a job related to forests in that year? (Source: www.statcan.gc.ca)

44. **[2.7.2]** Ukraine is one of the top ten countries with the biggest decline in population. The population of Ukraine in 2013 was 45.5 million. By 2050, the population is expected to be 33.4 million. What is Ukraine's expected percent of decrease in population? Round to the nearest percent. (*Source:* www.geography.about.com)

Quick Quiz 11.4

1. $A = \{(3, -4), (2, -6), (5, 6), (-3, 4)\}$

(a) Is A a function?

(b) Is A a one-to-one function?

(c) If the answers to **(a)** and **(b)** are yes, find A^{-1}.

2. Find the inverse function for $f(x) = 5 - 2x$. Graph f and its inverse. Graph the line $y = x$ as a dashed line.

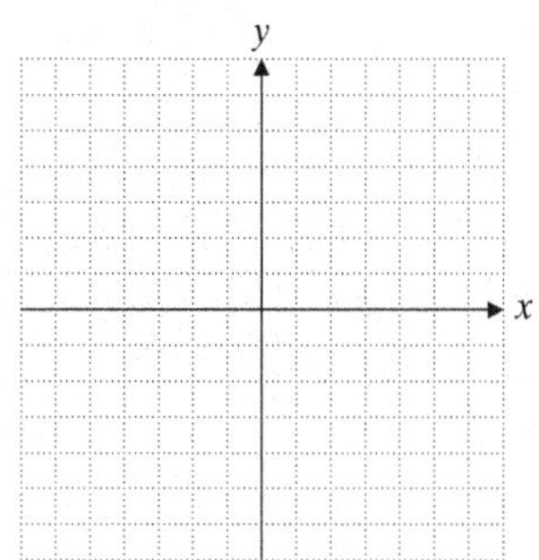

3. Find the inverse function for $f(x) = 2 - x^3$.

4. **Concept Check** If $f(x) = \frac{x - 5}{3}$, explain how you would find the inverse function.

Background: Criminal Justice Professional

Ross is a criminal justice professional who specializes in the development of crime-prevention programs. He works as a consultant to towns and cities seeking to begin such programs, and he also evaluates existing programs. In his work, he analyzes data and develops models that enlighten the work of local criminal justice and law enforcement professionals.

Currently, he is working on the evaluation of an established crime-prevention program in a large city. To evaluate its effectiveness, he has compiled and worked with data to create a few mathematical functions that can be evaluated thus providing information to everyone about the program's impact.

Facts

The function model for all crimes reported annually over the period from 2012 through 2015 is $C(x) = 1.4x^2 + 8.2x + 2170$, where x represents the number of years after 2012.

The function model for all petty crimes reported annually over the period from 2012 through 2015 is $T(x) = 0.27x + 2028$, where x represents the number of years after 2012.

The annual population function model for the city for the period 2012 through 2015 is $P(x) = 68.8x^2 + 1744.73x + 73{,}561$, where x represents the number of years after 2012.

Tasks

1. Ross wants to compute the number of crimes reported in 2012 and 2015. What are his results?

2. Using the population function, Ross computes the population for years 2012 and 2015. What are his results?

3. Ross now wants to express the crime rate per person as a percent rounded to the nearest hundredth for years 2012 and 2015. How can he use the functions and the evaluations he's completed to do this?

4. Using the petty-crime model, Ross can calculate some additional information. What are the total numbers of petty crimes in 2012 and in 2015? What are the petty crime rates per person for those years, expressed as percents rounded to the nearest hundredth?

5. Ross determines that a function model for violent crimes $V(x)$, can be generated by subtracting the petty crime function from the function for all crimes. What is the violent crime model, and what were the numbers of violent crimes in 2012 and 2015?

Chapter 11 Organizer

Topic and Procedure	Examples	You Try It

Finding function values, p. 680

Replace the variable by the quantity inside the parentheses. Simplify the result.

If $f(x) = 2x^2 + 3x - 4$, then we have the following:

(a) $f(-2) = 2(-2)^2 + 3(-2) - 4$
$= 8 - 6 - 4 = -2$

(b) $f(a) = 2a^2 + 3a - 4$

(c) $f(a + 2) = 2(a + 2)^2 + 3(a + 2) - 4$
$= 2(a^2 + 4a + 4) + 3(a + 2) - 4$
$= 2a^2 + 8a + 8 + 3a + 6 - 4$
$= 2a^2 + 11a + 10$

(d) $f(3a) = 2(3a)^2 + 3(3a) - 4$
$= 2(9a^2) + 3(3a) - 4$
$= 18a^2 + 9a - 4$

1. If $f(x) = -x^2 + x - 5$, find the following.
   (a) $f(1)$
   (b) $f(a)$
   (c) $f(a - 1)$
   (d) $f(4a)$

Vertical line test, p. 687

If any vertical line intersects the graph of a relation more than once, the relation is not a function. If no such line exists, the relation is a function.

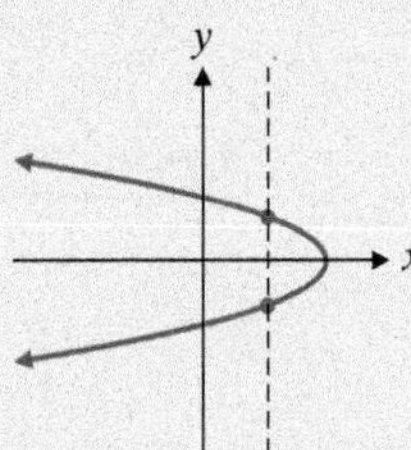

Does this graph represent a function?
No, because a vertical line intersects the curve more than once.

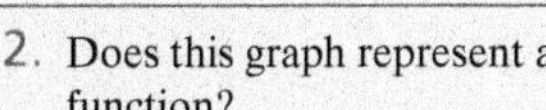

2. Does this graph represent a function?

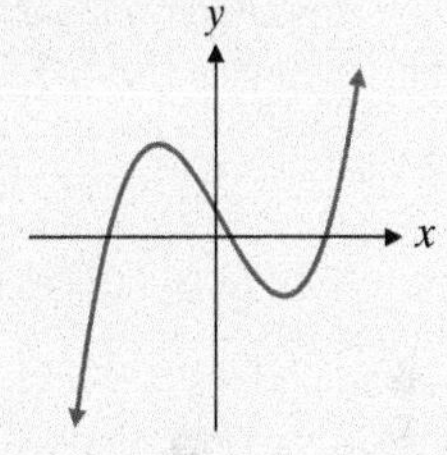

Vertical shifts of the graph of a function, p. 689

If $k > 0$:

1. The graph of $f(x) + k$ is shifted k units *upward* from the graph of $f(x)$.
2. The graph of $f(x) - k$ is shifted k units *downward* from the graph of $f(x)$.

(a) Graph $f(x) = x^2$ and $g(x) = x^2 + 3$.

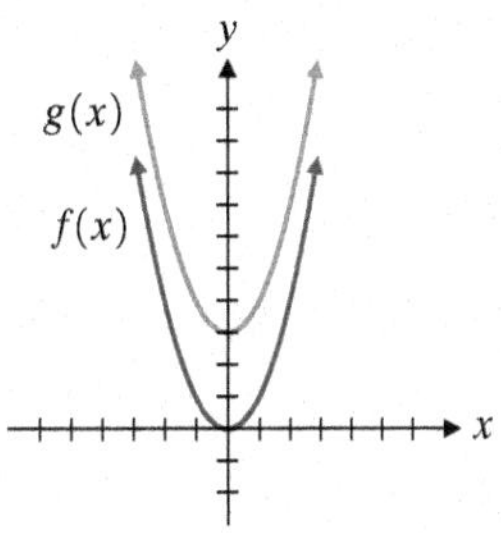

3. **(a)** Graph $f(x) = x^2$ and $g(x) = x^2 - 2$.

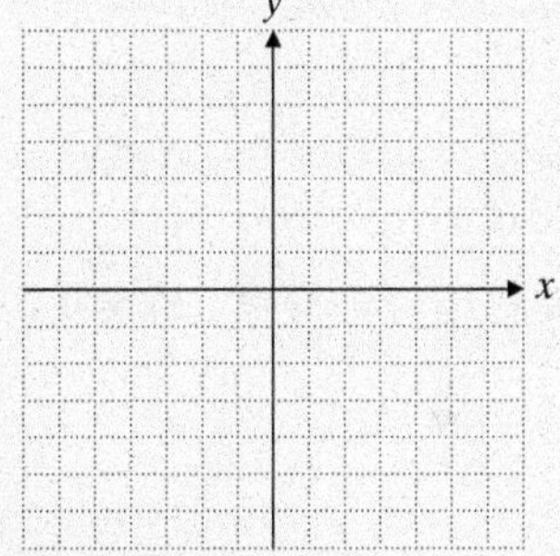

(b) Graph $f(x) = |x|$ and $g(x) = |x| - 2$.

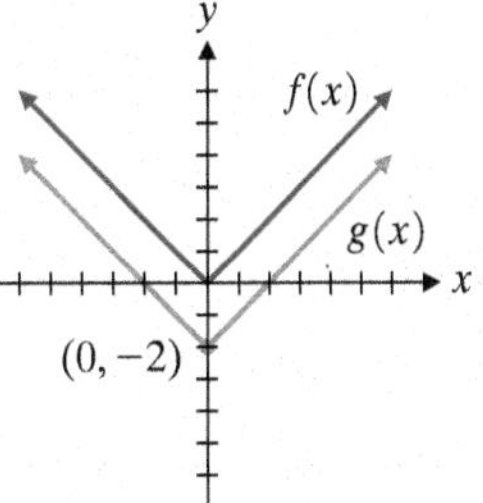

(b) Graph $f(x) = |x|$ and $g(x) = |x| + 4$.

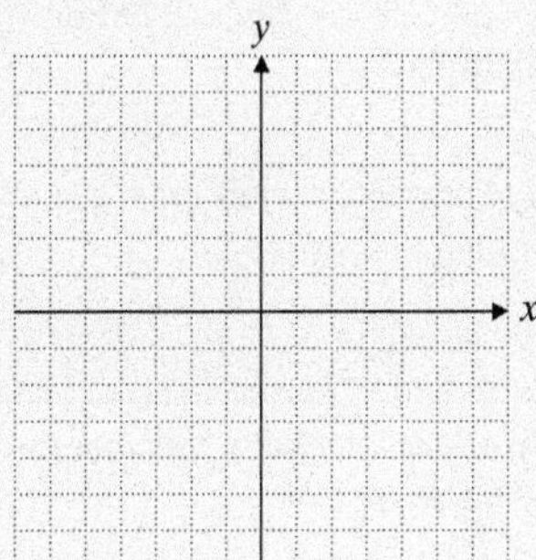

Topic and Procedure	Examples	You Try It

Horizontal shifts of the graph of a function, p. 690

If $h > 0$:

1. The graph of $f(x - h)$ is shifted h units to the *right* of the graph of $f(x)$.
2. The graph of $f(x + h)$ is shifted h units to the *left* of the graph of $f(x)$.

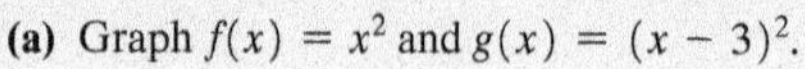

(a) Graph $f(x) = x^2$ and $g(x) = (x - 3)^2$.

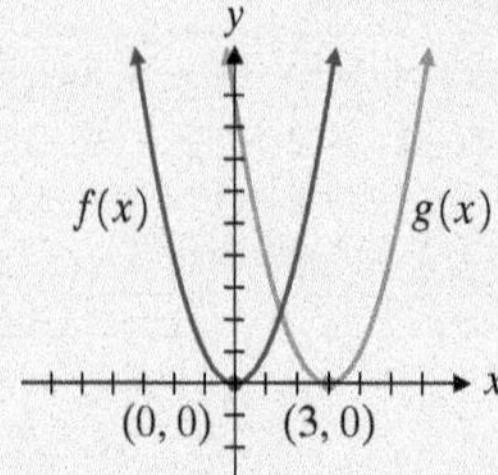

(b) Graph $f(x) = x^3$ and $g(x) = (x + 4)^3$.

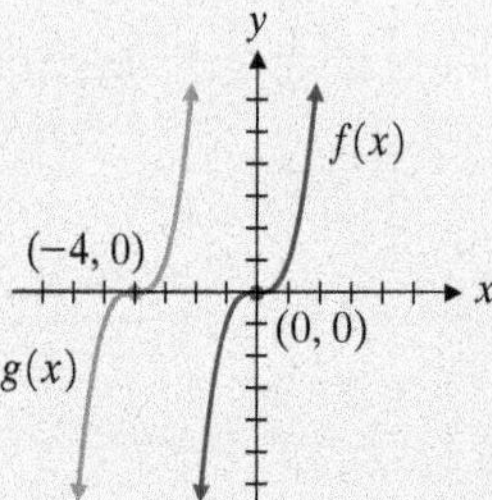

4. **(a)** Graph $f(x) = x^2$ and $g(x) = (x + 5)^2$.

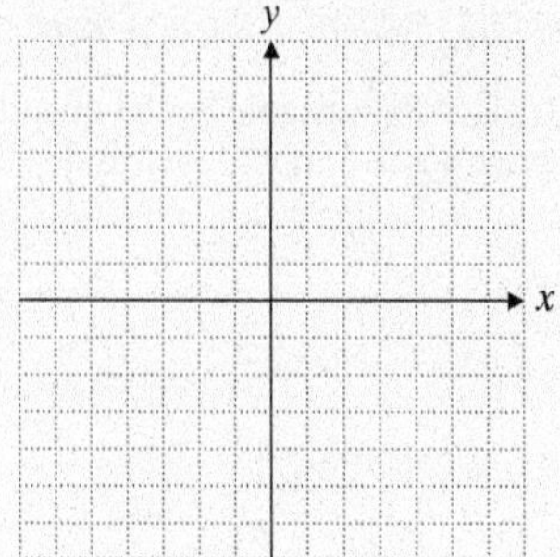

(b) Graph $f(x) = x^3$ and $g(x) = (x - 3)^3$.

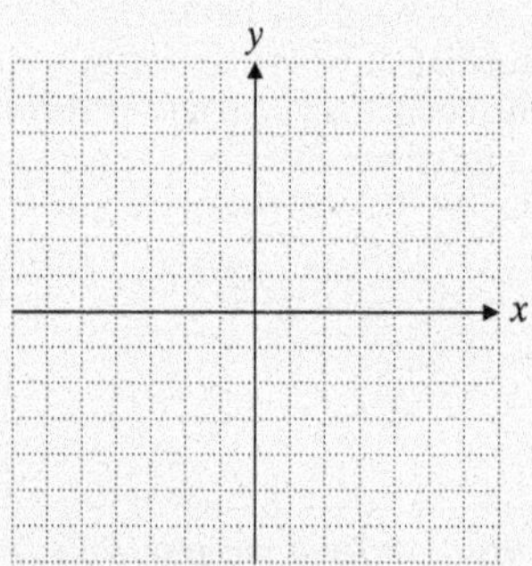

Sum, difference, product, and quotient of functions, p. 696

1. $(f + g)(x) = f(x) + g(x)$
2. $(f - g)(x) = f(x) - g(x)$
3. $(f \cdot g)(x) = f(x) \cdot g(x)$
4. $\left(\frac{f}{g}\right)(x) = \frac{f(x)}{g(x)}, g(x) \neq 0$

If $f(x) = 2x + 3$ and $g(x) = 3x - 4$, then we have the following:

(a) $(f + g)(x) = (2x + 3) + (3x - 4)$
$= 5x - 1$

(b) $(f - g)(x) = (2x + 3) - (3x - 4)$
$= 2x + 3 - 3x + 4$
$= -x + 7$

(c) $(f \cdot g)(x) = (2x + 3)(3x - 4)$
$= 6x^2 + x - 12$

(d) $\left(\frac{f}{g}\right)(x) = \frac{2x + 3}{3x - 4}, x \neq \frac{4}{3}$

5. If $f(x) = x^2 + 3x$ and $g(x) = x - 6$, find the following.

(a) $(f + g)(x)$
(b) $(f - g)(x)$
(c) $(f \cdot g)(x)$
(d) $\left(\frac{f}{g}\right)(x)$

Composition of functions, p. 698

The composition of functions f and g is written as $(f \circ g)(x) = f[g(x)]$. To find $f[g(x)]$ do the following:

1. Replace $g(x)$ by its equation.
2. Apply the formula for $f(x)$ to this expression.
3. Simplify the results.

Usually, $f[g(x)] \neq g[f(x)]$.

If $f(x) = x^2 - 5$ and $g(x) = -3x + 4$, find (a) $f[g(x)]$ and (b) $g[f(x)]$.

(a) $f[g(x)] = f(-3x + 4)$
$= (-3x + 4)^2 - 5$
$= 9x^2 - 24x + 16 - 5$
$= 9x^2 - 24x + 11$

(b) $g[f(x)] = g(x^2 - 5)$
$= -3(x^2 - 5) + 4$
$= -3x^2 + 15 + 4$
$= -3x^2 + 19$

6. If $f(x) = -x + 2$ and $g(x) = x^2 + 3$, find the following.

(a) $f[g(x)]$
(b) $g[f(x)]$

Relations, functions, and one-to-one functions, p. 704

A relation is any set of ordered pairs.

A function is a relation in which no ordered pairs have the same first coordinate.

A one-to-one function is a function in which no ordered pairs have the same second coordinate.

Is $\{(3, 6), (2, 8), (9, 1), (4, 6)\}$ a one-to-one function?
No, since $(3, 6)$ and $(4, 6)$ have the same second coordinate.

7. Is $[(-1, 0), (2, 3), (2, 5), (0, 4)]$ a one-to-one function?

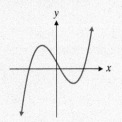

Topic and Procedure	Examples	You Try It
Horizontal line test, p. 704 If any horizontal line intersects the graph of a function more than once, the function is not one-to-one. If no such line exists, the function is one-to-one.	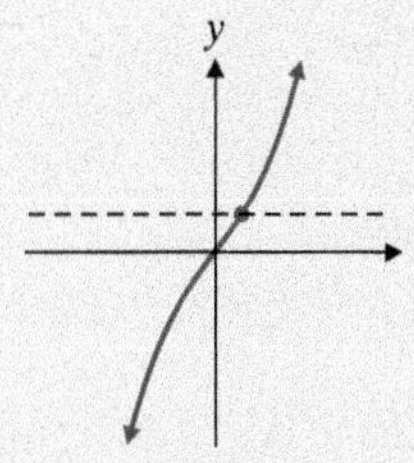 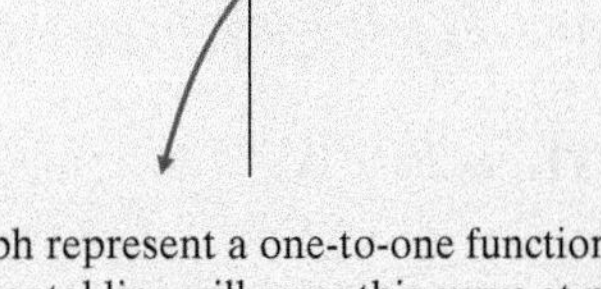 Does this graph represent a one-to-one function? Yes, any horizontal line will cross this curve at most once.	8. Does this graph represent a one-to-one function?
Finding the inverse of a function defined by a set of ordered pairs, p. 706 Reverse the order of the coordinates of each ordered pair from (a, b) to (b, a).	Find the inverse of $A = \{(5,6), (7,8), (9,10)\}$. $A^{-1} = \{(6,5), (8,7), (10,9)\}$	9. Find the inverse of $C = \{(3,-2), (0,4), (5,1)\}$.
Finding the inverse of a function defined by an equation, p. 707 Any one-to-one function has an inverse function. To find the inverse f^{-1} of a one-to-one function f, do the following: **1.** Replace $f(x)$ with y. **2.** Interchange x and y. **3.** Solve for y in terms of x. **4.** Replace y with $f^{-1}(x)$.	Find the inverse of $f(x) = -\frac{2}{3}x + 4$. $y = -\frac{2}{3}x + 4$ $x = -\frac{2}{3}y + 4$ $3x = -2y + 12$ $3x - 12 = -2y$ $\frac{3x - 12}{-2} = y$ $-\frac{3}{2}x + 6 = y$ $f^{-1}(x) = -\frac{3}{2}x + 6$	10. Find the inverse of $f(x) = \frac{x}{2} - 5$.
Graphing the inverse of a function, p. 708 Graph the line $y = x$ as a dashed line for reference. **1.** Graph $f(x)$. **2.** Graph $f^{-1}(x)$. The graphs of f and f^{-1} are symmetric about the line $y = x$.	$f(x) = 2x + 3$ $f^{-1}(x) = \frac{x - 3}{2}$ Graph f and f^{-1} on the same set of axes. 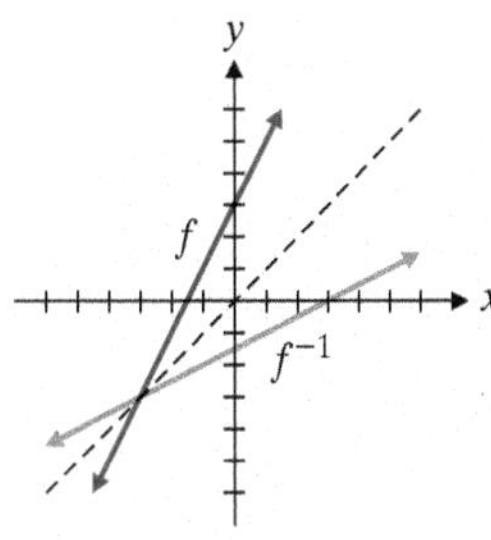	11. Graph $f(x) = \frac{x}{3} + 2$ and $f^{-1}(x) = 3x - 6$ on the same set of axes. 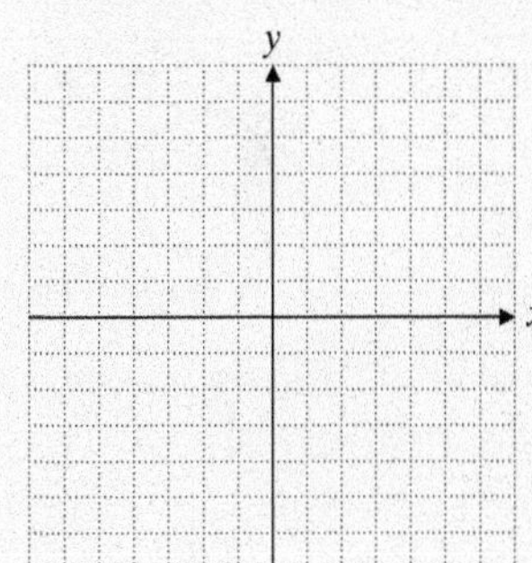

Chapter 11 Review Problems

For the function $f(x) = \frac{1}{2}x + 3$, *find the following:*

1. $f(a - 1)$

2. $f(a - 1) - f(a)$

3. $f(b^2 - 3)$

For the function $p(x) = -2x^2 + 3x - 1$, *find the following:*

4. $p(-3)$

5. $p(2a) + p(-2)$

6. $p(a + 2)$

For the function $h(x) = |2x - 1|$, *find the following:*

7. $h(0)$

8. $h\left(\frac{1}{4}a\right)$

9. $h(2a^2 - 3a)$

For the function $r(x) = \frac{3x}{x+4}$, $x \neq -4$, *find the following. In each case, write your answers as one fraction, if possible.*

10. $r(5)$

11. $r(2a - 5)$

12. $r(3) + r(a)$

Find $\frac{f(x+h) - f(x)}{h}$ *for the following:*

13. $f(x) = 7x - 4$

14. $f(x) = 6x - 5$

15. $f(x) = 2x^2 - 5x$

Examine each of the following graphs. **(a)** *Does the graph represent a function?* **(b)** *Does the graph represent a one-to-one function?*

16.

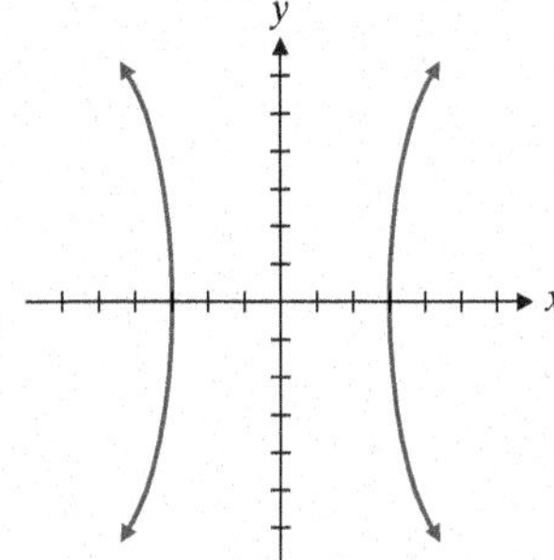

17.

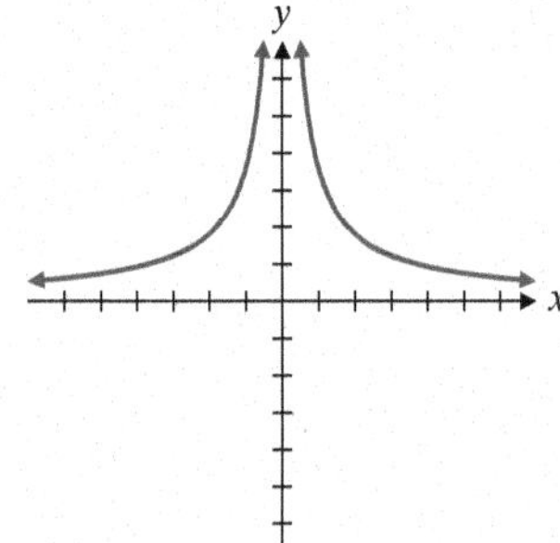

18.

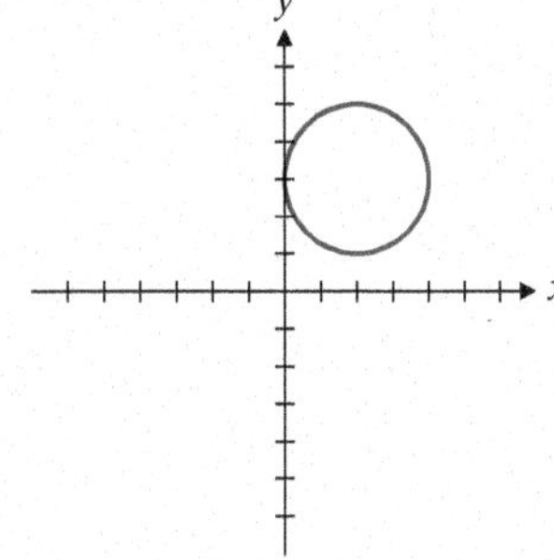

19.

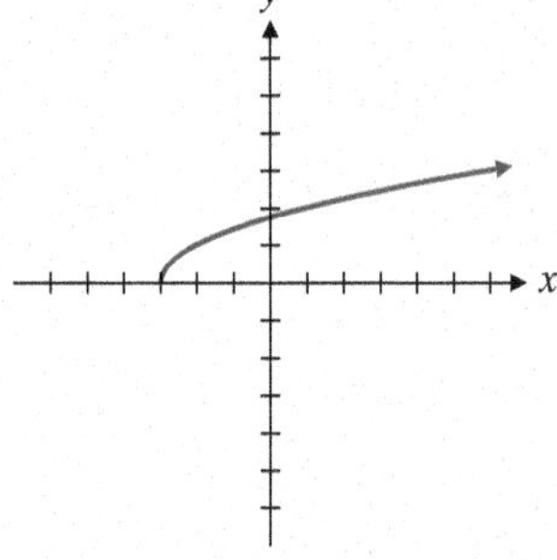

Graph each pair of functions on one set of axes.

20. $f(x) = x^2$
$g(x) = (x + 2)^2 + 4$

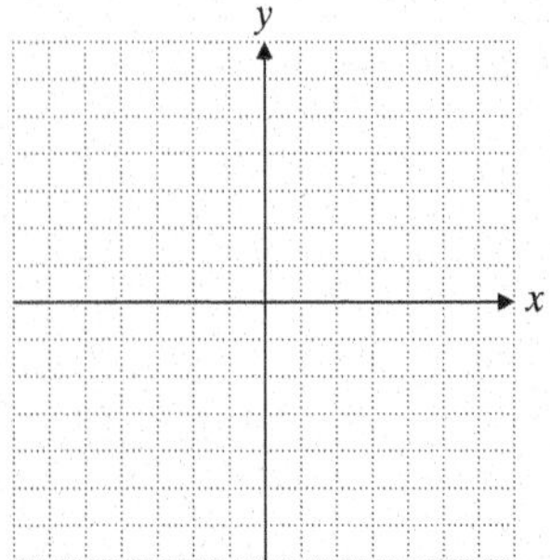

21. $f(x) = |x|$
$g(x) = |x - 4|$

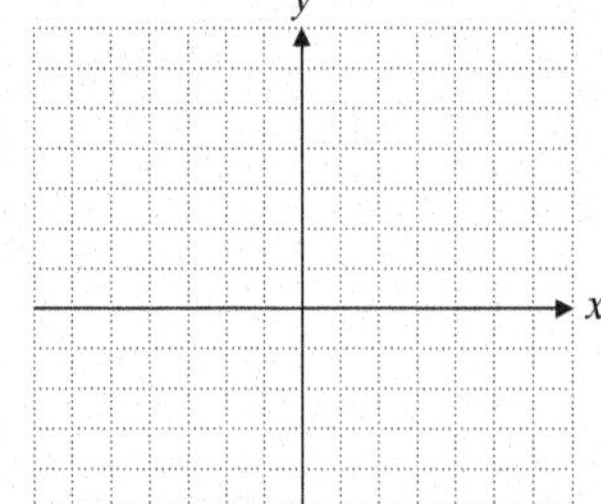

22. $f(x) = |x|$
$h(x) = |x| + 3$

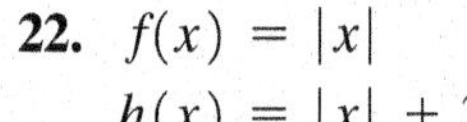

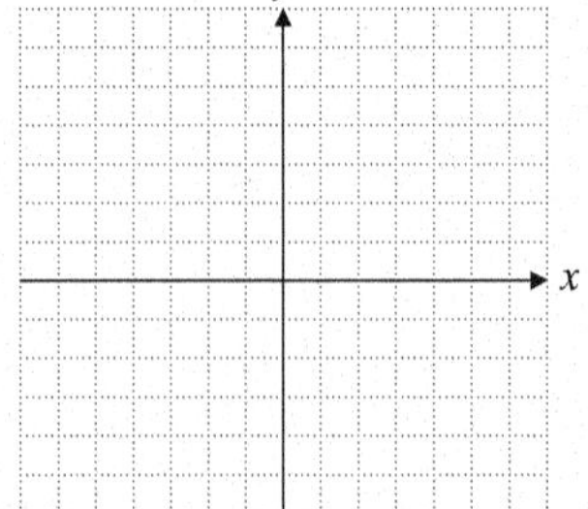

23. $f(x) = x^3$
$r(x) = (x + 3)^3 + 1$

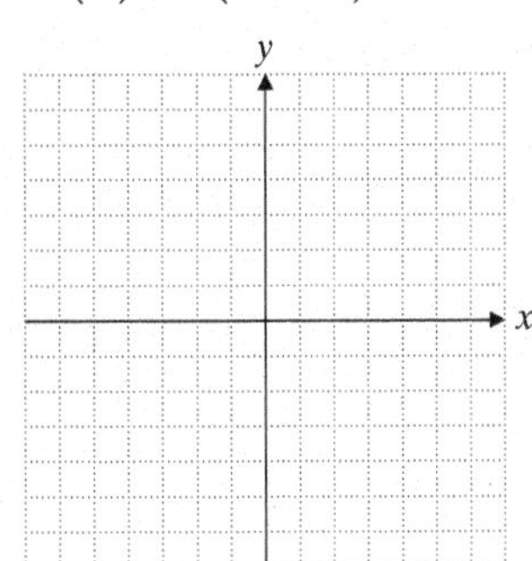

24. $f(x) = x^3$
$r(x) = (x - 1)^3 + 5$

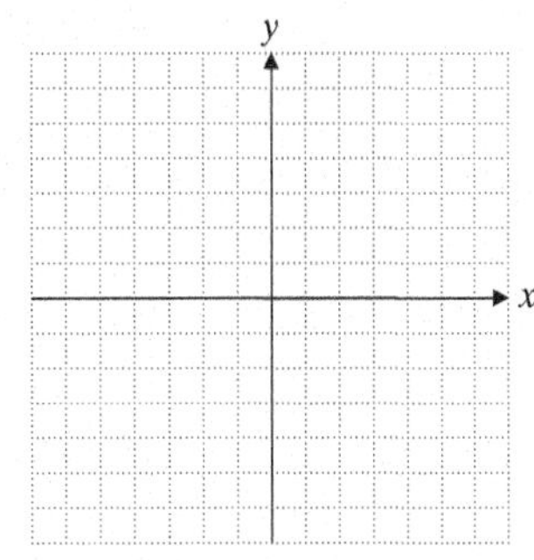

25. $f(x) = \frac{2}{x}, x \neq 0$
$r(x) = \frac{2}{x + 3} - 2, x \neq -3$

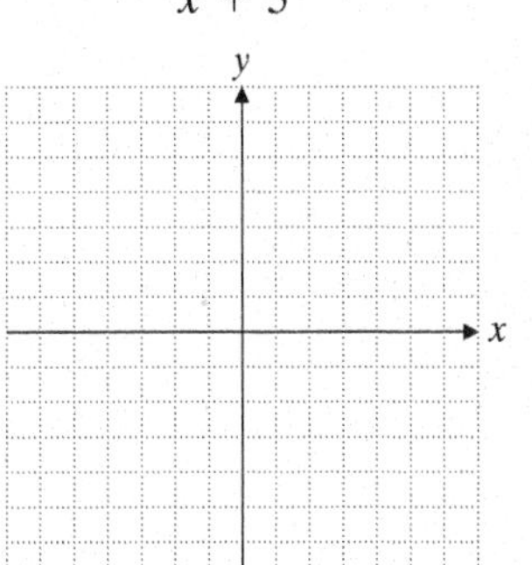

Use these functions for problems 26–35:

$$f(x) = 3x + 5; \quad g(x) = \frac{2}{x}, x \neq 0; \quad s(x) = \sqrt{x - 2}, x \geq 2;$$

$$h(x) = \frac{x + 1}{x - 4}, x \neq 4; \quad p(x) = 2x^2 - 3x + 4; \quad \text{and} \quad t(x) = -\frac{1}{2}x - 3,$$

find each of the following:

26. $(f + t)(x)$

27. $(p - f)(x)$

28. $(p - f)(2)$

29. $(fg)(x)$

30. $\left(\frac{g}{h}\right)(x)$

31. $\left(\frac{g}{h}\right)(-2)$

32. $p[f(x)]$

33. $s[p(x)]$

34. $s[p(2)]$

35. Show that $f[g(x)] \neq g[f(x)]$.

36. Given that $f(x) = \frac{2}{3}x + \frac{1}{2}$ and $f^{-1}(x) = \frac{6x - 3}{4}$, find $f^{-1}[f(x)]$.

For each set, determine **(a)** *the domain,* **(b)** *the range,* **(c)** *whether the set defines a function, and* **(d)** *whether the set defines a one-to-one function. (Hint: You can review the concept of domain and range in Section 3.5 if necessary.)*

37. $B = \{(3, 7), (7, 3), (0, 8), (0, -8)\}$

38. $A = \{(100, 10), (200, 20), (300, 30), (400, 10)\}$

39. $D = \left\{\left(\frac{1}{2}, 2\right), \left(\frac{1}{4}, 4\right), \left(-\frac{1}{3}, -3\right), \left(4, \frac{1}{4}\right)\right\}$

40. $F = \{(3, 7), (2, 1), (0, -3), (1, 1)\}$

Find the inverse of each of the following one-to-one functions.

41. $A = \left\{\left(3, \frac{1}{3}\right), \left(-2, -\frac{1}{2}\right), \left(-4, -\frac{1}{4}\right), \left(5, \frac{1}{5}\right)\right\}$

42. $f(x) = -\frac{3}{4}x + 2$

43. $g(x) = -8 - 4x$

44. $h(x) = \frac{6}{x + 5}$

45. $p(x) = \sqrt[3]{x + 1}$

46. $r(x) = x^3 + 2$

Find the inverse of each function. Graph the function and its inverse on one coordinate plane. Then on that same set of axes, graph the line $y = x$ as a dashed line.

47. $f(x) = \frac{-x - 2}{3}$

48. $f(x) = -\frac{3}{4}x + 1$

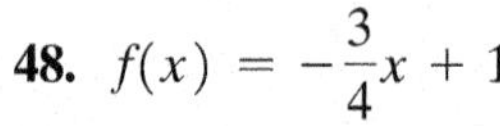

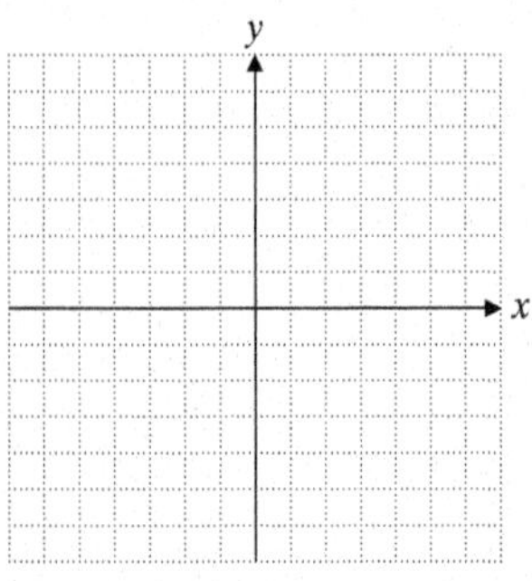

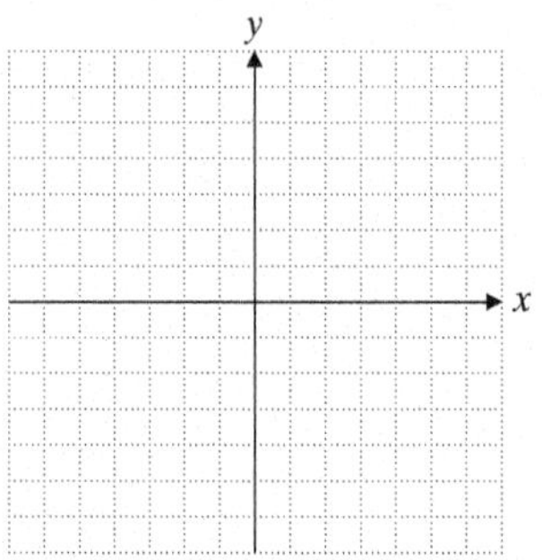

How Am I Doing? Chapter 11 Test

MATH COACH MyMathLab® YouTube

After you take this test read through the Math Coach on pages 721–722. Math Coach videos are available via MyMathLab and YouTube. Step-by-step test solutions in the Chapter Test Prep Videos are also available via MyMathLab and YouTube. (Search "TobeyBegInterAlg" and click on "Channels.")

For the function $f(x) = \frac{3}{4}x - 2$, *find the following:*

1. $f(-8)$ **2.** $f(3a)$ **3.** $f(a) - f(2)$

For the function $f(x) = 3x^2 - 2x + 4$, *find the following:*

4. $f(-6)$ MC **5.** $f(a + 1)$

6. $f(a) + f(1)$ **7.** $f(-2a) - 2$

Look at each graph below. **(a)** *Does the graph represent a function?* **(b)** *Does the graph represent a one-to-one function?*

8.

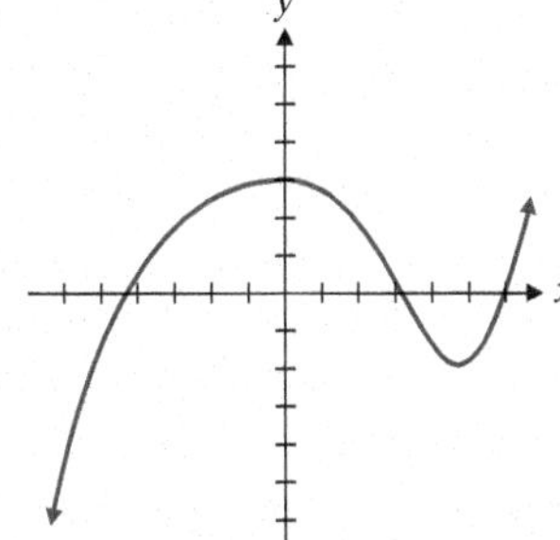

9.

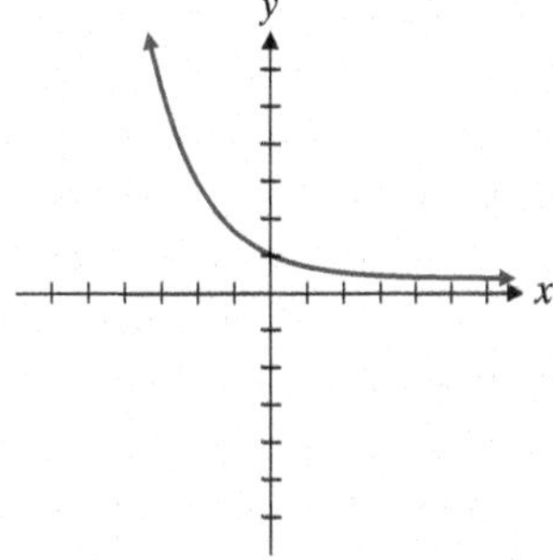

Graph each pair of functions on one coordinate plane.

MC **10.** $f(x) = x^2$
$g(x) = (x - 1)^2 + 3$

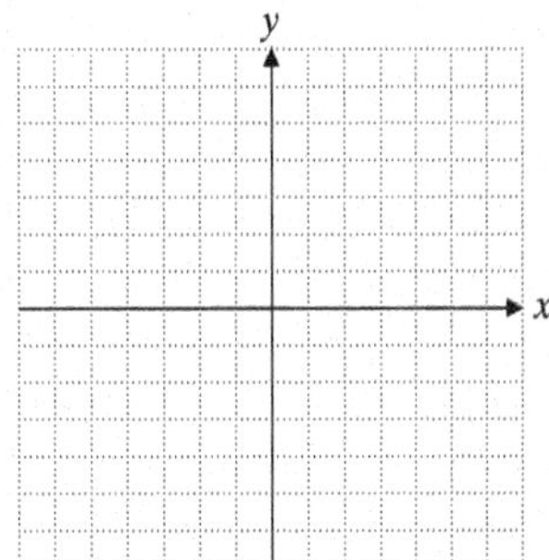

11. $f(x) = |x|$
$g(x) = |x + 1| + 2$

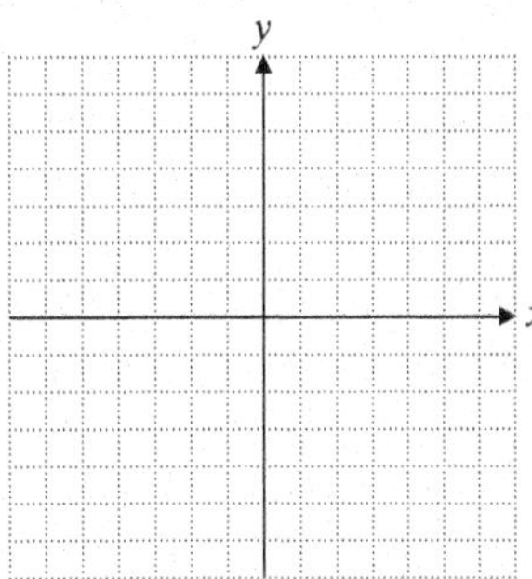

12. If $f(x) = 3x^2 - x - 6$ and $g(x) = -2x^2 + 5x + 7$, find the following:
(a) $(f + g)(x)$ **(b)** $(f - g)(x)$ **(c)** $(f - g)(-2)$

1.
2.
3.
4.
5.
6.
7.
8. (a)
(b)
9. (a)
(b)
10.
11.
12. (a)
(b)
(c)

13. (a) ______

(b) ______

(c) ______

14. (a) ______

(b) ______

(c) ______

15. (a) ______

(b) ______

16. (a) ______

(b) ______

17. ______

18. ______

19. ______

20. ______

Total Correct: ______

13. If $f(x) = \frac{3}{x}, x \neq 0$ and $g(x) = 2x - 1$, find the following:

(a) $(fg)(x)$ **(b)** $\left(\frac{f}{g}\right)(x)$ **(c)** $f[g(x)]$

14. If $f(x) = \frac{1}{2}x - 3$ and $g(x) = 4x + 5$, find the following:

MG **(a)** $(f \circ g)(x)$ **(b)** $(g \circ f)(x)$ **(c)** $(f \circ g)\left(\frac{1}{4}\right)$

Look at the following functions. **(a)** *Is the function one-to-one?* **(b)** *If so, find the inverse of the function.*

15. $B = \{(1, 8), (8, 1), (9, 10), (-10, 9)\}$

16. $A = \{(1, 5), (2, 1), (4, -7), (0, 7), (-1, 5)\}$

17. Determine the inverse of $f(x) = \sqrt[3]{2x - 1}$.

MG **18.** Find f^{-1}. Graph f and its inverse f^{-1} on one coordinate plane. Graph $y = x$ as a dashed line for a reference.

$$f(x) = -3x + 2$$

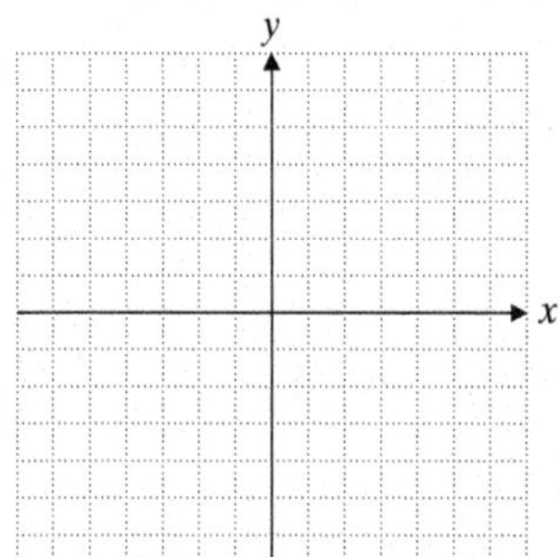

19. Given that $f(x) = \frac{3}{7}x + \frac{1}{2}$ and that $f^{-1}(x) = \frac{14x - 7}{6}$, find $f^{-1}[f(x)]$.

20. Find $\frac{f(x + h) - f(x)}{h}$ for $f(x) = 7 - 8x$.

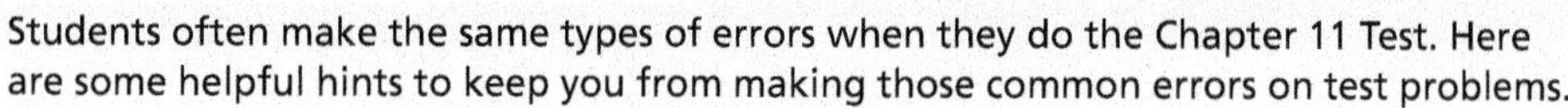

Mastering the skills you need to do well on the test.

Students often make the same types of errors when they do the Chapter 11 Test. Here are some helpful hints to keep you from making those common errors on test problems.

Chapter 11 Test, Problem 5 For the function $f(x) = 3x^2 - 2x + 4$, find $f(a + 1)$.

HELPFUL HINT The key idea is to replace every x by the expression $a + 1$ and then simplify the result. Use parentheses around the substitutions to avoid calculation errors.

Did you substitute $a + 1$ into the function to get $3(a + 1)^2 - 2(a + 1) + 4$?

Yes ______ No ______

Did you evaluate $(a + 1)^2$ to get $a^2 + 2a + 1$?

Yes ______ No ______

If you answered No to these questions, remember to replace every x with $a + 1$. Use parentheses to avoid calculation errors. Note that when you square a binomial, you must be sure to write down all the terms.

Next, did you simplify further to obtain $3a^2 + 6a + 3 - 2a - 2 + 4$?

Yes ______ No ______

If you answered No, remember to multiply all three terms of $a^2 + 2a + 1$ by 3. Multiply both terms of $a + 1$ by -2.

In your final step, collect like terms to write your answer in simplest form.

If you answered Problem 5 incorrectly, go back and rework the problem using these suggestions.

Chapter 11 Test, Problem 10 Graph each pair of functions on one coordinate plane. $f(x) = x^2$
$g(x) = (x - 1)^2 + 3$

HELPFUL HINT First graph $f(x)$. The graph of $f(x - h) + k$ is the graph of $f(x)$ shifted h units to the right and k units upward (assuming that $h > 0$ and $k > 0$).

Can you determine that the graph of $f(x) = x^2$ passes through the points $(0, 0), (1, 1), (-1, 1), (2, 4)$, and $(-2, 4)$?

Yes ______ No ______

If you answered No, try building a table of values in which you replace x with $0, 1, -1, 2$, and -2 and find the corresponding values of $f(x)$. Plot those points and connect the points with a curve to form the graph of $f(x) = x^2$.

Do you see that the function $g(x) = (x - 1)^2 + 3$ has the values of $h = 1$ and $k = 3$ when you apply the Helpful Hint?

Yes ______ No ______

Did you find that the graph of $g(x)$ is the graph of $f(x)$ shifted one unit to the right and 3 units up?

Yes ______ No ______

If you answered No to these questions, reread the Helpful Hint carefully. Notice that the values of h and k are both greater than zero.

Finish by graphing $g(x)$ on the same coordinate plane as your graph of $f(x)$.

If you answered Problem 10 incorrectly, go back and rework the problem using these suggestions.

Need more help? Watch the MATH COACH videos in MyMathLab® or on YouTube.

Chapter 11 Test, Problem 14(a) If $f(x) = \frac{1}{2}x - 3$ and $g(x) = 4x + 5$, find $(f \circ g)(x)$

> **HELPFUL HINT** First rewrite $(f \circ g)(x)$ as $f[g(x)]$. Most students find this expression more logical. Then substitute $g(x)$ for the value of x in $f(x)$.

First, did you rewrite the problem as $f[g(x)] = \frac{1}{2}(4x + 5) - 3$?

Yes ______ No ______

If you answered No, substitute the expression for $g(x)$ for the value of x in the expression for $f(x)$.

Next did you simplify the resulting expression to $f[g(x)] = 2x + \frac{5}{2} - 3$?

Yes ______ No ______

If you answered No, remember that $\frac{1}{2}(4x) = 2x$ and $\frac{1}{2}(5) = \frac{5}{2}$. As your final step, combine like terms to write your expression in simplest form.

If you answered Problem 14(a) incorrectly, go back and rework the problem using these suggestions.

Chapter 11 Test, Problem 18 Given $f(x) = -3x + 2$, find f^{-1}. Graph f and its inverse f^{-1} on one coordinate plane. Graph $y = x$ as a dashed line for reference.

> **HELPFUL HINT** Use the following four steps to find the inverse of a function:
> 1. Replace $f(x)$ with y.
> 2. Interchange x and y.
> 3. Solve for y in terms of x.
> 4. Replace y with $f^{-1}(x)$.

Did you substitute y for $f(x)$ to get $y = -3x + 2$ and then interchange x and y to get $x = -3y + 2$?

Yes ______ No ______

If you answered No, review the first two steps in the Helpful Hint and perform these steps again.

Did you solve the equation for y to get $y = -\frac{1}{3}x + \frac{2}{3}$?

Yes ______ No ______

If you answered No, remember to add $3y$ to each side. Next add $-x$ to each side and then divide each side of the equation by 3. As your last step in finding the inverse, replace y with $f^{-1}(x)$.

Remember to graph $f(x)$ and $f^{-1}(x)$ on the same coordinate plane. Add the graph of $y = x$ as a dashed line for reference.

If you answered Problem 18 incorrectly, go back and rework the problem using these suggestions.

Need more help? Look for section examples marked with MC to review.

CHAPTER 12

Logarithmic and Exponential Functions

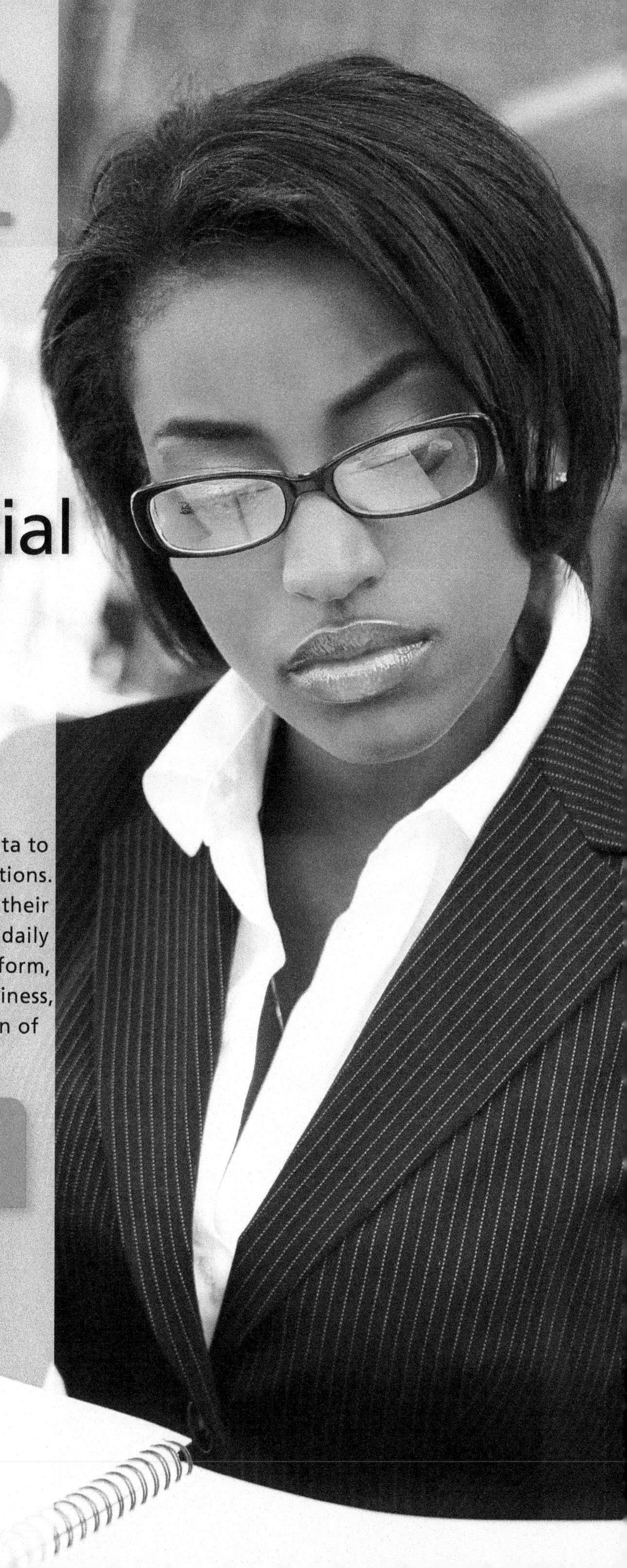

CAREER OPPORTUNITIES

Statistician

Statisticians collect, organize, and analyze data to help guide decision making and answer questions. Because so much of the world is data driven, their work impacts nearly every aspect of people's daily lives. Whether it is to solve problems or to inform, a statistician's work in sports, healthcare, business, and many other areas requires the application of mathematical principles.

To investigate how the mathematics in this chapter can help with this field, see the Career Exploration Problems on page 767.

12.1 The Exponential Function

Student Learning Objectives

After studying this section, you will be able to:

1. Graph an exponential function.
2. Solve elementary exponential equations.
3. Solve applications requiring the use of an exponential equation.

1 Graphing an Exponential Function

We have defined exponential equations $a^x = b$ for any rational number x. For example,

$$2^{-2} = \frac{1}{4},$$
$$2^{1/2} = \sqrt{2}, \text{ and}$$
$$2^{1.7} = 2^{17/10} = \sqrt[10]{2^{17}}.$$

We can also define such equations when x is an irrational number, such as π or $\sqrt{2}$. However, we will leave this definition for a more-advanced course.

We define an **exponential function** for all real values of x as follows:

DEFINITION OF EXPONENTIAL FUNCTION

The function $f(x) = b^x$, where $b > 0$, $b \neq 1$, and x is a real number, is called an **exponential function.** The number b is called the **base** of the function.

Now let's look at some graphs of exponential functions.

Example 1 Graph $f(x) = 2^x$.

Solution We make a table of values for x and $f(x)$.

$$f(-1) = 2^{-1} = \frac{1}{2}, \quad f(0) = 2^0 = 1, \quad f(1) = 2^1 = 2$$

Verify the other values in the table below. We then draw the graph.

x	f(x)
−2	$\frac{1}{4}$
−1	$\frac{1}{2}$
0	1
1	2
2	4
3	8

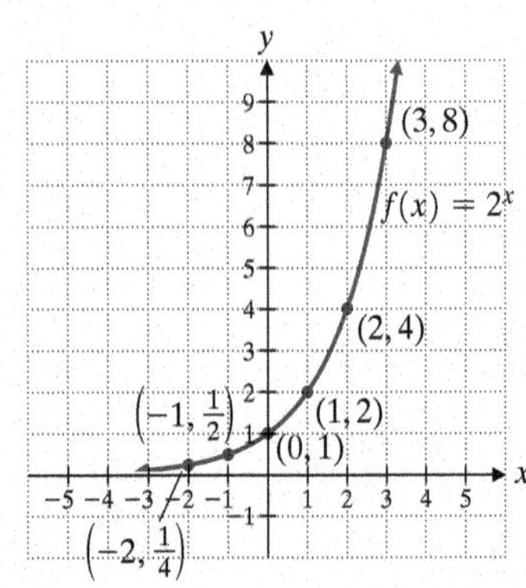

Notice how the curve of $f(x) = 2^x$ comes *very close to* the x-axis but *never touches* it. The x-axis is an **asymptote** for every exponential function of the form $f(x) = b^x$, where $b > 0$, $b \neq 1$, and x is a real number. You should also notice that $f(x)$ is always positive, so the range of f is the set of all positive real numbers (whereas the domain is the set of all real numbers). When the base is greater than one, as x increases, $f(x)$ increases faster and faster (that is, the curve gets steeper). □

Student Practice 1 Graph $f(x) = 3^x$.

Student Practice 1

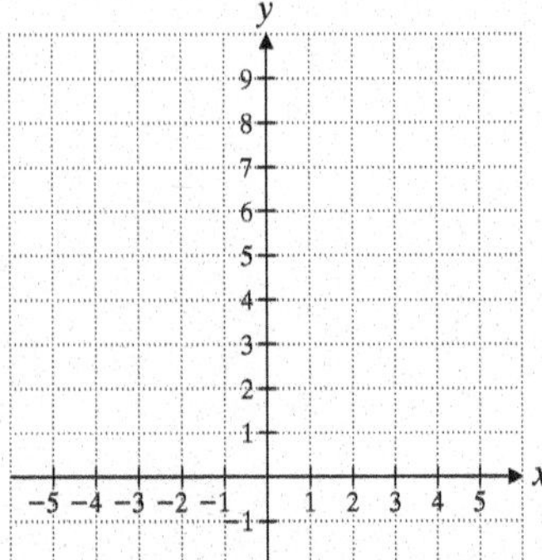

Example 2 Graph $f(x) = \left(\frac{1}{2}\right)^x$.

Solution We can write $f(x) = \left(\frac{1}{2}\right)^x$ as $f(x) = \left(\frac{1}{2}\right)^x = (2^{-1})^x = 2^{-x}$ and evaluate it for a few values of x. We then draw the graph.

x	f(x)
−3	8
−2	4
−1	2
0	1
1	$\frac{1}{2}$
2	$\frac{1}{4}$

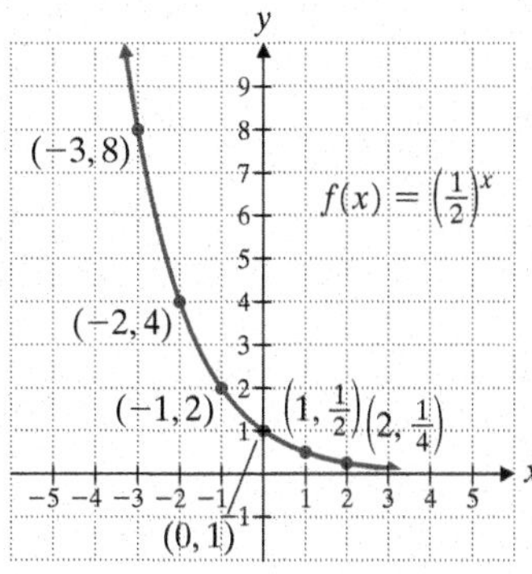

Note that as x increases, $f(x)$ decreases. □

Student Practice 2 Graph $f(x) = \left(\frac{1}{3}\right)^x$.

Student Practice 2

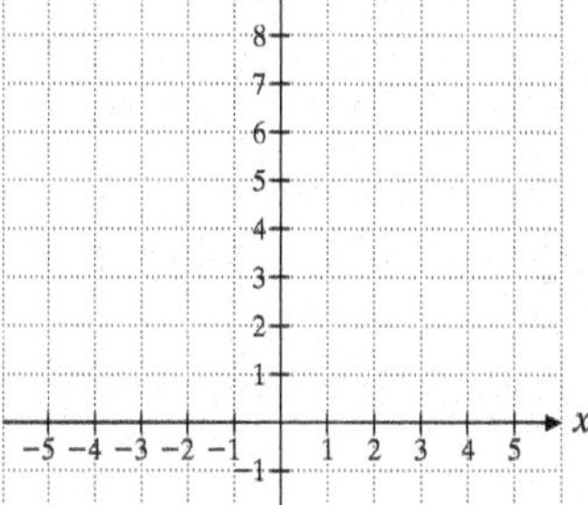

To Think About: *Comparing Graphs* Look at the graph of $f(x) = 2^x$ in Example 1 and the graph of $f(x) = \left(\frac{1}{2}\right)^x = 2^{-x}$ in Example 2. How are the two graphs related?

Example 3 Graph $f(x) = 3^{x-2}$.

Solution We will make a table of values for a few values of x. Then we will graph the function.

$$f(0) = 3^{0-2} = 3^{-2} = \frac{1}{3^2} = \frac{1}{9}$$
$$f(1) = 3^{1-2} = 3^{-1} = \frac{1}{3}$$
$$f(2) = 3^{2-2} = 3^0 = 1$$
$$f(3) = 3^{3-2} = 3^1 = 3$$
$$f(4) = 3^{4-2} = 3^2 = 9$$

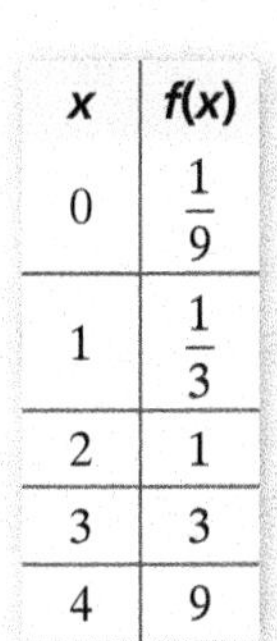

x	f(x)
0	$\frac{1}{9}$
1	$\frac{1}{3}$
2	1
3	3
4	9

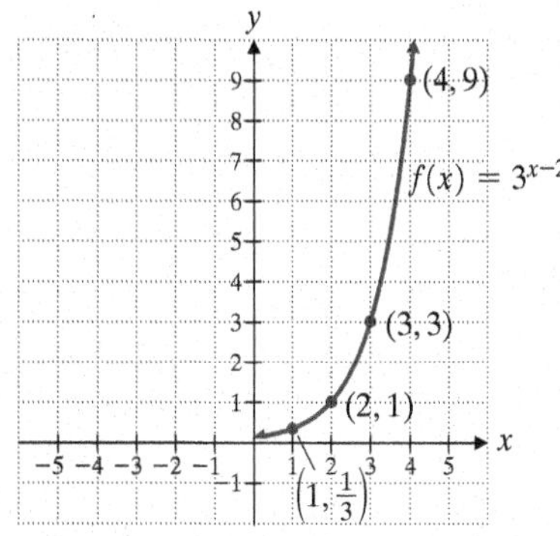

We observe that the curve is that of $f(x) = 3^x$ except that it has been shifted 2 units to the right. □

Student Practice 3

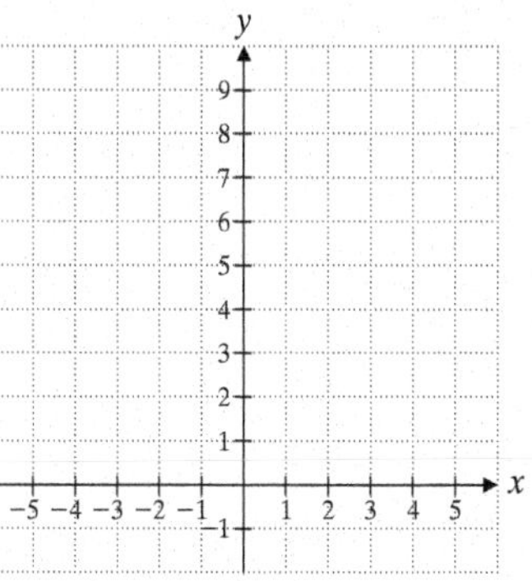

Student Practice 3 Graph $f(x) = 3^{x+2}$.

To Think About: *Graph Shifts* How is the graph of $f(x) = 3^{x+2}$ related to the graph of $f(x) = 3^x$? Without making a table of values, draw the graph of $f(x) = 3^{x+3}$. Draw the graph of $f(x) = 3^{x-3}$.

For the next example we need to discuss a special number that is denoted by the letter e. The letter e is a number like π. It is an irrational number. It occurs in many formulas that describe real-world phenomena, such as the growth of cells and radioactive decay. We need an approximate value for e to use this number in calculations: $e \approx$ **2.7183.**

The exponential function $f(x) = e^x$ is an extremely useful function. We usually obtain values for e^x by using a calculator or a computer. If you have a scientific calculator, use the $\boxed{e^x}$ key. (Many scientific calculators require you to press $\boxed{\text{SHIFT}}$ $\boxed{\ln}$ or $\boxed{\text{2nd F}}$ $\boxed{\ln}$ or $\boxed{\text{INV}}$ $\boxed{\ln}$ to obtain the operation e^x.) If you have a calculator that is not a scientific calculator, use $e \approx 2.7183$ as an approximate value.

Example 4 Graph $f(x) = e^x$.

Solution We evaluate $f(x)$ for some negative and some positive values of x. We begin with $f(-2)$.

To find $f(-2) = e^{-2}$ on a scientific calculator, we enter $-2\boxed{e^x}$ and obtain 0.135335283 as an approximation. (On some scientific calculators you will need to use the keystrokes 2 $\boxed{\text{2nd F}}$ $\boxed{\ln}$ or 2 $\boxed{\text{SHIFT}}$ $\boxed{\ln}$ or 2 $\boxed{\text{INV}}$ $\boxed{\ln}$.) Thus, $f(-2) = e^{-2} \approx 0.14$ to the nearest hundredth.

x	f(x)
−2	0.14
−1	0.37
0	1
1	2.72
2	7.39

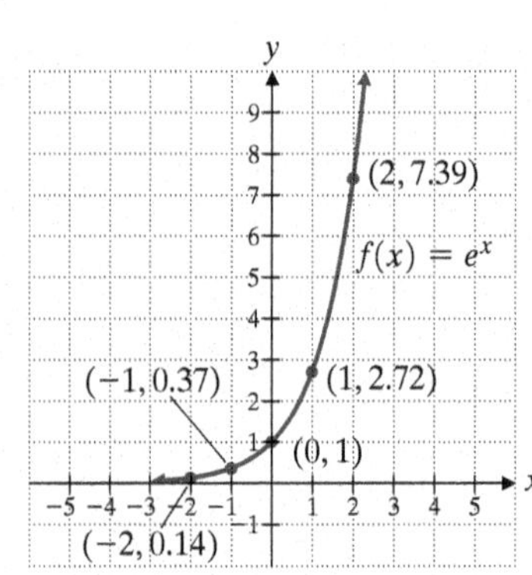

Student Practice 4 Graph $f(x) = e^{x-2}$.

Student Practice 4

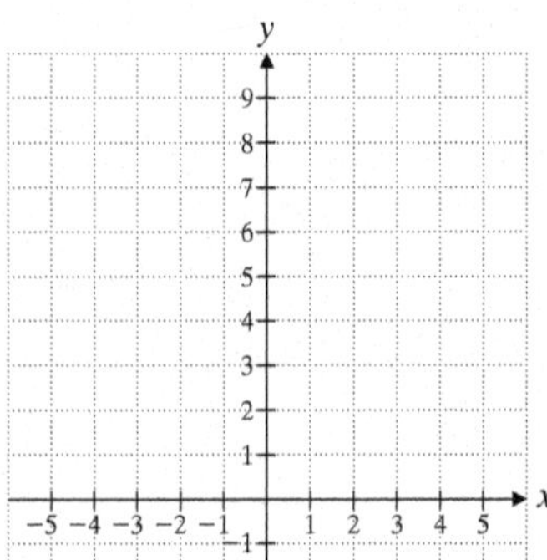

To Think About: *Graph Shifts* Look at the graphs of $f(x) = e^x$ and $f(x) = e^{x-2}$. Describe the shift that occurs. Without making a table of values, draw the graph of $f(x) = e^{x+3}$.

2 Solving Elementary Exponential Equations

All the usual laws of exponents are true for exponential functions. We also have the following important property to help us solve exponential equations.

> **PROPERTY OF EXPONENTIAL EQUATIONS**
>
> If $b^x = b^y$, then $x = y$ for $b > 0$ and $b \neq 1$.

Example 5 Solve. $2^x = \frac{1}{16}$

Solution To use the property of exponential equations, we must have the same base on both sides of the equation.

$$2^x = \frac{1}{16}$$

$$2^x = \frac{1}{2^4} \quad \text{Because } 2^4 = 16.$$

$$2^x = 2^{-4} \quad \text{Because } \frac{1}{2^4} = 2^{-4}.$$

$$x = -4 \quad \text{Property of exponential equations.}$$

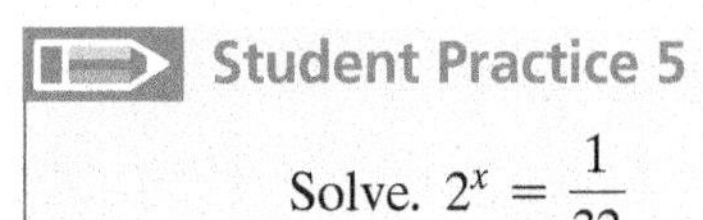

Student Practice 5

Solve. $2^x = \frac{1}{32}$

3 Solving Applications Requiring the Use of an Exponential Equation

An exponential function can be used to solve compound interest exercises. If a principal amount P is invested at an interest rate r compounded annually, the amount of money A accumulated after t years is $A = P(1 + r)^t$.

Example 6 If a young married couple invests \$5000 in a mutual fund that pays 10% interest compounded annually, how much will they have in 3 years?

Solution Here $P = 5000$, $r = 0.10$, and $t = 3$.

$$\begin{aligned} A &= P(1 + r)^t \\ &= 5000(1 + 0.10)^3 \\ &= 5000(1.10)^3 \\ &= 5000(1.331) \\ &= 6655.00 \end{aligned}$$

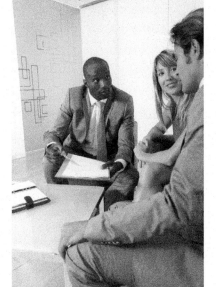

The couple will have \$6655.00.

If you have a scientific calculator, you can find the value of $5000(1.10)^3$ immediately by using the [×] key and the [y^x] key. On most scientific calculators you can use the following keystrokes.

5000 [×] 1.10 [y^x] 3 [=] 6655.00

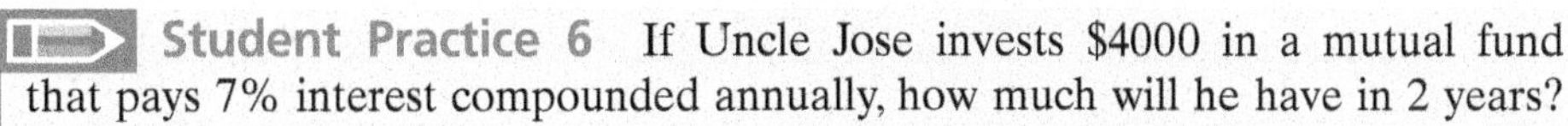

Student Practice 6 If Uncle Jose invests \$4000 in a mutual fund that pays 7% interest compounded annually, how much will he have in 2 years?

Interest is often compounded quarterly or monthly or even daily. Therefore, a more useful form of the interest formula that allows for variable compounding is needed. If a principal P is invested at an annual interest rate r that is compounded n times a year, then the amount of money A accumulated after t years is

$$A = P\left(1 + \frac{r}{n}\right)^{nt}.$$

Example 7 If we invest \$8000 in a fund that pays 6% annual interest compounded monthly, how much will we have after 6 years?

Solution In this situation $P = 8000$, $r = 6\% = 0.06$, and $n = 12$. The interest is compounded monthly or twelve times per year. Finally, $t = 6$ since the interest will be compounded for 6 years.

Continued on next page

$$
\begin{aligned}
A &= 8000\left(1 + \frac{0.06}{12}\right)^{(12)(6)} \\
&= 8000(1 + 0.005)^{72} \\
&= 8000(1.005)^{72} \\
&\approx 8000(1.432044278) \\
&\approx 11{,}456.35423
\end{aligned}
$$

Rounding to the nearest cent, we obtain the answer \$11,456.35. Using a scientific calculator, we could have found the answer directly by using the following keystrokes.

8000 [×] 1.005 [y^x] 72 [=] 11,456.35423

Depending on your calculator, your answer may contain fewer or more digits. □

Student Practice 7 How much money would Collette have if she invested \$1500 for 8 years at 8% annual interest if the interest is compounded quarterly? Round your answer to the nearest cent.

Exponential functions are used to describe radioactive decay. The equation $A = Ce^{kt}$ tells us how much of a radioactive chemical element is left in a sample after a specified time.

Example 8 The radioactive decay of the element americium 241 can be described by the equation

$$A = Ce^{-0.0016008t},$$

where C is the original amount of the element in the sample, A is the amount of the element remaining after t years, and $k = -0.0016008$, the decay constant for americium 241. If 10 milligrams (mg) of americium 241 is sealed in a laboratory container today, how much will theoretically be present in 2000 years? Round your answer to the nearest hundredth.

Solution Here $C = 10$ and $t = 2000$.

$$A = 10e^{-0.0016008(2000)} = 10e^{-3.2016}$$

Using a calculator, we have

$$A \approx 10(0.040697) = 0.40697 \approx 0.41 \text{ mg.}$$

The expression $10e^{-3.2016}$ can be found directly on some scientific calculators as follows:

10 [×] 3.2016 [+/−] [e^x] [=] 0.406970366

(Scientific calculators with no [e^x] key will require the keystrokes [INV] [ln] or [2nd F] [ln] or [SHIFT] [ln] in place of the [e^x].)

Thus, 0.41 milligrams of americium 241 would be present in 2000 years. □

Graphing Calculator

Exploration

Graph the function $f(t) = 10e^{-0.0016008t}$ from Example 8 for $t = 0$ to $t = 100$ years. Now graph the function for $t = 0$ to $t = 2000$ years. What significant change is there in the two graphs? From the graphs, estimate a value of t for which $f(t) = 5.0$. (Round your value of t to the nearest hundredth.)

Student Practice 8 If 20 milligrams of americium 241 is present in a sample now, how much will theoretically be present in 5000 years? Round your answer to the nearest thousandth.

12.1 Exercises MyMathLab®

Verbal and Writing Skills, Exercises 1 and 2

1. An exponential function is a function of the form ______________________.

2. The irrational number e is a number that is approximately equal to ________. (Give your answer with four decimal places.)

Graph each function.

3. $f(x) = 3^x$

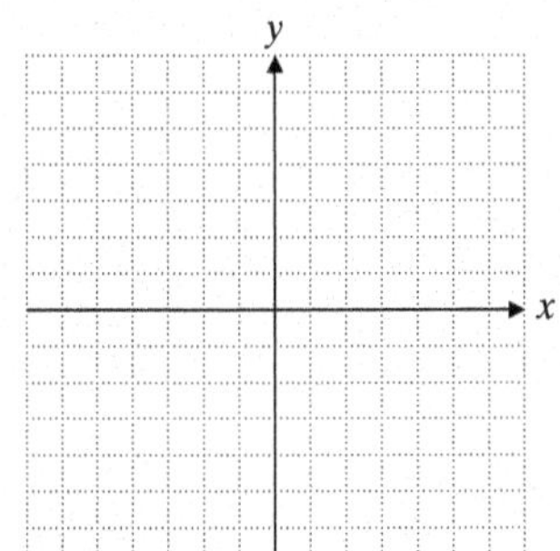

4. $f(x) = 2^x$

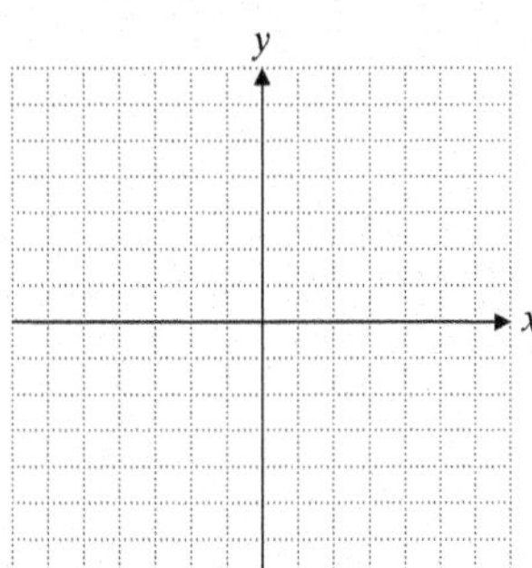

5. $f(x) = 2^{-x}$

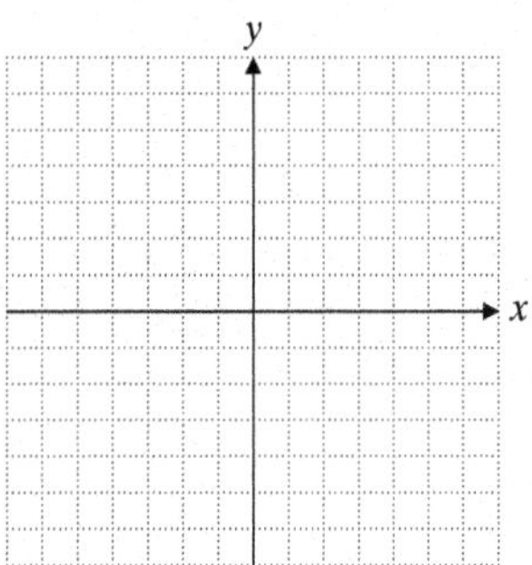

6. $f(x) = 5^{-x}$

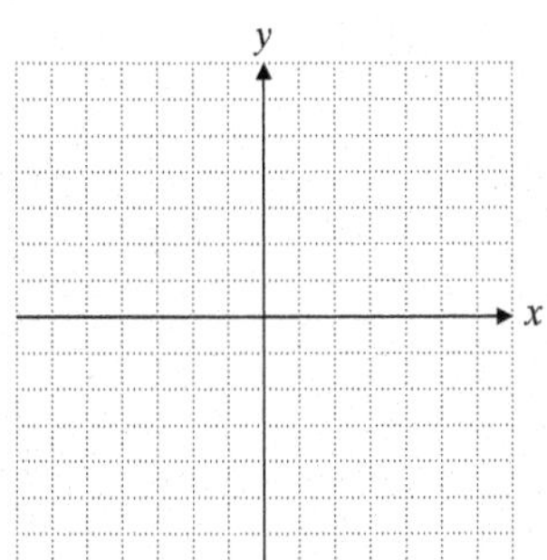

7. $f(x) = 3^{-x}$

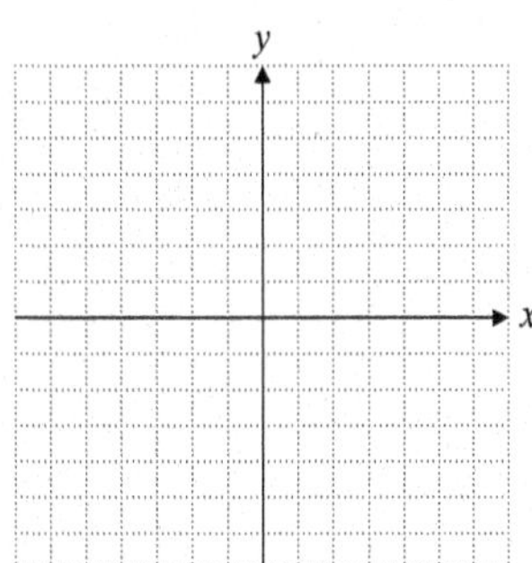

8. $f(x) = 4^{-x}$

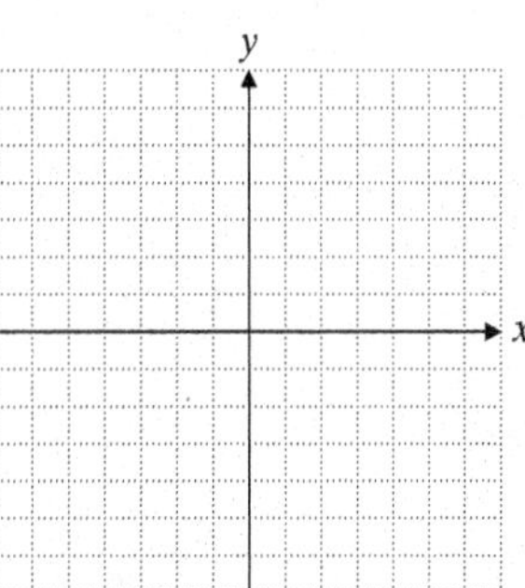

9. $f(x) = 2^{x+3}$

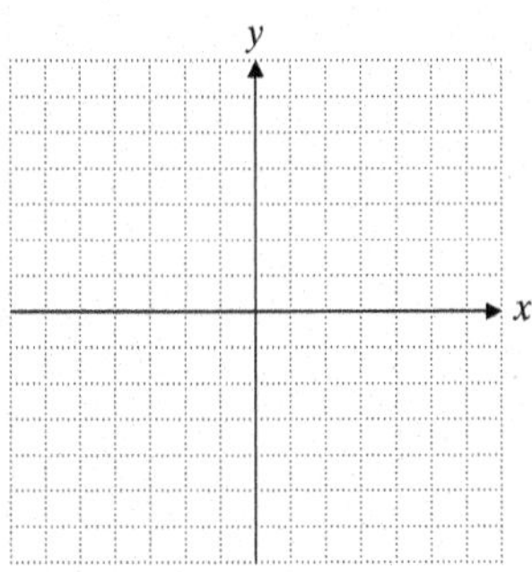

10. $f(x) = 2^{x-4}$

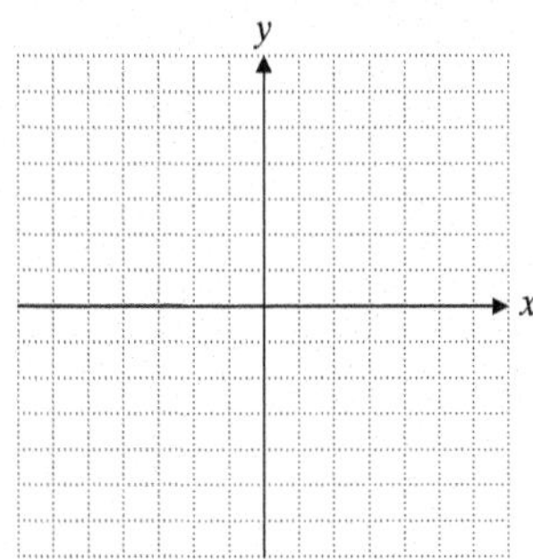

11. $f(x) = 3^{x-5}$

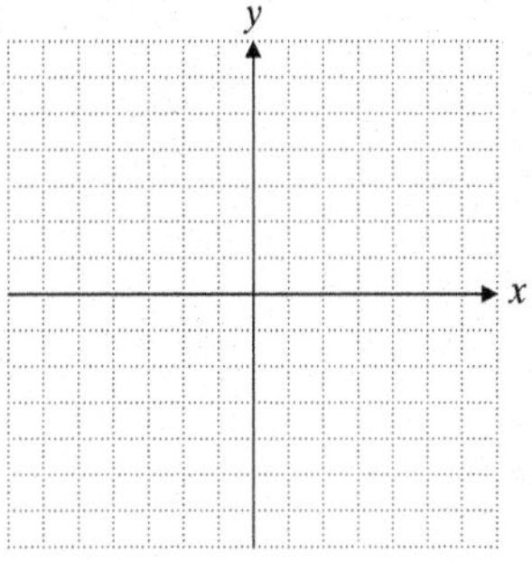

12. $f(x) = 3^{x+1}$

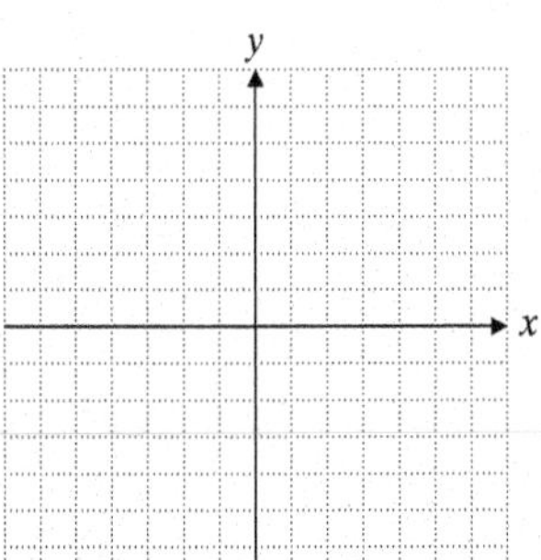

13. $f(x) = 2^x + 2$

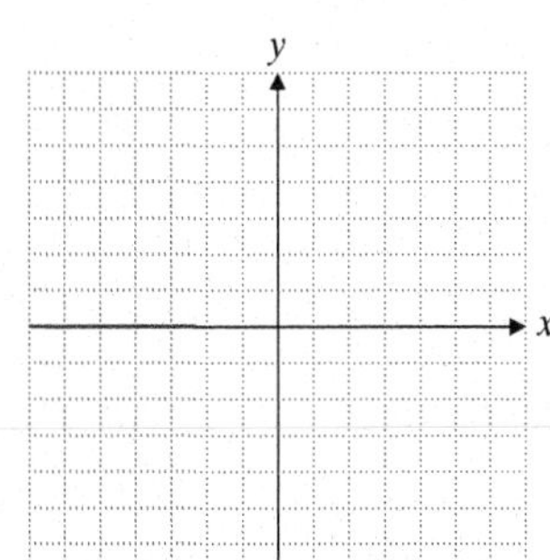

14. $f(x) = 2^x - 4$

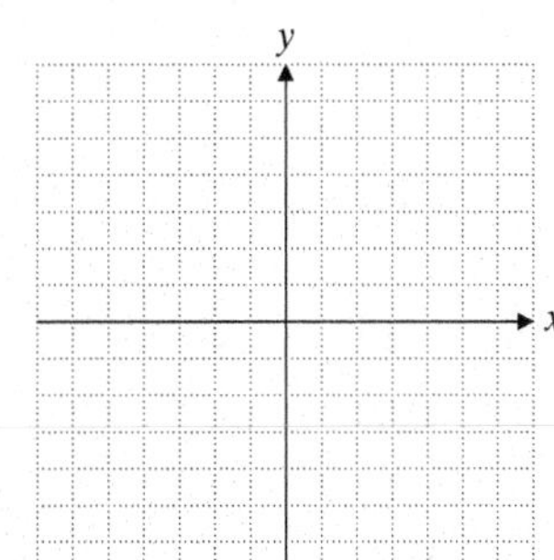

Graph each function.

15. $f(x) = e^{x-1}$

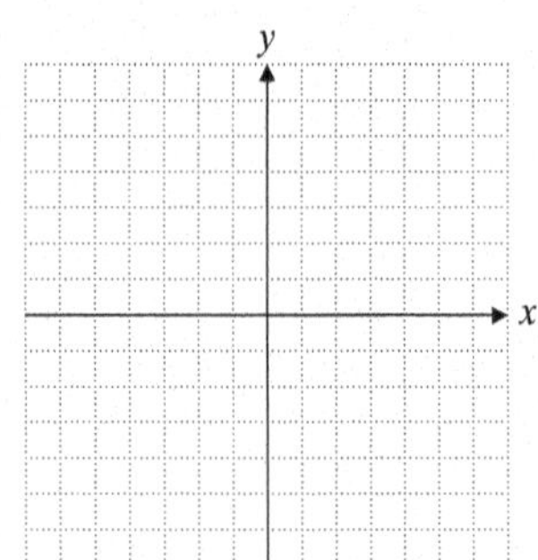

16. $f(x) = e^{x+1}$

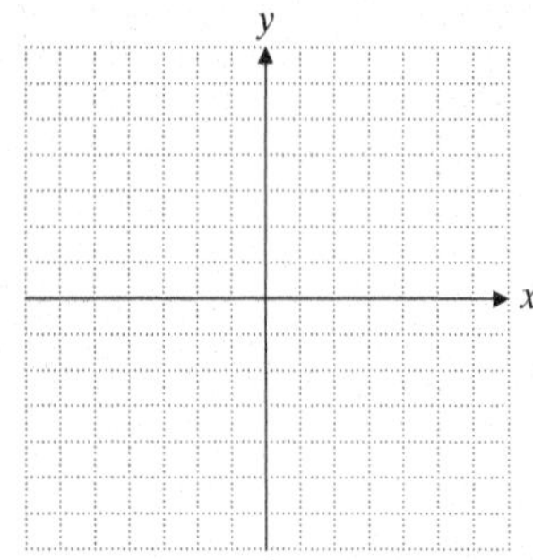

17. $f(x) = 2e^{x}$

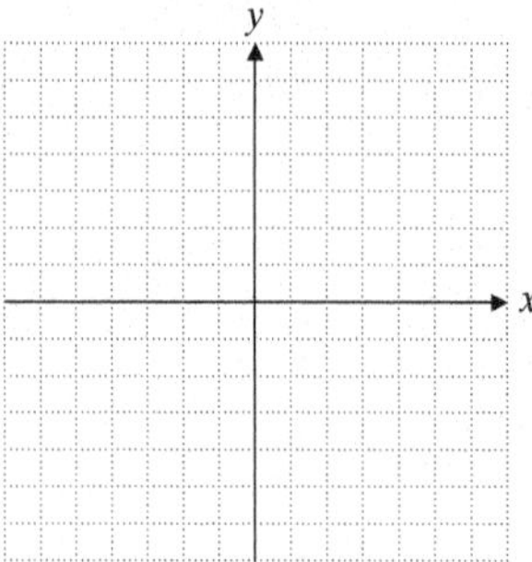

18. $f(x) = 3e^{x}$

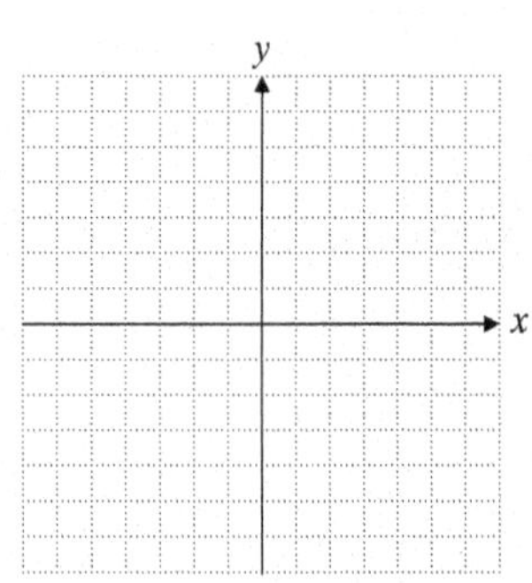

19. $f(x) = e^{1-x}$

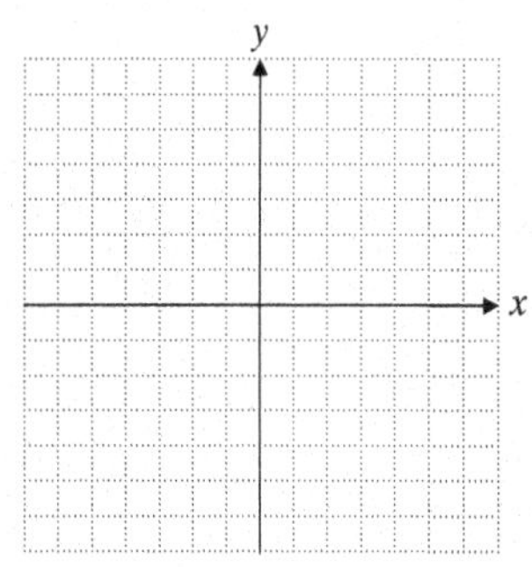

20. $f(x) = e^{2-x}$

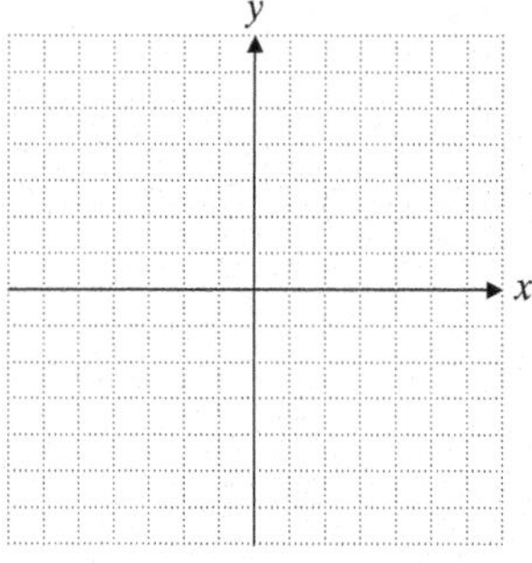

Solve for x.

21. $2^x = 4$

22. $2^x = 8$

23. $2^x = 1$

24. $2^x = 2$

25. $2^x = \frac{1}{2}$

26. $2^x = \frac{1}{64}$

27. $3^x = 81$

28. $3^x = 27$

29. $3^x = 1$

30. $3^x = 243$

31. $3^{-x} = \frac{1}{9}$

32. $3^{-x} = \frac{1}{3}$

33. $4^x = 256$

34. $4^{-x} = \frac{1}{16}$

35. $5^{x+1} = 125$

36. $2^{x-3} = 128$

37. $8^{3x-1} = 64$

38. $5^{4x+1} = 125$

To solve exercises 39–42, use the interest formula $A = P\left(1 + \frac{r}{n}\right)^{nt}$. *Round your answers to the nearest cent.*

39. ***Investments*** Alicia is investing \$2000 at an annual rate of 6.3% compounded annually. How much money will Alicia have after 3 years?

40. ***Investments*** Manza is investing \$5000 at an annual rate of 7.1% compounded annually. How much money will Manza have after 4 years?

41. ***Investments*** How much money will Isabela have in 6 years if she invests \$3000 at a 3.2% annual rate of interest compounded quarterly? How much will she have if it is compounded monthly?

42. ***Investments*** How much money will Waheed have in 3 years if he invests \$5000 at a 3.85% annual rate of interest compounded quarterly? How much will he have if it is compounded monthly?

43. ***Bacteria Culture*** The number of bacteria in a culture is given by $B(t) = 4000(2^t)$, where t is the time in hours. How many bacteria will grow in the culture in the first 3 hours? In the first 9 hours?

44. ***College Tuition*** Suppose that the cost of a college education is increasing 4% per year. The equation $C(t) = P(1.04)^t$ forecasts the tuition cost t years from now and is based on the present cost P in dollars. How much will a college now charging \$3000 for tuition charge in 10 years? How much will a college now charging \$12,000 for tuition charge in 15 years?

45. ***Diving Depth*** U.S. Navy divers off the coast of Nantucket are searching for the wreckage of an old World War II–era submarine. They have found that if the water is relatively clear and the surface is calm, the ocean filters out 18% of the sunlight for each 4 feet they descend. How much sunlight is available at a depth of 20 feet? The divers need to use underwater spotlights when the amount of sunlight is less than 10%. Will they need spotlights when working at a depth of 48 feet?

46. ***Sewer Systems*** The city of Manchester just put in a municipal sewer to solve an underground water contamination problem, and many homeowners would like to have sewer lines connected to their homes. It is expected that each year the number of homeowners who use their own private septic tanks rather than the public sewer system will decrease by 8%. What percentage of people will still be using their private septic tanks in 5 years? The city feels that the underground water contamination problem will be solved when the number of homeowners still using septic tanks is less than 10%. Will that goal be achieved in the next 25 years?

Use an exponential equation to solve each problem. Round your answers to the nearest hundredth.

47. ***Radium Decay*** The radioactive decay of radium 226 can be described by the equation $A = Ce^{-0.0004279t}$, where C is the original amount of radium and A is the amount of radium remaining after t years. If 12 milligrams of radium are sealed in a container now, how much radium will be in the container after 100 years?

48. ***Radon Decay*** The radioactive decay of radon 222 can be described by the equation $A = Ce^{-0.1813t}$, where C is the original amount of radon and A is the amount of radon after t days. If 3.5 milligrams are in a laboratory container today, how much was there in the container 1 day ago?

Atmospheric Pressure *Use the following information for exercises 49 and 50. The atmospheric pressure measured in pounds per square inch is given by the equation* $P = 14.7e^{-0.21d}$, *where d is the distance in miles above sea level. Round your answers to the nearest hundredth.*

49. What is the pressure in pounds per square inch on an American Airlines jet plane flying 10 miles above sea level?

50. What is the pressure in pounds per square inch experienced by a man on a Colorado mountain at 2 miles above sea level?

Professional Baseball Player *The average salary of an MLB player increased exponentially between 2000 and 2015. The average annual salary A (in millions) can be approximated by the equation* $A = 2.08e^{0.045t}$, *where t is the number of years since 2000. Round your answers to the nearest whole number.* (*Source:* www.cbssports.com)

51. Using the given equation, determine the average salary in 2000. In 2005. What was the percent increase from 2000 to 2005?

52. Using the given equation, determine the average salary in 2008. In 2015. What was the percent increase from 2008 to 2015?

World Population *Since the year 1750, the population of the world has been growing exponentially. The following table and graph contain population data for selected years and show a pattern of significant increases.*

Year	1750	1800	1850	1900	1950	2000	2050*
Approximate World Population in Billions	0.8	0.9	1.2	1.61	2.5	6.07	9

*estimated
Source: www.census.gov

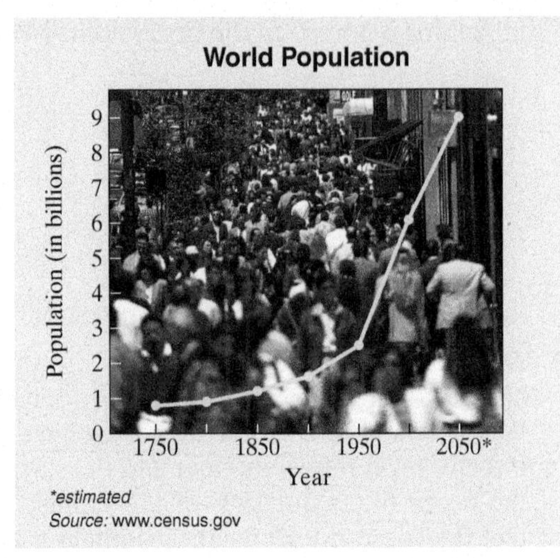

53. Based on the graph, in approximately what year did the world population reach three billion people?

54. Based on the graph, what was the approximate world population in 1900?

55. The growth rate of the world population peaked in 1963 at 2.1% per year, and growth continued at that rate through 1970. If that rate would have continued from 2000 to 2025, what would have been the world population in 2025? According to the graph, what will the population be in 2025?

56. Some leaders have reported that the growth rate of the world population in 2010 was 1.14% but continues to decrease. If this rate were to continue from 2000 to 2050, what would be the population in 2050? According to the graph, what will the population actually be in 2050?

Cumulative Review *Solve for x.*

57. **[2.3.3]** $5 - 2(3 - x) = 2(2x + 5) + 1$

58. **[2.4.1]** $\frac{7}{12} + \frac{3}{4}x + \frac{5}{4} = -\frac{1}{6}x$

Quick Quiz 12.1

1. Graph $f(x) = 2^{x+4}$.

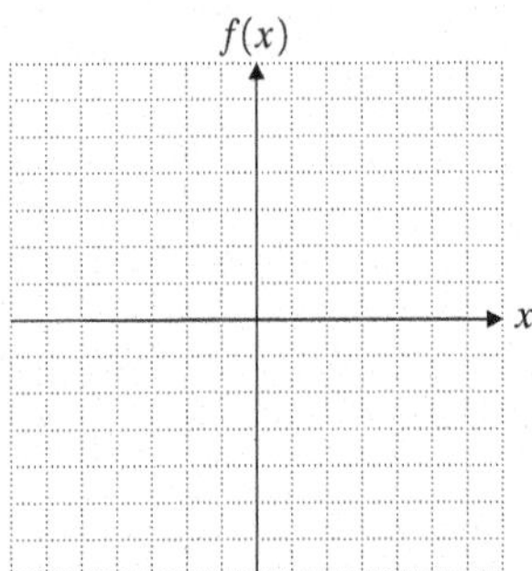

2. Dwayne invested $5500 at an annual rate of 4% compounded semiannually. How much money will Dwayne have after 6 years?

3. Solve for x. $3^{x+2} = 81$

4. **Concept Check** Explain how you would solve $4^{-x} = \frac{1}{64}$.

12.2 The Logarithmic Function

Logarithms were invented about 400 years ago by the Scottish mathematician John Napier. Napier's amazing invention reduced complicated exercises to simple subtraction and addition. Astronomers quickly saw the immense value of logarithms and began using them. The work of Johannes Kepler, Isaac Newton, and others would have been much more difficult without logarithms.

The most important thing to know for this chapter is that a logarithm is an exponent. In Section 12.1 we solved the equation $2^x = 8$. We found that $x = 3$. The question we faced was, "To what power do we raise 2 to get 8?" The answer was 3. Mathematicians have to solve this type of problem so often that we have invented a short-hand notation for asking the question. Instead of asking, "To what power do we raise 2 to get 8?" we say instead, "What is $\log_2 8$?" Both questions mean the same thing.

Now suppose we had a general equation $x = b^y$ and someone asked, "To what power do we raise b to get x?" We would abbreviate this question by asking, "What is $\log_b x$?" Thus, we see that $y = \log_b x$ is an equivalent form of the equation $x = b^y$.

The key concept you must remember is that a logarithm is an exponent. We write $\log_b x = y$ to mean that b to the power y is x. y is the exponent.

Student Learning Objectives

After studying this section, you will be able to:

1. Write exponential equations in logarithmic form.
2. Write logarithmic equations in exponential form.
3. Solve elementary logarithmic equations.
4. Graph a logarithmic function.

DEFINITION OF LOGARITHM

The **logarithm,** base b, of a *positive* number x is the power (exponent) to which the base b must be raised to produce x. That is, $y = \log_b x$ is the same as $x = b^y$, where $b > 0$ and $b \neq 1$.

Often you will need to convert logarithmic statements to exponential statements, and vice versa, to solve equations.

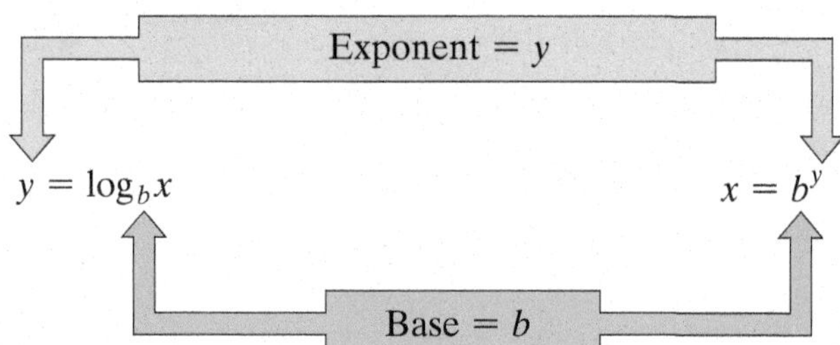

1 Writing Exponential Equations in Logarithmic Form

We begin by converting exponential statements to logarithmic statements.

Example 1 Write in logarithmic form.

(a) $81 = 3^4$ **(b)** $\dfrac{1}{100} = 10^{-2}$

Solution We use the fact that $x = b^y$ is equivalent to $\log_b x = y$.

(a) $81 = 3^4$

Here $x = 81$, $b = 3$, and $y = 4$. So $4 = \log_3 81$.

(b) $\dfrac{1}{100} = 10^{-2}$

Here $x = \dfrac{1}{100}$, $b = 10$, and $y = -2$. So $-2 = \log_{10}\left(\dfrac{1}{100}\right)$. □

Student Practice 1 Write in logarithmic form.

(a) $49 = 7^2$ **(b)** $\frac{1}{64} = 4^{-3}$

2 Writing Logarithmic Equations in Exponential Form 

If we have an equation with a logarithm in it, we can write it in the form of an exponential equation. This is a very important skill. Carefully study the following example.

Example 2 Write in exponential form.

(a) $2 = \log_5 25$ **(b)** $-4 = \log_{10}\left(\frac{1}{10{,}000}\right)$

Solution

(a) $2 = \log_5 25$

Here $y = 2$, $b = 5$, and $x = 25$. Thus, since $x = b^y$, $25 = 5^2$.

(b) $-4 = \log_{10}\left(\frac{1}{10{,}000}\right)$

Here $y = -4$, $b = 10$, and $x = \frac{1}{10{,}000}$. So $\frac{1}{10{,}000} = 10^{-4}$. □

Student Practice 2 Write in exponential form.

(a) $3 = \log_5 125$ **(b)** $-2 = \log_6\left(\frac{1}{36}\right)$

3 Solving Elementary Logarithmic Equations 

Many logarithmic equations are fairly easy to solve if we first convert them to an equivalent exponential equation.

Example 3 Solve for the variable.

(a) $\log_5 x = -3$ **(b)** $\log_a 16 = 4$

Solution

(a)
$$5^{-3} = x$$
$$\frac{1}{5^3} = x$$
$$\frac{1}{125} = x$$

(b)
$$a^4 = 16$$
$$a^4 = 2^4$$
$$a = 2$$
□

Student Practice 3 Solve for the variable.

(a) $\log_b 125 = 3$ **(b)** $\log_{1/2} 32 = x$

With this knowledge we have the ability to solve an additional type of exercise.

Example 4 Evaluate. $\log_3 81$

Solution Now, what exactly is the exercise asking for? It is asking, "To what power must we raise 3 to get 81?" Since we do not know the power, we call it x. We have

$$\log_3 81 = x$$
$$81 = 3^x \quad \text{Write an equivalent exponential equation.}$$
$$3^4 = 3^x \quad \text{Write 81 as } 3^4.$$
$$x = 4 \quad \text{If } b^x = b^y, \text{ then } x = y \text{ for } b > 0 \text{ and } b \neq 1.$$

Thus, $\log_3 81 = 4$. □

 **Student Practice 4** Evaluate. $\log_{10} 0.1$

4 Graphing a Logarithmic Function 

We found in Chapter 11 that the graphs of a function and its inverse have an interesting property. They are symmetric to one another with respect to the line $y = x$. We also found in Chapter 11 that the procedure for finding the inverse of a function is to interchange the x and y variables. For example, $y = 2x + 3$ and $x = 2y + 3$ are inverse functions. In similar fashion, $y = 2^x$ and $x = 2^y$ are inverse functions. Another way to write $x = 2^y$ is the logarithmic equation $y = \log_2 x$. Thus, the logarithmic function $y = \log_2 x$ is the **inverse** of the exponential function $y = 2^x$. If we graph the function $y = 2^x$ and $y = \log_2 x$ on the same set of axes, the graph of one is the reflection of the other about the line $y = x$.

Example 5 Graph $y = \log_2 x$.

Solution If we write $y = \log_2 x$ in exponential form, we have $x = 2^y$. We make a table of values and graph the function $x = 2^y$.

In each case, we pick a value of y as a first step.

If $y = -2$, $\quad x = 2^y = 2^{-2} = \frac{1}{2^2} = \frac{1}{4}$.

If $y = -1$, $\quad x = 2^{-1} = \frac{1}{2}$.

If $y = 0$, $\quad x = 2^0 = 1$.

If $y = 1$, $\quad x = 2^1 = 2$.

If $y = 2$, $\quad x = 2^2 = 4$.

x	y
$\frac{1}{4}$	-2
$\frac{1}{2}$	-1
1	0
2	1
4	2

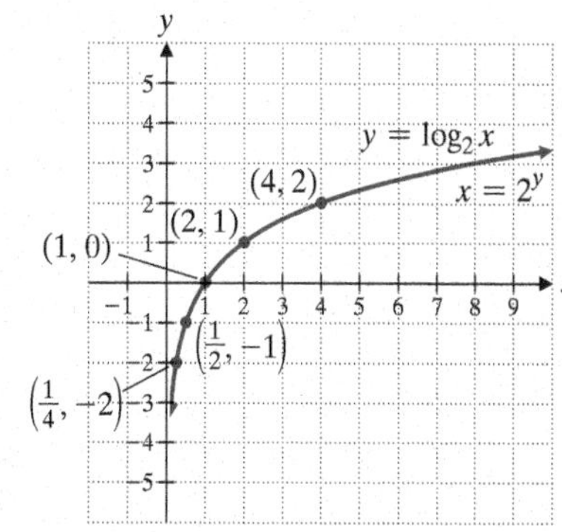

□

Student Practice 5 Graph $y = \log_{1/2} x$.

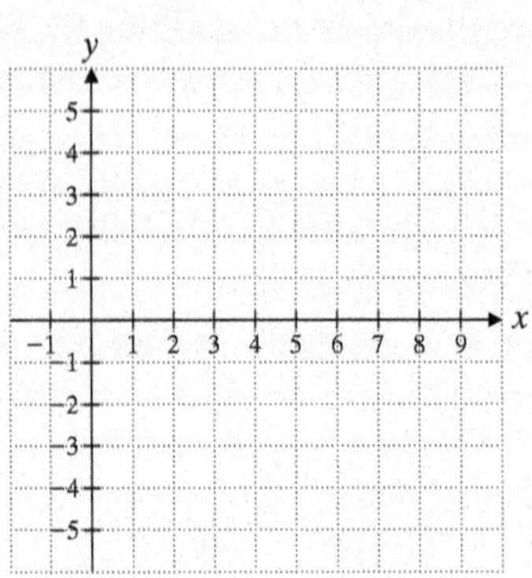

$f(x) = a^x$ and $f(x) = \log_a x$ are inverse functions. As such they have all the properties of inverse functions. We will review a few of these properties as we study the graphs of two inverse functions, $y = 2^x$ and $y = \log_2 x$.

Example 6 Graph $y = \log_2 x$ and $y = 2^x$ on the same set of axes.

Solution Make a table of values (ordered pairs) for each equation. Then draw each graph.

$y = 2^x$

x	y
-1	$\frac{1}{2}$
0	1
1	2
2	4

$y = \log_2 x$

x	y
$\frac{1}{2}$	-1
1	0
2	1
4	2

Coordinates of ordered pairs are reversed

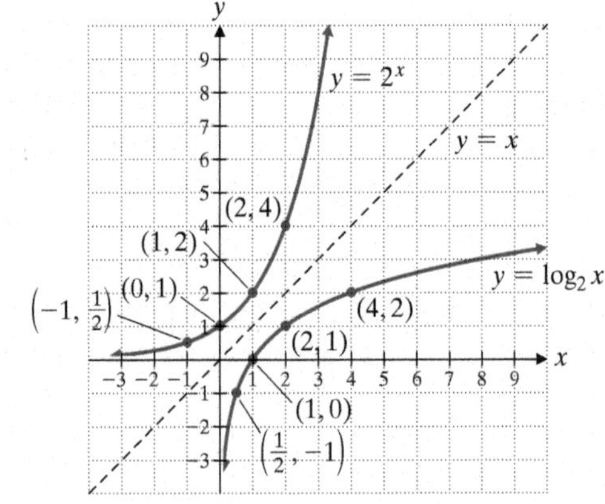

Note that $y = \log_2 x$ is the inverse of $y = 2^x$ because the ordered pairs (x, y) are reversed. The sketch of the two equations shows that they are inverses. If we reflect the graph of $y = 2^x$ about the line $y = x$, it will coincide with the graph of $y = \log_2 x$.

Recall that in function notation, f^{-1} means the inverse function of f. Thus, if we write $f(x) = \log_2 x$, then $f^{-1}(x) = 2^x$. □

Student Practice 6 Graph $y = \log_6 x$ and $y = 6^x$ on the same set of axes.

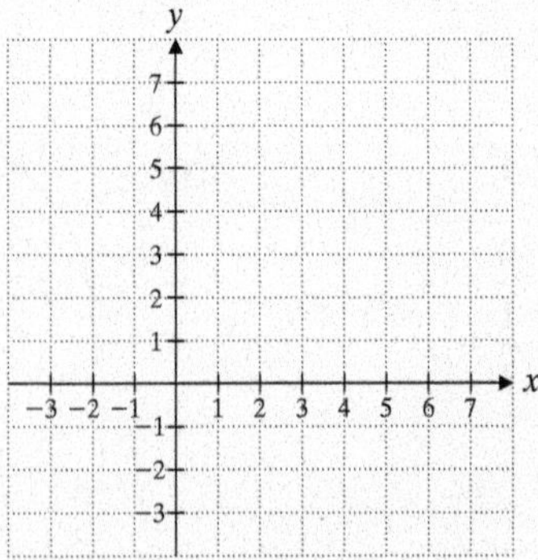

12.2 Exercises MyMathLab®

Verbal and Writing Skills, Exercises 1–4

1. A logarithm is an _______________.

2. In the equation $y = \log_b x$, the value b is called the _______________.

3. In the equation $y = \log_b x$, the domain (the set of permitted values of x) is _______________.

4. In the equation $y = \log_b x$, the permitted values of b are _______________.

Write in logarithmic form.

5. $49 = 7^2$

6. $216 = 6^3$

7. $128 = 2^7$

8. $100 = 10^2$

9. $0.001 = 10^{-3}$

10. $0.04 = 5^{-2}$

11. $\frac{1}{32} = 2^{-5}$

12. $\frac{1}{64} = 2^{-6}$

13. $y = e^5$

14. $y = e^{-12}$

Write in exponential form.

15. $2 = \log_3 9$

16. $2 = \log_2 4$

17. $0 = \log_{17} 1$

18. $0 = \log_{13} 1$

19. $\frac{1}{2} = \log_{16} 4$

20. $\frac{1}{2} = \log_{64} 8$

21. $-2 = \log_{10} 0.01$

22. $-1 = \log_{10} 0.1$

23. $-4 = \log_3\left(\frac{1}{81}\right)$

24. $-7 = \log_2\left(\frac{1}{128}\right)$

25. $-\frac{3}{2} = \log_e x$

26. $-\frac{2}{5} = \log_e x$

Solve.

27. $\log_2 x = 4$

28. $3 = \log_3 x$

29. $\log_{10} x = -3$

30. $\log_{10} x = -1$

31. $\log_4 64 = y$

32. $\log_7 343 = y$

33. $\log_8\left(\frac{1}{64}\right) = y$

34. $\log_3\left(\frac{1}{243}\right) = y$

35. $\log_a 121 = 2$

36. $\log_a 81 = 2$

37. $\log_a 1000 = 3$

38. $\log_a 100 = 2$

39. $\log_{25} 5 = w$

40. $\log_8 2 = w$

41. $\log_3\left(\frac{1}{3}\right) = w$

42. $\log_{37} 1 = w$

43. $\log_{15} w = 0$

44. $\log_{10} w = -4$

45. $\log_w 3 = \frac{1}{2}$

46. $\log_w 2 = \frac{1}{3}$

Evaluate.

47. $\log_{10} 0.001$

48. $\log_{10} 0.0001$

49. $\log_2 128$

50. $\log_5 125$

51. $\log_{23} 1$

52. $\log_{18} \frac{1}{18}$

53. $\log_6 \sqrt{6}$

54. $\log_3 \sqrt{3}$

55. $\log_{57} 1$

56. $\log_3 \frac{1}{27}$

Graph.

57. $\log_3 x = y$

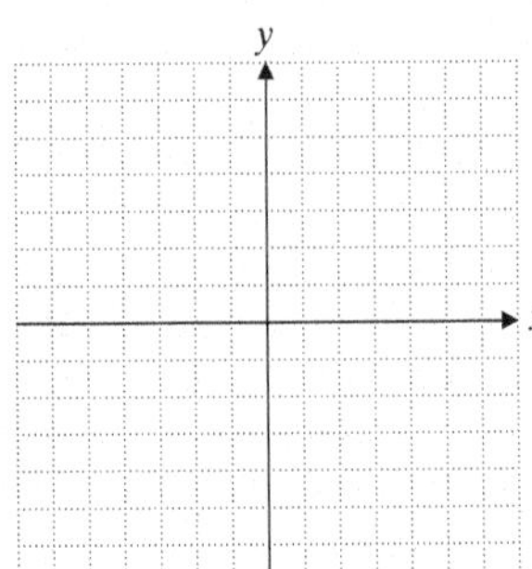

58. $\log_4 x = y$

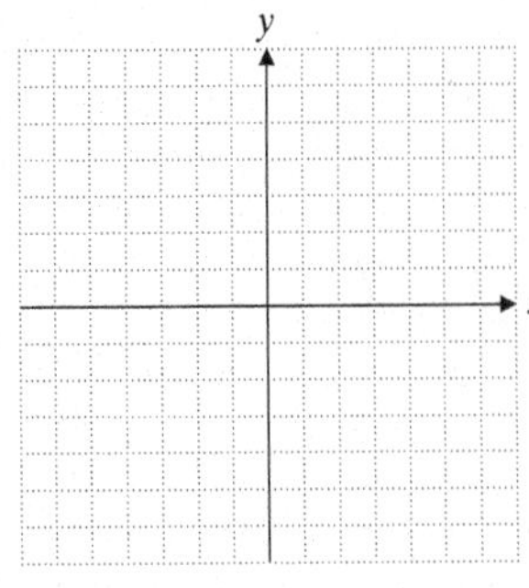

59. $\log_{1/4} x = y$

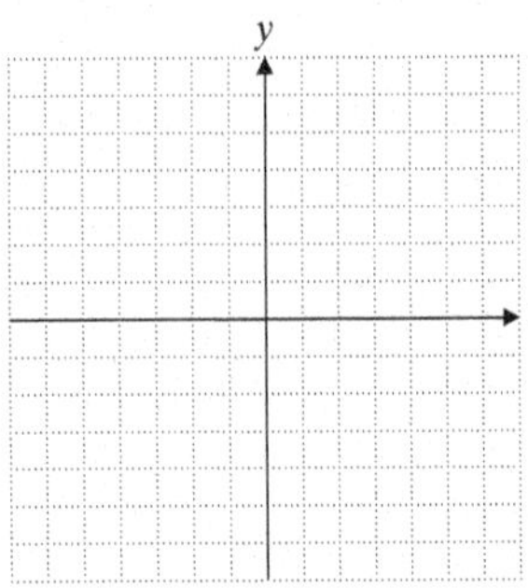

60. $\log_{1/3} x = y$

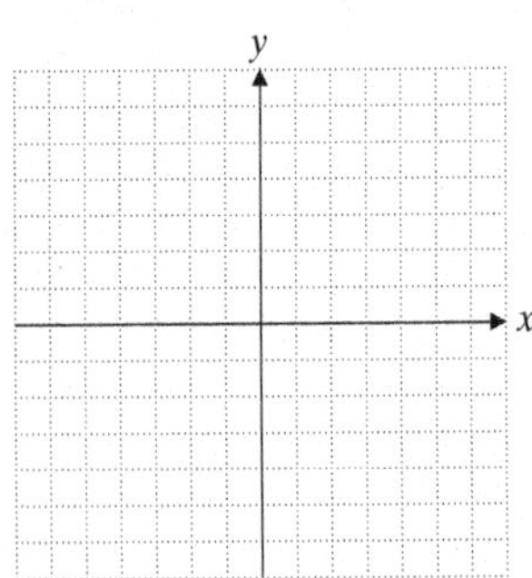

61. $\log_{10} x = y$

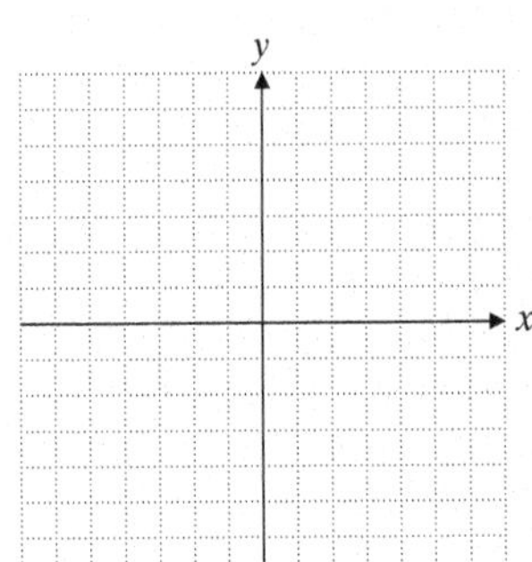

62. $\log_8 x = y$

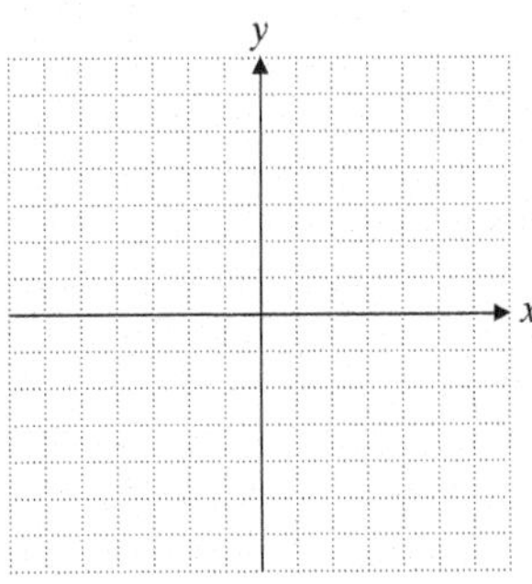

On one coordinate plane, graph the function f and the function f^{-1}. Then graph a dashed line for the equation $y = x$.

63. $f(x) = \log_3 x, f^{-1}(x) = 3^x$

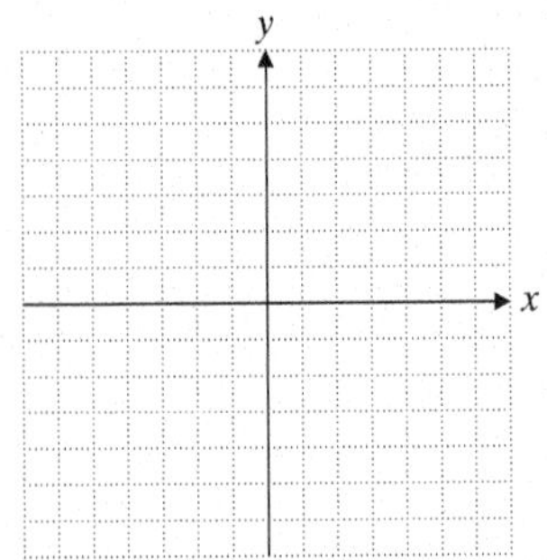

64. $f(x) = \log_4 x, f^{-1}(x) = 4^x$

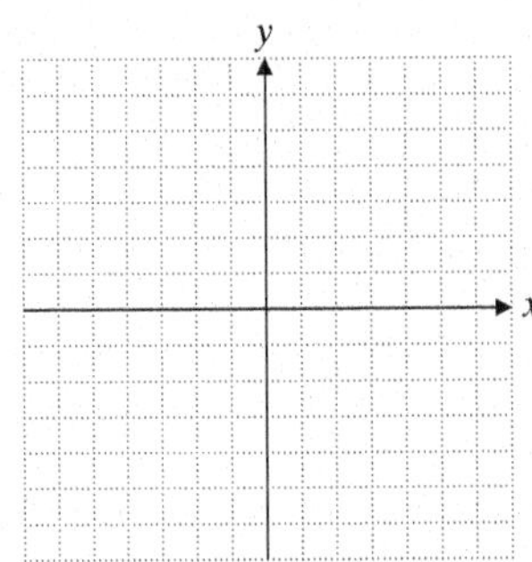

Applications

Chemist *To determine whether a solution is an acid or a base, chemists check the solution's* pH. *A solution is an acid if its* pH *is less than 7 and a base if its* pH *is greater than 7. The* pH *is defined by* $\text{pH} = -\log_{10}[\text{H}^+]$, *where* $[\text{H}^+]$ *is the concentration of hydrogen ions in the solution.*

65. The concentration of hydrogen ions in bleach is approximately $10^{-12.5}$. What is the pH of bleach?

66. The concentration of hydrogen ions in artichokes is approximately $10^{-5.6}$. What is the pH of artichokes?

67. A food scientist knows that for jelly to gel properly, its pH needs to be close to 3.5. Find the concentration of hydrogen ions in the jelly.

68. A lawn care employee tested a customer's soil and found that it had a pH of 7. Find the concentration of hydrogen ions in the soil.

69. The chef of The Depot Restaurant has prepared a special balsamic vinaigrette salad dressing. What is the pH of the dressing if the concentration of hydrogen ions is 1.103×10^{-3}? The logarithm, base 10, of 0.001103 is approximately -2.957424488. Round your answer to three decimal places.

70. The EPA is testing a batch of experimental industrial solvent with a pH of 9.25. Find the concentration of hydrogen ions in the solution. Give your answer in scientific notation rounded to three decimal places.

Marketing Manager *Jamie is the marketing manager for Easy Tax, a company that produces tax software for people who want to use their home computers to file their income tax returns. In the first two years of sales of the software, Jamie has found that for N sets of software to be sold, the company needs to invest d dollars for advertising according to the equation*

$$N = 1200 + 2500 \log_{10} d,$$

where d is always a positive number not less than 1.

71. How many sets of software were sold when they spent \$10,000 on advertising?

72. How many sets of software were sold when they spent \$100,000 on advertising?

73. They have a goal for two years from now of selling 18,700 sets of software. How much should they spend on advertising two years from now?

74. They have a goal for next year of selling 16,200 sets of software. How much should they spend on advertising next year?

Cumulative Review

75. **[3.4.4]** Find an equation of the line perpendicular to $y = -\frac{2}{3}x + 4$ that contains $(-4, 1)$.

76. **[3.3.1]** Find the slope of the line containing $(-6, 3)$ and $(-1, 2)$.

77. **[12.1.3]** ***Viral Culture*** The number of viral cells in a laboratory culture in a biology research project is given by $A(t) = 9000(2^t)$, where t is measured in hours.

(a) How many viral cells will be in the culture after the first 2 hours?

(b) How many viral cells will be in the culture after the first 12 hours?

78. **[12.1.3]** ***College Tuition*** Assume the cost of a college education is increasing at 4% per year. The equation $C(t) = P(1.04)^t$ forecasts the tuition cost t years from now, based on the present cost P in dollars.

(a) How much will a college now charging \$4400 for tuition charge in 5 years?

(b) How much will a college now charging \$16,500 for tuition charge in 10 years?

Quick Quiz 12.2

1. Evaluate $\log_3 81$.

2. Solve for x.

$$\log_2 x = 6$$

3. Solve for w.

$$\log_{27} 3 = w$$

4. Concept Check Explain how you would solve for x.

$$-\frac{1}{2} = \log_e x$$

12.3 Properties of Logarithms

Student Learning Objectives

After studying this section, you will be able to:

1. Use the property $\log_b MN = \log_b M + \log_b N$.
2. Use the property $\log_b\left(\frac{M}{N}\right) = \log_b M - \log_b N$.
3. Use the property $\log_b M^p = p \log_b M$.
4. Solve simple logarithmic equations.

1 Using the Property $\log_b MN = \log_b M + \log_b N$

We have already said that logarithms reduce complex expressions to addition and subtraction. The following properties show us how to use logarithms in this way.

> **PROPERTY 1: THE LOGARITHM OF A PRODUCT**
>
> For any positive real numbers M and N and any positive base $b \neq 1$,
>
> $$\log_b MN = \log_b M + \log_b N.$$

To see that this property is true, we let

$$\log_b M = x \quad \text{and} \quad \log_b N = y,$$

where x and y are any values. Now we write the statements in exponential notation:

$$b^x = M \quad \text{and} \quad b^y = N.$$

Then

$$MN = b^x b^y = b^{x+y} \quad \text{Laws of exponents.}$$

If we convert this equation to logarithmic form, then we have the following:

$$\log_b MN = x + y \quad \text{Definition of logarithm.}$$

$$\log_b MN = \log_b M + \log_b N \quad \text{By substitution.}$$

Note that the logarithms must have the same base.

Example 1 Write $\log_3 XZ$ as a sum of logarithms.

Solution By property 1, $\log_3 XZ = \log_3 X + \log_3 Z$. □

Student Practice 1 Write $\log_4 WXY$ as a sum of logarithms.

Example 2 Write $\log_3 16 + \log_3 x + \log_3 y$ as a single logarithm.

Solution If we extend our rule, we have $\log_b MNP = \log_b M + \log_b N + \log_b P$. Thus,

$$\log_3 16 + \log_3 x + \log_3 y = \log_3 16xy.$$ □

Student Practice 2 Write $\log_7 w + \log_7 8 + \log_7 x$ as a single logarithm.

2 Using the Property $\log_b\left(\frac{M}{N}\right) = \log_b M - \log_b N$

Property 2 is similar to property 1 except that it involves two expressions that are divided, not multiplied.

> **PROPERTY 2: THE LOGARITHM OF A QUOTIENT**
>
> For any positive real numbers M and N and any positive base $b \neq 1$,
>
> $$\log_b\left(\frac{M}{N}\right) = \log_b M - \log_b N.$$

Property 2 can be proved using a similar approach to the one used to prove property 1. We encourage you to try to prove property 2.

Example 3 Write $\log_3\left(\frac{29}{7}\right)$ as the difference of two logarithms.

Solution

$$\log_3\left(\frac{29}{7}\right) = \log_3 29 - \log_3 7$$ □

 **Student Practice 3** Write $\log_3\left(\frac{17}{5}\right)$ as the difference of two logarithms.

Example 4 Express $\log_b 36 - \log_b 9$ as a single logarithm.

Solution

$$\log_b 36 - \log_b 9 = \log_b\left(\frac{36}{9}\right) = \log_b 4$$ □

 **Student Practice 4** Express $\log_b 132 - \log_b 4$ as a single logarithm.

CAUTION: Be sure you understand property 2!

$$\frac{\log_b M}{\log_b N} \neq \log_b M - \log_b N$$

Do you see why?

3 Using the Property $\log_b M^p = p \log_b M$

We now introduce the third property. We encourage you to try to prove Property 3.

PROPERTY 3: THE LOGARITHM OF A NUMBER RAISED TO A POWER

For any positive real number M, any real number p, and any positive base $b \neq 1$,

$$\log_b M^p = p \log_b M.$$

Example 5 Write $\frac{1}{3}\log_b x + 2\log_b w - 3\log_b z$ as a single logarithm.

Solution First, we must eliminate the coefficients of the logarithmic terms.

$$\log_b x^{1/3} + \log_b w^2 - \log_b z^3 \quad \text{By property 3.}$$

Now we can combine either the sum or difference of the logarithms. We'll do the sum.

$$\log_b x^{1/3}w^2 - \log_b z^3 \quad \text{By property 1.}$$

Now we combine the difference.

$$\log_b\left(\frac{x^{1/3}w^2}{z^3}\right) \quad \text{By property 2.}$$ □

Student Practice 5 Write $\frac{1}{3}\log_7 x - 5\log_7 y$ as one logarithm.

Example 6 Write $\log_b\left(\dfrac{x^4y^3}{z^2}\right)$ as a sum or difference of logarithms.

Solution

$$\log_b\left(\frac{x^4y^3}{z^2}\right) = \log_b x^4y^3 - \log_b z^2 \quad \text{By property 2.}$$
$$= \log_b x^4 + \log_b y^3 - \log_b z^2 \quad \text{By property 1.}$$
$$= 4\log_b x + 3\log_b y - 2\log_b z \quad \text{By property 3.}$$

□

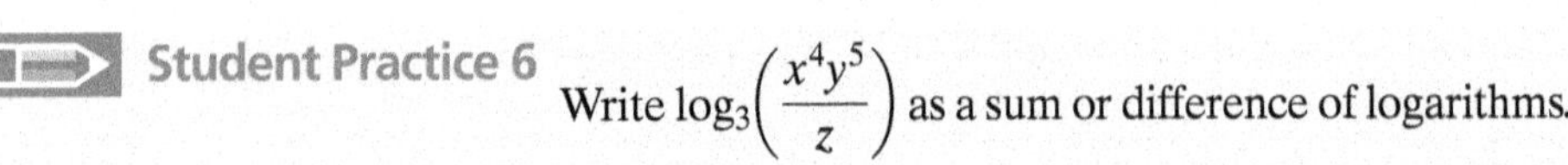

Student Practice 6 Write $\log_3\left(\dfrac{x^4y^5}{z}\right)$ as a sum or difference of logarithms.

4 Solving Simple Logarithmic Equations

A major goal in solving many logarithmic equations is to obtain a logarithm on one side of the equation and no logarithm on the other side. In Example 7 we will use property 1 to combine two separate logarithms that are added.

Example 7 Find x if $\log_2 x + \log_2 5 = 3$.

Solution

$$\log_2 5x = 3 \quad \text{Use property 1.}$$
$$5x = 2^3 \quad \text{Convert to exponential form.}$$
$$5x = 8 \quad \text{Simplify.}$$
$$x = \frac{8}{5} \quad \text{Divide both sides by 5.}$$

□

Student Practice 7 Find x if $\log_4 x + \log_4 5 = 2$.

In Example 8, two logarithms are subtracted on the left side of the equation. We can use property 2 to combine these two logarithms. This will allow us to obtain the form of one logarithm on one side of the equation and no logarithm on the other side.

MC **Example 8** Find x if $\log_3(x + 4) - \log_3(x - 4) = 2$.

Solution

$$\log_3\left(\frac{x+4}{x-4}\right) = 2 \quad \text{Use property 2.}$$
$$\frac{x+4}{x-4} = 3^2 \quad \text{Convert to exponential form.}$$
$$x + 4 = 9(x - 4) \quad \text{Multiply each side by } (x - 4).$$
$$x + 4 = 9x - 36 \quad \text{Simplify.}$$
$$40 = 8x$$
$$5 = x$$

□

Student Practice 8 Find x if $\log_{10} x - \log_{10}(x + 3) = -1$.

To solve some logarithmic equations, we need a few additional properties of logarithms. We state these properties now.

The following properties are true for all positive values of $b \neq 1$ and all positive values of x and y.

Property 4 $\log_b b = 1$

Property 5 $\log_b 1 = 0$

Property 6 If $\log_b x = \log_b y$, then $x = y$.

We now illustrate each property in Example 9.

Example 9

(a) Evaluate $\log_7 7$.

(b) Evaluate $\log_5 1$.

(c) Find x if $\log_3 x = \log_3 17$.

Solution

(a) $\log_7 7 = 1$ because $\log_b b = 1$. Property 4.

(b) $\log_5 1 = 0$ because $\log_b 1 = 0$. Property 5.

(c) If $\log_3 x = \log_3 17$, then $x = 17$. Property 6. □

Student Practice 9 Evaluate.

(a) $\log_7 1$ **(b)** $\log_8 8$ **(c)** Find y if $\log_{12} 13 = \log_{12}(y + 2)$.

We now have the mathematical tools needed to solve a variety of logarithmic equations.

Example 10 Find x if $2 \log_7 3 - 4 \log_7 2 = \log_7 x$.

Solution We can use property 3 in two cases.

$$2 \log_7 3 = \log_7 3^2 = \log_7 9$$
$$4 \log_7 2 = \log_7 2^4 = \log_7 16$$

By substituting these results, we have the following.

$$\log_7 9 - \log_7 16 = \log_7 x$$
$$\log_7\left(\frac{9}{16}\right) = \log_7 x \quad \text{Property 2.}$$
$$\frac{9}{16} = x \quad \text{Property 6.}$$

□

Student Practice 10 Find x if $\log_3 2 - \log_3 5 = \log_3 6 + \log_3 x$.

12.3 Exercises MyMathLab®

Express as a sum of logarithms.

1. $\log_3 AB$
2. $\log_{12} CD$
3. $\log_5(7 \cdot 11)$
4. $\log_6(13 \cdot 5)$
5. $\log_b 9f$
6. $\log_c 10a$

Express as a difference of logarithms.

7. $\log_9\left(\frac{2}{7}\right)$
8. $\log_{11}\left(\frac{23}{17}\right)$
9. $\log_b\left(\frac{12}{Z}\right)$
10. $\log_a\left(\frac{G}{7}\right)$
11. $\log_a\left(\frac{E}{F}\right)$
12. $\log_5\left(\frac{X}{Y}\right)$

Express as a product.

13. $\log_8 a^7$
14. $\log_3 c^8$
15. $\log_b A^{-2}$
16. $\log_a B^{-5}$
17. $\log_5 \sqrt{w}$
18. $\log_4 \sqrt{m}$

Mixed Practice *Write each expression as a sum or difference of logarithms of a single number or variable.*

19. $\log_8 x^2y$
20. $\log_3 ab^3$
21. $\log_{11}\left(\frac{6M}{N}\right)$
22. $\log_5\left(\frac{2C}{7}\right)$
23. $\log_2\left(\frac{5xy^4}{\sqrt{z}}\right)$
24. $\log_9\left(\frac{4x^2\sqrt[3]{y}}{z^3}\right)$
25. $\log_a\left(\sqrt[3]{\frac{x^4}{y}}\right)$
26. $\log_a\left(\sqrt[5]{\frac{y}{z^3}}\right)$

Write as a single logarithm.

27. $\log_4 13 + \log_4 y + \log_4 3$
28. $\log_5 8 + \log_5 m + \log_5 n$
29. $5\log_3 x - \log_3 7$
30. $3\log_8 5 - \log_8 z$
31. $2\log_b 7 + 3\log_b y - \frac{1}{2}\log_b z$
32. $4\log_b 2 + \frac{1}{3}\log_b z - 5\log_b y$

Use the properties of logarithms to simplify each of the following.

33. $\log_3 3$

34. $\log_7 7$

35. $\log_e e$

36. $\log_{10} 10$

37. $\log_9 1$

38. $\log_e 1$

39. $3 \log_7 7 + 4 \log_7 1$

40. $\frac{1}{2} \log_5 5 - 8 \log_5 1$

Find x in each of the following.

41. $\log_8 x = \log_8 7$

42. $\log_9 x = \log_9 5$

43. $\log_5(2x + 7) = \log_5 29$

44. $\log_{15} 26 = \log_{15}(3x - 1)$

45. $\log_3 1 = x$

46. $\log_7 1 = x$

47. $\log_7 7 = x$

48. $\log_5 5 = x$

49. $\log_9 27 + \log_9 x = 2$

50. $\log_{10} x + \log_{10} 5 = 1$

51. $\log_2 7 = \log_2 x - \log_2 3$

52. $\log_6 1 = \log_6 x - \log_6 9$

53. $3 \log_5 x = \log_5 8$

54. $\frac{1}{2} \log_3 x = \log_3 4$

55. $\log_e x = \log_e 5 + 1$

56. $\log_e x + \log_e 7 = 2$

57. $\log_6(5x + 21) - \log_6(x + 3) = 1$

58. $\log_2(3x + 8) - \log_2(x - 1) = 2$

59. It can be shown that $y = b^{\log_b y}$. Use this property to evaluate $5^{\log_5 4} + 3^{\log_3 2}$.

60. It can be shown that $x = \log_b b^x$. Use this property to evaluate $\log_7 \sqrt[4]{7} + \log_6 \sqrt[12]{6}$.

Cumulative Review

▲ **61.** **[1.8.2]** Find the volume of a cylinder with a radius of 2 meters and a height of 5 meters. Round your answer to the nearest tenth.

▲ **62.** **[1.8.2]** Find the area of a circle whose radius is 4 meters. Round your answer to the nearest tenth. Use 3.14 for π.

63. **[4.1.4]** Solve the system.

$$5x + 3y = 9$$
$$7x - 2y = 25$$

64. **[4.2.2]** Solve the system.

$$2x - y + z = 3$$
$$x + 2y + 2z = 1$$
$$4x + y + 2z = 0$$

At the end of 2010, a study ranked the top ten cities where households owed the highest percent of their average yearly income to credit card companies. Below is the information for Wilmington, North Carolina (ranked #1), and Toledo, Ohio (ranked #3). For each city, find the average credit card debt per household. Then find the percent of the average yearly household income that was owed to credit card companies. Round the debt to the nearest dollar and the percent to the nearest hundredth of a percent.

65. [0.5.4, 5.2.2] ***Wilmington, North Carolina***
Number of households: 1.53×10^5

Total credit card debt for all households: $\$1.117 \times 10^9$

Average yearly household income: $42,392

66. [0.5.4, 5.2.2] ***Toledo, Ohio***
Number of households: 2.56×10^5

Total credit card debt for all households: $\$1.9 \times 10^9$

Average yearly household income: $44,349

Quick Quiz 12.3

1. Express $\log_5\left(\frac{\sqrt[3]{x}}{y^4}\right)$ as a sum or difference of logarithms.

2. Express $3 \log_6 x + \log_6 y - \log_6 5$ as a single logarithm.

3. Find x if $\frac{1}{2}\log_4 x = \log_4 25$.

4. Concept Check Explain how you would simplify $\log_{10} 0.001$.

Use Math to Save Money

Did You Know? You can save money by purchasing store brand products.

Food and Rice Prices

Understanding the Problem:

Many large stores and chains have their own brands of products. Store brand products often cost much less than the equivalent name brand products. For example, a one-pound bag of rice from a national name brand can cost $2.50. A one-pound bag of the same rice from a store brand might cost only $1.00.

Making a Plan:

Lucy and her family enjoy rice as part of their dinner four times a week. Lucy has a family of four that consumes approximately 2/3 cup of rice with each meal. Lucy wants to calculate the cost of feeding her family and see if they can save money.

Step 1: Lucy needs to know how much rice her family consumes in a week.

Task 1: How many cups of rice does Lucy's family eat in a week?

Task 2: If each cup of rice weighs approximately 21 ounces, what is the weight in ounces of the rice Lucy's family eats in a week?

Task 3: There are 16 ounces in a pound. Find the number of pounds of raw rice Lucy's family eats each week?

Step 2: Lucy notices that her supermarket has its own brand of rice that is much less expensive than the brand she normally buys. She usually buys name brand rice for $2.66 per pound. The store brand is only 88 cents per pound.

Task 4: Find out how much Lucy spends per week on the name brand rice.

Task 5: Find out how much Lucy could save per week by buying the same amount of rice from the store brand.

Task 6: Find the percent savings that Lucy gets by buying the store brand rice instead of the name brand.

Finding a Solution:

Step 3: Lucy realizes that she can have this kind of savings each week on all the food she buys by choosing the store brand products over the name brand.

Task 7: If Lucy typically spends $162 per week on groceries by buying the name brand products, how much could she save in a year by purchasing all store brand products?

Task 8: Lucy finds that her family still prefers some of the name brand products. She continues to buy some of the name brand products and some of the store brands. She finds that her weekly grocery bill is reduced to around $130 per week. It this continues, how much will she save in the course of a year?

Applying the Situation to Your Life:

You should calculate how much you spend on groceries every week. Then see how you can reduce that by purchasing store brands. You may not change to the store brand for all your shopping, but for every case where you do, you can often save money.

12.4 Common Logarithms, Natural Logarithms, and Change of Base of Logarithms

Student Learning Objectives

After studying this section, you will be able to:

1. Find common logarithms.
2. Find the antilogarithm of a common logarithm.
3. Find natural logarithms.
4. Find the antilogarithm of a natural logarithm.
5. Evaluate a logarithm to a base other than 10 or e.

1 Finding Common Logarithms on a Scientific Calculator 

Although we can find a logarithm of a number for any positive base except 1, the most frequently used bases are 10 and e. Base 10 logarithms are called *common logarithms* and are usually written with no subscript.

DEFINITION OF COMMON LOGARITHM

For all real numbers $x > 0$, the **common logarithm** of x is

$$\log x = \log_{10} x.$$

Before the advent of calculators and computers, people used tables of common logarithms. Now most work with logarithms is done with the aid of a scientific calculator or a graphing calculator. We will take that approach in this section of the text. To find the common logarithm of a number on a scientific calculator, enter the number and then press the [log x] or [log] key.

Example 1 On a scientific calculator or a graphing calculator, find a decimal approximation for each of the following.

(a) log 7.32 **(b)** log 73.2 **(c)** log 0.314

Solution

(a) 7.32 [log] ≈ 0.864511081

(b) 73.2 [log] ≈ 1.864511081

(c) 0.314 [log] ≈ -0.503070352

Note that the only difference in the two answers is the 1 before the decimal point.

Note: Your calculator may display fewer or more digits in the answer.

Student Practice 1 On a scientific calculator or a graphing calculator, find a decimal approximation for each of the following.

(a) log 4.36 **(b)** log 436 **(c)** log 0.2418

To Think About: *Decimal Point Placement* Why is the difference in the answers to Example 1 **(a)** and **(b)** equal to 1.00? Consider the following.

$$\begin{aligned} \log 73.2 &= \log(7.32 \times 10^1) && \text{Use scientific notation.} \\ &= \log 7.32 + \log 10^1 && \text{By property 1.} \\ &= \log 7.32 + 1 && \text{Because } \log_b b = 1. \\ &\approx 0.864511081 + 1 && \text{Use a calculator.} \\ &\approx 1.864511081 && \text{Add the decimals.} \end{aligned}$$

2 Finding the Antilogarithm of a Common Logarithm on a Scientific Calculator

We have previously discussed the function $f(x) = \log x$ and the corresponding inverse function $f^{-1}(x) = 10^x$. The inverse of a logarithmic function is an exponential function. There is another name for this function. It is called an **antilogarithm.**

If $f(x) = \log x$ (here the base is understood to be 10), then $f^{-1}(x) = \text{antilog } x = 10^x$.

Example 2 Find an approximate value for x if $\log x = 4.326$.

Solution Here we are given the value of the logarithm, and we want to find the number that has that logarithm. In other words, we want the antilogarithm. We know that $\log_{10} x = 4.326$ is equivalent to $10^{4.326} = x$. So to solve this problem, we want to find the value of 10 raised to the 4.326 power. Using a calculator, we have the following.

$$4.326 \ \boxed{10^x} \approx 21{,}183.61135$$

Thus, $x \approx 21{,}183.61135$. (If your scientific calculator does not have a $\boxed{10^x}$ key, you can usually use $\boxed{\text{2nd F}}$ $\boxed{\log}$ or $\boxed{\text{INV}}$ $\boxed{\log}$ or $\boxed{\text{SHIFT}}$ $\boxed{\log}$ to perform the operation.) □

Student Practice 2 Using a scientific calculator, find an approximate value for x if $\log x = 2.913$.

Example 3 Evaluate antilog(-1.6784).

Solution Asking what is antilog(-1.6784) is equivalent to asking what the value is of $10^{-1.6784}$. To determine this on most scientific calculators, it will be necessary to enter the number 1.6784 followed by the $\boxed{+/-}$ key. You may need slightly different steps on a graphing calculator.

$$1.6784 \ \boxed{+/-} \ \boxed{10^x} \approx 0.020970076$$

Thus, antilog$(-1.6784) \approx 0.020970076$. □

Student Practice 3 Evaluate antilog(-3.0705).

Example 4 Using a scientific calculator, find an approximate value for x.

(a) $\log x = 0.07318$ **(b)** $\log x = -3.1621$

Solution

(a) $\log x = 0.07318$ is equivalent to $10^{0.07318} = x$.

$$0.07318 \ \boxed{10^x} \approx 1.183531987$$

Thus, $x \approx 1.183531987$.

(b) $\log x = -3.1621$ is equivalent to $10^{-3.1621} = x$.

$$3.1621 \ \boxed{+/-} \ \boxed{10^x} \approx 0.0006884937465$$

Thus, $x \approx 0.0006884937465$.

(Some calculators may give the answer in scientific notation as $6.884937465 \times 10^{-4}$. This is often displayed on a calculator screen as $6.884937465 \ -4$.) □

Student Practice 4 Using a scientific calculator, find an approximate value for x.

(a) $\log x = 0.06134$ **(b)** $\log x = -4.6218$

3 Finding Natural Logarithms on a Scientific Calculator 

For most theoretical work in mathematics and other sciences, the most useful base for logarithms is e. Logarithms with base e are known as *natural logarithms* and are usually written $\ln x$.

> **DEFINITION OF NATURAL LOGARITHM**
>
> For all real numbers $x > 0$, the **natural logarithm** of x is
>
> $$\ln x = \log_e x.$$

On a scientific calculator we can usually approximate natural logarithms with the $\boxed{\ln x}$ or $\boxed{\ln}$ key.

Example 5 On a scientific calculator, approximate the following values.

(a) $\ln 7.21$ **(b)** $\ln 72.1$ **(c)** $\ln 0.0356$

Solution

(a) $7.21\ \boxed{\ln} \approx 1.975468951$

(b) $72.1\ \boxed{\ln} \approx 4.278054044$

(c) $0.0356\ \boxed{\ln} \approx -3.335409641$

Note that there is no simple relationship between the answers to parts **(a)** and **(b)**. Do you see why these are different from common logarithms? □

Student Practice 5 On a scientific calculator, approximate the following values.

(a) $\ln 4.82$ **(b)** $\ln 48.2$ **(c)** $\ln 0.0793$

4 Finding the Antilogarithm of a Natural Logarithm on a Scientific Calculator

Example 6 On a scientific calculator, find an approximate value for x for each equation.

(a) $\ln x = 2.9836$ **(b)** $\ln x = -1.5619$

Solution

(a) If $\ln x = 2.9836$, then $e^{2.9836} = x$.

$$2.9836\ \boxed{e^x} \approx 19.75882051$$

(b) If $\ln x = -1.5619$, then $e^{-1.5619} = x$.

$$1.5619\ \boxed{+/-}\ \boxed{e^x} \approx 0.209737192$$

□

Student Practice 6 On a scientific calculator, find an approximate value for x for each equation.

(a) $\ln x = 3.1628$ **(b)** $\ln x = -2.0573$

An alternative notation is sometimes used. This is $\text{antilog}_e(x)$.

5 Evaluating a Logarithm to a Base Other Than 10 or e 

Although a scientific calculator or a graphing calculator has specific keys for finding common logarithms (base 10) and natural logarithms (base e), there are no keys for finding logarithms with other bases. What do we do in such cases? The logarithm of a number for a base other than 10 or e can be found with the following formula.

CHANGE OF BASE FORMULA

$$\log_b x = \frac{\log_a x}{\log_a b},$$

where a, b, and $x > 0$, $a \neq 1$, and $b \neq 1$.

Let's see how this formula works. If we want to use common logarithms to find $\log_3 56$, we must first note that the value of b in the formula is 3. We then write

$$\log_3 56 = \frac{\log_{10} 56}{\log_{10} 3} = \frac{\log 56}{\log 3}.$$

Do you see why?

Example 7 Evaluate using common logarithms. $\log_3 5.12$

Solution $\log_3 5.12 = \dfrac{\log 5.12}{\log 3}$

On a calculator, we find the following.

5.12 [log] [÷] 3 [log] [=] 1.486561234

Our answer is an approximate value with nine decimal places. Your answer may have more or fewer digits depending on your calculator. □

 **Student Practice 7** Evaluate using common logarithms. $\log_9 3.76$

If we desire to use base e, then the change of base formula is used with natural logarithms.

Example 8 Obtain an approximate value for $\log_4 0.005739$ using natural logarithms.

Solution Using the change of base formula, with $a = e$, $b = 4$, and $x = 0.005739$, we have the following.

$$\log_4 0.005739 = \frac{\log_e 0.005739}{\log_e 4} = \frac{\ln 0.005739}{\ln 4}$$

This is done on some scientific calculators as follows.

0.005739 [ln] [÷] 4 [ln] [=] −3.722492455

Thus, we have $\log_4 0.005739 \approx -3.722492455$.

Check. To check our answer we want to know the following.

$$4^{-3.722492455} \stackrel{?}{=} 0.005739$$

Using a calculator, we can verify this with the [y^x] key.

4 [y^x] 3.722492455 [+/−] [=] 0.005739 ✓ □

Student Practice 8 Obtain an approximate value for $\log_8 0.009312$ using natural logarithms.

Graphing Calculator

Graphing Logarithmic Functions

You can use the change of base formula to graph logarithmic functions on a graphing calculator.

To graph $y = \log_2 x$ in Example 9 on a graphing calculator, enter the function $y = \frac{\log x}{\log 2}$ into the Y = editor of your calculator.

Display:

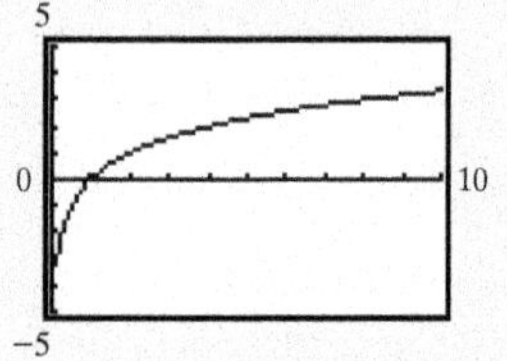

Example 9 Using a scientific calculator, graph $y = \log_2 x$.

Solution If we use common logarithms $(\log_{10} x)$, then for each value of x, we will need to calculate $\frac{\log x}{\log 2}$. Therefore, to find y when $x = 3$, we need to calculate $\frac{\log 3}{\log 2}$. On most scientific calculators, we would enter 3 [log] [÷] 2 [log] [=] and obtain 1.584962501. Rounded to the nearest tenth, we have $x = 3$ and $y = 1.6$. In a similar fashion we find other table values and then graph them.

x	$y = \log_2 x$
0.5	−1
1	0
2	1
3	1.6
4	2
6	2.6
8	3

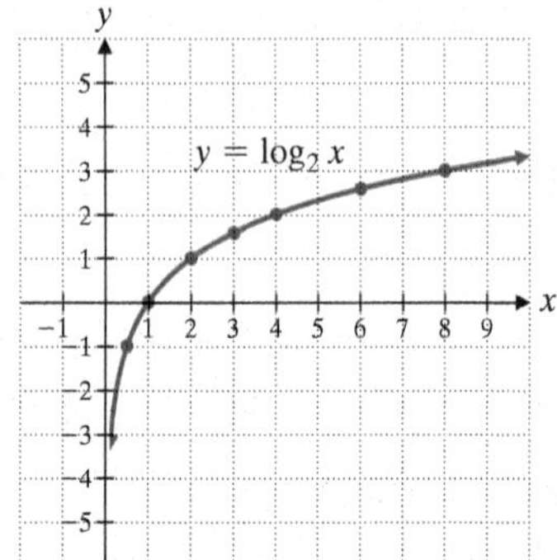

Student Practice 9 Using a scientific calculator, graph $y = \log_5 x$.

x	$y = \log_5 x$

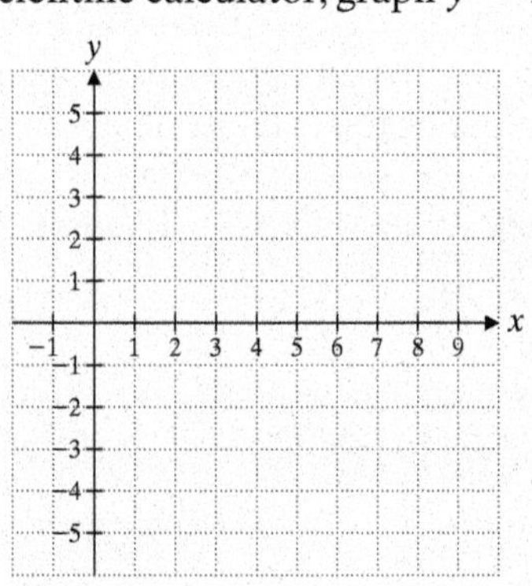

12.4 Exercises MyMathLab®

Verbal and Writing Skills, Exercises 1 and 2

1. Try to find $\log(-5.08)$ on a scientific calculator. What happens? Why?

2. Try to find $\log(-6.63)$ on a scientific or a graphing calculator. What happens? Why?

Use a scientific calculator to approximate the following.

3. $\log 12.3$

4. $\log 11.7$

5. $\log 25.6$

6. $\log 30.4$

7. $\log 15$

8. $\log 6$

9. $\log 125{,}000$

10. $\log 52{,}300$

11. $\log 0.0123$

12. $\log 0.95$

Find an approximate value for x using a scientific or a graphing calculator.

13. $\log x = 2.016$

14. $\log x = 2.754$

15. $\log x = -2$

16. $\log x = -3$

17. $\log x = 3.9304$

18. $\log x = 3.9576$

19. $\log x = 6.4683$

20. $\log x = 4.85$

21. $\log x = -3.3893$

22. $\log x = -4.0458$

23. $\log x = -1.5672$

24. $\log x = -1.2345$

Approximate the following with a scientific or a graphing calculator.

25. $\text{antilog}(7.6215)$

26. $\text{antilog}(4.3894)$

27. $\text{antilog}(-1.0826)$

28. $\text{antilog}(-2.0083)$

29. $\ln 5.62$

30. $\ln 8.81$

31. $\ln 1.53$

32. $\ln 7.35$

33. $\ln 136{,}000$

34. $\ln 129{,}000$

35. $\ln 0.00579$

36. $\ln 0.00134$

Find an approximate value for x using a scientific or a graphing calculator.

37. $\ln x = 0.95$

38. $\ln x = 0.55$

39. $\ln x = 2.4$

40. $\ln x = 4.4$

41. $\ln x = -0.05$

42. $\ln x = -0.03$

43. $\ln x = -2.7$

44. $\ln x = -3.8$

Approximate the following with a scientific or a graphing calculator.

45. $\text{antilog}_e(6.1582)$

46. $\text{antilog}_e(1.9047)$

47. $\text{antilog}_e(-2.1298)$

48. $\text{antilog}_e(-3.3712)$

Use a scientific or a graphing calculator and **common logarithms** *to evaluate the following.*

49. $\log_3 9.2$

50. $\log_2 6.13$

51. $\log_7 7.35$

52. $\log_9 9.85$

53. $\log_6 0.127$

54. $\log_5 0.173$

55. $\log_{15} 12$

56. $\log_{17} 18$

Use a scientific or a graphing calculator and **natural logarithms** *to evaluate the following.*

57. $\log_4 0.07733$

58. $\log_7 0.004462$

59. $\log_{21} 436$

60. $\log_{30} 913$

Mixed Practice *Use a scientific or a graphing calculator to find an approximate value for each of the following.*

61. $\ln 1537$

62. $\log 92.81$

63. $\text{antilog}_e(-1.874)$

64. $\log_6 0.5437$

Find an approximate value for x for each of the following.

65. $\log x = 8.5634$

66. $\ln x = 7.9631$

67. $\log_4 x = 0.8645$

68. $\log_3 x = 0.5649$

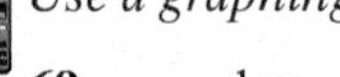

Use a graphing calculator to graph the following.

69. $y = \log_6 x$

70. $y = \log_4 x$

71. $y = \log_{0.4} x$

72. $y = \log_{0.2} x$

Applications

Median Age *The median age of people in the United States is slowly increasing as the population becomes older. In 1980 the median age of the population was 30.0 years. This means that approximately half the population of the country was under 30.0 years old and approximately half the population of the country was over 30.0 years old. By 2035 the median age is expected to peak at 39.1 years. An equation that can be used to predict the median age N (in years) of the population of the United States is* $N = 32.53 + 1.55 \ln x$, *where x is the number of years since 1990 and* $x \geq 1$. *(Source: www.census.gov)*

73. Use the equation to find the median age of the U.S. population in 2000 and in 2010. Round to the nearest hundredth. If this model is correct, by what percent did the median age increase from 2000 to 2010? Round to the nearest tenth of a percent.

74. Use the equation to find the median age of the U.S. population in 2008 and in 2018. Round to the nearest hundredth. If this model is correct, by what percent will the median age increase from 2008 to 2018? Round to the nearest tenth of a percent.

Seismologist *One aspect of a seismologist's job is to measure the magnitude of the many earthquakes that happen around the world each day. If an earthquake has a shock wave x times greater than the smallest shock wave that can be measured by a seismograph, then its magnitude R on the Richter scale is given by the equation* $R = \log x$*. An earthquake that has a shock wave 25,000 times greater than the smallest shock wave that can be detected will have a magnitude of* $R = \log 25{,}000 \approx 4.40$*. (Usually we round the magnitude of an earthquake to the nearest hundredth.)*

75. What is the magnitude of an earthquake that has a shock wave that is 56,000 times greater than the smallest shock wave that can be detected?

76. What is the magnitude of an earthquake that has a shock wave that is 184,000 times greater than the smallest shock wave that can be detected?

77. In April 2015, an earthquake of magnitude $R = 7.8$ occurred in Nepal. What can you say about the size of the earthquake's shock wave?

78. In April 2015, an earthquake of magnitude $R = 5.5$ occurred in Mexico. What can you say about the size of the earthquake's shock wave?

Cumulative Review *Solve the quadratic equations. Simplify your answers.*

79. **[9.2.1]** $3x^2 - 11x - 5 = 0$

80. **[9.2.1]** $2y^2 + 4y - 3 = 0$

Highway Exits *On a specific portion of Interstate 91, there are six exits. There is a distance of 12 miles between odd-numbered exits. There is a distance of 15 miles between even-numbered exits. The total distance between Exit 1 and Exit 6 is 36 miles.*

81. **[2.6.2]** Find the distance between Exit 1 and Exit 2. Find the distance between Exit 1 and Exit 3.

82. **[2.6.2]** Find the distance between Exit 1 and Exit 4. Find the distance between Exit 1 and Exit 5.

Quick Quiz 12.4 *Use a scientific calculator to evaluate the following.*

1. Find log 9.36. Round to the nearest ten-thousandth.

2. Find x if $\log x = 0.2253$. Round to the nearest hundredth.

3. Find $\log_5 8.26$. Round to the nearest ten-thousandth.

4. **Concept Check** Explain how you would find x using a scientific calculator if $\ln x = 1.7821$.

12.5 Exponential and Logarithmic Equations

Student Learning Objectives

After studying this section, you will be able to:

1. Solve logarithmic equations.
2. Solve exponential equations.
3. Solve applications using logarithmic or exponential equations.

1 Solving Logarithmic Equations 

In general, when solving logarithmic equations we try to collect all the logarithms on one side of the equation and all the numerical values on the other side. Then we seek to use the properties of logarithms to obtain a single logarithmic expression on one side.

We can describe a general procedure for solving logarithmic equations.

Step 1 If an equation contains some logarithms and some terms without logarithms, try to get one logarithm alone on one side and one numerical value on the other.

Step 2 Then convert to an exponential equation using the definition of a logarithm.

Step 3 Solve the equation.

Example 1 Solve. $\log 5 = 2 - \log(x + 3)$

Solution

$\log 5 + \log(x + 3) = 2$	Add $\log(x + 3)$ to each side.
$\log[5(x + 3)] = 2$	Property 1.
$\log(5x + 15) = 2$	Simplify.
$5x + 15 = 10^2$	Write the equation in exponential form.
$5x + 15 = 100$	Simplify.
$5x = 85$	Subtract 15 from each side.
$x = 17$	Divide each side by 5.

Check. $\log 5 \stackrel{?}{=} 2 - \log(17 + 3)$

$\log 5 \stackrel{?}{=} 2 - \log 20$

Since these are common logarithms (base 10), the easiest way to check the answer is to find decimal approximations for each logarithm on a calculator.

$$0.698970004 \stackrel{?}{=} 2 - 1.301029996$$

$$0.698970004 = 0.698970004 \checkmark$$

Graphing Calculator

Solving Logarithmic Equations

Example 1 could be solved with a graphing calculator in the following way. First write the equation as

$$\log 5 + \log(x + 3) - 2 = 0$$

and then graph the function

$$y = \log 5 + \log(x + 3) - 2$$

to find an approximate value for x when $y = 0$.

 **Student Practice 1** Solve. $\log(x + 5) = 2 - \log 5$

Example 2 Solve. $\log_3(x + 6) - \log_3(x - 2) = 2$

Solution

$\log_3\left(\dfrac{x + 6}{x - 2}\right) = 2$	Property 2.
$\dfrac{x + 6}{x - 2} = 3^2$	Write the equation in exponential form.
$\dfrac{x + 6}{x - 2} = 9$	Evaluate 3^2.
$x + 6 = 9(x - 2)$	Multiply each side by $(x - 2)$.
$x + 6 = 9x - 18$	Simplify.
$24 = 8x$	Add $18 - x$ to each side.
$3 = x$	Divide each side by 8.

Check.

$$\log_3(3+6) - \log_3(3-2) \stackrel{?}{=} 2$$
$$\log_3 9 - \log_3 1 \stackrel{?}{=} 2$$
$$2 - 0 \stackrel{?}{=} 2$$
$$2 = 2 \checkmark$$

□

Student Practice 2 Solve. $\log(x+3) - \log x = 1$

Some equations consist of logarithmic terms only. In such cases we may be able to use property 6 to solve them. Recall that this rule states that if $b > 0$, $b \neq 1, x > 0, y > 0$, and $\log_b x = \log_b y$, then $x = y$.

What if one of our possible solutions is the logarithm of a negative number? Can we evaluate the logarithm of a negative number? Look again at the graph of $y = \log_2 x$ on page 735. Note that the domain of this function is $x > 0$. (The curve is located on the positive side of the x-axis.) Therefore, the logarithm of a negative number is *not defined.*

You should be able to see this by using the definition of logarithms. If $\log(-2)$ were valid, we could write the following.

$$y = \log_{10}(-2)$$
$$10^y = -2$$

Obviously, no value of y can make this equation true. Thus, we see that **it is not possible to take the logarithm of a negative number.**

Sometimes when we attempt to solve a logarithmic equation, we obtain a possible solution that leads to the logarithm of a negative number. We can immediately discard such a solution.

Example 3 Solve. $\log(x+6) + \log(x+2) = \log(x+20)$

Solution

$$\log[(x+6)(x+2)] = \log(x+20)$$
$$\log(x^2 + 8x + 12) = \log(x+20)$$
$$x^2 + 8x + 12 = x + 20$$
$$x^2 + 7x - 8 = 0$$
$$(x+8)(x-1) = 0$$
$$x + 8 = 0 \qquad x - 1 = 0$$
$$x = -8 \qquad x = 1$$

Check. $\log(x+6) + \log(x+2) = \log(x+20)$

$$x = 1: \quad \log(1+6) + \log(1+2) \stackrel{?}{=} \log(1+20)$$
$$\log(7) + \log(3) \stackrel{?}{=} \log(21)$$
$$\log(7 \cdot 3) \stackrel{?}{=} \log 21$$
$$\log 21 = \log 21 \checkmark$$
$$x = -8: \quad \log(-8+6) + \log(-8+2) \stackrel{?}{=} \log(-8+20)$$
$$\log(-2) + \log(-6) \neq \log(12)$$

We can discard -8 because it leads to taking the logarithm of a negative number, which is not allowed. Only $x = 1$ is a solution. The only solution is 1. □

Student Practice 3 Solve $\log 5 - \log x = \log(6x - 7)$ and check your solution.

2 Solving Exponential Equations

You might expect that property 6 can be used in the reverse direction. It seems logical, for example, that if $x = 3$, we should be able to state that $\log_4 x = \log_4 3$. This is exactly the case, and we will formally state it as a property.

PROPERTY 7

If x and $y > 0$ and $x = y$, then $\log_b x = \log_b y$, where $b > 0$ and $b \neq 1$.

Property 7 is often referred to as "taking the logarithm of each side of the equation." Usually we will take the common logarithm of each side of the equation, but any base can be used.

Example 4 Solve $2^x = 7$. Leave your answer in exact form.

Solution

$$\log 2^x = \log 7 \quad \text{Take the logarithm of each side (property 7).}$$

$$x \log 2 = \log 7 \quad \text{Property 3.}$$

$$x = \frac{\log 7}{\log 2} \quad \text{Divide each side by log 2.}$$

□

Student Practice 4 Solve $3^x = 5$. Leave your answer in exact form.

When we solve exponential equations, it will often be useful to find an approximate value for the answer.

Example 5 Solve $3^x = 7^{x-1}$. Approximate your answer to the nearest thousandth.

Solution

$$\log 3^x = \log 7^{x-1}$$

$$x \log 3 = (x - 1)\log 7$$

$$x \log 3 = x \log 7 - \log 7$$

$$x \log 3 - x \log 7 = -\log 7$$

$$x(\log 3 - \log 7) = -\log 7$$

$$x = \frac{-\log 7}{\log 3 - \log 7}$$

We can approximate the value for x on most scientific calculators by using the following keystrokes.

7 [log] [+/−] [÷] [(] 3 [log] [−] 7 [log] [)] [=] 2.296606943

Rounding to the nearest thousandth, we have $x \approx 2.297$. □

Student Practice 5 Solve $2^{3x+1} = 9^{x+1}$. Approximate your answer to the nearest thousandth.

If the exponential equation involves e raised to a power, it is best to take the natural logarithm of each side of the equation.

Example 6 Solve $e^{2.5x} = 8.42$. Round your answer to the nearest ten-thousandth.

Solution

$$\ln e^{2.5x} = \ln 8.42 \quad \text{Take the natural logarithm of each side.}$$
$$(2.5x)(\ln e) = \ln 8.42 \quad \text{Property 3.}$$
$$2.5x = \ln 8.42 \quad \ln e = 1. \text{ Property 4.}$$
$$x = \frac{\ln 8.42}{2.5} \quad \text{Divide each side by 2.5.}$$

On most scientific calculators, the value of x can be approximated with the following keystrokes.

8.42 [ln] [÷] 2.5 [=] 0.85224393

Rounding to the nearest ten-thousandth, we have $x \approx 0.8522$. □

Student Practice 6

Solve $20.98 = e^{3.6x}$. Round your answer to the nearest ten-thousandth.

3 Solving Applications Using Logarithmic or Exponential Equations

We now return to the compound interest formula and consider some other exercises that can be solved with it. For example, perhaps we would like to know how long it will take for a deposit to grow to a specified goal.

Example 7 If P dollars are invested in an account that earns interest at 12% compounded annually, the amount available after t years is $A = P(1 + 0.12)^t$. How many years will it take for \$300 in this account to grow to \$1500? Round your answer to the nearest whole year.

Solution

$$1500 = 300(1 + 0.12)^t \quad \text{Substitute } A = 1500 \text{ and } P = 300.$$
$$1500 = 300(1.12)^t \quad \text{Simplify.}$$
$$\frac{1500}{300} = (1.12)^t \quad \text{Divide each side by 300.}$$
$$5 = (1.12)^t \quad \text{Simplify.}$$
$$\log 5 = \log(1.12)^t \quad \text{Take the common logarithm of each side.}$$
$$\log 5 = t(\log 1.12) \quad \text{Property 3.}$$
$$\frac{\log 5}{\log 1.12} = t \quad \text{Divide each side by log 1.12.}$$

On a scientific calculator we have the following.

5 [log] [÷] 1.12 [log] [=] 14.20150519

Thus, it would take approximately 14 years. Look at the graph on the next page to get a visual image of how the investment increases.

Continued on next page

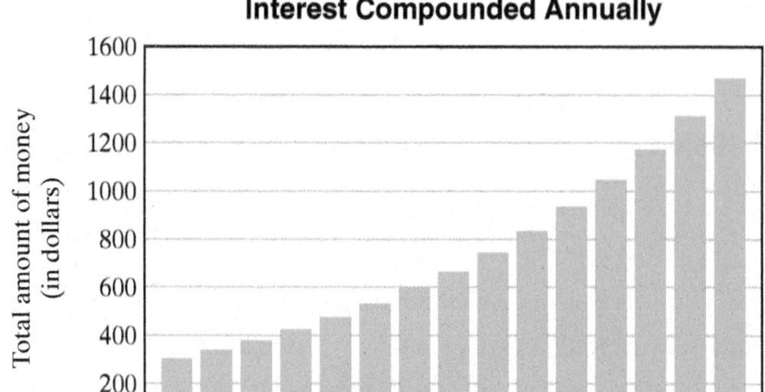

Student Practice 7 Mon Ling's father has an investment account that earns 8% interest compounded annually. How many years would it take for \$4000 to grow to \$10,000 in that account? Round your answer to the nearest whole year.

The growth equation for things that appear to be growing continuously is $A = A_0e^{rt}$, where A is the final amount, A_0 is the original amount, r is the rate at which things are growing in a unit of time, and t is the total number of units of time.

For example, if a laboratory starts with 5000 cells and they reproduce at a rate of 35% per hour, the number of cells in 18 hours can be described by the following equation.

$$A = 5000e^{(0.35)(18)} = 5000e^{6.3} \approx 5000(544.57) = 2{,}722{,}850 \text{ cells}$$

Example 8 At the beginning of 2011, the world population was seven billion people and the growth rate was 1.2% per year. If this growth rate continues, how many years will it take for the population to double to fourteen billion people?

Solution

$$A = A_0e^{rt}$$

To make our calculation easier, we write the values for the population in terms of billions. This will allow us to avoid writing numbers like 14,000,000,000 and 7,000,000,000. Do you see why we can do this?

$$14 = 7e^{0.012t} \quad \text{Substitute known values.}$$

$$\frac{14}{7} = e^{0.012t} \quad \text{Divide each side by 7.}$$

$$\ln\left(\frac{14}{7}\right) = \ln e^{0.012t} \quad \text{Take the natural logarithm of each side.}$$

$$\ln 2 = (0.012t)\ln e \quad \text{Property 3.}$$

$$\ln 2 = 0.012t \quad \text{Since } \ln e = 1. \text{ Property 4.}$$

$$\frac{\ln 2}{0.012} = t \quad \text{Divide each side by 0.012.}$$

Using our calculator, we obtain the following.

2 [ln] [÷] 0.012 [=] 57.76226505

Rounding to the nearest whole year, we find that the population will grow from seven billion to fourteen billion in about 58 years if the growth continues at the same rate.

Student Practice 8 The wildlife management team in one region of Alaska has determined that the black bear population is growing at the rate of 4.3% per year. This region currently has approximately 1300 black bears. A food shortage will develop if the population reaches 5000. If the growth rate remains unchanged, how many years will it take for this food shortage problem to occur?

Example 9 The magnitude of an earthquake (amount of energy released) is measured by the formula $R = \log\left(\frac{I}{I_0}\right)$, where I is the intensity of the earthquake and I_0 is the minimum measurable intensity. A 1964 earthquake in Anchorage, Alaska, had a magnitude of 8.4. A 1906 earthquake in Taiwan had a magnitude of 7.1. How many times more intense was the Anchorage earthquake than the Taiwan earthquake? (*Source:* www.noaa.gov)

Solution

Let I_A = intensity of the Alaska earthquake. Then

$$8.4 = \log\left(\frac{I_A}{I_0}\right) = \log I_A - \log I_0.$$

Solving for $\log I_0$ gives

$$\log I_0 = \log I_A - 8.4.$$

Let I_T = intensity of the Taiwan earthquake. Then

$$7.1 = \log\left(\frac{I_T}{I_0}\right) = \log I_T - \log I_0.$$

Solving for $\log I_0$ gives

$$\log I_0 = \log I_T - 7.1.$$

Therefore,

$$\log I_A - 8.4 = \log I_T - 7.1.$$

$$\log I_A - \log I_T = 8.4 - 7.1$$

$$\log \frac{I_A}{I_T} = 1.3$$

$$10^{1.3} = \frac{I_A}{I_T}$$

$$19.95262315 \approx \frac{I_A}{I_T} \quad \text{Use a calculator.}$$

$$20 \approx \frac{I_A}{I_T} \quad \text{Round to the nearest whole number.}$$

$$20I_T \approx I_A$$

The Alaska earthquake was approximately twenty times more intense than the Taiwan earthquake. □

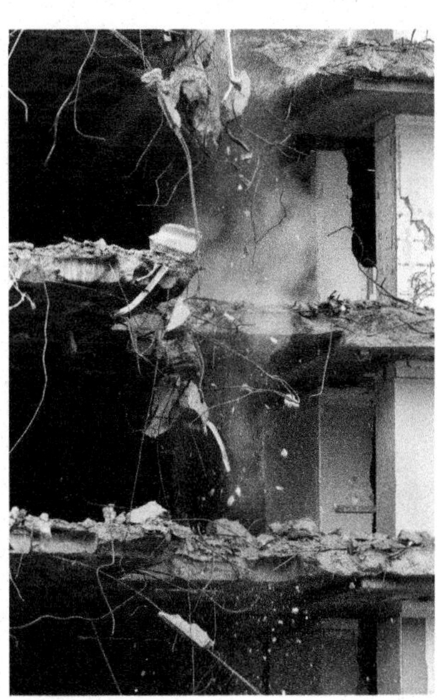

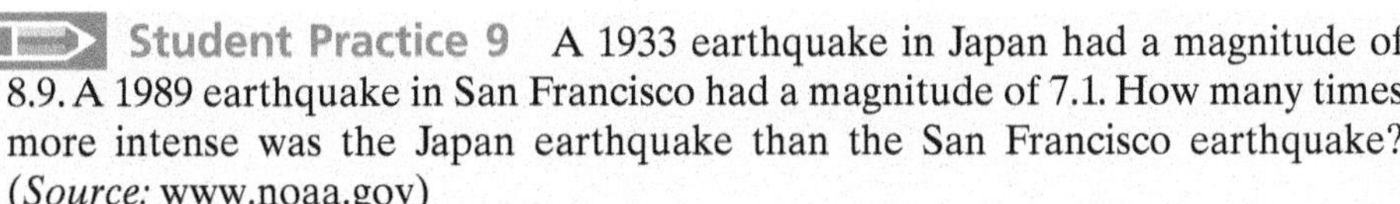

Student Practice 9 A 1933 earthquake in Japan had a magnitude of 8.9. A 1989 earthquake in San Francisco had a magnitude of 7.1. How many times more intense was the Japan earthquake than the San Francisco earthquake? (*Source:* www.noaa.gov)

12.5 Exercises MyMathLab®

Solve each logarithmic equation and check your solutions.

1. $\log_7\left(\frac{2}{3}x + 3\right) + \log_7 3 = 2$

2. $\log_5\left(\frac{3}{4}x - 2\right) + \log_5 4 = 2$

3. $\log_9 3 + \log_9(2x + 1) = 1$

4. $\log_2 4 + \log_2(x - 1) = 5$

5. $\log_2\left(x + \frac{4}{3}\right) = 5 - \log_2 6$

6. $\log_4(2x - 5) = 2 - \log_4 2$

7. $\log(30x + 40) = 2 + \log(x - 1)$

8. $1 + \log x = \log(9x + 1)$

9. $2 + \log_6(x - 1) = \log_6(12x)$

10. $\log_2 x = \log_2(x + 5) - 1$

11. $\log(75x + 50) - \log x = 2$

12. $\log_8(x + 1) = \log_8(x - 6) + 1$

13. $\log_3(x + 6) + \log_3 x = 3$

14. $\log_8 x + \log_8(x - 2) = 1$

15. $1 + \log(x - 2) = \log(6x)$

16. $\log_5(2x) - \log_5(x - 3) = 3\log_5 2$

17. $\log_2(x + 5) - 2 = \log_2 x$

18. $\log_5(5x + 10) - 2 = \log_5(x - 2)$

19. $2\log_7 x = \log_7(x + 4) + \log_7 2$

20. $\log x + \log(x - 1) = \log 12$

21. $\ln 10 - \ln x = \ln(x - 3)$

22. $\ln(2 + 2x) = 2\ln(x + 1)$

Solve each exponential equation. Leave your answers in exact form. Do not approximate.

23. $7^{x+3} = 12$

24. $4^{x-2} = 6$

25. $2^{3x+4} = 17$

26. $5^{2x-1} = 11$

Solve each exponential equation. Use your calculator to approximate your solutions to the nearest thousandth.

27. $8^{2x-1} = 90$

28. $15^{3x-2} = 230$

29. $5^x = 4^{x+1}$

30. $3^x = 2^{x+3}$

31. $28 = e^{x-2}$

32. $e^{x+2} = 88$

33. $88 = e^{2x+1}$

34. $3 = e^{1-x}$

Applications

Financial Planner *One part of the job of a financial planner is to help people invest their money appropriately. If a customer knows he or she will need a certain amount of money in 10 or 20 years, the financial planner can help the customer plan how much needs to be invested today and and at what rate in order to have that amount. When a principal P earns an annual interest rate r compounded yearly, the amount A after t years is* $A = P(1 + r)^t$. *Use this information to solve exercises 35–40. Round all answers to the nearest whole year.*

35. How long will it take \$1500 to grow to \$5000 at 8% compounded annually?

36. How long will it take \$1000 to grow to \$4500 at 7% compounded annually?

37. How long will it take for a principal to triple at 6% compounded annually?

38. How long will it take for a principal to double at 5% compounded annually?

39. What interest rate would be necessary to obtain \$6500 in 6 years if \$5000 is the amount of the original investment and the interest is compounded yearly? (Express the interest rate as a percent rounded to the nearest tenth.)

40. If \$3000 is invested for 3 years with annual interest compounded yearly, what interest rate is needed to achieve an amount of \$3600? (Express the interest rate as a percent rounded to the nearest tenth.)

World Population *The growth of the world population can be described by the equation* $A = A_0e^{rt}$, *where time* t *is measured in years,* A_0 *is the population of the world at time* $t = 0$, r *is the annual growth rate, and* A *is the population at time* t. *Use this information to solve exercises 41–44. Round your answers to the nearest whole year.*

41. In 1963, the world population was about three billion people and the growth rate was 2.1% per year. At that rate, how long would it have taken for the population to increase to seven billion?

42. At a growth rate of 2.1% per year, how long would it take a population to double?

43. By 2009, the world population was about six billion and the growth rate had slowed to 1.3%. At that rate, how long would it take for the population to increase to nine billion?

44. At a growth rate of 1.3% per year, how long would it take a population to double?

Physical Therapy Assistants *According to the Department of Labor, the physical therapy assistant field is one of the fastest-growing occupations for the time period 2012–2022. The number* N *of physical therapy assistants in the United States can be approximated by the equation* $N = 120{,}500(1.04)^x$, *where* x *is the number of years since 2012. Use this equation to answer the following questions. (Source: www.bls.gov)*

45. Approximately how many physical therapy assistants were there in 2014? Round to the nearest whole number.

46. Approximately how many physical therapy assistants will there be in 2020? Round to the nearest whole number.

47. In what year will the number of physical therapy assistants reach 175,000?

48. In what year will the number of physical therapy assistants reach 185,000?

Use the equation $A = A_0e^{rt}$ to solve exercises 49–54. Round your answers to the nearest whole number.

49. ***Population*** The population of Bethel is 80,000 people, and it is growing at the rate of 1.5% per year. How many years will it take for the population to grow to 120,000 people?

50. ***Population*** The 2010 population of Melbourne, Australia, was approximately four million people. The growth rate is 2% per year. In how many years will there be 4.5 million people?

51. ***Skin Grafts*** The number of new skin cells on a revolutionary skin graft is growing at a rate of 4% per hour. How many hours will it take for 200 cells to become 1800 cells?

52. ***Workforce*** The workforce in a state is increasing at the rate of 1.5% per year. During the last measured year, the workforce was 3.5 million. If this rate continues, how many years will it be before the workforce reaches 4.5 million?

53. ***Lyme Disease*** Unfortunately, some U.S. deer carry ticks that spread Lyme disease. The number of people who are infected by the virus is increasing by 6% every year. If 29,959 people were confirmed to have Lyme disease in 2009, how many are expected to be infected by the end of the year 2015? (*Source:* www.cdc.gov)

54. ***DVD Rentals*** In the city of Scranton, the number of DVD rentals is increasing by 7.5% per year. For the last year that data are available, 1.3 million DVDs were rented. How many years will it be before 2.0 million DVDs are rented per year?

To Think About

Earthquakes *The magnitude of an earthquake (amount of energy released) is described by the formula $R = \log\left(\frac{I}{I_0}\right)$, where I is the intensity of the earthquake and I_0 is the minimum measurable intensity. Use this formula to solve exercises 55–58. Round answers to the nearest tenth.*

55. October 17, 1989, brought tragedy to the San Francisco/Oakland area. An earthquake measuring 7.1 on the Richter scale and centered in the Loma Prieta area (Santa Cruz Mountains) collapsed huge sections of freeway, killing sixty-three people. Almost 6 years later, an earthquake measuring 8.2 on the Richter scale killed 190 people in the Kurile Islands of Japan and Russia. How many times more intense was the Kurile earthquake than the Loma Prieta earthquake? (*Source:* www.noaa.gov)

56. On January 17, 1993, in Northridge, California, residents experienced an earthquake that measured 6.8 on the Richter scale, killed sixty-one people, and undermined supposedly earthquake-proof steel-framed buildings. Exactly one year later near Kobe, Japan, an earthquake measuring 7.2 on the Richter scale killed more than 5300 people, injured more than 35,000, and destroyed nearly 200,000 homes, in spite of construction codes reputed to be the best in the world. How many times more intense was the Kobe earthquake than the Northridge earthquake? (*Source:* www.noaa.gov)

57. The 1906 earthquake in San Francisco had a magnitude of 8.3. In 1971 an earthquake in Japan measured 6.8. How many times more intense was the San Francisco earthquake than the Japan earthquake? (*Source:* www.noaa.gov)

58. A 1933 Japan earthquake had a magnitude of 8.9. In Turkey a 1975 earthquake had a magnitude of 6.7. How many times more intense was the Japan earthquake than the Turkey earthquake? (*Source:* www.noaa.gov)

Cumulative Review *Simplify. Assume that x and y are positive real numbers.*

59. [8.4.1] $(\sqrt{3} + 2\sqrt{2})(\sqrt{6} - \sqrt{2})$

60. [8.2.2] $\sqrt{98x^3y^2}$

61. [2.6.2] ***Spelling Bee*** 285 students competed in the 2015 Scripps National Spelling Bee. A total of seventy students were 12 years old. Eighteen more students were 13 years old than were 12 years old. Seventy-eight students were 14 or 15 years old. All the remaining students were younger. Only three students, the youngest in the competition, were 9 years old. There were eleven 10-year-old students. Twenty-four more students were 11 years old than were 10 years old. How many students were there in each age category (ages 9, 10, 11, 12, 13, and 14–15)?

62. [2.6.2] ***London Subway*** The London subway system has a staff of 19,000, and an average of 3.4 million passengers use the subway each weekday. Between 1993 and 1999, an expansion of the system was built that extended the system 16 kilometers and cost 3.5 billion British pounds. How many dollars per mile did this extension cost? (Use 1 kilometer = 0.62 mile and 1 U.S. dollar = 0.64 British pound.) (*Source*: en.wikipedia.org)

Quick Quiz 12.5 *Solve the equation.*

1. $2 - \log_6 x = \log_6(x + 5)$

2. $\log_3(x - 5) + \log_3(x + 1) = \log_3 7$

3. $6^{x+2} = 9$ (Round your answer to the nearest thousandth.)

4. Concept Check Explain how you would solve $26 = 52\,e^{3x}$.

CAREER EXPLORATION PROBLEMS

Background: Statistician

Caitlyn, a recent college graduate who majored in math, is now working as a statistician at the Internal Revenue Service. Her job responsibilities include designing data-collection surveys and reviewing their results. This process includes performing mathematical tests on samples of data taken from the surveys.

Caitlyn is studying trends in the rapidly growing occupational field of data information security. Her work involves summarizing and examining data from various surveys she has designed and administered. The mathematical testing she completed has led her to put together a few growth models that explain the growth seen in this field, and she wants to report her findings to her colleagues at their weekly staff meeting.

Facts

- Data from the surveys Caitlyn has conducted reveal that the number of employed data security analysts is approximated by the equation $N = 14{,}000(1.032)^x$, where x represents the number of years since 2013.
- One of Caitlyn's surveys also reveals that the population of data security analysts is approximated by the continuous growth equation $A = A_0e^{rt}$ where $A =$ the projected number of data security analysts, $A_0 =$ current number of data security analysts, $r =$ the annual growth rate for this occupation, and $t =$ the number of years. The approximate value for e is 2.7183.
- The equation $A = A_0(1 + r)^t$ is a general model for forecasting the average amount of debt owed by college graduates entering the workforce with degrees qualifying them as data security analysts. In this model, $A =$ the future average amount, $A_0 =$ the current average amount, $r =$ the annual growth rate, and $t =$ the number of years.

Tasks

1. Given that the current number of employed data security information analysts is approximated by $N = 14{,}000(1.032)^x$, what is the number of these analysts employed in 2017 rounded to the nearest whole number?

2. Caitlyn also wants to use the equation to determine by what year the number of these analysts will reach 18,000. What is her result?

3. Using the model $A = A_0e^{rt}$, Caitlyn again calculates, for comparison, by what year the number of analysts will reach 18,000, given an annual growth rate of 3.2%. What is her result?

4. The average amount of debt graduates of a four-year college owe when they pursue education leading to employment in this occupation is currently \$45,000. Caitlyn uses the model $A = A_0(1 + r)^t$, which assumes an annual inflation (growth) rate of 0.8%, and lets 2013 correspond to $t = 0$. What does she find is the average amount of debt for four-year school graduates in 2016, rounded to the nearest dollar?

Chapter 12 Organizer

Topic and Procedure	Examples	You Try It
Exponential function, p. 724 $f(x) = b^x$, where $b > 0, b \neq 1$, and x is a real number.	Graph $f(x) = \left(\frac{2}{3}\right)^x$. | x | $f(x)$ | | -2 | 2.25 | | -1 | 1.5 | | 0 | 1 | | 1 | $0.\overline{6}$ | | 2 | $0.\overline{4}$ | 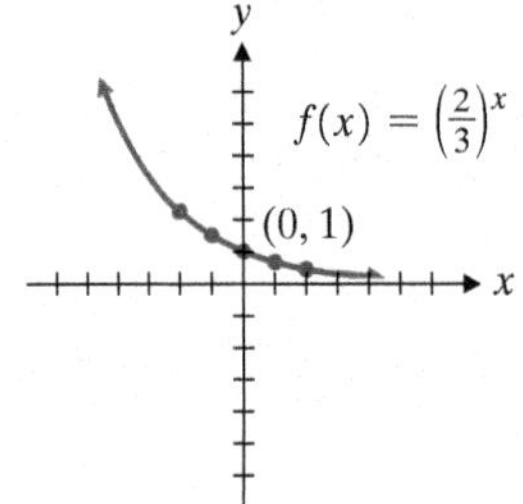	1. Graph $f(x) = \left(\frac{1}{4}\right)^x$. 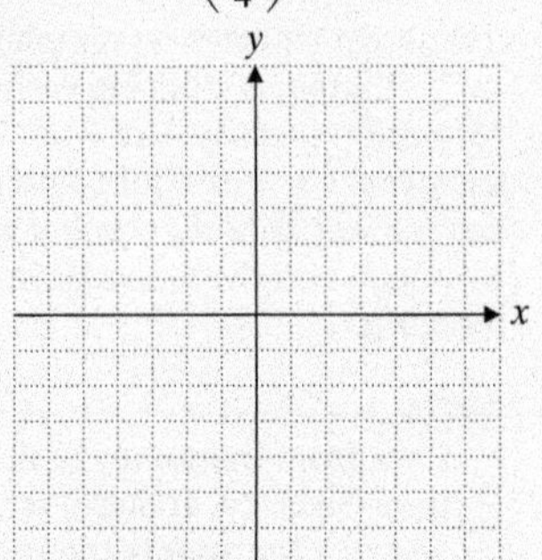
Property of exponential equations, p. 726 When $b > 0$ and $b \neq 1$, if $b^x = b^y$, then $x = y$.	Solve for x. $2^x = \frac{1}{32}$ $2^x = \frac{1}{2^5}$ $2^x = 2^{-5}$ $x = -5$	2. Solve for x. $3^x = \frac{1}{81}$
Definition of logarithm, p. 733 $y = \log_b x$ is the same as $x = b^y$, where $x > 0, b > 0$, and $b \neq 1$.	**(a)** Write in exponential form. $\log_3 17 = 2x$ $3^{2x} = 17$ **(b)** Write in logarithmic form. $18 = 3^x$ $\log_3 18 = x$ **(c)** Solve for x. $\log_6\left(\frac{1}{36}\right) = x$ $6^x = \frac{1}{36}$ $6^x = 6^{-2}$ $x = -2$	3. **(a)** Write in exponential form. $\log_4 9 = 0.5x$ **(b)** Write in logarithmic form. $28 = 5^x$ **(c)** Solve for x. $\log_4\left(\frac{1}{64}\right) = x$
Properties of logarithms, pp. 740–743 Suppose that $M > 0, N > 0, b > 0$, and $b \neq 1$. $\log_b MN = \log_b M + \log_b N$ $\log_b\left(\frac{M}{N}\right) = \log_b M - \log_b N$ $\log_b M^p = p \log_b M$ $\log_b b = 1$ $\log_b 1 = 0$ If $\log_b x = \log_b y$, then $x = y$. If $x = y$, then $\log_b x = \log_b y$.	**(a)** Write as separate logarithms of x, y, and w. $\log_3\left(\frac{x^2\sqrt[3]{y}}{w}\right)$ $= 2\log_3 x + \frac{1}{3}\log_3 y - \log_3 w$ **(b)** Write as one logarithm. $5\log_6 x - 2\log_6 w - \frac{1}{4}\log_6 z$ $= \log_6\left(\frac{x^5}{w^2\sqrt[4]{z}}\right)$ **(c)** Simplify. $\log 10^5 + \log_3 3 + \log_5 1$ $= 5\log 10 + \log_3 3 + \log_5 1$ $= 5 + 1 + 0$ $= 6$	4. **(a)** Write as separate logarithms of a, b, and c. $\log_2\left(\frac{a\sqrt{b}}{c^2}\right)$ **(b)** Write as one logarithm. $3\log_5 a + \frac{2}{3}\log_5 b - 2\log_5 c$ **(c)** Simplify. $\log_8 1 - \log 10^3 + \log_4 4$
Finding logarithms, pp. 748–750 On a scientific calculator: $\log x = \log_{10} x$, for all $x > 0$ $\ln x = \log_e x$, for all $x > 0$	**(a)** Find log 3.82. 3.82 [log] $\log 3.82 \approx 0.5820634$ **(b)** Find ln 52.8. 52.8 [ln] $\ln 52.8 \approx 3.9665112$	5. Find each logarithm. **(a)** log 16.5 **(b)** ln 32.7

Topic and Procedure	Examples	You Try It
Finding antilogarithms, pp. 748–750 If $\log x = b$, then $10^b = x$. If $\ln x = b$, then $e^b = x$. Use a calculator or a table to solve.	**(a)** Find x if $\log x = 2.1416$. $10^{2.1416} = x$ 2.1416 [10^x] ≈ 138.54792 **(b)** Find x if $\ln x = 0.6218$. $e^{0.6218} = x$ 0.6218 [e^x] ≈ 1.8622771	6. Find x. **(a)** $\log x = 2.075$ **(b)** $\ln x = 1.528$
Finding a logarithm to a different base, p. 751 Change of base formula: $\log_b x = \dfrac{\log_a x}{\log_a b}$, where a, b, and $x > 0$, $a \neq 1$, and $b \neq 1$.	Evaluate $\log_7 1.86$. $\log_7 1.86 = \dfrac{\log 1.86}{\log 7}$ 1.86 [log] [÷] 7 [log] [=] 0.3189132	7. Evaluate $\log_8 2.5$.
Solving logarithmic equations, p. 756 **1.** If some but not all of the terms of an equation have logarithms, try to rewrite the equation with one single logarithm on one side and one numerical value on the other. Then convert the equation to exponential form. **2.** If an equation contains logarithmic terms only, try to get only one logarithm on each side of the equation. Then use the property that if $\log_b x = \log_b y$, $x = y$. *Note:* Always check your solutions when solving logarithmic equations.	Solve for x. $\log_5 3x - \log_5(x^2 - 1) = \log_5 2$ $\log_5 3x = \log_5 2 + \log_5(x^2 - 1)$ $\log_5 3x = \log_5[2(x^2 - 1)]$ $3x = 2x^2 - 2$ $0 = 2x^2 - 3x - 2$ $0 = (2x + 1)(x - 2)$ $2x + 1 = 0 \quad x - 2 = 0$ $x = -\dfrac{1}{2} \quad x = 2$ *Check.* $x = 2$: $\log_5 3(2) - \log_5(2^2 - 1) \stackrel{?}{=} \log_5 2$ $\log_5 6 - \log_5 3 \stackrel{?}{=} \log_5 2$ $\log_5\left(\dfrac{6}{3}\right) \stackrel{?}{=} \log_5 2$ $\log_5 2 = \log_5 2$ ✓ $x = -\dfrac{1}{2}$: For the expression $\log_5(3x)$, we would obtain $\log_5(-1.5)$. You cannot take the logarithm of a negative number. $x = -\dfrac{1}{2}$ is not a solution. The solution is 2.	8. Solve for x. $-\log_6 3 + \log_6 8x = \log_6(x^2 - 1)$
Solving exponential equations, p. 758 **1.** See whether each expression can be written so that only one base appears on one side of the equation and the same base appears on the other side. Then use the property that if $b^x = b^y$, $x = y$. **2.** If you can't do step 1, take the logarithm of each side of the equation and use the properties of logarithms to solve for the variable.	Solve for x. $2^{x-1} = 7$ $\log 2^{x-1} = \log 7$ $(x - 1)\log 2 = \log 7$ $x \log 2 - \log 2 = \log 7$ $x \log 2 = \log 7 + \log 2$ $x = \dfrac{\log 7 + \log 2}{\log 2}$ (We can approximate the answer as $x \approx 3.8073549$.)	9. Solve for x. $3^{x+2} = 11$

Chapter 12 Review Problems

Graph the functions in exercises 1 and 2.

1. $f(x) = 4^{3+x}$

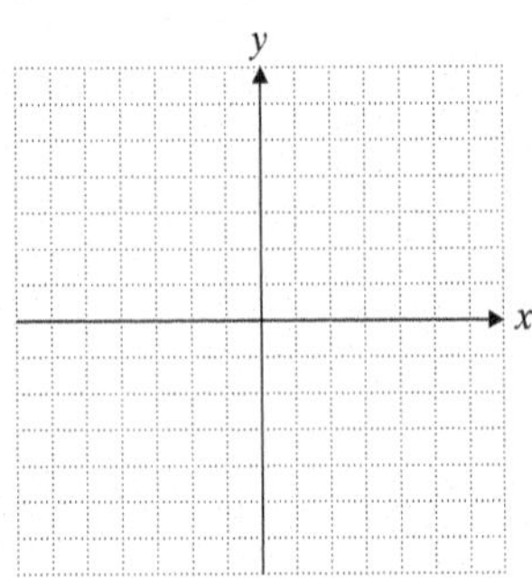

2. $f(x) = e^{x-3}$

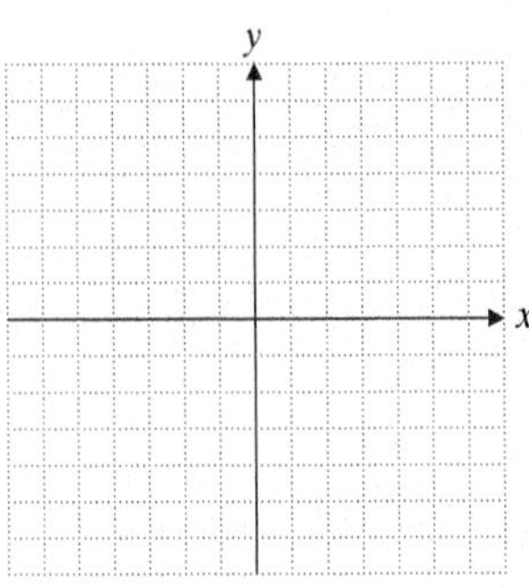

3. Solve. $3^{3x+1} = 81$

4. $2^{x+9} = 128$

5. Write in exponential form. $-2 = \log(0.01)$

6. Change to logarithmic form. $8 = 4^{3/2}$

Solve.

7. $\log_w 16 = 4$

8. $\log_8 x = 0$

9. $\log_7 w = -1$

10. $\log_w 64 = 3$

11. $\log_{10} w = -1$

12. $\log_{10} 1000 = x$

13. $\log_2 64 = x$

14. $\log_2\left(\frac{1}{4}\right) = x$

15. Graph the equation $y = \log_3 x$.

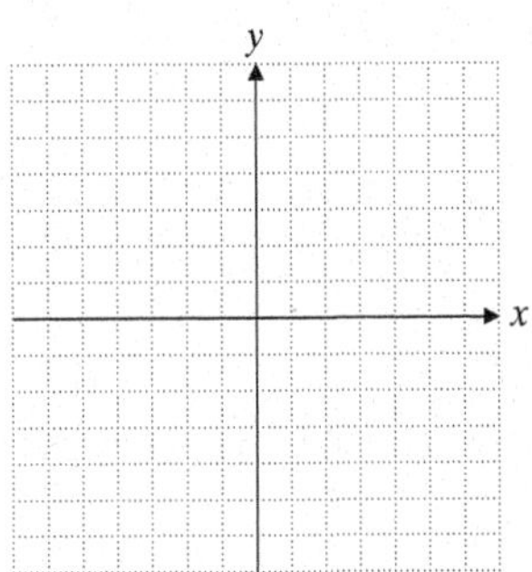

Write each expression as the sum or difference of $\log_2 5$, $\log_2 w$, $\log_2 x$, $\log_2 y$, *and* $\log_2 z$.

16. $\log_2\left(\frac{5x}{\sqrt{w}}\right)$

17. $\log_2 x^3\sqrt{y}$

Write as a single logarithm.

18. $\log_3 x + \log_3 w^{1/2} - \log_3 2$

19. $4 \log_8 w - \frac{1}{3}\log_8 z$

20. Evaluate $\log_e e^6$.

Find the value with a calculator.

21. $\log 23.8$

22. $\log 0.0817$

23. $\ln 3.92$

24. $\ln 803$

25. Find n if $\log n = 1.1367$.

26. Find n if $\ln n = 1.7$.

27. Evaluate. $\log_8 2.81$

28. Evaluate. $\log_4 72$

Solve each equation and check your solutions.

29. $\log_5 100 - \log_5 x = \log_5 4$

30. $\log_8 x + \log_8 3 = \log_8 75$

31. $\log_{11}\left(\frac{4}{3}x + 7\right) + \log_{11} 3 = 2$

32. $\log_8(x - 3) = -1 + \log_8 6x$

33. $\log(2t + 3) + \log(4t - 1) = 2 \log 3$

34. $\log(2t + 4) - \log(3t + 1) = \log 6$

Solve each equation. Leave your answers in exact form. Do not approximate.

35. $3^x = 14$

36. $5^{x+3} = 130$

37. $e^{2x-1} = 100$

Solve each equation. Round your answers to the nearest ten-thousandth.

38. $2^{3x+1} = 5^x$

39. $e^{3x-4} = 20$

40. $(1.03)^x = 20$

Compound Interest *For exercises 41 and 42, use $A = P(1 + r)^t$, the formula for interest that is compounded annually.*

41. How long will it take Frances to double the money in her account if the interest rate is 8% compounded annually? (Round your answer to the nearest year.)

42. How much money would Chou Lou have after 4 years if he invested \$5000 at 6% compounded annually?

Populations *The growth of many populations can be described by the equation $A = A_0e^{rt}$, where time t is measured in years, A_0 is the population at time $t = 0$, r is the annual growth rate, and A is the population at time t. Use this information to solve exercises 43 and 44. Round your answers to the nearest whole year.*

43. How long will it take a population of seven billion to increase to sixteen billion if $r = 2\%$ per year?

44. The number of moose in northern Maine is increasing at a rate of 3% per year. It is estimated in one county that there are now 2000 moose. If the growth rate remains unchanged, how many years will it be until there are 2600 moose in that county?

45. ***Gas Volume*** The work W done by a volume of gas expanding at a constant temperature from volume V_0 to volume V_1 is given by $W = p_0V_0 \ln\left(\frac{V_1}{V_0}\right)$, where p_0 is the pressure at volume V_0.

(a) Find W when $p_0 = 40$ pounds per cubic inch, $V_0 = 15$ cubic inches, and $V_1 = 24$ cubic inches.

(b) If the amount of work is 100 pounds, $V_0 = 8$ cubic inches, and $V_1 = 40$ cubic inches, find p_0.

46. ***Earthquakes*** An earthquake's magnitude is given by $M = \log\left(\frac{I}{I_0}\right)$, where I is the intensity of the earthquake and I_0 is the minimum measurable intensity. A 1964 earthquake in Anchorage, Alaska, had a magnitude of 8.4. A 1975 earthquake in Turkey had a magnitude of 6.7. How many times more intense was the Alaska earthquake than the Turkey earthquake? (*Source:* www.noaa.gov)

How Am I Doing? Chapter 12 Test

MATH COACH MyMathLab® YouTube

After you take this test read through the Math Coach on pages 773–774. Math Coach videos are available via MyMathLab and YouTube. Step-by-step test solutions in the Chapter Test Prep Videos are also available via MyMathLab and YouTube. (Search "TobeyBegInterAlg" and click on "Channels.")

1. ______
2. ______
3. ______
4. ______
5. ______
6. ______
7. ______
8. ______
9. ______
10. ______
11. ______
12. ______
13. ______
14. ______
15. ______
16. ______
17. ______

Total Correct: ______

1. Graph $f(x) = 3^{x+1}$.

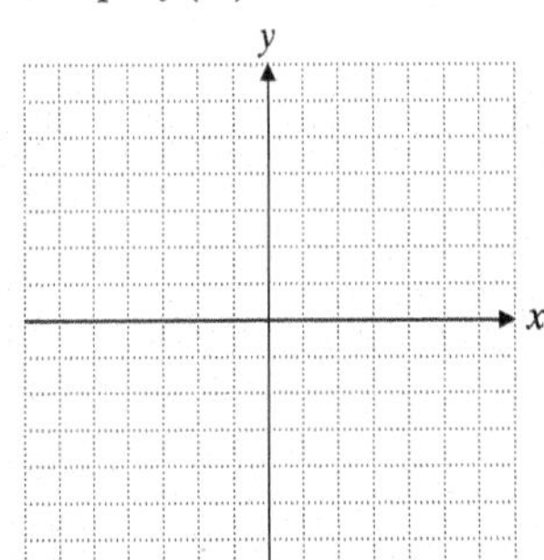

2. Graph $f(x) = \log_2 x$.

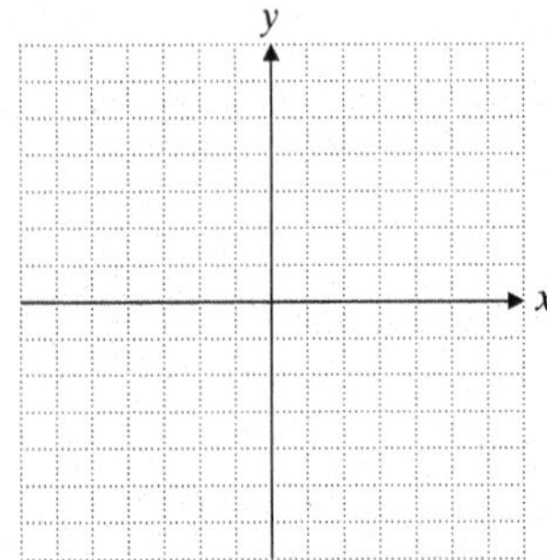

MC **3.** Solve. $4^{x+3} = 64$

In exercises 4 and 5, solve for the variable.

4. $\log_w 125 = 3$

5. $\log_8 x = -2$

MC **6.** Write as a single logarithm. $2 \log_7 x + \log_7 y - \log_7 4$

Evaluate using a calculator. Round your answers to the nearest ten-thousandth.

7. $\ln 5.99$

8. $\log 23.6$

9. $\log_3 1.62$

Use a calculator to approximate x.

10. $\log x = 3.7284$

11. $\ln x = 0.14$

Solve the equation and check your solutions for problems 12 and 13.

MC **12.** $\log_8(x + 3) - \log_8 2x = \log_8 4$

13. $\log_8 2x + \log_8 6 = 2$

MC **14.** Solve the equation. Leave your answer in exact form. Do not approximate. $e^{5x-3} = 57$

15. Solve. $5^{3x+6} = 17$ (Approximate your answer to the nearest ten-thousandth.)

16. How much money will Henry have if he invests $2000 for 5 years at 8% annual interest compounded annually?

17. How long will it take for Barb to double her money if she invests it at 5% compounded annually? Round to the nearest whole year.

MATH COACH

Mastering the skills you need to do well on the test.

Students often make the same types of errors when they do the Chapter 12 Test. Here are some helpful hints to keep you from making those common errors on test problems.

Chapter 12 Test, Problem 3 Solve. $4^{x+3} = 64$

HELPFUL HINT If one side of the equation is in exponential form, it is best to try to write the other side of the equation in exponential form. Then the procedure will be easier to complete.

Did you rewrite the equation as $4^{x+3} = 4^3$?

Yes _____ No _____

If you answered No, remember to write your equation in the form $b^x = b^y$. Note that $64 = 4^3$, and you want the base of the exponent on each side of the equation to be the same number, b, such that $b > 0$ and $b \neq 1$.

Did you use the property of exponential equations to write the equation $x + 3 = 3$?

Yes _____ No _____

If you answered No, remember that if $b^x = b^y$, then $x = y$ for any $b > 0$ and $b \neq 1$. Now solve the equation for x.

If you answered Problem 3 incorrectly, go back and rework the problem using these suggestions.

Chapter 12 Test, Problem 6 Write as a single logarithm. $2 \log_7 x + \log_7 y - \log_7 4$

HELPFUL HINT Try your best to memorize the three properties of logarithms in Objectives 12.3.1, 12.3.2, and 12.3.3. They are essential to know when working with logarithmic expressions.

Did you use property 3 to eliminate the coefficient of the first logarithmic term, $2 \log_7 x$, and rewrite that term as $\log_7 x^2$?

Yes _____ No _____

If you answered No, review property 3 in Objective 12.3.3 and complete this step again.

Did you use property 1 to combine the sum of the first two logarithmic terms and obtain the expression, $\log_7 x^2y - \log_7 4$?

Yes _____ No _____

If you answered No, review property 1 in Objective 12.3.1 and complete this step again.

In your last step, you will need to use property 2 in Objective 12.3.2 to combine the difference of two logarithms.

Now go back and rework the problem using these suggestions.

Chapter 12 Test, Problem 12

Solve the equation and check your solution. $\log_8(x + 3) - \log_8 2x = \log_8 4$

HELPFUL HINT Use the properties of logarithms to rewrite the equation such that one logarithmic term appears on each side of the equation.

First, did you combine the two logarithms on the left side of the equation and rewrite the equation as $\log_8\left(\frac{x+3}{2x}\right) = \log_8 4$?

Yes _____ No _____

If you answered No, use property 2 from Objective 12.3.2 to complete this first step again.

Next, did you rewrite the equation as $\frac{x+3}{2x} = 4$?

Yes _____ No _____

If you answered No, notice that you now have one logarithmic term on each side of the equation, which is the goal mentioned in the Helpful Hint. You can use property 6 from Objective 12.3.4 to evaluate these two logarithms.

In your final step, solve the equation for x. Check your solution by substituting for x in the original equation, evaluating the logarithms and then simplifying the resulting equation.

If you answered Problem 12 incorrectly, go back and rework the problem using these suggestions.

Chapter 12 Test, Problem 14

Solve the equation. Leave your answer in exact form. Do not approximate. $e^{5x-3} = 57$

HELPFUL HINT The first step is to take the natural logarithm of each side of the equation. Then you can simplify the equation further using the properties of logarithms.

Did you take the natural logarithm of each side of the equation and rewrite as $\ln e^{5x-3} = \ln 57$?

Yes _____ No _____

Next did you rewrite this equation as $(5x - 3)(\ln e) = \ln 57$?

Yes _____ No _____

If you answered No to these questions, review property 3 of logarithms from Objective 12.3.3 and complete this step again.

Did you rewrite the equation as $5x - 3 = \ln 57$?

Yes _____ No _____

If you answered No, remember that $\ln e = \log_e e = 1$. This combines the definition of natural logarithms or $\ln e = \log_e e$ and property 4 of logarithms from Objective 12.3.4, which indicates that $\log_e e = 1$.

In your final step, solve the equation for x without evaluating ln 57.

Now go back and rework the problem using these suggestions.

Need more help? Look for section examples marked with MC to review.

Practice Final Examination

Review Chapters 0–12 and then try to solve the problems in this Practice Final Examination.

Chapter 0

1. Add. $3\frac{1}{4} + 2\frac{3}{5}$

2. Multiply. $\left(1\frac{1}{6}\right)\left(2\frac{2}{3}\right)$

3. Divide. $\frac{15}{4} \div \frac{3}{8}$

4. Multiply. $(1.63)(3.05)$

5. Divide. $120 \div 0.0006$

6. Find 7% of 64,000.

Chapter 1

7. Evaluate $(4 - 3)^2 + \sqrt{9} \div (-3) + 4$.

8. Simplify. $5a - 2ab - 3a^2 - 6a - 8ab + 2a^2$

9. Simplify $-2x + 3y\{7 - 2[x - (4x + y)]\}$.

10. Evaluate if $x = -2$ and $y = 3$. $2x^2 - 3xy - 4y$

11. Find the Fahrenheit temperature when the Celsius temperature is $-35°$. Use the formula $F = \frac{9}{5}C + 32$.

Chapter 2

12. Solve for x. $\frac{1}{3}x - 4 = \frac{1}{2}x + 1$

13. Solve for y. $-4x + 3y = 7$

14. Solve for x and graph the resulting inequality on a number line.
$5x + 3 - (4x - 2) \leq 6x - 8$

▲**15.** A piece of land is rectangular and has a perimeter of 1760 meters. The length is 200 meters less than twice the width. Find the dimensions of the land.

▲**16.** Find the volume of a sphere with a diameter of 6 feet. Use 3.14 as an approximation for π.

Chapter 3

17. Find the intercepts and then graph the line $7x - 2y = -14$.

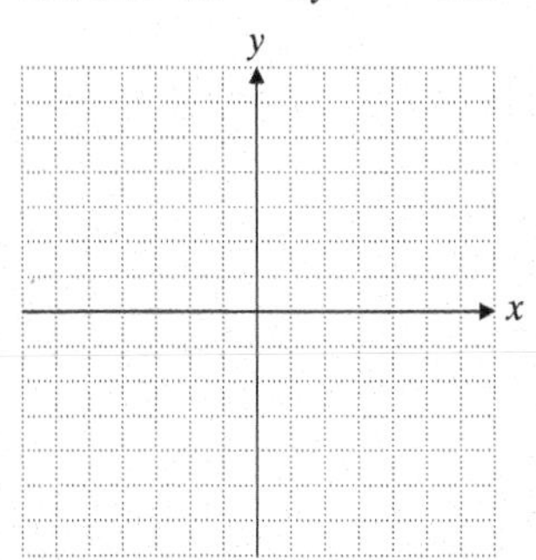

18. Graph the region $3x - 4y \leq 6$.

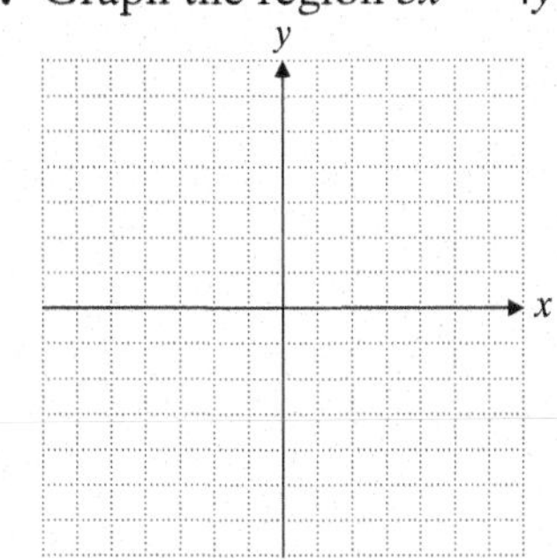

1. ______
2. ______
3. ______
4. ______
5. ______
6. ______
7. ______
8. ______
9. ______
10. ______
11. ______
12. ______
13. ______
14. ______
15. ______
16. ______
17. ______
18. ______

19. ______

20. ______

21. ______

22. ______

23. ______

24. ______

25. ______

26. ______

27. ______

28. ______

29. ______

30. ______

31. ______

32. ______

33. ______

34. ______

35. ______

36. ______

37. ______

38. ______

39. ______

19. Find the slope of the line passing through $(1, 5)$ and $(-2, -3)$.

20. Find the slope of the line that is parallel to $y = -\frac{2}{3}x + 4$.

Given the function defined by $f(x) = 3x^2 - 4x - 3$, *find the following.*

21. $f(3)$

22. $f(-2)$

23. Determine the domain and range of the relation. $\{(2, -7), (-1, 1), (3, 2)\}$

Chapter 4

24. Solve for x and y.

$$\begin{aligned} \frac{1}{2}x + \frac{2}{3}y &= 1 \\ \frac{1}{3}x + y &= -1 \end{aligned}$$

25. Solve for x and y.

$$\begin{aligned} 4x - 3y &= 12 \\ 3x - 4y &= 2 \end{aligned}$$

26. Solve for x, y, and z.

$$\begin{aligned} 2x + 3y - z &= 16 \\ x - y + 3z &= -9 \\ 5x + 2y - z &= 15 \end{aligned}$$

27. Solve for x, y, and z.

$$\begin{aligned} y + z &= 2 \\ x + z &= 5 \\ x + y &= 5 \end{aligned}$$

28. Graph the region.

$$\begin{aligned} 3y &\geq 8x - 12 \\ 2x + 3y &\leq -6 \end{aligned}$$

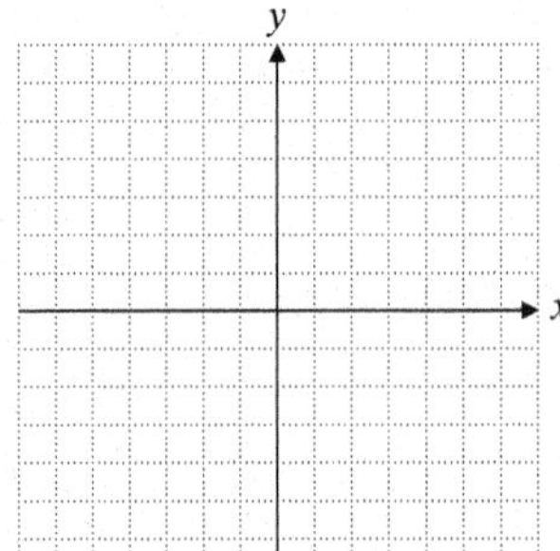

Chapter 5

29. Simplify. $(-3x^2y)(-6x^3y^4)$

30. Multiply and simplify your answer. $(3x - 1)(2x + 5)$

31. Divide. $(x^2 + 7x + 12) \div (x + 3)$

Chapter 6

Factor the following completely.

32. $9x^2 - 30x + 25$

33. $x^3 + 2x^2 - 4x - 8$

34. $2x^3 + 15x^2 - 8x$

35. Solve for x. $x^2 + 15x + 54 = 0$

Chapter 7

Simplify the following.

36. $\dfrac{9x^3 - x}{3x^2 - 8x - 3}$

37. $\dfrac{x^2 - 9}{2x^2 + 7x + 3} \div \dfrac{x^2 - 3x}{2x^2 + 11x + 5}$

38. $\dfrac{3x}{x + 5} - \dfrac{2}{x^2 + 7x + 10}$

39. $\dfrac{\dfrac{3}{2x + 1} + 2}{1 - \dfrac{2}{4x^2 - 1}}$

40. Solve for x. $\dfrac{x-1}{x^2-4} = \dfrac{2}{x+2} + \dfrac{4}{x-2}$

Chapter 8

41. Simplify $\dfrac{5x^{-4}y^{-2}}{15x^{-1/2}y^3}$.

42. Simplify $\sqrt[3]{40x^4y^7}$.

43. Combine like terms.
$5\sqrt{2} - 3\sqrt{50} + 4\sqrt{98}$

44. Rationalize the denominator.
$\dfrac{2\sqrt{3}+1}{3\sqrt{3}-\sqrt{2}}$

45. Simplify and add together.
$i^3 + \sqrt{-25} + \sqrt{-16}$

46. Solve for x and check your solutions.
$\sqrt{x+7} = x + 5$

47. If y varies directly with the square of x and $y = 15$ when $x = 2$, what will y be when $x = 3$?

Chapter 9

48. Solve for x. $5x(x+1) = 1 + 6x$

49. Solve for x. $5x^2 - 9x = -12x$.

50. Solve for x. $x^{2/3} + 5x^{1/3} - 14 = 0$

51. Solve $3x^2 - 11x - 4 \geq 0$.

52. Graph the quadratic function $f(x) = -x^2 - 4x + 5$. Label the vertex and the intercepts.

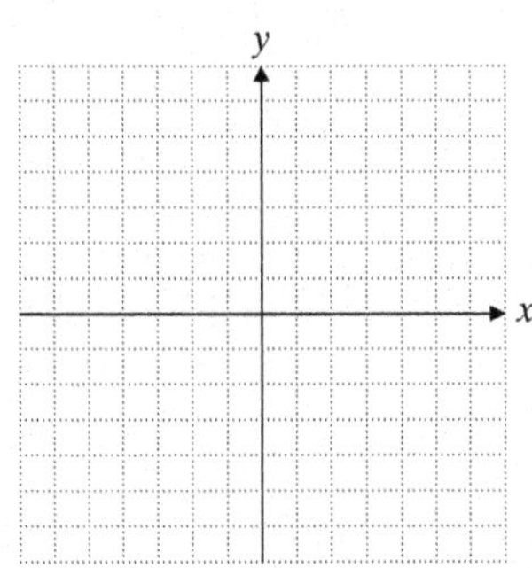

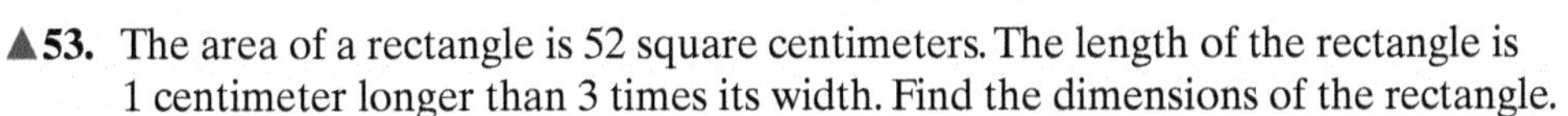

▲**53.** The area of a rectangle is 52 square centimeters. The length of the rectangle is 1 centimeter longer than 3 times its width. Find the dimensions of the rectangle.

54. Solve for x. $\left|\frac{2}{3}x - 4\right| = 2$

55. Solve the inequality. $|2x - 5| < 10$

Chapter 10

56. Place the equation of the circle in standard form. Find its center and radius.

$$x^2 + y^2 + 6x - 4y = -9$$

Identify and graph.

57. $\dfrac{x^2}{16} + \dfrac{y^2}{25} = 1$

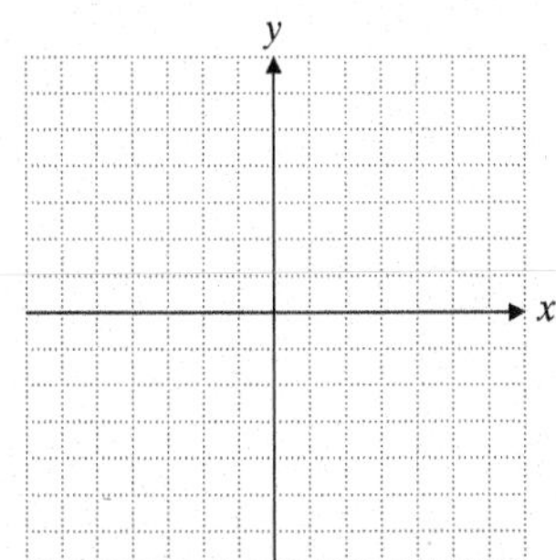

58. $\dfrac{x^2}{4} - \dfrac{y^2}{9} = 1$

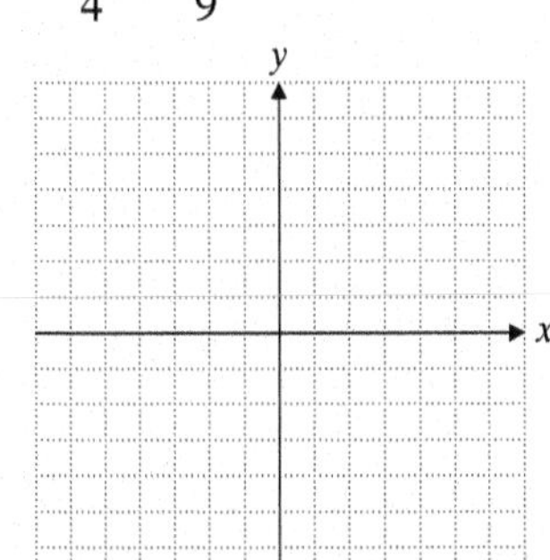

40. ______
41. ______
42. ______
43. ______
44. ______
45. ______
46. ______
47. ______
48. ______
49. ______
50. ______
51. ______
52. ______
53. ______
54. ______
55. ______
56. ______
57. ______
58. ______

59. ______

60. ______

(a)
(b)
61. (c) ______

62. ______

63. ______

64. ______

65. ______

66. ______

67. ______

68. ______

69. ______

59. $x = (y - 3)^2 + 5$

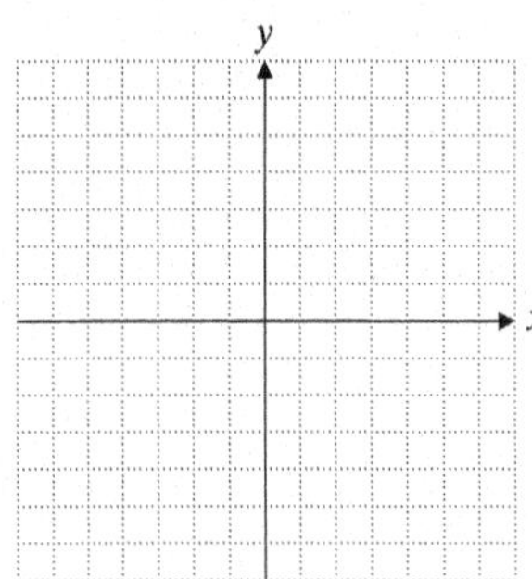

60. Solve the following system of equations.

$x^2 + y^2 = 16$
$x^2 - y = 4$

Chapter 11

61. Let $f(x) = 3x^2 - 2x + 5$.

(a) Find $f(-1)$.

(b) Find $f(a)$.

(c) Find $f(a + 2)$.

62. If $f(x) = 5x^2 - 3$ and $g(x) = -4x - 2$, find $f[g(x)]$.

63. If $f(x) = \frac{1}{2}x - 7$, find $f^{-1}(x)$.

64. Graph on one set of axes: $f(x)$, $f(x + 2)$, and $f(x) - 3$, if $f(x) = |x|$.

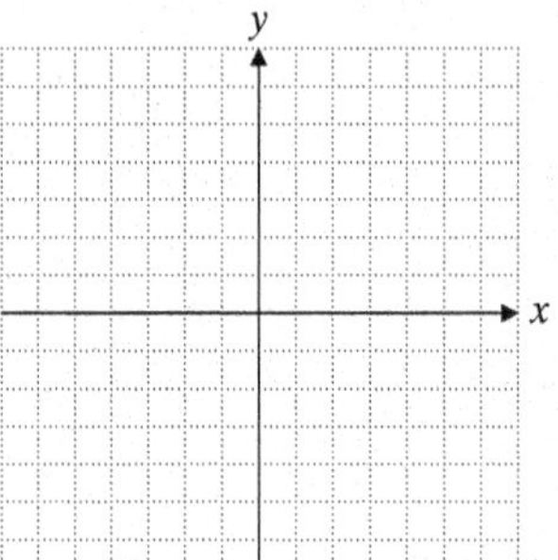

Chapter 12

65. Graph $f(x) = 2^{1-x}$. Plot three points.

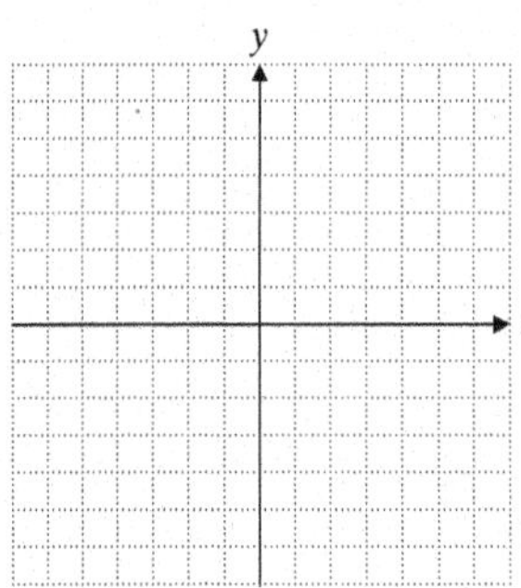

Solve for the variable.

66. $\log_5 x = -4$

67. $\log_4 (3x + 1) = 3$

68. $\log_{10} 0.01 = y$

69. $\log_2 6 + \log_2 x = 4 + \log_2 (x - 5)$

Appendix A Foundations for Intermediate Algebra: A Transition from Beginning to Intermediate Algebra

A.1 Integer Exponents, Square Roots, Order of Operations, and Scientific Notation

Student Learning Objectives

After studying this section, you will be able to:

1. Raise a number to a positive integer power.
2. Find square roots of numbers that are perfect squares.
3. Evaluate expressions by using the proper order of operations.
4. Rewrite expressions with negative exponents as expressions with positive exponents.
5. Use the product rule of exponents.
6. Use the quotient rule of exponents.
7. Use the power rules of exponents.
8. Express numbers in scientific notation.

1 Raising a Number to a Power

Exponents or powers are used to indicate repeated multiplication. For example, we can write $6 \cdot 6 \cdot 6 \cdot 6$ as 6^4. In the expression 6^4, 4 is the **exponent** or **power** that tells us how many times the **base**, 6, appears as a factor. This is called **exponential notation.**

EXPONENTIAL NOTATION

If x is a real number and n is a positive integer, then

$$x^n = \underbrace{x \cdot x \cdot x \cdot x \cdots}_{n \text{ factors}}$$

Example 1 Evaluate.

(a) $(-2)^4$ **(b)** -2^4 **(c)** 3^5 **(d)** $(-5)^3$ **(e)** $\left(\frac{1}{3}\right)^3$

Solution

(a) $(-2)^4 = (-2)(-2)(-2)(-2) = (4)(-2)(-2) = (-8)(-2) = 16$

Notice that we are raising -2 to the fourth power. That is, the base is -2. We use parentheses to clearly indicate that the base is negative.

(b) $-2^4 = -(2 \cdot 2 \cdot 2 \cdot 2) = -16$

Here the base is 2. The base is not -2. We wish to find the negative of 2 raised to the fourth power.

(c) $3^5 = 3 \cdot 3 \cdot 3 \cdot 3 \cdot 3 = 243$

(d) $(-5)^3 = (-5)(-5)(-5) = (25)(-5) = -125$

(e) $\left(\frac{1}{3}\right)^3 = \left(\frac{1}{3}\right)\left(\frac{1}{3}\right)\left(\frac{1}{3}\right) = \frac{1}{27}$ □

Student Practice 1 Evaluate.

(a) $(-3)^5$ **(b)** $(-3)^6$ **(c)** $(-4)^4$ **(d)** -4^4 **(e)** $\left(\frac{1}{5}\right)^2$

To Think About: *Raising Negative Numbers to a Power* Look at Student Practice 1. What do you notice about raising a negative number to an even power? To an odd power? Will this always be true? Why?

2 Finding Square Roots

We say that a square root of 16 is 4 because $4 \cdot 4 = 16$. You will note that, since $(-4)(-4) = 16$, another square root of 16 is -4. For practical purposes, we are usually interested in the nonnegative square root. We call this the **principal square root.** $\sqrt{}$ is the principal square root symbol and is called a **radical.**

$$\text{radical} \rightarrow \sqrt{9} = 3$$

radical → $\sqrt{9}$ = 3 ; radicand ↑ (9) ; ↑ principal square root (3)

The number or expression under the radical sign is called the **radicand.** Both 16 and 9 are called *perfect squares* because their square roots are integers.

> If x is an integer and a is a positive real number such that $a = x^2$, then x is a **square root** of a, and a is a **perfect square**.

Example 2 Find the square roots of 25. What is the principal square root?

Solutio Since $(-5)^2 = 25$ and $5^2 = 25$, the square roots of 25 are 5 and -5. The principal square root is 5.

Student Practice 2 What are the square roots of 49? What is the principal square root of 49?

We can find the square root of a positive number. However, there is no real number for $\sqrt{-4}$ or $\sqrt{-9}$. The square root of a negative number is not a real number.

Example 3 Evaluate.

(a) $\sqrt{0.04}$ **(b)** $\sqrt{\dfrac{25}{36}}$ **(c)** $\sqrt{-16}$

Solution

(a) $(0.2)^2 = (0.2)(0.2) = 0.04$. Therefore, $\sqrt{0.04} = 0.2$.

(b) We can write $\sqrt{\dfrac{25}{36}}$ as $\dfrac{\sqrt{25}}{\sqrt{36}}$, and $\dfrac{\sqrt{25}}{\sqrt{36}} = \dfrac{5}{6}$. Thus, $\sqrt{\dfrac{25}{36}} = \dfrac{5}{6}$.

(c) This is not a real number.

Student Practice 3 Evaluate.

(a) $\sqrt{0.09}$ **(b)** $\sqrt{\dfrac{4}{81}}$ **(c)** $\sqrt{-25}$

The square roots of some numbers are irrational numbers. For example, $\sqrt{3}$ and $\sqrt{7}$ are irrational numbers. We often use rational numbers to *approximate* square roots that are irrational. They can be found using a calculator with a square root key. To approximate $\sqrt{3}$ on most calculators, we enter the 3 and then press the  key. Using a calculator, we might get 1.7320508 as our approximation.

3 The Order of Operations of Real Numbers

Parentheses are used in numerical and algebraic expressions to group numbers and variables. When evaluating an expression containing parentheses, evaluate the numerical expressions inside the parentheses first. When we need more than one set of parentheses, we may also use brackets. To evaluate such an expression, work from the inside out.

When many arithmetic operations or grouping symbols are used, we use the following order of operations.

> **ORDER OF OPERATIONS FOR CALCULATIONS**
>
> 1. Combine numbers inside grouping symbols.
> 2. Raise numbers to their indicated powers and take any indicated roots.
> 3. Multiply and divide numbers from left to right.
> 4. Add and subtract numbers from left to right.

Example 4 Evaluate. $12 - 3[7 + 5(6 - 9)]$

Solution

$12 - 3[7 + 5(6 - 9)] = 12 - 3[7 + 5(-3)]$	Begin with the innermost grouping symbols.
$= 12 - 3[7 + (-15)]$	Multiply inside the grouping symbols.
$= 12 - 3[-8]$	Add inside the grouping symbols.
$= 12 + 24$	Multiply.
$= 36$	Add.

□

Student Practice 4 Evaluate.

(a) $6(12 - 8) + 4$ **(b)** $5[6 - 3(7 - 9)] - 8$

To Think About: *Importance of Grouping Symbols* Rewrite the expression in Example 4 without grouping symbols. Then evaluate the expression. Remember to first multiply from left to right and then add and subtract from left to right. Explain why the answers may differ.

A radical or absolute value bars group the quantities within them. Thus, we simplify the numerical expressions within the grouping symbols before we find the square root or the absolute value.

Example 5 Evaluate.

(a) $\sqrt{(-3)^2 + (4)^2}$ **(b)** $|5 - 8 + 7 - 13|$

Solution

(a) $\sqrt{(-3)^2 + (4)^2} = \sqrt{9 + 16} = \sqrt{25} = 5$

(b) $|5 - 8 + 7 - 13| = |-9| = 9$

□

Student Practice 5 Evaluate.

(a) $\sqrt{(-5)^2 + 12^2}$ **(b)** $|-3 - 7 + 2 - (-4)|$

A fraction bar acts like a grouping symbol. We must evaluate the expressions above and below a fraction bar before we divide.

Example 6 Evaluate. $\dfrac{2 \cdot 6^2 - 12 \div 3}{4 - 8}$

Solution We evaluate the numerator first.

$$\begin{aligned} 2 \cdot 6^2 - 12 \div 3 &= 2 \cdot 36 - 12 \div 3 && \text{Raise to a power.} \\ &= 72 - 4 && \text{Multiply and divide from left to right.} \\ &= 68 && \text{Subtract.} \end{aligned}$$

Next we evaluate the denominator.

$$4 - 8 = -4$$

Thus,

$$\frac{2 \cdot 6^2 - 12 \div 3}{4 - 8} = \frac{68}{-4} = -17$$

□

Student Practice 6 Evaluate.

$$\frac{2(3) + 5(-2)}{1 + 2 \cdot 3^2 + 5(-3)}$$

4 Rewriting Expressions with Negative Exponents as Expressions with Positive Exponents

Before we formally define the meaning of a negative exponent, let us look for a pattern.

On this side we decrease each exponent by 1 to obtain the expression on the next line.	↓		↓	On this side we divide each number by 3 to obtain the number on the next line.
		$3^4 = 81$		
		$3^3 = 27$		
		$3^2 = 9$		
		$3^1 = 3$		
		$3^0 = 1$		
		$3^{-1} = ?$		
		$3^{-2} = ?$		

What results would you expect on the last two lines? $3^{-1} = \frac{1}{3}$? Then $3^{-2} = \frac{1}{3^2} = \frac{1}{9}$. Do you see the pattern? Then we would have

$$3^{-3} = \frac{1}{3^3} = \frac{1}{27} \quad \text{and} \quad 3^{-4} = \frac{1}{3^4} = \frac{1}{81}.$$

Now we are ready to make a formal definition of a negative exponent.

DEFINITION OF NEGATIVE EXPONENTS

If x is any nonzero real number and n is an integer,

$$x^{-n} = \frac{1}{x^n}.$$

Example 7 Simplify. Do not leave negative exponents in your answers.

(a) 2^{-5} **(b)** w^{-6}

Solution **(a)** $2^{-5} = \dfrac{1}{2^5} = \dfrac{1}{32}$ **(b)** $w^{-6} = \dfrac{1}{w^6}$ □

Student Practice 7 Simplify. Do not leave negative exponents in your answers. **(a)** 3^{-2} **(b)** z^{-8}

5 The Product Rule of Exponents

Numbers and variables with exponents can be multiplied quite simply if *the base is the same.* For example, we know that

$$(x^3)(x^2) = (x \cdot x \cdot x)(x \cdot x).$$

Since the factor x appears five times, it must be true that

$$x^3 \cdot x^2 = x^5.$$

Hence we can state a general rule.

PRODUCT RULE OF EXPONENTS

If x is a real number and m and n are integers, then

$$x^m \cdot x^n = x^{m+n}.$$

This rule says that the exponents are integers. Thus, they can be negative.

Example 8 Multiply, then simplify. Do not leave negative exponents in your answer.

(a) $(8a^{-3}b^{-8})(2a^5b^5)$ **(b)** $(a+b)^2(a+b)^3$

Solution

(a) $(8a^{-3}b^{-8})(2a^5b^5) = 16a^{-3+5}b^{-8+5}$

$= 16a^2b^{-3}$

$= 16a^2\left(\dfrac{1}{b^3}\right) = \dfrac{16a^2}{b^3}$

(b) $(a+b)^2(a+b)^3 = (a+b)^{2+3} = (a+b)^5$ (The base is $a+b$.)

You may leave your answer in exponent form when the base is a binomial. □

Student Practice 8 Multiply, then simplify. Do not leave negative exponents in your answers.

(a) $(7xy^{-2})(2x^{-5}y^{-6})$ **(b)** $(x+2y)^4(x+2y)^{10}$

6 The Quotient Rule of Exponents

We now develop the rule for dividing numbers with exponents. We know that

$$\frac{x^5}{x^3} = \frac{x \cdot x \cdot \cancel{x \cdot x \cdot x}}{\cancel{x \cdot x \cdot x}} = x \cdot x = x^2.$$

Note that $x^{5-3} = x^2$. This leads us to the following general rule.

QUOTIENT RULE OF EXPONENTS

If x is a nonzero real number and m and n are integers,

$$\frac{x^m}{x^n} = x^{m-n}.$$

Example 9 Divide. $\dfrac{3x^{-5}y^{-6}}{27x^2y^{-8}}$

Solution Simplify your answer so there are no negative exponents.

$$\frac{3x^{-5}y^{-6}}{27x^2y^{-8}} = \frac{1}{9}x^{-5-2}y^{-6-(-8)} = \frac{1}{9}x^{-5-2}y^{-6+8} = \frac{1}{9}x^{-7}y^2 = \frac{y^2}{9x^7}$$ □

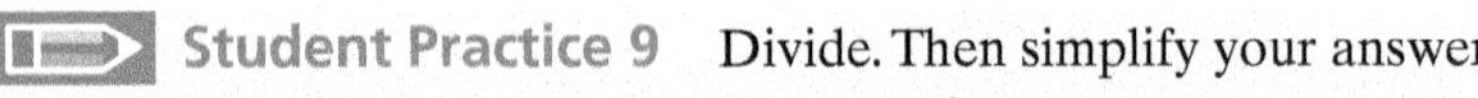

Student Practice 9 Divide. Then simplify your answer.

$$\frac{2x^{-3}y}{4x^{-2}y^5}$$

Our quotient rule leads us to an interesting situation if $m = n$.

$$\frac{x^m}{x^m} = x^{m-m} = x^0$$

But what exactly is x^0? Whenever we divide any nonzero value by itself we always get 1, so we would therefore expect that $x^0 = 1$. But can we prove that? Yes.

$$\text{Since} \quad x^{-n} = \frac{1}{x^n},$$

Then $x^{-n} \cdot x^n = 1$ Multiplying both sides by x^n.

$x^{-n+n} = 1$ Using the product rule.

$x^0 = 1$ Since $-n + n = 0$.

RAISING A NUMBER TO THE ZERO POWER

For any nonzero real number x, $x^0 = 1$.

Example 10 Divide, then simplify your answers. Do not leave negative exponents in your answers.

(a) $(3x)^0$ **(b)** $\dfrac{-150a^3b^4c^2}{-300abc^2}$

Solution

(a) $(3x)^0 = 1$ Note that the entire expression is raised to the zero power.

(b) $$\frac{-150a^3b^4c^2}{-300abc^2} = \frac{-150}{-300} \cdot \frac{a^3}{a^1} \cdot \frac{b^4}{b^1} \cdot \frac{c^2}{c^2} = \frac{1}{2} \cdot a^2 \cdot b^3 \cdot c^0 = \frac{a^2b^3}{2}$$

Remember that we don't usually write an exponent of 1. Thus, $a = a^1$ and $b = b^1$. □

Student Practice 10 Divide, then simplify your answers. Do not leave negative exponents in your answers.

(a) $\dfrac{30x^6y^5}{20x^3y^2}$ **(b)** $\dfrac{-15a^3b^4c^4}{3a^5b^4c^2}$ **(c)** $(5^{-3})(2a)^0$

For the remainder of this section, we will assume that for all exercises involving exponents, a simplified answer should not contain negative exponents.

7 The Power Rules of Exponents

Note that $(x^4)^3 = x^4 \cdot x^4 \cdot x^4 = x^{4+4+4} = x^{4 \cdot 3} = x^{12}$. In the same way we can show

$$(xy)^3 = x^3y^3$$

$$\text{and } \left(\frac{x}{y}\right)^3 = \frac{x^3}{y^3} \qquad (y \neq 0).$$

Therefore, we have the following rules.

POWER RULES OF EXPONENTS

If x and y are any real numbers and m and n are integers,

$$(x^m)^n = x^{mn}, \quad (xy)^n = x^ny^n, \text{ and}$$

$$\left(\frac{x}{y}\right)^n = \frac{x^n}{y^n}, \quad \text{if } y \neq 0.$$

Example 11 Use the power rules of exponents to simplify.

(a) $(x^6)^5$ **(b)** $(2^8)^4$ **(c)** $[(a+b)^2]^4$

Solution

(a) $(x^6)^5 = x^{6 \cdot 5} = x^{30}$

(b) $(2^8)^4 = 2^{32}$ Be careful. Don't change the base of 2.

(c) $[(a+b)^2]^4 = (a+b)^8$ The base is $a + b$. □

Student Practice 11 Use the power rules of exponents to simplify.

(a) $(w^3)^8$ **(b)** $(5^2)^5$ **(c)** $[(x-2y)^3]^3$

Example 12 Simplify.

(a) $(3xy^2)^4$ **(b)** $\left(\frac{2a^2b^3}{3ab^4}\right)^3$ **(c)** $(2a^2b^{-3}c^0)^{-4}$

Solution

(a) $(3xy^2)^4 = 3^4x^4y^8 = 81x^4y^8$

(b) $\left(\frac{2a^2b^3}{3ab^4}\right)^3 = \frac{2^3a^6b^9}{3^3a^3b^{12}} = \frac{8a^3}{27b^3}$

(c) $(2a^2b^{-3}c^0)^{-4} = 2^{-4}a^{-8}b^{12} = \frac{b^{12}}{2^4a^8} = \frac{b^{12}}{16a^8}$ □

Student Practice 12 Simplify.

(a) $(4x^3y^4)^2$ **(b)** $\left(\frac{4xy}{3x^5y^6}\right)^3$ **(c)** $(3xy^2)^{-2}$

We need to derive one more rule. You should be able to follow the steps.

$$\frac{x^{-m}}{y^{-n}} = \frac{\frac{1}{x^m}}{\frac{1}{y^n}} = \frac{1}{x^m} \cdot \frac{y^n}{1} = \frac{y^n}{x^m}$$

RULE OF NEGATIVE EXPONENTS

If m and n are positive integers and x and y are nonzero real numbers, then

$$\frac{x^{-m}}{y^{-n}} = \frac{y^n}{x^m}.$$

For example, $\frac{x^{-5}}{y^{-6}} = \frac{y^6}{x^5}$ and $\frac{2^{-3}}{x^{-4}} = \frac{x^4}{2^3} = \frac{x^4}{8}$.

Summary of Rules of Exponents when $x, y \neq 0$

1. $x^m \cdot x^n = x^{m+n}$
2. $\frac{x^m}{x^n} = x^{m-n}$
3. $x^{-n} = \frac{1}{x^n}$
4. $x^0 = 1$
5. $(x^m)^n = x^{mn}$
6. $(xy)^n = x^n y^n$
7. $\left(\frac{x}{y}\right)^n = \frac{x^n}{y^n}$
8. $\frac{x^{-m}}{y^{-n}} = \frac{y^n}{x^m}$

Example 13 Simplify.

(a) $\frac{3x^{-2}y^3z^{-1}}{4x^3y^{-5}z^{-2}}$ **(b)** $\left(\frac{5xy^{-3}}{2x^{-4}yz^{-3}}\right)^{-2}$

Solution

(a) First remove all negative exponents.

$$\frac{3x^{-2}y^3z^{-1}}{4x^3y^{-5}z^{-2}} = \frac{3y^3y^5z^2}{4x^3x^2z^1}$$ Only variables with negative exponents will change their position.

$$= \frac{3y^8z^2}{4x^5z^1}$$

$$= \frac{3y^8z}{4x^5}$$

(b) First remove the parentheses by using the power rules of exponents.

$$\left(\frac{5xy^{-3}}{2x^{-4}yz^{-3}}\right)^{-2} = \frac{5^{-2}x^{-2}y^6}{2^{-2}x^8y^{-2}z^6}$$

$$= \frac{2^2y^6y^2}{5^2x^8x^2z^6}$$

$$= \frac{4y^8}{25x^{10}z^6}$$ □

Student Practice 13 Simplify.

(a) $\frac{7x^2y^{-4}z^{-3}}{8x^{-5}y^{-6}z^2}$ **(b)** $\left(\frac{4x^2y^{-2}}{x^{-4}y^{-3}}\right)^{-3}$

8 Scientific Notation

Scientific notation is a convenient way to write very large or very small numbers. For example, we can write 50,000,000 as 5×10^7 since $10^7 = 10{,}000{,}000$, and we can write $0.0000000005 = 5 \times 10^{-10}$ since $10^{-10} = \dfrac{1}{10^{10}} = \dfrac{1}{10{,}000{,}000{,}000} = 0.0000000001$.

In scientific notation, the first factor is a number that is greater than or equal to 1, but less than 10. The second factor is a power of 10.

SCIENTIFIC NOTATION

A positive number written in **scientific notation** has the form $a \times 10^n$, where $1 \le a < 10$ and n is an integer.

Decimal form and scientific notation are just equivalent forms of the same number. To change a number from decimal notation to scientific notation, follow the steps below. Remember that the first factor must be a number between 1 and 10. This determines where to place the decimal point.

CONVERTING FROM DECIMAL NOTATION TO SCIENTIFIC NOTATION

1. Move the decimal point from its original position to the right of the first nonzero digit.
2. Count the number of places that you moved the decimal point. This number is the absolute value of the power of 10 (that is, the exponent).
3. If you moved the decimal point to the right, the exponent is negative; if you moved it to the left, the exponent is positive.

Example 14 Write in scientific notation.

(a) 7816 **(b)** 15,200,000 **(c)** 0.0123 **(d)** 0.00046

Solution

(a) $7816 = 7.816 \times 10^3$ We moved the decimal point three places to the left, so the power of 10 is 3.

(b) $15{,}200{,}000 = 1.52 \times 10^7$

(c) $0.0123 = 1.23 \times 10^{-2}$ We moved the decimal point two places to the right, so the power of 10 is -2.

(d) $0.00046 = 4.6 \times 10^{-4}$ □

Student Practice 14 Write in scientific notation.

(a) 128,320 **(b)** 476 **(c)** 0.0786 **(d)** 0.007

We can also change a number from scientific notation to decimal notation. We simply move the decimal point to the right or to the left the number of places indicated by the power of 10.

Calculator

Scientific Notation

Most scientific calculators can display only eight digits at one time. Numbers with more than eight digits are shown in scientific notation.

[1.12 E 08] means 1.12×10^8.

Note that the display on your calculator may be slightly different. You can use the calculator to compute large numbers by entering the numbers using scientific notation. For example, to compute

$(7.48 \times 10^{24}) \times (3.5 \times 10^8)$

on a scientific calculator, press these keys:

7.48 [EE] 24 [×]

3.5 [EE] 8 [=]

The display should read:

[2.618 E 33]

The label on the key used for entering the power of 10 will vary depending on the calculator.

Some scientific calculators have an [EXP] key.

Try the following:

(a) $35{,}000{,}000 + 77{,}000{,}000$

(b) $800{,}000{,}000 - 29{,}000{,}000$

(c) $(6.23 \times 10^{12}) \times (4.9 \times 10^5)$

(d) $(2.5 \times 10^7)^5$

(e) How many seconds are there in 1000 years?

Example 15 Write in decimal form.

(a) 8.8632×10^4 **(b)** 6.032×10^{-2} **(c)** 4.4861×10^{-5}

Solution

(a) $8.8632 \times 10^4 = 88{,}632$

Move the decimal point two places to the left.

(b) $6.032 \times 10^{-2} = 0.06032$

(c) $4.4861 \times 10^{-5} = 0.000044861$ □

Student Practice 15 Write in decimal form.

(a) 4.62×10^6 **(b)** 1.973×10^{-3} **(c)** 4.931×10^{-1}

Using scientific notation and the laws of exponents can greatly simplify calculations.

Example 16 Evaluate using scientific notation. $\dfrac{(0.000000036)(0.002)}{0.000012}$

Solution Rewrite the expression in scientific notation.

$$\frac{(3.6 \times 10^{-8})(2 \times 10^{-3})}{1.2 \times 10^{-5}}$$

Now rewrite using the commutative property.

$$\frac{\overset{3}{\cancel{(3.6)}}(2)(10^{-8})(10^{-3})}{\underset{1}{\cancel{(1.2)}}(10^{-5})} = \frac{6.0}{1} \times \frac{10^{-11}}{10^{-5}} \quad \text{Simplify and use the laws of exponents.}$$

$$= 6.0 \times 10^{-11-(-5)}$$

$$= 6.0 \times 10^{-6}$$ □

Student Practice 16 Evaluate using scientific notation

$$\frac{(55{,}000)(3{,}000{,}000)}{5{,}500{,}000}.$$

A.1 Exercises MyMathLab®

Verbal and Writing Skills, Exercises 1–6

1. In the expression a^3, identify the base and the exponent.

2. When a negative number is raised to an odd power, is the result positive or negative?

3. When a negative number is raised to an even power, is the result positive or negative?

4. Will $-a^n$ always be negative? Why or why not?

5. What are the square roots of 121? Why are there two answers?

6. What is the principal square root?

Evaluate.

7. 2^5

8. 7^3

9. $(-5)^2$

10. $(-4)^3$

11. -6^2

12. -3^4

13. -1^4

14. -3^2

15. $\left(-\frac{1}{4}\right)^4$

16. $\left(-\frac{1}{5}\right)^3$

17. $(0.7)^2$

18. $(-0.5)^2$

19. $(0.04)^3$

20. $(0.03)^3$

Find the value of each expression containing a principal square root.

21. $\sqrt{81}$

22. $\sqrt{121}$

23. $-\sqrt{16}$

24. $-\sqrt{64}$

25. $\sqrt{\frac{4}{9}}$

26. $\sqrt{\frac{1}{36}}$

27. $\sqrt{0.09}$

28. $\sqrt{0.25}$

29. $\sqrt{\frac{5}{36}+\frac{31}{36}}$

30. $\sqrt{\frac{1}{9}+\frac{3}{9}}$

31. $\sqrt{-36}$

32. $\sqrt{-49}$

Follow the proper order of operations to evaluate each of the following.

33. $-15 \div 3 + 7(-4)$

34. $16 \div (-8) - 6(-2)$

35. $-5^2 + 3(1 - 8)$

36. $-8^2 - 4(1 - 12)$

37. $5[(1.2 - 0.4) - 0.8]$

38. $-2[(3.6 + 0.3) - 0.9]$

39. $4(-6) - 3^2 + \sqrt{25}$

40. $9(-2) - 2^4 + \sqrt{81}$

41. $\frac{|2^2 - 5| - 3^2}{-5 + 3}$

42. $\dfrac{-2 + |2^3 - 10|}{3 - 4}$

43. $\dfrac{\sqrt{(-5)^2 - 3 + 14}}{|19 - 6 + 3 - 25|}$

44. $\dfrac{\sqrt{(-2)^2 - 3 + 3}}{6 - |3 \cdot 2 - 8|}$

45. $\dfrac{\sqrt{6^2 - 3^2 - 2}}{(-3)^2 - 4}$

46. $\dfrac{\sqrt{4 \cdot 7 + 2^3}}{3^2 - 5}$

Simplify. Rewrite all expressions with positive exponents only.

47. 3^{-2}

48. 4^{-3}

49. x^{-5}

50. y^{-4}

Multiply.

51. $(3x)(-2x^5)$

52. $(5y^2)(3y)$

53. $(-11x^2y^2)(-x^4y^7)$

54. $(-15x^4y)(-6xy^5)$

55. $4x^0y$

56. $-6a^2b^0$

57. $(3xy)^0(7xy)$

58. $-8a^2b^3(-6a)^0$

59. $(-6x^2yz^0)(-4x^0y^2z)$

60. $(5^0a^3b^4)(-2a^3b^0)$

61. $\left(-\dfrac{3}{5}m^{-2}n^4\right)(5m^2n^{-5})$

62. $\left(\dfrac{2}{3}m^2n^{-3}\right)(6m^{-5}n)$

Divide. Simplify your answers.

63. $\dfrac{2^8}{2^5}$

64. $\dfrac{3^{16}}{3^{18}}$

65. $\dfrac{2x^3}{x^8}$

66. $\dfrac{4y^3}{8y}$

67. $\dfrac{-15x^4yz}{3xy}$

68. $\dfrac{40a^3b}{-5a^3}$

69. $\dfrac{-20a^{-3}b^{-8}}{14a^{-5}b^{-12}}$

70. $\dfrac{-27x^7y^{10}}{-6x^{-2}y^{-3}}$

Use the power rules to simplify each expression.

71. $\left(\dfrac{x^2y^3}{z}\right)^6$

72. $\left(\dfrac{x^3}{y^5z^8}\right)^4$

73. $\left(\dfrac{3ab^{-2}}{4a^0b^4}\right)^2$

74. $\left(\dfrac{5a^3b}{-3a^{-2}b^0}\right)^3$

75. $\left(\dfrac{2xy^2}{x^{-3}y^{-4}}\right)^{-3}$

76. $\left(\dfrac{3x^{-4}y}{x^{-3}y^2}\right)^{-2}$

77. $(x^{-1}y^3)^{-2}(2x)^2$

78. $(x^3y^{-2})^{-2}(5x^{-5})^2$

79. $\dfrac{(-3m^5n^{-1})^3}{(mn)^2}$

80. $\dfrac{(m^4n^3)^{-1}}{(-5m^{-3}n^4)^2}$

Mixed Practice, Exercises 81–92

Simplify. Express your answers with positive exponents only.

81. $\dfrac{2^{-3}a^{2}}{2^{-4}a^{-2}}$

82. $\dfrac{3^{4}a^{-3}}{3^{3}a^{4}}$

83. $\left(\dfrac{y^{-3}}{x}\right)^{-3}$

84. $\left(\dfrac{z}{y^{-5}}\right)^{-2}$

85. $\dfrac{a^{0}b^{-4}}{a^{-3}b}$

86. $\dfrac{c^{-3}d^{-2}}{c^{-4}d^{-5}}$

87. $\left(\dfrac{14x^{-3}y^{-3}}{7x^{-4}y^{-3}}\right)^{-2}$

88. $\left(\dfrac{25x^{-1}y^{-6}}{5x^{-4}y^{-6}}\right)^{-2}$

89. $\dfrac{7^{-8}\cdot 5^{-6}}{7^{-9}\cdot 5^{-5}}$

90. $\dfrac{9^{-2}\cdot 8^{-10}}{9^{-1}\cdot 8^{-9}}$

91. $(9x^{-2}y)\left(-\dfrac{2}{3}x^{3}y^{-2}\right)$

92. $(-12x^{5}y^{-2})\left(\dfrac{3}{4}x^{-6}y^{3}\right)$

Write in scientific notation.

93. 38

94. 759

95. 1,730,000

96. 405,300,000

97. 0.83

98. 0.0654

99. 0.0008125

100. 0.0000048

Write in decimal notation.

101. 7.13×10^{5}

102. 4.006×10^{6}

103. 3.07×10^{-1}

104. 7.07×10^{-3}

105. 9.01×10^{-7}

106. 6.668×10^{-9}

Perform the calculations indicated. Express your answers in scientific notation.

107. $(3.1 \times 10^{-4})(1.5 \times 10^{-2})$

108. $(2.3 \times 10^{-4})(3.0 \times 10^{9})$

109. $\dfrac{3.6 \times 10^{-5}}{1.2 \times 10^{-6}}$

110. $\dfrac{9.3 \times 10^{-8}}{3.1 \times 10^{6}}$

Applications

111. ***Oxygen Molecules*** The weight of one oxygen molecule is 5.3×10^{-23} gram. How much would 2×10^{4} molecules of oxygen weigh?

112. ***Solar Probe*** The average distance from Earth to the sun is 4.90×10^{11} feet. If a solar probe is launched from Earth and travels at 2×10^{4} feet per second, how long would it take to reach the sun?

A.2 Polynomial Operations

Student Learning Objectives

After studying this section, you will be able to:

1 Identify types and degrees of polynomials.

2 Evaluate polynomial functions.

3 Add and subtract polynomials.

4 Multiply two binomials by FOIL.

5 Multiply two binomials $(a + b)(a - b)$.

6 Multiply two binomials $(a - b)^2$ or $(a + b)^2$.

7 Multiply polynomials with more than two terms.

1 Identifying Types and Degrees of Polynomials

A **polynomial** is an algebraic expression of one or more terms. A **term** is a number, a variable raised to a nonnegative integer power, or a product of numbers and variables raised to nonnegative integer powers. There must be no division by a variable. Three types of polynomials that you will see often are **monomials, binomials,** and **trinomials.**

1. A **monomial** has *one* term.
2. A **binomial** has *two* terms.
3. A **trinomial** has *three* terms.

Number of Variables	Monomials	Binomials	Trinomials	Other Polynomials
One Variable	$8x^3$	$2y^2 + 3y$	$5x^2 + 2x - 6$	$x^4 + 2x^3 - x^2 + 9$
Two Variables	$6x^2y$	$3x^2 - 5y^3$	$8x^2 + 5xy - 3y^2$	$x^3y + 5xy^2 + 3xy - 7y^5$
Three Variables	$12uvw^3$	$11a^2b + 5c^2$	$4a^2b^4 + 7c^4 - 2a^5$	$3c^2 + 4c - 8d + 2e - e^2$

The following are *not* polynomials.

$$2x^{-3} + 5x^2 - 3 \qquad 4ab^{1/2} \qquad \frac{2}{x} + \frac{3}{y}$$

To Think About: *Understanding Polynomials* Give a reason each expression above is not a polynomial.

Polynomials are also classified by degree. The **degree of a term** is the sum of the exponents of its variables. The **degree of a polynomial** is the degree of the highest-degree term in the polynomial. If the polynomial has no variable, then it has degree zero.

Example 1 Name the type of polynomial and give its degree.

(a) $5x^6 + 3x^2 + 2$ **(b)** $7x + 6$

(c) $5x^2y + 3xy^3 + 6xy$ **(d)** $7x^4y^5$

Solution

(a) This is a trinomial of degree 6.

(b) This is a binomial of degree 1.

(c) This is a trinomial of degree 4.

Remember that if a variable has no exponent, the exponent is understood to be 1.

(d) This is a monomial of degree 9. □

Student Practice 1 State the type of polynomial and give its degree.

(a) $3x^5 - 6x^4 + x^2$ **(b)** $5x^2 + 2$

(c) $3ab + 5a^2b^2 - 6a^4b$ **(d)** $16x^4y^6$

Some polynomials contain only one variable. A **polynomial in** x is an expression of the form

$$a_nx^n + a_{n-1}x^{n-1} + a_{n-2}x^{n-2} + \cdots + a_0$$

where n is a nonnegative integer and the constants $a_n, a_{n-1}, a_{n-2}, \ldots, a_0$ are real numbers. We usually write polynomials in **descending order** of the variable. That is, the exponents on the variables decrease from left to right. For example, the polynomial $4x^5 - 2x^3 + 6x^2 + 5x - 8$ is written in descending order.

2 Evaluating Polynomial Functions

A **polynomial function** is a function that is defined by a polynomial.
For example,

$$p(x) = 5x^2 - 3x + 6 \quad \text{and} \quad p(x) = 2x^5 - 3x^3 + 8x - 15$$

are both polynomial functions.

To evaluate a polynomial function, we use the skills developed in Section 3.6.

Example 2 Evaluate the polynomial function $p(x) = -3x^3 + 2x^2 - 5x + 6$ to find **(a)** $p(-3)$ and **(b)** $p(6)$.

Solution

(a)
$$\begin{aligned} p(-3) &= -3(-3)^3 + 2(-3)^2 - 5(-3) + 6 \\ &= -3(-27) + 2(9) - 5(-3) + 6 \\ &= 81 + 18 + 15 + 6 \\ &= 120 \end{aligned}$$

(b)
$$\begin{aligned} p(6) &= -3(6)^3 + 2(6)^2 - 5(6) + 6 \\ &= -3(216) + 2(36) - 5(6) + 6 \\ &= -648 + 72 - 30 + 6 \\ &= -600 \end{aligned}$$

□

Student Practice 2 Evaluate the polynomial function $p(x) = 2x^4 - 3x^3 + 6x - 8$ to find

(a) $p(-2)$ **(b)** $p(5)$

3 Adding and Subtracting Polynomials

We can add and subtract polynomials by combining like terms as we learned in Sections 1.7 and 5.3.

Example 3 Add. $(5x^2 - 3x - 8) + (-3x^2 - 7x + 9)$

Solution

$5x^2 - 3x - 8 - 3x^2 - 7x + 9$ We remove the parentheses and combine like terms.

$= 2x^2 - 10x + 1$

□

Student Practice 3 Add. $(-7x^2 + 5x - 9) + (2x^2 - 3x + 5)$

To subtract real numbers, we add the opposite of the second number to the first. Thus, for real numbers a and b, we have $a - (b) = a + (-b)$. Similarly for polynomials, to subtract polynomials we add the opposite of the second polynomial to the first.

Example 4 Subtract. $(-5x^2 - 19x + 15) - (3x^2 - 4x + 13)$

Solution

$(-5x^2 - 19x + 15) + (-3x^2 + 4x - 13)$ We add the opposite of the second polynomial to the first polynomial.

$= -8x^2 - 15x + 2$ □

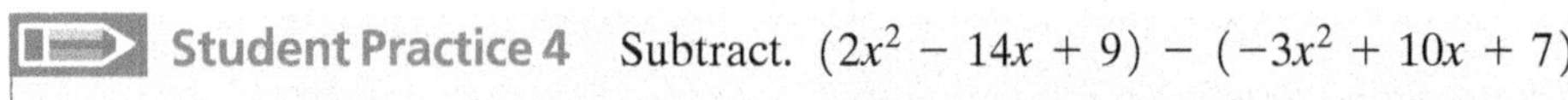

Student Practice 4 Subtract. $(2x^2 - 14x + 9) - (-3x^2 + 10x + 7)$

4 Multiplying Two Binomials by FOIL

The FOIL method for multiplying two binomials has been developed to help you keep track of the order of the terms to be multiplied. The acronym FOIL means the following:

F	**First**
O	**Outer**
I	**Inner**
L	**Last**

That is, we multiply the first terms, then the outer terms, then the inner terms, and finally, the last terms.

Example 5 Multiply. $(5x + 2)(7x - 3)$

Solution First Last First + Outer + Inner + Last

$$(5x + 2)(7x - 3) = 35x^2 - 15x + 14x - 6$$

Inner

Outer

$$= 35x^2 - x - 6$$ □

Student Practice 5 Multiply. $(7x + 3)(2x - 5)$

Example 6 Multiply. $(7x^2 - 8)(2x - 3)$

Solution First Last

$$(7x^2 - 8)(2x - 3) = 14x^3 - 21x^2 - 16x + 24$$

Inner

Outer

Note that in this case we were not able to combine the inner and outer products. □

Student Practice 6 Multiply. $(3x^2 - 2)(5x - 4)$

5 Multiplying $(a + b)(a - b)$

Products of the form $(a + b)(a - b)$ deserve special attention.

$$(a + b)(a - b) = a^2 - ab + ab - b^2 = a^2 - b^2$$

Notice that the middle terms, $-ab$ and $+ab$, equal zero when combined. The product is the difference of two squares, $a^2 - b^2$. This is always true when you multiply binomials of the form $(a + b)(a - b)$. You should memorize the following formula.

$$(a + b)(a - b) = a^2 - b^2$$

Example 7 Multiply. $(2a - 9b)(2a + 9b)$

Solution $(2a - 9b)(2a + 9b) = (2a)^2 - (9b)^2 = 4a^2 - 81b^2$

Of course, we could have used the FOIL method, but recognizing the special product allowed us to save time. □

Student Practice 7 Multiply. $(7x - 2y)(7x + 2y)$

6 Multiplying $(a - b)^2$ or $(a + b)^2$

Another special product is the square of a binomial.

$$(a - b)^2 = (a - b)(a - b) = a^2 - ab - ab + b^2 = a^2 - 2ab + b^2$$

Once you understand the pattern, you should memorize these two formulas.

$$(a - b)^2 = a^2 - 2ab + b^2 \qquad (a + b)^2 = a^2 + 2ab + b^2$$

This procedure is also called **expanding a binomial.** *Note*: $(a - b)^2 \neq a^2 - b^2$ and $(a + b)^2 \neq a^2 + b^2$

Example 8 Multiply.

(a) $(5a - 8b)^2$ **(b)** $(3u + 11v^2)^2$

Solution

(a) $(5a - 8b)^2 = (5a)^2 - 2(5a)(8b) + (8b)^2 = 25a^2 - 80ab + 64b^2$

(b) Here $a = 3u$ and $b = 11v^2$.

$$\begin{aligned}(3u + 11v^2)^2 &= (3u)^2 + 2(3u)(11v^2) + (11v^2)^2 \\ &= 9u^2 + 66uv^2 + 121v^4\end{aligned}$$ □

Student Practice 8 Multiply.

(a) $(4u + 5v)^2$ **(b)** $(7x^2 - 3y^2)^2$

7 Multiplying Polynomials with More Than Two Terms

The distributive property is the basis for multiplying polynomials. Recall that

$$a(b + c) = ab + ac.$$

We can use this property to multiply a polynomial by a monomial.

$$\begin{aligned}3xy(5x^3 + 2x^2 - 4x + 1) &= 3xy(5x^3) + 3xy(2x^2) - 3xy(4x) + 3xy(1) \\ &= 15x^4y + 6x^3y - 12x^2y + 3xy\end{aligned}$$

A similar procedure can be used instead of FOIL to multiply two binomials.

$$\begin{aligned}(3x+5)(6x+7) &= (3x+5)6x + (3x+5)7 \quad \text{We use the distributive property again.}\\ &= (3x)(6x) + (5)(6x) + (3x)(7) + (5)(7)\\ &= 18x^2 + 30x + 21x + 35\\ &= 18x^2 + 51x + 35\end{aligned}$$

The multiplication of a binomial and a trinomial is more involved. One way to multiply two polynomials is to write them vertically, as we do when multiplying two- and three-digit numbers. We then multiply them in the usual way.

Example 9 Multiply. $(4x^2 - 2x + 3)(-3x + 4)$

Solution

$$\begin{array}{rrrrl} & 4x^2 & -\ 2x & +\ 3 & \\ & & -3x & +\ 4 & \\ \hline & 16x^2 & -\ 8x & +\ 12 & \text{Multiply } (4x^2-2x+3)(+4).\\ -12x^3 & +\ 6x^2 & -\ 9x & & \text{Multiply } (4x^2-2x+3)(-3x).\\ \hline -12x^3 & +\ 22x^2 & -\ 17x & +\ 12 & \text{Add the two products.}\end{array}$$

□

Student Practice 9 Multiply. $(2x^2 - 3x + 1)(x^2 - 5x)$

Another way to multiply polynomials is to multiply horizontally. We redo Example 9 in the following example.

Example 10 Multiply horizontally. $(4x^2 - 2x + 3)(-3x + 4)$

Solution By the distributive law, we have the following:

$$\begin{aligned}(4x^2 - 2x + 3)(-3x + 4) &= (4x^2 - 2x + 3)(-3x) + (4x^2 - 2x + 3)(4)\\ &= -12x^3 + 6x^2 - 9x + 16x^2 - 8x + 12\\ &= -12x^3 + 22x^2 - 17x + 12.\end{aligned}$$

In actual practice you will find that you can do some of these steps mentally. □

Student Practice 10 Multiply horizontally. $(2x^2 - 3x + 1)(x^2 - 5x)$

STEPS TO SUCCESS Look Ahead to See What Is Coming

You will find that learning new material is much easier if you know what is coming. Take a few minutes at the end of your study time to glance over the next section of the book. If you quickly look over the topics and ideas in this new section, it will help you get your bearings when the instructor presents new material. Students find that when they preview new material it enables them to see what is coming. It helps them to be able to grasp new ideas much more quickly.

Making it personal: Do this right now: Look ahead to the next section of the book. Glance over the ideas and concepts. Write down a couple of facts about the next section. ▼

A.2 Exercises MyMathLab®

Name the type of polynomial and give its degree.

1. $2x^2 - 5x + 3$

2. $7x^3 + 6x^2 - 2$

3. $-3.2a^4bc^3$

4. $26.8a^3bc^2$

5. $\frac{3}{5}m^3n - \frac{2}{5}mn$

6. $\frac{2}{7}m^2n^2 + \frac{1}{2}mn^2$

For the polynomial function $p(x) = 5x^2 - 9x - 12$, *evaluate the following:*

7. $p(3)$

8. $p(-4)$

For the polynomial function $g(x) = -4x^3 - x^2 + 5x - 1$, *evaluate the following:*

9. $g(-2)$

10. $g(1)$

For the polynomial function $h(x) = 2x^4 - x^3 + 2x^2 - 4x - 3$, *evaluate the following:*

11. $h(-1)$

12. $h(3)$

Add or subtract the following polynomials as indicated.

13. $(x^2 + 3x - 2) + (-2x^2 - 5x + 1) + (x^2 - x - 5)$

14. $(3x^2 + 2x - 4) + (-x^2 - x + 1) + (-3x^2 + 5x + 7)$

15. $(7m^3 + 4m^2 - m + 2.5) - (-3m^3 + 5m + 3.8)$

16. $(5x^3 - x^2 + 6x - 3.8) - (x^3 + 2x^2 - 4x - 10.4)$

17. $(5a^3 - 2a^2 - 6a + 8) + (5a + 6) - (-a^2 - a + 2)$

18. $(a^5 + 3a^2) + (2a^4 - a^3 - 3a^2 + 2) - (a^4 + 3a^3 - 5)$

19. $\left(\frac{2}{3}x^2 + 5x\right) + \left(\frac{1}{2}x^2 - \frac{1}{3}x\right)$

20. $\left(\frac{5}{6}x^2 - \frac{1}{4}x\right) + \left(\frac{1}{6}x^2 + 3x\right)$

21. $(2.3x^3 - 5.6x^2 - 2) - (5.5x^3 - 7.4x^2 + 2)$

22. $(5.9x^3 + 3.4x^2 - 7) - (2.9x^3 - 9.6x^2 + 3)$

Multiply.

23. $(5x + 8)(2x + 9)$

24. $(6x + 7)(3x + 2)$

25. $(5w + 2d)(3a - 4b)$

26. $(7a + 8b)(5d - 8w)$

27. $(-6x + y)(2x - 5y)$

28. $(5a + y)(-2x - y)$

29. $(7r - s^2)(-4a - 11s^2)$

30. $(-3r - 2s^2)(5r - 6s^2)$

Multiply mentally. See Examples 7 and 8.

31. $(5x - 8y)(5x + 8y)$

32. $(2a - 7b)(2a + 7b)$

33. $(5a - 2b)^2$

34. $(4a + 3b)^2$

35. $(7m - 1)^2$

36. $(5r + 3)^2$

37. $(6 + 5x^2)(6 - 5x^2)$

38. $(8 - 3x^3)(8 + 3x^3)$

39. $(3m^3 + 1)^2$

40. $(4r^3 - 5)^2$

Multiply.

41. $2x(3x^2 - 5x + 1)$

42. $-5x(x^2 - 6x - 2)$

43. $-\frac{1}{2}ab(4a - 5b - 10)$

44. $\frac{3}{4}ab^2(a - 8b + 6)$

45. $(2x - 3)(x^2 - x + 1)$

46. $(4x + 1)(2x^2 + x + 1)$

47. $(3x^2 - 2xy - 6y^2)(2x - y)$

48. $(5x^2 + 3xy - 7y^2)(3x - 2y)$

49. $(3a^3 + 4a^2 - a - 1)(a - 5)$

50. $(4b^3 - b^2 + 2b - 6)(3b + 1)$

First multiply any two binomials in the exercise; then multiply the result by the third binomial.

51. $(x + 2)(x - 3)(2x - 5)$

52. $(x - 6)(x + 2)(3x + 2)$

53. $(a - 2)(5 + a)(3 - 2a)$

54. $(4 + 3a)(1 - 2a)(3 - a)$

Applications

▲ **55.** ***Geometry*** Ace Landscape Design makes large raised boxes for plantings. The height of the boxes they build is $(2x + 5)$ feet, and the area of the rectangular base measures $(2x^2 + x + 4)$ feet^2. If the box is filled with soil, what is the volume of soil needed?

▲ **56.** ***Geometry*** At Wilkinson Auditorium, $(3n + 8)$ rows of chairs can be arranged for large ceremonies. Each row can have $(3n^2 + 2n + 3)$ chairs. Find the number of chairs that can be arranged.

A.3 Factoring Polynomials

When two or more algebraic expressions (monomials, binomials, and so on) are multiplied, each expression is called a **factor.** We learned to multiply factors in Section 5.4 and Appendix A.2.

In this section, we will learn how to find the factors of a polynomial. **Factoring** is the opposite of multiplication and is an extremely important mathematical technique.

Student Learning Objectives

After studying this section, you will be able to:

1. Factor out the greatest common factor from a polynomial.
2. Factor a polynomial by the grouping method.
3. Factor trinomials of the form $x^2 + bx + c$.
4. Factor trinomials of the form $ax^2 + bx + c$.

1 Factoring Out the Greatest Common Factor

To factor out a common factor, we make use of the distributive property.

$$ab + ac = a(b + c)$$

The **greatest common factor** is simply the largest factor that is common to all terms of the expression.

It must contain

1. The largest possible common factor of the numerical coefficients and
2. The largest possible common variable factor

Example 1 Factor out the greatest common factor.

(a) $7x^2 - 14x$ **(b)** $40a^3 - 20a^2$

Solution

(a) $7x^2 - 14x = 7 \cdot x \cdot x - 7 \cdot 2 \cdot x = 7x(x - 2)$

Be careful. The greatest common factor is $7x$, not 7.

(b) $40a^3 - 20a^2 = 20a^2(2a - 1)$

The greatest common factor is $20a^2$.

Suppose we had written $10a(4a^2 - 2a)$ or $10a(2a)(2a - 1)$ as our answer. Although we have factored the expression, we have not found the *greatest* common factor. □

 **Student Practice 1** Factor out the greatest common factor.

(a) $19x^3 - 38x^2$ **(b)** $100a^4 - 50a^2$

How do you know whether you have factored correctly? You can do two things to verify your answer.

1. Examine the polynomial in the parentheses. Its terms should not have any remaining common factors.
2. Multiply the two factors. You should obtain the original expression.

In each of the remaining examples, you will be asked to **factor** a polynomial (i.e., to find the factors that, when multiplied, give the polynomial as a product). For each of these examples, this will require you to factor out the greatest common factor.

Example 2 Factor $6x^3 - 9x^2y - 6x^2y^2$. Check your answer.

Solution $6x^3 - 9x^2y - 6x^2y^2 = 3x^2(2x - 3y - 2y^2)$

Check:

1. $(2x - 3y - 2y^2)$ has no common factors. If it did, we would know that we had not factored out the *greatest* common factor.
2. Multiply the two factors.

$$3x^2(2x - 3y - 2y^2) = 6x^3 - 9x^2y - 6x^2y^2$$

Observe that we do obtain the original polynomial. □

Student Practice 2 Factor $9a^3 - 12a^2b^2 - 15a^4$. Check your answer.

The greatest common factor need not be a monomial. It may be a binomial or even a trinomial. For example, note the following:

$$5a(x + 3) + 2(x + 3) = (x + 3)(5a + 2)$$
$$5a(x + 4y) + 2(x + 4y) = (x + 4y)(5a + 2)$$

The common factors are binomials.

Example 3 Factor.

(a) $2x(x + 5) - 3(x + 5)$

(b) $5a(a + b) - 2b(a + b) - (a + b)$

Solution

(a) $2x(x + 5) - 3(x + 5) = (x + 5)(2x - 3)$ The common factor is $x + 5$.

(b) $5a(a + b) - 2b(a + b) - (a + b) = 5a(a + b) - 2b(a + b) - 1(a + b)$
$= (a + b)(5a - 2b - 1)$

The common factor is $a + b$.

Note that if we place a 1 in front of the third term, it makes it easier to factor. □

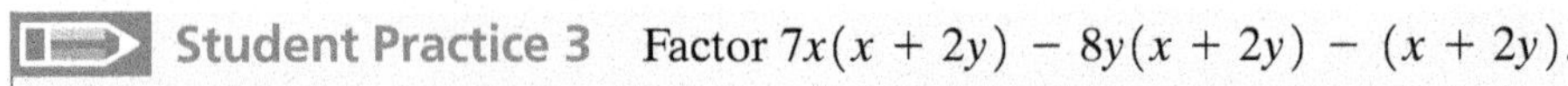

Student Practice 3 Factor $7x(x + 2y) - 8y(x + 2y) - (x + 2y)$.

2 Factoring by Grouping

Because the common factors in Example 3 were grouped inside parentheses, it was easy to pick them out. However, this rarely happens, so we have to learn how to manipulate expressions to find the greatest common factor.

Polynomials with four terms can often be factored by the method of Example 3(a). However, the parentheses are not always present in the original problem. When they are not present, we look for a way to remove a common factor from the first two terms. We then factor out a common factor from the first two terms and a common factor from the second two terms. Then we can find the greatest common factor of the original expression.

Example 4 Factor $ax + 2ay + 2bx + 4by$.

Solution

Remove the greatest common factor (a) from the first two terms.

$$ax + 2ay + 2bx + 4by = a(x + 2y) + 2b(x + 2y)$$

Remove the greatest common factor ($2b$) from the last two terms.

Now we can see that $(x + 2y)$ is a common factor.

$$a(x + 2y) + 2b(x + 2y) = (x + 2y)(a + 2b)$$ □

Student Practice 4 Factor $bx + 5by + 2wx + 10wy$.

Sometimes a problem can be factored by this method, but we must first rearrange the order of the four terms so that the first two terms do have a common factor.

Example 5 Factor $xy - 6 + 3x - 2y$.

Solution Rearrange the terms so that the first two terms have a common factor and the last two terms have a common factor.

$xy + 3x - 2y - 6$	Factor out a common factor of x from the first two terms and -2 from the second two terms.
$= x(y + 3) - 2(y + 3)$	Since we factor out a negative number, we have: $-2y - 6 = -2(y + 3)$.
$= (y + 3)(x - 2)$	Factor out the common binomial factor $y + 3$. □

Student Practice 5 Factor $xy - 12 - 4x + 3y$.

To Think About: *Example 5 Follow-Up* Notice that if you factored out a common factor of +2 from the last two terms, the resulting terms would not contain the same parenthetical expression: $x(y + 3) + 2(-y - 3)$. If the expressions inside the two sets of parentheses are not exactly the same, you cannot express the polynomial as a product of two factors!

Example 6 Factor $2x^3 + 21 - 7x^2 - 6x$. Check your answer by multiplication.

Solution

$2x^3 - 7x^2 - 6x + 21$	Rearrange the terms.
$= x^2(2x - 7) - 3(2x - 7)$	Factor out a common factor from each group of two terms.
$= (2x - 7)(x^2 - 3)$	Factor out the common binomial factor $2x - 7$.

Check:

$(2x - 7)(x^2 - 3) = 2x^3 - 6x - 7x^2 + 21$	Multiply the two binomials.
$= 2x^3 + 21 - 7x^2 - 6x$	Rearrange the terms.

The product is identical to the original expression. □

Student Practice 6 Factor $2x^3 - 15 - 10x + 3x^2$.

3 Factoring Trinomials of the Form $x^2 + bx + c$

If we multiply $(x + 4)(x + 5)$, we obtain $x^2 + 9x + 20$. But suppose that we already have the polynomial $x^2 + 9x + 20$ and need to factor it. In other words, suppose we need to find the expressions that, when multiplied, give us the polynomial. Let's use this example to find a general procedure.

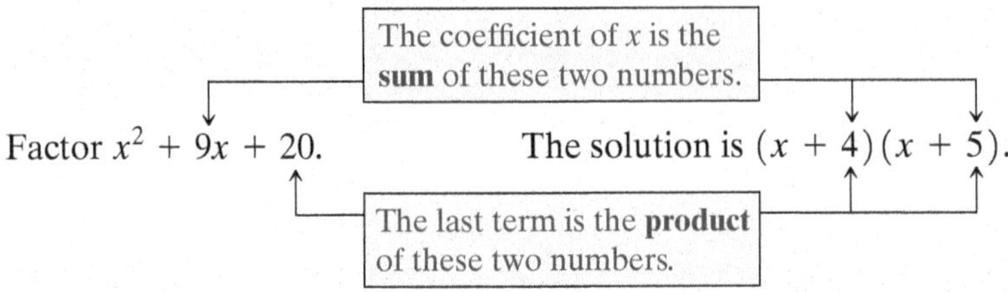
The coefficient of x is the sum of these two numbers.
Factor x² + 9x + 20.
The solution is (x + 4)(x + 5).
The last term is the product of these two numbers.

FACTORING TRINOMIALS OF THE FORM $x^2 + bx + c$

1. The answer has the form $(x + m)(x + n)$, where m and n are real numbers.
2. The numbers m and n are chosen so that
   (a) $m \cdot n = c$ and
   (b) $m + n = b$.

If the last term of the trinomial is positive and the middle term is negative, the two numbers m and n will be negative numbers.

Example 7 Factor $x^2 - 14x + 24$.

Solution We want to find two numbers whose product is 24 and whose sum is -14. They will both be negative numbers.

Factor Pairs of 24	Sum of the Factors
$(-24)(-1)$	$-24 - 1 = -25$
$(-12)(-2)$	$-12 - 2 = -14$ ✓
$(-6)(-4)$	$-6 - 4 = -10$
$(-8)(-3)$	$-8 - 3 = -11$

The numbers whose product is 24 and whose sum is -14 are -12 and -2. Thus,

$$x^2 - 14x + 24 = (x - 12)(x - 2).$$

Student Practice 7 Factor $x^2 - 10x + 21$.

If the last term of the trinomial is negative, the two numbers m and n will be opposite in sign.

Example 8 Factor $x^4 - 2x^2 - 24$.

Solution Sometimes we can make a substitution that makes a polynomial easier to factor. We need to recognize that we can write this as $(x^2)^2 - 2(x^2) - 24$. We can make this polynomial easier to factor if we substitute y for x^2.

Then we have

$$y^2 - 2y - 24.$$

The two numbers whose product is -24 and whose sum is -2 are -6 and 4.

Therefore, we have

$$y^2 - 2y - 24 = (y - 6)(y + 4).$$

But $y = x^2$, so our answer is

$$x^4 - 2x^2 - 24 = (x^2 - 6)(x^2 + 4).$$

 **Student Practice 8** Factor $x^4 + 9x^2 + 8$.

Note from Example 8 that c, the last term of the trinomial, is negative $(c = -24)$ and that m and n are opposite in sign $(m = -6$ and $n = 4)$. When we know the signs of both b and c, we then also know the signs of m and n as described in the box below.

FACTS ABOUT SIGNS

Suppose $x^2 + bx + c = (x + m)(x + n)$. We know certain facts about m and n.

1. m and n have the same sign if c is positive. (*Note*: We did *not* say that they will have the same sign as c.)
   (a) They are positive if b is positive.
   (b) They are negative if b is negative.
2. m and n have opposite signs if c is negative. The larger number is positive if b is positive and negative if b is negative.

If you understand these sign facts, continue on to Example 9. If not, review Examples 7 and 8.

Example 9 Factor.

(a) $y^2 + 5y - 36$ **(b)** $x^4 - 4x^2 - 12$

Solution

(a) $y^2 + 5y - 36 = (y + 9)(y - 4)$ The larger number (9) is positive because $b = 5$ is positive.

(b) $x^4 - 4x^2 - 12 = (x^2 - 6)(x^2 + 2)$ The larger number (6) is negative because $b = -4$ is negative.

 **Student Practice 9** Factor.

(a) $a^2 - 2a - 48$ **(b)** $x^4 + 2x^2 - 15$

Does the order in which we write the factors make any difference? In other words, if $x^2 + bx + c = (x + m)(x + n)$, then

$$x^2 + bx + c = (x + m)(x + n) = (x + n)(x + m).$$

Since multiplication is commutative, the order of the factors is not important.

We can also factor trinomials that have more than one variable.

Example 10 Factor.

(a) $x^2 - 21xy + 20y^2$ **(b)** $x^2 + 4xy - 21y^2$

Solution

(a) $x^2 - 21xy + 20y^2 = (x - 20y)(x - y)$

The last terms in each factor contain the variable y.

(b) $x^2 + 4xy - 21y^2 = (x + 7y)(x - 3y)$ □

 **Student Practice 10** Factor.

(a) $x^2 - 16xy + 15y^2$ **(b)** $x^2 + xy - 42y^2$

4 Factoring Trinomials of the Form $ax^2 + bx + c$

Using the Grouping Number Method. One way to factor a trinomial $ax^2 + bx + c$ is to write it as four terms and factor it by grouping as described earlier in this section (and also in Section 6.2). For example, the trinomial $2x^2 + 11x + 12$ can be written as $2x^2 + 3x + 8x + 12$.

$$\begin{aligned} 2x^2 + 3x + 8x + 12 &= x(2x + 3) + 4(2x + 3) \\ &= (2x + 3)(x + 4) \end{aligned}$$

We can factor all factorable trinomials of the form $ax^2 + bx + c$ in this way. Use the following procedure.

GROUPING NUMBER METHOD FOR FACTORING TRINOMIALS OF THE FORM $ax^2 + bx + c$

1. Obtain the grouping number ac.
2. Find the factor pair of the grouping number whose sum is b.
3. Use those two factors to write bx as the sum of two terms.
4. Factor by grouping.

Example 11 Factor $6x^2 + 7x - 5$.

Solution

1. The grouping number is $(a)(c) = (6)(-5) = -30$.

2. Since $b = 7$, we want the factor pair of -30 whose sum is 7.

$$\begin{aligned} -30 &= (-30)(1) & -30 &= (5)(-6) \\ &= (30)(-1) & &= (-5)(6) \\ &= (15)(-2) & &= (3)(-10) \\ &= (-15)(2) & &= (-3)(10) \end{aligned}$$

3. Since $-3 + 10 = 7$, use -3 and 10 to write $6x^2 + 7x - 5$ with four terms.

$$6x^2 + 7x - 5 = 6x^2 - 3x + 10x - 5$$

4. Factor by grouping.

$$\begin{aligned} 6x^2 - 3x + 10x - 5 &= 3x(2x - 1) + 5(2x - 1) \\ &= (2x - 1)(3x + 5) \end{aligned}$$ □

Student Practice 11 Factor $10x^2 - 9x + 2$.

If the three terms have a common factor, then prior to using the four-step grouping number procedure, we first factor out the greatest common factor from the terms of the trinomial.

Example 12 Factor $6x^3 - 26x^2 + 24x$.

Solution First we factor out the greatest common factor $2x$ from each term.

$$6x^3 - 26x^2 + 24x = 2x(3x^2 - 13x + 12)$$

Next we follow the four steps to factor $3x^2 - 13x + 12$.

1. The grouping number is 36.
2. We want the factor pair of 36 whose sum is -13. The two factors are -4 and -9.
3. We use -4 and -9 to write $3x^2 - 13x + 12$ with four terms.

$$3x^2 - 13x + 12 = 3x^2 - 4x - 9x + 12$$

4. Factor by grouping. Remember that we first factored out the factor $2x$. This factor must be part of the answer.

$$\begin{aligned} 2x(3x^2 - 4x - 9x + 12) &= 2x[x(3x - 4) - 3(3x - 4)] \\ &= 2x(3x - 4)(x - 3) \end{aligned}$$

□

Student Practice 12 Factor $9x^3 - 15x^2 - 6x$.

Using the Trial-and-Error Method. Another way to factor trinomials of the form $ax^2 + bx + c$ is by trial and error. This method has an advantage if the grouping number is large and we have to list many factors. In the trial-and-error method, we try different values and see which can be multiplied out to obtain the original expression.

If the last term is negative, there are many more sign possibilities. We will see this in the following example.

Example 13 Factor by trial and error $10x^2 - 49x - 5$.

Solution The first terms in the factors could be $10x$ and x or $5x$ and $2x$. The second terms could be $+1$ and -5 or -1 and $+5$. We list all the possibilities and look for one that will yield a middle term of $-49x$.

Possible Factors	Middle Term of Product
$(2x - 1)(5x + 5)$	$+5x$
$(2x + 1)(5x - 5)$	$-5x$
$(2x + 5)(5x - 1)$	$+23x$
$(2x - 5)(5x + 1)$	$-23x$
$(10x - 5)(x + 1)$	$+5x$
$(10x + 5)(x - 1)$	$-5x$
$(10x - 1)(x + 5)$	$+49x$
$(10x + 1)(x - 5)$	$-49x$

Thus, $10x^2 - 49x - 5 = (10x + 1)(x - 5)$.

As a check, multiply the two binomials to see if you obtain the original expression.

$$\begin{aligned} (10x + 1)(x - 5) &= 10x^2 - 50x + 1x - 5 \\ &= 10x^2 - 49x - 5 \end{aligned}$$

□

Student Practice 13 Factor by trial and error $8x^2 - 6x - 5$.

Example 14 Factor by trial and error $6x^4 + x^2 - 12$.

Solution The first term of each factor must contain x^2. Suppose that we try the following:

Possible Factors	Middle Term of Product
$(2x^2 - 3)(3x^2 + 4)$	$-x^2$

The middle term we get is $-x^2$, but we need its opposite, $+x^2$. In this case, we just need to reverse the signs of -3 and 4. Do you see why? Therefore,

$$6x^4 + x^2 - 12 = (2x^2 + 3)(3x^2 - 4).$$

Student Practice 14 Factor by trial and error $6x^4 + 13x^2 - 5$.

STEPS TO SUCCESS I Don't Really Need to Actually Read This Book, Do I?

Successful students find that they can get a lot of benefit from reading the book. Here are some suggestions they have made after they completed the course:

1. This book was written by faculty who have learned how to help you succeed in doing math problems. So take the time to read over the assigned section of the book. You will be amazed by the understanding you will acquire.
2. Read your textbook with a paper and pen handy. As you come across a new definition or a new idea, underline it. As you come across a suggestion that helps you understand things better, be sure to underline it.
3. Interact with the book. When you see something you don't understand put a big question mark next to it. Try reading it again. Find a similar example in the book and go over the steps. If that doesn't help, ask your instructor or a classmate to explain the procedure or idea.
4. Use your pen to follow each step that is worked out for each example. Underline the steps that you think are especially challenging. This will help you when you do the homework.

Making it personal: Which of these four suggestions do you think you would benefit from the most? Put a check mark by that suggestion. Now write down what you need to do in order to take full advantage of your textbook. Make a commitment to doing this for your next homework assignment.

A.3 Exercises MyMathLab®

Factor out the greatest common factor. For additional review and practice, see Section 6.1.

1. $80 - 10y$

2. $16x - 16$

3. $5a^2 - 25a$

4. $7a^2 - 14a$

5. $4a^2b^3 - 8ab + 32a$

6. $15c^2d^2 + 10c^2 - 60c$

7. $30y^4 + 24y^3 + 18y^2$

8. $16y^5 - 24y^4 - 40y^3$

9. $15ab^2 + 5ab - 10a^3b$

10. $-12x^2y - 18xy + 6x$

11. $10a^2b^3 - 30a^3b^3 + 10a^3b^2 - 40a^4b^2$

12. $28x^3y^2 - 12x^2y^4 + 4x^3y^4 - 32x^2y^2$

13. $3x(x + y) - 2(x + y)$

14. $5a(a + 3b) - 4(a + 3b)$

Hint for Exercises 15 and 16: Is the expression in the first parentheses equal to the expression in the second parentheses?

15. $5b(a - 3b) + 8(-3b + a)$

16. $4y(x - 5y) - 3(-5y + x)$

17. $3x(a + 5b) + (a + 5b)$

18. $2w(s - 3t) - (s - 3t)$

19. $2a^2(3x - y) - 5b^3(3x - y)$

20. $7a^3(5a + 4) - 2(5a + 4)$

21. $3x(5x + y) - 8y(5x + y) - (5x + y)$

22. $4w(y - 8x) + 5z(y - 8x) + (y - 8x)$

23. $2a(a - 6b) - 3b(a - 6b) - 2(a - 6b)$

24. $3a(a + 4b) - 5b(a + 4b) - 9(a + 4b)$

Factor by grouping. For additional review and practice, see Section 6.2.

25. $x^3 + 5x^2 + 3x + 15$

26. $x^3 + 8x^2 + 2x + 16$

27. $2x + 6 - 3ax - 9a$

28. $2bc + 4b - 5c - 10$

29. $ab - 4a + 12 - 3b$

30. $2m^2 - 8mn - 5m + 20n$

31. $5x - 20 + 3xy - 12y$

32. $3x - 21 + 4xy - 28y$

33. $9y + 2x - 6 - 3xy$

34. $10y + 3x - 6 - 5xy$

Factor each polynomial. For additional review and practice, see Section 6.3.

35. $x^2 + 8x + 7$ **36.** $x^2 + 12x + 11$ **37.** $x^2 - 8x + 15$ **38.** $x^2 - 10x + 16$

39. $x^2 - 10x + 24$ **40.** $x^2 - 9x + 18$ **41.** $a^2 + 4a - 45$ **42.** $a^2 + 2a - 35$

43. $x^2 - xy - 42y^2$ **44.** $x^2 - xy - 56y^2$ **45.** $x^2 - 15xy + 14y^2$ **46.** $x^2 + 10xy + 9y^2$

47. $x^4 - 3x^2 - 40$ **48.** $x^4 + 6x^2 + 5$ **49.** $x^4 + 16x^2y^2 + 63y^4$ **50.** $x^4 - 6x^2 - 55$

Factor out the greatest common factor from the terms of the trinomial. Then factor the remaining trinomial.

51. $2x^2 + 26x + 44$ **52.** $2x^2 + 30x + 52$ **53.** $x^3 + x^2 - 20x$ **54.** $x^3 - 4x^2 - 45x$

Factor each polynomial. You may use the grouping number method or the trial-and-error method. For additional review and practice, see Section 6.4.

55. $2x^2 - x - 1$ **56.** $3x^2 + x - 2$ **57.** $6x^2 - 7x - 5$ **58.** $5x^2 - 13x - 28$

59. $3a^2 - 8a + 5$ **60.** $6a^2 + 11a + 3$ **61.** $4a^2 + a - 14$ **62.** $3a^2 - 20a + 12$

63. $2x^2 + 13x + 15$ **64.** $5x^2 - 8x - 4$ **65.** $3x^4 - 8x^2 - 3$ **66.** $6x^4 - 7x^2 - 5$

67. $6x^2 + 35xy + 11y^2$ **68.** $5x^2 + 12xy + 7y^2$ **69.** $7x^2 + 11xy - 6y^2$ **70.** $4x^2 - 13xy + 3y^2$

Factor out the greatest common factor from the terms of the trinomial. Then factor the remaining trinomial.

71. $8x^3 - 2x^2 - x$ **72.** $9x^3 - 9x^2 - 10x$ **73.** $10x^4 + 15x^3 + 5x^2$ **74.** $16x^4 + 48x^3 + 20x^2$

A.4 Special Cases of Factoring

1 Factoring the Difference of Two Squares

We learned in Appendix A.2 and Section 5.5 how to multiply binomials using the special product formula: $(a + b)(a - b) = a^2 - b^2$. We can use it now as a factoring formula.

Student Learning Objectives

After studying this section, you will be able to:

1. Factor a binomial that is the difference of two squares.
2. Factor a perfect square trinomial.
3. Factor a binomial that is the sum or difference of two cubes.

> **FACTORING THE DIFFERENCE OF TWO SQUARES**
>
> $$a^2 - b^2 = (a + b)(a - b)$$

Example 1 Factor. $x^2 - 16$

Solution In this case $a = x$ and $b = 4$ in the formula.

$$\begin{array}{ccccc} a^2 & - & b^2 & = & (a + b)(a - b) \\ \downarrow & & \downarrow & & \downarrow \quad \downarrow \quad \downarrow \quad \downarrow \\ (x)^2 & - & (4)^2 & = & (x + 4)(x - 4) \end{array}$$

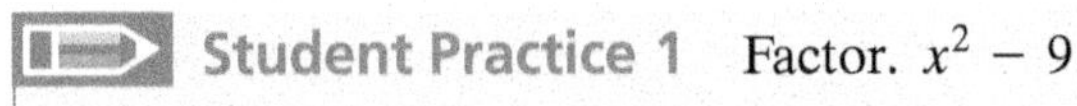

Student Practice 1 Factor. $x^2 - 9$

Example 2 Factor. $25x^2 - 36$

Solution Here we will use the formula $a^2 - b^2 = (a + b)(a - b)$.

$$25x^2 - 36 = (5x)^2 - (6)^2 = (5x + 6)(5x - 6)$$

Student Practice 2 Factor. $64x^2 - 121$

Example 3 Factor. $100w^4 - 9z^4$

Solution $100w^4 - 9z^4 = (10w^2)^2 - (3z^2)^2 = (10w^2 + 3z^2)(10w^2 - 3z^2)$

Student Practice 3 Factor. $49x^2 - 25y^4$

Whenever possible, a common factor should be factored out in the first step. Then the formula can be applied.

Example 4 Factor. $75x^2 - 3$

Solution We factor out the common factor 3 from each term.

$$\begin{aligned} 75x^2 - 3 &= 3(25x^2 - 1) \\ &= 3(5x + 1)(5x - 1) \end{aligned}$$

Student Practice 4 Factor. $7x^2 - 28$

2 Factoring Perfect Square Trinomials

Recall the formulas for squaring a binomial from Appendix A.2 and Section 5.5.

$$(a - b)^2 = a^2 - 2ab + b^2$$
$$(a + b)^2 = a^2 + 2ab + b^2$$

We can use these formulas to factor perfect square trinomials.

PERFECT SQUARE FACTORING FORMULAS

$$a^2 - 2ab + b^2 = (a - b)^2$$
$$a^2 + 2ab + b^2 = (a + b)^2$$

Recognizing these special cases will save you a lot of time when factoring. How can we recognize a perfect square trinomial?

1. The first and last terms are perfect squares. (The numerical values are 1, 4, 9, 16, 25, 36, . . . , and the variables have an exponent that is an even whole number.)
2. The middle term is twice the product of the values that, when squared, give the first and last terms.

Example 5 Factor. $25x^2 - 20x + 4$

Solution Is this trinomial a perfect square? Yes.

1. The first and last terms are perfect squares.

$$25x^2 - 20x + 4 = (5x)^2 - 20x + (2)^2$$

2. The middle term is twice the product of the value $5x$ and the value 2. In other words, $2(5x)(2) = 20x$.

$$25x^2 - 20x + 4 = (5x)^2 - 2(5x)(2) + (2)^2$$

Therefore, we can use the formula $a^2 - 2ab + b^2 = (a - b)^2$. Thus,

$$\begin{array}{ccccccccc} a^2 & - & 2ab & + & b^2 & = & (a & - & b)^2 \\ \downarrow & & \downarrow & & \downarrow & & \downarrow & & \downarrow \\ 25x^2 & - & 20x & + & 4 & = & (5x & - & 2)^2. \end{array}$$

□

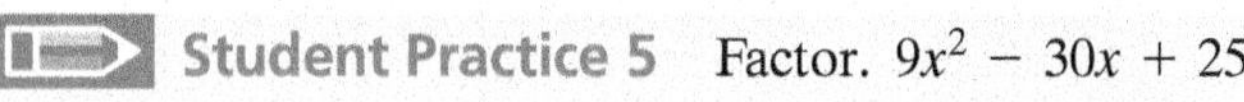

Student Practice 5 Factor. $9x^2 - 30x + 25$

Example 6 Factor. $200x^2 + 360x + 162$

Solution First we factor out the common factor 2.

$$200x^2 + 360x + 162 = 2(100x^2 + 180x + 81)$$
$$a^2 + 2ab + b^2 = (a + b)^2$$
$$2(100x^2 + 180x + 81) = 2[(10x)^2 + (2)(10x)(9) + (9)^2]$$
$$= 2(10x + 9)^2$$

□

Student Practice 6 Factor. $242x^2 + 88x + 8$

Example 7 Factor.

(a) $x^4 + 14x^2 + 49$ **(b)** $9x^4 + 30x^2y^2 + 25y^4$

Solution

(a) $$\begin{aligned} x^4 + 14x^2 + 49 &= (x^2)^2 + 2(x^2)(7) + (7)^2 \\ &= (x^2 + 7)^2 \end{aligned}$$

(b) $$\begin{aligned} 9x^4 + 30x^2y^2 + 25y^4 &= (3x^2)^2 + 2(3x^2)(5y^2) + (5y^2)^2 \\ &= (3x^2 + 5y^2)^2 \end{aligned}$$ □

Student Practice 7 Factor.

(a) $49x^4 + 28x^2 + 4$ **(b)** $36x^4 + 84x^2y^2 + 49y^4$

3 Factoring the Sum or Difference of Two Cubes

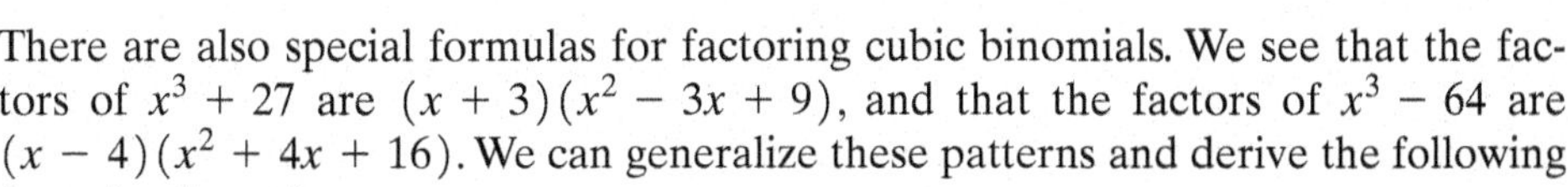

There are also special formulas for factoring cubic binomials. We see that the factors of $x^3 + 27$ are $(x + 3)(x^2 - 3x + 9)$, and that the factors of $x^3 - 64$ are $(x - 4)(x^2 + 4x + 16)$. We can generalize these patterns and derive the following factoring formulas.

SUM AND DIFFERENCE OF CUBES FACTORING FORMULAS

$$a^3 + b^3 = (a + b)(a^2 - ab + b^2)$$
$$a^3 - b^3 = (a - b)(a^2 + ab + b^2)$$

Example 8 Factor. $125x^3 + y^3$

Solution Here $a = 5x$ and $b = y$.

$$\begin{array}{rcl} a^3 + b^3 &=& (a \;\; + b)(a^2 \;\; - \;\; ab \;\; + b^2) \\ && \;\downarrow \quad\; \downarrow \;\; \downarrow \qquad\;\; \downarrow \qquad \downarrow \\ 125x^3 + y^3 = (5x)^3 + (y)^3 &=& (5x + y)(25x^2 - 5xy + y^2) \end{array}$$ □

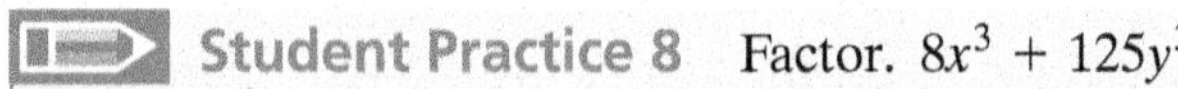

Student Practice 8 Factor. $8x^3 + 125y^3$

Example 9 Factor. $64x^3 + 27$

Solution Here $a = 4x$ and $b = 3$.

$$\begin{array}{rcl} a^3 + b^3 &=& (a \;\; + b)(a^2 \;\; - \;\; ab \;\; + b^2) \\ && \;\downarrow \quad\; \downarrow \;\; \downarrow \qquad\;\; \downarrow \qquad \downarrow \\ 64x^3 + 27 = (4x)^3 + (3)^3 &=& (4x + 3)(16x^2 - 12x + 9) \end{array}$$ □

Student Practice 9 Factor. $64x^3 + 125y^3$

Example 10 Factor. $125w^3 - 8z^6$

Solution Here $a = 5w$ and $b = 2z^2$.

$$a^3 - b^3 = (a - b)(a^2 + ab + b^2)$$

$$\downarrow \quad \downarrow \quad \downarrow \quad \downarrow \quad \downarrow$$

$$125w^3 - 8z^6 = (5w)^3 - (2z^2)^3 = (5w - 2z^2)(25w^2 + 10wz^2 + 4z^4)$$ □

Student Practice 10 Factor. $27w^3 - 125z^6$

Example 11 Factor. $250x^3 - 2$

Solution First we factor out the common factor 2.

$$250x^3 - 2 = 2(125x^3 - 1)$$
$$= 2(5x - 1)\underbrace{(25x^2 + 5x + 1)}$$
$$\uparrow$$

Note that this trinomial cannot be factored. □

Student Practice 11 Factor. $54x^3 - 16$

What should you do if a polynomial is the difference of two cubes *and* the difference of two squares? Usually, it's easier to use the difference of two squares formula first. Then apply the difference of two cubes formula.

Example 12 Factor. $x^6 - y^6$

Solution We can write this binomial as $(x^2)^3 - (y^2)^3$ or as $(x^3)^2 - (y^3)^2$. Therefore, we can use either the difference of two cubes formula or the difference of two squares formula. It's usually better to use the difference of two squares formula first, so we'll do that.

$$x^6 - y^6 = (x^3)^2 - (y^3)^2$$

Here $a = x^3$ and $b = y^3$. Therefore,

$$(x^3)^2 - (y^3)^2 = (x^3 + y^3)(x^3 - y^3).$$

Now we use the sum of two cubes formula for the first factor and the difference of two cubes formula for the second factor.

$$x^3 + y^3 = (x + y)(x^2 - xy + y^2)$$
$$x^3 - y^3 = (x - y)(x^2 + xy + y^2)$$

Hence,

$$x^6 - y^6 = (x + y)(x^2 - xy + y^2)(x - y)(x^2 + xy + y^2).$$ □

Student Practice 12 Factor. $64a^6 - 1$

Often the various types of factoring problems are all mixed together. We need to be able to identify each type of polynomial quickly. The following table summarizes the information we have learned about factoring in Appendix A.3 and A.4.

Many polynomials require more than one factoring method. When you are asked to factor a polynomial, it is expected that you will factor it completely. Usually, the first step is factoring out a common factor; then the next step will become apparent.

Carefully go through each example in the following **Factoring Organizer.** Be sure you understand each step that is involved.

Factoring Organizer

Number of Terms in the Polynomial	Identifying Name and/or Formula	Examples
A. Any number of terms	**Common factor** The terms have a common factor consisting of a number, a variable, or both.	$2x^2 - 16x = 2x(x - 8)$ $3x^2 + 9y - 12 = 3(x^2 + 3y - 4)$ $4x^2y + 2xy^2 - wxy + xyz = xy(4x + 2y - w + z)$
B. Two terms	**Difference of two squares** First and last terms are perfect squares. $a^2 - b^2 = (a + b)(a - b)$	$16x^2 - 1 = (4x + 1)(4x - 1)$ $25y^2 - 9x^2 = (5y + 3x)(5y - 3x)$
C. Two terms	**Sum and Difference of Cubes** First and last terms are perfect cubes. $a^3 + b^3 = (a + b)(a^2 - ab + b^2)$ $a^3 - b^3 = (a - b)(a^2 + ab + b^2)$	$8x^3 + 27 = (2x + 3)(4x^2 - 6x + 9)$ $8x^3 - 27 = (2x - 3)(4x^2 + 6x + 9)$
D. Three terms	**Perfect-square trinomial** First and last terms are perfect squares. $a^2 + 2ab + b^2 = (a + b)^2$ $a^2 - 2ab + b^2 = (a - b)^2$	$25x^2 - 10x + 1 = (5x - 1)^2$ $16x^2 + 24x + 9 = (4x + 3)^2$
E. Three terms	**Trinomial of the form $x^2 + bx + c$** It starts with x^2. The constants of the two factors are numbers whose product is c and whose sum is b.	$x^2 - 7x + 12 = (x - 3)(x - 4)$ $x^2 + 11x - 26 = (x + 13)(x - 2)$ $x^2 - 8x - 20 = (x - 10)(x + 2)$
F. Three terms	**Trinomial of the form $ax^2 + bx + c$** It starts with ax^2, where a is any number but 1.	Use trial-and-error or the grouping number method to factor $12x^2 - 5x - 2$. **1.** The grouping number is -24. **2.** The two numbers whose product is -24 and whose sum is -5 are -8 and 3. **3.** $12x^2 - 5x - 2 = 12x^2 + 3x - 8x - 2$ $= 3x(4x + 1) - 2(4x + 1)$ $= (4x + 1)(3x - 2)$
G. Four terms	**Factor by grouping** Rearrange the order if the first two terms do not have a common factor.	$wx - 6yz + 2wy - 3xz = wx + 2wy - 3xz - 6yz$ $= w(x + 2y) - 3z(x + 2y)$ $= (x + 2y)(w - 3z)$

A.4 Exercises MyMathLab®

Verbal and Writing Skills, Exercises 1–4

1. How do you determine if a factoring problem will use the difference of two squares?

2. How do you determine if a factoring problem will use the perfect square trinomial formula?

3. How do you determine if a factoring problem will use the sum of two cubes formula?

4. How do you determine if a factoring problem will use the difference of two cubes formula?

Use the difference of two squares formula to factor. Be sure to factor out any common factors.

5. $a^2 - 64$

6. $y^2 - 49$

7. $16x^2 - 81$

8. $4x^2 - 25$

9. $64x^2 - 1$

10. $81x^2 - 1$

11. $49m^2 - 9n^2$

12. $36x^2 - 25y^2$

13. $100y^2 - 81$

14. $49y^2 - 144$

15. $1 - 36x^2y^2$

16. $1 - 64x^2y^2$

17. $32x^2 - 18$

18. $50x^2 - 8$

19. $5x - 20x^3$

20. $49x^3 - 36x$

Use the perfect square trinomial formulas to factor. Be sure to factor out any common factors.

21. $9x^2 - 6x + 1$

22. $16y^2 - 8y + 1$

23. $49x^2 - 14x + 1$

24. $100y^2 - 20y + 1$

25. $81w^2 + 36wt + 4t^2$

26. $25w^2 + 20wt + 4t^2$

27. $36x^2 + 60xy + 25y^2$

28. $64x^2 + 48xy + 9y^2$

29. $8x^2 + 40x + 50$

30. $108x^2 + 36x + 3$

31. $3x^3 - 24x^2 + 48x$

32. $50x^3 - 20x^2 + 2x$

Use the sum and difference of cubes formulas to factor. Be sure to factor out any common factors.

33. $x^3 - 27$

34. $x^3 - 8$

35. $x^3 + 125$

36. $x^3 + 64$

37. $64x^3 - 1$

38. $125x^3 - 1$

39. $8x^3 - 125$

40. $27x^3 - 64$

41. $1 - 27x^3$

42. $1 - 8x^3$

43. $64x^3 + 125$

44. $27x^3 + 125$

45. $64s^6 + t^6$

46. $125s^6 + t^6$

47. $5y^3 - 40$

48. $54y^3 - 2$

49. $250x^3 + 2$

50. $128y^3 + 2$

51. $x^5 - 8x^2y^3$

52. $x^5 - 27x^2y^3$

Mixed Practice *Factor by the methods taught in both Appendix A.3 and A.4.*

53. $x^2 - 2x - 63$

54. $x^2 + 6x - 40$

55. $6x^2 + x - 2$

56. $5x^2 + 17x + 6$

57. $25w^4 - 1$

58. $16m^4 - 25$

59. $b^4 + 6b^2 + 9$

60. $a^4 - 10a^2 + 25$

61. $yz^2 - 15 - 3z^2 + 5y$

62. $ad^4 - 4ab - d^4 + 4b$

63. $ab^3 + c + b^2 + abc$

64. $25x^2z - 14y - 10yz + 35x^2$

65. $9m^6 - 64$

66. $144 - m^6$

67. $36y^6 - 60y^3 + 25$

68. $100n^6 - 140n^3 + 49$

69. $45z^8 - 5$

70. $2a^8 - 98$

71. $125m^3 + 8n^3$

72. $64z^3 - 27w^3$

73. $2x^2 + 4x - 96$

74. $3x^2 + 9x - 84$

75. $18x^2 + 21x + 6$

76. $24x^2 + 26x + 6$

77. $40ax^2 + 72ax - 16a$

78. $60bx^2 - 84bx - 72b$

79. $6x^3 + 26x^2 - 20x$

80. $12x^3 - 14x^2 + 4x$

81. $24a^3 - 3b^3$

82. $54w^3 + 250$

83. $4w^2 - 20wz + 25z^2$

84. $81x^4 - 36x^2 + 4$

85. $36a^2 - 81b^2$

86. $400x^4 - 36y^2$

87. $16x^4 - 81y^4$

88. $256x^4 - 1$

89. $125m^6 + 8$

90. $27n^6 + 125$

Try to factor the following four trinomials by using the formulas for perfect square trinomials. Why can't the formulas be used? Then factor each trinomial correctly using an appropriate method.

91. $25x^2 + 25x + 4$

92. $16x^2 + 40x + 9$

93. $49x^2 - 35x + 4$

94. $4x^2 - 25x + 36$

Applications

▲ **95.** ***Carpentry*** Find the area of a maple cabinet surface that is constructed by a carpenter as a large square with sides of $4x$ feet and has a square cut out region whose sides are y feet. Factor the expression.

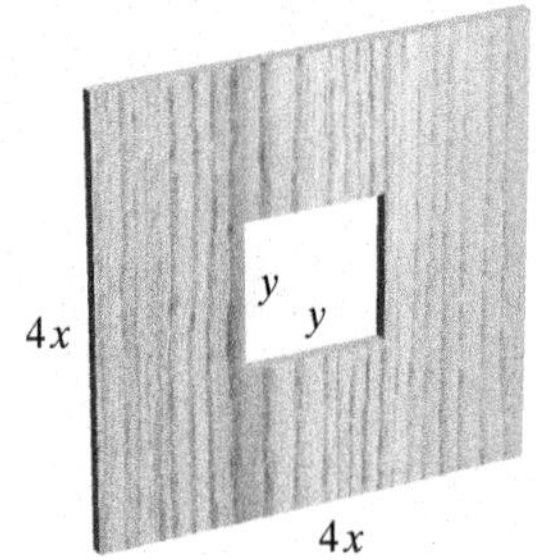

▲ **96.** ***Base of a Lamp*** A copper base for a lamp consists of a large circle of radius $2y$ inches with a cut out area in the center of radius x inches. Write an expression for the area of this copper base. Write your answer in factored form.

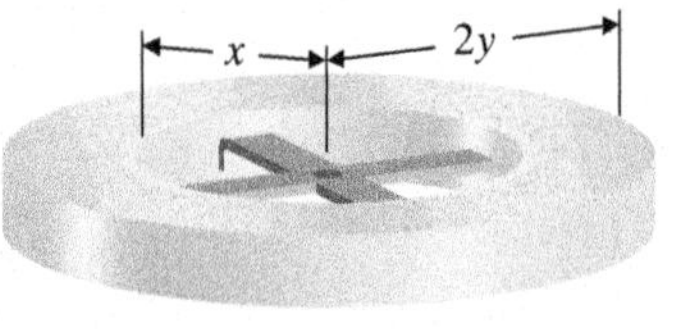

▲ **97.** ***Tree Reforestation*** A plan has been made in northern Maine to replace trees harvested by paper mills. The proposed planting zone is in the shape of a giant rectangle with an area of $30x^2 + 19x - 5$ square feet. Use your factoring skills to determine a possible configuration of the number of rows of trees and the number of trees to be placed in each row.

▲ **98.** ***Tree Reforestation*** A plan has been made in northern Washington to replace trees harvested by paper mills. The proposed planting zone is in the shape of a giant rectangle with an area of $12x^2 + 20x - 25$ square feet. Use your factoring skills to determine a possible configuration of the number of rows of trees and the number of trees to be placed in each row.

A.5 The Point–Slope Form of a Line 

1 Using the Point–Slope Form of the Equation of a Line

Student Learning Objectives

After studying this section, you will be able to:

1. Use the point–slope form of the equation of a line.
2. Write the equation of the line passing through a given point that is parallel or perpendicular to a given line.

Recall from Section 3.1 that we defined a linear equation in two variables as an equation that can be written in the form $Ax + By = C$. This form is called the **standard form of the equation of a line** and although this form tells us that the graph is a straight line, it reveals little about the line. A more useful form of the equation was introduced in Section 3.3 called the **slope–intercept form,** $y = mx + b$. This form immediately reveals the slope and y-intercept of a line, and also allows us to easily write the equation of a line when we know its slope and y-intercept. But, what happens if we know the slope of a line and a point on the line that is not the y-intercept? Can we write the equation of the line? By the definition of slope, we have the following:

$$m = \frac{y - y_1}{x - x_1}$$

$$m(x - x_1) = y - y_1$$

$$\text{That is, } y - y_1 = m(x - x_1).$$

This is the point–slope form of the equation of a line.

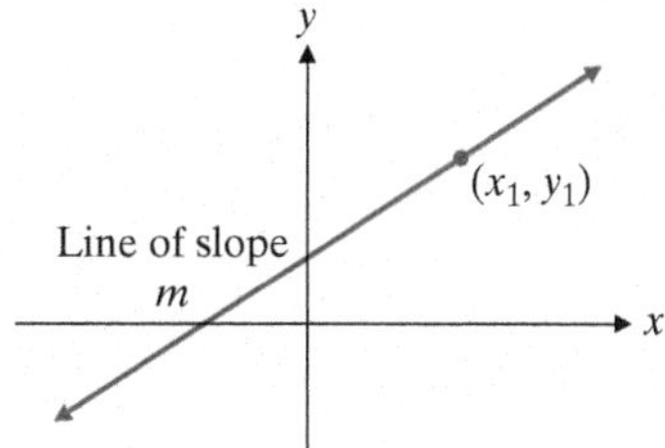

> **POINT–SLOPE FORM**
>
> The **point–slope form** of the equation of a line is $y - y_1 = m(x - x_1)$, where m is the slope and (x_1, y_1) are the coordinates of a known point on the line.

Write the Equation of a Line Given Its Slope and One Point on the Line

Example 1 Find an equation of the line that has slope $-\frac{3}{4}$ and passes through the point $(-6, 1)$. Express your answer in standard form.

Solution Since we don't know the y-intercept, we can't use the slope–intercept form easily. Therefore, we use the point–slope form.

$y - y_1 = m(x - x_1)$	
$y - 1 = -\frac{3}{4}[x - (-6)]$	Substitute the given values.
$y - 1 = -\frac{3}{4}x - \frac{9}{2}$	Simplify. (Do you see how we did this?)
$4y - 4(1) = 4\left(-\frac{3}{4}x\right) - 4\left(\frac{9}{2}\right)$	Multiply each term by the LCD 4.
$4y - 4 = -3x - 18$	Simplify.
$3x + 4y = -18 + 4$	Add $3x + 4$ to each side.
$3x + 4y = -14$	Add like terms.

The equation in standard form is $3x + 4y = -14$. □

(Continued on next page)

Student Practice 1 Find an equation of the line that passes through $(5, -2)$ and has a slope of $\frac{3}{4}$. Express your answer in standard form.

Write the Equation of a Line Given Two Points on the Line We can use the point–slope form to find the equation of a line if we are given two points. Carefully study the following example. Be sure you understand each step. You will encounter this type of problem frequently.

Example 2 Find an equation of the line that passes through $(3, -2)$ and $(5, 1)$. Express your answer in slope–intercept form.

Solution First we find the slope.

$$m = \frac{y_2 - y_1}{x_2 - x_1} = \frac{1 - (-2)}{5 - 3} = \frac{1 + 2}{2} = \frac{3}{2}$$

Now we substitute the value of the slope and the coordinates of either point into the point–slope equation. Let's use $(5, 1)$.

$$y - y_1 = m(x - x_1)$$

$$y - 1 = \frac{3}{2}(x - 5) \quad \text{Substitute } m = \frac{3}{2} \text{ and } (x_1, y_1) = (5, 1).$$

$$y - 1 = \frac{3}{2}x - \frac{15}{2} \quad \text{Remove parentheses.}$$

$$y = \frac{3}{2}x - \frac{15}{2} + 1 \quad \text{Add 1 to each side of the equation.}$$

$$y = \frac{3}{2}x - \frac{15}{2} + \frac{2}{2}$$

$$y = \frac{3}{2}x - \frac{13}{2} \quad \text{Add the two fractions and simplfy.}$$

□

Student Practice 2 Find an equation of the line that passes through $(-4, 1)$ and $(-2, -3)$. Express your answer in slope–intercept form.

Graphing Calculator

Using Linear Regression to Find an Equation

Many graphing calculators, such as the TI-84 Plus, will find the equation of a line in slope–intercept form if you enter the points as a collection of data and use the Regression feature. We would enter the data from Example 2 as follows:

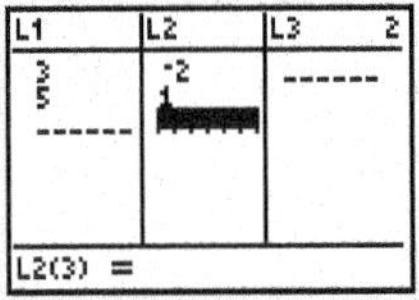

The output of the calculator uses the notation $y = ax + b$ instead of $y = mx + b$.

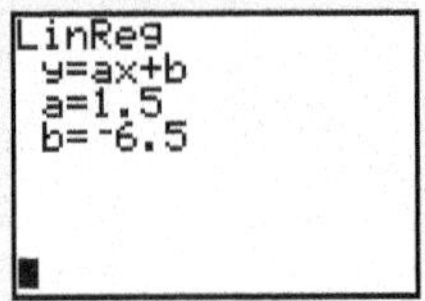

Thus, our answer to Example 2 using the graphing calculator would be $y = 1.5x - 6.5$.

Before we go further, we want to point out that these various forms of the equation of a straight line are just that *forms* for convenience. We are *not* using different equations each time, nor should you simply try to memorize the different variations without understanding when to use them. They can easily be derived from the definition of slope, as we have seen. And remember, you can *always* use the definition of slope to find the equation of a line. You may find it helpful to review Examples 1 and 2 for a few minutes before going ahead to Example 3. It is important to see how each example is different.

2 Writing the Equation of a Parallel or Perpendicular Line

Let us now look at parallel and perpendicular lines. If we are given the equation of a line and a point not on the line, we can find the equation of a second line that passes through the given point and is parallel or perpendicular to the first line. We can do this because we know that the slopes of parallel lines are equal and that the slopes of perpendicular lines are negative reciprocals of each other.

We begin by finding the slope of the given line. Then we use the point–slope form to find the equation of the second line. Study each step of the following example carefully.

Example 3 Find an equation of the line passing through the point $(-2, -4)$ and parallel to the line $2x + 5y = 8$. Express the answer in standard form.

Solution First we need to find the slope of the line $2x + 5y = 8$. We do this by writing the equation in slope–intercept form.

$$5y = -2x + 8$$

$$y = -\frac{2}{5}x + \frac{8}{5}$$

The slope of the given line is $-\frac{2}{5}$. Since parallel lines have the same slope, the slope of the unknown line is also $-\frac{2}{5}$. Now we substitute $m = -\frac{2}{5}$ and the coordinates of the point $(-2, -4)$ into the point–slope form of the equation of a line.

$$y - y_1 = m(x - x_1)$$

$$y - (-4) = -\frac{2}{5}[x - (-2)] \quad \text{Substitute.}$$

$$y + 4 = -\frac{2}{5}(x + 2) \quad \text{Simplify.}$$

$$y + 4 = -\frac{2}{5}x - \frac{4}{5} \quad \text{Remove parentheses.}$$

$$5y + 5(4) = 5\left(-\frac{2}{5}x\right) - 5\left(\frac{4}{5}\right) \quad \text{Multiply each term by the LCD 5.}$$

$$5y + 20 = -2x - 4 \quad \text{Simplify.}$$

$$2x + 5y = -4 - 20 \quad \text{Add } 2x - 20 \text{ to each side.}$$

$$2x + 5y = -24 \quad \text{Simplify.}$$

$2x + 5y = -24$ is an equation of the line passing through the point $(-2, -4)$ and parallel to the line $2x + 5y = 8$. □

Student Practice 3 Find an equation of the line passing through $(4, -5)$ and parallel to the line $5x - 3y = 10$. Express the answer in standard form.

An extra step is needed if the desired line is to be perpendicular to the given line. Carefully note the approach in Example 4.

Example 4 Find an equation of the line that passes through the point $(2, -3)$ and is perpendicular to the line $3x - y = -12$. Express the answer in standard form.

Solution To find the slope of the line $3x - y = -12$, we rewrite it in slope–intercept form.

$$-y = -3x - 12$$
$$y = 3x + 12$$

This line has a slope of 3. Therefore, the slope of a line perpendicular to this line is the negative reciprocal $-\frac{1}{3}$.

Now substitute the slope $m = -\frac{1}{3}$ and the coordinates of the point $(2, -3)$ into the point–slope form of the equation.

$$y - y_1 = m(x - x_1)$$

$$y - (-3) = -\frac{1}{3}(x - 2) \quad \text{Substitute.}$$

$$y + 3 = -\frac{1}{3}(x - 2) \quad \text{Simplify.}$$

$$y + 3 = -\frac{1}{3}x + \frac{2}{3} \quad \text{Remove parentheses.}$$

$$3y + 3(3) = 3\left(-\frac{1}{3}x\right) + 3\left(\frac{2}{3}\right) \quad \text{Multiply each term by the LCD 3.}$$

$$3y + 9 = -x + 2 \quad \text{Simplify.}$$

$$x + 3y = 2 - 9 \quad \text{Add } x - 9 \text{ to each side.}$$

$$x + 3y = -7 \quad \text{Simplify.}$$

$x + 3y = -7$ is an equation of the line that passes through the point $(2, -3)$ and is perpendicular to the line $3x - y = -12$. □

Student Practice 4 Find an equation of the line that passes through $(-4, 3)$ and is perpendicular to the line $6x + 3y = 7$. Express the answer in standard form.

A.5 Exercises MyMathLab®

Find an equation of the line that passes through the given point and has the given slope. Express your answer in slope–intercept form.

1. $(6, 4), m = -\frac{2}{3}$

2. $(4, 6), m = -\frac{1}{2}$

3. $(-7, -2), m = 5$

4. $(8, 0), m = -3$

5. $(6, 0), m = -\frac{1}{5}$

6. $(0, -1), m = -\frac{5}{3}$

Find an equation of the line passing through the pair of points. Write the equation in slope–intercept form.

7. $(-4, -1)$ and $(3, 4)$

8. $(7, -2)$ and $(-1, -3)$

9. $\left(\frac{1}{2}, -3\right)$ and $\left(\frac{7}{2}, -5\right)$

10. $\left(\frac{7}{6}, 1\right)$ and $\left(-\frac{1}{3}, 0\right)$

11. $(12, -3)$ and $(7, -3)$

12. $(4, 8)$ and $(-3, 8)$

Find an equation of the line satisfying the conditions given. Express your answer in standard form.

13. Parallel to $5x - y = 4$ and passing through $(-2, 0)$

14. Parallel to $3x - y = -5$ and passing through $(-1, 0)$

15. Parallel to $x = 3y - 8$ and passing through $(5, -1)$

16. Parallel to $2y + x = 7$ and passing through $(-5, -4)$

17. Perpendicular to $2y = -3x$ and passing through $(6, -1)$

18. Perpendicular to $y = 5x$ and passing through $(4, -2)$

19. Perpendicular to $x + 7y = -12$ and passing through $(-4, -1)$

20. Perpendicular to $x - 4y = 2$ and passing through $(3, -1)$

To Think About

Without graphing determine whether the following pairs of lines are (a) parallel, (b) perpendicular, or (c) neither parallel nor perpendicular.

21. $-3x + 5y = 40$
$5y + 3x = 17$

22. $5x - 6y = 19$
$6x + 5y = -30$

23. $y = -\frac{3}{4}x - 2$
$6x + 8y = -5$

24. $y = \frac{2}{3}x + 6$
$-2x - 3y = -12$

25. $y = \frac{5}{6}x - \frac{1}{3}$
$6x + 5y = -12$

26. $y = \frac{3}{7}x - \frac{1}{14}$
$14y + 6x = 3$

Appendix B Practice with Operations of Whole Numbers

Addition Practice

1. $\begin{array}{r} 23 \\ +\ 14 \\ \hline \end{array}$
2. $\begin{array}{r} 42 \\ +\ 33 \\ \hline \end{array}$
3. $\begin{array}{r} 50 \\ +\ 44 \\ \hline \end{array}$
4. $\begin{array}{r} 83 \\ +\ 16 \\ \hline \end{array}$
5. $\begin{array}{r} 51 \\ +\ 27 \\ \hline \end{array}$
6. $\begin{array}{r} 16 \\ +\ 13 \\ \hline \end{array}$
7. $\begin{array}{r} 32 \\ +\ 29 \\ \hline \end{array}$
8. $\begin{array}{r} 64 \\ +\ 17 \\ \hline \end{array}$
9. $\begin{array}{r} 327 \\ +\ 42 \\ \hline \end{array}$
10. $\begin{array}{r} 223 \\ +\ 54 \\ \hline \end{array}$
11. $\begin{array}{r} 463 \\ +\ 28 \\ \hline \end{array}$
12. $\begin{array}{r} 504 \\ +\ 96 \\ \hline \end{array}$
13. $\begin{array}{r} 739 \\ +\ 682 \\ \hline \end{array}$
14. $\begin{array}{r} 567 \\ +\ 485 \\ \hline \end{array}$
15. $840 + 60$
16. $364 + 37$
17. $915 + 796$
18. $420 + 899$
19. $213 + 46 + 30$
20. $326 + 21 + 52$
21. $132 + 441 + 16$
22. $671 + 204 + 12$
23. $139 + 61 + 222$
24. $524 + 73 + 195$
25. $701 + 166 + 24 + 11$
26. $439 + 365 + 45 + 81$

Subtraction Practice

1. $\begin{array}{r} 32 \\ -\ 11 \\ \hline \end{array}$
2. $\begin{array}{r} 87 \\ -\ 25 \\ \hline \end{array}$
3. $\begin{array}{r} 56 \\ -\ 34 \\ \hline \end{array}$
4. $\begin{array}{r} 73 \\ -\ 30 \\ \hline \end{array}$
5. $\begin{array}{r} 93 \\ -\ 25 \\ \hline \end{array}$
6. $\begin{array}{r} 21 \\ -\ 16 \\ \hline \end{array}$
7. $\begin{array}{r} 40 \\ -\ 11 \\ \hline \end{array}$
8. $\begin{array}{r} 60 \\ -\ 15 \\ \hline \end{array}$
9. $\begin{array}{r} 576 \\ -\ 45 \\ \hline \end{array}$
10. $\begin{array}{r} 294 \\ -\ 71 \\ \hline \end{array}$
11. $\begin{array}{r} 780 \\ -\ 54 \\ \hline \end{array}$
12. $\begin{array}{r} 208 \\ -\ 17 \\ \hline \end{array}$
13. $\begin{array}{r} 406 \\ -\ 28 \\ \hline \end{array}$
14. $\begin{array}{r} 100 \\ -\ 34 \\ \hline \end{array}$
15. $\begin{array}{r} 635 \\ -\ 126 \\ \hline \end{array}$
16. $\begin{array}{r} 375 \\ -\ 147 \\ \hline \end{array}$
17. $\begin{array}{r} 500 \\ -\ 244 \\ \hline \end{array}$
18. $\begin{array}{r} 200 \\ -\ 137 \\ \hline \end{array}$
19. $\begin{array}{r} 922 \\ -\ 739 \\ \hline \end{array}$
20. $\begin{array}{r} 646 \\ -\ 377 \\ \hline \end{array}$
21. $\begin{array}{r} 1729 \\ -\ 856 \\ \hline \end{array}$
22. $\begin{array}{r} 2382 \\ -\ 490 \\ \hline \end{array}$
23. $\begin{array}{r} 7806 \\ -\ 327 \\ \hline \end{array}$
24. $\begin{array}{r} 3024 \\ -\ 156 \\ \hline \end{array}$
25. $\begin{array}{r} 8200 \\ -\ 6134 \\ \hline \end{array}$
26. $\begin{array}{r} 2004 \\ -\ 1326 \\ \hline \end{array}$

Multiplication Practice

1. $\begin{array}{r}23\\ \times\ 3\\ \hline\end{array}$ **2.** $\begin{array}{r}13\\ \times\ 2\\ \hline\end{array}$ **3.** $\begin{array}{r}54\\ \times\ 7\\ \hline\end{array}$ **4.** $\begin{array}{r}67\\ \times\ 9\\ \hline\end{array}$ **5.** $\begin{array}{r}74\\ \times 21\\ \hline\end{array}$ **6.** $\begin{array}{r}53\\ \times 31\\ \hline\end{array}$

7. $\begin{array}{r}92\\ \times 40\\ \hline\end{array}$ **8.** $\begin{array}{r}70\\ \times 52\\ \hline\end{array}$ **9.** $\begin{array}{r}82\\ \times 95\\ \hline\end{array}$ **10.** $\begin{array}{r}69\\ \times 39\\ \hline\end{array}$ **11.** $\begin{array}{r}212\\ \times\ 43\\ \hline\end{array}$ **12.** $\begin{array}{r}341\\ \times\ 22\\ \hline\end{array}$

13. $\begin{array}{r}295\\ \times\ 41\\ \hline\end{array}$ **14.** $\begin{array}{r}419\\ \times\ 72\\ \hline\end{array}$ **15.** $\begin{array}{r}304\\ \times\ 68\\ \hline\end{array}$ **16.** $\begin{array}{r}620\\ \times\ 39\\ \hline\end{array}$ **17.** $\begin{array}{r}261\\ \times 144\\ \hline\end{array}$ **18.** $\begin{array}{r}124\\ \times 433\\ \hline\end{array}$

19. $\begin{array}{r}545\\ \times 522\\ \hline\end{array}$ **20.** $\begin{array}{r}634\\ \times 799\\ \hline\end{array}$ **21.** $\begin{array}{r}391\\ \times 609\\ \hline\end{array}$ **22.** $\begin{array}{r}817\\ \times 460\\ \hline\end{array}$ **23.** $\begin{array}{r}3844\\ \times\ 209\\ \hline\end{array}$ **24.** $\begin{array}{r}7409\\ \times\ 106\\ \hline\end{array}$

25. 72,499(683) **26.** 86,243(725)

Division Practice

1. $8\overline{)128}$ **2.** $3\overline{)168}$ **3.** $7\overline{)415}$ **4.** $6\overline{)287}$

5. $9\overline{)1116}$ **6.** $4\overline{)1184}$ **7.** $6\overline{)1404}$ **8.** $3\overline{)1701}$

9. $8\overline{)4174}$ **10.** $5\overline{)3697}$ **11.** $17\overline{)5468}$ **12.** $13\overline{)9795}$

13. $146\overline{)12,994}$ **14.** $163\overline{)14,833}$ **15.** $1728 \div 54$ **16.** $3813 \div 93$

17. $3701 \div 34$ **18.** $6052 \div 49$ **19.** $15,836 \div 74$ **20.** $23,256 \div 68$

21. $30,632 \div 27$ **22.** $85,069 \div 79$ **23.** $30,752 \div 248$ **24.** $49,878 \div 326$

25. $271,125 \div 241$ **26.** $546,924 \div 357$

Appendix C Determinants and Cramer's Rule

Student Learning Objectives

After studying this section, you will be able to:

1. Evaluate a second-order determinant.
2. Evaluate a third-order determinant.
3. Solve a system of two linear equations with two unknowns using Cramer's rule.
4. Solve a system of three linear equations with three unknowns using Cramer's rule.

1 Evaluating a Second-Order Determinant 

Mathematicians have developed techniques to solve systems of linear equations by focusing on the coefficients of the variables and the constants in the equations. The computational techniques can be easily carried out by computers or calculators. We will learn to do them by hand so that you will have a better understanding of what is involved.

To begin, we need to define a matrix and a determinant. A **matrix** is any rectangular array of numbers that is arranged in rows and columns. We use the symbol [ ] to indicate a matrix.

$$\begin{bmatrix} 3 & 2 & 4 \\ -1 & 4 & 0 \end{bmatrix}, \quad \begin{bmatrix} 4 & -3 \\ 2 & \frac{1}{2} \\ 1 & 5 \end{bmatrix}, \quad \begin{bmatrix} -4 & 1 & 6 \end{bmatrix}, \quad \text{and} \quad \begin{bmatrix} \frac{1}{4} \\ 3 \\ -2 \end{bmatrix}$$

are matrices. If you have a graphing calculator, you can enter the elements of a matrix and store them for future use. Let's examine two systems of equations.

$$\begin{aligned} 3x + 2y &= 16 \\ x + 4y &= 22 \end{aligned} \quad \text{and} \quad \begin{aligned} -6x \qquad &= 18 \\ x + 3y &= 9 \end{aligned}$$

We could write the coefficients of the variables in each of these systems as a matrix.

$$\begin{matrix} 3x + 2y \\ x + 4y \end{matrix} \Rightarrow \begin{bmatrix} 3 & 2 \\ 1 & 4 \end{bmatrix} \quad \text{and} \quad \begin{matrix} -6x \qquad \\ x + 3y \end{matrix} \Rightarrow \begin{bmatrix} -6 & 0 \\ 1 & 3 \end{bmatrix}$$

Now we define a determinant. A **determinant** is a *square* arrangement of numbers. We use the symbol | | to indicate a determinant.

$$\begin{vmatrix} 3 & 2 \\ 1 & 4 \end{vmatrix} \quad \text{and} \quad \begin{vmatrix} -6 & 0 \\ 1 & 3 \end{vmatrix}$$

are determinants. The value of a determinant is a *real number* and is defined as follows:

VALUE OF A SECOND-ORDER DETERMINANT

The value of the second-order determinant $\begin{vmatrix} a & c \\ b & d \end{vmatrix}$ is $ad - bc$.

Example 1 Find the value of each determinant.

(a) $\begin{vmatrix} -6 & 2 \\ -1 & 4 \end{vmatrix}$ **(b)** $\begin{vmatrix} 0 & -3 \\ -2 & 6 \end{vmatrix}$

Solution

(a) $\begin{vmatrix} -6 & 2 \\ -1 & 4 \end{vmatrix} = (-6)(4) - (-1)(2) = -24 - (-2) = -24 + 2 = -22$

(b) $\begin{vmatrix} 0 & -3 \\ -2 & 6 \end{vmatrix} = (0)(6) - (-2)(-3) = 0 - (+6) = -6$ □

Student Practice 1 Find the value of each determinant.

(a) $\begin{vmatrix} -7 & 3 \\ -4 & -2 \end{vmatrix}$ **(b)** $\begin{vmatrix} 5 & 6 \\ 0 & -5 \end{vmatrix}$

2 Evaluating a Third-Order Determinant

Third-order determinants have three rows and three columns. Again, each determinant has exactly one value.

VALUE OF A THIRD-ORDER DETERMINANT

The value of the third-order determinant

$$\begin{vmatrix} a_1 & b_1 & c_1 \\ a_2 & b_2 & c_2 \\ a_3 & b_3 & c_3 \end{vmatrix}$$

is

$$a_1b_2c_3 + b_1c_2a_3 + c_1a_2b_3 - a_3b_2c_1 - b_3c_2a_1 - c_3a_2b_1.$$

Because this definition is difficult to memorize and cumbersome to use, we evaluate third-order determinants by a simpler method called **expansion by minors.** The **minor** of an element (number or variable) of a third-order determinant is the second-order determinant that remains after we delete the row and column in which the element appears.

Example 2 Find **(a)** the minor of 6 and **(b)** the minor of -3 in the determinant.

$$\begin{vmatrix} 6 & 1 & 2 \\ -3 & 4 & 5 \\ -2 & 7 & 8 \end{vmatrix}$$

Solution

(a) Since the element 6 appears in the first row and the first column, we delete them.

$$\begin{vmatrix} \not{6} & \not{1} & \not{2} \\ -\not{3} & 4 & 5 \\ -\not{2} & 7 & 8 \end{vmatrix}$$

Therefore, the minor of 6 is

$$\begin{vmatrix} 4 & 5 \\ 7 & 8 \end{vmatrix}.$$

(b) Since -3 appears in the first column and the second row, we delete them.

$$\begin{vmatrix} \not{6} & 1 & 2 \\ -\not{3} & \not{4} & \not{5} \\ -\not{2} & 7 & 8 \end{vmatrix}$$

The minor of -3 is

$$\begin{vmatrix} 1 & 2 \\ 7 & 8 \end{vmatrix}.$$

□

Student Practice 2 Find **(a)** the minor of 3 and **(b)** the minor of -6 in the determinant.

$$\begin{vmatrix} 1 & 2 & 7 \\ -4 & -5 & -6 \\ 3 & 4 & -9 \end{vmatrix}$$

To evaluate a third-order determinant, we use expansion by minors of elements in the first column; for example, we have

$$\begin{vmatrix} a_1 & b_1 & c_1 \\ a_2 & b_2 & c_2 \\ a_3 & b_3 & c_3 \end{vmatrix} = a_1\begin{vmatrix} b_2 & c_2 \\ b_3 & c_3 \end{vmatrix} - a_2\begin{vmatrix} b_1 & c_1 \\ b_3 & c_3 \end{vmatrix} + a_3\begin{vmatrix} b_1 & c_1 \\ b_2 & c_2 \end{vmatrix}$$

Note that the signs alternate. We then evaluate the second-order determinants according to our definition.

Example 3 Evaluate the determinant $\begin{vmatrix} 2 & 3 & 6 \\ 4 & -2 & 0 \\ 1 & -5 & -3 \end{vmatrix}$ by expanding it by minors of elements in the first column.

Solution

$$\begin{vmatrix} 2 & 3 & 6 \\ 4 & -2 & 0 \\ 1 & -5 & -3 \end{vmatrix} = 2\begin{vmatrix} -2 & 0 \\ -5 & -3 \end{vmatrix} - 4\begin{vmatrix} 3 & 6 \\ -5 & -3 \end{vmatrix} + 1\begin{vmatrix} 3 & 6 \\ -2 & 0 \end{vmatrix}$$

$$\begin{aligned} &= 2[(-2)(-3) - (-5)(0)] - 4[(3)(-3) - (-5)(6)] + 1[(3)(0) - (-2)(6)] \\ &= 2(6 - 0) - 4[-9 - (-30)] + 1[0 - (-12)] \\ &= 2(6) - 4(21) + 1(12) \\ &= 12 - 84 + 12 \\ &= -60 \end{aligned}$$

□

Student Practice 3

Evaluate the determinant. $\begin{vmatrix} 1 & 2 & -3 \\ 2 & -1 & 2 \\ 3 & 1 & 4 \end{vmatrix}$

3 Solving a System of Two Linear Equations with Two Unknowns Using Cramer's Rule

We can solve a linear system of two equations with two unknowns by Cramer's rule. The rule is named for Gabriel Cramer, a Swiss mathematician who lived from 1704 to 1752. Cramer's rule expresses the solution for each variable of a linear system as the quotient of two determinants. Computer programs are available to solve systems of equations by Cramer's rule.

CRAMER'S RULE

The solution to

$$\begin{aligned} a_1x + b_1y &= c_1 \\ a_2x + b_2y &= c_2 \end{aligned}$$

is

$$x = \frac{D_x}{D} \quad \text{and} \quad y = \frac{D_y}{D}, \quad D \neq 0,$$

where

$$D_x = \begin{vmatrix} c_1 & b_1 \\ c_2 & b_2 \end{vmatrix}, \quad D_y = \begin{vmatrix} a_1 & c_1 \\ a_2 & c_2 \end{vmatrix}, \quad \text{and} \quad D = \begin{vmatrix} a_1 & b_1 \\ a_2 & b_2 \end{vmatrix}.$$

Example 4 Solve by Cramer's rule.

$$\begin{aligned} -3x + y &= 7 \\ -4x - 3y &= 5 \end{aligned}$$

Solution

$$D = \begin{vmatrix} -3 & 1 \\ -4 & -3 \end{vmatrix} \qquad D_x = \begin{vmatrix} 7 & 1 \\ 5 & -3 \end{vmatrix} \qquad D_y = \begin{vmatrix} -3 & 7 \\ -4 & 5 \end{vmatrix}$$

$$\begin{aligned} &= (-3)(-3) - (-4)(1) \\ &= 9 - (-4) \\ &= 9 + 4 \\ &= 13 \end{aligned} \qquad \begin{aligned} &= (7)(-3) - (5)(1) \\ &= -21 - 5 \\ &= -26 \end{aligned} \qquad \begin{aligned} &= (-3)(5) - (-4)(7) \\ &= -15 - (-28) \\ &= -15 + 28 \\ &= 13 \end{aligned}$$

Hence,

$$x = \frac{D_x}{D} = \frac{-26}{13} = -2$$

$$y = \frac{D_y}{D} = \frac{13}{13} = 1.$$

The solution to the system is $x = -2$ and $y = 1$. Verify this. □

Student Practice 4 Solve by Cramer's rule.

$$\begin{aligned} 5x + 3y &= 17 \\ 2x - 5y &= 13 \end{aligned}$$

4 Solving a System of Three Linear Equations with Three Unknowns Using Cramer's Rule

It is quite easy to extend Cramer's rule to three linear equations.

CRAMER'S RULE

The solution to the system

$$\begin{aligned} a_1x + b_1y + c_1z &= d_1 \\ a_2x + b_2y + c_2z &= d_2 \\ a_3x + b_3y + c_3z &= d_3 \end{aligned}$$

is

$$x = \frac{D_x}{D}, \quad y = \frac{D_y}{D}, \quad \text{and} \quad z = \frac{D_z}{D}, \quad D \neq 0,$$

where

$$D = \begin{vmatrix} a_1 & b_1 & c_1 \\ a_2 & b_2 & c_2 \\ a_3 & b_3 & c_3 \end{vmatrix}, \qquad D_x = \begin{vmatrix} d_1 & b_1 & c_1 \\ d_2 & b_2 & c_2 \\ d_3 & b_3 & c_3 \end{vmatrix},$$

$$D_y = \begin{vmatrix} a_1 & d_1 & c_1 \\ a_2 & d_2 & c_2 \\ a_3 & d_3 & c_3 \end{vmatrix}, \quad \text{and} \quad D_z = \begin{vmatrix} a_1 & b_1 & d_1 \\ a_2 & b_2 & d_2 \\ a_3 & b_3 & d_3 \end{vmatrix}.$$

Example 5 Use Cramer's rule to solve the system.

$$\begin{aligned} 2x - y + z &= 6 \\ 3x + 2y - z &= 5 \\ 2x + 3y - 2z &= 1 \end{aligned}$$

Solution We will expand each determinant by the first column.

$$D = \begin{vmatrix} 2 & -1 & 1 \\ 3 & 2 & -1 \\ 2 & 3 & -2 \end{vmatrix}$$

$$= 2\begin{vmatrix} 2 & -1 \\ 3 & -2 \end{vmatrix} - 3\begin{vmatrix} -1 & 1 \\ 3 & -2 \end{vmatrix} + 2\begin{vmatrix} -1 & 1 \\ 2 & -1 \end{vmatrix}$$

$$= 2[-4 - (-3)] - 3(2 - 3) + 2(1 - 2)$$

$$= 2(-1) - 3(-1) + 2(-1)$$

$$= -2 + 3 - 2$$

$$D = -1$$

Graphing Calculator

Copying Matrices

If you are using a graphing calculator to evaluate the four determinants in Example 5 or similar exercises, first enter matrix D into the calculator. Then copy the matrix using the copy function to three additional locations. Usually we store matrix D as matrix A. Then store a copy of it as matrices B, C, and D. Finally, use the Edit function and modify one column of each of matrices B, C, and D so that they become D_x, D_y, and D_z. This allows you to evaluate all four determinants in a minimum amount of time.

$$D_x = \begin{vmatrix} 6 & -1 & 1 \\ 5 & 2 & -1 \\ 1 & 3 & -2 \end{vmatrix}$$

$$= 6\begin{vmatrix} 2 & -1 \\ 3 & -2 \end{vmatrix} - 5\begin{vmatrix} -1 & 1 \\ 3 & -2 \end{vmatrix} + 1\begin{vmatrix} -1 & 1 \\ 2 & -1 \end{vmatrix}$$

$$= 6[-4 - (-3)] - 5(2 - 3) + 1(1 - 2)$$

$$= 6(-1) - 5(-1) + 1(-1)$$

$$= -6 + 5 - 1$$

$$D_x = -2$$

$$D_y = \begin{vmatrix} 2 & 6 & 1 \\ 3 & 5 & -1 \\ 2 & 1 & -2 \end{vmatrix}$$

$$= 2\begin{vmatrix} 5 & -1 \\ 1 & -2 \end{vmatrix} - 3\begin{vmatrix} 6 & 1 \\ 1 & -2 \end{vmatrix} + 2\begin{vmatrix} 6 & 1 \\ 5 & -1 \end{vmatrix}$$

$$= 2[-10 - (-1)] - 3(-12 - 1) + 2(-6 - 5)$$

$$= 2(-9) - 3(-13) + 2(-11)$$

$$= -18 + 39 - 22$$

$$D_y = -1$$

$$D_z = \begin{vmatrix} 2 & -1 & 6 \\ 3 & 2 & 5 \\ 2 & 3 & 1 \end{vmatrix}$$

$$= 2\begin{vmatrix} 2 & 5 \\ 3 & 1 \end{vmatrix} - 3\begin{vmatrix} -1 & 6 \\ 3 & 1 \end{vmatrix} + 2\begin{vmatrix} -1 & 6 \\ 2 & 5 \end{vmatrix}$$

$$= 2(2 - 15) - 3(-1 - 18) + 2(-5 - 12)$$

$$= 2(-13) - 3(-19) + 2(-17)$$

$$= -26 + 57 - 34$$

$$D_z = -3$$

$$x = \frac{D_x}{D} = \frac{-2}{-1} = 2; \qquad y = \frac{D_y}{D} = \frac{-1}{-1} = 1; \qquad z = \frac{D_z}{D} = \frac{-3}{-1} = 3$$ ◻

 **Student Practice 5** Find the solution to the system by Cramer's rule.

$$
\begin{aligned}
2x + 3y - z &= -1\\
3x + 5y - 2z &= -3\\
x + 2y + 3z &= 2
\end{aligned}
$$

Cramer's rule cannot be used for every system of linear equations. If the equations are dependent or if the system of equations is inconsistent, the determinant of coefficients (D) will be zero. Division by zero is not defined. In such a situation the system will not have a unique answer.

If $D = 0$, then the following are true:

1. If $D_x = 0$ and $D_y = 0$ (and $D_z = 0$, if there are three equations), then the equations are *dependent*. Such a system will have an infinite number of solutions.
2. If at least one of D_x or D_y (or D_z if there are three equations) is nonzero, then the system of equations is *inconsistent*. Such a system will have no solution.

Appendix C Exercises MyMathLab®

Evaluate each determinant.

1. $\begin{vmatrix} 5 & 6 \\ 2 & 1 \end{vmatrix}$

2. $\begin{vmatrix} 3 & 4 \\ 1 & 8 \end{vmatrix}$

3. $\begin{vmatrix} 2 & -1 \\ 3 & 6 \end{vmatrix}$

4. $\begin{vmatrix} -4 & 2 \\ 1 & 5 \end{vmatrix}$

5. $\begin{vmatrix} -\frac{1}{2} & -\frac{2}{3} \\ 9 & 8 \end{vmatrix}$

6. $\begin{vmatrix} 10 & 4 \\ -\frac{3}{2} & -\frac{2}{5} \end{vmatrix}$

7. $\begin{vmatrix} -5 & 3 \\ -4 & -7 \end{vmatrix}$

8. $\begin{vmatrix} 2 & -3 \\ -4 & -6 \end{vmatrix}$

9. $\begin{vmatrix} 0 & -6 \\ 3 & -4 \end{vmatrix}$

10. $\begin{vmatrix} -5 & 0 \\ 2 & -7 \end{vmatrix}$

11. $\begin{vmatrix} 2 & -5 \\ -4 & 10 \end{vmatrix}$

12. $\begin{vmatrix} -3 & 6 \\ 7 & -14 \end{vmatrix}$

13. $\begin{vmatrix} 0 & 0 \\ -2 & 6 \end{vmatrix}$

14. $\begin{vmatrix} -4 & 0 \\ -3 & 0 \end{vmatrix}$

15. $\begin{vmatrix} 0.3 & 0.6 \\ 1.2 & 0.4 \end{vmatrix}$

16. $\begin{vmatrix} 0.1 & 0.7 \\ 0.5 & 0.8 \end{vmatrix}$

17. $\begin{vmatrix} 7 & 4 \\ b & -a \end{vmatrix}$

18. $\begin{vmatrix} \frac{1}{4} & \frac{3}{5} \\ \frac{2}{3} & \frac{1}{5} \end{vmatrix}$

19. $\begin{vmatrix} \frac{3}{7} & -\frac{1}{3} \\ -\frac{1}{4} & \frac{1}{2} \end{vmatrix}$

20. $\begin{vmatrix} -3 & y \\ -2 & x \end{vmatrix}$

In the determinant $\begin{vmatrix} 3 & -4 & 7 \\ -2 & 6 & 10 \\ 1 & -5 & 9 \end{vmatrix}$,

21. Find the minor of 3.

22. Find the minor of −2.

23. Find the minor of 10.

24. Find the minor of 9.

Evaluate each of the following determinants.

25. $\begin{vmatrix} 4 & 1 & 2 \\ 3 & -1 & 0 \\ 1 & 2 & 3 \end{vmatrix}$

26. $\begin{vmatrix} 2 & 3 & 1 \\ -3 & 1 & 0 \\ 2 & 1 & 4 \end{vmatrix}$

27. $\begin{vmatrix} -4 & 0 & -1 \\ 2 & 1 & -1 \\ 0 & 3 & 2 \end{vmatrix}$

28. $\begin{vmatrix} 3 & -4 & -1 \\ -2 & 1 & 3 \\ 0 & 1 & 4 \end{vmatrix}$

29. $\begin{vmatrix} \frac{1}{2} & 1 & -1 \\ \frac{3}{2} & 1 & 2 \\ 3 & 0 & -2 \end{vmatrix}$

30. $\begin{vmatrix} 1 & 2 & 3 \\ 4 & -2 & -1 \\ 5 & -3 & 2 \end{vmatrix}$

31. $\begin{vmatrix} 4 & 1 & 2 \\ -1 & -2 & -3 \\ 4 & -1 & 3 \end{vmatrix}$

32. $\begin{vmatrix} -\frac{1}{2} & 2 & 3 \\ \frac{5}{2} & -2 & -1 \\ \frac{3}{4} & -3 & 2 \end{vmatrix}$

33. $\begin{vmatrix} 2 & 0 & -2 \\ -1 & 0 & 2 \\ 3 & 4 & 3 \end{vmatrix}$

34. $\begin{vmatrix} 7 & 0 & 2 \\ 1 & 0 & -5 \\ 3 & 0 & 6 \end{vmatrix}$

35. $\begin{vmatrix} 6 & -4 & 3 \\ 1 & 2 & 4 \\ 0 & 0 & 0 \end{vmatrix}$

36. $\begin{vmatrix} 7 & 0 & 3 \\ 1 & 2 & 4 \\ 3 & 0 & -7 \end{vmatrix}$

Optional Graphing Calculator Problems *If you have a graphing calculator, use the determinant function to evaluate the following:*

37. $\begin{vmatrix} 1.3 & 1.8 & 2.5 \\ 7.9 & 5.3 & 6.0 \\ 1.7 & 1.8 & 2.8 \end{vmatrix}$

38. $\begin{vmatrix} 0.7 & 5.3 & 0.4 \\ 1.6 & 0.3 & 3.7 \\ 0.8 & 6.7 & 4.2 \end{vmatrix}$

39. $\begin{vmatrix} -55 & 17 & 19 \\ -62 & 23 & 31 \\ 81 & 51 & 74 \end{vmatrix}$

40. $\begin{vmatrix} 82 & -20 & 56 \\ 93 & -18 & 39 \\ 65 & -27 & 72 \end{vmatrix}$

Solve each system by Cramer's rule.

41. $x + 2y = 8$
$2x + y = 7$

42. $x + 3y = 6$
$2x + y = 7$

43. $5x + 4y = 10$
$-x + 2y = 12$

44. $3x + 5y = 11$
$2x + y = -2$

45. $x - 5y = 0$
$x + 6y = 22$

46. $x - 3y = 4$
$-3x + 4y = -12$

47. $0.3x + 0.5y = 0.2$
$0.1x + 0.2y = 0.0$

48. $0.5x + 0.3y = -0.7$
$0.4x + 0.5y = -0.3$

Solve by Cramer's rule. Round your answers to four decimal places.

49. $52.9634x - 27.3715y = 86.1239$
$31.9872x + 61.4598y = 44.9812$

50. $0.0076x + 0.0092y = 0.01237$
$-0.5628x - 0.2374y = -0.7635$

Solve each system by Cramer's rule.

51. $2x + y + z = 4$
$x - y - 2z = -2$
$x + y - z = 1$

52. $x + 2y - z = -4$
$x + 4y - 2z = -6$
$2x + 3y + z = 3$

53. $2x + 2y + 3z = 6$
$x - y + z = 1$
$3x + y + z = 1$

54. $4x + y + 2z = 6$
$x + y + z = 1$
$-x + 3y - z = -5$

55. $x + 2y + z = 1$
$3x - 4z = 8$
$3y + 5z = -1$

56. $3x + y + z = 2$
$2y + 3z = -6$
$2x - y = -1$

Optional Graphing Calculator Problems

Round your answers to the nearest thousandth.

57. $10x + 20y + 10z = -2$
$-24x - 31y - 11z = -12$
$61x + 39y + 28z = -45$

58. $121x + 134y + 101z = 146$
$315x - 112y - 108z = 426$
$148x + 503y + 516z = -127$

59. $28w + 35x - 18y + 40z = 60$
$60w + 32x + 28y = 400$
$30w + 15x + 18y + 66z = 720$
$26w - 18x - 15y + 75z = 125$

Appendix D Solving Systems of Linear Equations Using Matrices

Student Learning Objective

After studying this section, you will be able to:

1 Solve a system of linear equations using matrices.

1 Solving a System of Linear Equations Using Matrices 

In Appendix C we defined a matrix as any rectangular array of numbers that is arranged in rows and columns.

$$\begin{bmatrix} 2 & 3 \\ 5 & 6 \end{bmatrix}$$ This is a 2×2 matrix with two rows and two columns.

$$\begin{bmatrix} 1 & -5 & -6 & 2 \\ 3 & 4 & -8 & -2 \\ 2 & 7 & 9 & -4 \end{bmatrix}$$ This is a 3×4 matrix with three rows and four columns.

A matrix that is derived from a system of linear equations is called the **augmented matrix** of the system. This augmented matrix is made up of two smaller matrices separated by a vertical line. The coefficients of each variable in the linear system are placed to the left of the vertical line. The constants are placed to the right of the vertical line.

The augmented matrix for the system of equations

$$\begin{aligned} -3x + 5y &= -22 \\ 2x - y &= 10 \end{aligned}$$

is the 2×3 matrix

$$\left[\begin{array}{cc|c} -3 & 5 & -22 \\ 2 & -1 & 10 \end{array}\right].$$

The augmented matrix for the system of equations

$$\begin{aligned} 3x - 5y + 2z &= 8 \\ x + y + z &= 3 \\ 3x - 2y + 4z &= 10 \end{aligned}$$

is the 3×4 matrix

$$\left[\begin{array}{ccc|c} 3 & -5 & 2 & 8 \\ 1 & 1 & 1 & 3 \\ 3 & -2 & 4 & 10 \end{array}\right].$$

Example 1 Write the solution to the system of linear equations represented by the following matrix.

$$\left[\begin{array}{cc|c} 1 & -3 & -7 \\ 0 & 1 & 4 \end{array}\right]$$

Solution The matrix represents the system of equations

$$\begin{aligned} x - 3y &= -7 \quad \text{and} \\ 0x + y &= 4. \end{aligned}$$

Since we know that $y = 4$, we can find x by substitution.

$$\begin{aligned} x - 3y &= -7 \\ x - 3(4) &= -7 \\ x - 12 &= -7 \\ x &= 5 \end{aligned}$$

Thus, the solution to the system is $x = 5$; $y = 4$. We can also write the solution as $(5, 4)$. □

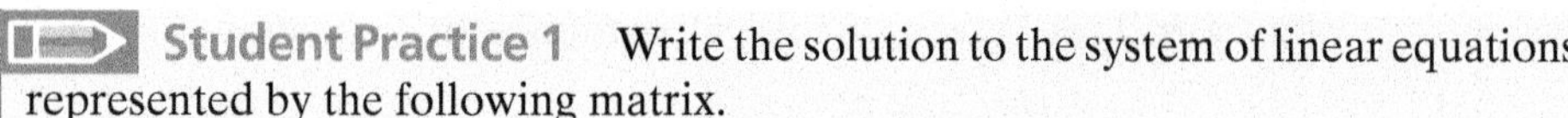

Student Practice 1 Write the solution to the system of linear equations represented by the following matrix.

$$\left[\begin{array}{cc|c} 1 & 9 & 33 \\ 0 & 1 & 3 \end{array}\right]$$

To solve a system of linear equations in matrix form, we use three row operations of the matrix.

MATRIX ROW OPERATIONS

1. Any two rows of a matrix may be interchanged.
2. All the numbers in a row may be multiplied or divided by any nonzero number.
3. All the numbers in any row or any multiple of a row may be added to the corresponding numbers of any other row.

To obtain the values for x and y in a system of two linear equations, we use row operations to obtain an augmented matrix in a form similar to the form of the matrix in Example 1.

The desired form is

$$\left[\begin{array}{cc|c} 1 & a & b \\ 0 & 1 & c \end{array}\right] \quad \text{or} \quad \left[\begin{array}{ccc|c} 1 & a & b & d \\ 0 & 1 & c & e \\ 0 & 0 & 1 & f \end{array}\right].$$

The last row of the matrix will allow us to find the value of one of the variables. We can then use substitution to find the other variables.

Example 2 Use matrices to solve the system.

$$\begin{aligned} 4x - 3y &= -13 \\ x + 2y &= 5 \end{aligned}$$

Solution The augmented matrix for this system of linear equations is

$$\left[\begin{array}{cc|c} 4 & -3 & -13 \\ 1 & 2 & 5 \end{array}\right].$$

First we want to obtain a 1 as the first element in the first row. We can obtain this by interchanging rows one and two.

$$\left[\begin{array}{cc|c} 1 & 2 & 5 \\ 4 & -3 & -13 \end{array}\right] \quad R_1 \longleftrightarrow R_2$$

Next we wish to obtain a 0 as the first element of the second row. To obtain this we multiply -4 by all the elements of row one and add this to row two.

$$\left[\begin{array}{cc|c} 1 & 2 & 5 \\ 0 & -11 & -33 \end{array}\right] \quad -4R_1 + R_2$$

Next, to obtain a 1 as the second element of the second row, we multiply each element of row two by $-\frac{1}{11}$.

$$\left[\begin{array}{cc|c} 1 & 2 & 5 \\ 0 & 1 & 3 \end{array}\right] \quad -\frac{1}{11}R_2$$

This final matrix is in the desired form. It represents the linear system

$$\begin{aligned} x + 2y &= 5 \\ y &= 3. \end{aligned}$$

Since we know that $y = 3$, we substitute this value into the first equation.

$$x + 2(3) = 5$$
$$x + 6 = 5$$
$$x = -1$$

Thus, the solution to the system is $(-1, 3)$. □

Student Practice 2 Use matrices to solve the system.

$$\begin{aligned} 3x - 2y &= -6 \\ x - 3y &= 5 \end{aligned}$$

Now we continue with a similar example involving three equations and three unknowns.

Example 3 Use matrices to solve the system.

$$\begin{aligned} 2x + 3y - z &= 11 \\ x + 2y + z &= 12 \\ 3x - y + 2z &= 5 \end{aligned}$$

Solution The augmented matrix that represents this system of linear equations is

$$\left[\begin{array}{ccc|c} 2 & 3 & -1 & 11 \\ 1 & 2 & 1 & 12 \\ 3 & -1 & 2 & 5 \end{array}\right].$$

To obtain a 1 as the first element of the first row, we interchange the first and second rows.

$$\left[\begin{array}{ccc|c} 1 & 2 & 1 & 12 \\ 2 & 3 & -1 & 11 \\ 3 & -1 & 2 & 5 \end{array}\right] \quad R_1 \longleftrightarrow R_2$$

Now, in order to obtain a 0 as the first element of the second row, we multiply row one by -2 and add the result to row two. In order to obtain a 0 as the first element of the third row, we multiply row one by -3 and add the result to row three.

$$\left[\begin{array}{ccc|c} 1 & 2 & 1 & 12 \\ 0 & -1 & -3 & -13 \\ 0 & -7 & -1 & -31 \end{array}\right] \quad \begin{array}{l} -2R_1 + R_2 \\ -3R_1 + R_3 \end{array}$$

To obtain a 1 as the second element of row two, we multiply all the elements of row two by -1.

$$\left[\begin{array}{ccc|c} 1 & 2 & 1 & 12 \\ 0 & 1 & 3 & 13 \\ 0 & -7 & -1 & -31 \end{array}\right] \quad -1R_2$$

Next, in order to obtain a 0 as the second element of row three, we add 7 times row two to row three.

$$\left[\begin{array}{ccc|c} 1 & 2 & 1 & 12 \\ 0 & 1 & 3 & 13 \\ 0 & 0 & 20 & 60 \end{array}\right] \quad 7R_2 + R_3$$

Finally, we multiply all the elements of row three by $\frac{1}{20}$. Thus, we have the following:

$$\left[\begin{array}{ccc|c} 1 & 2 & 1 & 12 \\ 0 & 1 & 3 & 13 \\ 0 & 0 & 1 & 3 \end{array}\right] \quad \frac{1}{20}R_3$$

From the final line of the matrix, we see that $z = 3$. If we substitute this value into the equation represented by the second row, we have

$$\begin{aligned} y + 3z &= 13 \\ y + 3(3) &= 13 \\ y + 9 &= 13 \\ y &= 4. \end{aligned}$$

Now we substitute the values obtained for y and z into the equation represented by the first row of the matrix.

$$\begin{aligned} x + 2y + z &= 12 \\ x + 2(4) + 3 &= 12 \\ x + 8 + 3 &= 12 \\ x + 11 &= 12 \\ x &= 1 \end{aligned}$$

Thus, the solution to this linear system of three equations is (1, 4, 3). □

 **Student Practice 3** Use matrices to solve the system.

$$\begin{aligned} 2x + y - 2z &= -15 \\ 4x - 2y + z &= 15 \\ x + 3y + 2z &= -5 \end{aligned}$$

We could continue to use these row operations to obtain an augmented matrix of the form

$$\left[\begin{array}{cc|c} 1 & 0 & a \\ 0 & 1 & b \end{array}\right] \quad \text{or} \quad \left[\begin{array}{ccc|c} 1 & 0 & 0 & a \\ 0 & 1 & 0 & b \\ 0 & 0 & 1 & c \end{array}\right].$$

This form of the augmented matrix is given a special name. It is known as the **reduced row echelon form.** If the augmented matrix of a system of linear equations is placed in this form, we would immediately know the solution to the system. Thus, if a system of linear equations in the variables x, y, and z has an augmented matrix that could be placed in the form

$$\left[\begin{array}{ccc|c} 1 & 0 & 0 & 7 \\ 0 & 1 & 0 & 32 \\ 0 & 0 & 1 & 18 \end{array}\right],$$

we could determine directly that $x = 7$, $y = 32$, and $z = 18$. A similar pattern is obtained for a system of four equations in four unknowns, and so on. Thus, if a system of linear equations in the variables w, x, y, and z has an augmented matrix that could be placed in the form

$$\left[\begin{array}{cccc|c} 1 & 0 & 0 & 0 & 23.4 \\ 0 & 1 & 0 & 0 & 48.6 \\ 0 & 0 & 1 & 0 & 0.73 \\ 0 & 0 & 0 & 1 & 5.97 \end{array}\right],$$

we could directly conclude that $w = 23.4$, $x = 48.6$, $y = 0.73$, and $z = 5.97$. Reducing a matrix to reduced row echelon form is readily done on computers. Many mathematical software packages contain matrix operations that will obtain the reduced row echelon form of an augmented matrix. A number of graphing calculators such as the TI-84 can be used to obtain the reduced row echelon form by using the **rref** command on a given matrix.

Graphing Calculator

Obtaining a Reduced Row Echelon Form of an Augmented Matrix

If your graphing calculator has a routine to obtain the **reduced row echelon form** of a matrix **(rref)**, then this routine will allow you to quickly obtain the solution of a system of linear equations if one exists. If your calculator has this capability, solve the following system.

$$\begin{aligned} 5w + 2x + 3y + 4z &= -8.3 \\ -4w + 3x + 2y + 7z &= -70.1 \\ 6w + x + 4y + 5z &= -13.3 \\ 7w + 4x + y + 2z &= 14.1 \end{aligned}$$

Ans:

$w = 3.1$, $x = 2.2$,

$y = 4.6$, $z = -10.5$

Appendix D Exercises MyMathLab®

Solve each system of equations by the matrix method. Round your answers to the nearest tenth when necessary.

1. $\begin{aligned} 2x + 3y &= 5 \\ 5x + y &= 19 \end{aligned}$

2. $\begin{aligned} 3x + 5y &= -15 \\ 2x + 7y &= -10 \end{aligned}$

3. $\begin{aligned} 2x + y &= -3 \\ 5x - y &= 24 \end{aligned}$

4. $\begin{aligned} x + 5y &= -9 \\ 4x - 3y &= -13 \end{aligned}$

5. $\begin{aligned} 5x + 2y &= 6 \\ 3x + 4y &= 12 \end{aligned}$

6. $\begin{aligned} -5x + y &= 24 \\ x + 5y &= 10 \end{aligned}$

7. $\begin{aligned} 3x - 2y + 3 &= 5 \\ x + 4y - 1 &= 9 \end{aligned}$

8. $\begin{aligned} 3x + y - 4 &= 12 \\ -2x + 3y + 2 &= -5 \end{aligned}$

9. $\begin{aligned} -7x + 3y &= 2.7 \\ 6x + 5y &= 25.7 \end{aligned}$

10. $\begin{aligned} x - 2y - 3z &= 4 \\ 2x + 3y + z &= 1 \\ -3x + y - 2z &= 5 \end{aligned}$

11. $\begin{aligned} x + y - z &= -2 \\ 2x - y + 3z &= 19 \\ 4x + 3y - z &= 5 \end{aligned}$

12. $\begin{aligned} 5x - y + 4z &= 5 \\ 6x + y - 5z &= 17 \\ 2x - 3y + z &= -11 \end{aligned}$

13. $\begin{aligned} x + y - z &= -3 \\ x + y + z &= 3 \\ 3x - y + z &= 7 \end{aligned}$

14. $\begin{aligned} 2x - y + z &= 5 \\ x + 2y - z &= -2 \\ x + y - 2z &= -5 \end{aligned}$

15. $\begin{aligned} 2x - 3y + z &= 11 \\ x + y + 2z &= 8 \\ x + 3y - z &= -11 \end{aligned}$

16. $\begin{aligned} 4x + 3y + 5z &= 2 \\ 2y + 7z &= 16 \\ 2x - y &= 6 \end{aligned}$

17. $\begin{aligned} 6x - y + z &= 9 \\ 2x + 3z &= 16 \\ 4x + 7y + 5z &= 20 \end{aligned}$

18. $\begin{aligned} 3x + 2y &= 44 \\ 4y + 3z &= 19 \\ 2x + 3z &= -5 \end{aligned}$

Optional Graphing Calculator Problems *If your graphing calculator has the necessary capability, solve the following exercises.*

19. $\begin{aligned} 5x + 6y + 7z &= 45.6 \\ 1.4x - 3.2y + 1.6z &= 3.12 \\ 9x - 8y + 22z &= 70.8 \end{aligned}$

20. $\begin{aligned} 2x + 12y + 9z &= 37.9 \\ 1.6x + 1.8y - 2.5z &= -20.53 \\ 7x + 8y + 4z &= 39.6 \end{aligned}$

21. $\begin{aligned} 6w + 5x + 3y + 1.5z &= 41.7 \\ 2w + 6.7x - 5y + 7z &= -21.92 \\ 12w + x + 5y - 6z &= 58.4 \\ 3w + 8x - 15y + z &= -142.8 \end{aligned}$

22. $\begin{aligned} 2w + 3x + 11y - 14z &= 6.7 \\ 5w + 8x + 7y + 3z &= 25.3 \\ -4w + x + 1.5y - 9z &= -53.4 \\ 9w + 7x - 2.5y + 6z &= 22.9 \end{aligned}$

Appendix E Sets

1 Writing a Set in Roster Form

Student Learning Objectives

After studying this section, you will be able to:

1 Write a set in roster form.

2 Write a set in set-builder notation.

3 Find the union and intersection of sets.

4 Identify subsets.

Set theory is the basis of several mathematical topics. Sorting and classifying objects into categories is something we do every day. You may organize your closet so all your sweaters are together. When you go through your mail, you may separate bills from junk mail.

A **set** is a collection of objects called **elements.** Numbers can be classified into several different sets. The natural numbers, for example, are the set of whole numbers excluding 0. Prime numbers make up another set. They are the set of natural numbers greater than 1 whose only natural number factors are 1 and itself. We can write these sets the following way.

$$N = \{1, 2, 3, 4, 5, \ldots\}$$
$$P = \{2, 3, 5, 7, 11, 13, \ldots\}$$

There are several things to notice. Capital letters are usually used to represent sets. When elements of a set are listed, they are separated by commas and enclosed by braces. When we list the elements of a set this way, we say the set is in **roster form.** The three dots in sets N and P indicate that the pattern of numbers continues.

To indicate an element is part of a set, we use the symbol $\in$. Since 17 is a prime number, we can write $17 \in P$. This is read "17 is an element of set P."

Example 1 Write in roster form.

(a) Set D is the set of Beatles.

(b) Set X is the set of natural numbers between 2 and 7.

(c) Set Y is the set of natural numbers between 2 and 7, inclusive.

Solution

(a) Writing set D in roster form, we have

$$D = \{\textit{John Lennon, Paul McCartney, George Harrison, Ringo Starr}\}.$$

(b) $X = \{3, 4, 5, 6\}$.

(c) The word inclusive means the numbers 2 and 7 are included.

$$Y = \{2, 3, 4, 5, 6, 7\}$$ □

Student Practice 1 Write in roster form.

(a) Set A is the set of continents on Earth.

(b) Set C is the set of natural numbers between 35 and 42.

(c) Set D is the set of natural numbers between 35 and 42, inclusive.

The sets in Example 1 are **finite sets**; we can count the number of elements. Set D in part (a) contains four elements and set Y in part (c) has six elements. The set of natural numbers $N = \{1, 2, 3, \ldots\}$ is an example of an **infinite set.** The list of numbers continues without bound; there are infinitely many elements. Some sets contain no elements and are called **empty sets.** The empty set is denoted by the symbol { } or $\varnothing$. The set of students in your class that are 11 feet tall is an empty set.

2 Writing a Set in Set-Builder Notation

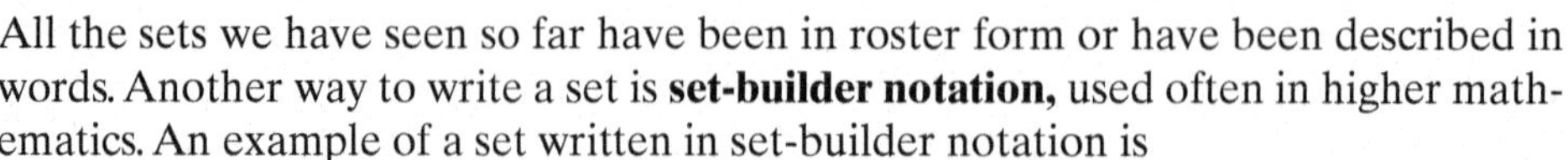

All the sets we have seen so far have been in roster form or have been described in words. Another way to write a set is **set-builder notation,** used often in higher mathematics. An example of a set written in set-builder notation is

$$A = \{x \mid x \text{ is a natural number greater than } 10\}.$$

We read this as "Set A is the set of all elements x such that x is a natural number greater than 10." We also could have written $A = \{x \mid x \in N \text{ and } x > 10\}$. $x \in N$ means x is an element of the natural numbers.

Let's look at each part of set-builder notation and its meaning.

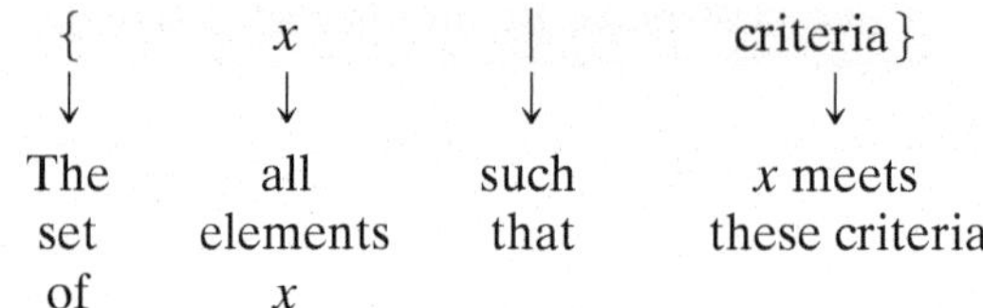

Example 2 Write set B in set-builder notation. $B = \{a, e, i, o, u\}$

Solution For an element to be in set B, it must be a vowel of the alphabet. We write $B = \{x \mid x \text{ is a vowel}\}$. Notice what is written to the right of the bar. We don't describe the set in words here. We simply indicate what criteria an element must meet to be in the set. $B = \{x \mid x \text{ is the set of vowels}\}$ is *not* correct. □

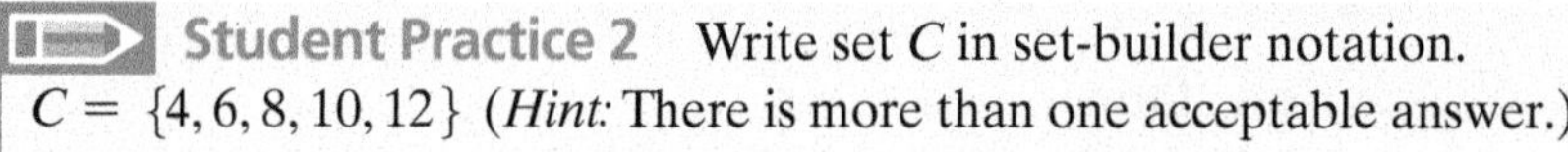

Student Practice 2 Write set C in set-builder notation. $C = \{4, 6, 8, 10, 12\}$ (*Hint:* There is more than one acceptable answer.)

3 Finding the Union and Intersection of Sets

Addition, subtraction, multiplication, and division are operations used on numbers. There are other operations used on sets. The two most common operations are union and intersection.

The **union** of two sets A and B, written $A \cup B$, is the set of elements that are in set A *or* set B.

Example 3 Find $A \cup B$ if $A = \{a, b, c, d, e\}$ and $B = \{a, c, d, g\}$.

Solution To find the set $A \cup B$, we combine the elements of A with those of B. We have $A \cup B = \{a, b, c, d, e, g\}$. □

Student Practice 3 Find $G \cup H$ if $G = \{!, *, \%, \$\}$ and $H = \{\$, ?, \wedge, +\}$

The **intersection** of sets A and B, written $A \cap B$, is the set of elements in set A *and* set B.

Example 4 Find $A \cap B$, if $A = \{a, b, c, d, e\}$ and $B = \{a, c, d, g\}$.

Solution The elements that are common to sets A and B are a, c, and d. Thus $A \cap B = \{a, c, d\}$. □

Student Practice 4 Find $G \cap H$ if $G = \{!, *, \%, \$\}$ and $H = \{\$, ?, \wedge, +\}$.

We have talked about one important set of numbers, the natural numbers. There are several other sets of numbers that we summarize below.

Sets of Numbers	
Real numbers	$\{x \mid x \text{ can be placed on the number line}\}$
Natural numbers	$\{1, 2, 3, 4, 5, \ldots\}$
Whole numbers	$\{0, 1, 2, 3, 4, 5, \ldots\}$
Integers	$\{\ldots, -2, -1, 0, 1, 2, \ldots\}$
Rational numbers	$\{x \mid x \text{ can be written as } \frac{p}{q} \text{ where } p \text{ and } q \text{ are integers, and } q \neq 0\}$
Irrational numbers	$\{x \mid x \text{ is a real number that is not rational}\}$

You are probably more familiar with the terms natural numbers, whole numbers, and integers than the others. Let's look at rational, irrational, and real numbers in more detail.

A **rational number** is a number that can be written as a fraction (with a denominator not equal to 0). Here are some examples of rational numbers:

$$-5, \quad 3.54, \quad \sqrt{9}, \quad \frac{1}{4}$$

The first three numbers can be written as $\frac{-5}{1}, \frac{354}{100}$, and $\frac{3}{1}$, respectively, and so are considered rational numbers. Every integer is rational since it can be written with 1 in the denominator. When a number is written in decimal form, we can easily determine whether or not it is a rational number. If the decimal repeats or terminates, it is a rational number.

$\frac{1}{3} = 0.3333\ldots = 0.\overline{3}$ and $\frac{13}{22} = 0.5909090\ldots = 0.5\overline{90}$ are repeating decimals

rational numbers

$\frac{3}{10} = 0.3$ and $-1\frac{9}{16} = -1.5625$ are terminating decimals

There are some numbers whose decimal representation is not a repeating or terminating decimal. For example, if we looked at the decimal forms of $\sqrt{2}$, $\sqrt{6}$, and $\sqrt{7}$, we would see that the decimal does not end and does not contain digits that repeat. These are **irrational numbers.** Pi (π) is another irrational number. When these numbers are used in calculations, we use approximations: $\sqrt{6} \approx 2.449$ and $\pi \approx 3.14$.

All the numbers we have discussed above can be placed on the number line. Some of them are shown below.

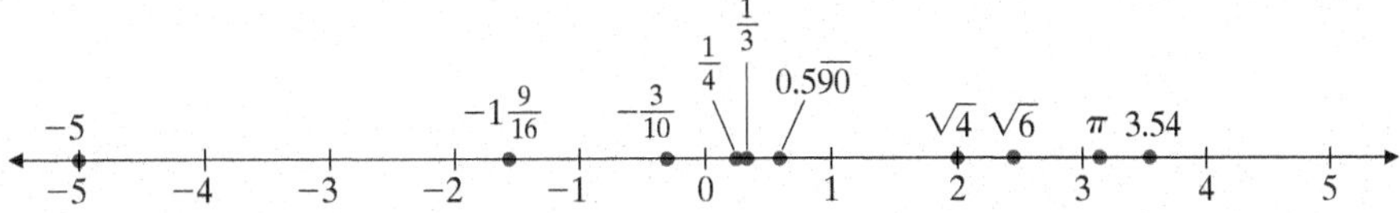

Any number that can be placed on the number line is a **real number.** The set of real numbers is the union of the rational and irrational numbers.

4 Identifying Subsets

We have seen that all integers are rational numbers and all rational numbers are real numbers. When all the elements of one set are contained in another set, we say that it is a **subset.** Here is a more formal definition.

Set A is a subset of set B, written $A \subseteq B$, if all elements in A are also in B.

Consider the sets $A = \{\text{Amy, Jack, Ron}\}$ and $B = \{\text{Amy, Harry, Lena, Jack, Ron}\}$. All three elements of set A are also in set B. Therefore, set A is a subset of set B, and we can write $A \subseteq B$.

Example 5 Determine if the statement is true or false. If false, state the reason.

(a) $A = \{t, v\}$ and $B = \{r, s, t, u, v, w\}$, so $A \subseteq B$.

(b) The set of integers is a subset of the natural numbers.

Solution

(a) True. All the elements of A are also in B.

(b) False. To see why, consider the number -3.
-3 is an element of the set of integers, but -3 is not a natural number. So the integers is not a subset of the natural numbers.

The natural numbers, however, is a subset of the set of integers. Do you see why? □

Student Practice 5 Determine if the statement is true or false. If false, state the reason.

(a) $C = \{a, b, c, d, e, f\}$ and $D = \{c, f\}$ so $C \subseteq D$.

(b) The set of whole numbers is a subset of the rational numbers.

The table below shows the relationship among the sets of numbers we have discussed.

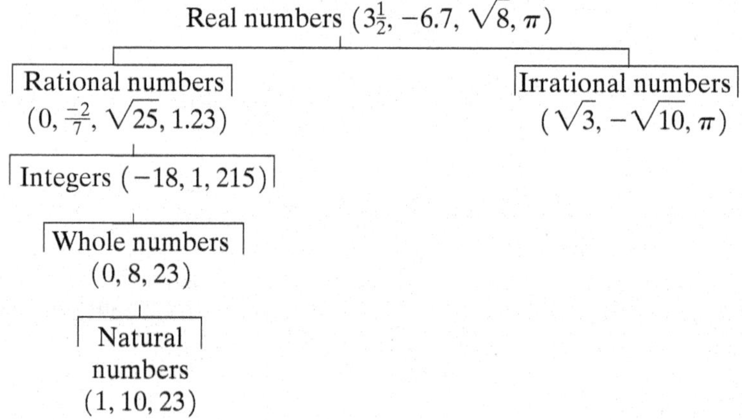

Appendix E Exercises MyMathLab®

Fill in the blank with appropriate word or words.

1. The objects of a set are called ________.

2. When the elements of a set are listed in braces, the set is in ________.

3. The ________ of two sets is the elements the sets have in common.

4. The symbol ________ means "intersection."

5. A set that contains no elements is called the ________.

6. If a set is ________, we can count the number of elements it contains.

In exercises 7–14, write the set in roster form.

7. The set of states in the United States that begin with the letter C.

8. $A = \{x \mid x \in N \text{ and } 4 < x < 10\}$

9. $C = \{x \mid x \in N \text{ and } x \text{ is odd}\}$

10. $M = \{x \mid x \in P \text{ and } 2 < x < 23\}$

11. $F = \{x \mid x \in N$ and x is a multiple of 5 between 10 and 50$\}$

12. $B = \{x \mid x$ is an integer between -3 and 2, inclusive$\}$

13. A is the set of integers less than 0.

14. The set of the last five months of the year

In exercises 15–20, write the set in set-builder notation.

15. $O = \{$Atlantic, Pacific, Indian, Arctic$\}$

16. $B = \{3, 6, 9, 12, 15\}$

17. $G = \{2, 4, 6, 8\}$

18. $K = \{31, 37, 41, 43\}$

19. $T = \{$scalene, isosceles, equilateral$\}$

20. C is the set of planets in our solar system

21. If $A = \{-1, 2, 3, 5, 8\}$ and $B = \{-2, -1, 3, 5\}$, find
(a) $A \cup B$ **(b)** $A \cap B$

22. If $C = \{d, e, f, g, h, i, j\}$ and $D = \{x, z\}$, find
(a) $C \cup D$ **(b)** $C \cap D$

Given sets A, B, and C, decide if the statements in exercises 23–34 are true or false. If it is false, give the reason.
$A = \{1, 2, 3, 4, 5, \dots\}$, $B = \{10, 20, 30, 40, \dots\}$, $C = \{1, 2, 3, 4, 5\}$

23. B is a finite set

24. $A \subseteq C$

25. $68 \in A$

26. $120 \in B$

27. $B \subseteq A$

28. $B \cap C = \{10, 20, 30, 40\}$

29. $A \cap C = \{\ \}$

30. $A \cup B = \{1, 2, 3, 4, 5, \dots\}$

31. A is the set of whole numbers

32. $B = \{x \mid x \in N$ and x is a multiple of 10 greater than $5\}$

33. $A \cup C$ is the set of natural numbers

34. $B \cap C = \{\ \}$

35. Give an example of a subset of A. $A = \{$Joe, Ann, Nina, Doug$\}$

36. Give an example of a set of which B is a subset. $B = \{$poodle, Irish setter, dachshund$\}$

Below is a table of the top 10 most popular boy and girl names for 1980 and 2000. Use the table to answer exercises 37 and 38. Source: www.cherishedmoments.com.

1980		2000	
Boy	**Girl**	**Boy**	**Girl**
1. Michael	Jennifer	Jacob	Emily
2. Jason	Jessica	Michael	Hannah
3. Christopher	Amanda	Matthew	Madison
4. David	Melissa	Joshua	Ashley
5. James	Sarah	Christopher	Sarah
6. Matthew	Nicole	Nicholas	Alexis
7. John	Heather	Andrew	Samantha
8. Joshua	Amy	Joseph	Jessica
9. Robert	Michelle	Daniel	Taylor
10. Daniel	Elizabeth	Tyler	Elizabeth

37. (a) Write set B in roster form. B is the set of the most popular boys' names in 1980 or 2000.

(b) Your answer to part (a) represents a(n) _____ of two sets.

38. (a) Write set G in roster form. G is the set of the most popular girls' names in 1980 and 2000.

(b) Your answer to part (a) represents a(n) __________ of two sets.

39. Decide which elements of the following set are whole numbers, natural numbers, integers, rational numbers, irrational numbers, or real numbers.

$$\left\{3.62, \sqrt{20}, \frac{-3}{11}, 15, \frac{22}{3}, 0, \sqrt{81}, -17\right\}$$

40. Is the set of whole numbers a subset of the real numbers? Explain.

41. Is the set of integers a subset of the whole numbers? Explain.

42. Which sets of numbers are subsets of the rational numbers?

43. List all sets of numbers of which the whole numbers are a subset.

Solutions to Student Practice Problems

Chapter 0 0.1 Student Practice

1. (a) $\frac{10}{16} = \frac{2 \times 5}{2 \times 2 \times 2 \times 2} = \frac{5}{2 \times 2 \times 2} = \frac{5}{8}$

(b) $\frac{24}{36} = \frac{2 \times 2 \times 2 \times 3}{2 \times 2 \times 3 \times 3} = \frac{2}{3}$

(c) $\frac{36}{42} = \frac{2 \times 2 \times 3 \times 3}{2 \times 3 \times 7} = \frac{2 \times 3}{7} = \frac{6}{7}$

2. (a) $\frac{4}{12} = \frac{2 \times 2 \times 1}{2 \times 2 \times 3} = \frac{1}{3}$ **(b)** $\frac{25}{125} = \frac{5 \times 5 \times 1}{5 \times 5 \times 5} = \frac{1}{5}$

(c) $\frac{73}{146} = \frac{73 \times 1}{73 \times 2} = \frac{1}{2}$

3. (a) $\frac{18}{6} = \frac{2 \times 3 \times 3}{2 \times 3 \times 1} = 3$ **(b)** $\frac{146}{73} = \frac{73 \times 2}{73 \times 1} = 2$

(c) $\frac{28}{7} = \frac{2 \times 2 \times 7}{7} = 2 \times 2 = 4$

4. 56 out of 154 $= \frac{56}{154} = \frac{2 \times 7 \times 2 \times 2}{2 \times 7 \times 11} = \frac{4}{11}$

5. (a) $\frac{12}{7} = 12 \div 7$ $\quad 7\overline{)12}$ quotient 1, -7, 5 Remainder

$\frac{12}{7} = 1\frac{5}{7}$

(b) $\frac{20}{5} = 20 \div 5$ $\quad 5\overline{)20}$ quotient 4, -20, 0 Remainder

$\frac{20}{5} = 4$

6. (a) $3\frac{2}{5} = \frac{(3 \times 5) + 2}{5} = \frac{15 + 2}{5} = \frac{17}{5}$

(b) $1\frac{3}{7} = \frac{(1 \times 7) + 3}{7} = \frac{7 + 3}{7} = \frac{10}{7}$

(c) $2\frac{6}{11} = \frac{(2 \times 11) + 6}{11} = \frac{22 + 6}{11} = \frac{28}{11}$

(d) $4\frac{2}{3} = \frac{(4 \times 3) + 2}{3} = \frac{12 + 2}{3} = \frac{14}{3}$

7. (a) $\frac{3}{8} = \frac{?}{24}$; $8 \times 3 = 24$ **(b)** $\frac{5}{6} = \frac{?}{30}$; $6 \times 5 = 30$

$\frac{3 \times 3}{8 \times 3} = \frac{9}{24}$ $\qquad \frac{5 \times 5}{6 \times 5} = \frac{25}{30}$

(c) $\frac{2}{7} = \frac{?}{56}$; $7 \times 8 = 56$

$\frac{2 \times 8}{7 \times 8} = \frac{16}{56}$

0.2 Student Practice

1. (a) $\frac{3}{6} + \frac{2}{6} = \frac{3 + 2}{6} = \frac{5}{6}$ **(b)** $\frac{3}{11} + \frac{8}{11} = \frac{3 + 8}{11} = \frac{11}{11} = 1$

(c) $\frac{1}{8} + \frac{2}{8} + \frac{1}{8} = \frac{1 + 2 + 1}{8} = \frac{4}{8} = \frac{1}{2}$

(d) $\frac{5}{9} + \frac{8}{9} = \frac{5 + 8}{9} = \frac{13}{9}$ or $1\frac{4}{9}$

2. (a) $\frac{11}{13} - \frac{6}{13} = \frac{11 - 6}{13} = \frac{5}{13}$

(b) $\frac{8}{9} - \frac{2}{9} = \frac{8 - 2}{9} = \frac{6}{9} = \frac{2}{3}$

3. The LCD is 56, since 56 is exactly divisible by 8 and 7. There is no smaller number that is exactly divisible by 8 and 7.

4. Find the LCD using prime factors.

$\frac{8}{35}$ and $\frac{6}{15}$

$35 = 7 \cdot 5$
$15 = \quad 5 \cdot 3$

LCD $= 7 \cdot 5 \cdot 3 = 105$

5. Find the LCD of $\frac{5}{12}$ and $\frac{7}{30}$.

$12 = 3 \cdot 2 \cdot 2$
$30 = 3 \cdot \quad 2 \cdot 5$

LCD $= 3 \cdot 2 \cdot 2 \cdot 5 = 60$

6. Find the LCD of $\frac{2}{27}$, $\frac{1}{18}$, and $\frac{5}{12}$.

$27 = 3 \cdot 3 \cdot 3$
$18 = \quad 3 \cdot 3 \cdot 2$
$12 = \quad\quad 3 \cdot 2 \cdot 2$

LCD $= 3 \cdot 3 \cdot 3 \cdot 2 \cdot 2 = 108$

7. From Student Practice 3, the LCD is 56.

$\frac{5}{7} = \frac{5 \times 8}{7 \times 8} = \frac{40}{56}$ $\qquad \frac{1}{8} = \frac{1 \times 7}{8 \times 7} = \frac{7}{56}$

$\frac{5}{7} + \frac{1}{8} = \frac{40}{56} + \frac{7}{56} = \frac{47}{56}$

$\frac{47}{56}$ of the farm fields were planted in corn or soybeans.

8. We can see by inspection that both 5 and 25 divide exactly into 50. Thus 50 is the LCD.

$\frac{4}{5} = \frac{4 \times 10}{5 \times 10} = \frac{40}{50}$ $\qquad \frac{6}{25} = \frac{6 \times 2}{25 \times 2} = \frac{12}{50}$

$\frac{4}{5} + \frac{6}{25} + \frac{1}{50} = \frac{40}{50} + \frac{12}{50} + \frac{1}{50} = \frac{40 + 12 + 1}{50} = \frac{53}{50}$ or $1\frac{3}{50}$

9. Add $\frac{1}{49} + \frac{3}{14}$

First find the LCD.

$49 = 7 \cdot 7$
$14 = \quad 7 \cdot 2$

LCD $= 7 \cdot 7 \cdot 2 = 98$

Then change to equivalent fractions and add.

$\frac{1}{49} = \frac{1 \times 2}{49 \times 2} = \frac{2}{98}$ $\qquad \frac{3}{14} = \frac{3 \times 7}{14 \times 7} = \frac{21}{98}$

$\frac{1}{49} + \frac{3}{14} = \frac{2}{98} + \frac{21}{98} = \frac{23}{98}$

10. $\frac{1}{12} - \frac{1}{30}$

First find the LCD.

$12 = 2 \cdot 2 \cdot 3$
$30 = \quad 2 \cdot 3 \cdot 5$

LCD $= 2 \cdot 2 \cdot 3 \cdot 5 = 60$

Then change to equivalent fractions and subtract.

$\frac{1}{12} = \frac{1 \times 5}{12 \times 5} = \frac{5}{60}$ $\qquad \frac{1}{30} = \frac{1 \times 2}{30 \times 2} = \frac{2}{60}$

$\frac{1}{12} - \frac{1}{30} = \frac{5}{60} - \frac{2}{60} = \frac{3}{60} = \frac{1}{20}$

11. $\frac{2}{3}+\frac{3}{4}-\frac{3}{8}$

First find the LCD.

$$\begin{array}{l} 3 = 3 \\ 4 = \;\; 2\cdot 2 \\ 8 = \;\; 2\cdot 2\cdot 2 \\ \quad\downarrow\;\downarrow\;\downarrow\;\downarrow \end{array}$$

LCD $= 3\cdot 2\cdot 2\cdot 2 = 24$

$\frac{2}{3}=\frac{2\times 8}{3\times 8}=\frac{16}{24}$ $\quad \frac{3}{4}=\frac{3\times 6}{4\times 6}=\frac{18}{24}$ $\quad \frac{3}{8}=\frac{3\times 3}{8\times 3}=\frac{9}{24}$

Now combine the fractions.

$\frac{2}{3}+\frac{3}{4}-\frac{3}{8}=\frac{16}{24}+\frac{18}{24}-\frac{9}{24}=\frac{25}{24}$ or $1\frac{1}{24}$

12. (a) The LCD of 3 and 5 is 15.

$1\frac{2}{3}=\frac{5}{3}=\frac{5\times 5}{3\times 5}=\frac{25}{15}$ $\quad 2\frac{4}{5}=\frac{14}{5}=\frac{14\times 3}{5\times 3}=\frac{42}{15}$

$1\frac{2}{3}+2\frac{4}{5}=\frac{25}{15}+\frac{42}{15}=\frac{67}{15}=4\frac{7}{15}$

(b) The LCD of 4 and 3 is 12.

$5\frac{1}{4}=\frac{21}{4}=\frac{21\times 3}{4\times 3}=\frac{63}{12}$ $\quad 2\frac{2}{3}=\frac{8}{3}=\frac{8\times 4}{3\times 4}=\frac{32}{12}$

$5\frac{1}{4}-2\frac{2}{3}=\frac{63}{12}-\frac{32}{12}=\frac{31}{12}=2\frac{7}{12}$

13.

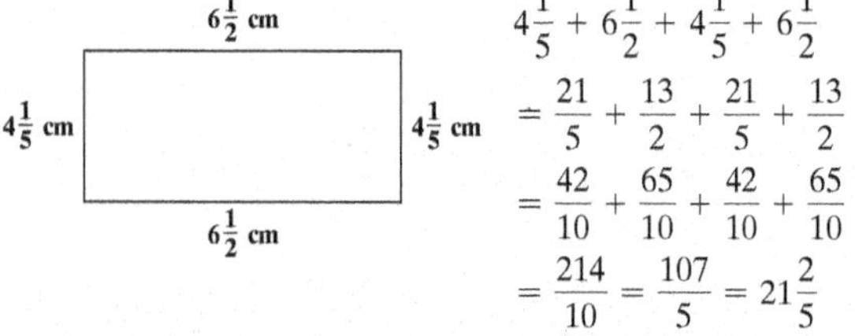

$4\frac{1}{5}+6\frac{1}{2}+4\frac{1}{5}+6\frac{1}{2}$

$=\frac{21}{5}+\frac{13}{2}+\frac{21}{5}+\frac{13}{2}$

$=\frac{42}{10}+\frac{65}{10}+\frac{42}{10}+\frac{65}{10}$

$=\frac{214}{10}=\frac{107}{5}=21\frac{2}{5}$

The perimeter is $21\frac{2}{5}$ cm.

0.3 Student Practice

1. (a) $\frac{2}{7}\times\frac{5}{11}=\frac{2\cdot 5}{7\cdot 11}=\frac{10}{77}$

(b) $\frac{1}{5}\times\frac{7}{10}=\frac{1\cdot 7}{5\cdot 10}=\frac{7}{50}$

(c) $\frac{9}{5}\times\frac{1}{4}=\frac{9\cdot 1}{5\cdot 4}=\frac{9}{20}$

(d) $\frac{8}{9}\times\frac{3}{10}=\frac{8\cdot 3}{9\cdot 10}=\frac{24}{90}=\frac{4}{15}$

2. (a) $\frac{3}{5}\times\frac{4}{3}=\frac{3\cdot 4}{5\cdot 3}=\frac{4\cdot 3}{5\cdot 3}=\frac{4}{5}$

(b) $\frac{9}{10}\times\frac{5}{12}=\frac{3\cdot 3}{2\cdot 5}\times\frac{5}{2\cdot 2\cdot 3}=\frac{3}{8}$

3. (a) $4\times\frac{2}{7}=\frac{4}{1}\times\frac{2}{7}=\frac{4\cdot 2}{1\cdot 7}=\frac{8}{7}$ or $1\frac{1}{7}$

(b) $12\times\frac{3}{4}=\frac{12}{1}\times\frac{3}{4}=\frac{4\cdot 3}{1}\times\frac{3}{4}=\frac{9}{1}=9$

4. Multiply. $5\frac{3}{5}$ times $3\frac{3}{4}$

$5\frac{3}{5}\times 3\frac{3}{4}=\frac{28}{5}\times\frac{15}{4}=\frac{4\cdot 7}{5}\times\frac{3\cdot 5}{4}=\frac{21}{1}=21$

The area of the field is 21 square miles.

5. $3\frac{1}{2}\times\frac{1}{14}\times 4=\frac{7}{2}\times\frac{1}{14}\times\frac{4}{1}=\frac{7}{2}\times\frac{1}{2\cdot 7}\times\frac{2\cdot 2}{1}=1$

6. (a) $\frac{2}{5}\div\frac{1}{3}=\frac{2}{5}\times\frac{3}{1}=\frac{6}{5}$ or $1\frac{1}{5}$

(b) $\frac{12}{13}\div\frac{4}{3}=\frac{12}{13}\times\frac{3}{4}=\frac{4\cdot 3}{13}\times\frac{3}{4}=\frac{9}{13}$

7. (a) $\frac{3}{7}\div 6=\frac{3}{7}\div\frac{6}{1}=\frac{3}{7}\times\frac{1}{6}=\frac{3}{7}\times\frac{1}{2\cdot 3}=\frac{1}{14}$

(b) $8\div\frac{2}{3}=\frac{8}{1}\div\frac{2}{3}=\frac{8}{1}\times\frac{3}{2}=\frac{2\cdot 4}{1}\times\frac{3}{2}=\frac{12}{1}=12$

8. (a) $\dfrac{\frac{3}{11}}{\frac{5}{7}}=\frac{3}{11}\div\frac{5}{7}=\frac{3}{11}\times\frac{7}{5}=\frac{21}{55}$

(b) $\dfrac{\frac{12}{5}}{\frac{8}{15}}=\frac{12}{5}\div\frac{8}{15}=\frac{12}{5}\times\frac{15}{8}=\frac{3\cdot 4}{5}\times\frac{3\cdot 5}{2\cdot 4}=\frac{9}{2}$ or $4\frac{1}{2}$

9. (a) $1\frac{2}{5}\div 2\frac{1}{3}=\frac{7}{5}\div\frac{7}{3}=\frac{7}{5}\times\frac{3}{7}=\frac{3}{5}$

(b) $4\frac{2}{3}\div 7=\frac{14}{3}\div\frac{7}{1}=\frac{14}{3}\times\frac{1}{7}=\frac{2\cdot 7}{3}\times\frac{1}{7}=\frac{2}{3}$

(c) $\dfrac{1\frac{1}{5}}{1\frac{2}{7}}=1\frac{1}{5}\div 1\frac{2}{7}=\frac{6}{5}\div\frac{9}{7}=\frac{6}{5}\times\frac{7}{9}=\frac{2\cdot 3}{5}\times\frac{7}{3\cdot 3}=\frac{14}{15}$

10. $64\div 5\frac{1}{3}=\frac{64}{1}\div\frac{16}{3}=\frac{64}{1}\times\frac{3}{16}=\frac{4\cdot 16}{1}\times\frac{3}{16}=\frac{12}{1}=12$

He can fill 12 jars.

11. $126\div 5\frac{1}{4}=\frac{126}{1}\div\frac{21}{4}=\frac{126}{1}\times\frac{4}{21}=\frac{6\cdot 21}{1}\times\frac{4}{21}=24$

The car got 24 miles per gallon.

0.4 Student Practice

1. (a) 1; $0.9=\frac{9}{10}$ = nine tenths

(b) 2; $0.09=\frac{9}{100}$ = nine hundredths

(c) 3; $0.731=\frac{731}{1000}$ = seven hundred thirty-one thousandths

(d) 3; $1.371=1\frac{371}{1000}$ = one and three hundred seventy-one thousandths

(e) 4; $0.0005=\frac{5}{10{,}000}$ = five ten-thousandths

2. (a) $\frac{3}{8}=0.375$

$$\begin{array}{r} 0.375 \\ 8\overline{)3.000} \\ \underline{2\,4} \\ 60 \\ \underline{56} \\ 40 \\ \underline{40} \\ 0 \end{array}$$

(b) $\frac{7}{200}=0.035$

$$\begin{array}{r} 0.035 \\ 200\overline{)7.000} \\ \underline{6\,00} \\ 1\,000 \\ \underline{1\,000} \\ 0 \end{array}$$

(c) $\frac{33}{20}=1.65$

$$\begin{array}{r} 1.65 \\ 20\overline{)33.00} \\ \underline{20} \\ 13\,0 \\ \underline{12\,0} \\ 1\,00 \\ \underline{1\,00} \\ 0 \end{array}$$

3. (a) $\frac{1}{6}=0.1666\ldots$ or $0.1\overline{6}$

$$\begin{array}{r} 0.166 \\ 6\overline{)1.000} \\ \underline{6} \\ 40 \\ \underline{36} \\ 40 \\ \underline{36} \\ 4 \end{array}$$

(b) $\frac{5}{11}=0.454545\ldots$ or $0.\overline{45}$

$$\begin{array}{r} 0.4545 \\ 11\overline{)5.0000} \\ \underline{4\,4} \\ 60 \\ \underline{55} \\ 50 \\ \underline{44} \\ 60 \\ \underline{55} \\ 5 \end{array}$$

4. (a) $0.8 = \frac{8}{10} = \frac{4}{5}$ **(b)** $0.88 = \frac{88}{100} = \frac{22}{25}$

(c) $0.45 = \frac{45}{100} = \frac{9}{20}$ **(d)** $0.148 = \frac{148}{1000} = \frac{37}{250}$

(e) $0.612 = \frac{612}{1000} = \frac{153}{250}$ **(f)** $0.016 = \frac{16}{1000} = \frac{2}{125}$

5. (a) $\begin{array}{r} 3.12 \\ +5.08 \\ \hline 8.20 \end{array}$ **(b)** $\begin{array}{r} 152.003 \\ -136.118 \\ \hline 15.885 \end{array}$ **(c)** $\begin{array}{r} 1.1 \\ 3.16 \\ +5.123 \\ \hline 9.383 \end{array}$ **(d)** $\begin{array}{r} 1.0052 \\ -0.1234 \\ \hline 0.8818 \end{array}$

6. (a) $\begin{array}{r} 0.0610 \\ 5.0008 \\ +1.3000 \\ \hline 6.3618 \end{array}$ **(b)** $\begin{array}{r} 18.000 \\ -0.126 \\ \hline 17.874 \end{array}$

7. 0.5 (one decimal place)
× 0.3 (one decimal place)
0.15 (two decimal places)

8. 0.12 (two decimal places)
× 0.4 (one decimal place)
0.048 (three decimal places)

9. (a) 1.23 (two decimal places)
× 0.005 (three decimal places)
0.00615 (five decimal places)

(b) 0.003 (three decimal places)
× 0.00002 (five decimal places)
0.00000006 (eight decimal places)

10. $6\overline{)31.56}$ = 5.26
30
1 5
1 2
36
36
0
Each box of paper costs $5.26.

11. $0.06_\wedge\overline{)1800.00_\wedge}$ = 300 00.
18
00000
Thus, 1800 ÷ 0.06 = 30,000.

12. $4.9_\wedge\overline{)0.0_\wedge1764}$ = 0.0036
147
294
294
0
Thus, 0.01764 ÷ 4.9 = 0.0036.

13. (a) 0.0016 × 100 = 0.16
Move decimal point 2 places to the right.
(b) 2.34 × 1000 = 2340
Move decimal point 3 places to the right.
(c) 56.75 × 10,000 = 567,500
Move decimal point 4 places to the right.

14. (a) $\frac{5.82}{10} = 0.582$ (Move decimal point 1 place to the left.)
(b) $\frac{123.4}{1000} = 0.1234$ (Move decimal point 3 places to the left.)
(c) $\frac{0.00614}{10{,}000} = 0.000000614$ (Move decimal point 4 places to the left.)

0.5 Student Practice

1. (a) 0.92 = 92% **(b)** 0.0736 = 7.36%
(c) 0.7 = 0.70 = 70% **(d)** 0.0003 = 0.03%

2. (a) 3.04 = 304% **(b)** 5.186 = 518.6%
(c) 2.1 = 2.10 = 210%

3. (a) 7% = 0.07 **(b)** 9.3% = 0.093
(c) 131% = 1.31 **(d)** 0.04% = 0.0004

4. Change the percents to decimals and multiply.
(a) 18% of 50 = 0.18 × 50 = 9
(b) 4% of 64 = 0.04 × 64 = 2.56
(c) 156% of 35 = 1.56 × 35 = 54.6

5. (a) 4.2% of 38,000 = 0.042 × 38,000 = 1596
His raise is $1596.
(b) 38,000 + 1596 = 39,596
His new salary is $39,596.

6. $\frac{37}{148}$ reduces to $\frac{37 \cdot 1}{37 \cdot 4} = \frac{1}{4} = 0.25 = 25\%$

7. (a) $\frac{24}{48} = \frac{1}{2} = 0.5 = 50\%$
(b) $\frac{4}{25} = 0.16 = 16\%$

8. $\frac{430}{1256} = \frac{215}{628} \approx 0.342 \approx 34\%$

9. Round 128,621 to 100,000. Round 378 to 400.
100,000 × 400 = 40,000,000

10. Round $12\frac{1}{2}$ feet to 10 feet. Round $9\frac{3}{4}$ feet to 10 feet.
First room: 10 × 10 = 100 square feet
Round $11\frac{1}{4}$ feet to 10 feet. Round $18\frac{1}{2}$ feet to 20 feet.
Second room: 10 × 20 = 200 square feet
100 + 200 = 300
The estimate is 300 square feet.

11. (a) Round 422.8 miles to 400 miles. Round 19.3 gallons to 20 gallons.
$20\overline{)400}$ = 20
Roberta's truck gets about 20 miles per gallon.
(b) Round 3862 miles to 4000 miles.
$20\overline{)4000}$ = 200
She will use about 200 gallons of gas for her trip.
Round $\$3.69\frac{9}{10}$ to $4.00.
200 × $4 = $800
The estimated cost is $800.

0.6 Student Practice

1. Mathematics Blueprint for Problem Solving

Gather the Facts	What Am I Solving for?	What Must I Calculate?	Key Points to Remember
Living room measures $16\frac{1}{2}$ ft by $10\frac{1}{2}$ ft. The carpet costs $20.00 per square yard.	Area of room in square feet. Area of room in square yards. Cost of the carpet.	Multiply $16\frac{1}{2}$ ft by $10\frac{1}{2}$ ft to get the area in square feet. Divide the number of square feet by 9 to get the number of square yards. Multiply the number of square yards by $20.00.	9 sq feet = 1 sq yard

$16\frac{1}{2}\text{ ft} \times 10\frac{1}{2}\text{ ft} = \frac{33}{2}\text{ ft} \times \frac{21}{2}\text{ ft} = \frac{693}{4}$ or $173\frac{1}{4}$ square feet

$173\frac{1}{4} \div 9 = \frac{693}{4} \div \frac{9}{1} = \frac{693}{4} \times \frac{1}{9} = \frac{77}{4} = 19\frac{1}{4}$

$19\frac{1}{4}$ square yards of carpet are needed.

$19\frac{1}{4} \times 20 = \frac{77}{4} \times \frac{20}{1} = 385$

It will cost a minimum of $385.

Check: Estimate area of room: $20 \times 10 = 200$ square feet
Estimate area in square yards: $200 \div 10 = 20$
Estimate the cost: $20 \times 20 = \$400.00$

This is close to our answer of $385.00. Our answer seems reasonable.

To Think About: 6.5 sq ft; 52.5 sq ft; $468

2. (a) $\frac{10}{55} \approx 0.182 \approx 18\%$ **(b)** $\frac{3{,}660{,}000}{13{,}240{,}000} \approx 0.276 \approx 28\%$

(c) $\frac{15}{55} \approx 0.273 \approx 27\%$ **(d)** $\frac{3{,}720{,}000}{13{,}240{,}000} \approx 0.281 \approx 28\%$

(e) We notice that 18% of the company's sales force is located in the Northwest, and they were responsible for 28% of the sales volume. The percent of sales compared to the percent of sales force is about 150%. 27% of the company's sales force is located in the Southwest, and they were responsible for 28% of the sales volume. The percent of sales compared to the percent of sales force is approximately 100%. It would appear that the Northwest sales force is more effective.

To Think About: 88; 35

Chapter 1

1.1 Student Practice

1.

	Number	Integer	Rational Number	Irrational Number	Real Number
(a)	$-\frac{2}{5}$		X		X
(b)	1.515151...		X		X
(c)	−8	X	X		X
(d)	π			X	X

2. (a) Population growth of 1259 is +1259.
(b) Depreciation of $763 is −763.
(c) Wind-chill factor of minus 10 is −10.

3. (a) The additive inverse (opposite) of $\frac{2}{5}$ is $-\frac{2}{5}$.
(b) The additive inverse (opposite) of −1.92 is +1.92.
(c) The opposite of a loss of 12 yards on a football play is a gain of 12 yards on the play.

4. (a) $|-7.34| = 7.34$
(b) $\left|\frac{5}{8}\right| = \frac{5}{8}$
(c) $\left|\frac{0}{2}\right| = 0$

5. (a) $37 + 19$
$37 + 19 = 56$
$37 + 19 = +56$

(b) $-23 + (-35)$
$23 + 35 = 58$
$-23 + (-35) = -58$

6. $-\frac{3}{5} + \left(-\frac{4}{7}\right)$
$-\frac{21}{35} + \left(-\frac{20}{35}\right)$
$\frac{21}{35} + \frac{20}{35} = \frac{41}{35}$ or $1\frac{6}{35}$
$-\frac{21}{35} + \left(-\frac{20}{35}\right) = -\frac{41}{35}$ or $-1\frac{6}{35}$

7. $-12.7 + (-9.38)$
$12.7 + 9.38 = 22.08$
$-12.7 + (-9.38) = -22.08$

8. $-7 + (-11) + (-33)$
$= -18 + (-33)$
$= -51$

9. $-9 + 15$
$15 - 9 = 6$
$-9 + 15 = +6$ or 6

10. $-\frac{5}{12} + \frac{7}{12} + \left(-\frac{11}{12}\right)$
$= \frac{2}{12} + \left(-\frac{11}{12}\right) = -\frac{9}{12} = -\frac{3}{4}$

11. $-6.3 + (-8.0) + 3.5$
$= -14.3 + 3.5$
$= -10.8$

12.
$$\begin{array}{rr} -6 & +5 \\ -7 & +5 \\ \underline{-2} & \underline{+3} \\ -15 & 13 \end{array}$$
$-6 + 5 + (-7) + (-2) + 5 + 3 = -15 + 13 = -2$

13. (a) $-2.9 + (-5.7) = -8.6$
(b) $\frac{2}{3} + \left(-\frac{1}{4}\right)$
$= \frac{8}{12} + \left(-\frac{3}{12}\right) = \frac{5}{12}$

1.2 Student Practice

1. $9 - (-3) = 9 + (+3) = 12$

2. $-12 - (-5) = -12 + (+5) = -7$

3. (a) $\frac{5}{9} - \frac{7}{9} = \frac{5}{9} + \left(-\frac{7}{9}\right) = -\frac{2}{9}$
(b) $-\frac{5}{21} - \left(-\frac{3}{7}\right) = -\frac{5}{21} + \frac{3}{7} = -\frac{5}{21} + \frac{9}{21} = \frac{4}{21}$

4. $-17.3 - (-17.3) = -17.3 + 17.3 = 0$

5. (a) $-21 - 9$
$= -21 + (-9)$
$= -30$

(b) $17 - 36$
$= 17 + (-36)$
$= -19$

(c) $12 - (-15)$
$= 12 + 15$
$= 27$

(d) $\frac{3}{5} - 2$
$= \frac{3}{5} + (-2)$
$= \frac{3}{5} + \left(-\frac{10}{5}\right) = -\frac{7}{5}$ or $-1\frac{2}{5}$

6. $350 - (-186) = 350 + 186 = 536$
The helicopter is 536 feet from the sunken vessel.

1.3 Student Practice

1. (a) $(-6)(-2) = 12$ **(b)** $(7)(9) = 63$
(c) $\left(-\frac{3}{5}\right)\left(\frac{2}{7}\right) = -\frac{6}{35}$ **(d)** $\left(\frac{5}{6}\right)(-7) = \left(\frac{5}{6}\right)\left(-\frac{7}{1}\right) = -\frac{35}{6}$ or $-5\frac{5}{6}$

2. $(-5)(-2)(-6) = (+10)(-6) = -60$

3. (a) positive; $-2(-3) = 6$
(b) negative; $(-1)(-3)(-2) = +3(-2) = -6$
(c) positive;
$-4(-\frac{1}{4})(-2)(-6) = +1(-2)(-6) = -2(-6) = +12$ or 12

4. (a) $-36 \div (-2) = 18$ **(b)** $49 \div 7 = 7$
(c) $\frac{50}{-10} = -5$ **(d)** $\frac{-39}{13} = -3$

5. (a) The numbers have the same sign, so the result will be positive. Divide the absolute values.
$1.8_\wedge\overline{)12.6_\wedge}$ quotient 7.; $12\,6$
$-12.6 \div (-1.8) = 7$
(b) The numbers have different signs, so the result will be negative. Divide the absolute values.
$0.9_\wedge\overline{)0.4_\wedge5}$ quotient 0.5; $4\,5$
$0.45 \div (-0.9) = -0.5$

6. $-\frac{5}{16} \div \left(-\frac{10}{13}\right) = \left(-\frac{5}{16}\right)\left(-\frac{13}{10}\right) = \left(-\frac{\cancel{5}^{1}}{16}\right)\left(-\frac{13}{\cancel{10}_{2}}\right) = \frac{13}{32}$

7. (a) $\dfrac{-12}{-\frac{4}{5}} = -\frac{12}{1} \div \left(-\frac{4}{5}\right) = \left(-\frac{\cancel{12}^{3}}{1}\right)\left(-\frac{5}{\cancel{4}_{1}}\right) = 15$

(b) $\dfrac{-\frac{2}{9}}{\frac{8}{13}} = -\frac{2}{9} \div \frac{8}{13} = -\frac{\cancel{2}^{1}}{9}\left(\frac{13}{\cancel{8}_{4}}\right) = -\frac{13}{36}$

8. (a) $6(-10) = -60$
The team lost approximately 60 yards with plays that were considered medium losses.
(b) $7(15) = 105$
The team gained approximately 105 yards with plays that were considered medium gains.
(c) $-60 + 105 = 45$
A total of 45 yards were gained during plays that were medium losses or medium gains.

1.4 Student Practice

1. (a) $6(6)(6)(6) = 6^4$
(b) $-2(-2)(-2)(-2)(-2) = (-2)^5$
(c) $108(108)(108) = 108^3$
(d) $-11(-11)(-11)(-11)(-11)(-11) = (-11)^6$
(e) $(w)(w)(w) = w^3$
(f) $(z)(z)(z)(z) = z^4$

2. (a) $3^5 = (3)(3)(3)(3)(3) = 243$
(b) $2^2 = (2)(2) = 4$
$3^3 = (3)(3)(3) = 27$
$2^2 + 3^3 = 4 + 27 = 31$

3. (a) $(-3)^3 = -27$ **(b)** $(-2)^6 = +64$
(c) $-2^4 = -(2^4) = -16$ **(d)** $-(6^3) = -216$

4. (a) $(\frac{1}{3})^3 = (\frac{1}{3})(\frac{1}{3})(\frac{1}{3}) = \frac{1}{27}$
(b) $(0.3)^4 = (0.3)(0.3)(0.3)(0.3) = 0.0081$
(c) $(\frac{3}{2})^4 = (\frac{3}{2})(\frac{3}{2})(\frac{3}{2})(\frac{3}{2}) = \frac{81}{16}$
(d) $3^4 = (3)(3)(3)(3) = 81$
$4^2 = (4)(4) = 16$
$(3)^4(4)^2 = (81)(16) = 1296$
(e) $4^2 - 2^4 = 16 - 16 = 0$

1.5 Student Practice

1. $25 \div 5 \cdot 6 + 2^3$
$= 25 \div 5 \cdot 6 + 8$
$= 5 \cdot 6 + 8$
$= 30 + 8$
$= 38$

2. $(-4)^3 - 2^6 = -64 - 64 = -128$

3. $6 - (8 - 12)^2 + 8 \div 2$
$= 6 - (-4)^2 + 8 \div 2$
$= 6 - (16) + 8 \div 2$
$= 6 - 16 + 4$
$= -10 + 4$
$= -6$

4. $\left(-\frac{1}{7}\right)\left(-\frac{14}{5}\right) + \left(-\frac{1}{2}\right) \div \left(\frac{3}{4}\right)^2$
$= \left(-\frac{1}{7}\right)\left(-\frac{14}{5}\right) + \left(-\frac{1}{2}\right)\left(\frac{16}{9}\right)$
$= \frac{2}{5} + \left(-\frac{8}{9}\right)$
$= \frac{2 \cdot 9}{5 \cdot 9} + \left(-\frac{8 \cdot 5}{9 \cdot 5}\right)$
$= \frac{18}{45} + \left(-\frac{40}{45}\right)$
$= -\frac{22}{45}$

1.6 Student Practice

1. (a) $3(x + 2y) = 3x + 3(2y) = 3x + 6y$
(b) $-2(a + 3b) = -2(a) + (-2)(+3b) = -2a - 6b$

2. (a) $-(-3x + y) = (-1)(-3x + y)$
$= (-1)(-3x) + (-1)(y)$
$= 3x - y$

3. (a) $\frac{3}{5}(a^2 - 5a + 25) = \left(\frac{3}{5}\right)(1a^2) + \left(\frac{3}{5}\right)(-5a) + \left(\frac{3}{5}\right)(25)$
$= \frac{3}{5}a^2 - 3a + 15$
(b) $2.5(x^2 - 3.5x + 1.2)$
$= (2.5)(1x^2) + (2.5)(-3.5x) + (2.5)(1.2)$
$= 2.5x^2 - 8.75x + 3$

4. $-4x(x - 2y + 3) = (-4)(x)(x) + (-4)(x)(-2)(y) + (-4)(x)(3)$
$= -4x^2 + 8xy - 12x$

5. $(3x^2 - 2x)(-4) = (3x^2)(-4) + (-2x)(-4) = -12x^2 + 8x$

6. $400(6x + 9y) = 400(6x) + 400(9y)$
$= 2400x + 3600y$
The area of the field in square feet is $2400x + 3600y$.

1.7 Student Practice

1. (a) $5a$ and $8a$ are like terms.
$2b$ and $-4b$ are like terms.
(b) y^2 and $-7y^2$ are like terms. These are the only like terms.

2. (a) $16y^3 + 9y^3 = (16 + 9)y^3 = 25y^3$
(b) $5a + 7a + 4a = (5 + 7 + 4)a = 16a$

3. $-8y^2 - 9y^2 + 4y^2 = (-8 - 9 + 4)y^2 = -13y^2$

4. (a) $1.3x + 3a - 9.6x + 2a = -8.3x + 5a$
(b) $5ab - 2ab^2 - 3a^2b + 6ab = 11ab - 2ab^2 - 3a^2b$
(c) There are no like terms in the expression $7x^2y - 2xy^2 - 3x^2y^2 - 4xy$, so no terms can be combined.

5. $5xy - 2x^2y + 6xy^2 - xy - 3xy^2 - 7x^2y$
$= 5xy - xy - 2x^2y - 7x^2y + 6xy^2 - 3xy^2$
$= 4xy - 9x^2y + 3xy^2$

6. $\frac{1}{7}a^2 - \frac{5}{12}b + 2a^2 - \frac{1}{3}b$

$\frac{1}{7}a^2 + 2a^2 = \frac{1}{7}a^2 + \frac{2}{1}a^2 = \frac{1}{7}a^2 + \frac{2 \cdot 7}{1 \cdot 7}a^2$

$= \frac{1}{7}a^2 + \frac{14}{7}a^2 = \frac{15}{7}a^2$

$-\frac{5}{12}b - \frac{1}{3}b = -\frac{5}{12}b - \frac{1 \cdot 4}{3 \cdot 4}b = -\frac{5}{12}b - \frac{4}{12}b$

$= -\frac{9}{12}b = -\frac{3}{4}b$

Thus, our solution is $\frac{15}{7}a^2 - \frac{3}{4}b$.

7. $5a(2 - 3b) - 4(6a + 2ab) = 10a - 15ab - 24a - 8ab$
$= -14a - 23ab$

1.8 Student Practice

1. $4 - \frac{1}{2}x = 4 - \frac{1}{2}(-8)$
$= 4 + 4$
$= 8$

2. **(a)** $4x^2 = 4(-3)^2 = 4(9) = 36$
(b) $(4x)^2 = [4(-3)]^2 = (-12)^2 = 144$

3. $2x^2 - 3x = 2(-2)^2 - 3(-2)$
$= 2(4) - 3(-2)$
$= 8 + 6$
$= 14$

4. $6a + 4ab^2 - 5 = 6\left(-\frac{1}{6}\right) + 4\left(-\frac{1}{6}\right)(3)^2 - 5$
$= -1 + 4\left(-\frac{1}{6}\right)(9) - 5$
$= -1 + (-6) - 5$
$= -1 + (-6) + (-5)$
$= -12$

5. Area of a triangle is $A = \frac{1}{2}ab$
altitude = 3 meters (m)
base = 7 meters (m)

$A = \frac{1}{2}(3\text{ m})(7\text{ m})$
$= \frac{1}{2}(3)(7)(\text{m})(\text{m})$
$= \left(\frac{3}{2}\right)(7)(\text{m})^2$
$= \frac{21}{2}(\text{m})^2$
$= 10.5\text{ m}^2$

6. Area of a circle is
$A = \pi r^2$
$r = 3$ meters
$A \approx 3.14(3\text{ m})^2$
$\approx 3.14(9)(\text{m})^2$
$\approx 28.26\text{ m}^2$

7. Formula
$C = \frac{5}{9}(F - 32)$
$= \frac{5}{9}(68 - 32)$
$= \frac{5}{9}(36)$
$= 5(4)$
$= 20°$ Celsius

The temperature is 20° Celsius or 20°C.

8. Use the formula.
$k \approx 1.61(r)$
$\approx 1.61(35)$ Replace r by 35.
≈ 56.35

The truck is traveling at approximately 56.35 kilometers per hour. It is violating the minimum speed law.

1.9 Student Practice

1. $5[4x - 3(y - 2)]$
$= 5[4x - 3y + 6]$
$= 20x - 15y + 30$

2. $-3[2a - (3b - c) + 4d]$
$= -3[2a - 3b + c + 4d]$
$= -6a + 9b - 3c - 12d$

3. $3[4x - 2(1 - x)] - [3x + (x - 2)]$
$= 3[4x - 2 + 2x] - [3x + x - 2]$
$= 3[6x - 2] - [4x - 2]$
$= 18x - 6 - 4x + 2$
$= 14x - 4$

4. $-2\{5x - 3[2x - (3 - 4x)]\}$
$= -2\{5x - 3[2x - 3 + 4x]\}$
$= -2\{5x - 3[6x - 3]\}$
$= -2\{5x - 18x + 9\}$
$= -2\{-13x + 9\}$
$= 26x - 18$

Chapter 2 2.1 Student Practice

1.
$x + 14 = 23$
$x + 14 + (-14) = 23 + (-14)$
$x + 0 = 9$
$x = 9$

Check. $x + 14 = 23$
$9 + 14 \stackrel{?}{=} 23$
$23 = 23$ ✓

2.
$17 = x - 5$
$17 + 5 = x - 5 + 5$
$22 = x + 0$
$22 = x$

Check. $17 = x - 5$
$17 \stackrel{?}{=} 22 - 5$
$17 = 17$ ✓

3.
$0.5 - 1.2 = x - 0.3$
$-0.7 = x - 0.3$
$-0.7 + 0.3 = x - 0.3 + 0.3$
$-0.4 = x$

Check. $0.5 - 1.2 = x - 0.3$
$0.5 - 1.2 \stackrel{?}{=} -0.4 - 0.3$
$-0.7 = -0.7$ ✓

4.
$x + 8 = -22 + 6$
$-2 + 8 \stackrel{?}{=} -22 + 6$
$6 \neq -16$ This is not true.

Thus $x = -2$ is not a solution. Solve to find the solution.
$x + 8 = -22 + 6$
$x + 8 = -16$
$x + 8 - 8 = -16 - 8$
$x = -24$

5.
$\frac{1}{20} - \frac{1}{2} = x + \frac{3}{5}$
$\frac{1}{20} - \frac{1 \cdot 10}{2 \cdot 10} = x + \frac{3 \cdot 4}{5 \cdot 4}$
$\frac{1}{20} - \frac{10}{20} = x + \frac{12}{20}$
$-\frac{9}{20} = x + \frac{12}{20}$
$-\frac{9}{20} + \left(-\frac{12}{20}\right) = x + \frac{12}{20} + \left(-\frac{12}{20}\right)$
$-\frac{21}{20} = x$
$x = -\frac{21}{20}$ or $-1\frac{1}{20}$

Check.
$\frac{1}{20} - \frac{1}{2} = x + \frac{3}{5}$
$\frac{1}{20} - \frac{1}{2} \stackrel{?}{=} -\frac{21}{20} + \frac{3}{5}$
$\frac{1}{20} - \frac{10}{20} \stackrel{?}{=} -\frac{21}{20} + \frac{12}{20}$
$-\frac{9}{20} = -\frac{9}{20}$ ✓

2.2 Student Practice

1.
$$\frac{1}{8}x = -2$$
$$8\left(\frac{1}{8}x\right) = 8(-2)$$
$$\left(\frac{8}{1}\right)\left(\frac{1}{8}\right)x = -16$$
$$x = -16$$

2. $9x = 72$
$$\frac{9x}{9} = \frac{72}{9}$$
$$x = 8$$

3. $6x = 50$
$$\frac{6x}{6} = \frac{50}{6}$$
$$x = \frac{25}{3} \text{ or } 8\frac{1}{3}$$

Check.
$$4\left(\frac{7}{2}\right) - \left(\frac{7}{2} + 3\right) \stackrel{?}{=} 12 - 3\left(\frac{7}{2} - 2\right)$$
$$14 - \frac{13}{2} \stackrel{?}{=} 12 - 3\left(\frac{3}{2}\right)$$
$$\frac{28}{2} - \frac{13}{2} \stackrel{?}{=} \frac{24}{2} - \frac{9}{2}$$
$$\frac{15}{2} = \frac{15}{2} \checkmark$$

4. $-27x = 54$
$$\frac{-27x}{-27} = \frac{54}{-27}$$
$$x = -2$$

5. $-x = 36$
$$-1x = 36$$
$$\frac{-1x}{-1} = \frac{36}{-1}$$
$$x = -36$$

6.
$$-51 = 3x - 9x$$
$$-51 = -6x$$
$$\frac{-51}{-6} = \frac{-6x}{-6}$$
$$\frac{17}{2} \text{ or } 8\frac{1}{2} = x$$

7. $16.2 = 5.2x - 3.4x$
$$16.2 = 1.8x$$
$$\frac{16.2}{1.8} = \frac{1.8x}{1.8}$$
$$9 = x$$

2.3 Student Practice

1.
$$9x + 2 = 38$$
$$9x + 2 + (-2) = 38 + (-2)$$
$$9x = 36$$
$$\frac{9x}{9} = \frac{36}{9}$$
$$x = 4$$

Check.
$$9(4) + 2 \stackrel{?}{=} 38$$
$$36 + 2 \stackrel{?}{=} 38$$
$$38 = 38 \checkmark$$

2.
$$13x = 2x - 66$$
$$13x + (-2x) = 2x + (-2x) - 66$$
$$11x = -66$$
$$\frac{11x}{11} = \frac{-66}{11}$$
$$x = -6$$

3.
$$3x + 2 = 5x + 2$$
$$3x + (-3x) + 2 = 5x + (-3x) + 2$$
$$2 = 2x + 2$$
$$2 + (-2) = 2x + 2 + (-2)$$
$$0 = 2x$$
$$\frac{0}{2} = \frac{2x}{2}$$
$$0 = x$$

Check. $3(0) + 2 \stackrel{?}{=} 5(0) + 2$
$$2 = 2 \checkmark$$

4.
$$-z + 8 - z = 3z + 10 - 3$$
$$-2z + 8 = 3z + 7$$
$$-2z + 2z + 8 = 3z + 2z + 7$$
$$8 = 5z + 7$$
$$8 + (-7) = 5z + 7 + (-7)$$
$$1 = 5z$$
$$\frac{1}{5} = \frac{5z}{5}$$
$$\frac{1}{5} = z$$

5. $4x - (x + 3) = 12 - 3(x - 2)$
$$4x - x - 3 = 12 - 3x + 6$$
$$3x - 3 = -3x + 18$$
$$3x + 3x - 3 = -3x + 3x + 18$$
$$6x - 3 = 18$$
$$6x - 3 + 3 = 18 + 3$$
$$6x = 21$$
$$\frac{6x}{6} = \frac{21}{6}$$
$$x = \frac{7}{2} \text{ or } 3\frac{1}{2}$$

6. $4(-2x - 3) = -5(x - 2) + 2$
$$-8x - 12 = -5x + 10 + 2$$
$$-8x - 12 = -5x + 12$$
$$-8x + 8x - 12 = -5x + 8x + 12$$
$$-12 = 3x + 12$$
$$-12 - 12 = 3x + 12 - 12$$
$$-24 = 3x$$
$$\frac{-24}{3} = \frac{3x}{3}$$
$$-8 = x$$

7. $0.3x - 2(x + 0.1) = 0.4(x - 3) - 1.1$
$$0.3x - 2x - 0.2 = 0.4x - 1.2 - 1.1$$
$$-1.7x - 0.2 = 0.4x - 2.3$$
$$-1.7x + 1.7x - 0.2 = 0.4x + 1.7x - 2.3$$
$$-0.2 = 2.1x - 2.3$$
$$-0.2 + 2.3 = 2.1x - 2.3 + 2.3$$
$$2.1 = 2.1x$$
$$\frac{2.1}{2.1} = \frac{2.1x}{2.1}$$
$$1 = x$$

8. $5(2z - 1) + 7 = 7z - 4(z + 3)$
$$10z - 5 + 7 = 7z - 4z - 12$$
$$10z + 2 = 3z - 12$$
$$10z - 3z + 2 = 3z - 3z - 12$$
$$7z + 2 = -12$$
$$7z + 2 - 2 = -12 - 2$$
$$7z = -14$$
$$\frac{7z}{7} = \frac{-14}{7}$$
$$z = -2$$

Check.
$$5[2(-2) - 1] + 7 \stackrel{?}{=} 7(-2) - 4[(-2) + 3]$$
$$5[-4 - 1] + 7 \stackrel{?}{=} 7(-2) - 4[1]$$
$$5(-5) + 7 \stackrel{?}{=} -14 - 4$$
$$-25 + 7 \stackrel{?}{=} -18$$
$$-18 = -18 \checkmark$$

2.4 Student Practice

1.
$$\frac{3}{8}x - \frac{3}{2} = \frac{1}{4}x$$
$$8\left(\frac{3}{8}x - \frac{3}{2}\right) = 8\left(\frac{1}{4}x\right)$$
$$\left(\frac{8}{1}\right)\left(\frac{3}{8}\right)(x) - \left(\frac{8}{1}\right)\left(\frac{3}{2}\right) = \left(\frac{8}{1}\right)\left(\frac{1}{4}\right)(x)$$
$$3x - 12 = 2x$$
$$3x + (-3x) - 12 = 2x + (-3x)$$
$$-12 = -x$$
$$12 = x$$

2.
$$\frac{5x}{4} - 1 = \frac{3x}{4} + \frac{1}{2}$$
$$4\left(\frac{5x}{4}\right) - 4(1) = 4\left(\frac{3x}{4}\right) + 4\left(\frac{1}{2}\right)$$
$$5x - 4 = 3x + 2$$
$$5x - 3x - 4 = 3x - 3x + 2$$
$$2x - 4 = 2$$
$$2x - 4 + 4 = 2 + 4$$
$$2x = 6$$
$$\frac{2x}{2} = \frac{6}{2}$$
$$x = 3$$

Check.
$$\frac{5(3)}{4} - 1 \stackrel{?}{=} \frac{3(3)}{4} + \frac{1}{2}$$
$$\frac{15}{4} - 1 \stackrel{?}{=} \frac{9}{4} + \frac{1}{2}$$
$$\frac{15}{4} - \frac{4}{4} \stackrel{?}{=} \frac{9}{4} + \frac{2}{4}$$
$$\frac{11}{4} = \frac{11}{4} \checkmark$$

3.
$$\frac{x+6}{9} = \frac{x}{6} + \frac{1}{2}$$
$$\frac{x}{9} + \frac{6}{9} = \frac{x}{6} + \frac{1}{2}$$
$$18\left(\frac{x}{9}\right) + 18\left(\frac{6}{9}\right) = 18\left(\frac{x}{6}\right) + 18\left(\frac{1}{2}\right)$$
$$2x + 12 = 3x + 9$$
$$2x - 2x + 12 = 3x - 2x + 9$$
$$12 = x + 9$$
$$12 - 9 = x + 9 - 9$$
$$3 = x$$

4.
$$\frac{1}{2}(x+5) = \frac{1}{5}(x-2) + \frac{1}{2}$$
$$\frac{x}{2} + \frac{5}{2} = \frac{x}{5} - \frac{2}{5} + \frac{1}{2}$$
$$10\left(\frac{x}{2}\right) + 10\left(\frac{5}{2}\right) = 10\left(\frac{x}{5}\right) - 10\left(\frac{2}{5}\right) + 10\left(\frac{1}{2}\right)$$
$$5x + 25 = 2x - 4 + 5$$
$$5x + 25 = 2x + 1$$
$$5x - 2x + 25 = 2x - 2x + 1$$
$$3x + 25 = 1$$
$$3x + 25 - 25 = 1 - 25$$
$$3x = -24$$
$$\frac{3x}{3} = \frac{-24}{3}$$
$$x = -8$$

Check. $\frac{1}{2}[(-8) + 5] \stackrel{?}{=} \frac{1}{5}[(-8) - 2] + \frac{1}{2}$
$$\frac{1}{2}(-3) \stackrel{?}{=} \frac{1}{5}(-10) + \frac{1}{2}$$
$$-\frac{3}{2} \stackrel{?}{=} -2 + \frac{1}{2}$$
$$-\frac{3}{2} \stackrel{?}{=} -\frac{4}{2} + \frac{1}{2}$$
$$-\frac{3}{2} = -\frac{3}{2} \checkmark$$

5.
$$2.8 = 0.3(x-2) + 2(0.1x - 0.3)$$
$$2.8 = 0.3x - 0.6 + 0.2x - 0.6$$
$$10(2.8) = 10(0.3x) - 10(0.6) + 10(0.2x) - 10(0.6)$$
$$28 = 3x - 6 + 2x - 6$$
$$28 = 5x - 12$$
$$28 + 12 = 5x - 12 + 12$$
$$40 = 5x$$
$$\frac{40}{5} = \frac{5x}{5}$$
$$8 = x$$

2.5 Student Practice

1. **(a)** $x + 4$ **(b)** $3x$ **(c)** $x - 8$ **(d)** $\frac{1}{4}x$

2. **(a)** $3x + 8$ **(b)** $3(x + 8)$ **(c)** $\frac{1}{3}(x + 4)$

3. Let a = Ann's hours per week.
Then $a - 17$ = Marie's hours per week.

4. $l = 2w + 5$; w

width $= w$
length $= 2w + 5$

5. Number of degrees in 1st angle $= s - 16$
Number of degrees in 2nd angle $= s$
Number of degrees in 3rd angle $= 2s$

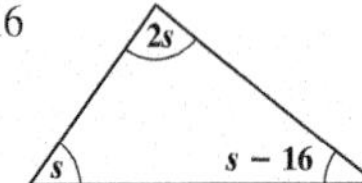

6. Let x = the number of students in the fall.
$\frac{2}{3}x$ = the number of students in the spring.
$\frac{1}{5}x$ = the number of students in the summer.

2.6 Student Practice

1. Let x = the unknown number.
$$\frac{3}{4}x = -81$$
$$4\left(\frac{3}{4}x\right) = 4(-81)$$
$$3x = -324$$
$$\frac{3x}{3} = \frac{-324}{3}$$
$$x = -108$$
The number is -108.

2. Let x = the unknown number.
$$3x - 2 = 49$$
$$3x - 2 + 2 = 49 + 2$$
$$3x = 51$$
$$\frac{3x}{3} = \frac{51}{3}$$
$$x = 17$$
The number is 17.

3. Let x = the first number.
Then the second number is $3x - 12$.
$$x + (3x - 12) = 24$$
$$4x - 12 = 24$$
$$4x = 36$$
$$x = 9$$
First number is 9. Second number $= 3(9) - 12 = 15$.

4. 132°; x; x
$$132° + x + x = 180°$$
$$132° + 2x = 180°$$
$$2x = 48°$$
$$x = 24°$$
Both angles measure 24°.

5. Let x = the measure of the first angle. Then $2x + 5$ = the measure of the second angle, and $\frac{1}{2}x$ = the measure of the third angle.
$$x + 2x + 5° + \frac{1}{2}x = 180°$$
$$\frac{7}{2}x + 5° = 180°$$
$$\frac{7}{2}x = 175°$$
$$x = \frac{2}{7}(175)°$$
$$x = 50°$$
The measure of the first angle is 50°, the measure of the second angle is $2(50°) + 5° = 105°$, and the measure of the third angle is $\frac{1}{2}(50°) = 25°$

6. **(a)**
$$d = rt$$
$$220 = 4r$$
$$55 = r$$
Leaving the city, her average speed was 55 mph.

(b)
$$d = rt$$
$$225 = 4.5r$$
$$50 = r$$
On the return trip, her average speed was 50 mph.

(c) She traveled 5 mph faster on the trip leaving the city.

7. Let $x =$ her final exam score. Since the final counts as two tests, divide her total score by 6.

$$\frac{78 + 80 + 100 + 96 + x + x}{6} = 90$$
$$\frac{354 + 2x}{6} = 90$$
$$6\left(\frac{354 + 2x}{6}\right) = 6(90)$$
$$354 + 2x = 540$$
$$2x = 186$$
$$x = 93$$

She needs a 93 on the final exam.

2.7 Student Practice

1. Let $m =$ the number of miles.

$$3(25) + (0.20)m = 350$$
$$75 + 0.20m = 350$$
$$0.20m = 275$$
$$\frac{0.20m}{0.20} = \frac{275}{0.20}$$
$$m = 1375$$

He can travel 1375 miles for \$350.

2. Let $x =$ the cost of the chef's knives he sold.

$$0.38x = 17{,}100$$
$$\frac{0.38x}{0.38} = \frac{17{,}100}{0.38}$$
$$x = 45{,}000$$

He sold \$45,000 worth of chef's knives last year.

3. Let $x =$ the price last year.

$$x + 0.07x = 19{,}795$$
$$1.07x = 19{,}795$$
$$\frac{1.07x}{1.07} = \frac{19{,}795}{1.07}$$
$$x = 18{,}500$$

A similar model would have cost \$18,500 last year.

4. $I = prt$

$$I = (7000)(0.12)(1)$$
$$I = 840$$

The interest charge on borrowing \$7000 for one year at a simple interest rate of 12% is \$840.

5. Let $x =$ the amount invested at 9%.
Then $8000 - x$ was invested at 7%.

$$0.09x + 0.07(8000 - x) = 630$$
$$0.09x + 560 - 0.07x = 630$$
$$0.02x + 560 = 630$$
$$0.02x = 70$$
$$\frac{0.02x}{0.02} = \frac{70}{0.02}$$
$$x = 3500$$
$$8000 - x = 8000 - 3500 = 4500$$

Therefore, she invested \$3500 at 9% and \$4500 at 7%.

6. Let $x =$ the number of dimes.
Then $x + 5 =$ the number of quarters.

$$0.10x + 0.25(x + 5) = 5.10$$
$$0.10x + 0.25x + 1.25 = 5.10$$
$$0.35x + 1.25 = 5.10$$
$$0.35x = 3.85$$
$$\frac{0.35x}{0.35} = \frac{3.85}{0.35}$$
$$x = 11$$
$$x + 5 = 11 + 5 = 16$$

Therefore, she has 11 dimes and 16 quarters.

7. Let $x =$ the number of dimes.
Then $2x =$ the number of nickels and
$x + 4 =$ the number of quarters.

$$0.05(2x) + 0.10x + 0.25(x + 4) = 2.35$$
$$0.10x + 0.10x + 0.25x + 1 = 2.35$$
$$0.45x + 1 = 2.35$$
$$0.45x = 1.35$$
$$\frac{0.45x}{0.45} = \frac{1.35}{0.45}$$
$$x = 3$$
$$2x = 2(3) = 6$$
$$x + 4 = 3 + 4 = 7$$

Therefore, the boy has 3 dimes, 6 nickels, and 7 quarters.

2.8 Student Practice

1. **(a)** $7 > 2$ **(b)** $-2 > -4$ **(c)** $-1 < 2$
(d) $-8 < -5$ **(e)** $0 > -2$ **(f)** $5 > -3$

2. **(a)** x is greater than 5

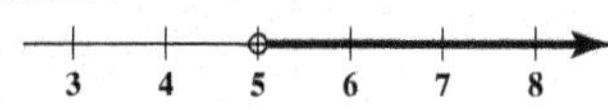

(b) x is less than or equal to -2

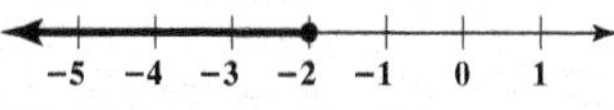

(c) x is less than 3 (or 3 is greater than x)

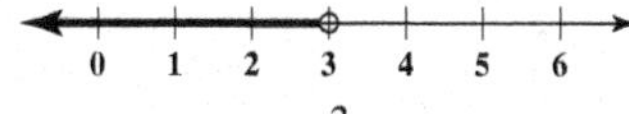

(d) x is greater than or equal to $-\frac{3}{2}$

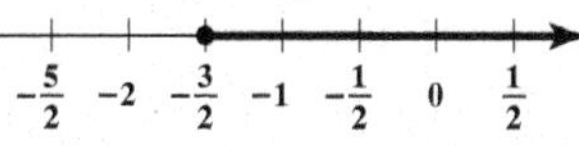

3. **(a)** Since the temperature can never exceed 180 degrees, then the temperature must always be less than or equal to 180 degrees. Thus, $t \le 180$.
(b) Since the debt must be less than 15,000, we have $d < 15{,}000$.

4. **(a)** $7 > 2$
$-14 < -4$
(b) $-3 < -1$
$3 > 1$
(c) $-10 \ge -20$
$1 \le 2$
(d) $-15 \le -5$
$3 \ge 1$

5.
$$8x - 2 < 3$$
$$8x - 2 + 2 < 3 + 2$$
$$8x < 5$$
$$\frac{8x}{8} < \frac{5}{8}$$
$$x < \frac{5}{8}$$

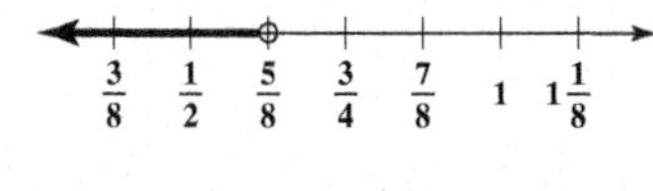

6.
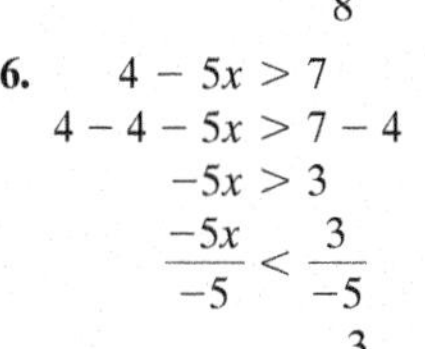

$$4 - 5x > 7$$
$$4 - 4 - 5x > 7 - 4$$
$$-5x > 3$$
$$\frac{-5x}{-5} < \frac{3}{-5}$$
$$x < -\frac{3}{5}$$

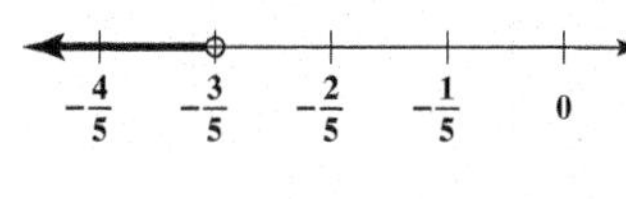

7.
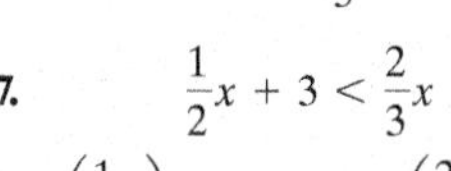

$$\frac{1}{2}x + 3 < \frac{2}{3}x$$
$$6\left(\frac{1}{2}x\right) + 6(3) < 6\left(\frac{2}{3}x\right)$$

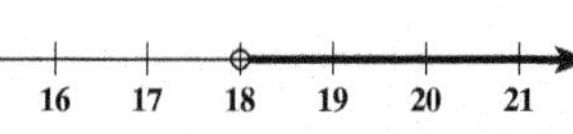

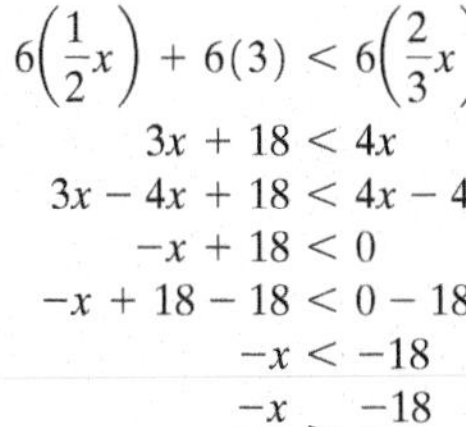

$$3x + 18 < 4x$$
$$3x - 4x + 18 < 4x - 4x$$
$$-x + 18 < 0$$
$$-x + 18 - 18 < 0 - 18$$
$$-x < -18$$
$$\frac{-x}{-1} > \frac{-18}{-1}$$
$$x > 18$$

8. $\frac{1}{2}(3 - x) \le 2x + 5$

$\frac{3}{2} - \frac{1}{2}x \le 2x + 5$

$2\left(\frac{3}{2}\right) - 2\left(\frac{1}{2}x\right) \le 2(2x) + 2(5)$

$3 - x \le 4x + 10$

$3 - x - 4x \le 4x - 4x + 10$

$3 - 5x \le 10$

$3 - 3 - 5x \le 10 - 3$

$-5x \le 7$

$\frac{-5x}{-5} \ge \frac{7}{-5}$

$x \ge -\frac{7}{5}$

9. $2000n - 700{,}000 \ge 2{,}500{,}000$

$2000n - 700{,}000 + 700{,}000 \ge 2{,}500{,}000 + 700{,}000$

$2000n \ge 3{,}200{,}000$

$\frac{2000n}{2000} \ge \frac{3{,}200{,}000}{2000}$

$n \ge 1600$

Chapter 3 3.1 Student Practice

1. Point B is 3 units to the right on the x-axis and 4 units up from the point where we stopped on the x-axis.

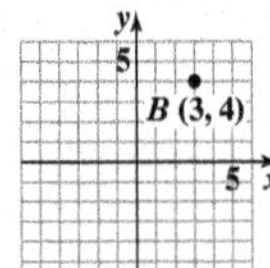

2. **(a)** Begin by counting 2 squares to the left, starting at the origin. Since the y-coordinate is negative, count 4 units down from the point where we stopped on the x-axis. Label the point I.
(b) Begin by counting 4 squares to the left of the origin. Then count 5 units up because the y-coordinate is positive. Label the point J.
(c) Begin by counting 4 units to the right of the origin. Then count 2 units down because the y-coordinate is negative. Label the point K.

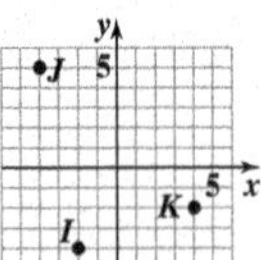

3. The points are plotted in the figure.

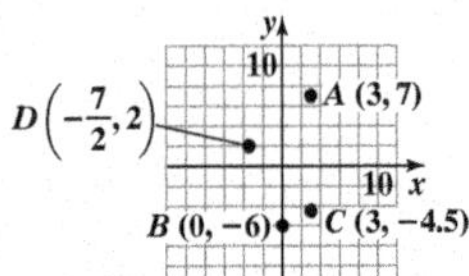

4. To find the point C move along the x-axis to get as close as possible to C. We end up at 2. Thus the first number of the ordered pair is 2. Then count 7 units upward on a line parallel to the y-axis to reach C. So the second number of the ordered pair is 7. Thus, point C is represented by (2, 7).

5. **(a)**

Number of motor vehicle deaths (in thousands)

Number of years since 1980

(b) Motor vehicle deaths were significantly high in 1980. During 1985–2005, the number of motor vehicle deaths was relatively stable. There was a significant decrease in the number of deaths in 2010.

6. **(a)** Replace x with 3 and y with -1.

$3x + 2y = 5$

$3(3) + 2(-1) \stackrel{?}{=} 5$

$9 - 2 \stackrel{?}{=} 5$

$7 = 5$ False

The ordered pair $(3, -1)$ is not solution to $3x + 2y = 5$.

(b) Replace x with 2 and y with $-\frac{1}{2}$.

$3x + 2y = 5$

$3(2) + 2\left(-\frac{1}{2}\right) \stackrel{?}{=} 5$

$6 + (-1) \stackrel{?}{=} 5$

$5 = 5$ ✓ True

The ordered pair $\left(2, -\frac{1}{2}\right)$ is a solution to $3x + 2y = 5$.

7. **(a)** The ordered pair $(-2, 13)$ is a solution if, when we replace x with -2 and y with 13 in the equation $x + y = 11$, we get a true statement.

$x = -2$ and $y = 13$, or $(-2, 13)$ $\quad x + y = 11$

$-2 + 13 = 11, \quad 11 = 11$ ✓

Since $11 = 11$ is true, $(-2, 13)$ is a solution of $x + y = 11$.

(b) There are infinitely many solutions. We can choose any two numbers whose sum is 11.

$x = -3$ and $y = 14$, or $(-3, 14)$ $\quad x + y = 11$

$-3 + 14 = 11, \quad 11 = 11$ ✓

$x = 2$ and $y = 9$, or $(2, 9)$ $\quad x + y = 11$

$2 + 9 = 11, \quad 11 = 11$ ✓

$x = 5$ and $y = 6$, or $(5, 6)$ $\quad x + y = 11$

$5 + 6 = 11, \quad 11 = 11$ ✓

Therefore $(-3, 14)$, $(2, 9)$, and $(5, 6)$ are solutions to $x + y = 11$.

8. **(a)** Replace x by 0 in the equation.

$3x - 4y = 12$

$3(0) - 4y = 12$

$0 - 4y = 12$

$y = -3$

The ordered pair is $(0, -3)$.

(b) Replace the variable y by 3.

$3x - 4y = 12$

$3x - 4(3) = 12$

$3x - 12 = 12$

$3x = 24$

$x = 8$

The ordered pair is $(8, 3)$.

(c) Replace the variable y by -6.

$3x - 4y = 12$

$3x - 4(-6) = 12$

$3x + 24 = 12$

$3x = -12$

$x = -4$

The ordered pair is $(-4, -6)$.

9. First solve the equation for y.

$$8 - 2y + 3x = 0$$
$$-2y = -3x - 8$$
$$y = \frac{-3x - 8}{-2}$$
$$y = \frac{-3x}{-2} + \frac{-8}{-2}$$
$$y = \frac{3}{2}x + 4$$

(a) Replace x with 0.

$$y = \frac{3}{2}(0) + 4$$
$$y = 4$$

(b) Replace x with 2.

$$y = \frac{3}{2}(2) + 4$$
$$y = 3 + 4$$
$$y = 7$$

Thus, we have the ordered pairs $(0, 4)$ and $(2, 7)$.

3.2 Student Practice

1. Graph $y = -3x - 1$.

Let $x = 0$.

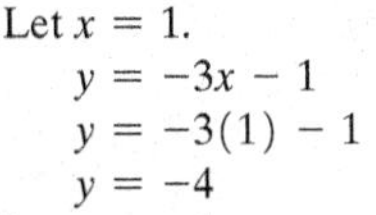

$$y = -3x - 1$$
$$y = -3(0) - 1$$
$$y = -1$$

Let $x = 1$.

$$y = -3x - 1$$
$$y = -3(1) - 1$$
$$y = -4$$

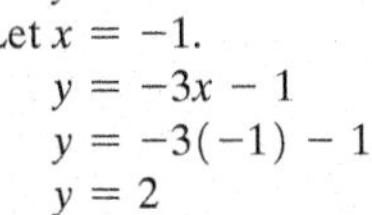

Let $x = -1$.

$$y = -3x - 1$$
$$y = -3(-1) - 1$$
$$y = 2$$

Plot the ordered pairs $(0, -1)$, $(1, -4)$, $(-1, 2)$

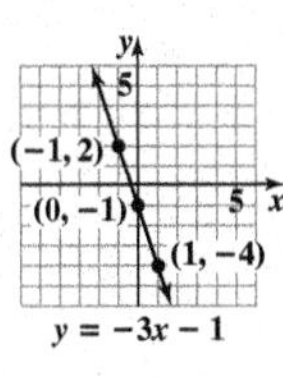

2.

$$7x + 3 = -2y + 3$$
$$7x + 3 - 3 = -2y + 3 - 3$$
$$7x = -2y$$
$$7x + 2y = -2y + 2y$$
$$7x + 2y = 0$$

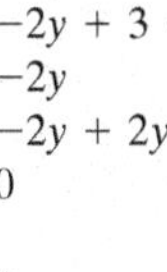

Let $x = 0$.

$$7(0) + 2y = 0$$
$$2y = 0$$
$$y = 0$$

Let $x = -2$.

$$7(-2) + 2y = 0$$
$$-14 + 2y = 0$$
$$2y = 14$$
$$y = 7$$

Let $x = 2$.

$$7(2) + 2y = 0$$
$$14 + 2y = 0$$
$$2y = -14$$
$$y = -7$$

Graph the ordered pairs $(0, 0)$, $(-2, 7)$, and $(2, -7)$.

3. $2y - x = 6$

Find the two intercepts.

Let $y = 0$.

$$2(0) - x = 6$$
$$-x = 6$$
$$x = -6$$

Let $x = 0$.

$$2y - 0 = 6$$
$$2y = 6$$
$$y = 3$$

The x-intercept is $(-6, 0)$, the y-intercept is $(0, 3)$.

Find a third point.

Let $y = 1$.

$$2(1) - x = 6$$
$$2 - x = 6$$
$$-x = 4$$
$$x = -4$$

Graph the ordered pairs $(-6, 0)$, $(0, 3)$, and $(-4, 1)$.

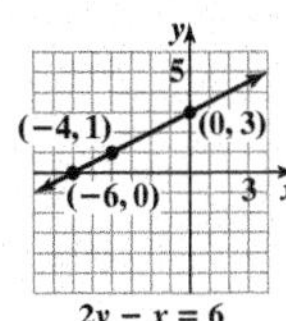

4. $y = 2$

This line is parallel to the x-axis. It is a horizontal line 2 units above the x-axis.

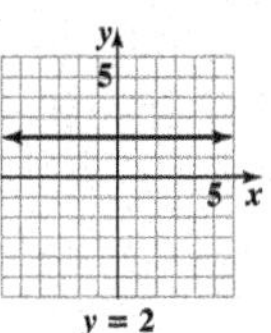

5. $3x + 1 = -8$

Solve for x.

$$3x + 1 - 1 = -8 - 1$$
$$3x = -9$$
$$x = -3$$

This line is parallel to the y-axis. It is a vertical line 3 units to the left of the y-axis.

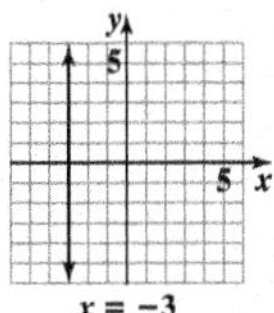

3.3 Student Practice

1. $m = \frac{y_2 - y_1}{x_2 - x_1} = \frac{-1 - 1}{-4 - 6} = \frac{-2}{-10} = \frac{1}{5}$

2. $m = \frac{y_2 - y_1}{x_2 - x_1} = \frac{1 - 0}{-1 - 2} = \frac{1}{-3} = -\frac{1}{3}$

3. **(a)** $m = \frac{3 - 6}{-5 - (-5)} = \frac{-3}{0}$

$\frac{-3}{0}$ is undefined. Therefore there is no slope and the line is a vertical line through $x = -5$.

(b) $m = \frac{-11 - (-11)}{3 - (-7)} = \frac{0}{10} = 0$

$m = 0$. The line is a horizontal line through $y = -11$.

4. Solve for y.

$$4x - 2y = -5$$
$$-2y = -4x - 5$$
$$y = \frac{-4x - 5}{-2}$$
$$y = 2x + \frac{5}{2} \qquad \text{Slope} = 2 \qquad y\text{-intercept} = \left(0, \frac{5}{2}\right)$$

5. **(a)** $y = mx + b$

$m = -\frac{3}{7}$ $\quad y$-intercept $= \left(0, \frac{2}{7}\right)$, $b = \frac{2}{7}$

$$y = -\frac{3}{7}x + \frac{2}{7}$$

(b)

$$y = -\frac{3}{7}x + \frac{2}{7}$$
$$7(y) = 7\left(-\frac{3}{7}x\right) + 7\left(\frac{2}{7}\right)$$
$$7y = -3x + 2$$
$$3x + 7y = 2$$

6. y-intercept $= (0, -1)$. Thus the coordinates of the y-intercept for this line are $(0, -1)$. Plot the point. Slope is $\frac{\text{rise}}{\text{run}}$. Since the slope for this line is $\frac{3}{4}$, we will go up (rise) 3 units and go over (run) 4 units to the right from the point $(0, -1)$. This is the point $(4, 2)$.

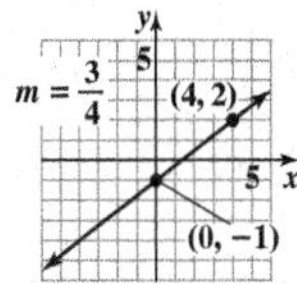

7. $y = -\frac{2}{3}x + 5$
The y-intercept is $(0, 5)$ since $b = 5$. Plot the point $(0, 5)$. The slope is $-\frac{2}{3} = \frac{-2}{3}$. Begin at $(0, 5)$, go down 2 units and to the right 3 units. This is the point $(3, 3)$. Draw the line that connects the points $(0, 5)$ and $(3, 3)$.

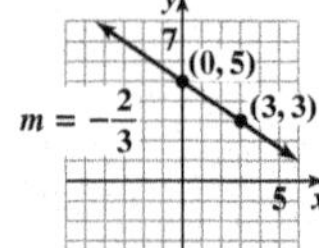

8. **(a)** Parallel lines have the same slope. Line j has a slope of $\frac{1}{4}$.
(b) Perpendicular lines have slopes whose product is -1.
$m_1m_2 = -1$
$\frac{1}{4}m_2 = -1$
$4\left(\frac{1}{4}\right)m_2 = -1(4)$
$m_2 = -4$
Thus line k has a slope of -4.

9. **(a)** The slope of line n is $\frac{2}{3}$. The slope of a line that is parallel to line n is $\frac{2}{3}$.
(b) $m_1m_2 = -1$
$\frac{2}{3}m_2 = -1$
$m_2 = -\frac{3}{2}$
The slope of a line that is perpendicular to line n is $-\frac{3}{2}$.

3.4 Student Practice

1. $y = mx + b$
$12 = -\frac{3}{4}(-8) + b$
$12 = 6 + b$
$6 = b$
An equation of the line is $y = -\frac{3}{4}x + 6$.

2. Find the slope.
$m = \frac{y_2 - y_1}{x_2 - x_1} = \frac{1 - 5}{-1 - 3} = \frac{-4}{-4} = 1$
Using either of the two points given, substitute x and y values into the equation $y = mx + b$.
$m = 1 \quad x = 3$ and $y = 5$.
$y = mx + b$
$5 = 1(3) + b$
$5 = 3 + b$
$2 = b$
An equation of the line is $y = x + 2$.

3. The y-intercept is $(0, 1)$. Thus $b = 1$. Look for another point on the line. We choose $(6, 2)$. Count the number of vertical units from 1 to 2 (rise). Count the number of horizontal units from 0 to 6 (run) $m = \frac{1}{6}$.
Now we can write an equation of the line.
$y = mx + b$
$y = \frac{1}{6}x + 1$

3.5 Student Practice

1. Graph $x - y \geq -10$.
Begin by graphing the line $x - y = -10$. Use any method discussed previously. Since there is an equals sign in the inequality, draw a solid line to indicate that the line is part of the solution set. The easiest test point is $(0, 0)$. Substitute $x = 0, y = 0$ in the inequality.
$x - y \geq -10$
$0 - 0 \geq -10$
$0 \geq -10$ true
Therefore, shade the side of the line that includes the point $(0, 0)$.

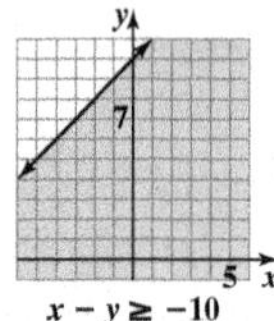

2. **Step 1** Graph $2y = x$. Since $>$ is used, the line should be a dashed line.
Step 2 The line passes through $(0, 0)$.
Step 3 Choose another test point, say $(-1, 1)$.
$2y > x$
$2(1) > -1$
$2 > -1$ true
Shade the region that includes $(-1, 1)$, that is, the region above the line.

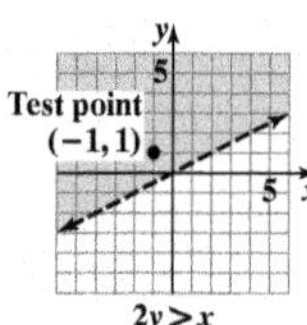

3. **Step 1** Graph $y = -3$. Since $\geq$ is used, the line should be solid.
Step 2 Test $(0, 0)$ in the inequality.
$y \geq -3$
$0 \geq -3$ true
Shade the region that includes $(0, 0)$, that is, the region above the line $y = -3$.

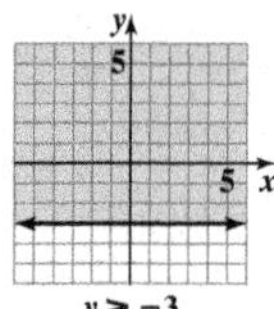

3.6 Student Practice

1. The domain consists of all the first coordinates of the ordered pairs. The domain is $\{-3, 0, 3, 20\}$. The range consists of all the second coordinates of the ordered pairs. The range is $\{-5, 5\}$.

2. **(a)** Look at the ordered pairs. No two ordered pairs have the same first coordinate. Thus this set of ordered pairs defines a function.
(b) Look at the ordered pairs. Two different ordered pairs, $(60, 30)$ and $(60, 120)$, have the same first coordinate. Thus this relation is not a function.

3. **(a)** Looking at the table, we see that no two different ordered pairs have the same first coordinate. The cost of gasoline is a function of the distance traveled.
Note that cost depends on distance. Thus distance is the independent variable. Since a negative distance does not make sense, the domain is {all nonnegative real numbers}.
The range is {all nonnegative real numbers}.
(b) Looking at the table, we see two ordered pairs, $(5, 20)$ and $(5, 30)$, have the same first coordinate. Thus this relation is not a function.

4. Construct a table, plot the ordered pairs, and connect the points.

x	$y = x^2 - 2$	y
-2	$y = (-2)^2 - 2 = 2$	2
-1	$y = (-1)^2 - 2 = -1$	-1
0	$y = (0)^2 - 2 = -2$	-2
1	$y = (1)^2 - 2 = -1$	-1
2	$y = (2)^2 - 2 = 2$	2

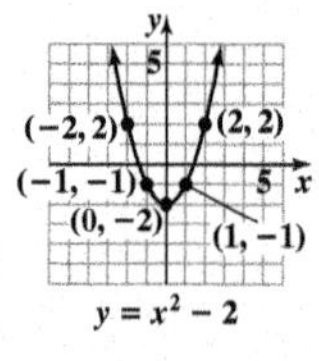

5. Select values of y and then substitute them into the equation to obtain values of x.

y	$x = y^2 - 1$	x	y
-2	$x = (-2)^2 - 1 = 3$	3	-2
-1	$x = (-1)^2 - 1 = 0$	0	-1
0	$x = (0)^2 - 1 = -1$	-1	0
1	$x = (1)^2 - 1 = 0$	0	1
2	$x = (2)^2 - 1 = 3$	3	2

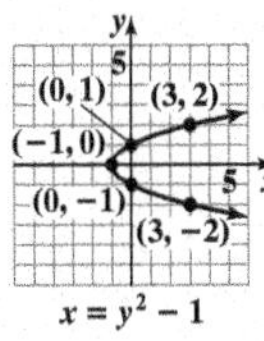

6. $y = \frac{6}{x}$

x	$y = \frac{6}{x}$	y
-3	$y = \frac{6}{-3} = -2$	-2
-2	$y = \frac{6}{-2} = -3$	-3
-1	$y = \frac{6}{-1} = -6$	-6
0	We cannot divide by 0.	
1	$y = \frac{6}{1} = 6$	6
2	$y = \frac{6}{2} = 3$	3
3	$y = \frac{6}{3} = 2$	2

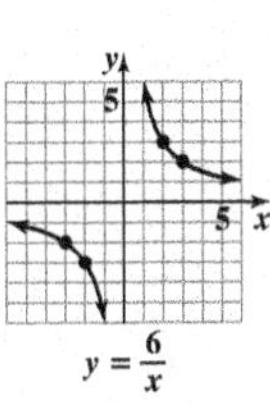

7. **(a)** The graph of a vertical line is not a function.
(b) This curve is a function. Any vertical line will cross the curve in only one location.
(c) This curve is not the graph of a function. There exist vertical lines that will cross the curve in more than one place.

8. $f(x) = -2x^2 + 3x - 8$

(a) $f(2) = -2(2)^2 + 3(2) - 8$
$= -2(4) + 3(2) - 8$
$= -8 + 6 - 8$
$= -10$

(b) $f(-3) = -2(-3)^2 + 3(-3) - 8$
$= -2(9) + 3(-3) - 8$
$= -18 - 9 - 8$
$= -35$

(c) $f(0) = -2(0)^2 + 3(0) - 8$
$= -2(0) + 3(0) - 8$
$= 0 + 0 - 8$
$= -8$

Chapter 4 4.1 Student Practice

1. Substitute $(-3, 4)$ into the first equation to see if the ordered pair is a solution.

$$2x + 3y = 6$$
$$2(-3) + 3(4) \stackrel{?}{=} 6$$
$$-6 + 12 \stackrel{?}{=} 6$$
$$6 = 6 \checkmark$$

Likewise, we will determine if $(-3, 4)$ is a solution to the second equation.

$$3x - 4y = 7$$
$$3(-3) - 4(4) \stackrel{?}{=} 7$$
$$-9 - 16 \stackrel{?}{=} 7$$
$$-25 \neq 7$$

Since $(-3, 4)$ is not a solution to both equations in the system, it is not a solution to the system itself.

2. You can use any method from Chapter 3 to graph each line. We will change each equation to slope-intercept form to graph.

$$3x + 2y = 10$$
$$2y = -3x + 10$$
$$y = -\frac{3}{2}x + 5$$
$$x - y = 5$$
$$y = x - 5$$

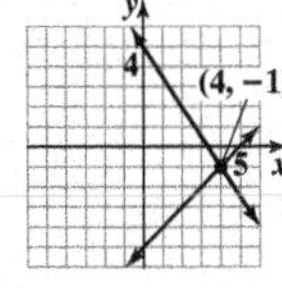

The solution is $(4, -1)$.

The lines intersect at the point $(4, -1)$. Thus $(4, -1)$ is the solution. We verify this by substituting $x = 4$ and $y = -1$ into the system of equations.

$$3x + 2y = 10$$
$$3(4) + 2(-1) \stackrel{?}{=} 10$$
$$12 - 2 \stackrel{?}{=} 10$$
$$10 = 10 \checkmark$$

$$x - y = 5$$
$$4 - (-1) \stackrel{?}{=} 5$$
$$5 = 5 \checkmark$$

3. $2x - y = 7$ [1]
$3x + 4y = -6$ [2]
Solve equation [1] for y.

$$-y = 7 - 2x$$
$$y = -7 + 2x \quad [3]$$

Substitute $-7 + 2x$ for y in equation [2].

$$3x + 4(-7 + 2x) = -6$$
$$3x - 28 + 8x = -6$$
$$11x - 28 = -6$$
$$11x = 22$$
$$x = 2$$

Substitute $x = 2$ into equation [1].

$$2(2) - y = 7$$
$$4 - y = 7$$
$$-y = 3$$
$$y = -3$$

The solution is $(2, -3)$.

4. $\frac{1}{2}x + \frac{2}{3}y = 1$ [1]

$\frac{1}{3}x + y = -1$ [2]

Clear both equations of fractions.

$$6\left(\frac{1}{2}x\right) + 6\left(\frac{2}{3}y\right) = 6(1)$$
$$3x + 4y = 6 \quad [3]$$
$$3\left(\frac{1}{3}x\right) + 3(y) = 3(-1)$$
$$x + 3y = -3 \quad [4]$$

Solve equation [4] for x.

$$x = -3 - 3y$$

Substitute $-3 - 3y$ for x in equation [3].

$$3(-3 - 3y) + 4y = 6$$
$$-9 - 9y + 4y = 6$$
$$-9 - 5y = 6$$
$$-5y = 15$$
$$y = -3$$

Substitute $y = -3$ into equation [2].

$$\frac{x}{3} - 3 = -1$$
$$\frac{x}{3} = 2$$
$$x = 6$$

The solution is $(6, -3)$.

5. $-3x + y = 5$ [1]

$2x + 3y = 4$ [2]

Multiply equation [1] by -3.

$$-3(-3x) + (-3)(y) = -3(5)$$
$$9x - 3y = -15 \quad [3]$$

Add equations [3] and [2].

$$9x - 3y = -15$$
$$2x + 3y = 4$$
$$11x = -11$$
$$x = -1$$

Now substitute $x = -1$ into equation [1].

$$-3(-1) + y = 5$$
$$3 + y = 5$$
$$y = 2$$

The solution is $(-1, 2)$.

6. $\frac{x}{4} + \frac{y}{5} = \frac{23}{20}$ [1]

$\frac{7}{15}x - \frac{y}{5} = 1$ [2]

Clear both equations of fractions.

$$20\left(\frac{x}{4}\right) + 20\left(\frac{y}{5}\right) = 20\left(\frac{23}{20}\right)$$
$$5x + 4y = 23 \quad [3]$$
$$15\left(\frac{7}{15}x\right) - 15\left(\frac{y}{5}\right) = 15(1)$$
$$7x - 3y = 15 \quad [4]$$

We now have an equivalent system.

$$5x + 4y = 23 \quad [3]$$
$$7x - 3y = 15 \quad [4]$$

Multiply equation [3] by 3 and equation [4] by 4.

$$15x + 12y = 69$$
$$28x - 12y = 60$$
$$43x = 129$$
$$x = 3$$

Now substitute $x = 3$ into equation [3].

$$5(3) + 4y = 23$$
$$15 + 4y = 23$$
$$4y = 8$$
$$y = 2$$

The solution is $(3, 2)$.

7. $4x - 2y = 6$ [1]

$-6x + 3y = 9$ [2]

Multiply equation [1] by 3 and equation [2] by 2.

$$3(4x) - 3(2y) = 3(6)$$
$$12x - 6y = 18 \quad [3]$$
$$2(-6x) + 2(3y) = 2(9)$$
$$-12x + 6y = 18 \quad [4]$$

When we add equations [3] and [4] we get $0 = 36$.
This statement is of course false. Thus, we conclude that this system of equations is inconsistent, so there is no solution.

8. $0.3x - 0.9y = 1.8$ [1]

$-0.4x + 1.2y = -2.4$ [2]

Multiply both equations by 10 to obtain a more convenient form.

$$3x - 9y = 18 \quad [3]$$
$$-4x + 12y = -24 \quad [4]$$

Multiply equation [3] by 4 and equation [4] by 3.

$$12x - 36y = 72$$
$$-12x + 36y = -72$$
$$0 = 0$$

This statement is always true. Hence these are dependent equations. There is an infinite number of solutions.

9. (a) $3x + 5y = 1485$

$x + 2y = 564$

Solve for x in the second equation and solve using the substitution method.

$$x = -2y + 564$$
$$3(-2y + 564) + 5y = 1485$$
$$-6y + 1692 + 5y = 1485$$
$$-y = -207$$
$$y = 207$$

Substitute $y = 207$ into the second equation and solve for x.

$$x + 2(207) = 564$$
$$x + 414 = 564$$
$$x = 150$$

The solution is $(150, 207)$.

(b) $7x + 6y = 45$

$6x - 5y = -2$

Use the addition method. Multiply the first equation by -6 and the second equation by 7.

$$-6(7x) + (-6)(6y) = (-6)(45)$$
$$7(6x) - 7(5y) = 7(-2)$$

$$-42x - 36y = -270$$
$$42x - 35y = -14$$
$$-71y = -284$$
$$y = 4$$

Substitute $y = 4$ into the first equation and solve for x.

$$7x + 6(4) = 45$$
$$7x + 24 = 45$$
$$7x = 21$$
$$x = 3$$

The solution is $(3, 4)$.

4.2 Student Practice

1. Substitute $x = 3, y = -2, z = 2$ into each equation.

$$2(3) + 4(-2) + 2 \stackrel{?}{=} 0$$
$$6 - 8 + 2 \stackrel{?}{=} 0$$
$$0 = 0 \ \checkmark$$
$$3 - 2(-2) + 5(2) \stackrel{?}{=} 17$$
$$3 + 4 + 10 \stackrel{?}{=} 17$$
$$17 = 17 \ \checkmark$$
$$3(3) - 4(-2) + 2 \stackrel{?}{=} 19$$
$$9 + 8 + 2 \stackrel{?}{=} 19$$
$$19 = 19 \ \checkmark$$

Since we obtained three true statements, the ordered triple $(3, -2, 2)$ is a solution to the system.

2. $x + 2y + 3z = 4$ [1]
$2x + y - 2z = 3$ [2]
$3x + 3y + 4z = 10$ [3]
Eliminate x by multiplying equation [1] by -2 (call it equation [4]) and adding it to equation [2].

$$\begin{aligned} -2x - 4y - 6z &= -8 \quad [4] \\ 2x + y - 2z &= 3 \quad [2] \\ \hline -3y - 8z &= -5 \quad [5] \end{aligned}$$

Now eliminate x by multiplying equation [1] by -3 (call it equation [6]) and adding it to equation [3].

$$\begin{aligned} -3x - 6y - 9z &= -12 \quad [6] \\ 3x + 3y + 4z &= 10 \quad [3] \\ \hline -3y - 5z &= -2 \quad [7] \end{aligned}$$

Now eliminate y and solve for z in the system formed by equation [5] and equation [7].

$$\begin{aligned} -3y - 8z &= -5 \quad [5] \\ -3y - 5z &= -2 \quad [7] \end{aligned}$$

To do this, multiply equation [5] by -1 (call it equation [8]) and add it to equation [7].

$$\begin{aligned} 3y + 8z &= 5 \quad [8] \\ -3y - 5z &= -2 \quad [7] \\ \hline 3z &= 3 \\ z &= 1 \end{aligned}$$

Substitute $z = 1$ into equation [8] and solve for y.

$$\begin{aligned} 3y + 8(1) &= 5 \\ 3y &= -3 \\ y &= -1 \end{aligned}$$

Substitute $z = 1$ and $y = -1$ into equation [1] and solve for x.

$$\begin{aligned} x + 2y + 3z &= 4 \\ x + 2(-1) + 3(1) &= 4 \\ x - 2 + 3 &= 4 \\ x &= 3 \end{aligned}$$

The solution is $(3, -1, 1)$.

3. $2x + y + z = 11$ [1]
$4y + 3z = -8$ [2]
$x - 5y = 2$ [3]
Multiply equation [1] by -3 and add the result to equation [2], thus eliminating the z terms.

$$\begin{aligned} -6x - 3y - 3z &= -33 \quad [4] \\ 4y + 3z &= -8 \quad [2] \\ \hline -6x + y &= -41 \quad [5] \end{aligned}$$

We can solve the system formed by equation [3] and equation [5].

$$\begin{aligned} x - 5y &= 2 \quad [3] \\ -6x + y &= -41 \quad [5] \end{aligned}$$

Multiply equation [3] by 6 and add the result to equation [5].

$$\begin{aligned} 6x - 30y &= 12 \quad [6] \\ -6x + y &= -41 \quad [5] \\ \hline -29y &= -29 \\ y &= 1 \end{aligned}$$

Substitute $y = 1$ into equation [3] and solve for x.

$$\begin{aligned} x - 5(1) &= 2 \\ x &= 7 \end{aligned}$$

Now substitute $y = 1$ into equation [2] and solve for z.

$$\begin{aligned} 4y + 3z &= -8 \\ 4(1) + 3z &= -8 \\ 3z &= -12 \\ z &= -4 \end{aligned}$$

The solution is $(7, 1, -4)$.

4.3 Student Practice

1. Let x = the number of baseballs purchased and y = the number of bats purchased.
Last week: $6x + 21y = 318$ [1]
This week: $5x + 17y = 259$ [2]
Multiply equation [1] by 5 and equation [2] by -6.

$$\begin{aligned} 30x + 105y &= 1590 \\ -30x - 102y &= -1554 \\ \hline 3y &= 36 \\ y &= 12 \end{aligned}$$

Substitute $y = 12$ into equation [2].

$$\begin{aligned} 5x + 17(12) &= 259 \\ 5x &= 55 \\ x &= 11 \end{aligned}$$

Thus 11 baseballs and 12 bats were purchased.

2. Let x = the number of small chairs and y = the number of large chairs. (*Note*: All hours have been changed to minutes.)

$$\begin{aligned} 30x + 40y &= 1560 \quad [1] \\ 75x + 80y &= 3420 \quad [2] \end{aligned}$$

Multiply equation [1] by -2 and add the results to equation [2].

$$\begin{aligned} -60x - 80y &= -3120 \\ 75x + 80y &= 3420 \\ \hline 15x &= 300 \\ x &= 20 \end{aligned}$$

Substitute $x = 20$ in equation [1] and solve for y.

$$\begin{aligned} 600 + 40y &= 1560 \\ 40y &= 960 \\ y &= 24 \end{aligned}$$

Therefore, the company can make 20 small chairs and 24 large chairs each day.

3. Let a = the speed of the airplane in still air in kilometers per hour and w = the speed of the wind in kilometers per hour.

	R	· T	= D
Against the wind	$a - w$	3	1950
With the wind	$a + w$	2	1600

From the chart, we have the following equations.

$$\begin{aligned} (a - w)(3) &= 1950 \\ (a + w)(2) &= 1600 \end{aligned}$$

Remove the parentheses.

$$\begin{aligned} 3a - 3w &= 1950 \quad [1] \\ 2a + 2w &= 1600 \quad [2] \end{aligned}$$

Multiply equation [1] by 2 and equation [2] by 3 and add the resulting equations.

$$\begin{aligned} 6a - 6w &= 3900 \\ 6a + 6w &= 4800 \\ \hline 12a &= 8700 \\ a &= 725 \end{aligned}$$

Substitute $a = 725$ into equation [2].

$$\begin{aligned} 2(725) + 2w &= 1600 \\ 1450 + 2w &= 1600 \\ 2w &= 150 \\ w &= 75 \end{aligned}$$

Thus, the speed of the plane in still air is 725 kilometers per hour and the speed of the wind is 75 kilometers per hour.

4. Let A = the number of boxes that machine A can wrap in 1 hour, B = the number that machine B can wrap in 1 hour, and C = the number that machine C can wrap in one hour.

$$\begin{aligned} A + B + C &= 260 \quad [1] \\ 3A + 2B &= 390 \quad [2] \\ 3B + 4C &= 655 \quad [3] \end{aligned}$$

Multiply equation [1] by -3 and add it to equation [2].

$$\begin{aligned} -3A - 3B - 3C &= -780 \quad [4] \\ 3A + 2B &= 390 \quad [2] \\ \hline -B - 3C &= -390 \quad [5] \end{aligned}$$

Now multiply equation [5] by 3 and add it to equation [3].

$$\begin{aligned} -3B - 9C &= -1170 \quad [6] \\ 3B + 4C &= 655 \quad [3] \\ \hline -5C &= -515 \\ C &= 103 \end{aligned}$$

Substitute $C = 103$ into equation [3] and solve for B.

$$\begin{aligned} 3B + 412 &= 655 \\ 3B &= 243 \\ B &= 81 \end{aligned}$$

Now substitute $B = 81$ into equation [2] and solve for A.

$$\begin{aligned} 3A + 162 &= 390 \\ 3A &= 228 \\ A &= 76 \end{aligned}$$

Machine A wraps 76 boxes per hour, machine B wraps 81 boxes per hour, and machine C wraps 103 boxes per hour.

4.4 Student Practice

1. The graph of $-2x + y \leq -3$ is the region on and below the line $-2x + y = -3$. The graph of $x + 2y \geq 4$ is the region on and above the line $x + 2y = 4$.

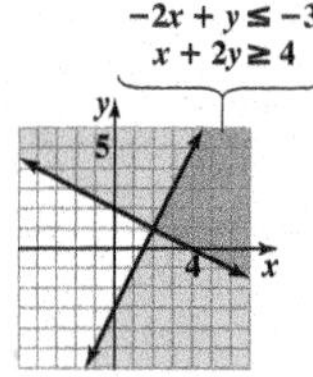

2. The graph of $y > -1$ is the region above the line $y = -1$, not including the line. The graph of $y < -\frac{3}{4}x + 2$ is the region below the line $y = -\frac{3}{4}x + 2$, not including the line.

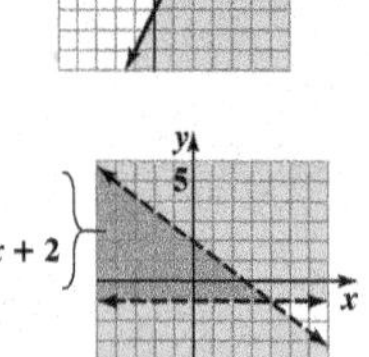

3. The graph of $x + y \leq 6$ is the region on and below the line $x + y = 6$. The graph of $3x + y \leq 12$ is the region on and below the line $3x + y = 12$. The lines $x + y = 6$ and $3x + y = 12$ intersect at the point $(3, 3)$. The graph of $x \geq 0$ is the y-axis and all the region to the right of the y-axis. The graph of $y \geq 0$ is the x-axis and all the region above the x-axis. The vertices of the solution are $(0, 0)$, $(0, 6)$, $(3, 3)$, and $(4, 0)$.

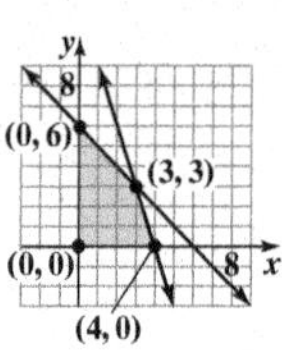

Chapter 5

5.1 Student Practice

1. (a) $a^7 \cdot a^5 = a^{7+5} = a^{12}$ (b) $w^{10} \cdot w = w^{10+1} = w^{11}$

2. (a) $x^3 \cdot x^9 = x^{3+9} = x^{12}$ (b) $3^7 \cdot 3^4 = 3^{7+4} = 3^{11}$
(c) $a^3 \cdot b^2 = a^3 \cdot b^2$ (cannot be simplified)

3. (a) $(7a^8)(a^4) = (7 \cdot 1)(a^8 \cdot a^4) = 7(a^8 \cdot a^4) = 7a^{12}$
(b) $(3y^2)(-2y^3) = (3)(-2)(y^2 \cdot y^3) = -6y^5$
(c) $(-4x^3)(-5x^2) = (-4)(-5)(x^3 \cdot x^2) = 20x^5$

4. $(2xy)\left(-\frac{1}{4}x^2y\right)(6xy^3) = (2)\left(-\frac{1}{4}\right)(6)(x \cdot x^2 \cdot x)(y \cdot y \cdot y^3) = -3x^4y^5$

5. (a) $\frac{10^{13}}{10^7} = 10^{13-7} = 10^6$
(b) $\frac{x^{11}}{x} = x^{11-1} = x^{10}$
(c) $\frac{y^{18}}{y^8} = y^{18-8} = y^{10}$

6. (a) $\frac{c^3}{c^4} = \frac{1}{c^{4-3}} = \frac{1}{c^1} = \frac{1}{c}$
(b) $\frac{10^{31}}{10^{56}} = \frac{1}{10^{56-31}} = \frac{1}{10^{25}}$
(c) $\frac{z^{15}}{z^{21}} = \frac{1}{z^{21-15}} = \frac{1}{z^6}$

7. (a) $\frac{-7x^7}{-21x^9} = \frac{1}{3x^{9-7}} = \frac{1}{3x^2}$
(b) $\frac{15x^{11}}{-3x^4} = -5x^{11-4} = -5x^7$
(c) $\frac{23b^8}{46b^9} = \frac{1}{2b^{9-8}} = \frac{1}{2b}$

8. (a) $\frac{r^7s^9}{s^{10}} = \frac{r^7}{s^{10-9}} = \frac{r^7}{s}$
(b) $\frac{12x^5y^6}{-24x^3y^8} = \frac{x^{5-3}}{-2y^{8-6}} = -\frac{x^2}{2y^2}$

9. (a) $\frac{10^7}{10^7} = 1$
(b) $\frac{12a^4}{2a^4} = 6\left(\frac{a^4}{a^4}\right) = 6(1) = 6$

10. (a) $\frac{-20a^3b^8c^4}{28a^3b^7c^5} = -\frac{5a^0b}{7c} = -\frac{5(1)b}{7c} = -\frac{5b}{7c}$
(b) $\frac{5x^0y^6}{10x^4y^8} = \frac{5(1)y^6}{10x^4y^8} = \frac{1}{2x^4y^2}$

11. $\frac{(-6ab^5)(3a^2b^4)}{16a^5b^7} = \frac{-18a^3b^9}{16a^5b^7} = -\frac{9b^2}{8a^2}$

12. (a) $(a^4)^3 = a^{4 \cdot 3} = a^{12}$
(b) $(10^5)^2 = 10^{5 \cdot 2} = 10^{10}$
(c) $(-1)^{15} = -1$

13. (a) $(3xy)^3 = (3)^3x^3y^3 = 27x^3y^3$
(b) $(yz)^{37} = y^{37}z^{37}$
(c) $(-3x^3)^2 = (-3)^2(x^3)^2 = 9x^6$

14. (a) $\left(\frac{x}{5}\right)^3 = \frac{x^3}{5^3} = \frac{x^3}{125}$
(b) $\frac{(4a)^2}{(ab)^6} = \frac{4^2a^2}{a^6b^6} = \frac{16a^2}{a^6b^6} = \frac{16}{a^4b^6}$

15. $\left(\frac{-2x^3y^0z}{4xz^2}\right)^5 = \left(\frac{-x^2}{2z}\right)^5 = \frac{(-1)^5(x^2)^5}{2^5z^5} = -\frac{x^{10}}{32z^5}$

5.2 Student Practice

1. (a) $x^{-12} = \frac{1}{x^{12}}$ (b) $w^{-5} = \frac{1}{w^5}$ (c) $z^{-2} = \frac{1}{z^2}$

2. (a) $4^{-3} = \frac{1}{4^3} = \frac{1}{64}$ (b) $2^{-4} = \frac{1}{2^4} = \frac{1}{16}$

3. (a) $\frac{3}{w^{-4}} = 3w^4$ (b) $\frac{x^{-6}y^4}{z^{-2}} = \frac{y^4z^2}{x^6}$ (c) $x^{-6}y^{-5} = \frac{1}{x^6y^5}$

4. (a) $(2x^4y^{-5})^{-2} = 2^{-2}x^{-8}y^{10} = \frac{y^{10}}{2^2x^8} = \frac{y^{10}}{4x^8}$
(b) $\frac{y^{-3}z^{-4}}{y^2z^{-6}} = \frac{z^6}{y^2y^3z^4} = \frac{z^6}{y^5z^4} = \frac{z^2}{y^5}$

5. (a) $78{,}200 = 7.82 \times 10{,}000 = 7.82 \times 10^4$
Notice we moved the decimal point 4 places to the left.
(b) $4{,}786{,}000 = 4.786 \times 1{,}000{,}000 = 4.786 \times 10^6$

6. (a) $0.98 = 9.8 \times 10^{-1}$
(b) $0.000092 = 9.2 \times 10^{-5}$

7. (a) $1.93 \times 10^6 = 1.93 \times 1{,}000{,}000 = 1{,}930{,}000$
(b) $8.562 \times 10^{-5} = 8.562 \times \frac{1}{100{,}000} = 0.00008562$

8. $30{,}900{,}000{,}000{,}000{,}000$ meters $= 3.09 \times 10^{16}$ meters

9. (a) $(56{,}000)(1{,}400{,}000{,}000) = (5.6 \times 10^4)(1.4 \times 10^9) = (5.6)(1.4)(10^4)(10^9) = 7.84 \times 10^{13}$
(b) $\frac{0.000111}{0.00000037} = \frac{1.11 \times 10^{-4}}{3.7 \times 10^{-7}} = \frac{1.11}{3.7} \times \frac{10^{-4}}{10^{-7}} = \frac{1.11}{3.7} \times \frac{10^7}{10^4} = 0.3 \times 10^3 = 3.0 \times 10^2$

10. $159 \text{ parsecs} = (159 \text{ parsecs})\dfrac{(3.09 \times 10^{13} \text{ kilometers})}{1 \text{ parsec}}$
$= 491.31 \times 10^{13}$ kilometers
$d = r \times t$
$491.31 \times 10^{13} \text{ km} = \dfrac{50{,}000 \text{ km}}{1 \text{ hr}} \times t$
$4.9131 \times 10^{15} \text{ km} = \dfrac{5 \times 10^4 \text{ km}}{1 \text{ hr}} \times t$
$\dfrac{4.9131 \times 10^{15} \text{ km } (1 \text{ hr})}{5.0 \times 10^4 \text{ km}} = t$
$0.98262 \times 10^{11} \text{ hr} = t$
It would take the probe about 9.83×10^{10} hours.

5.3 Student Practice

1. **(a)** This polynomial is of degree 5. It has two terms, so it is a binomial.
(b) This polynomial is of degree 7, since the sum of the exponents is $3 + 4 = 7$. It has one term, so it is a monomial.
(c) This polynomial is of degree 3. It has three terms, so it is a trinomial.

2. $(-8x^3 + 3x^2 + 6) + (2x^3 - 7x^2 - 3)$
$= [-8x^3 + 2x^3] + [3x^2 + (-7x^2)] + [6 + (-3)]$
$= [(-8 + 2)x^3] + [(3 - 7)x^2] + [6 - 3]$
$= -6x^3 + (-4x^2) + 3$
$= -6x^3 - 4x^2 + 3$

3. $\left(-\frac{1}{3}x^2 - 6x - \frac{1}{12}\right) + \left(\frac{1}{4}x^2 + 5x - \frac{1}{3}\right)$
$= \left[-\frac{1}{3}x^2 + \frac{1}{4}x^2\right] + [-6x + 5x] + \left[-\frac{1}{12} + \left(-\frac{1}{3}\right)\right]$
$= \left[\left(-\frac{1}{3} + \frac{1}{4}\right)x^2\right] + [(-6 + 5)x] + \left[-\frac{1}{12} + \left(-\frac{1}{3}\right)\right]$
$= \left[\left(-\frac{4}{12} + \frac{3}{12}\right)x^2\right] + [-x] + \left[-\frac{1}{12} - \frac{4}{12}\right]$
$= -\frac{1}{12}x^2 - x - \frac{5}{12}$

4. $(3.5x^3 - 0.02x^2 + 1.56x - 3.5) + (-0.08x^2 - 1.98x + 4)$
$= 3.5x^3 + (-0.02 - 0.08)x^2 + (1.56 - 1.98)x + (-3.5 + 4)$
$= 3.5x^3 - 0.1x^2 - 0.42x + 0.5$

5. $(5x^3 - 15x^2 + 6x - 3) - (-4x^3 - 10x^2 + 5x + 13)$
$= (5x^3 - 15x^2 + 6x - 3) + (4x^3 + 10x^2 - 5x - 13)$
$= (5 + 4)x^3 + (-15 + 10)x^2 + (6 - 5)x + (-3 - 13)$
$= 9x^3 - 5x^2 + x - 16$

6. $(x^3 - 7x^2y + 3xy^2 - 2y^3) - (2x^3 + 4xy - 6y^3)$
$= (x^3 - 7x^2y + 3xy^2 - 2y^3) + (-2x^3 - 4xy + 6y^3)$
$= (1 - 2)x^3 - 7x^2y + 3xy^2 - 4xy + (-2 + 6)y^3$
$= -x^3 - 7x^2y + 3xy^2 - 4xy + 4y^3$

7. **(a)** 1990 is 5 years later than 1985, so $x = 5$.
$0.16(5) + 20.7 = 0.8 + 20.7 = 21.5$
We estimate that the average light truck in 1990 obtained 21.5 miles per gallon.
(b) 2025 is 40 years later than 1985, so $x = 40$.
$0.16(40) + 20.7 = 6.4 + 20.7 = 27.1$
We estimate that the average light truck in 2025 will obtain 27.1 miles per gallon.

5.4 Student Practice

1. $4x^3(-2x^2 + 3x) = 4x^3(-2x^2) + 4x^3(3x)$
$= 4(-2)(x^3 \cdot x^2) + (4 \cdot 3)(x^3 \cdot x)$
$= -8x^5 + 12x^4$

2. **(a)** $-3x(x^2 + 2x - 4) = -3x^3 - 6x^2 + 12x$
(b) $6xy(x^3 + 2x^2y - y^2) = 6x^4y + 12x^3y^2 - 6xy^3$

3. $(-6x^3 + 4x^2 - 2x)(-3xy) = 18x^4y - 12x^3y + 6x^2y$

4. $(5x - 1)(x - 2) = 5x^2 - 10x - x + 2 = 5x^2 - 11x + 2$

5. $(8a - 5b)(3a - b) = 24a^2 - 8ab - 15ab + 5b^2$
$= 24a^2 - 23ab + 5b^2$

6. $(3a + 2b)(2a - 3c) = 6a^2 - 9ac + 4ab - 6bc$

7. $(3x - 2y)(3x - 2y) = 9x^2 - 6xy - 6xy + 4y^2$
$= 9x^2 - 12xy + 4y^2$

8. $(2x^2 + 3y^2)(5x^2 + 6y^2) = 10x^4 + 12x^2y^2 + 15x^2y^2 + 18y^4$
$= 10x^4 + 27x^2y^2 + 18y^4$

9. $A = (\text{length})(\text{width}) = (7x + 3)(2x - 1)$
$= 14x^2 - 7x + 6x - 3$
$= 14x^2 - x - 3$
There are $(14x^2 - x - 3)$ square feet in the room.

5.5 Student Practice

1. $(6x + 7)(6x - 7) = (6x)^2 - (7)^2 = 36x^2 - 49$

2. $(3x - 5y)(3x + 5y) = (3x)^2 - (5y)^2 = 9x^2 - 25y^2$

3. **(a)** $(4a - 9b)^2 = (4a)^2 - 2(4a)(9b) + (9b)^2$
$= 16a^2 - 72ab + 81b^2$
(b) $(5x + 4)^2 = (5x)^2 + 2(5x)(4) + (4)^2 = 25x^2 + 40x + 16$

4.
$$\begin{array}{r} 4x^3 - 2x^2 + x \\ x^2 + 3x - 2 \\ \hline -8x^3 + 4x^2 - 2x \\ 12x^4 - 6x^3 + 3x^2 \quad \\ 4x^5 - 2x^4 + x^3 \qquad \quad \\ \hline 4x^5 + 10x^4 - 13x^3 + 7x^2 - 2x \end{array}$$

5. $(2x^2 + 5x + 3)(x^2 - 3x - 4)$
$= 2x^2(x^2 - 3x - 4) + 5x(x^2 - 3x - 4) + 3(x^2 - 3x - 4)$
$= 2x^4 - 6x^3 - 8x^2 + 5x^3 - 15x^2 - 20x + 3x^2 - 9x - 12$
$= 2x^4 - x^3 - 20x^2 - 29x - 12$

6. $(3x - 2)(2x + 3)(3x + 2) = (3x - 2)(3x + 2)(2x + 3)$
$= [(3x)^2 - 2^2](2x + 3)$
$= (9x^2 - 4)(2x + 3)$
$$\begin{array}{r} 9x^2 - 4 \\ 2x + 3 \\ \hline 27x^2 + 0x - 12 \\ 18x^3 + 0x^2 - 8x \qquad \\ \hline 18x^3 + 27x^2 - 8x - 12 \end{array}$$
Thus we have
$(3x - 2)(2x + 3)(3x + 2) = 18x^3 + 27x^2 - 8x - 12.$

5.6 Student Practice

1. $\dfrac{15y^4 - 27y^3 - 21y^2}{3y^2} = \dfrac{15y^4}{3y^2} - \dfrac{27y^3}{3y^2} - \dfrac{21y^2}{3y^2} = 5y^2 - 9y - 7$

2.
$$\begin{array}{r} x^2 + 6x + 7 \\ x + 4 \overline{)\, x^3 + 10x^2 + 31x + 25} \\ \underline{x^3 + 4x^2} \qquad \qquad \\ 6x^2 + 31x \qquad \\ \underline{6x^2 + 24x} \qquad \\ 7x + 25 \\ \underline{7x + 28} \\ -3 \end{array}$$
Ans: $x^2 + 6x + 7 + \dfrac{-3}{x + 4}$

3.
$$\begin{array}{r} 2x^2 + x + 1 \\ x - 1 \overline{)\, 2x^3 - x^2 + 0x + 1} \\ \underline{2x^3 - 2x^2} \qquad \qquad \\ x^2 + 0x \qquad \\ \underline{x^2 - x} \qquad \\ x + 1 \\ \underline{x - 1} \\ 2 \end{array}$$
Ans: $2x^2 + x + 1 + \dfrac{2}{x - 1}$

4.
$$\begin{array}{r} 5x^2 + x - 2 \\ 4x - 3 \overline{)\, 20x^3 - 11x^2 - 11x + 6} \\ \underline{20x^3 - 15x^2} \qquad \qquad \\ 4x^2 - 11x \qquad \\ \underline{4x^2 - 3x} \qquad \\ -8x + 6 \\ \underline{-8x + 6} \\ 0 \end{array}$$
Ans: $5x^2 + x - 2$

Check. $(4x - 3)(5x^2 + x - 2)$
$= 20x^3 + 4x^2 - 8x - 15x^2 - 3x + 6$
$= 20x^3 - 11x^2 - 11x + 6$

Chapter 6 6.1 Student Practice

1. **(a)** $21a - 7b = 7(3a - b)$ because $7(3a - b) = 21a - 7b$.
   (b) $5xy + 8x = x(5y + 8)$ because $x(5y + 8) = 5xy + 8x$.
2. 4 is the greatest numerical common factor and a is a factor of each term. Thus, $4a$ is the greatest common factor.
   $12a^2 + 16ab^2 - 12a^2b = 4a(3a + 4b^2 - 3ab)$
3. **(a)** The largest integer common to both terms is 8.
   $16a^3 - 24b^3 = 8(2a^3 - 3b^3)$
   (b) r^3 is common to all the terms.
   $r^3s^2 - 4r^4s + 7r^5 = r^3(s^2 - 4rs + 7r^2)$
4. We can factor 9, a, b^2, and c out of each term.
   $18a^3b^2c - 27ab^3c^2 - 45a^2b^2c^2 = 9ab^2c(2a^2 - 3bc - 5ac)$
5. We can factor 6, x, and y^2 out of each term.
   $30x^3y^2 - 24x^2y^2 + 6xy^2 = 6xy^2(5x^2 - 4x + 1)$
   Check: $6xy^2(5x^2 - 4x + 1) = 30x^3y^2 - 24x^2y^2 + 6xy^2$
6. $3(a + 5b) + x(a + 5b) = (a + 5b)(3 + x)$
7. $8y(9y^2 - 2) - (9y^2 - 2) = 8y(9y^2 - 2) - 1(9y^2 - 2)$
   $= (9y^2 - 2)(8y - 1)$
8. $\pi b^2 - \pi a^2 = \pi(b^2 - a^2)$

6.2 Student Practice

1. $3y(2x - 7) - 8(2x - 7) = (2x - 7)(3y - 8)$
2. $6x^2 - 15x + 4x - 10 = 3x(2x - 5) + 2(2x - 5)$
   $= (2x - 5)(3x + 2)$
3. $ax + 2a + 4bx + 8b = a(x + 2) + 4b(x + 2)$
   $= (x + 2)(a + 4b)$
4. $6a^2 + 5bc + 10ab + 3ac = 6a^2 + 10ab + 3ac + 5bc$
   $= 2a(3a + 5b) + c(3a + 5b)$
   $= (3a + 5b)(2a + c)$
5. $6xy + 14x - 15y - 35 = 2x(3y + 7) - 15y - 35$
   $= 2x(3y + 7) - 5(3y + 7)$
   $= (3y + 7)(2x - 5)$
6. $3x + 6y - 5ax - 10ay = 3(x + 2y) - 5a(x + 2y)$
   $= (x + 2y)(3 - 5a)$
7. $10ad + 27bc - 6bd - 45ac = 10ad - 6bd - 45ac + 27bc$
   $= 2d(5a - 3b) - 9c(5a - 3b)$
   $= (5a - 3b)(2d - 9c)$
   Check: $(5a - 3b)(2d - 9c) = 10ad - 45ac - 6bd + 27bc$
   $= 10ad + 27bc - 6bd - 45ac$

6.3 Student Practice

1. The two numbers that you can multiply to get 12 and add to get 8 are 6 and 2. $x^2 + 8x + 12 = (x + 6)(x + 2)$
2. The two numbers that have a product of 30 and a sum of 17 are 2 and 15. $x^2 + 17x + 30 = (x + 2)(x + 15)$
3. The two numbers that have a product of $+18$ and a sum of -11 must both be negative. The numbers are -9 and -2.
   $x^2 - 11x + 18 = (x - 9)(x - 2)$
4. The two numbers whose product is 24 and whose sum is -11 are -8 and -3. $x^2 - 11x + 24 = (x - 8)(x - 3)$ or $(x - 3)(x - 8)$
5. The two numbers whose product is -24 and whose sum is -5 are -8 and $+3$. $x^2 - 5x - 24 = (x - 8)(x + 3)$ or $(x + 3)(x - 8)$
6. The two numbers whose product is -60 and whose sum is $+17$ are $+20$ and -3. $y^2 + 17y - 60 = (y + 20)(y - 3)$
   Check: $(y + 20)(y - 3) = y^2 - 3y + 20y - 60 = y^2 + 17y - 60$
7. List the possible factors of 60.

Factors of 60	Difference	
60 and 1	59	
30 and 2	28	
20 and 3	17	
15 and 4	11	
12 and 5	7	← Desired Value
10 and 6	4	

For the coefficient of the middle term to be -7, we must add -12 and $+5$. $x^2 - 7x - 60 = (x - 12)(x + 5)$
8. $a^4 = (a^2)(a^2)$
   The two numbers whose product is -42 and whose sum is 1 are 7 and -6.
   $a^4 + a^2 - 42 = (a^2 + 7)(a^2 - 6)$
9. $3x^2 + 45x + 150 = 3(x^2 + 15x + 50)$
   $= 3(x + 5)(x + 10)$
10. $4x^2 - 8x - 140 = 4(x^2 - 2x - 35)$
    $= 4(x + 5)(x - 7)$
11. $x(x + 1) - 4(5) = x^2 + x - 20$
    $= (x + 5)(x - 4)$

6.4 Student Practice

1. To get a first term of $2x^2$, the coefficients of x in the factors must be 2 and 1. To get a last term of 5, the constants in the factors must be 1 and 5. Possibilities:
   $(2x + 1)(x + 5) = 2x^2 + 11x + 5$
   $(2x + 5)(x + 1) = 2x^2 + 7x + 5$
   Thus $2x^2 + 7x + 5 = (2x + 5)(x + 1)$ or $(x + 1)(2x + 5)$.
2. The different factorizations of 9 are (3)(3) and (1)(9). The only factorization of 7 is (1)(7).

Possible Factors	Middle Term	Correct?
$(3x - 7)(3x - 1)$	$-24x$	No
$(9x - 7)(x - 1)$	$-16x$	No
$(9x - 1)(x - 7)$	$-64x$	Yes

The correct answer is $(9x - 1)(x - 7)$ or $(x - 7)(9x - 1)$.
3. The only factorization of 3 is (3)(1). The different factorizations of 14 are (14)(1) and (7)(2).

Possible Factors	Middle Term	Correct Factors?
$(3x + 14)(x - 1)$	$+11x$	No
$(3x + 1)(x - 14)$	$-41x$	No
$(3x + 7)(x - 2)$	$+x$	No (wrong sign)
$(3x + 2)(x - 7)$	$-19x$	No

To get the correct sign on the middle term, reverse the signs of the constants in the factors.
The correct answer is $(3x - 7)(x + 2)$ or $(x + 2)(3x - 7)$.
4. We list some of the options and check to see which one yields the middle term $+7x$.

$(2x \quad 2)(5x \quad 3)$: $10x$, $6x$ — No possible sum of $+7x$
$(2x \quad 3)(5x \quad 2)$: $15x$, $4x$ — No possible sum of $+7x$
$(2x \quad 1)(5x \quad 6)$: $-5x$, $+12x$, $+7x$ ✓
$(2x \quad 6)(5x \quad 1)$: $30x$, $2x$ — No possible sum of $+7x$

Only one of the possibilities yields $+7x$, so we can write the factors $(2x - 1)(5x + 6)$.
Check: $(2x - 1)(5x + 6) = 10x^2 + 7x - 6$ ✓
5. The grouping number is $3(-4) = -12$. The two numbers with a product of -12 and a sum of 4 are 6 and -2. Write $4x$ as the sum $6x + (-2x)$ or $6x - 2x$.
   $3x^2 + 4x - 4 = 3x^2 + 6x - 2x - 4$
   $= 3x(x + 2) - 2(x + 2)$
   $= (x + 2)(3x - 2)$
6. The grouping number is $9 \cdot 7 = 63$. The two numbers with a product of 63 and a sum of -64 are -63 and -1. Write $-64x$ as the sum $-63x + (-1x)$ or $-63x - 1x$.
   $9x^2 - 64x + 7 = 9x^2 - 63x - x + 7$
   $= 9x(x - 7) - 1(x - 7)$
   $= (x - 7)(9x - 1)$

7. $8x^2 + 8x - 6 = 2(4x^2 + 4x - 3)$
$= 2(2x - 1)(2x + 3)$

8. $24x^2 - 38x + 10 = 2(12x^2 - 19x + 5)$
$= 2(4x - 5)(3x - 1)$

6.5 Student Practice

1. $64x^2 - 1 = (8x + 1)(8x - 1)$ because $64x^2 = (8x)^2$ and $1 = (1)^2$.

2. $36x^2 - 49 = (6x + 7)(6x - 7)$ because $36x^2 = (6x)^2$ and $49 = (7)^2$.

3. $100x^2 - 81y^2 = (10x + 9y)(10x - 9y)$

4. $x^8 - 1 = (x^4 + 1)(x^4 - 1)$
$= (x^4 + 1)(x^2 + 1)(x^2 - 1)$
$= (x^4 + 1)(x^2 + 1)(x + 1)(x - 1)$

5. The first and last terms are perfect squares: $x^2 = (x)^2$ and $25 = (5)^2$.
The middle term, $10x$, is twice the product of x and 5.
$x^2 + 10x + 25 = (x + 5)^2$

6. $30x = 2(5x \cdot 3)$
Also note the negative sign. $25x^2 - 30x + 9 = (5x - 3)^2$

7. **(a)** $25x^2 = (5x)^2, 36y^2 = (6y)^2$, and $60xy = 2(5x \cdot 6y)$.
$25x^2 + 60xy + 36y^2 = (5x + 6y)^2$
(b) $64x^6 = (8x^3)^2, 9 = (3)^2$, and $48x^3 = 2(8x^3 \cdot 3)$.
$64x^6 - 48x^3 + 9 = (8x^3 - 3)^2$

8. $9x^2 = (3x)^2$ and $4 = (2)^2$, but $15x \neq 2(3x \cdot 2) = 12x$.
$9x^2 + 15x + 4 = (3x + 1)(3x + 4)$

9. $20x^2 - 45 = 5(4x^2 - 9)$
$= 5(2x + 3)(2x - 3)$

10. $75x^2 - 60x + 12 = 3(25x^2 - 20x + 4)$
$= 3(5x - 2)^2$

6.6 Student Practice

1. **(a)** $3x^2 - 36x + 108$
$= 3(x^2 - 12x + 36)$
$= 3(x - 6)^2$
(b) $9x^4y^2 - 9y^2$
$= 9y^2(x^4 - 1)$
$= 9y^2(x^2 + 1)(x^2 - 1)$
$= 9y^2(x^2 + 1)(x + 1)(x - 1)$
(c) $12x - 9 - 4x^2 = -4x^2 + 12x - 9$
$= -(4x^2 - 12x + 9)$
$= -(2x - 3)^2$

2. $5x^3 - 20x + 2x^2 - 8 = 5x(x^2 - 4) + 2(x^2 - 4)$
$= (5x + 2)(x^2 - 4)$
$= (5x + 2)(x + 2)(x - 2)$

3. **(a)** $x^2 - 9x - 8$
The factorizations of -8 are $(-2)(4)$, $(2)(-4)$, $(-8)(1)$, and $(-1)(8)$.
None of these pairs will add up to be the coefficient of the middle term. Thus the polynomial cannot be factored. It is prime.
(b) $25x^2 + 82x + 4$
Check to see if this is a perfect-square trinomial.
$2[(5)(2)] = 2(10) = 20$
This is not the coefficient of the middle term. The grouping number is 100. No factors add to 82. It is prime.

6.7 Student Practice

1. $10x^2 - x - 2 = 0$
$(5x + 2)(2x - 1) = 0$
$5x + 2 = 0 \qquad 2x - 1 = 0$
$5x = -2 \qquad 2x = 1$
$x = -\frac{2}{5} \qquad x = \frac{1}{2}$

Check: $10\left(-\frac{2}{5}\right)^2 - \left(-\frac{2}{5}\right) - 2 \stackrel{?}{=} 0 \qquad 10\left(\frac{1}{2}\right)^2 - \frac{1}{2} - 2 \stackrel{?}{=} 0$
$10\left(\frac{4}{25}\right) + \frac{2}{5} - 2 \stackrel{?}{=} 0 \qquad 10\left(\frac{1}{4}\right) - \frac{1}{2} - 2 \stackrel{?}{=} 0$
$\frac{8}{5} + \frac{2}{5} - 2 \stackrel{?}{=} 0 \qquad \frac{5}{2} - \frac{1}{2} - 2 \stackrel{?}{=} 0$
$\frac{10}{5} - \frac{10}{5} \stackrel{?}{=} 0 \qquad \frac{4}{2} - 2 \stackrel{?}{=} 0$
$0 = 0 \qquad 2 - 2 \stackrel{?}{=} 2$
$0 = 0$

Thus $-\frac{2}{5}$ and $\frac{1}{2}$ are both roots of the equation.

2. $3x^2 + 11x - 4 = 0$
$(3x - 1)(x + 4) = 0$
$3x - 1 = 0 \qquad x + 4 = 0$
$3x = 1 \qquad x = -4$
$x = \frac{1}{3}$
Thus $\frac{1}{3}$ and -4 are both roots of the equation.

3. $7x^2 + 11x = 0$
$x(7x + 11) = 0$
$x = 0 \quad 7x + 11 = 0$
$7x = -11$
$x = \frac{-11}{7}$
Thus 0 and $-\frac{11}{7}$ are both roots of the equation.

4. $x^2 - 6x + 4 = -8 + x$
$x^2 - 7x + 12 = 0$
$(x - 3)(x - 4) = 0$
$x - 3 = 0 \qquad x - 4 = 0$
$x = 3 \qquad x = 4$
The roots are 3 and 4.

5. $\frac{2x^2 - 7x}{3} = 5$
$3\left(\frac{2x^2 - 7x}{3}\right) = 3(5)$
$2x^2 - 7x = 15$
$2x^2 - 7x - 15 = 0$
$(2x + 3)(x - 5) = 0$
$2x + 3 = 0 \qquad x - 5 = 0$
$x = -\frac{3}{2} \qquad x = 5$
The roots are $-\frac{3}{2}$ and 5.

6. Let w = width, then
$3w + 2$ = length.
$(3w + 2)w = 85$
$3w^2 + 2w = 85$
$3w^2 + 2w - 85 = 0$
$(3w + 17)(w - 5) = 0$
$3w + 17 = 0 \qquad w - 5 = 0$
$w = -\frac{17}{3} \qquad w = 5$
The only valid answer is width = 5 meters
length = $3(5) + 2 = 17$ meters

7. Let b = base.

$b - 3$ = altitude.

$$\frac{b(b-3)}{2} = 35$$
$$\frac{b^2 - 3b}{2} = 35$$
$$b^2 - 3b = 70$$
$$b^2 - 3b - 70 = 0$$
$$(b+7)(b-10) = 0$$
$$b + 7 = 0$$
$$b = -7$$

This is not a valid answer.

$$b - 10 = 0$$
$$b = 10$$

Thus the base = 10 centimeters

altitude = 10 − 3 = 7 centimeters

8.
$$-5t^2 + 45 = 0$$
$$-5(t^2 - 9) = 0$$
$$t^2 - 9 = 0$$
$$(t+3)(t-3) = 0$$
$$t + 3 = 0 \qquad t - 3 = 0$$
$$t = -3 \qquad t = 3$$

$t = -3$ is not a valid answer.

Thus it will be 3 seconds before he breaks the water's surface.

Chapter 7

7.1 Student Practice

1. $$\frac{28}{63} = \frac{\cancel{7}\cdot 4}{\cancel{7}\cdot 9} = \frac{4}{9}$$

2. $$\frac{12x-6}{14x-7} = \frac{6\cancel{(2x-1)}}{7\cancel{(2x-1)}} = \frac{6}{7}$$

3. $$\frac{4x^2-9}{2x^2-x-3} = \frac{\cancel{(2x-3)}(2x+3)}{\cancel{(2x-3)}(x+1)} = \frac{2x+3}{x+1}$$

4. $$\frac{x^3-16x}{x^3-2x^2-8x} = \frac{x(x^2-16)}{x(x^2-2x-8)} = \frac{x(x+4)\cancel{(x-4)}}{x(x+2)\cancel{(x-4)}} = \frac{x+4}{x+2}$$

5. $$\frac{8x-20}{15-6x} = \frac{4\cancel{(2x-5)}}{-3\cancel{(-5+2x)}} = \frac{4}{-3} = -\frac{4}{3}$$

6. $$\frac{4x^2+3x-10}{25-16x^2} = \frac{(4x-5)(x+2)}{(5+4x)(5-4x)} = \frac{\cancel{(4x-5)}(x+2)}{-1\cancel{(-5+4x)}(5+4x)}$$
$$= \frac{x+2}{-1(5+4x)} = -\frac{x+2}{5+4x}$$

7. $$\frac{x^2-8xy+15y^2}{2x^2-11xy+5y^2} = \frac{(x-3y)\cancel{(x-5y)}}{\cancel{(x-5y)}(2x-y)} = \frac{x-3y}{2x-y}$$

8. $$\frac{25a^2-16b^2}{10a^2+3ab-4b^2} = \frac{\cancel{(5a+4b)}(5a-4b)}{\cancel{(5a+4b)}(2a-b)} = \frac{5a-4b}{2a-b}$$

7.2 Student Practice

1. $$\frac{6x^2+7x+2}{x^2-7x+10}\cdot\frac{x^2+3x-10}{2x^2+11x+5}$$
$$= \frac{\cancel{(2x+1)}(3x+2)}{(x-5)\cancel{(x-2)}}\cdot\frac{\cancel{(x+5)}\cancel{(x-2)}}{\cancel{(2x+1)}\cancel{(x+5)}} = \frac{3x+2}{x-5}$$

2. $$\frac{2y^2-6y-8}{y^2-y-2}\cdot\frac{y^2-5y+6}{2y^2-32}$$
$$= \frac{2(y^2-3y-4)}{(y-2)(y+1)}\cdot\frac{(y-3)(y-2)}{2(y^2-16)}$$
$$= \frac{\cancel{2}\cancel{(y-4)}\cancel{(y+1)}}{\cancel{(y-2)}\cancel{(y+1)}}\cdot\frac{(y-3)\cancel{(y-2)}}{\cancel{2}(y+4)\cancel{(y-4)}} = \frac{y-3}{y+4}$$

3. $$\frac{x^2+5x+6}{x^2+8x} \div \frac{2x^2+5x+2}{2x^2+x}$$
$$= \frac{x^2+5x+6}{x^2+8x}\cdot\frac{2x^2+x}{2x^2+5x+2}$$
$$= \frac{\cancel{(x+2)}(x+3)}{\cancel{x}(x+8)}\cdot\frac{\cancel{x}\cancel{(2x+1)}}{\cancel{(2x+1)}\cancel{(x+2)}} = \frac{x+3}{x+8}$$

4. $$\frac{x+3}{x-3} \div (9-x^2) = \frac{x+3}{x-3} \div \frac{9-x^2}{1} = \frac{x+3}{x-3}\cdot\frac{1}{9-x^2}$$
$$= \frac{\cancel{x+3}}{x-3}\cdot\frac{1}{\cancel{(3+x)}(3-x)}$$
$$= \frac{1}{(x-3)(3-x)}$$

7.3 Student Practice

1. $$\frac{2s+t}{2s-t} + \frac{s-t}{2s-t} = \frac{2s+t+s-t}{2s-t} = \frac{3s}{2s-t}$$

2. $$\frac{b}{(a-2b)(a+b)} - \frac{2b}{(a-2b)(a+b)}$$
$$= \frac{b-2b}{(a-2b)(a+b)}$$
$$= \frac{-b}{(a-2b)(a+b)}$$

3. $$\frac{7}{6x+21},\ \frac{13}{10x+35}$$
$$6x + 21 = 3(2x+7)$$
$$10x + 35 = 5(2x+7)$$
$$\text{LCD} = 3\cdot 5\cdot(2x+7) = 15(2x+7)$$

4. (a) $$\frac{3}{50xy^2z},\ \frac{19}{40x^3yz}$$
$$50xy^2z = 2\cdot \qquad 5\cdot 5\cdot x\cdot \qquad y\cdot y\cdot z$$
$$40x^3yz = 2\cdot 2\cdot 2\cdot 5\cdot \mid x\cdot x\cdot x\cdot y\cdot \mid z$$
$$\downarrow\ \downarrow\ \downarrow\ \downarrow\ \downarrow\ \downarrow\ \downarrow\ \downarrow\ \downarrow\ \downarrow\ \downarrow$$
$$2\cdot 2\cdot 2\cdot 5\cdot 5\cdot x\cdot x\cdot x\cdot y\cdot y\cdot z$$
$$\text{LCD} = 2^3\cdot 5^2\cdot x^3\cdot y^2\cdot z = 200x^3y^2z$$

(b) $$\frac{2}{x^2+5x+6},\ \frac{6}{3x^2+5x-2}$$
$$x^2 + 5x + 6 = (x+3)(x+2)$$
$$3x^2 + 5x - 2 = \qquad (x+2)(3x-1)$$
$$\downarrow \qquad \downarrow \qquad \downarrow$$
$$\text{LCD} = (x+3)(x+2)(3x-1)$$

5. LCD = abc

$$\frac{7}{a} + \frac{3}{abc} = \frac{7}{a}\cdot\frac{bc}{bc} + \frac{3}{abc} = \frac{7bc}{abc} + \frac{3}{abc} = \frac{7bc+3}{abc}$$

6. $a^2 - 4b^2 = (a+2b)(a-2b)$

LCD = $(a+2b)(a-2b)$

$$\frac{2a-b}{a^2-4b^2} + \frac{2}{a+2b}$$
$$= \frac{2a-b}{(a+2b)(a-2b)} + \frac{2}{(a+2b)}\cdot\frac{a-2b}{a-2b}$$
$$= \frac{2a-b}{(a+2b)(a-2b)} + \frac{2a-4b}{(a+2b)(a-2b)}$$
$$= \frac{2a-b+2a-4b}{(a+2b)(a-2b)}$$
$$= \frac{4a-5b}{(a+2b)(a-2b)}$$

7. $$\frac{7a}{a^2+2ab+b^2} + \frac{4}{a^2+ab}$$
$$= \frac{7a}{(a+b)^2} + \frac{4}{a(a+b)} \qquad \text{LCD} = a(a+b)^2$$
$$= \frac{7a}{(a+b)^2}\cdot\frac{a}{a} + \frac{4}{a(a+b)}\cdot\frac{(a+b)}{(a+b)}$$
$$= \frac{7a^2}{a(a+b)^2} + \frac{4(a+b)}{a(a+b)^2} = \frac{7a^2+4a+4b}{a(a+b)^2}$$

8. $\frac{x+7}{3x-9} - \frac{x-6}{x-3} = \frac{x+7}{3(x-3)} - \frac{x-6}{x-3}$

$= \frac{x+7}{3(x-3)} - \frac{x-6}{x-3}\cdot\frac{3}{3} = \frac{x+7-3(x-6)}{3(x-3)}$

$= \frac{x+7-3x+18}{3(x-3)} = \frac{-2x+25}{3(x-3)}$

9. $\frac{x-2}{x^2-4} - \frac{x+1}{2x^2+4x} = \frac{x-2}{(x+2)(x-2)} - \frac{x+1}{2x(x+2)}$

$= \frac{x-2}{(x+2)(x-2)}\cdot\frac{2x}{2x} - \frac{x+1}{2x(x+2)}\cdot\frac{x-2}{x-2}$

$= \frac{2x(x-2)-(x+1)(x-2)}{2x(x+2)(x-2)} = \frac{2x^2-4x-x^2+x+2}{2x(x+2)(x-2)}$

$= \frac{x^2-3x+2}{2x(x+2)(x-2)} = \frac{(x-1)\cancel{(x-2)}}{2x(x+2)\cancel{(x-2)}} = \frac{x-1}{2x(x+2)}$

7.4 Student Practice

1. $\dfrac{\frac{1}{a}+\frac{1}{a^2}}{\frac{2}{b^2}} = \dfrac{\frac{1}{a}\cdot\frac{a}{a}+\frac{1}{a^2}}{\frac{2}{b^2}} = \dfrac{\frac{a+1}{a^2}}{\frac{2}{b^2}} = \frac{a+1}{a^2} \div \frac{2}{b^2}$

$= \frac{a+1}{a^2}\cdot\frac{b^2}{2} = \frac{b^2(a+1)}{2a^2}$

2. $\dfrac{\frac{1}{a}+\frac{1}{b}}{\frac{x}{2}-\frac{5}{y}} = \dfrac{\frac{1}{a}\cdot\frac{b}{b}+\frac{1}{b}\cdot\frac{a}{a}}{\frac{x}{2}\cdot\frac{y}{y}-\frac{5}{y}\cdot\frac{2}{2}} = \dfrac{\frac{b+a}{ab}}{\frac{xy-10}{2y}} = \frac{b+a}{ab}\cdot\frac{2y}{xy-10} = \frac{2y(b+a)}{ab(xy-10)}$

3. $\dfrac{\frac{x}{x^2+4x+3}+\frac{2}{x+1}}{x+1} = \dfrac{\frac{x}{(x+1)(x+3)}+\frac{2}{x+1}\cdot\frac{x+3}{x+3}}{x+1}$

$= \dfrac{\frac{x+2x+6}{(x+1)(x+3)}}{(x+1)} = \frac{3x+6}{(x+1)(x+3)}\cdot\frac{1}{(x+1)}$

$= \frac{3(x+2)}{(x+1)^2(x+3)}$

4. $\dfrac{\frac{6}{x^2-y^2}}{\frac{1}{x-y}+\frac{3}{x+y}} = \dfrac{\frac{6}{x^2-y^2}}{\frac{1}{x-y}\cdot\frac{x+y}{x+y}+\frac{3}{x+y}\cdot\frac{x-y}{x-y}}$

$= \dfrac{\frac{6}{x^2-y^2}}{\frac{x+y}{(x+y)(x-y)}+\frac{3x-3y}{(x+y)(x-y)}} = \dfrac{\frac{6}{(x+y)(x-y)}}{\frac{4x-2y}{(x+y)(x-y)}}$

$= \frac{6}{\cancel{(x+y)(x-y)}}\cdot\frac{\cancel{(x+y)(x-y)}}{2(2x-y)} = \frac{\cancel{2}\cdot 3}{\cancel{2}(2x-y)} = \frac{3}{2x-y}$

5. The LCD of all the denominators is $3x^2y$.

$\dfrac{\frac{2}{3x^2}-\frac{3}{y}}{\frac{5}{xy}-4} = \dfrac{3x^2y\left(\frac{2}{3x^2}-\frac{3}{y}\right)}{3x^2y\left(\frac{5}{xy}-4\right)} = \dfrac{3x^2y\left(\frac{2}{3x^2}\right)-3x^2y\left(\frac{3}{y}\right)}{3x^2y\left(\frac{5}{xy}\right)-3x^2y(4)} = \frac{2y-9x^2}{15x-12x^2y}$

6. The LCD of all the denominators is $(x+y)(x-y)$.

$\dfrac{\frac{6}{x^2-y^2}}{\frac{7}{x-y}+\frac{3}{x+y}}$

$= \dfrac{(x+y)(x-y)\left(\frac{6}{(x+y)(x-y)}\right)}{(x+y)(x-y)\left(\frac{7}{x-y}\right)+(x+y)(x-y)\left(\frac{3}{x+y}\right)}$

$= \frac{6}{7(x+y)+3(x-y)}$

$= \frac{6}{7x+7y+3x-3y}$

$= \frac{6}{10x+4y}$

$= \frac{2\cdot 3}{2(5x+2y)}$

$= \frac{3}{5x+2y}$

7.5 Student Practice

1. $\frac{3}{x}+\frac{4}{5} = -\frac{2}{x}$ LCD $= 5x$

$5x\left(\frac{3}{x}\right)+5x\left(\frac{4}{5}\right) = 5x\left(-\frac{2}{x}\right)$

$15+4x = -10$

$4x = -10-15$

$4x = -25$

$x = -\frac{25}{4}$ or $-6\frac{1}{4}$ or -6.25

Check.

$\dfrac{3}{-\frac{25}{4}}+\frac{4}{5} \stackrel{?}{=} -\dfrac{2}{-\frac{25}{4}}$

$-\frac{12}{25}+\frac{4}{5} \stackrel{?}{=} \frac{8}{25}$

$-\frac{12}{25}+\frac{20}{25} \stackrel{?}{=} \frac{8}{25}$

$\frac{8}{25} = \frac{8}{25}$ ✓

2. LCD $= (2x+1)(x+2)$

$\frac{6}{2x+1} = \frac{2}{x+2}$

$(2x+1)(x+2)\left(\frac{6}{2x+1}\right) = (2x+1)(x+2)\left(\frac{2}{x+2}\right)$

$6(x+2) = 2(2x+1)$

$6x+12 = 4x+2$

$2x+12 = 2$

$2x = -10$

$x = -5$

Check.

$\frac{6}{2(-5)+1} \stackrel{?}{=} \frac{2}{(-5)+2}$

$\frac{6}{-10+1} \stackrel{?}{=} \frac{2}{-5+2}$

$\frac{6}{-9} \stackrel{?}{=} \frac{2}{-3}$

$-\frac{2}{3} = -\frac{2}{3}$ ✓

3. $\frac{x-1}{x^2-4} = \frac{2}{x+2}+\frac{4}{x-2}$

$\frac{x-1}{(x+2)(x-2)} = \frac{2}{x+2}+\frac{4}{x-2}$

$(x+2)(x-2)\left[\frac{x-1}{(x+2)(x-2)}\right]$

$= (x+2)(x-2)\left(\frac{2}{x+2}\right)+(x+2)(x-2)\left(\frac{4}{x-2}\right)$

$x-1 = 2(x-2)+4(x+2)$

$x-1 = 2x-4+4x+8$

$x-1 = 6x+4$

$-5x-1 = 4$

$-5x = 5$

$x = -1$

Check. $\frac{-1-1}{(-1)^2-4} \stackrel{?}{=} \frac{2}{-1+2}+\frac{4}{-1-2}$

$\frac{-2}{-3} \stackrel{?}{=} \frac{2}{1}+\frac{4}{-3}$

$\frac{2}{3} \stackrel{?}{=} \frac{6}{3}-\frac{4}{3}$

$\frac{2}{3} = \frac{2}{3}$ ✓

4. The LCD is $x + 1$.

$$\frac{2x}{x+1} = \frac{-2}{x+1} + 1$$
$$(x+1)\left(\frac{2x}{x+1}\right) = (x+1)\left(\frac{-2}{x+1}\right) + (x+1)(1)$$
$$2x = -2 + x + 1$$
$$2x = x - 1$$
$$x = -1 \quad \text{(but see the check)}$$

Check. $\frac{2(-1)}{-1+1} \stackrel{?}{=} \frac{-2}{-1+1} + 1$

$$\frac{-2}{0} \stackrel{?}{=} \frac{-2}{0} + 1$$

These expressions are not defined; therefore, there is **no solution** to this equation.

7.6 Student Practice

1. Let $x =$ the number of hours it will take to drive 315 miles.

$$\frac{8}{420} = \frac{x}{315}$$
$$8(315) = 420x$$
$$\frac{2520}{420} = x$$
$$6 = x$$

It would take Brenda 6 hours to drive 315 miles.

2. Let $x =$ the distance represented by $2\frac{1}{2}$ inches.

$$\frac{\frac{5}{8}}{30} = \frac{2\frac{1}{2}}{x}$$
$$\frac{5}{8}x = 30\left(2\frac{1}{2}\right)$$
$$\frac{5}{8}x = 30\left(\frac{5}{2}\right)$$
$$\frac{5}{8}x = 75$$
$$8\left(\frac{5}{8}x\right) = 8(75)$$
$$5x = 600$$
$$x = 120$$

Therefore $2\frac{1}{2}$ inches represents 120 miles.

3. $$\frac{13}{x} = \frac{16}{18}$$
$$13(18) = 16x$$
$$234 = 16x$$
$$\frac{234}{16} = x$$

Side x has length

$$\frac{117}{8} = 14\frac{5}{8} \text{ cm.}$$

4. $$\frac{6}{7} = \frac{x}{38.5}$$
$$6(38.5) = 7x$$
$$231 = 7x$$
$$x = 33$$

The height of the flagpole is 33 feet.

5. Let $x =$ the speed of train B. Then train A time $= \frac{180}{x+10}$ and train B time $= \frac{150}{x}$.

$$\frac{180}{x+10} = \frac{150}{x}$$
$$180x = 150(x+10)$$
$$180x = 150x + 1500$$
$$30x = 1500$$
$$x = 50$$

Train B traveled 50 kilometers per hour. Train A traveled $50 + 10 = 60$ kilometers per hour.

6.

	Number of Hours	Part of the Job Done in One Hour
John	6 hours	$\frac{1}{6}$
Dave	7 hours	$\frac{1}{7}$
John & Dave Together	x	$\frac{1}{x}$

LCD $= 42x$

$$\frac{1}{6} + \frac{1}{7} = \frac{1}{x}$$
$$42x\left(\frac{1}{6}\right) + 42x\left(\frac{1}{7}\right) = 42x\left(\frac{1}{x}\right)$$
$$7x + 6x = 42$$
$$13x = 42$$
$$x = 3\frac{3}{13}$$
$$\frac{3}{13}\text{ hour} \times \frac{60 \text{ min}}{1 \text{ hour}} = \frac{180}{13}\text{ min} \approx 13.846 \text{ min}$$

Thus, doing the job together will take 3 hours and 14 minutes.

Chapter 8 8.1 Student Practice

1. $$\left(\frac{3x^{-2}y^4}{2x^{-5}y^2}\right)^{-3} = \frac{(3x^{-2}y^4)^{-3}}{(2x^{-5}y^2)^{-3}}$$
$$= \frac{3^{-3}(x^{-2})^{-3}(y^4)^{-3}}{2^{-3}(x^{-5})^{-3}(y^2)^{-3}}$$
$$= \frac{3^{-3}x^6y^{-12}}{2^{-3}x^{15}y^{-6}}$$
$$= \frac{3^{-3}}{2^{-3}} \cdot \frac{x^6}{x^{15}} \cdot \frac{y^{-12}}{y^{-6}}$$
$$= \frac{2^3}{3^3} \cdot x^{6-15} \cdot y^{-12+6}$$
$$= \frac{8}{27}x^{-9}y^{-6} \text{ or } \frac{8}{27x^9y^6}$$

2. (a) $(x^4)^{3/8} = x^{(4/1)(3/8)} = x^{3/2}$

(b) $\dfrac{x^{3/7}}{x^{2/7}} = x^{3/7-2/7} = x^{1/7}$

(c) $x^{-7/5} \cdot x^{4/5} = x^{-7/5+4/5} = x^{-3/5}$

3. (a) $(-3x^{1/4})(2x^{1/2}) = -6x^{1/4+1/2} = -6x^{1/4+2/4} = -6x^{3/4}$

(b) $\dfrac{13x^{1/12}y^{-1/4}}{26x^{-1/3}y^{1/2}} = \dfrac{x^{1/12-(-1/3)}y^{-1/4-1/2}}{2}$

$= \dfrac{x^{1/12+4/12}y^{-1/4-2/4}}{2}$

$= \dfrac{x^{5/12}y^{-3/4}}{2}$

$= \dfrac{x^{5/12}}{2y^{3/4}}$

4. $-3x^{1/2}(2x^{1/4} + 3x^{-1/2}) = -6x^{1/2+1/4} - 9x^{1/2-1/2}$

$= -6x^{2/4+1/4} - 9x^0$

$= -6x^{3/4} - 9$

5. (a) $(4)^{5/2} = (2^2)^{5/2} = 2^{2/1 \cdot 5/2} = 2^5 = 32$

(b) $(27)^{4/3} = (3^3)^{4/3} = 3^{3/1 \cdot 4/3} = 3^4 = 81$

6. $3x^{1/3} + x^{-1/3} = 3x^{1/3} + \dfrac{1}{x^{1/3}}$

$= \dfrac{x^{1/3} \cdot 3x^{1/3}}{x^{1/3}} + \dfrac{1}{x^{1/3}}$

$= \dfrac{3x^{2/3} + 1}{x^{1/3}}$

7. $4y^{3/2} - 8y^{5/2} = 4y^{2/2+1/2} - 8y^{2/2+3/2}$

$= 4(y^{2/2})(y^{1/2}) - 8(y^{2/2})(y^{3/2})$

$= 4y(y^{1/2} - 2y^{3/2})$

8.2 Student Practice

1. (a) $\sqrt[3]{216} = \sqrt[3]{6^3} = 6$

(b) $\sqrt[5]{32} = \sqrt[5]{2^5} = 2$

(c) $\sqrt[3]{-8} = \sqrt[3]{(-2)^3} = -2$

(d) $\sqrt[4]{-81}$ is not a real number.

2. (a) $f(3) = \sqrt{4(3) - 3} = \sqrt{12 - 3} = \sqrt{9} = 3$

(b) $f(4) = \sqrt{4(4) - 3} = \sqrt{16 - 3} = \sqrt{13} \approx 3.6$

(c) $f(7) = \sqrt{4(7) - 3} = \sqrt{28 - 3} = \sqrt{25} = 5$

3. $0.5x + 2 \geq 0$

$0.5x \geq -2$

$x \geq -4$

The domain is all real numbers x where $x \geq -4$.

4.

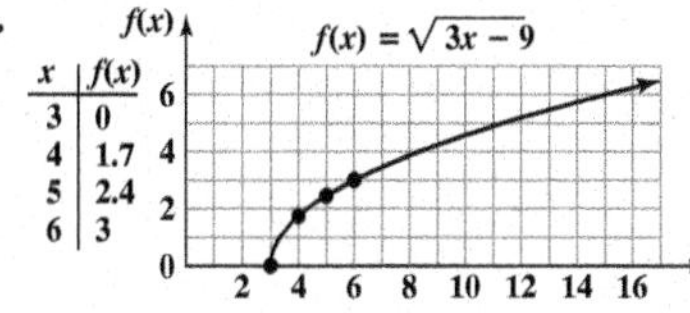

5. (a) $\sqrt[3]{x^3} = (x^3)^{1/3} = x^{3/3} = x^1 = x$

(b) $\sqrt[4]{y^4} = (y^4)^{1/4} = y^{4/4} = y^1 = y$

6. (a) $\sqrt[4]{x^3} = (x^3)^{1/4} = x^{3/4}$

(b) $\sqrt[5]{(xy)^7} = [(xy)^7]^{1/5} = (xy)^{7/5}$

7. (a) $\sqrt[4]{81x^{12}} = (3^4x^{12})^{1/4} = 3x^3$

(b) $\sqrt[3]{27x^6} = (3^3x^6)^{1/3} = 3x^2$

(c) $(32x^5)^{3/5} = (2^5x^5)^{3/5} = 2^3x^3 = 8x^3$

8. (a) $(xy)^{3/4} = \sqrt[4]{(xy)^3} = \sqrt[4]{x^3y^3}$ or $(xy)^{3/4} = \left(\sqrt[4]{xy}\right)^3$

(b) $y^{-1/3} = \dfrac{1}{y^{1/3}} = \dfrac{1}{\sqrt[3]{y}}$

(c) $(2x)^{4/5} = \sqrt[5]{(2x)^4} = \sqrt[5]{16x^4}$ or $(2x)^{4/5} = \left(\sqrt[5]{2x}\right)^4$

(d) $2x^{4/5} = 2\sqrt[5]{x^4}$ or $2x^{4/5} = 2\left(\sqrt[5]{x}\right)^4$

9. (a) $8^{2/3} = \left(\sqrt[3]{8}\right)^2 = 2^2 = 4$

(b) $(-8)^{4/3} = \left(\sqrt[3]{-8}\right)^4 = (-2)^4 = 16$

(c) $100^{-3/2} = \dfrac{1}{100^{3/2}} = \dfrac{1}{\left(\sqrt{100}\right)^3} = \dfrac{1}{10^3} = \dfrac{1}{1000}$

10. (a) $\sqrt[5]{(-3)^5} = -3$

(b) $\sqrt[4]{(-5)^4} = |-5| = 5$

(c) $\sqrt[4]{w^4} = |w|$

(d) $\sqrt[7]{y^7} = y$

11. (a) $\sqrt{36x^2} = 6|x|$

(b) $\sqrt[4]{16y^8} = 2|y^2| = 2y^2$

(c) $\sqrt[3]{125x^3y^6} = \sqrt[3]{5^3x^3(y^2)^3} = 5xy^2$

8.3 Student Practice

1. $\sqrt{20} = \sqrt{4 \cdot 5} = \sqrt{4}\sqrt{5} = 2\sqrt{5}$

2. $\sqrt{27} = \sqrt{9}\sqrt{3} = 3\sqrt{3}$

3. (a) $\sqrt[3]{24} = \sqrt[3]{8}\sqrt[3]{3} = 2\sqrt[3]{3}$

(b) $\sqrt[3]{-108} = \sqrt[3]{-27}\sqrt[3]{4} = -3\sqrt[3]{4}$

4. $\sqrt[4]{64} = \sqrt[4]{16}\sqrt[4]{4} = 2\sqrt[4]{4}$

5. (a) $\sqrt{45x^6y^7} = \sqrt{9 \cdot 5 \cdot x^6 \cdot y^6 \cdot y} = \sqrt{9x^6y^6}\sqrt{5y} = 3x^3y^3\sqrt{5y}$

(b) $\sqrt[3]{27a^7b^8c^9} = \sqrt[3]{27 \cdot a^6 \cdot a \cdot b^6 \cdot b^2 \cdot c^9}$

$= \sqrt[3]{27a^6b^6c^9}\sqrt[3]{ab^2}$

$= 3a^2b^2c^3\sqrt[3]{ab^2}$

6. $19\sqrt{xy} + 5\sqrt{xy} - 10\sqrt{xy} = (19 + 5 - 10)\sqrt{xy} = 14\sqrt{xy}$

7. $4\sqrt{2} - 5\sqrt{50} - 3\sqrt{98}$

$= 4\sqrt{2} - 5\sqrt{25}\sqrt{2} - 3\sqrt{49}\sqrt{2}$

$= 4\sqrt{2} - 5(5)\sqrt{2} - 3(7)\sqrt{2}$

$= 4\sqrt{2} - 25\sqrt{2} - 21\sqrt{2}$

$= -42\sqrt{2}$

8. $4\sqrt{2x} + \sqrt{18x} - 2\sqrt{125x} - 6\sqrt{20x}$

$= 4\sqrt{2x} + \sqrt{9}\sqrt{2x} - 2\sqrt{25}\sqrt{5x} - 6\sqrt{4}\sqrt{5x}$

$= 4\sqrt{2x} + 3\sqrt{2x} - 10\sqrt{5x} - 12\sqrt{5x}$

$= 7\sqrt{2x} - 22\sqrt{5x}$

9. $3x\sqrt[3]{54x^4} - 3\sqrt[3]{16x^7}$

$= 3x\sqrt[3]{27x^3}\sqrt[3]{2x} - 3\sqrt[3]{8x^6}\sqrt[3]{2x}$

$= 3x(3x)\sqrt[3]{2x} - 3(2x^2)\sqrt[3]{2x}$

$= 9x^2\sqrt[3]{2x} - 6x^2\sqrt[3]{2x}$

$= 3x^2\sqrt[3]{2x}$

8.4 Student Practice

1. $\left(-4\sqrt{2}\right)\left(-3\sqrt{13x}\right) = (-4)(-3)\sqrt{2 \cdot 13x} = 12\sqrt{26x}$

2. $\sqrt{2x}\left(\sqrt{5} + 2\sqrt{3x} + \sqrt{8}\right)$

$= \left(\sqrt{2x}\right)\left(\sqrt{5}\right) + \left(\sqrt{2x}\right)\left(2\sqrt{3x}\right) + \left(\sqrt{2x}\right)\left(\sqrt{8}\right)$

$= \sqrt{10x} + 2\sqrt{6x^2} + \sqrt{16x}$

$= \sqrt{10x} + 2\sqrt{x^2}\sqrt{6} + \sqrt{16}\sqrt{x}$

$= \sqrt{10x} + 2x\sqrt{6} + 4\sqrt{x}$

3. By FOIL: $\left(\sqrt{7} + 4\sqrt{2}\right)\left(2\sqrt{7} - 3\sqrt{2}\right)$

$= 2\sqrt{49} - 3\sqrt{14} + 8\sqrt{14} - 12\sqrt{4}$

$= 14 + 5\sqrt{14} - 24$

$= -10 + 5\sqrt{14}$

4. $\left(2 - 5\sqrt{5}\right)\left(3 - 2\sqrt{2}\right) = 6 - 4\sqrt{2} - 15\sqrt{5} + 10\sqrt{10}$

5. By FOIL: $\left(\sqrt{5x} + \sqrt{10}\right)^2 = \left(\sqrt{5x} + \sqrt{10}\right)\left(\sqrt{5x} + \sqrt{10}\right)$

$= \sqrt{25x^2} + \sqrt{50x} + \sqrt{50x} + \sqrt{100}$

$= 5x + 2\sqrt{25}\sqrt{2x} + 10$

$= 5x + 2(5)\sqrt{2x} + 10$

$= 5x + 10\sqrt{2x} + 10$

6. (a) $\sqrt[3]{2x}(\sqrt[3]{4x^2} + 3\sqrt[3]{y})$
$= (\sqrt[3]{2x})(\sqrt[3]{4x^2}) + 3(\sqrt[3]{2x})(\sqrt[3]{y})$
$= \sqrt[3]{8x^3} + 3\sqrt[3]{2xy}$
$= 2x + 3\sqrt[3]{2xy}$

(b) $(\sqrt[3]{7} + \sqrt[3]{x^2})(2\sqrt[3]{49} - \sqrt[3]{x})$
$= 2\sqrt[3]{343} - \sqrt[3]{7x} + 2\sqrt[3]{49x^2} - \sqrt[3]{x^3}$
$= 2(7) - \sqrt[3]{7x} + 2\sqrt[3]{49x^2} - x$
$= 14 - \sqrt[3]{7x} + 2\sqrt[3]{49x^2} - x$

7. (a) $\dfrac{\sqrt{75}}{\sqrt{3}} = \sqrt{\dfrac{75}{3}} = \sqrt{25} = 5$

(b) $\sqrt[3]{\dfrac{27}{64}} = \dfrac{\sqrt[3]{27}}{\sqrt[3]{64}} = \dfrac{3}{4}$

(c) $\dfrac{\sqrt{54a^3b^7}}{\sqrt{6b^5}} = \sqrt{\dfrac{54a^3b^7}{6b^5}} = \sqrt{9a^3b^2} = 3ab\sqrt{a}$

8. $\dfrac{7}{\sqrt{3}} = \dfrac{7}{\sqrt{3}}\cdot\dfrac{\sqrt{3}}{\sqrt{3}} = \dfrac{7\sqrt{3}}{\sqrt{9}} = \dfrac{7\sqrt{3}}{3}$

9. Method 1: $\dfrac{8}{\sqrt{20x}} = \dfrac{8}{\sqrt{4}\sqrt{5x}} = \dfrac{8}{2\sqrt{5x}}\cdot\dfrac{\sqrt{5x}}{\sqrt{5x}} = \dfrac{8\sqrt{5x}}{10x} = \dfrac{4\sqrt{5x}}{5x}$

Method 2: $\dfrac{8}{\sqrt{20x}} = \dfrac{8}{\sqrt{20x}}\cdot\dfrac{\sqrt{5x}}{\sqrt{5x}} = \dfrac{8\sqrt{5x}}{\sqrt{100x^2}} = \dfrac{8\sqrt{5x}}{10x} = \dfrac{4\sqrt{5x}}{5x}$

10. Method 1: $\sqrt[3]{\dfrac{6}{5x}} = \dfrac{\sqrt[3]{6}}{\sqrt[3]{5x}} = \dfrac{\sqrt[3]{6}}{\sqrt[3]{5x}}\cdot\dfrac{\sqrt[3]{25x^2}}{\sqrt[3]{25x^2}} = \dfrac{\sqrt[3]{150x^2}}{\sqrt[3]{125x^3}} = \dfrac{\sqrt[3]{150x^2}}{5x}$

Method 2: $\sqrt[3]{\dfrac{6}{5x}} = \sqrt[3]{\dfrac{6}{5x}\cdot\dfrac{25x^2}{25x^2}} = \sqrt[3]{\dfrac{150x^2}{125x^3}} = \dfrac{\sqrt[3]{150x^2}}{\sqrt[3]{125x^3}} = \dfrac{\sqrt[3]{150x^2}}{5x}$

11. $\dfrac{4}{2+\sqrt{5}} = \dfrac{4}{2+\sqrt{5}}\cdot\dfrac{2-\sqrt{5}}{2-\sqrt{5}}$
$= \dfrac{8-4\sqrt{5}}{2^2-(\sqrt{5})^2}$
$= \dfrac{8-4\sqrt{5}}{4-5}$
$= \dfrac{8-4\sqrt{5}}{-1}$
$= -(8-4\sqrt{5})$
$= -8+4\sqrt{5}$

12. $\dfrac{\sqrt{11}+\sqrt{5}}{\sqrt{11}-\sqrt{5}} = \dfrac{\sqrt{11}+\sqrt{5}}{\sqrt{11}-\sqrt{5}}\cdot\dfrac{\sqrt{11}+\sqrt{5}}{\sqrt{11}+\sqrt{5}}$
$= \dfrac{\sqrt{121}+2\sqrt{55}+\sqrt{25}}{(\sqrt{11})^2-(\sqrt{5})^2}$
$= \dfrac{11+2\sqrt{55}+5}{11-5}$
$= \dfrac{16+2\sqrt{55}}{6}$
$= \dfrac{2(8+\sqrt{55})}{2\cdot 3} = \dfrac{8+\sqrt{55}}{3}$

8.5 Student Practice

1. $\sqrt{3x-8} = x-2$
$(\sqrt{3x-8})^2 = (x-2)^2$
$3x-8 = x^2-4x+4$
$0 = x^2-7x+12$
$0 = (x-3)(x-4)$
$x-3=0$ or $x-4=0$
$x=3$ $\quad x=4$

Check.

For $x=3$: $\sqrt{3(3)-8} \stackrel{?}{=} 3-2$
$\sqrt{1} \stackrel{?}{=} 1$
$1 = 1$ ✓

For $x=4$: $\sqrt{3(4)-8} \stackrel{?}{=} 4-2$
$\sqrt{4} \stackrel{?}{=} 2$
$2 = 2$ ✓

The solutions are 3 and 4.

2. $-4+\sqrt{x+4} = x$
$\sqrt{x+4} = x+4$
$(\sqrt{x+4})^2 = (x+4)^2$
$x+4 = x^2+8x+16$
$0 = x^2+7x+12$
$0 = (x+3)(x+4)$
$x+3=0$ or $x+4=0$
$x=-3$ $\quad x=-4$

Check.

For $x=-3$: $-4+\sqrt{-3+4} \stackrel{?}{=} -3$
$-4+\sqrt{1} \stackrel{?}{=} -3$
$-3 = -3$ ✓

For $x=-4$: $-4+\sqrt{-4+4} \stackrel{?}{=} -4$
$-4+\sqrt{0} \stackrel{?}{=} -4$
$-4 = -4$ ✓

The solutions are -3 and -4.

3. $\sqrt{2x+5} - 2\sqrt{2x} = 1$
$\sqrt{2x+5} = 2\sqrt{2x}+1$
$(\sqrt{2x+5})^2 = (2\sqrt{2x}+1)^2$
$2x+5 = (2\sqrt{2x}+1)(2\sqrt{2x}+1)$
$2x+5 = 8x+4\sqrt{2x}+1$
$-6x+4 = 4\sqrt{2x}$
$-3x+2 = 2\sqrt{2x}$
$(-3x+2)^2 = (2\sqrt{2x})^2$
$9x^2-12x+4 = 8x$
$9x^2-20x+4 = 0$
$(9x-2)(x-2) = 0$
$9x-2=0$ or $x-2=0$
$9x = 2$
$x = \dfrac{2}{9}$ $\quad x=2$

Check. For $x = \dfrac{2}{9}$:

$\sqrt{2\left(\dfrac{2}{9}\right)+5} - 2\sqrt{2\left(\dfrac{2}{9}\right)} \stackrel{?}{=} 1$
$\sqrt{\dfrac{4}{9}+5} - 2\sqrt{\dfrac{4}{9}} \stackrel{?}{=} 1$
$\sqrt{\dfrac{49}{9}} - 2\left(\dfrac{2}{3}\right) \stackrel{?}{=} 1$
$\dfrac{7}{3} - \dfrac{4}{3} \stackrel{?}{=} 1$
$\dfrac{3}{3} \stackrel{?}{=} 1$
$1 = 1$ ✓

For $x=2$:
$\sqrt{2(2)+5} - 2\sqrt{2(2)} \stackrel{?}{=} 1$
$\sqrt{9} - 2\sqrt{4} \stackrel{?}{=} 1$
$3-4 \stackrel{?}{=} 1$
$-1 \neq 1$

The only solution is $\dfrac{2}{9}$.

4. $\sqrt{y-1}+\sqrt{y-4} = \sqrt{4y-11}$
$(\sqrt{y-1}+\sqrt{y-4})^2 = (\sqrt{4y-11})^2$
$(\sqrt{y-1}+\sqrt{y-4})(\sqrt{y-1}+\sqrt{y-4}) = 4y-11$
$y-1+2\sqrt{(y-1)(y-4)}+y-4 = 4y-11$
$2\sqrt{(y-1)(y-4)} = 2y-6$
$\sqrt{(y-1)(y-4)} = y-3$
$(\sqrt{y^2-5y+4})^2 = (y-3)^2$
$y^2-5y+4 = y^2-6y+9$
$y = 5$

Check. $\sqrt{5-1}+\sqrt{5-4} \stackrel{?}{=} \sqrt{4(5)-11}$
$\sqrt{4}+\sqrt{1} \stackrel{?}{=} \sqrt{9}$
$2+1 \stackrel{?}{=} 3$
$3=3$ ✓

The solution is 5.

8.6 Student Practice

1. (a) $\sqrt{-49}=\sqrt{-1}\sqrt{49}=(i)(7)=7i$
(b) $\sqrt{-31}=\sqrt{-1}\sqrt{31}=i\sqrt{31}$

2. $\sqrt{-98}=\sqrt{-1}\sqrt{98}=i\sqrt{98}=i\sqrt{49}\sqrt{2}=7i\sqrt{2}$

3. $\sqrt{-8}\cdot\sqrt{-2}=\sqrt{-1}\sqrt{8}\cdot\sqrt{-1}\sqrt{2}$
$=i\sqrt{8}\cdot i\sqrt{2}$
$=i^2\sqrt{16}$
$=-1(4)=-4$

4. $-7+2yi\sqrt{3}=x+6i\sqrt{3}$
$x=-7,\quad 2y\sqrt{3}=6\sqrt{3}$
$y=3$

5. $(3-4i)-(-2-18i)$
$=[3-(-2)]+[-4-(-18)]i$
$=(3+2)+(-4+18)i$
$=5+14i$

6. $(4-2i)(3-7i)$
$=(4)(3)+(4)(-7i)+(-2i)(3)+(-2i)(-7i)$
$=12-28i-6i+14i^2$
$=12-28i-6i+14(-1)$
$=12-28i-6i-14=-2-34i$

7. $-2i(5+6i)$
$=(-2)(5)i+(-2)(6)i^2$
$=-10i-12i^2$
$=-10i-12(-1)=12-10i$

8. (a) $i^{42}=(i^{40+2})=(i^{40})(i^2)=(i^4)^{10}(i^2)=(1)^{10}(-1)=-1$
(b) $i^{53}=(i^{52+1})=(i^{52})(i)=(i^4)^{13}(i)=(1)^{13}(i)=i$

9. $\dfrac{4+2i}{3+4i}=\dfrac{4+2i}{3+4i}\cdot\dfrac{3-4i}{3-4i}$
$=\dfrac{12-16i+6i-8i^2}{9-16i^2}$
$=\dfrac{12-10i-8(-1)}{9-16(-1)}$
$=\dfrac{12-10i+8}{9+16}$
$=\dfrac{20-10i}{25}$
$=\dfrac{5(4-2i)}{5\cdot 5}$
$=\dfrac{4-2i}{5}$ or $\dfrac{4}{5}-\dfrac{2}{5}i$

10. $\dfrac{5-6i}{-2i}=\dfrac{5-6i}{-2i}\cdot\dfrac{2i}{2i}$
$=\dfrac{10i-12i^2}{-4i^2}$
$=\dfrac{10i-12(-1)}{-4(-1)}$
$=\dfrac{10i+12}{4}$
$=\dfrac{2(5i+6)}{2\cdot 2}$
$=\dfrac{6+5i}{2}$ or $3+\dfrac{5}{2}i$

8.7 Student Practice

1. Let s = maximum speed,
h = horsepower.
$s=k\sqrt{h}$
Substitute $s=128$ and $h=256$.
$128=k\sqrt{256}$
$128=16k$
$8=k$
Now we know the value of k so $s=8\sqrt{h}$.
Let $h=225$.
$s=8(\sqrt{225})$
$s=8(15)=120$
The maximum speed is 120 miles per hour.

2. $y=\dfrac{k}{x}$
Substitute $y=45$ and $x=16$.
$45=\dfrac{k}{16}$
$720=k$
We now write the equation $y=\dfrac{720}{x}$.
Find the value of y when $x=36$.
$y=\dfrac{720}{36}$
$y=20$

3. Let w = weight,
h = height of column.
$w=\dfrac{k}{h^2}$
Substitute $w=2$ and $h=7.5$.
$2=\dfrac{k}{(7.5)^2}$
$112.5=k$
We now write the equation $w=\dfrac{112.5}{h^2}$.
Now substitute $h=3$ and solve for w.
$w=\dfrac{112.5}{(3)^2}$
$w=\dfrac{112.5}{9}=12.5$
The column can support 12.5 tons.

4. $y=\dfrac{kzw^2}{x}$
To find the value of k, substitute $y=20$, $z=3$, $w=5$, and $x=4$.
$20=\dfrac{k(3)(5)^2}{4}$
$20=\dfrac{75k}{4}$
$\dfrac{80}{75}=k$
$\dfrac{16}{15}=k$
We now substitute $\dfrac{16}{15}$ for k.
$y=\dfrac{16zw^2}{15x}$
We use this equation to find y when $z=4$, $w=6$, and $x=2$.
$y=\dfrac{16(4)(6)^2}{15(2)}=\dfrac{2304}{30}$
$y=\dfrac{384}{5}$

Chapter 9

9.1 Student Practice

1. $x^2 - 121 = 0$
$x^2 = 121$
$x = \pm\sqrt{121}$
$x = \pm 11$
The two roots are 11 and -11.

Check.
$(11)^2 - 121 \stackrel{?}{=} 0$ $\quad$ $(-11)^2 - 121 \stackrel{?}{=} 0$
$121 - 121 \stackrel{?}{=} 0$ $\quad$ $121 - 121 \stackrel{?}{=} 0$
$0 = 0$ ✓ $\quad$ $0 = 0$ ✓

2. $x^2 = 18$
$x = \pm\sqrt{18} = \pm\sqrt{9 \cdot 2}$
$x = \pm 3\sqrt{2}$
The roots are $3\sqrt{2}$ and $-3\sqrt{2}$.

3. $5x^2 + 1 = 46$
$5x^2 = 45$
$x^2 = 9$
$x = \pm\sqrt{9}$
$x = \pm 3$
The roots are 3 and -3.

4. $3x^2 = -27$
$x^2 = -9$
$x = \pm\sqrt{-9}$
$x = \pm 3i$
The roots are $3i$ and $-3i$.

Check.
$3(3i)^2 \stackrel{?}{=} -27$ $\quad$ $3(-3i)^2 \stackrel{?}{=} -27$
$3(9i^2) \stackrel{?}{=} -27$ $\quad$ $3(9i^2) \stackrel{?}{=} -27$
$3(-9) \stackrel{?}{=} -27$ $\quad$ $3(-9) \stackrel{?}{=} -27$
$-27 = -27$ ✓ $\quad$ $-27 = -27$ ✓

5. $(2x + 3)^2 = 7$
$2x + 3 = \pm\sqrt{7}$
$2x = -3 \pm \sqrt{7}$
$x = \dfrac{-3 \pm \sqrt{7}}{2}$
The roots are $\dfrac{-3 + \sqrt{7}}{2}$ and $\dfrac{-3 - \sqrt{7}}{2}$.

6. $x^2 + 8x + 3 = 0$
$x^2 + 8x = -3$
$x^2 + 8x + (4)^2 = -3 + (4)^2$
$x^2 + 8x + 16 = -3 + 16$
$(x + 4)^2 = 13$
$x + 4 = \pm\sqrt{13}$
$x = -4 \pm \sqrt{13}$
The roots are $-4 + \sqrt{13}$ and $-4 - \sqrt{13}$.

7. $2x^2 + 4x + 1 = 0$
$2x^2 + 4x = -1$
$x^2 + 2x = -\dfrac{1}{2}$
$x^2 + 2x + (1)^2 = -\dfrac{1}{2} + 1$
$(x + 1)^2 = \dfrac{1}{2}$
$x + 1 = \pm\sqrt{\dfrac{1}{2}}$
$x + 1 = \pm\dfrac{1}{\sqrt{2}}$
$x = -1 \pm \dfrac{\sqrt{2}}{2}$ or $\dfrac{-2 \pm \sqrt{2}}{2}$

9.2 Student Practice

1. $x^2 + 5x = -1 + 2x$
$x^2 + 3x + 1 = 0$
$a = 1, b = 3, c = 1$
$x = \dfrac{-3 \pm \sqrt{3^2 - 4(1)(1)}}{2(1)}$
$x = \dfrac{-3 \pm \sqrt{5}}{2}$

2. $2x^2 + 7x + 6 = 0$
$a = 2, b = 7, c = 6$
$x = \dfrac{-7 \pm \sqrt{7^2 - 4(2)(6)}}{2(2)}$
$x = \dfrac{-7 \pm \sqrt{49 - 48}}{4} = \dfrac{-7 \pm \sqrt{1}}{4}$
$x = \dfrac{-7 + 1}{4} = -\dfrac{6}{4}$ or $x = \dfrac{-7 - 1}{4} = -\dfrac{8}{4}$
$x = -\dfrac{3}{2}$ $\quad$ $x = -2$

3. $2x^2 - 26 = 0$
$a = 2, b = 0, c = -26$
$x = \dfrac{-0 \pm \sqrt{0^2 - 4(2)(-26)}}{2(2)}$
$x = \dfrac{\pm\sqrt{208}}{4} = \dfrac{\pm\sqrt{16}\sqrt{13}}{4} = \dfrac{\pm 4\sqrt{13}}{4} = \pm\sqrt{13}$

4. $0 = -100x^2 + 4800x - 52{,}559$
$a = -100, b = 4800, c = -52{,}559$
$x = \dfrac{-4800 \pm \sqrt{(4800)^2 - 4(-100)(-52{,}559)}}{2(-100)}$
$x = \dfrac{-4800 \pm \sqrt{23{,}040{,}000 - 21{,}023{,}600}}{-200}$
$x = \dfrac{-4800 \pm \sqrt{2{,}016{,}400}}{-200}$
$x = \dfrac{-4800 \pm 1420}{-200}$
$x = \dfrac{-4800 + 1420}{-200} = \dfrac{-3380}{-200} = 16.9 \approx 17$
or
$x = \dfrac{-4800 - 1420}{-200} = \dfrac{-6220}{-200} = 31.1 \approx 31$
For zero profit, they should manufacture 17 or 31 modems.

5. $\dfrac{1}{x} + \dfrac{1}{x - 1} = \dfrac{5}{6}$ $\quad$ The LCD is $6x(x - 1)$.
$6x(x - 1)\left(\dfrac{1}{x}\right) + 6x(x - 1)\left(\dfrac{1}{x - 1}\right) = 6x(x - 1)\left(\dfrac{5}{6}\right)$
$6(x - 1) + 6x = 5x(x - 1)$
$6x - 6 + 6x = 5x^2 - 5x$
$0 = 5x^2 - 17x + 6$
$a = 5, b = -17, c = 6$
$x = \dfrac{-(-17) \pm \sqrt{(-17)^2 - 4(5)(6)}}{2(5)}$
$x = \dfrac{17 \pm \sqrt{289 - 120}}{10} = \dfrac{17 \pm \sqrt{169}}{10} = \dfrac{17 \pm 13}{10}$
$x = \dfrac{17 + 13}{10} = \dfrac{30}{10} = 3,\quad x = \dfrac{17 - 13}{10} = \dfrac{4}{10} = \dfrac{2}{5}$

6. $2x^2 - 4x + 5 = 0$

$a = 2, b = -4, c = 5$

$$x = \frac{-(-4) \pm \sqrt{(-4)^2 - 4(2)(5)}}{2(2)}$$

$$x = \frac{4 \pm \sqrt{16 - 40}}{4} = \frac{4 \pm \sqrt{-24}}{4}$$

$$x = \frac{4 \pm 2i\sqrt{6}}{4} = \frac{2(2 + i\sqrt{6})}{4} = \frac{2 \pm i\sqrt{6}}{2}$$

7. $9x^2 + 12x + 4 = 0$

$a = 9, b = 12, c = 4$

$b^2 - 4ac = 12^2 - 4(9)(4) = 144 - 144 = 0$

Since the discriminant is 0, there is one rational solution.

8. (a) $x^2 - 4x + 13 = 0$

$a = 1, b = -4, c = 13$

$b^2 - 4ac = (-4)^2 - 4(1)(13) = 16 - 52 = -36$

Since the discriminant is negative, there are two complex solutions containing i.

(b) $9x^2 + 6x - 7 = 0$

$a = 9, b = 6, c = -7$

$b^2 - 4ac = 6^2 - 4(9)(-7) = 36 + 252 = 288$

Since the discriminant is positive but not a perfect square, there are two different irrational solutions.

9. $x = -10 \qquad x = -6$

$x + 10 = 0 \qquad x + 6 = 0$

$(x + 10)(x + 6) = 0$

$x^2 + 16x + 60 = 0$

10. $x - 2i\sqrt{3} = 0 \qquad x + 2i\sqrt{3} = 0$

$$(x - 2i\sqrt{3})(x + 2i\sqrt{3}) = 0$$

$$x^2 + 2ix\sqrt{3} - 2ix\sqrt{3} - (2i\sqrt{3})^2 = 0$$

$$x^2 - 4i^2(\sqrt{9}) = 0$$

$$x^2 - 4(-1)(3) = 0$$

$$x^2 + 12 = 0$$

9.3 Student Practice

1. $x^4 - 5x^2 - 36 = 0$

Let $y = x^2$. Then $y^2 = x^4$.

Thus, our new equation is

$y^2 - 5y - 36 = 0.$

$(y - 9)(y + 4) = 0$

$y - 9 = 0 \quad$ or $\quad y + 4 = 0$

$y = 9 \qquad y = -4$

$x^2 = 9 \qquad x^2 = -4$

$x = \pm\sqrt{9} \qquad x = \pm\sqrt{-4}$

$x = \pm 3 \qquad x = \pm 2i$

2. $x^6 - 5x^3 + 4 = 0$

Let $y = x^3$. Then $y^2 = x^6$.

$y^2 - 5y + 4 = 0$

$(y - 1)(y - 4) = 0$

$y - 1 = 0 \quad$ or $\quad y - 4 = 0$

$y = 1 \qquad y = 4$

$x^3 = 1 \qquad x^3 = 4$

$x = \sqrt[3]{1} = 1 \qquad x = \sqrt[3]{4}$

3. $3x^{4/3} - 5x^{2/3} + 2 = 0$

Let $y = x^{2/3}$. Then $y^2 = x^{4/3}$.

$3y^2 - 5y + 2 = 0$

$(3y - 2)(y - 1) = 0$

$3y - 2 = 0 \quad$ or $\quad y - 1 = 0$

$y = \frac{2}{3} \qquad y = 1$

$x^{2/3} = \frac{2}{3} \qquad x^{2/3} = 1$

$(x^{2/3})^3 = \left(\frac{2}{3}\right)^3 \qquad (x^{2/3})^3 = 1^3$

$x^2 = \frac{8}{27} \qquad x^2 = 1$

$x = \pm\sqrt{\frac{8}{27}} \qquad x = \pm\sqrt{1}$

$x = \pm\frac{2\sqrt{2}}{3\sqrt{3}} \qquad x = \pm 1$

$x = \pm\frac{2\sqrt{2}}{3\sqrt{3}} \cdot \frac{\sqrt{3}}{\sqrt{3}}$

$x = \pm\frac{2\sqrt{6}}{9}$

Check. $x = \frac{2\sqrt{6}}{9}$:

$$3\left(\frac{2\sqrt{6}}{9}\right)^{4/3} - 5\left(\frac{2\sqrt{6}}{9}\right)^{2/3} + 2 \stackrel{?}{=} 0$$

$$3\left(\frac{4}{9}\right) - 5\left(\frac{2}{3}\right) + 2 \stackrel{?}{=} 0$$

$$\frac{4}{3} - \frac{10}{3} + 2 \stackrel{?}{=} 0$$

$$0 = 0 \ ✓$$

$x = -\frac{2\sqrt{6}}{9}$:

$$3\left(-\frac{2\sqrt{6}}{9}\right)^{4/3} - 5\left(-\frac{2\sqrt{6}}{9}\right)^{2/3} + 2 \stackrel{?}{=} 0$$

$$3\left(\frac{4}{9}\right) - 5\left(\frac{2}{3}\right) + 2 \stackrel{?}{=} 0$$

$$\frac{4}{3} - \frac{10}{3} + 2 \stackrel{?}{=} 0$$

$$0 = 0 \ ✓$$

$x = 1$:

$3(1)^{4/3} - 5(1)^{2/3} + 2 \stackrel{?}{=} 0$

$3(1) - 5(1) + 2 \stackrel{?}{=} 0$

$0 = 0$ ✓

$x = -1$:

$3(-1)^{4/3} - 5(-1)^{2/3} + 2 \stackrel{?}{=} 0$

$3(1) - 5(1) + 2 \stackrel{?}{=} 0$

$0 = 0$ ✓

4. $3x^{1/2} = 8x^{1/4} - 4$

$3x^{1/2} - 8x^{1/4} + 4 = 0$

Let $y = x^{1/4}$. Then $y^2 = x^{1/2}$.

$3y^2 - 8y + 4 = 0$

$(3y - 2)(y - 2) = 0$

$3y - 2 = 0 \quad$ or $\quad y - 2 = 0$

$y = \frac{2}{3} \qquad y = 2$

$x^{1/4} = \frac{2}{3} \qquad x^{1/4} = 2$

$(x^{1/4})^4 = \left(\frac{2}{3}\right)^4 \qquad (x^{1/4})^4 = (2)^4$

$x = \frac{16}{81} \qquad x = 16$

Check. $x = \frac{16}{81}$: $\qquad x = 16$:

$3\left(\frac{16}{81}\right)^{1/2} \stackrel{?}{=} 8\left(\frac{16}{81}\right)^{1/4} - 4 \qquad 3(16)^{1/2} \stackrel{?}{=} 8(16)^{1/4} - 4$

$3\left(\frac{4}{9}\right) \stackrel{?}{=} 8\left(\frac{2}{3}\right) - 4 \qquad 3(4) \stackrel{?}{=} 8(2) - 4$

$\frac{4}{3} = \frac{4}{3}$ ✓ $\qquad 12 = 12$ ✓

5. $2x^{-2} + x^{-1} = 6$

$2x^{-2} + x^{-1} - 6 = 0$

Let $y = x^{-1}$. Then $y^2 = x^{-2}$.

$2y^2 + y - 6 = 0$

$(2y - 3)(y + 2) = 0$

$2y - 3 = 0$ or $y + 2 = 0$

$y = \frac{3}{2}$ $\quad y = -2$

$x^{-1} = \frac{3}{2}$ $\quad x^{-1} = -2$

$(x^{-1})^{-1} = \left(\frac{3}{2}\right)^{-1}$ $\quad (x^{-1})^{-1} = (-2)^{-1}$

$x = \frac{2}{3}$ $\quad x = -\frac{1}{2}$

Check. $x = \frac{2}{3}$: $\quad x = -\frac{1}{2}$:

$2\left(\frac{2}{3}\right)^{-2} + \left(\frac{2}{3}\right)^{-1} \stackrel{?}{=} 6$ $\quad 2\left(-\frac{1}{2}\right)^{-2} + \left(-\frac{1}{2}\right)^{-1} \stackrel{?}{=} 6$

$2\left(\frac{9}{4}\right) + \frac{3}{2} \stackrel{?}{=} 6$ $\quad 2(4) + (-2) \stackrel{?}{=} 6$

$6 = 6$ ✓ $\quad 6 = 6$ ✓

9.4 Student Practice

1. $V = \frac{1}{3}\pi r^2 h$

$\frac{3V}{\pi h} = r^2$

$\pm\sqrt{\frac{3V}{\pi h}} = r$ so $r = \sqrt{\frac{3V}{\pi h}}$

2. $2y^2 + 9wy + 7w^2 = 0$

$(2y + 7w)(y + w) = 0$

$2y + 7w = 0$ or $y + w = 0$

$2y = -7w$ $\quad y = -w$

$y = -\frac{7}{2}w$

3. $3y^2 + 2fy - 7g = 0$

Use the quadratic formula.

$a = 3, b = 2f, c = -7g$

$$y = \frac{-2f \pm \sqrt{(2f)^2 - 4(3)(-7g)}}{2(3)}$$

$$y = \frac{-2f \pm \sqrt{4f^2 + 84g}}{6}$$

$$y = \frac{-2f \pm \sqrt{4(f^2 + 21g)}}{6}$$

$$y = \frac{-2f \pm 2\sqrt{f^2 + 21g}}{6}$$

$$y = \frac{-f \pm \sqrt{f^2 + 21g}}{3}$$

4. $d = \frac{n^2 - 3n}{2}$

Multiply each term by 2.

$2d = n^2 - 3n$

$0 = n^2 - 3n - 2d$

Use the quadratic formula.

$a = 1, b = -3, c = -2d$

$$n = \frac{-(-3) \pm \sqrt{(-3)^2 - 4(1)(-2d)}}{2(1)}$$

$$n = \frac{3 + \sqrt{9 + 8d}}{2}$$

Since n is the number of sides of a polygon, $n > 0$ and we use only the principal root.

5. (a) $a^2 + b^2 = c^2$

$b^2 = c^2 - a^2$

$b = \sqrt{c^2 - a^2}$

Only the positive root is used since b is a length.

(b) $b = \sqrt{c^2 - a^2}$

$b = \sqrt{(26)^2 - (24)^2} = \sqrt{676 - 576} = \sqrt{100}$

$b = 10$

6. $x + (x - 7) + c = 30$

$c = -2x + 37$

$a = x, b = x - 7, c = -2x + 37$

By the Pythagorean Theorem,

$x^2 + (x - 7)^2 = (-2x + 37)^2$

$x^2 + x^2 - 14x + 49 = 4x^2 - 148x + 1369$

$-2x^2 + 134x - 1320 = 0$

$x^2 - 67x + 660 = 0$

Use the quadratic formula.

$a = 1, b = -67, c = 660$

$$x = \frac{-(-67) \pm \sqrt{(-67)^2 - 4(1)(660)}}{2(1)}$$

$$x = \frac{67 \pm \sqrt{1849}}{2} = \frac{67 \pm 43}{2}$$

$$x = \frac{67 + 43}{2} = 55 \text{ or } x = \frac{67 - 43}{2} = 12$$

The only answer that makes sense is $x = 12$; therefore,

$x = 12$

$x - 7 = 5$

$-2x + 37 = 13$

The longest boundary of the piece of land is 13 miles. The other two boundaries are 5 miles and 12 miles.

7. $A = \pi r^2$

$A = \pi(6)^2 = 36\pi$

The area of the cross section of the old pipe is 36π square feet.

Let $x =$ the radius of the new pipe.

(area of new pipe) $-$ (area of old pipe) $= 45\pi$

$\pi x^2 - 36\pi = 45\pi$

$\pi x^2 = 81\pi$

$x^2 = 81$

$x = \pm 9$

Since the radius must be positive, we select $x = 9$. The radius of the new pipe is 9 feet. The radius of the new pipe has been increased by 3 feet.

8. Let $x =$ width. Then $2x - 3 =$ the length.

$x(2x - 3) = 54$

$2x^2 - 3x = 54$

$2x^2 - 3x - 54 = 0$

$(2x + 9)(x - 6) = 0$

$x = -\frac{9}{2}$ or $x = 6$

We do not use the negative value.

Thus, width $= 6$ feet

length $= 2x - 3 = 2(6) - 3 = 9$ feet

9. Let $x =$ his speed on the secondary road. Then $x + 10 =$ his speed on the better road.

	Distance	Rate	Time
Secondary Road	150	x	$\frac{150}{x}$
Better Road	240	$x + 10$	$\frac{240}{x + 10}$
Total	390	(not used)	7

$$\frac{150}{x} + \frac{240}{x+10} = 7$$

The LCD of this equation is $x(x+10)$. Multiply each term by the LCD.

$$x(x+10)\left(\frac{150}{x}\right) + x(x+10)\left(\frac{240}{x+10}\right) = x(x+10)(7)$$

$$150(x+10) + 240x = 7x(x+10)$$

$$150x + 1500 + 240x = 7x^2 + 70x$$

$$0 = 7x^2 - 320x - 1500$$

$$0 = (x-50)(7x+30)$$

$$x - 50 = 0 \quad \text{or} \quad 7x + 30 = 0$$

$$x = 50 \qquad x = -\frac{30}{7}$$

We disregard the negative answer. Thus, $x = 50$ mph, so Carlos drove 50 mph on the secondary road and 60 mph on the better road.

9.5 Student Practice

1. $f(x) = x^2 - 6x + 5$
   $a = 1, b = -6, c = 5$

   Step 1 The vertex occurs at $x = \frac{-b}{2a}$. Thus, $x = \frac{-(-6)}{2(1)} = \frac{6}{2} = 3$

   The vertex has an x-coordinate of 3.
   To find the y-coordinate, we evaluate $f(3)$.

   $$f(3) = 3^2 - 6(3) + 5 = 9 - 18 + 5 = -4$$

   Thus, the vertex is $(3, -4)$.

   Step 2 The y-intercept is at $f(0)$.

   $$f(0) = 0^2 - 6(0) + 5 = 5$$

   The y-intercept is $(0, 5)$.

   Step 3 The x-intercepts occur when $f(x) = 0$.

   $$x^2 - 6x + 5 = 0$$
   $$(x-5)(x-1) = 0$$
   $$x - 5 = 0 \qquad x - 1 = 0$$
   $$x = 5 \qquad x = 1$$

   Thus, the x-intercepts are $(5, 0)$ and $(1, 0)$.

2. $g(x) = x^2 - 2x - 2$
   $a = 1, b = -2, c = -2$

   Step 1 The vertex occurs at

   $$x = \frac{-b}{2a}$$
   $$x = \frac{-(-2)}{2(1)} = \frac{2}{2} = 1$$

   The vertex has an x-coordinate of 1. To find the y-coordinate, we evaluate $g(1)$.

   $$g(1) = 1^2 - 2(1) - 2 = 1 - 2 - 2 = -3$$

   Thus, the vertex is $(1, -3)$.

   Step 2 The y-intercept is at $g(0)$.

   $$g(0) = 0^2 - 2(0) - 2 = -2$$

   The y-intercept is $(0, -2)$.

   Step 3 The x-intercepts occur when $g(x) = 0$. We set $x^2 - 2x - 2 = 0$ and solve for x. The equation does not factor, so we use the quadratic formula.

   $$x = \frac{-b \pm \sqrt{b^2 - 4ac}}{2a} = \frac{-(-2) \pm \sqrt{(-2)^2 - 4(1)(-2)}}{2(1)}$$
   $$= \frac{2 \pm \sqrt{12}}{2} = 1 \pm \sqrt{3}$$

   The x-intercepts are approximately $(2.7, 0)$ and $(-0.7, 0)$.

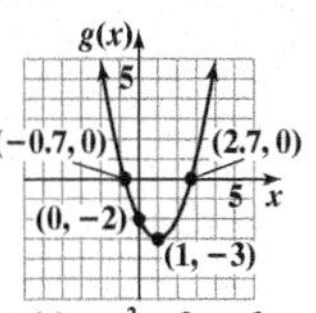

3. $g(x) = -2x^2 - 8x - 6$
   $a = -2, b = -8, c = -6$
   Since $a < 0$, the parabola opens downward.
   The vertex occurs at

   $$x = \frac{-b}{2a} = \frac{-(-8)}{2(-2)} = \frac{8}{-4} = -2.$$

   To find the y-coordinate, evaluate $g(-2)$.

   $$g(-2) = -2(-2)^2 - 8(-2) - 6 = -8 + 16 - 6 = 2$$

   Thus, the vertex is $(-2, 2)$.
   The y-intercept is at $g(0)$.

   $$g(0) = -2(0)^2 - 8(0) - 6 = -6$$

   The y-intercept is $(0, -6)$.
   The x-intercepts occur when $g(x) = 0$.
   Use the quadratic formula.

   $$x = \frac{-(-8) \pm \sqrt{(-8)^2 - 4(-2)(-6)}}{2(-2)}$$
   $$= \frac{8 \pm \sqrt{16}}{-4} = -2 \pm -1$$
   $$x = -3, x = -1$$

   The x-intercepts are $(-3, 0)$ and $(-1, 0)$.

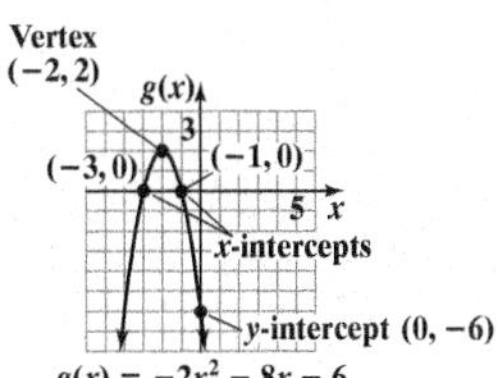

9.6 Student Practice

1. $-8 < x < -2$

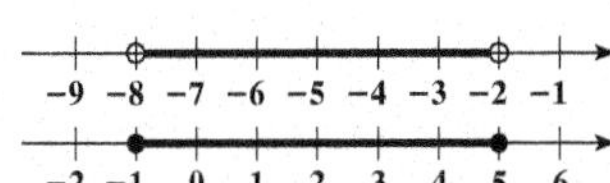

2. $-1 \le x \le 5$

3. "At least \$200" means $s \ge \$200$ and "never more than \$950" means $s \le \$950$.

4. $x < 8 \quad or \quad x > 12$

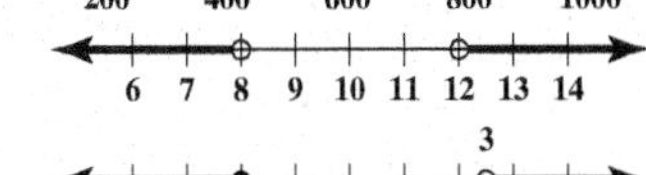

5. $x \le -6 \quad or \quad x > 3$

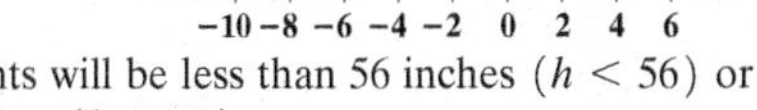

6. Rejected applicants' heights will be less than 56 inches $(h < 56)$ or will be greater than 70 inches $(h > 70)$.

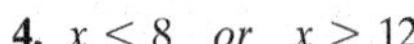
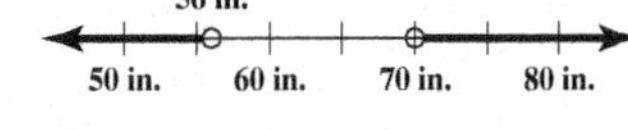

7. $3x - 4 < -1 \quad or \quad 2x + 3 > 13$
   $3x < 3 \qquad 2x > 10$
   $x < 1 \quad or \quad x > 5$

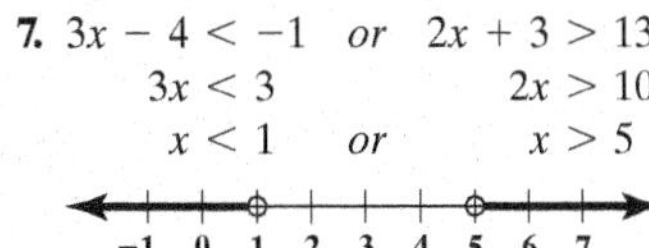

8. $-2x + 3 < -7 \quad \textit{and} \quad 7x - 1 > -15$

$-2x < -10 \qquad 7x > -14$

$x > 5 \quad \textit{and} \quad x > -2$

$x > 5$ *and* at the same time $x > -2$. Thus $x > 5$ is the solution to the compound inequality.

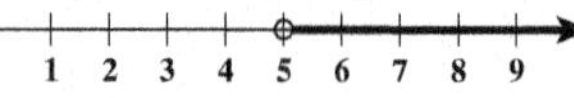

9. $-3x - 11 < -26 \quad \textit{and} \quad 5x + 4 < 14$

$-3x < -15 \qquad 5x < 10$

$x > 5 \quad \textit{and} \quad x < 2$

Now clearly it is impossible for one number to be greater than 5 *and* at the same time less than 2. There is no solution.

10. $x^2 - 2x - 8 < 0$

Replace the inequality symbol by an equals sign and solve.

$$x^2 - 2x - 8 = 0$$
$$(x + 2)(x - 4) = 0$$
$$x + 2 = 0 \quad \text{or} \quad x - 4 = 0$$
$$x = -2 \qquad x = 4$$

Region I $\quad x < -2$ Use $x = -3$

$(-3)^2 - 2(-3) - 8 = 9 + 6 - 8 = 7 > 0$

Region II $\quad -2 < x < 4$ Use $x = 0$

$0^2 - 2(0) - 8 = -8 < 0$

Region III $\quad x > 4$ Use $x = 5$

$(5)^2 - 2(5) - 8 = 25 - 10 - 8 = 7 > 0$

Thus, $x^2 - 2x - 8 < 0$ when $-2 < x < 4$.

11. $3x^2 - x - 2 \geq 0$

$3x^2 - x - 2 = 0$

$(3x + 2)(x - 1) = 0$

$3x + 2 = 0 \quad \text{or} \quad x - 1 = 0$

$x = -\frac{2}{3} \qquad x = 1$

Region I $\quad x < -\frac{2}{3}$ Use $x = -1$

$3(-1)^2 - (-1) - 2 = 3 + 1 - 2 = 2 > 0$

Region II $\quad -\frac{2}{3} < x < 1$ Use $x = 0$

$3(0) - 0 - 2 = -2 < 0$

Region III $\quad x > 1$ Use $x = 2$

$3(2)^2 - 2 - 2 = 12 - 2 - 2 = 8 > 0$

Thus, $3x^2 - x - 2 \geq 0$ when $x \leq -\frac{2}{3}$ or when $x \geq 1$.

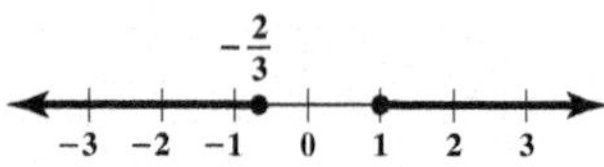

12. $x^2 + 2x < 7$

$x^2 + 2x - 7 < 0$

$x^2 + 2x - 7 = 0$

$$x = \frac{-2 \pm \sqrt{2^2 - 4(1)(-7)}}{2(1)} = \frac{-2 \pm \sqrt{4 + 28}}{2}$$
$$= \frac{-2 \pm \sqrt{32}}{2} = \frac{-2 \pm 4\sqrt{2}}{2} = -1 \pm 2\sqrt{2}$$

$-1 + 2\sqrt{2} \approx -1 + 2.828 \approx 1.828 \approx 1.8$

$-1 - 2\sqrt{2} \approx -1 - 2.828 \approx -3.828 \approx -3.8$

Region I $\quad$ Use $x = -5$

$(-5)^2 + 2(-5) - 7 = 25 - 10 - 7 = 8 > 0$

Region II $\quad$ Use $x = 0$

$(0)^2 + 2(0) - 7 = -7 < 0$

Region III $\quad$ Use $x = 3$

$(3)^2 + 2(3) - 7 = 9 + 6 - 7 = 8 > 0$

Thus, $x^2 + 2x < 7$ when $-1 - 2\sqrt{2} < x < -1 + 2\sqrt{2}$.

13. Approximately $-3.8 < x < 1.8$

9.7 Student Practice

1. $|3x - 4| = 23$

$3x - 4 = 23 \quad \text{or} \quad 3x - 4 = -23$

$3x = 27 \qquad 3x = -19$

$x = 9 \qquad x = -\frac{19}{3}$

Check.

if $x = 9$

$|3x - 4| = 23$

$|3(9) - 4| \stackrel{?}{=} 23$

$|27 - 4| \stackrel{?}{=} 23$

$|23| \stackrel{?}{=} 23$

$23 = 23$ ✓

if $x = -\frac{19}{3}$

$|3x - 4| = 23$

$\left|3\left(-\frac{19}{3}\right) - 4\right| \stackrel{?}{=} 23$

$|-19 - 4| \stackrel{?}{=} 23$

$|-23| \stackrel{?}{=} 23$

$23 = 23$ ✓

2. $\left|\frac{2}{3}x + 4\right| = 2$

$\frac{2}{3}x + 4 = 2 \quad \text{or} \quad \frac{2}{3}x + 4 = -2$

$2x + 12 = 6 \qquad 2x + 12 = -6$

$2x = -6 \qquad 2x = -18$

$x = -3 \qquad x = -9$

Check.

if $x = -3$

$\left|\frac{2}{3}(-3) + 4\right| \stackrel{?}{=} 2$

$|-2 + 4| \stackrel{?}{=} 2$

$|2| \stackrel{?}{=} 2$

$2 = 2$ ✓

if $x = -9$

$\left|\frac{2}{3}(-9) + 4\right| \stackrel{?}{=} 2$

$|-6 + 4| \stackrel{?}{=} 2$

$|-2| \stackrel{?}{=} 2$

$2 = 2$ ✓

3. $|2x + 1| + 3 = 8$

$|2x + 1| + 3 - 3 = 8 - 3$

$|2x + 1| = 5$

$2x + 1 = 5 \quad \text{or} \quad 2x + 1 = -5$

$2x = 4 \qquad 2x = -6$

$x = 2 \qquad x = -3$

Check.

if $x = 2$

$|2(2) + 1| + 3 \stackrel{?}{=} 8$

$|4 + 1| + 3 \stackrel{?}{=} 8$

$|5| + 3 \stackrel{?}{=} 8$

$5 + 3 \stackrel{?}{=} 8$

$8 = 8$ ✓

if $x = -3$

$|2(-3) + 1| + 3 \stackrel{?}{=} 8$

$|-6 + 1| + 3 \stackrel{?}{=} 8$

$|-5| + 3 \stackrel{?}{=} 8$

$5 + 3 \stackrel{?}{=} 8$

$8 = 8$ ✓

4. $|x - 6| = |5x + 8|$

$x - 6 = 5x + 8 \quad \text{or} \quad x - 6 = -(5x + 8)$

$x - 5x = 6 + 8 \qquad x - 6 = -5x - 8$

$-4x = 14 \qquad x + 5x = 6 - 8$

$x = \frac{14}{-4} = -\frac{7}{2} \qquad 6x = -2$

$x = \frac{-2}{6} = -\frac{1}{3}$

Check.

if $x = -\frac{7}{2}$

$\left|-\frac{7}{2} - 6\right| \stackrel{?}{=} \left|5\left(-\frac{7}{2}\right) + 8\right|$

$\left|-\frac{7}{2} - 6\right| \stackrel{?}{=} \left|-\frac{35}{2} + 8\right|$

$\left|-\frac{7}{2} - \frac{12}{2}\right| \stackrel{?}{=} \left|-\frac{35}{2} + \frac{16}{2}\right|$

$\left|-\frac{19}{2}\right| \stackrel{?}{=} \left|-\frac{19}{2}\right|$

$\frac{19}{2} = \frac{19}{2}$ ✓

if $x = -\frac{1}{3}$

$\left|-\frac{1}{3} - 6\right| \stackrel{?}{=} \left|5\left(-\frac{1}{3}\right) + 8\right|$

$\left|-\frac{1}{3} - 6\right| \stackrel{?}{=} \left|-\frac{5}{3} + 8\right|$

$\left|-\frac{1}{3} - \frac{18}{3}\right| \stackrel{?}{=} \left|-\frac{5}{3} + \frac{24}{3}\right|$

$\left|-\frac{19}{3}\right| \stackrel{?}{=} \left|\frac{19}{3}\right|$

$\frac{19}{3} = \frac{19}{3}$ ✓

5. $|x| < 2$

$-2 < x < 2$

6. $|x - 6| < 15$

$$-15 < x - 6 < 15$$
$$-15 + 6 < x - 6 + 6 < 15 + 6$$
$$-9 < x < 21$$

−12 −9 −6 −3 0 3 6 9 12 15 18 21 24

7. $|2 + 3(x - 1)| < 20$

$$|2 + 3x - 3| < 20$$
$$|-1 + 3x| < 20$$
$$-20 < -1 + 3x < 20$$
$$-20 + 1 < 1 - 1 + 3x < 20 + 1$$
$$-19 < 3x < 21$$
$$\frac{-19}{3} < \frac{3x}{3} < \frac{21}{3}$$
$$-\frac{19}{3} < x < 7$$
$$-6\frac{1}{3} < x < 7$$

$-6\frac{1}{3}$

−7 −6 −5 −4 −3 −2 −1 0 1 2 3 4 5 6 7

8. $|x| > 2.5$

$$x < -2.5 \quad or \quad x > 2.5$$

−3 −2 −1 0 1 2 3

9. $|-5x - 2| > 13$

$$-5x - 2 > 13 \quad or \quad -5x - 2 < -13$$
$$-5x > 15 \qquad -5x < -11$$
$$\frac{-5x}{-5} < \frac{15}{-5} \qquad \frac{-5x}{-5} > \frac{-11}{-5}$$
$$x < -3 \qquad x > \frac{11}{5}$$
$$x > 2\frac{1}{5}$$

$$x < -3 \; or \; x > 2\frac{1}{5}$$

$2\frac{1}{5}$

−4 −3 −2 −1 0 1 2 3 4

10. $\left|4 - \frac{3}{4}x\right| \geq 5$

$$4 - \frac{3}{4}x \geq 5 \quad or \quad 4 - \frac{3}{4}x \leq -5$$
$$4(4) - 4\left(\frac{3}{4}x\right) \geq 4(5) \qquad 4(4) - 4\left(\frac{3}{4}x\right) \leq 4(-5)$$
$$16 - 3x \geq 20 \qquad 16 - 3x \leq -20$$
$$-3x \geq 4 \qquad -3x \leq -36$$
$$\frac{-3x}{-3} \leq \frac{4}{-3} \qquad \frac{-3x}{-3} \geq \frac{-36}{-3}$$
$$x \leq -\frac{4}{3} \qquad x \geq 12$$
$$x \leq -1\frac{1}{3}$$

$$x \leq -1\frac{1}{3} \quad or \quad x \geq 12$$

$-1\frac{1}{3}$ 0 12

11.

$$|d - s| \leq 0.37$$
$$|d - 276.53| \leq 0.37$$
$$-0.37 \leq d - 276.53 \leq 0.37$$
$$-0.37 + 276.53 \leq d - 276.53 + 276.53 \leq 0.37 + 276.53$$
$$276.16 \leq d \leq 276.90$$

Thus the diameter of the transmission must be at least 276.16 millimeters, but not greater than 276.90 millimeters.

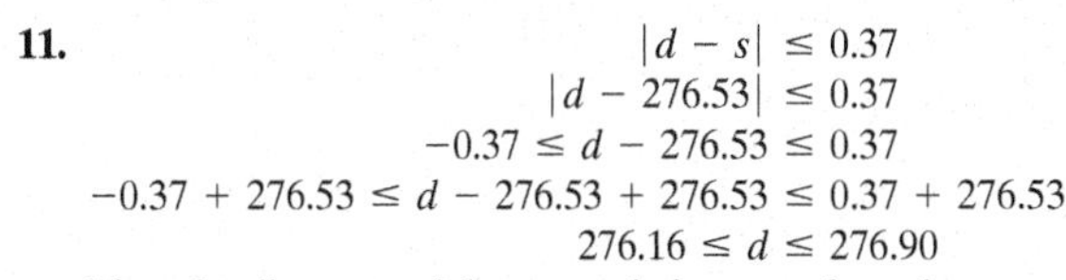

Chapter 10

10.1 Student Practice

1. Let $(x_1, y_1) = (-6, -2)$ and $(x_2, y_2) = (3, 1)$

$$d = \sqrt{(x_2 - x_1)^2 + (y_2 - y_1)^2}$$
$$= \sqrt{[3 - (-6)]^2 + [1 - (-2)]^2}$$
$$= \sqrt{(3 + 6)^2 + (1 + 2)^2}$$
$$= \sqrt{(9)^2 + (3)^2}$$
$$= \sqrt{81 + 9} = \sqrt{90} = 3\sqrt{10}$$

2. $(x + 1)^2 + (y + 2)^2 = 9$

To compare this to $(x - h)^2 + (y - k)^2 = r^2$, we write it in the form

$$[x - (-1)]^2 + [y - (-2)]^2 = 3^2.$$

Thus, we see the center is $(h, k) = (-1, -2)$ and the radius is $r = 3$.

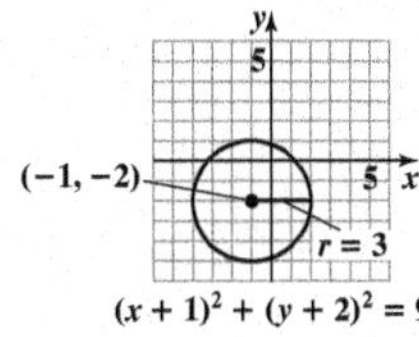

3. We are given that $(h, k) = (-5, 0)$ and $r = \sqrt{3}$. Thus, $(x - h)^2 + (y - k)^2 = r^2$ becomes

$$[x - (-5)]^2 + (y - 0)^2 = (\sqrt{3})^2$$
$$(x + 5)^2 + y^2 = 3$$

4. To write $x^2 + 4x + y^2 + 2y - 20 = 0$ in standard form, we complete the squares.

$$x^2 + 4x + \underline{\quad} + y^2 + 2y + \underline{\quad} = 20$$
$$x^2 + 4x + 4 + y^2 + 2y + 1 = 20 + 4 + 1$$
$$x^2 + 4x + 4 + y^2 + 2y + 1 = 25$$
$$(x + 2)^2 + (y + 1)^2 = 25$$

The circle has its center at $(-2, -1)$ and the radius is 5.

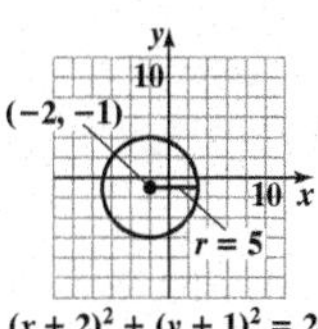

10.2 Student Practice

1. Make a table of values. Begin with $x = -3$ in the middle of the table because $(-3 + 3)^2 = 0$. Plot the points and draw the graph.

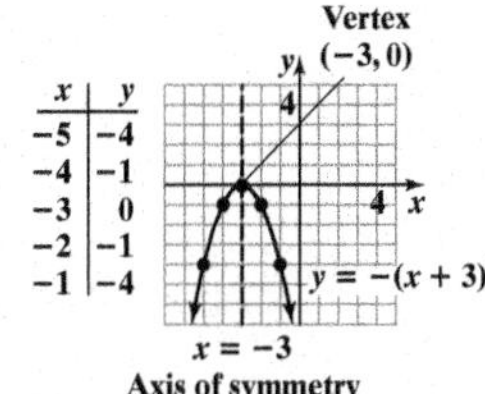

The vertex is $(-3, 0)$, and the axis of symmetry is the line $x = -3$.

2. This graph looks like the graph of $y = x^2$, except that it is shifted 6 units to the right and 4 units up.
The vertex is $(6, 4)$. The axis of symmetry is $x = 6$.
If $x = 0$, $y = (0 - 6)^2 + 4 = 36 + 4 = 40$, so the y-intercept is $(0, 40)$.

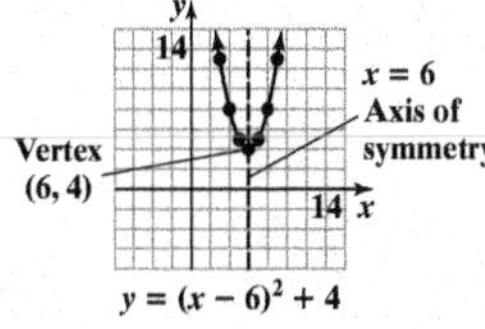

3. **Step 1** The equation has the form $y = a(x - h)^2 + k$, where $a = \frac{1}{4}, h = 2$, and $k = 3$, so it is a vertical parabola.
Step 2 $a > 0$; so the parabola opens upward.
Step 3 We have $h = 2$ and $k = 3$. Therefore, the vertex is $(2, 3)$.
Step 4 The axis of symmetry is the line $x = 2$. Plot a few points on either side of the axis of symmetry. At $x = 4, y = 4$. Thus, the point is $(4, 4)$. The image from symmetry is $(0, 4)$. At $x = 6, y = 7$. Thus the point is $(6, 7)$. The image from symmetry is $(-2, 7)$.
Step 5 At $x = 0$,

$$y = \frac{1}{4}(0 - 2)^2 + 3$$
$$= \frac{1}{4}(4) + 3$$
$$= 1 + 3 = 4.$$

Thus, the y-intercept is $(0, 4)$.

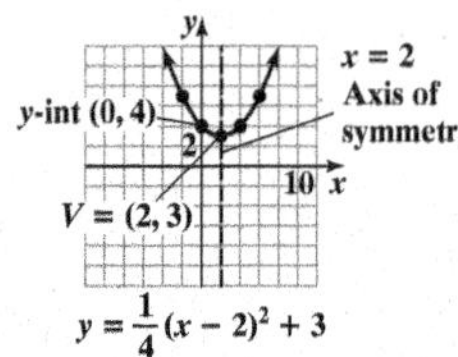

4. Make a table, plot points, and draw the graph. Choose values of y and find x. Begin with $y = 0$.

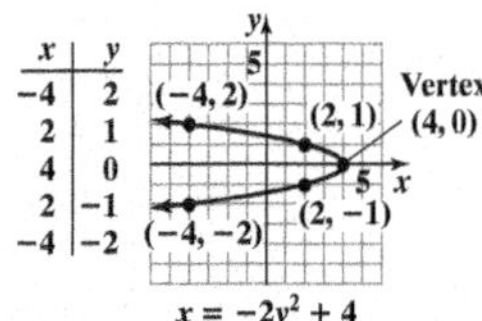

The vertex is at $(4, 0)$. The axis of symmetry is the x-axis.

5. **Step 1** The equation has the form $x = a(y - k)^2 + h$, where $a = -1, k = -1$, and $h = -3$, so it is a horizontal parabola.
Step 2 $a < 0$; so the parabola opens to the left.
Step 3 We have $k = -1$ and $h = -3$. Therefore, the vertex is $(-3, -1)$.
Step 4 The line $y = -1$ is the axis of symmetry. At $y = 0, x = -4$. Thus, we have the point $(-4, 0)$ and $(-4, -2)$ from symmetry. At $y = 1, x = -7$. Thus, we have the point $(-7, 1)$ and $(-7, -3)$ from symmetry.
Step 5 At $y = 0$,

$$x = -(0 + 1)^2 - 3$$
$$= -(1) - 3$$
$$= -1 - 3 = -4$$

Thus, the x-intercept is $(-4, 0)$.

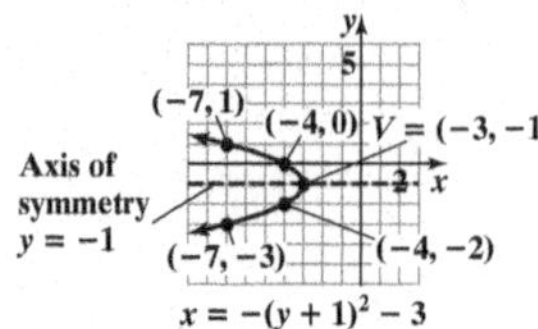

6. Since the y-term is squared, we have a horizontal parabola. The standard form is $x = a(y - k)^2 + h$.

$$x = y^2 - 6y + 13$$
$$= y^2 - 6y + \left(\frac{6}{2}\right)^2 - \left(\frac{6}{2}\right)^2 + 13$$
$$= (y^2 - 6y + 9) + 4 \qquad \text{Complete the square.}$$
$$= (y - 3)^2 + 4$$

Therefore, we know that $a = 1, k = 3$, and $h = 4$. The vertex is at $(4, 3)$. The axis of symmetry is $y = 3$. If $y = 0$, $x = 13$. So the x-intercept is $(13, 0)$.

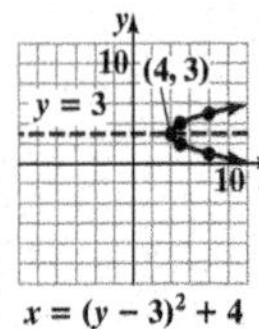

7. $y = 2x^2 + 8x + 9$
Since the x-term is squared, we have a vertical parabola. The standard form is

$$y = a(x - h)^2 + k$$
$$y = 2x^2 + 8x + 9$$
$$= 2(x^2 + 4x + \_\_\_\_) - \_\_\_\_ + 9$$
$$= 2(x^2 + 4x + 4) - 2(4) + 9 \qquad \text{Complete the square.}$$
$$= 2(x + 2)^2 + 1$$

The parabola opens upward. The vertex is $(-2, 1)$, and the y-intercept is $(0, 9)$. The axis of symmetry is $x = -2$.

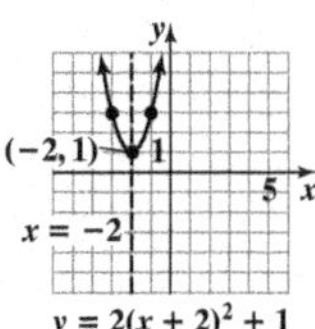

10.3 Student Practice

1. Write the equation in standard form.

$$4x^2 + y^2 = 16$$
$$\frac{4x^2}{16} + \frac{y^2}{16} = \frac{16}{16}$$
$$\frac{x^2}{4} + \frac{y^2}{16} = 1$$

Thus, we have:

$$a^2 = 4 \quad \text{so} \quad a = 2$$
$$b^2 = 16 \quad \text{so} \quad b = 4$$

The x-intercepts are $(2, 0)$ and $(-2, 0)$ and the y-intercepts are $(0, 4)$ and $(0, -4)$.

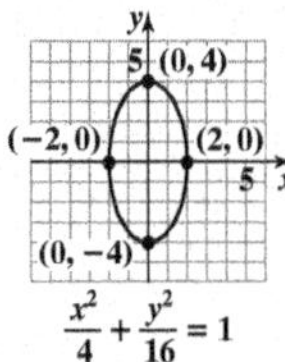

2. $\dfrac{(x - 2)^2}{16} + \dfrac{(y + 3)^2}{9} = 1$
The center is $(h, k) = (2, -3)$, $a = 4$, and $b = 3$. We start at $(2, -3)$ and measure to the right and to the left 4 units, and up and down 3 units.

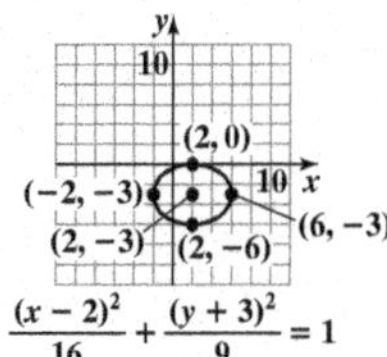

10.4 Student Practice

1. $\dfrac{x^2}{16} - \dfrac{y^2}{25} = 1$
This is the equation of a horizontal hyperbola with center $(0, 0)$, where $a = 4$ and $b = 5$. The vertices are $(-4, 0)$ and $(4, 0)$. Construct a fundamental rectangle with corners at $(4, 5)$, $(4, -5)$, $(-4, 5)$, and $(-4, -5)$. Draw extended diagonals as the asymptotes.

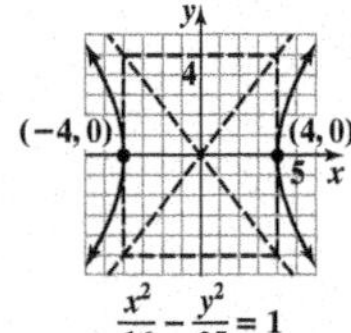

2. Write the equation in standard form.

$$y^2 - 4x^2 = 4$$
$$\frac{y^2}{4} - \frac{4x^2}{4} = \frac{4}{4}$$
$$\frac{y^2}{4} - \frac{x^2}{1} = 1$$

This is the equation of a vertical hyperbola with center $(0, 0)$, where $a = 1$ and $b = 2$. The vertices are $(0, 2)$ and $(0, -2)$.
The fundamental rectangle has corners at $(1, 2)$, $(1, -2)$, $(-1, 2)$, and $(-1, -2)$.

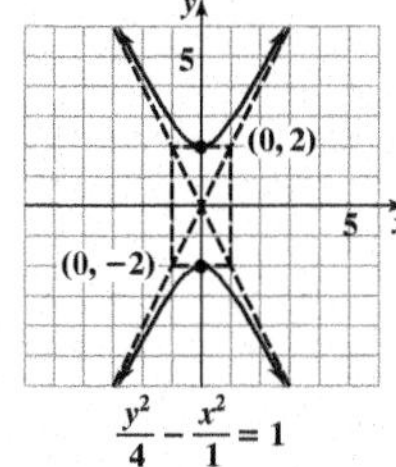

3. $\frac{(y + 2)^2}{9} - \frac{(x - 3)^2}{16} = 1$
This is a vertical hyperbola with center at $(3, -2)$, where $a = 4$ and $b = 3$. The vertices are $(3, 1)$ and $(3, -5)$. The fundamental rectangle has corners at $(7, 1)$, $(7, -5)$, $(-1, 1)$, and $(-1, -5)$.

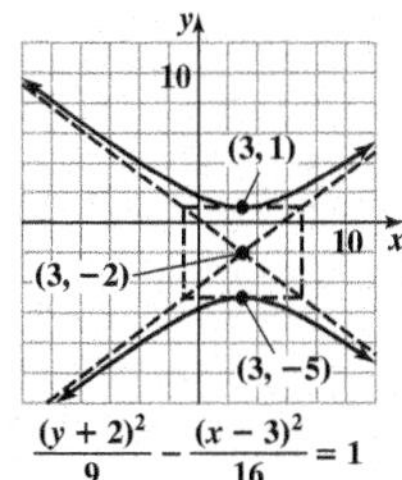

10.5 Student Practice

1. $\frac{x^2}{4} - \frac{y^2}{4} = 1$ $\qquad x + y + 1 = 0$
$x^2 - y^2 = 4$ (1) $\qquad y = -x - 1$ (2)
Substitute (2) into (1).

$$x^2 - (-x - 1)^2 = 4$$
$$x^2 - (x^2 + 2x + 1) = 4$$
$$x^2 - x^2 - 2x - 1 = 4$$
$$-2x - 1 = 4$$
$$-2x = 5$$
$$x = \frac{5}{-2} = -2.5$$

Now substitute the value for x in the equation $y = -x - 1$.
For $x = -2.5$: $y = -(-2.5) - 1 = 2.5 - 1 = 1.5$.
The solution is $(-2.5, 1.5)$.

2. $2x - 9 = y$ (1)
$xy = -4$ (2)
Equation (1) is solved for y.
Substitute (1) into equation (2).

$$x(2x - 9) = -4$$
$$2x^2 - 9x = -4$$
$$2x^2 - 9x + 4 = 0$$
$$(2x - 1)(x - 4) = 0$$
$$2x - 1 = 0 \quad \text{or} \quad x - 4 = 0$$
$$x = \frac{1}{2} \qquad x = 4$$

For $x = \frac{1}{2}$: $y = 2\left(\frac{1}{2}\right) - 9 = -8$
For $x = 4$: $y = 2(4) - 9 = -1$
The solutions are $(4, -1)$ and $\left(\frac{1}{2}, -8\right)$.

The graph of $y = \frac{-4}{x}$ is the graph of $y = \frac{1}{x}$ reflected across the x-axis and stretched by a factor of 4. The graph of $y = 2x - 9$ is a line with slope 2 passing through the point $(0, -9)$. The sketch shows that the points $(4, -1)$ and $\left(\frac{1}{2}, -8\right)$ seem reasonable.

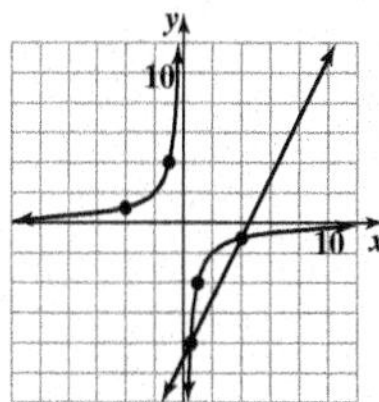

3. $x^2 + y^2 = 12$ (1)
$3x^2 - 4y^2 = 8$ (2)

$4x^2 + 4y^2 = 48$ Multiply (1) by 4.
$3x^2 - 4y^2 = 8$ Add the equations.
$7x^2 = 56$

$$x^2 = 8$$
$$x = \pm\sqrt{8}$$
$$x = \pm 2\sqrt{2}$$

If $x = 2\sqrt{2}$, then $x^2 = 8$. Substituting this value into equation (1) gives

$$8 + y^2 = 12$$
$$y^2 = 4$$
$$y = \pm\sqrt{4}$$
$$y = \pm 2$$

Similarly, if $x = -2\sqrt{2}$, then $y = \pm 2$.
Thus, the four solutions are $(2\sqrt{2}, 2)$, $(2\sqrt{2}, -2)$, $(-2\sqrt{2}, 2)$, $(-2\sqrt{2}, -2)$.

Chapter 11 11.1 Student Practice

1. (a) $g(a) = \frac{1}{2}a - 3$

(b) $g(a + 4) = \frac{1}{2}(a + 4) - 3$
$= \frac{1}{2}a + 2 - 3$
$= \frac{1}{2}a - 1$

(c) $g(a) = \frac{1}{2}a - 3$
$g(4) = \frac{1}{2}(4) - 3 = 2 - 3 = -1$
Thus, $g(a) + g(4) = \left(\frac{1}{2}a - 3\right) + (-1)$
$= \frac{1}{2}a - 3 - 1$
$= \frac{1}{2}a - 4$

2. (a) $p(-3) = -3(-3)^2 + 2(-3) + 4$
$= -3(9) + 2(-3) + 4$
$= -27 - 6 + 4$
$= -29$

(b) $p(a) = -3(a)^2 + 2(a) + 4$
$= -3a^2 + 2a + 4$

(c) $p(2a) = -3(2a)^2 + 2(2a) + 4$
$= -3(4a^2) + 2(2a) + 4$
$= -12a^2 + 4a + 4$

(d) $p(a - 3) = -3(a - 3)^2 + 2(a - 3) + 4$
$= -3(a - 3)(a - 3) + 2(a - 3) + 4$
$= -3(a^2 - 6a + 9) + 2(a - 3) + 4$
$= -3a^2 + 18a - 27 + 2a - 6 + 4$
$= -3a^2 + 20a - 29$

3. (a) $r(a + 2) = \dfrac{-3}{(a + 2) + 1}$
$= \dfrac{-3}{a + 3}$

(b) $r(a) = \dfrac{-3}{a + 1}$

(c) $r(a + 2) - r(a) = \dfrac{-3}{a + 3} - \left(\dfrac{-3}{a + 1}\right)$
$= \dfrac{-3}{a + 3} + \dfrac{3}{a + 1}$
$= \dfrac{(-3)(a + 1)}{(a + 3)(a + 1)} + \dfrac{3(a + 3)}{(a + 1)(a + 3)}$
$= \dfrac{-3a - 3}{(a + 1)(a + 3)} + \dfrac{3a + 9}{(a + 1)(a + 3)}$
$= \dfrac{-3a - 3 + 3a + 9}{(a + 1)(a + 3)}$
$= \dfrac{6}{(a + 1)(a + 3)}$

4. $g(x + h) = 2 - 5(x + h) = 2 - 5x - 5h$
$g(x) = 2 - 5x$
$g(x + h) - g(x) = (2 - 5x - 5h) - (2 - 5x)$
$= 2 - 5x - 5h - 2 + 5x$
$= -5h$

Therefore, $\dfrac{g(x + h) - g(x)}{h} = \dfrac{-5h}{h} = -5.$

5. (a) $S(r) = 16(3.14)r + 2(3.14)r^2 = 50.24r + 6.28r^2$

(b) $S(2) = 50.24(2) + 6.28(2)^2$
$= 50.24(2) + 6.28(4)$
$= 125.6 \text{ m}^2$

(c) $S(e) = 50.24(2 + e) + 6.28(2 + e)^2$
$= 50.24(2 + e) + 6.28(2 + e)(2 + e)$
$= 50.24(2 + e) + 6.28(4 + 4e + e^2)$
$= 100.48 + 50.24e + 25.12 + 25.12e + 6.28e^2$
$= 125.6 + 75.36e + 6.28e^2$

(d) $S(e) = 125.6 + 75.36e + 6.28e^2$
$S(0.3) = 125.6 + 75.36(0.3) + 6.28(0.3)^2$
$= 125.6 + 22.608 + 0.5652$
$\approx 148.77 \text{ m}^2$

Thus, if the radius of 2 meters was incorrectly measured as 2.3 meters, the surface area would be approximately $148.77 - 125.6 = 23.17$ square meters too large.

11.2 Student Practice

1. By the vertical line test, this relation is not a function.

2. $f(x) = x^2$ $\qquad h(x) = x^2 - 5$

x	f (x)
−2	4
−1	1
0	0
1	1
2	4

x	h (x)
−2	−1
−1	−4
0	−5
1	−4
2	−1

3. $f(x) = |x|$ $\qquad p(x) = |x + 2|$

x	f (x)
−4	4
−3	3
−2	2
−1	1
0	0
1	1
2	2

x	p (x)
−4	2
−3	1
−2	0
−1	1
0	2
1	3
2	4

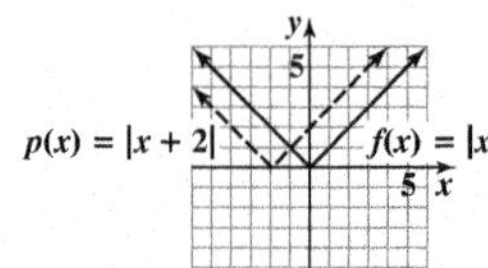

4. $f(x) = x^3$

x	f (x)
−2	−8
−1	−1
0	0
1	1
2	8

We recognize that $h(x)$ will have a similar shape, but the curve will be shifted 4 units to the left and 3 units upward.

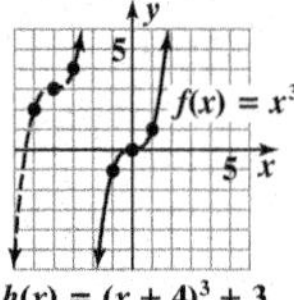

5. $f(x) = \dfrac{2}{x}$

x	f (x)
−4	$-\frac{1}{2}$
−2	−1
−1	−2
$-\frac{1}{2}$	−4
0	undefined
$\frac{1}{2}$	4
1	2
2	1
4	$\frac{1}{2}$

The graph of $g(x)$ is 1 unit to the left and 2 units below $f(x)$. We use each point on $f(x)$ to guide us in graphing $g(x)$.
Note: $g(x)$ is not defined for $x = -1$.

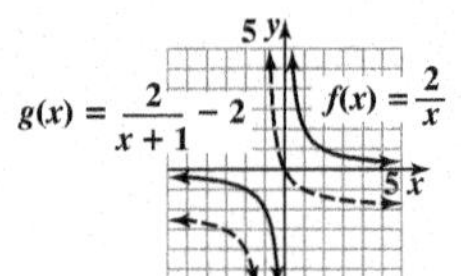

11.3 Student Practice

1. (a) $(f+g)(x) = f(x) + g(x)$
$= (4x+5) + (2x^2+7x-8)$
$= 4x + 5 + 2x^2 + 7x - 8$
$= 2x^2 + 11x - 3$

(b) Use the formula obtained in **(a)**.
$(f+g)(x) = 2x^2 + 11x - 3$
$(f+g)(4) = 2(4)^2 + 11(4) - 3$
$= 2(16) + 11(4) - 3$
$= 32 + 44 - 3$
$= 73$

2. (a) $(fg)(x) = f(x) \cdot g(x)$
$= (3x+2)(x^2-3x-4)$
$= 3x^3 - 9x^2 - 12x + 2x^2 - 6x - 8$
$= 3x^3 - 7x^2 - 18x - 8$

(b) Use the formula obtained in **(a)**.
$(fg)(x) = 3x^3 - 7x^2 - 18x - 8$
$(fg)(2) = 3(2)^3 - 7(2)^2 - 18(2) - 8$
$= 3(8) - 7(4) - 18(2) - 8$
$= 24 - 28 - 36 - 8$
$= -48$

3. (a) $\left(\frac{g}{h}\right)(x) = \frac{5x+1}{3x-2}$, where $x \neq \frac{2}{3}$

(b) $\left(\frac{g}{p}\right)(x) = \frac{5x+1}{5x^2+6x+1} = \frac{5x+1}{(5x+1)(x+1)} = \frac{1}{x+1}$,
where $x \neq -\frac{1}{5}, x \neq -1$

(c) $\left(\frac{g}{h}\right)(x) = \frac{5x+1}{3x-2}$
$\left(\frac{g}{h}\right)(3) = \frac{5(3)+1}{3(3)-2} = \frac{15+1}{9-2} = \frac{16}{7}$

4. $f[g(x)] = f(3x-4)$
$= 2(3x-4) - 1$
$= 6x - 8 - 1$
$= 6x - 9$

5. (a) $f[g(x)] = f(x+2)$
$= 2(x+2)^2 - 3(x+2) + 1$
$= 2(x^2+4x+4) - 3(x+2) + 1$
$= 2x^2 + 8x + 8 - 3x - 6 + 1$
$= 2x^2 + 5x + 3$

(b) $g[f(x)] = g(2x^2-3x+1)$
$= (2x^2-3x+1) + 2$
$= 2x^2 - 3x + 1 + 2$
$= 2x^2 - 3x + 3$

6. (a) $(g \circ f)(x) = g[f(x)]$
$= g(3x+1)$
$= \frac{2}{(3x+1)-3}$
$= \frac{2}{3x+1-3}$
$= \frac{2}{3x-2}, x \neq \frac{2}{3}$

(b) $(g \circ f)(-3) = \frac{2}{3(-3)-2} = \frac{2}{-9-2} = \frac{2}{-11} = -\frac{2}{11}$

11.4 Student Practice

1. (a) A is a function. No two pairs have the same second coordinate. Thus, A is a one-to-one function.

(b) B is a function. The pair $(1, 1)$ and $(-1, 1)$ share a common second coordinate. Therefore, B is not a one-to-one function.

2. The graphs of **(a)** and **(b)** do not represent one-to-one functions. Horizontal lines exist that intersect each graph more than once.

3. The inverse of B is obtained by interchanging x- and y-values of each ordered pair.
$B^{-1} = \{(2,1), (8,7), (7,8), (12,10)\}$

4. $g(x) = 4 - 6x$
$y = 4 - 6x$
$x = 4 - 6y$
$x - 4 = -6y$
$-x + 4 = 6y$
$\frac{-x+4}{6} = y$
$g^{-1}(x) = \frac{-x+4}{6}$

5. $f(x) = 0.75 + 0.55(x-1)$
$y = 0.75 + 0.55(x-1)$
$x = 0.75 + 0.55(y-1)$
$x = 0.75 + 0.55y - 0.55$
$x = 0.55y + 0.2$
$x - 0.2 = 0.55y$
$\frac{x-0.2}{0.55} = y$
$f^{-1}(x) = \frac{x-0.2}{0.55}$

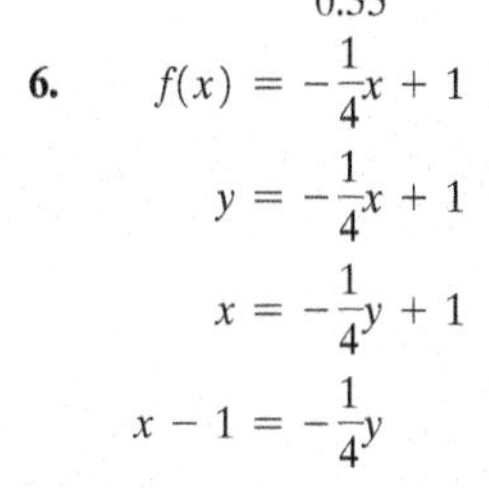

6. $f(x) = -\frac{1}{4}x + 1$
$y = -\frac{1}{4}x + 1$
$x = -\frac{1}{4}y + 1$
$x - 1 = -\frac{1}{4}y$
$-4x + 4 = y$
$f^{-1}(x) = -4x + 4$
Now graph each line.

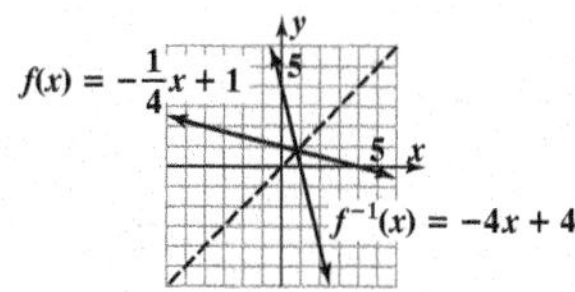

Chapter 12

12.1 Student Practice

1. $f(x) = 3^x$

$f(-2) = 3^{-2} = \left(\frac{1}{3}\right)^2 = \frac{1}{9}$

$f(-1) = 3^{-1} = \frac{1}{3}$

$f(0) = 3^0 = 1$

$f(1) = 3^1 = 3$

$f(2) = 3^2 = 9$

x	f(x)
−2	$\frac{1}{9}$
−1	$\frac{1}{3}$
0	1
1	3
2	9

$\left(-1, \frac{1}{3}\right)$ (2, 9) (1, 3) (0, 1)

$f(x) = 3^x$

2. $f(x) = \left(\frac{1}{3}\right)^x$

$f(-2) = \left(\frac{1}{3}\right)^{-2} = 3^2 = 9$

$f(-1) = \left(\frac{1}{3}\right)^{-1} = 3^1 = 3$

$f(0) = \left(\frac{1}{3}\right)^0 = 1$

$f(1) = \left(\frac{1}{3}\right)^1 = \frac{1}{3}$

$f(2) = \left(\frac{1}{3}\right)^2 = \frac{1}{9}$

x	f(x)
−2	9
−1	3
0	1
1	$\frac{1}{3}$
2	$\frac{1}{9}$

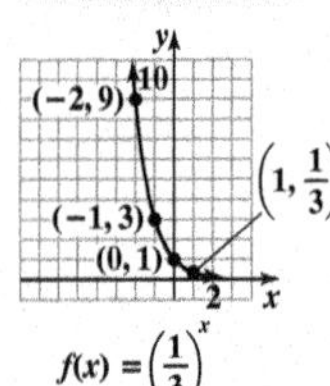

3. $f(x) = 3^{x+2}$

$f(-4) = 3^{-4+2} = 3^{-2} = \frac{1}{3^2} = \frac{1}{9}$

$f(-3) = 3^{-3+2} = 3^{-1} = \frac{1}{3}$

$f(-2) = 3^{-2+2} = 3^0 = 1$

$f(-1) = 3^{-1+2} = 3^1 = 3$

$f(0) = 3^{0+2} = 3^2 = 9$

x	f(x)
−4	$\frac{1}{9}$
−3	$\frac{1}{3}$
−2	1
−1	3
0	9

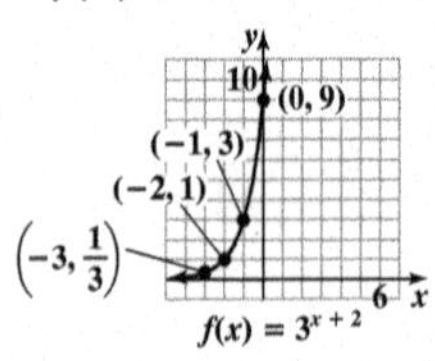

4. $f(x) = e^{x-2}$ (values rounded to nearest hundredth)

$f(4) = e^{4-2} = e^2 \approx 7.39$

$f(3) = e^{3-2} = e^1 \approx 2.72$

$f(2) = e^{2-2} = e^0 = 1$

$f(1) = e^{1-2} = e^{-1} \approx 0.37$

$f(0) = e^{0-2} = e^{-2} \approx 0.14$

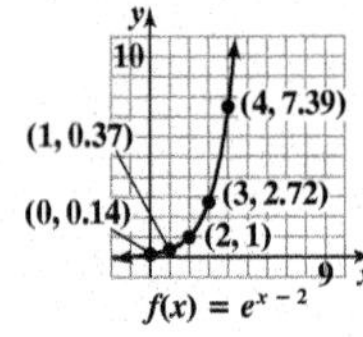

x	f(x)
4	7.39
3	2.72
2	1
1	0.37
0	0.14

5. $2^x = \frac{1}{32}$

$2^x = \frac{1}{2^5}$ Because $2^5 = 32$.

$2^x = 2^{-5}$ Because $\frac{1}{2^5} = 2^{-5}$.

$x = -5$ Property of exponential equations.

6. Here $P = 4000$, $r = 0.07$, and $t = 2$.

$A = P(1 + r)^t$

$= 4000(1 + 0.07)^2$

$= 4000(1.07)^2$

$= 4000(1.1449)$

$= 4579.6$

Uncle Jose will have \$4579.60.

7. Here $P = 1500$, $r = 8\% = 0.08$, $t = 8$, and $n = 4$.

$A = P\left(1 + \frac{r}{n}\right)^{nt}$

$= 1500\left(1 + \frac{0.08}{4}\right)^{(4)(8)}$

$= 1500(1 + 0.02)^{32}$

$= 1500(1.02)^{32}$

$\approx 1500(1.884540592)$

≈ 2826.81089

Collette will have \$2826.81.

8. Here $C = 20$ and $t = 5000$.

$A = 20e^{-0.0016008(5000)}$

$A = 20e^{-8.004}$

$A \approx 20(0.0003341) = 0.006682 \approx 0.007$

Thus, 0.007 milligram of americium 241 would be present in 5000 years.

12.2 Student Practice

1. Use the fact that $x = b^y$ is equivalent to $\log_b x = y$.

(a) Here $x = 49$, $b = 7$, and $y = 2$. So $2 = \log_7 49$.

(b) Here $x = \frac{1}{64}$, $b = 4$, and $y = -3$. So $-3 = \log_4\left(\frac{1}{64}\right)$.

2. (a) Here $y = 3$, $b = 5$, and $x = 125$. Thus, since $x = b^y$, $125 = 5^3$.

(b) Here $y = -2$, $b = 6$, and $x = \frac{1}{36}$. So $\frac{1}{36} = 6^{-2}$.

3. (a) $\log_b 125 = 3$; then $125 = b^3$

$5^3 = b^3$

$b = 5$

(b) $\log_{1/2} 32 = x$; then $32 = \left(\frac{1}{2}\right)^x$

$2^5 = \left(\frac{1}{2}\right)^x$

$\frac{1}{2^{-5}} = \left(\frac{1}{2}\right)^x$

$\left(\frac{1}{2}\right)^{-5} = \left(\frac{1}{2}\right)^x$

$x = -5$

4. $\log_{10} 0.1 = x$

$0.1 = 10^x$

$10^{-1} = 10^x$

$-1 = x$

Thus, $\log_{10} 0.1 = -1$.

5. To graph $y = \log_{1/2} x$, we first write $x = \left(\frac{1}{2}\right)^y$. We make a table of values.

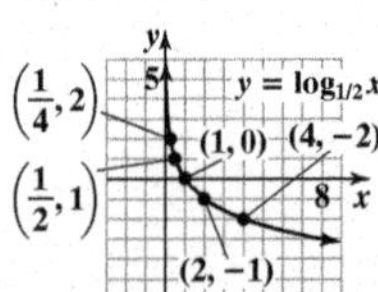

x	y
$\frac{1}{4}$	2
$\frac{1}{2}$	1
1	0
2	−1
4	−2

6. Make a table of values for each equation.

$y = \log_6 x$

x	y
$\frac{1}{6}$	-1
1	0
6	1
36	2

$y = 6^x$

x	y
-1	$\frac{1}{6}$
0	1
1	6
2	36

Ordered pairs reversed

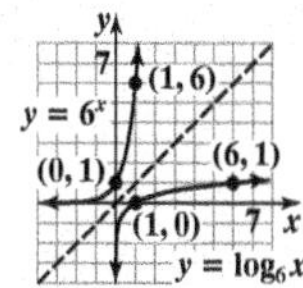

12.3 Student Practice

1. Property 1 can be extended to three logarithms.
$$\log_b MNP = \log_b M + \log_b N + \log_b P$$
Thus,
$$\log_4 WXY = \log_4 W + \log_4 X + \log_4 Y.$$

2. $\log_b M + \log_b N + \log_b P = \log_b MNP$
Therefore,
$$\log_7 w + \log_7 8 + \log_7 x = \log_7(w \cdot 8 \cdot x) = \log_7 8wx.$$

3. $\log_3\left(\frac{17}{5}\right) = \log_3 17 - \log_3 5$

4. $\log_b 132 - \log_b 4 = \log_b\left(\frac{132}{4}\right) = \log_b 33$

5. $\frac{1}{3}\log_7 x - 5\log_7 y = \log_7 x^{1/3} - \log_7 y^5$ By property 3.

$= \log_7\left(\frac{x^{1/3}}{y^5}\right)$ By property 2.

6. $\log_3\left(\frac{x^4y^5}{z}\right) = \log_3 x^4y^5 - \log_3 z$ By property 2.

$= \log_3 x^4 + \log_3 y^5 - \log_3 z$ By property 1.

$= 4\log_3 x + 5\log_3 y - \log_3 z$ By property 3.

7. $\log_4 x + \log_4 5 = 2$

$\log_4 5x = 2$ By property 1.

Converting to exponential form, we have

$4^2 = 5x$

$16 = 5x$

$\frac{16}{5} = x$

$x = \frac{16}{5}$

8. $\log_{10} x - \log_{10}(x + 3) = -1$

$\log_{10}\left(\frac{x}{x+3}\right) = -1$ By property 2.

$\frac{x}{x+3} = 10^{-1}$ Convert to exponential form.

$x = \frac{1}{10}(x + 3)$ Multiply each side by $(x + 3)$.

$x = \frac{1}{10}x + \frac{3}{10}$ Simplify.

$\frac{9}{10}x = \frac{3}{10}$

$x = \frac{1}{3}$

9. (a) $\log_7 1 = 0$ By property 5.

(b) $\log_8 8 = 1$ By property 4.

(c) $\log_{12} 13 = \log_{12}(y + 2)$

$13 = y + 2$ By property 6.

$y = 11$

10. $\log_3 2 - \log_3 5 = \log_3 6 + \log_3 x$

$\log_3\left(\frac{2}{5}\right) = \log_3 6x$ By property 1 and property 2.

$\frac{2}{5} = 6x$ By property 6.

$x = \frac{1}{15}$

12.4 Student Practice

1. (a) 4.36 [log] $\approx$ 0.639486489

(b) 436 [log] $\approx$ 2.639486489

(c) 0.2418 [log] $\approx$ -0.616543703

2. We know that $\log x = 2.913$ is equivalent to $10^{2.913} = x$.
Using a calculator we have

2.913 [10^x] $\approx$ 818.46479.

Thus, $x \approx 818.46479$.

3. To evaluate antilog(-3.0705) using a scientific calculator, we have

3.0705 [+/−] [10^x] $\approx 8.5015869 \times 10^{-4}$.

Thus, antilog$(-3.0705) \approx 0.00085015869$.

4. (a) $\log x = 0.06134$ is equivalent to $10^{0.06134} = x$.

0.06134 [10^x] $\approx$ 1.1517017

Thus, $x \approx 1.1517017$.

(b) $\log x = -4.6218$ is equivalent to $10^{-4.6218} = x$.

4.6218 [+/−] [10^x] $\approx 2.3889112 \times 10^{-5}$

Thus, $x \approx 0.000023889112$.

5. (a) 4.82 [ln] $\approx$ 1.572773928

(b) 48.2 [ln] $\approx$ 3.875359021

(c) 0.0793 [ln] $\approx$ -2.53451715

6. (a) If $\ln x = 3.1628$, then $e^{3.1628} = x$.

3.1628 [e^x] $\approx$ 23.636686

Thus, $x \approx 23.636686$.

(b) If $\ln x = -2.0573$, then $e^{-2.0573} = x$.

2.0573 [+/−] [e^x] $\approx$ 0.1277986

Thus, $x \approx 0.1277986$.

7. To evaluate $\log_9 3.76$, we use the change of base formula.

$$\log_9 3.76 = \frac{\log 3.76}{\log 9}$$

On a calculator, find the following.

3.76 [log] [÷] [9] [log] = 0.602769044.

8. By the change of base formula,

$$\log_8 0.009312 = \frac{\log_e 0.009312}{\log_e 8} = \frac{\ln 0.009312}{\ln 8}$$

On a calculator, find the following.

0.009312 [ln] [÷] 8 [ln] [=] -2.24889774

Thus, $\log_8 0.009312 \approx -2.24889774$.

9. Make a table of values.

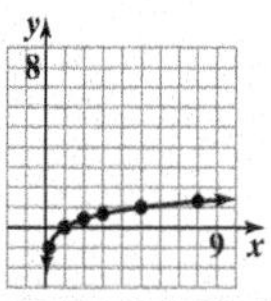

x	$y = \log_5 x$
0.2	-1
1	0
2	0.4
3	0.7
5	1
8	1.3

12.5 Student Practice

1.
$$\begin{aligned}\log(x+5) &= 2-\log 5\\ \log(x+5)+\log 5 &= 2\\ \log[5(x+5)] &= 2\\ \log(5x+25) &= 2\\ 5x+25 &= 10^2\\ 5x+25 &= 100\\ 5x &= 75\\ x &= 15\end{aligned}$$

Check.
$$\begin{aligned}\log(15+5) &\stackrel{?}{=} 2-\log 5\\ \log 20 &\stackrel{?}{=} 2-\log 5\\ 1.301029996 &\stackrel{?}{=} 2-0.698970004\\ 1.301029996 &= 1.301029996 \ \checkmark\end{aligned}$$

2. $\log(x+3)-\log x = 1$
$$\begin{aligned}\log\left(\frac{x+3}{x}\right) &= 1\\ \frac{x+3}{x} &= 10^1\\ \frac{x+3}{x} &= 10\\ x+3 &= 10x\\ 3 &= 9x\\ \frac{1}{3} &= x\end{aligned}$$

Check.
$$\begin{aligned}\log\left(\frac{1}{3}+3\right)-\log\frac{1}{3} &\stackrel{?}{=} 1\\ \log\frac{10}{3}-\log\frac{1}{3} &\stackrel{?}{=} 1\\ \log\left(\frac{\frac{10}{3}}{\frac{1}{3}}\right) &\stackrel{?}{=} 1\\ \log 10 &\stackrel{?}{=} 1\\ 1 &= 1 \ \checkmark\end{aligned}$$

3. $\log 5-\log x = \log(6x-7)$
$$\begin{aligned}\log\left(\frac{5}{x}\right) &= \log(6x-7)\\ \frac{5}{x} &= 6x-7\\ 5 &= 6x^2-7x\\ 0 &= 6x^2-7x-5\\ 0 &= (3x-5)(2x+1)\end{aligned}$$

$3x-5=0$ or $2x+1=0$

$3x=5$ $\quad 2x=-1$

$x=\frac{5}{3}$ $\quad x=-\frac{1}{2}$

Check. $\log 5-\log x = \log(6x-7)$

$x=\frac{5}{3}$: $\log 5-\log\left(\frac{5}{3}\right) \stackrel{?}{=} \log\left[6\left(\frac{5}{3}\right)-7\right]$
$$\begin{aligned}\log 5-\log\frac{5}{3} &\stackrel{?}{=} \log(10-7)\\ \log 5-\log\frac{5}{3} &\stackrel{?}{=} \log 3\\ \log\left[\frac{5}{\frac{5}{3}}\right] &\stackrel{?}{=} \log 3\\ \log 3 &= \log 3 \ \checkmark\end{aligned}$$

$x=-\frac{1}{2}$: $\log 5-\log\left(-\frac{1}{2}\right) \stackrel{?}{=} \log\left[6\left(-\frac{1}{2}\right)-7\right]$

Since logarithms of negative numbers do not exist, $x=-\frac{1}{2}$ is not a valid solution

The only solution is $\frac{5}{3}$.

4.
$$\begin{aligned}3^x &= 5\\ \log 3^x &= \log 5\\ x\log 3 &= \log 5\\ x &= \frac{\log 5}{\log 3}\end{aligned}$$

5. Take the logarithm of each side.
$$\begin{aligned}2^{3x+1} &= 9^{x+1}\\ \log 2^{3x+1} &= \log 9^{x+1}\\ (3x+1)\log 2 &= (x+1)\log 9\\ 3x\log 2+\log 2 &= x\log 9+\log 9\\ 3x\log 2-x\log 9 &= \log 9-\log 2\\ x(3\log 2-\log 9) &= \log 9-\log 2\\ x &= \frac{\log 9-\log 2}{3\log 2-\log 9}\end{aligned}$$

Use the following keystrokes.

[(] 9 [log] [−] 2 [log] [)] [÷] [(] 3 [×] 2 [log] [−] 9 [log] [)] [=] −12.76989838

Rounding to the nearest thousandth, we have $x \approx -12.770$.

6. Take the natural logarithm of each side.
$$\begin{aligned}20.98 &= e^{3.6x}\\ \ln 20.98 &= \ln e^{3.6x}\\ \ln 20.98 &= (3.6x)(\ln e)\\ \ln 20.98 &= 3.6x\\ \frac{\ln 20.98}{3.6} &= x\end{aligned}$$

On a scientific calculator, find the following.

20.98 [ln] [÷] 3.6 [=] 0.845436001

Rounding to the nearest ten-thousandth, we have $x \approx 0.8454$.

7. We use the formula $A=P(1+r)^t$, where $A=10{,}000$, $P=4000$, and $r=0.08$.
$$\begin{aligned}10{,}000 &= 4000(1+0.08)^t\\ 10{,}000 &= 4000(1.08)^t\\ \frac{10{,}000}{4000} &= (1.08)^t\\ 2.5 &= (1.08)^t\\ \log 2.5 &= \log(1.08)^t\\ \log 2.5 &= t(\log 1.08)\\ \frac{\log 2.5}{\log 1.08} &= t\end{aligned}$$

On a scientific calculator,

2.5 [log] [÷] 1.08 [log] [=] 11.905904.

Thus, it would take approximately 12 years.

8.
$$\begin{aligned}A &= A_0e^{rt}\\ 5000 &= 1300e^{0.043t} && \text{Substitute known values.}\\ \frac{5000}{1300} &= e^{0.043t} && \text{Divide each side by 1300.}\\ \ln\left(\frac{5000}{1300}\right) &= \ln e^{0.043t} && \text{Take the natural logarithm of each side.}\\ \ln\left(\frac{50}{13}\right) &= (0.043t)\ln e\\ \ln 50-\ln 13 &= 0.043t\\ \frac{\ln 50-\ln 13}{0.043} &= t\end{aligned}$$

Using a scientific calculator,

[(] 50 [ln] [−] 13 [ln] [)] [÷] 0.043 [=] 31.32729414

Rounding to the nearest whole year, the food shortage will develop in about 31 years.

9. Let I_J = intensity of the Japan earthquake.
Let I_S = intensity of the San Francisco earthquake.

$8.9 = \log\left(\frac{I_J}{I_0}\right) = \log I_J - \log I_0$

Solving for $\log I_0$ gives $\log I_0 = \log I_J - 8.9$

$7.1 = \log\left(\frac{I_S}{I_0}\right) = \log I_S - \log I_0$

Solving for $\log I_0$ gives $\log I_0 = \log I_S - 7.1$

Therefore, $\log I_J - 8.9 = \log I_S - 7.1$

$\log I_J - \log I_S = 8.9 - 7.1$

$\log\left(\frac{I_J}{I_S}\right) = 1.8$

$\frac{I_J}{I_S} = 10^{1.8}$

$\frac{I_J}{I_S} \approx 63.09573445$

$\frac{I_J}{I_S} \approx 63$

$I_J \approx 63I_S$

The earthquake in Japan was about sixty-three times more intense than the San Francisco earthquake.

Appendix A.1 Student Practice

1. (a) $(-3)^5 = (-3)(-3)(-3)(-3)(-3) = -243$
(b) $(-3)^6 = (-3)(-3)(-3)(-3)(-3)(-3) = 729$
(c) $(-4)^4 = (-4)(-4)(-4)(-4) = 256$
(d) $-4^4 = -(4 \cdot 4 \cdot 4 \cdot 4) = -256$
(e) $\left(\frac{1}{5}\right)^2 = \left(\frac{1}{5}\right)\left(\frac{1}{5}\right) = \frac{1}{25}$

2. Since $(-7)^2 = 49$ and $7^2 = 49$, the square roots of 49 are -7 and 7. The principal square root is 7.

3. (a) $(0.3)^2 = (0.3)(0.3) = 0.09$, therefore, $\sqrt{0.09} = 0.3$.
(b) $\sqrt{\frac{4}{81}} = \frac{\sqrt{4}}{\sqrt{81}} = \frac{2}{9}$
(c) This is not a real number.

4. (a) $6(12 - 8) + 4 = 6(4) + 4 = 24 + 4 = 28$
(b) $5[6 - 3(7 - 9)] - 8 = 5[6 - 3(-2)] - 8$
$= 5[6 + 6] - 8$
$= 5[12] - 8$
$= 60 - 8$
$= 52$

5. (a) $\sqrt{(-5)^2 + 12^2} = \sqrt{25 + 144} = \sqrt{169} = 13$
(b) $|-3 - 7 + 2 - (-4)| = |-3 - 7 + 2 + 4| = |-4| = 4$

6. $\frac{2(3) + 5(-2)}{1 + 2 \cdot 3^2 + 5(-3)} = \frac{6 - 10}{1 + 2 \cdot 9 + 5(-3)}$
$= \frac{-4}{1 + 18 + (-15)}$
$= \frac{-4}{4}$
$= -1$

7. (a) $3^{-2} = \frac{1}{3^2} = \frac{1}{9}$
(b) $z^{-8} = \frac{1}{z^8}$

8. (a) $(7xy^{-2})(2x^{-5}y^{-6}) = 14x^{1-5}y^{-2-6} = 14x^{-4}y^{-8} = \frac{14}{x^4y^8}$
(b) $(x + 2y)^4(x + 2y)^{10} = (x + 2y)^{4+10} = (x + 2y)^{14}$

9. $\frac{2x^{-3}y}{4x^{-2}y^5} = \frac{1}{2}x^{-3-(-2)}y^{1-5} = \frac{1}{2}x^{-1}y^{-4} = \frac{1}{2xy^4}$

10. (a) $\frac{30x^6y^5}{20x^3y^2} = \frac{30}{20} \cdot \frac{x^6}{x^3} \cdot \frac{y^5}{y^2} = \frac{3}{2}x^3y^3$ or $\frac{3x^3y^3}{2}$
(b) $\frac{-15a^3b^4c^4}{3a^5b^4c^2} = \frac{-15}{3} \cdot \frac{a^3}{a^5} \cdot \frac{b^4}{b^4} \cdot \frac{c^4}{c^2} = -5 \cdot a^{-2} \cdot b^0 \cdot c^2 = -\frac{5c^2}{a^2}$
(c) $(5^{-3})(2a)^0 = (5^{-3})(1) = \frac{1}{5^3} = \frac{1}{125}$

11. (a) $(w^3)^8 = w^{3 \cdot 8} = w^{24}$
(b) $(5^2)^5 = 5^{2 \cdot 5} = 5^{10}$
(c) $[(x - 2y)^3]^3 = (x - 2y)^{3 \cdot 3} = (x - 2y)^9$

12. (a) $(4x^3y^4)^2 = 4^2x^6y^8 = 16x^6y^8$
(b) $\left(\frac{4xy}{3x^5y^6}\right)^3 = \frac{4^3x^3y^3}{3^3x^{15}y^{18}} = \frac{64x^3y^3}{27x^{15}y^{18}} = \frac{64}{27x^{12}y^{15}}$
(c) $(3xy^2)^{-2} = 3^{-2}x^{-2}y^{-4} = \frac{1}{3^2x^2y^4} = \frac{1}{9x^2y^4}$

13. (a) $\frac{7x^2y^{-4}z^{-3}}{8x^{-5}y^{-6}z^2} = \frac{7x^2x^5y^6}{8y^4z^3z^2} = \frac{7x^7y^2}{8z^5}$
(b) $\left(\frac{4x^2y^{-2}}{x^{-4}y^{-3}}\right)^{-3} = \frac{4^{-3}x^{-6}y^6}{x^{12}y^9} = \frac{y^6}{4^3x^{12}x^6y^9} = \frac{1}{64x^{18}y^3}$

14. (a) $128{,}320 = 1.2832 \times 10^5$
(b) $476 = 4.76 \times 10^2$
(c) $0.0786 = 7.86 \times 10^{-2}$
(d) $0.007 = 7 \times 10^{-3}$

15. (a) $4.62 \times 10^6 = 4{,}620{,}000$
(b) $1.973 \times 10^{-3} = 0.001973$
(c) $4.931 \times 10^{-1} = 0.4931$

16. $\frac{(55{,}000)(3{,}000{,}000)}{5{,}500{,}000} = \frac{(5.5 \times 10^4)(3.0 \times 10^6)}{5.5 \times 10^6}$
$= \frac{3.0}{1} \times \frac{10^{10}}{10^6}$
$= 3.0 \times 10^{10-6}$
$= 3.0 \times 10^4$

A.2 Student Practice

1. (a) $3x^5 - 6x^4 + x^2$ This is a trinomial of degree 5.
(b) $5x^2 + 2$ This is a binomial of degree 2.
(c) $3ab + 5a^2b^2 - 6a^4b$ This is a trinomial of degree 5.
(d) $16x^4y^6$ This is a monomial of degree 10.

2. $p(x) = 2x^4 - 3x^3 + 6x - 8$
(a) $p(-2) = 2(-2)^4 - 3(-2)^3 + 6(-2) - 8$
$= 2(16) - 3(-8) + 6(-2) - 8$
$= 32 + 24 - 12 - 8$
$= 36$
(b) $p(5) = 2(5)^4 - 3(5)^3 + 6(5) - 8$
$= 2(625) - 3(125) + 6(5) - 8$
$= 1250 - 375 + 30 - 8$
$= 897$

3. $(-7x^2 + 5x - 9) + (2x^2 - 3x + 5)$
$= -7x^2 + 5x - 9 + 2x^2 - 3x + 5$
$= -5x^2 + 2x - 4$

4. $(2x^2 - 14x + 9) - (-3x^2 + 10x + 7)$
$= (2x^2 - 14x + 9) + (3x^2 - 10x - 7)$
$= 5x^2 - 24x + 2$

5. $(7x + 3)(2x - 5)$

$= 14x^2 - 35x + 6x - 15 = 14x^2 - 29x - 15$

6. $(3x^2 - 2)(5x - 4)$

$= 15x^3 - 12x^2 - 10x + 8$

7. $(7x - 2y)(7x + 2y) = (7x)^2 - (2y)^2$
$= 49x^2 - 4y^2$

8. (a) $(4u + 5v)^2 = (4u)^2 + 2(4u)(5v) + (5v)^2$
$= 16u^2 + 40uv + 25v^2$

(b) $(7x^2 - 3y^2)^2 = (7x^2)^2 - 2(7x^2)(3y^2) + (3y^2)^2$
$= 49x^4 - 42x^2y^2 + 9y^4$

9.
$$\begin{array}{r} 2x^2 - 3x + 1 \\ x^2 - 5x \\ \hline -10x^3 + 15x^2 - 5x \\ 2x^4 - 3x^3 + x^2 \\ \hline 2x^4 - 13x^3 + 16x^2 - 5x \end{array}$$

10. $(2x^2 - 3x + 1)(x^2 - 5x)$
$= (2x^2 - 3x + 1)(x^2) + (2x^2 - 3x + 1)(-5x)$
$= 2x^4 - 3x^3 + x^2 - 10x^3 + 15x^2 - 5x$
$= 2x^4 - 13x^3 + 16x^2 - 5x$

A.3 Student Practice

1. (a) $19x^3 - 38x^2 = 19 \cdot x \cdot x \cdot x - 19 \cdot 2 \cdot x \cdot x = 19x^2(x - 2)$

(b) $100a^4 - 50a^2 = 50a^2(2a^2 - 1)$

2. $9a^3 - 12a^2b^2 - 15a^4 = 3a^2(3a - 4b^2 - 5a^2)$

Check: $3a - 4b^2 - 5a^2$ has no common factors.
$3a^2(3a - 4b^2 - 5a^2) = 9a^3 - 12a^2b^2 - 15a^4$
This is the original polynomial. It checks.

3. $7x(x + 2y) - 8y(x + 2y) - (x + 2y)$
$= 7x(x + 2y) - 8y(x + 2y) - 1(x + 2y)$
$= (x + 2y)(7x - 8y - 1)$

4. $bx + 5by + 2wx + 10wy = b(x + 5y) + 2w(x + 5y)$
$= (x + 5y)(b + 2w)$

5. To factor $xy - 12 - 4x + 3y$, rearrange the terms. Then factor.
$xy - 4x + 3y - 12 = x(y - 4) + 3(y - 4)$
$= (y - 4)(x + 3)$

6. To factor $2x^3 - 15 - 10x + 3x^2$, rearrange the terms. Then factor.
$2x^3 - 10x + 3x^2 - 15 = 2x(x^2 - 5) + 3(x^2 - 5)$
$= (x^2 - 5)(2x + 3)$

7.

Factor Pairs of 21	**Sum of the Factors**
$(-21)(-1)$	$-21 - 1 = -22$
$(-7)(-3)$	$-7 - 3 = -10$ ✓

The numbers whose product is 21 and whose sum is -10 are -7 and -3. Thus, $x^2 - 10x + 21 = (x - 7)(x - 3)$.

8. $x^4 + 9x^2 + 8 = (x^2)^2 + 9(x^2) + 8$

Let $y = x^2$. Then $x^4 + 9x^2 + 8 = y^2 + 9y + 8$. The two numbers whose product is 8 and whose sum is 9 are 8 and 1, so $y^2 + 9y + 8 = (y + 8)(y + 1)$.
Thus, $x^4 + 9x^2 + 8 = (x^2 + 8)(x^2 + 1)$.

9. (a) $a^2 - 2a - 48 = (a + 6)(a - 8)$

(b $x^4 + 2x^2 - 15 = (x^2 + 5)(x^2 - 3)$

10. (a) $x^2 - 16xy + 15y^2 = (x - 15y)(x - y)$

(b) $x^2 + xy - 42y^2 = (x + 7y)(x - 6y)$

11. The grouping number is $(a)(c) = (10)(2) = 20$. The factor pairs of 20 are $(-20)(-1)$, $(-10)(-2)$, and $(-5)(-4)$. Since $-5 + (-4) = -9$, use -5 and -4.

$10x^2 - 9x + 2 = 10x^2 - 5x - 4x + 2$
$= 5x(2x - 1) - 2(2x - 1)$
$= (2x - 1)(5x - 2)$

12. $9x^3 - 15x^2 - 6x = 3x(3x^2 - 5x - 2)$
$= 3x(3x^2 - 6x + x - 2)$
$= 3x[3x(x - 2) + 1(x - 2)]$
$= 3x(x - 2)(3x + 1)$

13. The first terms of the factors could be $8x$ and x or $4x$ and $2x$. The second terms could be $+1$ and -5 or -1 and $+5$.

Possible Factors	**Middle Term of Product**
$(8x + 1)(x - 5)$	$-39x$
$(8x - 1)(x + 5)$	$+39x$
$(8x + 5)(x - 1)$	$-3x$
$(8x - 5)(x + 1)$	$+3x$
$(4x + 1)(2x - 5)$	$-18x$
$(4x - 1)(2x + 5)$	$+18x$
$(4x + 5)(2x - 1)$	$+6x$
$(4x - 5)(2x + 1)$	$-6x$

Thus, $8x^2 - 6x - 5 = (4x - 5)(2x + 1)$.

14. $6x^4 + 13x^2 - 5 = (2x^2 + 5)(3x^2 - 1)$

A.4 Student Practice

1. $x^2 - 9 = (x)^2 - (3)^2 = (x + 3)(x - 3)$

2. $64x^2 - 121 = (8x)^2 - (11)^2 = (8x + 11)(8x - 11)$

3. $49x^2 - 25y^4 = (7x)^2 - (5y^2)^2 = (7x + 5y^2)(7x - 5y^2)$

4. $7x^2 - 28 = 7(x^2 - 4) = 7(x + 2)(x - 2)$

5. $9x^2 - 30x + 25 = (3x)^2 - 2(3x)(5) + (5)^2 = (3x - 5)^2$

6. $242x^2 + 88x + 8 = 2(121x^2 + 44x + 4)$
$= 2[(11x)^2 + 2(11x)(2) + (2)^2]$
$= 2(11x + 2)^2$

7. (a) $49x^4 + 28x^2 + 4 = (7x^2)^2 + 2(7x^2)(2) + (2)^2 = (7x^2 + 2)^2$

(b) $36x^4 + 84x^2y^2 + 49y^4 = (6x^2)^2 + 2(6x^2)(7y^2) + (7y^2)^2$
$= (6x^2 + 7y^2)^2$

8. $8x^3 + 125y^3 = (2x)^3 + (5y)^3 = (2x + 5y)(4x^2 - 10xy + 25y^2)$

9. $64x^3 + 125y^3 = (4x)^3 + (5y)^3 = (4x + 5y)(16x^2 - 20xy + 25y^2)$

10. $27w^3 - 125z^6 = (3w)^3 - (5z^2)^3 = (3w - 5z^2)(9w^2 + 15wz^2 + 25z^4)$

11. $54x^3 - 16 = 2(27x^3 - 8)$
$= 2(3x - 2)(9x^2 + 6x + 4)$

12. Use the difference of two squares formula first.
$64a^6 - 1 = (8a^3)^2 - (1)^2$
$= (8a^3 + 1)(8a^3 - 1)$
$= [(2a)^3 + (1)^3][(2a)^3 - (1)^3]$
$= (2a + 1)(4a^2 - 2a + 1)(2a - 1)(4a^2 + 2a + 1)$

A.5 Student Practice

1.
$$y - y_1 = m(x - x_1)$$
$$y - (-2) = \frac{3}{4}(x - 5)$$
$$y + 2 = \frac{3}{4}x - \frac{15}{4}$$
$$4y + 4(2) = 4\left(\frac{3}{4}x\right) - 4\left(\frac{15}{4}\right)$$
$$4y + 8 = 3x - 15$$
$$-3x + 4y = -15 - 8$$
$$3x - 4y = 23$$

2. $(-4, 1)$ and $(-2, -3)$
$$m = \frac{y_2 - y_1}{x_2 - x_1} = \frac{-3 - 1}{-2 - (-4)} = \frac{-4}{-2 + 4} = \frac{-4}{2} = -2$$

Substitute $m = -2$ and $(x_1, y_1) = (-4, 1)$ into the point–slope equation.

$$\begin{aligned} y - y_1 &= m(x - x_1) \\ y - 1 &= -2[x - (-4)] \\ y - 1 &= -2(x + 4) \\ y - 1 &= -2x - 8 \\ y &= -2x - 7 \end{aligned}$$

3. First, we need to find the slope of the line $5x - 3y = 10$. We do this by writing the equation in slope–intercept form.

$$\begin{aligned} 5x - 3y &= 10 \\ -3y &= -5x + 10 \\ y &= \frac{5}{3}x - \frac{10}{3} \end{aligned}$$

The slope is $\frac{5}{3}$. A line parallel to this passing through $(4, -5)$ would have an equation

$$\begin{aligned} y - y_1 &= m(x - x_1) \\ y - (-5) &= \frac{5}{3}(x - 4) \\ y + 5 &= \frac{5}{3}x - \frac{20}{3} \\ 3y + 3(5) &= 3\left(\frac{5}{3}x\right) - 3\left(\frac{20}{3}\right) \\ 3y + 15 &= 5x - 20 \\ -5x + 3y &= -20 - 15 \\ 5x - 3y &= 35 \end{aligned}$$

4. Find the slope of the line $6x + 3y = 7$ by rewriting it in slope–intercept form.

$$\begin{aligned} 6x + 3y &= 7 \\ 3y &= -6x + 7 \\ y &= -2x + \frac{7}{3} \end{aligned}$$

The slope is -2. A line perpendicular to this passing through $(-4, 3)$ would have a slope of $\frac{1}{2}$, and would have the equation

$$\begin{aligned} y - y_1 &= m(x - x_1) \\ y - 3 &= \frac{1}{2}[x - (-4)] \\ y - 3 &= \frac{1}{2}(x + 4) \\ y - 3 &= \frac{1}{2}x + 2 \\ 2y - 2(3) &= 2\left(\frac{1}{2}x\right) + 2(2) \\ 2y - 6 &= x + 4 \\ -x + 2y &= 4 + 6 \\ x - 2y &= -10 \end{aligned}$$

Appendix C Student Practice

1. (a) $\begin{vmatrix} -7 & 3 \\ -4 & -2 \end{vmatrix} = (-7)(-2) - (-4)(3) = 14 - (-12) = 14 + 12 = 26$

(b) $\begin{vmatrix} 5 & 6 \\ 0 & -5 \end{vmatrix} = (5)(-5) - (0)(6) = -25 - 0 = -25$

2. (a) Since 3 appears in the first column and third row, we delete them.

$$\begin{vmatrix} 1 & 2 & 7 \\ -4 & -5 & -6 \\ 3 & 4 & -9 \end{vmatrix}$$

The minor of 3 is $\begin{vmatrix} 2 & 7 \\ -5 & -6 \end{vmatrix}$.

(b) Since -6 appears in the second row and third column, we delete them.

$$\begin{vmatrix} 1 & 2 & 7 \\ -4 & -5 & -6 \\ 3 & 4 & -9 \end{vmatrix}$$

The minor of -6 is $\begin{vmatrix} 1 & 2 \\ 3 & 4 \end{vmatrix}$.

3. We can find the determinant by expanding it by minors of elements in the first column.

$$\begin{aligned} \begin{vmatrix} 1 & 2 & -3 \\ 2 & -1 & 2 \\ 3 & 1 & 4 \end{vmatrix} &= 1\begin{vmatrix} -1 & 2 \\ 1 & 4 \end{vmatrix} - 2\begin{vmatrix} 2 & -3 \\ 1 & 4 \end{vmatrix} + 3\begin{vmatrix} 2 & -3 \\ -1 & 2 \end{vmatrix} \\ &= 1[(-1)(4) - (1)(2)] - 2[(2)(4) - (1)(-3)] \\ &\quad + 3[(2)(2) - (-1)(-3)] \\ &= 1[-4 - 2] - 2[8 - (-3)] + 3[4 - 3] \\ &= 1(-6) - 2(11) + 3(1) \\ &= -6 - 22 + 3 \\ &= -25 \end{aligned}$$

4. $D = \begin{vmatrix} 5 & 3 \\ 2 & -5 \end{vmatrix} = (5)(-5) - (2)(3) = -25 - 6 = -31$

$D_x = \begin{vmatrix} 17 & 3 \\ 13 & -5 \end{vmatrix} = (17)(-5) - (13)(3) = -85 - 39 = -124$

$D_y = \begin{vmatrix} 5 & 17 \\ 2 & 13 \end{vmatrix} = (5)(13) - (2)(17) = 65 - 34 = 31$

$$x = \frac{D_x}{D} = \frac{-124}{-31} = 4, \quad y = \frac{D_y}{D} = \frac{31}{-31} = -1$$

5. We will expand each determinant by the first column.

$$\begin{aligned} D = \begin{vmatrix} 2 & 3 & -1 \\ 3 & 5 & -2 \\ 1 & 2 & 3 \end{vmatrix} &= 2\begin{vmatrix} 5 & -2 \\ 2 & 3 \end{vmatrix} - 3\begin{vmatrix} 3 & -1 \\ 2 & 3 \end{vmatrix} + 1\begin{vmatrix} 3 & -1 \\ 5 & -2 \end{vmatrix} \\ &= 2[15 - (-4)] - 3[9 - (-2)] + 1[-6 - (-5)] \\ &= 2(19) - 3(11) + 1(-1) \\ &= 4 \end{aligned}$$

$$\begin{aligned} D_x = \begin{vmatrix} -1 & 3 & -1 \\ -3 & 5 & -2 \\ 2 & 2 & 3 \end{vmatrix} &= -1\begin{vmatrix} 5 & -2 \\ 2 & 3 \end{vmatrix} - (-3)\begin{vmatrix} 3 & -1 \\ 2 & 3 \end{vmatrix} + 2\begin{vmatrix} 3 & -1 \\ 5 & -2 \end{vmatrix} \\ &= -1[15 - (-4)] + 3[9 - (-2)] \\ &\quad + 2[-6 - (-5)] \\ &= -1(19) + 3(11) + 2(-1) \\ &= 12 \end{aligned}$$

$$\begin{aligned} D_y = \begin{vmatrix} 2 & -1 & -1 \\ 3 & -3 & -2 \\ 1 & 2 & 3 \end{vmatrix} &= 2\begin{vmatrix} -3 & -2 \\ 2 & 3 \end{vmatrix} - 3\begin{vmatrix} -1 & -1 \\ 2 & 3 \end{vmatrix} + 1\begin{vmatrix} -1 & -1 \\ -3 & -2 \end{vmatrix} \\ &= 2[-9 - (-4)] - 3[-3 - (-2)] + 1[2 - 3] \\ &= 2(-5) - 3(-1) + 1(-1) \\ &= -8 \end{aligned}$$

$$\begin{aligned} D_z = \begin{vmatrix} 2 & 3 & -1 \\ 3 & 5 & -3 \\ 1 & 2 & 2 \end{vmatrix} &= 2\begin{vmatrix} 5 & -3 \\ 2 & 2 \end{vmatrix} - 3\begin{vmatrix} 3 & -1 \\ 2 & 2 \end{vmatrix} + 1\begin{vmatrix} 3 & -1 \\ 5 & -3 \end{vmatrix} \\ &= 2[10 - (-6)] - 3[6 - (-2)] \\ &\quad + 1[-9 - (-5)] \\ &= 2(16) - 3(8) + 1(-4) \\ &= 4 \end{aligned}$$

$$x = \frac{D_x}{D} = \frac{12}{4} = 3; \; y = \frac{D_y}{D} = \frac{-8}{4} = -2; \; z = \frac{D_z}{D} = \frac{4}{4} = 1$$

Appendix D Student Practice

1. The matrix represents the equations $x + 9y = 33$ and $0x + y = 3$. Since we know that $y = 3$, we can find x by substitution.

$$\begin{aligned} x + 9y &= 33 \\ x + 9(3) &= 33 \\ x + 27 &= 33 \\ x &= 6 \end{aligned}$$

The solution to the system is $(6, 3)$.

2. The augmented matrix for this system of equations is

$$\left[\begin{array}{cc|c} 3 & -2 & -6 \\ 1 & -3 & 5 \end{array}\right].$$

Interchange rows one and two so there is a 1 as the first element in the first row.

$$\left[\begin{array}{cc|c} 1 & -3 & 5 \\ 3 & -2 & -6 \end{array}\right] \quad R_1 \leftrightarrow R_2$$

We want a 0 as the first element in the second row. Multiply row one by −3 and add this to row two.

$$\left[\begin{array}{cc|c} 1 & -3 & 5 \\ 0 & 7 & -21 \end{array}\right] \quad -3R_1 + R_2$$

Now, to obtain a 1 as the second element of row two, multiply each element of row two by $\frac{1}{7}$.

$$\left[\begin{array}{cc|c} 1 & -3 & 5 \\ 0 & 1 & -3 \end{array}\right]. \quad \frac{1}{7}R_2$$

This represents the linear system $\begin{aligned} x - 3y &= 5 \\ y &= -3. \end{aligned}$

We know $y = -3$. Substitute this value into the first equation.

$$\begin{aligned} x - 3(-3) &= 5 \\ x + 9 &= 5 \\ x &= -4 \end{aligned}$$

The solution to the system is $(-4, -3)$.

3. The augmented matrix is

$$\left[\begin{array}{ccc|c} 2 & 1 & -2 & -15 \\ 4 & -2 & 1 & 15 \\ 1 & 3 & 2 & -5 \end{array}\right]$$

$$\left[\begin{array}{ccc|c} 1 & 3 & 2 & -5 \\ 4 & -2 & 1 & 15 \\ 2 & 1 & -2 & -15 \end{array}\right] \quad R_1 \leftrightarrow R_3$$

$$\left[\begin{array}{ccc|c} 1 & 3 & 2 & -5 \\ 0 & -14 & -7 & 35 \\ 0 & -5 & -6 & -5 \end{array}\right] \quad \begin{array}{l} -4R_1 + R_2 \\ -2R_1 + R_3 \end{array}$$

$$\left[\begin{array}{ccc|c} 1 & 3 & 2 & -5 \\ 0 & 1 & \frac{1}{2} & -\frac{5}{2} \\ 0 & -5 & -6 & -5 \end{array}\right] \quad -\frac{1}{14}R_2$$

$$\left[\begin{array}{ccc|c} 1 & 3 & 2 & -5 \\ 0 & 1 & \frac{1}{2} & -\frac{5}{2} \\ 0 & 0 & -\frac{7}{2} & -\frac{35}{2} \end{array}\right] \quad 5R_2 + R_3$$

$$\left[\begin{array}{ccc|c} 1 & 3 & 2 & -5 \\ 0 & 1 & \frac{1}{2} & -\frac{5}{2} \\ 0 & 0 & 1 & 5 \end{array}\right] \quad -\frac{2}{7}R_3$$

We now know that $z = 5$. Substitute this value into the second equation to find y.

$$\begin{aligned} y + \tfrac{1}{2}(5) &= -\tfrac{5}{2} \\ y + \tfrac{5}{2} &= -\tfrac{5}{2} \\ y &= -5 \end{aligned}$$

Now substitute $y = -5$ and $z = 5$ into the first equation.

$$\begin{aligned} x + 3y + 2z &= -5 \\ x + 3(-5) + 2(5) &= -5 \\ x - 15 + 10 &= -5 \\ x - 5 &= -5 \\ x &= 0 \end{aligned}$$

The solution to this system is $(0, -5, 5)$.

Appendix E Student Practice

1. (a) A = {Africa, Antarctica, Asia, Australia, Europe, North America, South America}
(b) $C = \{36, 37, 38, 39, 40, 41\}$
(c) $D = \{35, 36, 37, 38, 39, 40, 41, 42\}$

2. Some possible answers:
$C = \{x|x \text{ is an even number between } 2 \text{ and } 14\}$
$C = \{x|x \text{ is an even number between } 4 \text{ and } 12, \text{inclusive}\}$
$C = \{x|x \text{ is even and } 2 < x < 14\}$
$C = \{x|x \text{ is even and } 4 \le x \le 12\}$
$C = \{x|x = 2n, \text{where } n = 2, 3, 4, 5, 6\}$

3. Combining elements of G and H, we have
$G \cup H$ = {!, *, %, $, ?, ^, +}.

4. The only element common to both G and H is $, thus
$G \cap H$ = {$}.

5. (a) False. All elements of C are not in D.
(b) True. Whole numbers are also rational numbers.

Answers to Selected Exercises

Chapter 0

0.1 Exercises **1.** 12 **3.** Answers may vary. When two or more numbers are multiplied, each number that is multiplied is called a factor. In 2×3, 2 and 3 are factors. **5.** 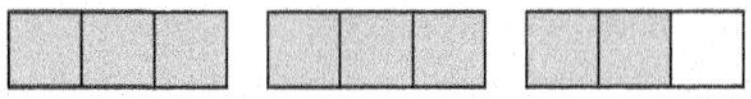 **7.** $\frac{3}{5}$ **9.** $\frac{1}{3}$ **11.** 5 **13.** $\frac{2}{3}$ **15.** $\frac{6}{17}$ **17.** $\frac{7}{9}$ **19.** $2\frac{5}{6}$ **21.** $9\frac{2}{5}$ **23.** $5\frac{3}{7}$ **25.** $20\frac{1}{2}$ **27.** $6\frac{2}{5}$ **29.** $12\frac{1}{3}$ **31.** $\frac{16}{5}$ **33.** $\frac{33}{5}$ **35.** $\frac{11}{9}$ **37.** $\frac{59}{7}$ **39.** $\frac{97}{4}$ **41.** 8 **43.** 24 **45.** 21 **47.** 12 **49.** 21 **51.** 15 **53.** 70 **55.** $23\frac{1}{2}$ **57.** $\frac{33}{160}$ **59.** $\frac{1}{4}$ **61.** $\frac{1}{2}$ **63.** $\frac{3}{5}$

Quick Quiz 0.1 *See Examples noted with Ex.* **1.** $\frac{21}{23}$ (Ex. 1) **2.** $\frac{75}{11}$ (Ex. 6) **3.** $4\frac{19}{21}$ (Ex. 5) **4.** See Student Solutions Manual

0.2 Exercises **1.** Answers may vary. A sample answer is: 8 is exactly divisible by 4. **3.** 36 **5.** 20 **7.** 54 **9.** 105 **11.** 120 **13.** 120 **15.** 546 **17.** 90 **19.** $\frac{5}{8}$ **21.** $\frac{2}{7}$ **23.** $\frac{25}{24}$ or $1\frac{1}{24}$ **25.** $\frac{31}{63}$ **27.** $\frac{11}{15}$ **29.** $\frac{35}{36}$ **31.** $\frac{2}{45}$ **33.** $\frac{1}{2}$ **35.** $\frac{53}{56}$ **37.** $\frac{3}{2}$ or $1\frac{1}{2}$ **39.** $\frac{11}{30}$ **41.** $\frac{1}{4}$ **43.** $\frac{1}{2}$ **45.** $7\frac{11}{15}$ **47.** $1\frac{35}{72}$ **49.** $4\frac{11}{12}$ **51.** $6\frac{13}{28}$ **53.** $5\frac{19}{24}$ **55.** $4\frac{3}{7}$ **57.** 9 **59.** $\frac{23}{24}$ **61.** $7\frac{9}{16}$ **63.** $\frac{10}{21}$ **65.** $2\frac{7}{10}$ **67.** $19\frac{11}{21}$ **69.** $1\frac{13}{24}$ **71.** $12\frac{1}{12}$ **73.** $33\frac{3}{7}$ **75.** $24\frac{2}{3}$ mi **77.** $2\frac{7}{12}$ hr **79.** $A = 12$ in.; $B = 15\frac{7}{8}$ in. **81.** $1\frac{5}{8}$ in. **83.** $\frac{9}{11}$ **84.** $\frac{133}{5}$

Quick Quiz 0.2 *See Examples noted with Ex.* **1.** $\frac{5}{3}$ or $1\frac{2}{3}$ (Ex. 8) **2.** $7\frac{8}{15}$ (Ex. 12a) **3.** $2\frac{5}{18}$ (Ex. 12b) **4.** See Student Solutions Manual

0.3 Exercises **1.** First, change each mixed number to an improper fraction. Look for a common factor in the numerator and denominator to divide by, and, if one is found, perform the division. Multiply the numerators. Multiply the denominators. **3.** $\frac{24}{25}$ **5.** $\frac{17}{30}$ **7.** $\frac{6}{25}$ **9.** $\frac{12}{5}$ or $2\frac{2}{5}$ **11.** $\frac{1}{6}$ **13.** $\frac{6}{7}$ **15.** $\frac{18}{5}$ or $3\frac{3}{5}$ **17.** $\frac{3}{5}$ **19.** $\frac{1}{7}$ **21.** 14 **23.** $\frac{8}{7}$ or $1\frac{1}{7}$ **25.** $\frac{7}{27}$ **27.** $\frac{7}{6}$ or $1\frac{1}{6}$ **29.** $\frac{15}{14}$ or $1\frac{1}{14}$ **31.** $\frac{8}{35}$ **33.** $1\frac{1}{3}$ **35.** $8\frac{2}{3}$ **37.** $6\frac{1}{4}$ **39.** $\frac{8}{15}$ **41.** 6 **43.** $\frac{5}{4}$ or $1\frac{1}{4}$ **45.** $54\frac{3}{4}$ **47.** 40 **49.** 28 **51.** $\frac{3}{16}$ **53.** **(a)** $\frac{5}{63}$ **(b)** $\frac{7}{125}$ **55.** **(a)** $\frac{7}{6}$ **(b)** $\frac{8}{21}$ **57.** 26 shirts **59.** 136 ft^2 **61.** 55 **62.** 49

Quick Quiz 0.3 *See Examples noted with Ex.* **1.** $\frac{5}{6}$ (Ex. 2) **2.** $14\frac{5}{8}$ (Ex. 4) **3.** $1\frac{8}{25}$ (Ex. 9a) **4.** See Student Solutions Manual

0.4 Exercises **1.** 10, 100, 1000, 10,000, and so on **3.** 3, left **5.** 0.875 **7.** 0.2 **9.** $0.\overline{63}$ **11.** $\frac{4}{5}$ **13.** $\frac{1}{4}$ **15.** $\frac{5}{8}$ **17.** $\frac{3}{50}$ **19.** $\frac{17}{5}$ or $3\frac{2}{5}$ **21.** $\frac{11}{2}$ or $5\frac{1}{2}$ **23.** 2.09 **25.** 10.82 **27.** 261.208 **29.** 131.79 **31.** 3.9797 **33.** 122.63 **35.** 30.282 **37.** 0.0032 **39.** 0.10575 **41.** 87.3 **43.** 0.0565 **45.** 2.64 **47.** 261.5 **49.** 0.508 **51.** 3450 **53.** 0.0076 **55.** 73,600 **57.** 0.73892 **59.** 14.98 **61.** 0.01931 **63.** 8.22 **65.** 16.378 **67.** 2.12 **69.** 768.3 **71.** 52.08 **73.** 1.537 **75.** 2.6026 L **77.** 21 hr; \$4 **79.** $\frac{2}{3}$ **80.** $\frac{1}{6}$ **81.** $\frac{93}{100}$ **82.** $\frac{11}{10}$ or $1\frac{1}{10}$

Quick Quiz 0.4 *See Examples noted with Ex.* **1.** 5.7078 (Ex. 5) **2.** 3.522 (Ex. 9) **3.** 28.8 (Ex. 11) **4.** See Student Solutions Manual

Use Math to Save Money **1.** \$100, \$300, \$2000, \$8000, \$8000, \$8000, \$12,000 **2.** $3 \times \$25 + \$50 + \$200 + 2 \times \$20 = \$365$ **3.** \$100 loan, \$300 loan, \$2000 car loan **4.** \$40 **5.** $\$2000 - \$400 = \$1600$; \$1600/\$240 is 7 months if we round to the nearest whole number. **6.** Answers will vary.

0.5 Exercises **1.** Answers may vary. Sample answers follow. 19% means 19 out of 100 parts. Percent means per 100. 19% is really a fraction with a denominator of 100. In this case it would be $\frac{19}{100}$. **3.** 79% **5.** 56.8% **7.** 7.6% **9.** 239% **11.** 360% **13.** 367.2% **15.** 0.03 **17.** 0.004 **19.** 2.5 **21.** 0.074 **23.** 0.0052 **25.** 1 or 1.00 **27.** 5.2 **29.** 13 **31.** 72.8 **33.** 150% **35.** 5% **37.** 6% **39.** 40% **41.** 85% **43.** \$4.92 tip; \$37.72 total bill **45.** 21% **47.** 540 gifts **49.** **(a)** \$29,640 **(b)** \$35,040 **51.** 240,000 **53.** 20,000,000 **55.** 240 **57.** 20,000 **59.** 0.1 **61.** \$4000 **63.** \$6400 **65.** 25 mi/gal **67.** 22 mi/gal **69.** 4.0 in.

Quick Quiz 0.5 *See Examples noted with Ex.* **1.** 96.9 (Ex. 4c) **2.** 15% (Ex. 7b) **3.** 20 (Ex. 9) **4.** See Student Solutions Manual

0.6 Exercises **1.** \$1287 **3. (a)** 76 ft **(b)** He should buy the cut-to-order fencing. He will save \$7.80. **5.** Jog $2\frac{2}{3}$ mi; walk $3\frac{1}{9}$ mi; rest $4\frac{4}{9}$ min; walk $1\frac{7}{9}$ mi **7.** Betty; Melinda increases each activity by $\frac{2}{3}$ by day 3 but Betty increases each activity by $\frac{7}{9}$ by day 3. **9.** $4\frac{1}{2}$ mi **11. (a)** 900,000 lb/day **(b)** 14.29% **13. (a)** \$48,635 **(b)** \$14,104.15 **15.** 18% **17.** 69%

Quick Quiz 0.6 *See Examples noted with Ex.* **1.** 350 stones (Ex. 1) **2.** 46% (Ex. 2) **3.** \$1215 (Ex. 1) **4.** See Student Solutions Manual

Career Exploration Problems **1.** \$1823.08 **2.** 43.4% **3. (a)** \$122.15 **(b)** \$242.47

You Try It **1. (a)** $\frac{2}{3}$ **(b)** $\frac{3}{4}$ **(c)** $\frac{2}{5}$ **2. (a)** $4\frac{1}{5}$ **(b)** $5\frac{2}{7}$ **3. (a)** $\frac{12}{5}$ **(b)** $\frac{55}{9}$ **4.** 15 **5. (a)** 60 **(b)** 100 **6. (a)** $\frac{11}{18}$ **(b)** $\frac{5}{12}$ **7. (a)** $5\frac{7}{9}$ **(b)** $2\frac{1}{2}$ **8. (a)** $\frac{10}{33}$ **(b)** $\frac{1}{2}$ **(c)** $\frac{18}{5}$ or $3\frac{3}{5}$ **9. (a)** $\frac{15}{28}$ **(b)** $\frac{5}{12}$ **10. (a)** $\frac{33}{8}$ or $4\frac{1}{8}$ **(b)** 2 **11.** 0.875 **12. (a)** $\frac{29}{100}$ **(b)** $\frac{7}{40}$ **13. (a)** 8.533 **(b)** 3.46 **14. (a)** 1.35 **(b)** 3.4304 **15.** 18.5 **16. (a)** 52% **(b)** 0.8% **(c)** 186% **(d)** 7.7% **(e)** 0.09% **17. (a)** 0.28 **(b)** 0.0742 **(c)** 1.65 **(d)** 0.0025 **18.** 13.8 **19. (a)** 83.3% **(b)** 125% **20.** 300 ft^2 **21.** \$154.06

Chapter 0 Review Problems **1.** $\frac{3}{4}$ **2.** $\frac{3}{10}$ **3.** $\frac{18}{41}$ **4.** $\frac{3}{5}$ **5.** $\frac{57}{8}$ **6.** $6\frac{4}{5}$ **7.** $26\frac{2}{3}$ **8.** 15 **9.** 5 **10.** 45 **11.** 22 **12.** $\frac{17}{20}$ **13.** $\frac{29}{24}$ or $1\frac{5}{24}$ **14.** $\frac{4}{15}$ **15.** $\frac{13}{30}$ **16.** $5\frac{23}{30}$ **17.** $6\frac{9}{20}$ **18.** $2\frac{29}{36}$ **19.** $1\frac{11}{12}$ **20.** $\frac{30}{11}$ or $2\frac{8}{11}$ **21.** $10\frac{1}{2}$ **22.** 50 **23.** $\frac{20}{7}$ or $2\frac{6}{7}$ **24.** $\frac{1}{16}$ **25.** $\frac{24}{5}$ or $4\frac{4}{5}$ **26.** $\frac{3}{20}$ **27.** 6 **28.** 7.201 **29.** 7.737 **30.** 29.561 **31.** 4.436 **32.** 0.03745 **33.** 362,341 **34.** 0.07956 **35.** 125.5 **36.** 0.07132 **37.** 1.3075 **38.** 90 **39.** 1.82 **40.** 0.375 **41.** $\frac{9}{25}$ **42.** 0.014 **43.** 0.361 **44.** 0.0002 **45.** 1.253 **46.** 0.25% **47.** 32.5% **48.** 90% **49.** 10% **50.** 120 **51.** 3.96 **52.** 95% **53.** 60% **54.** 13,480 students **55.** 75% **56.** 400,000,000,000 **57.** 2500 **58.** 300,000 **59.** 9 **60.** \$12,000 **61.** 25 **62.** \$320 **63.** \$800 **64.** 1840 mi; 1472 mi **65.** 1500 mi; 1050 mi **66.** \$349.07 **67.** \$462.80 **68.** \$1585.50 **69.** \$401.25

How Am I Doing? Chapter 0 Test **1.** $\frac{8}{9}$ (obj. 0.1.2) **2.** $\frac{4}{3}$ (obj. 0.1.2) **3.** $\frac{45}{7}$ (obj. 0.1.3) **4.** $11\frac{2}{3}$ (obj. 0.1.3) **5.** $\frac{15}{8}$ or $1\frac{7}{8}$ (obj. 0.2.3) **6.** $4\frac{7}{8}$ (obj. 0.2.4) **7.** $\frac{5}{6}$ (obj. 0.2.4) **8.** $\frac{4}{3}$ or $1\frac{1}{3}$ (obj. 0.3.1) **9.** $\frac{7}{2}$ or $3\frac{1}{2}$ (obj. 0.3.2) **10.** $1\frac{21}{22}$ (obj. 0.3.2) **11.** $8\frac{1}{8}$ (obj. 0.3.1) **12.** $\frac{7}{2}$ or $3\frac{1}{2}$ (obj. 0.3.2) **13.** 14.64 (obj. 0.4.4) **14.** 3.9897 (obj. 0.4.4) **15.** 1.312 (obj. 0.4.5) **16.** 73.85 (obj. 0.4.7) **17.** 230 (obj. 0.4.6) **18.** 263.259 (obj. 0.4.7) **19.** 7.3% (obj. 0.5.1) **20.** 1.965 (obj. 0.5.2) **21.** 6.3 (obj. 0.5.3) **22.** 6% (obj. 0.5.4) **23.** 18 computer chips (obj. 0.3.2) **24.** 100 (obj. 0.5.5) **25.** 700 (obj. 0.5.5) **26.** 65% (obj. 0.6.1) **27.** 60 tiles (obj. 0.6.1)

Chapter 1

1.1 Exercises **1.** Integer, rational number, real number **3.** Irrational number, real number **5.** Rational number, real number **7.** Rational number, real number **9.** Irrational number, real number **11.** −20,000 **13.** $-37\frac{1}{2}$ **15.** +7 **17.** −8 **19.** 2.73 **21.** 1.3 **23.** $\frac{5}{6}$ **25.** −15 **27.** −50 **29.** $\frac{3}{10}$ **31.** $-\frac{7}{13}$ **33.** $\frac{1}{35}$ **35.** −19.2 **37.** 0.4 **39.** −14.16 **41.** −6 **43.** 0 **45.** $\frac{9}{20}$ **47.** −8 **49.** −3 **51.** 59 **53.** $\frac{7}{18}$ **55.** $\frac{2}{5}$ **57.** −2.21 **59.** 12 **61.** −12 **63.** 15.94 **65.** \$167 profit **67.** −\$3800 **69.** 9-yd gain **71.** 3500 **73.** \$28,000,000 **75.** 18 **77.** $\frac{19}{16}$ or $1\frac{3}{16}$ **78.** $\frac{2}{3}$ **79.** $\frac{1}{12}$ **80.** $\frac{25}{34}$ **81.** 1.52 **82.** 0.65 **83.** 1.141 **84.** 0.26

Quick Quiz 1.1 *See Examples noted with Ex.* **1.** −34 (Ex. 5b) **2.** 0.5 (Ex. 11) **3.** $-\frac{13}{24}$ (Ex. 13b) **4.** See Student Solutions Manual

1.2 Exercises **1.** First change subtracting −3 to adding +3. Then use the rules for addition of two real numbers with different signs. Thus, $-8 - (-3) = -8 + 3 = -5$. **3.** −22 **5.** −4 **7.** −11 **9.** 8 **11.** 5 **13.** 0 **15.** 3 **17.** $-\frac{2}{5}$ **19.** $\frac{27}{20}$ or $1\frac{7}{20}$ **21.** $-\frac{19}{12}$ or $-1\frac{7}{12}$ **23.** −0.9 **25.** 4.47 **27.** $-\frac{17}{5}$ or $-3\frac{2}{5}$ **29.** $-\frac{14}{3}$ or $-4\frac{2}{3}$ **31.** −53 **33.** −73 **35.** 7.1 **37.** $\frac{35}{4}$ or $8\frac{3}{4}$ **39.** $-\frac{37}{6}$ or $-6\frac{1}{6}$ **41.** $-\frac{38}{35}$ or $-1\frac{3}{35}$ **43.** −8.5 **45.** $-\frac{29}{5}$ or $-5\frac{4}{5}$ **47.** 6.06 **49.** −5.047 **51.** 7 **53.** −48 **55.** −2 **57.** 0 **59.** 10 **61.** 1.1 **63.** 626 ft **65.** −21 **66.** −51 **67.** −19 **68.** 15°F **69.** $6\frac{2}{3}$ mi

Quick Quiz 1.2 *See Examples noted with Ex.* **1.** 7 (Ex. 2) **2.** −1.9 (Ex. 4) **3.** $\frac{51}{56}$ (Ex. 3b) **4.** See Student Solutions Manual

1.3 Exercises **1.** To multiply two real numbers, multiply the absolute values. The sign of the result is positive if both numbers have the same sign, but negative if the two numbers have opposite signs. **3.** −40 **5.** 0 **7.** 49 **9.** 0.264 **11.** −4.5 **13.** $-\frac{3}{2}$ or $-1\frac{1}{2}$ **15.** $\frac{9}{11}$

17. $-\frac{5}{26}$ **19.** 0 **21.** 6 **23.** 15 **25.** -12 **27.** -130 **29.** -0.6 **31.** -0.9 **33.** $-\frac{3}{10}$ **35.** $\frac{20}{3}$ or $6\frac{2}{3}$ **37.** $\frac{7}{10}$ **39.** 14 **41.** $-\frac{5}{4}$ or $-1\frac{1}{4}$ **43.** $-\frac{2}{3}$ **45.** -24 **47.** -72 **49.** 0 **51.** -0.3 **53.** $-\frac{8}{35}$ **55.** $\frac{2}{27}$ **57.** 9 **59.** 4 **61.** 17 **63.** -72 **65.** -1 **67.** He gave \$4.40 to each person and to himself. **69.** \$235.60 **71.** Approximately 20 yd gained **73.** Approximately 70 yd lost **75.** A gain of 20 yd **76.** -6.69 **77.** $-\frac{11}{6}$ or $-1\frac{5}{6}$ **78.** -15 **79.** -88

Quick Quiz 1.3

See Examples noted with Ex. **1.** $-\frac{15}{8}$ or $-1\frac{7}{8}$ (Ex. 1d) **2.** -120 (Ex. 2) **3.** 4 (Ex. 5) **4.** See Student Solutions Manual

1.4 Exercises

1. The base is 3 and the exponent is 4. Thus you multiply $(3)(3)(3)(3) = 81$. **3.** The answer is negative. When you raise a negative number to an odd power the result is always negative. **5.** If you have parentheses surrounding the -2, then the base is -2 and the exponent is 4. The result is 16. If you do not have parentheses, then the base is 2. You evaluate to obtain 16 and then take the negative of 16, which is -16. Thus $(-2)^4 = -16$ but $-2^4 = -16$. **7.** 6^5 **9.** w^2 **11.** p^4 **13.** $(3q)^3$ or 3^3q^3 **15.** 27 **17.** 81 **19.** 216 **21.** -27 **23.** 16 **25.** -25 **27.** $\frac{1}{16}$ **29.** $\frac{8}{125}$ **31.** 4.41 **33.** 0.00032 **35.** 256 **37.** -256 **39.** 161 **41.** -21 **43.** -128 **45.** 23 **47.** 1000 **48.** -19 **49.** $-\frac{5}{3}$ or $-1\frac{2}{3}$ **50.** -8 **51.** 2.52 **52.** \$1696

Quick Quiz 1.4

See Examples noted with Ex. **1.** 256 (Ex. 3) **2.** 3.24 (Ex. 4b) **3.** $\frac{27}{64}$ (Ex. 4c) **4.** See Student Solutions Manual

1.5 Exercises

1. $3(4) + 6(5)$ **3.** **(a)** 90 **(b)** 42 **5.** 10 **7.** 5 **9.** -29 **11.** 24 **13.** 21 **15.** 13 **17.** -6 **19.** 42 **21.** $\frac{9}{4}$ or $2\frac{1}{4}$ **23.** 0.848 **25.** $-\frac{7}{16}$ **27.** 5 **29.** $-\frac{23}{2}$ or $-11\frac{1}{2}$ **31.** 18.35 **33.** $\frac{29}{18}$ or $1\frac{11}{18}$ **35.** $1(-2) + 5(-1) + 10(0) + 2(+1)$ **37.** 2 over par **39.** 0.125 **40.** $-\frac{19}{12}$ or $-1\frac{7}{12}$ **41.** -1 **42.** $\frac{72}{125}$

Quick Quiz 1.5

See Examples noted with Ex. **1.** -77 (Ex. 1) **2.** 16.96 (Ex. 1) **3.** 11 (Ex. 3) **4.** See Student Solutions Manual

Use Math to Save Money

1. $(\$2500 \times 0.05) \times 12 = \$1500$ **2.** \$4500 **3.** \$3450 **4.** About 28 months or 2 years, 4 months **5.** About 14 months or 1 year, 2 months **6.** $(2500 + (5800/12)) \times 0.05 = \$149.17$ *a month* **7.** $(\$2500 + (\$5800/12)) \times 0.20 = \$596.67$ *a month* **8.** Answers will vary. **9.** Answers will vary. **10.** Answers will vary.

1.6 Exercises

1. variable **3.** Here we are multiplying 4 by x by x. Since we know from the definition of exponents that x multiplied by x is x^2, this gives us an answer of $4x^2$. **5.** Yes, $a(b - c)$ can be written as $a[b + (-c)]$.

$$3(10 - 2) = (3 \times 10) - (3 \times 2)$$
$$3 \times 8 = 30 - 6$$
$$24 = 24$$

7. $10x - 25y$ **9.** $-8a + 6b$ **11.** $9x + 3y$ **13.** $-8m - 24n$ **15.** $-x + 3y$ **17.** $-81x + 45y - 72$ **19.** $-10x + 2y - 12$ **21.** $10x^2 - 20x + 15$ **23.** $\frac{x^2}{5} + 2xy - \frac{4x}{5}$ **25.** $5x^2 + 10xy + 5xz$ **27.** $13.5x - 15$ **29.** $18x^2 + 3xy - 3x$ **31.** $-3x^2y - 2xy^2 + xy$ **33.** $-5a^2b - 10ab^2 + 20ab$ **35.** $2a^2 - 4a - \frac{5}{4}$ **37.** $0.36x^3 + 0.09x^2 - 0.15x$ **39.** $-1.32q^3 - 0.28qr - 4q$ **41.** $800(5x + 14y) = 4000x + 11{,}200y$ ft^2 **43.** $4x(3000 - 2y) = 12{,}000x - 8xy$ ft^2 **44.** -16 **45.** 64 **46.** 14 **47.** 4 **48.** 10

Quick Quiz 1.6

See Examples noted with Ex. **1.** $-15a - 35b$ (Ex. 1) **2.** $-2x^2 + 8xy - 16x$ (Ex. 4) **3.** $-12a^2b + 15ab^2 + 27ab$ (Ex. 4) **4.** See Student Solutions Manual

1.7 Exercises

1. A term is a number, a variable, a product, or a quotient of numbers and variables. **3.** The two terms $5x$ and $-8x$ are like terms because they both have the variable x with the exponent of one. **5.** The only like terms are $7xy$ and $-14xy$ because the other two have different exponents even though they have the same variables. **7.** $-31x^2$ **9.** $6a^3 - 7a^2$ **11.** $-5x - 5y$ **13.** $7.1x - 3.5y$ **15.** $-2x - 8.7y$ **17.** $5p + q - 18$ **19.** $5bc - 6ac$ **21.** $x^2 - 10x + 3$ **23.** $-10y^2 - 16y + 12$ **25.** $-\frac{1}{15}x - \frac{2}{21}y$ **27.** $\frac{11}{20}a^2 - \frac{5}{6}b$ **29.** $-2rs + 2r$ **31.** $\frac{19}{4}xy + 2x^2y$ **33.** $28a - 20b$ **35.** $-27ab - 11b^2$ **37.** $-2c - 10d^2$ **39.** $11x + 20$ **41.** $7a + 9b$ **43.** $14x - 8$ ft **45.** $-\frac{13}{12}$ or $-1\frac{1}{12}$ **46.** $-\frac{3}{8}$ **47.** $\frac{23}{50}$ **48.** $-\frac{15}{98}$

Quick Quiz 1.7

See Examples noted with Ex. **1.** $\frac{13}{6}xy + \frac{5}{3}x^2y$ (Ex. 6) **2.** $0.6a^2b - 4.4ab^2$ (Ex. 4a) **3.** $20x - 2y$ (Ex. 7) **4.** See Student Solutions Manual

1.8 Exercises

1. -7 **3.** -12 **5.** $\frac{25}{2}$ or $12\frac{1}{2}$ **7.** -26 **9.** -1.3 **11.** $\frac{25}{4}$ or $6\frac{1}{4}$ **13.** 10 **15.** 5 **17.** -24 **19.** -20 **21.** 9 **23.** 39 **25.** -2 **27.** 44 **29.** -2 **31.** -24 **33.** 15 **35.** 32 **37.** 32 **39.** $-\frac{1}{2}$ **41.** 352 ft^2 **43.** 1.24 cm^2 **45.** 32 in.^{2} **47.** 56,000 ft^2 **49.** ≈ 28.26 ft^2 **51.** $-78.5°$C **53.** \$2340.00 **55.** The coldest temperature was $-396.4°$F. The warmest temperature was $253.4°$F. **57.** 16 **58.** $-x^2 + 2x - 4y$

Quick Quiz 1.8

See Examples noted with Ex. **1.** 2 (Ex. 3) **2.** $\frac{9}{2}$ or $4\frac{1}{2}$ (Ex. 4) **3.** 33 (Ex. 4) **4.** See Student Solutions Manual

1.9 Exercises

1. $-(3x + 2y)$ **3.** distributive **5.** $4x + 12y$ **7.** $2c - 16d$ **9.** $x - 7y$ **11.** $8x^3 - 4x^2 + 12x$

13. $-2x + 26y$ **15.** $2x + 11y + 10$ **17.** $15a - 60ab$ **19.** $12a^3 - 19a^2 - 22a$ **21.** $3a^2 + 16b + 12b^2$ **23.** $-13a + 9b - 1$ **25.** $6x^2 + 30x - 30$ **27.** $12a^2 - 8b$ **29.** 1947.52°F **30.** 453,416 ft^2 **31.** 54 to 67.5 kg **32.** 4.05 to 6.3 kg

Quick Quiz 1.9

See Examples noted with Ex. **1.** $-14x - 4y$ (Ex. 1) **2.** $-6x + 15y - 36$ (Ex. 2) **3.** $-8a - 24ab + 8b$ (Ex. 4) **4.** See Student Solutions Manual

Career Exploration Problems

1. 21.6W per conductor; total power loss of 43.2W **2.** 172.8W **3.** 0.6A **4.** 168 $in.^3$

You Try It

1. (a) 5 (b) 1 (c) 0.5 (d) $\frac{1}{4}$ (e) 4.57 **2.** -14 **3.** (a) 6 (b) -6 **4.** 3 **5.** -1 **6.** (a) -54 (b) -8 (c) 6 (d) 21 **7.** (a) 81 (b) 2.25 (c) $\frac{1}{16}$ **8.** (a) -8 (b) 256 **9.** 67 **10.** (a) $8a - 12$ (b) $-25x + 5$ **11.** $-3a^2 - 3a + 8ab$ **12.** 103 **13.** 2500 ft^2 **14.** $28x + 8$

Chapter 1 Review Problems

1. -8 **2.** -4.2 **3.** -9 **4.** 1.9 **5.** $-\frac{1}{3}$ **6.** $-\frac{7}{22}$ **7.** $\frac{1}{6}$ **8.** $-\frac{1}{2}$ **9.** 8 **10.** 13 **11.** -33 **12.** 9.2 **13.** $-\frac{13}{8}$ or $-1\frac{5}{8}$ **14.** $\frac{11}{24}$ **15.** -22.7 **16.** -88 **17.** -3 **18.** 13 **19.** 32 **20.** $\frac{5}{6}$ **21.** $-\frac{25}{7}$ or $-3\frac{4}{7}$ **22.** -72 **23.** 60 **24.** -30 **25.** -243 **26.** 64 **27.** 625 **28.** $-\frac{8}{27}$ **29.** -81 **30.** 0.36 **31.** $\frac{25}{36}$ **32.** $\frac{27}{64}$ **33.** -44 **34.** 20.004 **35.** 1 **36.** $-21x + 7y$ **37.** $18x - 3x^2 + 9xy$ **38.** $-7x^2 + 3x - 11$ **39.** $-6xy^3 - 3xy^2 + 3y^3$ **40.** $-5a^2b + 3bc$ **41.** $-3x - 4y$ **42.** $-5x^2 - 35x - 9$ **43.** $10x^2 - 8x - \frac{1}{2}$ **44.** -55 **45.** 1 **46.** -4 **47.** -3 **48.** 0 **49.** 17 **50.** 15 **51.** \$810 **52.** 68°F to 77°F **53.** \$75.36 **54.** \$8580 **55.** 100,000 ft^2; \$200,000 **56.** 10.45 ft^2; \$689.70 **57.** $-2x + 42$ **58.** $-17x - 18$ **59.** $-2 + 10x$ **60.** $-12x^2 + 63x$ **61.** $5xy^3 - 6x^3y - 13x^2y^2 - 6x^2y$ **62.** $x - 10y + 35 - 15xy$ **63.** $-31a + 2b$ **64.** $15a - 15b - 10ab$ **65.** $-3x - 9xy + 18y^2$ **66.** $10x + 8xy - 32y$ **67.** -2.3 **68.** 8 **69.** $-\frac{22}{15}$ or $-1\frac{7}{15}$ **70.** $-\frac{1}{8}$ **71.** -1 **72.** -0.5 **73.** $\frac{81}{40}$ or $2\frac{1}{40}$ **74.** -8 **75.** 240 **76.** -25.42 **77.** \$600 **78.** 0.0081 **79.** -0.0625 **80.** 10 **81.** $-4.9x + 4.1y$ **82.** $-\frac{1}{9}$ **83.** $-\frac{2}{3}$ **84.** No: the dog's temperature is below the normal temperature of 101.48°F. **85.** $3y^2 + 12y - 7x - 28$ **86.** $-12x + 6y + 12xy$

How Am I Doing? Chapter 1 Test

1. -0.3 (obj. 1.1.4) **2.** 2 (obj. 1.2.1) **3.** $-\frac{14}{3}$ or $-4\frac{2}{3}$ (obj. 1.3.1) **4.** -70 (obj. 1.3.1) **5.** 4 (obj. 1.3.3) **6.** -3 (obj. 1.3.3) **7.** -64 (obj. 1.4.2) **8.** 2.56 (obj. 1.4.2) **9.** $\frac{16}{81}$ (obj. 1.4.2) **10.** 6.8 (obj. 1.5.1) **11.** -25 (obj. 1.5.1) **12.** $-5x^2 - 10xy + 35x$ (obj. 1.6.1) **13.** $6a^2b^2 + 4ab^3 - 14a^2b^3$ (obj. 1.6.1) **14.** $2a^2b + \frac{15}{2}ab$ (obj. 1.7.2) **15.** $-1.8x^2y - 4.7xy^2$ (obj. 1.7.2) **16.** $5a + 30$ (obj. 1.7.2) **17.** $14x - 16y$ (obj. 1.7.2) **18.** 122 (obj. 1.8.1) **19.** 37 (obj. 1.8.1) **20.** $\frac{13}{6}$ or $2\frac{1}{6}$ (obj. 1.8.1) **21.** ≈ 96.6 km/hr (obj. 1.8.2) **22.** 22,800 ft^2 (obj. 1.8.2) **23.** \$23.12 (obj. 1.8.2) **24.** 3 cans (obj. 1.8.2) **25.** $3x - 6xy - 21y^2$ (obj. 1.9.1) **26.** $-3a - 9ab + 3b^2 - 3ab^2$ (obj. 1.9.1)

Chapter 2

2.1 Exercises

1. equals, equal **3.** solution **5.** Answers may vary. **7.** $x = 7$ **9.** $x = 11$ **11.** $x = 17$ **13.** $x = -5$ **15.** $x = -13$ **17.** $x = 62$ **19.** $x = 15$ **21.** $x = 21$ **23.** $x = 0$ **25.** $x = 0$ **27.** $x = 21$ **29.** No; $x = 9$ **31.** No; $x = -14$ **33.** Yes **35.** Yes **37.** $x = -1.8$ **39.** $x = 0.6$ **41.** $x = 1$ **43.** $x = -\frac{1}{4}$ **45.** $x = -7$ **47.** $x = \frac{17}{6}$ or $2\frac{5}{6}$ **49.** $x = \frac{3}{7}$ **51.** $x = 5.2$ **53.** $x = 20.2$ **55.** $-2x - 4y$ **56.** $-2y^2 - 4y + 4$

Quick Quiz 2.1

See Examples noted with Ex. **1.** $x = 14.3$ (Ex. 2) **2.** $x = -3.5$ (Ex. 2) **3.** $x = -8$ (Ex. 3) **4.** See Student Solutions Manual

2.2 Exercises

1. 6 **3.** 7 **5.** $x = 48$ **7.** $x = -30$ **9.** $x = 80$ **11.** $x = -15$ **13.** $x = 4$ **15.** $x = 8$ **17.** $x = -\frac{8}{3}$ **19.** $x = 50$ **21.** $x = 15$ **23.** $x = -7$ **25.** $x = 0.2$ **27.** $x = 4$ **29.** No; $x = -7$ **31.** Yes **33.** $y = -0.03$ **35.** $t = \frac{8}{3}$ **37.** $y = -0.7$ **39.** $x = 3$ **41.** $x = -4$ **43.** $x = -36$ **45.** $x = 1$ **47.** $m = 2$ **49.** $x = 84$ **51.** $x = -10.5$ **53.** $9xy - 8y^2$ **54.** $x - 9$ **55.** 22.5 tons **56.** 27 earthquakes

Quick Quiz 2.2

See Examples noted with Ex. **1.** $x = -38$ (Ex. 4) **2.** $x = 14$ (Ex. 4) **3.** $x = -12$ (Ex. 6) **4.** See Student Solutions Manual

2.3 Exercises

1. $x = 9$ **3.** $x = 6$ **5.** $x = -4$ **7.** $x = 13$ **9.** $x = 3.1$ **11.** $x = 28$ **13.** $x = -27$ **15.** $x = 8$ **17.** $x = 3$ **19.** $x = \frac{11}{2}$ **21.** $x = -9$ **23.** Yes **25.** No; $x = -11$ **27.** $x = -1$ **29.** $x = 7$ **31.** $y = -1$ **33.** $x = 16$ **35.** $y = 3$ **37.** $x = 4$ **39.** $x = \frac{1}{4}$ or 0.25 **41.** $x = \frac{5}{2}$ **43.** $x = 6.5$ **45.** $a = 0$ **47.** $x = -\frac{16}{5}$ **49.** $y = 2$ **51.** $x = -4$

53. $z = 5$ **55.** $a = -6.5$ **57.** $x = -\frac{2}{3}$ **59.** $x = -0.25$ **61.** $x = 8$ **63.** $x = -1.5$ **65.** 42 **66.** -37 **67.** 21
68. \$1184.35 **69. (a)** \$629.30 **(b)** \$647.28

Quick Quiz 2.3 *See Examples noted with Ex.* **1.** $x = -\frac{4}{11}$ (Ex. 3) **2.** $x = 4$ (Ex. 1) **3.** $x = -\frac{8}{7}$ (Ex. 6)
4. See Student Solutions Manual

2.4 Exercises **1.** $x = -7$ **3.** $x = 1$ **5.** $x = 1$ **7.** $x = 12$ **9.** $y = 20$ **11.** $x = 3$ **13.** $x = \frac{7}{3}$ **15.** $x = -\frac{7}{2}$ or -3.5
17. Yes **19.** No **21.** $x = 1$ **23.** $x = 8$ **25.** $x = 2$ **27.** $x = -5$ **29.** $y = 4$ **31.** $x = -22$ **33.** $x = 2$
35. $x = -12$ **37.** $x = -\frac{5}{3}$ **39.** $x = \frac{10}{3}$ **41.** No solution **43.** Infinite number of solutions **45.** $x = 0$ **47.** No solution
49. $-\frac{52}{3}$ or $-17\frac{1}{3}$ **50.** $\frac{22}{5}$ or $4\frac{2}{5}$ **51.** 572 − 975 g **52.** 3173 seats

Quick Quiz 2.4 *See Examples noted with Ex.* **1.** $x = -\frac{7}{5}$ (Ex. 2) **2.** $x = \frac{19}{31}$ (Ex. 2) **3.** $x = \frac{11}{26}$ (Ex. 4) **4.** See Student Solutions Manual

Use Math to Save Money **1.** Shell: \$4.55; ARCO: \$4.88 **2.** Shell: \$13.65; ARCO: \$13.74 **3.** Shell: \$18.20; ARCO: \$18.17
4. Shell: \$45.50; ARCO: \$44.75 **5.** 3.75 gallons **6.** Shell **7.** ARCO **8.** Answers will vary. **9.** Answers will vary.
10. Answers will vary.

2.5 Exercises **1.** $x + 11$ **3.** $x - 12$ **5.** $\frac{1}{8}x$ or $\frac{x}{8}$ **7.** $2x$ **9.** $3 + \frac{1}{2}x$ **11.** $2x + 9$ **13.** $\frac{1}{3}(x + 7)$
15. $\frac{1}{3}x - 2x$ **17.** $5x - 11$ **19.** x = value of a share of AT&T stock; $x + 74.50$ = value of a share of IBM stock **21.** w = width; $2w + 7$ = length
23. x = number of boxes sold by Keiko; $x - 43$ = number of boxes sold by Sarah; $x + 53$ = number of boxes sold by Imelda **25.** Measure of 1st angle $= s - 25$; Measure of 2nd angle $= s$; Measure of 3rd angle $= 3s$ **27.** v = value of exports of Canada; $2v$ = value of exports of Japan
29. p = price of the Summer on the Beach Concert tickets; $\frac{1}{2}p$ = price of All Star Concert tickets **31.** x = number of men aged 16–24; $x + 82$ = number of men aged 25–34; $x - 25$ = number of men aged 35–44; $x - 110$ = number of men aged 45 and above **33.** $x = 7$ **34.** $x = -\frac{5}{2}$ or $-2\frac{1}{2}$

Quick Quiz 2.5 *See Examples noted with Ex.* **1.** $x + 10$ or $10 + x$ (Ex. 1) **2.** $2x - 5$ (Ex. 2) **3.** Measure of first angle: $x + 15$; Measure of second angle: x; Measure of third angle: $5x$ (Ex. 5)
4. See Student Solutions Manual

2.6 Exercises **1.** 1261 **3.** 2368 **5.** 182 **7.** 43 **9.** 9 **11.** -4 **13.** 12 **15.** 40 red tablet cases **17.** They each measure 72°.
19. first: 46°; second: 23°; third: 111° **21.** 4 items **23.** 11.5 sec **25.** 5 mi **27.** She traveled 52 mph on the mountain road. It was 12 mph faster on the highway route. **29.** The score on the final lab must be 92. **31. (a)** $F - 40 = \frac{x}{4}$ **(b)** 200 chirps **(c)** 77°F
32. $10x^3 - 30x^2 - 15x$ **33.** $-2a^2b + 6ab - 10a^2$ **34.** $-5x - 6y$ **35.** $-4x^2y - 7xy^2 - 8xy$

Quick Quiz 2.6 *See Examples noted with Ex.* **1.** 17 (Ex. 2) **2.** They each measure 55°. (Ex. 4) **3.** 74 (Ex. 7) **4.** See Student Solutions Manual

2.7 Exercises **1.** 38 coffee products **3.** 5 hr **5.** 12 weeks **7.** \$360 **9.** \$22,000 **11.** \$12,000 **13.** \$3000 at 7%; \$2000 at 5%
15. Conservative fund = \$250,000; Growth fund = \$150,000 **17.** \$12,000 **19.** 13 quarters; 9 nickels **21.** 18 nickels; 6 dimes; 9 quarters **23.** 10 boxes of beige paper; 20 boxes of white paper **25.** \$10 bills = 26; \$20 bills = 31; \$50 bills = 10
27. \$925,000 worth of furniture **29.** 12 **30.** -17 **31.** 25 **32.** -26

Quick Quiz 2.7 *See Examples noted with Ex.* **1.** 9 months (Ex. 1) **2.** \$11,600 (Ex. 3) **3.** \$2200 at 4%; \$2800 at 5% (Ex. 5)
4. See Student Solutions Manual

2.8 Exercises **1.** Yes, both statements imply 5 is to the right of -6 on a number line. **3.** > **5.** > **7.** < **9. (a)** < **(b)** >
11. (a) > **(b)** < **13.** < **15.** > **17.** > **19.** < **21.** < **23.** < **25.**

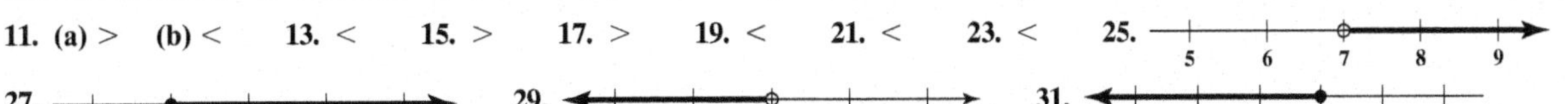

27. **29.** **31.**

33. **35.** $x \geq -\frac{2}{3}$ **37.** $x < -20$ **39.** $x \leq 3.7$ **41.** $c \geq 12$ **43.** $h \geq 48$

45. **47.** $x \leq -3$

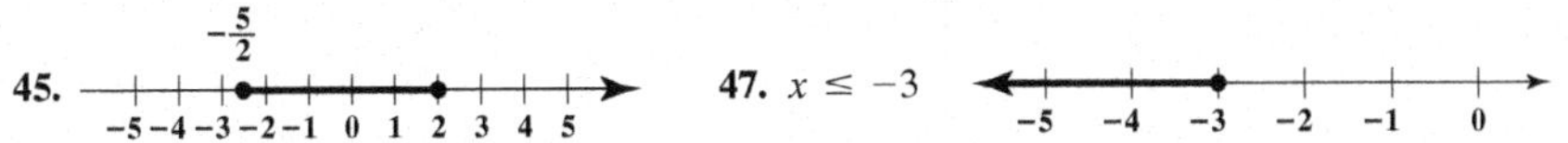

49. $x \leq 5$ **51.** $x > -9$

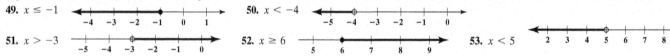

53. $x \geq 8$ **55.** $x < -12$

57. $x < -1$ **59.** $x < \frac{3}{2}$

61. $x > -6$ **63.** $x > \frac{1}{3}$

65. $3 > 1$ Adding any number to both sides of an inequality doesn't reverse the direction. **67.** $x < 3$ **69.** $x \geq 4$ **71.** $x < -1$ **73.** $x \leq 14$ **75.** $x < -3$ **77.** 76 or greater **79.** 8 days or more **81.** 6.08 **82.** 15% **83.** 2% **84.** 37.5%

Quick Quiz 2.8

See Examples noted with Ex. **1.** (Ex. 2)

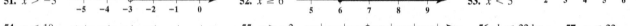

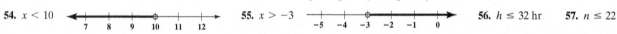

2. $x \leq 6$ (Ex. 5) **3.** $x > -5$ (Ex. 7)

4. See Student Solutions Manual

Career Exploration Problems

1. $\frac{BMR - 6.25H + 5A - 5}{10} = W; \frac{BMR - 10W + 5A - 5}{6.25} = H$

2. 74 inches or 6 feet 2 inches tall **3.** 125 pounds

You Try It

1. $x = -12$ **2.** $y = 4$ **3.** 10 and 50 **4.** \$600 at 2%; \$3000 at 4% **5.** $x \leq -4$

Chapter 2 Review Problems

1. $x = -7$ **2.** $x = -3$ **3.** $x = -3$ **4.** $x = 40.4$ or $40\frac{2}{5}$ **5.** $x = -2$ **6.** $x = -7$ **7.** $x = 3$ **8.** $x = -\frac{7}{2}$ **9.** $x = 5$ **10.** $x = 1$ **11.** $x = 20$ **12.** $x = \frac{2}{3}$ **13.** $x = 5$ **14.** $x = \frac{35}{11}$ **15.** $x = 4$ **16.** $x = -17$ **17.** $x = -7$ **18.** $x = 3$ **19.** $x = 4$ **20.** $x = 32$ **21.** $x = 0$ **22.** $x = -17$ **23.** $x = -32$ **24.** $x = -\frac{17}{5}$ **25.** $x + 19$ **26.** $\frac{2}{3}x$ **27.** $\frac{1}{2}x$ or $\frac{x}{2}$ **28.** $x - 18$ **29.** $3(x + 4)$ **30.** $2x - 3$

31. r = the number of retired people; $4r$ = the number of working people; $0.5r$ = the number of unemployed people **32.** $3w + 5$ = the length; w = the width **33.** b = the number of degrees in angle B; $2b$ = the number of degrees in angle A; $b - 17$ = the number of degrees in angle C **34.** a = the number of students in algebra; $a + 29$ = the number of students in biology; $0.5a$ = the number of students in geology **35.** 3 **36.** -7 **37.** 16 years old **38.** 88 **39.** 13.3 hr; 12.3 hr **40.** 1st angle = 32°; 2nd angle = 96°; 3rd angle = 52° **41.** 310 kilowatt-hours **42.** 280 mi **43.** \$3000 **44.** \$200 **45.** \$7000 at 12%; \$2000 at 8% **46.** \$2000 at 4.5%; \$3000 at 6% **47.** 18 nickels; 6 dimes; 9 quarters **48.** 24 nickels; 21 dimes; 26 quarters

49. $x \leq -1$ **50.** $x < -4$

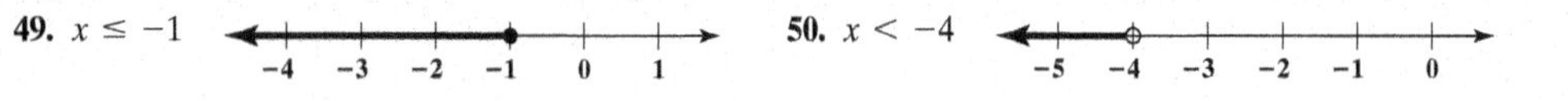

51. $x > -3$ **52.** $x \geq 6$ **53.** $x < 5$

54. $x < 10$ **55.** $x > -3$ **56.** $h \leq 32$ hr **57.** $n \leq 22$

58. $x = \frac{13}{6}$ **59.** $x = \frac{11}{8}$ **60.** $x = 0$ **61.** $x = 4$ **62.** $x < 2$

63. $x \leq -8$ **64.** $x \geq \frac{19}{7}$

65. $x \geq -15$

How Am I Doing? Chapter 2 Test

1. $x = 2$ (obj. 2.3.1) **2.** $x = \frac{1}{3}$ (obj. 2.3.2) **3.** $y = -\frac{7}{2}$ (obj. 2.3.3) **4.** $y = \frac{42}{5}$ (obj. 2.4.1) **5.** $x = 1$ (obj. 2.3.3) **6.** $x = -\frac{6}{5}$ or -1.2 (obj. 2.3.3) **7.** $y = 7$ (obj. 2.4.1) **8.** $y = \frac{7}{3}$ or $2\frac{1}{3}$ (obj. 2.3.3) **9.** $x = 13$ (obj. 2.3.3) **10.** $x = 20$ (obj. 2.3.3) **11.** $x = 10$ (obj. 2.3.3) **12.** $x = -4$ (obj. 2.3.3) **13.** $x = 12$ (obj. 2.3.3) **14.** $x = -\frac{1}{5}$ or -0.2 (obj. 2.4.1) **15.** $x = 3$ (obj. 2.4.1) **16.** $x = 2$ (obj. 2.4.1) **17.** $x = -2$ (obj. 2.4.1)

18. $x \leq -3$ (obj. 2.8.4) **19.** $x > -\frac{5}{4}$ (obj. 2.8.4)

20. $x < 2$ (obj. 2.8.4) **21.** $x \geq \frac{1}{2}$ (obj. 2.8.4)

22. 35 (obj. 2.6.1) **23.** 36 (obj. 2.6.1) **24.** 7 and 15 (obj. 2.6.1) **25.** 1 hour sooner (obj. 2.6.2) **26.** Second angle = 34°; first angle = 102°; third angle = 44° (obj. 2.6.1) **27.** 15 months (obj. 2.7.1) **28.** \$32,000 (obj. 2.7.2) **29.** \$1400 at 14%; \$2600 at 11% (obj. 2.7.3) **30.** 16 nickels; 7 dimes; 8 quarters (obj. 2.7.4)

Chapter 3 3.1 Exercises

1. 0 **3.** The order in which you write the numbers matters. The graph of $(5, 1)$ is not the same as the graph of $(1, 5)$. **5.** They are not the same because the x and y coordinates are different. To plot $(2, 7)$ we move 2 units to the right on the x-axis, but for the ordered pair $(7, 2)$ we move 7 units to the right on the x-axis. Then to plot $(2, 7)$ we move 7 units up, parallel to the y-axis, but for the ordered pair $(7, 2)$ we move 2 units up. **7.**

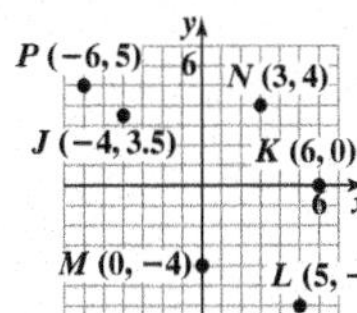

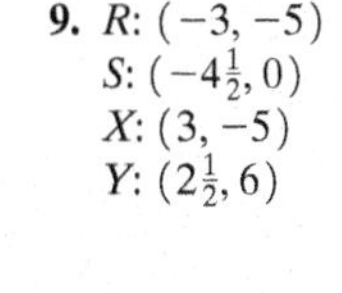

9. R: $(-3, -5)$ S: $(-4\frac{1}{2}, 0)$ X: $(3, -5)$ Y: $(2\frac{1}{2}, 6)$

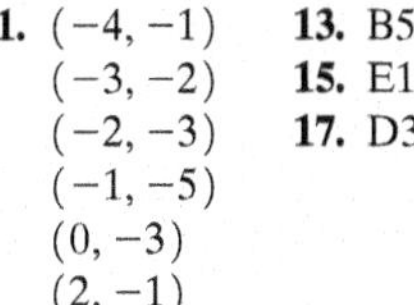

11. $(-4, -1)$ $(-3, -2)$ $(-2, -3)$ $(-1, -5)$ $(0, -3)$ $(2, -1)$ **13.** B5 **15.** E1 **17.** D3

19. (a)

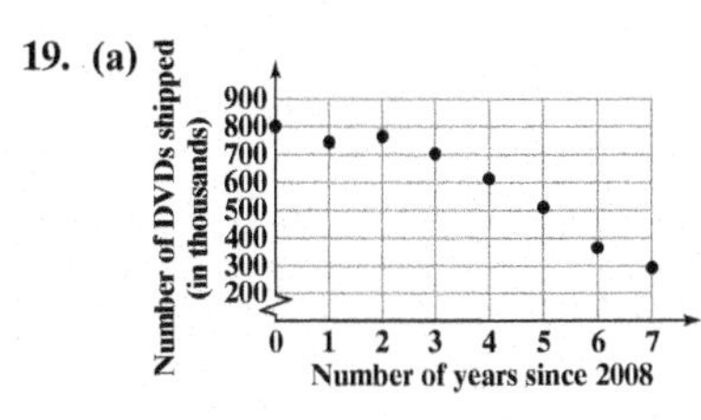

(b) The number of DVDs shipped decreased overall between 2008 and 2015, with a slight increase in 2010.

21. (a)

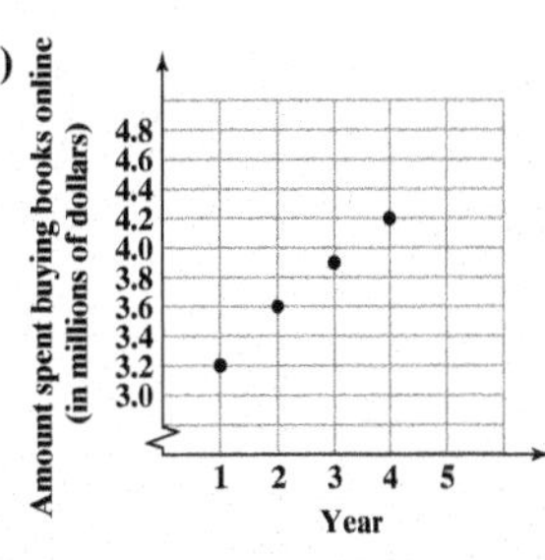

(b) An estimated \$4.5 billion will be spent buying books online in year 5.

23. No **25.** Yes **27.** Jon is right because for the ordered pair $(5, 3)$, $x = 5$ and $y = 3$. When we substitute these values in the equation we get $2(5) - 2(3) = 4$. For the ordered pair $(3, 5)$, when we replace $x = 3$ and $y = 5$ in the equation we get $2(3) - 2(5)$, which equals -4, not 4.

29. (a) $y = -\frac{2}{3}x + 4$ **(b)** $(6, 0)$ **31. (a)** $y = -4x + 11$ **(b)** $(2, 3)$ **33. (a)** $y = -\frac{2}{3}x + 2$ **(b)** $(3, 0)$ **35. (a)** $(0, 7)$ **(b)** $(2, 15)$

37. (a) $(-1, 11)$ **(b)** $(3, -13)$ **39. (a)** $(-3, -5)$ **(b)** $(5, 1)$ **41. (a)** $(-2, 0)$ **(b)** $(-4, 3)$ **43. (a)** $(7, 3)$ **(b)** $\left(-1, \frac{5}{7}\right)$

45. (a) $(2, 2)$ **(b)** $\left(\frac{3}{2}, 5\right)$ **47.** $\approx 1133.54\ \text{yd}^2$ **48.** 12

Quick Quiz 3.1

See Examples noted with Ex. **1.**

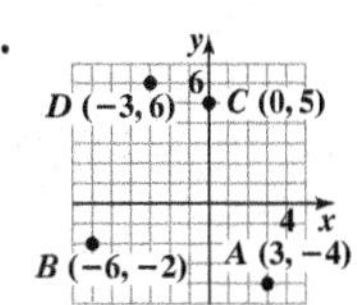

(Ex. 3)

2. (a) $(-2, 3)$ **(b)** $(3, -22)$ **(c)** $(0, -7)$ (Ex. 9) **3. (a)** $(3, 8)$ **(b)** $(-9, -8)$ **(c)** $(4.5, 10)$ (Ex. 8) **4.** See Student Solutions Manual

To Think About, page 218

By choosing multiples of 4 as replacements for x, we get integers as the corresponding y-values.

3.2 Exercises

1. No, replacing x by -2 and y by 5 in the equation does not result in a true statement. **3.** x-axis

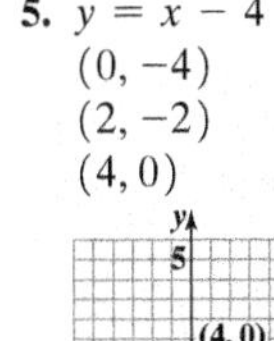

5. $y = x - 4$ $(0, -4)$ $(2, -2)$ $(4, 0)$

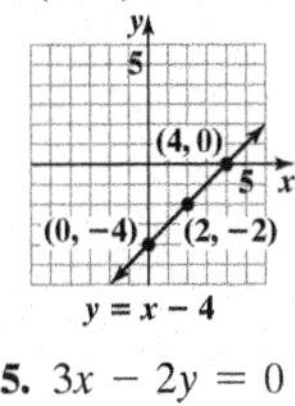

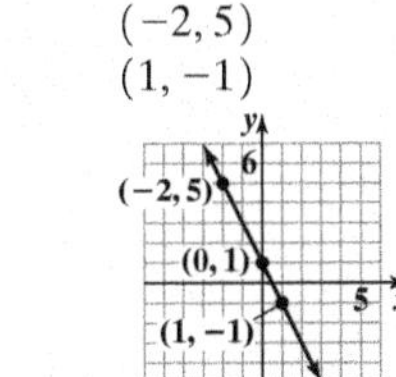

7. $y = -2x + 1$ $(0, 1)$ $(-2, 5)$ $(1, -1)$

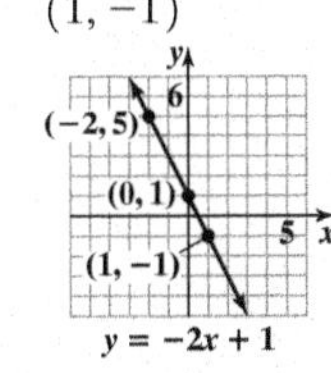

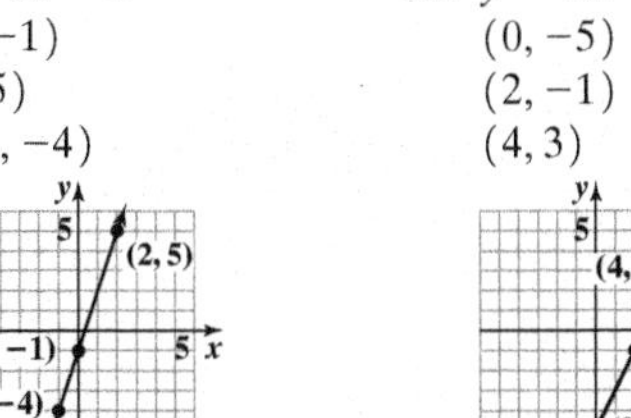

9. $y = 3x - 1$ $(0, -1)$ $(2, 5)$ $(-1, -4)$

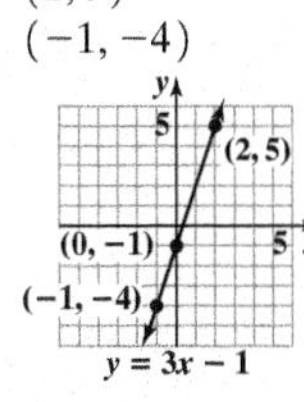

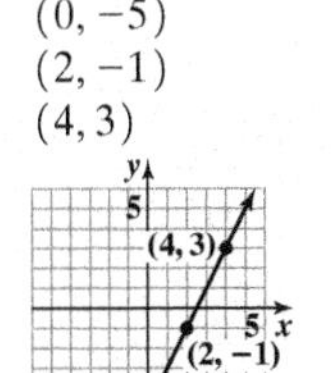

11. $y = 2x - 5$ $(0, -5)$ $(2, -1)$ $(4, 3)$

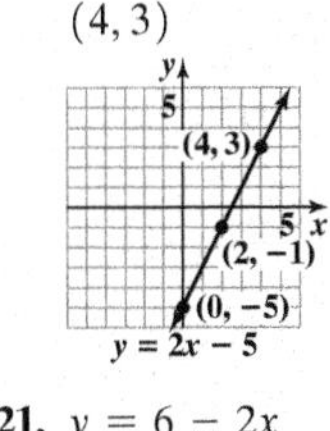

13. $y = -x + 3$

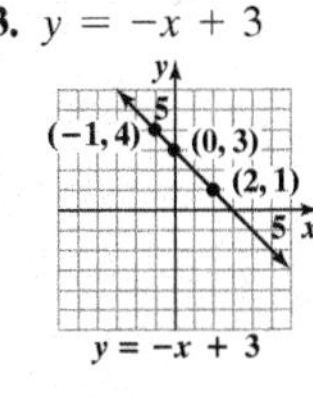

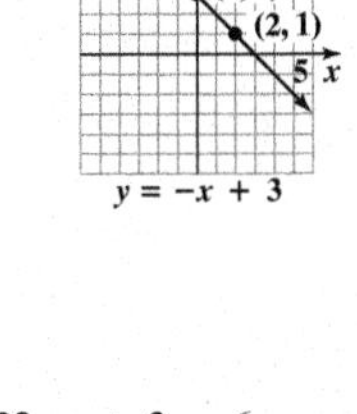

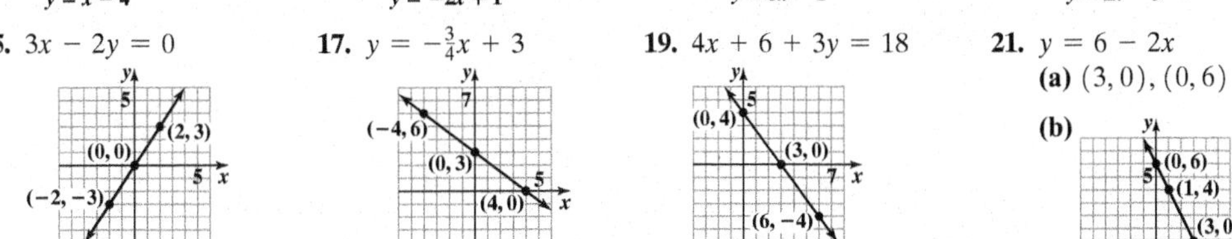

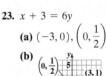

15. $3x - 2y = 0$

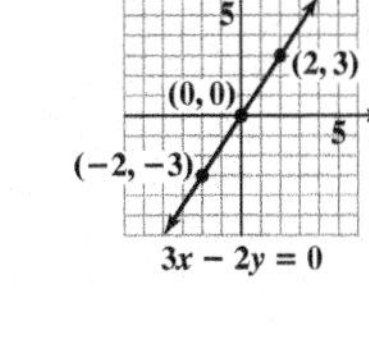

17. $y = -\frac{3}{4}x + 3$

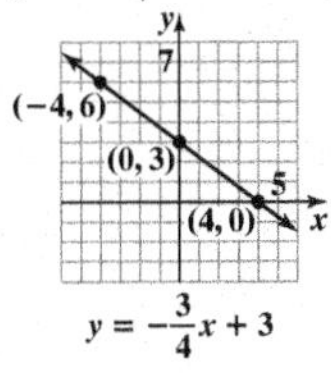

19. $4x + 6 + 3y = 18$

21. $y = 6 - 2x$ **(a)** $(3, 0)$, $(0, 6)$ **(b)**

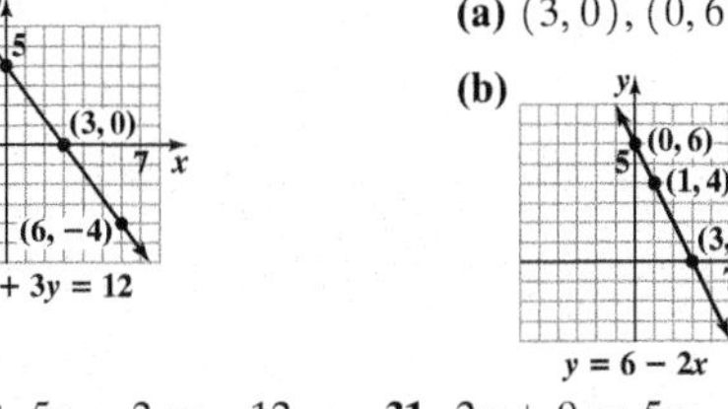

23. $x + 3 = 6y$ **(a)** $(-3, 0)$, $\left(0, \frac{1}{2}\right)$ **(b)**

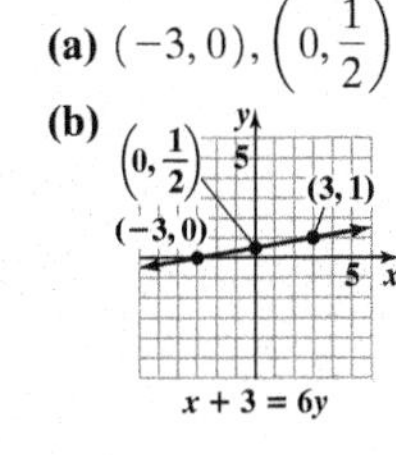

25. $x = 4$

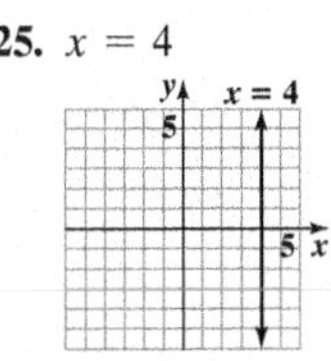

27. $y - 2 = 3y$

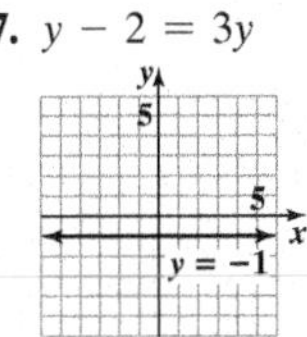

29. $2x + 5y - 2 = -12$

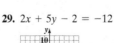

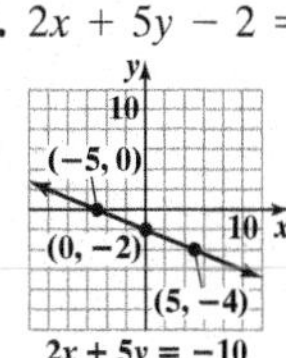

31. $2x + 9 = 5x$

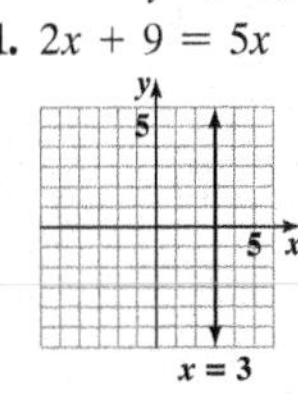

33.

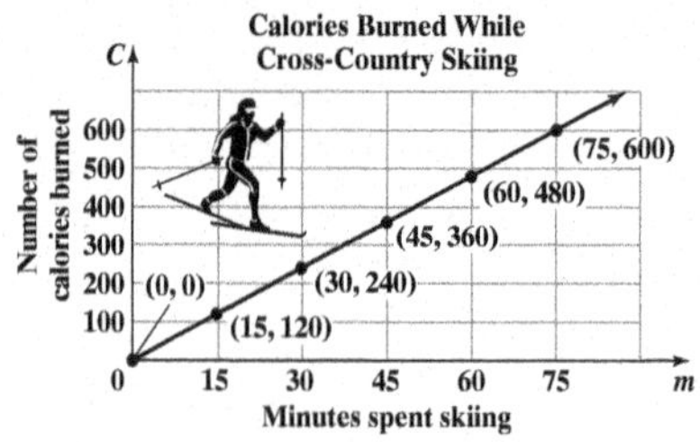

35.

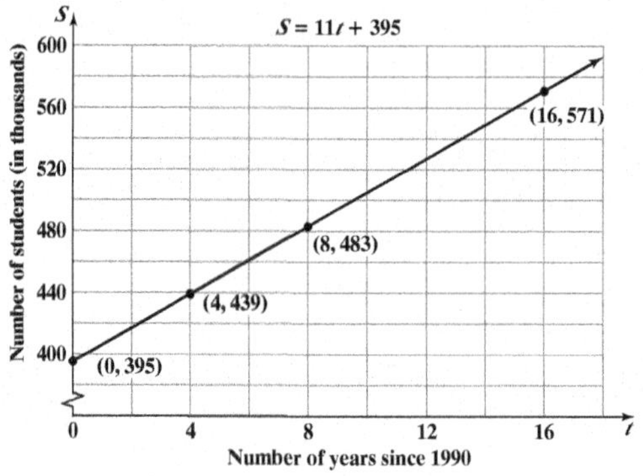

37. $x = -2$

38. $x \geq -\frac{14}{3}$

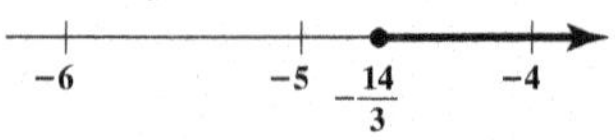

Quick Quiz 3.2

See Examples noted with Ex.

1.

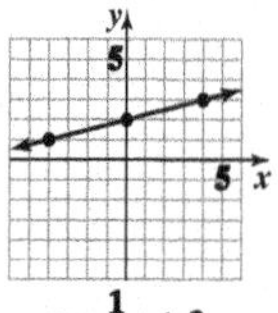

$y = \frac{1}{4}x + 2$ (Ex. 1)

2.

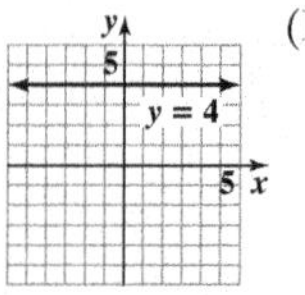

(Ex. 5)

3. (a) (2, 0), (0, 4) (Ex. 3)
(b)

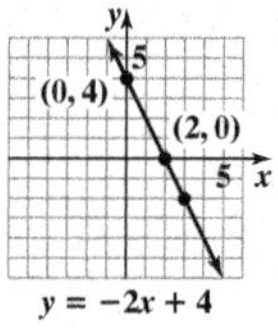

$y = -2x + 4$

4. See Student Solutions Manual

To Think About, page 228

Since the slope is negative, we would expect the y-values decrease as you go from left to right. In other words, the line will go down to the right, as we can see in the graph.

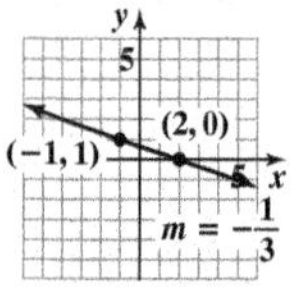

You Try It Graphing Organizer

Method 1:

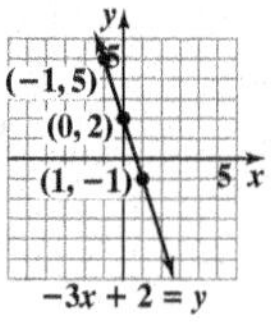

$-3x + 2 = y$

Method 2:

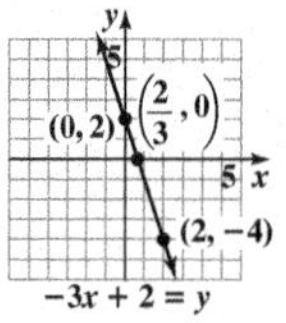

$-3x + 2 = y$

Method 3:

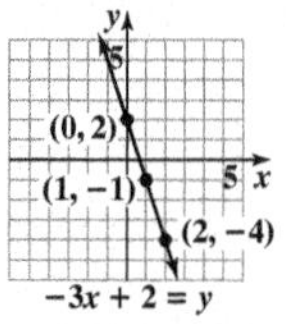

$-3x + 2 = y$

$m = -3;\ b = 2$

3.3 Exercises

1. No, division by zero is impossible, so the slope is undefined. **3.** 2 **5.** −5 **7.** $\frac{3}{5}$ **9.** $-\frac{1}{4}$ **11.** $-\frac{4}{3}$ **13.** 0 **15.** $-\frac{16}{5}$ **17.** $m = 8;\ (0, 9)$ **19.** $m = -3;\ (0, 4)$ **21.** $m = -\frac{8}{7};\ \left(0, \frac{3}{4}\right)$ **23.** $m = -6;\ (0, 0)$ **25.** $m = 0;\ (0, -2)$ **27.** $m = \frac{7}{3};\ \left(0, -\frac{4}{3}\right)$ **29.** $y = \frac{3}{5}x + 3$ **31.** $y = 4x - 5$ **33.** $y = -x$ **35.** $y = -\frac{5}{4}x - \frac{3}{4}$

37.

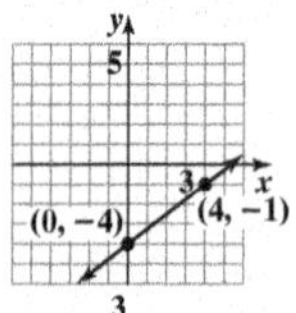

$y = \frac{3}{4}x - 4$

39.

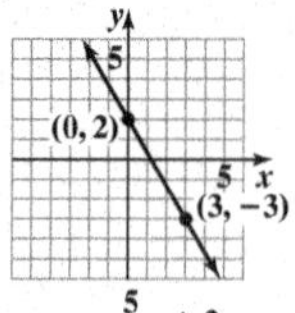

$y = -\frac{5}{3}x + 2$

41.

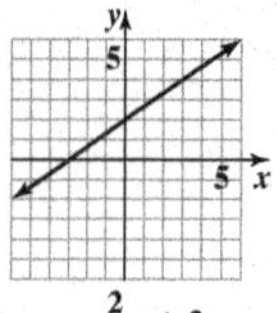

$y = \frac{2}{3}x + 2$

43.

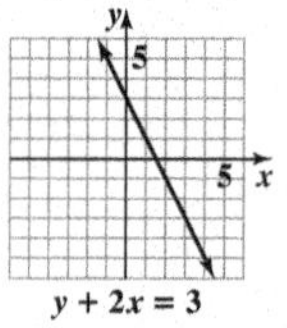

$y + 2x = 3$

45.

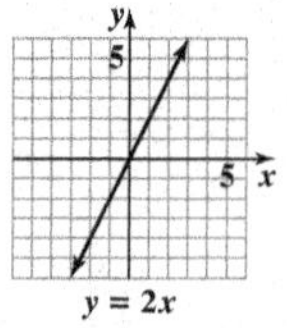

$y = 2x$

47. (a) $\frac{5}{6}$ **(b)** $-\frac{6}{5}$

49. (a) −8 **(b)** $\frac{1}{8}$ **51. (a)** $\frac{2}{3}$ **(b)** $-\frac{3}{2}$ **53.** Yes; $2x - 3y = 18$ **55. (a)** $y = 35x + 625$ **(b)** $m = 35;\ (0, 625)$
(c) The amount of increase (in thousands) in the number of cell phone accessories sold in the U.S. per year from 2010 to 2020.

57. $x < \frac{12}{5}$

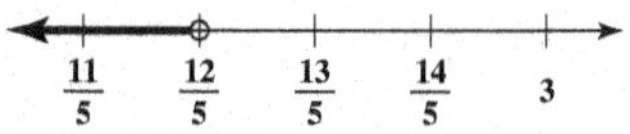

58. $x \leq 24$

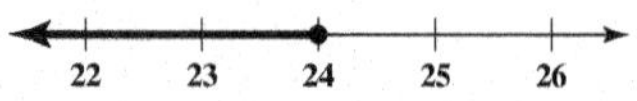

Quick Quiz 3.3

See Examples noted with Ex.

1. $\frac{1}{2}$ (Ex. 1) **2. (a)** $m = -3;\ (0, 2)$ (Ex. 4)

(b)

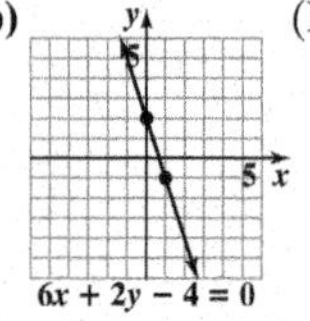

$6x + 2y - 4 = 0$

(Ex. 6) **3.** $y = -\frac{5}{7}x - 5$ (Ex. 5) **4.** See Student Solutions Manual

Use Math to Save Money **1.** $\$42{,}000/12 = \$3500$ **2.** $\$3500 \times 0.05 = \$175$ **3.** $FV = \$175 \times [1.0067^{480} - 1]/0.0067 \approx \$618{,}044$ **4.** $\$618{,}044 \times 0.04 \approx \$24{,}722$ **5.** $\$24{,}722/12 \approx \$2060$ **6.** $\$3500 - \$2060 = \$1440$ **7.** Yes, he needs to increase the amount.

3.4 Exercises **1.** $y = 3x + 6$ **3.** $y = -2x + 11$ **5.** $y = -3x + \frac{7}{2}$ **7.** $y = \frac{1}{4}x + 4$ **9.** $y = -2x - 6$ **11.** $y = -4x + 2$ **13.** $y = 5x - 10$ **15.** $y = \frac{1}{3}x + \frac{1}{2}$ **17.** $y = -3x$ **19.** $y = -3x + 3$ **21.** $y = -\frac{2}{3}x + 1$ **23.** $y = \frac{2}{3}x - 4$ **25.** $y = -\frac{2}{3}x$ **27.** $y = -2$ **29.** $y = -5$ **31.** $x = 4$ **33.** $y = \frac{1}{3}x + 5$ **35.** $y = -\frac{1}{2}x + 4$ **37.** $y = 2.4x + 227$ **39.** $x < -4$ **40.** $x \le 3$ **41.** \$61.20 **42.** 290 min

Quick Quiz 3.4 *See Examples noted with Ex.* **1.** $y = \frac{2}{3}x - 7$ (Ex. 1) **2.** $y = 6x + 19$ (Ex. 2) **3.** $x = 4$; the slope is undefined (Ex. 2) **4.** See Student Solutions Manual

3.5 Exercises **1.** No, all points in one region will be solutions to the inequality while all points in the other region will not be solutions. Thus testing any point will give the same result, as long as the point is not on the boundary line.

3.

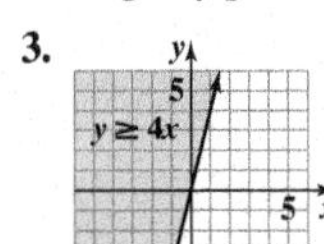

5. **7.**

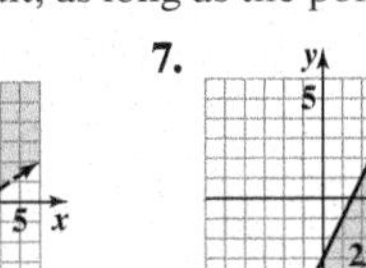

9.

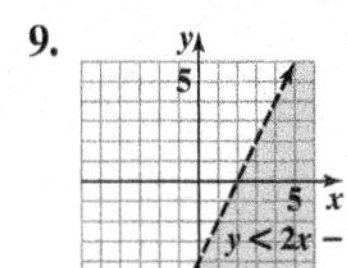

11.

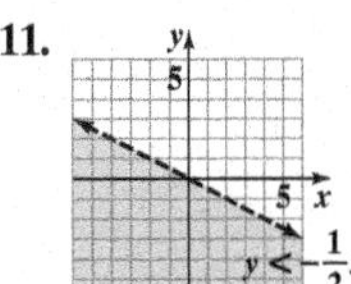

13.

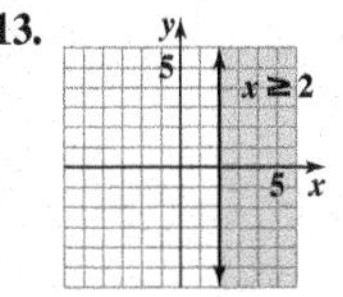

15.

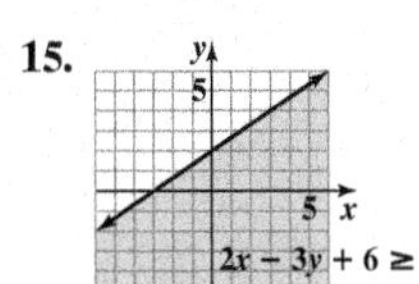

17.

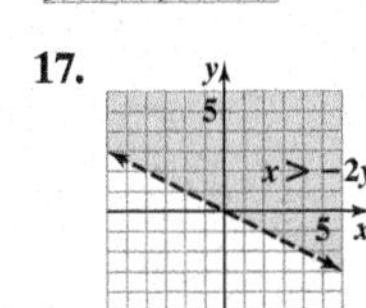

19.

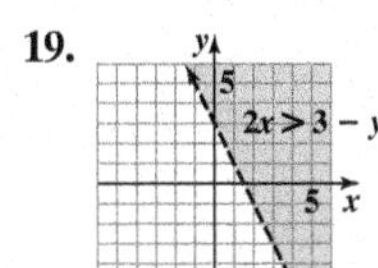

21.

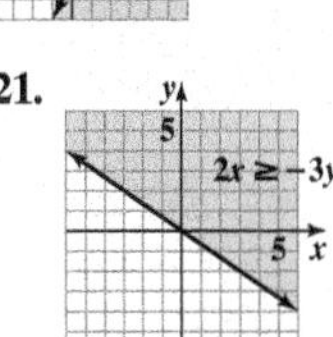

23. 7 **24.** −7 **25.** −28 **26.** −20 **27.** \$95.20 **28.** \$19,040

Quick Quiz 3.5 *See Examples noted with Ex.* **1.** Use a dashed line. If the inequality has a $<$ or a $>$ symbol, the points on the line itself are not included. This is indicated by a dashed line. (Ex. 1) **2.** (Ex. 2) **3.** (Ex. 1)

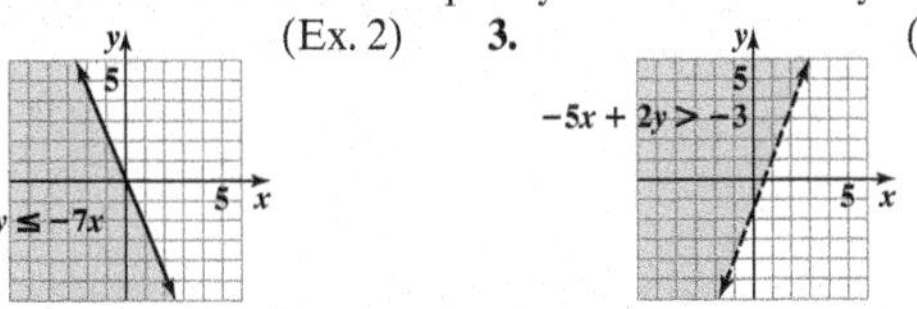

4. See Student Solutions Manual

To Think About, page 250 The relation is a function since for each time, there is exactly one bus stop. Since the location of the bus depends on the time, the bus stop is the dependent variable and time is the independent variable.

3.6 Exercises **1.** Using a table of values, an algebraic equation, or a graph **3.** possible values; independent **5.** If a vertical line can intersect the graph more than once, the relation is not a function. If no such line exists, then the relation is a function.

7. **(a)** Domain $= \left\{-3, \frac{3}{7}, 3\right\}$ Range $= \left\{-1, \frac{3}{7}, 4\right\}$ **(b)** Not a function **9.** **(a)** Domain $= \{0, 3, 6\}$ Range $= \{0.5, 1.5, 2.5\}$ **(b)** Function

11. **(a)** Domain $= \{1, 9, 12, 14\}$ Range $= \{1, 3, 12\}$ **(b)** Function **13.** **(a)** Domain $= \{3, 5, 7\}$ Range $= \{75, 85, 95, 100\}$ **(b)** Not a function

15.

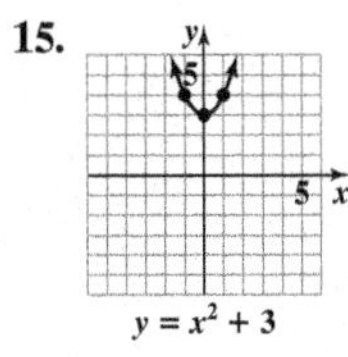

$y = x^2 + 3$

17.

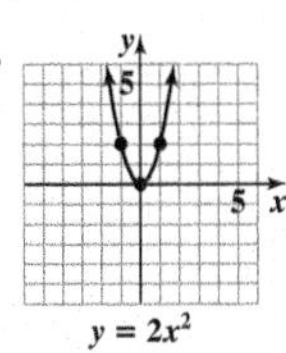

$y = 2x^2$

19. $x = -2y^2$

21.

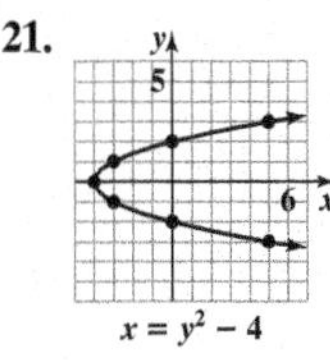

$x = y^2 - 4$

23.

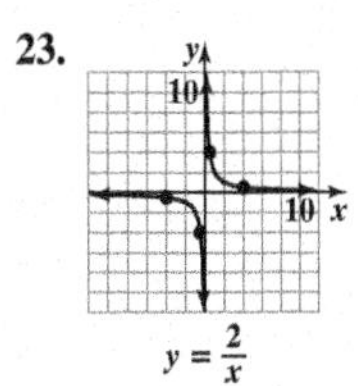

$y = \frac{2}{x}$

25.

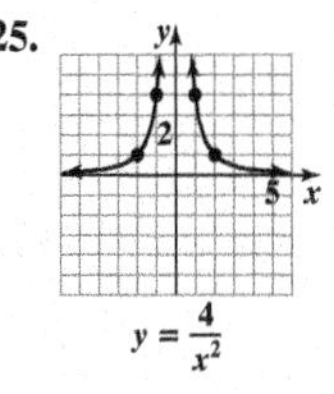

$y = \frac{4}{x^2}$

27. $x = (y + 1)^2$

29.

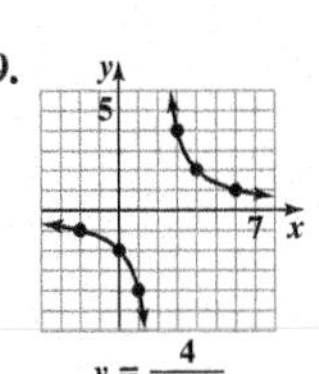

$y = \frac{4}{x - 2}$

31. Function **33.** Not a function **35.** Function **37.** Not a function **39.** **(a)** 26 **(b)** 2 **(c)** −4 **41.** **(a)** 3 **(b)** 24 **(c)** 9

43. $f(0) = 31.6, f(4) = 32.24, f(10) = 34.4$ The curve slopes more steeply for larger values of x. The increase in pet ownership is increasing as x gets larger.

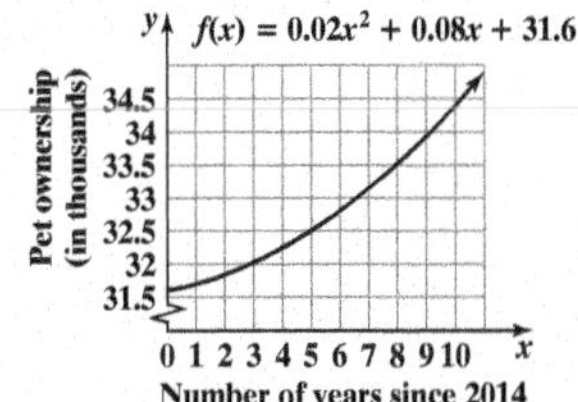

45. $-8x^3 + 12x^2 - 32x$ **46.** $5a^2b + 30ab - 10a^2$ **47.** $-19x + 2y - 2$ **48.** $9x^2y - 6xy^2 + 7xy$

Quick Quiz 3.6 *See Examples noted with Ex.*

1. No. Two different ordered pairs have the same first coordinate. (Ex. 2) **2. (a)** 41 **(b)** 34 (Ex. 8) **3. (a)** -7 **(b)** $-\frac{7}{8}$ (Ex. 8) **4.** See Student Solutions Manual

Career Exploration Problems

1.

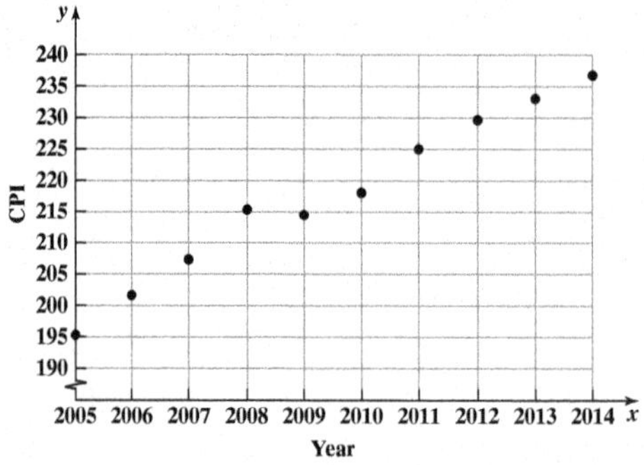

2. 2005–2008: \$6.67; 2008–2009: $-$\$0.76; 2010–2014: \$4.67 **3.** $y = 4.60x - 9027.66$ **4.** 2016: $y = 4.60(2016) - 9027.66$ $= \$245.94$; 2020: $y = 4.60(2020) - 9027.66$ $= \$264.34$

You Try It

1.

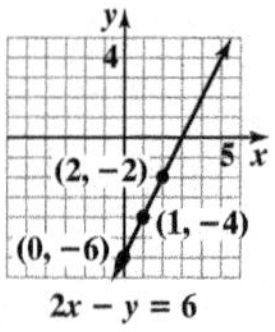

2. (a) x-intercept: $(4, 0)$ y-intercept: $(0, -2)$ **(b)**

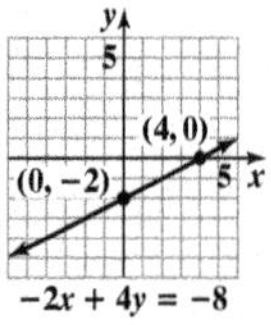

3. (a)

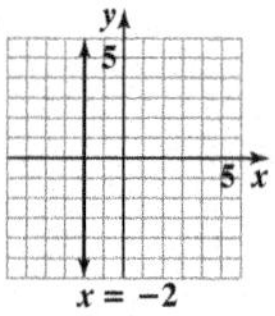

(b)

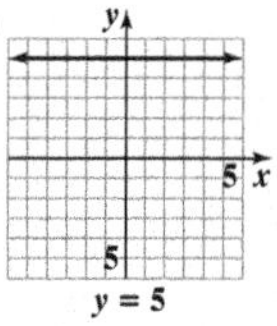

4. -2 **5.** Slope $= -\frac{5}{3}$; y-intercept $= (0, 3)$ **6.** $y = 2x - 3$

7.

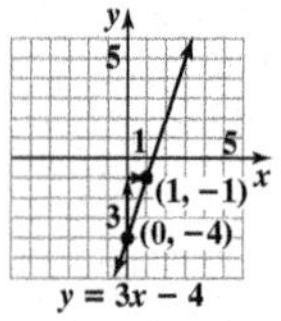

8. (a) 3 **(b)** $-\frac{1}{3}$ **9.** $y = -\frac{1}{2}x + \frac{7}{2}$ **10.** $y = \frac{1}{2}x + \frac{1}{2}$ **11.**

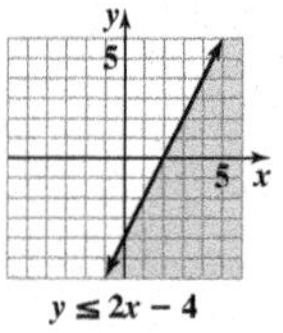

12. Yes **13.**

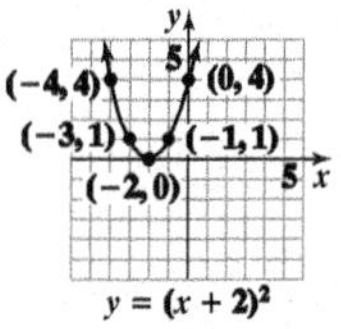

14. No **15. (a)** 20 **(b)** 6

Chapter 3 Review Problems

1.

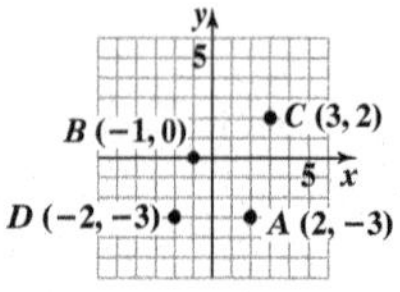

2.

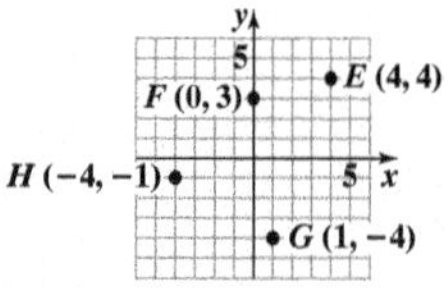

3. (a) $(0, 7)$ **(b)** $(-1, 10)$ **4. (a)** $(1, 2)$ **(b)** $(-4, 4)$ **5. (a)** $(6, -1)$ **(b)** $(6, 3)$

6.

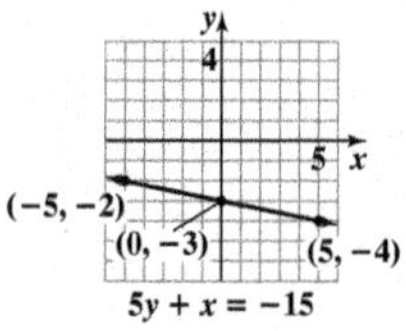

7.

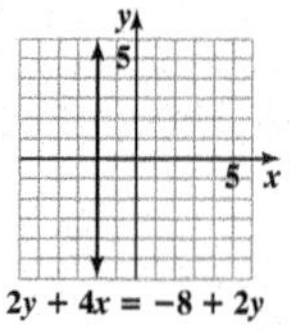

8.

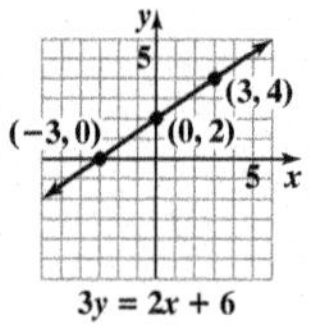

9. $m = -\frac{5}{6}$ **10.** $m = -\frac{5}{3}$ **11.** $m = \frac{9}{11}$; $\left(0, \frac{15}{11}\right)$ **12.** $y = -\frac{1}{2}x + 3$

13.

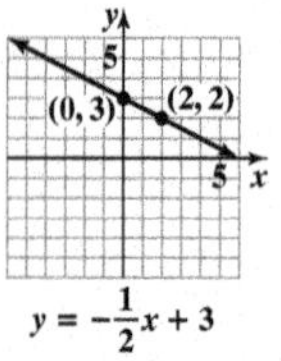

14.

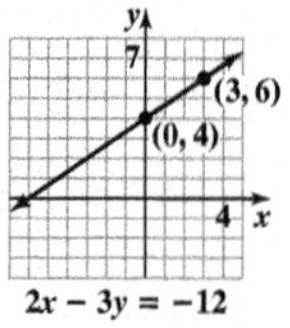

15.

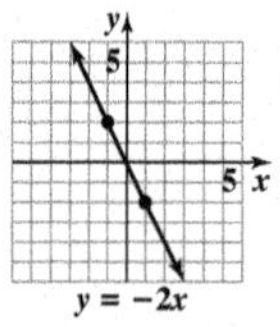

16. $y = -6x + 14$ **17.** $y = -\frac{1}{3}x + \frac{11}{3}$ **18.** $y = x + 3$ **19.** $y = 7$ **20.** $y = \frac{2}{3}x - 3$ **21.** $y = -3x + 1$ **22.** $x = 5$

23.

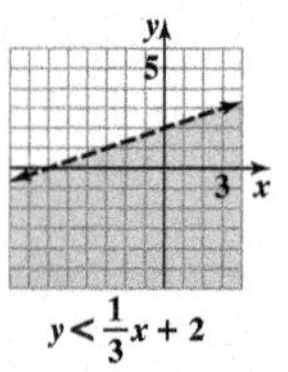

24.

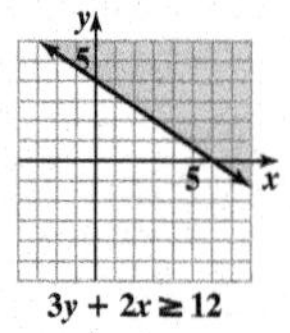

25.

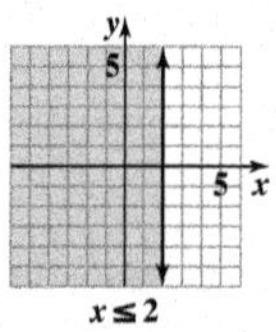

26. Domain: $\{-6, -5, 5\}$ Range: $\{-6, 5\}$ not a function **27.** Domain: $\{-2, 2, 5, 6\}$ Range: $\{-3, 4\}$ function **28.** Function **29.** Not a function **30.** Function

31. 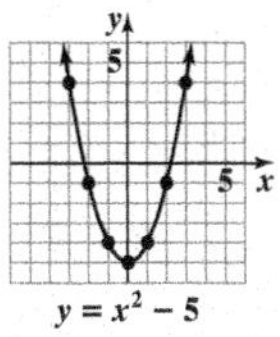 $y = x^2 - 5$ **32.** 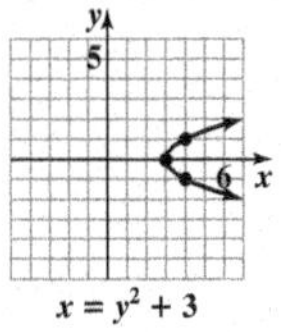 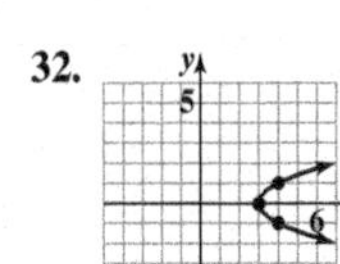 $x = y^2 + 3$ **33.** 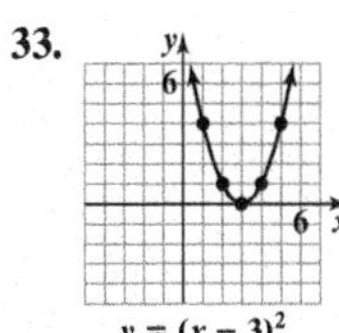 $y = (x - 3)^2$

34. (a) 7 **(b)** 31 **35. (a)** −1 **(b)** −5 **36. (a)** 1 **(b)** $\frac{1}{5}$ **37. (a)** 0 **(b)** 4

38. 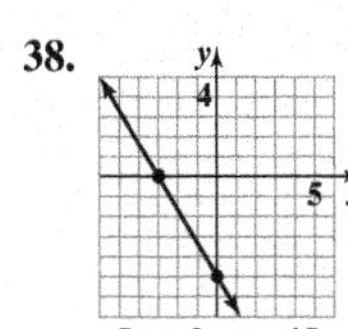 $5x + 3y = -15$ **39.** 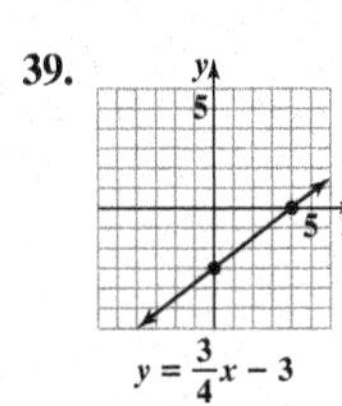 $y = \frac{3}{4}x - 3$ **40.** 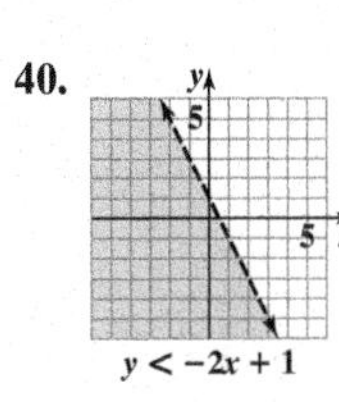 $y < -2x + 1$

41. $-\frac{2}{5}$ **42.** $m = -\frac{7}{6}$; y-intercept $= \left(0, \frac{5}{3}\right)$ **43.** $y = \frac{2}{3}x - 7$

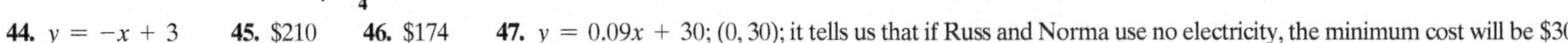

44. $y = -x + 3$ **45.** \$210 **46.** \$174 **47.** $y = 0.09x + 30$; $(0, 30)$; it tells us that if Russ and Norma use no electricity, the minimum cost will be \$30. **48.** $m = 0.09$; the electric bill increases \$0.09 for each kilowatt-hour of use. **49.** 1300 kilowatt-hours **50.** 2400 kilowatt-hours

51. 17,020,000 people in 1994
13,254,000 people in 2008
11,640,000 people in 2014

52. 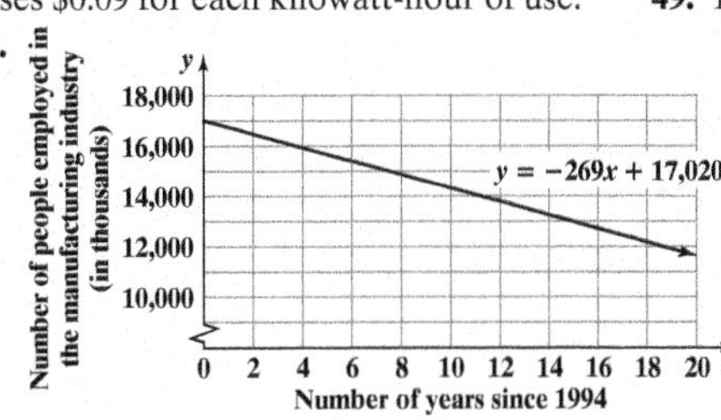

53. The slope is −269. The slope tells us that the number of people employed in manufacturing decreases each year by 269 thousand. In other words, employment in manufacturing goes down 269,000 people each year. **54.** The y-intercept is (0, 17,020). This tells us that in the year 1994, the number of manufacturing jobs was 17,020 thousand, which is 17,020,000. **55.** 2016 **56.** 2019

How Am I Doing? Chapter 3 Test

1. 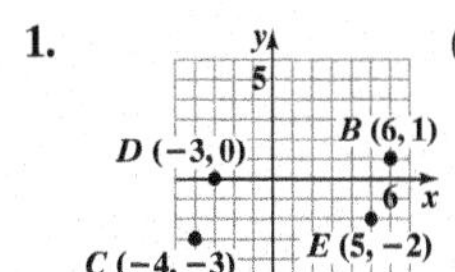

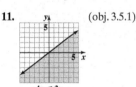

(obj. 3.1.1)

2. 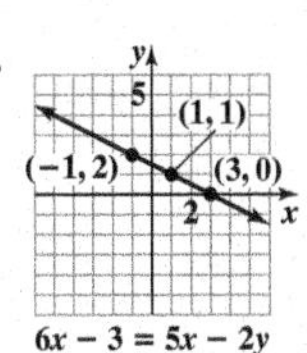

$6x - 3 = 5x - 2y$ (obj. 3.2.1)

3. 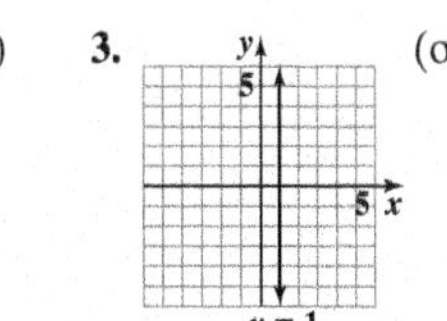 $x = 1$ (obj. 3.2.3)

4. 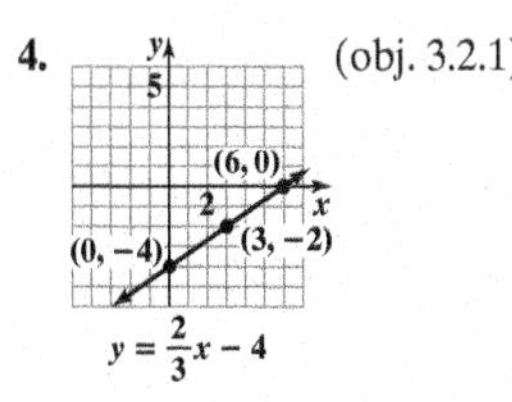

$y = \frac{2}{3}x - 4$ (obj. 3.2.1)

5. (a) x-intercept: $(-2, 0)$
y-intercept: $(0, -4)$

(b) 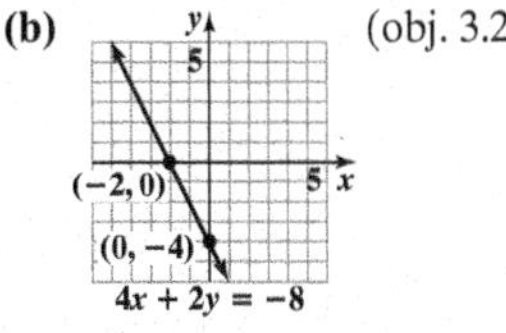

$4x + 2y = -8$ (obj. 3.2.2)

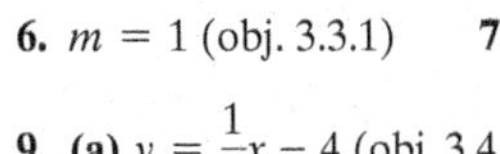

6. $m = 1$ (obj. 3.3.1) **7.** $m = -\frac{3}{2}$; $\left(0, \frac{5}{2}\right)$ (obj. 3.3.2) **8.** $y = \frac{3}{4}x - 6$ (obj. 3.3.3)

9. (a) $y = \frac{1}{2}x - 4$ (obj. 3.4.1) **(b)** −2 (obj. 3.3.5) **10.** $y = -\frac{3}{2}x + \frac{7}{2}$ (obj. 3.4.2)

11. $4y \leq 3x$ (obj. 3.5.1) **12.** 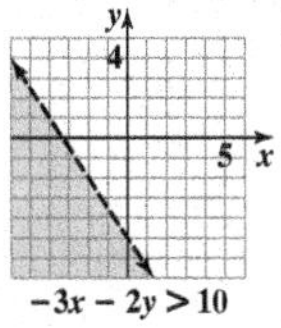 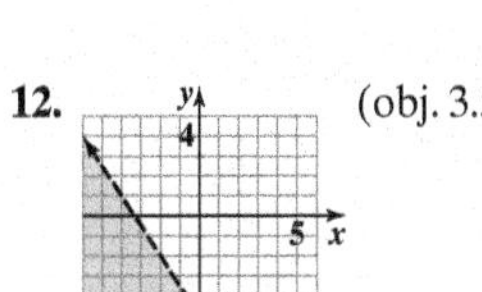 $-3x - 2y > 10$ (obj. 3.5.1)

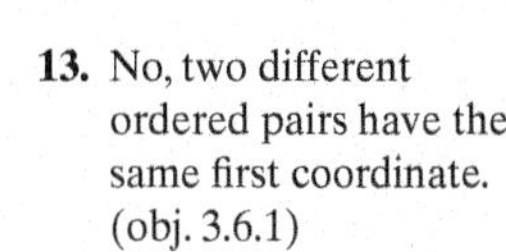

13. No, two different ordered pairs have the same first coordinate. (obj. 3.6.1)

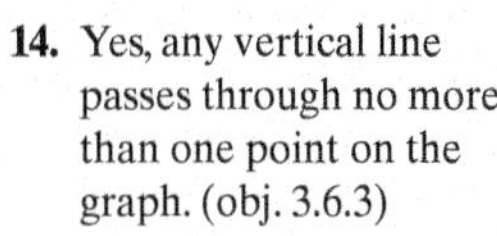

14. Yes, any vertical line passes through no more than one point on the graph. (obj. 3.6.3)

15. 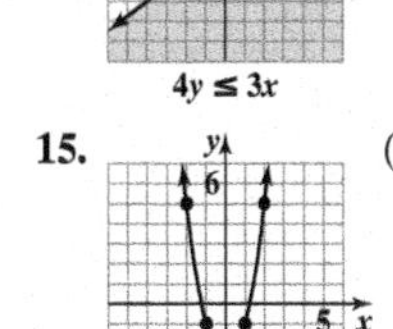 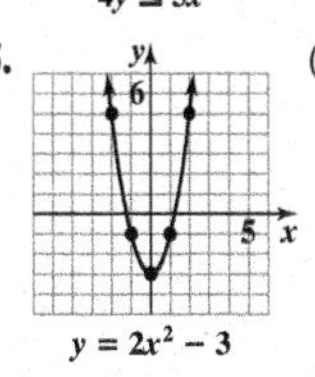 $y = 2x^2 - 3$ (obj. 3.6.2) **16. (a)** −3 **(b)** −3 (obj. 3.6.4)

Cumulative Test for Chapters 0–3

1. 3.69 **2.** $\frac{31}{24}$ or $1\frac{7}{24}$ **3.** −1.514 **4.** 8.52 **5.** −8.6688 **6.** $-34x + 10y - 20$

7. 21 **8.** 30 **9.** $-2st - 7s^2t + 14st^2$ **10.** $25x^2$ **11.** $6x - 40 + 48y - 24xy$ **12.** $x = \frac{5}{6}$ **13.** $x = 4$ **14.** $y = 1$

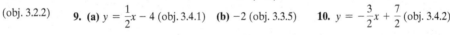

15. $x \leq 3$ (number line: 1 2 3 4 5) **16.** 28.26 in.^{2} **17.** $A = 162.5$ m^2; \$731.25 **18.** −7 **19.** 58 **20.** 92 **21.** $m = 0$

22. $y = -\frac{2}{3}x - 3$ **23.**

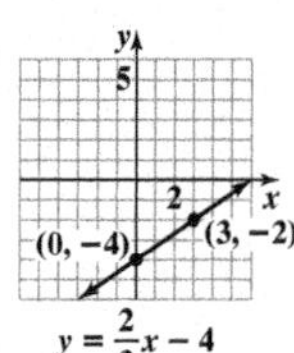

24.

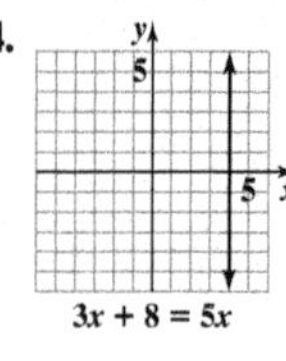

25.

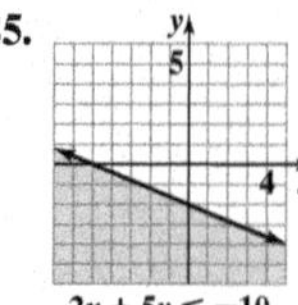
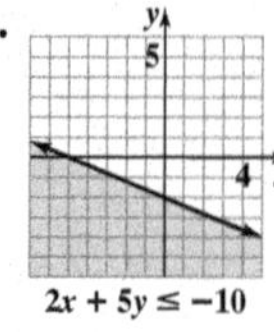

26. (a) 1 **(b)** 10

Chapter 4

4.1 Exercises

1. There is no solution. There is no point (x, y) that satisfies both equations. The graph of such a system yields two parallel lines. **3.** It may have one solution, it may have no solution, or it may have an infinite number of solutions. **5.** $\left(\frac{5}{2}, 2\right)$ is a solution to the system. **7.** $2x - y = 3$; $(1, -1)$; $3x + y = 2$ **9.** $3x - 2y = 6$; $(0, -3)$; $4x + y = -3$ **11.**

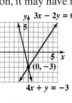

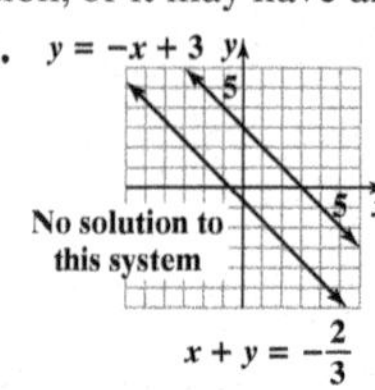

13.

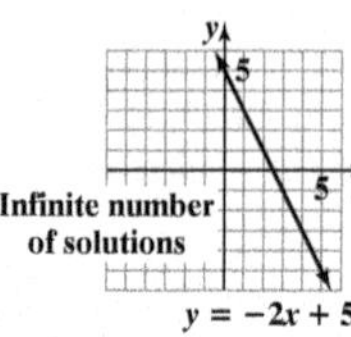

15. $(18, -10)$ **17.** $(2, -2)$ **19.** $(-1, 2)$ **21.** $(6, 0)$ **23.** $(0, 1)$ **25.** $\left(2, \frac{4}{3}\right)$ **27.** $(1, -3)$ **29.** $(3, -2)$ **31.** $(2, 8)$ **33.** $(-4, 5)$ **35.** $(6, -8)$ **37.** No solution; inconsistent system of equations **39.** Infinite number of solutions; dependent equations **41.** $(-4, 7)$ **43.** No solution; inconsistent system of equations **45.** $(12, 4)$ **47.** $(0, 2)$ **49.** Infinite number of solutions; dependent equations

51. (a) $y = 200 + 50x$ $y = 300 + 30x$

(b)

x	$y = 200 + 50x$
0	200
4	400
8	600

x	$y = 300 + 30x$
0	300
4	420
8	540

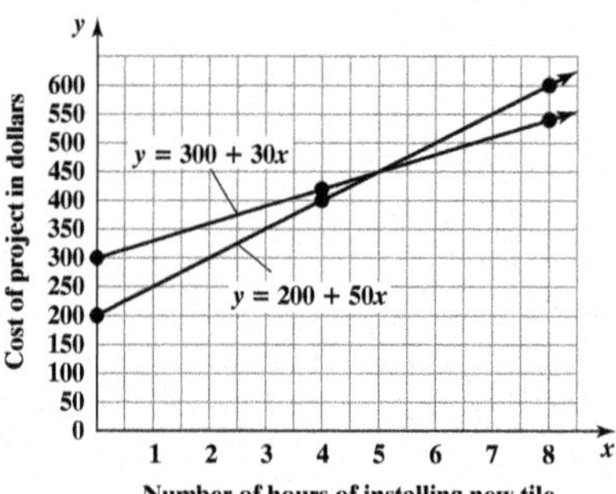

(c) The cost will be the same for 5 hours of installing new tile.
(d) The cost will be less for Modern Bathroom Headquarters.

53. $(2.46, -0.38)$

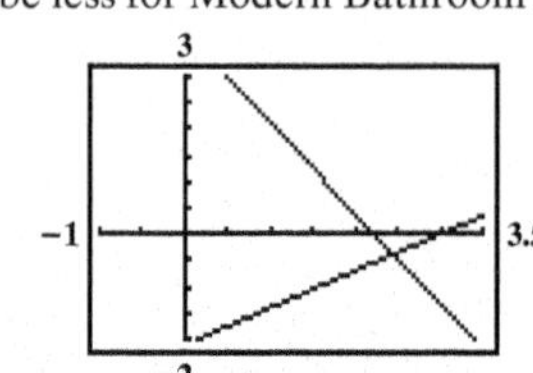

55. $(-2.45, 6.11)$

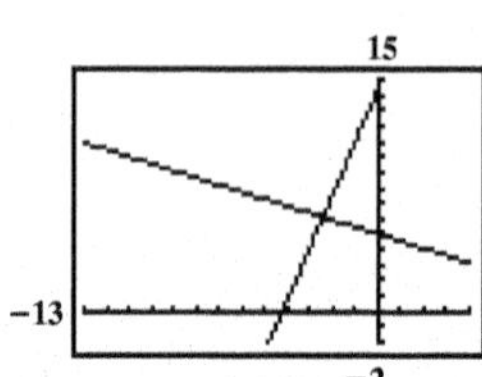

56. Approximately \$0.01/lb **57.** 341,889 cars

Quick Quiz 4.1

See Examples noted with Ex. **1.** $(4, 2)$ (Ex. 3) **2.** $(2, 2)$ (Ex. 6) **3.** $(-12, -15)$ (Ex. 6) **4.** See Student Solutions Manual

4.2 Exercises

1. Yes, it is. **3.** No, it is not. **5.** $(1, 3, -2)$ **7.** $(-3, 0, 2)$ **9.** $(0, -2, 5)$ **11.** $(1, -1, 2)$ **13.** $(2, 1, -4)$ **15.** $(3, -1, -2)$ **17.** $x = 1.10551, y = 2.93991, z = 1.73307$ **19.** $(4, -1, -3)$ **21.** $\left(\frac{1}{2}, \frac{2}{3}, \frac{5}{6}\right)$ **23.** $(1, 3, 5)$ **25.** $(-2, 6, -1)$ **27.** Infinite number of solutions; dependent equations **29.** No solution; inconsistent system of equations **31.** $(4, 0, 2)$ **33.** $x = 2$ **34.** $m = -\frac{9}{5}$ **35.** $y = \frac{1}{3}x + \frac{11}{3}$ **36.** $m = \frac{3}{2}$

Quick Quiz 4.2

See Examples noted with Ex. **1.** $(-1, 2, 3)$ (Ex. 2) **2.** $(1, 0, -1)$ (Ex. 2) **3.** $(2, 1, -1)$ (Ex. 3) **4.** See Student Solutions Manual

Use Math to Save Money

1. Job 1: \$50,310/52 = \$967.50; Job 2: \$20.50 × 30 = \$615 **2.** Job 1: \$967.50 × 0.15 ≈ \$145.13; Job 2: \$615 × 0.12 = \$73.80 **3.** Job 1: \$175 + \$75 + \$25 + \$50 + (\$0.35 × 45 × 5) = \$403.75; Job 2: \$150 + \$50 + \$25 + \$25 + (\$0.35 × 5 × 5) = \$258.75 **4.** Job 1: \$967.50 − (\$145.13 + \$403.75) = \$418.62; Job 2: \$615 − (\$73.80 + \$258.75) = \$282.45 **5.** Job 1: \$418.62/50 ≈ \$8.37; Job 2: \$282.45/31 ≈ \$9.11 **6.** She should take Job 2. **7.** She should take Job 1.

4.3 Exercises

1. 62 is the larger number; 25 is the smaller number **3.** 16 heavy equipment operators; 19 general laborers **5.** 51 tickets for regular coach seats; 47 tickets for sleeper car seats **7.** 25 experienced employees; 10 new employees **9.** 30 packages of old fertilizer; 25 packages of new fertilizer **11.** One scone costs \$2.75; one large coffee costs \$1.90 **13.** Speed of plane in still air is 216 mph; speed of wind is 36 mph **15.** Speed of boat is 14 mph; speed of current is 2 mph **17.** 28 free throws; 36 regular (2-point) shots **19.** 235 regular text messages;

80 multimedia text messages **21.** \$10,258 for a car; \$17,300 for a truck **23.** 7 pens, 5 notebooks, and 3 highlighters **25.** 80 adults, 100 high school students, and 120 children **27.** 800 senior citizens, 10,000 adults, and 1200 junior and high school students **29.** 5 small pizzas, 8 medium pizzas, and 7 large pizzas **31.** 3 of Box *A*, 4 of Box *B*, and 3 of Box *C* **32.** $x = \frac{26}{7}$ or $3\frac{5}{7}$ **33.** $x = \frac{7}{18}$ **34.** $y = \frac{5}{3}$ or $1\frac{2}{3}$ **35.** $x = -\frac{10}{11}$

Quick Quiz 4.3

See Examples noted with Ex. **1.** Speed of wind = 40 mph; speed of plane = 440 mph (Ex. 3) **2.** \$35/day; \$0.25/mi (Ex. 1) **3.** Drawings were \$5, carved elephants were \$15, and a set of drums was \$10. (Ex. 4) **4.** See Student Solutions Manual.

4.4 Exercises

1. They would be dashed. The boundary lines are not included in the solution of a system of inequalities whenever the system contains only < or > symbols. **3.** She could substitute $(3, -4)$ into each inequality. Since $(3, -4)$ does not satisfy the second inequality, we know that the solution does not contain that ordered pair. Therefore, that point would not lie in the solution region.

5. $y \geq 2x - 2$, $x + y \leq 4$

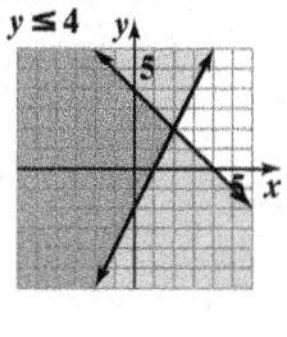

7. $y \geq -2x$, $y \geq 3x + 5$

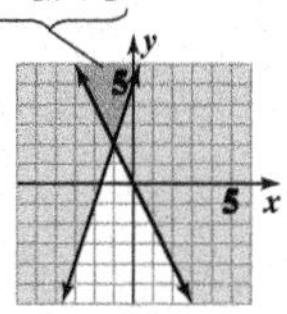

9.

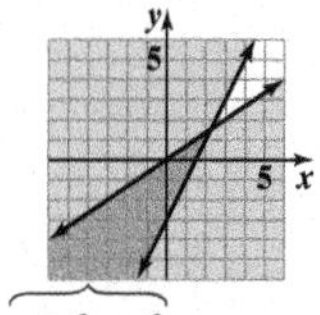

$y \geq 2x - 3$, $y \leq \frac{2}{3}x$

11. $x - y \geq -1$, $-3x - y \leq 4$

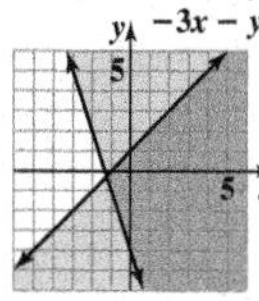

13. $x + 2y < 6$, $y < 3$

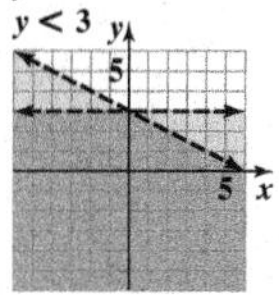

15. $y < 4$, $x > -2$

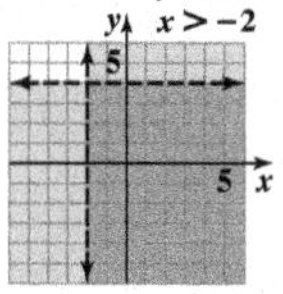

17.

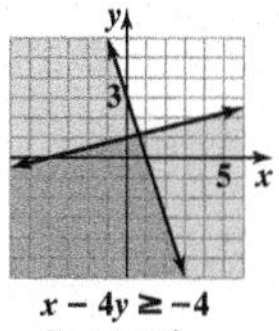

$x - 4y \geq -4$, $3x + y \leq 3$

19. $3x + 2y < 6$, $3x + 2y > -6$

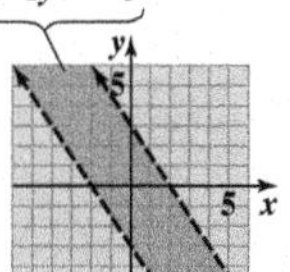

21. $x + y \leq 5$, $2x - y \geq 1$

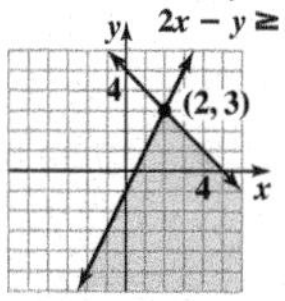

23. $x + 4y < -20$, $y \leq x$

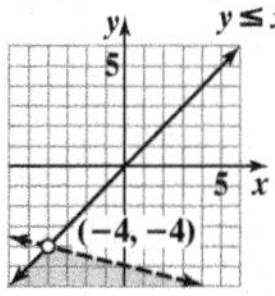

25.

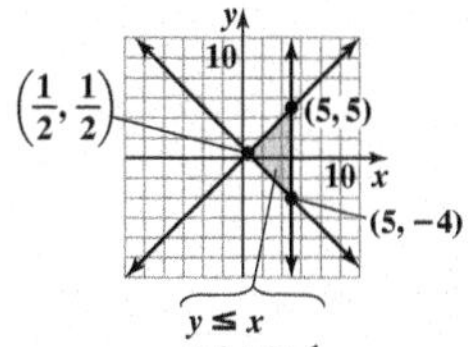

$y \leq x$, $x + y \geq 1$, $x \leq 5$

27.

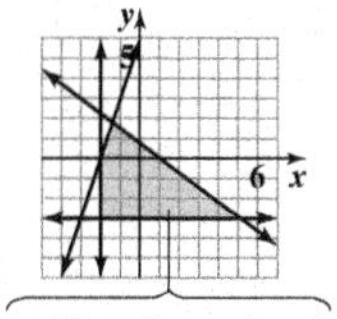

$y \leq 3x + 6$, $4y + 3x \leq 3$, $x \geq -2$, $y \geq -3$

29. (a)

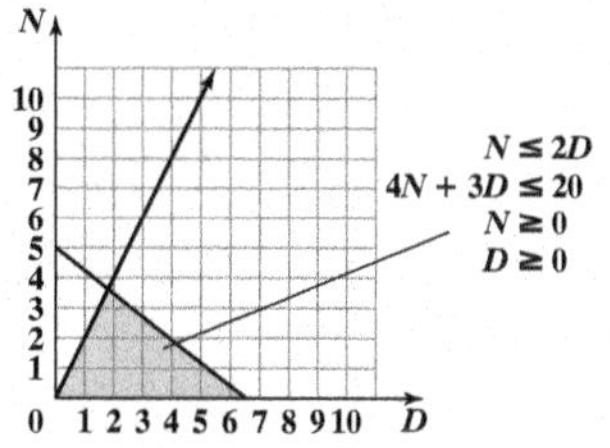

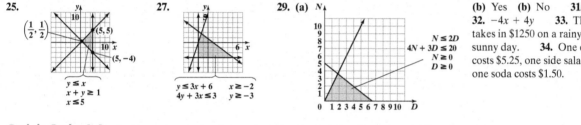

(b) Yes **(b)** No **31.** 13 **32.** $-4x + 4y$ **33.** The driving range takes in \$1250 on a rainy day and \$800 on a sunny day. **34.** One chicken sandwich costs \$5.25, one side salad costs \$3.75, and one soda costs \$1.50.

Quick Quiz 4.4

See Examples noted with Ex. **1.** Below the line (Ex. 2) **2.** Solid lines (Ex. 1) **3.** $3x + 2y > 6$, $x - 2y < 2$ (Ex. 2) **4.** See Student Solutions Manual.

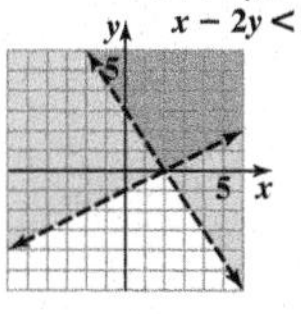

Career Exploration Problems

1. 1600 units **2.** 996 units **3.** 900 units

You Try It

1.

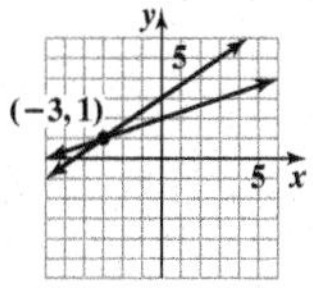

The solution is $(-3, 1)$. **2.** $(2, -1)$ **3.** $(10, 7)$ **4.** No solution **5.** Infinite number of solutions **6.** $(-4, -2, 5)$ **7.** $x - 3y < 6$, $2x + y > 5$

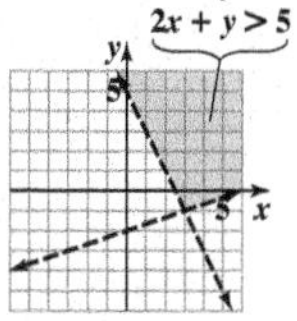

Chapter 4 Review Problems

1.

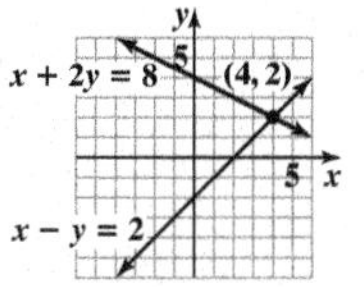

2.

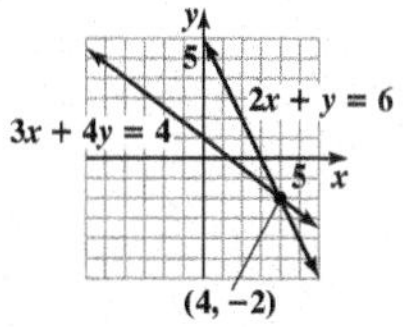

3. $(-1, 3)$ **4.** $(3, -2)$ **5.** $(-4, 0)$ **6.** $(1, -2)$ **7.** $(2, 3)$ **8.** $(5, -11)$ **9.** No solution; inconsistent system of equations **10.** Infinite number of solutions; dependent equations **11.** $(1, -5)$ **12.** $(0, 3)$ **13.** $\left(\frac{4}{3}, -\frac{1}{2}\right)$ **14.** $\left(0, \frac{2}{3}\right)$ **15.** No solution; inconsistent system of equations **16.** $(5, 2)$ **17.** $(1, 1, -2)$ **18.** $(1, -2, 3)$ **19.** $(5, -3, 8)$ **20.** $\left(7, \frac{1}{2}, -3\right)$ **21.** $(4, -2, 0)$ **22.** $(1, 2, -4)$ **23.** Speed of plane in still air = 264 mph; speed of wind = 24 mph **24.** 8 touchdowns; 2 field goals **25.** Laborers = 15; mechanics = 10 **26.** Children's tickets = 215; adult tickets = 115 **27.** Hats = \$7; shirts = \$15; pants = \$22 **28.** Jess scored 92; Nick scored 85; Chris scored 72 **29.** One jar of jelly = \$2.60; one jar of peanut butter = \$3.70; one jar of honey = \$2.20 **30.** Buses = 2; station wagons = 4; sedans = 3 **31.** $\left(\frac{4}{3}, \frac{1}{3}\right)$ **32.** $(-3, 2)$ **33.** $(0, 2)$ **34.** $(0, 12)$ **35.** $(2, 5)$

36. (11, 6) **37.** (3, 2) **38.** (−3, −2, 2) **39.** (3, −1, −2) **40.** (5, −5, −20)

41.

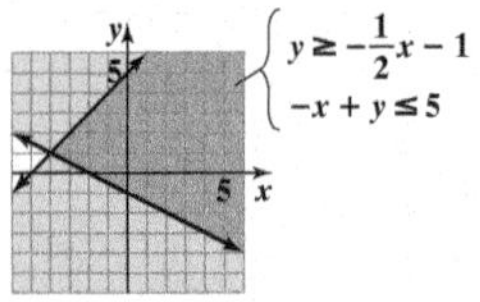

42.

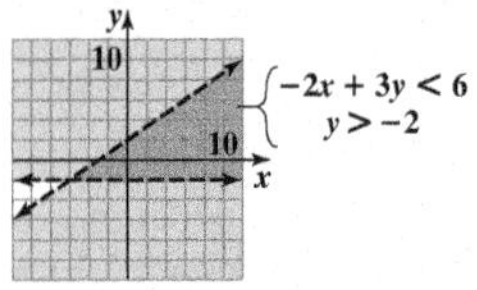

43.

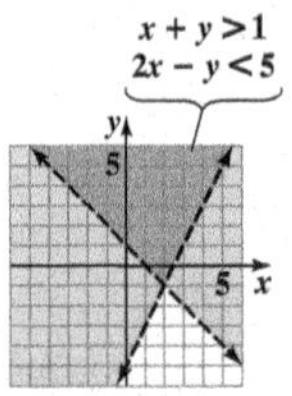

44.

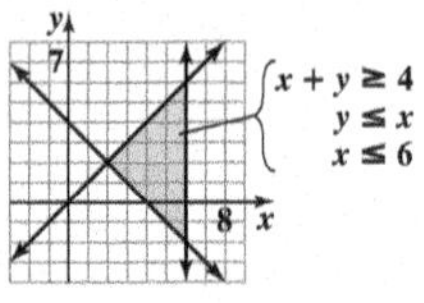

How Am I Doing? Chapter 4 Test

1. (10, 7) (obj. 4.1.3) **2.** (3, −4) (obj. 4.1.4) **3.** (3, 4) (obj. 4.1.6) **4.** $\left(\frac{1}{2}, \frac{3}{2}\right)$ (obj. 4.1.6) **5.** (1, 2) (obj. 4.1.6) **6.** No solution; inconsistent system of equations (obj. 4.1.5) **7.** (2, −1, 3) (obj. 4.2.2) **8.** (−2, 3, 5) (obj. 4.2.3) **9.** (−4, 1, −1) (obj. 4.2.2) **10.** Speed of plane in still air is 450 mph; speed of wind is 50 mph (obj. 4.3.1) **11.** Pen = \$1.50, mug = \$4.00, T-shirt = \$10.00 (obj. 4.3.2) **12.** \$30/day; \$0.20/mi (obj. 4.3.1) **13.**

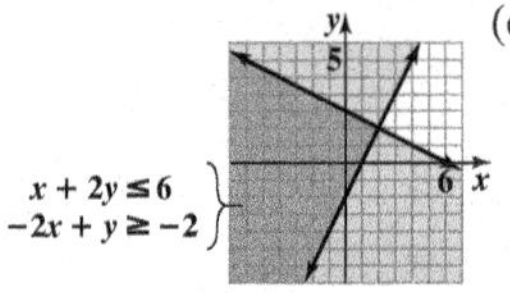

(obj. 4.4.1) **14.**

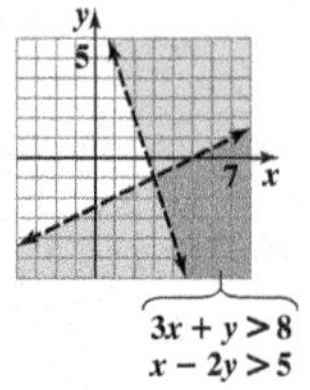

(obj. 4.4.1)

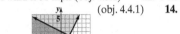

Chapter 5

5.1 Exercises

1. When multiplying exponential expressions with the same base, keep the base the same and add the exponents. **3.** $\frac{2^2}{2^3} = \frac{\not{2} \cdot \not{2}}{\not{2} \cdot \not{2} \cdot 2} = \frac{1}{2} = \frac{1}{2^{3-2}}$ **5.** 6; x and y; 11 and 1 **7.** 2^2a^3b **9.** $-5x^3y^2z^2$ **11.** 7^{10} **13.** 8^{21} **15.** x^{12} **17.** t^{16} **19.** $-20x^6$ **21.** $50x^3$ **23.** $18x^3y^8$ **25.** $\frac{2}{15}x^3y^5$ **27.** $-2.75x^3yz$ **29.** 0 **31.** $80x^3y^7$ **33.** $-24x^4y^7$ **35.** 0 **37.** $-24a^4b^3x^2y^5$ **39.** $-30x^3y^4z^6$ **41.** y^7 **43.** $\frac{1}{y^3}$ **45.** $\frac{1}{11^{12}}$ **47.** 2^7 **49.** $\frac{a^8}{4}$ **51.** $\frac{x^7}{y^9}$ **53.** $2x^4$ **55.** $-\frac{x^3}{2y^2}$ **57.** $\frac{f^2}{30g^5}$ **59.** $17x^4y^3$ **61.** $\frac{y^2}{16x^3}$ **63.** $\frac{3a}{4}$ **65.** x^{12} **67.** $x^{15}y^5$ **69.** r^6s^{12} **71.** $27a^9b^6c^3$ **73.** $9a^8$ **75.** $\frac{x^7}{128m^{28}}$ **77.** $\frac{25x^2}{49y^4}$ **79.** $81a^8b^{12}$ **81.** $-8x^9z^3$ **83.** $\frac{9}{x}$ **85.** $25a^5b^7$ **87.** $\frac{49}{a^{10}}$ **89.** $\frac{16x^4}{y^{12}}$ **91.** $\frac{7a^2c^3}{4b}$ **93.** $\frac{1}{3xy}$ **94.** −11 **95.** −46 **96.** $\frac{2}{25}$ **97.** −4 **98.** ≈64% **99.** 0.6%

Quick Quiz 5.1

See Examples noted with Ex. **1.** $-10x^3y^7$ (Ex. 3) **2.** $-\frac{4x^3}{5y^2}$ (Ex. 8) **3.** $81x^{12}y^{20}$ (Ex. 13c) **4.** See Student Solutions Manual

5.2 Exercises

1. $\frac{1}{x^4}$ **3.** $\frac{1}{81}$ **5.** y^8 **7.** $\frac{z^6}{x^4y^5}$ **9.** $\frac{a^3}{b^2}$ **11.** $\frac{x^9}{8}$ **13.** $\frac{3}{x^2}$ **15.** $\frac{1}{9x^2y^4}$ **17.** $\frac{3xz^3}{y^2}$ **19.** $4xy$ **21.** $\frac{b^3d}{ac^4}$ **23.** $\frac{1}{8}$ **25.** $\frac{z^8}{9y^4}$ **27.** $\frac{1}{x^6y}$ **29.** 1.2378×10^5 **31.** 6.3×10^{-2} **33.** 8.8961×10^{11} **35.** 3.42×10^{-6} **37.** 302,000 **39.** 0.00047 **41.** 983,000 **43.** 2.37×10^{-5} mph **45.** 149,600,000 km **47.** 6.3×10^{15} **49.** 1.0×10^1 **51.** 8.1×10^{-11} **53.** 4.5×10^5 **55.** 5.61×10^4 dollars **57.** 6.6×10^{-5} mi **59.** About 3.68×10^8 mi/yr **61.** 13.2% **63.** −0.8 **64.** −1 **65.** $-\frac{1}{28}$

Quick Quiz 5.2

See Examples noted with Ex. **1.** $\frac{3y^2}{x^3z^4}$ (Ex. 3c) **2.** $\frac{a^8}{2b}$ (Ex. 4b) **3.** 8.76×10^{-3} (Ex. 6) **4.** See Student Solutions Manual

5.3 Exercises

1. A polynomial in x is the sum of a finite number of terms of the form ax^n, where a is any real number and n is a whole number. An example is $3x^2 - 5x - 9$. **3.** The degree of a polynomial in x is the largest exponent of x in any of the terms of the polynomial. **5.** Degree 4; monomial **7.** Degree 5; trinomial **9.** Degree 6; binomial **11.** $-3x - 15$ **13.** $-2x^2 + 2x - 1$ **15.** $\frac{5}{6}x^2 + \frac{1}{2}x - 9$ **17.** $3.4x^3 + 2.2x^2 - 13.2x - 5.4$ **19.** $5x - 24$ **21.** $\frac{1}{15}x^2 - \frac{1}{14}x + 11$ **23.** $3x^3 - x^2 + 8x$ **25.** $-4.7x^4 - 0.7x^2 - 1.6x + 0.4$ **27.** $6x - 6$ **29.** $4x^2y^2 + 4xy - 10$ **31.** $x^4 - 3x^3 - 4x^2 - 24$ **33.** 743,000 **35.** 90,400 **37.** $3x^2 + 12x$ **39.** $m = \frac{8}{3}$; $\left(0, \frac{2}{3}\right)$ **40.**

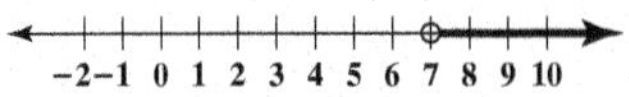

$x > 7$ **41.** $x = 7$ **42.** $x = 2$

Quick Quiz 5.3

See Examples noted with Ex. **1.** $-4x^2 - 11x + 5$ (Ex. 2) **2.** $6x^2 - 9x - 16$ (Ex. 5) **3.** $-3x + 8$ (Ex. 5) **4.** See Student Solutions Manual

Use Math to Save Money

1. (\$450 + \$425 + \$460)/3 = \$445 **2.** \$445 − (0.06 × \$445) − \$30 = \$388.30 **3.** \$388.30 − (\$388.30 × 0.2) = \$310.64 **4.** \$445 − \$310.64 = \$134.36; \$134.36/445 = 30.2% **5.** \$134.36 × 12 = \$1612.32

5.4 Exercises

1. $-12x^4 + 2x^2$ **3.** $24x^3 - 4x^2$ **5.** $-4x^6 + 10x^4 - 2x^3$ **7.** $x + \frac{3}{2}x^2 + \frac{5}{2}x^3$ **9.** $-2x^5y + 4x^4y - 5x^3y$ **11.** $9x^4y + 3x^3y - 24x^2y$ **13.** $3x^4 - 9x^3 + 15x^2 - 6x$ **15.** $-2x^3y^3 + 12x^2y^2 - 16xy$ **17.** $-28x^5y + 12x^4y + 8x^3y - 4x^2y$

19. $-6c^2d^5 + 8c^2d^3 - 12c^2d$ **21.** $12x^7 - 6x^5 + 18x^4 + 54x^3$ **23.** $-16x^6 + 10x^5 - 12x^4$ **25.** $x^2 + 12x + 35$ **27.** $x^2 + 8x + 12$ **29.** $x^2 - 6x - 16$ **31.** $x^2 - 9x + 20$ **33.** $-20x^2 - 7x + 6$ **35.** $2x^2 + 6xy - 5x - 15y$ **37.** $15x^2 - 5xy + 6x - 2y$ **39.** $20y^2 - 7y - 3$ **41.** $10x^4 + 23x^2y^3 + 12y^6$ **43.** The signs are incorrect. The result is $-3x + 6$. **45.** $20x$ **47.** $20x^2 - 23xy + 6y^2$ **49.** $49x^2 - 28x + 4$ **51.** $16a^2 + 16ab + 4b^2$ **53.** $0.8x^2 + 11.94x - 0.9$ **55.** $\frac{1}{4}x^2 + \frac{1}{24}x - \frac{1}{12}$ **57.** $6x^4 + 16x^2y^3 + 8y^6$ **59.** $10x^2 - 11x - 6$ **61.** $x = -10$ **62.** $w = -\frac{25}{7}$ or $-3\frac{4}{7}$ **63.** 9 twenties; 8 tens; 23 fives **64.** \$25 billion **65.** \$36.2 billion **66.** \$58.5 billion **67.** \$67.4 billion

Quick Quiz 5.4

See Examples noted with Ex. **1.** $8x^3y^4 - 12x^2y^3 + 16xy^2$ (Ex. 3) **2.** $6x^2 - x - 15$ (Ex. 4) **3.** $12a^2 - 26ab + 12b^2$ (Ex. 5) **4.** See Student Solutions Manual

5.5 Exercises

1. binomial **3.** The middle term is missing. The answer should be $16x^2 - 56x + 49$. **5.** $y^2 - 49$ **7.** $x^2 - 81$ **9.** $36x^2 - 25$ **11.** $4x^2 - 49$ **13.** $25x^2 - 9y^2$ **15.** $0.36x^2 - 9$ **17.** $4y^2 + 20y + 25$ **19.** $25x^2 - 40x + 16$ **21.** $49x^2 + 42x + 9$ **23.** $9x^2 - 42x + 49$ **25.** $\frac{4}{9}x^2 + \frac{1}{3}x + \frac{1}{16}$ **27.** $81x^2y^2 + 72xyz + 16z^2$ **29.** $49x^2 - 9y^2$ **31.** $9c^2 - 30cd + 25d^2$ **33.** $81a^2 - 100b^2$ **35.** $25x^2 + 90xy + 81y^2$ **37.** $x^3 - 4x^2 + 8x - 15$ **39.** $2x^4 + 7x^3 + x^2 + 7x + 4$ **41.** $a^4 + a^3 - 13a^2 + 17a - 6$ **43.** $3x^3 - 2x^2 - 25x + 24$ **45.** $2x^3 - x^2 - 16x + 15$ **47.** $2a^3 + 3a^2 - 50a - 75$ **49.** $24x^3 + 14x^2 - 11x - 6$ **51.** 19; 41 **52.** Width = 7 m; length = 10 m

Quick Quiz 5.5

See Examples noted with Ex. **1.** $49x^2 - 144y^2$ (Ex. 2) **2.** $6x^3 - x^2 - 19x - 6$ (Ex. 6) **3.** $15x^4 - 16x^3 - 8x^2 + 17x - 6$ (Ex. 5) **4.** See Student Solutions Manual

5.6 Exercises

1. $5x^3 - 3x + 4$ **3.** $y^3 + 7y - 3$ **5.** $9x^4 - 4x^2 - 7$ **7.** $8x^4 - 9x + 6$ **9.** $3x + 5$ **11.** $x - 3 - \frac{32}{x - 5}$ **13.** $3x^2 - 4x + 8 - \frac{10}{x + 1}$ **15.** $2x^2 - 3x - 2 - \frac{5}{2x + 5}$ **17.** $2x^2 + 3x - 1$ **19.** $2x^2 + 3x + 6 + \frac{23}{2x - 3}$ **21.** $y^2 - 4y - 1 - \frac{9}{y + 3}$ **23.** $y^3 + 2y^2 - 5y - 10 + \frac{-25}{y - 2}$ **25.** 2010: \$3.25; 2015: \$3.77 **26.** 170 and 171 **27.** **(a)** 7.5 **(b)** 7.3 **(c)** 6.3 **(d)** 2.7% decrease **(e)** 13.7% decrease

Quick Quiz 5.6

See Examples noted with Ex. **1.** $5x^3 - 16x^2 - 2x$ (Ex. 1) **2.** $4x^2 - 5x - 2$ (Ex. 4) **3.** $x^2 + 2x + 8 + \frac{13}{x - 2}$ (Ex. 3) **4.** See Student Solutions Manual

Career Exploration Problems

1. Transformation efficiency of first set: 1.4×10^6; transformation efficiency of second set: 3.28×10^4 **2.** **(a)** 50,100,000 **(b)** 0.00302 **(c)** $237 \times 10^6 = 237{,}000{,}000$ **3.** $0.25X^2 + 0.5Xy + 0.25y^2$

You Try It

1. **(a)** 2^{23} **(b)** $16a^8$ **(c)** $-3a^5b^4$ **2.** **(a)** $7x^3$ **(b)** $-\frac{1}{3x}$ **(c)** $\frac{b^6}{2a^2}$ **3.** **(a)** 1 **(b)** 1 **(c)** 1 **(d)** $6a$ **4.** **(a)** a^{20} **(b)** $4n^6$ **(c)** $\frac{27x^9}{y^3}$ **(d)** $25s^4t^{10}$ **(e)** $-a^{10}b^5$ **5.** **(a)** $\frac{1}{a^3}$ **(b)** x **(c)** $\frac{n^6}{m^9}$ **(d)** $\frac{1}{9}$ **6.** **(a)** 3.864×10^5 **(b)** 5.2×10^{-5} **7.** **(a)** 7.75×10^{10} **(b)** 3.04×10^4 **8.** $-6x^4 - 4x^3 + x^2$ **9.** $3 - 3x^2$ **10.** **(a)** $-6a^3 + 10a^2 - 2a$ **(b)** $-4x^3y + 12x^2y^2 - 12xy^3$ **11.** **(a)** $4a^2 - 25b^2$ **(b)** $4a^2 + 20ab + 25b^2$ **(c)** $4a^2 - 20ab + 25b^2$ **(d)** $6a^2 + 7ab - 5b^2$ **12.** **(a)** $12x^3 - 4x^2 + x + 3$ **(b)** $3x^3 - 17x^2 + 11x - 5$ **13.** $3x^3 + 14x^2 - 7x - 10$ **14.** $6a^2 - 3a + 1$ **15.** $x^2 + 5x - 1$

Chapter 5 Review Problems

1. $-18a^7$ **2.** 5^{23} **3.** $6x^4y^6$ **4.** $-14x^4y^9$ **5.** $\frac{1}{7^{12}}$ **6.** $\frac{1}{x^5}$ **7.** y^{14} **8.** $\frac{1}{9^{11}}$ **9.** $-\frac{3}{5x^5y^4}$ **10.** $-\frac{2a}{3b^6}$ **11.** x^{24} **12.** $\frac{2^4}{5^6b^{10}}$ **13.** $9a^6b^4$ **14.** $81x^{12}y^4$ **15.** $\frac{25a^2b^4}{c^6}$ **16.** $\frac{y^9}{64w^{15}z^6}$ **17.** $\frac{b^5}{a^3}$ **18.** $\frac{m^8}{p^5}$ **19.** $\frac{2y^3}{x^6}$ **20.** $\frac{x^{15}}{8y^3}$ **21.** $\frac{1}{36a^8b^{10}}$ **22.** $\frac{3y^2}{x^3}$ **23.** $\frac{4w^2}{x^5y^6z^8}$ **24.** $\frac{b^5c^3d^4}{27a^2}$ **25.** 1.563402×10^{11} **26.** 1.79632×10^5 **27.** 9.2×10^{-4} **28.** 1.74×10^{-6} **29.** 120,000 **30.** 6,034,000 **31.** 0.25 **32.** 0.0000432 **33.** 2.0×10^{13} **34.** 9.36×10^{19} **35.** 7.8×10^{-11} **36.** 7×10^8 dollars **37.** 7.94×10^{14} cycles **38.** 6×10^{12} operations **39.** $5.5x^2 - x - 2.3$ **40.** $7x^3 - 3x^2 - 6x + 4$ **41.** $\frac{1}{10}x^2y - \frac{13}{21}x + \frac{5}{12}$ **42.** $\frac{1}{4}x^2 - \frac{1}{4}x + \frac{1}{10}$ **43.** $-3x^2 - 15$ **44.** $15x^2 + 2x - 1$ **45.** $28x^2 - 29x + 6$ **46.** $20x^2 + 48x + 27$ **47.** $10x^3 - 30x^2 + 15x$ **48.** $-4x^2y^4 - 20x^2y^3 + 24xy^2$ **49.** $5a^2 - 8ab - 21b^2$ **50.** $8x^4 - 10x^2y - 12x^2 + 15y$ **51.** $16x^2 + 24x + 9$ **52.** $a^2 - 25b^2$ **53.** $49x^2 - 36y^2$ **54.** $25a^2 - 20ab + 4b^2$ **55.** $4x^3 + 27x^2 + 5x - 3$ **56.** $2x^3 - 7x^2 - 42x + 72$ **57.** $2y^2 + 3y + 4$ **58.** $6x^3 + 7x^2 - 18x$ **59.** $4x^2 - 6x + 8$ **60.** $3x - 2$ **61.** $3x - 7$ **62.** $2x^2 - 5x + 13 - \frac{27}{x + 2}$ **63.** $3x - 4 + \frac{3}{2x + 3}$ **64.** $x^2 + 3x + 8$ **65.** $2x^2 + 4x + 5 + \frac{11}{x - 2}$ **66.** About \$14.52 **67.** 1.847×10^9 **68.** 2.733×10^{-23} g **69.** 4.76×10^7 lb **70.** $3xy + 2x$ **71.** $2x^2 - 4y^2$

How Am I Doing? Chapter 5 Test

1. 3^{34} (obj. 5.1.1) **2.** $\frac{1}{25^{16}}$ (obj. 5.1.2) **3.** 8^{24} (obj. 5.1.3) **4.** $12x^4y^{10}$ (obj. 5.1.1) **5.** $-\frac{7x^3}{5}$ (obj. 5.1.2) **6.** $-125x^3y^{18}$ (obj. 5.1.3) **7.** $\frac{49a^{14}b^4}{9}$ (obj. 5.1.3) **8.** $\frac{3x^4}{4}$ (obj. 5.1.3) **9.** $\frac{1}{64}$ (obj. 5.2.1) **10.** $\frac{6c^5}{a^4b^3}$ (obj. 5.2.1) **11.** $3xy^7$ (obj. 5.2.1) **12.** 5.482×10^{-4} (obj. 5.2.2) **13.** 582,000,000 (obj. 5.2.2) **14.** 2.4×10^{-6} (obj. 5.2.2) **15.** $-2x^2 + 5x$ (obj. 5.3.2) **16.** $-11x^3 - 4x^2 + 7x - 8$ (obj. 5.3.3) **17.** $-21x^5 + 28x^4 - 42x^3 + 14x^2$ (obj. 5.4.1) **18.** $15x^4y^3 - 18x^3y^2 + 6x^2y$ (obj. 5.4.1) **19.** $10a^2 + 7ab - 12b^2$ (obj. 5.4.2) **20.** $6x^3 - 11x^2 - 19x - 6$ (obj. 5.5.3) **21.** $49x^4 + 28x^2y^2 + 4y^4$ (obj. 5.5.2) **22.** $25s^2 - 121t^2$ (obj. 5.5.1) **23.** $12x^4 - 14x^3 + 25x^2 - 29x + 10$ (obj. 5.5.3) **24.** $3x^4 + 4x^3y - 15x^2y^2$ (obj. 5.4.2) **25.** $3x^3 - x + 5$ (obj. 5.6.1) **26.** $2x^2 - 7x + 4$ (obj. 5.6.2) **27.** $2x^2 + 6x + 12$ (obj. 5.6.2) **28.** 3.77×10^9 barrels per year (obj. 5.2.2) **29.** 4.18×10^6 mi (obj. 5.2.2)

Chapter 6

6.1 Exercises **1.** factors **3.** No; $6a^3 + 3a^2 - 9a$ still has a common factor of $3a$. **5.** $8a(a + 1)$ **7.** $7ab(3 - 2b)$ **9.** $2\pi r(h + r)$ **11.** $5x(x^2 + 5x - 3)$ **13.** $4(3ab - 7bc + 5ac)$ **15.** $8x^2(2x^3 + 3x - 4)$ **17.** $7x(2xy - 5y - 9)$ **19.** $9x(6x - 5y + 2)$ **21.** $y(3xy - 2a + 5x - 2)$ **23.** $8xy(3x - 5y)$ **25.** $7x^2y^2(x + 3)$ **27.** $8x^2y(2x^2y - 3y - 1)$ **29.** $(x + 2y)(7a - b)$ **31.** $(x - 4)(3x - 2)$ **33.** $(2a - 3c)(6b - 5d)$ **35.** $(b - a^2)(7c - 5d + 2f)$ **37.** $(ab - 4)(3a - 5 - b)$ **39.** $(a - 3b)(4a^3 + 1)$ **41.** $(a + 2)(1 - x)$ **43.** $2\pi(x + y + z)$ **45.** 1,664,000 metric tons **46.** 2,752,000 metric tons **47.** About 41 lb/person **48.** About 30 lb/person

Quick Quiz 6.1 *See Examples noted with Ex.* **1.** $x(3 - 4x + 2y)$ (Ex. 2) **2.** $5x(4x^2 - 5x - 1)$ (Ex. 5) **3.** $(a + 3b)(8a - 7b)$ (Ex. 6) **4.** See Student Solutions Manual

6.2 Exercises **1.** We must remove a common factor of 5 from the last two terms. This will give us $3x(x - 2y) + 5(x - 2y)$. Then our final answer is $(x - 2y)(3x + 5)$. **3.** $(b - 4)(a + 6)$ **5.** $(x - 4)(x^2 + 3)$ **7.** $(a + 3b)(2x - y)$ **9.** $(3a + b)(x - 2)$ **11.** $(a + 2b)(5 + 6c)$ **13.** $(c - 2d)(6 + x)$ **15.** $(y - 2)(y - 3)$ **17.** $(9 - y)(6 + y)$ **19.** $(3x + y)(2a - 1)$ **21.** $(x + 4)(2x - 3)$ **23.** $(t - 1)(t^2 + 1)$ **25.** $(2x + 5y^2)(3x + 4w)$ **27.** We must rearrange the terms in a different order so that the expression in parentheses is the same in each case. We use the order $6a^2 - 8ad + 9ab - 12bd$ to factor $2a(3a - 4d) + 3b(3a - 4d) = (3a - 4d)(2a + 3b)$. **29.** $-\frac{15}{7}$ **30.** $\frac{2}{15}$ **31.** $-\frac{a}{5b^2}$ **32.** $4x^2 - 20x + 25$ **33.** Pharmacist: \$127,000; pharmacy technician: \$35,000 **34.** About 21,323.1 thousand barrels of oil each day

Quick Quiz 6.2 *See Examples noted with Ex.* **1.** $(7x + 12)(a - 2)$ (Ex. 5) **2.** $(y^2 + 3)(2x - 5)$ (Ex. 7) **3.** $(10y - 3)(x + 4b)$ (Ex. 6) **4.** See Student Solutions Manual

6.3 Exercises **1.** product, sum **3.** $(x + 4)(x + 4)$ **5.** $(x + 5)(x + 7)$ **7.** $(x - 3)(x - 1)$ **9.** $(x - 7)(x - 4)$ **11.** $(x + 8)(x - 3)$ **13.** $(x - 14)(x + 1)$ **15.** $(x + 7)(x - 5)$ **17.** $(x - 6)(x + 4)$ **19.** $(x + 12)(x + 3)$ **21.** $(x - 6)(x - 4)$ **23.** $(x + 3)(x + 10)$ **25.** $(x - 5)(x - 1)$ **27.** $(a + 8)(a - 2)$ **29.** $(x - 4)(x - 8)$ **31.** $(x + 7)(x - 3)$ **33.** $(x + 7)(x + 8)$ **35.** $(y + 9)(y - 5)$ **37.** $(x + 12)(x - 3)$ **39.** $(x + 3y)(x - 5y)$ **41.** $(x - 7y)(x - 9y)$ **43.** $4(x + 5)(x + 1)$ **45.** $6(x + 1)(x + 2)$ **47.** $5(x - 1)(x - 5)$ **49.** $3(x + 4)(x - 6)$ **51.** $7(x + 5)(x - 2)$ **53.** $5(x - 1)(x - 6)$ **55.** $(12 + x)(10 - x)$ **57.** $18a^6b^9$ **58.** $25y^{12}$ **59.** $\frac{x^6}{y^8}$ **60.** $8x^2 + 8xy - 6y^2$ **61.** 120 mi **62.** \$3800 **63.** 25°C **64.** June

Quick Quiz 6.3 *See Examples noted with Ex.* **1.** $(x + 7)(x + 10)$ (Ex. 2) **2.** $(x - 6)(x - 8)$ (Ex. 3) **3.** $2(x + 6)(x - 8)$ (Ex. 10) **4.** See Student Solutions Manual

6.4 Exercises **1.** $(4x + 1)(x + 5)$ **3.** $(5x + 2)(x + 1)$ **5.** $(4x - 3)(x + 2)$ **7.** $(2x + 1)(x - 3)$ **9.** $(3x + 1)(3x + 2)$ **11.** $(3x - 5)(5x - 3)$ **13.** $(2x - 5)(x + 4)$ **15.** $(4x - 1)(2x + 3)$ **17.** $(3x + 2)(2x - 3)$ **19.** $(5x - 1)(2x + 1)$ **21.** $(x - 2)(7x + 9)$ **23.** $(9y - 4)(y - 1)$ **25.** $(5a + 2)(a - 3)$ **27.** $(7x - 2)(2x + 3)$ **29.** $(5x - 2)(3x + 2)$ **31.** $(6x + 5)(2x + 3)$ **33.** $(6x + 1)(2x - 3)$ **35.** $(3x^2 + 1)(x^2 - 5)$ **37.** $(2x + 5y)(x + 3y)$ **39.** $(5x - 4y)(x + 4y)$ **41.** $5(2x + 1)(x - 3)$ **43.** $3x(2x - 5)(x + 4)$ **45.** $(5x - 2)(x + 1)$ **47.** $2(3x - 2)(2x - 5)$ **49.** $x(6x - 1)(2x - 3)$ **51.** $2(2x - 1)(2x + 7)$ **53.** $m = -\frac{2}{3}$ **54.** $x = \frac{7}{2}$ **55. (a)** 17,600,000 **(b)** Approximately 10.8% **56.** Approximately 27.4% **57. (a)** 1.8 million **(b)** Approximately 28.1% **58. (a)** 2.1 million **(b)** Approximately 35.6%

Quick Quiz 6.4 *See Examples noted with Ex.* **1.** $(2x + 3)(6x - 1)$ (Ex. 3) **2.** $(5x - 3)(2x - 3)$ (Ex. 4) **3.** $3x(2x - 5)(x + 2)$ (Ex. 7) **4.** See Student Solutions Manual

Use Math to Save Money **1.** \$5100 + \$3800 + \$3200 = \$12,100 **2.** \$5500 + \$4000 + \$3500 = \$13,000 **3.** \$12,100/\$13,000 ≈ 93% **4.** \$13,000 × 0.5 = \$6500 **5.** \$12,100 − \$6500 = \$5600 **6.** \$13,000 × 0.33 = \$4290, \$6500 − \$4290 = \$2210

6.5 Exercises **1.** $(10x + 1)(10x - 1)$ **3.** $(9x + 4)(9x - 4)$ **5.** $(x + 7)(x - 7)$ **7.** $(5x + 9)(5x - 9)$ **9.** $(x + 5)(x - 5)$ **11.** $(1 + 4x)(1 - 4x)$ **13.** $(4x + 7y)(4x - 7y)$ **15.** $(6x + 13y)(6x - 13y)$ **17.** $(10x + 9)(10x - 9)$ **19.** $(5a + 9b)(5a - 9b)$ **21.** $(3x + 1)^2$ **23.** $(y - 5)^2$ **25.** $(6x - 5)^2$ **27.** $(7x + 2)^2$ **29.** $(x + 7)^2$ **31.** $(5x - 4)^2$ **33.** $(9x + 2y)^2$ **35.** $(3x - 5y)^2$ **37.** $(4a + 9b)^2$ **39.** $(7x - 3y)^2$ **41.** $(8x + 5)^2$ **43.** $(12x + 1)(12x - 1)$ **45.** $(x^2 + 4)(x + 2)(x - 2)$ **47.** $(3x^2 - 4)^2$ **49.** No two binomials can be multiplied to obtain $9x^2 + 1$. **51.** 9; one answer **53.** $4(2x + 3)(2x - 3)$ **55.** $3(7x + y)(7x - y)$ **57.** $4(2x - 1)^2$ **59.** $2(7x + 3)^2$ **61.** $(x + 9)(x + 7)$ **63.** $(2x - 1)(x + 3)$ **65.** $3(2x + 3)(2x - 3)$ **67.** $(3x + 7)^2$ **69.** $9(2x - 1)^2$ **71.** $2(x - 9)(x - 7)$ **73.** $x^2 + 3x + 4 + \frac{-3}{x - 2}$ **74.** $2x^2 + x - 5$ **75.** 1.2 oz of greens; 1.05 oz of bulk vegetables; 0.75 oz of fruit **76.** 1.44 oz of greens; 1.26 oz of bulk vegetables; 0.9 oz of fruit

Quick Quiz 6.5 *See Examples noted with Ex.* **1.** $(7x + 9y)(7x - 9y)$ (Ex. 3) **2.** $(3x - 8)^2$ (Ex. 6) **3.** $2(9x + 10)(9x - 10)$ (Ex. 9) **4.** See Student Solutions Manual

6.6 Exercises **1.** $x(3x - 6y + 5)$ **3.** $(4x + 5y)(4x - 5y)$ **5.** Prime **7.** $(x + 5)(x + 3)$ **9.** $(3x + 2)(5x - 1)$ **11.** $(a - 3c)(x + 3y)$ **13.** $(y + 7)^2$ **15.** $(2x - 3)^2$ **17.** $(2x - 3)(x - 4)$ **19.** $(x - 10y)(x + 7y)$ **21.** $(a + 3)(x - 5)$ **23.** $4x(2 + x)(2 - x)$ **25.** Prime **27.** $3xy(z + 1)(z - 3)$ **29.** $3(x + 7)(x - 5)$ **31.** $5xy^3(x - 1)^2$ **33.** $-1(2x^2 + 1)(x + 2)(x - 2)$ **35.** Prime **37.** $5x(x + 2y)(x + 1)(x - 1)$ **39.** $5(x^2 + 2xy - 6y)$ **41.** $3x(2x + y)(5x - 2y)$ **43.** $2(5x - 3)(3x - 2)$ **45.** prime **47.** \$28,000 **48.** 372 live strains **49.** $(4, -5)$ **50.**

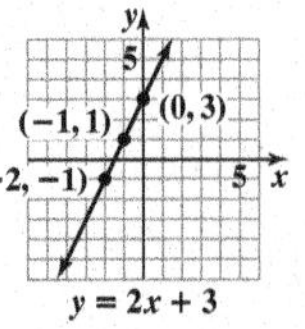

(−1, 1)
(0, 3)
(−2, −1)
$y = 2x + 3$

Quick Quiz 6.6 *See Examples noted with Ex.* **1.** $(2x - 3)(3x - 4)$ (Ex. 1b) **2.** $3(4x + 1)(5x - 2)$ (Ex. 1b) **3.** Prime (Ex. 3) **4.** See Student Solutions Manual

6.7 Exercises **1.** $x = -2, x = 6$ **3.** $x = -12, x = -2$ **5.** $x = \frac{3}{2}, x = 2$ **7.** $x = \frac{2}{3}, x = \frac{3}{2}$ **9.** $x = 0, x = -13$ **11.** $x = 3, x = -3$ **13.** $x = 0, x = 1$ **15.** $x = \frac{2}{3}, x = 2$ **17.** $x = -3, x = 2$ **19.** $x = -\frac{1}{3}$ **21.** $x = 0, x = -5$ **23.** $x = -8, x = -2$ **25.** $x = -\frac{3}{2}, x = 4$ **27.** You can always factor out x. **29.** $L = 14$ m; $W = 10$ m **31.** 66 groups **33.** 10 students **35.** 3 sec; 12 m above ground after 2 sec **37.** 2415 telephone calls **39.** 136 handshakes **41.** $-10x^5y^4$ **42.** $12a^{10}b^{13}$ **43.** $-\frac{3a^4}{2b^2}$ **44.** $\frac{1}{3x^5y^4}$

Quick Quiz 6.7 *See Examples noted with Ex.* **1.** $x = \frac{1}{5}, x = \frac{1}{3}$ (Ex. 1) **2.** $x = 3, x = -1$ (Ex. 4) **3.** $x = 3, x = -\frac{3}{4}$ (Ex. 4) **4.** See Student Solutions Manual

Career Exploration Problems **1.** Rectangular piece: width is 7 ft and length is 18 ft. Triangular piece: base is 5 ft and height is 4 ft. **2.** Base is 12 ft and height is 22 ft. **3.** $\frac{44}{3}$ or $14\frac{2}{3}$ yd^2

You Try It **1. (a)** $5a(a - 3)$ **(b)** $4x(x - 2y + 1)$ **(c)** $6x^2(x^2 - 3)$ **2.** $(3x^2 + 2)(a - 4)$ **3. (a)** $(x + 3)(x + 6)$ **(b)** $(x + 7)(x - 5)$ **(c)** $3(x + 1)(x - 4)$ **4.** $(2x + 3)(4x - 3)$ **5. (a)** $(3x + 4y)(3x - 4y)$ **(b)** $(9x^2 + 1)(3x + 1)(3x - 1)$ **(c)** $(4a + 3)^2$ **(d)** $(2x - 5y)^2$ **6. (a)** $4(x + 3)(x - 2)$ **(b)** $x(3x + 1)(x + 2)$ **(c)** $x(3x + 8)(3x - 8)$ **(d)** $3(4x - 1)^2$ **7. (a)** This is a sum of squares. **(b)** There are no two factors of 2 that add to 1. **8.** $x = \frac{3}{2}$ or $x = -1$ **9.** Length = 15 ft; width = 6 ft

Chapter 6 Review Problems **1.** $4x^2(3x - 5y)$ **2.** $5x^3(2 - 7y)$ **3.** $8x^2y(3x - y - 2xy^2)$ **4.** $3a(a^2 + 2a - 3b + 4)$ **5.** $(a + 3b)(2a - 5)$ **6.** $3xy(5x^2 + 2y + 1)$ **7.** $(2x + 5)(a - 4)$ **8.** $(a - 4b)(a + 7)$ **9.** $(x^2 + 3)(y - 2)$ **10.** $3(2x - y)(5a + 7)$ **11.** $(5x - 1)(3x + 2)$ **12.** $(5w - 3)(6w + z)$ **13.** $(x + 9)(x - 3)$ **14.** $(x + 10)(x - 1)$ **15.** $(x + 6)(x + 8)$ **16.** $(x + 3y)(x + 5y)$ **17.** $(x^2 + 7)(x^2 + 6)$ **18.** $(x^2 - 7)(x^2 + 5)$ **19.** $6(x + 2)(x + 3)$ **20.** $2(x - 6)(x - 8)$ **21.** $(4x - 5)(x + 3)$ **22.** $(3x - 1)(4x + 5)$ **23.** $(2x - 3)(x + 1)$ **24.** $(3x - 4)(x + 2)$ **25.** $(10x - 1)(2x + 5)$ **26.** $(5x - 1)(4x + 5)$ **27.** $2(x - 1)(3x + 5)$ **28.** $2(x + 1)(3x - 5)$ **29.** $2(2x - 3)(x - 5)$ **30.** $4(x - 9)(x + 4)$ **31.** $(4x + 3y)(3x - 2y)$ **32.** $(3x - 5y)(2x + 5y)$ **33.** $(7x + y)(7x - y)$ **34.** $4(2x + 3y)(2x - 3y)$ **35.** $(y + 6x)(y - 6x)$ **36.** $(3y + 5x)(3y - 5x)$ **37.** $(6x + 1)^2$ **38.** $(5x - 2)^2$ **39.** $(4x - 3y)^2$ **40.** $(7x - 2y)^2$ **41.** $2(x + 4)(x - 4)$ **42.** $3(x + 3)(x - 3)$ **43.** $7(2x + 5)^2$ **44.** $8(3x - 4)^2$ **45.** $(2x + 3y)(2x - 3y)$ **46.** $(x + 15)(x - 2)$ **47.** $(3x - 4)(3x + 1)$ **48.** $10x^2y^2(5x + 2)$ **49.** $3(x - 3)^2$ **50.** $x(5x - 6)^2$ **51.** $(4x + 3)(x - 4)$ **52.** $x^3a(3a + 4x)(a - 5x)$ **53.** $2(3a + 5b)(2a - b)$ **54.** $(11a + 3b)^2$ **55.** $(a - 1)(7 - b)$ **56.** $(3x + 5y)(x + 1)(x - 1)$ **57.** $(3b - 7)(6 + c)$ **58.** $(5b + 8)(2 - 3x)$ **59.** $(b - 7)(5x + 4y)$ **60.** $(x^2 + 9y^6)(x + 3y^3)(x - 3y^3)$ **61.** $(3x^2 - 5)(2x^2 + 3)$ **62.** $yz(14 - x)(2 - x)$ **63.** $x(3x + 2)(4x + 3)$ **64.** $3(2w - 1)^2$ **65.** $2y(2y - 1)(y + 3)$ **66.** $9(x^2 + 4)(x - 2)(x + 2)$ **67.** Prime **68.** Prime **69.** $4y(2y^2 - 5)(y^2 + 3)$ **70.** $(4x^2y - 7)^2$ **71.** $(2x + 5)(a - 2b)$ **72.** $(2x + 1)(x + 3)(x - 3)$ **73.** $x = -5, x = 4$ **74.** $x = -6, x = \frac{1}{2}$ **75.** $x = 0, x = \frac{5}{2}$ **76.** $x = 4, x = -3$ **77.** $x = -5, x = \frac{1}{2}$ **78.** $x = -8, x = -3$ **79.** $x = -5, x = -9$ **80.** $x = -\frac{3}{5}, x = 2$ **81.** $x = -3$ **82.** $x = -3, x = \frac{3}{4}$ **83.** $x = \frac{1}{5}, x = 2$ **84.** Base = 5 in.; altitude = 10 in. **85.** Width = 5 ft; length = 6 ft **86.** 6 sec **87.** 8 amperes, 12 amperes

How Am I Doing? Chapter 6 Test **1.** $(x + 14)(x - 2)$ (obj. 6.3.1) **2.** $(4x + 9)(4x - 9)$ (obj. 6.5.1) **3.** $(5x + 1)(2x + 5)$ (obj. 6.4.2) **4.** $(3a - 5)^2$ (obj. 6.5.2) **5.** $x(7 - 9x + 14y)$ (obj. 6.1.1) **6.** $(2x + 3b)(5y - 4)$ (obj. 6.2.1) **7.** $2x(3x - 4)(x - 2)$ (obj. 6.6.1) **8.** $c(5a - 1)(a - 2)$ (obj. 6.6.1) **9.** $(9x + 10)(9x - 10)$ (obj. 6.5.1) **10.** $(3x - 1)(3x - 4)$ (obj. 6.6.1) **11.** $5(2x + 3)(2x - 3)$ (obj. 6.6.1) **12.** Prime (obj. 6.6.2) **13.** $x(3x + 5)(x + 2)$ (obj. 6.6.1) **14.** $-5y(2x - 3y)^2$ (obj. 6.5.3) **15.** $(9x + 1)(9x - 1)$ (obj. 6.5.1) **16.** $(9y^2 + 1)(3y + 1)(3y - 1)$ (obj. 6.6.1) **17.** $(x + 3)(2a - 5)$ (obj. 6.6.1) **18.** $(a + 2b)(w + 2)(w - 2)$ (obj. 6.6.1) **19.** $3(x - 6)(x + 5)$ (obj. 6.6.1) **20.** $x(2x + 5)(x - 3)$ (obj. 6.6.1) **21.** $x = -5, x = -9$ (obj. 6.7.1) **22.** $x = -\frac{7}{3}, x = -2$ (obj. 6.7.1) **23.** $x = -\frac{5}{2}, x = 2$ (obj. 6.7.1) **24.** $x = 7, x = -4$ (obj. 6.7.1) **25.** Width = 7 mi; length = 13 mi (obj. 6.7.2)

Cumulative Test for Chapters 0–6 **1.** $\frac{1}{30}$ **2.** $-\frac{1}{3}$ **3.** $-\frac{6}{7}$ **4.** 110.55 **5.** 12,400 employees **6.** 66 **7.** $2x^2 - 13x$ **8.** $x = 15$ **9.** $x = -\frac{9}{2}$ **10.** $x < -3$ **11.** (24, 8) **12.** (5, 1) **13.** 1 mi/hr **14.** $m = \frac{1}{10}$ **15.** $y = -\frac{3}{4}x + \frac{13}{2}$ **16.**

y
5
5 x
$4x - 8y = 10$

17. $20x^2 - 21x - 5$ **18.** $9x^2 - 30x + 25$ **19.** $-20x^5y^8$ **20.** $16x^{12}y^8$ **21.** $\frac{9z^8}{w^2x^3y^4}$ **22.** 5.6×10^{-4} **23.** $(11x + 8y)(11x - 8y)$ **24.** $-4(5x + 6)(4x - 5)$ **25.** $x(4x + 5)^2$ **26.** $(4x^2 + b^2)(2x + b)(2x - b)$ **27.** $(2x + 3)(a - 2b)$ **28.** $x = -9, x = 4$

Chapter 7

7.1 Exercises **1.** 4 **3.** $\frac{6}{x}$ **5.** $\frac{3x+1}{1-3x}$ **7.** $\frac{a(a-2b)}{2b}$ **9.** $\frac{x+2}{x}$ **11.** $\frac{x-5}{3x-1}$ **13.** $\frac{x-3}{x(x-7)}$ **15.** $\frac{3x+1}{x+5}$ **17.** $\frac{3x-5}{4x-1}$ **19.** $\frac{x-5}{x+1}$ **21.** $-\frac{3}{5x}$ **23.** $-\frac{2x+3}{5+x}$ **25.** $\frac{3x+4}{3x-1}$ **27.** $\frac{3x-2}{4-x}$ **29.** $\frac{a-2b}{3a-b}$ **31.** $9x^2-42x+49$ **32.** $49x^2-36y^2$ **33.** $2x^2-5x-12$ **34.** $2x^3-9x^2-2x+24$ **35.** $\frac{23a^2}{7}+\frac{3b}{4}$ **36.** $-\frac{49}{6}$ or $-8\frac{1}{6}$ **37.** $1\frac{5}{8}$ acres **38.** 6 hr, 25 min

Quick Quiz 7.1 *See Examples noted with Ex.* **1.** $\frac{x}{x-5}$ (Ex. 4) **2.** $-\frac{2}{b}$ (Ex. 5) **3.** $\frac{2x-1}{4x+5}$ (Ex. 3) **4.** See Student Solutions Manual

7.2 Exercises **1.** factor the numerators and denominators completely and divide out common factors **3.** $\frac{4(x+5)}{x+3}$ **5.** $\frac{3x^2}{4(x-3)}$ **7.** $\frac{x-2}{x+3}$ **9.** $\frac{(x+6)(x+2)}{x+5}$ **11.** $x+3$ **13.** $\frac{3(x+2y)}{4(x+3y)}$ **15.** $\frac{(x+5)(x-2)}{3x-1}$ **17.** $\frac{-5(x+3)}{2(x-2)}$ or $\frac{5(x+3)}{-2(x-2)}$ or $-\frac{5(x+3)}{2(x-2)}$ **19.** 1 **21.** $\frac{x+4}{x+8}$ **23.** $x=-8$ **24.** $\frac{19}{28}$ **25.** $79.5(8981)\$x = \$713{,}989.5x$ **26.** Harold's was 6 ft by 6 ft. George's was 9 ft by 4 ft.

Quick Quiz 7.2 *See Examples noted with Ex.* **1.** $\frac{2(x+4)}{x-4}$ (Ex. 1) **2.** $\frac{x-25}{x+5}$ (Ex. 1) **3.** $\frac{2(x-3)}{3x}$ (Ex. 3)
4. See Student Solutions Manual

7.3 Exercises **1.** The LCD would be a product that contains each factor. However, any repeated factor in any one denominator must be repeated the greatest number of times it occurs in any one denominator. So the LCD would be $(x+5)(x+3)^2$. **3.** $\frac{2(2x+1)}{2x+5}$ **5.** $\frac{2x-5}{x+3}$ **7.** $\frac{2x-7}{5x+7}$ **9.** $3a^2b^3$ **11.** $90x^3y^5$ **13.** $18(x-3)$ **15.** $(x+3)(x-3)$ **17.** $(x+5)(3x-1)^2$ **19.** $\frac{7+3a}{ab}$ **21.** $\frac{3x-13}{(x+7)(x-7)}$ **23.** $\frac{y(5y-3)}{(y+1)(y-1)}$ **25.** $\frac{43a+12}{5a(3a+2)}$ **27.** $\frac{4z+x}{6xyz}$ **29.** $\frac{9x+14}{2(x-3)}$ **31.** $\frac{x+10}{(x+5)(x-5)}$ **33.** $\frac{3a+17b}{10}$ **35.** $\frac{-2(2x-17)}{(2x-3)(x+2)}$ **37.** $\frac{-7x}{(x+3)(x-1)(x-4)}$ **39.** $\frac{4x+23}{(x+4)(x+5)(x+6)}$ **41.** $\frac{4x-13}{(x-4)(x-3)}$ **43.** $\frac{11x}{y-2x}$ **45.** $\frac{-y(17y+10)}{(4y-1)(2y+1)(y-5)}$ **47.** $\frac{8x+3}{(2x+1)(x-3)}$ **49.** $x=-7$ **50.** $x=4$ **51.** $x>\frac{5}{7}$ **52.** $81x^{12}y^{16}$ **53.** At least 17 days **54.** 287,985 people

Quick Quiz 7.3 *See Examples noted with Ex.* **1.** $\frac{2x+7}{(x+2)(x-4)}$ (Ex. 6) **2.** $\frac{bx+by+xy}{bxy}$ (Ex. 8) **3.** $\frac{5x-1}{(x+3)(x-3)(x+4)}$ (Ex. 7)
4. See Student Solutions Manual

Use Math to Save Money **1.** $\$500 \times 0.15 = \$75$ **2.** $\$500 \times 0.25 = \$125$ **3.** $\$500/\$5000 = 10\%$ **4.** $\$9 \times 48 = \$432$

7.4 Exercises **1.** $\frac{5x}{4x+3}$ **3.** $\frac{4b+a}{5}$ **5.** $\frac{x(x-2)}{4+5x}$ **7.** $\frac{2}{3x+10}$ **9.** $\frac{1}{xy}$ **11.** $\frac{2x-1}{x}$ **13.** $\frac{x-6}{x+5}$ **15.** $\frac{3(a^2+3)}{a^2+2}$ **17.** $\frac{3(x+3)}{2x-5}$ **19.** $\frac{4x-1}{7-4x}$ **21.** No expression in any denominator can be zero because division by zero is undefined. So −3, 5, and 0 are not allowed. **23.** $\frac{5y(x+5y)^2}{x(x-6y)}$ **25.** $m=\frac{-5}{6}$; $\left(0, \frac{4}{3}\right)$ **26.** $x>-1$  **27.** 6 **28.** \$24,000

Quick Quiz 7.4 *See Examples noted with Ex.* **1.** $\frac{a(3a-4b)}{3(5a-16b)}$ (Ex. 2) **2.** ab (Ex. 2) **3.** $\frac{2}{x-1}$ (Ex. 4)
4. See Student Solutions Manual

7.5 Exercises **1.** $x=-12$ **3.** $x=2$ **5.** $x=15$ **7.** $x=10$ **9.** $x=-5$ **11.** $x=-\frac{14}{3}$ or $-4\frac{2}{3}$ **13.** $x=-2$ **15.** $x=8$ **17.** $x=-1$ **19.** No solution **21.** $x=-3$ **23.** $x=-9$ **25.** No solution **27.** $x=4$ **29.** No solution **31.** $x=-6, x=-\frac{1}{3}$ **33.** $(4x+1)(2x-1)$ **34.** $x=\frac{5}{2}$ or $2\frac{1}{2}$ or 2.5 **35.** Width = 10 in.; length = 12 in. **36.** Domain = $\{7, 2, -2\}$; Range = $\{3, 2, 0, -2, -3\}$; Not a function

Quick Quiz 7.5 *See Examples noted with Ex.* **1.** $x=\frac{5}{24}$ (Ex. 1) **2.** $x=2$ (Ex. 2) **3.** No solution (Ex. 4)
4. See Student Solutions Manual

7.6 Exercises **1.** $x=\frac{88}{5}$ or $17\frac{3}{5}$ or 17.6 **3.** $x=\frac{204}{5}$ or $40\frac{4}{5}$ or 40.8 **5.** $x=6.5$ **7.** $x=\frac{91}{4}$ or $22\frac{3}{4}$ or 22.75 **9. (a)** 650 New Zealand dollars **(b)** 75 New Zealand dollars less **11.** 56 mph **13.** 29 mi **15.** $n=\frac{377}{20}$ or $18\frac{17}{20}$ in. **17.** $y=200$ m **19.** $k=\frac{56}{5}$ or $11\frac{1}{5}$ ft **21.** $k=\frac{768}{20}$ or $38\frac{2}{5}$ m **23.** 48 in. **25.** 35 in. **27.** 61.5 mph **29.** Commuter airliner = 250 km/hr; Helicopter = 210 km/hr **31. (a)** \$0.17 **(b)** \$0.14 **(c)** \$5.48 **33.** $3\frac{3}{7}$ hr or 3 hr, 26 min **35.** 8.92465×10^{-4} **36.** 6,830,000,000 **37.** $\frac{w^8}{x^3y^2z^4}$ **38.** $\frac{27}{8}$ or $3\frac{3}{8}$

Quick Quiz 7.6 *See Examples noted with Ex.* **1.** $x = 55$ (Ex. 1) **2.** 24 ft (Ex. 4) **3.** 172 flights (Ex. 1) **4.** See Student Solutions Manual

Career Exploration Problems **1.** $C_{ave} = \dfrac{\$350n + \$22{,}000}{n}$ **2.** 8 hours **3.** \$98.75 per hour for 50 treadmills; \$71.25 per hour for 100 treadmills

You Try It **1.** $\dfrac{2(x-5)}{x-3}$ **2.** $\dfrac{x+y}{2x-y}$ **3.** $\dfrac{2x-5}{2}$ **4.** $\dfrac{x-2}{2(x-3)}$ **5.** 6 **6.** $x - 1$ **7.** No solution **8.** 15.75 ft

Chapter 7 Review Problems **1.** $\dfrac{x}{x-y}$ **2.** $-\dfrac{4}{5}$ **3.** $\dfrac{x}{x+3}$ **4.** $\dfrac{2x-3}{5-x}$ **5.** $\dfrac{2(x-4y)}{2x-y}$ **6.** $\dfrac{2-y}{3y-1}$ **7.** $\dfrac{x-2}{(5x+6)(x-1)}$ **8.** $2(2x+y)$ **9.** $\dfrac{x+6}{6(x-1)}$ **10.** $\dfrac{2y-1}{5y}$ **11.** $\dfrac{4y(3y-1)}{(3y+1)(2y+5)}$ **12.** $\dfrac{3y(4x+3)}{2(x-5)}$ **13.** $\dfrac{2(x+3y)}{x-4y}$ **14.** $\dfrac{3}{16}$ **15.** $\dfrac{4(5y+1)}{3y(y+2)}$ **16.** $\dfrac{3x^2+6x+1}{x(x+1)}$ **17.** $\dfrac{2(5x-11)}{(x+2)(x-4)}$ **18.** $\dfrac{(x-1)(x-2)}{(x+3)(x-3)}$ **19.** $\dfrac{2xy+4x+5y+6}{2y(y+2)}$ **20.** $\dfrac{(2a+b)(a+4b)}{ab(a+b)}$ **21.** $\dfrac{3x-2}{3x}$ **22.** $\dfrac{2x^2+7x-2}{2x(x+2)}$ **23.** $\dfrac{3}{2(x-9)}$ **24.** $\dfrac{1-2x-x^2}{(x+5)(x+2)}$ **25.** $-\dfrac{4}{9}$ **26.** $\dfrac{22}{5x^2}$ **27.** $w - 2$ **28.** 1 **29.** $-\dfrac{y^2}{2}$ **30.** $\dfrac{x+2y}{y(x+y+2)}$ **31.** $\dfrac{-1}{a(a+b)}$ or $-\dfrac{1}{a(a+b)}$ **32.** $\dfrac{-3a-b}{b}$ or $-\dfrac{3a+b}{b}$ **33.** $a = 2$ **34.** $a = 15$ **35.** $x = \dfrac{2}{7}$ **36.** $x = -2$ **37.** $x = \dfrac{1}{2}$ **38.** No solution **39.** $x = -\dfrac{12}{5}$ or $-2\dfrac{2}{5}$ or -2.4 **40.** $y = 2$ **41.** $y = -2$ **42.** No solution **43.** $y = \dfrac{5}{4}$ or $1\dfrac{1}{4}$ or 1.25 **44.** $y = 0$ **45.** $x = \dfrac{14}{5}$ or $2\dfrac{4}{5}$ or 2.8 **46.** $x = \dfrac{5}{4}$ or $1\dfrac{1}{4}$ or 1.25 **47.** $x = \dfrac{132}{5}$ or $26\dfrac{2}{5}$ or 26.4 **48.** $x = 6$ **49.** $x = 16$ **50.** $x = \dfrac{91}{10}$ or $9\dfrac{1}{10}$ or 9.1 **51.** 8.3 gal **52.** 167 cookies **53.** 240 mi **54.** Train = 60 mph, Car = 40 mph **55.** 182 ft **56.** 1200 ft **57.** $2\dfrac{2}{5}$ hr or 2 hr, 24 min **58.** 12 hr **59.** $-\dfrac{a+4}{3a^2}$ **60.** $\dfrac{4a^2}{2a+3}$ **61.** $\dfrac{x-2y}{x+2y}$ **62.** $\dfrac{x+6}{x}$ **63.** $\dfrac{x+6}{x}$ **64.** $\dfrac{(b-a)(b+a)}{ab(x+y)}$ **65.** $\dfrac{1}{3}$ **66.** $-\dfrac{6}{y^2}$ **67.** $\dfrac{y(x+3y)}{2(x+2y)}$ **68.** $x = 12$ **69.** $x = -32$ **70.** $b = -2$

How Am I Doing? Chapter 7 Test **1.** $\dfrac{2}{3a}$ (obj. 7.1.1) **2.** $\dfrac{2x^2(2-y)}{(y+2)}$ (obj. 7.1.1) **3.** $\dfrac{5}{12}$ (obj. 7.2.1) **4.** $\dfrac{1}{3y(x-y)}$ (obj. 7.2.1) **5.** $\dfrac{2a+1}{a+2}$ (obj. 7.2.2) **6.** $\dfrac{3a+4}{(a+1)(a-2)}$ (obj. 7.3.3) **7.** $\dfrac{x-a}{ax}$ (obj. 7.3.3) **8.** $-\dfrac{x+2}{x+3}$ (obj. 7.3.3) **9.** $\dfrac{x}{4}$ (obj. 7.4.2) **10.** $\dfrac{6x}{5}$ (obj. 7.4.1) **11.** $\dfrac{2x-3y}{4x+y}$ (obj. 7.1.1) **12.** $\dfrac{x}{(x+2)(x+4)}$ (obj. 7.3.3) **13.** $x = -\dfrac{1}{5}$ (obj. 7.5.1) **14.** $x = 4$ (obj. 7.5.1) **15.** No solution (obj. 7.5.2) **16.** $x = \dfrac{47}{6}$ (obj. 7.5.1) **17.** $x = \dfrac{45}{13}$ (obj. 7.6.1) **18.** $x = 37.2$ (obj. 7.6.1) **19.** 151 flights (obj. 7.6.1) **20.** \$368 (obj. 7.6.1) **21.** 102 ft (obj. 7.6.2)

Chapter 8

8.1 Exercises **1.** $\dfrac{81x^4}{y^4z^8}$ **3.** $\dfrac{8a^6b^6}{27}$ **5.** $\dfrac{y^3}{8x^6}$ **7.** $\dfrac{y^{10}}{9x^2}$ **9.** $x^{3/2}$ **11.** y^8 **13.** $x^{2/5}$ **15.** $x^{2/3}$ **17.** $a^{7/4}$ **19.** $x^{4/7}$ **21.** $a^{7/8}$ **23.** $y^{1/2}$ **25.** $\dfrac{1}{x^{3/4}}$ **27.** $\dfrac{b^{1/3}}{a^{5/6}}$ **29.** $\dfrac{1}{6^{1/2}}$ **31.** $\dfrac{3}{a^{1/3}}$ **33.** 243 **35.** 8 **37.** -32 **39.** 9 **41.** $xy^{1/6}$ **43.** $-14x^{7/12}y^{1/12}$ **45.** $6^{4/3}$ **47.** $2x^{7/10}$ **49.** $-\dfrac{4x^{5/2}}{y^{6/5}}$ **51.** $2ab$ **53.** $16x^{1/2}y^5z$ **55.** $x^2 - x^{13/15}$ **57.** $m^{3/8} + 2m^{15/8}$ **59.** $\dfrac{1}{2}$ **61.** $\dfrac{1}{16}$ **63.** 32 **65.** $\dfrac{3y+1}{y^{1/2}}$ **67.** $\dfrac{1+6^{4/3}x^{1/3}}{x^{1/3}}$ **69.** $2a(5a^{1/4} - 2a^{3/5})$ **71.** $3x(4x^{1/3} - x^{3/2})$ **73.** $a = -\dfrac{3}{8}$ **75.** Radius = 1.86 m **77.** Radius = 5 ft **79.** $x = -\dfrac{3}{2}$ **80.** $x = -3$

Quick Quiz 8.1 *See Examples noted with Ex.* **1.** $-12x^{5/6}y^{3/4}$ (Ex. 3) **2.** $2x^{10/3}$ (Ex. 3) **3.** $125x^{3/8}$ (Ex. 1 and 5) **4.** See Student Solutions Manual

8.2 Exercises **1.** A square root of a number is a value that when multiplied by itself is equal to the original number. **3.** $\sqrt[3]{-8} = -2$ because $(-2)(-2)(-2) = -8$ **5.** 10 **7.** 13 **9.** $-\dfrac{1}{3}$ **11.** Not a real number **13.** 0.5 **15.** 0, 2.8, 4, 4.9; all real numbers x where $x \geq -3$ **17.** 0, 1, 2, 2.2; all real numbers x where $x \geq 10$

19.

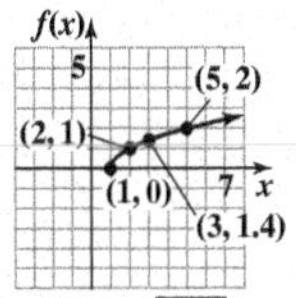

$f(x)$, 5, (5, 2), (2, 1), (1, 0), 7, x, (3, 1.4)

$f(x) = \sqrt{x-1}$

21.

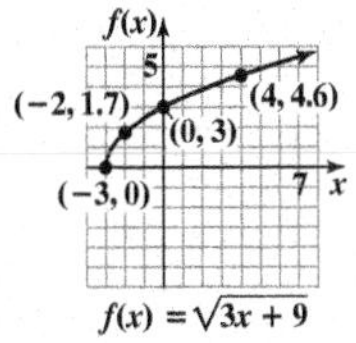

$f(x)$, 5, $(-2, 1.7)$, (4, 4.6), (0, 3), $(-3, 0)$, 7, x

$f(x) = \sqrt{3x+9}$

23. 4 **25.** -10

27. 2 **29.** 3 **31.** 8 **33.** 5 **35.** $-\frac{1}{2}$ **37.** $y^{1/3}$ **39.** $m^{3/5}$ **41.** $(2a)^{1/4}$ **43.** $(a+b)^{3/7}$ **45.** $x^{1/6}$ **47.** $(3x)^{5/6}$ **49.** 12

51. x^4y **53.** $6x^4y^2$ **55.** $2a^2b$ **57.** $\sqrt[7]{y^4}$ or $(\sqrt[7]{y})^4$ **59.** $\frac{1}{\sqrt[3]{49}}$ **61.** $\sqrt[4]{(a+5b)^3}$ or $(\sqrt[4]{a+5b})^3$ **63.** $\sqrt[5]{(-x)^3}$ or $(\sqrt[5]{-x})^3$

65. $\sqrt[5]{8x^3y^3}$ or $(\sqrt[5]{2xy})^3$ **67.** 27 **69.** $\frac{2}{5}$ **71.** 2 **73.** $\frac{1}{8x^2}$ **75.** $11x^2$ **77.** $12a^3b^{12}$ **79.** $5|x|$ **81.** $-2x^2$ **83.** $|x|y^6$ or $|xy^6|$

85. $|a^3b|$ **87.** $5x^6y^2$ **89.** $x = \frac{2y-6}{5}$ **90.** $y = \frac{3x-12}{2}$

Quick Quiz 8.2

See Examples noted with Ex. **1.** $\frac{8}{125}$ (Ex. 9) **2.** -4 (Ex. 1) **3.** $11x^5y^6$ (Ex. 7) **4.** See Student Solutions Manual

8.3 Exercises

1. $2\sqrt{2}$ **3.** $3\sqrt{2}$ **5.** $2\sqrt{7}$ **7.** $5\sqrt{2}$ **9.** $3x$ **11.** $2a^3b^3\sqrt{10b}$ **13.** $3xz^2\sqrt{10xy}$ **15.** 2 **17.** $2\sqrt[3]{5}$ **19.** $3\sqrt[3]{2a^2}$ **21.** $3ab^3\sqrt[3]{a^2}$ **23.** $2x^4y^4\sqrt[3]{5y}$ **25.** $3p^5\sqrt[4]{kp^3}$ **27.** $-2xy\sqrt[5]{y}$ **29.** $a = 4$ **31.** $12\sqrt{5}$ **33.** $4\sqrt{3} - 4\sqrt{7}$ **35.** $11\sqrt{2}$ **37.** $11\sqrt{3}$ **39.** $-5\sqrt{2}$ **41.** 0 **43.** $2\sqrt{6}$ **45.** $8\sqrt{3x}$ **47.** $19\sqrt{2x}$ **49.** $2\sqrt{11} - \sqrt{7x}$ **51.** $6x\sqrt{2x}$ **53.** $11\sqrt[3]{2}$ **55.** $4xy\sqrt[3]{x} - 3y\sqrt[3]{xy^2}$ **57.** 20.7846097 = 20.7846097 **59.** 7.146 amps **61.** 3.14 sec **63.** $x(4x-7y)^2$ **64.** $y(9x+5)(9x-5)$ **65.** $4x^2 - 9$ **66.** $4x^2 + 12x + 9$

Quick Quiz 8.3

See Examples noted with Ex. **1.** $2x^3y^4\sqrt{30x}$ (Ex. 5) **2.** $2x^5y^3\sqrt[3]{2y}$ (Ex. 5) **3.** $10\sqrt{3}$ (Ex. 7) **4.** See Student Solutions Manual

8.4 Exercises

1. $\sqrt{35}$ **3.** $-8\sqrt{21}$ **5.** $-24\sqrt{5}$ **7.** $-3\sqrt{5xy}$ **9.** $-12x^2\sqrt{5y}$ **11.** $15\sqrt{ab} - 25\sqrt{a}$ **13.** $-3\sqrt{2ab} - 6\sqrt{5a}$ **15.** $-a + 2\sqrt{ab}$ **17.** $14\sqrt{3x} - 35x$ **19.** $22 - 5\sqrt{2}$ **21.** $4 - 6\sqrt{6}$ **23.** $14 + 11\sqrt{35x} + 60x$ **25.** $\sqrt{15} + 3 + 2\sqrt{10} + 2\sqrt{6}$

27. $29 - 4\sqrt{30}$ **29.** $81 - 36\sqrt{b} + 4b$ **31.** $3x + 13 + 6\sqrt{3x+4}$ **33.** $3x\sqrt[3]{4} - 4x^2\sqrt[3]{x}$ **35.** $1 - \sqrt[3]{12} + \sqrt[3]{18}$ **37.** $\frac{7}{5}$

39. $\frac{2\sqrt{3x}}{7y^3}$ **41.** $\frac{2xy^2\sqrt[3]{x^2}}{3}$ **43.** $\frac{y^2\sqrt[3]{5y^2}}{3x}$ **45.** $\frac{3\sqrt{2}}{2}$ **47.** $\frac{2\sqrt{3}}{3}$ **49.** $\frac{\sqrt{5y}}{5y}$ **51.** $\frac{\sqrt{7ay}}{y}$ **53.** $\frac{\sqrt{3x}}{3x}$ **55.** $\frac{x(\sqrt{5}+\sqrt{2})}{3}$

57. $2y(\sqrt{6} - \sqrt{5})$ **59.** $\frac{\sqrt{6y} - y\sqrt{2}}{6-2y}$ **61.** $4 + \sqrt{15}$ **63.** $\frac{3x - 3\sqrt{3xy} + 2y}{3x-y}$ **65.** $11\sqrt{2}$ **67.** $-24 + \sqrt{6}$ **69.** $\frac{9\sqrt{2x}}{4x}$

71. $3 - \sqrt{5}$ **73.** 1.194938299; 1.194938299; yes; yes **75.** $-\frac{43}{4(\sqrt{2} - 3\sqrt{5})}$ **77.** \$79.58 **79.** $(x + 8\sqrt{x} + 15)$ mm^2

81. $(-3, 5)$ **82.** $(2, 0, -4)$ **83.** 48% **84.** 12,423,000 households **85.** 59,040,000 households

Quick Quiz 8.4

See Examples noted with Ex. **1.** $8 + \sqrt{15}$ (Ex. 3) **2.** $\frac{3\sqrt{3x}}{x}$ (Ex. 9) **3.** $\frac{14 + 9\sqrt{5}}{11}$ (Ex. 12) **4.** See Student Solutions Manual

Use Math to Save Money

1. \$545.75 **2.** \$578.06 **3.** He did not deposit enough money to cover the checks he wrote for May. But the \$300.50 he already had in the bank will help to cover his expenses for May. **4.** \$268.19 **5.** Eventually Terry will be in debt. **6.** Answers will vary. **7.** Answers will vary.

8.5 Exercises

1. Isolate one of the radicals on one side of the equation. **3.** $x = 3$ **5.** $x = 3$ **7.** $y = 2, y = 1$ **9.** $x = 3$

11. No solution **13.** $y = 7$ **15.** $y = -2, y = -3$ **17.** $x = 3, x = 7$ **19.** $x = 0, x = \frac{1}{2}$ **21.** $x = \frac{5}{2}$ **23.** $x = 7$ **25.** $x = 12$

27. $x = 0, x = 7$ **29.** $x = \frac{1}{4}$ **31.** $x = -3, x = -4$ **33.** $x = 0, x = 8$ **35.** No solution **37.** $x = 9$ **39.** $x \approx 3.3357, x \approx 0.9443$

41. **(a)** $S = \frac{V^2}{12}$ **(b)** 75 ft **43.** $x = 0.055y^2 + 1.25y - 10$ **45.** $c = 1$ **47.** $16x^4$ **48.** $\frac{1}{2x^2}$ **49.** $-6x^2y^3$ **50.** $-2x^3y$

51. Approximately 1.33 mph **52.** 16 dogs; 12 cats

Quick Quiz 8.5

See Examples noted with Ex. **1.** $x = 1, x = 4$ (Ex. 1) **2.** $x = 2$ (Ex. 2) **3.** $x = 5$ (Ex. 3) **4.** See Student Solutions Manual

8.6 Exercises

1. No. There is no real number that, when squared, will equal -9. **3.** No. To be equal, the real parts must be equal, and the imaginary parts must be equal. $2 \neq 3$ and $3i \neq 2i$ **5.** $5i$ **7.** $2i\sqrt{7}$ **9.** $\frac{5}{2}i$ **11.** $-9i$ **13.** $2 + i\sqrt{3}$ **15.** $-2.8 + 4i$

17. $-3 + 2i\sqrt{6}$ **19.** $-\sqrt{10}$ **21.** -12 **23.** $x = 5; y = -3$ **25.** $x = 1.3; y = 2$ **27.** $x = -6, y = 3$ **29.** $-5 + 11i$

31. $1 - i$ **33.** $1.2 + 2.1i$ **35.** -14 **37.** 42 **39.** $7 + 4i$ **41.** $6 + 6i$ **43.** $-10 - 12i$ **45.** $-\frac{3}{4} + i$ **47.** $-\sqrt{21}$

49. $12 - \sqrt{10} + 3i\sqrt{5} + 4i\sqrt{2}$ **51.** i **53.** 1 **55.** -1 **57.** $-i$ **59.** 0 **61.** $1 + i$ **63.** $\frac{1+i}{2}$ or $\frac{1}{2} + \frac{i}{2}$ **65.** $\frac{4+i}{17}$ or $\frac{4}{17} + \frac{i}{17}$

67. $\frac{-2-5i}{6}$ or $-\frac{1}{3} - \frac{5}{6}i$ **69.** $-2i$ **71.** $\frac{35+42i}{61}$ or $\frac{35}{61} + \frac{42}{61}i$ **73.** $\frac{11-16i}{13}$ or $\frac{11}{13} - \frac{16}{13}i$ **75.** $7i\sqrt{2}$ **77.** $9 - 8i$ **79.** $-12 - 14i$

81. $\frac{1-8i}{5}$ or $\frac{1}{5} - \frac{8}{5}i$ **83.** $Z = \frac{2-3i}{3}$ or $\frac{2}{3} - i$ **85.** 18 hr producing juice in glass bottles; 25 hr producing juice in cans, 62 hr producing juice in plastic bottles **86.** \$96,030

Quick Quiz 8.6 *See Examples noted with Ex.* **1.** $32 - 9i$ (Ex. 6) **2.** $\frac{-2 + 11i}{5}$ or $-\frac{2}{5} + \frac{11}{5}i$ (Ex. 9) **3.** i (Ex. 8) **4.** See Student Solutions Manual

8.7 Exercises **1.** Answers may vary. A person's weekly paycheck varies as the number of hours worked. $y = kx$, y is the weekly salary, k is the hourly salary, x is the number of hours. **3.** $y = \frac{k}{x}$ **5.** $y = 52$ **7.** 71.4 lb/in.^{2} **9.** 160 ft **11.** $y = 500$ **13.** 2993 gal **15.** 3.9 hr **17.** 400 lb **19.** Approximately 3.1 min **21.** $x = \frac{2}{3}, x = 2$ **22.** $x = 1, x = -8$ **23.** \$460 **24.** 55 gal

Quick Quiz 8.7 *See Examples noted with Ex.* **1.** $y = 4.5$ (Ex. 2) **2.** $y = 3$ (Ex. 4) **3.** 135.2 ft (Ex. 1) **4.** See Student Solutions Manual

Career Explorations Problems **1.** 174.2 feet and 79.5 feet; overall length is approximately 254 feet long **2.** 1078 yd^3 **3.** 20-foot light pole: 6093.75 lumens; 30-foot light pole: 2708.3 lumens **4.** 38.4 ft

You Try It **1. (a)** $x^{1/5}$ **(b)** $2^{1/3}a^{-1}b^{-1/12}$ **(c)** $\frac{5^{1/4}x^{-3/4}}{4^{-1/2}y^{1/2}}$ **2.** $-4a^{5/6}$ **3.** $-6x^{1/4}$ **4. (a)** $\frac{5}{a^3}$ **(b)** $\frac{a^4}{2}$ **(c)** $\frac{1}{32}$ **5.** $x^{7/6} - x$ **6.** 1 **7. (a)** 2 **(b)** -1 **(c)** Not a real number **8. (a)** $\sqrt[5]{x^4}$ or $(\sqrt[5]{x})^4$ **(b)** $v^{9/4}$ **(c)** 81 **9. (a)** $|x|$ **(b)** y **10.** $-4m^6$ **11. (a)** $2x^2\sqrt{6x}$ **(b)** $3rs^3\sqrt[3]{2r}$ **12.** $30\sqrt{2}$ **13. (a)** $3\sqrt{30}$ **(b)** $18\sqrt{2} - 9\sqrt{5}$ **(c)** $1 - \sqrt{15}$ **14.** $\frac{\sqrt[4]{3}}{2}$ **15. (a)** $\frac{\sqrt{6}}{2}$ **(b)** $-4\sqrt{2} - 4\sqrt{3}$ **16.** $x = 9$ **17. (a)** $10i$ **(b)** $2i\sqrt{6}$ **18. (a)** $11 - 6i$ **(b)** $-1 + 2i$ **19.** $9 - 7i$ **20.** 1 **21.** $\frac{3 - 14i}{5}$ or $\frac{3}{5} - \frac{14}{5}i$ **22.** $y = 36$ **23.** $w = 0.5$

Chapter 8 Review Problems **1.** $\frac{15x^3}{y^{5/2}}$ **2.** $4a^3b^{5/2}$ **3.** $3^{2/3}$ **4.** $\frac{x^{1/2}y^{3/10}}{2}$ **5.** $\frac{x}{32y^{1/2}z^4}$ **6.** $7a^5b$ **7.** $2a^2$ **8.** 16 **9.** $\frac{2x + 1}{x^{2/3}}$ **10.** $3x(2x^{1/2} - 3x^{-1/2})$ **11.** -4 **12.** -2 **13.** $-\frac{1}{5}$ **14.** 0.2 **15.** Not a real number **16.** $-\frac{1}{2}$ **17.** 16 **18.** 625 **19.** $7x^2y^5z$ **20.** $4a^4b^{10}$ **21.** $-2a^4b^5c^7$ **22.** $7x^{11}y$ **23.** $a^{2/5}$ **24.** $(2b)^{1/2}$ **25.** $(5a)^{1/3}$ **26.** $(xy)^{7/5}$ **27.** $\sqrt{m}$ **28.** $\sqrt[5]{y^3}$ or $(\sqrt[5]{y})^3$ **29.** $\sqrt[3]{9z^2}$ or $(\sqrt[3]{3z})^2$ **30.** $\sqrt[7]{8x^3}$ or $(\sqrt[7]{2x})^3$ **31.** 8 **32.** 9 **33.** $\frac{1}{3}$ **34.** 0.7 **35.** 6 **36.** $125a^3b^6$ **37.** $11\sqrt{2}$ **38.** $13\sqrt{7}$ **39.** $25\sqrt{3}$ **40.** $8x\sqrt{5x}$ **41.** $11\sqrt{2x} - 5x\sqrt{2}$ **42.** $-6\sqrt[3]{2}$ **43.** $90\sqrt{2}$ **44.** $-24x\sqrt{5}$ **45.** $12x\sqrt{2} - 36\sqrt{3x}$ **46.** $4\sqrt{3a} - 3a\sqrt{7}$ **47.** $2b\sqrt{7a} - 2b^2\sqrt{21c}$ **48.** $4 - 9\sqrt{6}$ **49.** $74 - 12\sqrt{30}$ **50.** $2x - \sqrt[3]{2xy} + 2\sqrt[3]{3x^2} - \sqrt[3]{6y}$ **51. (a)** $f(12) = 8$ **(b)** All real numbers x where $x \geq -4$ **52. (a)** $f(9) = 3$ **(b)** All real numbers x where $x \leq 12$ **53.** $\frac{y\sqrt{6x}}{x}$ **54.** $\frac{3\sqrt{5y}}{5y}$ **55.** $\sqrt{3}$ **56.** $2\sqrt{6} + 2\sqrt{5}$ **57.** $\frac{3x - \sqrt{xy}}{9x - y}$ **58.** $-\frac{\sqrt{35} + 3\sqrt{5}}{2}$ **59.** $\frac{2 + 3\sqrt{2}}{7}$ **60.** $\frac{\sqrt[3]{4x^2y}}{2y}$ **61.** $4i + 3i\sqrt{5}$ **62.** $x = \frac{-7 + \sqrt{6}}{2}; y = -3$ **63.** $-9 - 11i$ **64.** $-10 + 2i$ **65.** $21 + 9i$ **66.** $32 - 24i$ **67.** $-8 + 6i$ **68.** $-5 - 4i$ **69.** -1 **70.** i **71.** $\frac{13 - 34i}{25}$ or $\frac{13}{25} - \frac{34}{25}i$ **72.** $\frac{11 + 13i}{10}$ or $\frac{11}{10} + \frac{13}{10}i$ **73.** $\frac{-3 - 4i}{5}$ or $-\frac{3}{5} - \frac{4}{5}i$ **74.** $\frac{18 + 30i}{17}$ or $\frac{18}{17} + \frac{30}{17}i$ **75.** $x = 9$ **76.** $x = 3$ **77.** $x = 4$ **78.** $x = 5$ **79.** $x = 5, x = 1$ **80.** $x = 1, x = \frac{3}{2}$ **81.** $y = 16.5$ **82.** 22.5 g **83.** 3.5 sec **84.** $y = 0.5$ **85.** $y = 100$ **86.** 432 cm^3

How Am I Doing? Chapter 8 Test **1.** $-6x^{5/6}y^{1/2}$ (obj. 8.1.1) **2.** $\frac{7x^{9/4}}{4}$ (obj. 8.1.1) **3.** $8^{3/2}x^{1/2}$ or $16(2x)^{1/2}$ (obj. 8.1.1) **4.** $\frac{8}{27}$ (obj. 8.1.1) **5.** -2 (obj. 8.2.1) **6.** $\frac{1}{4}$ (obj. 8.2.1) **7.** 32 (obj. 8.2.1) **8.** $5a^2b^4\sqrt{3b}$ (obj. 8.3.1) **9.** $7a^2b^5$ (obj. 8.3.1) **10.** $3mn\sqrt[3]{2n^2}$ (obj. 8.3.1) **11.** $6\sqrt{6} + 2\sqrt{2}$ (obj. 8.3.2) **12.** $2\sqrt{10x} + \sqrt{3x}$ (obj. 8.3.2) **13.** $-30y\sqrt{5x}$ (obj. 8.4.1) **14.** $18\sqrt{2} - 10\sqrt{6}$ (obj. 8.4.1) **15.** $12 + 39\sqrt{2}$ (obj. 8.4.1) **16.** $\frac{6\sqrt{5x}}{x}$ (obj. 8.4.3) **17.** $\frac{\sqrt{3xy}}{3}$ (obj. 8.4.3) **18.** $\frac{9 + 7\sqrt{3}}{6}$ (obj. 8.4.3) **19.** $x = 2, x = 1$ (obj. 8.5.1) **20.** $x = 10$ (obj. 8.5.1) **21.** $x = 6$ (obj. 8.5.2) **22.** $2 + 14i$ (obj. 8.6.2) **23.** $-1 + 4i$ (obj. 8.6.2) **24.** $18 + i$ (obj. 8.6.3) **25.** $\frac{-13 + 11i}{10}$ or $-\frac{13}{10} + \frac{11}{10}i$ (obj. 8.6.5) **26.** $27 + 36i$ (obj. 8.6.3) **27.** $-i$ (obj. 8.6.4) **28.** $y = 3$ (obj. 8.7.2) **29.** $y = \frac{5}{6}$ (obj. 8.7.3) **30.** About 83.3 ft (obj. 8.7.1)

Chapter 9

9.1 Exercises **1.** $x = \pm 10$ **3.** $x = \pm\sqrt{3}$ **5.** $x = \pm 2\sqrt{10}$ **7.** $x = \pm 9i$ **9.** $x = \pm 4i$ **11.** $x = 3 \pm 2\sqrt{3}$ **13.** $x = -9 \pm \sqrt{21}$ **15.** $x = \frac{-1 \pm \sqrt{7}}{2}$ **17.** $x = \frac{9}{4}, x = -\frac{3}{4}$ **19.** $x = 1, x = -6$ **21.** $x = \pm\frac{3\sqrt{2}}{2}$ **23.** $x = -5 \pm 2\sqrt{5}$ **25.** $x = 4 \pm \sqrt{33}$ **27.** $x = 6, x = 8$ **29.** $x = \frac{-3 \pm \sqrt{41}}{2}$ **31.** $y = \frac{-5 \pm \sqrt{3}}{2}$ **33.** $x = \frac{-5 \pm \sqrt{31}}{3}$ **35.** $x = -2 \pm \sqrt{10}$ **37.** $x = 4, x = -2$ **39.** $x = \frac{1 \pm i\sqrt{11}}{6}$ **41.** $x = \frac{1 \pm i\sqrt{7}}{2}$ **43.** $x = \frac{3 \pm i\sqrt{7}}{4}$

45. $(-1+\sqrt{6})^2+2(-1+\sqrt{6})-5 \stackrel{?}{=} 0; 1-2\sqrt{6}+6-2+2\sqrt{6}-5 \stackrel{?}{=} 0; 0=0$ ✓ **47.** $x=12$ **49.** Approximately 1.12 sec **51.** 8 **52.** 7 **53.** 40 **54.** 4

Quick Quiz 9.1

See Examples noted with Ex. **1.** $x=\frac{3\pm 2\sqrt{3}}{4}$ (Ex. 5) **2.** $x=4\pm 2\sqrt{11}$ (Ex. 6) **3.** $x=\frac{-5\pm\sqrt{3}}{2}$ (Ex. 7) **4.** See Student Solutions Manual

9.2 Exercises

1. Place the quadratic equation in standard form. Find *a, b,* and *c*. Substitute these values into the quadratic formula. **3.** one real **5.** $x=\frac{-1\pm\sqrt{21}}{2}$ **7.** $x=\frac{-3\pm\sqrt{33}}{4}$ **9.** $x=0, x=\frac{2}{3}$ **11.** $x=1, x=-\frac{2}{3}$ **13.** $x=\frac{-3\pm\sqrt{41}}{8}$ **15.** $x=\pm\frac{\sqrt{6}}{2}$ **17.** $x=\frac{-1\pm\sqrt{3}}{2}$ **19.** $x=\pm 3$ **21.** $x=\frac{2\pm 3\sqrt{2}}{2}$ **23.** $x=2\pm\sqrt{10}$ **25.** $y=6, y=4$ **27.** $x=-2\pm 2i\sqrt{2}$ **29.** $x=\pm\frac{i\sqrt{22}}{2}$ **31.** $x=\frac{4\pm i\sqrt{5}}{3}$ **33.** Two irrational roots **35.** Two rational roots **37.** One rational root **39.** Two nonreal complex roots **41.** $x^2-15x+26=0$ **43.** $x^2+13x+42=0$ **45.** $x^2+16=0$ **47.** $2x^2-x-15=0$ **49.** $x\approx 1.4643, x\approx -2.0445$ **51.** 18 mountain bikes or 30 mountain bikes per day **53.** $-3x^2-10x+11$ **54.** $-y^2+3y$

Quick Quiz 9.2

See Examples noted with Ex. **1.** $x=\frac{9\pm 5\sqrt{5}}{22}$ (Ex. 1) **2.** $x=-\frac{8}{3}, x=1$ (Ex. 5) **3.** $x=\frac{5\pm i\sqrt{47}}{4}$ (Ex. 6) **4.** See Student Solutions Manual

9.3 Exercises

1. $x=\pm\sqrt{5}, x=\pm 2$ **3.** $x=\pm\sqrt{3}, x=\pm 2i$ **5.** $x=\pm\frac{i\sqrt{3}}{2}, x=\pm 1$ **7.** $x=2, x=-1$ **9.** $x=0, x=\sqrt[3]{3}$ **11.** $x=\pm 2, x=\pm 1$ **13.** $x=\pm\frac{\sqrt[4]{54}}{3}$; these are the only real roots. **15.** $x=-64, x=27$ **17.** $x=\frac{1}{64}, x=-\frac{8}{27}$ **19.** $x=81$ **21.** $\frac{1}{256}$ **23.** $x=32, x=-1$ **25.** $x=\sqrt[3]{7}, x=\sqrt[3]{-2}$ **27.** $x=-1, x=-2, x=-4, x=1$ **29.** $x=9, x=4$ **31.** $x=-\frac{1}{3}$ **33.** $x=\frac{1}{5}, x=-\frac{1}{2}$ **35.** $x=\frac{5}{6}, x=\frac{3}{2}$ **37.** $(-2,3)$ **38.** $\frac{15x+6}{7-3x}$ **39.** $3\sqrt{10}-12\sqrt{3}$ **40.** $6-2\sqrt{10}+6\sqrt{3}-2\sqrt{30}$ **41.** Less than H.S. diploma 31.6%; H.S. graduate 31.5%; some college or associate's 29.8%; Bachelor's degree or higher 37.0% **42.** An additional 15.5 weeks

Quick Quiz 9.3

See Examples noted with Ex. **1.** $x=\pm 4, x=\pm\sqrt{2}$ (Ex. 1) **2.** $x=\frac{1}{4}, x=-\frac{2}{5}$ (Ex. 5) **3.** $x=125, x=-1$ (Ex. 3) **4.** See Student Solutions Manual

Use Math to Save Money

1. The Gold plan **2.** The Silver plan **3.** The Silver plan **4.** The Bronze plan **5.** \$30 **6.** \$60 **7.** \$90 **8.** The Silver plan **9.** \$2319

9.4 Exercises

1. $t=\pm\frac{\sqrt{S}}{4}$ **3.** $r=\pm\frac{1}{3}\sqrt{\frac{S}{\pi}}$ **5.** $x=\pm\sqrt{\frac{15N}{2a}}$ **7.** $y=\pm\frac{\sqrt{7R-4w+5}}{2}$ **9.** $M=\pm\sqrt{\frac{2cQ}{3mw}}$ **11.** $r=\pm\sqrt{\frac{V-\pi R^2h}{\pi h}}$ **13.** $x=0, x=\frac{3a}{7b}$ **15.** $I=\frac{E\pm\sqrt{E^2-4RP}}{2R}$ **17.** $w=\frac{-5t\pm\sqrt{25t^2+72}}{18}$ **19.** $r=\frac{-\pi h\pm\sqrt{\pi^2h^2+S\pi}}{\pi}$ **21.** $x=\frac{-5\pm\sqrt{25-8aw-8w}}{2a+2}$ **23.** $b=2\sqrt{5}$ **25.** $a=\sqrt{15}$ **27.** $a=\frac{12\sqrt{5}}{5}, b=\frac{24\sqrt{5}}{5}$ **29.** $5\sqrt{2}$ in. **31.** Width is 0.13 mi; length is 0.2 mi **33.** Width is 7 ft; length is 18 ft **35.** Base is 8 cm; altitude is 18 cm **37.** Speed in rain was 45 mph; speed without rain was 50 mph **39.** 30 mi **41.** 2,263,000 incarcerated adults **43.** In the year 2015 **45.** $\frac{4\sqrt{3x}}{3x}$ **46.** $\frac{\sqrt{30}}{2}$ **47.** $\frac{3(\sqrt{x}-\sqrt{y})}{x-y}$ **48.** $-2-2\sqrt{2}$

Quick Quiz 9.4

See Examples noted with Ex. **1.** $y=\pm\sqrt{\frac{5ab}{H}}$ (Ex. 1) **2.** $z=\frac{-7y\pm\sqrt{49y^2+120w}}{12}$ (Ex. 3) **3.** Width = 7 yd; length = 25 yd (Ex. 8) **4.** See Student Solutions Manual

9.5 Exercises

1. $V(4,1)$; $(0,15)$; $(3,0), (5,0)$ **3.** $V(-4,25)$; $(0,9)$; $(-9,0), (1,0)$ **5.** $V(-2,-9)$; $(0,3)$; $(-0.3,0), (-3.7,0)$ **7.** $V\left(-\frac{1}{3},-\frac{17}{3}\right)$; y-intercept $(0,-6)$; no x-intercepts **9.** $V\left(-\frac{1}{2},-\frac{9}{2}\right)$; $(0,-4)$; $(1,0), (-2,0)$ **11.**

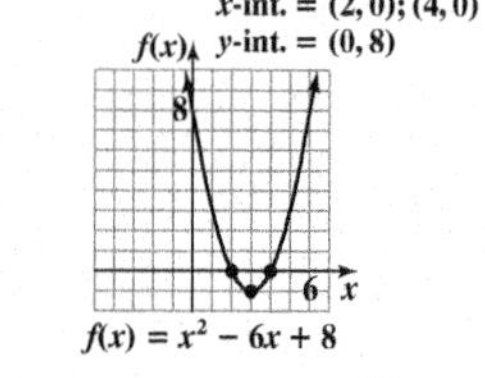

13. **15.** **17.** **19.** **21.**

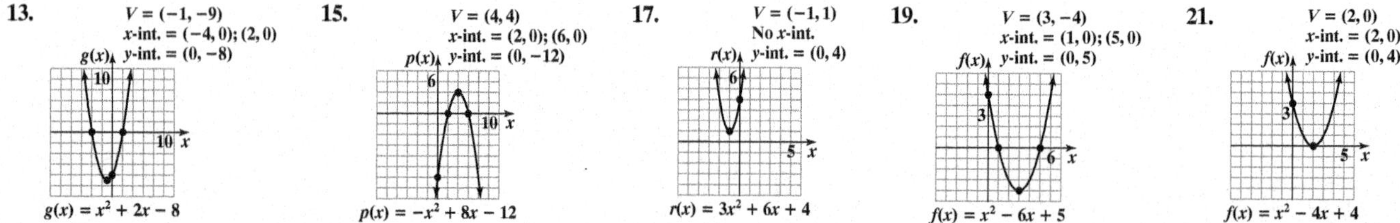

23. 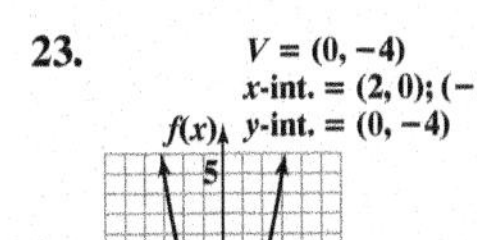

25. $N(10) = 1229.8$; $N(30) = 2686.6$; $N(50) = 5735.4$; $N(70) = 10{,}376.2$; $N(90) = 16{,}609$
27. $N(60) \approx 8000$. Approximately 8,000,000 people participate in boating and have a mean income of $60,000.
29. Approximately 68. This means that 10,000,000 people who participate in boating have a mean income of $68,000.
31. $P(18) = -220$; $P(20) = 0$; $P(28) = 480$; $P(35) = 375$; $P(42) = -220$ **33.** 30 tables per day; $500 per day
35. 20 tables per day or 40 tables per day **37.** 56 ft; about 2.9 sec

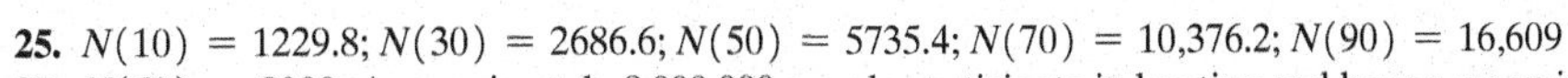
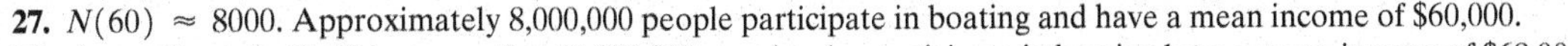

39. x-intercepts $(-0.3, 0)$ and $(2.6, 0)$

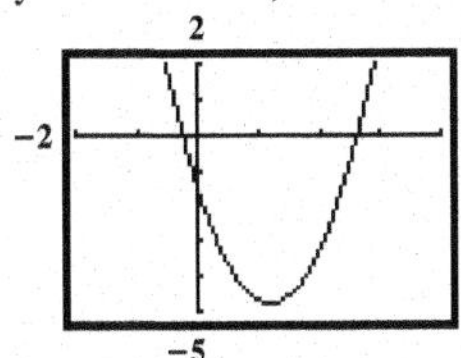

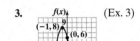

41. $a = 3, b = 8, c = -10$
43. $(3, 1, 2)$ **44.** $(3, -5, 8)$

Quick Quiz 9.5

See Examples noted with Ex. **1.** $(-1, 8)$ (Ex. 1) **2.** $(0, 6)$; $(-3, 0)$, $(1, 0)$ (Ex. 1) **3.** (Ex. 3)

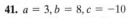

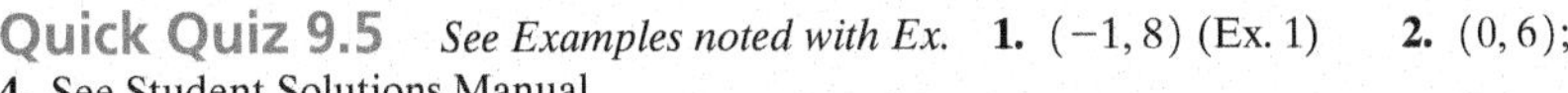

4. See Student Solutions Manual

9.6 Exercises

1. 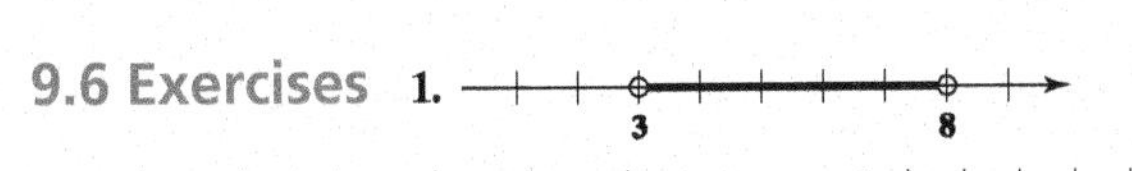

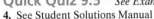

3.

5.

7. 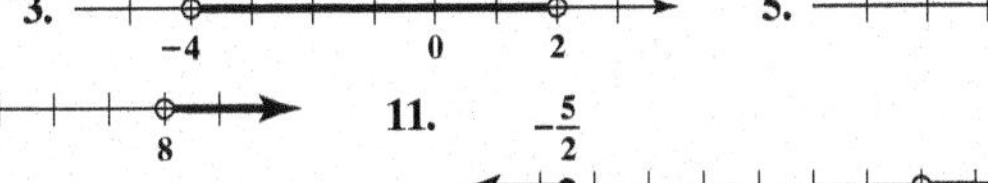

9.

11.

13. 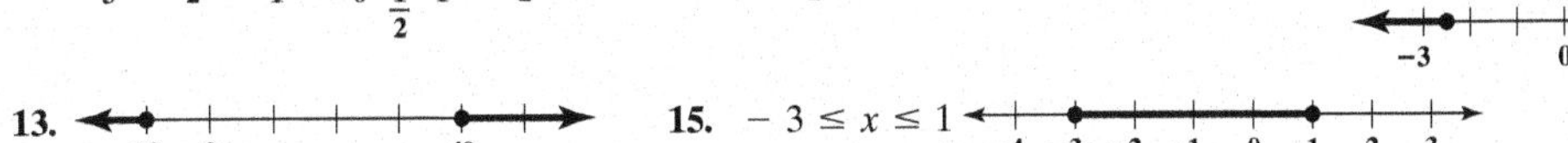

15. $-3 \le x \le 1$

17. $x < -6$ or $x > \frac{5}{2}$ 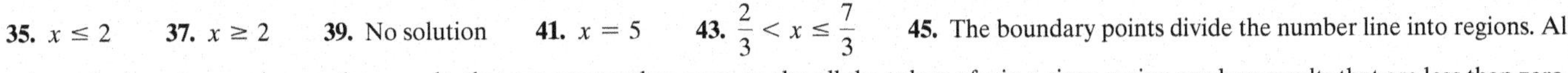

19. No solution **21.** $t < 10.9$ or $t > 11.2$ **23.** $5000 \le c \le 12{,}000$

25. $-4° \le F \le 51.8°$ **27.** $\$509.20 \le d \le \$593.24$ **29.** $-2 < x < 2$ **31.** $-3 \le x \le 1$ **33.** $x < -2$ or $x \ge 3$

35. $x \le 2$ **37.** $x \ge 2$ **39.** No solution **41.** $x = 5$ **43.** $\frac{2}{3} < x \le \frac{7}{3}$ **45.** The boundary points divide the number line into regions. All values of x in a given region produce results that are greater than zero, or else all the values of x in a given region produce results that are less than zero.

47. $-4 < x < 3$ 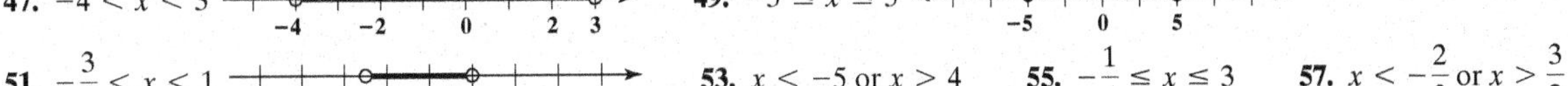

49. $-5 \le x \le 5$

51. $-\frac{3}{2} < x < 1$ 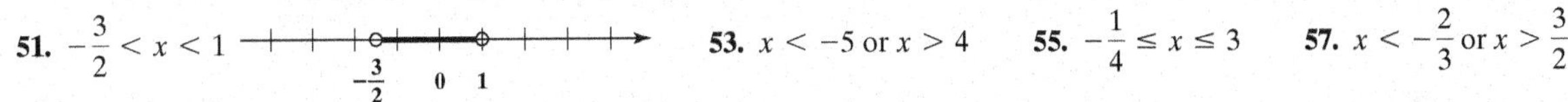

53. $x < -5$ or $x > 4$ **55.** $-\frac{1}{4} \le x \le 3$ **57.** $x < -\frac{2}{3}$ or $x > \frac{3}{2}$

59. $-10 \le x \le 3$ **61.** All real numbers **63.** $x = 2$ **65.** Approximately $x < -1.4$ or $x > 3.4$ **67.** Approximately $-0.8 < x < 1.8$ **69.** All real numbers **71.** No real number **73.** Greater than 15 sec but less than 25 sec **75. (a)** Approximately $9.4 < x < 190.6$ **(b)** $57,000 **(c)** $66,000 **77.** $400 for the 2-hour trip, $483 for the 3-hour trip **78.** 9 adults; 12 children; 2 seniors

Quick Quiz 9.6

See Examples noted with Ex. **1.** $-\frac{16}{3} < x < 2$ (Ex. 8) **2.** $x \le -8$ or $x \ge 1$ (Ex. 7) **3.** $x < 1$ or $x > 6$ (Ex. 10)
4. See Student Solutions Manual

9.7 Exercises

1. It will always have 2 solutions. One solution is when $x = b$ and one when $x = -b$. Since $b > 0$ the values of b and $-b$ are always different numbers. **3.** You must first isolate the absolute value expression. To do this you add 2 to each side of the equation. The result will be $|x + 7| = 10$. Then you solve the two equations $x + 7 = 10$ and $x + 7 = -10$. The final answer is $x = 3$ and $x = -17$. **5.** $x = 30, -30$

7. $x = 8, -22$ **9.** $x = 9, -4$ **11.** $x = 2, 7$ **13.** $x = 6, -10$ **15.** $x = -2, \frac{10}{3}$ **17.** $x = -9, 15$ **19.** $x = -\frac{1}{5}, \frac{13}{15}$

21. $x = 5, -1$ **23.** $x = -\frac{7}{3}, -1$ **25.** $x = 3, 1$ **27.** $x = 0, 12$ **29.** $x = 0, -8$ **31.** $x = \frac{5}{4}$ **33.** No solution **35.** $x = \frac{11}{15}, -\frac{1}{15}$

37. $-8 \le x \le 8$

39. $-9.5 < x < 0.5$

41. $-2 \le x \le 8$

43. $-\frac{11}{3} \le x \le 3$ **45.** $-5 < x < 15$ **47.** $-32 < x < 16$ **49.** $-4 < x < 8$ **51.** $-3\frac{1}{3} < x < 4\frac{2}{3}$ **53.** $x < -5$ or $x > 5$

55. $x < -7$ or $x > 3$ **57.** $x \le -1$ or $x \ge 3$ **59.** $x \le -\frac{1}{2}$ or $x \ge 4$ **61.** $x < 10$ or $x > 110$ **63.** $x < -9\frac{1}{2}$ or $x > 10\frac{1}{2}$

65. $-13 < x < 17$ **67.** $-5 < x < \frac{3}{2}$ **69.** $x < -2$ or $x > 2\frac{3}{4}$ **71.** $18.53 \le m \le 18.77$ **73.** $17.75 \le n \le 17.81$ **75.** $5xy^3$ **76.** $\frac{y^6}{9x^8}$

Quick Quiz 9.7

See Examples noted with Ex. **1.** $x = \frac{21}{2}, -\frac{27}{2}$ (Ex. 1) **2.** $-2 \le x \le 3$ (Ex. 5) **3.** $x < -1\frac{4}{5}$ or $x > 1$ (Ex. 9)
4. See Student Solutions Manual

You Try It

1. $x = \pm 2\sqrt{5}$ **2.** $x = \dfrac{-3 \pm \sqrt{15}}{2}$ **3.** $x = 4 \pm \sqrt{21}$ **4.** $2x^2 - 6x - 11 = 0$ **5.** $x = 1$ **6. (a)** $x = -\dfrac{y}{2}, x = 3y$ **(b)** $x = \pm 2\sqrt{a - y^2}$ **(c)** $x = \dfrac{w \pm \sqrt{w^2 + 36w}}{4}$ **7.** $b = 5\sqrt{3}$ **8.**

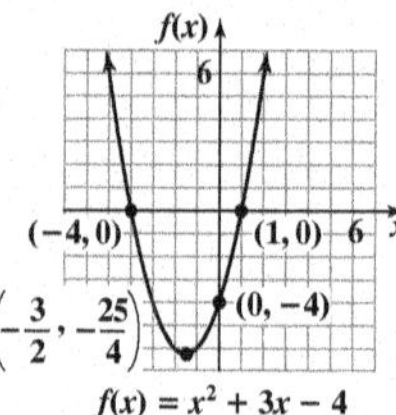

9. $x > -8$ *and* $x < 2$

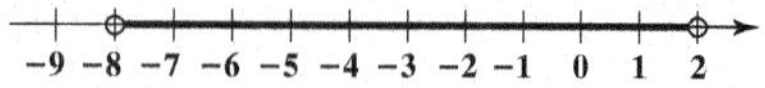

10. $x \leq -2$ *or* $x \geq 3$

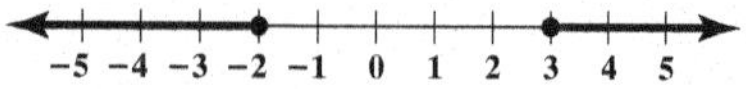

11. $-\dfrac{3}{2} < x < 3$ **12.** $x = 2, x = -\dfrac{16}{3}$ **13.** $-12 < x < 5$

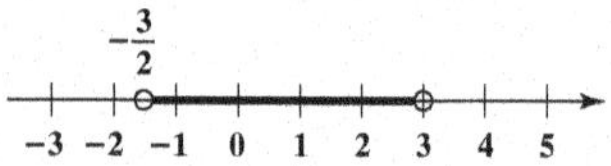

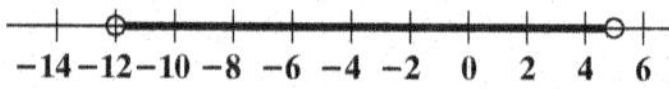

14. $x < -12$ *or* $x > -4$

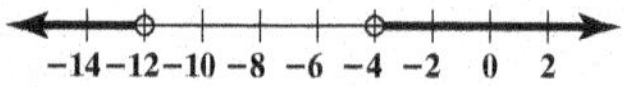

Chapter 9 Review Problems

1. $x = \pm 2$ **2.** $x = 1, -17$ **3.** $x = -4 \pm \sqrt{3}$ **4.** $x = \dfrac{2 \pm \sqrt{3}}{2}$ or $1 \pm \dfrac{\sqrt{3}}{2}$ **5.** $x = 2 \pm \sqrt{6}$ **6.** $x = \dfrac{2}{3}, 2$ **7.** $x = \dfrac{3}{2}$ **8.** $x = 7, -2$ **9.** $x = 0, \dfrac{9}{2}$ **10.** $x = \dfrac{5 \pm \sqrt{17}}{4}$ **11.** $x = \pm\sqrt{2}$ **12.** $x = \dfrac{2}{3}, 5$ **13.** $x = -1 \pm \sqrt{5}$ **14.** $x = \pm 2i\sqrt{3}$ **15.** $x = \dfrac{-5 \pm \sqrt{13}}{6}$ **16.** $x = -\dfrac{1}{2}, 4$ **17.** $x = \dfrac{-1 \pm i}{3}$ **18.** $x = -\dfrac{1}{4}, -1$ **19.** $y = -\dfrac{5}{6}, -2$ **20.** $y = -5, 3$ **21.** $y = 0, -\dfrac{5}{2}$ **22.** $x = -2, 3$ **23.** Two irrational solutions **24.** Two rational solutions **25.** One rational solution **26.** $x^2 - 25 = 0$ **27.** $x^2 + 9 = 0$ **28.** $8x^2 + 14x + 3 = 0$ **29.** $x = \pm 2, \pm\sqrt{2}$ **30.** $x = -\dfrac{\sqrt[3]{4}}{2}, \sqrt[3]{3}$ **31.** $x = 27, -1$ **32.** $x = \pm 1, \pm\sqrt{2}$ **33.** $A = \pm\sqrt{\dfrac{3MN}{2}}$ **34.** $t = \pm\sqrt{3ay - 2b}$ **35.** $x = \dfrac{3 \pm \sqrt{9 + 28y}}{2y}$ **36.** $d = \dfrac{x}{4}, -\dfrac{x}{5}$ **37.** $y = \dfrac{-2a \pm \sqrt{4a^2 + 6a}}{2}$ **38.** $x = \dfrac{2 \pm \sqrt{4 - 6y^2 + 3AB}}{3}$ **39.** $c = \sqrt{22}$ **40.** $a = 4\sqrt{15}$ **41.** Approximately 3.3 mi **42.** Base is 7 cm; altitude is 20 cm **43.** Width is 7 m; length is 29 m **44.** 20 mph for 80 mi, 10 mph for 10 mi **45.** 50 mph during first part, 45 mph during rain **46.** Approximately 2.4 ft wide **47.** 0.5 m wide **48.** Vertex $(3, -2)$; no x-intercepts; y-intercept $(0, -11)$ **49.** Vertex $(-5, 0)$; x-intercept $(-5, 0)$; y-intercept $(0, 25)$ **50.**

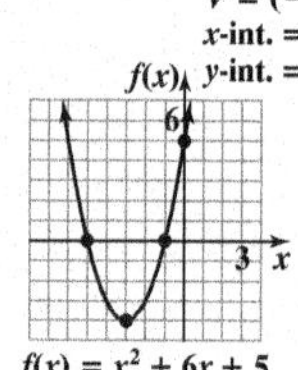

51.

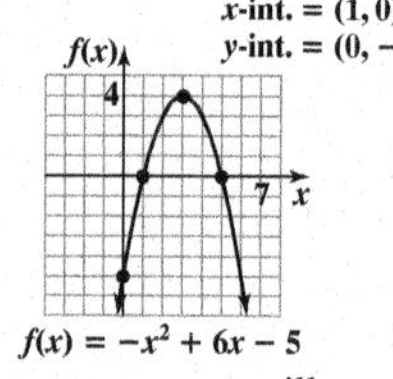

52. Maximum height is 2540 ft; about 25.1 sec for complete flight **53.** $R(x) = x(1200 - x)$; the maximum revenue will occur if the price is \$600 for each unit. **54.** **55.** 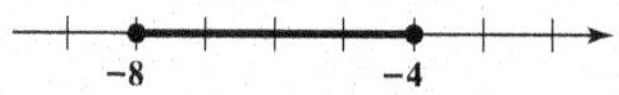 **56.** 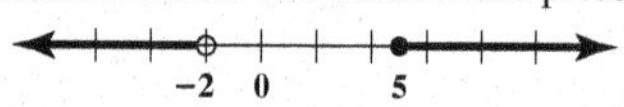 **57.** 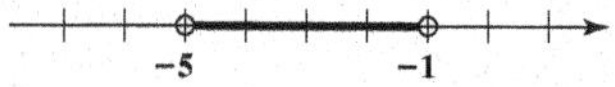 **58.** 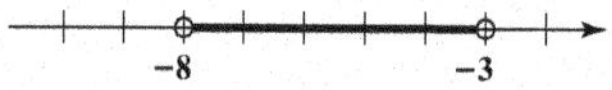 **59.**

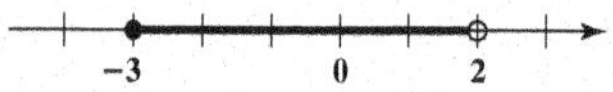

60. $x > 9$ *or* $x < -1$ **61.** No solution **62.** $-6 < x < 3$ **63.** $-4 \leq x \leq 1\dfrac{2}{3}$ **64.** $1\dfrac{4}{5} \leq x < 5$ **65.** All real numbers **66.** $-9 < x < 2$ **67.** $x < 4$ or $x > 5$

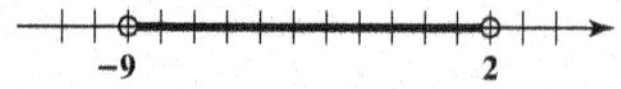

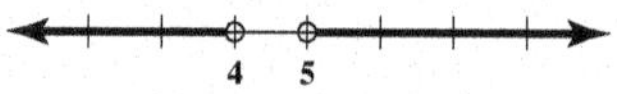

68. $-\dfrac{3}{2} \leq x \leq 2$ **69.** $\dfrac{4}{3} \leq x \leq 3$ **70.** $x < -\dfrac{2}{3}$ or $x > \dfrac{2}{3}$ **71.** $x \leq -6$ or $x \geq -2$ **72.** $x < -8$ or $x > 2$ **73.** $x < 1.4$ or $x > 2.6$ **74.** No real solution **75.** $x = 4, 3$ **76.** $-\dfrac{15}{4} \leq x \leq \dfrac{21}{4}$ **77.** $x < -3$ *or* $x > \dfrac{11}{5}$ **78.** $-22 < x < 8$ **79.** $-27 < x < 9$ **80.** $-7\dfrac{1}{2} < x < -\dfrac{1}{2}$ **81.** $x \leq -4$ *or* $x \geq 5$ **82.** $x \leq -\dfrac{1}{3}$ *or* $x \geq 1$ **83.** $x \leq 4$ *or* $x \geq 6$

How Am I Doing? Chapter 9 Test

1. $x = 0, -\dfrac{9}{8}$ (obj. 9.1.2) **2.** $x = \dfrac{3 \pm \sqrt{33}}{12}$ (obj. 9.1.2) **3.** $x = 2, -\dfrac{2}{9}$ (obj. 9.2.1) **4.** $x = -2, 10$ (obj. 9.2.1) **5.** $x = \pm 2\sqrt{2}$ (obj. 9.1.1) **6.** $x = \dfrac{7}{2}, -1$ (obj. 9.2.1) **7.** $x = \dfrac{3 \pm i}{2}$ (obj. 9.2.1) **8.** $x = \dfrac{3 \pm \sqrt{3}}{2}$ (obj. 9.2.1) **9.** $x = \pm 3, \pm\sqrt{2}$ (obj. 9.3.1) **10.** $x = \dfrac{1}{5}, -\dfrac{3}{4}$ (obj. 9.3.2) **11.** $x = 64, -1$ (obj. 9.3.2) **12.** $z = \pm\sqrt{\dfrac{xyw}{B}}$ (obj. 9.4.1) **13.** $y = \dfrac{-b \pm \sqrt{b^2 - 30w}}{5}$ (obj. 9.4.1) **14.** Width is 5 mi; length is 16 mi (obj. 9.4.3) **15.** $c = 4\sqrt{3}$ (obj. 9.4.2)

16. 2 mph during first part; 3 mph after lunch (obj. 9.4.3) **17.** $V = (-3, 4)$; y-int. $(0, -5)$; x-int. $(-5, 0)$; $(-1, 0)$ (obj. 9.5.2)

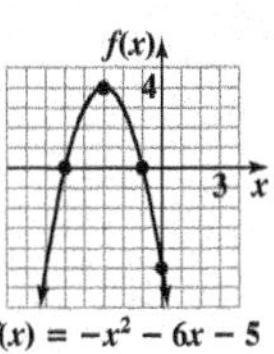

$f(x) = -x^2 - 6x - 5$

18. $-5 < x \leq -1$ (obj. 9.6.3) **19.** $x \leq -2$ *or* $x \geq 1$ (obj. 9.6.3) **20.** $x \leq -\frac{9}{2}$ *or* $x \geq 3$ (obj. 9.6.4) **21.** $-2 < x < 7$ (obj. 9.6.4)
22. $x < -4.5$ *or* $x > 1.5$ (obj. 9.6.5) **23.** $x = -7, \frac{39}{5}$ (obj. 9.7.1) **24.** $x = 6, -18$ (obj. 9.7.2) **25.** $-\frac{15}{7} \leq x \leq 3$ (obj. 9.7.4)
26. $x < -\frac{8}{3}$ *or* $x > 2$ (obj. 9.7.5)

Cumulative Test for Chapters 0–9

1. $\frac{81y^{12}}{x^8}$ **2.** $\frac{1}{4}a^3 - a^2 - 3a$ **3.** $x = 12$ **4.**

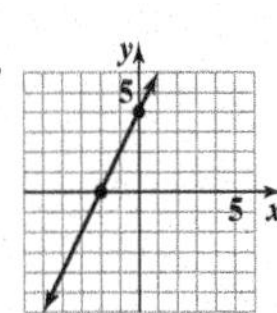

5. $m = -\frac{1}{2}$ **6.** $(2, -5)$ **7.** $(5x - 3y)(25x^2 + 15xy + 9y^2)$ **8.** $6xy^3\sqrt{2x}$ **9.** $4\sqrt{6} - 2\sqrt{2}$ **10.** $\frac{x\sqrt{6}}{2}$ **11.** $x = 0, \frac{14}{3}$
12. $x = \frac{2}{3}, \frac{1}{4}$ **13.** $x = \frac{3 \pm 2\sqrt{3}}{2}$ **14.** $x = \frac{2 \pm i\sqrt{11}}{3}$ **15.** $x = 4, -2$ **16.** $x = -27, -216$ **17.** $y = \frac{-5w \pm \sqrt{25w^2 + 56z}}{4}$
18. $y = \pm\frac{\sqrt{15w - 48z^2}}{3}$ **19.** $\sqrt{13}$ **20.** Base is 5 m; altitude is 18 m **21.** Vertex (4, 4); y-intercept $(0, -12)$; x-intercepts (2, 0); (6, 0)
22. [graph] $f(x) = -x^2 + 8x - 12$ **23.** $x \leq -9$ *or* $x \geq -2$ **24.** $x < -6$ *or* $x > 3$ **25.** $x = 5, x = -\frac{17}{3}$ **26.** $-20 \leq x \leq 12$ **27.** $x < -2\frac{1}{3}$ *or* $x > 5$

Chapter 10

10.1 Exercises

1. Subtract the values of the points and use the absolute value: $|-2 - 4| = 6$ **3.** Since the equation is given in standard form, we determine the values of h, k, and r to find the center and radius: $h = 1, k = -2$, and $r = 3$. Thus, the center is $(1, -2)$ and the radius is 3. **5.** $\sqrt{5}$ **7.** $2\sqrt{5}$ **9.** 10 **11.** $\frac{\sqrt{10}}{3}$ **13.** $\frac{2\sqrt{26}}{5}$ **15.** $5\sqrt{2}$ **17.** $y = 10, y = -6$ **19.** $y = 0, y = 4$
21. $x = 3, x = 5$ **23.** 4.4 mi **25.** $(x + 3)^2 + (y - 7)^2 = 36$ **27.** $(x + 2.4)^2 + y^2 = \frac{9}{16}$ **29.** $x^2 + \left(y - \frac{3}{8}\right)^2 = 3$
31. [graph] $x^2 + y^2 = 25$ **33.** [graph] $(x - 5)^2 + (y - 3)^2 = 16$ **35.** [graph] $(x + 2)^2 + (y - 3)^2 = 25$
37. $(x - 2)^2 + (y + 5)^2 = 25$; center $(2, -5)$, $r = 5$
39. $(x - 5)^2 + (y + 3)^2 = 36$; center $(5, -3)$, $r = 6$
41. $\left(x + \frac{3}{2}\right)^2 + y^2 = \frac{17}{4}$; center $\left(-\frac{3}{2}, 0\right)$, $r = \frac{\sqrt{17}}{2}$
43. $(x - 61.5)^2 + (y - 55.8)^2 = 1953.64$
45.

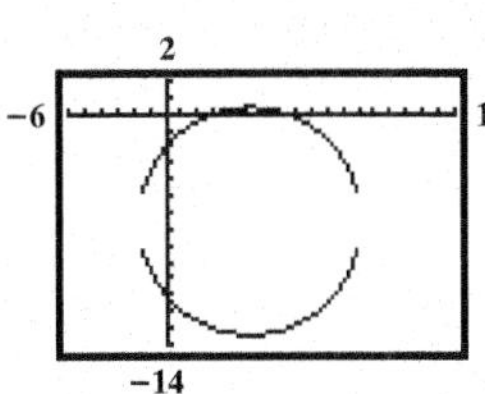

47. $x = \frac{-1 \pm \sqrt{5}}{4}$ **48.** $x = \frac{3 \pm 2\sqrt{11}}{5}$ **49.** Approximately 8.364×10^{10} ft^3 **50.** Approximately 81 sec

Quick Quiz 10.1

See Examples noted with Ex. **1.** $\sqrt{29}$ (Ex. 1) **2.** $(x - 5)^2 + (y + 6)^2 = 49$ (Ex. 3) **3.** $(x + 2)^2 + (y - 3)^2 = 9$; center $(-2, 3)$; $r = 3$ (Ex. 4) **4.** See Student Solutions Manual

10.2 Exercises

1. y-axis; x-axis **3.** If it is in the standard form $y = a(x - h)^2 + k$, the vertex is (h, k). So in this case the vertex is $(3, 4)$.
5. [graph] y-int (0, 0); $y = -4x^2$ **7.** [graph] (0, −2); y-int (0, −2); $y = x^2 - 2$ **9.** [graph] (0, −2); y-int (0, −2); $y = \frac{1}{2}x^2 - 2$ **11.**

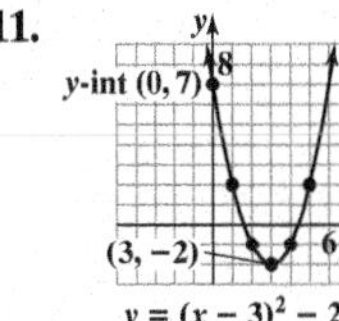

$y = (x - 3)^2 - 2$

13.

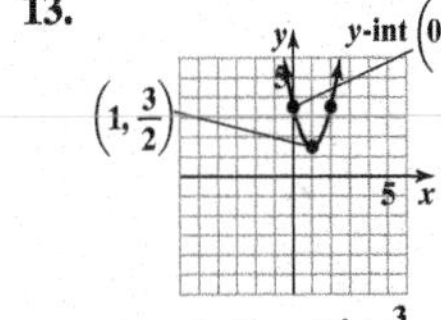

$y = 2(x - 1)^2 + \frac{3}{2}$

15. 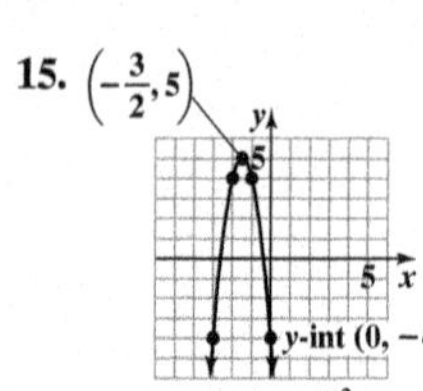

17. 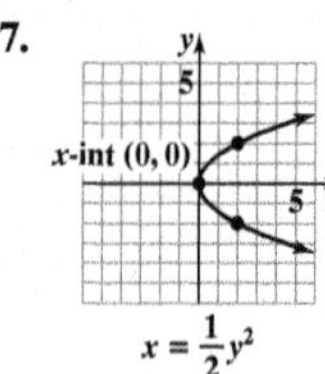

19. 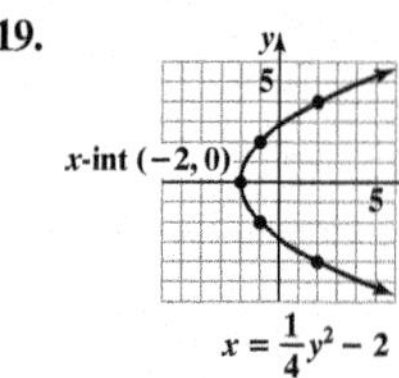

21. 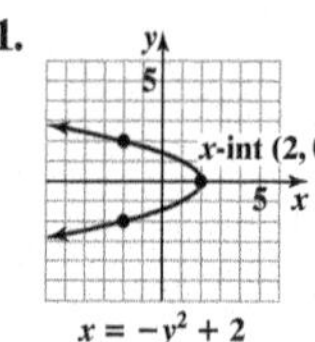

23. 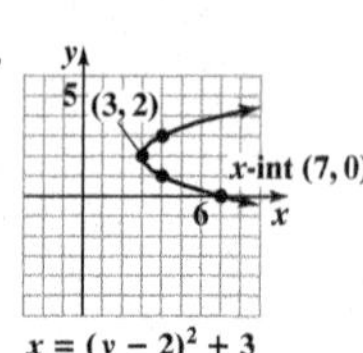

25. 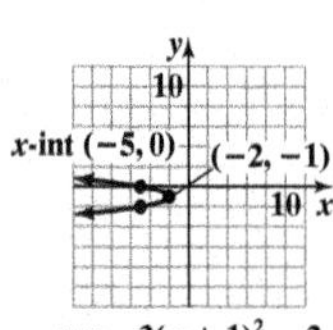

27. $y = (x - 2)^2 - 5$ **(a)** Vertical **(b)** Opens upward **(c)** $(2, -5)$ 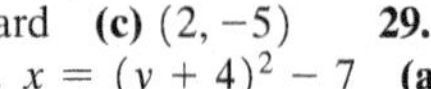 **29.** $y = -2(x - 3)^2 + 2$ **(a)** Vertical **(b)** Opens downward **(c)** $(3, 2)$ **31.** $x = (y + 4)^2 - 7$ **(a)** Horizontal **(b)** Opens right **(c)** $(-7, -4)$ **33.** $y = \frac{1}{20}x^2$ **35.** 5 in. **37.** Vertex $(-1.62, -5.38)$; y-intercept $(0, -0.1312)$; x-intercepts are approximately $(0.02012195, 0)$ and $(-3.260122, 0)$ **39.** Maximum profit = \$45,000; number of watches produced = 70 **41.** Maximum yield = 202,500 oranges; number of trees planted per acre = 450 **43.** $5x\sqrt{2x}$ **44.** $2xy\sqrt[3]{5y}$ **45.** $2x\sqrt{2} - 8\sqrt{2x}$ **46.** $2x\sqrt[3]{2x} - 20x\sqrt[3]{2}$ **47.** 7392 blooms **48.** Approximately 1098 buds

Quick Quiz 10.2

See Examples noted with Ex. **1.** Vertex $(-2, 5)$; y-intercept $(0, -7)$ (Ex. 2) **2.** $x = (y - 2)^2 + 3$ (Ex. 5)

3. 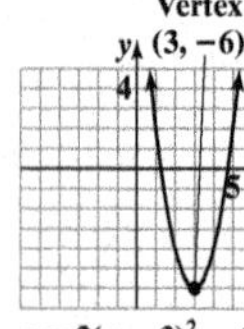

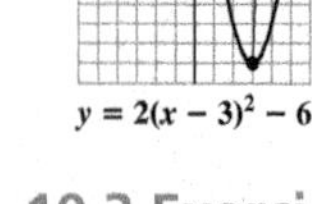

(Ex. 7) **4.** See Student Solutions Manual

10.3 Exercises

1. In the ellipse $\frac{(x - h)^2}{a^2} + \frac{(y - k)^2}{b^2} = 1$ the center is at (h, k). In this case, the center of the ellipse is $(-2, 3)$.

3. 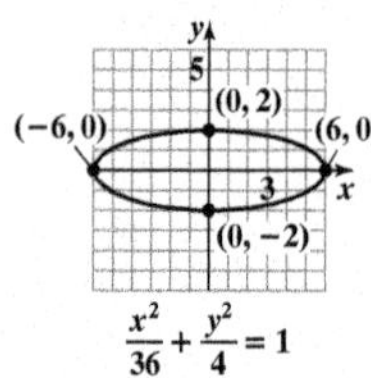

5. 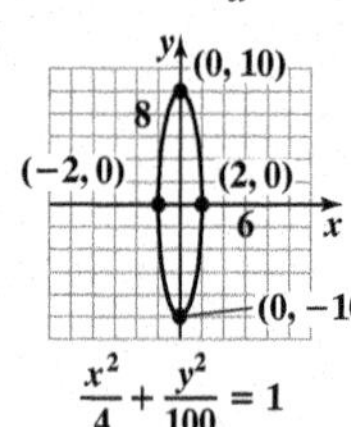

7. 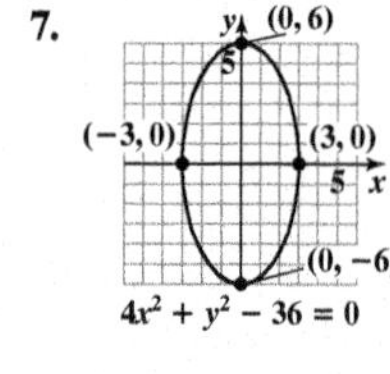

9. 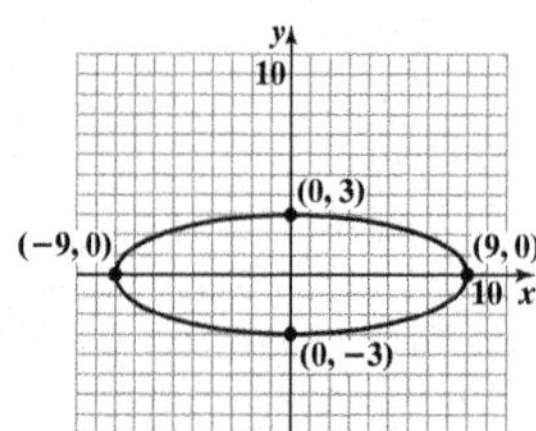

11. 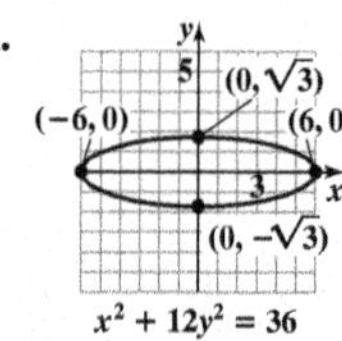

13. 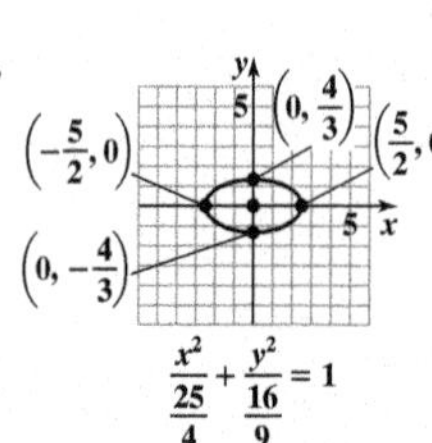

15. 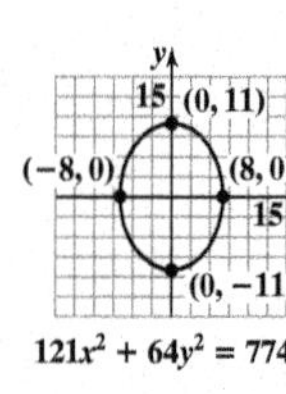
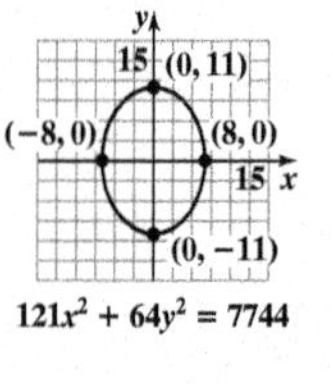

17. $\frac{x^2}{169} + \frac{y^2}{144} = 1$ **19.** $\frac{x^2}{36} + \frac{y^2}{48} = 1$ **21.** 142 million mi

23. 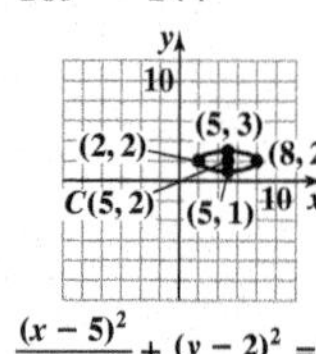

25. 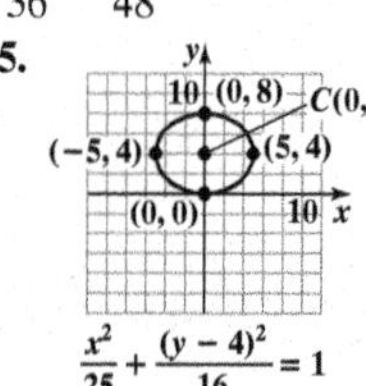

27. 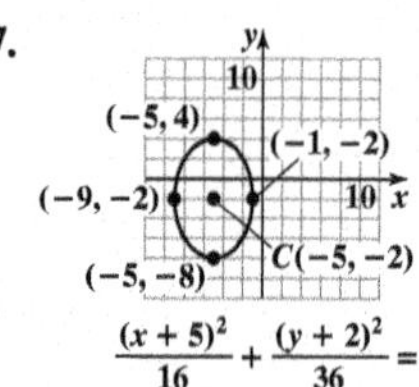

29. $(x - 5)^2 + \frac{(y - 1)^2}{4} = 1$ **31.** $\frac{(x - 30)^2}{900} + \frac{(y - 20)^2}{400} = 1$ **33.** $(0, 7.2768)$, $(0, 3.3232)$, $(4.2783, 0)$, $(2.9217, 0)$ **35.** $\frac{5(\sqrt{2x} + \sqrt{y})}{2x - y}$ **36.** $30\sqrt{2} - 2\sqrt{6} + 40\sqrt{3} - 8$ **37.** $22\frac{2}{3}$ weeks

Quick Quiz 10.3

See Examples noted with Ex. **1.** $\frac{x^2}{25} + \frac{y^2}{36} = 1$ (Ex. 1) **2.** $\frac{(x + 3)^2}{16} + \frac{(y + 2)^2}{9} = 1$ (Ex. 1)

3. 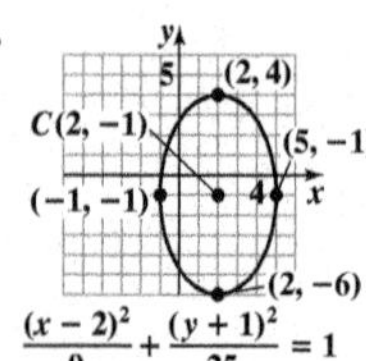

(Ex. 2) **4.** See Student Solutions Manual

Use Math to Save Money

1. Private loan: $I = (10{,}000)(0.0465)(20) = \$9300$. Total Amount = 10,000 + 9300 = \$19,300. Subsidized loan: $I = (10{,}000)(0.06)(20 - 4.5) = \$9300$. Total Amount = 10,000 + 9300 = \$19,300. **2.** The total amount that Alicia must pay back is the same for each loan. **3.** Total Amount ÷ Number of Payments = Monthly Payment. Private loan: 19,300 ÷ (12 months × 20 years) = 19,300 ÷ 240 = \$80.42. Subsidized loan: 19,300 ÷ (12 months × 15.5 years) = 19,300 ÷ 186 = \$103.76.

4. Private student loan **5.**

Type of Loan	Total Amount	Number of Monthly Payments	Monthly Payment Amount	When Payments Begin
Private loan	$19,300	240	$80.42	Immediately
Subsidized loan	$19,300	186	$103.76	6 months after graduation

6. Answers will vary.

10.4 Exercises

1. The standard form of a horizontal hyperbola centered at the origin is $\frac{x^2}{a^2} - \frac{y^2}{b^2} = 1$ with a and b being positive real numbers.
3. This is a horizontal hyperbola, centered at the origin, with vertices at $(4, 0)$ and $(-4, 0)$. Draw a fundamental rectangle with corners at $(4, 2)$, $(4, -2)$, $(-4, 2)$, and $(-4, -2)$. Extend the diagonals of the rectangle as asymptotes of the hyperbola. Construct each branch of the hyperbola passing through a vertex and approaching the asymptotes.

5.

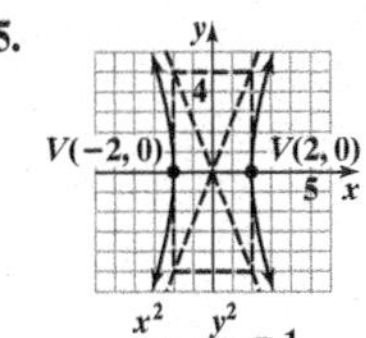

7.

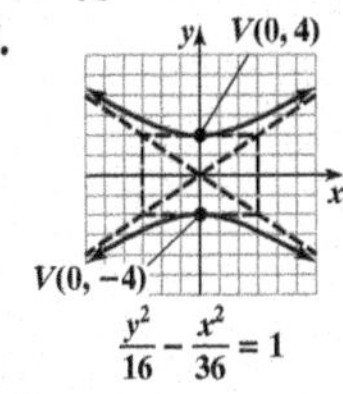

9.

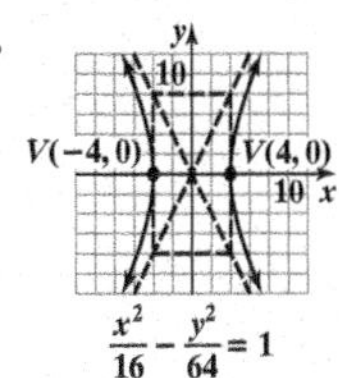

11.

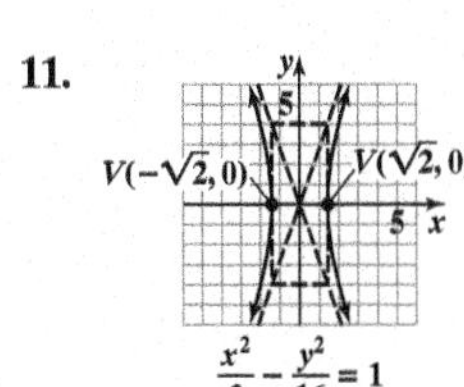

13.

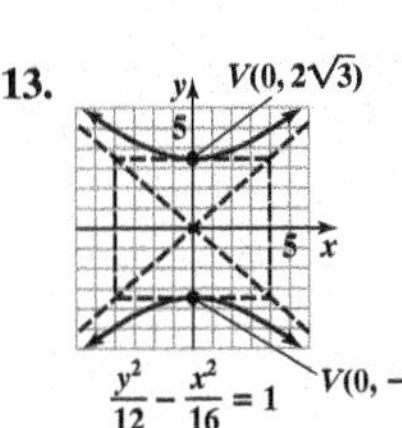

15. $\frac{x^2}{9} - \frac{y^2}{16} = 1$ **17.** $\frac{y^2}{121} - \frac{x^2}{169} = 1$

19. $\frac{x^2}{14{,}400} - \frac{y^2}{129{,}600} = 1$, where x and y are measured in millions of miles

21.

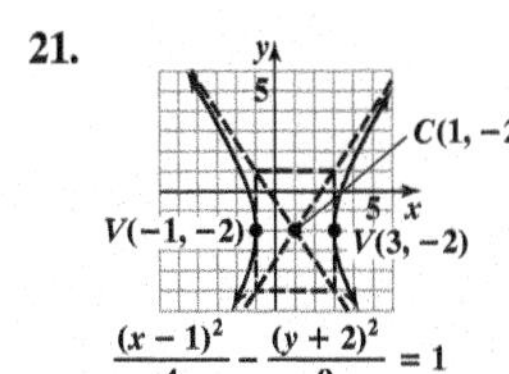

23.

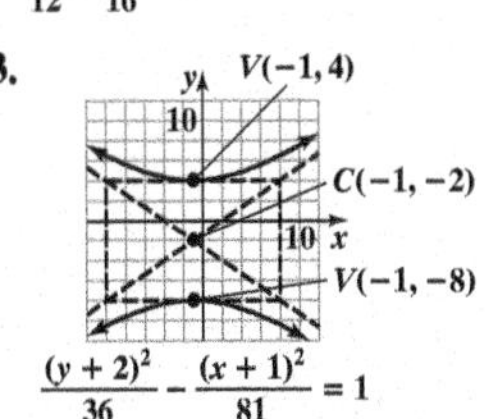

25. Center $(-6, 0)$; vertices $(-6 + \sqrt{7}, 0)$, $(-6 - \sqrt{7}, 0)$
27. $\frac{(y+7)^2}{49} - \frac{(x-4)^2}{16} = 1$ **29.** $y = \pm 9.055385138$
31. $\frac{5x}{(x-3)(x-2)(x+2)}$ **32.** $\frac{-x-6}{(5x-1)(x+2)}$ or $-\frac{x+6}{(5x-1)(x+2)}$
33. $123.4 million **34.** 88.2 million barrels

Quick Quiz 10.4

See Examples noted with Ex. **1.** $(0, 1)$, $(0, -1)$ (Ex. 2) **2.** $\frac{x^2}{16} - \frac{y^2}{25} = 1$ (Ex. 1) **3.**

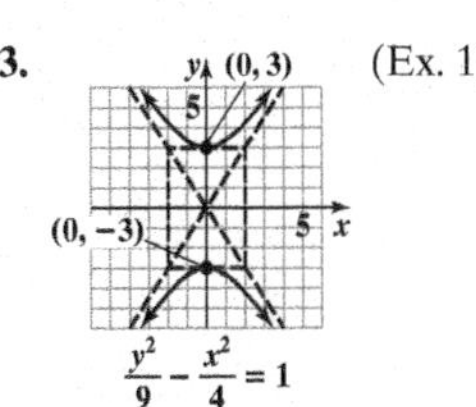

(Ex. 1)

4. See Student Solutions Manual

10.5 Exercises

1. $\left(\frac{1}{2}, 1\right)$, $(2, -2)$

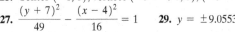

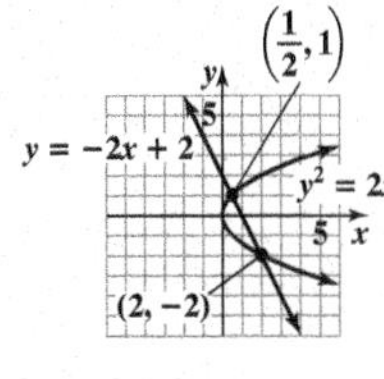

3. $(-4, 2)$, $(4, -2)$

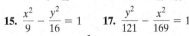

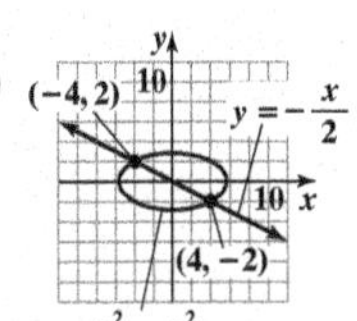

5. $(-2, 3)$, $(1, 0)$
7. $(-4, 0)$, $\left(-\frac{12}{5}, -\frac{16}{5}\right)$
9. $\left(\frac{2}{3}, \frac{4}{3}\right)$, $(2, 0)$ **11.** $\left(\frac{5}{2}, \frac{3}{2}\right)$

13. $(3, 2)$, $(-3, 2)$, $(3, -2)$, $(-3, -2)$

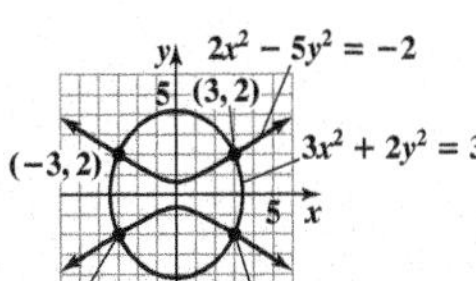

15. $(2, \sqrt{5})$, $(2, -\sqrt{5})$, $(-2, \sqrt{5})$, $(-2, -\sqrt{5})$
17. $\left(\frac{\sqrt{30}}{3}, \frac{\sqrt{21}}{3}\right)$, $\left(-\frac{\sqrt{30}}{3}, \frac{\sqrt{21}}{3}\right)$, $\left(\frac{\sqrt{30}}{3}, -\frac{\sqrt{21}}{3}\right)$, $\left(-\frac{\sqrt{30}}{3}, -\frac{\sqrt{21}}{3}\right)$
19. $(2, \sqrt{3})$, $(-2, \sqrt{3})$, $(2, -\sqrt{3})$, $(-2, -\sqrt{3})$ **21.** $\left(3, -\frac{2}{3}\right)$, $\left(-4, \frac{1}{2}\right)$

23. $(-3, 2)$, $(1, -6)$ **25.** No real solution **27.** Yes, the hyperbola intersects the circle; (3290, 2270) **29.** $x^2 - 3x - 10$ **30.** $\frac{2x(x+1)}{x-2}$

Quick Quiz 10.5

See Examples noted with Ex. **1.** $(4, 4)$, $(1, -2)$ (Ex. 1) **2.** $(0, -4)$, $(\sqrt{7}, 3)$, $(-\sqrt{7}, 3)$ (Ex. 3) **3.** $(-2, 4)$, $(1, 1)$ (Ex. 2) **4.** See Student Solutions Manual

Career Exploration Problems

1. About 8.5 feet **2.** About 26 feet **3.** Yes

You Try It

1. 5 **2.**

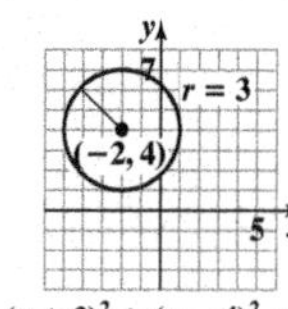

$(x + 2)^2 + (y - 4)^2 = 9$

3. Vertex $(-1, 3)$; $(0, 1)$

$y = -2(x + 1)^2 + 3$

4.

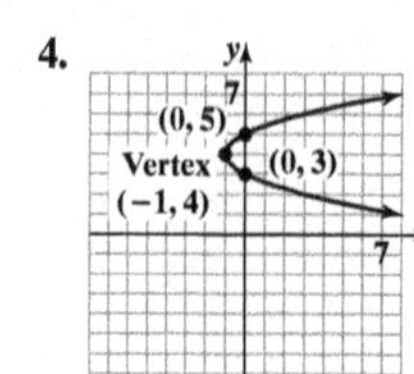

$x = (y - 4)^2 - 1$

5. $(-6, 0)$, $(0, 1)$, $(6, 0)$, $(0, -1)$

$\frac{x^2}{36} + y^2 = 1$

6.

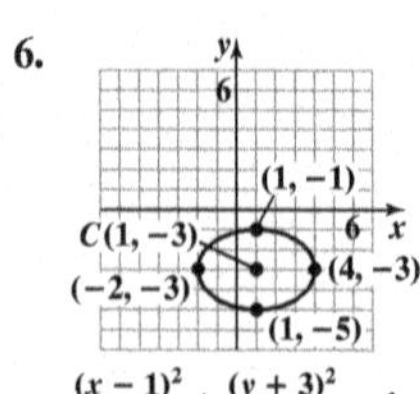

$\frac{(x - 1)^2}{4} + \frac{(y + 3)^2}{9} = 1$

7.

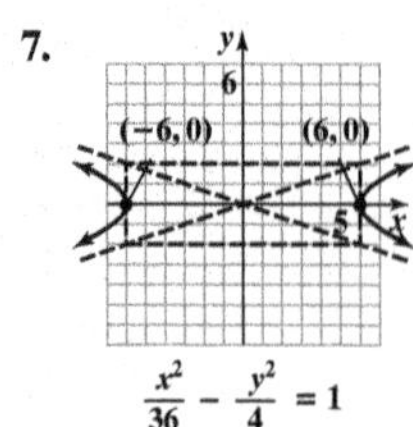

$\frac{x^2}{36} - \frac{y^2}{4} = 1$

8.

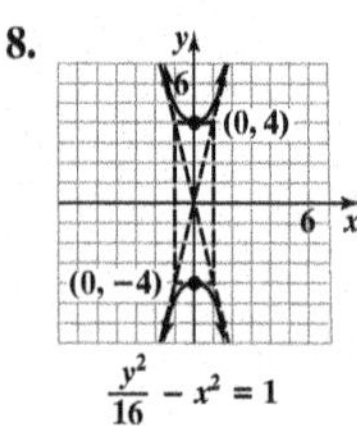

$\frac{y^2}{16} - x^2 = 1$

9.

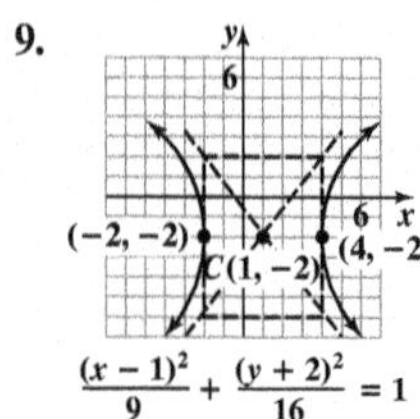

$\frac{(x - 1)^2}{9} + \frac{(y + 2)^2}{16} = 1$

10.

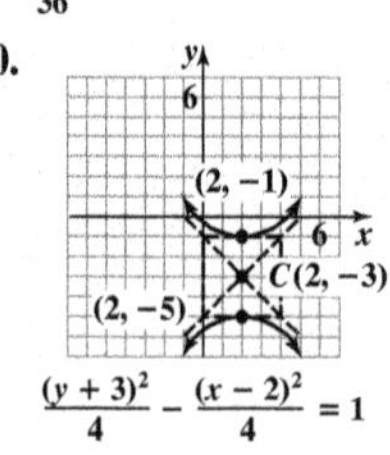

$\frac{(y + 3)^2}{4} - \frac{(x - 2)^2}{4} = 1$

11. $(0, -2), \left(\frac{9}{7}, \frac{13}{7}\right)$

Chapter 10 Review Problems

1. $\sqrt{73}$ **2.** $\sqrt{41}$ **3.** $(x + 6)^2 + (y - 3)^2 = 15$ **4.** $x^2 + (y + 7)^2 = 25$
5. $(x + 1)^2 + (y - 3)^2 = 5$; center $(-1, 3), r = \sqrt{5}$ **6.** $(x - 5)^2 + (y + 6)^2 = 9$; center $(5, -6), r = 3$

7.

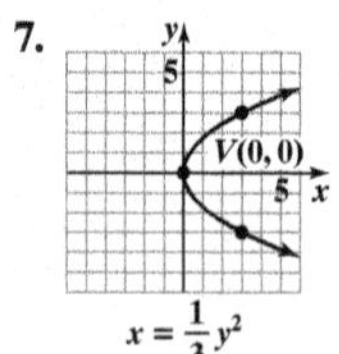

$x = \frac{1}{3}y^2$

8.

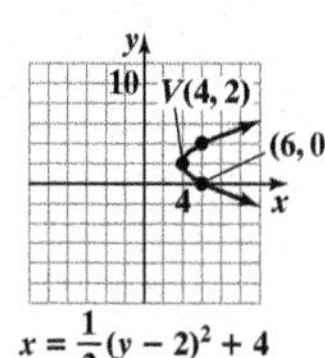

$x = \frac{1}{2}(y - 2)^2 + 4$

9.

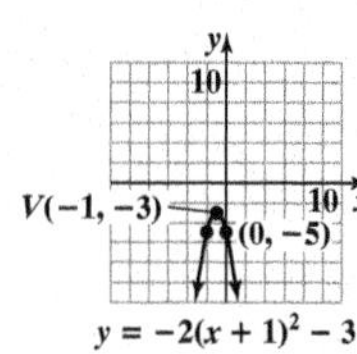

$y = -2(x + 1)^2 - 3$

10. $y = (x + 3)^2 - 5$; vertex at $(-3, -5)$; opens upward
11. $x = (y - 4)^2 - 6$; vertex at $(-6, 4)$; opens to the right

12.

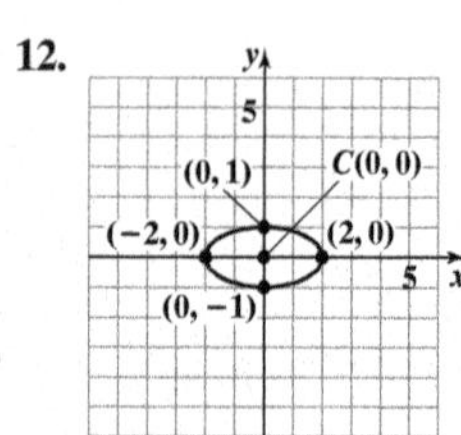

$\frac{x^2}{4} + y^2 = 1$

13.

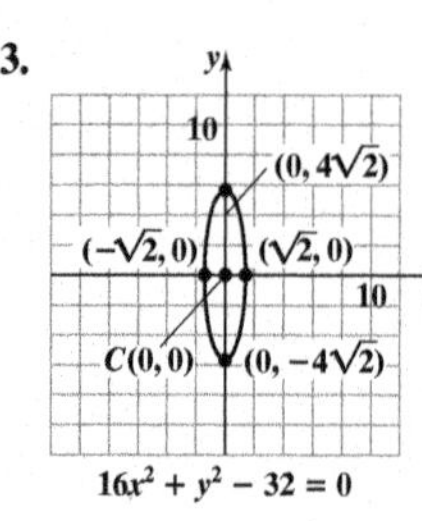

$16x^2 + y^2 - 32 = 0$

14. Center $(-5, -3)$; vertices are $(-3, -3), (-7, -3), (-5, 2), (-5, -8)$

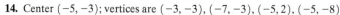

15.

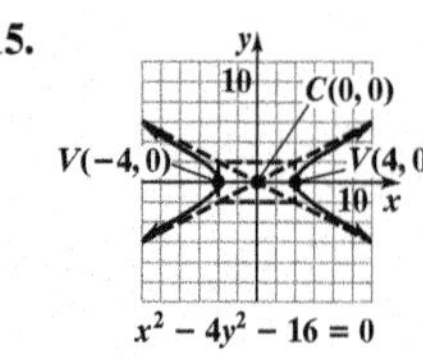

$x^2 - 4y^2 - 16 = 0$

16.

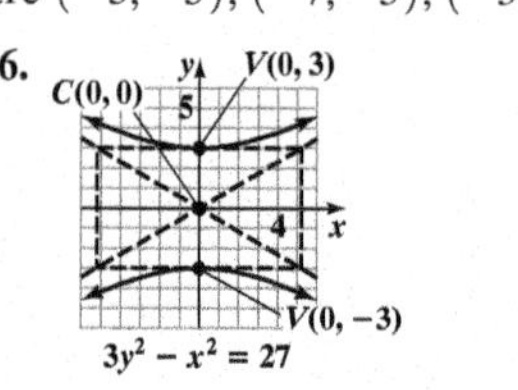

$3y^2 - x^2 = 27$

17. Vertices are $(0, -3), (4, -3)$; center $(2, -3)$ **18.** $(-3, 0), (2, 5)$ **19.** $(0, 2), (2, 0)$ **20.** $(2, 3), (-2, 3), (2, -3), (-2, -3)$
21. No real solution **22.** $(0, 1), (\sqrt{5}, 6), (-\sqrt{5}, 6)$ **23.** $(1, 4), (-1, -4), (2\sqrt{2}, \sqrt{2}), (-2\sqrt{2}, -\sqrt{2})$
24. $(1, 2), (-1, 2), (1, -2), (-1, -2)$ **25.** $(2, 2)$ **26.** 0.78 ft **27.** 1.56 ft

How Am I Doing? Chapter 10 Test

1. $\sqrt{185}$ (obj. 10.1.1) **2.** Parabola; vertex $(4, 3)$; $x = (y - 3)^2 + 4$ (obj. 10.2.3)

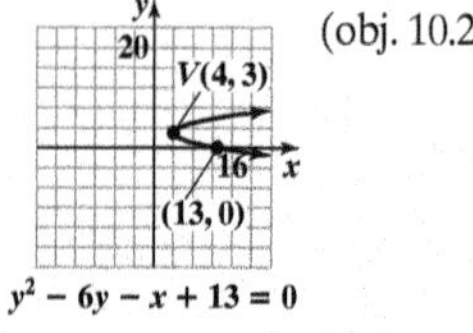

$y^2 - 6y - x + 13 = 0$

3. Circle; center $(-3, 2)$; $(x + 3)^2 + (y - 2)^2 = 4$ (obj. 10.1.4)

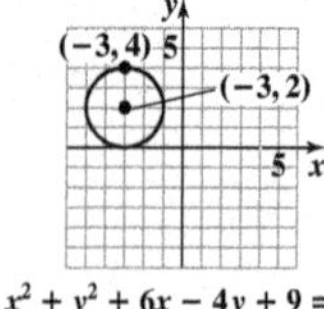

$x^2 + y^2 + 6x - 4y + 9 = 0$

4. Ellipse; center $(0, 0)$ (obj. 10.3.1)

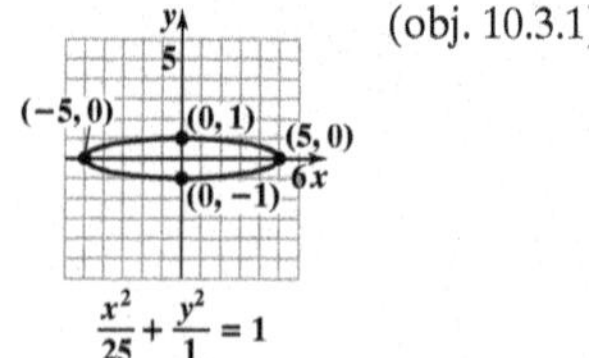

$\frac{x^2}{25} + \frac{y^2}{1} = 1$

5. Hyperbola; center $(0, 0)$ (obj. 10.4.1)

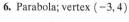

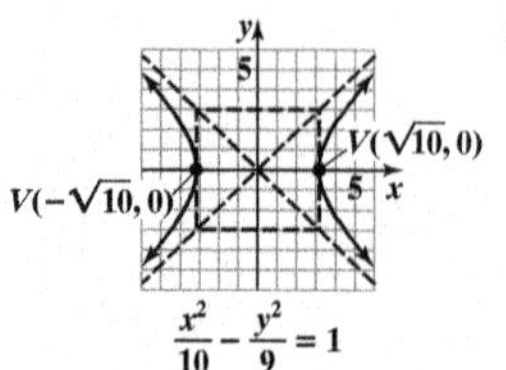

$\frac{x^2}{10} - \frac{y^2}{9} = 1$

6. Parabola; vertex $(-3, 4)$ (obj. 10.2.1)

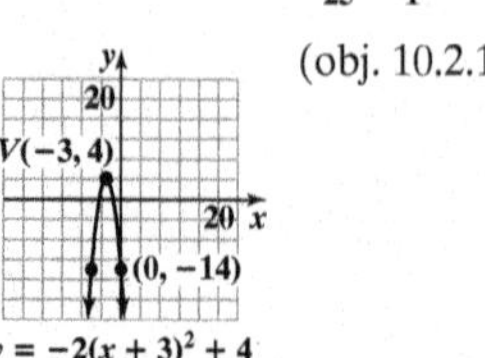

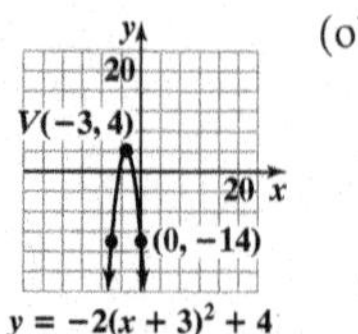

$y = -2(x + 3)^2 + 4$

7. Ellipse; center $(-2, 5)$ (obj. 10.3.2)

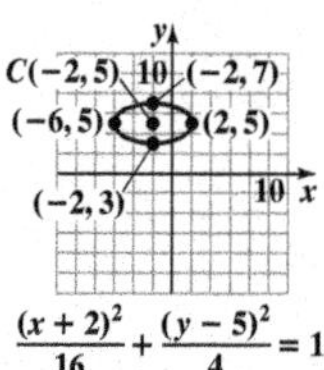

$\frac{(x + 2)^2}{16} + \frac{(y - 5)^2}{4} = 1$

8. Hyperbola; center $(0, 0)$ (obj. 10.4.1)

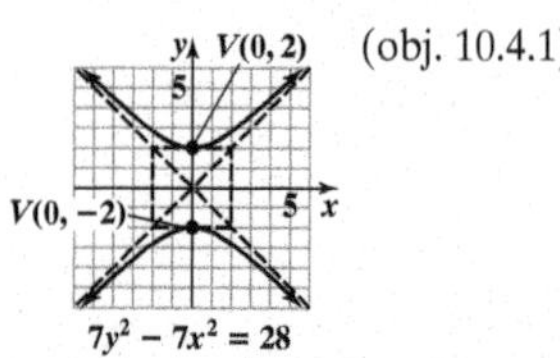

$7y^2 - 7x^2 = 28$

9. $(x-3)^2+(y+5)^2=8$ (obj. 10.1.3) **10.** $\frac{x^2}{9}+\frac{y^2}{25}=1$ (obj. 10.3.1) **11.** $x=(y-3)^2-7$ (obj. 10.2.3) **12.** $\frac{x^2}{9}-\frac{y^2}{25}=1$ (obj. 10.4.1) **13.** $(0,5),(-4,-3)$ (obj. 10.5.1) **14.** $(0,-3),(3,0)$ (obj. 10.5.1) **15.** $(1,0),(-1,0)$ (obj. 10.5.2) **16.** $(3,-\sqrt{3}),(3,\sqrt{3}),(-3,\sqrt{3}),(-3,-\sqrt{3})$ (obj. 10.5.2)

Chapter 11

11.1 Exercises

1. -7 **3.** $3a+10$ **5.** $\frac{1}{2}a-5$ **7.** $\frac{3}{2}a$ **9.** $a-5$ **11.** $\frac{1}{2}a^2-\frac{1}{5}$ **13.** 2 **15.** $\frac{3}{4}$ **17.** $3a^2+10a+5$ **19.** $\frac{4a^2}{3}-\frac{8a}{3}-2$ **21.** 3 **23.** $2\sqrt{3}$ **25.** $\sqrt{a^2+4}$ **27.** $\sqrt{-2b+5}$ **29.** $2\sqrt{a+1}$ **31.** $\sqrt{b^2+b+5}$ **33.** $\frac{7}{4}$ **35.** 14 **37.** $\frac{7}{a^2-3}$ **39.** $\frac{7}{a-1}$ **41.** $-\frac{7}{2}$ **43.** 2 **45.** $2x+h-1$ **47.** **(a)** $P(w)=2.5w^2$ **(b)** 1000 kilowatts **(c)** $P(e)=2.5e^2+100e+1000$ **(d)** 1210 kilowatts **49.** The percent would decrease by 13; 26 **51.** $0.002a^2-0.120a+1.23$ **53.** $x=-3$ **54.** $x=5$ **55.** Approximately 6.7 times greater **56.** Approximately 14.4 times greater

Quick Quiz 11.1

See Examples noted with Ex. **1.** $\frac{3}{5}a+\frac{9}{5}$ (Ex. 1) **2.** $\frac{8}{9}a^2-2a+4$ (Ex. 2) **3.** $\frac{3}{a-2}$ (Ex. 3) **4.** See Student Solutions Manual

11.2 Exercises

1. No, $f(x+2)$ means to substitute $x+2$ for x in the function $f(x)$. $f(x)+f(2)$ means to evaluate $f(x)$ and to evaluate $f(2)$ and then to add the two values. **3.** Up **5.** Not a function **7.** Function **9.** Function **11.** Not a function **13.** Function

15.

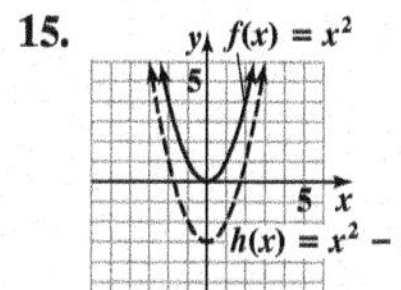

17.

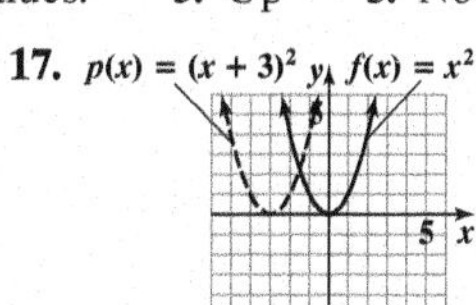

19.

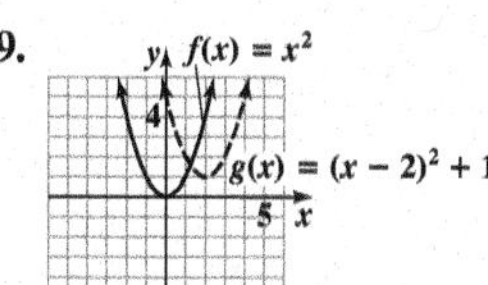

21.

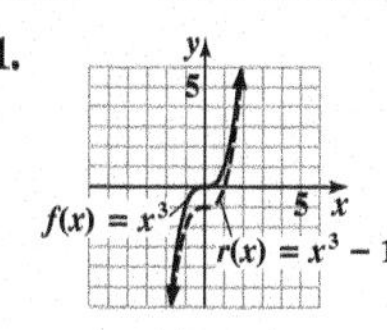

23.

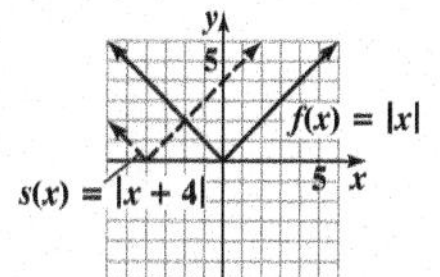

25.

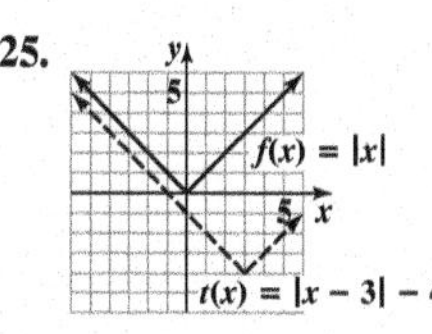

27.

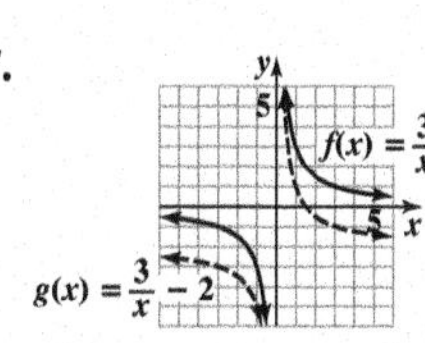

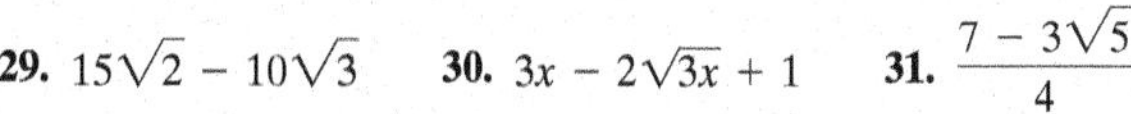

29. $15\sqrt{2}-10\sqrt{3}$ **30.** $3x-2\sqrt{3x}+1$ **31.** $\frac{7-3\sqrt{5}}{4}$

Quick Quiz 11.2

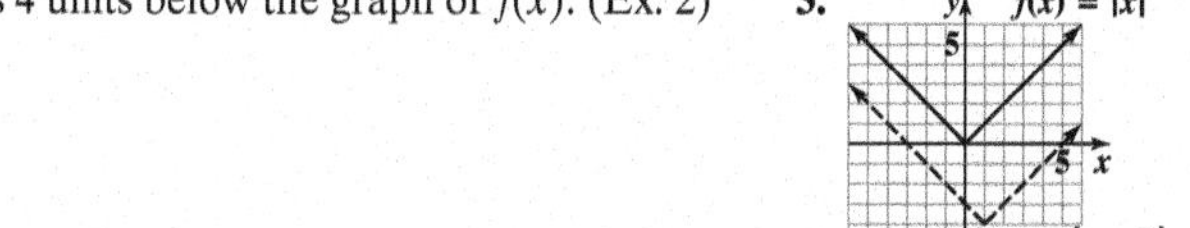

See Examples noted with Ex. **1.** The graph of $h(x)$ is 3 units to the right of the graph of $f(x)$. (Ex. 3) **2.** The graph of $g(x)$ is 4 units below the graph of $f(x)$. (Ex. 2) **3.** (Ex. 4)

4. See Student Solutions Manual

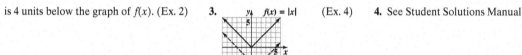

Use Math to Save Money

1. December \$295.00, November \$355.00 **2.** \$1775.00 **3.** 4° lower = 8% savings, \$1775.00 × 0.08 = \$142 **4.** \$355 = 20% of \$1775; 20% = 10 degrees lower, 72 degrees − 10 degrees = 62 degrees **5.** Answers will vary. **6.** Answers will vary.

11.3 Exercises

1. **(a)** $2x+5$ **(b)** $-6x+1$ **(c)** 9 **(d)** 7 **3.** **(a)** $2x^2+4x$ **(b)** $2x^2-4x-2$ **(c)** 16 **(d)** 4 **5.** **(a)** $x^3+\frac{1}{2}x^2+\frac{3}{4}x-5$ **(b)** $x^3-\frac{3}{2}x^2+\frac{5}{4}x+5$ **(c)** $\frac{13}{2}$ **(d)** $\frac{5}{4}$ **7.** **(a)** $3\sqrt{x+6}$ **(b)** $-13\sqrt{x+6}$ **(c)** $6\sqrt{2}$ **(d)** $-13\sqrt{5}$ **9.** **(a)** $-4x^3+11x-3$ **(b)** 72 **11.** **(a)** $\frac{2(x-1)}{x}, x\neq 0$ **(b)** $\frac{8}{3}$ **13.** **(a)** $-3x\sqrt{-2x+1}$ **(b)** $9\sqrt{7}$ **15.** **(a)** $\frac{x-6}{3x}, x\neq 0$ **(b)** $-\frac{2}{3}$ **17.** **(a)** $x+1, x\neq 1$ **(b)** 3 **19.** **(a)** $x+5, x\neq -5$ **(b)** 7 **21.** **(a)** $\frac{1}{x+2}, x\neq -2, x\neq \frac{1}{4}$ **(b)** $\frac{1}{4}$ **23.** $-x^2+5x+2$ **25.** $3x, x\neq 2$ **27.** -3 **29.** -3 **31.** $-6x-13$ **33.** $2x^2-4x+7$ **35.** $-5x^2-7$ **37.** $\frac{7}{2x+1}, x\neq -\frac{1}{2}$ **39.** $|2x+2|$ **41.** $9x^2+30x+27$ **43.** $3x^2+11$ **45.** 11 **47.** $\sqrt{x^2+1}$ **49.** $\frac{3\sqrt{2}}{2}+5$ **51.** $\sqrt{10}$ **53.** $K[C(F)]=\frac{5F+2297}{9}$ **55.** $a[r(t)]=28.26t^2$; 11,304 ft² **56.** $(6x-1)^2$ **57.** $(5x^2+1)(5x^2-1)$ **58.** $(x+3)(x-3)(x+1)(x-1)$ **59.** $(3x-1)(x-2)$

Quick Quiz 11.3

See Examples noted with Ex. **1.** $5x^2-9x-6$ (Ex. 1) **2.** $\frac{x^2}{4}-2x+1$ (Ex. 5) **3.** $\frac{1}{5}$ (Ex. 3) **4.** See Student Solutions Manual

11.4 Exercises

1. have the same second coordinate **3.** $y=x$ **5.** Yes, it passes the vertical line test. No, it does not pass the horizontal line test. **7.** Not one-to-one **9.** One-to-one **11.** Not one-to-one **13.** One-to-one **15.** Not one-to-one **17.** Not one-to-one

19. $J^{-1} = \{(2,8),(1,1),(0,0),(-2,-8)\}$ **21.** $f^{-1}(x) = \frac{x+5}{4}$ **23.** $f^{-1}(x) = \sqrt[3]{x+8}$ **25.** $f^{-1}(x) = -\frac{4}{x}$ **27.** $f^{-1}(x) = \frac{4}{x} + 5$ or $\frac{4+5x}{x}$ **29.** $f[f^{-1}(x)] = x$

31. $g^{-1}(x) = \frac{x-5}{2}$

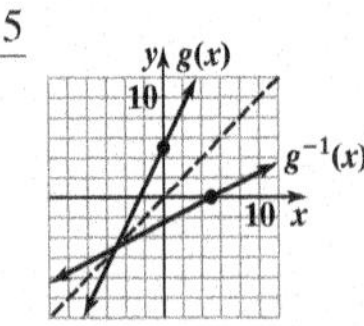

33. $h^{-1}(x) = 2x + 4$

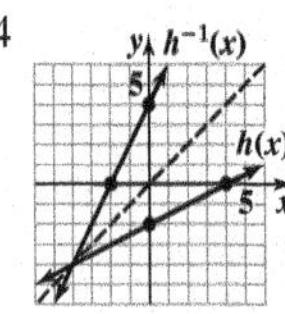

35. $r^{-1}(x) = -\frac{x+1}{3}$

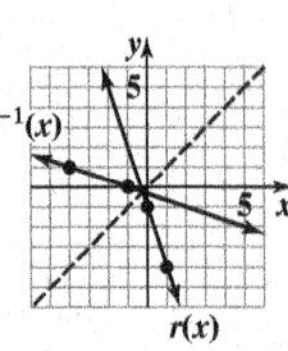

37. No, $f(x)$ is not one-to-one. **39.** $f[f^{-1}(x)] = 2\left(\frac{1}{2}x - \frac{3}{4}\right) + \frac{3}{2} = x$; $f^{-1}[f(x)] = \frac{1}{2}\left(2x + \frac{3}{2}\right) - \frac{3}{4} = x$ **41.** $x = 4$ **42.** $x = -64, x = -27$ **43.** 1,296,400 people **44.** 27%

Quick Quiz 11.4

See Examples noted with Ex. **1. (a)** Yes **(b)** Yes $A^{-1} = \{(-4,3),(-6,2),(6,5),(4,-3)\}$ (Ex. 1, 3)

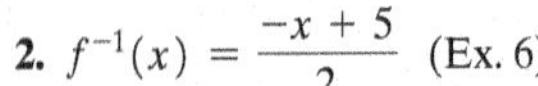

2. $f^{-1}(x) = \frac{-x+5}{2}$ (Ex. 6)

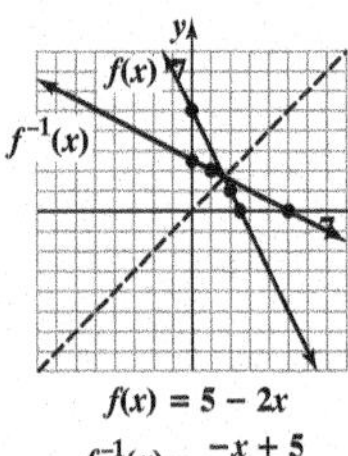

3. $f^{-1}(x) = \sqrt[3]{2-x}$ (Ex. 4) **4.** See Student Solutions Manual

Career Exploration Problems

1. In 2012, 2170 crimes were reported; in 2015, 2207 crimes were reported. **2.** The population in 2012 was 73,561; in 2015, it was 79,414. **3.** 2012: 2.95%; 2015: 2.78% **4.** 2012: 2.76%; 2015: 2.55% **5.** The violent crime model is $V(x) = 1.4x^2 + 7.93x + 142$. There were 142 violent crimes in 2012 and 178 violent crimes in 2015.

You Try It

1. (a) -5 **(b)** $-a^2 + a - 5$ **(c)** $-a^2 + 3a - 7$ **(d)** $-16a^2 + 4a - 5$ **2.** Yes **3. (a)**

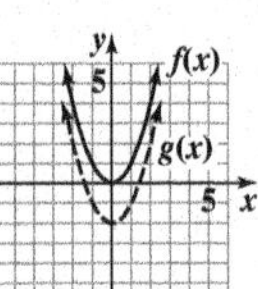

(b)

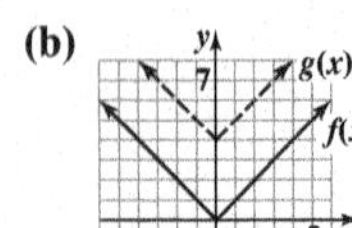

4. (a)

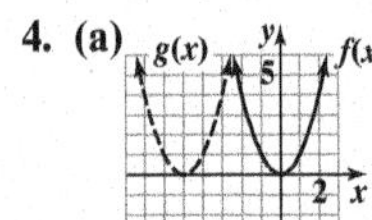

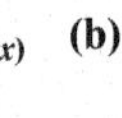

(b)

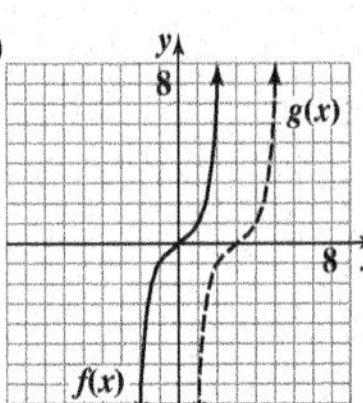

5. (a) $x^2 + 4x - 6$ **(b)** $x^2 + 2x + 6$ **(c)** $x^3 - 3x^2 - 18x$ **(d)** $\frac{x^2+3x}{x-6}, x \neq 6$ **6. (a)** $-x^2 - 1$ **(b)** $x^2 - 4x + 7$ **7.** No **8.** No **9.** $C^{-1} = \{(-2,3),(4,0),(1,5)\}$ **10.** $f^{-1}(x) = 2x + 10$ **11.**

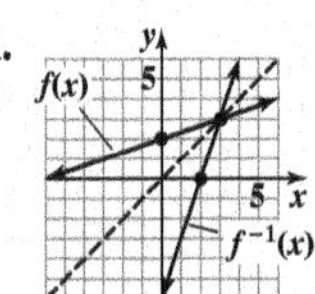

Chapter 11 Review Problems

1. $\frac{1}{2}a + \frac{5}{2}$ **2.** $-\frac{1}{2}$ **3.** $\frac{1}{2}b^2 + \frac{3}{2}$

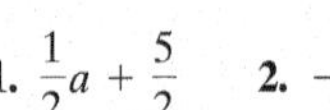

4. -28 **5.** $-8a^2 + 6a - 16$ **6.** $-2a^2 - 5a - 3$ **7.** 1 **8.** $\left|\frac{1}{2}a - 1\right|$ **9.** $|4a^2 - 6a - 1|$ **10.** $\frac{5}{3}$ **11.** $\frac{6a-15}{2a-1}$ **12.** $\frac{30a+36}{7a+28}$ **13.** 7 **14.** 6 **15.** $4x + 2h - 5$ **16. (a)** Not a function **(b)** Not one-to-one **17. (a)** Function **(b)** Not one-to-one **18. (a)** Not a function **(b)** Not one-to-one **19. (a)** Function **(b)** One-to-one

20.

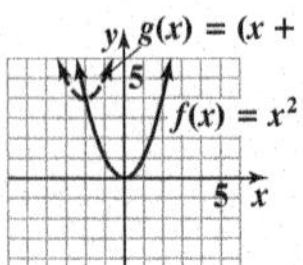

21.

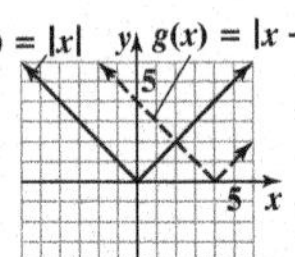

22.

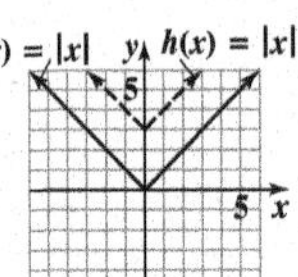

23.

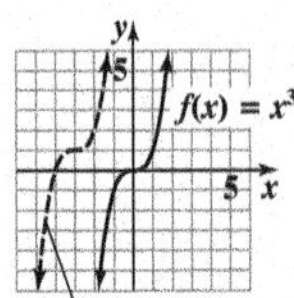

24.

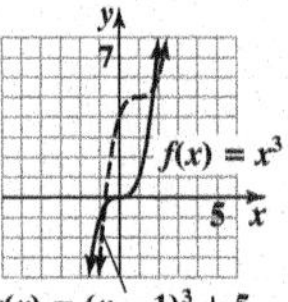

25.

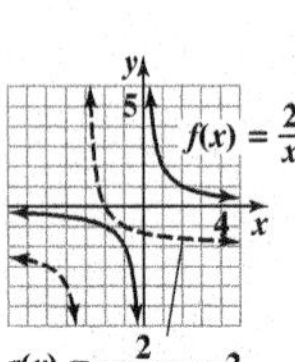

26. $\frac{5}{2}x + 2$ **27.** $2x^2 - 6x - 1$ **28.** -5 **29.** $\frac{6x+10}{x}, x \neq 0$ **30.** $\frac{2x-8}{x^2+x}, x \neq 0, x \neq -1, x \neq 4$ **31.** -6 **32.** $18x^2 + 51x + 39$ **33.** $\sqrt{2x^2 - 3x + 2}$ **34.** 2 **35.** $f[g(x)] = \frac{6}{x} + 5 = \frac{6+5x}{x}$; $g[f(x)] = \frac{2}{3x+5}$; $f[g(x)] \neq g[f(x)]$ **36.** $f^{-1}[f(x)] = x$ **37. (a)** Domain $= \{0, 3, 7\}$ **(b)** Range $= \{-8, 3, 7, 8\}$ **(c)** Not a function **(d)** Not one-to-one **38. (a)** Domain $= \{100, 200, 300, 400\}$ **(b)** Range $= \{10, 20, 30\}$ **(c)** Function **(d)** Not one-to-one **39. (a)** Domain $= \left\{-\frac{1}{3}, \frac{1}{4}, \frac{1}{2}, 4\right\}$ **(b)** Range $= \left\{-3, \frac{1}{4}, 2, 4\right\}$ **(c)** Function **(d)** One-to-one

40. **(a)** Domain $= \{0, 1, 2, 3\}$ **(b)** Range $= \{-3, 1, 7\}$ **(c)** Function **(d)** Not one-to-one **41.** $A^{-1} = \left\{\left(\frac{1}{3}, 3\right), \left(-\frac{1}{2}, -2\right), \left(-\frac{1}{4}, -4\right), \left(\frac{1}{5}, 5\right)\right\}$
42. $f^{-1}(x) = -\frac{4}{3}x + \frac{8}{3}$ **43.** $g^{-1}(x) = -\frac{1}{4}x - 2$ **44.** $h^{-1}(x) = \frac{6}{x} - 5$ or $\frac{6 - 5x}{x}$ **45.** $p^{-1}(x) = x^3 - 1$ **46.** $r^{-1}(x) = \sqrt[3]{x - 2}$
47. $f^{-1}(x) = -3x - 2$ **48.** $f^{-1}(x) = -\frac{4}{3}x + \frac{4}{3}$

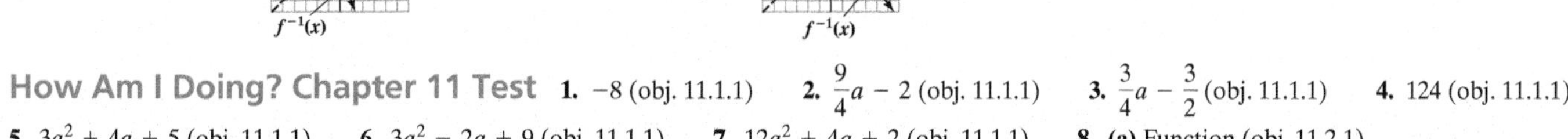

How Am I Doing? Chapter 11 Test

1. -8 (obj. 11.1.1) **2.** $\frac{9}{4}a - 2$ (obj. 11.1.1) **3.** $\frac{3}{4}a - \frac{3}{2}$ (obj. 11.1.1) **4.** 124 (obj. 11.1.1)
5. $3a^2 + 4a + 5$ (obj. 11.1.1) **6.** $3a^2 - 2a + 9$ (obj. 11.1.1) **7.** $12a^2 + 4a + 2$ (obj. 11.1.1) **8.** **(a)** Function (obj. 11.2.1)
(b) Not one-to-one (obj. 11.4.1) **9.** **(a)** Function (obj. 11.2.1) **(b)** One-to-one (obj. 11.4.1)
10. (obj. 11.2.2) **11.** (obj. 11.2.2) **12.** **(a)** $x^2 + 4x + 1$ (obj. 11.3.1) **(b)** $5x^2 - 6x - 13$ (obj. 11.3.1)

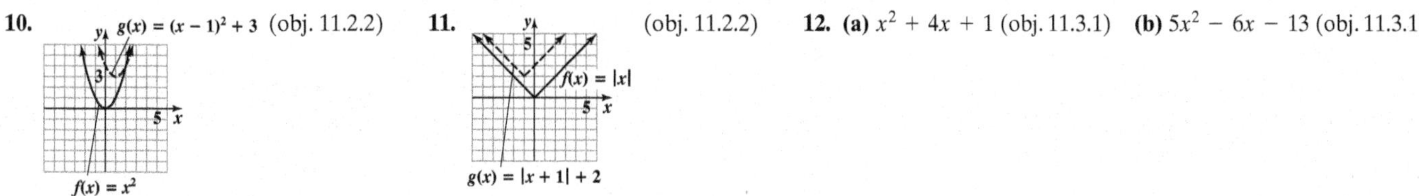

(c) 19 (obj. 11.3.1) **13.** **(a)** $\frac{6x - 3}{x}, x \neq 0$ (obj. 11.3.1) **(b)** $\frac{3}{2x^2 - x}, x \neq 0, x \neq \frac{1}{2}$ (obj. 11.3.1) **(c)** $\frac{3}{2x - 1}, x \neq \frac{1}{2}$ (obj. 11.3.2)
14. **(a)** $2x - \frac{1}{2}$ (obj. 11.3.2) **(b)** $2x - 7$ (obj. 11.3.2) **(c)** 0 (obj. 11.3.2) **15.** **(a)** One-to-one (obj. 11.4.1)
(b) $B^{-1} = \{(8, 1), (1, 8), (10, 9), (9, -10)\}$ (obj. 11.4.2) **16.** **(a)** Not one-to-one (obj. 11.4.1) **(b)** Inverse cannot be found (obj. 11.4.2)
17. $f^{-1}(x) = \frac{x^3 + 1}{2}$ (obj. 11.4.2) **18.** $f^{-1}(x) = -\frac{1}{3}x + \frac{2}{3}$ (obj. 11.4.3) **19.** $f^{-1}[f(x)] = x$ (obj. 11.3.2) **20.** -8 (obj. 11.1.1)

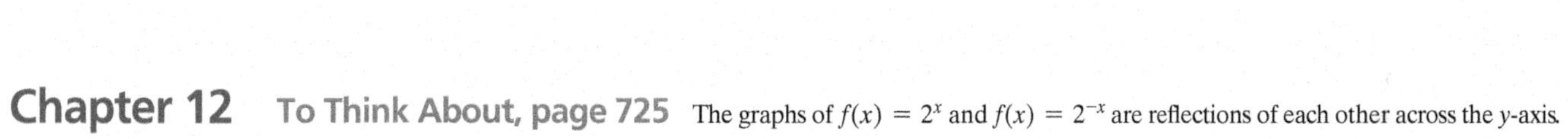

Chapter 12

To Think About, page 725 The graphs of $f(x) = 2^x$ and $f(x) = 2^{-x}$ are reflections of each other across the y-axis.

To Think About, page 726 **1.** To graph $f(x) = 3^{x+2}$, take the graph of $f(x) = 3^x$ and shift it two units to the left. **2.** The graph of $f(x) = e^x$ gets shifted two units to the right to obtain the graph of $f(x) = e^{x-2}$.

12.1 Exercises

1. $f(x) = b^x$, where $b > 0, b \neq 1$, and x is a real number. **3.** $f(x) = 3^x$ **5.** $f(x) = 2^{-x}$ **7.** $f(x) = 3^{-x}$

9. $f(x) = 2^{x+3}$ **11.** $f(x) = 3^{x-5}$ **13.** $f(x) = 2^x + 2$ **15.** $f(x) = e^{x-1}$ **17.** $f(x) = 2e^x$ **19.** $f(x) = e^{1-x}$

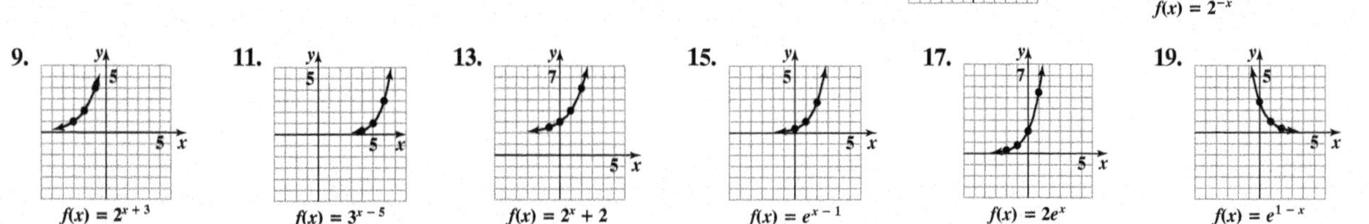

21. $x = 2$ **23.** $x = 0$ **25.** $x = -1$ **27.** $x = 4$ **29.** $x = 0$ **31.** $x = 2$ **33.** $x = 4$ **35.** $x = 2$ **37.** $x = 1$ **39.** \$2402.31
41. \$3632.24; \$3634.08 **43.** 32,000; 2,048,000 **45.** 37.1%; yes **47.** 11.50 mg **49.** 1.80 lb/in.^{2} **51.** \$2,080,000; about \$2,604,831; about 25%
53. 1955 **55.** 10.2 billion people; about 8 billion people **57.** $x = -6$ **58.** $x = -2$

Quick Quiz 12.1 *See Examples noted with Ex.* **1.** $f(x) = 2^{x+4}$ (Ex. 3) **2.** \$6975.33 (Ex. 7) **3.** $x = 2$ (Ex. 5)
4. See Student Solutions Manual

12.2 Exercises **1.** exponent **3.** $x > 0$ **5.** $\log_7 49 = 2$ **7.** $\log_2 128 = 7$ **9.** $\log_{10} 0.001 = -3$ **11.** $\log_2\left(\frac{1}{32}\right) = -5$ **13.** $\log_e y = 5$ **15.** $3^2 = 9$ **17.** $17^0 = 1$ **19.** $16^{1/2} = 4$ **21.** $10^{-2} = 0.01$ **23.** $3^{-4} = \frac{1}{81}$ **25.** $e^{-3/2} = x$ **27.** $x = 16$ **29.** $x = \frac{1}{1000}$ **31.** $y = 3$ **33.** $y = -2$ **35.** $a = 11$ **37.** $a = 10$ **39.** $w = \frac{1}{2}$ **41.** $w = -1$ **43.** $w = 1$ **45.** $w = 9$ **47.** -3 **49.** 7 **51.** 0 **53.** $\frac{1}{2}$ **55.** 0 **57.** $y = \log_3 x$ **59.** $y = \log_{1/4} x$ **61.** $y = \log_{10} x$ **63.** $f^{-1}(x) = 3^x$; $y = x$; $f(x) = \log_3 x$

65. 12.5 **67.** $10^{-3.5}$ **69.** 2.957 **71.** 11,200 sets **73.** \$10,000,000 **75.** $y = \frac{3}{2}x + 7$ **76.** $m = -\frac{1}{5}$ **77.** **(a)** 36,000 cells **(b)** 36,864,000 cells **78.** **(a)** \$5353.27 **(b)** \$24,424.03

Quick Quiz 12.2 *See Examples noted with Ex.* **1.** 4 (Ex. 4) **2.** $x = 64$ (Ex. 3) **3.** $w = \frac{1}{3}$ (Ex. 3) **4.** See Student Solutions Manual

12.3 Exercises **1.** $\log_3 A + \log_3 B$ **3.** $\log_5 7 + \log_5 11$ **5.** $\log_b 9 + \log_b f$ **7.** $\log_9 2 - \log_9 7$ **9.** $\log_b 12 - \log_b Z$ **11.** $\log_a E - \log_a F$ **13.** $7 \log_8 a$ **15.** $-2 \log_b A$ **17.** $\frac{1}{2}\log_5 w$ **19.** $2 \log_8 x + \log_8 y$ **21.** $\log_{11} 6 + \log_{11} M - \log_{11} N$ **23.** $\log_2 5 + \log_2 x + 4 \log_2 y - \frac{1}{2}\log_2 z$ **25.** $\frac{4}{3}\log_a x - \frac{1}{3}\log_a y$ **27.** $\log_4 39y$ **29.** $\log_3\left(\frac{x^5}{7}\right)$ **31.** $\log_b\left(\frac{49y^3}{\sqrt{z}}\right)$ **33.** 1 **35.** 1 **37.** 0 **39.** 3 **41.** $x = 7$ **43.** $x = 11$ **45.** $x = 0$ **47.** $x = 1$ **49.** $x = 3$ **51.** $x = 21$ **53.** $x = 2$ **55.** $x = 5e$ **57.** $x = 3$ **59.** 6 **61.** Approximately 62.8 m^3 **62.** Approximately 50.2 m^2 **63.** $(3, -2)$ **64.** $(-1, -2, 3)$ **65.** \$7301; 17.22% **66.** \$7422; 16.74%

Quick Quiz 12.3 *See Examples noted with Ex.* **1.** $\frac{1}{3}\log_5 x - 4 \log_5 y$ (Ex. 6) **2.** $\log_6\left(\frac{x^3 y}{5}\right)$ (Ex. 5) **3.** $x = 625$ (Ex. 10) **4.** See Student Solutions Manual

Use Math to Save Money **1.** 8/3 or 2 2/3 cups **2.** 56 oz **3.** 3.5 lb **4.** \$9.31 **5.** \$6.23 **6.** Lucy saves approximately 67%. **7.** \$162 × 52 = \$8424; \$8424 × 67% = \$5644.08 **8.** \$1664

12.4 Exercises **1.** Error. You cannot take the logarithm of a negative number. **3.** 1.089905111 **5.** 1.408239965 **7.** 1.176091259 **9.** 5.096910013 **11.** −1.910094889 **13.** $x \approx 103.752842$ **15.** $x = 0.01$ **17.** $x \approx 8519.223264$ **19.** $x \approx 2{,}939{,}679.609$ **21.** $x \approx 0.000408037$ **23.** $x \approx 0.027089438$ **25.** 41,831,168.87 **27.** 0.082679911 **29.** 1.726331664 **31.** 0.425267735 **33.** 11.82041016 **35.** −5.15162299 **37.** $x \approx 2.585709659$ **39.** $x \approx 11.02317638$ **41.** $x \approx 0.951229425$ **43.** $x \approx 0.067205513$ **45.** 472.576671 **47.** 0.1188610637 **49.** 2.020006063 **51.** 1.025073184 **53.** −1.151699337 **55.** 0.917599921 **57.** −1.84641399 **59.** 1.996254706 **61.** 7.337587744 **63.** 0.153508399 **65.** $x \approx 3.6593167 \times 10^8$ **67.** $x \approx 3.3149796$ **69.**

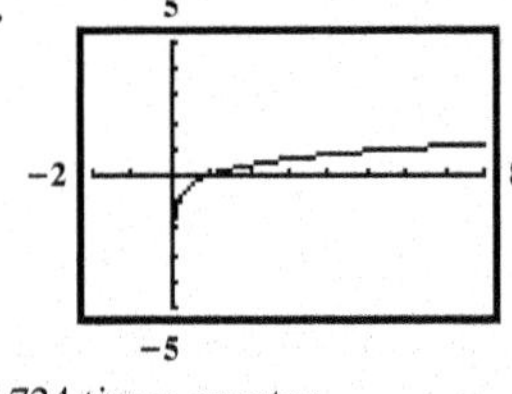

71. 6; −3; 8; −4

73. 36.10; 37.17; 3.0% **75.** $R \approx 4.75$ **77.** The shock wave was about 63,095,734 times greater than the smallest detectable shock wave. **79.** $x = \frac{11 \pm \sqrt{181}}{6}$ **80.** $y = \frac{-2 \pm \sqrt{10}}{2}$ **81.** 6 mi; 12 mi **82.** 21 mi; 24 mi

Quick Quiz 12.4 *See Examples noted with Ex.* **1.** 0.9713 (Ex. 1) **2.** $x \approx 1.68$ (Ex. 4) **3.** 1.3119 (Ex. 7) **4.** See Student Solutions Manual

12.5 Exercises **1.** $x = 20$ **3.** $x = 1$ **5.** $x = 4$ **7.** $x = 2$ **9.** $x = \frac{3}{2}$ **11.** $x = 2$ **13.** $x = 3$ **15.** $x = 5$ **17.** $x = \frac{5}{3}$ **19.** $x = 4$ **21.** $x = 5$ **23.** $x = \frac{\log 12 - 3 \log 7}{\log 7}$ **25.** $x = \frac{\log 17 - 4 \log 2}{3 \log 2}$ **27.** $x \approx 1.582$ **29.** $x \approx 6.213$ **31.** $x \approx 5.332$ **33.** $x \approx 1.739$ **35.** $t \approx 16$ yr **37.** $t \approx 19$ yr **39.** 4.5% **41.** 40 yr **43.** 31 yr **45.** 130,333 physical therapy assistants **47.** 2021 **49.** 27 yr **51.** 55 hr **53.** 42,941 people **55.** About 12.6 times more intense **57.** About 31.6 times more intense **59.** $3\sqrt{2} - \sqrt{6} + 4\sqrt{3} - 4$ **60.** $7xy\sqrt{2x}$ **61.** 9 years old, 3 students; 10 years old, 11 students; 11 years old, 35 students; 12 years old, 70 students; 13 years old, 88 students; 14 or 15 years old, 78 students **62.** Approximately \$551,285,282 per mi

Quick Quiz 12.5 *See Examples noted with Ex.* **1.** $x = 4$ (Ex. 1) **2.** $x = 6$ (Ex. 2 and 3) **3.** $x \approx -0.774$ (Ex. 5) **4.** See Student Solutions Manual

Career Exploration Problems **1.** 15,880 analysts **2.** 2021 **3.** 2021 (the same year) **4.** \$46,089

You Try It **1.**

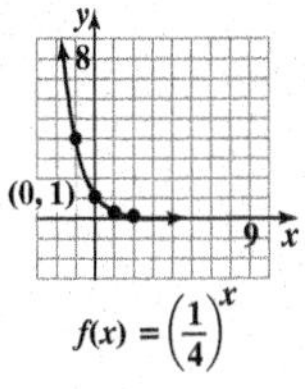

2. $x = -4$ **3. (a)** $4^{0.5x} = 9$ **(b)** $\log_5 28 = x$ **(c)** $x = -3$ **4. (a)** $\log_2 a + \frac{1}{2}\log_2 b - 2\log_2 c$ **(b)** $\log_5\left(\frac{a^3b^{2/3}}{c^2}\right)$ **(c)** -2 **5. (a)** Approximately 1.2174839 **(b)** Approximately 3.4873751 **6. (a)** $x \approx 118.8502227$ **(b)** $x \approx 4.6089497$ **7.** Approximately 0.4406427 **8.** $x = 3$ **9.** $x \approx 0.1826583$

Chapter 12 Review Problems **1.**

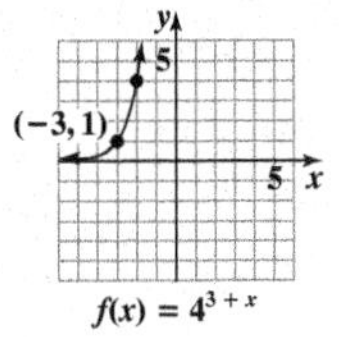

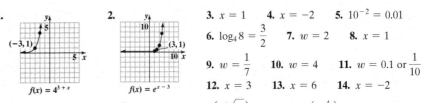

2.

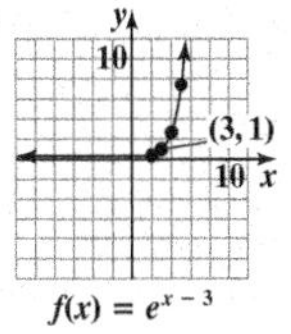

3. $x = 1$ **4.** $x = -2$ **5.** $10^{-2} = 0.01$ **6.** $\log_4 8 = \frac{3}{2}$ **7.** $w = 2$ **8.** $x = 1$ **9.** $w = \frac{1}{7}$ **10.** $w = 4$ **11.** $w = 0.1$ or $\frac{1}{10}$ **12.** $x = 3$ **13.** $x = 6$ **14.** $x = -2$

15.

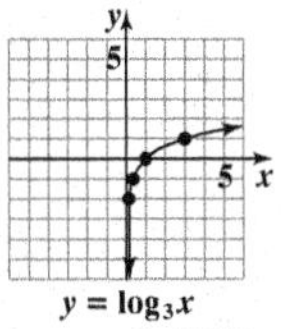

16. $\log_2 5 + \log_2 x - \frac{1}{2}\log_2 w$ **17.** $3\log_2 x + \frac{1}{2}\log_2 y$ **18.** $\log_3\left(\frac{x\sqrt{w}}{2}\right)$ **19.** $\log_8\left(\frac{w^4}{\sqrt[3]{z}}\right)$ **20.** 6 **21.** 1.376576957 **22.** -1.087777943 **23.** 1.366091654 **24.** 6.688354714 **25.** $n \approx 13.69935122$ **26.** $n \approx 5.473947392$ **27.** 0.49685671 **28.** 3.084962501 **29.** $x = 25$ **30.** $x = 25$ **31.** $x = 25$ **32.** $x = 12$ **33.** $t = \frac{3}{4}$ **34.** $t = -\frac{1}{8}$ **35.** $x = \frac{\log 14}{\log 3}$ **36.** $x = \frac{\log 130 - 3\log 5}{\log 5}$ **37.** $x = \frac{\ln 100 + 1}{2}$ **38.** $x \approx -1.4748$ **39.** $x \approx 2.3319$ **40.** $x \approx 101.3482$ **41.** 9 yr **42.** \$6312.38 **43.** 41 yr **44.** 9 yr **45. (a)** Approximately 282 lb **(b)** 7.77 lb/in.^3 **46.** 50.1 times more intense

How Am I Doing? Chapter 12 Test **1.**

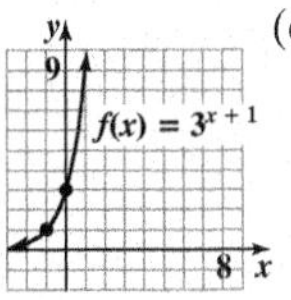

(obj. 12.1.1) **2.**

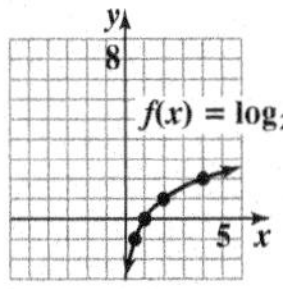

(obj. 12.2.4) **3.** $x = 0$ (obj. 12.1.2)

4. $w = 5$ (obj. 12.2.3) **5.** $x = \frac{1}{64}$ (obj. 12.2.3) **6.** $\log_7\left(\frac{x^2y}{4}\right)$ (obj. 12.3.3) **7.** 1.7901 (obj. 12.4.3) **8.** 1.3729 (obj. 12.4.1)

9. 0.4391 (obj. 12.4.5) **10.** $x \approx 5350.569382$ (obj. 12.4.2) **11.** $x \approx 1.150273799$ (obj. 12.4.4) **12.** $x = \frac{3}{7}$ (obj. 12.5.1)

13. $x = \frac{16}{3}$ (obj. 12.5.1) **14.** $x = \frac{3 + \ln 57}{5}$ (obj. 12.5.2) **15.** $x \approx -1.4132$ (obj. 12.5.2) **16.** \$2938.66 (obj. 12.5.3) **17.** 14 yr (obj. 12.5.3)

Practice Final Examination **1.** $5\frac{17}{20}$ **2.** $3\frac{1}{9}$ **3.** 10 **4.** 4.9715 **5.** 200,000 **6.** 4480 **7.** 4 **8.** $-a^2 - 10ab - a$

9. $-2x + 21y + 18xy + 6y^2$ **10.** 14 **11.** $F = -31$ **12.** $x = -30$ **13.** $y = \frac{4x + 7}{3}$ **14.** $x \geq 2.6$

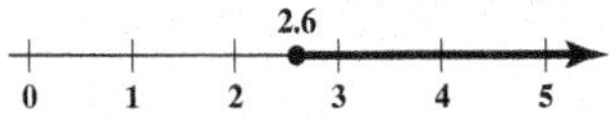

15. Length = 520 m; width = 360 m **16.** 113.04 cubic feet

17. x-intercept $(-2, 0)$; y-intercept $(0, 7)$

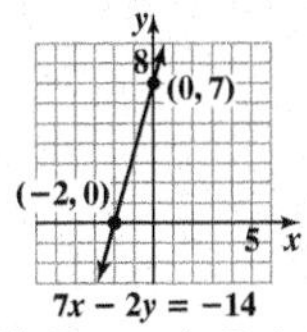

18.

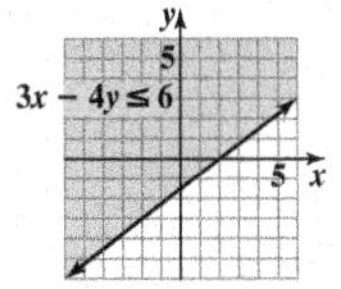

19. $m = \frac{8}{3}$ **20.** $m = -\frac{2}{3}$ **21.** $f(3) = 12$

22. $f(-2) = 17$ **23.** Domain $\{2, -1, 3\}$; Range $\{-7, 1, 2\}$ **24.** $(6, -3)$ **25.** $(6, 4)$ **26.** $(1, 4, -2)$ **27.** $(4, 1, 1)$

28.

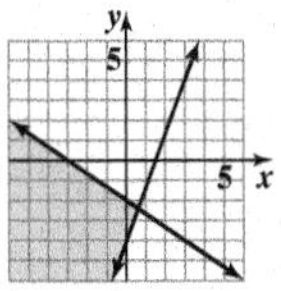

29. $18x^5y^5$ **30.** $6x^2 + 13x - 5$ **31.** $(x + 4)$ **32.** $(3x - 5)^2$ **33.** $(x + 2)(x + 2)(x - 2)$

34. $x(2x - 1)(x + 8)$ **35.** $x = -6, x = -9$ **36.** $\frac{x(3x - 1)}{x - 3}$ **37.** $\frac{x + 5}{x}$ **38.** $\frac{3x^2 + 6x - 2}{(x + 5)(x + 2)}$ **39.** $\frac{8x^2 + 6x - 5}{4x^2 - 3}$

40. $x = -1$ **41.** $\frac{1}{3x^{7/2}y^5}$ **42.** $2xy^2\sqrt[3]{5xy}$ **43.** $18\sqrt{2}$ **44.** $\frac{18 + 2\sqrt{6} + 3\sqrt{3} + \sqrt{2}}{25}$ **45.** $8i$ **46.** $x = -3$

47. $y = 33.75$ **48.** $x = \frac{1 \pm \sqrt{21}}{10}$ **49.** $x = 0, x = -\frac{3}{5}$ **50.** $x = 8, x = -343$ **51.** $x \leq -\frac{1}{3}$ or $x \geq 4$

52. $f(x) = -x^2 - 4x + 5$ **53.** Width = 4 cm; length = 13 cm **54.** $x = 3, x = 9$ **55.** $-\frac{5}{2} < x < \frac{15}{2}$

56. $(x + 3)^2 + (y - 2)^2 = 4$; center at $(-3, 2)$; radius = 2

57. $\frac{x^2}{16} + \frac{y^2}{25} = 1$; ellipse

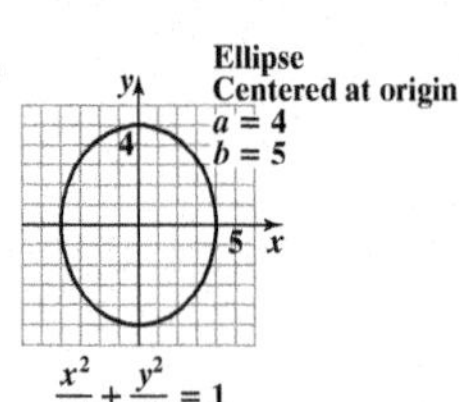

58. $\frac{x^2}{4} - \frac{y^2}{9} = 1$; hyperbola

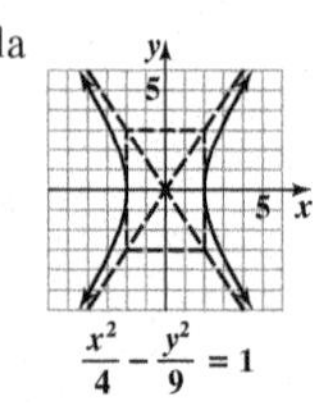

59. Parabola opening right

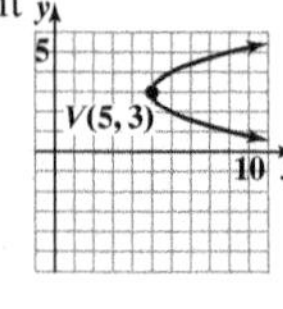

60. $(0, -4), (\sqrt{7}, 3), (-\sqrt{7}, 3)$ **61.** **(a)** $f(-1) = 10$ **(b)** $f(a) = 3a^2 - 2a + 5$ **(c)** $f(a + 2) = 3a^2 + 10a + 13$ **62.** $f[g(x)] = 80x^2 + 80x + 17$
63. $f^{-1}(x) = 2x + 14$ **64.**

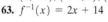

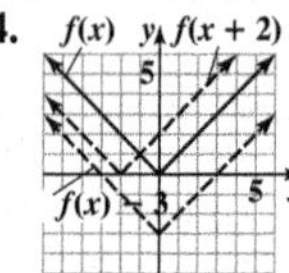

65. $f(x) = 2^{1-x}$

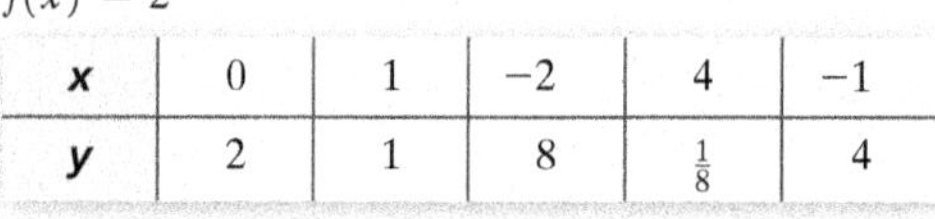

x	0	1	−2	4	−1
y	2	1	8	$\frac{1}{8}$	4

$f(x) = 2^{1-x}$

66. $0.0016 = x$ or $\frac{1}{625} = x$ **67.** $x = 21$ **68.** $y = -2$ **69.** $x = 8$

Appendix A

A.1 Exercises **1.** The base is a. The exponent is 3. **3.** Positive **5.** $+11$ and -11; $(+11)(+11) = 121$ and $(-11)(-11) = 121$ **7.** 32 **9.** 25 **11.** −36 **13.** −1 **15.** $\frac{1}{256}$ **17.** 0.49 **19.** 0.000064 **21.** 9 **23.** −4 **25.** $\frac{2}{3}$
27. 0.3 **29.** 1 **31.** Not a real number **33.** −33 **35.** −46 **37.** 0 **39.** −28 **41.** 4 **43.** $\frac{2}{3}$ **45.** 1 **47.** $\frac{1}{9}$ **49.** $\frac{1}{x^5}$
51. $-6x^6$ **53.** $11x^6y^9$ **55.** $4y$ **57.** $7xy$ **59.** $24x^2y^3z$ **61.** $-\frac{3}{n}$ **63.** 8 **65.** $\frac{2}{x^5}$ **67.** $-5x^3z$ **69.** $-\frac{10}{7}a^2b^4$ **71.** $\frac{x^{12}y^{18}}{z^6}$
73. $\frac{9a^2}{16b^{12}}$ **75.** $\frac{1}{8x^{12}y^{18}}$ **77.** $\frac{4x^4}{y^6}$ **79.** $-\frac{27m^{13}}{n^5}$ **81.** $2a^4$ **83.** x^3y^9 **85.** $\frac{a^3}{b^5}$ **87.** $\frac{1}{4x^2}$ **89.** $\frac{7}{5}$ **91.** $-\frac{6x}{y}$ **93.** 3.8×10^1
95. 1.73×10^6 **97.** 8.3×10^{-1} **99.** 8.125×10^{-4} **101.** 713,000 **103.** 0.307 **105.** 0.000000901 **107.** 4.65×10^{-6}
109. 3×10^1 **111.** 1.06×10^{-18} g

A.2 Exercises **1.** Trinomial, degree 2 **3.** Monomial, degree 8 **5.** Binomial, degree 4 **7.** 6 **9.** 17 **11.** 6 **13.** $-3x - 6$
15. $10m^3 + 4m^2 - 6m - 1.3$ **17.** $5a^3 - a^2 + 12$ **19.** $\frac{7}{6}x^2 + \frac{14}{3}x$ **21.** $-3.2x^3 + 1.8x^2 - 4$ **23.** $10x^2 + 61x + 72$
25. $15aw - 20bw + 6ad - 8bd$ **27.** $-12x^2 + 32xy - 5y^2$ **29.** $-28ar - 77rs^2 + 4as^2 + 11s^4$ **31.** $25x^2 - 64y^2$ **33.** $25a^2 - 20ab + 4b^2$
35. $49m^2 - 14m + 1$ **37.** $36 - 25x^4$ **39.** $9m^6 + 6m^3 + 1$ **41.** $6x^3 - 10x^2 + 2x$ **43.** $-2a^2b + \frac{5}{2}ab^2 + 5ab$ **45.** $2x^3 - 5x^2 + 5x - 3$
47. $6x^3 - 7x^2y - 10xy^2 + 6y^3$ **49.** $3a^4 - 11a^3 - 21a^2 + 4a + 5$ **51.** $2x^3 - 7x^2 - 7x + 30$ **53.** $-2a^3 - 3a^2 + 29a - 30$
55. $(4x^3 + 12x^2 + 13x + 20)$ ft³

A.3 Exercises **1.** $10(8 - y)$ **3.** $5a(a - 5)$ **5.** $4a(ab^3 - 2b + 8)$ **7.** $6y^2(5y^2 + 4y + 3)$ **9.** $5ab(3b + 1 - 2a^2)$
11. $10a^2b^2(b - 3ab + a - 4a^2)$ **13.** $(x + y)(3x - 2)$ **15.** $(a - 3b)(5b + 8)$ **17.** $(a + 5b)(3x + 1)$ **19.** $(3x - y)(2a^2 - 5b^3)$
21. $(5x + y)(3x - 8y - 1)$ **23.** $(a - 6b)(2a - 3b - 2)$ **25.** $(x + 5)(x^2 + 3)$ **27.** $(x + 3)(2 - 3a)$ **29.** $(b - 4)(a - 3)$
31. $(x - 4)(5 + 3y)$ **33.** $(x - 3)(2 - 3y)$ **35.** $(x + 7)(x + 1)$ **37.** $(x - 5)(x - 3)$ **39.** $(x - 6)(x - 4)$ **41.** $(a + 9)(a - 5)$
43. $(x - 7y)(x + 6y)$ **45.** $(x - 14y)(x - y)$ **47.** $(x^2 - 8)(x^2 + 5)$ **49.** $(x^2 + 7y^2)(x^2 + 9y^2)$ **51.** $2(x + 11)(x + 2)$
53. $x(x + 5)(x - 4)$ **55.** $(2x + 1)(x - 1)$ **57.** $(3x - 5)(2x + 1)$ **59.** $(3a - 5)(a - 1)$ **61.** $(4a - 7)(a + 2)$
63. $(2x + 3)(x + 5)$ **65.** $(3x^2 + 1)(x^2 - 3)$ **67.** $(3x + y)(2x + 11y)$ **69.** $(7x - 3y)(x + 2y)$ **71.** $x(4x + 1)(2x - 1)$
73. $5x^2(2x + 1)(x + 1)$

A.4 Exercises **1.** The problem will have two terms. It will be in the form $a^2 - b^2$. One term is positive and one term is negative. The coefficients and variables for the first and second terms are both perfect squares. So each one will be of the form 1, 4, 9, 16, 25, 36, and/or x^2, x^4, x^6, etc. **3.** There will be two terms added together. It will be of the form $a^3 + b^3$. Each term will contain a number or variable cubed or both. They will be of the form 1, 8, 27, 64, 125, and/or x^3, x^6, x^9, etc. **5.** $(a + 8)(a - 8)$ **7.** $(4x + 9)(4x - 9)$ **9.** $(8x + 1)(8x - 1)$
11. $(7m + 3n)(7m - 3n)$ **13.** $(10y + 9)(10y - 9)$ **15.** $(1 + 6xy)(1 - 6xy)$ **17.** $2(4x + 3)(4x - 3)$ **19.** $5x(1 + 2x)(1 - 2x)$
21. $(3x - 1)^2$ **23.** $(7x - 1)^2$ **25.** $(9w + 2t)^2$ **27.** $(6x + 5y)^2$ **29.** $2(2x + 5)^2$ **31.** $3x(x - 4)^2$ **33.** $(x - 3)(x^2 + 3x + 9)$

35. $(x + 5)(x^2 - 5x + 25)$ **37.** $(4x - 1)(16x^2 + 4x + 1)$ **39.** $(2x - 5)(4x^2 + 10x + 25)$ **41.** $(1 - 3x)(1 + 3x + 9x^2)$ **43.** $(4x + 5)(16x^2 - 20x + 25)$ **45.** $(4s^2 + t^2)(16s^4 - 4s^2t^2 + t^4)$ **47.** $5(y - 2)(y^2 + 2y + 4)$ **49.** $2(5x + 1)(25x^2 - 5x + 1)$ **51.** $x^2(x - 2y)(x^2 + 2xy + 4y^2)$ **53.** $(x - 9)(x + 7)$ **55.** $(3x + 2)(2x - 1)$ **57.** $(5w^2 + 1)(5w^2 - 1)$ **59.** $(b^2 + 3)^2$ **61.** $(y - 3)(z^2 + 5)$ **63.** $(ab + 1)(b^2 + c)$ **65.** $(3m^3 + 8)(3m^3 - 8)$ **67.** $(6y^3 - 5)^2$ **69.** $5(3z^4 + 1)(3z^4 - 1)$ **71.** $(5m + 2n)(25m^2 - 10mn + 4n^2)$ **73.** $2(x + 8)(x - 6)$ **75.** $3(3x + 2)(2x + 1)$ **77.** $8a(5x - 1)(x + 2)$ **79.** $2x(3x - 2)(x + 5)$ **81.** $3(2a - b)(4a^2 + 2ab + b^2)$ **83.** $(2w - 5z)^2$ **85.** $9(2a + 3b)(2a - 3b)$ **87.** $(4x^2 + 9y^2)(2x + 3y)(2x - 3y)$ **89.** $(5m^2 + 2)(25m^4 - 10m^2 + 4)$ **91.** $(5x + 4)(5x + 1)$ **93.** $(7x - 4)(7x - 1)$ **95.** $A = (4x + y)(4x - y)$ ft^2 **97.** One possibility is to have $6x + 5$ rows $5x - 1$ trees in each row. Another possibility is to have $5x - 1$ rows with $6x + 5$ trees in each row.

A.5 Exercises **1.** $y = -\frac{2}{3}x + 8$ **3.** $y = 5x + 33$ **5.** $y = -\frac{1}{5}x + \frac{6}{5}$ **7.** $y = \frac{5}{7}x + \frac{13}{7}$ **9.** $y = -\frac{2}{3}x - \frac{8}{3}$ **11.** $y = -3$ **13.** $5x - y = -10$ **15.** $x - 3y = 8$ **17.** $2x - 3y = 15$ **19.** $7x - y = -27$ **21.** Neither **23.** Parallel **25.** Perpendicular

Appendix B

Addition Practice **1.** 37 **2.** 75 **3.** 94 **4.** 99 **5.** 78 **6.** 29 **7.** 61 **8.** 81 **9.** 369 **10.** 277 **11.** 491 **12.** 600 **13.** 1421 **14.** 1052 **15.** 900 **16.** 401 **17.** 1711 **18.** 1319 **19.** 289 **20.** 399 **21.** 589 **22.** 887 **23.** 422 **24.** 792 **25.** 902 **26.** 930

Subtraction Practice **1.** 21 **2.** 62 **3.** 22 **4.** 43 **5.** 68 **6.** 5 **7.** 29 **8.** 45 **9.** 531 **10.** 223 **11.** 726 **12.** 191 **13.** 378 **14.** 66 **15.** 509 **16.** 228 **17.** 256 **18.** 63 **19.** 183 **20.** 269 **21.** 873 **22.** 1892 **23.** 7479 **24.** 2868 **25.** 2066 **26.** 678

Multiplication Practice **1.** 69 **2.** 26 **3.** 378 **4.** 603 **5.** 1554 **6.** 1643 **7.** 3680 **8.** 3640 **9.** 7790 **10.** 2691 **11.** 9116 **12.** 7502 **13.** 12,095 **14.** 30,168 **15.** 20,672 **16.** 24,180 **17.** 37,584 **18.** 53,692 **19.** 284,490 **20.** 506,566 **21.** 238,119 **22.** 375,820 **23.** 803,396 **24.** 785,354 **25.** 49,516,817 **26.** 62,526,175

Division Practice **1.** 16 **2.** 56 **3.** 59 R2 **4.** 47 R5 **5.** 124 **6.** 296 **7.** 234 **8.** 567 **9.** 521 R6 **10.** 739 R2 **11.** 321 R11 **12.** 753 R6 **13.** 89 **14.** 91 **15.** 32 **16.** 41 **17.** 108 R29 **18.** 123 R25 **19.** 214 **20.** 342 **21.** 1134 R14 **22.** 1076 R65 **23.** 124 **24.** 153 **25.** 1125 **26.** 1532

Appendix C

Exercises **1.** -7 **3.** 15 **5.** 2 **7.** 47 **9.** 18 **11.** 0 **13.** 0 **15.** -0.6 **17.** $-7a - 4b$ **19.** $\frac{11}{84}$ **21.** $\begin{vmatrix} 6 & 10 \\ -5 & 9 \end{vmatrix}$ **23.** $\begin{vmatrix} 3 & -4 \\ 1 & -5 \end{vmatrix}$ **25.** -7 **27.** -26 **29.** 11 **31.** -27 **33.** -8 **35.** 0 **37.** -3.179 **39.** 18,553 **41.** $x = 2$; $y = 3$ **43.** $x = -2$; $y = 5$ **45.** $x = 10$; $y = 2$ **47.** $x = 4$; $y = -2$ **49.** $x \approx 1.5795$; $y \approx -0.0902$ **51.** $x = 1$; $y = 1$; $z = 1$ **53.** $x = -\frac{1}{2}$; $y = \frac{1}{2}$; $z = 2$ **55.** $x = 4$; $y = -2$; $z = 1$ **57.** $x \approx -0.219$; $y \approx 1.893$; $z \approx -3.768$ **59.** $w \approx -3.105$; $x \approx 4.402$; $y \approx 15.909$; $z \approx 6.981$

Appendix D

Exercises **1.** $(4, -1)$ **3.** $(3, -9)$ **5.** $(0, 3)$ **7.** $(2, 2)$ **9.** $(1.2, 3.7)$ **11.** $(3, -1, 4)$ **13.** $(1, -1, 3)$ **15.** $(0, -2, 5)$ **17.** $(0.5, -1, 5)$ **19.** $(3.6, 1.8, 2.4)$ **21.** $(4.2, -3.6, 8.8, 5.4)$

Appendix E

Exercises **1.** Elements **3.** Intersection **5.** Empty set **7.** {California, Colorado, Connecticut} **9.** $\{1, 3, 5, 7, \ldots\}$ **11.** $\{15, 20, 25, 30, 35, 40, 45\}$ **13.** $\{\ldots, -3, -2, -1\}$ **15.** $O = \{x \mid x \text{ is an ocean}\}$ **17.** $G = \{x \mid x \text{ is even and } 2 \le x \le 8\}$ **19.** $T = \{x \mid x \text{ is a type of triangle}\}$ **21.** **(a)** $\{-2, -1, 2, 3, 5, 8\}$ **(b)** $\{-1, 3, 5\}$ **23.** False. B is an infinite set. **25.** True **27.** True **29.** False; $A \cap C = \{1, 2, 3, 4, 5\}$ **31.** False; the set of whole numbers includes 0. **33.** True **35.** Answers will vary. One possible answer is {Ann, Nina}. **37.** **(a)** {Andrew, Christopher, Daniel, David, Jacob, James, Jason, John, Joseph, Joshua, Matthew, Michael, Nicholas, Robert, Tyler} **(b)** Union **39.** Whole numbers: 15, 0, $\sqrt{81}$; natural numbers: 15, $\sqrt{81}$; integers: 15, 0, $\sqrt{81}$, -17; rational numbers: 3.62, $\frac{-3}{11}$, 15, $\frac{22}{3}$, 0, $\sqrt{81}$, -17; irrational numbers: $\sqrt{20}$; real numbers: 3.62, $\sqrt{20}$, $\frac{-3}{11}$, 15, $\frac{22}{3}$, 0, $\sqrt{81}$, -17 **41.** No. The set of integers contains negative numbers. Negative numbers are not part of the set of whole numbers. **43.** The set of whole numbers is a subset of the integers, the rational numbers, and the real numbers.

Beginning and Intermediate Algebra Glossary

Absolute value inequalities (9.7) Inequalities that contain at least one absolute value expression.

Absolute value of a number (1.1) The absolute value of a number x is the distance between 0 and the number x on the number line. It is written as $|x|$. $|x| = x$ if $x \geq 0$, but $|x| = -x$ if $x < 0$.

Altitude of a geometric figure (1.8) The height of the geometric figure. In the three figures shown the altitude is labeled a.

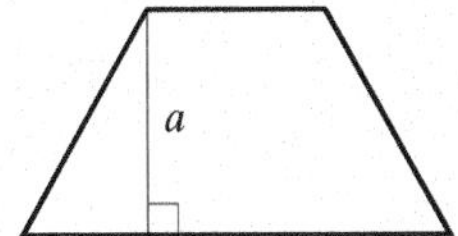

Altitude of a trapezoid

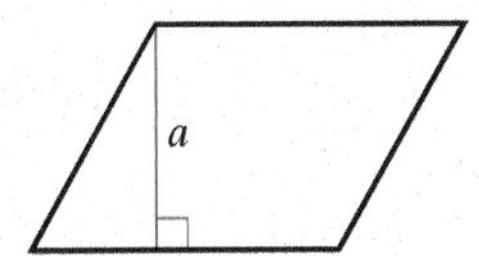

Altitude of a parallelogram

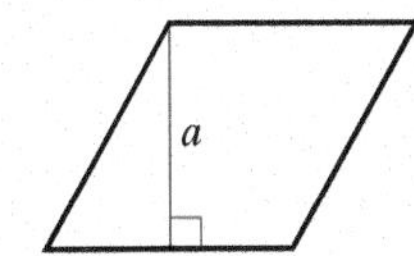

Altitude of a rhombus

Altitude of a triangle (1.8) The height of any given triangle. In the three triangles shown the altitude is labeled a.

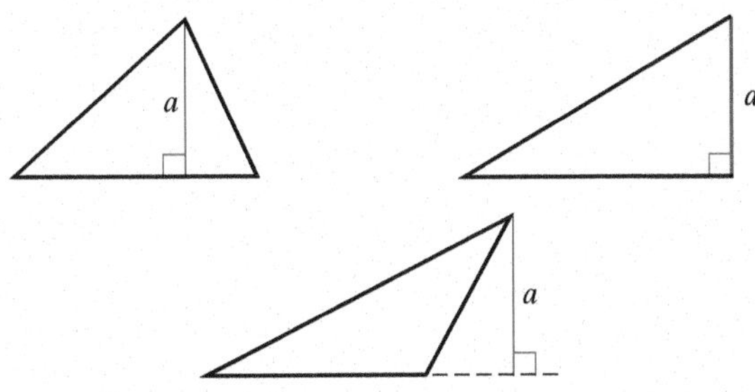

Associative property of addition (1.1) If a, b, and c are real numbers, then

$$a + (b + c) = (a + b) + c.$$

This property states that if three numbers are added it does not matter *which two numbers* are added first; the result will be the same.

Associative property of multiplication (1.3) If a, b, and c are real numbers, then

$$a \times (b \times c) = (a \times b) \times c.$$

This property states that if three numbers are multiplied it does not matter *which two numbers* are multiplied first; the result will be the same.

Asymptote (10.4) A line that a curve continues to approach but never actually touches. Often an asymptote is a helpful reference in making a sketch of a curve, such as a hyperbola.

Augmented matrix (Appendix D) A matrix derived from a linear system of equations. It consists of the coefficients of each variable in a linear system and the constants. The augmented matrix of the system $\begin{array}{rl} -3x + 5y = & -22 \\ 2x - \;\; y = & 10 \end{array}$ is the matrix $\left[\begin{array}{rr|r} -3 & 5 & -22 \\ 2 & -1 & 10 \end{array}\right]$. Each row of the augmented matrix represents an equation of the system.

Axis of symmetry of a parabola (10.2) A line passing through the focus and the vertex of a parabola, about which the two sides of the parabola are symmetric. See the sketch.

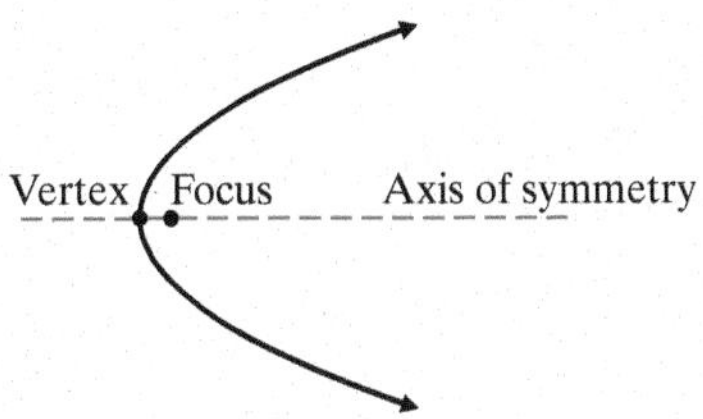

Base (1.4) The number or variable that is raised to a power. In the expression 2^6, the number 2 is the base.

Base of an exponential function (12.1) The number b in the function $f(x) = b^x$.

Base of a triangle (1.8) The side of a triangle that is perpendicular to the altitude.

Binomial (5.3) A polynomial of two terms. The expressions $a + 2b$, $6x^3 + 1$, and $5a^3b^2 + 6ab$ are all binomials.

Circumference of a circle (1.8) The distance around a circle. The circumference of a circle is given by the formula $C = \pi d$ or $C = 2\pi r$, where d is the diameter of the circle and r is the radius of the circle.

Coefficient (5.1) A coefficient is a factor or a group of factors in a product. In the term $4xy$ the coefficient of y is $4x$, but the coefficient of xy is 4. In the term $-5x^3y$ the coefficient of x^3y is -5.

Combined variation (8.7) Variation that depends on two or more variables. An example would be when y varies directly with x and z and inversely with d^2, written $y = \dfrac{kxz}{d^2}$ where k is the constant of variation.

Common logarithm (12.4) The common logarithm of a number x is given by $\log x = \log_{10} x$ for all $x > 0$. A common logarithm is a logarithm using base 10.

Commutative property for addition (1.1) If a and b are any real numbers, then $a + b = b + a$.

Commutative property for multiplication (1.3) If a and b are any real numbers, then $ab = ba$.

Complex fraction (7.4) A fraction that contains at least one fraction in the numerator or in the denominator or both. These three fractions are complex fractions:

$$\frac{7+\frac{1}{x}}{x^2+2}, \quad \frac{1+\frac{1}{5}}{2-\frac{1}{7}}, \quad \text{and} \quad \frac{\frac{1}{3}}{4}$$

Complex number (8.6) A number that can be written in the form $a + bi$, where a and b are real numbers and $i = \sqrt{-1}$.

Compound inequalities (9.6) Two inequality statements connected together by the word *and* or by the word *or*.

Conjugate of a binomial with radicals (8.4) The expressions $a\sqrt{x} + b\sqrt{y}$ and $a\sqrt{x} - b\sqrt{y}$. The conjugate of $2\sqrt{3} + 5\sqrt{2}$ is $2\sqrt{3} - 5\sqrt{2}$. The conjugate of $4 - \sqrt{x}$ is $4 + \sqrt{x}$.

Conjugate of a complex number (8.6) The expressions $a + bi$ and $a - bi$. The conjugate of $5 + 2i$ is $5 - 2i$. The conjugate of $7 - 3i$ is $7 + 3i$.

Constant (2.3) Symbol or letter that is used to represent exactly one single quantity during a particular problem or discussion.

Coordinates of a point (3.1) An ordered pair of numbers (x, y) that specifies the location of a point in a rectangular coordinate system.

Critical points of a quadratic inequality (9.6) In a quadratic inequality of the form $ax^2 + bx + c > 0$ or $ax^2 + bx + c < 0$, those points where $ax^2 + bx + c = 0$.

Degree of a polynomial (5.3) The degree of the highest-degree term of a polynomial. The degree of the polynomial $5x^3 + 2x^2 - 6x + 8$ is 3. The degree of the polynomial $5x^2y^2 + 3xy + 8$ is 4.

Degree of a term of a polynomial (5.3) The sum of the exponents of the variables in the term. The degree of $3x^3$ is 3. The degree of $4x^5y^2$ is 7.

Denominator (0.1) and (7.1) The bottom number or algebraic expression in a fraction. The denominator of

$$\frac{3x-2}{x+4}$$

is $x + 4$. The denominator of $\frac{3}{7}$ is 7. The denominator of a fraction may not be zero.

Dependent equations (4.1) Two equations are dependent if every value that satisfies one equation satisfies the other. A system of two dependent equations in two variables will not have a unique solution.

Determinant (Appendix C) A square array of numbers written between vertical lines. For example $\begin{vmatrix} 1 & 5 \\ 2 & 4 \end{vmatrix}$ is a 2×2 determinant. It is also called a *second-order determinant.* $\begin{vmatrix} 1 & 7 & 8 \\ 2 & -5 & -1 \\ -3 & 6 & 9 \end{vmatrix}$ is a 3×3 determinant. It is also called a *third-order determinant.*

Difference-of-two-squares polynomial (6.5) A polynomial of the form $a^2 - b^2$ that may be factored by using the formula

$$a^2 - b^2 = (a + b)(a - b).$$

Direct variation (8.7) When a variable y varies directly with x, written $y = kx$, where k represents some real number that will stay the same over a range of x-values. This value k is called the *constant of variation.*

Discriminant of a quadratic equation (9.2) In the equation $ax^2 + bx + c = 0$, where $a \neq 0$, the expression $b^2 - 4ac$. It can be used to determine the nature of the roots of the quadratic equation. If the discriminant is *positive,* there are two rational or irrational roots. The two roots will be rational only if the discriminant is a perfect square. If the discriminant is *zero,* there is only one rational root. If the discritninant is *negative,* there are two complex roots.

Distance between two points (10.1) The distance between point (x_1, y_1) and point (x_2, y_2) is given by the formula $d = \sqrt{(x_2 - x_1)^2 + (y_2 - y_1)^2}$.

Distributive property (1.6) For all real numbers a, b, and c, $a(b + c) = ab + ac$.

Dividend (0.4) The number that is to be divided by another. In the problem $30 \div 5 = 6$, the three parts are as follows:

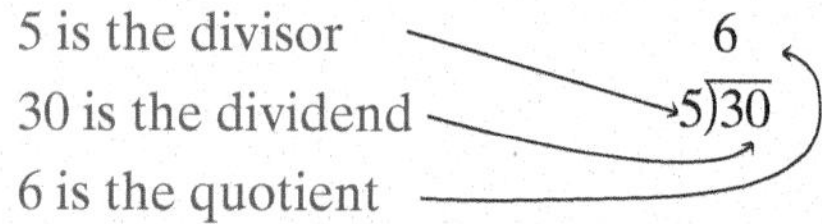

Divisor (0.4) The number you divide into another.

Domain of a relation (3.6) In any relation, the set of values that can be used for the independent variable is called its domain. This is the set of all the first coordinates of the ordered pairs that define the relation.

***e* (12.1)** An irrational number that can be approximated by the value 2.7183.

Ellipse (10.3) The set of points in a plane such that for each point in the set, the sum of its distances to two fixed points is constant. Each of the fixed points is called a *focus.* Each of the following graphs is an ellipse.

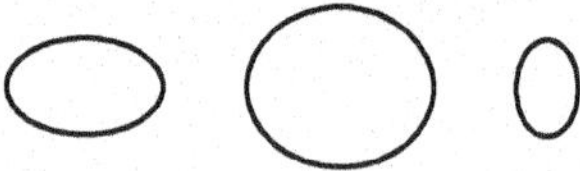

Equilateral hyperbola (10.4) A hyperbola for which $a = b$ in the equation of the hyperbola.

Equilateral triangle (1.8) A triangle with three sides equal in length and three angles that measure 60°. Triangle ABC is an equilateral triangle.

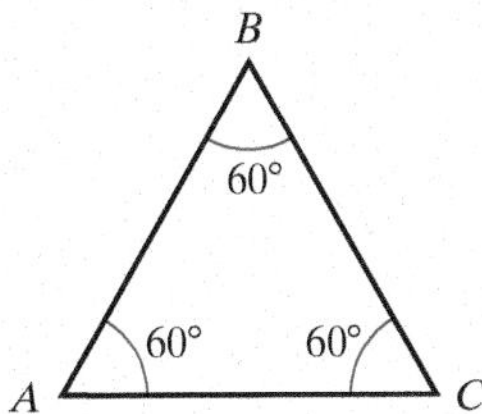

Even integers (1.3) Integers that are exactly divisible by 2, such as $\ldots, -4, -2, 0, 2, 4, 6, \ldots$.

Exponent (1.4) The number that indicates the power of a base. If the number is a positive integer, it indicates how many times the base is multiplied. In the expression 2^6, the exponent is 6.

Exponential function (12.1) $f(x) = b^x$, where $b > 0$, $b \neq 1$, and x is any real number.

Expression (5.3) A mathematic expression is any quantity using numbers and variables. Therefore, $2x$, $7x + 3$, and $5x^2 + 6x$ are all mathematical expressions.

Extraneous solution (7.5) and (8.5) An obtained solution to an equation that, when substituted back into the original equation, does *not* yield an identity. $x = 2$ is an extraneous solution to the equation

$$\frac{x}{x-2} - 4 = \frac{2}{x-2}$$

An extraneous solution is also called an *extraneous root.*

Factor (0.1) and (6.1) When two or more numbers, variables, or algebraic expressions are multiplied, each is called a factor. If we write $3 \cdot 5 \cdot 2$, the factors are 3, 5, and 2. If we write $2xy$, the factors are 2, x, and y. In the expression $(x - 6)(x + 2)$, the factors are $(x - 6)$ and $(x + 2)$.

Focus point of a parabola (10.2) The focus point of a parabola has many properties. For example, the focus point of a parabolic mirror is the point to which all incoming light rays that are parallel to the axis of symmetry will collect. A parabola is a set of points that is the same distance from a fixed line called the *directrix* and a fixed point. This fixed point is the focus.

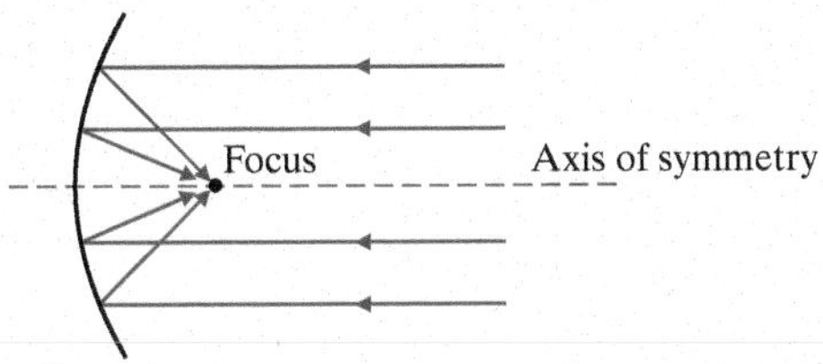

Fractions

Algebraic fractions (7.1) The indicated quotient of two algebraic expressions.

$$\frac{x^2 + 3x + 2}{x - 4} \quad \text{and} \quad \frac{y - 6}{y + 8}$$

are algebraic fractions. In these fractions the value of the denominator cannot be zero.

Numerical fractions (0.1) A set of numbers used to describe parts of whole quantities. A numerical fraction can be represented by the quotient of two integers for which the denominator is not zero. The numbers $\frac{1}{5}, -\frac{2}{3}, \frac{8}{2}, -\frac{4}{31}, \frac{8}{1}$, and $-\frac{12}{1}$ are all numerical fractions. The set of rational numbers can be represented by numerical fractions.

Function (3.6) A relation in which no two different ordered pairs have the same first coordinate.

Graph of a function (11.2) A graph in which a vertical line will never cross in more than one place. The following sketches represent the graphs of functions.

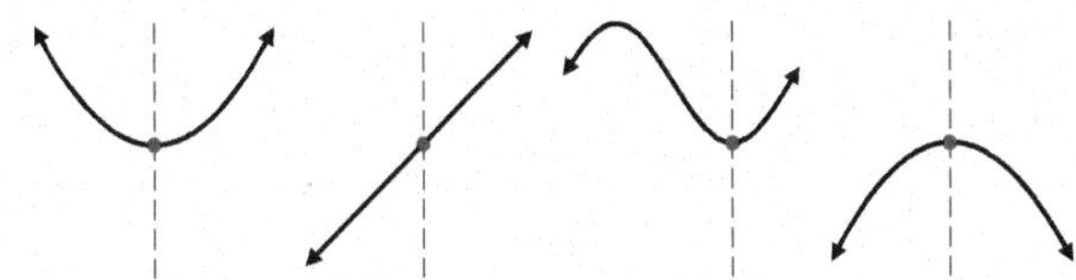

Graph of a one-to-one function (11.4) A graph of a function with the additional property that a horizontal line will never cross the graph in more than one place. The following sketches represent the graphs of one-to-one functions.

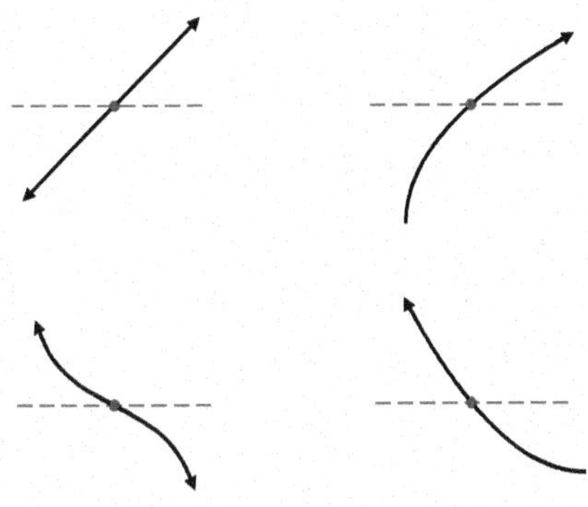

Higher-order equations (9.3) Equations of degree 3 or higher. Examples of higher-order equations are

$$x^4 - 29x^2 + 100 = 0 \quad \text{and} \quad x^3 + 3x^2 - 4x - 12 = 0.$$

Higher-order roots (8.2) Cube roots, fourth roots, and all other roots with an index greater than 2.

Horizontal parabolas (10.2) Parabolas that open to the right or to the left. The following graphs represent horizontal parabolas.

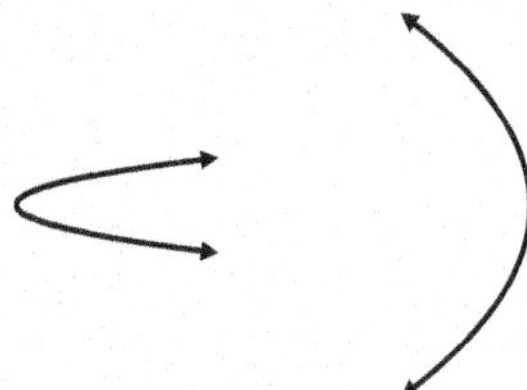

Hyperbola (10.4) The set of points in a plane such that for each point in the set, the absolute value of the difference of its distances to two fixed points is constant. Each of these fixed points is called a *focus*. The following sketches represent graphs of hyperbolas.

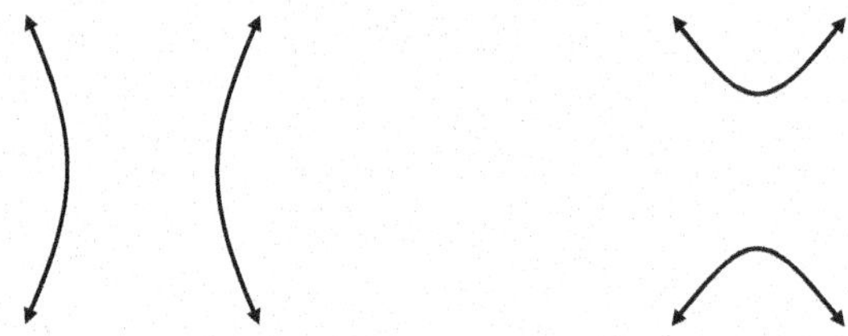

Hypotenuse of a right triangle (9.4) The side opposite the right angle in any right triangle. The hypotenuse is always the longest side of a right triangle. In the following sketch the hypotenuse is side c.

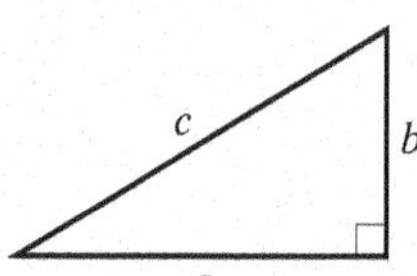

Imaginary number (8.6) i, defined as $i = \sqrt{-1}$ and $i^2 = -1$.

Improper fraction (0.1) A numerical fraction whose numerator is larger than or equal to its denominator. $\frac{8}{3}, \frac{5}{2}$, and $\frac{7}{7}$ are improper fractions.

Inconsistent system of equations (4.1) A system of equations that does not have a solution.

Independent equations (4.1) Two equations that are not dependent are said to be independent.

Index of a radical (8.2) Indicates what type of a root is being taken. The index of a cube root is $\sqrt[3]{x}$; the 3 is the index of the radical. In $\sqrt[4]{y}$, the index is 4. The index of a square root is 2, but the index is not written in the square root symbol, as shown: $\sqrt{x}$.

Inequality (2.8) and (3.5) A mathematical relationship between quantities that are not equal. $x \leq -3$, $w > 5$, and $x < 2y + 1$ are inequalities.

Integers (1.1) The set of numbers $\ldots, -5, -4, -3, -2, -1, 0, 1, 2, 3, 4, 5, \ldots$.

Intercepts of an equation (3.2) The point or points where the graph of the equation crosses the x-axis or the y-axis or both. (*See* x-intercept or y-intercept.)

Inverse function of a one-to-one function (11.4) That function obtained by interchanging the first and second coordinates in each ordered pair of the function.

Inverse variation (8.7) When a variable y varies inversely with x, written $y = \frac{k}{x}$, where k is the constant of variation.

Irrational number (1.1) A real number that cannot be expressed in the form $\frac{a}{b}$, where a and b are integers and $b \neq 0$. $\sqrt{2}, \pi, 5 + 3\sqrt{2}$, and $-4\sqrt{7}$ are irrational numbers.

Isosceles triangle (1.8) A triangle with two equal sides and two equal angles. Triangle ABC is an isosceles triangle. Angle BAC is equal to angle ACB. Side AB is equal in length to side BC.

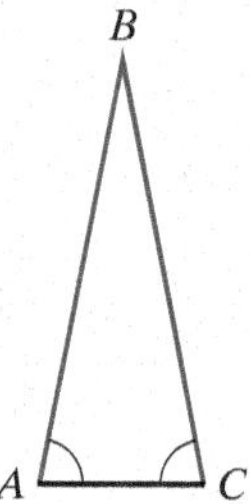

Joint variation (8.7) Variation that depends on two or more variables. An example would be when a variable y varies jointly with x and z, written $y = kxz$, where k is the constant of variation.

Least common denominator of numerical fractions (0.2) The smallest whole number that is exactly divisible by all denominators of a group of fractions. The least common denominator (LCD) of $\frac{1}{6}, \frac{2}{3}$, and $\frac{3}{5}$ is 30. The least common denominator is also called the *lowest common denominator.*

Leg of a right triangle (9.4) One of the two shorter sides of a right triangle. In the following sketch, sides a and b are the legs of the right triangle.

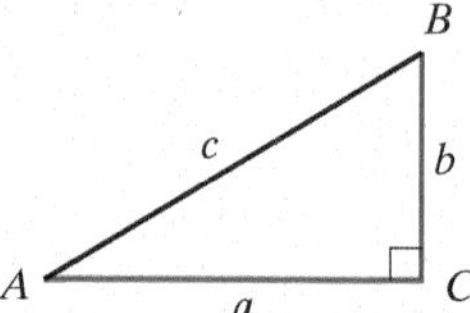

Like terms (1.7) Terms that have identical variables and exponents. In the expression $5x^3 + 2xy^2 + 6x^2 - 3xy^2$, the term $2xy^2$ and the term $-3xy^2$ are like terms.

Linear equation in two variables (3.1) An equation of the form $Ax + By = C$, where A, B, and C are real numbers. The graph of a linear equation in two variables is a straight line.

Logarithm (12.2) For a positive number x, the power to which the base b must be raised to produce x. That is, $y = \log_b x$ is the same as $x = b^y$, where $b > 0$ and $b \neq 1$. A logarithm is an exponent.

Logarithmic equation (12.2) An equation that contains at least one logarithm.

Magnitude of an earthquake (12.5) The magnitude of an earthquake is measured by the formula $M = \log\left(\frac{I}{I_0}\right)$, where I is the intensity of the earthquake and I_0 is the minimum measurable intensity.

Matrix (Appendix C) A rectangular array of numbers arranged in rows and columns. We use the symbol [ ] to indicate a matrix. The matrix $\begin{bmatrix} 3 & 4 & 5 \\ 6 & 7 & 8 \end{bmatrix}$ has two rows and three columns and is called a 2×3 *matrix*.

Minor of an element of a third-order determinant (Appendix C) The second-order determinant that remains after we delete the row and column in which the element appears. The minor of the element 6 in the determinant $\begin{vmatrix} 1 & 2 & 3 \\ 7 & 6 & 8 \\ -3 & 5 & 9 \end{vmatrix}$ is the second-order determinant $\begin{vmatrix} 1 & 3 \\ -3 & 9 \end{vmatrix}$.

Mixed number (0.1) A number that consists of an integer written next to a proper fraction. $2\frac{1}{3}$, $4\frac{6}{7}$, and $3\frac{3}{8}$ are all mixed numbers. Mixed numbers are sometimes called mixed fractions or mixed numerals.

Natural logarithm (12.4) For a number x, $\ln x = \log_e x$ for all $x > 0$. A natural logarithm is a logarithm using base e.

Natural numbers (0.1) The set of numbers 1, 2, 3, 4, 5, ... This set is also called the set of counting numbers.

Nonlinear system of equations (10.5) A system of equations in which at least one equation is not a linear equation.

Numeral (0.1) The symbol used to describe a number.

Numerator (0.1) The top number or algebraic expression in a fraction. The numerator of

$$\frac{x + 3}{5x - 2}$$

is $x + 3$. The numerator of $\frac{12}{13}$ is 12.

Numerical coefficient (5.1) The number that is multiplied by a variable or a group of variables. The numerical coefficient in $5x^3y^2$ is 5. The numerical coefficient in $-6abc$ is -6. The numerical coefficient in x^2y is 1. A numerical coefficient of 1 is not usually written.

Odd integers (1.3) Integers that are not exactly divisible by 2, such as . . . , −3, −1, 1, 3, 5, 7, 9, ...

One-to-one function (11.4) A function in which no two different ordered pairs have the same second coordinate.

Opposite of a number (1.1) Two numbers that are the same distance from zero on the number line but lie on different sides of it are considered opposites. The opposite of −6 is 6. The opposite of $\frac{22}{7}$ is $-\frac{22}{7}$.

Ordered pair (3.1) A pair of numbers presented in a specified order. An ordered pair is often used to specify a location on a graph. Every point in a rectangular coordinate system can be represented by an ordered pair (x, y).

Origin (3.1) The point (0, 0) in a rectangular coordinate system.

Parabola (10.2) The set of points that is the same distance from some fixed line (called the *directrix*) and some fixed point (called the *focus*) that is not on the line. The graph of any equation of the form $y = ax^2 + bx + c$ or $x = ay^2 + by + c$, where a, b, and c are real numbers and $a \neq 0$, is a parabola. Some examples of the graphs of parabolas are shown.

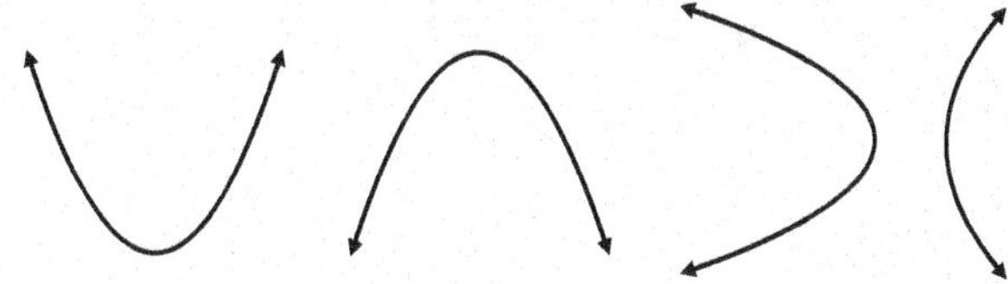

Parallel lines (3.4) and (4.1) Two straight lines that never intersect. The graph of an inconsistent system of two linear equations in two variables will result in parallel lines.

Parallelogram (1.8) A four-sided figure with opposite sides parallel. Figure $ABCD$ is a parallelogram.

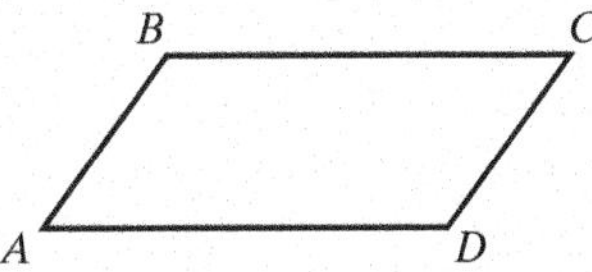

Percent (0.5) Hundredths or “per one hundred”; indicated by the % symbol. Thirty-seven hundredths $\left(\frac{37}{100}\right) = 37\%$ (thirty-seven percent).

Perfect square number (6.5) A number that is the square of an integer. The numbers 1, 4, 9, 16, 25, 36, 49, 64, 81, 100, 121, 144, . . . are perfect square numbers.

Perfect-square trinomial (6.5) A polynomial of the form $a^2 + 2ab + b^2$ or $a^2 - 2ab + b^2$ that may be factored using one of the following formulas:

$$a^2 + 2ab + b^2 = (a + b)^2$$

or

$$a^2 - 2ab + b^2 = (a - b)^2.$$

Perimeter (1.8) The distance around any plane figure. The perimeter of this triangle is 13. The perimeter of this rectangle is 20.

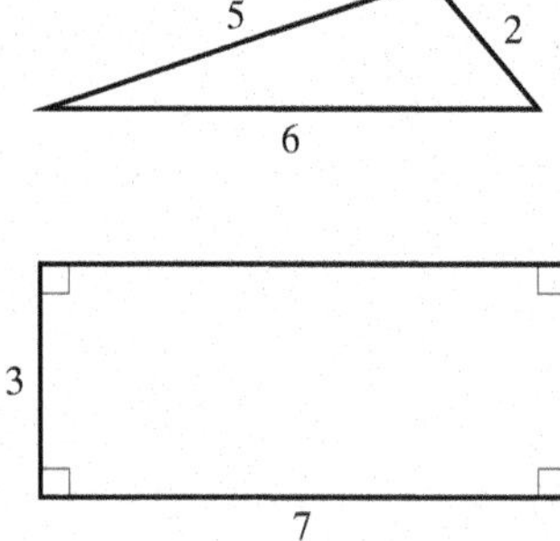

pH of a solution (12.2) Defined by the equation $\mathrm{pH} = -\log_{10}(\mathrm{H}^+)$, where H^+ is the concentration of the hydrogen ion in the solution. The solution is an acid when the pH is less than 7 and a base when the pH is greater than 7.

Pi (1.8) An irrational number, denoted by the symbol π, that is approximately equal to 3.141592654. In most cases 3.14 can be used as a sufficiently accurate approximation for π.

Point–slope form of the equation of a straight line (Appendix A.5) For a straight line passing through the point (x_1, y_1) and having slope m, $y - y_1 = m(x - x_1)$.

Polynomial (5.3) An expression that only contains terms with nonnegative integer exponents. The expressions $5ab + 6$, $x^3 + 6x^2 + 3$, -12, and $x + 3y - 2$ *are all polynomials*. The expressions $x^{-2} + 2x^{-1}$, $2\sqrt{x} + 6$, and $\frac{5}{x} + 2x^2$ *are not polynomials.*

Prime number (0.1) Any natural number greater than 1 whose only natural number factors are 1 and itself. The first eight prime numbers are 2, 3, 5, 7, 11, 13, 17, and 19.

Prime polynomial (6.6) A prime polynomial is a polynomial that cannot be factored by the methods of elementary algebra. $x^2 + x + 1$ is a prime polynomial.

Principal square root (8.2) The positive square root of a number. The symbol indicating the principal square root is $\sqrt{}$. Thus, $\sqrt{4}$ means to find the principal square root of 4, which is 2.

Proper fraction (0.1) A numerical fraction whose numerator is less than its denominator; $\frac{3}{7}, \frac{2}{5}$, and $\frac{8}{9}$ are proper fractions.

Proportion (7.6) A proportion is an equation stating that two ratios are equal.

$$\frac{a}{b} = \frac{c}{d} \quad \text{where } b, d \neq 0$$

is a proportion.

Pythagorean Theorem (9.4) In any right triangle, if c is the length of the hypotenuse and a and b are the lengths of the two legs, then $c^2 = a^2 + b^2$.

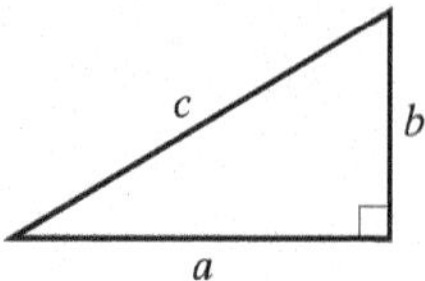

Quadratic equation in standard form (5.8) and (9.1) An equation of the form $ax^2 + bx + c = 0$, where a, b, and c are real numbers and $a \neq 0$. A quadratic equation is classified as a second-degree equation.

Quadratic formula (9.2) If $ax^2 + bx + c = 0$ and $a \neq 0$, then the roots to the equation are found by the formula

$$x = \frac{-b \pm \sqrt{b^2 - 4ac}}{2a}.$$

Quadratic inequality (10.5) An inequality written in the form $ax^2 + bx + c > 0$, where $a \neq 0$ and a, b, and c are real numbers. The $>$ symbol may be replaced by a $<$, $\geq$, or $\leq$ symbol.

Quotient (0.4) The result of dividing one number or expression by another. In the problem $12 \div 4 = 3$, the quotient is 3.

Radical equation (8.5) An equation that contains one or more radicals. The following are examples of radical equations.

$$\sqrt{9x - 20} = x \quad \text{and} \quad 4 = \sqrt{x - 3} + \sqrt{x + 5}$$

Radical sign (8.2) The symbol $\sqrt{}$, which is used to indicate the root of a number.

Radicand (8.2) The expression beneath the radical sign. The radicand of $\sqrt{7x}$ is $7x$.

Range of a relation (3.6) In any relation, the set of values that represents the dependent variable is called its range. This is the set of all the second coordinates of the ordered pairs that define the relation.

Ratio (7.6) The ratio of one number a to another number b is the quotient $a \div b$ or $\frac{a}{b}$.

Rational numbers (1.1) and (8.1) A number that can be expressed in the form $\frac{a}{b}$, where a and b are integers and $b \neq 0$. $\frac{7}{3}, -\frac{2}{5}, \frac{7}{-8}, \frac{5}{1}$, 1.62, and 2.7156 are rational numbers.

Rationalizing the denominator (8.4) The process of transforming a fraction that contains one or more radicals in the denominator to an equivalent fraction that does not contain any radicals in the denominator. When we rationalize the denominator of $\frac{5}{\sqrt{3}}$, we obtain $\frac{5\sqrt{3}}{3}$.

When we rationalize the denominator of $\frac{-2}{\sqrt{11} - \sqrt{7}}$, we obtain $-\frac{\sqrt{11} + \sqrt{7}}{2}$.

Rationalizing the numerator (8.4) The process of transforming a fraction that contains one or more radicals in the numerator to an equivalent fraction that does not contain any radicals in the numerator. When we rationalize the numerator of $\frac{\sqrt{5}}{x}$, we obtain $\frac{5}{x\sqrt{5}}$.

Real number (1.1) Any number that is rational or irrational. $2, 7, \sqrt{5}, \frac{3}{8}, \pi, -\frac{7}{5}$, and $-3\sqrt{5}$ are all real numbers.

Rectangle (1.8) A four-sided figure with opposite sides parallel and all interior angles measuring 90°. The opposite sides of a rectangle are equal.

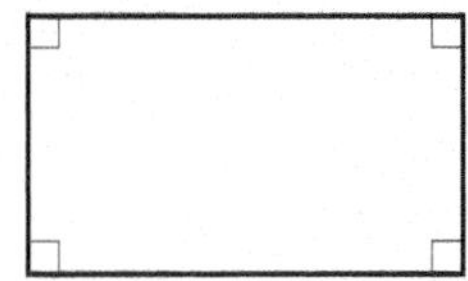

Reduced row echelon form (Appendix D) In the reduced row echelon form of an augmented matrix, all the numbers to the left of the vertical line are 1s along the diagonal from the top left to the bottom right. If there are elements below or above the 1s, these elements are 0s. Two examples of matrices in reduced row echelon form are

$$\left[\begin{array}{cc|c} 1 & 0 & 3 \\ 0 & 1 & 4 \end{array}\right] \quad \text{and} \quad \left[\begin{array}{ccc|c} 1 & 0 & 0 & 5 \\ 0 & 1 & 0 & 6 \\ 0 & 0 & 1 & 7 \end{array}\right].$$

Relation (3.6) A relation is any set of ordered pairs.

Rhombus (1.8) A parallelogram with four equal sides. Figure $ABCD$ is a rhombus.

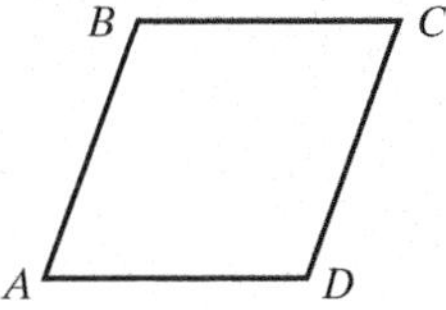

Right triangle (1.8) and (9.4) A triangle that contains one right angle (an angle that measures exactly 90 degrees). It is indicated by a small square at the corner of the angle.

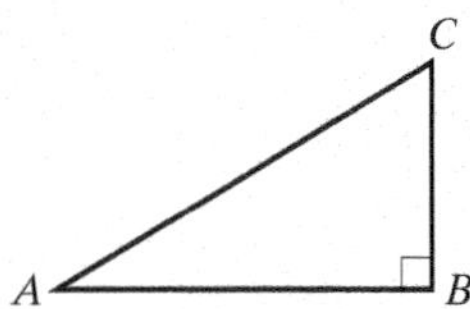

Root of an equation (2.1) and (6.7) A value of the variable that makes an equation into a true statement. The root of an equation is also called the solution of an equation.

Scientific notation (5.2) A positive number is written in scientific notation if it is in the form $a \times 10^n$, where $1 \le a < 10$ and n is an integer.

Similar radicals (8.3) Two radicals that are simplified and have the same radicand and the same index. $2\sqrt[3]{7xy^2}$ and $-5\sqrt[3]{7xy^2}$ are similar radicals. Usually similar radicals are referred to as *like radicals*.

Simplifying a radical (8.3) To simplify a radical when the root cannot be found exactly, we use the product rule for radicals, $\sqrt[n]{ab} = \sqrt[n]{a}\sqrt[n]{b}$ for $a \ge 0$ and $b \ge 0$. To simplify $\sqrt{20}$, we have $= \sqrt{4}\,\sqrt{5} = 2\sqrt{5}$. To simplify $\sqrt[3]{16x^4}$, we have $= \sqrt[3]{8x^3}\,\sqrt[3]{2x} = 2x\sqrt[3]{2x}$.

Simplifying imaginary numbers (8.6) Using the property that states for all positive real numbers a, $\sqrt{-a} = \sqrt{-1}\sqrt{a} = i\sqrt{a}$. Thus, simplifying $\sqrt{-7}$, we have $\sqrt{-7} = \sqrt{-1}\sqrt{7} = i\sqrt{7}$.

Slope–intercept form (3.3) The equation of a line that has slope m and the y-intercept at $(0, b)$ is given by $y = mx + b$.

Slope of a line (3.3) The ratio of change in y over the change in x for any two different points on a nonvertical line. The slope m is determined by

$$m = \frac{y_2 - y_1}{x_2 - x_1},$$

where $x_2 \ne x_1$ for any two points (x_1, y_1) and (x_2, y_2) on a nonvertical line.

Solution of an equation (2.1) A number that, when substituted into a given equation, yields an identity. The solution of an equation is also called the root of an equation.

Solution of an inequality in two variables (3.5) The set of all possible ordered pairs that, when substituted into the inequality, will yield a true statement.

Solution of a linear inequality (2.8) The possible values that make a linear inequality true.

Square (1.8) A rectangle with four equal sides.

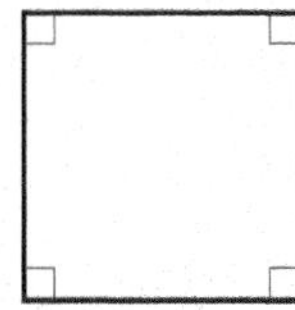

Square root (8.2) If x is a real number and a is a positive real number such that $a = x^2$, then x is a square root of a. One square root of 16 is 4 since $4^2 = 16$. Another square root of 16 is -4 since $(-4)^2 = 16$.

Standard form of the equation of a circle (10.1) For a circle with center at (h, k) and a radius of r,

$$(x - h)^2 + (y - k)^2 = r^2.$$

Standard form of the equation of an ellipse (10.3) For an ellipse with center at the origin,

$$\frac{x^2}{a^2} + \frac{y^2}{b^2} = 1, \qquad \text{where } a \text{ and } b > 0.$$

This ellipse has intercepts at $(a, 0), (-a, 0), (0, b)$, and $(0, -b)$.

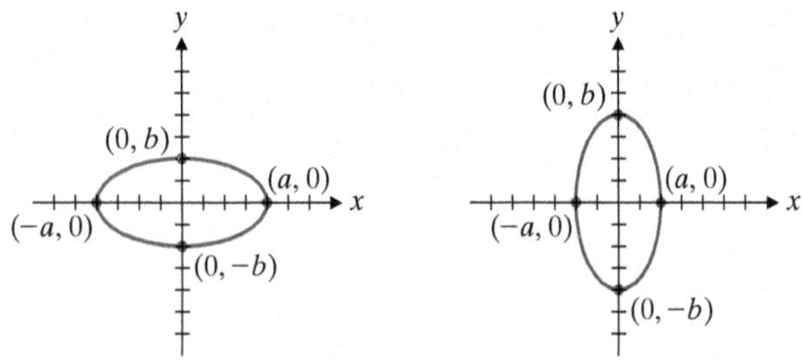

For an ellipse with center at (h, k),

$$\frac{(x-h)^2}{a^2} + \frac{(y-k)^2}{b^2} = 1, \qquad \text{where } a \text{ and } b > 0.$$

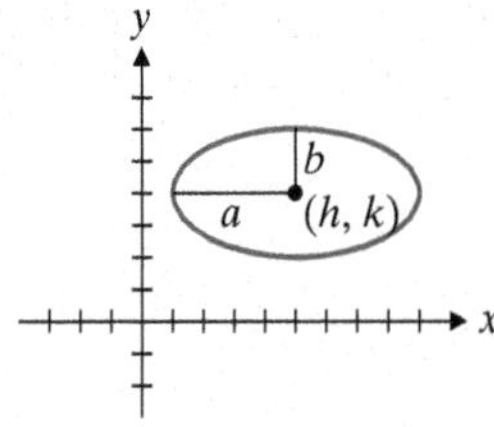

Standard form of the equation of a hyperbola with center at the origin (10.4) For a horizontal hyperbola with center at the origin,

$$\frac{x^2}{a^2} - \frac{y^2}{b^2} = 1, \qquad \text{where } a \text{ and } b > 0.$$

The vertices are at $(-a, 0)$ and $(a, 0)$.

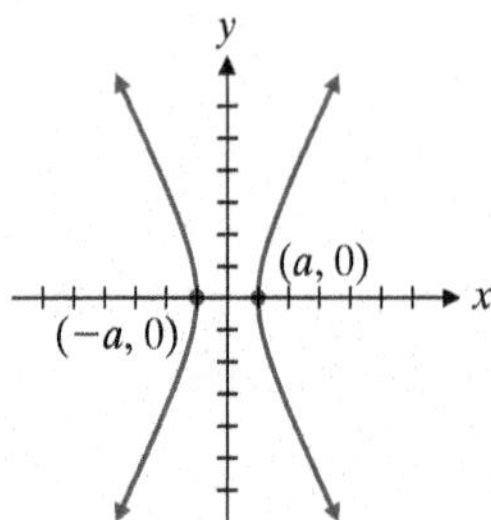

For a vertical hyperbola with center at the origin,

$$\frac{y^2}{b^2} - \frac{x^2}{a^2} = 1, \qquad \text{where } a \text{ and } b > 0.$$

The vertices are at $(0, b)$ and $(0, -b)$.

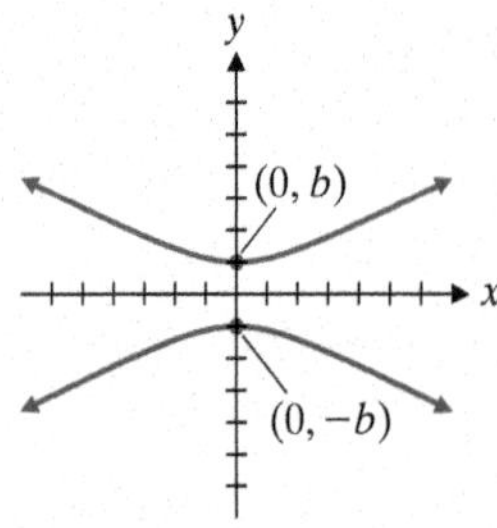

Standard form of the equation of a hyperbola with center at point (h, k) (10.4) For a horizontal hyperbola with center at (h, k),

$$\frac{(x-h)^2}{a^2} - \frac{(y-k)^2}{b^2} = 1, \qquad \text{where } a \text{ and } b > 0.$$

The vertices are at $(h - a, k)$ and $(h + a, k)$.

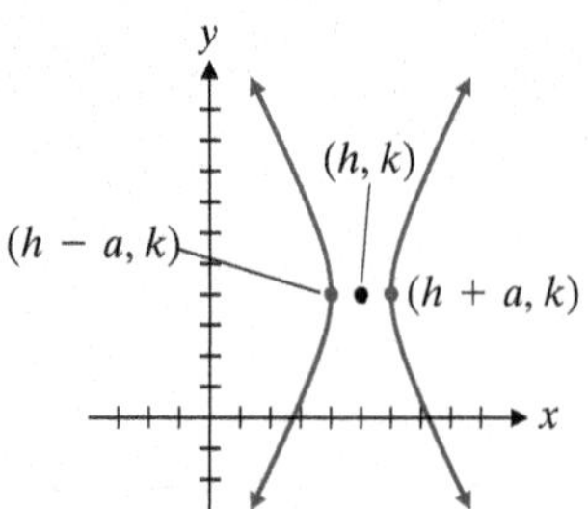

For a vertical hyperbola with center at (h, k),

$$\frac{(y-k)^2}{b^2} - \frac{(x-h)^2}{a^2} = 1, \qquad \text{where } a \text{ and } b > 0.$$

The vertices are at $(h, k + b)$ and $(h, k - b)$.

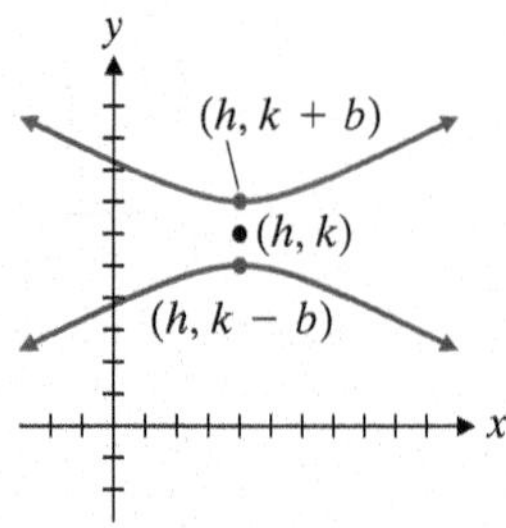

Standard form of the equation of a parabola (10.2) For a vertical parabola with vertex at (h, k),

$$y = a(x-h)^2 + k, \qquad \text{where } a \neq 0.$$

For a horizontal parabola with vertex at (h, k),

$$x = a(y-k)^2 + h, \qquad \text{where } a \neq 0.$$

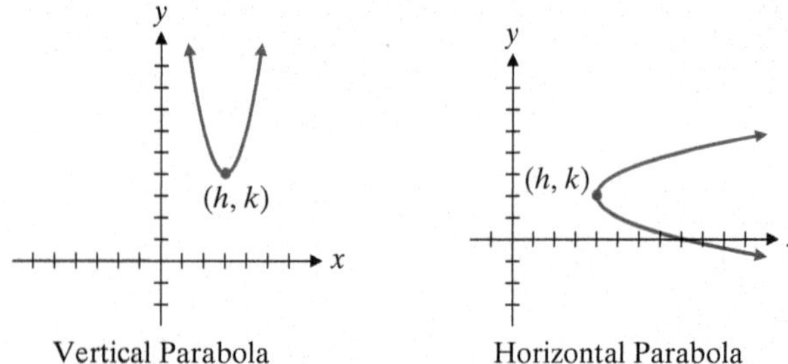

Vertical Parabola   Horizontal Parabola

Standard form of a quadratic equation (6.7) A quadratic equation that is in the form $ax^2 + bx + c = 0$.

System of equations (4.1) A set of two or more equations that must be considered together. The solution is the value for each variable of the system that satisfies each equation.

$$x + 3y = -7 \qquad 4x + 3y = -1$$

is a system of two equations in two unknowns. The solution is $(2, -3)$, or the values $x = 2, y = -3$.

System of inequalities (4.4) Two or more inequalities in two variables that are considered at one time. The solution is the region that satisfies every inequality at one time. An example of a system of inequalities is

$$y > 2x + 1 \qquad y < \frac{1}{2}x + 2.$$

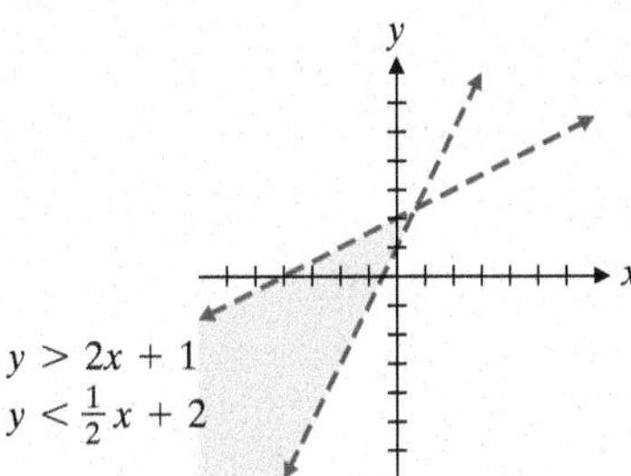

Term (1.7) A number, a variable, or a product of numbers and variables. For example, in the expression $a^3 - 3a^2b + 4ab^2 + 6b^3 + 8$, there are five terms. They are a^3, $-3a^2b$, $4ab^2$, $6b^3$, and 8. The terms of a polynomial are separated by plus and minus signs.

Trapezoid (1.8) A four-sided figure with two sides parallel. The parallel sides are called the bases of the trapezoid. Figure $ABCD$ is a trapezoid.

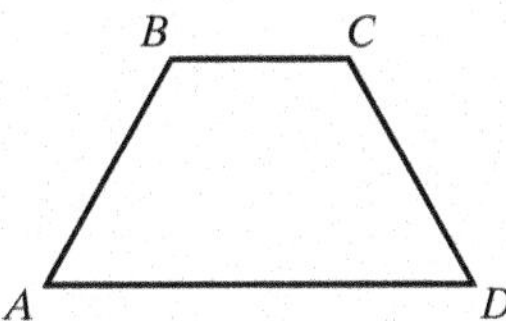

Trinomial (5.3) A polynomial of three terms. The expressions $x^2 + 6x - 8$ and $a + 2b - 3c$ are trinomials.

Value of a second-order determinant (Appendix C) For a second-order determinant $\begin{vmatrix} a & b \\ c & d \end{vmatrix}$, $ad - cb$.

Value of a third-order determinant (Appendix C) For third-order determinant $\begin{vmatrix} a_1 & b_1 & c_1 \\ a_2 & b_2 & c_2 \\ a_3 & b_3 & c_3 \end{vmatrix}$,

$a_1b_2c_3 + b_1c_2a_3 + c_1a_2b_3 - a_3b_2c_1 - b_3c_2a_1 - c_3a_2b_1$.

Variable (1.4) A letter that is used to represent a number or a set of numbers.

Variation (8.7) An equation relating values of one variable to those of other variables. An equation of the form $y = kx$, where k is a constant, indicates *direct variation.* An equation of the form $y = \frac{k}{x}$, where k is a constant, indicates *inverse variation.* In both cases, k is called the *constant of variation.*

Vertex of a parabola (10.2) In a vertical parabola, the lowest point on a parabola opening upward or the highest point on a parabola opening downward.

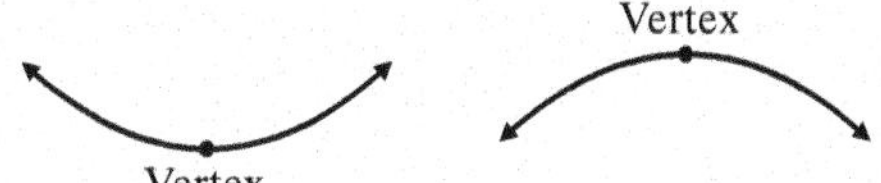

In a horizontal parabola, the leftmost point on a parabola opening to the right or the rightmost point on a parabola opening to the left.

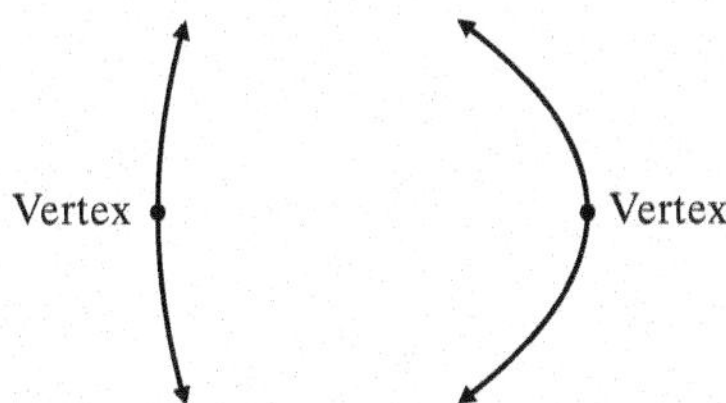

Vertical line test (3.6) If a vertical line can intersect the graph of a relation more than once, the relation is not a function.

Vertical parabolas (10.2) Parabolas that open upward or downward. The following graphs represent vertical parabolas.

Whole numbers (0.1) The set of numbers 0, 1, 2, 3, 4, 5, ...

***x*-intercept (3.2)** The ordered pair $(a, 0)$ is the x-intercept of a line if the line crosses the x-axis at $(a, 0)$. The x-intercept of line l on the following graph is $(4, 0)$.

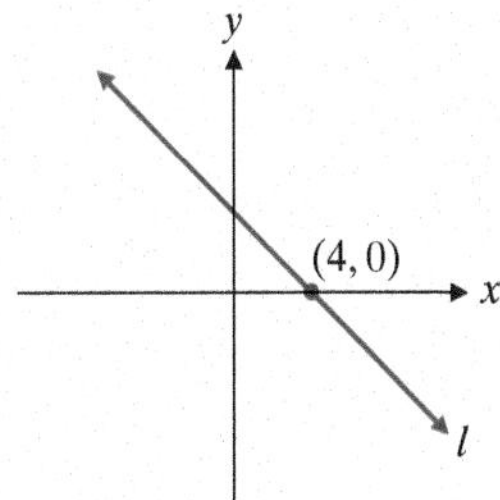

***y*-intercept (3.2)** The ordered pair $(0, b)$ is the y-intercept of a line if the line crosses the y-axis at $(0, b)$. The y-intercept of line p on the following graph is $(0, 3)$.

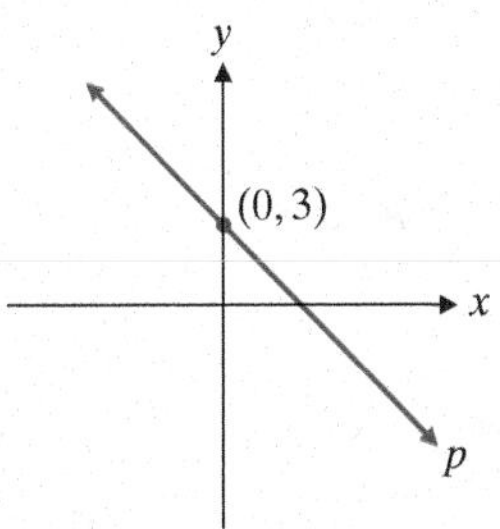

Photo Credits

CHAPTER 0 **CO** Monkey Business/Fotolia **p. 31** CandyBox Images/Fotolia **p. 39** Littlebloke/Fotolia **p. 40** Alex_Po/Fotolia **p. 62** John Tobey

CHAPTER 1 **CO** Wavebreak Media Ltd/123RF, Sergey Gavrilichev/123RF **p. 66** Jezper/Fotolia **p. 99** Kevin C. Cox/Getty Images **p. 100** Andres Rodriguez/Fotolia **p. 114** Tom McHugh/Science Source **p. 121** Jeanne White/Science Source **p. 128** Jeff Slater

CHAPTER 2 **CO** Michaeljung/Fotolia **p. 142R** Dmitrydesigner/123RF **p. 142L** Dale Walsh/E+/Getty Images **p. 157** VadimGuzhva/Fotolia **p. 172** Mattiaath/Fotolia **p. 188** Tyler Olson/123RF **p. 192** John Michael Evan Potter/Shutterstock **p. 201** Jamie Blair

CHAPTER 3 **CO** Andriy Popov/123RF **pp. 230, 233, 234, 252** Screenshots from Texas Instruments. Courtesy of Texas Instruments, Inc. **p. 238** Szefei/123RF **p. 269** Jenny Crawford

CHAPTER 4 **CO** Sjenner13/123RF **pp. 276, 287, 309** Screenshots from Texas Instruments. Courtesy of Texas Instruments, Inc. **p. 295** Tom Wang/Fotolia **p. 322** John Tobey

CHAPTER 5 **CO** WavebreakmediaMicro/Fotolia **p. 340** D. Calzetti/H. Ford/ Hubble Heritage (STScI/AURA)-ESA/Hubble Collaboration/NASA **p. 349** JohnKwan/Fotolia **p. 374** Jeff Slater

CHAPTER 6 **CO** Wavebreakmedia Ltd/123RF **p. 381** khamkula/Fotolia **p. 400** Rocketclips/Fotolia **p. 405** Santi Praseeratenang/123RF **p. 415** Corbis_infinite/Fotolia **p. 425** Jamie Blair

CHAPTER 7 **CO** Goodluz/123RF **p. 449** Laurence Mouton/PhotoAlto sas/Alamy **p. 476** Jenny Crawford

CHAPTER 8 **CO** Henryk Sadura/Fotolia **p. 510** Portra Images/Digital Vision/Getty Images **p. 526** Mik Lav/Shutterstock **p. 531** Jjava/Fotolia **p. 541** Jamie Blair

CHAPTER 9 **CO** Tyler Olson/123RF **p. 567** Nithid 18/Fotolia **p. 572** Singkamc/Fotolia **p. 574** Sergii Mostovy/Fotolia **p. 580** Screenshots from Texas Instruments. Courtesy of Texas Instruments, Inc. **p. 621** Jenny Crawford

CHAPTER 10 **CO** Gennadiy Poznyakov/Fotolia **p. 627** Screenshots from Texas Instruments. Courtesy of Texas Instruments, Inc. **p. 652** Songquan Deng/Shutterstock **p. 653** Tom Wang/Fotolia **p. 677** John Tobey

CHAPTER 11 **CO** LukaTDB/Fotolia **p. 695** lopolo/123RF **p. 698** Monkey Business/Fotolia **p. 703** Tom McHugh/Science Source **p. 721** Jeff Slater

CHAPTER 12 **CO** Stephen Coburn/Fotolia **pp. 708, 752** Screenshots from Texas Instruments. Courtesy of Texas Instruments, Inc. **p. 727** Goodluz/Fotolia **p. 747** Erik Isakson/Blend Images/Getty Images **p. 761** Lane Erickson/Fotolia **p. 773** Jamie Blair

APPENDIX **p. A-40** Screenshots from Texas Instruments. Courtesy of Texas Instruments, Inc.

Subject Index

Index of Applications

S

T

PROPERTIES OF THE REAL NUMBERS

If a, b, and c are real numbers:

Closure Properties	$a + b$ is a real number. ab is a real number.
Commutative Properties	$a + b = b + a$ $ab = ba$
Associative Properties	$a + (b + c) = (a + b) + c$ $a(bc) = (ab)c$
Identity Properties	$a + 0 = 0 + a = a$ $a \cdot 1 = 1 \cdot a = a$
Inverse Properties	$a + (-a) = -a + a = 0$ $a\left(\frac{1}{a}\right) = \left(\frac{1}{a}\right)a = 1$ (where $a \neq 0$)
Distributive Property	$a(b + c) = ab + ac$

QUADRATIC FORMULA

If $ax^2 + bx + c = 0$, where $a \neq 0$,

$$x = \frac{-b \pm \sqrt{b^2 - 4ac}}{2a}.$$

PYTHAGOREAN THEOREM

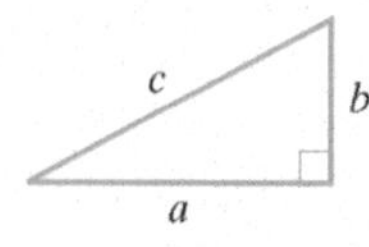

In a right triangle if c is the length of the hypotenuse and a and b are the lengths of the legs, then $c^2 = a^2 + b^2$.

PROPERTIES OF INEQUALITIES

If $a < b$, then $a + c < b + c$ and $a - c < b - c$.

If $a < b$ and c is a **positive** real number,

then $ac < bc$ and $\frac{a}{c} < \frac{b}{c}$.

If $a < b$ and c is a **negative** real number,

then $ac > bc$ and $\frac{a}{c} > \frac{b}{c}$.

If $|x| < a$, then $-a < x < a$.

If $|x| > a$, then $x > a$ or $x < -a$.

ABSOLUTE VALUE

$$|x| = \begin{cases} x & \text{if } x \geq 0 \\ -x & \text{if } x < 0 \end{cases}$$

EQUATIONS OF STRAIGHT LINES

Standard Form $Ax + By = C$
A, B, and C are real numbers.

Slope–Intercept Form $y = mx + b$
$m =$ slope $(0, b) = y$-intercept

Point–Slope Form $y - y_1 = m(x - x_1)$
(x_1, y_1) is a point on the line, and $m =$ slope

PROPERTIES ABOUT POINTS AND STRAIGHT LINES

The **distance between two points** (x_1, y_1) and (x_2, y_2), d, is $\sqrt{(x_2 - x_1)^2 + (y_2 - y_1)^2}$.

The **slope of a line**, m, passing through (x_1, y_1) and (x_2, y_2) is $\frac{y_2 - y_1}{x_2 - x_1}$, where $x_2 \neq x_1$.

Parallel lines have the same slope: $m_1 = m_2$

Nonvertical **perpendicular lines** have slopes whose product is -1. This may be written $m_1 = -\frac{1}{m_2}$.

Horizontal lines have a slope of 0.

Vertical lines have no slope. (The slope is not defined for a vertical line.)

FACTORING AND MULTIPLYING FORMULAS

$a^2 - 2ab + b^2 = (a - b)^2$	**Perfect Square Trinomial**
$a^2 + 2ab + b^2 = (a + b)^2$	**Perfect Square Trinomial**
$a^2 - b^2 = (a + b)(a - b)$	**Difference of Two Squares**
$a^2 + b^2$ cannot be factored.	
$a^3 - b^3 = (a - b)(a^2 + ab + b^2)$	**Difference of Two Cubes**
$a^3 + b^3 = (a + b)(a^2 - ab + b^2)$	**Sum of Two Cubes**